ENCYCLOPEDIA OF PRODUCTION AND MANUFACTURING MANAGEMENT

ENCYCLOPEDIA OF
PRODUCTION AND MANUFACTURING MANAGEMENT

Editor: Paul M. Swamidass, Auburn University

ENCYCLOPEDIA OF PRODUCTION AND MANUFACTURING MANAGEMENT

Editor

Paul M. Swamidass

Professor of Operations Management
College of Business
Auburn University
Auburn, AL 36849

and

Associate Director
Thomas Walter Center for Technology Management
Tiger Drive, Rm. 104
Auburn University
Auburn, AL 36849-5358

Kluwer Academic Publishers
Boston/Dordrecht/London

Distributors for North, Central and South America:
Kluwer Academic Publishers
101 Philip Drive
Assinippi Park
Norwell, Massachusetts 02061 USA
Telephone (781) 871-6600
Fax (781) 871-6528
E-Mail <kluwer@wkap.com>

Distributors for all other countries:
Kluwer Academic Publishers Group
Distribution Centre
Post Office Box 322
3300 AH Dordrecht, THE NETHERLANDS
Telephone 31 78 6392 392
Fax 31 78 6546 474
E-Mail <orderdept@wkap.nl>

Electronic Services <http://www.wkap.nl>

Library of Congress Cataloging-in-Publication

Encyclopedia of production and manufacturing management / editor, Paul M. Swamidass.
p. cm.
Includes bibliographical references and index.
ISBN 0-7923-8630-2 (alk. paper)
1. Production management – Encyclopedias. I. Swamidass, Paul. M.

TS9.E53 2000
658.5'03 21 – dc21 99-044533

Printed on acid-free paper.

Printed in the United States of America

To my parents:
Jacob M. and Gnanam Swamidass

Table of Contents

Alphabetical List of Major Articles

List of Authors, Contributors, and Reviewers

The following authors have written one or more long articles and several short pieces for this encyclopedia. Authors of longer articles are identified at the beginning of each article. Unless identified otherwise the institutions in the following list are located in the USA.

Gerald Aase, *Northern Illinois University.*

Hilary Bates, *University of Warwick, UK.*
Thomas J. Billesbach, *Northwest Missouri State University.*
JT Black, *Auburn University.*
Matthew C. Bloom, *University of Notre Dame.*
Steven F. Bolander, *Colorado State University.*
David E. Booth, *Kent State University.*
Kenneth K. Boyer, *DePaul University.*
Thomas F. Burgess, *The University of Leeds, UK.*

Pratap S. S. Chinnaiah, *Northeastern University.*
David Challis, *University of Melbourne, Australia.*
Chen H. Chung, *University of Kentucky.*
M. Bixby Cooper, *Michigan State University.*

Neil R. Darlow, *Auburn University.*
Edward W. Davis, *University of Virginia.*
Arnoud De Meyer, *INSEAD, France.*
Rommert Dekker, *Erasmus University Rotterdam, The Netherlands.*
Thomas W. Dougherty, *University of Missouri.*
George F. Dreher, *Indiana University.*
Rebecca Duray, *University of Colorado at Colorado Springs.*
S.P. Dutta, *University of Windsor, Canada.*

James R. Evans, *University of Cincinnati.*

Benito E. Flores, *Texas A&M University.*
Stanely E. Fawcett, *Brigham Young University.*
John W. Fowler, *Arizona State University.*

Larry C. Giunipero, *Florida State University.*
Noel P. Greis, *University of North Carolina—Chapel Hill.*
Hasan K. Gules, *Selcuk University, Turkey.*
Surendra M. Gupta, *Northeastern University.*

George C. Hadjinicola, *University of Cyprus, Cyprus.*
Robert B. Handfield, *Michigan State University.*
Jeffrey W. Herrmann, *University of Maryland.*

Sukran N. Kadipasaoglu, *University of Houston.*
Sagar V. Kamarthi, *Northeastern University.*
John D. Kasarda, *University of North Carolina—Chapel Hill.*
Corinne M. Karuppan, *Southwest Missouri State University.*
Moutaz Khouja, *University of North Carolina—Charlotte.*
Basheer M. Khumawala, *University of Houston—Houston.*
Elif Kizilkaya, *Northeastern University.*
Robert Klassen, *University of Western Ontario, Canada.*
Michael G. Kolchin, *Lehigh University.*
K. Ravi Kumar, *University of Southern California.*

R. Lawrence LaForge, *Clemson University.*
Kenneth D. Lawrence, *New Jersey Institute of Technology.*
Brian Leavy, *Dublin City University, Ireland.*
John Leschke, *University of Virginia.*
James P. Lewis, *The Lewis Institute, Inc.*
Christoph H. Loch, *INSEAD, France.*

Debasish N. Mallick, *Boston College.*
John K. McCreery, *North Carolina State University.*
Kathleen E. McKone, *University of Minnesota.*
Patrick R. McMullen, *Auburn University.*
Hokey Min, *University of Louisville.*
Amitava Mitra, *Auburn University.*
Kendra E. Moore, *ALPHATECH Inc.*

R. Nat Natarajan, *Tennessee Technological University.*

Kwan-Kee Ng, *National University of Singapore, Singapore.*
Hamid Noori, *Wilfrid Laurier University, Canada.*

O. Felix Offodile, *Kent State University.*
David L. Olson, *Texas A&M University.*
Rodney P. Parker, *University of Michigan.*
John N. Pearson, *Arizona State University.*
K. W. Platts, *The University of Cambridge, UK.*

Ramachandran Ramanan, *University of Notre Dame.*
Ranga Ramasesh, *Texas Christian University.*
Paul G. Ranky, *New Jersey Institute of Technology.*

Danny Samson, *University of Melbourne, Australia.*
Nada R. Sanders, *Wright State University.*
Marc J. Schniederjans, *University of Nebraska.*
Kristie Seawright, *Brigham Young University.*
S. Sengupta, *Oakland University.*
D. Scott Slocombe, *Wilfrid Laurier University, Canada.*
Timothy L. Smunt, *Wake Forest University.*
V. Sridharan, *Clemson University.*
Kathryn E. Stecke, *University of Michigan.*
Michael Suh, *Kent State University.*
Chee-Chuong Sum, *National University of Singapore, Singapore.*
Monte R. Swain, *Brigham Young University.*
Paul M. Swamidass, *Auburn University.*
Morgan Swink, *Michigan State University.*

Christian Terwiesch, *University of Pennsylvania.*
Sam G. Taylor, *University of Wyoming.*
David Twigg, *University of Brighton, UK.*

M. Michael Umble, *Baylor University.*
Elisabeth J. Umble, *Texas A&M University.*

Frank A. Van der Duyn Schouten, *Tilburg University, The Netherlands.*
Pitipong Veerakamolmal, *Northeastern University.*

Elliott N. Weiss, *University of Virginia.*
H. Joseph Wen, *New Jersey Institute of Technology.*
Gregory P. White, *Southern Illinois University—Carbondale.*
Richard E. White, *University of North Texas.*
James M. Wilson, *Glasgow Business School, UK.*

Enver Yücesan, *INSEAD, France.*

Manufacturers and Organizations Discussed in the Encyclopedia

One of the special features of this encyclopedia is that it gives several real life examples from manufacturing firms from around the world. The list below contains over 100 manufacturers, from the U.S., Japan, Europe, Korea and other nations. In addition, it includes trucking and transportation companies and leading retailers, who are used as examples in various articles. Some manufacturers such as Chrysler, GM, IBM, Motorola, and Toyota are discussed in several articles.

Manufacturers and organizations are listed on the left column. The articles in which the organization is mentioned are listed in the right column.

Organization	Article where discussed
3M	Environmental issues and operations mgmt.
	JIT evolution and use in the United States
	New product development through...
ABB Group	Environmental issues and operations mgmt.
ADAC Laboratories	Performance excellence: The Malcolm...
Aerotech Services Corporation	Virtual manufacturing
AIAG	JIT evolution and use in the United States
Albany International	Just-in-time manufacturing implications
Allegheny Ludlum Corporation	Core manufacturing competencies
Allen Edmonds Shoe Corporation	Supplier performance measurement
Allen-Bradley	Core manufacturing competencies and...
Allied-Signal	Purchasing: Acquiring the best inputs
Amdahl	Supplier partnership as strategy
Anderson Windows	Mass customization and manufacturing
Anheuser-Busch	Facilitics location decision
ANSI	Electronic data interchange in supply chain
	ISO 9000/QS 9000 quality standards
Apple	Supplier partnership as strategy
Applied Materials, Inc.	Virtual manufacturing
Applied Services Corporation	Virtual manufacturing
Army Ordnance	Quality: The implications of Deming's...
ASA	Quality: The implications of Deming's...
ASME	Quality: The implications of Deming's...
ASQC, ASQ	ISO 9000/QS 9000 quality standards
ASTM	Quality: The implications of Deming's...

Organization	Article where discussed
AT&T	Agile manufacturing
	Environmental issues: Reuse and recycling
Auto Air	Supplier performance measurement
AVO, Inc.	Virtual manufacturing
B.F. Goodrich	Facilities location decision
BAAN	Enterprise resource planning (ERP)
	MRP implementation
Bally Engineered Structures	Agile manufacturing
	Mass customization
Bausch & Lomb	Product design for global markets
Bechtel	Mass customization and manufacturing
Ben & Jerry's Ice Cream	Balanced scorecard
Bennetton	Agile Logistics
	Logistics: Meeting customer's real needs
	Supplier partnership as strategy
Black and Decker	Electronic data interchange in supply chain
	JIT evolution and use in the United States
	Mass customization and manufacturing
BMW	Environmental issues and operations mgmt.
Boeing	Learning curve analysis
	Manufacturing flexibility
	Product design for global markets
	Supplier performance measurement
Bose Corporation	Purchasing: The future
Boston Consulting Group	Learning curve analysis
Briggs and Stratton	Quality: The implications of Deming's...
Bristol-Myers Squibb	Bullwhip effect in supply chain
British Airways	Environmental issues and operations mgmt.
Burlington	Bullwhip effect in supply chain
CAM-I	Activity-based costing: An evaluation
	Capital investment in advanced technology
	New product development through...
Camtronics Medical Systems	Cost analysis for purchasing
Carolina Freight	Electronic data interchange in supply chain
Caterpillar	Logistics: Meeting customer's real needs
	Total quality management
CERES	Environmental issues and operations mgmt.
Chrysler	Agile manufacturing
	Concurrent engineering
	Electronic data interchange in supply chain
	Facilities location decision
	New product development through...
	Outsourcing of product design...

Organization	Article where discussed
	Product innovation
	Purchasing: Acquiring the best inputs
	Quality management systems: Baldridge...
	Supplier performance measurement
	Supplier relationships
	Virtual manufacturing
Coca-Cola	Mass customization and manufacturing
	Product design for global markets
Colgate	Product design for global markets
Compaq	Agile Logistics
	Human resource issues in manufacturing
	Mass customization and manufacturing
Consolidated Freightways	Electronic data interchange in supply chain
Copeland Corporation	Focused factory
Council of Logistics Management	Logistics: Meeting customer's real needs
C-TAD systems	Virtual manufacturing
Cummins Engine	Concurrent engineering
Curtis-Wright Corporation	Learning curve analysis
Custom Clothing Technology	Mass customization: Implementation
Custom Research Inc. (CRI)	Performance excellence: The Malcolm...
Daewoo (Korea)	Global manufacturing rationalization
Daihatsu Motors	JIT evolution and use in the United States
	Target costing
Daimler-Benz	Environmental issues and operations mgmt.
Dana Commercial Credit	Performance excellence: The Malcolm...
Dana Corporation	Total quality management
DARPA	Virtual manufacturing
Dell	Agile Logistics
	Bullwhip effect in supply chain
Digital Equipment Corporation	Supplier partnership as strategy
DMSO	Virtual manufacturing
Dodge	New product development through...
Du Pont	History of manufacturing management
	ISO 9000/QS 9000 quality standards
	Just-in-time manufacturing: Implications
	Total productive maintenance
Dun and Bradstreet	Purchasing: Acquiring the best inputs
Electro	Customer service, satisfaction and success
Eli-Lilly	Bullwhip effect in supply chain
	Environmental issues and operations mgmt.
Exxon	Facilities location decision
Federal Express	Agile Logistics
Fisher Technology	Purchasing: The future

Organization	Article where discussed
Florist Telegraph Service (FTD)	Agile Logistics
Ford	Agile manufacturing
	Assembly line design
	Concurrent engineering
	Electronic data interchange in supply chain
	Facilities location decision
	History of manufacturing management
	ISO 9000/QS 9000 quality standards
	Purchasing: Acquiring the best inputs
	Quality management systems: Baldridge...
	Quality: The implications of Deming's...
	Reengineering and the process view of...
	Statistical process control using control...
	Supplier performance measurement
	Total quality management
	U-shaped assembly lines
	Virtual manufacturing
Ford-VW	Outsourcing of product design...
Freightliner	ISO 9000/QS 9000 quality standards
Fuji	New product development through...
Fuji-Xerox	Product design for global markets
GAAP	Accounting system implications of TOC
	Activity-based costing
Gateway	Agile Logistics
GE	Facilities location decision
	JIT evolution and use in the United States
	Mass customization and manufacturing
	Purchasing: The future
	Total productive maintenance
	Total quality management
GE Fanuc	Agile manufacturing
	Mass customization and manufacturing
General Foods	Facilities location decision
Gillette	Mass customization and manufacturing
GM	Agile manufacturing
	Concurrent engineering
	Core manufacturing competencies
	Core manufacturing competencies and...
	Facilities location decision
	Global manufacturing rationalization
	Human resource issues in manufacturing
	ISO 9000/QS 9000 quality standards
	Logistics: Meeting customer's real needs

Organization	Article where discussed
	Manufacturing flexibility dimensions
	Manufacturing strategy
	Mass customization
	Quality management systems: Baldridge...
	Supplier partnership as strategy
	Supplier performance measurement
	Total quality management
	Virtual manufacturing
Graphic Enterprises	Quality: The implications of Deming's...
Harley-Davidson	Just-in-time manufacturing
Heineken	Product design for global markets
Heinz	Facilities location decision
Hermann Miller	Environmental issues and operations mgmt.
Hershey	Customer service, satisfaction and success
Hewlett Packard	Product design for global markets
	Concurrent engineering
	Environmental issues: Reuse and recycling
	International manufacturing
	JIT evolution and use in the United States
	Mass customization and manufacturing
	Virtual manufacturing
Hino	JIT evolution and use in the United States
Hitachi Seiki	Core manufacturing competencies
Home Depot	Supply chain management
Honda	Concurrent engineering
	Cost analysis for purchasing
	Product design for global markets
Honeywell	Agile manufacturing
	Facilities location decision
	JIT evolution and use in the United States
IBM	Agile Logistics
	Customer service, satisfaction and success
	Environmental issues: Reuse anrecycling
	JIT evolution and use in the United States
	Mass customization and manufacturing
	Total quality management
	Virtual manufacturing
Ikea	Product design for global markets
Imperial Chemical	Facilities location decision
Individual, Inc.	Mass customization and manufacturing
Institute of management Accountants	Activity-based costing: An evaluation
Intel	JIT evolution and use the United States
	Virtual manufacturing

Organization	Article where discussed
ISO	ISO 9000/QS 9000 quality standards
J.B. Hunt	Electronic data interchange in supply chain
J.C. Penny	Bullwhip effect in supply chain
J.D. Edwar	Enterprise resource planning (ERP)
JIPE	Total productive maintenance
JIPM	Total productive maintenance
John Crane Limited	Core manufacturing competencies
John Deere	Agile manufacturing
	Virtual manufacturing
Johnson and Johnson	Facilities location decision
JUSE	Quality: The implications of Deming's...
Kawasaki	Just-in-time manufacturing
Kia (Korea)	Bullwhip effect in supply chain
K-Mart	Bullwhip effect in supply chain
Kodak	Facilities location decision
	Environmental issues and operations mgmt.
Komatsu	Total quality management
KPMG	Balanced scorecard
Lawrence Livermore National Laboratory	Electronic data interchange in supply chain
Lawson Software	Enterprise resource planning (ERP)
	Activity-based costing
Lego	Mass customization: Implementation
	Product design for global markets
Levi-Strauss	Agile Logistics
	Human resource issues in manufacturing
	Mass customization: Implementation
	Product design for global markets
Lockheed Corporation	Learning curve analysis
Lutron Electronics Company	Mass customization and manufacturing
MacDonnell Aircraft/Aerospace	Manufacturing flexibility
	Virtual manufacturing
Mack Trucks	ISO 9000/QS 9000 quality standards
MANTECH	Virtual manufacturing
Marks and Spencer	Supplier partnership as strategy
Marriott	Product design for global markets
Mars	Product design for global markets
	Agile manufacturing
Massachusetts Institute of Technology	Flexible automation
Matlack	Electronic data interchange in supply chain
Matsushita	Target costing
	Agile manufacturing
	Mass customization and manufacturing
	Virtual manufacturing

Organization	Article where discussed
Mazda	JIT evolution and use in the United States
McDonalds	Product design for global markets
McKesson	Bullwhip effect in supply chain
Mercedes Benz	Mass customization and manufacturing
	Target costing
Metro Bank	Balanced scorecard
Mobil	Facilities location decision
Motorola	JIT evolution and use in the United States
	Mass customization and manufacturing
	Product design for global markets
	Purchasing: The future
	Statistical process control using control...
	Supplier relationships
	Total quality management
NAPM	Purchasing: Acquiring the best inputs
Nashua Corporation	Just-in-time manufacturing: Implications
	Quality: The implications of Deming's...
	Statistical process control using control...
National Association of Accountants	Capital investment in advanced technology
National Bureau of Standards	Virtual manufacturing
National Insurance	Balanced scorecard
National Semiconductor	Logistics: Meeting customer's real needs
Navistar International	ISO 9000/QS 9000 quality standards
NCPDM	Logistics: Meeting customer's real needs
Nestle	Facilities location decision
Nissan	JIT evolution and use in the United States
	Mass customization and manufacturing
NIST	Total quality management
Northern Telecom	Environmental issues and operations mgmt.
Nokia	U-shaped assembly lines
NUMMI	Core manufacturing competencies
	Just-in-time manufacturing
	Total quality management
Nuskin	Customer service, satisfaction and success
Olympus Optical Company	Target costing
Omark Industries	JIT evolution and use in the United States
	Setup reduction
Oracle	Enterprise resource planning (ERP)
	MRP implementation
Oxford	Bullwhip effect in supply chain
PACCAR, Inc.	ISO 9000/QS 9000 quality standards
Packard Electric	Manufacturing strategy
Panasonic	Agile manufacturing

Organization	Article where discussed
Peoplesoft	Enterprise resource planning (ERP)
	MRP implementation
Pininfarina, Italy	Outsourcing of product design...
Pizza Hut	Product design for global markets
Polaroid	Product design for global markets
Portsmouth Block Mill	History of manufacturing management
Pratt and Whitney	Supplier performance measurement
Procter and Gamble	Global manufacturing rationalization
	Supply chain management
Quaker Oats	Facilities location decision
RCA	Facilities location decision
Reebok	Mass customization and manufacturing
Renault	International manufacturing
RJR Nabisco	Electronic data interchange in supply chain
Roadway Express	Electronic data interchange in supply chain
Ross Controls	Mass customization
Ryan Transport Management Systems	Quality: The implications of Deming's...
Samsonite	Theory of constraints in manufacturing...
Samsung	International manufacturing
SAP	Activity-based costing
	Enterprise resource planning (ERP)
	MRP implementation
Schrader Bellows	Activity-based costing: An evaluation
Seagate Technologies	Facilities location decision
Siemens	Kanban-based manufacturing systems
Skoda	Manufacturing analysis using Chaos, ...
Sony	Virtual manufacturing
	Environmental issues and operations mgmt.
Space Electronics, Inc.	Mass customization and manufacturing
Standard Oil Company	Facilities location decision
Stanford University	Quality: The implications of Deming's...
Subway	Product design for global markets
Sun Microsystems	ISO 9000/QS 9000 quality standards
Sundstrand Corporation	Flexible automation
Surgical Focused Care	ISO 9000/QS 9000 quality standards
Taco Bell	Reengineering and process view of...
Target	Bullwhip effect in supply chain
Tektronics	New product development through...
Telco (India)	Mass customization and manufacturing
Telepad, Inc.	Virtual manufacturing
Tennant	JIT evolution and use in the United States
Texas Instruments	Concurrent engineering
	Learning curve analysis

Organization	Article where discussed
	Reengineering and process view of...
	Supplier performance measurement
The Gap	Logistics: Meeting customer's real needs
Thomas Register	Purchasing: Acquiring the best inputs
Thomson Consumer Electronics	Concurrent engineering
Timkin Company	Performance measurement
Toronto Plastics Ltd.	ISO 9000/QS 9000 quality standards
Toyo Ink	International manufacturing
Toyota	Core manufacturing competencies
	Customer service through system...
	JIT evolution and use in the United States
	Kanban-based manufacturing systems
	Lean manufacturing implementation
	Mass customization and manufacturing
	Product development and concurrent...
	Product innovation
	Supplier relationships
	Target costing
	Total quality management
Toys-R-Us	Supply chain management
Trident Precision Manufacturing	Performance excellence: The Malcolm...
Tyco	Product design for global markets
U.S. Air Force	Flexible automation
	Virtual manufacturing
U.S. Department of Defense	Electronic data interchange in supply chain
	Environmental issues: Reuse and recycling
U.S. Navy	History of manufacturing management
U.S. Postal Service	Performance excellence: The Malcolm...
UCAR Composites	Virtual manufacturing
UNEP	Environmental issues in operations mgmt.
Unilever	Logistics: Meeting customer's real needs
United Airlines	Supplier performance measurement
UPS	Electronic data interchange in supply chain
USCAR	Agile manufacturing
	Virtual manufacturing
Valic	Facilities location decision
Volkswagen	Outsourcing of product design...
	Purchasing: Acquiring the best inputs
Wal-Mart	Bullwhip effect in supply chain
	Electronic data interchange in supply chain
	Logistics: Meeting customer's real needs
	Purchasing: The future
	Supplier partnership as strategy

Organization	Article where discussed
	Supply chain management
	Virtual manufacturing
War Department	Quality: The implications of Deming's...
War Production Board	Quality: The implications of Deming's...
Watertown Arsenal	Scientific management
Western Textile Products	Theory of constraints in manufacturing...
Westinghouse	JIT evolution and use in the United States
Whirlpool	Activity-based costing: An evaluation
World Trade Organization (WTO)	Global facility location analysis
	Product design for global markets
Xerox	Customer service, satisfaction and success
	Environmental issues: Reuse and recycling
	Quality: The implications of Deming's...
	Supplier relationships
	Total quality management
Yamaha	Concurrent engineering
Yamazaki Machinery (Japan)	Capital investment in advanced technology
Yellow Freight	Electronic data interchange in supply chain
Yokohama Corporation (Japan)	Target costing

Foreword

In the last 20 years, we have witnessed the historic resurgence of manufacturing in the United States. This has been due, in part, to the power of emerging ideas in production and manufacturing management. The widespread use of these new developments has made American manufacturers lean, aggressive and successful competitors.

At the National Association of Manufacturers (NAM), we represent and promote the interests of 14,000 manufacturing firms and organizations. Having worked with manufacturing firms for decades, I am convinced that in this age of heightened global competition, the need for manufacturers to be informed and prepared for the challenges of the global economy is more acute and immediate than ever. The *Encyclopedia of Production and Manufacturing Management* is a valuable contribution by Dr. Paul Swamidass to that end. It provides timely, useful data to managers and manufacturing economists on issues ranging from industrial trends to recent progress in manufacturing management. I expect this single volume encyclopedia to become a standard reference work for managers and students of production and manufacturing management.

Dr. Paul Swamidass has collaborated with the National Association of Manufacturers since 1990 in studying the use of manufacturing technology for improving manufacturing processes in the United States. So far, in collaboration with the National Science Foundation, we have jointly completed three such studies, which are now available as the NAM's "Technology on the Factory Floor" series of reports. Industrial leaders and policy makers have used these reports for nearly ten years. Dr. Swamidass' work reflects a clear understanding of the needs of practicing managers and provides them with useful information that enhances their decision-making and professional performance.

Dr. Swamidass is exceptionally qualified to bring a well-rounded perspective to this complex subject. He holds a degree in mechanical engineering and has done graduate work in production management. He has seven years of industrial experience in production management and for decades has taught production management to students in American universities. I have long found that his strong academic background has complemented his real-world grasp of the problems of practicing managers.

This encyclopedia is an important work. It deserves widespread use by everyone who cares about American manufacturing at the dawn of the new century. I am pleased to commend it to persons who work in manufacturing management and to everyone who cares about the future of American industry.

Washington, D.C.
July 18, 1999

Jerry J. Jasinowski
President
The National Association of Manufacturers

Preface

The *Encyclopedia of Production and Manufacturing Management* is a specialized encyclopedia developed to serve as a basic reference resource for the practitioner, researcher, and student. Because of its specialized focus, this one-volume work is able to cover the entire field of production and manufacturing management. It contains factual and conceptual information for fundamental understanding while serving as a starting point for a deeper researched investigation. The material is state-of-the-art, covering the field of operations management and its exciting recent developments. These developments are covered extensively in this volume.

CONTENT AND PURPOSE: In the past twenty years, the field of production and operations management has grown with incredible speed, stretching its boundaries in all directions. For example, in the last two decades, production and manufacturing management absorbed in rapid succession several new production management concepts, including manufacturing strategy, focused factory, just-in-time manufacturing, concurrent engineering, total quality management, supply chain management, flexible manufacturing systems, lean production, and the list goes on. These considerable changes and developments in manufacturing highlight the critical need for an efficient, authoritative reference tool for manufacturing management students and practicing managers. Today's manufacturing managers are now expected to think more broadly than their counterparts two to three decades ago. The most notable change has been the need for manufacturing managers to think in technological, strategic and competitive terms. As a matter of record, technological and strategic developments have been the instrumental factors in the resurgence in manufacturing worldwide. The entries in this encyclopedia focus on these on-going technological and strategic changes in production and manufacturing management.

In addition to the technological advances in manufacturing, a special feature of this encyclopedia is the number of successful manufacturers and manufacturing organizations illustrated and examined in the encyclopedia. Throughout the volume, "real-world" examples are drawn from more than 100 international firms whose business operations originate from a variety of countries in Asia, America, and Europe. Moreover, the practices of an array of major manufacturers, including Chrysler, Ford, GM, and Toyota are examined in a number of the topical entries. To assist the reader in locating these operational examples, the table provides an easy-access list of the references of where these "real world" examples can be found in the encyclopedia.

Added to the encyclopedia's entries are two valuable appendixes, each with unique bibliographies. Appendix I is a bibliography organized alphabetically and includes writings on all topics covered by the encyclopedia. Appendix II is a topical bibliography covering 21 broad topics from Capacity Planning to Supply Chain. The second bibliography can greatly speed the search for publications on a given topic. These two appendixes should serve as valuable research tools.

ORGANIZATION: The encyclopedia's topical treatments vary in length. The longer articles on important concepts and practices range from three to ten pages. There are about 100 such articles written by nearly 100 authors from around the world. In addition, there are over 1000 shorter entries on concepts, practices and principles. The range of topics and depth of coverage is designed to fit the needs of students and professional managers. The shorter entries provide digests of unfamiliar and complicated subjects. Difficult subjects are made intelligible to the reader without oversimplification.

While some entries – because of their special nature – present a free-form discussion of the topic, the majority of the encyclopedia's entries are structured to provide the following: *introduction or description, historical context of the topic, seminal works, a strategic perspective, a technological perspective, significant analytical models, short examples or cases from real-life, implementation issues and references.* The material in these articles is arranged in sections using the titles below:

Content	Title of the Section
What is it?	Description or Introduction
How did it evolve and seminal works?	Historical perspective
Why is it important?	Strategic perspectives
How is this practiced or implemented?	Implementation
What technologies are essential?	Technology perspective
Problem solving with analytical models	significant analytical models
Effect on performance	Effect
When is it appropriate?	Timing
Where is it used or practiced?	Location
What results and problems to expect?	Results
Who uses them?	Cases
Caveats	Collective wisdom
References to Seminal works	References

A variety of entries examine a strategic perspective on the topic. *The Strategic Perspective* includes some or all of the following:

1. Long-term implications of the process/issue.
2. Its contribution to the organization's competitiveness.
3. The competitive advantages bestowed by the process or practice to the user.
4. Examples, short cases and illustrations.

There are topical entries that require a technological examination of the topic. *The Technological Perspective* includes some or all of the following:

1. The nature and magnitude of investments in specific technologies needed.
2. Historical evolution of the technology.
3. Recent advances in the technology.
4. Benefits of the technology, and illustrative cases.

Driving manufacturing competitiveness has been the strategic and technological responses of manufacturing firms to the increasing level of world wide competition. These two factors – increasing competition and the strategic implementation of manufacturing technology – are inextricably tied together. Hence strategy and technology are examined in the longer entries of the encyclopedia. In addition with the increasing use of hard and soft manufacturing technologies, examining the technological perspectives in appropriate topical entries provides an essential framework for understanding the "on-going technological revolution" in manufacturing and its future developments.

This encyclopedia is an organized summary of basic knowledge and important information in production and manufacturing management. One of the goals of this encyclopedia is to make the large volume of material easily accessible to its users. A network of cross-references enables the reader to start from a topic and move speedily to several related topics. Entries are arranged alphabetically (letter-by-letter) for ease of access with each entry ending with generous cross-references. An Index of all entries is provided at the beginning of the encyclopedia, followed by an alphabetical list of longer articles and their authors.

For those interested in pursuing a topic beyond the encyclopedia, longer articles contain several references to seminal and authoritative books and articles. All these references and additional reading materials are listed alphabetically in a comprehensive Bibliography in Appendix I. Appendix II is a topical bibliography covering 21 broad topics. Appendix III is an information resource of selected journals and periodical on production and manufacturing management.

The combination of my many years of teaching university level production and manufacturing management courses and my work with numerous manufacturing firms has underscored to me the need for a functional reference tool for the field. I hope that our efforts in producing the *Encyclopedia of Production and Manufacturing Management* represent an important step in satisfying this need.

CONTRIBUTING AUTHORS AND BOARD OF ADVISORS: The Advisory Board, composed of internationally known experts, provided guidance during the development of the encyclopedia. Collectively, the Board of Advisors for this encyclopedia represents hundreds of years of experience in the research, teaching and practice of production and manufacturing management. The Board's members have published numerous books on the subject, and they know the field intimately. Their input has measurably improved this volume. This distinguished Board of Advisors provided input in selecting the encyclopedia's topical treatment based on each topic's importance, usefulness and currency. Our objective was to be as comprehensive as possible within a one-volume framework. We have made choices realizing that this is a fast moving field and on its horizon there are new developments underway that will need to be addressed in future editions.

Through this encyclopedia, I am delighted to present to the reader the works of over one hundred experts and scholars. The authors of articles are international experts with a record of research, teaching and publications on the topic. The scholarship of the authors ensures the reliability of the entries. It was my privilege to work with each author.

It was a pleasure to work with Gary Folven, OR/MS publisher, who was also associated with the *Encyclopedia of Management Science and OR*. His experience with encyclopedia development and production provided me with the necessary assurance to undertake the project. Nevertheless, preparing an encyclopedia is a formidable undertaking but I was fortunate to have access to Linda Patillo's skills at Auburn, and the talents of Carolyn Ford and Kristin Piper at the editorial offices of Kluwer.

Paul M. Swamidass
Professor of Operations Management
Auburn University

3 "R"s

*R*educe, *R*euse and *R*ecycle are three actions that firms can use to manage the environmental impact of solid wastes. In particular, 3Rs lower the amount of solid and hazardous waste that requires disposal. First, *reduction* should be pursued, whereby the consumption of materials and energy is minimized and the generation of pollutants is lowered. Second, products should be *reused*, through repair or reconditioning if necessary, to lengthen their usable life. Finally, *recycling* should be pursued, whereby a product or material is collected, processed and utilized in another product or service. While recycling reduces the consumption of virgin materials, additional environmental and economic costs may be associated with waste collection and recycling.

See **Environmental issues and operations management.**

8-4-8-4 SCHEME

It is method of scheduling the plant to work with two eight hour shifts separated by two four hour time blocks for maintenance, tool change, restocking, long setups and so on.

See **Integrated preventive maintenance; Lean manufacturing implementation; Maintenance management models; Preventive maintenance; Total productive maintenance.**

ABC

See **Activity-based costing.**

ABC ANALYSIS OR ABC CLASSIFICATION

In marketing, since the 1920s, firms have sought to differentiate customers on the basis of selective grouping. ABC classification is one such attempt which is based on the well-known "Pareto Principle." The Italian Vilfredo Pareto observed in his analysis of countries and wealth that regardless of the country studied, a small percentage of the population controlled a large percentage of the wealth. This observation has been generalized into numerous settings and is frequently known as the 80-20 principle. In marketing, it has been observed that a small percentage (say 20%) of the customers account for a large percentage (80%) of an organization's sales volume, profitability or some other measure of activity/result. Application of the 80-20 rule is typically known as ABC classification by which three (or more) classes of customers are identified. For example, "A" customers may be identified as the top 10 percent of customers who contribute 70-80 percent of the firm's sales volume; "B" customers are the next 20 percent of customers who contribute approximately 10-15 percent of volume; "C" customers are the remaining 70 percent who contribute only 10-20 percent of volume. While the percentages may vary somewhat from company to company and the specific criteria may differ, the principle of ABC classification has wide acceptance because it allows management to concentrate on the areas of highest potential payoff.

The classification of the items in inventory in decreasing order of annual dollar volume (price multiplied by projected volume) or other criteria is employed in inventory management for the differential treatment of items in inventory. This array is then split into three classes, called A, B, and C. The A group usually represents 10% to 20% by number of items and 50% to 70% by projected dollar volume. The next grouping, B, usually represents about 20% of the items and about 20% of the dollar volume. The C class contains 60% to 70% of the items and represents about 10% to 30% of the dollar volume. The ABC principle states that efforts and money can be saved through the use of looser controls to the low-dollar-volume class of items than the controls applied to high-dollar-volume class of items. The ABC principle is applicable to inventories, purchasing, sales, etc.

See **ABC classification; Absolute lateness; Customer service; Safety stocks: Luxury or necessity; Sequencing rules.**

References

[1] Bowersox, D.J., Calantone, R.J., Clinton, S.R., Closs, D.J., Cooper, M.B., Droge, C.L., Fawcett, S.E., Frankel, R., Frayer, D.J., Morash, E.A., Rinehart,

L.M., and Schmitz, J.M. (1995). *World Class Logistics: The Challenge of Managing Continuous Change*. Council of Logistics Management, Oak Brook, IL.

[2] Leenders, M.R. and Fearon, H.E. (1998). *Purchasing and Supply Management* (11th ed.). Irwin, Inc., Homewood.

ABSORPTION COSTING

Manufacturing costs incurred in the production of goods can be divided into fixed costs and variable costs. Fixed costs remain the same when the volume of production changes. Variable costs vary directly with the level of production. Absorption costing is a cost accumulation and reporting method that treats all costs of manufacturing (direct materials, direct labor, variable overhead, and fixed overhead) as inventoriable or product costs. Fixed costs are most commonly allocated to units of output in multiples of direct labor hours, machine hours, or material costs. This cost allocation procedure is then called absorption costing (or full costing or traditional costing) because each unit produced absorbs a portion of all manufacturing costs incurred. It is the traditional approach to product costing and is used for external financial statements and tax returns. That is, under generally accepted accounting practices, fixed costs are allocated to the output. The result is, like variable costs, fixed manufacturing costs are treated as product costs.

See **Activity-based cost management; Activity-based costing: An evaluation.**

References

[1] Barfield, Jesse T., Cecily A. Raiborn, and Michael R. Kinney (1994). *Cost Accounting: Traditions and Innovations*, West Publishing Company, St. Paul, Minnesota.

[2] Henke, Emerson O. and Charlene W. Spoede (1991). *Cost Accounting: Managerial Use of Accounting Data*, PWS-Kent Publishing Company, Boston, Massachusetts.

ABSORPTION COSTING MODEL

Full absorption costing is the accounting approach required in the United States by Generally Accepted Accounting Principles (GAAP) in preparing public financial statements. Essentially, absorption costing is an accounting method that assigns all types of manufacturing costs (direct material, direct labor, fixed and variable overhead) to units produced. Absorption costing occurs as long as *all* costs related to production are allocated to (i.e., absorbed by) the units produced for inventory and eventual sale. Regardless of the specific accounting procedures employed, absorption costing generally takes places in both traditional unit-based costing systems and activity-based costing (ABC) systems.

See **Activity-based cost management; Activity-based costing; Unit cost model.**

ACCEPTABLE QUALITY

It refers to the quality which the customer (either internal or external) finds acceptable. What constitutes acceptable quality can only be determined by the customer. Often there is a formal agreement between the producer and customer as to specific quality criteria which must be adhered to for products produced or services provided. Quality criteria may include tolerance limits, operating characteristics, color clarity and sharpness, metal hardness, structural integrity, reliability, durability, and a host of other attributes deemed critical by the customer.

When a large number of products are shipped to a customer, acceptance or rejection of the shipment may be based on a statistical sample. Then, the acceptable quality level is often expressed as a percentage of defects in the sample. For example, a sample of 100 units may be permitted two or fewer defective items in order for the entire shipment to be accepted. The number of defects allowable depends to a great extent on the size of the sample and ultimately what the customer will accept.

See **Implications of JIT manufacturing; Quality assurance; Total quality management.**

ACCEPTANCE SAMPLING

Acceptance sampling is a quality control procedure, which uses the inspection of small samples instead of 100 percent inspection in making the decision to accept or reject much larger quantities, called a lot. This is a statistical procedure, which uses random samples, that is, each item in the lot has an equal chance of being a part of the sample that is inspected.

In its simplest form, If the sample from a larger lot has an acceptable level of defects, it will be accepted. If not, the entire lot will be rejected. If acceptance sampling is used prior to accepting goods from a supplier (i.e., incoming inspection), then the acceptable level of defects must be agreed between the supplier and customer because the supplier may have to take back the entire lot if it fails acceptance sampling. A sampling plan establishes the rules guiding the sampling and the criteria for accepting or rejecting the lot.

Making a decision about the entire lot based on an inspection of a smaller sample involves risks to the buyer and seller. The probability of rejecting an acceptable lot is called producer's risk, also called alpha or Type I error. The probability of accepting a lot that should have been rejected is called consumer's risk, also called beta or Type II error.

When acceptance sampling is used, the management decides on an Acceptable Quality Level, or AQL. It is the acceptable proportion of a lot that is defective. For example, AQL may be set as, two defects in a lot of 200, or one percent. Since 100 percent is not inspected, there will be lots that will not conform to AQL. The upper limit for such lots is agreed between the buyer and seller and it is called Lot Tolerance Percent Defective (LTPD). Sampling plans are created with AQL and LTPD in mind. The performance of a given sampling plan is represented by an Operating Characteristics Curve (OC). It plots the probability of accepting lots with all possible defect levels given a sample size (n) and acceptance level (c).

Acceptance sampling will pass on some defective items to the customer. It varies with the sampling plan selected. The expected number of defective items passed on to the customer is called Average Outgoing Quality (AOQ). The maximum value of AOQ (note, lower AOQ is better) for various values of defective proportions in the lot is called Average Outgoing Quality Limit (AOQL). If this limit is not acceptable to the management because it happens to be too high, the sampling plan should be changed until the AOQL is satisfactory.

The U.S. Department of Defense publishes acceptance sampling plans, commonly known as MIL-STD-105E. Others publishing such standards are the American National Standards Institute (ANSI) and American Society for Quality Control (ASQC).

See **All or none sampling; Operating characteristics curve; Statistical process control using control charts.**

Reference

[1] Montgomery, D.C. (1991). *Introduction to Statistical Quality Control*, 2nd edition, New York: Wiley and Sons.

ACCOUNTING SYSTEM IMPLICATIONS OF TOC

Monte R. Swain and Stanley E. Fawcett

Brigham Young University, Provo, Utah, USA

DESCRIPTION: The theory of constraints (TOC) is a management philosophy that identifies and leverages constraints in order to maximize the profit potential of a company. The intelligent management of a constraint (sometimes called *bottlenecks* in the case of resource constraints) involves the following five steps: (1) identify the system's constraint, (2) decide how to exploit the system's constraint, (3) subordinate everything else to the above decision, (4) elevate the constraint, and (5) if a constraint has been broken, go back to step 1 (Goldratt, 1990). Subordinating all other processes to the bottleneck is accomplished using a Drum-Buffer-Rope approach. Specifically, the "drum beat" of the bottleneck sets the production pace for the rest of the operation; processes upstream from the bottleneck are held in check by a "rope-generated" schedule of material releases; and a minimal "buffer" of inventory is kept in front of the bottleneck in order to keep the factory always running at full capacity (Goldratt and Fox, 1986), if demand requires it.

Throughput Accounting

Throughput accounting is specifically designed to support TOC-based management. This system has its own definition of some key accounting terms and its own profit and loss (P&L) statement. Throughput accounting defines throughput value as sales revenue less cost of raw materials expenditures (and out-of-pocket selling expenses, if any). Operations cost is the sum of all expenditures on production and administration activities other than expenditures on raw materials. Maximizing operating profit is primarily a function of maximizing throughput value per unit of the bottleneck consumed. However, TOC recognizes that profit is also a function of operations cost. Thus all costs in the system must be managed.

The throughput accounting view of profit management is significantly different from both financial accounting as defined by Generally Accepted Accounting Principles (GAAP) and the traditional managerial accounting approach, which emphasizes contribution margin analysis. GAAP-based P&L statements can promote overproduction of inventory. Contribution margin P&L

Exhibit 1. Contrasting assumptions of GAAP-based requirements, contribution margin accounting, and throughput accounting.

GAAP assumptions	Contribution margin assumptions	Throughput accounting assumptions
• All direct labor and direct material costs are specifically assigned to inventory.	• Only variable product costs (direct labor, direct material, and variable manufacturing overhead) are specifically assigned to inventory.	• No costs (including direct materials costs) are specifically assigned to inventory.
• All manufacturing overhead costs are allocated to inventory using a predetermined overhead application rate.	• Rather than allocating to inventory, fixed manufacturing overhead costs are immediately expensed to the income statement.	• All product costs are immediately expensed to the income statement.
• Inventory costs are not expensed to the income statement until the inventory is sold.	• Variable product costs are not expensed to the income statement until the inventory is sold.	

statements assume that variable costs such as direct labor and some overhead are avoidable within a 6 to 12 month period of time. Conversely, throughput accounting P&L statements function as tools for short-term (6 to 12 weeks) management of operations and assume that only expenditures for direct material are truly variable within this shorter time frame. Essential differences between these three accounting procedures are summarized in Exhibit 1.

Throughput accounting systems focus on the efficiency and capacity of the entire organization, rather than on individual processes and work centers. Only bottleneck processes are managed to maximize individual efficiency and output volume. The performance of non-bottlenecks is measured by how well they smoothly interface with the bottleneck to keep it operating efficiently without unnecessary buildup of work-in-process inventory. *Throughput accounting is intended as an operations management system*. Therefore, product costing and strategic cost management will require additional accounting systems, e.g., activity-based costing (ABC) systems.

HISTORICAL PERSPECTIVE: In 1984, TOC was introduced to the world in Eliyahu Goldratt's novel/textbook *The Goal*. Since then, Goldratt's ideas involving throughput, constraint management, and the Drum-Buffer-Rope scheduling system have been used by many companies. Since about 1988, the term "throughput accounting" has been used to identify the assemblage of accounting systems, reports, and performance measures used to support the implementation of TOC in an organization (Waldron, 1988). But throughput accounting does not necessarily replace more traditional managerial accounting methods. *Throughput accounting is used to support a very specific managerial view of an operation*. The TOC management technique concentrates on increasing the effective use of existing costs and asset structures to maximize throughput in the organization. The focus of throughput accounting is on the incremental value of more effective use of the constraint resource, rather than on the underlying cost or operation structures of non-constraint processes. Overall, *throughput accounting concentrates on extremely short-term relevant costs and capacity*. It does not necessarily provide data useful for other important management needs, such as cost reduction, innovation (particularly within non-bottleneck processes), and long-term marketing strategy (Mackay and David, 1993).

IMPLEMENTATION: When Goldratt first introduced TOC, he redefined three accounting terms: throughput, operations cost, and inventory (Noreen *et al.*, 1995). These new definitions, and how they differ from conventional definitions, must be understood in order to appreciate the distinction between throughput accounting and traditional cost accounting systems, and how new performance measures are created to support a TOC management model.

Throughput

The TOC version of throughput is the amount of money the system generates through sales (which is carefully distinguished from finished and unsold output). It is generally measured as sales revenue less direct materials. The TOC version of throughput is different from the traditional definition, which focuses generally on output in terms of units over time. Note that TOC throughput is closely related to the traditional measure of contribution margin: sales revenue less variable costs. Yet, there are important differences between throughput and contribution margin. The TOC throughput measure assumes that the only truly variable costs are direct materials (and perhaps some out-of-pocket selling costs such as sales commissions). The contribution margin measure, on the other hand, typically categorizes direct labor and some manufacturing overhead as variable costs. Time is the essential difference between these two measures (Holmen, 1995). As mentioned earlier TOC is very focused on maximizing short-term results – typically measured within a few weeks. The contribution margin definition of variable costs is based on a 6- to 12-month time frame. Generally, over two to six weeks period, management cannot or reluctant to vary the level of direct labor and manufacturing overhead costs in a plant. Only with careful planning backed by effort over time, it possible to adjust expenditures on direct labor and overhead without causing negative impacts on the organizations culture and strategic direction.

Operations Cost

Operations cost is all the money the organization spends to turn inventory into throughput. Essentially, operations cost is somewhat analogous to the traditional accounting definition of conversion costs: direct labor and manufacturing overhead. In addition, operation expense includes selling and general administrative expenses. Because of its short-term focus on effectively using constraints to maximize cash flow, TOC considers operations cost to be a fixed cost. In most manufacturing operations, the only costs that can vary significantly in the brief window of a few weeks are the costs of raw materials.

Inventory

The difference between conversion costs and operations cost is important to understand in light of how inventory value is measured in TOC. GAAP-based financial reporting guidelines require that conversion costs be allocated to both work-in-process and finished goods inventories. This allocation effectively keeps many costs off the P&L statement until products are actually sold. Conversely, under throughput accounting, all production-related costs consumed during the production period are expensed immediately to the P&L statement, regardless of whether the goods in production have been completed or sold. Inventory values in TOC consist solely of the original acquisition costs. No value, such as labor and overhead, is added during production. In fact, more extreme versions of throughput accounting will immediately expense all inventory acquisition costs to the income statement as well. To immediately expense all product costs (materials, labor, and overhead) regardless of inventory status violates GAAP. *TOC, however, does not emphasize accurate cost tracking for reporting purposes*. The emphasis of TOC and throughput accounting is on exploiting the system constraints to generate throughput. Given these GAAP violations, it is obvious that throughput accounting does not replace the organization's financial accounting system. As a result, many companies often run multiple types of information systems to support various needs of the organization (Constantinides and Shank, 1994).

Technology Perspective: Goldratt generally advocates simplified technology to manage a TOC-based operation. For example, Goldratt (1990) suggests labeling bottleneck output with gold ribbons because they must be handled "like gold." Nevertheless, the Drum-Buffer-Rope method of managing manufacturing operations can be implemented with some technological sophistication. For example, continually updated scheduling systems can release raw materials precisely (i.e., the rope) in order to keep an adequate level of stock (i.e., the buffer) in front of the bottleneck. Similar scheduling systems (i.e., the drum) can also be established to help operations downstream from the bottleneck adequately anticipate arrival of work-in-process and finished goods. Is should be understood that non-constraint operations in TOC theory are not typically schedule driven, unless they are a divergence point, an assembly operation, or a capacity-constraint resource.

Linear programming tools can be used as a first step to a solution for establishing a TOC-based operation. However, linear programming by itself is not an effective TOC management tool. Determining the mathematical solution to a constraint problem is one thing; implementing that solution into an actual organization is an entirely different matter. Using throughput accounting as its information system, TOC implements a solution and then manages the process containing one or more constraints using the Drum-Buffer-Rope system.

Exhibit 2. Comparing three different profit and loss statements.

GAAP Basis		Contribution Margin Basis		Throughput Basis	
Revenue	$500,000	Revenue	$500,000	Revenue	$500,000
Cost of Goods Sold	215,000	Variable Cost[a]	275,000	Direct Materials	110,000
Gross Margin	$285,000	Gross Margin	$225,000	Gross Margin	$390,000
Selling & General Administrative Exp.	250,000	Fixed Costs[b]	190,000	Operating Expense[c]	355,000
Operating Income	$35,000	Operating Income	$35,000	Operating Income	$35,000
First Priorities	↓Cost		↓Cost		↑Throughput
Second Priorities	↑Inventory		↑Throughput		↓Inventory
Third Priorities	↑Throughput		↓Inventory		↓Cost

[a]Direct Materials + Direct Labor + Variable Manufacturing Overhead and Variable Selling & General Administrative Expense.
[b]Fixed Manufacturing Overhead and Fixed Selling & General Administrative Expense.
[c]All costs in the organization other than Direct Materials.

EFFECTS: The implementation of throughput accounting can be a dramatic departure from classic standard cost systems in traditional manufacturing settings. The goal of a TOC system is to enable the bottleneck resource produce at full volume as effectively and efficiently as possible. This, in fact, is quite similar to a standard cost system. However, the goals for and management of non-bottleneck processes differ greatly in a throughput accounting system versus a standard cost accounting system.

Adjusted View of Standard Cost Systems

The TOC goal for a non-bottleneck process is to make it effectively fit into the overall organization by supporting smooth production flows through non-bottleneck processes that match the bottleneck's capacity. Since the bottleneck resource sets the capacity for all the other processes, throughput accounting has a system-wide perspective. It does not stress decentralized measures such as individual work center efficiency and volume variances for individual work center in the organization (Mackey and Davis, 1993). Since non-bottleneck processes are expected to operate at less than full capacity and at less than perfect efficiency, interpretation of both efficiency and volume variances of these processes are viewed as suspect.

Emphases of the P&L Statement

With new definitions of throughput and operations cost, throughput accounting provides a new P&L statement that is significantly different from the GAAP-approved income statement and the traditional contribution margin statement used by managerial accountants for the last several decades. Exhibit 2 contrasts these three P&L statements. Note that the revenues and operating incomes are the same across all three formats; only cost categorization is changed. It would seem that the differences between these three formats of the P&L statement are not that significant. What is important, though, is how these three formats affect the priority placed on managing costs, inventory, and throughput (Constantinides and Shank, 1994; Dugdale and Jones, 1996).

When production managers are oriented on GAAP-based P&L statements, their emphasis is first on decreasing cost, second on increasing inventory (in order to delay cost recognition on the P&L statement), and third on increasing throughput. Similarly, traditional contribution margin statements give top priority to cost reduction. Next, emphasizing contribution margin will naturally lead managers to focus on increasing the volume (i.e., throughput) moving through the plant. The third item of emphasis for managers using contribution margin statements will be inventory. However, in contrast to the GAAP orientation, the contribution margin orientation (as well as the throughput orientation) creates an appropriate emphasis on reducing (rather than increasing) inventory. Finally, throughput P&L statements encourage an emphasis first on increasing throughput, second on decreasing inventory, and third on decreasing costs. The different strategic priorities resulting from all three formats are summarized at the bottom of Exhibit 2.

Exhibit 3. Performance measures for throughput accounting (adapted from Galloway and Waldron 1988, 1989a, 1989b).

If:

Throughput return per factory hour $= \dfrac{\text{Product sales price} - \text{Product material cost}}{\text{Product's time on bottleneck process}}$

And:

Operations cost per factory hour $= \dfrac{\text{Total operations costs in factory}}{\text{Total factory time at the bottleneck process}}$

Then:

Throughput accounting (TA) ratio $= \dfrac{\text{Throughput return per factory hour}}{\text{Operations cost per factory hour}}$

(This ratio could be calculated per product item, product batch, or product line.)

Primary ratio (T/TFC) $= \dfrac{\text{Total sales revenue} - \text{Total material cost}}{\text{Total operations cost in factory}}$

(This ratio could be calculated either on a total factory basis or on a department basis.)

Cost per bottleneck minute ratio (C/PB) $= \dfrac{\text{Total operations cost in facility}}{\text{Bottleneck capacity (minutes)}}$

(This ratio is useful in a complex environment with multiple manufacturing facilities, each with its own bottleneck process.)

STRATEGIC PERSPECTIVE: As Exhibit 2 indicates, the strategic focus of TOC is on first increasing throughput, second on reducing inventory, and third on reducing cost. This contrasts with conventional accounting, which emphasizes costs reduction first. TOC's priority ranking is consistent with Just-in-Time (JIT) manufacturing which emphasize inventory reduction first (Dugdale and Jones, 1996). The success of a TOC operation requires that the throughput accounting system be developed and used to disseminate new measures of performance that focus on throughput. Galloway and Waldron (1988, 1989a, 1989b) suggest several measures to support a throughput emphasis in the plant. These measures are illustrated in Exhibit 3.

It is important to note that inappropriate use of the TOC performance measures illustrated in Exhibit 3 can lead to a throughput accounting system that actually emphasizes cost reduction over throughput increase. For example, improving the TA ratio, the T/TFC ratio, or the C/PB ratio requires that management avoid a typical strategy of first reducing operations costs in the factory. There is a limit to the improvement attainable using a cost reduction strategy. Further, some cost-reduction strategies may actually contribute over time to situation that limits or reduces throughput! On the other hand, throughput can, in principle, be increased without limit (Dugdale and Jones, 1996).

RESULTS: The bottleneck resource is the management focus in a TOC manufacturing operation. Because of its central importance to the organization's overall capability to make money, the bottleneck must necessarily receive the lion's share of management effort. Schedules must be well developed and controlled for optimal use of bottleneck resources. On the other hand, there is much more flexibility in the management of non-bottleneck resources. Operating budgets and schedules for non-bottleneck resources need not be nearly as precise. For example, making an error in the processing time on a non-bottleneck is not as large a concern as the same error on a bottleneck process. Further, in using the Drum-Buffer-Rope scheduling system, the traditional variances (such as labor efficiency and volume variances) do not apply to non-bottleneck processes.

COLLECTIVE WISDOM AND CAVEATS: Throughput accounting is strictly a horizontal system-wide view of the organization. The efficiency or effectiveness of any single employee or production process is not important, except to the extent that the employee or process supports efficient exploitation of the bottleneck to maximize throughput. Many organizations suffer from a strict functional view such that each production process is expected to generate as much high-quality goods or services as it is capable of producing. These organizations fail to ask questions such as "Are we producing the right products?" or "How effectively are the products moving through the production process?" Throughput accounting addresses these questions and allows the organization to design the management process around a constraint to maximize the system's capability to make money.

Use of Throughput Accounting for Product Costing

Throughput accounting will not provide the information management needed to make all decisions within the organization. Throughput accounting is not a cost system. It does not support the allocation of product costs (which an ABC system does do). Further, throughput accounting will not integrate with the GAAP-based general ledger system. Hence, if product costing and reconciliation with financial accounting standards are important to management, the organization will also need to invest in traditional costing or ABC systems too.

Excessive Budget Cuts

When improperly used, a throughput accounting system can create problems. For example, the excess of capacity in non-bottleneck processes may tempt management to make severe budget cuts within these processes, particularly in times of economic stress (Mackey and Davis, 1993). Understandably, employees in these non-bottleneck work centers may become concerned about their job security. Before workers are released and production capacity is scaled back, management needs to remember that in most operations it is impossible to balance individual process capacities. Every process has inherently different capacities, and bottlenecks are inevitable. Hence, releasing non-bottleneck workers in order to reduce capacity is non-bottleneck resource to match the bottleneck resource can remove the necessary flexibility in the enterprise to effectively implement a TOC system. Furthermore, employees are not likely to support a throughput accounting system if they fear the loss of their job.

Lack of Focus on Non-Bottlenecks

TOC implies that only by improving throughput through the constraint operation can the organization improve revenue and profits. Hence, throughput accounting focuses absolutely on management of bottleneck processes. The need to improve existing non-bottlenecks is also formally recognized in TOC. Under a Drum-Buffer-Rope scheduling system, the management of non-bottlenecks should be highly focused on ensuring optimal output at the bottleneck by emphasizing tightly scheduled, high-quality output at both upstream and downstream non-bottleneck operations. With proper attention to non-bottleneck operations, product quality and process productivity can be improved throughout the company. Therefore, a throughput system must include performance measures specific to non-bottlenecks that encourage continuous improvement of non-bottleneck operations. With a universal emphasis on improvement both bottleneck and non-bottleneck operations can be directly tied to performance that impacts the organization's competitive edge in its market.

Finally, external sources of constraints on revenue and profitability such as supplies of raw materials or demand for finished goods must be regularly evaluated and improvement opportunities must be considered.

See **Bottleneck resource; Contribution margin analysis; Cost per bottleneck minute (C/PB) ratio; Drum-buffer-rope; GAAP-based financial reporting guidelines; Operations cost; Primary ratio (T/TFC); Theory of constraints; Theory of constraints in manufacturing management; Throughput; Throughput accounting (TA) ratio.**

References

[1] Constantinides, K., and J.K. Shank (1994). "Matching accounting to strategy: One mill's experience," *Management Accounting* (USA, September), 32–36.

[2] Dugdale, D., and D. Jones (1996). "Accounting for throughput: Part 1 – the theory," *Management Accounting* (UK, April), 24–29.

[3] Galloway, D., and D. Waldron (1988). "Throughput Accounting: Part 2 – ranking products profitably," *Management Accounting* (UK, December), 34–35.

[4] Galloway, D., and D. Waldron (1989a). "Throughput Accounting: Part 3 – a better way to control labour costs," *Management Accounting* (UK, January), 34–35.

[5] Galloway, D., and D. Waldron (1989b). "Throughput Accounting: Part 4 – moving on to complex products," *Management Accounting* (UK, February), 34–35.

[6] Goldratt, E.M. (1990). *The Haystack Syndrome: Sifting Information out of the Data Ocean,*

North River Press, Croton-on-Hudson, New York.

[7] Goldratt, E.M., and J. Cox (1984). *The Goal: A Process of Ongoing Improvement*. North River Press, Croton-on-Hudson, New York.

[8] Holmen, J.S. (1995). "ABC vs. TOC: It's a matter of time," *Managament Accounting* (USA, January), 37–40.

[9] Mackey, J.T., and C.J. Davis (1993). "The Drum-Buffer-Rope application of the theory of constraints in manufacturing: What is it and what does it do to your accounting system?" *Proceedings of the 1993 American Accounting Association Annual Convention*, San Francisco, CA.

[10] Noreen, E., D. Smith, and J.T. Mackay (1995). *The Theory of Constraints and its Implications for Management Accounting*. North River Press, Great Barrington, MA.

[11] Waldron, D (1988). "Accounting for CIM: The new yardsticks," *EMAP Business and Computing Supplement* (February), 1–2.

ACTION BUCKET

This is part of material requirements planning (MRP) terminology. A bucket refers to a unit of planning period such as a week. In a *manufacturing planning and control* (MPC) system, planned order release data are produced according to gross requirement, scheduled receipt, and projected available data for the period, usually a week. A planned order is said to be in the action bucket if it is meant for the current or immediate period. An item in an action bucket requires an action by the planner. The actions may include, order release, rescheduling, or cancellation.

See **Bucketless material requirements; Bucketless system; Manufacturing planning and control; Manufacturing planning system; Material requirements planning (MRP); Planning order release.**

ACTION FLEXIBILITY

The capability of a manufacturing system with a high degree of action flexibility enables it to take a set of new actions to respond effectively (i.e., speedily and cost-efficiently) to changes in its internal and external operating environment. For example, a manufacturing system can effectively change the product or component design, the process, the materials used, etc. in response to changes in the environment.

See **Flexible automation; Manufacturing flexibility; Manufacturing flexibility types and dimension.**

ACTIVITY DRIVER

A cost driver is a factor that is presumed to have a direct cause-effect relationship to manufacturing related costs. For example, production volume has a direct effect on the total cost of raw material used and can be said to "drive" that cost. Thus, production volume could be considered to be a predictor of that cost. However, in most other cases, the cause-effect relationship is less clear since costs are commonly caused by multiple factors. For example, quality control costs are affected by a variety of factors such as production volume, quality of materials used, skill level of workers, and level of automation. Although it may be difficult to determine which factor actually caused a specific change in quality control cost, any one of these factors could be chosen to predict quality cost if confidence exists about the factor's relationship with quality cost changes. To be used as a predictor, it is only necessary that the predictor variable and the cost both change together in a predictable manner. Traditionally, a single cost driver has been used to allocate costs to cost objects, such as products or customers.

Activity-based costing is a cost accounting system that attempts to allocate costs recorded in the general ledger accounts to cost objects in a two-stage allocation system. In the first stage, costs are accumulated in activity center cost pools using first-stage cost drivers that reflect the appropriate level of cost incurrence. An activity center is a segment of the production process for which management wants separate cost collection and reporting for the activities performed. The costs from the activity center cost pools are then allocated to the appropriate cost objects using a second-stage cost driver that attempts to quantify the amount of the activity consumed by the cost object. This second-stage cost driver is known as an activity driver.

See **Activity-based costing; Activity-based costing: An evaluation; Cost drivers; Cost objects; Target costing; Two-stage allocation process.**

Reference

[1] Barfield, Jesse T., Cecily A. Raiborn, and Michael R. Kinney (1994). *Cost Accounting: Traditions and Innovations*, West Publishing Company, St. Paul, Minnesota.

ACTIVITY ON ARROW NETWORK (AOA)

See **Critical path network; PERT; Project management.**

ACTIVITY ON NODE NETWORK (AON)

See **Critical path network; PERT; Project management.**

ACTIVITY-BASED COST MANAGEMENT

It is a formal analysis and management of activities associated with a product or service. It's goal is to enhance activities that add value to the customer and reduce activities that do not. It attempts to satisfy customer needs while making smaller demands on costly resources.

See **Activity-based costing; Activity-based management; Balanced scorecard.**

ACTIVITY-BASED COSTING

Monte R. Swain and Stanley E. Fawcett

Brigham Young University, Provo, Utah, USA

DESCRIPTION: Activity-based costing was created in response to the problem Johnson & Kaplan noticed. Johnson and Kaplan (1987:1) argued that, today's management accounting information, driven by the procedures and cycle of the organization's financial reporting system, is too late, too aggregated, and too distorted to be relevant for managers' planning and control decisions.

The Traditional Cost Model

The classic product costing model, in harmony with Generally Accepted Accounting Principles (GAAP) of the United States, maintains that

cost of products = direct materials
+ direct labor
+ manufacturing overhead

Typically, the most difficult challenge facing cost accountants involves attributing the manufacturing overhead costs to the product. The traditional approach of assigning overhead costs to products uses a two-stage process. The first step allocates costs of manufacturing support centers to production centers, then combines all allocated and overhead costs in production centers to compute an overhead burden rate based on some input activity (such as direct labor hours or machine hours). The traditional model (Unit-based costing) may be characterized as:

Costs in Support Centers
→ Costs in Production Center
→ Costs assigned to Products

The ABC Model

The traditional two-stage allocation of overhead costs seems to indicate that products consume or incur costs. However, according to activity-based costing (ABC) theory, costs are consumed by activities and activities are necessary to manufacture products. In ABC, cost attribution process is accomplished in two stages. Costs incurred are first attributed to activities; then each activity is studied to determine its relationship to products (or other relevant cost objects such as customers, projects, geographic regions, office locations, etc.). Product costs are then determined on the basis of these activities. The flow of costs in the ABC model is characterized as

Costs of Resources
→ Cost Pools based on Activities
→ Cost assigned to Objects

Obviously, direct materials, direct labor, and manufacturing overhead do not make up the total set of costs in an organization. Yet, the costs allocated to products in the traditional GAAP-based model are strictly limited to these production-related costs (i.e., product costs). Other costs, not directly connected to any aspect of the production process, include selling and general administration costs. These costs, typically not included in the product cost category, fall under the heading "period costs." However, ABC does not distinguish between product and period costs. In an effort to provide better management of period costs, some selling and general administrative costs are in many cases attributed to products and other cost objects in ABC-based cost systems. Because of such handling of costs, output of the ABC system cannot be used for external financial statements. Thus, firms using ABC systems need multiple costs systems. (Kaplan, 1988).

The ABC system has a broader focus then the traditional system. Rather than focusing solely on measuring costs of products, ABC systems may identify a variety of cost objects for measurement, such as products, customers, sales representatives, operations, factory or office locations, geographic regions, etc.

Essentially, ABC proponents hold that the classic product cost view (cost of products = direct materials + direct labor + manufacturing overhead) is the wrong perspective (Turney, 1992). Further, ABC theorists view production costs to be the result of a hierarchy of activities: facility-support activities, product-sustaining activities, batch activities, and unit-level activities (Cooper, 1990). Identifying these activities and their relation to products produced by the organization is essential to the development of *multiple* activity drivers that can be used to create a comprehensive view of cost relationships affecting the product and other cost objects within the organization.

HISTORICAL PERSPECTIVE: The structure of traditional product cost systems was established during the nineteenth century by the textile mills, railroads, and steel mills of the United States and Great Britain (Johnson and Kaplan, 1987). It has changed little since then. However, the structure of actual product costs has changed dramatically over the last 80 years. Improvements in production processes through improved use of technology and less labor have resulted in important cost efficiencies and enhanced product quality. Consequently, direct labor costs continue to become a smaller percentage of total product costs, while overhead costs are becoming increasingly larger (Boer, 1994). In addition, in response to growing diversity in customer demands, as well as advances in production technology, many manufacturing organizations face increasing diversity in both their product mix and their production processes. Within these organizations, legacy (i.e., traditionally-focused and well-established) manufacturing accounting systems often struggle with the task of assigning a diversity of production costs on the basis of direct labor inputs, leading to grossly misstated cost assignments. Cost accountants and executives started becoming particularly sensitive to the difficulties caused by inadequate cost systems as a result of the seminal publication *Relevance Lost* by Thomas Johnson and Robert Kaplan (1987).

Cooper's Hierarchical Model of Activities

As a result of dissatisfaction with legacy systems, ABC applications gained popularity beginning in the late 1980s as product costing and product cost management alternatives to traditional costs systems (referred to as unit-based costing, or UBC, systems). Other individuals also made significant contributions to develop and disseminate the ABC model of manufacturing, service, and retail organizations. Professor Robin Cooper published a series of articles in 1988 and 1989 that initially codified the ABC model, identified the need for ABC information, and described issues involved in developing accurate activity drivers for the ABC system (Cooper, 1988a; 1988b; 1989a; 1989b). Subsequently, Cooper further refined the concept of activities with his hierarchical model of activities (Cooper, 1990). Cooper's hierarchy is shown in Exhibit 1.

One value of Cooper's hierarchy of activities is the insight it provides into selecting activity drivers useful for attributing activity costs to products. For example, unit-level activities have costs that vary in proportion to the number of units produced. Costs related to unit-level activities can be directly attributed to product units. On the other hand, costs of batch activities can only be attributed to batches of products, and product-sustaining activity costs are attributed to individual product lines. Perhaps most important, Cooper's hierarchy illustrates that it is difficult to intelligently attribute costs of facility-support activities to individual products, batches, or product lines.

Turney's Cross-View of the ABC Model

Peter Turney also expanded ABC theory with his cross-view of the ABC model that complements

Exhibit 1. Cooper's hierarchical model of activities (adapted from Cooper and Kaplan [1991] Figure 5.3).

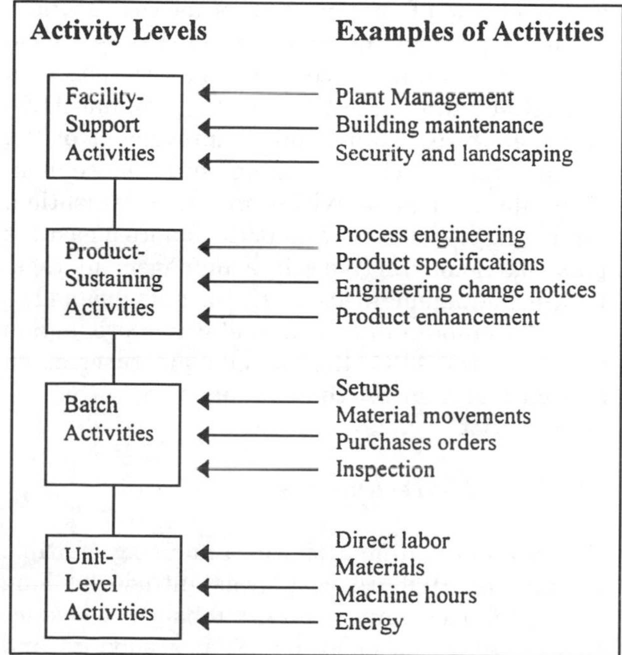

Exhibit 2. Turney's cross view of the ABC/ABM model (adapted from Turney [1992] Figure 5.2).

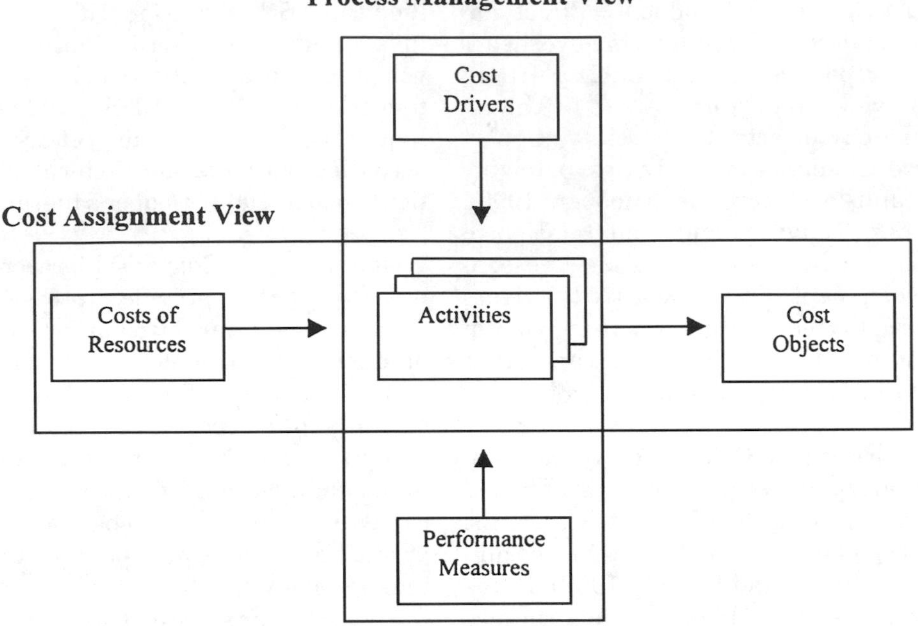

ABC with Activity-Based Management (ABM) as shown in Exhibit 2 (Turney, 1992).

ABM expands the basic ABC theory by separating the concept of product costing using activity-based relationships (the cost assignment view) from the task of directly managing activities to achieve cost efficiencies and to enhance quality in business processes (the process management view). In Turney's cross-view of ABC/ABM, opportunities for cost efficiencies are created by separately tracking multiple cost drivers for each critical activity. ABM provides a framework of cost measures as well as non-financial performance measures, that are used to support planning, control, and evaluation of critical activities identified in the original ABC-based analysis. Continuous improvement of the organization is achieved using performance measures for critical activities in the organization. Managing cost drivers and performance measures for a particular activity is a management task. Unless management uses the insights gleaned to change to more efficient and effective production processes, the ABC effort will be underused as an approach to *strategic* cost management within the organization.

The Need for Technology

The *idea* of defining activities and using multiple drivers to attribute costs was introduced long before Johnson and Kaplan published *Relevance Lost* in 1987. As early as 1921, academics and professionals published articles arguing that complex manufacturing costs were misrepresented by simplistic product costing systems (Harrison, 1921). ABC-type cost models have been promoted at least since the 1960s (see, for example, Drucker, 1963). Given the criticism of product cost systems, why has it taken so long for ABC, and other improved cost models, to emerge? The answer is twofold. First, in recent years, most markets have experienced increased competition as a result of the expanding global market, increasing consumer sophistication, and government trends to deregulate key industries. Increased competition, with attending tight profit margins, results in a *need* for tracking product and process costs better than ever before. Furthermore, as a result of the relatively recent proliferation of inexpensive computing power and the development of advanced database technology, firms have the *opportunity* to track cost data more cost-effectively than ever before. The most profound difference between ABC and traditional UBC models is the amount of data generated when an organization moves from single to multiple drivers of product costs. As Johnson and Kaplan were preparing their seminal text *Relevance Lost* (1987), the emergence of the Information Age was being heralded by cheaper, faster, and more sophisticated computer technology. As a result, the ability to access, store, and process large amounts of data quickly and inexpensively helps the implementation of sophisticated ABC models in manufacturing organizations. Essentially, ABC models finally took a foothold in manufacturing during the 1980s and continue to grow in popularity and application today.

STRATEGIC PERSPECTIVE: ABC has the potential to enhance strategic management by reducing cost distortions. Occasionally, product costs are used solely to establish prices. In other cases, however, open competitive markets effectively establish prices. In the latter more common context, the strategic use of product costs is twofold: first, to determine if the organization is able to compete effectively in the market, and second, to identify opportunities for target costing. Uncertainty or error in understanding the cost of providing a product to the customer may negatively affect profitability, organizations may miss market entrance and exit opportunities or miss opportunities for cost improvements. Hence, accurate understanding of product costs within a complex multi-production organization can provide a valuable competitive edge in a market of shrinking profit margins.

Increasing the accuracy of cost data via ABC also allows an organization to implement its market strategy more effectively. Another strategic use of the ABC model is its ability to trace activity costs to a number of different cost objects. This flexibility allows a company to determine the value to the organization of various cost objects. For example, it is not difficult to trace revenues generated by a geographic region, customer, or product group. By accurately tracing activities (and attendant costs) associated with cost objects, profits generated by various cost objects can be determined. More importantly, with this information, management develops insight into various cost objects, such as employees, customers, regions, or product groups, leading to better management of activity costs.

Identifying Cost Information Distortions

The traditional process of using overhead burden rates to allocate the costs of support departments to units produced can create serious cost distortions and cross-subsidization between groups of products. Essentially, product cost cross-subsidization is the result of using UBC that allocates average overhead costs across all types of products within the company. Hence, high-volume products may be forced to carry part of the costs of producing specialty products. In the effort to establish a new product, such subsidization might be a sensible strategy. However, in situations in which subsidy is unknown to the management, poor decisions may result. This undesirable cross-subsidization is the major result of six different information distortions created by the traditional UBC system (Cooper and Kaplan, 1991). The six distortions are explained below.

Allocating Unrelated Costs. The first potential distortion is the allocation of costs to unrelated products. For example, consider the situation in which costs related to creating engineering change notices (ECN) are included in the total pool of overhead costs, the overhead burden rate is on the basis of direct labor hours. As each product moves through the manufacturing process and uses direct labor hours, the overhead allocated to the product includes a measure of ECN costs. However, in a diversified product mix, some well-established products do not generate ECN activities. Nonetheless, under UBC, these standard products will carry part of the burden of this engineering activity, and new or specialized products will not carry their "fair share" of overhead costs.

Omitting Relevant Costs. The second source of distortion is the disregard of certain costs incurred by a particular product. This is likely the result of accounting systems established to satisfy external auditors. Financial accounting regulators allow only the inclusion of product costs (i.e., direct labor, direct materials, and manufacturing overhead) in the valuation of inventory and costs of goods sold. However, the movement of a product through the various phases of development, manufacturing, selling, and warranting incurs many more costs than those recognized by the traditional product cost model. Once manufactured and placed in inventory for sale, a product continues to require handling and movement, and security costs. Marketing expenses are those realized as the market for a particular product is developed and supported. Selling and delivery costs may vary with the product type. Finally, warranty costs are generated as products fail to perform for customers. These costs are as real as the traditional product cost components. Thus, for a company to maintain a strategic view of its market position, these costs must also be understood and included in the market price.

Omitting Cost Objects. The third cost distortion can be introduced by costing only a subset of the relevant cost objects of the organization. For example, when the output of an organization includes both tangible (i.e., manufactured) and intangible (i.e., service and support) products, traditional cost systems may assign costs only to tangible outputs. Obviously, this causes reported costs of tangible objects to be too high and does not support the effective management of costs of intangible outputs.

Erroneous Burden Rates. The fourth type of distortion is related to the first stage of the traditional UBC allocation model. In the first stage of developing the burden rate, overhead costs associated with support departments are allocated to production centers. If, for example, costs of a machine maintenance group are allocated to production centers based on the number of square feet in each production area, the overhead costs born by each production center may not truly reflect the maintenance costs actually incurred by each center. This data is then subsequently used to develop burden rates within the production center. Since distorted burden rates are used, faulty product costs will result.

Erroneous Allocation Bases. The fifth source of cost distortion is perpetuated within the second stage of the UBC model. The second stage uses the burden rate developed before a given production period in order to allocate overhead costs during the given production period to units produced. If the allocation basis is improperly selected, then overhead costs will be improperly assigned to individual products. Suppose, for example, that the costs to set up (e.g., clean and calibrate) a machine for a production run is the same regardless of the number of units scheduled for the production run. It would not be accurate, then, to allocate this overhead cost with all the other overhead costs in the production center based on any of the unit-based cost drivers (e.g., number of units, direct labor hours, machine hours).

Allocating Common Costs. The sixth cost distortion results from trying to allocate costs that simply cannot be sensibly assigned to specific products. These unallocatable costs, the common costs of the production facility, are typically related to activities defined by Cooper's (1990) hierarchical model as facility-support activities. Any attempt to allocate or attribute these costs to specific products cannot be done objectively. Consider, for example, the salary and office costs of the cost accountant assigned to track all production cost information. Clearly this individual is essential to the manufacturing activity. If the company chose to change from a manufacturing to a merchandising organization, this particular job description would be terminated. Therefore, the expense of having this accountant is manufacturing overhead by definition. However, the cost of the accountant may not be traceable to any particular product group. Attempts to allocate these costs will be arbitrary and may mislead the decision-maker trying to analyze individual product costs.

Alternative Views of Cost Objects

Rather than strictly focusing on products, consider specific customers as the cost objects. In the past few years, many companies have initiated customer rationalization programs in order to provide world-class service to their best customers. In a separate study we conducted, these firms had adopted a policy by which customers were rated based on their importance. The few "A" customers (most important; See ABC classification) received the greatest attention, with considerable resources being dedicated to customize products, orders, and deliveries normally not provided to lower-rated customers. By contrast, "C" customers were often poorly served and at times ignored. In a few cases, the firms indicated that they were no longer interested in the business of the "C" customers, since meeting their needs dissipated resources that were needed to keep the critical "A" customers satisfied. Firms discovered that some key accounts with important customers were actually losing money only after more rigorous and accurate ABC systems were implemented with customers as the cost object. Exhibit 3 illustrates the ABC attribution approach that led to this discovery. Much to the surprise of some companies, the new approach to costing revealed that many "C" customers whom the companies had previously disparaged as unimportant and too small were actually their most profitable accounts. When the costs of "extra effort" activities were averaged across all products and customers the companies found they were underpricing these special customers. Similar to the product cross-subsidization problem, these ABC analyses supported very important (though potentially painful) decisions to increase prices or discontinue some key account customers. This example illustrates well the strategic effect of ABC data on the management decision process.

IMPLEMENTATION:

The First Stage of ABC

Pragmatically, how does one specifically trace costs to cost objects in the ABC model? This is a two-stage process. The first step is to identify and cost the activities within the organization. Rather than thinking of the organization as a pool of departments (e.g., production, quality control, personnel, accounts payable, etc.), the organization is viewed as a pool of activities. Many of these

Exhibit 3. Costing the service requirements of selected customers.

Traditional UBC view		ABC view	
Salaries	$120,000	Receive Material	$86,600
Supplies	30,000		
Depreciation	20,000	Move Material	84,600
Overtime	15,000		
Space	30,000	Expedite Material	58,800
Other	5,000		
Total	$230,000		$230,000

The H.S. Sponge account represents 25% of units sold, yet consumes 25% of the receiving activities, 30% of the movement activities, and 40% of the expediting activities. How much does it cost to meet H.S. Sponge s service requirements?

Traditional UBC view:

Overhead		% of Units		Customer cost
$230,000	×	25%	–	$57,500

ABC view:

Activity costs		of Units		Customer cost
$86,600	×	25%	=	$21,650
$84,600	×	30%	=	$25,380
$58,800	×	40%	=	$23,520
				$70,550

activities will cut across departments, the departments often participating in many different activities. Since the ABC implementation team needs to orient itself by activities, rather than by departments, a good starting point is to understand that there are three essential cycles in a business: the Acquisition/Payment Cycle, the Conversion Cycle, and the Sales/Collection Cycle (Denna *et al.*, 1993). All activities in the company relate to the process of acquiring and paying for resources, converting resources into salable goods or services, or selling goods/services and collecting revenues. This view of the organization can be quite valuable in the processes of identifying and mapping critical activities for an ABC implementation. Attaching costs of resources to activities can be a rather imprecise process involving many estimates. In terms of materials resources (e.g., supplies), it is not as difficult for the organization to use the current accounting system to directly trace the cost of supplies required by a particular activity. On the other hand, labor and overhead costs are more challenging to trace to activities. It is possible to require employees to fill out detailed time records in order to track labor resources flowing to particular activities. However, such expensive and obtrusive record keeping is unlikely for most organizations. A more viable alternative than maintaining precise time sheets is to simply interview department managers for estimates of effort expended on specific activities. However, interviewing is not the only way to gather essential insight about activities. Other methods include simple observation of direct production and essential support activities, computerized or manual timekeeping systems that track work done and time spent on each activity, questionnaires, and storyboards (in order to discover consensus on specific activities among employees directly involved). These methods can be used alone or in conjunction with each other.

The Second Stage of ABC

Once the organization has traced all resource costs to the appropriate activities and established activity cost pools, the second ABC stage is to determine who or what requires each activity and the basis for the activity use. The output of this analysis is activity drivers that provide the link between activities and cost objects.

In analyzing for activity drivers, it is important to remember that products are not the only cost objects that consume activity costs. As described above, customers, market segments, and distribution channels all consume activities and can be identified as cost objects. The main idea of Cooper's (1990) hierarchical activity model is that costs and activities should be managed at the level at which they occur. Forcing all costs to behave as production unit costs is an artificial simplification of the complex reality of modern business. Additionally, activities themselves can also be cost objects. For example, many companies may find that their payroll activity requires computer maintenance activity.

In identifying activity drivers, Cooper's (1990) hierarchical model (Exhibit 2) provides valuable insight. For example, the "Quality Design" activity in many manufacturing organizations is required to support individual product types. Within Cooper's hierarchy, it would be classified a product-support activity and its costs may be assigned to overall product lines possibly based on number of engineering change notices or on the number of components within the product. In another example, investment and maintenance costs may be assigned to a "Production Machinery" activity. This activity is consumed by the production of various products or product components, and may be assigned to these cost objects on a per-unit basis. The "Purchase Order" activity, on the other hand, if required each time a new batch is initiated, should be assigned on a batch basis. The process of determining activity drivers is the same process the ABC implementation team follows in determining activity pools. Interviews, observations, database analysis, and sampling procedures may be used in this identification process. In fact, as part of the process of successfully developing activity pools, activity drivers become transparent as the consumers of each activity are identified.

TECHNOLOGY PERSPECTIVE: Most organizations need to reengineer their whole approach to information system design in order to take advantage of ABC. Current computerized accounting information systems are generally limited to financial measurements. This limitation affects the ABC model and its need to track nonfinancial cost drivers and performance measures. In accounting systems, preclassification in terms of debits, credits, assets, liabilities, and equity in very common. However, the ABC model is based on a set of activities, not on a chart of accounts. As a result, useful integration of financial and nonfinancial data necessary for the effective implementation of ABC does not exists in many organizations.

There are several proposed technology solutions to implementing ABC. Essentially, a disaggregated relational database could be developed that maintains detailed data for use in any model of decision support including ABC. Only a few years ago such enterprise-wide information warehouses were not possible. Today, with decreasing technology costs and modern advances in information theory, organizations can create a robust information environment conducive to ABC. Perhaps the best solution currently is an approach referred to as *events-driven system design* (Swain and Denna, 1998). Essentially, events-driven systems are designed around *all* the business events (or activities) of interest rather than shoehorning some events into an information artifact like the general ledger. For example, consider the "purchase inventory" activity. Why artificially separate pieces of this event into the accounts labelled "Inventory," "Purchases," and "Accounts Payable," and then fail to track other essential information like the time of delivery, who made the delivery, who placed the order, and the quality of the inventory? Events-driven systems integrate all these data into a single database (usually a relational database). This database becomes the organization's information warehouse – an asset that is becoming increasingly affordable as the costs of information technology continue to come down. Software companies like Lawson Software and SAP® provide off-the-shelf technology that incorporates many of these concepts.

TIMING AND LOCATION: Continued use of UBC systems that do not accurately portray complex production cost flows can lead to information distortions. However, implementing an ABC system is an expensive and time-consuming process. Organizations need to determine carefully the need for sophisticated cost information before beginning an ABC implementation project. Some production settings may be adequately served by a traditional UBC system.

Market Warning Signs

Competitive markets can be helpful in analyzing the need for ABC. Organizations can watch for specific warning signs in the market which may indicate that an ABC system may be justified (Cooper, 1987).

Confusing Profit Margins. First, management should evaluate honestly whether profit margins across the product mix make sense. One way to compare product groups within the overall product mix is to differentiate between high-volume standard products and low-volume customized products. Low-volume customized products indicate a desire to "delight the customer." These customized products are obviously more difficult to produce. Products are often customized to satisfy small pockets of customers.

However, management may be confused – though not unhappy – at the apparent ability of these customized products to show healthy profit margins with little or no markup on price. Most customized products logically either require significant price premiums or, in the effort to build customer loyalty, display acceptably tight or negative margins. If management cannot explicitly explain why the unexpectedly large profit margins exist, it is possible that reported margins are masking a problematic product mix relationship – low-volume customized products are being unknowingly subsidized by high-volume standard products.

Unexplained Market Barriers. Another indicator of the potential that high-volume products are subsidizing low-volume products is the inexplicable dominance of a market segment. Management may find itself in a market for one of its low-volume custom products with little or no competition. Again, management may be happy, but puzzled. Why do competitors not attempt to penetrate this market in order to share in the seemingly significant profits? A sophisticated ABC system may reveal what competition already knows – the company is pursuing phantom profits that do not actually exist!

Similarly, management may be frustrated at their inability to penetrate or hold the market for one of their high-volume standard products. The company may have developed a very efficient production system and should be able to compete. However, others may be bringing the product to the market at what seems to be an impossibly low price. Again, this may denote the inadvertent inclusion of the cost of producing low-volume products within the standard product's cost base. With ABC, the cross-subsidization may be removed from the high volume product to reveal previously unknown profit potential.

Bidding Mysteries. Bidding mysteries can also indicate a need for ABC. First, management should carefully evaluate their successes and failures in bidding for contract work. Regular loss of contracts for which bids are prepared with very tight profit margins, particularly for high-volume standard products, may indicate cross-subsidization. Additionally, if the company surprisingly wins contracts it didn't expect to win, this may indicate that cross-subsidization is taking place. Similar need for ABC may be seen in analyses that recommend outsourcing work which the company should be able to do itself profitably.

RESULTS AND COLLECTIVE WISDOM: Despite the introduction of ABC systems in the late 1980s, accountants and managers in many organizations today consider the tremendous investments in money and time required to bring online a new cost system to be a major impediment to implementation. Complicating the cost/benefit analysis of an implementation decision is the mixed success some organizations have had in adopting ABC as a product cost system (1995c). In a three-part series of articles, Player and Keys (1995a, 1995b, 1995c) discuss the difficulties involved in successfully launching an ABC system and provide important caveats to managers of implementation teams. These lessons are valuable to those who are considering ABC investment.

Beginning the Effort

Player and Keys (1995a) note that companies often fail to appreciate the effort required to initiate successfully an ABC system. Without the involvement of top management in the organization, ABC implementation is likely to stall. Everyone involved must clearly understand exactly why the organization needs a new cost system and what can be expected from making this investment. The lack of involvement across the organization is typically revealed by underfunded projects staffed by employees who are not adequately trained in the essential principles and technology of ABC. As a result of misunderstanding the commitment required, there are many industry examples of failed ABC projects. Further, ABC is not simply a financial or accounting project. The focus on activities requires that managers and employees from functional areas such as operations, engineering, and marketing must be involved. Otherwise, the new system will fail to deal effectively with operational or strategic issues. Finally, ABC is not an isolated, one-time project. To operate effectively as a strategic cost management system, ABC must be integrated with other strategic initiates that the organization is pursuing. For example, the cross-view of the ABC/ABM model as proposed by Turney (1992) in Exhibit 2 should be used to develop performance measurements and ideas for continuous improvement for the organization's just-in-time (JIT) or total quality management (TQM) initiatives.

Developing the ABC Pilot

"It is extremely difficult to implement a comprehensive [ABC] system without doing a pilot project first" (Player and Keys, 1995b, 20). By first launching an ABC pilot, the implementation team is able to generate greater support throughout the organization when the mainstream system is brought online. Further, the pilot program provides opportunities to train personnel in interviewing and observation methods necessary to identify critical activities and develop relevant drivers that link activity costs to cost objects. The pilot can also alert management to potential software problems or to difficulties in obtaining the necessary data required to support the ABC model. Finally, the organization can use the pilot to understand the critical balance between too much and too little detail in its proposed ABC system. As a result of a poorly designed ABC system, management within the organization can be frustrated by "information overload." The ABC design team must, therefore, be mindful of the effect of information detail provided to management. Too much information results in "information overload" and decision-makers, who are either confused by the information provided by the ABC system or who simply refuse to use it! On the other hand, the ABC pilot should incorporate enough new data to ensure that it supports insights into processes and costs that are significantly different from those provided by the legacy system.

Moving from the Pilot to Mainstream

Player and Keys (1995c) are careful to point out in their studies of ABC implementation projects that behavioral issues tend to be much more crucial than technical issues in successfully establishing an ABC system. Implementing ABC is a large undertaking. The implementation team should expect that both individuals and departments will naturally resist change until it is clearly understood how the information provided by the new system will impact jobs and relations within and outside of the organization. To move successfully from the pilot phase of the ABC project to full mainstream implementation, the organization must have in place formal procedures involving how and when reports will be developed and disseminated. Further, it must be clear to everyone how the organization plans to act on the ABC information and how those actions will affect individuals and profits. Finally, the decision to invest in the ABC system must be continually evaluated in terms of costs and benefits in order to justify the constant investments required to update the ABC system to reflect changing business processes within a dynamic organization.

SUMMARY: Dissatisfaction with traditional cost systems, coupled with the advent of inexpensive and sophisticated computer technology, has led to the emergence of ABC as a viable, strategic cost-management system. ABC can enhance an organization's product and market strategy by reducing cost distortions and by allowing flexibility in defining cost objects. Market events can provide valuable signals to help the organization evaluate the potential benefits of an ABC system. The process of designing and implementing an ABC is both costly and time-consuming, and some organizations have struggled in their launch of ABC systems. Most of the pitfalls that can undermine an ABC implementation project relate to behavioral, rather than technical issues of managing the change.

See **ABC classification; Activity-based management; Enterprise resource planning; Information warehouse; Target costing; Unit-based costing (UBC).**

References

[1] Boer, G. (1994). "Five modern management accounting myths," *Management Accounting* (USA, January), 22–27.

[2] Cooper, R. (1987). "Does your company need a new cost system?" *Journal of Cost Management* (Spring), 45–49.

[3] Cooper, R. (1988a). "The rise of activity-based costing – part one: What is an activity-based cost system?" *Journal of Cost Management* (Summer), 45–54.

[4] Cooper, R. (1988b). "The rise of activity-based costing – part two: When do I need an activity-based cost system?" *Journal of Cost Management* (Fall), 41–48.

[5] Cooper, R. (1989a). "The rise of activity-based costing – part three: How many cost drivers do you need, and how do you select them?" *Journal of Cost Management* (Winter), 34–46.

[6] Cooper, R. (1989b). "The rise of activity-based costing – part four: What do activity-based cost systems look like?" *Journal of Cost Management* (Spring), 38–49.

[7] Cooper, R. (1990). "Cost classifications in unit-based and activity-based manufacturing cost systems," *Journal of Cost Management for the Manufacturing Industry* (Fall), 4–14.

[8] Cooper, R. and R.S. Kaplan (1991). *The Design of Cost Management Systems*, Prentice Hall, Englewood Cliffs.

[9] Drucker, P.F. (1963). "Managing for business effectiveness," *Harvard Business Review* (May–June), 59–66.

[10] Denna, E.L., J.O. Cherrington, D.P. Andros, and A.S. Hollander (1993). *Event-Driven Business Solutions*, Business One Irwin, Homewood, Ill.

[11] Harrison, G.C. (1921). "What is wrong with cost accounting?" *N.A.C.A. Bulletin*.

[12] Johnson, H.T., and R.S. Kaplan (1987). *Relevance Lost! The Rise and Fall of Management Accounting*, Harvard Business School Press, Boston, MA.

[13] Kaplan, R.S. (1988). "One cost system isn't enough," *Harvard Business Review* (January–February), 61–66.

[14] Player, R.S., and D.E. Keys (1995a). "Lessons from the ABM battlefield: Getting off to the right start," *Journal of Cost Management* (Spring), 26–38.

[15] Player, R.S., and D.E. Keys (1995b). "Lessons from the ABM battlefield: Developing the pilot," *Journal of Cost Management* (Summer), 20–35.

[16] Player, R.S., and D.E. Keys (1995c). "Lessons from the ABM battlefield: Moving from pilot to mainstream," *Journal of Cost Management* (Fall), 31–41.

[17] Turney, P.B.B. (1992). *Common Cents: The ABC Performance Breakthrough*, Cost Technology, Portland, OR.

[18] Swain, M.R., and E.L. Denna (1998). "Making new wine: Useful management accounting systems," *International Journal of Applied Quality Management* (volume 1), 25–44.

ACTIVITY-BASED COSTING IN MANUFACTURING

Traditional accounting systems generally allocate all manufacturing overhead costs as a function of the number of products being produced or the amount of service being provided by the organization. This unit-based costing (UBC) works well in fairly simple settings. However, such an approach to allocation produces cost distortions as direct labor becomes a smaller percentage of product cost in automated environments. Activity-based costing (ABC), instead, identifies specific activities that incur costs and uses these "cost drivers" to allocate shared costs.

For example, purchasing department costs can be an overhead item. The specific cost driver used to allocate such costs to an individual product might be the number of purchase orders associated with components for that product. ABC is especially relevant to manufacturing because it allows companies to determine much more accurately the actual cost of making a product in today's manufacturing environment.

See **Activity-based costing; Activity-based costing: An evaluation; Unit-based costing.**

ACTIVITY-BASED COSTING: AN EVALUATION

M. Michael Umble

Baylor University, Texas, USA

Elisabeth J. Umble

Texas A&M University, Texas, USA

DESCRIPTION: Activity-based costing (ABC) systems allocate costs to products or customers in a way that theoretically reflects reality better than traditional standard cost systems (Unit-based Cost Systems). ABC systems are more sophisticated and complex than standard cost systems. Most standard cost systems, in a grossly oversimplified approach, allocate all overhead costs to individual products using a single allocation basis (such as direct labor, machine hours, or material dollars) that is highly correlated with unit volume. In contrast, ABC systems allocate costs through a two-stage process using multiple allocation bases, which may not be related to unit volume. ABC advocates claim that ABC can help identify opportunities for product mix changes, pricing changes, process changes, and organization restructuring. But, numerous critics have argued that ABC is simply standard costing warmed over, and therefore, does not eliminate the fundamental problems of standard costing. Perhaps the most severe critic of both standard cost and ABC cost systems is H.T. Johnson, who claims that trying to get relevant information from cost systems by implementing ABC is akin to "rearranging the deck chairs on the Titanic" (Johnson, 1992). In this article, we consider both sides of the argument.

THE CASE FOR ACTIVITY BASED COSTING:

The Problem

Shortly after World War II, a "manage-by-the-numbers" culture rapidly developed and almost universally captured the hearts and minds of business leaders in the U.S. The centerpiece of this movement was standard cost accounting systems that provided the "numbers" for a wide variety of decisions. Corporate managers generally assumed that their standard cost systems generated information that accurately reflected the costs of producing goods and services. But in general, the cost information grossly distorted reality. Hiding behind a cloak of precision and generally accepted accounting principles primarily designed to satisfy external reporting requirements, fatally flawed standard cost systems and their supporting performance measures routinely triggered poor decisions and dysfunctional organizational behaviors.

In the 1980s, standard costing systems first came under attack when Goldratt (1983) argued that "cost accounting is the number one enemy of productivity." A classic work (Kaplan and Johnson, 1987) described the shortcomings and effects of traditional costing systems in great detail. Managers soon began to realize that standard cost data distorts product cost and product margin estimates, causes poor product mix and product pricing decisions, results in poor resource allocation, and leads to improper outsourcing decisions. It also became evident that traditional absorption (standard costing; unit-based costing) costing encourages unneeded production in order to capture overhead in inventory rather than in cost of goods sold. This not only contributes to poor resource allocation, but leads to higher costs, longer production lead times, and lower quality. Interestingly, as the fallacies of standard cost systems became widely understood, most managers simply looked for a way to upgrade their traditional cost systems.

The ABC Approach

Implementing an ABC system is a two phase process. In the first phase, all key processes and activities in the company are charted. This normally involves a comprehensive interview and analysis process to identify the significant activities that are performed in the company and how all the pieces fit together. In the second phase, the key parameters of the ABC system – activities, resources, cost elements, resource drivers, activity cost pools, cost objects, and activity drivers – are analyzed, assumptions concerning activities, drivers, and capacities are made, and costs are allocated to products or customers.

ABC Benefits

The charting process performed in the first implementation phase routinely yields keen insights into organizational problems and often generates ideas on how to resolve those problems. The process of charting the key activities and business processes is probably responsible for many of the benefits reported in ABC implementations because it fosters a better understanding of these critical organizational elements. As the charting process proceeds, many redundancies and non-value added tasks are uncovered. In most companies, this preliminary activity-based analysis triggers immediate improvements as the "low-hanging fruit" is picked from the tree of potential improvements.

It should be noted that Activity analysis was not an accounting innovation. As early as 1950, W. Edwards Deming (1982: 4) stressed the value of diagramming and analyzing processes from a system perspective. And Peter Drucker (1954) described the need for activity analysis in organizations three decades before the term "activity-based cost" was coined. But it is clear that Deming and Drucker did not believe that activity analysis should assume a cost emphasis.

The second implementation phase emphasizes product cost data. Since ABC analyses can theoretically account for product volume, product complexity, or support service requirements, they can generate more realistic cost data than traditional standard cost systems. Thus, ABC can often identify cross-product subsidies (where one product subsidizes another by assuming a disproportionate share of allocated costs). In addition, ABC systems may be able to provide fair and more defensible cost allocations for transfer pricing situations and for firms that have "cost plus" pricing contracts.

The Whirlpool Experience with ABC

Managers of the Whirlpool plant in Evansville, Indiana were dissatisfied with their traditional cost system. The plant produced over 300 different refrigerator and vertical freezer models that varied greatly in production volume and assembly complexity. Management wanted more accurate cost data to perform product cost and profitability analyses, cost reduction analyses, investment analyses, and make/buy decisions. In 1990, a pilot ABC system was implemented in 16 weeks (Greeson and Kocakulah, 1997). Two plant employees with accounting and production backgrounds worked on the implementation full-time. In addition, three division/corporate level employees and an external consultant were all available on a part-time basis. To minimize the magnitude of the implementation, Whirlpool chose to develop a stand-alone ABC system that was maintained separately from the financial accounting system.

Analysts identified 237 activities, and all resource costs were allocated to one of the 237 activity cost pools according to their chosen resource drivers. The costs in the activity cost pools were then allocated to products using 23 activity drivers. In 1991, it was decided that the number of activity drivers could be reduced to 14 and still generate satisfactory results.

The ABC-generated cost data indicated that the old standard cost system had grossly distorted product costs. Analysts determined that the old system significantly undercosted low-volume and high-complexity products, and overcosted high-volume and low-complexity products. (Such findings are typical in ABC implementations). Many of Whirlpool's low-volume or high-complexity products received two to three times more overhead from the ABC system than they did from the standard cost system, while overhead allocations for many high-volume or low-complexity products were cut in half. One extremely complex product had its overhead allocation increased from $65.46 per unit under the old cost system to $728.92 under ABC.

ABC Implementation Problems:

Implementation Pitfalls

Companies have reported tremendous difficulties implementing ABC systems and many implementation attempts have failed. The best discussion of implementation problems is found in a book by Player and Keys (1995). This book describes the 30 most critical pitfalls encountered when implementing ABC systems. Most of these pitfalls involve a lack of top management or company-wide support, inadequate resource support, technical or data problems, system maintenance requirements, or organizational resistance to change.

Resource Requirements

Some of the earliest ABC systems were developed on a spreadsheet. However, for all but the most basic implementations, ABC system software should be considered a necessary tool. Another prerequisite is ABC expertise. Since most companies do not have such expertise, external consultants are often retained. It has been reported that a typical ABC "startup" engagement with one large consulting firm for a $100–$500 million company would require the active involvement of two consultants and two client employees with strong financial and operations background for about 12 weeks (Albright and Smith, 1996: 49). However, as noted by Player and Keys (1995), this limited amount of involvement, segregated from the rest of the company management is a classic recipe for failure.

Size Limitations

ABC has been implemented in a wide variety of firms, but the cost and complexity of activity-based systems may make implementations in small and midsize businesses difficult (Wiersema, 1996). Such firms exhibit three key characteristics that make ABC implementations exceedingly troublesome. One, they typically have unsophisticated cost systems that do not capture data about activities, and this limits the design choices for activity drivers. Two, there is a general lack of expertise necessary to develop and implement the system. Three, they generally have limited resources to spend on expensive improvement projects and have difficulty achieving information economies of scale. Not coincidentally, most companies that have reported "successful" activity-based implementations are relatively large firms.

ABC and Dysfunctional Decisions:

Standard costing systems can cause horrible distortions of reality and lead to poor decision making. Properly implemented, ABC can reduce some of the distortions caused by standard costing procedures. But ABC has been criticized for three reasons: (1) Despite its complexity, ABC still cannot provide accurate cost information; (2) The cost focus of ABC encourages poor decisions and dysfunctional behaviors; (3) ABC tends to keep managers' attention focused on cost to the detriment of the non-cost-oriented competitive elements necessary for success in today's global economy.

Inability to Provide Accurate Cost Information

Extreme care must be exercised in the use of any allocation-based cost data because true cost relationships are extremely complex and any cost allocation system will be comparatively simple and arbitrary. Since cost allocations are totally dependent upon numerous arbitrary decisions concerning activities, drivers, and capacities, ABC-generated numbers will always differ from traditional cost system numbers, but neither will reflect reality. There is no such thing as a "true unit cost" – it is a figment of the assumptions. And it is debatable whether ABC numbers can be accurate enough to provide a solid basis for organization-wide decision making.

Improper Cost Focus Encourages Dysfunctional Behaviors

Cost relationships are only part of the big picture. Decisions should not be made solely or even primarily on the basis of cost. This argument is easily illustrated. The natural tendency of users in any cost allocation system is to reduce the consumption of cost drivers. This has tremendous behavioral implications. For example, suppose that two activity driver candidates – number of purchase orders and number of unique parts – are being considered to allocate purchasing overhead to products. If the number of purchase orders is selected, users are likely to reduce the number of purchase orders generated, leading to larger order quantities and higher inventory levels. This conflicts with the nearly universal objective of inventory reduction. But, if the number of unique parts is the driver, the use of common parts is encouraged. This tends to reduce inventory requirements, engineering support requirements, and manufacturing complexity. Surely, encouraging appropriate future behaviors is important in the selection of cost drivers.

ABC Diverts Attention from Competitive Success Factors

A final criticism of ABC is that it is a cost-focused technique that requires considerable resource support. As such, it diverts attention away from the primary determinants of success in today's marketplace – quality, quick delivery, flexibility, and customer service (Johnson, 1992). Although cost control is important, organizations can no longer be successfully managed by focusing on costs. In fact, the reengineering philosophy has shown that the best way to cut costs is not by managing costs, but by improving processes. As a result, many believe that reliance on ABC systems is inconsistent with the proper focus of world-class competitors. Cooper and Kaplan (1991) suggest that the cost perspective is counterproductive if it diverts managers' attention away from appropriate process improvement actions.

The Schrader Bellows Case

The Schrader Bellows Company provides an interesting example of an ABC implementation (Montgomery and Cooper, 1986). In the early 1980's, bottom-line performance at Schrader Bellows was poor. Analysts from the parent Scovill corporation performed an ABC-type analysis using multiple cost drivers to reassign overhead to Schrader Bellows' complex array of products. The ABC analysis generated product cost data and identified a large number of "unprofitable" pneumatic control products that should be dropped from the product line. However, Schrader Bellows managers rationalized this number down to 250 unprofitable products that actually became candidates for being dropped. According to the analysis, dropping these products would cause projected sales revenue to decrease by about $1,178,000, but would supposedly increase net income by $750,000. However, a follow-up study indicated that by the end of 1984, only about 20 products had actually been dropped. The company continued to struggle and was eventually sold.

The classic argument is that the company continued to struggle because of its inability to overcome organizational resistance and successfully pare down its product line and cut costs. An alternative line of thinking is that Schrader Bellows had other significant problems that made them uncompetitive. And simply dropping selected products would not resolve those problems. Moreover, management may have had reservations about the validity of the ABC data and questioned whether costs would actually decrease enough to offset the loss of revenue. There are many examples of companies that inappropriately dropped products or outsourced production and were eventually swept up in a "death spiral" because the surviving products could not absorb all of the remaining overhead costs. It is impossible to ascertain whether significantly cutting the product line would have improved financial performance at Schrader Bellows.

RESULTS OF ABC IMPLEMENTATIONS: The impact of most ABC implementations on bottom-line profitability has been disappointing. Many firms have invested significant resources while implementing ABC, with little or no payback. Roberts and Silvester (1996) discuss two major surveys investigating the results of ABC implementations. According to a 1992 study sponsored by the Consortium for Advanced Manufacturing-International, (CAM-I) only one out of three firms could demonstrate a bottom-line improvement in profitability after adopting ABC. A survey published by the Institute of Management Accountants in 1994 also yielded interesting results. Approximately 59 percent of responding firms indicated they were using some method other than ABC to improve their management practices. Approximately 41 percent of the firms indicated that they were using an ABC-type system. Of these, only 29 percent (or 12 percent of the total respondents) were running their ABC system as a replacement for their old cost system. Moreover, of those companies running ABC systems, 81 percent reported no improvement in profitability.

COLLECTIVE WISDOM: Despite the criticisms of ABC, virtually everyone agrees that the process charting activities performed in the first phase of ABC are extremely powerful and worthwhile. As described earlier, this charting activity generates a better "big picture" understanding of how critical processes work and can play a vital role through continuous improvement. The valuable insights from process mapping, coupled with the less than spectacular success of ABC implementations, has led to an increasing focus on activity-based management (ABM). ABM is a more recent and much broader concept than ABC. It is a management philosophy that focuses on the planning, execution, and measurement of activities as the key to a firm's competitive advantage.

ABM downplays the importance of cost allocations that characterize ABC, and is more consistent with other methodologies that focus on improving critical processes. Today, ABM is perceived by some to be the salvation of the activity-based approach. If it is to succeed, it must be integrated within an overall program of continuous improvement that is aligned with the competitive realities that require firms to identify and deliver value to customers.

See **Activity-based costing; Activity-based management; Activity driver; Absorption costing; Cost accounting; CAM-I; Overhead; Product margin; Standard cost system; Standard costing; Unit-based costing; Unit cost; Value added tasks.**

References

[1] Albright, T.L. and T. Smith (1996). "Software for Activity-Based Costing," *Journal of Cost Management*, Summer, 47–53.

[2] Cooper, R. and R.S. Kaplan (1991). "Profit Priorities from Activity-Based Costing," *Harvard Business Review*, May–June, 130–135.

[3] Deming, W.E. (1986). *Out of the Crisis*, MIT Center for Advanced Engineering Study, Cambridge, Massachusetts.

[4] Drucker, P.F. (1954). *The Practice of Management*, Harper and Brothers, New York, 195–196.

[5] Greeson, C.B. and M.C. Kocakulah (1997). "Implementing an ABC Pilot at Whirlpool," *Journal of Cost Management*, March–April, 16–21.

[6] Goldratt, E.M. (1983). "Cost Accounting is Enemy Number One of Productivity," *International Conference Proceedings*, American Production and Inventory Control Society, October.

[7] Kaplan, R.S. and H.T. Johnson (1987). *Relevance Lost: The Rise and Fall of Cost Accounting*, Harvard Business School Press, Boston, Massachusetts.

[8] Johnson, H.T. (1992). *Relevance Regained: From Top-Down Control to Bottom-Up Empowerment*, Free Press, New York.

[9] Montgomery, J.E. and R. Cooper (1986). "Schrader Bellows (E) and (F)," *Harvard Business School Cases 186-054 and 186-055*, revised March and February.

[10] Player, S. and D. Keys, eds. (1995). *Activity-Based Management: Arthur Andersen's Lessons From the ABM Battlefield*, MasterMedia Limited, New York.

[11] Roberts, M.W. and K.J. Silvester (1996). "Why ABC Failed and How It May Yet Succeed," *Journal of Cost Management*, Winter, 23–35.

[12] Wiersema, W.H. (1996). "Implementing Activity-Based Management: Overcoming the Data Barrier," *Journal of Cost Management*, Summer, 17–21.

ACTIVITY-BASED MANAGEMENT

Identifying and using multiple cost drivers and performance measures to manage an activity is called activity-based management (ABM). Essentially, ABM expands the basic activity-based costing (ABC) model by separating the concept of product costing using activity-based relationships (Costs → Activities → Products) from the task of directly managing activities to achieve cost efficiencies and to enhance quality in business processes. Using ABM, opportunities for cost efficiencies are identified by separately tracking multiple cost drivers for each critical activity. Continuous improvement of the organization is then achieved using performance measures based on quality and timelines issues for strategically important activities in the organization. Hence, managing activity cost drivers and performance measures for a particular activity is a management task that is actually exclusive of the processes of tracking a single cost driver that allocates the activity's costs to products.

Activity-based management uses activity-based costing information about cost pools and drivers, and business processes to: (1) identify business strategies; (2) improve product design; and (3) remove waste from operations. Activity-based management focuses on the study of activities that are part of the production of goods to make improvements to the value received by a customer and improve profits. Activities are repetitive actions performed in the fulfillment of business functions. A primary component of activity-based management is activity analysis, which is the process of studying activities to classify them as either value-added activities or non-value-added activities and devising ways of minimizing or eliminating non-value-added activities and maximizing and encouraging value-added activities.

See **Activity-based costing; Activity-based costing: An evaluation; Activity driver.**

Reference

[1] Barfield, Jesse T., Cecily a Raiborn, and Michael R. Kinney (1994). *Cost Accounting: Traditions and Innovations*, West Publishing Company, St. Paul, Minnesota.

AD-HOC TEAMS

Ad-hoc teams are temporary groups created to address specific issues or problems. They do not have control over any work decisions and typically are not involved directly in production. These teams are formed for very limited periods of time to address narrow, short-term problems, and have a very limited scope of responsibilities. Usually ad hoc teams only offer ideas or suggestions related to the issue and have no role in decision making or implementation.

See **Team building; Teams: Design and implementation.**

ADAPTATION FLEXIBILITY

The capability of a manufacturing system that enables it to adapt rapidly and inexpensively to changes in its internal and external operating environment is called adaptation flexibility. For example, if a manufacturing system can make rapid and inexpensive implementation of engineering design changes for a particular product, it possesses adaptation flexibility.

See **Manufacturing flexibility implementation; Manufacturing flexibility types and dimension.**

ADAPTIVE CONTROL

In manufacturing processes, adaptive control is an advanced control logic that can react through feedback to changing conditions in the process itself and modify the process inputs accordingly. An example would be the ability of CNC machine tools to adjust the tool in response to tool wear without the need for operator intervention.

See **Flexible automation; Manufacturing system design.**

ADAPTIVE EXPONENTIAL SMOOTHING

The single exponential smoothing forecasting model requires the specification of a coefficient value, α, which determines the degree of model smoothing. Proper selection of the coefficient α is critical to the model's success. One approach to selecting this coefficient is called *adaptive exponential smoothing*, which allows the value of α to be modified as changes in the pattern of data occur.

See **Forecasting guidelines; Forecasting in manufacturing management.**

Reference

[1] S. Makridakis, S. Wheelwright and V. McGee (1983) *Forecasting: Methods and Applications* (2nd ed.), New York: John Wiley & Sons.

ADAPTIVE KANBAN CONTROL

Adaptive kanban control is a dynamic kanban control (DKC) policy the goal of which is to systematically add and remove kanbans so as to achieve improved system performance. Under a DKC policy, kanbans are added (released) when needed to improve system performance and removed (captured) when they are no longer needed or when their presence will result in lowered system performance. In general, we want additional kanbans when the benefits of their presence (e.g., reduced blocking and starvation, improved throughput) outweigh the costs (e.g., increased WIP and operating costs). In the adaptive kanban control policy, the release and capture thresholds are a function of demand and finished inventory levels. The thresholds are evaluated when a new demand arrives and when a part enters the finished inventory. The goal of this DKC policy is to reduce holding and inventory cost in the presence of extreme variations in demand and lead times.

See **Dynamic kanban control for JIT manufacturing; Flexible kanban control; JIT; Kanbans; Pull production systems; Reactive kanban control.**

ADVANCED COST MANAGEMENT SYSTEM (ACMS)

ACMS is a set of recently developed quantitative cost analysis approaches supporting operations and process control, product related decisions, and performance evaluation. Typically, four types of cost analysis systems comprise advanced cost management systems: (1) activity-based costing systems, (2) multiple performance measurement systems, (3) competitor cost analysis systems, and (4) product life cycle cost management systems. These systems are tyically integrated into an Enterprise Resource Planning system such as SAP, BAAN, Oracle, or People Soft.

See **Activity-based costing; Cost analysis for purchasing; ERP.**

ADVANCED MANUFACTURING TECHNOLOGIES (AMT)

When limited to the production equipment, AMT usually refers to various configurations of computer numerical control (CNC) machines, robots, direct numerical control (DNC), and flexible manufacturing systems (FMS). With CNC, a computer processor is built into the production machine controller. The controller guides the machine to perform its operations and may signal when equipment adjustments are needed. CNC robots are controlled like CNC machines and are usually used for assembly or material handling. DNC involves a remote central processor controlling several CNC machines and/or robots. Although multiple definitions of FMS's exist, the most common one used in research refers to a group of CNC machines/robots linked by an automated material handling system. This material handling system can be robots, automated conveyors, monorails, or automated guided vehicles. The main characteristic of the FMS is its ability to handle any member of similar families of parts in random order.

AMT may also include an integration of different manufacturing technologies such as robotics, machine vision, voice recognition, etc. and computer-based information technologies such as software programs and database management in manufacturing activities. The distinguishing characteristic of AMT is the potential for electronic integration which creates links among activities such as product design, process planning, production, materials handling, process control, etc.

See **Capital investments in advanced manufacturing technologies; CNC; Direct numerical control; FMS; Hard manufacturing technologies; Human resource issues and advanced manufacturing technology; Manufacturing technology.**

ADVERSARIAL BUYER-SUPPLIER RELATIONSHIP

Saunders (1997, p. 255) lists various factors that characterize the traditional (adversarial) nature of the relationship between the buyer and supplier in a supply transaction. Adversarial relationship is promoted when they operate at arm's length with communication carried out in a formal manner rather than by personal contact. In an adversarial relationship, gains by one partner are seen as being at the expense of the other. This contributes to the apparent lack of trust between the parties. The ensuing contracts are short-term. In general, the two parties are not ready to share information. Buyers are suspicious of the supplier's capabilities to produce quality goods and depend on inspection schemes for incoming goods.

See **Supplier development; Supplier relationships; Supply chain management: Competing through integration.**

Reference

[1] Saunders, M. (1997). *Strategic Purchasing and Supply Chain Management*. Pitman Publising, London.

AGGREGATE DEMAND

It is the sum of a factory's demands for all products over a specified period of time, commonly known as the planning horizon. The planning horizon is usually 6–18 months.

See **Aggregate planning; Capacity management in make-to-order production systems; Capacity planning: Long-range; Capacity planning: Medium- and short-range.**

AGGREGATE PLAN AND MASTER PRODUCTION SCHEDULE LINKAGE

Chen H. Chung

University of Kentucky, Kentucky, U.S.A.

Aggregate planning (AP) and master production scheduling (MPS) are two important functions in Manufacturing Resource Planning (MRP II).

Proper interfaces between the two functions will greatly enhance the effectiveness of the whole manufacturing planning and control (MPC) system. Many practical concerns need to be addressed. This article discusses practical issues such as the forecast horizons, the planning horizons, the planning time buckets and the replanning frequencies for both aggregate planning and master production scheduling. It also describes two procedures for interfacing these two planning activities. Illustrative examples and practical applications are presented.

INTRODUCTION: Aggregate planning (AP) and master production scheduling (MPS) are the front end of most production and operations planning and control systems (Vollmann *et al.*, 1996). In aggregate production planning, management is concerned with determining the aggregate levels of production, inventory and workforce to respond to fluctuating demands in the future. The aggregate plan provides an overall guideline for master production scheduling which specifies the timing and volume of the production of individual products. The master scheduler takes the aggregate decisions as targets and tries to achieve them as much as possible. Ideally, the sum of production (and inventory) quantities in the master schedule over a time period should equal the aggregate production (and inventory) quantities for that time period. However, deviations may occur due to factors such as capacity limitations at critical work-centers. The feedback from actual master scheduling performance provides information for improving future aggregate plans. To be feasible and acceptable, a master production schedule should meet the resource constraints at critical work centers. This feasibility check is called resource requirements planning (RRP) or rough-cut capacity planning (RCP). If the resource requirements planning indicates potential infeasibility at some critical operations, either the master schedule is revised to accommodate these capacity limitations or a decision is made to maintain the master schedule and implement adjustments to the capacities at relevant work centers. It may also require adjustments to the long-term capacity plan or the aggregate plan. After the feasibility check via resource requirements planning, the feasible and acceptable master schedule becomes an authorized master schedule and can be further broken down into detailed schedules with the use of lower level planning and scheduling systems such as Material Requirements Planing (MRP). This sequential production planning process is shown in Figure 1.

Obviously, each level of planning in the hierarchical planning process must be capacity sensitive. Also, the implementation of the planning process is iterative and should be repeated on a regular basis. In this respect, the interface between various levels of the planning hierarchy becomes an important concern for effective management. In the following sections, we will discuss various issues associated with the effective interface between aggregate planning and master scheduling. Methods for interfacing the two planning activities will also be discussed.

ISSUES IN INTERFACING AGGREGATE PLANS AND MASTER SCHEDULES: In a hierarchical production planning process, not only are the higher level decisions disaggregated, but also some planning parameters are broken down for detailed lower level planning. For example, different planning horizons, replanning frequencies, and planning time buckets may be used for aggregate plans and master schedules, respectively (Chung and Krajewski, 1984, 1986b). In this section, we discuss these and other issues which are relevant to the interface between aggregate planning and master production scheduling.

Forecast and Planning Horizons

A forecast horizon is the time period for which future demands are estimated. A planning horizon, on the other hand, determines how far into the future the production is planned. Since demand forecasts are part of the key inputs to the production planning process, the forecast horizon should be at least as long as the planning horizon for any planning activity. Furthermore, since aggregate plan will provide guidelines for master scheduling, the planning horizon for aggregate planning should be at least as long as that of master scheduling. On the other hand, a rule of thumb commonly used by practitioners is that the planning horizon for master scheduling should be at least as long as the cumulative manufacturing plus procurement lead-time (i.e., the critical path of the bill of material). Another rule of thumb is that the length of the planning horizon for master scheduling should be extended beyond the cumulative lead-time to provide "visibility" for planning lower level components and assemblies.

Table 1 shows the planning horizons for both aggregate planning and master scheduling in several companies as reported in Berry *et al.* (1979).

For effective planning, it is desirable to have a proper planning horizon for each level of planning. Chung and Krajewski (1984) found that the proper planning horizons for master scheduling are

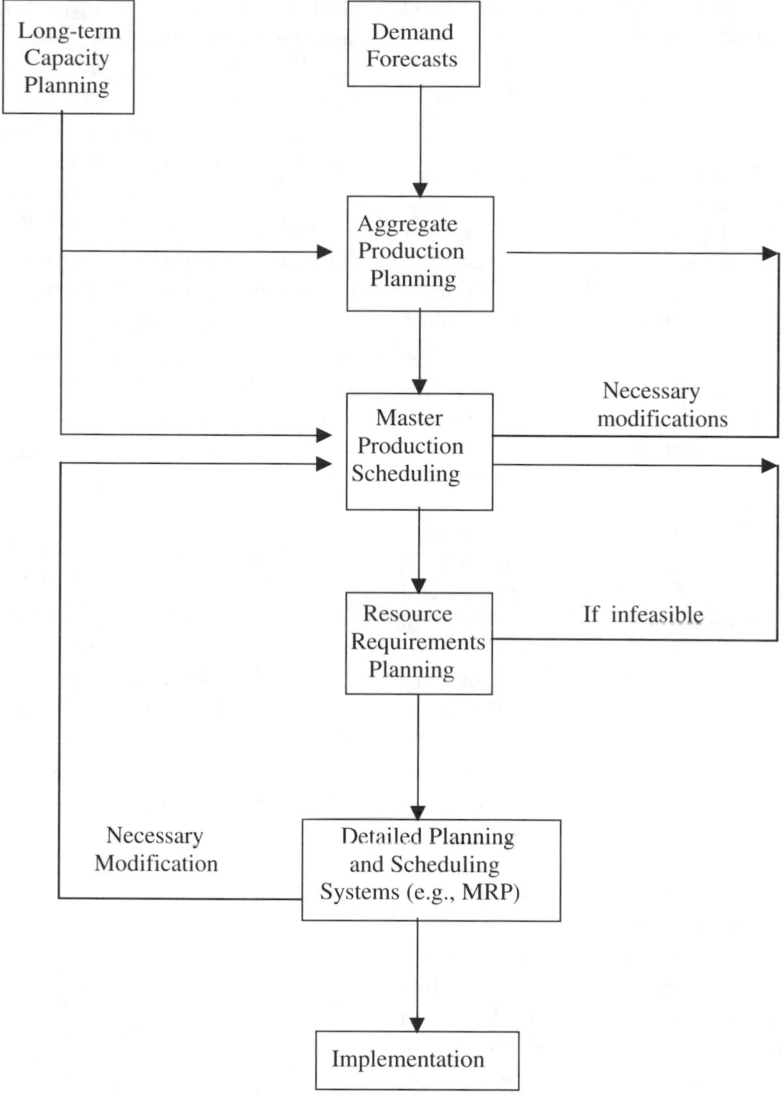

Fig. 1. A schematic diagram of sequential production planning process.

affected by the planning horizon of the aggregate plans. They also found that a firm's cost structure has an impact on the appropriate planning horizons for both aggregate planning and master scheduling. Some cost conditions allow for smaller master schedule horizons. The best horizon choice seems to be equal planning horizons for both aggregate planning and master scheduling.

Replanning Frequencies

The updating of production plans is typically achieved by rolling the planning horizon forward one period at a time. Intuitively, the more frequent the planning activities, the better the chance to correct the mistakes (if any) of the previous plan, and therefore results in better plans. However, there are at least two possible considerations which would go against frequent replanning. First, if the environment is relatively stable there is no significant benefit from frequent updating the plans. Second, there is a need to provide stability for planning components and assemblies by "freezing" a portion of the master production schedule. Typically, this frozen period corresponds to the maximum cumulative manufacturing and procurement lead-time of the products produced. It could be shorter if the company commits to having ample supplies of purchased items and/or lower level assemblies. Frozen period could be longer if resource requirements are stable.

Planning Time Buckets

It is a common practice to use different planning time buckets for aggregate planning and master scheduling. Since master scheduling is a more detailed level planning activity than aggregate planning, smaller time buckets are

Table 1. Planning horizons for aggregate planning and master scheduling in eight companies.

Company name	Planning horizon for aggregate planning	Planning horizon for master scheduling
Pfizer	1 year	1 quarter
Ethan Allen	6 months	18 weeks
Black & Decker	1 year	1 year
Abbot Lab	1 year	102 weeks[a] (20 weeks)
Dow Corning	5 quarters	26 weeks
Elliot	1 year	2 years
Hyser	1 year	1 year
Tennant	2 years	2 years

[a]Of the 102 weeks, 20 weeks of data are in weekly time buckets and the remaining 82 weeks are in monthly time buckets. The preparation of the master schedule is incorporated directly in the material requirements planning process as a part of the explosion of requirements. The first 20 weeks of data are the actual inputs to the MRP system. Thus, the actual planning horizon for master scheduling is 20 weeks.
Source: Summarized from Berry *et al.* (1979).

used. If monthly time buckets are used for aggregate planning while weekly time buckets are used for master scheduling, then one month's aggregate planning decisions will be broken down to four weeks in master schedules. Similarly, the available resources and demand forecasts at the aggregate planning level should be disaggregated and expressed in terms of the master schedule time buckets.

As shown in Table 1, it is possible to use different time buckets at the same level of planning. For example, weekly time buckets may be used for the firm or frozen portion while monthly time buckets may be used for the tentative portion of the same master schedule.

Resource Requirements Planning

Resource requirements planning or rough-cut capacity planning is an important but often overlooked step in master scheduling. Most RRP practices are basically a "trial and error" approach. If the resource requirements plan implied by the MPS turns out to be infeasible, the master schedule or even the aggregate plan has to be revised. It may take several iterations until a master production schedule can be finally authorized. The RRP function is directly linked to the available resources specified by the aggregate plans. Thus, RRP constitutes an important aspect of the interface between aggregate planning and master scheduling. Berry *et al.* (1982) propose three different methods for RRP. They are: (1) The Overall Factors Method, where the total resource requirements implied by a master schedule are allocated to critical work centers according to their historical usage percentages; (2) The Capacity Bills Method, where the resource requirements for the work centers are determined by the amount of worker-hours required by the products that are processed in these work centers; and (3) The Resource Profiles Method, where the resource requirements are determined by the time-phased operations setback charts of the products produced. Chung and Krajewski (1986b) propose a procedure, called Integrated Resource Requirements Planning (IRRP), which incorporates resource feasibility check in the preparation of the master production schedules. In this way, whenever the master schedule is developed, it is guaranteed to be a feasible schedule in terms of resource requirements. In their approach, the RRP function is included as a constraint of the master schedule.

METHODS FOR INTERFACING AP AND MPS:

Interfaces via a Goal-Stipulation/ Dual-Price Method

One approach to carrying out the between-level interface in a hierarchical management system is the goal-stipulation/dual-price method. Ruefli (1971) proposed a "generalized goal decomposition model" to represent decision-making in a three level hierarchical organization. The three levels – Central Unit, Management Units, and Operating Units – are actually similar to the "Anthony Taxonomy" (Anthony, 1965). The central unit coordinates decision activity by generating goals (and/or selecting resource levels) for a subordinate level of management units. Each management unit chooses activity levels based on the goals and resource levels set by the central unit. It also determines values for the dual variables (i.e., the opportunity costs) in its subunits and transmits these economic indicators to the central unit so that the latter can evaluate and improve its goal generating policies. The operating units are then responsible for generating project proposals for their respective management units.

The concept of goal decomposition by Ruefli (1971) can be used to interface aggregate planning and master production scheduling problems. The procedure starts with the solutions to the aggregate planning problem. Then a vector of goal stipulations is passed down to the maser scheduling problem which uses the goal stipulations in its objective funtion. After the master scheduling problem is solved, the vector of dual prices associated with the goal constraints is passed back to the aggregate planning model to determine a new set of goal stipulations. The process is repeated until the goal deviations of the master scheduling solutions are at a minimum and no further revisions are needed i.e., no further net decrease in the goal levels are expected. King and Love (1980) reported a successful real world application of interface procedures similar to those described above.

A Rolling-Horizon Feedback Procedure

As mentioned above, rolling horizons are commonly used in real world planning and problem-solving. A typical scenario of a rolling-horizon feedback procedure is as follows: Solve the planning problem and implement only the first period's decisions; for the following period, update the model to reflect information collected in the interim, re-solve the problem. Again implement only the imminent decision pending subsequent model runs (Baker, 1977). Thus, the practice of rolling horizons is to routinely update or revise the plans taking into consideration more reliable data as they become available.

To illustrate the interface between aggregate planning and master scheduling via a rolling-horizon feedback procedure, let us assume that the aggregate planning problem is to be done at the product family level. Master production scheduling determines the timing and sizing of the production of individual products (i.e., the end items) within product families. Since the aggregate production and inventory levels are determined by the aggregate plan, the master scheduler takes these aggregate decisions as targets and tries to achieve a master production schedule that adheres to the spirit of the aggregate production plan. The decisions made at master scheduling level are allowed to deviate (with some penalties) from the goals set by the aggregate plan. The sum of production (inventory) quantities of all items in each period are targets at the aggregate production (inventory) level for that period.

With the use of a rolling horizon strategy, the actual implementation of the first period or the first few periods of a master schedule will be reported back as input to the aggregate planning model for the next planning session. This feedback is not only useful but also necessary for improving aggregate planning (and therefore master scheduling) in the future.

Figure 2 outlines the solution procedure for interfacing the two planning activities via a rolling-horizon mechanism. The procedure begins with the solution of the aggregate plan for a given set of factor values. Each month's production and inventory quantities for each family, and the regular-time and overtime workforce levels are passed on to the master scheduling level. The master scheduling problem is then solved for the given planning horizon. The master scheduling decisions regarding the production, inventory, and backorder quantities for items in each product family are passed back to the aggregate planning level. Finally, the horizon is rolled forward one month and the procedure is repeated.

See Chung and Krajewski (1987) for a simulation study using the above procedure. Allen and Schuster (1994) report a practical application of the above aggregate planning and master scheduling models and the rolling-horizon feedback procedure for interfacing the two plans.

CONCLUSION: Aggregate planning and master scheduling, which form the front end of a manufacturing planning and control system, greatly influence the effectiveness of the whole manufacturing systems performance. Two methods for interfacing the two plans are described in this article. Additional issues such as planning horizons, replanning frequencies, planning time buckets, resource requirements planning, etc., are also important to the interface between aggregate plans and master production schedules.

We have illustrated two different mechanisms for dealing with the interfaces between the aggregate planning and master scheduling. Obviously, it is possible to combine the Goal-Stipulation/Dual-Pricing mechanism with the rolling-horizon practice. However, this becomes computationally cumbersome since goal stipulations and dual prices have to be passed back and forth between aggregate plans and master schedules for several iterations before the horizon is rolled forward. Decision support systems may be developed to facilitate such interface processes. See Chung and Luo (1996) for the design of a computer support system for decentralized production planning.

See **Aggregate planning; Bill of material; Capacity bills method; Capacity management in...; Capacity planning:...; Critical path of the bill of materials; Hierarchical planning and control; Hierarchical production planning; Integrated resource requirements planning (IRRP); Master production scheduling; Overall factors method; Resource profiles method; Resource requirements planning; Rough-cut planning.**

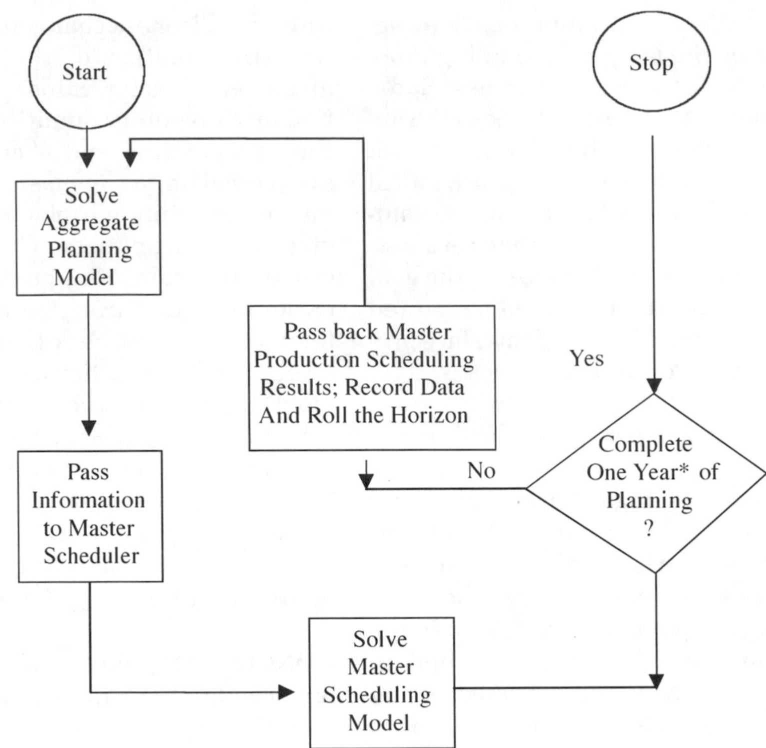

*Assuming that a one-year planning horizon is used.

Fig. 2. Solution procedure for a rolling-horizon schedule.

References

[1] Allen, S.J. and E.W. Schuster (1994). "Practical Production Scheduling with Capacity Constraints and Dynamic Demand: Family Planning and Disaggregation," *Production and Inventory Management*, fourth quarter, pp. 15–21.

[2] Anthony, R.N. (1965). *Planning and Control Systems: A Framework for Analysis*, Cambridge, Mass.: Harvard University Graduate School of Business Administration, Studies in Management Control.

[3] Baker, K.R. (1977). "An Experimental Study of the Effectiveness of Rolling Schedules in Production Planning," *Decision Sciences*, 8(1), pp. 19–27.

[4] Berry, W.L., T. Schmitt, and T.E. Vollmann (1982). "Capacity Planning Techniques for Manufacturing Control Systems: Information Requirements and Operational Features," *Journal of Operations Management*, 3(10), pp. 13–25.

[5] Berry, W.L., T.E. Vollmann, and D.C. Whybark (1979). *Master Production Scheduling*: Principles and Practices, American Production and Inventory Control Society.

[6] Chung, C.H. and L.J. Krajewski (1984). "Planning Horizons for Master Production Schedules," *Journal of Operations Management*, (4), pp. 389–406.

[7] Chung, C.H. and L.J. Krajewski (1986a). "Replanning Frequencies for Master Production Schedules," *Decision Sciences*, 17(2), pp. 263–273.

[8] Chung, C.H. and L.J. Krajewski (1986b). "Integrating Resource Requirements Planning with Master Production Scheduling," Lee, S.M., L.D. Digman and M.J. Schniederjans, eds., *Proceedings of the 18th Decision Sciences Institute National Meeting*, Hawaii, pp. 1165–1167.

[9] Chung, C.H. and L.J. Krajewski (1987). "Interfacing Aggregate Plans and Master Production Schedules via a Rolling Horizon Procedure," *OMEGA*, 15(5), pp. 401–409.

[10] Chung, C.H. and W.H. Luo (1996). "Computer Support of Decentralized Production Planning," *The Journal of Computer Information Systems*, 36(2), pp. 53–59.

[11] King, R.H. and R.R. Love, Jr. (1980). "Coordinating Decisions for Increased Profit," *Interfaces*, 10(6), pp. 4–19.

[12] Ruefli, T. (1971). "A Generalized Goal Decomposition Model," *Management Science*, 17(8), pp. 505–518.

[13] Vollmann, T.E., W.L. Berry, and D.C. Whybark (1996). *Manufacturing Planning and Control Systems*, (Fourth Edition), Richard D. Irwin, Homewood, Illinois.

AGGREGATE PLANNING

Aggregate planning is also called aggregate production planning or simply production planning. The goal of Aggregate planning is to determine the aggregate levels of production, inventory, and workforce to respond to fluctuating demand in the next 6–18 months. The term "aggregate" refers to some measure of output or input that permits aggregation across several products. Thus, the aggregate plan is usually stated in aggregate quantities such as hours, dollars, or units of product families. The aggregation process also implies the aggregation of resource and time. For example, in order to plan the use of various productive resources such as labor, it would be necessary to sum up or aggregate the resource requirements of various products. On the other hand, it is likely that bigger time buckets are used in developing aggregate plans while smaller ones are used for detailed level planning. In this sense, the aggregation process will also include the aggregation of time. Aggregate production planning problem may be viewed as the search for the optimal trade-off between the cost of changing production levels against the cost of avoiding production changes to meet demand changes during the plan period. The latter costs include the costs of inventories and the cost of changing workforce levels. There are many techniques that have been suggested in the literature for developing aggregate plans. The non-optimal solution methods include the graphical method, the tableau method, etc. The optimal solution methods include linear programming, the linear decision rules, etc. There are also some heuristic procedures such as the search decision rule, the management coefficient model, the production switch heuristic, etc.

See **Aggregate plan and master production schedule linkage; Capacity planning: Long-range; Time buckets.**

Reference

[1] Vollmann, T.E., W.L. Berry, and D.C. Whybark (1996). *Manufacturing Planning and Control Systems*, (Fourth Edition), Richard D. Irwin, Homewood, Illinois.

AGILE LOGISTICS (ENTERPRISE LOGISTICS)

Noel P. Greis and John D. Kasarda

University of North Carolina, North Carolina, U.S.A.

DESCRIPTION: Rapidly changing global markets for competitively-priced custom-configured goods have triggered major structural changes in how companies source, manufacture, and distribute component materials and assembled products (Greis and Kasarda, 1997; Council of Logistics Management, 1995). In this new architecture of production, networks of strategically-aligned companies joined together in an extended enterprise have replaced individual companies as the effective unit of competition, with supply chain competing against supply chain. Agile logistics is the integration of manufacturing, transportation and information across the extended enterprise (i.e. the entire supply chain) to provide competitive advantage through real-time flexible response to unique and changing customer requirements. It comprises the set of practices and technologies that enables these strategically-aligned suppliers, producers, and distributors to coordinate geographically and institutionally dispersed resources into a unified and effective extended enterprise with the following capabilities:

- Simultaneous information exchange across the enterprise
- Full integration of supply, manufacturing and delivery
- Seamless and uninterrupted flow of goods and materials
- 24-hour delivery almost anywhere in the world
- Customized supply chain reconfigurability
- Total resource visibility across the supply chain
- Real-time responsiveness to customer requirements

HISTORICAL PERSPECTIVE: Agile enterprises are different from traditional organizations in that they are dynamic and flexible, and are based on collaboration rather than competition Goldman and Preiss, 1996; Preiss, Goldman and Nagel, 1996). These strategically-aligned companies are focused on capturing specific market opportunities as they arise. To appreciate the role of logistics within the agile extended enterprise, it is useful to examine the functional relationship between manufacturing and logistics in mass production, lean and agile systems as shown in Figure 1 (Preiss, 1997).

1. ***Mass Production Logistics:*** Mass production emerged to manufacture large quantities of standard products in an environment where customer needs and competitive factors are relatively static or unchanging. In such a system, manufacturing processes are managed independently. Since processes are independent, information sharing is minimal and processes function on the basis of forecasts in a "push" system. Demand variability is buffered by inventory, either work-in-progress inventory within the factory or in warehouses and

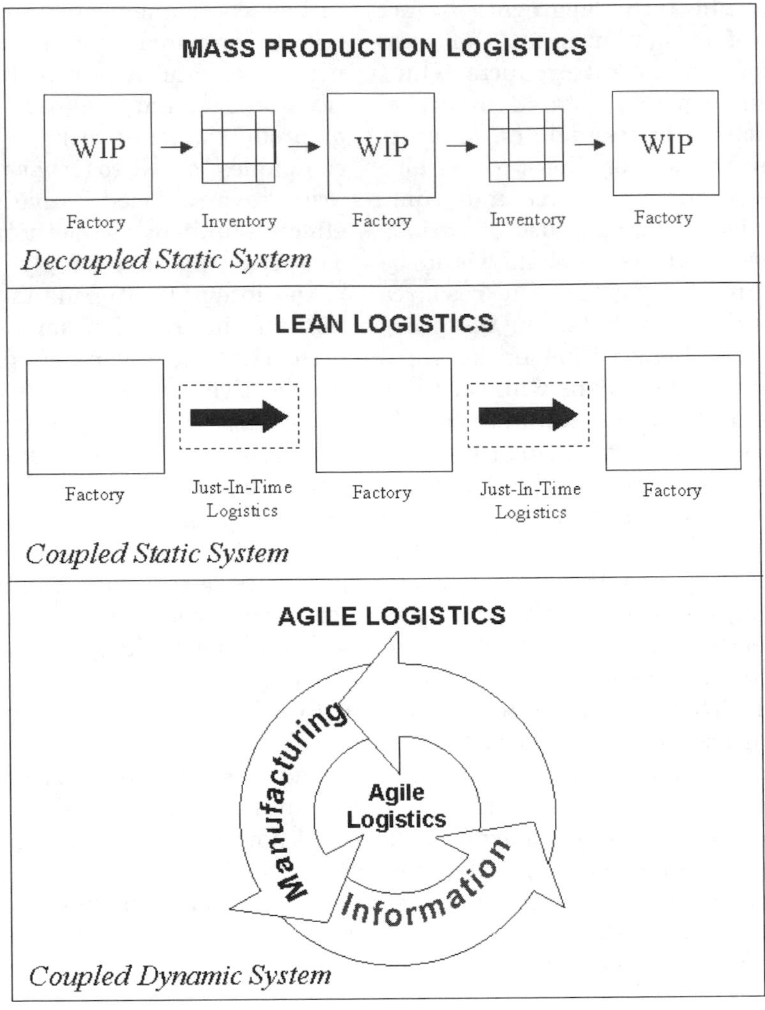

Fig. 1. Evolution of an agile logistical system.

distribution centers. Here, the function of logistics is to manage inventory, and to transport materials between decoupled processes in the system.

2. ***Lean Logistics:*** A distinguishing feature of lean systems that emerged in the early 1980s is closer alignment between manufacturing processes through logistics. Manufacturing processes at the upstream suppliers and downstream manufacturers are linked and dependent, which eliminates the need for warehouses and work-in-process inventories. The lean system succeeds by sharing information to assure the coordinated movement of materials from process to process on a just-in-time basis. Deliveries are frequent, small and tightly scheduled. In this static, but coupled system, the role of logistics is to link individual processes to their nearest neighbors so that lean manufacturing can operate as a "pull" system.
3. ***Agile Logistics:*** In agile systems, manufacturing and logistics are integrated into a single synchronized production system that spans the entire supply chain or extended enterprise. Unlike the mass production and lean systems, the agile system is designed to operate within a dynamic environment that is characterized by rapidly changing, and often unpredictable, customer needs. Agile system process linkages are so extensive that we can talk about a single extended enterprise-spanning logistics environment. The role of logistics is to manage the *total* process of product creation and delivery, *where response replaces inventory and flows replace stocks*. Because of its ability to span the multiple functions and boundaries of an organization, logistics becomes a critical knowledge center for coordinating the seamless and uninterrupted flow of materials and information across the agile enterprise.

STRATEGIC PERSPECTIVE: For companies like Wal-Mart, logistics has become the centerpiece of the retailer's entire competitive strategy, while for other companies logistics is a crucial enabler of their competitive strategy. Below, we describe key logistics-based elements of agile competitive strategy.

1. ***Customized Solutions:*** Customers are increasingly demanding products that are differentiated by feature, color, size, shape, handling and delivery requirements (St. Onge, 1997). Few companies today have the luxury of a stable and predictable customer base where shipments of standard products flow repeatedly along well-traveled supply and delivery paths. Individual consumers are demanding customized apparel, sporting equipment and consumer products. At the same time, mass merchandisers insist on increasing their control over aspects of design, manufacture and delivery. In response, consumer products companies like Proctor and Gamble have been forced to adjust their "one-size-fits-all" strategy to handle a range of unique supply chain needs from high-volume, low-priced needs of mass merchandisers like Kmart and Target who demand immediate responsiveness, to smaller "less-than-truckload" customers who get their products from distributors.
2. ***Changing Distribution Models – Direct to the Customer:*** Companies are adopting new logistics strategies that bypass traditional distribution channels – frequently moving products by air directly to the customer's premises. Non-store shopping – which eliminates the retailer altogether – includes home shopping channels, interactive shopping by computer, and direct mail ordering. Although a small phenomenon currently, Kurt Salmon estimates that direct or non-store sales will account for 55% of the more than $2 trillion worth of general merchandise sold in the United States by 2010. Companies such as Dell and Gateway 2000 have already shaken the personal computer industry by selling build-to-suit hardware direct to the customer, bypassing warehouses and retail outlets. Likewise, through a partnership with FedEx and leading growers of fresh cut flowers, Calyx and Corolla is able to bypass the Florists Telegraph Delivery to fulfill catalogue orders for floral arrangements.
3. ***Global Sourcing and Distribution:*** The dual trends of global sourcing and global product distribution place new demands on logistics practices and infrastructures to act as an integrated system that can move products large distances quickly and efficiently. The IBM personal computer that is assembled and sold in the United States contains electrical components manufactured in Thailand, glass screens from Taiwan, keyboards from Indonesia, disk drives from Singapore, integrated circuits from Japan and microprocessors from Korea. At the same time, "global product" strategies cannot work unless companies are able to provide the same high-quality service to all customers worldwide.
4. ***Logistics as a Source of Customer Value:*** As the requirements for speed and flexible response become greater due to rapidly-changing global markets and the increasing dispersion of productive resources worldwide, logistics becomes an increasingly attractive way for companies to differentiate their products (Daugherty, Sabath and Rogers, 1992). Companies that are able to add customer value to their product through better logistics – for example 24-hour global delivery – can carve out a distinctive niche in the marketplace. Increasingly, companies are required to provide value-added logistical services beyond speed and low cost. Companies may now be required to provide door-to-door service or special handling for sensitive, high-value or dangerous shipments, and even overnight replacement parts. Customer value can also be created through such logistical elements as customized handling, time-definite delivery, tracking and tracing, consistency of overall service, ease of placing orders, and convenient return through reuse logistics.

IMPLEMENTATION: Organizations today typically are not readily able to accommodate dynamic change in their logistical arrangements. They are less eager to create new competitive advantage by implementing agile logistics practices. Table 1 identifies eight agile change domains and logistics examples (Dove, 1995).

These change domains span the full range of practices that comprise agile logistics and that can be implemented to support the agile logistics strategies identified in the previous section. For example, agile change domains include operational capabilities with respect to changes in order size (i.e. doubling order size), delivery time (shipping one week early), and routing (change in customer location). These operational examples require flexibility within established supply chain configurations and relationships. At the opposite end of the spectrum, other change domains include the ability to completely reconfigure entire supply chains in response to new customer requirements, for example after the collapse of a third-party logistics relationship. Another

Table 1. Change domains for agile logistics.

Change domain	Description	Logistics application
Creation	Build something new	Customize delivery system for major customer
Capacity	Increase/Decrease Existing Resource Mix	Double shipment to customer
Capability	Add/Delete Resource Types	Ship to customer in new location
Reconfiguration	Change Relationships Among Modules	Fulfill order from another warehouse
Migration	Event-based Change of Fundamental Concepts	Implement postponement strategy
Performance	Real-time Operating Surprise	Ship one week early
Improvement	Continuous, Incremental Upgrade	Increase delivery speed for customer
Recovery	Reincorporate Corrected Failures or Alternatives	Recover from failure of third-party logistics provider

example is the ability to implement customized delivery systems so that each customer views the system as a private network.

TECHNOLOGY PERSPECTIVE: The ability to implement these agile practices requires a logistical system that is global, speed-based, collaborative and flexible. Five interlocking components of an agile logistical system are described below (Kasarda and Rondinelli, 1998; Jordan, 1997):

1. ***Global Transportation Network:*** In order to meet changing and geographically diverse customer requirements, all major modes of transportation (air, sea, truck and rail) must be integrated into a global transportation network capable of reaching members of the agile extended enterprise and its customers worldwide. As pressures for speed and agile response buildup, air freight will increasingly be used to move materials around the world. Companies must balance the cost of air freight against the cost of inventory and the strategic value of quick response, and not against the cost of alternative transportation modes alone.
2. ***Intermodal Capability:*** The speed and flexibility on which the agile extended enterprise depends requires the ability to switch modes of transport for shipments moving through the network – for example, from one rail route to another, or from sea to air. A manufacturer must be able to select from multiple transportation modes and must be able to mix modes optimally and seamlessly as inbound receipts and outbound shipments change. Transitions between modes must be quick, efficient, and transparent to the user. By increasing flexibility and reducing non-value-adding time, intermodal operations can significantly reduce order-to-delivery cycle times.
3. ***Shared Communications Infrastructure:*** Communication and information technologies must be in place to connect all members of the extended enterprise and its customers in real-time. Smooth transitions at the seams of the logistical system – when goods move from one transportation mode to another or from one nation to another – cannot occur without such information sharing. Although the use of electronic data interchange (EDI) is becoming increasingly common, these systems are typically customized and form dedicated links between trading partners without the ability to support the flexibility or integrated logistics of the extended agile enterprise. An agile communications infrastructure must be based on message standards and open architectures, with data separation capabilities to maintain security over those parts of the information system that are proprietary.
4. ***Tracking and Tracing:*** Shared communications throughout the extended enterprise enables the ability to track goods and materials across organizational and national borders. While many companies can monitor the movement of materials within their organizational walls, few have the ability to track their products or materials once they cross those walls. The agile extended enterprise requires not only up-to-date information about the location and disposition of assets, but also information linking the location of the asset with available transportation opportunities.

5. ***Shipment Integration Capability:*** Finally, the ability to deliver small batch sizes as cost effectively as large batch sizes – what is referred to as *shipment integration* – is essential to implement mass customization and just-in-time production. While a great deal of attention has been given to the problem of manufacturing low-cost customized products, relatively little attention has been given to interpreting this concept across the supply chain. The key to mass customized logistics is the ability to consolidate and bundle orders from different companies that are destined for the same geographic region or customer. Shipment integration, when combined with tracking and tracing, enables materials merger so that customized products can be produced efficiently and cost-effectively on a just-in-time basis.

CASES: Many companies are implementing elements of agile logistics within their supply chains:

1. ***Textile and Apparel:*** Benetton, the Italian sportswear manufacturer, distributes more than 50 million pieces of clothing worldwide through its Quick Response system (Dapiran, 1992). Benetton relies on a small number of manufacturers around the world to produce coordinated garments for its more than 6,000 small boutique-style shops. All distribution is orchestrated out of a single facility that air freights 230,000 pieces of clothing a day to 5,000 stores in 60 countries. Retail orders are transmitted to Benetton's main computer in Italy where production schedules are coordinated and distribution arranged. By linking retailers, production and warehousing, boutiques can expect to receive orders in only a few weeks while competitors take much longer. By integrating agile manufacturing with information and transportation, companies like Levi Strauss are able to deliver *customized* garments—directly to the customer's home (Holusha, 1996). Digitally scanned body measurements taken at the retail store are relayed electronically to Levi's factory in Tennessee for sewing, followed by express delivery to the customer's home. Using new manufacturing processes with express delivery, Levi Strauss has gone beyond quick response replenishment systems to provide a customized product for only $10 more than off-the-shelf prices.
2. ***Personal Computer Industry:*** Building computers only after customers order them has enabled companies like Dell and Gateway to avoid costly inventories of unsold, older-generation machines and components (McGraw, 1997). By selling directly to customers, thereby avoiding retailers and resellers, these companies are able to price their machines as much as 20% less than competitors. To challenge Dell, Compaq is implementing a modified direct-sales approach under which Compaq will build a portion of its machines only after an order is received from a reseller (Ramstad, 1997). To implement this strategy, Compaq opened a warehouse in Houston where suppliers store parts until required by the nearby Compaq assembly plant. Assemble-to-order strategies are still risky for manufacturers of high-end workstations since companies can still be left with expensive unused components when they misjudge product cycles. A manufacture-to-order strategy would solve these problems. Since most components are sourced from Asia, however, such a strategy would not be possible without a global logistics infrastructure that provides just-in-time express transportation directly from suppliers' manufacturing sites. Such an infrastructure must also feature an information system to support materials merger, and tracking and tracing for coordinated arrival at the assembly site.

GENERAL CONCLUSIONS: There is increasing recognition that established practices of moving goods are inadequate to meet the needs of manufacturers competing in the new speed-based competitive environment in which suppliers and markets are increasingly dispersed around the world. Companies have already reduced costs and improved productivity within the walls of their organizations through agile manufacturing practices and technologies. In the future, reduced costs, productivity increases, and improvements in customer responsiveness will come from better management of supply chain relationships, and more efficient information and material flows across organizational boundaries. Agile logistics has emerged in response to companies' concerns that streamlining and rationalization of their supply chains do not necessarily improve their ability to respond to customer needs for flexible and real-time response – and that new approaches and practices are needed. It is unlikely that the competitive forces that are driving companies to become fast and flexible will lessen in the future. Such pressures will continue to drive the integration of manufacturing, transportation and information – the hallmark of agile logistics. The implementation of agile logistics practices and infrastructures on a broad scale will no doubt be facilitated even further by continuing advances in new information and transportation technologies and infrastructures throughout the world.

See **Agile manufacturing; Council of logistics management, lean logistics, third party logistics relationship.**

References

[1] Bowersox, D.J. and P.J. Daugherty (1995). "Logistics Paradigms: The Impact of Information Technology," *Journal of Business Logistics*, 16(1), 65–80.

[2] Council of Logistics Management (1995). *World Class Logistics: The Challenge of Managing Continuous Change*, Council of Logistics Management, Illinois.

[3] Dapiran, P. (1992). "Benetton – Global Logistics in Action," *International Journal of Physical Distribution and Logistics Management*, 22(6).

[4] Daugherty P.J., R.E. Sabath and D.S. Rogers (1992). "Competitive Advantage Through Customer Responsiveness," *The Logistics and Transportation Review*, 28(3), 257–271.

[5] Dove, Rick (1995). "The Challenges of Change," *Production*, 107(2), 16–17.

[6] Evans, G.N., M.M. Naim, and D.R. Towill (1993). "Dynamic Supply Chain Performance: Assessing the Impact of Information Systems," *Logistics Information Management*, 6(4), 15–25.

[7] Goldman, S., R. Nagel and K. Preiss (1995). *Agile Competitors and Virtual Organizations*, Van Nostrand Reinhold, New York.

[8] Greis, N.P. and J.D. Kasarda (1997). "Enterprise Logistics in the Information Era." *California Management Review*, 39(4), 55–78.

[9] Holusha, J. (1996). "Producing Custom-Made Clothes for the Masses." *New York Times*, February 19, 1996.

[10] Jordan, J. (1997). "Enablers for Agile Virtual Enterprise Integration." *Agility and Global Competition*, 1(3), 26–46.

[11] Kanter, R.M. (1995). *World Class: Thriving Locally in the Global Economy*, Simon and Shuster, New York.

[12] Kasarda J.D. and D.A. Rondinelli (1998). "Innovative Infrastructure for Agile Manufacturing," *Sloan Management Review*, 39(2), 73–82.

[13] McGraw, D. (1997). "Shootout at PC Corral," *U.S. News and World Report*, June 23, 1997.

[14] Preiss, K., S. Goldman, and R. Nagel (1996). *Cooperate to Compete: Building Agile Business Relationships*. Van Nostrand Reinhold, New York.

[15] Preiss, K. (1997). "A Systems Perspective of Lean and Agile Manufacturing." *Agility and Global Competition*, 1(1), 59–75.

[16] Ramstad (1997). "Shift at Compaq May Push Down Prices for PCS." *Wall Street Journal*, March 31, 1997.

[17] St. Onge, A. (1997). "Personalized Production: Riding the Information Superhighway Back to the Future," *Supply Chain Management Review*, 1(1), 44–52.

AGILE MANUFACTURING

Pratap S.S. Chinnaiah and Sagar V. Kamarthi

Northeastern University, Boston, USA

INTRODUCTION: Business are witnessing unprecedented changes today. New products, new processes, new technologies, new markets, and even new competitors are appearing and disappearing within short periods of time. Historically, mass production has evolved into lean production. Now lean production is evolving into agile manufacturing. Until the 1950s companies focused on productivity improvement and in the 60s and 70s they concentrated on quality enhancement. In the eighties, while companies worked hard to achieve flexibility, in the 90s they are challenged by the need to increase agility. To review current issues in agility, this article describes the market forces that demand agility, the elements that constitute agility, agility enablers, and agility implementation.

ELEMENTS OF AGILITY: The central idea in agile manufacturing is that an enterprise should be built on the competitive foundations of continuous improvement, rapid response, quality improvement, social responsibility, and total customer focus.

Agility is the ability of producers of goods and services to thrive in rapidly changing, fragmented markets. Agility is a comprehensive response to the challenges posed by a business environment dominated by changes and uncertainty. For a company to be agile is to be capable of operating profitably in a competitive environment of continually and unpredictably changing customer opportunities. An agile enterprise must have broad change capability that is in balance across multiple dimensions. Would the enterprise be considered agile if a short-notice change was completed in time, but at a cost that eventually bankrupted the company? Would it be agile if the changed environment thereafter required the specialized wizardry and constant attention of one specific employee to keep it operational? Is it agile if change is virtually free and painless but out of synch with market opportunity timing? Is the enterprise agile if it can readily accommodate a broad category of changes that are no longer needed, or too narrow for the latest requirements? These questions lead to four principal change proficiency metrics to define agility: time, cost, robustness, and scope.

Completing a change in a timely manner is the only effective way to respond, while the *time* of the change alone does not provide a sufficient metric for agility. Virtually anything can be changed if *cost* is not a constraint, but change at any cost is not a viable solution. If the response to the change costs too much relative to a competitor's cost, it will limit shareholder profits. A change, quick and economical, is still not a sufficient profile of agility. If after the change, the modified system requires continual attention to remain functional, the change accommodation is insufficiently *robust*. A company may end up with a fragile result if standard procedures are bypassed in the process of changing quickly and economically. Finally, a system is considered to be agile precisely because it is able to thrive on change, but how much change can the system handle? The metric of *scope* addresses this question. Scope is the principal difference between flexibility and agility. Flexibility is the capability the deal with planned responses to anticipated contingencies. Agility, on the other hand, means the system can operate within a defined range of changes and it can be deconstructed and reconstructed to handle unplanned changes. Thus, for an enterprise to be agile, it must have a balanced response-to-change capability across the four change proficiency metrics: time, cost, robustness, and scope.

MARKET FORCES THAT DRIVE AGILITY: As the context of commercial competition changes dramatically, the conduct of business changes correspondingly. Today, the following market forces demand changes in the conduct of business.

Market Fragmentation

Markets of all kinds are being fragmented at an accelerating pace. Companies are segmenting customer groups and pricing the same goods or services differently depending on the circumstances of the transaction. For example, with virtually unlimited packaging possibilities, Mars company prices their M&M candies differently when they are sold in different packages of the same weight. Companies are segmenting markets according to function and are exploiting economies of scope which enables high variety and low unit volume production (Stipp, 1996).

Production in Smaller Lot Sizes

Many companies are already capable of making several products, that are different from each other, on a high volume line with little or no increase in production costs. Lot sizes can be as low as one. This type of production capability has revolutionized marketing (Kaplan, 1993).

Information Capacity to Treat Mass Customers as Individuals

More and more companies are discovering that they can produce customer-configured products, "to order" instead of "to forecast," and in doing so they can generate benefits far beyond elimination of inventories. The consumer initiates the interactive relationship with the company through which the product is jointly defined (Pine II, 1993). Massive customer databases and production equipment innovations permit effective production of smaller and smaller orders at significantly lower costs (Lau, 1995).

Shrinking Product Cycle Times

The decreasing product development cycle times, increasing proliferation of models, and accelerating pace of the introduction of new or improved models are among the most aggressive challenges of contemporary competition. Today, Panasonic's consumer electronics product development cycle time is three months. That is, the cycle time of any given model of CD player, TV, VCR, cassette deck, or stereo receiver is just 90 days. During that time its successor is being designed, tested, and put into production. The design, development, production, distribution, and marketing processes are continuous and overlapping.

Convergence of Physical Products and Services

The traditional distinction between goods and services and companies is vanishing. A direct result of this convergence is that manufacturers companies are acquiring the capability to create information and provide additional service to customers.

Global Production Networks

No market is domestic anymore, and no producer is restricted to domestic production only. The addition of high-capacity information and communication systems to the existing global transportation systems opens foreign markets to many producers.

Simultaneous Intercompany Cooperation and Competition

To a degree unprecedented in history, companies are entering into partnerships, joint ventures, and collaborations of every imaginable kind including the formation of virtual companies. The three U.S. Companies, GM, Ford, and Chrysler were unwilling to develop catalytic converters cooperatively in the 1970s. As a result, each company spent hundreds of millions of dollars developing essentially the same product. Today, the three firms have joined a multifaceted consortium, USCAR, which will permit the joint development of technologies, materials, and components ranging from structural plastics to electric vehicle-control systems.

Distribution Infrastructures for Mass Customization

Agile competition rests on individualized products and interactive customer relationships. Inevitably, new product distribution mechanisms are emerging as part of agile manufacturing. For example, direct marketing by producers of goods require customer-centered production.

Corporate Reorganization Frenzy

The US companies have been implementing a wide range of initiatives such as just-in-time, total quality management, flexible manufacturing systems, business process reengineering, ISO 9000 in order to improve their competitiveness and profitability.

ENABLING SYSTEMS FOR AGILITY: The change in a manufacturing system's behavior and structure needed for agility is enhanced and enabled by a number of items. Special skills are required of the people employed in agile systems. Because it is neither possible nor desirable to automate all activities in these systems, the people are required to be highly skilled, multiskilled craftsmen. Continuous improvement of the skills of the personnel will be an essential part of an agile manufacturing facility. The teams working on individual orders must operate in a frictionless manner and must stress the *process* rather than traditional *functions*. For example, at Bally Engineered Structures, Inc. factory floor workers are now responsible for planning and supervising their own work. The use of virtual teams consisting of product designers, tool planners, and production engineers located at far fling geographical locations is another example. The objective in agile manufacturing is to combine the organization, technology, and people into an integrated whole.

The implementation and operation of agile systems are possible with both low and high technologies. The more the complexity of the technology involved in an agile manufacturing system, the higher the cost of implementation and operation of the production system. Generally, highly flexible general-purpose technologies are preferred in agile manufacturing systems.

Several technologies such as computer-aided design (CAD), virtual reality, customer database, electronic kiosks, and multimedia communications may be used at the marketing stage for identifying individual customer requirements. Some technologies specifically required for meeting the customized orders in arbitrary lot sizes are configurators, stereolithography, internet, multimedia workstations, personalized smart cards, electronic data interchange (EDI), etc. A configurator is an expert system that assists in developing valid product and process descriptions to meet a customer's specific needs. Configurators create accurate drawings and designs, speed up proposal creation, and increase sales efficiency for mass customization. Many software vendors are now offering configurator modules integrated with Enterprise Resource Planning (ERP). These software systems permit integration of sales, order processing, and engineering. For example, a software known as PROSE (Product Offerings Expertise) is a knowledge-based engineering and order processing platform that supports sales and order processing at AT&T network systems. A sophisticated information management system called CDIN (Computer-Driven Intelligence Network) at Bally Engineered Structures, Inc. connects everyone in the company as well as independent sales representatives, suppliers, and customers together. Similarly, John Deere Harvester Works uses a configurator software called OptiFlex to configure the harvester features such as row count, or the nature of fertilizer it can handle. It is reported that literally thousands of configurations can be produced using OptiFlex.

Virtual reality can play an important role in the process of translating a customer's requirements into design specifications. For example, customers buying their kitchen units from Matsushita appliances can experience their kitchen environment with the help of data gloves and 3-D vision systems before placing an order. Such virtual reality tools are known as experience simulators. Similarly, close communication between manufacturers and customers can be realized with multimedia communications technology. Several geographically separated product development teams such as a customer group, a manufacturing group, and a design group can work together with the help of multimedia communication technology. Various teams can exchange documents, see each other on a tiled video display in a window on a computer screen, talk, exchange ideas, modify product specifications and solve manufacturing problems. Such facilities are labeled as virtual meeting rooms.

A new line of factory automation software called CIMPLICITY that facilitates mass customization is available from GE Fanuc. With the help of an FMS in its production shop, Honeywell Microswitch Corporation is able to drive product customization. Personalized smart cards that have details of personal features such as physical body dimensions and medical information are being visualized to make mass customization work faster. At the development stage, technologies such as personalized smart cards, electronic kiosks, internet, CAD, CAE, configurators, and concurrent engineering are useful in reducing design lead times. Some technologies such as rapid prototyping, web-aided design (or internet-aided design), virtual machining, virtual assembly & testing, cost & risk analysis models, and product realization process are being developed by TEAM (Technologies Enabling Agile Manufacturing) program established by the US Government (Herrin, 1996). At the production stage, in addition to flexible integrated manufacturing facilities, skilled people play a very important role. The technology should augment peoples' skills and abilities. However, technology is being used for automating certain tasks where productivity and safety can be improved considerably. Achieving seamlessness among the different phases of the value chain must be the goal of employing technology and skilled people together. The emerging production concept – agile manufacturing – offers strategies to integrate organization, people, and technology into a coordinated interdependent system.

CONSTRUCTING AGILE CAPABILITIES: The journey to agility is a never-ending quest to do better than the competition, even as the competitive environment constantly changes. The agile paradigm is concerned principally with an unpredictable change. To understand the kinds of change impacting an enterprise and analyze the enterprise's ability to respond, the change is decomposed into various domains. Building a model of *change domains* makes it easier to analyze potential agile characteristics. Dove *et al.* (Dove, 1996; Dove *et al.*, 1996) identified the following eight domains for analyzing and constructing agile capabilities: creation/deletion, expansion/contraction, addition/subtraction, reconfiguration, migration, variation, augmentation, and correction.

The last three change domains are common to both lean and agile systems. Table 1 shows the eight change domains and simple examples of how they might manifest themselves in four different areas of an enterprise. These eight change domains when combined with the four proficiency metrics of agility (time, cost, robustness, and scope) explained earlier gives an analytical tool for prioritizing problems and opportunities. For example, when considering alternatives, the creation/deletion change domain can be assessed on the four metrics of agility when by use of ratings on a scale of 1 to 10. Based on the rating the best alternatives can be chosen for increasing the agile capabilities to an enterprise.

IMPLENTATION ISSUES: Agile business practices may become core requirement for conducting business in future. Concepts of lean production, flexible manufacturing, mass customization, just-in-time are all operating strategies that a firm can implement to become more agile. Additionally, concepts such as reengineering, total quality management, statistical process control are the transformation tools that may be used to transform a company into a more competitive organization. There are five key issues that directly influence a firm's agility: communication connectedness, interorganizational participation, management involvement, production flexibility, and employee empowerment (DeVor *et al.*, 1997).

Communication Connectedness

The level of communication connectedness in the firm refers to: (1) the technical aspects of network linkages, computer systems compatibility, and hardware interface flexibility; and (2) the behavioral aspects of communication. These two aspects of communication connectedness refer to external connectedness with customer and supplier computer systems, as well as internal connectedness.

Interorganizational Participation

This refers to the level of business activities that cross boundaries of a company. The involvement

Table 1. Change domains of agility.

Domain	Production	Organizational structure	Information automation	Human resources
Creation/Deletion: Build something new or remove something completely	Build new production plant	Build new team with new people	Build information access and e-mail infrastructure	Hire all new people for the new facility
Expansion/ Contraction: Increase or decrease existing resource mix	Add similar production equipment	Add more people with similar skills to a team	Add acquired company to the network	Increase or decrease employee head count
Addition/ Subtraction: Add or delete resource types	Add different production equipment	Add more people with different skills to a team	Add access to a new database	Add people with new and different skills
Reconfiguration: Change relationships among modules	Convert a production line to different purpose	Abolish old teams and reform new teams	Change the network structure	Adjust dental vs. medical benefit mix
Migration: Event-based change of fundamental concepts	Convert to bid-based cellular scheduling	Institute self-direction in work teams	Full access to outside databases and e-mail	Institute on-the-job continuous learning
Variation: Ability to cope up with real-time operating surprise	Setup/changeover for unscheduled part	Function when team members are absent	Video traffic swamps the network	Deal with a union wildcat work shutdown
Augmentation: Continuous incremental upgradation	Daily control system upgrades	Continuous learning of team-work skills	Personal agents get smarter	Start monthly company communication sessions
Correction: Reincorporate corrected failures or alternatives	Return broken station to service	Fix dysfunction in a team structure	Route around a bad network node	Return to EEOC compliance

of suppliers and customers in the product design and process improvements is an example. In some cases, much tighter linkages are employed, which eliminate the need for quotations based on price, quality, delivery, etc. Then, purchasing processes can be made quicker. Sourcing and sales decisions involving personnel from product development, quality, and manufacturing functions increases agility. Such tightly-knit relationship will require much trust and dependence.

Production Flexibility

This issue refers to the ease and timeliness with which a product line or the entire factory can be re-configured to produce a new product or process. It also relates to the automation of the processes of a firm and their reconfigurability.

Management Involvement

This refers to the level of active participation of the management in the implementation of agile practices. Management involvement is a critical element in increasing the level of agility in a firm. Management provides physical and organizational resources in support of the creativity and initiative of the workforce. Management must be actively committed to continuous workforce education in the pursuit of the quality and capability of the workforce. It is management's role to implement business processes that lead to reconfigurability and flexibility.

The Practice of Agile Manufacturing: Practice of agility is highly context-dependent. Therefore what a company must do in order to become agile is based on its own understanding of its customers, markets, competitors, products, competencies, and resources. Agility is a continual process of managing change, and a constant adaptation of internal practices and external relationships to meet new opportunities.

The National Science Foundation (NSF) and the Defense Advanced Research Projects Agency (DARPA) have jointly established the Agile Manufacturing Research Institutes (AMRI) at three US Universities. The purpose of the AMRIs is to enhance the understanding of agile manufacturing enterprises, develop system performance measures based on quantitative data, structure a program of research to meet industry-defined needs, and move emerging agile technologies into the next stage where functional prototyping or proof-of-concepts test can take place (DeVor *et al.*, 1997). Many companies including Boeing, Bellcore, Chrysler, Caterpillar, Citicorp, Ford, Honda, Hitachi, IBM, Hyundai, Intel have realized the benefits of agility by applying the concepts developed at the AMRIs (Goldman *et al.*, 1995). The following case illustrates the agile competitive behavior (Goldman *et al.*, 1995).

The Case of Agility in the Apparel Industry

{TC^2}, the Textile/Clothing Technology Corporation is a consortium of 200 companies. This consortium demonstrated the evolving agility potential of the US apparel industry at the September 1994 Bobbin show held in Atlanta, Georgia. For the first time at a major trade show, a fully operational, integrated virtual apparel manufacturing capability was demonstrated. A blouse and coordinated skirt were designed, cut to customer order, printed, sewn, and distributed daily to customers at the show and at three other remote retail sites.

A designer located at the Fashion Institute of Technology (FIT) in New York City linked by two-way audio and video to {TC^2}'s booth at the Bobbin show in Atlanta designed the blouse print pattern interactively with the help of customers at the following locations (1) at the show; (2) at a retail store in Cincinnati, Ohio; (3) at EDS headquarters in Dallas, Texas; and (4) at the National Center for Manufacturing Sciences in Ann Arbor, Michigan. Approval from the customers for the print pattern, sizes, and colors was obtained by 3 p.m. on each day of the show. The FIT designer then transmitted the designs electronically to the Atlanta show booth of Computer Design, Inc. (CDI). CDI translated the designs into digital screen print color separation data and digital garment pattern design data. The former were transmitted to a screen-making facility in Atlanta, and the latter to computer-controlled laser cutting machines at the show. Concurrently, full-color posters displaying the designs being produced on the show floor were printed, as were custom garment tags that accompanied each order to fulfillment.

The screen print manufacturer in Atlanta produced the screens overnight and delivered them by 8 o'clock the next morning to the show booth of Precision Screening Machine (PSM) for off-site printing of that day's designs. Concurrently, laser cutting machines at {TC^2}'s booth used the digitized point-of-sale garment pattern and size data to cut garment pieces out of white fabric and deliver them to PSM for printing during the show hours. The printed pieces, collated by the customer orders for size, style, and design, were then shipped to two separate locations on the show floor for assembly by flexible, cross-trained manufacturing teams: blouses for assembly by Sunbrand, and skirts for assembly by Juki. When sewing was completed, the garments were then delivered to customers at the show and shipped overnight to the customers at the remote locations.

This demonstration which involved a partnership among 35 companies and the electronic coordination of four remote sites and five different locations at the Bobbin show, reveals how close the apparel industry is to realizing agile production, marketing, and distribution by producing customized products, "to order," for traditionally for mass-market customers. In this case, the firms leveraged resources to cut product cycle time by means of intensive interactive cooperation among companies, utilized people and information, and achieved flexibility, empowerment, and mutually advantageous sharing.

Furthermore, the limited Bobbin show demonstration is only a piece of {TC^2}'s larger agile manufacturing effort called "Apparel-on-Demand" demonstration project. In this project, {TC^2} members have participated in the development of three-dimensional body scanning hardware and software programs to translate this data into two-dimensional design data for printing fabric and cutting garment pieces using computer-controlled laser cutters, and high-speed computerized garment assembly machinery. Together with the software linking point-of-sale data to fiber, fabric, and garment manufacturers true agile clothing production moves to imminent reality.

POTENTIAL PROBLEMS AND BENEfiTS: There are many potential problems that act as barriers to the implementation of agility. The concepts of agility have yet to be incorporated into commercial systems, measures, laws, customs, and habits. The barriers to assimilating agility can be grouped into four categories: technological barriers, financial barriers, internal barriers, and external barriers. Many technologies such as intranets, internets, and rapid prototyping are still slow and costly. The commonly used financial accounting system is another barrier because it is reasonably accurate for classical mass production but is unsuitable for agile commerce. Internal barriers include performance measurement systems, budgeting procedures, dysfunctional organizations and information systems, lack of trust, and lack of knowledge of successful users of agility. External barriers include legal systems that are suitable for mass-production environment, tax laws, trade agreements and enterprise zone definitions, static and rigid requirements of public bodies (Goldman *et al.*, 1995). Despite these hurdles, some users experience benefits of agility such as increased customer satisfaction through customization, shortened cycle times for product development, and customer involvement in product specification and design. These outcomes of agility in conjunction with high quality and reliability promise improved competitive position, high market share, and profitability to the users.

CONCLUSIONS: Manufacturing evolved from mass production to lean production, and finally to agile manufacturing. Lean production practices that began in Japan at Toyota, Inc., in the 1950's deserve full credit to Japan's ascendancy in the automotive world. The lessons of lean production are extremely important for understanding agility (Kidd, 1994; Preiss, 1995a). *Lean* is a response to competitive pressures with limited resources. Whereas, *agile* is the response to complexity brought about by constant change. *Lean* is a collection of operational techniques focused on productive use of resources; *agile* is an overall strategy focused on thriving in an unpredictable environment. A very discernible difference surfaces when we look at the architectural roots of manufacturing paradigms. Craft production is based upon the comprehensive single unit: one person builds the entire rifle, or one team builds an entire car. Mass production introduced specialized work modules and sequential work flow. Lean production brought flexibility. Now agile manufacturing brings reconfigurable work modules and work environments. *Lean* is interested in those things that can be controlled; *agile* is interested in dealing with those things that cannot be controlled. Agile manufacturing may not solve all the problems, nor it is the correct approach for all things at all times. Agile manufacturing is a new option that needs to be understood and applied when relevant.

See **Agility change domains; Agility proficiency metrics; AMRI; CIMPLICITY; DARPA; Employee empowerment; Reconfigurability; TEAM; Virtual reality; Lean manufacturing; Manufacturing flexibility; Mass customization; Enterprise resource planning.**

References

[1] DeVor, R., Graves, R. and Mills, J.J. (1997). "Agile Manufacturing Research: Accomplishments and Opportunities," *IIE Transactions*, 29, 813–823.

[2] Dove, R. (1996). *Tools for Analyzing and Constructing Agile Capabilities*, Agility Forum, Bethlehem, PA.

[3] Dove, R., Hartman, S. and Benson, S. (1996). *An Agile Enterprise Reference Model with a Case Study of Remmele Engineering*, Agile Forum, Bethlehem, PA.

[4] Goldman, S.L., Nagel, R.N. and Presiss, K. (1995). *Agile Competitors and Virtual Organizations*, Van Nostrand Reinhold, New York, NY.

[5] Herrin, G.E. (1996). "Industry thrust areas of TEAM," *Modern Machine Shop*, 69, 146–147.

[6] Kaplan, G. (1993). "Manufacturing S La Carte – Agile Assembly Lines, Faster Development Cycles," *IEEE Spectrum*, 30, 24–34.

[7] Kidd, P.T. (1994). *Agile Manufacturing-Forging New Frontiers*, Addison-Wesley Publishing Company, Wokingham, England.

[8] Lau, R.S.M. (1995). "Mass Customization: The Next Industrial Revolution," *Industrial Management*, 37, 18–19.

[9] Pine II, J.B. (1993). *Mass Customization – The New Frontier in Business Competition*, Harvard Business School Press, Boston, MA.

[10] Preiss, K. (1995a). *Mass, Lean, and Agile as Static and Dynamic Systems*, Agility Forum, Bethlehem, PA.

[11] Preiss, K. (1995b). *Models of the Agile Competitive Environment*, Agility Forum, Bethlehem, PA.

[12] Richards, C.W. (1996). "Agile Manufacturing: Beyond Lean?," *Production and Inventory Management Journal*, 37, 60–64.

[13] Roos, D. (1995). *Agile/Lean: A Common Strategy for Success*, Agility Forum, Bethlehem, PA.

[14] Sheridan, J.H. (1993). "Agile Manufacturing: Beyond Lean Production," *Industry Week*, 242, 34–36.

[15] Stipp, D. (1996). "The Birth of Digital Commerce," *Fortune*, 134, 159–164.

AGILITY CHANGE DOMAINS

The agile manufacturing paradigm is concerned principally with unpredictable changes. To understand the kinds of change impacting an enterprise and analyze the enterprise's ability to respond, the "change" should be decomposed into various and interesting domains. Building a model of "change domains" gives a means for analyzing potential agile characteristics. The following eight domains for analyzing and constructing agile capabilities, (1) creation/deletion, (2) expansion/contraction, (3) addition/subtraction, (4) reconfiguration, (5) migration, (6) variation, (7) augmentation, and (8) correction, are now recognized.

See **Agile manufacturing.**

AGILITY PROFICIENCY METRICS

An agile enterprise must have broad "change capability" that is in balance across multiple dimensions. These dimensions are (1) time, (2) cost, (3) robustness, and (4) scope. Completing a change in a timely manner and economically is important. But a quick and economical change is not a sufficient profile of agility. The modified systems must remain functionally robust. The metric *scope* addresses the amount of change. Scope is the principal difference between flexibility and agility. Flexibility is the characteristic that is fixed at the time of specification considering the planned responses to the anticipated contingencies. But agility attempts to minimize the inhibitions to change in any direction. Rather than building an enterprise that anticipates a defined range of requirements, it is built so that it can be deconstructed and reconstructed as needed. Thus, for some element of an enterprise to be agile it must have a balanced response-to-change capability across the four change proficiency metrics: time, cost, robustness, and scope.

See **Agile manufacturing.**

AGV

Automated guided vehicles (AGV) are unmanned carriers or platforms that are controlled by a central computer that dispatches, tracks and governs their movements on guided loops. AGV systems can employ infrared, optical, inertial, embedded wire or ultrasonic methods for guidance. AGVs are primarily useful for material handling, where they deliver inventory from holding areas to production, or between work stations as a replacement for conventional forklifts and rigid transfer lines. Some AGVs are used in assembly systems, while others provide production platforms that support products such as automobiles and engines while work is performed.

See **Capital investment in advanced manufacturing technology; Flexible automation; Manufacturing flexibility implementation; Manufacturing technologies.**

ALL OR NONE SAMPLING

Quality expert, W. Edwards Deming showed that, in most practical cases, average total cost of inspection of incoming materials plus cost to repair and retest defective products downstream can be minimized by "all or none" sampling. Let k_1 be the cost to inspect one part and let k_2 be the cost to dismantle, repair, reassemble, and test an assembly that fails because a defective part was put into production. If the worst possible lot has fraction defective less than k_1/k_2, inspection cost is minimized with no inspection. If the best possible lot has fraction defective greater than k_1/k_2, cost is minimized with 100 percent inspection. Thus, for a process in control, cost is minimized with no sampling when proportion defective is less than the break-even proportion, and cost is minimized with 100 percent sampling when proportion defective exceeds the break-even proportion.

See **Quality: The implications of W. Edwards Deming's approach.**

References

[1] Deming, W. Edwards (1982). *Out of the Crisis*, Center for Advanced Engineering Study, Massachusetts Institute of Technology, Cambridge, Massachusetts.

[2] Deming, W. Edwards (1982). *Quality, Productivity, and Competitive Position*, Center for Advanced Engineering Study, Massachusetts Institute of Technology, Cambridge, Massachusetts.

ALLIANCE MANAGEMENT

Alliance management refers to the establishment of close working relationships between two or more firms. Alliances might be viewed as working partnerships or collaborations. Most alliances exist between a manufacturer and a supplier, a manufacturer and a customer, or a manufacturer and a service provider. However, alliances can exist between two competing firms that hope to gain mutual advantage through cooperation (the sharing of technology in exchange for greater market access). Indeed, the goal of most alliances is to use closer cooperation to gain competitive advantage. For example, the primary reasons for entering into a supply chain alliance include the following: to leverage capital, to access global markets, to access technology, to exploit a core competency, to enhance supply/demand stability, to increase customer involvement, to reduce inventory, to decrease lead times, and to improve quality.

Alliance management became much more important in the 1980s as relationships between buyers and suppliers transitioned from adversarial to cooperative. Firms of all kinds found that by sharing information and expertise, they could frequently achieve higher levels of joint competitiveness. At the same time, the long established tradition of vertical integration started to lose favor. Mangers found that vertical integration strategies were very difficult to implement successfully. Hoped for synergies were difficult to attain and integration distracted management from those tasks where value-added expertise existed. Because the loss of focus was difficult to overcome, many firms looked to develop alternatives to vertical integration. Alliance relationships offered what seemed to be the best of both worlds – the ability to focus on core competencies while reducing costs and improving control throughout the channel. Alliances also provided a greater degree of flexibility to meet the ever changing needs of today's dynamic global environment. The combination of these forces has led to an increased use of alliances.

Even as alliances have become more popular, many firms have experienced difficulty in managing alliances. Many alliances never yield the hoped for benefits while others seem to yield one-sided benefits. Because alliances can be difficult to manage, several key issues should be considered when evaluating alliance opportunities.

- View the alliance as the implementation of a strategic plan.
- Encourage the participants involved to consider their roles in terms of a value-added process.
- Seek an arrangement that achieves scale-economy benefits while spreading risk.
- Make sure that the needed information to function well over the long run is shared.
- Build trust between the organizations by setting unambiguous goals, establishing clear roles, laying down firm rules, and measuring performance rigorously.
- Consider the contingencies – even long-term alliances may not last forever.

See **Supplier partnership as strategy; Supply chain management: Competing through integration; Supplier relationships.**

Reference

[1] Bowersox, D.J., Calantone, R.J., Clinton, S.R., Closs, D.J., Cooper, M.B., Droge, C.L., Fawcett, S.E., Frankel, R., Frayer, D.J., Morash, E.A., Rinehart, L.M., and Schmitz, J.M. (1995). *World Class Logistics: The Challenge of Managing Continuous Change*. Council of Logistics Management, Oak Brook, IL.

ALLOCATION AND AVAILABILITY CHECKING

This term is associated with (MRP). Allocation refers to quantities of items assigned to specific orders but not yet released to production.

Allocation takes place before launching an order for production. Allocation requires checking that all necessary components of a product are available. If the availability check reveals that sufficient components are not available, the order will not be launched. In that case, the planner may consider releasing a partial order for production. In MRP systems, component availability is checked for any order to be launched. An order is created if the required component quantities are available. Once the order is created, the MRP system allocates the quantities of components to the shop order, and makes those components unavailable to any other shop order.

See **Material requirements planning (MRP); Order launching; Order release; Order release mechanism; Order release rule.**

References

[1] APICS Dictionary, 9th Edition, 1998.
[2] Vollmann, T.E., Berry, W.L., Whybark, D.C. Manufacturing Planning and Control Systems, Fourth Edition, Irwin/McGraw-Hill, New York, 1997.

AMERICAN PRODUCTION AND INVENTORY CONTROL SOCIETY (APICS)

See **Professional associations for manufacturing professionals.**

AMERICAN SOCIETY FOR QUALITY

See **Professional association for manufacturing professionals.**

AMRF

Automated Manufacturing Research Facility (AMRF) is a project created by the National Institute of Standards and Technologies (NIST) to address specific control problems in the design of automated manufacturing. This center provides a wide range of calibration services for mechanical artifact standards such as gage blocks, thread gages, and line scales. The AMRF allows research in measurement technology and provides a test bed where integrated manufacturing system measurement research can be performed.

See **Virtual manufacturing.**

AMRI

The National Science Foundation (NSF) and the Defense Advanced Research Projects Agency (DARPA) jointly established the Agile Manufacturing Research Institutes (AMRI) at three US universities. The Aerospace Agile Manufacturing Research Center (AAMRC) was established at the University of Texas at Arlington, the Electronics Agile Manufacturing Research Institute (EAMRI) at Rensselaer Polytechnic Institute, and the Machine Tool Agile Manufacturing Research Institute (MT-AMRI) at the University of Illinois at Urbaba-Champaign. The purpose of the AMRIs is to enhance the understanding of agile manufacturing enterprises, develop system performance measures based on quantitative data, structure a program of research to meet industry-defined needs, and move emerging technology which has a potential for impacting agile manufacturing into the next stage of functional prototyping or proof-of-concepts test beds.

See **Agile manufacturing.**

ANDON

Andon is a Japanese term signifying a signal board used on assembly lines to alert everyone to those stations on the line that may be having a problem or have been stopped.

See **Integrated quality control; Just-in-time manufacturing implications; Just-in-time manufacturing; Lean-manufacturing implementation.**

ANTICIPATED BUILD SCHEDULE

In MRP the master production schedule for end products is an anticipated build schedule. That is, it is a planning document that lays out what will be produced and when.

See **Master production schedule (MPS); Material requirements planning (MRP).**

AOQ

AOQ stands for Average Outgoing Quality.

See **Acceptance sampling; Operating characteristics curve; Statistical process control using control charts.**

AOQL

AOQL stands for Average Outgoing Quality Limit.

See **Acceptance sampling; Operating characteristics curve; Statistical process control using control charts.**

APICS

See **Professional associations for manufacturing professionals.**

APPRAISAL COSTS

Appraisal costs are a component of the cost of quality, and include those costs associated with maintaining the quality of a product or evaluating an organization's ability to maintain a desired level of quality in its products. Such costs generally include inspection and testing, but may also extend to costs of quality audits and even costs associated with quality certifications such as ISO 9000.

In the past, it was often believed that a company needs to increase its appraisal costs in order to improve quality. However, if appraisal includes only testing and inspection, then quality of the product reaching the customer may improve, but only because defective products are caught before they are shipped out. Today, it is generally realized that improvements in quality depend upon increased prevention of defects, and that appraisal costs can generally remain the same, or even decrease as the number of defects decreases due to prevention efforts.

See **Cost of external failure; Cost of quality; ISO 14000; ISO 9000/QS 9000 quality standards; ISO 9000; Performance measurement in manufacturing; Prevention costs.**

AQL

See **Acceptance sampling; Operating characteristics curve; Statistical process control using control charts.**

ARTIFICIAL INTELLIGENCE (AI)

Artificial Intelligence is that part of computer science that makes computers "think" and make "decisions" like humans. Computer Systems with AI can reason inductively and make decisions when data may change. It is beneficial to imbed AI in equipment when simple decisions have to made frequently by them.

See **Flexible automation.**

ASQ

See **Professional association for manufacturing professionals.**

ASQC

See **Professional associations for manufacturing professionals.**

ASSEMBLE-TO-ORDER PRODUCTION

In assemble-to-order production, assembly activity starts after receiving a customer order. The products such as computers that are assembled from modules come under this category. The module manufacturing shop is driven by manufacturing resources planning, and the product assembly activity is driven by customer orders. The assemble-to-order production can take two different variations. In one situation, the products are assembled from their modules exploiting the concept of modularity in product design. In another situation, after products are assembled from their modules, a small amount of manufacturing is employed at the delivery point.

See **Mass customization.**

ASSEMBLY LINE BALANCE

See **Assembly line design; U-shaped assembly line.**

ASSEMBLY LINE DESIGN

Patrick R. McMullen

Auburn University, AL, USA

INTRODUCTION: Assembly lines are a vital part high-volume production. Their proper implementation and management is of great importance to the competitive position of efficient, high-volume manufacturing firms.

The assembly line is a type of machine or work center layout that is dictated by the product. In an assembly line environment, the product being manufactured moves continuously through the line from one work center to the next. Automotive assembly is perhaps the most common application of the assembly line, also found in other high-volume products such as appliances and consumer electronics.

There are several reasons why manufacturers may prefer an assembly line layout to other layout approaches. These reasons are usually economic in nature. First of all, when an assembly line is used, the in-process jobs are in continual motion – usually by automated conveyance equipment. This results in relatively small material handling costs. Personnel assigned to various tasks on the line have a good deal of repetition in their jobs. This essentially means two things: attainment of a high level of skill is easy and quick and a low level of training requirement; these reasons translate into lower costs. These reasons were the impetus for Henry Ford's investment in the first ever assembly line for Ford Model T production in 1908. Prior to using an assembly line, a Model T required 728 hours to assemble. After the introduction of the assembly line, a Model T required a mere 93 minutes to assemble (Halberstam, 1986) – a strong reason for the assembly line's rapid popularity following Ford's adoption. The assembly line, however, does have its disadvantages. Perhaps the biggest disadvantage to using assembly lines is their low degree of flexibility in addition to their relatively high capital cost in some instances. To compare the advantages and disadvantages of the assembly line with alternative manufacturing layouts, see **job shops** and **manufacturing cell design**.

DESIGN OF THE ASSEMBLY LINE: As stated earlier, proper implementation and management of the assembly line is imperative for its success. This "design" has typically been in the form of Assembly Line Balancing. Assembly Line Balancing (ALB) is the practice of distributing definable units of work (referred to as tasks) into cohesive groups (referred to as work centers). It should be noted that the first attempt at addressing the ALB was made by Salveson (1955), and a comprehensive survey of seminal papers in ALB is provided by Baybars (1986).

Distribution of these tasks into groups is done in the hope of optimizing some objective. These objectives are usually in the form of one or more of the following: minimization of workers needed, minimization of equipment needed, minimization of resultant cycle time, minimization of work-in-process inventory, or minimization of workload discrepancy among workers. Despite these numerous types of objectives, there are typically two objectives that standout. The first is referred to as the Type I Assembly Line Balancing Problem. This is where it is intended to minimize the number of workers given a cycle time that is pre-specified by management. In other words, the design process attempts to use as few workers as possible while meeting a certain level of output. This type of problem exists when there is a demand for fast, repetitive production (such as auto assembly). The second is the Type II problem. In this case, it is desired to distribute the tasks among workers such that the cycle time is minimized given a pre-specified number of workers. This type of objective is relevant when the assembly line is placed in an area with physical limitations that restrict the number of workers. The challenge then is to distribute work among a given number of workers such that the cycle time of the assembly line is minimized.

For the purpose of brevity, a simple example will be provided that details how ALB can be done for the Type I problem. Consider an assembly with tasks needed for its completion provided in Table 1:

Table 1. Data for a type I ALB problem.

Task ID	Task description	Predecessor	Task duration (min)
A	Procure raw materials	None	.53
B	Inspect raw materials	A	.87
C	Prepare parts for assembly	B	.22
D	Subassembly #1	C	.83
E	Subassembly #2	C	1.22
F	Assemble chasis (#1 and #2)	D, E	1.33
G	Subassembly #3	C	.63
H	Assemble chasis and #3	F, G	2.31
I	Final assembly	H	3.81
J	Final inspection	I	.88

For the purpose of this example, it will be assumed that the cycle time established by management is .42 minutes per unit. This means that management expects a unit to be completed every .42 minutes (or, a unit will be completed every 25.2 seconds). The ten tasks detailed above describe the process for an assembly of a certain, generic product. Given are the identifying label (Task ID) for each task, its description, precedence requirements and the task duration in minutes. The challenge ALB is to organize the tasks into work centers in an attempt to minimize the necessary workers. A work center on the line may be required to complete one or more tasks. Notice from Table 1 that all task durations exceed the cycle time of .42 minutes. This means that multiple workers must work simultaneously on these tasks. This practice is referred to as paralleling, and is necessary whenever cycle time is less than individual task times.

One technique to perform ALB under such requirements is the Incremental Utilization Heuristic (Gaither and Frazier, 1998). This technique places tasks into work centers under the condition that the addition of a task results in an increase in the utilization of the work center (hence the name "incremental utilization"). The first work center is assigned the very first task in Table 1. If the addition of a task does not increase the utilization of the work center, the current work center is closed, and the task is added to a new work center in the sequence. This process is continued until all tasks have been asigned to a work center. The decision-maker must also be aware of precedence requirements – a task cannot be added to a work center until all its predecessors have been assigned to a work center. For purposes of this illustration, tasks will be added to work centers in a lexicographic fashion, which adheres

Table 2. Incremental utilization heuristic for example problem.

Work center	task(s)	Total task time (minutes)	Workers needed	Adjusted workers	Utilization of work
1	A	0.53	1.26	2	63.09%
	A, B	0.53 + 0.87 = 1.40	3.33	4	83.33%
	A, B, C	1.40 + 0.22 = 1.62	3.86	4	96.43%
	A, B, C, D	1.62 + 0.83 = 2.45	5.83	6	97.22%
	A, B, C, D, E	2.45 + 1.22 = 3.67	8.74	9	97.08%
2	E	1.22	2.90	3	96.83%
	E, F	1.22 + 1.33 = 2.55	6.07	7	86.73%
3	F	1.33	3.17	4	79.17%
	F, G	1.33 + .63 = 1.96	4.67	5	93.33%
	F, G, H	1.96 + 2.31 = 4.27	10.17	11	92.42%
4	H	2.31	5.5	6	91.67%
	H, I	2.31 + 3.81 = 6.12	14.57	15	97.14%
	H, I, J	6.12 + .88 = 7.00	16.67	17	98.04%

Note: Workers column is determined by dividing required time in work center by the cycle time (.42). Adjusted workers are adjusted to the nearest larger integer value.

Table 3. Summary of assembly line balancing solution.

Work center	Task(s)	Workers
1	A, B, C, D	6
2	E	3
3	F, G	5
4	H, I, J	17
Total		31

to all precedence requirements. Table 2 shows how Gaither and Frazier's Incremental Utilization Heuristic can be used to design an assembly line layout to assemble the product having tasks presented in Table 1.

The shaded rows in Table 2 represent situations where utilization decreased as a result of adding an additional task therefore, it signals the need to open the next work center. The tasks that cause the efficiency drop become the first task assigned to the newly created work center. For example, in Table 2, the addition of task E resulted in lowereing the efficiency in work center 1, therefore work center 2 was created and task E becomes its first task. Table 3 summarizes resulting line balance.

The resultant requirement of 31 workers is the fewest possible. The theoretical minimum number of required workers is determined by summing the task duration's for all tasks [12.63] and dividing by the target cycle time of .42 which = 30.07, rounded or adjusted to 31. The attainment of an optimal (minimum) number of workers is not guaranteed by this approach, but was obtained here nonetheless.

Figure 1 shows the details of how paralleling is performed. For example, Work Center 1 contains tasks A, B, C and D, and six workers are used. Each of these six workers performs tasks A, B, C, and D. As another example, Work Center 4 contains tasks H, I and J, and seventeen workers are used. Each of these seventeen workers performs tasks H, I, and J. All jobs are first processed in Work Center 1, then sequentially in Work Centers 2, 3, and 4.

The example detailed above illustrates one way to address the ALB problem. There are several other ways to address such problems, and problem objectives may vary. This example addresses a Type I problem that required paralleling of tasks where the objective is to minimize the number of workers required. For some insight into various objectives and various approaches to address them, the interested reader is referred to the work of McMullen and Frazier (1997, 1998).

OTHER ISSUES TO CONSIDER: A general introduction to basic Assembly Line Balancing problem has been presented here.

One of the more important considerations to make for problems of this type is the mixed-model type of problem. This is where two or more differing products must be assembled on the line simultaneously. This complicates the ALB process substantially. Despite its complexity, it is important not to ignore this challenge due to the fact that modern, Just-in-Time production applications frequently require differing products to be manufactured simultaneously.

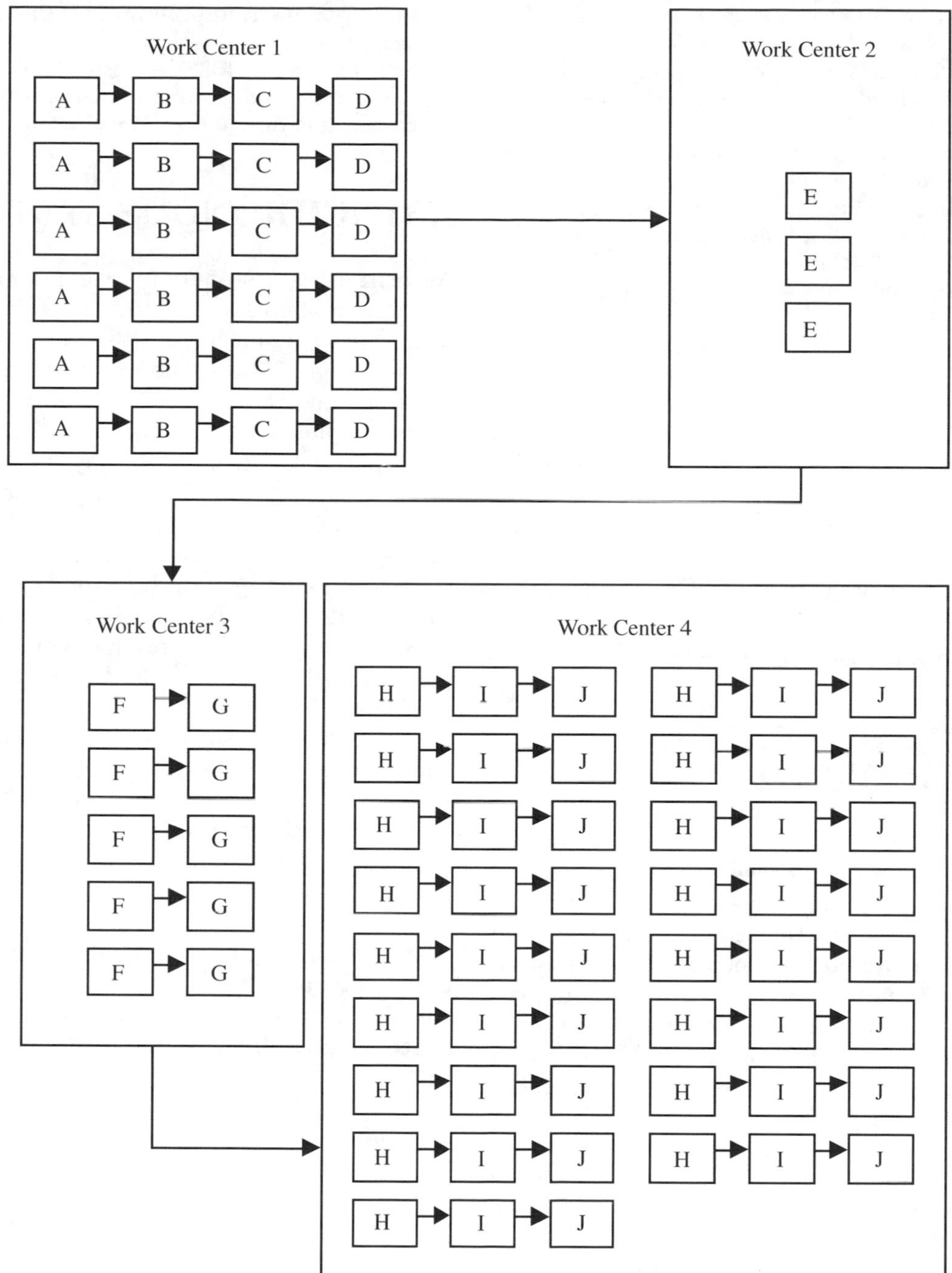

Fig. 1. Graphical illustration of paralleling of tasks.

Another consideration in addressing the ALB problem is when task durations exhibit stochastic (not known with certainty) durations. The example above, addresses task durations as deterministic (known with certainty). Stochastic task durations are more complicated than deterministic, but are frequently encountered in reality.

The work of McMullen and Frazier in the reference does provide some insight into these complicating factors of mixed-models and stochastic durations.

See **Assembly line balance; Cycle-time; Job shops; Manufacturing cell design; Manufacturing systems; U-shaped assembly lines.**

References

[1] Baybars, I. (1986). "A Survey of Exact Algorithms for the Simple Assembly Line Balancing Problem." *Management Science,* 32, 909–932.

[2] Gaither, N. and Frazier, G.V. (1998). Production and Operations Management (Eighth Edition), South-western Publishing, New York.

[3] Halberstam, D. (1986). *The Reckoning*. Morrow Publishers, New York.
[4] Hyer, N.L. and Wemmerlov, U. (1984). "Group Technology and Productivity." *Harvard Business Review*, 62, 140–149.
[5] McMullen, P.R. and Frazier, G.V. (1997). "A Heuristic Solution for Stochastic Assembly Line Balancing Problems with Parallel Processing for Mixed-Models." *International Journal of Production Economics*, 51, 177–190.
[6] McMullen, P.R. and Frazier, G.V. (1998). "Using Simulated Annealing to Solve a Multiobjective Assembly Line Balancing Problem with Parallel Work Stations." *International Journal of Production Research*, 10, 2717–2741.
[7] Salveson, M.E. (1955). "The Assembly Line Balancing Problem." *Journal of Industrial Engineering*, 6, 18–25.

ASSEMBLY STRUCTURE

See **U-shaped assembly lines.**

ASSIGNABLE CAUSES OF VARIATIONS

Assignable causes of variation are present in most production processes. These causes of variability are also called special causes of variation (Deming, 1982). The sources of assignable variation can usually be identified (assigned to a specific cause) leading to their elimination. Tool wear, equipment that needs adjustment, defective materials, or operator error are typical sources of assignable variation. If assignable causes are present, the process cannot operate at its best. A process that is operating in the presence of assignable causes is said to be "out of statistical control." Walter A. Shewhart (1931) suggested that assignable causes, or local sources of trouble, must be eliminated before managerial innovations leading to improved productivity can be achieved.

Assignable causes of variability can be detected leading to their correction through the use of control charts.

See **Quality: The implications of W. Edwards Deming's approach; Statistical process control; Statistical process control using control charts.**

References

[1] Deming, W. Edwards (1982). *Out of the Crisis*, Center for Advanced Engineering Study, Massachusetts Institute of Technology, Cambridge, Massachusetts.
[2] Shewhart, W.A. (1939). *Statistical Method from the Viewpoint of Quality Control*, Graduate School, Department of Agriculture, Washington.

ASYNCHRONOUS SYSTEMS

Asynchronous systems are systems in which sequences of events or activities occur in parallel (concurrently) and where there is no direct synchronization of the sequence of events. Synchronization may occur at certain points in the process (for example, the final assembly), but the activities that occur in parallel may not be synchronized. Job shops and flexible manufacturing systems frequently contain asynchronous processes. These asynchronous processes may occur when more than one type of product is being manufactured in different areas of the production facility, or when an order for a product generates multiple work orders which go to different parts of the production facility.

In contrast to asynchronous systems, in synchronous systems (such as automatic transfer lines) each stage in the production line is synchronized with the other stages in the line, and parts move from one stage to the next, at the same time.

See **Concurrent systems; Discrete-event systems; Manufacturing systems modeling using petri nets; Synchronous manufacturing systems; Synchronous manufacturing using buffers.**

References

[1] Browne, J., Dubois, D., Rathmill, K., Sethi, S., and Stecke, K.E. (1984), Classification of flexible manufacturing systems. *The FMS Magazine*: 114–117.
[2] Cassandras, C.G. (1993), *Discrete Event Systems: Modeling and Performance Analysis*. (Homewood, IL: Irwin).

ATP

See **Available-to-promise.**

AUTOMATED MANUFACTURING SYSTEM

An Automated Manufacturing System (AMS) is an interconnected system of material processing stations capable of automatically processing a wide variety of part types simultaneously under computer control. The system is not only interconnected by a material transport system, but also by a communication network for integrating all aspects of manufacturing. Such a system exhibits flexibility in parts routing, part processing, part handling, and tool changing. Additionally an automated manufacturing system exhibits the following characteristics: high degree of automation, high degree of integration, and high degree of flexibility. An automated manufacturing system may include several enabling technologies such as: computer aided design (CAD), computer aided process planning (CAPP), computer aided manufacturing (CAM), flexible assembly and flexible manufacturing, computer aided testing (CAT), production planning and control, process technologies, robotics and automated material handling. It should be noted that sometimes the terms Computer Integrated Manufacturing (CIM) and AMS are used interchangeably.

See **CAD; CAM; CAPP; CAT; CIM; Disaggregation in an automated manufacturing environment; Flexible automation; Manufacturing flexibility.**

References

[1] Pimentel, J.R. (1990). *Communication Networks for Manufacturing*, Prentice Hall, Englewood Cliffs, New Jersey.

[2] Ranky, P.G. (1986). *Computer Integrated Manufacturing*, Prentice Hall, Englewood Cliffs, New Jersey.

AUTOMATED TRANSFER LINES

See **Flexible automation; Manufacturing systems design.**

AUTOMOTIVE INDUSTRY ACTION GROUP (AIAG)

The Automotive Industry Action Group (AIAG) represents North American vehicle manufacturers and suppliers, 26200 Lahser Road, Suite 200-Southfield, MI 48034, www.aiag.org. AIAG coordinates work to improve industry-wide manufacturing through the development of new manufacturing technologies. In addition, AIAG develops standards for the automotive industry. The work of the AIAG is promoted and disseminated through publications, conferences, and expositions. Some examples of recent developments led by AIAG are:

Manufacturing Assembly Pilot (MAP). Completed in 1996, this pilot project investigated the use of electronic commerce (EC) technologies to improve communication throughout the supply chain. The study found that use of EC technology could save the industry approximately $1 billion annually, through reductions in information-flow lead time and business process reengineering.

AutoSTEP pilot. This three-phase project concerns the STandard for the Exchange of Product Model Data (STEP) and associated business processes. The project's goal is to use this STEP technology to gain reductions in capital investments and improve timing, cost and quality of data exchange within the supply chain.

AIAG assists in improving the automotive industry manufacturers by helping suppliers achieve QS-9000 certification.

See **ISO9000; ISO9000/QS9000; Quality management certifications and awards; Quality standards.**

AUTONOMATION

Autonomation means the autonomous control of quality and quantity. Taiichi Ohno, former vice-president of manufacturing for Toyota, was convinced that Toyota had to raise its quality to superior levels in order to penetrate the world automotive market. He wanted every worker to be personally responsible for the quality of the part or product that he/she produced.

Quite often, inspection devices are placed in the machines (inspection at the source) or in devices (called decouplers) between the machines, so inspection can be performed automatically. Inspection by a machine instead of a person can be faster, easier, and easily repeatable. This is called *in-process control inspection*.

Inspection is made part of the production process and does not involve a separate location or person to perform it. Parts are 100% inspected by devices which either stop the process if a defect is found or correct the process before the defect can occur. The machine shuts off automatically when a problem arises. This prevents mass production of defective parts. The machine may also shut off automatically when the necessary parts have been made. This contributes to inventory control.

See **Lean manufacturing implementation.**

AUTONOMOUS PROJECT ORGANIZATION

Organizational structures created for product development projects can be categorized based on how much they emphasize functional links opposed to project links. Based on functional links, three project organization have been recognized: lightweight project organizations, heavyweight project organization, and autonomous project organizations (Wheelwright and Clark 1992). For a description of light or heavy project manager See Clark and Fujimoto (1991).

In an autonomous project organization, the involved individuals give up – at least temporarily – the links to their function (e.g., to marketing, design, and manufacturing). This means that project members are fully dedicated to a single project, typically working in the same building or complex. In the extreme case of a "tiger team organization" or "skunk works", the project members are not only co-located, but purposely placed somewhere remote from the rest of the organization.

The project manager leading the autonomous project organization controls budgets, schedules, and the performance evaluation of the project members, all of whom report directly or indirectly to him/her. If the project is sufficiently large, it is not uncommon for the project leader to be a member of the executive board of the main organization.

The major advantage as well as disadvantage of the autonomous project organization over functional and matrix organizations is its detachment from the rest of the organization. On the one hand, this allows the autonomous project organization to resolve technical and organizational problems with no constraints from outside, which lead to fast decision making, innovative products and quick adaptation to changing conditions.

On the other hand, project members will have to – at least temporarily – give up their functional career and skill development. In fast changing technologies, leaving a functional field for more than one or two years can be an irreversible decision. Moreover, groups of people working towards the same goal, with little or no oversight from the outside, can get trapped in what is known as "group-think." Loch and Terwiesch (1996) illustrate how a pharmaceutical drug development team initially ignored and later downplayed negative information about side-effects of the drug, as they were obsessed by the idea of developing a highly successful drug.

As a result of these disadvantages, organizations frequently shy away from creating fully autonomous project organizations. The most common applications of autonomous project organizations include "tiger teams" and "skunk works," set up mostly by large organizations in search for radical innovations. The development of the Java programming language by Sun Corporation is one famous case. Small and relatively young one-product organizations by definition fall into this category.

See **Functional organization; Heavy weight project organization; Light weight project organization; Product development and concurrent engineering.**

References

[1] Clark, K.B. and T. Fujimoto (1991). *Product Development Performance: Strategy, Organization and Management in the World Auto Industry*. Harvard Business School Press, Cambridge.

[2] Loch, C.H. and C. Terwiesch (1996). "The Development of Nopane," INSEAD case 12/96 4661.

AUTONOMOUS STRUCTURE FOR DESIGN

See **Autonomous project organization; Product design and concurrent engineering.**

AUTONOMOUS WORK TEAMS

Autonomous work teams include all aspects of self-directed work teams, with the addition of significant human resource management responsibilities. These teams are given complete control over the hiring and termination of team members. They determine the proper allocation of rewards among team members, develop and implement performance management systems. They are given responsibility and authority over entire work processes which often includes all aspects of a team member's work conditions and duties. In many respects, autonomous work teams may be conceived as small organizations.

See **Design and implementation of manufacturing work teams.**

AVAILABLE-TO-PROMISE (ATP)

Available-to-promise (ATP) is a portion of a company's inventory that enables a manufacturer promise delivery by a certain date. It refers to uncommitted inventory balance and is part of the information in a master production schedule (MPS). The ATP quantity is usually calculated for each period in which a MPS receipt is scheduled.

The (ATP) value for a particular time period is calculated by subtracting the total orders before the next master production schedule from the total units of inventory on hand. The ATP value represents the number of that orders can be fulfilled if a delivery date is promised.

See **Master production schedule (MPS); Order promising.**

AVERAGE INVENTORY

The average inventory is the mean inventory on hand during a given time period. It is usually calculated as a simple arithmetic mean of inventory on hand at the beginning of the time period and the inventory on hand at the end of the time period. Average inventory is the basic measure used to calculate holding cost in conjunction with the holding cost rate. Inventory ordering policies where few orders are placed each year result in high average inventory versus policies where many small orders are placed each year. Ordering policies consider the impact of purchasing price, as well as the risk of stockout, and the implications of stocking out.

See **Inventory build and replenishment cycles; Inventory holding costs; Simulation of production problems using spreadsheet programs.**

AVERAGE JOB LATENESS

Average job lateness is one of many performance measures used in evaluating rules used for sequencing work orders awaiting processing at one or more work centers. Such rules are called job sequencing rules. Average job lateness criterion, given the jobs on hand and the machine resources available, attempts to maximize customer satisfaction by reducing late deliveries of orders. Other criterion employed for sequencing jobs may focus on improving operational efficiency of the system by reducing inventory.

See **Sequencing rules.**

AVERAGE LATENESS

See **Average job lateness; Priority scheduling rules; Priority sequencing rules.**

B

BACK FLUSHING

Also known as "postdeduct", back flushing is carried out in Just-in-Time (JIT) manufacturing. In this method, component inventory records are reduced once the products containing the components are received in finished goods. This technique carries the risk of the records not being totally accurate, but transaction processing is reduced due to the elimination of detailed work-in-process (WIP) accounting systems. To minimize problems in backflushing, accurate bills of materials (BOM) are necessary and all other records must be meticulously kept and updated in a timely manner. The BOM information is useful in updating inventory records when assembly of finished goods is completed.

See **Bill of materials; Just-in-time manufacturing; Work-in process inventory.**

BACK SCHEDULING

Back scheduling is a fundamental component of MRP. It is a technique for calculating operation start dates and due dates. To develop the schedule, start and due dates for each operation are calculated working backwards from the order due dates. Production schedule can be flexible if the processes are not on a product's critical path. While front scheduling starts individual processes as early as possible back scheduling starts each item on the production schedule as late as possible. The advantages of back scheduling are: Work in process (WIP) can be reduced by significant quantities, raw materials do not have to be committed to specific order numbers earlier than needed, and completed components are stored for shorter lengths of time. In order for back scheduling to work successfully, the production system must have accurate bill of materials (BOM) data and lead time estimates. In addition, the system must incorporate an effective method of tracking components and subassemblies.

See **Bill of materials; MRP; WIP inventory.**

BACK-END ACTIVITIES

Back end activities form a part of the Manufacturing Planning and Control (MPC) system. In this system, the front end handles production planning, using available resources to manage demand. This is achieved by master production scheduling. The heart of the MPC system is concerned with detailed capacity and materials planning. The back end of the system consists of execution systems for the shop floor and vendors. The nature of the shop floor systems will be partially dependent on the nature of the process. For example, Process-focused manufacturing cells are scheduled according to priorities, and product-focused cells use Just-in-Time (JIT) systems. Vendor scheduling information is provided by purchasing systems.

The interaction between all these levels of manufacturing planning and control systems requires integrated software systems in which each level of planning/scheduling provides the right input to the next level below.

See **JIT; Just-in-time manufacturing; Manufacturing cells; Manufacturing planning and control (MPC); Master production schedule.**

BACKFLOW

Materials or parts in the a manufacturing cell that moves in the opposite direction to the prescribed part flow.

See **Manufacturing cell design.**

BACKWARD INTEGRATION

Most businesses are individual links in a longer market chain. For example, the market chain that brings soft drinks to the consumer links extractors of metallic ores, canning companies to soft-drink concentrate producers and retailers. When a concentrate producer like Pepsi Cola decides to produce its own cans and enters the canning business, we call this development backward integration. In other words it is the extension of commercial activities to sectors further back from the end-consumer in the overall market chain.

There are many reasons why a company might want to expand by backwards integration. One is to increase its share of the profits available throughout the entire market chain. Another is to gain greater control over the availability, quality and cost of what were formally externally-sourced supplies. In many cases, the decision to backward integrate becomes attractive when the demand for the company's core product reaches a certain critical size. In the case of a beverage company, for example, this point would be at the minimum volume needed to support the profitable operation of a multi-million dollar high-speed canning operation. There are also risks involved in backward integration. For example, companies that integrate backward increase their risk and become even more vulnerable to unfavourable trends in the buying behaviour of the final consumer. Furthermore, even though the new activity is a link in the same market chain it might be a different business with different success criteria. For example, canning and beverage production are very different businesses. The former is a low-margin, production-oriented, business. The latter is a high-margin, marketing-oriented activity. The formula for being a successful soft-drink producer may not work in a canning operation.

See **Strategic supplier partnerships.**

BACKWARD SCHEDULING

Two dates are important in production scheduling – the present date and the date when the order is due. Within the period between these two points, the purpose of scheduling is to plan production of an order so that it will be completed by the due date. One approach to planning is to begin at the due date and work backward through time, scheduling each operation by working backward until the first operation is scheduled. In this way the job or order is planned for completion on its due date. An advantage of this approach is that it minimizes finished goods inventory. However, an order may go past its due date if unplanned delays occur because backward scheduling has little margin for error. In reality, scheduling usually combines both backward scheduling and forward scheduling.

See **Back scheduling; Forward scheduling; Medium- and short-range capacity planning.**

BALANCED SCORECARD

Ramachandran Ramanan

University of Notre Dame, Indiana, U.S.A.

DESCRIPTION: The **balanced scorecard** is a performance measurement system that uses a set of performance targets and results to show an organization's performance in meeting its objectives relating to its various stakeholders. It recognizes that organizations are responsible to different stakeholder groups, such as employees, suppliers, customers, business partners, the community and shareholders. This method of measuring performance focuses attention on achieving strategic organizational objectives relating to the above stakeholders. Further, the balanced scorecard provides a system to channel the energies, abilities, and specific knowledge held by people throughout the organization toward achieving these strategic objectives.

Sometimes different stakeholders have different wants. For example, employees depend on an organization for their employment. Customers expect a quality product with the shortest possible delivery time and price. Shareholders depend on an organization to maintain and grow their investment. The organization must balance these competing wants. For many years, organizations focused only on financial results, which reflected mainly the shareholders' interests. In recent years, organizations have shifted attention to customer issues, such as quality and service. They also pay attention to the needs of employees and the community. For example, Ben & Jerry's Ice Cream measures its social performance along with financial performance and has a social audit next to its financial audit in its annual report.

The concept of a balanced scorecard is to measure how well the organization is doing in view of competing stakeholder wants. A balanced scorecard could be used as a strategic management system. It promotes the balancing of the efforts of the organization among the financial, customer, internal business process, and learning and growth (innovation) objectives. For each objective, the balanced scorecard provides a framework for the firm to develop multiple measures of outcomes relating to the objective, to set target outcomes for a period and to identify key initiatives that would help in achieving the desired outcomes. A comparison of actual to the target at the end of the period can then be used for feedback, corrective actions and performance evaluation.

HISTORICAL PERSPECTIVE: The development of the balanced scorecard method can be traced back to 1990 when the Nolan Norton Institute, the research arm of the accounting firm KPMG, sponsored a one-year study, "Measuring Performance in the Organization of the Future." This study was motivated by a belief that existing performance measurement approaches relying on financial accounting measures, were becoming obsolete. David Norton, CEO of Nolan Norton, was the study leader and Robert Kaplan, Professor of Accounting, Harvard University was the academic consultant. Senior managers from 12 corporations participated in this year-long study. All of these managers shared a belief that exclusive reliance on summary performance measures was hindering organizations' abilities to create future economic value.

Participants of this group shared their experiences in bimonthly meetings and several ideas emerged in the group discussions over the year. The group examined various innovative performance measurement systems, such as shareholder value, productivity, quality improvements and compensation plans, were studied. The participants started focusing on multidimensional scorecards as offering the most promise for their needs. This led to the design of a rough "Corporate Scorecard." and evolved into what is now entitled "Balanced Scorecard." Kaplan and Norton (1996) state that, "...the name reflected the balance provided between short- and long-term objectives, between financial and non-financial measures, between lagging and leading indicators, and between external and internal performance perspectives." The scorecard measures organizational performance across four balanced perspectives: financial, customers, internal business processes, and learning and growth (innovation).

Seminal work on this topic is done by Kaplan and Norton, the recognized architects of balanced scorecard. They have published several articles since 1992 and a book in 1996. In their recently published book, the authors demonstrate how senior executives in industries such as banking, oil, insurance, and retailing are using the balanced scorecard both to guide current performance and target future performance. They show how to use measures in financial performance, customer knowledge, internal business processes and learning and growth to align individual, organizational and cross-departmental initiatives. They describe how these measures can be used to identify entirely new processes for meeting customer and shareholder objectives. The book also provides the steps that managers in any company can use to build their own balanced scorecard.

Interested readers could also gain useful insights from the various articles by these authors. For example, Kaplan and Norton (1992), summarizes the findings of their study group in 1990. In a subsequent article in 1993, the authors describe the importance of choosing measures based on strategic success and reflects their consulting experience with some corporations. More recently, Kaplan and Norton (1996), suggest how the use of balanced scorecard could be improved by extending it from a measurement system to a core management system. Based on their experiences of consulting partnerships with various organizations implementing this method, Kaplan and Norton (1996) posit that the balanced scorecards provides managers with the instrumentation they need to navigate to future competitive success. This approach to performance evaluation translates an organization's mission and strategy into a comprehensive set of performance measures that provides the framework for a strategic measurement and management system.

Strategic Perspective: An overemphasis on achieving and maintaining short-term financial results can cause companies to overinvest in short-term fixes and to under-invest in long-term value creation, particularly in the intangible and intellectual assets that generate future growth. To illustrate, Larry Brady, President of F.C. is quoted as saying: "*As a highly diversified company,... the return-on-capital-employed (ROCE) measure was especially important to us. At year-end, we rewarded division managers who delivered predictable financial performance. We had run the company tightly for the past 20 years and had been successful. But it was becoming less clear where future growth would come from and where the company should go for breakthroughs into new areas. We had become a high return-on-investment company but had less potential for further growth. It was also not at all clear from our financial reports what progress we were making in implementing long-term initiatives.*" Similarly, many senior executives realize that in several cases employees are not able to link the applicability of corporate mission and vision to their day-to-day activities.

Breakthroughs in performance require change, and change is required in the measurement and management systems as well. Balanced scorecard offers this framework. As argued by Kaplan and Norton in their book, balanced scorecard can help organizations in three identifiable ways. First, the scorecard describes the organization's vision of the future to the entire organization and promotes shared understanding. Second, the scorecard creates a holistic model of the strategy that allows all employees to see how they contribute to organizational success. Without such a linkage, individuals and departments can optimize their local performance but not contribute to achieving strategic objectives. Third, the scorecard also focuses attention on change efforts. If the right objectives and measures are identified, successful implementation will likely occur. If not, investments and initiatives will be wasted.

The book and articles by Kaplan and Norton describe how the balanced scorecard has been developed and used in many companies. Mostly, it has been used at the top management level where it supports the company's strategic management system. The authors observe that the balanced scorecard concept has also been helpful to both top and middle management to shape and clarify organizational goals and strategy. Moreover, it has been useful at the worker level, when the complex trade-offs implied by the balanced scorecard are translated into simple performance measures. For example, in the Analog Devices Case, the "Corporate Scorecard" contained, in addition to several traditional financial measures, performance measures relating to customer delivery times, quality and cycle times of manufacturing processes, and effectiveness of new product developments.

It is worth noting that this system provides an organization-wide framework to integrate the benefits of several improvement initiatives, such as total quality management, just-in-time production and distribution systems, time-based competition, lean production/lean enterprise, building customer-focused organizations, activity-based cost management, employee empowerment, and reengineering. Kaplan and Norton claim that the programs like those listed above, if fragmented and not linked to the organization's strategy, could and will lead to disappointing results.

IMPLEMENTATION: Kaplan and Norton (1993) provide a typical profile of the process of building a balanced scorecard. They point out that each organization is unique and so may follow its own path for building this system. They propose the following eight-step process that firms may adapt in their systematic efforts to develop a balanced scorecard:

1. *Preparation:* The organization must first define the business unit for which the top-level scorecard is to be developed. Typically a business unit would have its own customers, distribution channels, production facilities (if appropriate) and financial and performance measures.
2. *Interviews, First Round:* The balanced scorecard facilitator (either outside consultant or executive from the firm) conducts interviews of about 90 minutes each with senior managers to obtain their inputs on the company's objectives. Interviews with principal shareholders and key customers may also be conducted to gather inputs on their expectations.
3. *Executive Workshops, First Round:* The top management team is brought together with the facilitator to debate the proposed mission and strategy statements until consensus is reached. Subsequently the group brainstorms about the development of the measures. At this meeting, it is not necessary to reach consensus on the proposed measures as long as the group develops a menu of measures for each objective.
4. *Interviews, Second Round:* The facilitator consolidates the executive discussions and seeks further input from senior managers about the measures and the process of implementation.
5. *Executive Workshop, Second Round:* At this time more middle managers are involved in the group meeting in addition to the top management team. Participants work in groups, debate and develop the measures and start to develop an implementation plan. They could also be asked to propose preliminary targets for each objective and measure.
6. *Executive Workshop, Third Round:* The top management team is brought together to reach consensus on the balanced scorecard. This is the time for the group to develop an implementation plan, a communication plan, and an information system to support the scorecard.
7. *Implementation:* It may be beneficial to form a team for implementation. This team may also encourage the development of second-level metrics for decentralized units.
8. *Periodic Review:* Each quarter or month, top management may undertake reviews of the balanced scorecard information and discussions with managers of decentralized divisions and departments. The authors recommend that senior managers revisit the balanced scorecard metric annually as part of their strategic planning, goal setting and resource allocation processes.

DISCUSSION: Some general guidelines are offered below with regard to selecting measures for each of the four perspectives. First, the financial objectives have first to be related to the product life cycle (harvest, sustain or growth). Three typical themes encountered are revenue growth and mix, earnings improvement, asset utilization. Second, it is important that the customer-related objectives be identified with more specificity. For example, objectives relating to market share should not go unqualified; rather "market share of targeted customers" would be more appropriate to get across what actual measure the managers are seeking to define. Other common measures include percentage growth of business with existing customers to capture customer loyalty and the solicitation cost per new customer acquired to capture market development effort. Third, while traditional approaches may seek to improve existing internal business processes, balanced scorecard encourages identification of new processes and seeks to meet needs of future. In this context, product and service performance attributes that are likely to create value for the customer become critical in addition to measures such as response time, cost and quality. Finally, in the context of learning and growth, balanced scorecard provides a framework to reveals gaps between existing capabilities of people, systems, and procedures and those needed to meet the strategic goals of the organization.

In general, the authors recommend between 10 and 25 metrics in total as a benchmark and indicate that a typical firm takes about 16 weeks to develop the balanced scorecard system. These are carefully selected as strategic measures that focus on factors that will lead to competitive breakthroughs. Keeping the list short and manageable is key for top management to monitor progress without too much clutter.

Several case study examples are used to illustrate the concepts by the authors in their book and articles. *Metro Bank*, for example, developed a scorecard with measures that focused **directly** on achieving their strategy. In this case, the use of both "lead" and "lag" indicators was especially important, with " lead" performance driver measures signaling strategy to managers (e.g. hours spent with customer reinforced the importance of a new relationship-based sales approach) and "lag" outcome measures providing feedback on the process. For *National Insurance*, initially, only "lagging" core outcome measures could be identified. The importance of "lead" performance driver measures was obvious due to the severe "lag" of data in insurance industry. As a result of balanced scorecard thinking, the firm developed a survey to assess policyholder satisfaction and the results were leading indicators with regard to their effort in the area of customer acquisition and retention.

With regard to *technology*, Kaplan and Norton point out that although not essential it is helpful to have up-to-date corporate information systems that can give one the capability to have information quickly compiled. It is important to stress corporate-wide knowledge of balanced scorecard's implementation and use. Frms could make effective use of a "corporate billboard" displaying the balanced scorecarded in high-traffic areas such as meeting rooms, main thoroughfares, etc. and encouraging all the employees to periodically monitor performance in their functions.

In conclusion, the authors note that in order for the balanced scorecard approach to be effective, it must reflect the strategic vision of the senior executive group and have the full support of this group. They also indicate that benchmarking with other companies is traditionally *not* good since different competitive environments and different strategies typically exist from firm to firm. Further, if implementation takes too long, a loss of momentum will occur in the organizational shift. Accordingly, even if the measures are not perfect to begin with, the implementation should go on since balanced scorecard is meant to be a continually reviewed process with many opportunities for evolution.

See **Activity-based cost management; Financial accounting measures; Non-financial performance measures; Performance drivers; Performance excellence; Performance measurement in manufacturing.**

References

[1] Kaplan, R.S. and D.P. Norton, "The balanced scorecard-measures that drive performance" *Harvard Business Review*, January–February 1992.

[2] Kaplan, R.S. and D.P. Norton, "Putting the balanced scorecard to work" *Harvard Business Review*, September–October 1993.

[3] Kaplan, R.S. and D.P. Norton, *The Balanced Scorecard*, HBS Press 1996.

[4] Kaplan, R.S. and D.P. Norton, "Using the balanced scorecard as a strategic management system" *Harvard Business Review*, January–February 1996.

BALANCING THE LEAN MANUFACTURING SYSTEM

See **Leveling, balancing and synchronizing the lean manufacturing system; Lean manufacturing implementation.**

BALDRIDGE AWARD

See **Quality management systems; Total quality management.**

BAND WIDTH

Band width is an important aspect in the design of manufacturing processes for Just-in-Time (JIT) production systems. It is a measure of flexibility attributable to production capacity. A system is said to have a wide band width if it has the production capacity to manufacture a broad variety of products with some fluctuation in demand. Wide band width systems are capable of rapid changeovers, reduced throughput time, and reduced inventory. High band width is consistent with the philosophy of JIT production, which is to be as responsive as possible to as large a set of demands as necessary.

See **Just-in-time manufacturing; Just-in-time manufacturing implications.**

BASE KANBANS

Kanban is the Japanese word for "visible sign" or card. Kanbans are used to control the flow of materials, parts and products in just-in-time (JIT) or pull-production systems (Sugimori, *et al.*, 1977). During the 1990s, researchers developed a class of kanban control policies, known as dynamic kanban control policies, which systematically manipulate the number of kanbans in the production system. Some of these policies use the concept of maintaining a minimal or "base" number of kanbans in the system (Gupta and Al-Turki, 1997, and 1998). They may be augmented by additional kanbans as warranted; the extra kanbans may be removed according to the dynamic control policy. The base kanbans are always present in the system and represent the minimum number of kanbans in the system.

See **Adaptive kanban control; Dynamic kanban control for JIT manufacturing; Flexible kanban control; JIT; Kanbans; Pull production systems; Reactive kanban control.**

References

[1] Gupta, S.M., and Y.A.Y. Al-Turki (1997). "An algorithm to dynamically adjust .the number of kanbans in stochastic processing times and variable demand environment." *Production Planning and Control*, 8, 133–141.

[2] Gupta, S.M., and Y.A.Y. Al-Turki (1998). "The effect of sudden material handling system breakdown on the performance of a JIT system." *International Journal of Production Research*, 36, 1935–1960.

[3] Sugimori, Y., K. Kusunoki, F. Cho, and S. Uchikawa (1977). "Toyota production system and kanban system materialization of just-in-time and respect-for-human system." *International Journal of Production Research*, 15, 553–564.

BASE VALUE

Under experimental smoothing method of forecasting, the calculation of the base value is a procedural step in making trend-enhanced forecasts (TEF) of future demand. The formula for calculating the new base value is as follows:

$$\text{Base value}_t = \alpha(\text{Actual demand}_t) + (1-\alpha)(\text{Base value}_{t-1} + \text{Trend}_{t-1})$$

where:

α = Base value smoothing constant (between 0 and 1)

t = Current period (period for which most recent actual demand is known)

Base value_{t-1} = Base value computed one period previously (at the end of period $t-1$)

Trend_{t-1} = Trend value computed one period previously (at the end of period $t-1$)

See **Forecasting guidelines; Forecasting in manufacturing management; Forecasting models: Quantative.**

BATCH PRODUCTION

Batch production is distinguished by its intermittent manufacturing: a particular product is made in lots or batches that may range in size from a very few to a very many. The key feature is that these items are not made constantly but must instead fit in with the production of other products. Batch production may exist as regularly scheduled production, for example, when a month's sales requirements are made once each month. Batches also may be produced irregularly, as when demand is highly variable. Batch production is most frequently used when the general production capacity available greatly exceeds the demand for any specific product but the production of a number of different products fills the production capacity. In these cases, a specialized or dedicated process to make one product is generally not economically viable. There may be technical aspects of the production process that limit the size of batches.

See **Batch size; Job shop; Manufacturing systems.**

BATCH SIZE

A batch is a quantity either in production or scheduled to be produced. The concept of batch size is best defined in terms of two different concepts, the process batch and the transfer batch. A process batch is the quantity of a product processed at a work center before that work center is reset to produce a different product. A transfer batch is the quantity of units that move from one work center to the next. The transfer batch size need not, and in most cases, should not be equal to the process batch size; process batch can be equal or greater than transfer batch. In synchronized or lean manufacturing systems, the transfer batch size generally should be kept as small as possible in order to ensure a smooth and rapid flow of materials. The process batch size should be determined by the requirements of the system and should be allowed to be variable as needed over time. At bottleneck work centers, especially those with significant setup times, the small transfer batches should be used with relatively large, economical process batches.

See **Synchronous manufacturing using buffers.**

Reference

[1] Srikanth, M.L. and M.M. Umble (1997). *Synchronous Management: Profit-Based Manufacturing For The 21st Century, Volume One*, Spectrum Publishing, Guilford, Connecticut, 163.

BENCH MODEL

Bench model is a paper or cardboard model of a product used for assessing its market potential in the new product development process.

See **New product development: The role of suppliers; Product development and concurrent engineering; Sequential approach to product design.**

BENCHMARKING

Benchmarking is the search for industry best practices on the road to superior performance. Best practices refers to approaches that produce exceptional results. These practices are usually innovative in terms of the use of technology or human resources, and are recognized by customers or industry experts. Benchmarking includes performance benchmarking, which focuses on characteristics of competing products and services; process benchmarking, which centers on work processes of companies that perform similar functions, no matter what the industry; and strategic benchmarking, which examines how companies compete and seeks winning strategies that lead to competitive advantage and market success.

See **Managing for performance excellence: The Malcolm Baldridge national quality award criteria; Total quality management.**

References

[1] Bogan, C.E. and M.J. English. 1994. *Benchmarking for Best Practices*. New York: McGraw-Hill, Inc.

[2] Evans, James R. and W.M. Lindsay 1999. *The Management and Control of Quality, Fourth Edition*, Cincinnati, OH: South-Western Publishing Co.

[3] Walleck, A.S., J.D. O'Halloran, and C.A. Leader. 1991. "Benchmarking World-Class Performance." *The McKinsey Quarterly*, (1), 3–24.

BIAS IN FORECASTING

Lack of bias is desirable in any forecasting procedure; that is, the resulting forecasted value should not be consistently higher or lower than the actual value. The *mean error* is used to measure the bias of forecasts, as follows:

Mean error (bias)

$$= \sum_{i=1}^{n} (\text{Actual demand}_i - \text{Forecast demand}_i)/n$$

where:

i = Period number
n = Number of periods of data

The mean error is low when high and low forecast errors cancel each other out, but it may hide large errors. Therefore the magnitude of the errors is also important. The magnitude of forecast errors is addressed by the mean absolute deviation (MAD), as follows:

$$\text{MAD} = \sum_{i=1}^{n} |\text{Actual demand}_i - \text{Forecast demand}_i|/n$$

where:

i = Period number
n = Number of periods of data
|x| = Absolute value of x.

Forecasting results therefore are evaluated using a combination of bias and MAD.

See **Forecast accuracy; Forecast errors; Forecast evaluation; Forecasting guidelines; Forecasting in manufacturing management; Mean error.**

BIDDING PROTOCOL

A bidding protocol is a procedure that has been used in decision support systems for distributed and/or decentralized problem solving. According to this protocol, a project may be decomposed into several tasks or jobs to be distributed among a group of (cooperating) operating units through a contract negotiation process. The bidding process creates a market-like environment within which the operating units compete for the jobs that are most suitable for themselves. The whole system usually consists of a controlling unit, called manager, and a group of operating units which possess multiple task capabilities. The manager is responsible for the development, modification and implementation of the overall project. The manager ensures that the tasks or jobs are assigned to appropriate operating units so that certain performance measures are maximized. The operating units, on the other hand, are responsible for the assigned tasks which are under their control. They are required to optimize their performance by utilizing available resources. The bidding process furnishes a decentralized coordination mechanism and provides a communication protocol between the manager and operating units for task assignment. Furthermore, the bidding process serves as a means to facilitate the realization of objectives of both the manager and the operating units. The whole process can be briefly outlined as follows: First, the manager decomposes the project into a set of tasks that can be handled by at least one operating unit. Then, these tasks are broadcast to the operating units. After receiving the task announcement, each operating unit reviews all job opportunities and consider which job to bid. Each operating unit's objective is to obtain task assignments that would maximize its performance measures subject to local constrains such as time and capacities. Having received bids from the operating units, the manager begins the bid evaluation process. If a particular task did not secure sufficient number of bids from operating units, the manager may modify task specifications and announce it again. After choices are made, the manager informs the operating units of the outcomes. During the entire project duration, the manager continues to monitor the environment. Occasionally, if there are changes in the environment, new jobs will be announced and awarded on an ad-hoc basis. If an operating unit fails to complete a job according to agreement, the job has to be reassigned. The above bidding process can be facilitated by computerized support.

See **Disaggregation in an automated manufacturing environment.**

Reference

[1] Davis, R. and R.G. Smith (1983), "Negotiation as a Metaphor for Distributed Problem Solving," *Artificial Intelligence*, 20, pp. 63–109.

BILL OF LABOR

A bill of labor lists the number of standard hours required at each operation to produce a given item, usually a finished product. The following is an example of a bill of labor.

Bill of Labor for Product Z
(Standard Hours per Unit)

Work Center	**Std. Hrs. per unit of Product Z**
Drill	*.35*
Mill	*.40*
Assembly	*.25*
Total	**1.00 Hrs.**

This information is generally used in rough-cut capacity planning to estimate capacity requirements associated with the master production schedule (MPS). For example, if the MPS calls for producing 100 units of product Z in a given week, we can easily estimate the capacity that would be required at each work center by multiplying standard hours from the bill of labor by the planned production.

The bill of labor is a moderately accurate method for rough-cut capacity planning. It does not include the lead time information that would be present in a resource profile, but it is more detailed than the method of overall factors. However, developing a bill of labor does usually require that a company have detailed time study information.

See **Capacity planning; Capacity planning by overall factors; Resource profile; Long-range capacity planning; MRP; Resource profile; Rough-cut capacity planning; Time study.**

BILL OF MATERIALS

The bill of materials (BOM) for a product or subassembly is an engineering document that details all part numbers and their quantities that are components of the product. The BOM is an essential input to a *material requirements planning* (MRP) system. Bills of materials may exist at several levels. A single-level BOM is only concerned with the immediate components of a part, and does not additionally consider the components of the components; for example, a car requires five wheels, but the single-level BOM ignores rims, tires, and other wheel components. Other types of BOM exist, giving additional details of the lower tiers of components. Computerization of BOM leads to bill of materials files and product structure records. The BOM is used in conjunction with the *master production schedule* (MPS) to generate purchase orders and production orders for all required items.

See **Bill of materials, indented; Bill of materials structure; Master production schedule (MPC); Material requirements planning (MRP); Modular bill of materials.**

BILL OF MATERIALS CO-EFFICIENT

Bill of Materials (BOM) coefficient is defined as the percent ratio of the number of new components to be designed (for a specific order) to the total number of components in the product. In customizing an automobile, for example, with 5000 total number of components and 50 new components, the BOM coefficient is 1%.

See **Mass customization.**

BILL OF MATERIALS, INDENTED

An indented bill of materials (BOM) differs from a single-level bill of materials in that it includes the lower level components of the subassemblies of a product. The indented BOM lists all components of the product from the finished article to the raw materials.

The indented BOM is structured such that the highest level parents are closest to the left margin, with their components indented to the right. Each subsequent level of the BOM is indented further to the right. Components used in more than one parent component appear several times in the BOM. Example of an Indented BOM (Source: Vollmann, *et al.* 1997):

- 1605 Snow Shovel
  - 13122 Top Handle Assembly (1 required)
    - 457 Top Handle (1 required)
      - 082 Nail (2 required)
      - 11495 Bracket Assembly (1 required)
        - 129 Top Handle Bracket (1 required)
        - 1118 Top Handle Coupling (1 required)
- 048 Scoop-Shaft Connector (1 required)
- 118 Shaft (1 required)
- 062 Nail (4 required)
- —
- —

See **Bill of materials; Bill of materials structure; Modular bill of materials.**

Reference

[1] Vollmann, T.E., Berry, W.L., Whybark, D.C. *Manufacturing Planning and Control Systems*, Fourth Edition, Irwin/McGraw-Hill, New York, 1997.

BILL OF MATERIALS STRUCTURE

Bill of materials (BOM) structuring is the act of organizing bills of materials using computer files to fulfill specific applications. BOM structure defines the arrangement and format of bill of materials files. BOM files are computer records. The link between BOM structure and BOM files is maintained by computer software known as a BOM processor. It is usual for single level bills of materials to be linked together by the BOM structure as a substitute for indented bills of materials. Using the BOM processes, printed output can be had as single level or indented BOM, depending upon the purpose. The single level BOM provides component availability data for work order launching, whereas the indented BOM is useful for designing assembly processes and costing. The BOM structure permits these multiple uses of a single set of data.

See **Bill of materials; Bill of materials, indented; Material requirements planning (MRP); Modular bill of materials.**

BLOCKED OR STARVED MACHINE

A machine is in a "blocked upstream state", if the machine is idle and waiting for arrival of a part from the upstream machine for processing. This is also called starvation of a machine. A machine is in a "blocked down stream" state, if on completion of it work on a part it cannot deliver the part to the next machine.

See **Blocking; Kanaban-based manufacturing systems.**

BLOCKING

Blocking in a manufacturing system occurs when a machine cannot unload a part because its output buffer is full or there is no material handling equipment available to move the part to the next work center. Blocking can also occur for material handling equipment (e.g., automated guided vehicles (AGV) and robots) when they cannot move forward because another piece of equipment is in the way or the next operation is not ready. Blocking may either be temporary or permanent. In the former case, a machine is blocked until space opens up in the output buffer, or an AGV is blocked until the obstruction is gone. In the latter case, the machine or AGV may be deadlocked and require manual intervention.

See **AGV; Deadlock; Linked-cell manufacturing system (L-CMS); Manufacturing systems modeling using Petri nets; Synchronous manufacturing using buffers.**

Reference

[1] Desrochers, A.A., and Al-Jaar, R.Y., 1995, *Applications of Petri Nets in Manufacturing Systems: Modeling, Control, and Performance Analysis*. (Piscataway, NJ: IEEE Press.)

BOTTLENECK RESOURCE

See **Drum-buffer-rope synchronization; Theory of constraints; Theory of constrains in manufacturing management.**

BOTTOM-UP REPLANNING

Bottom-up replanning is used in computerized MRP systems to solve material availability problems. The planner, rather than the computer system, performs manual bottom-up replanning. The planner uses pegging data and evaluates effects of possible solutions. Pegging refers to the capability in MRP and MPS to identify the sources of an item's gross requirements (demand) or allocations. Possible replanning solutions could include lead time compression, reducing order quantities, substituting materials, or changing the master production schedule.

See **Master production schedule (MPS); Material requirements planning (MRP); Pegging.**

BOX-JENKINS METHODOLOGY

Box-Jenkins Mehtodology is a time series forecasting method popularized by George E. Box and G.M. Jenkins for applying ARIMA (Auto Regressive Integrated Moving Average) models to time series analysis, forecasting, and control. They put together in a comprehensive manner the relevant information required to understand and use univariate time series ARIMA models. The theoretical underpinnings are sophisticated but it is possible for the non-specialist to obtain an working understanding of the methodology. Several commercial softwares enable practitioners use this model for forecasting demand.

See **Forecasting guidelines; Forecasting in manufacturing management; Forecasting models: Causal.**

BREAK POINT

The break point, or break-even point, is the quantity of production (or sales volume) which yields neither loss nor profit. That is sales revenue completely covers all the cost without leaving a surplus. Graphically, it is represented at the intersection of total revenue and total costlines. (See figure next page). When sales exceeds the break point profits are realized; profits increase with volume.

See **Break-even analysis; Fixed cost; Variable cost.**

BREAK-EVEN ANALYSIS

See **Break point.**

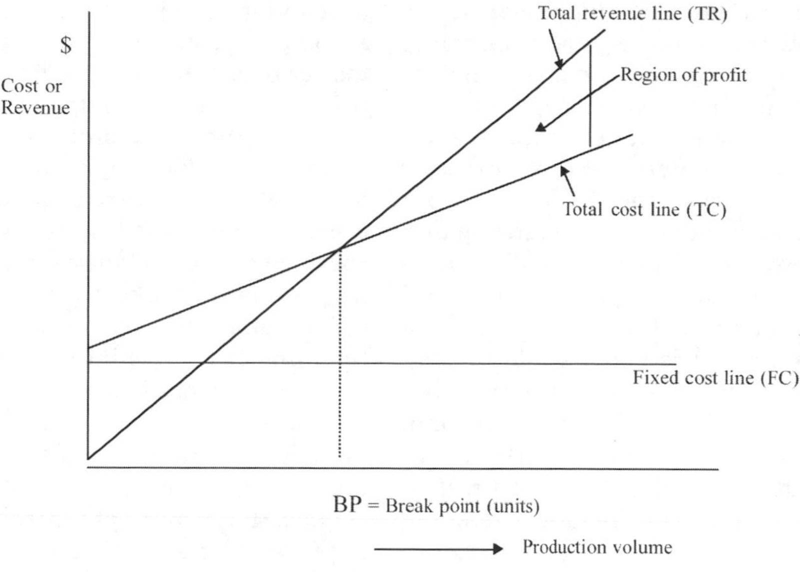

Let VC = Variable cost to produce 1 unit

Then: Total cost = Fixed cost + Units produced*(Variable cost)

Further: TC = FC+ BP(VC)

TC - FC = BP(VC)

Therefore: BP = (TC-FC)/VC

Also: Profit = Total revenue - Total cost

BREAK-EVEN POINT

See **Break point.**

BUCKETLESS MATERIAL REQUIREMENTS

The unit of time (usually a week) used in MRP are referred to as time buckets. The implications of using time buckets is the following convention: an order that is due in a particular week has to be ready at the beginning of that week, regardless of the exact due date. In theory, this problem could be reduced by using a greater number of smaller time buckets (example: time bucket = one day), but this approach increases storage costs, calculations, and paperwork. A bucketless MRP system uses exact due dates, rather than time periods, for controlling scheduled receipts and planned orders with greater precision. A further development is the real-time computer system, which uses bucketless concepts.

See **Material requirements planning (MRP); Real-time systems; Time buckets.**

BUCKETLESS SYSTEMS

A bucketless system is a manufacturing resources planning (MRP), distribution resources planning (DRP), or other time-phased system that does not use defined time unit such as one week (time buckets) to process, store, and display time-phased data. Instead, precisely dated records are used for planning; for example, an order release date of January 22nd will be used instead of the "week of January 20th," which is an example of a time bucket. This is in contrast to bucketed systems, where time-phased data are accumulated into buckets appropriate to the time period (for example, weekly buckets where the planning unit of time is one week).

See **Action bucket; Bucketless material requirements; DRP; MRP.**

BUFFER STOCK

Buffer stock is additional inventory over and above foreseen need that is kept on hand in the event of some likely but infrequent or unforeseen occurrence. This inventory is often referred to as safety stock. Buffer inventory provides a buffer to ensure that a manufacturing line does not stop for lack of parts or a store does not run out of an item too often. In a retail environment, buffer inventory may be kept for unanticipated surges in demand. In a manufacturing environment, buffer material and work-in-process inventory may be used to overcome machine breakdowns, operator absences, quality problems, line imbalances, and a host of other reasons. Because it is not possible to keep buffer stock for service activities, back up labor serves as a buffer for variations in service demand.

Statistics can be used to determine the likelihood of a stockout for varying levels of finished goods buffer stock. Accurate historical demand rates can reveal average demand and standard deviation for stated periods of time. (i.e., daily, weekly, monthly, quarterly, etc.) Using average demand, standard deviation of demand, and average inventory replenishment lead time, a manager can examine stockout possibilities for varying levels of buffer inventory. A statistical analysis of this type assists a manager in assessing the trade offs associated with carrying too much or too little inventory and the impact on stockouts.

See **Implications of JIT manufacturing; Safety stock; Safety stock: Luxury or necessity; Safety time; Simulation of production problems.**

Reference

[1] Shafer, S.M. and Meredith, J.R. (1998). *Operations Management*, John Wiley and Sons, Inc., New York.

BUILT TO ORDER

Manufacturers may produce goods under several different order-fulfillment methods. An order fulfillment system where products are manufactured upon the receipt of firm orders rather than in anticipation of orders is called built-to-order or make-to-order systems. In this system, inventory accumulation and forecasting errors are minimized.

See **Bullwhip effect in supply chain management; Make-to-stock production.**

BULK EXCHANGE OF TOOLS

Bulk Exchange is a tooling strategy which provides the machine a complete dedicated set of tools needed for each job or part visiting the machine. A complete set of tools are transferred from the tool crib to the machine for each different part being made on the equipment. At the completion of a part run, all tooling will return to the tool room to be replaced by a different dedicated set of tools for the next part to be manufactured. Bulk exchange offers the best control over tooling for a specific part. The disadvantages of this approach are extensive duplicate tooling, large tooling inventory and high level of tool handling. It limits the number of machines which can accept a part and creates more traffic due to the need to transfer many tools. That is, this strategy ignores the fact that sharing of tools can reduce the tooling inventory and tool handling. With Bulk Exchange, the problem of control is minimized at the expense of excessive tooling inventory. Bulk Exchange only becomes a sensible strategy with a high volume/low variety part mix application.

See **Disaggregation in an automated manufacturing environment; Flexible automation; Tools planning and management.**

BULK-BREAKING

Bulk-breaking refers to breaking down consolidated or bulky shipments into smaller, individual shipments for final delivery to local terminals.

See **Bullwhip effect in supply chain management.**

THE BULLWHIP EFFECT IN SUPPLY CHAIN MANAGEMENT

Hokey Min

University of Louisville, Louisville, KY, USA

INTRODUCTION: In today's fiercely competitive global marketplace, the increased efficiency and effectiveness within the individual function or organization alone will not give a firm comparative advantages over its rivals. Instead, the firm should create a series of value-added chains from the sourcing of raw materials to manufacturing, marketing, and distribution of finished goods to end customers. As such, the concept of supply chain management has grown in importance over the last decade. The International Center for Competitive Excellence in 1994 defined supply chain management (SCM) as "an integration of business processes from the end-customer through original suppliers that provides products, services and information that add value for customers." (Cooper, Lambert, and Pagh, 1997). The main thrust of SCM is to avoid sup-optimization of long-term business goals by strategically aligning with trading partners across the entire value chain and consequently leveraging the core competencies of each trading partner.

However, if the firm fails to achieve the SCM ideal, it may suffer from one of the most dramatic consequences called the *bullwhip (whiplash* or *whipsaw)* effect. The bullwhip effect (or phenomenon) is generally referred to as an inverse ripple effect of forecasting errors throughout the supply chain that often leads to amplified supply and demand misalignment where orders (perceived demand) to the upstream supply chain member (e.g., the supplier) tend to exaggerate the true patterns of end-customer demand since each chain member's view of true demand can be blocked by its immediate downstream supply chain member. The common symptoms of the bullwhip effect include delayed new product development, constant shortages and backorders, frequent order cancellations and returns, excessive pipeline inventory, erratic production scheduling, and chronic overcapacity problems. For example, in the $300 billion grocery industry, the bullwhip effect may account for the $75 to $100 billion worth of unproductive pipeline inventory, because 5 percent variation in end customer consumption can result in 300 percent to 400 percent variation in downstream suppliers' production (Artzt, 1993). In general, the bullwhip effect can increase total business operating costs by 12.5 percent to 25 percent (Lee, Padmanabhan, and Whang, 1997a). Metters (1997) also estimated that the elimination of the bullwhip effect could increase company profits by an average of 15 percent to 30 percent.

POTENTIAL CAUSES OF THE BULLWHIP EFFECT: The supply chain consists of boundary-spanning activities that transcend functional and organizational boundaries. The failure to recognize interdependence caused by boundary-spanning activities is often the main cause of the bullwhip effect. That is to say, uncoordinated demand creation and physical supply activities across functions and organizations can lead to supply chain oversupply with resultant inefficiency and ineffectiveness. In the following paragraphs, the various causes of the bullwhip effect are discussed:

1. ***Information Failure:*** An unprecedented number and variety of products competing in today's markets make it increasingly difficult for manufacturers, suppliers, retailers to predict and plan for orders and production volume (Fisher, Hammond, Obermeyer, and Raman, 1994). Coupled with such demand uncertainty, a firm's sales forecasting based on order history from its immediate downstream supply chain members can distort the true end-customer demand and thereby contribute to upstream order variability. In particular, during the period of continuous product shortages, overzealous downstream supply chain members tend to over-order and stockpile their shelves with inventory to catch up with previous demand, creating *"phantom demand"* which, in turn, triggers a series of over-forecast and the subsequent over-production in the supply chain.

   Lee *et al.* (1997a) argued that a lack of demand information sharing among supply chain members might contribute to the exaggeration of actual demand signals. Indeed, in the supply chain of pharmaceutical companies such as Eli Lilly, Bristol-Myers Squibb, and McKesson, Lee *et al.* (1997b) observed that information distortion had led the total pipeline inventory to exceed 100 days of supply, when the optimal pipeline inventory should have been much less than 30 days of supply. Reseller News estimates that 10 days of pipeline inventory was equivalent to a loss of one percent of profit (Austin, Lee, and Kopczak, 1997). Austin *et al.* (1997) further noted that the more complex a chain structure is or the more autonomous chain members are, the more demand information tends to get skewed by each chain member. In other words, supply chain members' failure to update demand information and their lack of coordinated communication escalates strong influences on the bullwhip effect.
2. ***Chain Complexity:*** In a typical chain structure, each chain member tends to specialize in certain functions. Manufacturers concentrate on production and national promotion, whereas distributors concentrate on product assortment, bulk-breaking, storing, and delivering. Although distributors are considered sales agents of manufacturers, they often represent multiple manufacturers with diverse products. As such, channel intermediaries such as distributors by nature cannot match the quantities and characteristics of products desired by ultimate end- customers to those of products provided by manufacturers. For this reason, the presence of channel intermediaries may contribute to the bullwhip effect. For

instance, Dell Computer successfully adopted a direct marketing plan called *"Built-to-Order"* to reinvent the personal computer (PC) industry's traditional supply chain by delivering its PCs directly to customers within 36 hours after the order (Austin *et al.*, 1997). With Build-to-Order, end-customer orders are manufactured on receipts, eliminating the need for demand forecasting. Therefore, a direct marketing plan which bypasses channel intermediaries can mitigate the bullwhip effect.

The chain complexity added by channel intermediaries can also increase order cycle time throughout the entire supply chain. Partly due to chain complexity, Kurt Salmon Associates discovered that apparel industry products averaged 66 weeks to reach its customers (Blackburn, 1991). Such time delays among supply chain links can add to forecasting difficulty and the subsequent bullwhip effect.

3. ***Product Proliferation:*** As many retailers such as Wal-mart, K-mart, and Target are demanding more diversified product lines and tailored services to meet the specific needs of different segments of customers, there has been explosive proliferation of products offered to customers in terms of color, design, and function. Such product proliferation, however, may backfire, because it can significantly increase the chain complexity and demand volatility, while dramatically shortening the product life cycle. With greater variations in product offerings, demand forecasting has become more arduous than ever before. In fact, Chaffee (1995) observed that it would be very difficult for the firm to forecast the demand for more than 20 different types of products with an acceptable degree of accuracy, and the resulting forecasting difficulty could create the supply and demand imbalance throughout the supply chain. For example, in the computer industry, which is typified by product proliferation, shortage and oversupply problems are endemic (Fisher *et al.*, 1994).
4. ***Sales Promotion:*** Today's shrewd customers tend to shop only when sales promotion in the form of coupons, rebates, and seasonal discounts is available. This shopping pattern often results in peaks and valleys of customer orders and correlates with frequent price changes. Although sales promotion is intended to benefit end-customers, many downstream distributors also tend to stockpile huge amount of promotional inventory with an expectation of future price increase. Thus, sales promotion is likely to create phantom demand throughout the supply chain, contributing to over-order, over-production, or temporary product shortages. Procter and Gamble once discovered such a bullwhip effect in its diaper sales and production. Procter and Gamble noticed that deep discount promotion accounted for more than 5 percent variation in customer orders because it encouraged its customers to buy considerably more than they could accommodate (Artzt, 1993). The same bullwhip effect was believed to be the important reason why annual supermarket sales declined by one percent in 1993 (Saporito, 1994).
5. ***Economies of Scale:*** To exploit cost saving opportunities resulting from economies of scale, many firms prefer to order on a periodic basis so that they can accumulate orders large enough for volume purchasing or freight consolidation. Such periodic ordering is likely to cause a *hockey stick* phenomenon (order surges at the end of month, quarter or year), which not only amplifies order variability, but also increases order cycle time. For instance, with the widening spread between truckload (TL) and less-than-truckload (LTL) rates in the wake of transportation deregulation in the U.S., a growing number of firms are prompted to pool small orders to create larger shipments weighing as little as 5,000 pounds; the maximum time that shippers could hold an order is about 30 days (Jackson, 1985). Freight consolidation also increases managerial time and responsibilities for sorting traffic, assembling loads, bulk-breaking, material handling and arranging prepayment. Therefore, freight consolidation can substantially increase order cycle time, exacerbating the bullwhip effect. Another form of economies of scale that may cause the bullwhip effect is *bundling* (providing the same package of products and services, regardless of the differences in customers' needs and preferences). Bundling forces downstream chain members to buy some of the items that they would not want, creating phantom demand which, in turn, becomes a source of the bullwhip effect.
6. ***Speculative Investment Behavior:*** In years of economic upturns, many firms tend to perceive product shortages as the loss of market share or revenue, thereby triggering a period of over-investment and production. In particular, a firm may overestimate the actual demand. For instance, encouraged by various forms of government incentives such as increased local content rules, preferential loans and government bailouts, many automakers in Asia substantially increased their production capacity in the early 90s and were left with huge amount of inventory as the growth rate

of auto sales declined. Such a speculative investment behavior may have contributed to the downfall of the Korean auto giant named Kia and the Indonesian national car-maker in the nineties. Similarly, manufacturer inducements such as liberal order return policies and "*monopoly rebates*" (dividends paid to Japanese *keiretsu* retailers for limiting their shelf space for competing brands) may increase downstream chain members' risk-taking behavior, thereby exaggerating the actual end-customer demand. As illustrated above, since a speculative investment behavior often impedes the firm's ability to match supply and demand, it aggravates the bullwhip effect.

MEASUREMENT OF THE BULLWHIP EFFECT IN SUPPLY CHAIN MANAGEMENT: Although there is no commonly accepted guideline for measuring bullwhip effects on supply chain performances, it is apparent that the lower the level of supply chain integration, the greater the likelihood of the bullwhip effect. Indeed, the results from the 1997 Integrated Supply Chain Benchmarking Study (ISC '97) reveal that improvements in supply chain performance are closely tied to the firm's ability to integrate its internal supply chains with a group of suppliers and customers (Helming and Mistry, 1998). Thus a key measure of the bullwhip effect in supply chain management is the level of supply chain integration. In a broad sense, supply chain integration is the integration of two or more independent organizational units' business processes through long-term relationships. Such relationship building is characterized by joint planning and investment, co-ownership of assets, and information and risk sharing. In effect, the level of supply chain integration may be measured in two ways: (1) *Breadth* (extent of integration), and (2) *Depth* (nature of integration).

Breadth represents the extent to which an organization has built the value-added relationship with its upstream and/or downstream supply chain partners. The potential breadth measures may include the percentage of supply chain members participating in joint product development and capacity planning, the percentage of supply chain members electronically connected to their supply chain-wide electronic network or information sharing systems, and the percentage of supply chain members utilizing an advanced synchronized order fulfillment methods such as vendor-managed inventory (VMI). *Depth* represents the permeability or compatibility of organizational boundaries with respect to the organization's functions, culture, technological advancement, and balance of power. The potential depth measures may include the frequency of information sharing and joint planning among supply chain members, the percentage of business functions (e.g., order processing, invoicing, payment) computerized or electronically transmitted through the supply chain electronic network, the degree of technological sophistication across the supply chain (e.g., a variety of information technology usages between supply chain members), and the level of supply chain members' dependency on one another in forecasting demand and controlling inventory.

MANAGERIAL IMPLICATIONS AND REMEDIES OF THE BULLWHIP EFFECT: To strive for value-added productivity throughout the supply chain, a firm should successfully overcome the bullwhip effect. The following strategic guidelines are proposed as potential remedies for the bullwhip effect.

1. ***Inter-organizational Information Sharing System:*** The interchange of information through open communication channels should form the basis of all supply chain activities including demand forecasting and planning. Thus one of the best solutions for the bullwhip effect is the combined use of an inter-organizational information sharing systems such as electronic commerce (EC) with a point-of-sale system (POS) and enterprise resource planning (ERP). To elaborate, the downstream retailer's POS captures customers' responses to products and services, permitting instant updating of its sales record and inventory position. Then POS data is passed along to the upstream distributors and manufacturers via the EC network (e.g., Electronic Data Interchange, the Internet) in such a way that they can update their order replenishment automatically. Using the data base built upon the POS data, ERP may be designed to integrate all of the key planning processes such as demand and time-phased inventory planning across the supply chain so that the firm can facilitate real-time responses to end-customer demand by tracking both forecasted and actual demand on an as-need basis.
2. ***Order Fulfillment Agility:*** Since a long lead-time delays adjustment to volatile orders at the downstream supply chain, a firm cannot mitigate the bullwhip effect without compressing order cycle-time. Thus, "pull" strategies such as Just-In-Time (JIT) and quick-response manufacturing may be an effective remedy for the bullwhip effect. The motivation for quick-response manufacturing is that cycle-time

reduction decreases the level of demand uncertainty and thus minimizes forecasting error and the subsequent bullwhip effect. Indeed, Blackburn (1991) observed that forecasting error grows exponentially as the time to forecast increases. He also reported that the U.S. apparel industry firms such as J.C. Penney, Oxford, and Burlington were able to reduce forecasting error by more than half and increase inventory turnover ratio by 90 percent through quick-response manufacturing.

3. ***The Postponement of Final Product Configuration:*** By delaying customization of finished products, a firm can minimize its finished good inventories and lessen the need to reconfigure finished good forecasts when end-customer demand shifts (Austin *et al.*, 1997). That is why a vendor-managed inventory (VMI) program has become increasingly popular among many firms. VMI is a customer-driven, synchronized inventory planning system that aims to reduce inventory and opportunity costs associated with unsold products by asking vendors (i.e., upstream supply chain members) to maintain ownership of all semi-configured products until they are actually demanded by end-customers. In other words, in VMI, customer order fulfillment is based on actual demand, not in anticipation of demand. Thus VMI, through postponement, can alleviate the bullwhip effect.
4. ***Value Pricing:*** Considering that short-term, deep-discounted promotional deals often contribute to erratic order variability and supply fluctuations, *value pricing* which offers customers better value at every stage of supply chain may be a sensible solution for the bullwhip effect. In fact, Proctor and Gamble is reported to have a great success with its value pricing policy such as everyday low pricing (EDLP). With EDLP, Proctor and Gamble charges a constant, everyday low price without temporary promotional deals or discounts and consequently increases its customers' confidence in price stability. Since such price stability makes customer orders more predictable, Proctor and Gamble can run its factories more efficiently and increase inventory turnover ratio. Proctor and Gamble also has reported a 3 percent to 5 percent decrease in cost of goods and the subsequent improvement in cash flows using EDLP (Saporito, 1994).

See **Built-to-order; Bulk-breaking; Downstream supply chain member; Electronic commerce; Enterprise resource planning; Point-of-sale system; Supplier partnership; Supply chain management; Vendor-managed inventory.**

References

[1] Artzt, E.L. (1993), "Customers Want Performance, Price, and Value." *Transportation and Distribution*, 32–34.

[2] Austin, T.A., Lee, H.L., and Kopczak, L. (1997), *Unlocking Hidden Value in the Personal Computer Supply Chain*, Andersen Consulting Report.

[3] Blackburn, J.D. (1991), "The Quick-Response Movement in the Apparel Industry: A Case Study in Time-Compressing Supply Chains." edited by J.D. Blackburn, *Time-Based Competition: The Next Battle Ground in American Manufacturing*, Irwin, Homewood, IL, 246–269.

[4] Chaffee, R.H. (1995), "Creating Fulfillment Agility in the Face of Product Proliferation." *Logistics: Mercer Management Consulting*, 20–23.

[5] Cooper, M.C., Lambert, D.M., and Pagh, J.D. (1997), "Supply Chain Management: More Than A New Name for Logistics." *International Journal of Logistics Management* 1, 1–13.

[6] Fisher, M.L., Hammond, J.H., Obermeyer, W.R., Raman, A. (1994), "Making Supply Meet Demand in an Uncertain World." *Harvard Business Review*, 83–93.

[7] Helming, B. and Mistry, P. (1998), "Making Your Supply Chain Strategic." *APICS: The Performance Advantage* 8, 54–56.

[8] Jackson, G.C. (1985), "A Survey of Freight Consolidation Practices." *Journal of Business Logistics* 6, 13–34.

[9] Lee, H.L., Padmanabhan, V., and Whang, S. (1997a), "Information Distortion in a Supply Chain: The Bullwhip Effect." *Management Science* 43, 546–558.

[10] Lee, H.L., Padmanabhan, V., and Whang, S. (1997b), "The Bullwhip Effect in Supply Chains." *Sloan Management Review*, 93–102.

[11] Metters, R. (1997), "Quantifying the Bullwhip Effect in Supply Chains." *Journal of Operations Management* 15, 89–100.

[12] Saporito, B. (1994), "Behind the Tumult at P&G." *Fortune*, 74–82.

BURDEN RATE

See **Two-stage allocation process; Unit-based costing.**

BUS MODULARITY

Bus modularity is a concept within mass customization. Bus modularity is most often exhibited in electrical and electronic products with busses, such as computers and circuit breaker systems.

Bus modularity is also exhibited in track lighting, where light fixtures can be placed at various locations on a track, and shelving systems, where the shelves can be moved to higher or lower location along the brackets installed in the wall or case. An important distinction between bus modularity and other forms of modularity is that bus modularity allows variation in the number and the location of the components in the system while the other forms of modularity allow only variation in the type of components used in identical product architecture.

See **Mass customization; Mass customization implementation.**

BUSINESS PROCESS REENGINEERING (BPR)

Business Process Reengineering (BPR), alternatively called reengineering, redesign, or process innovation, came into being in the 1990s. It proposes radical approaches to redefining work as processes, rather than tasks. The philosophy has been used successfully in a number of commercial organizations and the U.S. Department of Defense. BPR has been defined as "the analysis and design of workflows and processes within and between organizations" (Davenport and Short, 1990). Alternatively, Hammer and Champy (1993) define reengineering as "the fundamental rethinking and radical redesign of business processes to achieve dramatic improvements in critical, contemporary measures of performance, such as cost, quality, service, and speed." Davenport (1993) argues that one of the contributing factors to the success of Japanese organizations has been their ability to develop efficient business processes. However, BPR is not analogous to continuous improvement; it is appropriate where radical, rather than incremental, improvements in business processes are required. To emphasize this difference, BPR is sometimes referred to as process innovation (Davenport, 1993). Process reengineering improves responsiveness to customer demands.

WHAT ARE BUSINESS PROCESSES?: Davenport (1993) defines a process as "a specific ordering of work activities across time and place, with a beginning, an end, and clearly identified inputs and outputs: a structure for action." This structure produces an output for the customer or market.

Typically, a manufacturing firm will contain operational and management processes (Davenport, 1993). Operational processes include product development, customer acquisition, customer requirements identification, manufacturing processes, integrated logistics, order management, and after-sales service. Management processes include performance measurement, information management, asset management, human resource management, and planning and resource allocation.

IMPLEMENTATION: Davenport and Short (1990) suggest five steps for the implementation of BPR:

1. *Develop a business vision and process objectives.* BPR is driven by a business vision which implies specific business objectives such as cost reduction, time reduction, output quality improvement, quality of work life, learning and empowerment.
2. *Identify the processes to be redesigned.* Most firms use the "High-Impact" approach, which focuses on the most important processes or those that conflict most with the business vision. Lesser number of firms use the "Exhaustive" approach that attempts to identify all the processes within an organization and then prioritize them in order of redesign urgency.
3. *Understand and measure the existing processes.* This prevents the repeating of old mistakes and provides a baseline for future improvements.
4. *Identify information technology (IT) levers.* Awareness of IT capabilities can and should influence process design.
5. *Design and build a prototype of the new process.* A redesign of a process should not be viewed as the end of the BPR process. Rather, it should be viewed as a prototype, with successive iterations when necessary.

See **Quality of work-life; Quality: The implications of deming's approach; Reengineering and the process view of manufacturing.**

References

[1] Davenport, T.H. (1993). *Process Innovation: Reengineering Work through Information Technology*. Harvard Business School Press, Boston, MA.

[2] Davenport, T.H., and J.E. Short (1990). "The New Industrial Engineering: Information Technology and Business Process Redesign," *Sloan Management Review*, Summer, 11–27.

[3] Hammer, M., and J. Champy (1993). *Reengineering the Corporation: A Manifesto for Business Revolution*. Harper Business, New York.

[4] http://www.brint.com/papers/bpr.htm.

CAD

Computer-aided design (CAD) is a computer hardware and software combination with computer graphics, which allows engineers and designers to create, draft, manipulate and change product designs without the use of conventional drafting. CAD systems permit tremendous speed, precision and flexibility over traditional drafting systems. Being an electronic system, it permits the electronic transfer of design information to customers and machines equipped to receive and process electronic data. Thus, a product designed on a CAD system can be transferred through a LAN to shop floor CNC machines or CAM systems.

See **Advanced manufacturing technology; CAM; CNC machine tools.**

CALENDAR FILE

A **Material Requirements Planning** (MRP) database uses several different files for planning and control, Calendar File is one of them. The calendar file is used to convert the shop's day calendar to a conventional day/date/year calendar. This file accounts for vacations, shutdown periods, and so on.

See **Materials requirements planning (MRP).**

Reference

[1] Vollmann, T.E., Berry, W.L., Whybark, D.C. (1997) *Manufacturing Planning and Control Systems*, Fourth Edition, Irwin/McGraw-Hill, New York.

CALS

CALS is a computer program which was initiated by the U.S. Department of Defense and its contractors in an effort to facilitate electronic interchange of technical documents and to create product-related data bases accessible by both government agencies and their trading partners.

See **Electronic data interchange in supply chain management.**

CAM

CAM stands for Computer-aided manufacturing. CAM incorporates the use of computers to control and monitor several manufacturing elements such as robots, CNC machine tools, and automated guided vehicles. CAM implementation is often classified into several levels. At the lowest level, it represents an integrated system of programmable computerized machines controlled by a centralized computer. At the higher levels, large-scale integration of machines is controlled and supervised by multiple levels of computers. CAM systems are often integrated electronically with CAD systems to permit instantaneous transfer of electronic information between designers and shop floor production systems such as CAM.

See **Advanced manufacturing technologies; AGV; CAD; Computer-aided manufacturing; Flexible automation; Manufacturing flexibility.**

CAM-I

CAM-I stands for Computer-Aided Manufacturing International, an organization dedicated to the study of computer-aided manufacturing software, cost analysis, and manufacturing practices. A recent study conducted by Computer-Aided Manufacturing International (CAM-I) concluded that while the concept and design engineering phases of new product development incur only 5–8 percent of the total product development costs, these two activities commit or "lock in" 80 percent of the total cost of the product.

See **New product development: The role of suppliers.**

CAMPAIGN CYCLE LENGTH

The length of a production run consisting of two or more campaigns, where a campaign is the production run of a family of products that share a major setup. This is also known as a rotation schedule or a schedule wheel. For example, a company producing fiberglass material may produce three major product families; fiberglass material with a paper backing, a foil backing, and no backing. Within each product family a variety of widths and depths are produced. Furthermore, since a process line may produce all of the above products, a schedule must be established which cycles through the production of all product families and the variety of widths and depths produced in each family. It is also common to produce all products within a family while the equipment is setup to produce that product family. This avoids major setup or switchover costs typically associated with changing the family of products being produced. Thus, a campaign would be the production of all foil backed products and a campaign cycle would be the production run producing all product families. The campaign cycle length defines how long it would take to cycle through all product families and product varieties before the cycle is repeated.

See **Process industry scheduling.**

Reference

[1] Brown, R.G. (1967). *Decision Rules for Inventory Management*, Holt, Rinehart and Winston, New York.

CAPACITATED FACILITY LOCATION DECISION

The key consideration in facilities location decisions is the determination of the capacity of the facilities to be utilized. This has lead to the classification of facilities location decisions (FLD) as capacitated and uncapacitated. In the capacitated FLD, the capacity limit of each potential facility is set at an *a priori* capacity level; the decision as to which facilities should be utilized is made taking into consideration the *a priori* capacity levels for the potential facilities.

See **Facilities location decision.**

CAPACITY

Capacity is the maximum possible output over a specified period of time. In cases where the output is non-homogeneous, capacity may be measured in terms of the available machine hours.

See **Capacity available; Capacity management; Capacity planning; Manufacturing systems modeling using petri nets.**

CAPACITY AVAILABLE

The amount of capacity not already committed to specific customer orders is called Capacity Available. It is computed as the difference between the **total installed capacity** and the amount of capacity already committed to customer orders.

See **Managing capacity in make-to-order production systems.**

CAPACITY BILLS (CB) METHOD

The Capacity Bills (CB) method is one of the techniques that can be used for a company's **resource requirements planning** (RRP, i.e., **roughcut capacity planning**, RCP). Compared with the **overall factors** (OF) method, the CB method provides more direct linkage between individual end products in the **master production schedule** (MPS) and the capacity required for individual work centers. Consequently, it requires more data than the OF procedure. Bill of materials and routing data are required, and direct labor-hour or machine-hour data must be available for each operation. A bill of capacity indicates total standard time required to produce one end product in each work center required in its manufacture. The capacity requirements at individual work centers can be estimated by multiplying the capacity bills by the MPS quantities. Estimates obtained from the OF method are based on an overall historical ratio of work between work centers, whereas capacity bill estimates reflect the actual product mix planned for each period.

See **Capacity planning by overall factors; Linking aggregate plans and master production schedules; Master production schedule; Resource requirements planning; Roughcut capacity planning.**

Reference

[1] Berry, W.L., T.G. Schmitt, and T.E. Vollmann (1982). "Capacity planning Techniques for Manufacturing Control Systems: Information Requirements and Operational Features," *Journal of Operations Management*, 3(1), pp. 13–25.

CAPACITY INSTALLED

Capacity installed is the maximum output that a facility could attain under ideal conditions; also known as design capacity.

See **Capacity management in make-to-order production systems.**

CAPACITY MANAGEMENT IN MAKE-TO-ORDER PRODUCTION SYSTEMS

V. Sridharan
Clemson University, South Carolina, USA

DESCRIPTION: Make-to-order (MTO) manufacturing has been defined as "*a production environment where a product or service can be made after receipt of a customer's order. The final product is usually a combination of standard items and items custom designed to meet the special needs of customer* (APICS, 1995: 46)." Thus, MTO firms essentially sell the capability to custom design and produce a certain set of products. Some MTO firms experience less demand than capacity, on average. For such firms, order backlog management, operations scheduling, and lead-time determination constitute three primary areas of concern. Other MTO firms experience demand greater than capacity. Their major concern is capacity management. See Bertrand and Sridharan (1998) for specific examples of such firms. In some cases, capacity expansion is often not an economically viable option due to the already high fixed-to-variable cost ratio and/or the need for a highly skilled labor force. Further, when demand exceeds capacity, traditional approaches for adjusting capacity in the short-term such as scheduling overtime and offloading work to other machines are not feasible. Examples of firms that experience such an environment would include custom musical instrument manufacturers, low-volume component manufacturing firms, and high-end fashion apparel manufacturing firms. This article focuses on some creative approaches for managing capacity at such firms.

Difficulties in order promising and capacity management often create conflict and tension between manufacturing and marketing functions. Pressure for reduced levels of inventory and increased product variety coupled with an emphasis on on-time delivery exacerbates the problem. Central to the resolution of the conflict between manufacturing and marketing functions is the management of capacity. The chief issue is to ensure that the firm allocates the available capacity to satisfy customer demand in an efficient and effective fashion.

HISTORICAL PERSPECTIVE: Several alternative ways to ensure coordination between manufacturing and marketing have been prescribed. Most of these, however, focus on coordination at the strategic level. Very little work has been done on the operational coordination between the two functions. Whybark and Wijngaard (1994) argue that well coordinated marketing and manufacturing strategies cannot be executed effectively if the operational coordination is not effective. Effective operational coordination requires a set of guidelines, which govern operational decision making.

Most of the studies reported in the literature pertain to situations where the expected aggregate demand is assumed to be in the range of 80% to 85% (or less) of the available capacity (defined as installed capacity minus planned shut-downs for maintenance). As already mentioned, some make-to-order firms, however, face aggregate expected demand much higher than that. For example, empirical data pertaining to machine tool manufacturing firms indicates that, roughly a third of the firms from twelve different countries, report average capacity utilization levels in excess of 90% (Bertrand and Sridharan, 1998). A large fraction of firms also report capacity utilization in excess of 95%. These facts suggest that the average demand level is close to the available capacity or exceeds it.

Fransoo, Sridharan and Bertrand (1995) describe such a situation faced by firms in the semi-continuous process industry, where the actual aggregate demand level is between 95% and 100% of the available capacity. In other cases, such as low volume component manufacturing shops, the total demand is even greater than the available capacity; see in Bertrand and Sridharan (1998). A somewhat similar situation exists also in the high-fashion apparel manufacturing industry.

When the total demand is greater than the available capacity, and the objective is to maximize profit and delivery reliability, it may be tempting to conclude that a firm needs to do is to elect to satisfy demand for the most profitable product(s) and ignore the demand for less profitable products. However, often, it may be important from a strategic point of view for the firm to maintain a certain minimum service level (fill rate) for all products in its product mix. Otherwise, the firm would seriously limit its business. The alternative is to manage capacity at the tactical level by efficient allocation to the various products in the mix. In most real life situations, demand is stochastic and the capacity allocation decision must be made on a dynamic basis. Otherwise it would be trivial problem to allocate capacity between competing requirements.

The issue of dynamically allocating capacity between various competing sources of demand is neither new nor unique to MTO firms. Many service firms, such as airlines, hotels, and rental car companies, for example, routinely deal with the issue of allocating capacity between competing classes of customers (demand), commonly referred to as "yield management." The basic objective of yield management is to maximize revenue by effectively managing a perishable asset (i.e., capacity). Thus the focus of yield management in service firms has been on selling the right product to the right customer at the right price. The objective of the capacity allocation problem in make-to-order manufacturing firms is similar to that of yield management in service firms. In both cases, the main issue is allocating a known quantity of an asset (that has zero value at the end of the planning horizon) to demands from different classes of products (customers) to maximize the overall profit. This suggests that a tight management of demand and capacity could prove fruitful in manufacturing as they have been in many service operations. It is important to understand and learn from the characteristics of the environment faced by service firms, and how it is shaped largely by yield management practices.

In many services the customer is compelled to pay a higher price for the same product as the deadline for its use approaches; an example is airline tickets thus making them time sensitive. Due to the time sensitive nature of the price, the demands for different classes of service may be viewed as occurring at distinct time periods with no demand overlap. Finally, the sequence in which individual orders are received can be ignored in services if demand for different classes is non-overlapping. These characteristics of the different classes of service based on price help management simplify the capacity allocation problem.

In the case of manufacturing operations, it may be worthwhile to explore the possibility of adapting some or all of these characteristics from service capacity management. For example, charging a different price based on order lead-time. Such a scheme would facilitate a non-overlapping demand and avoid the need to either accept or reject each order as it is received. It reduces the magnitude of the need to sequence orders.

The demand for some products, may not be stationary. Consider, for example, the case of high-fashion apparel manufacturing firms. Such firms operate in an environment, where the market for products is at its peak at the beginning of the planning horizon but steadily declines over time and may reach zero at the end of the planning horizon.

In summary, the capacity allocation problem in manufacturing and the yield management problem in service operations have some similarities. However, significant differences exist between the two.

This article describes a few specific approaches for managing capacity in MTO firms with total demand in excess of capacity. In these cases, the manager is saddled with the difficult task of dynamically allocating the available capacity between competing classes of products and/or customers. In addition, the manager must ensure dependable delivery performance, and speed of delivery in order to stay competitive. At first glance, it may appear that, when demand is greater than capacity, the firm should consider expanding capacity to solve the problem. Although this could be a viable long-term option, it may not be a feasible option in the short term. If demand is greater than capacity including overtime and offloading of work to other machines, other alternatives, such as capacity rationing, coordination mechanisms, and subcontracting need to be considered to exploit the available flexibility.

CAPACITY RATIONING: The basic philosophy of rationing is to be discriminating so as to maximize a specified objective when faced with excess demand. Several authors have studied the issue of stock rationing in the context of make-to-stock manufacturing firms. They focus on allocating the available inventory to customers based on their priority, and recommend a stock rationing policy to discriminate between different customers (sources of demand), in the short-term. Stock rationing appears to be an effective approach when a sizable difference exists between the rewards associated with different customers.

Make-to-order manufacturing firms (and service firms), on the other hand, essentially hold capacity (rather than inventory) in stock. When such firms experience demands in excess of capacity, the manager is faced with the problem of allocating the available capacity to different priority products (or customers), resulting in the rejection of some orders for lower priority products or from lower priority customers. Clearly, there are rewards (e.g., increased profits) and penalties (e.g., long-term impact on market share) associated with accepting or rejecting orders based on capacity. Therefore, the manager has to carefully allocate capacity so that it achieves stated objectives such as maximizing overall profit, or meets prescribed fill rates for each product customer.

Recent studies suggest that a substantial and statistically significant increase in total profit may be obtained by systematically rationing capacity

(Balakrishnan, Sridharan, and Patterson, 1996). Capacity rationing, therefore, serves as an effective decision tool for managers in make-to-order manufacturing firms to discriminate between different classes of products or customers. Rationing, however, could sacrifice the service level (i.e., fill rate) for the lower priority class of products/customers. Under rationing, guaranteeing a minimum level of service for each class of customer (product) may prove to be tricky.

COORDINATION MECHANISMS: Efficient mechanisms for improving the operational coordination between marketing and manufacturing via improved communication have been reported in the literature. But, most of these studies focus on non-MTO firms. For example, the Hyundai Motor Company has been reported to have achieved effective coordination between the marketing and manufacturing functions via a master scheduling process in which communication between the two was enhanced by means of a centralized organizational structure and cross-functional teams (Hahn, Duplaga, and Kim, 1994). Erens and Hegge (1994) describes a product specification concept (via generic bill-of-materials) that allows improved communication between sales and manufacturing to improve logistic performance and customer satisfaction in an assemble-to-order firm. In each of these cases total demand was not very high compared to the available capacity.

Two forms of coordination between manufacturing and marketing have been considered: (1) the hierarchical approach and (2) the integrated approach (Ten Kate, 1994). The hierarchical approach involves the separation of order acceptance and production scheduling decisions, with a common restriction on the available capacity of the shared resource. In the integrated approach, both decisions are fully integrated, making a new and revised schedule upon every order arrival. Simulation results show that little difference between the two approaches, except in tight situations (short lead times, high utilization rate) where the integrated approach was found to marginally outperform the hierarchical approach. The difference between the two approaches, in practice, would be even smaller. Further, it would be reasonable to expect the integrated approach to produce considerably more nervousness in schedules compared to the hierarchical approach.

Fransoo, Sridharan, and Bertrand (1995) describe a two-level hierarchical planning and scheduling approach in which the control of product changeover times is the central feature in a factory where capacity is tightly constrained. Their hierarchical approach places demand management at the tactical level to coordinate order acceptance and production scheduling decisions. By avoiding uncontrolled changes in production cycle times, this approach attempts to limit the amount of productive capacity lost in changeovers. Further, it achieves stability in the shop floor via stable product cycle times. Simulation results suggest that the hierarchical approach could produce stable schedules and increased profit.

SUBCONTRACTING: When the total demand is greater than or nearly equal to the available capacity, subcontracting a portion of the orders may be an attractive alternative to ensure delivery speed and dependability. Bertrand and Sridharan (1998) investigate the subcontracting decision in the context of a low volume component manufacturing operation facing order arrival rate greater than the service rate. They develop four decision rules with varying informational needs and complexity to determine which orders should be subcontracted. They examine the performance of the policies over a range of environments. Their results suggest that rules based on a shop workload norm outperform simpler rules, which use only order lead time or order arrival rate information. The workload-based rules are also found to be robust for up to 50% error in system parameter estimates, whereas the non-workload-based rules are found to be extremely sensitive to estimation errors.

Their results are encouraging for the following reasons. They found order arrival rate to have very little impact on the performance of the rules. This implies that, coordination between marketing and manufacturing can be simplified to a great extent when subcontracting a portion of the demand is feasible. The actual utilization levels achieved are virtually identical to the target values for all rules under all conditions examined. This suggests that subcontracting could serve as a viable tool to achieve improved coordination at a low cost.

Although the more complex workload-based rules performed better, the average tardiness and percentage tardy values were high. Improving the estimation of job flow times and the setting of order due dates is akin to the demand forecasting problem. Flow time estimates may be significantly improved by taking account of current shop conditions.

A hybrid approach of subcontracting and flow time estimation may be fruitful in devising a sophisticated order promising and capacity management system. Such an approach would likely produce superior results not only in terms of cost minimization but also in terms of due date performance. This, however, remains to be seen.

CONCLUSION: Managing capacity in a tightly constrained MTO manufacturing system is a complicated problem. Capacity rationing, coordination mechanisms for improving communication between manufacturing and marketing functions, and subcontracting are three possible approaches for managing capacity. Evolving research exploring the viability of grouping products/customers into different classes and discriminating between them via rationing capacity suggests that this approach may be valuable in some make-to-order manufacturing environments. While the work done so far is limited to simple and specialized manufacturing environments and, hence, may not be of immediate practical value, increasingly realistic models can be developed for handling complex manufacturing situations.

Research focusing on coordination mechanisms suggests that hierarchical systems may prove to be robust, especially when the impact of schedule nervousness on cost and employee morale are considered.

See **Capacity planning: Long-range; Capacity planning: Medium- and short-range; Hierarchical production planning; Stock rationing.**

References

[1] APICS, 1995. *Apics Dictionary*, 8th Edition, Falls Church, VA.

[2] Balakrishnan, N., Sridharan, V., and Patterson, J.W. (1995). "Rationing Capacity Between Two Product Classes." *Decision Sciences*, 27-2, 185–215.

[3] Bertrand, J.W.M. and Sridharan, V. (1998). A Study of Simple Rules for Subcontracting in Make-to-Order Manufacturing. *European Journal of Operational Research*, Forthcoming.

[4] Erens, F.J. and Hegge, H.M.H. (1994). "Manufacturing and Sales Coordination for Product Variety." *International Journal of Production Economics*, 37-1, 83–100.

[5] Fransoo, J.C., Sridharan, V. and Bertrand, J.W.M. (1995). "A Hierarchical Approach for Capacity Coordination in Multiple Products Single-Machine Production Systems With Stationary Stochastic Demands." *European Journal of Operational Research*, 86-1, 57–72.

[6] Hahn, C.K., Duplaga, E.A., and Kim, K.Y. (1994). "Production/Sales Interface: MPS at Hyundai Motor." *International Journal of Production Economics*, 37-1, 5–18.

[7] Kate, H.A. ten (1994). "Towards a Better Understanding of Order Acceptance." *International Journal of Production Economics*, 37-1, 139–152.

[8] Powers, T.L., Sterling, J.U., and Wolter, J.F. (1988). "Marketing and Manufacturing Conflict: Sources and Resolution." *Production and Inventory Management Journal*, 29-4, 56–60.

[9] Shapiro, B.P. (1977). "Can Marketing and Manufacturing coexist?" *Harvard Business Review*, 55-5, 104–114.

[10] Weatherford, L.R., and Bodily, S.E., 1992. "A taxonomy and research overview of perishable-asset revenue management: Yield management, overbooking, and pricing." *Operations Research*, 40-5, 831–844.

[11] Whybark, W.C. and Wijngaard, J. (1994). "Editorial: Manufacturing-Sales Coordination." *International Journal of Production Economics*, 37-1, 1–4.

CAPACITY PLANNING BY OVERALL FACTORS

Capacity is a measure of the production capability of a company, a work center, a machine, or a person at any given point in time. It is an important consideration in the production planning process because if planned production exceeds available capacity, then the production plan probably will not be realized. In general, each step of the production planning process requires some calculation of the associated capacity requirements and, if necessary, a revision of the production plan to conform to capacity limitations.

When the master production schedule (MPS), or weekly planned production, is being developed, the associated capacity planning step is known as rough-cut capacity planning. It is called "rough-cut" because it represents a rough estimate of capacity needs. A more accurate estimate is determined later through capacity requirements planning (CRP) after the MPS has been used for material requirements planning (MRP).

The simplest, but least accurate, approach to rough-cut capacity planning is known as the method of overall factors (CPOF). This method begins with the total standard hours required to produce each item in the MPS, then multiplies those hours by the number of units planned for production in a given week. The resulting total hours required are then broken down for individual machines or work centers using the historical percentage of total time accounted for by each operation.

See **Capacity bills method; Capacity requirements planning; Capacity planning: Long range; Master production schedule; Material requirements planning.**

CAPACITY PLANNING: LONG-RANGE

Gregory P. White
Southern Illinois University at Carbondale, Illinois, U.S.A.

DESCRIPTION: Long-range capacity planning is the process of ensuring that sufficient production resources (facilities, people, equipment, and operating hours) are available to meet an organization's long-range production needs. The American Production and Inventory Control Society (APICS, 1996) considers sales and operations planning, and master production scheduling to be planning activities that are covered by long-range capacity planning.

As shown in Figure 1, different capacity planning methods are applied at different planning stages. For example, resource requirements planning is generally used as a means of ensuring sufficient capacity to execute the sales and operations plan, and rough-cut capacity planning deals with the capacity implications of the master production schedule. Capacity requirements planning is used in conjunction with material requirements planning (MRP). This article describes the techniques used for long-range capacity planning. For medium- and short-range capacity planning, see **Capacity Planning: Medium- and Short-Range.**

HISTORICAL PERSPECTIVE: Capacity planning, in one form or another, has probably been done for at least as long as manufacturing has

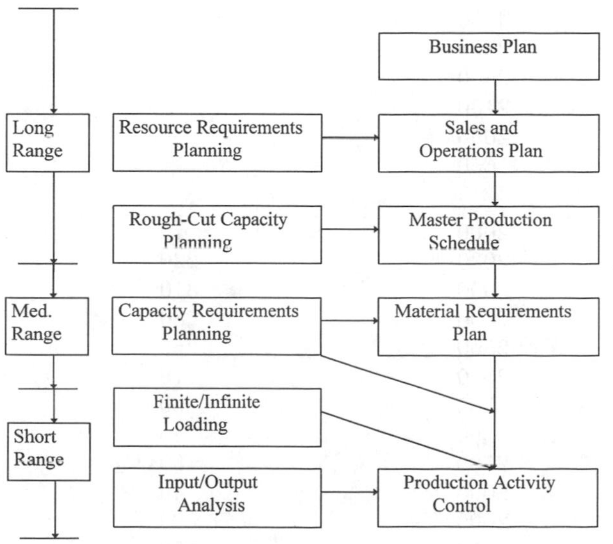

Fig. 1. The sequence of capacity planning steps.

been in existence (Wight, 1970). However, more formal approaches to capacity planning came into wider use with the advent of MRP systems (Berry *et al.*, 1979). As companies began utilizing these formal systems for material planning, it became not only possible to develop more formal systems for capacity planning, but also became necessary to do so (Blackstone, 1989).

Until recently, such systems have been largely iterative, requiring a trial-and-error approach. Some efforts have been devoted by academicians to developing mathematical optimization models for production planning, but these have not received much acceptance in industry (Vollmann *et al.*, 1997). Most recently, Theory of Constraints (TOC), described later in this article, has been receiving increased attention as an alternative to the trial-and-error capacity planning.

STRATEGIC PERSPECTIVES: Effective capacity planning can greatly enhance an organization's ability to reduce costs, meet delivery due dates, and shorten production cycle times. In the long range, capacity planning can involve decisions to either close or add facilities – decisions that can have far-reaching implications, both financially and competitively. Consequently, capacity planning really has to begin as a component of the organization's strategic planning process. Even in shorter-range time periods, capacity decisions will influence the company's ability to meet customer demand and to respond when demand shifts. Further, lack of capacity planning may force an organization into making short-term adjustments that can greatly increase costs. Therefore, the major purpose of long-range capacity planning is to ensure that both the sales and operations functions engage in planning, and the master production schedule is feasible with regard to the production capacity.

IMPLEMENTATION:

1. ***Resource Requirements Planning:*** The capacity planning activity that looks farthest into the future is resource requirements planning, also frequently referred to as resource planning. This activity, which can include facility planning, is linked directly to the sales and operations plan, also known as the production plan or aggregate plan. Broader, general resource planning may also take place earlier when the organization is developing its strategic plan and business plan. Because the sales and operations plan represents the first step in the production planning process, it is usually developed at higher management levels in the organization and, consequently, is frequently

stated either in dollar terms or in units of product groups. One month is usually the smallest planning time unit for the sales and operations plan.

Because the sales and operations plan is in broad, general terms, the resource requirements plan is also stated in broad, general capacity terms such as dollar output, labor hours, number of employees, or machines hours, among others. The purpose of resource requirements planning is to compare the amount of capacity that should be available against the amount of capacity that will be required by the sales and operations plan. Mismatches can highlight the need to plan overtime, to hire more employees, to buy more equipment or, in the longer run, to build new facilities. If none of these are feasible alternatives, then changes may be required in the sales and operations plan. The example in Table 1 shows two iterations of the resource requirements planning process, based on the given forecast for one product group.

The first trial plan shown in Table 1 utilizes what is known as the "chase strategy," in which planned production is varied to meet the demand forecast. Suppose that for capacity planning purposes, the company in this example estimates that five labor hours are required for every $1000 of production, resulting in the estimate of required labor hours as shown. If this level of, and amount of variation in, labor hours is acceptable to the company, then this plan might be finalized for the time being. Suppose, however, that a maximum of 2500 labor hours are available in any month, and that the company does not want to change its labor hour usage as frequently as this first trial plan would require. The second trial plan shows an alternative that uses more of a "level strategy" by allowing inventory to vary. At this point the plan might be finalized if it is acceptable. Alternatively, concerns about high levels of inventory in some months could require additional iterations. This example indicates that marketing, operations, and finance all must be involved in the resource requirements planning process to ensure that the objectives of each functional area are being met. Furthermore, the plan must be revisited on a regular basis, usually monthly, for updating as new information becomes available.

Table 1. Resource requirements planning in the sales and operations plan.

Month	Sales forecast ($000)	Planned production ($000)	Required labor hours	Planned inventory
First Trial Plan for Product Group A				
January	330	330	1650	
February	250	250	1250	
March	310	310	1550	
April	430	430	2150	
May	530	530	2650	
June	580	580	2900	
July	610	610	3050	
August	520	520	2600	
September	480	480	2400	
October	420	420	2100	
November	410	410	2050	
December	330	330	1650	
Second Trial Plan for Product Group A				
January	400	330	2000	70
February	400	250	2000	220
March	400	310	2000	310
April	400	430	2000	280
May	500	530	2500	250
June	500	580	2500	170
July	500	610	2500	60
August	500	520	2500	40
September	500	480	2500	60
October	400	420	2000	40
November	400	410	2000	30
December	400	330	2000	100

2. ***Rough-Cut Capacity Planning:*** The master production schedule (MPS) is developed from the sales and operations plan indicating planned production of individual products or component modules on a weekly basis. Consequently, capacity planning at this level, known as rough-cut capacity planning is more detailed. Although the end result of resource requirements planning performed earlier may have indicated that sufficient overall capacity should be available, capacity at some individual machines or work centers may not be sufficient. As a result, one objective of rough-cut capacity planning is to begin estimating capacity requirements at the machine level as soon as possible. Although the MPS does not provide sufficient information by itself to determine capacity needs at such a detailed level, we can combine it with other information to obtain an initial, or rough-cut, estimate of detailed capacity needs. Often, rough-cut capacity planning is needed only for work centers or machines that may have insufficient capacity; operations that are known to have ample capacity can frequently be ignored at this stage. The results from the rough-cut capacity plan may lead to revisions in the MPS.

Rough-Cut Capacity Planning by the Method of Overall Factors: Several approaches exist for performing rough-cut capacity planning, and each provides a different level of accuracy, but the approaches that provide more accurate information also require more detailed input. The first of these, the method of overall factors (CPOF), only requires information about total standard hours (either labor or machine hours) per unit of each product and the historical percentage of those hours by operation. Such information is often available as accounting data. An example below shows this approach. Because the CPOF method does not account for lead-time offset, it can be quite inaccurate if product mixes change frequently. However, CPOF can provide good capacity estimates when the production plan and product mix remain fairly constant.

Table 2 indicates the development of a rough-cut capacity plan given the master production schedules for two products. With the CPOF approach, total capacity required (direct labor

Table 2. Capacity planning using overall factors.

	Week							
Master production schedule (units)	1	2	3	4	5	6	7	8
Product X	100	100	100	100	80	80	80	80
Product Y	50	50	50	50	40	40	40	40

Total direct labor per unit (*standard hours*)

Product X	1.20
Product Y	2.50

	Week							
Total direct labor (standard hours)	1	2	3	4	5	6	7	8
Product X	120	120	120	120	96	96	96	96
Product Y	125	125	125	125	100	100	100	100
Total	245	245	245	245	196	196	196	196

Estimated capacity requirements (standard direct-labor hours)

	Historical percentage	Week							
Work center	of total	1	2	3	4	5	6	7	8
Drill	.3	73.5	73.5	73.5	73.5	58.8	58.8	58.8	58.8
Lathe	.6	147	147	147	147	117.6	117.6	117.6	117.6
Assembly	.1	24.5	24.5	24.5	24.5	19.6	19.6	19.6	19.6
Total	1.0	245	245	245	245	196	196	196	196

hours in this example) is available for each product. This number can be multiplied by the planned production of that product in a given week. By summing the results over all products in a week, total capacity requirements for that week can be estimated. The total for each week is then broken down based on the historical percentage of total capacity accounted for by each work center or machine. The estimated capacity requirements that result can be used to identify work centers for which future capacity problems may occur. If capacity cannot be added to meet estimated requirements, then some work might be offloaded to alternative machines. Otherwise, the MPS will have to be modified and capacity estimates recalculated until an MPS is obtained that does not require more capacity than available.

Rough-Cut Capacity Planning using the Bill of Labor: The second approach to rough-cut capacity planning is based on a Bill of Labor (BOL), also known as a capacity bill or resource bill. Just as a bill of materials can list all the components required to make a product, the bill of labor lists the capacity requirements, by work center or machine, to make that product. Such information, which is often available from engineering or accounting records, is applied to the MPS to estimate required capacity. Because the actual requirements by work center are known for each product, instead of being estimated from overall historical percentages, results of the BOL approach are usually more accurate than those obtained through CPOF. However, lead times still are not taken into account, so the method can be inaccurate if production output levels change frequently. An example of rough-cut capacity planning based on the bill of labor approach is shown in Table 3.

In this example, the BOL shows capacity usage broken down by work center for each product. Those figures, combined with the master production schedules, can be used to estimate capacity requirements for each week. It should be noted that the estimates produced in Table 3 are different than those in Table 2 because of

Table 3. Rough-cut capacity planning using the bill of labor.

Work Center	Bill of labor (Standard Hours per Unit) Product X	Product Y
Drill	.25	1.00
Lathe	.90	1.25
Assembly	.05	0.25
Total	1.20 Hrs.	2.50 Hrs.

Master Production Schedule (units)

	Week 1	2	3	4	5	6	7	8
Product X	100	100	100	100	80	80	80	80
Product Y	50	50	50	50	40	40	40	40

Estimated Capacity Requirements (Standard hours)

Work Center	Week 1	2	3	4	5	6	7	8
Drill	$.25 \times 100$ $+ 1.00 \times 50$ $= 75$	75	75	75	$.25 \times 80$ $+ 1.00 \times 40$ $= 60$	60	60	60
Lathe	$.90 \times 100$ $+ 1.25 \times 50$ $= 152.5$	152.5	152.5	152.5	$.90 \times 80$ $+ 1.25 \times 40$ $= 122$	122	122	122
Assembly	$.05 \times 100$ $+ 0.25 \times 50$ $= 17.5$	17.5	17.5	17.5	$.05 \times 80$ $+ .25 \times 40$ $= 14$	14	14	14
Total	245	245	245	245	196	196	196	196

the more detailed information provided by the BOL.

Rough-Cut Capacity Planning using the Resource Profile: The third approach to rough-cut capacity planning, and the most accurate, is the resource profile approach. This method is similar to the bill of labor approach, but includes information about lead time for each operation. This lead time information is used to "offset" the capacity requirements for each operation into earlier weeks, depending upon lead times. Although this approach can be quite accurate, it also assumes that actual production lots will correspond to the MPS quantities, which may be true only if lot-for-lot approach is used throughout the MRP system. Table 4 provides an example of rough-cut capacity planning using the resource profile approach.

In this example, final assembly occurs in the same week as indicated by the MPS, so assembly operations are shown with zero lead time. For Product X, the lathe operation will occur one week before final assembly so it is shown with a lead time of one week. The numbers within the resource profile represent capacity usage (again in direct labor hours for this example) per unit of product. These match the BOL information of Table 3, so the only additional information provided is lead-time offsets.

Capacity estimation by this method is a bit trickier than with the other two methods because the MPS value from a particular week must be used to calculate capacity requirements in different weeks based on the lead-time offset of each work center. For example, drill capacity for Product X in week 3 will be based on the MPS for week 5 because of the two-week lead time of the drill operation ($3 + 2 = 5$). However, the capacity estimate for the lathe time required to make Product X in that same week will be based on the MPS for week 4 because the lathe has a lead time of only one week

Table 4. Rough-cut capacity planning using resource profiles.

	Lead time (weeks)				Lead time (weeks)		
Product X Work center	2	1	0	Product Y Work center	2	1	0
Drill	.25			Drill		1.00	
Lathe		.90		Lathe	1.25		
Assembly			.05	Assembly			.25

Master production schedule (units)

	Week							
	1	2	3	4	5	6	7	8
Product X	100	100	100	100	80	80	80	80
Product Y	50	50	50	50	40	40	40	40

Estimated capacity requirements (standard hours) – abbreviated

	Week				
Work Center	1	2	3	4	5
Drill	$.25 \times 100$ $+1.00 \times 50$ $= 75$	75	$.25 \times 80$ $+1.00 \times 50$ $= 70$	$.25 \times 80$ $+1.00 \times 40$ $= 60$	60
Lathe	$.90 \times 100$ $+1.25 \times 50$ $= 152.5$	152.5	$.90 \times 100$ $+1.25 \times 40$ $= 140$	$.90 \times 80$ $+1.25 \times 40$ $= 122$	122
Assembly	$.05 \times 100$ $+0.25 \times 50$ $= 17.5$	17.5	17.5	17.5	$.05 \times 80$ $+.25 \times 40$ $= 14$
Total	245	245	227.5	199.5	196

Table 5. Use of TOC to develop an MPS for week 1.

	Market demand	Lathe Time (hrs./unit)	Profit per unit
Product X	100	.90	$450
Product Y	50	1.25	$500

MPS based on profit per unit

Product Y has highest profit per unit, so produce its market demand first.

Product Y: 50 units × 1.25 hrs./unit = 62.5 hours (77.5 hours remaining)

Product X: 77.5 hours ÷ .9 hrs./unit = 86 units

Total profit = 86 × $450 + 50 × $500 = $63,700

MPS based on profit per hour of constraint time

	Profit per Hour Of Lathe Time
Product X	$450/.9 = $500
Product Y	$500/1.25 = $400

Product X has highest profit per unit of constraint (lathe) time, so produce its market demand first.

Product X: 100 units × .9 hrs./unit = 90 hours (50 hours remaining)

Product Y: 50hours ÷ 1.25 hrs./unit = 40 units

Total profit = 100 × $450 + 40 × $500 = $65,000

Resulting MPS

Week	1
Product X	100
Product Y	40

(3 + 1 = 4). Otherwise, the calculations are essentially the same as the BOL approach.

3. ***Theory of Constraints:*** Theory of Constraints (TOC) develops a master schedule based on the constraint that limits an organization's ability to make money (Goldratt and Cox, 1986). If that constraint happens to be a particular machine, then the master schedule is developed to maximize profits given the time available on that bottleneck machine, so the master schedule is made feasible with existing capacity and an iterative approach is not used.

   Suppose in the examples of Tables 2, 3, and 4 that lathe capacity is strictly limited to 140 hours per week. In all three cases, the estimated requirements exceed that amount, which would necessitate iteratively replanning the MPS. However, TOC would begin with the knowledge, based on past experience, that the lathe is a constraint. An MPS based on that knowledge would be developed as shown in Table 5.

   Suppose that we know the profit per unit is $450 for Product X and $500 for Product Y. Further, suppose in week 1 that market demand for both products would require more capacity than is available on the lathe. Then, it would be desirable to determine how much of each product to produce within the capacity constraint. Ordinarily, we might choose to meet the market demand for Product Y because it has the highest profit per unit. However, that decision ignores the constraint on lathe time. By calculating each product's profit per hour of lathe time we realize that profits in week 1 will be maximized only if we produce as much Product X as the market demands because Product X has the highest profit per hour of constraint time. Remaining lathe time could be used to make Product Y. More details about TOC are described in Umble and Srikanth (1995)

TECHNOLOGY PERSPECTIVE: Long-range capacity planning often does not require the use of a computer, although some companies have found that computerizing the process enables them to explore more possibilities in less time. However, an increasing amount of commercial software is becoming available for capacity planning. For example, Enterprise Resource Planning (ERP) software now enables companies to plan for many different resource requirements, including equipment, material, labor, and capital, resulting in much better, and more comprehensive capacity planning.

SIGNIFICANT ANALYTICAL MODELS: Although analytical models have not been widely adopted by industry, some effort has been devoted by academicians to the development of such models for sales and operations planning, usually referred to in that literature as aggregate planning. One of the first and best-known of such efforts is the Linear Decision Rule (LDR) of Holt *et al.* (1960), which focused on minimizing costs. The model was actually used by a company that manufactured paint, but it never performed well. More recently, some researchers have been developing an approach called Hierarchical Production Planning (HPP) that prescribes procedures for successively disaggregating, or breaking down, the plan from one level in an organization to the next lower level. Often, mathematical models are used to disaggregate the plan subject to resource capacity constraints (Bitran *et al.*, 1982).

CASES: Some excellent company examples involving the use of capacity planning are provided in Wemmerlov (1984) and Berry *et al.* (1979). A summary of many of these cases, as well as others, is provided by Blackstone (1989).

See **APICS; Bill of labor; Chase strategy for capacity planning; Capacity planning by overall factors; Capacity planning: Medium- and short-range; Hierarchical production**

planning (HPP); Level strategy for capacity planning; Linear decision rule (LDR); Lot-for-lot procedure; Master production schedule; Material planning; MRP; Resource requirements planning; Rough-cut capacity planning; Sales and operations plan; Resource profile; Theory of constraints; Weekly planned production.

References

[1] APICS (1996). 1997 *CPIM Exam Content Manual.* APICS, the Educational Society for Resource Management, Virginia.

[2] Berry, W.L., T.E. Vollmann, and D.C. Whybark (1979). *Master Production Scheduling*, American Production and Inventory Control Society, Virginia.

[3] Bitran, G.D., E.A. Haas, and A.C. Hax (1982). "Hierarchical Production Planning: A Two-Stage System." *Operations Research*, March-April, 232–51.

[4] Blackstone, J.H. Jr. (1989). *Capacity Management*, South-Western, Cincinnati.

[5] Goldratt, E.M. and J. Cox (1986). *The Goal* (Rev. Ed.). North River Press, New York.

[6] Holt, C.C., F. Modigliani, J.F. Muth, and H.A. Simon (1960). *Planning Production, Inventories, and Workforce*. Prentice Hall, New York.

[7] Umble, M. and M.L. Srikanth (1995). *Synchronous Manufacturing*, Spectrum Publishing, Connecticut.

[8] Vollmann, T.E., W.L. Berry, and D.C. Whybark (1997). *Manufacturing Planning and Control Systems (4th ed.)*. Irwin/McGraw-Hill, New York.

[9] Wemmerlov, U. (1984). *Case Studies in Capacity Management*. American Production and Inventory Control Society, Virginia.

[10] Wight, O.W. (1970). "Input-output control, a real handle on lead time." *Production and Inventory Management*, 11, 9–31

CAPACITY PLANNING: MEDIUM- AND SHORT-RANGE

Gregory P. White

Southern Illinois University at Carbondale, Illinois, U.S.A.

DESCRIPTION: Capacity planning is the process of ensuring that sufficient production resources (facilities, people, equipment, and operating hours) are planned to be available to execute the organization's production plans. As Figure 1 shows, capacity planning is carried out for three different planning periods, with plans becoming more specific and detailed as the planning horizon becomes shorter. This article focuses on the methods used for capacity planning for medium and short time periods. Good descriptions of capacity planning can be found in Blackstone (1989) and Vollmann *et al.* (1997).

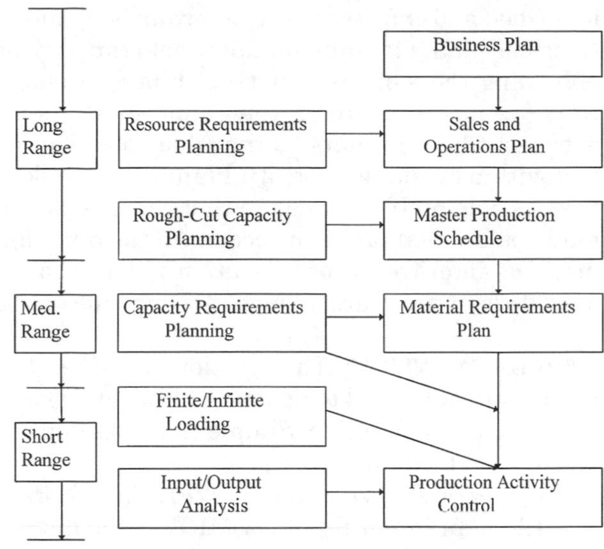

Fig. 1. The sequence of capacity planning steps.

HISTORICAL PERSPECTIVE: Historically, most capacity planning has been done for the long-range time period simply because the specific, detailed information needed for medium- and short-range capacity planning was not available. Any capacity planning done for shorter time frames was often done on an adhoc basis. However, the manual methods employed, such as scheduling boards, simply could not keep up with the complexity of capacity planning problems. It was only after the development of computerized material requirements planning (MRP) software that the necessary information and automated procedures for performing the required calculations became available. Consequently, the methods for medium-range capacity planning discussed in this article will be heavily based on MRP. In general, procedures for short-range capacity planning have not progressed as far as those for medium-range planning, although computerized procedures for finite capacity scheduling for the short-term, discussed later, are becoming more widely available.

STRATEGIC PERSPECTIVES: On-time delivery and short delivery lead times are becoming increasingly important in today's competitive environment. These competitive capabilities can depend heavily on an organization's ability to plan for adequate capacity. Medium-range capacity planning is designed to identify whether sufficient capacity should be available at individual work centers or machines to meet the material requirements plan. Possible problem areas can be identified and changes made accordingly, by modifying the plan, planning for additional capacity, or offloading the work to other machines. In short-range capacity planning, the emphasis shifts toward scheduling orders so that due dates will be met within available capacity. Planning tools identify possible problem areas so that changes can be made before real problems occur. All the planning methods improve an organization's ability to reduce its lead times and to meet delivery due dates.

IMPLEMENTATION: This section describes the most commonly used tools for medium- and short-range capacity planning. Simple examples demonstrate each technique.

1. ***Capacity Requirements Planning (CRP):*** CRP is an integral part of MRP, so the discussion here will assume that MRP calculations have been performed to obtain time-phased planned order releases for each level in the bill of materials. Of course, the MRP calculations are driven by a master production schedule (MPS), and techniques for long-range capacity planning should ensure that sufficient capacity would be available to make MPS feasible. However, as another article in this encyclopedia on long-range capacity planning indicates, the estimated capacity requirements obtained through rough-cut capacity planning may not be entirely accurate. But after MRP calculations are performed, very specific information becomes available about how many units of each component part will be started in production during each time period. This information can be combined with information from the routing sheet about the machines and work centers needed for that component part, and labor hours needed to determine very accurately the required capacity at each machine or work center during each time period. Table 1 provides an example of capacity requirements planning based on information generated from such MRP calculations.

   Table 1 shows the MPS that drives the MRP calculations. Those calculations, shown here for only one component part, must be completed for all components listed in the bill of materials. Because MRP includes both lead time and lot sizing information, the planned order release information is quite accurate with regard to both timing and quantity. Therefore, these planned order release quantities can be used to determine accurate capacity requirements in each time period.

   The routing sheet for each part would ordinarily indicate the processing steps required for that part. Setup and run time at each operation should be available from labor standards. It is quite common to assume that one setup will be required for each lot. Thus, for the part shown in Table 1, each planned order release of 1000 units will require one setup. For the drill, this means that a setup of one hour will be required and that each unit will require .05 hours of run time, thus requiring 51 hours of capacity on the drill in weeks 1 and 5. Of course, other parts will also be processed through the drill so their capacity requirements, calculated in a similar manner, would all be added to those just calculated in Table 1. This procedure generates overall capacity requirements for the drill in each week.

   The lathe is the second operation for part #3297, and there are two possible ways to estimate its capacity requirements. The first is to assume that the lathe operation will occur in the same week that the order is released, as was done for the drill. However, the second approach, which is to assume that one week will be required for each operation. Thus, at the bottom of Table 1, lathe capacity requirements are shown occurring one week later than those for the drill, assuming the drill operation occurred during the week the order was released, and that the second operation will occur in the next week. Other operations would be time-phased accordingly.

   The resulting capacity requirements can be used to identify possible problem areas. Such problems are usually dealt with by planning for additional capacity, such as the use of overtime, or offloading some work to other work centers. Also possible, but less common, is the possibility of going back to the material requirements plan to make changes in either timing or quantity of planned order releases to avoid capacity problems.

2. ***Finite/Infinite Loading:*** Capacity planning techniques discussed so far indicate that an iterative process allows for making changes in the plan when insufficient capacity exists. However, finite and infinite loading are techniques for sequencing orders at a point in time when planned order releases are usually fixed. Thus, the loading techniques focus sequencing orders in a specific time period (a week usually), and within available capacity. Table 2 shows examples of both finite and infinite loading.

   For purposes of simplicity, the examples in Table 2 assume that each machine can process only one order per day. Infinite loading ignores capacity limitations when sequencing orders. The top example in Table 2 indicates forward scheduling with infinite loading; all three orders are scheduled from the beginning of the planning period working forward (i.e.,

Table 1. Capacity requirements planning.

Master Production Schedule - Product X

1	2	3	4	5	6	7	8
100	100	100	100	80	80	80	80

Part #3297 - (**4 per Product X**), Lot size = 1000, Lead time = 2

Week		1	2	3	4	5	6	7	8
Gross Requirements		400	400	400	400	320	320	320	320
Scheduled Receipts		1000							
Projected ending inventory	100	700	300	900	500	180	860	540	220
Net Requirements				100			140		
Planned Receipts				1000			1000		
Planned order releases		1000			1000				

Routing Sheet -- Part #3297

Operation	Machine	Setup Time	Run Time
1	Drill	1 hr.	.05 hrs./unit
2	Lathe	2 hrs.	.20 hrs./unit

Drill Capacity Required (Standard Hours)

1	2	3	4	5	6	7	8
1+ .05x1000 = 51			1 + .05x1000 = 51				

Lathe Capacity Required (Assuming one week lead time per operation)

1	2	3	4	5	6	7	8
	2 + .2x1000 = 202			2 + .2x1000 = 202			

forward scheduling), and capacity limitations are ignored. As is often the case with infinite loading, capacity on each machine is exceeded on one or more days. Backward scheduling works backward from each order's due date. The second example indicates backward scheduling with infinite loading. Although the results are better, capacity on the mill has been exceeded on Wednesday. As can be seen, infinite loading can be fast and simple, but there is no provision to keep capacity from being overcommitted. Generally, capacity problems identified through infinite loading are solved either by manually rescheduling orders or by making short-term capacity adjustments through offloading or overtime.

Finite loading works within available capacity and schedules jobs within capacity limits, so that the resulting schedule is feasible. However, finite loading also requires that priorities be provided for each order so that conflicts can be resolved. Suppose in Table 2 the lower order number has higher priority (i.e., #100 has higher priority over #200). Example #3 shows backward scheduling with finite loading. Order #100 was scheduled first, then order #200, and no capacity problems were encountered. However, when order #300 was scheduled a conflict on the mill was encountered. Because order #300 has lowest priority, it was rescheduled so the mill work on #300 would be started earlier to avoid that conflict.

The examples of Table 2 are purposely kept simple. In real situations, the computational burden of finite loading can become overwhelming rather quickly. Consequently, finite loading has not been widely used in the past. Further, it is often argued that random events such as

Table 2. Infinite and finite loading.

Job #100 – due Wednesday		Job #200 – due Thursday		Job #300 – due Friday	
Operation	Time	Operation	Time	Operation	Time
Lathe	2 days	Mill	1 day	Mill	2 days
Drill	1 day	Lathe	1 day	Drill	1 day
				Lathe	1 day

Example #1, forward scheduling with infinite loading:

	Monday	Tuesday	Wednesday	Thursday	Friday
Job #100	Lathe	Lathe*	Drill*		
Job #200	Mill*	Lathe*			
Job #300	Mill*	Mill	Drill*	Lathe	

Example #2, backward scheduling with infinite loading:

	Monday	Tuesday	Wednesday	Thursday	Friday
Job #100	Lathe	Lathe	Drill		
Job #200			Mill*	Lathe*	
Job #300		Mill	Mill*	Drill	Lathe

Example #3, backward scheduling with finite loading:

	Monday	Tuesday	Wednesday	Thursday	Friday
Job #100	Lathe	Lathe	Drill		
Job #200			Mill	Lathe	
Job #300	Mill	Mill		Drill	Lathe

*Capacity exceeded.

machine breakdowns often end up invalidating the plan anyway. However, recent advances in computer hardware and software have greatly reduced the computational burden, resulting in a somewhat wider use of what is now referred to as finite capacity scheduling.

3. ***Finite Capacity Scheduling:*** Finite capacity scheduling (FCS) is a method of finite loading that takes advantage of advances in computer hardware and software to enable the rapid development of production schedules that efficiently utilize, without exceeding, available planned capacity (American Production and Inventory Control Society, 1996). In general, this approach assumes that a master production schedule has already been developed. However, some companies are reportedly using FCS software to develop master schedules by first sequencing orders. Because of its speed, FCS is also reported to be used for on-the-spot replanning after unforeseen events disrupt the original production schedule, which overcomes one of the common arguments against the use of finite loading.
4. ***Input/Output Analysis:*** Some variation will always exist between planned capacity and actual capacity due to unforeseen events. This variation will produce changes in the backlog, or queue, of orders waiting at a work center or machine. Companies in a make-to-order environment often try to maintain a desired amount of backlog in spite of these changes, and input/output analysis is one tool that can be used to maintain the backlog at a desired level (Wight, 1970). This approach can be iterative because input to a machine or work center is influenced by the schedule while its output is influenced by capacity. Thus, both factors may be adjusted to control the backlog. Table 3 provides an example of an input/output report showing both the planned and actual results.

Table 3. Input/Output report drill (standard hours) – as of end of week 4.

		Week			
		1	2	3	4
Planned Input		390	390	390	390
Actual Input		390	380	400	370
Cumulative Deviation		0	−10	0	−20
Planned Output		400	400	400	400
Actual Output		390	390	400	400
Cumulative Deviation		−10	−20	−20	−20
Planned Backlog (Hours)	80	70	60	50	40
Actual Backlog (Hours)	80	80	70	70	40

In this case, the company wanted to reduce its backlog at the drill to 40 hours of work. It had planned to do so by offloading some work to another machine, thus reducing planned input below planned output. However, a machine breakdown in week 1 reduced actual output below planned. In week 2, the company attempted to catch up with its plan by offloading additional work, but machine breakdowns persisted into that week. Unfortunately, offloading, as had been originally planned, was not available but machine output was brought back to expected levels. Finally, in week 4, the company was able to achieve its plan by offloading additional work from the drill. As this example indicates, input/output analysis provides a means for planning for short-range capacity changes. Because the planning horizon is so short, responses to variations from plan are generally quite limited

TECHNOLOGY PERSPECTIVE: MRP and capacity requirements planning almost certainly require computers and specialized software. For short-range scheduling and capacity planning, many companies still use manual scheduling boards, but the availability of finite capacity scheduling software is beginning to make such manual methods obsolete.

CASES: Some excellent case examples involving the use of capacity planning are provided by Wemmerlov (1984) and Berry *et al.* (1979). A summary of many of these cases, as well as others, is provided by Blackstone (1989). Many of these describe, in detail, actual capacity planning system implementations that have greatly improved each company's ability to minimize costs, decrease lead times, and meet delivery due dates.

See **Backward scheduling; Capacity planning: Long-range; Capacity requirements planning (CRP); Finite capacity scheduling; Finite loading; Forward scheduling; Infinite loading; Input/output analysis; Material requirements planning (MRP); Offloading; Planned order release; Production resource; Scheduling board.**

References

[1] American Production and Inventory Control Society (1996). "1996 finite capacity scheduling software and vendor directory." *APICS – the Performance Advantage*, 6, 72–82.

[2] Berry, W.L., T.E. Vollmann, and D.C. Whybark (1979). *Master Production Scheduling*, American Production and Inventory Control Society, Virginia.

[3] Blackstone, J.H. Jr. (1989). *Capacity Management*, South-Western, Cincinnati.

[4] Vollmann, T.E., W.L. Berry, and D.C. Whybark (1997). *Manufacturing Planning and Control Systems (4th ed.)*. Irwin/McGraw-Hill, New York.

[5] Wemmerlov, U. (1984). *Case Studies in Capacity Management*. American Production and Inventory Control Society, Virginia.

[6] Wight, O.W. (1970). "Input-output control, a real handle on lead time." *Production and Inventory Management*, 11, 9–31

CAPACITY REQUIREMENTS PLANNING

Capacity requirements planning (CRP) is a short-term capacity planning technique that utilizes **planned order releases** from MRP in combination with information from **routing sheets** to estimate capacity needs, usually in **standard hours**, at individual machines or work centers. CRP is the last capacity planning step prior to detailed production scheduling. As such, it is much more accurate than **rough-cut capacity planning** because it includes lead time offsetting and lot sizing information that were utilized when **planned order releases** were calculated in MRP. Although the resource profile approach to rough-cut capacity planning does include lead times, it does not take lot sizes into consideration.

Capacity requirements planning amounts to establishing, measuring, and adjusting production capacity levels. CRP determines in detail the resources needed for production. It may indicate that although the overall capacity is sufficient, capacity shortages could exist at specific times. Capacity requirements planning, in contrast to rough-cut capacity planning, takes account of capacity that already exists to calculate the extra capacity required to complete the **master production schedule** (MPS). The result is a more accurate timing of capacity requirements. To achieve this, substantial use is made of information from the material requirements planning (MRP) system. This information includes actual lot sizes, lead times and current status of scheduled receipts and planned orders, production capacity expressed as inventories of components and assemblies, work-in-process (WIP), and demands not accounted for by the MPS. In addition, bills of material, routing information, time standards, and lead times are required. The capacity requirements are given for each work center and time period.

Although CRP utilizes the best available information from **material requirements planning**

(MRP), the results still may not exactly match with reality. The main reason for this is that the planned lead times used in MRP and, consequently in CRP, are based on certain assumptions that may or may not be entirely accurate when the detailed schedule is developed. CRP is a standard component of commercial MRP software.

See **Capacity planning: Medium- and short-range; Capacity planning: Long-range capacity planning; Master production schedule; Material requirements planning (MRP); Planned order release; Routing sheets; Roughcut capacity planning.**

Reference

[1] Vollmann, T.E., Berry, W.L., Whybark, D.C. (1997) *Manufacturing Planning and Control Systems*, Fourth Edition, Irwin/McGraw-Hill, New York.

CAPITAL INVESTMENT IN ADVANCED MANUFACTURING TECHNOLOGY

Ranga Ramasesh

Texas Christian University, Fort Worth, Texas, USA

INTRODUCTION: Advanced Manufacturing Technologies (AMT) such as robots and Computer Numerically Controlled (CNC) machine tools, Flexible Manufacturing Systems (FMS), Computer Aided Design/Manufacture (CAD/CAM) etc., enable manufacturing firms to be competitive in today's dynamic environment characterized by shorter product life cycles, time-based competition, and demand for variety and high quality. Yet, the difficulties associated with the economic justification of the investment in AMT using the traditional capital budgeting techniques are well known. In the last decade, there has been a strong debate on the adequacy of the traditional financial techniques such as discounted cash flow (DCF) analysis for the justification of such investments. Although many researchers in the field of manufacturing strategy have argued that the DCF techniques are inadequate for AMT, researchers in finance have argued that the DCF techniques are basically sound and that these should not be abandoned in favor of *ad hoc* and subjective procedures (Kaplan, 1986 and Myers, 1984). A comprehensive review of the various approaches to planning and justifying new manufacturing technologies may be found in Swamidass and Waller (1990). The justification of capital investments in AMT warrants a broad multi-component framework that includes, among other things, (i) an economic or financial analysis to ascertain the commercial viability of the investment, (ii) a strategic analysis to ensure fit with the goals of the organization, and (iii) an organizational-behavior analysis to ensure conformity with the firm's social-, political- and technology-cultures (Ramasesh and Jayakumar, 1993). Economic analysis constitutes the most critical component of such a framework and this article reviews the key issues and the important findings pertaining to the economic analysis of investments in AMT.

HISTORICAL PERSPECTIVE: Not long ago, the only form of automation with high initial investments were highly dedicated systems which offered significant economies of scale. Savings in labor costs and high efficiencies in high volume, low-variety production were the main drivers that motivated capital investment in such automation. Traditional discounted cash flow (DCF) financial techniques such as the net present value (NPV) method and the internal rate of return (IRR) are admirably suited for the justification of investments in such traditional manufacturing technology. However, AMTs, that are being developed since the last two decades, are fundamentally different in that they offer significant economies of scope rather than economies of scale. Economies of scope come from flexibility rather than from savings in labor cost and the spreading of overheads over large production volumes. Traditional financial justification techniques, which are inadequate to capture the benefits arising out of flexibility, often indicate that the investment is not justified. A well-known example is that of the Yamazaki Machinery Company in Japan. The company invested $18 million in a flexible manufacturing system which resulted in a reduction in the number of machines from 68 to 18, the number of employees from 215 to 12, floor space from 103,000 square feet to 30,000 square feet, and the average processing time of parts from 35 days to 1.5 days (Kaplan, 1986). While these benefits seem impressive, the cost savings of $6.9 million reported over the first two years and an estimated savings of $1.5 million over the next 20 years, would result in an annual internal rate of return of less than 10%. Compared with the "hurdle rate" of about 15 percent or higher used by most American companies, this investment in AMT would not be justified based on the economic analysis alone. Clearly, the advent of modern day manufacturing technology calls for refinements to economic analysis to avoid the perils of being under invested in technology with the risk of being uncompetitive and unprofitable.

Strategic Perspective: The intriguing nature of the AMT-justification problem *vis-a-vis* the justification of traditional investment projects can be attributed to two key characteristics of AMT. First, the capital investment in AMT is generally very large and cannot be implemented in a piecemeal manner. As a result, the decision to adopt it involves a sizable all-or-nothing financial commitment. Second, many of the significant benefits from AMT-proposals, for example, improvements in flexibility, quality, manufacturing lead time etc., cannot be easily quantified in cash flow terms. Routine application of such techniques to AMT-proposals without recognizing the benefits of enhanced flexibility, quality, etc. generally results in an under valuation of the projects, and their rejection.

In this article, we identify four key issues in the economic analysis for the justification of AMT and present the important findings in the literature. These issues are:

1. How to refine the traditional DCF analysis to overcome its limitations and potential misapplication while justifying investments in AMT?
2. How to refine cost accounting to get reliable product cost information, and to assess the value of such benefits as reduced inventory, reduced scrap, and less floor space requirement?
3. How to incorporate the strategic, but difficult-to-quantify benefits arising out of the flexibility of AMT into the justification process? and,
4. How to value the strategic benefits of substantial future opportunities that can be harnessed only if the firm invested in the AMT under consideration?

Refinements to the DCF Analysis: The mechanics of the DCF analysis to determine the net present value (NPV) or the internal rate of return (IRR) of an investment proposal are well known.

1. ***Cash Flow Assumptions:*** The traditional accounting approach uses only cash flows which are based on the *incremental* sales from the current level of sales. It is implicitly assumed that the current competitive position and the current level of operating cash flows will remain unaltered if the proposed investment in AMT is not undertaken. However, this will be true only if the cost, quality, and delivery characteristics of the competitors will also remain unaltered. Therefore, if the investment in AMT is not undertaken, the firm may lose its competitive position and its market share may drop. As a result, the current level of operating cash flows will not be sustained and may drop either linearly or exponentially over time. The cash flow estimates should include not only the incremental cash flows but also the cash flows which would otherwise be lost if the proposed investment in ATM is not undertaken (Howell and Soucy, 1988).
2. ***Terminal value of the project:*** Traditionally, if the terminal value of the investment cannot be reliably estimated (as in the case of a AMT where a substantial portion of the investment pertains to system design, data base development and software), it is conservatively taken to be zero resulting in the under valuation of the terminal value (Howell and Soucy, 1988). A plausible approach would be to estimate a range of terminal values along with an estimate of the associated probabilities and derive an expected value. In the case of AMT much of the investment in data base and software development may have useful value in other projects. This value where appropriate, must be incorporated into the DCF analysis as the terminal value of the proposal. This will improve the NPV and IRR of the proposal, especially when the project life is small.
3. ***Treatment of Inflation:*** The discounting factor or the hurdle rate is usually increased to include a certain percentage for expected inflation to estimate cash flows using current selling prices. This is questionable because, in most cases, inflation would result in higher selling prices and larger cash flows in nominal terms. An alternative would be to consider the "real returns" and use a discounting factor or hurdle rate that is relevant for returns in real terms.
4. ***Treatment of Project Risks:*** Two sources of misapplication are encountered in the treatment of project risks: (i) excessive risk-adjustments and, (ii) the use of a large risk-adjusted factor throughout the project life. It is generally assumed that investments in AMT are highly risky and hence large risk-adjusted factors are used to discount the future cash flows. In a survey conducted by the National Association of Accountants and Computer Aided Manufacturing-International (Howell *et al.*, 1988), 36% of the manufacturers were using discounting rates between 13% and 17% and, more than 30% used rates over 19%. Such excessive risk adjustments are arbitrary which makes the justification process biased against justification. The use of a single risk-adjusted discounting factor *throughout the project life* implies that risks increase geometrically into the future. This is unrealistic because the project risk which is high in the preimplementation and start-up stages generally does not

remain the same throughout the project life. In fact, it would decrease significantly once the project is successfully commissioned and put on stream. Two approaches have been suggested to recognize the different levels of risk associated with the cash flows during the different phases of the project. One approach is to segment the project into different risk phases and then evaluating each phase sequentially using an appropriate discounting factor. Working backwards from the last phase is recommended, instead of applying a single large risk-adjusted rate through out the project life (Hodder and Riggs, 1984). Another approach is to use the concept of certainty equivalent (CEV). The CEV of a cash flow in a given year is simply its risk-adjusted value in that year. If all future cash flows are converted to CEV's, a single risk-free rate would be an adequate discounting factor (Sisk, 1986).

Refinements to Cost Analysis: The benefits of AMT may be broadly classified into two categories: The benefits in the first category are: (a) setup-time reduction; (b) process-time reduction; (c) labor reduction; (d) tooling-cost reduction; (e) scrap and rework reduction; and, (f) work-in-process reduction. It is relatively straight forward to quantify these benefits in dollar terms. The benefits in the second category are: (a) improvement in quality, which improves customer satisfaction; (b) reduction in manufacturing lead time; (c) reduction in the number of over-due orders; (d) reduction in the floor space requirement; (e) reduction in the design effort; and (f) reduction in the number of new shop-drawings made. Quantification of the savings in the second category needs some refinements to the conventional accounting procedures. For example, financial accounting systems do not provide a reliable estimate of the value of the space saved if the cost of building has been largely depreciated. Also, if there is no immediate use for the space released, conservative accounting practice would suggest a zero value based on opportunity valuation. A plausible approach might be to consider what it would cost to construct new space and compute the savings on an annualized basis. Although this approach violates the *principle of conservatism in accounting*, it is justifiable in that a growth-oriented company would find alternative uses for the floor space and assigning no value to it would unduly hurt the investment proposal (Cooper and Kaplan, 1988).

Some of the intangible benefits from quality improvements could be measured in dollar terms. For example, reduction in the warranty costs could be converted into an estimated stream of cash savings. Although reductions in the design and drawing efforts may not result in the reduction of staff level, a dollar value could be placed on the benefits from AMT by estimating the additional level of staffing or overtime that would have been required to handle the number of design modifications and product changes. In quantifying the cost savings from AMT it is easier to do so where activity-based cost accounting and cost accounting by goals and strategies are used because AMT benefits are better recognized in these systems.

Estimating the Benefits of Flexible Automation: Two types of models have been discussed extensively in the more recent literature to ascertain the benefits of flexibility provided by AMT under various factory-design scenarios. The first type of models use simulation and decision support system (DSS) analysis to generate measures of physical outputs such as number of products produced, quantities, system and transport utilization etc., from which financial measures such as NPV and IRR are calculated (Monahan and Smunt, 1989, Suresh, 1990). The second type of models use mathematical programming analysis to address the different issues related to flexible automation such as, (a) optimal technology acquisition policies for flexible automation subject to technological improvements over time; (b) the cost-flexibility tradeoffs in investing in product-flexible manufacturing capacity (Fine and Freund, 1990); and (c) the value of the flexibility of a single-product FMS (Kulatilaka, 1988). The use of these models is not widespread.

Valuing Future Opportunities As Real Options: Some of the most significant strategic benefits of AMT, which are difficult to quantify, are the capability for business survival and the future investment opportunities for growth. Such investment opportunities that are associated with today's investments are treated as analogous to the "call options" on securities in the financial markets and are called "real options" to distinguish them from the "financial options" (Kester, 1984). As an example, firms which had invested in the CNC machine tool technology that was emerging in the mid-1970s, may be viewed as having bought "options" to exploit huge gains in the 1980's from the breakthrough in microprocessor-based technology. Firms which did not invest in the CNC technology (perhaps because the benefits of future opportunities could not be recognized) were closed-out of the growth opportunities when the microprocessor-based technology became well established. The presence of real options enhances the value of the investment proposal although the valuation of such options in cash flow terms may be quite difficult and complex. In many situations, the nature of these options may be articulated using probabilistic statements at the time of evaluating the initial investment. More recently, researchers have extended the financial option-pricing models and applied the contingent claims analysis (CCA) to a variety of real options such as the flexibility of a production system, which can produce different combinations of products easily, or the option to abandon the project before its economic life expires (Triantis and Hodder, 1990). Option-pricing models are potentially powerful tools for the valuation of many of the strategic benefits of AMT. Their use is not yet widespread.

CONCLUSION: In the justification of large capital investments in advanced manufacturing technology, managers face a challenging decision with far reaching implications for their firms' long-term competitiveness and profitability. An analysis that includes economic, strategic, and organizational-behavior issues is appropriate and is likely to ensure a fit with the strategic goals and culture of the organization.

See **Activity-based costing; Certainty equivalent; Contingency claims analysis; Hurdle rate; Internal rate of returning financial justification; Manufacturing technologies; Net present value.**

References

[1] Cooper, R., and R.S. Kaplan (1988). "Measuring costs right make the right decisions." *Harvard Business Review*, 66, 96–103.

[2] Fine, C.H., and R.M. Freund (1990). "Optimal investment in product-flexible manufacturing capacity." *Management Science*, 13, 443–451.

[3] Hodder, J.E., and H.E. Riggs (1984). "Pitfalls in evaluating risky projects." *Harvard Business. Review*, 62 , 26–30.

[4] Howell, R.A., and S.R. Soucy (1987). "Major trends for management accounting." *Management Accounting*, 68, 21–27.

[5] Howell, R.A., J.D. Brown, S.R. Soucy, and A.H. Reed (1988). *Management Accounting in the New Environment*, National Association of Accountants, USA.

[6] Kaplan, R.S. (1986). "Must CIM be justified by faith alone." *Harvard Business Review*, 64, 87–95.

[7] Kester, W.C. (1984). "Today's options for tomorrow's growth." *Harvard Business Review*, 62, 153–160.

[8] Kulatilaka, N. (1988). "Valuing the flexibility of flexible manufacturing systems." *IEEE Transactions on Engineering Management*, 35, 250–257.

[9] Monahan, G.E., and T.L. Smunt (1987). "A multilevel decision support system for the financial justification of automated flexible manufacturing systems." *Interfaces*, 17, 29–40.

[10] Myers, S.C. (1984). "Finance theory and finance strategy." *Interfaces*, 114, 126–137.

[11] Ramasesh, R.V., and M.D. Jayakumar (1993). "Economic justification of advanced manufacturing technology." *OMEGA*, 21, 2819–306.

[12] Sisk, G.A. (1986). "Certainty equivalent approach to capital budgeting." *Financial Management*, Winter, 23–32.

[13] Suresh, N.C. (1990). "Towards an integrated evaluation of flexible automation investments." *International Journal of Production Research*, 28, 1657–1672.

[14] Swamidass, P.M. and M.A, Waller (1990). "A classification of approaches to planning and justifying new manufacturing technologies." *Journal of Manufacturing Systems*, 9, 3, 181–193.

[15] Triantis, A.J., and J.E. Hodder (1990). "Valuing flexibility as a complex option." *Journal of Finance*, XLV, 549–565.

CAUSAL MODELS

A causal model is a type of forecasting model. It is useful for long-term decision making. Examples of such situations are capital expansion projects, and new product development proposals.

See **Forecast evaluation; Forecast horizon; Forecasting guidelines; Forecasting models in manufacturing mangement.**

C-CHART

C-chart is one of many different visual charts used to control variability in production processes. The various visual charts used for the said purpose are called statistical process control charts. C- chart is used for controlling the number of nonconformities in samples of constant size. Typical applications are the nonconformities or defects in 100 square meters of paper in a paper mill.

The occurrence of nonconformities is assumed to follow a poisson distribution. Such a distribution is appropriate for modeling the number of events that occur over a specified time, space or volume.

See **Statistical process control using control charts.**

CELL MANUFACTURING (CELLULAR MANUFACTURING)

See **Linked-cell manufacturing system; Manufacturing cell design.**

CERES

Formerly CERES was known as the Valdez Principles. A U.S. based coalition of firms, environmental groups, religious organizations and socially responsible investors with a distinctive strategy of using shareholder resolutions to initiate discussions of environmental responsibility at the highest corporate levels. The ten principles espoused by the partnership are voluntarily; however, signatories are committed to responsible stewardship of the environment, ecologically sound operations and public environmental reporting. The principles emphasize the necessity to protect the biosphere, to use resources in a sustainable manner, to reduce wastes and conserve energy, to reduce any environmental, health and safety risks, and to audit and report environmental performance. These principles should be taken into consideration by environmentally responsible manufacturing firms.

See **Environmental issues and operations management.**

CERTAINTY EQUIVALENT (CEV)

A certain cash flow that has the same future value as an investment project's risky cash flow. Certainty equivalent is the *smaller, certain* payoff for which one would exchange the larger but riskier cash flows from a project. In investment decisions, use of the certainty equivalent is one approach to adjust for the project's risk.

See **Capital investments in advanced manufacturing technologies.**

CHAOTIC SYSTEMS

Chaos is popularly something formless, indescribable, unpredictable, or random. Chaotic systems exhibit unpredictable, non-repeating behavior over time. Technically, a chaotic system is one whose small-scale behavior can be described by usually simple, deterministic (non-random) rules or equations, but whose large-scale behavior is in fact so unpredictable as to appear random. This large-scale indeterminacy is a result of sensitive dependence on initial conditions (i.e. nonlinear interactions which amplify small differences very quickly) which produces chaotic, rather than periodic or even aperiodic dynamics. This implies that chaotic systems are never in exactly the same state twice. It also implies that if two identical systems are started from arbitrarily similar but not identical points, their states will quickly diverge and become wholly different. The difficulty of weather forecasting is largely a function of this kind of behavior which makes long-term prediction extremely difficult if not impossible, while short-term forecasts can be reasonably accurate.

See **Manufacturing analysis using chaos, fractals, and self-organization.**

CHASE STRATEGY FOR CAPACITY PLANNING

Many companies experience variable demand for their products. This is probably most noticeable in products that have seasonal demand patterns, such as snow blowers or lawn mowers, but demand for other products may vary appreciably throughout the year. In such cases, companies face several different options in determining how to meet demand. Generally, these options fall into two "pure strategies" – a chase strategy and a level strategy. Under the chase strategy, production is varied as demand varies. With the level strategy, production remains at a constant level in spite of demand variations.

The use of a chase strategy requires that a company have the ability to readily change its output level, which means that it must be able to readily change its capacity. In some industries where labor is the major determinant of capacity, and where additional labor is readily available, such changes may be feasible. However, other resources such as equipment and facilities are also often a determinant of output. Consequently, few companies follow a pure chase strategy, but usually combine it with some aspects of a level strategy.

See **Capacity planning: Long-range; Level strategy for capacity planning.**

CIM

See **Computer integrated manufacturing.**

CIMPLICITY

CIMPLICITY (brand name) is a new line of factory automation software from GE Fanuc. It facilitates mass customization. This software helps design engineers address the increasing demand for

mass customized goods. It handles enterprise-wide process monitoring, supervision and motion control. This helps in manufacturing planning and control of a wide variety of products in small quantities to meet increasing customer demands for unique products.

See **Agile manufacturing; Mass customization.**

C-KANBAN

C-Kanban is the signaling device used in factory floors that indicates the need for more material in Just-in-Time manufacturing (JIT) at a work center. It is often referred to as a kanban, which is a Japanese word meaning "card" or "visible marker". Some companies, such as Toyota, utilize two different types of kanban; one that authorizes the movement of material from one location to another, and one that authorizes the production of more material. In such a two-card system, the type of kanban that authorizes movement of material is often referred to as a C-kanban or conveyance kanban. Thus, if work center A produces a component part used by work center B, units of that component can be moved from A to B only when a C-kanban is available to indicate that more of the component is needed at work center B.

In general, the C-kanban will be a card that is attached to a container of parts. When a container is emptied so that the parts may be used by a downstream work center, the card is returned to the upstream location where those parts are kept (the work center or supplier producing the part) and placed on a full container. The C-kanban on that container authorizes its conveyance to the location at which the parts are used.

See **Just-in-time manufacturing; P-kanban.**

CLASS A MRP

Class A MRP **(Materials Requirements Planning)** is a basic tool for performing the detailed material planning function in the manufacture of component parts and their assembly into finished items. Class A MRP users employ MRP in a closed-loop mode. **Closed-loop MRP** is a system built around material requirements planning that includes the additional planning functions of sales and operations planning (production planning), **master production scheduling**, and **capacity requirements planning**. Once this planning phase is complete and the plans have been accepted as realistic and attainable, the execution functions come into play. These functions include the manufacturing control functions of input-output (capacity) measurement, detailed scheduling and dispatching, as well as anticipated delay reports from both the plant and suppliers, supplier scheduling, etc. The term *closed loop* implies not only that each of these elements is included in the overall system, but also that feedback is provided by the execution functions so that the planning can be kept valid at all times. In addition, a Class A MRP user is considered a sophisticated user who is able to exploit MRP to the fullest and hence able to reap many benefits. A key difference between Class A users and other users (Class B, C, and D users) is that Class A users have full support from top management.

See **Closed-loop MRP; Materials requirements planning (MRP); MRP implementation.**

Reference

[1] Vollmann, Thomas E., William L. Berry, D. Clay Whybark (1997). *Manufacturing Planning and Control Systems*, Fourth Edition, Irwin/McGraw-Hill, New York.

CLEAN TECHNOLOGIES

These technologies extract and use natural resources more efficiently, generate products with fewer harmful components, minimize pollutant releases to air, water and soil during manufacturing and product use, and produce more durable products that can be recovered or recycled to the greatest possible extent. Technology is broadly defined to include equipment, processes, techniques, and management practices and information. As such, pollution prevention is subsumed under this term. "Clean" is relative, as cleaner options might always be possible with better scientific understanding and further technical innovation.

See **Environmental issues and operations management.**

CLOSED-LOOP MATERIAL REQUIREMENTS PLANNING

Material Requirements Planing (MRP) develops plans and schedules for the production system. Due to uncertainty in lead-times, variations in the quantities of parts supplied or produced, inaccuracy in inventory status files, changes in design or customer orders, and inadequate production capacity, MRP plans and schedules are often disrupted. Without a system of feedback and MRP file updating, MRP plans and schedules tend to become invalid and do not reflect reality.

In closed-loop MRP, systematic and continuous feedback from the shop floor and other manufacturing entities keep the inventory status files, capacity and other files, that supply data to the MRP program, updated, accurate and valid. Timely updates and rescheduling, when necessary, keeps the schedules valid and achievable. In the Oliver Wight classification of MRP use, Class A companies are said to use closed-loop MRP.

See **Capacity requirement planning; Materials requirements planning.**

Reference

[1] Schonberger, R.J. and Knod, Jr., E.M. (1997). *Operations Management*, sixth edition. Chicago: Irwin.

CNC MACHINE TOOL

Computerized numerical control (CNC) machines are controlled by locally programmable, dedicated micro- or mini-computers. Computer programs resident in the memory of the computer guide the machine tool to cut or form metal to conform to the design and tolerance written into the program. In seconds or minutes, by switching the program or altering the program in the computer, the operator can direct the machine to work on a different part conforming to a different design and specifications. CNC provides great flexibility by allowing the machine to change setups quickly.

CNC allows computerized machines to be integrated with other complementary technologies such as computer-aided design (CAD), computer integrated manufacturing (CIM), and computer-aided manufacturing (CAM). CNC machines also serve as building blocks for flexible manufacturing systems (FMS).

See **CAD; CAM; CIM; FMS.**

COALITION FOR ENVIRONMENTALLY RESPONSIBLE ECONOMIES (CERES)

See **CERES; Environmental issues and operations management.**

CODING/CLASSIFICATION ANALYSIS (C/C)

Cellular manufacturing, which is based on the principles of **group technology**, is a powerful new manufacturing philosophy. Similar products or components forming a product family are manufactured in any given cell. A methodology for finding families of parts based on a part code number, which is often based on how the part is manufactured. Parts are then sorted based on the codes into classes of components, using a computer program. This is an important step in determining which components will be manufacturers in a given cell and it is a major step in designing manufacturing cells.

See **Cellular manufacturing; Group technology; Manufacturing cell design.**

COEFFICIENT-BASED CLUSTERING

The coefficient-based clustering or similarity coefficient method (SCM) refers to a set of clustering algorithms developed in the field of numerical taxonomy and has been applied to form machine cells for cellular manufacturing. In SCM, a measure of similarity between objects (similarity coefficient) is defined and used to group them into clusters. In machine cell formation, the similarity coefficient between two machines is defined as the number of parts visiting both machines divided by the number of parts visiting one of the two machines.

See **Manufacturing cell; Manufacturing cell design.**

COGNITIVE SKILLS OF OPERATORS

Cognitive skills refer to an individual's mental ability to understand abstract notions such as mathematics, digital programming, electronics, and CNC automation logic. These skills have become critical on the shop floors equipped with advanced manufacturing technologies.

See **Human resource issues and advanced manufacturing technology.**

COMBINATORIAL PROBLEM

This refers to class of mathematical problems useful in solving production problems. In these problems, the decision variables are discrete, the solution is a set or a sequence of integers and other discrete objects, and where the decision is to pick one optimal solution out of an enormous number of different alternatives. For example, in facility location problems, if three sites are to be selected from a large number of potential facilities (say 65), the total number of resulting combinations of facilities to choose from becomes extremely large. Other examples of combinatorial problems are Vehicle Routing, Scheduling, Project Management, Assembly Line Balancing, the Assignment Problem, the 0–1 Knapsack Problem, and the Set Covering Problem.

See **Facilities location decisions.**

Reference

[1] Gauss, Saul I. And Harris, Carl M. (eds.) (1996). *Encyclopedia of Operations Research and Management Science*, Kluwer Academic Press, Boston.

COMMON CAUSES OF VARIATIONS

Common causes of variation in production occur due to the natural variability of the production system. Walter A. Shewhart demonstrated how graphs can be used to distinguish common causes of trouble that occur due to the natural variability of the system from local sources of trouble, which Shewhart called assignable causes of variation. Local sources of trouble must be eliminated before managerial innovations leading to improved productivity can be achieved through removal of common causes. Examples of common cause variability include vibration in machines, slight variations in incoming raw materials, and human variability in taking measurements.

W. Edwards Deming argued that common cause variability is not necessarily acceptable, but it cannot be removed entirely by lower level employees attempting to do better. Problems generated by common cause variability can only be addressed by improving the production system, which is the responsibility of management. Deming believed that 85% of the variation in a process is due to common causes.

See **Quality: The implications of W. Edwards Deming's approach; Statistical process control; Statistical process control using control charts.**

References

[1] Deming, W. Edwards (1982). *Out of the Crisis*, Center for Advanced Engineering Study, Massachusetts Institute of Technology, Cambridge, Massachusetts.

[2] Shewhart, W.A. (1939). *Statistical Method from the Viewpoint of Quality Control*, Graduate School, Department of Agriculture, Washington.

COMMONALITY (PART, COMPONENT, etc.)

Commonality refers to a condition where parts, components and raw materials are used in multiple end products or parents. Part commonality results in standardization of parts or modularity. Where commonality is present, the same part may appear in the bill of materials for several different end products or may appear in several subassemblies of the same end product. Part commonality increases the volume of production for the part and consequently many cost saving measures can be instituted for the part. Today, the attainment of increased part commonality is a very desired goal in product design.

See **Bill of materials; Product design.**

COMPETITIVE ANALYSIS

It is an analysis of a firm's product or service features and value, compared to other comparable products or services currently available to customers on the market. A competitive analysis of an organization provides a mapping of internal strengths, weaknesses, opportunities, and threats so that it enables strategic decision-making in terms of pricing, marketing, and future improvements to the product or future generations of the product. In some cases, a competitive analysis may also cause an organization to discontinue a product or service, if it is felt that the competition is too strong to enable a foothold in the market. A good competitive analysis should test different market assumptions based on information secured on a competitor's cost structures. For instance, with information available from the public domain, such as selling price, labor pool wages, geographic transportation rates, and standard material costs, a user could overlay this information onto an internal cost structure for the competing product, thus revealing the relative cost advantages/disadvantages in a particular market.

See **Cost analysis for purchasing.**

COMPETITIVE PRIORITIES FOR MANUFACTURING

Competitive priorities address the objectives and the scope of a manufacturing organization (Wheelwright, 1984). Priorities may cover various aspects of manufacturing performance concerning product cost, quality, delivery, flexibility, innovation, or service. Each of these dimensions of manufacturing performance is multi-facetted. For example, many dimensions of flexibility can be identified. It is therefore important for manufacturing managers to develop and set precise and explicit performance goals based on priorities, because no single manufacturing operation can do all things well. It is imperative that a higher priority item receives more resources and attention than a lower priority item.

Competitive priorities convey the essential mission of a manufacturing operation. This mission derives from managers' understanding of customers' desires, how the firm differentiates its products, and the capabilities of the manufacturing function. Furthermore, competitive priorities govern choices regarding how resources are deployed, what skills are developed, and what initiatives are pursued. For a more detailed discussion of competitive priorities, see Garvin (1993).

See **International manufacturing; Manufacturing strategy.**

References

[1] Wheelwright, S.C. (1984). "Manufacturing Strategy: Defining the Missing Link," *Srategic Management Journal*, 5 (1), Jan/Mar, 77–91.

[2] Garvin, D.A., (1993), "Manufacturing Strategic Planning." *California Management Review*, Summer, 85–106.

COMPONENT COMMONALITY ANALYSIS

The Component Commonality Analysis System (CCAS) using a relational database software is a method for analyzing computerized bills of materials (BOM) to identify common and unique parts.

See **Bill of materials; Commonality; Product design.**

COMPONENT SHARING MODULARITY

In component sharing, the same basic component is used in different product families. In the automotive industry, the same brake shoes, alternators or spark plugs are used in several different product families of automobiles. Some of these components are even used across the product families of different manufacturers. Component sharing modularity is often associated with the idea of component standardization. Note that component sharing modularity and component swapping modularity are identical except that the "swapping" involves different components of the same basic products and "sharing" involves different basic products using the same components. The difference between them lies in how the basic product and components are defined in particular situations.

See **Standardization; Component swapping modularity; Mass customization; Mass customization and manufacturing.**

COMPONENT SWAPPING MODULARITY

Component swapping modularity occurs when two or more alternative types of a component can be paired with the same basic product creating different product variants belonging to the same product family. Examples of this type of modularity in the computer industry would be the possible matching of different hard disk types, monitor types, and keyboards with the same CPU. Component swapping modularity is often associated with product variety as perceived by the customer. Note that component sharing modularity and component swapping modularity are identical except that the "swapping" involves different components of the same basic products and "sharing" involves different basic products using the same components. The difference between them lies in how the basic product and components are defined in particular situations. (Ulrich and Tung, 1991)

See **Component sharing modularity; Mass customization; Product family.**

COMPUTER-AIDED DESIGN

See **Advanced manufacturing technologies; CAM; Flexible automation; Manufacturing flexibility.**

COMPUTER-AIDED MANUFACTURING

See **Advanced manufacturing technologies; CAD; Flexible automation; Manufacturing flexibility.**

COMPUTER ASSISTED ACQUISITION LOGISTICS SUPPORT (CALS)

CALS is computer program, which was developed at the initiative of the U.S. Department of Defense and its contractors in an effort to facilitate electronic interchange of technical documents and to create product-related data bases accessible by both government agencies and their trading partners.

See **Electronic data interchange in supply chain management.**

COMPUTER INTEGRATED MANUFACTURING (CIM)

CIM refers to total or near total integration of all computer systems in a manufacturing facility for free exchange of information and sharing of databases. The integration may extend beyond the confines of one factory into multiple manufacturing facilities and into the facilities of customers and vendors. Computer Integrated Manufacturing is an approach to the organization and management of a firm, in which all functions including order processing, design, manufacturing and production management, accounting, finance and computerized equipment on the shop floor are completely coordinated, through the use of multiple levels of computer and information/communication technologies. The idea is to form one large computer integrated system that connects all activities so that common information is shared on a real-time basis for decision making and control.

See **Advanced manufacturing technologies; CIM; Flexible automation; Kanaban-based manufacturing systems; Manufacturing flexibility.**

COMPUTER NUMERICAL CONTROL (CNC)

A Computer Numerical Control (CNC) machine is typically a stand-alone machine controlled by its own microprocessor. Stored programs are invoked to perform various operations on a variety of parts. The computer control simplifies very complex operations such as circular milling and the development of 3-D surface contours.

CNC machines are programmable, which means that an operator can reduce setup time to almost zero by calling for a different stored computer program from the machines computer while changing from job to a totally different job. CNC machines come in many forms including lathes, milling machines, metal cutting machines, and so on.

See **Advanced manufacturing technology; Flexible automation; Human resource issues and advanced manufacturing technology; Manufacturing flexibility.**

CONCURRENT ENGINEERING

Morgan Swink

Michigan State University, Michigan, USA

DESCRIPTION: Concurrent engineering (CE) is defined as the simultaneous design and development of all the processes and information needed to manufacture a product, to sell it, to distribute it, and to service it. Other terms sometimes used in place of CE include "simultaneous engineering," "design-for-manufacturability," and "integrated product development." Concurrent engineering represents an important evolution in new product development (NPD) practices. Two aspects that distinguish CE from conventional approaches to product development are cross-functional integration and concurrency. Conventional NPD programs execute concept exploration, product design, testing, and process design activities serially. Each of these development activities is typically under the control of one functional organization at a time (e.g., marketing, engineering, manufacturing). As one function completes its design and

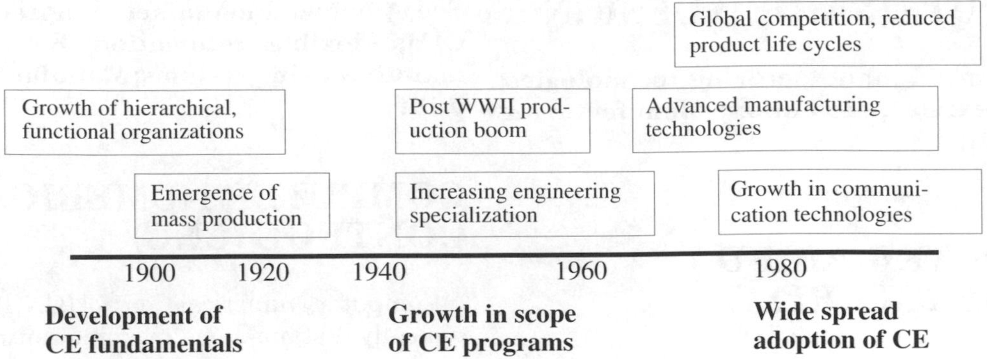

Fig. 1. Timeline of major events in the development and adoption of CE.

development tasks, it then hands over control and responsibility to the next organizational function. In the CE approach, integrated, multi-functional teams work together, simultaneously attacking all aspects of new product development. Control and responsibility are shared among functions and development activities overlap. For example, manufacturing process designers do not wait until product specifications are completed before developing tooling and processing equipment designs. Instead, they work closely with product designers to concurrently develop product and process concepts. In doing so, manufacturing personnel influence the product design in ways that make the product less costly or more producible. Other functions are similarly integrated into the design process so that their concerns may also be addressed.

HISTORICAL PERSPECTIVE: Figure 1 illustrates the development and growth of CE in the twentieth century. The undergirding principles of concurrent engineering have been around for a long time, and were discussed as early as the beginning of the twentieth century (see Smith, 1997 for an interesting history of CE). However, CE has grown to become a dominant product development management approach only in the past two decades. Smith (1997) suggests several reasons for this growth. First, the need for CE increased because engineering training became intensely specialized, emphasizing engineering science over engineering practice. Second, the advent of useful information and communication technologies has enabled and lowered the cost of implementing CE. Third, changes in the competitive environment have increased the importance of reducing product development lead time and improving product quality.

While the fundamentals of CE are not new, there has been a marked growth in the scope of CE. According to Nevins and Whitney (1989), early CE approaches focused only on identifying part fabrication issues early in the product development process. Over the years, this focus was expanded to include assembly issues and groups of parts in design decisions, until finally all production and product support processes were addressed. For example, the U.S. Department of Defense emphasizes "cradle-to-grave" considerations in CE programs with the primary objective of coordinating decisions between different engineering functions (MIL-HDBK-59A-CALS, 1995). Market-oriented advocates of CE also stress the need for integrating the "voice of the customer" and marketing strategies into design decisions, emphasizing information exchange between marketing and R&D personnel. Others suggest an even broader view of CE which addresses environmental and societal cost issues (Alting, 1993).

Performance goals associated with CE have grown as well. Early CE developments were aimed at improving quality or minimizing product acquisition costs, while more recent programs have emphasized reductions in product development time. The result of this growth in CE concepts is a more holistic, strategic view of CE.

STRATEGIC PERSPECTIVE: Recently, global competition has led to shorter product life cycles and increased technological sophistication. Products are becoming more complex due to rapid technological developments and increasing consumer demands for lower costs, greater variety, and greater performance. At the same time the proliferation of new technologies is rendering products obsolete at an increasingly rapid pace. These market and technology trends place great demands on the NPD process.

Several studies have shown that delivering a product to market on-time but over budget is more profitable than delivering the same product on

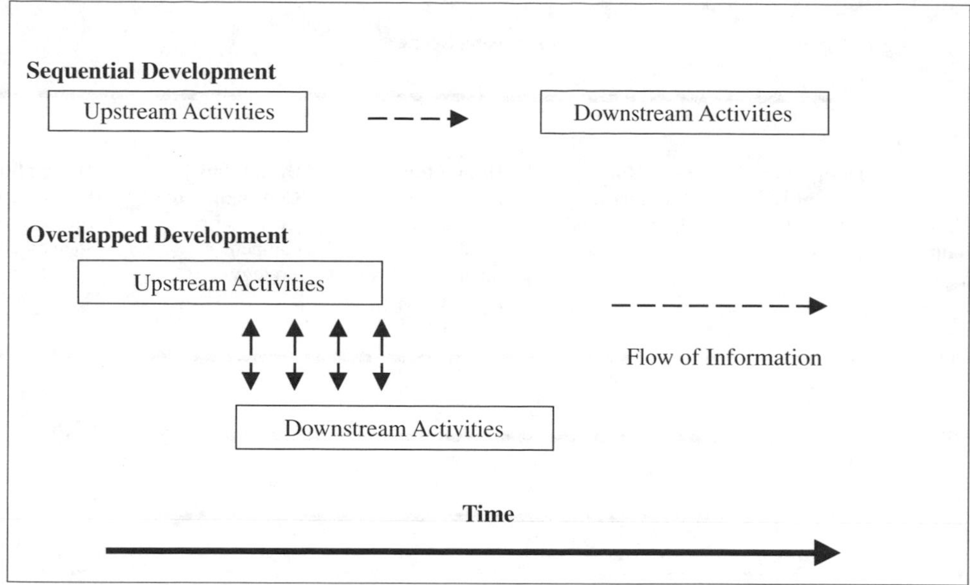

Fig. 2. Sequential and overlapped product development activities.

budget but six months late. In fact, a recent study estimated that a delayed new product entry reduces a firm's market value by $120 million on average (Hendricks and Singhal, 1997). On the other hand, getting new products to market quickly provides numerous competitive advantages. In many industries, product life cycles are currently only two years long or even less. First entrants in these markets typically capture large portions of market share and get a head start down the manufacturing learning curve. These advantages enable lowered manufacturing costs through economies of scale and improved manufacturing efficiency. Further, frequent new product introductions can help a firm dominate its markets. This was demonstrated in the motorcycle market in the 1980s. In an eighteen month period, Honda introduced eighty-one new models in Japan, while Yamaha could introduce only thirty-four. As a result Honda gained strategic advantage and market share at Yamaha's expense.

As we enter the new millennium, product development lead time is becoming an even more important basis for competition. The auto industry provides a good example. Pressure from Japanese firms has forced U.S. auto manufacturers to seek ways to reduce product development times. By offering new models more often, auto makers can more quickly incorporate newly developed technologies into product designs. In addition, they are able to more quickly respond to changes in market tastes or economic conditions. Because these abilities hold the promise of enormous competitive advantage, reducing product development time has become a key priority in many industries. Chrysler corporation recently announced its goal of developing a new car in as little as twelve months! Current development times in the auto industry vary from three to five years.

CE helps product development managers meet growing competitive challenges by producing higher quality products in less time. CE improves product quality in at least two dimensions. Design quality, the match between customers needs and product attributes, is improved by emphasizing customer needs in the concurrent execution of market research and product design activities. Product conformance quality is similarly improved. Manufacturing defects are reduced by considering manufacturing process capabilities while product design and process design activities are concurrently executed. Concurrent engineering efforts in various industries have improved process capability performance for critical product dimensions by as much as 100 percent, frequently producing C_{pk}[1] levels of 1.5 to 2.0 (Handfield, 1995). Development lead time is reduced by the overlap of NPD activities coupled with intense communication between upstream and downstream development activities (see Figure 2). Also, when CE is successful, fewer time-consuming manufacturing problems are encountered in the initial stages of production.

[1] C_{pk} is a commonly used measure of process capability that reflects the ratio of product specification limits (e.g., tolerances) to process variation limits. Values greater than 1.0 are desirable.

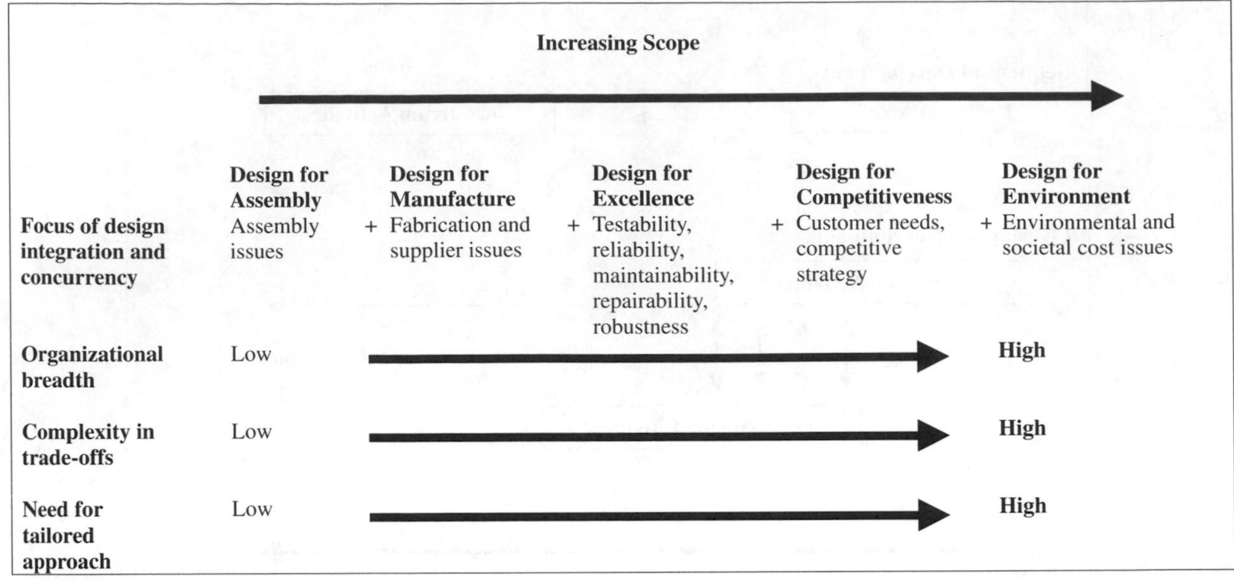

Fig. 3. The varying scope of concurrent engineering programs (Adapted from a figure in Clark (1992)).

IMPLEMENTATION: There are significant differences in the ways CE is conceived and implemented in different projects, companies, and industries. Some programs address only narrow product producibility issues (e.g., ease of assembly). More comprehensive CE programs address the impacts of product design decisions on competitive issues and product life-cycle considerations. Figure 3 illustrates the types of issues that add to the scope of a CE program. As the scope of CE increases, additional stakeholders (e.g., designers, manufacturers, suppliers, customers) must be involved early in design and development processes. As the group grows larger, the complexity of design decisions and related trade-offs increases, creating a greater need for a customized management to meet the specific needs of the NPD program. For example, managers can implement standard protocols across a wide variety of product development programs in order to insure that consistent product assembly issues are addressed early in product design. However, the protocols and approaches that managers use to address broad, strategic issues will differ, depending on the NPD program importance, the complexity and newness of the product, and the intended uses and markets for the product.

One of the difficulties of implementing CE is deciding what activities should be done concurrently and establishing where the most important points of integration are. Program priorities should drive these decisions. Customer desires and competitive threats influence the relative priorities placed on design quality, product costs, and product introduction speed. For example, lowering product cost is often a high priority in incremental product redesign projects. When new technologies or other product differentiation dimensions are absent, product cost becomes a primary basis for competition. Close interactions among designers, suppliers, and manufacturing personnel are important in this situation.

Comparable arguments can be made regarding other NPD outcomes in various circumstances. All functional groups are important although no single group contains all the knowledge needed to complete NPD. However, the degree of influence that each group exerts in NPD should reflect program priorities and project characteristics. Leadership roles should be played by functional representatives who are most able to reduce uncertainty in the areas of greatest need.

Concurrent execution of design and development activities stimulates needed cross-functional interactions. Figure 4 illustrates relationships among three different types of concurrency. *Product concurrency* is the overlap of separate but related new products requiring coordination between NPD programs. Product concurrency exists in the concurrent development of first generation and next generation products, or in the development of product variants. For a given product development effort, *project phase concurrency* involves simultaneously developing market concepts, product designs, manufacturing processes, and product support structures. Within any project phase, *design concurrency* involves the overlap of design disciplines (e.g., system, software, electrical, and mechanical engineering) so that system level and component level designs are

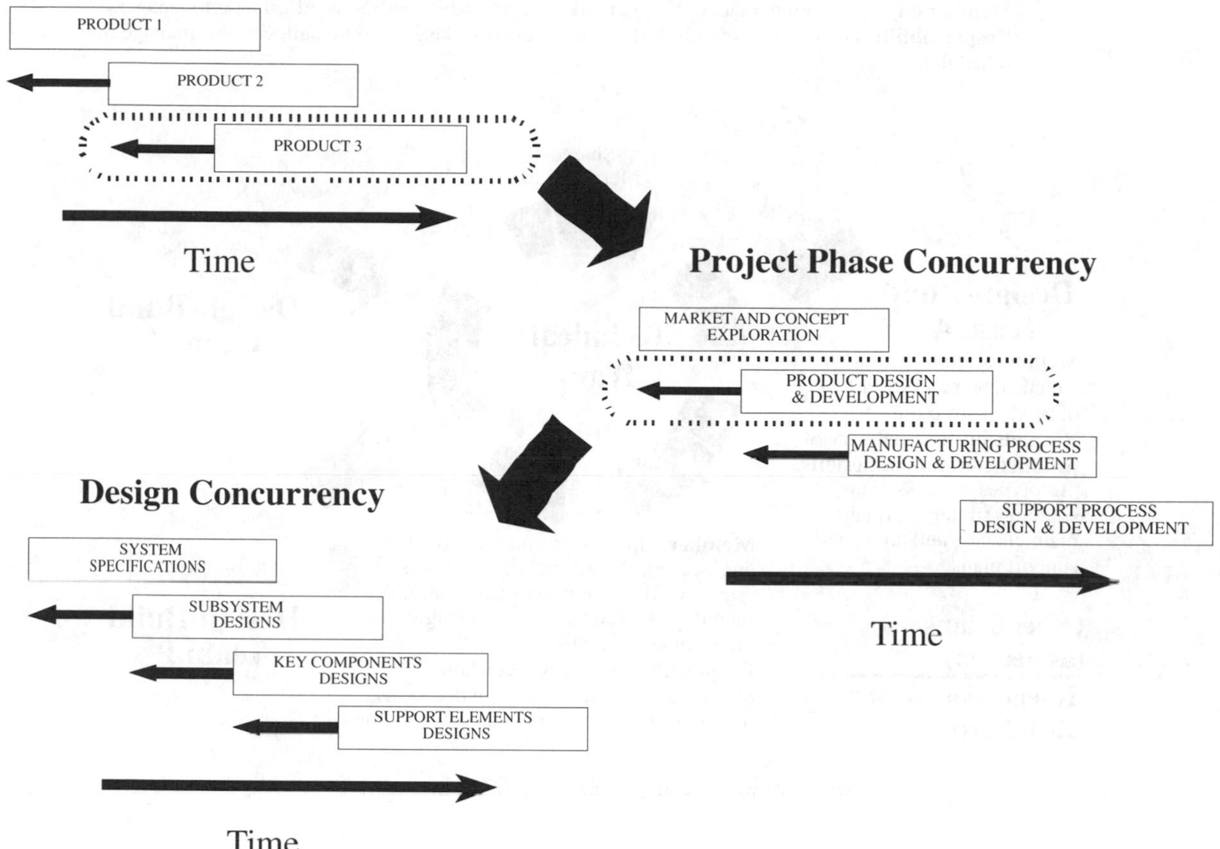

Fig. 4. Different types of concurrency.

produced concurrently. Greater the concurrency, greater the reduction in development lead time. However, greater concurrency also increases the number of functional relationships that must be managed and increases the severity of design risks undertaken.

While the scope and design of CE programs varies across firms and across NPD projects, two management initiatives remain central. First, project managers must organize personnel, policies, and procedures in ways that improve cross-functional integration and communication. Second, CE requires design analysis and decision making methods that foster design excellence.

1. ***Improving Cross-Functional Integration:*** In order to concurrently address the many decisions involved in NPD, persons from different organizational functions must be willing to collaborate, share information, and resolve conflicts quickly and effectively. Three management approaches have proven effective in this regard:

- Teamwork.
- Setting and analyzing goals.
- Directing and controlling integration.

Encouraging Teamwork: The use of cross-functional teams is a fundamental method for fostering communication in CE projects. Susman and Dean identify a number of integrative mechanisms that can be considered prerequisites to the success of CE teams (see Susman, 1992, chapter 12).

- Status parity: All members of the team must have equal perceived value and relevance. This also means having equal voting power in making decisions.
- Project focus: A project structure should be elevated over functional structures. The project manager should have real power and the project members should not be too tightly controlled by or rewarded only by functional department heads.
- Number of level: The firm should seek to minimize the number of organizational levels that exist between project-level personnel in design and manufacturing and the first manager who has authority over both.
- Frequency of rotation: Frequently rotating people among positions in design and manufacturing will improve CE.

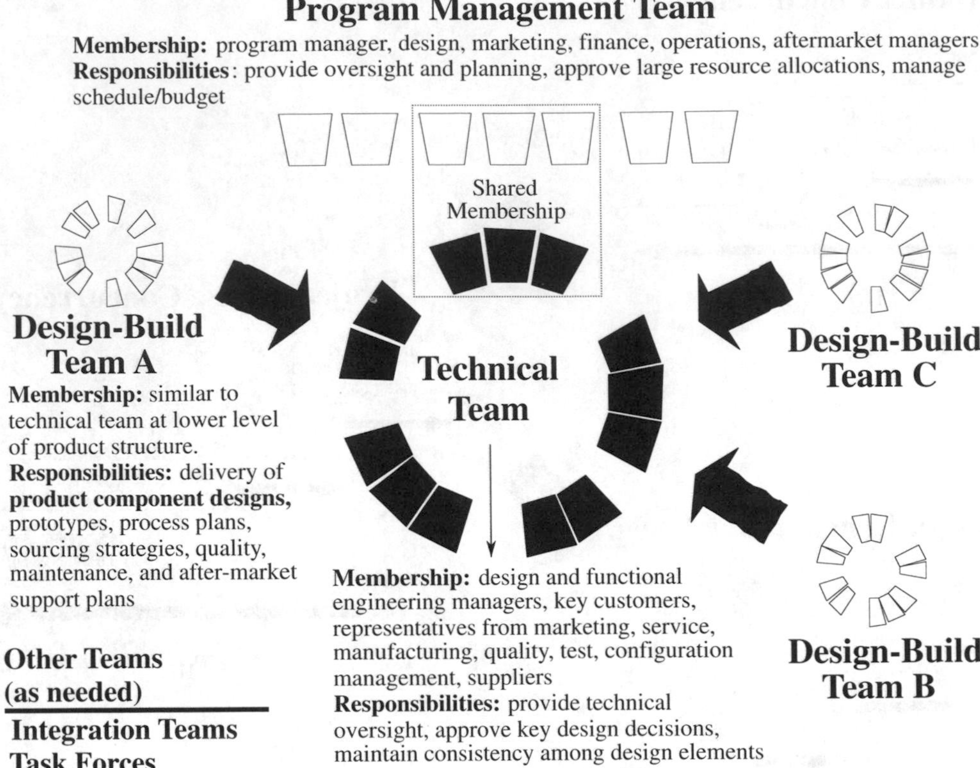

Fig. 5. Concurrent engineering team structure.

While team arrangements vary, three organizational levels of teams frequently appear in NPD programs: a program management team, a technical team, and design-build teams. Figure 5 illustrates relationships among the teams.

Program management teams typically include the program manager, marketing manager, finance manager, operations manager, aftermarket manager, and design managers. This group provides management oversight and planning, approves large resource allocations, approves and controls the project budget, and manages the project schedule.

The technical team reports to the program team and provides technical oversight, approves key design decisions, and maintains consistency between design elements. Engineering managers from design and functional support areas are typically members of the technical team, along with representatives from marketing, service, manufacturing, quality assurance, test engineering, drafting/documentation, and key customers and suppliers.

Membership of design-build teams replicates the technical project team with responsibility for components at the lower levels of the product structure. Each team is oriented around a particular product component, with responsibility for delivering designs, prototype hardware, process plans, sourcing strategies, quality engineering, maintenance plans, and after-market support plans. These teams are frequently co-led by design and manufacturing engineers who maintain high degrees of design authority and budgetry control. Suppliers and counterpart engineers from partnering companies also get involved at this level.

Managers often mention that it is difficult to get team members to work together as an actual team beyond a mere "information exchange group." The key to fostering teamwork is to vest the team with responsibility for its deliverables and the authority to obtain needed resources. Without these, most people will focus on their individual technical responsibilities and will view team membership as irrelevant to their "real jobs."

Performance measurement is another critical issue. Different functions typically emphasize different performance measures in NPD. Sometimes these differing priorities and their associated measures create conflicts. For example, product design engineers are frequently rewarded for reducing product material costs. On the other hand, operations personnel are usually rewarded for reducing labor costs. Choices made in product design can force trade-offs between these two priorities. Total product unit cost would be a better measure of performance, as this would encourage both functions to work together to produce an overall optimal design. Project managers should carefully consider how measures used in the development program might motivate collaboration or conflict within project teams.

Most firms use recognition as the primary incentive for motivating individuals on teams. For example, some firms hold team celebrations, competitions, or formally recognize teams at the corporate level. Others document product design improvements or cost-saving suggestions generated by teams and reward them with gifts or trophies. Team participation may be also assessed as part of the normal annual performance review process, addressing each employee's ability to participate in teams or to lead teams.

Another way companies encourage teamwork is by increasing each team member's sensitivity to the needs of the other team members. Larger projects establish liaison positions for personnel who act as go-betweens for major functional groups. Other approaches give team members a broader base of experiences by enlarging their job roles, or by rotating personnel through different job assignments and functional departments. In many cases, co-location is considered an important prerequisite to team success. Placing the work sites of personnel physically close to one another encourages team building and communication through daily contact and socialization.

Setting and Analyzing Goals: Clearly defined goals reduce nonconstructive conflict. High-level program goals also motivate cross-functional integration by providing rallying points for team members from different functional areas. For example, a product development goal of cutting product weight and cost by 30 percent motivates interactions among all affected functions (e.g., product design, manufacturing, purchasing, etc.). Clear, pre-specified goals help resolve conflicts between functional concerns because they focus the team's emphasis on goal achievement.

Furthermore, great benefits can be gained when project goals and product performance specifications are clearly linked to specific customer needs. Project team members are more likely to pull together to meet project timing and budget targets when they believe that customer satisfaction is on the line. In setting product specification goals, quality function deployment (QFD) is widely regarded as a useful tool for decomposing customer needs into engineering design parameters. Essentially, QFD requires a cross-functional team to generate and evaluate customer attributes and the requisite engineering specifications (Hauser and Clausing, 1988). The priority given to various customer needs and project goals can also influence both the degree of interaction between certain persons and the nature of their interactions. For example, setting aggressive product cost goals is likely to stimulate intense manufacturing-design interactions, since both parties are needed to generate and evaluate cost reduction ideas. Development project managers should carefully identify the greatest needs for functional integration and design program goals accordingly.

Directing and Controlling Integration: Project managers must also design procedures that direct and control functional interactions. Some firms formally document the major phases and general deliverables for any product development program in the form of a program contract. A "blank" program contract contains templates for establishing the "who, what, where, when, and how" of specific, key tasks on the project. Intense preprogram training may be connected with the development of the program contract including discussions that identify the most important project management aspects and product design requirements. The effectiveness of program contracts and training processes lies in the ability to adapt broad organizational rules and procedures to the specific needs of the program at hand.

Corporate initiatives such as quality and productivity improvement programs are often useful for directing integration. A number of NPD projects have effectively integrated key components of corporate initiatives into their program management plans. For example, one program manager leveraged a company-developed quality improvement procedure by establishing it as the basis for team interaction and problem solving on his project. The potential danger of relying heavily on corporate-wide initiatives is that project specific goals may be overshadowed by the initiative.

2. ***Improving Design Analysis and Decision making:*** An excellent product design maximizes customer satisfaction while effectively utilizing the company's production, sales, and service capabilities. The conventional way of insuring that good design practices are fol-

lowed is to give functional engineering representatives approval authority for all designs. A designer creates a part design, for example, that is then serially reviewed and revised by a number of functional experts. While this approach has merit, it is also inefficient because rework is institutionalized and very little learning is transferred to the original designer.

A more productive approach is for designers to incorporate good design practices in the initial process of generating the design. In order to minimize design rework, progressive firms are seeking innovative ways to train designers in good design practices. Training can include general design axioms, guidelines, or specific design rules pertaining to commonly used manufacturing processes. Design rules and best-practices can also be embedded into computer-aided-design (CAD) systems. These systems are discussed in more detail in the Technology Perspective section.

Another way to reinforce design rules and best practices is to create an internal consulting group, staffed by skilled and experienced engineers, and separated organizationally from other functional units. Representatives from the group can be assigned to NPD projects to assure that state-of-the-art design practices are followed and to serve as liaisons to functional experts. Descriptions of other tools and methodologies for facilitating manufacturing process design analyses are found in Nevins and Whitney (1989), Susman (1992), Kusiak (1992, 1993), Dorf and Kusiak (1994), and Ulrich and Eppinger (1995).

3. ***Summary:*** A summary of CE best practices is provided in Table 1. To maximize the effectiveness of CE, managers need to customize NPD processes to meet specific NPD program challenges, priorities, and product characteristics. Managers should consider company culture, the level of experience program personnel have with CE, the complexity of the product, the level of innovation attempted in the development effort, and the technical risk involved in the project. Product complexity is reflected by the numbers of people, functional specialties, and outside suppliers involved in NPD and by the degree of interdependency. Project teams and integration systems should be designed to ameliorate the negative effects of this complexity while dealing with the constraints and risks of innovation. For example, the design of radically new product components sometimes call for limited concurrency early in NPD due to an increased need for product testing and validation. At the same time, breakthrough technological developments increase the need for interactions among designers and team members inside and outside the firm who have the expertise needed to reduce technical uncertainties. (see Swink *et al.*, 1996a).

Table 1. Checklist of best practices in concurrent engineering.

- The design of the CE program is guided by a clear set of goals with attached priorities. These are communicated to team members in the form of a program contract.
- The design of the CE program takes into account the effects of the firm's organizational culture and prior experience with CE.
- Overlapped activities in NPD are supported by incentives, performance measures, technologies, and organizational arrangements that motivate and enable rich communications among the appropriate team members.
- Cross-functional teams are given clear objectives along with budget control, authority, and resources needed to meet those objectives.
- Barriers to CE are removed. These barriers can be created by functional silos, the sequential scheduling of activities, lack of access to information, and lack of co-location of personnel.
- Advanced technologies are used to support design, analysis, and communication. Managers plan for the additional time team personnel need to implement and become proficient in the use of new technologies.
- Project personnel are trained in the CE process. Each team member has a clear understanding of his/her role in achievement NPD program goals and priorities.
- Members of design-build teams have equal status and awareness of the requirements of other team members.
- The NPD project manager has greater power than that of functional heads.
- For any two team members, only two to three organizational levels must be bridged before reaching a common manager.
- A high level champion has helped to create a sense of urgency, uniqueness, or importance for the project.
- Methods have been established for capturing and conveying technical discoveries and learning gleaned from the various design and development activities.

TECHNOLOGY PERSPECTIVE: Technology can be a strong enabler of concurrent engineering. Many new computer applications improve the richness of communication among NPD team members, and improve the quality of product designs. These technologies can be broadly classified as follows:

- Physical facilities: The building layout and technologies facilitate co-location of personnel, team meetings, information display, and an "open" atmosphere.
- Communications Technologies: These technologies improve the speed, richness, and distance over which communications can occur.
- Computer-Aided-Design Systems and Databases: These systems provide digital design capabilities and a centralized source of design data that can be accessed electronically.
- Computer-Aided-Engineering and Analysis Systems: These systems provide simulation capabilities and easy access to engineering information that is useful to designers who are making choices among design alternatives.
- Group Technology and Coding Systems: These systems provide a scheme for classifying existing component designs and sources of components (e.g., suppliers), thereby reducing the need for creating new designs or finding new sources.

The physical layout of facilities used to house NPD team members can promote greater communication and teamwork. The Chrysler Corporation's new research and technical center has been highly touted for its ability to co-locate engineering and pilot production operations. Group discussions are enhanced by an open building concept that is sprinkled with numerous high-technology meeting rooms. Each floor of the circular building houses a single product development team.

Electronic mail and communications networks are powerful tools for rapidly communicating information and for providing wide audiences easy access to product and project data. Information can also be stored on centralized computer-aided-design (CAD) databases. Data captured in these systems can be accessed by persons located around the world for use in product design, process planning, and computer-aided manufacturing. Boeing made extensive use of these systems in the design of its 777 aircraft, the first plane to be completely "digitally" designed.

Computer-aided-engineering (CAE) tools are frequently linked to CAD systems in ways that reinforce good design practices. These sophisticated systems create and analyze three-dimensional models of parts and assemblies, reducing the need to build expensive and time consuming physical prototypes. For example, CAD/CAE systems can automatically analyze assembly designs to identify areas of potential interference between parts. Further, many CAD systems embed process information and design rules directly into the design software so that it may be linked to certain design features. For example, when a designer draws a hole, he can then select a pull-down window of information providing a list of processes that could create the hole, typical dimensional tolerances, defect rates associated with each process, and any other design rules related to the feature. Some companies have developed "expert systems" that aid the evaluation of design choices. In addition, numerous off-the-shelf CAE systems address stress and thermal analyses, mechanical assembly, printed circuit board design, and integrated circuit (IC) design.

In large organizations, designers often waste time and resources by unknowingly recreating existing designs. CAD systems can be linked with databases that contain information on preferred components, existing designs from other products, and suppliers of purchased items. Group technology-based classification and coding systems enable designers to easily search design databases for existing designs which meet their current needs. Similarly, databases which prioritize certain components and vendors can speed up a designer's search for suitable parts. Coding systems also allow manufacturing planners to identify "families" of parts that have similar design or processing characteristics. These approaches reduce design time and reap enormous manufacturing benefits because fewer unique parts must be fabricated and inventoried, less special tooling is needed, production scheduling is simplified, and less disruption is experienced. For more information on enabling technologies in CE, see Nevins and Whitney (1989), Kusiak (1992, 1993), Dorf and Kusiak (1994), and Ulrich and Eppinger (1995).

EFFECT: Concurrent engineering processes are sometimes viewed as expensive in the short term, requiring resources and levels of management and worker commitment that may not be readily available. However, when appropriately used, CE approaches can produce higher quality, more producible products, in less time. Further, CE produces organizational benefits that far exceed the profits associated with any single product. For example, some companies have documented savings in overall product development costs of 20 percent, reductions in development time of 50 percent, and reductions in engineering design changes of 45–50 percent (Swink *et al.*, 1996a). More importantly, successive CE development programs build upon one another. Improved design processes lead to improved products and services. CE experiences produce engineers and managers who are more aware of the company's goals and of the needs of other functional areas. Combining this type of organizational learning with effective CE processes can improve competitive advantage.

TIMING/LOCATION: Numerous cases of concurrent engineering success have been documented in a variety of industry and product development contexts. Even so, researchers continue to debate the appropriateness of CE in different situations. On-going research is investigating the dynamics of functional integration and concurrency by addressing such issues as to who should be integrated and when and how this is best achieved.

Some researchers and users of CE argue that it is especially important when there is stiff competition, when new manufacturing processes are being used, and when reducing development lead time is very important. Indeed, managers may have little choice but to implement CE when presented with a lucrative market opportunity or confronted by a particularly aggressive competitor. At the same time, there are indications that an over-emphasis on CE can hurt NPD performance outcomes in some circumstances. Heavy concurrency may increase development cost and lead time when NPD activities are risky with a high potential for failure. Because process design activities are begun earlier and with less complete information, there is an increased probability that some designs will need frequent rework.

Recent empirical research suggests that heavy manufacturing influences early in NPD can be detrimental to product innovation, and may unnecessarily increase lead time in NPD projects when they involve little new technology (Swink, 1997). For example, manufacturing personnel may sometimes be locked-in to the firm's current processing capabilities or remain unknowledgeable about new technologies or options outside the firm. In this case, their influence in NPD may work against the adoption of innovative new product features. In addition, R&D personnel sometimes complain that the inclusion of downstream process personnel early in NPD creates confusion and slows down decision making, especially at very early stages of NPD when design concepts are not well-defined (Gerwin, 1993). These findings are tentative and need to be confirmed. However, it is important to remember that the primary goals of NPD should govern the training and participation of various functional groups and the usage of cross-functional integration methods. Good communication among marketers, product designers, process designers and manufacturing personnel is always crucial. However, communications across certain functional groups should be prioritized according to the objectives established for the project (Swink *et al.*, 1996a). For example, the use of a important new manufacturing technology would necessitate rich communications among the designers using the technology, the designers and installers of the technology, and the users of the technology (i.e., manufacturing personnel). Further, each CE team member should realize that the strategic importance of certain NPD program outcomes (e.g., project timing, innovative features) might sometimes outweigh the importance of his or her own function's design guidelines.

CASES: Many well-known companies in different industries have adopted CE approaches including Thomson Consumer Electronics, Cummins Engine, Chrysler, Ford, Motorola, General Motors, Texas Instruments, Hewlett Packard, and Intel. Descriptions of these and other cases are found in Clark and Fujimoto (1991), Mabert *et al.* (1992), and Swink *et al.* (1996a).

Haddad (1996) provides an interesting account of a major CE implementation in an automotive company. The firm had experienced poor profitability and market share losses for a number of years. Executives at the firm realized that their Japanese competitors were able to bring new products to market in half the time. As a result, in 1990 the company launched a dramatic reorganization of engineering, purchasing, manufacturing, and planning activities into cross-functional, product-focused teams. The new arrangement required that personnel from these various functions, including marketing and first tier suppliers, be dedicated to a new product platform team. The overall platform team was then divided into groups known as "product action teams". Each group was headed by an executive engineer and was responsible for a portion of the platform (e.g., the interior of the automobile). These groups were further divided into departments responsible for components in the car (e.g., seats or controls). Groups averaged around 16 people in number. Groups were given significant authority and encouraged to make decisions by consensus.

The organizational change was supported by several changes in corporate standards, human resource policies, and technology. A new policy dictated that no more than five levels of organizational hierarchy should exist from vice presidents down through executive engineers, managers, and supervisors. Human resources managers developed training sessions and evaluation systems aimed at motivating team behavior. For example, the old, performance appraisal system was replaced by a new system that emphasized a broader set of criteria, which included the acquisition of skills in the areas of teamwork, communication, planning, process improvement, problem solving, and so on. Technological improvements included more extensive use of CAD, CAE, and electronic networking technologies. In addition, a new facility was built to co-locate vehicle engineering, design, pilot production, and training activities.

These changes by the automobile firm produced handsome returns. The first two new platform automobiles introduced using the new processes were brought to market in about three years. Prior models had required 5 years or more to develop. The company earned healthy profits in 1992 while its competitors showed losses for the same period. In 1993 and 1994 the company posted record earnings, and their stock increased sixfold in the period from 1991 to 1994. The changes brought high praise from market analysts, and workers inside the firm experienced many qualitative benefits of the change. Workers felt that the development projects had greater focus and less confusion than in the past. In addition, there appeared to be a higher level of worker commitment and buy-in.

While the change to CE produced great benefits for the company, it was not without its problems and challenges. Union relationships were strained due to decisions by the firm to increase the amount of outsourced design work. Some workers expressed feelings of greater stress and a lack of recognition for their contributions. The change to CE was hampered by shortage of design engineers who had meaningful manufacturing experience. Also, functional experts were spread across too many platform teams. Finally, newly hired engineers had difficulty knowing who the experts in other platform teams were, since they had no opportunity to work with them on prior projects.

This case study points to the challenges and benefits of implementing CE in a very large organization. However, similar benefits have been recorded in smaller firms who did not have the opportunity to make the kinds of sweeping organizational and technological changes described in this study. Regardless of the form CE takes, the key benefits are derived from the empowerment and teamwork of program personnel.

COLLECTIVE WISDOM: The involvement of corporate-level management is important to the success of CE, especially for initial attempts at using the approach. To increase the probability of success, managers must:

- Elevate the project.
- Elucidate goals.
- Eliminate barriers to integration.
- Elaborate CE processes (Swink *et al.*, 1996b).

1. ***Elevate the Project:*** For most firms, CE is an unconventional approach to product development, requiring personnel to behave in non-traditional ways that violate prevalent habits and management philosophies. Consequently, for early attempts at CE, it is vital to get people to view the project as "special." Creating a crisis environment is one way to break people out of conventional practices. A sense of urgency can encourage team members to overcome barriers formed by the corporate culture and standard operating procedures. Furthermore, when team members perceive that the project is unique, they often become more interested, take greater ownership, and are more willing to take risks.

   Top management commitment is also important. A high level champion for the project provides the energy and enthusiasm needed to get people to change. Committed top level managers are more willing to fight for necessary project resources. This is especially important since CE development programs can be initially expensive to implement. Initial programs may require resources dedicated to the development of CE concepts for the firm, extensive training for team members, and the purchase and implementation of communication systems. In addition, program managers should expect to spend more money early in the NPD effort since downstream personnel are asked to participate in up-front decision-making and analysis.
2. ***Elucidate Goals:*** In CE projects it is important to define goals that stimulate interactions between the right groups of people, at the right time, and to the right extent. Shared, globally-identifiable, customer-initiated goals produce a common language and sense of purpose for the project. In addition, team-based performance measures and feedback that are related to super-ordinate goals allow team members to share the success or failure of their efforts.
3. ***Eliminate Barriers to Integration:*** A fundamental barrier to integration in most large organizations is bureaucracy. Standard operating procedures often create functional barriers which must be broken down. In these situations, it may be useful to separate the project from the larger organization by dedicating resources and by realigning reporting relationships and reward structures, at least temporarily.

   The forming of teams can cause conflicts and upset well-established domains of power. Team members may be put in situations where they must choose between benefiting the project and benefiting their own home functions. Functional leaders may try to exert control over portions of the project that they consider to be within their domains. Project leaders should try to anticipate and manage these situations.

4. ***Elaborate Concurrent Engineering Processes:*** Team members require more than technical skills and information access to be effective players in a CE environment. They also need a clear understanding of the CE process itself. A well-defined process must be established and clearly articulated to all team members inside and outside the firm.

   Defining the process means developing a shared vision of the important dimensions of CE. Managers need to explore how CE approaches can be tailored to meet specific project needs. Furthermore, managers need to capture and leverage learning from CE experiences by peppering NPD projects with people who are experienced in the process. Sharing project documentation and post mortem project analyses across divisions within the firm can be helpful in disseminating learning.

CONCLUSION: Concurrent engineering can provide a powerful means for improving product development time, cost, and quality. To be successful, managers must develop a vision of CE that matches corporate capabilities and the objectives of the product development effort. In many cases, the implementation of CE will require major changes in NPD processes and corporate culture. There are a number of organizational and technological enablers that can be used to facilitate needed changes and to improve the quality of design and development activities. In the final analysis, however, teamwork is the source of the primary advantages offered by CE.

See **Cross functional team; Cross-functional integration; Design concurrency; Design-for-manufacturability (DFM); Functional organization; Integrated product development; Process capability; Product concurrency; Product life cycle; Project phase concurrency; Quality function deployment (QFD); Simultaneous engineering; Teams: Design and implementation.**

References

[1] Alting, L. (1993). "Life-Cycle Design of Products: A New Opportunity for Manufacturing Enterprises." in *Concurrent Engineering*, A. Kusiak (ed.). Wiley, New York.

[2] Clark, K.B. and T. Fujimoto (1991). *Product Development Performance*. Harvard Business School Press, Boston.

[3] Clark, K.B. (1992). "Design for Manufacturability at Midwest Industries: Teaching Note." Note no. 5-693-007, Harvard Business School Press, Boston.

[4] Dorf, R.C. and A. Kusiak (1994). *Handbook of Design, Manufacturing, and Automation*. Wiley, New York.

[5] Gerwin, D. (1993). "Integrating Manufacturing into the Strategic Phases of New Product Development." *California Management Review*, Summer, 123–136.

[6] Haddad, C.J. (1996). "Operationalizing the Concept of Concurrent Engineering: A Case Study from the U.S. Auto Industry." *IEEE Transactions on Engineering Management*, 43(2), 124–132.

[7] Handfield, R.R (1995). *Re-engineering for Time-Based Competition*, Quorum Books, Westport, CT.

[8] Hauser, J.R. and D. Clausing (1988). "The House of Quality." *Harvard Business Review*, May–June, 62–73.

[9] Hendricks, K.B. and V.R. Singhal (1997). "Delays in New Product Introductions and the Market Value of the Firm: The Consequences of Being Late to the Market," *Management Science*, 43(4), 422–436.

[10] Kusiak, A. (1992). *Intelligent Design and Manufacturing*. Wiley, New York.

[11] Kusiak, A. (1993). *Concurrent Engineering: Automation, Tools, and Techniques*. Wiley, New York.

[12] Mabert, V.A., J.F. Muth and R.W. Schmenner (1992). "Collapsing New Product Development Times: Six Case Studies." *Journal of Product Innovation Management*, 9, 200–212.

[13] MIL-HDBK-59A-CALS (1995). *Military Handbook on Computer-Aided Acquisition and Logistics Support*.

[14] Nevins, J.L. and D.E. Whitney (1989). *Concurrent Design of Products and Processes*. McGraw-Hill, New York.

[15] Susman, G.I. (1992). *Integrating Design and Manufacturing for Competitive Advantage*. Oxford University Press, New York.

[16] Ulrich, K.T. and S.D. Eppinger (1995). *Product Design and Development*. McGraw-Hill, New York.

[17] Smith, R.P. (1997). "The Historical Roots of Concurrent Engineering Fundamentals." *IEEE Transactions on Engineering Management*, 44(1), 67–78.

[18] Swink, M.L., J.C. Sandvig, and V.A. Mabert (1996a). "Customizing Concurrent Engineering Processes: Five Case Studies." *Journal of Product and Innovation Management*, 13(3), 229–244.

[19] Swink, M.L., J.C. Sandvig, and V.A. Mabert (1996b). "Adding Zip to Product Development: Concurrent Engineering Methods and Tools" *Business Horizons*, 39(2), 41–49.

[20] Swink, M.L (1997) "An Exploratory Study of Management Approaches and Their Relationships to Performance in Incremental, Radical and Breakthrough Product Development Projects." Working paper, Indiana University.

CONCURRENT SYSTEMS

The events and activities that occur in a manufacturing system may take place sequentially (one at a time) or concurrently (in parallel). In sequential systems, the events or activities take place one after the other. For example, the assembly of a computer may require the following sequence of steps: position the case; position the motherboard; secure the motherboard to the case with screws; insert and secure the specified components (hard drive, CD-ROM, floppy drive); attach the cables from the components to the motherboard; insert additional cards and attach their cables; and put on and secure the cover. In contrast, in concurrent systems, events and activities take place in parallel. For example, the components that are used in the assembly of the computer in the previous example, may themselves be produced for assembly in different parts of the plant, which operate concurrently.

Obviously, most systems are not entirely sequential or concurrent; rather, they are mixed, with some portions operating sequentially and some operating concurrently. At the most basic level, even when production of an individual product is a purely sequential process, there is likely to be more than one item undergoing production at the same time; i.e., the items are being produced concurrently. In more complex systems, an individual item may be produced in a mixed mode, where some of the production steps occur sequentially and others occur concurrently.

See **Asynchronous systems; Discrete-event systems; Manufacturing systems modeling using petri nets.**

References

[1] Browne, J., Dubois, D., Rathmill, K., Sethi, S., and Stecke, K.E. (1984), Classification of flexible manufacturing systems. *The FMS Magazine*: 114–17.

[2] Cassandras, C.G. (1993), *Discrete Event Systems: Modeling and Performance Analysis*. Homewood, IL: Irwin.

CONSTANT TIME ALLOWANCE

The constant time allowance is one of several ways of estimating manufacturing lead times. The same lead time is applied to all jobs; that is, a constant time allowance is added to the date of order receipt to compute the estimated due date for an order. This method is less demanding to use in terms of the information needed than alternative methods based on **slack time allowance** or **total work content allowance** for computing lead-time.

See **Slack time allowance; Total work content allowance.**

CONSTANT WORK-IN-PROCESS (WIP) METHOD

The constant **work-in-Process** (CONWIP) policy for **flow lines** is an **order release** mechanism. Spearman *et al.* (1990) proposed and also described by Hopp and Spearman (1996). CONWIP is a logical extention of **input/output control**. The CONWIP mechanism uses a set of (possibly electronic) cards. Each card accompanies a work order through the shop. After the shop completes the order, the card returns to the order release station, where it joins a new order. As the name implies, this mechanism uses a constant number of cards to maintain constant work in process in the shops. An order can be released only when a card is available. If there are multiple products, the production planning staff must sequence the new orders for release. The Earliest Release Time First sequencing rule is used to sequence orders pending at work centers. The number of cards should be the minimum number sufficient to avoid decreasing shop output.

See **Flow lines; Input/output control; Order release; Work-in-process (WIP) inventory.**

References

[1] Hopp, Wallace J., and Mark L. Spearman (1996). *Factory Physics*, McGraw-Hill, Boston, 1996.

[2] Spearman, Mark L., David L. Woodruff, and Wallace J. Hopp (1990). "CONWIP: a pull alternative to kanban," *International Journal of Production Research*, 28 (5), pp. 879–894.

CONSTRAINT MANAGEMENT

Constraint management lies at the heart of Goldratt's Theory of Constraints (TOC). Constraint refers to a bottleneck resource, which limits the output of a manufacturing system. TOC offers bottleneck managing principles such that bottleneck resources (such as work centers) are scheduled to maximize their throughput. Managing constraints in TOC is essentially a five-step process: (1) Identify the system's constraint(s); (2) Decide how to exploit the system's constraint(s);

(3) Subordinate everything else to the above decision; (4) Elevate the constraint(s); (5) If a constraint has been broken, go back to step 1.

See **Theory of constraints; Theory of constraints in manufacturing management.**

Reference

[1] E.M. Goldratt (1990). *The Haystack Syndrome: Sifting Information out of the Data Ocean* North River Press: Croton-on-Hudson, NY.

CONSTRAINT RESOURCE

A work center, machine or other resource may limit or constrain the output of a manufacturing system. Since the production outputs of all other work centers, operations, or machines in the system are constrained by the constraint resource (i.e., the constraint operation), according to the Theory of Constraints, management must focus attention on fully exploiting and eventually elevating the capacity of the constraint resource. According to TOC, the key to optimizing the moneymaking potential of a system lies in intimately understanding and managing the constraint resource. Synonyms for constraint resource in the context of TOC include "constraint operation" and "bottleneck." However, one needs to be careful in using the term "constraint resource" to refer to the bottleneck operation of a process. In some contexts, the term "capacity constraint resources" refers to a limited set of resources in the system that, if not management properly, can reduce output in the system.

See **Theory of constraints; Theory of constraints in manufacturing management.**

Reference

[1] E.M. Goldratt (1990). *The Haystack Syndrome: Sifting Information out of the Data Ocean* (North River Press: Croton-on-Hudson, NY.

CONSULTATIVE TEAMS

In the last ten years, the use of teams has increased in manufacturing firms. Consultative teams are fairly permanent teams that provide ideas, suggestions, or recommended solutions about ongoing performance and work process issues. Among the most well-known examples are quality circles. Consultative teams comprise employees whose normal work responsibilities are similar. Consultative teams do not possess decision-making authority. The team's major responsibility is to provide information and ideas to management.

See **Design and implementation of manufacturing work teams.**

CONTINGENCY CLAIMS ANALYSIS (CCA)

CCA is an approach based on financial options-valuation theory in modern finance to value future opportunities provided by today's investment projects, which offer future opportunities for survival and growth. The presence of future opportunities confers a valuable right (or a claim) to the investing firm to exploit them, but without an obligation to do so, contingent on how the future circumstances turn out. Such contingency claims, or claims whose value depends on the value of another asset, may be analyzed using the option pricing models in modern financial theory. Much like the *call options* on traded securities, which are claims on financial assets future opportunities are *real options* in that they represent claims on real assets. CCA provides a powerful technique to weigh the benefits of operating options such as flexibility embedded in many advanced manufacturing technologies against larger initial investment costs.

See **Capital investments in advanced manufacturing technologies.**

CONTINUOUS FLOW MANUFACTURING

See **Just-in-time; Lean manufacturing; Lean manufacturing implementation.**

CONTINUOUS IMPROVEMENT TEAMS

Continuous improvement is now well-known by the Japanese word, Kaizen. It is a constant endeavor to expose and eliminate problems at the root level. Factories now use teams of employees to achieve continuous improvement in all aspects of manufacturing. These teams are concerned with implementation of continuous improvement to enhance operations in three ways: waste reduction, performance improvement, and quality improvement. Internal performance measures such as rejection and rework rate are used to guide the work of continuous improvement teams.

See **Quality circle; Total quality management.**

CONTINUOUS REPLENISHMENT

The just-in-time revolution of the 1980s occurred predominantly in the manufacturing and industrial sectors of the economy. However, JIT's principle philosophies of eliminating waste and effectively using the human resource to reduce lead times, increase responsiveness, and establish a learning organization were viewed as highly attractive across a wide spectrum of industries. The core philosophies of JIT have thus been adopted and adapted to improve the competitiveness of apparel retailers, food marketers, and other retail and service-oriented companies. The names of these different strategies vary from quick response to continuous replenishment to efficient consumer response. But each of these strategies has at its core the elimination of waste (time and/or inventory) through 1) the better use and sharing of information, 2) the use of advanced logistics techniques, 3) the closer cooperation of channel members, and 4) the integration of people and technology systems. When these elements come together, not only are materials where they need to be when they need to be there (and in just the right quantities) but an agile and learning organization emerges.

Reference

[1] Bowersox, D.J., Calantone, R.J., Clinton, S.R., Closs, D.J., Cooper, M.B., Droge, C.L., Fawcett, S.E., Frankel, R., Frayer, D.J., Morash, E.A., Rinehart, L.M., and Schmitz, J.M. (1995). *World Class Logistics: The Challenge of Managing Continuous Change*. Council of Logistics Management, Oak Brook, IL.

See **Supply chain management: Competing through integration.**

CONTRIBUTION MARGIN ANALYSIS

Contribution margin, a traditional accounting term, is a dollar measure using the following formula:

$$\text{Sales Revenue} - \text{Variable Costs} = \text{Contribution Margin}$$

Variable costs in a manufacturing organization typically include costs of direct materials, direct labor, and some types of overhead, administration, and selling activities. Distinguishing variable costs from fixed costs in the organization allows analyses that support management decisions related to pricing of special orders, outsourcing, and cost-volume-profit analyses (sometime called "**breakeven analysis**). The challenge in contribution margin analysis lies in defining exactly what constitutes a "variable" cost. For example, given the strength of union contracts and the importance of employer/employee relations in modern production environments, direct labor costs are much more "fixed" than "variable" compared to business environments earlier in this century.

See **Accounting system implications of TOC; Breakeven analysis.**

CONTROL CHART

Control charts are integral part of statistical quality control programs. They are used to control variability in manufacturing process. These charts are graphic plots of process performance data on the background of predetermined control limits. The process performance data usually consist of groups of measurements selected in regular sequence of production that preserve the order. Control charts enable the detection of assignable causes of variation in the process as opposed to random variations. The removal of assignable causes of variations in the process improves the overall quality of the product.

See **Statistical process control using control charts; Total quality management.**

CORE COMPETENCY

Core competence is the combination of technologies and production skills that form the foundation of a company's competitive strength. The core competence of company, by definition, cannot be easily copied by competitors. Further, it enables the company to develop and market a line of competitive products or services. Core competencies represent the collective learning in the organization. Examples are the skills to coordinate complex production systems and integrate multiple streams of product technologies. For example,

Sony's core competence in miniaturization allows the company to make everything from the Sony walkman to videocameras to notebook computers. Canon's core competence in optics, imaging, and microprocessor controls have enabled it to enter seemingly diverse markets such as copiers, laser printers, cameras, and image scanners. Similarly Honda's core competence is in engines and power trains design, development and manufacturing.

See **Core manufacturing competencies; Virtual manufacturing.**

CORE MANUFACTURING COMPETENCIES

Morgan Swink

Michigan State University, Michigan, USA

DESCRIPTION: Recently, researchers have argued that "competencies" form the primary basis for competition between firms. It has been said that in the current business environment, the essence of strategy is to develop "...hard-to-imitate organizational capabilities that distinguish a company from its competitors in the eyes of its customers" (Stalk, Evans, and Schulman, 1992). Core competencies offered by a firm's manufacturing processes enable it to differentiate its products from competitors' products.

Core manufacturing competencies have not been well-defined. Hayes and Pisano (1996) suggest that competencies are *activities* that a firm can *do* better than its competitors. Further, competencies cannot be purchased. Competencies are organization specific; they are developed internally. The fact that they are difficult to imitate or transfer is what makes them valuable. Thus, competencies derive less from specific technologies or manufacturing facilities and more from manufacturing infrastructure: people, management and information systems, learning, and organizational focus. A manufacturing core competency is a very fundamental skill that is not easily copied.

HISTORICAL PERSPECTIVE: Over the years, several manufacturing strategic planning frameworks have been developed that involve the notions associated with manufacturing competencies. These frameworks have been very useful, yet they contain some inherent limitations. Strategic planning frameworks identify key manufacturing decision areas and stress the need for consistency among decisions affecting business strategy, competitive priorities, and manufacturing structure and infrastructure. The top portion of Figure 1 provides a highly summarized schematic, which is representative of frameworks advanced by Skinner (1969), Wheelwright (1978), Hill (1983), Fine & Hax (1985), and Schroeder, *et al.* (1986). This framework provide a powerful message, addressing manufacturing decisions of strategic nature and highlighting manufacturing's competitive value. However, these models are

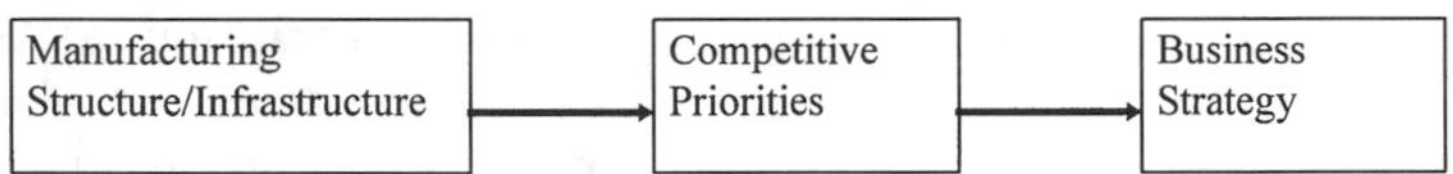

Historical Framework: General, Static, Priorities-based

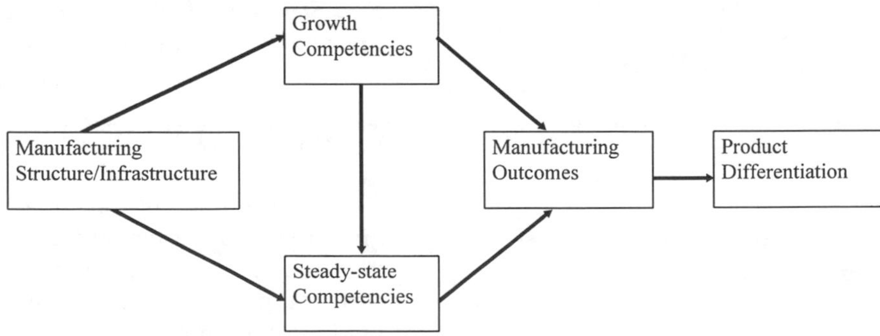

Revised Framework: Specific, Dynamic, Competencies-based

Arrows indicate supportive relationships necessary for effective strategy.

Fig. 1. Historical and revised manufacturing strategy linkage models.

also subject to two primary limitations: 1) they lack specificity, and 2) they do not directly address manufacturing competencies.

Making decisions using historical manufacturing strategy planning frameworks typically centered around "competitive priorities", such as cost, quality, dependability, flexibility, and service (Skinner, 1969; Wheelwright, 1978). These priorities are used to describe the content of a manufacturing strategy (Fine & Hax, 1985; Schroeder, Anderson, and Cleveland, 1986; Swamidass & Newell, 1987). Competitive priorities are too conceptually aggregated. Each priority is multifacetted and complex, making its interpretation very much dependent on the researcher, strategymaker, etc. Many different meanings and interpretations can be (and have been) attached to each term. For example, numerous aspects of flexibility have been discussed (Gerwin, 1993; Upton, 1994). Precise, disaggregated versions of competitive priorities are needed to translate them into specifics for manufacturing planning and decisionmaking.

Recognizing this need, researchers identified finer distinctions of competitive priorities (Skinner, 1992; Garvin, 1993). A complicating factor, however, lies in the perspective one employs when defining competitive priorities. An internal, manufacturing-based perspective leads one to consider priorities in terms of cost, product conformance (quality), and through-put lead time. An external, customer-based perspective defines priorities in terms of price, product performance, and delivery speed. While these two sets of priorities are related, there are frequently distinct differences between manufacturing outcomes and the product attributes that differentiate a product in the marketplace. For example, product pricing is almost certainly influenced by manufacturing costs, but other issues such as promotion and competition may exert even greater influences. Delivery speed may depend largely on manufacturing through-put time. Then again, delivery speed may be quite independent of through-put time if large finished goods inventories are held. Since most manufacturing operations are functionally buffered from customers, it is important to distinguish priorities related to manufacturing outcomes from priorities related to product differentiation. Making these distinctions enables the manufacturing function to more clearly define its strategic role in differentiating products (See ***Core Manufacturing Competencies: Implications***).

Another important limitation of competitive priorities literature is that it does not discriminate between manufacturing *competencies* and manufacturing *outcomes* (Corbett and Van Wassenhove, 1993). The ubiquitous list of manufacturing priorities: cost, quality, dependability, and flexibility, includes competencies and outcomes. Cost is a manufacturing outcome; flexibility is a manufacturing competency. The former construct refers to an *end*; the latter construct refers to a *means* to an end. Customer's view manufacturing outcomes and manufacturing competencies differently. As Penrose (1959) and McGrath, Tsai, Venkataraman, and MacMillan (1996) point out, customers do not purchase a firm's competencies, per se (e.g., flexibility). Customers desire and purchase product and service attributes (e.g., delivery speed) that a firm creates by deploying its competencies.

The strategic discussion of manufacturing is sometimes based on generic processing types. Some manufacturing strategy literature relate process forms (e.g., job shop, batch shop, assembly line) or technologies (e.g., FMS, transfer lines) to product and market characteristics (e.g., product life cycle stages) or to generic strategies (see Kim and Lee, 1993, for a review of this research). For example, a model developed by Kotha and Orne (1989) relates process complexity, product complexity, and organizational scope to business strategy, namely, cost leadership, differentiation, and market scope. It should be noted that, Kotha and Orne (1989) mention "differentiation" as a generic manufacturing strategy but many of product differentiation exist (Mintzberg, 1988), each requiring its own, potentially unique set of manufacturing structures.

STRATEGIC PERSPECTIVE: Numerous researchers have mentioned the ambiguity surrounding the concept of manufacturing strategy (Skinner, 1992; Gerwin, 1993; Swink and Way, 1995). A more specific and distinct terminology is required to resolve these ambiguities. Explicitly defining manufacturing competencies provides a step in this direction. In addition, a clear understanding of competencies should improve the implementation of manufacturing strategy models. Manufacturing competencies play three key roles in the formulation of strategy.

- Identifying important competencies clarifies differences between manufacturing outcomes and manufacturing means. Discussing competencies clarifies strategy formulation by not addressing *what* is needed but by addressing *how* it is delivered.
- An understanding of needed competencies clarifies the manufacturing objectives that undergird strategic manufacturing initiatives. A vision of needed competencies provides a dynamic basis for improvement that goes beyond simple strategic alignment and beyond static improve-

ment goals. A clear view of needed manufacturing competencies helps maintain a strategic direction over time (Garvin, 1993; Hayes and Pisano, 1994).

- Understanding manufacturing competencies provides deeper insights for translating manufacturing policies into product attributes that produce competitive advantages. Strategic manufacturing initiatives should seek to gain leverage from existing manufacturing competencies, or develop needed competencies that are currently lacking.

The framework pictured in the lower portion of Figure 1 illustrates links among manufacturing structure/infrastructure, manufacturing competencies, manufacturing outcomes, and product differentiation attributes. The framework is more specific than the manufacturing strategy framework based on "competitive priorities." Further, in the figure manufacturing outcomes are linked to specific dimensions of product differentiation. The framework explicitly addresses manufacturing competencies. Growth competencies enable a manufacturing operation to be dynamic and change over time. Steady state competencies represent the capabilities of a manufacturing operation if nothing changes, or at a given point in time. Further, the framework indicates that core competencies stem from the manufacturing structure and infrastructure.

SPECIFYING MANUFACTURING CORE COMPETENCIES: In order to identify specific manufacturing core competencies, we first have to distinguish them from manufacturing product outcomes.

1. ***Manufacturing Outcomes:*** Manufacturing outcomes are product attributes that reflect the cost, quality, and timing of production as well as the additional service provided by the operation (Corbett and Van Wassenhove, 1993; Chase *et al.*, 1992). These dimensions of manufacturing performance are corollaries to the familiar marketing dimensions of price, product, place, and promotion. A number of typical manufacturing outcomes are defined in Table 1.

   One of the outcomes in Table 1 is superior manufacturing technology. This construct refers not only to the characteristics of the technology itself but rather to the operating performance resulting from superior *uses* of technology. An example of a superior use of manufacturing technology is provided by R&R Engineering, Inc., a small manufacturer of bent bolts and custom made wire forms. This firm has perfected the use of various planetary thread rolling machines to make products at faster rates and with higher conformance quality than any of its competitors, who continue to use circular die rollers and flat die rollers. The complexity of the planetary threaders makes it

Table 1. Manufacturing outcomes

Development cost:	The cost to the manufacturer to design, test and develop production processes for a new product.
Production/transfer cost:	The cost to the manufacturer to make and deliver the product, including the cost to return or replace the item if necessary.
Product conformance:	Degree to which a product meets pre-established specifications, and the accuracy of the delivered order to the specified mix and quantities of products.
Superior mfg. technology:	Manufacturing processes which are unique or superior to competitors' manufacturing processes.
Order status information:	The availability and accuracy of real-time information about the location and status of an order shipment.
Mfg. process information	The availability and accuracy of data regarding manufacturing performance or process parameters.
Order processing time:	The time required for the customer and the supplier to communicate and agree upon the order specifications and to place the order (i.e., ease of ordering).
Development time:	The time required to create, design, and introduce a new product into manufacturing, including the time to develop and ramp-up needed manufacturing processes.
Production/transfer time:	The time required to produce and transfer the complete contents of the customer's order, including the time to return or replace defective products (i.e., the availability of the product as a function of time).
Lead time variance:	The variance between the scheduled delivery date and the actual, agreed upon, date.

very difficult to calibrate and adjust the machines for peak performance. Realizing this, the firm's managers invested a great deal of time and effort into experimenting with the equipment and understanding its capabilities. Competitors have also purchased planetary threaders, but have been unable to use them as effectively. Because R&R managers have a unique understanding of the equipment's competencies and have successfully integrated them into their operations, they have distinct quality advantages and can deliver products in half the time required by competitors.

2. ***Manufacturing Competencies:*** While manufacturing competencies span a wide range of attributes, seven *core* competencies address steady state and growth aspects of manufacturing performance. Steady state competencies can be measured at any given point in time and are indicated by superior manufacturing outcomes. Growth competencies enable changes in manufacturing outcomes over time or the development of new steady state competencies. The various components of each of manufacturing competency are described in Table 2.

   The framework in Table 2 views growth in manufacturing effectiveness as stemming from three core competencies for change: improvement, innovation, and integration. Steady state competencies stem from acuity, control, agility, and responsiveness. The seven components of manufacturing competing are elaborated below.

Improvement: Improvement relates to the ability to steadily increase the efficiency and productivity of existing manufacturing resources over time. The NUMMI plant, established as a joint venture between General Motors and Toyota, provides an example of superlative improvement competencies. Adler (1993) documents the steady performance improvement that resulted at the plant as a result of increased worker motivation, learning and problem solving, waste reduction, and work standardization.

Innovation: Innovation refers to the ability to radically improve manufacturing performance through the creation and implementation of new resources, methods, or technologies (Schroeder, Scudder, and Elmm, 1989). Innovation stems from the awareness of technological developments, plus the abilities to adapt and apply technology in ways that meet needs or create opportunities. Innovation competencies are vividly illustrated in the early development of flexible manufacturing systems by Hitachi Seiki (Hayes, 1990). The company cultivated the intellectual assets and organizational skills it needed to successfully apply a burgeoning micro-processor technology to machine tool systems. Technical and R&D expertise were important to the success of these developments. However, equally crucial were the contributions of manufacturing experts who understood the myriad production issues that needed to be addressed (e.g., scheduling, materials handling, etc.). Numerous other stories in the popular press recount the advantages manufacturing firms enjoy due to innovative developments of superior, often proprietary, processing competencies.

Integration: Integration is the ability to easily expand an operation to incorporate a wider range of products or process technologies. Upton (1994) discussed the ability to quickly manufacture new designs at John Crane Limited. The company's proficiency at introducing custom mechanical seal designs into an existing mix of manufactured components greatly enhanced its ability to meet unique customer needs. Related abilities to quickly introduce and utilize new processes or equipment are also important, especially for firms that compete in dynamic environments involving rapidly changing process technologies.

Acuity: Acuity refers to the insights of operations managers regarding process competencies and performance. These insights derive from high quality operations data and from abilities to translate internal or external customer needs into manufacturing specifications. Allegheny Ludlum Corporation, a specialty steel manufacturer, provides an example of superior acuity in manufacturing (March, 1985). The company has developed acuity through extensive process modeling and experimentation. Information systems rapidly and frequently provide in-depth data regarding productivity, utilization, yields, rejects, and operating variances. In addition, manufacturing personnel have close ties to key customers. Allegheny Ludlum uses its extensive process information coupled with a keen understanding of customer needs for evaluating strategic alternatives. In doing so, the company has achieved a high level of financial success even during periods of industry-wide recession.

Control: Control is the ability to direct and regulate operating processes. A necessary requirement for control is feedback, a property that permits comparisons of actual output values to desired output values. An illustration of control is found in statistical process control techniques. Statistical process control tools are used to analyze and understand process variables, to determine a process's capability to perform with respect to those variables (Gitlow, Gitlow, Oppenheim, and Oppenheim, 1989). In a larger sense, control refers to managers' abilities to understand

Table 2. Core manufacturing competencies.

Improvement:	**The ability to incrementally increase manufacturing performance using existing resources.**
Motivation:	The ability to impel human resources to higher levels of effort and effectiveness.
Learning:	The ability to increase and apply process understanding.
Waste reduction:	The ability to identify and remove non-value-adding activities.
Innovation:	**The ability to create and implement unique manufacturing processes that radically improve manufacturing performance.**
Scanning:	The ability to identify problems, process needs, or useful technological developments inside and outside the manufacturing organization.
Creativity:	The ability to generate and evaluate new ideas which meet organizational objectives.
Ingenuity:	The ability to apply new technologies or methods to solve problems.
Integration:	**The ability to incorporate new products or processes into the operation.**
Product intro. flexibility:	The ability to introduce and manufacture new products quickly.
Process ramp-up flexibility:	The ability to quickly learn new skills and adopt new processes.
Modification flexibility:	The ability to easily adjust processes to incorporate product design changes or special needs.
Aggregate change flexibility:	The ability to adjust smoothly to changes in product mix over the long term.
Acuity:	**The ability to understand customers' needs and to acquire, develop, and convey valuable information and insights regarding products and or processes.**
Consulting:	The ability to assist both internal groups and customers in problem solving (e.g., in new product development, design for manufacturability, quality improvement, etc.).
Information sharing:	The ability to furnish critical data on product performance, process parameters, and cost to internal groups and to external customers.
Showcasing:	The ability to enhance sales and marketing by exhibiting technology, equipment, or production systems in a way that conveys the value or quality of manufacturing competencies.
Control:	**The ability to direct and regulate operating processes.**
Process understanding:	The ability to understand manufacturing process capability limits and sources of variation.
Feedback:	The ability to monitor process outputs and to compare them with desired outputs.
Adjustment:	The ability to determine the causes of adverse effects and remedy undesired variations in manufacturing outcomes.
Agility:	**The ability to easily move from one manufacturing state to another.**
Volume flexibility:	The ability to efficiently produce wide ranges in the demanded volumes of products.
Mix flexibility:	The ability to manufacture a variety of products, over a short time span, without modifying facilities.
Responsiveness:	**The ability to react to changes in inputs or output requirements in a timely manner.**
Material flexibility:	The ability to accommodate raw material substitutions or variations.
Rerouting flexibility:	The ability to change product sequencing/loading in response to machine/equipment problems.
Sequencing flexibility:	The ability to rearrange the order in which parts are fed into the manufacturing process, because of changes in parts and raw material deliveries or changes in customer delivery requirements.
Shipment flexibility:	The ability to expedite or reroute shipments to accommodate special circumstances without loss of time.

and reduce sources of unwanted variation in a process.

Agility: Agility is the ability to move from one manufacturing state to another with very little cost or penalty. Manufacturing may be required to produce a wide range of products using a fixed set of resources. Agile processes are able to switch process set-ups quickly and efficiently, so that non-value-added time is minimized and so that smaller production runs are economical. Referring once again to John Crane Limited, Upton (1994) relates agility (which he names "mobility") to the ability of the company to maintain a wide offering of products without having to maintain a large finished goods inventory. Agility also involves the ability to produce a wide range of products in a range of quantities.

Responsiveness: Responsiveness refers to the ability to quickly adjust manufacturing processes to deal with changes in inputs, changes in resources, or changes in output requirements. For example, a responsive process can accommodate variations in the quality of raw materials or the uptime of equipment. Similarly, responsive processes can shift work schedules, job sequences, or physical routings to deal with unexpected changes in customer needs. In these ways, responsive processes are robust under conditions of input or demand variations.

IMPLEMENTATION: This article presents an inventory of manufacturing competencies. It is up to manufacturing managers to identify those competencies that are most important for their businesses. In formulating a strategic view of manufacturing competencies, manufacturing managers should consider the following:

- Multiple information sources will be required to supply adequate data and perspectives concerning manufacturing outcomes and manufacturing competencies.
- Measurement of some manufacturing competencies may be difficult. Table 2 suggests the salient dimensions for each competency. For the more vague concepts (e.g., creativity), managers may have to rely on perceptual measures rather than objective measures that could be difficult to define or obtain.
- Managers should consider inter-relationships among manufacturing competencies. Certain competencies probably build upon other competencies.

See **Core manufacturing competencies: Implications; Manufacturing flexibility; Manufacturing strategy.**

References

[1] Adler, P.S. (1993), "Time-and-Motion Regained." *Harvard Business Review*, January-February, 97–108.

[2] Chase, R.B.K., R. Kumar and W.E. Youngdahl (1992), "Service-Based Manufacturing: The Service Factory." *Production and Operations Management*, 1(2), 175–184.

[3] Corbett, C. and L. Van Wassehnove (1993), "Trade-Offs? What Trade-Offs? Competence and Competitiveness in Manufacturing Strategy." *California Management Review*, Summer, 107–122.

[4] Fine, C.H. and A.C. Hax (1985), "Manufacturing Strategy: A Methodology and an Illustration." *Interfaces*, 15(6), 28–46.

[5] Garvin, D.A. (1993), "Manufacturing Strategic Planning." *California Management Review*, Summer, 85–106.

[6] Gerwin, D., (1993), "Manufacturing Flexibility: A Strategic Perspective," *Management Science*, 39(4), 395–410.

[7] Gitlow, H., S. Gitlow, A. Oppenheim, and R. Oppenheim (1989), *Tools and Methods for the Improvement of Quality*, Irwin: Homewood, Il.

[8] Hayes, R. (1990), "Hitachi Seiki (Abridged"). Case No. 9-690-067, *Harvard Business School*, Boston: MA.

[9] Hayes, R. and G.P. Pisano (1994), "Beyond World Class: The New Manufacturing Strategy." *Harvard Business Review*, January—February, 77–86.

[10] Hill, T., (1983), "Manufacturing's Strategic Role." *Journal of the Operational Research Society*, 34(9), 853–860.

[11] Kim, Y. and J. Lee (1993), "Manufacturing Strategy and Production Systems: An Integrated Framework." *Journal of Operations Management*, 11(1), 3–15.

[12] Kotha, S. and D. Orne (1989), "Generic Manufacturing Strategies: A Conceptual Synthesis." *Strategic Management Journal*, 10(3), 211–231.

[13] March, A. (1985), "Allegheny Ludlum Steel Corporation", Case No. 9-686-087, *Harvard Business School*, Boston: MA.

[14] McGrath, R.G., M. Tsai, S. Venkataraman, I.C. MacMillan (1996), "Innovation, Competitive Advantage and Rent: A Model and Test." *Management Science*, 42(3), 389–403.

[15] Mintzberg, H. (1988), "Generic Strategies: Toward a Comprehensive Framework." *Advances in Strategic Management*, 5, 1–67.

[16] Penrose, E. (1959), *The Theory of the Growth of the Firm*, Wiley, New York.

[17] Porter, M.E. (1980), *Competitive Strategy: Techniques for Analyzing Industries and Competitors*, Free Press, New York.

[18] Schroeder, R.G., J.C. Anderson and G. Cleveland (1986), "The Content of Manufacturing Strategy: An Empirical Study." *Journal of Operations Management*, 6(3–4), 405–415.

[19] Schroeder, R.G. G.D Scudder, and D.R. Elmm

(1989), "Innovation in Manufacturing." *Journal of Operations Management*, 8(1), 1–15.

[20] Skinner, W. (1969), "Manufacturing–Missing Link in Corporate Strategy." *Harvard Business Review*, 47 (May–June), 136–145.

[21] Skinner, W. (1992), "Missing the Links in Manufacturing Strategy." in: C. Voss (Ed)., *Manufacturing Strategy: Process and Content*, Chapman-Hall, London, 13–25.

[22] Stalk, G., P. Evans and L.E. Schulman (1992), "Competing on Capabilities: The New Rules of Corporate Strategy." *Harvard Business Review*, March–April, 58–69.

[23] Swamidass, P.M. and W.T. Newell (1987), "Manufacturing Strategy, Environmental Uncertainty and Performance: A Path Analytic Model." *Management Science*, 33(4), 509–524.

[24] Swink, M. and M. Way (1995), "Manufacturing Strategy: Propositions, Current Research, Renewed Directions." *International Journal of Operations and Production Management*, 15(7), 4–27.

[25] Upton, D.M. (1994), "The Management of Manufacturing Flexibility", *California Management Review*, Winter, 72–89.

[26] Wheelwright, S.C. (1978), "Reflecting Corporate Strategy in Manufacturing Decisions." *Business Horizons*, 21(1), 57–66.

CORE MANUFACTURING COMPETENCIES AND PRODUCT DIFFERENTIATION

Morgan Swink

Michigan State University, Michigan, USA

DESCRIPTION: It is easy to make the case that business and manufacturing strategies should be "linked" (Skinner, 1969; Schendel & Hofer, 1979; Hill, 1983; Wheelwright, 1984; Corbett and Van Wassenhove, 1993; Garvin, 1993). Researchers have empirically verified the positive effects on performance that result from consistency in operations and marketing strategies (Swamidass and Newell, 1987; Deane, McDougall, and Gargeya, 1991). According to Skinner (1996, p. 12), inconsistency between functional strategies is "the first and most serious problem, and... the main weakness in MCS" (MCS stands for manufacturing in the corporate strategy).

In order to make manufacturing and business strategies mutually supportive, it is important to understand the core manufacturing competencies in the firm (see, ***Core Manufacturing Competencies***). In this article the link between core manufacturing competencies and product differentiation is discussed. Certain manufacturing competencies are more important than others to attain the various dimensions of product differentiation.

HISTORICAL PERSPECTIVE: Many researchers have specified that typical strategic business unit (SBU) can compete effectively using "generic strategies" (Buzzell, Gale, and Sultan, 1975; Utterback & Abernathy, 1975; Hofer & Schendel, 1978; Miles & Snow, 1978; Porter, 1980; Galbraith & Schendel, 1983; Mintzberg, 1988). Porter (1980) recommended two generic strategies, "cost leadership" and "differentiation." These strategies can be used to target an entire industry (i.e., a "breadth" strategy) or a market segment (i.e., a "focus" strategy). Although many other business strategy typologies have been proposed, Porter's model has arguably had the greatest influence on manufacturing strategy thinking. Product differentiation strategy has many implications for manufacturing.

TYPES OF PRODUCT DIFFERENTIATION: Mintzberg (1988) and others have discussed various types of product differentiation. Some basic dimensions of product differentiation are provided in Table 1. These are defined from a customer's perspective. For example, Mintzberg argued that competition on the basis of cost leadership (a generic strategy proposed by Porter, 1980) is a form of price differentiation. Mintzberg also recognized three dimensions of product quality-based differentiation: performance, uniqueness, and image. In Table 1 it is important to note that information-oriented customer support has one form of impact on manufacturing that is different from the impact of product delivery related customer support.

Non-manufacturing functions in the firm bear a big share of responsibility for delivering certain types of differentiation. For example, marketing frequently has a significant impact on product image than manufacturing.

CORE MANUFACTURING COMPETENCIES AND PRODUCT DIFFERENTIATION: Figure 1 indicates how certain manufacturing competencies support different dimensions of product differentiation. For a discussion on core manufacturing competencies are see, ***Core Manufacturing Competencies***. While many manufacturing outcomes and competencies may be important for a given product, Figure 1 includes only those manufacturing attributes that are essential competitive advantage through differentiation. Product cost, for example, is almost always an

Table 1. Types of product differentiation.[a]

Purchase price:	The expenditure of resources required of the customer to acquire the product, including costs of return or replacement.
Performance/uniqueness:	Product attributes that exceed comparable attributes in competitors' products or are unique in terms of greater reliability, greater durability, are and/or superior performance. These include relative measures of the product's fitness for customer use.
Product image:	Perceived differences for products that in fact do not differ in dimensions of performance. Image a measure of the impact of a product or company name, reputation, advertising, etc. on the customer's evaluation of the product.
Product information:	Instruction or data which accompany the sale, delivery, or use of the product and which augment the customer's value satisfaction.
Delivery speed:	The expenditure of time required of the customer to acquire the product or to receive replacement of defective product or replenishment of stocks.
Delivery reliability:	The customer's level of confidence that delivery will occur upon an agreed date.

[a] There are certainly other possible types of product differentiation. This list is limited to only those dimensions that are potentially influenced by manufacturing.

important product attribute. However, lower product cost is a source of differentiation only in cases in which product cost is a primary determinant of price. The linkages indicated in Figure 1 are based on examples, logic, and related theory.

1. ***Price Differentiation:*** Low product cost is often a highly prioritized manufacturing outcome when purchase price is a primary source of competitive differentiation. The total cost to produce and deliver the product to the customer determines the lower bound for profitable pricing. Price differentiation likely strategy to be found in predictable, stable markets.

Stable markets and technologies allow firms to establish well-designed organizational hierarchies and measurement systems. In this environment, (1) operations managers can establish efficient routines and performance

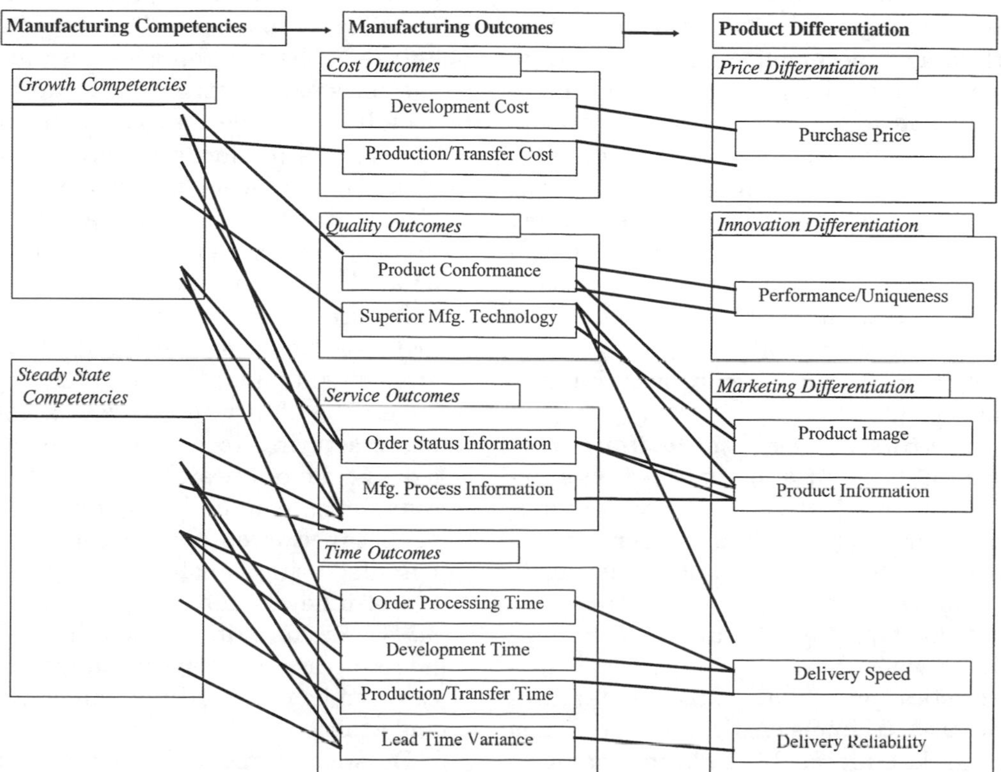

Fig. 1. Supportive manufacturing competencies for product differentiation.

standards, providing greater abilities to analyze and reduce variances and waste; (2) suggestions for job design, standards, and improvements can be generated by line workers who become experts as they get used to the process in a stable environment; and (3) repeated production of standardized products enables operations personnel to concentrate on process refinements that produce steady performance improvement over time. In conclusion, when product pricing is largely a function of production and delivery costs, price differentiation is realized through a manufacturing strategy which emphasizes improvement and control.

As the degree of product customization increases, product design costs comprise a larger share of total product cost. This is especially true for low volume orders. While many activities outside the realm of manufacturing contribute to product design costs, manufacturing resources are often heavily involved in product design and development. Further, manufacturing characteristics contribute to efficient product customization processes. Consider the example of a supplier firm that designs and produces metal castings which are used in low volume products such as medical equipment or weapon systems. The firms ability to differentiate itself on the basis of price is related to manufacturing's abilities to clearly communicate metal casting capability to product designers early during development activities. High quality communications serve to reduce product design and tool rework in latter development stages. In addition, manufacturing's ability to efficiently integrate new products, tools, and processing requirements into the existing production environment reduces process development costs. *Summary:* When product pricing is largely a function of product design costs, price differentiation is realized through a manufacturing strategy that emphasizes integration and acuity.

2. ***Innovation-Based Differentiation:*** Innovation-based differentiation is based on superior or unique product performance, features, reliability, durability, serviceability, or aesthetics (Garvin, 1987). For example, McDougall, Deane and D'Souza (1992) found that new venture firms typically offer a narrower range of products which have superior performance characteristics or unique product features and patented technologies. Businesses often exploit market niches by tailoring products and processes to the unique needs of the niche.

In order to deem a product superior or unique, customers implicitly or explicitly compare product attributes to competing product attributes or to their own expectations. In the case of competing products that have similar features, product conformance to design specifications is a necessity. In addition, the degree of conformance to specifications may greatly influence the reliable performance of the product. High conformance quality requires process capabilities that are consistent with design specifications. Conformance can be improved through greater process controls. Further, greater conformance also results from more appropriate allocation of design tolerances and specifications. If manufacturing personnel are able to relate process capability to the needs of product designers, this improves designers' abilities to specify design requirements that are producible. In this way, manufacturing acuity leads to greater product conformance.

Many unique or superior products are born of process and customer acuity and process innovation. For example, Allegheny Ludlum Steel, a producer of stainless and specialty steels, provides unique product solutions to its customers by leveraging its ability to translate customer needs into process refinements. Environments of high growth, research, and a large number of competitors provide a fertile ground for process innovation. The changing, dynamic nature of the environment described by these variables requires that operations personnel be acutely aware of changes in customer needs, competitors' products, and processing technologies. Further, operations personnel need to be able to distill and translate this information in ways that guide manufacturing innovations and improvement initiatives.

Reduced formality, increased delegation, and increased technocratic influence free-up creative talents to generate and evaluate new ideas. The ability to innovate a valuable new manufacturing technology or to use technology in a unique way, coupled with abilities to communicate an understanding of the technology's superior or unique capabilities, makes manufacturing capable of supporting a strategy of innovation-based differentiation. *Summary:* Product innovation differentiation is more successful when the manufacturing strategy emphasizes process innovation, control, and acuity.

Drastically reduced product life cycles in many industries testify to the fleeting nature of competitive advantages provided by innovation differentiation. In environments of dynamic product change, rapid and accurate input from manufacturing is important for reducing product development time. When many new or revised products are introduced in a manufacturing plant, product prototyping, testing, and production ramp-up activities compete with existing production activities for the use of manufacturing resources. Manufacturing managers are challenged to quickly adjust operating resources to integrate these new activities while maintaining acceptable levels of cost and efficiency.

Summary: In environments of dynamic product change, product innovation differentiation is more successful when the manufacturing strategy emphasizes integration of several activities in addition to process innovation, control, and acuity.

3. ***Marketing Differentiation:*** Marketing differentiation can take on a number of different forms, such as superior product promotion, service, delivery speed and reliability, packaging, installation, maintenance, etc.

Marketing differentiation is built on formal controls and procedures linked with high degrees of internal coordination, support, and communication. Superior informational support and/or product delivery by marketing requires a keen understanding of customer needs coupled with the corporate infrastructure required to meet those needs, including support staff, coordinating units, communication systems, policies, and procedures.

Chase and Garvin (1989) and Chase *et al.* (1992) identified the service roles manufacturing can play in improving the information and image characteristics of product differentiation. To enhance external customer satisfaction, "service factories" seek to make their products more attractive by offering customers easy access to manufacturing information and consultation, and by making inputs into design and service. For example, manufacturing functions at firms like Hewlitt Packard and Digital Equipment Corporation provide quality data sheets, video tapes, and equipment demonstrations for potential customers. Manufacturing's ability to share useful information with customers increases their total satisfaction and loyalty, and repeat business. Numerous suppliers provide greater access to product ordering and replenishment information to customers by electronically linking their scheduling systems to those of their customers. *Summary:* Product information differentiation is more successful when the manufacturing strategy emphasizes acuity.

Image differentiation seeks to shape customers' perceptions and expectations through promotions or other communications (Davidow and Uttal, 1989). Manufacturing contributes to a product's positive image by providing product or processing information which presents the firms' competencies as unique or superior to competitor's manufacturing competencies. Manufacturing's showcasing abilities make customers aware of product differences and process superiorities.

The value of manufacturing's showcasing abilities is closely tied to the customer's perception of the uniqueness or superiority of manufacturing processes. General Motors' Saturn plant provides a widely visible example. General Motors has promoted the plant to customers as a highly innovative operation. Its work systems and technologies are advertised in the media, public tours are frequently conducted, and Saturn owners are regularly invited to visit the plant. This marketing differentiation approach leverages manufacturing's abilities to successfully innovate superlative processing technologies and infrastructures (e.g., work environment, planning and control procedures, etc.). Customers' perceptions of these manufacturing superiorities influence their perceptions of product image. *Summary*: Product image differentiation is more successful when the manufacturing strategy emphasizes innovation and acuity, and customers can see product quality enhanced by manufacturing.

Product delivery timing is an important element of competition. Suppliers are increasingly asked to support just-in-time production requirements. In addition, consumers are continually pressing for faster product delivery. Time-oriented manufacturing outcome measures include product lead time and the variance in lead times. Three components of lead time are: 1) the time required to place the order, 2) the time required to develop the product and associated processes, and 3) the time required to produce and deliver the product.

For non-customized products, the time required to produce and deliver the items constitutes the majority of lead time, since a new product design is not required and order placement is routine. Control is the basis for process reliability, which in turn is essential for delivery speed (Ferdows & DeMeyer, 1990). Reduction in production/delivery lead time can also result from advances in technology use which increase processing speed (e.g., some machines run faster than others, some transport modes are faster than others). However, many products spend a majority of the lead time waiting for processing. Manufacturing's ability to efficiently vary product volume and variety over a wide range, can reduce production/delivery time. Agile systems reduce non-value-added time. An example of agility is provided by Allen-Bradley's World Contractor Facility. The plant has been touted for its ability to produce up to seven varieties of contractors and relays, with more than one-thousand different customer specifications, in small or large lot sizes, within a lead time of twenty-four hours. *Summary:* For standard product designs, delivery speed differentiation is more successful when the manufacturing strategy emphasizes control and agility.

By definition, new or customized products require design and development. Therefore, order time and development time make up a large proportion of total lead time for customized products. Product development lead time may be mostly due to activities which occur outside the manufacturing function. However, development lead time can be improved by manufacturing's ability to understand product performance requirements, to communicate process design information to designers, to participate in design activities, and to quickly integrate new products or processes into the existing manufacturing structure. *Summary:* When development time makes up a large proportion of lead time, delivery speed differentiation is more successful when the manufacturing strategy emphasizes integration and acuity.

Lead time variance results from unplanned events, from customer-initiated changes, or from inaccurate requirement forecasts. Unexpected downtime, late supplier deliveries, changes in customers' due dates, variations in processing times, and changes in demand also contribute to lead time variance. Consequently, delivery reliability is largely a function of manufacturing's abilities to predict, control, and respond. Better ability to predict resource capacities and resource requirements provides more accurate forecasts of production lead time. This ability comes from the acuity gained through improved communication, scanning, and analysis.

Greater control reduces process and schedule variations that contribute to lead time variance. Several of the structural correlates such as formal procedures, formal authority, and organizational differentiation increase organizational control.

Responsiveness is the flexibility to react to schedule variations and changes. Responsiveness is a function of support staff size and power, coordination mechanisms, and task forces. Greater support staff resources provide greater competencies to devise work-arounds and alternatives for meeting schedule requirements. In addition, a larger, more powerful support staff makes a firm more likely to be able to dedicate resources and focused attention to production problems that threaten on-time delivery. Coordination committees and task forces facilitate this type of problem solving. *Summary:* Delivery reliability differentiation is more successful when the manufacturing strategy emphasizes acuity, control, and responsiveness.

IMPLEMENTATION: The foregoing discussion addresses "pure" dimensions of product differentiation in the context of manufacturing outcomes and competencies. In practice, individual product differentiation strategies are rarely employed in isolation. For example innovation differentiation strategies often go with efforts to elevate product image. Marketing differentiation strategies often emphasize a combined package of superior delivery speed, dependability, and information. Value-based strategies seek to provide superior product performance at competitive prices. The recent move by many manufacturers toward mass-customization reflects the ultimate in combined differentiation strategies, linking low prices with high product uniqueness (customization) and quick and reliable delivery.

Overall, managers should work toward gaining a clear understanding of customer desires, differentiation strategy if the company and needed manufacturing competencies. This understanding can then be used to guide product development, market targeting, and investment decisions.

See **Core manufacturing competencies; Manufacturing strategy; Mass customization; Process capability.**

References

[1] Buzzell, R.D., B.T. Gale and R.G. Sultan (1975), "Market share: a key to profitability." *Harvard Business Review*, 54(1), 97–106.

[2] Chase, R.B.K., R. Kumar and W.E. Youngdahl (1992), "Service-Based Manufacturing: The Service Factory." *Production and Operations Management*, 1(2), 175–184.

[3] Chase, R.B. and D.A. Garvin (1989), "The Service Factory." *Harvard Business Review*, 67(4), 61–69.

[4] Corbett, C. and L. Van Wassehnove (1993), "Trade-Offs? What Trade-Offs? Competence and Competitiveness in Manufacturing Strategy." *California Management Review*, Summer, 107–122.

[5] Davidow, W.H. and B. Uttal (1989), "Service Companies: Focus or Falter." *Harvard Business Review*, 67(4), 77–85.

[6] Deane, R.H., P.P. McDougall and V.B. Gargeya (1991), "Manufacturing and Marketing Interdependence in the New Venture Firm: An Empirical Study." *Journal of Operations Management*, 10(3), 329–343.

[7] Dess, G.G. and P.S. Davis (1984), "Porter's (1980) generic strategies as determinants of strategic group membership and organizational performance." *Academy of Management Journal*, 27(3), 467–488.

[8] Ferdows, K., and A. De Meyer (1990), "Lasting Improvements in Manufacturing Performance: In Search of a New Theory." *Journal of Operations Management*, 9(2), 168–184.

[9] Galbraith, C. and D. Schendel (1983), "An empirical analysis of strategy types." *Strategic Management Journal*, 4(2), 153–173.

[10] Garvin, D.A. (1987), "Competing on the Eight Dimensions of Quality." *Harvard Business Review*, November-December, 101–109.

[11] Garvin, D.A. (1993), "Manufacturing Strategic Planning." *California Management Review*, Summer, 85–106.

[12] Goldhar, J.D. and M. Jelinek (1983), "Plan for Economies of Scope." *Harvard Business Review*, 61(6), 141–148.

[13] Hambrick, D.C. (1983), "High Profit Strategies in Mature Capital Goods Industries: A Contingency Approach." *Academy of Management Journal*, 26(4), 687–707.

[14] Hill, C. (1988), "Differentiation versus low cost or differentiation and low cost: A contingency framework." *Academy of Management Journal*, 13(3), 401–412.

[15] Hill, T. (1983), "Manufacturing's Strategic Role." *Journal of the Operational Research Society*, Vol. 34(9), 853–860.

[16] Hofer, C. and D. Schendel (1978), *Strategy Formulation: Analytical Concepts*, West, St. Paul.

[17] Kim, L. and Y. Lim (1988), "Environment, Generic Strategies, and Performance in a Rapidly Developing Country: A Taxonomic Approach." *Academy of Management Journal*, 31(4), 802–826.

[18] McDougall, P.P., R.H. Deane and D.E. D'Souza (1992), "Manufacturing Strategy and Business Origin of New Venture Firms in the Computer and Communications Equipment Industries." *Production and Operations Management*, 1(1), 53–69.

[19] Miles, R. and C. Snow (1978), *Organizational Strategy, Structure, and Process*, McGraw-Hill, New York.

[20] Miller, D. (1986), "Configurations of strategy and structure: Towards a synthesis." *Strategic Management Journal*, 7(3), 233–249.

[21] Miller, D. (1987), "The structural and environmental correlates of business strategy." *Strategic Management Journal*, 8(1), 55–76.

[22] Miller, D. (1988), "Relating Porter's business strategies to environment and structure: Analysis and performance implications." *Academy of Management Journal*, 31(3), 280–308.

[23] Miller, D. and P.H. Friesen (1984), *Organizations: A quantum view*, Prentice Hall, Englewood Cliffs, NJ.

[24] Mintzberg, H. (1988), "Generic Strategies: Toward a Comprehensive Framework." *Advances in Strategic Management*, 5, 1–67.

[25] Porter, M.E. (1980), *Competitive Strategy: Techniques for Analyzing Industries and Competitors*, Free Press, New York.

[26] Schendel, D. and C. Hofer (1979), *Strategic Management: A new view of Business Policy and Planning*. Little, Brown & Co., Boston, MA.

[27] Skinner, W. (1969), "Manufacturing–Missing Link in Corporate Strategy." *Harvard Business Review*, 47 (May–June), 136–145.

[28] Skinner, W. (1996), "Manufacturing Strategy on the "S" Curve." *Production and Operations Management*, 5(1), 3–14.

[29] Swamidass, P.M. and W.T. Newell (1987), "Manufacturing Strategy, Environmental Uncertainty and Performance: A Path Analytic Model." *Management Science*, 33(4), 509–524.

[30] Utterback, J.M. and W.J. Abernathy (1975), "A dynamic model of process and product innovation." *OMEGA*, 3(6), 639–656.

[31] Wheelwright, S.C. (1984), "Manufacturing Strategy: Defining the Missing Link." *Strategic Management Journal*, 5(1), 77–91.

CORE-PRODUCT APPROACH

This is principle used to develop customized products for various markets/countries around the globe. Under this approach, a standard product is designed which is capable of accepting a diverse number of attachments or components. The standard and central product is referred to as the core product. Depending on the choice of attachments used on the core product, customized products are created that meet the needs of various markets/countries.

See **Product design for global markets.**

CORPORATE BILLBOARD

Corporate billboard is a visual display used to promote awareness within the firm of key themes such as the mission of the firm, key strategies, key performance measures, key results, etc. These could be displayed in high-traffic areas such as meeting rooms, main thoroughfares etc. to get the attention of employees. It improves communication and may serve as a motivating force towards a common goal.

See **Balanced scorecard.**

CORRECTIVE OR BREAKDOWN MAINTENANCE

Corrective (or breakdown) maintenance refers to activities that are meant to bring a failed system or equipment back to operation. In evaluating the costs associated with corrective maintenance one usually includes the cost of failure consequences, like lost production time or worthless output that needs to be discarded.

See **Maintenance management decision models.**

COSMETIC CUSTOMIZERS

Cosmetic customizers are manufacturers who present a standard product differently to different customers, for instance, with changes in packaging, or with minor modification of some selected specifications. The cosmetic approach is appropriate when all customers use a product in the same manner but respond differently to the way it is marketed or presented. For example, the product may be displayed in more than one way with its attributes and benefits advertised in different ways.

See **Mass customization.**

COST ACCOUNTING

Cost accounting is the process of accumulating and accounting for the flows of costs in a business. It is defined as a technique or method for determining the cost of a project, process, or thing through direct measurement, arbitrary assignment, or systematic and rational allocation. It is concerned with the development of systems for relating costs recorded for the resources and services given up in the acquisition or production of resources by a company involved in the provision of products or services. The appropriate method of determining cost often depends on the circumstances that generate the need for information.

See **ABC analysis; Absorption costing; Absorption costing model; Activity-based costing; Activity-based costing: An evaluation.**

References

[1] Barfield, Jesse T., Cecily A. Raiborn, and Michael R. Kinney (1994). *Cost Accounting: Traditions and Innovations*, West Publishing Company, St. Paul, Minnesota.

[2] Henke, Emerson O. and Charlene W. Spoede (1991). *Cost Accounting: Managerial Use of Accounting Data*, PWS-Kent Publishing Company, Boston, Massachusetts.

COST ANALYSIS FOR PURCHASE NEGOTIATION

Robert B. Handfield

Michigan State University, Michigan, USA

ABSTRACT: For many years, purchasing professionals were accused of being too "price-conscious" in their dealings with suppliers, (at the expense of quality, delivery, technology, and service.) As organizations seek to achieve "low cost producer" status, however, the price paid for materials becomes increasingly important. Because the cost of materials in the majority of manufacturing enterprises is generally over 50% of the total cost of goods sold, price is nevertheless an important issue in negotiations with key suppliers. In this article, we present some key practices in the area of managing material costs jointly with suppliers. The approaches discussed here seek to find ways to reduce "non-value-added costs" considered part of overhead, while leaving suppliers' profit margins relatively intact. We also present some recent innovative approaches to "Total Cost Management" being applied by organizations, which seeks to assess the cost of non-conformance that is part of the total cost of doing business with a supplier.

DESCRIPTION: A major responsibility of purchasing is to make sure that the price paid for an item is fair and reasonable. (Monczka, Trent, and Handfield, 1997). How a purchasing professional evaluates a supplier's cost versus the price the buyer pays is an on-going challenge within many industries. The need to control costs requires that a buyer focus on the costs associated with the supplier producing an item or service versus simply analyzing final price. In fact, innovative pricing approaches require the identification of seller's cost before a buyer and seller even agree on final price.

The ultimate goal of the cost/price analysis process is to determine whether prices charged by a supplier are fair to both the purchaser and the supplier. A key part of this process is that purchasers must understand the difference between price and cost. The basic equation that underlies this concept is:

PRICE = COST + PROFIT

Once the implications of this equation are understood, insights into a major part of the procurement function can be derived. Price analysis refers to the process of comparing supplier prices against external price benchmarks. In contrast, cost analysis is the process of analyzing each individual element of cost (i.e. material, labor hours and rates, overhead, general and administrative costs, and profit), assuming these are not confidential.

Before discussing cost analysis techniques, it is important to review and define the major types of costs.

Fixed Costs: Fixed costs are those that remain constant or stable over different levels of volume, at least during the short term. For a firm to stay in business over the long term, it must price at a level that covers fixed costs and make some contribution to profit. Examples of traditional fixed costs include property taxes, depreciation, salaries of some employees, insurance payments, interest on outstanding loans, etc.

Variable Costs: Variable costs change proportionally with output. In particular, variable costs vary directly and proportionally with the quantity of production. As production increases, the cost increases; as production decreases, so do the costs. Examples of variable costs include the materials used directly in production, some types of labor, transportation, material ordering costs, and packaging.

Semi-Variable Costs: These costs have a fixed and variable component (and are therefore often called mixed costs). For example, the base salary of an executive manager is a fixed expense while an earned bonus is a variable expense. Telephone expenses are another example of a semi-variable cost. Regardless of usage, a customer still receives a fixed charge each month for basic phone service. As long-distance usage increases, costs increase. A third example is a basic equipment rental or lease (such as a car rental or lease). A firm must pay the monthly payment regardless of volume. As equipment usage increases, however, so do the operating expenses associated with the equipment. A semi- variable cost is present at some level regardless of volume but also increases as volume or usage increases.

Total Cost: Total cost is the sum of the variable, fixed, and semivariable costs. The total costs increase as the volume of production or service increases, while the cost to produce each unit decreases. Specifically:

Total fixed costs do not change with volume
Unit fixed costs decrease as volume increases
Total variable costs increase with volume
Unit variable costs generally do not change with volume

STRATEGIC PERSPECTIVE: Firms in competitive industries are experiencing unrelenting pressure to achieve continuous cost reductions. Because manufacturers commit a large percentage of their sales dollar to purchased goods and services, purchasing has an opportunity to contribute to a firm's competitiveness through aggressive cost analysis. Cost analysis is the *formal review of the detailed components comprising a producer's cost structure*. Ideally, this analysis identifies the actual cost to produce an item so a buyer and seller can: (1) determine a fair and reasonable price; and (2) develop plans to achieve future cost reductions. Price analysis, on the other hand, *focuses simply on a seller's price with little or no consideration given to the actual cost of production*. It is important to differentiate when and how pricing and cost analysis tools and tactics should be employed. Although both techniques are related, they are driven by fundamentally different principles.

The difference between price and cost analysis is significant. Buyers have historically focused on a quoted price from a seller to make sure it was not excessive compared to the prices of other producers. Buyers conducted some type of price analysis for almost all purchased items. Price analysis consists of comparing quotations from a number of suppliers or using historical prices as a basis to evaluate current prices. Another price analysis technique involves comparisons with published catalog prices. Cost-volume curves, escalation tables, and wage-rate projections are among the price- analysis tools used to develop prices for products and services. These same tools can also be used by purchasers to help evaluate their suppliers' pricing structure and to some degree predict pricing changes and strategies (White, 1995). Unfortunately, price analysis typically ignores the actual cost components of a purchased price. As a result, a buyer usually becomes a price taker even when he or she has the ability to influence a seller's cost and price. For this reason, price analysis is used extensively when the buyer has little control over the elements affecting cost.

A buyer's ability to perform a detailed cost analysis for an item is a function of several variables. First, the relationship between a buyer and seller strongly influences the quality of a cost analysis. Suppliers who have a close relationship with the purchaser will show a greater willingness to share cost information. When a buyer estimates cost data from external sources, this affects data reliability. Secondly, the ability of a buyer to understand traditional and progressive approaches to cost analysis is crucial. Today, buyers must practice innovative approaches to identifying costs and establishing prices. As buyers and sellers develop longer-term relationships, a joint emphasis on cost becomes more likely.

IMPLEMENTATION: A major responsibility of purchasing is to make sure that the price paid for an item is fair and reasonable. How a purchasing professional evaluates a supplier's cost versus the actual price paid is an on-going challenge within many industries. The need to control costs requires that a buyer focus on the costs associated with producing an item or service versus simply analyzing final price.

A buyer can use a variety of techniques to analyze the overall cost structure of a supplier or of a supplier's particular production process. These techniques are important because price is a major factor in a purchase decision. A supplier must be cost competitive even though the supplier provides superior quality, delivery, and customer responsiveness. A purchaser analyzes costs for several important reasons. First, cost analysis allows a buyer to evaluate whether a quoted price is fair and reasonable. A buyer should perform cost analysis even with competitively bid items to ensure the quoted price does not provide excessive profit to a supplier. Second, cost analysis allows a buyer to identify the most efficient production level at a supplier for a given item. As we will illustrate, purchase volume should match a supplier's process capability. A mismatch between a buyer's requirements and a supplier's capability can create higher per unit costs. Third, detailed cost analysis can help identify high cost areas that can benefit from cooperative improvement efforts. Cost data is a major area for information sharing in a collaborative relationship between a buyer and seller.

The ability to perform a cost analysis is a direct function of the quality and availability of information. If a buyer and seller maintain a distant purchase relationship, cost data will be more difficult to identify due the lack of support from the seller. A buyer may have to: (1) use internal engineering estimates about costs to produce an item; (2) rely on historical experience and judgment to estimate costs; or (3) review public financial documents to identify key cost data about the seller. The third approach works best with smaller suppliers with limited product lines. Financial documents allow estimation of a supplier's overall cost structure. The drawback is that these documents do not provide much information about breakdown of cost by product or product line. Also, if a supplier is a privately held company, cost data becomes even more difficult to obtain or estimate.

Another approach that can help in obtaining required cost data is to require a detailed production cost breakdown when a seller submits a purchase quotation. A buyer's ability to enforce this requirement relates to the relative size of the purchaser compared to the supplier, the volume of the purchase, and the type of product being purchased. A buyer must also consider the reliability of self-reported cost data. Again, closer relationships make it easier to obtain cost information. A third option involves the joint sharing of cost information. Joint cost savings initiatives involve the sharing of cost information between buyers and suppliers. A cross-functional team composed of engineers and manufacturing personnel from both company may then meet to identify potential areas of the supplier's process (or buyer's requirements) that can potentially reduce costs. The team then "brainstorms" possible solutions for cost reduction. Possibilities may include changing specifications, reducing setup, providing advance forecasts to reduce inventory, training supplier personnel in quality improvement techniques, and many other possible solutions.

TOTAL COST OF OWNERSHIP: In assessing and evaluating supplier performance, many firms have now begun to implement a new approach to measurement which includes the cost of nonconformance to performance expectations. Total Cost of Ownership (TCO) systems are often developed based on initial models of commodity cost behavior which are subsequently enhanced and refined as new data is developed (Handfield, 1995). While the effort, time and resources required to identify these costs are often enormous, the payback in terms of better decision-making, cost reduction, and improved time-based performance often outweigh them for particularly important critical items.

Many leading firms such as Eastman Kodak, General Motors, Sematech, and Intel are using TCO as a basis for leveraging information and assumptions regarding competitive and supply base cost performance. The philosophy underlying TCO is that organizations must understand the total cost of using critical suppliers in order to evaluate alternative sourcing strategies and maximize value added by the purchasing function.

However, the TCO concept extends beyond purchasing. Moreover, TCO can be viewed as an extension of cost-driven accounting systems such as Activity Based Costing (ABC), which seeks to allocate costs to appropriate cost drivers according to their value-added component. In this respect, TCO extends ABC to not only external sources of supply, but also to other value-adding activities both upstream and downstream from the organization. Armed with better information about costs, such systems allow managers to make better decisions regarding procurement strategy, and market and pricing strategies. New and relevant performance metrics, which provide true assessments of total cost can be developed to permit an integrated approach to supply chain management.

In order to develop a strategic business unit (SBU)-specific TCO model, purchasing managers must have a framework within which to work, as well as a cohesive decision support system for implementation. Such a system can help change the way managers think about measuring different attributes of supply chain activities, and can help focus their attention on metrics which reduce total cost, improve leadtime performance, and increase profits.

The steps required to implement TCO are as follows:

1. Identify the need.
2. Determine the purchase of interest.
3. Form a TCO development team: A team approach utilizes the expertise of different functional areas. In working with suppliers and members of the financial and marketing department, greater insights into cost components can be developed.
4. Identification of relevant costs: At this stage, the team can begin to identify and select those cost "drivers" which are believed to have the greatest impact on total cost.
5. High level design: Documentation of a conceptual model of TCO to facilitate analysis of numerous commodities. The design should identify data structures, user interfaces, and solution algorithms.
6. Detailed design: Preparation of specifications from the high level design which are sufficient for the actual programming effort.
7. Application, development and testing: Coding, testing, and verifying the model. The output of this stage is a working decision support tool with tested logic and user interfaces.
8. Testing and implementation of the model: A pilot test of the TCO model is necessary to determine whether the data is accurate, and also provide a real world test of the assumed data components and solution techniques.
9. Fine tuning: modification and refinement of data structures, equations, and an assessment of "lessons" learned is an important step in deriving a better cost model.
10. Linkage of TCO to other systems: The final step involves linking TCO modules to other financial and accounting reporting systems. Ideally, TCO should be a "visible system" which is accessible to all functions within the organization. This process is clearly a long-term goal.

Having developed a TCO system, there are a number of ways that such a system can be used:

- Internal costing: to create "what if" scenarios to conduct reverse cost analysis. For example, given an end cost, the system could breakdown this cost into labor costs, material, transportation, and overhead, as well as more detailed sub-categories. The tool would permit in-depth analysis of these costs, and allow assessment of value-added vs. non-value-added costs.
- Competitive analysis: the tool would permit the user to test assumptions based on the information he is able to secure on competitor's cost structures. For instance, with information available from the public domain, such as selling price, labor pool wages, geographic transportation rates, and standard material costs, the user could overlay this information onto an internal cost structure for the competing product, thus revealing the relative cost advantages/disadvantages in a particular market. An evolutionary model which provides a structure within which to accumulate information over time could then be developed.
- Supplier evaluation and negotiation: given a supplier's cost, the tool would permit the user to break down the supplier's costs and assess relative profit margins and performance (Monczka and Trecha, 1988). This tool could then be an important driver for improved negotiating leverage, particularly in long term contract and price discussions. As more information is obtained from the supplier or suppliers, the cost model could be updated and refined over time. Other uses in this category include target pricing for new product sourcing requirements, and the estimation of the effects of learning curves on total cost.

Total cost of ownership models can be as complex as the user desire them to be. A framework for development of such a tool is shown in Figures 1 and 2. Note that there is both an internal and external component to the system. The external component is used primarily for modeling the *supplier's price* and breaking it down into it's different components. The internal component is used for measuring the *total cost of using this particular supplier*, of which the supplier's price is only one component. This type of analysis can be very revealing. For instance, one manager using a TCO approach found that a supplier with a high price but excellent quality had a lower total cost of

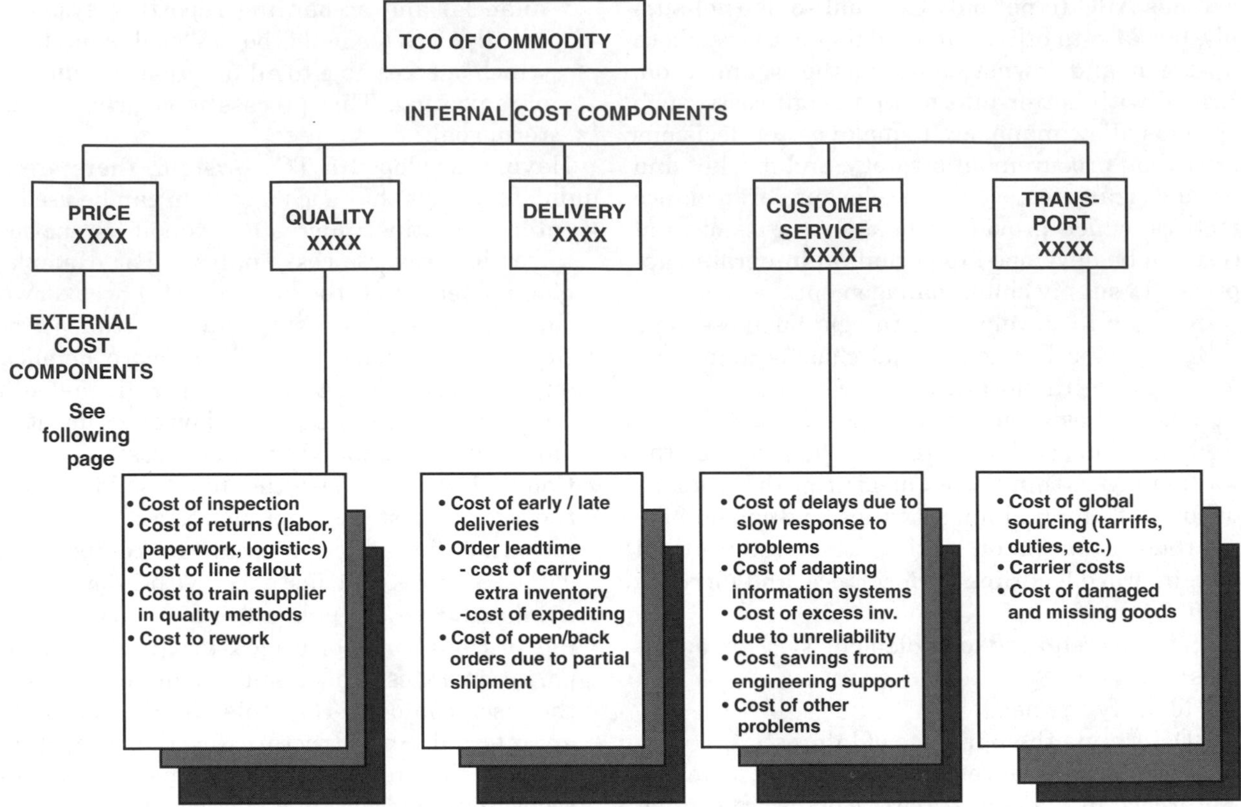

Fig. 1. Total cost of ownership (TCO) model internal components.

ownership than the other suppliers quoting a lower price.

The types of costs shown in Figures 1 and 2 can be extended to cover several commodities, or several suppliers. It can also be extended downward. For example, under materials cost, a set of subcategories might include:

- associated raw material costs
- usage rates and scrap rates
- tariff and transportation costs (global vs. domestic sourcing)
- inventory costs (for the assessment of JIT procurement strategies).

The evaluation system could therefore act both as an analysis engine, which allows for interactions which may exist between the categories of costs, and a structure for data collection by the purchasing manager. The real potential of such systems is that they allow the user to test assumptions regarding uncertain information. This can lead to better decisions regarding total cost and the elimination of non-value-added activities within the supply chain.

Collective Wisdom: Traditional cost accounting systems do not support the growing need for operational and product cost data (Kaplan, 1988). A firm must focus on product and process costs to drive out waste from its system. Also, dramatic changes in manufacturing technology (i.e., JIT purchasing and manufacturing) has presented cost accountants with new challenges. Historically, the primary function of a firm's cost accounting system was to value inventory for financial reporting requirements. The system did not assign costs (particularly overhead costs) to a unique operation or individual product. As a result, the true cost to produce an item is often unknown or reported inaccurately. The shortcomings of current accounting cost systems have become increasingly obvious within purchasing. For example, short comings become evident during attempts to develop total cost of ownership systems, when pursuing cost-based pricing with suppliers, or when analyzing internal cost components during make or buy analysis. A modern firm requires a cost accounting system(s) with the following three critical capabilities (Kaplan, 1988):

- Inventory valuation for financial and tax statements by allocating periodic production costs between goods sold and goods in stock
- Operational control by providing feedback on the amount of resources consumed in a timely manner
- Cost measurement at the individual product level

Modern manufacturing firms require more than one type of accounting and cost reporting system.

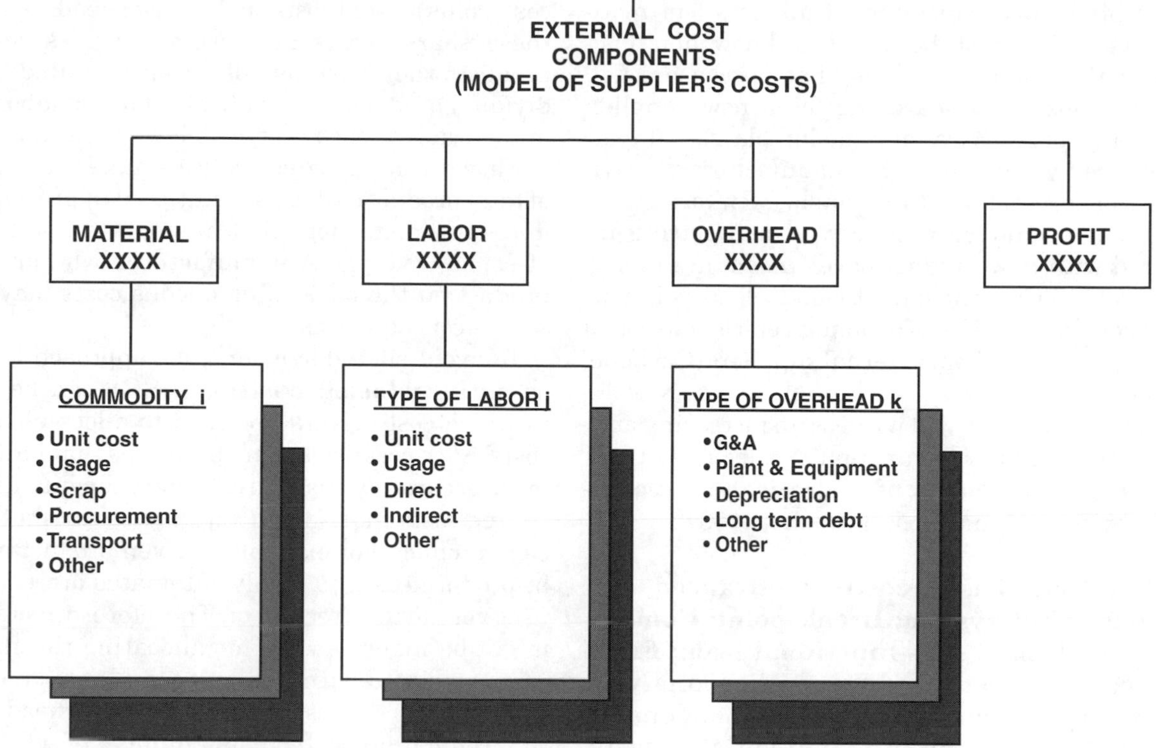

Fig. 2. Total cost of ownership (TCO) model external components. Note: Other cost classifications are possible.

Advanced cost management systems (ACMS) refers to recently developed quantitative cost analysis approaches supporting operational and process control, product related decisions, and managerial performance evaluation. Four types of systems comprise advanced cost management systems: (1) activity-based costing systems; (2) multiple performance measurement systems; (3) competitor cost analysis systems; and (4) product life-cycle cost management systems.

Activity-Based Costing Systems focus on the specific activities required to produce a product. Each product's usage of specific activities during the manufacturing process serves as the basis of the product's cost. The allocation bases, or cost drivers, are direct measures of the activities performed during the production of an item. The basis of the cost allocation approach is that products do not directly consume resources – they consume activities during production. Each activity has a cost associated with it that can be allocated directly to a product. An activity-based costing system, (1) provides an understanding of the overhead or indirect costs attributable to a specific item; (2) controls and manages costs by understanding the events and activities that create costs; and (3) includes all relevant costs in the decision making process.

CASES: Camtronics Medical Systems has developed an innovative cost analysis method, known as "should cost," to determine a supplier's cost. Camtronics obtains an evaluation unit from its supplier, tests it for systems compatibility, verifies materials, and then applies estimated overhead and margins. Analysts estimate a final cost and use this cost during negotiation. This approach also gives Camtronics a good understanding of the product from a manufacturing and service standpoint. A "should cost" analysis enabled Camtronics to realize a $10,000 per unit savings from a supplier's list price for a piece of electronic equipment. This resulted in a total savings of $600,000 over the 60 units purchased (Miller, 1995).

Similarly, today's auto industry buyer is focusing on the cost to manufacture a part instead of simply purchase price. This shift recognizes that price is simply an outcome affected by the *process* that produces an item. The focus on cost requires purchasing and supplier involvement earlier in the product development process. Early design phases of product development, for example, commit more than 80% of a final product's cost. By working with engineers and utilizing suppliers' expertise, purchasers can reduce product development time and keep costs low. Suppliers now propose designs according to a set of parameters or guidelines. The purchaser evaluates these design ideas, and then selects the final supplier. Suppliers are selected based on experience and the merits of their design proposal. Selected suppliers receive contracts covering the life of the product and work closely with the design team to develop final product specifications. Suppliers feel motivated to reduce costs because they know their business will be long term and that purchasers will not be constantly searching for a new supplier. At Honda of America, purchasing plays an important role by managing the manufacturing activities that are external to Honda. Monitoring and improving supplier productivity and getting involved with new product development are important roles for purchasers at Honda. This is important considering 80% of a Honda vehicle consists of parts purchased from external suppliers (Carbone, 1995).

In the future, we will witness an increasing ability of the purchasing function to effectively track total costs as a function of innovative cost accounting systems (Monczka and Trecha, 1988).

See **Activity-based costing; Advanced cost management system; Break point; Competitive analysis; Cross-functional team; Fixed costs; Low cost producer; Price analysis; Semi-variable costs; Total cost management; Total cost of ownership (TC); Variable costs.**

References

[1] Burt, D.N., Norquist, W., and Anklesaria, J. (1990). *Zero Base Pricing: Achieving World Class Competitiveness Through Reduced All-in-Cost*, Probus Publishing, Chicago.

[2] Carbone, J. (1995). "Lessons from Detroit," *Purchasing* (October 19), pp. 38–42.

[3] Hirst, Mark (1990). "Accounting education and advanced cost management systems," *Australian Accountant*, 60(11), 59–65.

[4] Kaplan, R.S. (1988). "One Cost System Isn't Enough," *Harvard Business Review*, January/February, 61–68.

[5] Miller, J. (1995). "Innovative cost savings," *NAPM Insights*, (September), pp. 30–32.

[6] Monczka, R., Trent, R. and Handfield, R. (1997). *Purchasing and Supply Chain Management*, Cincinnati, OH: Southwestern Publishing, College Division, Cincinnati.

[7] Monczka, R., and Trecha, S. (1988). "Cost-based supplier performance evaluation," *International Journal of Purchasing and Materials Management* (Spring), pp. 2–7.

[8] Schmidgall, R.S. (1986). *Managerial Accounting*, The Educational Institute, East Lansing.

[9] White, Jeffrey, "Learning Pricing Moves and Strategies," *NAPM Insights*, September 1995, pp. 16–19.

COST DRIVERS

The cost of producing a product has been traditionally attributed to direct labor and materials costs, plus shared costs such as overhead. Because these shared costs are incurred across several products they have usually been allocated to individual products as a multiple of direct labor cost attributable to each product. However, as direct labor has become progressively a smaller proportion of total product cost, this traditional approach produces cost distortions that misrepresent the costs of some products. A manufacturer who prices a product on the basis of erroneous costs may face severe consequences.

To avoid such distortions, an approach known as **activity-based costing** (ABC), can be used in which cost drivers are used to allocate shared costs. Such cost drivers are activities that indicate, more accurately than direct labor, the quantity of shared costs that should actually be attributed to each product. For example, a given product might be produced using a highly automated process that uses very little direct labor. Therefore, direct labor might be inappropriate for allocating the shared cost associated with this process. However, setup time for this process might be an overhead item that could be allocated using number of setups for each product as the cost driver.

In other words, cost drivers are shared activities that incur costs. Examples of cost drivers in manufacturing include order processing, issuing of engineering change orders, changing of production schedules, and production stoppage to change machine settings.

See **Activity-based costing; Activity-based management; Direct labor costs; Direct labor hours; Performance measurement in manufacturing; Shared costs; Target costing.**

COST LEADERSHIP STRATEGY

Cost leadership is one of three *generic* business strategies discussed by Porter in his well-known book, *Competitive Strategy* (1980). Essentially, a firm that follows a cost leadership strategy attempts to earn higher returns and competitive advantages through offering products or services at the lowest prices in the industry. Cost leadership requires the vigorous pursuit of cost minimization techniques. Costs may be reduced through improved operating efficiencies, production learning or scale economies, unique access to raw material, or special relationships with suppliers, distributors, or customers. Cost leaders are often vertically integrated or integrated into high value added, proprietary components and services. This capability allows them to be the most efficient processors in at least one stage of the value-added chain. Furthermore, a cost leader firm frequently owns high relative market share, enabling various scale advantages in purchasing or manufacturing due to large scale operations.

See **Manufacturing strategy.**

Reference

[1] Porter, M.E. (1980), *Competitive Strategy*, The Free Press: New York.

COST MANAGEMENT

It is the analysis and management of activities (and hence costs) necessary to make a product or deliver a service. The focus of cost management is to strengthen and increase activities that add value to the customer and reduce activities that do not.

See **Target costing.**

COST OBJECTS

In the language of activity-based costing (ABC), cost objects are the core aspects of the business, which serve as revenue-generating sources. Ideally, all activities within the organization should be structured directly or indirectly to support cost objects. Examples include products, customers, employees, sales outlets, sales regions, product components, etc. In an ABC accounting system, cost objects are identified for cost assignment.

See **Activity-based cost management; Activity-based costing; Activity-based costing: An evaluation.**

COST OF CAPITAL

See **Capital investment in advanced manufacturing technology; Robot selection.**

COST OF EXTERNAL FAILURE

External failure costs are one component of the cost of quality, and are incurred if a defective product reaches the customer and fails during use. The most common components of this cost are warranty work and returns. However, lawsuits from customers may also be a component. Some individuals also believe that a measure of the cost of lost goodwill should be included as a component of external failure cost. In truth, it may be difficult to accurately estimate the real cost of external failures.

Reducing external failure costs usually depends on increased prevention of defects and appraisal of quality during manufacture. Both of these may involve additional costs, but it is generally agreed that any increase in prevention and appraisal costs will be more than offset by decreases in failure costs. Although companies have often believed that these cost tradeoffs were subject to diminishing returns, and that total costs would be minimized at some non-zero level of defects, several individuals such as Crosby (1979) have argued that total costs are minimized only when zero defects are produced.

See **Cost of internal failure; Cost of quality; Performance measurement in manufacturing; Quality assurance; Total quality management.**

Reference

[1] Crosby, P.B. (1979). *Quality is Free*. McGraw-Hill, New York.

COST OF INTERNAL FAILURE

Cost of internal failure is one of the costs of ensuring product quality, and is associated with defects that are found at the manufacturer's site before the product reaches the customer. These costs are associated with either reworking the product or component to correct the defect, or scrapping that item if rework is not possible. Such costs can actually be quite substantial, although they are often overlooked because they usually are erroneously treated as a normal part of the manufacturing cost. Some companies have found that, when they separated out costs of internal failure, such costs were a significant percentage of cost of goods sold. A company can often bring about a large increase in profits just by reducing internal failure costs.

Reduction of internal failure costs rests on defect or error prevention, which generally requires increased employee training or the redesign of processes or products. Of course, such changes will incur some additional costs, but those costs are expected to be offset by reductions in failure costs.

See **Cost of external failure; Cost of quality; Performance measurement in manufacturing; Total quality management.**

COST OF QUALITY

As quality has become more important in the competitive environment, companies have searched for ways to capture the cost of quality as part of their traditional financial accounting systems. Because those systems use monetary values, and focus on cost and profits, the measurement of quality must be in similar terms. Thus, a cost of quality approach has evolved. All the costs associated with quality are divided into four categories: prevention, appraisal, internal failure, and external failure.

Prevention costs are associated with activities, such as employee training, that are designed to prevent defects from occurring. **Appraisal costs** include inspection and testing of the product. **Internal failure costs** are associated with defects that are found before the product reaches the customer. These costs usually include the cost of scrapping defective parts or products that cannot be corrected, or reworking ones that can be salvaged. **External failure costs** are incurred if a defective product reaches the customer and fails in during use. Such costs usually include warranty and return costs, although costs associated with lawsuits for damages, and even lost goodwill, may also be part of external failure costs.

See **Cost of external failure; Performance measurement in manufacturing; Quality assurance; Total quality management.**

COST PER BOTTLENECK MINUTE RATIO (C/PB)

The Cost per Bottleneck Minute (C/PB) ratio is one of a number of performance measures used in the implementation of the Theory of Constraints (TOC). The calculation for the C/PB ratio is Total Facility Cost/Bottleneck Capacity (in minutes). This ratio is useful in a complex production environment with multiple manufacturing facilities, each with its own bottleneck process (Galloway and Waldron, 1988, 1999).

See **Accounting system implications of TOC; Primary ratio; Theory of constraints; Throughput accounting ratio.**

References

[1] Galloway, D. and Waldron, D. (1988). "Throughput Accounting: Part 2 – ranking products profitably," *Management Accounting*, (UK) December, pp. 34–35.

[2] Galloway. D. and Waldron, D. (1999) "Throughput Accounting: Part 3 – a better way to control labour costs," *Management Accounting* (UK), January, pp. 34–35.

COST PLANNING

It is a form of financial planning of estimated results of future operations; frequently used to help control future operations.

See **Target costing.**

COST-BASED PRICING

The pricing of a product that is computed from the total cost of making a product plus a desired profit is called cost-based pricing.

See **Target costing.**

COST-OF-GOODS-SOLD

Cost-of-goods-sold is an accounting measure, which includes the cost associated with manufactured goods. It includes the direct costs of materials, labor and allocated overhead. It does not include markups or discounts.

See **Activity-based cost management; Activity-based costing; Target costing.**

COUNCIL OF LOGISTICS MANAGEMENT

The Council of Logistics Management (CLM) is the premier professional organization for logistics practitioners in the United States. CLM provides leadership in developing, defining, understanding, and enhancing the logistics process on a worldwide basis and provides a forum for the exchange of concepts and best practices among logistics professionals. CLM supports research that advances knowledge and leads to enhanced customer value and supply chain performance. The organization also provides education and career development programs to build career opportunities in logistics management. CLM currently has more than

15,000 members worldwide, and holds an annual conference each year.

See **Agile logistics (enterprise logistics); Professional associations for manufacturing professionals.**

CRITICAL PATH METHOD (CPM)

Considering a project as a network of activities assists with scheduling and execution. It is necessary to order the activities since some may have to be completed before others may be started; other, independent activities may be carried out concurrently. The project schedule is represented as a network, following one of two conventions: (1) activity-on-node (AON), in which the circles, or nodes, represent activities and the arrows between the nodes represent the sequence of activities in the schedule, (2) activity-on-arrow (AOA), where the activities are placed on the arrows and the intervening nodes represent events in the project. Activities use the resources of time, money and labor; events do not (See *Project Management*). Typically, the AON notation is used in the Critical Path Method (CPM), whereas AOA is used in Program Evaluation and Review Technique (PERT) (Meredith, 1992).

PROJECT SCHEDULING TECHNIQUES: Once the network has been constructed, the schedule of the project can be outlined. There are several tools for scheduling projects. The two main approaches are PERT (Program Evaluation and Review Technique) and CPM (Critical Path Method). Both of these techniques appeared during the late 1950s. The techniques of CPM and PERT have become largely similar over the years. The chief difference between the two is that PERT uses probability principles to introduce the chance of variability into activity times. Computer software can assist in the scheduling of large-scale projects using PERT or CPM.

Application of the Critical Path Method

After planning the project in terms of activities and constructing a network, the time required for each activity is added to the network diagram. The times inputted to the network are estimates of the time it should take to complete activities. Network scheduling techniques depend upon realistic estimates of activity duration. Hence, the techniques must be supported by the knowledge of the project team. After the activity times have been added, the *critical path* is identified. This is the longest continuous chain of activities in the network, and is equal to the total duration of the project. This critical path is of importance in scheduling and rescheduling the project, because activities that are not on the critical path may be delayed or brought forward by some extent. This leeway on non-critical path activities is known as 'float' or 'slack.' During project execution, the schedule is monitored against progress and may be changed by: (1) altering the start or finish times of such activities, (2) expediting activities on the critical path by trading off time and the cost of resources, a process known as 'crashing'.

Hence, in scheduling a project, there are two types of activities to consider: those on the critical path, and those on other paths that have float. Activities on the critical path have no float. For the remaining activities in the project, it may be advantageous to delay the commencement of some in order to distribute resources more evenly. The amount of float for the activities not on the critical path can be calculated from the time estimates given on the network diagram. This requires finding the early start (ES), late start (LS), early finish (EF) and late finish (LF) times for each activity. These refer to the earliest or latest time that an activity can take place to not delay the total project, given that the activity being considered and its predecessors and successors take the expected amount of time.

For an activity of length t,

$$EF = ES + t \quad \text{and} \quad LS = LF - t.$$

Hence, float = LS − ES or LF − EF.

These ES, LS, EF and LF times are calculated by going through the network first from left to right (start to finish), then from right to left. These procedures are known as the forward pass and backward pass. An example of a project network, using AOA notation, is given in Figure 1.

Forward Pass: Assuming the early start time (ES) of the first activity is 0, then its $EF = ES + t_1$. The ES of the second activity in the network can be set to the EF of its predecessor, as there is only one preceding activity. If more than one predecessor of an activity n exists, then ES_n is set equal to the *latest* EF of the predecessors. This procedure continues until the final activity to the right. The early start and finish dates are then known. The EF of the final activity is the EF of the project.

Backward Pass: From the right of the network, set the LF of the final activity = EF of the project, the LS for each activity is $LF - t$. $LF_n = LS_{n+1}$ if there is only one successor activity, otherwise LS = the earliest LS of the successors.

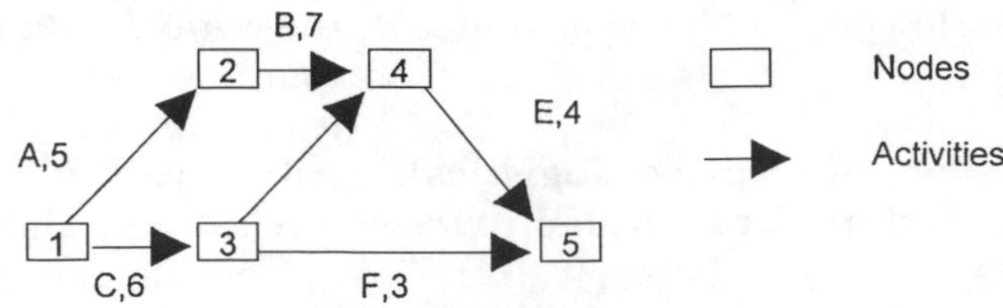

Given			Computed					
Activity	Preceded by	Activity duration	ES	LS	EF	LF	Slack or Float (LS-ES) or (LF-EF)	Critical Activities*
A	-	5	0	0	5	5	0	yes
B	A	7	5	5	12	12	0	yes
C	-	6	0	3	6	9	3	
D	C	3	6	9	9	12	3	-
E	B, D	4	12	12	16	16	0	yes
F	C	3	6	13	9	16	7	-

*Critical path is A, B, E; it has no slack.

Fig. 1. Critical path network.

Either of the float formulas then gives the float for each activity in the project. Critical path activities have no float.

Crashing a Project by Trading Time and Cost of Resources

CPM may be used to evaluate the effects of reducing activities by various lengths of time to reduce total project costs (due to reduced daily overhead) or to complete the project on time without vastly increased costs. Such calculations use time-cost curves and can involve complex computations for the costs of resources, especially in large projects or when large reductions in duration are being considered. Fortunately the process may be simplified by the availability of computer software for project management.

See **Activity on arrow network (AOA); Project evaluation and review technique (PERT); Project management.**

References

[1] Dilworth, J.B. (1992). *Operations Management: Design, Planning and Control for Manufacturing and Services*. McGraw-Hill.

[2] Meredith, J.R. (1992). *The Management of Operations: A Conceptual Emphasis*. Wiley, New York.

CRITICAL PATH OF THE BILL OF MATERIALS

A **bill of materials** (BOM) can be narrowly defined as an engineering document that specifies the ingredients or subordinate components required for the manufacture of a part, assembly or final product. Thus, a BOM is a product structure record showing all components, parts and subassemblies and quantities of each required for the manufacture one unit of the product. A **time-phased** BOM is a product structure added with the information about the unit production time for each item in the BOM. Thus, a time-phased BOM resembles a project network. The longest path of the network is the critical path for the project. The critical path of the BOM is the duration of cumulated manufacturing lead-time for making one unit. A simple example of a critical path of a BOM is shown below. In the diagram, the cumulated manufacturing lead time of end item A is 10. That is the critical path which consists of the production times of items E, B, and A.

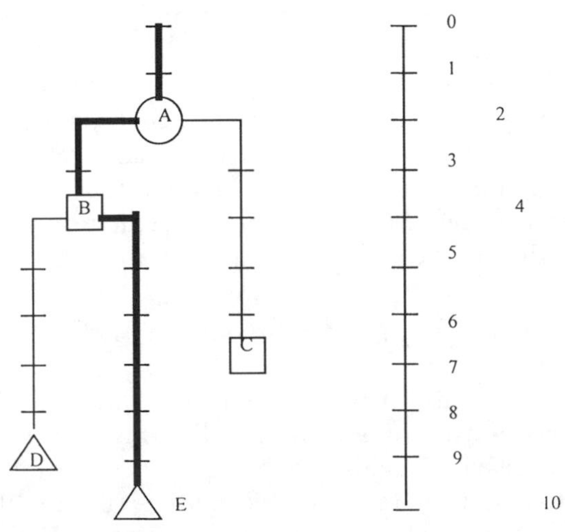

See **Bill of materials; Bill of materials structuring; Aggregate plans and master production schedule linkage.**

Reference

[1] Vollmann, T.E., W.L. Berry, and D.C. Whybark (1996). *Manufacturing Planning and Control Systems*, (Fourth Edition), Richard D. Irwin, Homewood, Illinois.

CRITICAL RATIO

The critical ratio (CR) is a rule used in priority sequencing of work waiting for processing at a work center. The CR priority index number is created by the formula (due date – today's date) divided by the total shop time remaining. If this rule is used for sequencing, the job with the lowest CR is scheduled first. A CR less than 1 indicate that the job will be late, and a CR greater than 1 indicates that the job can be completed ahead of the due date if no unexpected delay occurs.

See **Priority sequencing rules.**

CROSS TRAINING

Cross training refers to operator skill development such that each operator has the ability to perform multiple tasks. A workforce with more than one skill per worker is more flexible, less prone to product shortages when individual workers are absent. Such operators are generally less likely to fall a prey to boredom on the job due to repetitive work. Cross training is limited to a small number of jobs (less than ten) since learning a large number of jobs can be difficult. Job rotation is one method used to facilitate cross training, as is the use of on-the-job training such as mentor-apprentice relationships. Rewarding employees for learning new jobs is critical to the development of a motivated workforce. Studies have shown that cross training is most effective when designed in chains. For example, worker A can do jobs 1 and 2, worker B can do jobs 2 and 3, and worker C can do jobs 3 and 1. This chain allows workers to be shifted up and down the line as needed, while simultaneously keeping the number of jobs each worker must learn to a reasonable number.

See **Flexibility in manufacturing; Manufacturing cell; Manufacturing flexibility.**

CROSS-FUNCTIONAL INTEGRATION

Cross-functional integration refers to the close collaboration of personnel who represent different functional organizations through cross-functional teams or other organizational vehicles. A cross-functional team contains members who possess various essential perspectives and expertise required for reaching a common team goal.

No single functional organization is the repository of all the knowledge required to successfully complete activities such as designing a new product. Consequently, it is thought that activities and decisions rendered by cross-functional teams are superior to those rendered by individuals or groups who represent only one functional viewpoint. For example, a product design team that includes representatives from marketing, engineering, manufacturing, distribution, sales, purchasing, suppliers, customers and after-sales support functions is more likely to create a design that simultaneously meets the varied requirements of the different organizational functions and customers. An important benefit of such product development teams is that new products can be brought to market much faster, a definite competitive advantage in some markets. By working together, discussing different points of view, and cooperating, cross-functional teams can avoid biased or myopic decisions.

See **Concurrent engineering; Cross-functional team; New product development.**

CROSS-FUNCTIONAL TEAM

Cross-functional teams (CFT), also known as interfunctional teams, *consist of personnel from different functions and possibly even customers and suppliers brought together to complete specific tasks*. This includes tasks such as product design or supplier selection decisions on an "as needed" or continuous basis. The tasks assigned to cross-functional teams can also be very broad. For example, a team may have the responsibility of reducing purchased item cost. Teams often focus on complex tasks requiring members with diverse skills.

Firms have found that the increased use of cross-functional teams is critical for competitive success. When managed properly the team approach can break down communication barriers between functions. Improved and faster communication between entities represented in CFTs result in greater cooperation between departments and increased effectiveness.

Firms commit the time and energy to form teams to realize some specific performance benefits. The use of teams, however, is not without cost. Members on a cross-functional team often perform team assignments in addition to their regular duties. Team members usually meet on an "as needed" or regularly scheduled basis in addition to carrying out their normal assignments.

The cost of team maintenance has not slowed the trend towards the increased use of cross-functional teams. Cross-functional teams, when they meet their performance objectives, can provide benefits that far outweigh their costs. For example, the team approach can result in lower total cost for a new product. The team approach may also result in improved decision quality and easier implementation of a team's decision. One of the most advertised benefit of cross-functional product teams is the ability to develop and introduce new products quickly, which can be a very powerful competitive advantage. In some hi-tech industries, it is estimated that a new product price drops about five percent each month. At this rate, all things remaining equal, a product introduced three months after a competitor can command only 85% of the price charged three months ago by the competitor.

See **Concurrent engineering; Cost analysis for purchasing; Cross-functional integration; Product design.**

CUMULATIVE AVERAGE COST MODEL

This is one type of **learning curve analysis**, which projects or estimates the average cost assuming that the cost of manufacture declines progressively by a certain percentage as the number of units produced doubles. Typically, learning curve models are applied to direct costs of production.

See **Learning curve analysis; Learning rate.**

CURRENT RATIO

Current ratio is indicative of a company's ability to meet its short-term cash requirements, (including such expenses as accounts payable, short-term payroll costs, and short-term operating expenses). This ratio is calculated as:

Current ratio = Current assets/current liabilities.

The quick ratio is often used as a "red flag" while monitoring suppliers. A low ratio is an indication that the supplier is not being paid on time, or is overextended in terms of debt. A supplier with a current ratio less than 1.0 may be in danger of shutting down due to an inability to meet short term debts.

In cases when the supplier is a publicly traded company on the stock market, this and other financial ratios can be obtained from sources such as 10-K reports, Moody's Industrials, and Dunn and Bradstreet Reports available in most libraries. These ratios can also be calculated using information shown in income statements and balance sheets provided in company annual reports.

See **Supplier performance measurement.**

CUSTOMER ORDER DECOUPLING POINT

The **value chain** of mass-customization consists of marketing, development, production, and delivery in that order. The chain starts with the customer and ends with the customer. Customers can be anonymous as in the case of make-to-stock products, or the customer can be a specific individual as in the case of make-to-order production. In the value chain, marketing propels the development and development drives the production, and production leads the delivery to customers. The Customer-Order Decoupling Point (CODP) refers to the point in the value chain of mass customization at which a customer order triggers the production activities. All the activities ahead of the CODP are driven by forecast based planning rather than by customer orders.

See **Mass customization; Value chain.**

CUSTOMER RATIONALIZATION

Customer rationalization is founded on a basic and important proposition: not all customers are equal. Put another way, some customers are more important than others. This fundamental proposition simply suggests that managers evaluate closely the relationship that the firm maintains with current customers as well as the relationships that are likely to be established with future customers. If customers are classified on the basis of "**ABC Analysis**," "A" customers are those with the greatest importance to the firm. "A" customers are the firm's "customers of choice" and are the largest contributors to sales/profit. Because of their importance, the firm works closely with them to ensure that they are both satisfied and successful. "B" customers are of moderate importance to a firm because of their moderate contribution to sales/profit. "C" customers seldom merit special attention because their contribution to sales is very low and the firm cannot meet all the needs of these customers profitably.

For example, one grocery chain decided that printing and mailing of circulars with special advertisements to the general public was no longer profitable. Management realized that a large majority of its customers were "cherry picking" the loss leaders in the advertisement without contributing to additional sales. By developing a data base that tracked purchase behavior of key customers, the chain was able to back away from it general advertisements and more closely target its steady customers. In fact, specialized advertisements and coupons were mailed to best customers to reward them for their patronage. This approach eliminated much of the "cherry picking" while increasing the volume purchased by the chain's best customers. The bottom line is that most firms lack the resources to be all things to all customers and therefore need to build better, more profitable relationships with a fewer important customers.

See **ABC analysis; Supply chain management: Competing through integration.**

CUSTOMER SATISFACTION

Customer satisfaction should be the goal of every business. It is the judgment that a product or a service provided (or is providing) a pleasurable level of consumption-related fulfillment (Oliver, 1996: p. 11). Customer satisfaction, therefore, focuses on the customer and the customer's feeling of being fulfilled in a consumption experience. Fulfillment can only be judged with reference to some expected standard of performance, which forms a basis for comparison. While theories concerning the performance standards are numerous, the most commonly accepted theory is Expectancy Theory which holds that customers are satisfied when they perceive that a product or service performance is at least equal to their prior predictions (expectations) of how that product or service will perform. Customer satisfaction is the key to retaining customers and profitability.

See **Customer service, satisfaction and success; Customer service through value chain integration; Expectation theory.**

Reference

[1] Oliver, Richard L. (1996). *Satisfaction: A Behavioral Perspective on the Consumer*. McGraw-Hill, New York.

CUSTOMER SERVICE, SATISFACTION, AND SUCCESS

Stanley E. Fawcett

Brigham Young University, Utah, USA

M. Bixby Cooper

Michigan State University, Michigan, USA

DESCRIPTION: Meeting customers' real needs is the key to competitive success (Ohmae, 1988). Unfortunately, comparatively few companies excel at meeting and exceeding customer requirements. In fact, several recent articles in the business press have indicated that despite extensive efforts to provide ever higher levels of service, customers remain dissatisfied and demonstrate little loyalty (Stewart, 1995; Fierman, 1995; Stewart, 1997). The bottom line is that the concepts of customer service and satisfaction are frequently misunderstood and often poorly defined – even by leading companies. The result is that customers continue to complain about the poor service that they receive.

Creating true customer satisfaction and achieving sustained loyalty remains a persistent challenge. To meet this challenge, today's manufacturers need to reconsider their customer satisfaction strategies so that they can keep customers from defecting to competitors. The objective of any new customer-service paradigm must be to clearly define and describe the basic characteristics and likely outcomes of various service activities. As this is done, everyone within an organization comes to better understand exactly what needs to be done to profitably provide customer satisfaction. When everyone within the firm acts on this understanding, the firm is better able to meet customers' real needs. The ultimate goal is to provide truly superior products and services that will help key customers enhance their own competitiveness.

HISTORICAL PERSPECTIVE: For most of this century, firms have focused on enhancing the performance of internal service-related activities or processes with the expectation that internal excellence would somehow translate into satisfied customers. This approach frequently left customers disenchanted; however, a lack of adequate alternatives often meant that customers returned to sellers despite their dissatisfaction. Only as new alternatives arrived in the marketplace – in response to global competition – did companies begin to perceive the gaps that existed between the product/service packages that they offered and the desires of their customers. When the gaps became large enough, customers opted for competitors' products. Haas (1987) referred to this point of customer departure as a strategic breakpoint.

As competitive pressures increased during the 1980s, customers throughout the supply chain became more vocal in expressing their displeasure with existing service levels. As customers demanded higher levels of service, more and more companies began to pursue customer satisfaction. By 1990, the term customer satisfaction had all but replaced customer service in the trade press as well as in the managerial lexicon. Expectation theory emerged as the guiding principal in companies' efforts to provide higher levels of satisfaction (Zeithaml *et al.*, 1996). Customers were satisfied when their actual experience with a supplier's product/service package met apriori expectations. Any gap between expectations and experience led to dissatisfaction. The key to achieving high levels of satisfaction was to understand customers' needs so that the firm could develop and deliver distinctive product/service packages that would meet those needs.

Unfortunately, many initiatives designed to provide real customer satisfaction degenerated into efforts to convince customers that the company was truly interested in them and their level of satisfaction. One important aspect of communicating this concern to customers was to get them involved in sharing expectations as well as feelings about their experience with the company and its products. By eliciting customer feedback, companies generated expectations that the feedback would be used to improve both products and the buying experience. From the customers perspective, companies appeared to systematically ignore the feedback they had received. This resulted in cynical, dissatisfied, and increasingly vocal customers.

Further, even for those companies that assiduously sought to incorporate customer feedback into the way they conducted business, the challenge of meeting expectations was magnified by the fact that competitors were constantly evolving and improving. New products and services are constantly introduced in today's global marketplace, which has the effect of obsoleting "old" technologies and processes with little or no notice. To combat this challenge, many companies embarked on the quest to provide high levels of customer delight. The new mission was to exceed customer expectations. Yet, going beyond customer expectations can be both difficult and expensive. Many companies have found that exceeding customer expectations have been unprofitable (Bowersox *et al.*, 1995). Today, the focus of leading companies is to mesh internal excellence with the ability to deliver valued satisfactions. For these firms, customer success, the ability to improve the customer's competitive advantage, is the ultimate strategic objective.

The distinctive characteristics of customer service, satisfaction, and success are defined below:

- Customer Service: Customer service focuses on what the firm can do; that is, customer service measures internal service levels. The firm hopes that by performing well along these internal measures customers will be pleased.
- Customer Satisfaction: The measures of customer satisfaction are externally oriented. The firm seeks feedback from its key customers and uses this feedback to design its measurement system.
- Customer Success: Customer success focuses on helping customers succeed. Customer success requires an intimate understanding of the value-added potential of the entire supply chain. The firm uses this knowledge to help its customers meet the needs of customers further down the supply chain.

Strategic Perspective: The strategic benefit of customer success initiatives is that they move the firm closer to the position of indispensability. The critical component of these efforts is a knowledge of the entire supply chain, which can be used to help key customers improve their competitive position. For example, one manager pointed out that a particular supplier was practically indispensable because of the tremendous value provided by the supplier. The manager noted that the supplier knew more about the industry than his firm, and more importantly, actively shared that knowledge to help his firm be more successful. By promoting state-of-the-art strategic initiatives, sharing operational tools, and facilitating new merchandising techniques, the supplier had helped its customer compete more effectively against larger rivals. Loyalty to the supplier was a natural result. Customer success strategies recognize the importance of knowledge, and they use knowledge to increase switching costs through unique product/service offerings.

Of course, using knowledge to turn customers into winners can be a difficult process both in gaining the requisite knowledge and expertise as well as in translating that expertise into competitive advantage for the customer. The firm must obtain vital information about downstream requirements that can then be used to tailor the firm's product/service packages to deliver exceptional value. At the same time, the firm takes on the role of consultant to its customers, educating them in areas where they lack needed skills or knowledge. When a good relationship exists between the firm and its customers, this educational role is not only natural and straightforward but also genuinely appreciated. When close relationships do not exist, informing customers that they "are not always right" is difficult and can be highly uncomfortable. It is, however, better to turn down a customer request for a product that will not meet the customer's real needs than to deliver a product that will ultimately lead to dissatisfaction. The short-term struggle to explain how a product might be inadequate as well as what product/service package might better fulfill the customer's requirements is preferable to selling a product that dissatisfies the customer later. It is very important, however, not to use customer success as an excuse to needlessly oversell customers on higher margin products.

Finally, customer success strategies must recognize that the buyer/supplier relationship should yield competitive advantage, or profit, to both firms. Mutual benefit is particularly important given the experience of many firms that adopted "customer-delight-at-any-cost" approaches to keeping customers satisfied. These firms established selective relationships with key accounts that proved to be highly unprofitable because of the expense of meeting "excessive" service requests. The fact that many such relationships are cost ineffective can be discovered only if activity based costing systems are adopted. The bottom line is that being a of preferred supplier is only beneficial when it can be done at a reasonable profit.

Despite the fact that competitive leverage has shifted toward the final customer in recent years, losing money to keep customers happy is not a realistic long-term strategy.

IMPLEMENTATION: A key to the design and implementation of an appropriate customer service strategy is to understand the benefits and challenges inherent in different service strategies. Table 1 highlights some of the pertinent issues related to the implementation of service, satisfaction, and success strategies. The primary benefit of traditional customer service strategies is that they are relatively easy to implement since they are internally oriented. Objectives, measures, and procedures all originate and reside within the firm. This internal emphasis often meant that managers did not really understand customers' needs and desires. For example, Does a firm's internal measure of quality match the customer's definition of quality? It is not uncommon for a manufacturer and a customer to measure quality differently or to have different quality standards. One manufacturer was disappointed at the number of shipments that were returned as unacceptable by a sister division of the same firm. The two divisions used different quality standards in their internal operations. When firms do not understand

Table 1. Customer service, satisfaction, and success fundamentals.

Approach	Focus	Issues
Customer Service	Meet internally defined standards.	• Fail to understand what customers value. • Expend resources in wrong areas. • Measure performance inappropriately. • Fail to deliver more than mediocre service. • Operational emphasis leads to service gaps.
↓	↓	
Customer Satisfaction	Meet customer-driven expectations.	• Ignore operating realities while overlooking operating innovations. • Constant competitor benchmarking leads to product/service proliferation and inefficiency. • Maintain unprofitable relationships. • Vulnerable to new products and processes. • Focus on historical needs of customer does not help customer meet new market exigencies.
↓	↓	
Customer Success	Help customers meet their competitive requirements.	• Limited resources require that "customers of choice" be selected; that is, customer success is inherently a resource intensive strategy

customer requirements, they frequently "dissipate resources on things viewed as unimportant by customers." (Stock and Lambert, 1992). Traditional service strategies can easily lead to service gaps, which represent an invitation for competitors to enter the market and "steal" valuable customers (Parasuraman *et al.*, 1985).

The main benefit of customer satisfaction strategies is that they incorporate customer input into both product and process designs in order to remove service gaps. Typical approaches to gathering customer feedback include surveys, panels, focus groups, in-depth personal interviews, and shadowing. In recent years, senior executives have begun spending more time with key customers to gain a better understanding of customer requirements. Because these approaches are costly in terms of both time and money, relatively few companies systematically collect reliable and valid data from customers. Even fewer really incorporate the data that they do collect into product and process design decisions. Other pitfalls of satisfaction strategies is that they often lead to unnecessary product proliferation. A leading consumer products manufacturer recently asked shoppers to try and find a given product on the store shelves. Even when given a sample of the product to facilitate the search, many shoppers were unable to find the product and indicated that they were confused by the multitude of similar products on the shelf. The manufacturer decided that it needed to simplify its product lines. Customer satisfaction strategies can also lead to myopic, "me too" product/service strategies, preventing important innovation. Customer satisfaction strategies attempt to use information to bridge the gap between the firm and its customers (Stewart 1997).

The principal benefit of customer success strategies is that they are forward looking. That is, they consider the needs and requirements of the entire supply chain. This perspective enables a firm to actually help its customers become more competitive. Unfortunately, customer success strategies are inherently resource intensive and can only be effectively implemented with a few "key accounts" or "customers of choice." The challenge of implementation is twofold: first, to identify attractive opportunities where the firm's capabilities match the market needs of key customers and second, to coordinate the firm's value-added efforts so that the firm can efficiently deliver increased value to the selected customers. Even as management begins to evaluate market opportunities, it must consider marketing and operations implications to determine whether the needs of specific customers can really be met efficiently and effectively. This evaluation process requires that managers realize that not all customers are equal and that individual customers have different needs that require different resource commitments.

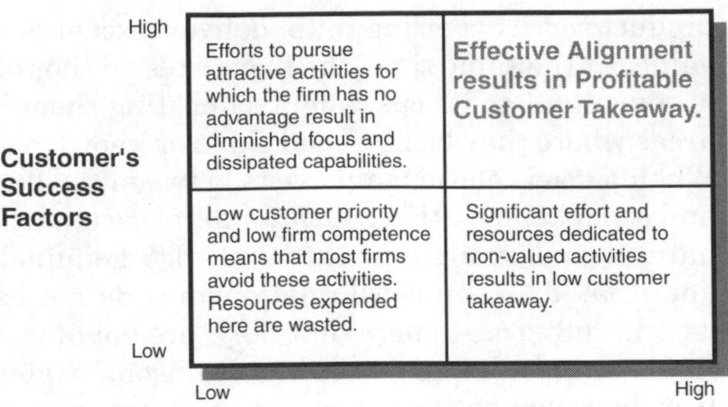

Fig. 1. The alignment matrix.

Figure 1 illustrates a matrix alignment approach that is useful in helping the firm identify opportunities where a customer success strategy would be appropriate. The vertical axis represents the important success factors of potential customers of choice. These success factors are determined by evaluating the entire supply chain to determine what imperatives drive the success of the customer. The key is to find what does the customer's customer define as important? For firm's at the end of the supply chain, the question translates into, how can we improve the operations or lives of key customers so that they feel not just satisfied but feel more successful. The horizontal axis focuses on the firm's distinctive capabilities; that is, what does the firm do particularly well? In the matrix, where the customer's success factors and the firm's distinctive capabilities come together, an opportunity for the successful implementation of a customer success strategy exists.

Having identified key customers and what constitutes real value to them, the task is to use the diverse value-added activities of the firm to profitably create and deliver the sought after value. To do this, "value-added" must be clearly defined and communicated, and each part of the organization must understand: (1) its role in the value-added process; (2) the value that is added by other areas of the firm; and (3) how decisions in one area of the firm impact the other areas as well as the overall value-added process. The following questions can help the firm explicitly assess where it is in its quest to achieve customer success. These questions also help direct implementation efforts.

- Does the firm still operate from a service orientation that is driven by a set of internal measures of activity performance?
- If the firm has adopted a true customer satisfaction orientation, are the systems and measures

in place to really know what customers want and need?

- Does the firm really understand the supply chain and the competitive needs of the different members of the supply chain? What imperatives are changing the competitive dynamics of the supply chain? Is the firm providing value to other supply chain members that will enhance the overall competitiveness of the supply chain? Can the firm's product/service package be provided by another member of the supply chain in a way that threatens the firm's active participation through role shifting?
- Does the firm know how profitable different customer relationships are? Where is the value created that leads to profitable customer success? How could communication and coordination be improved to deliver more profitable customer value?

TECHNOLOGY PERSPECTIVE: The impact of customer success initiatives depends heavily on the attitude of management and the training and measurement systems that are put into place. Technology is not as critical or central to these programs. However, the more the firm moves toward a success-driven strategy, the greater the opportunity to utilize technology in the development and delivery of tailored product/service packages. For example, a primary impediment to more effective customer satisfaction initiatives is the inability to accurately and inexpensively capture customer perceptions in a timely manner. Information technologies such as electronic data interchange (EDI) provide a direct, low-cost, and real-time vehicle for sharing customer satisfaction information both in terms of expectations as well as actual performance levels. Internet "chat rooms" and e-mail can also facilitate the low-cost, real- time sharing of information between a firm and its key customers. The key to getting real value from these information technologies is to motivate people to use them on a frequent and consistent basis. These technologies only work when the proper incentives and training are provided early in the implementation of the customer success strategy.

Information technologies can also play an important role in the design and management of supply chain alliances, providing the firm with a better opportunity to analyze and understand supply chain dynamics. The knowledge gained from this analysis is the key to accurately identifying customer success factors. Moreover, better enterprise resource planning (ERP) and inventory tracking systems help the firm reduce cycle times and make more accurate delivery promises. That is, when the managers knows exactly what products are available in inventory as well as what current production schedules look like, they can make realistic delivery promises based on information and not the guesswork of a sales person. Companies such as IBM, Hershey, and Nuskin have all been able to improve customer responsiveness through the use of ERP systems. Hershey has also been able to standardize its customer order and invoicing procedures via its ERP system to provide more customer friendly interactions. Bar coding and satellite positioning systems also provide more accurate information regarding expected and actual delivery times. These technologies allow firms to track, expedite, and reroute shipments in response to customer inquiries and requests. Each of these technological advances allows the firm to be more creative in the products and services that it offers. They also help the firm get closer to customers than ever before. This connectivity helps to eliminate surprises and allows better responsiveness when they do occur. In effect, technology is altering both the product/service packages that can now be offered to important customers and the relationship with them.

TIMING AND LOCATION: The foremost timing issue related to customer satisfaction strategies is the determination of which type of strategy a firm will employ with each customer in its client portfolio. Quite simply, service excellence is essential in today's intensely competitive global marketplace. Remember, any service gap is an open door for the competition, existing and potential, to enter the market and capture market share. Efforts to go beyond service excellence require greater resource intensity and should therefore be evaluated carefully with particular attention being paid to understanding customers' needs. The choice of a customer service, satisfaction, or success strategy depends on the following issues: customer expectations, customer leverage, the value of the relationship, the number and diversity of the customers, the firm's resource base, and the breadth of firm's product line. An "ABC" classification of customers might be useful. "A" customers would be likely candidates for implementation of a customer success strategy – the intensity of these relationships almost assures that a firm will lack the resources to offer dedicated, tailored product/services to more than a few non-"A" customers. Likewise, most "B" customers would be potential targets for focused satisfaction strategies where customer input is actively sought and utilized to meet expressed customer requirements. Finally, the sheer number of "C" customers generally means providing high levels of service excellence is the best a

Table 2. Location of decision making authority.

Decision	Activities/Processes	Functional Areas
Relationship	• Defines firm's customer strategy and allocates resources	• Top Management
	• Identify and evaluate customers	• Marketing leads cross-functional effort
	• Understand supply chain imperatives	• Marketing, Operations, Logistics, Purchasing
	• Define customer success factors	• Marketing leads cross-functional effort
	• Communicate success factors throughout the firm	• Top management
	• Select customers of choice	• Top Management with cross-functional input
	• Build relationships	• Top Management with cross-functional help
Competencies	• Recognize the firm's distinctive capabilities	• Top Mangement leads cross-functional effort
	• Balance market pull and technology push pressures in developing distinctive capabilities	• Engineering provides technical expertise while Marketing provides market awareness
	• Focus on selected capabilities and respective processes	• Operations takes lead; purchasing supports
	• Convert raw materials and component inputs into finished product/service package	• Operations creates the form
	• Make products and services available when customers want them	• Purchasing provides materials on time; Operations meets production due-dates; Logistics stages and moves products; and Marketing interacts with customer
	• Make products and services available where customers want them	• Logistics and Marketing

firm can do with constrained resources. As data capturing technologies improve and firms become more adept at using database mining software, "C" level customers may become candidates for more tailored satisfaction strategies.

Regarding location, the primary issue is who within the firm is responsible for making important decisions that will lead to the implementation of an appropriate and successful customer strategy. Table 2 identifies the primary decision areas, the key processes that must be performed, and the focus of responsibility for each process. The two major decision areas involve the determination of the type of relationship to be established and the competencies to be developed. Top management plays the most pervasive role in defining the firm's customer philosophy and making final decisions regarding the type of relationship to be established with each customer. Much of the customer evaluation process must be cross-functional; however, given most firms' organizational structures, marketing is likely to lead these efforts. The competency selection decision is led by top management with individual activities and processes directed by different functional areas. As a firm places more emphasis on customer satisfaction and success, the more integrated the management effort must be – customer success is by nature a cross-functional initiative.

RESULTS AND CASES: The rationale for devoting resources to achieve high levels of customer satisfaction has always been the belief that satisfaction and loyalty move in tandem – satisfied customers are loyal customers. Existing evidence, however, refutes this assumption for all but the most satisfied customers (Jones and Sasser, 1995). In fact, Xerox found that largely satisfied customers – those who marked a four on a five-point scale where five is equal to completely satisfied – are six times more likely to defect to the competition than those who are completely satisfied. Generally speaking, the reality is that customers in the 1990s are fickle and really do not trust the companies that promise to meet their needs. Only when a company truly excels at something that customers value can it expect to achieve some semblance of loyalty, which leads to profitable repeat business.

Figure 2 shows the relationship between satisfaction and loyalty. When a company's service offering is markedly lower than the competition, customers simply do not come back. Repeat

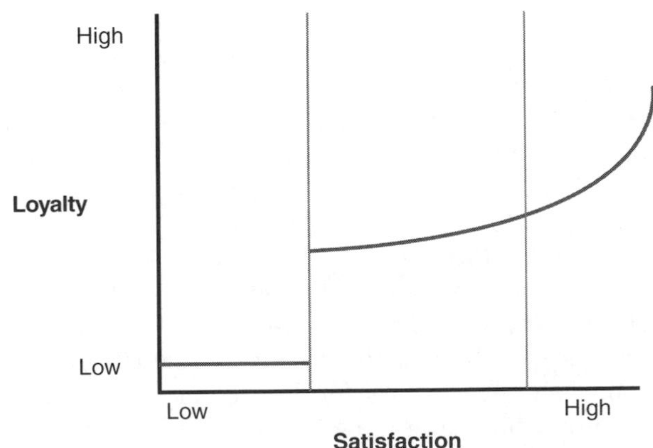

Fig. 2. The satisfaction/loyalty conundrum.

business and loyalty begin to appear as the service offering enters the zone where it is comparable to competitors' offerings. However, customers will defect if it is convenient to do so. Only when the firm achieves an intrinsically higher level of service or a truly distinctive product do customers exhibit strong loyalty. The challenge is that most products and services can be replicated relatively easily by the competition within a short period of time – frequently less than one year. Thus, most distinctive advantages are transitory if they are not constantly upgraded. Firms must learn to make their own product/service offerings obsolete with improved product/service offerings.

This predicament underlies the emergence of customer success as a viable strategy. As one CEO notes, "We turn our customers into winners. Their success is cash in our bank. Our customer is our most important partner in cooperation – his customer benefits from this as well" (Ginsburg and Miller, 1992). Unfortunately, relatively few companies are willing to invest the needed time, effort, brainpower, and capital necessary to develop truly distinctive capabilities and match them to customers' important success factors. In reality, among all firms, three out of four still operate using traditional customer service definitions – regardless of the management rhetoric concerning the importance of customers. Another 20 percent have truly implemented customer satisfaction programs that utilize a systematic approach to gathering and incorporating customer feedback. Only a very small percentage of today's firms understand that customer success is an investment in long-term profitability. The end result is that customer success companies achieve levels of loyalty and profitability that are generally envied by the competition.

The following case story illustrates the perspective and power of a well-implemented customer success strategy. Electro, a manufacturer of electrical components used in the assembly of consumer durable products such as household appliances and automobiles, adopted a customer success philosophy a little over three years ago. The firm's customer orientation was brought to life and communicated through the slogan, "Pride in helping customers compete!" The phrase was quickly disseminated throughout the company and can now be seen hanging on walls in reception areas, corporate offices, and throughout the firm's manufacturing facilities. More importantly, Electro provided the needed resources and training to make the slogan a reality. As a result, Electro is now a preferred supplier to the largest appliance manufacturer in the U.S. as well as to two of the largest automobile manufacturers in the world. One corporate official summarized Electro's philosophy as follows,

> "We feel it is no longer adequate to simply provide our customer good service since that has become the standard in our industry. It isn't even adequate to satisfy them if that means meeting their expectations since their past experience may lead them to expect certain problems to arise with their suppliers. Instead, we feel that the most appropriate thing for us to do is to do our best in making our customers better competitors in their industries. If they are more successful due to our ability to provide them with better products, more timely delivery, lower total costs, or whatever, then they will gain market share and grow. Of course, when they grow, we grow!"

COLLECTIVE WISDOM: All too often, customer service initiatives are viewed with tremendous skepticism by the firm's employees and customers alike. Employees consider such programs to be among top management's favorite "flavor-of-the-month" programs – if they can avoid making any big changes for a while, the initiative will go away, having been replaced by the next management fad. Customers, likewise, are inundated with statements that "customers are our reason for being" or "no one ever won an argument with a customer" while at the same time experiencing poor customer service time and time again. The evidence suggests that a relatively small set of impediments prevent the majority of firms from providing the high levels of customer service, satisfaction, and success that they desire. These impediments begin with a lack of true commitment from top management. Once real managerial commitment is in place, the other impediments, which follow below, can be addressed.

- Adversarial worker/manager relations that get translated to the company's products and customers.
- Inadequate training, especially with respect to how each job impacts customer success.
- Incongruent performance measures that do not reinforce appropriate customer success behavior.
- Lack of information flow to employees to help them learn; i.e., what worked and why!
- Compensation and reward systems that fail to recognize the customer relationship as critical.

As commitment to customers increases and as managers recognize and overcome the principal impediments to the implementation of an effective customer fulfillment strategy, their firms will be better positioned for long-term success. As noted by one executive, customer success is the new competitive imperative, "The only sure way to grow is to share in the growth of our customers. If we are the preferred supplier and they are their customer's choice, we can jointly identify breakthrough opportunities. Jointly, we can outperform the competition."

See **ABC classification; Customer satisfaction; Customer success; Customers of choice; Expectation theory; Preferred supplier; Product proliferation; Service orientation.**

References

[1] Bowersox, D.J., Calantone, R.J., Clinton, S.R., Closs, D.J., Cooper, M.B., Droge, C.L., Fawcett, S.E., Frankel, R., Frayer, D.J., Morash, E.A., Rinehart, L.M., and Schmitz, J.M. (1995). *World Class Logistics: The Challenge of Managing Continuous Change*. Council of Logistics Management, Oak Brook, IL.

[2] Fawcett, S.E. and Fawcett, S.A. (1995). "The Firm as a Value-Added System: Integrating Logistics, Operations, and Purchasing." *International Journal of Physical Distribution and Logistics Management*, 25(3), 24–42.

[3] Fierman, J. (1995). "Americans Can't Get No Satisfaction." *Fortune* (December 11), 186–194.

[4] Ginsburg, I. and Miller, N. (1992). "Value-Driven Management." *Business Horizons* (May-June), 23–27.

[5] Haas, E.A. (1987). "Breakthrough Manufacturing." *Harvard Business Review* (March-April), 75–81.

[6] Jones, T.O. and Sasser, W.E. Jr. (1995). "Why Satisfied Customers Defect." *Harvard Business Review* (November-December), 88–99.

[7] Ohmae, K. (1988). "Getting Back to Strategy." *Harvard Business Review*, 66(6), 149–156.

[8] Parasuraman, A., Zeithaml, V.A., and Berry, L.L. (1985). "A Conceptual Model of Service Quality and Its Implications for Future Research." *Journal of Marketing*, 49(4), 41–50.

[9] Stewart, T. (1995). "After all you've done for you customers, why are they still NOT HAPPY?" *Fortune* (December 11), 178–18.

[10] Stewart, T.A. (1997). "A Satisfied Customer Isn't Enough." *Fortune*, 136 (July 21), 112–113.

[11] Stock, J. and Lambert, D. (1992). "Becoming a "World Class" Company With Logistics Service Quality." *International Journal of Logistics Management*, 3(1), 73–80.

[12] Zeithaml, V.A., Berry, L.L., and Parasuraman, A. (1996). "The Behavioral Consequences of Service Quality." *Journal of Marketing*, 60, 31–46.

CUSTOMER SERVICE LEVEL

Customer service has many aspects; however manufacturers' service level is often measured in terms of the speed of delivery and inventory reliability. Speed of delivery is often measured by order cycle time, which is the elapsed time from customer order to order delivery. Speed of delivery is most important in make-to-order and assemble-to-order businesses. In contrast, inventory reliability is most important in make-to-stock environments. Inventory reliability is often measured by the fill rate, which is the percent of demand filled from stock. The fill rate is sometimes referred to as the customer service level. An alternative measure of the customer service level is the percent of orders shipped on time and complete. This customer service measure includes elements of both speed of delivery and inventory reliability.

It should be noted that customer service level may be measured in terms other than speed of delivery depending on the nature of the business and what is critical to the customer.

See **Make-to-order production; Make-to-stock production; Process industry scheduling; Safety lead time; Safety stock; Safety time.**

Reference

[1] Greene, J.H., ed. (1997). *Production and Inventory Control Handbook*, third Edition, McGraw Hill, New York.

CUSTOMER SERVICE THROUGH VALUE CHAIN INTEGRATION

Richard E. White

University of North Texas

John N. Pearson

Arizona State University

DESCRIPTION: Robert Hall and co-authors (1993a, 1993b, & 1993c) to describe a vision for globally competitive manufacturing organizations. These articles used examples of future manufacturing scenarios (three-day car & Sekisui's three-day house) and descriptive lists to paint a picture of the required characteristics for globally competitive manufacturers of the future. Generally, the characteristics included: increased flexibility, supplier partners, mass customization, customer interaction open communication, continuous improvement, and integrative relationships among all activities in the material pipeline. Several names were used throughout the series of articles to portray this abstract vision (i.e., agile manufacturing, virtual organizations, world-class, organizations, etc.). The last article in the series presented a new term to represent this vision, "total enterprise manufacturing".

Another approach to describing this vision of globally competitive future manufacturing organizations starts with Porter's (1985) value chain. Porter describes the value of the product in the marketplace is the vehicle that creates differentiation in contrast to competitors' products. The creation of effective product value involves not only developing effective value added activities, but also developing value in purchased inputs. In other words, even if a firm is effective in performing its internal activities (activities associated with the traditional factory floor), the high costs associated with purchased inputs may reduce the differentiation of a company's products from the competitors' products.

Value System. Porter's value system (1985) is much like Hall's total enterprise manufacturing; it provides a perspective of the expanded boundaries that must be considered to maximize an organization's performance. The value system is composed of a pipeline of businesses where each business in the value system has its own value chain. Porter provides generic descriptions of the activities that make up this larger stream of activities, that make up this larger stream of activities, the value system. For a manufacturer to gain and sustain competitive advantage, it must understand not only all the activities in its value chain, but, also how its value chain fits into the value system. This article discusses how the value chain integration is achieved through just-in-time (JIT) practices.

TOOLS FOR INTEGRATION:

JIT

Japanese manufacturers have demonstrated superior skills in applying value system perspective to the management of their organizations. The Japanese have been very successful in improving their productivity and global competitiveness by developing technologies oriented toward continuous improvement of whole value chain. This management approach is referred to by many names (i.e., world class, short cycle manufacturing, lean production, etc.), the name used in this paper is just-in-time (JIT) manufacturing. JIT is a holistic manufacturing system in which associated management practices are interrelated (Burnham, 1985; Schneider & Leatherman, 1992). White and Ruch (1990) identified ten management practices associated with JIT. These ten management practices include:

1. focused factory (FF),
2. group technology (GT),
3. reduction of setup times (SU),
4. total productive maintenance (TPM),
5. multi-function employees (MFE),
6. uniform workload (UW),
7. Kanban (KAN),
8. just-in-time purchasing (JITP),
9. total quality control (TQC), and
10. quality circles (QC).

Satisfying customer demand at each stage in the pipeline becomes the common goal for all employees in JIT manufacturing.

Value Chain Perspective

Porter (1985) categorizes the activities in the value chain into primary and support activities. The primary activities consist of inbound logistics, operations, outbound logistics, marketing and sales, and service activities. The support activities consist of procurement, technology development, human resource management, and firm infrastructure (see Figure 1).

The value chain and value system provide a framework for improving customer service. Integration of the primary and support activities

Primary Activities

Inbound Logistics
Operations
Outbound Logistics
Marketing and Sales
Service Activities

Support Activities

Procurement
Technology Development
Human Resource Management
Firm Infrastructure

Fig. 1. Value chain (Porter, 1985).

across the value chain can be achieved by implementing JIT systems.

Primary Activities. The inbound logistics activities are those activities involving receiving, storing, and disseminating inputs to the operations. The management practices of just-in-time purchasing (JITP) and Kanban (KAN) are important elements of JIT systems that can be used to integrate procurement and inbound logistics. Just-In-Time Purchasing (JITP) links the supplier to the shop floor by pulling needed materials from the supplier. The inbound logistics refers to the transportation of JIT purchased parts from both external and internal suppliers (units or divisions of the organization treated as suppliers that are physically located away from the main transformation process).

An organization in a global competitive environment must be concerned about reliable delivery of materials to its manufacturing transformation process; therefore, reliability (delivery of materials on time) of the transportation mode is a key criterion in mode selection for inbound logistics. Often, JIT purchasing activities are integrated with inbound logistics activities and milk runs (dedicated contract carriers with regular routes, typically motor carriers are used for short hauls) are commonly employed in JIT systems for transporting incoming materials from local suppliers.

Kanban (KAN) is a manual restocking system that evolved through refinement of the production system at Toyota Motor Corporation. Kanban is the visual sign, the indication, to a supplier (external and internal) of the need for more materials. The Kanban system tries to eliminate the traditional "push" system of material flow and develops a "pull" system where customer demand creates the initial pull to start material flow. For total system responsiveness, the total system must use Kanban throughout for maximum system flexibility.

Operations Activities: Operations activities include those activities that use machinery/equipment and their maintenance. Customer service is enhanced by the ability to process unanticipated demand, which requires system responsiveness (flexibility). JIT management practices that contribute to the system's flexibility include: reduced setup times (SU), group technology (GT), total productive maintenance (TPM), and uniform workload (UW). These practices develop greater efficiency by streamlining existing shop activities. Reduced setup times (SU) is an on-going program that continuously attempts to reduce setup time and costs. JIT approach does not accept traditional set up costs and carrying costs as given. By reducing set up times, the EOQ is reduced, thus making it more economical to produce in smaller and smaller batch sizes until it is economical to produce in a batch size of one. The less time is required for set ups, more time will be available for production.

Total Productive Maintenance (TPM) emphasizes effective utilization of machines. This JIT practice sets up and maintains a very vigorous and routine preventative maintenance and replacement program. Precise records maintained on the maintenance performed on the machines. This information with the production records of the machines allow for developing the routine and necessary preventative tasks. Total productive maintenance also involves getting the machine operator actively participating in minor machine maintenance functions. "The basic idea behind TPM is that production workers should take 'ownership' of their work area" (Turbide, 1993, p. 33).

Uniform Workload (UW) is a management practice that tries to stabilize and smooth the production workload through improved scheduling techniques. The variation in workload does not exceed the scheduled plan by more than plus or minus 10 percent during a 90-day period. Plans that exceed 90 days are routinely updated. In the short term, variations in demand would not change the number of different products produced each day. Variations in demand are satisfied by varying the frequency of the products within the daily product mix. Because the system is far more flexible, it does not require inventories to fulfill unexpected demand. The opportunities for improving efficiency derive from the repetitive work and the constant level of the workload. The concept of uniform workload when extends beyond the operations activities to all system activities results in system stability.

Outbound Logistics: Outbound logistics include those activities associated with collecting, storing, and physically distributing products to the market place. The distribution of materials to the market place is complex and involves multi-faceted decisions in both controlling and planning. Traditionally, these decisions often involve tradeoffs between costs and service. An organization's policy on customer service becomes an important criterion for the selection of transportation mode. Since customer service requirements usually vary over the life of a product, customer service policies should be derived from the firm's operating policies in relationship to product life cycles. As the product goes through the various stages of its life, e.g., introduction, growth, saturation and decline, decisions about mode selection moves from an emphasis on flexibility to handle unusual shipments to an overriding concern for selecting the mode for distribution based on cost. Reliability of the transportation mode also plays a significant role in selecting the mode in the later stages of the product life cycle.

Total Quality Control (TQC) or Total Quality Management (TQM) enables an organization to achieve its most important business objective; the objective to provide a quality product to the customer. With total quality control, system design efforts are centered around developing integrative mechanisms for solving and preventing quality problems, and thus, providing better customer service. Quality Circles (QC) play an important part in the continuous improvement effort.

Concurrent engineering is an example of an integrative mechanism used to improve organizational effectiveness in designing and developing new products. Today, multifunction teams are used in the product design and development phase. This team approach to new product management is also called "simultaneous engineering". This team approach emphasizes the simultaneous design of the product and production process; it brings together all relevant functions, such as design engineering, manufacturing, marketing (customers), purchasing (suppliers), quality, and accounting into an integrated team that simultaneously designs and develops new and existing products.

Conventional Service Activities: Customer service is often described as those activities that occur after production; this typically includes physical distribution activities. Customer service serves as the link that matches and integrates the supplier system with the customer system, i.e., the pipeline of material flow from the supplier to the manufacturer, through the factory and physical distribution activities. Since customer service is the driver of the total value system, early customer involvement is critical in the new competitive environment. Focus groups and other customer involvement activities provide mechanisms for involving the customer early.

Support Activities. Support activities of the value chain are also enhanced by JIT practices. Just-In-Time Purchasing (JITP) is used in the procurement of resources. It focuses on efficiency and flexibility upstream from the manufacturer and enables a manufacturer to work more closely with its suppliers. The objective of JITP is to develop a long term relationship with suppliers based on trust. By reducing the number of suppliers and increasing the quantity of orders and the number of parts purchased from each supplier, the manufacturer commits to a long-term relationship with suppliers. The purchasing manager is responsible for developing suppliers and facilitating the implementation of JIT practices in suppliers' facilities.

Technology Development: Technology is embodied in every value adding activity in the system (Porter, 1985); it includes the know-how, procedures, or the technology associated with process equipment. In addition, technology may be used to support the primary activities. Development of these technologies provides for a competitive advantage for organizations. The continuous improvement effort associated with the Japanese management approach and quality circles throughout the system serves as the foundation for this development. In this paper, technology is addressed in two categories: product and process technologies, and information technology.

Information technology can be pervasive throughout the system; it may involve the use of computer information systems and/or integrative mechanisms (e.g., concurrent engineering and quality circles) that provide for open communication between entities. The information flow effectively link activities and the competitive strategy. Information technology can be developed to provide real-time information throughout the system, and thus, can be used to integrate the secondary and primary activities of the value chain. For example, Womack *et al.* (1991) describe a potential future scenario for using information technology to integrate the distribution system with the production system in a Japanese automobile system. Other examples, pointed out by Hall *et al.* (1993a, 1993b) are used to show how customers through CAD can have direct input into the design of the product they purchase. The CAD data is then used for product design documentation, order entry, processing, etc.

Flatter organizational structures provide for more efficient information flows vertically in the organization. Horizontal informational flows are also necessary to achieve the degree of employee involvement and commitment necessary to develop the continuous improvement effort necessary in a world-class manufacturing environment. The integrative mechanisms provided with JIT practices are structural changes conducive to improving communication flows.

Human Resource Management: In JIT, various manufacturing practices serve as a means to accomplish two principles: elimination of all wastes and full utilization of human resources. Elimination of wastes includes all wastes that pertain to time, materials, parts, and equipment. In JIT systems, respect for people includes not only treating employees as human beings, but also making full use of their capabilities, physical and mental.

Multi-function employees (MFE) and quality circles (QC) allow for full utilization of employees and provide a foundation for eliminating wastes and/or continuous improvement.

Multi-Function Employees (MFE) are created by formal cross-training of employees on several different machines and in several different functions. Employees are compensated according to the skills that they learn, thus are encouraged towards self-development. The job rotation is another self development program that improves the employee's motivation and commitment. Multi-function employees are better prepared to fill in for other employees when needed, and an employee's input to the problem solving process is enhanced. By becoming more knowledgeable of other activities and functions, employees are better prepared to suggest improvements for their functions and other activities of the organization.

JIT practices and the appropriate technology can be applied to other business units in the system as described for the manufacturer's value chain. Integration occurs through the common goal of satisfying the customer. Integrative mechanisms can be developed across business units to support these efforts. These integrative mechanisms can provide for better employee involvement, communication, and interdependent relationships.

CONCLUSION: It is the integration of the activities in the value chain that leads to maximization of total enterprise performance. Many approaches to developing customer service have been proposed for an organization. The most appropriate approach is one that provides for maximizing the total system rather than optimization of functional areas or subsystems. This article describes how JIT activities can be utilized to integrate the value chain to improve customer service. JIT success depends on a long term plan and commitment to the plan. Other requirements are top management commitment and appropriate competitive strategy.

Top Management: The responsibility for maximization of the macro system cannot be left to middle and lower management. Top management must lead the effort for change through the value chain/system. A vision must be developed and communicated by the top management to others in the value chain (supplier, channel etc.).

Competitive Strategy: The organizational changes necessary to achieve the long-range objectives and competitive strategy must be understood and given organizational support. Customer service strategy must be matched with each business unit's strategy.

Organizations Mentioned: Toyota.

See **CAD; Focused factory; Group technology; Hard manufacturing technology; Just-in-time manufacturing; Just-in-time manufacturing implementation; Kanban; Lean manufacturing implementation; Multi-functional employee; Pull production system; Push system; Quality circles; Soft manufacturing technology; TQM; Value chain; World-class manufacturing.**

References

[1] Blois, K.J. (1986). "Marketing Strategies and the New Manufacturing Technologies," *International Journal of Operations and Production Management*, 6, 1, pp. 34–41.

[2] Burnham, J.M. (1985). "Improving Manufacturing Performance: A Need for Integration," *Production and Inventory Management*, 26, 4, pp. 51–59.

[3] Ellram, L.M. (1994). "The 'What' of Supply Chain Management," *NAPM Insights*, 5, 3, pp. 26–27.

[4] Hall, R.W. (1993a). "The Challenges of the Three-Day Car," *Target*, 9, 2, March/April, pp. 21–29.

[5] Hall, R.W. and Y. Yamada, (1993b). "Sekisui's Three-Day House," *Target*, 9, 4, July/August, pp. 6–11.

[6] Hall, R.W. (1993c). "AME's Vision of Total Enterprise Manufacturing," *Target*, 9, 6, November/December, pp. 33–38.

[7] Poist, R.F. (1986). "Evolution of Conceptual Approaches to Designing Business Logistics Systems," *Transportation Journal*, Fall, pp. 55–64.

[8] Porter, M.E. (1985). *Competitive Advantage: Creating and Sustaining Superior Performance*, New York: The Free Press.

[9] Schneider, J.D. and Leatherman, M.A. (1992). "Integrated Just-In-Time: A Total Business Approach," *Production and Inventory Management Journal*, 33, 1, pp. 78–82.

[10] Shycon, H.N. and C.R. Spraque (1975). "Put a Price Tag on Your Customer Servicing Levels," *Harvard Business Review*, 53 (July–August), pp. 71–78.

[11] Stobaugh, R. and P. Telesio (1983). "Match Manufacturing Policies and Product Strategy, *Harvard Business Review*, (March–April), pp. 113–120.

[12] Turbide, D.A. (1993). "Total Productive Maintenance Finds Its Place in the Sun," *APICS – The Performance Advantage*, August, pp. 32–34.

[13] White, R.E. and Ruch, W.A. (1990). The composition and scope of JIT. *Operations Management Review*, 7(3&4), 9–18.

[14] Womack, J.P., Jones, D.T., and D. Roos (1991). *The Machine That Changed the World*, New York: Harper Perennial.

CUSTOMER SUCCESS

Customer Success is an emerging concept in the field of marketing and logistics with strong implications for manufacturing. It basically holds that a company's marketing and logistics efforts should be not only be devoted to satisfying customer expectations but should be oriented toward helping customers succeed in meeting their objectives in their own competitive environment. While **customer satisfaction** implies "meeting or exceeding customer expectations," customer success implies more intimate knowledge of customers and a focus on their true stated or unstated interests rather than their expectations of supplier performance.

The principle of customer success moves the organization to a higher level of relevancy to customers by trying to satisfy the critical drivers of the customer's business. The resultant commitment to customers can be demonstrated by such activities as joint supplier-customer business teams, joint research by the customer and supplier, and general expansion of knowledge and understanding of what motivates the customer. Since no business has the resources to focus on all customers with equal zeal, it is necessary to carefully evaluate and select from the total customer base those who deserve to receive such special attention.

See **Customer satisfaction; Customer service, satisfaction and success.**

Reference

[1] Bowersox, D.J., Calantone, R.J., Clinton, S.R., Closs, D.J., Cooper, M.B., Droge, C.L., Fawcett, S.E., Frankel, R., Frayer, D.J., Morash, E.A., Rinehart, L.M., and Schmitz, J.M. (1995). *World Class Logistics: The Challenge of Managing Continuous Change*. Council of Logistics Management, Oak Brook, IL.

CUSTOMERS OF CHOICE

Customers of choice (also referred to as "key accounts") refer to the subset of potential customers that a firm selectively targets for its most intensive marketing and distribution efforts. Traditionally, many firms chose a mass marketing approach, which implies that all potential customers were viewed as being essentially equal. A somewhat more selective approach chosen by many organizations defines several different target market segments in a manner such that the customers within each segment have needs and requirements that were similar to one another but different from the needs and requirements of customers in other market segments. Due to constraint posed by resources, marketing may choose to focus on individual accounts or market segments with individualized marketing and logistics strategies for the enhanced satisfaction of these customers. Those accounts, which are targeted for such efforts are Customers of Choice, and receive the highest levels of service delivered by the firm.

See **Customer satisfaction; Customer service; Customer success.**

CUSTOMIZATION

Customization in products involves the design and manufacture of a product for a single customer's unique needs. This typically involves a set of engineering specifications that define the relevant product characteristics, materials, dimensions, manufacturing processes, etc. to satisfy a specific customer. Prior to the development of mass production virtually all products were custom-made, their uniqueness was dictated either by the customer's personal requirements or by the variations in the craftsmanship or materials used.

There are degrees of customization that range from a fully customized product that is unique in every detail, to quasi-customized ones that share a set of common design elements with some variations in the products manufactured for different customers.

See **Mass customization; Product design.**

CUT-TO-FIT MODULARITY

With cut-to-fit modularity, product variation is achieved through the use of one or more standard components with one or more infinitely variable add-on components. Most frequently, the variation is associated with the physical dimensions that can be modified (e.g., length). Although the concept also applies to components that can be infinitely varied by any simple production process. A common example of cut-to-fit modularity can be seen in products that are left unfinished until requested by the customer. For example, suit pants are not hemmed in production. The length of the pant is determined and hemmed once the customer purchases the garment. Another example being aluminum rain gutters installed on new homes. Aluminum rain gutters are formed and sold in standard lengths but are cut to specific lengths during installation. In cut-to-fit modularity, the standard product dimensions are altered to suit a specific customer. This is also known as fabricate- to-fit modularity.

See **Mass customization.**

CYCLE COUNTING

Cycle counting is an audit technique to bring accuracy to inventory records. The process involves physical counting and reconciliation of inventory records on a cyclical basis; more frequent counts for high volume or high value items. Each item in inventory is counted at a specified frequency to identify and correct errors, if any, in the records. Accurate inventory records are essential for effective production planning and scheduling and on-time delivery of orders.

See **Material planning; Material requirements planning.**

CYCLE STOCK

Cycle stock exists where the quantity of an item produced or received from a supplier exceeds the amount of the item required for use or shipment immediately. In the long run, cycle stock is not a surplus. It occurs because the lot size for production or supply of the item is larger than the immediate demand. For example, economies of scale may encourage producing monthly for sales that are shipped weekly. It gradually reduces as customer orders are received, and cycle stock builds up cyclically when replenishment is received from production or suppliers.

See **Inventory build and depletion cycles; Inventory management.**

CYCLE TIME

Cycle time refers to the interval of time between successively completed units that come off the end of an assembly line. While balancing assembly lines, the cycle time is determined from the required production rate of the line. For example, if the required production rate is 30 units per hour, then the cycle time would be 60 minutes divided by 30 = 2 minutes. The cycle time also represents the time allowed at each workstation to complete the assigned tasks on one unit before it moves to the next workstation.

See **Assembly line design; Manufacturing cell; Total quality management; U-shaped assembly lines.**

DARPA

DARPA which stands for Defense Advanced Research Projects Agency is the central research and development organization for the US Department of Defense (DoD). It manages and directs selected basic and applied research and development projects for DoD, and pursues research and technology where risk and payoff are both very high and where success may provide dramatic advances for traditional military roles and missions and dual-use applications. DARPA was established in 1958 as the first US response to the Soviet launching of Sputnik. Since that time DARPA's mission has been to assure that the US maintains a lead in applying state-of-the-art technology for military capabilities and to prevent technological surprise from the US adversaries. The DARPA organization was as unique as its role, reporting directly to the Secretary of Defense and operating in coordination with but completely independent of the military research and development establishment. Strong support from the senior DoD management has always been essential since DARPA was designed to be an anathema to the conventional military and research and development structure and, in fact, to be a deliberate counterpoint to traditional thinking and approaches.

See **Agile manufacturing.**

DEADLOCK

In a manufacturing system, a deadlock occurs when more than one job requests the same resource. For example, consider the case of a machine that has a common input/output buffer with a capacity of one. Assume that an incoming job is placed in the machine's buffer, and after the machine removes the job to work on, another incoming job is placed in the buffer. When the machine completes work on the first job, it has nowhere to put the completed part, since the buffer is full. In this case, the machine, the worked part, and the incoming part are all deadlocked. In many real systems, the only way to recover from this situation is to manually intervene to release the shared resource (e.g., a buffer). In some cases, this may necessitate shutting down a portion or all of the system. (See Desrochers and Al-Jaar, 1995; Murata, 1989; and Peterson, 1981.)

See **Blocking; Manufacturing systems modeling using petri nets.**

References

[1] Desrochers, A.A., and Al-Jaar, R.Y., 1995, *Applications of Petri Nets in Manufacturing Systems: Modeling, Control, and Performance Analysis.* (Piscataway, NJ: IEEE Press).

[2] Murata, T., 1989, Petri nets: properties, analysis and applications. *Proceedings of the IEEE*, 77(4): 541–580.

[3] Peterson, J.L., 1981, *Petri Net Theory and the Modeling of Systems.* (Englewood Cliffs, NJ: Prentice-Hall).

DEBT RATIO

Debt ratio is a numerical figure that is indicative of over-leveraged firms incapable of paying their long-term debts. This ratio is calculated as: Debt Ratio = Total Liabilities/Total Assets. The debt ratio indicates how much debt the supplier has incurred relative to assets. Acceptable figure will vary by industry, and also according to how "new" the company is. Generally speaking, companies that have just started up have higher debt ratios.

In cases when the supplier is a publicly traded company on the stock market, specific financial ratios can be obtained from sources such as 10-K reports, Moody's Industrials, and Dunn and Bradstreet Reports.

See **Supplier performance measurement.**

DECENTRALIZED PRODUCTION PLANNING

See **Aggregate plan and master production schedule linkage.**

DECISION HORIZON

Many manufacturing system control policies attempt to generate plans to cover some future period of time. This period of time is known as the decision horizon for the policy; it is the period of time for which the plan is active. Typically, the policy will be reevaluated prior to the end of the decision horizon, as new data become available; in this case, a new plan is implemented for the specified decision horizon. For example, some dynamic kanban control (DKC) policies plan the release and capture of kanbans over a specified period of time (Gupta and Al-Turki, 1997; 1998)

See **Capacity planning: Medium- and short-range; Dynamic kanban control; Dynamic kanban control for JIT manufacturing; Flexible kanban control; Kanbans; Reactive kanban control.**

References

[1] Gupta, S.M., and Y.A.Y. Al-Turki (1997). "An algorithm to dynamically adjust the number of kanbans in stochastic processing times and variable demand environment." *Production Planning and Control*, 8, 133–141.

[2] Gupta, S.M., and Y.A.Y. Al-Turki (1998). "Adapting just-in-time manufacturing systems to preventive maintenance interruptions." *Production Planning and Control*, 9, 349–359.

DECISION RULES IN INVENTORY CONTROL

Independent demand inventories are managed according to two decisions: order size and order timing. Four decision rules for inventory control are used in making these decisions. The four rules are formed by the combinations of fixed or variable order frequency and fixed or variable order quantity, as follows:

- Variable order frequency (R), fixed order quantity (Q): Q, R rule
- Fixed order frequency (T), fixed order quantity (Q): Q, T rule
- Variable order frequency (R), variable order quantity (S): S, R rule
- Fixed order frequency (T), variable order quantity (S): S, T rule

where:

Q = Order a fixed quantity (Q)
S = Order up to a fixed opening inventory quantity (S)
R = Place an order when the inventory balance drops to (R)
T = Place an order every (T) periods.

The Q, R rule is commonly known as the order point rule, where an order for a fixed quantity is placed when stock levels reach a reorder point.

See **Inventory flow analysis; Inventory holding cost.**

Reference

[1] Vollmann, T.E., Berry, W.L., Whybark, D.C. *Manufacturing Planning and Control Systems*, Fourth Edition, Irwin/McGraw-Hill, New York, 1997.

DECISION SUPPORT SYSTEM

It is a computer system that assists the decision maker in selecting a certain desired strategy among multiple alternative courses of action by storing, processing, and retrieving large amount of data and knowledge relevant to the selection decision.

It often is a software package to support management decision making. The package contains methods to handle data from information systems, or mathematical models to derive conclusions from these data. A typical characteristic of a decision support systems is that the user can interact with the system, for example by asking questions of "what-if" type.

See **Electronic data interchange in supply chain management; Maintenance management decision models.**

DECOUPLERS

Decoupler is an element or device in a manufacturing cell that assists part movement and makes the cell design more flexible. Decouplers can have multiple functions including part transportation and inspection.

See **Autonomation; Linked-cell manufacturing system; Manufacturing cell design; Manufacturing systems.**

DECOUPLING INVENTORIES

If inventories are positioned between process stages, those inventories decouple the flow of materials being received at a work station from an upstream process. This allows the two process stages to be scheduled somewhat independently of one another. This decoupling process allows for more effective and efficient scheduling of each process stage. While the ideal plant would like to achieve a continuous flow of material through the entire process train, this synchronization is often difficult to achieve. Where synchronization is not possible, it becomes advisable to decouple process stages.

In some production environments, equipment utilization can be improved by accumulating an intermediate product in inventory to satisfy peak demand requirements from a downstream unit. If process stage A feeds process stage B then process stage A must have sufficient capacity to meet the peak requirements from process stage B. If a decoupling inventory is positioned between these process stages, process stage A could continuously produce product. The decision involves a trade-off between the cost of extra, under-utilized capacity for the upstream stage and the cost of carrying intermediate inventory. If capacity is inexpensive relative to the cost of inventory, then excess capacity would be chosen. If capacity is expensive relative to the cost of inventory, then decoupling inventory should be used.

A similar decoupling need arises when two sequential process stages require different product sequencing strategies. Suppose that the first process stage produces slabs of 10 different metal alloys, each of which has a different chemical composition. The production sequence is based on alloy compatibility to avoid adjacent production of incompatible alloys that may lead to off-specification material during alloy transitions. Thus, there are strong economic incentives to follow a specified alloy sequence.

Suppose further that the second stage is a rolling mill that makes 20 widths of metal coils. Rolling the metal alloys leaves a groove on the roller at a coil's edge. If a wider coil is rolled next, the groove in the roller will impart a groove to the coil. This problem is minimized by sequencing production from wide to narrow coils.

Thus, if we have two sequential operations, each requiring its own technologically determined sequence. Which sequence should dominate? The answer is neither. By using a decoupling inventory of alloys slabs, the desired sequences for both operations can be maintained.

Inventories may also be used to decouple production processes from customer orders; thereby shortening delivery times to the customer. Any inventory that is positioned to allow each process stage to be scheduled independent of an up-stream or down-stream process stage becomes a decoupling inventory.

See **Process industry scheduling.**

Reference

[1] Taylor, S.G. and S.F. Bolander (1994). *Process Flow Scheduling: A Scheduling Systems Framework for Flow Manufacturing*, APICS, Falls Church, VA.

DEDICATED SYSTEMS

See **Flexible automation.**

DEMAND CHAIN MANAGEMENT

Demand chain management is an evolution of what has traditionally been called supply chain management. As the term implies, the focus shifts from working forward from the manufacturer or supplier, to working back from the end customer. This customer focus requires all members of the chain to continually improve goods and services for the benefit of the chain. In the process unnecessary transactions and costs can be eliminated.

See **Supply chain management.**

Reference

[1] Vollmann, T.E., Berry, W.L., Whybark, D.C. *Manufacturing Planning and Control Systems*, Fourth Edition, Irwin/McGraw-Hill, New York, 1997.

DEMAND FENCE

To achieve stability in master production schedules (MPS), common techniques include firm planned order treatment, frozen time periods, and time fencing. The time fencing technique assists in achieving a stable MPS, which in turn means stable component schedules and high productivity. However, some changes to the MPS are desirable to accommodate the needs of customers and to avoid building up needless inventory.

Time fencing establishes guidelines specifying permitted levels of MPS changes in certain time periods (see, Time fence). Commonly used fences include the demand fence, or demand time fence, and planning fence. The demand fence is the shorter of these. Company policies may make it very difficult to change the MPS within the demand fence, since actual customer orders are used in calculations rather than the forecast.

See **Master production schedule; MRP; Time-fence.**

Reference

[1] Vollmann, T.E., Berry, W.L., Whybark, D.C. *Manufacturing Planning and Control Systems*, Fourth Edition, Irwin/McGraw-Hill, New York, 1997.

DEMAND MANAGEMENT TECHNIQUES

The three master production scheduling (MPS) environments of make-to-stock (MTS), assemble-to-order (ATO), and make-to-order (MTO) utilize different demand management techniques. In MTS production, highly detailed, item level forecasts are required, by location and time period. The difficulty of forecasting individual items can be dealt with by using ratios or percentages of aggregated forecasts as a surrogate for the individual units. This is appropriate in firms with steady product mix ratios and allows management to devote time to forecasting overall sales. In the ATO environment, the key issue is accurate order promise dates, which calls for a stable MPS. This is achieved by using time fencing and related methods. In MTO environments, control of customer orders in the production system is important. Usually, there is much uncertainty concerning future orders, which may require extensive new design and engineering.

See **Demand fence; Make-to-order production; Make-to-stock production; Manufacturing systems; Master production schedule; Time-fence.**

Reference

[1] Vollmann, T.E., Berry, W.L., Whybark, D.C. *Manufacturing Planning and Control Systems*, Fourth Edition, Irwin/McGraw-Hill, New York, 1997.

DEMING CYCLE (PDCA)

The Deming cycle (Plan, Do, Check, Act) is a methodology for continuous improvement. This methodology, originally called the Shewhart cycle, was developed by Walter A. Shewhart. It was renamed the Deming cycle by the Japanese in 1950. W. Edwards Deming suggested that this procedure should be followed for the improvement of any stage of production and as a procedure for finding a special cause of variation indicated by statistical signals. The Plan stage involves studying the current situation, gathering data, and planning for improvement. The Do stage consists of implementing the plan on a trial basis. The Check stage is designed to determine if the trial plan is working and to see if any further problems or opportunities have been discovered. The Act stage consists of implementing the final plan. This leads back to the Plan stage for further diagnosis and improvement. The cycle is never ending. Deming used this Plan, Do, Check, Act cycle as the basis for accomplishing his 14th point for management: Put everyone in the company to work to accomplish the transformation to a competitive, prosperous organization.

See **Deming's 14 points; Quality: The implications of W. Edwards Deming's approach; Statistical process control using control charts.**

References

[1] Deming, W. Edwards (1982). *Out of the Crisis*, Center for Advanced Engineering Study, Massachusetts Institute of Technology, Cambridge, Massachusetts.
[2] Shewhart, W.A. (1939). *Statistical Method from the Viewpoint of Quality Control*, Graduate School, Department of Agriculture, Washington.

DEMING PRIZE (OR AWARD)

First awarded in 1951, Japan's Deming Prize marked the beginning of the modern quality movement. It recognizes successful efforts in instituting company wide quality control (CWQC) principles.

See **Quality: The implications of Deming's approach; Quality management systems; Total quality management.**

DEMING'S 14 POINTS

W. Edwards Deming offered 14 Points for management as the basis for keeping companies in business and making them more competitive. It assumes that 85% of the problems in any company are the responsibility of management. The 14 Points are:

1. Create constancy of purpose toward improvement of product and service with a plan to become competitive and stay in business.
2. Adopt the new philosophy of quality.
3. Cease dependence on inspection to achieve quality. Require instead statistical evidence that quality is built in.
4. End the practice of awarding business on the basis of price tag. Consider quality as well as price. Move toward a single supplier for an item based on a long-term relationship of trust.

5. Improve constantly and forever the system of planning, production, and service, to improve quality and productivity and constantly decrease costs.
6. Institute modern methods of training on the job.
7. Institute modern methods of supervision that emphasize quality rather than sheer numbers.
8. Drive out fear.
9. Break down barriers between departments.
10. Eliminate slogans, exhortations, and targets for the work force.
11. Eliminate numerical quotas for the work force and numerical goals for management.
12. Remove barriers that rob workers (and management) of their right to pride of workmanship.
13. Institute a vigorous program of education and retraining.
14. Put everyone in the company to work to accomplish the transformation.

See **Quality: The implications of W. Edwards Deming's approach; Quality management systems; Total quality management.**

References

[1] Deming, W. Edwards (1982). *Out of the Crisis*, Center for Advanced Engineering Study, Massachusetts Institute of Technology, Cambridge, Massachusetts.
[2] Deming, W. Edwards (1982). *Quality, Productivity, and Competitive Position*, Center for Advanced Engineering Study, Massachusetts Institute of Technology, Cambridge, Massachusetts.

DEPENDENT DEMAND

Dependent demand is the demand for items directly derived from the demand for another item usually higher up in the bill of materials (BOM) structure. Dependent demand derives from the concept of gross to net explosion in MRP systems. It occurs where the demand for components depends upon the demand for other subassemblies, or products. Dependent demand differs from independent demand in that dependent demand can be exactly calculated for demand for other products; it does not need to be forecast.

See **Bill of materials; Gross to net explosion; MRP; MPS.**

DESIGN CONCURRENCY

Within any new product development phase, product design usually involves a number of different design tasks (e.g., system design, software design, electrical design, mechanical design). Design concurrency refers to the degree of overlap in the execution of these tasks. For example, in a concurrent design effort, engineers might begin designing structural parts before the overall system specifications are completed. Design concurrency also indicates that product subsystems are simultaneously designed. For example, the drive train and cockpit of a new car might be designed at the same time. If the functions of the subsystems being concurrently designed are closely related, this calls for tight coordination of activities and frequent communications between personnel involved in the separate design efforts.

See **Concurrent engineering; Product design; Product development and concurrent engineering.**

DESIGN FOR CUSTOMIZATION

See **Design for localization; Product design for global markets.**

DESIGN FOR THE ENVIRONMENT (DFE)

In DFE, the environmental impact of products, services and processes is explicitly evaluated and minimized during their initial design. Environmental impact can occur during any stage in the product or service life-cycle, including raw material extraction, fabrication, delivery, use and post-use (i.e., reuse, recycle or disposal). Historically, firms have optimized product or service design around specific customer needs, such as cost or quality, without explicitly considering the environmental degradation that might result during the entire "cradle to grave" life-cycle. Ideally, a firm can use the principles of DFE in anticipation of new environmental demands of customers and regulators and thus proactively reduce any negative impacts.

One input to DFE is Life-Cycle Assessment (LCA). This increasingly sophisticated tool assists in accounting for the environmental impact or burden of a product or service across all life stages. At the simplest level, LCA catalogs all the materials and energy consumed, and all the direct air, water and solid waste pollution generated throughout a product's life-cycle. Any indirect burden, such as emissions associated with energy generation, also need to be considered. As the analysis becomes increasingly sophisticated, different aspects of any environmental burden, such as toxicity, are examined and synthesized to construct a holistic assessment of the product or service.

See **Environmental issues and operations management; Life-cycle assessment (LCA).**

DESIGN FOR LOCALIZATION

It refers to the design processes, which focuses on creating a product that meets the needs of local consumers in various countries, that is easy to manufacture (design for manufacturability), and that is easy to customize, distribute and deliver.

See **Product design for global markets.**

DESIGN FOR MANUFACTURABILITY (DFM)

It refers to the design process that simplifies the product to increase the quality and reduce the cost of manufacture.

In its simplest form, DFM is the philosophy and practice of designing a product to optimize the fit between product specifications and the manufacturing system. The philosophy of DFM encourages design of products that are easy to produce in the quantities needed (i.e., products that have high "producibility"). Depending on the industry and process technologies involved, DFM analyses might focus on improving process yields, fabrication, assembly, product test, or packaging. The practice of DFM usually involves technologies, procedures, and incentives aimed at improving communication and integration between product and process designers. Technologies that help DFM are, computer-aided-engineering and process-planning systems, group technology systems, shared design databases, and enhanced electronic communications. Procedures that help DFM are, requiring manufacturing authorization of product designs, and design teams that include production personnel. (Yousseff, 1994).

See **Concurrent engineering; Product design; Total quality management.**

Reference

[1] Yousseff, M.A. (1994). "Design for Manufacturability and Time-to-Market, Part 1: Theoretical Foundations," *International Journal of Operations and Production Management*, 14(12), 6–21.

DETERMINISTIC PROCESSING TIME

Processing time is the estimated or actual duration to complete a task (work element, activity or work unit). Deterministic processing times indicate that the duration is fixed and that it is known with certainty.

See **U-shaped assembly lines.**

DETERMINISTIC PRODUCTION ENVIRONMENTS

A deterministic production environment is one in which the demand, production times, maintenance times, and resource availability are predictable and have no variability. Such an environment is ideal from a planning perspective. In any event, plant supervisors strive to minimize the internal sources of variation (e.g., machine breakdown, maintenance schedules, staff turnover, etc.), even when they cannot completely control the sources of external variation (e.g., demand).

See **Dynamic kanban control for JIT manufacturing; Statistical process control using control charts.**

DEVELOP-TO-ORDER PRODUCTION

In develop-to-order production, a customer requests the develop-to-order company to carry out research and development, prototype development, process development, and subsequently manufacture and delivery of the product. Typically research and development institutions undertake these types of jobs. A customer order may conceptually specify the functions expected of the product. To satisfy the customer requirements, intense research and development is carried out, usually over several years to develop the product and in some situations the production process too. For example, the design and development of a spacecraft comes under this category.

See **Mass customization.**

DIAGNOSTIC SKILLS OF OPERATORS

Diagnostic skills refer to an individual's ability to identify a particular problem and define it. These skills are acquired through formal training, practice, and experimentation.

See **Human resource issues and advanced manufacturing technology**

DIFFERENTIATION STRATEGY

A differentiation strategy is one of three *generic* business strategies proposed by Porter in his well-known book, *Competitive Strategy* (1980). A differentiation strategy seeks to develop and offer products or services that are perceived as being unique or superior in some way. Approaches to differentiation can take many forms: product design, brand image, technology, product features, distribution and service characteristics, or other dimensions. Catepillar Tractor, a manufacturer of heavy construction equipment, provides an example of product differentiation. Catepillar is well-known as a producer of very high quality, reliable machinery. In addition, their dealer network and excellent spare parts availability make their products even more attractive to buyers, who suffer heavy expenses when equipment are unavailable. Pursuit of a differentiation strategy usually requires investments in product or process development, new technologies, market research, and image building such as promotion or advertising. Differentiation is based on a good understanding of customer needs or desires, and the ability to offer products or services that meet some specific need or desire in a unique or superior way.

Reference

[1] Porter, M.E. (1980). *Competitive Strategy*, The Free Press: New York.

DIGITAL SIGNATURE

It is the legal infrastructure that is designed to minimize frauds such as electronic forgeries by not only identifying the person who purportedly sent the electronic message and verifying that the electronic message has not been altered or otherwise damaged during transmission.

See **Electronic commerce; Electronic data interchange in supply chain management.**

DIRECT LABOR COSTS

Direct labor costs is that portion of costs associated with a product that are attributable to the price of labor directly consumed in producing the product. These costs can be directly traced and allocated to an item. The figure is calculated by multiplying the amount of time that workers actually spend manufacturing the product by their hourly wage. For instance, if a product requires 5 hours of labor to produce, and workers are paid an average of $20 an hour, the direct labor cost is $100 per unit of product. Direct labor costs are of interest when we need to calculate item-specific costs.

See **Activity-based costing; Cost analysis for purchasing; Supplier performance measurement.**

DIRECT LABOR HOURS

Direct labor hours is time spent by individuals who work specifically on manufacturing a product or performing a service. In contrast, indirect labor hours is the time spent by individuals involved in supervising or supporting the process of manufacturing a product or performing a service. For example, in a car manufacturing plant, direct labor is defined as those individuals who directly work on the car products (i.e., physically use the production equipment to make cars). Indirect labor in a car manufacturing plant would involve line supervisors, machine maintenance staff, plant security personnel, etc. Traditionally, the direct labor hours measure is often used to determine the cost of support overhead activities to assign to products.

See **Activity-based costing; Cost analysis for purchasing; Direct labor costs.**

DIRECT LABOR SHRINKAGE

An important trend in manufacturing is the direct labor cost as a percent of total manufacturing cost is shrinking. This has important impact upon costing and new equipment justification

See **Activity-based costing; Capital investment in advanced manufacturing technology; Direct labor costs; Direct labor hours; Hidden factory.**

DIRECT NUMERICAL CONTROL

A Direct Numerical Control (DNC) system operates with a remote central processor controlling several CNC machines via communication lines. The host computer dynamically downloads numerical data "on-the-fly" as CNC machines execute operations. When the system is equipped with adaptive control, feedback from the CNC machine tool allows for the evaluation of and possible intervention to the manufacturing process.

See **Advanced manufacturing technologies; Human resource issues and advanced manufacturing technology; Manufacturing flexibility implementation.**

DISAGGREGATION IN AUTOMATED MANUFACTURING ENVIRONMENT

Chen H. Chung

University of Kentucky, Kentucky, U.S.A.

A disaggregation process in manufacturing planning and control (MPC) converts a higher-level plan into lower-level detailed plans and schedules. The high level plan is usually too general and thus incomplete. It needs to be complemented by and implemented in detailed schedules of operations. The disaggregation process also reflects the real practice of hierarchical planning and control.

Disaggregation, as a class of manufacturing problems, was first identified by Krajewski and Ritzman (1977). They defined disaggregation as the total process of going from aggregate plans to more detailed plans. While the disaggregation process in the traditional manufacturing systems has been well recognized, there is a relative dearth in the investigation of the disaggregation issues in the automated manufacturing system (AMS) environment. For example, decisions which are crucial to the success of AMS such as parts family grouping, loading and routing, and tools planning and management need to be incorporated into MPC systems.

One of the key concerns in automated manufacturing over the years is the issue of integrating manufacturing decisions and activities. At the end of the 1970s, the manufacturing industries recognized that independent computerization of individual function resulted in "islands of automation." As a result, in the 1980s, efforts were focused in two areas: (1) the integration of computer-aided design (CAD) and computer-aided manufacturing (CAM); and (2) the integration of manufacturing system. Over the years, researchers have proposed various ways of integrating AMS. Below (1987) suggests that integration occurs at seven levels: physical, data, schedule, functions, attitudes, principles, and purpose. Singhal *et al.* (1987) use a life cycle approach to classify various research and models for AMS.

DISAGGREGATION IN TRADITIONAL MANUFACTURING: According to Krajewski and Ritzman (1977), disaggregation decisions in a manufacturing organization may exist in one or more of the following three forms:

1. Given aggregate decisions on output and capacity, determine the timing and sizing of specific final product production quantities over the time horizon (sometimes referred to as a master schedule).
2. Given the timing and sizing of final product production quantities, determine the timing and sizing of manufactured (or purchased) component quantities.
3. Given the timing and sizing of component quantities, determine the short-term sequences and priorities of the jobs (orders) and the resource allocation to individual operations.

The above three levels of disaggregation decisions are the essence of the traditional manufacturing planning and control (MPC) systems (Vollmann *et al.*, 1996). Production planning determines the aggregate levels of production, inventory, and workforce to respond to the fluctuating demands in the future. The master production schedule (MPS) is a disaggregate version of the production plan. It specifies the timing and sizing of the production of individual end products. Based on the master production schedule, the second level disaggregation is to develop the detailed schedules for manufacturing or purchasing the components and parts. The third level disaggregation specifies the plans for the most detailed operations. It determines the sequences of

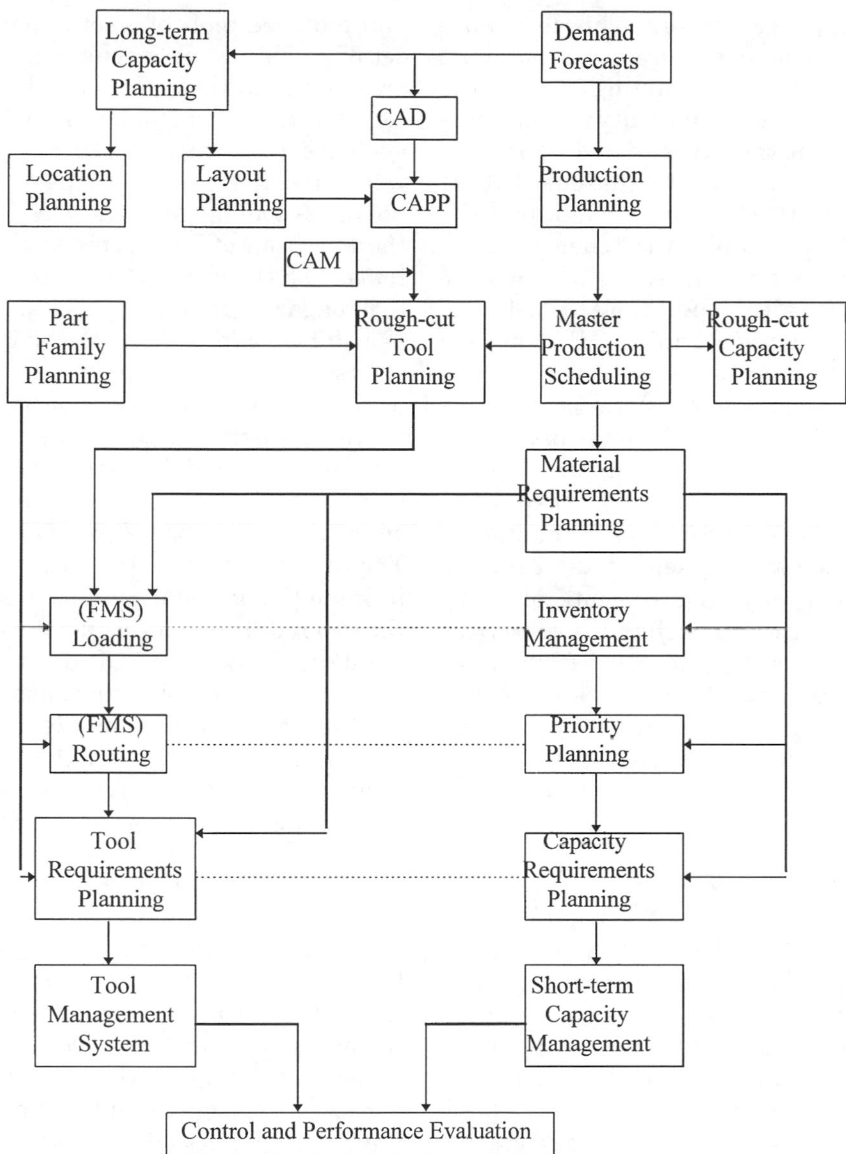

Fig. 1. A framework for disaggregation decisions in automated manufacturing environments.

jobs (orders) and relevant capacity and inventory decisions.

Most disaggregation issues in the traditional MPC system are applicable to the automated manufacturing environment. For example, Krajewski and Ritzman (1977) categorized the disaggregation problems according to the dimensions of single vs. multiple stage systems, linear vs. nonlinear assembly trees, single vs. multiple products, etc. These dimensions are also applicable to the automated manufacturing environment. However, there are many additional issues that are unique to the AMS environment that need to be addressed.

A Framework of Disaggregation Decisions in Automated Manufacturing: Vollmann *et al.* (1996) present a comprehensive framework for traditional manufacturing planning and control systems. Figure 1 extends the MPC framework by incorporating various AMS-related decisions such as parts family grouping, loading and routing, tool planing and management, etc.

The right half of the diagram basically consists of the traditional MPC systems. This portion is also known as a manufacturing resource planning (MRP II) system. It reflects the disaggregation process in the traditional manufacturing environment. Production planning determines the aggregate levels of production, inventory, and workforce to respond to the fluctuating demands in the future. Given the aggregate plan, the MPS specifies the timing and sizing of the production of individual end products. To be feasible and acceptable, an MPS should meet the resource constraints at key workcenters. This feasibility check is called resource requirements planning (RRP) or rough-cut capacity planning (RCP). If the RCP indicates potential infeasibility at some critical operations, then the MPS should be revised to accommodate the capacity limitations of these key facilities. After the feasibility check by RCP, the feasible and acceptable MPS then becomes an authorized MPS and serves as an input to the lower level detailed operations scheduling. A material requirements planning (MRP) system determines the timing and size of the production of components and parts. It also serves as the basis for detailed-level inventory management (e.g., procurement and materials control), priority planning (e.g., sequencing and scheduling), and capacity planning. In RCP, the estimation of capacity requirements is based on the MPS information. The capacity requirements planning (CRP) activity, on the other hand, utilizes the information generated by the MRP system, including open orders and planned order releases. CRP also takes into consideration the current status of all work-in-process inventories in the shop. Thus, CRP provides a more accurate account of the capacity requirements than RCP does.

The left half of Figure 1 deals with important decisions in an AMS environment. Computer-aided design uses computers in product design and analysis. Computer-aided manufacturing applies computers and microprocessors to direct activities of manufacturing. Computer-aided process planning (CAPP) bridges the gap between CAD and CAM. Basically, CAPP determines how to convert a product design to the final form of the manufacturing product. Different volume representations and decompositions would entail different cutting paths, machine tools, setup and machining costs, etc. Thus, the actual manufacturing activities and related tool requirements are greatly affected by the CAPP.

Quite often, it would be desirable to pool operations with common tool requirements in AMS. If a small set of tools can perform many different operations, tool utilization can be increased. Thus, parts family grouping (PFG) is desirable for the purpose of loading flexible manufacturing systems (FMS) and planning tool requirements. In fact, Chakravarty and Shtub (1984) approach the FMS loading problem by first forming workpiece-tool groups and then assigning the groups to machines. The grouping of workpiece-tools is similar to the machine-component grouping problem in group technology (GT) (King and Nakoronchai, 1982; Kusiak, 1985; Wemmerlov and Hyer, 1986). The FMS loading decision allocates the operations and required tools of a set of part types among a set of machines (or machine groups). Many modern FMSs are equipped with the capability of automated tool interchange (ATI). With ATI, partial system set-ups will not cause any loss of machine work time. If a machine needs a specific tool that is not on its tool magazine, the ATI device can bring the tool from other machines or tool storage systems (TSSs). On the other hand, if the machine's tool magazine is full, the tool(s) not needed in the near future can be transported to other machines or TSSs.

The routing decision is concerned with the order in which a part type visits a set of machines. The routing decision differs from the scheduling problem in that the latter is concerned with the order in which the operations of parts are to be performed. The final (i.e., optimal) routing decisions will determine the actual machines used and, therefore, the tool requirements. Thus, the loading and routing decisions will have direct or indirect impacts upon the actual tool requirements.

Rough-cut tool planning (RTP) is meant to provide a rough estimate of the tool requirements implied by the MPS. RTP serves two purposes. First, it performs a feasibility check for the MPS in the context of tool availability. Such availability checks are applied to major tool types only. Second, it provides early planning for tool requirements implied by the production schedule. Figure 1 shows that tool requirements planning (TRP) has CRP as its counterpart in the traditional MPC system. However, a direct application of the CRP logic to TRP activities may not be adequate. In most automated manufacturing environments, the actual tool requirements are quite often dictated by the tooling policies in tool management systems (TMSs). For example, Hankins and Rovito (1984) listed four different tooling policies: bulk exchange, sharing tools in a frozen production schedule, migration at the completion of a part type, and resident tools. Different tooling policies result in different levels of tool inventories. Also, with different tooling policies, the relationships between TRP and other subsystems in Figure 1 would be different (Chung, 1991).

COMPUTER SUPPORT FOR THE DISAGGREGATION PROCESS: Most computer softwares developed to support the traditional MPC system or MRP II are not adequate to support the disaggregation process in an AMS environment. An AMS usually consists of many production units which have multiple manufacturing capabilities. In order to achieve manufacturing flexibility in an AMS, production planning must take advantage of production units' multi-functionality and overlapping capability. Although many issues addressed in conventional MPC systems are still important to the disaggregation in the AMS environment, we have to take into account the latter's unique characteristics. First, automated manufacturing requires shorter planning horizon and easier adjustments of the production plans. Second, an AMS production plan should take into account the versatility of production units. Third, an AMS production plan may involve not only production units within a firm but also interorganizational partners.

See **Automated manufacturing system; Capacity planning: Medium- and short-term; Hierarchical planning and control; Hierarchical production planning; Manufacturing planning and control; Manufacturing planning system; Planning horizon; Rough-cut capacity planning; Tool magazine; Tool planning and management.**

References

[1] Allen, S.J. and E.W. Schuster (1994). "Practical Production Scheduling with Capacity Constraints and Dynamic Demand: Family Planning and Disaggregation," *Production and Inventory Management*, fourth quarter, pp. 15–21.

[2] Below, L.J. (1987). "The Meaning of Integration," *Proceedings of the 4th European Conference on Automated Manufacturing*, Birmingham, UK, May, pp. 345–352.

[3] Chakravarty, A.K. and A. Shtub (1984). "Selecting Parts and Loading Flexible Manufacturing Systems," *Proceedings of The 1st ORSA/TIMS FMS Conference*, Ann Arbor, MI.

[4] Chen, I.J. (1989). "Interfacing the Loading and Routing Decisions for Flexibility and Productivity of Flexible Manufacturing Systems," Ph.D. Dissertation, University of Kentucky, Lexington, Kentucky.

[5] Chung, C.H. (1991). "Planning Tool Requirements for Flexible Manufacturing Systems," *Journal of Manufacturing systems*, 10(6), pp. 476–483.

[6] Chung, C.H. and W.H. Luo (1996). "Computer Support of Decentralized Production Planning," *The Journal of Computer Information Systems*, 36(2), pp. 53–59.

[7] George, J.F. (1991). The Conceptual and Development of Organizational Decision Support System," *Journal of Management Information Systems*, 8(3), pp. 109–125.

[8] Gershwin, S.B. (1986). "An Approach to Hierarchical Production Planning and Scheduling," in Jackson, R.H.F. and A.W.T. Jones, eds., *Real-Time Optimization in Automated Manufacturing Facilities*, U.S. Department of Commerce, National Bureau of Standards, pp. 1–14.

[9] Hankins, S.L. and V. P. Rovito (1984). "The Impact of Tooling in Flexible Manufacturing Systems," Paper presented in International Machine Tool Technical Conference.

[10] King, J.R. and V. Nakornchai (1982). "Machine-Component Group Formation in Group Technology: Review and Extension," *International Journal of Production Research*, 20(2), pp. 117–133.

[11] Krajewski, L.J. and L.P. Ritzman (1977). "Disaggregation in Manufacturing and Service Organizations: Survey of Problems and Research." *Decision Sciences*, 8(1).

[12] Kusiak, A. (1985), "The Part Families Problem in Flexible Manufacturing Systems," *Annals of Operations Research*, 3, pp. 279–300.

[13] Sarin, S.C. and S.K. Das (1994). "Computer Integrated Production Planning and Control," in Dorf, R.C. and A. Kusiak, eds., *Handbook of Design, Manufacturing, and Automation*, New York: John Wiley & Sons.

[14] Shaw, M.J. and A.B. Whinston (1988). "A Distributed Knowledge-Based Approach to Flexible Automation: The Contract Net Framework," *Flexible Manufacturing Systems*, pp. 85–104.

[15] Singhal, K., C. Fine, J.R. Meredith, and R. Suri (1987). "Research and Models for Automated Manufacturing," *Interfaces*, 17(6), November-December, pp. 5–14.

[16] Stecke, K.E. (1986). "A Hierarchical Approach to Solving Machine Grouping and Loading Problems of Flexible Manufacturing Systems," *European Journal of Operational Research*, 24, pp. 369–378.

[17] Veeramani, D., J.J. Bernardo, C.H. Chung, and Y.P. Gupta (1995). "Computer-Integrated Manufacturing: A Taxonomy of Integration and Research Issues," *Production and Operations Management*, 4(4), pp. 360–380.

[18] Vollmann, T.E., W.L. Berry, and D.C. Whybark (1996). *Manufacturing Planning and Control Systems*, (Fourth Edition), Richard D. Irwin, Homewood, Illinois.

[19] Wemmerlov, U. and N.L. Hyer (1986). "The Part Family/Machine Group Identification Problem in Cellular Manufacturing," *Journal of Operations Management*, 6(2), pp. 125–147.

DISASSEMBLY

Disassembly is the process of systematic removal of desirable constituent parts from an assembly for reuse or recycling. In addition, disassembly provides the opportunity to recover valuable materials or isolate poisonous substances from scrap by means of manual or automatic separation process. It is an alternative to traditional disposal method like incineration and landfill. Besides it adds value to the existing product because, instead of throwing it away, the products can be disassembled for refurbishment, reuse, or recycling.

There are essentially two types of disassembly, viz., non-destructive and destructive. Non-destructive disassembly removes desirable parts from an assembly while ensuring that there is no impairment to the parts. Destructive disassembly, on the other hand, is the process of separating similar metals and materials from an assembly in order to sort each material type for recycling (Gupta and McLean, 1996).

See **Environmental issues: Reuse and recycling in manufacturing systems.**

Reference

[1] Gupta, S.M. and McLean, C.R., "Disassembly of Products", *Computers and Industrial Engineering*, Vol. **31**, 1996, 225–228.

DISCOUNTED CASH FLOW (DCF)

Given the time value of money, in DCF analysis, cash flows generated by an investment project in future time periods are reduced by applying appropriate discounting factors to denote their reduced present value (PV) so that the future cash flows could be compared with the investment costs incurred in the present. The discounting factors represent the present value of $1 received at a future date. PV depend on when the payment is received and the discounting rate, which usually depends on the opportunity cost of capital.

See **Capital investments in advanced manufacturing technologies.**

DISCRETE EVENT SYSTEMS

This term is sued with mathematical or simulation models of systems. A discrete-event system is one in which events occur at discrete points in time, there exist a discrete set of states, and the states change at discrete points in time (Cassandras, 1993; Law and Kelton, 1991). For example, an automobile manufacturing facility is a discrete-event system. The following events occur at discrete points in time: orders arrive, a production step begins, a production step ends, a machine breaks down, a machine comes back on line after repair, and so on. Similarly, at any point in time, there are a discrete number of orders pending, machines operating, and automobiles in production.

Examples of discrete-event systems in manufacturing include production lines, flow and assembly lines, job shops, flexible manufacturing systems, automated manufacturing systems, and so on (Desrochers and Al-Jaar, 1995). Other examples of discrete-event systems include any queueing type of system (e.g., a bank, grocery store, or toll booth), computer systems, communications systems, and traffic systems.

In contrast in continuous systems, the system variables change continuously with respect to time. Examples of continuous systems in the manufacturing domain include chemical manufacturing (e.g., batch processing and mixing), annealing processes, and other processes which involve heating or cooling, and chemical reactions.

See **Manufacturing systems modeling using petri nets.**

References

[1] Cassandras, C.G., 1993, *Discrete Event Systems: Modeling and Performance Analysis.* (Homewood, IL: Irwin).

[2] Desrochers, A.A., and Al-Jaar, R.Y., 1995, *Applications of Petri Nets in Manufacturing Systems: Modeling, Control, and Performance Analysis*. (Piscataway, NJ: IEEE Press).

[3] Law, A.M., and Kelton, W.D., 1991, *Simulation Modeling and Analysis*. (New York: McGraw-Hill).

DISCRETE LOT SIZES

In material requirements planning (MRP), occasionally it is desirable to combine requirements for different time periods into a single lot, in order to avoid repetitious machine set-ups for small lots of components. Where lot sizes are not fixed, lot sizes can equal exactly the net requirements over multiple time periods. Such lot sizes are called discrete lot sizes. The advantage of discrete lot sizes or fixed lot sizes is that they can reduce machine setups.

See **Capacity planning: Medium- and short-range; Lot-for-lot procedures; Material requirements planning.**

DISPATCHING DECISION

In a job shop scheduling and control decisions in manufacturing includes dispatching. The dispatching decision in a job shop identify the sequence in which pending jobs will be processed at the work center. Priority sequencing rules are used to make such decisions.

See **Input/output control; Manufacturing systems; Priority sequencing rules.**

Reference

[1] Vollmann, T.E., Berry, W.L., Whybark, D.C. *Manufacturing Planning and Control Systems*, Fourth Edition, Irwin/McGraw-Hill, New York, 1997.

DISSIPATIVE SYSTEMS

A dissipative system is one that uses up, "dissipates", energy, matter and/or information in the process of maintaining its (non-equilibrium) structure and organization. The term comes from the work of chemist Ilya Prigogine and colleagues (Prigogine, 1980; Prigogine & Stengers, 1984). Most complex structures in the real world are dissipative systems, from living organisms to manufacturing firms, as they all use up and degrade energy, in maintaining themselves. Complex, dissipative systems tend to experience periods of instability, bifurcation, and re-organization. Thus dissipative systems are the foundation for theories of self-organization.

See **Manufacturing analysis using chaos, fractals, and self-organization.**

References

[1] Prigogine, I. (1980). *From Being to Becoming: Time and Complexity in the Physical Sciences*, W.H. Freeman & Co., San Francisco, CA.

[2] Prigogine, I. & I. Stengers. 1984. *Order out of Chaos: Man's New Dialogue with Nature*. Bantam, New York.

DISTRIBUTION REQUIREMENT PLANNING (DRP)

DRP applies MRP principles to distribution inventories (Martin, 1983). DRP is a time-phased, backward scheduling technique that integrates the planning of distribution inventories with manufacturing planning to ensure that the right materials are available when needed at the right location. The DRP system provides information that identifies need dates, replenishment dates, and order dates for material requirements. This information can be unified within the distribution system to provide one set of numbers to be used by everyone.

See **Process industry scheduling; Resource planning: MRP to MRP II and ERP; Supply chain management.**

Reference

[1] Martin, Andre J. (1983). *Distribution Resource Planning: Distribution Management's Most Powerful Tool*, Oliver Wight Limited Publications.

DOABLE MASTER PRODUCTION SCHEDULE

Optimized production technology (OPT) when used as a scheduling tool performs finite loading of bottleneck resources. This results in better utilization of the bottleneck capacity. OPT enhances master production scheduling (MPS) by using the data from the MPS to establish how doable the MPS is.

See **Bottleneck resource; Finite loading; Master production scheduling; Optimized production technology; Theory of constraints in manufacturing.**

DOWNSIZING

Downsizing are the actions by an organization to reduce the number and scope of activities performed. Removing activities are those in which the firm does best in relation to its competitors. Downsizing often leads to a reduction in personnel and is associated with reductions in the number of employees within a firm.

See **Business process reengineering (BPR); The future of purchasing.**

DOWNSTREAM SENSITIVITY

Many tasks in product development projects are sequentially dependent. That is, a "downstream" task requires information from the "upstream" task, but not *vice versa*. In spite of the sequential relationship of the tasks, the quest to reduce time to market may justify conducting the tasks in an overlapped fashion. Task overlapping leads to a direct time saving and is one important aspect of concurrent engineering.

However, overlapping carries a risk. In a fully sequential process, downstream starts with finalized information from upstream, whereas in an overlapping process it has to rely on preliminary information which is prone to change. If upstream information changes substantially (as the task progresses), or if a large part of downstream work is made obsolete by an upstream change, overlapping may cause so much downstream rework that the project may be delayed rather than accelerated in addition to increased costs.

See **Concurrent engineering; Product development and concurrent engineering.**

DOWNSTREAM SUPPLY CHAIN MEMBER

It refers to a trading partner or supplier at the latter end of a supply chain or a distribution channel, who is closer to end-customers.

See **Bullwhip effect in supply chain management.**

DRP

See **Distribution requirements planning.**

DRUM-BUFFER-ROPE SYNCHRONIZATION

Eliyahu M. Goldratt, author of the Theory of Constraints (TOC), uses the idea of drums, buffers, and ropes to implement TOC in the typical operation where the constraint resource (or bottleneck) is somewhere other than the beginning of the production operation. A drum sets the tempo or pace of the work for the plant, and is particularly useful for downstream operations after the bottleneck in anticipating output from the bottleneck. Buffer inventories in front of the constraint assure that the bottlenecked operation works at its maximum output and is not idle waiting for another upstream operation (operation before the bottleneck). Ropes restrain the upstream operations from overloading the bottleneck with too much work-in-process inventory.

See **Theory of constraints in manufacturing management.**

Reference

[1] E.M. Goldratt and J. Cox (1984). *The Goal: A Process of Ongoing Improvement* (North River Press: Croton-on-Hudson, NY).

DRY SORTING

Dry sorting is a method of separating a mixture of different materials by distinguishing the differences in material properties, such as size or density. Materials with different particle sizes can be sorted by screening the smaller ones from the larger ones.

See **Environmental issues: Reuse and recycling in manufacturing systems.**

DYNAMIC CAPACITY ALLOCATION

The process of allocating the available capacity to new customer orders as they are received. The process may result in changes to existing schedules.

See **Capacity management in make-to-order production systems.**

DYNAMIC DUE DATE

Many changes can occur within a manufacturing environment that require the frequent revision of due dates for open orders. Dynamic due date (DDD) maintenance is a responsibility of the MRP planner. The advantage of dynamic due dates is that a high level of customer service can accrue from using timely, accurate dispatching information. Arguments against implementing dynamic due dates include the effects of nervousness on scheduling performance, causing instability, and overloading the production planner with rescheduling information for processing. The concept of dynamic due dates is still under debate. Dynamic due dates are calculated as follows:

$$\text{DDD} = \text{Current date} + (\text{On hand inventory} - \text{balance Buffer stock})/\text{Average daily usage}$$

See **Dispatching decision; Nervousness in material requirements.**

Reference

[1] Vollmann, T.E., Berry, W.L., Whybark, D.C. *Manufacturing Planning and Control Systems*, Fourth Edition, Irwin/McGraw-Hill, New York, 1997.

DYNAMIC KANBAN CONTROL

Dynamic kanban control (DKC) is the general name for a class of kanban control policies whose aim is to systematically manipulate the number of kanbans in a just-in-time (JIT) system in order to achieve certain performance goals. DKC policies have been developed to cope with variations in demand, sudden equipment breakdown, and processing time variations.

See **Adaptive kanban control; Dynamic kanban control for JIT manufacturing; Flexible kanban control; JIT; Reactive kanban control.**

DYNAMIC KANBAN CONTROL FOR JIT MANUFACTURING

Kendra E. Moore

ALPHATECH, Inc., Burlington, Massachusetts, USA

Elif Kizilkaya and Surendra M. Gupta

Northeastern University, Boston, Massachusetts, USA

DESCRIPTION: During the past two decades, the desire to achieve manufacturing excellence has become the driving force behind various innovative manufacturing techniques. One of the best known is the just-in-time (JIT) technique. The primary objective of JIT is to produce the right quantity of product in the right place at the right time, while maintaining minimal work-in-process (WIP) and minimal stock of completed work. JIT systems are "pull" production systems in which the product is manufactured in response to a specific demand. Control of JIT systems is maintained using kanbans (Japanese for "visible sign" or card). JIT has been enormously successful in Japan, where it was developed in the 1970s (Sugimori, *et al.*, 1977). Since then, many manufacturing companies outside of Japan have either implemented or are considering implementing JIT systems.

JIT systems were originally designed for deterministic production environments with a smooth and stable demand and constant processing times; their performance is optimum in that environment. Once implemented, however, JIT systems face the uncertainties inherent in any manufacturing system, including variations in processing time and demand, equipment malfunctions, as well as known or planned interruptions such as preventive maintenance. These variations can lead to blocking and starvation (Berkley, 1990). A workstation (WS) is blocked when all of its kanbans are attached to full containers in its output buffer; a WS is starved when its machine is idle and at least one of its kanbans is waiting for a container at the preceding WS's kanban post. The result may be lowered production throughput, decreased machine utilization, increased order completion time, and greater backlogs or increased overtime requirements. Increasing the number of kanbans can help smooth out the variations in processing time by providing more buffer between workstations. This, in turn, can improve production flow, lower order completion time, and increase throughput, at the cost of increased WIP.

The overall goal of the JIT production philosophy is to reduce or eliminate the variations that can lead to these problems. The ability to minimize variations may be effected by a variety of factors. These include cultural and economic influences, as well as distributor and supplier relationships. In Japan, where JIT was first introduced, there exist strong cultural factors which support the teamwork approach advocated in the JIT philosophy. This is less true in other countries including the United States. In addition, changing economic conditions and greater mobility in the work force may lead to higher turnover when employees leave for better paying jobs. The geographical distribution of suppliers and distributors in the United States, also leads to greater variation in supply and demand. Although JIT manufacturers should work to minimize all sources of variation, they may be limited in their ability to control some of these exogenous factors. As a result, even those manufacturers who benefit from a JIT are likely to experience the problems associated with variations.

As Hall, (1983): Stevenson (1996) reported, JIT production supervisors vary the number of kanbans on an *ad hoc* basis in response to internal and external system conditions (shortages or excess inventory). Unfortunately, this *ad hoc* approach may result in adding or subtracting kanbans too early or too late to achieve the desired results. By not adding extra kanbans soon enough or removing them too early, the system will continue to experience blocking and starvation; by adding extra kanbans too early or removing them too late, the system will have greater than desired WIP. Selecting the wrong time to increase or decrease the number of kanbans may lead to higher operating costs. In response to this situation, a number of researchers have recently begun to explore dynamic kanban control (DKC) policies; DKC policies have been referred to as flexible, adaptive, or reactive kanban control policies.

The goal of DKC policies is to systematically add and remove kanbans so as to achieve improved system performance. Under a DKC policy, kanbans are added when needed to improve system performance and removed when they are no longer needed or when their presence will result in lowered system performance. In general, we want additional kanbans when the benefits of their presence (e.g., reduced blocking and starvation, improved throughput) outweighs the costs (e.g., increased WIP and operating costs). The point at which kanbans are added is referred to as the *release threshold*; the point at which they are removed is the *capture threshold*. Specifying the release and capture thresholds, and the number of kanbans to release or capture, for different operating scenarios is the focus of research in DKC.

Classifying Dynamic Kanban Control Policies

DKC policies can be classified according to several criteria: objectives, release/capture threshold, release/capture method, decision periodicity, decision horizon, model structure, and type of kanban system. Each of these criteria is described below.

1. *Objectives:* In general, the objective of a DKC policy is to improve the performance of a JIT system which is operating under some type of variation (e.g., processing, demand, or supply variations). Possible objectives include: reducing order completion time and backorders, increasing throughput, increasing utilization, and reducing WIP. A DKC policy may address one or more of these objectives.
2. *Thresholds:* What are the release and capture thresholds and how are they calculated? For example, the release/capture thresholds may be a function of the current inventory (WIP), current demand, or expected demand. Related issues include the locality of the threshold calculations (local to an individual workstation or global), and the type of information available for the threshold calculation.
3. *Release/Capture Method:* How are the extra kanbans added/removed? Possibilities range from all at once to one at a time. Are they added/removed locally (one workstation at a time) or globally (at all workstations)?
4. *Decision Timing:* When are the threshold calculations made? When is the current situation reviewed? Possibilities include: daily, hourly, when new demand arrives, and when a product is generated.
5. *Decision Horizon:* For what length of time does the control policy apply? Possibilities range from a multi-day planning horizon to an immediate planning horizon
6. *Model Structure:* What is the underlying model structure? Possibilities include queuing system, simulation, etc.
7. *Type of Kanban System:* Is the system a one- or two-card kanban system? Does it handle single or multiple products? Are kanbans transferred according to a fixed-quantity or fixed-cycle withdrawal? (Berkley, 1992)

HISTORICAL PERSPECTIVE: DKC is a new area of research with three known research approaches to studing DKC policies. In the following paragraphs, we describe these emerging DKC studies.

Gupta *et al.* (1995) and Tardif *et al.* (1996) both distinguish between *base* kanbans and *extra* kanbans. *Base* kanbans are the kanbans that are always present in the system. The minimum number of kanbans is bounded by the number of base kanbans. *Extra* kanbans are additional kanbans that may be added or removed from the system, based on the control policy. The maximum number of kanbans is bounded by the number of base kanbans plus the number of extra kanbans. In contrast, Takahashi *et al.* (1996) use DKC policies that do not establish minimum or maximum levels of kanbans. Rather, the minimum and maximum are effectively controlled by the demand which is bounded.

Gupta *et al.* (1995) evaluate a DKC policy for a two-card, single-product, fixed-quantity withdrawal kanban system, where demand arrives once per day (in a batch order). The release and capture thresholds are a function of the current demand and production levels at each workstation. The release threshold is evaluated once per day, when the daily demand arrives. Extra kanbans are immediately added to each workstation whose release threshold has been exceeded. The capture threshold is evaluated at each workstation when a kanban is detached from a container. The extra kanbans are individually removed from the workstations as they meet their production goals for the day. The objective of this kanban control policy is to reduce backlog and order completion times, with minimal increase in WIP, in the presence of variations in processing times. This control policy utilizes an immediate planning horizon. The policy is evaluated for a range of parameters using a simulation model. Moore and Gupta (1999) extend this approach to a multi-product system modeled using stochastic, colored Petri nets.

Gupta and Al-Turki (1997) evaluate a DKC policy for a two-card, single-product, fixed-quantity withdrawal kanban system. The release and capture thresholds are a function of the demand, processing times, and available production time. Demand is assumed to be known for a fixed number of days in advance (planning horizon). The algorithm sets the number of extra kanbans, and schedules their addition and removal during the planning horizon; extra kanbans can be added or removed to workstations globally (all at once) or locally (one workstation at a time). The control policy is reevaluated once each day, when the future demand information arrives and the current status is updated. The goal of the control policy is to minimize order completion time and backlog, while keeping WIP to a minimum.

Gupta and Al-Turki (1998b) evaluate a DKC policy for a two-card, single-product, fixed-quantity withdrawal kanban system with sudden material handling system (MHS) breakdowns, where only a portion of the line depends on the MHS. The goal of the policy investigated is to minimize order completion time and backlog, while keeping WIP to a minimum. Simulation is used to evaluate the performance of the DKC policy.

Tardif and Maaseidvaag (1997) propose a DKC policy for a single-stage pull control system. The system is a single-card, single-product, fixed-quantity withdrawal kanban system. Orders arrive randomly according to an exponential distribution. The release and capture thresholds are a function of demand and finished inventory levels. The thresholds are evaluated when a new demand arrives and when a part enters the finished inventory; hence, the control policy has an immediate time horizon. The goal of this DKC policy is to reduce holding and inventory cost in the presence of extreme variations in demand and lead times.

Takahashi and Nakamura (1996), Takahashi and Nakamura (1997a) Takahashi and Nakamura (1997b) study a reactive kanban control policy for a single-card, single-product, fixed-quantity withdrawal kanban system with unstable demand changes (changes in the mean interarrival time of demand). The release and capture thresholds are a function of future demand. Unstable demand changes are detected by monitoring the time series of the demand, filtering the time series using exponential smoothing, and comparing the filtered demand information with an exponentially weighted moving average. When an unstable change in the mean interarrival time is detected, the number of kanbans is increased or decreased according to the size of the unstable change in the demand. The goal of the control policy is to achieve a desired order completion time and WIP. Threshold calculations are performed for all stations when unstable demand changes are detected. Each station's kanbans are adjusted locally when the thresholds are reached. All three papers use simulation to evaluate the proposed DKC policies.

STRATEGIC PERSPECTIVE AND TIMING: DKC policies provide a means for improving system performance in terms of reduced order completion time and backlog, without drastically increasing WIP. In fact, DKC policies can actually reduce WIP (vs. traditional kanban control), without increasing order completion time or backlog. DKC policies are suitable for use in systems which experience variations in processing time, demand, lead time, and so forth. Within these systems, DKC policies are most effective in those systems which have a high demand with short individual workstation processing times.

IMPLEMENTATION: Although analytic methods may be used to analyze small systems (e.g., Tardif and Maaseidvaag, 1997), the complexity introduced by varying the number of kanbans dynamically during a production cycle requires the use of simulation. Eventually, these simulation models could be turned into control software and used online to monitor system behavior and control the release and capture of extra kanbans.

RESULTS AND CASES: Work in the area of DKC is currently in the research stage. To date, a number of research studies have shown that this approach outperforms the traditional kanban control methods. Studies have shown that DKC results in lower backlog and order completion time without significantly increasing the average WIP, when compared to static kanban control. Further, the WIP which is carried over from one production period to another, is actually lower for this DKC policy than for the static policy. The greatest benefits for DKC are achieved when the volume is high and the individual processing times are low.

Based on early case studies, DKC appears to be a promising approach for improving the performance of kanban systems in the presence of system variations.

See **Adaptive kanban control; Captive threshold; Dynamic kanban control; Flexible kanban control; JIT; Just-in-time manufacturing implementation; Just-in-time manufacturing; Kanban-based manufacturing**

systems; Kanban; Reactive kanban control; Release threshold; Two-card system.

References

[1] Berkley, B.J. (1990). "Analysis and approximation of a JIT production line: a comment." *Decision Sciences*, 21, 660–669.

[2] Berkley, B.J. (1992). "A review of the kanban production control literature." *Production and Operations Management*, 1, 393–411.

[3] Gupta, S.M., and Y.A.Y. Al-Turki (1997). "An algorithm to dynamically adjust the number of kanbans in stochastic processing times and variable demand environment." *Production Planning and Control*, 8, 133–141.

[4] Gupta, S.M., and Y.A.Y. Al-Turki (1998a). "Adapting just-in-time manufacturing systems to preventive maintenance interruptions." *Production Planning and Control*, 9(4), 349–359.

[5] Gupta, S.M., and Y.A.Y. Al-Turki (1998b). "The effect of sudden material handling system breakdown on the performance of a JIT system." *International Journal of Production Research*, 36(7), 1935–1960.

[6] Gupta, S.M., Y.A.Y. Al-Turki, and R.F. Perry (1995). "Coping with processing times variation in a JIT environment". *Proceedings of Northeast Decision Sciences Institute Conference.*

[7] Hall, R.W. (1983). *Zero Inventories*. Dow Jones-Irwin, Homewood, Illinois.

[8] Moore, K.E., and S.M. Gupta (1995). "Stochastic, colored Petri net models of flexible manufacturing systems: materials handling systems and machining." *Computers and Industrial Engineering*, 29, 333–337.

[9] Moore, K.E., and S.M. Gupta (1999). "Stochastic, colored Petri net (SCPN) models of traditional and flexible kanban systems." *International Journal of Production Research*, 37(9), 2135–2158.

[10] Sepehri, M. (1986). "Just-in-time, not just in Japan: case studies of American pioneers in JIT implementation." *Proceedings of American Production and Inventory Control Society.*

[11] Stevenson, W.J. (1996). *Production/Operations Management* (5th Ed.). Irwin, Homewood, Illinois.

[12] Sugimori, Y., K. Kusunoki, F. Cho, and S. Uchikawa (1977). "Toyota production system and kanban system materialization of just-in-time and respect-for-human system." *International Journal of Production Research*, 15, 553–564.

[13] Takahashi, K., and N. Nakamura (1996). "Reactive JIT ordering systems for unstable demand changes". *Proceedings of Advances in Production Management Systems*, Information Processing Society of Japan.

[14] Takahashi, K., and N. Nakamura (1997a). "Reactive logistics in a JIT environment." *Proceedings of 3rd International Symposium on Logistics*, SGEditoriali.

[15] Takahashi, K., and N. Nakamura (1997b). "A study on reactive kanban systems." *Journal of Japan Industrial Management Association (in Japanese)*, 48, 159–165.

[16] Tardif, V., and L. Maaseidvaag (1997). Adaptive single-stage pull control systems. University of Texas at Austin, Graduate Program in Operations Research and Industrial Engineering, Austin, Texas.

DYNAMIC ROUTING

The routing and scheduling of MCFMS can be considered at three levels according to the structure of MCFMS. The top level problem is to schedule the material handling system and route the parts among cells. The second level is to scheduling the material handling device in each cell and route the parts among the machines in the cell. The bottom level is to sequence parts on each individual machine and determine the tool changing sequence. Using conventional analytical methods, such as array-based clustering, similarity coefficient-based clustering, and mathematical programming, each part can only belong to one part family. The purpose of dynamic routing of part families among FMCs is to prevent a workload imbalance among FMCs or an imbalance of machine loads within each FMC. These imbalances can cause excessive flowtimes and tardiness and improper machine utilization. However, it is difficult, if not impossible, to implement dynamic routing of part families among FMCs using a conventional 0-1 part-family incidence matrix that assumes part families are mutually exclusive.

See **Dynamic routing in flexible manufacturing systems.**

DYNAMIC ROUTING IN FLEXIBLE MANUFACTURING SYSTEMS

H. Joseph Wen

Kenneth D. Lawrence, New Jersey Institute of Technology, Newark, NJ, USA

Traditionally, production facilities have two conflicting objectives: flexibility and productivity. Flexibility refers to producing a number of distinct products in a job shop environment where opportunities for production variability exist. Productivity, on the other hand, refers to a high rate production. Studies have demonstrated

Table 1. 0-1 Part-family incidence matrix.

	Part numbers										
	P1	P2	P3	P4	P5	P6	P7	P8	P9	P11	P12
Cell 1 = Part Family 1	0	1	1	0	0	1	0	0	0	0	1
Cell 2 = Part Family 2	1	0	0	1	0	0	1	1	0	0	0
Cell 3 = Part Family 3	0	0	0	0	1	0	0	0	1	1	0

that the productivity in a job shop environment is low as compared to flow shop (Hutchinson, 1984). Increasing job shop productivity while maintaining production flexibility has been a desired but difficult goal. Flexible manufacturing systems (FMS) attempt to fulfill this goal.

MCFMS: A multicell flexible manufacturing system (MCFMS) is a type of FMS which consists of a number of flexible manufacturing cells (FMCs) connected by an automated material handling system (MacCarthy and Liu, 1993). The routing in a MCFMS is at three levels. The top-level problem is to route the parts among cells, and the second-level problem is to route parts among the machines in a cell. The third-level problem is to sequence parts on each individual machine and to determine the tool changing sequence.

Fixed and Dynamic Routing: A conventional manufacturing cell is capable of processing only a fixed number of part types; thus, conventional part-family formation methods are not appropriate for FMCs. The conventional methods, such as array-based clustering (King, 1980; King and Nakornchai, 1982), similarity coefficient-based clustering (McAuley, 1972; Seifoddini and Wolfe, 1986), and mathematical programming (Kusiak, 1987; Gunasingh and Lashkari, 1989), can assign a part to only one machine cell. Thus, each part type has a fixed route through the system, as shown in Table 1. For example, in Table 1, Part P1 is processed in Cell 2 and P2 in Cell 1. When each part is limited to a fixed route through the system, the performance of an MCFMS is diminished because the inherent flexibility of the FMS is not fully utilized (Chen and Chung, 1991).

Table 2 illustrates that a fuzzy part-family method that assumes part families are not mutually exclusive allows for dynamic routing of part families among FMCs. The dynamic routing method provides priorities for all the possible routes among FMCs. Route priority is based on the "degree of membership" of a part in the various part families associated with cells. For example, the first routing priority for part P1 is Cell 2 because it has the highest degree of membership (.7) associated with Family 2 and the second (membership = .2) and third (membership = .1) routing priorities are Cell 1 and Cell 3, respectively.

It is important to consider the part routing problem in a MCFMS production system with machine breakdowns. In most FMS scheduling and control problems discussed in the literature, a machine is assumed to be continuously working. However, this is often an unrealistic assumption because unpredictable machine breakdowns occur from time to time. When the performance of the dynamic routing method using fuzzy clustering and certainty factors techniques is compared with a traditional fixed routing method based on three different performance criteria, the authors found interesting results.

Conventional research involving part-family formation has included integer programming (Kusiak, 1987; Gunasingh and Lashkari, 1989; Bruyand *et al.*, 1988), the generalized assignment model (Shtub, 1989), a graph-theoretic approach (Rajagopalan and Batra, 1975), Rank Order Clustering (ROC) (King, 1980; Chandrasekharan and Rajagopalan, 1986), the Similarity Coefficient Method (SCM) (McAuley, 1972; Seifoddini and Wolfe, 1986), and Ideal Seed Non-hierarchical

Table 2. Fuzzy part-family incidence matrix.

	Part numbers										
	P1	P2	P3	P4	P5	P6	P7	P8	P9	P11	P12
Cell 1 = Part Family 1	.2	.6	.9	.3	.2	.7	.4	.2	0	.2	.9
Cell 2 = Part Family 2	.7	0	.1	.6	0	.3	.5	.7	.3	0	0
Cell 3 = Part Family 3	.1	.4	0	.1	.8	0	.1	.1	.7	.8	.1

Clustering (ISNC) (Chandrasekharan and Rajagopalan, 1986). One common weakness of these conventional approaches is their implicit assumption that disjoint part families exist in the data; therefore, a part can only belong to one part family and families are static. In practice, it is clear that some parts definitely belong to certain part families; however, there exist parts that may belong to more than one family. Recent studies suggest that fuzzy logic approaches offer a special advantage over conventional clustering (Chu and Hayya, 1991). Since fuzzy clustering allows for a degree of membership of a part in more than one part family, users have flexibility in assigning a part to a family, which permits workload balancing among machine cells.

Dynamic routing problems have been addressed by a few researchers. For example, Maimon and Choong (1987) studied a dynamic routing problem in which the dynamic routing decisions were dependent on the state of the local station. Yao (1985) and Kumar (1987) took an information theoretic approach to dynamic routing problems. They suggested a measure of flexibility, where parts moved to nearby machines based on reduction of entropy. Their models, however, are not driven by uncertain conditions such as varying demand, and they stop short of showing numerical results for system improvement. Previous work on optimal dynamic routing between two machines has been done by researchers including Ephremides, Varaiya, and Walrand (1980), and Hajek (1984). However, in the MCFMS setting, the routing problem is extremely complicated. There are only some very simple routing problems that can be solved optimally by using integer programming. Accordingly, it is not practical, even for moderate-sized systems, to pursue an optimal solution to the routing problem. Wen, Smith, and Minor (1996) used computer simulation is to compare fuzzy-based performance of the dynamic routing method with the performance of the fixed routing method under a variety of conditions. The system studied was a 22-part, 11-machine, hypothetical MCFMS similar to the one studied by Seifoddini (1989) and Wen *et al.* (1996).

SYSTEM PERFORMANCE: Several criteria have been used in FMS research to measure shop performance. Due date performance and reduction of average inventory level are commonly stressed in today's manufacturing environment. Mean flowtime, mean tardiness, and mean absolute lateness of part orders are often used as FMS performance measurers. Lower mean flowtimes indicate reduction in average inventory. Mean absolute lateness measures reliability in completing orders near the due date by penalizing both early and late completion. Mean tardiness has been the most popular measure of due date performance in the published literature, indicating the average time beyond the due date for order completion.

Wen, Smith and Minor (1996) found that the dynamic routing method performs much better than the fixed routing method on mean flowtime, mean tardiness and mean absolute lateness at high system utilization levels. However, when the cells are at low utilization, the fixed routing method performs better. The dynamic routing method shows better flowtime, tardiness, and absolute lateness when machine breakdowns are considered.

Despite the advantages of the dynamic routing method, further research is warranted. It was expected that the dynamic routing method would work better than the fixed routing method at all system utilization levels. However, research findings do not support this. While studies have examined dynamic routing among FMCs at the system level, future research should also examine the impact of the dynamic routing method within the FMCs at the cell level. It would be valuable to study the effect of combining dynamic routing among FMCs and within the FMCs, that is, both at the system and the cell levels. The findings of such a study can assist as take full advantage of the inherent flexibility of an MCFMS.

See **Coefficient-based clustering; Flexibility in manufacturing; Flexible automation; Flexible manufacturing systems; Fuzzy logic part family formation; Manufacturing flexibility dimensions; Manufacturing flexibility; Mathematical programming methods.**

References

[1] Bruyand, A., Garcia, H., Gateau, D., Molloholli, L., and Mutel, B. (1989), Part families and manufacturing cell formation, *Prolamat's Proceedings*, Dresden, GDR.

[2] Chandrasekharan, M.P., and Rajagopalan, R. (1986), An ideal-seed non-hierarchical clustering algorithm for cellular manufacturing, *International Journal of Production Research*, 24, 451–554.

[3] Chen, I.J., and Chung, C.H. (1991), Effects of loading and routing decisions on performance of flexible manufacturing systems, *International Journal of Production Research*, 29, 2209–2225.

[4] Chu, C.H., and Hayya, J.C. (1991), A fuzzy clustering approach to manufacturing cell formation, *International Journal of Production Research*, 29, 1475–1487.

[5] Ephremides, A., Varayia, P., and Walrand, J. (1980), A simple dynamic routing problem, *IIEE Transactions on Automatic Control*, 690–693.

[6] Gunasingh, K.R., and Lashkari, R.S. (1989), Machine grouping problem in cellular manufacturing systems: an integer programming approach, *International Journal of Production Research*, 27, 1465–1473.

[7] Hajek, B. (1984), Optimal control of two interacting service stations, *IEEE Transactions on Automatic Control*, 29, 491–499.

[8] King, J.R. (1980), Machine-component grouping in production flow analysis: an approach using a rank order clustering algorithm, *International Journal of Production Research*, 18, 231–237.

[9] King, J.R., and Nakornchai, V. (1982), Machine-component group formation in group technology - review and extension, *International Journal of Production Research*, 20, 117–133.

[10] Kumar, V. (1987), Entropy measures of measuring manufacturing flexibility, *International Journal of Production Research*, 25, 957–966.

[11] Kusiak, A. (1987), The production equipment requirements problem, *International Journal of Production Research*, 25, 319–325.

[12] Maccarthy, B.L., and Liu, J. (1993), A new classification scheme for flexible manufacturing systems, *International Journal of Production Research*, 31, 299–309.

[13] Maimon, O.Z., and Choong, Y.F. (1987), Dynamic routing in reentrant flexible manufacturing systems, *Robotics and Computer-Integrated Manufacturing*, 3, 295–300.

[14] Mcauley, J. (1972), Machine grouping for efficient production, *Production Engineering*, 2, 53–57.

[15] Rajagopalan, R., and Batra, J.L. (1975), Design of cellular production systems: a graph-theoretic approach, *International Journal of Production Research*, 13, 567–579.

[16] Seifoddini, H.K. and Wolfe, P. (1986), Application of the similarity coefficient method in group technology, *AIIE Transactions*, 18, 271–277.

[17] Seifoddini, H.K. (1989), Single linkage versus average linkage clustering in machine cells formation, *Computers Ind. Engng.*, 16, 419–426.

[18] Shtub, A. (1989), Modeling group technology cell formation as a generalized assignment problem, *International Journal of Production Research*, 27, 775–782.

[19] Wen, H.J., Smith, C.H., and Minor, E.D. (1996), Formation and dynamic routing of part families among flexible manufacturing cells, *International Journal of Production Research*, 34, 2229–2245.

[20] Yao, D.D. (1985), Material and information flows in flexible manufacturing systems, *Material Flow*, 2, 143–149.

DYNAMIC SCHEDULING

Dynamic scheduling refers to scheduling problems where new jobs are added, in contrast to static scheduling where jobs are fixed. Further, the processing times for the jobs are often stochastic. Different dispatching rules may be used at the work centers. Simulation is appropriate for dynamic scheduling situations.

See **Dynamic Kanban control for JIT manufacturing; Dynamic routing in flexible manufacturing systems; Simulation analysis of manufacturing and logistics systems.**

EARLIEST DUE DATE RULE

The earliest due date (EDD) rule is one of many sequencing rules used in sequencing jobs in a job shop. It prioritizes jobs according to their due date, the job with the earliest due date being scheduled first. The rule is appropriate when the goal is to reduce/minimize job lateness.

See **Dispatching decision; Priority sequencing rules.**

ECONOMIC ANALYSIS

Economic analysis assesses the impact of any decision, such as a decision to invest in advanced manufacturing technologies, in economic terms by determining the dollar values for costs and benefits associated with the decision.

See **Capital investments in advances manufacturing technologies; Industrial robot selection.**

ECONOMIC DOLLAR VALUE ORDER METHOD

The economic dollar value order method may be used when ordering multiple items from the same vendor. Instead of ordering as and when an item reaches an order point, all items from the vendor are ordered to meet demand over a fixed period. An example would be to order all 10 items supplied by a given vendor to meet two weeks of demand.

See **Economic order quantity; Economic time between orders; Inventory flow analysis; Inventory holding cost.**

Reference

[1] Vollmann, T.E., Berry, W.L., Whybark, D.C. *Manufacturing Planning and Control Systems*, Fourth Edition, Irwin/McGraw-Hill, New York, 1997.

ECONOMIC GLOBALIZATION

Economic globalization refers to the increased integration of economic activity across national borders. One area where economic globalization is readily seen is in the amount of foreign direct investment that has occurred in recent years. Foreign direct investment (FDI) typically refers to an investment in a value-added asset such as a company or a facility and generally implies a controlling interest. Approximately 40,000 companies worldwide have foreign direct investments that encompass every type of business activity from agricultural and mining to manufacturing and distribution to services. The 1995 value of foreign direct investments approximated $2.7 trillion. These investments generated over $8 trillion in sales, a number considerably larger than the total gross domestic product of the North American Free Trade Area. Foreign direct investments have become the foundation of many countries' industrialization efforts. Indeed, attracting FDI is critical to the economic development of countries from Brazil to Bangladesh and Mexico to Malaysia and has influenced policy decisions in countries around the world. When local investment capital is a scarce commodity, countries compete vigorously for FDI.

Another indicator of the importance of globalization is the amount of goods exported each year. In 1995, the total worldwide exports of goods and services exceeded $6 trillion – an amount equal to the gross domestic product of the United States. Simply stated, economic globalization has led to much greater economic interdependence among the countries of the world.

From a micro or company-level standpoint, economic globalization has become a dominant influences in strategic decision making. The key questions are, where should the company produce its various products? and where should it sell these products? These questions are especially important as companies face fierce global competition and look for market opportunities as their own home markets become saturated. The vast majority of all firms are involved in some form of global economic activity. That is, they either produce products outside their home countries, sell products in foreign countries, or buy components from worldwide suppliers. Among large companies, global operations are almost a given. Caterpillar, for example, operates 74 manufacturing plants, with 50 percent of these facilities located outside the U.S. As a result, Caterpillar generates 51% of its revenues from international markets (the goal is to increase this percentage to 75% by 2008). Similarly, McDonalds generates 50% of its sales and 60% of its profits in international markets. While the sales and profit numbers vary across Fortune 500 companies, global markets are becoming increasingly important for most firms.

To summarize, economic globalization is a phenomenon that is here to stay. Countries have increasingly linked their economic futures through free trade and other economic treaties (the European monetary union is an example.

See **Global manufacturing rationalization.**

ECONOMIC ORDER PERIOD MODEL

See **Fixed time-period inventory model.**

ECONOMIC ORDER QUANTITY (EOQ) MODEL

The economic order quantity (EOQ) model is the most basic and oldest of all mathematical lotsizing models. Indeed, its roots can be traced to the early twentieth century during the early stages of the movement toward scientific management. Its objective is to determine the optimal (i.e., lowest cost) production or purchase order quantity based on the tradeoff between setup and holding costs.

The EOQ is based on the total production costs for a single item for a period of time such as one year. The key assumptions include: constant and continuous demand rate (D) which is known with certainty; infinite production rate resulting in immediate supply when the order is placed; setup cost (A); unit production cost (v); and a per unit per period holding charge (r = fraction of v). The total relevant costs (TRC) for the period include total inventory holding costs (average inventory Q/2 multiplied by the per unit per period holding charge) and total setup costs (cost per setup multiplied by the number of setups). The formulation is as follows:

$$TRC(Q) = \frac{Qvr}{2} + \frac{AD}{Q}$$

Q is the order quantity. The Q/2 in the first term represents the average inventory level over the period (note that, if the demand rate is continuous and constant, the maximum inventory level will equal the order quantity and the minimum will equal zero, hence the average inventory is one-half the order quantity). The D/Q in the second term represents the setup frequency (note that the demand rate per period divided by the order quantity yields the number of orders per period). Solving this equation yields an optimal order quantity $Q^* =$ square root of (2AD/vr). The optimal order quantity represents the recommended quantity to be ordered with each setup (or purchase) based on these assumptions.

It is critical to recognize Q^* is optimal *only based on above assumptions*. A close look at the assumptions reveals that they are very restrictive and somewhat unrealistic. Thus, managers need to be aware of the model's limitations and maintain realistic expectations. To address these very narrow assumptions, a tremendous body of research has accumulated involving less restrictive models and many more realistic assumptions and scenarios. Nevertheless, despite its simplicity and naivete, the EOQ model captures the bulk of the critical tradeoffs and yields reasonable results in many situations. In general, the effect of more advanced mathematical lotsizing models is to fine tune the initial recommendation of the EOQ model.

See **Just-in-time manufacturing; Lean manufacturing; Marginal analysis; Mathematical lotsizing models; Optimal lot size; Setup reduction; SMED.**

ECONOMIC TIME BETWEEN ORDERS

The economic time between orders (TBO) which is expressed in weeks is a useful inventory control tool. It can be computed by dividing economic order quantity (EOQ) by average weekly usage as shown below:

TBO = EOQ/Average weekly usage rate

TBO serves as an estimate of the ordering frequency or time between inventory reviews where EOQ is practiced.

See **Economic order quantity; Inventory flow analysis.**

Reference

[1] Vollmann, T.E., Berry, W.L., Whybark, D.C. *Manufacturing Planning and Control Systems*, Fourth Edition, Irwin/McGraw-Hill, New York, 1997.

ECONOMIES OF COOPERATION

The notion of partnering is intuitively appealing to most people. However, in a commercial context, the practice of supplier partnering will be adopted only if it is more economical than alternative procurement approaches. In this context, the ability of partnering to create new commercial benefits is important. Such potential benefits can be referred to as 'economies of cooperation.' For example, whereas we use the term 'economies of scale' to refer to unit cost reduction made possible by operating at higher volumes, we can use the term 'economies of cooperation' to refer to unit cost reduction made possible by higher degrees of cooperation through supplier partnering relationship.

The most obvious economies created through the supplier partnering approach come from improved scheduling and from joint efforts at product and process cost reduction. These economies result from the reductions in inventory and production planning and control costs. However, additional intangible benefits are shared learning, which can result in an increased capacity for innovation in product and process development.

See **Strategic supplier partnerships; Supply chain management: Competing through integration; Supplier relationships.**

ECONOMIES OF SCALE

This refers to productivity increases made possible through the high-volume production larger is a given facility.

See **Flexibility in manufacturing; Flexible automation; Learning curve analysis; Manufacturing flexibility.**

ECONOMIES OF SCOPE

See **Flexible automation; Flexibility in manufacturing; Manufacturing flexibility.**

EFFECTIVENESS

Effectiveness refers to successful goal attainment. Where there is no goal, effectiveness cannot be measured. If on-time delivery is a goal, then effectiveness may be measured in terms on-time delivery.

See **Manufacturing strategy.**

EFFICIENCY

Efficiency is a measure of the outputs with reference to inputs. It can be expressed on the ratio (outputs) (inputs). In an efficient system, this ratio will be larger than in less efficient systems. When inputs are measured in terms of costs, it is common to express efficiency as cost efficiency, where cost is the denominator in the above ratio.

See **Effectiveness; Performance measurement in manufacturing.**

E-COMMERCE/ELECTRONIC COMMERCE

EC is an inter-organizational information system that facilitates business-to-business electronic communication, data exchange, and transaction support through a web of either public access or private networks; also known as e-commerce.

See **Bullwhip effect in supply chain management; Electronic data interchange in supply chain management.**

ELECTRONIC DATA INTERCHANGE (EDI)

EDI refers to the direct computer-to-computer exchange of business information in a standard communication format. EDI allows transaction documents, such as purchase orders, invoices, and shipping notices, to be transmitted electronically and entered directly into the supplier's computer or to a third-party network for processing.

See **Electronic data interchange in supply chain management; The future of purchasing.**

ELECTRONIC DATA INTERCHANGE IN SUPPLY CHAIN MANAGEMENT

Hokey Min

University of Louisville, Louisville, KY, USA

Introduction: Electronic data interchange (EDI) is an inter-organizational information system that is intended to facilitate computer-to-computer exchange of data among multiple business trading partners in standard, machine readable formats through a web of business-to-business communication network. EDI is useful for efficient and effective supply chain management, because EDI can help reduce lead time, save documentation processing cost, eliminate procurement errors, clarify inventory status information, and enhance strategic alliances throughout the supply chain. In a broader perspective, EDI plays three important roles in supply chain management. These roles are classified as: (1) electronic integration; (2) information diffusion and sharing; and (3) electronic marketplaces.

Electronic integration is referred to as the integration of two or more independent organizational units' business processes by exploiting the capabilities of computers and communication technologies (Venkatraman and Zaheer, 1990). The establishment of a common computer network via EDI will lead to an integration of communication and information processing activities at all of the supply chain locations and among the different business functions such as design, procurement, manufacturing, distribution and marketing. Increased integration of efforts across supply chains, in turn, enables firms to compress the time to bring a product to the market. Also, electronic integration formed by EDI can help reduce manufacturing cost and enhance product quality by involving suppliers in the product design and development process. Furthermore, Piore and Sable (1984) observed that electronic integration via EDI might assists the shift in from mass production to flexible, specialized production due to increasing outsourcing opportunities.

EDI allows more information to be communicated across the supply chain in the same amount of time through faster electronic transmission of data. Thus EDI is likely to reduce information search cost associated with acquisition of information about product offerings and market prices, thereby expedites information diffusion. EDI linkages also increase interdependence across the supply chain and enable sustainable long-term relationships and risk sharing.

EDI can serve as intermediaries connecting the buyers and the suppliers in a vertical supply chain thereby creating an electronic marketplace. An electronic marketplace is defined as an inter-organizational information exchange mechanism that allows the participating buyers and the suppliers to access market information in a more efficient and timely fashion and subsequently makes comparison shopping much easier (e.g., Bakos, 1991). With abundant market information available from the electronic marketplace, EDI participants can compare a fairly large number of product offerings in a short period of time. Therefore, EDI may lead to a high degree of product differentiation.

STRATEGIC IMPLICATIONS OF EDI: Currently, there is no commonly accepted guideline for measuring the strategic significance of EDI to the firm's competitiveness. Nevertheless, EDI can be seen as a competitive weapon because EDI affects the bargaining power of buyers and suppliers, the extent of barriers to market entry, and the degree of strategic alliances among trading partners. To elaborate, EDI enhances the buyer's ability to gather information concerning the competing suppliers' product prices and offerings. That is to say, with the help of EDI, the buyer will be better informed about the particular supplier's price advantages and product availability relative to other competing suppliers. Therefore, EDI is likely to promote price competition among the suppliers and consequently increases the bargaining power of the buyer. On the other hand, if EDI technology is too proprietary or costly for the non-EDI suppliers to adopt, EDI may diminish the possibility of replacing the existing EDI suppliers with the more favorable non-EDI suppliers. Thus, EDI can increase the bargaining power of such EDI suppliers.

By the same token, the exclusive and proprietary nature of EDI network between the buyer and his/her existing EDI suppliers may create deterrents (so-called network externalities) for new market entrants and consequently limit the potential pool of competing suppliers. Whereas, EDI may build barriers to entry, EDI can also solidify strategic alliances among EDI participants through electronic integration of their communication and information processing activities. Furthermore, EDI increases synchronization across the vertical supply chain where the upstream unit's output becomes the input to its downstream units. In particular, an improved information link via EDI allows better control of buffer inventories throughout the entire supply chain and facilitates "pull" manufacturing strategies.

BENEFITS AND RISKS OF EDI: In addition to its strategic gains described above, EDI provides numerous operating benefits. Some of these benefits are documentation cost savings resulting from reduced paper transactions, shorter order cycle time resulting from speedy transmission of purchase data, and simplified purchase payment resulting from electronic funds transfer. Additional operating benefits of EDI include reduced mailing expenses and better cash management. Despite substantial benefit potentials, not every practicing manager is ready to embrace EDI because some serious risks of EDI include a host of security, legal, and compatibility problems. To overcome such risks, the following practical guidelines are suggested.

Security: EDI encompasses the transmission of business-related proprietary information across intra/inter-organizational boundaries through electronic media. Due to EDI necessitating the integration of the firm's internal information system to its trading partner's external information systems, it often poses the risk of information interception by unauthorized hackers not directly involved with the transaction. In addition to the risk of unauthorized data access, contents of the electronic messages can be deliberately altered and mis-sequenced during the electronic transmission. Therefore, without systematic control over the access of information, EDI is shunned by many wary firms. There are, however, some remedies that may alleviate security problems involving privacy, authentication, integrity, and disinformation by protecting data transfers through EDI. These methods are: (1) Digital signature; (2) Access control; (3) Encryption; and (4) Value-added network (VAN).

Digital signature can be used to not only identify and validate who sent the EDI message originally, but also verify that no unauthorized changes in EDI message have been made after a contract was signed. Thus digital signature can provide message authenticity and integrity by ensuring "non-repudiation", protecting a recipient against a sender who denies that he/she is the source of the message. The *access control* mechanism may include user identification, passwords, gatekeepers and firewalls that limit or deny unauthorized access to the EDI network through dedicated safeguards. *Encryption* is intended to preserve information integrity by developing ways to hide (encode) information and then recover (decrypt) it by only the intended recipient with the proper software (decryption) key. Such methods include scrambling of data that prevents the intruder from deciphering and understanding data during electronic transmission. *VAN* can work as a safety buffer which ensures an accurate, seamless, and secure connection between the message sender and the receiver via electronic mailboxes set up by the third-party network service provider. VAN can often increase data security by requiring both a specific mailbox identification code and a password.

Legality: Contracts involving the purchase of goods are governed by the uniform commercial code (UCC) which is based on written agreements between the buyer and the supplier. In particular, the UCC mandates contracts for the purchase of goods priced at $500 or more to be evidenced by a writing, a signature, and a hard copy of written documents (Murray, 1996). Since the purchasing contract via EDI may not leave any paper trails of written agreements, it can be repudiated by either the buyer or the supplier and invalidated under the UCC. Such repudiation may stem from three types of fraud: (1) *Supplier fraud* with a false claim that the sale was not authorized; (2) *Merchant fraud* when a buyer never authorizes a transaction; and (3) *Third-party fraud* committed through the unauthorized use of an account (Cooper *et al.*, 1996). In addition, EDI may increase the risk of overpayments to suppliers with subsequent kickbacks and bribery. Furthermore, with the advent of world-wide EDI, buyers and suppliers from different countries under different jurisdictions may face disputes over purchasing contracts and shipment claims. For instance, international trade laws may vary on what constitutes an order confirmation and acknowledgment. Considering the legal complexity of EDI, trading partners of EDI should agree upon specific legal definitions, responsibilities, and obligations prior to final purchasing contracts via EDI.

Compatibility: In general, compatibility may be defined as the degree to which EDI is perceived as being consistent with the existing values, past experiences, and needs of the potential adopter (Rogers, 1983). In the context of EDI, there are two additional important compatibility issues: organizational and technical (Premkumar *et al.*, 1994). *Organizational compatibility* problems may arise in a form of organizational resistance to EDI when EDI adoption leads to radical organizational restructuring and extensive re-engineering of work procedures. For example, paperless transaction would reduce clerical and administrative works and thereby redefine the job responsibility of some employees. Organizational compatibility, therefore, is directly related to organizational readiness in terms of the EDI user's or the potential adopter's willingness to change, degree of computerization, and financial capacity. *Technical compatibility* problems often stem from non-uniformity of EDI standards. Since EDI standards can shorten the length of messages and reduce data entry errors, EDI requires a standardized data format that allows the computer to read, process, and interpret transmitted data. Thus the buyer's hardware and software supporting EDI should be consistent with those of his/her trading partners. Otherwise, the use of incompatible hardware platforms, transaction standards, and communication protocols can hamper seamless transfer of data between the computers.

Incompatibility can pose a serious problem because EDI currently has the three different levels of standards: industry-specific, national, and international standards. Examples of industry-specific standards include uniform communication standards (UCS) used by the grocery industry, and transportation data coordinating committee standards (TDCCS) used by the transportation industry. As a national standard, the American National Standards Institute (ANSI) ASC X12 standard supports cross-industry business transactions in both the U.S. and Canada. To facilitate international trade, international standards such as the United Nations/Electronic Data Interchange for Administration, Commerce, and Transport (UN/EDIFACT) were developed in 1986. Despite the progress, a lack of widely-accepted EDI standards within a particular industry and across industries can increase the barrier of entering into EDI linkages with trading partners.

ILLUSTRATIONS OF EDI APPLICATIONS: Supply chain management (SCM) is concerned with coordinated linkage, design and control of material and information flows across the vertical value chain involving procurement, production, and distribution. Its major aim is to increase joint problem solving and concerted cost reduction efforts among different organizational units or trading partners through information and risk sharing. section illustrates selected cases of successful EDI implementation as well as potential EDI applications for SCM.

1. ***EDI in procurement:*** Due to the large amount of documentation requirements, EDI is frequently used to promote paperless transactions in purchasing. For instance, Ford, Chrysler, Wal-Mart, RJR Nabisco and Black & Decker electronically generate purchasing documents such as requests for quote (RFQ), purchase requisitions, purchase orders, and invoices through EDI. They reported tremendous savings of time and cost during the preordering and ordering process. Also, EDI can support purchasing bids. In fact, the Department of Defense launched a pilot military procurement project allowing suppliers to bid electronically on relatively small government purchases under $25,000 (Messmer, 1992). EDI suppliers can participate in government bids through a VAN hub located at Lawrence Livermore National Laboratory. An important benefit of EDI to purchasing is that EDI frees purchasing professionals from repetitive clerical responsibilities and subsequently enhances their professionalism by allowing them to concentrate on their core professional and managerial duties.
2. ***EDI in production:*** Since EDI can help closely track and monitor the flow of material from the suppliers and their downstream manufacturer, EDI may lower the buffer inventory level in the production process. Buffer inventory is also reduced by EDI due to its ability to reduce order fulfilment cycle and order filling errors. For example, Navistar reported a reduction $167 million in inventory in the first 18 months it was involved in EDI (Milbrandt, 1987). Reduced inventory not only improves inventory turnover, but also increases the awareness to quality failures and consequently helps improve product quality. A recent field study conducted by Kekre and Mukhopadhyay (1992) verified that EDI transactions had favorable impacts on quality and inventory measures. Furthermore, Walton (1996) discovered that EDI was used to facilitate a vendor managed inventory (VMI) system which can provide suppliers with more control of their products on the shelf, to develop production schedules to reflect actual replenishment needs, and to tie downstream manufacturers to sales via point-of-sale (POS) data transmission. In light of the above, EDI contributes to JIT manufacturing principles that require real-time access to supplier information, increased responsiveness to trading partners, and better control of multi-echelon inventories along the supply chain.

   In addition to JIT applications, EDI information can be exploited to better forecast customer demand and plan more precise production schedules. Similarly, EDI can be used to share Computer Aided Design (CAD) information with the suppliers for timely and seamless product development. Despite its enormous application potentials, Premkumar *et al.* (1994) observed that the extent of EDI applications to production areas such as inventory control and production planning has been minimal.
3. ***EDI in Distribution:*** As competition in the distribution industry intensified in the aftermath of deregulation and globalization,

distribution service providers attempted to differentiate themselves from one another by providing superior customer services. Since superior customer services cannot be sustained without accurate, real-time information on material movements in the distribution pipeline, EDI is widely utilized in distribution operations. For example, EDI is used to transmit the information concerning advanced shipping notice, advanced bookings, shipment status, and carrier routing guides directly to the customer so that the customer can schedule delivery, prevent critical delays and waiting time at the customer's (shipper's) dock. Thus EDI can reduce distribution bottlenecks.

EDI applications in the transportation sector include automation of freight invoicing and billing, remittance advices, rate audit services and bill of lading exchanges. Typical EDI users in the motor carrier industry include many leading freight carriers such as Consolidated Freightways, Roadway Express, Carolina Freight, United Parcel Service, J.B. Hunt, Matlack, and Yellow Freight (Anderson, 1992).

Current EDI usage, however, is not limited to traditional traffic operations. Notable emerging EDI applications include intermodal EDI and port-to-port EDI systems in global distribution environments. *Intermodal EDI* deals with the interchange of information on bookings, shipment status, cargo tracking, and billing that is necessary to facilitate multi-modal movements, particularly information required to locate or interchange the container between different modes such as ocean and motor carriers (LeRose, 1993). With the passage of the Customs Modernization Act of 1993 that reduces the mandate for paper-based import/export documents, *port-to-port EDI* has become more prevalent in customs clearance procedures. Port-to-port EDI may replace or reduce the traditional function of customs brokers through communication and documentary supports. For instance, international ports such as the Singapore Port Authority and the Australian Customs Office provided automatic customs clearance as well as on-line computer access for consignment status, container arrival notice, cargo advice, rail requirements, dangerous goods information and statistical trade information via port-to-port EDI (Branch, 1996).

EDI AND THE INTERNET: EDI in its simplest form may be regarded as nothing more than electronic-mail service or an electronic document management system. However, with the advent of multi-media concepts and digital communication technology, EDI's role as the mere automation of messaging and documentation is gradually diminishing. In particular, EDI's survival has been challenged by the rapidly emerging Internet system which has already become dominant communication tools integrating supply chains by linking tens of millions of computers worldwide. The Internet is a global, non-proprietary, and inter-organizational communication network linked by telephone lines, digital cables, optical fiber, and satellite transmissions. The Internet is used to support inter-organizational decision making, provide e-mail, distribute software, display web pages, and offer universal access to information and documents via its global backbone of communication links. Unlike EDI, the Internet does not require highly structured, machine-readable data and consequently does not necessitate translator application software packages. Another difference between the Internet and EDI is that the Internet promotes "open" communication with any users including the general pubic, whereas EDI is primarily intended for exchanges of proprietary information among the "closed" community of trading partners.

In general, the Internet is cheaper and easier-to-use than EDI, because there is no central authority managing the Internet. However, the former is less secure than the latter. Despite such differences, the Internet still can perform a variety of functions similar to EDI. For example, the Internet can be used to place purchase orders, make on-line payments, transmit paper-less invoicing, and peruse pre-sale technical product information. Considering the increasing vulnerability of EDI to alternative forms of electronic commerce, a stand-alone EDI system seems to be no longer viable as an innovative technology or a new way of doing business. Thus there is a growing need for integrating EDI with other types of electronic commerce (e.g., Internet, electronic catalogs) and computerized systems such as bar-coding, CALS (Computer Assisted Acquisition and Logistics Support), POS (Point-of-Sale System), DSS (Decision Support Systems), and CAD/CAM.

In particular, due to considerable cost savings and faster response time, the use of Internet-based EDI is on the upswing. As a matter of fact, approximately 7% of companies using EDI is reported to have exchanged transaction sets over the Internet in mid-1996 (Turban and Aronson, 1998). The major benefit of Internet-based EDI stems from the fact that it can replace a costly VAN with the Internet. Such a replacement eliminates various charges related to VAN services. These charges typically include base billing on the document transaction, mailbox fees, and data transmission fees to peak and off-peak rates. Internet-based EDI can also provide real-time information to trading partners anywhere in the world due to its "ubiquity." Considering these synergistic benefits, Internet-based EDI may facilitate the use of EDI among small organizations that, otherwise, would not have financial and technical resources to embrace EDI.

In light of the above discussion, EDI's future lies in its ability to incorporate non-character electronic media such as graphics, sound, video, and animation which are available on the Internet. To remain its strategic value within the next generation virtual organization, EDI needs to be more flexible enough to embrace other forms of electronic commerce such as the Internet.

IMPLEMENTATION GUIDELINES: Realizing enormous competitive advantages of EDI, an increasing number of firms have attempted to formulate a viable EDI implementation strategy. A successful EDI implementation strategy should address the following key issues.

1. ***Trading partner agreements:*** Due to the nature of EDI that inherently promotes interorganizational communication and information transfer, the EDI initiating firm (or a "hub" company) may not gain the full benefit of EDI without the support of its trading partners ("followers"). For instance, if trading partners are forced into the EDI network too far in advance of their technical capability to embrace EDI, the return-on-investment (ROI) from EDI would not be great for both the EDI initiating firm and its trading partners. Consequently, if implemented hastily, EDI may weaken trading partnerships among EDI participants rather than strengthen them. Considering such a negative impact, the EDI initiating firm needs to identify appropriate EDI partners who fully understand EDI technology, represent a high volume of repetitive transactions, have access to the same VAN, adopt compatible EDI standards, and best fit the EDI initiating firm's long-term business strategy.
2. ***Management support:*** Regardless of the potential EDI benefits, unconvinced management would lack commitment to EDI investment and would be less inclined to provide adequate financial and personnel resources required for company-wide EDI implementation. It would be difficult to manage transition from traditional paper-based systems to EDI programs without senior management commitment. Senior management must be educated on specific strategic gains that can be obtained from EDI, processes that are affected by EDI, the cost of establishing an EDI infrastructure, and how EDI can re-shape corporate culture.
3. ***Performance measurement:*** To justify EDI investments, a firm should conduct an objective cost-benefit analysis of the proposed EDI program covering its life cycle. Such an analysis may begin with cost comparisons among third-party service alternatives, existing program modification, and in-house development. Typical EDI costs are incurred in relation to the purchase and maintenance of hardware (e.g., computers, modems), software (e.g., translation and communication software, software upgrade, mapper integration, technical support), communication (e.g., per transmission VAN charge, mailbox cost), employee and trading partner training, and personnel hiring. On the other hand, as discussed earlier, EDI can bring numerous operating benefits along with some intangible benefits such as enhanced trading partnerships, higher employee morale, improved customer loyalty, and increased competitive advantages. These benefits, however, may not be realized in the short-term.

See **Computer assisted acquisition and logistics support; Decision support system; Hub company; Digital signature; Electronic market place; Encryption; Intermodal EDI; Paperless transactions; Supply chain management; Virtual manufacturing; Virtual organization.**

References

[1] Anderson, D.R. (1992), *Motor Carrier/Shipper Electronic Data Interchange*, The American Trucking Associations, Inc., Alexandria, VA.

[2] Auguston, K. (1995), "How EDI Helps us Clear the Dock Every Day." *Modern Material Handling*, 46–47.

[3] Bakos, J.Y. (1991), "A Strategic Analysis of Electronic Marketplaces." *MIS Quarterly* 15, 295–310.

[4] Branch, A.E. (1996), *Elements of Shipping*, seventh edition, Chapman and Hall, London, UK.

[5] Cooper, L.K., Duncan, D.J., and Whetstone, J. (1996), "Is Electronic Commerce Ready for The Internet?: A Case Study of a Financial Service Provider." *Information Systems Management* 13, 25–36.

[6] Kekre, S., and Mukhopadhyay, T. (1992), "Impact of Electronic Data Interchange Technology

on Quality Improvement and Inventory Reduction Programs: A Field Study." *International Journal of Production Economics* 28, 265–282.
[7] LeRose, J.W. (1993), "Intermodal EDI." *EDI World* 3, 22–24.
[8] Messmer, E. (1992), "EDI Suppliers to Test Fed Purchasing Net." *Network World* 9, 43–44.
[9] Milbrandt, B. (1987), *Electronic Data Interchange: Making Business More Efficient*, Automated Graphic Systems, White Plains, MD.
[10] Murray, Jr., J.E. (1996), "Electronic Purchasing and Purchasing Law." *Purchasing* (February, 15), 29–30.
[11] Piore, M.J., and Sabel, C.F. (1984), *The Second Industrial Divide*, Basic Books, New York, N.Y.
[12] Premkumar, G., Ramamurthy, K., and Nilakanta, S. (1994), "Implementation of Electronic Data Interchange: An Innovation Diffusion Perspective." *Journal of Management Information Systems* 11, 157–186.
[13] Pye, C. (1997), "Another Option for Internet-Based EDI: Electronic Communication." *Purchasing Today* 8, 38–40.
[14] Rogers, E.M. (1983), *Diffusion of Innovation*, Free Press, New York, NY.
[15] Turban, E., and Aronson, J. (1998), *Decision Support Systems and Intelligent Systems*, fifth edition, Prentice-Hall Inc., Upper Saddle River, New Jersey.
[16] Venkatraman, N., and Zaheer, A. (1990), "Electronic Integration and Strategic Advantage: A Quasi-experimental Study in the Insurance Industry." *Information Systems Research* 1, 377–393.
[17] Walton, L.W. (1996), "The ABC's of EDI: The Role of Activity-Based Costing (ABC) in Determining EDI Feasibility in Logistics Organizations." *Transportation Journal* 36, 43–50.

ELECTRONIC MARKETPLACE

Electronic marketplace is an interorganizational information exchange mechanism that allows the participating buyers, suppliers, and third parties to access market or product-related information in an efficient and timely fashion. It makes comparison shopping easier and quicker.

See **Electronic data interchange in supply chain management.**

EMPLOYEE EMPOWERMENT

Employment empowerment is the practice of giving non-managerial employees the responsibility and the power to make decisions regarding their jobs or tasks. It is a form of decentralizing managerial responsibility to lower-level employee. Empowerment allows the employee to take on responsibility for tasks normally associated with management or staff specialists. Examples of such decisions include scheduling, quality, process design, or purchasing decision.

Employee empowerment helps in attaining business goals such as zero defects, zero break downs, and supplier coordination, which are essential in an agile business environment. Employee empowerment is very important in implementing the concepts such as business process reengineering, total quality management, statistical process control and just-in-time manufacturing.

See **Agile manufacturing; Business process reengineering; Just-in-time manufacturing; Manufacturing cell design; Total quality management.**

EMPLOYEE TURNOVER

Employee turnover in a company refers to the movement of workers in and out of employment. It disrupts planning and production. Employee turnover comes in several forms: voluntary separations, layoffs, discharges for disciplinary reasons, retirement, or permanent disability. Employee turnover may be defined as the ratio of the number of employees hired to replace those who have left, to the total number employees.

See **Dynamic kanban control for JIT manufacturing; Human resource issues in manufacturing.**

EMPOWERED WORKFORCE

See **Employee empowerment.**

ENCRYPTION

Encryption is the technical method used to protect sensitive information transmitted through potentially unsafe electronic network by hiding information and then permitting its recovery by the intended recipient with a proper deciphering device.

See **Electronic data interchange in supply chain management.**

END-OF-PIPE TECHNOLOGIES

These technologies remove pollutants or toxic substances from the waste byproducts of a process before they reach the natural environment. Examples of end-of-pipe technologies include sulfur dioxide flue gas scrubbers and sewage treatment plants.

See **Environmental issues and operations management.**

ENGINEERING CHANGE ORDERS

See **Engineering changes.**

ENGINEERING CHANGES

Engineering changes are revisions to drawings and designs that modify a part or assembly. These changes are formalized by the engineering department but may have been requested by another department, a customer, or a supplier.

Engineering changes are often necessary but they have adverse effect upon all functions of a company. Engineering changes cause updates to manufacturing planning and control (MPC) systems, routings and bills of materials (BOM). Change transactions are very expensive compared to most other transactions. The goal of manufacturing systems should be eliminate or reduce engineering changes.

See **Concurrent engineering; Design for manufacturing; Product design and concurrent engineering.**

ENGINEERING DESIGN

Engineering Design is the process that ensures the technical feasibility of a product concept and the required "functions" of the product. In engineering design, engineers select the process, specify materials, and determine shapes to satisfy technical requirements and to meet performance specifications.

See **Concurrent engineering; Product design; Product development and concurrent engineering.**

ENGINEERING MODEL

It is used in new product development. It is a prototype that embodies the exterior and internal components, and materials used.

See **Concurrent engineering; Product development and concurrent engineering; Product development through supplier integration.**

ENGINEER-TO-ORDER PRODUCTION

In engineer-to-order production, products are designed from scratch, then manufactured and delivered to a customer. The customer initiates the design activity and may even participate in product development process. Job shops, which can design, produce, and deliver according to customers' specifications and requirements fall into this category. For example, an equipment manufacturer, after receiving a customer order, designs, manufactures, and then delivers to the customer.

See **Mass customization.**

ENTERPRISE RESOURCE PLANNING (ERP)

DEFINITION: Enterprise Resource Planning (ERP) is a term associated with a multi-module software for managing and controlling a broad set of activities that helps a manufacturer or other businesses. ERP can assist in, data capturing, storage, product planning, parts purchasing, inventory control, purchasing, customer service, and order tracking. ERP also includes application modules for finance and human resources management. The deployment of an ERP system involves considerable business process reengineering, and retraining.
(Source: http://whatis.com/erp.htm, February 1999)

ERP SYSTEMS-EVOLUTION: In the 1980's the concept of MRP-II (Manufacturing Resources Planning) evolved, which was an extension of Material Requirements Planning (MRP) software. MRP-II software covers data collection and storage for planning, operation and control purposes in Manufacturing, Marketing, Engineering, Finance, Human Resources, Project Management etc., i.e., the complete range of activities within any business enterprise. Eventually, the term ERP

(Enterprise Resource Planning) emerged to capture the true scope of the software.

SCOPE OF APPLICATION: ERP solutions are meant to provide comprehensive and seamless integration of the information system covering manufacturing, distribution, accounting, finance, project management, service, maintenance, etc. An ERP system can allow the quick flow of information across the supply chain, i.e. from the customer to manufacturers to supplier: ERP softwares provide centralized or de-centralized accounting functions with the ability to do activity-based costing.

Software Evaluation Criteria

According to industry sources, some important items to be evaluated while investing in ERP software:

1) Functional fit with the company's business processes
2) Degree of integration between the various components of the ERP system
3) Flexibility and scalability
4) Complexity; user friendliness
5) Quick implementation; shortened ROI period
6) Ability to support multi-site planning and control
7) Technology; client/server capabilities, database independence, security
8) Availability of regular upgrades
9) Customization
10) Local support infrastructure
11) Availability of reference sites
12) Total costs, including cost of license, training, implementation, maintenance, customization and hardware requirements.

(Source: http://www.expressindia.com/newads/bsl/advant.htm, February 1999)

IMPLEMENTATION: ERP implementation takes time. One mid-sized U.S. plant with 950 employees begins ERP implementation in January 1999 and expects completion in five years at the expense of about $3 million. Two approaches to implementation are recognized by industry (source: http://www.expressindia.com/newads/bsl/advant.htm, February 1999):

1) *Comprehensive Implementation:* Here the focus is on business improvement. This approach is expensive and requires software customization. It is suitable when improvements in business processes are required, a high level of integration with other systems is required, and multiple sites are involved.
2) *Compact Implementation:* Here, first establish a strong technical base during implementation with business improvements coming at a later stage. This approach is suitable when improvements in business processes can wait, and a single site is involved.

SOME CURRENT ERP SOFTWARE:

SAP

SAP is the leading supplier of ERP software in the U.S. in 1998. SAP products empower companies of all sizes and all industry sectors to respond quickly and decisively to dynamic market conditions, helping businesses achieve and maintain a competitive advantage. Its strategic product architecture links all areas of an enterprise. Crucial to making sure that SAP products, legacy systems, and third-party products all work together as seamlessly as possible are two efforts: One is the development of open interfaces to link disparate systems and provide enormous scalability and flexibility. The other is the TeamSAP certification program. SAP, in collaboration with its hundreds of complementary software partners and world-class hardware providers, has created this program to ensure the use of its standard implementation methodologies; and realize the solution in an accelerated time frame at least cost to the customer. SAP R/3 applications cover accounting and controlling, production and materials management, quality management and plant maintenance, sales and distribution, human resources management, and project management.
(Source: http://www.sap-ag.de/products/index.htm, February 1999)

Baan

The company's family of products is designed to help corporations maintain a competitive advantage in the management of critical business processes by having a product architecture that lends itself to fast implementations and ease of change. Baan IV is an integrated family of manufacturing, distribution, finance, transportation, service, project and orgware modules. The solution offers a new concept in business management software that incorporates and goes beyond Enterprise Resource Planning (ERP). Using the principle of Dynamic Enterprise Modeling (DEM) implemented via its Orgware capabilities, Baan IV enables a company to match their specific business processes and their organization model with the extensive functionality of the BAAN applications. BAAN IV is specially designed to meet the needs

of key vertical markets. Furthermore, BAAN also extends supply chain support beyond the boundaries of an organization to support trading partner management. BAAN IV has a client-server architecture, runs on most popular environments and is Internet enabled to provide customers with the operational flexibility they need to be the leaders in their marketplace.
(Source: http://www.baanbbs.com/solutions/Default.htm, February 1999)

J.D. Edwards

OneWorld allows you to build highly flexible workflow solutions and execute both predefined and ad hoc processes in your organization. And you can change your processes quickly to take advantage of new – and unexpected – opportunities. You can even reassign tasks from a central location to avoid costly bottlenecks. With OneWorld, your ability to learn, implement, and maintain workflow at all levels of your organization is simplified. The Company's software operates in multiple computing environments, and is both Java™ and HTML enabled. J.D. Edwards enables Idea to Action with ActivEra™, a suite of products and technologies that allows companies to change enterprise software after implementation as their business needs demand. ActivEra extends the capabilities of the Company's existing OneWorld™ and J.D. Edwards SCOREx products.
(Source: http://www.jdedwards.com/, February 1999)

Lawson

- Lawson's commitment to and understanding of the retail industry is reflected in its development of retail-specific functionality within the LAWSON INSIGHT. Retail Business Management System.

(Source: http://www.lawson.com/, February 1999)

Peoplesoft

Peoplesoft Enterprise Performance Management includes the following software products:

- *Peoplesoft Balanced Scorecard.* Functioning as the central nervous system of the Enterprise Performance Management system, the Balanced Scorecard allows senior management to communicate the strategy of the organization to all employees through a system of key performance indicators, including financial measures, customer knowledge, internal business processes, learning and growth. The Balanced Scorecard allows executives to express to employees how their individual efforts contribute to the business' success, and it lets employees tell executives what day-to-day realities affect their progress.
- *Peoplesoft WorkBench.* Peoplesoft WorkBench will enable organizations to put PeopleSoft analytic functionality on every desktop. A tool to enable strategic, operational and tactical decision-making at every level, the WorkBench is a predefined template, configured by role, functional area and industry, that allows the user to view, analyze and share information. PeopleSoft WorkBench in role-based versions includes Finance WorkBench, Human Resources WorkBench, and Retail WorkBench.
- *Peoplesoft Analytic Applications.* These rules-based applications transform data from enterprise resource planning (ERP) systems and external sources, stored in the Enterprise Warehouse, into new and relevant information to aid decision-making in such areas as customer or product profitability, planning and simulation, and customer acquisition and retention. PeopleSoft Analytic Applications include: Activity Based Management (ABM), Funds Transfer Pricing (FTP), Risk Weighted Capital (RWC), Total Compensation Management (TCM), and Asset Liability Management (ALM).
- *Peoplesoft Enterprise Warehouse.* The main repository of enterprise information, the PeopleSoft Enterprise Warehouse draws from PeopleSoft applications as well as other ERP and legacy systems to stage, store and make information available for analysis.

(Source: http://www.peoplesoft.com/, February 1999)

Oracle

Now more than ever, businesses face the challenge of simplifying and lowering the cost of managing the systems that run on their corporate, interconnected networks. Effective systems management for the Oracle environment requires that teams of administrators have the ability to centrally view their managed topology; associate and organize disparate-but-linked services, such as databases and applications; and effectively monitor and respond to the health of these systems 24 hours a day. Oracle Enterprise Manager is the comprehensive management framework built on Internet standards, providing a robust console, a rich set of tools, and the extensibility to detect, solve, and simplify the problems of any managed environment. Oracle Enterprise Manager includes the following features:

- *Manage from Anywhere* – Oracle Enterprise Manager is fully accessible through thin clients or Web browsers providing extraordinary flexibility to administrators in any environment.
- *"Set and Forget" Management* – Oracle Enterprise Manager enables any size organization to cut operational costs by minimizing the manual care and feeding of enterprise information systems.
- *Manage Everything* – Oracle Enterprise Manager provides a consistent and universal system management framework to support the end-to-end management of your Oracle solutions. Having an automated system that routes documents, information, and tasks from user to user sounds pretty good. But one that doesn't lock you into prescribed processes – and out of opportunities – sounds even better. And one that results in increased efficiency on the bottom line is better still.

(Source: http://www.oracle.com/, February 1999)
Note: Softwares are described in the words of the vendors from company sources referenced.

See **MRP; Resource planning: MRP to MRP II and ERP.**

References

[1] http://whatis.com/erp.htm, February 1999.
[2] http://www.expressindia.com/newads/bsl/advant.htm, February 1999.
[3] http://www.sap-ag.de/products/index.htm, February 1999.
[4] http://www.baanbbs.com/solutions/Default.htm, February 1999.
[5] http://www.jdedwards.com/, February 1999.
[6] http://www.lawson.com/, February 1999.
[7] http://www.peoplesoft.com/, February 1999.
[8] http://www.oracle.com/, February 1999.

ENTREPRENEURSHIP

Entrepreneurship is the process of transforming an idea into a commercial product or service. Resources must be assembled in order to commercialize an idea. This means that entrepreneurship involves putting together human resources and financial resources in particular.

See **Process innovation; Product innovation.**

ENVIRONMENTAL ISSUES AND OPERATIONS MANAGEMENT

Robert D. Klassen

University of Western Ontario, London, Ontario, Canada

INTRODUCTION: The management of the natural environment is becoming increasingly important within operations as customers, suppliers and the public demand that manufacturers minimize any negative environmental effects of their products and operations. Managers play a critical role in determining the environmental impact of manufacturing operations through choices of raw materials used, energy consumed, pollutants emitted and wastes generated. Over the past three decades, conceptual thinking on environmental issues have slowly expanded from a narrow focus on pollution control to include a large set of management decisions, programs and technologies. Pressures to apply the concept of sustainable development to manufacturing underscore the need to think strategically about environmental issues. In this context, sustainable development translates into the integration of environmental management into manufacturing design and technology decisions. However, at this point, the implications of environmental management for manufacturing performance appear mixed. Limited empirical research and anecdotal evidence point to benefits such as reduced waste, and new markets.

DESCRIPTION: Management of the natural environment has become increasingly important to manufacturing firms as regulatory requirements tighten, customer demands change, employee concerns multiply, and public scrutiny increases. In its broadest sense, management of environmental issues encompasses a diverse set of management decisions, programs, tools, and technologies that respond to or integrate environmental issues into all aspects of business competitiveness. While much research is underway across a variety of other disciplines, including public policy, "green" marketing, corporate social performance, economics and environmental engineering, this article is limited to the area of production and manufacturing management. Because of the rapid pace of change in this area, managerial practice often has been forced to lead or at least match research.

At first glance, environmental issues may appear to concern only a small set of heavily polluting industries. The popular press is filled with examples of companies that demonstrate poor environmental management – oil spills, explosions, leaking storage tanks or contaminated ground water – and to a lesser extent, examples of firms that exhibit exemplary performance. In virtually all cases, operations managers played a critical role in developing management systems and implementing decisions related to these outcomes. However, at this time, manufacturers outside historically polluting industries also must concern themselves with environmental issues. Product design and process technology typically determine the types of pollutants emitted, the solid and hazardous wastes generated, resources harvested and energy consumed. In addition, upstream and downstream supplier partnerships, transportation and logistics, and customer relationships further magnify or attenuate environmental risks related to production.

At a more general level, public and private concerns over the environment recently have been embodied in the concept of sustainable development, with its proposition that economic growth can occur while simultaneously protecting the environment (World Commission on Environment and Development). Other organizations, including World Business Council for Sustainable Development (WBCSD), United Nations Environmental Programme (UNEP) and Coalition for Environmentally Responsible Economies (CERES), also advocate principles that promote sustainable development. In response, there has been significant international movement toward standardized certification of environmental management systems, an example being ISO 14000 (comparable to ISO 9000 for quality certification). Increasingly, international agreements also have implications for operations systems and their target markets. One recent example is the Montreal Protocol, which eliminates the use of chloro-fluorocarbons (CFCs) by the year 2000.

HISTORICAL PERSPECTIVE: Historically firms have viewed environmental management as a narrowly defined corporate legal function, responsible for minimizing the impact of environmental legislation on the firm and for ensuring regulatory compliance. Not surprisingly, environmental regulation, which for the most part only began to seriously develop in the late 1960's and 1970's, was the primary driver for environmental improvement. As a result, investment in pollution abatement expenditures in many developed nations showed a corresponding increase as manufacturing firms attempted to control their pollutant emission into air and water and onto land (Lanjouw and Mody, 1996). This limited view has evolved and expanded recently because of intense pressures from many different stakeholder directions: governments, typically through regulation, customers, public interest groups, plant neighbors and employees.

Paralleling this evolution, early research and conceptual thinking on environmental issues focused only on the concept of pollution control (Bragdon and Marlin, 1972). More recent research has broadened this basic view to integrate the consideration of environmental impacts into and across functional areas with implications for strategy and functional area decisions. Research and practice can be structured along at least two dimensions: degree of specificity; and intended level of impact for operations (Figure 1). The first dimension, degree of specificity, captures the relative level of focus, which ranges from narrowly defined tools to broad, directional philosophies. In contrast, the second dimension considers the broadest level at which implications are felt or changes occur in operations. While other dimensions could be proposed, these two provide a conceptual map for identifying the relative position of various management initiatives in operations on environmental issues. For illustrative purposes, several management initiatives are plotted in Figure 1, with the caveat that specific initiatives by any individual firm may vary somewhat from the figure.

STRATEGIC PERSPECTIVE: The management of environmental issues draws on elements of strategy, capabilities and performance. At the strategic level, product design, process technologies and managerial systems are major determinants in the environmental performance of manufacturing firms. For example, if complex multi-layer plastics are used for packaging, closed-loop recycling of materials becomes very difficult, if not impossible. Choices of particular process technologies, such as the kraft bleaching process in the pulp and paper industry, demand long-term commitments to particular raw materials and waste streams that often are regulated, either now or in the future. The kraft process requires chlorine or chorine dioxide for bleaching, which leads to the emission of chlorinated toxins.

Many initiatives to better manage environmental issues within operations are directed at improving productivity, raising resource efficiency and fostering innovation (Porter and Van der Linde, 1995). Moreover, adept management of environmental issues can establish barriers to

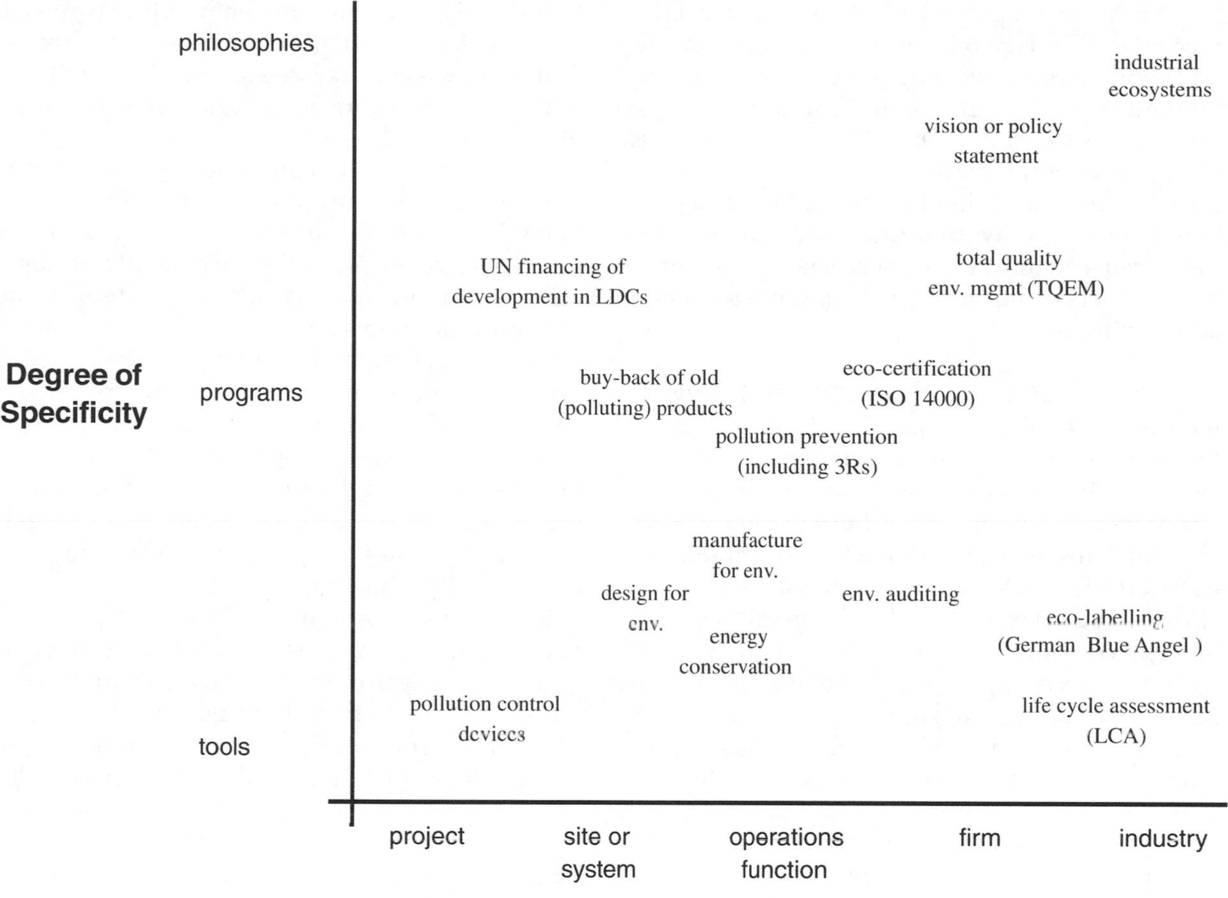

Fig. 1. Management initiatives on environmental issues.

entry for competitors as new regulation is introduced. For example, regulatory agencies continue to tighten acceptable emission and exposure standards, citing technology that is used by industry leaders, thereby forcing other firms to invest in new technology. Manufacturing firms often must evaluate difficult trade-offs between acquiring or developing new technology, buying pollution credits, or developing new in-house management and control systems. Unfortunately, the risks associated with each of these alternatives often are difficult to predict and quantify.

Research is continuing on the difficult theoretical task of integrating manufacturing strategy concepts, including structural and infrastructural investments, with the current policies, programs and tools for managing environmental issues noted in Figure 1. As a first step, the complex interactions between multiple stakeholders, including customers, government agencies and environmental interest groups, must be explicitly recognized (Klassen, 1993). The most obvious linkages are to elements of operations strategies that leverage product design (Zhang *et al.*, 1997), process technologies (Gupta, 1995; Sarkis, 1995), and product recovery (Brennan *et al.*, 1994; Thierry *et al.*, 1995). Newman and Hanna (1996) report empirical evidence that expanding manufacturing's strategic role in the firm tends, although not necessarily, to favor more proactive management of environmental issues.

Empirical studies characterizing approaches to the management of environmental issues align along two general perspectives: strategic choice; and developmental progression. The first perspective proposes that manufacturing firms make distinct choices from among a set of alternative positions, typically on a spectrum from reactive to proactive (Newman and Hanna, 1996; Smith and Melnyk, 1996). In contrast, the developmental progression captures the notion that technical understanding, skills and cross-functional integration build over time, thereby allowing firms to move from a naïve, crisis-oriented beginner to an enlightened, proactive innovator (Petulla, 1987;

Hunt and Auster, 1990). While both perspectives recognize that capabilities change over time, the strategic perspective suggests that operations *might* use increasingly sophisticated environmental capabilities to defend their current levels of environmental performance. In contrast, the second perspective implies a subtle linkage between values/motivation and capabilities, where increased capabilities are associated with additional efforts to improve environmental leadership and performance.

TECHNOLOGICAL PERSPECTIVE: Environmental technologies can be broadly defined to include any design, equipment, method or systems that conserve resources, minimize environmental burden, or otherwise protect the natural environment **Error! Bookmark not defined**. Technologies are often further broadly classified as (1) *pollution abatement*, alternatively labeled end-of-pipe or pollution control, and (2) *pollution prevention*. Other related terms are clean technologies, design for environment, 3R (reduce, reuse, and recycle) technologies or green manufacturing, to list just a few terms (OECD, 1995). In general, much of the environmental management literature, usually in the context of proactive environmental management, espouses the use of pollution prevention as the most effective means by which to improve environmental performance. In contrast, pollution abatement merely tries to contain or control wastes and pollutants.

Much has been written about barriers to greater development and implementation of pollution prevention (for more complete reviews, see Ashford, 1993; and OECD, 1995). These barriers are broadly summarized as:

- *public policies* which favor command-and-control regulations, are inflexible, and enact incremental reductions;
- *financial constraints* in firms which focus on short term profitability and limit R&D expenditures to develop new technologies;
- *workforce related concerns* which resist new technological change that may displace jobs or require significant retraining;
- *technological capabilities and risk* which may hurt current manufacturing processes optimized for quality, cost and efficiency; and
- *managerial attitudes and structures* which cause inertia, poor communication, education and redirection of resources elsewhere.

Of these, the two barriers that receive the greatest amount of attention are public policy and technology. Of necessity, environmental regulation is a critical concern for most operations managers. Collectively, and often incrementally, regulations develop at the local or national level to form very different legislative systems, with potentially very different impacts on technological choices by manufacturers.

For example, U.S. regulations tend to limit the amount of pollutants released into the environment based on the availability of existing pollution control technologies, commonly termed a command-and-control regulatory system. Efforts to take risks and to innovate with environmental technologies generally are not rewarded. In contrast, while Germany certainly has not ignored the emission of pollutants, much greater emphasis has been placed on regulating the total product system, through environmental consumer labeling (i.e. "Blue Angel" eco-label) and mandated requirements to take back and recycle high percentages of packaging.

In response, firms in the U.S. have focused on end-of-pipe technologies installed at the end of the manufacturing process (Lanjouw and Mody, 1996), whereas German firms have been pushed to more actively pursue process redesign, material recycling, and product disassembly (Cairncross, 1992). Broader recognition of this barrier and additional research into the managerial implications of public policies is gradually encouraging the adoption of more flexible regulation, such as tradable emission permits and multi-media (air, water and land) regulations (Ashford, 1993).

EFFECTS & RESULTS: Two main theoretical arguments link the management of environmental issues to manufacturing and firm performance. First, if traditional economic analysis is made, where social (public) costs of pollution are simply transferred back to the (private) firm, manufacturers that improve their environmental performance are at an economic and competitive disadvantage. Alternatively, if the costs of improved environmental performance are overshadowed by other benefits, including increased productivity, reduced liabilities, access to new markets and better public reputation, firms could be in a better competitive position. However, critics contend that new environmental regulations and specific actions to improve environmental performance often impose significant financial burdens and risk, and only a small fraction are likely to offer any competitive benefit. Furthermore, these "easy wins" – low risk, easily implemented solutions with positive financial returns – may have already been implemented in most companies (Walley and Whitehead, 1994).

Much research in fields such as finance, corporate social performance, economics, accounting and environmental management has been devoted to testing the relationship between environmental and economic performance (see Klassen and McLaughlin, 1996). Environmental performance, such as emission levels or reputation, is typically regressed against some form of past, present or future economic performance. Unfortunately, the results generally have been ambiguous, possibly because any direct effects were masked at such an aggregate level. Yet, based on firm- and industry-level studies, a growing number of researchers strongly advocate that an integrative approach coupled with pollution prevention can be competitively advantageous (Bragdon and Marlin, 1972; Cairncross, 1972; Schmidheiny, 1992; Porter and Van der Linde, 1995).

More specifically within manufacturing, case examples cite both positive and negative outcomes from investments in environmental technologies. Several cases of positive outcomes, related to pollution prevention, are noted in the next section. On the negative side, at the plant-level, analysis of confidential U.S. Bureau of Census data indicates that, for at least the paper, petroleum and steel industries, a $1 increase in compliance costs at the plant-level translates into higher total factor productivity costs of $3 to $4 (Gray and Shadbegian, 1993). However, this study, like others, only captured direct, immediate costs, and failed to quantify potential benefits such as new market opportunities or reduced operating risk. More research is needed to clarify the broad-scale performance implications of pollution prevention technologies and regulatory flexibility, and to provide operations managers with much clearer guidance for action.

CASES: Many examples of strong or weak management of environmental issues are part of the public record. Rather than emphasize poor actors, a few notable positive examples are highlighted here. As a general indication of the perceived importance of environmental performance, an increasing number of manufacturing firms are now publishing annual environmental reports, similar to financial reports. Multinational companies, covering a range of industries, include: ABB Group, British Airways, Daimler-Benz, Eastman Kodak, Eli Lilly & Co., Northern Telecom and Sony.

One firm with a notable track record is 3M, with its Pollution Prevention Pays program (now called 3P Plus) which claims to have saved over $790 million through 4,500 projects over a 20-year period. As implied by its name, the program is very pragmatic. Employees attempt to explicitly link environmental improvements with product or technological improvements that provide business advantages. Managers reported three primary lessons: strong commitment is essential from senior management, including having a senior manager responsible for environment; environmental issues need to be integrated into business plans; and employees must be involved in achieving environmental goals which can be best supported through formal recognition (for further discussion of this case, and other examples of benefits arising from improved environmental practice, see Schmidheiny (1992) and Shrivastava (1995)).

Examples of manufacturers taking a systems-oriented perspective on environmental management include BMW and Hermann Miller, with their emphasis on taking responsibility for their products, automobiles and office furniture, respectively, at the end of their useful life (Cairncross, 1992; Smith and Melnyk, 1996). Historically, while much of the metal and many high value components have been recovered from old automobiles, virtually none of the plastic material, accounting for 10% of the weight of an automobile, was recovered. By building on existing recycling capabilities, BMW is focusing on developing new product designs and production technologies that will meet their recycling target of 80% for plastics. A plant has been constructed to pilot vehicle disassembly, an increasing number of parts are being refurbished, and designs have been modified to reduce the use of complex plastics, to label the composition of parts and to use more recycled materials (Gupta, 1995; Thierry *et al.*, 1995).

Thus manufacturers seeking practical approaches to integrate environmental issues into operations can turn to real examples of firms that have, in fact, introduced environmentally responsive management systems, process technology and product design. While some of these initiatives carry significant costs and risks, these firms have generally discovered that insightful and proactive management of environmental issues represent more than addressing regulatory constraints, but also pursuing new market opportunities and cost savings.

Organizations Mentioned: 3M; ABB Group; BMW; British Airways; CERES; Daimler-Benz; Eli-Lilly and Co.; Hermann Miller; Kodak; Northern Telecom; OECD; Sony; UNEP.

See **3 "R"s; Clean technologies; Coalition for environmentally responsible economies (CERES); Design for the environment (DFE); End-of-pipe technologies; Environmental**

issues: Reuse and recycling in manufacturing systems; ISO 9000; ISO 14000; Manufacturing strategy; Montreal protocol; Pollution abatement; Pollution prevention; Total quality management; United nations' environment programme (UNEP).

References

[1] Ashford, N.A. (1993). "Understanding Technology Responses of Industrial Firms to Environmental Problems: Implications for Government Policy." In K. Fischer & J.Schot (Eds.), *Environmental Strategies for Industry*. Washington, DC: Island Press, 277–307.

[2] Bragdon, J., and J. Marlin (1972). "Is Pollution Profitable?" *Risk Management*, 1972 (April), 9–18.

[3] Brennan, L., S.M. Gupta, and K. Taleb (1994). "Operations Planning Issues in an Assembly/Disassembly Environment." *International Journal of Production and Operations Management*, 14(9), 57–67.

[4] Cairncross, F. (1992). *Costing the Earth*. Boston, MA: Harvard and Business School Press.

[5] Gray, W.B., and R.J. Shadbegian (1993). *Environmental Regulation and Manufacturing Productivity at the Plant Level* Report No. CES 93-6. Washington, DC: Center for Economics Studies, U.S. Bureau of the Census.

[6] Gupta, M.C. (1995). "Environmental Management and Its Impact on the Operations Function." *International Journal of Operations and Production Management*, 15(8), 34–51.

[7] Hunt, C.B., and E.R. Auster (1990). "Proactive Environmental Management: Avoiding the Toxic Trap." *Sloan Management Review*, 31(2), 7–18.

[8] Klassen, R.D. (1993). "Integration of Environmental Issues into Manufacturing." *Production and Inventory Management Journal*, 34(1), 82–88.

[9] Klassen, R.D., and C.P. McLaughlin (1996). "The Impact of Environmental Management on Firm Performance." *Management Science*, 42(8), 1199–1214.

[10] Lanjouw, J.O., and A. Mody (1996). "Innovation and the International Diffusion of Environmentally Responsive Technology." *Research Policy*, 25, 549–571.

[11] Newman, W.R., and M.D. Hanna (1996). "An Empirical Exploration of the Relationship Between Manufacturing Strategy and Environmental Management." *International Journal of Operations and Production Management*, 16(4), 69–87.

[12] OECD (1995). *Technologies for Cleaner Production and Products*. Paris, France: OECD.

[13] Petulla, J.M. (1987). "Environmental Management in Industry." *Journal of Professional Issues in Engineering*, 113(2), 167–183.

[14] Porter, M.E., and C. van der Linde (1995). "Green and Competitive: Ending the Stalemate." *Harvard Business Review*, 73(5), 120–134.

[15] Sarkis, J. (1995). "Manufacturing Strategy and Environmental Consciousness." *Technovation*, 15(2), 79–97.

[16] Schmidheiny, S. (1992). *Charging Course : A Global Business Perspective on Development and the Environment*. Cambridge, MA: MIT Press.

[17] Shrivastava, S. (1995). "Environmental Technologies and Competitive Advantage." *Strategic Management Journal*, 16(3), 183–200.

[18] Smith, R.T., and S.A. Melnyk (1996). *Green Manufacturing: Integrating the Concerns of Environmental Responsibility with Manufacturing Design and Execution*. Dearborn, MI: Society for Manufacturing Engineering.

[19] Thierry, M., M. Salomon, J. Van Nunen and L. Van Wassenhove (1995). "Strategic Issues in Product Recovery Management." *California Management Review*, 37(2), 114–135.

[20] Walley, N., and B. Whitehead (1994). "Its Not Easy Being Green." *Harvard Business Review*, 72(3), 46–52.

[21] World Commission on Environment and Development (1987). *Our Common Future*. New York: Oxford University Press.

[22] Zhang, H.C., T.C. Kuo, H. Lu, and S.H. Huang (1997). "Environmentally conscious design and manufacturing: A state-of-the-art survey." *Journal of Manufacturing System*, 16(5), 352–371.

ENVIRONMENTAL ISSUES: REUSE AND RECYCLING IN MANUFACTURING SYSTEMS

Surendra M. Gupta and Pitipong Veerakamolmal

Northeastern University, Boston, MA, USA

DESCRIPTION: The rapid development and improvement of products have resulted in shortened lifetime of most products. This in turn has increased the quality and quantity of used products scrapped. The bulk of the scrap comes from automobiles, household appliances, consumers electronic goods, and computers. There is an urgency to find alternative ways to dispose of products because of the alarming rate at which the landfills are being used up. In addition, environmental awareness and recycling regulations have been putting pressure on many manufacturers and consumers, forcing them to produce and dispose of products in an environmentally responsible manner. In many parts of the world (and specially in Europe), the regulations are becoming more stringent and manufacturers are urged to take-back and recycle their products at the end of their useful lives. Furthermore, companies are also urged to use recycled materials whenever possible.

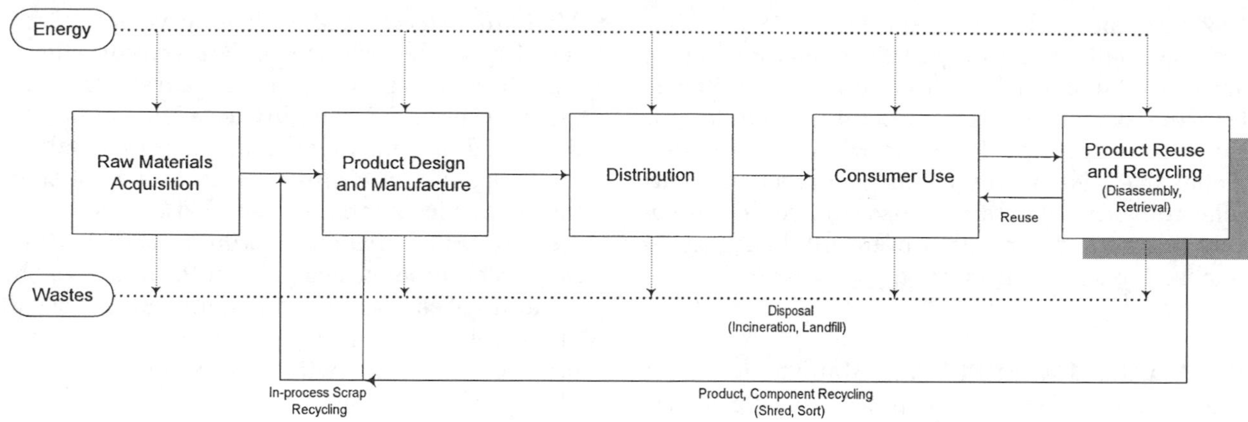

Fig. 1. Product life-cycle.

Product manufacturers need to be responsible for promoting the utilization of recycled materials while entertaining the possibility for taking the products apart at the end of life (EOL) stage and reusing their components. To reuse components in new products is a step towards conserving precious natural resources (e.g. raw materials, energy) and also preventing the degradation of the environment by reducing land and atmospheric emissions (Graedel and Allenby, 1995). Economically justified, products made with reused components are sometimes not only cheaper but also better, since parts would have survived the "burn-in" period.

An assessment technique that is used to monitor the movement of products and their components is called *Life-cycle Assessment* (LCA) (Fava *et al.*, 1991). The phases in a products' life-cycle range from raw materials acquisition, product design and manufacturing, distribution, usage, and disposal (Figure 1). Essentially, the cost of reusing components (which involves cost of disassembly, retrieval and recondition etc.) must not be too high compared to other disposal alternatives, and the overall benefit must outweigh the cost of using new components.

Environmentally benign alternatives to conventional disposal methods are *remanufacturing, refurbishing*, and *recycling*, all of which require disassembly first. Since a disassembly process is usually labor intensive, products that are specifically designed for disassembly are likely yield cheaper reusable component. For example, products built with permanent fixtures and joints can be costly to disassemble.

HISTORICAL PERSPECTIVE: Since the beginning of the industrial revolution, human activities involving production, manufacturing, and distribution of goods have had an impact on environmental quality. Along with industrial development, growing human population has led to an escalation of this negative impact which often effects the quality of our environment. At first, the environmental impact was limited and concentrated in highly populated areas. Industrial development, along with other factors, contribute to an ever accelerating rate of pollution on earth, air and water.

The most prominent evidence of our environmental problem comes from the growing need for waste disposal. Originally, the majority of our municipal wastes were landfilled. However, throughout 1980s and 1990s, the shortcomings of our reliance on landfilling has become evident: they pose unacceptable environmental risks because of their location or they pose hazardous risks to human health through ground water contamination and toxic air emissions. As a result, numerous landfills, especially in big cities, where an enormous amount of waste is generated everyday, have closed down. While new landfills are being built at a relatively slower rate, they are located further away, which increases the costs of hauling waste. (Denison and Ruston, 1990). Furthermore, U.S. Congress passed a toxic wastes cleanup bill known as "Superfund" in 1981, requiring the costs of cleaning up of contaminated waste sites to be shared among those who caused it. Hence, the growing expense of solid waste management has fueled the growth of recycling and reuse. Starting with materials of simpler forms – papers, glasses, plastics and metals – an increasing number of products are currently being collected, processed, and diverted from the waste stream to be reused productively.

Government and non-profit organizations in America have assumed a leadership role in the establishment of environmental standards. Germany introduced the "Blue Angel" eco-labeling program in 1977, that is the oldest system in Europe to identify environmentally harmful product groups (Klassen and Greis, 1993). During the 1980s, the European Commission set up an Environment Directorate that, later, helped lead the standardization of pollution regulations throughout Europe (Fava *et al.*, 1991). Lately, global attention on environmental problems is accelerating the efforts to reuse and recycle products to reduce the number of harmful substances released to the environment.

STRATEGIC PERSPECTIVE: Manufacturing for reuse is concerned with the processing of products at the design and manufacturing phases. Most products have not been manufactured with disassembly and reuse in mind. Except for automobile parts that can be used as spares, large scale application of disassembly is yet to be proven in terms of cost and benefit.

There are many advantages of manufacturing reusability in products. They include: the reduction of work needed in recovering recyclable parts and materials (due to uniformity and predictability in product configuration), simple and rapid disconnecting operations, ease in handling of removed parts, ease in separation and treatment of removed materials and residuals, and reduction in component variability (Gupta and McLean, 1996).

Manufacturers have to take into consideration the product specific characteristics with regard to disassembly and reuse. For each component in a product assembly, there is a different life-time expectancy. In mechanical products, moving parts, for example, are known to have a higher failure rate and need frequent replacements. On the contrary, modern electronic components tend to have a higher life expectancy and can be reused in other products. In some cases, computer chips are known to posses even higher reliability than their newly manufactured counterparts. This allows the use of recovered components as spares, or in the manufacture of new products.

There are many issues that make manufacturing for reuse a cost effective solution. The following is a partial list of cost-effective approaches that can help save the effort required in disassembly and reuse (Brennan *et al.*, 1994):

- *Ease of separation:* This refers to components designed to separate easily. Important factors that may influence the ease of separation are the special separation tool needed, and the orientation of components layout in the assembly.
- *Easy-to-separate fasteners:* Replace screws, glues and welds with new two-way snap-fit or velcro fasteners. Taking apart a snap-fitted or velcro-fitted product is easier, and requires less energy than taking apart a welded product.
- *Materials selection:* Minimize the variety of material types in order to realize volume advantage from large and efficient plants for reuse and recycling. Use of highly recyclable materials such as aluminum and thermoplastics should be encouraged, while minimizing the use of thermosets, which cannot be recycled.
- *Parts consolidation and modular product structure:* The product design should minimize the varieties of components and materials. This will allow easy sorting of the various components for recycling, and will require fewer assembly/disassembly steps.

IMPLEMENTATION: Products that can no longer function and cannot be restored or modified to perform their original function can be disassembled to retrieve reusable parts and materials. Some of these reusable parts, especially in the case of electronic components, may still reliably function in new products even though they have been in service for years. To implement the recovery system efficiently and effectively, this section discusses product disassembly and retrieval techniques (for reuse and recycling), including their merits and limitations.

Disassembly scheduling seeks to enhance the efficiency in recovery operations by means of planing and scheduling. Disassembly scheduling can be classified into two planning levels, each accounting for scheduling efforts for a different planning horizon. On a longer planning horizon, disassembly scheduling focuses on the aggregation of a disassembly plan, resulting in the total number of products to be disassembled in a given planning horizon (Gupta and Taleb, 1994; Taleb *et al.*, 1997). It takes into consideration demand constraints and calculates the number of products (root items) to disassemble in order to fulfill the demand. On a shorter planning horizon, disassembly scheduling looks at the sequencing of disassembly tasks, and creating a disassembly process plan (Veerakamolmal and Gupta, 1998). Disassembly process planning looks into possible disassembly sequences.

There are two types of disassembly methods: destructive and nondestructive. Destructive disassembly shreds assemblies into a pool of mixed-material chunks and uses sophisticated separation technique to separate materials. Techniques of materials separation include: dry sorting (e.g. riddle, wind sifter and gas cyclone); wet sorting where liquid is used as the separating medium (e.g. float-sink technique and hydrocyclone); and electrical and magnetic sorting to separate ferrous from non-ferrous materials (de Ron and Penev, 1995).

The nondestructive method disassembles a product into subassemblies, and then into parts, breaking fixtures and/or connecting mechanisms, instead of crushing the whole subassemblies into mixed-material chunks. This way, reusable parts within the product structure can be retrieved.

In practice, a destructive disassembly is much more economical than its nondestructive counterpart in terms of labor costs of separating and sorting subassemblies. However, to accommodate the retrieval of reusable components, thus capturing more benefits from reuse, a nondestructive disassembly is preferred. Therefore, to maximize the benefit of the overall operation, a combination of both methods is often used.

Issues Associated with Disassembly Operations

Disassembly systems may exist in a variety of configurations, degrees of complexity, and sizes. Individual parts recovered from a disassembly system range in size from small precision components to very large structural components. The following issues are relevant to disassembly operations (Brennan *et al.*, 1994):

- *Accumulations:* This problem is currently most severe in the automotive industry which generates huge numbers of items, such as tires and windscreens, with no market value and high disposal cost.
- *Location:* Plant location decisions are influenced by the transportation cost of raw materials. However, when dealing with disassembly and recycling, manufacturers will have to plan the locations of new assembly, disassembly or recycling plants appropriately.
- *Logistical Network:* In a disassembly environment there will be potentially three separate entities: the assembly plant, the disassembly plant and the recycling plant. Operations therefore have to be planned for all three entities.
- *Resource availability and allocation:* Disassembly requires additional work, and resources are likely to become frequent constraints. This is especially true in the initial transition periods when plants are experimenting with disassembly and may be using the same location for production as well as disassembly.
- *Buffer stock location(s):* In a disassembly environment, the optimal inventory policies will alter in terms of the level and location of buffer stocks.
- *Alternative disassembly sequences:* Disassembly may be more efficient if the steps are tailored to accommodate different modes of reuse.
- *Implementation and transition:* While companies may have a clear idea of where they want to be, they may not know what is the most efficient way to get there. Transitional and implementation problems, when going from manufacturing of a component to its disassembly and reuse, require very careful scrutiny since they have the potential of adding significant costs.

LOCATION: America, Canada and some countries in Europe and Asia have already established their environmental programs (Klassen and Greis, 1993). The well established programs and their eco-labels, are: German "Blue Angel", Swedish "White Nodic Swan", Dutch "Stichting Milieukeur", Canadian "Environmental Choice", United States' "Green Seal", and European Union "Flower with twelve star-petals". Countries in Asia that have environmental programs include: Japanese "Eco-mark" and Singapore's "Green Label".

Germany submitted legislation for enactment in 1994 which obligated producers and distributors of complex electronic products to take back their products after use and have them recycled. Complex electronic products are: electric toys, laboratory equipment, household appliances, lightening equipment and computers. The take-back ordinance, also known as Dual System Deutschland (DSD), requires that: products must be manufactured from environmentally compatible and recyclable materials; products must be designed and manufactured for easy disassembly and repair; used equipment collection centers must be in reach of the end-user; and parts deemed nonrecyclable must be disposed of safely and in a documented way.

More stringent regulations in Europe and America have affected their trade partners across the globe, especially in developing nations, to produce more environmentally benign products.

Legislatures in Europe are trying to enforce companies to declare the life-cycle assessment of their products, which will be evaluated and certified. In the US, companies are trying to make some economic comparisons using the environmental impact assessment model. To prevent governmental intervention in businesses, US companies internally develop their assessment methods so that they will benefit from making environmental related decisions.

RESULTS: From an economic point of view, disassembly operations can be managed in conjunction with manufacturing operations to obtain the highest benefit. The following is a partial list of implications of reuse and recycling in manufacturing (Brennan *et al.*, 1994):

- *Production cost:* Manufacturers will have to make better quality and more reliable products that accommodates reuse. In return, to offset the added costs, manufacturers have to facilitate product take-back to reclaim and reuse valuable components.
- *Product maintenance:* Maintenance and servicing of products are essential ingredients in ensuring longer product lives. Often these activities require partial disassembly in order to replace or repair parts that are embedded deep in the product structure. The inclusion of disassembly considerations in the product designing process can greatly facilitate maintenance and service activities.
- *Financial decisions:* The capital budgeting process for developing and manufacturing a new product considers a planning horizon that is about equal to its product life-cycle. This process will be more difficult under "manufacturing for reuse" and "recycling" because they require a much longer horizon with increased uncertainties. In the future, when manufacturers have the capacity to facilitate product take-back, leasing may be cheaper as opposed to buying.
- *Capacity and storage requirements:* A difficult variable to forecast will be the distribution of the returns of used products over the planning horizon. Governmental regulations are expected to force manufacturers to take back their used products when consumers decide to return them (Hentschel, 1993). Manufacturers may be faced with unexpected return patterns that will depend on their product success in the market and competing products.

CASES: Major US companies, especially those producing products to sell in the international market (e.g. XEROX and AT&T), have foreseen the urgent need for designing products for disassembly and reuse. Disassembly is a relatively fast growing trend in manufacturing. BMW, the German car manufacturer, has already opened a dismantling plant in Orlando, Florida, and a number of plants in Europe. The big three U.S. automobile makers have formed the Vehicle Recycling Partnership as part of their effort towards disassembly and recycling. Computer and peripheral makers such as IBM, Xerox, and Hewlett Packard are also working on disassembly. Another current application of disassembly is in disposing of unneeded weapons. It is no longer environmentally acceptable to discard weapons and ammunitions by blowing them up, burning them or dumping them in the ocean. The U.S. Defense Department is now hiring military contractors to disassemble and recycle unwanted weapons.

CONCLUSIONS: To make manufacturing for reuse and recycling possible, product manufacturers have to engage in the practice of environmental manufacturing, or Green Manufacturing, at the earliest stage of a product life-cycle. Design and manufacturing can reduce the use of materials and manufacturing processes that foster environmental emissions. Careful selection of materials, placement of components layout, and documentation of reuse methodologies are required to facilitate disassembly scheduling at the product's end-of-life. The life cycles of most products can be significantly prolonged by extensive reuse. Proper planning of disassembly processes ensures the quality of reusable products before being put into use again.

See **Disassembly; Environmental issues and operations management; Life cycle assessment.**

References

[1] Brennan, L., S.M. Gupta and K. Taleb (1994). "Operations Planning Issues in an Assembly/Disassembly Environment." *International Journal of Operations and Production Management*, 14(9), 57–67.

[2] Denison, R.A., J. Ruston (1990). *Recycling and Incineration*. Island Press, Washington D.C.

[3] de Ron, A. and K. Penev (1995). "Disassembly and Recycling of Electronic Consumer Products: An overview." *Technovation*, 15(6), 365–374.

[4] Doos, B.R. (1994). "The State of the Global Environment: Is there a need for 'An orderly retreat?" *Environment Cooperation in Europe: The political dimension*, Holl, O. (Eds.), Westview Press, Boulder, Colorado, 33–50.

[5] Fava, J.A., R. Denison, B. Jones, M.A. Curran, B. Vigon, S. Selke, and J. Barnum (Ed.) (1991). *A Technical Framework for Life-Cycle Assessments*. Washington, DC: Society of Environmental Toxicology and Chemistry (SETEC).

[6] Graedel, T.E. and B.R. Allenby (1995). *Industrial Ecology*. Prentice Hall, New Jersey.

[7] Gupta, S.M. and C.R. McLean (1996). "Disassembly of Products." *Computers and Industrial Engineering*, 31, 225–228.

[8] Gupta, S.M. and K. Taleb (1994). "Scheduling Disassembly." *International Journal of Production Research*, 32(8), 1857–1866.

[9] Handfield, R.B., S.V. Walton, L.K. Seegers and S.A. Melnyk (1997). "'Green' Value Chain Practices in the Furniture Industry." *Journal of Operations Management*, 15(4), 293–315.

[10] Hentschel, D.C. (1993). "The Greening of Products and Production – A New Challenge for Engineers."

Advances in Production Management Systems, Elsevier Science, North-Holland, 39–46.

[11] Keoleian, G.A. and D. Menerey (1994). "Sustainable Development by Design: Review of life cycle Design and Related Approaches." *Journal of the Air and Waste Management Association*, 44(5), 645–668.

[12] Klassen, R.D. and N.P. Greis (1993). "Managing Environmental Improvement through Product and Process Innovation: Implications of Environmental Life Cycle Assessment." *Industrial and Environmental Crisis Quarterly*, 7(4), 293–318.

[13] Meadows, D.H., D.L. Meadows, J. Randers and W.W. Behrens, III (1974). *The Limits to Growth*. Universe Books, New York.

[14] Taleb, K., Gupta, S.M. and L. Brennan (1997). "Disassembly of Complex Products with Parts and Materials Commonality." *Production Planning and Control*, 8(3), 255–269.

[15] Veerakamolmal, P., and S.M. Gupta (1998). "Disassembly Process Planning." *Engineering Design and Automation*, (forthcoming).

ENVIRONMENTAL SCANNING

Environmental scanning involves both unstructured and structured consideration of the external and internal factors that impact an organization and its future. It may involve considering the market influences, regulatory forces, competitor actions and strategies, labor and financial markets and the impact of new technology on the company. From an environmental scanning activity, the analyst may generate one or more scenarios of the future as input for long-term planning.

See **Process innovation; Product innovation.**

EQUAL ALLOCATION RULE (EAR)

A Kanban based production system typically operates with a specified number of Kanbans. Rules are used to allocate available Kanbans. EAR implies that available Kanbans are allocated equally to different part types without assigning any priority to a part based on one or more of its attributes. This simple Kanban allocation rule is easy to implement though there are other more efficient allocation rules.

See **Kanaban-based manufacturing systems.**

ERP

See **Enterprise resource planning.**

EXPECTATION THEORY

Expectation theory (also commonly known as Expectancy-Disconfirmation Theory) is the most widely accepted theory concerning customer satisfaction processes. The theory holds that satisfaction/dissatisfaction results from a customer's comparison of performance (of a product or service) with predetermined standards of performance. According to the view, the predetermined standards are the customer's predictive expectations. Three possible outcomes of the comparison are possible. Positive disconfirmation occurs when performance is perceived to be better than the predetermined expectations. In this scenario, customer is delighted. Zero disconfirmation occurs when performance is perceived to be exactly equal to expectations–customers are likely to be satisfied. Finally, negative disconfirmation occurs when performance is lower than expectations. Of course, negative disconfirmation leads to dissatisfied or unhappy customers. The need to achieve positive disconfirmation has been popularized by the familiar phrase: "Meet or exceed customer expectations." Substantial evidence in support of the theory has been accumulated over a number of years.

See **Customer service, satisfaction and success.**

Reference

[1] Oliver, Richard L. (1996). *Satisfaction: A Behavioral Perspective on the Consumer*. McGraw-Hill, New York.

EXPEDITING

See **Purchasing: Acquiring the best inputs.**

EXPERIENCE CURVE

See **Learning curve analysis.**

EXPERT SYSTEMS

Expert systems are a new generation of computer software that can make decisions in place of human experts. Often expert systems are used to delegate decision-making to non-experts or machines.

See **Capital investment in advanced manufacturing technology; Decision support**

system; Reengineering and the process view of manufacturing.

EXPLOSION (BILL OF MATERIALS)

Explosion refers to the break down of a bill of materials into its lower level components, parts and materials. Explosion is performed from the end product right down to the raw materials and/or purchased parts. This comprehensive process calculates the quantities of all components needed to meet end product requirements while taking into account any inventory on hand or scheduled receipts.

See **Bill-of-materials; MRP.**

EXPONENTIAL ARRIVALS

While studying a production system using simulation models, it is important to mimic the arrival of jobs to a production system. The exponential distribution is appropriate for events that recur randomly over time including jobs arriving in a system. A key property of the exponential distribution is that it is memoryless, that is the current time has no effect on future outcomes. For instance, if machine failure was exponentially distributed, the length of time between machine failures has the same distribution no matter how long the machine has been running. The mean of the exponential distribution is equal to its standard deviation, reflecting a relatively high variance, and means that the exponential distribution can be fully described by one parameter (the mean). The exponential distribution is commonly observed in the distribution of time between arrivals in waiting line systems. The existence of exponentially distributed times should not be assumed, but should be tested by chi-squared tests, or other means of identifying probability distributions.

See **Simulation of production problems; Waiting line model.**

EXPONENTIAL SERVICE TIMES

While studying waiting lines, where jobs wait for processing or service, the model must account for random variation in the processing/service time. The exponential distribution is often a good approximation of service time. If so, the variance of exponential service times is usually much wider than other distributions, such as the normal distribution. The existence of exponentially distributed times should not be assumed, but should be tested by chi-squared tests, or other statistical methods for identifying probability distributions.

See **Exponential arrivals; Simulation of production problems using spreadsheet programs; Waiting line model.**

EXPONENTIAL SMOOTHING

Exponential smoothing methods of forecasting provide forecasts using weighted averages of past values and forecast errors. In these methods, recent observations are given more weight in forecasting than older observations. In exponential smoothing, there is a single smoothing parameter that is used to weight the data. The parameter is called alpha. Exponential smoothing methods are commonly used in inventory control systems where many items are to be forecast and low forecast cost is a primary concern.

See **Forecasting guidelines and methods; Forecasting in manufacturing management.**

EXPONENTIAL SMOOTHING CONSTANT

See **Exponential smoothing; Forecasting in manufacturing management.**

EXPONENTIAL SMOOTHING MODEL

See **Forecasting guidelines and methods; Forecasting in manufacturing management.**

EXTERNAL CONSTRAINTS

See **Internal constraints; Manufacturing management using the theory of constraints.**

EXTERNAL PERFORMANCE MEASUREMENT

External performance measures are used to assess an organization's performance in terms of its externally observable outputs. For a manufacturing organization these will be items such as price, delivery lead time, customer returns, warranty claims, etc. These are essentially output measures which relate directly to the competitive priorities of the organization's products. As such they are measures that can usually be compared directly against equivalent measures of competitors and can provide a basis for benchmarking against competitors.

See **Performance measurement in manufacturing; Process approach to manufacturing strategy development.**

EXTERNAL SETUP OPERATION

External setup operation refers to change over activity machine which can be performed while the machine is still running and producing. Identifying, differentiating and separating external from internal setup operations is a critical step in reducing setup costs and times. By performing all activities that can be performed, while the machine is running, downtime due to setup. Since lost productive capacity and degree of flexibility are determined by the length of down time due setup, controlling and managing external setup time is a less critical than controlling internal setup time. Nevertheless, reducing external setup times reduces total cost by decreasing the time, labor and cost associated with setups.

See **Internal setup operation; Setup reduction; SMED.**

FACILITIES LOCATION DECISIONS

Basheer M. Khumawala and Sukran N. Kadipasaoglu

University of Houston, Houston, TX, USA

INTRODUCTION: Facility Location decisions are typically long-term and are crucial in terms of a firm's profitability and success. In the last two decades, due to the opening of economies, the end of cold war and communism, reduced trade barriers, the proliferation of trade agreements, and advances in communication technology, markets have become global and the number of multinational firms has increased tremendously. Globalization of markets and businesses has necessitated the locations of manufacturing plants and distribution centers globally, in order to obtain advantages of customer proximity and low labor costs, reduction in transportation and distribution time and costs, and other benefits.

Many companies have spread a variety of their operations such as fabrication, assembly, warehousing and distribution to different parts of the world, creating a network of operations, which is now commonly referred to as the global supply chain. Consequently, location decisions, an integral and vital component of the optimal configuration of the global supply chain, have become all the more significant.

CLASSIFYING LOCATION DECISIONS: In general, Facilities Location Decisions (FLD) are classified into two broad categories. The first one is where the location of the facility is not limited to pre-determined potential sites. The assumption here is that a facility can be established at any point in a given area/region. The second one is where a pre-determined set of potential sites is first identified and then, selection of facilities from this set is made. Location decisions of facilities such as manufacturing plants, distribution centers, warehouses in the manufacturing sector as well as offices in the service sector fall in the second category. As such, the focus of this paper is the location decisions to be made from a pre-determined set of potential sites. The traditional objective in location decisions of this type has been to minimize the total relevant distribution costs, which is the sum of the inbound and outbound transportation costs to and from the facility, and the operating costs of the facility. These costs will of course consist of fixed, semi-variable and variable components. In addition to the traditional cost minimization objective in FLD, other objectives have been used such as maximizing customer service, maximizing profitability, establishing a pre-determined number of facilities within a given budget, or a combination of these.

A wide variety of practical considerations have led to a number of other classifications of the FLD. One consideration concerns the decision about the capacity of the facilities. There are two possibilities: capacitated FLD and uncapacitated FLD. In the capacitated FLD, each potential site is identified with its *a priori* capacity level. Whereas in the uncapacitated FLD, the problem solution determines the ideal capacity level for the selected sites. FLD models have been developed for single and multiple products. The degree of certainty with which future demand is known and the volatility of costs/prices give rise to deterministic or stochastic FLD.

The scope of facilities location decisions has evolved considerably along with the growth in international trade. This has given rise to International FLD. Since international location decisions are an integral part of the optimal configuration of a firm's global supply chain the strategic impact of location decisions is now greater than ever, and this trend is expected to continue in the future. One such large-scale published study on global supply chain management is that of Digital Equipment Corporation. As a result of the study, Digital reduced its number of plants from 33 to 12 in the three major trade regions, Pacific Rim, the Americas, and Europe, reducing manufacturing cost by $500 million and logistics costs by $300 million, while dramatically increasing output (Arntzen *et al.*, 1995).

GLOBALIZATION OF LOCATION DECISIONS: The shift to globalization location decisions has occurred due to several reasons: among these are the need to: (1) acquire raw materials and lower labor costs; (2) secure a presence in foreign markets to increase market share and profitability; and (3) achieve proximity to customers. The competitive benefits from proximity to customers vary from: (1) better understanding the customer needs, and; (2) a reduction in transportation/

distribution costs. Survey results from 224 companies with multi-national operations show the following as the key factors in international location decisions: availability of transportation, labor attitudes, ample space for future expansion, proximity to markets, and availability of suitable plant sites (Tong and Walter, 1980). Several other considerations are included in international location decisions. Some of them are: cost of labor vs. automation at home country, challenges facing technology transfer, the proximity to technical expertise, the availability of telecommunications, the proximity to support services to maintain the high technology equipment, the availability of utilities, the availability and reliability of different modes of transportation, environmental regulations, government incentives, and the possibility of local partnerships.

Three regions where the US-owned multinational companies have expanded their operations are Mexico, Far East and Europe. The primary incentive to US firms to locate in Mexico was inexpensive labor, its proximity to US, and the Mexican government policy of allowing 100% foreign-owned establishments under certain conditions such as the maquiladoras. Major US companies such as Ford, General Motors, Chrysler, RCA, Honeywell, General Electric and Johnson and Johnson have established assembly plants (maquiladoras) in Mexico. Cheap labor and proximity of these plants to the US reduced production and distribution costs while giving easy access for management and technical personnel to oversee the Mexican operations. In addition to US Companies, Japan, Germany and other European companies also have strong presence in Mexico.

Beginning in late 1970's companies from South Korea, Hong Kong, Taiwan, Singapore, Malaysia, Thailand and Indonesia (the Newly-Industrialized Countries, NICs) have become prominent in labor intensive manufactured goods including semi-conductors, computers, machine tools, audio and video equipment, garments, toys and household appliances. The NICs, due their supply of low-cost skilled and unskilled labor, lower material costs, reliable suppliers, adequate infrastructure and government incentives have become a major attraction for multinational firms from developed counties (Lim and Fong, 1988). With the opening of their economies to more foreign interests, China and India have also become attractive to multi-national companies. While lower costs are still a major factor for locating in Asian countries, the changing nature (the trend is that these countries are rapidly becoming consumer economies with a great appetite for western products) and market size and wealth (Japan's per capita income is now the highest in the world (MacCormack *et al.* 1994)) of these markets make them attractive to globally-oriented manufacturers.

Another major international market is the European Community whose collective income has now surpassed that of North America. The major attraction of Europe is its size, wealth, educated work force, and favorable governmental relationships (Thurow, 1992). Several major US companies have facilities in Europe, and more recently in Eastern Europe, formerly communist countries. These countries have a special advantage in terms of their proximity to industrialized European nations, with a supply of low cost labor and land, and a highly educated work force. Thus, these eastern European countries attract investments from the Western European nations as well as from US and Japan.

FUTURE: It is now well understood that global competition affects a firm's business strategy. Facilities Location decisions are an integral part of the optimal configuration of the global supply chain. Therefore, location decisions cannot be made based on quantitative analysis and cost trade-offs alone. These decisions should support the firm's global business strategy and must be in line with its competitive priorities.

The global trade dynamics, trade agreements, and the non-tariff barriers greatly influence location decisions. In Europe, for example, rules of origin and local content requirements have been increasing. This penalizes an export-based strategy and requires foreign firms to locate parts of their manufacturing in Europe (MacCormack *et al.*, 1994). In the Pan-Pacific region, the NICs and the ASEAN nations, have been enjoying rapid growth, which has already attracted lots of foreign firms, notably from US, Europe and Japan. This trend appears to continue in the near future. In the American continent, in addition to NAFTA, other trade agreements are in the works. This has certainly created tremendous shifts in the establishment of manufacturing and distribution facilities across borders. Notable among these are the new facilities established by GM and Ford in Brazil.

An opposing trend is also evident. The JIT systems require frequent and close interactions with suppliers and, thus, result in greater utilization of the local suppliers with supporting communication and transportation systems. For example, maquiladoras in Mexico are unable to obtain input materials from Mexican firms due to poor transportation and communication. The lack of entrepreneurship and managerial skills on the Mexican side further exacerbate the problem.

For any firm, competitive dimensions include quality, cost, delivery and flexibility. The traditional cost minimization objective for location decisions will not be viable for companies competing on these dimensions. Competing on flexibility, delivery or new product innovation requires close proximity to customers, strong coordination between design and manufacturing and strong local supplier relationships. Bartmess and Cerny (1993) refer to this as the "capability-focused" approach to facility location decisions. This approach is dynamic and targets a long-term strategic advantage. Bartmess and Cerny (1993) give examples of several firms including Kodak Business Imaging Systems Division and Seagate Technology Inc., who made erroneous moves by locating their manufacturing facilities in Asian countries based on the overemphasis on low direct labor costs but endangered their long-term competitiveness.

LOCATION DECISION MODELS: Selecting a set of locations from a pre-determined set of sites is a difficult combinatorial problem; it becomes increasingly difficult to solve as the problem size (number of potential sites) is increased. In the past, finding the guaranteed best (optimum) solution to practical size problems was not feasible with the then prevalent computing technology. Two limitations of the computing technology were data storage capability and computation time needed to find the best solution; these limitations have decreased with the progress in computer technology. Many researchers have developed techniques for improving the quality of the FLD (as measured by its proximity to the optimal solution). The techniques generally are of the following types: (1) heuristic approaches, (2) mathematical programming techniques, and (3) simulation (Khumawala and Whybark, 1971).

The first noteworthy study in this area was by Kuehn and Hamburger (1963) which dealt with the development of what is now well known as the "Add Heuristic". Subsequently, the researchers at Exxon (at the time called Esso Research) developed the "Drop Heuristic" as a complement to the Add Heuristic (Feldman *et al.*, 1966). During this period, variants of the Branch and Bound method were also proposed (Khumawala, 1972). Parallel branch and bound application on the capacitated FLD was first formulated by Davis and Ray (1969) and was implemented by Akinc and Khumawala (1977). The Add and Drop Procedures as well as the Branch and Bound method have been used at several companies. Among the known ones are those at Anheuser Busch, H.J. Heinz, Nestle, Exxon, B.F. Goodrich, Standard Oil Company, Mobil, Quaker Oats, Imperial Chemical, General Foods, and Valic. Some of these companies reported considerable savings from these applications (Geoffrion and Graves, 1974; Gelb and Khumawala, 1984). The above mentioned studies dealt primarily with the minimization of total distribution costs. At the time of these studies globalization was still at its infancy, therefore, global location factors were not as predominant in these decision models.

The explosion in computing technology resulted in a shift in the FLD solution methodology. It is now possible to formulate and solve a more comprehensive FLD, i.e., multi-product, multi-period, capacitated, with the increased complexities. As discussed earlier, the International Facilities Location Decisions (IFLD) requires the inclusion of several additional considerations. It should be noted that the IFLD literature to date is classified into two categories: Descriptive and Quantitative. The descriptive literature identifies the importance of IFLD, factors to be considered in IFLD, and how to make successful IFL decisions (Canel, 1992). The quantitative literature includes the mathematical models developed for solving the IFLD (Haug, 1985; Hodder and Jucker, 1985; Canel and Khumawala, 1997).

More recently, solution techniques such as "Tabu Search", "Genetic Algorithms" and "Simulated Annealing" have now been applied to these complex FLD and tested for their effectiveness. These techniques are general heuristics that can be applied to all variants of FLD and IFLD. One large-scale study compared the application of Simulated Annealing, Generic Algorithms, and Tabu Search techniques to the capacitated, multi-period, and multi-product FLD (Arostegui, 1997). His results demonstrated the robustness of the Tabu Search procedure over Simulated Annealing and Genetic Algorithms.

Developing models that truly represent the global supply chain in its entirety is extremely challenging. Part of this challenge is to incorporate the myriad of qualitative and intangible factors involved into the formulation.

CONCLUSION: Although the strategic importance of location decisions has always been recognized, their significance increased tremendously along with the globalization of businesses.

See **Capacitated facility location decision; Combinatorial problem; Generic algorithm; International manufacturing; Manufacturing strategy; Simulated annealing; Supply chain management: Competing through integration; Tabu search.**

References

[1] Akinc, U. and B.M. Khumawala (1977). "An Efficient Branch and Bound Algorithm for Capacitated Warehouse Location Problem", *Management Science*, 23, pp. 585–594.

[2] Arostegui, M.A. (1997). "An Empirical Comparison of Tabu Search, Simulated Annealing, and Genetic Algorithms for Facilities Location Problems", *Unpublished Doctoral Dissertation*, University of Houston.

[3] Bartmess, A. and K. Cerny (1993). "Building Competitive Advantage Through a Global Network of Capabilities", *California Management Review*, pp. 78–102, Winter 1993.

[4] Arntzen, B.C., G.G. Brown, T.P. Harrison, and L.L. Trafton (1995). "Global Supply Chain Management At Digital Equipment Corporation", *Interfaces*, 25(1), pp. 69–93, Jan.–Feb.

[5] Canel, T.C. (1992). "International Facilities Location: Issues and Solutions", *Unpublished Doctoral Dissertation*, University of Houston.

[6] Canel, T.C. and B.M. Khumawala (1997). "Multi-Period International Facility Location: An Algorithm and Application", *International Journal of Production Research*, 35(7), pp. 1891–1910.

[7] Davis, P.S. and T.L. Ray (1969). "A Branch and Bound Algorithm for the Capacitated Facilities Location Problems", *Naval Research Logistics Quarterly*, 16(3), Sept.

[8] Feldman E., F.A. Lehrer, and T.L. Ray (1966). "Warehouse Locations Under Continuous Economies of Scale", *Management Science*, 12, May.

[9] Gelb, B.D. and B.M. Khumawala (1984). "Reconfiguration of an Insurance Company's Sales Regions", *Interfaces*, 14, pp. 87–94.

[10] Geoffrion, A.M. and G.W. Graves (1974). "Multicomodity Distribution System Design by Bender's Decomposition", *Management Science*, 12, pp. 822–844.

[11] Haug, P. (1985). "A Multiple Period, Mixed Integer Programming Model for Multinational Facility Location", *Journal of Management*, 11(3), pp. 83–96.

[12] Hodder, J.E. and J.V. Jucker (1985). "International Plant Location under Price and Exchange Rate Uncertainty", *Engineering Costs and Production Economics*, 9, pp. 225–229.

[13] Keuhn, A.A. and M.J. Hamburger (1963). "A Heuristic Program for Locating Warehouses", *Management Science*, 9, July.

[14] Khumawala, B.M. and D.C. Whybark (1971). "A Comparison of Some Recent Warehouse Location Techniques", *The Logistics Review*, 7(31), pp. 3–19.

[15] Khumawala B.M. (1972). "An Efficient Branch and Bound Algorithm for the Warehouse Location Problem", *Management Science*, 18(12), pp. 718–731.

[16] Khumawala, B.M. (1974). "An Efficient Heuristic Procedure for the Capacitated Warehouse Location Problem", *Naval Research Logistics Quarterly*, 21, pp. 609–623.

[17] Lim, L.Y.C. and P.E. Fong (1988). "High-Tech and Labor in the Asian NICS", *Working Paper #547*, University of Michigan.

[18] MacCormack, A.D., L.J. Newmann, III and D.B. Rosenfield (1994). "The New Dynamics of Global Manufacturing Site Location", *Sloan Management Review*, 35(4), pp. 69–80, Summer.

[19] Tong, H. and C.K. Walter (1980). "An Empirical Study of Plant Location Decisions of Foreign Manufacturing Investors in the United States", *Columbia Journal of World Business*, pp. 66–73, Spring.

[20] Thurow, L.C. (1992). "Who Owns the Twenty-First Century?", *Sloan Management Review*, pp. 5–17, Spring.

FACILITY LAYOUT

This refers to the relative location of equipment and/or work centers on the factory floor. Facility layout affects the flow good and materials through the factory. It has a strong effect on manufacturing lead-time, work-in-process inventory, and other performance measures.

See **Assembly line design; Flexible automation; Functional layout; History of manufacturing management; Linked-cell manufacturing system; Manufacturing cell design; U-shaped assembly lines.**

FACTOR DRIVEN STRATEGIES

Factor-driven strategies are strategies designed to help a firm gain access to the best factor inputs available worldwide. Most often factor-driven strategy seeks low-cost labor (other factors are raw materials and purchased components). Companies from industrialized countries with high labor costs have built facilities all over the world to reduce their total labor costs. For German, Japanese, or U.S. companies that face manufacturing wages that average 15 to 30 dollars an hour, the lure of low labor costs of under two dollars an hour can be very attractive (in some countries, wage rates can be as low as 20 cents an hour). Some of the countries that have attracted tremendous foreign direct investment because of the low-cost labor available include Brazil, the Czech Republic, China, Indonesia, Malaysia, Mexico, Poland, among many others.

Of course, the low-cost labor country selected for manufacturing is selected for many reasons including the labor rate. The size of the local as well as surrounding markets is a very important factor. Further, any firm looking to establish a production operation in a foreign country in order to take advantage of a lower cost structure should follow a rigorous facility location decision process that looks at a number of issues including political and economic stability, currency issues, tax laws, productivity and quality concerns, logistics costs, and market conditions. Many companies have made investment decisions based simply on wage rates and never achieved any cost savings because they: (1) did not achieve needed productivity or quality levels; (2) experienced high indirect labor costs; (3) discovered that logistics and other costs outweighed the production cost saving; (4) experienced difficulties in working with the foreign government; and (5) were never able to integrate the operation into their remaining productive system. For these, and many other, reasons, managers should be very careful in evaluating factor driven opportunities to be certain that they really are strategically sound.

One other potential problem in today's world is the appearance of exploitation that often accompanies the use of 50 cents an hour labor by multinational firms. Labor unions in of multi-national firms dislike it. Despite all of the potential difficulties, factor-driven strategies can provide excellent returns, and can even mean the difference between competitive success and diminishing market share/profitability.

See **Global manufacturing rationalization; International manufacturing.**

FAMILY OF PARTS OR PRODUCTS

Family of parts refers parts having the same raw material and the same sequence of operations (processing steps) or similar geometry. The concept of family of parts in the foundation upon which principles of group technology and manufacturing cells are built.

See **Group technology; Linked-cell manufacturing system; Manufacturing cell design.**

FILL RATE

Fill rate is a measure of delivery performance of finished goods, usually expressed as a percentage. In make-to-stock company, this percentage usually represents the number of items or dollars (on one or more customer orders) that were shipped on schedule for a specific time period, compared with the total that were supposed to be shipped in that time period. In a make-to-order company, it is usually some comparison of the number of jobs or dollars shipped in a given time period (e.g., a week) compared with the number of jobs or dollars that were supposed to be shipped in that time period. (APICS Dictionary, 9th ed., APICS)

See **Capacity management in make-to-order production systems; Make-to-stock production; Safety stocks: Luxury or necessity.**

FILTERING PROCEDURES

Filtering procedures are used in rescheduling selected due date changes. Three alternative filters are suggested for rescheduling based on the ability to change, magnitude of change and time horizon.

With the ability filter, only due date changes that are doable are sent to production.

Magnitude and horizon filters use a threshold value to eliminate trivial due date changes. The magnitude filter changes only those due dates where the difference between the old and revised due dates exceeds the threshold value. The horizon filter does not consider changes to due dates that are still a long way into the future.

See **Capacity planning: Medium- and short-range; Master production schedule; Material requirements planning.**

FINAL ASSEMBLY SCHEDULE (FAS)

This refers to a schedule of finished goods in a make-to-order or assemble-to-order environment. It is sometimes referred to as the finishing schedule. The FAS is prepared to suit customer orders as constrained by the availability of material and capacity. FAS schedules the operations required to complete the product from subassemblies and components that are stocked to the end-item level. FAS may be stated in terms of customer orders or end product items.

See **Capacity planning: Medium- and short-Range; Master production schedule.**

FINANCIAL ACCOUNTING MEASURES

These are performance metrics derived from the financial accounting system of the firm. Typical examples include net income, return on equity, return on assets, book-to-market ratio, and price-earnings ratio.

See **Balanced scorecard; Performance measurement in manufacturing.**

FINANCIAL ACCOUNTING SYSTEM

All organizations measure their performance. Traditionally, such measurement has been done in financial terms that would be useful to outside groups, such as financial markets and stockholders. To ensure uniformity across all companies, predetermined standards have been established that indicate what should be measured and how it should be measured. These standards are developed and administered by the Financial Accounting Standards Board (FASB) in the USA. Because the standards deal essentially with accounting functions, companies have historically placed responsibility for performance measurement in the hands of their accounting departments. Consequently, most performance measures within the financial accounting system focus on costs, profits, and returns to stockholders.

Recently, organizations have found that they must compete on the basis of other criteria, such as quality, innovativeness, or delivery speed. Such criteria have not traditionally been considered as part of the financial accounting system and, consequently, were not been measured. Several different approaches have been taken by companies to remedy this oversight. One approach is to broaden the financial accounting system so that costs associated with these other criteria can be included. A second approach involves adding non-cost performance measures to those collected and reported by the accounting function. A third approach is to spread responsibility for performance measurement throughout the entire organization, making various functional areas responsible for collecting and reporting some of the non-cost measures.

See **Balanced scorecard; Performance measurement in manufacturing.**

FINANCIAL JUSTIFICATION

The process of comparing the costs and benefits of an investment decision by determining the inflows and outflows associated with the investment decision in cash flow terms to see if the decision is economically viable. Financial justification is an important aspect of the investment decision. Discounted cash flow analyses such as the net present value method and internal rate of return are traditionally used in financial justification.

See **Capital investments in advances manufacturing technologies; Internal rate of return (IRR); Net present value (NPV).**

FINISHED-GOODS STOCK BUFFER

Finished goods are items that are subject to a customer order or sales forecast and may include products that are sold as a completed item or repair parts. A finished-goods buffer, also known as a finished-goods stock buffer, are finished goods that are kept in inventory in order to satisfy customer demand. Examples of finished-goods stock buffers include products kept on the shelves of a retail store or end item products kept at the manufacturer's warehouse. A finished-goods buffer is needed when it is necessary or highly desirable to fill customer orders immediately or in the very near future. It may also be needed when manufacturing lead time is relatively long, making it difficult or impossible to manufacture products in time to meet customer requirements. Other categories of stock buffers include raw-material stock buffers and work-in-process stock buffers. Generally, stock buffers should include only products that are relatively standardized with high demand. This is especially true for finished-goods stock buffers.

See **Buffer stock; Synchronous manufacturing using buffers.**

FINITE CAPACITY SCHEDULING

The concept of finite loading machines and work centers requires that machines be loaded only upto their capacity limits. However, actual use of the concept has been hindered by the computational difficulties. Recent advances in computer speed and better software have now made finite loading a feasible scheduling procedure. This approach is now called finite capacity scheduling (FCS).

FCS has some significant advantages over traditional approaches to scheduling, especially ones that utilize infinite loading. First, FCS allows organizations to avoid overloading. Second, FCS enables an organization to consider different alternative schedules that might achieve different objectives. Third, FCS software offers simulation capabilities so that a company can perform "what if" analysis to evaluate different possible scenarios. Finally, companies can promise more realistic due dates to customers in the absence of excessive overloading of workcenters.

See **Finite loading; Infinite loading; Medium- and short-range capacity planning.**

Reference

[1] Kirchmier, B. (1998). "Finite capacity scheduling methods," *APICS*: *The Performance Advantage*, 8, 38–40.

FINITE LOADING

When developing production schedules, there are two general approaches to dealing with production resource limitations. One approach, known as infinite loading, ignores capacity limitations when developing the schedule. The second approach, finite loading, takes into account the actual amount of each resource that will be available. Historically, this second approach could not be used in practice because it required excessive amounts of computer time to develop a schedule. Furthermore, many individuals have argued that all the time and effort required to develop a schedule with finite loading could end up being wasted because of unforeseen changes to the plan.

Recently, improvements in computer speed, and developments in software, have made finite loading a reasonable alternative for computerized scheduling. This approach, which is now referred to as finite capacity scheduling, enables companies to quickly determine a production schedule using finite loading, then to adjust that schedule on the fly as conditions change during execution of the schedule.

See **Finite capacity scheduling; Infinite loading; Medium- and short-range capacity planning.**

FIRM PLANNED ORDERS (FPO)

A planned order in material requirements planning (MPS) can be frozen (i.e., fixed) in quantity and time. Once declared FPO, it is not allowed to change automatically when conditions change. However, either the quantity or timing of a FPO can be changed by managerial action. FPO has several benefits. It helps the system to respond to material and capacity problems by firming up selected planned orders.

Before a FPO could be changed, the design maker must consider alternatives to solve the problem causing the change in FPO.

See **Capacity planning: Medium- and short-range; Master production schedule; Material requirements planning.**

FIRST COME/FIRST SERVED RULE (FCFS)

This is one of many dispatching rules for sequencing jobs at a work center. Under FCFS rule jobs are sequenced by their arrival times. FCFS also refers to a method of inventory valuation for accounting purposes. In this accounting practice the oldest inventory (first in) is the first to be used (first out). While the physical use of inventory is not constrained by FCFS rule.

While serving people, this rule is 'fair' in that people are processed in the order they arrive at the work center.

See **Capacity planning: Medium- and short-range; Sequencing rules.**

FIRST-TIER SUPPLIER

See **Supply chain management: Competing through integration.**

FISHBONE DIAGRAM

See **Integrated quality control; Total quality management.**

FIVE S's

The five routine activities or duties of the operator include proper arrangement, orderliness, cleanliness, cleanup and discipline – These 5 activities all begin with the letter S in Japanese where this term orginated.

See **Implementing lean manufacturing.**

FIXED ASSET TURNOVER

It is a performance measure that is indicative of how effectively a company is utilizing its assets. This ratio is calculated as:

Fixed asset turnover = Sales/Fixed Assets.

The fixed asset turnover ratio is often used as a "red flag" while evaluating companies. Low values are symptomatic of an underlying problem. If this measure is used in evaluating a supplier, the buyer may investigate the following: Is the weakness a result of poor quality or delivery performance? Is it a result of wasteful spending by management? Has the supplier assumed too much debt? Can the supplier's bank provide a statement of good financial health?

In cases when the supplier is a publicly traded company on the stock market, specific financial ratios can be obtained from sources such as 10-K reports, Moody's Industrials, and Dunn and Bradstreet Reports. Ratios can also be calculated using information shown in income statements and balance sheets.

An assessment of a potential supplier's financial condition almost always occurs during the initial evaluation process. Some firms view the financial assessment as a screening process or preliminary condition that the supplier must pass before a detailed evaluation can begin. A firm can use a financial rating service to help analyze a supplier's financial condition. If the supplier is a publicly held company, other financial documents will be readily available. Because buyers are relying on fewer suppliers today to support their purchase requirements, it is important to reduce risk by selecting financially sound suppliers who are expected to remain in business for the foreseeable future.

Selecting a supplier in poor financial condition presents a number of risks. First, there is the risk that the supplier will go out of business. This can present serious problems if there are not other sources of supply readily available. Second, suppliers who are in poor financial condition may not have the resources to invest in plant, equipment or research which are essential to longer-term technological or other performance improvements. A purchaser must feel confident that a supplier can commit the resources necessary to support continuous improvement.

See **Performance measurement in manufacturing; Supplier performance measurement.**

FIXED COSTS

Fixed costs are those that remain constant or stable over different levels of production volume, at least during the short term. For a firm to stay in business over the long term, it must price its products at a level that covers fixed costs, variable costs and profit. Examples of traditional fixed costs include:

Property taxes
Depreciation expense
Salaries of some employees
Pension obligations
Rent
Insurance payments
Interest on outstanding loans
Equipment lease payments and charges

Technically, these costs are present even if a firm does not produce any material output. However, a firm can partially reduce its fixed costs, during slower economic conditions. For example, a company can close a section of a plant to realize lower insurance premiums due to an idle condition. Avoidable costs are the portion of fixed costs reduced during slower economic periods.

Most firms attempt to maintain fixed costs at the lowest level possible because these costs represent financial obligations. During slower business conditions, some firms take on even marginal business simply to increase the utilization of production facilities and cover some fixed costs. The logic behind this is that the fixed cost per unit becomes lower as sales increase. For example, if a seller has $10,000 in fixed costs and a current sales volume of 5,000 units, the average fixed cost per unit is $2.00 (i.e., $10,000/5000 units = $2 per unit). If volume doubles to 10,000 units, the average fixed cost per unit lowers to $1.00 ($10,000/10,000 units = $1 per unit). This illustrates why firms are sometimes willing, in the short run, to take on business that actually appears unprofitable. Then, the seller's motivation would be to recover variable costs while making at least some contribution to fixed costs.

See **Break-even analysis; Cost analysis for purchasing.**

FIXED ORDER-QUANTITY INVENTORY MODEL

This is a method of managing inventory for an item with independent demand. According to this model, a fixed amount Q is reordered every time the inventory level drops to a predetermined reorder point R. This inventory model is termed as the Q/R model states that "order Q units, when inventory level reaches a predetermined level = R." This system of inventory management is also termed as, fixed order quantity system, lot-size system.

See **Economic order quantity model; Fixed time period inventory model; Inventory flow models; Safety stocks: Luxury or necessity.**

FIXED TIME-PERIOD INVENTORY MODEL

This is an inventory management system in which inventory is counted and reorder is placed in predetermined intervals, such as a week or a month. Placing orders on a periodic basis is desirable in situations where vendors make routine visits to customers and take orders for their complete line of products, or when buyers want to combine orders to save transportation costs. Fixed-Time Period models generate order quantities that vary from period to period depending on the usage rates. This system of inventory management requires a higher level of safety stock than a fixed-order quantity system.

In the system, the order quantity is not fixed. If M is the maximum inventory, and if the inventory on hand at the time of periodic review is B, then the order quantity will be M-B. This system is also called, periodic-review system, and fixed-interval order system.

See **Economic order quantity model; Fixed order quantity inventory model; Inventory flow model; Safety stocks: Luxury or necessity.**

FIXED WITHDRAWAL CYCLE

This refers to a manufacturing system using Kanbans. In a two-card Kanban system either fixed quantity variable withdrawal cycle or variable quantity fixed withdrawal cycle may be used. Under the fixed withdrawal cycle system, the period between material movements is fixed and the quantity moved (withdrawal Kanban) depends on the usage over the withdrawal cycle. In other works, material is withdrawn at fixed intervals from the output buffer of one workstation and moved (or transported) to the input buffer of the next workstation. Fixed withdrawal cycle systems are often used to control inventories between supplier and assembly plants when delivery trucks run on a fixed schedule.

See **Dynamic kanban control for JIT manufacturing; Kanban; Kanban-based manufacturing systems; Two-card system.**

FLEXIBILITY DIMENSIONS

See **Manufacturing flexibility dimensions.**

FLEXIBILITY IN MANUFACTURING

Kenneth K. Boyer

DePaul University, Chicago, IL, USA

DESCRIPTION: Manufacturing companies have two basic alternatives for addressing the challenge posed by variable demand: (1) build manufacturing plants with excess capacity and/or stock excess goods in inventory to smooth fluctuations in demand, or (2) increase the flexibility of their manufacturing plants so that production volume and variety can be varied more easily to match changes in demand. Numerous manufacturers have pursued this second option of increased flexibility over the past decade. The widespread popularity of manufacturing initiatives such as just-in-time inventory management and total quality management has served to underscore the benefits of flexibility as an alternative to wasted resources in the form of excess capacity and inventory. This article provides a review of various types of flexibility and factors which facilitate the development of increased flexibility.

Manufacturing flexibility is an important element of a firm's manufacturing strategy, providing the capability to respond quickly to shifts in market requirements. Flexibility is one of the key factors which determine the competitive position of the firm and is one of the four competitive priorities (cost, quality, flexibility and delivery) which comprise a firm's manufacturing strategy (Hayes and Wheelwright, 1984; Leong *et al.*, 1990). In uncertain, rapidly changing and dynamic environments, the value of flexibility is magnified due to its ability to alleviate problems associated with such chaos (Swamidass and Newell, 1987). See Table 1 for examples of flexibility types.

TYPES OF FLEXIBILITY: There are several different ways to classify flexibility types. The framework developed by Gerwin (1987, 1993) is widely accepted and will therefore be examined here. Other frameworks that provide good guidance and insight include those by Browne *et al.* (1984), Sethi and Sethi (1990) and Slack (1987). All these authors agree that flexibility is a multi-dimensional concept with different types and that there are important interactions amongst types. A brief description of individual flexibility types is provided here.

1. **Mix:** Mix flexibility is the ability of a system to handle a range of products or variants with fast setups. The faster a plant can change from the production of one product to another, the more efficiently that plant can support a larger number of product offerings. Plants with high mix flexibility can combine greater product variety with short lead times without sacrificing low costs which traditionally have been associated with standardized products with high volumes (Stalk and Hout, 1990).
2. **Changeover:** Changeover flexibility is the ability to quickly substitute new products for current offerings. The range of changeover flexibility measures the variety of product changes that can occur (i.e., 3 new products). Temporal aspect of changeover flexibility deals with the percentage of new product introduction time that is attributable to necessary changes in manufacturing. For example, automobile manufacturers in recent years have enlarged the number of new models offered, while simultaneously reducing the time to introduce new models from 5 years to less than 3 (Womack *et al.*, 1990). A major component of offering new models quickly is the time it takes for the manufacturing operations to be switched from the old model to accommodate the new. This changeover time is currently a major source of competition among automobile manufacturers.
3. **Modification:** Modification flexibility deals with relatively minor changes to existing products. The range of flexibility deals with the number of different changes possible, and the speed with which changes can be made.
4. **Volume:** The ability to adjust the aggregate volume of production allows companies to manage demand fluctuations. Many production systems have fixed capacities which can only be increased by lumpy capital investment for stepwire increase. In the face of such volume inflexibility, manufacturers forecast demand and stockpile inventory in anticipation of future sales. Such an approach has a large degree of risk. In contrast, companies with high volume flexibility can readily adjust their aggregate production volume to match demand.
5. **Material:** This is the ability to deal with variation in the composition and dimensions of parts or materials being processed. Material flexibility also allows manufacturers to handle more than a single type of material, either for the same component or for different components.
6. **Sequencing:** Sequencing flexibility allows manufacturers to rearrange the sequence in which different orders are released to production. The need for this type of flexibility arises from uncertain delivery times for parts and raw materials. As the reliability and quality of suppliers increase, often due to just-in-time or total quality management programs, there is less need for sequencing flexibility.

METHODS TO IMPROVE FLEXIBILITY: There are three basic approaches to increasing the flexibility of manufacturing processes. First, new programmable automation technologies such as computer aided design and flexible manufacturing systems allow manufacturers to combine economies of scale and scope. Second, a capable, well trained and empowered workforce provides the underlying skills necessary to increase flexibility. Finally, it is essential that the process be designed to facilitate greater flexibility. These approaches for improving flexibility are highly synergistic and should be implemented together for maximum benefit (Boyer *et al.*, 1997). Each method is briefly discussed below.

1. **Programmable Automation:** Traditional manufacturing technologies rely primarily on *economies of scale* in seeking improvements in efficiency. The essence of economies of scale is that fixed costs can be spread out over large volumes of production, thus reducing the per unit cost. The drawback to achieving economies of scale through dedicated equipment (such as an assembly line) is that flexibility is sacrificed. In contrast to traditional manufacturing technologies, programmable automation such as computer numerical control (CNC) machine tools, flexible manufacturing systems (FMS) or robotics derive many of their benefits from *economies of scope*. Economies of scope are defined as "efficiencies wrought by variety, not volume" (Goldhar and Jelinek, 1983) and encompass advantages such as extreme flexibility, rapid response to changes in demand and product design, greater control and repeatability of processes, reduced waste and faster throughput. When a single process or piece of equipment can be used for two or more products, economies of scope are realized by spreading fixed costs. Programmable automation facilitates flexibility by embedding machine control in computers.

   Figure 1 depicts the traditional trade-off between economies of scale and scope. Low-volume production facilities (job shops) typically employ general purpose equipment which provides great flexibility, but at high per unit cost. High-volume production facilities (line flows) employ dedicated equipment which

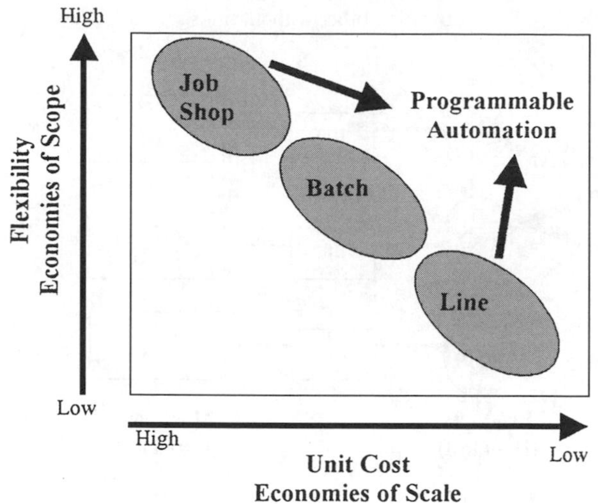

Fig. 1. Trade-offs between economies of scale and scope.

provide low unit costs at the cost of flexibility. Figure 1 also illustrates that programmable automation recaps the benefits of both kinds of economies by making possible high flexibility simultaneously with low per unit cost.

2. **Workforce Development:** Although many experts considered the development of programmable automation as the beginning of the "factory of the future" in which a lot size of one would be produced as efficiently as large volume orders, the reality is that none of these advantages can be realized without a well trained, empowered workforce. Flexibility is enhanced through the use of programmable automation, but the implementation of programmable automation must be supported by other, "softer" improvement programs such as quality leadership, training, worker empowerment, and the use of small, cross-functional teams in order to be successful. This view is express by Ettlie (1988), who asserts that technological innovation must be tempered with administrative innovation in order to achieve success through synchronous innovation.

Empowering workers to make decisions regarding the best method of capturing the benefits of new technologies is critical to achieving increased flexibility (Boyer *et al.*, 1997). Flexibility demands greater knowledge in workers and machines, and thus makes the operating environment more complex. Cross training, or job rotation through a series of assignments, allows workers to develop a greater array of skills

Table 1. Flexibility examples.

Flexibility type	Example of low flexibility	Example of high flexibility
Mix	Radio producer that only offers 2 models of radios in either blue or black.	Radio producer that offers 10 different models in choice of 8 colors.
Changeover	Ford Motor Co. had to shut down production for an entire year when changing from production of the Model T to the Model A in 1926.	Allen-Bradley's Milwaukee plant can changeover from one type of contactor to another essentially instantaneously, thus making lot sizes of one economically feasible.
Modification	If a customer wants a particular brand or type of car stereo, they must go to an after-market stereo store rather than to the automobile manufacturer that does not possess the capability to provide such a feature.	Levi Strauss can mass customize jeans according to individual customer specifications for certain key dimensions. These jeans are sold at a slightly higher price than "standard" jeans.
Volume	A typical automobile assembly is limited in terms of volume expansion. Increases in production can typically only be accomplished by adding/dropping shifts.	Fast food restaurants (McDonald's or Burger King) are typically designed to handle wide fluctuations in demand. Production can be adjusted through a variety of levels to meet high/low demand.
Material	A high volume bread manufacturer (such as Wonder Bread) is limited in the range of materials which can be incorporated due to equipment designed for a very specific type of bread.	A neighborhood bakery with more general purpose ovens has a much higher ability to bake breads from a variety of ingredients.
Sequencing	A typical dedicated assembly line in which a product is made sequentially in a fixed sequence.	A group technology or cellular format for manufacturing the same product allows for easier rearrangement of steps.

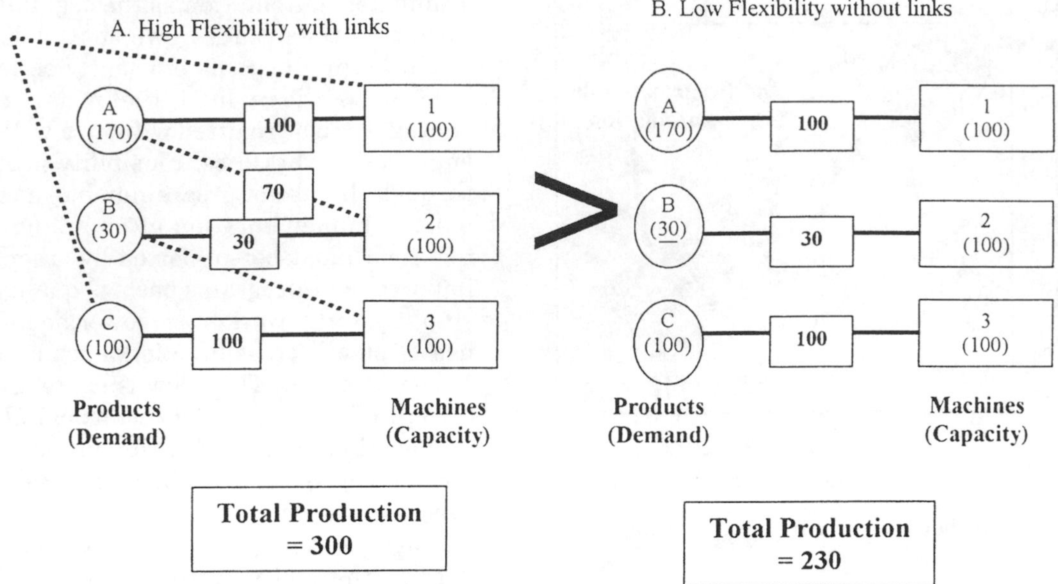

Fig. 2. Linked chains for flexibility.

which help facilitate greater flexibility (Jaikumar, 1986). The use of small teams also allows more frequent and spontaneous communication. In short, workforce development through cross-training, empowerment and job enlargement is a critical prerequisite for increasing flexibility. Without a skilled workforce, the most advanced technology is similar to a Ferrari with an untrained driver – it looks nice, but will not win the race!

3. **Process Design:** A third approach to increasing flexibility involves process design. Group technology in an approach to manufacturing where all the equipment required to make a family of products are placed in a cell. Allocating equipment to a family of products with similar characteristics allows some of the flexibility in a low volume job shop to be realized along with the low per unit cost of a larger volume line production. For example, a family of hubcaps could share essential characteristics: all may have round holes in the middle, may have round boltholes for mounting, and may be made from the same metal. A cell for manufacturing these hubcaps would be appropriate.

Adding machines or plants in a linked chain is another method for increasing flexibility. Figure 2 illustrates a situation with three machines (1, 2 and 3), each with capacity shown in parentheses and three products (A, B and C) with demand shown in parentheses. In Figure 2B, each machine can only make one product (i.e. machine 1 makes product A). Figure 2A illustrates product-machine links added (the dashed lines) to form a linked chain. This type of linked chain allows production to be shifted efficiently among machines. The amount produced at each plant is shown in bold letters on the rectangle associated with each product-machine link. In Figure 2B, the maximum overall production is 230, since only 100 of the demanded units of product A can be produced at machine 1 while machine 2 is under-utilized with only 30 units of production for product B. In contrast, the total production for the situation in Figure 2A is 300. This is possible because the added links (dashed lines) allow machine 2 to produce the 70 units of product A, which can not be done if the process is de-linked as in 2B. Thus, adding the ability to handle multiple products greatly increases the flexibility of the aggregate system. The most efficient manner to add flexibility is in the form of a chain as in Figure 2A.

See **Cross training; Economies of scope; Flexible automation; Group technology; Lean manufacturing; Manufacturing cell design; Manufacturing flexibility dimensions; Manufacturing flexibility; Manufacturing flexibility; Programmable automation; Teams: Design and implementation.**

References

[1] Boyer, K.K. and Leong, G.K., 1996. "Manufacturing Flexibility at the Plant Level", *Omega: The International Journal of Management Science*, Vol. 24, No. 5, pp. 495–510.

[2] Boyer, K.K., Leong, G., Ward, P.T. and Krajewski, L., "Unlocking the Potential of Advanced Manufacturing Technologies", *Journal of Operations Management*, Vol. 15, No. 4, 1997, pp. 331–347.

[3] Browne, J., Dubois, D., Rathmill, K., Sethi, S.P. and Stecke, K.E., 1984. "Classification of Flexible Manufacturing Systems", *The FMS Magazine*, pp. 114–117.

[4] Ettlie, J.E., 1988., *Taking Charge of Manufacturing*, Jossey-Bass Publishers, San Francisco.

[5] Gerwin, D., 1987. "An Agenda for Research on the Flexibility of Manufacturing Processes", *International Journal of Operations and Production Management*, Vol. 7, pp. 38–49.

[6] Gerwin, D., 1993. "Manufacturing Flexibility: A Strategic Perspective", *Management Science*, Vol. 39, pp. 395–410.

[7] Goldhar, J.D. and Jelinek, M., "Plan for Economies of Scope", *Harvard Business Review*, Vol. 61, 1983, pp. 141–148.

[8] Hayes, R.H. and S.C. Wheelwright, 1984. *Restoring Our Competitive Edge: Competing Through Manufacturing*, John Wiley & Sons, New York.

[9] Jaikumar, R., 1986. "Postindustrial Manufacturing", *Harvard Business Review*, Vol. 64, No. 6, November–December, pp. 69–76.

[10] Leong, G.K., D. Snyder and P. Ward, 1990. "Research in the Process and Content of Manufacturing Strategy", *Omega*, Vol. 18, pp. 109–122.

[11] Sethi, A.K. and Sethi, S.P., 1990. "Flexibility in Manufacturing: A Survey", *International Journal of Flexible Manufacturing Systems*, Vol. 2, pp. 289–328.

[12] Slack, N., 1987. "The Flexibility of Manufacturing Systems", *International Journal of Operations and Production Management*, Vol. 7, pp. 35–45.

[13] Stalk, G. and Hout, T., 1990. *Competing Against Time*, The Free Press, New York.

[14] Swamidass, P.M. and W.T. Newell, 1987. "Manufacturing Strategy, Environmental Uncertainty and Performance: A Path Analytic Model", *Management Science*, Vol. 33, No. 4, pp. 509–524.

[15] Womack, J., Jones, D., and Roos, D., 1990. *The Machine That Changed the World*, Harper Perennial, New York.

FLEXIBILITY IN WORK ASSIGNMENTS

See **Lean manufacturing; Manufacturing cell design; U-shaped assembly lines.**

FLEXIBILITY TYPES

See **Flexibility in manufacturing; Manufacturing flexibility; Manufacturing flexibility dimensions.**

FLEXIBLE AUTOMATION

Kathryn E. Stecke and Rodney P. Parker

The University of Michigan, Ann Arbor Michigan, USA

DESCRIPTION: Flexible automation (FA) is a type of manufacturing automation which exhibits some form of "flexibility". Most commonly this flexibility is the capability of making different products in a short time frame. This "process flexibility" allows the production of different part types with overlapping life-cycles. Another type of flexibility that comes with flexible automation is the ability to produce a part type through many generations. Clearly, there are several other manifestations of flexibility.

Flexible automation allows the production of a variety of part types in small or unit batch sizes. Although FA consists of various combinations of technology, flexible automation most typically takes the form of machining systems, that is, manufacturing systems where material is removed from a workpiece. The flexibility comes from the programmability of the computers controlling the machines. Flexible automation is also observed in assembly systems. The most prominent form of flexible assembly is observed in the electronics industry, where flexible machines (automated surface mount technologies) are used to populate printed circuit boards with integrated circuits and other componentry. In this instance, manufacturers have found the machines' far superior accuracy and reliability to be sufficient to warrant the significant investment. Overall, however, manufacturers tend to use automation for fabrication, and leave assembly to human operators who can adapt to a greater variety of changing circumstances more rapidly and easily than machines. In this article, the discussion of flexible automation is primarily focused on machining systems.

The building block of flexible automation is the computer numerical controlled (CNC) machine tool which is typically augmented by automated materials handling systems, centralized controlling computers, automated storage and retrieval systems, and human operators. The variety of installations of flexible automation are numerous. Some typical configurations are discussed below. A CNC machine tool is a self-contained machine, where the tool cutting movements, spindle speeds, tool exchange, and other operations are controlled by a part program executed by the computer controller based at the machine tool. The spindle is a spinning device which holds the tool used to cut into the workpiece.

Conventional machine tools (e.g., lathes, drill presses, milling machines) are not computer controlled. The operation of conventional tools is typically done by skilled craftsmen. There can be variation to dimensions on parts made on a conventional tool, whereas this variation is decreased on CNC machine tools. The elimination of this variation is one objective (benefit) of automating the discrete part production process.

Additional benefits include a reduction in required floorspace, reduced delivery and production leadtimes, higher utilization, increased quality, and smoother implementation of changes and improvements in product design. Another significant benefit is the ability to mass produce with a machine tool that is able to produce identical parts and with the ability to switch production between part types of different designs. This latter capability comes from the part programs stored in the local machine memory, or downloaded from a centralized storage device when needed. Effectively, this is a step towards gaining the benefits of both mass production and job shop customization, commonly known as mass customization. Mass customization refers to the practice of producing single parts or small batches to custom modifications to a part design. Flexible automation helps to achieve mass customization.

Flexible automation is created when the CNC machine tools are augmented by ancillary equipment such as automated materials handling systems, automated inspection, and central controllers. The materials handling systems are responsible for loading and unloading parts from the machines, transporting parts between machines in the system, and handling work-in-process inventory storage. These materials handling systems could consist of robot arms, conveyors, automated guided vehicles, and gravity feed chutes. Most commonly, a combination of these technologies are used with human operators introducing unworked parts into the system and for removing finished parts from the system.

Another aspect of flexible automation is the usage of multiple cutting tools to perform each operation on a part and the automatic changing of cutting tools at each CNC machine tool in the system. A magazine containing cutting tools is located at each machine, and cutting tools are automatically changed (i.e., without human intervention) as the part program dictates. Typical tool magazines can hold 30–90 tools. The selection of which tools should be loaded into which magazine is known as the loading problem, which is one of several production planning problems (see Stecke, 1983) associated with flexible automated machining system. Occasionally, centralized tool magazines permit the sharing of tools between various machine tools, potentially reducing the total tooling cost. However, this requires additional tool transportation devices and a more difficult coordination activity by the central computer system.

The central computer system is another feature of flexible automation systems. The central computer system differs from the local computer controller that resides at an individual machine in the system. It has an integrative role of managing the overall operation of the system. The responsibilities of the central computer can be divided into off-line and real-time activities. Typical off-line activities include effective planning and scheduling for the most productive use of the system during a given production period. Typical real-time activities include monitoring the operation of the system, adapting the schedule and production plans when problems arise, and alerting personnel when catastrophic failure occurs. The degree of 'intelligence' and automatic control in the computer system varies greatly across systems, with levels of human intervention. When the central computer system is also responsible for downloading the part programs to the workstations, the system is known as distributed numerical control.

Other ancillary equipment that is contained in flexible automation systems is some form of automated inspection system that checks the location of either the raw part, the specifications of the finished part, or some intermediate version of the part, or several of these. Video cameras and automated gauges can verify whether the part adheres to some pre-determined quality standard and alert the central computer when a part falls outside of the specifications.

Another form of flexible automation is seen primarily in the semi-conductor industry using surface mount technologies. These machines are used to 'populate' printed circuit boards (PCBs) with integrated circuits and other componentry. Typically the components in question are presented to the machine on large reels which are loaded onto the surface mount machine. As the PCBs pass alongside or though the machine, a component is extracted from the reel, and a gantry arm places the component into a specified location on the PCB. Sometimes the PCB itself is attached to the gantry arm and is moved to a location where the component is inserted into its correct position. The board is then moved along a conveyor to the next component loading position. The component locations are stored in computer memory, and the CNC gantry and inserter locations are controlled by this information. Different PCB designs can follow one another through the surface mount machine without disrupting the machine as long as there is sufficient commonality of components or capacity to load different component reels.

Due to the level of automation and presumed consistency, when a single part fails to meet specifications, it is a signal that some problem that could affect many parts could exist. This could be a faulty fixture, materials handling device, or worn or damaged tool. How the central computer deals with the situation depends upon the level of autonomy granted to it. Most systems will merely alert their human overseers. However, others may take action, checking various possible sources for the problem. The amount of artificial intelligence (AI) built into most industrial systems today is fairly low, largely limited to image processing and monitoring activities rather than the management of contingencies. These machine vision systems typically consist of a video camera, lighting, computer-based artificial intelligence to analyze and filter the image into a recognizable form, and a monitor to display the image and process status (Cohen and Apte, 1997). The use of AI is likely to expand as the adaptive control hardware technology improves. In the future, opportunities will exist for problems to be prevented before they occur. For example, tool wear can be monitored and new tools substituted before a frail tool has the opportunity to damage a part in production. Limited applications of these are beginning to appear in practice.

While most of the discussion so far is concerned with flexible *machining* systems because of their prevalence, there are other forms of flexible automation. Industrial robots can be used for more than part handling. If equipped with the appropriate monitoring and sensing devices, they can perform quite sophisticated and dexterous functions such as welding, inspection, and assembly. For example, a spot-welding robot arm on an automotive production line can move more quickly across an entire vehicle, performing more consistent and rapid welds than a human operator. The flexible robot can recognize the vehicle type by some sensing technology (e.g., by identifying the fixture type) or from sequencing information from the central computer, and adapt the weld location and sequence from car model to car model. Such robots are usually trained by an operator manually moving the robot arm in a learning mode, where the x-y-z Cartesian coordinates of the robot arm's position, the various arm-segment angles, and joint rotations are recorded by the robot's controller for use in real production.

Flexible automation can be configured in several different ways to achieve different production objectives. For example, a flexible cell is typically a single CNC machine tool possibly with automated materials handling system, and can make many part types at low volumes, sometimes one-off prototypes of products. Another form is where CNC machine tools can be lined serially with a conveyor for part movement in a 'transfer line' arrangement to get a high throughput of a limited number of part types. A flexible manufacturing system is typically more elaborate in design, involving several CNCs doing different sets of operations, linked together logically by computer communications and physically by materials handling devices. These systems can be operated in different ways. For example, sometimes several machines may perform identical operations for reasons of system balance and/or redundancy during periods of machine failure. This then allows for different routes for parts going through the system.

HISTORICAL PERSPECTIVE: Flexible automation is a form of manufacturing technology which is the culmination of a long evolution in production automation. Most of the development in industrial automation, as we know it, has largely occurred during the twentieth century.

Automation has long been the dream of engineers and scientists, whereby the simple, dirty, repetitive, and dangerous tasks traditionally done by people, could be undertaken by machines. More recently, this vision has been extended to complex physical, computational, and analytical tasks. Advances in computer technologies allowed this vision to recent automate several aspects of personal, administrative, industrial, and logistic activities is now a reality.

Initially, automation was *fixed*, that is, it could perform a single task, or small set of tasks, efficiently and effectively but changing this task set was difficult, costly, or impossible. This feature common to earlier of fixed automation is called "rigidity." Fixed automation commonly can do a small set of well-defined tasks particularly well, but has trouble doing anything else, without significant and time-consuming intervention from human operators.

Historically, automation as a substitute for general human activity. Automation where mechanical, electronic, or computational apparatus were substituted for human activity in an organized industrial context can be traced to the 18th century.

Scale economics through fixed automation dominated the industry for much of the 20th century until competitive pressures, primarily from Japan, forced the American auto industry to change their focus to one of product variety and quick response to market needs. The emergence of the new flexible automation technologies enabled a more agile approach to cater to these pressures.

The birth of numerical control (NC), resulting from the combination of conventional machine tools and computers in the 1940s, is credited to John Parsons (Chang, Wysk, and Wang, 1998). Further development occurred at MIT, funded by the U.S. Air Force. The first fully fledged NC machine tool, which could machine complex shapes was developed in 1952 at MIT. Parson used punched cards containing programs which delivered instructions to hardwired machine tools. The hardwired controller was succeeded by an NC controller and the punched cards gave way to paper tape. These machine tools evolved into NC machine centers that could drill, bore, and mill. Developments during the 1960s included automated tool changers and indexing work tables (Viswanadham and Narahari, 1992).

Parallel developments in computing technologies resulted in much progress in the NC controller, allowing a centralized controller to issue commands to numerous numerical control machine tools. This *direct numerical control* (DNC) was appropriate when the available computing technology was bulky and expensive. However, as electronics and computers became miniaturized it became possible to place computer controllers within each machine tool, with a central controller responsible for a smaller array of operations, mostly real-time monitoring at the system level.

One of the earliest full-fledged flexible manufacturing systems was developed by the Sunstrand Corporation in 1965. It involved eight NC machine tools with a computer automated roller conveyor. Although it did not have much process flexibility, it marked the advent of flexible automation where part programs for different part types could now be loaded quickly into local microprocessors and production could switch between different part types without significant setup time. Automated movement was done using relay switches. Developments since then have been mostly refinements in the technologies and the variety of machine tools covered. Today, flexible automation technology is far more robust and cost significantly less.

STRATEGIC PERSPECTIVE: Flexible automation was hailed as a remedy for the competitive challenges that modern manufacturing was encountering through rising quality standards, shortened product life cycles, and greater demand for product variety (Hill, 1994). Some disappointment resulted from these great expectations. Some commentators suggested that the problem lay in the strategic mis-use of the systems. For example, Hayes and Jaikumar (1988) suggested that managers using these new technologies in the same manner in which they used their previous conventional technologies were destined for disappointment. They stressed the need for a new mindset to experience the 'revolutionary' benefits these new flexible systems promised. Hayes and Clark (1986) observe that productivity can fall for significant periods after the introduction of new production technologies, but this can be prevented by enlightened management and reorganization. Jaikumar (1986) observed a difference in the early usage (late 1970s, early 1980s) of these technologies between certain Japanese and American manufacturers. He noticed that the flexible systems in Japan were used more for their flexible benefits than in the U.S. He also noticed that Japanese managers introduced more products every year than their American counterparts. Consequently, Japanese manufacturers also had fewer problems financially justifying these flexible technologies than American manufacturers.

Hill (1994) suggests that a great deal of disappointment resulted from managers investing in 'flexible' equipment believing that the possession of new flexible technologies would result in a 'strategic response' to competitive pressures. One lesson appears to be that flexible automation is appropriate when its capabilities (e.g., producing multiple part types in medium volumes) are aligned with the company's needs and defined manufacturing and technology strategies.

Much has been written about the economic justification of flexible technologies (see Son, 1992). There has been evidence to suggest conventional justification techniques are inappropriate for flexible automation (Kaplan, 1986) and much activity has been directed at attempting to capture the more elusive benefits of flexible technologies. Foremost among these benefits is the flexibility of making multiple products simultaneously, and many authors have attempted to capture and characterize this flexibility through mathematical programming models (e.g., Fine and Freund, 1990), real option models (Trigeorgis, 1996), and empirical studies (e.g., Upton, 1995b).

TECHNOLOGY PERSPECTIVE: There is an obvious difference between flexible automation and the conventional equipment. Not so obvious is the change in management practice required to secure the benefits of the new technologies. This need was not fully appreciated initially, and early performance of flexible automation in America was lacklustre. Necessary changes extend to the planning processes needed to operate flexible automation. Stecke (1983) identified five production planning issues necessary for effective operation of flexible manufacturing systems. Much subsequent research into FMSs has addressed one or more of these issues, which are grouping machines, selecting part types, choosing relative mixes of products, allocating system resources to part types, and determining appropriate tool magazine loading strategies. These are unique challenges faced by production managers of flexible automation that are driven by technology.

With all the advantages that existing flexible automation offer, a legitimate question is to ask, why all manufacturing is not done on such equipment. One reason is that dedicated equipment is generally faster, operation by operation, than flexible automation, and more appropriate in high-volume environments. Another reason is that there is a cost premium in the acquisition and operation of flexible automation over dedicated systems. Also, for all the tumult about the 'agility' of flexible automation, the ability to *easily* modify the systems to accommodate entirely new part types is limited. Therefore, the next phase of flexible automation appears to be the development of reconfigurable manufacturing systems (RMSs), where the technology (Koren and Ulsoy, 1997: 1) will be "designed for rapid adjustment of production capacity and functionality, in response to new circumstances, by rearrangement or change of its components". An example of a reconfigurable machine is one that, has milling and drilling capabilities but currently has no capability for turning. But a "reconfigurable" machine can easily, quickly, and cheaply be reconfigured to acquire the new turning capability also. Although, RMS technology does not currently exist, newly constructed hardware and software tools offer the capability to produce newly introduced part types. Another example of an RMS hardware is a milling machine where with room for the addition of several spindles that can be arranged in numerous configurations. The development of the hardware, the software, and the science of reconfiguration is ongoing.

See **CAD; CAM; CIM; Dynamic routing in flexible manufacturing systems; FMS; Flexibility in manufacturing; Manufacturing cell design; Manufacturing flexibility; Manufacturing strategy; Mass customization.**

References

[1] Chang, T.-C., R.A. Wysk, and H.-P. Wang (1998). *Computer-Aided Manufacturing* (2nd. Ed.), Prentice Hall, New Jersey.

[2] Cohen, M.A. and U.M. Apte (1997). *Manufacturing Automation*, Irwin, Illinois.

[3] Fine, C.H. and R.M. Freund (1990). "Optimal Investment in Product-Flexible Manufacturing Capacity." *Management Science*, 36(4), 449–466.

[4] Hayes, R.H. and K.B. Clark (1986). "Why Some Factories are More Productive Than Others." *Harvard Business Review*, September–October, 66–73.

[5] Hayes, R.H. and R. Jaikumar (1988). "Manufacturing's Crisis: New Technologies, Obsolete Organizations." *Harvard Business Review*, January–February, 77–85.

[6] Hayes, R.H., G.P. Pisano, and D.M. Upton (1996). *Strategic Operations*, Free Press, New York.

[7] Hill, T. (1994). *Manufacturing Strategy* (2nd. Ed.), Irwin, Massachusetts.

[8] Hopp, W.J. and M.L. Spearman (1996). *Factory Physics.* Irwin, Illinois.

[9] Jaikumar, R. (1986). "Postindustrial Manufacturing." *Harvard Business Review*, November–December, 69–76.

[10] Kaplan, R.S. (1986). "Must CIM be Justified by Faith Alone?" *Harvard Business Review*, March-April, 87–95.

[11] Koren, Y. and G. Ulsoy (1997). "Reconfigurable Manufacturing Systems." ERC Technical Report #1, The University of Michigan, Ann Arbor, Michigan.

[12] Son, Y.K. (1992). "A Comprehensive Bibliography on Justification of Advanced Manufacturing Technologies." *Engineering Economist*, 38(1), 59–71.

[13] Stecke, K.E. (1983). "Formulation and Solution of Nonlinear Integer Production Planning Problems for Flexible Manufacturing Systems." *Management Science*, 29(3), 273–288.

[14] Stecke, K.E. and I. Kim (1991). "A Flexible Approach to Part Type Selection in Flexible Flow Systems Using Part Mix Ratios." *International Journal of Production Research*, 29(1), 53–75.

[15] Trigeorgis, L. (1996). *Real Options: Managerial Flexibility and Strategy in Resource Allocation.* The MIT Press, Massachusetts.

[16] Upton, D.M. (1995a). "What Really Makes Factories Flexible." *Harvard Business Review*, July–August, 74–79.

[17] Upton, D.M. (1995b). "Flexibility as Process Mobility: The Management of Plant Capabilities for Quick Response Manufacturing." *Journal of Operations Management*, 12, 205–224.

[18] Viswanadham, N. and Y. Narahari (1992). *Performance Modeling of Automated Manufacturing Systems*. Prentice Hall, New Jersey.

FLEXIBLE DESIGN TECHNOLOGIES/ ENGINEERING CHANGE ORDERS

In many product development projects, a substantial portion of development time is spent on iterating previously established solutions to include the latest available information. In practice, such iterations are known as engineering change orders (ECOs). They result from the fact that engineering is an iterative rather than a purely linear process, and they are traditionally targeted toward correcting mistakes, integrating components, or the fine tuning of a product. ECOs are also an outcome of the growing level of parallelity in today's development processes, where information-absorbing downstream activities are often started prior to the completion of information-supplying upstream activities and thus have to rely initially on tentative information (Terwiesch and Loch, 1998).

ECOs may consume one-third to one-half of engineering capacity and represent 20–50% of tool costs (see Terwiesch and Loch (1998) and the references there). Thus, reducing the cost of iterations provides substantial leverage in reducing overall development time and cost. One way toward achieving these benefits lies in the application of flexible design technologies (Thomke, 1996). Flexible design technologies are development tools that allow incorporation of design changes quickly and cheaply.

The three most common examples include rapid prototyping, computer simulation, and soft tooling. The idea behind rapid prototyping is to use simplified prototypes specifically designed for a given experiment. Such prototypes are easier and faster to produce than traditional, fully functional ones. For example, in evaluating the fit of an automobile component into the engine compartment, it is sufficient to have a non-functional three dimensional representation of the component, which can be rapidly produced from existing CAD data by using stereolithography.

Computer simulations can replace a large portion of physical prototypes. As they represent "virtual prototypes," they are relatively inexpensive to develop and, in particular, inexpensive and fast to change. Quick and inexpensive changes are also the rationale for using soft tools. Soft tools do not have the durability required for large-scale manufacturing. For example, in the case of stamping dies, this allows the usage of softer steel variants. The softer steel is easier to produce and, easier to be modified.

Projects using flexible design technologies are more efficient than projects using inflexible technologies. The difference stems partly from the lower cost of direct iterations (ECOs), and partly from the reduced need for costly resource investments to reduce the risk of design changes.

See **Concurrent engineering; Product development and concurrent engineering.**

References

[1] Terwiesch, C., and C.H. Loch (1998). "Managing the Process of Engineering Change Orders," INSEAD Working Paper #98/31/TM, forthcoming in the *Journal of Product Innovation Management*, 1998.

[2] Thomke, S.H. (1996). "Managing Experimentation in the Design of New Products and Processes," Harvard Business School Working Paper 96-037, forthcoming in *Management Science*.

FLEXIBLE KANBAN CONTROL

Flexible kanban control is a dynamic kanban control (DKC) policy. The goal of DKC is to systematically add and remove kanbans so as to achieve improved system performance. Under a DKC policy, kanbans are added when needed to improve system performance and removed when they are no longer needed or when their presence will result in lowered system performance. In general, we want additional kanbans when the benefit of their presence (e.g., reduced blocking and starvation, improved throughput) outweighs the cost (e.g., increased WIP and operating costs). The goal of these DKC policies is to minimize backlog, while holding down WIP, in the face of variations in processing time and demand, and sudden equipment breakdown.

See **Adaptive kanban control; Dynamic kanban control for JIT manufacturing; JIT; Kanbans; Pull production systems; Reactive kanban control.**

FLEXIBLE KANBAN SYSTEM

In Kanban-based manufacturing systems, the number of Kanbans is generally held fixed during a production cycle. It is, however, well known that supervisors, from time to time, on an *ad hoc* basis, increase or decrease the number of Kanbans depending on whether the system is experiencing shortages (starvation) or inventory buildup (blocking). In fluctuating processing time and variable demand environments, it is beneficial to adjust the number of Kanbans during the production cycle. This type of system, which allows changes in the number of Kanbans, is termed as a Flexible Kanban System (FKS). Recent research, shows how to change the number of Kanbans in order to offset blocking and starvation caused by uncertainties during a production cycle.

See **Blocking; Dynamic kanban control for JIT manufacturing; Kanban-based manufacturing systems; Kanbans; Starvation.**

FLEXIBLE MACHINING SYSTEMS

See **Flexible automation.**

FLEXIBLE MANUFACTURING SYSTEM (FMS)

The development of CNC and DNC led to the first U.S. implementation of a Flexible Manufacturing System (FMS) at Caterpillar Tractor in the mid-1970's. The typical configuration of an FMS is a group of CNC machines linked together by an automated material handling system. Many of these systems use robots or automated guided vehicles for loading and unloading parts. The operation of the FMS is integrated by supervisory computer control. The major advantage of an FMS is its ability to produce in random order a variety of products as well as new products on the same machine and accommodate design changes.

See **Flexible automation; Human resource issues and advanced manufacturing technology.**

FLEXIBLE TRANSFER LINES

See **Flexible automation; Robot selection.**

FLOW LINES

Flow lines are composed of processes or operations laid out in a linear fashion with each processes operated by one worker. Work flows from one end of the line to the other. Processes are laid out linearly according to the sequence of operations dictated by the product without much flexibility.

See **Assembly line design; Manufacturing systems.**

FLOW MANUFACTURING

Flow manufacturing emphasizes the smooth flow of materials at a relatively constant rate through the manufacturing process. This manufacturing approach offers many advantages, including low work-in-process inventories and simplified planning and control. However, companies that produce a wider range of products or make products that require complex assembly operations have historically used batch production methods in which large batches of material flow intermittently through the factory. Large batch production results in long lead times, large inventory levels and complex scheduling problems.

As companies move from large-batch production to flow manufacturing, they must usually implement the following:

- Setup time reduction
- Small lot sizes
- Total Productive maintenance
- U-shaped manufacturing cells
- Pull system, such as kanban
- Mixed-Model Sequencing
- Statistical Process Control
- Just-in-Time Manufacturing or lean Manufacturing enable the successful implementation of flow manufacturing.

See **Just-in-time manufacturing; Just-in-time manufacturing implementation; Lean manufacturing.**

Reference

[1] Hirano, H. (1988). *JIT Factory Revolution*. Productivity Press, Massachusetts.

FLOW SHOP

See **Flow-lines; Manufacturing systems.**

FLOWTIME

Flowtime is defined to be the time that a job spends in the shop.

See **Dynamic routing in flexible manufacturing systems.**

FMS

See **Flexible manufacturing systems.**

FOCUSED FACTORY

Paul M. Swamidass and Neil R. Darlow

Auburn University, Alabama

The concept of the focused factory argues that problems arise when a plant attempts to mix too many manufacturing processes and/or serve too many markets at the same time. A focused factory is more competitive than an unfocused factory which is bogged down by an unmanageable set of processes and/or products. The focused factory concentrates the plant resources on a limited manufacturing task that is closely tied to the business strategy; the manufacturing task defines what is demanded of manufacturing in terms of costs, deliveries, lead times, quality levels, and reliability.

THE ORIGIN OF FOCUSED FACTORY: The focused factory (FF) concept was first described by Skinner (1974). He argued the case for factory focus at a time when U.S. industry was experiencing problems with high costs and low efficiency. U.S. labor was expensive compared to that used by foreign competitors, and U.S. manufacturers were slipping in their competitiveness. Focused factory shifts the emphasis from being concerned solely with productivity to the issue of how the plant can be made more competitive.

The focused factory became the basis of Japanese competition in the mid-1960s (Stalk, 1988). From 1945 until the early 1960s, the success of Japanese manufacturing was largely attributed to low labor costs, made possible by the devaluation of the yen against the dollar. The Japanese strategy was to compete in industries in which low labor costs offset the effects of low productivity, such as textiles, shipbuilding, and steel. During the early 1960s rising Japanese labor costs and a more favorable exchange rate encouraged a change in strategy. Increased capital investment and labor productivity enabled the Japanese to compete through economies of scale. A quest for improvement led to the use of focused factories, making products otherwise unavailable in the world, or high-volume, low-variety products. Reducing the variety of products made led to increased productivity, reduced cost and increased return on investment.

The unfocused factory attempts to accomplish many different and conflicting manufacturing goals at the same time. In an unfocused plant, a wider range of products require a diversity of processes, and product volumes may differ widely. Further, unfocused plants serve markets with a diversity of needs. This may be the result of a belief that full and complete utilization of the assets of the factory, made of machines and people, should be given a higher priority in pursuing productivity.

An unfocused 'under one roof' strategy introduces the need for compromises in delivery times, quality, and cost and consequently all three may suffer rather than simultaneously improve. As production grows more complex, more management resources will be required. Complex, unfocused factories with these characteristics can find themselves unable to compete with focused plants that structure their manufacturing strategies in such a way as to concentrate on a sensible combination of products, processes and markets for maximum effectiveness.

THE PRINCIPLES BEHIND FOCUSED FACTORY: The rationale behind focused manufacturing are: (1) Low-cost production is not the only means of gaining competitive advantage, nor is it always the most important. Innovation in product features, quality, technology, delivery and other criteria may be equally important; (2) The multiple goals of a factory need prioritization in order for the factory to excel. Focus forces prioritization of goals and enables tradeoffs when and where needed; and (3) Competence in the specified manufacturing task is encouraged by simplicity and dedication. However, simplicity does not imply that all focused plants are small or that all unfocused plants are large. In larger facilities it is possible to adopt a plant within a plant (PWP) approach to factory layout, where several focused 'factories' operate under one roof with separate and distinct facilities, people and objectives.

The five characteristics of focused plants are: (1) Generally, a portfolio of tried and tested processes are sufficient to produce the products; (2) Excellence is attained in a small set of market-oriented criteria such as cost, quality or delivery; (3) Production volumes are consistent, and special orders are separated from the main production such that low-volume products do not disrupt high-volume production; (4) An overall quality philosophy is stressed in the plant encompassing all aspects of production and service; and (5) The competitive manufacturing task at the plant consists of one set of achievable criteria that are consistent with each other and the business strategy, which is the overall long-term direction of the company.

The manufacturing task can be considered as a statement of manufacturing philosophy. It must relate ends and means and link them with the business strategy. The process of developing this statement could be a series of discussions in which the manufacturing task is gradually clarified. The end statement should have the following characteristics:

1. It must be written in sentences and paragraphs, not merely outlined.
2. It must explicitly state the demands and constraints on manufacturing derived from the

competitive strategy of the business, and marketing, financial, technological and economic perspectives.
3. It must state what manufacturing must accomplish to be a competitive weapon and, conversely, how performance can be judged.
4. It must identify likely difficulties caused by prevailing industry problems. Accomplishing a difficult task will create a unique competitive advantage.
5. It must explicitly state priorities (tradeoffs).
6. It must explicitly state the impact on and change requirements for the manufacturing infrastructure, that is, the control and organization aspects.
7. The essence of the manufacturing task could be summarized as a memorable slogan or picture.

TYPES OF MANUFACTURING FOCUS: Hayes and Schmenner (1978) offer guidance on how to define the manufacturing task and create a plant focused on accomplishing it. They emphasize the importance of consistency within manufacturing decisions and between them and the business strategy. Creating a set of 'plants-within-a-plant' (PWP), each with its own manufacturing task, requires careful planning and organization. Hayes and Schmenner discuss two approaches for focus: product focus or process focus. A plant can be either product focused or process focused, or a hybrid. The warnings concerning the use of a hybrid plant are: a mixed focus will create confusion and be detrimental to the consistency of manufacturing tasks in a plant. If two forms of focus must exist within an organization, then they should each have their own central staff to ensure as much separation as possible.

Product Focus

In a product focused organization, each plant, or unit within a plant, is responsible for the production of one complete product using several manufacturing processes. With product focus, the organization can be made flexible and authority can be decentralized to the lower levels. In this scenario, the plant managers are more responsible for new product introduction. The role of corporate staff lies in coordinating product development teams, managing corporate human resources, training, and the allocation of capital. Performance measures for the product focused plant can center around profit and return on investment. That is, they can operate as profit centers. Product focus is generally appropriate for companies that view themselves as catering to several individual markets, rather than producing commodities without market or product distinctions.

Product focus is essential for an organization that competes through time (Stalk, 1988). The goal of time-based manufacturing is to get newer products to customers more quickly than competitors. Product focus enables the processes necessary for designing and making the product to be located closely together, so that there is minimum delay between processes.

Hayes and Pisano (1996), while confirming the utility of the focused factory concept, provide some warnings. A possible danger of focusing a plant on products is that the collective expertise on manufacturing processes will become dispersed. Each plant will develop varying levels of competence in the same processes.

Process Focus

The process focused organization consists of plants or units specializing in a limited set of manufacturing processes. Each of these units may be involved in the production of parts for a variety of products. Responsibilities therefore center around parts of the manufacturing cycle rather than on product lines. Performance measures typically relate to the manufacturing processes themselves: the quality and efficiency of production, scrap rates, and so on. In process focused plants, minimizing the cost per unit of the dominant processes may provide critical competitive advantage. In such cases, the process focused plant is unsuitable for treatment as a profit center; it is usually a cost center.

Process focus is appropriate for those companies that specialize in a particular material or technology that is applied to many products. New product introduction demands overall changes in the plant, since all the processes involved may be affected. Because plant managers must gain familiarity with the process for which they are responsible, they must acquire an in-depth technical perspective. Production of the end product of the Strategic Business Unit (SBU) may require extensive logistics between the various process-focused units.

Warnings have also been given about possible negative effects of adopting process focus (Hayes and Pisano, 1996). Plants will accumulate substantial knowledge about particular manufacturing processes, but the organization may be weaker at introducing new products when required in order to compete. It is therefore advised that in both product and process-focused plants, competences in processes necessary to competitiveness should be nurtured.

A logical extension of product or process focus is Group Technology (GT), the basis for cellular manufacturing. Facilities within a plant are grouped according to product or process layout. The individual machines in a manufacturing cell are linked with conveyors, automated guided vehicles (AGV), or robots to transfer parts between process operations. Such cells may be developed into Flexible Manufacturing Cells (FMC) for processing a variety of parts, and Flexible Manufacturing Systems (FMS) to produce various products in small batches with a high degree of automation.

IMPLEMENTING THE FOCUSED FACTORY: Focusing a plant may require minor to major reorganization and/or investment. The steps to focusing a plant are:

1. Develop an explicit, brief statement of corporate objectives and strategy. Typically, this describes what the company hopes to achieve in five years' time in terms of satisfying the customers.
2. Translate this statement into "what this means to manufacturing," that is, the manufacturing task. This addresses what the factory must become especially good at in order to support the business strategy.
3. Make a careful examination of each element of the production system. This audit assesses the key capabilities of the system and how they must change to meet the manufacturing task that has been defined.
4. Reorganize the elements of structure to produce a congruent focus. The plant structure consists of the hardware, that is, the machinery and processes and the capacity available. The key is that the plant structure and management are grouped to focus on the manufacturing task. This may involve reorganizing the plant into units based on product, process or customer focus.

The end result is a plant that has a clear sense of direction, rather than one containing an unmanageable mixture of different product volumes, manufacturing processes, and customer requirements. The measure of the success of the focused plant should not be productivity or economies of scale but the ability to compete.

The focused plant can more easily change to suit revisions in the business strategy. Consider the impact of company growth and resulting strategy changes on a plant. Four types of growth are possible: (1) a broadening of the products or product lines being offered; (2) vertical integration, that is an extension of the span of the production for existing products with the intention to increase value added; (3) an increased product acceptance within an existing market area; and (4) expansion of the geographic sales territory serviced by the company.

In the first case, where more products are to be made, a product-focused organization would find it easier to implement the strategy by adding additional product-focused plants. However, where vertical integration increases, additional process-focused plants will be needed. In the third case, as a product matures, the company generally will face cost-based competition. This will require low-cost manufacturing processes, which process focus can provide. This may call for moving from product focus to process focus depending on the volume. This idea is commonly called Product-Process Dynamics. Geographic expansion could prove more problematic. New, separate manufacturing facilities may be installed close to the new markets if they are product focused and the local demand justifies it. However, if plants are currently process focused, logistics costs must be traded off against setting up a new product focused plant near the new market. The product focused plant will be less efficient than a process focused plant, yet its output may remain competitive because of reduced logistics costs for moving products to markets far away from production facilities.

CASE: The experiences of the Copeland Corporation from 1975 to 1982 described by March and Garvin (1986) are helpful in illustrating factory focus. The case examines both the creation of new focused factories for certain products, and the refocusing of an existing complex plant. The points made by this classic case are highly relevant to manufacturing plants today.

In the late seventies, Copeland manufactured a range of compressors and condensers for commercial refrigeration systems at its Sidney, Ohio plant, which was unfocused. Compressors accounted for 90 percent of sales, and Copeland produced several types in various sizes. The two main types were: Copelametics, a semi-hermetic design introduced in the 1940s, and the fully-sealed Copelawelds, first made in the 1950s. The plant catered to a sizable parts and remanufacturing business too.

A New Focused Factory at Hartselle, Alabama

Copeland first used the term 'focused factory' to describe a new plant to be set up at Hartselle, Alabama, that was specifically built to manufacture a new range of compressor, the CR. The CR was to be more energy efficient and more reliable than the Copelaweld, and to compete through skilled high-volume manufacturing. This product was designed with little compromise. At this plant, quality was to be increased through three-stage reliability testing; energy efficiency was a key goal even higher than cost reduction; fewer models were planned, and standardized components were used. The investment was about $30 million.

The plant manager described Hartselle as "a focused plant, where the organization, manufacturing processes, and facilities are all concentrated on a single product." The manufacturing task was to support marketing effort through the use of processes capable of making a high quality compressor. The plant became a showcase for the company and its customers were frequently given plant tours. The plant needed a longer start-up time than planned, but eventually the company gained a 25% market share in a few years.

New Focused Factory for Remanufacturing at Rushville, Indiana

Copeland's remanufacturing business was initially done in five locations. Three of these were independent companies remanufacturing under license, and a fourth was a company bought by Copeland. The remaining remanufacturing operations took place at the Sidney, Ohio plant. The Sidney remanufacturing facility was entangled with the original compressor production to the extent that remanufactured parts were sometimes accidentally used in the manufacture of new compressors. This problem was addressed with the introduction of separate accounting and materials systems. Further separation was achieved by relocating remanufacturing to another building close to the Sidney plant. However, the labor union rules permitted workers to move between original manufacturing and remanufacturing operations, so the problems of separation still remained to a degree.

Remanufacturing workers required special skills: the ability to recognize parts by sight, both for currently produced models and for older, discontinued products; and to detect defects and assess their severity. In order to preserve these employee attributes in remanufacturing and to prevent the workers from migrating to original manufacturing, a decision was made to relocate remanufacturing to a totally new site at Rushville, Indiana.

Flexibility was required at the plant due to the ten different semi-hermetic designs to be rebuilt. Quality was achieved through the use of dedicated machines and gauges to ensure good tolerances. Workers were made responsible for the quality of parts that they remanufactured, and were trained to inspect their own as well as others' work. The remanufacturing facility was designated a division of manufacturing in its own right.

A New Focused Factory at Shelby, Ohio

Further reduction of the range of products made at Sidney took place when the larger models of the Copelametic design were moved to a new 'Large Cope' focused factory at Shelby, Ohio. There were several reasons for this move. Firstly, these large compressors had higher failure rates. Secondly, a design change was proposed which incorporated a new valve into the design to make the compressor 15% more energy efficient. These issues required the improvement of manufacturing for the large compressor products and a new focused factory was born at Shelby.

Focusing Remaining Production at the Sidney Plant

After building the three focused factories and moving some of Sidney's products into the new plants, Sidney manufactured eight Copelametic and five Copelaweld models. These older products were not expected to be competitive on cost and quality in the future. Therefore Sidney plant closure was considered. The idea was dropped because the plant was the company's headquarters and a sense of responsibility was felt towards the long-serving staff and the local community. The workers at Sidney were a valuable resource because of their in-depth product knowledge.

The company was faced with a need to restructure the plant to improve production. After a year of discussion with an external consultant, four strategic goals were defined: (1) improved quality for the Copelametic products, (2) reduced costs for Copelawelds, (3) improvement in material systems, and (4) simplified product designs. In an old, complex plant such as Sidney, the decision to refocus on products or processes is more complex, especially because existing equipment and workforce must be considered.

Two types of factory focus were evaluated for feasibility. Focusing the plant on process would result in two factories within the factory; one for machining operations and another for assembly. Product focus would create two plants also; one for Copelametics and the other for Copelawelds. Both types of foci would necessitate physical changes: in the case of process focus, some large and heavy machines, which were difficult to align, needed relocation; whereas in product focus, a solid wall was to be built between the two factories. The workforce would be affected in either case: different types of people gravitated towards assembly and machining jobs: they were different in their age and skill. Assembly workers were concerned with efficiency, whereas those in machining focused on quality. Copelametics attracted workers who liked to tinker with general purpose machines, tended to be older and preferred a more stable working environment. Copelaweld production, in contrast, attracted younger workers with its fast pace and high turnover.

The management chose product focus for the Sidney plant. Thus, an old plant was revitalized by implementing the focused factory concept in multiple steps.

See **Cellular manufacturing; Competitive strategy; Product-process dynamics; Manufacturing task; Strategic business unit (SBU).**

References and Further Reading

[1] Hayes, R.H., and G.P. Pisano (1996). "Manufacturing strategy: at the intersection of two paradigm shifts", *Production and Operations Management* 5 (1), 25–41.

[2] Hayes, R.H., and R.W. Schmenner (1978). "How should you organize manufacturing?" *Harvard Business Review* (January–February), 105–117.

[3] Hayes, R.H., and S.C. Wheelwright (1984). *Restoring Our Competitive Edge: Competing Through Manufacturing*, Wiley.

[4] March, A., and D.A. Garvin (1986). Copeland Corporation: *Evolution of a Manufacturing Strategy*, Harvard Business School Case 9-686-088.

[5] Skinner, W. (1974). "The focused factory". *Harvard Business Review*, (May–June) 113–121.

[6] Skinner, W. (1978). *Manufacturing in the corporate strategy*, Wiley.

[7] Skinner, W. (1996). "Manufacturing strategy on the "S" curve". *Production and Operations Management* 5 (1) 3–14.

[8] Stalk, G., Jr. (1988). "Time – The Next Source of Competitive Advantage". *Harvard Business Review* (July–August), 41–51.

FOLLOW THE LEADER CAPACITY EXPANSION STRATEGY

See **Capacity planning: Long range; Global facility location.**

FORECAST ACCURACY

Forecast accuracy has to do with the closeness of a forecasting model's predictions to the actual data. One of the basic principles of forecasting is that *forecasts are rarely perfect*. This is because forecasting future events involves uncertainty and perfect prediction is almost impossible. Forecasters know that they have to live with a certain amount of error. The goal in forecasting is to have a good forecast performance on the average over time. Forecast accuracy is also affected by the type of data being forecast. In general, forecasts are more accurate for groups of items rather than for individual items because group data tends to be less variable. Also, forecasts are more accurate for shorter than longer time horizons. The shorter the time horizon of the forecast, the lower the uncertainty. One way to improve forecast accuracy is through careful selection of the forecasting model. The forecasting model should be selected considering the type and amount of available data, the length of the forecast horizon, and the type of patterns present in the data. Forecast accuracy can be measured and monitored by using measures such as bias, mean absolute deviation (MAD) and tracking signal.

See **Bias in forecasting; Forecasting guidelines and methods; Forecasting in manufacturing management; Mean absolute deviation; Tracking signal.**

FORECAST ERRORS

Forecast error is the difference between the actual and the forecast for a given period. Forecast error is a measure forecast accuracy. There are many different ways to summarize forecast errors in order to provide meaningful information to the manager. Forecast errors can be separated into *standard* and *relative error measures*. Standard error measures typically provide error in the same units as the data. Relative error measures are based on percentages and make it easier for managers to understand the quality of the forecast. One of the most popular relative error measure is MAPE, which is the average of the sum of all the percentage errors for a given data without regard for sign. Bias, mean absolute deviation (MAD), and tracking signal are tools to measure and monitor forecast errors.

See **Bias in forecasting; Forecasting guidelines and methods; Forecasting in manufacturing management; Mean absolute deviation; Tracking signal.**

FORECAST EVALUATION

See **Forecast accuracy; Forecasting guidelines and methods; Forecasting in manufacturing management.**

FORECAST HORIZON

It is the length of time into the future covered by forecasts. The decision of which horizons to use depends on the managerial need for the forecast. For long term decision making, such as investments in capital equipment, the horizon may be between five to twenty years. For medium term needs such as aggregate production planning, the forecast horizon may be one to two years. For short term production scheduling, the horizon may be 1–4 months.

See **Capacity planning; Forecasting guidelines and methods; Forecasting in manufacturing management; Master production schedule.**

FORECASTING EXAMPLES

Benito E. Flores

Texas A&M University, College Station, TX, USA

Read articles, **Forecasting Guidelines and Methods** and **Forecasting in Manufacturing Management** before readings this one.

The simplest forecasting method is the Naïve method. It states that the forecast of demand for the next period is the value of the demand for the current period t. This method is simple and may be used to compare all forecasting methods. Any method chosen to forecast a time series should do no worse than the Naïve method.

EXAMPLE OF FORECASTS FOR STATIONARY DATA: A stationary (i.e., without trend) time series model can be represented as:

Actual Demand = Pattern + Error

$$Y_t = \theta + \varepsilon_t \tag{1}$$

Where, the value θ (theta) remains constant over time and there is noise in the data represented by ε. Y_t is the actual value of the demand for period t.

MOVING AVERAGES: Forecasting process estimates theta. A simple way of estimating the value of a stationary pattern (the value of theta) is to use the average of historical values. Simple Moving Averages (SMA) is one method. This methodology should be used only when the data displays a stationary (no-trend) form and no seasonality.

Example, assume a monthly time series data where the most recent value available is the sales for the month of November. A three period average can be calculated by using November sales and the previous two months sales.

$$\text{SMA}(3) = (\text{Sept} + \text{Oct} + \text{Nov})/3$$

This 3-period moving average can be used to forecast sales in December. Note that the implication of the above is that all three values are weighted equally, one-third each. If it were desired to forecast the sales for January from the same data (end of November), the forecast would be the same as December's forecast. The reason being the pattern is assumed stationary. So, the forecast into the future will be horizontal (flat), until additional data comes in.

The next moving average can be calculated when actual sales become known for December. The forecast for January would be the average of the actual sales for October, November, and December.

WEIGHTED MOVING AVERAGE: A simple variation to this model is the weighted moving average. As it was mentioned above the Simple Moving Average model weights all values the same. The same 3-period moving average can be changed to:

$$\text{WMA}(3) = (1^*\text{Sept} + 2^*\text{Oct} + 3^*\text{Nov})/6 \tag{2}$$

In this model, the sales figure for the month of November is weighted by the factor 3, October by the factor 2 and September stays the same. This weighting states that the more recent the sales values, the more they impact the moving average and thus the forecast. So the more recent the information, the larger the weight.

The denominator value of 6 in equation 2 is the sum of the three weights $(1 + 2 + 3 = 6)$. In all weighted schemes, the sum of the weighting coefficients is the value of the denominator. The selection of how many periods to use in the moving average is done by trial and error. The recommendation is to use the one that yields the lowest forecast error. For an example of forecasts using a moving average, view Table 1.

SIMPLE EXPONENTIAL SMOOTHING: Simple (or Single) Exponential Smoothing (SES) is a forecasting method that applies exponential weights to the historical sales values to create a moving average. The unequal weighting is obtained by using a smoothing constant alpha that determines how the weight is allocated to each observation. The scheme provides the most recent observation with more weight and lesser weight for all successively older observations. For a more detailed description, see Bowerman *et al.* (1993). The model is written as:

$$S_t = \alpha Y_t + (1 - \alpha)S_{t-1} \tag{3}$$

Table 1. Two period moving average tabular data. Output generated using MINITAB.

Period	Revenue	MA(2)	Forecast	Error
1	332.1	*	*	*
2	408.7	370.40	*	*
3	394.7	401.70	370.40	24.30
4	324.3	359.50	401.70	−77.40
5	334.0	329.15	359.50	−25.50
6	361.2	347.60	329.15	32.05
7	375.1	368.15	347.60	27.50
8	303.6	339.35	368.15	−64.55
9	320.3	311.95	339.35	−19.05
10	368.5	344.40	311.95	56.55
11	365.3	366.90	344.40	20.90
12	292.0	328.65	366.90	−74.90
13	296.8	294.40	328.65	−31.85
14	353.0	324.90	294.40	58.60
15	371.2	362.10	324.90	46.30
16	277.0	324.10	362.10	−85.10
17	313.4	295.20	324.10	−10.70
18	378.7	346.05	295.20	83.50
19	384.6	381.65	346.05	38.55
20	292.9	338.75	381.65	−88.75
21	315.2	304.05	338.75	−23.55
22	408.9	362.05	304.05	104.85
23	421.9	415.40	362.05	59.85
24	340.8	381.35	415.40	−74.60
25	357.2	349.00	381.35	−24.15
26	465.0	411.10	349.00	116.00
27	438.1	451.55	411.10	27.00
28	358.7	398.40	451.55	−92.85
29	378.3	368.50	398.40	−20.10
30	483.8	431.05	368.50	115.30
31	462.0	472.90	431.05	30.95
32	393.8	427.90	472.90	−79.10
33	415.8	404.80	427.90	−12.10
34	519.3	467.55	404.80	114.50
35	485.6	502.45	467.55	18.05
36	413.4	449.50	502.45	−89.05
37	431.8	422.60	449.50	−17.70
38	514.7	473.25	422.60	92.10
39	509.8	512.25	473.25	36.55
40	453.5	481.65	512.25	−58.75

Row	Period	FORECAST	LOWER LIMIT	UPPER LIMIT	Accuracy measures
					MAPE: 14.35
					MAD: 54.56
1	41	481.65	357.214	606.086	MSD: 4030.68
2	42	481.65	357.214	606.086	

Where, alpha, the smoothing constant, is usually assumed to exist in the range between zero and one.

One can describe the model by stating that the smoothed value is calculated by combining the weighted estimate of the (stationary) mean S_{t-1} for period $t-1$ with the weighted value of most recent sales, Y_t for period t. Equation 3 shows how α is used to weight Y_t and k S_{t-1}.

The forecasts for k periods into the future using this model are given by:

$$F_{t+k} = S_t \quad \text{for } k = 1, 2, \ldots \tag{4}$$

Equation 4 says that the forecast for k periods into the future are all equal to most recent estimate of the man S_t. Where:

$$S_t = S_{t-1} + \alpha(Y_t - S_{t-1}) \tag{5}$$

$$F_{t+1} = S_t \tag{6}$$

The term in parentheses in equation 5 is the forecast error (actual − forecast) for period t. One can describe the method as one that uses the previous forecast plus a fraction (determined by alpha) of the forecast error. The key to using the simple exponential smoothing relies on the estimation of the parameter alpha and an estimate of theta (given by S_t).

A practical way is to assume that the initial value (given by S_0) is given by the first data point (sales in period one). Another alternative is to average the first few terms of the time series. For example, the MINITAB software uses 6 time periods. Thus,

$$S_0 = Y_1 \tag{7}$$

Alpha is estimated by using part (or all) of the time series to test quality of the forecasts for different values of alpha. The quality of the forecasts can be evaluated by forecast error. The forecast error for period t is:

$$e_t = Y_t - F_t (\text{Actual} - \text{Forecast}) \tag{8}$$

An error metric can be used to compare the different forecasts using different values of alpha. A commonly used error metric is the Mean Squared Deviation, which is defined as

$$\underset{t=1}{\overset{n}{\text{MSD}}} = \left[\sum e_t^2\right] / n \tag{9}$$

The MSD calculated with a value of alpha, say 0.1, is compared with the one calculated with the alpha of, say 0.2. If the first MSD is smaller, then an alpha of 0.1 is preferable. This comparison is done throughout the range of values for alpha (between zero and one) and the alpha that yields the lowest MSD is chosen. The increment (0.1) in the value of alpha can be modified as desired (making it smaller or larger).

Once the best value of alpha is found, then the SES model can be used to forecast the demand for future time periods. Many software packages have an automatic feature to calculate the best value of alpha. The SES method has been found to be quite robust. It performs reasonably well under a wide range of time series. An example of single exponential smoothing is shown in Table 2.

If the data contains trend or seasonality, the SES model is inadequate. Read the article, Forecasting Guidelines and Methods for adaptations of the simple exponentially smoothed forecasts, when trend or seasonality are present in the data.

See **ABC analysis; Forecasting guidelines; Forecasting in manufacturing management.**

Table 2. Single exponential smoothing tabular data. Output generated using MINITAB.

Smoothing constant alpha: 0.274871

Period	Revenue	Smooth	Forecast	Error
1	332.1	351.309	358.591	−26.491
2	408.7	367.084	351.309	57.391
3	394.7	374.675	367.084	27.616
4	324.3	360.828	374.675	−50.375
5	334.0	353.454	360.828	−26.828
6	361.2	355.583	353.454	7.746
7	375.1	360.948	355.583	19.517
8	303.6	345.185	360.948	−57.348
9	320.3	338.345	345.185	−24.885
10	368.5	346.633	338.345	30.155
11	365.3	351.764	346.633	18.667
12	292.0	335.337	351.764	−59.764
13	296.8	324.744	335.337	−38.537
14	353.0	332.511	324.744	28.256
15	371.2	343.145	332.511	38.689
16	277.0	324.964	343.145	−66.145
17	313.4	321.785	324.964	−11.564
18	378.7	337.430	321.785	56.915
19	384.6	350.395	337.430	47.170
20	292.9	334.592	350.395	−57.495
21	315.2	329.261	334.592	−19.392
22	408.9	351.152	329.261	79.639
23	421.9	370.598	351.152	70.748
24	340.8	362.408	370.598	−29.798
25	357.2	360.976	362.408	−5.208
26	465.0	389.569	360.976	104.024
27	438.1	402.909	389.569	48.531
28	358.7	390.757	402.909	−44.209
29	378.3	387.333	390.757	−12.457
30	483.8	413.849	387.333	96.467
31	462.0	427.084	413.849	48.151
32	393.8	417.935	427.084	−33.284
33	415.8	417.348	417.935	−2.135
34	519.3	445.372	417.348	101.952
35	485.6	456.429	445.372	40.228
36	413.4	444.602	456.429	−43.029
37	431.8	441.083	444.602	−12.802
38	514.7	461.318	441.083	73.617
39	509.8	474.644	461.318	48.482
40	453.5	468.832	474.644	−21.144

Period	FORECAST	LOWER LIMIT	UPPER LIMIT	Accuracy measures
				MAPE: 10.93
				MAD: 42.17
41	468.832	365.513	572.152	MSD: 2428.99
42	468.832	365.513	572.152	

References

[1] Bowerman, B., and R. O'Connell, *Forecasting and Time Series*, Duxbury Press, Belmont CA, 3rd edition.

[2] Makridakis *et al.* "The Accuracy of Extrapolation (Time Series) Methods: Results of a Forecasting Competition", *Journal of Forecasting*, 1982, pp. 111–153.

FORECASTING GUIDELINES AND METHODS

Nada R. Sanders

Wright State University, Dayton, Ohio, USA

INTRODUCTION: Planning for future events is an integral aspect of operating any business. Planning allows actions to be taken that will meet lead time requirements and create a competitive operation. The process of planning, however, assumes that forecasts of the future are readily available. Manufacturers must anticipate future demand for products or services and plan to provide capacity and resources necessary to meet that demand. Forecasting is the first step in planning. It is one of the most important tasks, as many other organizational decisions are based on a forecast of the future. The quality of these decisions can only be as good as the quality of the forecast upon which they are based.

In business, forecasts are made in virtually every function and at every organizational level. For example, a bank manager might need to predict cash flows for the next quarter or a marketing manager might need to predict consumer response. Forecasting in operations and manufacturing is different from forecasting in other functional areas because the needs and uses of that forecast are different. In operations and manufacturing, forecasts are used to make numerous decisions that include machine and labor scheduling, production and capacity planning, inventory control, and many others. It is the forecast that drives production schedules and inventory requirements. Forecasting in manufacturing use approaches that primarily extrapolate history into the future. This is different from marketing approaches that are oriented toward analysis of the environment and future needs of customers.

Though each forecasting situation is unique, there are certain general principles that are common to almost all forecasting problems. Also, there is a large variety of forecasting methodologies available to practitioners to choose from, ranging in degree of complexity, cost, and accuracy. Understanding the basic principles of forecasting and the forecasting options available is the first step toward generating good forecasts. We begin with a brief historical overview and a discussion of these basic principles. Then we discuss techniques of interest to manufacturing managers and develop collective wisdom for applying these forecasts.

Forecasting models can be classified into two groups: *quantitative* and *qualitative models*. Quantitative forecasting models are approaches based on mathematical or statistical modeling. Qualitative methods, on the other hand, are subjective in nature. They are based on judgment, intuition, experience and personal knowledge of the practitioner. Quantitative models are more appropriate for manufacturing management.

QUANTITATIVE VERSUS QUALITATIVE MODELS: Much research has been done to compare the forecast accuracy of quantitative versus qualitative models. Quantitative models, based on mathematics, are objective, consistent, and can consider much information at a time. Qualitative models, based on judgment, are subject to numerous biases that include limited processing ability, lack of consistency, and selective perception. The biases found in human judgment cannot be disputed and can indeed contribute to poor forecast accuracy. Though quantitative and qualitative forecasts have usually been viewed in the literature as mutually exclusive, recent studies have shown that combining their strengths may lead to improved forecast accuracy.

COMBINING QUANTITATIVE AND QUALITATIVE APPROACHES TO FORECASTING: The sophistication of many quantitative forecasting models has made them superior in fitting historical data. However, this sophistication has not necessarily provided superiority in their ability to forecast the future. This shortcoming has been demonstrated to show that simpler models often outperform the more sophisticated ones (Makridakis *et al.*, 1982).

In the search for ways to improve forecast accuracy, researchers have found that combining forecasts from different models in the form of an average forecast can lead to significant improvements in forecast accuracy (Clemen, 1989). For example, averaging the forecasts from a few different quantitative models to get a final forecast should improve accuracy. Further, combining is more effective when the forecasts combined are not highly correlated. As qualitative and quantitative forecasts usually do not have a high correlation, their combination has shown to be quite competitive (Blattberg, 1990). Combining qualitative and quantitative forecasts shows particular promise for practitioners who inevitably adjust quantitative forecasts with their own judgment (Sanders and Manrodt, 1994).

BASIC PRINCIPLES OF FORECASTING: Forecasting future events involves uncertainty and perfect prediction is almost impossible. Forecasters know that they have to live with a certain amount of error. Our goal in forecasting is to have good forecast performance on the average over time, and minimize the chance of very large errors.

Another principle of forecasting is that *forecasts are more accurate for groups of items rather than for individual items themselves*. Due to pooling of variances, the behavior of group data can have very stable characteristics even when individual items in the group exhibit high degrees of randomness. Consequently, it is easier to obtain a high degree of accuracy when forecasting groups of items rather than individual items themselves.

The last principle of forecasting is that *forecasts are more accurate for shorter than longer time horizons*. The shorter the time horizon of the forecast, the lower the uncertainty of the future. There is a certain amount of inertia inherent in the data, and dramatic pattern changes typically do not occur over the short run. As the time horizon increases, however, there is a much greater likelihood that a change in established patterns and relationships may occur. Therefore, forecasters cannot expect to have the same degree of forecast accuracy for long range forecasts as they do for shorter ranges.

SELECTING A FORECASTING MODEL: There are a number of factors that influence the selection of a forecasting model. The first determining factor to consider is the *type* and *amount of available data*. Certain types of data are required for using quantitative forecasting models and in the absence of these qualitatively generated forecasts may be the only option. Also, different quantitative models require different amounts of data. The amount of data available may preclude the use of certain quantitative models reducing the pool of possible techniques.

Another important factor to consider in model selection is the *degree of accuracy required*. Some situations only require crude forecasts, whereas others require great accuracy. Increasing accuracy, however, usually raises the costs of data acquisition, computer time, and managerial involvement. A simpler but less accurate model may be preferable over a complex but highly accurate one. As a general rule, it is best to use as simple a model as possible for the conditions present and data available.

A third factor to consider is the *length of the forecast horizon*. Forecasting methods vary in their appropriateness for different time horizons, and short-term versus long-term forecasting methods differ greatly. It is important to select the correct forecasting model for the forecast horizon being used. For example, a manufacturing manager should use a vastly different model when forecasting product sales for the next 3 months as opposed to forecasting capacity requirements for the next two years.

Finally, an important criteria in selecting an appropriate method is to consider the *types of patterns present* in the data. Forecasting models differ in their ability to handle different types of data patterns. One of the most important factors in forecasting is to select a model that is appropriate for the pattern present in the data pattern. Four basic types of data patterns can be distinguished:

1. **Level or Stationary** – A level pattern exists when data values fluctuate around a constant mean. In forecasting such a data is "stationary" in its mean. An example of this would be a product whose sales do not increase or decrease over time. This type of pattern is frequently encountered when forecasting demand for products in the mature stage of their life cycle.
2. **Trend** – We can say that a trend is present when there is a steady increase or decrease in the demand over time.
3. **Seasonality** – Any pattern that regularly repeats itself and is of constant duration is considered a seasonal pattern. This pattern frequently exists when the data are influenced by the events associated with a given month or quarter of the year, as well as day of the week. An example of this would be the peaking of sales during Christmas.
4. **Cycles** – When data are influenced by longer-term economic changes, such as those associated with the business cycle, a cyclical pattern is present. The major distinction between a seasonal and cyclical pattern is that a cyclical pattern varies in length and magnitude. Consequently, cyclical factors are typically more difficult to forecast than other patterns.
5. **Randomness** – randomness are changes in the data caused by chance events that cannot be forecast. Randomness can frequently obscure the true patterns in the data making it difficult to forecast. In forecasting, we try to predict the other data patterns while trying to 'smooth out' as much of the randomness as possible. As stated in the second principle of forecasting, group data has less randomness present due to 'pooling of variances.'

Any one or more of these patterns can be present in the data.

FORECASTING MODELS USED IN OPERATIONS: Manufacturing managers most frequently need short- to medium-term demand forecasts to make decisions in production control, inventory control, and work force scheduling. Given the frequency with which many of these forecasts are made, automating as much of the process is desirable. Efficiency is gained by relying on forecasting models that are easy to use, easy to understand, and require little data storage. Qualitative forecasts of practitioners can be combined with quantitative forecasts during specific time periods, as will be discussed later in this chapter. For the most part, reliance should be placed on quantitative models to gain efficiency and to keep costs down.

Quantitative models can be divided into two categories: *time series models and causal models*. Time series models are based on the assumption that data representing past demand can be used to obtain a forecast of the future. Causal models, by contrast, assume that demand is related to some underlying factor or factors in the environment. Generating a forecast with causal models typically requires the forecaster to build a model that captures the relationship between the variable being forecast and the other factor (s) in the environment.

Time series models, on the other hand, are typically much easier to use, provide more expedient forecasts, and require smaller amounts of data. Further, time series models are available on numerous forecasting software packages that do not require much user expertise and involvement. Some of the most common of these models are described below.

The Mean

One of the simplest forecasting models available is the mean, or the simple average. Given a data set covering N time periods, $X_1, X_2, \ldots, X_n$, the forecast for next time period $t+1$ is given as:

$$F_{t+1} = \sum_{i=1}^{T} X_i/T$$

Though this model is only useful for stationary data patterns, it has the advantage of being simple and easy to use. As the mean is based on a larger and larger historical data set, forecasts become more stable. One

Simple Moving Average

When using the mean to forecast, one way to control the influence of past data can be limited by the number of observations included in the mean. This process is described by the term *moving average* because as each new observation becomes available, the oldest observation is dropped and a new average is computed. The number of observations used to compute the average is kept constant and always includes the most recent observations. This model is particularly good at removing randomness from the data. Like the simple mean, this model is only good for forecasting stationary data and is not suitable for data with trend or seasonality.

Given N data points and a decision to use T observations for each average, the simple moving average is computed as follows:

Time	Forecast
T	$F_{t+1} = \sum_{i=1}^{T} X_i/T$
T + 1	$F_{t+2} = \sum_{i=2}^{T+1} X_i/T$
T + 2	$F_{t+3} = \sum_{i=3}^{T+2} X_i/T$

The decision on the number of periods to include in the moving average is important and there are several conflicting effects that must be considered. In general, the greater the number of observations in the moving average, the greater the smoothing of the random elements. However, if there is a change in data pattern, such as a trend, the larger the number of observations in the moving average, the greater the propensity to lag the forecast will lag this pattern. For an example of the moving average forecasting model, see **Forecasting Examples**.

Single Exponential Smoothing

Single exponential smoothing belongs to a general class of models called exponential smoothing models. These models are based on the premise that the importance of past data diminishes as the past becomes more distant. As such, these models place exponentially decreasing weights on progressively older observations. Exponential smoothing models are the most used of all forecasting techniques and are an integral part of virtually all computerized forecasting programs. They are widely used for forecasting in practice, particularly in production and inventory control environments. There are many reasons for their widespread use. First, these models have been shown to produce accurate forecasts under many conditions. Second, model formulation is relatively easy and the user can understand how the model works. Finally, little computation is required to use the model and computer storage requirements are quite small.

Single exponential smoothing, SES, is the simplest of the exponential smoothing models. Forecasts using SES are generated as follows:

$$F_{t+1} = \alpha X_t + (1-\alpha)F_t \tag{1}$$

where F_{t+1} and F_t are next period's and this period's forecasts, respectively. X_t is this period's actual observation and α is a smoothing constant that can theoretically vary between 0 and 1. It can be seen that next period's forecast is actually a weighted average of this period's forecast and this period's actual.

An alternative way of writing equation (1) is as follows:

$$F_{t+1} = F_t + \alpha(X_t - F_t)$$

$$F_{t+1} = F_t + \alpha e_t \tag{2}$$

where e_t is the forecast error for period t. This provides another interpretation of SES. It can be seen that the forecast provided through SES is simply the old forecast plus an adjustment for the error that occurred in the last forecast. When α is close to 1, the new forecast includes a large adjustment for the error. The opposite is true when α is close to 0; the new forecast will include very little adjustment. These equations show that SES has a built-in self adjusting mechanism. The past forecast error is used to correct the next forecast in a direction opposite to that of the error.

SES is only appropriate for stationary data and is not appropriate for data containing trend as the forecasts will always lag the trended data. However, there are many versions of the basis SES model that have been adjusted to accommodate trend and/or seasonality. These are discussed later.

How much weight is placed on each of the components depends on the value of α. The proper selection of α is a critical component in generating good forecasts with exponential smoothing. High values of α will generate forecasts that react quickly to changes in the data, but do not offer much smoothing of the random components. On the other hand, low α values will not allow the model to respond rapidly to changes in data pattern, but will smooth out much of the randomness. A good rule to consider is that small values of α are best if the demand is stable in order to minimize the effects of randomness. If the demand pattern is volatile and changing, a large value of α should be selected in order to keep up with the changes.

Another option to consider when selecting α is to use what is known as a tracking signal for α, a technique sometimes called *adaptive exponential smoothing*. Here, the tracking signal is defined as the exponentially smoothed actual error divided by the exponentially smoothed absolute error. As α is set equal to the tracking signal, its value automatically changes from period to period as changes occur in the data pattern. This allows α to adapt to changes in the data pattern and helps automate the process. For example of the SES model, see, **Forecasting Examples**.

Holt's Two-Parameter Model

Holt's two-parameter model, also known as linear exponential smoothing, is one of many models applicable for forecasting data with a trend pattern (Makridakis, Wheelwright, and McGee, 1983). As mentioned earlier, stationary models will generate forecasts that will lag the data if there is a trend present. All trend models are simply level models that have an additional mechanisms that allows them to track the trend and adjust the level of the forecast to compensate for that trend. Holt's model does this through the development of a separate trend equation that is added to the basic smoothing equation in order to generate the final forecast. The series of equations is as follows:

1. Overall smoothing

$$S_t = \alpha X_t + (1-\alpha)(S_{t-1} + b_{t-1}) \tag{3}$$

2. Trend smoothing

$$b_t = \gamma(S_t - S_{t-1}) + (1-\gamma)b_{t-1} \tag{4}$$

3. Forecast

$$F_{t+m} = S_t + b_t m \tag{5}$$

In equation (3) $(S_{t-1}b_{t-1})$ is forecast for period t adjusted for trend, b_{t-1} to generate next period's S_t. This is the technique that helps adjust the value of S_t for trend and eliminate lag. The trend itself is updated over time through equation (4), where the trend is expressed as the difference between the last two smoothed values. The form of equation (4) is the basic single smoothing equation applied to trend. Similar to α, the coefficient γ is used to smooth out the randomness in the trend and can vary between 0 and 1. Finally, equation (5) is used to generate the final forecast. The trend, b_t, is multiplied by *m*, the number of periods ahead and added to the base value S_t to forecast several period into the future. To forecast the next period, set $M = 1$.

Winters' Three-Parameter Trend and Seasonality Model

Winters' model is just one of several models appropriate for use with data exhibiting either a seasonal pattern or a pattern containing a combination of trend and seasonality. It is available on most software packages and is often considered the model of choice for these data patterns. Winters' model is based on three smoothing equations; one for forecasting the level (St) one for trend (b_t), and one for seasonality (I_t). The equations of this model are similar to Holt's model, with an added equation to deal with seasonality. The model is described as follows:

1. Overall smoothing

$$S_t = \alpha X_t / I_{t-L} + (1-\alpha)(S_{t-1} + b_{t-1}) \quad (6)$$

2. Trend smoothing

$$b_t = \gamma(S_t - S_{t-1}) + (1-\gamma)b_{t-1} \quad (7)$$

1. Seasonal smoothing

$$I_t = \beta X_t / S_t + (1-\beta)I_{t-L} \quad (8)$$

1. Forecast

$$F_{t+m} = (S_t + b_t m)I_{t-L+m} \quad (9)$$

Equations (6) to (9) are similar to Holt's with a few exceptions. Here, L is the length of seasonality, such as the number of months or quarters in a year, and I_t is the corresponding seasonal adjustment factor. As in Holt's model, the trend component is given by b_t. Equation (8) is the seasonal smoothing equation which is comparable to a seasonal index that is found as a ratio of the current values of the series X_t, divided by the current single smoothed value for the series, S_t. When X_t is larger than S_t, the ratio is greater than 1. The opposite is true when X_t is smaller than S_t, when the ratio will be less than 1. It is important to understand that S_t is a smoothed value of the series that does not include seasonality. The data values X_t, on the other hand, do contain seasonality which is why they are deseasonalized in equation (6). Like parameters α and γ, β can theoretically vary between 0 and 1.

As with other smoothing models, one of the problems in using Winters' method is determining the values of parameters α, β and γ. The approach for determining these values is the same as the selection of parameters for other smoothing procedures. Trial and error on historical data is one approach that can be used. Most software packages have the option of automatically selecting the parameters for the given data set to minimize some criterion described later.

MEASURING FORECAST ERROR: Regardless of which forecasting model is used, a critical aspect of forecasting is evaluating forecast performance. Unfortunately, there is little consensus among forecasters as to the best and most reliable forecast error measures (Armstrong and Collopy, 1992). Complicating the issue is in fact that different error measures often provide conflicting results. The challenge for managers is to know which forecast error measure to rely on in order to expediently assess forecast performance.

Common Forecast Error Measures

Measures of forecast accuracy are most commonly divided into *standard* and *relative* error measures, which are fundamentally different from each other. In this section the most common measures in these categories are shown. The next section provides some specific suggestions for their use.

If X_t is the actual value for time period t and F_t is the forecast for the period t, the forecast error for that period can be computed as the difference between the actual and the forecast: $e_t = X_t - F_t$. When evaluating forecasting performance for multiple observations, say *n*, there will be *n* error terms. The following are *standard* forecast error measures:

1. Mean Error:

$$ME = \sum_{t=1}^{n} e_t/n$$

2. Mean Absolute Deviation:

$$MAD = \sum_{t=1}^{n} |e_t|/n$$

3. Mean Squared Error:

$$MSE = \sum_{t=1}^{n} (e_t)^2/n$$

4. Root Mean Squared Error:

$$RMSE = \left[\sum_{t=1}^{n} (e_t)^2/n\right]^{1/2}$$

Some of the most common *relative* forecast error measures are:

1. Mean Percentage Error:

$$MPE = \sum_{t=1}^{n} PE_t/n$$

where

$$PE = [(X_t - F_t)/X_t](100)$$

2. Mean Absolute Percentage Error:

$$\text{MAPE} = \sum_{t=1}^{n} |PE_t|/n$$

Standard Versus Relative Forecast Error Measures

Standard error measures, such as mean error (ME) or mean squared error (MSE), typically provide the error in the same units as the data. Because of this the true magnitude of the error can be difficult for managers to comprehend. For example, a forecast error of 50 units has a completely different level of gravity if the actual forecast for that period was 1000 as opposed to 100 units. Also, a forecast error of 50 dollars is vastly different from a forecast error of 50 cartons. In addition, having the error in actual units of measurement makes it difficult to compare accuracy's across time series. For example, in inventory control some series might be measured in dollars, others in pallets or boxes.

Relative error measures, which are unit-free, do not have these problems. They are significantly better to use, and provide an easier understanding of the quality of the forecast. As managers easily understand percentages, comparisons across different time series or different time intervals are easy and meaningful. For example, it is easy to understand the differences in the quality of forecasts between 12% and 48% error. One of the most popular of the relative error measures is MAPE.

Error Measures Based on Absolute Values

Error measures that use absolute values, such as the mean absolute deviation (MAD), provide valuable information. Because absolute values are used, these measures do not have the problem of errors of opposite signs canceling themselves out. For example, a low mean error (ME) may mislead the manager into thinking that forecasts are doing well when in fact very forecasts and very low forecasts may be canceling each other out. This problem is avoided with absolute error measures which provide the average total magnitude of the error.

One shortcoming of some absolute error measures is that they assume a symmetrical loss function. That is, the organizational cost of overforecasting is assumed to equal to the cost of underforecasting when these values are summed. The manager is then provided with the total magnitude of error but does not know the true bias or direction of that error. When using an error measure like MAD it is useful to also compute a measure of forecast bias. This may be mean error (ME) or mean percentage error (MPE). ME or MPE, by contrast to MAD, provide the direction of the error which is the systematic tendency to over- or underforecast. It is very common for managers and sales personnel to have a biased forecast in line with the organizational incentive system, such as performance against a sales quota. Information on forecast bias is useful as the forecast can then be adjusted to compensate for the bias. The two pieces of information, the total forecast error and forecast bias, can work to complement each other and provide a more complete picture for the manager.

HOW TO HANDLE OUTLIERS: A frequent problem that can distort forecasts are outliers, or unusually high or low data points. The problem can also arise due to a mistake in recording the data, promotional events or just random occurrences. Outliers can make it particularly difficult to compare performance across series. Some error measures, such as the mean square error (MSE), are especially susceptible due to the squaring of the error term which can make overall error appear unusually high. This can distort true forecast accuracy.

One method which is immune to outliers is to compare forecast performance against a comparison model which can serve as a baseline. Then a computation can be made of the percentage of forecasts for which a given method is more accurate than the baseline. This is called the Percent Better. A good model to use as a baseline is the Naive Model, where one period's actual is assumed to the next period's forecast. Finally, extreme values can be replaced by more typical values, as keeping outliers in the data can distort future forecasts (Armstrong and Collopy, 1992).

THE EFFECT OF ZERO VALUES AND MISSING OBSERVATIONS: Another problem that is often encountered in practice, especially in inventory control, is the existence of zero or near zero values. This is often an issue for items that experience 'lumpy' demand where sales do not occur everyday. Relative error measures, because they are based on a ratio, cannot handle a zero in the denominator as the error would become infinite. In such a case, the results from using either the percentage error (PE), mean percentage error (MPE) or the mean absolute percentage error (MAPE) would become infinite and evaluation would be impossible. There are a couple of ways to handle this problem. One is to replace the missing value with a more typical value that can be obtained through a moving average. Another option, suggested by practitioner Bernie Smith (Smith, 1984) is simply to define the error as 100% for all data points equal to zero.

Developing an Operations Forecasting System: There are certain requirements manufacturing managers need to consider for any forecasting system used in an operations environment. These are described below:

- the forecasting models used should generate forecasts for expected demand in a *timely* fashion. This means that forecasts should be generated to allow for ample time for decisions to be made. Typically, short range forecasts could be made in actual product units, such as by SKU's. Medium to longer range forecasts should be generated by an aggregate unit measure, such as dollars.
- in addition to a *point estimate of demand*, the system should also generate *a confidence interval*, providing an estimate of forecast error. This information is particularly important for computing inventory and safety stock levels.
- the forecasting system should provide past forecast errors, allowing for regular monitoring of forecast performance. Summary error statistics should be regularly updated, outliers recorded, and exception reports generated for large errors.
- the forecasting system should provide convenient presentation in both tabular and graphical format for easy viewing.
- the database should allow data presentation by different dimensions, such as by product, product group, geographic region, customer, distribution channel, and by time period.
- the forecasting system should allow for forecasts to be updated frequently, allowing decisions to be revised in the light of new information.
- the system should allow human judgment to override quantitatively generated forecasts and keep track of managerial adjustments. Managers occasionally become aware of events that can significantly impact the forecast. Keeping track of these adjustments allows for more accurate monitoring of forecast accuracy and discourages unjustified adjustments.

There are numerous forecasting software packages available to manufacturing managers and the most sophisticated is not necessarily the best. It should be kept in mind that simplicity and ease of use are key elements for regular usage. A review of available software is provided by Yurkiewicz (1993).

Getting the Best From Quantitative and Qualitative Forecasts: The newest thinking on forecasting acknowledge that a combination of both quantitative and qualitative forecasts may be best under certain conditions (Goodwin and Wright, 1993; Webby and O'Connor, 1996).

In practice, the easiest and most common way to combine qualitative and quantitative forecasts is through a qualitative adjustment of quantitatively generated forecasts (Sanders and Manrodt, 1994). However, this practice can harm the forecast more often than it can improve it. Managers need to exercise extreme caution when overriding the system.

As discussed earlier, quantitative models are objective and can process much information at a time. However, quantitative models are based on the assumption that past patterns will continue into the future. As such, they do not work well during periods of change. Qualitative techniques subject to biases of human judgment, but they may perform better during periods of change. The reason is that practitioners can sometimes bring information to the forecasting process not available to the quantitative model. This means that primary reliance should be placed on quantitative forecasts in order to maximize efficiency. However, managerial judgment may be used to adjust the quantitative forecasts when management becomes aware of events and information that is not available to the quantitative model.

To adjust quantitative forecasts successfully, the managerial adjustment ought to come from experienced practitioners, who are knowledgeable. The managerial adjustments should be based on *specific information* that the practitioner becomes aware of. For example, it is appropriate to adjust the quantitative forecast in anticipation of increased demand from an advertising campaign that is under way.

The cost of managerial involvement necessary to make qualitative adjustments to quantitative forecasts is often quite high. Therefore, the benefits in forecast accuracy need to be cost justified, as managerial time may be better served elsewhere. Managerial involvement should be restricted to a small number of important forecasts. Majority of forecasts should be strictly automated and the accuracy of all the forecasts needs to be evaluated on a regular basis.

Finally, it must be kept in mind that managerial adjustment is subject to biases of human judgment. All adjustments should be documented. Keeping accurate records can help managers see which adjustments are most effective and which are not.

See **Adaptive exponential smoothing; Forecast accuracy; Forecasting examples; Forecasting in manufacturing management;**

Forecasting models - causal; Forecasting models - qualitative; Forecasting models - quantitative, Holt's forecasting model; Mean absolute deviation (MAD); Mean percentage error (MPE); Forecast errors; Winter's forecasting model.

References

[1] Armstrong, J.S. (1995). *Long-range Forecasting: From Crystal Ball to Computer*, New York: John Wiley and Sons.

[2] Armstrong, J.S. and F. Collopy (1992). "Error measures for generalizations about forecasting methods: Empirical comparisons with discussion." *International Journal of Forecasting*, 8, 69–80.

[3] Blattberg, R.C. and S.J. Hoch (1990). "Database models and managerial intuition: 50% model and 50% manager." *Management Science*, 36, 889–899.

[4] Clemen, R.T. (1989). "Combining forecasts: A review and annotated bibliography" *International Journal of Forecasting*, 559–584.

[5] Edmundson, R., Lawrence, M. and M. O'Connor (1988). "The use of non-time series information in sales forecasting: a case study." *Journal of forecasting*, 7, 201–211.

[6] Goodwin, P. and G. Wright (1993). "Improving judgmental time series forecasting: A review of the guidance provided by research." *International Journal of Forecasting*, 9, 147–161.

[7] Hogarth, R. (1987). *Judgment and Choice*, John Wiley & Sons, NY.

[8] S. Makridakis, S. Wheelwright and V. McGee (1983). *Forecasting: Methods and Applications* (2nd ed.), New York: John Wiley & Sons.

[9] S. Makridakis, A. Andersen, R. Carbone, R. Fildes, M. Hibon, R. Lewandowski, J. Newton, E. Parzen and R. Winkler (1982). "The accuracy of extrapolation (time series) methods: Results of a forecasting competition." *Journal of Forecasting*, 1, 111–153.

[10] Mentzer, J. and K. Kahn (1995). "Forecasting techniques familiarity, satisfaction, usage and application," *Journal of Forecasting*, 14, 465–476.

[11] Sanders, N.R. and K. B. Manrodt (1994). "Forecasting practices in US corporations: survey results." *Interfaces*, 24 (2), 91–100.

[12] Sanders, N.R. and L.P. Ritzman (1992). "The need for contextual and technical knowledge in Judgmental forecasting." *Journal of Behavioral Decision Making*, 5, 39–52.

[13] Webby, R. and M. O'Connor (1996). "Judgmental and statistical time series forecasting: A review of the literature." *International Journal of Forecasting*, 12, 91–118.

[14] Yurkiewicz, J. (1993). "Forecasting software: Clearing up a cloudy picture." *OR/MS Today*, 2, 64–75.

FORECASTING IN MANUFACTURING MANAGEMENT

Benito E. Flores

Texas A&M University, College Station, TX, USA

INTRODUCTION: If every forecaster had his wish, all the forecast values would be perfectly accurate. Alas, this is not likely, but that does not mean that the operations forecaster should give up. Operations forecaster should engage in continuous search for lower forecast errors.

The amount of managerial attention to the forecasting process should be in direct proportion to the importance of the product/service being forecast. The more important the product for the company, the more care should be given to the process to make sure that everything possible has been done to improve the accuracy of the forecast.

Thus, the forecasting effort ought to be patterned after ABC analysis. For the important few items (type A – top 10–20% in dollar value), specialized care should be given to the forecast of these items. From all the forecasting methods available, a method should be selected, and carefully fitted to the product demand history. On the other hand, for the other items, especially the C items (type C – lower 80%), a simple generic method should be selected that can handle large volumes of data and perform adequate forecasting.

The collection of historical data for forecasting is called a time series. Time series are values of the variable(s) of interest collected over some time frame and at regular time intervals (days, weeks, months, quarters, years). Of the many types of variables collected for forecasting, sales data is the most common one. Demand forecasting is often done using past sales data.

DEFINITIONS: A simple definition of forecasting is that it is a process that has as its objective the prediction of a future event. Business forecasting includes the analysis of historical information with the purpose of identifying the characteristics relevant for forecasting. Using these characteristics, the model that best fits these properties is selected. Once this is done, forecasts of the variable in question (sales, etc.) are made. Assuming that the time series properties will remain in effect, the extrapolation of the historical values will provide a forecast with a reasonable error.

Operations Forecasts therefore are: Projections of sales volumes (shipments or demand) for a product or product line, usually measured in units, based on historical information, estimated by week, month, or year, and limited by the forecast horizon. It is bounded on the lower side by zero and can be limited by production capacity.

The forecaster in a business environment is part of the management staff. In order to have an impact on the organization, the operations forecaster is required to sell the forecast, and the process of developing it to management. This may require defining with managers what the forecast should be, when it should be prepared, how accurate should it be, and the form of the presentation.

The forecaster should be knowledgeable about the forecasting process, the best methods to use, the problems associated with the data, and the use of the forecast. Most of all, the forecaster should be immersed in the management process and preferably is the user.

HISTORICAL DEVELOPMENTS: Many of the forecasting methods that are in use today have been around for some time. For example, the exponential smoothing models were popularized in the fifties, others like the Box-Jenkins methodology in the sixties, and some other methods like Neural Networks have been developed in the last few years. The evolution of forecasting methods has been characterized by the continuous development of ever more (statistically) sophisticated methods. The sophistication has been driven by the use of statistical techniques.

Makridakis *et al.* (1982) showed that no single method (or family of methods) of forecasting was best for several time series data . In addition, his study showed that simple methods performed just as well as more complicated ones. Similar studies have been carried out later and the results have been basically the same.

Experience and intuitive reasoning reveal that future events are not solely a function of historical trends. In the business world, goods and services are purchased by companies for many diverse reasons. Thus, in some cases the subjective opinion of the business executive must be brought to bear on the forecasting process. Subjective forecasts or the subjective modification of forecasts can be included in the forecasting process.

OPERATIONS FORECASTING: The only stable paradigm in forecasting is that there will always be a forecast error. The future will never be known with certainty. No matter how sophisticated the forecasting method is, forecast errors are the norm. But, forecasting is still necessary. Forecasting tries to reduce uncertainty not eliminate it. The forecast is one of the major inputs to the decision making process.

It should be remembered that as the forecast horizon lengthens, the accuracy will probably decline. Forecast for a family of products will tend to be better than the forecasts for individual products.

The forecasting systems used in operations generate short to medium term forecasts that are input information to such things as inventory management systems, production planning and control, labor and production scheduling, aggregate planning, inventory replenishment, materials contracting and Materials Requirement Planning (MRP).

Forecasts drive the Master Production Schedule (MPS) which in turn drives both MRP and other systems. Safety stocks and inventory level cannot be minimized without a good forecast. Distribution Requirements Planning (DRP) also needs a forecast.

THE FORECASTING PROCESS: Forecasting is a process. As soon as the forecasting process is completed for a said month (or whatever time bucket is used), the next cycle begins. Forecasting processes should be flexible and should take into account the realities of the business world. It may be said that it is both an art and a science. The best possible science (statistical forecast) should be used but, in addition, human judgement (subjective forecast or subjectively adjusted forecast) should be added to make it hopefully more accurate.

A simplified forecasting process would consist of the following steps for each major items such as the A items (See, ABC Analysis):

1. Identify, from the user, the purpose of the forecast.
2. Establish, with the user, the time horizon and the accuracy needed.
3. Gather, and clean (i.e. remove outliers, irrelevant data) the appropriate data
4. Select a forecasting method.
5. Forecast using the chosen method.
6. Define, with the user, the presentation of the output of the final forecast.
7. Present the output to the forecast users.
8. Have the user provide input and feedback about the forecasts. If the method is not providing good enough results, return to Step 4 for a method change.
9. Evaluate and Monitor the forecasts for error or error patterns.
10. If errors become too large return to Step 1.
11. For every forecast period, update all the historical records.

The control of forecast is quite important. Monitoring the forecast values for errors and determining whether the process is performing in a satisfactory manner is absolutely indispensable for the prolonged functioning of the process.

For the B and C items (See, ABC Analysis), because they are not as important, another option could be used by the forecaster. The suggested process is to use a fairly general model of forecasting or an automated expert system. The forecasting model should be simple and robust (that is, it should be able to handle the presence and/or absence of trend, or seasonality or both).

The forecasting process can be set up in a couple of ways:

1. Obtain a software package that can process all the time series data for the B and C items. The software performs a 'contest' among all available forecasting models, and the 'best' (in terms of lower forecast error metric) is chosen. The selected model is used to forecast the time series.

A general model or a model contest, on average, may provide forecasts good enough for B and C items.

2. A different process is to use an 'expert system.' The expert system can use the same set of forecasting models as the previously explained but in addition, it can have 'rules' that can identify and eliminate data that is considered irrelevant, use available information concerning the forces that drive the demand, or identify and eliminate outliers. An expert system like this can be run in an automatic form or can have input from the forecaster. An example of this methodology is the RBF (Rule Based-Forecasting) developed by Collopy and Armstrong (1992).

Another commonly used process (that can include the one mentioned above) is called the pyramidal forecasting process. In this process, the forecast information, however generated, flows upward in the organization pyramid. As the information flows upward, at every organizational level the information is reconciled with the forecasts generated at that level. Once it reaches the top, senior management may or may not accept the aggregated forecast. If it is not accepted, the information flows downward and forecasts are adjusted. The cycle is repeated.

SOFTWARE: It is always a challenge to decide on a software package for operations forecasting. There are many good ones available. What is needed is to find the one that provides a happy medium between sophistication and ease of use (Rycroft, 1993).

The forecasting packages should forecast each active Stock Keeping Unit (SKU). They should provide point estimates of demand for each time period in the forecast horizon with a confidence interval. Further, they should provide estimates of forecast error and variance for safety stock estimation. The database used by the forecasting software may be structured using the structure shown below.

Level of detail:

- Geographic region,
- Industry served,
- Distribution channel.

Hierarchies:

- Product dimension:
  - Product group level,
  - Product line level,
  - SKU level, (Stock Keeping Unit)
- Geographic dimension:
  - National,
  - Sales territory,
  - Customer,
- Time period dimension:
  - Annual,
  - Quarterly,
  - Monthly,
  - Weekly, etc.

A multidimensional database management system can retrieve actual demand, forecast, and adjusted data different products, markets, and time periods. The requirements for a good forecasting software are listed below. It should:

1. Show past performance and next year's forecast at a glance.
2. Display average forecast errors (MAD, MAPE, MSE) to help set safety stock levels.
3. Show earlier management adjustments.
4. Display actual vs. forecast graphics.
5. Display Data from several models on one screen.
6. Provide summary error statistics.
7. Identify outliers.
8. Detect early irrelevant data.
9. Allow for forecast override using managerial judgment.
10. Computer tracking signal print in an exception report.

FORECASTING METHOD SELECTION FOR "A" ITEMS: In a classic paper, Georgoff and Murdick (1986) reviewed the forecasting field and proposed a system to be used by managers in order to select the most appropriate forecasting technique. They start by acknowledging the fact that managers are faced with rapidly changing and uncertain environments. If anything, the speed of change in today's world has increased. The most important considerations according to them are:

- Time
- Technical Sophistication
- Cost
- Data Availability
- Variability and Consistency of Data
- Amount of Detail Necessary
- Accuracy
- Turning Points, and
- Form.

Time refers to the amount of time required to generate a forecast. It can be divided into development and execution time. Development time refers to the acquisition time of the software, data gathering activities, modification of software, report generation, start up of the system, etc. The execution time refers to the time it takes to produce a forecast.

Technical sophistication refers to the fact that many techniques require statistical know how in the use and interpretation of the information.

Cost is the developmental cost. In the execution phase, the costs per forecast drop especially because computer costs continue to drop. The most important cost may be the cost of the inaccuracy of the forecast. The forecast error cost is difficult to measure but, in time, it may be more important than any other cost.

Data availability and quality is a very important issue. In addition, it is important to decide which data are important for you. Is it sales data? Is it shipments? Is it demand data? How can you insure that there are no inaccuracies in the collection and storage of the data? How much does it cost to store the data and keep it up to date? What should be the frequency of data collection? Weekly, monthly or quarterly? Nowadays, with the advent of point of sale data collection even daily data can be captured.

Variability and consistency of data should be examined periodically to establish if the relationships between the variables have remained stable. Are there any significant changes that make the assumptions invalid? Is information being tracked to identify changes?

Amount of detail necessary. How fine should the data be divided? What should be the time period for collection purposes? How should the data be broken down geographically? What is the frequency and segmentation required by the firm to make good use of the forecast information?

Accuracy. accuracy in the forecast is the most important feature. Since total accuracy is out of the question, it is important that management define what is the accuracy that the firm can live with. Can a forecast with an average error of five percent be good enough for management purposes?

Turning Points. The presence of turning points in the demand curve may give the company to consider threats and/or opportunities. Can a significant shift be forecast? Is a permanent shift in demand in the offing?

Form. The forecasts should be presented as point estimates but should include a range of possible values. If a probability can be attached to the forecasts, the better they are. This provides the manager with additional information for decision making. Provide a confidence interval for a forecast where possible.

In a simplified selection process, accuracy is the overriding factor. The method that generates the lowest forecast fitting error should be chosen.

SUBJECTIVE MODELS: Subjective models are popular and can be quite effective under some circumstances. For instance, they are appropriate for A items, where there is either no historical data or limited data (such as a new product), or when the data are changing drastically (due to the entry of a new competitor into the market). In this case, historical information is not of much use. The business use of subjective models is described by Dalrymple (1987), Sanders and Mandrot (1994), and Sanders (1997).

Another scenario that can make a subjective model essential is when there are drastic changes in the economic or political conditions. When this happens, models based on historical data become less relevant and subjective ones become relevant.

This form of forecasting makes use of information from consumer surveys, executive opinion, sales force composites, etc. Sometimes, sources such as these can provide an additional insight into the forecast (which may not be present in the time series data). In addition to the statistical components of the data, there may exist other factors that are not readily identifiable. In such a case, the forecast can be adjusted by an expert who will provide a modification of the value. This would be called subjective adjustment of an objective forecast. There is no consensus as to whether this practice improves or worsens a forecast. A major decision factor concerning the subjective adjustment of a forecast is whether there is domain information available as described by Armstrong and Collopy (1993). That is, is there external information available that can help improve the accuracy of the forecast?

Executive opinions are gathered by getting together groups of upper level managers to collectively develop a sales forecast. It has the advantage of bringing together the considerable talents of top management people. A danger, is that the responsibility of the forecast may be diffused among the entire group, with lost accountability and control.

Sales force composites from sales personnel can be a good source of information for forecasting. They are aware of customers' future plans. Of course, what a customer says and what actually happens may not be one and the same. If the information collected from sales personnel is used to establish sales quotas it will make the forecast more conservative.

Consumer surveys are useful if they truly elicit customer demand information. If all customers can be contacted then the information can be very reliable. A good source of information for these and other models is the work of Lilien and Kotler (1983).

DATA CONSIDERATIONS: The operations forecaster should study and consider the characteristics of the data that will be used in forecasting. The major question is whether the data are available in a form fit for forecasting.

The forecasting process is a major user of time series data. Mostly, the data is internal, i.e., inventories, costs, prices, sales or demand values. Sometimes, the information may come from government documents (Bureau of Labor Statistics, Federal Reserve Bank System, Census data, etc.). In other cases, the data may be gathered via market research or through the acquisition of databases from private sources.

The frequency of data collection depends on the business needs. For planning purposes, the frequency of data collection may be quarterly, semi-annually or annually. For re-stocking retail outlets from warehouses, data collection is done daily.

DATA TYPE: Before the data collection process begins, data definition must be made. Data is defined by:

- forecast horizon (immediate, short, medium, or long term).
- frequency of the forecast (daily, weekly, monthly, quarterly, yearly).
- the geographic segmentation of the information (city, county, state, region, national) be
- aggregation level (individual product or family of products, etc.).

The type of data to be collected should be in agreement with the forecasting task. For instance, if the horizon required for the forecast is defined to be monthly, data collection must be in monthly interval or shorter.

Companies that handle thousands of items usually store between 2–5 years of historical data. A frequently quoted number is three years of information. Nowadays, with the advent of point of sale equipment, the amount of data storage requirements has exploded since data can and is collected at more frequent intervals.

As was discussed earlier, the items in the database should be classified as either important (A items; see, ABC Analysis), intermediate (B items), or trivial (C items). The care with which data are collected and stored should be consistent with this classification.

DATA QUALITY: Another important consideration is the quality of the information. Data quality controls should cover data storage, data currency, data accuracy, etc. Procedures should exist to check the information (data audits) and the way it is stored. Identify outliers as part of the data handling methodology. There are no standard ways of identifying outliers, but in a normal distribution, they are located more than three standard deviations away from the mean. If an outlier is identified, it should be investigated. Was it a random event? Is it possible that the value will reoccur? If it can happen again, it is not an outlier, and cannot be ignored.

GRAPHICAL ANALYSIS OF DATA: Analysis of time series data involves visually searching for patterns in the historical information. One of the most explicit and powerful methods, even though it is the simplest, is to display the data visually. Data, that in tabular form will not reveal specific patterns or forms. Patterns are easier to see when data is graphically presented. The data can be classified as follows based on data characteristics:

- Stationary in the mean,
- Linear Trend, and
- Seasonal.

The most commonly used graphical analysis in forecasting is the Time Series plot. This figure has, as its horizontal axis (X axis), the time variable (independent). In the vertical axis (Y axis), the dependent variable, usually sales or demand, is graphed. This graph can help identify demand patterns.

Stationary in the Mean

Data which exhibits a level (no-trend) pattern can be defined as a time series that is stationary. The term stationary, in this article, refers to the mean value. It refers to a series whose values will

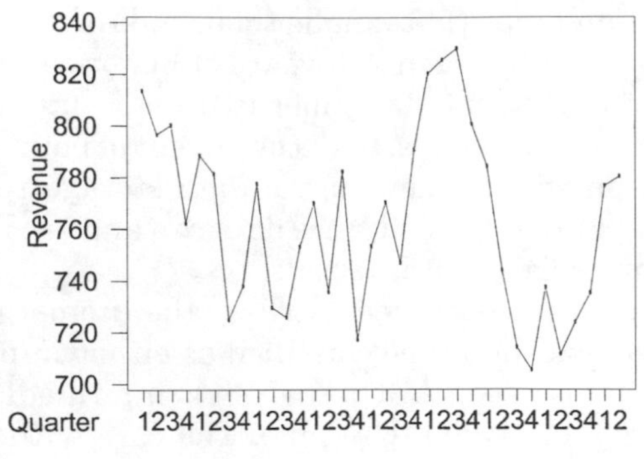

Fig. 1. Stationary time series.

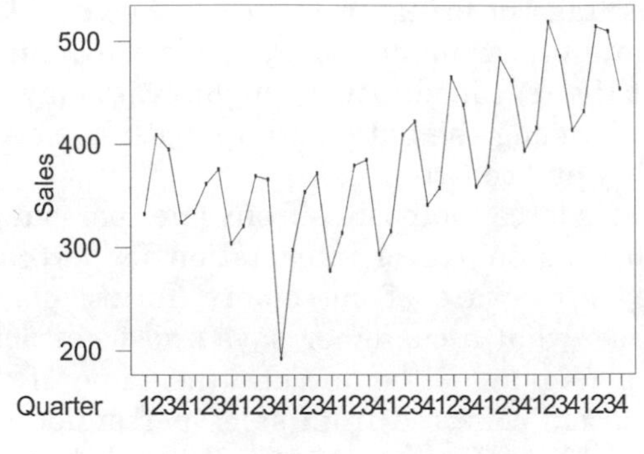

Fig. 3. Time series with seasonality.

fluctuate randomly around a mean value that does not change. This is the behavior of many time series data.

For instance, in Figure 1 observe the sales pattern for a company for the 1989–1997 period. There seem to be no apparent growth or downward trend in sales.

Trend

The trend in a time series data is the component that describes the long term pattern of growth or decline of an economic variable under study. Sometimes it is also called secular trend.

The forces that drive trends are population growth (in terms of mortality and birth rates), price changes in commodities, divorce rates, inflation rates, etc. Some of these forces create linear trends and, in some cases, exponential growth or decline.

The behavior of many time series departs from the stationary pattern. This is shown in Figure 2. As can be seen in the graph, the data have a linear trend. For this company, sales seem fairly flat in the first few years. Later, there is growth that shows incremental gains and a linear growth for an extended period of time.

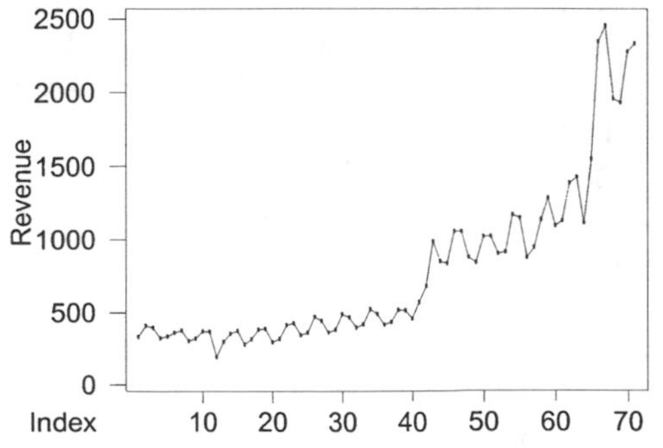

Fig. 2. Time series with trend.

Seasonality

Seasonality is the component of demand associated with the periodic fluctuations of economic data. Common seasonal influences are climate, human habits, holidays, etc. Factors such as summer or winter temperatures drives the production and sale of lawn mowers, snow blowers, and different types of garments. Other seasonal factors may have to do with Christmas, which drives demand for many items.

The pattern of sales (Figure 3) for the company for the 1980–1989 period shows growth and seasonality within each year. In Figure 3, Quarters 2 and 3 have high or sales as compared to quarters 1 and 4.

There are analytical methods to determine trend. For analytical methods the reader is referred to books by Hanke *et al.* (1995), or DeLurgio *et al.* (1991, 1998).

Random

After trend and seasonality have been accounted for in a time series, what remains is the random component. Random time series are the result of many influences that act independently to yield non-systemic and non-repeating patterns around some mean value. As this component is random, it can not be predicted.

The term noise is sometimes used to name this component. If the behavior of the noise has no pattern, then the noise is called White Noise.

See **ABC analysis; Forecasting examples; Forecasting guidelines and methods.**

References

[1] Adya, M., F. Collopy, and M. Kennedy, "Critical Issues in the Implementation of Rule-Based Forecasting: Evaluation, Validation, and Refinement," Working Paper, University of Maryland Baltimore County, Baltimore, MD, 1997.

[2] Armstrong, J., and F. Collopy, "Causal Forces: Structuring Knowledge for Time Series Extrapolation," *Journal of Forecasting*, Vol. 12, pp. 103–115, 1993.

[3] Bowerman, B., and R. O'Connell, *Forecasting and Time Series*, Duxbury Press, Belmont CA, 3rd edition.

[4] Box, G., and G. Jenkins, *Time Series Analysis: Forecasting and Control*, Holden-Day, San Francisco, 1976.

[5] Collopy, F., and J.S. Armstrong, "Rule Based Forecasting: Development and Validation of an Expert System Approach to Combining Time Series Extrapolations," *Management Science*, Vol. 38, No. 10, pp. 1394–1414, 1992.

[6] Dalrymple, D., "Sales Forecasting Practices – The Results of a US Survey," *International Journal of Forecasting*, Vol. 3, No. 3, pp. 379–391, 1987.

[7] DeLurgio, S., *Forecasting Principles and Applications*, Irwin/McGraw Hill, 1998.

[8] DeLurgio, Stephen A., and Carl D. Bahme, *Forecasting Systems for Operations Management*, Business One Irwin, Homewood, IL, 1991.

[9] Gardner, E., and E. McKenzie, "Forecasting Trends in Time Series," *Management Science*, V. 31, No. 10, October, 1985, pp. 1237–1246.

[10] Gardner, E., "Exponential Smoothing: The State of the Art," *Journal of Forecasting*, Vol. 4, pp. 1–38, 1985.

[11] Georgoff, David M., and Robert G. Murdick, Manager's Guide to Forecasting, *Harvard Business Review*, Jan.–Feb., 1986, pp. 110–120.

[12] Hanke, John E., and Arthur G. Reitsch, *Business Forecasting*, Fifth Edition, Prentice Hall, 1995.

[13] Lilien, G. and P. Kotler, *Marketing Decision Making – A Model-Building Approach*, Harper & Row Publishers, New York, 1983.

[14] Makridakis *et al.* "The Accuracy of Extrapolation (Time Series) Methods: Results of a Forecasting Competition," *Journal of Forecasting*, 1982, pp. 111–153.

[15] Mentzer, J, K. Kahn, and C. Bienstock, Sales Forecasting Benchmarking Study, The University of Tennessee, Knoxville, 1996.

[16] McKenzie, J., R. Schaefer, E. Farber, *The Student Edition of MINITAB for Windows*, Addison-Wesley, Reading, Mass., 1995.4

[17] Rycroft, R., "Microcomputer Software of Interest to Forecasters in Comparative Review: An Update," *International Journal of Forecasting*, Vol. 9. No. 4, December 1993, pp. 531–575.

[18] Sanders, N., "The Status of Forecasting in Manufacturing Firms," *Production and Inventory Management Journal*, Vol. 38, No. 2, 2nd Quarter, pp. 32–35, 1997.

[19] Sanders, N., and K. Manrodt, "Forecasting Practices in US Corporations: Survey Results," *Interfaces*, Vol. 24, No. 2, pp. 92–100, 1994.

FORECASTING MODELS: CASUAL

Causal forecasting models assume that demand is related to some underling factor or factors in the environment. The relationship between various productive factors and demand is mathematically modeled using historical data. One of the best known and simplest of the causal models is *linear regression*. Multiple regression models can estimate demand using multiple independent variables. For example, using regression, demand for an item can be forecasted given historic annual data on household income, and prevailing interest rates.

See **Forecasting guidelines and methods; Forecasting in manufacturing management.**

FORECASTING MODELS: QUALITATIVE

Qualitative forecasting models are those based on judgment, intuition, and informed opinions. These models are subjective in nature and rely on the unique knowledge of the forecaster. As different forecasters can typically arrive at different forecasts using the same data, qualitative forecasting results may not be consistent. These models are also subject to many biases of human decision making, such as optimism, wishful thinking, and regency effects. However, there are many situations when qualitative forecasts are the only option, such as when data from the past cannot be quantified. This is often the case in strategic decisions and while forecasting demand for a brand new product.

See **Forecasting guidelines and methods; Forecasting in manufacturing management.**

FORECASTING MODELS: QUANTITATIVE

Quantitative forecasting models are models noted in mathematics and statistics. Quantitative models are objective and will produce the same forecast for the same set of data regardless of who does the forecasting. These models have the advantage of being able to consider large amounts of data and many variables at one time. However, to be able to use quantitative forecasting models, information about the past must be available in a quantitative form.

See **Forecasting guidelines; Forecasting in manufacturing.**

FORWARD INTEGRATION

Most businesses are individual links in a more extensive market chain that often stretches from basic material extraction to end-product consumption. For example, the soft-drink market chain links the extractors of basic metals to sheet metal producers to canning companies to cola concentrate manufacturers to retailers to consumers. When a basic metal producer like Reynolds Metals, the aluminum company, decides to enter the canning business, we call such a development forward integration. Forward integration, in other words, is a strategy for growth based on expanding beyond current activities to include some of those that are closer to the consumer in the overall market chain.

There are many reasons why any company might want to expand by forward integration. One is to allow the company to increase its share of the overall profit margin available throughout the market chain. Another is to secure greater control over the market further along the chain. For example, one of the reasons why Reynolds Metals entered the canning industry when it did was to stimulate the switch from tin-plated steel to aluminum cans further along the market chain in order to increase the overall demand for aluminum production. Intel integrated forward into the production of personal computers and other microprocessor based products for similar reasons. However, there are also risks involved in forward integration. Companies that choose to integrate forward increase their exposure to a particular market chain and become more vulnerable to any unfavorable trends in the purchase patterns of its end consumers.

See **Manufacturing strategy; Supplier partnerships as strategy; Supply chain management.**

FORWARD SCHEDULING

The production schedule of a job or order is usually determined by two specific dates-the present time or the time when the order is due. Within the period between these two dates, the purpose of scheduling is to plan production of the job so that it will be completed by the due date. One approach to planning is to begin work on the order at the present time and work forward through time by, scheduling operations in sequence until the last operation is scheduled. In this way, order should be completed before its due date if there is sufficient time between now and the due date. This approach can allow for unforeseen delays, but can have the disadvantage of generating excess finished goods inventory if orders are consistently completed before their due dates. If excessive inventory build up due to forward scheduling is undesirable, use backward scheduling. In practice, forward scheduling is usually combined with backward scheduling.

See **Backward scheduling; Capacity planning: Medium- and short-range.**

FREEZING THE MASTER PRODUCTION SCHEDULE

The Master Production Schedule (MPS) is often prepared for several weeks; 26 to 52 weeks into the future are common. In order to prevent costly disruptions in the entire system, it is a common practice to bar any changes to the firs few weeks of the MPS. The period that is barred from changes to the MPS is the frozen period.

See **Capacity planning: Medium- and short-range; Master production schedule; Material requirements planning; Scheduling stability.**

FUNCTIONAL BARRIERS

For many years, companies have been organized along functional lines; viz., marketing, finance, operations, logistics, etc. Employees in each functional area developed a notion of what role they need to play in helping the firm succeed. Over time, Managers and employees of a function tend to become somewhat territorial with respect to the "tasks" for which they feel responsible. Unfortunately, this functional approach segments the firm and creates conflicts. For example, marketing may prefer to meet customers' needs by having large inventories located close to the customer. This inventory strategy facilitates high levels of availability and happy customers. However, it also creates tremendous inventory and warehousing costs for the logistics function, which is often managed as a cost center and therefore tries to minimize inventories. A similar conflict exists between marketing and operations–marketing may promise delivery dates that are unrealistic from a production scheduling perspective. As a result of these conflicts, tension may rise and overall firm performance may deteriorate as communication and coordination break down. The "turf" mentality, policies, procedures, measures, training, and organizational culture of functional alignment prevent cross-functional interaction and integration. These functional barriers have become highly visible in recent years as firms have endeavored to reengineer key value-added processes.

See **Business process reengineering; Concurrent engineering; Design; Reengineering and the process view of manufacturing; Supply chain management: Competing through integration; Teams: Design and implementation.**

FUNCTIONAL LAYOUT

This is also called process-oriented layout. A factory that has a functional layout is one in which manufacturing operations are grouped together on the basis of the function, technology or equipment used. A factory that uses this approach would install all equipment of a particular type in one area with separate areas assigned each type of equipment. Thus, a manufacturer using lathes would put all of them in one area. One alternative to a functional layout is a product-oriented layout, in which the equipment are placed in the exact sequence the operations will be carried out. Functional layouts are common for two reasons. First, functional layouts reflect traditional factory organization and as workshops grew into factories it was easy for their owners to simply lodge additional equipment alongside similar existing machines. This then lead to a *de facto* segregation of equipment by function and type, and became the traditional model for factory layout. Second, such functional layouts may be easily justified in many factories where there are no clear or consistent flows of work that allow identifiable sequences of operations. Advantages are, (1) workers using particular types of equipment could be better trained, supervised and supported if they were concentrated in one area; and (2) specialized tooling could be kept in proximity to the machines.

See **Manufacturing systems design; Scientific management.**

FUNCTIONAL ORGANIZATION

The term "functional organization" refers to departmental organization of firms based on functional groups. Examples of functions in organizations include marketing, sales, manufacturing, research and development, customer service, distribution, procurement, quality assurance, etc. Most large firms divide personnel into functional groups in order to encourage productivity and effectiveness. Dividing labor in this way enables concentrated learning in each discipline. A hazard associated with functional organizations is that persons working within a single function might lose sight of how their activities affect the activities of other functions. Since the goals of different functions are sometimes in conflict, functional organization is not effective for reducing lead-times and for improved product quality.

Team-based organizations are becoming more popular at the expense of rigid functional organizations.

See **Concurrent engineering; Functional barriers; Reengineering and the process view of manufacturing; Teams: Design and implementation.**

FUNCTIONALLY-ORIENTED ORGANIZATIONS

These are organizations that are structured around functions such as operations, finance, marketing, engineering, etc. as compared to those that are structured around processes such as order fulfillment, customer attainment, product development, manufacturing capability development, etc.

See **Functional organization; Reengineering and the process view of manufacturing; Teams: Design and implementation.**

FUZZY LOGIC PART-FAMILY FORMATION

Part-family concept is integral to group technology and cellular production. Grouping parts into a family of similar parts is the first step in cellular manufacturing. The conventional part-family formation methods implicitly assume part families are mutually exclusive and collectively exhaustive. The conventional methods assume that a part can only belong to one part-family. However, very often part families are not completely disjoint; part-families overlap. The fuzzy logic approach offers a special advantage over conventional part-family clustering. It not only reveals the specific part family that a part belongs to, but also provides the degree of membership of a part with different part-families. The fuzzy logic clustering algorithm provides extra information that is not available in conventional algorithms. This information could permit managers to make better-informed dynamic routing decisions, and allows for more flexible assignment of parts to cells.

See **Dynamic routing in flexible manufacturing systems; Group technology; Manufacturing cell design.**

GAAP-BASED FINANCIAL REPORTING GUIDELINES

Generally Accepted Accounting Principles (GAAP) are promulgated by the U.S. Financial Accounting Standards Board (FASB). GAAP represents a set of standards that companies use to publish financial statements, including the Income Statement, the Balance Sheet, and the Statement of Cash Flows. Companies registered with the Securities and Exchange Commission (SEC) in the USA are required to adhere to GAAP. In addition, many other companies are similarly required by creditors, investors, industry regulatory bodies, and employee unions to follow GAAP. Hence, GAAP has a tremendous influence on the accounting systems in all business sectors. One aspect of GAAP important to manufacturing organizations is that all production-related costs must be allocated to products for purposes of inventory valuation on the Balance Sheet, and for the calculation of income in the Income Statement. However, such all-inclusive allocation procedures can impede many management analysis techniques such as contribution margin analysis, throughput accounting for Theory of Constraints (TOC) management systems, and activity-based costing (ABC). Hence, many organizations maintain multiple accounting systems in order to satisfy FASB requirements and to provide valuable management accounting data that are outside the scope of GAAP.

See **Accounting system implications of TOC; Activity-based costing.**

GANTT CHART

GANTT CHART FOR A PROJECT:

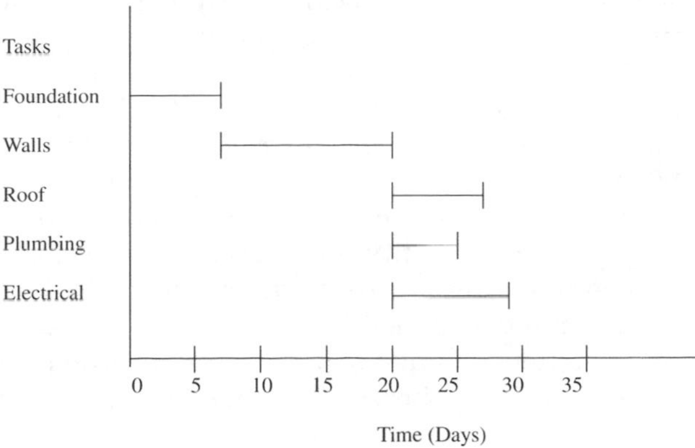

Reading the Schedule

According to the Gantt chart, this project will be completed in 29 days, when all its activities end. The plumbing task begins on the 20th day after the walls are completed. The plumbing activity takes five days to complete.

See **Critical path method; PERT; Process industry scheduling; Project management.**

GENERAL-PURPOSE LANGUAGE

Computer languages designed to support general computer operations are called general-purpose languages. FORTRAN, one of the original computationally oriented computer languages, is a general-purpose language. Other general-purpose languages often encountered in simulation include various versions of BASIC, Pascal, and C. Once the logic of how a system operates is developed, the most flexible means of developing simulation models is with general-purpose languages. But general-purpose languages require more coding than special simulation languages. Languages oriented toward conducting simulations include general.

See **Simulation of production problems using spreadsheet programs.**

GENERIC STRATEGIES

The concept of generic strategies refers to a few basic types of strategies for competition found in many different industrial settings. Several researchers have described and empirically studied generic strategies. Porter (1980) proposed that most business units follow a cost leadership strategy, a differentiation strategy, or a focus strategy. Stuck in the middle firms pursue less effective combination of strategies or no strategy at all.

Numerous researchers have studied and debated the question of whether or not "pure" generic strategies really exist in practice, and empirical findings have been mixed. Nevertheless, the discussion and contrast of strategic types has aided the formulation of strategy in many firms. Concepts associated with generic strategies have produced major impacts on strategic thinking over the last twenty years. See Miller (1992) for a helpful discussion of the merits and pitfalls of thinking in terms of generic strategies.

See **Cost leadership strategy; Differentiation strategy; Focused factory.**

References

[1] Miller, D. (1992). "The Generic Strategy Trap", *The Journal of Business Strategy*, 13(1), Jan/Feb, 37–41.

[2] Porter, M.E. (1980). *Competitive Strategy*, The Free Press: New York.

GENETIC ALGORITHMS

Genetic algorithms are computer programs which mimic the behavior of biological evolution in finding a solution to a complex problem. This procedure uses an evolutionary, "survival of the fittest" strategy where new solutions (children) are generated by combining portions of two prior solutions (parents). Genetic algorithms originated in John Holland's (1992) work, and have been steadily refined over the last fifteen years. In practice they are computationally intensive and may produce what some consider to be inelegant, difficult-to-follow solutions. Even so, they are gaining increasing application in search, optimization, learning and related problems.

See **Facilities location decision; Manufacturing analysis; Manufacturing analysis using Chaos, Fractals, and self-organization.**

References

[1] Goldberg, D.E. 1989. *Genetic Algorithms in Search, Optimization and Machine Learning*. Addison-Wesley, Reading, Mass.

[2] Holland, J.H. 1992. *Adaptation in Natural and Artificial Systems: An Introductory Analysis with Applications to Biology, Control, and Artificial Intelligence*. MIT Press, Cambridge.

GLOBAL COMMON DENOMINATORS

This is an approach used to develop standardized products for global markets. Under this approach, an international enterprise identifies a segment of consumers of relatively homogeneous tastes around the world. Using the preferences of the consumers in this segment, the company designs a standardized product which is marketed around the world. This approach presupposes that the international enterprise has carefully collected data on a global scale so as to identify the international segment. Often though, it is hard to isolate truly global market segments. Thus, companies tend to focus on sub-global markets segments, usually in the more affluent western countries.

See **Product design for global markets.**

GLOBAL FACILITY LOCATION ANALYSIS

Marc J. Schniederjans

University of Nebraska-Lincoln, Nebraska, USA

DESCRIPTION: *Domestic facility location analysis* is a process by which an organization operating in one country selects a location for an additional production or manufacturing facility within the same country. *International facility location analysis* is a process by which an organization operating in one country selects a location for a manufacturing facility in a different country. *Global facility location analysis* is a process by which a location for a production facility determined in the global context. Global facility location must also consider the interactivity and interconnectivity of a global lattice of facilities, and how a new site, fits within the existing lattice of facilities. The focus here is on selecting a location that will maximize the interconnectivity of a global system of facilities to satisfy the needs of a global market.

All of the facility location criteria used in a domestic or international location analyses can also used in a global facility location analyses. What guides these analyses are the identification of *critical success factors* or criteria that establishes success. Typical domestic location criteria like nearness to customers, nearness to raw materials, local costs (i.e., taxes, labor, etc.), local labor sources and a lack of local governmental restrictions can be important location factors in global facility location analyses also. Further, international location criteria such as political stability, legal aspects (i.e., export restrictions, product liability laws, etc.), social norms (e.g., work ethic), cultural customs (i.e., similarity in language), and economic stability (i.e., currency, transportation systems, etc.) are also important location factors in a global facility location analysis. Evaluation criteria unique to a global production facility network and, therefore, global facility location analysis would include the establishment of a *global supply-chain*, the ability to reach unique local foreign markets, the ability to learn from the people of a new country, and the ability to attract global talent.

A *global operation* (i.e., a production operation that consists of a network of facilities whose goal is to service a global market) will operate many facilities in an almost autonomous mode in a variety of foreign countries.

Almost autonomous global operation is a critical success factor in the global location decision. This autonomy permits the global operation's access to "home" markets in foreign countries that externally based operations would not have. Also the autonomy permits the global organization an opportunity to extensively acquire and use foreign country personnel in their own home markets. This permits the operation to learn how best to serve the local markets. It also permits the global operation to acquire local talent. So, locating in countries where human resource talent exists. is an important and unique critical success factor in global location analysis.

HISTORICAL PERSPECTIVE: Global facility location analysis follows the same logical progression that business organizations follow as they grow and acquire more markets for their products. Virtually all companies start in a domestic market. If successful, these firms dominate or at least experience maturation in the domestic market. To maintain market growth, they may choose to expand by going after markets in foreign countries (i.e., going international). In turn, these international markets will mature and a firm may go after more foreign markets in until, by default, it will become a global organization. Some firms are forced by the nature of their products to directly seek a global market. This is particularly true for extremely small market items like luxury automobiles.

STRATEGIC PERSPECTIVE: We have seen in the 1980's and 1990's, the growth in mergers and acquisitions of production facilities have continually increased. *North American Free Trade Agreement* (i.e., or NAFTA is an international agreement the eliminates much of the tariffs between the U.S., Canada and Mexico) and the expansion of the *World Trade Organization* (i.e., or WTO is an international trade organization that was created to act as a third party in international trade disputes and to develop world trade guidelines) have encouraged global operations. Business organizations in most countries are now being encouraged by their respective governments to go "global" as a means of rectifying their import/export imbalances. Advances in telecommunications and computer systems have contributed to global operations. Further, the development of the Internet and the various technologies to support it is allowing firms of all sizes to reach the global market. The enhanced order taking capacity that the Internet provides may create a demand for global manufacturing capacity to satisfy the global customer.

SIGNIFICANT ANALYTICAL MODELS: The complexity of global facility location analysis demands modeling methodologies that can incorporate both objective (i.e., dollars, miles, etc.) and subjective (i.e., cultural, attitudes, etc.) decision making criteria. There is a wide range of methodologies that can be used in global facility location analysis from fairly simplistic *rating/scoring methods* to complex *mathematical programming* and *genetic algorithm* methods.

Rating/scoring methods can be used in determining specific facility sites, for region selection, country selection, or continent selection. These methods require users to first identify critical success factor criteria for the operation. Once identified, the criteria are rated or scored on a scale of numbers to represent the relative rank of a specific location with other possible locations. The rating scale could be 1 to 5, where 1 represents a highly preferred rating 5 represents an undesirable rating. Once the ratings for each location are determined, they are summed up and a total score is determined upon which the location decision is based.

This procedure assumes that each of the individual critical success factors are equally important. A common extension of this approach is to weight the relative importance of the criteria. The weighting factor can be expressed as a mathematical weight ranging from 0 to 1, the sum of the weights for all critical success factors = 1. The weighted total score may be used for facility location.

Mathematical programming methods permit users to model the global facility location problem and use algorithms to arrive at an optimal solution. Mathematical programming methods can be used to optimize very complex location problems that might require the consideration of dozens of critical success factors. Some of the more typical mathematical programming methods that can be used in global facility location analysis include: *shortest-route problem, minimal spanning-tree problem, transportation problem, assignment problem, dynamic programming, linear programming* and *goal programming* (see Gass and Harris, 1996).

Shortest-route problem are actually a methodology that can be used in determining where to locate an additional production facility in a global network of facilities to achieve the shortest route from an origin to a destination through the network. This methodology is focused on seeking the location of the facility that minimizes the distance for traveling along a supply-chain network.

Minimal spanning-tree problem is a methodology that is used to connect all the network facilities in such way that the total supply-chain branch lengths can be configured to achieve minimal distance or time. This methodology can be used to locate a facility that minimizes the distance in the entire supply-chain network.

Transportation problem is a collection of methodologies that can be used to minimize distance between a series of supply sources (i.e., manufacturing plants) and demand destinations (i.e., markets). It can be used to optimize the choice of location for a global facility based on time, distance or cost/profit. The methodology requires as input data the total available production capacity at each of the supply sources and the total demand at the demand destinations.

Assignment problem is a set of methodologies that can be used to optimally assign a set of facility locations to service a set of customer markets. This methodology is limited to a one-to-one assignment but permits the user to optimize distance, time, or dollars (i.e., input in to the model). This methodology can be used in network location problems where an existing facility is to be added or one is to be re-located to a new location within a network of facilities.

Dynamic programming is a methodology that can be used to explore multistage facility location decisions (i.e., a sequential decision process). This optimization methodology allows users to model complex global problems where the first stage of a location will alter the possible payoffs (i.e., cost or profit).

Linear programming is a multi-variable, constrained optimization methodology that can simultaneously consider multiple critical success factors (representing input constraints and parameters in the model) and multiple facility locations (representing variables in the model) and arrive at an optimal facility choice. This methodology provides unique economic tradeoff information that can reveal opportunities and opportunities lost in making a global facility location decision.

Goal programming is an extension of linear programming that can be used to perform the same type of analysis but adds the unique feature of ranking or prioritizing multiple critical success factors. Goal programming is quiet robust permitting solutions to be obtained even when conflicting critical success factors prevent a lesser model such as linear programming from reaching a solution.

More recently, *genetic algorithms* have been suggested for use in global facility location problems (Gen and Cheng, 1997, pp. 359–379). Genetic algorithms (see Gass and Harris 1996, pp. 250–252) are more appropriate for global facilities analysis than other mathematical methods.

CASES: While numerous studies of global manufacturing and the need for global facility location analysis exist in the literature (see Flaherty, 1996, and Locke, 1996), most have focused on a single aspect of global facility location analysis, such as establishing a supply-chain network. Research on the concepts (see Dreifus, 1992 and Gates, 1994) and methodology (see Schniederjans, 1998, 1999) continues to explore new ideas and new problems in global location analysis. For new applications, based on genetic algorithms, see Gong *et al.* (1998).

COLLECTIVE WISDOM: Global facility location analysis represents a high-risk decision. Review of the current literature of applied global facility location analysis shows that the overwhelming majority of applications are based on ideas and methodologies from domestic and international facility location analyses.

See **Mathematical programming methods; NAFTA and manufacturing in the U.S.; WTO.**

References

[1] Dreifus, S.B., ed. (1992). *Business International's Global Management Desk Reference*, McGraw-Hill, New York.

[2] Flaherty, M.T. (1996). *Global Operations Management*, McGraw-Hill, New York.

[3] Gass, S.I. and Harris, C.M. (1996). *Encyclopedia of Operations Research and Management Science*, Kluwer, Boston.

[4] Gates, S. (1994). *The Changing Global Role of the Foreign Subsidiary Manager*, The Conference Board, New York.

[5] Gen, M. and Cheng, R. (1997). *Genetic Algorithms and Engineering Design*, Wiley and Sons, New York.

[6] Gong, D., Gen, M., Yamazaki, G. and Xu, W. (1998). "Hybrid Evolutionary Method for Capacitated Location-Allocation Problem," *Engineering Design and Automation* (to appear).
[7] Hoffman, J.J. and Schniederjans, M.J. (1994). "A Two-Stage Model for Structuring Global Facility Site Selection Decisions: The Case of the Brewing Industry," *International Journal of Operations and Production Management*, 14:4, pp. 79–96.
[8] Hoffman, J.J. and Schniederjans, M.J. (1990). "An International Strategic Management/Goal Programming Model for Structuring Global Expansion Decisions in the Hospitality Industry," *International Journal of Hospitality Management*, 9:3, 175–190.
[9] Hoffman, J.J., Schniederjans, M.J. and Sirmans, G.S. (1991). "A Strategic Value Model for Inter national Property Appraisal," *The Journal of Real Estate Appraisal and Economics*, 5:1, 15–19.
[10] Locke, D. (1996). *Global Supply Management*, Irwin, Chicago.
[11] Marquardt, M. and Reynolds, A. (1994). *The Global Learning Organization*, Irwin, Burr Ridge, IL.
[12] Miller, J.G., De Meyer, A. and Nakane, J. (1992). *Benchmarking Global Manufacturing*, Irwin, Homewood, IL.
[13] Schniederjans, M.J. and Hoffman, J.J. (1992). "Multinational Acquisition Analysis: A Zero-One Goal Programming Model," *European Journal of Operational Research*, 62, 174–185.
[14] Schniederjans, M.J. (1999). *Global Facility Location and Acquisition*, New York, Quorum Books.
[15] Schniederjans, M.J. (1998). *Operations Management: A Global Context*, New York, Quorum Books.

GLOBAL MANUFACTURING NETWORK

As firms seek to take advantage of worldwide resources, incorporating low-cost labor and materials into their productive systems, they establish fabrication and assembly facilities in different countries around the world. These geographically dispersed productive facilities combine to form the firm's global manufacturing network. Of course, the efficiency and effectiveness of these global networks can be evaluated through answers to several questions.

Is the network designed in a way that makes sense strategically? Many companies make individual facility location decisions on an ad hoc or reactive basis without evaluating how the new facility will fit within the larger network. How will the firm support the new facility logistically? Are suppliers in position to provide the needed materials on time? Are service providers readily available to provide adequate delivery to and from the facility? Is the national logistics infrastructure adequate to support the facility? Does the firm have the human resources needed to make the most of the new facility? Are superior local managers available? Will expatriates be needed? and does the firm have the resources to provide them? Finally, does the firm have the policies and procedures in place to create a cohesive and seamless network?

When a firm can answer these basic but essential questions, it is in a position to evaluate the competitive impact of its global manufacturing network.

See **Global manufacturing rationalization; International manufacturing.**

GLOBAL MANUFACTURING RATIONALIZATION

Stanley E. Fawcett and Kristie Seawright

Brigham Young University, Utah, USA

DESCRIPTION: Coordinated global manufacturing has become an important strategic issue for companies throughout the industrialized world (Sapsford, 1994; Roth, 1994). Coordinated global manufacturing strategies depend on the firm's ability to rationalize its productive activities; that is, the firm must assign each value-added activity to an appropriate area of the world so that the greatest cumulative competitive advantage is achieved. The appropriate area of the world can be defined in terms of either comparative advantage or market access.

Indeed, most global manufacturing endeavors can be classified as either factor-driven or market-driven (see Figure 1). Factor-driven strategies seek to gain access to worldwide resources – particularly inexpensive labor – in order to improve the firm's overall cost position. In the U.S., the classic example of manufacturing rationalization involves the electronics firm that moves assembly and production activities of labor intensive components to a foreign country with abundant and relatively inexpensive labor such as Malaysia or Mexico. Knowledge and capital intensive activities are maintained in the U.S. The objective of market-access strategies is to increase the firm's global market share. They do this by overcoming the physical and emotional barriers that impede international market entry and development. For example, Honda, Nissan, and Toyota established assembly facilities in the U.S. in order to establish a local presence and thereby circumvent the quotas that had been established in the early 1980s

Factor-Input Network:

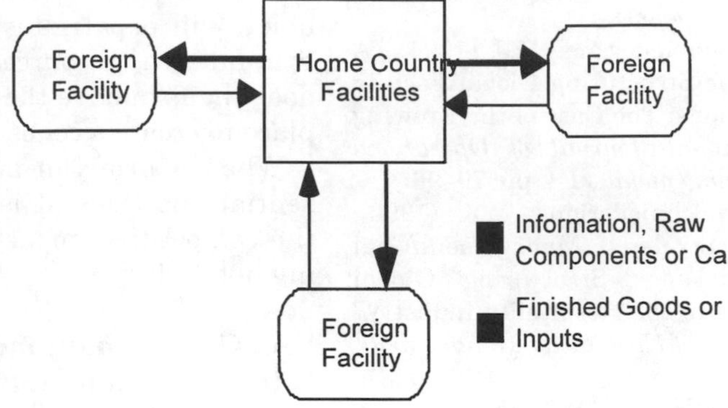

Market-Access Network:

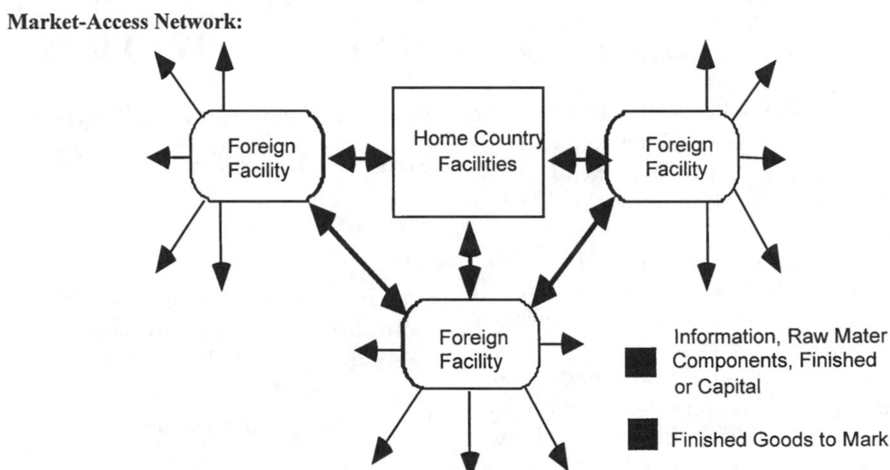

Fig. 1. Factor-input and market-access strategies.

via the voluntary restraint agreements. In the longer term, these Japanese automakers hoped to mitigate the negative feelings created by the loss of U.S. jobs. In today's complex and dynamic marketplace, most firms incorporate aspects of both strategies in their worldwide operations – using a "best mix" of worldwide resources to deliver distinctive products to worldwide customers.

Historical Perspective: The advent of coordinated global manufacturing can be closely tied to the emergence of today's global economy (Porter, 1986). Advances in technology as well as greater efforts to overcome political barriers have essentially made the world smaller – more accessible to companies that are willing to view the world as a single large marketplace. The transition toward economic globalization has effectively been moderated by three factors: the ability to exchange information, the availability of reliable transportation, and the amount of government regulation. First, the growth in information capability has enabled coordination among production entities, suppliers, and customers that was previously impossible. Second, transportation methods, reliability, and availability have all inccreased, removing some of the guess work from the management of globally dispersed operations. The increased capabilities of thrid-party logistics providers has also made if possible for many manufacturers to effectively outsource much of their global logistics requirements. Third, government resricctions, though still restricctive in some ccountries, have generally diminished as politicians have realized that economic growth is often driven by international trade. Of course, a fourth factor – deeply rooted cultural differences – also complicates the transition to a truly integrated global economy; however, the astute company has always been able to successfully overcome cultural differences to achieve success in diverse geographic and cultural regions of the world.

Improvements in the flow of information and the development of a variety of dependable modes of transportation have actually become the prime drivers of economic globalization (Ohmae, 1989). With much greater access to information and enhanced international travel experiences, worldwide consumers have come to expect higher levels of performance with respect to quality, cost, and innovation (Levitt, 1983). The fall of the Berlin Wall has even been attributed to heightened global awareness and consumerism among the people of Central Europe. With the barriers created by information and transportation diminishing, the greatest impediment to global economic integration is nationalistic tendencies and the resulting government intervention. Forces other than technology will be needed to reduce pervasive protectionistic tendencies. The following forces promise to keep the pressure on for continued reductions in trade restrictions, enhancing companies' ability to take more complete advantage of international opportunities.

- Consumers worldwide have expressed a strong desire for better access to a greater variety of goods and services at lower prices.
- Governments are increasingly willing to trade greater market access for the direct foreign investment that is needed to achieve greater industrialization and national competitiveness.
- Governments now view foreign competitors as an important motivating force to drive improved efficiency among domestic producers.
- Governments use lower trade restrictions to induce other countries to lower their barriers to international trade. (Daniels and Radebauagh, 1998)
- Governments are adopting a more liberal view of trade as critical driver of economic growth.

STRATEGIC PERSPECTIVE: Several implications of economic globalization make global manufacturing rationalization an important component of competitive strategy. The foremost implication of globalization is intensified competition. By the end of the 1980s it was estimated that 70 percent of all U.S. industry faced intense global competition. The forecast for the year 2000 is that nearly 100 percent of all currently existing U.S. industry will be challenged by competitive international rivals. The existence of formidable world competitors makes access to the best mix of resources – labor, materials, and technology – imperative. A second vital implication is the creation of worldwide markets and the resultant need to capture worldwide market share. Effectively building global market share eventually requires that companies establish operations in local markets around the world. By targeting similar market segments across different countries, a firm can sell more of a given product than it could if it only sold the product in its home country. This allows a company to increase its production volume and take advantage of significant scale economies. Viewing the world as a single market also helps a firm to justify and recover the huge research and development investments that are needed to bring to market the high-technology products that consumers desire (Ohmae, 1987). Further, as a company is faced with the challenge of protecting its home markets from aggressive foreign rivals, it only makes sense to keep the pressure on these rivals by competing in their home markets. To summarize, the dual requirements of meeting the challenge of foreign rivals and penetrating world markets makes global manufacturing rationalization a key component in many firms' competitive arsenals.

IMPLEMENTATION: Successful coordinated global manufacturing rationalization depends on the appropriate design and the effective management of the firm's global facility network. The facility network design determines both the deployment and productivity of a firm's value-adding resource and the proximity to key customers. Issues involved in the design and management of global manufacturing networks must therefore be well understood. Four issues – compatibility, configuration, coordination, and control – are of particular importance in the development of a competitive facility network. The fact that very different and distinct factors often motivate the decision to rationalize operations raises the first issue that should be evaluated to assess the attractiveness of a given country for manufacturing investment. Each manufacturing investment decision should enhance the overall network's competitiveness; therefore, each investment must be compatible to the firm's strategy as well as with the existing network. For instance, if the firm is seeking enhanced comparative advantage in the form of reduced manufacturing costs, then a country that offers abundant low-cost, semi-skilled labor would pass the initial test of compatibility. Likewise, a firm seeking greater worldwide market share would emphasize market potential, which is typically measured in terms of population size and wealth per capita.

The two issues of configuration and coordination are of particular importance in the development of a competitive global facility network (Porter, 1986). Configuration deals with the location of facilities and the allocation of productive activities among the facilities. Coordination involves the linking or integration of activities into a unified system. It is interesting to note that configuration and coordination issues should be addressed for

Table 1. Configuration and coordination issues.

	Configuration	Coordination
Product development	Number and location of R&D centers	• Interchange among centers • Degree of product customization • Sequence of product introduction
Purchasing	Location of purchansing function	• Managing dispersed suppliers • Transferring market knowledge • Taking advantage of commonalities
Operations	Location and allocation of production activities	• Transferring technology and know-how • Networking different facilities
Logistics	Number and location of warehouses and service suppliers	• Sharing information and know-how • Moving materials to point of need
Marketing	• Market select • Product selection	• Brand development • Account management • Coordination of pricing and channels
Service	Location of service organization	Development of service standards and procedures

See: (Porter, 1986).

all of the functional areas of the firm to achieve the most effective global facility network (see Table 1). Further, configuration and coordination issues should be considered simultaneously for the firm to achieve true global synergies. This contrasts with traditional practice, in which configuration issues typically play a preeminent role in global network design. Unfortunately, experience has shown that network performance declines when configuration issues take precedence in the design process (Fawcett, 1990; McDonald, 1986).

Most configuration decisions rely heavily on traditional facility location analyses, which unfortunately considers only a small number of relevant factors such as the location of inputs, the location of demand centers, and the cost of transportation. The greatest emphasis has been on production-related and facility costs while logistics (primarily transportation) costs have been approximated to reduce the complexity of location and allocation (Aikens, 1985; Daskin, 1985; Brandeau and Chiu, 1989). The emphasis on financial considerations and labor costs has been even more pronounced in actual international location decisions. However, cost tradeoffs must be carefully considered. the low-cost location of today will probably not be the low-cost location of tommorow. While production costs, especially labor costs, can change in relatively short periods of time, coordination and control costs remain as a part of the total global network costs. These costs must therefore play a greater role in the original design of the network. A sensitivity analysis of potential alterations in cost scenarios should be included with the up-front financial analysis.

Table 2 lists various criteria that should be considered by a firm as it decides where in the world it is going to do business. Looking at overall economic and political conditions, the investment climate, value-added capabilities, and market potential among other factors provides a firm a balanced comparison of the opportunities and challenges that should be considered in global network rationalization.

Table 2. Criteria to consider in country evaluations.

Overall economic and political conditions
- Inflation rate
- Unemployment rate
- Debt load
- Government protection of state-owned enterprises
- Political stability

Foreign investment climate
- Ownership regulations – land and company
- Currency convertibility
- Profit repatriation
- Bankruptcy and other commercial laws
- Domestic content restrictions
- Availability of joint-venture partners
- Tax rates

Value-added capabilities
- Labor rates
- Managerial and technical skills
- Quality and production control
- Productivity levels
- Communications and transportation infrastructure
- Access to materials

Market potential
- Country population
- Household income/Gross Domestic Product per capita
- Income distributions
- Access to other regional countries
- Access to worldwide economies
- Level of existing competition

When only a few important but narrow issues dominate the decision-making process, productive activities are often located in diverse regions of the world in such a way that it is very difficult to coordinate the geographically dispersed operations. That is, the basic and necessary coordinating processes of sharing information and transporting goods among worldwide facilities can be hindered by decisions made in the initial configuration of international networks. Part of the challenge stems from the fact that these basic coordinating processes are often taken for granted in domestic operations; yet, they can present serious impediments to the effective integration of global operations. For example, buying components from suppliers in the Pacific Basin to be used in Mexican assembly operations to build a product that will be sold primarily in Europe and the U.S. involves much more complicated operations than buying, assembling, and selling within a single nation's borders. Transborder channels introduce a higher level of uncertainty into the decision-making process because many more things can go wrong (e.g., customs checks and exchange rate fluctuations).

Longer international supply lines also reduce managerial control – once a shipment is loaded on a containership for movement across the Pacific Ocean, there is little that can be done to expedite delivery. Further, communicating product specifications and/or changes in desired delivery times can become a complicated process because of time zone differences, breakdowns in infrastructure (the phone might not work), or a variety of cultural/language differences. The end result is that longer transborder channels lead to a more uncertain operating environment in which managers often have less control than they are used to in their domestic operations. While the role of coordination is to deal with uncertainty and enhance managerial control, substantial performance tradeoffs result when coordination occurs after configuration decisions have been made. The concurrent assessment of how manufacturing operations in a country will complement existing worldwide operations (configuration) as well as how easily they can be integrated into the firm's overall global value-added system (coordination) is thus vital to network design.

A final issue that merits consideration as a firm designs its global network is that of day-to-day operating control. Indeed, the ultimate success of a firm's international operations is heavily dependent on its ability to obtain high levels of operational performance on a daily basis. That is, even when strategic compatibility exists and the operation is well positioned from a configuration and coordination perspective, poor on-site control can lead to disappointing results. Moreover, international operations are generally more difficult to manage than domestic operations because of cultural and other environmental differences. For instance, differences in language, measurement and reward systems, workforce relations, and infrastructure are among the many challenges firms must deal with daily to achieve sustainable competitive advantage. For this reason, firms should aggressively assess the challenges inherent in managing the day-to-day operations that they are likely to encounter in a specific country. To summarize, the key to the effective implementation of a coordinated global manufacturing strategy is to systematically assess compatibility, configuration, coordination, and control issues. An analysis of current and future tradeoffs must be included as a vital part of the assessment proccess. By so doing, a firm gains the understanding necessary to design and manage its global operations for competitive success.

TECHNOLOGY PERSPECTIVE: Technological advances have played a key role in making truly coordinated and therefore effective global manufacturing rationalization strategies possible. The most important advances have come in the communications and analysis arenas; however, important changes have also occurred in transportation. The use of electronic data interchange (EDI) and satellite technology makes communication and information sharing among all of the firm's geographically dispersed facilities possible. It is common practice to locate a satellite dish next to a manufacturing facility in Southeast Asia or Mexico so that communication with corporate headquarters can be maintained in real time. Improvements in the internet promise even better communication. More powerful computers are also being combined with data warehouses and systems analysis software to make better location and allocation decisions. Today's enterprise resource planning software is also helping to close the gaps that have long existed among a firm's geographically dispersed activities. Advances in communication and computer technologies are making global manufacturing decisions much more straightforward, inexpensive, and effective.

In the area of transportation, satellite tracking and bar-coding systems work together to track and expedite international shipments. These systems can pinpoint the location of a shipment anywhere in route to within a couple of football fields. Information systems are also being used to preclear customs, which reduces the time needed to cross national borders. Fast ships are also being developed that reduce the time required to ship

Table 3. Logistics techniques used as facilitators of global coordination.

- Buying and shipping items in container or smaller lot sizes on a periodic or systematic schedule.
- Developing partnership relationships with providers of transportation services.
- Developing partnership relationships with domestic and foreign suppliers of sourced components.
- Pre-clearance through customs.
- Greater use of air freight for regular shipments.
- Reliance on third-party transportation companies.
- Increased use of intermodal transportation including sea-air and double stack.
- Use of advanced information systems including EDI to track and/or expedite shipments.
- Use of local third-party warehousing to buffer global and JIT purchasing.

containers across the Atlantic Ocean by half or more. Such improvements in transportation service reliability help logistics managers remove uncertainty and increase coordination in international operations to both reduce costs and increase service levels. Table 3 lists several logistics practices that are commonly used to make coordination of global operations truly possible. In each of these instances, technology and better logistics management enhance network coordination.

RESULTS: Depending on the firm's strategic intent, global manufacturing rationalization can help the firm dramatically reduce production costs, increase market share, or both. For example, U.S. manufacturing firms that have established production sharing facilities in Mexico have been able to reduce their overall manufacturing costs by approximately 36%. Of note, the costs of coordination increased somewhat to offset the production cost reduction; even so, total costs reductions often exceed 20–25%. Likewise, manufacturers that have been willing to establish localized operations in order to penetrate foreign markets are often able to capture significant market share above and beyond what was available through export penetration.

Because of its potential impact on the firm's competitive position, a rationalization strategy can be implemented any time that the firm has the financial and managerial resources necessary to evaluate the issues related to compatibility, configuration, coordination, and control. The increased availability of third-party service providers makes rationalized manufacturing more assessable to firms of all sizes. Further, global sourcing enables firms to develop a virtual manufacturing network to take advantage of regional economies (and at a lower level of risk than a direct foreign investment network). A firm should seriously consider global manufacturing rationalization when its cost competitiveness is threatened or when attractive foreign markets are identified. Given the financial investment and managerial commitment inherent in manufacturing rationalization, compatibility, configuration, and coordination decisions should be managed centrally by top management. Day-to-day control issues must be managed on site at each manufacturing facility; however, consistent policies and procedures should be established to facilitate coordination of value-added activities.

CASES: General Motors exemplifies a company that has used manufacturing rationalization to improve its global competitiveness. GM operates production facilities in over 30 countries, maintains product research activities in approximately 60 countries, and markets its automotive products in over 140 countries around the world. As North America's market share leader, GM has established production facilities in Canada, Mexico, and the U.S. In fact, GM is the single largest private employer in both Canada and Mexico. Two types of operations have been established in Mexico. First, full-scale auto assembly operations have been located in Mexico to circumvent high tariffs (historically 20 percent or more) and thus target the growing Mexican market. Second, GM has located many labor-intensive auto parts operations in Mexico to take advantage of Mexico's abundant low-cost labor. For example, the majority of all wiring harnesses produced for cars manufactured in the U.S. are assembled in Mexico. Moreover, GM has found its Mexican workforce to be highly productive and capable of producing at the highest quality levels.

Through its Adam Opel division, General Motors has also established itself as a leading auto manufacturer throughout Europe. Not only does the Opel division provide manufacturing capacity and market access throughout Europe but it has been used as GM's primary platform for entry into the emerging Central European market. In addition to providing greater scale economies for parts divisions, extensive European operations provide market diversification. Further, technological and model innovations can be shared across divisions. For example, GM's new Cadillac Catera is based on an Opel design. While GM has long held strong positions in the U.S. and European auto markets, the newly emerging Asian countries have represented a serious weak link in GM's global network. To combat its weakness, GM has entered into joint ventures with several Japanese automakers and is currently evaluating the potential of investing in

Korea's Daewoo. GM has also targeted China and established production facilities, including kit assembly operations, in Southeast Asian countries. Despite the recent economic and financial crisis in the region, GM has indicated that it will maintain its newly established production presence in the region. Through manufacturing rationalization, General Motors has gained access to markets around the while simultaneously taking advantage of low-cost regionalized inputs.

Procter and Gamble is another company that has utilized manufacturing rationalization for global competitive success. P&G's entry into China illustrates some of the opportunities and challenges inherent in global manufacturing. P&G's decision to produce in China was driven by the mission statement to "produce products of superior value that enhance the life of the world's consumers." Entry into China also provided the dual advantages of access to the world's most populous market and access to extremely low-cost labor. While P&G has adequate worldwide capacity to currently supply the few products being sold in the People's Republic of China, import duties would elevate costs unnecessarily – thus justifying the initial investment in production operations. It is interesting to note that the low labor costs also provided a strong incentive for local Chinese production. However, while these labor costs are considerably lower that in many other locations in P&G's global production network, they are rising consistently with China's economic growth.

Within China, P&G operates production facilities in a variety of widely dispersed locations. This geographic dispersion provides excellent market coverage in the world's third largest country and provides P&G access to China's 1.2 billion consumers. Unfortunately, it has created several supply chain challenges that result primarily from China's weak but developing transportation infrastructure as well as the need to find suitable, high-quality suppliers near each of P&G's operations. That is, the lack of transportation infrastructure creates long lead times as products are moved throughout China. Indeed, manager's at some firms that import products from the U.S. note that the lead time from the U.S. to China is considerably shorter than for internal shipments within China. This fact impacts the sourcing decision, placing added emphasis on finding proximate suppliers for each facility. Historically, many inputs have been sourced from Asian suppliers outside China; however, a growing percentage of materials are now being purchased from Chinese suppliers. As the infrastructure improves and as managers continue to gain experience that will enable them to circumvent logistical problems, the coordination challenges of operating in China will diminish.

The new Xiqing plant, located in the industrial city of Tianjin, is an example of P&G's ability to work around some of the challenges encountered in Chinese operations. The shampoo plant is oversized to accommodate future market growth as well as to facilitate automation as labor rates continue to increase. The loading docks have been designed to be especially flexible to facilitate the loading and unloading of the non-standardized vehicles used for transportation within China. The facility is also located near a deep sea port to ease import and export shipping. Further, new lease agreements will give the facility access to a rail connection. The forethought that has gone into the site location decision as well as the actual facility design has helped P&G integrate the Xiqing facility into its Chinese and worldwide manufacturing networks.

COLLECTIVE WISDOM: Rationalized manufacturing offers opportunities to reduce production costs and access diverse consumer markets around the world; however, the process of establishing global operations is not without costs. Building an effective global operating network requires both extensive managerial time and expertise as well as substantial financial investment. The decision to either produce or market internationally should therefore be preceded by a thorough cost/benefit analysis. For example, reductions in production costs achieved by shifting manufacturing activities to different worldwide locations are at least somewhat offset by the logistics costs required to support the geographically dispersed operations (Scully and Fawcett, 1993). Bowersox *et al.* (1989) noted that, "Task complexity, increased lead times, market-unique product and service requirements and foreign business environmental factors combine to create a challenging and potentially costly environment for international logistics." Further, in today's business environment, non-cost issues such as flexibility and quality can be adversely affected by poorly planned global operations. The tradeoffs between the benefits and burdens of global manufacturing should therefore be considered to determine the value of global manufacturing to the firm. The following points should be explored.

- Decreasing global transportation and communication costs have made global operations available to firms around the world. As competitors establish global operations, a firm might need to look at global operations as a defensive strategy to prevent market erosion.

- Cost advantages are frequently short lived, co-ordination must therefore lead to higher quality and better speed/agility to create superior customer value.
- Global operations can allow firms to cross-subsidize key markets to increase market share (Kogut, 1985a). That is, profits earned in one geographic region can be used to support operations in another area of the world.
- Long-term competitive ability based on global operations requires strategic investments in a support infrastructure to create a unique value-added process (Stalk *et al.*, 1992). For example, significant investment in information systems is needed because global operating decisions are information intensive.
- Constant changes in comparative cost advantages result from exchange rate fluctuations and the opening of new labor markets. Thus, the low-cost location of today probably will not be the low-cost choice of tomorrow. Further, innovation, scale economies, and learning are all approaches a firm can use to neutralize a competitor's location-based cost advantage (Kogut, 1985b; Ohmae, 1987; Porter, 1990).
- As products become more capital intensive, the value-added advantages from direct labor savings become less significant and the advantages from co-ordination activities such as logistics become more important (Ohmae, 1989).
- International operating environments are different. from those that managers are familiar with. Cultural, political, financial, and legal environments vary greatly from country to country. Thus, doing business internationally requires a high degree of organizational adaptability combined with a willingness to invest in new skills and knowledge.

See **Economic globalization; Factor-driven strategies; Global manufacturing network; Production sharing facilities; Scale econo mies; Sustainable competitive advantage; Transborder channels; Virtual manufacturing network.**

References

[1] Aikens, C.H. (1985). "Facility Location Models of Distribution Planning." *European Journal of Operations Research*, 263–279.

[2] Bowersox, D., Daugherty, P., Droge, C., Rogers, D., and Wardlow, D. (1989). *Leading Edge Logistics Competitive Positioning for the 1990's*. Council of Logistics Management, Oakbrook, IL.

[3] Brandeau, M.L. and Chiu, S.S. (1989). "An Overview of Representative Problems in Location Research." *Management Science* (June), 645–678.

[4] Daniels, J.D. and Radebaugh, L.H. (1998). *International Business Environments and Operations* (8th ed.). Addison Wesley, Reading, Mass.

[5] Daskin, M.S. (1985). "Logistics: An Overview of the State of the Art and Perspectives on Future Research." *Transportation Journal*, 19A(5/6), 383–398.

[6] Fawcett, S.E. (1990). "Logistics and Manufacturing Issues in Maquiladora Operations." *International Journal of Physical Distribution and Logistics Management*, 20(4), 13–21.

[7] Kogut, B. (1985a). "Designing Global Strategies: Comparative and Competitive Value-Added Chains." *Sloan Management Review*, 26(4), 15–28.

[8] Kogut, B. (1985b). "Designing Global Strategies: Profiting from Operational Flexibility." *Sloan Management Review*, 27(1), 27–38.

[9] Levitt, T. (1983). "The Globalization of Markets." *Harvard Business Review*, 61(3), 92–102.

[10] McDonald, A.L. (1986). "Of Floating Factories and Mating Dinosaurs." *Harvard Business Review* (November–December), 81–86.

[11] Ohmae, K. (1987). "The Triad World View." *Journal of Business Strategy* (Spring), 8–19.

[12] Ohmae, K. (1989). "Managing in a Borderless World." *Harvard Business Review*, 67(3), 151–161.

[13] Porter, M. (1986). "Changing Patterns of International Competition." *California Management Review* (Winter), 9–40.

[14] Porter, M. (1990). "The Competitive Advantage of Nations." *Harvard Business Review*, 68(2), 73–93.

[15] Roth, T. (1994). "Global Report Finds U.S. Has Replaced Japan as Most Competitive Economy." *Wall Street Journal*.

[16] Sapsford, J. (August 31, 1994). "Unemployment In Japan Hits a 7-Year High." *Wall Street Journal*.

[17] Scully, J. and Fawcett, S.E. (1993). "Comparative Logistics and Production Costs for Global Manufacturing Strategy." *International Journal of Operations and Production Management*, 13(12), 62–78.

[18] Stalk, G., Evans, P., and Schulman, L.E. (1992). "Competing on Capabilities: The New Rules of Corporate Strategy." *Harvard Business Review*, 70(2), 57–69.

GLOBAL MARKETING CONCEPT

Global Marketing Concept is the overall marketing framework which an international enterprise adopts to design, introduce, distribute, promote and maintain its products in the global arena. Dimensions of the global marketing concept include the product design, the nature of packaging and labeling, the choice of a brand name, the presence or absence of warranty, the advertising strategy in each country, and the after-sales service support. For each dimension, the international enterprise must choose the degree of standardization and customization. For example, whether the advertising strategy will include a standardized advertisement in all countries or customized advertisements for each country.

See **Product design for global markets.**

GLOBAL SOURCING

See **Purchasing: acquiring the best inputs.**

GLOBAL SUPPLY CHAIN

This includes the flow and transformation of goods (as well as the flow of the associated information) from the raw materials stage to the end user, including the supplier's supplier and the customer's customer. This flow of goods and information may encompass several different facilities (plants, warehouses, sales and distribution centers) belonging to several different business entities located in various parts of the globe.

See **Facilities location decisions; Supplier partnership as strategy; Supply chain management: competing through integration.**

GLOBALIZATION

This refers to the increasing integration of business activities on a worldwide scale. Increasing international trade has linked consumers and producers across the world and has created an economic environment in which companies compete on a worldwide scale. This has a significant impact on pricing, and cost.

See **Global manufacturing rationalization; International manufacturing; Product design for global markets; Supplier partnership as strategy.**

GLOBALLY RATIONALIZED MANUFACTURING

Globally rationalized manufacturing in a company focuses on the assignment of each production activity to the appropriate plant in the world so that the greatest cumulative competitive advantage is achieved. The appropriate area of the world can be defined in terms of either comparative advantage or market access. The best location for labor intensive, but relatively low-skill assembly work may be in a country with low labor costs. The best location for a research and development office may be in the country where the product is sold. Often, comparative advantage and market access interact in the design of a global manufacturing network. For example, hundreds of production and assembly facilities are located along the 2,000 mile United States/Mexico border. The intent is to use low-cost Mexican labor for basic manufacturing and assembly, while gaining access to the affluent consumer market in the U.S. as well as the emerging Mexican market. Most major U.S. manufacturers have located manufacturing operations in Brazil to gain access to the large Brazilian market. Commercial offices with light assembly and distribution facilities are located in Argentina and other South American countries to gain access to these country's markets while they are still to small to support large scale manufacturing.

The key to success in global manufacturing rationalization is to take advantage of regionalized factor inputs while obtaining scale economies through large scale manufacturing. By selectively locating facilities, a firm can rationalize manufacturing and gain access to important markets at the same time. For example, a local presence can be very important to penetrating European markets; however, the countries with the largest markets – France and Germany – also have some of the highest labor costs in the world. The rational solution may be to manufacture in one of the lower-cost European countries such as Ireland, Spain, Greece, or England and export to the other members of the European Union. This example illustrates that rationalization can occur at both the global and regional levels. Ultimately, designing a globally rationalized facility network requires that several priorities including market access, comparative advantage, and scale economies be balanced in the location decision. At the same time, issues involved in coordinating the flow of both information and materials among geographically dispersed production facilities must be seriously considered. When these configuration and coordination decisions are carefully addressed, a firm can use the "best mix" of worldwide resources to access the "best" markets around the globe.

See **Global manufacturing rationalization; Logistics: meeting customers' real needs.**

References

[1] Porter, M. (1986). "Changing Patterns of International Competition." *California Management Review* (Winter), 9–40.

[2] Fawcett, S.E. and Smith, S.R. (1993). "Manufacturing Rationalization Under North American Free Trade." *Production and Inventory Management Journal*, 34(3), 58–63.

GRAVITY FEED CHUTES

See **Flexible automation.**

GREEN MANUFACTURING

The term 'green' is commonly used to denote an environmental friendly activity. The objective of green manufacturing is to design and produce products in an environmentally benign manner. The characteristics of an environmentally benign product ranges from the use of recyclable materials or recycled materials wherever possible, the use of fixtures and/or components that can be easily taken apart, or the design of product structure to allow ease of disassembly (Graedel and Allenby, 1995).

See **Environmental issues and operations management; Environmental issues: reuse and recycling.**

Reference

[1] Graedel, T.E. and Allenby, B.R., *Industrial Ecology*, Prentice Hall, Englewood Cliffs, NJ, 1995

GROSS PROFIT MARGIN

Gross profit margin is indicative of a company's profitability. This ratio is calculated as:

$$\text{Gross profit margin} = (\text{Profit in dollars after taxes})/(\text{Sales in dollars}).$$

This margin may vary according to the number of other companies in the industry, as well as the research intensity of the industry. For instance, certain types of companies, such as semiconductor and pharmaceutical manufacturers, have very high margins, because of the high level of research and development required in these industries. The companies in these industries must continually reinvest their profits back into the organization to update their products and ensure that they remain "state of the art".

While evaluating a supplier, a purchasing manager must use good judgement and industry averages to determine what is a "fair and reasonable" profit margin for a given supplier. When the supplier is a publicly traded company, financial ratios can be obtained from sources such as 10-K reports, Moody's Industrials, and Dunn and Bradstreet Reports to gain insights into the supplier's financial health. Profit margins can be calculated using information shown in income statements and balance sheets.

Once a contract is awarded to a supplier, a prudent purchasing manager must monitor this ratio after the contract is awarded. If profit margins increase, the supplier may be overcharging the buyer. If they decrease, they may have underbid the contract, and performance of the supplier may suffer in other ways.

See **Performance measurement in manufacturing; Supplier performance measurement.**

GROSS REQUIREMENTS

Gross requirements refers to the total of independent and dependent demand for a component in material requirements planning.

See **Master production schedule; Material requirements planning.**

GROUP INCENTIVE PAY PLANS

This is a wage incentive system whereby employees are paid in part or in whole based upon the successful attainment of group performance measures.

See **Learning curve analysis; Teams: Design and implementation.**

GROUP REORDER POINTS METHODS

This is an inventory reordering system where a group of items are ordered together. In this system, a reorder point is established for the group of items. If 10 items form the group, the joint reorder point may be the sum of individual reorder points. When the combined inventory for the 10 items drops below the joint reorder point, a joint order will be released. Alternatively, the reorder point can be computed using a joint inventory's dollar value.

See **Safety stock: Luxury or necessity.**

GROUP TECHNOLOGY (GT)

It refers to grouping machines to process families of components with similar if not exact sequences of operations. The family of parts dictate the processes used to form a specific manufacturing cell.

Machines and equipment are grouped toge ther into cells for producing the family of parts/ products. Grouping characteristics may include size, shape, manufacturing or routing requirements and demand. The objective is to group products with similar processing requirements in order to minimize machine setup or changeover time. Advantages of group technology include lower setup time, lower work-in-process inventory, reduced materials handling and decreased cycle time. Group technology cells generally seek to combine the advantages of batch or job shop manufacturing with assembly line type layouts.

See **Dynamic routing in flexible manufacturing systems; Flexibility in manufacturing; Manufacturing cell design.**

HARD AUTOMATION

In this form of automation is not easy to change, or reconfigure, or reprogram the equipment for a new product or new design.

See **Flexible automation; Hard manufacturing technology; Manufacturing systems; Soft manufacturing technology.**

HARD MANUFACTURING TECHNOLOGY

See **Manufacturing technologies.**

HEAVY WEIGHT STRUCTURE FOR DESIGN

See **Concurrent engineering; Product development and concurrent engineering.**

HEAVY WEIGHT PROJECT MANAGER

See **Concurrent engineering; Heavy weight project organization; Product development and concurrent engineering.**

HEAVYWEIGHT PROJECT ORGANIZATION

Organizational structures of product development projects can be categorized based on how much functional links are emphasized as opposed to project links. It is common to distinguish between functional organization, lightweight project organizations, heavyweight project organization, and autonomous project organizations (Wheelwright and Clark, 1992). In distinguishing between the four structures, it is important to understand that they should be seen as points on a continuum rather than discrete categories. An index for measuring how light or how heavy a project manager is can be found in Clark and Fujimoto (1991).

In a heavyweight project organization, the involved individuals have a stronger link to the project than to their function (e.g., to marketing, design, and manufacturing). Project members are typically working for one single project at a time, and might be co-located with the project members from other functions. The *heavyweight project manager* has full control over budgets and has a substantial influence on the performance evaluation of the project members. He or she frequently has a dedicated staff group that directly report to him/her and is of (at least) similar status in the overall organization as the functional managers. The heavyweight project organization is a special form of a matrix organization with an emphasis on the project links.

The major advantage of the heavyweight project organization over a functional organization and a lightweight project organization is its increased focus on speed and integration. Whereas the lightweight project manager can only act as a coordinator, the heavy weight project manager can create his or her own organizational structures, which ensures effective information exchange between project members. He or she also has the authority to resolve technical and organizational problems, which has been shown to reduce product development time and to increase the technical integrity of the product (Clark and Fujimoto, 1991).

Similar to the lightweight project organization, the heavyweight project organization does allow the involved project members to continue their regular day-to-day work in the corresponding functional organization. This feature is an advantage over the autonomous project organization, where the functional links of the project members are substantially weakened, maintenance of functional expertise can be difficult and future career development within the function might not be possible. However, many companies do value the broader scope of project work as an important experience for long-term career development in the overall organization.

Heavyweight project organization may be weak in the coordination between multiple projects and may be driven by short-run goals. As the heavyweight project organization is primarily set-up to achieve its own project goals, it is likely to ignore opportunities for other ongoing or future projects. For example, the sharing of components between projects headed by heavyweight project managers may be hard to achieve. Similarly, heavyweight project organizations have little incentives to research technologies and develop components that go beyond the lifetime of their project. Probably the largest problem of the approach is the extremely high demand it places on the heavyweight project manager. Even if the other two disadvantages are carefully managed, the success of the project will ultimately depend on the individual characteristics of the heavyweight project manager.

In summary, the heavyweight project organization provides an effective way of organizing development projects in environments that call for rapid product development. The automotive industry provides numerous success stories of its application (e.g., Clark and Fujimoto, 1991). Other good examples can be found in the electronics and software development industry.

See **Concurrent engineering; Product development and concurrent engineering.**

References

[1] Clark, K.B. and T. Fujimoto (1991). *Product Development Performance: Strategy, Organization and Management in the World Auto Industry*. Harvard Business School Press, Cambridge.

[2] Wheelwright, S.C., and K.B. Clark (1992). *Revolutionizing Product Development*. The Free Press, New York.

HEURISTIC

Heuristic is a "rule-of-thumb" that identifies a feasible solution for a problem, but does not necessarily produce an optimal solution. Heuristic solution procedures are used to solve many operations management problems including but not limited to: facility location, inventory, job scheduling, project scheduling, and line balancing. Some of the more common heuristic rules include first-come-first-served (FCFS), shortest processing time (SPT), and earliest due date (EDD) when sequencing jobs, or ranked positional weight, longest operation time, and most followers, when solving the **Line Balancing Problem**.

See **Assembly line design; Sequencing rules; U-shaped assembly lines.**

HIDDEN FACTORY

The 'hidden factory' refers to that part of a manufacturing firm that is not part of the direct manufacturing processes, but comprises the paperwork and computer transactions that support manufacturing. This paperwork is expensive and contribute to overhead cost. The Just-in-Time (JIT) production philosophy aims to eliminate much of this overhead by simplifying and quickening the flow of material batches, and by simplifying or removing the need for work orders. Hidden factory cost can be reduced by stabilizing operations and automating transactions using various technologies such as bar coding.

See **Just-in-time manufacturing.**

HIERARCHICAL MODEL OF ACTIVITIES

The Hierarchical Model of Activities is an activity-based costing (ABC) accounting model, originally proposed by Cooper (1990), which identifies an alternative view of product costs that reclassifies products in terms of the activities they support. Hence, rather than defining product costs as being either direct materials, direct labor, or manufacturing overhead (i.e., the traditional accounting model of product costs), the Hierarchical Model of Activities breaks product costs into four classifications as follows: Unit-Level Activities, Batch Activities, Product-Sustaining Activities, and Facility-Support Activities.

Unit-Level Activities within the Hierarchical Model of Activities are directly associated with manufacturing a single product unit such as a hubcap or a pair of shoes. Examples of unit-level activities include direct labor, raw materials, energy costs to operate stamping machines, etc. Batch Activities are those associated with manufacturing an entire batch of products. These activities are generally not related to the number of product units in the batch. Examples of batch-level activities include material movements, inspection sampling, purchase and receipt, set-ups, etc. Product-Support Activities are those that provide the capability to produce a particular product. However, these activities are not directly related to the actual production of units or batches of that product. Examples of Product-Support Activities include engineering change notices (ECNs); investments in product, production, or market enhancements; cost-of work-in-process inventory; etc. Finally, Facility-Support Activities within the Hierarchical Model of Activities are directly associated with maintaining the production capacity. These activities are not directly connected with any particular product type in the plant. Examples of facility-support activities include plant security, plant custodians, cost accounting, general administration (such as VP of manufacturing), etc. Because there exists little substantive relationship between facility-support activities and products, most ABC accounting systems are careful *not* to allocate these costs to products.

See **Activity-based costing: An evaluation.**

Reference

[1] Cooper, Robin (1990). "Cost classifications in unit-based and activity-based manufacturing cost systems," *Journal of Cost Management for the Manufacturing Industry* (Fall), 4–14.

HIERARCHICAL PLANNING AND CONTROL

Organizational activities can usually be classified into several levels. Anthony (1965) proposed a framework which classifies organizational decisions into three categories (i.e., three levels): strategic planning, tactical planning and operational control. These three categories of decisions form a hierarchical planning and control process in an organization. When applied to production and operations management, the process is called a hierarchical production planning and control process, which is the essence of a manufacturing planning and control (MPC) system. Strategic planning is the process of making decisions concerning the objectives of the organization, changes to objectives, the resources used to attain these objectives, and concerning the policies that are to govern the acquisition, use and disposition of these resources. Strategic decisions have long-lasting effects. Thus strategic planning requires a long planning horizon perspective. Also, these decisions are usually made by the upper levels of the management hierarchy.

Tactical planning is the process by which managers assure that resources are obtained and used effectively and efficiently in the accomplishment of the organization's objective. These decisions usually involve the consideration of a medium range time horizon. Typical decisions to be made in the context of production planning are the utilization of regular and overtime workforce, allocation of aggregate capacity resources to product families, build-up of seasonal inventories, etc.

Operational control is the process of assuring that specific tasks are carried out effectively and efficiently. Typical decisions at this level are the assignment of customer orders to individual machines, the sequencing of these orders in the work shop, inventory accounting and inventory control activities, dispatching, expediting and processing of orders, vehicular scheduling, and so on.

See **Capacity planning: Medium- and short-range; Disaggregation in an automated manufacturing environment; Manufacturing planning and control.**

References

[1] Anthony, R.N. (1965). *Planning and Control Systems: A Framework for Analysis*, Harvard University, Graduate School of Business Administration, Boston, Massachusetts.

[2] Hax, A.C. and G.R. Bitran (1979). "Hierarchical Planning Systems – A Production Application, "in Ritzman, L.P., L.J. Krajewski, W.L. Berry, S.H. Goodman, S.T. Hardy and L.D. Vitt, eds., *Disaggregation: Problems in Manufacturing and Service Organizations*, Martin Nijhoff, Boston, Massachusetts.

HIERARCHICAL PRODUCTION PLANNING (HPP)

Hierarchical production planning (HPP) is a mathematical approach to production planning based on the concept of disaggregation. Disaggregation breaks the production planning problem down into a series of smaller problems, each corresponding to a level in the organizational hierarchy.

See **Hierarchical planning and control; Disaggregation in an automated manufacturing environment; Linear decision rule; Capacity planning: Long-range.**

HISTORY OF MANUFACTURING MANAGEMENT

James M. Wilson

Glasgow University, Glasgow, Scotland, U.K.

> ...the man who aspires to fortune or fame by new discoveries must be content to examine with care the knowledge of his contemporaries, or to exhaust his efforts in inventing again, what he will most probably find has been better executed before. (Charles Babbage, 1835, *On the Economy of Machinery and Manufactures*: §327)

THE BEGINNINGS OF MANUFACTURING MANAGEMENT: The Industrial Revolution brought with it new processes, new products and new management problems. Before the industrial revolution most production occurred in small workshops employing few skilled people using hand tools and craft methods (Chandler, 1977). With the industrial revolution, workshops evolved into integrated factories using power-driven machinery attended by large numbers of lower-skilled workers performing increasingly specialized tasks. As the scale and complexity of manufacturing increased management was needed to provide direction and control. Manufacturing management grew in the industrial revolution: it stimulated the development of production processes as Hounshell (1984) shows, and evolved with them. Modern production systems and thinking about them has been influenced by a long series of earlier developments – developments whose effects have not always been recognized in spite of their influence.

TECHNOLOGICAL PERSPECTIVE: A HISTORICAL FRAMEWORK OF PRODUCTION PROCESSES: Textbooks conventionally discuss operations management by focusing on the production processes managed. Different processing technologies impose varied requirements on management; and, although general principles unite the discipline there are significant adaptations to different environments. The most common perspective provides a cross-section of the current interests of manufacturing management by differentiating between: project management, batch production, repetitive manufacture, assembly lines and processing plants. This view can be strengthened through an analysis of their origins and

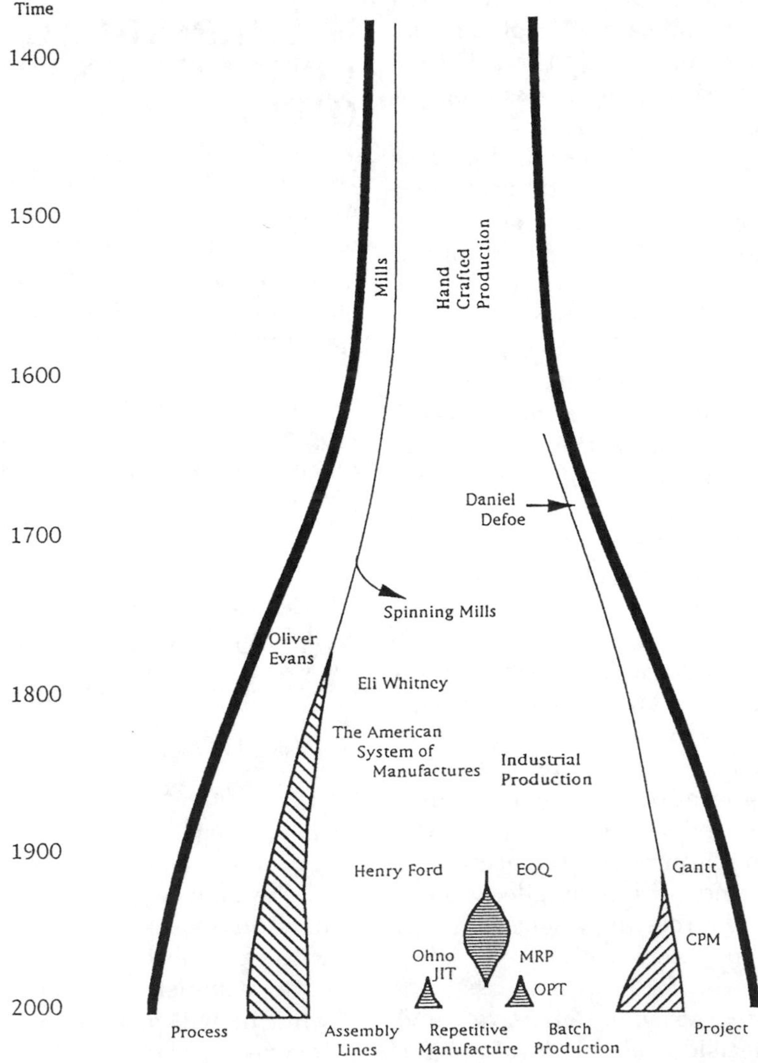

Fig. 1. Developmental tree. Source: *Production & Inventory Management Journal*, 1995, Vol. 36, Figure 1: 62, Reprinted with permission of the APICS – The Educational Society for Resource Management, Falls Church, VA.

development. Figure 1 shows an evolutionary tree summarising historic technical and managerial innovations and their impact. In the distant past there were no specialized production processes. We now recognize various categories and these distinctions signify specialized management technique. Branching points at particular times indicate when each process became distinct in its development. These branching points often denote new conceptualizations of production systems that are embodied in archetypal physical production systems. These archetypes and their underlying concepts provide distinct models or patterns for subsequent developments to improve: they represent the genesis of new modes of production.

PROCESSING AND PROCESS MANAGEMENT: Processing plants are typically large, expensive facilities dedicated to producing large volumes of generic products in a continuous flow. Technical requirements often dominate design and operational considerations with high levels of automation with monitoring and control systems. Relatively few people are employed in processing plants with vital roles in planning, monitoring and maintaining the facility's operations. Processes were some of the first activities transformed during the industrial revolution. The introduction of powered machinery to processing allowed continuous operation where manual methods of materials handling and product transformation had permitted only intermittent and irregular operation. The most significant of the new processing plants were the spinning mills. By the mid-1700's water-powered, mechanized mills were used and quickly became common.

Case 1: An Archetypal Processing Plant

The distinctly modern feature of processing plants: automatic operation, was applied in 1783 by Oliver Evans (1834) in a water-powered, fully mechanized flour mill. The mill could be left to run for short periods without anyone actively controlling its operations. Evans' Mill was a system of inter-dependent machines designed to operate in sequence. Figure 2 illustrates the mill's layout.

It shows the processing activities (milling, cooling, screening, etc.) along with the materials handling devices, storage buffers and facilities in their proper relationships. The flow of wheat from delivery through the whole process to its packing can be traced by following the numbers in the Figure, much as with a modern process chart. Evans' mill was an advance on earlier mills that had virtually all the same machinery, in much the same layout. Evans recognized that machines could be linked to the materials handling devices that brought their inputs and carried away their outputs. People were not involved with machine tending. The machines operated more steadily while producing a more consistent quality and volume of product. One feature of continuous operation is its need for balanced capacity and flows throughout the facility. Any inconsistencies in the materials flows or processing would have either starved or overloaded other processes and demanded rectification. The technical constraints of flour milling thus provided design parameters for the mill.

There were no automatic control or feedback devices in Evans' Mill. Problems in one operation would not automatically generate compensatory adjustments to other operations, either up-stream or down. Nevertheless, once the mill was set running properly, it could be left to continue running; at least until some problem arose, then the miller needed to respond and set it running properly again. People provided the control elements in Evans' mill, just as they ultimately do in current processing plants. Milling is a relatively straightforward processing operation using robust equipment to effective crush wheat into flour. It was eminently suited to early mechanization where more complex and delicate processes were not. English fabric mills were heavily mechanized at an early period but were not interlinked as completely as Evan's mill – the processing of cotton and wool involved finer operations that the mechanical systems could not provide until later in the nineteenth century. It is notable too that Evan's explicitly recognized a need to "save" labor that English mill owners did not.

ASSEMBLY LINES AND FLOW MANUFACTURING MANAGEMENT: Flow manufacturing arises from the sub-division of production into a number of tasks. These more narrowly defined tasks may then be the focus for specialized production techniques using equipment, tools or staff with particular skills or capabilities for those tasks. Even very small scale production can benefit from specialization, and its benefits were recognized very early (Wild, 1974). In discussing craftsmen, Plato (Lee, 1986: 61) observes that "Quantity and quality are therefore more easily produced when each man specializes appropriately on a single job for which he is naturally suited, and neglects all others." Adam Smith (1776) noted the great benefits further specialization brought to pin-making and Charles Babbage (1835) recognized the balance-delay problem in pin making. Its tasks could be optimally spread between five workers, and capacity expansion would be most effective if a multiple of five were used. Although Babbage noted

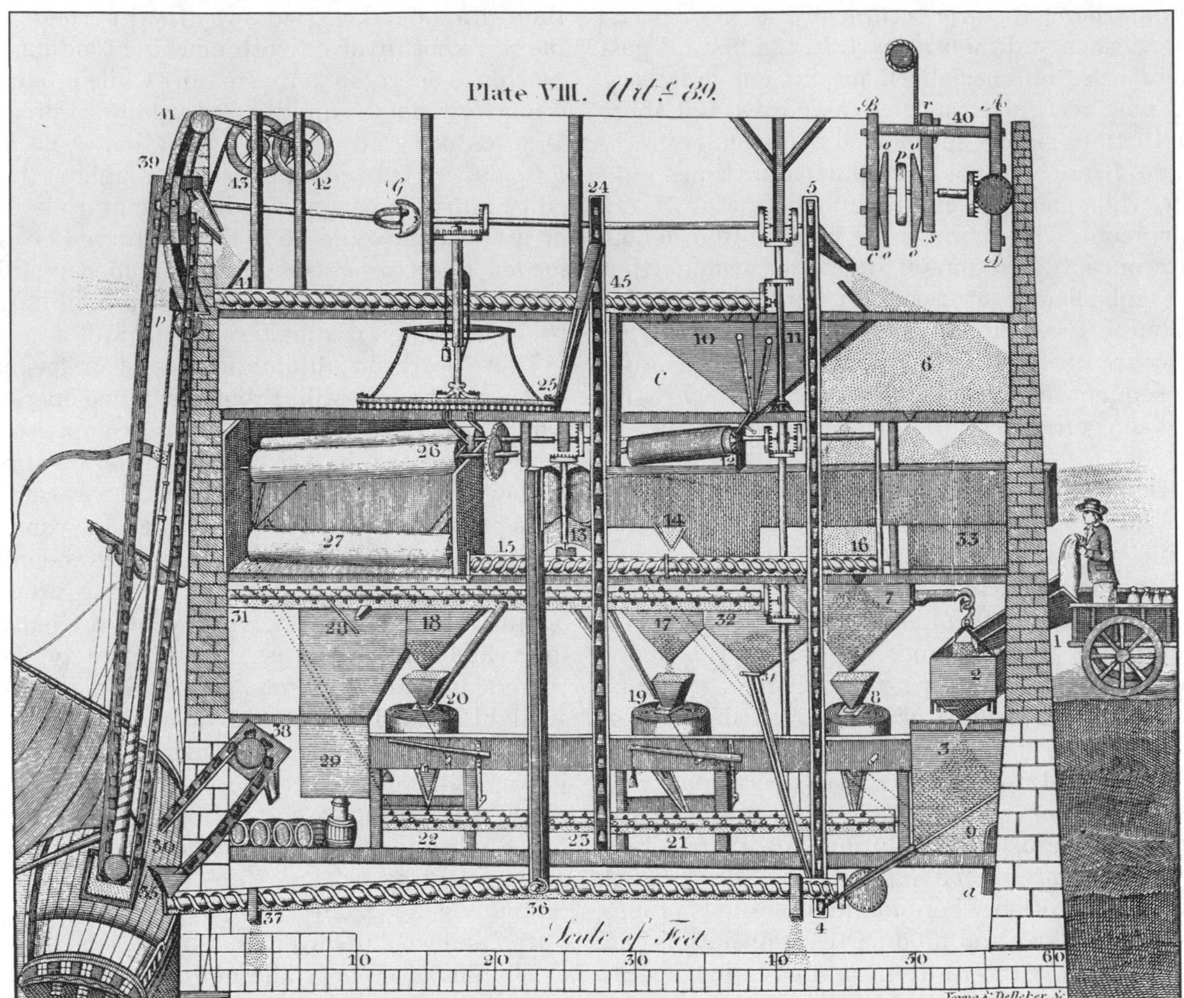

Fig. 2. Evan's mill. This diagram shows the flow of wheat through Evans' mill: it enters at the right unloaded from a wagon at 1; or unloaded from ships by 35, 36; and is weighed at 2 and then temporarily stored at 3. Then it is moved upstairs by an elevator 4, 5 where it is stored at 6 and 7. The wheat then goes through a cleaning process with shelling under the millstone 8 and cleaning and filtering at 9, 10, 11, 12 and 13. It is then stored at 6 and 7 (with horizontal conveyors to 17, 18) that then feed the millstones 8, 19 and 20. The milled flour is then conveyed by 21, 22 to an elevator 23, 24 that moves it upstairs where it is spread out to dry and cool by a mechanical spreader at 25. The cooling process gradually moves the flour to a point where it is feed into filtering screens at 26, 27. Flour of acceptable fineness then falls through to be stored at 28 and 29 before being packed into barrels or loaded onto ships, while flour that is too coarse is conveyed by 31, 32 to the millstones to be reground. (Reprinted with permission). (Evans, 1834: Plate VIII and Article 89, p. 216–218.)

that different numbers would be less efficient he did not determine task assignments that would minimize the inefficiencies.

The ideal of "progressive manufacture" arose early in the industrial revolution, and was avidly pursued. Wright (1880: 540) comments that as early as 1814 "... all the processes involved in the manufacture of goods ... were carried on in one establishment by successive steps, mathematically considered, under one harmonious system." Hounshell (1984) shows the development of the American System of Manufactures during the first half of the nineteenth century relied on sub-dividing work and increasing mechanization. The American System (Rosenberg, 1969) grew from governmental support for interchangeable parts; this gave manufacturers a stable, long term demand that justified investment and technological development. Interchangeability redefined quality: fit and function became more important than a fine finish. Major difficulties in achieving interchangeability continued throughout the 1800's. It took until the early 1900's before advances in equipment, tooling, materials,

Fig. 3. Ford's first assembly line. The flywheel magneto assembly line in 1914. This was the first of the Model T's subassemblies to be organized as a manually driven line process. This improved throughput times from 20 to 13 minutes. It is notable that Ford did not "design" this is any formal way. They tried various arrangements until the most effective one was found. From the collection of the Henry Ford Museum and Greenfield Village, The Edison Institute Negative No. 833-167 (Reprinted with permission).

engineering documentation and measuring devices allowed interchangeability. Interchangeability allowed efficient processing: the factory used a sequence of machines run by specialists, with assembly by non-craftsmen. Volume justified flow processing.

Case 2: Ford's Assembly Line at Highland Park

Ford's development of the moving assembly line at its Highland Park factory in 1913 represents the culmination of a century's developments in progressive manufacturing. Before the assembly line was introduced cars were manufactured on stands. All materials and workmen moved to the cars in the sequence required, this created a complex logistics problem. Nevertheless, Ford was able to assemble cars in volume. Further productivity and volume increases depended on process improvements. The first assembly line was installed at Ford's Highland Park plant on April 1, 1913 to manufacture flywheel magnetos. This is illustrated in Figure 3. It was a manually driven line, with employees doing their task and then pushing the assembly on to the next work station. Ford did not design this line in any meaningful sense: they just set up tables and experimented to see what worked best. The speed of the line was varied, its height was raised and lowered, and other variations were tried in a trial-and-error approach to design. There was no work study, though experience soon revealed where bottlenecks arose and what task assignments worked best (Ford and Crowther, 1926, 1928). Arnold and Faurote (1916) note that the first manual line reduced average labor time from 20 minutes per unit to 13 minutes. When a chain driven line was introduced for flywheel magnetos this further reduced the time per unit to 5 minutes. The use of lines allowed phenomenal productivity improvements – as much as 400% where they were used. The assembly line spread rapidly throughout Ford's factory and dominated public perceptions of mass production and factory life.

Ford's Model T factory was a highly specialized and focused manufacturing system (Wilson, 1996). Ford was committed to a single product, the Model T automobile. The system was designed around it, constrained by the capabilities of the workforce and existing technology. Quality was a critical prerequisite for Ford: the line could not run steadily at speed without consistently good components. On time delivery was also critical for Ford: the desire to keep men and machines busy with materials flowing constantly made scheduling critical. Ford's objective was to minimize costs; this was extremely profitable, but it also allowed Ford to fulfil his desire to make a car for "everyman" and to pay his workers well. Product, processes, material, logistics and people were well integrated and balanced in the design and operation of the Highland Park factory.

REPETITIVE MANUFACTURE: Repetitive manufacturing involves the concentration of manufacturing capacity for the high volume production of discrete units. High volumes of standardized products justify using flow production methods while smaller volumes of more customized products do not. However, these smaller volumes may be produced more efficiently if flow principles can be introduced into their manufacture. Current practices originated with Toyota as it considered producing automobiles. This plainly differs from Ford's "Continuous" manufacturing, but the underlying ideal of a continuous flow of single units is common to both. Ohno alludes to a possible Fordist influence on the origins of JIT systems. He quotes Toyoda Kiichiro saying "We shall learn production techniques from the American method of mass production. But we shall not copy it as is. We shall . . . develop a method that suits our own company's situation." Ohno (1988: 91) then says "I believe that this was the origin of Toyoda Kiichiro's idea of just-in-time." Eiji Toyoda (1987: 57) described Kiichiro's approach in 1938 as converting operations from batch to flow production with a balanced throughput. This was the beginning of the Toyota production system. It was totally dismantled at the start of the Second World War, and laid dormant for several years until revived by Ohno and refined over the next three decades. The wide usefulness of repetitive manufacturing principles can be appreciated by recognizing that they too were discovered early during the Industrial Revolution. There are no direct links between Toyota and the Portsmouth Block Mill, but the two operations illustrate the possibility that people and organizations confronting similar problems can develop similar solutions.

Case 3: The Portsmouth Block Mill

The Portsmouth Block Mill is an early implementation of a repetitive manufacturing system. Beginning in 1805 it produced rigging blocks for the British Navy that required large numbers of these blocks. At its peak the Mill produced over 130,000 a year; in 200 different variants to irregular schedules dictated by the outfitting requirements of new construction and recommissioning; and to meet the on-going maintenance requirements of an operational fleet. (Cooper, 1981–2, 1984) Although the overall production volumes might have justified mass production methods it was comprised of a variety of models, with no single one required in sufficient volumes to make an individual line viable. The Block Mill remained in operation until the early 1900's, and some equipment was used as late as the mid-1960's. It long life reflects the basic nature and stable design of the product and the circumstances of its construction. The Block Mill was built at the on-set of the Napoleonic Wars and seems to have been designed to match expected demands arising from them. The following century of relative peace in Europe combined with a shift to powered ships did not create any demand requirements outwith the Mills' capacity. It was then well suited to supplying its captive market.

Blocks are simple products with few components or moving parts and have relatively liberal tolerances compared with more sophisticated products (such as the guns that Eli Whitney was then attempting to produce without much success). This allowed interchangeability within a batch to be readily achieved. The high demand made mass production attractive, but the variety of blocks required was a complication. The mechanical power distribution systems were restrictive: the layout was static and could not adapt to shifting demand patterns. Figure 4 illustrates the general, sequential flows of work within the Mill. Figure 5 shows the factory in operation with the wheeled bin in the foreground used to moved roughly a day's output between the various machines for processing. The Block Mill mechanized the manufacture of Blocks and co-ordinated the machining and material handling processes to achieve significant productivity improvements. The Mill with 10 staff could produce as much as 110 people did previously

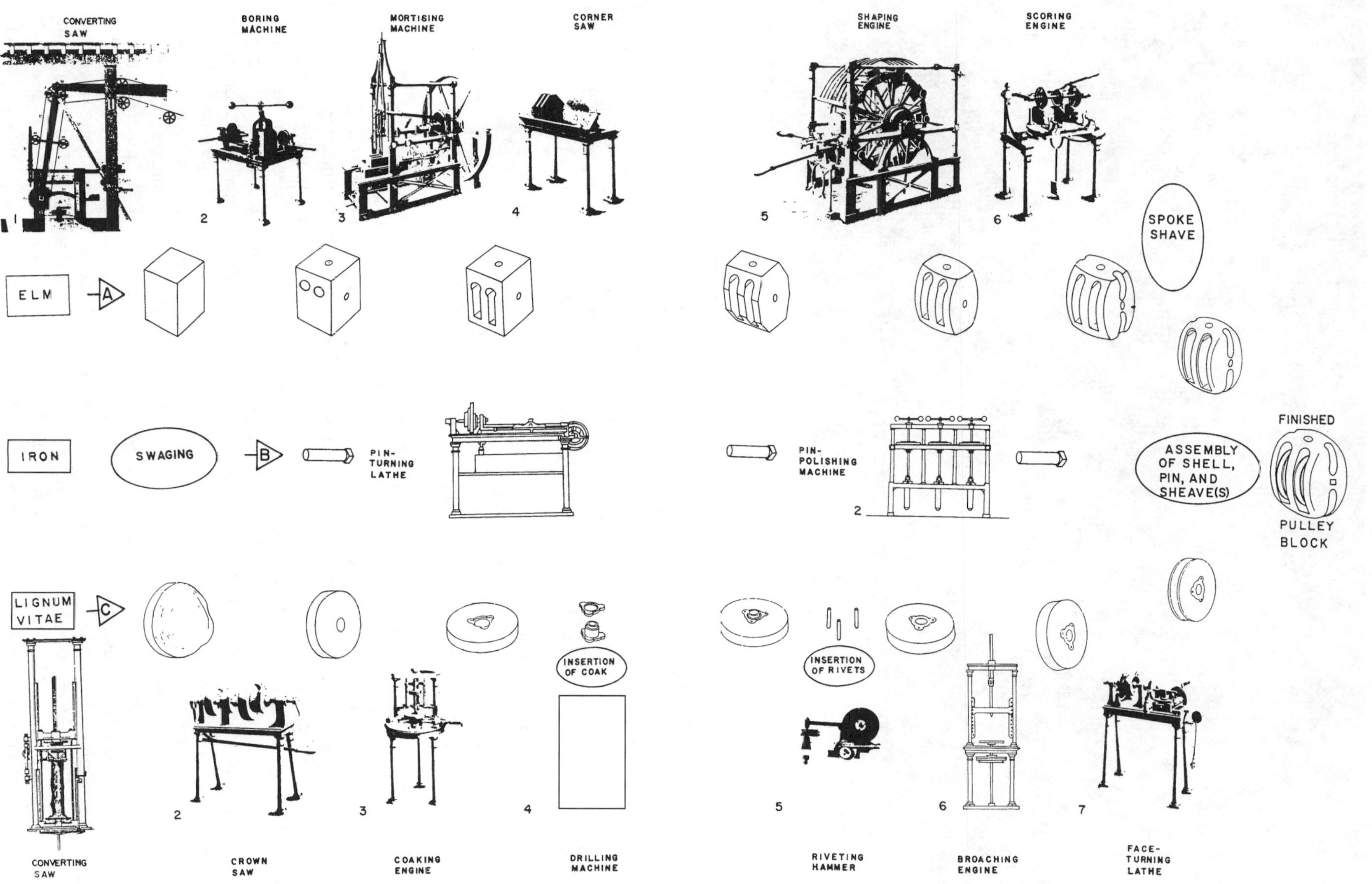

Fig. 4. Portsmouth block mill process flow. Source: Cooper, 1984: (Figure 6, p. 200–201). Figure used with the permission of John Hopkins University Press.

Fig. 5. Portsmouth block mill operations. This photograph is used with the permission of the Portsmouth Naval Base Property Trust, Portsmouth, England.

(Gilbert, 1965). Under the circumstances, the Luddite fear of mechanization seems well justified.

BATCH PRODUCTION AND ITS MANAGEMENT: In batch production a wide variety of parts are made intermittently in small quantities. This precludes using automation or narrowly specialized labor. Flexibility is essential for effective operation and dictates using general purpose machinery with highly skilled staff, typically organized in functional layouts. Gallagher (1978) speculates that increasing demand lead to such layouts simply when manufacturers simply added new machines next to their existing ones. Unless there was clear backtracking in materials flows such an approach is sensible in small facilities. F.W. Taylor also noted advantages of functional layouts in more complex operations. Technical problems were more easily managed: mechanical power distribution was easier, commonly used jigs or tools were readily available and specialist foremen could better supervise less skilled operatives. The labor supply was relatively lacking in industrial skills and the objective was to make best use of the available labor. Batch production allowed higher productivity: set-up costs were an important factor and, where demand was limited, a batch size could easily comprise the quantities required in a full order. Batch production also allowed better quality and manufacturability: it was far easier to obtain interchangeability of parts within a batch than between batches. Batch production was an effective approach to manufacturing and many historical systems were no less complex than modern ones.

Case 4: Frederick W. Taylor and the Planning Office

Frederick Taylor is best known for advocating work study, but it was one element in a larger production planning and control system. (Copley, 1923) Taylor recognized that managing the workman's productivity was futile unless the shop as a whole functioned in an orderly manner. Production planning and control was essential if his innovations in work study and mechanical engineering were to succeed. Taylor introduced many elements that have since been re-discovered and incorporated in materials requirements planning systems. (Thompson, 1917) The "order of work" was established and served as a master schedule – it was created in consultation with the sales department and linked the shop's activities to demand. Taylor also introduced routings to indicate where specific jobs were to go in the shop; and their movement was closely controlled and monitored using "move tickets" so that shop managers could readily tell where jobs were in the shop, and what work remained to be done. Taylor also used modern stock control methods, quantities on hand were monitored along with quantities on order. Then, whenever incoming orders were projected to cause shortages, replenishment stock could be ordered. Though Taylor relied upon fixed order quantities he developed his system before economic order quantities became prevalent and order quantities remained subjective. Taylor created a planning office responsible for planning, monitoring and controlling the shop and it anticipated the functions of modern planning and control systems.

PROJECTS AND THEIR MANAGEMENT: The Pyramids, Stonehenge and innumerable Castles, Cathedrals and other great edifices of the past were projects. There is little information on their construction management. Several factors may have made management in a modern sense difficult: the projects were political or religious, and often ordered built by authoritarian regimes so that non-economic motives may have made modern management practices unworkable. The backward state of accounting, economics and the behavioral sciences would also have created practical difficulties for managers. The first clear recognition of projects as a specific type of product can be found in Daniel Defoe's *Essay Upon Projects* in the late 1600's. He comments that "projecting" began in the 1640's and became popular in the 1680's. Projects then were somewhat disreputable: Defoe uses the term "Project" much as we today would use the term "Scheme", and "Projector" carried with it the same connotations that "Schemer" does today. No distinction was drawn between commercial projects and engineering or industrial ones. Some large commercial projects were fraudulent and this image affected all projects. However, even large engineering projects were susceptible to image problems of their own, as Defoe recognized: "... the true definition of a project, according to the modern [1693] acceptation, is, as said before, a vast undertaking, too big to be managed, and therefore likely enough to come to nothing." (Defoe, 1697: 30)

Although Defoe recognized the special nature of projects it does not appear that any specialized techniques were developed for their management until Gantt Charts were applied during World War I. Gantt charts were developed during the 1890's as a method for controlling production activities in general (Gantt, 1903). In the late 1950's computers allowed more detailed network based approaches to develop.

Case 5: The First Networks

The nearly simultaneous development of network methods for project management by Du Pont and the U.S. Navy shows an independent genesis of similar solutions to common problems. The management challenge present by developing nuclear submarines was recognized in mid-1956. Experience during 1956 lead managers to recognize that a detailed planning and control system was needed and outside consultants were engaged to help model projects using operations research techniques. PERT began development early in 1958, with Dr. C.E. Clark demonstrating a network representation (activity on node) of a project to his colleagues in February (Fazar, 1962: 618). There was great apprehension about implementing PERT in an ongoing project, but strong support from the program's top management ensured PERT's survival. The full-scale use of PERT in October, 1958 followed. Similarly, the development of CPM at Du Pont followed from their recognition that the scale of activities had grown too large for effective management. In late 1956 Walker (Kelley and Walker, 1989: 7) was given the broad remit to "See what you can do about scheduling." A parametric linear programming model was subsequently developed; but was known to be impractical for real projects. The similarities of this formulation with transportation models suggested a network and algorithmic solution procedures. A test case was identified and a network drawn for it, dated July 24, 1957. This test was successful and development work proceeded. CPM was applied to a real project beginning in December, 1957, with further refinement and de-bugging of the system. CPM was presented to the public in March, 1959 and its commercial success guaranteed its future.

STRATEGIC PERSPECTIVE: THE FUTURE AND THE PAST: It is almost impossible to consider contemporary developments without being struck by historic antecedents. Recent developments all have parallels with earlier systems and practices.

1. Interchangeability and Total Quality Management: Total quality management can be compared with the century long crusade to achieve interchangeability. Eli Whitney made the first concerted efforts at achieving interchangeability in an industrial setting in the late 1790's but was not successful. Since then production processes have developed significantly and so too have inspections and quality management methods. Ford's mass production system was based on achieving "total quality"

through product and process design. Ford had a fine understanding of the principles of design for manufacture and a key element was setting achievable quality standards. This involved both rigorous specifications and thorough inspection. In a continuous process with a single product (the Model T) there were no batches to be periodically inspected – Ford undertook to inspect their output more-or-less continuously with shop-based inspectors that monitored output and corrected any faults on the run. The design, production methods and inspection policies all combined to yield a high quality product at a competitive price.

2. Business Process Reengineering and Systems Redesign: Business process reengineering too represents a revival of past practices when systems redesign was fundamental to progress. The industrial revolution itself was based on just such fundamental re-appraisals of operations. They lie at the root of every archetypal system described. Hounshell's discussion (Hounshell, 1985) of the American System of Manufactures shows such development as critical in the evolution of manufacturing systems. Process reengineering in the 1800's was more readily appreciated: companies were smaller with fewer processes and employees, and more readily visible impediments to flows of information and materials. Such leanness and simplicity was the norm in the 1800's. (Chandler, 1977) The sub-division of labor and separation of responsibilities was in many cases an objective that promoted productivity, not an impediment. In sub-dividing work managers often found significant improvements followed from redefining the products and processes. Such an overall perspective is vital to improving production processes. In the 1800's and early 1900's specialization and mechanization were essential for improving productivity: particularly because traditional craft methods were closely scrutinized to eliminate and simplify as much of the work as possible. The ASM departed from craft methods and relied on a fundamental reconsideration of product design, production activities and facility operation.
3. Outsourcing and Internal Contracting: Parallels between outsourcing and internal contracting are also apparent. Outsourcing involves contracting with outsiders to provide either non-core activities that otherwise distracts management from their mainstream business; or core activities with sophisticated skills or technology not easily understood by management.

The objectives for many organizations not familiar with a particular area are to obtain access to its best technologies and talent cheaply using mechanisms that allow effective control. In the 1800's similar problems in production management were resolved through internal contracting. (Chandler, 1977) Management then was unfamiliar with production technologies and shop management, as with many information technologies today. The solution was to produce items on a contractual basis: the shop or work center contractor bid on work – offering to produce some item against specifications and delivery schedule at a mutually agreed price. This effectively removed the organization's management from the shop floor and allowed specialists free rein. The contractor determined the most appropriate production techniques; hired, fired and allocated labor as required, and established wage rates for specific jobs; the company's management provided the materials and equipment used in house, and paid for the output as agreed. The contractor was responsible for managing the work: scheduling its performance, ensuring quality standards were met, controlling materials and work flow, etc. So long as management was unfamiliar with rapidly changing production technologies and shop management the contractors were an effective solution. In the later 1800's these conditions began to change. Production technologies became more stable and less dependent on specialist knowledge. Contractors also came to be seen as unfairly exploiting the relationship: holding contract prices steady while reducing wage rates was one source of income for contractors and industrial relations suffered. Contractors were also thought to establish prices based on out-moded practices and then introduce new modes once contracts had been agreed, thus taking all the savings from new technologies for themselves. Further problems arose from management's inability to effectively monitor and control the contractors: although management provided materials and equipment to contractors their usage was often obscured so management was never clear whether all the resources provided were used as intended. Such misgivings over souring industrial relations and fears that contractors were building inefficiencies into contracts that they could then exploit led managers to assert more control over production management by bringing contractors in-house as foremen and supervisors. Managers then assumed the reduced risks of managing production and the former contractors were paid wages and assumed a lesser role.

The past can provide many insights into the present. All too often the difficulties arise because managers familiar with contemporary problems know relatively little about former industrial systems and practices; and historians and technologists familiar with such practices seem to know little about contemporary concerns, developments and issues.

CONCLUSIONS: Modern problems have often been faced in the past, concerns with economically producing high quality products to schedule are not new. Managers may believe that modern changes have rendered historical systems irrelevant but this is a delusion. It is quite easy to believe that customers and markets today are different from those that existed in the past, but we should be careful not to exagerate the differences. Change too is much more common today than in the past, but this platitude rather overlooks the significance of past developments. Fundamental improvements in transportation and communications systems have altered the business environment substantially. It is easy for us today to look at the onslaught of minor changes and assume that life was so much easier in the past. Perhaps there were fewer changes – that is not at all certain – but the known changes were substantial and far-reaching in their impact. Telegraphy, telephony and radio altered communications systems beyond all earlier recognition. The impact of mobile telephones and even e-mail pale by comparison. Similarly the development of railways, steamships, and motor transport greatly affected the business environment – the mobility of goods, services and people has been enhanced beyond belief. If a key problem for modern managers is coping with change their position seems little different than that faced by earlier managers. Production managers in the past confronted many of the same basic problems faced today, and historic solutions have often been remarkably "modern" in their conception. The history of operations management reveals many examples of well considered and implemented production systems. The past certainly cannot be used to predict the future but it can be very useful in understanding and coping with the present. Many promoters of new management techniques often denigrate "traditional" management when the distinction should more properly be drawn between sound and poor management practices.

Organizations Mentioned: DuPont; Ford; Portsmouth Block Mill; Toyota; U.S. Navy.

See **Assembly line design; Business process reengineering; Critical path method; Gantt chart; Human resource management in manufacturing; Just-in-time manufacturing; Manufacturing systems; Project management; Scientific management; Total quality management.**

References

[1] Babbage, Charles (1835). *On the Economy of Machinery and Manufactures*, Charles Knight, London, England.

[2] Chandler, Alfred (1977). *The Visible Hand: The Managerial Revolution in American Business.* Belknap Press, Cambridge, Mass.

[3] Cooper, Carolyn C. (1981–2). The Production Line at Portsmouth Block Mill. *Industrial Archaeology Review.* 6, 28–44.

[4] Cooper, Carolyn C. (1984). The Portsmouth System of Manufactures. *Technology and Culture.* 25, 182–225.

[5] Copley, Frank Barkley (1923). *Frederick W. Taylor: Father of Scientific Management.* Reprinted 1969, Augustus Kelley, New York, New York.

[6] Defoe, Daniel (1697). An Essay Upon Projects. Reprinted 1887. Cassell & Co., London, England.

[7] Evans, Oliver (1834). The Young Mill-wright & Miller's Guide. Reprinted 1989. Arno Press. New York, New York.

[8] Faurote, Fay Leone and Harold Lucien Arnold (1916). The Ford Methods and the Ford Shops. *Engineering Magazine* Co., New York, New York.

[9] Fazar, Willard (1962). The Origins of PERT. *The Controller*. Vol. 8, 598–602, 618–621.

[10] Ford, Henry with Samuel Crowther (1926). *My Life and Work*. William Heineman, Ltd., London.

[11] Ford, Henry with Samuel Crowther (1928). *Today and Tomorrow*. William Heineman, Ltd., London.

[12] Gallagher, Colin (1978). Batch Production and Functional Layout. *Industrial Archaeology*. 13, 157–161.

[13] Gantt, H.L. (1903). A Graphical Daily Balance in Manufacture. *ASME Transactions*, 24, 1322–1336.

[14] Gilbert, K.R. (1965). The Portsmouth Block-Making Machinery. Her Majesty's Stationery Office. London, England.

[15] Hounshell, David. (1984). *From the American System to Mass Production*, 1800–1932. The John Hopkins University Press, Baltimore, Maryland.

[16] Kelley, James E., Jr. and Morgan R. Walker (1989). *The Origins of CPM: A Personal History. PMNetwork*, 3, 7–22.

[17] Plato, (translated by Desmond Lee). (1987). *The Republic*. Penguin Classics, London, England.

[18] Ohno, Taiichi (1988). *Toyota Production System*. Productivity Press, Cambridge, Mass.

[19] Rosenberg, Nathan (1969). *The American System of Manufactures*. Edinburgh University Press, Edinburgh, Scotland.

[20] Thompson, Clarence B. (1917). *The Taylor System*

of Scientific Management. A.W. Shaw Co. Chicago, Illinois.
[21] Toyoda, Eiji (1987). *Toyota: Fifty Years in Motion*. Kodansha International, Tokyo, Japan.
[22] Wild, Ray (1974). The Origins and Development of Flow-Line Production. *Industrial Archaeology Review*. 11, 43–55.
[23] Wilson, James M. (1995). An Historical Perspective on Operations Management. *Production & Inventory Management Journal*. 36, 61–66.
[24] Wilson, James M. (1996). Henry Ford: A Just-in-Time Pioneer. *Production & Inventory Management Journal*. 37, 26–31.
[25] Wren, Daniel (1994). *The Evolution of Management Thought*. John Wiley & Sons, New York, New York.

HMMS PRODUCTION PLANNING MODEL

See **Capacity planning: Long-range; Linear decision rule.**

HOLDING COST RATE

Inventory has value at least theoretically equal to the price paid for its purchase. Inventory is far less liquid than cash and therefore the firm incurs a cost of holding the asset. The financial aspect of holding cost consists of the marginal cost of capital, which is equal to what the firm foregoes in the form of income that could have been generated using the money tied up in inventory. Holding cost also reflects the cost of storing inventory (such as warehousing expense, and possibly costs of refrigeration, and other expenses), and losses due to pilferage or obsolescence. Holding cost rate is expressed in terms of the cost to hold one unit for one year. Holding costs increase with the size of inventory on hand.

See **Inventory flow analysis; Simulation of production problems using spreadsheet programs.**

HOLT'S FORECASTING MODEL

Holt's two-parameter model, also known as linear exponential smoothing, is a popular smoothing model for forecasting data with trend. Holt's model has three separate equations that work together to generate a final forecast. The first is a basic smoothing equation that directly adjusts the last smoothed value for last period's trend. The trend itself is updated over time through the second equation, where the trend is expressed as the difference between the last two smoothed values. Finally, the third equation is used to generate the final forecast. Holt's model uses two parameters, one for the overall smoothing and the other for the trend smoothing equation. The method is also called double exponential smoothing or trend-enhanced exponential smoothing.

See **Forecasting guidelines and methods; Forecasting in operations management.**

HOSHIN KANRI

In the Japanese language it means breakthrough planning. It refers to a Japanese strategic planning process in which a company develops up to four vision statements that indicate where the company should be in the next five years. Company goals and work plans are derived from the vision statements. Periodic audits monitor the progress.

See **Total quality management.**

HRM

See **Human resource management in manufacturing.**

HUB COMPANY

An Electronic Data Interchange (EDI) initiating company which represents an accumulation center around which trading partners can build electronic networks can be built.

See **Electronic data interchange (EDI) in supply chain management.**

HUMAN FACTOR DESIGN

It is the process of ensuring a "fit" between a product and its user during product design. Human Factor Design (HFD) uses scientific analysis of human performance and behavior with respect to the environment influenced by new products and technology. It is especially important for complex products where human interventions and interface are high.

See **Product design.**

HUMAN RESOURCE ISSUES AND ADVANCED MANUFACTURING TECHNOLOGY

Corinne M. Karuppan

Southwest Missouri State University, Missouri, USA

STRATEGIC ISSUES: Shorter product life cycles, new market opportunities, and fierce competition have made flexibility one of the top competitive priority. Technology enables flexibility. However, while some manufacturers are successful, not all manufacturers who invest in advanced technologies such as Computer Numerical Control (CNC) machines, Direct Numerical Control (DNC), and Flexible Manufacturing Systems (FMS) are satisfied. Some manufacturers' inability to benefit from advanced manufacturing technology (AMT) can often be traced to human resource problems. Under pressure to get a new system up and running, management often gives a low priority to the time, cost, and expertise needed to ensure a fit between automated systems and the workforce. Brief bursts of training occur initially when new systems are adopted and have, at best, a short-term effect. Indeed most companies do not develop comprehensive human resource strategies to support their investment in technology (Fallik, 1988).

When the costs of wages, benefits, recruiting, selecting, training and placement, sick leave, etc., are considered people are already the most expensive resource in an organization. However, these costs do not include the lost productivity due to operator/staff performance problems that are AMT-related. It is thus clear that AMT plans should consider its effects on users and include comprehensive employee development plans.

BACKGROUND ON TECHNOLOGY: Even though the impetus to AMT was the commercial availability of computers in the mid-1950's, the relevent concepts such as group technology and robotics emerged decades ago. The early work on industrial robots started at the end of World War II with the development of controlled mechanical manipulators to handle radioactive materials and magnetic control devices (Fu, Gonzalez, and Lee, 1987). Also during the 1940's, Eckert and Mauchley built the ENIAC, the first fully functional electronic calculator, and numerical control for machine tools was introduced in the airplane industry. In the 1950's, a computer-driven manipulator permitted robots to be reprogrammed easily and thus provided the flexibility to perform a variety of manufacturing tasks. Further advances in these and other technologies took place in the 1960's. In the mid-1970's, the first FMS was implemented at Caterpillar Tractor. The rapid proliferation of personal computers in the 1980's and, later, local area network technology, enabled the linkages of manufacturing devices to mainframes, minicomputers, and personal workstations (Goldstein, 1986).

AMT configurations on the shop floor include the popular CNC machines, robots, DNC, and FMS configurations. A CNC machine or CNC robot is typically a stand-alone machine controlled by its own microprocessor. In a DNC environment, a remote central processor controls several CNC machines. Despite a lack of international congruence on the meaning of an FMS, a common definition refers to a group of CNC machines linked by an automated material handling system and controlled by a host computer (Wan, 1993). The versatility of an FMS lies in its ability to produce a variety of products as well as new products on the same machine and accommodate design changes.

THE IMPACT OF AMT ON PEOPLE: Despite rapid and substantial progress in the development of these technologies, they are still characterized by varying degrees of: 1) hardware and software unreliability; 2) unpredictability; and 3) system-imposed pacing, which occurs when technology rather than the operator determines the beginning and length of the work cycle. Moreover, typical features of AMT-related jobs include: 1) interdependence among tasks and functions; 2) one-person cells; 3) management's expectations of increased output rates; and 4) skill obsolescence (Karuppan, 1997). Table 1 explains how both technology and AMT work characteristics create the following pressures: quantitative overload, psychological demands, lack of social support, role ambiguity, and loss of control.

Individuals face quantitative overload when the quantity of work to be performed within a certain time frame exceeds their capabilities. Psychological demands refer to operators' fear of making errors as well as the attention, concentration, and endurance imposed by technology-based work. The notion of social support evokes the availability of help, with others, and the sense of being supported by others. Workers face role ambiguity when the expectations associated with their jobs, the methods for carrying out known expectations, and/or the outcomes of their performance are not clearly defined. People perceive that they have control when they believe that their own response can alter or neutralize an adverse situation. A large body of field research has linked the above pressures to: 1) psychological strains (e.g., job

Table 1. AMT characteristics and impact on operators.

Technological characteristics	Nature and demands of the technology	Impact on operators
Hardware and software unreliability	Machine failures, loss of data, loss of program, and software glitches require intense vigilance and concentration. Moreover, the operator may fear that these failure conditions will produce costly errors.	Psychological demands, i.e., amount of vigilance/concentration, and cost of errors (Dwyer and Ganster, 1991)
Unpredictability	– Errors occur at random times, are often complex, and require prompt corrective action.	Quantitative overload, i.e., too much work to do within a given time (Frankenhaeuser and Gardell, 1976)
	– Random, complex errors can make intervention and performance criteria ambiguous to workers (Weick, 1990)	Role ambiguity, i.e., lack of clarity in a role, the methods to carry out role expectations, and consequences of role performance (Kahn *et al.*, 1964).
Pacing	– Some technologies determine the beginning and the duration of the work cycle – Program and equipment speed are beyond the operator's control	Loss of timing control, i.e., the opportunity to determine the timing of one's work behavior (Corbett, 1987)
AMT Work Characteristics		
Interdependence among tasks and functions	– Traditional, individual-based reward systems foster competition rather than collaboration among employees – Lack of trust in other departments' data heightens the state of alertness – Tight interdependence affects the pacing and quality of operations – Planning and control systems reduce the need for supervisors	Psychological demands Loss of timing and method control, i.e., the choice in how to complete one's tasks (Corbett, 1987) Lack of social support from supervisor and colleagues, i.e., absence of help (Gore, 1987)
One-Person cells	– More vigilance is needed – Responsibility for lost output is placed on a single individual – Fewer opportunities to interact with fellow workers and obtain their collaboration	Psychological demands Lack of social support from colleagues
Pressure to increase output rates	To justify AMT investments, work loads are increased and there is pressure to run machines at top speeds	Quantitative overload
Skill obsolescence	Workers need to learn new material continuously while keeping up with their regular workload	Quantitative overload

dissatisfaction, anxiety, depression, somatic complaints, etc.); 2) physiological strains (e.g., increased heart rate, higher blood pressure, higher cholesterol levels, etc.); and 3) behavioral strains (poor performance, low motivation, absenteeism, tardiness, intent to quit, etc.). The prolonged experience of such strains, in turn, may affect the individual's health. It also has repercussions on the organization's bottom line due to the costs of poor performance on the job, sick days, turnover, accidents, etc. (Ivancevich and Matteson, 1987).

The pressures described above exist in all work environments but exacerbated by the new technologies. A study comparing AMT and non-AMT workers showed that quantitative overload, lack of social support by the supervisor, role ambiguity, and lack of control contributed to the AMT group's job dissatisfaction. Quantitative overload, psychological demands, and lack of control were associated with somatic complaints. Further, psychological demands in workers were related to intent to quit the organization (Karuppan, 1995). Workers experiencing such symptoms are unlikely to demonstrate the levels of commitment needed for flexibility. Workers commitment is key to the key ingredient in their ability to switch tasks and functions, which is essential to manufacturing flexibility. In the next section, technological features and work design policies are suggested to create a more harmonious, supportive, and flexible AMT environment.

IMPLEMENTATION: As technology evolves, new equipment and software upgrades have the potential to significantly reduce the level of stress induced in workers. For example, improved stop devices, improved automated error detection capabilities, improved software editing options for faster and easier error correction, recoverability of equipment after an error, equipment pauses in the cycle where errors are most likely, switches in equipment allowing operators to stop the machine upon error detection, interactive help about equipment capabilities, and more efficient technical support services can significantly lessen psychological demands, quantitative overload, and system-imposed pacing on the operator. However, technology enhancements alone are insufficient to guarantee successful AMT implementations. Rather, management must create an appropriate infrastructure that facilitates and promotes the operator's decision making authority. The foundation of this infrastructure are enhanced operator control, continuous learning, compensation, and social support.

1. ***Operator Control:*** As mentioned earlier, the primary objective of AMT technologies is flexibility. In an environment where errors occur randomly, flexibility requires the operator to restore the hardware and/or software to a working condition promptly or, better, anticipate these errors and prevent them. To do so, the operator must have the authority (and of course knowledge) to engage in problem diagnosis and solving, when he/she deems appropriate. Furthermore, when machine failures threaten system flexibility, the operator should have the authority to re-route jobs if necessary. The policies for expanding the operator's functions beyond mere monitoring to problem solving and decision making give more control back to the operator. This policy also benefits the organization by minimizing lost production due to errors. But, to delegate problem solving authority to line personnel, management must trust their competence. The next section provides a blueprint for training AMT problem solvers, not merely AMT operators.
2. ***Continuous Learning:*** AMT workers are knowledge workers. For knowledge-based work, Frese (1989) has shown that a holistic approach to training is more effective than a step-by-step approach, which is typical of on-the-job training. In the holistic training approach, workers develop a mental model of the system. If users can develop an internal model of the system functions, they will be better prepared to solve complex problems or anticipate them. Moreover, a holistic approach is better suited to provide the extensive skill set necessary in AMT work, i.e., cognitive, social, diagnostic, and tacit skills and pave the ground for continuous learning.
   1. *Cognitive Skills Development:* AMT skills refer to mental ability rather than manual ability. AMT work requires the acquisition of abstract knowledge in mathematics, digital programming, and CNC automation logic, and its application to machine operation and production planning and control. To fully understand the system, operators be taught the evolution of AMT from the basic system to today's complex architectures. As each stage of development is described, trainees should be capable of identify each stage's benefits as well as its shortcomings.
   2. *Social Skills Development:* Given the tight interdependence of AMT systems, there is a vital need for workers performing related tasks to engage in collective problem solving. This kind of social support largely depends on the individual's willingness to establish social "networks" with colleagues, trust their judgment, listen to their opinions, and communicate his/her opinions clearly. Training should promote teamwork in all phases and aspects of the program: knowledge acquisition, mental model construction, machine operation, problem diagnosis, and problem solving. Cross-training with job rotation is in fact a powerful tool to promote not only flexibility, but also interaction with other workers and broader understanding of the production system.
   3. *Diagnostic Skills Development:* Diagnostic skills depend not only on the cognitive skills, but also on the practical knowledge of the various machine commands, and the functions of machine time, planning, scheduling, and quality control. To refine their diagnostic skills, operators should be taught hypotheses formulation about machine status and try out solutions during the training phase. However, AMT is increasingly complex and requires the acquisition of a vast knowledge base a single operator to single operator is limited in what he/she can do. Therefore, team problem solving is a desirable way of reinforcing each other's diagnostic skills.
   4. *Tacit Skills Development:* Flexibility is better enhanced by fault prevention than equipment recovery after a fault/failure. Tacit skills involve learning to avoid mistakes

to overcome imperfections in the system (Wood, 1990) based on implicit knowledge. The development of implicit knowledge is gained only through active involvement in system control, not through formal training. However, it has been shown that Total Quality Management (TQM) tools and techniques, and an organizational environment that emphasizes continuous learning and appropriate compensation schemes foster the development of tacit skills.

3. ***Compensation Schemes:*** Since AMT environments require team work, team-based incentives or gainsharing should be key elements of compensation packages. Accordingly, the reward is pegged to group outcomes or collective skills accumulated over time. One of the potential drawbacks is the "free riding" effect in which some team members carry most of the burden for the team. This is why team-based plans must be supplemented with some form of individual plan. In the AMT work environment, pay-for-knowledge compensation is an obvious form of individual compensation. More than 60% of Fortune 1000 firms use some form of pay-for-knowledge compensation scheme (Ledford, 1995). An overwhelming majority of these firms have seen worker productivity, motivation, and flexibility to adapt to changing production needs increase. However, in AMT environments, the change in compensation plans must be synchronized with technological change. Human resource specialists must therefore realize the need to change pay on a "just-in-time" basis. In this fast-paced environment, this means that rapid, incremental changes are more effective than dramatic, periodic changes.
4. ***Peer and Supervisory Support:*** One-person cells promote task identity as operators identify with the completion of an entire operation or piece of work, but they are not conducive to productive, work-related discussions among coworkers. The fact is, coworkers are often the preferred source of information to help define and solve problems. This problem is minimized when the opening of a cell coincides with that of a contiguous cell, and when one-person cells are placed closer together, where dialog among peers can take place. In this AMT environment, the role of the supervisor also changes. Even if planning and control systems have reduced the number of supervisors needed, it is also true that the number of operators has also decreased. As a result, there might be more personal contacts between employees and supervisors, and the supervisor can be more involved in technical training, in advising workers, and in compiling a documentation of problems and solutions for operators.

CONCLUSION: The man-machine interface must be examined from many different angles. This article provides a quick overview of how AMT work can affect the individuals that operate it and vice versa. Too often, AMT operators have been considered less important than the equipment. Management must view line employees as a work*force* capable of enhancing machine effectiveness and flexibility. The implementation guidelines presented here are intended to help build such a workforce.

Despite an increasing interest in this domain, there remains a strong need to conduct more research. Academics and practitioners have emphasized the need for worker participation, especially in process quality management. More research should focus on the benefits and pitfalls of making AMT operators participants in the work design process.

See **Advanced manufacturing technology (AMT); Cognitive skills of operators; Computer numerical control (CNC); Diagnostic skills of operators; Direct numerical control (DNC); Flexible manufacturing system (FMS); Job stress; Physiological strains of operators; Psychological strains of operators; Quantitative overload; Role ambiguity; Skill obsolescence; System-imposed pacing; Tacit skills of operators.**

References

[1] Corbett, J.M. (1987). "A Psychological Study of Advanced Manufacturing Technology: the Concept of Coupling." *Behaviour and Information Technology*, 6, 4, 441–453.

[2] Dwyer, D.J., and D.C. Ganster (1991). "The Effects of Job Demands and Control on Employee Attendance and Satisfaction." *Journal of Organizational Behaviour*, 12, 7, 595–608.

[3] Fallik, F. (1988). *Managing Organizational Change: Human Factors and Automation*. Taylor & Francis, Philadelphia, 65.

[4] Frankenhaeuser, M., and B. Gardell (1976). "Underload and Overload in Working Life: Outline of a Multidisciplinary Approach." *Journal of Human Stress*, 2, September, 35–46.

[5] Frese, M. (1989). "Human Computer Interaction within an Industrial Psychology Framework." *Journal of Applied Psychology: An International Review*, 38, 1, 29–44.

[6] Fu, K.S., R.C. Gonzalez, and C.S.G. Lee (1987). *Robotics: Control, Sensing, Vision, and Intelligence*. McGraw-Hill Book Company, New York.

[7] Goldstein, L.J. (1986). *Computers and Their Applications*. Prentice-Hall, Englewood Cliffs, 307–308.

[8] Gore, S. (1987). "Perspectives on Social Support and Research on Stress Moderating Processes." In *Job Stress: From Theory to Suggestion*, J.M. Ivancevich and D.C. Ganster (eds.), The Haworth Press, New York, 85–101.

[9] Ivancevich, J.M., and M.T. Matteson (1987). "Organizational Level Stress Management Interventions: a Review and Recommendations." In *Job Stress: From Theory to Suggestion*, J.M. Ivancevich and D.C. Ganster (eds.), The Haworth Press, New York, 229–248.

[10] Kahn, R.L., D.M. Wolfe, R.P. Quinn, J.D. Snoek, and R.A. Rosenthal (1964). *Occupational Stress: Studies in Role Conflict and Ambiguity*, John Wiley & Sons, New York.

[11] Karuppan, C.M. (1997). "AMT and Stress: Technology and Management Support Policies." *International Journal of Technology Management*, 14, 2/3/4, 254–264.

[12] Karuppan, C.M. (1995). "How Stressful Is the Automated Shop Floor?" *Benchmarking for Quality Management & Technology*, 2, 4, 27–40.

[13] Ledford, G.E., Jr. (1995). "Paying for the Skills, Knowledge, and Competencies of Knowledge Workers." *Compensation & Benefits Review*, 27, 4, 55–62.

[14] Swamidass, P.M. (1998). *Technology on the Factory Floor III: Technology Use and Training in the United States Manufacturing*. Manufacturing Institute, National Association of Manufacturers, Washington, D.C., August 1998.

[15] Wan, C.A. (1993). *Foreign Industry Analysis: Flexible Manufacturing Systems*. U.S. Department of Commerce, Bureau of Export Administration, Office of Foreign Availability, BXA/OFA 93–05.

[16] Weick, K.E. (1990). "Technology as Equivoque: Sensemaking in New Technologies." In *Technology and Organizations*, P.S. Goodman, L.S. Sproull, and Associates (eds.), Jossey-Bass, San Francisco, 1–44.

[17] Wood, S. (1990). "Tacit Skills, the Japanese Management Model and New Technology." *Applied Psychology: an International Review*, 39, 2, 169–190.

HUMAN RESOURCE MANAGEMENT IN MANUFACTURING

Thomas W. Dougherty

University of Missouri-Columbia, MO, U.S.A.

George F. Dreher

Indiana University-Bloomington, IN, U.S.A.

WHAT IS HUMAN RESOURCE MANAGEMENT?: Human resource management (HRM) can be defined as "planning, implementing, and evaluating programs for the people in an organization, in a fashion consistent with the organization's strategies and values." Human resource management includes decisions about a number of activities involving the employees of the organization. The major activities comprising HRM systems can be classified as **staffing** (e.g., employee recruiting, selection, promotion, termination), **reward** (e.g., pay, incentives, performance management), and **development** (e.g., employee training) activities. Human resource management decisions and activities have been linked to a number of **outcomes** for the organization, including employee performance, satisfaction, retention, attendance, and product/service quality.

HRM has recently been recognized as an important source of sustained competitive advantage for a firm, a source of advantage that, unlike (for example) technology, cannot be easily duplicated (Pfeffer, 1995). Effective HRM systems must be "internally congruent" and aligned with business objectives, and sensitive to the changing nature of the firm's labor markets and legal environments. Some of these issues, along with examples, will be discussed in more detail later. It is important to acknowledge that different business settings call for different prescribed HRM systems. The proper mix of HRM practices depends upon how work processes are organized and the unique environmental circumstances surrounding a given firm.

Human resource activities are designed and implemented by both line managers (e.g., operations managers) and by HRM staff specialists, working as partners in a coordinated fashion. HRM specialists located in the Human Resources or Personnel Departments possess detailed technical knowledge to assist line managers in human resource management activities. HRM specialists also contribute to maintaining consistency and fairness in how HRM decisions and systems are applied across various parts of the organization. Thus, to insure effective HRM systems, the typical line manager (e.g., operations manager) needs to be an "informed consumer" of the services provided by HRM staff specialists or by consulting firms.

EVOLUTION OF HUMAN RESOURCE MANAGEMENT: HISTORICAL PERSPECTIVE: The role of human resource management has changed over the years. One current textbook in HRM suggests that a useful way to categorize historical periods would be the 1800s, the period from 1911–1930, from 1930 to 1970, and the modern period (Noe, Hollenbeck, Gerhart, & Wright, 1997). During the 1800s the U.S. economy was dominated by agriculture and small family businesses.

Not surprisingly, HRM practices were conducted by owners and senior employees. New employees learned their jobs through apprenticeship to more senior employees. Hiring relatives and friends was quite typical, and compensation often included food and housing in addition to wages. During this period HRM could be described as casual, informal, and even arbitrary and inconsistent.

The Industrial Revolution stimulated the growth of more formal HRM practices. Large numbers of people were hired and trained for specific skills involved in operating machines performing specialized operations. Managers specializing in human resources were needed to train and schedule these workers. The development of "scientific management" in 1911 led to an emphasis on identifying employees with the appropriate skills for performing each job, administering incentives to employees for their productivity, providing employees with rest breaks, and studying jobs to determine the "best way" for performing each job (Noe *et al.*, 1997). Most organizations were bureaucratic entities, organized into a hierarchy of authority.

These changes brought on by the Industrial Revolution and scientific management promoted the growth of the "Personnel" department in organizations from 1911 to 1930. The major focus of this department was administrative activities, such as keeping track of employee records, administering payroll, processing new hires, and administering terminations.

The period from 1930 to around 1970 saw a recognition of the link between employee participation and involvement in decision making and outcomes such as employee morale, attendance, turnover, and efforts to unionize. A new "employee-oriented" philosophy was taking hold in the more progressive organizations, a philosophy reflected in terms such as Theory Y management and the Human Relations approach to management. The famous Hawthorne Studies conducted at Western Electric in the late 1920s also emphasized that treating employees in a considerate and positive manner is a possible key to productivity.

This period also saw the growth of the use of tests in employee selection. Testing had first been used during World War I, when group intelligence tests were used to select officer candidates from the enlisted ranks in the military. The development of testing continued during World War II. Following the war, private industry eagerly adopted a large variety of paper and pencil tests for employee selection, which continues to this day. These tests measured many different aptitudes, abilities, and interests of job applicants.

A key development during this period was the need for Personnel departments to maintain firms' compliance with legislation regulating employment practices. The most significant piece of legislation was the Civil Right Act of 1964's Title VII, which made it illegal for employers to treat people differently on the basis of attributes such as race and sex. The role of the Personnel department began to expand significantly as a result of the above developments, and Personnel departments began to take on new names, such as Human Resources or Employee Relations. The role of HRM now expanded beyond administrative work, to include such activities as applicant testing, conducting attitude surveys, negotiating labor contracts, and compliance with the expanding employment legislation. As Noe *et al.* (1997) explain it, "The major role of employee relations departments was to ensure that employee-related issues did not interfere with the manufacturing, selling, or development of goods and services. HRM tended to be reactive – this is, it occurred in response to a business problem or need (p. 4)."

Finally, in recent years companies have begun to view HRM systems as linked to the firm's strategy and as a source of competitive advantage. HR managers and specialists in progressive organizations are often considered to be "strategic partners" in management of the firm. They are expected to promote HRM practices that enhance the operations of the other functions of the business and enhance firm profitability.

A STRATEGIC PERSPECTIVE FOR HRM: Current ideas regarding the design of HRM systems are very strategic. These ideas are about aligning HRM practices with a company's business strategy. At a very general level, most strategic perspectives begin with a framework for classifying firm types. For instance, Gubman (1998) builds linkages between three strategic business styles and three distinct kinds of workforce strategies. The business styles are represented by **operations**, **customers**, and **products** companies. Operations companies (and these are heavily represented by manufacturing firms) focus on delivering the best quality, cost/value they can to customers. They tend to focus on continuously improving work processes in an attempt to reduce manufacturing costs and improve product quality. Gubman argues that with *efficiency*, *order*, and *process* as its foundation, operations companies tend to:

- Compete by building teams to deliver high-value, low-cost processes
- Place heavy emphasis on creating motivation and esprit de corps

- Measure process- and results-oriented outcomes and reward what is measured
- Build a continuous improvement mindset and environment

Companies that strategically focus on customer service or product design and innovation require very different workforce strategies. For instance, products companies would need to foster innovation and risk taking among key employees. Staffing, reward, and development practices required to encourage and maintain a high level of innovation will differ from the practices needed to encourage a customer-service orientation or a process-improvement orientation.

The Process of Aligning HRM and Business Strategy

The explicit link between business strategy and HRM system design is shown in Figure 1. This represents a more fine-tuned approach than that of Gubman (1998). This is necessary because it is possible for there to be HRM system variation among firms within one of Gubman's categories. That is, as will be subsequently illustrated, two operations companies may require different approaches to managing their workforces.

As shown in Figure 1, the first step in the alignment process is to carefully consider the business strategy and manufacturing technologies that characterize a particular business. These attributes of the business environment will dictate, in large part, how work needs to be organized and structured. How work processes are ordered, in turn, will lead to a determination of the **key employee behaviors** that will be required to successfully accomplish specified business objectives. Only by focusing on these specific roles and behavioral requirements is it possible to begin to make informed decisions about the design of an appropriate HRM system. As shown in the figure, decisions about the HRM system come directly out of the analysis of the behavioral requirements of the job. Of course, after considering the behavioral requirements of the situation, informed decisions about HRM practices must take the conditions of the labor market and relevant labor/employment laws and regulations into account. Finally, while not explicitly shown in the figure, the costs associated with implementing and maintaining the HRM system must be compared to the likely returns. To be useful, the rate of return associated with implementing and maintaining an HRM system must meet or exceed the rate of return of other investment options. To illustrate this process consider a company like Compaq Computer Corporation. Historically, Compaq pursued an aggressive market-share strategy (known for high-end personal computers) by continuously reducing manufacturing costs and improving product quality. In

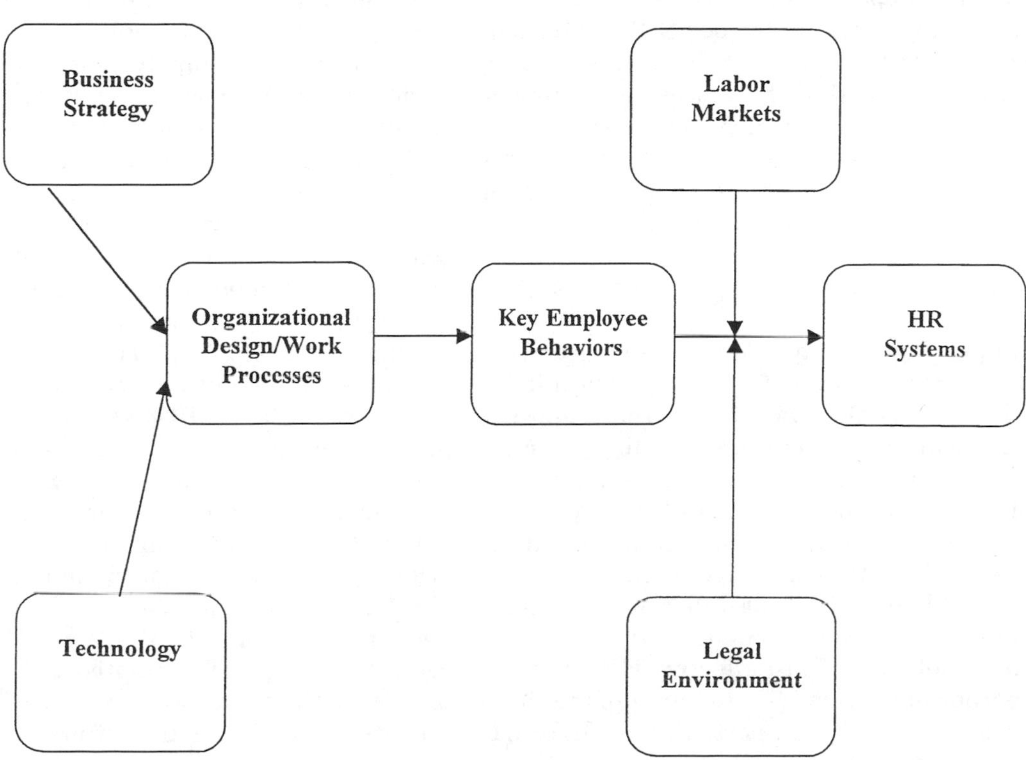

Fig. 1. Human resource system.

the summer of 1966, Compaq modified its strategy and unveiled a plan to segment the PC market (Blumenthal, 1996). This plan called for the building of computers all along the price continuum. Having the capability to build machines across many price levels was deemed critical because Compaq needed to be ready to meet consumer demand wherever it might develop. The available technology for forecasting customer demand had been notoriously poor in this industry – creating pressures to build machines to customer order. By being able to adjust production schedules to take advantage of changing customer preferences, the company also would be able to reduce costs related to inventory control, handling, freight, and the processing of unsold and returned goods. All of this was a precursor to manufacturing in a direct-sales environment in which individual customers place phone or electronic-commerce orders with the manufacturer.

This type of business environment creates the need for a very high level of manufacturing flexibility. To operate in this environment, Compaq developed a sophisticated assembly process relying on assembly cells (Levin, 1994). Workers in three- or four-person assembly cells, unlike traditional assembly line workers, perform multiple tasks. Each team assembles an entire computer (and has the capability of building machines across the entire price/feature continuum). One person prepares the subassemblies, a second team member installs the components into the computer's external case, and a third member of the team conducts tests of the circuits and electronics. It, also, is possible to multiskill the process even further by training each team member to be able to do the tasks of the other team members. This would further enrich the jobs and create a higher degree of flexibility. In addition to the assembly process, it is common to assign teams additional tasks. For example, given the focus on cost cutting and production efficiency, teams can be charged with continuously improving the way PCs are assembled. This would require monitoring for quality, determining the causes for quality variability, and proposing and evaluating changes in the assembly process.

As will be noted in a subsequent section, this type of assembly process requires that a company be able to attract, motivate, and retain individuals who have (or have the capability of acquiring) a set of skills and knowledge that differs from the skill profile required of workers in other types of manufacturing settings. Thus, effective HR practices will appropriately vary as a function of the business strategy. Finally, as portrayed in Figure 1, final decisions about staffing, reward, and development systems will depend upon prevailing labor market conditions and the legal environment surrounding the specified business setting. For example, a shortage of critical skills needed to implement the assembly cell approach could render the approach as inappropriate or require moving the manufacturing facility to a more suitable setting.

HRM SYSTEMS AND THE NEED FOR CONGRUENCE: The last section emphasized the aligning of HRM systems with business strategy. It is also important that these HRM systems prescribed for a particular firm be internally **congruent**: these systems should be integrated and mutually supportive. This need for congruency has often been overlooked in HRM in organizations. In the previous section of this chapter we asserted that the strategic perspective for HRM includes determination of the "key employee behaviors" for the work in a particular job class. Managers must then design an HRM system to promote these employee behaviors – in a congruent fashion.

What is included in the notion of "congruent" HRM practices? We believe that there are three critical components. First, HRM systems should consist of focussing powerful forces on the target behaviors desired of employees. Another way of saying this is to **use multiple levers** available to promote the behavior. For example, assume that "teamwork" is a key employee behavior for a job class in a firm. The firm would probably want to **select** employees with the capacity and motivation for teamwork, **develop** (train) employees to be more effective as team members, and **reward** employees on the basis of team performance. Managers would want to "pull as many levers as possible" toward teamwork. Second, the firm must insure that staffing, reward, and development practices do not *conflict*. For example, managers who wish to promote employee teamwork would want to avoid a set of HRM practices that included (for example): (1) individual performance measurement, along with (2) team training in group interaction skills, along with (3) individual employee bonuses for achieving quality goals. This HRM system would lack congruence because some activities promote teamwork, while some actually work *against* teamwork by focusing on individual employee performance.

Finally, the third component of congruence in HRM systems is to insure that if interrelated, HRM practices are supportive of each other. Managers must understand that some HRM practices require the support of related HRM practices in order to maintain a targeted class of employee behaviors. For example, the ultimate success of certain reward practices may depend on the use of certain interrelated selection practices. Likewise, the success of certain training initiatives may depend on the application of complementary reward programs. More specifically, consider a firm's quality goals in the production process. Managers might implement cash **bonuses** for employees (or teams) who reach their goals. But these efforts may fail if the firm does not also provide quality improvement **training** to equip employees with the necessary skills to earn the bonuses.

In summary, it is becoming increasingly recognized that HRM systems must be aligned with the firm's business strategy – but that managers must also pay careful attention to the internal congruence of these HRM systems.

THE ENVIRONMENT OF HRM: LABOR MARKETS AND LEGAL CONSTRAINTS: Business decisions and human resource management systems are also influenced and constrained by a number of factors in the environment of HRM. In this section we will discuss labor markets and legal constraints (see Figure 1). Both domestic and international **labor markets** play a role here. For example, certain strategic options may be ruled out if sufficient talent (from inside or outside the firm) is not available. And a firm's comparative strengths in terms of the number and quality of employees available in various functional areas (e.g., marketing, operations) can influence a firm in taking new strategic directions.

General Motors provides an illustration of how labor markets can influence business decisions. In 1997 GM initiated a "four-plant international expansion" strategy by opening four new assembly plants in Argentina, Poland, China, and Thailand (Blumenstein, 1997). When building these new plants, their designs were influenced by attributes of local labor markets. A very successful new plant in Eisenach, Germany is 95 percent automated. In the new Rosario, Argentina plant, the company will depend much more on *workers* (using a 45 percent automation ratio) – taking advantage of lower labor costs. This contributes to an estimated $350 million price tag for the facility – one of the lowest ever for a new GM plant (Bluemenstein, 1997).

In the U.S., while the competition mounts for skilled workers, less skilled workers are losing their jobs. In November, 1997, Levi Strauss & Company announced that it would close 11 U.S. plants and lay off 6,400 workers (Associated Press, 1997). The company, along with other clothing makers, has had a history of moving operations offshore to take advantage of lower labor costs.

It is also important to note that basic workforce demographics and other characteristics can dramatically influence business decisions and HRM systems. Major workforce trends that relate to worker age, gender, and ethnic diversity have direct consequences for the way work gets done in a firm. Understanding the nature of these "internal labor market" characteristics is now central to designing sound HRM systems. The degree of change is profound. For example, in the U.S., over 45 percent of all net additions to the labor force between 1992 and 2005 will likely be non-white, especially first-generation immigrants from Asian and Latin countries, and almost two-thirds will be female (Cox, 1994). These demographic changes can lead to conflicts related to value and temperament differences among various groups. As a result, many see the successful management of diversity as an increasingly important dimension of leadership effectiveness (Cox, 1994). Taking full advantage of all available talent in a firm must certainly be linked to higher levels of firm performance.

A related environmental constraint for human resource management is legal constraints, especially **equal employment opportunity** (EEO) laws and government guidelines. In the U.S., for example, all aspects of a firm's human resources activities, such as hiring, promotions, discipline, termination, transfers, training, and compensation are to some extent influenced and constrained by the equal employment environment. These legal constraints do vary across countries, such that global managers must grapple with a complex web of laws, regulations, and levels of enforcement in formulating and implementing human resource strategies.

Managers in the U.S. are constrained in HRM activities by a large number of laws and guidelines. Probably the most significant law is Title VII of the Civil Rights Act of 1964. This forbids discrimination in all aspects of employment on the basis of race, color, religion, sex, and national origin. Over the last thirty years, this law has transformed the workplace by opening employment opportunities previously unavailable to many people, especially women and ethnic minorities. Additional laws protect workers over the age of 40 and those with mental or physical disabilities from employment discrimination. Under Title VII, managers must first avoid **disparate treatment**, meaning overtly treating people differently or restricting employment opportunities because of race, sex, etc. Second, employers create a **disparate impact** on a protected group of people if the firm engages in a (apparently neutral) practice which *differentially affects* the employment of those in a protected group (see Ledvinka & Scarpello, 1991). For example, if a firm uses a "cutoff score" on a test for screening applicants for jobs, and that cutoff score screens out significantly more African Americans compared to whites, this screening process might be determined to have a disparate impact on African Americans. Disparate impact in itself is not illegal. But when disparate impact occurs, the "burden of proof" will shift to employers to defend their employment practices.

Thus, managers must be careful to avoid any overt differential treatment of people based on race, color, religion, sex, or national origin. We might make special note of **sexual harassment** here – unwanted behavior of a sexual nature that interferes with one's work performance (a type of sex discrimination under Title VII). As a proactive approach, many firms have issued explicit policies forbidding sexual harassment, provided training to managers, and set up internal procedures for investigating and dealing with harassment. But managers must also be prepared to defend *any* practice (e.g., employment test, experience/education requirement, etc.) that may have a differential impact on certain groups. Each year a large number of discrimination complaints are lodged with the enforcement agency, the Equal Employment Opportunity Commission. Many of these complaints are settled out of court, but a large number do make it to court (usually a federal district court). Needless to say, these complaints and lawsuits can be very costly to the firm, in terms of time, money, and morale. Thus, it is important for all managers to have a basic understanding of equal employment opportunity issues.

MANUFACTURING LAYOUT DECISIONS AND ALTERNATIVE WORK ROLES: A firm's business strategy ultimately leads to choices about the mix of products and services offered to its customers. This, in turn, leads to decisions about how work is organized. An effective manufacturing layout facilitates the flow of materials, people, and information to meet the systems' stated objectives. To meet various manufacturing objectives, a variety of layouts have been developed. In order to show the possible variety of approaches, consider these components of a framework presented by Render and Heizer (1997, pp. 267–286). These authors offer the following:

- Fixed-position layout: Typically used in the manufacture of large, bulky projects such as ships and buildings.
- Process-oriented layouts: Typically used for either low-volume or high-variety production where highly flexible manufacturing processes are required.
- Product-oriented layout: Typically used when the best personnel and machine utilization is sought for continuous or repetitive manufacturing cycles (e.g., assembly lines for building one type of television or automobiles with limited feature variation).

Each layout is appropriate for a particular business strategy. When Compaq decided to build PCs all along the price and feature continuum, they needed to move from traditional product-oriented manufacturing layouts to process-oriented layouts. A special case of the process-oriented layout is the work cell. The work cell is built around the product and, according to Render and Heizer (1997, p. 275) provides the following advantages:

- Reduced work-in-process inventory
- Less floor space
- Heightened sense of employee participation
- Reduced direct labor costs
- Reduced investment in machinery and equipment
- Reduced raw material and finished goods inventories

Of course, the personnel requirements of cellular production will be vastly different from the requirements associated with something like traditional assembly-line production. The skills, interests, and work-orientation of highly effective assembly cell members will all converge on being able to quickly learn new procedures, adopt and be flexible, and enjoy the problem solving associated with improving a work process (i.e., modifying the form of the assembly cell). The ability to hone a particular (but narrow) assembly skill and then being able to stay motivated and focused when repeating the same short-cycle operation throughout the work shift (as in assembly-line work) requires a different set of employee attributes. It is for these differing workforce requirements that we will now consider some alternative HR practices.

HR PRACTICE/MANUFACTURING SYSTEM CONGRUENCE: One very important place to begin when considering HR practice alternatives is to focus on task dependence versus independence. Do workers operate as members of teams (as they do when organized as assembly cells) or do they work independently of others? Differences at this simple starting point often lead to vastly different prescriptions regarding appropriate HR practices. For example, effective assembly-cell or semi-autonomous work team manufacturing will typically require staffing, reward, and development practices that look like this:

- The use of above market entry-level base pay to attract individuals who not only possess the physical stamina and agility to assembly products over long hours but who also possess the basic reasoning skills, interpersonal communication skills, and team-oriented interests needed in this environment.
- The use of team-oriented contingent pay (e.g., gainsharing) where the pay-off rule is at the team level. This is necessary because there often is a high level of task interdependence among team members (making it difficult to measure

individual-level performance) and the reward system needs to encourage cooperation, a willingness to train newcomers, and other behaviors required to improve team efficiency.

- The use of a flexible and well funded benefit plan that works to provide employees with the benefits they most value. Receiving non-wage benefits are typically contingent upon staying with a current employer (i.e., not seeking employment elsewhere). It is for this reason that they can be so important to companies competing in a tight labor market for employees possessing unique skills. The costs of turnover for these types of companies can be quite high.
- The use of both results- and process-oriented performance measurement and monitoring systems. If teams are charged with not only producing goods but also with reducing manufacturing costs and modifying procedures to improve product quality, they need timely information about how they are achieving their results (e.g., information about quality variation for product components, information about subassembly cycle times, measures related to the abuse of equipment).
- The extensive use of both formal "classroom" training and innovative new training methods that can provide "just-in-time" instructional material to operational assembly cells.
- The use of exit practices that create stability and loyalty by creating a reasonably high level of job security. This is because replacing workers can be so costly and because workers need to be encouraged to take appropriate risks in the design of improved work processes. Few workers will strive to improve a work process if their job security is threatened.

When work is organized around repetitive, narrow, and independent tasks (the flow of work moves the product from one work station to the next), HR system design must be adjusted. Using the dimensions previously discussed for assembly-cell manufacturing, appropriate HR practices likely will take the following form.

- Pay above market entry level wages may not be required because the skill profile needed in this environment is less likely to be in short supply.
- The use of performance-contingent pay may or may not become an issue. If the pace of work is controlled by the speed of the assembly line, workers not able to maintain the required pace become so disruptive that they may need to be replaced. Receiving any pay level will be contingent upon meeting production standards. If workers control output levels (i.e., their effort and perseverance directly relates to the enhanced production of goods and services) there may be a case for the use of performance-contingent rewards – but due to task independence, the reward system would need to be at the level of the individual. Here something like a traditional piece-rate approach may be appropriate. Under a piece-rate system, workers typically receive a minimum hourly base rate and then are paid based on the number of pieces made.
- The need for a well funded and flexible benefit plan will likely depend on the nature of the relevant labor market. If job candidates are heterogeneous with respect to interests and values and there is a shortage of talent, flexible and well funded benefit plans may be the best way to attract and retain needed employees.
- Performance management systems will primarily address measurable outcomes. Employees working in this type of environment are typically not charged with improving the production process. Monitoring employee performance is more likely to be used to allocate individual bonuses (to stimulate high and continuous levels of effort) or to encourage attendance and adherence to work schedules.
- Training will typically be informal and of the on-the-job variety. If the needed skill set is more complex and multidimensional, more extensive and formal classroom training may be reqquired.
- If the needed skills are in abundance in the labor market, there will be a tendency to aggressively retain only employees capable of meeting production standards. A piece-rate system often leads to the natural removal of below standard workers because these individuals are not able to command wages that meet their minimum personal needs. This, of course, all changes if employees are represented by unions that bargain for stringent security provisions.

The point of these brief illustrations is to argue that the design and implementation of high-quality HR systems is contingent upon the unique business circumstances confronting a company. The HR practices needed to promote a high level of creative and innovative thinking will be vastly different from the practices need to foster great customer service or the practices needed to energize production workers controlled by the pace of an assembly line. Sound HR practices will also be very sensitive to the values and interests represented among job holders. The HR systems needed to manage television manufacturing plants in midwestern states of the U.S. will likely not be as effective when used in other countries and cultures. Referring back to Figure 1, sound HR system design must take into account business strategy, relevant technology, labor market attributes, and prevailing employment and labor laws. HR practices can vary on many dimensions-making informed choices among the many alternatives requires a careful analysis of the relevant setting.

CONCLUSION: Human resource management involves decisions and activities in the areas of staffing, developing, and rewarding employees. HRM has recently been recognized as a potentially important source of sustained competitive advantage, not easily duplicated by other firms. HRM has evolved from an emphasis on record-keeping and maintaining good employee relations to the current emphasis on linkages to business strategy and firm performance. Not surprisingly, firms that focus on different business styles, such as operations, customers, or products typically need to implement different workforce strategies.

HRM systems should be carefully aligned with business strategy. We argue that this alignment takes into consideration the firm's business strategy and manufacturing technologies, how work is organized and structured, and the resulting key employee behaviors required to accomplished the firm's business objectives. Only after the behavioral requirements (for employees in specific job classes) have been specified can appropriate HRM systems be designed. Decisions about HRM systems should also take into consideration the conditions of the labor market and relevant employment laws and regulations. The costs and likely returns of implementing particular HRM activities should also be assessed.

These HRM systems should be designed to be internally congruent. In maintaining congruence, managers focus "multiple levers" toward the target employee behaviors, choosing from among many options for staffing, reward, and development activities. Managers should also insure that HRM activities do not conflict with each other, and that if interrelated, HRM practices support each other.

Manufacturing managers use a variety of manufacturing layouts to facilitate the flow of materials, people, and information to meet their objectives. For example, firms may use a fixed-position layout, process-oriented layout, or product-oriented layout, to fulfill particular business strategies. The workforce requirements may vary considerably for these differing approaches, and call for alternative HRM practices. We concluded this chapter by providing an example of the likely nature of staffing, reward, and development practices when using assembly-cell or semi-autonomous work team manufacturing. We then illustrated how the HRM system might be different when work is organized around repetitive, narrow, and independent tasks.

Organization mentioned: Compaq; General Motors; Levi Strauss and Co.

See **History of manufacturing management; Manufacturing cell design; Manufacturing flexibility; Manufacturing system design teams: Design and implementation.**

References

[1] Associated Press (1997). "Levi Strauss plans closings, layoffs." *Bloomington Herald-Times*, November 4, B1.

[2] Bluemenstein, R. (1997). "Global strategy: GM is building plants in developing nations to woo new markets." *Wall Street Journal*, August 4, A1, A4.

[3] Blumenthal, K. (1996). "Compaq is segmenting the home-computer market." *Wall Street Journal*, July 16, A3.

[4] Cox, T. (1994). *Cultural Diversity in Organizations: Theory, Research and Practice*. Berret-Koehler, San Francisco.

[5] Gubman, E.L. (1998). *The Talent Solution: Aligning Strategy and People to Achieve Extraordinary Results*. New York: McGraw-Hill.

[6] Ledvinka, J., & Scarpello, V.G. (1991). *Federal Regulation of Personnel and Human Resource Management (2nd ed.)*. PWS-Kent, Boston.

[7] Levin, D.P. (1994). "Compaq storms the PC heights from its factory floor." *New York Times*, November 13, C5.

[8] Noe, R.A., Hollenbeck, J.R., Gerhart, B., & Wright, P.M. (1997). *Human Resource Management: Gaining a Competitive Advantage (2nd ed.)*. Irwin, Chicago.

[9] Pfeffer, J. (1995). "Producing sustainable competitive advantage through the effective management of people." *Academy of Management Executive*, 9, 55–69.

[10] Render, B., & Heizer, J. (1997). *Principles of Operations Management*. Prentice-Hall, Upper Saddle River, New Jersey.

HURDLE RATE: Hurdle rate is the minimum acceptable rate of return for a proposed project to merit investment. Usually, the hurdle rate is the opportunity cost of capital or the best return the firm could get by investing the capital elsewhere. To be financially justifiable, a project's internal rate of return must exceed the hurdle rate set by the firm for such projects. Conversely, a project's cash flows must yield a positive net present value when discounted at the hurdle rate to be financially justifiable.

See **Capital investments in advances manufacturing technologies**

I

IDEAL SEED NON-HIERARCHICAL CLUSTERING

Chandrasekharan and Rajagopalan (1986) developed a non-hierarchical algorithm for grouping parts into families and machines into cells. This algorithm consists of three stages: 1) Non-hierarchical clustering of columns and rows is carried out with the help of graph theory and clustering analysis. 2) Diagonalization is performed because the output form the first stage cannot be used directly. The result is useful only if the part families are assigned to their respective machine cells. The objectives of the assignment are to maximize utilization of machines and minimize inter-cell movement. A group efficiency factor is defined for the assignments. 3) The final stage further improves the output from the second stage in terms of both utilization and inter-cell movement, using ideal seeds which improve defects caused by the initial disposition of the matrix.

Later, Chandrasekharan and Rajagopalan (1987) developed an improved approach called ZODIAC (Zero-One Data Idea Seed Algorithm of Clustering). In this, the choice of seeds in the first stage was not restricted to the arbitrary choice of classical clustering analysis but different seeding methods were introduced and tried for different data sets. There included arbitrary, artificial, representative and natural seeds. The ideal seed was used only in the last phase of ZODIAC. The algorithm establishes that there exist a limit for grouping efficiency, depending on data structure and cluster membership. This led to the concept of relative efficiency to act as a stopping criterion. Subsequently, Chandrasckharan and Rajagopalan (1989) proposed a groupability concept to determine whether the 0-1 incidence matrix can be diagonalized. To measure the groupability, they used Jaccard similarity as a characteristic and examined the Jaccard similarity distribution and its relation to groupability using a number of data sets.

See **Dynamic routing in flexible manufacturing systems.**

IDLE TIME

Idle time refers to the period when a resource is not in use. It is sometimes expressed as a percent of total time. Idle time is the inverse of resource utilization.

See **Assembly line design; Manufacturing cell design; Manufacturing systems modeling using petri nets.**

IMPLOSION

Implosion uses the bill of materials to determine the where-used relationship for a given component. Single-level implosion show only the parents of the component on the next higher level, multilevel implosion show the ultimate top-level parent.

See **Bill of materials; Explosion; Where-used data.**

INCREMENTAL INNOVATION

Incremental innovation is small but distinct progress in either a product or a service. It signifies a significant step forward form the previous state without being a major breakthrough. For example, a new feature added to an established product that enhances it but does not change its basic utility would be an incremental innovation. A new and more effective tread pattern for an automobile tire would be as an incremental innovation.

See **Process innovation; Product innovation.**

INCREMENTAL INVENTORY COSTS

Only certain costs are used in computing inventory costs and in making inventory management decision. Inventory costs are those that represent an actual out-of-pocket expenditure or a forgone profit. For example, item cost used in calculating inventory carrying cost includes item's variable material, labor, and overhead costs.

See **Inventory flow analysis.**

INDEPENDENT DEMAND

It refers to the demand for a product that is independent of demand for other items. The demand for finished goods or assemblies are examples of independent demand. In contrast the demand for components such as a automobile steering wheel is dependent on the demand for assembled cars. The demand for steering wheels is called dependent demand.

See **Dependent demand; Material requirements planning.**

INDIRECT LABOR COSTS

Indirect labor costs are not directly associated with the production of a item. They are indirectly required to ensure that the product or order is completed. For example, consider a material handler, who supports two separate machines producing multiple items. It is impractical, however, to assign the material handler's labor to a specific product. In this case, the cost accounting system usually allocates the material handling expense based on some allocation formula. Other allocated costs include the energy required to operate the machine, managerial and administrative expenses, and shipping containers.

Some firms account for indirect or non-traceable costs by simply assigning a single expense, often called a burden expense, to the item, when calculating the total cost of production. The burden expense is typically some percentage of the direct costs to produce an item. In a recently developed accounting system called "Activity-based Costing," indirect labor costs are handled differently.

See **Activity-based costing; Supplier performance measurement; Unit-based cost; Unit cost.**

INDUSTRIAL DESIGN

Industrial design is the process of providing "form" and "appeal" during the design and development of a product. Industrial design process adds appearance or styling to a product. It may also influences the functionality of a product, such as reducing the footprint of a computer, or by making the product more user friendly. Industrial design takes a holistic new of the product and user, and therefore can enhance the esthetics as well as the functionality of the product.

See **Product design.**

References

[1] Bridger, R.S. (1995). *Introduction to Ergonomics*, McGraw-Hill, NY.

[2] Lorenz, C. (1986). *The Design Dimension: The New Competitive Weapon for Business*, Basil Blackwell, NY.

[3] Ulrich, K.T., and S.D. Eppinger (1995). *Product Design and Development*, McGraw-Hill, NY.

INFINITE LOADING

In developing production schedules by computer, there are two possible approaches to dealing with production resource limitations. One approach, known as infinite loading, ignores any such limitations when developing the schedule. The second approach, finite loading, takes into account the actual amount of each resource that will be consumed. Although, the second approach would appear more desirable, it requires excessive computer processing. Thus, infinite loading the most prevalent. Although this method of scheduling might appear unrealistic, it can often work reasonably well in many circumstances. This is because, as overloaded resources are identified, the overloading can often be dealt with through manual methods, such as offloading of work to equivalent resources. With increased inexpensive computing power, finite loading practices are on the rise.

See **Medium- and short-range capacity planning; Finite loading.**

INFORMATION TECHNOLOGY

The technologies that constitutes computers, telecommunications, and other devices that permit business to integrate data, equipment, personnel, and problem-solving methods are called information technologies. Information technology is useful for collecting, storing, encoding, processing, analyzing, transmitting, receiving and printing data, text, audio, or video information. These activities are essential for the success of businesses in the technology intensive world.

See **Electronic data interchange; Safety stocks: Luxury or necessity.**

INITIALIZATION BIAS IN SIMULATION

One of the considerations in the simulation analysis of a system is to decide whether the simulation run would have a finite horizon (terminate at the end of a specified period) or reach a steady state before termination. While steady-state simulations are useful for long-term problems such as capacity planning, most tactical simulations tend to be for finite horizon. For instance, a printed circuit board production line never attains steady-state conditions as new models are continuously introduced into the line. In fact, most manufacturing systems producing a wide variety of products, each in moderate volumes, will exhibit similar behavior.

The ideal scenario is to start a simulation in steady state; however, this is rarely feasible. Hence, initial conditions are set arbitrarily as a matter of convenience, contaminating the data with the impact of the starting conditions. This is typically referred to as *initialization bias.*

For example, consider the simulation analysis of a manufacturing cell with the objective of estimating the long-run average cycle time for parts processed at this cell. This is a steady-state simulation experiment. For convenience, suppose that the simulation run is initiated with the system empty and with all workcenters idle. As a consequence, the first few parts going through the cell will experience unusually low cycle times, causing the estimate for the average cycle time to be biased low. In fact, until the cell reaches its long-run congestion level, the cycle time experienced by the incoming parts will be biased low. Our estimate will therefore be tainted with initialization bias.

See **Simulation analysis of manufacturing and logistics systems; Simulation of production problems using spreadsheet programs.**

INNOVATION

Innovation usually refers to the process of taking a new or creative idea and producing from it a commercially viable product, a service, or an useful organizational process or activity. An innovation may not be entirely new but if it is new in its application to a company or market as a result of an adaptation, then it can be considered an innovation.

See **Process innovation; Product innovation.**

INPUT/OUTPUT ANALYSIS

Companies working in make-to-order environments usually operate with order backlogs. Likewise, individual machines or work centers within those companies generally have a backlog, or a queue of jobs waiting to be processed. This queue is actually beneficial because it can often provide greater flexibility in scheduling and prevent machines and/or operators for idling. However, if the queue becomes too long, jobs may be delayed and work-in-process inventory could increase.

Input/output analysis represents a method for maintaining queue lengths that are neither too short nor too long. Input and output, usually in standard hours, are planned for each work center or machine, usually on a week by week basis. By adding the planned input for a given week to the planned backlog from the previous week, then subtracting planned output, an estimate of ending backlog for each week can be determined. This weekly backlog number can be varied by changing planned input, planned output, or both.

Of course, the actual input and actual output will often be different than planned. Actual input and output may be used to determine actual backlog at the end of the work.

See **Capacity planning: Medium- and short-range; Order release.**

INPUT/OUTPUT CONTROL

Input/output control is another name for order release, which controls a manufacturing system's input. It is also known as job release, order review/release, or just input control. Input/output control also refers to a specific order release mechanism (Wight, 1970; Fry and Smith, 1987) that keeps the input to a shop equal to the output during a given time period. After measuring the system's actual output over some time period, the order release mechanism releases this number of orders each period. If the system output increases, the order release mechanism will release more. If the system output decreases, the order release mechanism will release less.

Fry and Smith (1987) describe input/output control at a tool manufacturer and present a six-step procedure for implementing input/output control. In the first step, management must begin using system-wide performance measures such as throughput. The second step is to identify the bottlenecks that limit system throughput. The third step is to set maximum inventory levels for each station. When inventory reaches this level the previous operation must stop production. The fourth step is to reduce production lot sizes. The fifth step is to prioritize orders correctly so that no station hinders system performance. The sixth step is to match the input rate to the output rate. The authors report that, after implementing input/output control, manufacturer reduced work-in-process inventory (WIP) by 42%, shrank the order backlog by 93%, reduced customer lead times by over 50%, and increased customer service to above 90%.

See **Capacity planning: Medium- and short-range; Order release.**

References

[1] Fry, Timothy D., and Allen E. Smith, "A procedure for implementing input/output control: a case study," *Production and Inventory Management Journal*, Volume 28, Number 3, pages 50–52, 1987.

[2] Wight, O.W., "Input/output control: a real handle on lead times," *Production and Inventory Management*, Volume 11, pages 9–30, 1970.

INSOURCING

It is the decision to acquire a task or service internally from within the company. This is the opposite of outsourcing which refers to the acquisition of goods and services from sources outside of the company.

See **Supply chain management; The future of purchasing.**

INTEGRATED LOGISTICS

Strategic integrated logistics are intended to support the firm's competitive strategy and to deliver high levels of customer satisfaction. To do this successfully, the many diverse logistics activities such as inventory management, transportation, order processing, demand forecasting, documentation, packaging, parts support, warehouse management, and reverse distribution must be managed in a cohesive and systematic manner. Ad hoc management of these diverse activities will increase the effort expended to achieve mediocre results.

Integrated logistics views all the different logistics activities as interrelated components of a single value-added system that is comprised of facilities, equipment, people, and operating policies. It covers the flow of goods and related information from the acquisition of raw materials through production and distribution to the customer. It is important to note that all the related movement and storage activities may not be under the control of a single manager, or even a single management group. Indeed, different firms organize their logistics activities differently to meet situation specific needs. For example, customer demands and expectations, the magnitude of logistics costs incurred by the firm, and the firm's historical approach to logistics all influence the firm's view of logistics as a value-added function and thus impact the way the firm organizes its logistics activities. However, regardless of the organization, the integrated logistics concept provides a unified vision of what logistics must accomplish to help the firm succeed and provide the impetus for coordinating decisions across decision areas. (See: Logistics: Meeting Customer's Real Needs)

See **Logistics: Meeting customers real needs.**

References

[1] Ballou, R.H. (1992). *Business Logistics Management* (3rd ed.). Prentice Hall, Englewood Cliffs, New Jersey.

[2] Lambert, D.M. and Stock, J.R. (1993). *Strategic Logistics Management* (3rd ed.). Richard D. Irwin, Inc., Homewodd, IL.

INTEGRATED MANUFACTURER

Many industries, such as the automobile, household appliance and computer equipment industries, involve the manufacture of products with multiple levels of assembly, ranging from sheet metal or silicon wafer production to complex to sub-assemblies and product assemblies. In such industries, end-product assemblers have a range of choices as to the number of levels of subassembly and component production to own. When a company chooses to be involved in most levels of product or system assembly, it is called an integrated manufacturer. In contrast, companies that choose to concentrate on one or few levels are known as focused producers.

Integrated manufacturers tend to make many of the components and assemblies that more focused producers of the same products choose to buy from outside suppliers. For example, till the 1980's, leading American automobile producers tended to be much more highly integrated than their main Japanese rivals. U.S. manufacturers manufactured 50% of their components in-house as opposed to 25% made by their Pacific Rim competitors. More recently, the trend throughout the industry in the U.S. has been to move towards the more focused approach.

This trend has been mirrored in many other industries, most notably in computing, where Sun Microsystems in an example of a focused manufacturer, and IBM and Digital Equipment Corporation are integrated manufacturers. As modern automobile and computer equipment products change frequently and are more complex, the technologies needed at the various levels of assembly have become more complex, so that it has become increasingly difficult for any one producer to excel in integrated manufacturing. However, we now know that many of the benefits associated with being an integrated manufacturer can be secured by focused producers through strategic supplier partnerships. The partners, while remaining individually focused, can achieve high degrees of collective integration.

Organizations Mentioned: Compaq; IBM; Sun Microsystems.

See **Strategic supplier partnerships.**

INTEGRATED PLANTS

In a multiplant company, integrated plants have a lot of interaction (both in terms of information exchange and supply of goods) with the other plants in the network.

See **International manufacturing.**

INTEGRATED PREVENTIVE MAINTENANCE

Under integrated preventive maintenance, multifunctional operators are trained to perform routine machine tool maintenance. Adding lubricants (oiling the machine), checking for wear and tear, replacing damaged nuts and bolts, routinely changing and tightening belts and fasteners, and listening for telltale whines and noises that signify impending failures can do wonders for machine tool reliability. For this to occur, the maintenance department must teach the operators on how to carryout these functions and help operators use a routine check list for machine maintenance. The machine operators must also be responsible for keeping their areas clean and neat. Thus, maintenance and housekeeping gets integrated into the manufacturing system.

Naturally, machines still need attention from the experts in the maintenance department, just as the airplane is taken out of service periodically for major engine overhaul and maintenance. One alternative is to use two eight-hour shifts separated by two four-hour time blocks for machine maintenance, tooling changes, restocking, long setups, overtime, earlytime, and so on. This is called the 8-4-8-4 scheme.

The main advantage that equipment has over people is low variability, but to be effective, machines must be reliable and dependable. Smaller machines are simpler and easier to maintain and therefore are more reliable. Small machines in multiple copies add to the flexibility of the system as well. The linked-cell manufacturing system permits certain machines in the cells to be slowed down to run longer without breakdown. Consequently, many observers of the JIT manufacturing system get the feeling that production machines are "babied." In reality, they are run at the pace needed to just meet the demand without overproducing.

See **8-4-8-4 Scheme; Just-in-time manufacturing; Lean manufacturing implementation; Linked-cell manufacturing system (L-CMS); Manufacturing cell design; Manufacturing systems; Multi-functional employees; Total productive maintenance.**

Reference

[1] Nakajima, Seiichi (1988). *TPM, Introduction to TPM: Total Productive Maintenance*. Productivity Press.

INTEGRATED PRODUCT DEVELOPMENT

See **Concurrent engineering.**

INTEGRATED QUALITY CONTROL

When management and production workers trust each other, it is possible to implement an integrated quality control program. Since the Japanese treat their co-workers as extended family this mutual trust is natural. Japan was started on the road to superior quality with the visits of W. Edward Deming, an American, to Japan in the late 1940's and early 1950's. The Japanese were desperate to learn about quality, and therefore statistical quality control techniques were readily accepted by them when introduced by Deming. They wrongly believed that everyone in America used statistical process control techniques. Thus, the Japanese readily accepted quality control concepts, and then they did something that even the few American companies using SQC did not do. Japanese taught the techniques and concepts to everyone, including the top management and the production workers.

At Toyota, under the leadership of Taiichi Ohno, a new idea took hold, quite different in concept from western product inspection philosophy. First, every worker was an inspector responsible for quality. The workers on the assembly line could stop the line if they found something wrong. Pull cords were installed on the assembly lines to stop the lines if anything went wrong. Workers were obligated to stop a line, (1) if workers found defective parts; (2) if they could not keep up with production; (3) if production line was going too fast; or (4) if a safety hazard was found. The problem causing the line stoppage received immediate attention. Meanwhile, while the line was stopped, the other workers maintained their equipment, changed tools, swept the floor, or practiced setups; but the line would not move until the problem was solved. The root cause of the problem was always sought after and permanently resolved.

When every worker is responsible for quality, he/she is trained to use the seven basic tools of quality control (see, *Lean Manufacturing Implementation*). With workers assuming responsibility for quality, the number of inspectors on the plant floor is reduced, and products that fail to conform to specifications are immediately uncovered. The seven tools of quality control are: Pareto Analysis or ABC Analysis; flow diagram, histogram; scatter diagram; run chart; control chart and fishbone diagram. See *Lean Manufacturing Implementation* for more on these tools.

For manual assembly line work, an *andon* system is used to prevent defective work. An *andon* in the Japanese language is actually an electric light board which hangs high above the conveyor assembly lines so that everyone can see it. When a worker on the line needs help, he can turn on a yellow light. Nearby (multifunctional) workers who have finished their jobs within the allotted cycle time move to assist workers having problems (This is called mutual assistance).

If the problem cannot be solved within the cycle time, a red light turns on within 10 seconds. When the problem is solved, the next cycle begins and a green light comes on and all processes on the line start back up together. The name for this practice in the Japanese language is *Yo-I-don*, which literally means "ready, set, go." All the stages of the assembly line are synchronized to start and stop at the same time. Such systems are built on teamwork and a cooperative spirit among the workers fostered by a management philosophy based on harmony and trust. Line shutdowns in lean factories are encouraged to protect quality. Management exercises confidence in the individual worker to allow him/her to shut down a line.

Contrast this with the way things often operate in a job shop. In a job shop, the length of time it takes to find a problem, to convince somebody that it is a problem, to get the problem solved, and to get the fix implemented may be too long. Many defective parts may be produced in the meantime.

Another useful technique in integrated quality control is quality circles. The Japanese call them Small Group Improvement Activities (SGIA) and they are a key part of company-wide quality control efforts in Japan (it is also part of Total Quality Management, TQM). A quality circle is a group of employees who meet on a scheduled basis (daily, weekly) to discuss production and quality problems to try and devise a solution to the problems, or to propose solutions to their management. The group may be led by a supervisor or by a production worker. It usually includes people from a given discipline, or a given production area. They are basically a technique to identify causes of problems and to generate ideas and suggestions to solve problems on the shop floor level. The seven basic tools of quality control (see, *Lean Manufacturing Implementation*) can be used beneficially by quality circles.

See **ABC analysis; *Andon*; Control chart; Just-in-time manufacturing; Lean manufacturing implementation; Quality assurance; Quality circles; Statistical process control using charts; Total quality control; Total quality management (TQM); *Yo-I-Don*.**

INTEGRATED RESOURCE REQUIREMENT PLANNING (IRRP)

Resource Requirements Planning (RRP), or Rouch-Cut Capacity Planning (RCP) checks the resource availability at critical work centers based on prospective master production schedules (MPS). Various techniques, such as the overall factors method, the capacity bills method, and the resource profiles method, can be used for RRP. Traditionally, these methods are used in a "trial and error" fashion.

A resource requirements plan is developed *after* the MPS is developed. If the resource requirements plan implied by the MPS turns out to be infeasible (i.e., exceeding the available resources), then the MPS is revised. These revisions would continue until either a schedule satisfying all resource limitations is found or it is determined that such a feasible schedule does not exist. In that case, the aggregate production plan may be revised to either adjust the production requirements or increase the authorized resources. Since the RRP is done in a trial-and-error manner, it may take several iterations before a feasible MPS can be finalized.

Integrated Resource Requirements Planning (IRRP) offers an alternative approach to the traditional methods of RRP. IRRP integrates the function of developing a resource requirements plan with the process of aggregate planning and master production scheduling. The estimate of capacity requirements obtained this way will be more accurate than those obtained from traditional methods.

See **Aggregate plans and master production schedule linkage; Capacity bills method; Capacity planning: Medium- and short-range; Overall factors method; Resource profiles method; Resource requirement planning.**

References

[1] Berry, W.L., T.G. Schmitt, and T.E. Vollmann (1982). "Capacity planning Techniques for Manufacturing Control Systems: Information Requirements and Operational Features," *Journal of Operations Management*, 3(1), pp. 13–25.

[2] Chung, C.H. and Krajewski, L.J. (1986), "Integrating Resource Requirements Planning with Master Production Scheduling," *Proceedings of The 18th Annual Meeting of Decision Science Institute*, Hawaii, pp. 128–130.

INTEGRATED SUPPLY CHAIN

An integrated supply chain is a set of firms – including multiple tiers of materials suppliers, service providers, the reference firm, and multiple tiers of customers – that work together as a cohesive team to provide a high level of value to the end user. The terminology that is often used to describe a supply chain is "from the supplier's supplier to the customer's customer." The word "integrated" simply highlights the need to manage processes across firm boundaries. In effect, an integrated supply chain has consistent and aligned objectives, well-defined roles, effective and frequent communication among all members, consistent and proactive performance measurement, appropriate alliance management techniques, and a cross-functional mentality that promotes flexibility. The combination of these characteristics creates a "team of companies" that can effectively compete against other supply chain teams to meet the real needs of final customers.

See **Supply chain management: Competing through integration.**

Reference

[1] Bowersox, D.J. and Closs, D.J. (1996). *Logistical Management, the Integrated Supply Chain Process.* McGraw Hill, New York.

INTERCHANGEABILITY

Interchangeability is a characteristic of components and subassemblies that go into modern manufactured products. The basic principle is that every unit made should be exactly like every other unit produced. All units made must conform to design specifications within a narrowly identified allowable range of variation. Typically, these specifications pertain to every characteristic of the component that is relevant to the manufacturing process and/or the customer's requirement.

There are two objectives that make interchangeability desirable. First, the of quality of the finished product is function of a manufacturing processes that is consistent. Historically, consistency in meeting customer requirements was impossible without rigorous specifications guiding the manufacturing processes.

Secondly, the interchangeability principle benefits maintenance. Interchangeability is vital for effective maintenance, it allows worn or faulty components to be readily replaced without time-consuming and expensive hand crafting or customized fitting of replacement parts. This point was a major motivation for the U.S. government when it first contracted with Eli Whitney to produce military muskets using interchangeable parts. More significantly, the use of interchangeable parts allows assembly processes to function more efficiently and rapidly. It allows parts to be assembled easily with no need for modifications to fit them togetherAs parts deviate from their specifications they become less interchangeable. The quality of the resulting assembly may suffer, and the effort required during assembly increases significantly. Interchangeability is an essential prerequisite for high volume mass production systems such as assembly lines.

See **History of manufacturing management; Scientific management.**

References

[1] Green, Constance McL. 1956. *Eli Whitney and the Birth of American Technology*. Little, Brown & Co., Boston, Mass.
[2] Hounshell, David A. 1988. *The Same Old Principles in the New Manufacturing. Harvard Business Review*, Vol. 66, No. 6, 54–61.
[3] Hounshell, David. 1983. *From the American System to Mass Production*. University of Delaware Press, Wilmington, Delaware.
[4] Howard, Robert A. 1978. *Interchangeable Parts Reexamined: The Private Sector of the American Arms Industry on the Eve of the Civil War. Technology and Culture*, Vol. 19, No. 4, 633–649.
[5] Lord, Chester B. 1921. *Fundamentals of Interchangeable Manufacture. ASME Transactions*, Vol. 43, 421–427.
[6] Woodbury, Robert S. 1960. *The Legend of Eli Whitney and Interchangeable Parts. Technology and Culture*, Vol. 2, Summer, 235–253.

INTERFERENCE BETWEEN OPERATORS

See **U-Shaped assembly lines.**

INTERMEDIATE JIG

See **Setup reduction; SMED.**

INTERMODAL EDI

See **Electronic data interchange in supply chain management.**

INTERNAL CONSTRAINTS

Constraints in a manufacturing organization can be classified as either internal constraints (internal to the organization) or external constraints (outside of the organization). Examples of internal constraints include process and policy constraints. Process constraints occur when a given process or operation in the company has insufficient capacity to fully satisfy market demand. Policy constraints occur when management or employee unions enforce a rule that limits an organization's operation abilities or restricts its flexibility (e.g., a freeze on overtime or hiring, or a restriction on purchasing direct materials).

External constraints include material constraints and market constraints. Material constraints occur when an outside source of material becomes restricted. Market constraints occur when there is inadequate market demand to fully utilize a company's production capacity.

See **Theory of constraints in manufacturing management.**

References

[1] Atwater B. and M.L. Gagne. (1997). "The theory of constraints versus contribution margin analysis for product mix decisions," *Journal of Cost Management*, January/February, 6–15.
[2] Fawcett, S.E. and J.N. Pearson. (1991). "Understanding and applying constraint management in today's manufacturing environment," *Production and Inventory Management Journal* Third Quarter, 46–55.

INTERNAL CUSTOMER

In Total Quality Management (TQM), everybody in the organization, i.e. the total organization, takes on the responsibility for delivering a quality product or service. The organization is perceived as a large quality chain linking customers and suppliers together by mutual interactions (Oakland, 1993, p. 8). This applies not just to customers and suppliers external to the organization but also to individuals and departments within the organization. The people and departments within the organization are linked to others by transformation processes. Oakland (1993, p. 14) defines a process that transforms a set of inputs to include actions, methods, and operations. The input is transformed into outputs that satisfy customer needs and expectations, in the form of products, information, services or – generally – results.

For each process at least one customer and one supplier can be identified. For example, a secretary can be considered an "internal" supplier to his/her principal, who is the "internal" customer. Every process must be capable of doing the job correctly and must continue to do the job correctly!

See **Supplier partnerships as strategy; Supplier relationships; Supply chain management: Competing through integration.**

Reference

[1] Oakland, J.S. (1993). *Total Quality Management: The Route to Improving Performance*. Butterworth-Heinemann, Oxford.

INTERNAL PERFORMANCE MEASUREMENT

Internal performance measures are used to assess and monitor the internal operation of an organization. These are essentially measures of processes. For a manufacturing organization, they could include measures such as equipment utilization, set up times, work in progress levels, queues, etc. As these measures are often directly related to specific activities they can be used for identifying opportunities for improvement. Internal performance measurers do not always capture the competitive position of a company. For example, set up time does not by itself is not an indicator of the competitive position of a company, but reducing set up time may enable a company to offer faster deliveries which may improve its competitive position. However, reduced setup times may result in an improvement in the external performance measure, 'delivery lead time'.

See **A process approach to manufacturing strategy development.**

INTERNAL RATE OF RETURN (IRR)

IRR refers to the discount rate applied to the future cash flows from an investment such that the present value of the investment equal the cost of the investment. Alternatively, it is the discount rate that gives an investment project a zero net present value (NPV). IRR represents an investment project's rate of return. If the opportunity cost of capital for financing an investment project, i.e., the discount rate, is less than the IRR, then the investment project will have a positive NPV. Projects with larger NPVs are financial more attractive.

Companies establish a minimum IRR for capital investments. Projects with IRR less than the acceptable rate are often rejected or reevaluated.

See **Capital investments in advances manufacturing technologies; Net present value.**

INTERNAL SETUP OPERATION

Internal setup operation refers to a setup activity that can be performed *only* while the machine or workcenter is turned off and production is stopped. Identifying, differentiating and separating internal from external setup operations is a critical step in reducing setup costs and times. External setup activities are performed while the machine is running. To reduce setup time and costs, either modify internal setup operations into external setup operation so as to be completed while the machine is running, or streamline internal setup operations so they may be completed more quickly. Since lost productive capacity and the degree of machine flexibility are determined by the length of time a work center is idled by setup, controlling and minimizing internal setup time is critical to productivity and flexibility.

See **External setup operation; Setup reduction; SMED.**

Reference

[1] Shingo, S. (1985). *SMED: A Revolution in Manufacturing*, Productivity Press, Cambridge, MA.

INTERNAL SUPPLIER

The people and departments within the organization are linked to others by processes. Oakland (1993, p. 14) defines a process as a transformation of inputs using actions, methods, and operations. Each process has at least one customer and one supplier. For example, a secretary can be considered an "internal" supplier to his/her principal who is the "internal" customer.

See **Supplier relationships; Total quality management.**

Reference

[1] Oakland, J.S. (1993). *Total Quality Management: The Route to Improving Performance*. Butterworth-Heinemann, Oxford.

INTERNATIONAL MANUFACTURING

Arnoud De Meyer

INSEAD, Fontainebleau, France

INTRODUCTION: The globalisation of the economic activity has dramatically increased the extent to which manufacturing is carried out and managed on an international scale. Managing international manufacturing poses challenges similar to the ones faced by international management in general. Issues such as cultural differences, coordination problems, global efficiencies, and local responsiveness are relevant to manufacturing as well as to other functions.

International manufacturing, however pose some specific challenges. These challenges are:

1. What determines the architecture of the plant network: how many plants does one need? What kind of charter should the different plants have? What should be their capacity?
2. Where should one build the factories?
3. How does one determine the contributions that a specific plant must make to the overall network and the business organisation to which it belongs? In particular what should be its contribution in terms of product portfolio and how should it contribute to the knowledge creation of the organisation?
4. How does one manage the flows in the network of plants: networks are characterised by flows of people, knowledge, goods and capital. How should these be managed?
5. How does one manage the dynamics of the network of plants?

The Architecture of the Network: Multiple plants are integrated in an international network of subsidiaries of the company. One legitimate question is whether the different plants should be clones of the original 'mother' plant, or whether international plants have a different role to play. The case for the idea that all plants should be copies of the original plant is not without merit. Indeed the original plant is probably successful, since the company has been able to grow and internationalise. Repeating that success in another part of the world may make sense. Yet, the idea of standardisation of all plants according to some original blueprint is intuitively not attractive, because it negates the need for local responsiveness.

A useful model of the architecture of networks, and the roles that plants can play in the overall network, is the one proposed by Ferdows (1997). He distinguished plants along two dimensions: the primary strategic reason for the site, and the depth of technical activities on the site.

As primary strategic reasons he proposes the access to market, the cost of production input factors, and the access to technological ideas. Though many other authors have proposed different strategic reasons for manufacturing abroad, the categories offered by Ferdows are general enough to include most other proposals. The one strategic reason that deserves to be added to his list is the need to learn about international management. But, this can be partially included in the access to markets or technological resources. It could be a fairly short term reason. Ferdows himself mentions two other reasons that may occur, but are not sufficiently important to be included in his model: pre-emption of the competition, and the control and amortisation of technological assets through economies of scale.

The second dimension, that is, the extent of technical activities at the particular site, refers to the value added activities in the plant. Ferdows notes that the depth of technical activities increase progressively from simple production and assembly, to planning, quality management, process adaptation and development, to product adaptation and development. Based on these two dimensions his map of the network of plants describes six types of plants (Figure 1):

- *Offshore plants* with a limited extent of activities and located in an area of low cost production.
- *Source plants* located in an area where the cost of production factor is still low, but have a charter to source a larger region of the world by serving as a focal point for a specific component, product or process.
- *Server plants* serve a specific market with a relatively low level of value addition.
- *Contributor plants* serve a particular market which has a strong local industrial network.
- *Outpost plants* tap into local technological resources, and their main function is to gather and process information
- *Lead plants* tap into local technological resources and are also major producers and diffusers of ideas and information.

This architecture as outlined by Ferdows enables us to propose a few assertions about good and less than good performing architectures. Ferdows claims that superior manufacturers have a larger proportion of their global factories in the upper part of Figure 1: source, contributor and lead plants. These plants provide their companies with a formidable strategic advantage. As a corollary, it can be argued that better performing networks are characterised as follows (De Meyer and Vereecke, 1996):

- Better performing networks have a higher degree of variety in the portfolio of plants; in other words they have less clones, but adjust each plant's charter to its specific strategic objective.
- Plants on the different locations in Figure 1 require different management systems and performance metrics. For example, an offshore plant has to be measured on its cost efficiency, a source plant on its market responsiveness, and a lead plant on producing and diffusing ideas throughout the network. (Contrast this with the prevailing practice in many large

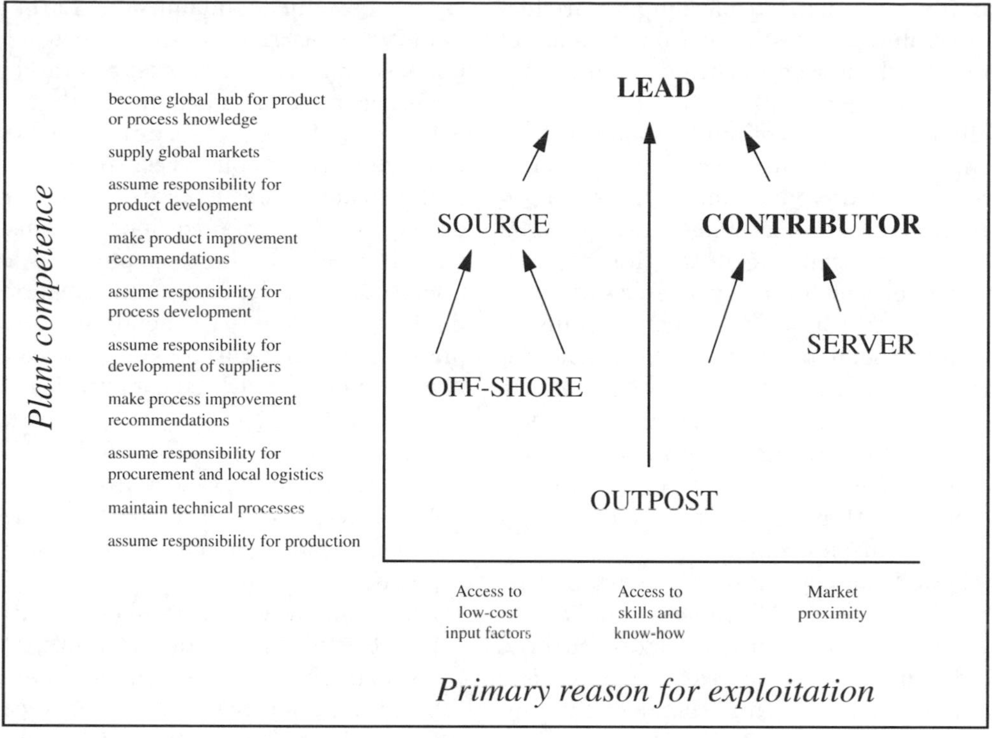

Fig. 1. Strategic role of the plant: Ferdows' model.

multinationals to use standardised management accounting systems and performance measures.) Since objectives and measurement systems differ, management of the various typed of plants must be different too.

- Plants on the upper end of the map entail more assets and thus more power and status in the organisation. Therefore, plant managers have a tendency to move their plants up in Ferdows' map. Plants on the upper end are more difficult to close down: the impact on the local business community of the closure of a contributor plant can be quite high, and may be severely hindered by social or political influences. The case of the closure in 1997 of the Renault plant in Belgium, which was a clear example of a contributor plant, illustrates this.

For over 75 years, Renault had an assembly plant for private cars in Belgium. This plant was originally created to serve the local and northern European market. Over the years it had developed into one of the most productive plants in the Renault network, and was heavily integrated into the local network of suppliers. The plant was seen as an essential element in the industrial fabric of Vilvoorde, the city in which it was located. However the rising labour costs in Belgium were not sufficiently compensated by the high level of productivity, and when Renault was forced to rationalise its worldwide production capacity, this plant was one of the first to be closed. The closure caused a lot of political and social tensions, not only because 3000 people lost their jobs, but because a whole region lost its key industrial player. One of the consequences of this friction was that over a period the Belgian population bouycotted all Renault products.

Having most plants in the upper end of the map may be a characteristic of superior companies according to Ferdows, but they also limit the flexibility to manage the dynamics of the plant architecture. The activities of local managers to move the plant up in the map will require entrepreneurial behaviour, which may well benefit the company. The case of Hewlett-Packard in Singapore (Thill and Leonard-Barton, 1994) illustrates this entrepreneurial behaviour of local management to move from electronic assembly to process development, and product adaptation to product development for the Asia Pacific market. This provided the company with a breakthrough in the Japanese market.

Location Decisions: This is probably the most extensively researched area in international manufacturing. Both economist and students of manufacturing have contributed to this literature. Traditionally, the location problem has been mathematically modelled, originally to optimise locations with minimal transport distances between supply, production, distribution and consumption. In more recent work, there is more interest in the optimization of complex production and distribution cost functions. Some of these models are discussed in Brandeau and Chiu (1989).

An equally large stream of publications emphasises the more qualitative aspects of site location decisions. Subjective location decisions based on the preferences of top management, political risk analysis of a particular country, the link with other functions of the company, the example set by competitors, the competitive move against a competitor, have been used extensively. Anecdotal evidence says that site selection based on the availability of a golf course, or more relevant perhaps, the living conditions – housing, schools, safety, etc. for expatriates? They emphasise and document the point that site location is not merely a rational optimisation problem.

There are two important issues concerning this location problem. First, the choice is usually a multi-stage problem: after having determined the strategic reason for creating a new site, one chooses a region or country, and then a site within that region or country. The first two steps should be determined with the company's long term strategy as the most important guiding factor. The third step is often influenced by short term perspectives, e.g. tax optimisation or personal preferences. This is most probably not so bad as long as the first two steps are decided with the long term interests of the company as the major criterion.

Secondly, most of the location models do not capture well the dynamics of a site during selection. The advantage of a site (in terms of market attractiveness or cost advantages) evolves over time. Some advantages may disappear, new advantages may arise. A successful plant will be one that can adapt itself to these changing conditions. Site selection processes should consider the evolution of both the strategy of the company and the characteristics of the site.

THE CONTRIBUTION OF A PLANT TO THE NETWORK:

> Within the network each plant plays a significant and different role. In addition to the Ferdows model, two other models are useful to interpret the contribution of a plant to a network of plants.

Hayes and Schmenner (1978) describe two extremes of plant type: the product focused and the process focused plant. In the product focused plant, each plant produces an end product for a particular market. In the process focused plant, the plant focuses on a particular process, and produces components or intermediary goods for a range of products and markets. Later on, Schmenner (1982) recogmozed other types of focii: general purpose plants, and market oriented plants.

Vereecke (1997) has empirically tested the idea of focus and finds that focus improves performance relative to their objectives. In particular, product and market focused plants do better than unfocused complex organisations. In the same study she proposes four types of plants, which differ in terms of their role, focus, and coordination mechanisms tying them to the headquarters and to other plants in the network:

- *Isolated plants* which communicate little or not at all with the rest of the network and do not exchange innovative knowledge throughout the network. In many cases these isolated plants are also physically isolated and they do not depend on other plants for components or intermediary goods.
- *Blueprint receivers* are plants which get the innovative ideas from other plants, and which implement them. There is little other communication, and usually the managerial capabilities of these plants are relatively less developed.
- *Integrated plants* are plants which have a lot of interaction (both in terms of information exchange and supply of goods) with the other plants in the network. The integration is often a consequence of the long history of these plants.
- *True network players* are plants with a high level of capabilities and play a proactive role in the management of the network. They share innovative knowledge and are highly integrated in the flow of goods and information.

Isolated plants and blueprint receivers play a limited strategic role and are vulnerable in the medium term. They do not contribute to the strategic development of the organisation but are useful in the current network as a provider of capacity. They may evolve into the two other types of plants with the appropriate support. The true network players are essential to the competitive evolution of the organisation. The integrated plants are often plants with a long history and which are, almost by default, in the position of a network player. But they may underperform compared to what the company expects from them. If true, they represent a lost opportunity for the organisation. Vereecke found that many 'mother' plants are actually of this nature. which suggests that the original plant from which the company's activity started may remain in a position of 'satisfactory underperformance'.

MANAGEMENT OF THE NETWORK: Managing international networks has been the topic of many studies in international management (see for example Doz, 1986 and Bartlett and Ghoshal, 1989). The biggest challenge in international manufacturing is the coordination and sharing of learning.

In an integrated international network of manufacturing there are many forms of flows: there is the flow of goods, human resources, administrative information and actionable process knowledge, and capital. The coordination of these flows requires special attention.

Martinez and Jarillo (1989) make a distinction between formal and informal coordination mechanisms. Coordination is not an exclusive activity of international corporations. In any multi-site corporation, and even in larger one site corporations, the coordination issue is of utmost importance. But coordination becomes more complicated in the case of a multi-country network.

The five formal coordination mechanisms described by Martinez and Jarillo are: grouping of organisational units; the centralisation of decision making; formalisation and standardisation of procedures; planning; and output and behaviour control. In addition, the three informal coordination mechanisms they describe are: lateral or cross departmental informal relations through direct contacts (throught temporary teams and taskforces, committees and integrative departments); informal communication; and socialisation (i.e., building an organisational culture of shared values and objectives by education, transfer of personnel, and appropriate performance appraisal and reward systems).

Technological and process knowledge is the most difficult flow to manage in a network of plants. The international transfer of explicit knowledge is a challenge, but can be planned and implemented through information systems, procedures, and rules about updating of standard operating procedures. But the transfer of tacit knowledge requires the interaction between human beings through face-to-face contacts (Nonaka and Takeuchi, 1994). That means a lot of travelling!

Sophisticated communication technology could overcome the need for geographical closeness (De Meyer, 1991). It is true that sophisticated videoconferencing systems and computer conferencing can, to some extent, overcome the need for face-to-face interactions. But two lessons emerge from research:

- In order to transfer tacit knowledge effectively through electronic media one needs to trust the person at the other side. Creating that trust still needs personal interaction. Once that trust is created, it enables people to collaborate over large geographical distances. But that trust will gradually disappear once both parties are separated. Misunderstandings could arise, time zone differences could lead to lack of availability precisely when a quick response is needed from the other side, etc. This may gradually lead to a breakdown of trust.
- Electronic media does not remove the decay of trust but decreases the rate of decay.

The transfer of tacit knowledge could be enhanced considerably by using 'ambassadors', i.e. representatives from one site who go and work for a considerable time (one to two years) at the other site, and who know that one of their major tasks is to build up a network of contacts. Once they return back home, the networ, so created, network could be used to keep sharing tacit knowledge.

MANAGING THE DYNAMICS OF THE NETWORK: The roles of factories will evolve over time. First, the architecture of the network in Figure 1 evolves over time because the strategy of the organisation evolves. Secondly, Vereecke pointed out that the isolated or blueprint factory is vulnerable over time, and thus must evolve towards more integration, or will risk closure.

How does one manage this transition? It is clear that the strategy of the organisation should be the driving element: network architecture and plants' roles have to follow the intended strategic evolution of the firm. But how does one manage that evolution over time? An interesting parallel can be found in a company which has to integrate the factories of an acquired organisation. Two case studies illustrate this point.

First case is the acquisition of a plant in Berlin by Samsung (De Meyer and Pycke, 1996), the second one is the acquisition by Toyo Ink of a plant in France (De Meyer and Probert, 1998). These two acquisitions were quite different from each other. In one case there was a massive upfront investment to turn around the plant in a few months, in the other case, considerable time was allowed for observing the plant and its capabilities before a few limited and gradual changes were implemented. The choice of the speed of integration was determined by how the acquiring company evaluated the competencies of the acquired company. In the first case, there was an attitude that, apart from the buildings and the technical competence of the workers, there was little of value to the acquirer. In the second case, the acquiring company considered the investment in France as an opportunity to learn about international manufacturing and about Europe's markets. But, as different as they were, they had a few common characteristics:

- In both cases there were attempts to instill in the acquired company the values of the acquirer. But, in both cases, there was considerable opposition and success came out of the newly created culture.

- In both cases there was a similar sequence in the programs that were implemented. The first programs were focused on 'cleaning up' (from literally improving the factory floor to cleaning up obsolete or unadapted procedures and traditions). The second wave of programs phased in company specific quality management procedures and approach. And the third wave of programs were aimed at turning the newly acquired factory into a 'true network player'.
- Success in both cases can be attributed to congruency of goals between the acquiring company, and the management and employees of the acquired company.

These two case studies illustrate the management of the dynamics in an international network of plants. Moving a plant from one role to another requires four conditions to be met:

- Management and employees of the plant must understand and accept the need for change. This is very much in line with the procedural justice or perceived fairness of all successful change programmes as described in Kim and Mauborgne (1997). The key issue here is communication.
- Moving to a new position requires new values, metrics, management systems, etc. It may be better to first 'clean up' the existing procedures and then implement new ones, as opposed to adding a new layer onto the existing procedures. Clean up, create new procedures, and then let the factories become network players.
- Speed can be important, but is not essential.
- Goal congruency between the company and the plant management is important.

Organizations Mentioned: Hewlett-Packard; Renault; Samsung; Toyo Ink.

See **Facilities location decisions; Global facility location analysis; Global manufacturing rationlization; Globalisation; Integrated plants; Isolated plants; Mother plant; Offshore plants; Process focused plants; Product focused plants; Production input factors; Server plants; Source plants.**

References

[1] Bartlett, C. and S. Ghoshal (1989). *Managing Across Borders: The Transnational Solution*. Hutchinson Business Books.

[2] Brandeau, M.L. and S.S. Chui (1989). "An overview of representative problems on location research." *Management Science*, 35:645–74.

[3] Collins, R. and R.W. Schmenner (1995). "Taking manufacturing advantage of Europe's single market." *European Management Journal*, 13:257–68.

[4] De Meyer, A. (1991). "Tech Talk: How managers are stimulating global R&D communication." *Sloan Management Review*, 49–58.

[5] De Meyer, A. and B. Pycke (1996). *Samsung (Berlin)*. INSEAD EAC case study 03/97-4672.

[6] De Meyer, A. and A. Vereecke (1996). "International Operations" in *International Encyclopedia of Business and Management*. (Ed M Werner) Routledge.

[7] De Meyer, A. and J. Probert (1998). *Francolor Pigments: A Toyo Ink Acquisition*. INSEAD EAC case study 05/98-4756.

[8] Doz, Y. (1986). *Strategic Management in Multinational Companies*. Pergamon Press, Oxford.

[9] Ferdows, K. (1997). "Making the most of foreign factories." *Harvard Business Review*, 73–88.

[10] Hayes, R.H. and R.W. Schmenner RW (1978). "How should you organise manufacturing?" *Harvard Business Review*, 105–19.

[11] Kim, W.C. and R. Mauborgne (1997). "Fair Process: Managing in the Knowledge Economy." *Harvard Business Review*, Jul/Aug.

[12] Martinez, J.I. and J.C. Jarillo (1989). "The evolution of research on coordination mechanisms in multinational corporations." *Journal of International Business Studies*, 20:489–514.

[13] Nonaka, I. and H. Takeuchi (1995). *The Knowledge Creating Company*. Oxford University Press, New York.

[14] Schmenner, R.W. (1982). *Making Business Location Decisions*. Prentice Hall.

[15] Thill, G. and D. Leonard-Barton (1994). *Hewlett-Packard: Singapore (A), (B) and (C)*. Harvard Business School Case Study.

[16] Vereecke, A. (1997). *The determinants of the creation of an optimal international network of production plants: an empirical study*. (Unpublished doctoral dissertation) The Vlerick School of Management, University of Ghent, Belgium.

INTERNATIONAL ORGANIZATIONS FOR STANDARDIZATION (ISO)

See **ISO 9000; ISO 14000.**

INTRANET

Intranet is web-based computer communication systems within an organization, as compared to the Internet, which allows access to and from users outside the organization.

See **Reengineering and the process view of manufacturing.**

INTRAPRENEURSHIP

Intrapreneurship refers to the innovative activities of a mature, possibly large organization. This term drawn from companies such as 3M, which tried to energize creative activities as if they were small entrepreneurs. Intrapreneurs act as if they are free of the bureaucratic restrictions that often exist in large companies.

See **Process innovation; Product innovation.**

INVENTION

Invention refers to the discovery or conception of a new idea for a product or service. When inventions is commercialized or otherwise used, it is called innovation.

See **Process innovation; Product innovation.**

INVENTORY BUILD AND DEPLETION CYCLES

See **Inventory flow analysis.**

INVENTORY CARRYING COSTS

The terms inventory carrying costs or holding costs are interchangeable. Holding cost are often defined as a percentage of the dollar value of inventory for one year.

The cost of holding inventory in a function of inventory quantity, item value, and length of time the item is held in inventory since it takes capital to acquire inventory, a firm forgoes use of these funds for other and cannot invest the funds. Therefore, a cost of capital, which is expressed as an annual interest rate, is one component of inventory carrying costs.

In addition to the cost of capital, costs of maintaining the inventory as taxes and insurance, obsolescence, spoilage, and storage space are included in carrying costs. They vary from 10% to 35% annually, for a typical item in inventory.

See **Inventory flow analysis; Inventory holding costs. Safety stocks: Luxury or necessity?**

INVENTORY CONTROL

Inventory control is an important activity to ensure the timely and inexpensive availability of materials, parts, components, subassemblies and finished goods. It is the organized, formal effort within manufacturing firms to manage and limit inventory levels and associated costs. It includes decisions affecting the following:

1. Which items to carry in inventory?
2. How much to order?
3. When to order?
4. Where to hold inventory?

Without inventory control, a firm's inventory can become excessive and result in excessive costs. On the other hand, poor inventory control can result in stockouts of raw materials or components, which may bring productive activities to halt. In the case of finished goods stockout, shipment to customers may be delayed, which may lead customer dissatisfaction.

Inventory control decisions concerning how much to order and when to order are often solved using optimizing, mathematical models. These models are incorporated in computerized inventory control software, which automatically decide when to order and how much to order, and generate purchase orders or place orders electronically through computers with predetermined suppliers.

See **Fixed order-quantity inventory model; Fixed time-period inventory model; Inventory flow analysis; Inventory holding costs; Inventory management; Just-in-time manufacturing; Lean manufacturing.**

INVENTORY FLOW ANALYSIS

Edward W. Davis

Darden Graduate School of Business, University of Virginia, Virginia, USA

INTRODUCTION: Inventory flow models are graphic representations of inventory behavior over the manufacturing lead time. They are a powerful tool that enables manufacturing managers to evaluate how effectively they add value to inventory. Step-by-step construction of a simple inventory flow diagram is shown in Figure 1, which illustrates how these diagrams show the distribution of the various categories of inventory (e.g., Raw Materials, Work in Process, Finished Goods)

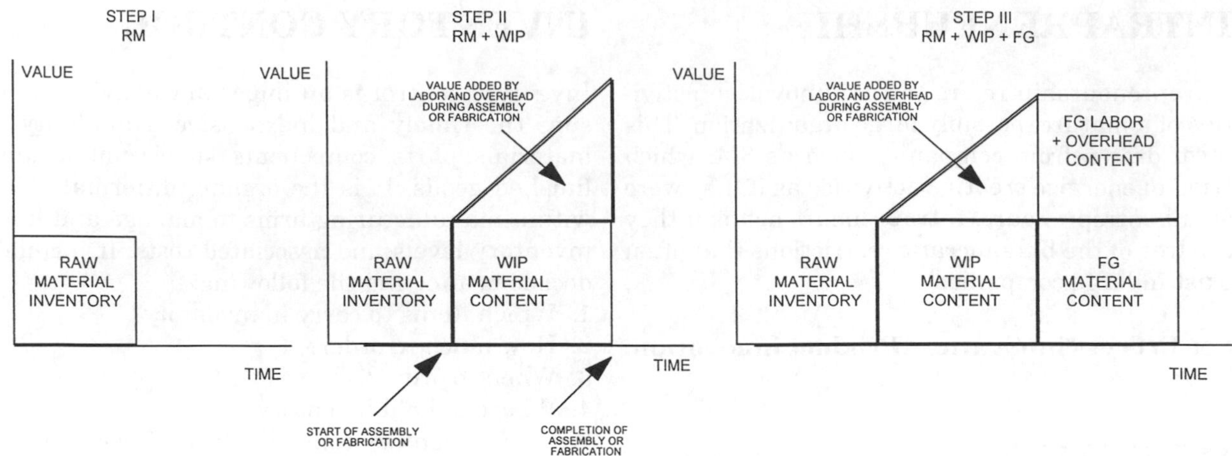

Step I: The value of the raw material inventory is graphed in periods of supply.

Step II: The value of the work in process (WIP) inventory is added, broken down into components of material content and value added by labor and overhead during assembly (or fabrication).

Step III: The finished goods (FG) inventory, in terms of periods of supply, is added to Graph III to complete the inventory flow diagram.

Fig. 1. Construction of an inventory flow diagram.

in terms of weeks of supply. They also show inventory build and depletion rates, in terms of dollars-per-week. Thus, they provide information about both the investment and time which inventories represent.

Inventory flow models are not new. While the exact origin of the concept is uncertain, the approach is known to have been used at Raytheon Corporation, Hughes Aircraft Company and Westinghouse Corporation in the mid-1960's. However the first publicly available document describing the approach did not appear until the early 1970's (Davis and Whitelaw, 1972). Other articles and case studies describing variations of the concept as used at IBM, Bendix and other companies appeared during the mid-late 1970's and early-mid 1980's (Sirianni, 1975, 1979, 1972; Colley, 1977; Wantuck, 1978; Clark, 1984; Davis and Schaefer, 1985). More recently, the approach has been described in connection with cycle time reduction and Business Process Reengineering (Cherukuri, Nierman and Sirianni, 1995).

Inventory flow diagrams show inventory investment at a point in time. Total inventory investment, equal to the total area under the diagram, is therefore shown as a "snapshot," in much the same way that the balance sheet is a financial snapshot. With this purely descriptive snapshot of inventory investment, the relationships and flow of parts, material and product between the various categories of inventory can be better understood. Inventory flow models such as shown in Figure 1 can also be developed for normative purposes, to set inventory objectives, and answer questions like "how much inventory *should* we have?" One such normative approach will be outlined later after first illustrating the application of the concept in a purely descriptive sense.

An Example of the Graphical-Flow Model ("Descriptive" Approach): This example shows how the inventory position in a hypothetical firm in the electronics industry can be represented by an inventory flow model, or diagram. The example company fabricates parts from raw materials, puts them together with purchased parts, and assembles equipment for sale to a variety of customers. The firm's assumed income statement and the inventory portion of its balance sheet are shown in Table 1.

Table 1. Sample financial data.

Income statement ($ in millions)		Inventories ($ in millions)	
Sales	$120.0	Raw materials and purchased parts	$5.28
Cost of sales	100.0		
Gross margin	20.0		
SG&A expenses	8.0	Work in process	7.52
Pretax profit	12.0		
Federal taxes	6.0	Finished goods	8.00
After-tax profit	$6.0	Total	$20.80

The overall inventory turnover rate is: $100M ÷ $20.8M[a] = 4.8.
[a]M denotes millions.

It is assumed that raw material and purchased parts are 60% of the cost of goods sold, with labor and factory overhead constituting 40%. With this information and the data given in Table 1, we could convert the three inventory categories on the balance sheet into "weeks-of-stock". For example, assuming 50 weeks of operation per year, the firm is shipping an average of \$2 million/week at cost (\$100 million cost of sales ÷ 50 weeks). The finished-goods inventory represents the equivalent of 4 weeks of sales at the given sales rate (\$8 million inventory ÷ \$2 million cost of sales/week). Similarly, the total raw-material and purchased-parts inventory of \$5.28 million could be expressed as 4.4 weeks of stock (\$5.28 million inventory balance ÷ \$1.2 million material and parts usage/week = 4.4 weeks, where 60% of \$2M cost of sales/week = \$1.2M/week). However, a more detailed model provides more useful results when the data are available, as described next.

Table 2 shows the composition of the inventory balances of Table 1 as determined from the company's accounting or operations data.

If the sales rate varied during the year, a relatively stable inventory level would represent a variable coverage in "weeks of stock." On the other hand, if the firm were to maintain controls to keep a predetermined level of "weeks-of-stock," the inventory balances would vary with the sales or production rate. For the sake of simplicity, this example is based on the assumption that the *sales rate is constant during the year*.

Assume also that, from the accounting records of this firm, we find that 70% of the purchased material used is raw material, which is fabricated prior to assembly, and 30% is purchased parts, which are used directly in assembly. Also, assume that 70% of the factory labor and overhead is consumed in fabrication and 30% in assembly.

The relevant inventory flow rates calculated from this information are summarized in Table 3.

By dividing the inventory balances in Table 2 by the flow rates in Table 3 we arrive at the equivalent weeks-of-stock for raw materials and for purchased parts, separately.

Table 2. Detailed inventory balances (\$ in millions).

Raw materials	\$3.57	
Purchased parts	1.71	
Materials and parts		\$5.28
Fabrication WIP	4.12	
Assembly WIP	3.40	
Work in process		7.52
Finished goods		8.00
Total inventory		\$20.80

Table 3. Detailed inventory flow rates.

Materials and parts @ 60% of	
CGS = \$60M/year	= \$1.2M/week
Fabrication raw materials @ 70%	= \$840,000/week
Purchased parts @ 30%	= \$360,000/week
Labor plus overhead @ 40% of	
CGS = \$40M/year	= \$800,000/week
Fabrication (L + OH) @ 70%	= \$560,000/week
Assembly (L + OH) @ 30%	= \$240,000/week

Raw materials

\$3.57M (in inventory) ÷ \$840,000/week
= 4.25 weeks

Purchased parts

\$1.71M (in inventory) ÷ \$360,000/week
= 4.75 weeks

Finished-goods stocks remain at 4 weeks as determined previously. However, the conversion of the work in process (WIP) inventory balance into equivalent weeks-of-stock requires additional information. The WIP inventory balance reflects the fact that some orders are just beginning production, others are almost finished, and thus it is assumed that the average order is half completed. Applying the accounting convention that values each order in work in process inventory as half completed from a labor and overhead standpoint, orders in process are valued at 100% of the material value and 50% of the labor and factory overhead, since the overhead is usually absorbed to the job on the basis of the labor costs or hours incurred. This implies that the buildup of labor and overhead in work in process is linear.[1] Using this assumption the equivalent weeks-of-stock for Fabrication and Assembly work-in-process are calculated as follows:

Fabrication work in process

Materials + Labor + Overhead
= Fabrication WIP Inventory Balance

\$840,000/wk. (X) + 1/2(\$560,000/wk.) (X)
= \$4.12M (from Table 2)

(\$840,000/wk.) (X) + (\$280,000/wk.) (X)
= \$4.12M

(\$1,120,000/wk.) (X) = \$4.12M

X = 3.68 weeks

[1] The assumption of linearity used in this example is not a necessary constraint of the methodology. Non-linear assumptions are possible if the relevant area under the diagram can be calculated, or estimated.

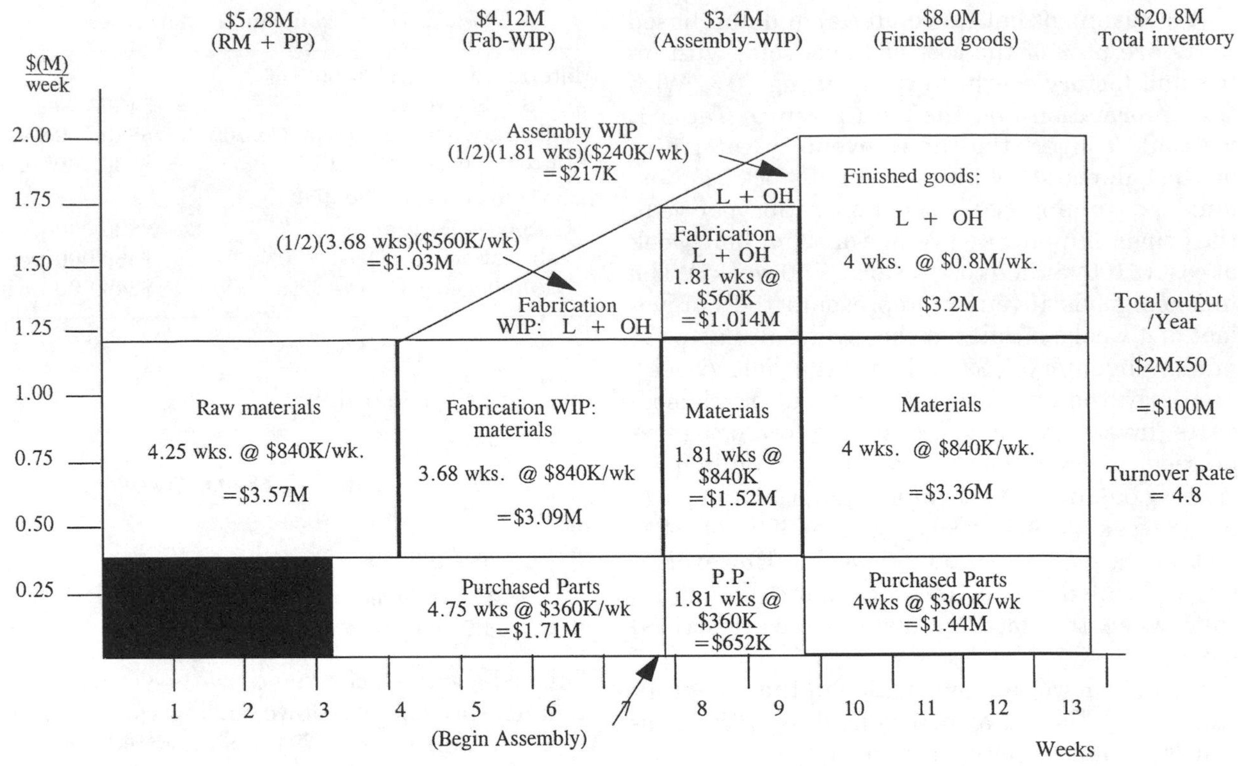

Fig. 2. Detailed inventory flow diagram.

Assembly work in process

Purchased parts + Fabricated raw materials
+ Fabrication labor & overhead
+ Assembly labor plus overhead
= Assembly WIP balance

($360,000/wk.) (X) + ($840,000/wk.) (X)
+ ($560,000/wk.) (X) + 1/2($240,000/wk.) (X)
= $3.40M (from Table 2)

($1.88M/wk.) (X) = $3.40M

X = 1.81 weeks

We can now graph the various stock balances we have converted to weeks-of-stock. The cycle times or throughput times are different for raw materials and purchased parts (4.25 weeks vs. 4.75 weeks) and there is a 3.68 week fabrication work-in-process cycle time. Figure 2, shows the complete and detailed flow diagram.

From this inventory flow model we see that a typical raw-material item arrives in stores almost eight weeks prior to its use in assembly. Four and a quarter of the eight weeks are spent in inventory and 3.68 weeks in fabrication work in process. A typical purchased part is received 4.75 weeks prior to beginning assembly.

COMMENTS ON "DESCRIPTIVE MODEL": The example given thus far is one form of the inventory flow model, constructed primarily from financial data. The advantages of this "descriptive" modeling approach include:

- it is easily constructed from a minimum amount of data;
- it is an easily understood graphic representation of the inventory assets;
- it graphically emphasizes those portions of inventory that have potentially high leverage;
- it is a useful graphic tool for year-to-year or company vs. company comparisons.

Disadvantages of this "descriptive" approach include the fact that the inventory balances are a "snapshot" in time, usually at fiscal year-end, and may not be representative of the average inventory position. This can be remedied by using average inventory values, or by developing multiple flow diagrams for different segments of the year to handle seasonal fluctuations. Another disadvantage is that the model shown does not allow for product line differences in inventory and cost of goods sold, but this can be overcome by constructing models for individual representative product lines, if the necessary data are available.

The most serious disadvantage of the "descriptive" model is that it simply shows inventory levels as they currently *are*. But because the inventory flow diagram shows the various inventory balances as weeks of stock, it can be a useful tool for management inquiry, even in a "descriptive-only" sense. For example, why must the firm stock 4.25 weeks' usage of raw materials? Could a better scheduling system shorten the fabrication work-in-process cycle? Could the firm maintain adequate customer service with 3 weeks of finished goods rather than 4 weeks? What happens to the inventory balances if sales increase by 20% next year and the same equivalent weeks of supply are maintained, assuming the cost structure does not change? The direct financial effect of such operational changes in weeks-of-stock, or in sales levels can be readily calculated, and the resulting impact on the Inventory diagram easily shown.

"NORMATIVE" INVENTORY FLOW MODELS FROM MANUFACTURING LEAD TIME AND COST DATA: An alternative, and powerful, use of the inventory flow model is in a *normative sense*; i.e., to investigate how much inventory *should* there be? For example, instead of starting with inventory dollar balances and converting into equivalent weeks, the procedure can start by determining the average cycle or lead times required for parts, material, and product to move through the existing manufacturing process. This information (along with other necessary data as shown here) can then be used to construct the inventory flow diagram. The inventory dollar balances which can be calculated from this diagram represent levels which *should* be possible for this particular manufacturing process based on the way the business is currently being run (existing lead times and costs). These levels can then be compared with actual balances to determine potential opportunities for improvement. A detailed description of such an approach in case study format is provided by Davis and Schaefer (1985) and also in Clark's (1984) APICS presentation. The steps involved in this approach are listed in Figure 3. Variations of the normative approach are presented by Sirianni (1975) (1979) and (1981) and Wantuck (1978).

When working with the normative inventory flow model there are additional levels of detail that can be added to enhance the accuracy and use of the model. One example, originally described by Wantuck (1978) and also shown by Clark (1984) is the scheduling of raw material for fabrication, is shown in Figure 4.

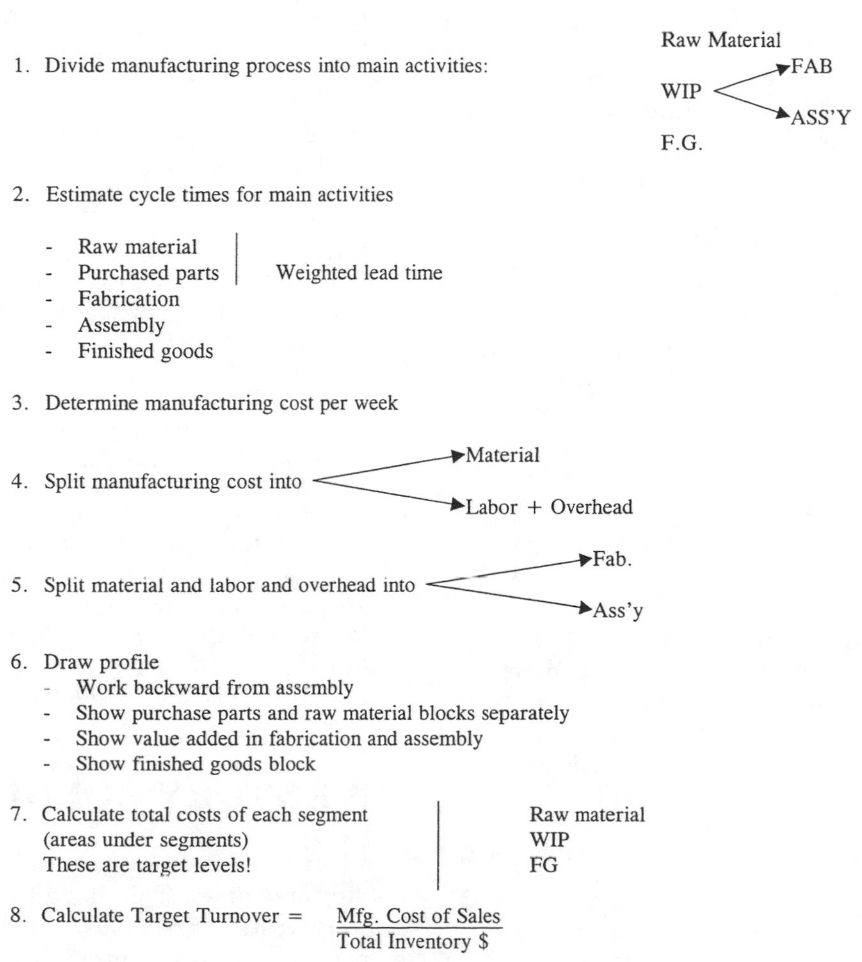

Fig. 3. Inventory flow model (normative approach) steps in process.

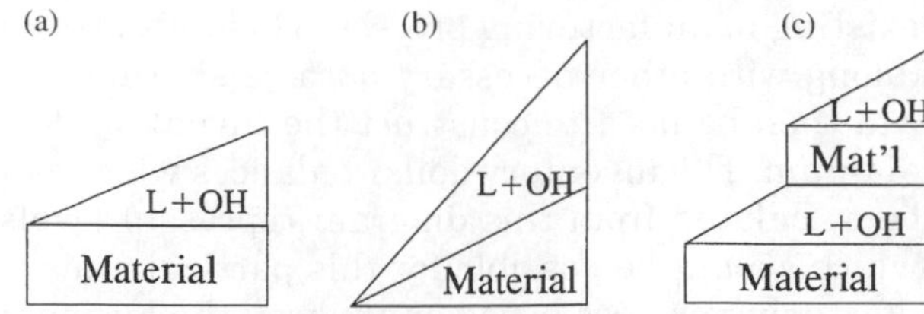

Fig. 4. Alternative models of WIP inventory.

Figure 4(a) depicts the assumption that all raw material is available before the application of labor and overhead, as in our first example here. However if material were added at a constant rate like labor, the model would be as shown in 4(b). Alternatively, Figure 4(c) represents the situation where 50% of the material arrives at the beginning of fabrication and the remaining 50% arrives halfway through. The use of inventory flow models for evaluating such production scheduling alternatives in production planning and master production scheduling is described by Clark, as well as the extension of these models "upstream" in the supply chain to include vendors and "downstream" to include physical distribution, to provide top management a useful view of their logistics system.

The advantages of developing "normative" inventory flow models from manufacturing costs and lead time data include:

Table 4. Inventory flow model uses.

In descriptive sense	In normative sense
"How Much Inventory Do We Have?"	"How Much Inventory Should We Have?"
Given: Company Financial Data	*Given:* MFG Cycle Times and MFG Costs
e.g., COGS $ Inventory $ Raw Matl. WIP FG	e.g., Raw Mtl : 8 weeks WIP : 2 weeks FG : 1 week
Calculate: Mfg. Cycle Times Days of Supply Of Inventory	*Calculate:* Inventory $ Target Turnover
Uses: – Comparison with other Companies – Another way of measuring inventory – To help find areas for improvement	*Uses:* – To set inventory standards – To help find areas for improvement

- they portray what the inventory should be for the way the business is currently being managed;
- they highlight potential areas of improvement;
- they provide an easy-to understand model of the business;
- they permit simulation to evaluation the probable results of different courses of action; and
- they can be presented in varying levels of detail to facilitate different levels of analysis.

A summary of the two major approaches described here for using the inventory flow model concept is outlined in Table 4.

See **Cost-of-goods-sold; Finished goods stock; Inventory build and depletion cycles; Weeks-of-stock.**

References

[1] Cherukuri, S.S., R.G. Nierman and N.C. Sirianni (1995). "Cycle Time and The Bottom Line", *Industrial Engineering*, March, 1995.

[2] Clark J.T. (1984). "Inventory Flow Models", *Proceedings, 27th Annual Conference, American Production and Inventory Control Society*.

[3] Colley, J.L. (1977). "Inventory Flow Models", Technical Note OM-120, Darden Graduate School of Business, University of Virginia, Charlottesville, Virginia.

[4] Davis, E.W. and S. Whitelaw (1972). "Leisure Products Company", case study 9-673-036, Harvard Business School of Business, Boston, Massachusetts.

[5] Davis, E.W. and P.G.D. Schaefer (1985). "Sullivan Products Company", case study OM-580, Darden Graduate School of Business, University of Virginia, Charlottesville, Virginia.

[6] Sirianni, N.C. (1975). "Calculating Inventory Objectives", Presentation, 1975 Inventory Workshops, Westinghouse Electric Corporation.

[7] Sirianni, N.C. (1979). "Inventory: How Much Do You Really Need?", *Production Magazine*, November, 1979.

[8] Sirianni, N.C. (1981). "Right Inventory Level: Top Down or Bottom UP", *Inventories and Production Magazine*, May–June, 1981.

[9] Wantuck, K.A. (1978). "Calculating Optimum Inventories", *Proceedings 21st Annual Conference, American Production and Inventory Control Society*.

INVENTORY HOLDING COSTS

It refers to all costs associated with carrying or holding inventory. This includes both direct and indirect costs. Direct costs include the cost of money tied up in inventory. This varies by company but includes the cost of capital, or the interest the company pays for borrowing money to pay for inventory. It also includes the cost of storage space, insurance, special containers to hold inventory, and inventory handling equipment. Indirect costs include those costs associated with managing the inventory, taking physical inventory counts, theft, obsolescence, and damage. In addition, opportunity costs linked to money tied up in inventory versus using it for other activities is an indirect holding cost. All costs associated with logistics support including equipment such as hardware and software to track and monitor inventory and the personnel who perform those duties are indirect costs of holding inventory.

See **Implications of JIT manufacturing; Inventory flow analysis; Safety stocks: Luxury or necessity.**

INVENTORY MANAGEMENT

Inventory refers to physical items stored for future use. Inventory is created through purchase or through production. There are generally four kinds of inventory:

1. Raw materials, and purchased part and components.
2. Work-in-process.
3. Finished goods in the factory or in the distribution system.
4. Suppliers.

On the average, materials cost represent 50 percent of a manufactured product's cost. Therefore, the management of materials is a important function in companies engaged in manufacturing and distribution.

The goal of inventory management are:

1. Minimize the cost attributes to inventory, and
2. Ensure the availability of materials, parts and components, when need for production.

Inventory management must deal effectively with the following costs:

1. Inventory carrying or holding costs.
2. Inventory order costs – the cost of procuring inventory.
3. The cost of stockouts.
4. The cost of the item.

There is often a trade off between holding costs and ordering costs. That is, if one increases, the other decreases. For example, if inventory is ordered in small lots of 10 everyday, it will keep the holding cost very low but will increase the number of orders and associated cost of ordering. Therefore, as part of good inventory management, companies use mathematical models to compute optimal order quantities to balance the two costs.

The nature of demand for an item has an effect on how its inventory is managed. The demand for some items is independent demand items is different from the inventory management for dependent demand items. Dependent demand items are those usually components, whose demand is dependent on the demand for the end item. An example, the demand for auto steering wheel is dependent on the demand for assembled automobiles (one steering wheel demanded for every car assembled). Materials requirements planning is appropriate for dependent demand situations. When to order decisions are time-phased and discrete qualities are ordered based on forecast of demand, lead time for procurement, on-hand quantity, etc.

Key inventory decisions in independent demand situations concern the following:

1. When to order, and
2. How much to order.

When to order decisions are mad using, (1) Order points; or (2) periodic reordering rules. In order point systems, an order is generated when inventory levels drops to a predetermined value such as 20 units, which is adequate to cover the demand during the lead time to get the order. Periodic reordering requires reordering every week or month or some prespecified interval. The question of how much to order is determined in independent demand situations using simple or complex mathematical models. Simple fixed order quantity and variable quantity models (Fogarty, Blackstone and Hoffman, 1991) can be used to compute the order quantity. Modern developments in factory management such as Just-in-Time manufacturing reduce the need for inventory through close relationships with suppliers, and the use of Kanban-based manufacturing systems, which reduce work-in-process and finished goods inventories.

See **Fixed order-quantity inventory model; Fixed time-period inventory model; Inventory flow analysis; JIT purchasing; Just-in-time manufacturing; Material requirements planning; Safety stocks: Luxury or necessity.**

References

[1] Fogarty, D.W., J.H. Blackstone, and T.R. Hoffman (1991). *Production and Inventory Management*, second edition. Cincinnati, OH: South-Western Publishing Co.

[2] McLeavey, D.W. and S.L. Narasimhan (1985), *Production Planning and Inventory Control*. Boston, MA: Allyn and Baeon, Inc.

INVENTORY TURNOVER RATIO

Inventory turnover or turns has become a common and important measure of inventory performance. Inventory turnover expresses inventory level as a function of sales volume. Inventory turnover can be computed easily as annual sales in dollars divided by the dollar value of average on-hand. For example, an average inventory of $4 million and an annual cost of sales of $32 million would mean that inventory turned over eight times.

AS a manufacturing system performs more and more efficiently, its inventory turns would increase. Just-in-time and lean manufacturing principles, when implemented, tend to increase the inventory turns substantially. Studies by Swamidass (1996, 1998) shows that the discrete products manufacturers in SIC classificatins 34 to 38, the inventory turns changed from 8.0 to 9.7 from 1993 to 1997. Over 1000 U.S. manufacturing plants provided the data for the study. It is a good estimate of the average inventory turns for the industries studied.

See **Inventory flow analysis; Manufacturing technologies and their use; Performance measurement in manufacturing.**

References

[1] Swamidass, P.M. (1996). Benchmarking Manufacturing Technology use in the United States in the *Handbook of Technology Management*, edited by G.H. Gaynor. New York: McGraw-Hill, p. 37.1 to 37.37.

[2] Swamidass, P.M. (1998) *Technology on the Factory Floor III*. Washington, D.C.: The Manufacturing Institute National Association of Manufacturers.

INVENTORY TURNS

See **Inventory turnover ratio.**

INVOLUNTARY PRODUCT ADAPTATION

This refers to changes/modifications that international enterprises are obliged to incorporate in their products for overseas markets in order to comply with legal requirements of foreign governments. Such requirements, often enforced by law, impose restrictions on how a product must be packed, labeled, and restrict the type of material used (especially in food products).

See **Product design for global markets.**

ISO 14000

ISO14000 is a voluntary, multi-part series of standards to guide an organization's environmental management system established by the International Organization for Standardization. In many ways, this series mirrors ISO 9000 (quality management systems), and was preceded by a British Standard, BS 7750. As they are not a product-based standard, no particular level of environmental performance is either mandated or assured. Instead, ISO 14000 specifies the way in which manufacturing or service operations must be managed. Explicit, systematic organizational structures, practices, activities and processes must be in place to minimize a firm's impact on the natural environment through monitoring, control and improvement. Firms set their own performance targets based on self-identified considerations, including the demands of customers, regulators, communities, lenders or environmental groups. Collectively, the series of standards provide a framework for firms to develop plans to meet their own environmental performance targets and to monitor progress. At present, ISO 14000 standards of varying specificity are available for environmental auditing, performance evaluation and life-cycle assessment.

See **Environmental issues and operations management; Environmental issues: Reuse and recycling.**

ISO 9000

A set of five individual but related international standards on quality management and quality assurance developed to help companies effectively document the quality system elements by International Organization for Standardization of the company's quality system. The standards, initially published in 1987, are not specific to any particular industry, product, or service. The standards were developed by the International Standards Organization (ISO), an international agency for standardization composed of the national standards bodies of over 90 countries.

See **ISO 9000/QS 9000 quality standards; Total quality management.**

ISO 9000/QS-9000 QUALITY STANDARDS

James R. Evans
University of Cincinnati, Cincinnati, Ohio, USA

DESCRIPTION: The ISO 9000 family of International Standards are *quality system standards* that guide a company's performance of specified requirements in the areas of design/development, production, installation, and service. The standards define three levels of quality assurance:

- *Level 1* (ISO 9001) provides a model for quality assurance in firms that design, develop, produce, install, and service products.
- *Level 2* (ISO 9002) provides a quality assurance model for firms engaged only in production and installation.
- *Level 3* (ISO 9003) applies to firms engaged only in final inspection and test.

Two other standards, ISO 9000 and ISO 9004, define the basic elements of a comprehensive quality assurance system and provide guidance in applying the appropriate level. ISO 9000 describes the principal concepts of quality assurance, such as the objectives and responsibilities for quality, stakeholder expectations, the concept of a process, the role of processes in a quality system, the roles of documentation and training in support of quality improvement, and how to apply the different standards. ISO 9004 guides the development and implementation of a quality system. It examines each of the elements of the quality system in detail and can be used for internal auditing purposes. Together, these five standards are referred to as the ISO 9000 series.

Late in 1994, the three major U.S. automobile manufacturers – Ford, Chrylser, and General Motors – released QS-9000, an interpretation and extension of ISO 9000 for automotive suppliers. QS-9000 is a collaborative effort of these firms to standardize their individual quality requirements while drawing upon the global ISO standards. Truck manufacturers – Mack Trucks, Freightliner, Navistar International, PACCAR Inc, and Volvo GM – also participated in the process. QS 9000 applies to all internal and external suppliers of production and service parts and materials. Chrysler, Ford, GM, and truck manufacturers require suppliers to establish, document, and implement quality systems based on these standards according to individual customers' timing.

HISTORICAL PERSPECTIVE: As quality became a major focus of businesses throughout the world, various organizations developed standards and guidelines. Terms such as *quality management, quality control, quality system*, and *quality assurance* acquired different, and sometimes conflicting meanings from country to country, within a country, and even within an industry. As the European Community moved toward the European free trade agreement, which went into effect at the end of 1992, quality management became a key strategic objective. To standardize quality requirements for European countries within the common market and those wishing to do business with those countries, a specialized agency for standardization, the International Organization for Standardization (ISO), founded in 1946 and composed of representatives from the national standards bodies of 91 nations, adopted a series of written quality standards in 1987, which were revised in 1994. The standards have been adopted in the United States by the American National Standards Institute (ANSI) with the endorsement and cooperation of the American Society for Quality Control (ASQC). The U.S. standards are called the ANSI/ASQC Q9000-1994 series. The standards are recognized by about 100 countries. In some foreign markets, companies will not buy from non-certified suppliers. Thus, meeting these standards is becoming a requirement for participation in international trade and competition. The standards are intended to apply to all types of businesses, including electronics and chemicals, and to services such as health care, banking, and transportation.

STRATEGIC PERSPECTIVE: ISO 9000 standards are based on the premise that certain generic characteristics of management practices can be standardized, and that a well-designed, well-implemented, and carefully managed quality system provides confidence that the outputs will meet customer expectations and requirements. The standards prescribe documentation for all processes affecting quality and suggest that compliance through auditing leads to continuous improvement. Thus, the standards have five objectives:

1. Achieve, maintain, and seek to continuously improve product quality (including services) in relationship to requirements.
2. Improve the quality of operations to continually meet customers' and stakeholders' stated and implied needs.

3. Provide confidence to internal management and other employees that quality requirements are being fulfilled and that improvement is taking place.
4. Provide confidence to customers and other stakeholders that quality requirements are being achieved in the delivered product.
5. Provide confidence that quality system requirements are fulfilled.

The specific goal of QS-9000 was to develop fundamental quality systems that provide for continuous improvement, emphasizing defect prevention, and the reduction of variation and waste in the supply chain – issues not explicitly addressed in the ISO 9000 standards. This standardized quality is expected to reduce the cost of doing business with suppliers and enhance the competitive position of the automakers and suppliers alike.

The most important reasons why companies seek ISO 9000 or QS-9000 certification include:

- To meet contractual obligations. Some customers now require certification of all their suppliers. Automotive manufacturers have mandated compliance. Suppliers that do not pursue registration will eventually lose customers.
- To meet trade regulations. Many products sold in Europe, such as telecommunication terminal equipment, medical devices, gas appliances, toys, and construction products require product certifications to assure safety. Often, ISO certification is necessary to obtain product certification.
- To market goods in Europe. ISO 9000 is widely accepted within the European Community. It is fast becoming a de facto requirement for doing business within the trading region.
- To gain a competitive advantage. Many customers use ISO registration as a basis for supplier selection. Companies without it may be at a market disadvantage.

IMPLEMENTATION: ISO 9000 standards focus on 20 key requirements. To illustrate the scope of the requirements, consider the first one, *Management Responsibility*. The standards require that:

- The company establishes, documents, and publicizes its policy, objectives, and commitment to quality.
- The company designates a representative with authority and responsibility for implementing and maintaining the requirements of the standard.
- The company provides adequate resources for managing, performing work, and verifying activities including internal quality audits.
- The company conducts in-house verification and review of the quality system. These reviews should consider the results of internal quality audits, management effectiveness, defects and irregularities, solutions to quality problems, implementation of past solutions, handling of nonconforming products, results of statistical scorekeeping tools, and the impact of quality methods on actual results.

A brief summary of the basic requirements for the remaining elements of ISO 9001 are summarized below.

- *Quality system.* The company must write and maintain a quality manual that meets the criteria of the applicable standard (9001, 9002, or 9003), which defines conformance to requirements; effectively implement the quality system and its documented procedures; and prepare quality plans for determining how requirements will be met.
- *Contract review.* The company must review contracts to assess whether requirements are adequately defined and whether the capability exists to meet requirements.
- *Design control.* The company must verify product design to ensure that requirements are being met and that procedures are in place for design planning and design changes. This includes plans for each design and development activity, defining organizational and technical interfaces, validating outputs against design input requirements, and design verification and validation procedures.
- *Document and data control.* The company must establish and maintain procedures for controlling documentation and data through approval, distribution, change, and modification.
- *Purchasing.* The company must have procedures to ensure that purchased products conform to requirements. This includes evaluating subcontractors, clearly-written purchasing documents, and verifying purchased products.
- *Control of customer-supplied products.* Procedures to verify, store, and maintain items supplied by customers must be established.
- *Product identification and traceability.* The company must identify and trace products during all stages of production, delivery, and installation.
- *Process control.* The company must carry out production processes under controlled conditions. The processes must be documented and monitored, and workers must use approved equipment and have specified criteria for workmanship.
- *Inspection and testing.* The company must maintain records at all stages of inspection and testing to verify that requirements are met. This includes receiving, in-process, and final inspection and testing.

- *Control of inspection, measuring, and test equipment.* The company must establish procedures to control, calibrate, and maintain equipment used to demonstrate conformance to requirements.
- *Control of nonconforming product.* Procedures should ensure that the company avoids inadvertent use of nonconforming product. This includes how nonconforming product is reviewed, and reinspection of repaired or reworked product.
- *Corrective and preventive action.* The company should investigate causes of nonconformance and take action both to correct the problems and to prevent them in the future. Corrective action includes handling customer complaints, investing causes of nonconformities, and applying appropriate controls. Preventive action includes detecting, analyzing, and eliminating potential causes of nonconformities and initiating preventive actions.
- *Handling, storage, packaging, preservation, and delivery.* The company should develop procedures for properly handling, storing, packaging, preserving, and delivering products.
- *Control of quality records.* The company should identify, collect, index, file, and store all records relating to the quality system.
- *Internal quality audits.* The company must establish a system of internal audits to verify whether its activities comply with requirements and to evaluate the effectiveness of the quality system.
- *Training.* The company must establish procedures for identifying training needs and provide for training of all employees performing activities that affect quality.
- *Servicing.* The company must develop procedures to ensure that service is performed as required by its contracts with customers.
- *Statistical techniques.* Procedures should identify statistical techniques used to control processes, products, and services and how they are implemented.

ISO requires that all published standards be reviewed on a periodic basis. Many of the changes in the 1994 revision were improvements in language, such as clarifying a "product" to be "hardware, software, processed materials, or services," reducing the manufacturing focus of the standards (a college in England achieved ISO registration in 1994, and Surgical Focused Care, a New York City orthopedic practice received registration in 1996). The current set of complete standards can be purchased through the American Society for Quality (http://www.asq.org).

REGISTRATION: The ISO 9000 standards originally were intended to be advisory in nature and to be used for two-party contractual situations (between a customer and supplier) and for internal auditing. However, they quickly evolved as a criterion for companies who wished to "certify" their quality management or achieve "registration" through a third-party auditor, usually a laboratory or some other accreditation agency (called a *registrar*). This process began in the United Kingdom. Rather than a supplier being audited for compliance to the standards by each customer, the registrar certifies the company, and this certification is accepted by all of the supplier's customers.

The registration process includes document review by the registrar of the quality system documents or quality manual; preassessment, which identifies potential noncompliance in the quality system or in the documentation; assessment by a team of two or three auditors of the quality system and its documentation; and surveillance, or periodic re-audits to verify conformity with the practices and systems registered.

Recertification is required every three years. Individual sites – not entire companies – must achieve registration individually. All costs are borne by the applicant, so the process can be quite expensive. A registration audit may cost anywhere from $10,000 to over $40,000 while the internal cost for documentation and training may exceed $100,000.

EFFECT: As of early 1993, only about 550 company sites in the United States were certified. In contrast, some 20,000 companies were certified in the United Kingdom. During the first nine months of 1993, registrations grew by 70 percent to about 45,000 worldwide. This evidence of the growing global interest in the standards is driven primarily by marketplace demands. By June of 1994, the United States had almost 4000 registrations, and the United Kingdom had more than 36,000. Over the last several years, growth has been considerable. Diverse organizations such as schools, physician's offices, and even a ski resort in France have achieved ISO 9000 registration.

The Big Three automakers have mandated that their Tier 1 suppliers receive QS-9000 certification or risk losing business. Chrysler had set a July 31, 1997 deadline; GM's deadline was December 31, 1997. Ford does not require formal QS registration; however to be Q1 certified, suppliers must meet QS requirements. Shortly before the deadline, GM stated that suppliers who are not certified by the deadline will not be eligible for any business with General Motors. Chrysler, however, did not make such a strong statement.

RESULTS: The rigorous documentation standards help companies uncover problems and improve their processes. At DuPont, for example, ISO 9000 has been credited with increasing on-time delivery from 70 to 90 percent, decreasing cycle time from 15 days to 1.5 days, increasing first-pass yields from 72 to 92 percent, and reducing the number of test procedures by one-third. Sun Microsystems' Milpitas plant was certified in 1992, and managers believe that it has helped delivery, and improved quality and service. In Canada, Toronto Plastics, Ltd. reduced defects from 150,000 per million to 15,000 after one year of ISO implementation. Thus, using ISO 9000 as a basis for a quality system can improve productivity, decrease costs, and increase customer satisfaction.

COLLECTIVE WISDOM: Many misconceptions exist about what ISO 9000. The standards do not specify any target measure for quality performance; specific product quality levels are set by the company. The standards only require that the supplier have a verifiable process in place to ensure that it consistently produces what it says it will produce, thus providing confidence to customers and company management that certain principles of good management are followed. The standards emphasize documenting conformance of quality systems to the company's quality manual and established quality system requirements. As one consultant explained it, "Document it, and do it like you document it. If it moves, train it. If not, calibrate it." A supplier can comply with the standards and still produce a poor quality product – as long as it does so consistently! In addition, ISO 9000 does not consider activities such as leadership, strategic planning, or customer relationship management.

ISO 9000 has generated considerable controversy. Many now are questioning its usefulness. The European Union has called for de-emphasizing ISO 9000 registration, citing that companies are focused more on "passing a test" than focusing their energies on quality processes. New quality policies are being debated at this time. The Australian government has stopped requiring it for government contracts. The Australian "Business Review Weekly" noted that "Its reputation among small and medium businesses continues to deteriorate. Some small businesses have almost been destroyed by the endeavor to implement costly and officious quality assurance ISO 9000 systems that hold little relevance to their businesses."

Despite such criticisms, the standards are useful to many organizations. They provide a set of good common practices for quality assurance systems and are an excellent starting point for companies with no formal quality assurance program. Many companies find that their current quality systems already comply with most of the standards. For companies in the early stages of formal quality programs, the standards enforce the discipline of control that is necessary before they can seriously pursue continuous improvement. The requirements of periodic audits reinforce the stated quality system until it becomes ingrained in the company.

Organizations Mentioned: ANSI; ASQC; DuPont; Ford; Freightliner; GM; International Organization for Standardization (ISO); Mack Trucks; Navistar International; PACCAR Inc.; Sun Microsystems; Surgical Focused Care; Toronto Plastics Ltd.

See **Quality management systems: Baldridge, ISO, and QS-9000; Quality: The implications of deming's approach;**

References

[1] AT&T Corporate Quality Office (1994). *Using ISO 9000 to Improve Business Processes.*

[2] Eckstein, A.L.H. , and J. Balakrishnan (1993). "The ISO 9000 Series: Quality Management Systems for the Global Economy," *Production and Inventory Management Journal* 34, 66–71.

[3] Evans, J.R. and W.M. Lindsay (1999). The Management and Control of Quality, Fourth Edition, Cincinnati, OH: South–Western.

[4] n.a. (1996). ISO 9000 Update, *Fortune*, September 30, 134[J].

[5] n.a. (1997). "QS-9000 Deadlines Approach for Big Three Suppliers," *Quality Digest*, 9.

[6] Timbers, M.J. (1992). "ISO 9000 and Europe's Attempts to Mandate Quality," *The Journal of European Business*, 14–25.

[7] Webster, S.E. (1997). "ISO 9000 Certification, A Success Story at Nu Visions Manufacturing," *IIE Solutions*, 18–21.

[8] Zuckerman, A., "ISO/QS-9000 Registration Issues Heating Up Worldwide," *The Quality Observer*, June 1997, 21–23.

ITEM MASTER FILE

A material planning and control system based on Material Requirements Planning (MRP) database uses several key computer files.

Two files are commonly used to store part related data. Some data concerning parts will remain the same, or very similar, from period to period, and these data are stored in the Item Master File. Information on part status, which is more dynamic, is stored in the Subordinate Item Master File. The item master file contains all the static data necessary to give a complete description of each part. The data in the item master file are used for MRP, purchasing, costing, and so on. The data on each consist of part number, part name, low-level code, unit of measure, engineering change number, drawing reference, release date, planner code, order policy code, lead time, safety stock, standard costs, and links to routing and bill of material (BOM) files. Th e subordinate item master file contains data pertaining to current shop order numbers, time-phased scheduled receipts with order numbers, time-phased gross requirements, planned orders, firm planned orders, pegging data, and links to the item master file for the part.

See **Bill of materials; Material requirements planning.**

J

JAPANESE BUYER/SUPPLIER RELATIONSHIP

The recent drive for collaboration in supply chains has been fuelled by Japanese practice. Japanese collaborative relationship is the opposite of an adversarial relationship which used to very common in the West. Saunders (1997, p. 255) highlights a number of factors that characterize this collaborative style of partnership (1) the partners communicate with each other frequently using both formal and informal channels; (2) personnel from the partners work together co-operatively, and relationships are based on mutual trust; (3) negotiations are focused on solving the supply problem by minimizing overall costs; (4) contracts are negotiated for the long term; (5) information is shared openly between the two partners; (6) suppliers are involved in the design of new products; and (7) the quality of supply is assured by joint efforts with an emphasis on defect prevention.

See **Just-in-time manufacturing; Purchasing: Acquiring the best inputs; Supplier relationships; Supply chain managing.**

Reference

[1] Saunders, M. (1997). *Strategic Purchasing and Supply Chain Management*. Pitman Publising, London.

JIGS AND FIXTURES

Jigs and fixtures are specially designed attachments used to facilitate easier and faster set up in manufacturing operations, especially machining and fabrication. Jigs are attachments that locate tools (e.g., drill bits) accurately so that multiple operations (e.g., drilling of multiple holes) may be repeated easily on a number of jobs without costly set up. Fixtures are attachments that hold the jobs and rotate them as necessary to facilitate fabrication or machining. For example, if the fabrication of a job involves placing welds at several different locations, fixtures facilitate holding and rotating the job with speed and accuracy.

See **Manufacturing flexibility dimensions; SMET; Setup reduction.**

JIT

See **Just-in-time manufacturing.**

JIT AND MRP

See **Just-in-time manufacturing; Material requirements planning.**

JIT DELIVERY

See **Just-in-time manufacturing.**

JIT EVOLUTION AND USE IN THE UNITED STATES

Richard E. White

University of North Texas, Denton, TX, USA

DESCRIPTION: Since the Japanese manufacturers have demonstrated their ability to simultaneously achieve high levels of productivity and quality while producing a wide variety of products, Japanese management practices are often adopted by U.S. manufacturers as progressive alternatives to their traditional management practices. Much of the Japanese manufacturers' success has been attributed to a management philosophy and a set of management practices initially referred to as "Just-In-Time" manufacturing. The creation of Just-In-Time (JIT) manufacturing is generally credited to Taiichi Ohno, a Vice President of Toyota Motor Corporation (Hall, 1983).

1. ***JIT Philosophy:*** The philosophy associated with JIT has been described in a variety of ways. Common terms used in describing it include: simplicity, rationality, flexibility, and continuous improvement. The philosophy underlying JIT manufacturing involves continual improvement in cost, productivity, quality, and flexibility in the pursuit of excellence in all phases of the manufacturing cycle. Early literature in this area (Monden, 1983; Shingo, 1981) suggests that JIT manufacturing was developed under a simple philosophy supported by principles of efficiency (elimination of all

wastes), and the development and full utilization of the capabilities of people (respect for people).

(a) *Elimination of All Wastes:* Elimination of all wastes includes waste that pertains to time, materials, parts, and equipment (inventory is a reflection of these underlying wastes) within the industrial cycle. Shingo (1981) originally identified seven different types of wastes in the Toyota Production System (TPS) including waste from processing, delay, defective products, motion, overproduction, stock, and transportation. These wastes were referred to as 'unnecessaries' by Monden (1983), and it was through the attack and elimination of these 'unnecessaries' that the Toyota Production System evolved.

(b) *Respect for People:* Respect for people includes making full use of their capabilities, both mental and physical. Traditional U.S. management practices have been built around the principle of division of labor and an adversarial relationship between labor and management; thus, efforts to develop employees beyond a specialized skill contradicted the division of labor principle. In addition, developing "respect for people" in the U.S. manufacturing environment includes breaking down the traditional adversarial barriers that exist between labor and management, changing management's attitude toward workers, integrating workers in the company decision making process, and utilizing small group improvement activities.

The JIT philosophy redefines the way we look at business practices and involves an implementation process that is "...the creation of a flexible environment in which everyone seeks to eliminate waste and to keep things simple so as to effect a continuous improvement in overall business" (Schneider and Leatherman, 1992: 78).

2. ***Confusion about JIT:*** When first introduced in the U.S. in the early 1980s, there was much confusion about JIT. Early JIT literature was often inconsistent in describing the concept and scope of JIT and typically emphasized that Just-In-Time manufacturing evolved from a repetitive production process, the Toyota Production System (Monden, 1983; Shingo). Many early studies presented JIT as just the material flow aspect of the Toyota Production System. This created confusion about the scope of JIT. These early studies presented a limited scope of JIT because quality-improvement and people-involvement activities were ingrained in the Japanese organizational culture; consequently, these activities were taken for granted by Japanese writers and not included in the early descriptions of JIT. Common misunderstandings created among U.S. managers about JIT included: JIT and Kanban are synonymous; JIT is a system only for materials movement aimed at reducing inventories; and JIT is applicable primarily to repetitive production processes. Moreover, confusion about JIT was created because the names given to JIT in the early literature were inconsistent, i.e., stockless production, short cycle manufacturing, just-in-time production, Kanban, and Japanese manufacturing techniques.

The level of confusion about JIT manufacturing among U.S. business people has decreased since the early 1980s, in part, because writers have developed a better understanding of JIT. In addition, it has become increasingly apparent that practices associated with good quality systems are integrally tied into JIT systems. Consequently, later descriptions of JIT systems typically involved broadened perspectives that consist of practices associated with efficient material flow, good quality, and increased employee involvement systems (Hall, 1987; Hay, 1988; Schonberger, 1986). Later writers more accurately describe JIT as a comprehensive management system that continuously strives for improvement; however, the names given JIT are still inconsistent, i.e., total enterprise manufacturing, world-class manufacturing, and lean production.

In reference to the names used for the philosophy of this broad conceptual system, Hall stated "None of the names often used for this philosophy suggest its total power and scope... and none are universally used" (Hall, 1987: 23). Even though the term, "JIT" has been misused and may be less than adequate to describe this broad conceptual system, it is the term used here to describe this broad based Japanese management system that strives to achieve excellence.

3. ***Just-In-Time Manufacturing:*** JIT manufacturing is a broad based manufacturing management system that covers all activities of the production system, from the design of the product, through production, to delivery to the customer, and centers on the elimination of all wastes throughout the manufacturing cycle. JIT manufacturing consists of three subsystems, systems for efficient material flow, good quality, and increased employee involvement. Each of these subsystems contributes to the

development of a conceptual system that resembles a pipeline of material flow from the raw material stage to the end customer (Hall, 1987).

Final assembly, for the JIT manufacturer, is the control point for pulling materials through the pipeline and the only products produced are those needed to satisfy immediate demand. Each operation in the pipeline serves as the customer for the preceding operation's products. The intent of a JIT manufacturer is to create a sense of serving the customer in each operation along the pipeline; therefore, good information flow among the operations is a critical aspect of JIT. With JIT, inventories are eliminated and interdependence among operations is intensified; thus, the tighter linkages make the operations more responsive to one another. The pull system (Kanban) of JIT allows for reciprocal interdependence, and the "... reciprocal interdependence encourages a wider perspective and a product-wide (if not plant- or company-wide) optimization driven by customer needs" (Arogyaswamy & Simmons, 1991: 57).

4. ***JIT Systems:*** After attending a 1983 APICS advanced study mission of various Japanese manufacturers, Burnham emphasized that the systems aspect was an important feature of the Japanese manufacturers' approach to management. He stated,

> "...the most impressive aspect of the visit was the always present evidence of integrative, systematic thinking. Systems appeared in everything. And balance, practicality, appropriate choices of equipment and people tasks, tie-ins between computers, materials, people, schedules, and equipment occurred seemingly without effort" (Burnham, 1985: 53).

Each JIT management practice provides some benefit for a production system, but the application of each practice may involve only certain areas in the organization, and unless a systems perspective is employed, the areas optimize locally, rather than at the organization level. Consequently, the potential synergistic benefits cannot be fully realized until all management practices in the system are functioning with a common goal in mind. Satisfying demand for this pipeline of material flow becomes the common goal for the employees in JIT manufacturing. Since the application of each management practice involves only certain groups of people, it is critical to understand the linkages among these different groups of people to effectively implement JIT. Suzaki (1987) points out that an organization must be viewed as a whole since problems in organizations often are found at organizational boundaries; therefore, integration of the JIT practices requires working through the boundaries formed between the different groups.

Based on their review of recent JIT literature, Goyal and Deshmukh (1992) state that understanding the concept of JIT requires a systems perspective.

TECHNOLOGY PERSPECTIVE: Although JIT manufacturing systems are presented as holistic integrated systems, in U.S. manufacturing it may consist of any number of different management practices that contribute to eliminating wastes in the system and continuous improvement of the system.

Based on a comprehensive survey of the JIT literature (articles, books and/or studies that address JIT from a broad-based perspective), White and Ruch (1990) synthesized the literature and identified ten management practices typically associated with JIT manufacturing systems and feasible for implementation by U.S. manufacturers. The ten management practices include the following: **focused factory, reduction of set-up times, group technology, total productive maintenance, multi-function employees, uniform workload, just-in-time purchasing, Kanban, total quality control, and quality circles**. White and Ruch proposed that the individual JIT management practices can be introduced into U.S. manufacturers as separate programs for improving performance. However, the researchers note that the greatest gains achieved by organizations implementing JIT may be the synergistic benefits derived from several JIT practices operating as an integrated system.

HISTORICAL PERSPECTIVE:

1. ***Toyota Production System (TPS):*** After World War II, in an attempt to catch up and compete with automotive industries in the Western nations, Toyota focused on improving their production system and making it more competitive. Initially, Toyota along with many other Japanese companies began the development of quality systems that focused on the customers and involved lower level employees in the quality improvement process. Early in the 1950s, Deming, a quality expert from the U.S., went to Japan teach statistical quality control techniques. Subsequently, Juran, another U.S. quality expert, who was asked to review the progress of the quality movement in Japan, noted that Japanese companies were too focused on just statistical techniques and

lacked the overall management of quality. However, later, Toyota successfully integrated many of the quality techniques and built the cultural and organizational framework to support an all around production improvement.

Due to scarcity of resources, the improvement effort at Toyota was built on a "non-cost principle" (Shingo, 1981), where, cost reduction was emphasized and improvement was achieved by attacking and eliminating all 'unnecessaries' in the production system (Monden, 1983). The foundation of the TPS had its roots in Henry Ford's original mass production system. Because Ford's production system was the largest, most efficient system in the world, Toyota, the largest postwar Japanese producer, studied Ford's production system and adopted many of Ford's principles of mass production. Toyota's managers believed they could apply Ford's principles of mass production to their high volume manufacturing environment and improve on the Ford system (Womack, Jones and Roos 1991). The approach for developing Toyota's Production System was different from that of Ford's; Toyota's effort centered not only around improving system efficiency, but also on improving system flexibility.

As Toyota's Production System evolved, the management practices were extended upstream to include Toyota's suppliers as well as their suppliers. Then, other large Japanese auto producers, Hino, Daihatsu, Mazda, and Nissan copied the management practices introduced at Toyota in the 1950s and 1960s. With diffusion of these management practices to other Japanese companies and a dedicated effort to pursue perfection by participating companies, Japanese Just-In-Time manufacturing evolved.

Kanban, the management practice created by Ohno, is the mechanism that connects organizational activities in a JIT environment. It is the key to attacking and eliminating inventories, and for identifying other unnecessaries to eliminate. It was Toyota's continual effort to improve the system that led to the integration of their improved practices, and subsequently, resulted in the world-class production system that involves new and improved manufacturing practices.

2. ***Diffusion to the U.S.:*** In the early 1980s, when they were losing market share to the Japanese, the U.S. automotive companies began sending teams of employees to Japan to study their methods of operation. Then, in 1983, the Automotive Industry Action Group (AIAG) was formed to help communicate the concepts of Japanese production methods to employees and suppliers, and assist automotive companies in the implementation of JIT manufacturing throughout the industry. AIAG defined 14 different practices that made up JIT manufacturing systems, and categorized them under the two principles of JIT: elimination of wastes and respect for people. Six of these practices were categorized under the principle of elimination of wastes (i.e., minimize set-up times, quality at the source, Kanban, group technology, uniform plant loading, and just-in-time production), and were promoted as being immediately applicable to U.S. automotive manufacturing plants. Those practices that were categorized under the principle of respect for people (i.e., consensus decision-making, quality circles, lifetime employment, company union structure, and management attitude) were promoted as practices not immediately applicable, or which required other changes for implementation of JIT in U.S. automotive companies. The practices categorized as immediately applicable have subsequently been implemented in many U.S. automotive companies.

Although diffusion of JIT to the U.S. began in the auto industry, later JIT implementations extended far beyond the auto industry. Then, other major U.S. industries using JIT are represented by companies, such as Westinghouse, Hewlett-Packard, Omark Industries, Intel, 3M, Tennant, General Electric, Black and Decker, IBM, Honeywell, and Motorola (Schonberger, 1986; Suzaki, 1987).

Trial and Error Effort: JIT, the offspring of Toyota, was the first in a series of management philosophies that U.S. businesses used to emulate the success achieved by Japanese manufacturers. Each focused on adopting new management practices that lead to improved performance by elimination of wastes and full utilization of human resources. Often, U.S. manufacturers adopted these Japanese management practices without a clear working knowledge of the practices and their applicability, and with little understanding of pertinent implementation issues. In addition, many organizations frequently encountered difficulties during implementation; this resulted in costly trial and error activities during implementation. Unfortunately, once an organization tried its version of the new practice and found limited success, it was difficult to regain employee support and the organizational momentum necessary to retry a refined version of the same practice again.

Many U.S. manufacturers had already implemented parts of Japanese management practices when it became apparent that JIT was more than just an inventory system, and that TQM principles provide the foundation necessary to make JIT successful (Shingo, 1981). World-Class Manufacturing (Schonberger, 1986) and Lean Production (Womack *et al.* 1991). Which are based on JIT are sought after by struggling organizations to achieve global competitiveness. Today, JIT implementations are common in U.S. manufacturing; however, often, without realizing the full potential of JIT systems. Since a series of trial and error type efforts were used to apply the Japanese manufacturing practices in the U.S., many of the issues associated with JIT implementations are still not fully understood by U.S. managers.

IMPLEMENTATION: Many U.S. manufacturers appear to have whole-heartedly adopted these new management practices; however, the adoption process typically begins by adopting those aspects of JIT that are easy to implement. Because the adoption process involves substantive changes in management systems that are well imbedded in the infrastructure of U.S. organizations, a high level of success is not easily achieved; thus, the gain is often less than anticipated.

Today, JIT implementations are more common and more advanced (longer period of time since implementation began) in large U.S. manufacturers, less so in small manufacturers. Eighty-two percent of larger plants report JIT use as opposed to 61 percent of small plants with less than 100 employees (Swamidass, 1996). Large manufacturers have reported substantial successes in improving organizational performance. Diffusion of JIT to small U.S. manufacturers may not be as advanced when compared to that of large manufacturers, but evidence suggests that implementation of JIT may be just as beneficial in small U.S. manufacturers as it is in large manufacturers.

Research on JIT implementations by U.S. manufacturers generally indicates: (1) not all organizations implement the same JIT management practices in their organization; (2) the type of production process and size of the manufacturer influences the specific JIT practices adopted and the number of practices adopted by an organization; (3) substantial benefits can be achieved from implementing JIT management practices; and (4) the time since the start of implementation of JIT influences the success achieved from JIT.

1. ***Partial JIT Implementations:*** Most JIT systems implemented by U.S. manufacturers are partial; the manufacturers typically do not implement all the JIT management practices. Only the practices that are easier to implement are adopted. Consequently, many U.S. adopters of JIT systems have not realized the full potential of JIT; many have not achieved the synergistic benefits from the JIT management practices operating as an integrated cohesive production system.

   One research study about JIT implementations indicates the implemented, in descending order of frequency are: (1) reduced setup times, (2) multi-function employees, (3) total quality control, (4) just-in-time purchasing, (5) Kanban, (6) tied-total productive maintenance and quality circles, (8) focused factory, (9) group technology, and (10) uniform workload (White, 1993).
2. ***Production Process Type:*** The type of production process employed by an organization influences the adoption of JIT management practices in the U.S. Early in the diffusion in the JIT to U.S., JIT manufacturing was generally understood to apply to repetitive production processes. However, subsequent research has shown that JIT systems are applicable and provide opportunities for improving productivity and quality in both repetitive and nonrepetitive production processes. Due to the characteristics of nonrepetitive production processes, implementation of JIT systems may require modifications.

   By implementing just the JIT practices of multi-function employees, reduced setup times, just-in-time purchased parts, group technology, and total quality control, nonrepetitive manufacturers can obtain substantial improvements in processing throughput time, productivity, improved quality levels and inventory reductions. Research indicates that U.S. manufacturers employing repetitive production processes typically implement the various JIT practices more often than organizations employing nonrepetitive production processes, except for the use of multi-function employee, which may be easier to implement in a nonrepetitive production environment.
3. ***Manufacturer Size:*** The results of research in this area indicate that manufacturer size influences the number of JIT practices adopted (Gilbert 1990; White, 1993). Since greater task specialization, more elaborate organizational structures (more levels in the hierarchy with sharper administrative division of labor), and more formalization of behavior (more operating policies for decision-makers, rules, procedures, norms of behavior, job descriptions, etc.) typically exist in larger organizations,

implementation of JIT in larger U.S. manufacturers face different deterrents during implementation compared to an those faced by smaller manufacturers. Large U.S. manufacturers typically have a structure with characteristics contrary to the structure of Japanese manufacturers.

Compared to small U.S. manufacturers, large manufacturers have more clout over their suppliers to support the implementation of the purchasing aspects of JIT. Large U.S. manufacturers may also have more financial resources to assure proper implementation of JIT systems when compared to that of small manufacturers. Research indicates that larger U.S. manufacturers typically implement the various JIT management practices more often than small manufacturers, except for the use of multi-function employees, which may not require dramatic cultural change in small manufacturing firms.

4. ***Implementation Sequence:*** Researchers are investigating the best sequential implementation of JIT practices. A given JIT practice may influence the implementation of another JIT practice. For example, implementing quality circles first by manufacturers can provide the organizational mechanism for involving employees in problem solving and decision making, establishing cooperative work relationships, and developing the necessary informal lines of communication (horizontally and vertically). Then, quality circles can serve as a facilitator for implementing other JIT practices.

   Since the organization's characteristics influence the JIT management practices implemented, the appropriate implementation sequence will vary dependent on the organization. By choosing the first JIT management practice to be implemented based on the highest probability of success confidence within the organization for JIT implementation can be developed quickly. Since each success attained during the implementation of the JIT practices can potentially modify the existing organization characteristics, a reevaluation of applicability and implementation priority may be warranted following the successful implementation of each JIT management practice.

5. ***Time Since the start of JIT Implementation Influences the Success Achieved:*** Research indicates that length of time since JIT implementation began influences the success achieved from the JIT system. Toyota took over forty years to implement JIT practices, refine the practices and integrate them into a cohesive operating system. Still today, Toyota continues to develop and refine their production system. U.S. organizations must have a long term commitment to the JIT implementation process. Ultimately, with time U.S. manufacturers should obtain the synergic benefits from JIT as a cohesive integrated system.

   Quality circles is the management practice that has typically been in operation the longest in the U.S. Although, not necessarily as an integral part of a JIT system, but more as a motivational technique. After quality circles, total quality control has been in the U.S. the longest, and total productive maintenance is the JIT practice most recently adopted by U.S. manufacturers. Typically, all the JIT management practices have been in operation longer in large manufacturers than in small, and in operation longer in manufacturers.

COLLECTIVE WISDOM: With the implementation of JIT, the organizational structure must be redesigned to create coordination mechanisms that increasingly involve employees in problem solving. Researchers studying JIT implementations in traditional U.S. manufacturers suggest that effective implementation requires a sense of shared vision by all employees and a long term commitment by the organization. It is extremely important that the various functional members of an organization all have the same vision, or common goals to be accomplished when implementing JIT. The increased interdependencies created among organizational subareas with the implementation of JIT require more open communication (informal lines of communication vertically and horizontally) and cooperative work relationships across functional areas and levels of management hierarchy.

Since the production orientation of U.S. workers and managers differs from that of JIT's birthplace (Womack *et al.*, 1991), implementation of JIT requires substantial training to create employee involvement. The decision-making authority with JIT implementation must be moved down the organizational levels to provide for responsive and continuous improvement of the system. This requires the development of new employee skills (i.e., communication skills, conflict resolution skills, negotiation, etc.). Senge (1990), in a related discussion, suggests that developing a common vision and shared mental model allows for more effective adoption of new approaches to management in organizations. He states, "...new insights fail to get put into practice because they conflict with deeply held internal images of how the world works, images that limit us to familiar ways of thinking and acting" (Senge, 1990: 174).

The effective implementation of JIT in U.S. manufacturers must begin with top management's commitment. Management must be dedicated to supporting the long term effort required for JIT implementation. Developing a formal implementation plan will further enhance the effectiveness of the implementation process. In addition to goals and desired completion dates, the formal implementation plan should involve training to provide all employees with the necessary skills, understanding, and vision to achieve the desired success.

Total quality control is a fundamental building block of JIT systems; it should be in place to provide the foundation for implementing the remaining JIT practices. In addition, manufacturers must develop "respect for people" early in the implementation process. Schonberger (1982) points out that lack of increased respect for humans and people involvement will fail to set the foundation for the advanced stages of JIT implementation. Implementing quality circles early will allow the organization to begin it's movement toward "respect for people".

JIT purchasing is the practice that allows manufacturers to extend JIT upstream to the suppliers; it should be implemented in the later stages of the implementation process. Researchers suggest that manufacturers should implement JIT internally, before implementing JIT purchasing. A successful implementation of JIT internally provides an example for the supplier in addition, the manufacturer is much more knowledgeable about what they want from the supplier and how to help the supplier.

EFFECT: Overall, findings of pertinent research in this area suggest that implementation of JIT provides improved performance for U.S. manufacturers. The results generally indicate that JIT provides both tangible and intangible benefits. The tangible benefits typically include improvements in the following areas: throughput time, quality levels, labor productivity, employee relations, inventory levels (raw materials, work-in-progress, finished goods), and manufacturing costs. The intangible benefits typically include perceived improvement in: problem solving capability, control of engineering changes, worker enhancement, and communication between operations.

A synthesis of various studies on JIT implementation indicates that benefits can be achieved in a variety of manufacturing situations. Hay (1988), a management consultant who specializes in JIT implementation, reported a range of benefits achieved from the implementation of JIT by his clients, who represent a number of different

Table 1. Percent improvement in benefits from JIT implementation.

Benefits	Percent improvement
Productivity Increase	
Direct Labor	5–50
Indirect Labor	21–60
Cost of Quality Reduction	26–63
Purchased Material	
Price Reduction	6–45
Inventory Reduction	
Purchased Material	35–73
Work In Process	70–89
Finished Goods	0–90
Setup Reduction	75–94
Lead Time Reduction	83–92
Space Reduction	39–80

Note: Adapted from *The Just-In-Time Breakthrough: Implementing the New Manufacturing Basics* (p. 23) by E.J. Hay, 1988, New York: John Wiley & Sons.

industries (see Table 1). Overall, research indicates that JIT manufacturing is beneficial to all U.S. manufacturers regardless of size or type of process employed.

Organizations Mentioned: 3M; AIAG; Black and decker; Daihatsu; General electric; Hewlett-Packard; Hino; Honeywell; IBM; Intel; Mazda; Motorola; Nissan; Omark industries; Tennant; Toyota; Westinghouse.

See **Focused factory; Group technology; Just-in-time manufacturing; Just-in-time purchasing; Kanban; Lean manufacturing implementation; Manufacturing all design; Multi-function employees; Quality circles; Reduction of set-up times; SMED; Total productive maintenance; Total quality control; Uniform workload.**

References

[1] Arogyaswamy, B. and R.P. Simmons (1991). "Thriving on independence: The key to JIT implementation." *Production and Inventory Management Journal*, 32, 3, 56–60.

[2] Automotive Industry Action Group (AIAG) (1983). *The Japanese Approach to Productivity*. Videotape Series.

[3] Burnham, J.M. (1985). "Improving manufacturing performance: A need for integration." *Production and Inventory Management*, 26, 4, 51–59.

[4] Gilbert, J.P. (1990). "The state of JIT implementation and development in the USA." *International Journal of Production Research*, 28, 1099–1109.

[5] Goyal, S.K. and S.G. Deshmukh (1992). "A critique of the literature on just-in-time Manufacturing."

International Journal of Operations & Production Management, 12, 1, 18–28.
[6] Hall, R.W. (1983). *Zero Inventories*. Dow-Jones-Irwin, Homewood, IL.
[7] Hall, R.W. (1987). *Attaining Manufacturing Excellence*. Dow Jones-Irwin, Homewood, IL.
[8] Hay, E.J. (1988). *The Just-In-Time Breakthrough: Implementing the New Manufacturing Basics*. John Wiley & Sons, New York: NY.
[9] Monden, Y. (1983). *Toyota Production System: A Practical Approach to Production Management*. Industrial Engineers and Management Press, Norcross, GA.
[10] Schneider, J.D. and Leatherman, M.A. (1992). "Integrated just-in-time: A total business approach." *Production and Inventory Management Journal*, 33, 1, 78–82.
[11] Schonberger, R.J. (1982). *Japanese Manufacturing Techniques: Nine Hidden Lessons in Simplicity*. The Free Press, New York.
[12] Schonberger, R.J. (1986). *World Class Manufacturing: The Lessons of Simplicity Applied*. The Free Press, New York.
[13] Senge, P.M. (1990). *The Fifth Discipline: The Art and Practice of the Learning Organization*, Doubleday Currency, New York: NY.
[14] Shingo, S. (1981). *Study of Toyota Production System From Industrial Engineering Viewpoint*. Japan Management Association, Tokyo.
[15] Suzaki, K. (1987). *The New Manufacturing Challenge: Techniques for Continuous Improvement*. The Free Press, New York.
[16] Swamidass, P.M. (1996). "Benchmarking Manufacturing Technology Use in the United States." *Handbook of Technology Management*, Gerard H. "Gus" Gaynor (Ed.), 37.1–37.40, McGraw-Hill Inc., New York.
[17] White, R.E. (1993). "An empirical assessment of JIT in U.S. manufacturers." *Production and Inventory Management Journal*, 34, 2, 38–42.
[18] White, R.E. and W.A. Ruch (1990). "The composition and scope of JIT." *Operations Management Review*, 7, 3&4, 9–18.
[19] Womack, J.P., D.T. Jones, and D. Roos (1991). *The Machine That Changed the World: The Story of Lean Production*. Harper Perennial, New York.

JIT II

JIT II refers to a program used to enhance the flow of information between the buyer and seller by having a supplier representative located in the buyer's facility. The supplier representative is responsible for meeting delivery schedules with quality goods. The seller gains access to important scheduling information to better manage and schedule their facility.

See **JIT; Purchasing: The future.**

JIT PHILOSOPHY

Just-in-Time (JIT) is not just a manufacturing technique but a philosophy of manufacturing that influences a company's relationship with its suppliers, customers, and employees. The two basic underpinnings of this philosophy are elimination of anything that does not add value for the customer, and continuous improvement. Thus, the emphasis is on efficient utilization of resources, where resources can include time, material, and people. JIT activities include setup and lead time reduction, minimization of inventory, employee involvement in the decision making process, cooperative arrangements with suppliers, and a focus on meeting the needs of the customer.

The JIT philosophy fosters an environment, where continuous improvements are sought in waste reduction and quality. Another important aspect of the JIT philosophy is that improves the relationships with employees, and employees are given broad problem-solving and decision-making authority.

See **Just-in-time manufacturing; Just-in-time manufacturing: Implications.**

References

[1] Schonberger, R.J. (1982). *Japanese Manufacturing Techniques: Nine Hidden Lessons in Simplicity*. The Free Press, New York.
[2] Sepehri, M. (1986). *Just-in-Time, Not Just in Japan*. American Production and Inventory Control Society, Virginia.

JIT PURCHASING STRATEGY

Just-in-time (JIT) purchasing and total quality control (TQC) concepts are not limited to the shop floor but extend back to a company's suppliers. JIT purchasing arrangements often result in long-term contracts and in the co-location of facilities. That is, suppliers and buyers come to depend on each others' business to the point that they build facilities next to each other to make the logistics of frequent, small deliveries easier. In some instances, managers from the supplier actually work at the buyer's facility and manage the purchasing process for all of the materials purchased from that supplier. In essence, the supplier manages both the inventory and the purchasing for the buyer/supplier relationship. This approach has come to be known as JIT II.

In almost all instances, JIT purchasing requires a reduction in the total number of suppliers. For example, Xerox reduced its supply base from almost 5,000 suppliers down to fewer than 500 as it implemented a JIT purchasing philosophy. As the supply base is reduced, the buyer has more time to devote to the remaining suppliers to build better, stronger relationships, where information and expertise are exchanged and shared. Indeed, as the number of suppliers is reduced, and more particularly, as sole sourcing becomes the norm, it is very important to develop the very best set of suppliers possible.

When the firm no longer has at least two suppliers for every purchased part, it must make sure that its sole supplier is reliable and competitive. For this reason, supplier certification is a common practice in JIT purchasing. The philosophy can be stated as follows: traditional purchasers believe in "never putting all their eggs in one basket," after JIT purchasers believe in "putting all their eggs in one basket but making sure its a good one" and then they watch it closely."

Establishing JIT purchasing is not a simple task even where close geographic proximity simplifies the relationship and logistical challenges implicit in JIT purchasing. The difficulty arises in developing the close relationship needed to make JIT purchasing mutually beneficial for both the buyer and the supplier. However, when the characteristics of JIT purchasing – larger volumes, more stability, information sharing, better training, shared risks and rewards – are closely examined, it becomes apparent that the interests of both buyer and supplier are indeed served. Ultimately, JIT purchasing arrangements are characterized by the integration of buyer-supplier activities. The degree of integration and the specific arrangements depend on the strength of the relationship between the buyer and the supplier.

Many significant benefits are achieved through the integration of buyer and supplier activities including the simplification of the administrative work involved in the purchasing function through the use of long-term, blanket contracts, which greatly reduce paperwork. In addition, by working closely with the supplier and by simplifying production specs, quality can be improved to the point that in-bound inspections can be eliminated. Supplier expertise can also be more rapidly incorporated into the new product development process, leading to better products developed with shorter concept-to-market cycle times. The end result is a reduction of overall costs, an increase in productivity, and an improvement in product quality.

Organizations Mentioned: Xerox corporation.

See **Just-in-time manufacturing; JIT II; Logistics: Meeting customers real needs; Total quality control; Total quality management.**

References

[1] Schonberger, R.J. (1986). *World Class Manufacturing*. The Free Press, New York.

[2] Fawcett, S.E. and Birou, L.M. (1993). "Just-In-Time Sourcing Techniques: Current State of Adoption and Performance Benefits." *Production and Inventory Mangement Journal*, 34(1), 18–24.

JIT VERSUS MRP

See **Just-in-time manufacturing; Kanban-based manufacturing systems.**

JOB ENRICHMENT

Job enrichment expands the scope of an operator's job to include more diverse tasks or activities. It results in the expansion of operator skills. It reduces operator boredom and has the potential for greater job satisfaction. It has been found that satisfied workers are more productive and effective.

See **Human resource issues and advanced manufacturing technology; Human resource management in manufacturing; Manufacturing cell design.**

JOB INVOLVEMENT

Job involvement expands the responsibility of the worker to include more decision making.

See **Human resource management in manufacturing; Manufacturing cell design.**

JOB RELEASE

Job release is another name for order release, which controls a manufacturing system's input. It is also known as order review/release, input/output control, or just input control.

See **Input/output control; Order release.**

JOB ROTATION

Job rotation involves moving of employees into different job assignments. The objective is to encourage broad skill development and a cross-fertilization of ideas. Employees can be rotated through different jobs in a number of ways. Long term programs may involve assignments of six months to two years in three or four different departments or functional areas over the course of three to five years. Shorter term programs can last from a few days to a few months. Job rotation helps expand the skill set of employees, develop new viewpoints and encourages more cross-functional communication. For example, manufacturing engineers placed in temporary positions with a marketing or sales team generally can develop new skills for selling products, as well as an appreciation of the challenges involved in marketing. Similarly, marketing professionals temporarily assigned to a production team can develop new skills and an understanding of the challenges involved in production.

See **Flexibility in manufacturing; Human resource issues in manufacturing.**

JOB SEQUENCING RULES

In a job shop, it is common for several jobs to wait in a queue for processing at each work center. The order in which the waiting jobs are processed can have a profound effect on the number of jobs waiting in the plant, the number of jobs that are delivered on-time, the time the jobs spend in the plant, the work-in-process inventory, and the average lateness of deliveries to customers. Therefore, the sequencing of jobs at work centers in an important decision. Since thousands of sequences must be determined every week in an average plant, it is important to develop simple and practical rules for job sequencing.

Job sequencing rules, or dispatching rules are used for sequencing jobs at a work center. A few of the common rules are:

1. Earliest due date rule (EDD) – job that is due first is started first. Used when companies are sensitive to due date changes.
2. Shortest processing time rule (SPT) – this gets the most work done rapidly and minimizes work-in-process (WIP). A drawback of this rule is that jobs with very long processing time may get delayed. To prevent this, periodically the rule is changed to longest processing time rule.
3. Longest processing time rule (LPT) – this rule increases the WIP and may cause many short jobs to miss their due dates.
4. First come, first served rule (FCFS) – it is the fairest rule, especially when people are to be sequenced at a service facility. It does not perform as well as the SPT rule.
5. Critical ratio rule (CR) – this rule combines the due date and processing time. CR is computed by the ratio:

$$CR = \frac{\text{time until due date}}{\text{processing time remaining}}$$

   Jobs with the lowest CR are schedules first. If CR is less than 1, the job will be past due. This rule schedules first those jobs which have the greatest chance of missing the due date.

In selecting a rule for use, one or more performance criteria are employed. A few of the common performance criteria used in evaluating sequencing rules are:

1. Average time to produce the item (called average flow time, which includes waiting and processing time).
2. Average WIP inventory.
3. Average lateness of jobs (a job is late if it is completed after a known due date).

Often, a simulation can be performed to assess the most appropriate rule to use in offer to meet one or more of the performance criteria specified by management.

See **Manufacturing systems; Setup reduction; Schedule stability; Simulation analysis of manufacturing logistics systems; Simulation of production problems using spreadsheet programs.**

JOB SHOP

See **History of manufacturing; Manufacturing systems.**

JOB STRESS

Objective work characteristics, i.e., stressors, can lead to perceptions of stress in employees. These perceptions are influenced by individual factors such as occupation, sex, and experience, which may act as potential moderators of stress. Individual perceptions of stress trigger job-related strains. Strains commonly refer to physiological, behavioral, and psychological responses to stressors. Prolonged experience of strains may affect the individual's health and may also have repercussions on the individual's performance on the job. Positive or negative situations can lead to stress. However, empirical research has shown that negative events had stronger effects on strain.

See **Human resource issues and advanced manufacturing technology.**

JUST-IN-TIME (JIT)

See **JIT philosophy; JIT purchasing; Just-in-time manufacturing; Safety stocks: Luxury or necessity; Total quality management.**

JUST-IN-TIME MANUFACTURING

Gregory P. White

Southern Illinois University at Carbondale, Illinois, USA

DESCRIPTION: Just-in-Time (JIT) is a philosophy of operation that seeks to utilize all resources in the most efficient manner by eliminating anything that does not contribute value for the customer. In this philosophy, resources include – but are not limited to – equipment, facilities, inventory, time, and human resources. Because of this broad definition of resources, JIT includes many different components. Some of those components emphasize the efficient use of material resources; other components of JIT focus on the efficient use of human resources. In fact, experts don't even agree completely on the factors that should be included as components of JIT. To make matters worse, some components of JIT are also important components of other techniques, such as Total Quality Management (TQM).

One possible way to define JIT that avoids the problem of identifying specific components is to define JIT as "flow manufacturing." The model for this definition is a continuous flow type of process such as an oil refinery or a paper mill. Such processes generally operate very efficiently, with material flowing rapidly from raw material to finished product. The objective of JIT is to make manufacturing processes resemble such flow processes in as many ways as possible.

HISTORICAL PERSPECTIVE: Although Henry Ford purportedly used some JIT concepts to produce the Model T (Womack *et al.*, 1990), Just-in-Time as a complete philosophy was developed by the Japanese company Toyota and was used only in Japan until the late 1970s, when some U.S. companies began to notice that certain Japanese competitors were producing better products for lower cost. Subsequently, delegations from U.S. manufacturers visited Japanese plants. What most impressed those delegations was the drastically reduced inventory level with which most Japanese plants functioned. Other, more important results of JIT, that were not as noticeable, did not originally receive much attention. As a result, JIT was known early on in the U.S. under such names as "zero inventory" (Hall, 1983) or "stockless production", or even as "kanban", which, as will be described later, is just one means of controlling production and inventory levels under the JIT system of manufacturing.

As more individuals studied JIT, they eventually realized that it involves more than just an inventory control system (Schonberger, 1982; Monden, 1981; Monden, 1983). In fact, many organizations today believe that other components of JIT are even more important than inventory reduction, and that inventory reduction is really just a by-product of other changes brought about by JIT. Keller and Kazazi (1993) provide a good overview of the JIT literature.

STRATEGIC PERSPECTIVES: (Krajewski *et al.*, 1987) used a computer to simulate various approaches to manufacturing, including the use of MRP and JIT. Surprisingly, they found that the use of a particular methodology was not nearly as important as the implementation of certain procedures for improving manufacturing performance. Such procedures include reduced setup times, small lot sizes, worker flexibility, and preventive maintenance, all of which are components of JIT. Thus, it does not appear that a company must necessarily adopt the entire JIT philosophy to obtain performance improvements; the use of key components that enable flow manufacturing may be sufficient. However, giving lip service to JIT by requesting smaller lot sizes from suppliers without reducing one's own lot sizes or setup times, as some companies have reportedly done, probably will generate only marginal, if any, performance improvement.

IMPLEMENTATION: Sakakibara *et al.*, (1993) identified sixteen "core JIT components" and divided them into six categories. Those categories and the components of each are described briefly below.

1. ***Production Floor Management:*** Production floor management refers those to components of JIT that are directly applied to control production activities at the shop floor level with the objective of maintaining a smooth flow of materials. Those components are:
   - Setup time reduction.

- Small lot sizes.
- Preventive maintenance.
- Kanban (if applicable).
- Pull system support (if applicable).

Small Lot Sizes and Setup Time Reduction: Most companies find that significant changes must be made in shop floor procedures as they move toward the flow manufacturing goal of JIT. One way to achieve smooth, continuous flow is by using small lot sizes with the objective of eventually achieving a lot size of one. Of course, if such small lot sizes are to be feasible, then setup times must be reduced close to zero, or to the minimum level possible. A considerable body of knowledge has been developed concerning setup time reduction (Shingo, 1985), and some companies have achieved dramatic results, often going from setups that required hours to than ten minutes (single-minute setups; see SMED).

Preventive Maintenance: While reduced lot sizes and setup times are important first steps toward achieving flow manufacturing, companies implementing JIT must also take steps to ensure that their manufacturing equipment will be able to maintain the requisite smooth flow of material. Thus, preventive maintenance is used extensively to avoid unplanned down time. Often, a predetermined portion of each day, or even each shift, is set aside for preventive maintenance. Gupta and Al-Turki (1998) describe approaches to preventive maintenance in a JIT system.

Kanban and Pull System Support: Batch-oriented manufacturing relies on an extensive paperwork system – including shop orders and dispatch lists – to control and monitor production. However, when small lot sizes are used, the usual paperwork system would become unmanageable. Thus, kanban is often used as a system for controlling the production and movement of material through a JIT manufacturing process. Kanban is a "pull" system in which material movement, and often production, is based on *actual need* for more materials at downstream work centers, not on *planned need* as in an MRP system, which is a "push" system. The kanban system operates by using a signaling device to indicate the need for materials. Although the signaling device is often a card, it can be any visible or audible signal. For example, Harley-Davidson uses empty containers for the pull signal while one Kawasaki plant has used colored golf balls. The important point is that the number of such devices are strictly limited so that excess inventory cannot be built up. Further, the number of units that can go into a container is predetermined. Thus, two rules that apply when using a kanban system are: (1) produce or move material only when authorized by a kanban; (2) fill containers only with the predetermined number of parts.

Figure 1 depicts the basic flow in a two-card kanban system, in which one kanban controls conveyance (or withdrawal) of material from storage to the work center using that material while another kanban controls production of the material. Often, only the *conveyance kanban* is used, with production based on a predetermined schedule. A conveyance kanban (*C-kanban*) can be returned for more materials only when the level of

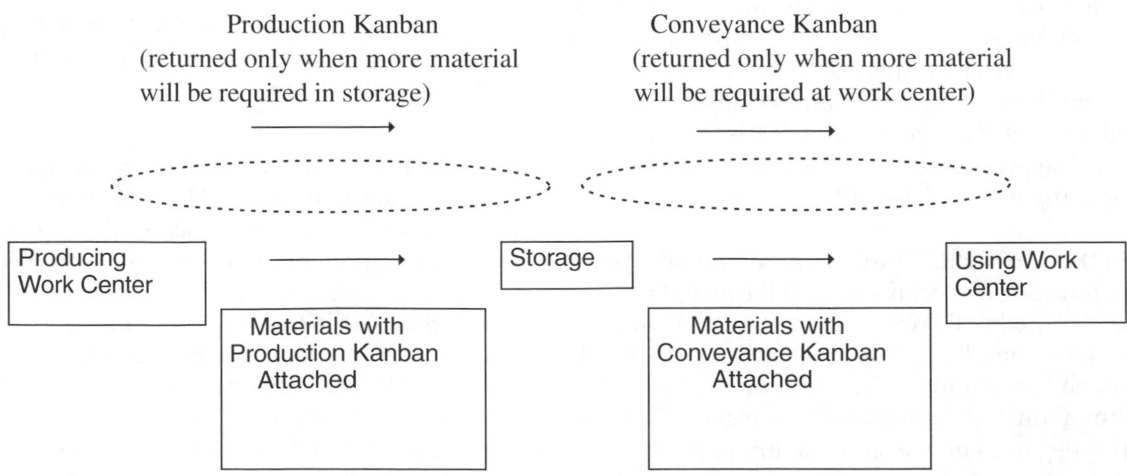

Fig. 1. Flow of kanban cards.

materials at the using work center has dropped to a predetermined level. Likewise, *a production kanban (P-kanban)* can be returned to the producing work center to authorize more production only when a container has been removed by a downstream work center from storage using a C-kanban. In this way, inventory build up is strictly limited by the number of kanban cards.

2. ***Scheduling:*** Experience tends to indicate that JIT works best in an environment in which the production schedule remains fairly stable. Thus, two components of JIT related to scheduling are:
   - Repetitive master schedule.
   - Daily schedule adherence.

   Companies using JIT often develop a master schedule that is repeated each day over an extended period of time, usually for at least several months. Each day's production of a given product is based on the average daily demand for that product, thus minimizing the amount of excess inventory at the end of the day. Ideally, this "rate-based" production schedule should produce each product in the smallest possible lot sizes. As a result, instead of producing each product in large batches that may extend over several days or weeks, small batches of each product are produced each day. This is referred to as "uniform load scheduling" or "mixed model sequencing." Producing in this manner not only minimizes inventory buildup of any product, but also levels the requirements for all component parts and for the resources used to produce those components. For example, if a company manufactures three products (A, B, and C) and the average daily demand is 5 units for product A, 10 units for product B, and 15 units for product C, then an "ideal" mixed model sequence could be CBCBCA, which would be repeated five times per day. Of course, setup times or other considerations might require that larger lot sizes be used for certain products, but the basic goal is to produce every product every, day if possible.

3. ***Process and Product Design:*** As companies implement JIT, they often find that flow manufacturing is facilitated through modifications to both the process and the product. Thus, the following are two components of JIT that influence design of processes and products:
   - Equipment layout.
   - Product design simplicity.

   Equipment Layout: With the JIT focus on smooth material flow and minimal inventory, equipment layouts are often changed under JIT so that work centers are close together. In fact, manufacturing cells are frequently used to produce a family of parts or products. These manufacturing cells group machines together in close proximity so that material handling times and workforce requirements are minimized. By working only on a family of parts or products that have similar machining requirements, setup times in these manufacturing cells can be reduced or eliminated.

   Product Design Simplicity: By simplifying product design to reduce the number of different parts, and even the number of parts required, inventory is reduced. Products are designed so that they are easy to manufacture and materials flow smoothly through production.

4. ***Work Force Management:*** One significant change under JIT, which is also being used widely by companies that do not use JIT, is a different approach to managing human resources. With JIT the focus is on fully utilizing a person's capabilities, including mental as well as physical capabilities, through the following components:
   - Multi-functional workers.
   - Small group problem solving.
   - Training.

   Employees are trained to perform several different tasks or jobs so they can move around as needed, and can help other employees if problems occur. Further, workers' problem-solving capabilities are fully utilized by allowing them to work in self-directed groups to solve production problems. Finally, to ensure that employees can work effectively in groups, possess know the how to solve problems, and can perform various jobs without errors, a significant amount of time is devoted to training.

5. ***Supplier Management:*** The emphasis on reducing inventory and achieving a smooth flow of materials involves suppliers. However, this step should usually be delayed until the customer company has its JIT house in order and has gained substantial experience working with JIT. At that time, a company may consider including the following JIT components:
   - JIT delivery.
   - Supplier quality level.

   After a company has implemented JIT in its own operations, it should have a level production schedule that can be transmitted to suppliers, who can then work to provide JIT deliveries. However, if deliveries occur just as material is needed, then there is no time for inspection and no room for quality problems.

Consequently, the supplier must also be able to achieve a consistently high level of quality so that inspection of incoming material is not necessary because very few, if any, incoming components are defective.

6. ***Information System:*** Because JIT brings about significant changes in manufacturing processes, it also produces major alterations to the information systems associated with those processes. Two major areas of change are:
   - MRP adaptation to JIT.
   - Accounting adaptation to JIT.

   Although it may appear that MRP and JIT are mutually exclusive, that is not true. In fact, many companies are now combining MRP and JIT, although doing so does require adaptations to the MRP system. These adaptations are prompted generally by the smaller lot sizes and shorter lead times of JIT. Thus, companies may find that *bucketless MRP* systems are more appropriate for JIT use. While this modification can help in adapting to reduced lead times and increased rate of material flow, it does not necessarily alleviate the paperwork problems. MRP systems are based on the usual approach of shop orders, allocation, and picking tickets. Applying this approach to JIT can quickly bog the system down in paperwork. To avoid that outcome, many companies utilize "backflushing" to enter material changes into the system. With backflushing, material usage is calculated from the production of final products. If a certain number of units of final product were produced, then it can be assumed that an appropriate number of units of each component part were also used. This same backflushing approach can also be used to adapt cost accounting systems to a JIT environment.

7. ***Supporting JIT Components:*** The preceding six categories have been identified as "core JIT components." However, there is also a large set of "supporting JIT components." This latter group is discussed briefly in the following paragraphs.

   Top Management Support: JIT, because it requires some rather drastic organizational changes, must have top management support to motivate those changes and keep everyone on track through the periods of turmoil that are nearly inevitable with change. Further, JIT often involves functions other than just manufacturing and, thus, needs top management support for cross-functional cooperation.

   Quality Management: The need for a high level of quality from suppliers has already been mentioned. However, the same – or even higher – quality level will also be needed from internal manufacturing processes. Achieving such high quality often requires strong management leadership in the area of quality, implementation of methodologies such as statistical process control, and an orientation toward meeting customer needs and expectations.

   Human Resource Management: The use of human resources in JIT frequently require changes throughout the entire organization. For example, the need for employees who can work in groups, make their own decisions, and perform multiple jobs will influence the new-employee recruitment and selection process. With employees working in groups the employee reward and compensation programs may need to be modified. Likewise, the role of managerial personnel may change from that of a decision maker to a facilitator.

   Technology Management: The need for smooth material flow and consistently high quality under JIT can often lead to changes in manufacturing technology such as the use of flexible manufacturing systems and robotics. Similarly, simplified product designs and the utilization of common parts are increased through the use of computer aided design systems.

   Manufacturing Strategy: Changes that occur with JIT will often lead to changes in a company's manufacturing strategy. For instance, JIT companies often develop distinctive competencies in terms of product quality, delivery speed, and delivery reliability. These competencies can be used advantageously as part of the company's competitive strategy. Further, the JIT philosophy helps companies to develop a plant-wide manufacturing philosophy and a long-range orientation, both of which are useful characteristics for developing and implementing competitive strategies.

   Company-Wide Continuous Improvement and Problem-Solving Activities by Workers, Engineers, and Management: An important part of the JIT philosophy is that waste, in some form, can always be found, and that problem-solving efforts can be used to find ways to eliminate that waste. Thus, continuous improvement is both a component and a result of JIT. Because JIT spreads across functional boundaries, the continuous improvement and problem-solving activities eventually become company-wide across all levels of the organization.

TECHNOLOGY PERSPECTIVE: JIT does not require any particular technology. In fact, the simplification that results from using JIT can often lead to a reduction in the use of technology. For example, companies frequently find that sophisticated material handling systems are no longer needed as workers are moved closer together so they can easily pass parts, by hand, from one to the other. However, this is not to say that technology is not used with JIT. Many companies find that as they simplify and integrate processes, a point is reached at which those processes can be automated. Automation is not seen merely as a way of eliminating people, but as a way to make a process more consistent and reliable. The Japanese word "jidoka" is often used to describe the use of technology to monitor automated processes so that the process will be stopped automatically if a problem occurs (Hirano, 1988).

Computer technology has also been applied to planning and control in JIT. This can involve either the integration of JIT with computerized MRP or actual stand-alone computer software for JIT (Vollmann *et al.*, 1997).

ANALYTICAL MODELS: A considerable amount of research effort has been expended by academicians to develop analytical models that are designed to optimize the number of kanban cards (Berkley, 1992). Many of these models are very complex mathematically. However, it is questionable whether such effort is really worthwhile as the formula reportedly used by Toyota (Hall, 1983) is the simple one shown below.

$$\text{Number of kanban cards} = [(\text{Demand rate}) \times (\text{Lead time}) \times (1 + \text{safety factor})]/(\text{container size}).$$

In practice, efforts are always under way to reduce the number of cards and, consequently, reduce the level of inventory.

EFFECT: Plossl (1985) reports the results of a group of companies that implemented JIT. Within a five-year period, those companies experienced reductions in manufacturing cycle times in the 80 to 90 percent range, inventory reductions of 35 to 90 percent, labor cost reductions of ten to 60 percent, and quality cost reductions of 25 to 60 percent.

PREREQUISITES: Although MRP and JIT are sometimes thought of as being at different ends of a spectrum, they are both actually similar in many ways. One similarity is that both are formal systems, and both require discipline to for success. Thus, companies have found quite often that once they have established the necessary discipline to use MRP effectively, it is not so difficult to implement JIT. On the other hand, a company that has no discipline and is operating with informal systems will often have a much more difficult time of implementing JIT. Thus, the implementation of JIT appears to proceed most smoothly when a company an has established a formal manufacturing system such as MRP.

Most of the early successes with JIT were experienced by automobile companies or other repetitive manufacturers. Consequently, it was believed for a while that JIT was only applicable in such industries. Today, we have learned that many of the JIT components can be applied in nearly every manufacturing environment, and even in service industries. It is important to realize that individual components of JIT can be implemented without adopting the entire JIT philosophy. Some companies may focus only on the worker flexibility aspects of JIT while others may implement only setup and lot size reduction. Thus, JIT, in at least some form, is applicable in any manufacturing situation.

RESULTS: A common analogy used for the inventory reduction that results from JIT implementation uses a stream of water in a channel strewn with rocks. Water represents the inventory level and rocks are problems which are uncovered as the inventory is reduced. This means that as inventory is reduced, problems will be uncovered and must be solved before smooth operations can resume. Unfortunately, some companies tend to avoid the problems by "canoeing around the rocks." Doing so goes against the basic concept of JIT, which is continuous improvement. This means that companies implementing JIT will not always have smooth sailing, but will often be faced with new problems that must be solved. However, solving them will eventually lead to further improvements in performance.

CASES: One of the best known JIT success stories is Harley-Davidson, the U.S. motorcycle manufacturer. In the early 1980s Harley was the last remaining company of its kind in the U.S., and from all indications its days were numbered as Japanese-made motorcycles with higher quality and a lower price tag were rapidly stealing Harley's market share. Harley-Davidson management decided that, to fight back, it would have to learn its competitors' secret weapon, which turned out to be JIT. Over several years Harley successfully implemented JIT and was able to drastically reduce its inventory levels and production costs while significantly increasing product quality. Harley-Davidson today is a strong proponent of JIT as demand for its motorcycles outstrips production capacity and the product sells throughout the world, even in Japan.

Harley-Davidson and other success stories are presented in an excellent collection of cases developed by (Sepehri, 1986). Those cases describe the use of JIT by companies in the United States, including Toyota USA and NUMMI, which is a joint venture between Toyota and General Motors, located in Fremont, California. Schonberger (1987) has developed a similar collection that includes companies using JIT, TQM, or both.

COLLECTIVE WISDOM: Implementation JIT is a complex, lengthy task. Fortunately, all components need not be implemented simultaneously, so companies have generally found it best to begin with a few and add others progressively. Hallihan *et al.* (1997) and Sohal *et al.* (1993) describe models for implementing JIT based on surveys of industry experience. The following points summarize some wisdom gained from experience:

- Top-management commitment and support are essential for successful implementation.
- Everyone in the organization must be educated about the importance of JIT and what they can expect from JIT.
- Begin with a small pilot project and get it operating smoothly before expanding the use of JIT throughout the organization.
- Focus on an application with a high probability of success as the first step. Rearranging an operation for improved product flow can often produce visible results quickly.
- Do not involve suppliers until your own operation is working smoothly and a stable production schedule has been achieved.

Organizations Mentioned: Harley-Davidson; Kawasaki; NUMMI

See **C-kanban; JIT philosophy; Kanban-based production systems; Just-in-time manufacturing: Implications; Lean manufacturing; Manufacturing cell design; Mixed model sequencing; P-kanban; Rate-based production schedule; Repetitive production; SMED; Total productive maintenance; Uniform load scheduling.**

References

[1] Berkley, B. (1992). "A review of the kanban production control research literature." *Production and Operations Management*, 1(4), 393–411.

[2] Gupta, S.M. and Y.A.Y. Al-Turki (1998). "Adapting just-in-time manufacturing systems to preventive maintenance interruptions." *Production Planning and Control*, 9(4).

[3] Hall, R.W. (1983). Zero Inventories. Dow Jones-Irwin, Illinois.

[4] Hallihan, A.P. Sackett, and G.M. Williams (1997). "JIT manufacturing: The evolution to an implementation model founded in current practice." *International Journal of Production Research*, 35(4), 901–920.

[5] Hirano, H. (1988). *JIT Factory Revolution*. Productivity Press, Massachusetts.

[6] Keller, A.Z. and A. Kazazi (1993). "Just-in-time manufacturing systems: A literature review." *Industrial Management and Data Systems*, 93(7), 1–32.

[7] Krajewski, L.J., B.E. King, L.P. Ritzman, and D.S. Wong (1987). "Kanban, MRP, and shaping the manufacturing environment." *Management Science*, 33, 39–57.

[8] Monden, Y. (1981). "What makes the Toyota production system really tick?" *Industrial Engineering*, 13, 36–46.

[9] Monden, Y. (1983). *Toyota Production System*. Industrial Engineering and Management Press, Georgia.

[10] Plossl, G.W. (1985). *Just-in-Time: A Special Roundtable*. George Plossl Educational Services, Georgia.

[11] Sakakibara, S., B.B. Flynn, and R.G. Schroeder (1993). "A framework and measurement instrument for just-in-time manufacturing." *Production and Operations Management*, 2, 177–194.

[12] Schonberger, R.J. (1982). *Japanese Manufacturing Techniques: Nine Hidden Lessons in Simplicity*. The Free Press, New York.

[13] Schonberger, R.J. (1987). *World Class Manufacturing Casebook*. The Free Press, New York.

[14] Sepehri, M. (1986). *Just-in-Time, Not Just in Japan*. American Production and Inventory Control Society, Virginia.

[15] Shingo, S. (1985). A Revolution in Manufacturing: *The SMED [Single-Minute Exchange of Die] System*. The Productivity Press, Massachusetts.

[16] Sohal, A.S., L. Ramsay, and D. Samson (1993). "JIT manufacturing: Industry analysis and a methodology for implementation." *International Journal of Operations and Production Management*, 13(7), 22–56.

[17] Vollmann, T.E., W.L. Berry, and D.C. Whybark (1997). *Manufacturing Planning and Control Systems* (4th ed.). Irwin/McGraw-Hill, New York.

[18] Womack, J.P., D.T. Jones, and D. Roos (1990). *The Machine That Changed the World*. R.A. Rawston Associates, New York.

JUST-IN-TIME MANUFACTURING: IMPLICATIONS

Thomas J. Billesbach

Northwest Missouri State University, Missouri, USA

INTRODUCTION: The concept of Just-in-Time (JIT) was pioneered by Toyota in the 1950's and since then has been expanded and modified by companies all over the world. Originally, managers viewed JIT as an inventory reduction technique (Hall, 1993). This narrow focus centered on small batches of product delivered just in time for downstream consumption. In today's progressive organizations the meaning of JIT embodies a wide variety of concepts and principles. (Kupanhy, 1995; Lee, 1984; Lubben, 1988; Schonberger, 1982; Schonberger, 1984; Schonberger, 1996).

EXPANDED/BROAD VIEW: A broader and more contemporary definition of JIT is the right amount of goods, services, and/or information, on time, and of acceptable quality. This definition takes into account both the service and administrative side of business as well as the manufacturing of products. There are essentially three fundamental components of this definition. The first component deals with definition of the right amount of goods, services, and/or information. In essence, it means no more or less than what the customer specifies. If a customer orders 500 parts, the delivery should be exactly 500 parts, not 500 plus or minus 10%. Often, suppliers indicate that a shipment contains a certain amount plus or minus a small percent. Customers might complain if they received only 480 parts but often say nothing if they received 520. If we only have to pay for 500 and received 520, they believe they have received 20 free parts. However, the cost is likely to be imbedded in the price paid. If a company cannot accurately control the amount of goods they ship to a customer, they need to focus on process control. *True JIT is based on process control.*

The second component of the definition deals with on-time delivery which means a product should not be late, or early. Most of us recognize the consequences of being late but are often rewarded for early delivery. Early delivery of parts crowds the production floor, ties up working capital, and gets in the way of production. A very small window of time for delivery is important for a JIT system. For many companies this means multiple deliveries daily directly to the production line.

The third component of the definition focuses on quality. JIT is wedded to quality. If one receives the right amount of goods on time but of poor quality, the JIT will breakdown. Therefore, without quality, which is acceptable to the customer, one does not have a true JIT system. A customer can be external or internal to the organization. An internal customer is likely to be the downstream work station, while the external customer is typically the end user of the product.

COSMETIC JIT – A FALSE SYSTEM: In the 1970's and 80's many United States automotive manufacturers required vendors to deliver a predetermined amount of supplies several times a day to the plant. They no longer wanted to have a huge stockpile of inventory on the plant floor, that is, they wanted a JIT system. Rather than choosing to ship the right amount of products as they are needed on a periodic basis, some vendors chose to build a warehouse next to the plant and ship large quantities to the warehouse on an infrequent basis. From the warehouse, the correct amount is be sent to the plant several times a day, if needed. On the surface, this system looks like JIT, however, it is not. True JIT marries production to consumption. If consumption is 200 units per day, production should be 200 units per day. It is acceptable to carry a small buffer stock but to produce and store the entire annual demand for release in small amounts is not JIT. Vendors, who send large amounts of inventory to a warehouse incur additional working capital costs in holding that inventory. While the automotive manufacturers lowered their costs from having very little inventory in their plants, those vendors engaged in stockpiling parts saw their working capital costs rise. As a result, the total cost of inventory in the system remained the same or in some cases increased.

Another cosmetic JIT uses the consignment approach. Managers of some companies believe they have a JIT system because they do not pay for parts until they use them. For example, assume 100 units are stored on the buyer's site for use by the buyer. The buyer does not own them or pay for them until they are used in production. From an accounting perspective, it appears as if the consumer of goods has a JIT system because needed parts are delivered on time. However, no actual inventory reduction in the system occurs, carrying costs remain the same. Hence, one has a false JIT system.

KANBAN AND PULL SYSTEMS: In the United States, managers tend to view JIT as more of a management philosophy versus a simple set of techniques (Billesbach, Harrison, and Croom-Morgan, 1993). In today's business environment, JIT encompasses many terms and concepts. These include kanban, pull systems, lean production, elimination of non-value added activities, and a host of related concepts designed to swiftly move a product, service, or information through the system.

Many companies implementing JIT often utilize pull systems which incorporate kanban. Pull systems allow the downstream work station to control the flow of the work station. In a traditional company, product is completed at one work station and pushed to the next downstream work station. Push systems can generate a tremendous amount of work-in-process inventory resulting in delayed feedback of quality problems and disconnecting the maker of parts from the user of parts. Kanban limits the amount of product that can accumulate in front of a work station. In essence, kanban acts as a governor controlling the amount of inventory within a system. With limited inventory between work stages, feedback regarding quality is immediate. Pull systems and kanban help synchronize production and consumption. This reduces formal scheduling within a plant as kanban signals the need for production to begin. JIT does not necessarily occur because one has kanban signals. It is possible to have numerous kanban signals resulting in large amounts of inventory. What kanban does is limit the maximum amount of inventory between any two work stations or within a system. If you set the limit excessively high, you do not have a JIT system. While using kanban, operators at a work station cannot produce a product unless a kanbansignal is received.

Because acceptable quality is imperative in a JIT operation, operators are given "line stop" authority to prevent unacceptable quality from being produced. Operators must feel they are in control of the process. In a ten-stage manufacturing process, which limits the amount of inventory between any two stations, if one work station cannot keep up the quality and quantity, eventually, all stations must shut down. Attention must then be directed to fixing the problem.

In order for many organizations to implement a JIT system, set-up times often must be reduced so that small batches of product can be produced as needed. With long set-up times it becomes too costly to produce a small batch of parts. When implementing JIT in manufacturing, either quick set-ups, or equipment which are dedicated to making specific parts are necessary.

BEHAVIORAL IMPLICATIONS: JIT has many powerful behavioral implications. Consider for a moment what happens when there is very little buffer between two work stations, and parts are rejected for lack of quality. If a work station produces only what is needed but poor in quality, and sends them immediately to the down stream work station, the down stream stations will be forced to shut down. Carrying a large buffer of inventory is not the solution. In order to keep carrying costs down, only what is needed should be produced. To keep a production line running when there is very little buffer stock between stations, workers must pay close attention to the quality of their work.

Large batch sizes often result in workers rationalizing that a few bad parts in a large lot is acceptable. Two rejects in 100 parts seems fine. However, with a small batch amount of 5 units, 2 rejects is unacceptable. Small JIT batches encourage workers to produce a quality product each time and every time. Large batch sizes result in a quality disconnect between the producer and consumer of goods. If a work station produces a batch size of 500 parts and moves them to the down stream work station, there is the potential of all 500 parts to be reworked or scrapped. With JIT, quality problems are noticed very quickly and feedback can be immediate. Quick identification and feedback of quality problems leads to quicker diagnosis of the true cause and resolution of the problem. With large production batches there is a long time lag between production of goods and the identification and feedback of a problem. Long time lags increase the difficulty in determining the real cause of the problem, if it can be identified at all.

BENEFITS OF JUST-IN-TIME: Perhaps the greatest measurable impact of adopting JIT is inventory reduction. A study of JIT and its impact on selected aspects of the balance sheet found statistically significant improvements in managing inventories over a ten year period. This longitudinal study of 28 companies which adopted JIT principles indicates significant improvements in turnover as evidenced by Figure 1. Twenty five of the 28 companies listed had improved inventory turnover. For most organizations, JIT can have a significant impact on the company's financial statements (Courtis, 1995).

Drastic reductions in both raw material, work-in-process, and finished goods inventory can be realized through JIT. Although JIT seems most appropriate for repetitive operations it can be applied to both a make-to-order business and process industries. Albany International Corporation's Perth, Canada facility reduced throughput time at the plant from 8 weeks to approximately

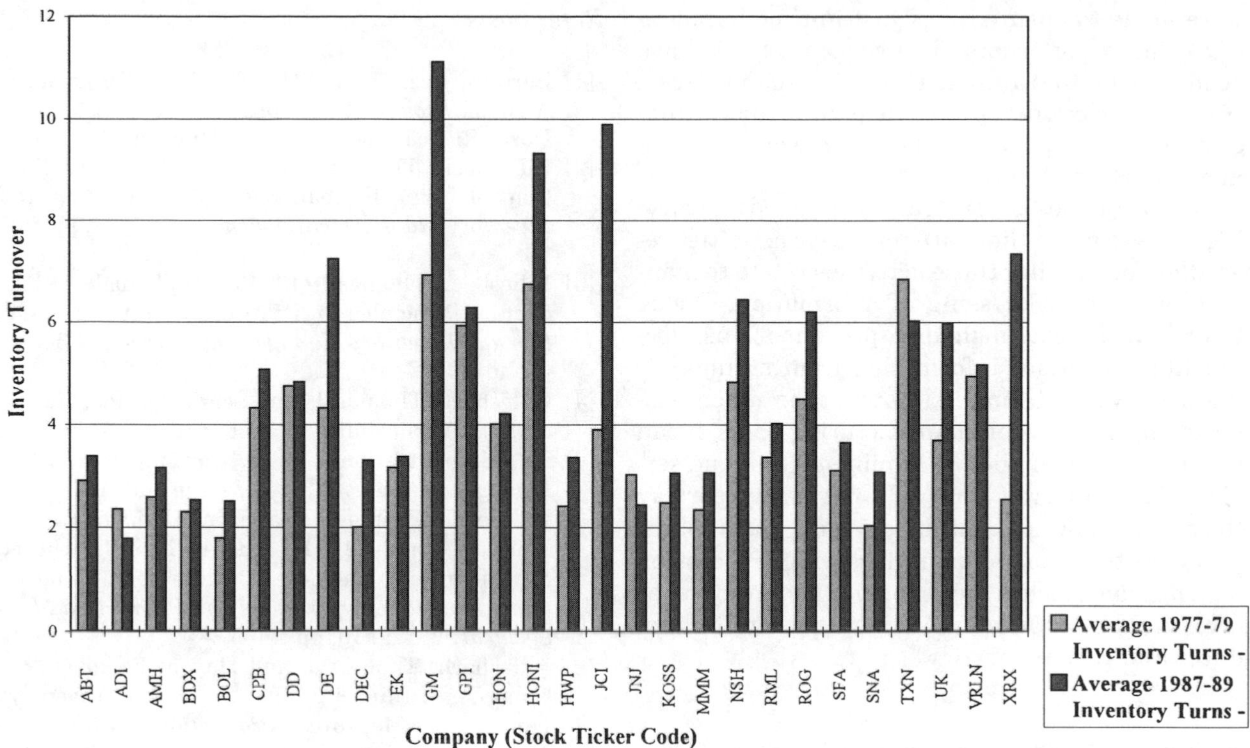

Fig. 1. Inventory turnover comparison (Ballesbach and Heyen, 1994: 65).

3.5 weeks and reduced inventory by $1.5 million. This was accomplished in a make-to-order facility with no negative impact on customer service. At the corporation level, JIT efforts resulted in inventory reductions of over $43 million dollars (Billesbach, 1994a). Applying lean production principles to DuPont's May Plant in Camden, SC, resulted in an astounding 96% reduction in work-in-process inventory.(Billesbach, 1994b) Effective JIT often results in one or more of the following benefits:

- Less raw material, work-in-process, and finished goods inventory
- Less potential rework and scrap
- Problems are detected quicker
- Problem diagnosis is much easier
- Communication and feedback between work stations is improved
- Improved customer response time
- Shorter cycle time
- Reduced number of outside suppliers
- Factory floor space requirements decline as less inventory occupies the plant floor

One other noteworthy benefit of JIT is the system itself, which creates an environment that influences the "right" kinds of behaviors.

ORGANIZATIONAL IMPLICATIONS: The effects of JIT on an organization can be dramatic. With less inventory, fewer inventory control specialists are needed. Tracking systems and information needs are considerably less. Implementing a pull system using kanban results in very little need for scheduling as kanban is the signal to produce. As a result, fewer production control specialists and schedulers are required. If operators are empowered to halt production for quality related issues and problem diagnosis is easier to accomplish, less supervision is needed. In a mature JIT system, operators are no longer the traditional "parts maker" but the manager of the process. Pull systems with kanbans link operators into a team working together to focus on the process rather than one's own work station. The result is less overhead and supervision.

ADMINISTRATIVE/SERVICE APPLICABILITY: Most applications of JIT have been in repetitive manufacturing settings with some job-shops and process industry implementations. An early pioneer in applying JIT was Nashua Corporation. After tremendous success in changing their manufacturing operations, the focus shifted to other parts of the value chain. Utilizing the principles of JIT, order entry cycle time was cut from approximately 8 days to 1 hour (Billesbach and Schnierderjans, 1989). Rather than allowing information to stay in a queue for batch processing, changes were made to ensure quick just-in-time handling of all incoming orders. The concept of JIT have been applied to training. This approach utilizes a train-do cycle whereby participants learn a concept and then apply it to their own work environment (Bertain, 1993).

The principles of JIT can be applied in varying degrees when there are repetitive activities regardless of whether those activities relate to manufacturing, or processing of information. Queue limits can be established to prevent forms, documents, and other information from accumulating at a work station. Although it is much easier to apply JIT when manufacturing parts, it can be successfully applied in administration and service. While piles of paper do not easily portray the amount of work contained within, the implications are the same as manufacturing. Processing information quickly and expeditiously minimizes errors and strengthens the link between the producer and consumer.

FORWARD LOOK: Managers, today, often do not call their change efforts JIT. Instead, they embark on cycle time reduction, aggressively focus on becoming lean and mean, and strive to be a seamless organization in the delivery of goods and service. The labels have changed, and yet the underlying concepts and principles of JIT remain fundamentally the same.

Companies no longer produce products only in their native country but manufacture parts all over the world. This globalization provides new challenges in managing inventories and the distribution of products. Managers today talk of global supply chain management utilizing the same fundamental principles of JIT that were used at the plant level.

A few repetitive manufacturing organizations first adopted JIT because of strong competitive pressures. Once significant benefits were realized, organizations in other industries followed. Eventually, job shops and process industries started utilizing the same principles. Those manufacturing organizations that benefited greatly by JIT eventually implemented these principles and concepts in the administrative side of their business. The greatest gains yet to be made lie in the adaptation and application of JIT to the administrative and service side of business. Quick delivery of information and service provides a distinct competitive advantage. Therefore, in the future we will see more organizations implementing JIT in the administrative and service side of the business and may label it something different.

References

[1] Bertain, Leonard (1993). *The New Turnaround: A Breakthrough for People, Profits, and Change*, North River Press, Croton-on-Hudson, New York.

[2] Billesbach, Thomas J. (1994a). "Simplified Flow Control Using Kanban Signals", *Production and Inventory Management Journal*, V. 35, #2, pp. 72–75.

[3] Billesbach, Thomas J. (1994b). "Applying Lean Production Principles to A Process Facility", *Production and Inventory Management Journal*, V. 35, #3, pp. 40–44.

[4] Billesbach, Thomas J., and Schniederjans, Marc J. (1989). "Applicability of Just-In-Time Techniques in Administration", *Production and Inventory Management Journal*, V. 30, #3, pp. 40–45.

[5] Billesbach, Thomas J., Harrison, Alan, and Croom-Morgan, Simon (1991). "Just-in-Time: A United States-United Kingdom Comparison", *International Journal of Operations and Production Management*, V. 11, #10, pp. 44–57.

[6] Billesbach, Thomas J. and Hayen, Roger (1994). "Long-Term Impact of Just-In-Time on Inventory Performance Measures", *Production and Inventory Management Journal*, V. 35, #1, pp. 62–67.

[7] Courtis, John K. (1995). "JIT's Impact on a Firm's Financial Statements", *International Journal of Purchasing and Materials Management*, V. 31, #1, Winter, pp. 46–51.

[8] Hall, R.W. (1983). *Zero Inventories*, Homewood, IL, Irwin Company.

[9] Kupanhy, Lumbidi (1995). "Classification of JIT Techniques and Their Implications", *Industrial Engineering*, V. 27, #2, Feb., pp. 62–66.

[10] Lee, S., and Ebrahimpour, M. (1984). "Just-In-Time Production System: Some Requirements for Implementation", *International Journal of Operations and Production Management*, V. 4, #1, pp. 3–11.

[11] Lubben, Richard T. (1988). *Just-In-Time: An Aggressive Manufacturing Strategy*, McGraw-Hill, New York.

[12] Schonberger, Richard J. (1982). *Japanese Manufacturing Techniques: Nine Hidden Lessons in Simplicity*, The Free Press, New York.

[13] Schonberger, Richard J. (1984). "Just-In-Time Production System: Replacing Complexity with Simplicity in Manufacturing Management", *Industrial Engineering*, Oct., pp. 52–63.

[14] Schonberger, Richard J. (1996). *World Class Manufacturing: The Next Decade*, The Free Press, New York.

See **JIT evolution and use in the USA; Just-in-time manufacturing.**

JUST-IN-TIME PURCHASING

See **JIT II; JIT philosophy; JIT purchasing; Just-in-time manufacturing.**

JUST-IN-TIME SHOP FLOOR SYSTEM

JIT shop-floor scheduling systems seek minimal flow times for the entire product. The scheduling emphasis is on products. Where cellular manufacturing techniques are employed, Kanban cards, containers, and other signals of downstream needs for components serve as the authorization to produce. In this environment, sequencing is not an issue. Order tracking is nonexistent in JIT environments because work-in-process is minimal, and material moves through the factory quickly. Thus, JIT can result in a paperless shop-floor system.

See **JIT evolution and use in the United States; JIT philosophy; Just-in-time manufacturing; Just-in-time manufacturing: Implications.**

KAIZEN

Kaizen is the Japanese term for improvement. It is part of JIT manufacturing, TQM and other efforts originating in Japan. Kaizen promotes continuous improvement for eliminating waste in machinery, labor, or production methods.

See **Just-in-time manufacturing; Lean manufacturing; Total quality management.**

KANBAN

Kanban is the Japanese word for "visible sign" or card. Kanbans are used to control the flow of product through a production facility; they are typically used in just-in-time (JIT) or pull-production systems (Sugimori *et al.*, 1977). Kanbans are attached to the products or to containers of products that flow through the system. A kanban may contain information identifying the part, the workstation, and the order. By keeping track of the kanbans, the system can keep track of the work in process (WIP).

In kanban-based production systems, Kanban cards in manufacturing represents a type of signal to begin production of a part. Without this signal, production cannot occur. The number of Kanban cards thus limits the amount of work-in-process inventory that can accumulate between work stations or within a cell. It can also be used to limit the amount of raw material and finished goods. Many things such as cards, containers, taped spaces, etc., can be used as kanban signals.

See **Adaptive kanban control; Dynamic kanban control for JIT manufacturing; Flexible kanban control; Just-in-time manufacturing; Just-in-time manufacturing implications; Lean manufacturing; Manufacturing cell design; Manufacturing performance measures; Pull production systems; Reactive kanban control.**

Reference

[1] Sugimori, Y., K. Kusunoki, F. Cho, and S. Uchikawa (1977). "Toyota production system and kanban system materialization of just-in-time and respect-for-human system." *International Journal of Production Research*, 15, 553–564.

KANBAN-BASED MANUFACTURING SYSTEMS

S. Sengupta

Oakland University, Rochester, Michigan, U.S.A.

S.P. Dutta

University of Windsor, Windsor, Ontario, Canada

DESCRIPTION: The **Kanban** system is a multistage production scheduling and inventory control system. It is motivated by the concept of just-in-time production and aims at reducing inventory levels within the system to a minimum. The system was originally conceived at the Toyota Motor Corporation in Japan in the 1950s to reduce production costs through elimination of various sources of waste such as excessive inventory and excessive workforce. Today it is used as an information system to keep tight control over inventory through which a just-in-time production system can be managed.

Just-in-Time (JIT) production systems have received considerable attention in recent years because of their potential to improve competitiveness in a repetitive manufacturing environment. The goal of a JIT system is to produce the appropriate items in the necessary quantity at the right time. The JIT operating philosophy is an integral part of the Toyota Production System (TPS), which strives to achieve an all around improvement in the economics of manufacturing operations by eliminating wastes of all forms, inventory included. The current TPS has extended the JIT approach to Computer Integrated Manufacturing and Strategic Information Systems and has incorporated continuous improvement activities to maintain the integrity of the overall process and to increase worker morale. A flexible workforce and the involvement of workers to achieve continuous improvement are two essential features of the TPS.

A JIT system operates in a ***pull*** mode rather than in a ***push*** mode. The push mode has been the dominant mode of operation in North America until recently. In a push mode, jobs are released to the first stage of manufacturing, which, on completion, pushes it to the next stage. Planning is carried out with the assumption that the demand rate is known and does not show any variability. Changes in the demand rate call for duplication in planning and lead to a waste of time and resources, increase in lead time and the generation of inventory. By comparison, in a pull mode, only the last stage of manufacturing interfaces with a demand pattern, and information flows back through the system. Also, in a pull system, the preceding stage must produce the exact quantity withdrawn by the subsequent manufacturing stage. To achieve this goal, flow of information is controlled by the flow of **Kanbans** connecting two adjacent manufacturing stages. A Kanban is a tag-like card that contains information about the type and quantity of units to be produced by the manufacturing system. In other words, a Kanban is an authorization for the previous stage to produce a specified quantity of units. A Kanban is released when the succeeding stage withdraws units from the output buffer of the preceding stage. The two types of kanban mainly used on the shop-floor are the *withdrawal Kanban* and the *production ordering Kanban*. A withdrawal Kanban conveys information about the quantity which the succeeding stage should withdraw from buffer storage, while a production ordering Kanban specifies the quantity which the preceding stage must produce. Kanban cards are therefore used as production triggers. In a *single card system*, a card is attached to each standard container of a specified type of part when these are produced.

When the container is moved to a consumption point, the card is removed and serves to authorize the production of a replacement container of parts. Production may be delayed if there is a queue of cards waiting their turn to be processed. In a *two card system*, a second type of card is used to authorize movement of containers from production to consumption workstations. The single card concept is similar to a Re-order Point (ROP) system–removal of a container triggers the order for its replacement. Since Material Resources Planning (MRP) was adopted partly as a cure for the perceived limitations of the ROP system–in particular, the lack of visibility of true need–how can the ROP system work in a JIT environment? For it to work, two conditions must exist:

1. output of the finished product must be level for a reasonably long time horizon, and
2. A mixed model final assembly must be practiced.

If the conditions exist, component inventory depletes at a constant rate, and so the problem of lumpy demand, which MRP attempts to solve, is avoided. Quick setups are also needed for small lot production, but these are not essential to the ROP system itself (Keaton, 1995).

Another classification of Kanbans is based on function which includes primary kanban, supply kanban, procurement kanban and the auxiliary kanban (Huang and Kusiak, 1996).

The following sections identify design and control issues of a Kanban controlled production system, and highlight those issues which need to be resolved in the future. A number of case studies are also presented to convey the nature of the operating environment suitable for Kanban use.

HISTORICAL PERSPECTIVE: The kanban system was pioneered by Toyota Motor Corporation in Japan and subsequently adopted by numerous other Japanese companies. Over the last two decades, it has grown roots in manufacturing organizations worldwide, with automotive industries leading the way. Today it is used as a production control tool in nearly every type of discrete manufacturing facility, from electronics to aerospace to automotive to consumer goods sectors. As the concept has evolved, many forms of kanban production control have been examined. These have been reported in English language production and operations management literature for nearly twenty years. It has become evident during this period that, as a system of material queues, kanban systems are fascinatingly complex. Historically, practitioners have addressed many questions to improve the efficacy of kanban based JIT systems. The most important of these are:

- Setting batch and container sizes
- Setting the number of kanban
- Sequencing rules to overcome blocking by part type
- Adaptation to job shops and other untypical manufacturing environments
- Advantages of kanban driven systems over materials requirements planning driven production control.

Kanban controlled JIT system have been studied using simulation, mathematical programming and stochastic modeling. A majority of simulation studies investigate system parameters such as types and number of kanbans in use, number of processing stages, set up time, variability in demand pull rate, container size, machine failure and repair patterns. System throughput is the most widely used measure of performance. Other measures of performance include service level and the amount of work-in-progress. In a simulation-based study, Krajewski *et al.* (1987) report that simultaneous reduction of set up time and lot size is the single most effective way to cut inventory levels and improve customer service. The list of factors with potential impact on performance include worker flexibility, degree of product standardization, and yield rate. Though it is difficult to generalize a simulation based result, knowledge gained from simulation based studies is expected to provide valuable guidance to the practitioners who design and operate a JIT system.

A number of studies have considered mathematical relationships to determine the number of kanbans under different types of material flow such as multi-stage production, multi-stage production with assembly and multi-item single stage production with dynamic demand. Philipoom *et al.* (1996) present a cost based model and Tayur (1993) reports that if the goal is to maximize the throughput with a given number of kanbans, then all of the machines should be placed in one cell.

The following section summarizes reports about some real world implementation of Kanban-based systems. The operating environment, the implementation problems and the benefits realized are presented for each application. The purpose of this section is to make readers aware of the types of problems to be encountered in implementing a kanban-controlled system and the nature of benefits that it may generate.

CASE STUDIES ON IMPLEMENTATION OF KANBAN SYSTEMS: Gravel *et al.* (1988) describes a small firm manufacturing high-quality outdoor clothing grouped into summer and winter categories. The firm operated as a job shop and dealt with a seasonal demand pattern. Traditionally the orders were grouped into a production run that included all sizes and colors. The plant operated with a production lot size equal to the order size, and scheduling of jobs to machines was done manually by the supervisor using the First-in-First-out (FIFO) rule. The study reports extensive use of simulation to identify the important operating parameters for a Kanban-controlled system. The number of production kanbans at each operation, the production lot size, as well as the scheduling rule were selected based on simulation results.

Based on the results of a simulation study, two production Kanbans were used for each operation and a Kanban lot size of 10 units was selected for production. Scheduling at each operation was based on the Shortest Processing Time (SPT) rule. The results reported include, i) improvement in quality, ii) less work-in-progress , iii) reduction in production lead time, iv) better floor space utilization due reduction in Work in Process (WIP), and reduction in the layout space (40%–50%) and v) improvement in worker morale arising from job enrichment. The study also pointed out that a number of existing practices must be improved to ensure better use of a Kanban system. These include: better estimates of processing time, better training, better coordination with the purchase, and maintaining the correct process flow for each product. The study reports that success of a pilot study at this company led to the implementation of Kanban-based production for other products in the company.

The study by Schwarzendahl (1996) describes a Kanban system recently introduced in a Siemens manufacturing plant in Florida, USA. The company manufactures telecommunication switching equipment. The study describes the conversion of the production flow from a push system to a pull system. The product reported in the study requires assembly of three different types of components: racks, modules and cables. A two-bin Kanban system was installed to control the flow of racks from the stockroom to the production floor. A three-bin Kanban system was installed for modules and a multiple – bin buffer was installed for cables. The size of each kanban bin was estimated primarily based on the demand pattern, the time for a replenishment cycle, and a safety factor to take into account the variation in replenishment cycles and demand rate. Several training sessions were held to communicate the concept of Kanban controlled flow and requirements for its successful implementation. The conversion from a push-system to a pull-system was carried out in a phased manner. The following benefits are reported:

1. 32% increase in throughput,
2. 19% reduction in work-in-progress,
3. 50% reduction in overtime, and
4. better availability of material, and a better awareness of the status of inventory.

The paper summarizes two major benefits from a successful Kanban system. A Kanban system provides an efficient tool for the control of material flow. It also provides a mechanism to expose systematically the problem areas in a system and thereby help a continuous improvement drive.

A paper by Crawford *et al.* (1988) report a number of implementation as well as operating problems related to implementation of a Kanban controlled JIT system. The study is based on a survey of 39 companies, who implemented JIT systems. The participants include automotive parts manufacturers, manufacturer of electronic products, air-conditioning, diesel engines, etc. The paper summarizes the common implementation problems and the benefits realized. The implementations problems fall into two broad categories-people problems and technical problems.

1. People Problems can be further categorized into cultural resistance to change, lack of top management support, and lack of organizational communication. Cultural resistance to change includes lack of acceptance by unions, by middle managers and by different functional areas of the organization and a general resistance to change. Ability to accept a failure and learn from it is cited as a major requirement for successful implementation of a JIT system. The lack of senior management understanding regarding the magnitude of change necessary, and lack of adequate communication among the various functional areas such as manufacturing, engineering, accounting etc., are cited as major impediments.
2. Technical Problems include the lack of resources, absence of appropriate measures of performance, and other system related problems. The lack of resources, in most cases, refer to lack of education or training in the principles of JIT. The lack of training led to mis-communication on the shop floor and with other functional areas. Performance measures based on individual incentive is ineffective in a JIT system. Other *system* related problems include change in layout, interfacing with existing MRP systems, maintaining quality and line balancing.

The *operating* problems are as follows:

1. Inability to meet a schedule,
2. Poor quality,
3. Lack of vendor support in the form of poor quality or missed delivery schedule,
4. Poor demand forecasting,
5. Data accuracy, and
6. Machinery breakdowns.

The major benefits reported are reduction of manufacturing cost, reduction of inventory, reduction of lead time, improvement in quality and improvement of worker morale. More recently, Aigbedo and Monden (1997) have described a multi-attribute approach to the installation of a Kanban system. This is outlined in the following section.

Factors Inhibiting the Effectiveness of a Kanban System:

Limitations of Pull Systems

The Kanban policy is often described as a "pull" policy because demands propagate from the downstream end of the production line. The last work station provides a finished product when a demand arrives; this frees a Kanban, which acts as demand for the previous work station, and so on. Now, if there were inventory after the last workstation, orders would be met immediately. **In reality, the only way to guarantee this would be to have a very large inventory at the last stage**. Alternatively, inventory can be distributed throughout the system, but still, the smaller the inventory, the longer the delay between receiving and meeting orders. In general, there is no easy way to calculate the performance of a Kanban system, and easy ways of determining the best distribution of cards other than by experimentation. Hence, if orders arrive in nearly equal intervals and operation times are nearly deterministic, the production output can be made to match the demand. Then, little inventory is needed. The more variability there is in the system, the more inventory is needed to cope with it. Thus, a Kanban system described earlier, has some disadvantages:

1. There must be only a few rather high volume products in the manufacturing process, so that maintaining permanent work-in-process (WIP) for each product in front of each machine is practical.
2. The volumes of the products must be fairly stable over time, to keep demand surges from swamping the smooth operation of the system.
3. The system will not tolerate even a brief failure of machines; this will bring production to a halt.

In recent years, the goal has been to reduce inventory buffers between various production stages in the Kanban based system. When inventory buffers are reduced, there will inevitably be disturbances experienced in the production system, as explained earlier. The practical choices available to Kanban system implementers to cope with these disturbances are: (a) to reduce variabilities at their source; and (b) to increase variability handling mechanisms within production. Practically speaking the second choice, i.e., (b), can be achieved by developing a multi-functional workforce through training, and by increasing manufacturing flexibility through acquisition of high performance numerically controlled machine tools; this is a capital-intensive situation. Hence, it is more sensible to first attempt to reduce variability at the source, i.e., alternative (a), which minimizes disturbances, rather than cope with them afterwards. Manufacturers now recognize that one of the ways of achieving this goal is to institute a Total Quality Management (TQM) system throughout the manufacturing operation, and the supply chain as well, to reduce rejects and rework any at every stage of production.

Where Inappropriate

It is therefore considered that Kanban systems ***will not be effective*** if:

(a) Completely customized orders are handled in large numbers by the production system.
(b) An order or design needs modifications repeatedly during manufacture.
(c) Overall production requirements are unstable and quite lumpy.
(d) The company does not have the elements of Total Quality Management and Total Productive Maintenance in place to reduce rework and increase machine reliability.
(e) The labour force is not multi-functional.
(f) The variety of products being manufactured is greater than 10 to 15.
(g) Setup times are high and attempts have not been made to reduce these times by 50–70%, at the least, in order to offset the increased number of setups, which accompany the introduction of Kanban systems

Recent Trends in Application of Kanban-Based Systems: The key to monitoring and analysis of the performance of a Kanban-based manufacturing system is the identification of the sources of "variability" and their management. Variability can be viewed as the degree of compliance to contractually binding delivery schedules. The major contributing sources of such variability include:

- Variation in cycle time – greater variability is expected if the task has a significant manual component. A completely machine controlled cycle has little or no variability.
- Variation in the availability of resources due to random machine failure and repair patterns – a machine is likely to fail during an operation in an unpredictable manner, and the corresponding repair time may depend on a host of factors such as complexity of the failure, diagnostic capability built into the machine, skill level of the repair person etc.
- Variation in demand pattern created by the customer – this is a major impediment to the successful operation of Kanban based manufacturing systems. The demand pattern must be "levelled" through negotiations with the customer.
- Material shortages forcing one or more subsystems (or work centres) out of operation from time to time – in a large and complex manufacturing facility, it is possible to fail to produce/procure a part on time.

The impact of variability, both internal and external to a manufacturing system can be severe on the entire system because output of one system (or manufacturing stage) is used as input to another system (or stage). The impact of variation propagates from one stage to the next with increasingly adverse effects on production planning and control activities.

By far the most common strategy for "managing variability" in manufacturing has been to increase the ***flexibility*** of manufacturing systems through Computer Integrated Manufacturing (CIM) and Flexible Manufacturing Systems (FMS). Recent investigations (Sengupta *et al.*, 1998) have indicated that, a ***Common Pool*** of Kanbans for a group of parts manufactured on the same line (or within a FMS) provides more flexibility in terms of adherence to schedules/due dates than ***Equal Allocation***. The Equal Allocation rule specifies that the total number of Kanbans at any manufacturing stage is to be divided equally among the various part types within a group, whereas the Common Pool rule does *not* specify that a particular number of Kanbans should be dedicated to each part type. The number of Kanbans at any processing stage beyond a specific quantity does not improve the performance of the FMS in terms of smoothening the effects of variation in demand at the final stage of manufacturing.

It is well known that, in traditional Kanban systems, since the number of Kanbans is fixed, there are times when intermediate stations can either be blocked or starved. The blocking of a station occurs when all its production Kanbans are attached to full containers in its output buffer. Similarly, the starvation of a station occurs when at least one production Kanban is at its input buffer waiting for a container while a machine at that station is idle. One of the main reasons for such blocking and starving of stations is the variability in processing times, as well as the limited number of Kanbans available at each processing station. In order to overcome these deficiencies, two possible techniques have been suggested.

Savsar (1996) reported on the effects of **two different Kanban withdrawl policies** on the performance of JIT systems with variable processing times at different workstations. The *variable withdrawal cycle – fixed order quantity* (FQ) policies appear to improve throughput rate and average machine/workstation utilisation, while *fixed withdrawal cycle – variable order quantity* (FC) policies appear to reduce total work-in-process (WIP). A method was suggested to determine an optimum Kanban level which stabilizes throughput rate while keeping total WIP below some predetermined limit. Gupta & El-Turki (1997) have proposed the concept of the **Flexible Kanban System** (FKS) to counteract the effects of blocking and starvation due to high processing time variations.

"Hybrid" production control mechanism is essentially a combination of the traditional

Kanban system and the so called CONWIP (Constant Work-in-Process) production control system (Bonvik *et al.*, 1997, Gstettner and Kuhn, 1996). In the CONWIP system, information about a product demand flows directly from the final buffer to the first station. The hybrid model is essentially a "combination" pull system and seems to reduce the inventory in the internal buffers almost to zero under decreasing end point demand conditions.

A "pure" traditional Kanban system ensures the overall production rate by making a few parts ahead of time and storing them in buffers. These parts can then be used to let downstream machines continue working even if there is a disruption *upstream*. The number of Kanbans allocated to each workstation regulate the size of these buffers, and the production rate is regulated by buffers filling up and blocking production. The hybrid policy, on the other hand, regulates the production rate by holding the release rate to the system equal to the demand rate. This keeps the internal buffers almost empty. The empty space is then used to let upstream machines continue working, even if there is a disruption *downstream*. This achieves the same boost in production rate by partially decoupling the machines as if parts were stockpiled, but without the inventory cost. Further, in the hybrid control system, **the demand information is sent to the first workstation only**.

It has been reported that a manufacturing line controlled by such a hybrid policy can adjust to significant demand rate changes with minimal blockage or starvation. This is achieved, as explained earlier, by simply routing Kanbans from the finished goods buffer to the first production stage instead of the last, and adjusting the single (or constant) WIP limit (CONWIP) for the system. Of course, the price paid for this level of performance adjustment is that, in exceptional cases, the overall WIP for the hybrid system may be higher than the regular Kanban system; however, throughput will certainly be higher in the case of the hybrid system.

It may therefore be stated that steps are being taken to use Kanban-based JIT pull systems, and their modifications, to "shape" manufacturing environments with more uniform workflows and flexibility to adjust to changing capacity requirements.

CONCLUDING REMARKS: A Kanban-controlled production system improves the economics of operation through a number of benefits described earlier. A Kanban-controlled system is relevant not only for high volume repetitive manufacturing but also for a job shop. Based on the research reported in the literature, it may be concluded that a Kanban controlled system is not suitable for a system with variable demand where the variability cannot be controlled through planning. A Kanban controlled system is also not suitable for a very low volume demand.

Based on a representative sample of real world applications, the following guidelines are proposed for future implementations of Kanban systems.

1. Use group technology to form part families and machine groups within a job shop.
2. Allocate adequate resources for training to overcome the resistance to change to a Kanban controlled system and cross-train the workers.
3. Make every effort to reduce set-up times.
4. Focus on Total Quality Control and Total Preventive Maintenance
5. Involve as many functional departments as feasible to communicate the organizational needs for successful implementation of a Kanban system.
6. Compute the number of production and withdrawal Kanbans carefully with due consideration to relevant system parameters which should include, at the least, the demand rate, replenishment cycle time and their variations.
7. Use Kanban control of material flow as a tool for continuous improvement.

Readers should be also aware of the recent trends in the adaptation of the conventional Kanban based production control system to manufacturing environments with variable demand, small order sizes, and relatively short cycle times. Kanban based systems are continuously evolving and adapting to the need for continuous reduction in manufacturing costs as well as logistics and distribution costs.

Organizations Mentioned: Siemens; Toyota.

See **Blocked or starved station; Dynamic kanban control for JIT manufacturing; Equal allocation rule; Fixed withdrawal cycle; Flexible automation; Flexible kanban system; Just-in-time manufacturing; Material resources planning; Production ordering kanban; Single card system; Total productive maintenance (TPM); Two card system; Variable withdrawal cycle; Withdrawal kanbans.**

References

[1] Aigbedo, H. and Y. Monden (1997). "Scheduling for Just-in-Time Mixed Model Assembly Lines-A Parametric Procedure for Multi-Criterion Sequence", *International Journal of Production Research*, 35, 2543–2564.

[2] Bonvik, A.M., C.E. Couch and S.B. Gershwin (1997). "A Comparison of Production Line Control Mechanisms," *International Journal of Production Research*, 35, 789–804.

[3] Crawford, K.M., J.H. Blackstone and J.F. Cox (1988). "A Study of JIT Implementation and Operating Problems," *International Journal of Production Research*, 26, 1561–1568.

[4] Deleersnyder, J.L., T.J. Hodgson, H. Muller and P.J. O'Grady (1989). "Controlled Pull Systems: An Analytical Approach," *Management Science*, 35, 1079–1091.

[5] Draper, C. (1984). *Flexible Manufacturing Systems Handbook*, Automation and Management Systems Division, The Charles Stark Draper Laboratory Inc., Noyes Publication.

[6] Gravel, M. and W.L. Price (1988). "Using the Kanban in a Job-shop Environment," *International Journal of Production Research*, 26, 1105–1118.

[7] Gstettner, S. and H. Kuhn (1996). "Analysis of Production Control Systems Kanban and CONWIP," *International Journal of Production Research*, 34, 3253–3273.

[8] Gupta, S.M. and Y.A.Y. Al-Turki (1997). "An Algorithm to Dynamically Adjust the Number of Kanbans in Stochastic Processing Times and Variable Demand Environment," *Production Planning and Control*, 8, 133–141.

[9] Huang, C.C. and A. Kusiak (1996). "Overview of Kanban Systems," *International Journal of Computer Integrated Manufacturing*, 9, 169–189.

[10] Keaton, M. (1995). "A New look at the Kanban Production Control System," *Production and Inventory Management Journal*, Third Quarter, 71–78.

[11] Krazewski, L.L., B.E., King, L.P. Ritzman and D.S. Wong (1987). "Kanban, MRP and Shaping the Manufacturing Environment," *Management Science*, 33, 39–57.

[12] Philipoom, P.R., L.P. Rees and B.W. Taylor (1996). "Simultaneously Determining the Number of Kanbans, Container Sizes and the Final Assembly Sequence of Products in a Just-in-Time Shop," *International Journal of Production Research*, 34, 51–69.

[13] Savsar, M. (1996). "Effects of Kanban Withdrawal Policies and other Factors on the Performance of JIT Systems – A Simulation Study," *International Journal of Production Research*, 34, 2879–2899.

[14] Schwarzendahl, R. (1996). "The Introduction of Kanban Principles to Material Management in EWSD Production," *Production Planning and Control*, 7, 212–221.

[15] Sethi, A.K. and S.P. Sethi (1990). "Flexibility in Manufacturing: A Survey," *International Journal of Flexible Manufacturing Systems*, 2, 289–321.

[16] Sengupta, S., F. Sharief and S.P. Dutta (1998). "Determination of Optimal Number of Kanbans and Kanban Allocation in a FMS: A Simulation Based Study," *Production Planning and Control*, under review.

[17] Siha, S. (1994). "The Pull Production System: Modelling and Characteristics," *International Journal of Production Research*, 32, 933–950.

[18] Tayur, S.R. (1993). "Structural Properties and a Heuristic for Kanban-Controlled Serial Lines," *Management Science*, 39, 1347–1368.

LABOR ASSIGNMENT/ REASSIGNMENT

Good labor scheduling practices are essential to the success of cellular manufacturing and just-in-time manufacturing. Flexible labor scheduling practices at work centers are used to vary capacity as work loads fluctuate. Using flexibility in assigning people to work centers, we can improve manufacturing performance as measured by reduced flow times, better delivery, and decreased work-in-process inventory. Flexibility in making labor assignments depends on cross-training of the work force. A cross-trained work force reduces the costs of shifting people between work centers.

See **Just-in-time manufacturing; Lean manufacturing implementation; Linked-Cell manufacturing systems; Manufacturing cell design.**

LATENESS (TARDINESS)

Lateness is defined as delivery date minus due date. Lateness may be positive or, in the case of early jobs, negative. While evaluating priority sequencing rules, lateness is used to compare various priority sequencing rules such as, first-come-first served, or shortest processing time rule. Late jobs will have positive lateness.

See **Dynamic routing in flexible manufacturing systems; Priority sequencing rules; Simulation analysis manufacturing and logistics systems.**

LAYOUT

In manufacturing, there are several options for the layout of machines on the shop floor. Layout determines the flow of materials, parts and products through a facility. Layout has significant impact on the efficiency of the production system, and the lead-time for manufacturing.

The different types of factory layouts are: fixed position layout; process-oriented layout or job shops; product-oriented layout or production/assembly lines; and manufacturing cells. The decision to choose one type of location over the other depends on:

- The design
- Volume of production
- Nature of production equipment
- Production capacity
- Quality of life issues for the production workers.

Good factory layout is a function of several key factors, which are: capacity and space requirements; material handling; flow of material and information; and cost of moving materials and people between workstations.

The general principles for selecting one layout over the other are: for low volume, high variety production, use job shops with general purpose machines; for high volume production with very little product variety, use line production; and use manufacturing cells for conditions in between the two. For extensive discussion of the various layouts, see *Manufacturing Systems*.

Job shop layouts can be done using several software packages such as CRAFT, SPACECRAFT, CORELAP, ALDEP, COFAD, and FADES.

See **Assembly line design; History of operations management; Manufacturing cell design; Manufacturing systems.**

References

[1] Apple, J.M. (1977). *Plant Layout and Material Handling*. New York: Wiley.

[2] Fisher, E.L. (1986). An AI based methodology for factory design. *AI Magazine*, Vol. 3 (4), Fall, pp. 72–85.

[3] Muther, R. (1976). *Systematic Layout Planning*. 2nd edition, Boston: Cahners.

[4] Baybars, K (1986). A survey of exact algorithms for the simple assembly line balancing problem. *Management Science*, Vol. 32 (8), August, pp. 909–932.

LEAD-TIME

Lead-time is a very important performance measure. It refers to the time taken to complete a task. The task may be purchasing of raw materials, in which case, lead-time refers to the difference in time between the time an order is placed and the time the material is delivered. Manufacturing lead-time in a factory could mean the elapsed time between order receipt and order shipment to customer. In flexible manufacturing facilities, lead-times are generally lower than in inflexible systems.

See **Manufacturing cell design; Manufacturing systems; Lean manufacturing implementation; Just-in-time manufacturing.**

LEAD-TIME OFFSETTING

It is a technique used in MRP to set the release date for an order earlier than planned order receipt by an amount of time equal to the lead time for the item. Example: If planned order receipt time period is week 8, lead time is 3 weeks, then order release period should be (8 – 3) week 5.

See **Material requirements planning (MRP).**

LEAN LOGISTICS

Lean logistics is the application of lean management principles to supply chain operations. The underlying principle of lean management is to eliminate any part of the system that does not add value, such as costly inventories or warehousing. The introduction of lean manufacturing, first introduced by Toyota in the late 1980s, represented a major shift away from mass production, and depended on new logistics approaches to make it successful. Specifically, the shift to smaller lot sizes and decreased inventories required continual delivery of parts and components on a just-in-time basis to feed the assembly line. Frequent delivery of parts eliminated the need for costly inventory but made it necessary for suppliers to be located within easy trucking distance of the assembly site. Lean systems were successful because suppliers and customers shared information about real-time customer orders and production schedules on a regular basis. Overall, lean systems have shown significant productivity and quality gains over their mass production counterparts.

See **Agile logistics (enterprise logistics); Just-in-time manufacturing; Lean manufacturing implementation; Supplier relationships; Supply chain management: Competing through integration.**

Reference

[1] Cusunamo, M. (1994). "The Limits of Lean," *Sloan Management Review*, 35(4), 27–32.

LEAN MANUFACTURING

See **Just-in-time manufacturing; Lean manufacturing implementation; Linked-cell manufacturing; Manufacturing cell design; Manufacturing cells, linked.**

LEAN MANUFACTURING CELL

It is a manufacturing cell designed with dissimilar machine tools for use in cellular manufacturing in a lean production system.

See **Just-in-time manufacturing; Lean manufacturing implementation; Manufacturing cell; Manufacturing cell design; Manufacturing systems.**

LEAN MANUFACTURING IMPLEMENTATION

J.T. Black

Auburn University, AL

Many companies have implemented lean production in the last decade. Lean production describes a manufacturing system that uses less resources to make a company's products. The term lean production was first used in the book, *The Machine that Charged the World* (Womack, Jones and Roos, 1990), to describe the Toyota Production System. Lean manufacturing systems are called by various names in the United States. Most commonly used names for lean manufacturing are listed in Table 1.

Figure 1 outlines the ten key steps to converting a factory from a job shop/flow shop manufacturing system to a linked-cell manufacturing system in achieving a true lean production system. The requirements for the successful implementation of lean manufacturing are:

1. All levels in the plant, from the production worker (the internal customer) to the president must be educated in lean production philosophy and concepts.
2. Top management must be totally committed to this venture and provide necessary leadership. Everyone must be involved in the change, and the internal customer must be empowered to play a vital role in this evolutional process.

Table 1. Various names for lean manufacturing system.

System name	Name used by
LEAN PRODUCTION	MIT group of authors, researchers
TOYOTA PRODUCTION SYSTEM – (TPS)	Toyota Motor Company
OHNO SYSTEM	Taiichi Ohno, Inventor of TPS
INTEGRATED PULL MANUFACTURING SYSTEM	AT&T
MINIMUM INVENTORY PRODUCTION SYSTEM – (MIPS)	Westinghouse
MATERIAL NEEDED – (MAN)	Harley Davidson
JUST-IN-TIME/TOTAL QUALITY CONTROL (JIT/TQC)	Schonberger
WORLD CLASS MANUFACTURING – (WCM)	Schonberger
ZERO INVENTORY PRODUCTION SYSTEMS – (ZIPS)	Omark and Bob Hall
QUICK RESPONSE – OR MODULAR MANUFACTURING	Apparel Industry
STOCKLESS PRODUCTION	Hewlett Packard
KANBAN SYSTEM	many companies in Japan
THE NEW PRODUCTION SYSTEM	Suzaki
ONE PIECE FLOW	Sekine
CONTINUOUS FLOW MANUFACTURING – (CFM)	Chrysler
INTEGRATED MANUFACTURING PRODUCTION SYSTEMS – (IMPS)	Black

3. Everyone in the plant must understand that cost, not price determines profit. Since the customer determines price, customer satisfaction is the key.
4. Everyone must be committed to the elimination of waste. This is fundamental for becoming lean.
5. The concept of standardization must be taught to everyone and applied to documentation, methods, processes as well as system metrics.

The steps for implementing lean manufacturing identified in Figure 1 are elaborated step by step.

Step 1. Form U-Shaped Cells – Restructure the Factory Floor: In the linked-cell manufacturing system (L-CMS), cells replace the job shop. The first task is to restructure and reorganize the basic manufacturing system into manufacturing cells that manufacture families of parts. This prepares the way for a systematically created linked-cell system designed for one-piece movement of parts within cells and for small-lot movement between cells. Creating cells is the first step in designing a manufacturing system in which production control, inventory control, quality control, and machine tool maintenance are integrated.

Step 2. Rapid Exchange of Tooling and Dies: Everyone on the plant floor must be taught how to reduce setup time using SMED (single-minute exchange of dies) principles. A setup

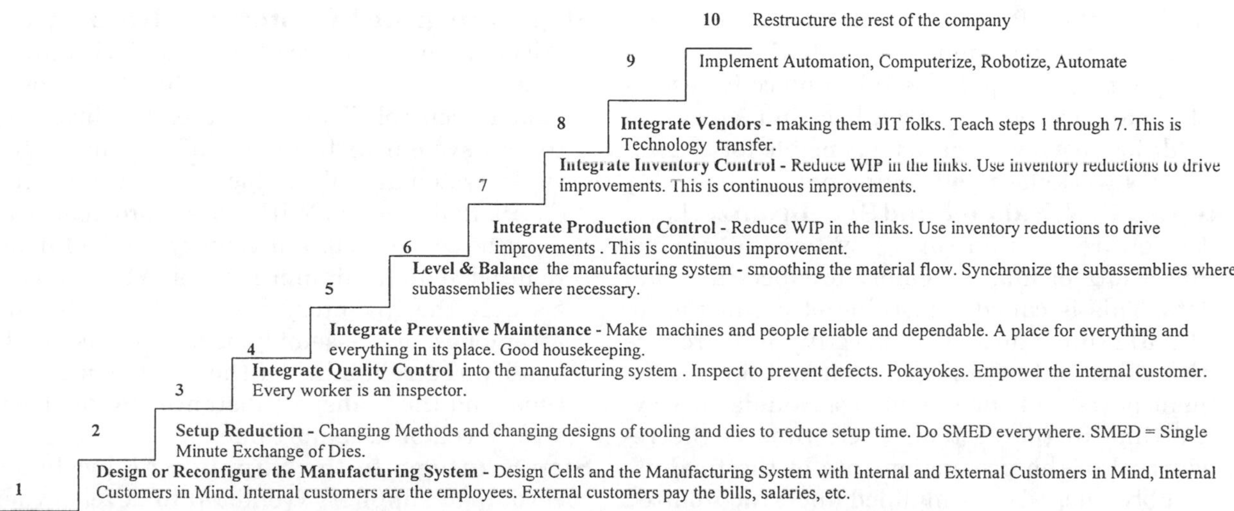

Fig. 1. Black's ten steps to lean production, or how to design a linked-cell manufacturing system.

reduction team acts to facilitate the SMED process for production workers and supervisors. The key is that everyone must get involved in setup reduction. The team may attack the plant's worst setup problem as a demonstration project. Reducing setup time is critical to reducing lot size.

Step 3. Integrate Quality Control: *A multiprocess* worker can run more than one kind of manufacturing process. A *multifunctional* worker can do more than operate machines; he/she is also an inspector, who understands process capability, quality control, and process improvement. In lean production, every worker has the responsibility to make the product right the first time and every time, and has the authority to stop the process when something goes wrong. This integration of quality control into the manufacturing system markedly reduces defects while eliminating roaming inspectors and associated costs. Cells provide a natural environment for the integration of quality control. The fundamental idea here is to use inspection to *prevent defects from occurring.* The cells do not release defective parts to downstream subassembly areas.

Step 4: Integrate Preventive Maintenance: To make machines operate reliably, begin with the installation of an integrated preventive maintenance (IPM) program by giving workers the training and tools to maintain equipment properly. The extra or additional processing capacity obtained by reducing setup time allows operators to reduce equipment speeds or feeds and to run processes at less than full capacity, thus reducing the pressure on workers and the wear and tear on tools. Reducing pressure on workers and processes fosters in workers a drive to produce near perfect quality. A key to IPM is housekeeping. Simply put, there is a place for everything and everything is put back in its place. In addition, each worker is responsible for cleanliness of workplace and equipment.

Step 5. Level, Balance and Synchronize: Level the entire manufacturing system by producing a mix of final assembly products in small lots. This is called smoothing of production in the JIT literature. The objective is to reduce the effect of lumpiness of demand for component parts and subassemblies. Standardize cycle time in the system. Cycle time is the reciprocal of the production rate of the final assembly line. Use a simplified and synchronized system to produce the proper number of components everyday, as needed. Some companies begin at mixed-model final assembly, and work backward through subassembly and manufacturing cells. Each process, cell, and subassembly tries to match the daily build quantity and cycle time of final assembly. This is called *balancing*. If the time for the manufacture of the subassembly is matched to that of final assembly, then the system is synchronized. Only through the effective use of JIT manufacturers can accomplish this.

The cycle time for the final assembly line is determined by the demand rate for parts according to the following calculations. Let CT = cycle time; PR = production rate.

$$\text{Daily Demand} = \frac{\text{monthly demand (forecast plus customer orders)}}{\text{number of days in month}}$$

$$\text{CT} = \frac{1}{\text{PR}} \quad \text{where}$$

$$\text{PR} = \frac{\text{Daily Demand (in units)}}{\text{available hours in day (hrs)}}$$

This incredibly simple approach highlights the method used by JIT companies calculate cycle time. Manufacturing becomes simpler when a linked-cell system is installed.

Step 6. Use Linked Cells: The simplification and integration of production control is materialized by linking the cells, subassemblies, and final assembly elements by utilizing the kanban subsystem. By connecting process elements with kanban links, the need for route sheets is eliminated. All the cells, processes, subassemblies, and final assemblies are connected by kanban links that pull materials to final assembly. This achieves the integration of production control into the manufacturing system, forming a linked-cell manufacturing system (see Figures 2 and 3).

Step 7. Integrate Inventory Control: People on the plant floor can directly control the inventory levels in their areas through the utilization of kanban control. This integrates the inventory control system with the manufacturing system while systematically reducing work-in-process (WIP). Reduction of WIP exposes problems that must be solved before inventory can be further reduced. The minimum level of WIP (which is actually the inventory *between* the cells, subassemblies, and assembly) is determined by the quality, the reliability of the equipment, setup time, and the transport distances to the downstream cell or assembly line.

Step 8. Integrate the Suppliers: Educate and encourage suppliers (vendors) to develop their own lean production system for superior quality, low cost, and rapid on-time delivery. They must be able to deliver parts to the customer when needed and where needed without incoming

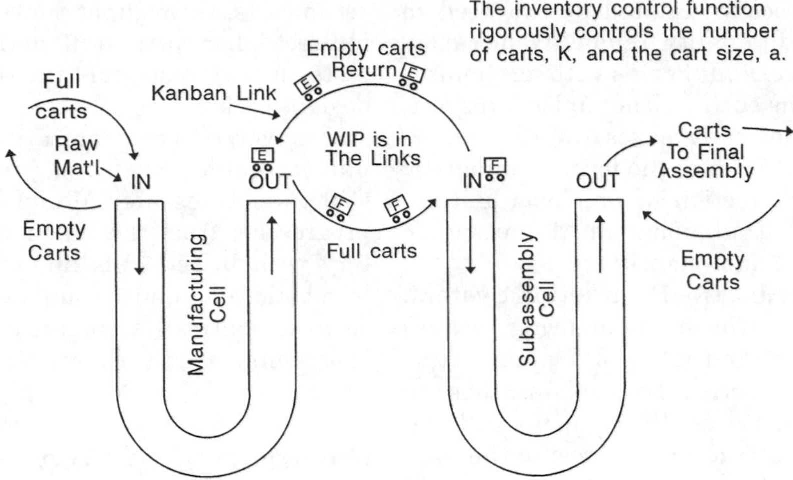

Fig. 2. Cells are linked by a Kanban system. The arrival of an empty container at the manufacturing cell is the signal to produce more parts.

inspection. The linked-cell network ultimately should include every supplier. Thus, suppliers become remote cells in L-CMS.

After these eight steps have been completed, and the manufacturing system has been redesigned and infused (integrated) with the critical production functions of quality control, inventory control, production control, and machine tool maintenance, the autonomation of the integrated system is the next step.

Step 9. Institute Autonomation: Converting manned cells to unmanned cells is an

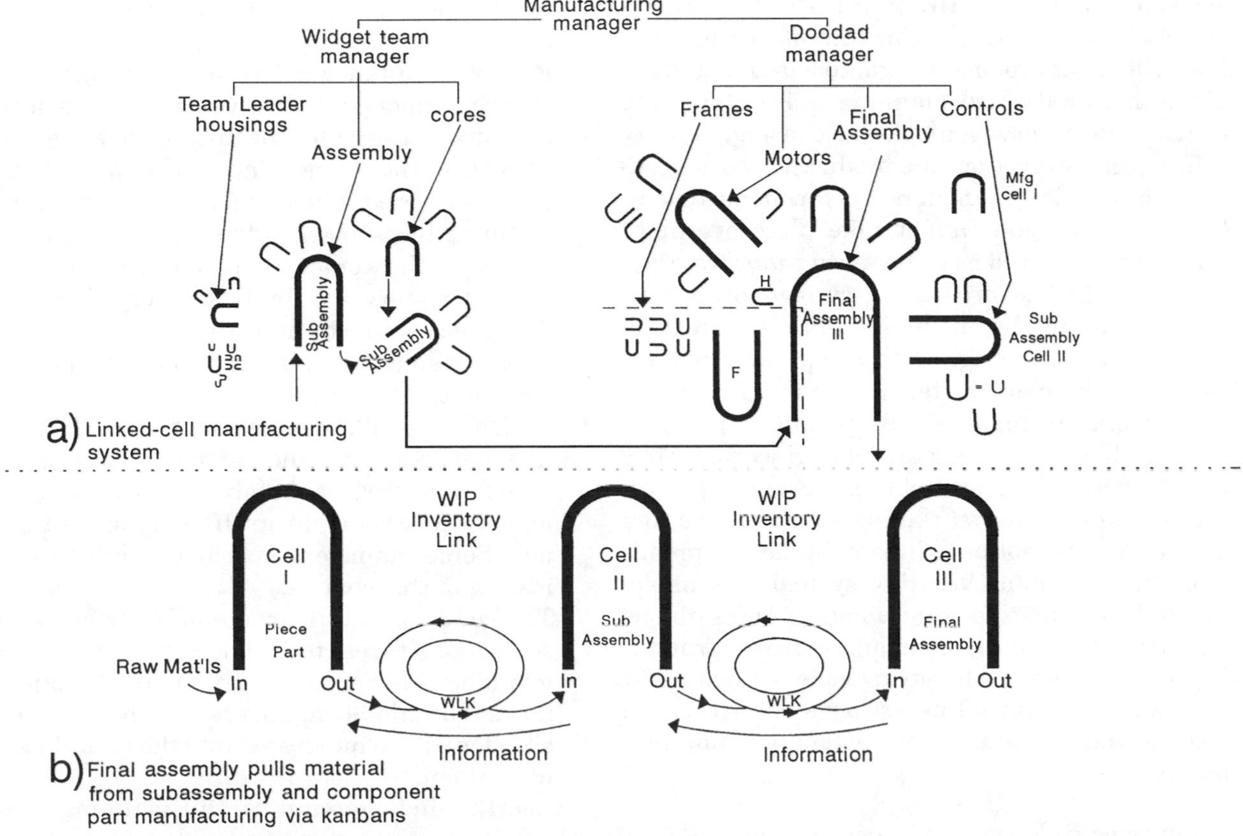

Fig. 3. In the linked-cell manufacturing system, the cells are links with controllable inventory buffers call kanban links or loops.

evolutionary process initiated by the need to solve problems in quality, reliability, or capacity (i.e., bottlenecks). It begins with mechanization of operations such as load, unload, inspect, and clamp, and logically moves toward more advanced mechanization in the form of automatic detection and correction of problems and defects. This is called autonomation, the automatic control of quality and quantity.

Step 10. Restructure the Production System: Once the factory (the manufacturing system) has been restructured into a JIT manufacturing system, and the critical control functions are integrated and matured, the company will find it expedient to restructure the rest of the company. Production system will require removing or reducing the functional orientation of the various departments and forming cross-functional teams, often along product lines. The implementation of concurrent engineering teams decrease the time needed to bring new products to market. This movement is gaining ground in many companies and is often called *Business Process Reengineering* (BPR); it is basically the restructuring of the plant operation to be as waste free and efficient as the manufacturing system.

RESTRUCTURING THE REST OF THE BUSINESS: Changing to a Lean Manufacturing system will affect product design, tool design and engineering, production planning (scheduling) and control, inventory control, purchasing, quality control and inspection, the production worker, supervisors, middle managers and top management. Such a conversion cannot take place overnight and must be viewed as a *long-term transformation* from one type of *production system* to another. This effort often begins with building product realization teams designed to bring new products to the marketplace faster. In the automotive industry, one example of this is called, platform teams, which enable concurrent engineering. They are composed of personnel from design, engineering, manufacturing, marketing, sales, finance, and so on. As the notion of team building spreads and the lean manufacturing system gets implemented, a leaner factory/company takes shape. Unfortunately, many companies are restructuring their non-manufacturing side without first completing the initial necessary steps 1 through 8 to get the manufacturing system lean and productive.

BENEFITS OF CONVERSION: The conversion to lean manufacturing results in significant cost savings over a two- to three-year period. Specifically, manufacturing companies report significant reduction in raw materials, in-process inventories, setup costs, throughput times, direct labor costs, indirect labor costs, staff, overdue orders, tooling costs, quality costs, and the cost of bringing new designs on line.

However, this reorganization has a greater and immeasurable benefit. It prepares the way for Computer Integrated Manufacturing (CIM). The progression from the functional shop to the factory with linked cells and, perhaps, ultimately to robotic cells under computer control gradually turns the entire system into a CIM system, which is accomplished in several logical steps.

CONSTRAINTS TO CONVERSION: A major effort on the part of a business is required to undertake the conversion to lean manufacturing. The hurdles to implementation are:

1. The top management person (or the real leader) does not totally buy into the conversion.
2. Systems changes are inherently difficult to implement. Changing the entire manufacturing production system is a huge task.
3. Companies spend freely for new manufacturing processes (i.e., machines) but not for the conversation to a new manufacturing system. It is easier to justify new hardware than to implement a new manufacturing system (i.e., linked cells).
4. Fear of the unknown. Decision making is often choosing among alternatives in the face of uncertainty; the greater the uncertainty, the more likely that the "do-nothing" option will prevail.
5. Faulty criteria for investment decisions. Manufacturing decisions should seek to enhance the ability of the company to compete (on items such as quality, reliability, delivery time, flexibility for product change or volume change) rather than merely minimize the cost of the investment.
6. Lack of blue-collar involvement in the decision-making process of the company. Getting the production workers involved in the decision-making process is, in itself, a significant change. Some managers may have problems adjusting to this change.
7. The conversion to lean manufacturing represents a real threat to middle managers. In lean manufacturing systems, some of the functional tasks that middle managers have been responsible for will be integrated into the manufacturing system.

Clearly, education and careful planning at all levels of the enterprise is needed to overcome these constraints. The attitudes of management and workers must change. By the next century, we may see significantly fewer workers on the plant floor. These workers will be far better educated

Table 2. Managing the lean production system (LPS).

The LPS includes the linked-cell manufacturing system (cells linked by a kanban pull system), the 5 S's (see below), standard operation, the seven tools of QC and other key elements.

Kanban Pull System

It is the production process that uses a card system, standard container sizes, and pull versus push production to accomplish just-in-time production.

5 S's (Seiri, Seiton, Seiketsu, Seiso, Shitsuke)

The five S's in Japanese language are proper arrangement, orderliness, cleanliness, cleanup, and discipline.

Standard Operation in Manufacturing Cells

The components of standard operation include: cycle time, work sequence, and standard stock on hand in the cells.

Morning Meeting

A daily meeting is held for the purpose of sharing production and safety information, quite often by a quality circle.

Process Sheets

The process sheets which are visually posted at each work station detail the work sequence and most critical points for performing the tasks.

Change-Over and Setup

The machine setup that must take place when an assembly line changes products.

Seven Tools of Quality

The seven tools of quality are: Pareto diagrams, check sheets, histograms, cause and effect diagrams, run charts for individuals, control charts for samples, and scatter diagrams. (See Integrated Quality Control.)

Production Behavior

Rules which include information on personal safety, safety equipment, clothing, restricted areas, vehicle safety, equipment safety, and housekeeping.

Visual Management

Each production line in the plant has a complete set of charts, graphs, or other devices like *Andons* (An electronic line stop indication board) for reporting the status and progress in the area.

---

and more productive than today's workers. They will be involved in: (1) solving daily production problems, (2) working to improve the entire system, and (3) making decisions to improve their job, the processes, and the manufacturing system. Table 2, obtained from a first-tier supplier to Toyota, summarizes some of the key managerial aspects of the Lean Production system.

Table 2 mentions the use of the seven tools of quality in managing the lean production system. The seven tools of quality are: flow diagram; histogram; pareto chart or ABC analysis; scatter diagram; fishbone diagram; run chart; and control chart. These tools are displayed in Figure 4. In the figure, Histogram, Pareto Chart, Scatter Diagram, Run Chart and Control Chart use quantitative or numerical data to understand, and/or identify and explain a problem.

A Histogram identifies the more frequently occurring events or problems. A scatter Plot visually shows the nature of relationship between two variables of interest. The shape of Histograms may reveal if the process is skewed to the left or right.

Run Chart and Control Chart use data for statistical Process control. A Run chart is used with small batches to detect process shift. Each item is inspected and plotted on the Run Chart. A sequence of plots with upward or downward trend suggest a process that is shifting and may go out of control, if not corrected. Run charts are bounded by USL and LSL, which are upper and lower specification limits. Plots outside the limits indicate the failure to meet specifications by the product; it should be reworked or discarded and the process should be adjusted.

For extensive description of control charts, See: *Statistical Process Control Using Control Charts*. Fishbone Diagram is effective for use in a team for the identification of root cause to a problem. It is also called "cause effect" diagram, which is a tool for systematically identifying and investigating all causes, subcauses, and so on.

Organizations Mentioned: Toyota.

See **ABC analysis; Autonomation; Business process reengineering (BPR); Concurrent engineering; Integrated quality control; Island of automation; Just-in-time manufacturing; Kanban-based manufacturing; Linked-cell manufacturing system (L-CMS); Manufacturing systems; Platform teams; Statistical quality control using control charts.**

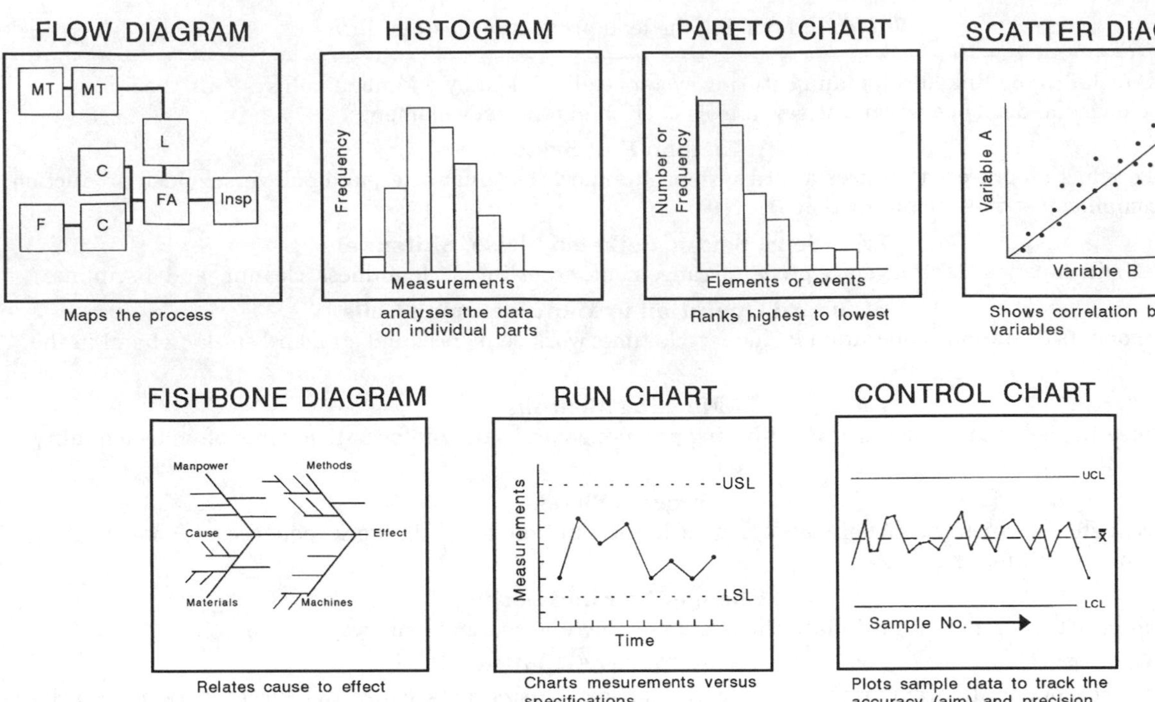

Fig. 4. The seven tools of quality control.

References

[1] Black, J.T. (1991). *The Design of The Factory With A Future*. McGraw Hill.

[2] Hall, Robert W. (1983). *Zero Inventories*. Dow Jones-Irwin.

[3] Harmon, Roy L., Peterson, Leroy D. (1990). *Reinventing the Factory: Productivity Breakthroughs In Manufacturing Today*. The Free Press.

[4] Ishikawa, Kaouru (1972). *Guide to Quality Control*. Asian Productivity Organization.

[5] Monden, Yasuhiro (1983). *Toyota Production System*. Industrial Engineering and Management Press, IIE.

[6] Nakajima, Seiichi (1988). *TPM, Introduction to TPM: Total Productive Maintenance*. Productivity Press, 1988.

[7] Ohno, Taiichi (1988). *Toyota Production System: Beyond Large-Scale Production*. Productivity Press.

[8] Schonberger, Richard J. (1986). *World Class Manufacturing*. The Free Press.

[9] Sekine, Kenichi (1990). *One-Piece Flow: Cell Design for Transforming the Production Process*. Productivity Press.

[10] Schonberger, Richard J. (1982). *Japanese Manufacturing Techniques: Nine Hidden Lessons in Simplicity*. The Free Press.

[11] Shingo, Shigeo (1985). *A revolution in Manufacturing: The SMED System*. Productivity Press.

[12] Shingo, Shigeo (1986). *Zero Quality Control: Source Inspection and the Poka-Yoke System*. Productivity Press.

[13] Shingo, Shigeo (1988). *Non-Stock Production: The Shingo System for Continuous Improvement*. Productivity Press.

[14] Suzaki, Kioyoski (1987). *The New Manufacturing Challenge*. The Free Press.

[15] Womack, J.P., D.T. Jones and D. Roos. *The Machine that Changed the World*. New York: Harper Perenial, 1990.

LEAN PRODUCTION

This is a term coined by the authors of the book *The Machine That Changed the World* (Womack, Jones and Roos, 1991). It is known by other names such as Just-in-Time Production, World Class Manufacturing, and Linked-Cell Manufacturing Systems.

See **Just-in-time manufacturing; Just-in-time manufacturing implementation; Linked-cell manufacturing; Manufacturing cells, linked.**

LEAN SUPPLY CHAIN

Many companies go beyond lean manufacturing to transform the entire company into 'lean organization'. The next logical step is to develop a 'lean supply chain' so that the linkages between a company's suppliers and customers are also based on 'lean' manufacturing principles such as waste elimination, cooperation, and trust. The synergy among organizations forming the whole supply chain, from the suppliers' suppliers to the customers' customers, and to the end consumer is maximized. This requires a new approach to supply chain management (SCM). The benefits are: Supply chain cost reduction and improved customer satisfaction.

See **Customer service; Logistics: Meeting customer's real needs; Supplier partnership as strategy; Supplier relationships; Supply chain management: Competing through integration.**

LEARNING CURVE ANALYSIS

Timothy L. Smunt

Wake Forest University, Winston-Salem, NC, USA

DESCRIPTION AND HISTORICAL PERSPECTIVE: The learning curve concept was formally introduced by T.P. Wright (Wright, 1936) through his pioneering empirical study on the productivity trends of the production of aircraft. Since that time, and especially following World War II, the learning curve concept has been extensively researched and applied in numerous settings. While the most publicized applications are related to the production of aircraft and defense products, other products have also experienced similar learning trends.

The most popular learning curve model that is used today is the same one that was proposed by Wright, i.e., that production costs decrease by a certain percentage each time the quantity produced doubles. It is sometimes referred to as the *cumulative average cost* model, or simply as the Wright model.

The cumulative average cost model was the first learning curve model to be published and used in practice. Wright first discovered the learning curve phenomenon in 1922 during his analysis of airplanes produced by the Curtis-Wright Corporation. He noticed that as the cumulative quantity produced doubled, the cumulative average cost for this quantity decreased by a constant ratio known as the learning curve percentage. The cumulative average cost curve is formulated as:

$$Y_{ca}(x) - (a)(x^{-b}) \tag{1}$$

where

$Y_{ca}(x)$ = cumulative average cost for the first **x** units
a = first unit cost

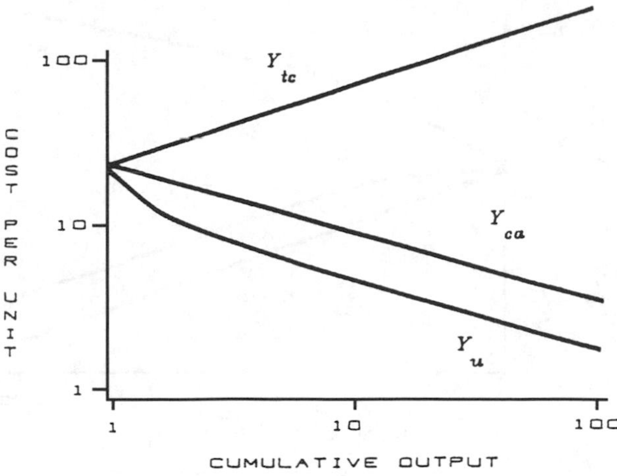

Fig. 1. Cumulative average model.

x = cumulative unit number
b = learning curve factor
$$= -\frac{\log(\text{learning rate }\%)}{\log(2.0)}$$

The total cost for the first x units is then,

$$Y_{tc}(x) = (a)(x^{1-b}) \tag{2}$$

The unit cost in the cumulative average model is a discrete function based on the total cost through any given point in question less the cost of prior units:

$$Y_u(x) = Y_{tc}(x) - Y_{tc}(x-1) \tag{3}$$

Graphed on a logarithmic scale, these relationships are illustrated in Figure 1. For simple examples illustrating the various models, see a later section titled, significant Analytical Models.

Determination of the learning curve parameters, **a** and **b**, is usually accomplished by regression analysis and requires the use of actual cumulative average costs. Projecting unit costs from the cumulative average cost function can become cumbersome since the unit cost function is discrete in nature. Also, if cost data are missing certain values, cumulative costs cannot be calculated; and therefore, the cumulative average model cannot be used to determine learning curve parameters unless missing data are estimated.

A Lockheed employee, J.R. Crawford, developed the *unit* cost model in the mid-1940s. Interestingly, this model became known as the "Boeing Curve," since Boeing was the prime user of this concept and published tables of factors for using this learning curve. Crawford chose to establish the unit cost function as the logarithmic function instead of the cumulative average function. In this model the unit cost is calculated as:

$$Y_u(x) = (a)(x^{-b}) \tag{4}$$

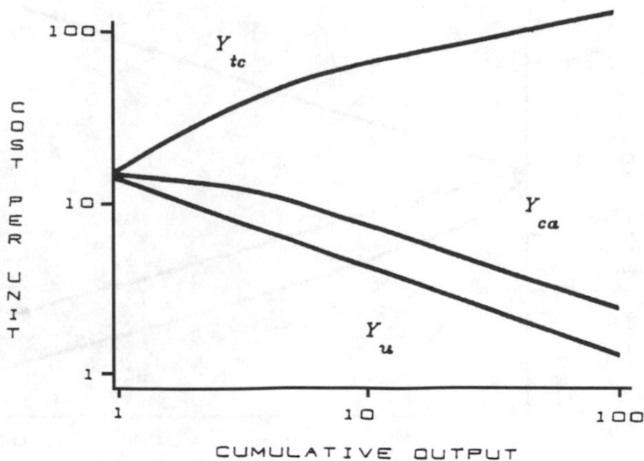

Fig. 2. Unit cost model.

This necessitates two discrete cost functions;

$$Y_{tc}(x) = \sum Y_u \tag{5}$$

and

$$Y_{ca}(x) = \frac{\sum Y_u}{x} \tag{6}$$

The unit average model (Figure 2) illustrates that the discrete nature of the total cost function necessitates either the use of a large set of conversion factors for each learning rate percentage, or substantial amounts of computation time to individually add all unit costs in question. See the later section, Significant Analytical Models for simple illustrations of these models.

This model became widely utilized, however, due to the often missing cost values of certain units, as stated previously, missing unit cost values negate the possibility of calculating cumulative average costs. Another major advantage of using the unit cost model is that changes in the learning rate become identifiable more quickly. The use of cumulative average costs, on the other hand, introduces considerable lag in detecting the change of the learning trend.

It should also be noted that the learning rate determined by regression analysis of unit costs is not appropriate for use in the cumulative average cost function. The cumulative average cost has a flatter slope since it includes all past cost data. Although the unit cost curve and the cumulative average cost curve become asymptotic to each other in either model, large projection errors are possible if similar slopes are assumed for both the cumulative average costs and unit costs, especially in the early units of production.

STRATEGIC AND TECHNOLOGY PERSPECTIVE: Learning curve analysis can be used in practice, not only for projecting and analyzing direct labor costs, but also for total costs, including materials, purchased items and overheads. When applied to costs beyond those caused by labor, it is sometimes referred to as the "experience curve." While the Boston Consulting Group is often mentioned as the catalyst of implementing the experience curve concept in industry, the notion of using learning curves for strategic purposes can be traced to the original paper by Wright (1936). In this paper he states:

> "We have the usual circle of relationships wherein price can be reduced most effectively by increasing quantity but wherein quantity production can only be obtained through market possibilities brought about by cheaper prices."

This notion of the pricing relationship to competitive advantage was reported to be implemented by Texas Instruments, most notably in the pricing of their early hand-held calculators. They priced their calculators based upon *expected* price reductions that would occur due to the increasing production experience caused by gaining additional market share due to their relatively lower prices.

Other sources of the experience effect include redesigning the product and its standardization, improving production processes, obtaining economies of scale, implementing concurrent engineering activities, and providing stability in product offerings. The intention of these activities is to provide the environment for stable, high quantity production. Research has shown that any interruptions in the production of most items will cause either a step up to a higher cost level before the learning curve again begins to take hold (top line of Figure 3) or an overall reduction in the

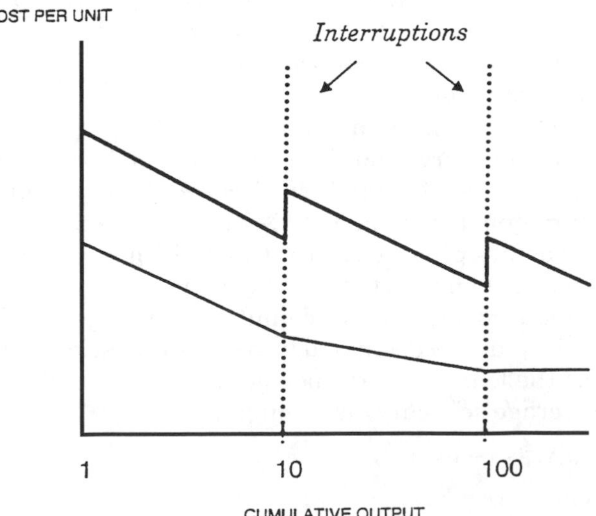

Fig. 3. Production interruption-step-up effect and lower learning rate effect.

learning rate, i.e., a flatter slope (bottom line of Figure 3). The first condition is noticeable when a few or less major interruptions occur (see, for example, Conway and Schultz, 1959). The second condition is apparent when the system moves from a stable learning environment to one where additional variance is introduced into the system vis-à-vis a high labor turnover rate, the reduction or elimination of proper engineering support, or a reduction of applicable training for direct labor.

One of the key issues that management must confront is the way in which they plan to compete for market share. Firms compete generally on different competitive dimensions such as cost, quality, flexibility and timeliness. Firms that compete more on the cost dimension would appear to have the most to gain by explicitly incorporating the learning curve concept into their planning and control procedures. On the other hand, firms that compete relatively more strongly on the flexibility dimension will choose to increase investments in processes that can be set up quickly to provide responsiveness to changes in demand volumes or product designs. While these flexible types of processes are chosen to reduce the costs associated with changes, it is unlikely that the overall cost of production will be smaller than that of a firm that chooses to maintain a fairly stable product design, and low prices.

The strategic use of learning curves also provides a basis for any firm to analyze the breadth of the product line offered and the types of processes used. Companies should consider the life cycle stage of a product, the types and amount of resources available, and their relative position with relation to costs. For example, firms that typically compete during the introduction stage of a product's life cycle, e.g., R&D firms, would be less likely to find particular advantage in pursuing the learning curve concept for managing production costs. On the other hand, large manufacturers of appliances or other consumer goods will probably find that investments in increasing the learning rates of their production processes will yield a good return on appropriate investments.

IMPLEMENTATION: A number of factors contribute to both the learning rate of an individual worker or the organization as a whole. The factors that contribute to individual learning are:

- training
- previous experience of the worker
- methods design
- individual incentive pay plans
- lighting and environmental conditions
- raw material and purchased part quality
- layout of the workplace

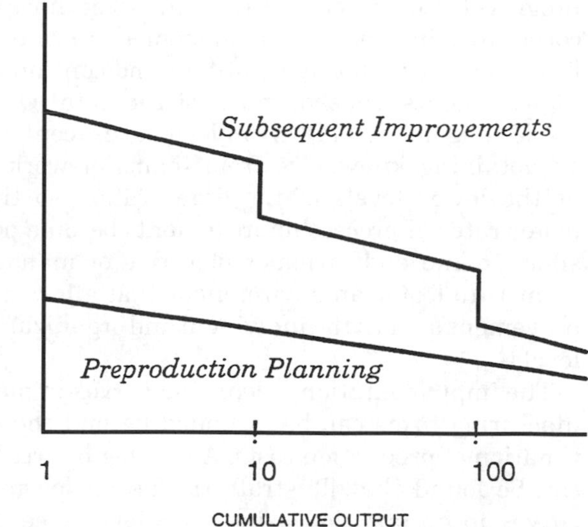

Fig. 4. Effect of preproduction and subsequent efforts.

Factors that contribute to organizational learning include:

- product stability
- capital equipment specialization
- process rationalization
- group incentive pay plans
- overall education level of employees
- level of vertical integration
- degree of planning and design efforts

Many of the above factors that contribute to learning can be implemented either before production activities begin, called "preproduction efforts," or after production has commenced, called "subsequent efforts." Interestingly, the amount of time and money spent on preproduction design and planning activities lower overall learning rate during full production. (See Figure 4 for an illustration of this effect.) But, appropriate preproduction activities reduce the cost of the first units of production (the y-axis intercept of the learning curve plot). Indeed, this could very well be the best approach, since changes made to the system during full production can be very disruptive and costly in a number of ways, including delayed shipments and poor quality. At the extreme, with large amounts of preproduction planning activities, the first unit cost will be very low and will remain almost at the same level throughout its life cycle. Note, however, that properly forecasting the exact design and volume requirements from the customer can be extremely difficult. In most cases, firms will find that there is an appropriate mix of efforts both before and after the start of production activities that maximizes the return to the shareholders.

A number of the preproduction and subsequent progress factors discussed above are sometimes incorporated into broader management initiatives focused on stimulating creativity and continuous improvements. Indeed, much of the Total Quality Management (TQM) philosophy is centered on obtaining knowledge from teams of workers at the lower levels of the organization so that faster rates of process improvements become possible. In the end, a major objective of management is to foster an environment that allows improvement at both the individual and organization levels.

The implementation of learning curves in manufacturing firms can be extended beyond the estimation of production costs. A number of articles can be found that illustrate the use of learning curves in optimal lot sizing techniques (see, for example, Smunt and Morton (1985), Mazzola and McCardle (1996) and Pratsini *et al.* (1994)). In addition, learning curves can be incorporated into the capacity requirements planning (CRP) modules of material requirements planning (MRP) systems, or rough cut capacity planning (See, Smunt (1986a) and Smunt (1996) for discussions on the use of learning curve analysis in capacity planning). Hiller and Shapiro (1986) illustrate how learning curve analysis can be extended to the aggregate production planning problem. Finally, it should be noted that learning curve analysis is often used as a basis for pricing in long-term purchasing contracts. It is especially appropriate to require a supplier to base future deliveries on a negotiated aggregate learning curve when the buyer grants a single source relationship. Since the power of competition is reduced in this situation, the learning curve phenomenon provides the incentive for continuous improvement.

In the following sections, the methods for projecting manufacturing costs with learning curves are illustrated.

SIGNIFICANT ANALYTICAL MODELS: The first set of examples will use the *unit cost* model since it is the simplest form for basic calculations. Let us assume that our past experience indicates we can expect a 90% learning curve to apply to the manufacture of a new product. Let us also assume that we have determined the first unit cost will be 100 hours of manufacturing time. Without the use of any formula, we can easily estimate the future cost of units 2, 4, 8, etc. since these unit numbers correspond to a doubling of production. We estimate the manufacturing time in Table 1 and Table 2 in the following way:

Table 1. Unit cost model calculations for doubling unit experience.

Unit number	Calculation	Cost
1		100.0
2	100 × .9	90.0
4	90 × .9	81.0
8	81 × .9	72.9

In order to calculate manufacturing costs for units that are not associated with doubling of output, we first need to calculate **b**, the learning curve factor. From equation (1) we see that

$$\mathrm{b} = \frac{-\log(\text{learning rate }\%)}{\log(2.0)} \tag{7}$$

Thus,

$$\mathrm{b} = \frac{-\log(.9)}{\log(2.0)} = \frac{.04576}{.30103} = .152$$

In order to project costs to units 1 through 8, we use equation (4) in the following manner:

Table 2. Unit cost model calculations for incremental unit numbers.

Unit number	Calculation	Unit cost	Cumulative cost	Cumulative average
1		100.00	100.00	100.00
2	$100\,(2^{-.152})$	90.00	190.00	95.00
3	$100\,(3^{-.152})$	84.62	274.62	91.54
4	$100\,(4^{-.152})$	81.00	355.62	88.91
5	$100\,(5^{-.152})$	78.30	433.92	86.78
6	$100\,(6^{-.152})$	76.16	510.08	85.01
7	$100\,(7^{-.152})$	74.39	584.47	83.50
8	$100\,(8^{-.152})$	72.90	657.37	82.17

Note that each unit cost must be individually calculated in order to determine the cumulative (total) cost or cumulative average for any experience level using this simple model. The cost for any batch of units can be calculated by summing the appropriate unit costs, e.g., the cost of a batch of units from unit 3 to unit 5 would equal 243.92 (84.62 + 81.00 + 78.30). While this model can be used easily for low production volumes, it becomes cumbersome when estimating manufacturing costs for unit numbers in the thousands and tens of thousands.

The *cumulative average* model is often used to estimate batch costs and analyze empirical learning rates, especially when manufacturing costs are gathered by batch rather than by individual unit (typical in practice). To estimate the costs of units 1 through 8 in Table 3, we have:

Table 3. Cumulative average cost model calculations for incremental unit numbers.

Unit number	Calculation	Cumulative average	Cumulative cost	Unit cost
1		100.00	100.00	100.00
2	$100\,(2^{-.152})$	90.00	180.00	95.00
3	$100\,(3^{-.152})$	84.62	253.86	91.54
4	$100\,(4^{-.152})$	81.00	324.00	88.91
5	$100\,(5^{-.152})$	78.30	391.50	86.78
6	$100\,(6^{-.152})$	76.16	456.95	85.01
7	$100\,(7^{-.152})$	74.40	520.77	83.50
8	$100\,(8^{-.152})$	72.90	583.20	82.17

It is important to note that the unit cost estimates are lower in the cumulative average model than in the unit cost model. The same formula is used to calculate the cumulative average and unit costs and, thus, the unit costs for the cumulative average model will always be lower for any unit number greater than one (see Figure 1 for the graphical illustration). We should note that neither of the two model's assumptions is inherently incorrect. The important consideration, however, is that regression analyses of actual data must explicitly consider the source of the data. That is, if the data is collected as unit costs, then the resulting regression equation will be the estimate for the unit cost model. Conversely, if total costs to date are used and converted to cumulative average costs, then the model will be the estimate for cumulative average costs.

The main advantage of the cumulative average cost model is the ability to easily estimate future total batch costs. For example, the estimated batch cost of units 5 through 8 can be calculated as the total costs through unit 8 less the total costs through unit 4 or 259.2 (583.20 − 324.00 = 259.2). This calculation can be accomplished directly by using the formula:

$$Y_{batch}(y - y') = Y_{tc}(y') - Y_{tc}(y) \tag{8}$$

where

y = the first unit number of the batch
y' = the last unit number of the batch
$Y_{tc}(y') = (a)(x^{1+b})$

To illustrate, consider that the first unit cost and learning rate remains the same and that we wish to estimate the cost of producing a batch from unit 11 to unit 20. In this case, the batch cost is

$$\begin{aligned} Y_{batch}(11 - 20) &= Y_{tc}(20) - Y_{tc}(10) \\ &= (100)20^{1.152} - (100)10^{1.152} \\ &= 1268.45 - 704.69 \\ &= 563.76 \end{aligned}$$

In order to estimate the learning rate from a set of data points, we use the relationship that states that a slope of a line is equal to the "Rise" divided by the "Run," or more specifically, the change in the cost divided by the change in the quantity. Since the learning curve models use a logarithmic relationship, the learning curve factor can be estimated as

$$b = \frac{\log y}{\log x} \tag{9}$$

where

x and y = the first unit number of two different batches
x' and y' = the last unit number of two different batches

As an illustration, assume that we have the following data points in Table 4 in our manufacturing cost data base:

Table 4. Historical cost data.

Unit number	Cumulative average
10	100
20	80
50	60
200	40
1000	25

Using the above formula for **b**, we can calculate the learning rate factor between any two points. For example, application of this formula for unit numbers 10 and 20 results in **b** calculated as .3219. Applying the formula for the unit numbers 50 and 1000 results in a **b** calculated as .2922, for unit numbers 10 and 1000 results in a **b** calculated as .30103. The average learning rate factor from these three calculations is .30504. In order to convert **b** to a learning curve, we use the following formula:

$$\begin{aligned} &= 2^{-.30504} \\ &= .8094 \text{ or approximately an 81\% learning curve} \end{aligned}$$

Of course, the practical application of this technique would typically utilize a log linear least squares regression analysis using any statistical analysis software such as SAS, SPSS, JMP or Systat.

In the above examples, either the cumulative average or unit costs were calculated using the log-linear learning curve function. If one assumes that learning can take place continuously, then a slight modification to the log-linear function allows both the cumulative average cost and unit cost curves to be parallel, and allows each to be calculated with a simple formula (See Smunt (1997) for a detailed description of this method and examples of its use).

The ability to calculate accurate learning curves is constrained by the quantity and quality of available data. In many companies, data gathering is problematic due to either input errors or a lack of detailed data. Ideally, data should be gathered immediately at the source at the most detailed level possible. The use of bar code technology and other automated data input devices provides an excellent means of insuring a meaningful database creation for future analysis. When detailed data is not available for learning curve analysis, it is sometimes possible to find aggregate data by department or product line. Such data can be found in capacity utilization reports (for departments) and in accounting and financial reports (for product lines).

CASES: A vast number of empirical studies have shown that the learning curve exists in a wide variety of production environments. While many of these empirical studies were performed on aircraft and other defense products soon after World War II, there have been recent studies of the learning curve effect in commercial production. Some recent studies have investigated the interaction effects of engineering changes and other variables with production experience.

In a study of semiconductor manufacturing plants by Bohn (1995), it was found that process variability had a large impact on the rate of learning. Bohn obtained detailed die yield data from five plants and found that process variability was high in most. Due to high process "noise levels," these plants found it difficult to run successful controlled experiments that lead to improvements in the process design. Due to the noise, it was estimated that more than one quarter of the potential information of the experiment was lost. Interestingly, Bohn found that learning caused increased yields, and the level of process noise (process variability) seemed to decrease with experience.

Argote *et al.* (1990) found that knowledge in an organization that is acquired during production depreciates rapidly. This is an important finding since production experience itself can overestimate the amount of learning that can be expected in practice. Organizations do not benefit from learning at other organizations (shared learning) once production commences. Knowledge can depreciate for a number of reasons, including labor turnover, technological obsolescence, or lost/inadequate organizational records. One published example of the inadequate record problem (Lenehan, 1982) indicates that Steinway, the piano manufacturer, had great difficulty putting a discontinued model back into production due to poorly documented production methods.

Adler (1990) studied several departments of an electronic equipment manufacturer that had production facilities located globally. He found that shared learning occurred in three significant forms. The first was sharing between Development and Manufacturing, the second was sharing between the production plants from the initial location to plants that start up later, and the third was ongoing sharing across all plants after start-up. Adler concludes that the role of manufacturing engineering may be critical to the continual progress of the learning function of a particular manufacturing organization.

Epple *et al.* (1996) studied an automotive assembly plant that switched from a one-shift operation after two years of production to a two-shift operation. In this study, they not only identified the strong existence of learning in this high volume operation, but also that the knowledge gained in the first shift operation carried over virtually in total to the second shift instantaneously. However, they also found that the rate of learning slowed to half of the previous rate after the second shift was implemented. The authors suspect that this was due to the fact that indirect, support efforts did not increase substantially even though a second shift was added. Thus, managers should be aware that the learning rate can be significantly affected by the amount of continuing managerial and other support efforts.

While the above cases came from very recent studies, it should be noted that the early studies by Conway and Schultz (1959) provide case studies that illustrate a number of important variables that should be considered when managing vis-à-vis learning curves. In their seminal paper, issues are discussed concerning low versus high volume production, the impact of inflation on reported manufacturing costs, problems of aggregation, interruptions during production, and the effect of incentive systems.

While each firm presents a unique set of environmental variables that affect learning rate potential, empirical data provides a good starting point for learning rate expectations. Empirical results from Nadler and Smith (1963) and Smunt and Watts (1997) indicate that batch production learning rates vary, based mostly on the degree of labor intensity. Table 5 illustrates approximate learning rates that can be expected for a variety of processes.

CAVEATS: Most learning does not take place without conscious and active management involvement. While workers by themselves will

Table 5. Typical learning rates for batch production processes.

Process	Expected learning rate range
Milling	85–90%
Drill Press	85–95%
Lathe	80–85%
Deburring	90–95%
Bench Inspection	85–90%
Bench Assembly	80–90%
Line Assembly	85–95%
Welding	90–98%

"learn by doing" to some extent, appropriate incentives must be provided for them to fully implement their potential for individual productivity improvement. In addition, most productivity improvements that occur with increased experience come mainly from capital investments. For example, Hayes *et al.* (1988) found that 75% of productivity improvements were related to additions of capital stock. Thus, managers need to recognize that while the use of learning curves can be a powerful planning and control tool in most organizations, learning is derived from a variety of sources other than worker learning. A learning environment must be created in the organization so that it can be responsive to changes in competitors' moves and to customers' expectations.

Finally, it should be noted that productivity improvements could sometimes be estimated using simpler means than a log-linear regression analysis. In Smunt (1986b), it is shown how moving average calculations can sometimes provide as good or better predictions of productivity improvements, especially when the reported production data exhibits a high degree of noise. In the end, however, the inclusion of productivity trends in some way can greatly enhance both production cost and production planning accuracy. The use of the log-linear learning curve function has been shown to provide such improvements in accuracy in a number of firms and should be considered a valuable managerial tool.

Organizations Mentioned: Boeing; Boston Consulting Group; Curtis-Wright Corporation; Lockheed Corporation; Steinway; Texas Instruments.

See **Business process reengineering (BPR); Capital investment in advanced manufacturing technology; Concurrent engineering; Cost analysis for purchasing; Manufacturing flexibility; Manufacturing strategy; Performance measurement is manufacturing.**

References

[1] Adler, Paul. S. (1990). "Shared Learning," *Management Science*, Vol. 36, No. 8, 938–957.

[2] Argote, L., S. Beckman and D. Epple (1990). "The Persistence and Transfer of Learning in Industrial Settings." *Management Science* (February), Vol. 36, No. 2, 140–154.

[3] Bohn, Roger E. (1995). "Noise and Learning in Semiconductor Manufacturing," *Management Science*, Vol. 41, No. 1, 31–42.

[4] Conway, R.W. and Andrew Schultz, Jr. (1959). "The Manufacturing Progress Function," *Journal of Industrial Engineering*, Vol. 10, No. 1, 39–54.

[5] Epple, Dennis, Linda Argote and Kenneth Murphy (1996). "An Empirical Investigation of the Microstructure of Knowledge Acquisition and Transfer Through Learning by Doing," *Operations Research*, Vol. 44, No. 1, 77–86.

[6] Hayes, R., S. Wheelwright, and K. Clark (1988). *Dynamic Manufacturing: Creating the Learning Organization*, The Free Press, New York, 161–184.

[7] Hiller, Randall S. And Jeremy F. Shapiro (1986). "Optimal Capacity Expansion Planning When There Are Learning Effects," *Management Science*, Vol. 32, No. 9, 1153–1163.

[8] Mazzola, Joseph B. And Kevin F. McCardle (1996). "A Bayesian Approach to Managing Learning-Curve Uncertainty," *Management Science*, Vol. 42, No. 5, 680–692.

[9] Nadler, G. and D. Smith (1963), "Manufacturing Progress Functions for Types of Processes," *International Journal of Production Research*, Vol. 2, No. 2, 115–135.

[10] Pratsini, Eleni, Jeffrey D. Camm and Amitabh S. Raturi (1993). "Effect of Process Learning on Manufacturing Schedules," *Computers and Operation Research*, Vol. 20, No. 1, 15–24.

[11] Smunt, Timothy L. and Thomas E. Morton (1985). "The Effects of Learning on Optimal Lot Sizes: Further Developments on the Single Product Case," *IIE Transactions*, Vol. 17, No. 1, 3–37.

[12] Smunt, Timothy L. (1986a). "Incorporating Learning Curve Analysis into Medium-Term Capacity Planning Procedures: A Simulation Experiment," *Management Science*, Vol. 17, No. 4, 475–495.

[13] Smunt, Timothy L. (1986b). "A Comparison of Learning Curve Analysis and Moving Average Ratio Analysis for Detailed Operational Planning," *Decision Sciences*, Vol. 17, No. 4, 475–495.

[14] Smunt, Timothy L. (1996). "Rough Cut Capacity Planning in a Learning Environment," *IEEE Transactions on Engineering Management*, Vol. 43, No. 3, 334–341.

[15] Smunt, Timothy L. (1997). "Log-Linear and Non-Log-Linear Learning Curve Models for Production Research and Cost Estimation," *Babcock Graduate*

School of Management, Wake Forest University working paper, No. 97–009.

[16] Smunt, Timothy L. and Charles A. Watts (1997), "An Empirical Investigation of Learning at the Component Part Production Level, *Babcock Graduate School of Management, Wake Forest University working paper*, No. 97–007.

[17] Wright, T. P. (1936) "Factors Affecting the Cost of Airplanes," *Journal of Aeronautical Sciences*, Vol. 3, No. 4, 122–128.

LEARNING RATE

Learning rate is a measure of the ability to reduce manufacturing costs, manufacturing cycle times, defect rates, and equipment down times in repetitive production. In learning curve analysis, a learning rate of .9 (or 90%) means that, when output doubles, the cost per unit will drop to 90% of the current unit. For example, if it takes 10 hours to produce the 4th unit, the time taken to make the 8th unit (double of 4) will be 9 hours, if the learning rate is 90%.

See **Learning curve analysis; Mass customization.**

LEVEL STRATEGY

While planning production over a period of a year, it becomes necessary to deal with fluctuation in demand month to month. One approach to planning is to change production output to match sales (i.e., chase sales). In level strategy, output is kept constant, with buildup of inventory when demand is low and inventory depletions when demand is greater than the production output. Constant production output incurs inventory holding and backorder costs while a chase strategy incurs costs associated with changes in output.

See **Capacity planning: Long-range; Capacity planning: Medium- and short-range; Level strategy for capacity planning.**

LEVEL STRATEGY FOR CAPACITY PLANNING

Many companies experience variable demand for their products. This is probably most noticeable in products that have seasonal demand patterns, such as snow blowers or lawn mowers, but demand for other products may vary appreciably throughout the year. In such cases, companies face several different options meeting demand. Generally, these options fall into two "pure strategies" – a chase strategy and a level strategy. Under the chase strategy, production is varied as demand varies. With the level strategy, production remains at a constant level in spite of demand variations.

The use of a level strategy means that a company will produce at a constant rate regardless of the demand level. In companies that produce to stock, this means that finished goods inventory levels will grow during low demand periods and decrease during high demand periods. In make-to-order or assemble-to-order environments the backlog of orders will increase when demand is high and decrease when demand is low. In any environment, having either inventories or backlogs at a high level may be undesirable. Consequently, few companies follow a pure level strategy, they usually combine it with some aspects of a chase strategy.

See **Capacity planning: Long-range; Chase strategy for capacity planning.**

LEVELING, BALANCING AND SYNCHRONIZING THE LEAN MANUFACTURING SYSTEM

A lean production system requires the *smoothing of the manufacturing system*. Simply stated, the idea is to eliminate variation or fluctuation in production quantities in the feeder processes that supply the assembly process. This is done by eliminating fluctuations in final assembly. This is also called leveling final assembly. Here is a simple example to show the basic idea of leveling.

Cycle time (CT) is calculated as follows: The monthly demand is divided by the number of days in the month to compute the daily demand (DD). If DD is 240 cars per day, and 480 production minutes are available (60 minutes × 8 hours/day) for production each day, cycle time (CT) = 2 minutes (i.e., 480/240 = 2). That is every 2 minutes a car rolls off the line. Cycle time for final assembly is called takt time because it paces the entire system. Suppose that the production mix for a certain day is as given in Table 1.

Say that the subprocesses that feed the two-door fastback (with rear hatch) assembly are controlled by the cycle time of 4.8 minutes for this model. Every 4.8 minutes, the rear hatch line will produce a rear hatch for the fastback version per Table 1. Every 4.8 minutes, two doors for the fastback are made on a line that also makes doors for the two door coupe and front doors for the sedan and wagon. This is accomplished by separate lines for the left front and right front doors, each making

Table 1. Example of mixed model final assembly line that determines the cycle time by model.

Daily prodution Q	Car mix for line model	Cycle time by model (min)	Production minutes by model
50	Two-door coupe	9.6	100
100	Two-door fastback	4.8	200
25	Four-door sedan	18.2	50
65	Four-door wagon	7.7	130
240 (total)	cars/8 hr;	CT = 480 min/240 = 2 min per car	

doors at the rate of 240 per eight hours. Thus individual lines are synchronized with the final assembly. Initially, the doors are painted with the body of the cars, then the doors are removed and sent through synchronized subassembly lines, and later returned to the final assembly line for reinstallation on the same car.

Let us further say that, every car, regardless of model type, has the same engine. Each engine needs four pistons; therefore every 2.0 minutes, four pistons are produced. Piston production, is balanced to engine production which is balanced to car production. Parts and assemblies are produced in their minimum lot sizes and delivered to the next process, all under the control of kanban. Let us look at a specific cell in this example. Suppose that two disk brake hubs are used in every car. Therefore the CT for the cell is about one minute (actually 55 seconds) so the cell produces two hubs in two minutes. The extra five seconds per cycle provides for some spare capacity for making spare hubs for replacement parts. *Thus, balancing makes the output of parts from cells equal to the necessary demand for the parts downstream.*

When the car body exits from painting, an order for its seats is issued to the seat supplier. The seats are made in the same order as the cars on the assembly line. They are made and delivered in the same amount of time as it takes the car to get from painting to the point on the line where, seats are installed (about four hours). That is, seat manufacturing is synchronized, and the seats are sequenced in the same order as the cars on the line. Thus, the system is leveled, balanced and synchronized.

See **Just-in-time manufacturing; Kanban-based manufacturing systems; Lean manufacturing implementation; Line balancing; Manufacturing cell; Manufacturing systems design; Takt time.**

References

[1] Monden, Yasuhiro (1983). *Toyota Production System*. Industrial Engineering and Management Press, IIE.

[2] Schonberger, Richard J. (1982). *Japanese Manufacturing Techniques: Nine Hidden Lessons in Simplicity*. The Free Press.

[3] Black, J.T. (1991). *The Design of The Factory With A Future*. McGraw Hill.

[4] Hall, Robert W. (1983). *Zero Inventories*. Dow Jones-Irwin.

[5] Harmon, Roy L., Peterson, Leroy D. (1990). *Reinventing the Factory: Productivity Breakthroughs In Manufacturing Today*. The Free Press.

[6] Schonberger, Richard J. (1986). *World Class Manufacturing*. The Free Press.

[7] Shingo, Shigeo (1988). *Non-Stock Production: The Shingo System for Continuous Improvement*. Productivity Press.

[8] Suzaki, Kioyoski (1987). *The New Manufacturing Challenge*. The Free Press.

LEVELING PRODUCTION

While demand may vary from one period to the other, a level output may be maintained using a stable production rate while allowing inventory levels to vary. In this approach to production planning, finished goods inventories are used to deal with seasonal demand. In JIT production, a level schedule (usually constructed monthly) ideally means planning to produce a full mix of models each day (or week or some other short period). The production of a full-mix of products in a short interval reduces inventory buildup for each model. As demand conditions change, a full-mix schedule can respond more quickly to the changes.

The process of leveling production during production planning results in a level production schedule.

See **Just-in-time manufacturing; Level strategy for capacity planning.**

LIFE CYCLE ASSESSMENT (LCA)

According to the Society of Environmental Toxicology and Chemistry, Life Cycle Assessment is defined as (SETAC, 1991) "An objective process to evaluate the environmental burdens associated with a product, process or activity by identifying and quantifying energy and materials used and wastes released to the environment, to assess the impact of those energy and materials uses and releases on the environment, and to evaluate and implement opportunities to affect environmental improvements. The assessment includes the entire life cycle of the product, process or activity, encompassing extraction and processing of raw materials, manufacturing, transportation and distribution, use/reuse/maintenance, recycling and final disposal."

See **Environmental issues and operations management; Environmental issues: Reuse and recycling in manufacturing systems.**

Reference

[1] SETAC (Society of Environmental Toxicology and Chemistry) (1991). "A Technical Framework for Life-Cycle Assessment", January.

LIGHT WEIGHT STRUCTURE FOR DESIGN

See **Heavyweight project organization; Lightweight project organization.**

LIGHTWEIGHT PROJECT ORGANIZATION (LIGHTWEIGHT PROJECT MANAGER)

Organizational structures of product development projects can be categorized based on how much they emphasize functional links opposed to project links. It is common to distinguish between functional organization, lightweight project organizations, heavyweight project organization, and autonomous project organizations (Wheelwright and Clark, 1992). In distinguishing between the four structures, it is important to understand that they should be seen as points on a continuum rather than discrete categories. An index for measuring how "light" or "heavy" a project manager is can be found in Clark and Fujimoto (1991).

In a lightweight project organization, the functional links (e.g., to marketing, design, and manufacturing) dominate the project links. As a result, the lightweight project manager is largely a coordinator and an administrator. He or she has little authority over budgets and people, which primarily remain under the control of the corresponding functional managers. The lightweight project organization is a special form of a matrix organization with an emphasis on the functional links.

The major advantage of the lightweight project organization over a functional organization is its increased focus on coordination between functions. The lightweight project manager facilitates the information exchange between parts of the organization that typically have few – if any – explicit channels for information flow. He or she can achieve this by calling meetings, by providing information platforms (e.g., telephone lists, intranet links), by developing and updating schedules, and by bringing unresolved problems and conflicts to the attention of upper management. At the same time, the lightweight project organization allows the project members involved to continue their regular day-to-day work in the corresponding functional organization. This ensures the maintenance of functional expertise, and they must seek further career development within the functional organization.

Like any matrix organization, the lightweight project organization poses a conflict between its functional links and its project links. The project member working for a light weight project organization receives work from the lightweight project manager, but is ultimately evaluated (and potentially promoted or penalized) by his or her functional manager. This frequently causes substantial ambiguity and stress to the project member who has to choose priorities and has to allocate time and effort between functional duties and project duties. It also puts the lightweight project manager in a position where he or she is largely dependent on the cooperation and good will of the functional managers. Other advantages and disadvantages are discussed in Wheelwright and Clark (1992).

The lightweight project organization has been the traditional form for automobile development in the U.S. and Europe during the 1970s and 1980s. Recently, however, prompted by the disadvantages mentioned above, the industry has moved towards heavyweight project organizations.

See **Concurrent engineering; Product development and concurrent engineering.**

References

[1] Clark, K.B. and T. Fujimoto (1991). *Product Development Performance: Strategy, Organization and Management in the World Auto Industry*. Harvard Business School Press, Cambridge.

[2] Wheelwright, S.C., and K.B. Clark (1992). *Revolutionizing Product Development*. The Free Press, New York.

LINE BALANCING

Line balancing enables efficient high volume production or assembly on a line. It involves assigning all production/assembly tasks required for making one unit to a series of workstations so that each workstation has no more work assigned to it than can be done within a fixed time known as the **cycle time**. The problem is complicated by strict **precedence relationships** among tasks imposed by product design and process technologies. Generally, the objective of the line balancing problem is to assign all tasks to the fewest number of workstations.

See **Assembly line design; Cycle time, Manufacturing systems; U-shaped assembly lines.**

LINE STOP AUTHORITY

In just-in-time manufacturing systems, authority may be granted to production line employees to stop the production line for quality related problems. Once the line is stopped, the quality problem is fixed before restarting the line again. This is a principle that originated in Japan and is associated with the just-in-time manufacturing system and its variations.

See **Just-in-time manufacturing; Kanban-based production systems; Lean manufacturing implementation.**

LINEAR DECISION RULE (LDR)

The Linear Decision Rule is a mathematical approach to production planning developed by Holt *et al.* (1960). It was first proposed at a time when there was a strong belief among academics that problems of production planning for a period such as a year could be solved using mathematical models. LDR seeks to meet demand while minimizing the costs of hiring, firing, holding inventory, backorders, payroll, overtime, and undertime. The primary application of this method was in a paint factory, but that application failed because employees would not implement the plan developed by the model (Gordon, 1966). This approach to production planning is also know as the HMMS production planning model (Hold, Modigliani, Muth and Simon, 1960.)

There are several reasons why such mathematical models are not more widely used. The major one is that most real decisions are not represented accurately by mathematical formulas. Thus, such methods are limited to a small number of applications that can be modeled realistically, such as oil refineries or feed mixing. A second reason for the limited use of such models is that managers are reluctant to utilize methods they do not understand. As more complex mathematical models are needed to accurately represent real problems, it becomes less likely that managers either understand them or use them for decision making purposes.

See **Aggregate planning and master production schedule linkage; Capacity planning: Long-range.**

References

[1] Gordon, J.R.M. (1966). "A Multi-Model Analysis of an Aggregate Scheduling Decision." Ph.D. Dissertation, Sloan School of Management, MIT.

[2] Holt, C.C., F. Modigliani, J.F. Muth, and H.A. Simon (1960). *Planning Production, Inventories, and Workforce*. Prentice Hall, New York.

LINEAR PROGRAMMING

Linear programming is a class of mathematical models, which are used to find solutions to problems by minimizing or maximizing a linear function subject to constraints. Linear programming models have been used to provide solutions to production problems such as aggregate production planning; lot sizing, etc., where cost minimization or revenue maximization is the objective.

See **Capacity planning: Long-range; Capacity planning: Medium- and short-range.**

LINKED-CELL MANUFACTURING SYSTEM (L-CMS)

JT Black

Auburn University, Auburn, AL, USA

A manufacturing system composed of linked cells is the newest innovation in manufacturing systems. In manufacturing cells, processes are grouped according to the sequence of operations

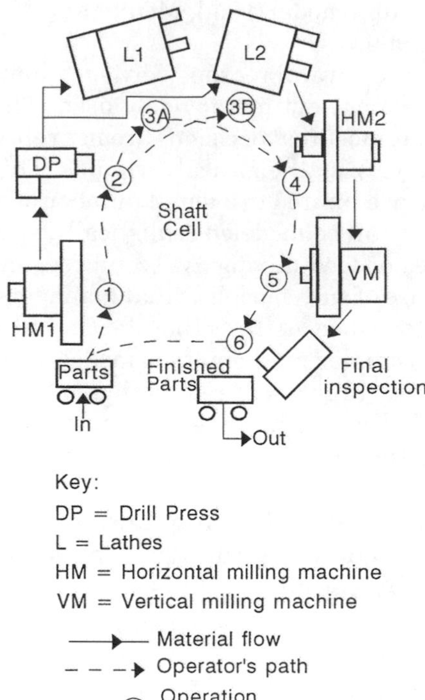

Work Sequence	Name of Operation	Time Manual	Time Walking	Time Machine
①	Mill ends HM1	12"	5"	30"
②	Drill hole DP	15"	5"	20"
3A or 3B	Turn, bore L1 or L2	13"	5" 8"	180"
④	Mill flats HM2	12"	8" 5"	20"
⑤	Mill steps VM	13"	7"	30"
⑥	Final inspect	10"	5"	—
		75"	35"	280"

Cycle Time = 75sec + 35 sec = 110 sec
Longest Machining Time = 180 sec
Total Machining Time = 280 sec per part

All machines in the cell are capable of processing unattended while the operator is are doing manual operations (unload, load, inspect, and deburr)and walking from machine to machine. The time to change tools and workholders (perform setup) is not shown.

Fig. 1. Example of an interim manned cell part of an CMS. This cell operates with six machines and one multi-functional worker. The lathe operation, #3, is duplicated.

needed to make a product, much like a flow shop but the L-CMS is designed for flexibility. Manufacturing and assembly cells are designed in U-shape so that the workers can move from machine to machine, loading and unloading parts; Figure 1 shows an example of a manned cell. The cell has one worker who can make a walking loop around the cell in 110 seconds; see the breakdown of time in the table to the right. The machines in the cell are usually single cycle automatics, so that once the operator has started the machining cycle, the machine can complete the desired processing untended and turn itself off, when completed with a machining cycle.

The operator comes to a machine, unloads a part, checks the part, loads a new part into the machine, and starts the machining cycle. The cell usually includes all the processing needed for a complete part or subassembly. The table shows the typical manual times for the operator at each machine plus the walking time between machines. The sum of the manual and walking time is the cell cycle time (CT). The cell is designed such that the machining time (MT) for any part on any machine in the cell is less than the cycle time, CT. That is, $MT_{ij} < CT$ so that all MT's should be less than 110 seconds.

However, note that the machining time for the third operation is greater than the necessary cycle time. That is 180 seconds >110 seconds. One solution to this problem is to duplicate this lathe and have the operator alternate between the two lathes, visiting each lathe every other trip around the cell. This makes the average MT = 180/2 = 90 sec.

Cells are typically manned, but unmanned cells are beginning to be designed with a robot replacing a worker. Such a robotic cell design is shown in Figure 2 with one robot and three CNC machines. For the cell to operate autonomously, all machines must have adaptive control capability. This kind of automation is recommended only after manned cell has been developed and optimized within the L-CMS.

In some cells, *decouplers* are placed between the processes, operations, or machines to provide flexibility, part transportation, inspection for defect prevention (*pokayoke*) and quality control, and process delay for the manufacturing cell. For example, referring to Figure 3, notice the *pokayoke* arrangement for M3 mill and decoupler 3. The decoupler allows inspection of the part for a critical dimension and feeds-back adjustments to the machine to prevent the machine from making oversize parts as the mill's cutter wears. A process delay decoupler would delay the part movement to the next operation to allow the part to cool down, heat up, cure, or whatever is necessary for a period of time greater than the cycle time for the cell. Decouplers and flexible fixtures

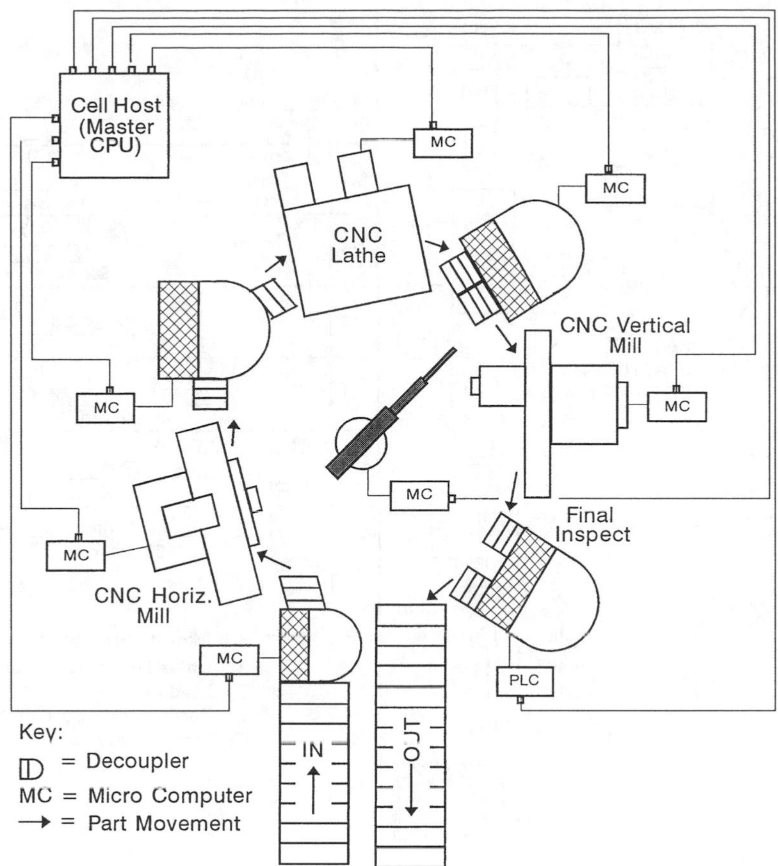

Fig. 2. Unmanned robotic cells will have CNC machine tools, robot for material handling, and decouplers for flexibility and capability.

are vital parts of both manned and unmanned cells.

Cells have many features that make them unique and different from other manufacturing systems. Parts move from machine to machine *one at a time* within the cell. For material processing, the *machines are typically* capable of completing a machining cycle initiated by a worker. The "U" shape puts the start and finish points of the cell next to each other. Every time the operator completes a walking trip around the cell, a part is completed.

The cell is designed so that this cycle time (CT) is equal to or less than the "necessary cycle time" established by the final assembly line, and it is called the takt time. The cycle time is determined by downstream requirements since each cell, like all the others in the plant, is geared to produce parts as needed, when needed, by the downstream process.

The machining time (MT) for each machine needs only to be less than the time it takes for the operator to complete the walking trip around the cell. As shown in Figure 3, *machining processes are overlapping and need not be equal (i.e., balanced)* as long as no MT is greater than the CT. In Figure 3b, the cycle time for the operator is 150 seconds while the total machine time for each part is 100 + 80 + 60 + 80 + 80 = 400 seconds. Certainly a large CNC machining center (MC) capable of performing the five machining operations could be used to replace the cell. However, the cycle time for a part would jump from 150 seconds to over 400 seconds because combining the processes into one machine prevents *overlapping of the machining times*. The MC processes in series, while the cell processes in parallel. The CT for the cell can readily be altered by adding (a portion of) an additional worker to the cell. The objective in designing the cell is to minimize the resources within the system that do not depreciate – direct material and labor.

Note also that the machining speeds and feeds can be relaxed to extend the tool life of the cutting tools and reduce the wear and tear on the machines, so long as the MT does not equal or exceed the CT.

Figure 3 is an example of a standard operations routine sheet used to plan the manufacture of one of the parts in the family produced in the cell. The plan shows the relationships between the manual operations performed by the worker,

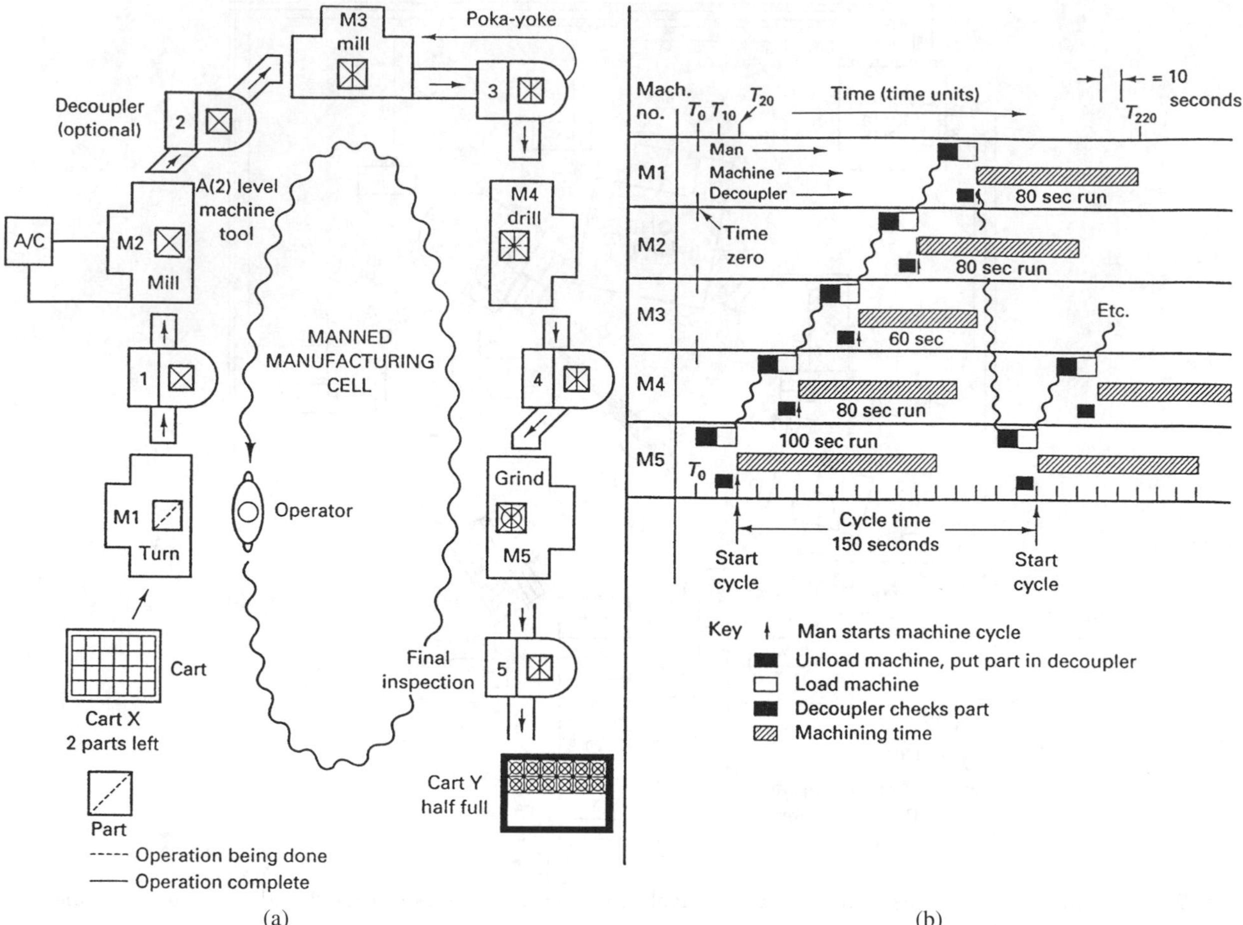

Fig. 3. (a) Layout and (b) time bar diagram of a five-machine, U-shaped cell with decouplers, operator upstreaming, and variable machine times (MT).

the machining operations performed by the machine, and the time spent by the worker walking from machine to machine. The manual operations include loading and unloading of machines, checking for quality, deburring, and cleaning chips out of fixtures.

In the cell, there is no need to balance the MTs. It is necessary only that no MT be greater than the required CT. If the MT for a process is greater than the cycle time, the process is duplicated and the worker alternates machines on his trips through the cell. Duplicating machines is not the ideal solution as it can increase the variability in the product.

Each worker in a cell can operate more than one kind of process (or multiple versions of the same process) and can also perform inspection and machine maintenance duties. Cells eliminate the job shop concept of one person/one machine and thereby greatly increase worker productivity and utilization. The restriction of the cell to one part family makes setup reduction in the cell easier. Fixtures are designed to hold a family of parts and they permit rapid changeover from one part to another. The fixtures are designed for easy load/unload and so that parts cannot be loaded incorrectly. Also fixturing can be designed to reject defective parts from being loaded.

FORMING CELLS AND LINKED CELLS: Cells are designed to manufacture specific groups (families) of parts. To form a linked-cell manufacturing system, the first step is to restructure portions of the job shop, converting it in stages into manned cells (see Figures 1 and 3). At the same time, the flow line portions of the factory need to be reconfigured into U-shaped cells as well (see Figure 4). The flow shop manufacturing within the plant can be redesigned to make these systems operate on the basis of one piece flow (Sekine, 1990). To do this, the long setup times typical in flow lines must be vigorously attacked and reduced so that the flow lines can be changed quickly from making one product to another. The need to line balance the flow line must be eliminated. This is accomplished by conversion to manufacturing

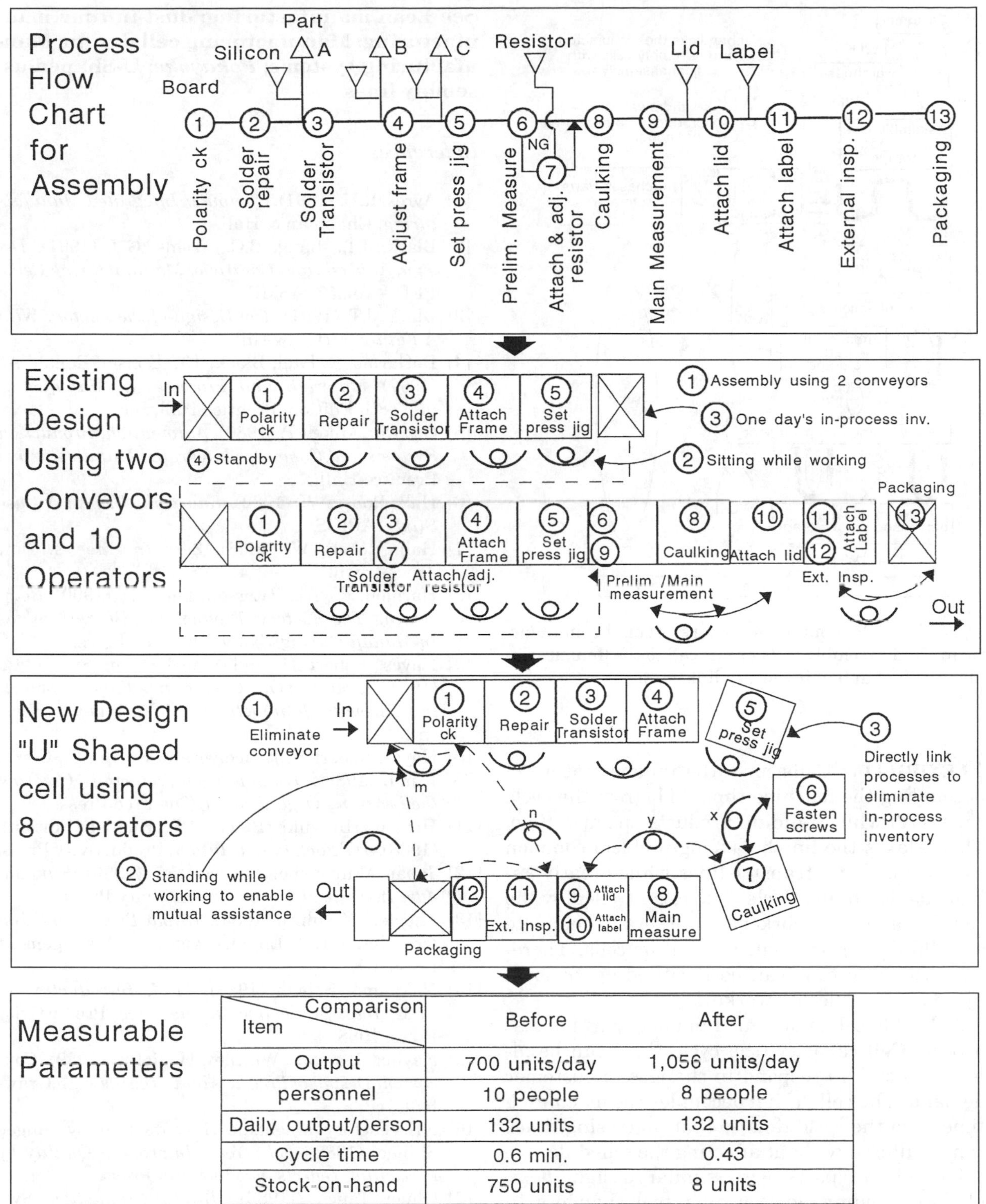

Item \ Comparison	Before	After
Output	700 units/day	1,056 units/day
personnel	10 people	8 people
Daily output/person	132 units	132 units
Cycle time	0.6 min.	0.43
Stock-on-hand	750 units	8 units

Fig. 4. Assembly cell redesigned into U-shaped cell with standing walking workers. (Sekine)

cell and using standing, walking workers who are capable of doing multiple operations. Assembly cells should be flexible and compatible with the cells designed to make piece parts and with the subassembly lines and final assembly lines. Cells are linked either *directly* to each other, or to subassembly points *indirectly* by a pull system using kanbans for material control.

In Figure 5, a linked-cell plant is shown with manufacturing cells to produce components, subassembly, and final assembly, all connected with kanban links.

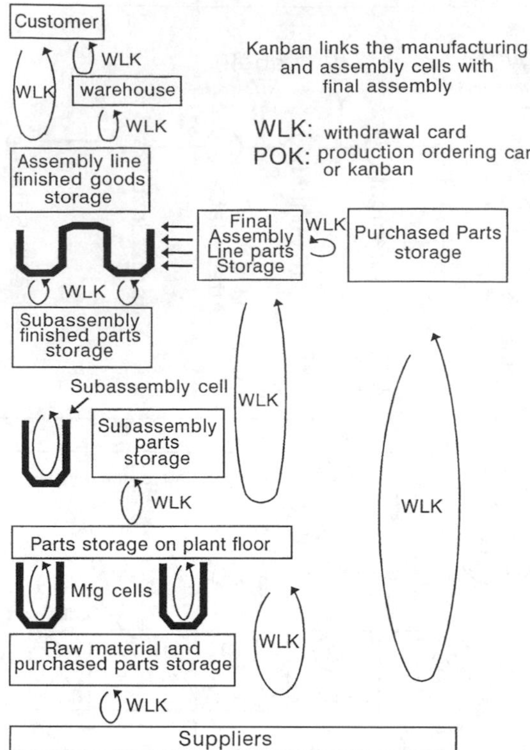

Fig. 5. In the lean manufacturing system, the manufacturing and assembly cells are linked to the final assembly area by kanban inventory links or loops.

FLEXIBILITY: Cells have the unique feature of decoupling the machine time (MT) from the cycle time (CT) while building products one at a time. This relaxes the line-balancing problem common to flow lines and transfer lines while greatly enhancing flexibility. Cycle time is controlled by the time it takes the worker or workers to complete a walking loop through the cell or cells. Therefore, the cycle time can be throttled up or down by adding or deleting workers.

A flexible cell is designed to make parts one at a time. Cell capacity (the cycle time) can be altered quickly to respond to changes in customer demand. The cell design decouples the machining time from the cycle time so that processing times can be altered without affecting the schedule.

Families of parts with similar design, flexible workholding devices, and tool changers in programmable machines allow rapid changeover from one component to another. Rapid changeover means that quick or one-touch setup is employed, often as quickly as flipping a light switch. Significant inventory reductions between the cells are possible, and the inventory level can be directly controlled. Quality is controlled by operators and the equipment within the cell is also maintained routinely by cell operators.

See **Lean manufacturing; Just-in-time manufacturing; Manufacturing cell design; Manufacturing systems; *Pokayoke*; U-Shaped assembly lines.**

References

[1] Ayres, R.U. (1991). *Computer Integrated Manufacturing*. Chapman & Hall.

[2] Black, J.T., Jiang, B.C., Wiens, G.J. (1991). *Design, Analysis and Control of Manufacturing Cells*. PED – Vol. 53, ASME.

[3] Black, J.T. (1991). *The Design of the Factory With A Future*. McGraw Hill.

[4] DeGarmo, E. Paul, Black, J.T., Kohser, Ronald A., (1997). *Materials and Processes in Manufacturing* – 8th Edition. Prentice-Hall, Inc.

[5] Groover, Mikell P. (1980). *Automation, Production Systems and Computer-Integrated Manufacturing*. Prentice-Hall.

[6] Hall, Robert W. (1982). *Kawasaki U.S.A.: A Case Study*. APICS.

[7] Hall, Robert W. (1983). *Zero Inventories*. Dow Jones-Irwin.

[8] Harmon, Roy L., Peterson, Leroy D. (1990). *Reinventing The Factory: Productivity Breakthroughs In Manufacturing Today*. The Free Press.

[9] Hayes, Robert H. and Wheelwright, Steven C., (1984). *Restoring Our Competitive Edge – Competing Through Manufacturing*. John Wiley & Sons, Inc.

[10] Hayes, Robert H., Wheelwright, Steven C., Clark, Kim B. (1988). *Dynamic Manufacturing – Creating the Learning Organization*. The Free Press.

[11] Hirano, Hiroyuki (Black, J.T., English Edition). (1988). *JIT Factory Revolution*. Productivity Press.

[12] Japan Management Association (1985). *Kanban: Just-In-Time at Toyota*. Productivity Press.

[13] Monden, Yasuhiro (1983). *Toyota Production System*. Industrial Engineering and Management Press, IIE.

[14] Nakajima, Seiichi (1988). *TPM, Introduction to TPM: Total productive Maintenance*. Productivity Press, 1988.

[15] Nevins, James L. Whitney, Daniel E. (1989). *Concurrent Design of Products and Processes*. McGraw-Hill, Inc.

[16] NKS/Factory Magazine (Introduction by Shigeo Shingo) (1988). *Poka-Yoke: Improving Quality by Preventing Defects*. Productivity Press.

[17] Ohno, Taiichi (1988). *Toyota Production System: Beyond Large-Scale Production*. Productivity Press.

[18] Schonberger, Richard J. (1982). *Japanese Manufacturing Techniques: Nine Hidden Lessons in Simplicity*. The Free Press.

[19] Schonberger, Richard J. (1986). *World Class Manufacturing*. The Free Press.

[20] Schonberger, Richard J. (1987). *World Class Manufacturing Casebook: Implementing JIT and TOC*. The Free Press.

[21] Schonberger, Richard J. (1996). *World Class Manufacturing: The Next Decade*. The Free Press.
[22] Sekine, Kenichi (1990). *One-Piece Flow: Cell Design for Transforming the Production Process.* Productivity Press.
[23] Sekine, Kenichi and Arai, Keisuke (1992). *Kaizen for Quick Changeover*. Productivity Press.
[24] Shingo, Shigeo (1985). *A Revolution in Manufacturing: The SMED System*. Productivity Press.
[25] Shingo, Shigeo (1986). *Zero Quality Control: Source Inspection and the Poka-Yoke system*. Productivity Press.
[26] Shingo, Shigeo (1988). *Non-Stock Production: The Shingo System for Continuous Improvement*. Productivity Press.
[27] Shingo, Shigeo (1989). *A Study of the Toyota Production System*. Productivity Press.
[28] Suzaki, Kioyoshi (1987). *The New Manufacturing Challenge*. The Free Press.
[29] Womack, James P., Jones, Daniel T., Roos, Daniel (1991). *The Machine That Changed The World*. Harper Perennial.

LOAD LEVELING

Load leveling is one of the steps of the order release process. The first step is order preparation. The second step is order review and evaluation. Load leveling has received much attention from researchers who feel that it can significantly improve shop performance. The process that performs the load leveling step is called an order release mechanism.

See **Order release.**

LOADING

In a production system, the term load refers to the work planned for a work center, the work released (sent) to a work center or assembly area in a given period of time such as a week. In a job shop, loading of work centers with various orders needs careful consideration of several items such as the due date for the order, size of the order in terms of labor/machine hours, batch size, capacity of the work center, existing backlog at the work center, as so on.

Loading has several shades of meaning. One use of the term loading is in the context of machine scheduling. Infinite loading occurs when jobs or orders are released for production at a work center or facility without considering the existing load or backlog on the work center. Under infinite loading, the most recently released job will join the existing queue of jobs waiting at the work center. Overloading a facility can occur with infinite loading.

Finite loading takes into account the finite capacity at each work center while releasing work to the center. A process known as input/output control is practiced to manage the workflow and queue lengths in manufacturing facilities.

See **Capacity planning: Medium- and short-range; Order release.**

LOCATION FILE

An MRP system includes a Material Requirements Planning (MRP) database. This database includes several files.

The location file of an MRP database contains data about the precise locations of stored parts. The file record also contains a part number and location information such as department, row, bay, tier, quantities, and so on.

See **MRP.**

LOGISTICS

Logistics refers to the process of planning, implementing, and controlling the efficient, effective flow and storage of goods, services, and related information from point of origin to point of consumption while meeting customer requirements.

See **Logistics: Meeting customer's real needs; Purchasing: Acquiring the best inputs; Purchasing: The future.**

LOGISTICS INFRASTRUCTURE

Logistics infrastructure exist at the national and company levels. The national logistics infrastructure consists of a nation's air, motor, rail, and shipping systems. This infrastructure consists of miles of improved highways, miles of railroad tracks, miles of navigable waterways, operating ports complete with adequate loading and unloading equipment, miles of gas and oil pipeline, and commercial airports in operation. The real importance of a well-established national logistics infrastructure is that it allows both people and materials to travel from point to point at relatively low cost. It enables farmers to readily get their crops to market; lumber, minerals, and other raw materials to be easily transported for processing or refining; components and parts to be shipped to manufacturers, and manufactured goods to be shipped at reasonable cost to the market. In today's global marketplace, it is important to have adequate ports of entry to bring in imports and ship out exports. As a general rule, countries with sound logistics infrastructures tend to be more highly industrialized and their citizens experience a higher standard of living than countries without an adequate logistics infrastructure.

Traditionally, a firm's logistics infrastructure consisted of a set of equipment and facilities used to move and store materials. The fleet of trucks, rail cars, and shipping containers represent the majority of firm's transportation equipment.

Wal-Mart uses a private fleet to achieve replenishment times that are the envy of the industry. However, other firms have decided that they are in the manufacturing or retail business, and not the shipping business, and therefore have outsourced their shipping requirements. In some instances, alliance relationships with third-party service providers could actually outperform a dedicated, private fleet. From a storage perspective, the firm's abilities depend on its network of warehouses and distribution centers complete with the materials handling equipment used to move materials in, through, and out of the storage facilities. In recent years, time has become a more important competitive dimension and has led firms to invest in automated storage and retrieval systems as well as cross docking facilities. These logistical solutions take advantage of better information systems to increase the velocity with which materials pass through the storage facility.

In reality, a firm's information systems and logistical policies are now an important part of the logistics infrastructure because of their impact on the way the rest of the logistics infrastructure performs. Improvements in both national and firm-level logistics infrastructures has led to higher standards of living for the citizens of many countries around the world.

Organizations Mentioned: Wal-Mart.

See **Logistics: Meeting customers real needs.**

References

[1] Sampson, R.J., Farris, M.T., and Schrock, D.L. (1990). *Domestic Transportation: Practice, Theory, and Policy* (6th ed.). Houghton Mifflin Company, Boston.

[2] Lambert, D.M. and Stock, J.R. (1993). *Strategic Logistics Management* (3rd ed.). Richard D. Irwin, Inc., Homewodd, IL.

LOGISTICS SUPPORT

The basic logistics activities such as transportation, inventory management, materials handling, packaging, warehousing, and order processing have always been important to the success of manufacturing and retail companies. These activities are essential to deliver high levels of customer satisfaction. Logistics support functions perform those activities which move and store incoming materials, work in process, and finished goods in a way that supports the firm's strategy and provides superior customer service. Basic logistics activities shorten cycle times, increase customer responsiveness, reduce costs, and provide differentiated services.

See **Logistics: Meeting customer's real needs.**

Reference

[1] Lambert, D.M. and Stock, J.R. (1993). *Strategic Logistics Management* (3rd ed.). Richard D. Irwin, Inc., Homewodd, IL.

LOGISTICS: MEETING CUSTOMERS' REAL NEEDS

Stanley E. Fawcett

Brigham Young University, Provo, Utah, USA

DESCRIPTION: Logistics has been described as the art and science of moving things from here to there and storing them along the way. Specifically, logistics management can be defined as the management of the movement and storage activities performed by the firm. Over time, and in different organizations various names have been used to describe the management of logistics activities. Some of the terms used include physical distribution, rochrematics, materials management, distribution engineering, business logistics, logistical management, and supply-chain management (Ballou, 1992). Today, logistics management is the term most commonly used to identify and describe a firm's movement and storage activities. The Council of Logistics Management (CLM), the leading professional association of logistics managers and educators, has defined logistics as follows:

> Logistics is the process of planning, implementing, and controlling the efficient, cost-effective

> flow and storage of raw materials, in-process inventory, finished goods and related information from point of origin to point of consumption for the purpose of conforming to customer requirements.

Every firm, manufacturing or service, must effectively manage movement and storage activities to support production and to provide adequate levels of customer service. Most casual observers quickly grasp the role of logistics in supporting manufacturing operations – raw materials and component parts must be transported to the manufacturing facility and finished goods must be shipped to the location where they will ultimately be purchased by consumers. However, the fact that logistics management is a cornerstone of the competitive strategy of such well-known service companies as Wal-Mart, The Gap, and Benetton is often overlooked. Logistics support is vital to most organizations in our modern society.

HISTORICAL PERSPECTIVE: As a formal area of management study, logistics is relatively new. The activities performed in logistics have, however, been around for a very long time. That is, logistics activities go back to the beginning of economic trade. Formalized logistics activities date back at least to the City State of Venice, which used logistics effectively to increase wealth creation before the thirteenth century. Going back to the 1800s, the most recognized logistics' activity was traffic management or the purchasing of transportation services. The traffic manager's job in the U.S. before 1887, was the negotiation of prices with the railroads. With the passage of The Act to Regulate Commerce in 1887, the role of traffic managers in the United States changed since the act prohibited price discrimination between places, shippers, and commodities. Transportation rates had to be published and filed with the U.S. Interstate Commerce Commission, and shippers were required to pay the published rate. Negotiations were no longer required. Further, the railroad was the only mode of inland transportation for much of the country. In effect, traffic managers became rate clerks, responsible for finding the lowest published rate for a given shipment. This "clerk" mentality permeated the logistics management until World War II. During the war, scarcity and the need for timely materials movement led to the recognition that by analyzing logistics on a total activity basis, costs could be reduced and service improved.

Since World War II, business logistics has undergone a slow but continual evolution. A major step in the development of logistics occurred as a result of a 1956 study on the economics of air freight. The study pointed out that because air freight is the most expensive form of transportation, it is relegated to emergency shipments. However, the use of air freight reduces other logistics costs such as inventory and warehousing costs. The study concluded that to understand the true impact of air freight on a firm's costs, all of the logistics costs should be totaled. The twin notions of tradeoff and total cost analysis were firmly introduced as foundation elements of logistics management. Subsequently, many changes in the operating and competitive environments have brought attention to logistics, helping to elevate logistics to a strategic level. In fact, one of the first textbooks dedicated to logistics appeared in print in 1961. Shortly thereafter, the National Council of Physical Distribution Management was formed in 1963 to develop the theory of logistics and promote the "art and science" of managing logistics systems.

The next major change to influence the development of logistics was the deregulation of transportation services which began in the 1970s. Deregulation created a very dynamic transportation environment with greater competition among carriers, greater pricing freedom, and the introduction of a variety of new transportation services. New skills including carrier selection and management, performance measurement, and negotiation were needed to manage within this rapidly changing environment. Overall, deregulation increased the opportunity for logistics to add value and support companies' efforts to increase customer service levels. The 1980s also introduced new competitive strategies such as just-in-time manufacturing and the globalization of manufacturing networks that placed a premium on better logistics management. In effect, logistics became a critical coordinating activity able to provide a firm with a unique and hard-to-replicate capability.

STRATEGIC PERSPECTIVE: Logistics' ability to reduce costs and enhance customer service has helped it emerge as an important strategic weapon. The value-added objectives of logistics have been summarized in terms of the "seven rights" of logistics – to deliver the **right** product in the **right** quantity and the **right** condition to the **right** place at the **right** time for the **right** customer at the **right** price (Gecowets, 1979). When logistics works effectively by fulfilling these seven rights, time and place utility are created, customers are satisfied, and the firm is more competitive and society is benefited.

The modern business world would be brought to a stand still without adequate logistics. Almost every aspect of the lives of citizens in industrialized countries is influenced by the performance of modern logistics systems. Imagine going to the store to buy groceries and finding the shelves empty. Consider the challenges that today's manufacturers would have if they could not get materials delivered on time. What would most consumer products cost if the most competitive products could not be sold in different markets around the world? What would happen if a major transportation company were not able to operate for a week? How much business would be lost? The important point is that we seldom have to worry about these situations because logistics systems generally do a fine job of supporting the needs of a modern economy. Unfortunately, the efficiency and effectiveness of logistics creates a significant problem – logistics issues are visible only when something goes wrong and expectations are not met. Logistics' strategic value can been seen by evaluating three specific issues:

- Logistics plays a key role in economic development, global growth and global competitive strategies.
- Logistics costs represent a large percentage of not only a company's operating costs but also a nation's gross domestic product.
- Effective logistics management enhances the firm's ability to serve customers and build a unique competitive advantage.

Let us consider the three items in greater detail:

1. ***Logistics's Role in Economic Development:*** Logistics' foremost influence on economic development comes from the fact that logistics facilitates economic trade. A product's competitiveness in a given market depends on its total landed cost – the sum of the production, transportation, and other costs. The better the logistics system, the more markets a product can be sold in. Martin Farris (Sampson *et al.*, 1990) has stated that the extent of the market is controlled by the cost of logistics. Logistics then is not only a *necessary* function in any organized economy but it is a *limiting* factor in the overall economic development of a country or region.

   The development of extensive logistics infrastructures and advanced logistics management techniques has enabled large portions of the world to experience the benefits of commercial trade, which are evidenced in the high standards of living found in industrialized countries and the recent reductions in poverty found in third-world countries (Lachica, 1997). Further, well-managed and efficient logistics systems allow firms to use the principle of comparative advantage to acquire the best resources available worldwide. Low-cost Mexican or Malaysian labor can be combined with U.S. or Japanese technology to build a product that is then sold in the European Union. Thus, many firms have developed globally rationalized manufacturing and marketing networks. For example, General Motors operates production facilities in over 28 countries and markets its vehicles in over 140 countries. Likewise, Unilever operates in 90 countries and sells in 160 countries. Without modern logistics, these extensive global networks would be untenable. Moreover, the direct foreign investment of large multinational companies in these global production and marketing operations has provided a catalyst for economic growth in many developing nations. Developing nations compete vigorously to attract this foreign investment because it brings the jobs and technology that are vital to economic development. For example, in Mexico, over 500,000 workers are employed in production sharing facilities operated by foreign (mostly U.S.) firms.
2. ***Logistics' Impact on Costs:*** Logistics activities can be very costly, especially when they are not managed carefully. The data in Table 1 show that in 1994, the total logistics costs for the U.S. were equal to 10.8 percent of Gross

Table 1. A breakdown of logistics costs in the U.S. in 1994.

Source of cost	$ Billions
Inventory Carrying Costs	
Interest	53
Taxes, Obsolescence, Depreciation, Insurance	161
Warehousing	63
Subtotal	**277**
Transportation Costs	
Motor Carriers:	
Public and for hire	96
Private and for own account	109
Local freight services	128
Subtotal	**333**
Other Carriers:	
Railroads	33
Water Carriers	22
Oil Pipelines	10
Air Carriers	17
Forwarders and others	5
Subtotal	**87**
Shipper Related Costs	5
Distribution Administration	28
Total Logistics Costs (10.8% of GDP)	**730**

Domestic Product (GDP). U.S. industry spent an estimated $420 billion dollars on freight transportation and another $277 billion on business inventory in 1994. These two activities – transportation and inventory – are far and away the most costly logistics activities in the U.S. economy. The data in the table, however, does not tell the complete story since the costs shown are only the direct costs. They do not include investment in transportation equipment, highways, port facilities, airways, airports, and miscellaneous costs. For example, when all of the investment in freight and passenger transportation equipment (excluding highways and other physical infrastructure) is totaled, transportation costs by themselves account for 18 per cent of the GDP. Because logistics costs represent a sizable component of the U.S. GDP, they influence not only product availability but also interest rates, inflation, and productivity. Improvements in the management of logistics therefore promise to have a positive impact on economic growth and consumer living standards.

From a company perspective, logistics costs are a substantial percent of both the cost of goods sold and the final sales dollar. Depending on the industry, total logistics costs can represent up to 40 or 50 percent of the value added. For most firms, outbound logistics costs account for somewhere between 7 and 10 percent of each sales dollar. Further, the dollars spent on logistics are roughly the same as those spent on direct labor. In fact, for firms in high-tech industries, logistics costs often exceed direct labor costs. This is particularly true in global manufacturing. Moreover, whereas production costs generally decrease in rationalized global manufacturing, logistics costs typically increase, creating a cost tradeoff that is often overlooked in the design of global networks.

While logistics costs directly impact the firm's bottom line performance, they also greatly influence the firm's ability to take advantage of scale economies, which reduce per unit production costs. That is, large-scale production requires sound transportation to bring raw materials and component parts together where production occurs and then to distribute the outputs to customers located in many different market areas. Without efficient logistics systems, mass production would be impossible. Logistics impacts product costs in at least two additional ways. First, logistics allows companies to effectively operate globally dispersed and rationalized multi-plant networks by coordinating materials flows and allowing all of the different operations to be combined into a single production system. Second, well-developed logistics support is needed to implement advanced purchasing strategies such as global and just-in-time purchasing. In fact, well-managed logistics activities make the integration of global and JIT purchasing strategies possible (see Table 2). These purchasing strategies reduce the total cost of purchased inputs, which often represent 50 to 80 percent of the cost of goods sold. Thus, logistics influences a firm's operating costs both directly and indirectly.

Table 2. Logistics techniques used to facilitate integrated use of global and JIT sourcing.

Facilitating practices	Percent usage
Buying and shipping in container-sized lots	59
Developing partnership relationships with providers of transportation services	57
Pre-clearance of customs	43
Use of local third-party warehousing to buffer global and JIT sourcing	43
Use of EDI to track and/or expedite shipments	31
Reliance on air freight for regular shipments	26
Reliance on third-party logistics services	26
Use of intermodal transportation	26

3. ***Logistics' Impact on Customer Service:*** Though logistics costs are substantial, cost is much less important than the competitive leverage a firm can gain when the logistics system is well-managed and supports the overall company strategy. That is, the real competitive potential of logistics lies in its ability to help the firm differentiate itself from the competition. The most visible arena in which logistics provides a differential advantage is in the delivery of high levels of customer service. Logistics derived customer service helps companies differentiate their product, keeps customers loyal, increases sales and improves profits. (LaLonde *et al.*, 1988; Bowersox *et al.*, 1989). Discussing logistics' customer service, Bowersox suggested that because delighting customers through "world-class" logistics service is important to long-term success, firms should consciously focus on those activities that their customers value, including the following:
   - Customers want to receive exactly what they ordered: no more, no less, no substitutions, no defects, no breakage, no spoilage.

- Customers want delivery of their perfectly filled orders at the agreed-upon time.
- Customers want to pay as little as possible.

Focusing on what customers really value is particularly important since many companies invest tremendous amounts of time and money in providing logistics services that are not valued by customers (Stock and Lambert, 1992). At the same time, they neglect the things that customers truly value–order accuracy, consistent lead times, and the ability to expedite emergency shipments. In the past, companies could get by with mediocre customer service because customers did not expect their desires to be fulfilled. Customers had become accustomed to relatively poor customer service and felt powerless to demand higher service levels. Today, channel power has shifted toward the final customer such that retailers and other customers possess greater leverage and use it to demand better service. In this environment, service gaps between what customers really value and what companies actually deliver are opportunities for competitors to "steal valuable customers." Efficient, customer-oriented logistics systems can be a very important and proprietary source of competitive advantage. A senior financial manager at a Fortune 500 firm recently noted, "We've done most of what we need to do to be competitive. We've changed the way we develop products, manufacture, market, and advertise. The one piece of the puzzle we haven't addressed is logistics. It's the next source of competitive advantage. The possibilities are just astounding." (Henkoff, 1994)

IMPLEMENTATION: Logistics can be viewed as a single value-added system that is comprised of facilities, equipment, people, and operating policies that make possible the flow of goods and related information from the acquisition of raw materials through production and distribution into the hands of the customer. To effectively manage the flow of materials from point of origin to point of ultimate consumption requires that numerous related decisions be made. Some of the important logistics decisions focus on transportation, inventory, warehousing, order processing, packaging, forecasting, and customer service. To facilitate understanding of the decisions that must be made in the management of each of these logistics activities, Table 3 identifies and describes many of the activities typically managed as part of the logistics system. It is important to understand that all of these related movement and storage activities may not be under the control of a single manager, or even a single management group. Indeed, different firms organize their logistics activities differently to meet situation specific needs. Several factors combine to influence organizational decisions. For example, customer demands and expectations, the magnitude of logistics costs incurred by the firm, as well as the firm's historical approach to logistics all influence the firm's view of logistics as a value-added function and thus impact the way the firm organizes its logistics activities.

Despite the fact that companies tend to adopt somewhat unique organizational structures to manage their logistics activities, it is widely accepted that logistics consists of two primary areas – *materials management* on the inbound side and *physical distribution management* on the outbound side. Materials management is concerned with the procurement, transportation, and storage of raw materials, purchased components, and subassemblies entering the manufacturing process along with the flow of goods within and through the manufacturing process. The central objectives of materials management are to insure that the manufacturing function has the necessary inputs at the right time and place, and that the flow within the manufacturing system takes place in a timely and efficient manner. Physical distribution focuses on the transportation and storage of finished products from point of manufacture to where the customers wish to acquire them. The key to physical distribution is to meet or exceed the satisfaction expectations of customers at the lowest possible costs. Logistics' ability to bridge the spatial and temporal gaps between suppliers, the firm, and customers has made logistics an important weapon of an increasing number of company's competitive arsenals. The following sections discuss some of the relevant issues in implementing logistics strategies for competitive success.

1. ***Integrated Logistics and the Total Cost Concept:*** While each logistics activity can be used to add value, world-class performance comes through the integration of the different logistics activities. Recognizing that integrated decision making is critical to successful logistics management is perhaps the single most important implementation issue. Central to the integration effort is the notion that decisions for each activity described in Table 3 be made only after considering the decision's impact on the other logistics activities. For instance, by selecting a higher-priced transportation carrier that provides higher levels of on-time delivery, the amount of inventory that must be held to protect against late delivery can be reduced. The cost savings that come from lower inventory levels should be factored into

Table 3. Logistics Activities.

Activity	Basic Functions and Roles
Customer Service	Customer service activities focus on understanding what customers want and measuring logistics performance against these customer requirements. Customer service drives all of the logistics activities.
Demand Forecasting	Logistics decisions are typically made under conditions of uncertainty. This situation makes it very difficult to allocate resources and provide high levels of service at minimal costs. Therefore, forecasts – estimates of demand – must be developed to help plan other logistics activities.
Documentation	Shipments are almost always accompanied by documents that identify ownership and provide other details about the shipment. The bill of lading is the most common document. Accurate documents are important, especially in international shipments, to assure that the product gets to the customer in a timely manner.
Information Mgmt.	Tracking all of the items that are moved and stored by the typical company requires extensive amounts of information. Data on carriers, customers, and inventories must be turned into information that is useful for decision making. Information is increasingly being substituted for inventory in today's logistics systems.
Inventory Mgmt.	Inventories must be available to meet both production requirements and customer demand. If purchased inputs are not available, the production process will be forced to shut down. If finished goods are not available where and when customers want them, then sales are lost, hurting the firm's customer service reputation. However, keeping inventories on hand is also expensive. Inventories take up space, require capital investment, and divert managerial attention from other value-added activities. Successful inventory control must therefore support high levels of customer service with as little inventory as possible.
Material Handling	The term materials handling refers primarily to the movement or flow of materials within a plant or warehouse. Because the handling of materials costs money leads to damage, efforts are made to reduce the total amount of required handling. and often This reduction in handling means that goods are moved fewer times and that move distances are minimized. Materials handling requirements play an important role in the design of plants and warehouses.
Order Processing	When an order is received, it must be processed before work on the order begins. Order processing lets the firm know what work needs to be done and thus initiates the value-added process. Today, many orders are transmitted directly from the buyer's computer to the supplier's computer. This electronic processing can improve the speed and accuracy of the order fulfillment process.
Packaging	Packaging serves two important functions. First, packaging protects the product. Packaging can make materials handling easier by reducing the number of times the actual product must be touched. Also, the shape of the package can be easier to handle than the actual product. Second, packaging performs a promotional role by conveying information about the product and by presenting an attractive appearance.
Parts and Service Support	Logistics must support after sales efforts by assuring the availability of needed spare and replacement parts. Customer satisfaction can be greatly influenced by making spare parts available when a product fails. Caterpillar built a strong service reputation by promising delivery of needed replacement parts anywhere in the world within 48 hours.
Plant & Warehouse Location	The location decision influences both access to inputs to the transformation process and customer service levels. Strategic facility location is especially important in international operations. For example, by building a production facility in a low labor cost country, a firm can reduce its production costs. Likewise, by establishing operations in a given consumer market, the firm can gain good will by establishing an image as a local player.
Return Goods Handling	While the vast majority of products flow from point-of-origin toward the final consumer, some items – defective products and inaccurate orders – flow in the opposite direction. Developing an effective approach to handle these "reverse logistics" movements is very important to achieving high levels of customer satisfaction.

Table 3. (Continued).

Activity	Basic Functions and Roles
Reverse Logistics	Handling excess materials is often overlooked. However, this is an important logistics activity, especially when hazardous materials or recyclable items must be handled.
Traffic Mgmt.	Transportation of materials from point of origin to point of consumption is the most visible logistics activity. Five modal options exist: rail, truck, air, water, and pipeline. After choosing a transportation mode, the traffic manager must select a specific carrier to make a given move. Tracking the carrier performance and working to improve the efficiency of the transportation system are other important traffic issues.
Warehouse/DC Mgmt.	Warehouses allow firms to overcome spatial and temporal distances that separate production from delivery. When products are not consumed immediately after they are produced, they must be held in inventory until they are needed by a customer. This storage activity is the basic role of warehousing. Combining different items into a single shipment for a specific customer is the standard role of a distribution center.

the carrier selection decision. Numerous interactions – some highly visible and some quite hidden – exist among the different logistics activities. The interactive nature among logistics activities makes it unwise to make decisions in isolation.

Two tools are used to make integrated logistics decisions. The first tool is process mapping, which involves a careful diagramming of product, information, and financial flows so that they can be understood and analyzed. Mapping also provides a focal point for discussion, allowing managers from different areas to work through problems on a realistic, factual basis. With the "guess work" removed, communication is enhanced. For example, in one firm, a "war room" was constructed to map out the order fulfillment process (Shapiro *et al.*, 1992). Walls made of poster board were constructed and "coated with color-coded sheets of paper and knitting yarn that graphically charted the order flow from the first step to the last, highlighting problems, opportunities, and potential action steps." This process map was 200 feet long and portrayed an accurate view of the process that was then used to provide a clear and shared vision of the order system. Mapping's real value comes from its ability to make interrelationships that are not obvious but that affect overall logistics performance visible so that they can be managed effectively.

The second tool is total-cost analysis. Total cost analysis recognizes that different activities' cost patterns often display characteristics that put them in conflict with one another. The conflict is managed by balancing the activities so they are collectively optimized in terms of the firm's strategic goals. The classic example of tradeoff analysis involves the modal choice for transportation services in which inventory costs are traded off against transportation costs. When total costs are calculated and compared, the lowest cost transportation alternative is not always the best overall choice. Real-life experience reveals that even this basic level of tradeoff analysis is not always performed. In one firm, the transportation manager approached the logistics manager with a new rail contract he had just negotiated. According to the transportation manager's calculations, the new contract would save several hundred thousand dollars in transportation costs. After indicating that he would need some time to evaluate the new contract, the logistics manager carefully analyzed the tradeoffs in inventory and customer service responsiveness (tradeoff analysis should be performed for all relevant performance objectives). This analysis revealed that the transportation cost savings would cost the logistics group over one million dollars. Other typical tradeoffs are noted below:

- In an ideal world, a company would be able to increase logistics service levels to 100 percent; however, improvements in service are almost always accompanied by increases in costs. More specifically, as customers receive a higher customer service level, fewer customers are lost as a result of inventory out-of-stock situations, slow and unreliable deliveries, and inaccurate order filling. Expressed another way, the cost due to lost sales (opportunity cost) decreases with improved service. Further, the increase in customer service may increase the quantity sold. Conversely, the cost for maintaining a given level of service increases. Improved service usually (but not in all cases) means that more must be paid for transportation, order processing, and inventories. The main point here is that all these considerations must be taken into account when making logistics decisions. In this example, the cost of lost sales is traded off against the increased transportation, order processing, and inventory costs required to support higher service levels.

- The issue of how many warehouses to build and where to locate them focuses on tradeoffs among customer service, inventory cost, transportation cost, and warehouse costs. Fewer warehouses require more transportation to deliver products to geographically dispersed customers but a lower total amount of inventory is needed. More warehouses located closer to consumers decrease transportation requirements but increase inventory costs. As the number of warehouses increases, the level of customer service also goes up. The extreme example would be to place a warehouse in every city where a customer was located; however, the costs of supporting this type of system would be very high.
- Tradeoff analysis extends across disciplines. For example, production scheduling creates logistical tradeoffs. The longer the production run (the more items produced without interruption) the more efficient the production process and the lower the production costs. However, inventory carrying and warehousing costs increase as production runs get larger.

To summarize, by mapping the logistics system and by performing rigorous tradeoff analysis a clearer and more precise picture of not only how value is added but also what each logistics activity must do to achieve the desired levels of customer service would emerge. This understanding provides the foundation required to successfully integrate value-added logistics activities.

2. ***Integrated Logistics and Order Fulfillment:*** The primary value created by logistics is in the form of time and place utilities; that is, having orders where they are needed when they are needed whether the orders are to be used as inputs in a subsequent value-added process or to be purchased and consumed by an end user. Order fulfillment is one of the major responsibilities of logistics and should be a major driver in the design and implementation of logistics strategies. Figure 1 shows the key activities that comprise the order cycle, which can be defined as the elapsed time from recognition of need until the product is available for use. By definition, the order cycle begins and ends with the buyer. That is, the buyer must recognize a need and prepare an order that when transmitted to the seller will trigger a response. The seller must then process the order, which involves several steps: 1) order entry, 2) production of the order in a make-to-order setting, or picking of the order in a make-to-stock setting, and 3) preparation and packaging for shipment. The order is then shipped from the seller to the buyer. Once the order arrives at the buyer's receiving dock, it must be inspected and accepted before it is moved to the point where it will be used.

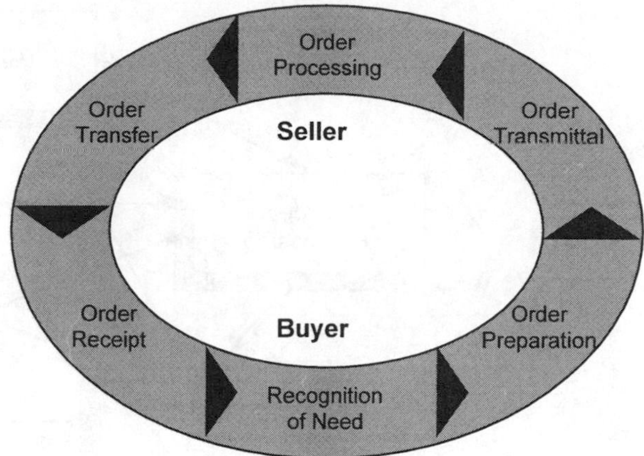

Fig. 1. The order cycle.

Just as in other areas of logistics performance, order fulfillment depends on how well the firm manages all of the activities that comprise the order cycle. Thus, the order fulfillment process should be carefully mapped and analyzed via tradeoff and total cost techniques. In fact, given the nature of order fulfillment, the total order performance should be analyzed. Total order performance considers quality, delivery, responsiveness, service and other performance dimensions. Figure 2 shows the five primary activities that must be managed well to improve order fulfillment. Each of these activities is discussed briefly below.

- By placing facilities in the right location and designing them well with the appropriate process technologies, the firm can reduce the combined production and delivery time.
- By carrying the right quantity and mix of inventory, the firm can more easily fill orders rapidly and completely.
- By streamlining the order processing system to eliminate unnecessary steps and assure that orders are entered in an accurate and

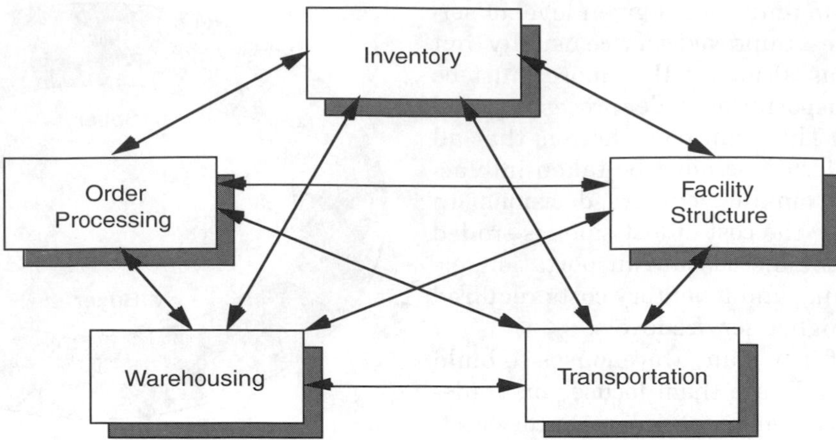

Fig. 2. Principal logistics components that facilitate order fulfillment.

timely manner, delays can be eliminated and orders filled to specification.

- By developing good relationships with quality transportation service providers, the firm can reduce transit time and increase on-time delivery performance. While each of the previous components of the order fulfillment process tend to be cross functional, transportation issues fall primarily under the domain of logistics management.
- By adopting emerging technologies and implementing innovative materials handling processes, today's firm can greatly increase the amount of material that flows quickly and efficiently through warehouses to meet customer demands.

As noted above, decisions made in these five areas create many tradeoffs, making it important to look at how each of the components in the order cycle work together to fill customers' orders. A thorough analysis of the order cycle can lead to reduced lead times and competitive advantage, especially in today's time-driven marketplace; after all, for most firms, time is money. National Semiconductor, for instance, found that it was not competitive in delivering products to key customers and was therefore losing market share. After careful analysis of its order cycle and each of the activities used in order fulfillment, National Semiconductor embarked on an ambitious experiment. To improve customer service through better order fulfillment, six warehouses in different areas of the world were shut down. The inventory in the warehouses was consolidated into a single 125,000 square-foot distribution center in Singapore. This move greatly reduced the total amount of inventory that needed to be held while increasing the probability that National Semiconductor could deliver complete orders. However, the only way to provide acceptable delivery lead times to customers worldwide from a single distribution center was to switch transport modes to air freight. Unfortunately, air freight is the highest cost mode of transportation, costing at least six times more than ocean container shipments. However, by combining air freight with consolidated inventory holdings in a single distribution center, National Semiconductor reduced delivery time 47% while lowering total distribution costs by 2.5%. The impact on customer satisfaction was remarkable, and national Semiconductor increased sales by 34% over a two-year time period. Only by looking at the entire order fulfillment picture was National Semiconductor able to make the right tradeoffs as it designed its new distribution system.

3. ***Integrated Logistics and Supply Chain Management:*** For logistics to have maximum impact on firm competitiveness, the concept of integration must be extended beyond the firm to include both suppliers and customers. The notion of working closely with other members of the distribution channel is now called supply-chain management. The move to more integrated supply-chain management is a result of greater competitive intensity in many industries. In particular, economic globalization has made it very difficult for a single company in one country to compete with strong rivals from other countries that work more closely with a network of allied companies. As Harold Sirkin has stated, "As the economy changes, as competition becomes more global, it's no longer company vs. company but supply chain vs. supply chain (Henkoff, 1994).

Logistics' role in supply chain management is as a central coordinating mechanism that links operations among members of the supply chain. Coordinated working relationships among supply-chain members can occur in many different logistics activities. For example, through integrated plant and warehouse location decisions, a firm and its suppliers might decide to locate operations in the same city or even in the same industrial complex. Likewise, two firms in closely linked supply chain relationship might decide to place some of their workers on site with each other. This sharing of workers is at the core of Just-In-Time II strategies in which the supplier locates one of its workers on-site at the buyer's location. This individual then manages all of the supplier's materials used by the buyer, placing orders and managing inventories almost as if she were an employee of the buyer. Logistics integration is also facilitated by technology. By using electronic data interchange, a buyer and supplier can communicate more frequently and share the information that is needed for the two to work more closely together. This information might include real-time inventory levels as well as production and delivery schedules. As logistics integration occurs, a seamless supply-chain team capable of providing higher service levels at lower total costs emerges.

TECHNOLOGY PERSPECTIVE: Even as the competitive and operating environments have changed in recent years, technological advancements have helped logistics meet the new challenges; in fact, technology has turned many challenges into opportunities. The most important technological factor has been the development of computer technology. Logistics management is very detail oriented, requiring the ability to gather, track, and process large quantities of information. Computers, data warehouses, and algorithm-based software allow for effective and efficient data analysis that leads to much better decision making. Indeed, complex, computer dependent decision making algorithms exist for many logistics decisions including facility location and design, vehicle loading and routing, and inventory management. The increasingly popular practice of cross-docking could not be performed without powerful, data-crunching computer systems.

Other vital technologies that play an increasingly visible role in logistics management include communications technologies, satellite tracking systems, bar-coding applications, and automated materials handling systems. Wal-Mart's distribution prowess comes from integrating these diverse technologies. Wal-Mart uses modern communication systems to tie all of its stores and distribution centers together. Likewise, Wal-Mart's prime suppliers are also electronically plugged in. By using bar-coding and computer-based inventory management systems, inventory levels are monitored in real time and orders are placed automatically. Product needs are consolidated at the distribution centers and orders are relayed to suppliers – this consolidation allows for full-truckload orders. Shipments are then tracked via satellite as they move from supplier to distribution center where they are cross-docked (most products flow through the distribution center in less than 24 hours). Cross-docking assures that full truck loads containing product from many different suppliers are then shipped to individual retail stores. By combining intelligent procedures with appropriate technology Wal-Mart has achieved the highest on-shelf stock percentage and the lowest distribution costs in the industry. Technology has made integrated logistics management truly possible.

EFFECT: Well-managed logistics can improve the firm's competitiveness by providing high levels of responsive and cost effective customer service. The examples of National Semiconductor and Wal-Mart discussed above highlight the specific competitive impacts that can be expected. These competitive advantages are listed below:

- Lower total landed costs.
- Shorter order fulfillment cycle times.
- Consistent on-time delivery of complete orders.
- Innovative product/service packages.
- Tailored services to meet individual customer's needs (Fuller *et al.*, 1993).
- More responsive after-sales service (e.g., Caterpillar's worldwide parts support system).
- Coordination of dispersed global operations.

When these benefits are combined, they provide a powerful competitive weapon for the firm.

TIMING: For many years, firms have managed logistics as a cost center such that most logistics decisions have been made reactively. Globalization, intensified competition, and technological change have altered the competitive landscape, placing much greater emphasis on the strategic and integrated management of logistics. Indeed, meeting the needs and demands of today's customers requires a proactive stance on logistics. Thus, the question, When is it appropriate to pay particular attention to the implementation of more fully integrated logistics management? In fact, the primary conclusion of the most extensive study ever done in the field of best practice logistics management is that the key to success is continual change

(Bowersox *et al.*, 1995). Firms with world-class levels of logistics performance are constantly evolving new logistics practices designed to enhance integrated decision-making and thereby provide customers with innovative, hard-to-replicate service.

LOCATION: Logistics activities take place wherever materials are moved or stored. From an organizational perspective, a variety of alternative organizations are used to manage the logistics processes. Some firms place responsibility for all logistics activities in a single integrated logistics department. Others separate out one or more major activities such as transportation into a separate traffic management department. Perhaps the most common and traditional organizational structure breaks activities into inbound and outbound categories which are typically referred to as materials management and physical distribution respectively. The trend today is toward a greater degree of centralization in an effort to more effectively employ information and other technologies and thereby achieve greater internal integration and interorganizational coordination.

RESULTS: Anecdotal evidence abounds to support logistics impact on firm competitiveness. For example, Fuller *et al.* (1993) and Henkoff (1994) discuss several examples of companies using logistics to reduce cycle times and offer distinctive service offerings that actually help customers succeed. Companies with world-class logistics capabilities also tend to have lower logistics costs as a percent of sales than their competitors. Large-scale empirical studies that validate the anecdotal evidence remain rare; however, a recent study demonstrated that firms that possess a superior logistics capability significantly outperform their counterparts (Fawcett *et al.*, 1997).

Unfortunately, numerous studies suggest that few companies actually manage logistics for competitive advantage. Bowersox *et al.* (1989) found that a very small percentage of U.S. firms – approximately 16 percent – have in place world-class logistics practices. Another study reported that the average U.S. company could improve its logistics productivity by 20 percent (Lambert and Stock, 1993). A third study found that logistics activities have not historically received the same level of management attention that has been given to other activities such as financial and strategic planning (Fawcett and Smith, 1995). The bottom line is that while logistics possesses tremendous potential to improve firm competitiveness, relatively few firms actively exploit this potential.

COLLECTIVE WISDOM: The recent and positive attention given to logistics belies the fact that for a large percentage of all firms, logistics is still managed reactively as a satisficing support activity. This attitude is a vestige of the fact that logistics has always been managed as a cost center. As a result, many opportunities to use logistics to expand markets and create customer loyalty are overlooked. This attitude also makes justifying investments in logistics training and infrastructure difficult. Other caveats that companies must be aware of as they strive to establish world-class logistics and tend to focus on information and performance measurement systems. First, U.S. firms have invested heavily in logistics-related information technologies; yet, managers still do not have access to the information that they need to make the best logistics decisions. Too frequently, technology investments are incompatible or are not supported with appropriate managerial procedures. With respect to total cost analysis, which is extremely information intensive, few firms have access to more than four or five relevant cost categories. For example, a recent study found that only 38% of leading companies had ready access to the cost of backorders (Bowersox *et al.*, 1995). The lack of information availability has thus led to a rather narrow definition of total cost despite its preeminent role in logistics system design and management. Second, even when the strategic intent is to pursue a legitimate logistics competence, it is not uncommon for firms to find that performance measures and strategic objectives are not aligned. As a result, tremendous effort – emotional and managerial – is expended without the early tangible results that are so important to building momentum and establishing the track record that will capture top management's attention and attract the resources needed to achieve world-class logistics performance. To use integrated logistics as a strategic weapon, firms must improve and align performance measurement, enhance information technologies, adopt better alliance management techniques, and understand that logistics can and does impact performance, for better or worse, depending on management's attitude toward logistical excellence.

Organizations Mentioned: Benetton; Caterpillar; Council of Logistics Management (CLM); General Motors; National Council of Physical Distribution Management; National Semiconductor; The Gap; Unilever; Wal-Mart.

See **Council of logistics management (CLM); Globally rationalized manufacturing; Integrated logistics; JIT purchasing strategy; Just-in-time manufacturing; Logistics infrastructure; Logistics support; Physical distribution management; Principle of comparative advantage; Production sharing facilities; Total cost analysis.**

References

[1] Ballou, R.H. (1992). *Business Logistics Management* (3rd ed.). Prentice Hall, Englewood Cliffs, New Jersey.

[2] Fawcett, S.E., Stanley, L.E. and Smith, S.R. (1997). "Developing a Logistics Capability to Improve the Performance of International Operations," *Journal of Business Logistics*, 18(2), 101–128.

[3] Bowersox, D., Daugherty, P., Droge, C., Rogers, D., and Wardlow, D. (1989). *Leading Edge Logistics Competitive Positioning for the 1990's*. Council of Logistics Management, Oakbrook, IL.

[4] Bowersox, D.J., Calantone, R.J., Clinton, S.R., Closs, D.J., Cooper, M.B., Droge, C.L., Fawcett, S.E., Frankel, R., Frayer, D.J., Morash, E.A., Rinehart, L.M., and Schmitz, J.M. (1995). *World Class Logistics: The Challenge of Managing Continuous Change*. Council of Logistics Management, Oak Brook, IL.

[5] Fawcett, S.E. and Smith, S.R. (1995). "Logistics Measurement and Performance for United States-Mexican Operations under NAFTA." *Transportation Journal*, 34(3), 25–34.

[6] Fuller, J.B., O'Conor, J., and Rawlinson, R. (1993). "Tailored Logistics: The Next Advantage." *Harvard Business Review*, 71, 87–98.

[7] Gecowets, G.A. (1979). "Physical Distribution Management." *Defense Transportation Journal*, 35, 11.

[8] Henkoff, R. (1994). "Delivering the Goods." *Fortune* (November 28), 64–78.

[9] Lachica, E. (1997, August 27). "East, Southeast Asia Make Huge Strides In Reducing Poverty, World Bank Says." *Wall Street Journal*, A8.

[10] LaLonde, B.J., Cooper, M.C., and Noordewier, T.G. (1988). *Customer Service: A Management Perspective*. Council of Logistics Management, Oak Brook, IL.

[11] Lambert, D.M. and Stock, J.R. (1993). *Strategic Logistics Management* (3rd ed.). Richard D. Irwin, Inc., Homewood, IL.

[12] Sampson, R.J., Farris, M.T., and Schrock, D.L. (1990). *Domestic Transportation: Practice, Theory, and Policy* (6th ed.). Houghton Mifflin Company, Boston.

[13] Shapiro, P., Rangan, V., and Sviokla, J. (1992). "Staple Yourself to an Order." *Harvard Business Review* (July–August), 113–122.

[14] Stock, J. and Lambert, D. (1992). "Becoming a "World Class" Company With Logistics Service Quality." *International Journal of Logistics Management*, 3(1), 73–80.

LOT-FOR-LOT PROCEDURE

There are several alternatives for computing the size of a batch or lot that goes into production at a given period. Lot-for-lot is a lot sizing technique, where orders produced are the exact amount needed. In a materials requirement planning (MRP) environment, lot or run size is equal to the net requirements for that period, or demand for that period. This lot sizing technique "eliminates inventory holding costs for parts carried over to other periods . . . If setup costs can be significantly reduced, this approach may approximate a minimum-cost lot size." (Stevenson, 1993: 669)

See **Capacity planning: Medium- and short range; Customer service through system integration; Lot-sizing models; Master production schedule; Mathematical lotsizing models; MRP; Schedule stability.**

Reference

[1] Stevenson, W.J. *Production/Operations Management*, 4th Ed., Richard D. Irwin, Inc., 1993.

LOT SIZE

Lot size refers to the quantity to be ordered or produced. This is an important decision for any manufacturing facility. Lot sizes generally vary with the type of manufacturing process used. In job shops, the lot sizes tend to be much smaller; a lot size of one unit is not uncommon. In line production, the lot sizes could be much larger.

Many mathematical models have been developed to compute optimal or near optimal lot sizes to minimize cost or maximize revenue or to provide a desired level of service. Generally, the models take into account the cost of holding inventory if what is produced cannot be consumed or shipped to customers immediately. If lot sizes become very small, then the need for frequent setup of production facilities or the need for placing several orders with suppliers increases. This may lead to increased setup or order costs. Mathematical models balance these and other costs to compute an optimal model.

Under just-in-time production, or lean production, lot sizes tend to be small and do not exceed the immediate demand. This philosophy prevents inventory buildup and costs associated with inventory holding and inventory management.

See **Capacity management in make-to-order production systems; Capacity planning: Medium- and short-range; Economic order quantity (EOQ) model; Inventory flow analysis; Just-in-time manufacturing; Kanban-based manufacturing systems; Lean manufacturing; Lot-for-lot procedure; Manufacturing order quantity; MRP; Order release.**

References

[1] Buffa, W.S. and J.G. Miller (1979). *Production and Inventory Systems*. 3rd edition, Homewood, IL: Irwin.

[2] Fogarty, D.W., J.H. Blackstone, and T.R. Hoffman (1991). *Production and Inventory Management*. Cincinnati, OH: South-Western Publishing Co.

[3] McLeavey, D.W. and S.L. Narasimhan (1985). *Production Planning and Inventory Control*. Boston: Allyn and Bacon.

[4] Vollman, T.E., W.L. Berry and D.C. Whybark (1997). *Manufacturing Planning and Control Systems*. New York: Irwin/McGraw-Hill.

LOT-SIZING MODELS

See **Lot size.**

LOW COST PRODUCER

A low cost producer is a firm within a specific industry with the lowest cost-of-goods-sold. Cost of goods generally consists of fixed and variable costs of production and logistics, and manufacturing and corporate overhead. The total cost-of-goods-sold therefore includes direct and indirect costs associated with producing and delivering a product or service. Only one, or very few organization within an industry enjoy this status. This serves as a competitive advantage. The low cost producer within an industry is able to price products and services lower than competitors and yet achieve a competitive return on assets.

Alternatively, low cost producers may price their product at the same level as competitors to reap a higher return on assets, which enables them to invest greater amounts into research and development, capital improvements, etc. In most manufacturing industries, 55–60% of total cost-of-goods-sold is comprised of material costs. Therefore, most low cost producers focus on purchasing and sourcing strategies to minimize the cost of procured materials and services.

See **Cost analysis for purchasing; Manufacturing strategy; Purchasing: Acquiring the best inputs.**

LOW-LEVEL CODE NUMBERS

The level code is a number starting from zero assigned to each level in a bill of materials. Low-level code identifies the lowest level in a bill of materials based on the part's use in various products. Gross requirements for a part are computed progressively from one level code to the next as part of MRP processing.

See **Bill of materials; Bill of materials indented; Explosion; Material requirements planning.**

MACHINE HOURS

Machine hours is the time in hours a product or component spends on a particular piece of production equipment while being processed. This measure is sometimes used to determine the direct cost of manufacturing and overhead costs to be assigned to the part.

See **Activity-based costing; Activity-based costing: An evaluation; cost analysis for purchasing.**

MACHINE RELIABILITY

Machine reliability is the probability of a machine operating without failure. This may be stated as a percentage, calculated by dividing actual operating time (total scheduled operating time minus unscheduled down time) by total scheduled operating time. Thus, a machine with a higher reliability will be expected to operate for a larger percentage of its scheduled operating time.

Machine reliability can be an important determinant of delivery speed and delivery dependability because unexpected machine down time will not only increase lead time but also disrupt the production plan. Such disruptions can be detrimental to a Just-in-Time (JIT) manufacturing environment. Thus, companies using JIT often emphasize preventive maintenance of equipment using an approach known as Total Preventive Maintenance that places responsibility for most machine maintenance with the person who operates that machine.

See **Performance measurement in manufacturing; Just-in-time manufacturing; Total productive maintenance.**

MACHINING CENTER

Shaping, drilling, turning, milling, sawing, broaching, and abrasive machining are seven basic machining processes. The following machines have been developed to carry out these basic machining operations: drill presses; boring machines, shapers/planers, saws, broaches and grinders. Lately, with the advent of computers and computerized numerical control (CNC) machines, it is now possible to design a single machining center that could combine several of the operations previously carried out by several individual machines. Among the many advantages is the need for a single setup before the machining center could carry out several different operations on the job.

See **Flexible automation; Manufacturing cell design; Manufacturing systems.**

MAD

MAD stands for Mean Absolute Deviation. It is a measure of forecast accuracy.

See **Forecasting guidelines and methods; Forecasting in operations management; Mean absolute deviation.**

MAINTENANCE AND CONTROL SYSTEM

Maintenance procedures can be categorized into two types: periodic maintenance and irregular maintenance. Periodic maintenance can be carried out according to any of the following guidelines:

- Regular intervals: by calendar date or clock time
- Logged usage: measured in time or number of cycles
- Regular inspection: inspect and repair when necessary, at regular intervals

Periodic maintenance may be part of a Total Productive Maintenance (TPM) program. Irregular maintenance includes equipment installation and relocation, repairs, overhauls, and maintenance upon suspicion of machine malfunction.

Maintenance systems management is served well using a maintenance database and maintenance inventories. Detailed database records of every problem occurring and maintenance procedures carried out enables the prediction of component failure rates, and in devising effective maintenance schedules.

See **Just-in-time management; Lean manufacturing; Maintenance management decision models; Total productive maintenance; Total quality management.**

MAINTENANCE MANAGEMENT DECISION MODELS

Frank A. Van der Duyn Schouten

Tilburg University, Tilburg, The Netherlands

Rommert Dekker

Erasmus University Rotterdam, Rotterdam, The Netherlands

INTRODUCTION: Maintenance management encompasses all activities aimed at the proper functioning of equipment and installations. As a management activity, the importance of maintenance in industry gained renewed importance during the last few decades in industry, where it is now realized that a good maintenance program can have a major impact on quality and reliability. While only a few decades ago, maintenance was considered as a necessary evil, today it is an important business function. This change in perception has been influenced by the increased level of automation, the higher complexity of products and equipment, and more importantly, the evolution of new production philosophies, such as JIT (just-in-time) manufacturing, in which unexpected malfunction of systems becomes less and less acceptable.

Maintenance management covers strategic, tactical and operational issues. At the strategic level, the issues are, the planning of the timely replacement of capital intensive production units, the outsourcing of maintenance activities, and the organizational relationship between production and maintenance departments. Typical decisions at the tactical level are the choice of a maintenance concept, the determination of maintenance frequencies, the size of the maintenance workforce and the location and size of spare parts stocks. At the operational level, one is concerned with the short term day-to-day scheduling of maintenance, like priority setting of jobs and work orders. Scientific models are used at the three levels for decision making. At the strategic level the most efficient tool is scenario analysis, while on the operational level, maintenance planning and scheduling models have shown their usefulness. At the tactical level, stochastic models are the most appropriate.

HISTORICAL PERSPECTIVE: Intense scientific interest in problems related to maintenance management originated only a few decades ago. The basis for the scientific support of maintenance management is found in reliability engineering. The book *Mathematical Theory of Reliability* (Barlow and Proschan, 1965) indicates the start of scientific interest in maintenance problems. In the early stages, maintenance was almost exclusively interpreted as replacement of components. The development of some simple but insightful models describing aging phenomena of technical components, as well as the choice of appropriate actions to cope with these phenomena, showed that a scientific approach can be useful. In the seventies and early eighties, the basic models were extended in several directions. A substantial part of this work was primarily motivated by mathematical curiosity, rather than practical need.

In the eighties, scientists made a step forward in bridging the gap between the analytical models and practice by a systematic analysis of multi-component systems. Multi-component systems are much closer to reality than the classical single-component systems. Since maintenance actions in practice quite often boil down to the replacement of one or more components in a multi-component system, the study of multi-component systems was the adequate way to bring reliability studies closer to practical maintenance management. Moreover, similar to many other areas of applied (stochastic) optimization, the major advances in information and computer technology brought the use of quantitative analytical models within computational reach.

During the last ten years substantial progress has been made in developing quantitative decision support systems for maintenance management of complex systems. Sophisticated decision support systems are in operation nowadays in the oil industry (both for maintenance of refineries as well as offshore installations), road maintenance, and electric power generation. The number of companies that make use of some kind of quantitative tools to support inspection and replacement of equipment is increasing rapidly.

STRATEGIC PERSPECTIVE: Many companies now a days consider the use of (quantitative) decision support in maintenance management as a strategic issue. At present, maintenance is responsible for a substantial part of companies' labor requirements. Employment in production has stabilized or decreased over the last two decades, while the number of people employed in maintenance increased gradually, and, in the process industry, even substantially. Management is aware that substantial savings in manpower costs can be realized by improving maintenance processes. Due to the stochastic character of the demand for maintenance, one has to hold a larger maintenance workforce as compared to the situation in which this work is plannable. Therefore a strategic question is, to what extent unplanned maintenance (maintenance actions based on failures, i.e., "fire fighting") can be replaced by plannable (i.e., preventive) maintenance. To support managers in their decision making about the size of a maintenance department or whether or not to contract out part of their maintenance activities, one could use quantitative tools both at a strategic, tactical as well as operational level. Most of the commercial decision support tools are offered at the tactical level. Decision support systems at the operational level are scarce and usually company specific.

IMPLEMENTATION: Tactical decision support systems for maintenance management appeared for the first time in the eighties. Initially, commercial packages were developed and marketed by academics, who were active in studying the underlying models. Later on, more packages appeared on the market, under names like MAINOPT and MACRO (see Woodhouse, 1997). Such packages claim substantial savings (up to 30%) on maintenance costs. MAINOPT was the first commercial PC-based decision support system for maintenance optimization. It makes use of standard age- and block-replacement, and efficiency models (see later) as developed in the sixties. The main strength of the package is that these basic models are embedded in a user-friendly environment, enabling engineers to formulate the necessary input to these models. MAINOPT has been successfully applied in several companies, with savings of millions of dollars.

We refer to Dekker, Van Rijn and Smit (1994) for a report on the implementation of MAINOPT within the Royal/Dutch Shell Group. An essential success factor of MAINOPT is the combination of technical reliability aspects with economic considerations, i.e., costs and revenues. An often heard objection against the use of quantitative models, is that these require input data, which are not (exactly) known. However, the use of the models in practical situations reveals that the decision support tools help managers to get a better feeling for the accuracy of the data that is actually required. The decision support system itself can be used to show that, in many situations, the outcome of the model is not strongly affected by the accuracy of the data; that is, the model is insensitive to data accuracy. Moreover, the opportunity to use the model as a laboratory, enables the maintenance manager to obtain better insight into additional data needs. Finally, the opportunity to use quantitative models to answer "what-if" questions, has substantially contributed to the satisfaction of its users.

Besides the quantitative based packages, there are qualitative approaches to improve maintenance efficiency, like Reliability Centered Maintenance (RCM) and Total Productive Maintenance (TPM).

RCM is a structured way to evaluate the maintenance requirements of complex systems (Smith, 1993), while TPM (Nakajima, 1986) is based on the Japanese philosophy, that maintenance, in the first place, should be the responsibility of the production personnel, and that problems should be solved in small teams with both production and maintenance expertise.

The supremacy of the qualitative models over the quantitative or the other way around has been the subject of a strong debate in the literature. This debate is now converging into the consensus, that both approaches should complement each other. For example, when a qualitative TPM analysis concludes that maintenance and production should be dealt with under joint responsibility, additional quantitative models are needed to translate this conclusion into operational production/maintenance plans.

ANALYTICAL MODELS: The most important developments in maintenance model building over the last three decades are summarized here. For a recent overview of the state of the art in reliability and maintenance modeling, see Özekici (1996). We will briefly discuss six models of increasing complexity. Most basic are the *age-and block replacement model*. The *modified block-replacement model* shows how certain drawbacks of basic models can be circumvented by model extensions. When opportunities for preventive maintenance are not always present, *opportunity based maintenance models* are relevant. Many models are based on the assumption of costless information about the state of the system. When this assumption is untenable, because of costly inspections, *delay time models* are appropriate. Finally, recent developments in *multi component* replacement modeling are discussed.

Age Replacement Model

In this model a single system in isolation is considered. This system can represent both a basic component or a complex system. In the latter case, however, the mutual influence of the composing components is not taken into account explicitly. It is assumed that the system is subject to failure according to a stochastic process. Moreover, we assume that the system under consideration has many functioning replica, so that, ample statistical data about its failure behavior is available. More precisely, we assume that the maintenance manager has a fairly accurate knowledge about the distribution of the time to failure. The time to failure represents the period until the first failure, starting with a perfect system. Upon failure there will be an immediate repair, which takes a negligible small amount of time. The possibility of preventive repair exists, meaning that the system can be taken out of operation for repair, while it is still functioning satisfactorily. Both the corrective and preventive repairs are assumed to be perfect in the sense that both types of maintenance bring the system back into perfect condition.

The advantage of a preventive repair over a corrective repair is that its associated cost is lower, reflecting the fact that a corrective repair always comes unannounced, requires more expensive action or may cause higher production losses.

Let the time until failure be a random variable with distribution function $F(.)$. The cost of a preventive replacement is c_p, while the cost of a corrective repair is denoted with c_f. The objective now is to find a critical age (i.e., running time) at which a preventive repair has to be started such that the expected average costs per unit time are minimized. Using the fact that the same problem repeats itself after every repair (corrective or preventive) it is fairly straightforward to conclude that, using a critical age t, the expected average costs per unit of time, $C(t)$, is given by:

$$C(t) = \frac{c_p + (c_f - c_p)F(t)}{\int_0^t (1 - F(s))\,ds} \tag{1}$$

The general shape of the function $C(t)$ is that of a unimodal function. This property can be exploited in finding the optimal value of t. Apart from finding the location of the optimal value of t, the formula for $C(t)$ can also be used to do sensitivity analysis, i.e., to investigate what effects small disturbances in the input data (the distribution $F(t)$, c_p, and c_f) may have on the optimal critical age. For example, an immediate conclusion from the formula for $C(t)$ is that the optimal value of the critical age does not depend on the values of c_p and c_f separately, but only on the ratio c_f/c_p, which eliminates the problem of scaling.

Suppose now that the age replacement model is used to control the replacement of bearings in a large number of identical machines. A major drawback of the age replacement policy is that after some time the ages of the bearings in various machines are no longer synchronous. Once the bearing in a certain machine has been replaced correctively, this machine will be out of step with the others. The stochastic nature of the failure occurrences implies that, after some time, every machine will have its own replacement time, implying that coordination between replacement activities is no longer possible. Also, application of the age replacement policy requires the record of the last replacement time of each individual bearing. The block replacement model has been developed to cope with these disadvantages.

Block Replacement Model

The assumptions underlying the block replacement model are the same as those for the age replacement model. The only difference is the replacement regime. Instead of setting a critical *age* for the system, here the *time between two subsequent preventive replacement actions* is fixed. In the age replacement system the time between two preventive maintenance actions remains only fixed as long as no corrective replacements occur. The class of block replacement policies is characterized by choosing a fixed time interval between two subsequent preventive maintenance actions, independent of the number of corrective replacements in between. Using an interval length of t, the resulting expected average costs per unit of time, $C(t)$ is given by:

$$C(t) = \frac{c_p + c_f M(t)}{t} \tag{2}$$

In this expression, $M(t)$ denotes the expected number of corrective replacements over a time interval of length t in the absence of preventive replacements. The precise form of this function depends on the form of the distribution function $F(t)$. For most cases the function $M(t)$ can only be computed or approximated numerically. Using the expression of $C(t)$ one can easily find the optimal length of the replacement interval by minimizing $C(t)$ as a function of t.

The formulas for the average cost for both the age replacement and block replacement model can be generalized to one generic form. This form holds good for most maintenance optimization models and allows for a generic mathematical analysis. For details, see Aven and Dekker (1997).

For the example of the bearings, the block replacement policy requires preventive replacement of all bearings simultaneously. This provides economies of scale as far as labor costs are concerned. There are, however, some drawbacks. The average cost under the optimal block replacement policy is in general higher than the average cost under the optimal age replacement regime. Under the block replacement policy it may happen that the system fails and is correctively replaced shortly before a planned preventive replacement.

It is obvious that such a close sequence of replacement actions is far from optimal. Hence several relaxation's of the block replacement policy have been proposed, like the modified block replacement policy.

Modified Block Replacement Model

In the modified block replacement policy we assume that on the occurrence of a failure a short time before a planned preventive replacement a "minimal repair" is executed instead of a complete corrective replacement. A minimal repair is a description for a maintenance action which does not bring the system back into the perfect condition, but into the condition it was immediately before the occurrence of the failure. So, the system is repaired temporarily in a minimal way such that it will hopefully survive the small time interval to the next planned preventive replacement. In this way the basic structure of the block replacement policy is maintained, but an obvious disadvantage is removed.

Opportunity-Based Replacement Model

All models presented so far are based on the assumption that preventive maintenance activities can be executed without any physical restrictions. This assumption is quite often too optimistic. For example, in very time-critical production situations, one might have to restrict the opportunities for preventive maintenance to low demand periods. Sometimes a production planner is only prepared to do preventive maintenance on non-failed units, when these are out of operation due to failures of other units. These approaches are called opportunity-based replacement models.

Both the age replacement model as well as the block replacement model have been extended to allow for the situation when opportunities for preventive maintenance actions may be limited. These extended models mostly assume that the process according to which the maintenance opportunities arise is random and independent of the aging process of the systems under consideration. Sometimes, this assumption is justified by the physical situation (like weather conditions in case of offshore maintenance activities). Sometimes, however, this assumption is made as an approximation of reality in order to reduce the analytical complexity of the model.

Delay Time Model

All previous models assume that the failure behavior of the system is dependent on the age (or running hours) of the system. This implies that the decision to do a preventive repair or replacement is based on the (observable) elapsed age. For many systems, however, the entrance of the system into a critical stage prior to failure is not obvious, but can be revealed by inspections only.

The *delay time model*, first proposed by Christer and Waller (1984), assumes that a failure is always preceded by an invisible fault. During the presence of a fault the system still functions properly, but the existence of a fault announces the occurrence of a failure in the near future and therefore summons a preventive repair. Under these circumstances the planning of the (costly) inspections constitutes an important decision. The time until failure is composed of two stochastically independent stages, the *delay time* and the *incubation period*. The delay time is the time until the occurrence of the fault, while the incubation period represents the time between the occurrence of the fault and the actual failure. Various statistical techniques as well as expert opinion methods have been proposed to determine the distribution of the delay time. More than 10 successful cases using the delay time model are reported by Baker and Christer (1994), but the delay time model has not yet been widely used in decision support systems.

Multi-Component Maintenance Systems

The weakness of the single component maintenance models described above is that dependencies between components are not taken into account explicitly. Generally, three type of dependencies can be distinguished: *statistical dependence, structural dependence and economic dependence*. Statistical dependence refers to the situation that the time to failure of various components can no longer be considered independent stochastic variables. This can be caused by the common environment in which the components are operating, implying that the failure of one component provides information on the residual time to failure of the others. Another situation in which statistical dependence occurs is when the failure of one component causes a higher stress on other components, which can have an impact on their time to failure.

Components exhibit structural dependence when they are part of a larger system. For example, in a series system, a failed component has to be repaired immediately, while in a parallel system a failed component can be left in a failed condition for a while without preventing the system as a whole from proper functioning. Economic dependence refers to the situation that combination of maintenance activities yields lower costs compared to multiple individual activities. Another type of economic dependence occurs in a system, where the unavailability of more than one unit is to be avoided. For a recent overview of multi-component maintenance models, see Dekker *et al.* (1997).

EFFECT: Effects of maintenance management on the performance of companies can be very large, but are not always easy to quantify. The costs of a maintenance program are the cost associated with organizing and executing the program, while the benefits are to be found in cost savings. Costs can be separated into *tangible costs* and *hidden costs*. Tangible costs include costs of the use of manpower and material, and are usually not hard to specify. In production environments, downtime is an important hidden cost factor, which cannot be extracted directly from the financial records.

The wide acceptance of new concepts in production management such as MRP, JIT and OPT has substantially contributed to a better understanding of maintenance cost and cost savings. For example, a good notion of the role of a bottleneck machine helps in realizing the costs and potential cost savings due to failures and preventive maintenance, respectively. The production loss caused by a machine failure depends on the impact of the failure on the production process as a whole, and on the relation between demand, production capacity and inventory position. Consider, for example, a plant facing a demand that is, on average, just below the available production capacity. One hour of unscheduled downtime due to a failure of a bottleneck machine will usually result in one hour loss of production, which in turn leads to lost sales.

Preventive maintenance also effects production output due to the unavailability of the bottleneck machine, but its impact can be minimized by scheduling of maintenance in periods of low demand or during of high inventories. For non-bottleneck machines, the idle times are natural opportunities for maintenance activities. Reasoning along these lines enables management to make hidden costs visible and quantifiable. Opportunity-based preventive maintenance models can play a role in analyzes of this kind.

It should be noted that, in most companies, a good effectiveness analysis (which machine should be improved and how) and efficiency analysis (how the profit contribution of a particular machine may be improved) is still in its infancy. This is partly due to the fact that the necessary cost figures cannot be extracted directly from the existing accounting systems. Moreover, traditional cost accounting methods which are developed mainly for external reporting purposes, can be misleading for management decision making. For an extensive discussion of decision making in maintenance see Pintelon and Gelders (1992). A detailed discussion as well as a case study on effectiveness and efficiency of maintenance activities can be found in Vanneste and Van Wassenhove (1995).

EXAMPLE: Some kind of maintenance management is used almost everywhere in private and public organizations. In most cases, however, these management activities are based purely on technical instructions or safety requirements, while optimization hardly plays a role. One area where the application of optimization methods have been used successfully is described below.

Off-Shore Oil Platform Maintenance

In the oil-industry, the relative share of the labor costs associated with maintenance has increased over the last two decades. It is now at the same level as the share of production and sometimes even higher. Hence the awareness of the crucial role of maintenance optimization for better company performance is large in this industry. The maintenance of gas turbines on oil platforms consists of a number of activities, ranging from checks and adjustments of instruments to replacements of individual components like filters. The execution of these activities requires the shut down of the turbine.

During the year, unexpected shutdowns of one of the turbines may be necessary due to other reasons like oil-well maintenance. These situations provide opportunities for turbine maintenance. These opportunities arise at random and are of limited duration. To take advantage of these opportunities, a prioritization in advance of the work is necessary using a decision support system so that certain maintenance activities may be permanently grouped into packages. This grouping of activities can be based on engineering judgement in combination with analytical model analysis. Activities should be divided into safety related and non-safety related work; for the former, a fixed time schedule will usually be required. In order to build a priority schedule for the non-safety related activities opportunity-based replacement models can be used. Such models can also estimate the additional costs of deferring the execution of a maintenance package from one opportunity to the next. This information can then be used to provide the maintenance manager with information on the cost effectiveness of the execution of each maintenance package.

Collective Wisdom

In maintenance management research two directions can be distinguished; qualitative versus quantitative approaches. Actual developments indicate that both approaches can claim substantial successes.

Successful application of a quantitative analysis requires well organized data collection and analysis as well as application of the latest information technology. The use of models in a laboratory type environment can provide insight into the actual need for failure data. The motivation for data collection can be improved substantially when their use can be foreseen. In many production environments, the reliability of failure data is low due to the lack of motivation on the part of those who are responsible for their collection.

In the education of engineers, there is little attention paid to maintenance. Traditionally, engineering education focuses on the design and not on the maintenance of systems.

See **Activity based costing; Corrective or breakdown maintenance; Decision support system; Maintenance frequencies; Preventive maintenance; Reliability centered maintenance; Reliability; Time-to-failure; Total productive maintenance (TPM).**

References

[1] Aven, T. and R. Dekker (1997). "A useful framework for optimal replacement models." *Reliability Engineering and System Safety*, 58, 61–67.

[2] Baker, R.D. and A.H. Christer (1994). "Review of delay-time OR modeling of engineering aspects of maintenance." *European Journal of Operational Research*, 73, 407–422.

[3] Barlow, R.E. and F. Proschan (1965). *Mathematical theory of reliability.* John Wiley, New York.

[4] Christer, A.H. and W.M. Waller (1984). "Reducing production downtime using delay-time analysis." *Journal of the Operational Research Society*, 35, 499–512.

[5] Dekker, R., C.F.H. Van Rijn and A.C.J.M. Smit (1994). "Mathematical models for the optimization of maintenance and their application in practice." *Maintenance*, 9, 22–26.

[6] Dekker, R., F.A. Van der Duyn Schouten and R.E. Wildeman (1997). "A review of multi-component maintenance models with economic dependence." *Mathematical Methods of Operations Research*, 45, 411–435.

[7] Nakajima, S. (1986). "TPM-challenge to the improvement of productivity by small group activities." *Maintenance Management International*, 6, 73–83.

[8] Özekici, S. (ed.) (1996). *Reliability and maintenance of complex systems*. NATO ASI Series, vol. 154, Springer, Berlin.

[9] Pintelon, L.M. and L.F. Gelders (1992). "Maintenance management decision making." European *Journal of Operational Research*, 58, 301–317.

[10] Smith, A.M. (1993). *Reliability centered maintenance*. McGraw-Hill, New York.

[11] Vanneste, S.G. and L.M. Van Wassenhove (1995). "An integrated and structured approach to improve maintenance." *European Journal of Operational Research*, 82, 241–257.

[12] Woodhouse, J. (1997). "The MACRO project: cost/risk evaluation of engineering and management decisions." In: C. Guedes Soares (ed.), *Advances in Safety and Reliability*. Pergamon Press, Oxford, UK, 237–246.

MAKE OR BUY DECISION

See **Cost analysis for purchase negotiation; Outsourcing product design and development; Purchasing: Acquiring the best inputs; Supply chain management; Value analysis.**

MAKESPAN

Makespan is the total time required to process all the jobs in an order or in a batch. Makespan is measured beginning from the time the first job in an order enters the manufacturing process through the time the last job in the order is completed.

See **Cycle time; Lean manufacturing; Manufacturing systems; Manufacturing systems Modeling using petri nets.**

MAKE-TO-ORDER (MTO) PRODUCTION

In make-to-order production, manufacturing of the product starts after receiving a customer order. Therefore in this production system, customer orders drive the activities of component manufacturing, their assembly into products, and the delivery of products. For example, a chemical equipment supplier produces equipment such as reactors, centrifuges, dryers, and filters after receiving customer orders.

See **Mass customization; Mass customization and manufacturing.**

MAKE-TO-STOCK (MTS) PRODUCTION

In make-to-stock production, standardized products, which may have a provision for easy customization, are produced in large volumes with long production runs and kept in inventory. Make-to-stock production of standard parts or products permits mass production. If the parts or modules can be easily customized, then MTS may be used with mass-customized products.

See **Manufacturing systems; Mass customization; Mass customization and manufacturing.**

MALCOLM BALDRIDGE NATIONAL QUALITY AWARD

The U.S. Congress, in 1987, established the Malcolm Baldridge National Quality Award to raise awareness to quality management and to recognize U.S. companies that have implemented successful quality management systems. Annually awards are give in each of three categories: manufacturing company, service company, and small business. The award is named after the late Secretary of Commerce, Malcolm Baldridge, a proponent of quality management. The U.S. Commerce Department's National Institute of Standards and Technology (NIST) manages the award, and American Society for Quality (ASQ) administers it.

See **ASQ; ISO 9000 and QS-9000; NIST; Quality management systems: Baldridge, ISO 9000 and QS 9000; Total quality management.**

MANAGEMENT COEFFICIENTS MODEL

This approach to aggregate production planning was proposed by Bowman. The purpose of his model is to determine the production rate given forecast of demand, inventory on hand, and other aggregate production planning variables.

$$P_t = aW_{t-1} - bI_{t-1} + cF_{t+1} + K$$

where:

Pt = The production rate set for period t

W_{t-1} = The work force in the previous period

I_{t-1} = The ending inventory for the previous period

F_{t+1} = The forecast of demand for the next period

and a, b, c, and K are constants.

To use the model, first gather historical data for P, W, I, and F. Then, using the data through regression analysis, estimate the values of a, b, c, and K. A notable aspect of the model is that it lacks explicit cost functions. Linear programming models for aggregate planning can incorporate cost and revenue.

See **Aggregate plan and master schedule linkage; Aggregate planning; Capacity planning: Long-range; Linear decision rule; Linear programming.**

MANNED CELLULAR MANUFACTURING

Manufacturing cells can be either manned or unmanned. In the case of the latter, a robot functions as an 'operator.' One or more operators staff a manned cell. The responsibilities of operators in a cell are: material handling; set up; trouble shooting; preventive maintenance; and implementing production plan or schedule. In a cell, operators move around to attend to machines and materials.

See **Linked-cell manufacturing system; Manufacturing cell design.**

MANTECH

MANTECH is an offshoot of the US Department of Defense (DoD). The US Air Force MANTECH (Manufacturing Technology) group along with Defense Advanced Research Projects Agency (DARPA) and Defense Modeling and Simulation Office (DMSO) has been promoting the development of Virtual Manufacturing (VM) technology. Their objective is to develop VM simulation and control models for manufacturing. These models will have the capability to evaluate if a product is manufacturable. They can analyze various production and supplier scenarios. The models are intended to enhance manufacturing operations by providing timely answers to the following questions. Can we make the product? How much will it cost? What is the best way to produce the product? And, what are the alternatives?

The government-sponsored Agile Manufacturing Initiative (AMI) is a joint DARPA and National Science Foundation (NSF) program. Its purpose is to develop, demonstrate, and evaluate advanced design, manufacturing, and business transaction processes.

See **Agile manufacturing; DARPA; Virtual manufacturing.**

MANUFACTURABILITY

See **Concurrent engineering; Product design.**

MANUFACTURING ANALYSIS USING CHAOS, FRACTALS, AND SELF-ORGANIZATION

Hamid Noori and D. Scott Slocombe

Wilfrid Laurier University, Waterloo, ON, Canada

DESCRIPTION: Many business process activities, including manufacturing processes in the broadest sense, are what systems scientists call complex, middle-number systems (Weinberg 1975): too many parts to be deterministically predicted, too few to be statistically forecasted. Such systems are also the prime exhibitors of complex, chaotic and fractal patterns in time and space. They also exhibit unpredictability, uncertainty and even self-organizing, emergent behaviors. Chaos, fractals, and self-organization are ways of describing and explaining the source of the patterns of behavior of complex systems with many interacting parts.

These kinds of systems and behaviors are the subject of the recent interest in the sciences of complexity, or complex adaptive systems (CAS) (Gleick, 1987; Holland, 1995; Kauffman, 1995). CAS draws on many different empirical and theoretical roots. Some of these roots have histories going back as much as a century to origins of thermodynamics and dynamical systems theory. Recent roots are in intelligent agents and artificial life.

The idea of fractals emerges from the recognition of the effects of hierarchy and scale, which help us to recognize that patterns may be different depending on the scale of focus. To the extent one system is embedded within another, larger patterns may be determined by processes operating at a range of space and time scales. *Fractals* are extremely complex structural patterns, considered to have nonintegral dimensionality, i.e. between points (0), lines (1), areas (2), or volumes (3). Self-similarity, the other distinguishing feature of fractals, is the recognition that some phenomena, from coastlines to trees to mathematical constructs, show broadly similar patterns which repeat at every scale of examination.

Chaos and self-organization are both dynamic processes, often with structural implications. A chaotic system is one whose small-scale behavior can be described by usually simple, deterministic (non-random) rules or equations, but whose large-scale behavior is, in fact, so unpredictable as to appear random. A self-organizing system exhibits spontaneous generation of new organization from information and processes wholly internal to the system. Both chaos and self-organization emerge from nonlinear processes, implying factors that interact multiplicatively or exponentially, rather than additively. Feedbacks, and many other behaviours of complex systems, are key factors in their development. Such processes often create path-dependence or irreversibility: a system's current state is a result of the amplification of past microscopic fluctuations by nonlinear feedbacks. Which fluctuation ends up "winning" is largely a matter of chance as a tiny, random advantage of one may be amplified enough that others cannot catch up. For example the "locking-in" of the QWERTY typewriter keyboard as the dominant format, winning out over many other keyboard designs not because it was the best design but because an early initial lead was amplified (Arthur 1989). The success of the VHS video format over Beta is similar. The lesson is, the current state is path-dependent because, if you started from the same place again, the end result would almost certainly be different.

This is one place where chance creeps into understanding of complex systems. Historically, natural science has believed that the evolution of systems is governed by deterministic (non-probabilistic) equations and laws of nature, whether we know them or not. This has changed today, although when scientists speak of *chaos* in relation to systems such as the atmosphere and weather, they are usually speaking properly of *deterministic chaos*. The equations that are believed to govern the evolution of the system are deterministic; they have no probabilistic elements. The unpredictability of the system after so many iterations of the equations is the result of nonlinear feedbacks and amplifications of microscopic differences.

Much attention is being given to the significance of the edge of chaos in a wide range of systems today. It is a particular state of a complex system, between order and chaos, where system dynamics are particularly prone to fluctuation, amplification, and sudden state changes or bifurcation's. Many writers use analogies from physical systems and phase transitions, e.g. from water to steam, principally to highlight the long-range order and qualitative transition in the character (structure and organization) of the system.

Another way of looking at the edge of chaos is to do so in terms of dynamic non-equilibrium. Complex, adaptive and chaotic systems are generally not in equilibrium with their environments. They maintain their identity and distinction from their environment by dissipating energy and resources that flow (actively or passively) across their boundaries. Such dissipative systems are the classic examples of *self-organizing systems*, systems that generate order internally, often exhibiting sudden bifurcations in state that are not, however, chaotic in origin. Their time evolution often seems directional, if not directed. The directional behavior of complex social systems such as the industrial revolution is a classic example.

HISTORICAL PERSPECTIVE: The origins of fractal, chaotic, and self-organizing ideas are found in natural sciences: in physics, chemistry, and biology (e.g., Abraham & Shaw, 1992; Nicolis & Prigogine, 1989), meteorology, mathematics and physical geography (Mandelbrot, 1983). The initial studies of dynamic systems that identified the first chaotic patterns of behavior and analysis of their form can be traced to Henri Poincaré in the late nineteenth and early twentieth century. But the elaboration of dynamic systems awaited the development of computers, numerical simulation and solution of equations that could not be solved analytically. Edward Lorenz in meteorology and Robert May in population biology in the 1960s are commonly cited pioneers.

Fractals can trace their origins to empirical studies on coastlines in the early twentieth century, and pure mathematics work in the nineteenth century, but their modern origins were in the 1960s and 1970s in Benoit Mandelbrot's conceptual and empirical work. Fractals have been applied in the 1980s and 1990s to a wide range of natural and mathematical phenomena. For example, fractals generate realistic computer animation sequences for movies. Fractal models of soot formation help engineers design more effective pollution control processes. The success of these highly varied applications suggests that fractals do indeed describe a real underlying phenomenon (Kiel & Elliot, 1996; Mandelbrot, 1983).

Ideas of self-organization can be traced to thermodynamics in the late nineteenth century, but especially to the post World War II work of Ilya Prigogine in the chemistry and physics of nonequilibrium systems (Nicolis & Prigogine 1989). Through the late 1970s and the 1980s a range of similar ideas were developed by others in biology and physics, e.g. synergetics and autopoiesis. Wider applications by Prigogine's colleagues and other scientists such as Herman Haken and Peter Allen included evolution of urban form, transportation modelling, population ecology, social evolution, social ecology, and many others.

The literature on complex systems such as chaos, fractals, and self-organization, is vast (cf. Slocombe 1997). Some disciplines such as economics and a few other social sciences, have seen a veritable explosion in the use of complex systems analyses (see, for example, Anderson *et al.*, 1988; Kiel & Elliot, 1996).

STRATEGIC PERSPECTIVE: One fundamental challenge of management has always been how business system structures and processes need to change to adapt to new and changing environmental conditions – environment is always changing. There is a perception today, right or wrong, that change has become more frequent and faster, and that organizations must change and adapt more quickly. This is frequently set in the context of globalization trends such as increased competition, reduced product lifecycles, fragmented markets, greater uncertainty in forecasting demand and other critical variables, and increased innovation and diffusion of new technology.

Not only is the environment of business becoming more dynamic, complex, and uncertain, but internal structures and processes of business are also becoming more complex and dynamic. Business processes which transform "selected resources into defined products through a series of discrete tasks or activities performed by an integrated series of people and equipment" (Noori & Radford 1995: 249) are most likely to change in response to increasing complexity, dynamics, and uncertainty. In the context of manufacturing there are at least three potential applications of complex systems ideas.

Integrated Product Introduction: There are multiple stages in product and service development and introduction, from strategic planning to management of the product life cycle. The relevance of notions such as turbulence, nonlinearity, and adaptive learning are increasingly recognized in this domain (e.g., Iansiti, 1995; Phillips & Kim, 1996). In terms of innovation and product development, application of complex systems ideas should enhance our understanding of the process, the discontinuities and difficulties surrounding it.

Aggregate Planning and Production Scheduling: The aggregate production plan specifies how a firm will provide the capacity to meet demand in the intermediate term. Aggregate planning is a difficult task since production requirements vary widely from one period to the next. In general, the complexity of the operational planning system will increase as companies experience larger product lines and multiple production facilities. This can be viewed as a quantitative, quasi-optimization problem with strategic elements. In this problem, critical complex systems elements include spatial resource deployment complexity, time lags and uncertainties, and issues of coordination and adaptive learning (Cohen & Lee, 1989). Tharumarajah and Wells (1997) have used a simulation to demonstrate the utility of adaptive, learning agent-based approaches (see below) to shop-floor scheduling.

Supply Chain Management: Supply chain management includes production processes, suppliers, and markets to develop the most efficient flow of supplies and products through it. Trends toward just-in-time manufacturing and supplies are based on an understanding of the dynamics of the supply chain to avoid disruptions while minimizing inventories and lead-times. Supply chain is a quantitative, dynamical system, whose complex systems elements include time lags and uncertainties, product flows, and the interaction of various agents in the process, which can potentially produce unpredictable dynamics (Levy, 1994, 1997).

IMPLEMENTATION: The application of complex systems ideas in manufacturing and elsewhere depends, most of all, on the knowledge of the theories and concepts themselves, and on several enabling approaches and ideas as well. Critical supporting knowledge includes systems analysis and progress in computation including artificial intelligence and artificial life.

Systems Analysis

Systems analysis is not easy to define. The term includes several diverse approaches that are fundamentally incompatible. The most basic distinction is between hard systems approaches, derived especially from engineering and computer sciences, and soft systems analysis derived from biology and social sciences. Hard systems analysis has its historical origins in the design of computers and other machines where the fundamental problem is one of predicting and controlling system behavior and communication to achieve a specific goal. These approaches are systemic in the simplest structural sense, and are strongly reductionist, quantitative, predictive, and design oriented. Soft systems analysis is quite different. It has a long history in biology and social science, aimed at understanding the behavior of existing systems rather than designing new ones. These approaches are relatively qualitative in approach and, more or less explicitly, challenge the traditional scientific methods of reductionism, linear cause and effect, predictability, and controlled experimentation.

At the most basic level, systems approaches include: (1) challenging and extending naive reductionism; (2) seeking understanding through analogy, comparisons, and case-studies; (3) aiming for a practical holism that studies entities as interacting wholes; and (4) focusing on system structure and organization that includes connections and interactions.

The definition of a system is somewhat arbitrary. It is determined by the minimum entity of interest to the observer that is clearly identifiable and coherent. This leads to identification of system boundaries and processes that maintain system identity and integrity. The system itself is usually divided into subsystems, often organized hierarchically.

Emergent Computation and Artificial Intelligence

Artificial intelligence (AI) has many definitions. Its roots go way back, even to classical times, in philosophy, mathematics, psychology, computing, and linguistics. Broadly one can identify four different conceptions of AI: systems that think like humans, systems that act like humans, systems that think rationally, and systems that act rationally (Russell & Norvig, 1995). They differ in their relative emphasis on replicating thought processes and reasoning versus implementing particular behaviours.

Four major applications of AI have emerged: natural language processing, knowledge representation, automated reasoning, and machine learning. Appplications of AI have led to a wide range of tools and demonstration programs, and a few commercial applications. The most famous are probably expert systems and neural networks. Common tools include methods of search, pattern matching, inference and logics, and neural networks and genetic algorithms.

Artificial life has emerged as a means for the study of evolutionary and ecological processes in computer environments (simulations). Intelligent agents are an evolution of artificial life to applications such as information retrieval and searching, and robotics and machine control, where intelligent, autonomous software "agents" or programs operate independently to perform tasks.

Genetic algorithms are related programs, which use processes of variation and selection to "evolve" new programs or solutions to problems through a repetitive, computation-intensive method. This permits us to think of the systems in terms of independent agents operating with simple rules, flows of resources, sources of variation, means of identification, and simple internal models that can serve as building blocks for complex behaviours and responses (Holland 1995).

Boolean and other networks take a slightly different approach. These are often conceptualized as *networks* of connected elements that can take on particular states (if only on/off states are possible then it is a Boolean network). A state space of alternatives is often conceived as a *fitness landscape*, when the entities that are evolving can be assessed in terms of some fitness function such as dollars earned or units produced or offspring produced. Then the key to success, as defined, is to get to the state represented by the highest fitness peak in the landscape (state space) of possibilities.

These and related approaches to working with complex systems depend on the practice of simulation modelling which has been substantially boosted by both recent advances in computing power, and the unexpected results from emergent computation exercises (see Casti 1997).

TECHNOLOGY PERSPECTIVE: Planning and management must address self-organization processes including the spontaneous emergence of order from disorder and its links to chaotic processes and fractal structures. The management and manufacturing-oriented literature in complex systems is still relatively small. The works of Phelan (1995) and Stacey (1996) are particularly relevant here. These studies are generally qualitative and focus on issues of adaptation, learning, expecting the unexpected, flexibility, and organizational design to facilitate dealing with turbulent environments.

A few writers are undertaking in-depth analyses of the dynamics and management of business processes to look for evidence of chaotic, fractal and self-organising patterns (Stacey, 1996; Warnecke, 1993), or parts of the manufacturing process such as supply chains (Levy, 1994). And a few others are exploring specifically fractal ideas in the layout and design of manufacturing processes (Jirasek, *et al.*, 1997), including the use of agents and adaptive modelling to better design processes (Tharumarajah & Wells, 1997). Kauffman (1995; esp. p. 265–66) comments on the relevance of his work on boolean network models and fitness landscapes, and optimal patchiness for studying and planning manufacturing facilities, particularly in the context of determining the optimum level for breaking down the process and fostering flexibility. There are some analyses of emergence and self-organization, which help us understand new product and technology emergence (Iansiti, 1995; Phillips & Kim, 1996; Silverberg, *et al.*, 1988).

There is a resurgence of interest in simulation modelling in the light of interest in complex systems, not the least because many such systems are not analytically tractable (e.g. Casti, 1997). The late 1980's saw several initial explorations of complex systems ideas in forecasting and understanding social systems, including organizations. Newer approaches based on object-oriented ideas (Khoshafian & Abnous 1995) utilizing artificial intelligence, artificial life, agents, and emergent computation are trying to more closely simulate the dynamics of complex organizations (Holland, 1995; Russell & Norvig 1995).

COLLECTIVE WISDOM: It is early still in the application of these ideas to manufacturing to say anything definitive about their effects. Detailed, quantitative case-studies and applications of these ideas are still largely lacking. The one exception is Levy's (1994) empirical and simulation work on an international supply chain as a complex dynamic system. This work better identified the potential volatility and even chaotic behavior to be seen in such systems. And it underscored the potential of changing the structure of the system by reducing the number of suppliers to reduce volatility. His studies suggest that reducing delivery times, increasing inventory levels, and reducing demand uncertainty can reduce the chaotic dynamics often found in inventory and demand.

Fractal concepts, and particularly their self-similar nature, provide a tool for the analysis of modular and cellular manufacturing approaches to the development of factories within factories, with different firms working together in a spatially integrated network. As fractal structures extend to multiple levels, coordination, and information and materials flow become more complex and chaotic, and self-organizing dynamics will have to be more explicitly considered. This may already be seen in the human resources problems and issues faced by "Škoda in Czechoslovakia (Jirasek *et al.*, 1997).

On the subject of new product development, firms which take an essentially nonlinear, flexible, semi-structured, and self-organizing approach to multiple product innovation are much more successful at identifying and bringing to market new products than those that take a linear, structured approach (Brown & Eisenhardt, 1997). This is an example of being on edge of chaos.

There are some grounds for suspecting that the quantitative applicability of chaotic and self-organizing mathematics to management may be limited. Phillips and Kim (1996), in explaining why earlier analyses of marketing and new product diffusion data do not show chaotic dynamics suggest that, in practice, most businesses change too fast for chaotic dynamics to emerge. However, there can be little doubt of the qualitative usefulness of these ideas in understanding complex systems in manufacturing processes, supply chains, and innovation and product development.

Organizations Mentioned: Skoda.

See **Chaotic systems; Dissipative systems; Genetic algorithms; Lean manufacturing; Neural networks; Self-organizing systems; Supply chain management; System dynamics.**

References

[1] Abraham, R.H. & C.D. Shaw (1992). *Dynamics: The Geometry of Behavior*, 2nd ed. Addison-Wesley, Redwood City, CA.

[2] Anderson, P.W., K.J. Arrow, & D. Pines, eds. (1988). *The Economy as an Evolving Complex System*. Addison-Wesley, Redwood City, CA.

[3] Arthur, W.B. 1989. "Competing technologies, increasing returns, and lock-in by historical events." *The Economic J.*, 99: 116–31.

[4] Brown, S.L. & K.M. Eisenhardt (1997). "The art of continuous change: linking complexity theory and time-paced evolution in relentlessly shifting organizations." *Administrative Science Quarterly*, 42: 1–34.

[5] Casti, J.L. (1997). *Would-Be Worlds: How Simulation is Changing the Frontiers of Science*. John Wiley & Sons, New York.

[6] Cohen, M.A. & H.L. Lee (1989). "Resource deployment analysis of global manufacturing and distribution networks." *Journal of Manufacturing & Operations Management*, 2: 81–104.

[7] Gleick, J. (1987). *Chaos: Making a New Science*. Viking, New York.

[8] Holland, J.H. (1995). *Hidden Order: How Adaptation Builds Complexity*. Addison-Wesley, Reading, Mass. 185 pp.

[9] Iansiti, M. (1995). "Shooting the rapids: managing product development in turbulent environments." *California Management Review*, 38(1): 37–58.

[10] Jirasek, J., H. Noori, and B. Moroz (1997). "Fractal Production at Škoda, automobilová a.s." Case published by Canadian International Development Agency, Ottawa. 13 pp.

[11] Kauffman, S. (1995). *At Home in the Universe: The Search for Laws of Self-Organization and Complexity*. Oxford University Press, New York.

[12] Kiel, L.D. & E. Elliott, eds. (1996). *Chaos and the Social Sciences: Foundations and Applications*. University of Michigan Press, Ann Arbor.

[13] Levy, D. (1994). "Chaos theory and strategy: theory, application, and managerial implications." *Strategic Management Journal*, 15: 167–78.

[14] Levy, D. (1997). "Lean production in an international supply chain." *Sloan Management Review*, 38(2): 94–102.

[15] Mandelbrot, B.B. (1983). *The Fractal Geometry of Nature*, W.H. Freeman & Co., San Francisco, CA.

[16] Nicolis, G. & I. Prigogine (1989). *Exploring Complexity: An Introduction*. W.H. Freeman & Co., San Francisco, CA.

[17] Noori, H. & R. Radford (1995). *Production and Operations Management: Total Quality and Responsiveness*. McGraw-Hill, Inc., New York.

[18] Phillips, F. & N. Kim (1996). "Implications of chaos research for new product forecasting." *Technological Forecasting and Social Change*, 53: 239–61.

[19] Russell, S. and P. Norvig (1995). *Artificial Intelligence: A Modern Approach*, Prentice-Hall, Englewood Cliffs, NJ.

[20] Silverberg, G., G. Dosi, & L. Orsenigo (1988). "Innovation, diversity and diffusion: a self-organization model." *The Economic Journal*, 98: 1032–54.

[21] Slocombe, D.S. (1997). *Chaos, Complexity and Self-Organization – An Annotated Bibliography and Review Essay on Management and Design in Complex Systems from Organizations to Ecosystems*. WLU Dept. of Geography & School of Business, Waterloo.

[22] Stacey, R.D. (1996). *Complexity and Creativity in Organizations*. Berrett-Koehler, San Francisco.

[23] Tharumarajah, A. and A.J. Wells (1997). "A behaviour-based approach to scheduling in distributed manufacturing systems." *Integrated Comput. Aided Eng.*, 4: 235–49.

[24] Warnecke, H.J. (1993). *The Fractal Company*, Springer-Verlag, New York, N.Y.

[25] Weinberg, G.M. (1975). *An Introduction to General Systems Theory*. Wiley, New York.

MANUFACTURING CELL

Manufacturing cells are one of the most powerful manufacturing innovations to take hold since the 1980's. They capture the benefits of a job shop as well as line production without the drawbacks of these systems. A cell is a layout of machine tools, usually single-cycle automatics, arranged in a U-Shape.

See **Manufacturing cell design; Manufacturing systems; U-shaped assembly lines.**

MANUFACTURING CELL DESIGN

J.T. Black

Auburn University, Auburn, AL, USA

Conversion of a functional manufacturing system into a flexible, linked-cell system is a manufacturing system design task. Table 1 outlines the some of the steps and methodologies by which manufacturing cells can be formed.

Companies that are trying to implement lean manufacturing say they want a system to deliver products on-time (have short lead times), to have the lowest possible unit cost, to have the best quality, and to be flexible (change to accommodate customer demand or design changes). It is difficult for a system to be able to achieve all four of these requirements simultaneously but manufacturing cells or a linked-cell manufacturing systems have the potential to meet these goals.

A manufacturing cell produces a family of part. A family of parts is defined by the sequence of processes needed to produce the individual components. A cell to produce the rack for a rack-and-pinion steering gear for a car can be cited as an example. A single model of a car may have 10 variations of racks (right-hand drive vs. left-hand drive, power steering vs. manual steering, etc.). The operations in the cell will include several metal cutting and grinding operations, heat treating, inspection, assembly, and superfinishing, all carried out in steps in one cell designed to make all rack variations for a car model.

Most companies "design" (layout) their first cell by trial-and-error for expediency in gaining experience in cellular manufacturing.

Sometimes, a key production machine is used to identify a family of parts. All the support equipment needed to make the family of parts are gathered around the key machine as the focus to form a cell. Often the key machine is a machining center capable of doing many different operations and therefore its cycle-time (CT) could be rather long.

Cell design and analysis using *digital simulation* is gathering wider usage with the advent of newer, more versatile languages. This technology can utilize virtual reality to simulate the workplace and the ergonomic involvement of the operators with the machines. Virtual reality can replace another technique called *physical simulation* which used small instructional grade robots and scaled-down versions of machine tools (mini-machines) to emulate real-world systems. The small machines can employ essentially the same control computers and software as the full-scale systems.

Group technology (GT) offers a systems solution to the reorganization of the functional system and restructuring of the job shop into manufacturing

Table 1. Multiple approaches to forming cellular manufacturing systems.

Lean manufacturing systems are based on linked-cell manufacturing cells. Knowing how to design the cells to be flexible is the key to successful manufacturing.

I. Use Group Technology methodology
   A. Production Flow Analysis (PFA) finds families and defines cells.
   B. Coding/classification is more complete and expensive.
   C. Other GT methods, including eyeball or tacit judgments.
      1. Find the key machine, often a machining center, and declare all parts going to this machine a family. Move secondary machines needed to complete all parts in a family around the key machine.
      2. Build cells around common sets of components such as gears, splines, spindles, rotors, hubs, shafts, etc.
      3. Build the cell around a common sequence of processes, such as drill, bore, ream, keyset, chamfer holes.
      4. Build the cell around a set of parts that eliminates the longest (most time consuming) element in setups between parts being made in the cell.

II. Simulation
   A. Digital simulation.
   B. Physical simulation.
   C. Object-oriented and graphical simulation.
   D. Virtual manufacturing simulation.

III. Pick a product or products.
   Beginning with the final assembly line, convert final assembly to mixed model, and move upstream through the subassembly to component parts and suppliers.

cells. Because of the magnitude of the changes, careful planning and full cooperation from everyone involved are essential.

The application of GT to a manufacturing facility results in the grouping of units or components into families where the components have similar geometry or manufacturing sequences. Machines are then collected into groups or cells (manufacturing cells) to process the parts family. The arrangement of the machines in the cells is defined by the manufacturing sequence of the family of parts.

The entire factory will not be able to convert into families immediately. Therefore, during transition, the manufacturing systems will be a mix of cellular and noncellular systems, evolving toward a complete linked-cell system over time. This will create scheduling problems because in-process times for components made by cells will be vastly different from those made under the traditional job shop environment. However, as the volume in the job shops decreases, the total system will become less chaotic and more productive.

PRODUCTION FLOW ANALYSIS (PFA): PFA is a popular method for conducting GT analysis. This methodology uses information available on route sheets. The idea is to sort through all the components and group them with a matrix analysis using product-routing information. This method is more analytical than tacit judgement, but not as comprehensive as the coding/classification method described below. PFA can be used as an up-front analysis, a sort of "before the fact" analysis that will yield some cost/benefit information. The following kinds of information can be generated through PFA: (1) what can the company expect in terms of the percentage of their product that could be made by cellular methods; (2) what would be a good "first cell" to undertake; (3) what coding/classification system would work best and; (4) how much money the organization might have to invest. In short, PFA can greatly reduce the uncertainty in making the decision on the reorganization to cellular production. As part of this technique, an analysis of the flow of materials in the entire factory can be performed to lay the groundwork for the new linked-cell layout for the entire factory.

CODING/CLASSIFICATION ANALYSIS (C/C): Many companies converting to a cellular system have used a *coding/classification* method. A classification step sorts components into classes or families based on their similarities. Computer software uses part codes to accomplish the classification. Coding is the assignment of symbols (letters or numbers or both) to specific components based on differences in shape, function, material, size, processes, operations, and so on. There are design codes, manufacturing codes, and codes that cover both design and manufacturing.

Once a C/C systems is selected, the software for doing C/C should be tailored to the particular company. The system should be a simple as possible so that everyone understands it. It is not necessary that old part numbers be discarded, but every component will have to be coded prior to finding families of parts. This coding procedure can be costly and time consuming, but most companies that choose this conversion understand the necessity of performing this analysis.

Using a classification system part families are formed, which leads to the design of cells but cell design is not automatic. It is the critical step in the reorganization and must be carefully planned. Different families of parts will not be the same with regard to their material flow, and, therefore, will require different cell designs (layouts). In some families, every part will go to every machine in exactly the same sequence without skipping a machine and without back flow. This is, of course, the purest form of a cellular system. Other families may require that some components skip some machines and that some machines be duplicated. However, back flow should always be avoided.

Many companies begin with a pilot cell so that everyone can see how cells function. It will require time and effort to train the cell operators who have to adjust to the standing, walking routine in cellular manufacturing. A company should proceed with developing manned pilot cells without waiting until all the parts have been coded. To start with, select a product or group of products that seem most logical. *Operators must be involved in designing the cell to get them to take over ownership of the cell.*

The pilot cell will show everyone not only how cells operate but also how setup time is reduced on each machine. Machines in cells will not be utilized 100 percent. Machine utilization rate usually improves with cell use but may not be what it was in the functional layout where *overproduction* is allowed. *The objective in manned cellular manufacturing is to utilize the people fully*, enlarging and enriching their jobs by allowing them to become fully functional.

One of the inherent results of the conversion to a cellular system is that workers become multifunctional. Operators learn to operate many different kinds of machines and perform tasks that include inspection and quality control, machine tool maintenance, setup reduction, and continuous improvement. In unmanned cells and systems, the utilization of the equipment becomes more important because the most flexible and smartest element in the cell, the operator, is removed and "replaced" by a robot.

The manned cellular system provides the worker with a natural environment for job enlargement. Greater job involvement enhances job enrichment possibilities and clearly provides an ideal arrangement for improving quality. Part quality is checked between each step in the cell and the resulting system is called: MO, CO, MOO – make one, check one, move one on.

See **Group technology; Human resource issues and advanced manufacturing technology; Job enrichment; Just-in-time manufacturing; Kanban-based manufacturing systems; Lean manufacturing; Linked-cell manufacturing system; Manufacturing systems; Route sheets**.

MANUFACTURING CELLS, LINKED

Manufacturing and assembly cells can be linked to each other in a pull production system using *kanban*. Kanban is a visual (physical) control system that works well in lean manufacturing systems with linked cells. Kanbans control the route that the parts must take (while doing away with the route sheet) kanban controls the timing and the amount of material flowing between any two work stations. To accomplish this, two kinds of kanban are used: *withdrawal* (or conveyance) *kanban* (WLK) and *production-ordering kanban* (POK). One can think of kanban as a link connecting the output side of one cell with the input point of the next cell. The link uses carts or containers that hold parts in specific numbers. Every cart is designed to hold the same number of parts.

See **JIT evolution and use in the U.S.; Just-in-time manufacturing; Just-in-time manufacturing implementation; Kanban-based manufacturing systems; Lean manufacturing implementation; Linked-cell manufacturing system; Manufacturing cell design; Manufacturing systems.**

References

[1] Monden, Yasuhiro (1983). *Toyota Production System*. Industrial Engineering and Management Press, IIE.

[2] Schonberger, Richard J. (1982). *Japanese Manufacturing Techniques: Nine Hidden Lessons in Simplicity*. The Free Press.

MANUFACTURING ENGINEERING

The main objective of manufacturing engineering is to develop and design manufacturing processes used in the transformation of inputs (material, labor, machines, etc.) into products sold to customers. Manufacturing Engineering function designs and develops and acquires production machinery, fixtures, material handling equipment, and all hardware and software for manufacturing.

See **Process innovation; Reengineering and the process view of manufacturing.**

MANUFACTURING FLEXIBILITY

Paul M. Swamidass

Auburn University, Auburn, AL USA

INTRODUCTION: Everyday, manufacturers are investing in computer aided design (CAD), just-in-time (JIT) manufacturing, computer-aided manufacturing (CAM), and a dozen other technologies. Although the above technologies are diverse, one thing common to all hard and soft technologies (Swamidass, 1998) is that they add flexibility to manufacturing operations. Hard technologies such as CAD, CAM, robots, and CIM are hardware and software intensive, whereas, soft technologies such as JIT, SQC and TQM are driven by know-how and techniques.

It is notable that over 70 percent of U.S. manufacturers reported that they experienced cycle time reduction due to the use of 17 different technologies (Swamidass, 1998). Reduced cycle time means improved flexibility. Figure 1 shows the use of 17 different manufacturing technologies and their use in the United States (Swamidass, 1998). The data in Figure 1 were gathered from 1025 U.S. manufacturing plants from the following industries: SIC 34: metal fabrication; SIC 35: machinery including computers; SIC 36: electrical; SIC 37: transportation; SIC 38: Instruments and photo goods. These industries employ more than 40% of all manufacturing employment in the U.S.

What is Manufacturing Flexibility?

Manufacturing flexibility (MF) could refer to the capacity of a manufacturing system to adapt

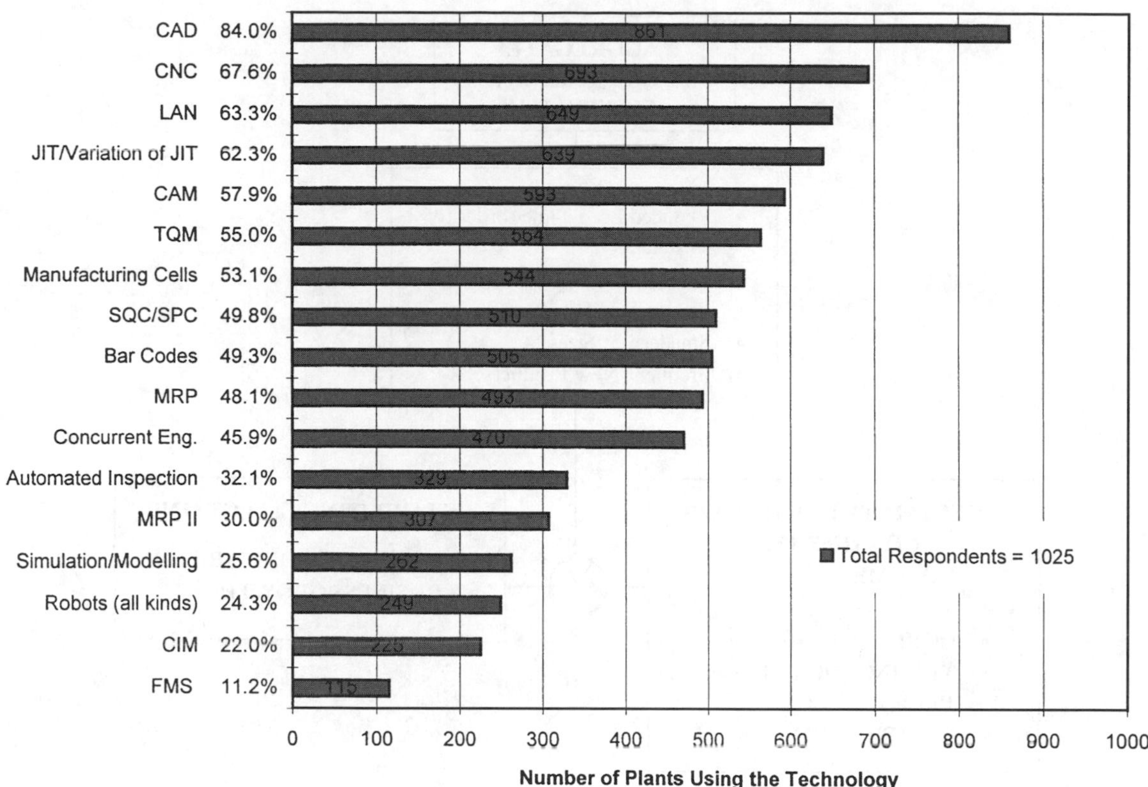

Fig. 1. Technology users.

successfully to changing environmental conditions as well as changing product and process requirements. It could refer to the ability of the production system to cope successfully with the instability induced by the environment. Flexibility, provides the manufacturing plant the ability to maintain customer satisfaction and profitability under conditions of change and uncertainty.

For superior understanding of the term Manufacturing flexibility, we must learn to distinguish clearly between flexibility offered by a single machine as opposed to the flexibility of an entire plant. Machine level flexibility is predominantly technology based, but plant level manufacturing flexibility is a complex blend of several ingredients including, (1) hard technologies, (hardware, software, and equipment), (2) soft technologies, (i.e., know-how, procedures, organizations and techniques) (3) design, and (4) manufacturing infrastructure.

Competitive Value of Manufacturing Flexibility

The environment that influences manufacturers is in a constant state of flux. However, the rate of change occurring in the environment and the extent of changes occurring may vary from one firm to another. Further, not all changes occurring in the environment may affect a firm's operation, its viability, and its success. But when the changing environment adversely affects a manufacturing firm, it disturbs or robs the firm of its ability to satisfy customers and operate profitably in a sustained manner. Under such circumstances, the firm needs flexibility to successfully cope with the changes in the environment.

In a changing environment, the primary challenge to manufacturing arises from the uncertainty associated with demand. The capability of manufacturing to respond appropriately to demand uncertainty will determine the stability and growth of the business unit's profitability. The competitive value of MF lies in the extend to which it can neutralize the effects of demand uncertainty. This strategic value of MF is captured in Figure 2. It shows that under conditions of demand uncertainty, MF could help maintain or stabilize the profitability of the business. It implies that the greater the demand uncertainty created by competitors, changing technology, and so on, the greater the need for MF in preserving sales growth and profitability of the business.

Manufacturing flexibility has been used strategically for offensive as well as defensive purposes as summarized in Figure 3. For example, in a given period of time, an ability to launch larger numbers of new models than one's competition could

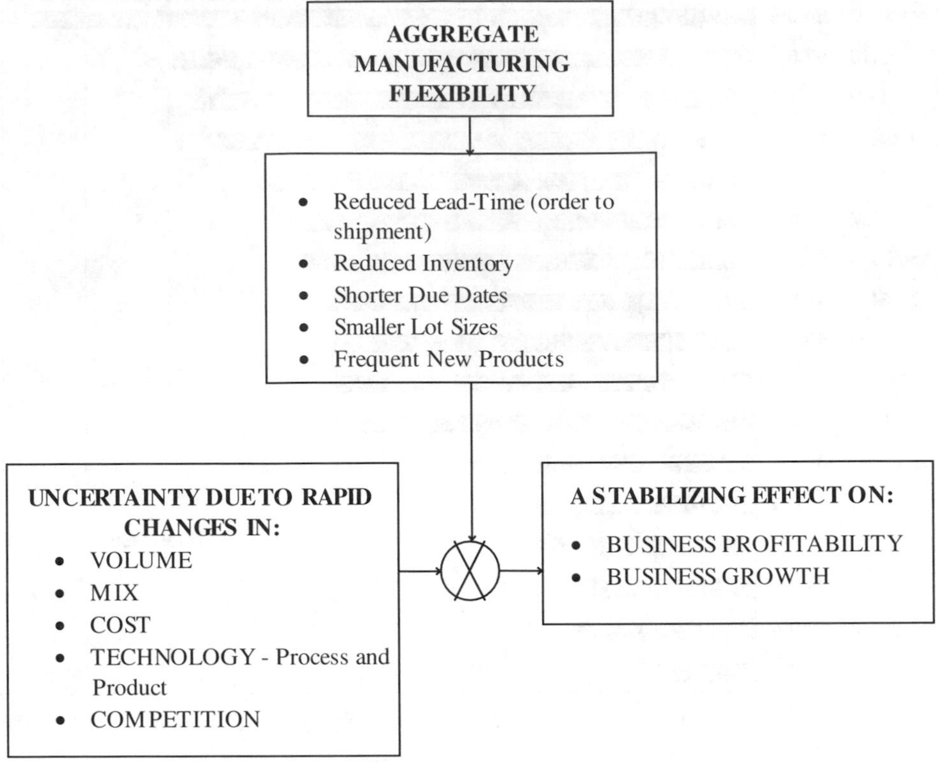

Fig. 2. The strategic role of manufacturing flexibility.

be a formidable competitive weapon in certain markets. An illustration of this is the well-documented Honda-Yamaha "war" in the early 1980's when Honda's number one position in market share for motorcycles was threatened by Yamaha. To eliminate the threat, within eighteen months, "Honda introduced 81 new (motorcycle) models and discontinued 32 models for a total of 113 changes in its product line. Yamaha retired only 3 models and introduced 34 new models for a total of 37 changes" (Abegglen & Stalk, 1985). By January 1983, about a year after Honda's successful counter attack, President Koske of Yamaha admitted defeat and declared, "... I would like to end the Honda-Yamaha war..."

MANUFACTURING FLEXIBILITY TYPES: Is manufacturing flexibility a homogeneous entity? Are there various types of flexibilities? Swamidass (1988) and Upton (1995b, 1997) support the view

IN OFFENSE	IN DEFENSE	IN OFFENSE AND DEFENCE
Responding to Opportunities	Desensitizing the System to Adverse Changes	Increasing Efficiency
1. Ability to introduce large number of new models into the market (Honda vs. Yamaha motorcycles. 2. Time required to change entire product line reduced (3 years in Japan vs. 10 years in U.S. air-conditioning industry).	1. Less susceptible to changes in demand, supply and tastes due to a broader range of product mix. 2. Enables the manufacturer to cope with uncertainties caused by changes in the external environment —demand, mix, material.	1. Better utilization of capacity through wider range of product mix. 2. Reduction or elimination of setup time or change-over time. 3. Better use of capacity of the production of counter cyclical products.

Fig. 3. Competitive value of manufacturing flexibility.

that manufacturing flexibility is not homogenous. Major hurdles to the understanding of manufacturing flexibility has been a homogeneous view of manufacturing flexibility, and a lack of consensus on the terms used to describe it.

Terms Associated with Manufacturing Flexibility

Most terms used in connection with manufacturing flexibility in discrete products industries are listed in Table 1. The table illustrates some of the problems hindering our understanding of MF. First, the scope of flexibility related terms used by various authors overlap considerably. Second, some flexibility terms are aggregates of other flexibility terms used. Finally, identical flexibility related terms used by more than one writer do not necessarily mean the same thing. In addition to the terms in Table 1, Upton (1995a, 1997) defines flexibility in process industries in terms of "operational mobility" and "process range."

Table 1. List of terms associated with manufacturing flexibility in OM literature[a].

1. Action flexibility
2. Adaptation flexibility
3. Application flexibility
4. Assembly system flexibility
5. Demand flexibility
6. Design flexibility
7. Dispatch flexibility
8. Job flexibility
9. Machine flexibility
10. Machining flexibility
11. Material flexibility
12. Mix flexibility
13. Modification flexibility
14. Process flexibility
15. Program flexibility
16. Product flexibility
17. Production flexibility
18. Routing flexibility
19. State flexibility
20. Volume flexibility

[a]Adapted from Swamidass (1988).

Process Contingencies

Consider Figure 4 which has been popularized in various forms by many writers on automation (Jelinek and Goldhar, 1983) in non-process industries. Typically, many authors have used similar figures, in one form or the other, to demonstrate the appropriateness of various manufacturing processes to selected manufacturing situations. In Figure 4, the two-dimensional space represents the entire range of manufacturing possibilities or process contingencies. A comprehensive discussion on MF should be in the context of the complete range of manufacturing possibilities or process contingencies represented by Figure 4; after all, MF broadens a manufacturer's

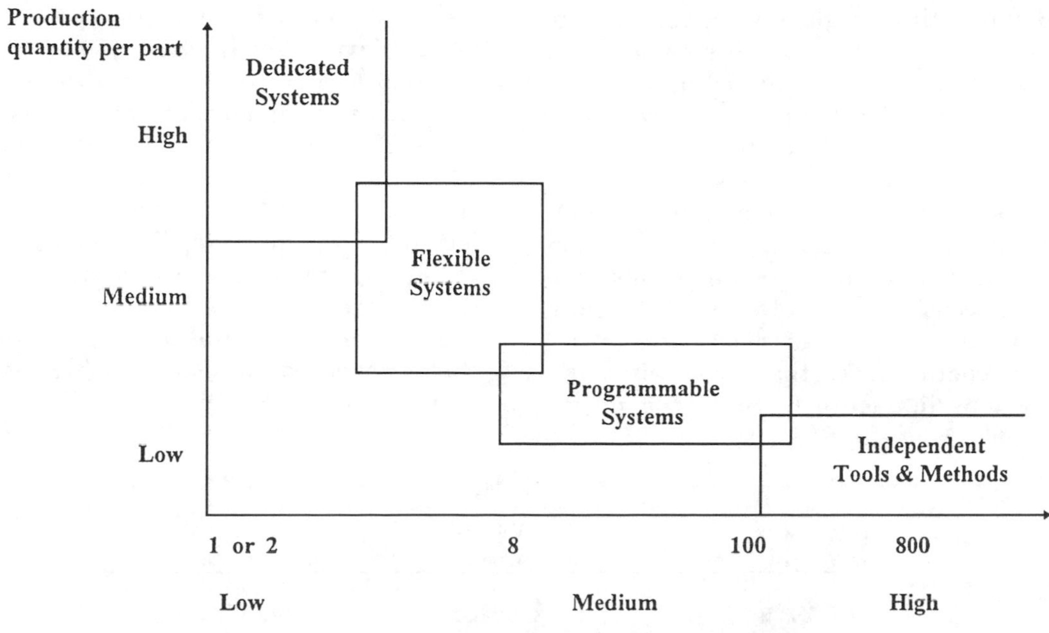

Fig. 4. Manufacturing contingencies spectrum [Adapted from Jelinek and Golhar (1983)].

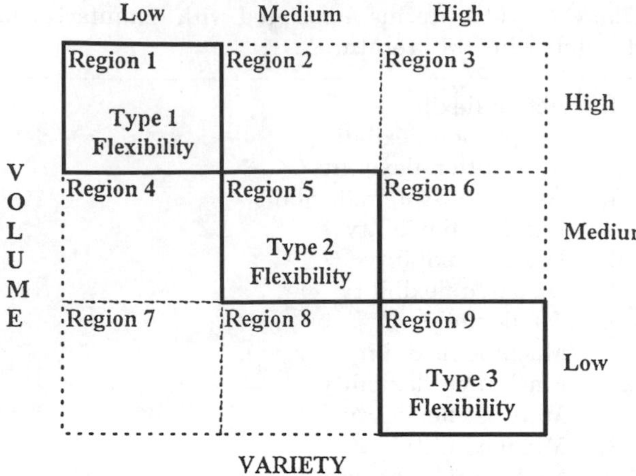

Fig. 5. Three regions of the manufacturing contingency space that represent almost all manufacturing firms [Swamidass (1988)].

capability to deal with a wider spectrum of process contingencies.

Figure 5 is an extension of Figure 4 where the two-dimensional space of manufacturing contingencies is partitioned into nine regions. In Figure 5, the three regions along the diagonal represent almost all manufacturing firms: the remaining regions are redundant. If all the redundant regions are dropped, Figure 5 reduces to the manufacturing flexibility continuum shown in Figure 6.

Three Manufacturing Flexibility Types

The manufacturing flexibility continuum in Figure 6 shows three regions with very distinct flexibility needs. For the sake of simplicity and ease of identification, the manufacturing flexibility associated with the three regions are labeled Type 1, Type 2, and Type 3 flexibilities in the context of discrete products industries.

The first step towards an understanding of MF requires our accepting multiple types of manufacturing flexibilities to cover all possible manufacturing contingencies. Different types of manufacturing flexibilities are needed to cope with the process contingencies that arise within the diverse regions in Figure 6. Since the process contingencies within the regions are different, MF to deal with the contingencies in these regions are also different. Failure to recognize the differences in process contingencies among the various regions leads to erroneous assumptions about the nature and usefulness of MF. A major consequence of such erroneous assumptions is the difficulty faced by practitioners when justifying investments in flexibility enhancing systems.

Examples of Flexibility Types

The three types of flexibilities introduced here are made more obvious to the reader through the following examples.

Type 1 Flexibility: An example of Type 1 flexibility is the Honda Motorcycle Division, whose head-to-head competition with Yamaha, was described earlier. In this example, a mass producer of motorcycles was able to introduce 81 new models and discontinue 32 models for a total of 113 changes in its product line in about 18 months using Type 1 flexibility.

Another good example of Type 1 flexibility is then IBM's (now Lexmark) automated Selectric typewriter/printer plant at Lexington, Kentucky. During the period 1981–1986, this plant underwent a major 350 million dollar renovation to implement flexible automation which enabled the plant to manufacture almost any product that can be assembled "from above" (to permit assembly by robots) within an envelop of size 18″ × 22″ × 28″. This case is described in greater detail in a later section.

Type 2 Flexibility: In 1983, General Electric's Series 8 locomotive plant in Erie, Pennsylvania machines a family of motor frames and gear boxes. The plant installed a Gidding and Lewis FMS with a capacity of 5,000 motor frames of sizes up to 4′ × 4′ × 5′ with over 100 machining surfaces. The system included two vertical milling machines, three horizontal machining center, three heavy horizontal boring mills and one medium horizontal mill. The machines included robotic or automatic tool changers with over 500 cutting tools. Machining time was

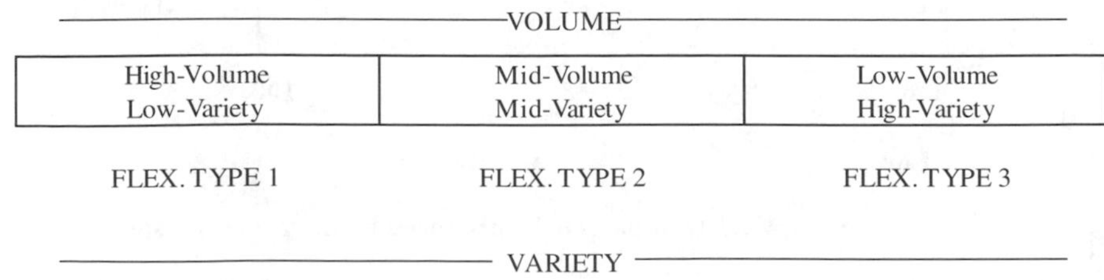

Fig. 6. Volume variety continuum and flexibility types*.

Table 2. Examples of the three types of flexibilities.

Type 1: Flexibility in automation	Type 2: Flexibility in manufacturing	Type 3: Flexibility in design and manufacturing
IMB Lexington, Kentucky: Selectric System 200 typewriters and printers plant. Annual production about 1,000,000 units. Capability to produce an infinite number of models a low cost. Lead time for new product introduction reduced to 18 months. (as of 1985)	GE's series 8 locomotive plant, Erie, Pennsylvania. Gidding and Lewis FMS for machining a family of motor frames and gear boxes. Yearly capacity 5000 motor frames of sizes up to $4' \times 4' \times 5'$ with over 100 machining surfaces. The system includes two vertical milling machines, three horizontal machining centers, three heavy horizontal machining centers, three heavy horizontal boring mills and on medium horizontal mill. The machines include robot or automated tool changers with over 500 cutting tools. Machining time reduced from 16 days to 16 hours per frame. (as of 1983)	Ingersoll Milling Machine Co., a very special machinery producer with a CIM system which includes CAD/CAM. A typical lot is one or two pieces; seldom builds a duplicate (as of 1984).

reduced from 16 days to 16 hours per frame with the installation of the FMS.

Type 3 Flexibility: Ingersoll Milling Machine Co., a special machinery producer, uses a CIM system which includes CAD/CAM. Orders for special machines come in lots of one and the firm seldom builds a duplicate of a machine, thus each order individually goes through the various stages of design, engineering, and manufacturing. The foregoing examples illustrating the three types of manufacturing flexibilities are summarized in Table 2.

AGENTS OF MANUFACTURING FLEXIBILITY: Manufacturing flexibility is the result of several contributing factors. In order to make substantial gains in MF there has to be a coordinated effort to address most of the contributing factors of flexibility. According to Figure 7, process, product design, and infrastructure induce MF.

In recent years, technological developments such as Flexible Manufacturing Systems (FMS), Computer Integrated Manufacturing (CIM), Computer Aided Manufacturing (CAD), and other technologies come to a manager's mind when considering manufacturing flexibility. It may turn out to be a serious error to consider manufacturing flexibility only in terms of process hardware. Only by blending flexibility in manufacturing hardware, product design, and infrastructive can one assure plant level flexibility and ensure the effective use of investments in flexibility.

Infrastructure is vital to integrating the benefits provided by process technology and product design. Manufacturing infrastructure is the framework made up of people, procedures, information systems, and decision making systems (Skinner, 1978) which enables process flexibility and product design flexibility to prosper.

Figure 8 is a matrix that presents the four flexibility types in the columns and five agents of flexibility in rows. In the figure, flexibility types 1 to 3 are found in non-process industries whereas type 4 flexibility is found in process industries. The cells in Figure 8 (i.e., intersections of columns and rows) contain selected examples of agents appropriate to the attainment of the type of flexibility unique to the column using a method that is unique to the row.

Figure 8 shows that while ***five agents*** are used in attaining all three types of flexibility, their use differs across flexibility types. For example, whereas product design is an agent of flexibility, the contribution of design to flexibility depends upon the type of flexibility sought. While Type 1 flexibility requires a few, stable designs for its success, extreme variability in design, from one order to the other, and an ability to rapidly accommodate customer initiated changes to design are essential to the success of Type 3 flexibility. The reader is encouraged to extend to other cells the above example of how Figure 8 is to be interpreted.

IMPLICATIONS OF MANUFACTURING FLEXIBILITY TYPES: There are several powerful implications of the three types of flexibilities proposed in Figure 8. These implications are discussed under the following headings:

- The strategic goals associated with flexibility types are different.
- Certain aspects of manufacturing flexibility types are mutually conflicting.

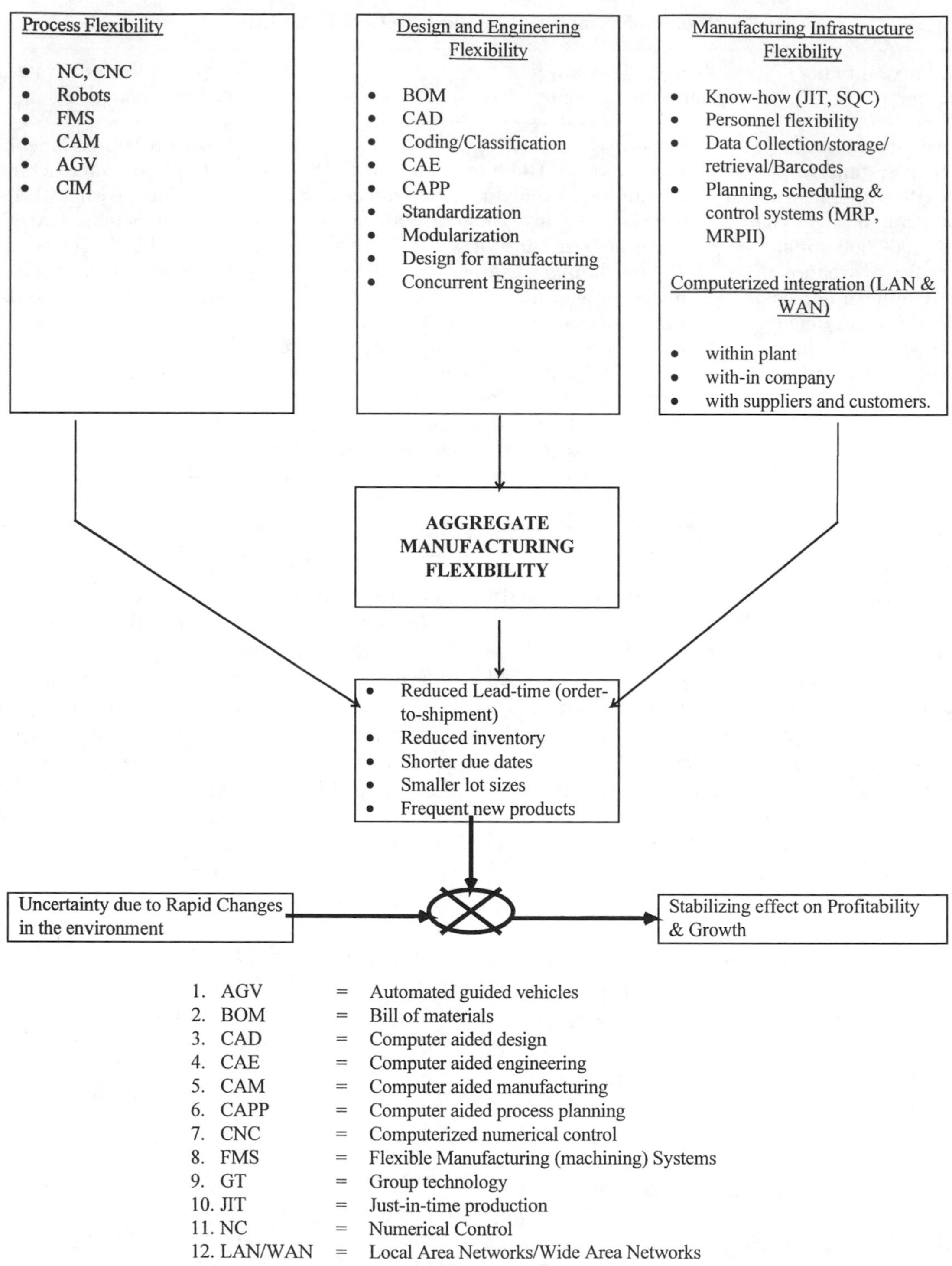

1. AGV = Automated guided vehicles
2. BOM = Bill of materials
3. CAD = Computer aided design
4. CAE = Computer aided engineering
5. CAM = Computer aided manufacturing
6. CAPP = Computer aided process planning
7. CNC = Computerized numerical control
8. FMS = Flexible Manufacturing (machining) Systems
9. GT = Group technology
10. JIT = Just-in-time production
11. NC = Numerical Control
12. LAN/WAN = Local Area Networks/Wide Area Networks

See: Appropriate Entries in the Encyclopedia for a description of most items above.

Fig. 7. Ingredients of aggregate manufacturing flexibility.

- Product design is integral to the attainment of MF in non-process industries.
- Justification procedures for investment in MF should account for the differences in flexibility types.
- An advanced information system is essential to the attainment of manufacturing flexibility.

Strategic Choice and Flexibility Types

In Figure 8, each flexibility type is defined by a unique set of strategic choices or goals. Each set of goals are made of a unique ***mix*** of the following: (a) variety of models produced, (b) stability of design, (c) number of units produced per model,

		Flexibility Context			
		A. Flexibility In Automated Line **(Type 1)**	B. Flexibility In Manufacturing **(Type 2)**	C. Flexibility in Design and Manufacturing **(Type 3)**	D. Process Industry **(Type 4)**
	Strategic Choices	1. Low-cost high-volume 2. New product introduction 3. Multiple models	1. High variety, mid volume 2. Different configurations 3. Different routings	1. Custom design 2. Very low volume 3. Design change frequent	1. Low Cost Change over 2. Range of Process 3. Process Mobility
Flexibility Agents	1. Design	A few stable frozen designs	Variety of moderately stable designs	Extremely variable custom designs	not relevant
	2. Process/ Technology/Flow	• Flexible automation, Robotics, AS/RS • Continuous flow	• FMS/AGV, CAD, CAM • Automated flow, AGV	FMS, NC/CNC. CAD/CAM, CIM Intermittent flow	Fixed at Installation
	3. Infrastructure	• JIT • A few dependable vendors • Flexible employees	• GT, Cells • MRP/JIT • Flexible multi-skilled employees	• GT • Flexible PPC • Alternate schedules and routings	• Flexible employers • Computer Control
	4. Computerized Integration among Design, Process, and Infrastructure	• Enhances planning, scheduling flexibility. • Reduces inventory • Reduces lead-time	• Enhances mix, schedule and routing flexibility • Reduces lead-time	• Enables Concurrent engineering, • Reduces design time • reduces design changes • Enables easier design changes.	• Improves scheduling • Decision making at the plant floor
	5. Computerized Integration with vendors and suppliers	• Smoothes production scheduling • Reduces Inventory • Reduces lead-time	• Reduces inventory • Reduces lead-time	• Improves concurrent engineering • Reduces lead-time	• Reduces inventory • Reduces lead-time • Customer Responsiveness • Better planning and forecasting

Fig. 8. A manufacturing flexibility typology.

(d) capability for custom design and production, (e) routing flexibility, and (f) lot size.

The goal of Type 1 flexibility is: low cost, high volume production of very few models of stable design on a common line with ease of new product introduction. While low cost, high volume production has always been the result of automated lines, modern technologies such as robots provide the additional capability to introduce new models on the line with ease and more rapidly – i.e., greater flexibility. Such lines have mixed model production capability and can handle lot sizes as small as one.

The goal of Type 2 flexibility is: moderate variety, mid-volume production of different configurations in lot sizes as small as one. In Type 2 environments, although product design change capabilities exist, the goal is to minimize disruptions due to design changes by concentrating on the production of relatively stable designs. This type of flexibility may be found most frequently in the manufacture of components or subassemblies requiring several machining operations. There is a moderate level of routing flexibility in Type 2 flexibility.

The goal of Type 3 flexibility is: custom design and manufacturing of an infinite variety of very low volume (often one-of-a-kind) product design and redesign. The design, engineering, and manufacturing systems are properly matched to cope with frequent minor as well as major changes to design. There is a high level of routing flexibility in Type 3 environments.

The goal of Type 4 flexibility is: increased process mobility and process range, and lower changeover cost and time.

Flexibility Types Can Be Mutually Conflicting

As can be inferred from the discussions above, MF is heterogeneous. The heterogeneity of MF results in some mutual incompatibilities among the three types of flexibilities. For example, the methods used for attaining Type 3 flexibility may be at cross purpose to the methods used for attaining Type 1 flexibility. For example, while it is extremely important for the hardware and infrastructure in Type 3 environments to have the capability to frequently deal with minor and major changes in the design and manufacture of components, such changes are very disruptive and incompatible with the low cost manufacturing objectives that invariably go with Type 1 environments – see IBM's example described later in this paper.

The Important Role of Design in Flexibility

The typology proposed here recognizes that, regardless of the type of flexibility, design plays an important role in the attainment of the three types of flexibilities, although as shown in Figure 8, its role varies substantially from one type of flexibility to the other. The message of Figure 8 is clear and simple – product design should be taken into account when developing manufacturing flexibility.

Economic Justification of Investments in Flexibility

The thorny issue of justifying investments in automation technologies has occupied many researchers and practitioners. One reason it confounds those working on the topic is the difficulty associated with the evaluation of flexibility for use in investment analyses. Any progress in evaluating the benefits of MF must be preceded by an improved understanding of the concept of MF.

In three recent instances of actual investments in automated manufacturing by manufacturers, the author found that the firms made their investments with either inadequate, or no justification. Kaplan (1986) calls this practice of justification of investment in manufacturing technology, "...justification by faith." Kaplan also reminds his readers that the practice of justification of investment in manufacturing technologies is difficult because there exist only extremely weak methods for analyzing the value of flexibility to the manufacturer.

The heterogeneous view of MF proposed here should improve investment analyses because it encourages the consideration of different as well as narrow and specific types of flexibilities in justifying investments in manufacturing technology. For example, increasing flexibility in Type 1 environments may mean increasing the number of models that can be produced on an existing line or the ease with which new models can be introduced on an existing line – none of which may be relevant to Type 2 and Type 3 flexibilities. Therefore, analyses of investments in manufacturing flexibility with explicit recognition of the three flexibilities can bring greater focus to the analyses and thus reduce the difficulties associated with their justification.

The Need for Computerized Integration

As shown in Figure 8, all three types of flexibilities require integration of design, technology, and infrastructure. Further, for the manufacturing system to be highly responsive to changing needs, setup time between changes in design, technology, and infrastructure must tend towards zero. This would make massive data and information processing an imperative. Furthermore, given that today's manufacturing technologies are invariably computer controlled and need computerized data-bases and knowledge-based systems, the need for real-time, on-line processing of huge centralized or distributed data-bases becomes inevitable. For example, real-time computer integrated manufacturing today integrates activities all the way from customers, suppliers, production planning to real-time control/monitoring of thousands of operations on the factory floor. The instantaneous flow of manufacturing information to relevant users reduces manufacturing response time to changes and increases flexibility.

ILLUSTRATIVE CASES: Beginning around 1980, Manufacturing flexibility became a new goal aggressively pursued by manufacturers. In the last two decades, while manufacturers have explored and exploited several new techniques for increasing manufacturing flexibilities, the basic principles of attaining manufacturing flexibility have not changed since early 1980's. This section, details the experience of two different manufacturers in the U.S. who took very different, yet successful, approaches to attaining manufacturing flexibility. These classic examples of enhancing manufacturing flexibility provide generalizable lessons that are valid today just as they were more than a decade ago (Swamidass, 1988).

Case 1: IBM's Former Lexington Plant (Now Lexmark)

An example of systematically developing a plant for increased manufacturing flexibility for competitive advantage is the investment at the former facility of IBM's Information Products Division at Lexington, Kentucky now operating under Lexmark (Swamidass, 1988). This factory, which is one of the largest automated factories in the world, manufactured typewriters and printers in a million square feet of floor space.

The most noteworthy aspect of 350 million dollar project was the extent of flexibility that was planned and achieved along with a high degree of automation. The top down planning and execution of this project (lasting almost seven years) exemplified two things. First, it illustrated the strategic value of flexibility in the changed competitive environment, and second, it demonstrated the comprehensive approach needed to achieve meaningful manufacturing flexibility throughout the plant. Figure 9 summarizes chronologically the

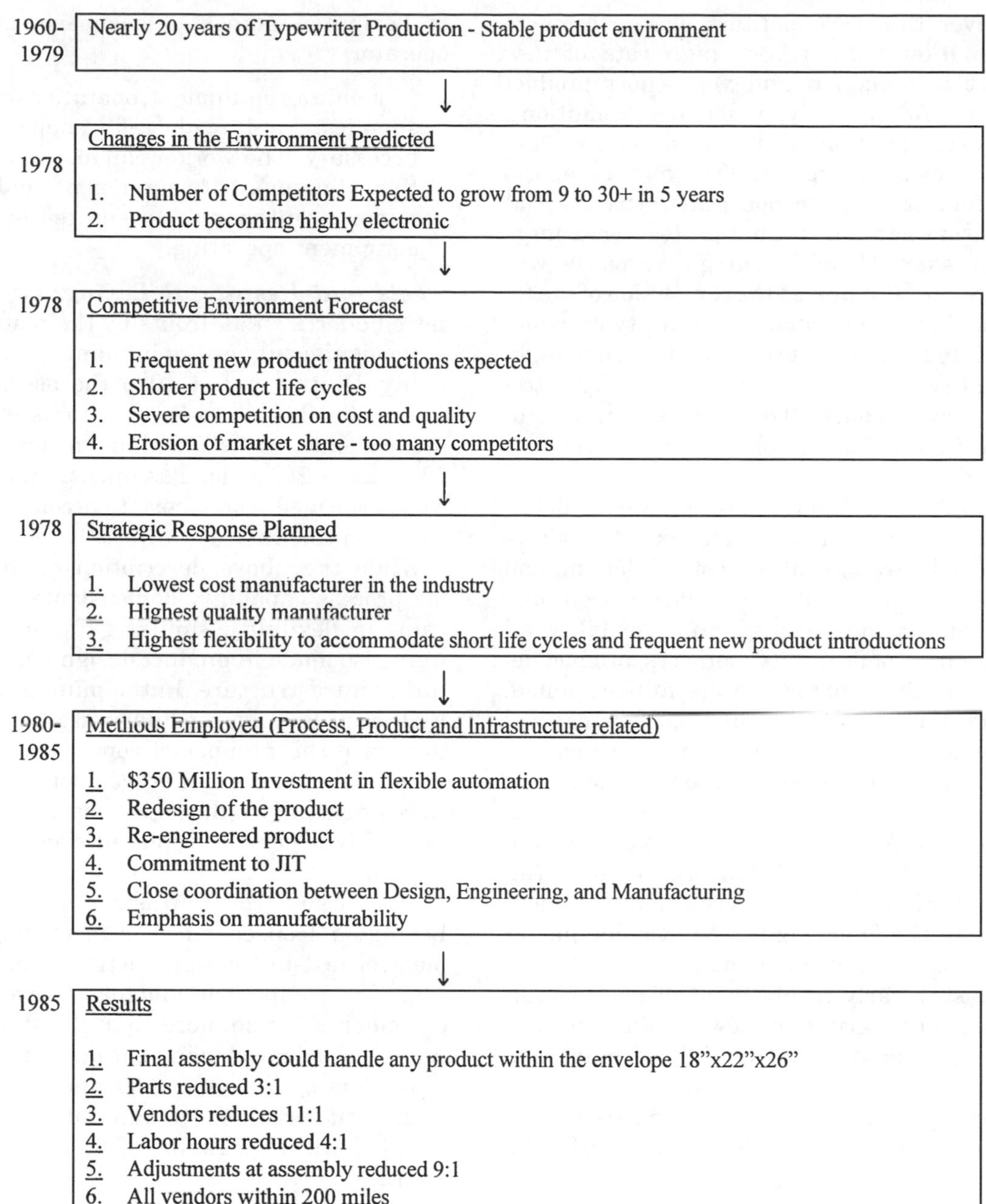

Fig. 9. A chronological study of Lexington plant.

steps taken at Lexington leading to the planning and execution of a manufacturing system very highly biased towards flexibility and automation at the same time.

Changing Competitive Environment: Beginning around 1960, the plant manufactured typewriters for nearly 20 years in a relatively stable environment characterized by very slow changes to the product and very few competitors. In 1978, IBM's planners recognized the signs of major change in the competitive environment. Primarily, market forces were shaping typewriters into electronic products as opposed to the traditional electro-mechanical machines they used to be. The transformation of typewriters into electronic products led IBM to expect several giant manufacturers of electronics products to enter this product market. IBM estimated that the number of competitors in the market would grow from nine in 1978 to over thirty by 1985.

Adoption of New Goals: For the IBM plant, the full impact of the ongoing changes in the market place was not difficult to see. The history of electronic product markets showed that the consequence of making an electronic product with over thirty competitors was to compete in a market with a very high rate of new product introduction and very short product life cycles. Additionally, under these conditions, intense competition on the basis of cost and quality was to be expected. The above scenario called for a strategic response on the part of IBM if the firm was serious about protecting its large market share. IBM's strategic response was very specific. It required the conversion of the facility at Lexington to manufacture typewriters and related products at the lowest cost and highest quality with maximum flexibility for introducing new products. These goals were instrumental in shaping the plant in the years that followed.

Implementation: What followed was a determined commitment to implement the above mentioned strategy at a cost of 350 million dollars over a period of five years beginning in 1980. In attaining maximum flexibility, all three sources of flexibility – process, product design, and infrastructure – were fully exploited. Lexington plant's actions implicitly recognized that major gains in flexibility, quality, and cost reduction could not be attained without coordinated changes in process, product, and infrastructure. According to Reichenback (*Modern Materials Handling*, 1985), Lexington's Automation Manager, this recognition is demonstrated in the following ten-point guideline for automation used at the plant:

1. Ensure early involvement of manufacturing engineering in new product designs to keep products compatible with automation.
2. Set up modular manufacturing areas.
3. Make products for shipment without finished goods storage.
4. Closely schedule and monitor manufacturing operations. Provide just-in-time parts delivery from vendors.
5. Minimize the work-in-process inventory.
6. Maximize automation of inplant manufacturing operations.
7. Require zero defects in manufacturing and vendor parts supply.
8. Minimize engineering changes.
9. Streamline information handling in all areas.
10. Train people to take on new responsibilities as "owner-operators" of equipment areas within production lines.

The use of the owner-operator concept at Lexington is another example of how investment in equipment alone was not sufficient to achieve higher levels of MF. According to Lexington plant's Reichenbach, an owner-operator,

> . . . monitors equipment, operations, and parts supplies, and will refill hoppers when necessary. The worker can also make corrective adjustments to equipment, and perform minor maintenance as required to keep the equipment operating.

Targets and Results at Lexington: The concerted efforts on all fronts by the plant yielded some significant results in manufacturing flexibility. By the end of 1985 the assembly line was so flexible as to be able to assemble any product that can fit within an envelope size 18″ × 22″ × 28″ as long as the assembly process was performed from above to accommodate the robots on the line.

While the above description of the assembly process capability demonstrates enormous gains in flexibility, similar gains in flexibility were also made in product design and manufacturing infrastructure. In the pursuit of flexibility, the product was modularized and redesigned to reduce the number of components by a ratio of about 3:1. Further, component varieties were drastically reduced. For example, a single type of ball bearing replaced over 60 sizes or varieties.

In pursuit of infrastructure flexibility, Lexington focused on ever increasing attainment of just-in-time production. This involved targeting component and parts inventory requirements for no more than 2.4 shifts usage. Additionally, purchasing cut down the number of vendors by a ratio of 11:1 and set targets to restrict the purchase of materials from vendors situated within a radius of 60 miles from the plant.

The plant at Lexington which used to manufacture a single model prior to 1978, by 1985 was making more than seven different models with almost unlimited capability for making many more models. The lead time for introducing a new product dropped from about four years to less than 18 months.

Trade-offs in Flexibility: In recent years, in the pursuit of manufacturing flexibility by trial and error, manufacturers have come to realize that certain flexibilities have to be surrendered to gain other more important flexibilities.

This trade-off between various forms of flexibility is evident in Lexington plant's actions. For example, the new system at Lexington consciously compromised the flexibility of indiscriminate design change and the flexibility to use multiple vendors. Before the new system, minor design changes were implemented by quick and informal agreement between the designer and manufacturing. In the new system, the procedure for even minor design changes involved careful study of the technical and cost implications of the change on engineering, purchasing, manufacturing, and production planning and control.

If the freedom to use many vendors is a form of flexibility, then IBM chose to sacrifice this form of flexibility too. The number of vendors was reduced from over 700 to about 60 in a period of five years from 1980–85.

Thus, some flexibilities were sacrificed to gain greater flexibility in the rate of new product introduction, in the variety of models that can be simultaneously manufactured, and in the ability to deal with shorter product life cycles without sacrificing the ability to be one of the lowest cost manufacturers. At the Lexington plant, the preeminent flexibility among the various competing flexibilities was never in doubt. A prioritization of flexibilities facilitated the resolution of trade-offs between conflicting flexibilities.

Case 2: McDonnell Aircraft Company (MAC)

This case pertains to the Automated Wing Drilling System installed at the firm's St. Louis plant. The system described here went into operation in November 1986 (Swamidass, 1988).

Process Description: In the manufacture of the advanced version Harrier II jet combat aircraft, the manufacture of the wing is a major task. Drilling about 6,500 holes in each wing structure for mounting its skin is an important and substantial part of wing manufacturing.

Since the surface of the wing is highly contoured, unique set ups of the drill in three dimensional space are mandated for each of the 6,500 holes to ensure absolutely perpendicular drilling. Further, different versions of the wing used in the British and the U.S. Marine models of the aircraft add to the variety of drilling set ups. Additionally, modifications to the wing or occasional rework may cause variations in drilling. Finally, the drilling operation also accommodates normal variability that occurs during the fabrication of the wing before it reaches the drilling station.

Corporate Goals: Corporate or divisional goals called for 90 percent improvement in quality as measured by rework and rejects, 40 percent reduction in cost, and 25 percent reduction in cycle time over a five year period starting in 1985.

The Need for Flexible a Automated Drilling System: A major cost component of wing drilling is associated with the three dimensional set up of the drill thousands of times on the highly contoured wing surface. MAC realized that substantial cost reductions could be made by automating the slow and difficult set up operation, and that major reductions in set up time would require flexible technology for wing drilling. Thus, the implication of corporate goals for the wing drilling operation was flexibility in addition to cost and quality.

Project Initiation: The acquisition of a flexible automated drilling systems (ADS) at MAC was the result of a bottom up process in response to the corporate goals described above. The project to acquire ADS was initiated by the Manufacturing Equipment and Process Engineering groups. In contrast, at Lexington the project that resulted in great increases in flexibility at the plant was the result of top down planning. The scope of the project at MAC was much smaller than that at the IBM; one covered an entire plant, while the other covered one major operation.

Benefits of Flexible ADS: The investment of ADS which amounted to 4.5 million dollars required careful justification. The benefits of the system could be categorized into three groups. First, cost related benefits were very obvious. Labor was reduced by 250 hours per wing. The wing drilling operation, which formerly took nine days for completion, reduced to just three days.

Second, quality related benefits include a 50 percent reduction in rework and rejection in hole drilling. Positioning accuracy increased from $+/=1/64''$ (about .016) to a range between $+/-.003$ to $+/-.007''$.

Third, the flexible ADS permits the drilling of any configuration which can fit within an envelope that is 108′ long, 14′ high, and 11′ wide. The benefits of flexibility include the random order in which different wing configurations can be drilled with little effect on set up time. The cost of retooling for new wings yet to be designed will be substantially lower than the system replaced by the new ADS. Further, the lead time required for retooling and start up for an entirely new wing will be much shorter than the older system.

Justification of the Investment: The strictly cost-based justification of the equipment revealed three-year pay back period. The anticipated economic life of the equipment was estimated to be about 10 years due to its ability to accommodate wings (or fuselages) of all configurations that do not exceed the envelope's dimensions.

Table 3. Comparisons between the Lexington and MAC projects.

	Item	Lexington	MAC
1.	Scope of the flexibility project	Entire plant, used technology, design and infrastructure	Major assembly, used only technology
2.	Flexibility Type	Type 1	Type 2
3.	Corporate goals	Lowest cost, highest quality, most flexible plant	90% improvement in quality, 40% less cost in 5 years
4.	Design	New products, CAD Introduced	No change, CAD already in use
5.	Process	Flexible lines	Massive drilling station
6.	Volume (units)	About 1 million/year	About 100/year
7.	Investment	$350 million	$4.5 million
8.	Infrastructure	Total change JIT, etc.	Little change
9.	Effect	Low cost, high quality very flexible, competitive plant	60% cost reduction, 15% quality gains, very flexible process, cycle time 60% less
10.	New product Introduction	Increased frequency, faster	Faster
11.	Project Initiation	Top-down	Bottom-up

Benefits due to reduced rejection and rework were not included in the analysis. Nor was reduction in work-in-process due to the reduction in cycle time included in the justification analysis. The investment was justifiable on labor cost savings alone.

Supporting Infrastructure: Unlike the experience of the Lexington plant described above, MAC did not acquire flexibility in wing drilling as part of a master plan for enhancing plant-wide flexibility. While Lexington altered the entire manufacturing system, infrastructure and all, MAC took a more localized approach in its acquisition of the flexible ADS.

Relatively speaking, the supporting infrastructure at MAC was little disturbed by the introduction of flexible ADS. If anything, the ADS had to meet existing infrastructure demands. The design function was already flexible through the use of a CAD system. Therefore, an important requirement for the ADS was its ability to interact with the existing CAD, and its ability to access design parameters directly from the CAD system and automatically convert them into drilling instructions.

Production planning and control, and inventory management, which are part of the infrastructure, were treated as though there would be minimal disturbance to them due to the acquisition of the flexible ADS. Actually, ADS fits into the existing system with little change to the infrastructure. The most significant interface involved software development to tie together a supervisory computer, CAD system, drilling system, a vision system, and parts program. A total of 12 months was required for the development of this interface at a cost of about six to seven man-years.

IMPLIMENTING MANUFACTURING FLEXIBILITY PROJECTS:

Comparing the Experiences of Lexington and MAC

The two cases described here are both successful yet very dissimilar in nature. The experiences of the two firms are compared in Table 3. There are some notable items in this comparison. First, Lexington needed Type 1 flexibility and MAC needed Type 2 flexibility. Second, Lexington employed several agents to gain flexibility while MAC essentially employed hardware (i.e. process technology) to gain flexibility. The various strategies one could use in implementing flexibility are explored below.

Strategies for Implementation

The examples from Lexington and MAC and the discussions throughout this paper should be of use to those interested in taking on projects intended to increase manufacturing flexibility. In prior discussions, we saw that flexibility can be enhanced through technology, design, and infrastructure dimensions. Flexibility enhancing projects may emphasize one or more of these three dimensions to attain flexibility. If we scale the emphasis placed on these dimensions to gain flexibility as either high, or low, then, as shown in Figure 10, there are eight possible implementation strategies for manufacturing flexibility projects.

Lexington's implementation strategy shown in Figure 10 is an all out strategy, which is high on technology, high on design, and high on infrastructure changes. It is a strategy that calls for large investments, a long implementation time-table,

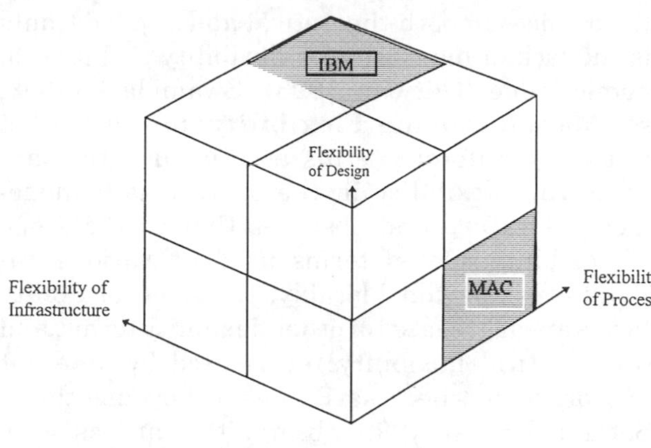

Fig. 10. Strategies for implementing manufacturing flexibility (Comparison between IBM* and MAC** cases).

and maximum disruption during implementation – but it results in the greatest benefits.

The hardware only approach used by MAC shown in Figure 9 is a good example of the implementation of localized flexibility projects with limited investments and equally limited results. Localized projects make sense if they provide logical progression from one project to the other. For example, in the MCA case, the addition of ADS to existing CAD system exploited the flexibility of the drilling system fully while expanding the benefits of the CAD system already in place. While MAC's implementation of ADS was preceded by a CAD system implementation, the two implementations were the result of two separate decisions.

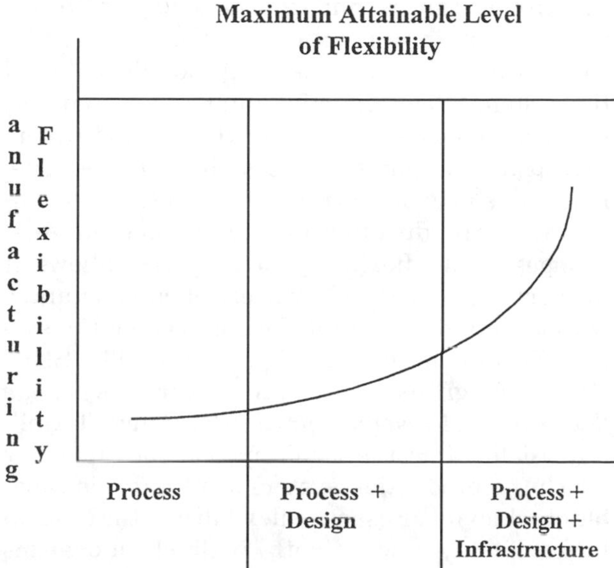

Fig. 11. Importance of process, design and infrastructure flexibilities in achieving manufacturing flexibility.

Figure 11 shows that total MF results from a buildup of the results from the employment of flexible technology, design, and infrastructure. Several conclusions can be made based on Figure 11 including the following: only so much flexibility can be gained in a manufacturing system without infrastructure flexibility. The same can be said about technology and design.

Organizations Mentioned: Boeing; MacDonnell Aircraft Corporation.

See **Capital investment in advanced manufacturing technology; Flexible automation; Flexibility in manufacturing; Just-in-time manufacturing; Lean manufacturing; Manufacturing flexibility dimensions; Manufacturing technology use and benefits; Robot selection; TQM.**

References

[1] Abegglen, J.C. and G. Stalk (1985). *The Japanese Corporation*, New York: Basic Books, Inc.

[2] Gerwin, D. (1993). Manufacturing flexibility: A Strategic Perspective. *Management Science*, Vol. 39, 4, 395–409.

[3] Gerwin, D. and Kolodny, H. (1992). *Management of Advanced Manufacturing Technology*, New York: Wiley-Interscience.

[4] "IBM's Automated Factory – A Giant Step Forward" (1985). *Modern Materials Handling*, March.

[5] Jelinek, M. and J. Goldhar (1983). "The Interface Between Strategy and Manufacturing Technology," *Columbia Journal of World Business*, Spring.

[6] Kaplan, Robert S. (1986). "Must CIM be Justified by Faith Alone?" *Harvard Business Review*, March-Vol. 86, 2, March–April, pp. 87–95.

[7] Skinner, W. (1978). *Manufacturing in Corporate Strategy*, New York: John Wiley and Sons.

[8] Stalk, G., Jr. (1988). Time – The Next Source of Competitive Advantage. *Harvard Business Review*, July–August, 41–51.

[9] Swamidass, Paul M. (1986). Manufacturing Strategy: Its Assessment and Practice, *Journal of Operations Management*, Vol. 6 No. 4, (combined issue) August.

[10] Swamidass, Paul M. (1988). *Manufacturing Flexibility*. Monograph #2, Operations Management Association, Plano, TX.

[11] Swamidass, Paul M. (1996). "Benchmarking Manufacturing Technology use in the United States." In *Handbook of Technology Management* edited by Gus Gaynor. New York: McGraw-Hill.

[12] Swamidass, Paul M. (1998). *Technology on the Factory Floor III: Technology use and Training in the United States*. Washington, D.C.: Manufacturing Institute of the National Association of Manufacturers.

[13] Upton, David M. (1995a). Flexibility as process mobility: The management of plant capabilities for quick response manufacturing. *Journal of Operations Management*, Vol. 12, 3&4, June, 205–224.

[14] Upton, David M. (1995b). What really makes factories flexible? *Harvard Business Review*, July–August, 74–84.

[15] Upton, David M. (1997). Process range in manufacturing: An empirical study of flexibility. *Management Science*, Vol. 4, 8, August, 1079–1092.

[16] U.S. Dept. of Commerce, Washington, D.C. (1987). *A Competitive Assessment of the U.S. Flexible Manufacturing System Industry*, July.

[17] Venketesan, R. (1990). Cummins flexes its Factory. *Harvard Business Review*, Vol 90, 2, March–April, 120–127.

[18] Voss, C.A. (1986). *Managing New Manufacturing Technologies*, Monograph No. 1, Operations Management Association, September.

Note: This piece is adopted from Swamidass (1988).

MANUFACTURING FLEXIBILITY DIMENSIONS

Ranga Ramasesh

Texas Christian University, Fort Worth, Texas, USA

INTRODUCTION: Manufacturing flexibility is generally construed as the ability of the manufacturing system to respond effectively to changes in the environment. Environmental changes include both internal disturbances (equipment breakdown, variable task times, queuing delays, rejects, rework etc.) and external uncertainties (level of product demand, availability of resources, technological innovations, etc.). Flexibility has emerged as a key competitive aspect of today's manufacturing systems in both traditional smoke-stack and completely new industries. The economic future of industrially-advanced nations may very well lie in flexible production systems, which can respond to the competitive threat from developing countries by offering a wide variety of highly customized and technologically superior products.

A clear understanding of different notions of manufacturing flexibility and their measurement is therefore of vital importance to researchers and practitioners in manufacturing management. However, such an understanding is complicated by the lack of a homogeneous view of manufacturing flexibility and the lack of a consensus on terms used to describe it (Swamidass, 1988). It has also been pointed out that the main stumbling block to advances in both theoretical and applied fronts is the lack of measures for flexibility and its economic value (Gerwin, 1993). Swamidass (1988; see **Manufacturing Flexibility**) provides a list of twenty different terms associated with manufacturing flexibility in the operations management literature and observes that (i) the scope of flexibility-related terms used by various authors overlaps considerably; (ii) some flexibility terms are aggregates of other flexibility terms, and (iii) identical flexibility terms used by different authors do not necessarily mean the same thing. Sethi and Sethi (1990) observe that at least fifty different terms are used to describe various types of flexibility, and several terms are used to refer to the same type of flexibility.

The plethora of concepts and notions of manufacturing flexibility are best understood at two levels: first, there are a few aggregate broadly defined types of flexibility; second, there are a large number of specific more narrowly defined attributes or dimensions of flexibility. In this article, the broader concepts or the aggregate notions of manufacturing flexibility are called "*flexibility types*," and the specific attributes or constituent notions of manufacturing flexibility are called "*flexibility dimensions*."

FLEXIBILITY TYPES:

State Flexibility and Action Flexibility

Mandelbaum (1978) characterized manufacturing flexibility into two different forms: *action flexibility* or "the capacity for taking new action to meet new circumstances" and *state flexibility* or "the capacity to continue functioning effectively despite changes in the environment." Action flexibility of the system allows effective interaction from outside and makes it possible for the present system to respond to a change in the environment, e.g., demand for a new product line, by being able to handle the new product line albeit with some marginal changes. State flexibility of a system allows it to operate under a wide variety of circumstances without any need for interaction outside the system. Within the notion of state flexibility, Slack (1987) recognizes two distinct variants: *range flexibility* and *response flexibility*. Range flexibility is defined as the total envelope of capability which the production system or resource is capable of achieving. Response flexibility is the ease (in terms of cost, time, or both) with which changes can be made within the capability envelope.

Buzacott (1984) observes that state flexibility is the primary feature of computer-aided production systems that are classified as "flexible manufacturing systems," and recognizes two distinct types: *Job flexibility* or "the ability of the system to cope with changes in the jobs to be processed by the system," and *Machine flexibility* or "the capacity of the system to cope with disturbances and changes at the machines and work stations." Taking a slightly different approach, Zalenovic (1982) identifies two types of manufacturing flexibility: *design adequacy*, which is related to the concept of job flexibility, and *adaptation flexibility*, which refers to the time required to switch from one type of job to another and thus is related to the notion of action flexibility.

Process Flexibility, Product Flexibility, and Program Flexibility

It is useful (especially in the context of flexible manufacturing systems) to recognize three types of flexibility at the system level: *product flexibility, process flexibility, and program flexibility*. Process flexibility subsumes the notion of machine flexibility and includes flexibility due to different routings, machine redundancy, tool redundancy, improved speed of operations, etc. Program flexibility subsumes the notion of adaptive flexibility and includes the ability of a system to run unattended for extended periods of time. Product flexibility is similar to job flexibility and design adequacy.

Aggregate Flexibility including Process, Product Design, and Infrastructure Flexibility

By recognizing the multidimensional nature of manufacturing flexibility, Swamidass (1988) develops the concept of an aggregate manufacturing flexibility with three ingredients – process flexibility, product design flexibility, and infrastructure flexibility. Sources of process flexibility include advanced manufacturing technology such as numerical control machines, computer aided manufacturing systems, automated material handling, storage, and retrieval systems, etc. Sources of product design flexibility include computer aided design, part commonality in the bill of materials for items, modular design, etc. Infrastructure flexibility is derived from cross-functional training of workers, just-in-time systems, etc. Ramasesh and Jayakumar (1991) provide a value-based approach to the measurement of aggregate manufacturing flexibility.

FLEXIBILITY DIMENSIONS:

Product Mix Flexibility

Product mix flexibility pertains to the responsiveness of the manufacturing system to changes in the market's needs for products over a period of time. It represents the ability of the manufacturing system to handle a range of products or variants efficiently. Mix flexibility comes from versatile machines and work force, cell manufacturing and CIM (computer integrated manufacturing) which allow quick changeovers at low costs. Suggested measures for mix flexibility can be measured by (i) the number of different outputs from the system such as the number of parts produced; (ii) the number of new products introduced per year; (iii) the ratio of the total set up costs incurred per period to the value of total output; and (iv) the incremental value of the new products that can be produced for a specified cost of new fixtures and tools.

Volume Flexibility

Volume flexibility refers to the capability of a manufacturing system to operate efficiently over a wide range of output levels by adjusting production levels upwards or downwards primarily to respond to variations in the demand. Volume flexibility comes from modular production facilities, multiple machines, interchangeable workforce, etc. Volume flexibility is measured by: (i) the ratio of average volume of production to total production capacity; (ii) the stability of the average cost of production over a wide range of production volume as indicated by the slope of the average cost curve; and (iii) the smallest profitable level of production volume.

Operation Flexibility

Operation flexibility pertains to the ability of a manufacturing system to produce a particular item in many different ways by interchanging the order of the different operations necessary to produce it. Operation flexibility derives from the design of the item and its manufacturing process. Operations flexibility is usually measured by the number of different ways in which an item can be produced.

Material Flexibility

Material flexibility pertains to the capability of the manufacturing system to employ and process alternative raw materials or grades of the same raw material in the production of parts that meet the desired functional requirements. Sources of material flexibility include versatile machines and robust part design. Material flexibility is measured by (i) the average number of alternative material that may be used for making a product; and

(ii) the percentage decrease in the throughput because of non-availability of the specified materials.

Labor Flexibility

Labor flexibility pertains to the capability of a manufacturing system to alter size and composition of the work force efficiently across the different production stages/work stations, or over the different production time periods. Labor flexibility to the number of workers, functions performed by the workers, financial implications of changing the labor force, and the grouping of workers in a work station. Labor flexibility is measured by, (i) the average number of different types of tasks performed by the workers; (ii) the ease with which a work group handles breakdowns and changes in job routings in a work station; and (iii) the cost associated with the change in work force level.

Machine Flexibility

Machine flexibility pertains to the different types of operations that a machine can perform without incurring high costs (labor and machine down time) in switching from one operation to another. Sources of machine flexibility include numerical control, rule-based languages and easily accessible programs, sophisticated part and tool changing devices, automatic chip removal, automatic diagnostics and control, etc. Measures of machine flexibility include, (i) the number of different operations that the machine can perform for a fixed level of switching cost; (ii) the cost of switching from one operation to another; and (iii) the number of tools or the number of programs the machine can use.

Material Handling Flexibility

Material handling flexibility pertains to the ability to move different items of production efficiently between different machines or work stations. This flexibility is derived from (i) versatile transportation devices such as conveyors, forklift trucks, monorails, push carts, robots, and automated guided vehicles coupled with a layout providing multiple paths for movement of the materials. Material handling flexibility can be measured by, (i) a rank ordering of the material handling equipment in terms of the range of their application; (ii) a weighted score based on relative importance of speed of movement, etc.; and an estimate of improvement in the output performance of the system that would result from changes to the material handling system relative to the existing system.

Routing Flexibility

Routing flexibility (also referred to as *scheduling flexibility)* pertains to the capability of a manufacturing system to handle breakdowns by using alternative processing routes to make different products. It is distinguished from operation flexibility, which pertains to the capability to use alternate sequencing of operations in making the same part. In routing flexibility, the system may use the same sequence of operations but different machine routes. Routing flexibility is derived from multi-purpose machines, multiple machines, system control software, multi-skilled workforce, etc. Routing flexibility is measured by, (i) the average number of routes available for making a product; (ii) percentage decrease in the throughput because of non-availability of some machines; (iii) cost of lost production due to rescheduling of a job to keep the delivery deadline; and (iv) the ratio of the number of paths available for routing certain parts between machines to the number of all possible paths between machines in the system.

Modification Flexibility

Modification flexibility pertains to the capability of the manufacturing system to respond to engineering change orders requiring modifications to the initial product/process designs and production and delivery plans. Modification flexibility depends on other flexibilities such as operation, material, and routing.

Modification flexibility can be measured by (i) overall cost and time needed to incorporate a certain number of modifications; and (ii) the number of design changes or part or material substitutions that can be handled by the system for specific products.

Expansion Flexibility

Expansion flexibility of a manufacturing system pertains to the ease with which its capacity (i.e., the output per unit time) can be increased. Expansion flexibility is derived from modular production facilities such as cells, special purpose machinery that do not require heavy foundations, etc. Expansion flexibility is measured by, (i) overall cost and time needed to expand the capacity by a certain amount such as doubling the current level of capacity; and (ii) the upper bound on capacity expansion.

STRATEGIC INSIGHTS:

Interactions Between Different Flexibility Types

The different notions of flexibility are not independent of each other. For example, improvements in process flexibility can have impact on product flexibility and with the other forms of flexibility. Several studies have observed that strong interactions exist between the different dimensions of flexibility.

Different Types of Measures and the Relationship between them

Measures of several different flexibility types have been used for flexibility. Among these, three (physical, value, and ratio) types of measures are most commonly used and are easily derived. *Physical measures* such as number of part types handled by the system, average changeover time to switch between two parts, and average number of different routings available have been used to assess specific dimensions of flexibility. *Value measures* such as shadow prices, or incremental net optimal revenues have been commonly used for the broader or aggregate types of flexibility, and these are generally derived from appropriate mathematical programming models of the manufacturing system. *Ratio measures* such as the ratio of part type to part families, the ratio of part families to changeover time, and part numbers scheduled per unit changeover time, have been used to recognize the impact of multiple products and their joint production (Ettlie and Penner-Hahn, 1994).

Choice Among the Flexibility Types

Uncertainties in the internal or external operating environments may be dealt by a flexible manufacturing system endowed with many flexibility dimensions. While guidelines for choosing the optimal mix of flexibility are not available, research into the benefits of manufacturing process flexibility and its application within General Motors Corporation has yielded some principles that are applicable to other manufacturing operations (Jordan and Graves, 1995). These are (a) a small amount of flexibility added *in the right way* can yield virtually all the benefits of nearly total flexibility; (b) the right way to add flexibility is to create fewer, longer plant-product chains; (c) there is not a single flexibility plan that optimizes the benefits of flexibility; and (d) benefits of adding more and more flexibility diminish rapidly.

Flexibility and Other Performance Criteria

Another issue of significant importance, which must be considered is the relationship between flexibility and other manufacturing performance criteria such as productivity and quality, especially because all three are considered essential for success in today's competitive environment (Gerwin, 1993). There is much speculation that a positive association exists between flexibility and productivity in computer-controlled flexible manufacturing systems. However, in a conventional manufacturing system, flexibility is considered to be negatively associated with productivity. Likewise, while flexibility is viewed as an unqualified good for manufacturing systems, counter arguments do exist that increasing flexibility may increase waste. A better understanding of the indirect costs and the overall economics of flexibility is necessary to select the appropriate quantity and types of flexibility in any manufacturing system.

Organizations Mentioned: General Motors.

See **Flexible automation; Lean manufacturing; Linked-cell manufacturing system; Manufacturing Cell Design; Manufacturing Flexibility; Manufacturing with flexible systems.**

References

[1] Browne, J., Dubois, D., Rathmill, K., Sethi, S., and Stecke, K. (1984) "Classification of flexible manufacturing systems." *The FMS Magazine*, April.

[2] Buzacott, J. (1982). "The fundamental principles of flexibility in manufacturing systems." *Proceedings of 1st Intl. Conference on Flexible Manufacturing Systems*, North Holland, UK.

[3] Ettlie, J.E. and J.D. Penner-Hahn (1994). "Flexibility ratios and manufacturing strategy." *Management Science*, 40, 1444–1454.

[4] Gerwin, D. (1993). "Manufacturing flexibility: a strategic perspective." *Management Science*, 39, 395–410.

[5] Jordan, W.C. and S.C. Graves (1995). "Principles on the benefits of manufacturing process flexibility." *Management Science*, 41, 577–594.

[6] Mandelbaum, M. (1978). "Flexibility in decision making: an exploration and unification." *Ph.D. Dissertation*, Department of Industrial Engineering, University of Toronto, Toronto, Canada.

[7] Ramasesh, R.V. and M.D. Jayakumar (1991). "Measurement of manufacturing flexibility: a

value-based approach." *Journal of Operations Management*, 10, 1–8.

[8] Sethi, A.K. and S.P. Sethi (1990). "Flexibility in manufacturing: a survey." *International Journal of Flexible Manufacturing Systems*, 2, 289–328.

[9] Slack, N. (1987). "The flexibility of manufacturing systems." *International Journal of Operations and Production Management*, 7, 35–45.

[10] Swamidass, P.M. (1988). *Manufacturing Flexibility*. Operations Management Association Monograph No. 2, Naman and Schneider Associates Group, Texas. See **Manufacturing Flexibility.**

[11] Zelenovic, D.M. (1982). "Flexibility – a condition for effective production systems." *International Journal of Production Research*, 20, 319–337.

MANUFACTURING LEAD TIME

Manufacturing lead time is defined as the total time required to manufacture an item, exclusive of the purchasing lead time for materials or component parts. For make-to-order products, it is the length of time between the release of an order to production and shipment to the final customer. For make-to-stock products, it is the length of time between the release of an order to the production process and receipt into finished goods inventory. Included here are order preparation time, queue time, setup time, run time, move time, inspection time, and put-away time.

In a production system, the manufacturing lead time for an order that does not receive special priority is roughly proportional to the amount of work-in-process inventory in the system. The degree to which the relationship is truly proportional is dependent upon the relative size of total processing time compared to total queue time. Manufacturing lead times can be shortened by reducing the amount of work-in-process inventory.

See **Cycle time; Just-in-time manufacturing; Lean manufacturing implementation; MRP; Synchronous manufacturing using buffers.**

References

[1] Cox, J.F., J.H. Blackstone, and M.S. Spencer (eds.) (1995). *APICS Dictionary*: Eighth Edition, American Production and Inventory Control Society, Falls Church, Virginia, 47–48.

[2] Srikanth, M.L. and M.M. Umble (1997). *Synchronous Management: Profit-Based Manufacturing For The 21st Century, Volume One*, Spectrum Publishing, Guilford, Connecticut, 13.

MANUFACTURING ORDER QUANTITIES

The amount of a certain product or part to be manufactured in a time period is called manufacturing order quantity. In MRP environments, the MRP logic determines quantities and timing for planned orders (i.e. the quantity to be produced) using inventory on hand, the gross requirements data, and other data in MRP files. In MRP systems, order size or quantity may be as needed, or computed (lot-for-lot) using decision rules and optimizing procedures. Lot-sizing procedures in MRP are primarily based on requirements data. Net requirements can vary substantially from period to period and often have periods with no requirements. In MRP net requirements are stated on a period-by-period basis (time-phased), rather than as a rate (e.g. as an average demand per month or year).

In computing manufacturing order quantity in MRP, the following assumptions are used. First, all requirements for each time period (i.e. week 6) must be available at the beginning of the period. Second, all requirements for future periods must be met with no provision for back ordering. Third, manufacturing orders are released at regular time intervals (e.g. daily or weekly). Fourth, manufacturing lead time is taken into account in planned orders.

See **Economic order quantities (EOQ); Inventory flow analysis; MRP; Part-period balancing (PPB); Periodic order quantities (POQ); Safety stock: Luxury or necessity; Simulation experiments; Wagner-Whitin (WW) algorithm.**

MANUFACTURING PERFORMANCE MEASURES

See **Performance measurement in manufacturing.**

MANUFACTURING PLANNING AND CONTROL (MPC)

A Manufacturing Planning and Control (MPC) system plans and controls the manufacturing process (including materials, machines, people and

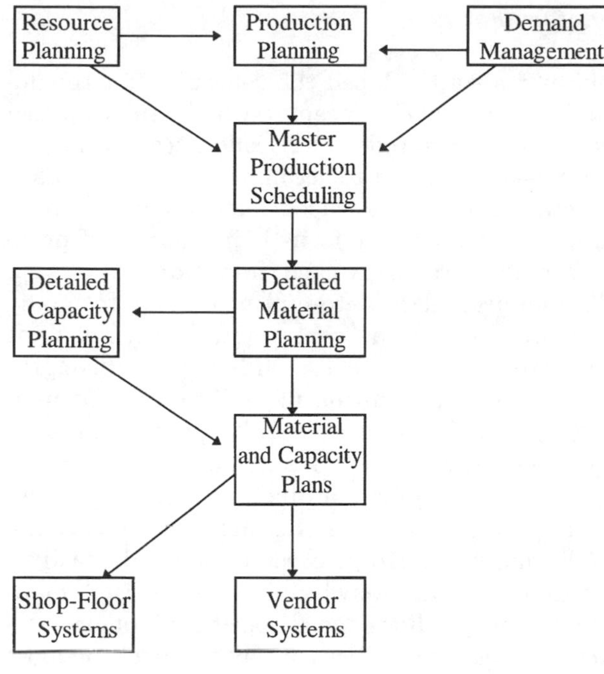

Fig. 1.

suppliers). Basically the MPC system provides information to efficiently manage the flow of materials, effectively utilize people and equipment, coordinate internal activities with those of suppliers, and communicate with customers about market requirements. Figure 1 shows a simplified schematic of a modern MPC system:

See **Capacity planning: Medium- and short-range; Disaggregation in an automated manufacturing environment; MRP.**

Reference

[1] Vollmann, T.E., W.L. Berry, and D.C. Whybark (1996). *Manufacturing Planning and Control Systems*, (Fourth Edition), Richard D. Irwin, Homewood, Illinois.

MANUFACTURING PLANNING SYSTEM

A manufacturing planning system consists of procedures that perform production planning, master production scheduling, and material requirements planning. This system is responsible for processing accepted customer orders and creates the required work orders.

See **MRP; Order release; Resource planning: MRP to MRP II and ERP.**

MANUFACTURING STRATEGY

Paul M. Swamidass and Neil R. Darlow

Auburn University, Auburn, AL, USA

DEFINITION: Strategy is the long-term direction of an organization in which its resources are matched to the economic climate and the expectations of clients and business owners. Manufacturing strategy is defined as "a sequence of decisions that, over time, enables a business unit to achieve a desired manufacturing structure, infrastructure, and set of specific capabilities" (Hayes and Wheelwright, 1984). Moreover, the pattern of decisions actually made constitutes realized manufacturing strategy, which may differ from the intended manufacturing strategy. The terms 'structure', 'infrastructure', and 'capabilities' will be subsequently explained. The primary aim of a manufacturing strategy is to ensure that all long-term manufacturing developments are congruent with the overall business strategy or competitive priorities of the firm.

BACKGROUND: Skinner (1969) introduced the concept of manufacturing strategy and its implementation in the 1960s. The concept derives from corporate strategy literature of the early 1960s. From its ancient military origins, the word 'strategy' came to be defined and conceptualized in the 1960s in business literature, which often treated strategy in terms of long-term planning. The domain of strategy includes the following components:

Mission: Overriding premise in line with the values or expectations of stakeholders.

Goal: General statement of aim or purpose.

Strategies: Broad categories or types of action to achieve objectives.

Tasks: Individual steps to implement strategies.

There are three levels of strategy of interest to a manufacturing company. At each level, the strategy contributes to the strategy immediately above it. The key questions that define strategies in descending order of scope are:

Corporate strategy: What set of businesses should we be in?

Business strategy: How should we compete in a given business?

Manufacturing strategy: How can the manufacturing function contribute to the competitive advantage of the business?

Much of the literature on manufacturing strategy assumes the hierarchy shown above. Hayes and Wheelwright (1984) classify the role played by manufacturing strategy in firms into four categories or stages: internally neutral (Stage 1), externally neutral (Stage 2), internally supportive (Stage 3) and externally supportive (Stage 4). In Stage 1, manufacturing strategy seeks to minimize its negative contribution to the business, whereas in Stage 2, capital investments assist manufacturing to catch up with the industry norm. In Stage 3, manufacturing strategy and business strategy are made consistent by making investments in manufacturing in the light of business goals, and longer-term manufacturing developments are systematically addressed. Finally, in Stage 4, manufacturing strategy results in world class manufacturing (WCM), where manufacturing is a competitive weapon.

THE ORIGIN OF MANUFACTURING STRATEGY THINKING: With the rise of business strategy thinking, Skinner reasoned that if manufacturing assumed a subordinate, reactive role in the corporate strategy, U.S. manufacturers will be ultimately uncompetitive (Skinner, 1969). He found that top management tended to distance themselves from the manufacturing function with the result that production management became an area for specialists with a limited view of the business. This led to unrealized competitive potential of the manufacturing function in such firms, and a lack of manufacturing people with a voice in top management decisions.

Skinner proposed a top-down approach to production management in which the manufacturing function and its managers are linked to business strategy and corporate strategy. He recognized the existence of 'tradeoffs' in manufacturing decisions that are unique to each business. Although tradeoffs remain an area of debate, Hayes and Pisano (1996) point out that competitive emphasis has changed over long periods of time, from low-cost production in the 1960s to the Japanese-led quality emphasis of the 1980s without sacrificing cost. In the nineties, cost, quality and flexibility have become collectively attainable strategic goals. To implement manufacturing strategy, Skinner recommended that manufacturing should be guided by a manufacturing task, which is a set of targets for productivity, service, quality and return on investment. These tasks reduce the plant's manufacturing strategy to a simple and practical set of goals and guidelines for implementation.

Factory Focus

Skinner also developed the reasoning for the 'focused factory' (*FF*) concept (Skinner, 1974), which is widely used today. FF concentrates the resources of the plant to accomplish a limited manufacturing task. A focused factory may be large and complex, but the focus is in the mix of products made, processes used, and markets served. Skinner argued that attempting to serve too many purposes within the same manufacturing plant or system could lead to conflicts and incongruent objectives, resulting in inefficiency and ineffectiveness. There are several aspects of focus to consider (Hayes and Wheelwright, 1984). First, a factory can focus on manufacturing products in either low or high volumes. Second, focus may relate to the manufacturing processes used in the facility. Third, the facility may be customer focused, manufacturing products for a specific client or market. The focused factory concept is one of the most widely used principles in large scale factory reorganization for more than a decade. An extension of the principle can be found in cell manufacturing which is revolutionizing manufacturing everywhere. Manufacturing cells permit 'lean', 'agile' and 'flexible' manufacturing.

WHAT IS MANUFACTURING STRATEGY?: The *content* of manufacturing strategy sets priorities for a plant in five major areas so that the plant's products and processes are competitive, and these priorities guide key manufacturing decisions and tradeoffs. The areas are: (1) *Differentiation*: technological sophistication and product features; (2) *Flexibility* to modify products to suit customers, or to modify delivery volumes; (3) Manufacturing *cost*; (4) *Timeliness*: lead time, delivery reliability; and (5) *Quality*: conformance or perceived quality in regards to product performance, reliability, durability, etc. Hill (1985) recommends further clarifying these priorities as 'order-qualifying' or 'order-winning'. Order qualifiers refer to those attributes of a product that are required in order for entry to a market, whereas order winners differentiate a product by excelling in particular areas or offering features not available from the competition. Hill's approach allows for different industries having different order qualifiers and an order winner at one point in time may become an order qualifier at a later date.

Manufacturing priorities of a plant defined by its manufacturing strategy have a major impact on decisions concerning all 'hard' and 'soft' manufacturing issues. The 'hard' or structural issues affected are: (1) *Capacity*: amount, timing, subcontracting; (2) *Facilities*: size, location, focus; (3) *Process technology*: equipment, automation, configuration; and (4) *Vertical integration*: make-or-buy, supplier policies. The infrastructural, or 'soft' issues affected are: (1) *Organization*: management structure; (2) *Quality policy*: monitoring, intervention, assurance; (3) *Production control*: decision rules, material control; (4) *Human resources*: skills, wage, management style; (5) *New products*: design for manufacture, flexibility; and (6) *Performance measurement and reward*: financial and non-financial systems.

The structural and infrastructural decisions play a dynamic role in shaping the development of specific long-term manufacturing capabilities. The decisions drive continual manufacturing improvement and reflect changes in manufacturing strategy. The 'hard' areas require capital investments in machinery and processes. The 'soft' areas require employee training and retraining.

HOW MANUFACTURING STRATEGY IS DEVELOPED: Manufacturing strategy process is concerned with how manufacturing strategy decisions are made and implemented. In the business strategy literature it is recognized that a company's strategy may not be completely planned, but may comprise various proportions of deliberate planning and emergent components (Mintzberg, 1978). However, planning plays a key role in manufacturing strategy making. A number of approaches to manufacturing strategy formulation have appeared in the literature, for example see Fine and Hax (1985), Miller (1988), Mills, Platts, Neely, Richards, Gregory and Bourne (1996), and Schroeder and Lahr (1990). A typical manufacturing strategy development and implementation process is illustrated in Figure 1.

The chief stages of the process in Figure 1 are:

1. Define corporate objectives: A top-down approach to manufacturing strategy is often taken, where manufacturing strategy is derived from corporate strategy. In this way, top management owns the manufacturing strategy. These corporate objectives may include a proliferation of new products, low costs, or a high level of product customization.
2. Select product families: Products made at the plant may be grouped in terms of their competitive requirements, followed by the formulation of a manufacturing strategy for each group. Manufacturing strategy for each group will be distinctive to suit the firm's business interests.
3. Audit of the external conditions: This audit addresses market requirements and competition in order to determine what is required from the firm's products to meet each of the strategic goals. This stage uses input from the marketing function of the company in order to view the competitive environment. Benchmarking may assist in understanding who the competitors are and what threat they represent.
4. Audit of internal capabilities: The internal audit considers the manufacturing capability of the firm in the context of the existing manufacturing strategy and the priorities outlined earlier, i.e. differentiation, flexibility, cost, timeliness, and quality. This stage assesses the state of the current manufacturing facilities, technology and infrastructure with respect to the intended manufacturing strategy.
5. Analyze gap between actual and desired performance: Gap analysis is a well-recognized concept from the business strategy literature. A comparison is carried out between the internal and external audits in order to assess how the manufacturing capabilities of the firm need to change to meet the competitive criteria, whilst remaining congruent with the corporate objectives and business strategy of the firm. This stage ensures a role for the manufacturing function in the competitive strategy.
6. Prioritize key issues and define objectives of manufacturing: After gap analysis, the manufacturing task is defined. This step uses the results of the gap analysis to translate the shortcomings of the manufacturing system (relative to the competitive strategy) into a set of tangible priorities and objectives.
7. Choose manufacturing strategies: The priorities and objectives from the previous stage are subject to specific action plans. This stage uses creativity techniques, such as brainstorming or mind mapping, to develop a set of action plans for manufacturing strategies that could be deployed. If multiple options are developed, one among them is chosen for implementation. The evaluation of alternatives may be done subjectively or by using computer models.

IMPLEMENTATION OF MANUFACTURING STRATEGY: Many consultants thrive in developing and assisting businesses in implementing a manufacturing strategy. Management processes and tools such as those used by Mills, Platts, Neely, Richards, Gregory and Bourne (1996) can be used as an aid to choosing and building appropriate manufacturing capabilities that will enhance and perhaps influence the competitive strategy of the company. Some plants use a steering group sponsored at a high level for developing and implementing manufacturing strategy. All key functions of the company including the manufacturing department should be involved in

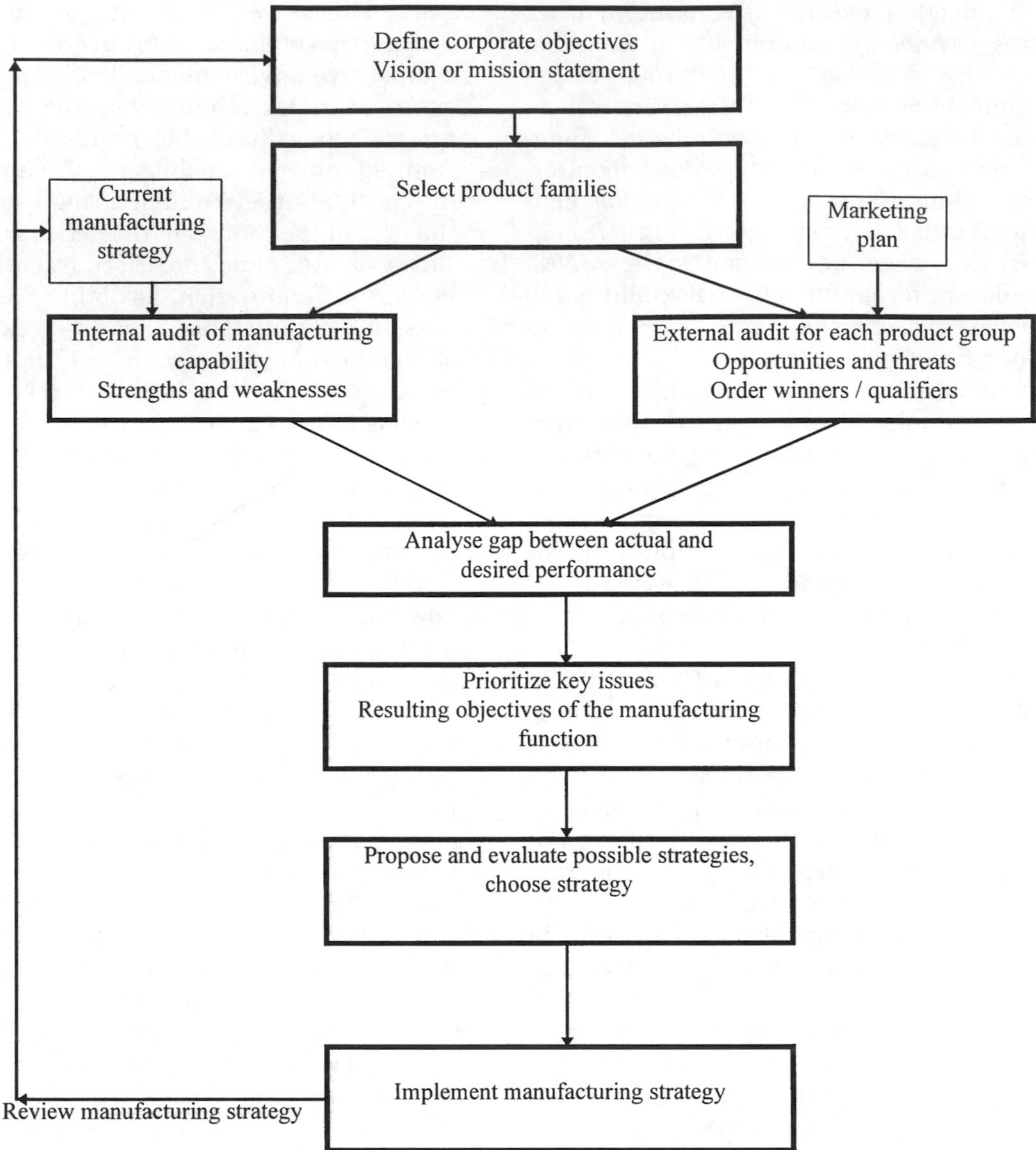

Fig. 1. A process for developing manufacturing strategy.

the process. Marketing, finance, human resources functions, and others should be part of the process.

The following cases illustrate some aspects of the implementation of manufacturing strategy. Case 1 shows how cooperation between top management and cross-functional teams helped in identifying shortcomings in the manufacturing strategy, and then developed and implemented a revised one in several phases. In Case 2, a systematic process for developing manufacturing strategy leads to the development of implementation programs for the various areas of the plant.

Case 1: Specialty Paper Producer (Bennigson, 1996)

This is a case of a company that perceived a need to change its manufacturing strategy due to a change in the competitive situation. For several years, it had occupied a niche in the market thanks to a unique production process developed internally, which enabled the company to manufacture a unique paper product. However, competitors were attracted to its market segment for its high profit margin potential, and the company began to lose its differentiation advantage. It then competed chiefly on price, but this strategy had limited success when other companies could offer better delivery timeliness with similar high quality. A steering group was initiated to investigate the company's manufacturing strategy. Members of manufacturing, sales, marketing, finance, management information systems, and research and development departments were involved in the steering group, and cross-functional teams were formed to carry out investigations.

The teams, together with top management, formulated the manufacturing strategy in the following steps. Firstly, the competitive challenge was clarified. Price was no longer an appropriate competitive criterion. Delivery reliability was the factor that was damaging the company's competitive edge, therefore customer service was identified as a problem area that required improvement. With this focus, machine scheduling was made the key issue. This was approached from the total company perspective, not simply as a manufacturing issue. Management found that many issues were impacted by machine scheduling. These were investigated from an external perspective, paying close attention to competitors.

Secondly, the current manufacturing strategy was identified and questioned. Management began to see clearly how the current manufacturing strategy had evolved and how it needed to be changed in the light of the new competitive challenge. Alternative approaches were examined for feasibility. Many 'sacred cows' were abandoned, such as scheduling to minimize set-ups rather than set-up time. Thirdly, analytical studies of the proposed changes to focused manufacturing facilities were done. These showed that by using 'focused factory' approaches and employing better machine scheduling, cost savings could be made whilst improving customer service. A goal was set to have the capability to improve delivery reliability performance by 300%. To work towards this goal, the cross-functional teams were reorganized to address ten tasks for implementation of the new customer-focused manufacturing strategy. These tasks included both the 'hard' issues, such as machine layout, and 'soft' issues such as information systems. The implementation of the manufacturing strategy took place in four phases, each of six months duration. Over a period of two years, improvement was achieved in the areas of delivery timeliness, inventory costs, and capital requirements, strengthening the competitive position of the company. The change in manufacturing strategy worked: They beat a chief competitor in the areas of low cost and volume flexibility, thus gaining market share.

The case highlights the following factors relevant to manufacturing strategy development: (1) A market perspective on the business rather than an overtly technical view of the manufacturing system; (2) Cross-functional cooperation in the analysis; (3) Understanding by and involvement of all levels in the company; and (4) Good communication in the manufacturing strategy development process. These are contrasted with the pitfalls that can damage a manufacturing strategy review: (1) Internal, technical view of manufacturing strategy without an external, competitive view; (2) Involvement of only the manufacturing function in consultation and decisions; and (3) Lack of communication outside of the steering group.

Case 2: Packard Electric (Fine and Hax, 1985)

This case describes the process of manufacturing strategy development in a division of a supplier of automotive parts. A strategic business unit manufacturing wire and cable, comprising three plants, was the focus of Fine and Hax's study. The division decided to develop a new manufacturing strategy when the chief customer, General Motors, required reduced costs, improved quality and better product development. Using the cost-delivery-quality-flexibility framework and the 'hard' and 'soft' issues described earlier, the first step in developing the manufacturing strategy was to look at the strategy hierarchy. Business strategy follows from corporate goals; the functional strategies are derived from business strategy. The manufacturing requirements to successfully implement the action programs of the business strategy were outlined. The requirements were: flexibility in capacity and processes for future demand, a changeover to new electronics technologies, and the development of a new packaging capability.

The current manufacturing strategy was then examined, broken down into the ten areas outlined earlier, and strengths and weaknesses of each policy were discussed. Product grouping was done using the *product-process life cycle* matrix of Hayes and Wheelwright (1984). The products were grouped according to their position on the product life cycle, from newly-introduced to mature, and additionally according to the market they served. This information was collected by the use of custom-developed forms about the products. A product-process matrix was again used to detect the degree of product or process focus at each of their three plants. A product was removed from a plant and reallocated if another plant was deemed more suitable to manufacture it. Next, given the capabilities of the current manufacturing system, long-term and short-term objectives were developed for the manufacturing function in each of the 'hard' and 'soft' decision areas. These objectives were developed into action programs for each decision category, such as quality management. The action programs were prioritized, with estimated cost and resources requirements and schedules. Responsible persons were identified for each program. Finally, the implementation programs included the development of new performance measurement systems, simulation modeling of capacity planning, and the introduction of a standardized decision process for make-or-buy decisions.

Organizations Mentioned: General Motors; Packard Electric.

See **Competitive priorities for manufacturing; Core manufacturing competencies; Core manufacturing competencies and product differentiation; Focused factory; Product-process dynamics.**

References

[1] Bennigson, L.A. (1996). "Changing manufacturing strategy". *Production and Operations Management*, 5 (1) 91–102.

[2] Fine, C.H. and A.C. Hax (1985). "Manufacturing strategy: a methodology and an illustration". *Interfaces*, 15 (6), 28–46.

[3] Hayes, R.H. and S.C. Wheelwright (1984). *Restoring our competitive edge: competing through manufacturing*, Wiley, New York.

[4] Hayes, R.H. and G.P. Pisano (1996). "Manufacturing strategy: at the intersection of two paradigm shifts". *Production and Operations Management*, (Spring), 25–41.

[5] Hill, T.J. (1985). *Manufacturing strategy*, Macmillan, London.

[6] Kinni, T.B. (1996). *America's Best: Industry Week's guide to world-class manufacturing plants*, Wiley, New York.

[7] Leong, G.K. and P.T. Ward (1995). "The six Ps of manufacturing strategy." *International Journal of Operations and Production Management*, 15 (12) 32–45.

[8] Miller, S.S. (1988). *Competitive manufacturing*, Van Nostrand Reinhold.

[9] Mills, J., K.W. Platts, A.D. Neely, H. Richards, M. Gregory, and M. Bourne (1996). *Creating a winning business formula*, Findlay, London.

[10] Mintzberg, H. (1978). "Patterns in strategy formation." *Management Science*, 24 (9), 934–948.

[11] Schroeder, R.G. and T.N. Lahr (1990). "Development of manufacturing strategy: a proven process." *Proceedings of the Joint Industry University Conference on Manufacturing Strategy*, 8–9 Jan., Ann Arbor, Michigan, 3–14.

[12] Skinner, W. (1969). "Manufacturing – the missing link in corporate strategy." *Harvard Business Review*, 47 (3), 136–145.

[13] Skinner, W. (1974). "The focused factory." *Harvard Business Review*, 52 (3), 113–119.

[14] Spring, M. and R. Boaden (1997). "One more time: how do you win orders?: a critical reappraisal of the Hill manufacturing strategy framework." *International Journal of Operations and Production Management*, 17 (8) 757–779.

[15] Womack, J., D. Jones, and D. Roos (1990). *The machine that changed the world*, Rawson Associates, New York.

MANUFACTURING SYSTEM, FLEXIBLE

Flexible means *"the ability to be bent."* According to Dr. Nigel Seaul, Oxford University, UK, *"flexibility is a useful concept which translates into the production context as the ability to take up different positions or alternatively the ability to adopt a range of state"*. This means that, one design or production system is more flexible than another if it is capable of exhibiting a wider range of states or behaviors; for example, if it can design and/or make a greater variety of products, manufacture at different output levels, achieve a range of alternative quality levels or delivery lead times, and so on.

The range of states a production system can adopt does not totally describe its flexibility. The ease with which a system moves from one state to another, in terms of cost or organizational disruption, is also important. A production system which moves smoothly and at a low cost from *State A* to *State B* should be considered more flexible than a system, which can only achieve the same change at a great cost and/or organization disruption.

In discussing this cost element of a system's flexibility one should be careful to distinguish between the cost of:

- making the change itself, and
- providing the capability to change.

It is the cost of making the change, which is an element of flexibility rather than the cost of providing the capability to make the change. A system which can move quickly from a State A to State B is more flexible than one, which takes longer to make the same change.

The time and cost elements of flexibility are clearly related. Not only is cost often a consequence of the time taken to make a change, the two elements may also be substitutable. The time to change from one state to another may be shortened at extra cost, and conversely, the cost of change may be reduced by extending the time given for the change. Both cost and time can be regarded as the "friction" elements of the flexibility of a system.

See **Manufacturing with flexible systems.**

Reference

[1] Ranky, P.G. (1998). *Flexibility Manufacturing Cells and System in CIM*, CIMware Ltd, http://www.cimwareukandusa.com (Book and CD-ROM combo).

MANUFACTURING SYSTEMS

J.T. Black
Auburn University, Auburn, AL, USA

In a factory, *manufacturing processes* are assembled together to form a *manufacturing system* (MS) to produce a desired set of goods. The manufacturing system takes specific inputs, adds value and transforms the inputs into products for the customer. It is important to distinguish between the production system which includes the manufacturing system and services it.

As elaborated in Figure 1, the production system services the manufacturing system using all the other functional areas of the plant for information, design, analysis, and control. These subsystems are connected to each other to produce either goods or services or both.

PRODUCTION SYSTEM: A production system includes all aspects of business and commerce including manufacturing, sales, advertising, and distribution.

MANUFACTURING SYSTEMS: A manufacturing system refers to a smaller part of the production systems; the collection of operations and processes used to produce a desired product(s) is called a *manufacturing system*. In Figure 2, the manufacturing system is defined as the complex *arrangement of the manufacturing elements* (physical

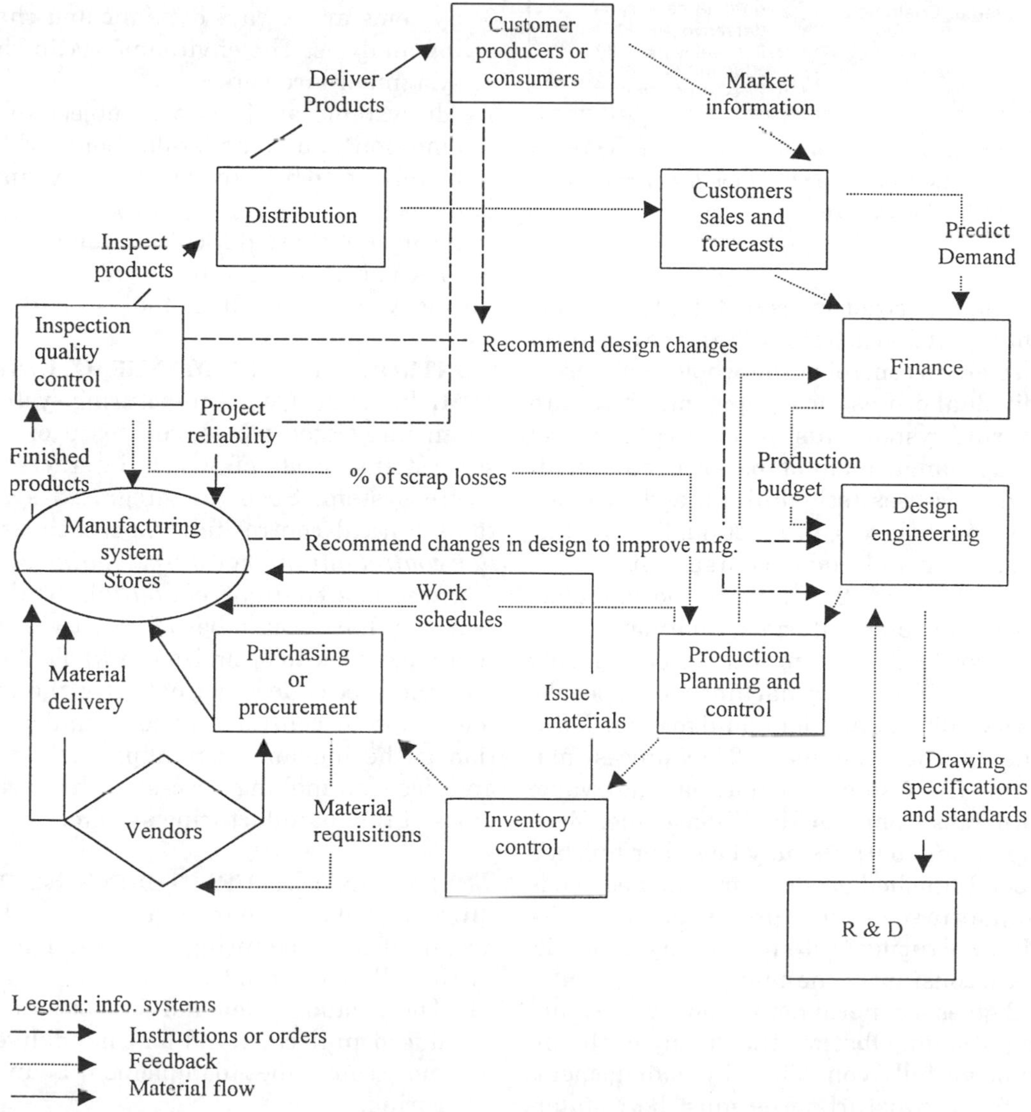

Fig. 1. The production system includes the manufacturing system. The former is composed of many functional departments connected to one another by formal and informal information systems designed to service the manufacturing system that "makes the product".

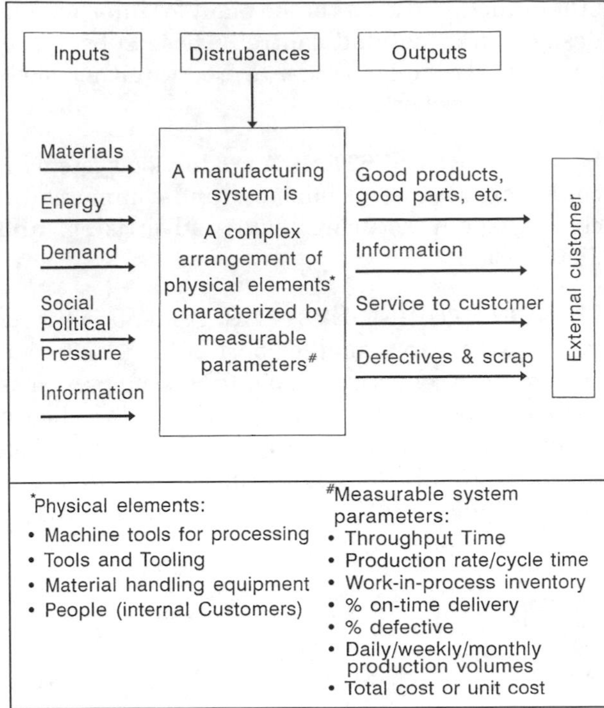

Fig. 2. Definition of a manufacturing system with its inputs and outputs (from *The Design of the Factory with a Future*, by J.T. Black).

elements) characterized and controlled by measurable parameters (Robenstein, 1975). The control of a system refers to control of the whole, not merely of the individual process or equipment. The entire manufacturing system must be subject to control to enable the management of material movement, people and processes (scheduling), and to reach goals in inventory levels, product quality, production rates, throughput and of course, cost.

As shown in Figure 2, inputs to the manufacturing system include materials, information and energy. The system is a complex set of elements that includes machines (or machine tools), people, materials handling equipment, and tooling. Workers are the internal customers. They process materials within the system and materials gain value as they progress from machine to machine. Manufacturing system outputs may be either finished goods or semi-finished goods; semi-finished goods serve as inputs to some other process at another location. Manufacturing systems are very dynamic, i.e., they are constantly changing.

Figure 2 gives a general definition for any manufacturing system. Observe that many of the inputs cannot be fully controlled by management, and the effect of disturbances must be counteracted by manipulating the controllable inputs or the system itself. Controlling material availability and/or predicting demand fluctuations may be difficult. The national economic climate can cause shifts in the business environment which can seriously change any of these inputs. In other words, in manufacturing systems, not all inputs are fully controllable. In order to understand and design manufacturing systems, their modeling and analysis are desirable, but it can be a difficult task because:

1. Manufacturing systems are very complex.
2. System objectives are difficult to define. Goals may have mutual conflict.
3. The data or information may be difficult to secure, inaccurate, conflicting, missing, or even too abundant to digest.
4. Relationships may be awkward to express in analytical terms, and interactions may be nonlinear; thus many analytical tools cannot be applied with accuracy.
5. System size may inhibit analysis.
6. Systems are always dynamic and change during analysis. The environment can change the system, or vice versa.
7. All systems analysis are subject to errors of omission and commission. Some of these will be related to breakdowns or delays in feedback elements.

Because of these difficulties, digital simulation has become an important technique for manufacturing systems modeling, analysis and design.

Control of the Manufacturing System: In general, a manufacturing system should be an integrated whole, composed of integrated subsystems, each of which interacts with the entire system. Such an integrated system contains critical control functions such as, *production control, inventory control, quality control, and machine tool control (reliability).* While the system may have a number of objectives, its operation must seek to optimize the whole. Optimizing bits and pieces does not optimize the entire system. System control function requires information gathering and communication capabilities, and decision-making processes that are integral parts of the manufacturing system.

Trends in Manufacturing System Design: Significant changes are taking place in the design of manufacturing systems. This is fueled by the following trends:

1. The customer demand for better quality, reduced unit cost, and on-time delivery forces more companies to implement Lean Manufacturing.
2. The spread of simplified and Lean Manufacturing systems globally as auto producers take this know-how worldwide.

3. The implementation of manufacturing cells as the first step in Just-In-Time manufacturing and Lean Production.
4. The proliferation of variety in products, resulting in a decrease in quantities (lot size). Variety increases through the use of more flexible manufacturing systems.
5. The increased variety in *materials*, including the growth in the use of composite materials with widely diverse properties, causes a need for the proliferation of new manufacturing processes. Manufacturing systems are designed so that new materials and processes can be readily introduced.
6. The continued decrease in direct labor used in manufacturing results in the cost of materials (including materials handling and energy) being the major part of the total product cost. Meanwhile, direct labor has dropped to 5 to 10% of the total manufacturing cost.
7. The increase in product reliability is in response to the increasing cost of product liability.
8. The decrease in the time between design of a product and its manufacture (called agile manufacturing by some) is reflected in shorter time-to-market for new products.
9. The increase in the concern for the environment forces companies to consider cleaner processes and manufacturing systems.
10. The continued growth in the use of computers for product and system design include the use of simulation. This phenomenon leads to virtual manufacturing.

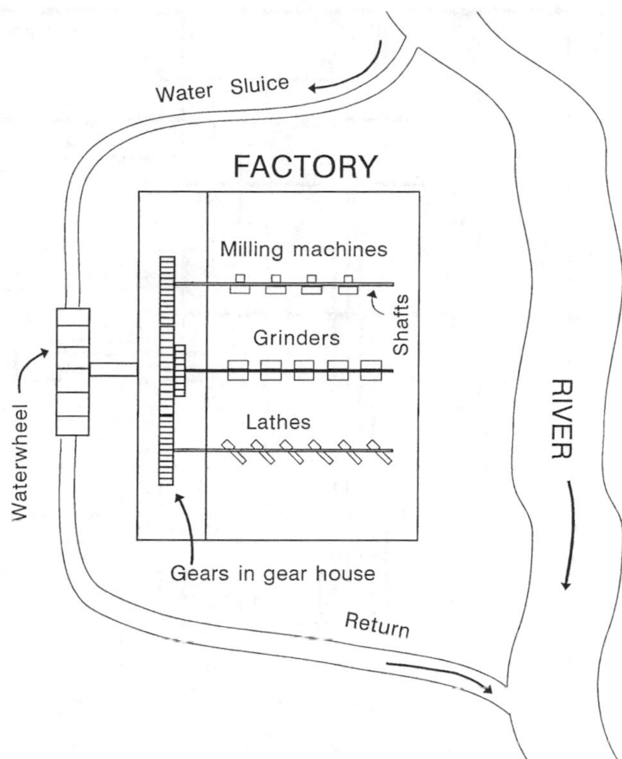

Fig. 3. Early factories used water power to run the machines.

A History of Manufacturing System Design: During the early stages of the industrial revolution, basic machine tools were invented and developed and provided mechanization and automation. Factories needed to focus resources (materials, workers, and processes) at the site where power for processes was available. Early plants were placed beside rivers where water was used to turn waterwheels that turned large gears connected to overhead shafts, which ran the length of the factory (see Figure 3). Belts from the main shaft powered individual machines, so similar machines were lined up under shafts that ran at a particular speed. The grouping of similar machines that needed to run at about the same speed was logical and expedient. Machines were arranged functionally because they were an extension of some human capability or human attribute. Machinists developed different skills depending on the type of machine they operated. The processes were departmentalized according to the kinds of skills needed to operate the processes. Steam engines and, later, electric motors replaced other types of machine power and greatly increased manufacturing system flexibility. Eventually, each machine acquired its own motor but the functional arrangement persisted and became known as the *job shop*. Thus the design of the first manufacturing system was based on functionality. The historians called it the *American Armory System* since so many of the early job shops produced guns.

In the early 1900's, as product complexity increased and the factory grew larger, separate departments evolved for product design, accounting (bookkeeping), and sales. In the scientific management era of Taylor and Gilbreth (the founders of Industrial Engineering), departments for production planning, work scheduling, and methods improvement were added to what became the *production system*, which serviced the manufacturing system. Because the production system was composed of functional areas to support the functionally designed job shop, it naturally evolved with a functional structure.

Job Shop Characteristics: The job shop's distinguishing feature is its functional design. In the job shop, a variety of products are manufactured; this results in small manufacturing lot sizes, often one of a kind. Job shop manufacturing

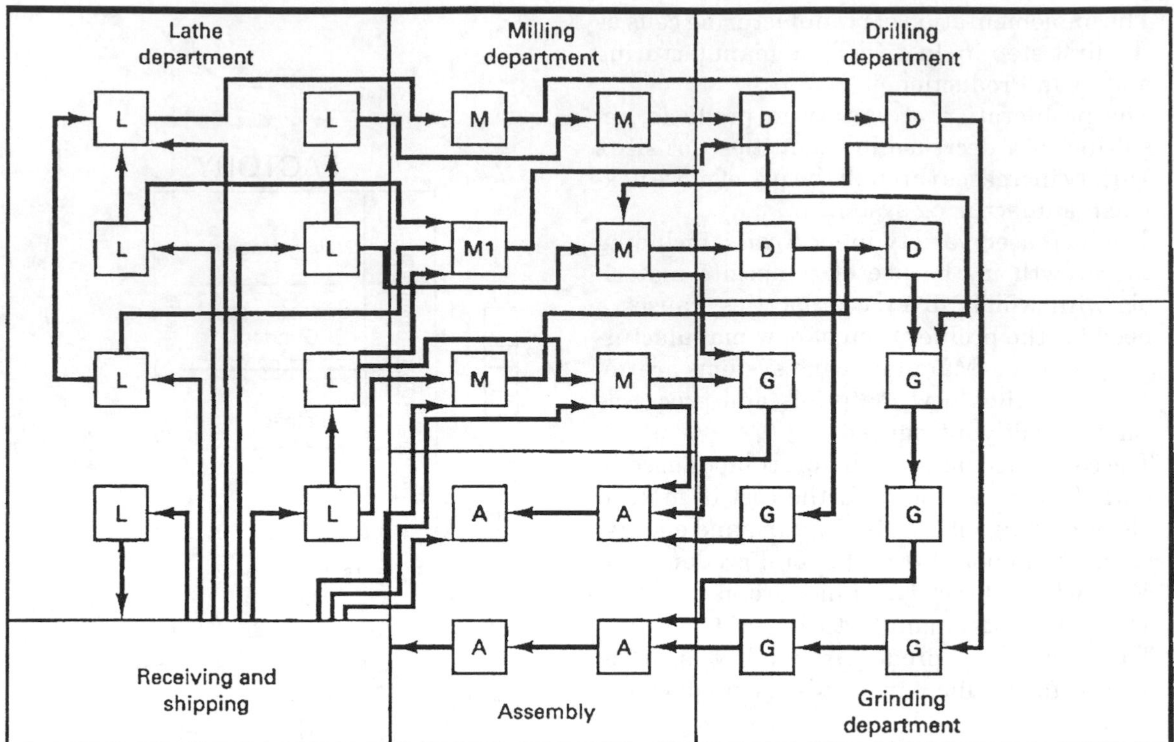

Fig. 4. Schematic layout of a job shop where processes are gathered functionally into areas or departments. Each square block represents a manufacturing process.

is commonly done to specific customer order, but in truth, many job shops produce finished-goods inventories. Because the job shop must perform a wide variety of manufacturing processes, general-purpose production equipment is adequate. Workers must have relatively high skill levels to perform a range of different work assignments. Job shop products include space vehicles, aircraft, machine tools, special tools, and equipment. Figure 4 depicts the functionally arranged job shop. Production machines are grouped according to the general type of manufacturing processes. The lathes are in one department, drill presses in another, plastic molding in still another, and so on. The advantage of this layout is its ability to make a wide variety of products. Each part requiring its own unique sequence of operations can be routed through the respective department in the proper order. *Route sheets* are used as the production control device to control the movement of the material through the manufacturing system. Forklifts and handcarts are used to move materials from one machine to the next. As the company grows, the job shop evolves into a *production job shop*, a combination of a job shop and a flow shop.

The Production Job Shop (PJS) becomes extremely difficult to manage as it grows larger. This results in long product throughput times and very large in-process inventory levels. In the job shop, parts spend 95% of the time waiting in queue, or being transported, and only 5% of the time on the machine. Of the time during which parts are on the machine, only 20 to 30% involves machining (see Figure 5). Thus, the time spent actually adding value to the part may only be 2 or 3% of the total time spent in the shop.

The PJS manufacturing system produces large volumes of products in flow lines but still builds them in lots or batches, usually medium-sized lots of 50 to 200 units. The lots may be produced only once, or they may be produced at regular intervals. The purpose of batch production is often to satisfy continuous customer demand for an item. This system usually operates in the following manner. Because the production rate can exceed customer demand rate, the shop builds an inventory of an item, then makes changes to the machines to produce other products to fill other orders. This involves tearing down the setups on machines and resetting them for new products. When the stock of an item becomes depleted, production is initiated to build inventory, and the cycle repeats.

Industrial equipment, furniture, textbooks, and components for many assembled consumer products (household appliances, lawn mowers, etc.) are made in production job shops. Such systems are called by various names including; machine shops,

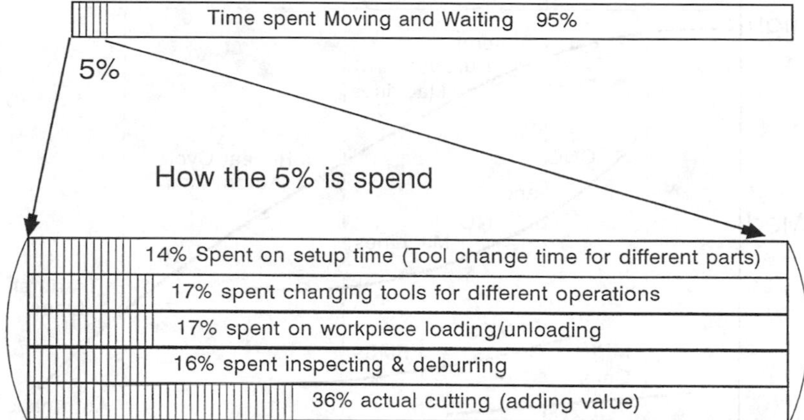

Fig. 5. In the job shop, 95% of the time the parts are waiting for processing. Only about 36% of the time that the part is on the machine are chips being made. (After Charles Carter, Cincinniti Milicron).

foundries, plastic molding factories, and press-working shops. As much as 75% of all piece-part manufacturing is estimated to be in lot sizes of 50 pieces or fewer. Hence, production job shops constitute an important portion of the total manufacturing activity.

FLOW SHOP: In the early 1900's a new manufacturing system design began to appear, now called the flow shop. The flow shop has a product-oriented layout composed mainly of flow lines. This kind of manufacturing system became famous in 1913 when Henry Ford's manufacturing engineers developed the moving assembly line for the assembly of cars. This dramatically increased the rate of car production to about one car/minute and reduced the cost per unit significantly. However, there was no flexibility in the processes, and no variety in the product.

When the volume gets very large in an assembly line it reaches *mass production*. This kind of system can have (very) high production rates. Specialized equipment, dedicated to the manufacture of a particular product, is used. The entire plant may often be designed exclusively to produce the particular product or family of products using special-purpose rather than general-purpose equipment. The investment in specialized machines and specialized tooling is high. Many production skills are transferred from the operator to the machines so that the manual labor skill level in a flow shop tends to be lower than in a production job shop. Items are made to "flow" through a sequence of operations by automated materials-handling devices such as conveyors, moving belts, and transfer devices. The items move through the operations one at a time. The time an items spends at each station or location is fixed and equal (i.e. a balanced line).

Many factories are mixtures of the job shop with flow lines. Obviously, the demand for products can precipitate a shift from batch to high-volume production. Subassembly lines and final assembly lines are further extensions of the flow line but the latter are usually much more labor intensive.

In the flowline manufacturing system, the processing and assembly facilities are arranged in accordance with the product's sequence of operations. Work stations or machines are arranged in line with only one work station of a type, except where duplicates are needed for equalizing; i.e. balancing the time products take at each station. The line can be sequenced to make a single product or a regular mix of products. A hybrid form of the flow line produces batches of products moving through clusters of work stations or processes organized by product flow. In most cases, the setup times to change from one product to another are long, difficult and often complicated.

Various approaches and techniques have been used to develop machine tools that would be highly effective in large-scale manufacturing. Their productivity is closely related to the degree to which the design of the products is standardized and the duration the design remains stable or fixed.

If a part or product is highly standardized, and its demand is very high, then machines that can produce the parts with a minimum of skilled labor are developed for use in a line. These machines can be single machine units or large collections

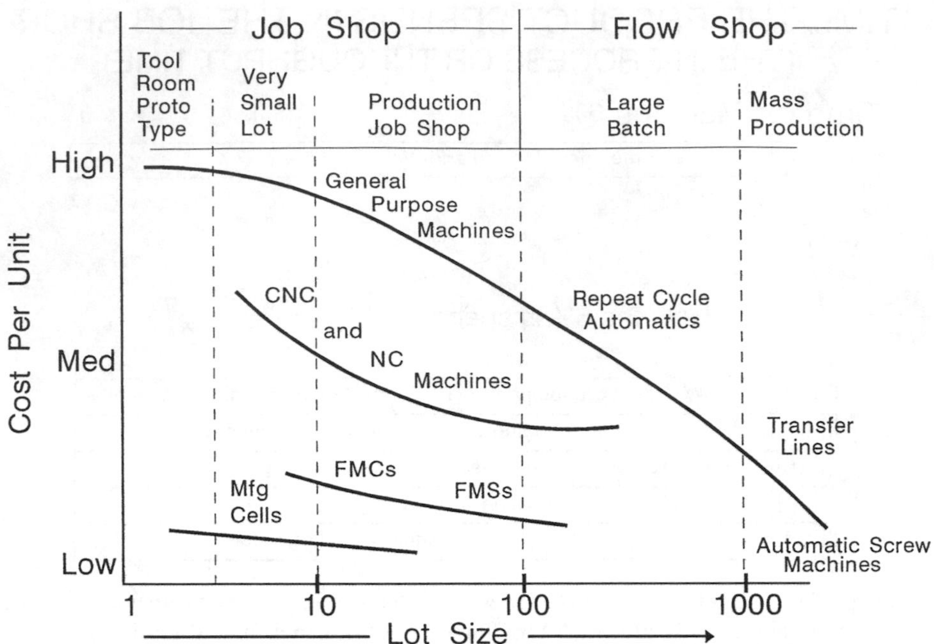

Fig. 6. General relationship between different manufacturing systems.

of machines called transfer lines. These machines are not vary flexible and are often referred to as hard automation systems. In order to make transfer lines more flexible, the machines are constructed from basic building blocks, or modular units that accomplish a function rather than produce a specific part. The production machine modules are combined to produce the desired system of machines for making the product. These units are called self-contained power-head production machines. Such machines are then connected by automatic transfer devices to handle the material movement automatically from one process to the next.

The incorporation of Programmable Logic Controllers (PLCs) and feedback control devices have made these machines more flexible. Modern PLCs have the functional sophistication to perform virtually any control task. These devices are rugged, reliable, easy to program, and economically competitive with alternative control devices, and they have replaced conventional hard-wired relay panels in many applications. Relay panels are hard to reprogram, where as PLCs are very flexible. Relays have the advantage of being well understood by maintenance people and are invulnerable to electronic noise, but construction time is long and tedious. PLCs allow for mathematical algorithms to be included in the closed-loop control system and are being widely used for single-axis, point-to-point control; which is typically required in straight-line machining.

PLC's do not, at this time, challenge computer numerical controls (CNCs) used on multiaxis contouring machines. However, PLCs are used for monitoring temperature, pressure, and voltage on such machines. PLCs are used on transfer lines to handle complex material movement problems, gaging automatic tool setting, on-line tool wear compensation, and automatic inspection. PLCs provide these systems with flexibility that they never had before.

In the 1970's the transfer line was combined with Numerical Control (NC) machines (now, Computer Numeric Control CNC machines) to form what were called Flexible Manufacturing Systems (FMS). The different manufacturing systems can be generally compared to each other in terms of unit cost and lot sizes as shown in Figure 6.

PROJECT SHOP: In the typical project manufacturing system, a product remains in a fixed position or location during manufacturing because of its size and/or weight. Materials, machines, and people used for fabrication are brought to the site. Locomotive manufacturing, commercial jetliners, large machine tools, and shipbuilding use fixed-position *layout*. Fixed-position *fabrication* is also used in construction jobs (buildings, bridges, and dams). Fixed-position fabrication is similar to fixed-position layout, the product is large, and the construction equipment and workers are brought to the site. When the job is completed, the people and equipment move to the next construction site. At the site, a project shop invariably has a job shop/flow shop manufacturing system for producing all the components for the large complex project, and, thus, uses a functionalized production system.

Continuous Processes: In continuous process manufacturing, the product physically flows continuously without a break. Oil refineries, chemical processing plants, and food processing operations are examples of this system. This system is sometimes called continuous flow production when applied to the manufacture of either complex single parts (such as a canning operation or automotive engine blocks) or assembled products (such as television sets). However, these are not continuous processes but simply high-volume dedicated, flow lines. In continuous processes, the products really do flow because they are liquids, gases, or powders. Continuous processes are the most productive but least flexible type of manufacturing systems. They usually have the leanest, simplest production systems, have the least work-in-process (WIP), and are the easiest to control.

The examination of many different industries reveals four kinds of classical manufacturing system designs (or hybrid combinations thereof) as shown in Figure 7. The classical systems are the *job shop*, the *flow shop*, the *project shop*, and the *continuous process*. The *linked-cell* is a new kind of manufacturing system associated with lean production. This design is based on *manufacturing cells* as shown in Figure 8, which is a three-cell reconfiguration of the job shop shown in Figure 4. See **Linked-Cell Manufacturing** for details of this system.

Table 1 summarizes the discussion on manufacturing systems by comparing different systems based on their production rate and product flexibility; that is, the number of different parts the system can handle. The project shop and continuous processes are not shown. The widely publicized Flexible Manufacturing System (FMS) is often operated like a job shop, wherein a random order of part movement is permitted and therefore scheduling of parts and machines within the FMS is necessary, just as it is in the job shop. This feature can make FMS designs difficult to link to the rest of the manufacturing system.

See **Autonomation; BPR; Concurrent engineering; Five S's; Flexible automation; History of manufacturing; Island of automation; Just-in-time manufacturing; Lean manufacturing; Linked-cell manufacturing; Manufacturing flexibility; Platform teams; Production systems.**

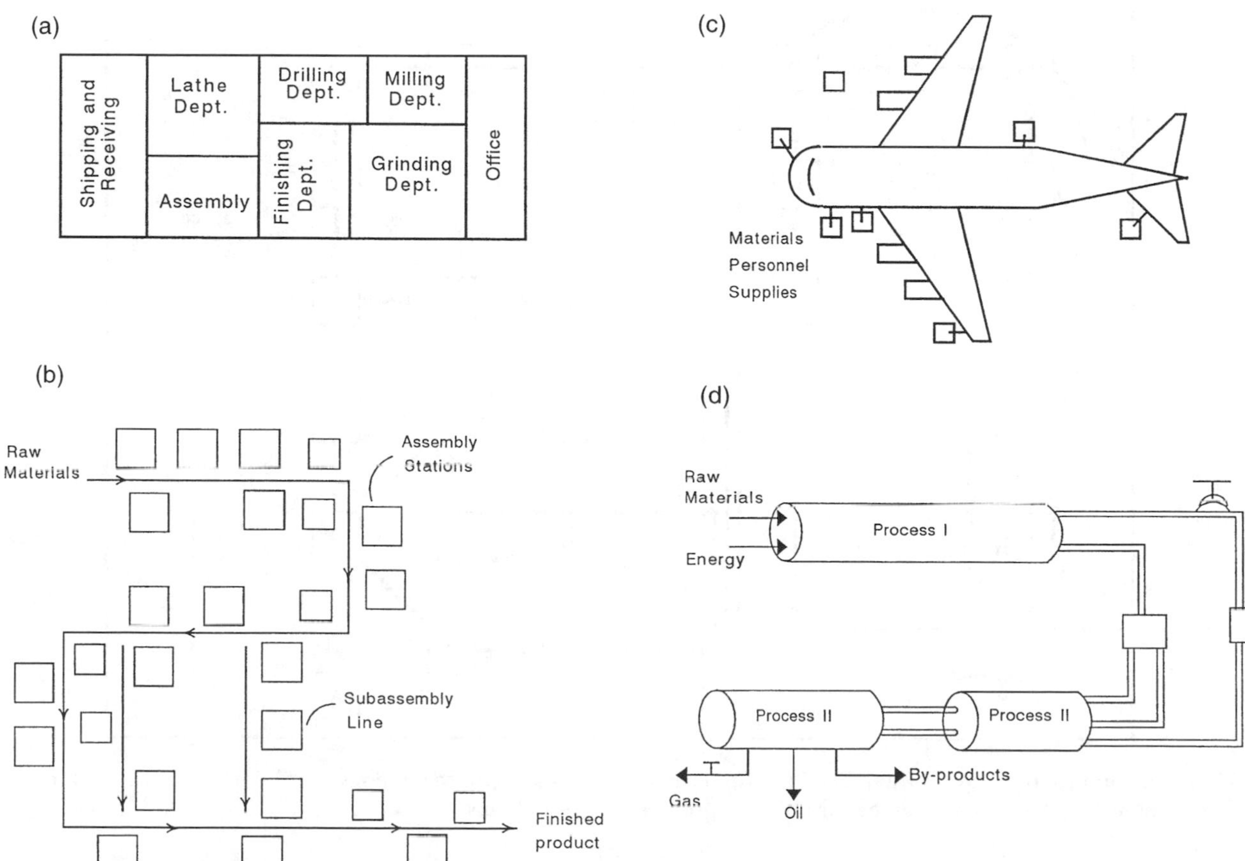

Fig. 7. Schematic layouts of four classical manufacturing systems: (a) job shop, (b) flow shop, (c) project shop, and (d) continuous process.

Table 1. Characteristics of basic manufacturing systems.

Characteristics	Job shop	Flow shop	Project shop	Continuous process	Linked cells
Types of machines	Flexible, general purpose some specialization	Single purpose, single function	General purpose; mobile, manual	Specialized, high technology	Simple, customized single-cycle automation
Design of processes	Functional or process	Product flow layout	Project or fixed-position layout	Product	U-shaped sequenced to flow for family of parts, overlapping
Setup time	Long, variable, frequent	Long and complex	Variable, every job a new setup	Rare and expensive	Short, frequent, one touch
Workers	Single-functioned; highly skilled; one an-one machine	One function, lower-skilled, one man–one machine	Specialized highly skilled	Skill level varies	Multifunctional, respected
Inventories (WIP)	Large inventory to provide for large variety	Large to provide buffer storage	Variable, usually large	Very small	Small
Lot sizes	Small to medium	Large lot	Small lot	Very large	Small
Manufacturing lead time	Long, variable	Short, constant	Long, variable lead time	Very fast, constant	Short, constant

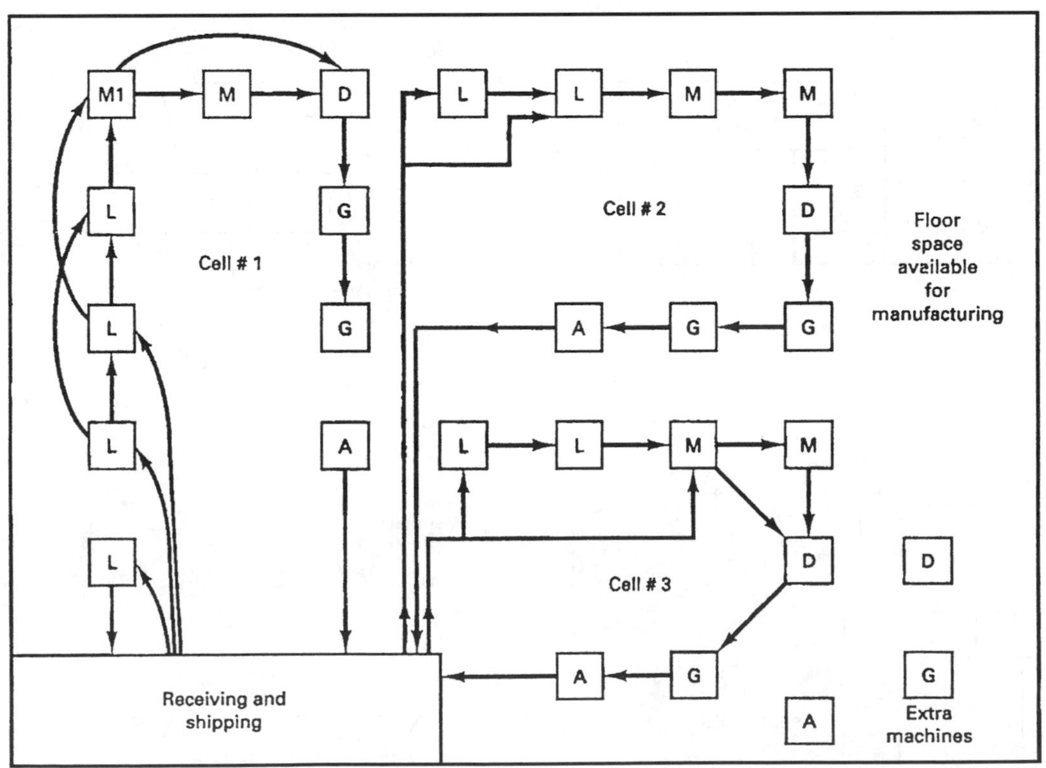

Fig. 8. Schematic layout of a linked-cell manufacturing system. Group technology can be used to restructure the factory floor by grouping processes into cells to process families of parts.

References

[1] Black, J.T. (1991). *The Design of The Factory With A Future*. McGraw Hill.

[2] Hall, Robert W. (1983). *Zero Inventories*. Dow Jones-Irwin.

[3] Harmon, Roy L. and Peterson, Leroy D. (1990). *Reinventing the Factory: Productivity Breakthroughs In Manufacturing Today*. The Free Press.

[4] Ishikawa, Kaouru (1972). *Guide to Quality Control*. Asian Productivity Organization.

[5] Monden, Yasuhiro (1983). *Toyota Production System*. Industrial Engineering and Management Press, IIE.

[6] Nakajima, Seiichi (1988). *TPM, Introduction to TPM: Total Productive Maintenance*. Productivity Press, 1988.

[7] Ohno, Taiichi (1988). *Toyota Production System: Beyond Large-Scale Production*. Productivity Press.

[8] Schonberger, Richard J. (1986). *World Class Manufacturing*. The Free Press.

[9] Sekine, Kenichi (1990). *One-Piece Flow: Cell Design for Transforming the Production Process*. Productivity Press.

[10] Shconberger, Richard J. (1982). *Japanese Manufacturing Techniques: Nine Hidden Lessons in Simplicity*. The Free Press.

[11] Shingo, Shigeo (1985). *A revolution in Manufacturing: The SMED System*. Productivity Press.

[12] Shingo, Shigeo (1986). *Zero Quality Control: Source Inspection and the Poka-Yoke System*. Productivity Press.

[13] Shingo, Shigeo (1988). *Non-Stock Production: The Shingo System for Continuous Improvement*. Productivity Press.

[14] Suzaki, Kioyoski (1987). *The New Manufacturing Challenge*. The Free Press.

[15] Womack, J.P., D.T. Jones, and D. Roos. *The Machine that Changed the World*. New York: Harper Perenial, 1990.

MANUFACTURING SYSTEMS MODELING USING PETRI NETS

Kendra E. Moore

ALPHATECH, Inc., Burlington, MA, USA

Surendra M. Gupta

Northeastern University, Boston, MA, USA

DESCRIPTION: Petri nets (PNs) have recently emerged as a promising approach for modeling manufacturing systems. Systems modeled range from individual manufacturing elements (workstations, machines, machine centers, assembly stations) to subsystems (material handling systems, automated storage/retrieval systems, and control systems) to complete manufacturing systems (assembly lines, transfer lines, flowshops, jobs shops, and automated and flexible manufacturing systems). PNs can be used to develop control logic, test system design approaches, and conduct performance analyses. Recent survey results document the application of PNs in manufacturing (D'Souza and Khator, 1994; Zurawski and Zhou, 1994; Desrochers and Al-Jaar, 1995; Moore and Gupta, 1996). PNs are a graphical and mathematical modeling technique developed in the early 1960s (Petri, 1962) to model concurrent computer system operations. Since then, they have been extended and applied to a wide variety of systems (Murata, 1989). PNs are useful for modeling systems which can be characterized as concurrent, asynchronous, distributed, parallel, nondeterministic, and/or stochastic, and in which operations share multiple resources.

PNs provide a number of advantages for modeling manufacturing systems. These include the following:

- PNs provide a natural way to model conflicts, blocking, deadlock, buffers, synchronization, sequencing, sharing of common resources, and priorities. These properties may be difficult to model using queueing networks.
- Once a PN model is developed, exploring alternative levels of resources or job requirements requires only minor modifications to the PN. PN structural complexity increases only slightly with respect to system complexity. This is in contrast to Markov chains, whose structure increases exponentially with system complexity, and where the addition of resources or job requirements means redefining all of the states. As a result, simple PN models can be used to represent complex manufacturing systems.
- PNs can be used to perform behavioral and temporal analysis. Behavioral analysis can be used to determine liveness, blocking, deadlocks, mutual exclusion, and buffer overflow. Temporal analysis can be used to determine a variety of performance measures for a system including throughput rate, resource utilization, work-in-process, and makespan.
- PNs provide a straightforward way to visualize complex systems and to communicate system design. The PN's graphical form can be related to real states in a discrete manufacturing system, and hence, to key design and control issues.
- PN models can be built in a modular, hierarchical fashion. It is possible to build valid subsystem models and link the subsystems together in a way that maintains their behavioral properties of the larger system.

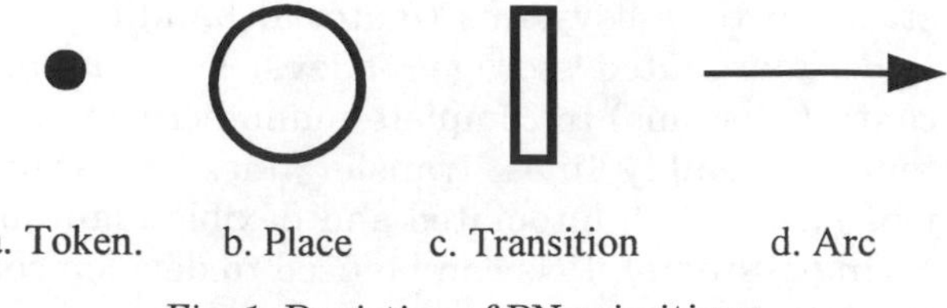

Fig. 1. Depiction of PN primitives.

- PNs can be automatically converted into simulation or control software code. Hence, it is possible to use the same model for systems design and analysis, simulation, planning, and control.

DEFINITIONS: PNs consist of four primitive elements (tokens, places, transitions, and arcs) and the rules that govern their operation (Figure 1). PNs are based on a vision of tokens moving around an abstract network. Tokens are conceptual entities; they appear as small solid dots and model the objects that move in a real network. Places are shown as circles and represent the locations where objects await processing or the conditions (states) that objects are in. Transitions appear as bars or rectangles and represent processes, events, or activities. Arcs represent the paths of objects through the system. Arcs connect places to transitions and transitions to places; the direction of the path is indicated by an arrowhead at the end of the arc. Each transition has some number of input and output places. From a modeling perspective, these input and output places can represent the pre-conditions and post-conditions of an event (i.e., transition), or the resources required and released by an event.

PETRI NET ANALYSIS TECHNIQUES: Two primary techniques are used to determine the behavior of PNs: the reachability tree and the matrix-equations method. The reachability tree is an enumeration of all reachable markings in a PN; it can be used to determine if the PN is bounded, safe, live, and reversible, if M_n is reachable from M_0, and if M_n coverable. Unfortunately, this method is exhaustive and cannot be applied to large nets. The matrix method takes advantage of the linear algebraic structure of the PN graph and manipulates the incidence matrix to determine reachable markings. The matrix method can be used to generate place and transition invariants (*p*- and *t*-invariants), which describe the behavior of the net. Though not exhaustive, the matrix method is restricted to pure PNs (PNs with no self-loops) and loses some information in the problem formulation. A third technique, known as reduction, can be used to transform larger, more complex nets into smaller, simpler nets, while maintaining the behavioral properties of boundedness, liveness, and safeness (Lee-Kwang and Favrel, 1985; Lee-Kwang *et al.*, 1987; Murata, 1989). When analytic techniques cannot be applied, due to model size or assumptions, simulation and other methods, including occurrence graphs, may be employed.

PETRI NET EXTENSIONS: Performance analysis requires adding temporal elements to PNs. *Timed PNs* (TPNs) are PNs with deterministic transition times (Ramchandani, 1974; Ramamoorthy and Ho, 1980). TPNs can be analyzed to determine the maximum performance of the system by finding the elementary circuits in the PN. *Stochastic PNs* (SPNs) are PNs with random transition times (Florin and Natkin, 1982; Molloy, 1982). The reachability graph of a bounded SPN is isomorphic to a finite Markov chain. The Markov chain can be analyzed to determine system behavior.

While classic PNs offer enormous modeling power, managing large-scale PN models is difficult due to the fact that tokens are indistinguishable from one another. The only way to distinguish between tokens representing different entities is to construct different subnets for each entity and connect the subnets to each other via the resources that they share. For example, consider the PN model shown in Figure 2a for three different jobs where one job is processed by resources A and B, one by resources A, B, and C, and the last by B and C (Moore and Gupta, 1996).

A class of PNs known as colored PNs (CPNs) provide a method for distinguishing between tokens in places by allowing the tokens to carry attributes or data structures; these data structures are referred to as "color sets" (Jensen and Rozenberg, 1991). Additional logic (arc expressions and guards) controls token flow. With CPNs it is possible to develop a single net to represent a complex system (see Figure 2b) (Moore and Gupta, 1996).

HISTORICAL PERSPECTIVE: PNs were originally developed to characterize concurrent operations in computer systems (Petri, 1962). Over the years they have been extended to include a wide variety of discrete-event systems, including manufacturing systems. Early work in applying PNs to manufacturing include (Dubois and Stecke, 1983; Alla *et al.*, 1984; Archetti and Sciomachen, 1989; Hillion and Proth, 1989). During the 1990s there has been an explosion of interest in applying PNs to manufacturing systems. Survey papers by (D'Souza and Khator, 1994; Zurawski and Zhou, 1994; Desrochers and Al-Jaar, 1995; Moore and Gupta, 1996) document the application of PNs to flexible and automated manufacturing, kanban

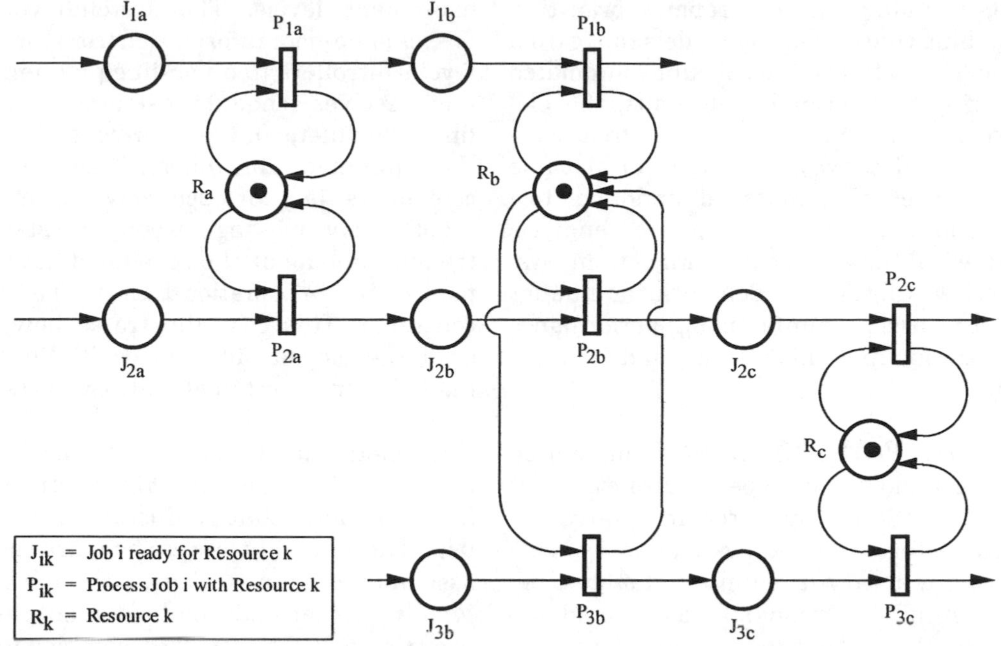

a. Classic PN Model of Three Job Types with Different Processing Requirements.

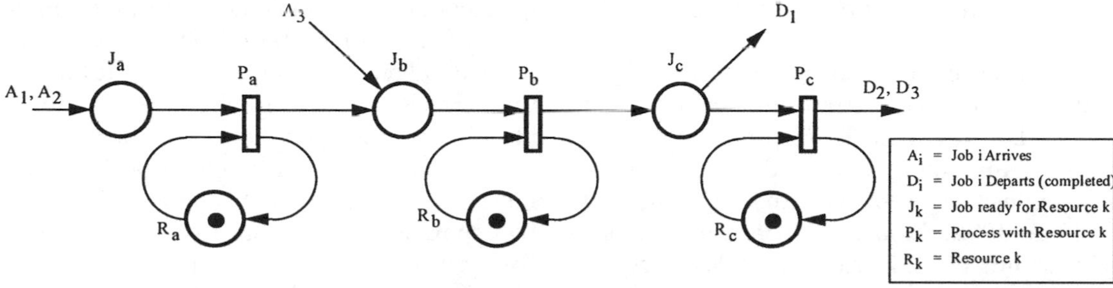

b. HLPN Model of Three Job Types with Different Processing Requirements.

Fig. 2. Comparison of classic PNs vs. HLPNs in terms of net size.

systems, assembly/disassembly, and manufacturing control.

In addition to the advantages of PNs (described above), two developments have led to this increased interest in applying PNs to manufacturing systems. First, techniques have been developed to support the synthesis of large-scale models from smaller modules which have desirable behavioral properties (e.g., liveness, boundedness, reversibility, etc.). The synthesis rules preserve these properties in the larger PN. Second, a number of software tools have been developed which support PN model development, analysis, and simulation (see below).

IMPLEMENTATION: All of the closed form methods applied to PNs require exponential amount of time and space to solve. Decomposition methods have helped to reduce the computational burden; however, these approaches are still limited by special requirements of the structure of the model and the parameters. Where closed form and decomposition methods are not feasible, occurrence graphs and simulation are the methods of choice. The former tends to be exhaustive, while the latter requires multiple runs to attain statistically valid results.

A number of PN and PN-based modeling tools have been developed to support PN modeling and analysis. Tools for modeling SPNs include DSPNexpress (Lindemann, 1998) and GreatSPN (Chiola *et al.*, 1995). PN-based simulation tools include ALPHA/Sim (ALPHATECH, 1995), Artifex (ARTIS, 1989), and DesignCPN (META, 1993). The World of Petri Nets web site maintains a comprehensive list of PN tools (www.daimi.aau.dk/ PetriNets/).

APPLICATIONS: PNs have been used to model a wide range of manufacturing systems and system elements including workstations, material handling, jobs, storage, other resources, and other constraints including failures, repairs, priorities, and dispatching rules. Existing models range from simple flow shops to complex flexible/automated manufacturing and assembly systems; the elements that can be represented range from classic machines and conveyor systems to machine centers, machine cells, material handling, tool handling, and storage. The types of analysis performed with these models range from system correctness (absence of deadlocks and buffer overflows) to classic manufacturing performance measures (throughput, makespan, and resource utilization).

When to Use Petri Nets: PNs can be used for modeling almost any type of discrete-event manufacturing system. They are particularly useful when modeling complex systems or systems which can have a varying number of states (depending on initial conditions). One of the advantages to using PNs is that the same model can be used for systems design and analysis, simulation, planning, and control. Hence, PNs are a good candidate for use when designing a system from the ground up.

Current research is focused on improving or developing qualitative techniques for large-scale PN models. This includes developing improved algorithms for analyzing PNs and developing improved composition and decomposition methods (Lin and Lee, 1992; Zhou and DiCesare, 1993; Cao and Sanderson, 1994; Zurawski, 1994). A second area of research is in the development of generic models of individual manufacturing elements that can be rapidly assembled to support the design and reconfiguration of real manufacturing systems. The goal of this research is to develop models that can be used to model a wide variety of systems by simply changing their parameters, rather than their structure (Zurawski, 1994).

Examples: Moore and Gupta (1999) use CPNs to model a Just-In-Time (JIT) production facility. The model is shown to be live and bounded, and incorporates an algorithm to dynamically adjust the number of kanbans in the system. Simulation experiments demonstrate that the dynamic control policy results in lowered WIP, while improving system performance (decreased shortages). This case illustrates how PNs can be used to conduct performance analysis of a kanban system and test various kanban control strategies to select the proper numbers of kanbans to have in the system.

Cho (1998) uses PNs to design logical controllers for a flexible manufacturing system. PN-based controllers operate at the system and equipment levels. The system-level controller passes shop-floor information from the equipment-level controllers to an on-line planner and scheduler, receives schedule instructions from the on-line scheduler, and passes control messages to the equipment controllers. The equipment-level controllers (one for each conveyor, machine, and robot) receive messages from the system-level controller, implement the scheduled actions, and report status information back to the system-level controller. This case illustrates how controllers can be developed and verified with PNs; the resulting PNs can be automatically converted to control code.

Ezpeleta and Martínez (1992) use PNs to model the assembly operations at an electric oven manufacturing plant that produces four different products. The assembly process has three steps: 1) pre-assembly at four stations, one for each product type where each station has an output buffer capacity of two; 2) select pre-assembled items to load in the assembly workshop; and 3) the assembly process. The assembly system consists of 30 workstations, a loading station, an unloading station, and a transport system. Each workstation has a capacity of one, with input and output buffers having capacity of two. Each assembly has a work plan listing the operations to be performed, while each workstation is capable of performing one or more operations. The PN model focuses on the transport control system to route the assemblies through the system in accordance with their work plan, and the workstation capabilities and capacity constraints. This case illustrates how PNs can be used to model a large-scale manufacturing system and, specifically, the transport control system.

See **Blocking; Capital investment in advanced manufacturing technology; Deadlock; Discrete-event systems; Dynamic kanban control; Dynamic kanban control for JIT manufacturing; JIT; Liveness; Makespan; Manufacturing analysis using chaos, Fractals and self organization; Manufacturing performance measures; Manufacturing systems; Simulation analysis of manufacturing and logistics systems; Simulation of production problems using spreadsheet programs; System correctness.**

References

[1] Alla, H., P. Ladet, J. Martínez, and M. Silva (1984). "Modeling and validation of complex systems by colored Petri nets application to a flexible manufacturing system." G. Rozenberg (ed.), *Advances*

in Petri Nets 1984, 15–31, Springer-Verlag, New York.

[2] ALPHATECH (1995). *ALPHA/Sim User's Guide, version 1.0.* ALPHATECH, Inc., Burlington, MA.

[3] Archetti, F. and A. Sciomachen (1989). "Development, analysis and simulation of Petri net models: an application to AGV systems." F. Archetti, M. Lucertini, and P. Serafini (eds.), *Operations Research Models in Flexible Manufacturing Systems*, 91–113, Springer-Verlag, New York.

[4] ARTIS (1989). *Artifex Tutorial Version 4.1.* ARTIS, S.r.l., Torino, Italy.

[5] Cao, T., and A.C. Sanderson (1994). "Task decomposition and analysis of robotic assembly task plans using Petri nets." *IEEE Transactions on Industrial Electronics*, 41, 620–630.

[6] Chiola, G., G. Franceschinis, R. Gaeta, and M. Ribaudo (1995). "GreatSPN 1.7: graphical editor and analyzer for timed and stochastic Petri nets." *Performance Evaluation*, 24, 47–68.

[7] Cho, H. (1998). "Petri net models for message manipulation and event monitoring in an FMS cell." *International Journal of Production Research*, 36, 231–250.

[8] D'Souza, K.A., and S.K. Khator (1994). "A survey of Petri net applications in modeling controls for manufacturing systems." *Computers in Industry*, 24, 5–16.

[9] Desrochers, A.A., and R.Y. Al-Jaar (1995). *Applications of Petri Nets in Manufacturing Systems: Modeling, Control, and Performance Analysis.* IEEE Press, Piscataway, NJ.

[10] Dubois, D., and K.E. Stecke (1983). "Using Petri nets to represent production processes." *Proceedings of 22nd IEEE Conf. Decision and Control*, IEEE Press.

[11] Ezpeleta, J., and J. Martińez (1992). "Petri nets as a specification language for manufacturing systems." J.-C. Gentina, and S. G. Tzafestas (eds.), *Robotics and Flexible Manufacturing Systems*, 427–436, North-Holland, New York.

[12] Florin, G., and S. Natkin (1982). "Evaluation based upon stochastic Petri nets of the maximum throughput of a full duplex protocol." C. Girault, and W. Reisig (eds.), *Application and Theory of Petri Nets*, 280–288, Springer-Verlag, New York.

[13] Hillion, H.P., and J.M. Proth (1989). "Performance evaluation of job-shop systems using timed event-graphs." *IEEE Transactions on Automatic Control*, 34.

[14] Jensen, K., and G. Rozenberg (eds.) (1991). *High-Level Petri Nets: Theory and Application.* Springer-Verlag, New York.

[15] Lee-Kwang, H., and J. Favrel (1985). "Hierarchical reduction method for analysis and decomposition of Petri nets." *IEEE Transactions on Systems, Man, and Cybernetics*, SMC-15, 272–280.

[16] Lee-Kwang, H., J. Favrel, and P. Baptiste (1987). "Generalized Petri net reduction method." *IEEE Transactions on Systems, Man and Cybernetics*, SMC-17, 297–303.

[17] Lin, J.T., and C.-C. Lee (1992). "A modular approach for the modelling of a class of zone-control conveyor system using timed Petri nets." *International Journal of Computer Integrated Manufacturing*, 5, 277–289.

[18] Lindemann, C. (1998). *Performance Modelling with Deterministic and Stochastic Petri Nets.* John Wiley.

[19] META (1993). *Design CPN Tutorial for X-Windows.* Meta Software Corporation, Cambridge, MA.

[20] Molloy, M.K. (1982). "Performance analysis using stochastic Petri nets." *IEEE Transactions on Computers*, C-31, 913–917.

[21] Moore, K.E., and S.M. Gupta (1996). "Petri net models of flexible and automated manufacturing systems: a survey." *International Journal of Production Research*, 34, 3001–3035.

[22] Moore, K.E., and S.M. Gupta (1999). "Stochastic, colored Petri net (SCPN) models of traditional and flexible kanban systems." *International Journal of Production Research*, 37(9), 2135–2158.

[23] Murata, T. (1989). "Petri nets: properties, analysis and applications." *Proceedings of the IEEE*, 77, 541–580.

[24] Petri, C.A. (1962). *Kommunikation mit automaten.* PhD thesis. Bonn: Institut für Instrumentelle Mathematik. English translation: Holt, A.W., trans., Communication with automata. Technical Report, RADC-TR-65-377, vol. 1, Suppl. 1. New York: Griffiss Air Force Base, 1966.

[25] Ramamoorthy, C.V., and G.S. Ho (1980). "Performance evaluation of asynchronous concurrent systems using Petri nets." *IEEE Transactions on Software Engineering*, SE-6, 440–449.

[26] Ramchandani, C. (1974). *Analysis of Asynchronous Concurrent Systems by Timed Petri Nets.* Doctoral Dissertation. Massachusetts Institute of Technology.

[27] Zhou, M., and F. DiCesare (1993). *Petri Net Synthesis for Discrete Event Control of Manufacturing Systems.* Kluwer, Boston.

[28] Zurawski, R. (1994). "Systematic construction of functional abstractions of Petri net models of flexible manufacturing systems." *IEEE Transactions on Industrial Electronics*, 41, 584–591.

[29] Zurawski, R., and M. Zhou (1994). "Petri nets and industrial applications: a tutorial." *IEEE Transactions on Industrial Electronics*, 41, 567–583.

MANUFACTURING TASK

The manufacturing task defines the scope of the manufacturing mission of a plant. Without a clearly defined manufacturing task, a factory may contain a proliferation of products and processes; in other words, it would lack focus. The manufacturing task can be considered as a practical statement of manufacturing philosophy. It relates ends and means and links them with the business strategy. A manufacturing task statement should have the following characteristics:

1. It must explicitly state the demands and constraints on manufacturing derived from the competitive strategy of the business, and marketing, financial, technological and economic perspectives.
2. It must state what manufacturing must accomplish to be a competitive weapon and, conversely, how performance is to be judged.
3. It must explicitly state priorities (tradeoffs).
4. The essence of the manufacturing task could be summarizable as a memorable slogan or picture.

An example of manufacturing task, which is an adaptation of a task for a company described by Skinner (1978) is given below.

1. Manufacturing's role is to be predictable, make clean commitments in regards to cost, quality, schedule and return on assets, and then meet them.
2. Grow from a controlled base only after meeting the constraints below:
   a. Cost: maintain competitive product costs at +5% of the best competitors.
   b. Quality/reliability: Maintain quality at +5% of industry norms.
   c. Delivery Performance: 95 percent on time.
   d. Employee relations: employees must judge us to be among the top quintile of companies to work for.

SUMMARY: we must be predictable, and we will sacrifice growth to be no worse than the minimums above.

See **Core manufacturing competence; Focused factory; Manufacturing strategy; Product-process dynamics.**

Reference

[1] Skinner, W. (1978). *Manufacturing in the Corporate Strategy*, Wiley, New York.

MANUFACTURING TECHNOLOGY USE IN THE U.S. AND BENEFITS

Paul M. Swamidass

Auburn University, Auburn, AL, USA

INTRODUCTION: Industrial revolution over the last few centuries was made possible by the progressive use of more and more manufacturing technologies for the purpose of replacing manual labor, and for improving quality and quantity of output. The process of new manufacturing technology introduction goes on even today. Today, manufacturing technologies are more diverse and complex than they used to be.

Table 1. A partial list of hard and soft technologies.

	The technology	Explanation
	Hard technologies	
1.	Automated inspection	
2.	CAD ...	Computer aided design
3.	CAM ...	Computer aided manufacturing including programmable automation of single or multi-machine systems.
4.	CIM ...	Computer integrated manufacturing.
5.	CNC ...	Machines with computerized numerical control
6.	LAN ...	Local area networks
7.	FMS ...	Flexible manufacturing systems; automated multi-machine systems linked by an automated material handling system.
8.	Robots	All kinds of robots.
	Soft technologies	
1.	Bar Codes	
2.	Concurrent Engineering	
3.	JIT ...	Just-in-time manufacturing
4.	Manufacturing cells	
5.	MRP ...	Material requirements planning
6.	MRP II ...	Manufacturing resource planning
7.	SQC ...	Statistical quality control
8.	Simulation and Modeling	
9.	TQM ...	Total quality management.

Manufacturing technologies may be classified as either hard or soft technologies. Table 1 is a list of hard and soft technologies included in a recent study by Swamidass (1998). Hard technologies are hardware intensive with associated software. Soft technologies are manufacturing principles, practices, techniques and know-how that may be enhanced by hardware and software.

The annual growth rate of U.S. manufacturing productivity (output per work hour) is at about three percent since early 1980s (Bureau of

Labor Statistics; The Manufacturing Institute). U.S. manufacturing productivity has increased 285 percent since 1960. Due to higher productivity growth in manufacturing, inflation rates in manufacturing have been lower than the whole U.S. economy – the inflation rate in manufacturing between 1995–1997 was 1.2 percent per year while the overall inflation rate was 2 percent (Department of Commerce; Manufacturing Institute). Manufacturing productivity in the U.S. exceeds that of other major industrial nations including Japan and Germany; 1996 Japan's productivity was 78 percent of the U.S., and Germany's productivity was at 82 percent of the U.S. (Manufacturing Institute; Van Ark and Pilat, 1993). The productivity of the U.S. manufacturing sector enables it to hold its share of world exports; U.S. share of world merchandise exports has risen from 11.4 percent in 1986 to 12.9 percent in 1997.

A SURVEY OF TECHNOLOGY USE IN THE U.S.: Given the impressive performance of U.S. manufacturing in recent years, the use of manufacturing technology in the U.S. can be revealing. Figures 1 and 2 are based on responses from 1025 manufacturing plants to a survey on manufacturing technology use in the United States; the study was sponsored by the National Association of Manufacturers (NAM) and the National Science Foundation, both of the USA (Swamidass, 1998).

Figure 1 shows the extent of manufacturing technology use in the United States. According to the figure, 84 percent of U.S. manufacturers use CAD; 67.6 percent use CNC and so on. Figure 2 shows the benefits of technology use reported by U.S. manufacturers. According to Figure 2, more than 2/3, or more than 67 percent of the manufacturers report that the benefits of technology use included the following:

1. decreased manufacturing cycle-time
2. decreased manufacturing cost
3. Increased product line
4. Increased return on investment

The benefits listed above are more than mere cost reduction from the use of technology; decreased cycle-time provides strategic and competitive advantages. This study of technology use (Swamidass, 1998) found that the average three-year return on investment was at 16.85 percent in 1997 compared to 12.99 percent in 1993 (Swamidass, 1996). Partial explanation for this performance is the judicious use of the various manufacturing technologies in Table 1.

Additionally, the evidence from the 1998 study leads one to infer that plants are making substantial gains in manufacturing flexibility and agility through the increased use of computerized integration, and manufacturing cells. These benefits provide competitive advantage to manufacturers.

Table 2. A comparison of selected 1993 and 1997 data.

	1993 sample	1997 sample
1. Sample	1042	1025
2. Sales ($ million)	47.2	34.5
3. Employment	228	168
4. Sales per employee ($000)	133	147
5. Inventory turns	8.0	9.7
6. Direct labor	18.3%	19.8%
7. Rejection and rework	4.0%	3.5%
8. Lead time (weeks)	7.2	7.4
9. ISO certified plants	4%	20%
10. Foreign-owned (foreign ownership >50%)	3.4%	6.3%
11. No sales to defense department	54%	64.8%
12. Percent reporting cycle time reduction	66%	76%
13. Return on investment	13.0%	16.8%

Improved Operational Excellence and Technology Use

Table 2 compares key operational measures for U.S. Manufacturers participating in the 1993 and 1997 studies (Swamidass, 1996; 1998). The following are notable trends in the table.

Inventory Turns Increase. Inventory turns are an index of manufacturing health; larger the better. Inventory turns have increased to 9.7 since 1993 when it was 8.0 – i.e., the average plant in 1997 had 1.23 months of inventory while it was 1.5 months of inventory in 1993.

Sales per Employee Increases. Sales per employee are an index of productivity. Sales per employee in 1997 were $147K as opposed to $133K in 1993.

Rejection and Rework Rate Decreases. Rejection and rework rate is an index of the ability of the manufacturing process to produce quality products; lower the rate, the better the quality. Rejection and rework rate has decreased from 4 percent in 1993 to 3.5 percent 1997.

Cycle Time and Manufacturing Cost Reduction. In 1997, 76 percent of all manufacturers reported reduction in cycle time as a result of technology use and 75 percent reported reduction in manufacturing costs. Other benefits of technology use and the percent of

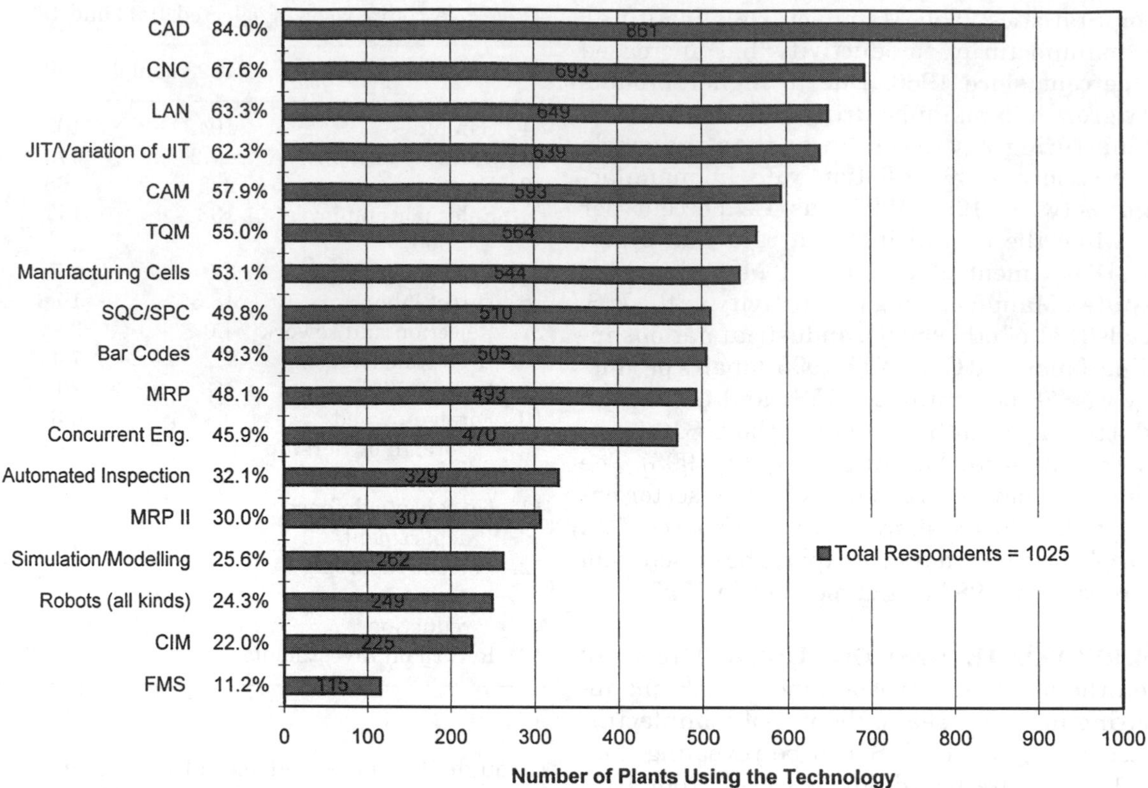

Fig. 1. Technology users.

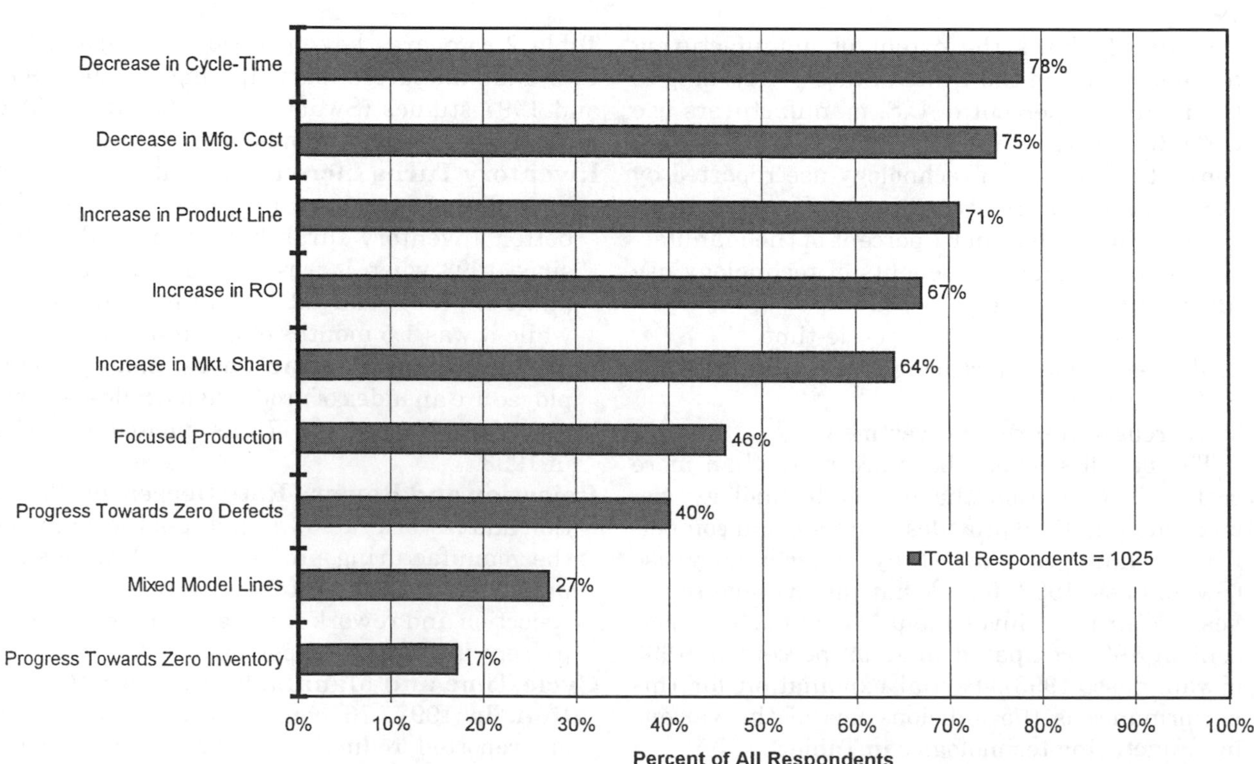

Fig. 2. Reports of benifits due to technology use.

manufacturers reporting these benefits are displayed in Figure 2.

Extremely Skilled Use of Technology Has Superior Pay Off. Not all manufacturers use manufacturing technologies with extreme skill. Extremely skilled users of JIT and its variations report the best inventory turns (19.4), lowest manufacturing lead times (5.7 weeks), and one of the highest return on investment (20.8%). Extremely skilled users of manufacturing cells report the highest return on investment (21.8%). Extremely skilled use of technologies requires a thorough understanding of the technology being used as well as constant training and retraining.

Network Technology Use Increase

The users of local area network (LAN) have grown more than the users of any other technology since 1993; from 49 percent of respondents to 63.3 percent. LAN enables the integration of various computerized technologies in the factory, and it is essential to the integration between factories and their customers or vendors. LAN also permits the expanded and better use of existing technologies on the factory floor through the synergy generated by the integration of several hard and soft technologies.

The 1997 study found that 58 percent of all transactions between the shop floor and production/materials planning are now computerized, and 41 percent of all transactions design and customers is now computerized. For the extent of computerized integration among other units internal and external to the factory, see Swamidass (1998).

Workforce Training

Modern manufacturing technologies need educated and trained workforce. Training and retraining of the workforce and management ensures the best use of investment in technologies. The average training budget in 1997 was five percent of payroll (Swamidass, 198). The training and retraining needs of manufacturing industries using technologies will remain high in the U.S. because (1) the training needs of manufacturing cell users increases because of the need to train operators to handle multiple machines in cells; (2) manufacturers experiencing down-sizing and reengineering face the need to train and retrain employees that are retained; (3) manufacturers hiring employees leaving a downsized facility need to retrain incoming employees; (4) the extremely skilled use of technologies requires frequent training and retraining of the employee; and (5) flexibility and agility in operations requires more frequent retraining of employees.

On-the-job training (OJT) is the most commonly used training technique for training operators in the U.S., although it takes more time to train an employee (8.3 months). Training by vendors is the quickest training method (5.3 months). Manufacturers report that the availability of trained operators delays the skilled use of technologies by as much as 4.9 Months.

LESSONS:

Increased LAN Use Means Increased Computerized Integration

The secret to understanding one of the trends in technology use in manufacturing plants lies in our ability to interpret the increased use of LAN technology. Clearly, by 1997, the use of some technologies had reached saturation levels – this is made evident by the time-series data gathered by the Swamidass (1996, 1998). Manufacturers normally begin to invest in several different technologies in the form of "islands of automation." LAN technology enables the integration of these "islands" and taps the synergistic benefits that flow from the integration of several technologies. Thus, one trend in technology use in the nineties is the computerized *integration of the various factory floor technologies*.

Swamidass (1998) also found computerized integration among internal and external units of a plant. For example, in 1997, 48 percent of all transactions between design and shop floor was compurterized, and 41 percent of all transactions between design and customers was computerized. While the growth in the use of individual technologies may taper off, growth is to be expected in the integration of technologies through the computerization of transactions between internal and external units of manufacturing plants.

Increased Agility

Computerized integration of units inside and outside a factory contributes to agile manufacturing by enhancing the speed of information flow and the ability of manufacturing systems to respond to changes. Swamidass (1998) found that more plants used manufacturing cells in 1997 than in 1993. Manufacturing cells require operators to be skilled in the use of multiple tasks, which adds to the agility of the plant. Through the increased use of LAN and manufacturing cells in the U.S., the flexibility of the average plant is being enhanced.

Table 3. The average plant in the study by size.

	Small plants*	Larger plants	Very large plants
Sample	570	428	69
Average employment	45	329	995
Average sales ($ million)	5.9	59.6	188.7
Sales per employee	$135 K	$163 K	$190.5 K
Rejection and rework (% of manufacturing costs)	3.4%	3.6%	3.2%
Direct labor cost (% of sales)	23%	15.3%	11.8%
Average manufacturing lead time	6.2 weeks	8.3 weeks	6.9 weeks
Inventory turns	9.5	10.1	12.9
LAN users (% of all plants)	51.9%	79.2%	97.1%
Cell users (% of all plants)	38.4%	72.9%	88.4%
Age of the plant	28 years	30 years	33 years
Exporters (percent of all manufacturers)	56.8%	82.2%	78.3%
On-the-job training time for skilled employees	9.4 months	7.1 months	5.1 months
Vendors' training time for skilled employees	5.3 months	5.2 months	4.6 months
Training budget (% of payroll)	5.6%	4.3%	3.5%
Extremely skilled operators (% of all operators)	43%	36.7%	34.5%
Percent of all transactions computerized between shops and production/materials planning	49%	71%	78%
Plants with 90%+ computerized transactions between shops and production/materials planning (% of all plants)	20%	42.1%	55.1%
Delay in skilled technology use due to the non-availability of skilled labor (months)	5.5 months	4.4 months	3.7 months
Plants reporting cycle time reduction as a result of technology use (% of all plants)	73%	85%	84.1%
Return on investment	15.6%	18.3%	23.3%

- Small plants: employees <100.
- Larger plants: employees >99.
- Very large plants: employees >499 (included in column 2).

Systemic Changes brought on by Soft Technology Use

Since 1993, slightly fewer plants are reporting the use of JIT, TQM and SQC techniques; however, the benefits associated with the use of these soft technologies are on the rise. For example, inventory turns and rejection and rework rates have improved to 9.7 and 3.5 percent, respectively (Table 2). This is interpreted to mean that, since mid 1980s, essential features of JIT, TQM and SQC are becoming generic and ingrained manufacturing practices in the U.S. without being associated with any specific technique. These systemic changes bring permanence to manufacturing improvements, and the continuous improvement theme underlying these practices should continue to improve manufacturing performance in the U.S.

Transportation Industry is Leader in Inventory Turns

A remarkable finding of the 1997 survey is the inventory turns of 19.4 reported by the transportation industry (SIC 37), which includes the auto industry; this is twice that of the national average at 9.7. It appears that the transportation industry has become a mature user of lean manufacturing principles and the industry is reaping the benefits of long-term and consistent use of such practices.

The auto industry, which forms a sizable portion of the transportation industry, has been one of the earliest industry groups to adopt soft technologies such as JIT, TQM and SQC in the mid-1980's because of severe competition from Japan and the rapid erosion of their market share. The big three automakers instituted standards and certification for their suppliers based on world-class manufacturing principles. In the process, the industry has proved to be a good training ground for world-class manufacturing techniques. Individuals moving from this industry to other industries may have contributed to the spread of such techniques to other industries.

Training Methods

OJT is the most commonly used training technique but it is also the most time consuming method. Vendors provide the quickest training; if vendors can provide training, small plants have more to gain by using vendors than larger plants – small plants can save more than four months while larger plants can save less than two months by going to technology vendors for training. Each plant should evaluate if the added cost of vendor-provided training justifies the savings in training time.

More larger Plants Use Technologies than Small Plants

Swamidass (1998) found the unmistakable effect of size on technology use. He categorized the plants into three groups to study the effect of plant size (employees); the sample had 570 small plants with less than 100 employees, 428 larger plants with more than 99 employees, and a subset of 69 very large plants with more than 499 employees. See Table 3 for extensive comparison across size.

The following were found to increase with size: (1) sales per employee; (2) inventory turns, (3) LAN users; (4) cell users; (5) percent of transactions computerized between shops and production planning; (6) percent of plant with 90% + computerization between shops and production planning; and (7) percent of plants reporting cycle time reduction as a result of technology use; and (3) return on investment.

The following decrease with size: (1) on-the-job training time; (2) training budget as a percent of sales; (3) percent of extremely skilled operators in the plant; and (4) the delay in the skilled use of technologies for want of skilled workers.

The Needs of Small Plants

The data raises the questions, why small companies do not use technologies as often as larger plants? While not all technologies may be appropriate for small plants, the following are some of the reasons:

- They may need assistance to understand the use and benefits of several new technologies. Large companies have in-house experts to investigate and evaluate investments in new technologies.
- The training expense associated with the use of technologies may be holding back small manufacturers from using technologies more aggressively.
- Funds for investments may be more difficulty to obtain for smaller manufacturers.

See **Agile manufacturing; Bar codes; CAD; CAM; CIM; CNC; Concurrent engineering; Flexible automation FMS; JIT; LAN; Lean manufacturing; Manufacturing cell design; Manufacturing flexibility dimensions; Manufacturing flexibility; MRP II; Robots; Simulation models; SQC; TQM.**

References

[1] Swamidass, P.M. (1996). Benchmarking Manufacturing Technology use in the United States, in *The Handbook of Technology Management*, edited by Gus Gaynor, New York, NY: McGraw-Hill.

[2] Swamidass, P.M. (1998) *Technology on the Factory Floor III: Training and Technology Use in the U.S.*, Manufacturing Institute of the National Association of Manufacturers, Washington, DC.

[3] The Manufacturing Institute. *The Facts About Modern Manufacturing*, fourth edition. Washington, D.C.

[4] Van Ark, B. and D. Pilat (1993). "Productivity levels in Germany, Japan and the United States: Differences and Causes," in *Brookings Papers on Economic Activity, Microeconomics*, 2, pp. 1–49.

MANUFACTURING WITH FLEXIBLE SYSTEMS

Paul G. Ranky

New Jersey Institute of Technology
Newark, NJ, USA

INTRODUCTION: The design and management of a Flexible Manufacturing/Assembly System incorporates vast amounts of analysis and system development work.

It is often done by a team of product and process engineers, industrial engineer, manufacturing systems and system design engineers as well as data processing staff, headed by an experienced team leader who understands not only the data management problems but also the operation control and process planning issues.

Furthermore, today, the system development team must have expertise in internet-based interactive engineering multimedia for effective knowledge communication between distributed systems and their users.

Flexible Manufacturing System Design Steps

A Flexible Manufacturing System, or FMS, is a highly automated, distributed and feed-back controlled system. It is a system of data, information and physical processors, such as computer and manually controlled machine tools, robots and others in which decisions have to be taken real-time (Figure 1). In such systems information processors

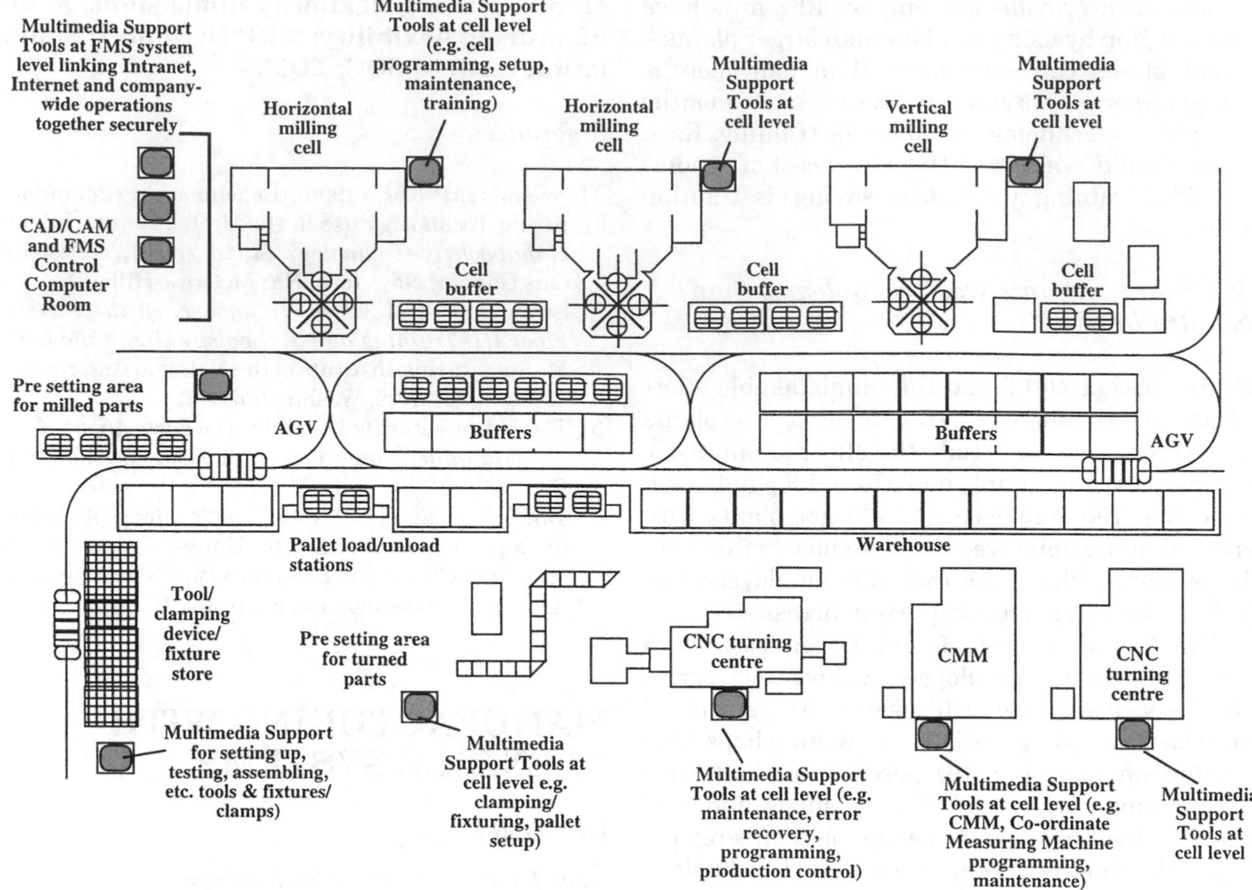

Fig. 1. An example of a distributed, networked and dynamically scheduled flexible manufacturing and design facility.

(including the human resources of such systems) are "well informed," meaning that they have the right information at the right time, in the format and mode they need it to take responsible decisions within given time constraints.

When designing the a Flexible Manufacturing/ Assembly System, or FMS/FAS, the design team should consider the following steps:

1. Collect all current and possible future user and system requirements.
2. Analyze the system (i.e., the data processing and the FMS/ FAS hardware and software constraints).
3. Design an appropriate data structure and database for describing processors and their resources, such as machines, robots, tools (and/or robot hands, probes, sensory-based inspection and assembly tools, etc.).
4. Specify and design programs, and query routines and dialogues that are capable of accessing this data base and communicating with the real-time production planning and control system of the FMS/FAS.
5. Design and integrate the system with the rest of the hardware and software, including on-line manuals, education and training packages, preferably in interactive, engineering multimedia format.
6. Maintain the system and continuously learn for the benefit of the existing as well as future system designs.

Probably the most important question to be answered before starting to design such a system and a data base is "Who" is going to use the data, "When" and "For what" purposes?

As an example, consider that tooling data in FMS are typically going to be used by several sub-systems as well as by human beings. These are as follows:

- The production planning sub-system.
- Process control.
- Part programming.
- Tool preset and tool maintenance.
- Tool assembly (manual or robotized).
- Stock control and material storage.

By employing the above listed sub-systems, the production planning system is kept informed in real-time about the availability of tools in stock, as well as about the current contents of the tool magazines of the machine tools (in the case of

Flexible Assembly Systems, the robot hands in the End-of-Arm-Tool magazines) otherwise it will not be able to generate a proper production schedule (Ranky, 1982; 1985b; Nof, Barash and Sholbert, 1979; Schriber and Stecke, 1986; and Eilon, 1967).

It must be noted that the real-time aspect is important because tools are changed in the magazines of machines, (or cells) not only because they wear, but also because different part programs may need different sets of tools. The actual tool changing operation is done in most cases by manipulators or by robots. The tool magazine loading/unloading procedure is performed mostly by human operators, sometimes by robots or special purpose mechanisms, such as a tool shuttle.

Both the process control and the production planning systems have to update any changes, and must act in real-time, otherwise the operation of the system can be disrupted.

From the FMS/FAS tooling and tool management points of view, one must emphasize the links between the CAD system in which the parts are designed (using design for manufacturing principles), and the CAM system, where the FMS part programs reside. Typically, an FMS part programmer analyses the CAD output (i.e., the design drawings of the component to be manufactured on the FMS), the fixturing, the different setup (i.e., work mounting) tasks, as well as the necessary operations, their alternatives, the required tools and finally a precedence list of the resources (i.e. the possible candidates of processing stations, or cells, or machines). Note that the actual selection of these candidates is done real-time by the dynamic scheduler.

The real-time data bases and software systems are also important since they provide the reports and status information that are needed for the smooth operation of the FMS.

A Flexible Manufacturing Model Integrated with Design

The output of the CAM (Computer Aided Manufacturing) system is a production rule base. This is the knowledge the FMS needs to produce each part. In this production rule base, amongst others, tools are assigned to each operation. The tool codes are selected by the FMS process planner, or automatically assigned by a process planning system, and are obtained from the tool data base.

On the basis of the requested tools, a list is sent via the network to the Tool Preparation Facility, or Station, where the actual tools are prepared (i.e., assembled and preset) and stored in an appropriate way, such that the material handling system of the FMS – typically AGVs, or a special purpose tool shuttle can pick them up.

The tool preparation station also deals with other activities, among which, the most important are as follows:

- Tool service and maintenance.
- Tool assembly to orders (as it is necessary to replace worn tools).
- Tool preset, tool inspection and adjustment.
- Real-time tool pickup and tool transportation organized to serve the needs of the real-time FMS.

The tool preparation station receives its orders, initially originated by the CAD data processing system via the FMS network and technically specified by the CAM system in the form of a production rule base. Order data arriving at the tool preparation station includes:

- Part orders (consisting of part codes and quantities). Note that this is a very important data set for the real-time FMS dynamic scheduler too.
- Notification of when the parts are physically available for FMS processing, which represents a due date for tool preparation.
- A priority order (note that this can change because of some real-time changes in the system, thus, the station must be able to cope with this task too).
- The portion of the production rule base which, describes the requirements regarding tool preparation.

The tool preparation station keeps in touch with the real-time FMS system, as well as with the rest of the system by feeding back important tooling system-related data relating to:

- Stock reviews (regarding tools).
- FMS status report (regarding tools).
- Part priority status reports (in case dynamic changes must be performed in the FMS, which have an effect on tooling needs and tool preparation due dates).

Real-time Operation Control: The real-time part of the FMS operation control and management system must deal with the following tasks:

- It must handle the application of tools for a variety of processes as defined in the production rule base, and assigned on a real-time basis to the FMS/FAS resources by the dynamic scheduler.
- It must provide data to control the transportation of tools and tool magazines within the FMS.
- It must provide information to perform and supervise tool changes and tool magazine changes at all levels.

- It must be notified of tool inspection results (e.g., in case it finds a worn-out tool as a result of an inspection procedure it must generate a command that would instruct the tool magazine update the system to change the tool in question in the appropriate tool magazine).
- It must provide information in the case of emergency, e.g., when a tool breaks.
- It must provide the necessary interfaces and data to perform diagnostic/recovery operations, preferably using diagnostic expert systems.

Finally, let us underline an important feedback loop starting at the real-time system, and ending at the tool preparation station, which contains the real-time tool status, wear and part priority information. This data is often useful to those people and/or software systems that deal with the generation of the production rule base. It is also a very useful data set for FMS designers (Ranky, 1982; 1998a; Suri and Desiraju, 1977).

The tool preset station in the tool preparation unit must be able to inform the process control system about tool preset and offset data, preferably via a digital tool preset unit linked directly to the data processing network of the FMS. Note that the final adjustments, i.e. tool length correction, have to be performed at the tool preset station or at the machine, if the machine is capable of performing this operation. Length and diameter values are often set at the cell prior to the first operation using that tool, and again between operations using the real-time operated tool in order to account for tool adjustments and tool wear (Ranky, 1998b).

When writing FMS part programs, one must know the actual sizes of tools and their characteristics and behavior in different conditions. Tool geometry data is also used when checking for tool collision by graphics simulation in the CAD/CAM environment.

The most important operation control activities in FMS/FAS may use three levels of simulation and optimization, prior to, or during FMS/FAS part manufacturing. The three levels are as follows:

1. The factory level, or business level handled by the business system of CIM, or even broader, the Enterprise Resource Management/Planning System.
2. The off-line scheduling, simulation and optimization activities prior to loading a batch or a single component on the FMS, (handled sometimes by the CAM system, sometimes by the FMS part programming computer).
3. The real-time control handled by the FMS/FAS operation control system, a dynamic scheduler with integrated tool management and multimedia support.

Due to its complexity, a truly integrated approach is required in designing a production rule base to provide the job description for the FMS dynamic scheduler. This is because the dynamic system relies heavily on the knowledge base as represented by the rule base, and an overly restrictive rule base will lead to inefficient, at times even wrong, decisions. This turns out to be a difficult task (Ranky 1998b and 1998c).

It should be underlined, that the application of multimedia at this level is extremely beneficial in terms of part program preparation, teaching/training operators on setting up parts, fixtures, tools, machines, for troubleshooting, for regular maintenance, CNC level programming, quality control, maintenance and other tasks.

Most FMSs have some part buffering capability (see Figure 1). This may not be for scheduling reasons, but for technological, i.e. process planning reasons (e.g., the part must cool down before an accurate inspection procedure is performed). Some level of buffering is useful and necessary because of reliability reasons. The actual number of buffer store locations should be established on the basis of simulation and experience.

CONCLUSIONS: One should realize that FMS scheduling cannot be studied, analyzed and designed in isolation from the distributed, lean and flexible CIM environment, because it simply would not work in practice. In other words, it must be part of an overall production control and management strategy based on fundamentally JIT oriented, "pull" versus "push type" production control methods, capable of considering all affected resources (i.e. machines, tools, fixtures, human operators, etc.) and changing dynamically according to real-time requests.

See **Flexible automation; Manufacturing cell design; Manufacturing system, flexible; Part programming.**

References

[1] Caprihan, R. and S. Wadhwa (1977). The Impact of Routing Flexibility on the Performance of an FMS – A simulation study, *The International Journal of Flexible manufacturing Systems* 9 (1977), p. 273–298.

[2] Eilon, S., *et al.* (1967). Job shop scheduling with due date. *Int. J. Production Res.* 6(1), 1–13.

[3] Nof, S.Y., M.M. Barash and J.J. Solberg (1979). Operational control of item flow in versatile

manufacturing systems, *International Journal of Production Research*, 17, 479–489.

[4] Ranky, P.G. (1985). *FMS in CIM (Flexible Manufacturing Systems in Computer Integrated Manufacturing)*, ROBOTICA, Cambridge University Press, Vol 3, pp. 205–214.

[5] Ranky, P.G. (1982a). *The Design and Operation of FMS* (IFS (Publications) Ltd. and North Holland).

[6] Ranky, P.G. (1985b). *Computer Integrated Manufacturing* (Prentice-Hall International).

[7] Ranky, P.G. (1998a). *An Introduction to Total Quality and the ISO90001 Standard An Interactive Multimedia Presentation on CD-ROM with off-line Internet support,* CIMware, http://www.cimwareukandusa.com (Book CD combo).

[8] Ranky, P.G. (1998b). *Flexible Manufacturing Cells and Systems in CIM*, CIMware Ltd, http://www.cimwareukandusa.com (Book and CD-ROM combo).

[9] Ranky, P.G. (1998c). *An Introduction to Concurrent/Simultaneous Engineering*, CIMware Ltd, http://www.cimwareukandusa.com (Book and CD-ROM combo).

[10] Schriber, T.J. and K.E. Stecke (1986). Machine utilization and production rates achieved by using balanced aggregate FMS production ratios in a simulated setting. *Proceedings of the Second ORSA/TIMS Conference on Flexible Manufacturing Systems: Operations Research Models and Applications*, (Amsterdam: Elsevier Science Publishers B.V.), pp. 405–416.

[11] Suri, R. and R. Desiraju (1977). Performance analysis of Flexible Manufacturing Systems with a Single Discrete Material – Handling Device, *The International Journal of Flexible manufacturing Systems* 9 (1977), pp. 223–249.

MAPE

See **Mean absolute percentage error.**

MARKET PULL MODEL OF INNOVATION

The market pull model of innovation assumes that articulated customer requirements should drive new product development decisions. In other words, new ideas in markets and goods looking for new technologies represents market pull view of innovation. While this approach has merits, in many industries it is possible to go beyond customers' knowledge and a pure market pull approach to innovation can be severely limiting. Optimal innovation efforts often result from a balance between market pull and technology push. Technology push model of innovation refers to new ideas from research labs that seek applications and markets. Customer is not the source of ideas in technology push innovation.

See **Process innovation; Product innovation; Technology push model innovation.**

MARKET SEGMENT OR FOCUS STRATEGY

A market segment or focus strategy is one of three *generic* business strategies discussed by Porter in his well-known book, *Competitive Strategy* (1980). A firm using a focus strategy concentrates on a particular buyer group, product line segment, or geographic market. The objective of a focus strategy is to serve this particular target more efficiently or effectively than competitors who are competing more broadly. A successful focus strategy either achieves differentiation by better serving the needs of the target market, or it achieves lower costs in serving this target, or both. Numerous examples of firms who pursue focus strategies are found in industrial markets. These firms choose to serve only a particular group of product users as opposed to attempting to serve all industrial and consumer markets. Porter (1980) provides the example of a large food distributor that serves only eight fast-food chains. In doing so, this firm gains a better understanding of its relatively few customers, and it builds close business relationships that are difficult for more broad-based competitors to match.

See **Core manufacturing competence; Focused factory; Manufacturing strategy.**

Reference

[1] Porter, M.E. (1980). *Competitive Strategy*, The Free Press: New York.

MASS CUSTOMIZATION

Rebecca Duray

University of Colorado at Colorado Springs CO, USA

INTRODUCTION: Mass customization has become popular in the nineties. Companies are striving to provide customized products and services to their customers at low costs. No industry is immune from the desire to produce low cost customized products. Mass customizers can be found in both consumer and industrial markets. As the concept of mass customization gains popularity, more companies may venture into the uncharted territory of mass customization without a clear idea about the nature of mass customization and without the appropriate manufacturing system to support this new marketing concept. Looking at companies that have successfully implemented mass customization provides little guidance. The diversity of the products and production system for mass customization does not lead to easily generalizable concepts.

The practice of mass customization does not fit the mindset associated with conventional manufacturing methods. Historically, companies chose to produce either customized, crafted products or mass produced, standardized products. This traditional practice means that customized products are made using low volume production processes that cope well with a high variety of products. Similarly, mass production process is chosen for making standardized products in a high volume, low cost environment. Mass customization, as defined by Davis (1987) provides a one-of-a-kind product manufactured on a large scale. Customers are able to purchase a customized product for the cost of a mass produced item. Thus, mass customization represents an apparent paradox by combining customization and mass production, offering unique products in a mass produced, low cost, high volume production environment.

When known mass customizers are examined, it is difficult to determine commonalties in manufacturing systems because each company uses a manufacturing systems appropriate to their product, process and industry. For example, Dell Computers manufactures a mass customized product with an assemble-to-order process. Each base computer is fitted with the appropriate features selected by the customer from a common list of options. In contrast, Levi Strauss provides personalized jeans for women through a process that cut a new pattern for each pair of customized jeans. Both manufacturers should be considered mass customizers, yet their manufacturing systems are very different. Mass customization includes a vast array of products made with appropriate manufacturing systems to deal with the mass customization paradox.

Although mass customization is usually thought of in terms of marketing or competitive strategy, mass customization should be viewed as a competitive capability that resides in marketing, manufacturing, and engineering functions. The integration of these functional areas through product design is the key to breaking the paradox of mass customization. A mass customizer must identify a market for low cost customization, determine the customizable features required by that market, design a product that can provide customization with mass production, and manufacture the product in a cost effective manner. This view of mass customization fits the resource-based view of business strategy (Wernerfelt, 1984; Barney, 1991) and suggests that the resources needed to provide a mass customization strategy reside in the integration of marketing, manufacturing, and engineering strategies.

DEFINING MASS CUSTOMIZATION: The essence of mass customization, as distilled from literature (Pine, 1993; Pine, Victor and Boyton, 1993; Kotha, 1995) is providing personalized products at reasonable prices. Although this description invokes the essence of mass customization, it does not possess the specificity required to identify companies as mass customizers. Literature on mass customization labels a broad range of companies as mass customizers, but the diversity of these companies further clouds the process of identification. The variety of the attributes possessed by the example companies requires a definition that establishes rigorous conceptual boundaries of mass customization and presents a means to distinguish among the vast array of mass customizers.

The boundaries of mass customization can be more clearly established by delineating two issues. The first issue is the nature of customization. When should a product be considered customized, and not just another version of product proliferation? The second issue for exploration is the "mass" in mass customization. How can unique products be developed in a mass production fashion? How can high volume, low cost customization be implemented? Pine (1993) suggested that modularity is the key to achieving mass customization. Modularity can be viewed as critical for gaining scale or "mass" in mass customization. The issues of customization and modularity are discussed below to provide a basis for a more comprehensive definition of mass customization.

Customization Issues

Customized products are uniquely produced for each customer, which requires that the customer be involved in the design process. Mintzberg's (1988) typology defined customization as taking three forms: *pure, tailored* and *standardized*. The role of the various parts of the value chain and the degree of uniqueness of the product differ in the three forms of customization.

A *pure* customization strategy furnishes a product developed from scratch for each unique customer. This type of customization affects the entire chain, from design, fabrication, assembly and delivery, and provides a highly unique product. Construction projects provide many good examples of pure customization: an architecturally designed house, an office building, or a highway overpass.

Tailored customization alters a basic design to meet the specific needs of the customer. The customer affects the value chain at the fabrication level, where standard products are changed. Eyeglasses are a good example of tailored customization. Customer select from standard frames, but each lens is fabricated to the unique optical prescription of the customer.

In *standardized* customization, a final product is assembled from standard components. Here, the customer penetrates the assembly and delivery processes through the selection of the desired features from a list of standard options. A familiar example of standardized customization can be seen when one orders a pizza. By selecting items on a pizza from a prescribed list, the customer enters the assembly stage of production which produces a standardized custom pizza.

Mintzberg's (1988) definitions showed that the type of customization represents different levels of customer involvement in the design and production process.

Mass Production Issues

Mass customization requires that unique products be provided in a cost effective manner, in essence, it requires volume production. To address the issues of mass production, mass customizers must adopt some form of modularity in order to provide both economies of scope and economies of scale simultaneously. This is true for all types of manufacturing systems. Even in group technology, family groupings are determined by component commonality which incorporates some form of modularity providing both economies of scope and scale. The production of standardized modules is the key to high volume "mass" customization. For mass customization, Ulrich and Tung (1991) recognized several types of modularity shown in Figure 1.

Component sharing modularity uses common components in the design of a product. Products are uniquely designed around a base unit of common components. *Component swapping* is the ability to switch options on a standard product; modules are selected from a list of options to be added to a base product. *Sectional modularity* is similar

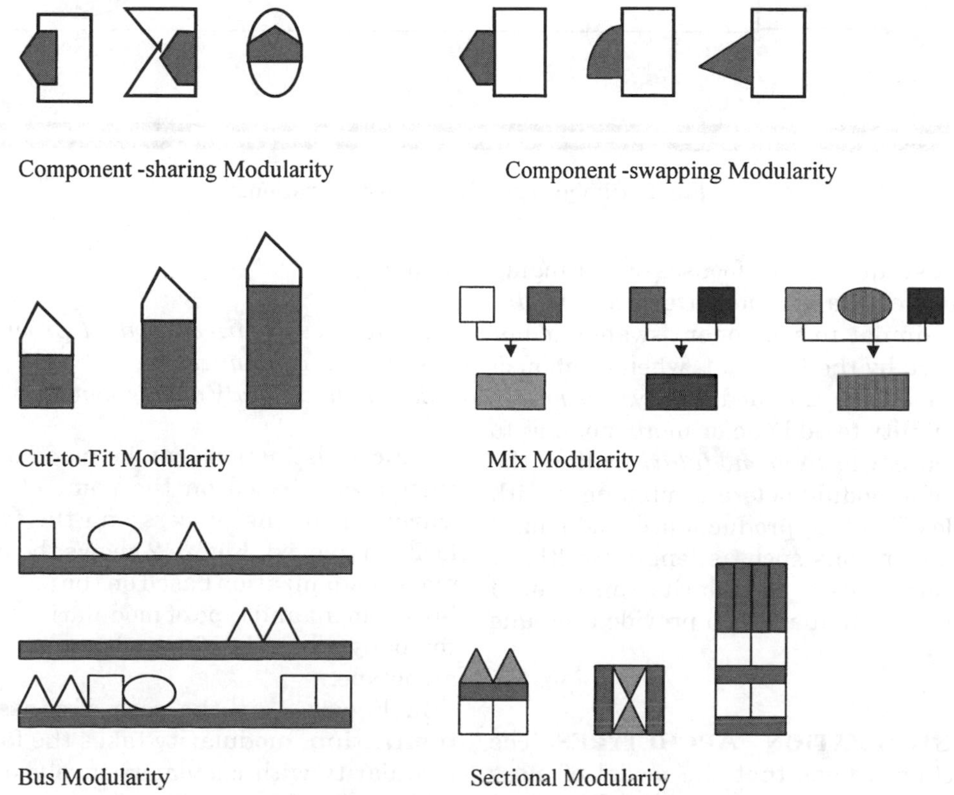

Fig. 1. Modularity types.

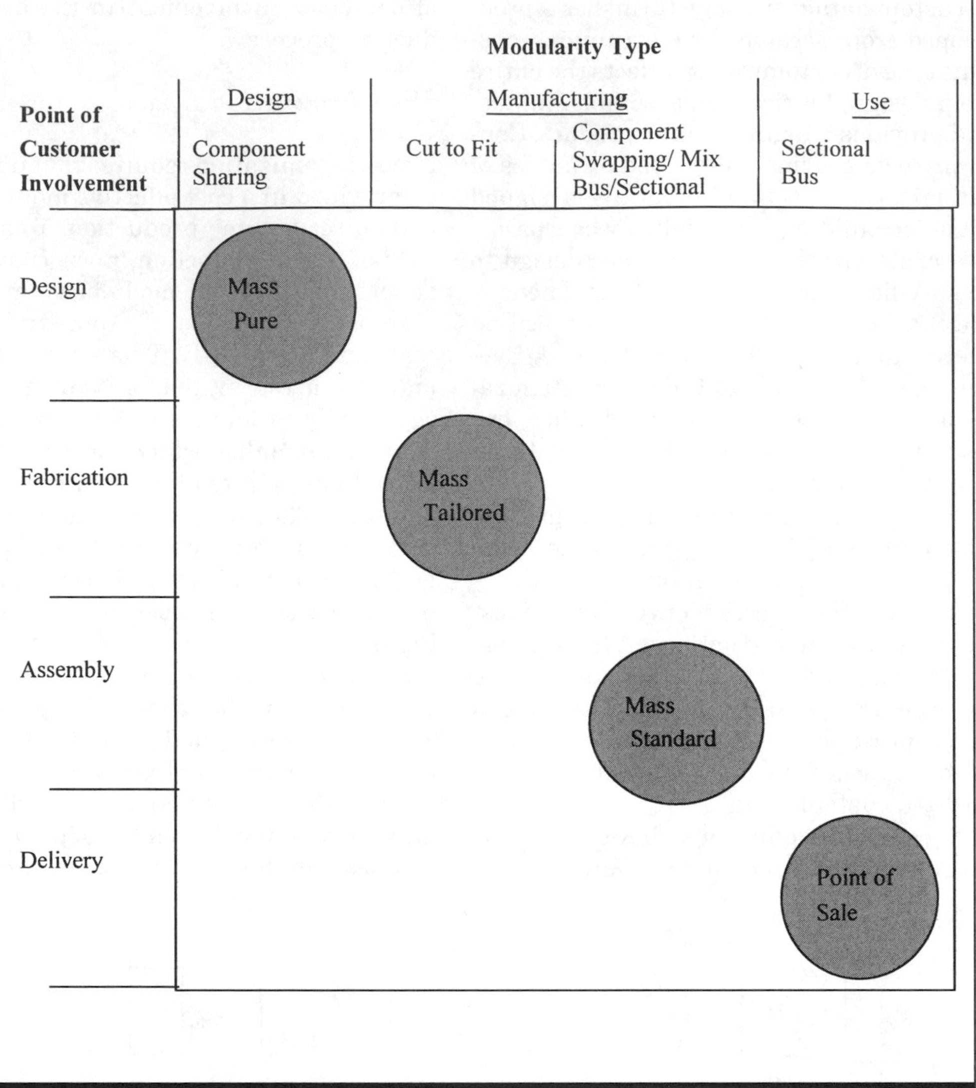

Fig. 2. Dimensions of mass customization.

to component swapping, but focuses on arranging standard modules in a unique pattern. *Mix modularity* also is similar to component swapping, but is distinguished by the fact that, when combined, the modules lose their unique identity. *Bus modularity* is the ability to add one or more modules to an existing base. *Cut-to-fit modularity* alters the dimensions of a module before combining it with other modules. That is, products are customized by varying dimensions such as length, width, or height. Various types of modularity can be used separately or in combination to provide a unique end product.

MASS CUSTOMIZATION ARCHETYPES: The previous section argues that the model of mass customization that emerges from the literature uses two critical definers: customization and modularity. Therefore, mass customization can be defined as:

> *A mass customized product provides end-user specified customization achieved through the use of modularity of components.*

Practical distinctions can be made between mass customizers based on the point of customer involvement in the process and the type of modularity employed. Figure 2 shows the archetypes of mass customization based on the point of customer involvement and type of modularity. Table 1 shows company examples for each mass customization archetype.

In Figure 2, in the case of **mass pure customization**, modularity takes the form of design modularity with customers involvement from an early stage of design. Core components are easily shared, and custom built features are designed around the common modules. For example, a

Table 1. Case examples of mass customization types.

Mass pure	Mass tailored	Mass standard	Point of sale
Sign-tic (Kk)	**Bally Engineered Systems** (PVB, P, PP)	**Motorola** (PVB, P, Do)	**Personics** cassettes (P, SP)
Ross Controls valve systems (PPR)	**Lenscrafter** (P)	**TWA Getaway Vacations** (P)	Electronic components – Stereo – Computers (D, GK)
	Andersen Windows (PPR)	**Lutron** lighting fixtures (SP)	**Ikea** Modular Furniture (D)
	National Bicycle (Ko)	**Hallmark/American Greetings** (PPR)	**Legos** (P)
		Peapod Virtual Supermarket (PPR)	
		R.R. Donnelley/ Farm Journal tailored to subscriber (PPR)	

D	Davis (1987)	P	Pine (1993)
Do	Donlon (1993)	PP	Pine and Pietrocini (1993)
GK	Garud and Kumaraswamy (1993)	PPR	Pine, Peppers, and Rogers (1995)
Ko	Kotha (1995)	PVB	Pine, Victor, Boyton (1993)
Kk	Kubiak (1993)	Sp	Spira and Pine (1993)

common engine type may be shared in specialty machines, such as a crane or mining equipment, but the actual application may require special features. Ross Controls provides mass customized valve systems for General Motors using a common platform that is individually customized for each stamping press. These valve systems, which are mass customized, perform better than the systems they replaced at one-third the price (Pine, Peppers, and Rogers, 1995).

Both **mass tailored** and **mass standard** customizers are modular in production; standard components are assembled to create unique products. However, **mass tailored** customizers use cut-to-fit modularity, where standard modules are altered to fit unique customer specifications. Customers are involved at the fabrication stage of production, where modules must be altered to customer specifications. Products that are produced to customer specified size or length fit this category. Bally Engineered Structures manufactures walk-in coolers, refrigerated rooms, and clean rooms. Product modules are cut-to-fit specific dimensions provided by the customer to make unique rooms manufactured from modular components (Pine, Victor, and Boyton, 1993).

Mass standard customization also occurs at the production stage, and requires modular components that can be joined in production to make unique end items. Component swapping, mix, sectional, and bus modularity could all be applied in this customization configuration. The unique nature of the product is developed from the many possible combinations of the components adapted to suit the unique needs of the customer. Customers are involved in the assembly stage of production as options are chosen from a limited set of module choices. Motorola pagers, are the result of "mass standard" customization. Pagers are designed to a customer's specification from a wide range of options that are introduced at the production phase (Pine, 1993; Donlon, 1993).

Point of Sale mass customization uses completed modules or standard products that are combined by the consumer to create a unique product. Customers provide the customization through their use and selection of modular components. Stereo components (receivers, CD players, cassette deck and speakers) are combined by the consumer to provide a customized audio system. Each customer selects and combines the required modules to form unique products. Modular furniture and bookshelves are also point of sale customized products. Bus modularity also may be employed as finished modules can be added to an existing structure. Track lighting (Pine, 1993) is a familiar example of bus modularity. Different lighting fixtures are added to an electrical track making track units unique. A child's yard play-set with a base frame to which swings, bars, horses, or rings are added can be considered a point of sale mass customized product

MISCLASSIFICATION: Only those manufacturers that employ modularity and involve the customer in the design process for specific units should be considered mass customizers. It is interesting to note that some companies used as examples of mass customizers do not fit this mass customization typology. Davis (1978) uses Cabbage Patch dolls as an example of mass customizers that gained a competitive edge from the unique nature of its products. Each doll is a unique end item that customers select at a retail outlet. Although customers do select a doll, they cannot have a doll made to their specifications such as, eye or hair color. Customers do not participate in the design process; therefore, the Cabbage Patch doll is not an example of mass customization. With a similar argument, Pine (1993) misclassifies Swatch Watch as an example of mass customization. Swatches are unique products, and the wide selection gives customers many choices. Although this provides great variety, customers do not have the ability to request changes. Therefore, this product is not an example of a mass customized product.

SUMMARY: Notions of customer involvement and modularity coupled with modularity types provide a basis for classifying mass customizers.

Organizations Mentioned: Bally Engineered Structure; Cabbage Patch Dolls; General Motors; Ross Controls.

See **Bus modularity; Component sharing modularity; Component swapping modularity; Cut-to-fit modularity; Manufacturing flexibility; Mass customization and manufacturing; Mass standard customizers; Mass tailored customizers; Mix modularity; Pure customization; Sectional modularity; Standardized customization; Tailored customization.**

References

[1] Barney, J. (1991). "Firm Resources and Sustained Competitive Advantage," *Journal of Management*, Vol. 19, No. 1.

[2] Davis, S.M. (1987). *Future Perfect*, Reading, Massachusetts, Addison-Wesley Publishing Company.

[3] Donlon, J.P. (1993). "The Six Sigma Encore; Quality Standards at Motorola." *Chief Executive*, November 1993, No. 90.

[4] Duray, R. (1997). Mass Customization Configurations, Ph.D. Thesis, The Ohio State University, Columbus, Ohio.

[5] Garud, R. and Kumaraswamy A. (1993). "Technological and Organizational Designs for Realizing Economies of Substitution." *Strategic Management Journal*, Vol. 16, pp. 93–109.

[6] Kotha, S. (1995). "Mass Customization: Implementing the Emerging Paradigm for Competitive Advantage." *Strategic Management Journal*.

[7] Kubiak, J. (1993). "A Joint Venture in Mass Customization." *Planning Review*, July 1993, Vol. 21, No. 4, pg. 25.

[8] Mintzberg, H. (1988). "Generic Strategies: Toward a Comprehensive Framework." *Advances In Strategic Management*, Vol. 5, pp. 1–67.

[9] Pine II, B.J. (1993). "Mass Customizing Products and Services." *Planning Review*, July 1993, Vol. 21, No. 4, pg. 6.

[10] Pine II, B.J. and Pietrocini, T.W. (1993). "Standard Modules Allow Mass Customization at Bally Engineered Structures." *Planning Review*, July 1993, Vol. 21, No. 4, pg. 20.

[11] Pine II, B.J., Peppers, D., and Rogers M. (1995). "Do You Want to Keep Your Customers Forever?" *Harvard Business Review*, pp. 103–114.

[12] Pine II, B.J., Victor B., and Boyton, A.C. (1993). "Making Mass Customization Work." *Harvard Business Review*, 71, pp. 108–119.

[13] Spira, J.S. and Pine II, B.J. (1993). "Mass Customization." *Chief Executive*, March 1993, No. 83, pg. 26.

[14] Ulrich, K. and Tung, K. (1991). "Fundamentals of Product Modularity." *Proceedings of the 1991 ASME Winter Annual Meeting Symposium on Issues in Design/Manufacturing Integration*. Atlanta.

[15] Wernerfelt, B. (1984). "A Resource-based View of the Firm." *Strategic Management Journal*, Vol. 5, pp. 171–180.

MASS CUSTOMIZATION AND MANUFACTURING

Pratap S.S. Chinnaiah and Sagar V. Kamarthi

Northeastern University, Boston, MA, USA

INTRODUCTION: The concept of mass production was introduced in the 1800s and reached its full development in the 1920s. However, craft production, where artisans built products to a customer's specifications continued catering to some select market niches. In the 1970s global competition intensified among producers and at the same time the variety of products proliferated. As a result, lean production which combines the advantages of craft and mass production while avoiding the high cost of the former and the inflexibility of the latter was born (Womack *et al.*, 1990). In the 1990s, while some markets are reaching their saturation limits for products, and customers are becoming more demanding, we see the rise of mass-customized production (Anderson, 1998; Davis, 1987; ICM Conferences Inc., 1996; Kotha, 1995; Lau, 1995; Preiss, 1995; Roos, 1995).

The concept of mass customization (MC) focuses on satisfying individual customer's unique needs with the help of technologies such as agile manufacturing, flexible manufacturing systems, computer integrated manufacturing, and information and communication systems. In MC, the needs of an individual customer are translated into design, accordingly produced, and delivered to the customer. Such mass-customized production (MCP) systems are also referred to as one-of-a-kind production (OKP) systems. MC draws from other production strategies such as personalized production, focused job shop, agile manufacturing, and virtual enterprise (Elantora, 1992; Wortmann, 1992).

Changing patterns of customer demands are beginning to give MC the competitive edge over lean production. For many companies, MC is becoming an inevitable means for survival in the market place. An observation of companies engaged in MC indicates that most of them have evolved over an extended period of time into this production scenario rather than by systematically implementing a well planned and well documented strategy. Lutron Controls is a good example of a company that entered into mass customization as a matter of survival. The company found that, by providing products according to each customer's unique specifications, they could survive and grow in the competitive market of lighting controls (ICM Conferences Inc., 1996).

Characteristics of Production Systems: Table 1 illustrates the differences between craft, mass, lean and MCP systems. A mass production system focuses on producing a single product for a mass market. A lean production system focuses on producing a finite number of variants of a single product or a finite set of standard products designed to meet demands of segmented markets. Lean production can offer a considerably large variety of products but this variety is attained without individual customer involvement. The Japanese method of fast paced introduction of an array of products sometimes has a downside. Toyota announced that it would reduce the varieties of the corolla model from 11 to 6. Nissan has a plan to reduce its variety of engines by 40% in five years (Lau, 1995). GE Fanuc has discovered that customers don't want more choice but want what they need. While lean production focuses on the elimination of waste in processes, MC aims at the elimination of waste in products by eliminating the features unwanted by customers. An MCP system attempts to produce just as many variants of a finite set of products as customers indicate they want.

When a production system produces only one product (e.g. GE's electric bulb, Coke's soda can, Gillette's Sensor razor), we can call it a mass production system. A mass production system is not usually designed to respond to a variety of customer demands. The Telco 1010 truck production line in India can be considered as a typical example of a real-world mass production system. On this production line thousands of trucks are produced with no variation in design. When a production system is capable of producing several variants of a product (e.g. the present range of automobiles available in the market, often produced on the mixed-model lines), then such a system can be called a lean production system. In fact, the variety of products that are requested by innumerable customers may be unlimited but the production system tries to satisfy their demands by providing a finite variety of products. A company that can manufacture shoes of any model, any size, any profile, with the user's choice of material can be considered an MCP system. That means, an MCP system can almost (if not completely) satisfy customer needs for a selected group of products.

Mass customization is defined as *a business strategy for profitably providing customers with anything they want, anytime, anywhere, any volume, in any way*. A simple intuitive definition of MC is captured in the 5A's of Nissan's vision for the year 2000: anybody, anything, any volume, any time, and anywhere (ICM Conferences Inc., 1996). However, the ideal mass customization represented by this definition is unachievable by even the most dedicated firms that hope to realize the ideal mass customization. A more realistic definition is that *mass customization is the use of flexible processes and organizational structures to produce varied and often individually customized products and services at a price of standardized, mass-produced alternatives*.

To understand the differences among craft, mass, lean production, and MCP systems, the concept of *customer sacrifice* must be understood. Customer sacrifice is the gap between what each customer truly wants or needs and what the company can supply. Companies have to go beyond aggregate customer-satisfaction and seek individual

Table 1. Comparison of craft, mass, lean, and mass-customized production systems.

Characteristic	Craft prod.	Mass prod.	Lean prod.	MCP
Anybody	Yes	Yes	Yes	Yes
Anything (Any product, Any design)	Yes	No	No	Yes
Any where (in the world)	No	Yes	Yes	Yes
Any time	No	Yes	Yes	Yes
Any volume	No	No	No	Yes
Main theme	Individualized product at unavoidably high cost.	Products for all at lowest possible cost.	One product for each market segment possibly at mass production cost.	One product for each customer possibly at mass production cost.
Nature of products	Nonstandard goods with a lot of variety.	Standard goods of negligible variety with expected tolerance for defects.	High quality standard goods with considerable variety.	High quality, individualized products and services with unlimited but manageable variety.
Productivity aspect	Human labor and skills.	Plant and equipment utilization at the expense of maintaining extra supplies, space, manpower, inventories, etc.	Elimination of nonvalue adding process and operations (process waste) in the value chain.	Increasing the value of product offered to customer by eliminating nonvalue adding features (product waste).
Technologies	Skill based technologies.	Special purpose fixed automation equipment. Narrowly skilled operators.	Programmable and flexible automation equipment: CNC, FMS, Robots, AS/RS, CAD, CAM, CIM, LAN, WAN.	Highly flexible and integrated automation equipment. Internet, ERnet, EInet, CAD, CAM, Digital & Rapid prototyping, FMS, CIM, Flexible tooling/fixturing, EDI, Configurators, Multimedia workstations, Electronic shopping, Electronic kiosks, Personalized cards, Virtual reality, Experience simulators, Portable computers.
Markets	Rich class one-to-one market.	Mass market.	Segmented markets driven by customer surveys.	Mass one-to-one market.

customer satisfaction (Pine II *et al.*, 1995). For example a woman tries, on average, thirteen pairs of jeans before making a purchase; and even after this considerable amount of effort, approximately 30% of the women do not buy pants at all because they can not find the pair that fit them. The cost of customers' time and their disappointment or dissatisfaction of having to live with imperfect pants is called, customer sacrifice. Levi-Strauss is capitalizing on such differences between customer needs and product functionalities by mass customizing blue jeans for women using the technology supplied by Custom Clothing Technology Corporation. After a woman customer has her measurements taken in a store, the information is then sent to the factory for prompt production.

In the process of manufacturing products according to customer requirements, Ford's truck plant in Kentucky offers 2.5 million variants of trucks, and Motorola offers 29 million varieties of pagers (Henricks, 1996). When a company advertises, *any design, anybody, any volume, anywhere, anything or whatever, whenever, or wherever*, it is an indication that the company has the capability to undertake mass customization (Pine Ii *et al.*, 1993). Even though craft producers could meet the challenges of *anybody* and *anything* but they would not claim *any volume, anywhere and any time*. Crafts producers made in small volumes, costs were very high. In contrast, MC attempts to give customers what they want at the cost of mass production.

CLASSIFICATION OF MASS CUSTOMIZATION PRODUCTION (MCP) SYSTEMS: A careful review and analysis of several products produced under the MC protocol demonstrates that, while some products are designed, produced and delivered as customized products, other products are assembled from modules to achieve customization. Some products are produced with embedded technology, which will allow the customers to customize product features to satisfy their individual requirements. In some cases, a small amount of manufacturing or customization is done at the delivery end to suit customers' unique needs. In this way, several opportunities for customization of products exist. They can be classified into five MC production systems depending on customer requirements and customer involvement in the production system.

Type 1 MCP: Make-to-stock MCP (mass-produced and mass-customized products)

Type 2 MCP: Assemble-to-order MCP (with or without a small amount of manufacturing at the delivery end)

Type 3 MCP: Make-to-order MCP (manufacture to order)

Type 4 MCP: Engineer-to-order MCP (design and manufacture to order)

Type 5 MCP: Develop-to-order MCP (R&D, design, and manufacture to order)

This classification has some similarities with the generic types of process choices such as continuous processing, line, batch, jobbing, and project. Products with built-in customizable technology are produced by Type-1 MCP systems. Production systems, which assemble products from a fixed set of pre-manufactured modules, are classified under Type-2 MCP systems. A Type-3 MCP system encompasses production systems, which manufacture products after receiving a customer order. Products that involve design and manufacturing are produced by Type-4 MCP systems. If products involve both research and development in addition to usual design and manufacturing activities, they are delivered by Type-5 MCP systems. Each one of these five MCP systems is discussed in detail with several examples in the following sections.

Type 1: Make-to-Stock MCP

In a make-to-stock MCP system, a few standardized products which have a provision for easy customization are produced to stock in large volumes and with long production runs. Products with embedded technology that enable customization by the customer to suit his or her needs are outcomes of make-to-stock MCP. To a large extent, and for all practical purposes, these customizable products are mass produced. Therefore, the products of this kind are known as mass-produced and mass-customized products. This is the simplest kind of MC production because such a production system derives the benefits of mass production and at the same time meets the unique needs of individual but anonymous customers. The following are some of the examples of mass-produced and mass-customized products (Pine Ii, 1993a):

1. A pair of Reebok shoes with an air pump: These shoes contain a pocket of air which can be pressurized or depressurized by the wearer to get the desired amount of cushion. Reebok shoes with air pump are mass produced but the embedded technology satisfies each customer's unique needs.
2. A Mercedes-Benz model with 13 different environmental controls: These built-in controls allow the driver to adjust the car's environment to suit his or her personal comforts and preferences.

What is common in the above examples is that the customers have the convenience of modifying or customizing the products to suit their needs by virtue of the products' embedded technology. This kind of MC requires a careful study of customer requirements to embed matching technologies at the product design stage.

Type 2: Assemble-to-Order MCP

In this type of production system, assembly activity starts after receiving a customer order. These products are assembled from modules that were already manufactured based on product demand forecasts. Therefore a module manufacturing shop is driven by manufacturing resources planning, and the module assembly activity is driven by customer orders. The assemble-to-order MCP can take two different variations. In one situation, the products are assembled from their modules exploiting the concept of modularity in product design. In another situation, after products are assembled from modules, a small amount of manufacturing is employed at the delivery point. These situations are discussed below.

Products Assembled from Modules: Modularity in product design helps to minimize costs and maximize customizability. Low product costs are achieved through economies of scale of the product modules rather than the products themselves. There are several examples of products that are constructed from modules.

1. Automobiles built with options like all-wheel drive, anti-lock brakes, air bags or navigation systems.
2. Newscasts on the Internet that are assembled together to suit individuals by sections on sports, travel, auto, real estate, or fashion. For example, Individual, Inc. provides published news stories selected to fit the specific and ever changing interests of each client.

From the above examples it can be seen that in such MCP systems, a customer's order typically initiates the assembly of modules into final products.

Products Assembled from Modules Plus a Small Amount of Manufacturing: In this case, the products are first assembled from the standardized modules at the production site and a small amount of manufacturing is carried out at the delivery end for incorporating a cosmetic, or an ergonomic feature into the products. The following are some examples of this type of MC.

1. Watches: The dial, hands, case, and strap are the appearance parts in a watch. If these visual features are customized, a customer can feel that his or her watch is unique. With CAD/CAM installed at dealer show rooms, such a features can be extended to customers.
2. Tennis Rackets: They are mass produced at central factories but can be subjected to a final customization step before their delivery to customers to give an individualized comfortable grip.

Type 3: Make-to-Order MCP

In this class of MC, manufacturing of the product starts after receiving a customer order. Therefore in this production system, customer orders drive component manufacturing, their assembly into products, and the delivery of products. For example,

1. Chemical equipment suppliers produce equipment such as reactors, centrifuges, dryers, filters, etc. after receiving the customer orders.
2. A general-purpose machine tool builder may start manufacturing machine tools only after receiving customer orders.

The production of specialty chemicals and fertilizers which are custom-blended for each hectare of a farm according to the type of soil, salinity, and nitrogen content, calls for redesign, movement, and relocation of process subsystems for each new order (Wortmann, 1992). IBM currently runs 95% of its business on a make-to-order basis. Hewlett-Packard conducts 80% of its business, and Compaq runs almost 100% of its business in a similar fashion (Sullivan, 1995).

The concept of evolving products also comes under this category. Products such as computers have traditionally been of the evolving type. In this type of product, both current and future requirements of the customers are considered together. Now this concept is being applied to other products as well. For example, Nissan has explored the concept of the "evolving car" that owners could bring to the dealer for the add-on latest innovations, such as improved exhaust system, or restyling features every a few years. Design of these types of products can be very complex because the future needs and expectations of the customers have to be charted in advance. These future needs and expectations must be within the capabilities of the value chain.

Type 4: Engineer-to-Order MCP

In this type of MCP system, products are designed from scratch, then manufactured and delivered to a customer. A customer order initiates the design and manufacturing. A job shop which can design, produce, and deliver according to customers' specifications and requirements falls into this category. The main building-blocks of this type of job shop are general-purpose machine tools, CAD, CNC, FMS, CIM, and a highly skilled workforce. Examples are:

1. An FMS supplier, after receiving a customer order, designs, manufactures, and then delivers a flexible manufacturing system to the customer.
2. Building construction and refurbishment.

Following two years of reengineering, Bechtel established an organizational structure, work processes, and information systems necessary to implement MC which enabled Bechtel to design and construct engineer-to-order power generation plants at a lower cost. Some percentage of orders of Lutron Electronics company are of the engineer-to-order type. Space electronics, Inc. designs and builds one-of-a-kind satellite electronic systems based on orders.

Type 5: Develop-to-Order MCP

In a develop-to-order MCP system, a customer requests that the company carry out research and development, prototype development, process development, and subsequently manufacture and delivery of the product. Typically, research and development institutions undertake these types of jobs. A customer order may conceptually specify the functions expected of the product. Then, intense research and development is carried out, usually over several years, to develop the product and in some situations the production methods also. Each order is treated as a project. Examples are:

1. Spacecraft design and development.
2. A project related to the development of a new and innovative packaging method. It may require a research and development effort, prototype development, testing, and manufacturing process development.

FACTORS THAT INFLUENCE AN MCP SYSTEM: As discussed in the previous section, MC takes on many different forms. In the following sections, a set of factors, which influence different types of MCP systems are discussed. The important factors that influence MCP systems are customer involvement, product variety, production volume, bill-of-material coefficient, tooling and process changes, quality system changes, MIS modifications, learning rate, customer-order decoupling point, manufacturing at delivery site, personnel, advanced technology, the basis for production planning and control, order promising, and the handling of demand uncertainty. The studies of Terry Hill (Hill, 1994) have identified these factors. An understanding of these factors will be most useful for the companies that are adopting a mass customization philosophy or restructuring themselves to achieve an effective MC.

Customer Involvement

In make-to-stock and assemble-to-order MCP systems, thorough research and development efforts are required to finalize a product's design. In some cases there is no customer involvement at the design and manufacturing stages but such an involvement is required at the point of sales or assembly stage when an order is placed. In other cases, customer involvement is required at the design stage or the manufacturing stage. In order to acquire customer specifications and requirements, an MCP system must have facilities which might call for an additional investment in technology, time, infrastructure and human skills. For example, Matsushita Bicycle Company, which manufactures unique bicycles, has ergonomic frames at each store for measuring physical dimensions of customers (Moffat, 1990). Companies also learn and capture the information about customer requirements using special software. Customer involvement is low or distant in Type-1 (make-to-stock) MCP and it is very high in Type-5 (develop-to-order) MCP.

Product Variety

In the case of make-to-stock and assemble-to-order MCP systems, the product variety, from the customer's point of view, is low compared to make-to-order, engineer-to-order, and develop-to-order MCP systems. In an ideal MC situation the ratio of the product variety to the number of customers is 1:1. The product variety varies from low in Type-1 (make-to-stock) and Type-2 (assemble-to-order) MCP systems to high in Type-5 (develop-to-order) MCP systems.

Production Volume

In make-to-stock and assemble-to-order MCP systems the production volumes and lot sizes can be very high and in other MCP systems the customer to lot size ratio tends toward 1:1. Thus, the lot sizes are truly small in make-to-order, engineer-to-order and develop-to-order MCP systems. At Lutron's mass customization plant over 95% of its products have annual shipments of less than 100 units. Therefore, the average production volume and lot size varies from very large in Type-1 (make-to-stock) MCP to 'close to one' in Type-5 (develop-to-order) MCP.

Bill-of-Material (BOM) Coefficient

Bill-of-Material (BOM) coefficient is defined as the ratio of the number of new components to be designed (for a specific order) to the total number of components in the product expressed as a percent. In customizing an automobile, for example, with 5000 total number of components and 50 new components, the BOM coefficient is 1%. Changes in the bill of materials or engineering can be quite expensive to the company.

When a product is designed and manufactured from scratch, the value of this variable can approach 100%. The number of new (added/modified) components manufactured for building a customized product can influence the cost of order fulfillment. For each and every order, some components could be unique. It can be generalized that the more the BOM coefficient the more the production cost. The BOM coefficient will be "close to zero" for Type 1 (make-to-stock) MCP systems and will increase as one progresses to other types of MC systems.

Cost of Production System Changes

A change in the production system requires changes in the tooling and process, quality system, and management information system. While the requirements of several MCP systems can be met by general-purpose tools, there are some MC situations which generally require a new set of tools or modification of the existing tools.

In several MCP systems, and particularly in engineer-to-order MCP, changes in quality assurance systems (changes in hardware and software) can be very complicated and costly. The higher the magnitude of the changes the greater the additional cost of manufacturing and quality control.

In MCP systems, in addition to the product and the process databases, a customer database has to be maintained. Modifications to management information systems, changes in tooling and processes, and changes in quality systems are expected to be minimal in Type-1 (make-to-stock) MCP, and are generally very high in Type-5 (develop-to-order) MCP.

Learning Rate

In mass production, a learning curve with a steep slope is possible because of continuous activity on a single repeating product. In contrast, the learning efforts in MCP systems can be time consuming and complex as each order may be executed only once. Therefore, learning rate varies from very high in the Type-1 (make-to-stock) MCP to very low in the Type-5 (develop-to-order) MCP.

Customer-Order Decoupling Point (CODP)

The Customer-Order Decoupling Point (CODP) refers to the point in the value chain at which a customer order drives or shapes the production activities. All the activities downstream of the CODP are driven by forecast based planning rather than by customer orders (Wortmann, 1992). As shown in Figure 1, the value chain consists of marketing, development, production, and delivery. The chain starts with the customer and ends with the customer. Customers can be anonymous as in the case of a mass-produced and mass-customized products, or the customer can be a specific individual as in the case of make-to-order MCP. As can be easily seen in the value chain, marketing drives the development, and development drives the production, and production drives the delivery of products. The CODP is 'delivery' in Type-1 (make-to-stock) and Type-2 (assemble-to-order) MCP systems and it moves backwards to 'production,' 'development', and 'marketing' in the value

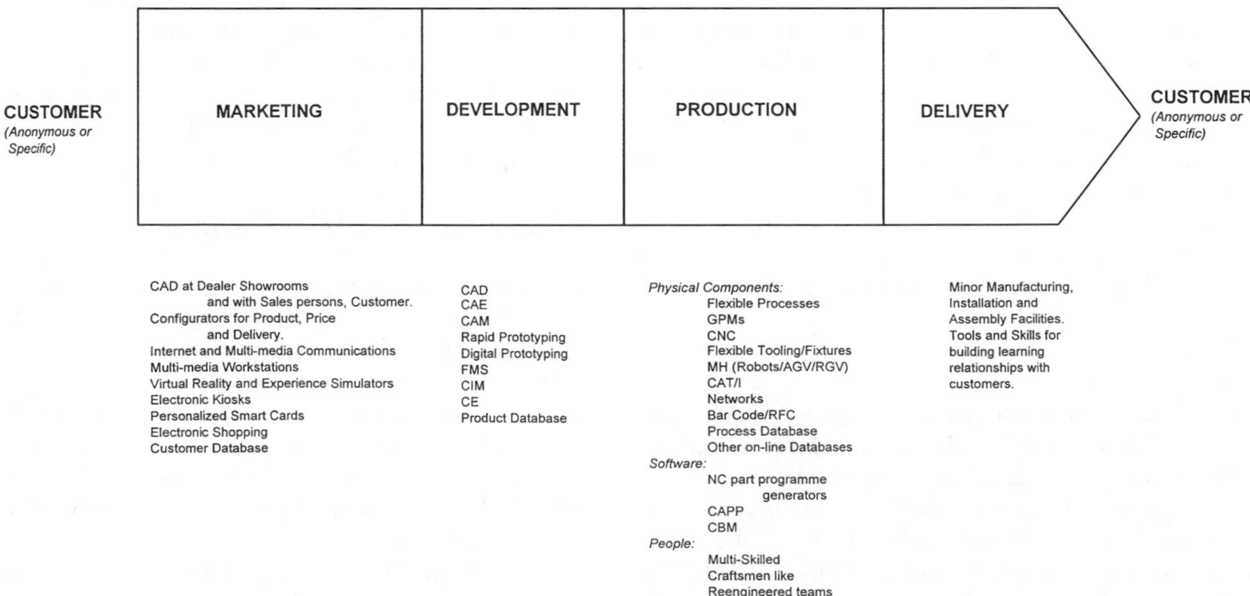

Fig. 1. Value chain and its supporting components for mass customization.

chain in Type-3 (make-to-stock), Type-4 (engineer-to-order), and Type-5 (develop-to-order) MCP systems, respectively.

Manufacturing at the Delivery End

In some MCP systems some manufacturing activity is required at the point of sales. The complexity of such a provision in terms of technology, human skills, and logistics should be worked out when developing an MC strategy. Extra manufacturing at the delivery end of the value chain might call for the installation of additional manufacturing facilities at the point of sales, and also training of sales personnel in those manufacturing operations. Obviously manufacturing at the delivery site doesn't exist in Type-1 (make-to-stock) MCP systems.

Personnel

In most MCP systems, special skills are required of the people employed to manage and operate these systems. Because it is neither possible nor desirable to automate all activities in MCP systems, the people are required to be highly skilled, multiskilled craftsmen. Continuous improvement of the skills of the personnel an essential part of the MC philosophy. The teams working on individual orders must operate in a frictionless manner to make MC possible. At Bally Engineered Structures, Inc. the workers themselves are now responsible for planning and supervising their own work (Pine II and Pietrocini, 1993). The concept of virtual teams consisting of product designers, tool planners, and production engineers placed at different geographical locations is advocated by many experts. As in agile manufacturing, the objective here should be to combine the organization, technology, and people into an integrated whole.

Advanced Technologies Used

As shown in Figure 1, the implementation and operation of MCP systems are possible with both low and high technologies. The more the complexity of the technology involved in MC, the higher the cost of implementation and operation of the production system. Except in the production of mass-produced and mass-customized products, highly flexible general-purpose technologies will be required to operate MCP systems.

To operate an MC business, several technologies such as CAD, virtual reality, customer database, electronic kiosks, and multimedia communications may be required at the marketing stage for identifying a customer's requirement. For example, Motorola's pager sales representatives use laptop computers when they work with a customer. Once the design specifications are acquired from the customer, the pager sales representative hooks up the laptop to a telephone line and transmits the design specifications to the factory. In a very short time a bar code identifying the customer is created. From that point onwards, a flexible manufacturing system takes over to manufacture the customized pager.

Sales representatives from Anderson Windows use a software system on a workstation that features 50,000 possible window components to help customers design their own windows. Once the customer submits the desired window design, the system automatically generates error-free quotations and manufacturing specifications. At the development stage, technologies such as personalized smart cards, electronic kiosks, internet, CAD, CAE, configurators, and concurrent engineering design tools will be useful in reducing the design lead times. As listed in Table 1, some other technologies specifically required for mass customization are configurators, stereolithography, internet, multimedia workstations, personalized smart cards, virtual reality, groupware, multimedia communications, electronic data interchange (EDI), flexible fixturing, etc.

Production Planning and Control (PPC)

Production Planning and Control (PPC) is based on demand forecasts for make-to-stock and assemble-to-order MCP. For example, Lego blocks are mass produced based on forecasts. Whereas in the other types of MCP systems, assembly, manufacturing, engineering, and development are driven by customer orders.

Order Promising

Order promising in a make-to-stock MCP system is based on the inventory of the finished goods, and in an assembled- to-order MCP system, it is based on the inventory of modules out of which the products are assembled. In other MCP systems, it is based on their production system capabilities. For example, a company manufacturing industrial equipment must quickly produce equipment designs according to customer requirements, translate the design into a product, and deliver it. Meeting the due dates on the orders depends on the rapid response and flexible technologies of the production system.

Handling of Demand Uncertainty

Demand uncertainty is handled by safety stocks in make-to-stock and assemble-to-order MCP systems. For example, in the Reebok's pump case, the safety stocks are calculated based on qualitative and quantitative forecasting models. In other MC schemes it is difficult, or almost impossible to forecast demand because there is no standard product for which forecasts can be prepared. Therefore the demand uncertainty in the make-to-order, engineer-to-order, and develop-to-order MCP systems is handled based on the capacity, capabilities, and the rapid responsiveness of the MCP systems.

STAGES IN MC IMPLEMENTATION: For any company, the first phase in planing for MC is to assess whether to adopt the MC philosophy and its associated operational requirements. Joseph Pine discussed a method consisting of five stages for implementing mass customization (Pine II, 1993b). In this method a company in the first stage aims at customizing services around the existing standardized products. In the second stage, it tries to mass produce customized services or products that customers can easily adjust to their individual needs. In the third stage, the company adjusts its production strategy to provide point of delivery customization. At the fourth stage, the company tries to provide quick response production customization, and at the fifth stage it tries to modularize building-blocks to customize its end products and services.

Some large companies can effectively produce products for multiple markets. However, for such an operation to be successful, a company must separate and organize its manufacturing facilities to meet the needs of each product and then develop sales volumes that are large enough to make those manufacturing units competitive. For example, Lutron Electronics has over 11,000 products which can be delivered anywhere in the world within 48 hours. Lutron has achieved this kind of customization with four levels of customization. In level 1, the customer configures the product from modules. In level 2, cosmetic customization (of color or surface finish) is allowed. In level 3 customization, some product specific manufacturing is carried out. Level 4 customization is meant for totally new designs. The organization of the company must be capable of handling all levels of customization into account.

Four distinct approaches to mass-customization are presented by Gilmore and Pine II (Gilmore and Pine II, 1997) that are helpful in redesigning a product, process, or business unit. These are named as collaborative, adaptive, cosmetic, and transparent customization strategies. In collaborative customization, the manufacturer conducts a dialogue with individual customers to help them articulate their needs, identify the product specifications, and make customized products for them. Adaptive customizers offer one standard product with embedded technology, which users can alter themselves. Cosmetic customizers present a standard product differently to different customers, for instance, with changes in packaging, or with minor modification of some selected specifications. Transparent customizers provide individual customers with unique goods or services without letting them know explicitly that those products have been customized for them (Henricks, 1996).

The degree of collaboration between a manufacturer and a customer can vary in different types of MC situations. Hewlett-Packard Company implemented mass-customization effectively by adapting three organizational principles (Feitzinger and Lee, 1997). These three principles are, modularity in the product, modularity in the process, and supply network modification to provide customization at the last step in the value chain, that is, the task of differentiating a product for a specific customer is postponed until the latest possible point in the value chain.

INCREASE MODULARITY: Products should be designed employing the concept of modularity to achieve customization. Modular components can be configured into a wide variety of end products easily and inexpensively. Black & Decker provides an excellent example of how component standardization can yield low costs and greater variety. As a result of standardizing, Black & Decker found that it could produce an entire line of 122 basic tools – jigsaws, trimmers, circular saws, grinders, polishers, sanders, and so forth – from a relatively small set of standardized components (Meyer, 1997).

There are six types of product modularity: component-sharing modularity, component-swapping modularity, cut-to-fit modularity, mix modularity, bus modularity, and sectional modularity (Pine II, 1993a). For a discussion of the types of modularities, see **Mass Customization**.

Organizations Mentioned: Anderson Windows; Bechtel; Black and Decker; Coca-Cola; Compaq; Custom Clothing Technology Corporation; GE Fanuc; GE; Gillette; Hewlett-Packard; IBM; Individual, Inc.; Lego; Levi-Strauss; Lutron Controls; Lutron Electronics co.; Matsushita Bicycle Company;

Mercedes Benz; Motorola; Nissan; Reebok; Space Electronics, Inc.; Telco, India; Toyota.

See **Agile manufacturing; Manufacturing flexibility; Mass customization.**

References

[1] Anderson, D.A. (1998). *Agile Product Development for Mass Customization, Niche Markets, JIT, Built-to-Order and Flexible Manufacturing*, Irwin, Chicago, IL.

[2] Davis, S.M. (1987). *Future Perfect*, Addision-Wesley, Reading, Massachusetts.

[3] Elantora, E. (1992). "The Future Factory: Challenge for One-of-a-Kind Production," *International Journal of Production Economics*, 28, 131–142.

[4] Feitzinger, E. and Lee, L.H. (1997). "Mass Customization at Hewlett-Packard: The Power of Postponement," *Harvard Business Review*, 75, 116–121.

[5] Gilmore, H.J. and Pine II, J.B. (1997). "The Four Faces of Mass Customization," *Harvard Business Review*, 75, 91–101.

[6] Handfield, R.B. (1995). *Re-engineering for Time-based Competition-Benchmarks and Best Practices for Production, R&D, and Purchasing, Quorum Books*, Westport, Connecticut.

[7] Henricks, M. (1996). "Mass Appeal," *Entrepreneur*, 24, 68.

[8] Hill, T. (1994). *Manufacturing Strategy, Text & Cases*, Richard D. Irwin Inc., Burr Ridge, Illinois.

[9] ICM Conferences Inc. (1996). *Mass Customization, A Two Day Conference*, ICM Conferences Inc., Chicago, IL.

[10] Kotha, S. (1995). "Mass Customization: Implementing the Emerging Paradigm for the Competitive Advantage," *Strategic Management Journal*, 16, 21–42.

[11] Lau, R.S.M. (1995). "Mass Customization: The Next Industrial Revolution," *Industrial Management*, 37, 18–19.

[12] Meyer, M.H. (1997). "Revitalize your Product Lines Through Continuous Platform Renewal," *Research Technology Management*, 40, 17–28.

[13] Moffat, S. (1990). "Japan's New Personalized Production," *Fortune*, 122, 132–135.

[14] Pine II, J.B. (1993a). *Mass Customization – The New Frontier in Business Competition*, Harvard Business School Press, Boston, MA.

[15] Pine II, J.B. (1993b). "Mass Customizing Products and Services," *Planing Review*, 21, 6–13.

[16] Pine II, J.B., Peppers, D., and Rogers, M. (1995). "Do You Want to Keep Your Customers Forever?," *Harvard Business Review*, 73, 103–114.

[17] Pine II, J.B. and Pietrocini, T.W. (1993). "Standard Modules allow Mass Customization at Bally Engineered Structures," *Planning Review*, 21, 20–22.

[18] Pine II, J.B., Victor, B. and Boynton, A.C. (1993). "Making Mass Customization Work," *Harvard Business Review*, 71, 108–119.

[19] Preiss, K. (1995). *Mass, Lean, and Agile as Static and Dynamic Systems*, Agility Forum, Bethlehem, PA.

[20] Roos, D. (1995). *Agile/Lean: A Common Strategy for Success*, Agility Forum, Bethlehem, PA.

[21] Sullivan, L. (1995). "Custom-made," *Forbes*, 156, 124–125.

[22] Womack, J.P., Jones, D.T., and Roos, D. (1990). *The Machine that Changed the World*, Rawson Associates, New York, NY.

[23] Wortmann, J.C. (1992). "Production Management Systems for One-of-a-Kind Products," *Computers in Industry*, 19, 79–88.

MASS PRODUCTION

Mass production refers to very high volume, low cost production using specialized equipment and labor specialization. This type of production frequently occurs on assembly lines, where repetitive work is performed on the jobs moved by an assembly line.

See **Assembly line design; History of operations management; Linked-cell manufacturing systems; Manufacturing systems; Moving assembly lines.**

MASS STANDARD CUSTOMIZERS

Mass standard customizers provide low cost custom products to customers by using modular components that can be custom assembled to provide unique end items. Component swapping, mix, sectional, and bus modularity could all be realized through mass standard customization. Customers are "involved" in the assembly stage of production as they choose options from a limited set of modules. Motorola pagers, a recognized leader in mass customization, can be considered a "mass standard" customizer. Pagers can be designed to a customer's specification from a wide range of options that are added at the production phase.

See **Mass customization; Mass customization and manufacturing.**

MASS TAILORED CUSTOMIZERS

Mass Tailored Customizers are one form of mass customizers. Mass tailored customizers use cut-to-fit modularity, where standard modules are altered to suit unique customer specifications to make low cost custom products. Customer inputs are received at the fabrication stage of production where modules are altered to customer specifications. Products that are finished to customer specifications, such as size or length, are examples of mass tailored customization. Bally Engineered Structures manufactures walk-in coolers, refrigerated rooms, and clean rooms. Product modules are cut-to-fit specific dimensions to suit the customer, unique rooms are manufactured from modular components. Mass tailored customization is similar to mass standard customization in that both use standard components to create unique products. However, mass tailored customizers alter the dimensions of the components (using cut-to-fit modularity) to achieve product uniqueness.

See **Mass customization; Mass customization and manufacturing.**

MASTER PRODUCTION SCHEDULE (MPS)

A master production schedule (MPS) is an anticipated build schedule for manufacturing end products or product options. It is an important component of production planning systems such as material requirements planning systems (MRP). It is a statement of production, not a statement of market demand. That is, the MPS is not a forecast. However, the sales forecast is a critical input into process that is used for determining the MPS. It represents what the company plans to produce in terms of models, quantities, and dates. It takes into account the demand forecast, the aggregate production plan, backlog, availability of material, and capacity.

In fact, a company's demand management activities (including forecasting, order entry, order promising etc.) should be closely coordinated with the MPS on an ongoing basis. The MPS takes into account capacity limitations. The MPS forms the basic communication link between marketing and manufacturing. The MPS provides the basis for resource requirements planning (RRP, i.e., Rough-cut Capacity Planning, RCP), which perform a feasibility check at critical work centers in terms of the resource required to carry out the MPS. The MPS translates the company's aggregate production plan into specific schedule for production activities. Thus, the aggregate plan constrains the MPS in that the sum of the MPS quantities must equal the totals in the production plan.

The MPS drives the "engine" and the "back end" of a manufacturing planning and control (MPC) system. That is, after the feasibility check via RRP, the feasible and acceptable MPS becomes an authorized MPS which is the key input to detailed-level planning such as material requirements planning (MRP).

See **Aggregate plan and master production schedule linkage; Capacity planning: Long-range; Capacity planning: Medium- and short-range; MRP; Safety stocks: Luxury or necessity.**

Reference

[1] Vollmann, T.E., W.L. Berry, and D.C. Whybark (1996). *Manufacturing Planning and Control Systems*, (Fourth Edition), Richard D. Irwin, Homewood, Illinois.

MATERIAL DOMINATED SCHEDULING

Any scheduling technique that schedules materials before processors (i.e., equipment/capacity) is called material dominated scheduling in the process industry; i.e., chemicals and commodities production. Material dominated scheduling (MDS) initially schedules production based on material need dates (material pull scheduling) or material availability (material push scheduling). This facilitates the development of schedules that minimize inventories. MDS is typically used in make-to-order environments where customer requirements directly define schedules (material pull scheduling). Capacity is checked after the material schedule has been created to ensure schedule feasibility. In some situations where the raw material is perishable or very costly, the arrival of raw material input drives the schedule (material push scheduling).

See **Process industry scheduling.**

Reference

[1] Taylor, S.G. and S.F. Bolander (1994). *Process Flow Scheduling: A Scheduling Systems Framework for Flow Manufacturing*, APICS, Falls Church, VA.

MATERIAL PLANNING

A major issue in manufacturing is to ensure that raw materials, component parts, and subassemblies are available when needed, where needed and in the correct quantities. This is accomplished several processes such as order releases, schedules, and inventory control – activities that fall under the heading of material planning. In certain manufacturing environments, material planning is covered by of material requirements planning (MRP). MRP is based on the concept of dependent demand. Any item that is a component part of another intermediate item, or an end item is a dependent demand item because its demand depends on the demand for the end item. Thus, component part needs to be planned based on the planned production of any item in which that part is used.

See **Capacity planning: Long-range; Capacity planning: Medium and short range; MRP; Resource planning: The evolution from MRP II and ERP.**

MATERIAL REQUIREMENTS PLANNING (MRP)

See **MRP**

MATERIALS MANAGEMENT

Materials management covers activities whose goal is to ensure the smooth flow of materials, from purchasing, through production and inventory, to distribution, while ensuring that raw materials, components and finished goods are available when needed, where needed, and in correct quantities at minimum cost to the organization.

See **Material planning.**

MATHEMATICAL LOTSIZING MODELS

Mathematical lotsizing model use mathematics to make relevant tradeoffs among production and inventory costs to determine an appropriate lot size (i.e., the quantity to produce) when production is in batches. Consider a product that is produced and sold repeatedly. There are costs associated with setting up each production run for the product, and there are inventory costs associated with holding finished products in inventory until they are sold or used. At one extreme, it would minimize setup costs if the entire amount needed in a year is produced in one lot using one setup; however inventory costs would be unacceptably high.

At the other extreme, inventory holding costs can be minimized if the production line is set up and produced each time it is needed; however, setup costs would be unacceptably high and the time spent repeadly in set ups may reduce production capacity drastically. Therefore, there is a tradeoff between the costs of carrying excess inventory and setting up frequently. The purpose of mathematical lotsizing models is to quantify these costs and model these tradeoffs so as to derive a production or order quantity (i.e., lotsize) that is economical.

A substantial body of research exists on this issue. The most basic model being the economic order quantity (EOQ) model which is based on a trade off between holding and setup costs. Advanced models extend this basic model to include other factors (e.g., variability in demand, degree of predictability of future demand, degree of risk associated with a stockout, time value of money, rate of production, quality levels, rate of obsolescence, and interactions among multiple products).

One caution when applying these models is that these models are static. That is, they derive an order quantity based on the estimated costs available at the time. If the costs change, the tradeoffs change. This caution is particularly relevant in terms of setup cost estimates. Since setup times (and costs) can be dramatically reduced with simple methods, it is important for managers to recognize that lotsizes recommended by these models can be significantly affected. Research has shown that this effect can be substantial and dramatic.

See **Economic order quantity (EOQ) model; Inventory flow analysis; Just-in-time manufacturing; Mathematical programming methods; Safety stocks: Luxury or necessity; Setup reduction; SMED; Wagner-Whitin algorithm.**

MATHEMATICAL PROGRAMMING METHODS

Mathematical programming is a term associated with problem formulation using mathematical symbols and their solution. For example, a linear programming problem is formulated showing relationship among multiple variables, constraints, and an objective. Mathematical programming may be used to formulate and save problems in engineering, education, government, business, or economics.

These methods and techniques usually include linear and non-linear models that seek an optimal or near optional solution. Books on Operations Research or Management Science are good sources for various mathematical programming methods.

See **Facility location decisions; Global facility location analysis; Simulation.**

References

[1] Ackoff, R.L. (1962). *Scientific Method: Optimizing applied Research Decisions*. New York, NY: John Wiley and Sons.
[2] Gass, S.I. and C.M. Harris (1996). *Encyclopedia of Operations Research and Management Science.* Boston, MA: Kluwer Academic Publishing.
[3] Grayson, C.J. (1973). "Management Science and Business Practice." *Harvard Business Review*, vol. 51.
[4] Kwak, N.K. and M.J. Schniederjans (1987). *Introduction to Mathematical Programming*, Krieger Publishing Co., Malabar, FL.
[5] Von Neremann, J. and O. Morgenstern (1944). *Theory of Games and Economic Behavior*.

MEAN ABSOLUTE DEVIATION (MAD)

This is a statistical measure used in evaluating forecasts. MAD is the average of the absolute values of the error, or deviations of observed (or actual) values from expected or forecasted values. In other words Mean absolute deviation (MAD) is the mean or average of the absolute values of all the *errors*. Because MAD is based on absolute errors, it provides the average magnitude of error regardless of the direction of errors. Smaller the MAD the better the forecast.

See **Forecasting examples; Forecasting guidelines; Forecasting in operations management; Mean absolute percentage error (MAPE); Safety stocks: Luxury or necessity.**

MEAN ABSOLUTE PERCENTAGE ERROR (MAPE)

The mean absolute percentage error (MAPE) is the mean or average of the absolute percentage errors of forecasts. Error is defined as actual or observed value minus the forecasted value. Percentage errors are summed without regard to sign to compute MAPE. This measure is easy to understand because it provides the error in terms of percentages. Also, because absolute percentage errors are used, the problem of positive and negative errors canceling each other out is avoided. Consequently, MAPE has managerial appeal and is a measure commonly used in forecasting. The smaller the MAPE the better the forecast.

See **Bias; Forecasting guidelines; Forecasting in operations management.**

MEAN ERROR

This is a measure of forecast accuracy. Error is defined as actual or observed value minus forecasted value. Forecast errors (could be either +ve or −ve) are summed and averaged over a period of time to develop, mean error. The smaller the mean error the better the forecast.

See **Forecasting examples; Forecasting guidelines; Forecasting in operations management.**

MEAN LATE TIME

Mean late time is a measure of performance to evaluate production or service systems. It is the average difference between the actual completion time of the job and its due date. If the job is finished after its due date, this value is positive, and the job is considered *tardy*. If the job is finished prior to its due date, this is a negative value. If an early job cannot be shipped to customers, it adds to inventory costs.

See **Manufacturing systems modeling using petri nets; Sequencing rules; Simulation analysis of manufacturing and logistics systems; Simulation of production problems using spreadsheet programs.**

MEAN PERCENTAGE ERROR (MPE)

This is one of the measures used to evaluate forecasts using forecast errors. Forecast error is defined as actual observation minus forecast. The mean percentage error (MPE) is the average or mean of all the percentage errors. The averaging process sums up positive and negative percentage errors allowing them to cancel each other out. Consequently, the mean error can mislead the manager into thinking that the average error is low, when in fact high forecasts and low forecasts may be canceling each other out. This problem is avoided with absolute error measures which provide the average based on total error regardless of sign. However, the mean percentage error does have value as a measure of forecast bias, telling the manager whether the forecasting model is over or under predicting.

See **Bias in forecasting; Forecasting guidelines; Forecasting in operations management; Mean absolute deviation (MAD); Mean absolute percentage error (MAPE).**

MEAN QUEUE TIME

In a job shop, work orders or jobs wait at work centers or buffers for processing. The average total time jobs spend waiting in lines or queues is used as a measure of performance or effectiveness.

See **Manufacturing systems modeling using petri nets; Simulation analysis of manufacturing and logistics systems; Simulation of production problems using spreadsheet programs; Waiting line model.**

MEAN SQUARED DEVIATION (MSD) ERROR

MSD is one of several measures for evaluating forecasts accuracy. It is calculated by squaring the individual forecast deviation (error) for each period and then finding the average or mean value of the sum of squared errors. Forecast error is actual observation for a period minus the forecast for the period. The mean squared error is used because, by squaring the error values, the resulting values are all positive. Mean squared deviations are easier to handle mathematically, and so it is often used in statistical optimization. The only problem with using this metric is that it is not easy to interpret by management. In addition, the squaring of individual large forecast deviations (errors) can exaggerate their importance for a time series.

See **Bias in forecasting; Forecasting guidelines and methods; Forecasting in operations management; Mean absolute deviation (MAD).**

METHODS DESIGN (OR ENGINEERING)

It refers to the process of developing production tasks and steps associated with the manufacturer of a component, part or product.

See **Learning curve analysis; Process planning.**

MINIMUM INVENTORY PRODUCTION SYSTEM (MIPS)

Many U.S. manufacturers use Just-in-Time manufacturing or world-class manufacturing systems but call such practices by names unique to their company. MIPS is the name of the system used in Westinghouse Corporation which is based on Just-in-Time or World Manufacturing Systems.

See **Just-in-time manufacturing; Linked-cell manufacturing; Manufacturing cells, linked; Total quality management.**

MISTAKE-PROOFING

See **Lean manufacturing implementation; Poka-Yoke; Total quality management.**

MIX MODULARITY

Mix modularity is a form of modularity employed in implementing mass customizatoin. When modules are combined, the modules lose their unique identity. A common example of mix modularity can be seen in custom blended paint for home interiors. A wide variety of paint colors can be mixed from a few base colors and a collection of primary tints. Paint colors can be made to match exactly a color supplied by the customer.

See **Mass customization; Mass customization and manufacturing.**

MIXED-MODEL SEQUENCING

Two objectives of JIT manufacturing are to reduce inventory levels and achieve flow manufacturing. Neither one of these can be achieved if products are manufactured intermittently in large batches. Large-batch production results in a large inventory buildup with associated costs. Large-batch production of final assemblers also creates uneven demand for component parts. When a large batch of one product is being produced, demand for components not used in the product in current production will be zero.

To reduce inventory levels and create a more level demand for component parts, batch sizes for final assemblies must be reduced. To achieve smooth flow, ideally, each product should be produced in a quantity that approximately matches the daily demand for that product. To maintain a smooth flow of component parts, these products are made in a sequence so that individual product batch sizes come as close to one as possible. For example, suppose a company sells three products, X, Y, and Z, and that the average daily demand for X is 15, demand for Y is 10, and demand for Z is 5. An ideal mixed-model sequence would be XYXYXZ, repeated 5 times each day. This sequence is considered ideal because the production of each product is spread evenly throughout the day. However, setup considerations might require a different sequence such as XXXYZ. This would still match daily production to average daily demand and spread the production of each product evenly throughout the day.

See **Just-in-time manufacturing; Uniform load scheduling; Lean manufacturing implementation; Linked-cell manufacturing system; Manufacturing cell design; manufacturing systems.**

MODEL VALIDATION AND VERIFICATION

Simulation is an increasingly important tool for analysis and study of complex manufacturing systems. Once the simulation model has been implemented and populated by an appropriate data set, it must be validated and verified. *Model validation* is the process of ensuring that the model accurately represents the actual system *for the purposes of the simulation study*. If a model is not valid, then the conclusions derived from the model will be of questionable value. If a model is valid, it may be expected to behave like the real system, and decisions affecting the real system can be made with confidence based on the results of the simulation model.

See **Simulation analysis of manufacturing and logistics systems; Simulation of production problems using spreadsheet programs.**

MODULAR ARCHITECTURE

Modular architecture is a product design and development related concept. A modular architecture for a product is characterized by groups of components (called *modules*) which interact only through *de-coupled interfaces*. De-coupled interfaces allow changes in one module without affecting the other (Ulrich 1995). For example, the video cassette is a module in the video player system. As long as the interfaces with the video player are unaffected (dimensions of the casing, position of the axles, and position and direction of the magnetic tape), the inner workings of the cassette are irrelevant for the system to work.

Architectural modularity has several advantages (Ulrich 1995, Baldwin and Clark 1998): it increases the range of "manageable" complexity by limiting the scope of interactions among subparts (modules) of the design. It allows higher concurrency of the design process, as the modules can be designed in parallel and independently of one another (as long as the interfaces, or design rules, are respected). It allows "mixing and matching," and thus higher part commonality across products, implying a higher manageable product line variety. Finally, modularity accommodates uncertainty, as module designs are reversible and can be changed later in response to changed demands, without having to re-design the whole system. This may lead to competition-driven innovation at the level of the modules: in the PC industry, for example, strong competition drives dynamic innovation in separate component industries of disk drives, microprocessors, DRAM memories, and monitors.

However, modularity comes at a cost. Setting design rules, and checking whether the separate modules really do work together (system integration), adds overhead activities to design. Moreover, modularity imposes design constraints and limits performance.

See **Concurrent engineering; Mass customization; Mass customization and manufacturing; Product development and concurrent engineering.**

References

[1] Baldwin, C.Y. and K.B. Clark. *Design Rules: The Power of Modularity*. MIT Press 1998.

[2] Ulrich, K. (1995). "The Role of Product Architecture in the Manufacturing Firm." *Research Policy* 24, 419–440.

MONTE CARLO SIMULATION

Simulation is a tool for studying and analyzing manufacturing systems for decision making and for design of manufacturing systems.

Monte Carlo simulation (MCS) is a method of simulation, which is appropriate when one or more elements of the system being investigated exhibit chance in their behavior. MCS uses random sampling to study through experimentation the probabilistic (chance) elements of the system.

Von Neumann developed Monte Carlo Simulation during WWII at the Los Alamos Scientific laboratory. In the 1960's computer programs such as GPSS and SIMSCRIPT were created to handle large simulation problems.

Later, in the 1980's simulation programs such as Xcell, SLAM, Witness and MAP/1 were developed to handle manufacturing systems design and problem solving.

Typical examples of variables that are simulated as part of Monte Carlo Simulation of production systems are: inventory demand, time between machine breakdowns, processing time at the work center and time between arrivals at a work center.

An example of a simple Monte Carlo Simulation is presented by Ronder and Stair (1992), where a hardware store investigates an inventory policy consisting of an order quantity of 10 and a reorder point of 5. With the help of MCS, the example shows that the following annual costs can be estimated: ordering cost, holding cost, shortage cost, and total annual inventory cost. The example also shows that through experimentation using MCS, a better inventory policy is an order quantity of 10 with a reorder point of 6.

See **Safety stocks: Luxury or necessity; Simulation analysis of manufacturing and logistics systems; Simulation of production problems using spreadsheet programs.**

References

[1] Render, B. and R.M. Stair, Jr. (1992). *Management Science*. Boston, MA: Allyn and Bacon.

MONTREAL PROTOCOL

Montreal Protocol concerns substances that deplete the Ozone Layer in the atmosphere. Initially, signed by 27 countries in 1987 in Montreal, Canada, and subsequently amended in 1990 and 1992, this document addressed the depletion of stratospheric Ozone that occurs through the use of industrial compounds such as chorofluorocarbons (CFCs), halons, carbon tetrachloride, and methyl chloroform. These substances, once emitted to the atmosphere, could significantly deplete the stratospheric ozone layer that shields the planet from damaging UV-B radiation. The production and consumption of specific compounds are to be phased out according to established schedules. For example, CFCs were phased out by developed nations by 1996, with a "grace period" of 10 years. The final objective of the Protocol is the elimination of all ozone depleting substances. By 1998, over 165 countries had signed the agreement.

See **Environmental issues and operations management.**

MOTHER PLANT

In plant location studies, mother plant refers to the original plant from which a company evolved. It often is the leading plant in technology development and a source of managerial capabilities for other plants of the corporation.

See **International manufacturing.**

MOVING ASSEMBLY LINE

It is a manufacturing system for high volume production. The idea here is to subdivide assembly operations into a number of smaller tasks that are assigned to workers placed sequentially in a fixed order, and the product is moved from one worker to the other by conveyor. Prior to the development of the assembly line, Ford Motor Co. built cars on fixed assembly stands with workers and the components moving to cars on fixed locations. A great deal of time and effort was taken up by the movement of workers and parts throughout the factory.

With moving assembly lines, productivity increased significantly since the time and effort spent on non-productive movement was eliminated. Workers no longer had to move, the cars came to them. Since specific activities were always performed in one spot, material flows could be regularized with single delivery points and storage spaces. The moving assembly line was so-named because it did not stop: workers had to perform tasks on the moving assembly. The assembly line was mechanically driven and paced so that workers had tightly controlled spaces (and times) in which to complete their work.

The moving assembly line was developed at Ford Motor Co. in 1913 as a means for increasing production volume and productivity. Although Henry Ford himself took credit for the innovation this claim has been contested by some.

Organizations Mentioned: Ford Motor Co.

See **Flow lines; History of operations management; Manufacturing systems.**

References

[1] Ford, Henry and Samuel Crowther (1924). *My Life and Work*. William Heinemann, Ltd., London, England.

[2] Ford, Henry (1926). Mass Production. *Encyclopaedia Britannica*, 13th Edition, Supplemental Volume 2, 821–823.

[3] Hounshell, David (1983). *From the American System to Mass Production*. University of Delaware Press, Wilmington, Delaware.

MOVING AVERAGE FORECASTS

This is simple forecasting method. The moving average is the average (mean) value of a set of most recent m successive observations in a time series. The calculated value is used as a forecast for the next time period. As time advances the oldest observation used in computing the average is dropped and the newest one is included. For details regarding the forecasting method and other methods, see the cross-referenced items below.

See **Forecasting examples; Forecasting guidelines and methods; Forecasting in operations management.**

MPC

See **Manufacturing planning and control; MRP.**

MPS (MASTER PRODUCTION SCHEDULE)

See **Master production schedule; MRP; MRP implementation; Aggregate plan and master production schedule linkage.**

MRP (MATERIAL REQUIREMENTS PLANNING)

V. Sridharan and R. Lawrence LaForge

Clemson University, Clemson, SC, USA

INTRODUCTION: What, how much, and when to produce are questions at the core of every manufacturing operation. Answers to these three questions constitute the essence of manufacturing planning and, in turn, establish the basis for material planning. Until the early 1970s the planning of material requirements for manufacturing was based on reorder point models. For independent demand items, reorder point models are used for stock replenishment. The depletion of items in inventory is monitored, and a replenishment order for purchase or production is released in periodic intervals, or when inventory reaches a predetermined level called reorder level. Order quantities are determined by considering the trade-off among related costs, demand forecast and demand fluctuations. The chief draw back of these models has been their lack of forward visibility. Furthermore, reorder point models, while well suited for a distribution environment where demand for individual items may be viewed as independent, failed to recognize the dependence between inventory items in a manufacturing environment.

Recognition of dependent demand in manufacturing and the need for forward visibility led to the development of a framework for making material replenishment decisions in the context of a manufacturing operation. It is this framework that is often called the material requirements planning (MRP) framework. It is important to recognize here that the framework facilitates production and materials related decision making by providing timely and appropriate information for decision making, and about implications of decisions. In this sense, an MRP system may be best viewed as a management information system rather than a decision-making system. Likewise, it is also important to recognize that the material requirements planning activity is a part of the manufacturing planning and control system. Figure 1 presents a schematic of a manufacturing planning and control system and serves to illustrate the role of material requirements planning (MRP). In this article we will briefly introduce the reader to the MRP-framework.

BASIC IDEA: The MRP framework views time in discrete intervals (eg., a week), and uses three basic inputs to compute the time-phased material requirements for all items (parts, components, and sub assemblies) needed to make the final product, commonly referred to as the end-item. The basic MRP inputs are: (1) Master Production Schedule (MPS); (2) Bill of Material (BOM); and (3) Inventory Status (IS).

The master production schedule is a time-phased plan that stipulates the completion dates for end-item production. In other words, the MPS states what the company is planning to make, and when it will make then. Thus, for each period,

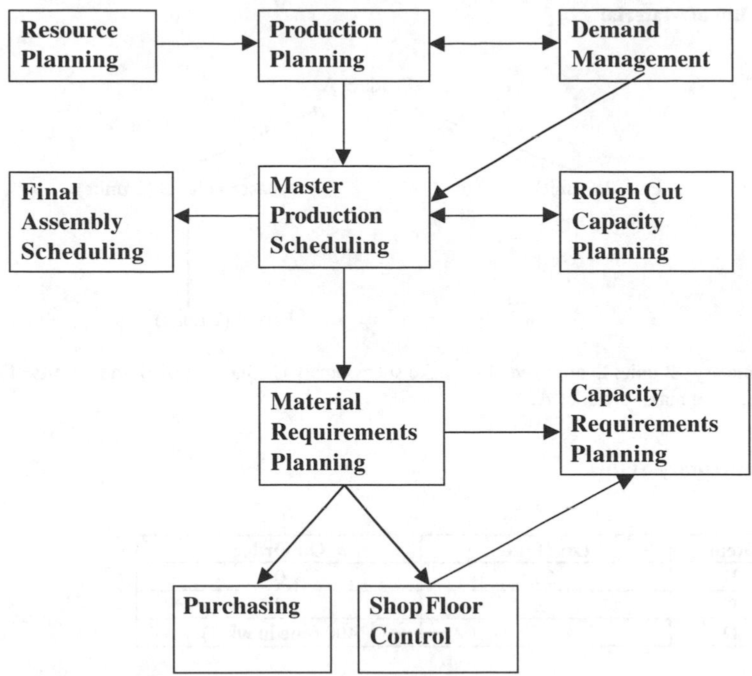

Fig. 1. Manufacturing planning and control system [adapted from Vallmann, Berry, and Whybark (1997)].

it specifies the anticipated production quantities of each kind of finished product, and serves as a key input for determining the requirements of all purchased parts, components, and sub-assemblies needed to make the finished products. The MPS is often expressed on a weekly basis, and is stated for three or more months (known as planning horizon). When several variations of the end-item are offered the master production schedule is accompanied by a final assembly schedule (FAS). Where FAS is maintained for a specific product configuration (such as, Model A or Model B), the MPS is maintained at the common sub-assembly level for options.

The thrust of material planning is to determine component item requirements based on the MPS over the planning horizon. This obviously requires information about components needed, quantity needed, assembly process, and lead-time to obtain each component. The bill-of-material (BOM) provides such information by describing the ingredients (components, purchased parts, and sub-assemblies) needed to produce the end-item. Thus, it may be viewed as the recipe for making the finished product. It may contain routing and lead-time information. However, in practice, the routing and lead time information is often maintained separately in what is called the "item master file."

Note that the determination of material requirements cannot be divorced from the information on how many units of each component item are already on hand and how many are on order. This information is reflected in the inventory record of the item and is frequently referred to as "inventory status".

In summary, once we know what end items (and when) we want to make them, what (and when) we need to make them, and what end items and components we have, the next logical step is to determine what materials, parts and components (and when) we need to get. This is exactly what material requirements planning is all about. MRP systems take the aforementioned three inputs and calculate the exact time-phased net requirements for all component items. In turn, this serves as the basis for authorizing the commencement of production for manufactured parts, and release of purchase orders for purchased parts.

AN ILLUSTRATIVE EXAMPLE: In this section we present a simple example to illustrate the mechanics of how the three basic inputs are processed to obtain time-phased net requirements for component items. The example serves also to demonstrate the two basic steps of the MRP logic – (a) gross-to-net explosion and (b) lead time offsetting. Answers to how much and when to produce (or procure) are determined using these two steps of MRP logic. While the first step helps deduce how much to order, the second step determines the timing of the orders for each item.

The MPS for a fictitious finished product A is depicted in Table 1. It calls for starting the assembly of 80 units of end-item A in week 5. Figure 2 presents the BOM for end-item A, and the inventory status information. The BOM shows

A. Bill of Material

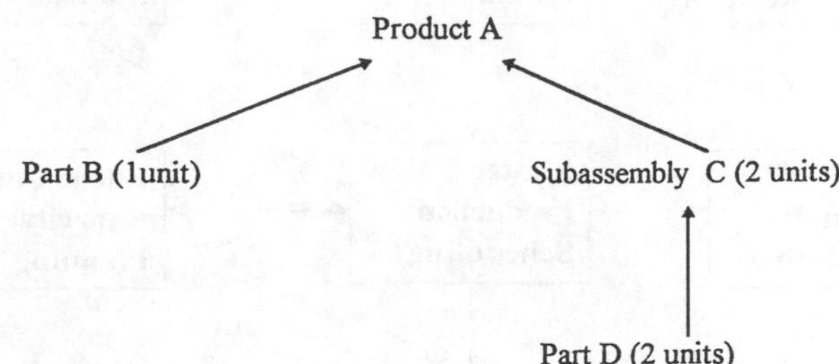

Summary : 2 units D are needed to make subassembly C. One unit of B and 2 units of C are needed to make product A.

B. Inventory Status.

Item	On Hand	On Order
B	50	0
C	0	0
D	0	400 (due in wk.3)

Fig. 2. Bell of material and inventory system.

that end-item A is made by assembling one unit of item B and two units of items C. Each unit of item C is made from two units of item D. Items B and D are purchased parts. The lead times needed for each item are also shown (in parenthesis) in Figure 2. The inventory status information shows, for each item, the on-hand and on-order quantities and their due dates.

Gross-to-Net Explosion: Table 2 depicts completed MRP records for all items in the BOM for product A. The record shown for product A reflects the decision made by the master scheduler about what to produce. This decision drives the material planning process.

In Table 2, consider the MRP record for item B. The first row, labeled Gross Requirements, shows the quantity that must be on hand at the beginning of each week in order to meet the production schedule stipulated by the MPS. Since 1 unit of B is needed for each unit of end item A, the gross requirement for B is 80 units in week 5. The second row indicates the scheduled receipts, which are the result of orders already issued. Any quantities shown in the Scheduled Receipts row are *expected to be received at the beginning of the week indicated.* In the record for B, no order is

Table 1. Master production schedule for end item A.

Weeks	1	2	3	4	5
MPS Release					80

Table 2. MRP records

Week	1	2	3	4	5
MPS Releases for End Item A					
MPS Releases	0	0	0	0	80
MRP Record for Item B					
Gross Requirements	0	0	0	0	80
Scheduled Receipts					
On Hand 50	50	50	50	50	0
Net Requirements					30
Planned Order Release				30	
MRP Record for Item C					
Gross Requirements	0	0	0	0	160
Scheduled Receipts					
On Hand 0	0	0	0	0	0
Net Requirements					160
Planned Order Release			160		
MRP Record for Item D					
Gross Requirements	0	0	320		
Scheduled Receipts			400		
On Hand 0	0	0	80	80	80
Net Requirements					
Planned Order Release					

outstanding and the scheduled receipts are zero in each week.

The third row shows the quantity projected to be on-hand at the end of each week. Based on the inventory record, which shows that, presently 50 units of item B are on hand, 50 units are shown to be available at the beginning of week 1. Since, the gross requirement during the first four weeks is zero, these 50 units will remain on hand until the end of week 4. The gross requirement in week five is 80. Clearly, the 50 units on hand can be used to meet part of the gross requirement. However, an additional 30 units are needed to fully meet the requirement. Thus, the net requirement in week 5 is 30 units. This information is shown in the row titled Net Requirements. Thus, the derivation of the numbers in the Net Requirements row is based on the application of the gross-to-net MRP logic.

LEAD-TIME OFF-SETTING: A work order needs to be released to the shop floor for the 30 units of item B that are needed. Since the lead time in the BOM for item B is one week, to have 30 units completed by week 5, the order must be released in week 4. This is reflected in the row titled, Planned Order Release, and illustrates the lead-time off-setting of MRP logic.

A similar logic is used to complete the MRP record for item C, which has a gross requirement of 160 units in week 5 (since two units of C are needed for each unit of A). Based on MRP logic, once the start date of the work order for making item C is determined as week 3, the plan for item D can be made as follows. Since 2 units of item D are needed for each unit of item C, 320 units of item D are needed in week 3. Therefore, as shown in the MRP record for item D, the gross requirement for item D is 320 units in week 3. Inventory record shows an order for item D in the amount of 400 units is scheduled to be received at the beginning of week 3. Thus, the projected on-hand inventory at the end of week 3 is 80. Consequently, the net requirement is zero and no order is planned for release for this item.

The two step of MRP logic, as demonstrated in this example, are netting of the requirements for each item after taking into account on-hand and on-order quantities, and time-phasing the order releases to meet the net requirements.

BOM Explosion also demonstrated by the above example is the conversion of the order release date and order quantities for an higher level item (e.g., B) into gross requirements for a lower level component (e.g., C). This process is referred to as the BOM explosion. Thus, as a result of the BOM, explosion the MRP system produces the planned order schedule for manufactured items, and purchase orders.

MRP SYSTEM NERVOUSNESS: Once the MRP records are processed for a planning horizon (which is predetermined), the first period's decisions indicated by the MRP documents in Table 2 are implemented. Replanning is performed at the beginning of the next period to take into account the actual outcome of period 1 (based on one week replanning frequency). The frequency of replanning is influenced by the dynamics of the operating environment and by the computer processing power. With the ever-increasing processor speed it has become easier to update MRP records frequently.

Frequent replanning keeps MRP records up-to-date. However, replanning too frequently is not desirable because it could result in constantly-changing production schedules. Changes in demand and the master production schedule result in due-date changes for open orders, and changes in quantity and timing for planned orders of end items. Since planned orders for the end-items translate into gross requirements for component items, the due dates for open orders and planned orders (timing or quantity) for components also change. This phenomenon is referred to as system nervousness, and could be an obstacle to the successful implementation of MRP systems.

BENIFITS OF MRP: A company can expect to obtain several benefits from the successful implementation of an MRP system. A successful implementation brings a good match between demand and supply. Through a closer match between demand and supply, finished goods inventories can be reduced. Better planning of priorities and improved scheduling of work orders help reduce work-in-process inventory. Accurate timing of vendor deliveries brings about a reduction in raw material inventories. Overall, inventory investment and cost, and obsolescence are reduced. The integration of production and capacity planning improves the management of human resources, equipment, and customer delivery promise performance. Accurate MRP information on inventory records, shop floor routings, and lead times reduce accounting adjustments and hidden overhead costs.

Successful implementation requires more than an MRP information system software. The accuracy of the input information determines the validity of the material plans. Thus, ideally, the various databases should be integrated. Furthermore, they should be updated on a regular and frequent basis. Training all MRP users and creating a proper culture to use the system are also important. Meeting these prerequisites clearly has costs. However, in many cases, benefits from successful implementation outweigh the preparation costs.

WEAKNESS OF MRP: Despite the potential benefits, it is important to recognize the limitations of the MRP framework. A central weakness concerns the modus operandi of sequential, independent processing of information as used by MRP to plan requirements level by level of the bill of material. MRP does this in an attempt to "divide and conquer" due to the complexity of the issues involved. This is highlighted by the process of first planning material requirements at one level, and then the planning process moves on to address the utilization of resources and capacity at lower levels. This could result in a delayed discovery that the plan is infeasible due to capacity limitations.

Another weakness is the use of planned lead times, which are management parameters and represent the amount of time budgeted for orders to flow through the factory. The lead times used are based on average behavior, but actual lead times are highly variable, and depend on capacity utilization, lot sizes, and product mix. A third weakness is the absence of a formal feedback mechanism to handle deviations from the plan.

SUMMARY: Since late 1960's the MRP framework has revolutionized the practice of material planning in manufacturing. Its use is widespread. A vast majority of small and large manufacturing companies, around the world, have made significant and substantial investment in MRP systems and, hence, continue to use MRP-based systems for manufacturing planning and control. Evidence also indicates that about 50% of U.S. companies covering a wide spectrum of manufacturing industries use an MRP-based systems for manufacturing planning and control (Swamidass, 1998).

See **Capacity planning: Medium- and short-range; Manufacturing technology use and benefits; MRP implementation; Resource planning: MRP to MRP II and ERP.**

References

[1] Orlicky, J. (1975). *Material Requirements Planning*, McGraw-Hill.

[2] Plossl, G.W. (1994). *Orlicky's Material Requirements Planning*, 2nd edition, McGraw-Hill.

[3] Swamidass, P.M. (1998). *Technology on the Factory Floor III*. Washington, DC: manufacturing Institute of the National Association of Manufacturers.

[4] Ptak, C.A.M. (1997). *MRP and Beyond*, Richard D. Irwin.

[5] Vollmann, T.E., W.B. Berry, and D.C. Whybark (1997). *Manufacturing Planning and Control Systems*, 4th edition, Irwin/McGraw Hill.

MRP IMPLEMENTATION

Chee-Chung Sum and Kwan-Kee Ng

National University of Singapore, Singapore

INTRODUCTION: MRP (Material Requirements Planning) is the most popular computer-based manufacturing management system in the world. The term "MRP" is generally used to include all versions of MRP systems, namely, MRPI, closed-looped MRP, and manufacturing resource planning (MRPII). The MRP literature indicates that while MRP benefits can be substantial, the success rate of MRP implementation is small. The realization of MRP benefits is heavily dependent on management of the MRP implementation process. MRP implementation process, and critical success factors for MRP implementation are presented here.

BASICS OF MRP: The challenge faced by manufacturing companies is knowing what to make and when to make. This challenge can be addressed by an effective planning and scheduling system (Ang *et al.*, 1995). MRP, which is used for this purpose is the most widely used and successful computer-based production planning and inventory control system.

From its beginnings in the 1950's as an inventory re-ordering tool, MRP has evolved into a method for effectively planning all the resources in a company. In its highest form (class A MRP), it links a variety of functions, including business planning, production planning, purchasing, inventory control, shop floor control, cost management, capacity, and logistics management. MRP also integrates operational planning (in units of production), and financial planning (in dollars), and has the ability, using simulation capability, to answer "what if" questions.

The physical workings of an MRP system have been extensively documented in the operations management literature. The benefits derived from a successful MRP system have also been well documented. Surveys of MRP users have shown that a successful MRP system generates both tangible and intangible benefits. Tangible ones include better inventory and raw material control, reduced clerical support, and reduced lead times, while improved communication and better integration of planning have been cited as examples of intangible benefits.

While computer hardware and software are essential elements of MRP, it is important to realise that it is not just a computer system. Many of the determinants of its success (like handling initial resistance to change and degree of organisational commitment) are directly related to the human factor. Implementing an MRP system involves major changes to the formal and informal systems within the company.

SUCCESSFUL MRP IMPLEMENTATION AND CRITICAL SUCCESS FACTORS (CSFs): Key managers' attention can be focused on the most important aspects of MRP implementation. The literature on MRP implementation can be broadly categorized into case studies and theoretical modelling studies. Case studies are descriptive accounts of implementation experiences at one or a few companies (Bevis 1976, Brown 1994, Grubbs 1994, Hall and Vollman 1978, Ormsby *et al.* 1990, Sheldon 1991, Sheldon 1994, Sum *et al.* 1997, Works 1998). Most studies presented testimonials of successful MRP implementations (Sheldon 1994, Works 1998), while others provided accounts of difficult implementations and unsuccessful efforts (Andrews 1978, DeSantis 1977).

Bevis (1976) provided one of the earliest descriptions of a successful MRP implementation at Tennant Company. He emphasized the importance of commitment during the implementation process. Subsequent studies have identified other important implementation variables; top management commitment and involvement (Anderson and Schroeder 1984, Beddick 1983, Brown 1994, Hall and Vollman 1978); MRP education and training (Anderson and Schroeder 1984, Beddick 1983, Sheldon 1991); dedicated formal project team (Anderson and Schroeder 1984, Beddick 1983); data integrity (Beddick 1983, Hall and Vollman 1978); interdepartmental co-operation and support (Appleton 1997, Tadjer 1998); and clear goals and plans (Appleton 1997, Sheldon 1991).

MRP practitioners and consultants have also consolidated their implementation experiences into prescriptive advice and a roadmap for MRP implementation (Ptak 1996, Toomey 1996, Scott 1994, Wallace 1990, Wight 1984). Some of the recommendations of these "MRP gurus" include a phased implementation approach, an implementation time frame of three months to two or more years depending on the scope and scale of implementation, the appointment of an operations person as project leader, the use of education as an agent to accomplish behaviour change, and the proper use of benchmarks and performance measurements to keep the MRP project on track and avoid costly mistakes.

The MRP implementation process is an organizational change process (Blasingame and Weeks 1981, Brown 1994, Cooper and Zmud 1989, 1990, White 1980). Several studies argue that the problems of MRP implementation are basically behavioural, stemming from a natural resistance to change on the part of individuals. Thus, MRP implementation is likened to managing a change process within the company. These ideas were based on Lewin-Schein's (Lewin 1947, Schein 1964) change theory of unfreezing, change and refreezing stages. The central idea of this theory is that in order to learn something new, a person must first "unlearn/unfreeze" old ideas and behaviours and experience a "felt need" to change. Brown (1994) argued that effective management of technological change such as MRP implementation requires transformational leadership and elaborate and dramatic planned sets of activities or "rites" are needed to promote change in individuals at both the psychological and behavioural level.

Contextual variable studies have focused on a variety of individual, organizational and technological variables, which were important to MRP implementation (Burns and Turnipseed 1991, Duchessi *et al.* 1988, Schroeder *et al.* 1981, Sum *et al.* 1995, Sum *et al.* 1998, White *et al.* 1982). Most of these empirical studies used statistical modelling techniques such as regression, discriminant analysis, and chi-square tests to examine the correlates of successful MRP implementation.

White *et al.* (1982) examined the problems encountered during the MRP implementation process. The study found that data accuracy, top management support, marketing management support, and in-house training and education were all predictors of successful MRP implementation.

Schroeder *et al.* (1981) performed a regression analysis to relate MRP benefits and several independent variables. The findings indicated that company size, data accuracy, and the implementation method affected the level of MRP benefits achieved.

Duchessi *et al.* (1988) empirically identified the determinants of success in implementing MRP systems. The findings confirmed several variables identified by White *et al.* (1982) as important to successful MRP implementation. Support from top management, marketing management and production management were particularly important for successful MRP implementation.

Burns and Turnipseed (1991) used chi-square tests of independence to identify variables significant to MRP implementation success. The findings indicated that the use of a project team and data accuracy were associated with successful MRP implementation.

Sum *et al.* (1995 and 1998) identified determinants of specific outcomes of MRP implementation such as operational efficiency, customer service, and co-ordination. Data accuracy, people support, company size, degree of integration were important determinants of the outcome of MRP implementation outcome.

SIGNIFICANT REULTS AND MANAGERIAL IMPLICATIONS: The studies by Sum *et al.* (1995 and 1998) have significant implications for both MRP managers and users in managing the MRP implementation process successfully. The key findings and implications are as outlined below.

1. **Data accuracy** is an important pre-requisite to successful MRP implementation. Data accuracy appears to be critical in affecting operational efficiency, customer service, and coordination benefits. This is expected because data accuracy affects the quality of work flow and decisions made. High data accuracy tends to raise the users' level of confidence in and acceptance of the system. Unless the system is perceived to be credible, users will be hesitant to exploit the full capabilities of MRP. The importance of data accuracy increases with the size of the company. This implies that it becomes more crucial for larger companies to have a higher level of data accuracy in their MRP systems so as to generate integrity and credibility against a background of increasing operational complexity and functional linkages. Managers and users, especially those in larger companies, must expend extra effort to maintain a high level of data accuracy in order to attain a high level of implementation success.
2. When production support problems are severe, data accuracy could still enhance performance, when users utilise the accurate data from the formal system to perform their job duties. Therefore, managers must carefully monitor the production support problems in relation to data accuracy.
3. **Size** can have a negative impact on efficiency.
4. **Proper training and education** are important for successful MRP implementation. In addition, the level of training and education provided becomes more crucial as the size of the company increases. This is particularly important for larger companies with complex operations. Managers in larger companies have to expend extra effort to ensure that proper training and education are provided to all those being affected by the MRP implementation process.
5. **Support from both top management and production** is essential for successful MRP implementation. Top management, therefore, should periodically review its own commitment level.
6. **People support** was found to be crucial to achieving efficiency and coordination benefits. When people problems reach a critical level, users abandon the formal system for informal backup systems to accomplish their work. The managerial implication is that people problems must be monitored closely. Managers must also understand that informal systems can play a positive role in the presence of extreme people problems. Managerial judgement must be exercised to approve temporary informal systems while efforts are made to dissipate people conflicts and tensions.
7. **Having clear goals and objectives** for the MRP implementation effort is a pre-requisite to the system's success. Therefore, top management should have clear goals and objectives for its implementation efforts before embarking on the MRP project.
8. The degree of integration has a non-linear effect on efficiency and coordination benefits. Where highly integrated information is needed, **total or near total integration** is essential if substantial benefits are to be realised.
9. Managers should monitor and promote usage among different departments. As it might be difficult to monitor the usage pattern in different departments, the profiles of technical complaints of the departments could provide valuable information on the usage pattern. For example, if Department A registers a large number of technical and system problems, and there is a notable absence of similar complaints in Department B, it could mean that that users in Department B are not utilising the system. A low incidence of technical complaints should be carefully interpreted as it could mean either an efficient system, or that the system is not being used.

ADVANCES IN MRP AND THE FUTURE: As organisations adopt a more global posture, distributed manufacturing will become more and more common. The latest version of MRP systems called Enterprise Resource Planning (ERP) can plan and mobilize geographically dispersed enterprise resources that manufacture, ship, distribute, and account for customer orders. ERP differs primarily from MRPII in technical requirements such as graphical user interface, relational database, use of fourth-generation language, client-server architecture and open- system portability. Most ERP products (eg., SAP, Baan, Oracle, PeopleSoft) which cost millions of dollars have added functions such as logistics, supply chain management, and distribution to allow organisations to exchange plans and implement actions across geographical boundaries.

More research needs to be conducted on the design, implementation, and management of ERP systems. Training, cross-cultural implementation, project management, organisational structure, and data management are some critical issues that would affect the success of these systems. Another research direction pertains to the use of MRP to gain competitive advantage. Many organisations still perceive MRP systems as a tool for incremental improvement. More research needs to be directed at identifying how successful MRP users are using MRP as a strategic weapon to gain competitiveness. The MRP implementation process is very complex involving both technical and organisational factors. A new stream of research could be directed at understanding the complexities of the process, especially the interaction of various technical, organisational and behavioural variables.

Organizations Mentioned: BAAN; Oracle; Peoplesoft; and SAP.

See **Aggregate plan and master production schedule linkage; ERP; Manufacturing technology use and benefits; MRP; Order release; Resource planning: MRP to MRP II and ERP.**

References

[1] Anderson, J.C., Schroeder R.G., Tupy S.E. and White E.M., "Material requirements planning systems: the state of the art", *Production and Inventory Management*, Vol. 23, No. 4, 1982, pp. 51–66.

[2] Andrews, C.G., "Leadership and motivation: the people side of systems implementation", *In Proceedings of the American Production and Inventory Control Society Conference*, 1978, pp. 352–387.

[3] Ang, J.S.K., Sum, C.C., and Chung, W.F., 1995. Critical success factors in implementing MRP and government assistance: A Singapore Context, *Information & Management*, Vol. 29, pp. 63–70.

[4] Appleton, E.L., "How to survive ERP", *Datamation*, Vol. 43, No. 3, 1997, pp. 50–53.

[5] Beddick, J.F., "Elements of success – MRP implementation", *Production and Inventory Management*, Vol. 24, No. 2, 1983, pp. 26–30.

[6] Bevis, G.E., "A management viewpoint on the implementation of a MRP system", *Production and Inventory Management*, Vol. 16, No. 1, 1976, pp. 105–116.

[7] Blasingame, J.W. and Weeks, J.K., "Behavioural dimensions of MRP change; assessing your organisation's strengths and weaknesses", *Production and Inventory Management*, Vol. 22, No. 1, 1981, pp. 81–95.

[8] Brown, A.D., "Implementing MRP: leadership, rites and cognitive change", *Logistics Information Management*, Vol. 7, No. 2, 1994, pp. 6–11.

[9] Burns, O.M. and Turnipseed, D., "Critical success factors in manufacturing resource planning implementation", *International Journal of Operations and Production Management*, Vol. 11, No. 4, 1991, pp. 5–19.

[10] Cooper, R.B. and Zmud, R.W., "Materials requirements planning system infusion", *OMEGA International Journal of Management Sciences*, Vol. 17, No. 5, 1989, pp. 471–481.

[11] Cooper, R.B. and Zmud, R.W., "IT implementation research: a technological diffusion approach", *Management Science*, Vol. 36, No. 2, 1990, pp. 123–139.

[12] DeSantis, G.F., "Implementation considerations in system design", In *Proceedings of the American and Production Inventory Control Society Conference*, 1977, pp. 210–217.

[13] Duchessi, P., Schaninger, C.M., Hobbs, D.R. and Pentak, L.P., "Determinants of success in implementing materials requirements planning (MRP)", *Journal of Manufacturing and Operations Management*, Vol. 1, No. 3, 1988, pp. 263–304.

[14] Grubbs, S.C., "MRPII: successful implementation the hard way", *Hospital Material Management Quarterly*, Vol. 15, No. 4, 1994, pp. 40–47.

[15] Hall, R.W. and Vollman, T.E., "Planning your requirements planning", *Harvard Business Review*, Vol. 56, No. 2, 1978, pp. 105–112.

[16] Lewin, K., "Group decision and social change", In *Readings in Social Psychology*, New York; Holt, Rinehart and Winston, 1947, pp. 197–211.

[17] Ormsby, J.G., Ormsby, S.Y. and Ruthstrom, C.R., "MRPII implementation: a case study", *Production and Inventory Management Journal*, Vol. 31, No. 4, 1990, pp. 77–81.

[18] Ptak, C.A., "MRP and Beyond: A Toolbox for Integrating People and Systems," *Irwin Professional Publications* 1996.

[19] Schein, E.H., "The mechanism of change", In *Interpersonal Dynamics*, 1964, pp. 199–236. Homewood, IL: The Dorsey Press.

[20] Schroeder, R.G., Anderson, J.C., Tupy, S.E. and White, E.M., "A study of MRP benefits and costs", *Journal of Operations Management*, Vol. 2, No. 1, 1981, pp. 1–9.

[21] Scott, B., *Manufacturing Planning Systems*, McGraw Hill, 1994.
[22] Sheldon, D., "MRPII – what it really is", *Production and Inventory Management Journal*, Vol. 32, No. 3, 1991, pp. 12–15.
[23] Sheldon, D., "MRPII implementation: a case study", *Hospital Material Management Quarterly*, Vol. 15, No. 4, 1994, pp. 48–52.
[24] Skinner, W., "The Focused Factory." *Harvard Business Review*, Vol. 52, No. 3, 1974, pp. 113–121.
[25] Sum, C.C., Yang, K.K., Ang, J.S.K. and Quek, S.A., "An analysis of material requirements planning (MRP) benefits using alternating conditional expectation (ACE)", *Journal of Operations Management*, Vol. 13, No. 1, 1995, pp. 35–58.
[26] Sum, C.C., Ang, J.S.K. and Yeo, L.N., "Contextual elements of critical success factors in MRP implementation", *Production and Inventory Management Journal*, Vol. 38, No. 3, 1997, pp. 77–83.
[27] Sum, C.C., Quek, S.A. and Lim, H.E., "Analysing interaction effects on MRP implementation using ACE", to appear in *International Journal of Production Economics*.
[28] Tadjer, R., "Enterprise resource planning", *InternetWeek*, No. 710, 1998, pp. 40–41.
[29] Toomey, J.W., *MRP II: Planning for Manufacturing Excellence*, Chapman and Hall, New York. 1996.
[30] Wallace, T.F., "MRPII: Making It Happen", *Oliver Wight Limited Publications*, Inc. 1990.
[31] White, E.M., "Implementing an MRP system using the Lewin-Schein theory of change", *Production and Inventory Management*, Vol. 21, No. 1, 1980, pp. 1–12.
[32] White, E.M., Anderson, J.C., Schroeder, R.G. and Tupy, S.E., "A study of the MRP implementation process", *Journal of Operations Management*, Vol. 2, No. 3, 1982, pp. 145–153.
[33] Wight, O.W., *MRPII – Unlocking America's Productivity Potential*, Oliver Wight Limited Publications, Inc. 1984.
[34] Works, M., "GenCorp Aerojet says time is right for SAP in the aerospace and defense industries", *Automatic I.D. News*, Vol. 14, No. 2, 1998, pp. 50–52.

MRP II

Manufacturing Resource Planning (MRPII) is a method for the effective planning of all resources of a manufacturing company. Because of its database and information processing needs, it is always implemented with the help of massive software. Ideally, it addresses operational planning in units, financial planning in dollars, and has a simulation capability to answer "what if" questions. The software covers a variety of functions, each linked together: business planning, sales and operations planning, production planning, master production scheduling, material requirements planning, capacity requirements planning, order processing, and capacity planning. Further the system is integrated with accounting and finance subsystems. Output from these systems produce financial reports such as the business plan, purchase commitment report, shipping budget, and inventory projections in dollars.

See **ERP; MRP implementation; Resource planning: MRP to MRP II and ERP.**

MRP LOT-SIZING

Several different rules may be used to compute lot sizes in MRP systems. One, method is to combine planned orders in an MRP record to reduce the number of orders (setups). The number of planned orders to combine into a single production lot may be a function of the trade off between the cost of ordering and inventory carrying costs.

See **MRP; Order release; Schedule stability.**

MRP REGENERATION

Plans and schedules prepared by an MRP system using an MRP logic fail due to a number of reasons including machine breakdown and processing time variability. When MRP plans are revised one approach is to replan by the recalculation of all MRP records; this called MRP regeneration.

See **MRP; MRP implementation; Net change processing; Order release; Schedule stability.**

MRP REPLANNING

Material requirements planning (MRP) generates production and purchase orders subject to many assumptions concerning production and purchase lead times, capacity, etc. Periodic re-evaluation of these plans are made through replanning to accommodate changing market and plant conditions.

See **MRP; MRP implementations; Schedule stability.**

MRP WITHOUT TIME BUCKETS

See **Bucketless material requirements.**

MULTI-ATTRIBUTE DECISION MODELS

Several quantitative models have been developed to assist decision-makers in industry. Multi-attribute models are a class of multi-criteria decision models in which all objectives and preference of the decision maker are unified under a super-function termed the decision maker's utility, which is maximized by the solution provided by the model. Such models are used in selecting one among several alternatives.

See **Capital investment advanced manufacturing technology; Industrial robot selection; Mathematical programming; Multi-criteria decision models.**

MULTICELL FLEXIBLE MANUFACTURING SYSTEM

A multi-cell flexible manufacturing system (MCFMS) is a type of FMS which consists of a number of flexible manufacturing cells (FMCs) all connected by an automated material handling system. One example of MCFMS contains three cells. Each cell contains two machines. The system manufactures nine part types. Part #8 can be processed in either cell 1 or cell 2 with different operation times and Part #9 can be processed in any of the three cells. Such MCFMS focus on cutting down the proportion of time spent on non-productive activities like material handling and waiting relative to the actual machining time.

See **Dynamic routing in flexible manufacturing systems; Flexible automation; FMS linked-cell manufacturing system; Manufacturing cell design.**

MULTI-CRITERIA DECISION MODELS

Certain decisions in industry must satisfy multiple criteria. Multi-criteria decision models are mathematical models which aide a decision maker select among alternatives based on the multiple criteria and the decision maker's preferences. Goal programming is one multi-criteria optimizing model that allows the decision maker to rank the various criteria used in making the decision.

See **Capital investment in advanced manufacturing technology; Industrial robot selection; Mathematical programming.**

MULTI-FUNCTIONAL EMPLOYEES

Multi-functional employees play a vital role in supporting just-in-time management practices. It is accomplished through the extended training of employees on several different machines and in various functions. Multi-functional employees are rarely idle, provide flexibility for adaptation to demand changes, and improve productivity and problem-solving capabilities on the shop floor. Multi-functional employees are essential for implementing manufacturing cells using group technology principles. Further, they are essential for implementing total quality control, quality circles, total productive maintenance, and uniform workload practices of JIT manufacturing. (Monden, 1983; Schonberger, 1982).

See **Customer service through system integration; Just-in-time manufacturing; Lean manufacturing implementation.**

References

[1] Monden, Y. (1983). *Toyota Production System: A Practical Approach to Production Management*. Industrial Engineers and Management Press, Norcross, GA.

[2] Schonberger, R.J. (1982). *Japanese Manufacturing Techniques: Nine Hidden Lessons in Simplicity*. The Free Press, New York.

MULTI-FUNCTIONAL TEAM

See **Cross-functional team.**

MULTI-ITEM INVENTORY MANAGEMENT

See **Multiple criteria ABC analysis.**

MULTI-OBJECTIVE DECISION MODELS

See **Multi-criteria decision models.**

MULTIPLE CRITERIA ABC ANALYSIS

ABC Analysis arbitrarily separates inventory into three groups on the basis of annual dollar usage (cost of a unit x annual usage of the item). Typically, 20 percent of items in inventory account for 80 percent of the dollar usage. These are called A items. About 30 percent of items account for about 15 percent of the dollar usage; these are B items. Remaining 50 percent of the items account for about five percent of dollar usage. For better use of inventory resources A items receive the most attention, and C items receive the least amount of attention. ABC analysis ensures that inventory control procedures have the greatest impact on the inventory cost and inventory performance. By adding the noncost criteria to the dollar usage criterion in ABC analysis, a multiple-criteria ABC Analysis can be developed. Lead-time and criticality are examples of noncost criteria that can be used in multiple-criteria ABC Analysis.

If we arbitrarily set three criticality classifications (i.e., highly critical, moderately critical, not critical) and combine them with a conventional ABC Analysis, it will create nine classification of inventory. Then nine different inventory policies can be adopted to suit the various classifications based on dollar usage and criticality. The different inventory policies for different classifications improve the overall effectiveness of inventory control using scarce inventory management resources.

See **ABC analysis.**

MULTIPROCESS WORKER

The implementation of manufacturing cells, just-in-time manufacturing, lean manufacturing and total productive maintenance require operators who can operate several different machines. They are called multiprocess workers. Such workers are developed through extensive training and returning.

They provide flexibility to the entire production system. Labor Unions in U.S. have accepted this practice. In the past unions did not approve work rules that required an operator to work on different machines. The recent success of U.S. manufacturers is to a good part due to the growth in the number of multiprocess workers. The idea originated in Japan.

See **Manufacturing cell design; Lean manufacturing implementation.**

NAFTA AND MANUFACTURING IN THE U.S.

NAFTA (North American Free Trade Agreement) is a 1993 US legislative act whose major objective is to phase out trade barriers during the following 15 years among the North American countries, Canada, Mexico and the United States. It also seeks to promote conditions of fair competition, increase investment opportunities and provide for protection of certain intellectual property rights of individuals and businesses. Also, the NAFTA agreement is designed to for expansion to include other countries in South America.

NAFTA has a significant impact on human resource management. NAFTA permits citizens from Canada and Mexico to come to the United States much more easily to work in the manufacturing sector. Under NAFTA, as long as the Canadian or Mexican citizen maintains the citizenship in the country of origin, they and groups of laborers can be brought into the United States to work on projects. The agreement requires businesses in each of the NAFTA member countries to recognize and accept as equivalent baccalaureate degrees when considering the credentials of job applicants for professional jobs. NAFTA agreement allows Mexican and Canadian professionals to seek jobs in the US job market in accounting, economics, engineering, dentistry, management consulting, registered nursing, science and teaching.

NAFTA impacts bidding and procurement in the manufacturing sector. Under NAFTA purchasing agents should (and in some cases must by law) consider all of North America as a possible source of bids and supplies. Even the US Government contracts to build roads or bridges must be offered to all eligible construction firms in Mexico and Canada.

See **Global facility location analysis; International manufacturing.**

Reference

[1] *North American Free Trade Agreement*, (1993). Volumes I and II, Superintendent of Documents, US Government Printing Office, Washington, D.C.

NAM

See **National association of manufacturers.**

NAPM

See **Professional associations for manufacturing professional.**

NATIONAL ASSOCIATION OF MANUFACTURERS (NAM)

The National Association of Manufacturers (NAM) is the oldest and largest broad-based industrial trade organization in the United States. Its more than 14,000 member companies and subsidiaries, including approximately 10,000 small manufacturers, are in every state and produce about 85 percent of U.S. manufactured goods. Through its member companies and affiliated associations, the NAM represents every industrial sector and more than 18 million employees.

NAM can be reached at 1331 Pennsylvania Avenue, NW; Washington, D.C., 20004-1790; phone 202-637-3000; manufacturing@nam.org.

NATURAL VARIABILITY

Natural variability is a quality control related concept. All manufacturing or service exhibit a certain amount of "natural" variability. (Such variability is also known as chance or random variation.) This natural variability is the cumulative effect of several minor factors. The extent of natural variability inherent in a process differs from process to process, and changes over time. For instance, older machines will generally exhibit a higher degree of natural variability than newer machines, partly because of worn parts and partly because newer machines may incorporate design improvements that reduce the variability of their output.

Ideally, only random variation (or natural variability) should be present in a process, because it represents the acceptable amount of variation. Improving the system itself can reduce natural variability. A process that is operating with only natural variability is said to be "in a state of statistical control."

Walter A. Shewhart (1931) promoted the use of graphs, such as control charts, to identify variations in processes and to determine when a process is exhibiting more than natural variability or random variation. The graph helped identify common causes of trouble which he called assignable causes of variation. A process is normally stopped and the assignable cause of variation is fixed before restarting the process.

See **Integrated quality control; Lean manufacturing implementation; Performance excellence: The Malcolm Baldridge award criteria; Quality: The implications of W. Edwards Deming's approach; Statistical process control using control charts; TQM.**

Reference

[1] Shewhart, W.A. (1939). *Statistical Method from the Viewpoint of Quality Control*, Graduate School, Department of Agriculture, Washington.

NC MACHINE TOOL

See **CNC machine tool; Flexible automation; Manufacturing technology use and benefits; Numerical control.**

NET CHANGE PROCESSING

Net Change processing is an MRP ability. MRP plans and records need changes as business conditions change. Changes may be needed in requirements, inventory status, bill of material, lead time, or the MRP plan.

Changes can be made by processing some or the all MRP records and files to incorporate current information. If all the records are processed at the same time in one run while making changes and updates, it is called regeneration. During regeneration the records for all part numbers are completely reconstructed. Net change processing of MRP records is an alternative to regeneration. In net change processing only the records affected by the change or new information are reconstructed. Net change processing takes less computer time for processing and is used to make daily or more frequent updates. All transactions may be accumulated daily and then processed overnight using the net change approach. Net change users must do periodic regeneration to clean up all possible records.

See **MRP; MRP implementation.**

NET PRESENT VALUE (NPV)

Investment options may be evaluated using several criteria. Net present value is one such investment evaluating tool. Its is the present value of an investment's future cash flows minus the initial cost of the investment. The present value is today's value of a future cash flow while recognizing that a dollar today is worth more than a dollar tomorrow. Present values of future cash inflows or revenue are determined by the discounted cash flow (DCF) analysis. If an investment project has a positive NPV, its discounted cash inflows are greater than the costs associated with the project.

See **Capital investments in advanced manufacturing technologies; Discounted cash flow (DCF); Robot selection.**

NEURAL NETWORKS

Neural networks permit more complete modeling of complex, multivariate, non-linear manufacturing systems. A neural network is particularly attractive for modeling a complex manufacturing system because *neural networks do not require any relationships between the numerous variables in the model to be defined or hypothesized a priori*. This is because of the unique distributed architecture of neural networks making them somewhat model-free estimators. Once trained using a large number of real-life observations, a neural network can mimic a complex real system. An axiom to remember is that, the larger and more diverse the number of observations used in training, the more accurate or valid the resulting trained neural model.

Neural networks possess the potential for quantifying complex mapping characteristics in a compact and elegant manner. Such representations have been shown to be very useful in modeling extremely complex systems encountered in areas such as economic analysis, pattern recognition, speech recognition and synthesis, medical diagnosis, seismic signal recognition, and control systems. Interest in neural networks is, in part, due to powerful neural models such as the multilayer perceptron (McCord-Nelson and Illingworth, 1991), and neural network learning methods such as back-propagation. Further, it is also due to the rapid recent developments in hardware design that have brought within reach the realization of neural networks with a very large number of nodes, i.e., networks of very large complex systems.

A neural network is a system with inputs and outputs and is composed of many simple and similar processing elements called neurons. These processing elements, or neurons, have a number of internal parameters called weights. Changing the weights of an element will alter the behavior of the element and, therefore, will also alter the behavior of the entire network. The goal here is to choose the weights of the network elements to achieve a desired input/output relationship. The desired input-output relationship is the empirically observed relationship between input variables and output variables included in the model.

Neural networks permit the modeling of systems, where generalizations concerning interactions between and among variables are not transparent, or indeterminate. Further, neural networks tolerate some raw data that is incomplete or susceptible to noise. They have properties such as content addressability, graceful degradation, retrieval from a partial description, default assignment and spontaneous generalization, which make them robust models or dependable mimics of complex systems; these properties are described in a number of basic texts on neural networks. The property of **spontaneous generalization** is particularly interesting and meaningful to the management scientist engaged in modeling complex systems defined by numerous interconnected variables. For example, neural networks do not require the specification of **a priori** generalizations, but the networks are capable of drawing generalizations inspired by the data used to train the network. *The authors also report a novel extension of the neural modeling concept to parametric sensitivity analysis in a later section.*

Neural Networks Versus Common Linear Models

Neural network models represent a massive system of distributed non-linear regression, as such, they permit us to model a wider variety of real-life complex systems without the unrealistic, and limiting assumptions of linearity. Linear multivariate regression models, as we know them, are mere subsets of non-linear regression models. Further, the ability to model many inputs and outputs, sets neural nets farther apart form linear multivariate regression models. Therefore, neural network is finding applications in modeling complex systems even though it is lot more expensive and time consuming to develop/train a neural network compared to a linear multivariate regression model.

See **Manufacturing analysis using chaos, fractals, and self-organization; Manufacturing system modeling using petri nets; Simulation analysis of manufacturing and logistics systems.**

References

[1] Bau, E.B., and D. Haussler, "What Size Net Gives Valid Generalization," *Neural Computation,* vol. 1, (1989) pp. 151–160.

[2] Finhoff, W., Hergert, F., and Zimmermann, "Improving Model Selection by Nonconvergent Methods," *Neural Networks*, 6, (1993), pp. 771–783.

[3] Fogel, D., "An Information Criterion for Optimal Neural Network Selection," *IEEE Transactions on Neural Networks*, vol. 2, no. 5, (1991), pp. 490–497.

[4] Hopfield, J.J. and D.W. Tank, "Neural Computation of Decisions in Optimization Problems," *Biological Cybernetics*, vol. 52, (1985) pp. 141–152.

[5] Hornik, K., M. Stinchcombe, and H. White, "Multilayer Feedforward Networks Are Universal Approximators", *Technical Report*, Dept. of Economics, University of California, San Diego, CA (June 1988).

[6] Kosko, B. 1992. *Neural Networks and Fuzzy Systems: A Dynamical Systems Approach to Machine Intelligence*. Prentice Hall, Englewood Cliffs, NJ.

[7] Lenard, M.J., Alam, P., and Madey, G. "An Application of Neural Networks and Qualitative Response Model to the Auditor's Going Concern Uncertainty Decision," *Decision Sciences*, 26, 2 (1995), pp. 209–228.

[8] Mistry, S.I., and Nair, S.S., "Nonlinear HVAC Computations Using Neural Networks," *ASHRAE Transactions*, 99, 1 (1993), 775–784.

[9] Mistry, S.I. and Nair, S.S., "Identification and Control Experiments Using Neural Designs," *IEEE Control Systems Magazine*, 14, 3 (1994), 48–57.

[10] McCord-Nelson, M. and W.T. Illingworth, *A Practical Guide to Neural Nets*, Addison-Wesley Publishing Company, Inc., (1991).

[11] Salchenberger, L.M., Cinar, E.M. and Lash, N.A. "Neural Networks: A New Tool for Predicting Failures," *Decision Sciences*, 23, 4 (1992), pp. 899–916.

[12] Tam, K.Y. and Kiang, Y. "Managerial Applications of Neural Networks: The Case of Bank Failure Prediction," *Management Science*, 38, 7 (1990), pp. 926–947.

NEW PRODUCT DEVELOPMENT THROUGH SUPPLIER INTEGRATION

Robert B. Handfield

Michigan State University, East Lansing, MI, USA

DESCRIPTION: In striving to reduce development time and increase new technology introduction into the market place, organizations are seeking to closely involve their material suppliers into the new product development cycle. In the past, organizations relied on developing detailed specifications, which were used as a basis for bids submitted by several suppliers. Although this technique is still utilized to some extent, a growing number of organizations are beginning to harness the intellectual capital, technology, and capabilities of suppliers through close interaction with them through new product and process design teams. Some of the resulting benefits include reduced product development time, lower costs, and innovative products and services, which would otherwise be impossible under the bidding process.

The extent of involvement of an organization with key suppliers may range from simple consultation with suppliers on design ideas to making suppliers fully responsible for the design of components or systems they supply.

STRATEGIC PERSPECTIVE: Within the last decade, the rapid rate of technological change, shortened product life cycles, and increasing global competition have made new product development a critical concern of US manufacturers. In a recent study, it was reported that procurement managers expect that over the next five years, competitive pressures will require manufacturers to reduce costs by 5–8 percent per year (after inflation), and continue to improve product quality, while simultaneously reducing time to market by 40–60 percent (Monczka and Trent, 1997).

In this competitive environment, suppliers are an increasingly important resource for manufacturers. Across all US manufacturers, purchased materials account for over 50 percent of the cost of goods sold. In addition, suppliers have a large and direct impact on cost, quality, technology, speed, and responsiveness of their companies. Effective integration of suppliers into the product value/supply chain is a key factor for some manufacturers to remain competitive.

A study conducted by Computer-Aided Manufacturing International (CAM-I) concluded that, while the concept and design engineering phases of new product development account for only 5–8 percent of the total product development costs, these two activities commit or "lock in" 80 percent of the total cost of the product. Decisions made in the design process have a significant impact on the resulting product quality, cycle time, and cost (Handfield, 1994). As the development process continues, it becomes increasingly difficult and costly to make design changes. It is crucial, therefore, for firms to bring to bear as much product, process, and technical expertise as possible early in the development process.

There is intuitive appeal to the idea that using the knowledge and expertise of suppliers to complement internal capabilities can help reduce concept-to-customer cycle time, costs, quality problems, and improve the overall design effort. Reports in the popular press indicate that leading companies in a variety of industries have made successful efforts at involving suppliers in the new product development process, and that interest in such efforts is growing.

Though not necessarily a strategic alliance, supplier integration into new product/process/service development (NPD) efforts are often pursued by strategically allied partners. Important factors such as equity sharing, trust, co-location, asset specificity and information sharing strengthens such alliance (Lewis and Wiegert, 1984).

Littler *et al.* (1995) examined the key success factors for collaborative new product development efforts in 106 UK firms in which the collaborative partner could be a supplier, customer, or competitor. They concluded that success came through frequent inter-company communication, trust building, equity partnership, and by ensuring that parties contribute as expected, and employing a product or collaboration champion.

Dyer and Ouchi (1993) suggest that the length of a buyer/supplier relationship positively impacts product development efforts. The supplier's existing knowledge of the buying firm's internal processes and objectives enables the supplier to plan for future product development efforts of customers and to develop, in advance, the capabilities to meet those needs.

Although these studies identify the potential benefits of supplier integration, the processes which lead to successful vs. unsuccessful deployment of this strategy has, until recently, been poorly understood.

IMPLEMENTATION: The decision to integrate suppliers in new product development processes is typically made by a cross-functional sourcing team. Cross-functional supplier selection (or "sourcing") teams consist of members from different departments assembled to evaluate a supplier's capability in a number of areas. Purchasing plays a critical role in this type of team for obvious reasons. The use of sourcing teams became popular during the latter part of 1980s as firms reduced the number of suppliers. Supplier selection has now become a critical activity with strategic implications. Purchasing often coordinates or leads the selection process and shares

Exhibit 1. The "Sashimi" concurrent development schedule.

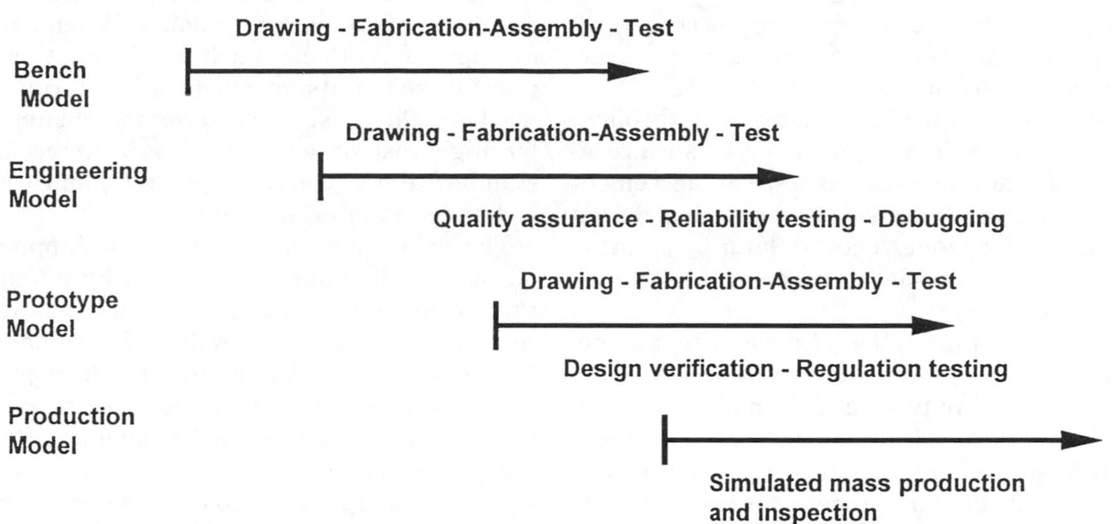

the decision making authority with other team members during the final selection decision.

Sourcing teams often make decisions related to new product development, which requires close interaction and communication between functions to speed the development of new products. Purchasing can actively support new product development teams by evaluating existing and potential suppliers, developing clear and concise material specifications, and developing long-term material plans (often referred to as "technology roadmaps").

The use of teams often generates innovative new product ideas and speeds the development and marketing of a new product. While the use of teams is popular, there are other methods that can help reduce the time it takes to develop a new product.

When implemented properly, the team approach can be an effective way to meet new product goals. The traditional new product development approach does not support the development of innovative new products in a short time. This "business as usual" approach excludes purchasing, supplier and manufacturing input during the product design phase. Under this scenario, product designers rarely communicated with process engineers, who rarely communicated with manufacturing, and innovative ideas that often take years to develop may never reach the customer.

Product designers can mistakenly assume that process engineers and manufacturing can develop the process capability required for a new product. Early cooperation and communication in teams can eliminate some of the problems in this area.

The most important way that cross-functional teams achieve reduced new product cycle times is through a concurrent versus sequential approach to new product development. *Concurrent engineering* is defined by the Institute for Defense Analysis (IDA) as: "the systematic approach to the integrated concurrent design of products and related processes including manufacture and support. This approach is to cause the developers, from the outset, to consider all the elements of product lifecycle from conception through disposal including quality, cost, and schedule and user requirements." (Kusiak and Belhe, 1992).

As shown in Exhibit 1, in concurrent engineering, activities in the product development cycle are scheduled in parallel at the interface of different phases. The benefits of concurrent engineering at Fuji included faster development, increased flexibility, and greater sharing of information. This system (which has been referred to as "Sashimi" development by Fuji because it resembles the manner in which slices of raw fish overlap one another on a plate) also increases shared responsibility of project members. Members of sales, research and development (R&D), manufacturing, purchasing and suppliers all work together as a team to solve the problems occurring in new product design (Handfield, 1995). Some of the disadvantages of the Sashimi system include the amplification of ambiguity, tension, and conflict in the group (Imai *et al.*, 1989).

If we compare the "Sashimi" and "sequential" approaches to product design, it is apparent that the former involves the compression of development lead times. All of the stages in product development are carried out by both systems. While the total time (in man-hours) spent on development may be equal using either approach, the "Sashimi" concurrent system compresses development time through overtime and parallel development activities. In addition, in the concurrent system, the organization compresses learning into a shorter time period.

What is it about a concurrent approach that reduces product development cycle times? A concurrent approach (Monczka and Trent, 1994):

- Requires cross-functional agreement throughout the product development process concerning critical development issues. This agreement can reduce costly and time-consuming design and product change requests later by creating agreed product specifications early in the development process.
- Supports the interaction of a competent group of professionals that should result in better decisions related to product design and manufacturability. Better decisions should require less time to implement and require less "rework" later on, particularly, when the team has the authority to make key decisions.
- Supports early supplier involvement. Suppliers can provide insight into design and manufacturing requirements that can save time and cost.
- Accelerates learning throughout the new product development process. Functional team members learn together simultaneously rather than separately as a project "passes" from one group or function to another. A team can spend a greater proportion of total development time on "value-added" activities rather than separate learning across functions.
- Requires establishment of customer-oriented and competitive goals by the cross-functional team.

IMPLEMENTATION – THE ROLE OF PURCHASING: Purchasing should play a key role in new product development. As the main contact with suppliers, purchasing is in a unique position to include suppliers early in the design process and perform early evaluation of supplier capabilities.

A major responsibility of purchasing is to provide information to suppliers. Sharing of information can help avoid unwelcome surprises throughout the life of a project, particularly if suppliers are brought in to the parts design process early in the concept stage. If purchasing selects capable and trustworthy suppliers, it will be able to share product information early in the development process. For example, if a component for a new part requires a specific production process, it is important to make sure that the supplier has the required process capability. Suppliers have delayed the launch of new products because they were unable to meet production requirements. Early visibility to product requirements allows purchasing to share critical information with suppliers that can avoid delays.

Purchasing's early involvement allows purchasing to determine at an earlier point the material requirements for a new product. Purchasing has an opportunity to provide input during the design phase based on its knowledge of material supply markets. Purchasing can recommend substitutes for high cost or volatile materials, recommend standard items wherever possible, and evaluate longer-term material trends.

The evaluation and selection of suppliers requires a major time commitment by purchasing. Purchasing must evaluate and select suppliers regardless of the new product development approach used. The team approach allows purchasing to anticipate product requirements earlier so that it can identify the most capable suppliers. Many purchasing departments are now developing global supplier databases for use by design engineers in new product designs. When a design engineer determines the need for a particular material or commodity during the design phase, he or she could consult the database to determine the preferred supplier for that commodity class. These databases can avoid duplication of effort, promote greater leverage, and assure that only state-of-the-art suppliers are used for sourcing new materials.

A process of pre-approving suppliers for commodity families or part numbers before they are used is often known as "pre-sourcing". For its compact sedans, the Chrysler Cirrus and Dodge Stratus, Chrysler Corporation presourced 95 percent of the parts required for production. That is, Chrysler used a team approach and chose the suppliers before the parts were even designed, which meant virtually eliminating traditional supplier bidding (Bennet, 1995).

Purchasing should always monitor and anticipate activity in its supply markets. For example, purchasing should forecast long-term supply and prices for its basic commodities. It should monitor technological innovations that impact its primary materials or make substitute materials economically attractive. It should evaluate not only its existing suppliers but other potential suppliers. Because team participation provides timely visibility to new product requirements, purchasing can monitor and forecast change on a continuous basis.

Purchasing can also develop contingency supply plans, including end-of-life strategies to phase out obsolete components to avoid being "stuck" with their inventory. This is particularly important in cases when component life cycles are much shorter than the actual product life cycle. It is not uncommon for high tech firms who anticipate end-of-life situations to plan phase-out of the soon to be obsolete part after manufacture of the product. The firm typically notifies customers that support will be phased out, and one-time buys of the now obsolete component will be possible until some specified date. Finally, the component is written off and disposed. Possible forms of disposition of the obsolete component include sale incentives for finished goods, disassembling and restocking non-obsolete components, returning to the supplier for credit, donation for a tax break, or scrap. In almost all of these cases, some loss will be incurred (Handfield and Pannesi, 1994).

Purchasing can assess whether project timing is realistic as it applies to a new part's sourcing requirements. If the timing is not realistic, purchasing should reevaluate the timing requirements, or come up with plans to meet the proposed time frame.

Bringing suppliers into the product development process is different from simply sharing information. This involves the including of important suppliers early in the design process of a new product, perhaps even as part of the new product team. The benefits of early supplier involvement include gaining a supplier's insight into the design process, allowing early comparisons of proposed production requirements against a supplier's existing capabilities, and allowing a supplier to begin pre-production work early. A supplier can bring to the development process a fresh perspective and new ideas.

If given the opportunity, suppliers can have a major impact on the overall timing and success of a new product. The type of involvement can also vary. At one extreme, the supplier is given blueprints and told to produce them. At an intermediate level, often described as a "gray box" design, the supplier's engineers work cooperative with the buying company's engineers to jointly design the product. At the highest level of supplier involvement known as "black box" design, suppliers are provided with functional specifications and are asked to complete all technical specifications, including materials to be used, blueprints, etc.

An important function of purchasing is the development of clear material and component specifications. The team approach allows purchasing to work directly with design and product engineers during the development of new product requirements. This helps purchasing develop a clear set of product specifications for suppliers. This enables building in quality at the source. Also, purchasing can work to overcome a major complaint that engineering has about purchasing – its lack of product knowledge. The team approach supports a closer working relationship between purchasing and engineering. This should support the development of clear material specifications that satisfy engineers and suppliers.

EFFECTS: Using a team approach, which involves suppliers communication, barriers during product development can be removed. Removing barriers allows a team to identify potential problems early in the design process. A team goal should be an average of 30–50 percent reduction in total development cycle time compared to traditional product development times. The team approach should also contribute to lower total product cost through creative ways to design and manufacture products.

Advantages

In a recent study funded by the U.S. National Science Foundation (NSF), companies reported important benefits of supplier integration in the areas of purchased material cost and quality, and reduced product development time (Ragatz *et al.*, 1997). Benefits also stem from better access to and application of technology. Particularly in industries employing complex product and/or process technologies, no single company can master all the relevant technologies.

The ability to tap the knowledge and expertise of suppliers, for design as well as manufacturing, is likely to lead to better technology decisions and ultimately, better designs. Furthermore, closer involvement with suppliers gives a buyer company an opportunity to influence the direction of suppliers' technology development efforts in order to align those efforts with the buyer company's needs. A number of the companies responding to the NSF sponsored study reported sharing technology roadmaps with suppliers, and holding formal discussions about aligning customer-supplier technology plans. At a more basic level, regardless of the technology issue, design involvement by suppliers can help in identifying potential problems and solutions earlier, reducing both time and cost of the design effort.

Drawbacks

Although the potential benefits are substantial, integrating suppliers into new product development efforts is a relatively new and sometimes uncomfortable way of doing business for some companies. It appears that a variety of internal and external barriers must be overcome to make such integration work well.

The barriers result from a variety of issues and concerns. First, there may be resistance at a number of levels within the organization to sharing proprietary information with suppliers. The usual concern is that such information may be revealed, intentionally or unintentionally, to competitors. This may be a particularly critical issue, when the supplier company itself is also a competitor. Second, in many organizations, a "not invented here" culture poses a challenge to the acceptance of ideas coming from suppliers, and buyer company designers and engineers may resist giving up any control over design decisions. Finally, there may be resistance within the supplier's organization as well based on concerns about revealing proprietary information or technologies. Particularly, a supplier dealing with a more-powerful buyer may worry about inequitable treatment, and loss of control.

Methods to Overcome Barriers

Two important conceptual themes regarding how companies can overcome or minimize these barriers emerge from the study mentioned above. First theme is relationship structuring. These factors include shared education and training, formal trust development processes/practices, formalized risk/reward sharing agreements, joint agreement on performance measurements, top management commitment from both companies, and confidence in the supplier's capabilities. Relationship structuring factors break down barriers, expand the boundaries of the relationship between the buying and supplying companies, open communication channels, build mutual trust, and define the expectations of both parties.

The second theme is asset allocation. *Intellectual* assets are: customer requirements, technology information, and direct cross-functional inter-company communication. *Physical* assets are: common and linked information systems, technology, and shared plant and equipment. *Human* assets are: supplier participation on the project team, and co-location of personnel.

Relationship structuring practices facilitate integration and sharing of assets, but do not directly affect the speed, cost and quality of NPD. The asset sharing practices more directly influence the results of the NPD process. In this sense, the relationship structuring factors are "enablers" for the asset allocation practices. For example, top management commitment at both companies helps to change and align the cultures of the organizations by indicating that it is "okay" to share information and that resources will be made available to support the integration effort. Shared education and training, formal trust development processes, and formalized risk/reward sharing agreements help minimize concerns that one party may take advantage of the other. Joint agreement on performance measures provides clear directions and expectations for the project and allows each party to identify the mutual benefits of the integration effort.

The allocation of intellectual assets provides the detailed knowledge base to facilitate better decision making, and to see that customer requirements are met. Physical asset allocation ensures that the necessary technical tools and resources are available to both parties to perform coordinated design and development activities. Human asset allocation leads to interaction between buyer and supplier personnel, and it encourages joint problem identification and resolution.

CASES:

The Tektronix Experience

Tektronix successfully uses a cross-functional team approach for new product development (Masimilian and Pedro, 1990). Tektronix, a manufacturer of high technology equipment, recognized that it needed to improve efficiency, lower manufacturing costs, and beat the competition to the marketplace with new products. To accomplish this, new product development required an entirely new organizational approach. The traditional approach relied on a sequential process where one function did not begin work on a new product until the previous function had finished. This resulted in longer product development times and reduced communication.

The company selected a new product idea – a portable oscilloscope – for development by a cross-functional team. The team completed product development in only twelve months – half the usual time. Besides working hard, the team attributed its success to a number of factors:

- Extraordinary Teamwork
  This required the development of an "us" mentality instead of "us versus them" mentality. It required the team to recognize success or failure as a team – not as individuals.
- Effective Communication
  The organizational barriers to free and open communication were removed and was replaced with constructive exchange between team members.
- Innovation
  The team questioned all past practices. Traditional product development methods would not have allowed the team to meet its project goals. The team had to create innovative new methods.

- Removal of Obstacles
  Instead of allowing obstacles to slow down progress, the team attacked all obstacles head on until it found solutions.
- Coaching
  Failure of even one team member meant failure for the entire team. The members took it upon themselves to coach one another to make sure the entire team succeeded.

The 3M Approach

3M is a diversified multibillion dollar company selling products worldwide (McKeown, 1990). It is a company committed to product and technology innovation as a way of life to remain a world leader. The company also commits itself to a goal of having at least 25 percent of its total sales revenue come from newer products. Each of 3M's 40 operational units are responsible for managing its own global growth of sales and profit. This case if about the 3M Industrial Specialties Division.

The Industrial Specialties Division uses cross-functional teams to manage families of products. These teams always include purchasing personnel and suppliers, who are actively involved in each stage of the process. In addition, the teams are responsible for developing new products from new technologies and developing product extensions from existing products and technologies. The teams follow a broad four-step procedure designed to generate new product ideas and streamline the new product development process. The four steps are:

- opportunity identification
- new idea prioritization
- product development, and
- review of the development process.

Opportunity Identification: This step involves the actual generation of new product ideas. The team, with its different functional backgrounds, identifies ideas requiring new technologies as well as extensions of existing products and technology. In addition, a reward system encourages personnel to submit new product ideas for the appropriate team to review. The cross-functional team accumulates the new product ideas for further consideration during the second step of the process.

New Idea Prioritization: This step involves prioritizing the new ideas. The team uses a subjective scale to evaluate ideas against a number of variables to arrive at an initial ranking of the ideas. For a potential new product, the team rates the potential customer need for the product, the absence of competitors producing the product, the ability to harness existing technology capabilities within 3M, 3M's ability to manufacture the product, and the ability to meet competitors pricing and performance features. These scores provide a good indication of the likclihood of new product success. Products with higher ratings are more attractive for initial development. A second approach involves the development of a ratio comparing the expected five-year cumulative sales to projected costs. 3M has found a strong correlation between the probability of technical success and the probability of marketing success. Each variable is estimated and placed into the opportunity rating equation:

$$\text{Opportunity Rating (in \$\$)} = (CS \times TS \times MS)/(LB \times OC)$$

where: CS = Five-year cumulative sales
TS = Probability of technical success
MS = Probability of marketing success
LB = Lab costs
OC = Other costs

The two rating systems allow the team to evaluate and compare new product ideas. From this, the team identifies the ideas that move to the product development phase.

Product Development: This phase involves the physical development of the new product. To assist the team in its development efforts, the division uses an expediting system called FAS-TRACK. Special stickers identify all documents and activities relating to a new product. Stickers alerts all personnel to the special priority. Special priority for selected items does not mean the neglect of other items. This system simply serves as a notice that streamlining the product development process is a high priority at 3M.

Review of the Development Process: The last phase is the periodic review of the new product development process. If a product does not meet its timing or market objectives, the team determines the reasons for the shortfall. The objective of the review process is to improve the product development process in a non-threatening and non-punitive environment. 3M recognizes that a punitive process only stifles creativity. The most common reasons for missing a product's development objectives are poor planning, technical difficulties, and poor support from outside functions.

The point of these two examples is that firms are using the team approach to streamline and improve the product development process. This is important for firms that rely on innovative new products for their continued success.

Purchasing has a key role to play on these teams. Their role involves helping to select suppliers for inclusion in the process, advising engineering personnel of suppliers' capabilities, and helping to negotiate contracts once the product team has selected a supplier. Purchasing also acts as a liaison throughout this process, in facilitating supplier participation at team meetings, and helping to resolve conflicts between the supplier and the team. Purchasing may also be involved in developing a target price for the supplier to shoot for in developing the component or system, and helping the supplier to meet this target price. Finally, purchasing may also be involved in developing nondisclosure and confidentiality agreements, when technology sharing occurs.

Organizations Mentioned: 3M; CAM-I; Chrysler; Dodge; Fuji; Tektronics

See **Bench model; CAM-I; Engineering model; NPD; Outsourcing of product design and development; Product champion; Prototype model; Purchasing: Acquiring the best inputs; Purchasing: The future; Risk/reward sharing agreements; Sashimi approach to product design; Sequential approach to product design; Supply chain management; Supplier partnership as strategy; Supplier relationships; Technology roadmap**.

References

[1] Bennet, J. (1994). "Detroit Struggles to Learn Another Lesson From Japan," *The New York Times*, June 19, 1994, section F, 5.

[2] Dyer, J.H. and W.G. Ouchi (1993). Japanese-style partnerships: giving companies a competitive edge. *Sloan Management Review* 35 (1), 51–63.

[3] Handfield, R. (1994). Effects of concurrent engineering on make-to-order products. *IEEE Transactions on Engineering Management* 41 (4), 1–11.

[4] Handfield, Robert and Ronald Pannesi (1994). "Managing component life cycles in dynamic technological environments", *International Journal of Purchasing and Materials Management (Spring)*. 20–27.

[5] Ken-ichi Imai, Ikujiro Nonaka, and Hirotaka Takeuchi, (March 27–29, 1989). "Managing the new product development process: How japanese companies learn and unlearn," Colloquium on Productivity and Technology, Harvard Business School.

[6] Kusiak, A. and U. Belhe (1992). "Concurrent engineering: A design process perspective," *Proceedings of the American Society of Mechanical Engineers*, PED 59, 387–401.

[7] Lewis, J. and A. Wiegert (1985). Trust as a social reality. *Social Forces* 63, 967–988.

[8] Littler, D., F. Leverick, and M. Bruce (1995). Factors affecting the process of collaborative product development: a study of UK manufacturers of information and communications technology products. *The Journal of Product Innovation Management* 12 (1), 16–23.

[9] Mabert, Vincent A., John F. Muth, and Roger W. Schmenner (1992). Collapsing new product development times: six case studies. *The Journal of Product Innovation Management* 9 (3), 200–212.

[10] Massimilian, R.D. and L. Pedro (1990). "Back from the future: Gearing up for the productivity challenge," 79 (2), *Management Review*, 41–43.

[11] McAllister, D. (1995). Affect- and cognition-based trust as foundations for interpersonal cooperation in organizations. *Academy of Management Journal* 38 (1), 24–59.

[12] McKeown, R. (1990). "New products from existing technologies", *The Journal of Business and Industrial Marketing*, 5 (1).

[13] Monczka, R., R. Trent, and R. Handfield (1997). *Purchasing and Supply Chain Management*, Cincinnati, OH: Southwestern Publishing, College Division.

[14] Monczka, R.M. and R.J. Trent (1997). "Purchasing and sourcing 1997: trends and implications," forthcoming by the *Center for Advanced Purchasing Studies* (CAPS), Tempe, Arizona,.

[15] Monczka, R., D. Frayer, R. Handfield, G. Ragatz, and R. Trent (1998). *Supplier Integration into New Product/Process Development: Best Practices*, Milwaukee: ASQC.

[16] Monczka, R. and Robert J. Trent (1995). Global sourcing: a development approach. *International Journal of Purchasing and Materials Management* 11 (2), 2–8.

[17] Ragatz, G., R. Handfield, and T. Scannell (1997). "Success factors for integrating suppliers into new product development", *Journal of Product Innovation Management*, 7–20.

NIST

The National Institute of Standards and Technology was established by Congress "to assist industry in the development of technology... needed to improve product quality, to modernize manufacturing processes, to ensure product reliability... and to facilitate rapid commercialization... of products based on new scientific discoveries." (www.NIST.gov).

It is an agency of the U.S. Department of Commerce's Technology Administration. NIST's primary mission is to strengthen the U.S. economy and improve the quality of life by working with industry to develop and apply technology, measurements, and standards. It carries out this mission through a portfolio of four major programs:

1. Measurement and Standards Laboratories that provide technical leadership for vital components of the nation's technology infrastructure needed by U.S. industry to continually improve its products and services;
2. The Advanced Technology Program, accelerating the development of innovative technologies for broad national benefit through R&D partnerships with the private sector;
3. A grassroots Manufacturing Extension Partnership (MEP) with a network of local centers offering technical and business assistance to smaller manufacturers; and
4. A highly visible quality outreach program associated with the Malcolm Baldrige National Quality Award that recognizes business performance excellence and quality achievement by U.S. manufacturers, service companies, educational organizations, and health care providers.

MEP is a nationwide network of more than 70 not-for-profit Centers whose sole purpose is to provide small and medium-sized manufacturers with the help they need to succeed. The Centers, located in all 50 states and Puerto Rico, are linked together through the Department of Commerce's National Institute of Standards and Technology. MEP makes it possible for even the smallest firms to have access to more than 2,000 knowledgeable manufacturing and business specialists.

MEP has assisted more than 62,000 firms. MEP provides assistance in the following areas:

1. quality management systems
2. business management systems
3. human resource development
4. market development
5. materials engineering
6. plant layout
7. product development
8. energy audits
9. environmental studies
10. financial planning
11. CAD/CAM/CAE
12. electronic commerce/EDI

MEP can be reached in the United States by calling 1-800-MEP-4MFG.

See **Quality management systems: Baldridge, ISO 9000, and QS 9000.**

Reference

[1] *www.NIST.org.* (the above materials are derived from this site).

NEXT QUEUE RULE

In a job shop, more than one job may wait in line for processing at the various work centers. To satisfy an efficiency or effectiveness criterion, items in the line are sequenced using an appropriate sequencing rule. Next queue (NQ) rule is based on a goal of maximizing machine utilization. The queues at each of the succeeding work centers to which the jobs will go are considered before sequencing the job at the current work center. The next job selected for processing is the one with the smallest queue at the next work center.

See **Order release; Sequencing rules.**

NO TOUCH EXCHANGE OF DIES (NOTED)

Setup reduction is at the heart of new manufacturing philosophies such as just-in-time manufacturing or lean manufacturing. NOTED, as defined by S. Shingo is a very rapid setup or change over where the tool or die is not removed from the machine.

See **Just-in-time manufacturing, lean manufacturing implementation; Setup reduction, SMED.**

Reference

[1] Shingo, Shigeo (1985). *A revolution in Manufacturing: The SMED system*. Productivity Press.

NON-COST CRITERIA IN INVENTORY MANAGEMENT

Normally, cost minimization drives inventory management decisions. However, lead time, obsolescence, availability, substitutability, and criticality are non cost items that also influence inventory management decisions. Non cost items may override cost considerations. For example, criticality may override cost issues in maintenance items. It considers the impact of parts when machines break down. Generally in ABC analysis, A items receive most weight in inventory management. A items are usually the top 15 to 20% of the items in inventory based on annual dollar usage. If criticality is taken into account in inventory management, criticality may weigh more heavily than dollar usage. Criticality is then a non-cost criterion based on the impact of running out of an item on production, and the effect of delivery delays on important customers. A low-cost item may be assigned 'A' status on the basis of its criticality measure.

See **ABC analysis; Multiple criteria ABC analysis.**

NON-FINANCIAL PERFORMANCE MEASURES

Production systems are managed to attain specific company goals by monitoring one or more performance measures. Non-financial performance metrics such as delivery, response time and quality are used in addition to financial performance measures to move the production system to higher levels of performance in the areas of delivery and quality, which are important to customers.

See **Balanced scorecard; Performance measurement in operations management.**

NONREPETITIVE JUST-IN-TIME SYSTEM

JIT was bred and developed for high-volume repetitive manufacturing. Low-volume high variety manufacturers have stayed away from JIT manufacturing because high-volume flow lines dedicated to a few products and level loading are unsuitable for their environment. In order to satisfy high customer need for high variety, high-volume repetitive manufacturers are learning to reduce the repetitiveness in their manufacturing system and yet continue to use JIT principles. At the other end of the spectrum, the low-volume job shop manufacturers are learning to adapt JIT principles.

See **Load leveling; Just-in-time manufacturing; Manufacturing systems.**

NON-VALUE ADDED ACTIVITIES

Not all manufacturing activities in a factory add value to the product. These activities are referred to as waste because they do not directly enhance the value of the product. Even if these activities are necessary, they do not directly add any value to the product. In the production of kitchen cabinets, cutting lumber to the right size directly adds value to the finished product while sharpening the saw is non-value added. Sharpening the saw may be necessary for efficient cutting of lumber but does not directly enhance the value of the finished product. Typical non-value added activities include scheduling, moving work-in-process from point to point, setting up equipment, recording time spent on a particular job, inspecting a part, and billing a customer. Typical valued added activities include assembling a product, fabricating a part, painting, grinding, cutting, coating, and a host of other activities necessary to complete a product.

See **Lean manufacturing implementation; Just-in-time manufacturing; Just-in-manufacturing implications.**

NORMAL DISTRIBUTION

See **Safety stocks: Luxury or necessity; Statistical process control using control charts.**

NPD

NPD stands for New Product Development. It refers to the process an organization use to conceive, design, test, and manufacture new products for introduction into the market. Product development lead time is a critical determinant of successful innovation and market competitiveness. The benefits of reducing development lead time through the use of an effective NPD process include increased market share, heightened barriers to new competitors, and higher sales farther into the future. Companies with shorter new product development cycles enjoy high profit margins. They progress down the manufacturing learning curve and ahead of others. Research has shown that the ability to develop new products quickly provides a window of opportunity for competitors to seize industry leadership at the expense of current leaders (Stalk and Hout, 1990). The incurred costs in the early stage of product development are low, yet in real terms, 40%–80% of the costs of the product are frozen during this stage in the form of material and process choice decisions. Changes made to the product are very inexpensive at this stage. Changes to the product made during final production may be as high as 10,000-fold more expensive. In the production stage, resources are already committed, processes established, and therefore any changes to the product require long man hours and expensive re-tooling to implement.

To become fast-to-market, many time-based competitors are embracing "concurrent engineering". Concurrent engineering suits a "time-based" corporate culture that places a greater reliance on external supplier design capabilities and greater reliance on automated design technologies.

See **Concurrent engineering; Process innovation; Product development and concurrent engineering; New product development through supplier integration.**

Reference

[1] Stalk, G.S. and Hout, T.M. 1990 *Competing Against Time*, The Free Press, New York.

NUMERICAL CONTROL

In the last fifty years, machine tools have been automated more and more. Machine tool automation takes many forms. In machine tool automation using numerical control, the machine is controlled by coded numerical instructions. The role of the operator in running system is reduced or eliminated by the numerical instructions.

See **CNC machine tool; Flexible automation; Manufacturing technology use and benefits; NC machine tool; Programmable automation; Programmable logic controllers (PLCs).**

OBSOLESCENCE COSTS

Inventory management recognizes several costs, one of them being obsolescence costs. Obsolescence costs are incurred when an item in inventory becomes obsolete before it is sold or used. Inventory may become obsolete due to product design changes, changes in customer demand, or by remaining unsold after an acceptable shelf life. Obsolescence costs include the labor and materials consumed in producing the original product and the cost of disposal (e.g., identifying, transporting and disposing obsolete inventory). Reduced setup times result in smaller lot sizes and less inventory, which reduces the risk of obsolescence.

See **Inventory flow analysis; Safety stocks: Luxury or necessity; Setup reduction.**

OC CURVE IN ACCEPTANCE SAMPLING

It is not always possible to inspect every piece received from a supplier. The common reasons being that the testing process destroys the product, and or the cost of testing each piece could be prohibitive. Under such circumstances, it is common to use acceptance sampling, where samples of batches (or lots) of products are tested, and the quality of the sample is used to judge the quality of all items in the lot. It controls the quality of incoming goods.

Acceptance sampling uses a "sampling plan" which specifies that, take a random sample if n units from an incoming lot, and if there are less than or equal to c defective units in the sample, the whole lot is to be accepted; otherwise reject the entire lot, or subject it to 100 percent inspection. In this plan c is the pre-specified acceptable number of defects.

The OC (Operating Characteristics) Curve shows how well the acceptance sampling plan works, that is, it shows the probability of accepting lots coming in with various amount of defective items. The use of sampling plans involves risks to the buyer and seller. There is producer's risk (alpha) associated with Type I error, it is the risk of rejecting a good lot. The buyer's risk (beta) associated with Type II error, it is the risk of accepting a bad lot. Producer's risk of 5 percent and buyer's risk of 10 percent are frequently used in industry.

These risks are managed by pre-selecting *acceptable quality level* (**AQL**) and *lot tolerance percent defective* (**LTPD**). AQL specifies the acceptable quality level; example, 2 percent defects in the incoming lot. LTPD defines the level of quality that is not acceptable; for example, 5 percent or more defects in the lot may be predefined as unacceptable.

Once the producer's risk, buyer's risk, AQL and LTPD are selected by managers, the sampling plan (n, c) can be derived or can be obtained from published tables and charts (Messina, 1987).

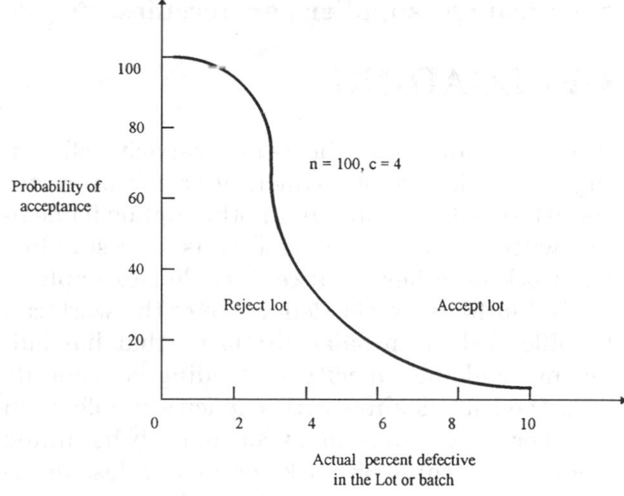

See **Statistical process control using control charts; Sampling in quality control.**

Reference

[1] Messina, W.S. (1987). *Statistical Quality Control for Manufacturing Managers*, New York: John Wiley and Sons.

OEM – ORIGINAL EQUIPMENT MANUFACTURER

Many products such as automobiles, household appliances, computer equipment and consumer electronics are complex and involve many levels of assembly. Often the product that reaches the final consumer has other products nested within them. For example, when General Radio developed a complex tester for printed circuit boards, its product had a Digital Equipment Corporation (DEC) computer in its core, a model that Digital sold separately to many customers. To the company supplying the core product, in this case Digital Equipment Corporation, Genreral Radio is an OEM (original equipment manufacturer) customer.

In some cases the OEM customer does little more than change the label and market the supplier's product under OEM's brand name and model number. This practice is quite common in the computer equipment and automobile industries. For example, some Mazda products reach the market in certain regions of the world under the label of its partner Ford Motor Company.

Organizations Mentioned: Cannon; Digital Equipment Corp; Ford Motor Company; General Radio; Mazda; Strategic Supplier Partnerships.

See **Strategic supplier partnerships.**

OFFLOADING

When a facility lacks short-term capacity, offloading the work to another facility or subcontractor could be used as an option. Another option for dealing with an overloaded facility is to reschedule the work on other work centers. For example, if a CNC lathe is overloaded, some of the work can be offloaded to an older NC lathe that has sufficient available capacity. Offloading is generally a method for dealing with capacity problems in the short run. If work must consistently be shifted from one machine or work center to a less desirable one, then this may indicate that permanent increase in capacity is needed. Offloading is generally done manually in conjunction with a scheduling system that is based on infinite loading. Methods that use finite loading, such as finite capacity scheduling, will generally handle the offloading decisions automatically.

See **Capacity planning: Medium and short-range; Capacity planning in make-to-order production; Finite capacity scheduling; Finite loading; Infinite loading; Order release.**

OFFSHORE PLANTS

Offshore plants are factories built in low cost environments, with the aim of exporting their outputs to other parts of the world. Very often these plants have a minimum management, and value adding activities are limited to those, which exploit the advantages of the low cost environment. Since the costs in a particular environment may change rapidly, offshore plants are very often temporary commitments.

See **Global facility location analysis; Global manufacturing rationalization; International manufacturing.**

OFF-THE-SHELF DEMAND

Off-the-shelf demand occurs when customer demand or orders for a product require immediate shipment upon receipt of the order. The order is therefore filled from existing inventory stocks, that is, finished-goods stocks which are already "on-the-shelf." Retail stores with shelves stocked with products and manufacturers warehouses filled with finished products can meet off-the-shelf demand. The need to satisfy "off-the-shelf" demand gives rise to a make-to-stock production environments, where products are manufactured before the receipt of a customer order. In this environment, customer orders are normally filled from existing stocks, and production orders are issued to replenish those inventory stocks.

See **Synchronous manufacturing using buffers; Theory of constraints in manufacturing management.**

Reference

[1] Srikanth, M.L. and M.M. Umble (1997). *Synchronous Management: Profit-Based Manufacturing For The 21st Century, Volume One*, Spectrum Publishing, Guilford, Connecticut, 224.

OHNO SYSTEM

See **Lean manufacturing implementation.**

ONE PIECE FLOW

See **Lean manufacturing implementation; Linked-cell manufacturing system; Manufacturing cell; Manufacturing cell design.**

ONE TOUCH EXCHANGE OF DIES (OTED)

Setup times are non-value-adding activities. Lean manufacturing attacks setup times until they are reduced to zero or near zero. OTED refers to setup or change over accomplished by the operator in one quick handling of the dies or tooling, usually in 10-15 seconds.

See **Lean manufacturing implementation; Just-in-time manufacturing; Just-in-time manufacturing implications; No touch exchange of dies; Setup reduction; SMED.**

OPEN ORDER FILE

A Material Requirements Planning (MRP) database consists of several files. Open order files in MRP are a set of files that hold records of scheduled receipts, or open order, in the MRP system. Scheduled receipts include purchase orders and shop orders. Data pertaining to the purchase orders include open purchase orders, open quotations, vendor master file, vendor performance data, alternate sources, and price/quantity. Data in the open order files include open orders, routings, work centers, employees, shifts, tooling, and labor/performance reporting.

See **MRP; MRP implementation.**

Reference

[1] Vollmann, T.E., Berry, W.L., Whybark, D.C. *Manufacturing Planning and Control Systems*, Fourth Edition, Irwin/McGraw-Hill, New York, 1997.

OPERATING CHARACTERISTICS CURVE

See **OC curve in acceptance sampling.**

OPERATIONS COST

According to the Theory of Constraints (Goldrath, 1984), the terms "operation costs", "operational expense", and "operational costs" are all synonyms. Operations cost is a catch-all description that includes all costs in the organization other than costs of direct materials. Essentially, it is all the money an organization spends in order to turn direct materials and work-in-process inventory into throughput. Operations cost includes traditional production costs such as direct labor and manufacturing overhead, as well as selling and general administrative expenses. TOC considers operations cost to be a fixed cost.

See **Accounting system implications of TOC; Theory of constraints in manufacturing management.**

Reference

[1] Goldratt, E.M. and J. Cox (1984). *The Goal: A Process of Ongoing Improvement*. North River Press, Croton-on-Hudson, N.Y.

OPERATIONS FORECASTING

See **Forecasting guidelines and methods; Forecasting in operations management.**

OPTIMAL LOT SIZES

Optimal lot size refers to the lot size (i.e., production or purchase order quantity) resulting in the lowest possible operating costs. Specifically, it would be the lot size recommended by a mathematical lotsizing model, which accounts for trade-offs between selected operating costs, relevant operating conditions and key constraints. In most cases, optimum refers to lowest possible cost, but, depending on the objective of the model, it may refer to other goals (e.g., maximum service level, shortest average delivery time or lowest cost within given constraints).

See **Inventory flow analysis; Mathematical lotsizing models; Safety stocks: Luxury or necessity; Setup reduction.**

OPTIMIZED PRODUCTION TECHNOLOGY (OPT)

Optimized Production Technology (OPT) is a predecessor to a production management approach known as the Theory of Constraints (TOC).

See **Accounting system implications of TOC; Theory of constraints in operations management.**

ORDER BACKLOG

See **Order release; Capacity planning in made-to-order production systems; MRP.**

ORDER LAUNCHING

Order launching is the process by which shop orders are released to the shop or purchase orders are released to vendors. In MRP systems, this process is triggered when a planned order release is in the action bucket, or current time period. Order launching is preceded by availability checking and allocation. Order launching creates open orders, that is, it converts planned orders into scheduled receipts in planning documents. Open orders are taken up for production. When the production is completed and the product is received in stockrooms, the order is closed.

See **Action bucket; Allocation and availability checking; MRP; MRP implementation; Order release.**

Reference

[1] Vollmann, T.E., Berry, W.L., Whybark, D.C. *Manufacturing Planning and Control Systems*, Fourth Edition, Irwin/McGraw-Hill, New York, 1997.

ORDER POINT

Timing of replenishment orders for items in inventory is sometimes based on the reorder point (R) or on order point. Where order points are used, inventory is under continuous monitoring(review), and, when the stock level drops to recorder point, a replenishment order for a fixed order quantity (Q) is generated. Order point is a function of demand rate, lead time to make or buy, uncertainty in the demand rate, and the company policy regarding the stockout (i.e., the level of customer service).

If there is uncertainty in an item's demand rate or lead time, a safety stock is required. Safety stocks protect against stockout while waiting for replenishment. The replenishment lead time for manufactured products can vary because it is a function of machine breakdowns, absenteeism, material shortages, etc., in the factory and distribution systems.

See **Inventory control; Inventory flow analysis; Inventory management; Safety stocks: Luxury or necessity; Simulation of production problems using spread-sheet programs.**

ORDER PREPARATION

Order preparation is the first step of the order release process. The second step is order review and evaluation. The third step is load leveling. During order preparation, the production planner creates and collects the documentation that the shop floor personnel need (drawings, control documents, process plans and the material requisitions) to process the shop order. If something is unavailable, then the planner cannot release the order.

See **MRP; MRP implementation; Order launching; Order release.**

ORDER PREPARATION COSTS

An order is an authorization to initiate production. An order requires preparatory work before releasing it for production. The costs of order production include variable clerical costs, paperwork, and material handling costs incurred in transporting goods from warehouse to the shop floor. Determining order preparation costs could be difficult. Work measurement techniques are often recommended to assess the cost of order preparation. Some costs of filling and preparing documents are fixed. Companies incur large costs for computer files with information relevant to order preparation.

See **MRP; MRP implementation; Order release; Order launching.**

ORDER PROMISING

Many products cannot be delivered to their customers immediately; rather, they are subject to production for future delivery. The customer is promised delivery on a certain date (promise date), decided by the customer's preferences or practical delivery date. Many orders therefore await production and delivery at any one time. Order promising is a commitment to sales or customer when delivery can take place, using the available-to-promise (ATP) concept. Order promising lets the company maintain lower inventory levels, since it reduces uncertainties in demand. The more accurate the order promising, the closer the actual shipments are to the master production schedule (MPS).

See **Available-to-promise (ATP); Master production schedule; MRP; MRP implementation; Order release.**

Reference

[1] Vollmann, T.E., Berry, W.L., Whybark, D.C. *Manufacturing Planning and Control Systems*, Fourth Edition, Irwin/McGraw-Hill, New York, 1997.

ORDER QUANTITY

Order quantity is an important concept in production. Order quantity is a variable that is a function of the customer order, lot-sizing models, Kanban, backlog, and MRP net requirements.

See **Economic order quantity; Just-in-time manufacturing; Kanban-based manufacturing systems; Manufacturing order quantity; MRP; Net requirement; Order release.**

ORDER RELEASE

Jeffrey W. Herrmann
University of Maryland, College Park, Maryland, USA

DESCRIPTION: Order release controls a manufacturing system's input. It may be known as job release, order review/release (Melnyk & Ragatz, 1989), input/output control (Wight, 1970), or just input control.

Order release connects the manufacturing planning system and the shop floor. Figure 1 shows a typical manufacturing operation. The large arrows represent orders that flow through the system. On the left-hand side, the arrows represent customer orders. On the right-hand side, the arrows represent work orders.

Typically, the manufacturing planning will process the accepted customer orders and create the required work orders. On an order's approval date, which arrives before the order's due date, manufacturing planning approves it for release so that it can be completed on-time. The lead-time is the predetermined time between an order's approval date and its due date. When approved, the order moves into the order release pool. The order release pool contains all approved orders awaiting release.

The order release process has three steps: order preparation, order review and evaluation, and load leveling. The first two steps are almost always performed. Although it is not required, the third step has received much attention from researchers who feel that it can significantly improve shop performance. In this article, any method that performs the load leveling step is called an order release mechanism.

During order preparation, the production planner creates and collects the documentation that the shop floor personnel need (drawings, control documents, process plans). If something is unavailable, then the planner cannot release the order.

During order review and evaluation, the production planner verifies that sufficient inventory is available to begin production, and that the needed tools are available. Again, if something is missing, then the planner cannot release the order.

Finally the production planner performs load leveling by using the order release mechanism to consider work-in-process (WIP), and releases orders if the WIP is low. If the WIP is too high, the production planner releases no orders, because WIP level would indicate that the shop has no available capacity.

The production planner sends each released order and its documentation to its first operation. The shop floor personnel begin performing the operations necessary to complete the order. The released orders and partially completed orders form the WIP. When complete, the order leaves the shop floor. The time between release and completion is an order's cycle time.

HISTORICAL PERSPECTIVE: Manufacturers have always used some order release function to

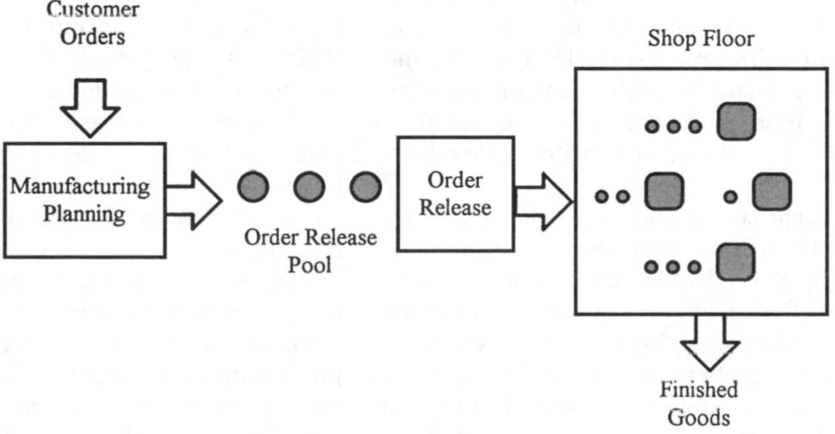

Fig. 1. The order release link.

prepare orders and release them to the shop. Over time, many practitioners began to view order release mechanisms as an important way to control the shop. Wight (1970) was one of the first to describe an order release mechanism. He proposed input/output control, which sets the shop input equal to the shop output.

Computer-based shop floor control systems made it possible to measure a shop's workload. This led to more sophisticated order release mechanisms, some developed for the complex flow shops in semiconductor manufacturing and some for more general shops. Starvation avoidance (Glassey and Resende, 1988) and workload regulation (Wein, 1988) are order release methods for semiconductor manufacturing. Spearman *et al.* (1990) proposed the constant WIP (CONWIP) order release policy for flow lines (also described by Hopp and Spearman, 1996). Bechte (1988) developed the load-oriented order release mechanism for job shop manufacturing.

Since then, researchers have developed order release mechanisms for various manufacturing systems and have evaluated how well they improve manufacturing system performance, if at all. Melnyk & Ragatz (1988) provide a survey of the field and the major findings. Bergamaschi *et al.* (1997) review and classify order release policies for job shop manufacturing.

STRATEGIC PERSPECTIVES: Order release is a critical function, and it is an unavoidable step in the manufacturing process, and the order release pool links the shop floor to the manufacturing planning systems. Improper order release starts too many orders that move to the shop floor and then wait for processing. This increases WIP and cycle times without increasing throughput. Resulting congestion complicates production, and personnel spend too much time seeking orders, expediting late orders, and scheduling resources. Long cycle times can increase loss and damage to waiting jobs if the products can be damaged while they wait for processing. This is an especially important issue in semiconductor manufacturing, where increased time in a clean room increases the chance of contamination and reduces the yield.

Reducing cycle times and WIP without reducing throughput is a universally important manufacturing objective. Proper order release limits the level of WIP, prevents congestion, reduces cycle time, and decreases the cycle time variance. More predictable cycle times will yield better on-time deliveries, less expediting, and simplify the planning and scheduling process. Lower WIP reduces congestion on the shop floor. Shorter cycle times reduce the chance that orders will be lost or damaged. Retaining or withholding an order in the order release pool has other benefits too: a cancellation or change causes fewer disruptions and less waste.

Proper order release can work better than capacity planning because counting inventory is easier than calculating capacity, which requires the measurement of various (and unpredictable) factors: setups, failures, repair, preventive maintenance, breaks, operator efficiency, yield losses, rework, and starvation. Typical planning and scheduling techniques do not adequately incorporate these uncertain events, so schedules soon become invalid and require constant revision. Planned lead-times and actual cycle times often differ, but reliable due date planning can occur only if the lead-times accurately predict the cycle time. Proper order release helps a manufacturer control cycle times and thus achieve better on-time delivery.

IMPLEMENTATION: The order preparation, review, and evaluation steps reflect the shop's particular documentation needs. Job shops typically require more documentation because each order has a different sequence of operations and a different set of components. Some industries use documentation to satisfy traceability requirements.

The order release mechanism makes two decisions: when to release an order (the timing decision) and which order to release (the order release rule). The timing decision can be periodic (the beginning of each day or each shift) or continuous (at any time). The order release mechanism may use information about the orders in the order release pool and about conditions in the shop and at each workcenter. The order release mechanism often measures the WIP and compares it to some threshold. The WIP may be the total WIP in the shop, or it may be the WIP at prespecified bottlenecks (the workcenters that limit shop throughput). The threshold may be based on the current WIP, or it may include the orders that will be released in future periods.

Many order release mechanisms ignore non-bottleneck workcenters although managing the bottlenecks is the key to managing the shop. This simplification reduces the data collection and processing effort.

The order release mechanism can measure WIP as the number of orders if all orders require the same amount of processing, or as the number of parts if all parts require the same amount of processing. Otherwise, it can measure WIP as the amount of time that a workcenter would require to process the remaining orders.

Input/output control (Wight, 1970; Fry and Smith, 1987) governs order release so that, during a given time period, the input to a shop equals the output. After measuring the system's actual output over some time period, the order release mechanism releases an amount equal to the actual output each period. If the system output increases, the order release mechanism will release more. If the system output decreases, the order release mechanism will release less.

Constant WIP, or CONWIP (Spearman *et al.*, 1990), logically extends input/output control. The CONWIP mechanism uses a set of (possibly electronic) cards. Each card accompanies an order through the shop. After the shop completes the order, the card returns to the order release step, where it joins a new order. As the name implies, this mechanism uses a constant number of cards. The number of cards should be the minimum number sufficient to avoid decreasing shop output. An order can be released only when a card is available. From the list of waiting jobs, each operation in the shop processes the order with the earliest release time first.

Starvation avoidance (Glassey and Resende, 1988) method of order release constantly measures the virtual inventory W for a bottleneck workcenter and compares that to the bottleneck's replenishment lead time L. If $W > L$, then a newly released order would reach the bottleneck before the bottleneck starved (became idle due to lack of work). It is unwise to starve a bottleneck work center. To account for manufacturing uncertainties, the starvation avoidance order release mechanism has an adjustable threshold; it triggers an order release whenever $W < \alpha L$ (where $\alpha > 0$). L is the total processing time needed for a newly released order to reach the bottleneck. (L does not include queue time. Ignoring the queue time simplifies the calculations.) The virtual inventory contains all released orders that require less than L time units of processing before reaching the bottleneck (this may include orders approaching the bottleneck for a return visit in a reentrant flow line). The value W for the bottleneck work center is the virtual inventory's total processing time, and the time needed to repair any down machines at the bottleneck workcenter, divided by the total number of machines.

Load-oriented order release (Bechte, 1988) is a periodic order release mechanism for job shops. Each period, the production planner first identifies the urgent orders that are, or will be approved during the time period. Then the production planner prioritizes the orders by approval date and releases orders while no workcenters exceed their load limits. A workcenter's load limit is the maximum allowable ratio of inventory (measured as hours of work) to the length of the time period. If urgent orders cannot be released, the production planner can use the load limits to calculate the needed production capacity during the time period. Using the load limits will keep actual cycle times close to planned lead times at each workcenter.

Herrmann *et al.* (1996) developed a system for bottleneck starvation avoidance in a large job shop that has multiple bottleneck workcenters. Each bottleneck workcenter has a WIP threshold measured in hours of work. This order release mechanism gathers information from the shop floor control system and examines the previously released orders at that workcenter or headed to that workcenter. The production planner may release orders if the current workload is less than the threshold. The order release system also collects information from the MRP system so that the production planner can examine the approved orders, and see the additional workload that they would add to the bottlenecks. Thus, the production planner can select orders for release while trying to level the load at bottleneck workcenters and maintain shop throughput with reduced WIP.

TECHNOLOGY PERSPECTIVE: Advanced order release mechanisms require information about the approved orders from the manufacturing planning system and information about the released orders from the shop floor control system. Like other shop floor control procedures, the order release function depends upon timely, accurate information. Poor information will lead to poor performance, even if the order release mechanism compensates for the information system constraints. Therefore, many order release mechanisms must be able to immediately and accurately acquire the necessary data.

Thus, the shop floor control system must collect and store information in real-time (or near real-time). This usually implies that each workcenter must have a terminal where the operator can update the shop floor control system, when an order starts, an order completes, a machine fails, a setup changes, or any other event occurs. In more automated manufacturing systems, the equipment itself may notify the shop floor control system.

At the minimum, the production planner must be able to retrieve data from the manufacturing planning and shop floor control systems. A more integrated approach automatically collects the required data from the manufacturing planning and shop floor control systems, and processes it to identify the approved orders that form the order release pool, calculate the WIP, signal if an order release is necessary, and identify the order that should be released. The production planner can verify that necessary inventory and tooling exist, select the order for release, create the required documentation, and send the order documentation electronically to the order's first operation. Bechte (1988) describes such an integrated system (see the Cases section).

SIGNIFICANT ANALYTICAL MODELS: One can use analytical models to evaluate an order release mechanism, and to find good parameter settings. Because manufacturing systems are so complex, simulation models are a popular technique. Some researchers (see, for example, Wein, 1988) have approximated complex systems with queueing networks and used these networks to evaluate order release mechanisms.

EFFECT: Proper order release reduces the inventory and cycle time within the shop and reduces the variability of these measures. However, they may increase the total manufacturing lead time because orders have to wait in the order release pool.

Order release mechanisms can reduce shop throughput if the production planner is not careful. The managers must set thresholds appropriately. Thresholds that are too small reduce WIP too much, which could starve the bottlenecks and reduce throughput. However, thresholds that are too large lead to unnecessary inventory and longer cycle times.

RESULTS: Order release mechanisms maintain low inventory levels and acceptable shop throughput by monitoring inventory and the bottleneck workcenters closely, and by releasing just enough orders to keep them busy. When a machine fails, order release is halted to prevent excessive inventory and congestion. If some orders require less time than expected, order release should release more work to avoid starving the bottleneck, which would reduce throughput.

An order release mechanism can reduce shop WIP and cycle time without degrading throughput. Since it also reduces performance variability, the shop should be able to set accurate leadtimes and then obtain better on-time delivery. Order release decreases the time-in-shop but increases the time-in-system because it shifts queue time from the shop to the pool. Since an order release mechanism synchronizes production to the bottlenecks, other workcenters may become idle. Managers must avoid the keep-busy attitude and realize that workers can use idle time productively by performing ancillary tasks. In addition, a well-implemented order release mechanism may simplify the order release task and allow production planners to perform other tasks. Finally, most order release mechanisms need good information, as discussed above. Implementing an order release mechanism may identify information system weaknesses and require changes.

CASES: Some authors have described the implementation of advanced order release procedures in manufacturing plants. Lozinski and Glassey (1988) describe a graphical tool that is used for order release and bottleneck starvation avoidance. The bottleneck starvation indicators are used daily for controlling high-volume semiconductor manufacturing. The indicators plot cumulative inventory versus cycle time for the bottleneck, and compare that to a target line that shows the inventory needed to achieve desired production levels. If the actual indicator line drops below the target, then the bottleneck resource is in danger of starving at some point in the future. Releasing more work may be necessary. However, if the actual indicator line is above the target, then the system has too much WIP, and no additional order release is necessary at that time.

Fry and Smith (1987) describe input/output control at a tool manufacturer and present a six-step procedure for implementing input/output control. In the first step, management must begin using system-wide performance measures such as throughput. The second step is to identify the bottlenecks that limit system throughput. The third step is to set maximum inventory levels for each station. When inventory reaches this level the previous operation must stop production. The fourth step is to reduce production lot sizes. The fifth step is to prioritize orders correctly so that no station hinders system performance. The sixth step is to match the input rate to the output rate. The authors report that, after implementing input/output control, the manufacturer reduced WIP by 42%, shrank the order backlog by 93%, reduced customer lead times by over 50%, and increased customer service to above 90%.

Bechte (1988) describes load-oriented order release at a European plastics processing firm. The load-oriented manufacturing control system includes a mainframe-based data management system, which provides the data that the PC-based planning and control systems need. A key component is a transaction data file that decribes all current orders and all orders completed in the last six months. From this data the mainframe can perform the large number of calculations needed to prepare data for the planning and control systems.

The planning system uses this data to perform order entry, order release, and order sequencing at each workcenter. The control system uses this data to generate reports that describe system performance.

COLLECTIVE WISDOM: Order release will not fix everything. Improper planning can still send too many orders to the pool. Order release methods assume that capacity is fixed. But order release succeeds only when it works with a capacity planning system and reacts to the dynamic shop floor conditions. If the order release pool increases or decreases, it signals that the shop floor capacity requires adjustment by shifting personnel or adding resources. The order release mechanism should support the system characteristics that most impact system performance: bottlenecks, batch processing machines, and shop flexibility. When choosing an order release mechanism, a planner should consider many issues, including the type of shop (flow lines or job shop), the due date setting policies, the dispatching rules, and the procedures for accepting customer orders. Additionally, the timing of release (continuous versus periodic) is a significant factor. In low-volume shops, complex order release rule may not be important. Sequencing rules such as SPT and FIFO rules should work well.

Organizations Mentioned:

See **Capacity management in make-to-order production systems; Capacity planning; FCFS; Input/output control; Just-in-time manufacturing; Order release mechanism; MRP; Theory of constraints in manufacturing management; Resource planning: MRP to MRP II and ERP; Sequencing rules; SPT.**

References

[1] Ashby, James R., and Reha Uzsoy (1995). "Scheduling and order release in a single-stage production system," *Journal of Manufacturing Systems*, Volume 14, Number 4, pages 290–306.

[2] Bechte, Wolfgang (1988). "Theory and practice of load-oriented manufacturing control," *International Journal of Production Research*, Volume 26, Number 3, pages 375–395.

[3] Bergamaschi, D., R. Cigolini, M. Perona, and A. Portioli (1997). "Order review and release strategies in a job shop environment: a review and a classification," *International Journal of Production Research*, Volume 35, Number 2, pages 399–420.

[4] Fry, Timothy D., and Allen E. Smith (1987). "A procedure for implementing input/output control: a case study," *Production and Inventory Management Journal*, Volume 28, Number 3, pages 50–52.

[5] Glassey, C. Roger, and Mauricio G.C. Resende (1988a). "Closed-loop job release for VLSI circuit manufacturing," *IEEE Transactions on Semiconductor Manufacturing*, Volume 1, Number 1, pages 36–46.

[6] Hendry, L.C., and B.G. Kingsman (1991). "Job release: part of a hierarchical system to manage manufacturing lead times in make-to-order companies," *OR: the Journal of the Operational Research Society*, Volume 42, Number 10, pages 871–883.

[7] Herrmann, Jeffrey W., Ioannis Minis, Murali Narayanaswamy, and Philippe Wolff (1996). "Work order release in job shops," pp. 733–736, *Proceedings of the 4th IEEE Mediterranean Symposium on New Directions in Control and Automation*, Maleme, Krete, Greece.

[8] Hopp, Wallace J., and Mark L. Spearman (1996). *Factory Physics*, McGraw-Hill, Boston.

[9] Lingayat, Sunil, John Mittenthal, and Robert M. O'Keefe (1995). "Order release in automated manufacturing systems," *Decision Sciences*, Volume 26, Number 2, pages 175–205.

[10] Lozinski, Christopher, and C. Roger Glassey (1988). "Bottleneck starvation indicators for shop floor control," *IEEE Transactions on Semiconductor Manufacturing*, Volume 1, Number 4, pages 147–153.

[11] Melnyk, Steven A., and Gary L. Ragatz (1989). "Order review/release: research issues and perspectives," *International Journal of Production Research*, Volume 27, Number 7, pages 1081–1096.

[12] Philipoom, P.R. and T.D. Fry (1992) "Capacity-based order review/release strategies to improve manufacturing performance," *International Journal of Production Research*, Volume 30, Number 11, pages 2559–2572.

[13] Roderick, Larry M., Don T. Phillips, Gary L. Hogg (1992). "A comparison of order release strategies in production control systems," *International Journal of Production Research*, Volume 30, Number 3, pages 611–626.

[14] Spearman, Mark L., David L. Woodruff, and Wallace J. Hopp (1990). "CONWIP: a pull alternative to kanban," *International Journal of Production Research*, Volume 28, Number 5, pages 879–894.

[15] Wein, Lawrence M. (1988). "Scheduling semiconductor wafer fabrication," *IEEE Transactions on Semiconductor Manufacturing*, Volume 1, Number 3, pages 115–130.

[16] Wein, Lawrence M., and Philippe B. Chevalier (1992). "A broader view of the job shop scheduling problem," *Management Science*, Volume 38, Number 7, pages 1018–1033.

[17] Wight, O.W. (1970). "Input/output control: a real handle on lead times," *Production and Inventory Management*, Volume 11, pages 9–30.

ORDER RELEASE MECHANISM

Order release sends work orders to the shop floor for production. An order release mechanism is a process for performing load leveling. Load leveling is the third step of the order release process. The first step is order preparation. The second step is order review and evaluation. A production planner performs load leveling by using the order release mechanism to release orders if the WIP is low. If the WIP is too high, the production planner releases no orders, because high WIP indicates all available capacity is fully occupied and several orders are waiting to be processed.

The order release mechanism makes two decisions: when to release an order (the timing decision) and which order to release (the order release rule). The timing decision can be periodic (the beginning of each day or each shift) or continuous (at any time). The order release mechanism may use information about the orders waiting for release in the order release pool, and about work load conditions in the shop and at each workstation. The order release mechanism often measures the WIP and compares it to some threshold. The WIP may be the total WIP in the shop, or it may be the WIP at prespecified bottlenecks (the workstations that limit shop throughput). It may be the current WIP, or it may include the orders that are expected to be released in future periods. Although many order release mechanisms ignore non-bottleneck workstations, managing the bottlenecks is important to managing the shop. This simplification reduces the data collection and processing effort.

The order release mechanism can measure WIP as the number of orders if all orders require the same amount of processing, or as the number of parts if all parts require the same amount of processing. Otherwise, it can measure WIP at a workstation as the amount of time that a workstation would require to process the remaining orders.

See **MRP; Order release; Order release rule; Schedule stability; Theory of constraint in operations management; Synchronous manufacturing using buffers.**

ORDER RELEASE RULE

An order release means sending a work order to the shop floor manufacturing. Not all orders that are ready for manufacturing are sent or released to the shop floor for production. An order release mechanism decides when and which order to release. An order release rule is the algorithm that an order release mechanism uses when deciding which order to release for production.

See **MRP; Order release; Order release mechanism; Schedule stability; Sequencing rules; Theory of constraint in operations management.**

ORDER REVIEW AND EVALUATION

Order review and evaluation is the second step of the order release process. The first step is order preparation. The third step is load leveling. During order review and evaluation, the production planner verifies that sufficient inventory materials and parts are available to begin processing the order, and that the needed tools are available. If something is lacking, the planner cannot release the order for production.

See **Order release; Order release mechanism; Order release rules; MRP; Schedule stability; Theory of constraints in operations management.**

ORDER-BASED FREEZING

Master Production Schedule (MPS) is a plan and an input to MRP. MPS is a plan for end items and major subassemblies. Its shows how many of each major item is needed and when it is needed to meet the delivery commitment. Changes to MPS entries can have very significant effect on planned orders, purchase orders and open orders already in production. Freezing of parts of MPS is done to prevent potentially disruptive changes . Order based freezing specifies that a certain number of orders cannot be changed.

See **Capacity planning in a make-to-order production; Master production schedule; MRP; Schedule stability.**

OUTSOURCING

Outsourcing refers to the purchase of raw materials, parts, subassemblies, end items or services from outside the organization. Companies outsource for various reasons including lack of capacity, lack of know-how or technology, or because the company strategy does not require it to be made internally.

See **Purchasing; Supply chain management; Supplier partnership as strategy; Supplier relationships.**

OUTSOURCING OF PRODUCT DESIGN AND DEVELOPMENT

Hilary Bates

University of Warwick, Coventry, United Kingdom

David Twigg

University of Brighton, Falmer, United Kingdom

INTRODUCTION: The buyer-supplier relationship has undergone substantial transformation over the past 30 years, from a traditional relationship focused largely on the buyer to a current focus on partnership and lean supply (Lamming, 1993). The emerging relationship has meant greater involvement in design, and product technology responsibility for many suppliers. It is based upon dialogue, free exchange of information, and long-term cost awareness.

Commentators such as Womack *et al.* (1990) and Lamming (1993) have studied the Japanese automotive industry, which, both in productivity and structure, provided stark contrast with the European and North American car industries. They see the benefits of externalising elements of the value chain through large scale outsourcing of aspects of the business, which were traditionally kept in-house. The benefits inherent in working with a small and very closely integrated supplier base as opposed to owning the supply base are seen as increased quality and productive efficiency, reduced inventory and costs. The arrival of the lean enterprise has focused attention on how companies could manage their supply chains in a manner, which would derive the benefits the Japanese have enjoyed over their competitors.

Lamming (1993) defines the lean supply model as comprising of:

- Fewer, larger and more talented companies who are the central provider of complete component systems.
- A small group of preferred suppliers with responsibility for co-ordination and control of their own supply base.
- R&D Collaboration between the vehicle manufacturer and preferred suppliers with early supplier involvement.
- Stronger vertical and horizontal relationships to share expertise (higher interdependence).
- Global sourcing and operations, and multi-market presence.
- Competitive advantage based on best practices, continual improvement, and the ability to collaborate.

Long term business relationships provide many potential benefits. In fact, close relationships between buyer and supplier can confer significant competitive advantage, and where this advantage is "invisible" to other competitors, it can create a barrier to entry (Jap and Weitz, 1996). On a more prosaic level, it is less costly for a firm to maintain and develop an existing client relationship than to attract a new one (Berry, 1983: 23; Grönroos, 1990: 5).

THE OUTSOURCING DECISION: Some of the reasons that manufacturers might consider for increased outsourcing include: the benefits of increased economies of scale; access to specialist expertise (product or service) in the supply base; short and long-term financial advantage; sale economics; strategic realignment; and better focus on core operations (Behara *et al.*, 1995). Quinn (1992) advocates the use of strategic outsourcing for greater flexibility and decreased product design cycles, especially where rapidly developing new technologies, or highly complex systems are involved. His reasons are:

> "Each supplier can (1) have a greater depth in knowledge about the design of its particular technologies, and (2) support more specialised facilities to produce higher quality than the co-ordinating company could possibly achieve alone." (Quinn, 1992: 48)

A further advantage Quinn notes is the ability to spread the risks for component development across a number of suppliers. The consideration of the outsourcing of product design activities requires attention to many other issues that go beyond what is commonly termed the 'make or buy decision'. Make or buy decisions fail to distinguish those services that are purely knowledge/information related. Elfring and Baven (1994) examine this decision for knowledge intensive industries, and conclude that the development of interfacing skills is perhaps more important. They argue that the expertise necessary to manage the external provision of knowledge intensive services (such as product design) is different from the sourcing of components for production. Hence, a firm that is likely to outsource, say for example component design, will need co-ordinating, strategic, and conceptual skills (interfacing skills), as well as the capacity to manage contractual relationships.

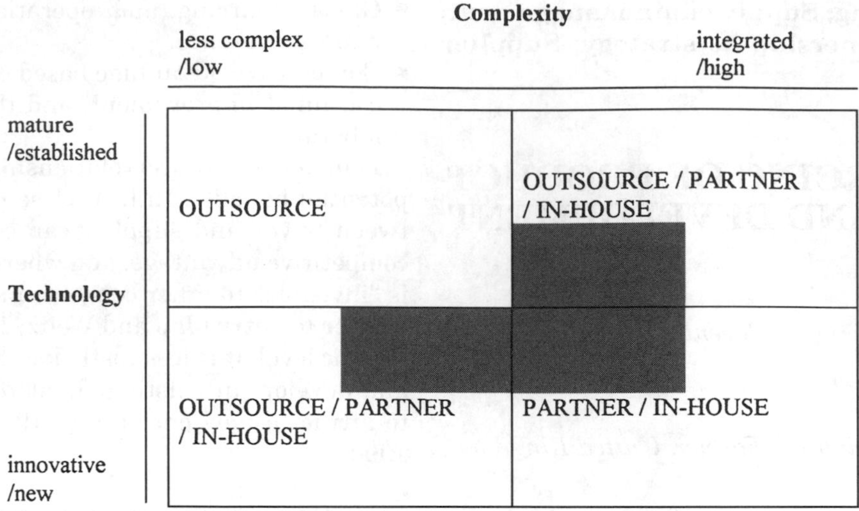

Fig. 1. The outsourcing decision.

Whilst the decision to outsource may be made along lines similar to the make or buy decision, additional mechanisms must be in place to ensure adequate control of the supplier relationship. Outsourcing of design requires the consideration of, *inter alia*: the reliability of suppliers to undertake the design work; the quality of their service; the supplier's access to the latest/advanced technology and expertise; and the ability to return to in-house operations if required. Supplier assessment/relationship programs of major companies are substantially detailed to ensure most of these requirements.

Figure 1 shows the relative position of various sourcing decisions that a company might consider. The figure considers the type of technology embedded within the component or system, and the level of complexity or the degree of interfacing it may have within an assembly. Where a component is mature and generally not complex(NW quadrant in Figure 1), the design work would normally be outsourced. Specifications of the product would be made by the manufacturer, but the detailed drawing would be the responsibility of the supplier. In the SE quadrant of Figure 1, one would expect components to be highly complex, perhaps with a large number of interfacing parts; design is preferred in this case, or an option would be to form a partnership with a supplier. Thus, the type of technology, the product and the interface requirements must be considered in all sourcing options.

The shaded areas of Figure 1 indicate critical areas in which manufacturers might outsource too much. But, a key threat to outsourcing is the fear of internal loss of knowledge in a particular technology. Two options exist for manufacturers to counteract this threat. First, customers may wish to co-locate supplier design engineers within their project team. Guest, or resident, engineers provide one means of achieving outsourcing of design, whilst maintaining close understanding of the design operations for a particular component. Alternatively, a customer may choose to locate their engineers with suppliers (Twigg and Slack, 1998). If competent engineers are co-located within the buyer-supplier network, then the groundwork is set for both improved communication and better long-term relationships.

The Case of the Automotive Industry: There are several important challenges facing the automotive industry:

- Worldwide over capacity in a mature market.
- Demanding customers.
- Increased global competition.
- The perceived need for a global presence.
- Fast cycle product development.

The competitive challenge has led vehicle manufacturers to consider the structure of their business. This has, in turn, led to a significant rationalisation of the automotive component supply base, through:

- Dual or single, rather than multi-sourcing of parts.
- Devolution of design and development of non-core components to the supply base, particularly the first-tier suppliers.
- Development of modular supply. (Modular supply uses suppliers, who have considerable design expertise and can deliver the complete sub-assemblies of the vehicle, rather than just individual components).

The trend towards increasing devolution of responsibility from vehicle manufacturer to the first tier component suppliers is, in part, an acknowledgement of the under utilised knowledge and expertise of suppliers and, in part, a way for vehicle manufacturers to reduce both costs and complexity within the organisation.

Vehicle manufacturers want to surround themselves with a small number of "approved and preferred suppliers". This has led to significant reductions in the number of suppliers customers. In 1986, for example, Chrysler used approximately 3,000 suppliers. By 1996 this number had fallen to 950, with 150 supplying 70% of the company's purchases. Only part of this reduction is due to the move from multi-sourcing to dual-sourcing, or single-sourcing. These suppliers will be certified as meeting predetermined quality specifications, they will be responsible for co-ordinating the work of suppliers in the lower tiers, and they will undertake the majority of the design and development work for any component with which they are involved, even if they do not manufacture that component.

The emergence of the vehicle "corner", or module, presents an additional challenge for component suppliers. Although the average vehicle requires somewhere in the region of 27,000 individual components, vehicle manufacturers can exercise choice as to how they organise the supply chain. For example, the 27,000 components could be supplied by 3000 suppliers, supplying individual components or small assemblies of components, or alternatively, 25 companies delivering larger more complex sub-assemblies or, conceivably, 5 suppliers delivering large modules or corners of the vehicle.

As suppliers of modules, they incur the task of integrating unfamiliar product technologies, the responsibility for the management of a wider supply chain than they have been accustomed to, and the task of managing a radically different relationship with vehicle assemblers. This opportunity to pass the responsibility, not merely for the design and development of a complete sub-assembly of a vehicle, but also for supplier co-ordination reduces the complexity for vehicle manufacturers.

There is evidence that closer integration with independent suppliers may prove more successful to vehicle manufacturers in developing new products and new technologies than similar relationships with wholly owned subsidiaries, or even between departments of the same company (Clark, 1989).

THE NATURE OF RELATIONSHIPS: At the concept and product planning stages, external expertise to vehicle styling may be sought from design and engineering consultants (such as Pininfarina of Italy) or suppliers of body trim may contribute manufacturing advice to model stylists. At the product engineering stage, multinational component system suppliers may take responsibility for proprietary parts. In the process engineering stage, manufacturing knowledge is paramount, and again external expertise may be sought. Toolmakers are particularly important here, especially for body aesthetic items. Equipment manufacturers, raw material suppliers, or process specialists (such as plastic injection moulding) also can provide expertise. These inputs are increasingly used further upstream in concept design and product engineering, such that toolmakers, raw material suppliers and equipment suppliers influence product specifications.

In aggregate terms, the level of supplier involvement in design and development activities may be more than half of the total procurement cost of engineering. In automotive engineering, for example, Clark and Fujimoto (1991) calculate that 10% of engineering procurement costs is for supplier proprietary (off the shelf items such as batteries or tyres), 40% is for black box items (systems or modules designed and developed to customer specifications by primary suppliers), and the remaining 50% is designed and developed in-house by vehicle manufacturers. What these figures do not demonstrate, however, is the increasing 'grey box' element where suppliers may sit in with a vehicle manufacturer and provide process knowledge for product design work (Twigg, 1997).

Firms may choose to collaborate for many different reasons. The trend towards the increased use of 'corner modules' will require new collaborative links between previously unconnected suppliers. The sharing of design responsibility will become increasingly necessary between assembler and supplier (vertical quasi-integration). Modular design of autos will require suppliers to collaborate with other suppliers who have complementary assets to complete the module. Sleigh (1993) noted that Volkswagen, for example, was looking for suppliers who could progress from being mere parts suppliers to suppliers of component systems. Volkswagen has been one of the foremost companies in utilising modular assembly. They have been using modular assembly in the Ford-VW joint venture plant in Portugal (Auto Europa), and in a more advanced form in their truck plant in Resende, Brazil. At the Resende plant, the vehicle has been divided into seven large modules, or "vehicle corners" based around: Chassis; Tires and Wheels; Axle and Suspension; Engine; Trim; Cabin Frame and Painting; and the Final test. Each corner is designed, developed, co-ordinated and assembled by a supplier. Apart from sales and distribution, Volkswagen is only responsible for the final module, which is overall vehicle testing and quality.

The sharp reductions in design and development costs provide Volkswagen with a much lower cost base than is the norm in the industry, which should leave them with plenty of funds to invest in research and development, or further globalisation. This example represents a complete devolution of design and development from manufacturer to supplier. Some vehicle manufacturers are concerned about the potential for losing control over total vehicle integrity when so much power is passed on to suppliers. There is only a fine line between delegation of responsibilities and complete devolution of power. This observation raises several questions:

- When manufacturers devolve design responsibility to supplier companies, do they intend to retain in-house knowledge about the component?
- Do manufacturers fully assess the features that contribute distinctiveness to their products?
- Are these features tied to specific components?
- Are there core 'design' capabilities that should not be outsourced?

If these questions are not addressed, vehicle manufacturers may indulge in injudicious outsourcing, which could result in an amputation of their technical knowledge base leaving them with limited future prospects as mere assemblers, and vendors of other people's design and production expertise.

Summary of the Auto Industry case: The automotive industry case indicates that:

- Rationalisation of the supply chain and outsourcing are actively undertaken with the supply base taking on more responsibility for the delivery of ever-larger modules.
- Not all suppliers exhibit the expected eagerness to become co-ordinators of major component modules, and some are sceptical about how outsourcing of the design and development of complete component systems will function.
- The control of design within the supply network is a major concern. Even if a vehicle manufacturer does not design every component, vehicle manufacturers perceive it as important to have more than a passing acquaintance with each component.
- Collaboration on design is already commonplace and functions well.
- Component suppliers perceive vehicle manufacturers as being ever more demanding on cost, quality and delivery.
- Detail design work is expected from suppliers as part of the service before firm orders are given.
- Vehicle manufacturers are aware of the risks inherent in over-enthusiastic outsourcing of design but few have reached conclusions within their own companies as to what components make up the critical core of the vehicle.
- Product liability and management of relationships are increasingly important issues for both component and vehicle manufacturers.

Conclusions: In recent years, companies have placed increasing emphasis on a 'core business' focus. This focus has resulted in growing interest in the core competencies of the firm, with a subsequent drive to concentrate internal resources on those tasks that support the strategic architecture and maintain the firm's competitive advantage.

There is a wide selection of alternative design capabilities available to firms in managing the design and development process. At one extreme, a firm may retain all the design capabilities inhouse, and until recently, this was indeed the scenario in most car companies. At the other extreme, a firm may outsource design work and act as a focal point for the co-ordination of the design process.

In general, vehicle manufacturers believe that their competitive advantage lies in those identifiable properties of the vehicle that mark it out as distinctive, and valued by the customer. In some cases, this advantage may be related solely to product design, but in many cases it is a mixture of both product and process design.

The decision to fully outsource the design and development of these features that give competitive advantage will diminish a firm's intangible assets of knowledge and learning. Before embarking on large-scale outsourcing, companies should identify the strategic components and component systems. Outsourcing routes must enhance, rather than diminish, their long-term competence. Whatever the sourcing decision, the ability to manage (or control) these external inputs is a key competence. This requires not only suitable coordination mechanisms but also individuals with the competencies to oversee the necessary process and relationships. Without them, the extinction of specific design knowledge within the organisation may lead to significant loss of competitiveness.

Within the automotive industry, the merger and acquisition activity among component makers is resulting in very large companies with the potential to produce whole subassemblies. These component companies are better prepared to confront the challenges of increased responsibility. Component manufacturers may grow to be significantly bigger than their customers.

Organisations Mentioned: Chrysler; Pininfarina, Italy; Ford-VW; Volkswagen.

See **Lean manufacturing implementation; New product development through supplier integration; Preferred supplier; Product development and concurrent engineering; Supply chain management; Supplier partnerships; Supplier relationships.**

References

[1] Behara, R.S., Gundersen, D.E. and E.A. Capozzoli (1995). "Trends in Information Systems Outsourcing." *International Journal of Purchasing and Materials Management*, 31 (2), 46–51.

[2] Berry, L.L. (1983). "Relationship Marketing." in L.L. Berry, G.L. Shostack and G.D. Upah (eds.) *Emerging Perspectives on Services Marketing*. AMA, Chicago, 25–28.

[3] Clark, K.B. (1989). "Project Scope and Project Performance: The Effect of Parts Strategy and Supplier Involvement on Product Development." *Management Science*, 35 (10), 1247–1263.

[4] Clark, K.B. and T. Fujimoto (1991). *Product Development Performance*. Harvard Business School Press, Boston, MA.

[5] Elfring, T. and G. Baven (1994). "Outsourcing Technical Services: Stages of Development." *Long Range Planning*, 27 (5), 42–51.

[6] Grönroos, C. (1990). "Relationship Approach to Marketing in Service Contexts: The Marketing and Organizational Behaviour Interface." *Journal of Business Research*, 20, 3–11.

[7] Jap, S. and B.A. Weitz (1996). *Achieving Strategic Advantages in Buyer-Seller Relationships*. Marketing Science Institute, Working Paper Report No. 96–108.

[8] Lamming, R. (1993). *Beyond Partnership: Strategies for Innovation and lean Supply*. Prentice Hall, London.

[9] Quinn, J.B. (1992). *Intelligent Enterprise*. Free Press, New York.

[10] Sleigh, P. (1993). *The world automotive components industry: A review of leading manufacturers and trends*, Research Report, 1–4, Economist Intelligence Unit, London.

[11] Twigg, D. (1997). "A Typology of Supplier Involvement in Automotive Product Development." *Warwick Business School Research Paper*, No. 271.

[12] Twigg, D. and N. Slack (1998). "Lessons from Using Supplier Guest Engineers in the Automotive Industry." *International Journal of Logistics, Research and Applications*, 1 (2), 181–192.

[13] Womack, J.P., Jones, D.T. and D. Roos (1990). *The Machine that Changed the World*. Macmillan Publishing, New York.

OVERHEAD

Overhead is a classification for certain types of costs in an organization. Overhead costs are incurred in the operation of a business but cannot be directly related to individual goods or services produced. That is, overhead costs are incurred indirectly in making the product. Overhead costs includes all indirect costs incurred in or by the production area, such as factory supervisors' salaries, electricity, depreciation of automated machinery, and insurance of the production facilities. Overhead does not include direct labor or direct materials, where direct labor refers to the time spent by workers in manufacturing the product and direct materials refers to materials used in the product.

See **Accounting system implications of TOC; Activity-based costing; Activity-based costing: An evaluation; Target costing; Unit cost; Unit cost model.**

Reference

[1] Barfield, Jesse T., Cecily A. Raiborn, and Michael R. Kinney (1994). *Cost Accounting: Traditions and Innovations*, West Publishing Company, St. Paul, Minnesota.

OVERSTATED MASTER PRODUCTION SCHEDULE

An MPS is overstated if the quantities included in schedule exceed the capacity to provide within the time stipulated in MPS. It can happen when past due quantities are included. An overstated MPS is made feasible before the MRP run. The overstated MPS causes MRP to fail and erodes belief in the system.

See **Capacity management in make-to-order production management systems; Capacity planning: Medium and short range; Master production schedule; MRP.**

PAPERLESS TRANSACTIONS

In paperless transactions, electronic, computer transactions and records are used instead of paper-based records and transactions.

See **Electronic data interchange in supply chain management.**

PARETO CHART

See **ABC analysis.**

PART PROGRAMMING (OF AN FMS)

In a manually controlled machine, manufacturing instructions were typically incomplete in the past, due to the fact that the skilled machine operators had the necessary knowledge to fill-in the gaps between what was provided on the worksheets that the production engineering department issued and what was actually needed at the shop- floor.

Obviously with computer control, such as CNC (Computer Numerical Control), the machine executes exactly what is "part programmed," meaning that the instructions fed into the machine must be precise and complete. Such instructions are created based on the drawing of the part, part geometry, tolerances and material, the tools needed, the tool-path designed, the setup, and machine selected.

In comparison to part programming a single machine, such as a machine tool or a robot, an FMS is a highly automated, distributed and feedback controlled system of data, information and physical processors in which decisions are made real-time. Consequently, in the case of FMS programming, not just an individual machine, but an entire system must be be programmed. This opens up the opportunity for various optimization techniques that can yield major economic benefits.

See **Manufacturing with flexible systems.**

Reference

[1] Ranky, P.G. (1998). *Flexible Manufacturing Cells and Systems in CIM*, CIMware, Ltd., cimware@cimwareukandusa.com (Book and CD-ROM).

PART-FAMILY FORMATION METHODS

One of the most important problems faced in designing manufacturing cells is called cell formation. It groups parts with similar geometry, function, material and processing into part families, and the corresponding machines to produce the family are parts go into machine cells. This grouping philosophy, which can simplify the manufacturing planning, scheduling, and control tasks, has been widely used in FMS or FMC (Kumar *et al.*, 1986) and in Just-In-Time (JIT) production (Schonberger, 1982).

See **Dynamic routing in flexible manufacturing systems; Just-in-time manufacturing; Kanban-based manufacturing systems; Linked-cell design; Manufacturing cell design.**

PART-FAMILY GROUPING

Each Flexible Manufacturing System (FMS) is limited to produce certain shapes and parts with certain technological properties. To meet these limitations, parts made in FMS are grouped into families. Parts grouping in FMSs has advantages such as: simplification of planning and scheduling, simplification of parts and tools flow, reduction of setup costs, reduction of production costs, etc. One of the basic tools used in group technology (GT) to accomplish parts grouping is the coding/classification analysis or system. This uses the description of parts based on their geometrical shape, dimensional accuracy, material, etc. This set of information is usually stored in the computer aided design (CAD) database. Various approaches can be used to solve the parts family grouping problem, including pattern recognition techniques. Various techniques for clustering analysis can also be used to formulate and solve the parts family grouping problem.

See **Coding/classification analysis; Disaggregation in an automated manufacturing environment; Dynamic routing in flexible manufacturing systems; Flexible automation; Group technology.**

Reference

[1] Kusiak, A. (1985), "The Part Families Problem in Flexible Manufacturing Systems," *Annals of Operations Research*, 3, pp. 279–300.

PART-PERIOD BALANCING (PPB)

PPB is a dynamic lot-sizing technique which computes the production lot size after considering the next or the previous period's demands to determine if it would be economical to include them in the current lot.

Thus, the part period balancing (PPB) procedure uses more information in the requirements schedule than other procedures. The PPB procedure allows both lot size and time between orders to vary. For example, smaller lot sizes and longer time intervals between orders are used in periods of low demand to reduce inventory costs.

See **Inventory floor analysis; MRP; Order release; Safety stocks: Luxury or necessity; Schedule stability; Setup reduction.**

P-CHART

A p-chart is used to control the proportion of nonconforming (or defective) items. A quality characteristic that does not meet specifications is said to be a nonconformity. A product with one or more nonconformities that is unable to function as required is defined to be a nonconforming item or a defective.

In the construction of p-charts, random samples are selected from the process and each item is classified as nonconforming or not. A sample proportion of nonconforming items is then calculated, which is similarly repeated for at least 25 samples. The center line on the p-chart represents the average nonconformance rate of the process. Control limits are usually placed at three standard deviations of the proportion nonconforming on either side of the center line. Based on observations from the process, the control limits can be calculated using the average nonconformance rate and the sample size. Note that, as the sample size increases, the control limits are drawn closer to the center line.

The choice of the sample size for a p-chart is important. It should be large enough to allow nonconforming items to be observed in the sample. For example, if a process has an average nonconformance rate of 1%, a sample size of 50 is not sufficient because the average number of nonconforming items per sample is only 0.5. Since no nonconforming items would be observed for many samples, misleading inferences could be made. It is possible to attribute a better nonconformance rate to the process that what actually exists. A sample size of 300 might be sufficient since the average number of nonconforming items per sample would be 3.

The p-chart is based on the binomial distribution. For the binomial distribution, the probability of occurrence of nonconforming items is assumed to be constant for each item and the items are assumed to be independent of each other with respect to meeting specifications. The latter assumption may not be valid if products are manufactured in groups and the same control chart is used for different groups. For instance, in steel manufacturing, suppose a batch produced in a cupola does not have the correct proportion of an additive. If a sample steel ingot chosen from the batch is found to be nonconforming, it is likely that other samples from the same batch are also likely to be nonconforming, so the samples are not independent of each other.

Sometimes, management may set a standard value for the proportion nonconforming, in which case the center line and control limits will be computed based on this standard. For observations from the process that fall above the upper control limit, special causes should be investigated and remedial actions taken for their elimination. However, if after the elimination of special causes, observations are still outside the upper control limit and no further special causes can be identified, management should question whether the imposed standards can be met by the process. It may be that the process is not capable of meeting the imposed standards.

See **Statistical process control using control charts.**

PDCA

See **Plan-do-check-act.**

PERCENT OF LATE JOBS

One of the criteria used to evaluate sequencing rules is "percent of late jobs." It refers to the number of jobs that are completed late divided by the total number of jobs completed.

See **Manufacturing systems modeling using Petri nets; Sequencing rules; Simulation analysis of manufacturing and logistics systems.**

PERFORMANCE DRIVERS

Performance drivers are underlying activities that cause performance to improve or deteriorate. For example, number of hours spent with customers is an activity likely to influence customer satisfaction, a performance measure.

See **Activity-based costing; Activity-based costing: An evaluation; Balanced scorecard.**

PERFORMANCE EXCELLENCE: THE MALCOLM BALDRIGE NATIONAL QUALITY AWARD CRITERIA

James R. Evans

University of Cincinnati, Cincinnati, Ohio, USA

DESCRIPTION: The Malcolm Baldrige National Quality Award (MBNQA) recognizes U.S. companies that excel in quality management practices and performance excellence. The Baldrige Award does not exist simply to recognize product excellence, nor does it exist for the purpose of "winning." Its principal focus is on promoting high-performance management practices throughout the economy that lead to customer satisfaction and business results. The award promotes the awareness of quality as an increasingly important element in competitiveness, an understanding of the requirements for performance excellence, and sharing of information on successful quality management practices and the benefits derived from implementation of these strategies. Up to two companies can receive a Baldrige Award in each of the categories of manufacturing, small business, and service. In 1995, pilot programs in education and health care were instituted.

HISTORICAL PERSPECTIVE: Recognizing that U.S. productivity was declining, President Reagan signed legislation mandating a national study/conference on productivity in October 1982. The American Productivity and Quality Center (formerly the American Productivity Center) sponsored seven computer networking conferences in 1983 to prepare for an upcoming White House Conference on Productivity. The final report on these conferences recommended that "a National Quality Award, similar to the Deming Prize in Japan, be awarded annually to those firms that successfully challenge and meet the award requirements. These requirements and the accompanying examination process should be very similar to the Deming Prize system to be effective." The Baldrige Award was signed into law (Public Law 100–107) on August 20, 1987. The award is named after President Reagan's Secretary of Commerce who was killed in an accident shortly before the Senate acted on the legislation. Malcolm Baldrige was highly regarded by world leaders, having played a major role in carrying out the administration's trade policy, resolving technology transfer differences with China and India, and holding the first Cabinet-level talks with the Soviet Union in seven years that paved the way for increased access for U.S. firms in the Soviet market. The Award is a public-private partnership, funded primarily through a private foundation.

PURPOSE: The purposes of the Award are to:

- Help stimulate American companies to improve quality and productivity for the pride of recognition while obtaining a competitive edge through increased profits
- Recognize the achievements of those companies that improve the quality of their goods and services and provide an example to others
- Establish guidelines and criteria that can be used by business, industrial, governmental, and other enterprises in evaluating their own quality improvement efforts
- Provide specific guidance for other American enterprises that wish to learn how to manage for high quality by making available detailed information on how winning enterprises were able to change their cultures and achieve eminence

The award examination is based upon a rigorous set of criteria designed to encourage companies to enhance their competitiveness through efforts toward dual, results-oriented goals:

1. Delivery of ever-improving value to customers, resulting in improved marketplace success
2. Improvement of overall company performance and capabilities

The primary emphasis of the Baldrige program is not on winning awards; but rather on educating American business.

IMPLEMENTATION: The award criteria are built upon a set of core values and concepts that are described below.

Customer-Driven Quality: Quality is judged by customers. All product and service features and characteristics that contribute value to customers and lead to customer satisfaction and preference must be a key focus of a company's management system. Customer-driven quality is directed toward customer retention, market share gain, and growth. It demands constant sensitivity to emerging customer and market requirements, and measurement of the factors that drive customer satisfaction and retention. It also demands awareness of developments in technology and of competitors' offerings, and rapid and flexible response to customer and market requirements. In addition, the company's success in recovering from defects and errors ("making things right for the customer") is crucial to building customer relationships and to customer retention.

Leadership: A company's senior leaders set direction and create customer orientation, clear and visible values, and high expectations. Reinforcement of the values and expectations requires personal commitment and involvement. The leaders need to guide the creation of strategies, systems, and methods for achieving excellence and building capabilities. Through their personal involvement in planning, communications, review of company performance, and employee recognition, the senior leaders serve as role models, reinforcing the values and building leadership and initiative throughout the company.

Continuous Improvement and Learning: Achieving the highest levels of performance requires a well-executed approach to continuous improvement. The term "continuous improvement" refers to both incremental and "breakthrough" improvement. Improvements may be of several types: (1) enhancing value to customers through new and improved products and services; (2) reducing errors, defects, and waste; (3) improving responsiveness and cycle time performance; (4) improving productivity and effectiveness in the use of all resources; and (5) improving the company's performance in fulfilling its public responsibilities and serving as a corporate citizenship role model. Continuous improvement must contain cycles of planning, execution, and evaluation with a basis for assessing progress and for deriving information for future cycles of improvement.

Employee Participation and Development: Companies need to invest in the development of the work force through education, training, and opportunities for continuing growth. Increasingly, training, development, and work organizations need to be tailored to a diverse work force and to more flexible, high performance work practices. Major challenges in the area of work force development include: (1) integration of human resource management selection, performance, recognition, training, and career advancement; and (2) aligning human resource management with business plans and strategic change processes.

Fast Response: Success in competitive markets increasingly demands ever-shorter cycles for new or improved product and service introduction. Also, faster and more flexible response to customers is now a more critical requirement. Response time improvements often drive simultaneous improvements in organization, quality, and productivity.

Design Quality and Prevention: Business management should place strong emphasis on design quality and waste prevention achieved through building quality into products and services and efficiency into production and delivery processes. Increasingly, design quality also includes the ability to incorporate information gathered from diverse sources and data bases, that combine factors such as customer preference, competitive offerings, marketplace changes, and external research findings and developments. Consistent with the theme of design quality and prevention, improvement needs to emphasize interventions "upstream" at early stages in processes.

Long-Range View of the Future: Pursuit of market leadership requires a strong future orientation and a willingness to make long-term commitments to all stakeholders-customers, employees, suppliers, stockholders, the public, and the community. Planning needs to anticipate many types of changes including those that may affect customers' expectations of products and services, technological developments, changing customer segments, evolving regulatory requirements, community/societal expectations, and thrusts by competitors.

Management by Fact: Modern business management systems depend upon measurement, data, information, and analysis. Measurements must derive from the company's strategy and encompass all key processes and the outputs and results of those processes. Analysis refers to extracting larger meaning from data to support evaluation and decision making at all levels within the company. Facts, data, and analysis support a variety of company purposes, such as planning, reviewing company performance, improving operations, and comparing company performance with competitors' or with "best practices" benchmarks. A system of measures or indicators tied to customer and/or company performance requirements represents a clear basis for aligning all activities with the company's goals.

Partnership Development: Companies should seek to build internal and external partnerships to better accomplish their overall goals. Internal partnerships might include those that promote labor-management cooperation, such as agreements with unions. External partnerships might be with customers, suppliers, and education organizations for a variety of purposes, including education and training. Internal and external partnerships should seek to develop longer-term objectives, thereby creating a basis for mutual investments.

Corporate Responsibility and Citizenship: A company's management should stress corporate responsibility and encourage corporate citizenship. Corporate responsibility refers to basic expectations of company – business ethics and protection of public health, safety, and the environment. Inclusion of public responsibility areas within a performance system means meeting all local, state, and federal laws and regulatory requirements. It also means treating these and related requirements as areas for continuous improvement "beyond mere compliance." This requires that appropriate measures be created and used in managing performance. Corporate citizenship refers to leadership and support of publicly important purposes, including areas of corporate responsibility.

Results Orientation: A company's performance system needs to focus on results. Results should be guided by and balanced by the interests of all stakeholders – customers, employees, stockholders, suppliers and partners, the public, and the community. To meet the sometimes conflicting and changing aims that balance implies, company strategy needs to explicitly address all stakeholder requirements to ensure that actions and plans meet the differing needs and avoid adverse impact on any stakeholders. The use of a balanced composite of performance measures offers an effective means to communicate requirements, to monitor actual performance, and to marshal support for improving results.

The Criteria consist of a hierarchical set of *categories items*, and *areas to address*. The seven categories form the organization of the Criteria as illustrated in Figure 1. Leadership, Strategic Planning, and Customer and Market Focus represent the "leadership triad," and suggest the importance of integrating these three functions. Human Resource Development and Management and Process Management represent how the work in an organization is accomplished and lead to business results. These functions are linked to the leadership triad. Finally,

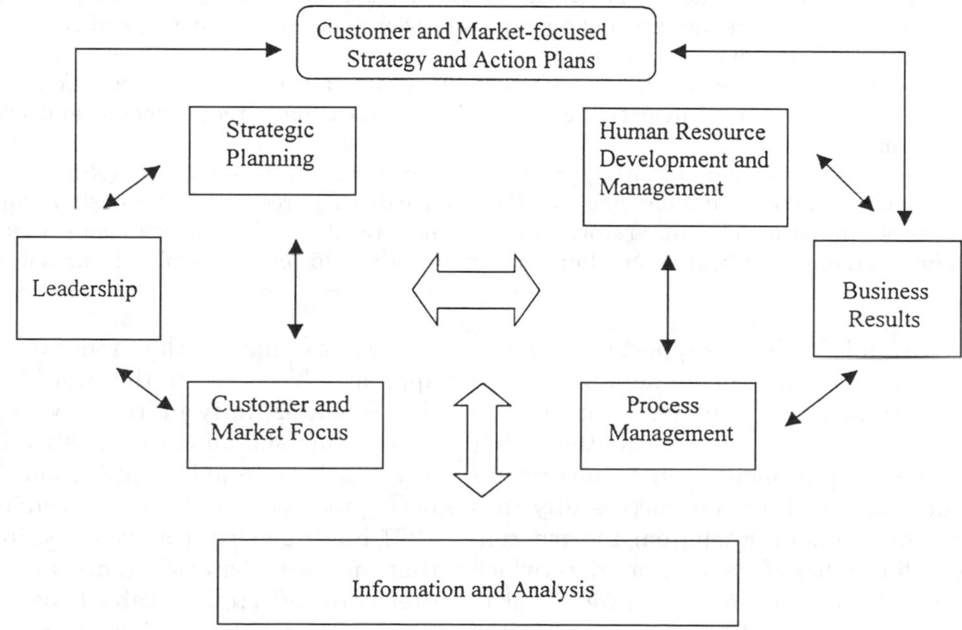

Fig. 1. Malcolm Baldrige national quality award framework.

Table 1. Key management processes reflected in the 1997 Baldrige criteria.

Leadership

1. How senior leadership creates and sustains values, sets company directions, and develops, sustains, and improves an effective leadership system.
2. How the company addresses impacts on society of its products, services, facilities and operations.
3. How the company practices good community citizenship.

Strategic Planning

4. The process of developing strategy, taking into account customers and markets, competitors, risks, company capabilities, and supplier/partner capabilities.
5. The process of translating strategy into action plans and tracking performance.

Customer and Market Focus

6. The process of determining longer-term requirements, expectations, and preferences of key customer groups and/or market segments, and evaluating and improving listening and learning strategies.
7. Approaches to providing access and information to enable customers to seek assistance, conduct business, and voice complaints to strengthen relationships and improve products and services.
8. The process of determining customer satisfaction and satisfaction relative to competitors.

Information and Analysis

9. The process of selecting, managing, and using information and data to support key company processes and improve company performance.
10. The process of selecting, managing, and using comparative information to improve overall performance and competitive position.
11. The process of integrating and analyzing customer, operational, competitive, and financial/market performance data.
12. The process of reviewing company performance and capabilities to assess progress and determine improvement priorities.

Human Resource Development and Management

13. The process of designing work and jobs to encourage all employees to contribute effectively to achieving the company's performance and learning objectives.
14. The design of compensation and recognition systems to reinforce work systems and performance.
15. The process of designing, delivering, evaluating, and improving education and training to build knowledge and capabilities and contribute to improved employee performance and development.
16. The design of a safe and healthful work environment.
17. The support of employee well-being, satisfaction, and motivation via services, facilities, and opportunities.
18. The process of determining employee well-being, satisfaction, and motivation, and using the results to identify improvement activities.

Process Management

19. The process of designing products, services, and production delivery processes to incorporate changing customer requirements, meet quality and operational performance requirements, and ensure trouble-free introduction and delivery of products and services.
20. The processes of managing, evaluating, and improving key product and service production/delivery processes to maintain process integrity, meet operational and customer requirements, and achieve better performance.
21. The process of evaluating and improving product and service production/delivery processes.
22. Approaches for managing and improving supplier and partnering processes, relationships, and performance.
23. The process of evaluating a balanced scorecard of business results, including customer satisfaction, financial/market, human resource, supplier/partner, and other efficiency and effectiveness measures.

Information and Analysis supports the entire framework by providing the foundation for performance assessment and management by fact.

Each category consists of several items that focus on the requirements that businesses should understand to function successfully in a competitive environment. Each item, in turn, consists of a small number of areas to address, which seek specific information on *approaches* used to ensure and improve competitive performance, the *deployment* of these approaches, or *results* obtained from such deployment.

Areas to address that request information on approach begin with the word "how"; that is, they fundamentally address key management processes that should form an effective system for any quality-minded organization. The key business processes that are embedded within the 1997 Baldrige Criteria are listed in Table 1. Note that not all Criteria items and areas to address are reflected in Tabe 1, because some deal explicitly with key information or results, and do not specifically address management processes.

Some key characteristics of the criteria are as follows:

1. **The criteria are directed toward business results:** The criteria focus principally on seven key areas of business performance: customer satisfaction and retention; market share and new market development; product and service quality; financial indicators, productivity, operational effectiveness, and responsiveness; human resource performance and development; supplier performance and development; and public responsibility and corporate citizenship. Improvements in these seven areas contribute significantly to overall company performance – including financial performance. Emphasis on results balances strategies in such a way that they avoid inappropriate tradeoffs among important stakeholders or objectives.
2. **The criteria are nonprescriptive:** The criteria are a set of basic interrelated, results-oriented requirements, but they do not prescribe specific quality tools, techniques, technologies, systems, or starting points. Companies are encouraged to develop and demonstrate creative, adaptive, and flexible approaches to meeting basic requirements.
3. **The criteria are comprehensive:** The criteria address all internal and external requirements of the company, including those related to fulfilling public responsibilities. Accordingly, all processes of all company work units are tied to these requirements.
4. **The criteria include interrelated learning cycles:** Learning takes place via feedback among the process and results elements. A learning cycle has four stages:
   - planning, selection of indicators, and deployment of requirements
   - execution of plans
   - assessment of progress
   - revision of plans based upon assessment findings
5. **The criteria emphasize alignment:** The criteria call for improvement at all levels and in all parts of the company. Such improvement is achieved via interconnecting and mutually reinforcing measures and indicators, derived from overall company requirements. These measures and indicators tie directly to customer value and operational performance and provide a communications tool and a basis for deploying consistent customer and operational performance requirements to all work units.
6. **The criteria are part of a diagnostic system:** Using the criteria provides a profile of strengths and areas for improvement that directs attention to processes and actions that contribute to business success.

The Baldrige Award Scoring System: Each examination item is assigned a point value that can be earned during the evaluation process. It is based upon three evaluation dimensions: approach, deployment, and results. Approach refers to the methods the company uses to achieve the requirements addressed in each category. The factors used to evaluate approaches include:

- The appropriateness of the methods, tools, and techniques to the requirements
- The effectiveness of methods, tools, and techniques
- The degree to which the approach is systematic, integrated, and consistently applied
- The degree to which the approach embodies effective evaluation/improvement cycles
- The degree to which the approach is based upon quantitative information that is objective and reliable
- Evidence of unique and innovative approaches, including the significant and effective new adaptations of tools and techniques used in other applications or types of businesses

Deployment refers to the extent to which the approaches are applied to all relevant areas and activities addressed and implied in each category. The factors used to evaluate deployment include:

- Appropriate and effective use of the approach in key processes
- Appropriate and effective application of the approach in the development and delivery of products and services
- Appropriate and effective use of the approach in all interactions with customers, suppliers of goods and services, and the public

Results refer to the outcomes and effects in achieving the purposes addressed and implied in the criteria. The factors used to evaluate results include:

- Current performance levels
- Performance levels relative to appropriate comparisons and benchmarks
- Rate of performance improvement
- Breadth and importance of performance improvements
- Demonstration of sustained improvement or sustained high-level performance

Examiners evaluate the applicant's response to each examination item, listing major strengths and areas for improvement relative to the criteria. Strengths demonstrate an effective and positive response to the criteria. Areas for improvement do not prescribe specific practices or examiners' opinions on what the company should be doing, but rather how management can better address the criteria. Based on these comments, a percentage score is given to each item. Each examination item is evaluated on approach/deployment or results.

Like quality itself, the specific award criteria are continually improved each year. The initial set of criteria in 1988 had 62 items with 278 areas to address. By 1991, the criteria had only 32 items and 99 areas to address. The 1995 criteria were streamlined significantly to 24 items and 54 areas to address. More significantly, the word quality has been judiciously dropped throughout the document. For example, until 1994, Category 3 had been titled "Strategic Quality Planning." The change to "Strategic Planning" signifies that quality should be a part of business planning, not a separate issue. Throughout the document, the term performance has been substituted for quality as a deliberate attempt to recognize that the principles of TQM are the foundation for a company's management system, not just the quality system. In 1997, the criteria were further streamlined and reorganized, with the goal being to develop the shortest list of key requirements necessary to compete in today's marketplace, improve the linkage between process and results, and make the criteria more generic and user-friendly. Only a few minor changes were made to the 1998 Criteria. As Curt Reimann, director of the Baldrige Award Program noted, "The things you do to win a Baldrige Award are exactly the things you'd do to win in the marketplace. Our strategy is to have the Baldrige Award criteria be a useful daily tool that simulates real competition."

A single free copy of the criteria can be obtained from the National Institute of Standards and Technology. Contact the Malcolm Baldrige National Quality Award, National Institute of Standards and Technology, Route 270 & Quince Orchard Road, Administration Building, Room A537, Gaithersburg, MD 20899 (301) 975-2036, FAX (301) 948-3716, or visit the web site at http://www.quality.nist.gov/

THE BALDRIGE AWARD EVALUATION PROCESS: The Baldrige evaluation process is rigorous. In the first stage, each application is thoroughly reviewed by up to 15 examiners chosen from among leading quality professionals in business, academia, health care, and government (all of whom are volunteers). The scores are reviewed by a panel of nine judges without knowledge of the specific companies. The higher scoring applications enter a consensus stage in which a selected group of examiners discuss variations in individual scores and arrive at consensus scores for each item. The panel of judges then reviews the scores and selects the highest scoring applicants for site visits. At this point, six or seven examiners visit the company for up to a week to verify information contained in the written application and resolve issues that are unclear. The judges use the site visit reports to recommend award recipients. Final contenders each receive more than 400 hours of evaluation. All applicants receive a feedback report that critically evaluates the company's strengths and areas for improvement relative to the award criteria. The feedback report, frequently 30 or more pages in length, contains the evaluation team's response to the written application. It includes a distribution of numerical scores of all applicants and a scoring summary of the individual applicant. This feedback is one of the most valuable aspects of the Baldrige Award program.

EFFECT: The Baldrige Award has generated an incredible amount of interest in quality, both within the United States and internationally. Winning companies have made thousands of presentations describing their quality management approaches and practices. Many states have instituted awards similar to the Baldrige and based on the Baldrige Criteria. The award criteria have been used to train hundreds of thousands of people and as a self-assessment tool within organizations. In assessing the applications submitted for the Baldrige Award, several key strengths and common weaknesses are evident. Most companies that apply have strong senior management leadership and are driven by the needs of customers and the marketplace. These firms have aggressive goals and high expectations. Companies have strong information systems that provide an excellent basis for assessing the state of quality and that link external customer satisfaction measurements with internal measurements such as process quality and employee satisfaction. Most companies have invested heavily in human resource development, and employee involvement is continuing and expanding.

Common weaknesses among firms that do not score well include

- Weak information systems
- Delegation of quality responsibility to lower levels of the company
- A partial quality system, for example, strong in manufacturing but weak in support services
- An unclear definition of what quality means in the organization
- A lack of alignment among diverse functions within the firm; that is, not all processes are driven by common goals or use the same approaches
- Failure to use all listening posts to gather information critical to decision making

Among the more disappointing results found by Baldrige Award administrators is that relatively few organizations have integrated systems. Many lack a quality vision or do not effectively translate this vision into a business strategy. Many companies still emphasize the negative side of quality – defect reduction, rather than customer focus. Finally, the gaps between the best and the average companies is quite large, indicating that much opportunity and hard work remains.

RESULTS: Baldrige Award-winning companies have shown excellent financial results. The Commerce Department reported that a person who would have invested $1000 in common stock in each publicly traded winning company on the first business day in April of the year they won the Baldrige Award (or the date when they began public trading, if it is later); or for subsidiaries, $1,000 times the percent of the whole company's employee base the subunit represented at the time of its application, would have outperformed the S&P 500 by approximately 3 to 1, achieving a 324.9% return. The group of five whole company winners outperformed the S&P 500 by greater than 3.5 to 1, achieving a 380.2% return.

A hypothetical sum was also invested in each 1990–1995 publicly-traded, site visited company. The 48 publicly-traded site visited applicants, as a group, outperformed the S&P 500 by 2 to 1, achieving a 167.5% return compared to a 83.3% return for the S&P 500. The group of 35 site visited applicants without the winners outperformed the S&P 500 by 68%, achieving a 138.3% return compared to a 82.0% return for the S&P 500. (Names of Baldrige applicants are kept confidential.)

Table 2. Baldrige award winners through 1997.

ADAC Laboratories (1996)
Ames Rubber Corporation (1993)
Armstrong World Industries (1995)
AT&T Consumer Communications Services (1994)
AT&T Network Systems Group, Transmission Systems Business Unit (now part of Lucent Technologies) (1992)
AT&T Universal Card Services (1992)
Cadillac Motor Car Company (1990)
Corning Telecommunications Products Division (1995)
Custom Research Inc. (1996)
Dana Commercial Credit Corporation (1996)
Eastman Chemical Company (1993)
Federal Express Corporation (1990)
Globe Metallurgical Inc. (1988)
Granite Rock Company (1992)
GTE Directories Corporation (1994)
IBM Rochester (1990)
Marlow Industries (1991)
Merrill Lynch Credit Corporation (1997)
Milliken & Company (1989)
Motorola Inc. (1988)
Solectron Corporation (1991, 1997)
Texas Instruments Defense Systems & Electronics Group (1992)
The Ritz-Carlton Hotel Company (1992)
3M Dental Products Division (1997)
Trident Precision Manufacturing Inc. (1996)
Wainwright Industries (1994)
Wallace Co. Inc. (1990)
Westinghouse Electric Corporation Commercial Nuclear Fuel Division (1988)
Xerox Business Services (1997)
Xerox Corporation Business Products & Systems (1989)
Zytec Corporation (1991)

CASES: A list of winners through 1997 is given in Table 2. In 1997, Solectron became the first company to receive the Award a second time. The 1996 winners included one manufacturing company, one service company, and two small businesses. ADAC Laboratories produces high-technology healthcare products for diagnostic imaging and healthcare information systems. Their management system based on Baldrige helped transform the company into a world leader after successfully coming out of a turnaround in the mid-1980s. Between 1990 and 1995, overall efficiency improvements resulted in an increase in revenue per employee from $200,000 to $330,000, 65% better than its competitors. ADAC consistently brings products to market faster than its larger competitors. A 1995 independent survey rated the first-month reliability of ADAC cameras as best in the industry. If a system breaks down, technicians will have it back up within an average of 17 hours after receiving a customer's call, less than a third of the time it took in 1990. On eight measures of customer service satisfaction, ADAC was the sole leader in five categories and tied for the top spot in the remaining ones.

Dana Commercial Credit Corporation provides leasing and financing services to a broad range of commercial customers. A collection of quality-linked "scoring processes" assesses how the company is progressing in its pursuit of continuous improvement goals set for all key areas of the business. Since 1994, customer-satisfaction scores for the Capital Markets Group have exceeded four on a five-point scale, and in 1995, topped the industry average by almost two points. The Dealer Products Group has scored between eight and nine on a 10-point scale, or nearly three points higher than the industry average. Rates of return on equity and assets have increased more than 45% since 1991.

Trident Precision Manufacturing, Inc. is a 167-person manufacturer of precision sheet metal components, electro-mechanical assemblies, and custom products. Quality is its basic business plan, focused on five key business drivers: customer satisfaction, employee satisfaction, shareholder value, operational performance, and supplier partnerships. Employee turnover has declined from 41% in 1988 to 5% in 1994 and 1995. Defect rates have fallen to a point where the company offers a full guarantee against defects for its custom products, and delivery performance has improved from 87% to 99.94 percent from 1990 to 1995.

Custom Research Inc. is a national marketing research firm, employing only about 100 people. With an intense focus on customer satisfaction, a team-oriented workforce, and information technology, CRI pursues individualized service and satisfies customers through a Surprise and Delight strategy. In 1996, most CRI employees received over 134 hours of training. CRI "meets or exceeds" clients expectations on 97% of its projects, and more remarkably, "exceeds expectations" on 70% of projects.

COLLECTIVE WISDOM: The Baldrige Award criteria form a blueprint for quality improvement in any organization. Hundreds of thousands of applications are distributed each year, yet only a handful of completed applications are received. Many companies are using the award criteria to evaluate their own quality programs, set up and implement TQM programs, communicate better with suppliers and partners, and for education and training. Even the U.S. Postal Service has decided to use the Baldrige Criteria as a basis to reestablish a quality system by identifying the areas that need the most improvement and providing a baseline to track progress. Using the award criteria as a self-assessment tool provides an objective framework, sets a high standard, and compares units that have different systems or organizations. It is also being used as a basis for giving awards within companies and at the local, state, and federal levels.

Many different philosophies and quality improvement programs exist. Organizations just getting started in quality improvement often have problems defining the quality system and setting objectives. The Baldrige Award addresses the full range of quality issues and can help those setting up new systems to get a complete picture of TQM.

The Baldrige Award criteria assist companies with internal communications, communications with suppliers, and communications with other companies seeking to share information. The criteria provide a focus on what to communicate and a framework for comparing strategies, methods, progress, and benchmarks. Finally, the Baldrige Award examination is being used for training and education, particularly for management, because it summarizes major issues that managers must understand. It draws the distinction between excellence and mediocrity.

Organizations Mentioned: ADAC Laboratories; Dana Commercial Credit Corporation; Trident Precision Manufacturing; Custom Research, Inc (CRI); U.S. Postal Service; See Table 2.

See **Benchmarking; The implications of Deming's approach; Integrated quality control; Just-In-time manufacturing; Kanban-based manufacturing system; Lean manufacturing implementation; Quality; Return on equity; Strategic quality planning.**

References

[1] Bemowski, K. (1996). "Baldrige Award Celebrates Its 10th Birthday With a New Look," *Quality Progress*, 29, 49–54.

[2] DeCarlo, N.J., and W.K. Sterret (1990). "History of the Malcolm Baldrige National Quality Award," *Quality Progress*, 23, 21–27.

[3] Evans, J.R. (1996). "Something old, something new: a process view of the Baldrige criteria," *International Journal of Quality Science*, 1, 62–68.

[4] Evans, J.R. and M.W. Ford (1997). "Value-Driven Quality," *Quality Management Journal*, 4(4), 19–31.

[5] Evans, J.R. and W.M. Lindsay (1998). *The Management and Control of Quality* (4th Edition), South-Western, Cincinnati.

[6] Neves, J.S. and B. Nakhai (1994). "The Evolution of the Baldrige Award," *Quality Progress*, 27, 65–70.

[7] U.S. Department of Commerce (1997). *Malcolm Baldrige National Quality Award Criteria For Performance Excellence.*

[8] U.S. Department of Commerce (undated). *Malcolm Baldrige National Quality Award, Profiles of Winners.*

PERFORMANCE MEASUREMENT IN MANUFACTURING

Gregory P. White

Southern Illinois University at Carbondale, Illinois, U.S.A.

DESCRIPTION: Performance measurement determines how well the manufacturing operation is succeeding in meeting specified organizational requirements. There are two important aspects of this definition. First, we must be able to measure the performance characteristic of interest in a meaningful way. Second, there must be some desired standard level of performance against which comparisons can be made.

Performance measurement ends up directing an organization; what gets measured is what gets attention. If an organization believes that labor utilization is an important measure, then labor utilization will get attention – sometimes to the detriment of more important criteria that are not being measured. This means it is very important that an organization's manufacturing performance measurement system be linked directly to its strategic plan. Areas that are of competitive importance to a company's strategic plan must be included in the performance measurement system for manufacturing.

Nearly every manufacturing organization measures costs; the financial accounting system requires cost measurement and costs have a direct effect on profitability. However, changes in the competitive environment, such as an increased emphasis on quality, and the use of Just-in-Time manufacturing, have highlighted the problems inherent in the use of traditional financial accounting systems for performance measurement. Efforts to develop new performance measurement systems are constantly evolving. This article provides a brief overview of the current state of such efforts.

HISTORICAL PERSPECTIVE: Historically, performance measurement in manufacturing has been based on traditional financial accounting systems. Such systems were developed originally during the late 1800s when manufacturing and the competitive environment were much different than today. In such manufacturing systems direct labor was a major component of product cost, and long production runs of a few standard products were common, so it made sense to allocate overhead and shared costs to each product as a proportion of its direct labor content. Today, with increased automation, short production runs, and greater product customization, allocation on the basis of direct labor often results in cost distortions. Furthermore, many companies find that financial measures such as cost or earnings per share do not reflect competitive strategy, and non-financial concerns such as delivery speed or customer service. As a result, academics and practitioners alike have expressed general dissatisfaction with traditional performance measurement systems. Efforts to change those systems fall into two major categories.

1. ***Modifying the Financial Measures:*** One approach to fixing performance measurement systems has been to change the way costs are measured. Activity-based costing (ABC) (Cooper *et al.*, 1992) is one approach that has been widely adopted. In ABC, "cost drivers" are associated with activities that utilize resources, and the cost of a product is determined based on these cost drivers. For example, "minutes of testing time" could be a cost driver for a product inspection activity. More recently, ABC has been broadened to constitute an entire approach to managing resources, known as activity-based cost management (ABCM), which can be applied to improvement programs. For example, dividing the cost of an activity by its output measure produces an indicator of efficiency, which can then be used to identify areas for process improvement (Gupta *et al.*, 1997).

Goldratt (1990) has proposed another approach to modifying the financial measurement system. Known as theory of constraints (TOC), it uses a completely different set of financial measures. TOC views the goal of a business as making more money. In measuring progress toward that goal, TOC uses three measures. Throughput is defined as the rate at which the system generates money through sales. Inventory is defined as all the money used to buy resources (including plant and equipment, as well as materials) that the system uses to generate throughput. Operating expense is defined as the money spent on turning inventory to throughput. More familiar financial measures can be derived from these three. For example, net profit is simply throughput minus operating expense, and return on investment is calculated as net profit divided by inventory. The focus in TOC is on identifying constraints that limit a system's ability to achieve its goal. Effort is devoted toward managing those constraints and continuously improving the system's performance. Although it might seem that TOC and ABC are completely different, the two can actually be used together to identify the product mix that maximizes profit per hour of constrained resource time (Hall *et al.*, 1997).

A third approach to changing the financial measurement system involves cost measures for non-cost aspects of an organization's competitive strategy. An excellent example of this approach is the concept of cost of quality (Crosby, 1979). Under this approach, all costs associated with quality are identified and explicitly measured, generally in four categories: costs of internal failure (e.g., scrap and rework), costs of external failure (e.g., warranty costs), appraisal costs (e.g., inspection), and prevention costs (e.g., employee training). Quite a few companies use this particular measure, which has the advantage that it can be integrated into existing cost measurement systems, whether a company is using traditional cost accounting or some other approach.

2. ***Performance Measurement Beyond Cost:*** A criticism of the preceding approaches is that they do not include measures that may be crucial to a company's competitive strategy, but which are not easily stated in financial terms, such as customer satisfaction or market share.

   Probably the best known among such systems is Balanced Scorecard proposed by Kaplan and Norton (1992). The scorecard includes four perspectives: financial (how do we look to shareholders?), customer (how do customers see us?), internal/business process (what must we excel at?), and learning and growth (can we continue to improve and create value?). Each perspective has its own goals and measures. However, there are linkages among the four perspectives so that each is influenced by and/or influences others in a cause-effect relationship. For example, a financial perspective measure might be quarterly sales growth. One driver of quarterly sales growth could be sales to new customers. It may turn out that one feature that attracts new customers is the company's excellent on-time delivery performance. Thus, delivery performance could become a customer perspective measure. The internal factors that influence delivery performance are identified as components of the internal/business process perspective. These might include schedule attainment, and machine reliability. Finally, the components of the learning and growth perspective may include employee training.

   The balanced scorecard has the advantage that it formalizes four important perspectives and ties them together. However, other models have also been developed. Lockamy and Cox (1994) developed a model that links together the finance function, customer functions, and resource functions based on a study of systems at six world-class companies. That study found that, although each company had a own competitive strategy, all had established linkages among the three functional areas through its performance measurement system.

   White and Ponce de Leon (1998) also present a cause-effect model that is used as a framework to study seven highly successful small to medium size companies. They found that, to varying degrees, each company perceived its performance measurement system as an important aspect of ensuring that activities on the shop floor are making contributions toward realizing the company's competitive strategy.

STRATEGIC PERSPECTIVES:

1. ***Linking Performance Measurement to the Strategic Plan:*** The primary requirement of a manufacturing performance measurement system is that it be linked to the organization's strategic plan. Dixon *et al.* (1990) have developed a procedure, called gap analysis, that involves determining what is important to the company's competitive strategy and comparing that to the performance measurement system. Their approach uses a questionnaire to determine how employees rate the importance of various performance factors to the firm, then to indicate the amount of emphasis placed on measuring each factor. These performance factors range from items such as overall market share to quality of after-sales service. The objective is to identify gaps between what is important to the organization competitively, and what is stressed in the performance measurement system. Gaps identify opportunities to change the performance measurement system. Although Dixon *et al.* propose "tinkering" with the cost accounting system as a first step, they definitely favor eventual separation of the performance measurement system from the cost accounting system. Dixon *et al.* suggest that a good measurement system has five attributes:
   - It is consistent with the business's operating goals, objectives, critical success factors, and programs.
   - Conveys information through as few and as simple a set of measures as possible.
   - Reveals how effectively customers' needs and expectations are satisfied. Focuses on measures that customers can see.
   - Provides a set of measurements for each organizational component, allowing all members of the organization to understand how their decisions and activities affect the entire business.
   - Supports organizational learning and continuous improvement.
2. ***Connections from the Top Floor to the Shop Floor:*** As the preceding list indicates, it is important that the performance measurement system be relevant to all parts of the organization. Unfortunately, different language and terminology at each level of the organization makes it difficult to achieve this

objective. Lynch and Cross (1995) have developed a model, called SMART, that can provide some guidance in going from top management to middle management, to the shop floor. The SMART model proposes that financial measures be used at the top organizational levels, but that these be translated as one moves down through the organization, with market and financial measures being used at the second level, while detailed measures related to quality, process time, and cost are used at the lowest levels of the organization.

Kaplan and Norton (1996) have developed four steps that tie an organization's strategy in with the balanced scorecard. Those steps are:

- Translating the vision (clarifying the vision, gaining consensus).
- Communicating and linking (communicating and educating, setting goals, linking rewards to performance measures).
- Business planning (setting targets, aligning strategic initiatives, allocating resources, establishing milestones).
- Feedback and learning (articulating the shared vision, supplying strategic feedback, facilitating strategy review and learning).

IMPLEMENTATION: Eccles (1991) notes that, as yet, there is no predetermined process to follow when implementing a new performance measurement system. However, he lists five areas of activity that must be addressed:

- Developing an information architecture.
- Putting the technology in place to support this architecture.
- Aligning incentives with the new system.
- Drawing on outside resources.
- Designing a process to ensure that the other four activities occur.

TECHNOLOGY PERSPECTIVE: Although Eccles (1991) implies that technology is a requirement, White and Ponce de Leon (1998) found that there is no particular technology that is essential for a manufacturing performance measurement system. Some companies have used networks of computer terminals and bar code scanners to gather data from individual machines or work centers, but others have achieved the same end using paper forms to collect data. A few companies that emphasize employee involvement are moving toward computer terminals, located throughout the plant, that can be used by any employee to access all performance measurement information available within the company – on a real-time basis.

RESULTS: White and Ponce de Leon (1998) found those companies that were most successful – financially and in terms of customer satisfaction – were also the ones that did the best job of tying their performance measurement system to the competitive strategy, deploying that measurement system throughout the entire organization, and then rewarding employees based on performance measures. Such companies have won numerous quality awards from customers and are experiencing sales growth and ROI rates well above the industry average.

CASES: A few collections of case studies on performance measurement for manufacturing have been written. White and Ponce de Leon (1998) studied the performance measurement systems of seven small-to-medium size companies with the objective of determining how such systems are developed, how they evolve, and how they are used to identify opportunities for improving the organization. Lockamy and Cox (1994) provide case studies of organizational performance measurement systems for large world-class manufacturers, some of which have multiple divisions. Both sets of studies provide good overviews of different possible approaches to performance measurement, highlighting best practices that were observed.

Kaplan and Norton (1996) describe a company that implemented the balanced scorecard approach by utilizing three layers of management. The senior executive group developed goals for the financial and customer perspectives, such as on-time delivery. The next two management layers developed goals for the internal/business process perspective and the learning and growth perspective. The goals developed at the lower levels were concerned with order processing, scheduling and fulfillment, and employee training.

Hall, Galambos, and Karlsson (1997) describe an application of TOC and ABCM in a division of the Timken Company. Based on TOC analysis of profit per hour of constrained resource time, the division determined that one product line should be emphasized while another should be outsourced and a third eliminated altogether. The resultant changes increased operating income by 20 percent.

Organizations Mentioned: Timkin Company.

See **Activity-based costing; Activity-based costing: Activity-based management; Accounting implications of TOC; An evaluation; Appraisal costs; Balanced scorecard; Cost drivers; Cost of external failure; Cost**

of internal failure; Cost of quality; Financial accounting system; Machine reliability; Prevention costs; Shared costs; SMART; Target costing.

References

[1] Cooper, R., R.S. Kaplan, L.S. Maisel, E. Morrissey, and R. M. Oehm (1992). *Acitivty-Based Cost Management: Moving from Analysis to Action*. The Institute of Management Accountants, New Jersey.

[2] Crosby, P.B. (1979). *Quality is Free*. McGraw-Hill, New York.

[3] Dixon, J.R., A.J. Nanni Jr., and T.E. Vollmann (1990). *The New Performance Challenge: Measuring Operations for World-Class Competition*. Business One Irwin, Illinois.

[4] Eccles, R.G. (1991). "The performance measurement manifesto." *Harvard Business Review*, 69, 131–137.

[5] Goldratt, E. (1990). *Theory of Constraints*. North River Press, New York.

[6] Gupta, M., S. Baxendale, and K. McNamara (1997). "Integrating TOC and ABCM in a health care company." *Journal of Cost Management*, 11, 23–33.

[7] Hall, R., N.P. Galambos, and M. Karlsson (1997). "Constraint-based profitability analysis: stepping beyond the theory of constraints." *Journal of Cost Management*, 11, 6–10.

[8] Kaplan, R.S. and D.P. Norton (1992). "The balanced scorecard – measures that drive performance." *Harvard Business Review*, 70, 71–79.

[9] Kaplan, R.S. and D.P. Norton (1996). "Using the balanced scorecard as a strategic management system." *Harvard Business Review*, 74, 79–85.

[10] Lockamy, A. III and J.F. Cox III (1994). *Reengineering Performance Measurement*. Irwin, New York.

[11] Lynch, L.L. and K.F. Cross (1995). *Measure up: Yardsticks for Continuous Improvement*. Basil Blackwell, Massachusetts.

[12] White, G.P. and J.A. Ponce de Leon (1998). *Performance Measurement Systems: A Model and Seven Case Studies*. APICS Education & Research Foundation, Virginia.

PERIOD COSTS

Period costs are general administrative and selling-related costs that are charged as an expense to the income statement in the *period* in which the costs occurred. All costs in an organization are considered by accountants to be either period or product costs. In a manufacturing setting, product costs include costs of materials and labor directly used in the manufacture of products. Product costs also include manufacturing overhead costs that are allocated to products using predetermined overhead application rates. As a result, product costs become the costs of product inventory on the balance sheet until the product is sold. Once sold, product costs become costs-of-goods-sold on the income statement. Period costs, on the other hand, cannot be connected to specific products or services in a sensible or economically feasible manner. As a result, period costs are not part of product inventory on the balance sheet, but are immediately expensed on the income statement.

See **Activity-based costing; Activity-based costing: An evaluation; Unit cost.**

PERIOD-BASED FREEZING

In order to achieve schedule stability in MRP environment parts of the master production schedule may be frozen, that is, it is not subject to changes or revision. In Period-based freezing, the master production schedule (MPS) is frozen for a specified number of periods.

See **Master production schedule; MRP; Order release; Schedule stability.**

PERIODIC ORDER QUANTITY (POQ)

Economic order quantities may require ordering large quantities resulting in light inventory costs. One way of overcoming this is to order smaller quantities periodically. Each order covers the demand for a fixed period of time such as a week or a month. The amount so ordered changes with the demand. Quantity so ordered is called periodic order quantity (POQ). In POQ procedures, orders for replenishment occur at fixed intervals. An example would be to order exactly the quantity needed every two weeks.

See **Economic order quantity (EOQ); Fixed order quantity inventory model; Fixed time period inventory model; Inventory flow analysis; MRP.**

PERT (PROGRAM EVALUATION AND REVIEW TECHNIQUE)

Activity Scheduling Techniques in Project Management: Before reading this section, the reader is encouraged to read Critical Path Method (CPM). The two main techniques for project scheduling are PERT (Program Evaluation and Review Technique) and CPM (Critical Path Method). Both of these techniques appeared during the late 1950s. The chief difference between the two is that PERT uses probability principles to account for variability into activity times (Dilworth, 1992). Network scheduling techniques depend upon realistic estimates of activity duration; however, the use of PERT can introduce the chance of delay into time estimates. PERT generally uses activity-on-arrow (AOA) notation (see **Critical Path Method**).

The Use of Probability with PERT

Although the times given for activity duration in the critical path method (CPM) are estimates, it is possible to better approximate a real-life situation by introducing the notion of chance in time estimates. Average activity time estimates in PERT are based on the most likely time (m), and optimistic (a) and pessimistic (b) times. It has been found that activity times tend to fit a beta distribution better than any other distribution. The average and standard deviation for each activity are approximated as shown below:

Average (i.e., expected) activity time
$= t_e = (a + 4m + b)/6;$

Standard deviation of activity time
$= (b - a)/6$: variance $= ((b - a)/6)^2$

While individual activity time conform to a beta distribution, the critical path time (see, Critical Path Method), which is the sum of the activity times on the critical path is normally distributed.

Probability of Project Completion

The variance of the expected duration of the project can be calculated by summing the individual variances of the activities on the critical path. A Z-score from a normal distribution table is then used to calculate the probability of completing the project in any desired length of time:

$Z =$ (desired completion time
− mean completion time)/$\surd$(variance)

EXAMPLE : The following data pertains to a small project, which is shown in the PERT chart below.

Activity	Activity Designation	Optimistic Time (a)	Pessimistic Time (b)	Most Likely Time (m)	Activity Time (a + 4m + b)	$\sigma = (b-a)/6$
A	1–2	3	9	6	6	1
B	1–3	2	8	5	5	1
C	2–4	2	6	4	4	.67
D	3–4	2	10	3	4	1.33
E	3–5	1	11	3	4	1.67
F	4–6	4	8	6	6	.67
G	5–6	1	15	5	6	2.33

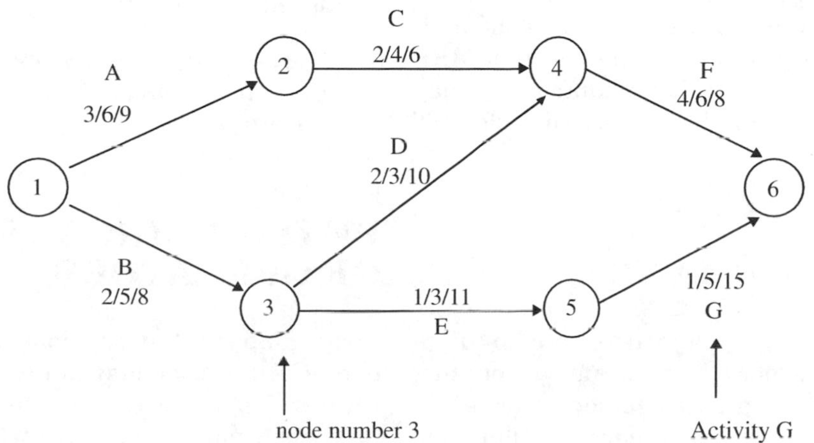

Path 1–2–4–6	16 weeks	(This path takes most time to complete; i.e., critical path)
Path 1–3–4–6	15 weeks	(non-critical path)
Path 1–3–5–6	15 weeks	(non-critical path)

Path 1-2-4-6 is critical or A-C-F is critical.

Standard deviation of the critical path (CP):

$$= \sigma_{CP} = \sqrt{((\sigma_A)^2 + (\sigma_C)^2 + (\sigma_F)^2)}$$

$$\sigma_{CP} = \sigma_{CP}\sqrt{((1)^2 + (0.67)^2 + (0.67)^2)} = \sqrt{1.9} = 1.38$$

To find the probability of completing this project in 18 days first find Z, where Z = (Desired completion time – CP time)/(Standard deviation of critical path time).

$$Z = (18 - 16)/1.38 = 1.45$$

From Z table, a Z value of 1.45 corresponds to 0.926. Thus, the probability of completion of the project in 18 weeks is 92.6%.

See **Critical path method (CPM); Project management.**

References

[1] Dilworth, J.B. (1992). *Operations Management: Design, Planning and Control for Manufacturing and Services*. McGraw-Hill.

[2] Meredith, J.R. and S.J. Mantel, Jr. (1985). *Project Management: A Managerial Approach*. New York: John Wiley & Sons.

PHANTOM BILLS

Some items called transient subassemblies are not stocked in inventory before their use in the next step in assembly. A phantom bill of material represents items that are physically built, but never stocked. Phantom bills are used in plants using MRP and JIT. Where a JIT cell exists in the middle of the process, the product moves from MRP to JIT and back to MRP. The phantom bill may ignore parts and assemblies performed under JIT Scheduling.

PHANTOM DEMAND

Nonexistent, illusory demand are created by overorders during the period of continuous product shortages. It is caused by customer's reaction to shortages. That is, they may order more than they need thereby creating a phantom demand.

See **Bullwhip effect in supply chain management.**

PHYSICAL DISTRIBUTION MANAGEMENT

Logistics activities have long been categorized into two primary areas – *materials management* on the inbound side, and *physical distribution management* on the outbound side. Materials management is concerned with the procurement, transportation, and storage of raw materials, purchased components, and subassemblies entering the manufacturing process, and manages the flow of goods within and through the manufacturing process. Physical distribution focuses on the transportation and storage of finished products from point of manufacture to where the customers wish to acquire them. The key to physical distribution is to meet or exceed the service expectations of customers at the lowest possible costs. Thus, the traditional key to successful physical distribution management has been to match inventory and transportation management to customer needs. The tradeoffs between inventory/transportation costs and customer service drive physical distribution decisions. These decisions include, (1) the number of distribution centers to operate and the technology to be employed in each warehouse; (2) the amount of inventory to hold; (3) where to hold it; and (4) how to ship it. When managed appropriately, physical distribution helps the firm bridge the spatial and temporal gaps between the firm and customers.

See **Logistics: Meeting customers real needs.**

References

[1] Ballou, R.H. (1992). *Business Logistics Management* (3rd ed.). Prentice Hall, Englewood Cliffs, New Jersey.

[2] Lambert, D.M. and Stock, J.R. (1993). *Strategic Logistics Management* (3rd ed.). Richard D. Irwin, Inc., Homewodd, IL.

PHYSIOLOGICAL STRAINS OF OPERATORS

Stress impacts the physiological systems or organs. Cardiovascular symptoms (e.g., higher heart rate and blood pressure), biochemical systems (e.g., abnormal levels of uric acid, cortisol, cholesterol, and catecholamines), and gastrointestinal symptoms (e.g., peptic ulcers) are all physiological indices of the degree of stress experienced by an individual.

See **Human resource issues and advanced manufacturing technology; Human resource issues in manufacturing.**

PIPELINE INVENTORIES

Inventory that are in the transportation network, the distribution system, and intermediate stocking points are called pipeline inventories. The higher the time for the materials to move through the pipeline the larger the pipeline inventory. The flow time through the pipeline is a function of the time required for order processing, shipping, transportation, stocking, etc.

See **Inventory flow analysis; Safety stocks: Luxury or necessity.**

P-KANBAN (POK)

The signaling device that indicates need for material in Just-in-Time manufacturing (JIT) is often referred to as a kanban, which is a Japanese word meaning "card" or "visible marker." Some companies, such as Toyota, utilize two different types of kanban; one that authorizes the *movement* of material from one location to another, and one that authorizes the *production* of more material. In such a *two-card system*, the type of kanban that authorizes production of more material is often referred to as a P-kanban or production kanban. Thus, if Work Center A produces a component part A-12 used by other work centers, more units of that A-12 can be produced at work center A only when a P-kanban is available to authorize that production.

In general, the P-kanban will be a card that is attached to a container of parts available at the location where those parts were made (the work center or supplier ofthe part). When an available C-kanban authorizes movement of the container to a location where the parts are needed, then the P-kanban is removed from that container and the C-kanban is placed in the container. The P-kanban removed from the container then authorizes production of one more container of those parts.

See **Just-in-time manufacturing; C-kanban; Dynamic control for JIT manufacturing; Kanban-based manufacturing systems.**

PLAN-DO-CHECK ACT (PDCA)

This is a multi-step process for quality improvement promoted by Shewhart and later by Deming. In the first step, a plan to make improvement is developed (plan). The plan is carried out (do), and the effects of the plan are observed (check). The results are studied for the purpose of learning and prediction (act). These steps are called the the Shewhart cycle, or Deming cycle.

See **Total quality management; Quality: The implications of Deming's approach.**

PLANNED ORDER RELEASE

The objective of material requirements planning (MRP) is to ensure that component parts and raw materials will be available when needed to carry out the production according to the master production schedule. Consequently, MRP determines when orders for these items should be released, and in what quantities. This information in MRP documents is called planned order releases. All order releases are planned for the future. Thus, they may be changed or even eliminated as conditions change. Only when a planned order release enters the current time period (known as the "action bucket") is it converted from a planned order to an open order and actually released to the shop floor for production, or to a supplier.

Two major issues determine planned order releases. The first is the lead for production, which determines how far in advance of actual need the order must be released. This "planned" lead time may not match the actual lead time. Second, many different factors are used in selecting the "lot-sizing rule" that determines how many units are ordered at one time when an order is released.

See **Master production schedule; Capacity planning: Mcdium- and short-range; MRP; MRP implementation; Order release; Schedule stability.**

References

[1] Orlicky, J. (1975). *Material Requirements Planning*. McGraw-Hill, New York.

[2] Vollmann, T.E., W.L. Berry, and D.C. Whybark (1997). *Manufacturing Planning and Control Systems* (4th *ed.*). Irwin/McGraw-Hill, New York.

PLANNING HORIZON

A planning horizon is the length of time (i.e., the number of weeks or months) into the future for which plans are made. An optimal plan should take into consideration all the information relevant to future events. However, this is not only impossible but also unnecessary. Employing a finite horizon in plans is justified partly on the supposition that far distant events have negligible effect on the current decisions and partly on the realization that the forecasted data far into the future tend to be inaccurate. In other words, if the time horizon for a planning model is too long, the amount of required input data – most of which often may be guesswork – tend to be very high and incur high computational costs. To spend large amount of time and money using data of questionable reliability could result in unnecessary expense and unsatisfactory results. If, on the other hand, the planner chooses too short a planning horizon, he or she may have ignored important events just past the horizon and, may consequently makes poor decisions. Thus, to achieve better plans, an optimal or proper planning horizon is both desirable and necessary.

See **Linking aggregate plans and master production schedules; Master production schedule; MRP.**

PLATFORM TEAMS

Automotive companies used the term platform teams to refer to concurrent engineering teams.

See **Concurrent engineering; Lean manufacturing implementation; Product development and concurrent engineering; Teams: design and implementation.**

PLC (PRODUCT LIFE CYCLE)

See **Product life cycle.**

POINT-OF-SALE SYSTEM (POS)

It is an information system that keeps track of end-customers' response to products, pricing, and promotions through on-line links to sales transactions.

See **Bullwhip effect in supply chain management.**

POKA-YOKE (*POKAYOKE*)

It is a Japanese term for mistake-proofing manufacturing or setup activities. It is intended to prevent errors in the process. The production of different products is prevented by the system.

See **Integrated quality control; Lean manufacturing implementation; Setup reduction; SMED; Quality: The implications of Deming's approach; Total quality management.**

POLLUTION ABATEMENT

The use of any processes, practices and materials to *capture* and/or *treat* pollutants, residuals or wastes prior to their release into the natural environment. These pollutants result from the production or consumption of a product or service. Rather than prevent the generation of pollutants (pollution prevention), pollution abatement emphasizes limiting the extent of any negative environmental impact. Activities related to collection, transport and monitoring of the pollutants and wastes also included in pollution abatement. In addition, the remediation of past environmental damage can be included, such as cleaning up oil spills, mine tailings or leaking storage vessels. Pollution abatement often controls the release of pollutants to air or water through the use of end-of-pipe technologies that mechanically or chemically convert pollutants into another form, which in turn can result in a solid waste by-product.

See **Environmental issues and operations management; Environmental issues: Rinse and recycling.**

POLLUTION PREVENTION

It refers to the use of any processes, practices, materials, products or energy that avoids or minimizes the *creation* of pollutants and waste, and thereby reduces overall risk to human health or the environment. Pollution prevention shifts the emphasis from controlling pollution once it has been generated (pollution abatement) to preventing generation. 3M's Pollution Prevention Pays (3P) program, launched in 1975, has been credited with popularizing both the term and its economic benefits. Examples of actions include material substitution (i.e., hazardous materials are replaced by less hazardous materials), process modifications (i.e., equipment is substituted or the sequence of operations altered), and better operating practices (e.g., better housekeeping, monitoring of material flows, or waste segregation).

See **Environmental issues and operations management; Environmental issues: Reuse and recycling.**

PREAPPROVAL OF SUPPLIERS

See **Product development through supplier integration.**

PRECEDENCE RELATIONSHIPS

Precedence relationship specifies the order in which tasks must be performed to complete a product or a project. For example, an operator in an assembly process would not package a product before assembly is completed. Precedence relationships are formally considered in project scheduling and **Line Balancing Problems.**

See **Assembly line design; CPM; Pert; Project management; U-shaped assembly lines.**

PREFERRED SUPPLIER

Most contemporary purchasing organizations actively evaluate the capabilities and performance of their suppliers. Using these evaluations, firms typically categorize suppliers according to their performance. "Preferred suppliers" are those who are able to meet all of the buyer's operational needs, and some or most of their strategic needs. Purchasers develop integration with the preferred supplier to avoid unnecessary duplication to speed up transactions. The "preferred" supplier is contrasted with other categories of supporters which include:

- "Unacceptable" suppliers are those who are unable to meet the buyer's operational or strategic needs. Discontinuing purchases from unacceptable suppliers is the normal course of action.
- "Acceptable" suppliers are those who meet operational needs as required by a contract. They provide performance which other suppliers could easily match.
- "Good" suppliers are somewhat better than acceptable suppliers in that they offer the potential to obtain value-added services in addition to products.

Purchasers actively seek to develop relationships with preferred suppliers who are judged capable of substantially enhancing the purchaser's abilities. Many suppliers actively seek the status of preferred supplier in order to capture a larger share of the purchaser's material (or service) requirements.

See **Customer service: Satisfaction and success; Purchasing: Acquiring the best inputs; Supplier partnerships as strategy; Supplier performance measurement; Supplier relationship; Supply chain management.**

Reference

[1] Leenders, M.R. and Fearon, H.E. (1998). *Purchasing and Supply Management* (11th ed.). Irwin, Inc., Homewood.

PREMIUM PROTOTYPE

Premium Prototyping is used to develop standardized products for global markets. Under this approach, a product is designed that meets the needs of the most demanding customers. This group of customers can be in the home market of the international enterprise or overseas. Furthermore, the product may be designed to operate in the most severe product use conditions. In other words, the standardized product is characterized by its high quality and by features that in some markets, may be viewed superfluous. Superfluous designs imply higher production costs. As a result, some consumers may be unwilling to pay a higher price for a product with features that they do not need, even though they may recognize the value of the product.

See **Concurrent engineering; New product development through supplier integration; Product design for global markets; Product development and concurrent engineering.**

PREVENTION COSTS

Prevention costs are incurred preventing product defects from occurring, and are a component of the cost of quality. Prevention costs usually include employee training and supplier certification. In general, the greatest improvements in product quality are associated with preventing defects. Although such prevention does incur additional costs, such costs are usually more than offset by savings associated with a reduced number of defects.

Historically, it was often believed that total costs of quality (prevention, appraisal, internal failure, and external failure) would be minimized at some non-zero number of defects. However, that belief has been widely replaced by one that advocates the goal of zero defects. This new belief is based on the fact that small increases in prevention costs often produce large decreases in failure costs.

See **Appraisal costs; Cost of external failure; Cost of internal failure; Performance measurement in manufacturing.**

Reference

[1] Crosby, P.B. (1979). *Quality is Free*. McGraw-Hill, New York.

PREVENTIVE MAINTENANCE

Preventive maintenance refers to maintenance of a system that is still performing its function. It is meant either to prevent the system from failing during operation or to restore the efficiency of the system. Preventive maintenance can be motivated by safety regulations (e.g., airplanes) or by economical considerations (preventing high costs of unscheduled process interruptions).

See **Implementation; Just-in-time manufacturing; Lean manufacturing; Manufacturing cell design; Optimizing models for maintenance management; Total production maintenance; Total quality management.**

PRICE ANALYSIS

Price analysis refers to the process of comparing supplier prices against external price benchmarks. Price analysis *focuses simply on a seller's price with little or no consideration given to the actual cost of production*. In conducting a price analysis, a number of factors should be considered by a buyer. Awareness of these factors allows a buyer to evaluate a quoted price, and possibly identify the major determinants of the seller's price structure.

A supplier's *market structure* has a major influence on price. Market structure refers to the number of competitors in an industry, the relative closeness of their products, and existing barriers of entry for new competitors. The different competitive market structures represent a continuum ranging from pure competition at one end to pure monopoly at the other. In between the two extremes is monopolistic competition and oligopolistic competition. The market structure of a seller's industry can directly affect how a supplier determines price.

Perfect competition occurs when identical products (standardized) are sold with minimal barriers for new suppliers to enter the market. Then, Price is a function of the forces of supply and demand. No single seller or producer controls enough of a market to affect the market price. Of course, a seller could reduce the price with the hope of selling more. In the long run, however, this simply results in lost revenue.

An industry is oligopolistic when only a few, large firms compete. The market and pricing strategies of one firm can directly influence other firms within the industry. Examples of oligopolies in the U.S. include the steel, automobiles, appliances, and large scale computer manufacturing industries. Within an oligopolistic industry, a firm may assume the role of a price leader and raise or lower prices. This can result in all other firms changing their prices or choosing to maintain existing price levels.

Monopolies represent the other extreme from pure competition. While we often think of monopolies as large producers who face no competition, few pure monopolies actually exist. Government or quasi-government agencies regulate the natural monopolies, such as utilities that are permitted to exist.

Buyers have the least amount of price leverage in purely competitive and monopolistic markets. In the former, the forces of supply and demand determine price while in the latter the producer controls price. Fortunately, the vast majority of industrial or commercial pricing occurs between the two extremes. The more knowledgeable a buyer is about a supplier's competitive market structure, the better prepared he or she can be when developing pricing and negotiating strategies.

The economic conditions within an industry often determines whether it is a sellers or buyers market. When capacity utilization at supply firms is high (supply is tight) and demand for output is strong, supply and demand factors combine to create pricing conditions favorable to the seller. When this occurs, buyers often attempt to keep price increases below the industry average. When an industry is in a decline, buyers can take advantage of this to negotiate favorable supply arrangements.

See **Cost analysis for purchasing.**

PRICE-BASED COSTING

This is a method of determining cost targets or allaowable costs for products and services. Allowable costs that are derived from the target/allowable price for a product/service plus a desired profit. The target or allowable price is determined from what the market will bear given the competition. In essence, allowable costs are worked backwards from the price.

See **Activity-based costing; Target costing; Unit cost.**

PRIMARY RATIO (T/TFC)

The Primary (T/TFC) ratio is one of a number of performance measures potentially useful in the effort to implement a system of management based on the Theory of Constraints (TOC). The Primary ratio = (Total Sales Revenue – Total Direct Material Costs)/Total Factory Costs. Note that this ratio can be calculated on a total factory basis or on a department basis.

See **Accounting system implications of TOC; Activity-based costing; Target costing; Theory of constraints (TOC) in manufacturing management; Throughput; Throughput accounting ratio;**

References

[1] Galloway, D., and D. Waldron (1988a). "Throughput Accounting: Part 2 – ranking products profitably," *Management Accounting* (UK, December), 34–35.

[2] Galloway, D., and D. Waldron (1988b)."Throughput Accounting: Part 3 – a better way to control labour costs," *Management Accounting* (UK, January), 32–33.

PRINCIPLE OF COMPARATIVE ADVANTAGE

The principle of comparative advantage was first postulated by the David Ricardo, an English economist, as an extension of the theory of absolute advantage. Absolute advantage simply stated that because different countries can produce some goods more efficiently than other countries, they should specialize in and export those items, and exchange them for products that they produce less efficiently. In essence, the absolute advantage pointed out that, under certain conditions, trade increases wealth.

Comparative advantage expanded the conditions under which trade would increase total wealth. Because the advantage ratio between two nations is likely to be different across product segments, specialization in product categories where the greatest "comparative advantage" existed coupled with trade would increase the wealth of both nations. Within today's global economy, comparative advantage generally refers to the fact that different countries possess different inherent advantages such as low-cost labor, advanced technology, or abundant natural resources. If a firm can identify and utilize the best resources – either through purchasing, global manufacturing, or global market penetration – available worldwide, then it can use the comparative advantages found in different countries to achieve a competitive advantage in both home and global markets.

See **Global facilities location; International manufacturing; Logistics: Meeting customers' real needs.**

Reference

[1] Daniels, J.D. and Radebaugh, L.H. (1998). *International Business Environments and Operations* (8th ed.). Addison Wesley, Reading, Mass.

PRIORITY SCHEDULING RULES

Priority sales are used to decide which job will be processed next at work center, where several jobs are waiting to be processed. The jobs waiting for processing are sequenced using one of many priority sequencing rules. It is assumed that the work center can process only one job at a time. A large number of sequencing rules are used in research and in practice to sequence the jobs waiting for processing at a work center. The common ones are:

FCFS (first come first served). Jobs are processed in the order they arrive at the work center.

SPT (shortest processing time). The job with the shortest processing time is processed first. This rule reduces work-in-process inventory, average job completion (flow) time, and average job lateness.

EDD (earliest due date). This rule is useful when the goal is to reduce job lateness. The job with the earliest due date is processed first.

CR (critical ratio). A priority index is computed using the formula: CR = (due date – today's date)/(manufacturing lead time remaining). Jobs with the smaller critical ratio are given priority over those with larger critical ratios.

FOR (fewest operations remaining). This is a variation of the SPT rule. The job with the fewest remaining operations is processed first.

ST (slack time). This is a variant of EDD rule. Sum the setup times and run times for all remaining operations, subtract this from the time remaining (now until the part due date), and call the remainder slack. The job with the smallest amount of slack gets top priority.

NQ (next queue). NQ increases machine utilization. It considers queues at each of the succeeding work centers where the jobs are headed, and the job going to the smallest queue gets top priority.

See **Order release; Queuing theory; Waiting line model.**

PRIORITY SEQUENCING RULES

See **Priority scheduling rules.**

PRISM SOFTWARE

PRISM is a software package used in for manufacturing planning and control (MPC) in the process industries. Production resources such as machine-hours, labor-hours, materials, tools, etc., are covered by the system.

See **Manufacturing planning and control; Process industry scheduling.**

PROCESS APPROACH TO MANUFACTURING STRATEGY DEVELOPMENT

K.W. Platts

University of Cambridge, Cambridge, UK

INTRODUCTION: The formulation of manufacturing strategy is one of the key tasks for operations managers, yet the process is extremely complex and not well understood. To date, the majority of reported work on the process of strategy formulation is in the area of corporate and business strategy. Much of this is based on the rational view of the strategy formulation process developed in the 1960s and 70s by, among others, Ansoff (1965) and Andrews (1971). The approach is essentially prescriptive and was succinctly summarized by Hofer and Schendel's seven stage model (Hofer and Schendel, 1978) (see Figure 1). These seven stages lay out the steps which organizations might undertake. The process is essentially analytical and rational. It specifies how strategies should be consciously formulated.

Strategy Identification	assessment of current strategy
Environmental Analysis	identification of opportunities and threats
Resource Analysis	assessment of principal skills and resources available to close the gaps identified in the next step
Gap Analysis	comparison of the organisations objectives, strategy and resources against the environmental opportunities and threats to determine the extent of change required in the current strategy
Strategic Alternatives	identification of the options upon which a new strategy may be built
Strategy Evaluation	evaluation of the strategic options to identify those that best meet the values and objectives of all stakeholders, taking into account the environmental opportunities and threats and the resources available
Strategic Choice	selection of the options for implementation

Fig. 1. The seven stages of prescriptive strategy formulation. (Source: Hofer and Schendel, 1978).

However, when the process of strategy formation is observed in practice other modes of the strategy making process are evident. Mintzberg (1973) identified two additional modes, the entrepreneurial mode, where a strong leader controls the organization, and the adaptive mode, where the organization makes adaptations in small, disjointed steps. Mintzberg (1978) observed the processes of strategies emerging rather than being deliberately planned. Strategies can form as well as be formulated, they can emerge in response to evolving situations. He introduced the definition of realized strategy as a pattern of actions – "strategies as ex post facto results of decisional behavior." This view of strategy formation is in marked contrast to the previous formalized planning view.

Quinn (1978) reconciled these two extremes to some extent. He proposed a process of 'logical incrementalism,' proceeding flexibly and experimentally from broad concepts towards specific commitments as late as possible. His rationale was that strategy dealt more with the unknowable than the uncertain, therefore one could not predict events in a probabalistic way, and should delay decisions as long as possible to ensure that the best possible information was available before the decision. However, within this process, he acknowledged a role for the classical, formal planning techniques because they:

- Provide a discipline, which forces managers to take a careful look ahead periodically
- Require rigorous communications about goals, strategic issues and resource allocations
- Stimulate longer term analyses
- Generate a basis for evaluating and integrating short term plans
- Lengthen time horizons
- Create an information framework.

However, there is little in the work referred to above that can enable manufacturing managers to operationalize the process of strategy development in a practical way. The remainder of this article is devoted to describing a process approach to manufacturing strategy development that has been well tested and found to be feasible, usable and useful.

A Manufacturing Strategy Process:

Defining Strategy

Before looking at a Manufacturing Strategy Process, it is necessary to understand what a manufacturing strategy needs to encompass. A review of the literature shows that most writers see manufacturing strategy as consisting of a set of manufacturing objectives which support business objectives, and a set of policies, decisions and actions aimed at achieving those objectives. More formally:

> A manufacturing strategy is defined by a pattern of decisions, both structural and infrastructural, which determine the capability of a manufacturing system and specify how it will operate in order to meet a set of manufacturing objectives which are consistent with the overall business objectives.

Outline of a Manufacturing Process

A strategy process needs to guide its users through the steps of determining the objectives for manufacturing, assessing the extent to which these are being met, and then developing courses of action to close the gaps between current and desired performance. This is similar to the Hofer and Schendel model referred to earlier.

There is an implicit process in the strategy frameworks of Skinner (1969), Wheelwright (1978), Hayes and Wheelwright (1984) and their derivatives. This implicit process depends on breaking manufacturing down into a number of decision areas, and making the goals of manufacturing explicit in terms of a number of performance criteria. The steps of identifying these criteria, prioritizing them and relating the decision areas to them form the basis of the process.

This process is explicit and is illustrated by a flow chart in Figure 2. Worksheets are used at each stage of the process to facilitate its progress.

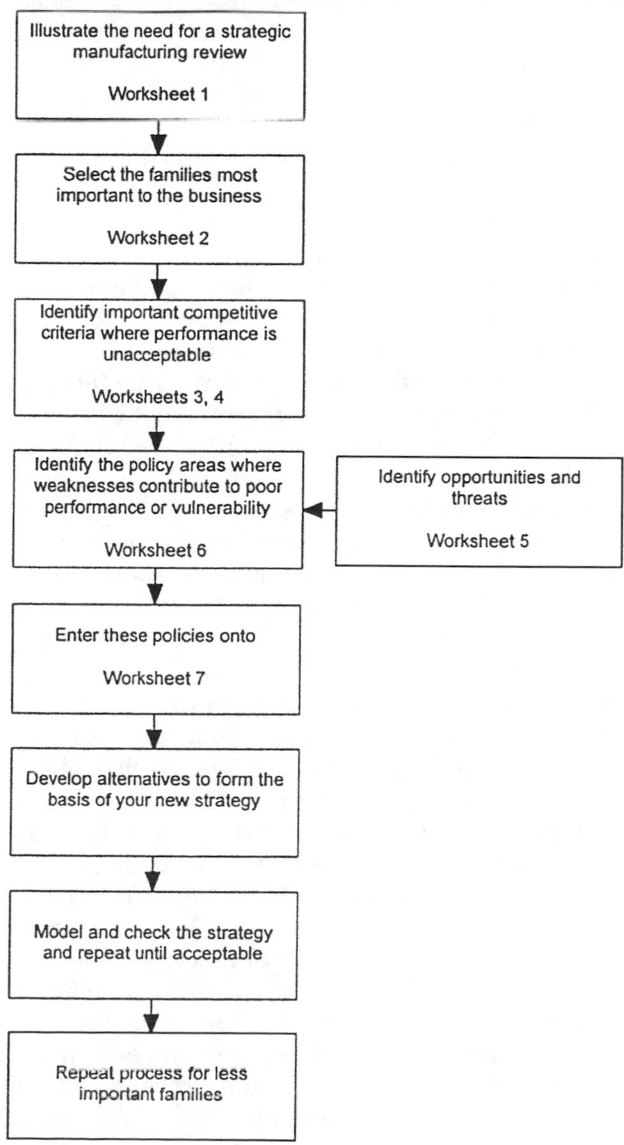

Fig. 2.

The use of worksheets is most important as it requires users to interact formally with the process rather than simply reviewing a "checklist." The worksheets also provide "traceability;" the logic and data for the analytical part of the process are recorded and can be revisited periodically.

Characteristics of a Process

However research has shown that a successful process is more than just a sequence of steps, i.e., the procedure. It needs other characteristics. Platts (1994) has termed these, Participation, Project Management and Point of entry. Together, the four characteristics considered below are: Procedure; Participation; Project Management; and Point of entry. below. These findings are based on the investigation of the manufacturing strategy processes used by three very large multi-national corporations and four consultancy organizations.

Procedure

There should be:

- A well defined procedure to progress through the stages of
  - Gathering information
  - Analyzing information
  - Identifying opportunities for improvement
- Simple and easily understood tools and techniques for use within the procedure
- A written record of the results at each stage.

This is the fundamental requirement of a process, it specifies the steps to be taken. The requirements for well defined procedures, and simple tools and techniques are significant. Operational managers feel more comfortable when they can see the overall structure of the process and appreciate how the individual pieces fit together. They like to understand any techniques that are used. They tend to be uncomfortable with "black box" tools. Managers like to see the stages of the process explicitly stated. This requires going to a greater level of detail than the four or five box flow chart so commonly found in the area. Much of the literature, where it addresses process at all, tends to define procedures at a very abstract level (see the Hofer and Schendel model in Figure 1) and the provision of tools within the procedures is rare.

The requirement for a written record of the process is to ensure that data and assumptions can be revisited at future dates. This will be useful both at subsequent formal strategy reviews and, more importantly, as a strategic management tool that can be used to assess the likely impact of changes in the business environment or incremental policy changes. As discussed in the introduction, descriptive studies have shown that strategies can evolve incrementally, yet formal planning followed by implementation, as suggested by the traditional prescriptive strategy models, does not fully take account of this.

Participation

There should be:

- Individual and group participation to achieve enthusiasm, understanding, and commitment.
- Workshop style interpretation meetings to collectively agree on objectives, identify problems, and develop improvements, and catalyze involvement
- A decision making forum leading to action.

The traditional view of strategy formulation embraces the idea of the brilliant strategist, working alone, contriving a grand plan in the manner of great army commanders of history. Indeed the very word "strategy" is derived from the Greek word for "the art of the General." This is not an appropriate model for manufacturing strategy formulation. Because manufacturing strategy is an integral part of a business strategy and because successful implementation is more likely if the strategy is widely accepted, there needs to be involvement throughout the strategy formulation process. The use of group working, particularly involving multi-functional groups is important as it offers many benefits; some are discussed briefly in the following paragraphs.

It provides a forum where misconceptions and errors in data collection can be detected at an early stage. Most strategy formulation methodologies are based on data to answer questions such as "where are we now?" Because this data is fundamental to the outcome of the process it is necessary to build in a crosscheck early in the methodology. The use of multi-functional groups provides such a mechanism. The diverse range of knowledge and access to information of such a group usually enables many misconceptions and data errors to be identified. It will not, of course, help in cases where there are company-wide, or even industry-wide, assumptions that may be questionable.

It provides inputs from different functions within a company and supplies expertise that can be pooled to aid the entire group. This point is linked to the previous one. Within any company there is usually a vast pool of knowledge on all aspects of the business. It is not however resident in the head of one person. If a strategy formulation methodology is to be successful it needs to tap in to this pool. Participative group working provides one way of achieving this.

It enables a wide range of opinions to be presented and discussed relatively quickly and provides a means of reaching consensus at each stage before progressing to the next stage. The use of structured group working ensures that key issues are adequately discussed and their implications explored before decisions are made. By seeking to achieve consensus at each stage of the process, participants are taken logically towards a conclusion based upon information and discussion. The requirement for absolute consensus is not proven. Of course, one always has to be wary of 'apparent consensus,' where one strong personality imposes his/her views on the rest of the group. Such a 'consensus' is ephemeral and should be avoided.

It ensures that company personnel are closely involved throughout the process and will therefore 'own' the strategy developed. This is one of the most important aspects of participation in the methodology. A strategy only shows effects when it is implemented, and a sense of ownership within the participants facilitates downstream implementation activity. This does not guarantee that the implementation will necessarily be straightforward, but that it is more likely to progress smoothly if those who will implement the strategy were closely involved in its formulation.

Project Management

There should be:

- Adequate resourcing; a managing, supporting and operating group should be identified.
- An agreed timetable.

These are typical project management issues of ensuring that the project is adequately resourced and works to a timetable. The resourcing comprises of three types: managing, the company personnel taking responsibility for ensuring that the project progresses; supporting, the internal or external personnel providing the process 'expertise;' and operating, the personnel who will be directly involved in carrying out the process. These roles may overlap considerably but it appears to be essential that all roles are adequately filled. The **managing group** needs to ensure that the project is adequately resourced and has the necessary profile within the company. In successful projects, the overall exercise is instigated by, or at the very least has the full backing of the Chief Executive for the business under review. This appears to be important for three main reasons;

- The importance of the exercise within the organization needs to be fully recognized.
- The exercise generally requires the cooperation of many different functions, and the Chief Executive provides the point of integration.
- This type of exercise (and its subsequent impact on strategy) is one of the Chief Executive's main roles.

The CEO may set up a core steering team, often the executive team, which oversees the conduct of the process. The exact composition of the team and the responsibility of individual managers obviously vary from company to company, and so it is difficult to be dogmatic about which job titles should be included. However a typical core team is likely to include senior representatives of the following functions: Product Design, Marketing, Manufacturing, Personnel, Finance. The functional focus may change as the exercise progresses, but the consistent involvement of senior management appears to favor success.

The **supporting group** supplies the "expertise" in the process of strategy formulation. The role of the supporting group embraces the following tasks and responsibilities.

- Moving the process through the various stages
- Arranging meetings (and guiding the chairman within meetings)
- Ensuring that the process is adequately recorded.
- Guiding and progressing the actions between meetings.
- Providing a "devil's advocate" role.

In many cases, the supporting group is actually one person, often called a facilitator. Such a person needs to be able to work closely with top management, and will need to have both the personality and the technical competence to interact effectively. He/she needs to be a critical, independent thinker, able to question the accepted wisdom, and yet needs to be acceptable to managers, and democratic in style.

The role of the facilitator is a difficult one to fill. Within larger companies, it is feasible that a senior person, maybe a manager or a professionally qualified employee could be found for this task. In smaller companies, it is less likely that such individuals are available. The choice for the smaller company seems to be between an outside consultant or a member of the managing team described above. The problem with using a team member for the facilitation role is that his/her personal views could dominate the team; there is also less opportunity for him to play the roles of "neutral intermediary" and "devil's advocate." One problem with using an outside consultant is finding one who has the necessary capability, and who is prepared to act in this role. The use of outsiders may also reduce the opportunity to develop people within the organization. The issues surrounding 'facilitators' are also considered by Rhodes (1991), who suggests that academics could undertake the role.

The **operating group** comprises of the people who do the bulk of the work: collecting and analyzing the data; assessing the requirements of the business; considering alternative policies; etc. The composition of the operating group may change during the process. During the early stages it is likely that the group members are market oriented, later on the focus is likely to move more towards manufacturing. Although the balance may change, the group should always remain multifunctional.

The timetable should be discussed and agreed at the outset of the exercise. If this is not done there is grave danger of the project never reaching conclusions, there will always be more data that need to be obtained, or different options that need to be explored. The well known "paralysis of analysis" could result, and rather than arriving at a new strategy, the company may stagnate, people may become disillusioned, and the whole exercise could lose credibility. Experience has shown that task-force style teams need to maintain a certain momentum to retain enthusiasm and productivity. The companies interviewed suggested that setting a tight but achievable timetable was advisable.

Point of Entry

There should be:

- A way of reaching agreement in the managing group
- A way of establishing commitment of the managing and operating groups
- Clearly defined expectations of what the process involves

It is necessary to provide a mechanism for introducing the strategy formulation process into an organization. There needs to be clear view of what the process involves and what type of results will be obtained. There needs to be the full agreement of the top management of the company about proceeding with the exercise. This goes beyond simply ensuring that everyone "knows what is going on."

The process may be formally documented and this could provide managers with the necessary understanding of the procedure; however, simply having a logical procedure and presenting it formally does not ensure that it will be used. The process might appear to be extremely time consuming, and show little direct benefit to operational managers who are often more concerned with day-to-day output than with standing back to review the overall rationale of what they are doing. There needs to be some way of demonstrating to the company the necessity of proceeding with the full process. The process needs to be "sold" to the personnel who are to be intimately involved in it.

The "Point of Entry" feature of the methodology needs to establish a common understanding within the company of what manufacturing strategy is, and what might be the outcome of the strategy process. It is important that management expectations are brought out into the open and discussed. These expectations will cover both the effort required (cost), and the benefits likely to be achieved. Managers are very accustomed to thinking in terms of cost/benefit analysis, and enabling them to put a strategy formulation project into this framework is beneficial. It must be recognized that the main aim of the "Point of Entry" stage is to obtain the commitment of managers.

THE MANUFACTURING STRATEGY PROCESS IN DETAIL: Having explained the characteristics of successful processes a specific process for manufacturing strategy is now presented. This is essentially a market-based approach following a procedure, which is described in more detail in a workbook (DTI, 1988). The main stages of the procedure are summarized below.

Illustrating the Need (Worksheet 1)

Worksheet 1 with profiles 1–3 (Figure 3) provides a way of illustrating the need for a strategic review of manufacturing by graphically illustrating the differences between market requirements and manufacturing performance. In this technique, a marketing person is asked to identify a product family and then sketch a profile to represent the market requirements in such terms as lead time, reliability, cost, etc. A manufacturing person is independently asked to sketch a profile of how manufacturing is performing. Both profiles are copied onto transparencies and overlaid. Any mismatches between requirements and performance become readily apparent and often the gaps are startling enough to encourage embarkation on, and commitment to, the whole process. Another revealing way to use the profiles is to ask managers from different functions to sketch their perceptions of the requirements of a particular product family. When overlaid these are likely to show dramatic differences and a lack of common agreement on requirements. And without agreement on requirements how can objectives and strategies be determined?

Profile 1
Market requirements

Product Family

Delivery Lead-time
Ex-stock — Not significant

Delivery Reliability
Into stock point — Critical: Project delay

Product Features
Fit for purpose — Many features/ High absolute level

Quality
Acceptable at price — Total reliability essential

Design Flexibility
Standard range only — All designs customer specified

Volume
Stable market little variation required — Highly cyclic variable market

Price
Price competition dominant — Non-price competition dominant

Profile 2
Achieved performance

Product Family

Delivery Lead-time
Short — Long

Delivery Reliability
variable — Good

Product Features
Few features — Many features

Quality
Acceptable — High

Design Flexibility
Standard only — All products customised

Volume
Volume variations low

Price
Low — High

Profile 3 (Composite of Profiles 1 and 2)

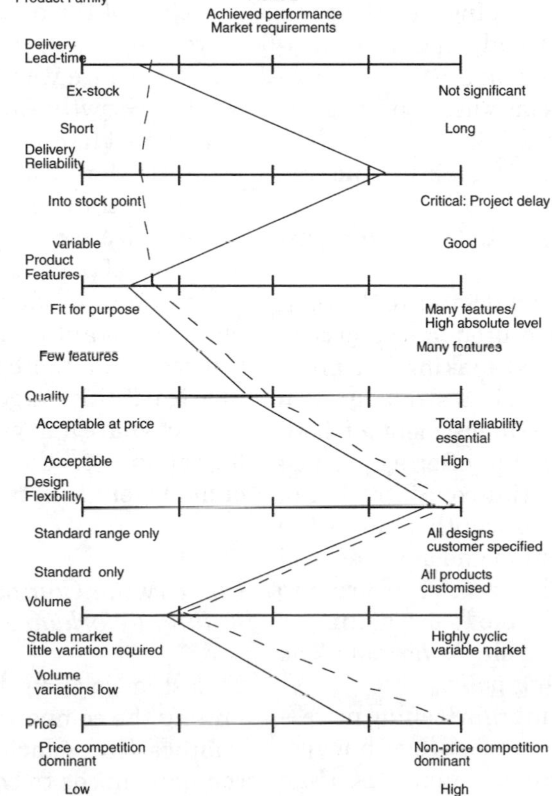

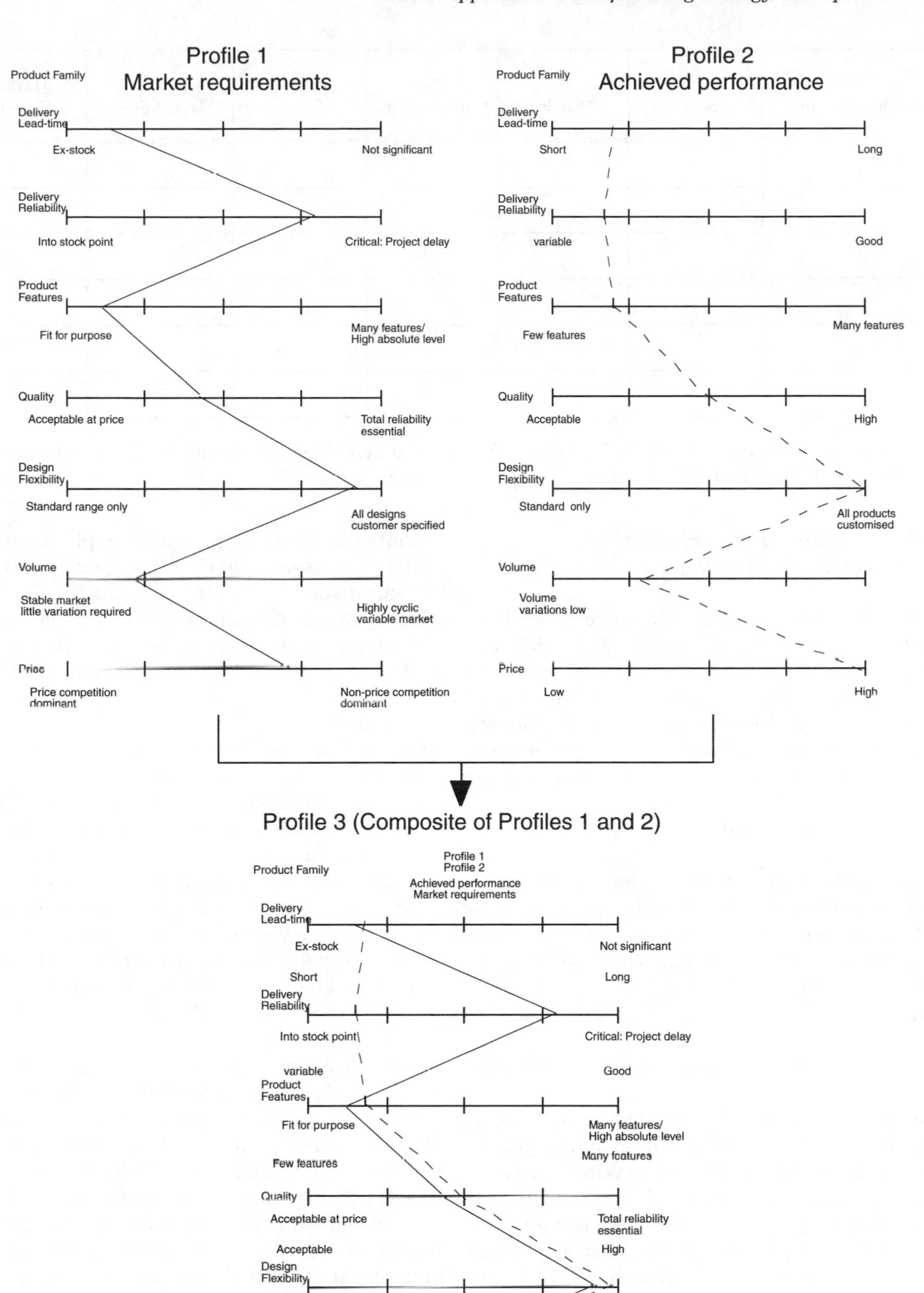

Fig. 3.

Product Families	Sales	% Sales	% Contribution.	Market Share	Growth of Market Share	Market Growth

Fig. 4. Worksheet 2 (basic product family data). For growth, if actual figures are not available, use scale of −2 (declining rapidly) to +2 (growth rapidly).

Identifying Basic Product Families, (Worksheet 2 – Figure 4)

It is necessary to identify different product families because it is highly unlikely that all the products within a company compete in the same way. A product family is defined as a grouping of technologically similar products, which compete within the market place in identical ways. Possible starting points when identifying product families are:

- Existing sales groupings (but not simply by geographical market);
- Existing manufacturing groupings (on the basis that the company was already operating logically). These could be based on the manufacturing processes used, or on classification s such as make-to-order or make-to-stock;
- Products at the same point in their product life cycle;
- Existing groupings of products for costing purposes.

Not all families will be of equal value to the company. Some might be mature products subject to intense price competition, and making low profit contributions, others might be in a strong growth phase requiring substantial investment of both money and managerial effort but offering promising by high rewards. An initial concentration of effort on the key families significantly reduces the effort required in data collection and analysis, and facilitates the execution of the exercise. In order to define key families it is desirable to use a number of measures rather than just a single variable. The following form a good starting point:

Contribution: Contribution margin (selling price – direct costs) is an indirect measure of how potentially profitable a particular family is. High contribution will generally imply high profitability assuming that the family does not incur disproportionate overheads;

Market Share: A high market share, particularly if the product is a cost-leader, reveals very strong position. Economies of scale and learning curve effects will further enhance competitive ability.

Growth of Market Share: If market share is increasing, this is one indicator that the product is competing successfully. On its own, of course, it has limited value. It is significant in the light of the overall market share, and the growth or decline of the overall markets. For example, increasing share in a growing market, where the initial share was high, is significant.

Market Growth: An indication of the potential for a product family. A growing market offers opportunities for future profits if a dominant position can be attained.

The objective of this worksheet is to obtain on one sheet of paper a sufficiently detailed picture of a company's products and markets to enable the most important families to be identified as an initial focus for the exercise. Problems most likely to arise at this stage are related to lack of data on market share and growth, and it may be necessary for reliance on the use of the subjective opinions of managers in establishing the market position.

Identify the Competitive Criteria for Each Family (Worksheet 3 – Figure 5)

This stage of the process is concerned with determining the competitive factors for each of the key families. These factors essentially capture what a company needs to be good at to win orders for this

Product Family	Features	Quality	Delivery Lead Time	Delivery Reliability	Flexibility Design	Flexibility Volume	Price

Fig. 5. Worksheet 3 (competitive criteria). Allocate 100 points among criteria.

product family in the market. There are of course many factors, both within and outside the scope of manufacturing. A typical list of manufacturing related competitive criteria includes:

Product features: Adding capability to the product, or choice for the customer.

Quality: Producing a product that performs per specification.

Delivery lead-time: Delivering the product within a short lead-time.

Delivery reliability: Always delivering to schedule.

Design flexibility: Having the ability to produce products to customer specification.

Volume flexibility: Having the ability to deal with demand volume without compromising leadtime.

Price: Selling at a competitive (may be lowest) price.

One way of identifying competitive criteria is by the allocation of 100 points among the competitive factors in such a way as to reflect the perceived market requirements. For example, a product family may compete 20 per cent on price and 80 per cent on delivery speed. Here it may become apparent if the initial choice of product families is appropriate. If it is relatively straightforward to assign the points in the competitive factor ratings then the original choice is sound. If, however, there is substantial difficulty then it is likely that the initial choice of family groups is suspect. A common reason for the inability to allocate points consistently across the factors is that there may be a number of products within a group which are 'exceptions' to the family being considered. This is often easily solved by sub-dividing the family into two or more groups, which can then be easily assessed. There are various methods of completing the worksheets that fall into two categories, individual views and consensus views.

The problem with individual views is that a large spread of responses is likely. This is not adequately dealt with by averaging as this loses the opportunity to explore the reasons for the differences which provides insights into the market. Discussions leading to consensus could be revealing. The method found to be the most productive is individual assessment of the requirements followed by a group discussion in an attempt to reach a consensus. Here all views get exposed for discussion, whereas group discussions without prior individual assessment can suppress some individuals' views.

In Worksheet 3 there may be difficulties in defining the competitive criteria. For example, 'flexibility' criteria especially 'volume flexibility,' could pose a problem. The problem is often caused by searching for the competitive criteria using the question: 'How do products win orders in the market?' This is interpreted by companies as; 'Why does a customer buy our product?' The argument then might proceed; "A customer doesn't buy a product because of 'volume flexibility', he buys it because the supplier can offer the delivery that he wants. How the supplier achieves that delivery is an internal feature, not a market criterion." Confusing 'the customer' with 'the market' causes this. Although the 'flexibility' criteria may be redundant from the viewpoint of a single customer, they are extremely important in specifying the manufacturing requirements imposed by the market. It is the latter, which defines the objectives for manufacturing, and it is the latter which the worksheet is designed to obtain. Thus a better question to ask is; "What are the requirements imposed upon manufacturing by the way products win orders in the market?"

In carrying out this analysis it is often necessary to distinguish between maintaining existing business, and winning new business. Hill's concept of "order winning" and "order qualifying" factors may prove useful here (Hill, 1985). The maintenance of existing business generally means achieving adequate performance in the qualifying factors. These qualifying factors however are not necessarily the same that win the business. Factors which were key to winning the business initially may become less so in maintaining that business.

Assessing Current Performance (Worksheet 4 – Figure 6)

The objective of this worksheet is to obtain an assessment of the current performance of manufacturing against the criteria identified as important Worksheet 3. Three issues need to be considered at this stage of the process: internal vs. external performance measures; factual vs subjective performance measures; and standards for comparison.

External vs. Internal Performance Measurement: External measures provide an assessment of the performance of the company as perceived by the market, while internal measures reveal how the various manufacturing sub-systems are performing. In many cases, existing performance measurement systems will not provide the necessary data. Thus it may be necessary to conduct a specific exercise to obtain this. Minimizing the effort in data collection will be an important factor if the approach is to be acceptable to the busy manager. Relatively small samples and concentration on one or two representative products will often provide a good indication of overall performance. Some factors are more easily assessed than others. Internal and external measures of quality and lead times are readily obtained. The price (an external measure) is also easily found but it is not, a useful indicator of manufacturing performance as it is affected by too many extraneous factors. Price is fixed more by the market than by the performance of the company. The internal measure, cost, is more useful but it is often not easily obtainable in a form which can be useful. The problems arise from costing systems which are designed primarily for the valuation of inventory (which is needed for financial statements) and not for use in exercises such as these. It may be more relevant to replace the 'cost' given by the accounting system with alternative internal cost

Product Family	**Features**	**Quality**	**Delivery Lead Time**	**Delivery Reliability**	**Flexibility Design**	**Volume**	**Cost**

Fig. 6. Worksheet 4 (achieved performance). Rate performance on a scale of −2 (Performance gives us a strong disadvantage vs. our competitors) to +2 (Performance gives us a strong advantage vs. our competitors).

indicators such as utilizations, efficiencies, stock levels, staffing etc.

Factual vs. Subjective Performance Measures: This issue is closely related to the section addressing market requirements. To what extent is it necessary to measure performance in an objective way? When it comes to filling in Worksheet 4, the requirement is to assess performance against the competition but many companies have little factual data on their competitors. In this case, the assessment might be done by the subjective judgements of the management team. This is considered in the following section.

Standards of Comparison: Worksheet 4 is structured around comparisons with competitors but such data is not easy to acquire. There are many sources of data at an industry level in the form of benchmark reports, financial statistics etc., but these are usually too aggregated. To be useful the data needs to be from actual competitors. One way is to use market intelligence. On the internal side, other techniques are available. For example, to assess costs, buying competitors products and reverse engineering them can be helpful. Non-competitive benchmarking can be another useful tool.

Assessing Opportunities and Threats (Worksheet 5 – Figure 7)

The objective of this worksheet is to identify opportunities and threats that might affect manufacturing in the future. It is designed as a 'catch-all' to pick up items not revealed on the other worksheets. Porter's (1980) framework for the basis of industrial competition provides a useful guide to typical factors to think about.

Assessing the Manufacturing Operations (Worksheet 6 – Figure 8)

The objective of this stage is to identify the current manufacturing strategy and to assess what effects this had on the achievement of the objectives established in the previous stages. The manufacturing strategy is expressed in terms of structural and infrastructural policy areas.

Facilities: The factories, their number, size, location, focus.

Capacity: The maximum output of the factory.

Span of Process: The degree of vertical integration.

Processes: The transformation processes (metal cutting, mixing, assembly, etc.) and most critically the way in which they are organized.

Human resources: All the people-related factors, including both at the personal and the organizational level.

Quality: The means of ensuring that product, processes and people operate to specification.

Control Policies: The control policies and philosophies of manufacture.

Suppliers: The methods of obtaining input materials at the right time, price and quality.

New Product: The mechanisms for coping with new product introduction, including links to design.

For each policy area: 1) Identify and describe the current practice; and 2) Determine the perceived effects of this practice on the achievement of the competitive criteria. Worksheet 6 (Figure 8) provides a way of structuring and presenting this data in a matrix format. The perceived effects are summarized by using a scale of −2 (The practice is extremely detrimental to achieving the required performance) to +2 (The practice provides strong

Product Family	Opportunities	Threats

Fig. 7. Worksheets 5 (opportunities and threats).

Policy Area	Features	Quality	Delivery Lead Time	Delivery Reliability	Flexibility Design	Flexibility Volume	Price
Facilities							
Capacity							
Span of Process							
Processes							
Human Resources							
Quality							
Control Policies							
Suppliers							
New Products							

Fig. 8. Worksheets 6 (existing practices). This worksheet is diagrammatic. Each of the policy areas would be expanded to allow the entry and rating of several practices.

support to achieving the required performance). This process ensures that, as far as possible, the influences of the existing practices on the achievement of the competitive criteria are understood and agreed. An important point to note is that the audit here must identify what is actually done, not what is meant to happen, or what the "policy" is.

The following procedure has been found to be a successful way of completing this stage of the process.

1. Blank worksheets are circulated to a number of personnel in manufacturing related roles. They are asked to identify the current practices.
2. The Facilitator collates the results into a master summary sheet. This may require selected interviewing of the respondents to clarify practices.
3. Copies of the summary master sheet are circulated to the manufacturing personnel who are asked to assess the effects of the practices.
4. A workshop is held involving the manufacturing personnel and the Facilitator where the worksheet is discussed, often at great length, and a consensus view on the practices and their effects are reached.

Developing the New Strategy (Worksheet 7 – Figure 9)

This is the stage at which all the strands of the process are pulled together. Briefly the worksheet provides a way of summarizing the major mismatches from the earlier worksheets. Identifying from Worksheet 6 the manufacturing practices which are most in need of revision. The process provides a clear identification of the manufacturing problems from which action plans for solutions can be developed. These action plans form the basis for a revised manufacturing strategy.

The problem identification stage is conducted by inspection of the worksheets. Worksheets 3 and 4 together reveal where performance is unsatisfactory, and worksheet 6 is then used to identify the practices that contribute to poor performance.

Policy Area	Weakness/Required Strength	Possible Actions/Strategic Choices

Fig. 9. Worksheet 7 (action worksheet).

Generally, a visual check of the worksheets is all that is required. The process then requires the creation of action plans that can remedy the deficient practices and build on the supportive practices. The creation of action plans (the basis of the new strategy) tend to follow relatively easily following problem definition, which results from the previous stages of the process.

CONCLUSION: The process described above has been extensively tested and has been found to provide a useable way for companies to reassess and develop their manufacturing strategies. However it must be clearly understood that the process does not prescribe strategies, it allows companies to develop their own strategies and plans that are the basis for competitive advantage. The benefits of a structured approach are that it ensures a much greater understanding of market requirements, competitive position and existing strategy than would normally be the case.

See **Core manufacturing competence; Manufacturing strategy.**

References

[1] Andrews, K.R. (1971). *The Concepts of Corporate Strategy*. Dow Jones-Irwin, Homewood, Illinois.

[2] Ansoff, H.I. (1965). *Corporate Strategy*. McGraw-Hill, New York, NY.

[3] DTI (1988). *Competitive Manufacturing: A Practical Approach to the Development of a Manufacturing Strategy*. IFS, Bedford.

[4] Hayes, R.H. and S.C. Wheelwright (1984). *Restoring our Competitive Edge*. John Wiley, New York, NY.

[5] Hill, T.J. (1985). *Manufacturing Strategy: The Strategic Management of the Manufacturing Function*. Macmillan, Basingstoke

[6] Hofer, C.W. and D. Schendel (1978). *Strategy Formulation: Analytical Concepts*. West Publishing, St Paul, MN.

[7] Mintzberg, H. (1973). "Strategy-making in three modes." *California Management Review*, 16, 2, 44–53.

[8] Mintzberg, H. (1978). "Patterns in strategy formation." *Management Science*, 24, 9, 934–48.

[9] Platts, K.W. (1994). "Characteristics of methodologies for manufacturing strategy formulation." *Computer Integrated Manufacturing Systems*, 7, 2, 93–9.

[10] Porter, M.E. (1980). *Competitive Strategy – Techniques for Analyzing Industries and Competitors*. Macmillan Free Press, New York.

[11] Quinn, J.B. (1978). "Strategic change: logical incrementalism." Sloan Management Review, 1, 20.

[12] Rhodes, D.J. (1991). "The facilitator – an organizational necessity for the successful implementation of IT and operations strategies." *Computer Integrated Manufacturing Systems*, 4, 2, 109–13.

[13] Skinner, W. (1969). "Manufacturing – the missing link in corporate strategy." *Harvard Business Review*, May–June.

[14] Wheelwright, S.C. (1978). "Reflecting corporate strategy in manufacturing decisions." *Business Horizons*, February.

PROCESS CAPABILITY

Process capability indicates the magnitude of process variation with respect to product design

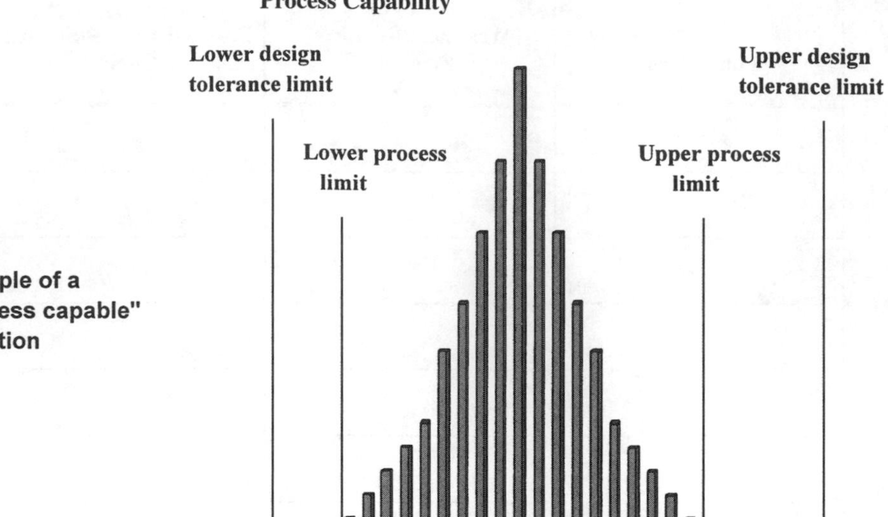

specifications. If the limits of variation produced by a manufacturing process are within the boundaries specified by product design tolerances, the process is deemed to be "process capable." Consider the example illustrated in the figure "Process Capability." The distribution indicates the variation of the process output along some given dimension (e.g., the diameter of a hole). A process capable condition exists when most of the output (more than 99%) lies within the boundaries established by the upper and lower design limits for the product. Three standard deviations on each side of the process mean include about 99.7 percent of the output between the Upper and Lower Process Limits, which are ±3 standard deviations from the mean, lie between Upper and Lower Tolerance Limits, then the process is considered "process capable."

See **Concurrent engineering; Just-in-time manufacturing; Statistical process control using control charts; Total quality management; Variability in manufacturing.**

PROCESS CONTROL

Process control ensures uniformity and consistency of output. All production processes have some natural variation. Deviation beyond the natural variation indicates that the production process is out of control, which may result in defective outputs. Process control ensures that the production process stays in control, and stops the process to identify and make adjustments when the process is out of control. Process control is often accomplished using statistical process control where certain quality characteristics are monitored on a periodic basis using statistical control charts. Control charts have both an upper and lower control limit. On a periodic basis, a sample of output is examined and the results plotted on a control chart. If a sample falls within the control limits, the process is likely in control, producing acceptable output. If a sample falls outside the control limits, the process is likely out of control, and should be stopped and corrected.

See **Integrated quality control; Just-in-time manufacturing; Quality: The implications of Deming's approach; Total quality management; Statistical process control using control charts.**

PROCESS FLOW CHART

Process flow chart results from the process mapping process, which is often a part of business process reengineering.

See **Business process reengineering; Reengineering and the process view of manufacturing.**

PROCESS FLOW SCHEDULING (PFS)

PFS is a scheduling framework that provides an integrative structure for planning and scheduling systems in a wide variety of high-volume process industries. The framework: (1) links the scheduling practices of different firms and different industries; (2) unifies planning and scheduling techniques, such as lot sizing methods and safety stock theory, into an integrated framework, and; (3) enhances communication between practitioners, software vendors, and researchers of process industry scheduling systems. Process Flow Scheduling (PFS) is characterized by the following three principles:

1. PFS uses the process structure to guide scheduling calculations.
2. Process clusters are scheduled using material-dominated or processor-dominated scheduling techniques.
3. Cluster schedules are linked together using a reverse-flow, forward-flow, or mixed-flow scheduling strategy.

See **Process industry scheduling; Processor-dominated scheduling; Material-dominated scheduling.**

References

[1] Taylor, S.G. and S.F. Bolander (1994). *Process Flow Scheduling: A Scheduling Systems Framework for Flow Manufacturing*, APICS, Falls Church, VA.

[2] Taylor, S.G. and S.F. Bolander (1991). "Process Flow Scheduling Principles," *Production and Inventory Management Journal*, Vol. 32, no. 1, pp. 67–71.

PROCESS FOCUSED PLANTS

See **Focused factory; Manufacturing strategy.**

PROCESS INDUSTRIES

Process industries which a standardized output or commodities on a continuous linear flow process. These are often referred to as continuous process industries. Generally, process industries tend to be highly automated. Typical process industries include those producing liquid products such as paint, petroleum, or other types of solvents, sugar refineries, flour mills, steel producers, and chemical producers.

See **Implications of JIT manufacturing; Process industry scheduling.**

PROCESS INDUSTRY SCHEDULING

Sam G. Taylor

University of Wyoming, Laramie, Wyoming, USA

Steven F. Bolander

Colorado State University, Fort Collins, Colorado, USA

INTRODUCTION: Process industry firms are found in the petroleum, chemical, paper, food, beverage, plastics, rubber, cement, glass, primary metals, and similar high-volume process industries. Process industry plants generally have large quantities of materials flowing through process trains. These process trains are, in turn, divided into process stages and within the process stages there are one or more process units. For example a process train may be divided into stages for feed preparation, reaction, and packaging. In turn, the packaging stage may be divided into one or more packaging lines, each of which represents a process unit.

Production schedules specify what is to be produced on each process unit in a plant with sufficient detail. Process operators execute the schedule. A production schedule typically covers a time horizon of several shifts to several weeks. Production schedules specify the lot sizes (or production run lengths) and sequences of items produced on process units. These schedules are the authorization for unit operators to produce specific intermediate and finished products. In contrast, production plans cover a time horizon ranging from several months for short-range plans to twenty or more years for strategic production plans. Thus, production plans have less detail than production schedules.

This article presents a framework for process industry scheduling called "process flow scheduling" and abbreviated with the acronym "PFS." While much has been written about the optimization of production plans in process industries, very little has been written about production scheduling in the process industries. In addition, many articles and books have been written about the planning and scheduling functions of MRP II systems. However most process industry firms have rejected the planning and scheduling modules of MRP II systems (Foley, 1988) and the MRP II literature generally does not apply to the high-volume, flow manufacturers in process industries. Like MRP II, the PFS framework is quite simple and can model a wide variety of plant operations. As with MRP II systems, there are significant differences among commercial PFS software packages.

The origins of process flow scheduling are obscured. Many process industry firms have used process flow scheduling concepts in their proprietary scheduling systems for decades. The first commercial PFS system was marketed in 1982 and today there are at least seven PFS software vendors whose systems are used in over 800 plants around the world. The framework integrating these scheduling practices was initially identified in 1988 and documented in several journal articles and presentations at APICS conferences by Taylor and Bolander (1994; 1997). One of their recent articles reviews the past, present, and future of PFS and gives references to earlier articles (Taylor and Bolander, 1997). The most complete documentation of PFS is their 1994 book (Taylor and Bolander, 1994).

An excellent, complementary view of process industry scheduling is given by Shobrys (1995). His article focuses on scheduling information requirements and the integration of production scheduling with production planning and process control.

We begin our overview of process industry scheduling by examining scheduling information flows. We then examine alternative procedures for scheduling process units and trains. We conclude by summarizing the benefits of implementing PFS systems.

SCHEDULING INFORMATION FLOWS: Scheduling information flows are shown in Figure 1. Inputs to the scheduling process are the production plan, scheduling data, and scheduling guidelines. The production plan determines the allocation of demand among plants, seasonal production plans, aggregate production rates for process trains, long term critical raw material contracts, and company-wide capacity strategies. The production plan also integrates operating plans with marketing plans, financial budgets, human resource plans, purchasing plans, and maintenance plans. Mathematical programming is frequently employed to optimize production plans. Camm *et al.* (1997), Bender, Northup, and Shapiro (1981), and Bradley, Hax, and Magnanti (1977) give examples of optimizing production plans.

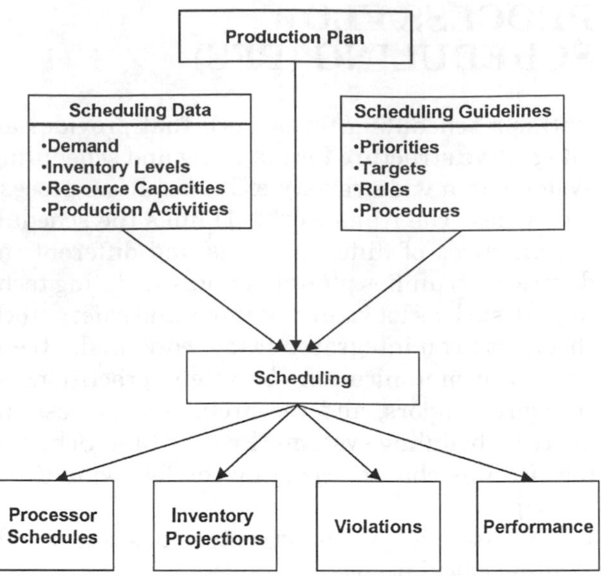

Fig. 1. Scheduling information flows.

Scheduling Data

Scheduling data must be detailed and is often voluminous. Demand forecasts (and customer orders in many cases) are needed for each finished product to be scheduled. Inventory levels are also required for each finished product, intermediate stock, and raw material to be scheduled. Capacities for individual process units, labor, and any other resource that may constrain the schedule are needed.

Production activity data for each alternative way process units may be scheduled are required. For example, production activity data may show that 0.3 pounds of A, 0.7 pounds of B, a box, and 0.0002 machine hours are required to make a one-pound box of product X. Although production activity data is similar to the bill of material in an MRP II system, there are two significant differences. First, production activity data includes resource requirements. Second, production activity data is limited to a single process unit; whereas as a bill of material shows all the parts and subassemblies required to make one unit of an end item. Thus, production activity data use a single-level, process oriented approach and bills of material use a multi-level, product oriented approach.

Scheduling Guidelines

The last group of scheduling inputs are guidelines. Guidelines are required for priorities, targets, rules, and procedures. Priorities establish the relative importance of customer service, cost, inventory, and quality objectives. For example when an order is received for an out-of-stock item, a firm with a high priority on customer service will break a least cost scheduling sequence and incur the costs of an expensive product changeover. On the other hand a firm with a high priority on low prices and costs would delay or decline this order and maintain their least cost production sequence. Schedulers need information on priorities to make decisions that are consistent with the business strategy.

Targets are based on economic and physical factors that constrain the production schedule. Targets include (1) minimum and maximum inventory levels, (2) production run lengths, (3) campaign cycle lengths, (4) production sequences, and (5) target customer service levels. Minimum inventory targets provide a safety stock or buffer for variations in requirements as well as for interruptions in supply. Traditional safety stock models (Brown, 1967) are quite useful for setting minimum inventory targets for finished goods; however, more research guidance is needed for setting raw material and intermediate inventory targets.

Maximum inventory targets may be either *hard* or *soft* constraints. A hard maximum inventory constraint, such as a tank maximum, must be observed. On the other hand soft constraint maximums are based on economic, rather than physical constraints, and can readily be violated. For example, a soft constraint for maximum inventory might be equal to the safety stock plus the peak cycle stock. In this situation a production run length might be extended for customer service, or perhaps quality considerations, and the maximum target inventory exceeded at a slight increase in inventory carrying costs. Note that an inventory might have both hard and soft constraints for the maximum target inventory.

Another target concerns production run lengths or lot sizes for individual items. Here again we can have hard and soft constraints on both minimum and maximum run lengths. A hard constraint on minimum run lengths/lot sizes may be derived from batch reactor sizes or from product quality considerations. A soft constraint may specify an optimal run length based on an economic lot size model (Brown, 1967).

Targets for campaign cycle lengths are similar in scope to targets for individual products discussed above. When multiple products are produced on the same process train, campaign cycles establish target production run lengths, product switching strategies, and production sequences. Hax and Candea (1984: 147) discuss procedures for determining lot sizes (production run lengths) for multiple items sharing the same equipment. Typically target sequences are based upon product characteristics such as light-to-dark, wide-to-narrow, low viscosity to high viscosity, etc. If process technology does not dictate a natural production sequence then a traveling salesman algorithm (Hoffman and Padberg, 1996) can be used to determine the production sequence. Alternatively a genetic algorithm might be employed (Rubin and Ragatz, 1994).

Rules are guidelines that are set by management and may have a significant impact on other functional areas. Examples of rules are (1) no overtime will be scheduled on Sundays, (2) never run product X after product A, (3) always produce product Z on day shift, and (4) production schedules are frozen for 2 days unless the vice-president of operations approves the change.

Procedures specify how schedules will be developed and communicated throughout the organization. Examples of procedures are (1) every Thursday schedules are issued for a six-week horizon, (2) only schedulers may revise the schedule, and (3) operators will record reasons for schedule deviations.

Scheduling Outputs

Production scheduling outputs are also shown in Figure 1. The processor schedules specify the schedule in sufficient detail that operators can execute the schedule. In most PFS systems the schedule is the authority to produce and work orders are not used. Inventory projections often show a plot of inventory over a specified time horizon. In addition reports may be used to show critical data such as days-of-supply (or run-out date) and available-to-promise quantities. Violations warn schedulers when a trial schedule breaks a target or a rule. Violations can be displayed graphically, by warning messages, or in reports.

The relative performance of alternative, trial schedules is useful for schedule evaluation. In some cases a set of schedule performance indicators are aggregated into a single, pseudo-cost number. In other cases, firms prefer to consider multiple scheduling objectives using factors such as the cost of changeovers, inventory-carrying cost, number of stockouts, bottleneck machine utilization, etc.

This section has presented a general framework for scheduling system information flows. This framework provides a starting point for designing a scheduling system for a particular process industry plant. However the framework should be modified as required to fit the needs of a particular plant environment. In the following sections we examine alternative techniques for developing process unit and process train schedules.

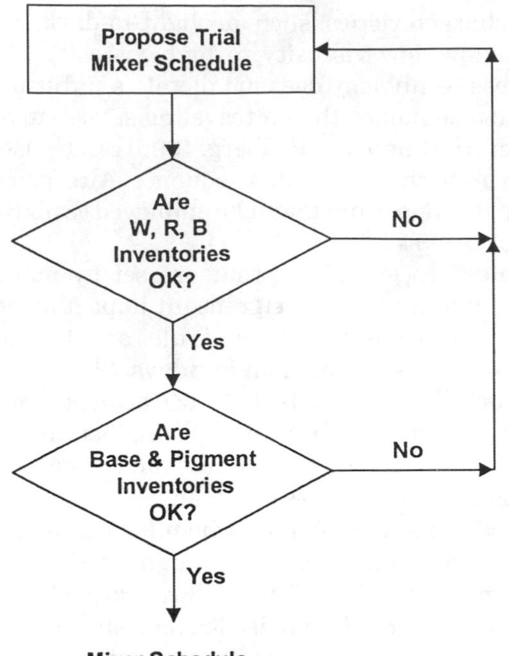

Fig. 2. Processor dominated scheduling flowchart.

PROCESS UNIT SCHEDULING: A process unit is a group of equipment that is scheduled as a single entity. For example a reactor, a bank of heat exchangers, and a set of distillation towers may all be scheduled as a single entity and designated as the reactor process unit. Process units can be scheduled using processor dominated scheduling (PDS) or material dominated scheduling (MDS.)

Processor Dominated Scheduling

Processor dominated scheduling first schedules capacity and then checks materials. PDS is commonly used when capacity must be utilized efficiently or when there is a dominating, least-cost production sequence. Let us illustrate PDS with a simple example. Suppose that a color mixer must be scheduled in the repeating, light-to-dark sequence: white (W), red (R), and black (B). No cleanouts are needed for the W-R or the R-B changeovers; however, a major changeover involving labor costs, material cost, and lost production time is required for the B-W product switch. A capacitated, multi-product lot sizing algorithm can be used to determine the target production run lengths (Hax and Candea, 1984: 147).

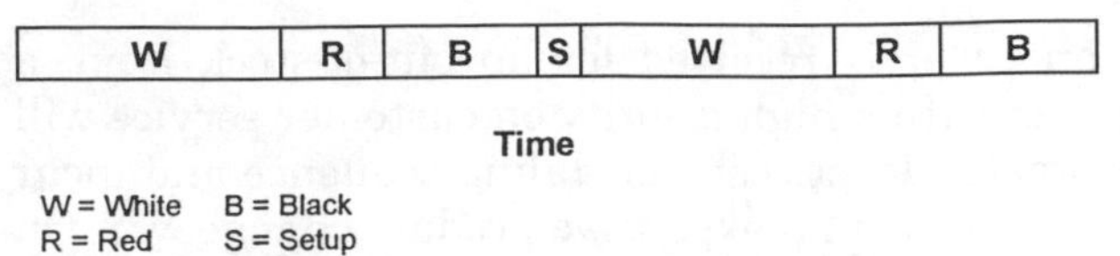

Fig. 3. Mixer Gantt chart.

After establishing target sequences, target run lengths, and obtaining data for the on-hand balances and demands of W, R, and B, we are ready to begin the scheduling process shown in Figure 2. First we propose a trial schedule by finite forward scheduling the mixer in the target sequence and run lengths. This trial schedule may be displayed as a Gantt chart as shown in Figure 3. The inventories are then checked for each product. Figure 4 shows the simple, spreadsheet calculations for projecting W inventories. The values in the production row are obtained from the trial production schedule and the demands come from a combination of demand forecasts and customer orders. The inventory in period t is calculated by adding the inventory in period t – 1 to the production in period t and then subtracting the demand for period t. The resulting inventory can be plotted as shown in Figure 5.

The next step, as shown by the flowchart in Figure 2, is to check the inventories of W, R, and B against their respective target minimum and maximum levels. If one or more of these inventory targets are violated, then the trial mixer schedule should be adjusted and the inventory targets checked again. The impact of these changes on changeover costs, inventory levels, and on-time shipments can be evaluated. Alternative, feasible schedules can be proposed by the scheduler. Generally the goal of scheduling is to obtain a satisfactory compromise that minimizes both schedule violations and finished product inventory violations.

The final step shown in Figure 2 is checking the availability of raw materials for the base material and pigments. If the raw material inventories are adequate the schedule is finalized. If raw material supplies are insufficient the raw material schedules must be revised to increase supplies. Alternatively, the mixer schedule can be revised to consume less of the unavailable materials. This completes the simple processor dominated scheduling example shown

Period	0	1	2	3	4	5	6	7	8	9	10
Production		100	100				100	100			
Demand		40	40	40	40	40	40	40	40	40	40
Inventory	20	80	140	100	60	20	80	140	100	60	20

Fig. 4. Sample white inventory calculations.

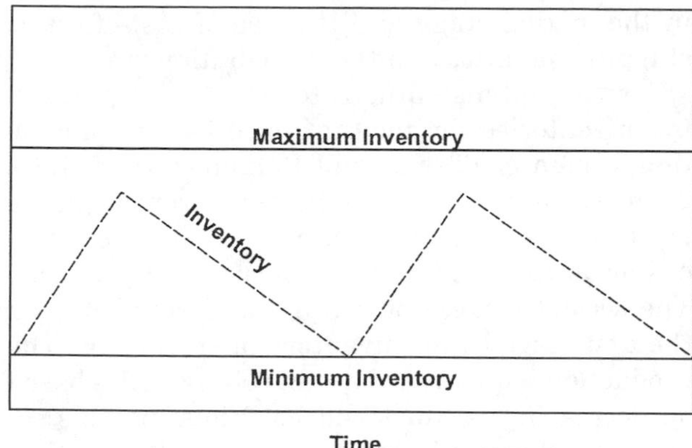

Fig. 5. White inventory projection.

in Figure 2. Let us now examine a simple example for the alternative – material dominated scheduling.

Material Dominated Scheduling

Material dominated scheduling (MDS) first schedules materials and then checks processor capacities. Suppose we extend our previous example and sell finished products from a field warehouse as well as the plant warehouse. Figure 6 presents a flowchart for MDS calculations.

Scheduling begins by proposing a trial material schedule. Figure 7 shows sample calculations for one product at the warehouse. The safety stock of 10 is the minimum, planned inventory target for this item at the warehouse. The lead time of 3 is the time for all activities between the release of an order at the warehouse and the receipt of the replenishment shipment. The lot size of 30 is the target replenishment lot size and is based on a full pallet quantity. The MDS spreadsheet is quite similar to MRP II and DRP records (Vollmann, Berry, and Whybark, 1997). The demand row is obtained from a combination of the demand forecast and customer orders. The inventory for period t is calculated by subtracting the demand in period t from the inventory in period t – 1. When the projected inventory falls below the safety stock level

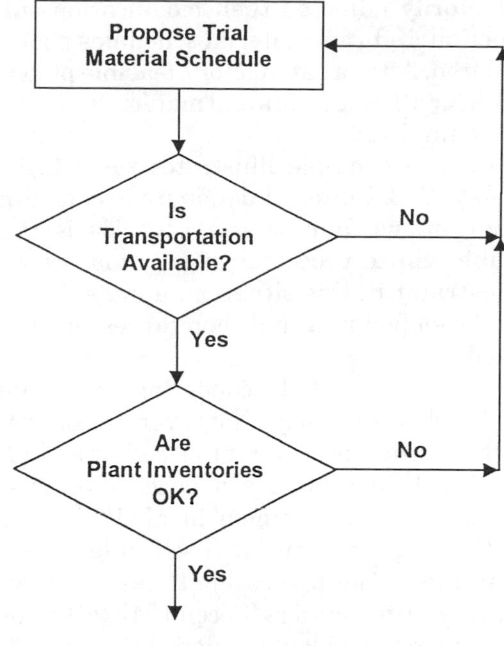

Fig. 6. Material dominated scheduling flowchart.

in period 4, a replenishment receipt is scheduled. A corresponding order release is then scheduled 3 periods earlier in period 1. The receipt in period 4 increases the projected inventory to 35 and prevents inventory from falling below its target minimum (safety stock) level. Subsequent periods are scheduled in a similar manner.

Having scheduled materials, we then check for the availability of transportation as shown in Figure 6. Since common carriers are used, transportation equipment (the processors in this example) is usually available. However in those rare situations when transportation shortages occur, the material schedule can be revised as shown by the feedback loop in Figure 6. Finally when the trial material schedule and the trial transportation schedule have been developed, the plant inventories are checked. If problems exist, then the trial warehouse and transportation schedules must be adjusted. If these schedules cannot be

Safety Stock: 10
Lead Time: 3
Lot Size: 30

Period	0	1	2	3	4	5	6	7	8	9	10
Demand		5	5	5	5	5	5	5	5	5	5
Inventory	25	20	15	10	35	30	25	20	15	10	35
Receipts					30						30
Order Releases		30						30			

Fig. 7. Trial material schedule.

satisfactorily adjusted then modifications in the production and raw material schedules should be considered. After a satisfactory, feasible plan is developed for all processors and materials, the schedules are finalized.

The above example illustrates the scheduling logic for MDS. Material dominated scheduling is appropriate when processor capacity is readily available. Since processor capacity is not a major constraint in this situation, a material schedule is developed first and then processor capacity checked.

Material dominated scheduling is similar to MRP II scheduling logic; however, there are two important differences. First the lead times in MDS schedules do not include a large allowance for queue times as is common in MRP II systems. Thus the lead times in MDS schedules are quite close to the actual processing times and have, at most, small time buffers. Second MRP II scheduling logic first develops a material plan for the entire manufacturing operation and then checks capacity. In contrast, MDS reconciles material and capacity for a stage before scheduling material and capacity for the next stage. Thus MDS uses a local, stage-by-stage reconciliation of material and capacity schedules while MRP II uses a global material plan followed by a global capacity plan.

This section has developed two alternatives for scheduling process units, PDS and MDS. These process units may be grouped into stages that are decoupled from other stages in a process train by inventories. Let us now examine some reasons for these decoupling inventories.

DECOUPLING INVENTORIES: Process trains are formed by a sequence of process stages that are decoupled by inventories. Because of the decoupling inventories, stages can be scheduled somewhat independently from preceding or succeeding stages. Ideally a process train would have a single stage which would tend to minimize work-in-process inventories. While some plants are able to achieve this single stage ideal, many plants require decoupling inventories.

The reasons for multiple stages are varied. Let us examine a few of these reasons and explore some examples. A common reason for multiple stages is the need for different sequences and lot sizes in two successive operations. The example of the previous section had a color mixing stage followed by a distribution stage. The mixing stage requires production in a light-to-dark sequence and in economic lot sizes. In contrast the distribution stage schedules uses full pallet lot sizes. Thus the decoupling inventory facilitates color sequencing in the mixing stage and the use of cost-effective shipping quantities in the distribution stage.

A steel mill also illustrates the use of decoupling inventories for facilitating different production sequences (Taylor and Bolander, 1994: 150). In the steel-making stage the production sequence is based on the amounts of metal alloys from preceding products allowed in succeeding products. The second stage, hot rolling, is decoupled from the first stage by an inventory of steel slabs. The production sequence in the hot rolling mill is based on decreasing product widths. Violating this sequence will result in grooves in the rolled steel or a very costly setup involving a roll change. Thus process technology dictates that a slab inventory be used to decouple the alloy-based sequence of the steel-making stage from the width-based sequence of the hot-rolling stage.

A second reason for decoupling inventories is different production rates in successive stages. A chemical plant provides an example of the need for rate-based decoupling inventories (Taylor and Bolander, 1994: 39). The first stage in this chemical plant continuously separates six intermediate products from a single feedstock. The second stage is a single reactor that processes one intermediate product at a time. Thus an inventory is required between stages to decouple the continuous production of an item in the separation stage from the intermittent consumption in the reactor stage.

A third reason for decoupling inventories is to protect a bottleneck. A plant with an extremely expensive film coater insures high utilization of the coater by maintaining an inventory of film and coating materials (Taylor and Bolander, 1994: 43). This buffer stock also permits production of the coating materials in a different sequence than the film oater uses them.

The fourth reason for multiple stages within a process train is to decouple the production of components from blending and packaging operations. For example, consider a brewery making seven brews that are packaged into several hundred finished products (Taylor and Bolander, 1994: 46). The seven brews require several months to produce and are made to stock. These brews are then finished and packaged to customer orders in a few days. A brew inventory decouples the production of brews from the packaging schedules and customer orders. This is much like an assemble-to-order strategy for assembly operations.

This section has illustrated several reasons for the use of decoupling inventories in the design of process trains. Let us now examine alternatives for creating process train schedules in a multi-stage manufacturing environment.

PROCESS TRAIN SCHEDULING ALTERNATIVES: Process flow scheduling uses the process structure to guide scheduling calculations. Process trains can be scheduled using reverse-flow scheduling, forward-flow scheduling, or mixed-flow scheduling. These process train scheduling alternatives are illustrated in Figure 8. Reverse-flow scheduling begins by scheduling the last stage in the train and progressively schedules upstream process units. The scheduling process continues until all upstream stages are scheduled. Each process stage in the train may be scheduled using processor dominated scheduling (PDS) or material dominated scheduling (MDS). Reverse flow scheduling is fairly common and employs elements of a "demand pull" scheduling philosophy. Taylor and Bolander (1994: 39, 49, 145, 147, 153) give several examples of reverse-flow scheduling.

Forward-flow scheduling begins by scheduling the arrival and processing of raw materials in the first process stage. The scheduling process continues in the same direction as the material flow until the last stage is scheduled. As with all process train scheduling strategies, each process stage in the train may be scheduled with PDS or MDS. Forward-flow scheduling is used in environments where an emphasis must be placed on quickly processing a perishable or extremely expensive raw material. Forward-flow scheduling often employs a "raw material push" scheduling philosophy. Taylor and Bolander (1994: 41) give an example of forward-flow scheduling.

Mixed-flow scheduling has several variations. One important variation is the "inside-out" strategy. The inside-out, mixed-flow scheduling strategy is used in process trains with an internal bottleneck. The bottleneck is scheduled first and then upstream and downstream process stages are scheduled. This scheduling strategy, like the Theory of Constraints scheduling philosophy, emphasizes efficient utilization of the bottleneck. Two examples of the inside-out, mixed-flow scheduling strategy are given by Taylor and Bolander (1994: 43, 148).

Another mixed-flow scheduling variation is the "outside-in" strategy. This strategy schedules from the first stage and last stages to an intermediate stocking point. Forward-flow scheduling is employed upstream of the intermediate stocking point and reverse-flow scheduling is used downstream from the intermediate stocking point. The intermediate stocking point is where the upstream "push" is reconciled with the downstream "pull." The brewery referenced in a previous section illustrates this scheduling strategy. The outside-in strategy is quite similar to an "assemble to-order" philosophy. This strategy provides a wide variety of products in a shorter lead time than a pure reverse-flow scheduling strategy.

BENEFITS: PFS provides a framework that describes the scheduling practices of many firms that have not embraced the MRP II framework. The potential benefits of PFS are like those obtained from any scheduling framework. PFS can reduce confusion and improve communication between vendors, practitioners, consultants, and researchers. The reduction in confusion can shorten software evaluations for practitioners, shorten sales cycles for PFS vendors, expand software availability by creating a demand for more software that employs PFS principles, and provide a foundation for

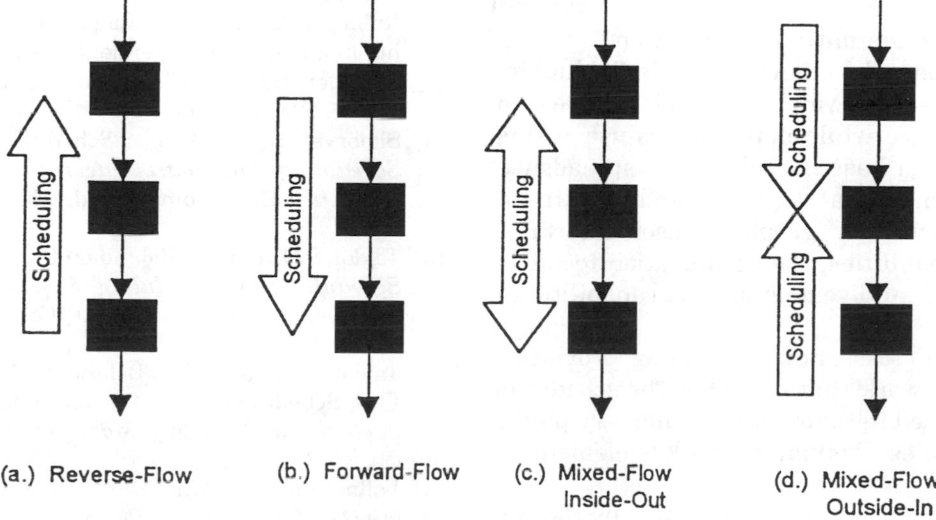

Fig. 8. Process train scheduling alternatives.

scheduling researchers to communicate new technologies.

Specific benefits for a plant implementing PFS are similar to benefits of other scheduling systems. Software firms report that over 800 plants have implemented PFS systems. Several of these firms have quantified the benefits of implementing PFS software and publicly released their information (Taylor and Bolander, 1997). These benefits include higher throughputs, lower operating costs, lower inventories, better customer service and faster schedule development. A European brewer reports a four percent increase in capacity utilization and dramatic reductions in inventory. A food processor reports fewer changeovers, higher equipment utilization, and reduced overtime while achieving an average customer service level of 99.5 percent. Two other food processors report rapid evaluation of alternative schedules, which has resulted in increased throughput or lower inventories. A carpet manufacturer reduced changeovers which, in turn, resulted in higher machine utilization and lower labor requirements.

Caveats: While PFS has proven helpful for many firms, it is not a universal scheduling tool and requires an organizational commitment for successful implementations. PFS is limited to use in flow manufacturing operations. It is not appropriate for job shops. PFS applies only in those plants where operations can be described using process units, stages, and trains. PFS, like MRP II, requires (1) intelligent, trained schedulers, (2) good communication between operating personnel and scheduling personnel, (3) operators who will follow good schedules, (4) support from the information technology personnel, (5) support from senior management, and (6) a large amount of accurate and timely data. A successful PFS implementation requires not only good software but also a capable, committed organization.

PFS is a general framework and individual implementations vary widely. Some plants use commercial software while many others use custom systems – often based on electronic spreadsheet software. Commercial software products are differentiated by their graphical user interfaces, modeling capabilities, the optimization technologies employed, and vendor support capabilities.

SUMMARY: Process flow scheduling provides a flexible framework that describes the scheduling procedures used in many process industry plants. Three principles summarize the key elements of PFS:

1. Scheduling calculations are guided by the process structure.
2. Process units are scheduled using processor dominated scheduling or material dominated scheduling.
3. Process trains are scheduled using reverse-flow scheduling, forward-flow scheduling, or mixed-flow scheduling.

See **Batch size; Distribution requirements planning, make-to-stock production; Management science and operations research in manufacturing; MRP; Optimization in manufacturing; Production scheduling; Resource planning: The evolution from MRP to MRP II and ERP; Safety stock; Scheduling issues; Supply chain management.**

References

[1] Bender, P.S., W.D. Northup, and J.F. Shapiro (1981). "Practical Modeling for Resource Management." *Harvard Business Review*, 59(2), 163–173.

[2] Bradley, S.P., A.C. Hax, and T.L. Magnanti (1977). *Applied Mathematical Programming*. Addison-Wesley, Reading, Massachusetts.

[3] Brown, R.G. (1967). *Decision Rules for Inventory Management*, Holt, Rinehart and Winston, New York.

[4] Camm, J.D., T.E. Chorman, F.A. Dill, J.R. Evans, D.J. Sweeney, and G.W. Wegryn (1997). "Blending OR/MS, Judgment, and GIS: Restructuring P&G's Supply Chain." *Interfaces*, 27(1), 128–142.

[5] Foley, M.J. (1988). "Post-MRP II: What Comes Next." Datamation, December 1, 1988, 24–36.

[6] Hax, A.C. and D. Candea (1984). *Production and Inventory Management*, Prentice Hall, Englewood Cliffs, NJ.

[7] Hoffman, K.L. and M. Padberg (1996). "Traveling Salesman Problem." *Encyclopedia of Operations Research and Management Science*, S.I. Gass and C.M. Harris, ed., Kluwer, Boston.

[8] Rubin, P.A. and G.L. Ragatz (1995). "Scheduling in a Sequence Dependent Setup Environment with Genetic Search." *Computers & Operations Research*, 22 (1), 85–99.

[9] Shobrys, D.E. (1995). "Scheduling." *Planning, Scheduling, and Control Integration in the Process Industries*, C.E. Bodington, ed., McGraw-Hill, New York.

[10] Taylor, S.G. and S.F. Bolander (1994). *Process Flow Scheduling: A Scheduling Systems Framework for Flow Manufacturing*, APICS, Falls Church, VA.

[11] Taylor, S.G. and S.F. Bolander (1997). "Process Flow Scheduling: Past, Present, and Future." *Production and Inventory Management Journal*, 38 (2), 21–25.

[12] Vollmann, T.E., W.L. Berry, and D.C. Whybark (1997). *Manufacturing Planning and Control Systems* (4th ed.), Irwin/McGraw-Hill, New York.

PROCESS INNOVATION

Danny Samson and David Challis
University of Melbourne, Parkville, Australia

INNOVATION: The literature abounds with definitions of innovation and the innovation process. The definition by Souder (1987) that summarizes the basic concept: "Innovation is a high risk idea that is new to the sponsoring organization. The innovation process is any system of organized activities that transforms a technology from idea to commercialization."

The terms innovation, entrepreneurship, invention, discovery, R&D and intrapreneurship are often used (and confused) interchangeably. Whereas innovation refers to new products, processes and services, entrepreneurship involves both the identification and exploitation of opportunities to innovate. Invention and discovery refer to beginnings in the innovation process and we refer to R&D as the formalized process for pursuing innovative ideas and bringing them to fruition. Intrapreneurship usually refers to processes of innovation within organizational boundaries, whereas entrepreneurship can include the creation and growth of new enterprises.

Specifically, innovation can be categorized into three types:

- Product innovation (e.g., new or improved products, new materials)
- Process innovation (e.g., new manufacturing technology, new distribution logistics)
- Managerial and systems innovation (e.g., Total Quality Management, Just In Time)

Discussion in this article is confined substantially to the topic of process innovation. Technological innovation is a key feature of today's manufacturing environment and as a consequence it has drawn significant attention from researchers. In this article we will review what is generally known about the nature of innovation and various models used to describe the innovation process.

THE NATURE OF INNOVATION: It has been long recognized by researchers in this field that the attitudes and values of organizations that existed during the 1950's and 1960's were not conducive to innovation. During this period, firms operated in relatively stable environments and adopted product development strategies characterized by incremental improvements rather than new directions, a slow pace of change, and functional championing of improvement effort. As economic, market and technological conditions become more dynamic, organizations recognized that to maintain competitive advantage, different strategies needed to be employed and, to be implemented effectively, different organizational characteristics were required. Risk taking, a tolerance for ambiguity, an increased customer and market orientation, a high degree of employee motivation and commitment, teamwork, effective horizontal communication channels, vesting decision making authority with entrepreneurs (Betz 1993) have become hallmarks of the firms that prospered in this environment. Innovative products and processes have been the outcome.

Conventional wisdom has consistently argued that innovation is fostered, or at least related to, firm size and industry structure. There are two schools of thought. On the one side are the Schumpeterians, arguing that developing markets and large firms are the driving forces behind innovation (Schumpeter, 1934). Their logic is based on the belief that the more competitive the market structure the smaller are the gains to innovation. Only by having the carrot of large potential profits can firms be induced to innovate. This logic is the basis of almost all existing patent and copyright law.

The opposite view (Robertson, 1971) holds that it is the fight to enter and compete in competitive markets that fosters innovation. The driving force to innovate is to survive and prosper as others encroach on your market, and you strive to keep up or stay ahead. According to this viewpoint it is the stick of annihilation that drives innovation. Today, we know that both mechanisms are valid under certain conditions.

Many authors have attempted to identify those organizational characteristics that differentiate "an innovative firm." Continuous and intensive organizational collaboration and interaction, management of uncertainty, a recognition of the cumulative nature of technological capabilities, and the differentiated nature of technological skills characterize superior innovative performance (Betz, 1993; Robertson, 1971; Steele, 1989).

FOSTERING PROCESS INNOVATION: Saleh and Wang (1993) have developed one of the more comprehensive views of innovation, which identifies three key organizational attributes for effective innovation:

1. Entrepreneurial strategies consist of:
   - Risk Taking: In such a dynamic business environment risk taking becomes part of doing business, as firms need to make strategic decisions with incomplete information.

- Proactive Approach: Successful firms are proactive and anticipate, rather than react to, change. Consequently they regularly scan environments (market, product and technology) and act accordingly.
- Management Commitment: Top management is expected to make a sustained commitment to policies and practices that may fly in the face of conventional wisdom.

2. Organizational structure and functional arrangements are made of:
   - Flexible Structures: The relationship between innovation and organic structural forms has long been recognized (Burns and Stalker, 1961).
   - Synthesis: Innovative firms are able to effectively integrate activities across organizational boundaries. Kanter (1985) has also indicated that there is significant evidence that many of the best ideas are interdisciplinary and inter-functional in origin.
   - Collective Orientation: Synthesizing and integrating activities between departments, groups and individuals is underpinned by a common collective understanding of the goals of the business. A number of authors believe that managing innovation requires an ability to manage interdependence and that this requires common organizational norms and beliefs, and not just reward and recognition systems.
3. Organizational climate, which involves:
   - Open Climate: Openness in exchanging information not only facilitates effective innovation but also trust and respect between employees.
   - Collegiality: In a collegial climate, authority and power are shared equally amongst colleagues and this characteristic has been found to be positively correlated with innovativeness in organizations (Souder, 1987).
   - Reward Systems: Well-planned reward systems have been found to be effective tools to reinforce expected behaviors and to shape the development of the desired climate (Baulkin 1988). Kanter (1985) emphasizes the importance of what she calls a "culture of pride" that expects and rewards high levels of achievement and assumes that investments in people pay off.

Ettlie (1990) studied the innovative practices of 50 manufacturing plants in the US. The focus of this work was to discover if the functional experience of general managers had any relationship to the genesis and success of manufacturing innovation. Notable findings included:

- firms that have CEO's with manufacturing experience are significantly more likely to implement an aggressive manufacturing technology policy.
- the company's commitment to training and workforce development is higher during modernization when senior executives have manufacturing experience.
- direct labor savings from modernization (as distinct from improvement in operating characteristics, efficiencies etc.) is more likely to be emphasized by traditional senior managers.

The study also found that over half the plants visited had implemented a mechanism such as a technology agreement with a union, or team based work structures in order that technological and/or organizational innovation would promote competitiveness. This finding further supported the linkage between an effective innovative capability and modern manufacturing practices.

Ettlie (1990) studied the mechanisms for the successful adoption of new process technology. He concluded that since firms buy most technologies for manufacturing operations it is difficult for them to use these technologies for competitive advantage because they are readily available to competitors. He suggests however that firms gain advantage from making process innovation a unique occasion for organizational restructuring and for adopting new patterns of behavior. These findings support the idea that redesigned organizational structures, improved coordination between design and manufacturing, and greater supplier cooperation positively affect the productivity of new manufacturing systems. Ettlie further posits that forming new customer alliances positively affects new system flexibility.

New process technology may simply be incremental, like replacing an old car with a new car capable of doing the same things faster or more efficiently. Alternatively, it may be more radical, like replacing a car with a helicopter (Hayes and Jaikumar, 1998). This latter change is not just one of 'degree' but of 'kind,' requiring new thinking, bringing higher potential rewards but also accompanied by higher risk of failure.

Other than for minor process change, it is clearly advisable to plan and execute a series of accompanying changes when a major process innovation is implemented. Human resource skills needs may be different, working arrangements may change, and the size of the direct and indirect workforces may change as a result of a process innovation. Some illustrative examples are:

- An automaker who brought in a robotic welding line then found that it needed far less direct labor, but more 'high-tech' support staff.
- A textile manufacturer who invested $65 million in state-of-the-art spinning and weaving

equipment without spending any money on upgrading its workforce skills. Without a change in skills and culture, the new equipment took far too long to commission and did not deliver the anticipated improvements in productivity or quality. The company essentially wasted two years making the accompanying changes before benefits accrued.

- a tire manufacturer who brought in leading edge tire building machines with the involvement of manufacturing shop floor staff form the inception of the process. This company carefully selected operators to run the new equipment based on skills and aptitude, openly communicated its business need and purpose for the new machines, and involved shop floor staff in design and selection. Implementation was smooth and successful. The ramp up of benefits in this instance went exactly according to schedule, even though the new process was a radical departure from the old process, and much less direct labor was needed. This was seen to be the result of a meticulous plan that brought together business strategy, technology strategy and human resource strategy elements together in an open environment.

CHARACTERISTICS OF INNOVATIVE FIRMS: Notwithstanding the fact that success in process innovation does not appear to have a general formula or prescription, the following are some common factors that correlate with success:

- Top managements' technological and business backgrounds, and hands-on involvement in technological issues.
- Technological capabilities focused toward supporting business strengths.
- Organizational culture, infrastructure, coordination mechanisms, organizational structures, reward systems etc., that are supportive of the innovation process T.
- Tolerance for failure and an understanding that innovation is accompanied with risk taking, ambiguity and a degree of "organized chaos."
- A customer and market orientation.
- A high degree of employee motivation and commitment.
- Use of flexible team-based work structures.
- Effective horizontal communication mechanisms.
- Decision making authority vested with entrepreneurs or project champions.
- Emphasis on technical skills development.
- Continuous and intensive organizational collaboration and cooperation.
- A recognition of the cumulative nature of technological capabilities.
- Relating technological skills to strategic intent.
- Using metrics that reflect process innovation.
- The senior R & D executive reporting directly to the CEO.
- Reward system that recognize entrepreneurial behavior, change and improvement.
- Establishing proprietary positions in respect to technological achievements.
- Leverage existing technological strengths.
- Clearly relating process technology strategy to product strategy.
- Institutionalized environmental scanning.
- Senior management commitment.
- Technological competency of executive management.
- Openness to sharing information.
- Egalitarian working environment.
- Manufacturing excellence.
- Intense communication about the new process.

See **Capital investment in advanced manufacturing technology; Flexible automation: Robot justification; Manufacturing technology use and benefits.**

References

[1] Baulkin (1988). "Reward Policies That Support Entrepreneurship." *Compensation Benefits Review* Vol. 20.

[2] Betz F. (1993). "Strategic Technology Management." New York, Mc Graw-Hill.

[3] Burns T. and Stalker G. (1960). "The Management Of Innovation." London, Tavistock.

[4] Ettlie J. (1990). "What makes a Manufacturing Firm Innovative?." Academy Of Management Executive, Vol. 4, No 4.

[5] Hayes, R.H. and J. Jaikumar (1988). Manufacturing's crisis: new technologies, obsolete organizations, Harvard Business Review, September, pp 77–85.

[6] Kanter R. (1985). "The Change Masters: Innovation For Productivity In The American Corporation." New York, Simon and Schuster.

[7] Robertson T. (1971). "Innovative Behavior and Communication." New York, Holt, Rinehart and Winston.

[8] Saleh S. and Wang C. (1993)."The Management Of Innovation: Strategy Structure and Organizational Climate." *IEEE Transactions On Engineering Management*, V40, No 1.

[9] Schumpeter J. (1934.) "The Theory Of Economic Development" Cambridge MA, Harvard University Press.

[10] Souder W. (1987). *Managing New Product Innovations*, Lexington MA, Lexington Books.

[11] Steele L. (1989). "Managing Technology: The Strategic View", New York, McGraw-Hill.

ABC Company Routing Sheet (sample, incomplete)			
Name of Part:	Turned rod		
Part Number	B324		
Quantity to be produced:	102		
Material Used:	SAE 1040		
Operations Number	Description of Operation	Equipment	Tooling
1	Turn ¾"	Turret Lathe	#24
2	Cut off 1 ½"	Turret Lathe	#2
3	Degrease	Degreaser	
4.	…	…	…

Fig. 1. Routing sheet.

PROCESS MAPPING

Process napping captures tasks and work and information flows using symbols; also called flow charting.

See **Business process reengineering; Reengineering and the process view of manufacturing.**

PROCESS PLANNING

Process planning is a preparatory step before manufacturing, which determines the sequence of operations or processes needed to produce a part or an assembly. This step is more important in job shops, where one-of-a-kind products are made or the same product is made infrequently.

The outputs of this step are: route sheet and operations sheet. A route sheet is a document which lists the exact sequence of operations needed to complete the job. The route sheet is intended to accompany parts moving individually or in batches. Route sheets provide the information to the material handlers to help them move materials or partly completed items from one work center to another, until the finished part or assembly reaches the shipping department. Route sheets are useful for Production planning. An example of a route sheet is in Figure 1.

An operations sheet is a document that lists all details of the operations needed to complete the part or assembly. It includes information on the machine to be used, the speed of cut, depth of cut, etc., when the process involves metal cutting. In the case of heat treatment operations, it will specify the furnace, the temperature, heating or cooling rate, and the cooling medium (air or oil). An example of an operation sheet is in Figure 2.

ABC Company Operation Sheet (sample, incomplete)						
Part Name:		Shaft				
Part Number:		C343				
Operation Number	Name of Operation	Machine	Cutting Tool	Cutting Speed	Feed	Remarks
1	Face end	Lathe		110 ft/min	400 rpm	Universal chuck, hand feed.
2…	…	….	…	….	….	….

Fig. 2. Operation sheet.

The purpose of operations sheet is to take the guesswork out of the process. The operation sheet tells the operator precisely what needs to be done. Without delay, an operator can proceed with the specified operation. However, an operator may fine-tune the recommended operations to suit changed conditions on the shop floor.

Today, process planning is increasingly done on the computer, it is termed Computer Aided Process Planning (CAPP). However, the challenge of computerizing process planning for all parts made in a job shop is very significant. Therefore, a significant portion of process planning in job shops may continue to be a manual operation.

See **Capacity planning; Capacity planning in made-to-order production systems; Capacity planning: Medium- and short-range; MRP.**

PROCESS RATIONALIZATION

The redesign of a process in order to simplify is called process rationalization.

See **Learning curve analysis; Reengineering and the rpocess view of manufacturing.**

PROCESS TRAIN SCHEDULES

In the process industry making chemicals and commodities, a process train is a sequence of process stages that produce an intermediate or finished product. The process train schedule defines the method by which this sequence of process stages is scheduled. Three methods are used to schedule a process train: reverse-flow scheduling, forward-flow scheduling, and mixed-flow scheduling.

Reverse-flow scheduling builds the process train schedule by starting with the last process stage and proceeds backward (countercurrent to the process flow) until all process stages are scheduled. Forward-flow scheduling builds the process train schedule by starting with the first process stage and proceeding sequentially through the process structure until all process stages are scheduled. Mixed-flow scheduling builds the process train schedule by using combinations of both forward and reverse-flow scheduling approaches. For example, a constraining process in the middle of the process train maybe scheduled first, then the process train schedule is completed by forward scheduling down-stream stages, and reverse flow scheduling up-stream stages.

See **Process industry scheduling.**

Reference

[1] Taylor, S.G. and S.F. Bolander (1994). *Process Flow Scheduling: A Scheduling Systems Framework for Flow Manufacturing*, APICS, Falls Church, VA.

PROCESSOR DOMINATED SCHEDULING

It refers to any scheduling technique that schedules processors (equipment/capacity) before materials in process industry, which produces chemicals and commofities. Processor dominated scheduling (PDS) initially schedules processors in low cost sequences or in economic run lengths. This results in the efficient use of processor capacity and in some cases improves quality. Material availability is checked after the processor is scheduled to ensure schedule feasibility. PDS is typically used as a scheduling concept when the efficient use of the processing equipment is critical to the financial success of the company. PDS is typically used more in make-to-stock environments, where finished goods inventory buffer the operation of process equipment from customer orders.

See **Process industry scheduling.**

References

[1] Taylor, S.G. and S.F. Bolander (1994). *Process Flow Scheduling: A Scheduling Systems Framework for Flow Manufacturing*, APICS, Falls Church, VA.

[2] Bolander S.F. and Taylor, S.G. (1987). "Processor Dominated Scheduling", *30th Annual International APICS Conference Proceedings*, APICS, Falls Church, VA.

PROCESS-ORIENTED FACTORY

A process-oriented factory has a process-oriented layout. The best examples of this approach to layout can be found in the processing industry. In most of these plants there is a single, unambiguous sequence of processes that a product goes through. The equipment used in each of these processes is then positioned so that it reflects the ordering and placement of the process for which it is used. For example, the production of beer from water, hops and other ingredients follows a clear and well understood sequence. This ordering then imposes a rigid layout on the transformation processes and the movement and storage of materials between them. In manufacturing generally, it is beneficial to use such linear arrangements where they are appropriate. Significant economies may be achieved through increased productivity and reduced material movement and work in process stocks.

See **History of operations management; Process industry scheduling; Scientific management.**

PRODUCT ARCHITECTURE

Product architecture is the scheme by which the desired functions of a product are assigned to physical components (Ulrich, 1995). Product Architecture consists of three elements: functions, components and their mapping to the functions, and interactions among components.

The *functional elements of a product* are the operations and transformations that together represent the performance of the product. Consider a standard VHS video cassette as an example: its *primary functions* are, to hold picture and sound information and to reproduce this information. At a more detailed level, the primary function of the cassette can be translated to moving the storage medium (magnetic tape) well-aligned along the magnetic head of the video player. *Supporting functions* are to hold the tape tight (so it does not tangle), and to protect the tape from dirt and mechanical stress.

The *physical components* of the cassette are magnetic tape, hubs, various pins around which the tape moves, a hub lock mechanism, and a casing with a door. The components can be mapped to the cassette's function: the tape stores information. The hubs with their inner holes that snugly fit the moving axles of the video player, together with the guiding pins, ensure movement of the tape along the magnetic head. The hub lock mechanism, itself a small subassembly of several parts, ensures that the tape can only move one way and thus stays tight. The casing provides mechanical protection. We see that some components map one-to-one to functions (tape to storage, hub lock mechanism to prevent tangling), and some components map many-to-one function (hubs and pins to tape transport).

The architecture gives an explicit description of the product's structure and its functions. The architecture has an important impact not only on the product's performance, but also on the concurrency of the design process used, on parts commonality and thus cost, and on the subsequent evolution of the design (Ulrich, 1995). A *modular* architecture is characterized by (groups of) components interacting only through well-defined interfaces, which allows them to develop and to change independently. In an integrated architecture, the product's components are coupled and interact. This makes the product more complex, but may have cost, space and/or performance advantages.

See **Concurrent engineering; Mass customization; Mass customization implementation; Product development and concurrent engineering.**

Reference

[1] Ulrich, K. (1995). "The Role of Product Architecture in the Manufacturing Firm." *Research Policy* 24, 419–440.

PRODUCT CHAMPION

Product Champion is an executive or manager who provides for a new product or a product development team at the upper levels of the organization. Such an individual acts as a "sponsor" for the team, and may help the team overcome barriers such as a lack of resources, lack of participation, or lack of assistance from other functional areas. Littler *et al.* (1995) examined the key success factors for collaborative new product development efforts in 106 UK firms in which the collaborative partner could be a supplier, customer, or competitor. They concluded that frequent inter-company communication, trust, partnership equity, and a product or collaboration champion increased the likelihood of success of the new product development.

See **Concurrent engineering; New product development through supplier integration; Product development and concurrent engineering; Supply chain management.**

Reference

[1] Littler, D., Leverick, F., and Bruce, M. (1995). Factors affecting the process of collaborative product development: a study of UK manufacturers of information and communications technology products. *The Journal of Product Innovation Management* 12(1): 16–23.

PRODUCT CONCEPT

Product concept is a concise description of how the product will satisfy customer needs. It is a proposed solution for matching customers' needs with a product for satisfying those needs. It describes "the form, function, and features of a product and is usually accompanied by a set of specifications, and an analysis of competitive products, and an economic justification of the project" (Ulrich and Eppinger, 1995).

See **Concurrent engineering; Mass customization; Product design for global markets; Product development and concurrent engineering.**

PRODUCT CONCURRENCY

Product concurrency refers to the simultaneous execution of design and development activities for two separate but related new products. For example, product concurrency exists when the development of a next generation product is begun before the development of the first generation product is complete. Product concurrency of this type is typically needed when the product life cycle is short relative to the time it takes to develop and market the new product. Product concurrency also applies to the simultaneous development of multiple product variants. Different product variants may be designed to meet the distinct needs or preferences of different market segments. While separate design teams may be assigned to develop each product line, it is usually important to create some means of integrating the separate efforts. In this way each team may benefit from other teams' experience and learning.

See **Concurrent engineering; Mass customization; Mass customization implementation; New product development through supplier integration; Product development and concurrent engineering.**

PRODUCT CUMULATIVE LEAD TIME

It is the total time required to make a product, from raw material acquisition through final assembly.

See **Capacity planning in make-to-order production; Order release; Scheduling stability; Theory of constraints in manufacturing management.**

PRODUCT DESIGN

Debasish N. Mallick

Boston College, Chestnut Hill, MA, USA

INTRODUCTION: Product design is one of the most important corporate activities that can be managed strategically for gaining competitive advantage in the marketplace. However, unlike other mainstream business functions such as manufacturing, marketing, and finance, management of product design, as it is practiced today, is more of an art than a science. The area lacks systematic and logical knowledge that can be generalized, and rarely do the executives responsible for managing the business understand it. One of the major difficulties faced by business executives in understanding product design is caused by the absence of a clear and concise definition of what "product design" is. The lack of a precise definition of "product design" often leads to dilution of focus and increased confusion in managing product design (Mallick and Kouvelis, 1992).

While the importance of design in gaining competitive advantage is being recognized slowly by academia and the business world, there is no general agreement on the meaning of the word "design." The word "design" is often used in a too narrow or a too broad context. As the design historian Jeffrey Meikle (1989) points out, at the one extreme it is "identified with the most superficial elements of style" such as "designer hairstyle," "designer jeans" etc., and at the opposite extreme it is related, "at least by implication, to all scientific and technological advancements." However, neither of these approaches can be used to define design in a way that is beneficial to both business and the general public.

Successful firms must design products to satisfy the needs and wants of their customers, communicate how these products will meet or exceed their expectation, and be able to manufacture and deliver these products in a profitable manner. Thus, the word design is embodied in products, communication, processes, and facilities used by a firm for carrying out their business. Products are "anything that can be offered to someone to satisfy a need or want."

In the context of product design, the word design has different meanings and interpretations. Engineers believe that product design and development is one of the major functions of their profession. According to the Industrial Design Society of America (IDSA): "Industrial Design is the professional service of creating and developing concepts and specifications that optimize the function, value and appearance of products and systems for the mutual benefit of both users and manufacturer." At the same time, US design patent laws refer only to the appearance of a product. These various views illustrate the ambiguity in the definition of product design. A precise definition is essential for not only understanding a product design problem but also for developing strategies for managing it.

According to the dictionary (Simpson and Weiner, 1989) the word "design" can either be used as a noun or as a verb. It can be used as a noun to refer to a "state" such as a preliminary plan of a product or an established version of a product. For example, we can say that "this is the design of a radio I wish to make" or we can point to an existing radio and referred to it as "a well designed product." It can also be used as a verb to describe the "process" of developing a product. For example, we may say that "we are designing a boat" to describe the process of developing a new boat. However, we would like to manage the product design (e.g., process) to arrive at a product design (e.g., state). Thus, from a managerial perspective we are interested in both. The importance of the distinction between design as a state and design as a process for understanding design was pointed out by Jay Doblin (1987). In this paper we extend his ideas to define product design in a way that is useful to business executives.

PRODUCT DESIGN AS A STATE: The state of a product design is defined by three elements: function, form, and fit. *Function* refers to the primary purpose of a product. For example, the primary purpose of a pencil is to write on paper. The primary purpose of a dishwasher is to clean dirty dishes. *Form* refers to the appearance of a product. It is defined by the shape, size, color, texture, and other visual attributes of a product. It is often influenced by the parts and components used in a product to achieve functionality. But it can also be manipulated to improve the attractiveness of a product. For example, the appearance of an automobile has often been a key factor in a customer's buying decision. Finally, the user must interact with the product to use it. The ease with which a user can interact with a product is defined by the product-user interface and it is referred to as *fit*. For example, the thickness of a steering wheel of an automobile affects the ease with which it can be steered.

Thus, a product must perform its intended *functions*, conform to an appropriate but aesthetically pleasing *form*, and achieve a *fit* to ensure its ease of use. Take, for example, the telephone instrument. The primary function of a telephone is to send and receive encoded voice messages across geographically dispersed locations. In order to perform its primary function, that is to send and receive encoded voice messages across distance, the telephone needs a transducer that transforms sound into electronic signals and another transducer that transforms electronic signals into audible sounds. The early telephone designs were influenced by their functions. It was made of two separate pieces: a mouthpiece to transmit message and an earpiece to receive the message. The user had to use both hands to use these telephones. The ease of use or fit improved significantly when a linkage was added between the earpiece and the mouthpiece for making one-handed operation possible. While the basic form of the telephone is influenced by its function and fit, they are produced in a variety of shapes, sizes and colors to appeal to different tastes.

Good design calls for a careful balancing of these three elements of design, i.e., function, form, and fit. Often it is helpful to classify a product based on its dominant design element to gain an understanding of such a balance. Products can be classified as performance products, appearance products, or convenience products. Performance products are dominated by their functions. For example, products such as paper clips, mechanical fasteners etc. are dominated by their function and can be classified as performance products.

Appearance products are dominated by their forms. For example, ornaments and decorative items are dominated by their form and are classified as appearance products. Convenience products are dominated by their ease-of-use or fit. For example, Post-it notes, disposable diapers, paper napkins etc. are used primarily because of their ease-of-use and can be classified as convenience products. These are some examples of products which clearly fall in one category or another. However, the majority of products that we encounter in our man-made world would probably occupy a three dimensional space defined by these three elements of design, i.e., function, form, and fit.

Most products are not purely performance products or purely appearance products or purely convenience products. Yet, in such cases, it is possible to identify if the design is dominated by function, by form, or by fit. Household appliances such as dishwashers, refrigerators etc. are dominated by their functional requirements, i.e., a dishwasher must clean dirty dishes and a refrigerator must keep food at a safe temperature. Even though a dishwasher or a refrigerator that blends well with the rest of the kitchen is preferable, we rarely buy appliances for their appearance alone to fill our already crowded kitchen space.

Decorative and style products such as flower vases, sunglasses etc. are dominated by their form: shape, color, texture etc. We buy them because of their appearance. We rarely keep flowers in empty plastic bottles for decorating our living room even though plastic bottles are more functional than a vase made of delicate glass or porcelain. At the same time, products such as telephone handsets and computer keyboards are dominated by the fit or human factor consideration. In a complex product, often different parts and components are dominated by different elements of design. For example, in an automobile, the engine is dominated by the function or performance requirements, the exterior body work is dominated by the form or appearance considerations, and the interior, especially the driver's seat and the controls, are dominated by the fit or convenience of the driver. Similarly, for a personal computer, the internal electronics are dominated by the function, the housing is dominated by the form, and the keyboard which allows a user to interact with the machine is dominated by fit.

PRODUCT DESIGN AS A PROCESS: The process view of product design recognizes the activities and actions necessary to achieve the balance between function, form, and fit. Three elements of a design process are engineering design, industrial design, and human factor design. These three elements of a design process are needed to achieve the function, form, and fit needed to define a design state. Product design as a process begins with a product concept: existing or new. "Product concept is a concise description of how the product will satisfy the customer needs (Ulrich and Eppinger, 1995)."

Engineering design is the process that validates the technical feasibility of a product concept and determines how to provide the required "functions" for a product. In designing a product, the engineers select the method, specify materials, and determine shapes to satisfy technical requirements and to meet performance specifications. *Industrial design* refers to the process of providing "forms" to these functions. Sometimes industrial design may involve simple appearance design or styling. Yet industrial design often influences the functionality of a product, such as arranging a component set in a personal computer to achieve a reduced foot print, and it should not be confused with the cheap frills that are often present in a product (Lorenz, 1985). However, "the conception of visual form, has become progressively separated from the act of making" a product and it is this process that is least subject to rational analysis and often most advantageous in commercial competition (Heskett, 1980).

Human factor design, also known as ergonomics, is the process of ensuring the "fit" between a product and its user. It involves the application of scientific analysis of human performance and behavior with respect to the environment influenced by new products and technology. It is specially important for complex products requiring human interventions for performance and control (Bridger, 1995). A good example is the simple four row by three column layout of push-buttons on a modern telephone. It was developed after extensive ergonomic study to ensure quick and error free dialing. Similarly, the design and placement of controls and gauges in the cockpit of an aircraft are specified by the human factor designers for easier operation.

MANAGERIAL IMPLICATION: The recognition of this distinction between the state and the process is helpful for identifying the *what* and *how* of a product design problem. Three elements (i.e., function, form, and fit) are used for defining the state of a product design. Recognition of these three elements in a product design problem will not only provide executives with a structure for understanding a design problem better but will also help them in identifying processes appropriate for addressing a design problem (see Figure 1). This is because there is a one-to-one relationship between the three elements of the product design state and the three elements of the product design process (i.e., engineering design, industrial design, and human factor design).

It is possible to identify many products with a good balance of function, form, and fit, which were designed entirely by engineers or by industrial designers. However, executives must recognize that the three elements of a product design process call for distinctly different types of expertise in terms of knowledge, skills, and abilities. A careful comparison of university programs and curriculum will reveal that the education and training of engineers, industrial designers, and ergonomists are significantly different from each other (Marshal and Lawrence, 1991). An understanding of the potentials and limitations of the technology used for achieving the functionality of a product is essential for engineering design. Thus, individuals trained in electrical engineering are necessary for designing the functionality of a computer. Industrial designers are trained for

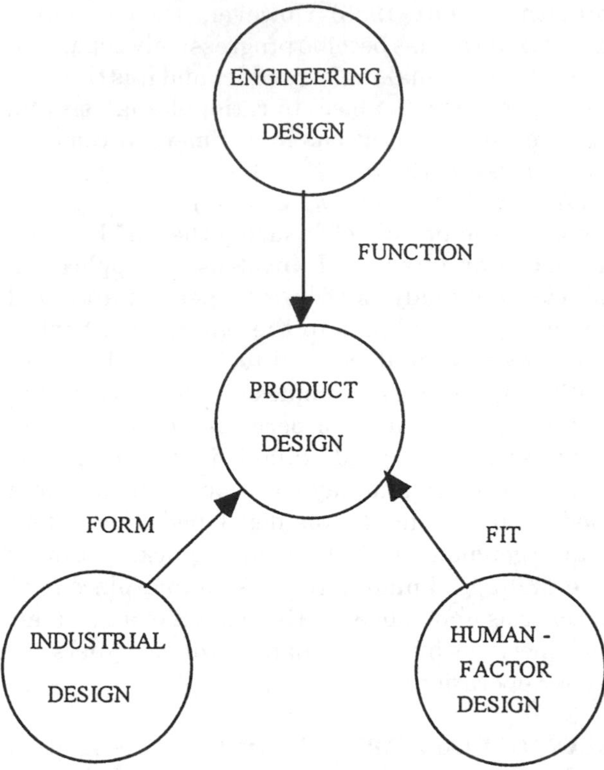

Fig. 1. Elements of product design.

addressing the unique qualities of external form. Thus, they are better prepared for designing the look and feel of a computer, which not only affects the aesthetic appeal but also the functionality and interface with the user. Ergonomists or the human factor specialists have the specialized knowledge about human anatomy, physiology, and psychology which are essential for ensuring safety and suitability of a product. Thus, the design of the keyboard, monitor and other parts of a personal computer requires ergonomists or human factor specialists. Recognition of the dominant element in a design problem, therefore, is essential for addressing a design problem well.

Although the proposed framework emphasizes the distinction between function, form, and fit to enhance understanding, executives must recognize that a good design is defined by a balance of function, form, and fit (Caplan, 1985). Product design is the entire process involving engineering design, industrial design, and human factor design to arrive at a balanced state of function, form and fit as specified by the product concept. Thus, significant interaction among engineering design, industrial design, and human-factor design functions is essential for a successful product design.

See **Engineering design; Concurrent engineering; Human factor design; Industrial design; Product concept; Product design for fit; Product design for form; Product design for function; Product design for global markets; Product development and concurrent engineering: Product innovations.**

References

[1] Bridger, R.S. (1995). Introduction to Ergonomics, McGraw-Hill, NY, p. 1–30.

[2] Caplan, R. (1985). "Good Design Is," Across the Board, 22, May 1985, p. 23–26.

[3] Davenport, T.H. (1993). *Process Innovation: Reengineering Work through Information Technology*, Harvard Business School Press, Boston, Massachusetts, p. 5–9.

[4] Doblin, J. (1987). "A Short Grandiose Theory of Design," *STA Design Journal.*

[5] Fitzsimmons, James A., P. Kouvelis, and D.N. Mallick (1991). "Design Strategy and Its Interface with Manufacturing and Marketing: A Conceptual Framework", *Journal of Operations Management*, Vol. 10, No. 3, August 1991, p. 398–415.

[6] Heskett, J. (1980). Industrial Design, Thames and Hudson, London, p. 7.

[7] Kotler, P. (1984). Marketing Management: Analysis, Planning, and Control, (Fifth edition), Prentice-Hall, New Jersey.

[8] Lorenz, C. (1986). The Design Dimension: The New Competitive Weapon for Business, Basil Blackwell, NY.

[9] Mallick, D.N. and P. Kouvelis (1992). "Management of Product Design: A Strategic Approach," In Intelligent Design and Manufacturing, ed. A. Kusiak, Wiley, New York.

[10] Marshal, J. and P. Lawrence (1991). "Design as A Strategic Resource: A Business Perspective," *Proceedings of the Design Leadership Symposium*, Corporate Design Foundation, Boston, MA.

[11] Meikle, J. (1989). *Design in the Contemporary World: A paper prepared from the proceedings of the Stanford Design Forum 1988*. ed. Stanford Design Forum, Pentagram Design, AG, p. 15.

[12] Olins, W. (1989). *Corporate Identity: Making Business Strategy Visible through Design*, Harvard Business School Press, Boston, Massachusetts, 1989.

[13] Simpson, J.A. and E.S.C. Weiner (1989). *The Oxford English Dictionary*, Second Edition, Clarendon Press, Oxford, UK. p. 519.

[14] Tompkins, J.A., J.A. White, Y.A. Bozer, E.H. Frazelle, J.M.A. Tanchoco, and J. Trevino (1996). Facilities Planning, Wiley, NY, p. 1–17.

[15] Ulrich, K.T., and S.D. Eppinger (1995). Product Design and Development, McGraw-Hill, NY, p. 77–127.

[16] Wasserman, A.S. (1989). "Redesigning Xerox: A Design Strategy Based on Operability" in *Ergonomics: Harness the Power of Human Factors in Your Business*, ed Klemmer, E.T., Norwood, N.J.

PRODUCT DESIGN FOR FIT

The user interacts with a product to use it. The ease with which a user can interact with a product is defined by the product-user interface and it is referred to as *fit*. For example, the early telephone designs had two separate units: a mouth piece to transmit message and an ear piece to receive the message. The user had to use both hands to use these telephones. The ease of use or fit improved significantly when the ear piece and the mouth piece were joined together in design making one hand free operation possible.

See **Concurrent engineering; Mass customization; Product design; Product design for form; Product design for function; Product design for global markets; Mass customization.**

PRODUCT DESIGN FOR FORM

Appearance of a product is defined by the shape, size, color, texture, and other visual attributes of a product. It is often influenced by the parts and components used in a product to achieve functionality. It can also be manipulated to improve the attractiveness of a product. For example, while the basic form of the telephone is influenced by its function and fit, phones are produced in a variety of shapes, sizes and colors to appeal to different tastes.

See **Concurrent engineering; Mass customization; Product design; Product design for fit; Product design for function; Product design for global markets.**

PRODUCT DESIGN FOR FUNCTION

The term function in product design refers to the primary purpose of a product. For example, the primary function of a telephone is to send and receive voice messages across geographically dispersed locations. In order to perform its primary function, that is to send and receive voice messages, two transducers must be designed; one to transform sound into electronic signals and another to transform electronic signals, into audible sounds. The early telephone designs were dominated by these functional requirements. It was made of two separate pieces: a mouth piece to transmit message and an ear piece to receive the message. Later designs improved the form and fit of the design without compromising the function.

See **Concurrent engineering; Mass customization; Product design; Product design for fit; Product design for form; Product design for global markets.**

PRODUCT DESIGN FOR GLOBAL MARKETS

K. Ravi Kumar

University of Southern California, Los Angeles CA, USA

George C. Hadjinicola

University of Cyprus, Nicosia, Cyprus

INTRODUCTION: Contemporary business firms are facing increased globalization of markets. Companies that have traditionally dominated their domestic markets, recognize that they cannot survive unless they establish their presence in all major markets. Various factors, termed globalization drivers, are forcing companies to compete in the global market place. Market factors include the consumers with similar income levels, spending habits, educational background, and life-style. Cost factors, stemming from the rising cost of developing a new product, force firms to enter major markets so as to achieve production and distribution economies of scale. Environmental factors contributing to globalization include the removal of nationalistic barriers and protectionism, the establishment of the General Agreement of Tariffs and Trade (GATT) and World Trade Organization (WTO) that promote free trade, the formation of economic communities such as the European Union and the North American Free Trade Agreement. Competitive factors make the need for a global market an imperative since firms need to disseminate their new product ideas worldwide before competitors mimick their in other markets.

Two of the strategic decisions that International Enterprises (IEs) operating in the global market have to make are the type of product policy to adopt, and the management of the design process for the global marketplace. Product policy encompasses a number of issues of managerial concern such as physical product design, packaging, labeling, brand, warranty, and after sales service. Research on product policies in a global environment originated in the early sixties, when the issue was raised as to whether IEs should provide customers around the world the same standardized product or a country-tailored product. The management of the process for designing new products in a global context has certain idiosyncrasies.

PRODUCT POLICIES IN A GLOBAL ENVIRONMENT: Levitt (1983) noted that advances in technology and more specifically, improvements in communication, transport, and travel, are making consumers around the world to be aware of, and demand the same set of products. Facing this phenomenon termed "globalization of markets," Levitt states that "companies must learn to operate as if the world were one large market – ignoring superficial regional and national differences," and as a result, IEs should "sell the same things in the same way everywhere." Offering a standard product design worldwide embraces the belief of the existence of homogeneous markets across countries. This belief has been challenged by the traditional paradigm that consumers across countries have different needs and therefore, demand different sets of products (Boddewyn *et al.*, 1986; Douglas and Wind, 1987). The two viewpoints of offering a standardized product versus offering a custom-tailored product seem to hold their own. Both sides can report examples of companies who have succeeded or failed when they adopted either viewpoint. Such examples can be found in Levitt (1983) and Kashani (1989). Walters (1986) provides an extensive discussion on the product customization versus standardization debate.

In general, the issue of product policies across countries has become a debate of polar opposites, namely complete uniformity versus full localization of the product policy. Some researchers, though, have pointed out that the advocacy of extreme points of view should not be the focus of the debate. The effort should be directed in finding ways to adapt the global marketing concept to suit the company's needs and provide it with a sustainable competitive advantage (Takeuchi and Porter, 1986; Quelch and Hoff, 1986). Product uniformity is simply one strategy in an array of strategic choices an IE may pursue. The crucial point is to find the product policy that suits the needs of the organization in the environment in which it operates.

Which types of product policy is appropriate? Takeuchi and Porter (1986) identified three basic types of international product policies: *the universal, the country-tailored, and the modified.* Two of the above three product policies, the universal and the country-tailored product policies, represent the polar positions in the product standardization versus customization debate. The third policy, modified product, deals with products that are relatively similar in overseas markets, but minor adaptations are made to these products to meet basic market needs or customer idiosyncracies. We provide below a more detailed discussion of the above three international product policies.

Universal Product Policy

The first international product policy identified by Takeuchi and Porter (1986) is the universal product policy. A universal product, also referred to as standardized product or uniform product, is physically identical in all overseas markets in which it is sold, with the exception of such elements as labeling and the language used in the manuals. Studies by Sorenson and Wiechmann (1975), and Hill and Still (1984) provide evidence that product policy is an area where IEs have the highest propensity for standardization. Products such as basic materials, components, high technology products, industrial products, steel, chemicals, plastics, ceramic castings used in memory chips, aircraft turbine engines, and cameras can be considered as universal products. Jain (1989) states that industrial and high technology products are more suitable for standardization than consumer products.

The primary benefits derived from product standardization are cost savings. Five main reasons contribute to these cost savings. First, offering a standardized product across countries implies higher production volumes that enable the IE to reduce the unit cost due to economies of scale. Fixed costs such as wages, insurance costs on the plant and equipment, property taxes, and interest costs are distributed over a larger number of products causing the unit cost to decrease. Such economies of scale are referred to as short-term economies of scale.

Second, cost savings can be derived from the design of a single product instead of a number of products. Third, the production of a standardized product enables the IE to increase the volume of the purchased raw materials and components. This assists the IE to increase its bargaining power over its suppliers, and eventually enjoy better purchasing contracts. Fourth, the production of a standardized product may allow the IE to centralize the production of the product to few specialized facilities. The construction of fewer but bigger facilities results in the cost savings. Lastly, promoting a standardized product across countries results in economies of scale in marketing. Advertising the standardized product may require the same type of campaign in each country so as to maintain a common global image of the product. The commonality in the advertising campaigns results in cost savings. In general, savings from standardization allow the company to position itself as a low-cost producer.

The higher production volume of the standardized product, as discussed above, leads to low production costs that eventually ensure lower prices for the customers. Nevertheless, this benefit may come at the risk that the standardized product may not meet the needs of customers in a specific country. The universal product approach can result in overstandardization of a product, and may prove to be catastrophic for some companies. Kashani (1989) describes the negative experiences of overstandardization in the Danish toy-maker Lego. Lego was marketing around the world its educational toys in the same fashion. In the United States, Lego faced fierce competition after Tyco, its major competitor, started marketing its toys in plastic buckets that could store the toys. Lego's universal approach to packaging was the elegant transparent boxes, which was not appealing to American parents, who preferred the functional packaging approach of toys in buckets. Lego's eroding market position alarmed the U.S. management but headquarters refused the idea of introducing toys in buckets, claming that this could harm Lego's reputation for high quality. Nevertheless, massive losses forced Lego to design its own bucket for storing toys. The result reversed its eroding market.

Walters and Toyne (1989) describe three methods that an IE can follow to develop standardized products. The first method is *product extension* under which the product developed for a single market, usually the home market of the IE, is sold to foreign countries. Companies that adopt this method have an ethnocentric orientation such as companies that are at the first stages of their internationalization and have started exporting. Textile and fitted carpet manufacturers are examples of such companies. Fuji-Xerox designed and targeted its high performance plain paper copiers exclusively for the Japanese market. Subsequent expansion to foreign markets was simply limited to exporting the copiers designed for the Japanese market. The probability of success increases when the foreign markets are "similar" to the home market of the enterprise.

The second method is referred to as *premium prototype*. The universal product developed for this method meets the needs of the most demanding customers whether the group is overseas or in the home market. The product is also designed to withstand the most severe conditions of use. Such a high quality product can be appealing to customers in foreign markets, but it may often include superfluous design features that are not necessary for the average foreign customer. The high quality design and, perhaps, high production cost due to the enhanced nature of the product may result in high prices to discourage a group of potential buyers.

The third method, known as *global common denominators*, strives to identify a *global segment* of international customers with homogeneous preferences and conditions of product use. After identifying such a segment, a product is developed to meet the common needs of customers in that segment. The difficulty with this approach is the existence of global segments. Douglas and Wind (1987), for example, argue that substantial heterogeneity exists not only among customers across countries but also within countries, making the existence of a global segment almost utopic. The new hotel chain Courtyard by Marriott introduced by Marriott is an example of the effort of a company to satisfy a global segment. Marriott tries to identify the attributes that travelers value most in a hotel in designing the hotel of their preference.

Country-Tailored Product Policy

A country-tailored product is designed and produced to meet the idiosyncratic needs of the country (or group of countries) for which it is intended for. Czinkota and Ronkainen (1993) present various factors that encourage the design of products that are exclusively targeted for certain countries. Such factors include: (1) different conditions of use depicted by different climatic conditions; (2) government regulations that have be to met in each country, (3) different consumer behavior patterns; and (4) an attempt to meet local competition when competing firms customize their products to the local market. The primary benefit obtained from designing products specifically for certain countries or regions is that this customized product fully meets the needs of local consumers. This leads to higher demand, higher market share, and eventually higher profitability.

The case of Fuji Xerox and its high performance plain paper copiers is an example of a product designed exclusively for the Japanese market. Bausch & Lomb developed 25 new products in 1986 with only one being custom-tailored for foreign markets. Since then, Baush & Lomb has changed its strategy. In 1991 more than half of its products were developed for foreign markets. In Europe, Bausch & Lomb's Ray-Ban glasses are made flashier, more avantgarde, and are costlier, in order to accommodate the high fashioned European consumer. Ray-Ban glasses in Asia have been redesigned to better suit the Asian face with its flatter bridge and higher cheekbones (Fortune Magazine, 1992: 76).

Consider the case of Mars producing its traditional chocolate bar in Europe and the United States. The two chocolate bars, even though they have the same brand name, have different tastes, texture, and packaging in order to meet the different tastes of consumers. Colgate toothpaste offers a more spicy toothpaste, which has been exclusively designed for the Middle East market. In fact, many consumer products, especially food and household products, are country-tailored products because of their strong culture dependency.

Designing a customized product that meets the consumer preferences in a specific or region has the drawback that the cost of the making product may be pushed higher. The higher cost is caused primarily by two reasons. First, the design of localized products significantly affects the new product identification and development costs. Costs increase because market research has to be conducted in each country or region and in addition, more funds must be allocated to design more than one product. As a result, the customized product designed for a specific country has to absorb all the developmental costs. If the product was to be marketed in several countries, this cost would have been distributed over more units demanded by the various countries. Second, since each customized product is sold in one country, the smaller production volumes may prevent the company from exploiting economies of scale.

Another drawback of the country-tailored product policy is that the logistic operations for distributing the customized products may be inefficient. Take for example the experience of Hewlett-Packard and its deskjet printers. Initially, Hewlett-Packard was producing and packaging printers for the United States and European countries. Customization involved the installation of the appropriate power supply, software, and manuals in the specific language. Inaccurate sales predictions left the distribution centers with inventories of printers meant for some countries while suffering shortages in some others. The distribution centers were forced to rework the packages and printers to customize products for the countries where demand was not satisfied.

Modified Product Policy

A modified product is substantially similar in each country in which it is sold but minor alterations/adaptations are made on the basic product to conform to local needs and regulations. Modified products are considered products whose adaptations represent a "modest" percentage of the total cost (Takeuchi and Porter, 1986). The modified product policy enables an IE to reap the benefits of standardization and centralized production, and at the same time, respond to local needs. Modified products include industrial products such as CT scanners, broadcast video cameras, mainframe computers, copiers, precision testing equipment, and construction equipment. In consumer goods, modified products include cars, motorcycles, calculators, and microwave ovens.

Product adaptations are classified as *involuntary* or *voluntary*. Involuntary adaptations are made to meet legal requirements of the foreign government on packaging, labeling, metric system, and also local climatic and economic conditions. Examples include the use of different paper trays for copiers in Japan, the United States, and Europe since all three regions use papers of different sizes. Similarly, car manufacturers must modify the speedometer on cars sold in the United States to display in miles per hour where as car speedometers in Europe provide the reading in kilometers per hour. Manufacturers of electronic appliances must install the proper power supply module in order to comply with the country's voltage requirements.

Voluntary adaptations are those made as part of the company's marketing strategy to meet local needs and consumer idiosyncracies such as color, the product's size, and its accessories. For example, the color of the package especially on consumer products is very important since different colors are perceived differently across countries. Packages in the Middle East tend to include the green color that is preferred by consumers in this region.

McDonalds, besides the standard products that it serves, tries to adapt to local needs by selling wine in France, vodka in Russia, and beer in Germany. Besides the world famous sandwiches that it serves, Subway prepares sandwiches that contain cheeses and ham made in the country of operation. Similarly, Pizza Hut prepares pizzas not only with the toppings that have traditionally been used, but also uses toppings that include cheese and processed meat found in the local country. As such, these two companies modify their overall products to meet local tastes. Heineken, the Dutch beer manufacturer, offers its beers in bottles of various sizes depending on the country of operation. Finally, an example of product modification involving car accessories is the cup-holder, which is almost a standard feature in cars in the

United States but not so commonly found in cars destined for the European countries.

Two approaches that allow an international enterprise to apply a modified product policy are the *modular approach*, and *core-product approach*. Adoption of either approach permits the firm to reap the benefits of standardization, and at the same time, respond to local needs and differences in conditions of product use across countries (Walters and Toyne, 1989).

Modular Approach: The notion of modular design was pioneered by Starr (1964: 137) who suggested that products should be made of interchangeable modules to supply "consumers with apparent variety even though the production output is based on the concepts of mass production." The modular approach entails the design and production of a range of standard parts that can be assembled in a variety of combinations yielding a series of products. These products can be assembled from the standard components to meet, as closely as possible, the preferences of local consumers in a particular foreign market. Ikea, the Swedish furniture manufacturer, is an IE that has adopted the modular approach. Specifically, Ikea is selling furniture that can be assembled from a standard collection of pieces.

Cost savings can be achieved when the components are mass-produced in a single facility, and the number of variant product combinations is limited when the number of components is limited. Nevertheless, under the modular approach, one must consider the tradeoffs between the number of product variants that enhance the attractiveness of the product to local markets (due to customization) and the number of components produced that promote economies of scale. Apparently, such a solution resides in the specifics of each firm.

Core-Product Approach: The core-product approach is based on the same principles as the modular approach. Under this approach, a uniform "central/core" product is designed that can accept a number of standard attachments, parts or components. The combination of attachments to the core product allows it to meet performance criteria and local consumer preferences in a country. The core product is of higher value relative to the total value of the finished product, whereas the components in the modular approach are of lower value relative to the total value of the finished product. As such, the core product represents a significant proportion of the total value of the finished product.

The production of a standard core product enables a firm to reap the benefits of economies of scale and thus enjoy lower production costs. On the other hand, the flexibility allowed by the attachment of components to the core product helps a firm localize its product and increase its attractiveness. Yip (1989) argues that "in a global strategy, the ideal is a standardized core product that requires minimal local adaptation." An additional advantage of the core product approach is that it facilitates delayed product differentiation, which enables marketers to react to market demand.

An example of an IE adopting the core-product approach is a manufacturer of agricultural machinery. Across countries, the IE was confronted with differences in climatic conditions, topography, and customer needs. To overcome these differences, the IE produced a robust and standard product that could operate in a variety of conditions. This core-product was also able to accept a number of attachments that gave the basic product different performance characteristics to meet local requirements and conditions of use (Walters and Toyne, 1989). In a similar fashion, car manufacturers employ the core-product approach by using a basic product, e.g., the chassis of the car, and then attach to it features in order to customize it for specific countries and even consumer needs. Such attachments include bigger engines, air conditioning units, electric windows, and other accessories. The core product is also referred to as platform product in the automobile industry or as a generic product in the electronics industry (Lee *et al.*, 1993). McGrath and Hoole (1992) state that each new product development should revolve around the design of a core product.

Honda recognizes the need for localized products since customers in different countries use them in different ways. To suit the needs of their customers and conform to the socio-economic conditions of the diverse marketplace, Honda uses its common basic technology to develop different types of motorcycles for different regions. In fact, the core product concept reduced the cost of Honda's products, which "cheaply built its new car lines from the platforms of existing cars" (Fortune Magazine, 1996: 32). Yip (1989) describes the product modification made by Boeing on its 737 in order to revive leveling sales. Boeing recognized that its current design (early 1970) was not appealing to developing countries because the shortness and softness of their runways, combined with the lower technical skills of their pilots made the planes bounce upon landing. Bouncing resulted in brake ineffectiveness. Boeing modified its 737 design by adding thrust to the engines, redesigning the wings and landing gear, and installing lower pressure tires.

Lee *et al.* (1993) describe Hewlett-Packard (HP)'s adoption of the concept of generic/core product. Specifically, HP is currently applying the principles of "postponement in product differentiation" on its deskJet printers. Postponing the point of product differentiation means delaying, as much as possible, the stage at which value is added to the work-in-process inventory before the final product is completed. Under this strategy, a generic/core product is manufactured at Vancouver, Canada, and Singapore that is shipped to the distribution centers in the U.S., Far East, and Europe. The distribution centers perform the product customization by inserting into the packaging boxes the language-specific manuals, power supply modules, and software. HP has even adopted the concept of packaging postponement under which final packaging is performed at the distribution center and not at the manufacturing facilities. The above concept requires the design of a generic/core product that can accommodate and facilitate the postponement of the customization process. Lee *et al.* (1993) refer to this product design as *design for localization* or *design for customization*.

FACTORS AFFECTING THE CHOICE OF A PRODUCT POLICY IN A GLOBAL ENVIRONMENT: The need to compete globally is becoming an imperative for firms aspiring to be key players in their industry and enjoy the benefits of globalization such as entering new markets, and exploiting economies of scale. Competing globally requires IEs to choose one of the product policies previously described. This choice for an international product policy is influenced by various factors that can be broadly categorized into market-related, production-related, and international economic and political factors. These factors must be analyzed before reaching a decision on the type of international product policy to adopt. We provide below a partial list of such factors and their repercussions on the product policy choice.

Market-Related Factors

- When the "similarity" of customer preferences across countries is significant and creates a homogeneous global segment, IEs tend to produce standardized products. Levi's jeans, Coca-Cola, and Colgate toothpaste are often cited as universal products (Levitt, 1993). Nevertheless, there are still substantial differences in consumer preferences across countries (Boddewyn *et al.*, 1986). Cultural compatibility also promotes standardized products (Jain, 1989).
- The nature of the product is important since industrial products are more suited for standardization than consumer products (Boddewyn *et al.*, 1986; Jain, 1989; Cavusgil and Zou, 1993).
- Where the product is marketed; in a developed or in a lesser-developed country (Hill and Still, 1984).
- The size of the market in various countries and their potential for growth.
- The level of experience of the IE in the global marketplace. That is, whether the IE is simply an exporting company, or has a multinational or global orientation (Cavusgil and Zou, 1993).
- The presence and capabilities of international and local competitors (Jain, 1989).

International Economic and Political Factors

- Economic recession in the countries where the IE promotes its products (Boddewyn *et al.*, 1986). The IE may not design a custom-tailored product and simply export the ones designed for other countries to countries in a recession.
- The exchange rate and its fluctuations of the currency of the various countries where the IE operates (Hadjinicola and Kumar, 1997).
- Quotas and tariffs set by local governments.
- The presence of nationalistic feelings among consumers.
- National government regulations that products must conform to.
- The political stability and similarity of the legal environments of the countries the IE operates.

Manufacturing Factors

The international product policy adopted by an IE is closely related to its manufacturing *configuration* and *coordination* (Porter, 1990). Configuration refers to the degree of concentration/dispersion of activities in the value-added chain. On the other hand, coordination of the activities around the globe deals with information sharing among the activities, the allocation of responsibility, and alignment of effort. Each of the three product policies described above requires the support of a distinct manufacturing strategy, and accordingly, the right configuration and coordination. The universal product policy is facilitated by centralized production, whereas the country-tailored product policy is often promoted through localized production. A modified product policy calls for the adoption of manufacturing rationalization, implying factory specialization and closely integrated flow of materials among facilities in various countries. Some manufacturing factors that affect the choice of a product policy are presented below.

- The cost of setting up facilities in the various countries. When the cost of setting up product facilities is high, the IE will tend to centralize production in a small number of countries. This may lead to the design and production of standardized products.
- The level of centralization/decentralization of the production function. Centralization tends to result in less country-tailored products (Jain, 1989).
- The expected production volume and the savings derived from economies of scale. If the expected benefits from high production volumes are significant, the IE may tend to design standardized products (Hadjinicola and Kumar, 1997).
- The coordination and the cost of the new product development process. High development costs for new products may force the IE to design a standardized product.
- High transportation costs may lead the IE to set up production facilities in each country, which eventually may lead to the design of customized products.
- Maintaining a few global suppliers to ensure consistency in the components may lead the IE to design standardized products. On the other hand, maintaining multi-regional suppliers with different capabilities improves relations and may lead the IE to design country-tailored or modified products.

More discussion and examples of IEs whose product supply was influenced by the factors above can be found in Czincota and Ronkainen (1993).

THE PRODUCT DESIGN PROCESS IN A GLOBAL ENVIRONMENT: Developing new products in a global environment often requires the various collaborating teams to be located in different parts of the world. Reasons for dispersed product design activities include: 1) need to tap into the skilled human resources of specific regions in some countries; 2) take advantage of financial incentives offered by the host countries; 3) exploit cost benefits from local labor; 4) be close to academic and business centers; 5) exploit reduced costs in equipment acquisition and building installations; and 6) take advantage of the technology and infrastructure of the foreign country. Having research and development facilities dispersed around the world allows the IE to roll out the product in less time since the design process go on for 24 hours a day. Of course the choice of locating a new design center also depends on the competitive advantage acquired by locating a design center in that region. An inhibiting factor for designing a product in various locations around the world is the need for transferring technology abroad. Companies are often reluctant to transfer their research and development technology abroad to prevent leakage to competitors.

The management of the design process in a global context is truly a challenging one since it involves the coordination of people in different locations, organizations, and cultures. The design process relies heavily on advanced communication technology since face-to-face interactions between the dispersed team members cannot be frequent. Differences in language, leadership/management styles, risk aversion, propensity to meet deadlines, organization and planning are only a few of the cultural issues that need to be resolved in order to ensure the success of the new product design process in a displaced environment. Rosenthal (1992) provides an extensive discussion on the impact of cultural differences on the effectiveness of new product design teams.

A CASE: The case of Motorola and the development of the Keynote pocket pager highlight some of the issues raised above concerning the design process in a global environment (Rosenthal, 1992). Motorola decided to split the design activity of the Keynote pager between Singapore and Boynton Beach in Florida. Boynton Beach was responsible for the conceptual design and factory introduction, while Singapore was assigned such activities as the detailed design, the sourcing of parts in Asia, and pilot runs. Three basic reasons led Motorola to create a design center to Singapore. First, it was meant to off-load work from the design center in Florida. Second, the design center was to be located close to the manufacturing facility in Singapore. As a result, management believed that the close contact with production and process people would assist the design process. Third, from the strategic point of view, Motorola wanted to establish a design center in Asia, thus preparing for global market, where products had to be designed in the vicinity of the customer.

As it turned out, the biggest challenge in the design of the Keynote pager was the management of the geographically separated teams. The 12-hour time difference meant that during normal working hours there no was communication between the members of the two teams. Issues that arose had to be resolved using the company's electronic mail system. This meant a delay in resolving even the minor problems that appeared in the design process. Language was also a barrier for the two teams since Singapore has four official languages.

Another problem that arose was the difference in priorities and perceptions of the two design teams in Florida and Singapore. Working in a manufacturing environment, the priorities of the people in Singapore were meeting delivery times and controlling the budget. There was a natural aversion towards new products since they perturb the system. People in Singapore were trying to balance daily operations with the new product design. As a result, the design team in Florida perceived that management in Singapore was not fully committed to the design process. Furthermore, designers in Florida were apprehensive of the idea of delegating work to offshore sites, for fear that their work would disappear.

From Motorola's experience, two key factors ensure the success of the design process. First the selection of design teams with prior experience, and success in designing new products. If possible, teams should have some experience on design collaborations with teams located in other countries. Second, the choice of the manager with necessary technical and personality traits to lead the design process, acting as a coordinator and cohesive agent among the dispersed teams. Other ways to increase the probability of success of the design process in a global context is to allow designers from different locations to visit other research facilities so as to assimilate the organizational culture and establish personal relationships with the rest of the team members.

De Meyer (1991) describes six broad approaches that IEs are adopting to solve the communication problem among the dispersed laboratories. Such approaches include socialization efforts through temporary assignments and travelling, the establishment of rules and procedures to enhance formal communication, the presence of people serving as a liaison among the laboratories, the centralized control of the design activity, and the extensive use of electronic communication.

CONCLUSION: The success of an international product policy does not only lie in the thorough analysis of the factors discussed above but also in the way products are designed. Consider the experience of Polaroid, when the company was planning to introduce a camera that had a voice chip to simulate the photographer saying "cheese" or "smile" in order to get the target to smile. In this case, in which language should the company program its chip given that the camera was going to be distributed around the world? Initially, Polaroid introduced customized cameras with chips programmed with various languages. The number of different cameras led to inventory problems, similar to those experienced by HP described before, forcing the company to offer a standardized product with three phrases/sounds. In other words, Polaroid alleviated its operational problems, at the expense of complete customization by offering a standardized product. An intermediate solution could have been to think of the camera as the core-product and allow customers themselves to record the phrase/joke in their own language. This idea has also been used in dolls that allow customers to record the sounds that they would like the dolls to produce. As such, the above example illustrates the importance of the design process that can allow the IE to reap the benefits of customization and at the same time exploit operational efficiencies.

Organizations Mentioned: Bausch & Lomb; Boeing; Coca-Cola; Colgate; Fuji-Xerox; Heineken; Hewlett-Packard; Honda; Ikea; Lego; Levi; Marriott; Mars; McDonalds; Motorola; Pizza Hut; Polaroid; Subway; Tyco.

See **Core-product approach; Design for customization; Design for localization; Global common denominators; Global manufacturing rationalization; Global marketing concept; Implementation; Involuntary product adaptations; Mass customization; Premium prototype; Product development through concurrent engineering; Product design; Product extension; Universal product policy; Voluntary product adaptations.**

References

[1] Boddewyn, J.J., R. Soehl, and J. Picard (1986). "Standardization in international marketing: Is Ted Levitt in fact right?" *Business Horizons*, November–December, 69–75.

[2] Czinkota, M.R. and I.A. Ronkainen (1993). *International Marketing*. Dryden Press, Fort Worth.

[3] De Meyer, A. (1991). "Tech tack: How managers are stimulating global R&D communication." *Sloan Management Review*, Spring, 49–58.

[4] Douglas, S.P. and Y. Wind (1987). "The myth of globalization." *Columbia Journal of World Business*, Winter, 19–29.

[5] Hadjinicola, G.C. and K.R. Kumar (1997). "Factors affecting international product design." *Journal of the Operational Research Society*, 48, 1131–1143.

[6] Hill, J.S. and R.R. Still (1984). "Adapting products to LDC tastes." *Harvard Business Review*, March–April, 92–101.

[7] Jain, S.J. (1989). "Standardization of international marketing strategy: Some research hypotheses." *Journal of Marketing*, 53, 70–79.

[8] Kashani, K. (1989). "Beware of the pitfalls of global marketing." *Harvard Business Review*, September–October, 91–98.
[9] Lee, H.L., C. Billington, and B. Carter (1993). "Hewlett-Packard gains control of inventory and service through design for localization." *Interfaces*, 23, 1–11.
[10] Levitt T. (1983). "The globalization of markets." *Harvard Business Review*, May–June, 92–102.
[11] McGrath, M.E. and R.W. Hoole (1992). "Manufacturing's new economies of scale." *Harvard Business Review*, May–June, 94–102.
[12] Porter, M.E. (1990). *The Competitive Advantage of Nations*. The Free Press, New York.
[13] Quelch, J.A. and E.J. Hoff (1986). "Customizing global marketing." *Harvard Business Review*, May–June, 59–68.
[14] Rosenthal, S.R. (1992). *Effective Product Design and Development*. Business One Irwin, Homewood.
[15] Sorenson, R.Z. and U.E. Wiechmann (1975). "How *multinationals* view marketing standardization." *Harvard Business Review*, May–June, 38–54.
[16] Starr, M.K. (1965). "Modular production – a new concept." *Harvard Business Review*, November–December, 137–145.
[17] Takeuchi, H. and M.E. Porter (1986). "Three roles on international marketing global strategy." In M.E. Porter (Ed.) *Competition in Global Industries*. Harvard Business School Press, Boston, 111–146.
[18] Walters, P.G. (1986). "International marketing policy, a discussion of the standardization construct and its relevance for corporate policy." *Journal of International Business Studies*, 17, 55–69.
[19] Walters, P.G. and B. Toyne (1989). "Product modification and standardization in international markets: Strategic options and facilitating policies." *Columbia Journal of World Business*, Winter, 37–44.
[20] Yip, G.S. (1989). "Global strategy . . . in a world of nations?" *Sloan Management Review*, 31, 29–41.

PRODUCT DEVELOPMENT AND CONCURRENT ENGINEERING

Christoph Loch

INSEAD, Fontainebleau, France

Christian Terwiesch

University of Pennsylvania, Philadelphia, Pa, USA

DESCRIPTION AND HISTORICAL PERSPECTIVE: Concurrent Engineering (CE) represents a key trend in product development over the last decade. It has changed academic and industrial approaches of looking at the product development process. It can be defined as an integrated "new product development process to allow participants, who make upstream decisions to consider downstream and external requirements." Central characteristics of a concurrent development process are activity overlapping, information transfer in small batches, and the use of cross-functional teams (Gerwin and Susman, 1996).

Many elements of CE have been recognized since the first half of this century (Smith, 1997). They include the formation of cross-functional teams, an important role for manufacturing process design in product development, and a focus on fast time-to-market. The term CE, however, was not adopted until the 1980s. Coming from manufacturing engineering, for example, Nevins and Whitney (1989) were among the earliest to demand concurrent design of the product and the manufacturing system. Others emphasize the cross-functional aspect by observing a ". . . need to transfer decision making authority from managers to teams" (Gerwin and Susman, 1996).

Concurrent engineering became popular in operations management literature with the work by Imai *et al.* (1985) and Takeuchi and Nonaka (1986). These studies contrasted two practices in product development: on the one hand, most Western organizations followed a more sequential process, similar to running a relay race with one specialist passing the baton to the next. This approach has also been referred to as "batch processing" (Blackburn, 1991; Wheelwright and Clark, 1992). In contrast, some high performing development organizations, most of them in Japan, followed a "rugby team" approach with a strong emphasis on cross-functional integration (a cross-functional team running in a staggered fashion). These early examples were motivated mainly from the camera and automobile industries (Takeuchi and Nonaka, 1986).

In a landmark study on the automobile industry, Clark and Fujimoto (1991) extended this pioneering work, and provided a detailed empirical analysis of product development processes, their impact on development performance and their differing use across regions. This study operationalized many important variables of CE, including measures for task overlapping and cross-functional communication. Clark and Fujimoto observed that companies that succeeded in short time-to-market were found to combine activity overlap with intensive information transfer, a practice they referred to as "integrated problem solving." These ideas were refined in further studies, e.g., Wheelwright and Clark (1992).

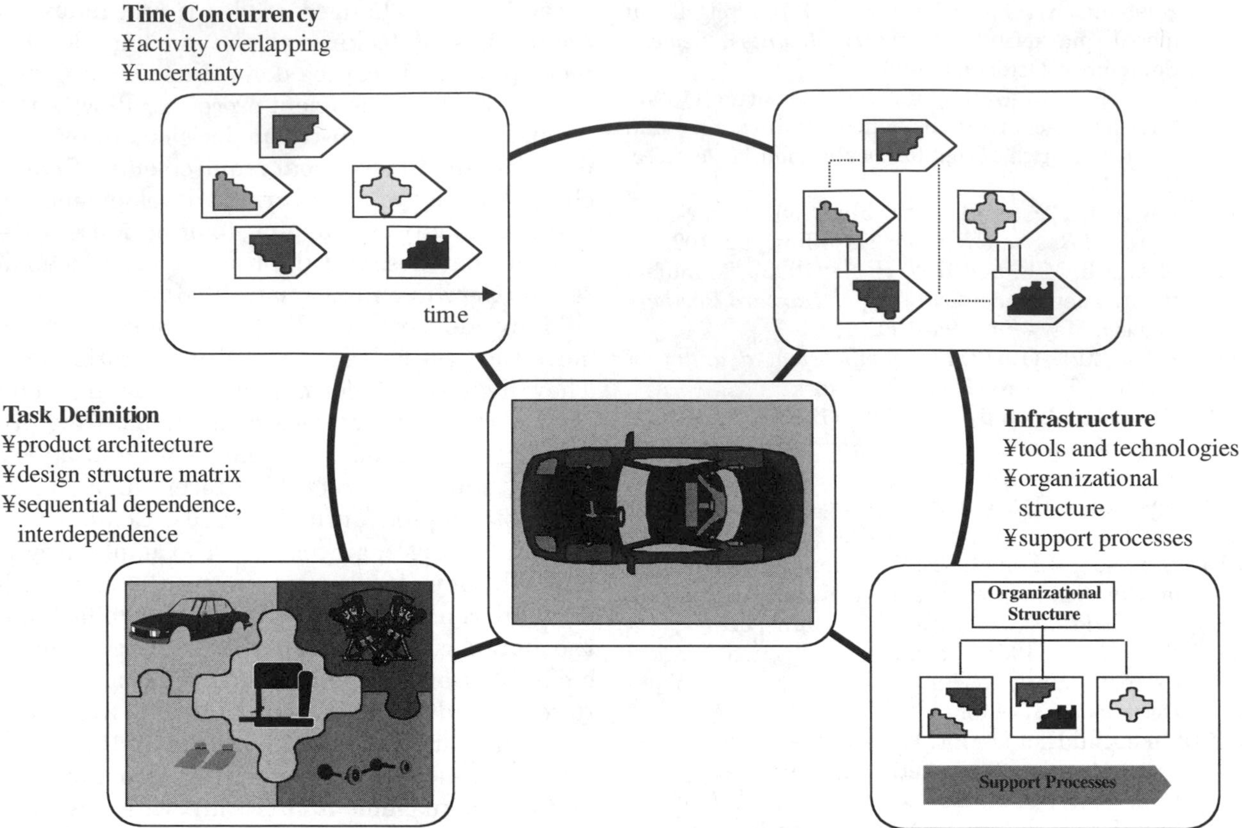

Fig. 1. Operations management issues in concurrent engineering.

The managerial framework of CE pose four key managerial challenges (see Figure 1).

- Defining development tasks based on the product architecture.
- Timing of activities with emphasis on task overlapping.
- Defining coordination and integration mechanisms among development tasks, which ensure adequate information exchange.
- Establishing support processes and an appropriate organizational context.

TASK DEFINITION IN COMPLEX DESIGN PROBLEMS: When faced with product development on a large scale, often hundreds of people work simultaneously on separate portions of the overall development effort, organized in many small teams (Eppinger *et al.*, 1994). The division of the complex overall design task into smaller, separately manageable portions depends on couplings among these portions, which are largely determined by the product architecture.

The *product architecture* identifies the basic building blocks, their relationships to the functionality of the product ("what they do"), and the interfaces among them. A modular architecture is one with standardized interfaces and a one-to-one correspondence between building blocks and functions: a module roughly corresponds to a function. The other extreme is an integrated architecture, with complex interfaces and a one-to-many correspondence between blocks and functions: functions are spread out over several building blocks, or several building blocks share one function (Ulrich, 1995).

In terms of the final product, a modular architecture offers the possibility of component standardization, easily achievable product variety, and add-ons. An integrated architecture, on the other hand, may increase the compactness of the product, or allow better adaptation to specific customer needs, higher performance, or longer life through the tight integration of its building blocks. Moreover, the architecture heavily influences the product development process, since modularity reduces interactions among product components, and thus complexity (Fitzsimmons *et al.*, 1991). In a modular architecture, product components can be developed in a decomposed and parallel manner (Ulrich, 1995).

The goal of task definition is to minimize interactions among activities, or at least make them explicit so as to clearly understand the information transfer necessary. Two types of couplings exist, defined by information flows among activities. First, *sequential dependence* refers to a one-way information flow from an upstream to a downstream activity. In this case, one activity is the information supplier and the other the information receiver. Second, task interdependence refers to several tasks requiring information input from one another.

The process of decomposing the overall design problem can be supported by operations management methods. The most important support tool is the design structure matrix (DSM), which groups activities so as to minimize the interdependencies among the groups (Steward, 1981; Eppinger *et al.*, 1994; Smith and Eppinger, 1997). A brief example of the application of this tool is below.

Based on the product architecture and the resulting component interfaces, the DSM captures interdependencies among the development tasks (corresponding to components), in the sense that tasks need input (physical or informational) from other tasks in order to be completed.

As is shown in Figure 2, information-receiving tasks are listed along the columns, and information-supplying tasks along the rows. Crosses (X) mark information dependencies. Task B is *sequentially dependent* of task A, as the information flow goes only one way. Tasks B and C are *independent*. Tasks C and D require mutual information input and are *coupled* (interdependent). The matrix suggests a plan for the order of the tasks: A, then B in parallel with C and D, the latter two being performed in a closely coordinated way.

Significant Analytical Models: Eppinger *et al.* (1994) extend the DSM concept to include task completion times (marked on the diagonal) and strength levels of dependencies. By grouping the tasks with the strongest couplings together, the DSM can thus also be used to suggest design team formation; a team is best for coupled tasks. They demonstrate the application of the method in a semiconductor design project.

TIME CONCURRENCE AND ACTIVITY OVERLAPPING: Task overlapping originated in the automotive (notably, Toyota and Honda), camera (e.g., Canon), and aerospace industries (Takeuchi and Nonaka 1986). It has also been adopted in electronics (Krishnan *et al.*, 1997) and the software industry, and recently even in making movies (*New York Times*, May 5, 1997: D1).

Once development tasks have been defined in the overall design problem, their execution over time must be planned. While it is natural to execute independent tasks in parallel, it is easiest to perform sequentially dependent tasks in their logical sequence. This results, however, in a lengthy sequential process similar to a relay race. To reduce the time needed, CE attempts to, at least partially, overlap the sequentially dependent activities.

In the context of classical project planning terms, overlapping proposes, in effect, to shorten the critical path of a project by "softening" precedence relationships, and by conducting sequential activities in parallel. This offers a fundamental time advantage, but also has drawbacks. In a fully sequential process, downstream starts with finalized information from upstream, whereas in an overlapping process, it has to rely on preliminary information. This approach can be risky if the outcome of the upstream activity is too uncertain to be accurately predicted. Under these conditions, overlapped activities create uncertainty for the downstream activity. The uncertainty would not exist in a sequential process. Thus, a trade-off exists between time gained from parallel execution and rework caused by uncertainty in the project. An optimal balance between parallelity and rework has been derived via analytical models and confirmed by several empirical studies showing that concurrence benefits do decrease with increasing project uncertainty.

	A	B	C	D	
Task A	**A**				A-B: sequential
Task B	x	**B**			B-C: independent
Task C			**C**	x	C-D: coupled
Task D	x		x	**D**	A-D: sequential

Fig. 2. Design structure matrix (Ulrich and Eppinger, 1995: 262).

Significant Analytical Models

Several models have been put forward to address possible trade-offs in applying CE. Some are reviewed here. Ha and Porteus (1995) investigate a situation in which two development tasks are inherently coupled, and must be carried out in parallel to avoid quality problems. They model the "how frequent to meet" question as a dynamic program. If one design activity proceeds without incorporating information from the other, design flaws and corresponding rework result. Thus, parallel development together with design reviews save time and rework. Similar to a quality inspection problem in production, these gains have to be traded off with the time spent on review meetings. The main question addressed in the study is: how often to communicate?

Krishnan *et al*. (1997) developed a framework for concurrence in case of sequentially dependent activities. The authors model preliminary information passed from an upstream to a downstream activity in the form of an interval. A parameter, e.g., the depth of a car door handle, is initially known only up to an interval, which narrows over time as the design becomes final. In this framework, two concepts determine the overlap trade-off. "Evolution" is defined as the speed at which the interval converges to a final upstream solution. "Downstream sensitivity" is defined as the duration of a downstream iteration to incorporate upstream changes associated with the narrowing of the interval. If upstream information is frozen before the interval has been reduced to a point value, a design quality loss occurs.

The authors formulate the problem as a mathematical program and show when overlapping (and thus preliminary information) should be used, and when upstream information should be frozen early. The authors solve an application example numerically and suggest that "generally, a fast evolution and low sensitivity situation is more favorable for overlapping. "The concept is illustrated for a door handle, a pager, and parts of a dashboard (Krishnan, 1996; Krishnan *et al*., 1997).

Loch and Terwiesch (1998) conceptualize uncertainty resolution as the distribution of engineering changes (ECs) over the course of the project: the more uncertain the upstream activity, the more ECs are likely to arise. ECs become more difficult to implement the later they occur. The authors investigate the trade-off between time gained from overlapping, and the downstream rework caused by implementing ECs. Sensitivity analysis on the optimal overlap level shows that gains from overlapping activities are larger if ECs can be avoided, if dependence among activities can be reduced, and if uncertainty (the rate of ECs) can be reduced early in the process.

Implementation

The above-cited models suggest that overlap in product development is not equally applicable in all situations. This is supported by several empirical studies. In their study of the world computer industries, Eisenhardt and Tabrizi (1995) identify substantial differences across different market segments. For the stable and mature segments of mainframes and microcomputers, the authors find that overlapping development activities significantly reduces time-to-market. However, in rapidly changing markets such as printers and personal computers, overlapping is no longer found to be a significant accelerator. Eisenhardt and Tabrizi argue that compressing the development process through activity overlaps only yields a time reduction if the market environment is stable and predictable.

Terwiesch and Loch (1996) find that project uncertainty in general (caused by the market or the technology) reduces the benefits from overlapping. Their study is based on data from 140 completed development projects across several global electronics industries. Although overlap helps to reduce project completion time overall, it is less helpful in projects with late uncertainty reduction than it is in projects with early uncertainty reduction. Intensive testing and frequent design iterations emerge as effective measures to reduce completion times in projects with late uncertainty reduction.

The effects of *uncertainty* in CE are compounded by product *complexity*. Complexity of a product refers to the number of elements in the system and the level of interactions among these elements (Fitzsimmons *et al*., 1991). The higher the complexity, the more interactions among subsystems exist in the design structure matrix, and the more components are affected by uncertainty manifested in ECs. This makes intensive coordination among the parallel tasks even more important, which is discussed next.

INFORMATION CONCURRENCE: COORDINATION AND INTEGRATION: When the subtasks of the overall product design problem are executed, information exchange is required to keep the tasks coordinated. In simple cases, the information exchange may occur prior to the start of the downstream activity. Such *ex-ante* coordination can be achieved by, for example, design rules or preferred parts lists, which are often used in the context of design-for-manufacture. These methods help to anticipate problems in the manufacturing process as early as possible, but they are not sufficient when the design problem is complex (Wheelwright and Clark, 1992). In this case, ongoing coordination is required during task execution. This is consistent with Clark and Fujimoto's (1991) finding that successful task overlapping is supported by intensive information exchange.

Coordination Frequency

Coupled activities are almost always performed in parallel (see the earlier discussion of the DSM), and the key question is how intensively they should be coordinated. Ha and Porteus (1995) offer a model where this coordination is analyzed in terms of the frequency of communication between the two activity teams. The more significant the coupling (i.e., the more quality problems exist and the more severe the impact of rework per quality problem), the more frequently the task teams should communicate.

Loch and Terwiesch (1998) show that the same principle holds in the case of sequentially dependent tasks: the more significant the number of changes from upstream and the dependency of the downstream task, the more intensive communication is required. High communication capabilities are thus an important enabler of task overlapping. If shared problem solving activities are performed before the actual design tasks begin, uncertainty levels during development may be significantly reduced in routine development projects.

These analytical results are consistent with empirical evidence: Morelli *et al.* (1995) examined communication patterns during the development of electrical computer board connectors, and they found that most communication took place among activities that were heavily interdependent. In a more general context, the relationship among dependencies, uncertainty, and communication has been extensively studied in the organizational sciences (e.g., Thomson, 1967).

Coordination Format

In addition to the frequency of communication, the format of information exchanged between concurrent activities is important. Ward *et al.* (1995) report that Toyota in its product development process pursues a seemingly excessive number of parallel prototypes and "communicates ambiguously." They conclude that Toyota uses "set-based" concurrence, where whole sets of possible design solutions are explored in parallel, and over time, the set of solutions considered is narrowed down (as a rule, not broadened). For example, key body dimensions are not fixed until late in the development process. However, when a fixed solution is reached, it no longer changes unless absolutely necessary. Ward *et al.* contrast the set-based approach to "iterative" or "point-to-point" design, where design teams fix solutions for their respective components and then coordinate among teams by iteration.

One cannot, however, conclude that a set-based approach is always superior. Terwiesch and Loch (1997) examined climate control system development of an automobile manufacturer and encountered both approaches, set-based and iterative, the choice based on the nature of the component developed. The trade-offs involved in the choice are summarized in Figure 3, where the two

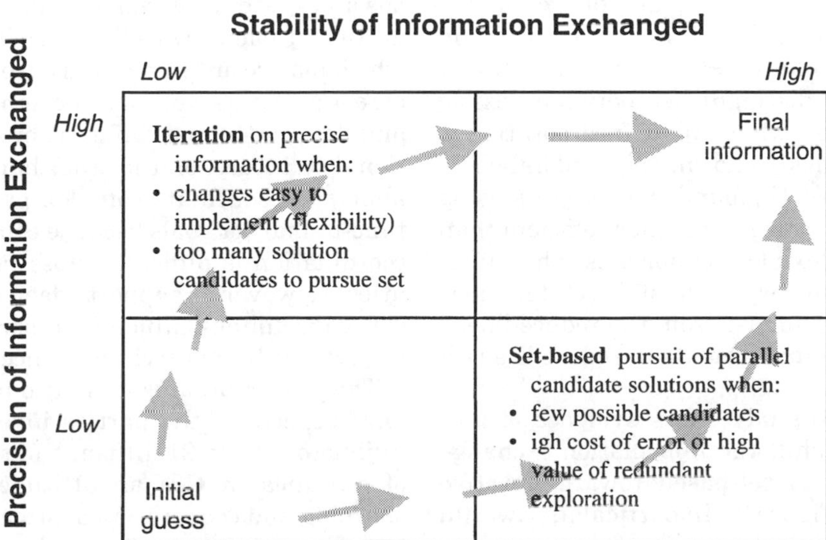

Fig. 3. Set-based and iterative approach to communication.

communication approaches are classified according to the nature of the information exchanged. Information precision refers to whether it communicated as a range or a precise value, while stability refers to whether the information tends to remain stable afterwards, or whether it is subject to change.

If the cost of making the wrong choice, and if the number of candidate solutions that must be pursued is relatively small, then the set-based approach offers advantages. On the other hand, if the development of a component is very flexible, i.e., engineering changes are very easy to incorporate, then the iterative approach is preferable.

Thus, the choice between set-based and iterative concurrence depends on the nature of the component developed, and often a combination is used. The choice is influenced by the technological and process capabilities of the organization. This is examined further in the next Section.

SUPPORTING INFRASTRUCTURE AND IMPLEMENTATION: The three aspects of CE (task definition, time and information concurrence) must be viewed within the context of a supporting infrastructure (Fitzsimmons *et al.*, 1991). Three infrastructure components are discussed below: technologies, project organization, and support processes.

Technology and Tools

New problem solving technologies can play an important role not only in diminishing the cost of iteration, but also in anticipating and eliminating project uncertainty in the first place. Examples of such new technologies are CAD, simulation tools, and rapid prototyping. These technologies have the potential of fundamentally changing the trade-offs involved in time and information concurrence. Thomke (1996) studied integrated circuit design and distinguished between flexible design technologies, which allow incorporation of design changes quickly and cheaply, and inflexible design technologies. He found that projects using flexible design technologies are more efficient than projects using inflexible technologies. The difference stems from the lower cost of direct iterations (design changes), and also from the reduced need for costly resource investments to reduce the risk of design changes.

Thomke's (1996) study offers evidence of how new technologies shift the information exchange trade-off away from set-based toward iterative concurrence (see Figure 3). In particular, powerful CAD systems allow the frequent information sharing and coordination necessary for activity overlapping in complex projects (a widely quoted example is the Boeing 777 jet, which was developed with heavy CAD usage).

Communication technologies also have the potential of influencing CE practices (e.g., Allen 1986). E-mail, portable phones, and shared databases all have the effect of making communication more time-efficient and independent of personal meetings between busy parties. However, even without considering implementation problems of incompatible data formats and protocols, communication technologies do not make communication instantaneous and painless.

Project Organization

Often, the structure of the project organization mirrors the logical structure of the design problem: the architecture determines communication patterns and physical proximity.

Commonly, four main types of project organizations are distinguishedable (e.g., Wheelwright and Clark 1992, Chapter 8). In a *functional structure*, people are grouped by discipline, and projects are coordinated via *a priori* agreed-upon specifications complemented by occasional meetings. In a *lightweight structure*, each function has a representation in a coordinating committee, led by a project coordinator. A *heavy weight structure* makes the coordinator a real project manager, with control over budgets and direct supervision of the functional representatives. Finally, the *autonomous structure* pulls the whole project outside of the regular organizational structure, and the project manager becomes akin to a line manager, who controls all resources and evaluates all project employees.

The strength of the autonomous structure is focus and speed. However, it often suffers from integration problems with the rest of the organization, which may result in rejected output and disrupted career paths, as well as lost commonality (of parts and designs) across different products. The functional structure, on the other hand, is weak in its ability to coordinate complex projects with tight time-to-market goals because communication and coordination is difficult across functional boundaries. However, it supports deep expertise in specialty disciplines, which is important in research projects with high technical uncertainty.

The project organization also influences behavioral aspects of the parties involved. Clark and Fujimoto (1991: 212f.) point out the importance of *attitudes* in CE: an attitude of information shareing and cooperation is preferred. Hauptman and Hirji (1996) confirm this observation in a multi-industry study of development projects using CE. They find that CE overcome the differences resulting from functional specialization: the parties involved must understand and trust one another sufficiently to be willing to share information. Once two-way communication happens, it tends to improve trust and employee satisfaction. For example, misunderstandings may arise from two engineering groups speaking different technical languages and having different perceptions of the same problem (e.g., Terwiesch and Loch, 1998).

In addition, Hauptman and Hirji find that CE requires a tolerance for releasing and using preliminary information, information that is imprecise and/or unreliable. Workers tend to minimize the risk of having to re-do tasks by suppressing information, procrastinating, *etc*. Active management is necessary to overcome such risk-avoiding behavior in implementing CE.

Support Processes

CE requires a number of effective support processes, interfaces with suppliers, engineering change orders, and interfaces among multiple development projects.

Clark and Fujimoto (1991) observed that close *integration with suppliers* in product development is necessary for applying CE. This was further studied by Liker *et al.* (1995), who identified important success patterns.

1. Simplify the coordination problem by outsourcing self-contained chunks.
2. Communicate clearly and often (e.g., by having resident engineers a the supplier's site).
3. Use a simple, stable development process that is well understood by both sides.
4. For suppliers participating in CE to develop their capability of solving technical problems without external help ("full service capability").

A general principle is that the investments and open information sharing required for involving suppliers in CE only work in long-term relationships with mutual commitment that prevents abuse and cheating. Such mutual commitment can be supported by investments in common infrastructure and systems (such as CAD or order management) and by establishing a mutually agreed-upon and understood development process. The process should describe who is responsible for which deliverables, what information must be exchanged, and what milestones must be met.

Engineering change orders are an almost unavoidable consequence of CE, as many design tasks started on the basis of preliminary information that is subject to change. ECs arise also during component integration, and during fine-tuning of the design to improve market success. EC's often consume significant resources and therefore deserve careful management. Five principles can help make the EC management process effective (Terwiesch and Loch, 1998):

- Give clear responsibilities and avoid unnecessary handoffs in the process.
- Manage capacity in order to reduce congestion. Technical personnel often have multiple responsibilities, which can result in congestion and delays.
- Set-ups occur every time an engineer switches from one activity to another. To reduce set-ups, engineers (just as in manufacturing) match batch their work, which can lead to substantial delays for jobs in the queue.
- A complex product with a highly integrated architecture makes EC management more difficult, as ECs will have a "snowball" effect on other components. In this situation, it is important to educate all personnel involved about the key interactions among components, and give strong incentives for fast problem communication and resolution.

In many companies, engineering work is often performed in functional organizations of specialists, who serve several projects at the same time. On the one hand, projects compete for the same resources, leading to congestion and interference with the intensive communication and coordination within the project (Adler *et al.*, 1995). This requires a clear prioritization of projects, not only when projects are initiated, but also during execution, when resource conflicts arise. Clear rules are necessary (e.g., degree of lateness) to settle conflicts speedily. On the other hand, the pursuit of multiple projects in parallel builds expertise in people. Nobeoka and Cusumano (1995) find that overlapping development projects for consecutive product generations, rather than sequential development projects, facilitate improved sharing of specifications and the transfer of knowledge.

OTHER BENEFITS OF CE: The operations management literature on CE has largely emphasized efficiency of product development, seeking to reduce time-to-market or development costs without mentioning product quality.

Nobeoka and Cusumano (1995) find that parallel execution of whole projects is advantageous not only for efficiency reasons, but to ensure effective knowledge transfer from one project to the next. Pisano (1997) reports that, in the pharmaceutical industry, the overlap between product and manufacturing process development is necessary for learning purposes. Overlapping may be necessary to experiment with the manufacturing process during development to detect major problems that cannot otherwise be foreseen. This is called "learning by doing."

Organizations Mentioned: Toyota.

See **Autonomous teams; Concurrent engineering; Convergence rate of a project; Downstream sensitivity; Flexible design technologies; Heavy weight project manager; Light weight project manager; Mass customization; Mass customization implementation; Modular architecture; Product architecture; Product design; Product design for global markets; Product innovation; Time to market; Work transformation matrix.**

References

[1] Adler, P.S., A. Mandelbaum, V. Nguyen, and E. Schwerer. "From Project to Process Management: An Empirically Based Framework for Analyzing Product Development Time," *Management Science*, 41, 1995, 458–484.

[2] Allen, Th. J. (1986). "Organizational Structure, Information Technology, and R&D Productivity." *IEEE Transactions on Engineering Management*, EM 33, 212–217.

[3] Blackburn, J.D. (1991). *Time Based Competition*, Homewood: Business One Irwin.

[4] Clark, K.B. and T. Fujimoto (1991). *Product Development Performance: Strategy, Organization and Management in the World Auto Industry*. Harvard Business School Press, Cambridge.

[5] Eisenhardt, K.M., and B.N. Tabrizi (1995). "Accelerating Adaptive Processes: Product Innovation in the Global Computer Industry." *Administrative Science Quarterly*, 40, 84–110.

[6] Eppinger, S.D., D.E. Whitney, R.P. Smith, and D.A. Gebala (1994). "A Model-Based Method for Organizing Tasks in Product Development." *Research in Engineering Design*, 6, 1–13.

[7] Fitzsimmons, J.A., P. Kouvelis, and D.N. Mallik. "Design Strategy and Its Interface with Manufacturing and Marketing: a Conceptual Framework," *Journal of Operations Management*, 10, 398–415.

[8] Gerwin, D., and G. Susman (1996). "Special Issue on Concurrent Engineering." *IEEE Transactions on Engineering Management*, 43, 118–123.

[9] Gulati, R.K., and S.D. Eppinger (1996). "The Coupling of Product Architecture and Organizational Structure Decisions." MIT Working Paper 3906.

[10] Ha, A.Y., and E.L. Porteus (1995). "Optimal Timing of Reviews in Concurrent Design for Manufacturability." *Management Science*, 41, 1431–1447.

[11] Hauptman, O., and K.K. Hirji (1996). "The Influence of Process Concurrency on Project Outcomes in Product Development: an Empirical Study with Cross-Functional Teams." *IEEE Transactions in Engineering Management*, 43, 153–164.

[12] Imai, K., I. Nonaka, and H. Takeuchi (1985). "Managing the New Product Development Process: How the Japanese Companies Learn and Unlearn." In: Clark, K.B., R.H. Hayes, and C. Lorenz (eds.). *The Uneasy Alliance*. Harvard Business School Press, Boston.

[13] Jaikumar, R. and R. Bohn (1992). "A Dynamic Approach to Operations Management: An Alternative to Static Optimization." *International Journal of Production Economics*.

[14] Krishnan, V., S.D. Eppinger, and D.E. Whitney (1997). "A Model-Based Framework to Overlap Product Development Activities." *Management Science*, 43, 437–451.

[15] Liker, J.K., R.R. Kamath, S.N. Wasti, and M. Nagamachi (1995). "Integrating Suppliers into Fast-Cycle Product Development." In J.K. Liker, J., J.E. Ettlie, and J.C. Campbell (eds.). *Engineered in Japan: Japanese Technology Management Practices*. Oxford University Press, Oxford.

[16] Loch, C.H., C. Terwiesch (1998). "Communication and Uncertainty in Concurrent Engineering," *Management Science* 44, August.

[17] Morelli, M.D., S.D. Eppinger, and R.K. Gulati (1995). "Predicting Technical Communication in Product Development Organizations." *IEEE Transactions on Engineering Management*, 42, 215–222.

[18] Nevins, J.L., and D.E. Whitney (1989). *Concurrent Design of Products and Processes*. McGraw Hill, New York.

[19] Nobeoka, K., and M.A. Cusumano (1995). "Multiproject Strategy, Design Transfer, and Project Performance." *IEEE Transactions on Engineering Management*, 42, 397–409.

[20] Pisano, G. (1997). *The Development Factory*. Harvard Business School Press, Boston.

[21] Smith, R. (1997). "The Historical Roots of Concurrent Engineering Fundamentals." *IEEE Transactions on Engineering Management*, 44, 67–78.

[22] Smith, R.P., and S.D. Eppinger (1997). "Identifying Controlling Features of Engineering Design Iteration." *Management Science*, 43, 276–293.

[23] Steward, D.V. (1981). *Systems Analysis and Management: Structure, Strategy and Design*. Petrocelli Books, New York.

[24] Takeuchi, H., and I. Nonaka (1986). "The New Product Development Game." *Harvard Business Review*, 64, 137–146.

[25] Terwiesch, C., C.H. Loch, and M. Niederkofler (1996). "Managing Uncertainty in Concurrent Engineering," Proceedings of the 3rd EIASM Conference on Product Development, 693–706.

[26] Terwiesch, C., and C.H. Loch (1997). "Management of Overlapping Development Activities: a Framework for Exchanging Preliminary Information." Proceedings of the 4th EIASM Conference on Product Development, 797–812.

[27] Terwiesch, C., and C.H. Loch (1998). "Managing the Process of Engineering Change Orders: The Case

of the Climate Control System in Automobile Development." Forthcoming in the *Journal of Product Innovation Management*, 1998.

[28] Thomke, S. (1996). "The Role of Flexibility in the Design of New Products: An Empirical Study." Harvard Business School Working Paper 96-066.

[29] Ulrich, K. (1995). "The Role of Product Architecture in the Manufacturing Firm." *Research Policy*, 24, 419–440.

[30] Ulrich, K., and S.D. Eppinger (1995). *Product Design and Development*. McGraw Hill, New York.

[31] Ward, A., J.K. Liker, J.J. Cristiano, D.K. Sobek II (1995). "The Second Toyota Paradox: How Delaying Decisions Can Make Better Cars Faster." *Sloan Management Review*, Spring, 43–61.

[32] Wheelwright, S.C., and K.B. Clark (1992). *Revolutionizing Product Development*. The Free Press, New York.

PRODUCT DIFFERENTIATION

Product differentiation refers to products that have unique attributes, which effectively distinguish one product from another in form, function, reliability, etc. In manufacturing, product differentiation is created at divergence points, where a material or component can be processed into a two or more distinctly different products. One environment, plants may process a single material into a variety of unique end items. For example, in the textile industry, yarn can be processed into a variety of different fabrics. In another type of environment, in assembly plants, where manufactured or purchased components are used as common parts to assemble a variety of different products. For example, in the electronics industry, a number of common components are used to assemble a wide variety of distinct end products.

See **Manufacturing strategy; Mass customization; Mass customization implementation; Product design; Product design for global markets; Synchronous manufacturing using buffers.**

References

[1] Cox, J.F., J.H. Blackstone, and M.S Spencer (eds.) (1995). *APICS Dictionary*, Eighth Edition, American Production and Inventory Control Society, Falls Church, Virginia, 65.

[2] Umble, M.M. and M.L. Srikanth (1997). *Synchronous Management: Profit-Based Manufacturing For The 21st Century, Volume Two, Implementation Issues and Case Studies*, Spectrum Publishing, Guilford, Connecticut.

PRODUCT EXTENSION

Product Extension is used to develop standardized products for global markets. Under this approach, a product designed for a single market, usually the home market of the international enterprise, is sold to other countries. In a way, the international enterprise is "projecting" a product released in a single market to overseas markets. Product extensions may ignore the idiocyncracies of the international environment. Companies that adopt the product extension approach are usually companies that are entering their internationalization process by beginning their exporting activities. In such cases, the company may not have the resources or the know-how to customize its products for overseas markets. Furthermore, the product extension approach is adopted when the market for which the product was originally designed for sets the trend in consumer demand. The product extension approach has a higher probability of being successful when the foreign markets in which the standardized product is sold are "similar" (in terms of consumer tastes) to the market that the product was originally designed for.

See **Mass customization; Mass customization implementations; Product design; Product design for global markets.**

PRODUCT INNOVATION

Danny Samson

University of Melbourne, Parkville, Australia

DESCRIPTION: New product development is vital to the survival of most organizations with the exception of pure commodity products. Once an organization has its basic infrastructure in place it has value to be gained by renewing its product and service offerings. Organizations that do not invest or that under-invest in new products are in danger of losing their customers' interest. Hence, product innovation is a strategic item for competitive advantage.

In some industries, such as the automobile industry, it is not enough to produce a quality product at a good price in an efficient manufacturing process. To ensure that sales and profits continue, product innovation essential. In many industries, cost, quality and delivery reliability are 'qualifiers' that keep a company alive in the short term but do not guarantee longevity.

Product innovation is becoming the real order-winner in more and more industries. As an illustration, consider the rebirth of Chrysler over the past decade. The cornerstone of their revitalization was the marriage of manufacturing process improvement with a series of successful product innovations. Their new models captured the imagination of consumers.

SUCCESS FACTORS: Many researchers have studied product innovation and produced checklists of success factors. The work of McQuater *et al.* (1998), Cooper and Kleinschmidt (1987) and Barczak (1985) can be synthesized to produce the following success factors:

- Product leader
- Innovation/development protocols and well-defined disciplines and processes
- Technological strength
- Research and pre-development investments
- Marketing expertise
- Technological and marketing synergy and interfaces
- Appropriate timing
- Product champions
- cross functional teams
- Idea generation and screening processes
- Tolerance of risk and senior management support
- Product innovation strategy

In an international study of innovative practices Quinn (1985) found that successful large firms, like many small entrepreneurial companies, accept the essential chaos of innovation and product development. According to Quinn, they pay close attention to their users' needs and desires, avoid detailed early technical or marketing plans, and allow entrepreneurial teams to pursue competing alternatives within a clearly conceived framework of goals and limits. Downside risks are minimized not by detailed controls but by spreading risks among multiple projects, and by keeping early costs low. Quinn also noted that entrepreneurial firms adopt different practices than traditional firms to project justification, decision making and incentive schemes.

Schewe (1994) studied the question of whether innovative success is a function of firm related variables (e.g., innovation experience, market and customer knowledge, distribution channels) or project related variables (e.g., technical skills, development costs, development time etc.). He concluded that, above all, innovative success is determined by the overall capabilities of the innovative firm.

Steele (1989) has identified some of the hard-won lessons about risks and success in innovation, which can be applied to process and product innovation:

- Unless a new technology offers real value to the customer in terms of functionality or performance it will not succeed as an innovation
- Technologies are systems, and unless the total system is managed, success cannot be assured
- Although the customer or user determines the ultimate success or failure of an innovation, the criteria that the customer uses may be multidimensional and vary with time
- A radically new technology usually requires a new business organization for success.

Betz (1993) identified speed and "correctness" as the essential criteria for evaluating a product development process, and indicated the difficulty in achieving them due to the uncertainties, variations and changes that need to be accommodated in the product or process development cycle. Bower and Hout (1988) illustrate the importance of these attributes by analyzing Toyota's fast cycle capability. They found that Toyota aimed to introduce new processes and products ahead of competitors. Through constantly trying out new models, Toyota studied customer choices. This enabled the company to keep up with changing market trends and facilitate the development of a "customer oriented culture."

For successful technological innovation to occur, the process of innovation needs to be carefully managed. Many authors have stressed the complex nature of the innovation process. An appreciation for the complexity of the innovation process can be gained by reviewing some of the many models that have been developed to help describe it. These range from simple "pipeline" or "black box" models to more complicated models. Some models focus on consumer or industrial product innovation, others are concerned with technical process innovation, and still others attempt to have a more universal application. It is generally agreed that no model appears to be capable of being a general model of innovation (Saren, 1984). Models may be catalogued (Forest, 1991) as:

Stage Models: Innovation is treated principally as a sequential activity where the process of innovation is viewed as a series of discrete stages. The unit of analysis within each stage is generally activity or functional responsibility. Examples of stage nodels are found in the works of Utterback (1971) and Saren (1984). Although useful, these models overlook the non-linearity of the innovation process. Kelly and Krantzberg (1978) observed that these models provide a poor representation of the complexity of the innovation process.

Conversion Models (Technology Push/Market Pull Models): Technological innovation has also been described as a conversion process with inputs being converted through a number of steps to products or processes. Twiss (1980) developed the simplest form of model, which is a black box model where the inputs of materials and knowledge are converted to outputs. While a number of models have been developed based on technology push or market pull, pure technology push and market pull models are viewed as extreme examples of innovation models.

Integrative Models: These models attempt to integrate process and product innovation in respect to the technology lifecycle (birth, prosperity, maturity). In the model developed by Utterback and Abernathy (1978), the dynamic nature of the innovation process was stressed, and the process was represented as changing over time as the industry evolved and matured. In this life cycle view, the initial focus was on new process or product development, and later it changed to one where the focus was on process optimization and cost reduction. Although appealing and extremely useful, the model failed to adequately represent the innovation process when tested over a range of industries (Utterback and Abernathy, 1981).

Decision Models: Throughout the process of innovation, many decisions need to be made, and many models of the innovation process approach product innovation in terms of these decision points. Some point to the evolutionary nature of the decision making process, and show that decisions take place between the various activities involved in the process of innovation. Various authors however have commented on the lack of generalizability of these models, and indicated that the constructs used and approaches taken limit their use to particular industry segments.

Other models have also been developed which fall outside these categories. For example Kanter (1985) conceptualizes the innovation process as "newstream" activities that must be autonomous of the mainstream value adding processes such as production, distribution and marketing as they are fundamentally different in nature and form. The author develops descriptions of the relationships that need to exist between the newstream and mainstream in order for innovation processes to be effective. This approach is powerful in recognising that investment in research and development must provide a business return, and that return is gained through the increase in competitiveness of the mainstream of the business.

Forest (1991) makes the following comments in respect of the degree to which these models are representative of the innovation process:

- Most models overlook the complexity of the pre-innovation stage: idea generation and screening.
- Although useful in conceptualizing the process, most models have not been of value to the strategic planning processes.
- Environmental considerations (political, economic, socio-cultural etc.) which can have a significant impact on the innovation process are generally excluded.
- The "human element" is overlooked. The impact of champions of innovation (organizational agents, governments, trade unions, employer bodies, etc.) has often not been considered.
- Strategic alliances, which are generally regarded as a key component in developing an innovative capability, have not been considered.
- Models fail to recognize the cumulative and somewhat chaotic nature of innovation. Any construct or process applied to manage the innovation process, needs to have adequate feedback mechanisms and flexibility to address the nature of the innovation process which is by it's very nature both evolutionary and revolutionary, deductive and inductive. Einstein once stated that "I never came upon any of my discoveries through a process of rational thinking."

Hayes *et al.* (1988) provide a comprehensive guide to good practice in product development, pointing out the need for speed in development cycle time, and the importance of coordinating product and process development.

Organizations Mentioned: Chrysler; Toyota.

See **Product design; Product design for global markets; Product development and concurrent engineering; Concurrent engineering.**

References

[1] Barczak, G. (1995). New Product Strategy, Structure, Process and Performance in the Telecommunications Industry, Journal of *Product Innovation Management*, V12, N3, pp. 374–391.

[2] Betz F. (1993). "Strategic Technology Management" Mc Graw-Hill, New York.

[3] Bower J. and Hout T. (1988). "Fast Cycle Capability For Competitive Power." *Harvard Business Review, Nov/Dec.*

[4] Cooper, R.G., E.J. Kleinschmidt (1987). New Products: What Separates Winners from Losers?, *Journal of Product Innovation Management*, V4, N3, pp. 169–184.

[5] Forest J. (1991). "Models For The Process Of Technological Innovation." *Technology Analysis And Strategic Management* V3, No 4.
[6] Hayes, R.H., S.C. Wheelwright, K.B. Clark (1988). *Dynamic Manufacturing*, Free Press, NY.
[7] Kelly P. and Krantzberg M. (1978). "Technological Innovation: A Critical Review Of Current Knowledge" San Francisco Press, USA.
[8] McQuater, R.E., A.J. Peters, B.G. Dale, M. Spring, J.H. Rogerson, E.M. Rooney (1998). The Management and Organisational Context of New Product Development, *International Journal of Production Economics*, V55, pp. 121–131.
[9] Quinn J. (1985). "Managing Innovation: Controlled Chaos." *Harvard Business Review*, May/June.
[10] Saleh S. and Wang C. (1993). "The Management Of Innovation: Strategy Structure and Organisational Climate" *IEEE Transactions On Engineering Management*, V40, No. 1.
[11] Saren A. (1984). "A Classification And Review of Models of The Intra-Firm Innovation Process," R&D Management, 14.
[12] Schewe G. (1994). "Successful Innovation Management: An Integrative Perspective." *Journal Of Engineering and Technology Management*, 11.
[13] Steele L. (1989). "Managing Technology: The Strategic View," McGraw-Hill, New York.
[14] Twiss, B.C. (1980). "Managing Technological Innovation" 2nd ed., Longman Group, London.
[15] Utterback J. (1972). "The Process Of Technological Innovation Within Firms," Academy Of Management Journal.
[16] Utterback J.M. and Abernathy W.J. (1981). "Multivariant Models for Innovation/Looking at the Abernathy and Utterback Model with Other Data," OMEGA, 9.

PRODUCT LIFE CYCLE

Products pass through distinct stages of demand. Typically, these stages include product introduction, growth, maturity, and decline. It is understood that each of these stages requires different marketing, financial, manufacturing, purchasing, and human resource strategies. Sometimes the term "product life cycle" is used to indicate the length of time between product introduction and product obsolescence. In recent years, many products have continually shrinking product life cycles due to increased competition, rapid changes in technologies, and consumer tastes. The figure below is a graphical representation of product life cycle. It shows that HD TV is in the Introduction Phase, Internet business is in Growth Phase, and so on.

See **Concurrent engineering; New product development through supplier integration; Product development and concurrent engineering; Product-process dynamics.**

PRODUCT MARGIN

Product margin is the difference between selling price and product cost. Product cost is the cost of producing a product or service that generates revenues through sales. This cost includes the cost of direct materials, direct labor, and overhead. Any readily identifiable part of product (such as the

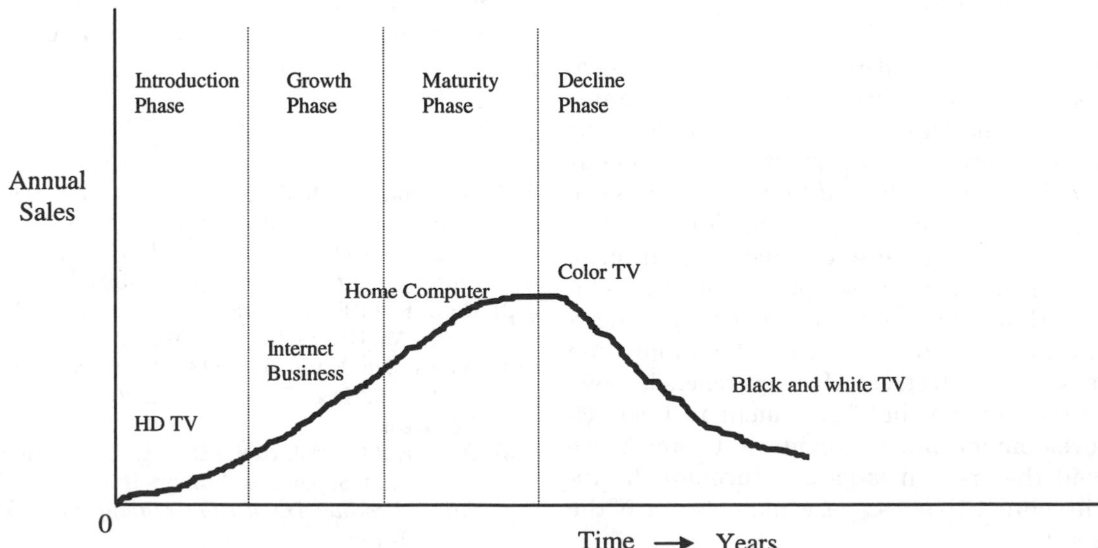

wood used in the production of a table) constitutes direct material. Direct labor refers to the time spent by employees, who work specifically on manufacturing the product or performing the service. Indirect costs, or overhead, include any factory or production cost that is indirect to the product or service. Examples of overhead costs include costs of management and supervision, depreciation of equipment, and insurance.

See **Activity-based costing; Activity-based costing: An evaluation; Target costing; Unit cost.**

Reference

[1] Barfield, Jesse T., Cecily A. Raiborn, and Michael R. Kinney (1994). *Cost Accounting: Traditions and Innovations*, West Publishing Company, St. Paul, Minnesota.

PRODUCT PRODUCIBILITY

The term, producibility, expresses how easily a product can be reliably produced in the desired volume using the existing manufacturing capabilities of a firm. Producibility can be measured in a number of ways, including the time or cost that must be expended to produce the item. Process capability, production yields, and production speed are other attributes frequently used to measure producibility. These measures may be estimated quantities (during product design phases) or actual outcomes (once the product has reached volume production levels). It is important to remember that product producibility is mostly a function of product design. Once the design is solidified, it is difficult to improve producibility.

See **Concurrent engineering; Lean manufacturing implementation; Mass customization; Product design; Product design for global markets; Product development and concurrent engineering.**

PRODUCT PROLIFERATION

Product proliferation is a charge frequently leveled against some organizations that they are marketing so many new products that economic resources are wasted, or consumers are confused. The new products themselves are frequently minor modifications or extensions of existing products in the companies' product lines, and are based on insignificant or trivial differentiation from existing products. Examples include minor alterations to product packaging, new sizes; new colors or flavors of a product; and the addition of unimportant features to a product.

The dangers of such additions to a product line are, first, that they may require considerable incremental resources from the firm which may not be justified by the incremental increase in sales generated. In fact, they may take sales away from successful products already existing in the firm's product line. Secondly, consumer's may be unable to distinguish any real differences in the new products. If consumers fail to benefit from the new products, is the effort justified?

See **Customer service, satisfaction and success; Mass customization; Mass customization implementation.**

Reference

[1] Kotler, Philip (1997). *Marketing Management: Analysis, Planning, Implementation, and Control* (9th ed.). New Jersey: Prentice-Hall.

PRODUCT REALIZATION TEAMS

Concurrent engineering teams are sometimes called product realization teams.

See **Concurrent engineering; Lean manufacturing implementation; New product development through supplier integration; Product development and concurrent engineering, teams: Design and implementation.**

PRODUCT THROUGHPUT TIME

Production throughput time is the time between the start of production until it is done. The shorter the better. In a job shop, partly finished products spend considrable time waiting before processing. It has been found that when products move through production swiftly, the efficiency of the plant is high and its product quality improves.

See **Cycle time; Manufacturing system design; Theory of constraints in manufacturing management; Synchronous manufacturing using buffers.**

PRODUCTION ACTIVITY CONTROL

Several activities keep production going smoothly. Production activity control is carried out by the production planning and control functions. They manage routing, shop and vendor scheduling, dispatching, order tracking, and input/output control.

See **Input/Output control; Manufacturing planning and control; MRP; Order release; Schedule stability**

PRODUCTION FLOW ANALYSIS (PFA)

Group Technology (GT) principles are used in designing manufacturing cells where a family of similar parts or products are produced. Group technology uses PFA methodology for finding families of parts and groups of machines based on routing information to develop manufacturing cells. Manufacturing cells have revolutionized manufacturing.

See **Group technology; Linked-cell manufacturing system; Manufacturing cell design.**

PRODUCTION INPUT FACTORS

To manufacture goods, several inputs are essential. Manufacturing transforms inputs into value for the customer. Traditionally one considers three types of input factor: labour, capital, materials and components. In more recent times, management systems and information are considered as additional factors.

See **International manufacturing production kanban; Production ordering kanban.**

PRODUCTION ORDERING KANBAN (POK)

Kanban is a Japanese term which refers to a card used in the pull inventory and production control system to signal to an upstream process to begin to make the appropriate parts listed in the card. This card is used only in the two card Kanban system.

A production-ordering Kanban specifies the type and quantity of product, which a work center must produce at the request of the next work center in the sequence of operations. A Production-ordering Kanban is also called an in-process Kanban or a production Kanban.

See **Just-in-time manufacturing; Just-in-time manufacturing implementation; Kanbans; Kanban-based manufacturing systems; Linked-cell manufacturing systems.**

PRODUCTION PLANNER

A production planner is an individual who plans, schedules, and monitors production. A manufacturing facility may have many production planners, each responsible for a particular portion of the factory or a subset of the products in the factory. Production planners create production plans, construct master production schedules, perform material requirements planning, release orders to the shop, and create schedules for the shop.

Most factories use a manufacturing planning software to assist the planner perform production planning, master production scheduling, and material requirements planning. The production planner uses the system and approves plans, schedules, and order releases. Such a system can make a production planner more productive by automating tedious work. It can also allow the planner to make better decisions by supplying more accurate and up-to-date information.

See **Capacity planning: Medium- and short-term; Master production schedule; MRP; MRP implementations; Production planning and control; Order release; Resource planning.**

Reference

[1] Vollmann, Thomas E., William L. Berry, D. Clay Whybark. (1997). *Manufacturing Planning and Control Systems*, Fourth Edition, Irwin/McGraw-Hill, New York.

PRODUCTION PLANNING AND CONTROL

See **Capacity planning: Manufacturing planning and control; Medium- and short-range; MRP.**

PRODUCTION REPORTING

Periodically on a daily, weekly, and monthly basis, a statement of output of a factory is produced and distributed to decision makers. Such a production report includes type and quantity of products provided, efficiency by work center or department, labor hours, overtime used, and so on. The report serves as a feedback and permits corrective action and maintenance.

See **MRP; Order release; Production planning and control; Schedule stability.**

PRODUCTION RESOURCE

Any resource used to make a product is considered a production resource. Such resources can include machine time, labor, or facilities. In general, a production resource will be available in limited quantities, thus planning activities must be undertaken to ensure that sufficient quantities of all resources are available when needed. For example, suppose a given product requires one hour of assembly time per unit in an assembly facility which has a total of eight hours capacity per day. If the company plans to produce more than eight units per day, then such a plan can be realized only if the assembly facility capacity is increased beyond eight hours per day.

Additional amounts of most resources can be obtained in the long run. Long-range capacity planning considers such resource additions through new construction. In the shorter term, capacity planning often must deal with available facilities: additional shifts, overtime, etc. are options for capacity increase in the short term.

See **Capacity planning: Long range; Capacity planning medium- and short-range.**

PRODUCTION RULE BASE

A production rule base contains all information needed by the flexible manufacturing system (FMS) to produce a part according to the quality, time and cost expectations. The production rule base feeds the dynamic scheduler with instructions, typically in the form of "if-then-else" and/or "in case of" structures, explaining exactly what needs to be done on the FMS (Flexible Manufacturing System) with "what part," in "what setup," on "which machines," using "which part program," and in "what order."

An FMS is a highly automated, distributed and feed-back controlled system of data, information, and physical processors, computers, machine tools, robots and others. The production rule base enables complex decisions of the FMS to be made by the dynamic scheduler on a real-time basis.

See **Flexible automation; Flexible manufacturing systems; Manufacturing flexibility; Manufacturing with flexible systems.**

Reference

[1] Ranky, P.G. (1998). *Flexible Manufacturing Cells and Systems in CIM*, CIMware Ltd, http://www.cimwareukandusa.com (Book and CD-ROM combo).

PRODUCTION RUN

When a component or product is produced in a factory, the production run normally refers to the batch size or quantity of a specific component or product produced before resetting the machine to make a batch of a different product. Every production run normally requires a variety of setup activities such as equipment preparation, cleaning, adjusting, and tools and fixtures change. Therefore, every production run normally has some associated setup time and setup costs. A production run size that is supposed to minimize all relevant costs including carrying costs and setup costs may be determined by various formulas to compute an economic run quantity. Some contemporary production systems, such as kanban or drum-buffer-rope systems, require that the production run size should be variable and determined by the requirements of the overall system.

See **Safety stocks: Luxury or necessity; Setup reduction; Synchronous manufacturing using buffers.**

References

[1] Stevenson, W.J. (1996). *Production/Operations Management*, Richard D. Irwin, Chicago, Illinois, 543–544.

[2] Srikanth, M.L. and M.M. Umble (1997). *Synchronous Management: Profit-Based Manufacturing For The 21st Century, Volume One*, Spectrum Publishing, Guilford, Connecticut.

PRODUCTION SCHEDULING

See **Manufacturing planning and control; MRP; Order release; Production planning and control; Schedule stability.**

PRODUCTION SHARING FACILITIES

International production sharing has become a popular cost-reduction strategy for companies throughout the industrialized world. By taking advantage of worldwide resources – particularly inexpensive labor – firms in Europe, Japan, and the U.S. are able to compensate for high domestic wage rates and improve their overall cost position. At the same time, these firms begin to establish a local presence in the emerging markets, where they perform the low-cost manufacturing. In essence, production sharing can potentially improve firm performance by locating productive activities in different countries to take advantage of inherent factor and market advantages.

Production sharing, also known as co-production, is generally a part of a broader manufacturing strategy – globalized manufacturing rationalization. Perhaps the best known production sharing situation in the United States involves the so-called "maquiladoras" of Mexico. These maquiladora facilities are low-cost manufacturing and assembly operations established in Mexico south of the U.S.-Mexico border to take advantage of low-cost Mexican labor, and the U.S. market. The key to successfully using production sharing is to use a thorough total cost analysis before making the facility location decision. All relevant manufacturing, logistics, international trade, supply base, marketing, and engineering issues should be considered in the analysis.

See **Facility location decisions; Global facility location; Global manufacturing rationalization; International manufacturing; NAFTA.**

References

[1] Fawcett, S.E. (1990). "Logistics and Manufacturing Issues in Maquiladora Operations." *International Journal of Physical Distribution and Logistics Management*, 20(4), 13–21.

[2] Davis, E. (1992). "Global Oursourcing: Have U.S. Managers Thrown the Baby Out With the Bath Water?" *Business Horizons* (July–August), 58–65.

PRODUCT-PROCESS DYNAMICS

Paul M. Swamidass and Neil R. Darlow

Auburn University, Auburn, Alabama, USA

The product-process matrix in Figure 1, proposed by Hayes and Wheelwright (1979a, 1979b, 1984), is a tool for analyzing the links between the maturation of products and the processes used in their manufacture. The path taken by the products and processes of individual companies over time can be explained using the matrix. The product-process concept has a strong impact on a company's *manufacturing strategy* and its *focus*. Linking product and process considerations brings manufacturing managers and marketing managers together in the strategy making process.

There are at least five uses for the product-process matrix:

1. Determining the appropriate mix of manufacturing facilities, setting the key manufacturing objectives for each plant.
2. Reviewing investment decisions for plant and equipment in terms of their consistency product and process plans.
3. Determining the direction and timing of major changes in a company's production processes.
4. Evaluating product and market opportunities in the light of the company's manufacturing capabilities.
5. Selecting an appropriate process and product combination for entry into a new market.

PRODUCT LIFE CYCLE: The product life cycle is a long-established concept from business strategy literature. A product passes through phases similar to a living organism, which are, birth, rapid growth, maturation, decline, and death. The life of a product is the length of time from its market introduction to discontinuation. The market for a new product begins slowly and creates interest in the marketplace. Once accepted, its sales grow rapidly. Later, the product matures and is accepted as standard, and sales level off. Finally, in the last stage of the life cycle, the product may lose its popularity or lose its markets to competitors' products, and sales subsequently dwindle. At some point thereafter, it is wise to remove it from the company's portfolio of products.

A product life cycle is easy to understand and visualize, and can be applied to an individual product, a company, or an industry. However, the simple product life cycle model does not address

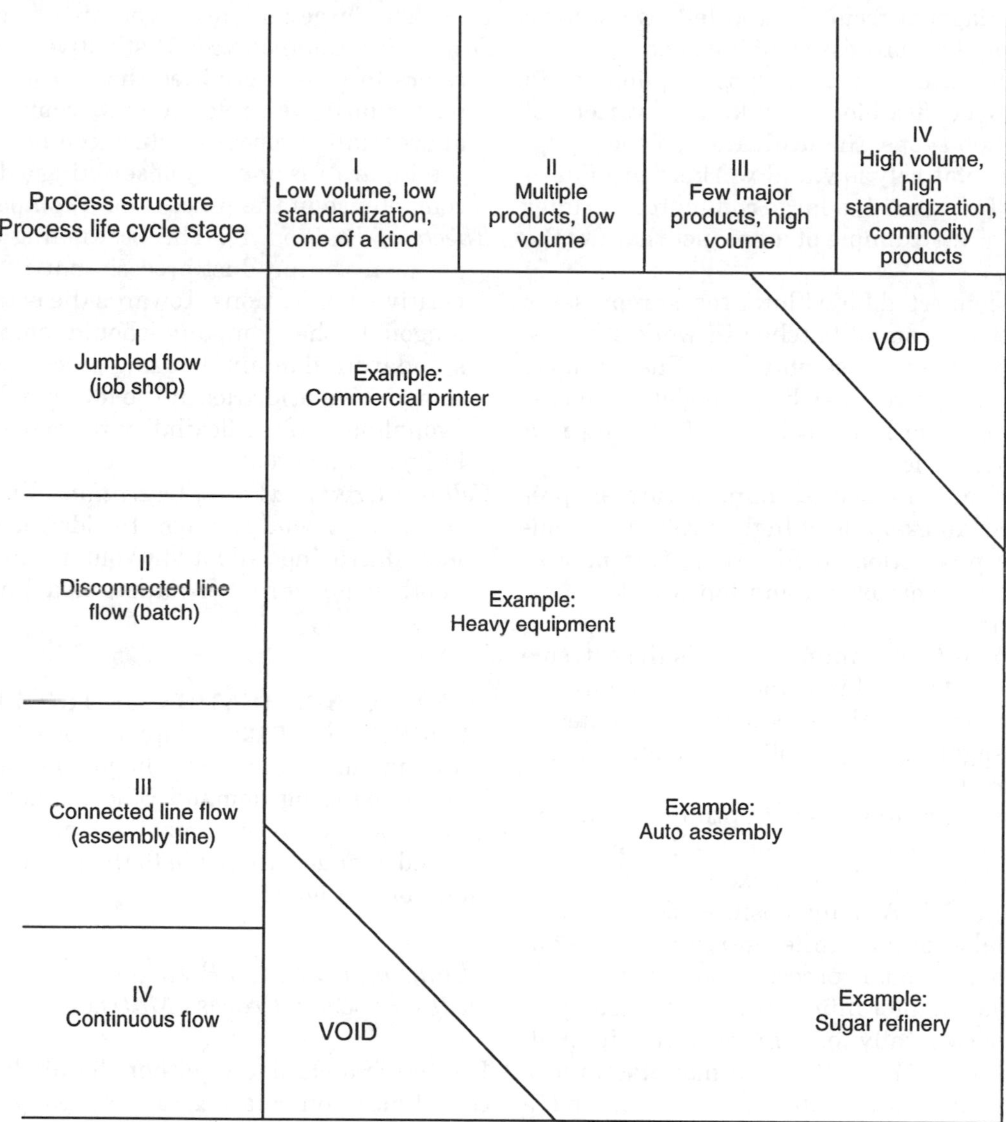

Fig. 1. The product-process life cycle matrix. (Source: Hayes and Wheelwright, 1984.)

the implications for the manufacturing function, which has to cope with demand changes and accompanying process changes during the different life cycle stages of a given product.

PROCESS LIFE CYCLE: The process life cycle addresses the evolution in production processes as a product's volume requirements change over its life cycle. For a new product, a highly flexible manufacturing facility is usually used, where the product is produced in small batches. Later, as the product enters high volume production, automation in production increases, with a resulting loss of manufacturing flexibility. Product and process life cycles should be matched appropriately for a company to remain competitive as a product progresses through its life cycle. Strategic positioning of a manufacturer is possible by the appropriate choice of process as the product goes through its life cycle. This phenomenon is illustrated using the product-process matrix in Figure 1.

PRODUCT-PROCESS MATRIX: The rows in the matrix in Figure 1 represent four types of manufacturing processes that could be used, and the columns in the matrix represent increasing volume of production as the product progresses through the product life cycle. The examples of manufacturers given on the diagonal represent appropriate product-process combinations. The top right and bottom left corners are void because they represent industries that are infeasible: job shop manufacture of commodities, and repetitive manufacture of one-of-a-kind products. Sliding down the diagonal from the top left, we see the following product-process combinations:

1. Job shops use general purpose equipment in order to be flexible in making a variety of jobs. In job shops, the utilization of the equipment is relatively low and workers are multi-skilled. In Figure 1, a large commercial printer is cited as an example of a product that fits the process.
2. The Disconnected Line Flow process represents production in small batches in work stations, or in low-volume assembly line. The products that fit this process are basic models with customization; machine tool manufacturing is a typical example.
3. Traditional automobile manufacture is presented as an example of higher volume assembly line production in Figure 1, this process uses high levels of automation and less customization.
4. Continuous flow manufacturing is the extreme example of high volume and high standardization employed in the production of commodities. A sugar refinery is offered as an example in Figure 1.

The diagonal represents the optimal production process for a given demand volume. However, not all companies are positioned exactly on the diagonal in Figure 1. A plant positioned away from the diagonal allows for differentiation and variation on product and/or process: moving vertically from the diagonal for differentiation through process, and horizontally for differentiation in product. For example, Rolls-Royce, a manufacturer of exclusive luxury automobiles, can command a price premium by manufacturing in an environment more typical of a job shop than an assembly line. "Hand-made" low volume cars such as the Rolls-Royce command a premium price, because the manufacturer positions production vertically above the diagonal where typical high volume automobile production should be in Figure 1. However, if the intention is not to gain such differentiation, a company must be wary of deviating too far from the diagonal, because the company risks uncompetitive production. That is, such a move could leave the company without markets and/or vulnerable to attack from its competitors.

USING THE MATRIX: The product-process matrix can be used in at least three different ways: (1) to identify the distinctive competence of the organization; (2) to understand the competitive implications of selecting a product-process combination; and (3) to be the basis for organizing the factory into manufacturing cells, each focusing on one exclusive manufacturing task and market need. The three uses are elaborated below.

Distinctive Competence: Distinctive competence refers to those activities that a company does particularly well relative to its competitors. An appropriate process-product combination chosen Figure 1 is one way of selecting a distinctive competence in the process for a company.

Effects of Positioning: The positioning of a company on the product-process matrix has competitive implications. Towards the bottom of the diagonal, the company should employ more standardized manufacturing processes, and the competitive priorities at the lower end of the diagonal shift from flexibility to increased reliability and low costs.

Cellular Organization of Operations: Product-process matrix would support the idea of a separate manufacturing cell for low volume models while another cell for higher volume models within the same factory.

DYNAMIC APPLICATION OF THE PRODUCT-PROCESS MATRIX: Figure 1 shows how a company may move along the diagonal or away from it with changing demand. The competitive edge and/or focus of an organization can be lost if ill-advised changes are made to the plant's processin violation of Figure 1.

Changing a Plant's Position on the Product-Process Matrix

If there is a change in either the product or process dimension relating to a company without a corresponding move along the other axis, it may have marketing and manufacturing implications, which should be addressed. An example of a problem is the automation of manufacturing processes without considering the accompanying product strategy. Adding additional incompatible manufacturing processes as a response to increased markets may cause problems. Subjecting the manufacturing function to piecemeal improvements that evolve and expand can result in the attendant loss of plant focus. As the manufacturing processes change, the company may no longer be competitive due to diluted focus.

Instead of simply expanding either the process or product, the core competences of the plant should be nurtured, protected and enlarged by making coordinated changes in process and product dimensions. Thus, the product-process matrix could be used when planning for company growth. This issue is addressed in greater detail below.

GROWTH AND THE PRODUCT-PROCESS MATRIX: A company's growth may take many different forms. Four generic types of growth that a company may pursue are:

Type 1: Simple growth of sales volume within an existing product line and market.

Type 2: Expansion of the product line within a single market, using an existing process structure (often called product proliferation).

Type 3: Expansion of the process (usually termed vertical integration).

Type 1 Growth: Incremental changes to both product and process may be necessary as product volume increases in existing markets. The average company in the industry may progress down the diagonal of the matrix from top left to bottom right. Exceptions are those industries where extreme standardization is resisted by the customer, as in the case of home construction. An individual company's path may diverge from the diagonal, reflecting the fact that companies usually change the product alone or the process alone followed by a matching change in the process or product at a later date. Thus, the progression down the diagonal line usually takes place in a series of alternating horizontal and/or vertical steps, determined by product maturation and process innovation. In another growth scenario, the path traced by a growing company may parallel the diagonal, i.e., be consistently to the right or left of the diagonal.

A path Left of the diagonal indicates that the plant has the capability to be flexible in making rapid changes to production volumes and processes, but may be vulnerable to competition on several other dimensions: cost, delivery reliability and possibly conformance quality. Moving along a path consistently right of the diagonal gives greater economies of scale at the expense of flexibility, but this strategy requires relative stability in product design, volumes and requires additional investment in processes. Whether the path is maintained to the left or the right of the diagonal, as the distance between the company's path and the diagonal path increases, the firm becomes increasingly dissimilar from its competitors; it then occupies a niche.

The product-process matrix incorporates the principles of the learning curve: i.e., a percent reduction in unit costs with each doubling of production volume. The matrix gives credit to the fact that standardized processes or increased production volumes provide opportunities for cost reduction. The diagonal path down benefits from both product and process improvements that lead to cost reduction. Companies which choose to move along a path to the left of the diagonal sacrifice some cost savings per unit increase in volume but maintain process flexibility.

Type 2 Growth: In this process case, growth occurs from broadened product line. The addition of more standardized products while maintaining other less standardized products in the product line is one case of growth. Marketing and manufacturing functions often find it difficult to resist this form of product proliferation, since marketing sees itself as offering a full range of products and manufacturing sees the additional products as a vehicle to distribute fixed production costs. However, the production processes may not easily adjust to this change in product strategy; without appropriate organizational support, quality and delivery reliability may suffer. The addition of special features to existing standard products constitutes a shift to the left of the diagonal, in other words, to an earlier stage of the product life cycle. This can create problems associated with loss of focus if the standardized production process lacks the flexibility to handle product design changes.

Type 3 Growth: Vertical integration involves the manufacture of components or products that were previously bought from a supplier. The effect is similar to product proliferation, with attendant process proliferation. The creation of separate focused factories or cells to address the expansion in business is one way of overcoming potential loss of focus and related problems.

UPDATING THE NATURE OF THE PRODUCT-PROCESS MATRIX: During the 1980s and 1990s, the development of advanced manufacturing philosophies such as Just in Time, Flexible Manufacturing Systems, Lean Production, and Agile Manufacturing resulted in changes to the tradeoffs and relationships implied in the product-process matrix. The changes occurring in manufacturing as a result of these advances are: smaller batch sizes have become economical; reduction in inventory levels produces cost savings; and 'mature' products such as the automobile can be customized. Therefore the discussions surrounding the matrix in Figure 1 need updating. Mainly, greater cost efficiencies are now possible in low volume production and greater customization is now possible in high volume production. Further, product life cycles have been drastically reduced, and new products are introduced more frequently. The effect of all this is that a plant can move vertically or horizontally in Figure 1 within reason without a compelling need to make a matching move along the other axis to remain competitive. That is, plants are now inherently more flexible across the board. For another discussion of the implications of these changes in manufacturing, see Spencer and Cox (1995). Pisano (1997) demonstrates the continuing importance of the link between product and process innovations in a discussion of developments in the pharmaceuticals industry.

See **Activity-based costing; Assembly line design; Concurrent engineering; Focused factory; History of manufacturing; Just-in-time manufacturing; Lean manufacturing implementation; Learning curves; Manufacturing core competencies; Manufacturing flexibility; Manufacturing strategy; Manufacturing systems; Mass customization; Mass customization implementation; Product development and concurrent engineering.**

References

[1] Hayes, R.H. and S.C. Wheelwright (1979a). "Link manufacturing process and product life cycles". *Harvard Business Review*, (January–February) 133–140.

[2] Hayes, R.H. and S.C. Wheelwright (1979b). "The dynamics of process-product life cycles". *Harvard Business Review*, (March–April) 127–136.

[3] Hayes, R.H. and S.C. Wheelwright (1984). *Restoring Our Competitive Edge: Competing Through Manufacturing*, Wiley, New York.

[4] Pisano, G.P. (1997). *The Development Factory*, Harvard Business School Press.

[5] Spencer, M.S., and J.F. Cox (1995). "An analysis of the product-process matrix and repetitive manufacturing". *International Journal of Production Research*, 33(5), 1275–1294.

PROFESSIONAL ASSOCIATIONS FOR MANUFACTURING PROFESSIONALS AND MANAGERS

APICS (The Educational Society for Resource Management): The Educational Society for Resource Management is a not-for-profit international educational organization respected throughout the world for its education and professional certification programs. With more than 70,000 individual members worldwide, APICS is dedicated to using education to improve the business bottom line. APICS is recognized globally as:

- the source of knowledge and expertise for manufacturing and service industries across the entire supply chain in such areas as materials management, information services, purchasing and quality
- the leading provider of high-quality, cutting-edge educational programs that advance organizational success in a changing, competitive marketplace
- a successful developer of two internationally recognized certification programs, Certified in Production and Inventory Management (CPIM) and Certified in Integrated Resource Management (CIRM)
- a distribution center for hundreds of business management publications and educational materials
- a source of solutions, support, and networking local chapters, workshops, symposia, and the annual APICS International Conference and Exposition.

APICS publishes *The Performance Advantage* (monthly), *The Production and Inventory Management Journal* for practitioners and researchers. The APICS Education and Research Foundation sponsors the research journal, *Journal of Operations Management*.

Certification

Certified in Integrated Resource Management (CIRM): CIRM certified professionals are prepared to meet the challenges of today's cross-functional workplace by teaching them techniques to abolish the walls that traditionally have separated people based on departments, divisions, functions, disciplines, and culture.

Certified in Production & Inventory Management (CPIM): CPIM certified professionals have in-depth knowledge of the key aspects of production and inventory management. (Source: www.apics.org)

American Society for Quality (ASQ, ASQC): ASQ is a society of individual and organizational members dedicated to the ongoing, development, advancement, and promotion of quality concepts, principles, and techniques. It was once known as American Society for Quality Control (ASQC). ASQ has products, services, and – most of all – information to help people in all walks of life grapple with perplexing issues like total quality management, benchmarking, productivity, and more. ASQ was founded in 1946 with the merger of several local quality societies scattered across the United States. These groups were formed to share information about statistical quality control after classes on that subject were held during World War II to improve and maintain the quality of defense materials.

Certification

Certification is formal recognition by ASQ that an individual has demonstrated a proficiency within and a comprehension of a specified body of knowledge at a point in time. It is peer recognition and not registration or licensure.

Quality Engineer Certification: Designed for those who understand the principles of product and service quality evaluation and control.

Quality Auditor Certification: Designed for those who understand the standards and principles of auditing and the auditing techniques of examining, questioning, evaluating, and reporting to determine quality systems adequacy.

Reliability Engineer Certification: Designed for those who understand the principles of performance evaluation and prediction to improve product/systems safety, reliability, and maintainability.

Quality Technician Certification: Designed for those who can analyze quality problems, prepare inspection plans and instruction, select sampling plan applications, and apply fundamental statistical methods for process control.

Mechanical Inspector Certification: Designed for those who, under professional direction, can evaluate hardware documentation, perform laboratory procedures, inspect products, measure process performance, record data, and prepare formal reports.

Quality Manager Certification: Designed for those who understand quality principles and standards in relation to organization and human resource management.

Software Quality Engineer Certification: Designed for those who have a comprehensive understanding of software quality development and implementation; have a thorough understanding of software inspection and testing, verification, and validation; and can implement software development and maintenance processes and methods. (Source: www.asq.org)

National Association of Purchasing Management: Founded in 1915, the National Association of Purchasing Management, Inc. (NAPM) is one of the most respected professional organizations in the United States. NAPM is a communication link with more than 44,000 purchasing and supply management professionals. NAPM is a progressive association with a mission to provide national and international leadership in purchasing and materials management, particularly in the areas of education, research and standards of excellence. Through various resources and a network of more than 180 affiliated organizations, NAPM provides opportunities for expansion of professional skills and knowledge. A not-for-profit association, NAPM offers a wide range of educational products and programs.

Certification

The C.P.M. (Certified Purchasing Manager) Program was originated by NAPM in 1974, and is the first nationally accepted standard of competence and knowledge for the purchasing and supply management field. Those who earn the title of C.P.M. join a select but fast growing professional group, widely recognized by management and colleagues to be among the most knowledgeable in today's competitive world of purchasing and materials management. Applicants for C.P.M. certification must pass all four modules of the C.P.M. exam. In addition, the applicant must (a) have five years of full-time professional purchasing and supply management experience, or (b) have a four-year degree from an accredited institution and three years of full-time professional purchasing and supply management experience.

The Accredited Purchasing Practitioner (A.P.P.) program was established in 1996 for (a) entry-level buyers who are primarily engaged in the tactical and operational side of purchasing, and (b) persons who work outside the organization's purchasing/supply management department, but nevertheless have definite procurement responsibilities. Both the A.P.P. and the C.P.M. are designed for individuals working in any of the major private, public, or non-profit work sectors. (Source: www.napm.org)

The Society of Manufacturing Engineers (SME): SME, headquartered in Dearborn, Michigan, U.S.A., is an international professional society dedicated to serving its members and the manufacturing community through the advancement of professionalism, knowledge, and learning. Founded in 1932, SME has nearly 65,000 members in 70 countries. The society also sponsors some 275 chapters, districts, and regions, as well as 240 student chapters worldwide.

Certification

Certification through SME's Manufacturing Engineering Certification Institute (MECI/SME) is a program of professional documentation and recognition of an individual's manufacturing-related knowledge, skills and capabilities.

Certified Manufacturing Technologist (CMfgT), recognizing competence in the fundamentals of manufacturing, including mathematics, applied sciences, computer applications and engineering design.

Certified Manufacturing Engineer (CMfgE), recognizing a more comprehensive knowledge of manufacturing processes and practices, including product and tool design, facility operations, floor operations and material handling.

The Certified Enterprise Integrator (CEI) exam is an experience-based test created in collaboration with recognized experts. Questions are based on their experience, following proven professional test development procedures and were evaluated to assure relevancy and fairness. Although a wide range of topics is covered, its 250 multiple choice questions are drawn from the six fundamental elements of enterprise integration: Customer Focus explores the role of an enterprise mission to align all work toward meeting customer expectations. After all, marketing, design, manufacturing, and support must be aligned to meet customer needs. People and Teamwork in the organization are included as the means to organize, hire, train, motivate, measure and communicate, ensuring teamwork and cooperation. Shared Knowledge and Systems is an element that covers both manual and computer tools to aid research, analysis, innovation, documentation, decision making and control of every process in the enterprise. Processes addresses processes/product process definition, manufacturing and customer support. Resources and Responsibilities combine two crucial areas. Resources (input) include capital, people, materials, management, information, technology and suppliers. Responsibilities (output) include employee, investor, and community relations as well as regulatory, ethical, and environmental obligations. Infrastructure because a company's success depends on customers, competitors, suppliers and other factors in the manufacturing environment.

Licensure is one way in which Professional Engineers (PEs) can demonstrate their credibility within the engineering profession and to the public. PE licensure is important to manufacturing engineers. Because of the significant role engineering plays in society, the profession regulates itself by setting high standards for Professional Engineers (PEs). Manufacturing Engineers can benefit from becoming a Professional Engineer through job opportunities and promotions and respect from peers. Manufacturing engineers are encouraged to take the Principles and Practice of Engineering (P&P/PE) Examination. SME supports the PE licensing process by writing the examination problems for the Manufacturing Engineering discipline.

The Principles & Practice of Engineering Examination (P&P/PE) is designed to document: (1) General knowledge of the manufacturing industry, and (2) The ability to apply that general knowledge of the engineering principles to some problems of practice in the manufacturing industry. (Source: www.sme.org)

SOURCES:

http://www.apics.org/
http://www.asq.org/
http://www.napm.org/
http://www.sme.org/

PROGRAMMABLE AUTOMATION

It is a form of automation, where machines and processes are controlled by programmable computers. Computers instead of people control machines that transform parts or products. Programmable automation can lead to reduced setup time, greatly increased flexibility, and economies of scope. Whereas human operators previously positioned parts on the machining equipment, loaded appropriate fixtures and tools, and monitored the machining process, now programmable computers can be used to control machines. Programmability of computers enables the change over from the production of one part to another by merely changing the program in the computer that controls the machine. This could be done in a few seconds, which represents the enormous flexibility offered by programmable automation.

See **Flexible automation; Flexibility in manufacturing; Manufacturing flexibility; Manufacturing with flexible systems.**

PROGRAMMABLE LOGIC CONTROLLERS (PLCs)

PLC's are micro-processor-based devices which control automated equipment or processes. They are built hardy enough to build into shop floor hardware to work in harsh environment.

See **Flexible automation; Manufacturing systems; Manufacturing with flexible systems.**

PROJECT CONTROL

Project control activity ensures that each task on the project as well as the whole project is completed on time and within cost. Project management techniques such as PERT and CPM are useful tools for planning, scheduling and control of projects.

See **Critical path method; PERT; Project management.**

PROJECT EVALUATION AND REVIEW TECHNIQUE

See **PERT; Project management.**

PROJECT MANAGEMENT

James P. Lewis

The Lewis Institute, Inc., Vinton, VA, USA

Managing projects can easily be said to date back at least 4500 years to the building of pyramids and stone henges, yet the role of project manager is only recently becoming recognized as a discipline in its own right. Universities are now offering courses in project management, and many of them also offer a Master of Project Management degree, as an option to the conventional M.B.A. This response is a recognition that project management is different from general management, uses specialized tools and techniques. Naturally, there are many points of overlap as well. In recent years, the development of scheduling software that runs on personal computers has made the scheduling of relatively large projects more or less routine.

WHAT IS A PROJECT?: There are all kinds of projects – projects to develop new products, to develop a marketing plan, to build a large office building, to remodel a home, to landscape one's lawn, to develop a new vaccine, and so on. The possibilities are almost endless, making project management a nearly universal discipline. One appropriate definition of a project is shown below:

> **A project is a one-time job, that has definite starting and ending points, clearly defined objectives, scope, and (usually) a budget.**

The key words are *one-time*, *definite starting* and *ending points*, *budget*, and *objectives/scope*. Thus, a project is differentiated from repetitive activities such as production, order processing, and so on.

Even for such seemingly repetitive activities as building a number of homes using the same design, the terrain will be different for each, the weather will vary, and the work group will probably change, so each construction job will be unique. It is the uniqueness of each project that makes special demands on project managers and simultaneously makes project management an exciting discipline.

Note also the part of the definition that says a project has a definite ending point. Some projects never seem to end. This is because the parties involved have failed to properly define what will constitute *finished*. Worse yet, they may have failed to define exactly what was to be done in the first place, thus allowing the project to grow, taking on a life of its own. This is commonly called "scope creep" since the magnitude of the job continues to grow.

A second definition of projects that has merit is offered by Dr. J.M. Juran, the quality expert. A project is a problem scheduled for solution. This means that a project is always conducted to solve a problem for the organization. The importance of this definition lies in forcing us to recognize that we are solving problems.

Note, however, that there are two connotations to the word *problem*. One is that something is messed up. That is not the definition we are using here. For our purposes, a problem is a gap between where you are and where you want to be, and that obstacles prevent easy movement to close the gap. For example, you would like to develop a new product. If you could wave a magic wand and have the product appear, you would have no problem, only a desired result. The obstacles that prevent easy movement to close the gap are many: the need to fully define requirements, solve design problems, manufacture it, and so on. Finding ways to deal with these obstacles constitutes problem solving.

WHAT IS PROJECT MANAGEMENT?: First, it might be good to say what project management *is not*. It is not just scheduling. With the growth in popularity of personal-computer-based scheduling software, many people think that if they simply buy the software and apply it, they will be doing sound project management. Then they find that they really do not know how to apply the software.

As a matter of fact, this is putting the "cart before the horse," since it is very difficult to know how to fully utilize the software unless you first have a good grasp of how to manage projects and exactly what you are going to do with the software. I could buy a superb accounting software package and be unable to use it, as I know very little about accounting practice. The software is simply a tool that will help me do the job better, assuming that I understand good project management practice in the first place.

For many years it has been customary to say that project management is the planning, scheduling, and controlling of project activities to achieve *performance*, *cost*, and *time* objectives, for a given *scope* of work, while using resources efficiently and effectively. These have been referred to as PCT objectives. They are also commonly called *good*, *fast*, and *cheap*. These more colorful terms capture the essence of what a project manager must achieve.

The important issue is that the three objectives must be met *while using resources efficiently and effectively*. This is a key point in project management, and one that is too often overlooked. Every organization has limited resources, and unless the project manager can deal successfully with the resource allocation problem, she will not be successful. Experience shows that in many environments, failure to properly manage resources is one of the most common causes of project failure.

The relationship of the four variables to each other is given by the following expression:

$$C = f(P, T, S)$$

In words, the expression says, "Cost is a function of Performance, Time, and Scope." Ideally, an equation could be written prescribing the actual relationships precisely. In practice, however, we never know that precise relationship. We have to estimate times and costs. What is important to note is that you must make tradeoffs between the variables in many situations. We often have more to do (scope) than we can afford (cost) in the time allowed, at the performance level desired.

This leads to an important question: in an equation containing four variables, to how many can you assign values? Naturally, the answer is three. You cannot dictate all four of them. They simply won't go together (unless you just get lucky).

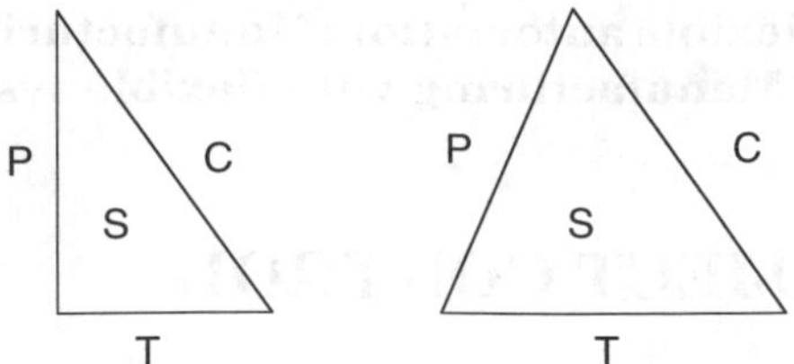

Fig. 1. The relationships of P, T, C, and S.

Once you pick three of them, the fourth one will be whatever the equation says it will be. One way to see this is to visualize the relationship as forming a triangle, as shown in Figure 1. Performance, Cost, and Time represent the sides of the triangle, while Scope is the area. If I choose values for three sides, that will determine the area (scope) of the project. Or, I can pick values for the scope and any two sides, and that will determine the length of the third side. The values are determined by the equation for calculating the area of a triangle.

Of course, in a project, we have no exact equation, as was stated above. We are estimating values. Once estimates have been established for three of the variables, it is okay to estimate the fourth variable and let that number be a *target*, but we recognize that it will probably not come out exactly as figured, since all of the variables on which it is based are themselves estimates.

It is interesting to examine the relationship between cost and time alone. This is shown in Figure 2. Note that there is some optimum duration for the project, at which costs are a minimum. As the duration is extended, costs rise because of inefficiency. As the duration is reduced (we say the project is being *crashed*), costs rise because you generally have to apply premium labor rates to reduce time, and you also reach a point of diminishing returns. In addition, there is a duration below which you cannot go. No matter how many resources are applied, the project duration cannot be further reduced.

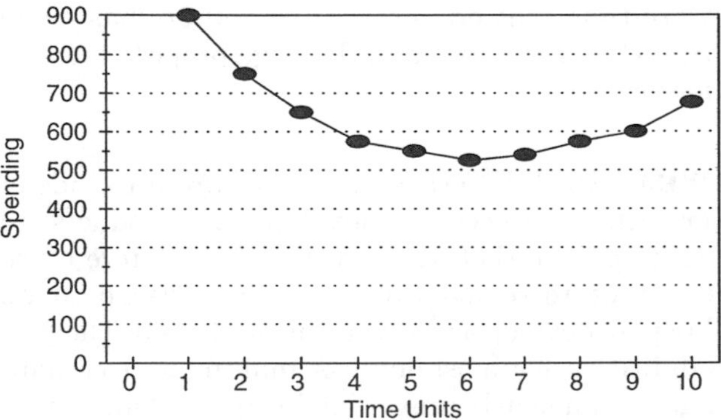

Fig. 2. Cost versus time curve.

THE PROJECT MANAGEMENT BODY OF KNOWLEDGE: Project Management Institute (PMI), which is the professional association for project managers, has identified nine areas of knowledge that a project manager needs to have some familiarity with in order to be effective. In order to be certified by PMI as a *Project Management Professional*, an individual must pass an examination that covers all nine of these areas, which are defined as follows:

- *Schedule/Time Management:* Managing the project to meet time objectives. This usually involves the use of Critical Path Method (CPM), Project Evaluation and Review Technique (PERT) and/or Gantt schedules.
- *Cost Management:* The methods used to keep a project within its budget.
- *Quality Management:* The tools and techniques of Total Quality Management and related disciplines are applied to projects to ensure that work is done properly.
- *Human Resources:* Managing the human aspect of projects involves dealing with the behavior of contributors as well as the administrative component, which includes pay and benefits, government compliance with EEO requirements, and so on.
- *Contract/Procurement:* The project manager must be able to deal with the acquisition of materials and supplies. In addition, it may be necessary to administer contracts when outside contractors perform part of the work on a project.
- *Communications:* This involves having the proper skills to communicate to the right people at the right time, using the appropriate process. Sometimes written communications are appropriate, and at other times only face-to-face communication should be used.
- *Scope Management:* The scope of a project is the magnitude of the work being done. Scope changes that are uncontrolled usually result in cost and schedule overruns. Project managers must know how to manage the scope of a job to avoid such consequences.
- *Risk Management:* Murphy's much-quoted law says that "Whatever can go wrong will go wrong." This is certainly true in projects, and failure to manage risks can lead to disasters.
- *Integration Management:* This area means that the organization must examine project portfolios to ensure that all of them have a strategic fit with overall business objectives.

To cover all nine of these areas in detail would require much more space than I have available, so I must refer the interested reader to the references at the end of this section.

THE PROJECT LIFE CYCLE: While the nature of projects will vary somewhat with the content of the work being done, there are a number of similarities that are independent of content. Most projects have a life cycle that consists of four to six phases. When six phases are involved, they are called *Concept, Definition and Planning, Design, Development or Construction, Application, and Post-Completion*. No matter how many phases are involved, the following points apply:

a. The character of the program changes in each life-cycle phase (see Figure 3).
b. The uncertainty regarding the ultimate time and cost diminishes as each phase is completed.

Project Life Cycle

Concept	Definition	Design	Development or Construction	Application	Post-Completion
• Marketing input • Investigation of technology, feasibility studies, etc. • Survey of competition	• Specify objectives • Establish PCT targets • Quality Assurance procedures • Set up control system • Establish project organization • Set up project notebook	• Architectural, engineering • Design reviews • Assessment reports • Revise cost & performance targets	• First units • Begin sales campaigns • Quality control procedures	• Install and field test • Begin de-staffing • Advertising begins • De-bug and redesign	• Final de-staffing • Post-mortem analysis • Final reports • Closeout

Fig. 3. A typical project life-cycle.

Figure 3 illustrates the life-cycles as they should be in the typical project.

THE LIFE-CYCLE PHASES:

Concept Phase

The *concept phase* of a project is the point at which someone identifies a need that must be met. It can be for a new product or service, including buildings or renovations, a move from one location to another, a new banking service, advertising campaign, research program, and so on. At the concept phase, there is a very fuzzy definition of the problem to be solved. In fact, note that a *feasibility study* may have to be conducted in order to clarify the definition of the project, before proceeding to the next phase.

Definition and Planning Phase

It is the *definition and planning phase* of the project that is intended to tie down exactly what is to be done. A problem statement is developed, which is meant to concretely define the problem to be solved. This is, of course, the ideal way of doing it. There are some projects in which the project definition will continue to evolve as work progresses, simply because neither the customer nor the project team is able to fully define what the customer needs.

In any case, once a suitable definition has been written, objectives are developed, strategies for achieving them are selected, and detailed working plans are constructed to insure that those objectives are achieved.

During this stage, an organization is established by identifying those individuals who will participate in the project, deciding to whom they will report, and defining the limits of their authority, responsibility, and accountability. Note also that a control system is developed, quality assurance procedures are set up (if they do not already exist), and detailed planning is completed.

When the plan is finished, it is placed in a *project notebook*, which then serves as the controlling document throughout the life of the project. The notebook is a very useful organizing device for project management, and may actually be a number of ring binders for very large projects.

Design Phase

The *design* phase may not exist (as in a construction project) or may be combined with the development phase. If it does exist, note that there is provision for revising cost and performance targets. Although this idea is "heresy" to some managers, like it or not, one often finds during design of new products or services that original cost, schedule, or performance targets cannot be met.

Winston Churchill once remarked that we must have the right to be smarter today than we were yesterday. Because no one can forecast with 20/20 vision, changes to original targets must be made as new information becomes available.

Development or Construction Phase

The *development or construction phase* is the point at which completed products and services can be tested to see if they meet all performance targets. In the case of new product development, sales campaigns usually begin here. The danger, of course, is that if any serious glitches are discovered in the product after the sales campaign is launched, the organization may be placed in the very difficult position of not being able to meet promised deliveries, that can result in loss of good will from customers.

Application Phase

Application of the product or service can also yield some unhappy results if problems are discovered; therefore, it is important that products be field tested before being shipped to customers. Software development companies usually ask some of their customers to beta-test their products, so that glitches can be uncovered and corrected before any advertising campaign is begun to the general public.

Horror stories abound, however, of some major companies that have fallen into the trap of shipping products before thorough testing was completed, only to incur considerable expenses repairing or replacing defective product later.

Post-Completion Phase

The *post-completion phase* sounds contradictory at first glance. If the project is complete, how can this be a phase? In fact, this phase is *aborted* in many projects. Note that it involves a *lessons-learned* analysis of the project. The project manager should look back at the project and ask what was done well and what was not. Those things that were done well should be repeated in future projects, and conversely, problems that occurred should be avoided. If the lessons-learned review is not conducted, it is likely that the mistakes of yesterday will be repeated.

One suggestion about lessons-learned reviews: they should be performed at each major milestone

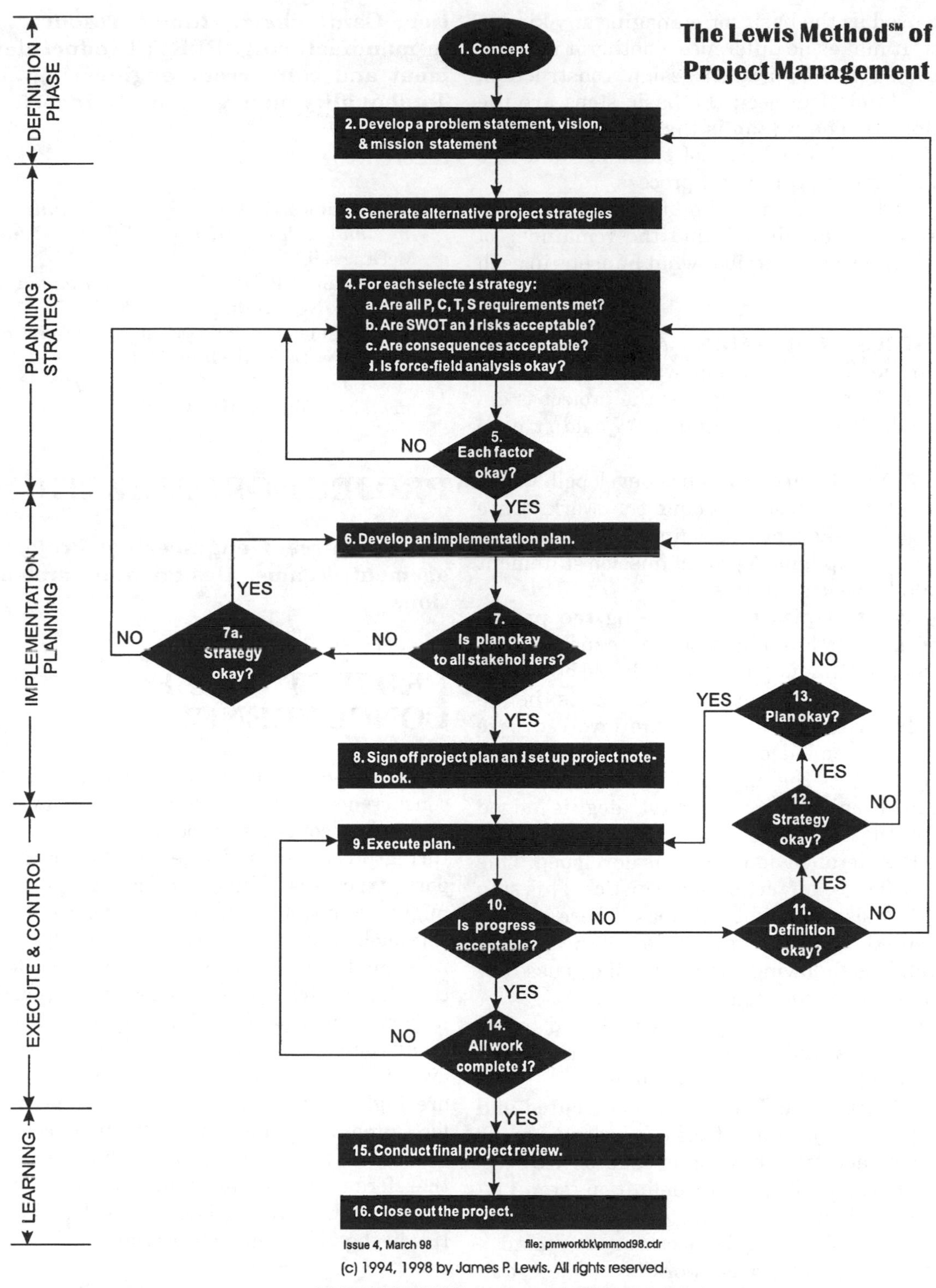

Fig. 4. A model for managing projects.

in the project, so that you can take advantage of what you learn for the current project, as well as for future ones.

A MODEL FOR MANAGING PROJECTS: If you are new to managing projects, the first question you might ask is, "Where do I start? In what order do I do things?" To answer those questions, the flow chart in Figure 4 could be used as a model for managing projects. It is called the Lewis Method[SM] by its author, and it is based on best practice of organizations throughout the U.S., and the model can be used as the basis for managing any kind of project. It makes no difference whether it is an information systems, product design, construction, or other kind of project; the basic steps are the same for all. The reason is that project management is a disciplined way of thinking, and this model outlines that thought process.

The model is shown in Figure 4. Each step in the process is numbered, and the remainder of this section briefly describes what happens in each step.

The Steps in Managing a Project: The steps in the model are as follows:

- Step 1. A general concept for the project is developed. "We need something." "Wouldn't it be nice if..."
- Step 2. A problem definition is developed, which states the purpose of doing the work. Care should be taken not to define the problem in terms of symptoms. A project mission statement should be written in this step.
- Step 3. Strategies for conducting the project work are listed in a brainstorming manner. All approaches that might be valid should be stated.
- Steps 4 & 5. Strategies are assessed. Is the approach feasible? What risks are involved? Are there unintended consequences? Can we live with them? Is the approach acceptable in this organization's culture? Which strategy is judged best of those available?
- Step 6. A detailed action plan is developed. This plan will develop tactics and logistics. This step itself consists of eight substeps. These will be presented following the overview of the general model. The following chapters will discuss how the plan is constructed.
- Steps 7, 7a, & 8. The plan is reviewed by all stakeholders and signed, signifying that it is considered acceptable and feasible.
- Steps 9, 10, & 14. The project is executed and results are monitored. If the desired outcome is not being achieved, control procedures are exercised. Is the fault with the definition (step 11)? Is project strategy correct (step 12)? Is the plan okay (step 13)? It may be necessary to replan.
- Step 15. When all project work is complete, a final post-mortem should be conducted. What was done well? What could have been done better?
- Step 16. Project is closed out and final reports filed for future reference. The project notebook containing *all* reports and documentation is placed in a central file for use by managers in planning subsequent projects.

See **Cost management; Cost planning; Critical path method; Design and implementation; Gantt chart; Human resource issues in manufacturing; PERT; Product development and concurrent engineering; Teams; Total quality management; Unit cost.**

References

[1] Lewis, James P. (1995). *Project planning, scheduling, and control, Revised Edition*. New York: McGraw-Hill.

[2] Lewis, James P. (1993). *The project manager's desk reference*. New York: McGraw-Hill.

[3] Lewis, James P. (1998). *Mastering project management*. New York: McGraw-Hill.

[4] Lewis, James P. (1997). *Team-based project management*. New York: AMACOM.

PROJECT ORGANIZATION

See **Concurrent engineering; Project management; Teams: Design and implementation.**

PROJECT PHASE CONCURRENCY

In a new product development effort, project phase concurrency means overlapping the development of market concepts, product designs, manufacturing processes, and after-market product support structures. Thus, product design activities might begin well before market concepts are fully established, manufacturing process design activities might begin before the product design is completed, etc. Project phase concurrency is an important method for collapsing product development times and improving design quality. However, the various stages of product development are highly interdependent; market planning affects product design, which affects process design, and so on. Successful project phase concurrency therefore requires close communications between personnel who are executing development activities in the different project phases.

See **Concurrent engineering; New product development through supplier integration; Product development and concurrent engineering.**

PROJECT PLANNING

The goal of project management is to complete the project on time without cost overruns. Project planning, schedule and control is the key to on-time and completion within budget. Project scheduling techniques such as CPM and PERT are useful tools for project scheduling and control.

See **Critical path method (CPM); PERT; Project management.**

PROJECTED AVAILABLE BALANCE

This term is used in materials Requirements Planning (MRP). It refers to inventory balance of an item projected into the future. It is computed in MRP records using on-hand inventory minus requirements plus scheduled receipts and planned orders. In MRP, records, it is computed each period using end of the period inventory status.

See **Master production schedule; MRP; MRP implementation; Order release.**

PROTOTYPE MODEL

See **New product development through supplier intergration.**

PSYCHOLOGICAL STRAINS OF OPERATORS

Psychological strains refer to an individual's affective, cognitive, and somatic experience of stress and they manifest themselves through job dissatisfaction, tension, anxiety, boredom, psychological fatigue, apathy, poor self-esteem, depression, etc.

See **Human resource issues and advanced manufacturing technology; Human resource issues in operations management.**

PULL PRODUCTION SYSTEM

Production systems may be characterized as "push" or "pull" systems (Hall, 1983; Stevenson, 1996). In push production systems, raw materials and parts are pushed through the production system, and the finished product is stocked to meet predicted demand. Even when production occurs in response to an order, the order triggers a release from the raw stock, which is then pushed through the production system. In pull production systems, the product is manufactured in response to a specific demand. The order is used to trigger a pulling action from the end of the production line (e.g., from the last workstation). If that workstation cannot fill the order, it requests additional units from the preceding workstation. This action continues with each following workstation requesting units from its predecessor workstation. Hence, the product is pulled through the system. When properly implemented, pull production systems result in less work in process (WIP) than push production systems, which in turn reduces warehousing and investment costs. Pull production systems are also known as Just-in-Time (JIT) systems or Kanban systems. Pull production systems are controlled using Kanbans. The Kanbans are used to signal the pulling action from one workstation to another (Sugimori *et al.*, 1977).

See **Dynamic kanban control for JIT manufacturing; Just-in-time manufacturing; Just-in-time manufacturing implementation; Kanban-based manufacturing systems; Kanbans; Lean manufacturing implementations; Manufacturing performance measures; Manufacturing systems.**

References

[1] Hall, R.W. (1983). *Zero Inventories*. Dow Jones-Irwin, Homewood, Illinois.

[2] Stevenson, W.J. (1996). *Production/Operations Management* (5th Ed.). Irwin, Homewood, Illinois.

[3] Sugimori, Y., K. Kusunoki, F. Cho, and S. Uchikawa (1977). "Toyota production system and kanban system materialization of just-in-time and respect-for-human system." *International Journal of Production Research*, 15, 553–564.

PURCHASING RATING

This refers to an assessment of a particular supplier's performance by the purchasing department (Monczka, Trent and Handfield, 1997). This type of rating is typically developed using historical information on the supplier's performance, which is often stored in a database. Some of the most common measures of supplier performance include:

1. *Ontime delivery* – Percentage of time the supplier delivered within the promised delivery window. Percentage of orders that arrived and were found to be of the correct quantity and mix.
2. *Delivery cycle time* – The average delivery time from when the order was placed to receipt by the purchaser.

3. *Pricing history* – The supplier's average price compared to one or more reference such as the "producer price index" published by the Bureau of Labor Statistics.
4. *Quality history* – Percentage of orders received that met acceptable levels for incoming defects and/or damage.

In addition to a Purchasing Rating, other ratings of a supplier's performance can be made other functions within the organization including the following:

1. Receiving Rating (good packaging, delivers per routing instructions, good service)
2. Accounting Rating (invoices correctly, issues credit memos punctually, does not ask for special financial consideration)
3. Engineering Rating (reliability of products, technical ability, accepts responsibility for defects and reacts promptly)
4. Quality Rating (quality of material, furnishes certification, replies with corrective action)

See **Performance measurement in manufacturing; Purchasing: Acquiring the best inputs; Supplier performance measurement.**

Reference

[1] Monczka, R., Trent, R., and Handfield, R., *Purchasing and Supply Chain Management*, Southwestern Publishing, 1997.

PURCHASING: ACQUIRING THE BEST INPUTS

Stanley E. Fawcett

Brigham Young University, Provo, Utah, USA

DESCRIPTION: Every organization whether it is a hospital, a city government, a toy manufacturer, or a retail store buys materials, supplies, and services from outside suppliers to support its internal operations. This buying process is managed by the organization's purchasing group. Purchasing's primary role is to buy inputs and bring them together at the firm, where they are used to support the production process. Purchased inputs represent the single largest cost element of most organizations. Today's firms purchase not only raw materials and basic supplies but complex fabricated components and finished goods. In addition, many firms outsource customized services, too.

Purchasing managers have developed the following statement to concisely define the objectives of purchasing: to obtain the **right materials** at the **right time** in the **right quantity** for delivery to the **right place** from the **right supplier** with the **right service** at the **right price**. This statement contains the "seven rights" of purchasing. First, purchasing must work to identify the most appropriate type of material to be used for a given task. This process involves working with the person or department that needs the material: to 1) identify and compare alternative materials; and 2) clearly state specifications. Second, the materials must be available when needed – neither too early nor too late.

Information sharing and good relationships, with both suppliers and internal customers, is needed to assure proper materials availability. Third, based on the internal customer's forecast of needs and a knowledge of the materials' cost and availability, purchasing must determine how much should be ordered at any given time. Fourth, purchasing must arrange for delivery to the correct location – materials available at the Chicago warehouse do very little good if they are needed at the Detroit plant. Fifth, the key to meeting all of the other objectives is to find and develop high-capability suppliers. Supplier selection is the single most important decision addressed by purchasing. Sixth, acquiring the right service is increasingly important because the service component today can be quite diverse, and include financing, after sales support, installation, warranty service, technical assistance, and finally participation in new product design. Getting the best available product/service package is vital. Finally, materials must be purchased at the best overall price possible. Once, this criterion meant purchasing at the lowest unit price. Now, best overall price is synonymous with the total cost of ownership; i.e., all of the costs related to buying the materials over the life of the contract.

Constant tradeoffs arise in trying to meet these basic objectives. For example, the lowest-price supplier might not be the right supplier if the supplier cannot meet delivery schedules, or if the materials fail to meet performance specifications. Likewise, a supplier that can provide timely and consistent delivery but does not provide adequate technical assistance might not be the appropriate choice. Further, if an item is needed immediately, the purchaser might have no alternative but to buy from a high-price supplier. Finding or developing a supplier that performs well in all of the "rights" of purchasing is not an easy endeavor. The key is to balance the often conflicting objectives while working with suppliers to overcome specific performance deficiencies.

HISTORICAL PERSPECTIVE: Before World War II, most firms looked at purchasing as a necessary but certainly not a glamorous activity. For a firm to produce a product it had to have materials to use in the transformation process, but materials were widely available and labor represented the most important and most visible cost of doing business. As a result, purchasing was viewed as a clerical activity – orders needed processing to trigger the inflow of materials. However, during the war, a general situation of scarcity prevailed. The materials and services that were plentiful before were now difficult to obtain, requiring greater effort to secure a steady and continuous flow materials to keep production running smoothly. With time, more attention was given to the organization and procedures that governed the purchasing process.

For the next two decades, relatively little change occurred in the way firms viewed purchasing. Most firms recognized that planning was essential to purchasing success; however, most planning was driven by events outside purchasing. Purchasing operated on a largely reactive basis. In effect, purchasing had attained managerial status, but it was relegated to the managerial backwaters.

The materials shocks of the 1970s – including the Middle East oil embargo in the summer of 1973 – brought renewed interest to purchasing. Material scarcity combined with dramatic price increases provided purchasing an opportunity to show that it could have a positive impact on firm operations. Purchasing had to identify new sources of supply and manage increasingly complex buyer-supplier relationships to make a profit in a very challenging competitive climate. Purchasing was becoming a strategic function.

The perception of purchasing as a strategic weapon continued to emerge gradually throughout the 1980s and into the 1990s as purchased inputs became a major part of operating cost. The transformation of purchasing into a strategic function was promoted by two important developments. First, better trained and more competent managers began to enter the purchasing arena. Second, the advent of new information technologies made it possible to store and track the large amounts of information needed to strategically manage buyer-supplier relationships. The combination of more highly trained managers, better technology, and greater organizational support allowed purchasing to take on a more proactive role in helping firms meet new competitive challenges. Purchasing is now recognized as a critical value-added activity capable of impacting a firm's competitive position.

STRATEGIC PERSPECTIVE: Purchased inputs from raw materials to overnight mail now represent the single largest cost category for most firms. Of course, purchased inputs' percentage of costs varies considerably across different industries depending on the mix of inputs used in the value-added process. For example, industries that require large amounts of engineering and extensive assembly work will have a lower percentage of purchased inputs. The data in Table 1 show that purchased inputs represent a sizable cost element in most manufacturing industries and average 55% of sales across manufacturing industries (Tully, 1995). This translates to approximately 60 to 80 percent of a firm's operating expenses. By comparison, direct labor costs have decreased to 10 percent of total operating costs in many industries (in some cases, direct labor accounts for as little as 2 percent of operating costs) (Drucker, 1986; Leenders and Blenkhorn, 1988). Thus, purchasing decisions play an important role in determining the relative cost position of manufacturing firms (purchasing also influences the cost position of service, public, and non-profit organizations).

The value-added component of purchased goods and services has also increased in recent years (Burt, 1989; Birou and Fawcett, 1993). Firms today not only purchase raw materials but also buy complex fabricated components and finished goods to which the company's brand name is added. A desire to focus on a limited set of operations has also led many firms to outsource customized services such as inventory management and distribution services. The impact of purchasing on firm competitiveness in terms of quality, dependability, flexibility, and innovation is also significant. For example, quality expert Philip Crosby claims that up to 50% of a firm's quality problems can be traced to defective purchased materials. Long-term success now depends on managing purchasing more strategically to assure that the best mix of inputs are obtained at the lowest total cost of ownership.

IMPLEMENTATION: Establishing a successful purchasing program begins by clearly defining the scope of purchasing's responsibilities. Once the role and responsibilities of purchasing are understood, the basic steps of the purchasing process must be carefully managed to assure that purchasing contributes to firm competitiveness. Effective implementation requires appropriate training coupled with adequate systems and procedures. These critical implementation issues are discussed below.

1. ***Scope of Purchasing Responsibilities:*** Purchasing must be given responsibility and authority over several key decision areas. These basic prerogatives are: 1) the right to select suppliers; 2) the right to determine the most

Table 1. Importance of purchased items – purchased inputs as a percent of sales, 1992.

Standard industrial code	Industry	Cost of materials (millions)	Capital expenditures, new (millions)	Total material and capital expenditures (millions)	Value of industry shipments (millions)	Material sales ratio	Total purchase sales ratio
20	Food and kindred products	247,203	9,933	257,136	403,836	61	64
21	Tobacco products	7,994	389	8,383	35,137	22	24
22	Textile mill products	41,049	2,387	43,436	70,694	58	61
23	Apparel and other textile products	35,606	996	36,602	71,617	50	51
24	Lumber and wood products	48,637	1,767	50,404	81,798	59	62
25	Furniture and fixtures	20,916	840	21,756	43,688	48	50
26	Paper and allied products	73,182	7,814	80,996	132,954	56	61
27	Printing and publishing	53,965	5,506	59,471	167,284	32	36
28	Chemicals and allied products	141,510	16,526	158,036	305,761	46	52
29	Petroleum and coal products	125,891	6,677	132,568	149,961	84	88
30	Rubber, miscellaneous plastics products	55,256	4,820	60,076	113,544	49	53
31	Leather and leather products	5,207	133	5,340	9,677	54	55
32	Stone, clay, glass products	27,977	2,467	30,444	62,479	45	49
33	Primary metal industries	85,904	5,368	91,272	138,333	62	66
34	Fabricated metal products	82,992	4,448	87,440	167,015	50	52
35	Industrial machinery and equipment	125,120	8.063	133,183	258,273	48	52
36	Electronic, electric equipment	95,590	9,001	104,591	217,906	44	48
37	Transportation equipment	234,777	10,882	245,659	401,214	59	61
38	Instruments and related products	44,787	4,595	49,382	135,479	33	36
39	Miscellaneous manufacturing	17,729	1,045	18,774	39,626	45	47

Source: U.S. Bureau of the Census, 1992 Census of Manufactures (Washington, D.C.; U.S. Government Printing Office).

appropriate pricing method; 3) the right to question input specifications (the user makes the final decision), and 4) the right to manage contacts with suppliers (Ruch *et al.*, 1992). Empowered with these four prerogatives, purchasing can begin to help the firm build a sustainable competitive advantage.

Purchasing's responsibilities, above and beyond these four prerogatives, vary greatly from firm to firm. In some cases, purchasing does nothing more than react to internal demands for purchased inputs. In these firms, purchasing's responsibilities are limited to selecting suppliers, contracting for purchased items, and managing the documentation associated with purchases. Most organizations, however, involve purchasing more actively in the strategic management of materials. Typical responsibilities now include many of the following:

- Analysis of make versus buy decisions
- Evaluation and execution of capital-equipment purchases
- Inspection of incoming materials
- Involvement in the new-product-development process
- Management of inbound transportation, and sometimes outbound transportation
- Management of inventories

- Management of value analysis/value engineering activities
- Management of third-party logistics service providers
- Participation in strategic planning
- Developing economic forecasts
- Scrapping and disposal of surplus equipment and materials

The trend today is to consolidate the activities related to planning and controlling the flow of materials into and through the transformation process. Consolidating the materials-related activities provides better access to information and facilitates the trade-off analysis that is so important to the strategic management of material flows.

2. ***Steps in the Purchasing Process:*** The steps in the standard purchasing process are familiar to almost everyone. As consumers, we go through the procedural steps found in most purchasing systems every time a major purchase is made. For example, whether we are buying a new car or planning a dream vacation, we go through each of the following steps: 1) we identify or recognize a need; 2) we clearly describe just exactly what we need to buy; 3) we shop around to find the best place to buy the desired product; 4) we decide what we are willing to pay; 5) we make the purchase; 6) we make sure that we get our new purchase when we need it; 7) we receive our new purchase and carefully inspect it; 8) we write a check for our purchase; and 9) we keep our receipts and other records – at least until the warranty expires.

   Of course, the actual purchasing process varies depending on what we are buying. We typically take more time and use more caution when buying a house than we do when buying donuts and milk. The same is true for companies – different companies perform the above nine steps at varying levels of sophistication depending on the items being purchased, the skill of the purchasing managers, and the resources dedicated to the purchasing process. The nine steps discussed below.

1. *Recognition of Need:* The purchasing process begins when someone within the firm identifies a need to acquire some input – materials, supplies, or services – for use in production or in the general support of the firm's day-to-day business activities. Recognizing that a need exists is the responsibility of the user of the item to be purchased. The well-managed firm typically uses a purchasing policy or procedure handbook to establish guidelines for the interaction between users within the firm and the purchasing group. Today, many standard production needs are communicated automatically through a computerized production control system that monitors inventory levels and reorder points. However, a clear and precise set of guidelines must be in place to facilitate communications regarding new, unique, or one-time purchases.
2. *Description of Need:* To assure that purchasing buys the right items as they are needed, the user must clearly specify what is needed. This specification process generally takes place as a purchase requisition is created. The purchase requisition contains the information needed to purchase most items, including the item description, requisitioning department, authorizing signature, purchase quantity, and delivery date and location. The information should be stated clearly to assure that the purchasing process goes smoothly. For example, the delivery date should be specific to the dateand time, and not "as soon as possible." If the item isn't needed for four weeks, the user should indicate four weeks and not two weeks just to make sure the item is available on time. The two-week cushion might require a higher price than necessary to get the item delivered "on-time."

   Item description is particularly critical since purchasing must have a clear understanding of what is needed to identify the appropriate supplier and then communicate the needs to the supplier. The actual description of the item can be performed in several ways, depending on the item as well as the amount of experience purchasing has in buying the item. Some items can be described in a few words, with a brand name or via an established part number; others require listing extensive physical and performance characteristics. The less experience purchasing has with an item, the more clear and specific the description must be. The following six approaches are widely used to describe items to purchasing and to potential suppliers: a brand name, an established market grading system, a word picture that describes the item, a blueprint, a sample or prototype, a set of performance expectations.
3. *Selection and Development of Suppliers:* Once purchasing understands the needs of the user, the next step is to find the best – capable and committed – supplier possible. It has been said that, "the success of the purchasing department depends on its skill in locating or developing suppliers, analyzing vendor capabilities, and then selecting the appropriate vendor." (Leenders and Blenkhorn, 1988).

   Since purchasing's objectives include providing the highest-quality product at the lowest total cost supported with the best service, a supplier should be selected based on its ability to meet these requirements. However, many additional issues should be considered in selecting a supplier. A comprehensive definition of a good supplier is:

> A good supplier is one who is at all times honest and fair in his dealings with the customers, his own employees, and himself; who has adequate plant facilities, and know-how so as to be able to provide materials which meet the purchaser's specifications, in the quantities required, and at the times promised; whose financial position is sound; whose prices are reasonable both to the buyer and to himself; whose management policies are progressive; who is alert to the need for continued improvement in both his products and his manufacturing processes; and who realizes that, in the last analysis, his own interests are best served when he best serves his customers. (Dobler, 1996)

The actual process used to identify good suppliers varies a great deal depending on the importance of the item being purchased. Suppliers of maintenace and repair order (MRO) items do not receive a lot of attention while suppliers of major components such as engines or flat panel displays are carefully analyzed and developed by the purchasing department. In general, the higher the dollar value, or the greater the impact of the purchased item on the end product's performance, the more important the supplier selection process.

The standard supplier selection process can be characterized by four stages: identification, evaluation, approval, and monitoring. Identification involves making a list of every supplier that could possibly meet the firm's needs for the item of interest. The list-making process usually begins with a review of the buying firm's purchasing database, which contains information on all suppliers used in the past several years. Supplier names also often come from contacts – purchasing and sales managers – within and outside the firm. Additional sources of information include trade publications and supplier directories such the *Thomas Register of American Manufacturers*, which claims to include all American manufacturers by type of product. Such directories do not contain any performance information; nor do they include international suppliers outside the USA. In today's global marketplace, the best suppliers are often scattered all over the world.

Determining the performance potential of the companies on the supplier list is the primary task of the evaluation stage. This stage really takes place in two separate steps. First, a list of supplier selection criteria is put together. The possibilities are almost endless; however, experience shows that a relatively small number of criteria receive high priority. These criteria include quality, price, delivery dependability, capacity (current and future), service responsiveness, technical expertise, managerial ability (attitude, skills, and talent), and financial stability. The second step in the evaluation stage is to gather performance information that can be used to evaluate possible suppliers. Among the many sources of information are the following:

- The firm's own past experience in working with specific suppliers. Recent experiences are almost always a good indication of what can be expected in the future.
- The experience of others. Purchasing managers are often quite willing to share insight gained from their past experiences (as long as price is not under discussion).
- The suppliers themselves. Initial efforts should focus on telephone or other low-cost interviews. Suppliers that are considered viable after initial contacts should be visited. Plant visits should be carefully planned with key written questions. This helps make sure that consistent information is obtained from each supplier visited.
- Public information. Trade publications contain valuable information on suppliers in specific industries. Financial information can be obtained from Dun & Bradstreet Credit Reports.

After carefully evaluating potential suppliers, purchasing must reduce the list to an approved list of suppliers. Suppliers that remain on the approved list are eligible to receive an order if their price and other terms are competitive at the time the order is being placed. If the firm has a policy of sourcing from multiple suppliers to assure competition in the hope of getting lower prices, then several suppliers will be maintained on the approved list. If the firm seeks to single source from only one supplier, then the approved list must be reduced to a single supplier. This sole source is usually determined through the negotiation of a partnership relationship.

The final stage in the supplier selection and development process is to monitor the performance of the firm's suppliers. Monitoring performance does not actually help the firm in the up-front supplier selection process, but it does help the firm determine whether future orders should be placed with current suppliers. Further, monitoring performance provides the information needed to help suppliers improve. By working with suppliers to help them improve, a company can develop and maintain a world-class supply base. Further, monitoring supplier performance helps the buying firm evaluate and improve its selection process.

4. *Determination of Price:* Purchasing can take one of three primary approaches to determine an item's purchase price: buy the item at list price, use competitive bidding, or negotiate a price/service agreement. For many items, the best approach is to buy the item at the published asking price. Lower-volume or lower-valued items often do not merit sufficient purchasing attention to obtain a lower-than-list price. These items are often classified as C items using ABC analysis. ABC analysis recognizes that, of all the items purchased by a typical company, 10 to 20 percent represent 80 to 90 percent of the purchase dollars (these are the A and B items). For C items, the cost of getting a better price often exceeds the price reduction. Sometimes it is possible to group together a number of related C items so that the total dollar value is large enough to make spending the time needed to obtain lower-than-list prices worthwhile.

   Competitive bidding relies on market forces to get prospective suppliers to offer a lower price. Because the order typically goes to the supplier with the lowest reasonable bid, competitive bidding has often been viewed as the most efficient means of obtaining a fair price. Competitive bidding requires:

   - The dollar value of the order must justify the expense of the bidding process.
   - A minimum of two qualified suppliers must be actively interested in the order.
   - The specifications must be sufficient to assure that the bids are comparable.
   - The bidding must be honest – collusion must be avoided.
   - An adequate lead time is required. The bidding process requires considerable time to request bids, prepare bids, compare bids, and award the contract.

   Planning is the key to the successful use of competitive bidding. Most of the real decision making actually occurs in the up-front planning process. For example, during the planning process specifications are set, potential suppliers are identified and invited to offer bids, order quantities are established, delivery schedules are set, and the timing of the entire process is considered. When this initial planning is done properly, evaluating the bids and selecting the best supplier is relatively straightforward. If all of the bids are all considered high, or if other problems arise (for example, collusion occurred or a different approach such as negotiation is determined to be more appropriate), a buyer can notify the bidders that no bid will be accepted.

   Finally, negotiation is used when the dollar value of the purchase is large, and one or more of the required conditions for effective competitive bidding is missing. Negotiation generally brings the buyer and the supplier together face-to-face to work out the details of a buyer-supplier relationship. Negotiation is more commonly used today because of an increased use of long-term, sole-sourcing relationships. Negotiation in these partnership relationships goes beyond price determination to develop an overall agreement that is mutually beneficial to both negotiating parties. Issues typically included in a negotiation include price, delivery, product specifications, warranty, and financial terms. Almost any issue can be discussed in a negotiation. Ultimately, a win-win situation should result from most negotiations. Key negotiation objectives include the following.

   - A fair and reasonable price for the specified quality.
   - Full performance of the purchasing agreement, especially in terms delivery performance.
   - Adequate control over the manner in which the agreement is carried out.
   - Supplier cooperation throughout the length of the relationship.
   - A platform for building a continuing relationship with an increasingly competent supplier.

   When all else is equal, the side with the best data or most knowledge is in the strongest negotiating position. That is, some attributes such as access to proprietary technology or sheer size provide a strong natural advantage during negotiation. Where no clear natural advantage exists, the foundation of successful negotiation is built on information and trust. Most often, the supplier holds the information advantage because of its intimate knowledge of production and other costs. A special research burden is placed on the buying firm to accurately estimate production and other costs. Negotiation success depends on the amount and quality of planning which has been done by both sides.

5. *Preparation of Purchase Orders:* After a need has been recognized and described, a supplier selected, and all of the necessary details regarding quantity, quality, price, delivery, etc. determined, a purchase order (PO) is issued. A PO is specifies the terms and conditions of the purchase agreement and initiates supplier action. The actual preparation of the PO is relatively easy compared to the work that has been done up to this point. Even so, it is important to make sure that every PO is filled out correctly because of the potentially high costs of fixing problems that might result from an inaccurate PO. For example, the time and effort required to make up for mistakes such as filling in the wrong part number, or specifying the wrong delivery time and place increase the purchasing costs, and can negate the benefits of careful supplier selection and astute negotiating.

   Efforts to streamline the purchasing process have focused on reducing the paperwork associated with POs and have led to the use of blanket orders. A blanket order specifics the overall terms of agreement for a given time period – often one year – and covers the entire quantity to be purchased even though smaller quantities will be delivered periodically during the agreement. Once the blanket order is on file, the buying firm issues a delivery release to trigger a new shipment. Because a blanket order promises the supplier future business, the buyer often qualifies for a quantity discount. Today, POs can be sent electronically using electronic data interchange (EDI) by either an end user or the purchasing department. The use of corporate procurement cards is another approach to reducing paperwork-costs.
6. *Follow-up and Expediting:* Purchasing must routinely follow-up with suppliers to make sure that the they live up to the terms found in the purchase agreement. Regular follow-up with suppliers helps identify quality or delivery problems as they arise. When problems come up, purchasing might: 1) increase the frequency and noise level of follow-up efforts ("the squeaky wheel gets the grease"); 2) send personnel to work with the supplier to overcome the problem; 3) use higher-cost emergency transportation; 4) arrange for delivery from an alternative supplier; or 5) alter the buying firm's production schedule. Purchasing should take into account the problem type, the importance of the order, and the nature of the buyer-supplier relationship before deciding what action to take.

   Expediting refers to efforts to speed up the delivery of an order. Expediting occurs under two scenarios: first, an order is behind schedule yet it is needed as originally promised, and second, the buying firm needs the order to be delivered ahead of schedule. The typical expediting request involves either an appeal for special consideration, or the threat of order cancellation or loss of future business. A well-managed firm expedites only a small percentage of total shipments. If purchasing performs well, only suppliers capable of delivering as promised will be selected, and if production adequately plans its materials requirements, requests for early shipments should be rare. To avoid late delivery problems, many companies now build penalty clauses into their purchasing agreements. One company charges $5,000 for the first late delivery; $50,000 for the second late delivery; and removes the supplier from the approved list for a year after the third late delivery.
7. *Receipt and Inspection:* When an order arrives at the buying firm, it must pass through a receiving process. The receiving process requires matching the invoice to the order contents which involves a physical count and quality inspection. The objective is to make sure that the purchased inputs are fit for use. In most firms, the receipt and inspection is actually managed by a separate organizational unit to assure an accurate and objective inspection. Purchasing gets involved when an incoming shipment fails to pass inspection. Two primary reasons for failure are that the count is off – either too much or not enough was shipped – or that the inputs do not meet quality standards. When an order does not meet specification, purchasing must work with the supplier to correct the problem.

   At many companies, the use of long-term relationships with preferred suppliers has changed the receiving process. Purchasing works closely with key suppliers to help them achieve certification. Most certification programs focus primarily on improving supplier's ability to produce high-quality products so that quality inspections can be eliminated. Making sure the supplier has the procedures and skills in place to deliver the right quantities on time is also a primary consideration in the certification process. Once the supplier achieves certification, incoming inspection is eliminated and incoming orders are moved immediately to incoming storage or directly to the production area where they will be used. The benefits of certifying suppliers and streamlining receiving process are improved quality and reduced cycle times.

8. *Invoice Clearance and Supplier Payment:* Suppliers must be paid in a timely fashion to maintain good relationships with capable suppliers. Actual payment is done by accounts payable, but purchasing should design a system that clears and submits the invoice to accounts payable as quickly as possible. The traditional invoice clearance process required four documents be compared to assure that no discrepancies exist. The four documents are: 1) the purchase order; 2) the supplier invoice, which acts as the bill and is the supplier's record of shipment; 3) the receiving report, which documents quantities received; and 4) the inspection report, which verifies the order's quality. When agreement exists within specified limits (e.g., ±5 percent regarding quantity or $25 of PO price), accounts payable issues payment to the supplier. Any discrepancy must be resolved by purchasing.

   Supplier certification programs combined with information technology can effectively streamline the invoice clearance and payment process. By allowing orders from certified suppliers to bypass incoming inspection, the inspection report and invoice can be eliminated from the payment process. In fact, when Ford redesigned its accounts payable department, it adopted "invoiceless processing." When an order arrives at receiving, a computer database is checked to make sure that a corresponding order is expected. If record of the order is found on the database, the order is accepted and accounts payable issues payment. This example shows how computer technology, and improved purchasing procedures can be combined to improve efficiency.

9. *Maintenance of Records:* The final step in the purchasing process is to update the purchasing database so that the firm will have accurate records for future decision making. At least four types of information should be maintained in the database: 1) the current status of all POs; 2) performance information for the firm's selected evaluation criteria for key suppliers; 3) information regarding each part type or commodity; and 4) information regarding all contracts and relationships. If a volume contract is in effect that commits the firm to a total annual volume, purchasing needs to know how much has been ordered to date. If a contract stipulates that the supplier will reduce the item price by 3 percent a year over the life of the relationship, purchasing needs to know whether or not the supplier is living up to its obligation. Care should be given to designing and maintaining a database capable of providing accurate, relevant, and timely information.

TECHNOLOGY PERSPECTIVE: Technology is dramatically improving purchasing effectiveness in three areas. First, the advent of EDI greatly increases the firm's ability to communicate in real time with suppliers. Requests for proposals or bids, purchase orders, invoices, etc. can be sent via EDI or other electronic means (e-mail or the world wide web). This electronic communication reduces the paper trail and the associated bureaucracy while reducing lead times and increasing responsiveness.

Second, better information databases/warehouses makes supplier performance evaluation much easier and more effective. Worldwide suppliers can be tracked using database technologies. An innovative firm can use database and communication technologies in tandem to create a cost effective and flexible organizational structure, especially where global production and sourcing are used. Third, expert systems are now often employed to help make supplier selection decisions (Yacura, 1994). Expert systems combine the knowledge of "experts" with decision-making logic to help managers take a systematic approach to decisions such as supplier selection.

A supplier selection expert system would contain not only the appropriate list of selection criteria but also relative weightings for each. After the purchasing manager collects the necessary information for each supplier, the expert system provides an evaluation for the suppliers being considered. While the essential steps in the purchasing process have not been altered by technology, the process is streamlined and made much quicker and efficient.

EFFECT: Purchasing's potential to contribute to a firm's competitive performance is often discussed in terms of the following two impact areas:

- Purchasing's impact on efficiency and effectiveness
- Purchasing's impact on financial performance

1. ***Purchasing's Impact on Efficiency and Effectiveness:*** The single most important decision made in purchasing is supplier selection because a supplier's performance affects, directly or indirectly, the firm's performance along each of five important competitive dimensions – cost, quality, delivery, flexibility, and innovation.

Table 2. The profit impact of purchasing improvements.

Beginning position:	Sales	$100,000,000	
	Purchases (55%)	55,000,000	
	Labor (15%)	15,000,000	
	Other (22%)	22, 000, 000	
	Pre-tax profit (8%)	8,000,000	
	Reduce purchase costs by 10%	**Increase sales by 68%**	**Reduce labor costs by 36%**
Sales	$100,000,000	$168,000,000	$100,000,000
Purchases	49,500,000	92,400,000	55,000,000
Labor	15,000,000	25,200,000	9,600,000
Other	22,000,000	36,960,000	22,000,000
Pre-tax profit	13,500,000	13,440,000	13,400,000

Cost is the competitive dimension most visibly influenced by purchasing. In fact, purchasing's job is often identified as finding suppliers capable of delivering inputs at the lowest total cost of ownership. The total cost of ownership includes not just the purchase price but also the cost of managing the buyer-supplier relationship, the cost of poor quality, future engineering costs, and all other costs associated with the use of the purchased items over the life of the buyer-supplier relationship. By carefully selecting and then working with key suppliers to improve their operations, purchasing can achieve dramatic cost reductions.

Purchasing's impact on the other competitive dimensions is equally important. Given that up to 50% of quality problems are the result of poor supplier performance, purchasing possesses tremendous potential to improve quality. Similarly, if a supplier fails to meet a promised delivery date, then extra costs are incurred in expediting the needed materials, buying from a back-up supplier, or rescheduling production. In the worst-case scenario, a production line or entire facility might be forced to shut down. The buyer's ability to meet its own promised delivery dates is also hampered by a supplier's poor delivery performance. By selecting well-managed suppliers with adequate capacity, a firm can increase the flexibility of its production scheduling to meet unexpected demands. Early supplier involvement in product development can lead to more innovative products and processes and can reduce development cycle times dramatically. When purchasing does its job well, the firm can better position itself to efficiently and effectively compete in an open marketplace.

2. ***Purchasing's Impact on Financial Performance:*** Purchasing's influence on the firm's bottom-line performance can be demonstrated in two ways. First, every dollar saved on the cost of purchased materials by better purchasing practices goes straight to pre-tax profit. This relationship between the cost of purchased inputs and profit is known as the profit-leverage (Leenders and Blenkhorn, 1988). Consider the following example involving a firm that has $100 million in sales (see Table 2). Assume that purchased inputs account for 55% of each sales dollar while labor costs are 15% of sales, and that the firm has a pre-tax profit margin of 8%. If the firm were able to reduce the cost of purchased materials by 10%, pre-tax profit would be increased by $5.5 million. The same increase in pre-tax profit would require a 68% increase in sales. Likewise, a 36% reduction in labor costs would be required. For most firms, reducing the cost of purchased inputs by 10% is much more feasible than increasing sales by 68% or decreasing labor costs by 36%.

Second, reductions in materials costs achieved through better purchasing management have a large positive influence on the firm's return on assets (ROA). Because return on assets is a popular measure of operating performance, purchasing's effect on ROA can be useful in demonstrating purchasing's valuable contribution to bottom-line performance. Figure 1 shows the important factors that contribute directly to the calculation of ROA. The example in Figure 1 uses the same numbers that were used in the profit-leverage example above and assumes that total assets are equal to 50% of sales and that inventory accounts for 20% of total assets. By reducing purchase costs 10%, the average inventory asset base would also be reduced 10%. The reduction in inventory carries over to a reduction in total assets and therefore increases asset turnover. A more frequent asset turnover combined with a larger profit margin

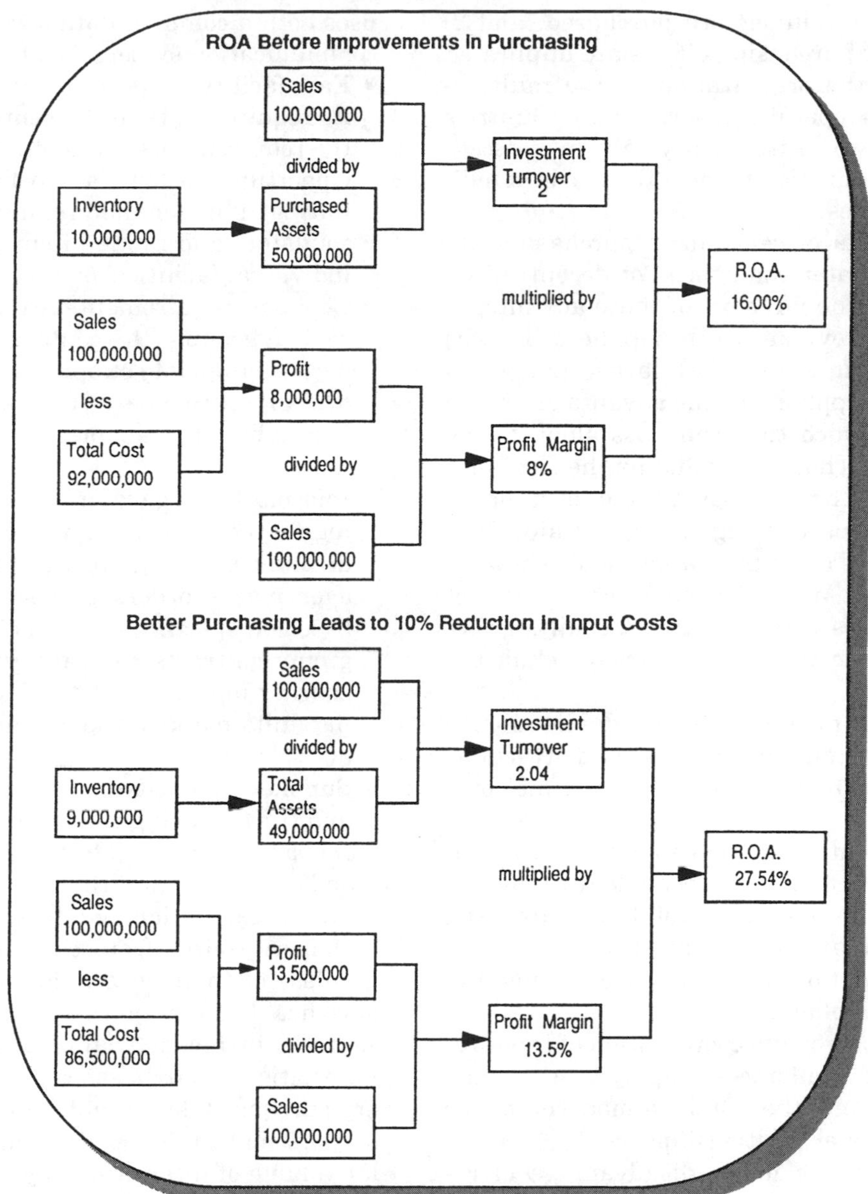

Fig. 1. Purchasing's impact on the firm's return on assets.

yields a substantial improvement in ROA. In fact, the ROA goes up by over 70 percent. The conclusion is clear – strategic management of the purchasing process can provide dramatic bottom-line performance benefits.

LOCATION: Two location issues affect purchasing: where should purchasing decisions be made? and where should suppliers be located?

1. ***Centralization versus Decentralization:*** In the single-facility organization, the question is whether decisions should be made by department managers on a decentralized basis or by a central purchasing department? In the multiple-facility organization, the question is whether decisions should be made at the plant or division level or by a central purchasing group based at headquarters? Pros and cons exist for centralized and decentralized purchasing in both the single-facility and multiple-facility settings.

The primary benefits of decentralization are: 1) increased department or plant control over purchases and 2) better responsiveness to department or plant needs. The fact is that no one knows the needs of the unit better than managers in the unit. A third benefit exists for the organization that operates multiple, geographically dispersed purchasing. This benefit is the more effective use of local suppliers. The principal drawbacks of decentralized purchasing are: 1) the loss of leverage with suppliers since smaller volumes are purchased, and 2) the fact that purchasing efforts are duplicated throughout the organization. Large multinationals with operations located in countries around the world frequently rely on a decentralized organization to meet the needs of individual facilities.

The benefits of centralized purchasing are generally greater than those of decentralized purchasing. The foremost of these advantages is increased leverage with suppliers because purchase volumes are much larger. Large volumes allow suppliers to take advantage of scale economies, which they can pass along to the buying firm. Thomas Stallkamp, head of procurement at Chrysler, noted the importance of high-volume purchasing when he said, "The advantages of decentralization don't show up in purchasing. We centralize purchasing to get the best prices from suppliers." (Tully, 1995). Other advantages of centralization include the following:

- A reduction of administrative duplication.
- Greater managerial focus – no distractions created by the need to be a "part-time" purchaser.
- Development of expertise in purchasing techniques and commodity/technology areas.
- Reduction in the number of items purchased through greater standardization.
- Common database of purchased items and capable suppliers.
- Greater control and coordination of input requirements enhances supply-base reduction and and the establishment of closer buyer/supplier relationships.

Centralization's major disadvantage of reduced responsiveness is mitigated, if not completely overcome, by these advantages. Thus, since the early 1980s, the organizational trend has been toward centralization. Among single-facility organizations, all but the smallest have centralized their purchasing organizations. Multi-facility organizations, even those with operations located in different countries around the world, are also actively pursuing the benefits of centralization. The challenge encountered by the multi-facility companies has been one of coordinating and consolidating requirements without disrupting the smooth inflow of materials to each facility. Modern information technology is greatly reducing this challenge. As a result, larger, multi-facility companies are implementing a hybrid organization designed to achieve the efficiencies of centralization and the responsiveness of decentralization. The hybrid organization, which uses both a common database and a real-time communication system, is outlined below.

- Each facility or operating unit communicates its requirements to the central group where the requirements are compared.
- Opportunities for standardization are carefully sought, common requirements are consolidated, and unique items assigned to the individual facilities' purchasing managers.
- Common requirements are managed by the central group – the center group selects the most appropriate supplier and negotiates an overall contract based on the total volume required by all of the organization's operating units.
- Information regarding the blanket purchasing contract is communicated back to the individual units where the purchasing manager places orders and schedules receipts based on specific needs. In effect, the central group contracts for capacity and the purchasing managers at the individual operating units manage the actual flow of materials.
- Supplier performance is constantly updated and tracked using the common database.

2. ***Domestic versus Global Sourcing:*** The mandate to acquire the best inputs for use by the organization requires that purchasing managers source from the best suppliers regardless of their geographic location. The result has been a marked increase in global sourcing in recent years. In the early 1980s, international purchases represented 5% of every purchased dollar and 2% of each sales dollar; raw materials represented the largest dollar volume of internationally sourced material (Monczka and Giunipero, 1984). International purchasing was also viewed as something only large firms with lots of resources could do. Today, the typical firm spends about 13% of every purchase dollar internationally, and larger firms with more than 5,000 employees acquire 17% of their total purchases internationally. The value-added nature of items sourced internationally has also increased substantially. By the early 1990s, fabricated components were the most frequently sourced items followed by raw materials and finished goods. Firms of all sizes also participated in international purchasing activities. Small firms have been greatly assisted by new information technologies and the assistance of third-party service providers (Scully and Fawcett, 1994).

Interestingly, while lower cost is the primary reason firms initiate international purchasing efforts, quality and availability become more important decision criteria as firms gain international purchasing experience (Min and Galle, 1991; Birou and Fawcett 1993). Increased experience with international sourcing also leads to more proactive global sourcing – global sourcing often becomes a vital strategic weapon that allows a firm to efficiently use resources found in countries around the world without incurring the costs or risks that come with establishing overseas operations. From this strategic perspective, the coordination of a firm's global purchasing activities has become very important. In fact, the term coordinated global sourcing now "refers to the integration and coordination of procurement requirements across worldwide business units, looking at common items, processes, technologies, and suppliers" (Monczka and Trent, 1991).

Today, competitive pressures, expanding markets, advancing information technologies, and better logistics services have made it possible for firms to effectively find, develop, and integrate worldwide suppliers. Thus, the question is no longer one of domestic versus global sourcing but rather, where can the best suppliers be found, and how can they be integrated into the firm's competitive strategy?

Firms that are considering international purchasing as a competitive option must remember that success requires substantial resource dedication and considerable management effort. Management must clearly define the firm's purchasing needs and objectives before venturing into the complex international environment. A follow-the-leader mentality almost always leads to poor performance and costly mistakes. Well-developed certification standards along with a program to qualify international suppliers must also be put in place as part of the firm's international purchasing foundation. In fact, just as supplier selection is the single most important issue in domestic purchasing, developing good international suppliers is central to the success of any international purchasing effort. Two final steps in establishing the foundation for international purchasing success are 1) to obtain top management support early in the strategy formulation process and 2) to couple a long-term perspective with intensive skill building (Fawcett and Scully, 1997).

RESULTS: The typical firm that has not made the management of purchased inputs a strategic priority can expect to achieve an initial reduction in overall purchase costs of approximately 10%. Once early cost reductions are achieved, firms continue to seek cost reductions via closer linkages with suppliers that lead to greater cooperation and communication. For instance, Allied Signal's approach to component cost reduction requires an initial price reduction, when a contract is signed with a new supplier as well as an agreement by the supplier to reduce the component's total cost by 6% per year (Tully, 1995). Contract provisions of this kind are relatively common and generally call for year-to-year cost reductions of 3–5%. By managing the purchasing function strategically – using just-in-time and global sourcing as well as supply chain alliances, supplier involvement in product development, and other advanced purchasing practices – year-to-year quality improvements of 10%, and reductions in product development times of 40% or more can be expected. As buyer/supplier relations continue to evolve in the current environment, where supply chain management is a common strategic theme, purchasing can yield even greater benefits. For instance, Volkswagen's new automobile plant in Brazil calls for suppliers to perform a large percentage of the actual car assembly. The objective is to substantially reduce costs while increasing quality and reducing lead times. The bottom line is that progressive purchasing can and does lead to improved competitiveness, but only when the firm recognizes the benefits, requirements, and challenges inherent in establishing a world-class purchasing program (see Table 3).

COLLECTIVE WISDOM: Purchasing has become a valuable contributor to firm performance and is recognized as such by an increasing number of leading companies. Top corporate managers now realize that purchasing managers must possess both highly developed skills and substantial managerial judgment to build and manage a world-class supply base, and work closely with other areas within the firm. Indeed, to be effective, purchasing managers must: 1) recognize the many sourcing alternatives that are available in today's marketplace; 2) assess a wide range of tradeoffs at many different levels in order to make good decisions; and 3) balance both the short- and long-term requirements of the organization. Developing a managerial base that possesses these skills remains a challenge.

Formal training remains somewhat difficult as only a small number of universities in the U.S.A. offer programs in purchasing and supply chain management. The National Association of Purchasing Management (NAPM) in the U.S.A. is, however, dedicated to helping develop high- caliber purchasing managers. NAPM offers skill building seminars throughout the U.S., sponsors local affiliates where purchasing managers can learn from each other, and runs a certification program. The designation C.P.M. denotes that

Table 3. Purchasing's impact on the firm's return on assets (benefits, requirements, and challenges of world-class purchasing).

Benefits	Requirements	Challenges
Reduced inventories	Top management support	Lack of top management support
Reduced costs	Supply-base reduction	Supplier reticence
Improved quality	Supplier development programs	Employee reticence
Increased productivity	Supplier certification	Inadequate engineering support
Improved supplier relations	Long-term buyer/supplier relationships	Not a panacea
Shorter delivery lead times	Quality at the source	Poor product quality
Enhanced responsiveness	Buyer training programs	Poor communication
Better management focus	Performance-based product specification	Relationship building
Enhanced production scheduling	Computerized data warehouses	Logistics support
Greater design innovativeness	Real-time communication	Production schedule stability
Enhanced product technology	Blanket orders	Finding qualified suppliers
Reduced product development cycles	Minimize variability of receipt quantity	Employee and manager training
Better customer service	Minimize administrative paper work	Long-term attitude
Improved competitive position	Management of inbound freight	

a manager is Certified in Purchasing Management and therefore possesses the skills to make good decisions. In addition, more companies are providing in-house training for their purchasing managers.

Now that purchasing has become established as a strategic value-added activity, its future evolution will depend on how it manages emerging trends, including the following:

- The reduction of the supplier base
- Supply-chain partnering
- Expanded range of purchased items
- Early supplier involvement in new product development programs

Each of these issues is influencing the management of purchasing activities by changing both what purchasing must do as well as how well purchasing must perform to help the firm achieve a sustainable competitive advantage.

Organizations Mentioned: Thomas Register; Dun and Bradstreet; Ford; Chrysler; Allied Signal; ABC analysis; NAPM; Volkswagen.

See **Competitive bidding; CPM – certified purchasing manager; Expediting; Global sourcing; MRO items; Multiple criterion in ABC analysis; NAPM; New product development through supplier integration; Supplier certification programs; Supply chain management; Supplier partnership as strategy; Supplier performance measurement; Supplier relationships; Thomas register of American manufacturers; Value analysis.**

References

[1] Birou, L.M. and S.E. Fawcett (1993). "International Purchasing: Benefits, Requirements, and Challenges." *International Journal of Purchasing and Materials Management*, 29(2), 28–37.

[2] Burt, D.N. (1989). "Managing Suppliers Up To Speed." *Harvard Business Review*, 67(4).

[3] Dobler, D.W., and D.N. Burt (1996). *Purchasing and Materials Management*. McGraw-Hill Book Company, New York.

[4] Drucker, P. (1986). "The Changed World Economy." *Foreign Affairs (Spring)*, 768–791.

[5] Fawcett, S.E. and J.I. Scully (1997). "The Impact of Experience and Facilitators on the Performance of International Sourcing Strategies," submitted to *Production and Inventory Management Journal*.

[6] Leenders, M. and D. Blenkhorn (1988). *Reverse Marketing – The New Buyer Seller Relationship*. The Free Press, New York.

[7] Min, H. and W.P. Galle (1991). "International Purchasing Strategies of Multinational U.S. Firms." *International Journal of Purchasing and Materials Management* (Summer), 9–18.

[8] Monczka, R.M. and L.C. Giunipero (1984). "International Purchasing: Characteristics and Implementation." *Journal of Purchasing and Materials Management* (Autumn), 2–9.

[9] Monczka, R.M. and R.J. Trent (1991). "Global Sourcing: A Development Approach." *International Journal of Purchasing and Materials Management* (Spring), 27.

[10] Ruch, W.A., H.E., Fearon, and C.D. Wieters (1992). *Fundamental of Production/Operations Management* (5th ed.). West Publishing Company, St. Paul.

[11] Scully, J.I. and S.E. Fawcett (1994). "International Procurement Strategies: Opportunities and

Challenges for the Small Firm." *Production and Inventory Management Journal* 35(2), 39–46.
[12] Tully, S. (1995). "Purchasing's New Muscle." *Fortune* (February 20), 75–83.
[13] Yacura, J.A. (1994). "The Expert System: Purchasing's Tool of the Future." *NAPM Insights* (October), 32–33.

PURCHASING: THE FUTURE

Larry C. Giunipero

Florida State University, Tallahassee, FL, USA

Michael G. Kolchin

Lehigh University, Bethelehem, PA, USA

BRIEF HISTORY OF THE PURCHASING FUNCTION: Purchasing as an activity is as old as civilization itself. Individuals realized that it was impossible to produce everything themselves and that they could enjoy greater efficiencies and wealth by specializing. This specialization created a need to exchange one's goods for those produced by others. When this exchange process began buying and selling became integral parts of all societies. The majority of buying and selling activities were both individual and local in nature until the Industrial Revolution of the late 19th and early 20th centuries. Industrial output was dramatically increased through highly mechanized approaches to production. Increased output lead to larger business sizes, and required more complex organizational structures. Managing these complex organizations required specialized functions in order to serve national markets. They also required purchasing specialists who sourced various equipment, products, and services to meet the organization's needs.

Although good buying has always been a part of any successful business, purchasing is most likely to come to management's attention during periods of raw material scarcity (Heinritz, Farrell *et al.*, 1991). In the United States, prior to World War I purchasing was regarded as a clerical function at most firms (Leenders and Fearon, 1997). However, after World War II pent up consumer demand caused by years of war production required the procurement of scarce items efficiently. The purchasing function's efficiency in acquisition translated into substantial profits for their firms since during this time period most organizations could sell all the products they could manufacture.

In the 1970s and 1980s sharply rising prices of materials again brought purchasing departments into the spotlight. During the late 1980s and into the 1990s purchasers were asked to obtain quality products and services while keeping costs under control. As markets began to mature and international competitors (of the United States) improved the variety and quality of their product offering, U.S. firms had to become more competitive to retain their market share. The transition to becoming competitive was not easy. Some firms closed entire plants while others downsized by reducing the number of employees.

All firms were asking more of their purchasing organizations. Purchasers were challenged to cut costs from their supply base, yet not quality. Meeting these increased expectations required the training, development and selection of a more sophisticated and skilled purchaser. The successful professional purchaser of the future will possess skills different from their counterparts of the past. Purchasing is now recognized as a key contributor to corporate profits by most senior management, thus, the function needs to become an innovator and not just a clerical order placer. How well they succeed in accomplishing this task will depend of their ability to anticipate future changes and develop proactive responses.

Data for this article is based on several studies done by the Center for Advanced Purchasing Studies located in Tempe, Arizona. One such study analyzed purchasing trends and practices and was based on responses from 131 chief purchasing officers of large U.S. based firms (Kolchin and Giunipero, 1993), and in-depth interviews conducted with 25 top purchasing professionals. These high level practitioners were asked to present future trends in the purchasing field for the year 2000.

The major trends are identified in Table 1 and are supplemented with other current research and technological changes affecting the purchasing function.

Table 1. Purchasing's future trends – 2000.

1. Fewer sources of supply will be used.
2. Purchasers will be more concerned with final customer satisfaction.
3. Purchasers will manage supplier relations.
4. Purchasers will drive shorter cycle times.
5. Supply Chain Management will receive greater emphasis.
6. Design engineers and buyers will be part of sourcing teams.
7. Global sourcing will increase.
8. Order releasing will be relegated to users.
9. Teams will make sourcing decisions.
10. Single sourcing will increase.

Source: Purchasing education and training, Kolchin and Giunipero.

MAJOR PURCHASING TRENDS:

1. ***Supplier Relations:*** Purchasers will be continually using fewer sources but spending more time managing the relations with these key suppliers. This represents a fundamental change from previous philosophies where the purchaser's philosophy was that the greater the number suppliers, the greater the competition, and therefore the better price and service one could obtain. However, by managing fewer suppliers purchasers can concentrate on managing the total cost and building a relationship with the supplier. Total cost starts with the suppliers selling price but also includes the supplier's: 1) managerial and operational capabilities; 2) quality focus; and 3) ability to schedule and deliver on time. While these costs are not as visible as price they have a significant impact on a firm's profits.

   Dealing with fewer suppliers allows for the building of closer relationships with suppliers. Such stronger relationships provide each party with a set of goals that foster continuous improvement and greater efficiencies for both parties. Purchasers must concern themselves with total cost since they will be more involved in understanding and satisfying their firm's customers. World class organizations require their people to understand not only their particular function but the process of the organization in which they work.

2. ***Reengineering Processes – Supply Chain Management:*** Since speed is a factor in everything the organization does, purchasers are asked to "think outside the box." "Thinking outside the box," is a term popularized by Messrs Hammer and Champy in their book on reengineering. It challenges people to identify better ways to accomplish a task. If a task doesn't add value to a process it should be eliminated. Tasks which add value need to be performed expediently, efficiently, and effectively. Reengineering is defined as the fundamental rethinking and radical redesign of business processes to achieve dramatic improvements in cost, quality service, and speed.

   One of the newer concepts used by reengineered firms that have been adopted by progressive organizations is the concept of Supply Chain Management. The growth of this concept is expected to spread quite rapidly in the future as it was ranked as the fifth most significant trend in purchasing. The International Center for Competitive Excellence defines Supply Chain Management (SCM) as "the integration of business processes from end users through original suppliers that provide product, services, and information that add value for customers." This definition of SCM is the broadest implying that SCM must go beyond logistics and focus on making business processes more effective and efficient. Previous definitions limited the scope of SCM.

   Early researchers defined SCM as the total flow of goods from supplier to end users and the linkage of each element of production and supply in the chain (see Jones and Riley, 1985). Next, writers discussed treating SCM as a philosophy that should be managed and analyzed as a total system and includes not only flow of goods but also flow of information (see Cooper and Ellram, 1978). Another stream of research defined SCM as a series of activities (e.g., sourcing and marketing) which must be effectively managed to provide the best customer value. It was postulated that achieving value would require building a relational philosophy with suppliers and customers (Cavinato, 1992). In its broadest form SCM is a strategic management tool to enhance customer satisfaction and improve a firm" competitive position and profits.

   SCM requires involvement with both customers and suppliers. The customer is "downstream" from the organization while the supplier is "upstream." One of purchasing's future job responsibilities is to challenge suppliers to improve and streamline their processes to facilitate the exchange of goods in a manner which provides competitive pricing and outstanding services. Good supply chain management has a direct impact on improving customer service. Perhaps one of the best examples of an efficient supply chain is found at WalMart. WalMart's real time system feeds data continually from the point of sale (the cashier upon customer checkout) to the warehouse (for timing on when to reorder) to the suppliers. Thus suppliers know when WalMart's warehouse will require a replacement order. Adopting a supply chain management system will require purchasers to coordinate supplier linkages in a much more efficient manner.

3. ***Supply Chain Management and Purchasing:*** Previous research has shown that purchasers interpret supply chain management (SCM) narrowly to mean building good relationships with suppliers (Giunipero and Brand, 1996). Many purchasers today are working with their suppliers to build mutually beneficial long-term relationships with a key group of suppliers. Future purchasers will be expected to expand this relational concept of supply chain management to their supplier or second

and third tier suppliers. They will begin to monitor the upstream supply chain and look for ways to streamline the process and also insure it is cost effective and competitive. The firm with the most efficient supply chain will win and delight its customers and gain market share. Thus firms will be using their supply chain to gain competitive advantage and purchasers will need to play a large role. There is already some evidence of this evolution to second tier supplier management. For example, one large automotive firm expects its first tier suppliers to pass their quality standards down to the second tier suppliers.

There is evidence to indicate purchasers will be in position to manage the supply chain linkages. Purchasers in the study discussed mentioned earlier very committed to sustaining the relationship they had with their largest supplier. They were also strongly committed to the relationship, placed a high value on the relationship, and were committed to developing future arrangements with the chosen supplier.

Multiple benefits were realized from utilizing this supply chain concept. These benefits included: (1) improved coordination from supplier to customer; 2) reduced lead times; 3) greater productivity; 4) lower inventories; and 5) more reliable deliveries. The movement to supply chain management will force purchasers to put more emphasis on continually reducing costs. These reduced costs will not come by purchasers acting alone but as part of teams.

4. ***Cross Functional Teams:*** The trends data indicated future purchasers will be part of a team when making decisions. Design engineers and buyers will be part of a sourcing team and teams will make sourcing decisions. One of the reasons that teams include design engineers is the need to reduce the total time from design or concept to actual product available for purchase. Previous research has shown that the firm that is first in getting a quality product to the market will capture the greater market share and profits.

In the race to cut product development cycle time, purchasing has started involving trusted suppliers earlier during the design phase. This early use of suppliers prior to specification development is termed Early Supplier Involvement. Brainstorming with suppliers early in the process provides a forum to evaluate alternatives and costs prior to specification writing. The final written specification will have considered suppliers latest technology and manufacturability. Attacking these product development and design issues during design phase assures smoother transition from initial design to finished product in a shorter time.

While the benefits to effective teams are well known, team performance will vary greatly in organizations. In certain organizations, teams act as surrogates for traditional decision making while in others teams work as facilitators. In successful organizations teams are the preferred way of conducting business. Much of the team's success comes from the structure and support teams receive. These teams are rewarded by joint profit sharing and/or given formal recognition. Benefits generally associated with team decision making include: 1) greater skill and knowledge; 2) consideration of a wider range of alternatives; and 3) increased participation by individuals in the decision process.

Typically, purchasing teams are either project teams (with a beginning and an end), type or ongoing team. The project or task team is brought together for a special purpose then disbanded once the project is completed. On-going type teams work on continuing projects. A sourcing team is an example of an ongoing type team. Purchasers belong to a variety of teams including those dealing with: sourcing, cost improvement, inventory reduction, supplier certification, and standardization teams. Currently, the largest single drawback associated with team involvement is the reward structure. Many organizations expect employees to work as part of teams yet reward is individual-based. Future reward structures aimed at promoting teamwork include team bonuses, division-wide profit sharing, and stock option plans for professional and clerical employees.

5. ***Electronic Commerce:*** While teams will continue to play a vital role in the purchasing function of the future, purchasers will still be responsible for issuing purchase orders and contracts. Managing the purchase order process efficiently will require a movement away from paper-based systems. Additionally, purchasers will control the majority of expenditures through contracts and then allow users to contact suppliers and order against these prenegotiated contracts. Two systems now widely used to reduce paperwork include the procurement card and electronic data interchange (EDI). In the future, purchasers will utilize other forms of electronic commerce, in addition to EDI. These include the Internet, electronic search agents, and other electronic linkages.

The procurement card system provides users a credit card similar to their personal Visa, Mastercard, or American Express. The purchaser sets up a Master Agreement with select suppliers and users simply call up and give the supplier salesperson their card number along with a list of items required. The user must verify quantity received and keep all paperwork and packing slips for accounting and audit verification. The card company pays the supplier, then later bills the buyer's firm. For small dollar items the procurement card is an excellent way to lower transaction costs, provide excellent service and obtain the user competitive pricing.

Fax and electronic data interchange provide two other means by which purchasers can reduce transaction costs, increase data accuracy and provide enhanced service to users. Fax machines permit instaneous transmittal of information such as quotes and purchase orders, and reduce cycle time and cost.

Electronic Data Interchange (EDI) is the electronic transfer of data between buying and selling firms. There are a variety of methods utilized to transfer data electronically. In the early 1980s, seller firms would put a terminal in the buyer's office. Buyers would enter their requirements and check stock status on these terminals. However, they limited electronic transfer to one supplier. In order to deal with multiple suppliers the buying firm would have to own a terminal for each supplier. Recognizing a need for supplier/buyer interface, third party providers were formed. These third party providers would perform the electronic interface between the buying and selling company for a fee. After several years of work, the American National Standards Institute (ANSI) developed a standard communication format for EDI technology. This standard (ANSI X12) has provided firms with the capability to go computer to computer, and created a growth in EDI usage.

The future purchaser will be very much involved with electronic commerce. Electronic commerce involves doing business electronically. It can be performed exchanging structured messages (e.g., EDI), unstructured messages (e-mail), or through supplier home pages and data bases accessible on the Internet (World Wide Web). The phenomenal growth in the acceptance and use of the Internet provides purchasers with new tools for transacting business. Preliminary research done by one of the authors indicates that, currently, use of the Internet is to: 1) keep up with business news; 2) visit supplier home pages to find product information and offerings; and 3) obtain supplier financial status.

Due to security issues, a very small percentage of purchaser orders are placed on the internet. Once effective security measures are in place growth, in order transactions over the Internet will occur. Since much work has been done developing the ANSI X12 standard for data transmission it may well become the format used by firms when ordering via the internet. Several firms have begun to develop secure on-line networks that facilitate the exchange of buyer/seller information. Currently, these firms offer buyers a secure web browser that will allow purchasers to directly access preferred suppliers. For example, the Fisher Technology Group located in Pittsburgh offers a secure internet service. The service called Cornerstone allows organizational buyers to access the products of their preferred suppliers via their web browsers.

Buying firms are beginning to use the Internet as a way to collect competitive bids online. Meanwhile, other companies are offering purchasers the opportunity to outsource the bidding process. These third party sourcers prepare the bid package and conduct the bid online in an auction format. Future tools available to purchasers will include intelligent agents. These intelligent agents are programmed computer search vehicles, when given a particular commodity, they will search for the best price and terms on a worldwide basis.

ENTERPRISE RESOURCE PLANNING SYSTEMS: Enterprise Resource Planning Systems (ERP) integrates the business processes of a company. These ERP systems provide the upper management with one set of computer databases to run the business. They can help drive a firm towards improving its business practices and attaining best in class status. They integrate all key elements of a business from customer service, financial management, purchasing, manufacturing, engineering, and marketing. Firms can quickly see the results of their reengineering, Total Quality Management and cost reduction efforts. ERP systems provide the decision support power necessary to improve customer service, reduce costs, and lower inventory.

OUTSOURCING: Outsourcing is the process of having a supplier firm handle tasks that were performed in-house. A recent study used a similar definition "the transfer of responsibility to a third party of activities which used to be performed internally" (Maltz and Ellram, 1997). Outsourcing could be compared to an older concept, "make or buy parts." However, outsourcing usually is much broader than make or buy, it covers any activity, not just parts. For example, a firm may outsource its entire customer service or order entry functions. Many firms have outsourced their electronic data processing operations while others their logistics and warehousing operations. Reasons given for outsourcing include:

1. Allows management to focus on core competencies.
2. Cost savings.
3. Lower investment in facilities.
4. Gaining access to latest technology.
5. Increased operating effectiveness.
6. Improved service (Maltz and Ellram, 1997).

Future purchasers can expect to see a continuation of the outsourcing trends. Purchasing must continue to impress top management that it does add value or it could be outsourced. For there are examples of cases where the entire purchasing function was outsourced. As with all programs outsourcing has disadvantages including: a loss of control; poorer than expected long term service; extra costs; loss of competitive information; and a reduction in internal skills. Current trends favor outsourcing to continue and they include management's desire to keep employee headcount low, and the desire to focus on core competencies.

INSOURCING: Insourcing refers to the practice of bringing outside contractors and suppliers into the buyer's plant to perform duties normally carried out by full time employees. The popular term for insourcing is JIT II. The JIT II was developed and popularized by Bose Corporation of Framingham, Massachusetts. Bose has various suppliers in its plants working side by side with its buyers managing the day to day requirements for its products. The suppliers are called in-house representatives and have access to data personnel and facilities. Insourcing replaces both the buyer and salesperson. The inplant representative is responsible for keeping the inventory to a minimum while keeping the plant running. The in-plant representative attends design and new products meetings, which involve his/her products (Dixon, 1994). Purchasers can expect to see insourcing used as a way of keeping a firm competitive since inplants provide a seamless flow of information between buyers and sellers.

Table 2. Future skills and abilities required of purchasers – 2000.

1. Interpersonal communication
2. Customer focus
3. Ability to make decisions
4. Negotiation
5. Analytical
6. Managing change
7. Conflict resolution
8. Problem solving
9. Influencing and persuasion
10. Computer literacy

Source: Purchasing education and training, Kolchin and Giunipero.

FUTURE SKILLS AND ABILITIES REQUIRED: The most critical skills identified by purchasers as important in the future are shown in Table 2. What this ranking of most important skills reflects is, perhaps, the change in how sourcing decisions are being made. As noted in the trends section above, purchasing personnel will be involved in developing closer relationships with fewer sources, and the sourcing decisions will be made by teams. The skills rated as important for the year 2000 place high value on interpersonal skills, a necessary ingredient in team success and relationship building.

Many of these skills are already required for success in the purchasing function; however, the increased importance of customer focus, change management, and computer literacy are very evident in the responses. Evidence that purchasing will be working through teams is particularly evident in the required skill sets of: 1) negotiation; 2) problem solving; 3) conflict resolution; and 4) influencing/persuasion.

Negotiation is the most popular training course for purchasers. However, most training applies to dealings with suppliers. Negotiating in a team environment within the firm must be added to purchasing training. Corporations such as Motorola and General Electric provide negotiation training for their purchasers as well as supplier personnel. This allows their suppliers to become more efficient. Future corporations will begin to do more of this training with their key tier I suppliers.

The skills required by purchasers in the future could be categorized into the following three areas: 1) enterprise (that is, having a good understanding of the overall business); 2) interpersonal; and 3) technical.

These enterprise skills will involve market analysis, negotiation with partners, effectively managing internal and external relations and change. Interpersonal skills include risk taking, written and oral communications, leadership, and cultural awareness. Lastly, technical skills include cost analysis, product knowledge, and computer literacy. What these responses suggest is that purchasing professionals in the future must be technically and interpersonally competent and must have a good grasp of the total enterprise.

KNOWLEDGE REQUIRED: Participants in the interviews and respondents to the survey were asked what knowledge areas would be most important to purchasing professionals in the future. The responses could be categorized into the same three areas: enterprise, interpersonal, and technical. Interviews with the purchasing executives who participated in the study led to the development of a list of 53 subjects that purchasing professionals should know if they were to be successful in the next decade. Respondents were then asked to rate which subjects were most critical. The ten most important future areas of purchasing knowledge are shown in Table 3. The knowledge areas illustrate the importance of quality, supplier relations/development and supply chain management. As buyers enter into closer relationships with sellers, it is the purchasing function that must provide the knowledge on how to insure the best value obtained through these relationships. Purchasing must provide the means to achieve lower costs that were previously provided by the competitive marketplace.

Renaming Purchasing: Survey respondents were also asked if the purchasing function would still be called purchasing in the future. Sixty-four percent of this sample of purchasing executives said no. The three most cited alternatives for the purchasing function in the future were: (1) Supply Management; (2) Sourcing Management; and (3) Logistics.

Table 3. Future knowledge required of purchasers.

1. Total quality management
2. Cost of poor quality
3. Supplier relations
4. Analysis of suppliers
5. Lowest total cost
6. Price/cost analysis
7. Source development
8. Quality assurance
9. Supply chain management
10. Competitive market analysis

Source: Purchasing education and training, Kolchin & Giunipero.

Organizations Mentioned: Bose Corporation; Fisher Technology; General Electric; Motorola; Wal-Mart.

See **Downsizing; Electronic data interchange (EDI); Electronic data interchange in supply chain management; ERP; Evolution and from MRP to MRP II and ERP; Planning, the; Insourcing; JIT II; Logistics; Outsourcing; Reengineering; Reengineering and the process view of manufacturing; Sourcing; Supplier partnership; Supplier performance measurement; Supplier relationships; Supply chain management.**

References

[1] Cavinato, Joseph L. *A Total Cost/Value Model for Supply Chain Competitiveness, Journal of Business Logistics*, 13(2), pp. 285–301.

[2] Cooper, Martha C. and Lisa M. Ellram (1993). "Characteristics of Supply Chain Management and the Implications for Purchasing & Logistics Strategy," *International Journal of Logistics Management*, 4(2), pp. 10–15.

[3] Dixon, Lance and Anne Miller Porter (1994). "JIT II – Revolution in Buying and Selling," Cahner's Publishing Company, Newton, MA.

[4] Giunipero, L. and J. Vogt (1997). "Empowering the Purchasing Function: Moving to Team Decisions," *International Journal of Purchasing & Materials Management*, Winter, pp. 8–15.

[5] Giunipero, Larry C. and Richard R. Brand (1996). "Purchasing's Role in Supply Chain Management," *International Journal of Logistics Management*, Vol. 7, Number 1, pp. 25–38.

[6] Hammer M., and J. Champy (1993). "Reengineering the Corporation: A Manifesto for Business Revolution," Harper, New York, NY.

[7] Heinritz, S., P. Farrell, L. Giunipero, and M. Kolchin (1991). "Purchasing Principles and Applications," Prentice Hall, Englewood Cliffs, N.J.

[8] Jones, Thomas C. and D.W. Riley (1985). "Using Inventory for Competitive Advantage Through Supply Chain Management," *International Journal of Physical Distribution and Materials Management*, 15(5), pp.16–26.

[9] Kolchin, M. and L. Giunipero (1993). "Purchasing Education and Training," Center for Advanced Purchasing Studies, Tempe, AZ.

[10] Leenders, M. and H. Fearon (1997). "Purchasing and Supply Management," R.D. Irwin, Homewood, IL.

[11] Maltz, Arnald and Lisa Ellram (1997). "Outsourcing: Implications for Supply Management," CAPS, p. 140.

[12] Monczka, R.M. and R.J. Trent (1993). "Cross-Functional Sourcing, Team Effectiveness," Center for Advanced Purchasing Studies, Tempe, AZ.

PURE CUSTOMIZATION

This is a form of mass customization. In pure customization, a product is developed from scratch for each customer. Pure customization infiltrates the entire operating chain. The product is designed for the original customer, assembled and fabricated to order and delivered in a personalized way. Pure customization is an expensive proposition, but there is a surprising amount of activity that fits this category. Many construction projects can be considered to be pure customization, such as, an architect designed house, or a company headquarters' building. In these projects, an unique design is conceived by an architect for the specific client and building site. Many industrial products are also manufactured for one specific customer or specific task; examples are special purpose machines or instrumentation. Pure customization does not only pertain to physical goods, but can also be found in organizations assembled for specific tasks, such as, the Olympics Organizing Committee or the Apollo moon project. (Mintzberg, 1988)

See **Concurrent engineering; Mass customization; Mass customization implementation; New product development through supplier integration; Production design.**

PUSH SYSTEM

See **Just-in-time manufacturing; Lean manufacturing implementation; Manufacturing systems; MRP; Order release; Pull system.**

QS 9000

See **Quality management systems: Baldridge, ISO 9000, and QS 9000.**

QUALITY ASSURANCE

According to ASQC (American Society for Quality Control) Standard A3-1971, quality assurance is making sure that quality is what it should be. Quality assurance ensures product conformance with quality specifications. The producer employs procedures to ensure the quality of its products and services, and undertakes to inform the consumer about that level of quality. This information serves two different purposes:

1. To assure the consumer that the product is fit for use, the manufacturing process is behaving normally, or proper procedures are being followed.
2. To provide the consumer with an early warning that all is not well. The consumer is then placed in a position to take preventive action to avert disaster.

Quality assurance normally requires documented evidence of the procedures used in production to ensure the attainment of quality. Two major forms of assurance which companies use are summary reports of laboratory tests, factory tests, field performance data, and quality audits, which are independent evaluations of various aspects of quality performance.

See **ASQ; Integrated quality control; Professional associations for manufacturing professionals and managers; Quality: The implications of W. Edwards Deming's approach; Statistical quality control using control charts; Total quality management.**

Reference

[1] Crosby, Philip B. (1979). *Quality is Free*, McGraw Hill, New York.

QUALITY CIRCLE

Quality circle originated in Japan.

See **Integrated quality control; Lean manufacturing implementation; Quality: The implications of deming's approach; Teams: Design and implementation; Total quality management.**

QUALITY CONTROL

See **Integrated quality control; Just-in-time manufacturing; Statistical quality control using control charts; Total quality management.**

QUALITY FUNCTION DEPLOYMENT (QFD)

QFD is a formal method employed in product design to ensure that all customer requirements are identified and met by the resulting product design. QFD serves as a tool that translates consumer/customer requirements into design specifications, product functions and features.

See **Product design for global markets; Product design; Target costing; Total quality management.**

QUALITY MANAGEMENT

See **Integrated quality control; Just-in-time manufacturing; Statistical quality control using control charts; Total quality management.**

QUALITY MANAGEMENT SYSTEMS: BALDRIGE, ISO 9000, AND QS 9000

James R. Evans

University of Cincinnati, Cincinnati, Ohio, USA

DESCRIPTION: With all the publicity surrounding the Malcolm Baldrige National Quality Award of the U.S., ISO 9000, and QS-9000, many misconceptions about them have arisen. Two common misconceptions are that the Baldrige Award and ISO 9000 registration cover similar requirements, and that both address improvement and results. In reality, the Baldrige Award and ISO 9000 are distinctly different instruments that can reinforce one another when properly used. QS-9000 is an auto industry standard, which is something of a hybrid of the two, encompassing all of ISO 9000 but not the full scope of Baldrige. The reader is referred to the articles *Performance Excellence and ISO 9000/QS-9000 Quality Standards* for further details on each topic.

BALDRIGE VERSUS ISO 9000: The Baldrige Criteria for Performance Excellence consist of seven categories that reflect key management practices deemed important to achieve success in today's competitive global marketplace. Within each category there are one or more items that focus on an important requirement within that category. Some items may consist of one or several areas to address, which define more specific requirements related to that item. The seven categories are:

1. *Leadership.*
2. *Strategic Planning.*
3. *Customer and Market Focus.*
4. *Information and Analysis.*
5. *Human Resource Focus.*
6. *Process Management.*
7. *Business Results.*

The Baldrige Criteria have been revised and continuously improved since their inception, and undergo a major evaluation and revision every two years.

ISO 9000 standards are based on the premise that certain generic characteristics of management practices can be standardized, and that a well-designed, well-implemented, and carefully managed quality system provides confidence that the outputs will meet customer expectations and requirements. The standards prescribe documentation for all processes affecting quality and suggest that compliance through auditing leads to continuous improvement.

ISO 9001 consists of the following elements:

- *Management responsibility.*
- *Quality system.*
- *Contract review.*
- *Design control.*
- *Document and data control.*
- *Purchasing.*
- *Control of customer-supplied products.*
- *Product identification and traceability.*
- *Process control.*
- *Inspection and testing.*
- *Control of inspection, measuring, and test equipment.*
- *Control of nonconforming product.*
- *Corrective and preventive action.*
- *Handling, storage, packaging, preservation, and delivery.*
- *Control of quality records.*
- *Internal quality audits.*
- *Training.*
- *Servicing.*
- *Statistical techniques.*

The differences between the ISO 9000 standards and the Baldrige Criteria are substantial. Table 1 shows a brief summary of the key differences; a more complete discussion can be found in Reimann and Hertz (1993).

BALDRIGE VERSUS QS-9000: QS 9000 is based on ISO 9000 and includes all ISO requirements. However, QS 9000 goes well beyond ISO 9000 standards by including additional requirements such as continuous improvement, manufacturing capability, and production part approval processes. QS 9000 not only states *what* must be done, but often *how* to do it. Many of the concepts in the Malcolm Baldrige National Quality Award Criteria for Performance Excellence are reflected in QS 9000. In many cases, the wording is almost identical to that found in the Baldrige Award criteria.

We will describe the portions of QS-9000 that are not included in ISO 9000, but which relate directly to the Baldrige Criteria. All references to "elements" refer to the QS-9000 standard; the other references (e.g. "2.2a") refer to Baldrige items or areas to address in the 1998 Criteria.

Element 4.1 – *Management Responsibility* – incorporates a variety of areas from the Baldrige Criteria. The additional language and requirements in this element are the most pertinent to moving the QS-9000 standard toward the world-class management practices reflected in the Baldrige.

- *Strategic Planning*. The company must utilize a formal, documented, comprehensive business plan and develop both short- and longer-term goals and plans based on the analysis of competitive products and benchmarking information (2.2a). It is also required to track, update, revise, and review the plan and communicate it throughout the organization (2.2a).
- *Customer and Market Focus*. The standard requires methods to determine current and future customer expectations along with an objective and valid process to collect the information (3.1). In addition, it shall have a documented process for determining customer satisfaction, including the frequency of determination and how objectivity and validity are assured (3.2b).
- *Information and Analysis*. Trends in data and information should be compared with progress toward overall business objectives and translated into actionable information to support prompt solutions to customer-related problems

Table 1. Contrasts between Baldrige and ISO 9000.

	Baldrige award program	ISO 9000 registration
Focus	Competitiveness; customer value and operational performance	Conformity to practices specified in the registrant's own quality system
Purpose	Educational; shares competitiveness learning	To provide a common basis for assuring buyers that specific practices conform with the providers' stated quality systems
Quality Definition	Customer-driven	Conformity of specified operations to documented requirements
Improvement/results	Heavy dependence on results and improvement	Does not assess outcome-oriented results or improvement trends
Role in the marketplace	A form of recognition, but not intended to be a product endorsement or certification	Provides customers with assurances that a registered supplier has a documented quality system and follows it
Nature of assessment	Four-stage review process or comprehensive internal self-assessment	Evaluation of quality manual and working documents and site audits to ensure conformance to stated practices
Feedback	Diagnostic feedback on approach, deployment, and results focusing on strengths and areas for improvement	Audit feedback on discrepancies and findings related to practices and documentation
Criteria improvement	Annual revisions	Revisions of 1987 document issued in 1994, focusing on clarification
Responsibility for information sharing	Winners required to share their successful strategies	No obligation to share information
Service quality	Service excellence a principal concern	Standards focused on repetitive processes, without a focus on critical service quality issues such as customer relationship management and human resource development
Scope of coverage	All operations and processes of all work units; includes all ISO 9001 requirements	Covers only design/development, production, installation, and servicing; address less than 10% of Baldrige criteria
Documentation requirement	Not spelled out in criteria	A central audit requirement
Self-assessment	Principal use of criteria in the area of improvement practices	Standards primarily for contractual situations or other external audits

and determination of key customer-related trends and correlations to support status review, decision making, and longer-term planning. These address areas 4.1a, 4.3a, and 4.3b in the Baldrige framework.

- *Business Results*. The company shall document trends and current levels in quality and operational performance for key product and service features, and compare them with competitors and/or appropriate benchmarks (7.5). In addition, it is required to document trends in customer satisfaction and key dissatisfaction indicators, compared to competitors or appropriate benchmarks (7.1).

Two QS-9000 elements deal specifically with Societal Responsibilities (1.2a). Element 4.6 – *Purchasing* – requires that all materials used in part manufacture satisfy current governmental and safety constraints on restricted, toxic, and hazardous materials, as well as environmental, electrical, and electromagnetic considerations. Element 4.9 – Process Control – requires a process for compliance with all applicable government safety and environmental regulations.

Some human resource issues related to Work Systems (5.1) appear in various QS-9000 elements. Element 4.2 – *Quality System* – states that companies should convene internal cross-functional teams to prepare for production of new or changed products. The sector-specific Manufacturing Capabilities element also requires the use of cross-functional teams for developing facilities, processes, and equipment plans. This element also addresses some issues associated with 5.3a, the Work Environment. Element 4.9 – *Process Control* – requires that documented process monitoring and operator instructions are accessible at the workstation. Finally, Element 4.18 – Training – states that training should be viewed as a strategic issue affecting all of the company's personnel, and that training effectiveness be periodically evaluated (5.2). The relationship between training and strategy also addresses Baldrige area 2.2a.

The majority of the requirements new to QS-9000 fall into the Baldrige category of Process Management. A summary of Baldridge items adopted by QS-9000 is given below.

6.1a-Design Processes The company shall investigate and confirm the manufacturing feasibility of proposed products prior to contracting to produce them, and consider all special characteristics in process failure mode effects analysis (FMEA) (Element 4.2, QS-9000).

The design process should include efforts to simplify, optimize, innovate and reduce waste; consider cost/performance/risk trade-offs; and use feedback from testing, production, and in the field (Element 4.4). Preliminary process capability studies are required for each company- or customer-designated special characteristic for new processes and to meet customer requirements (Element 4.9). Mistake-proofing methods should be used during the planning of processes, facilities, equipment and tooling, as well as during problem resolution (Sector-Specific Requirement 3).

6.1b-Baldridge Production and Delivery Processes The company shall develop control plans at the system, subsystem, component and/or material level as appropriate (Element 4.2). Appropriate resources for machine/equipment maintenance shall be provided along with an effective, planned total preventive maintenance system. Documented process monitoring and operator instructions for all employees having responsibilities for process operations shall be prepared. Process performance requirements are defined by the customer (Element 4.9).

All process activities should be directed towards defect prevention rather than defect detection (Element 4.10). Evidence is required that appropriate statistical studies have been conducted to analyze variation in measuring and test equipment (Element 4.11). Disciplined problem solving methods shall be used when internal or external nonconformances occur (Element 4.14).

The company shall establish a goal of 100% on-time shipments for its customers. When this goal is not met, the supplier shall implement a system to improve delivery performance, including communication of delivery problem information to the customer (Element 4.15). A procedure for communicating information on service concerns to manufacturing, engineering, and design activities shall be established and maintained (Element 4.19). A comprehensive continuous improvement philosophy shall be fully deployed throughout the organization, and specific action plans for continuously improving processes that are the most important to the customer shall be developed. The company shall identify opportunities for quality and productivity and implement appropriate improvement projects (Sector-Specific Requirement 2).

6.2-Baldridge Management of Support Processes Suppliers should extend continuous improvement philosophy to all business processes and support services (Sector-Specific Requirement 2).

6.3-Management of Supplier and Partnering Relationships Appropriate customer engineering and quality personnel should approve control plans unless waived (Element 4.2).

Subcontractor (i.e., supplier) quality system development is a fundamental quality system requirement. Assessments should occur at appropriate frequencies. Companies shall require 100% on-time delivery performance from suppliers and implement a system to monitor delivery performance (Element 4.6).

In addition, registration to QS 9000 requires demonstration of effectiveness in meeting the intent of the standards, rather than simply the "do it as you document it" philosophy. For instance, while ISO 9000 requires "suitable maintenance of equipment to ensure continuing process capability" under process control, QS-9000 requires suppliers to identify key process equipment and provide appropriate resources for maintenance, and develop an effective, planned total preventive maintenance system. The system should include a procedure that describes the planned maintenance activities, scheduled maintenance, and predictive maintenance methods. Also, extensive requirements for documenting process monitoring and operator instructions and process capability and performance requirements are built into the standards.

Finally, additional requirements pertain specifically to Ford, Chrysler, and GM suppliers. Thus, registration under QS 9000 standards will also achieve ISO 9000 registration, but ISO-certified companies must meet the additional QS 9000 requirements to achieve QS certification.

Organizations Mentioned: Ford; Chrysler; GM.

See **Implications of Deming's approach; Integrated quality control; ISO-9000/QS-9000; Malcom Baldridge award criteria; Performance excellence; Quality standards; Quality; Total productive maintenance.**

References

[1] Evans, James R. "Beyond QS-9000" *Production and Inventory Management Journal*, 38, 3, Third Quarter, 1997, 72–76.

[2] *Quality System Requirements QS-9000*, International Organization for Standardization, Chrysler Corporation, ford Motor Company, and General Motors Corporation, August 1994.

[3] Reimann, C.W. and H.S. Hertz, "The Malcolm Baldrige National Quality Award and ISO 9000 Registration," *ASTM Standardization News*, November 1993, 42–53.

[4] U.S. Department of Commerce, *Malcolm Baldrige National Quality Award 1998 Criteria for Performance Excellence*, Washington, D.C., GPO, 1998.

QUALITY OF WORK LIFE INITIATIVES

Quality of work life (QWL) initiatives include any actions designed to make work more intrinsically rewarding, increase job satisfaction, or remove or diminish negative aspects of the job. Many QWL initiatives are directed at improving worker health or safety. Examples include job rotation and job enlargement (adding additional responsibilities or tasks to a job) and initiatives to assist workers balance work and non-work aspects of life (e.g., on-site day care, flexible schedules). Employee participation systems may be considered QWL initiatives if employees value the opportunities to offer ideas, opinions, or become involved in work-related decision making.

See **Human resource issues and advanced manufacturing technology; Human resource issues in manufacturing; Teams: Design and implementation.**

QUALITY: THE IMPLICATIONS OF DEMING'S APPROACH

Elisabeth J. Umble

Texas A&M University, College Station, Texas, USA

DESCRIPTION: W. Edwards Deming (1900–1993) was a statistician, physicist, consultant, and teacher. He is best known for developing a system of Statistical Quality Control, although his contributions far exceed the introduction of those techniques. He contended that quality is the responsibility of top management and that high quality costs less than poor quality. He also championed the idea that quality must be built into the product at all stages of development and production in order to achieve excellence.

Deming was one of the earliest well-known proponents of quality, and his philosophy concerning the managerial changes necessary to achieve quality changed management practices in both manufacturing and service industries worldwide. His work illuminated and extended that of earlier proponents of quality control and productivity. Deming's many contributions to quality improvement and the corresponding managerial changes that are a prerequisite for quality include his famous 14 Points, his development of ideas in Statistical Process Control, and his System of Profound Knowledge.

HISTORICAL PERSPECTIVE: Quality has always been important in manufacturing. Before the industrial revolution, skilled craftsmen served both as manufacturers and inspectors, and their pride of workmanship inspired them to build quality into their products. By the turn of the twentieth century, mass production techniques had become firmly entrenched in manufacturing. In 1911, Frederick Taylor published *Principles of Scientific Management*, substantially changing the nature of quality assurance. By focusing on production efficiency and decomposing jobs into small work tasks, the assembly line destroyed the holistic nature of manufacturing. This necessitated the development of independent "quality control" departments that used inspection to ensure that products were manufactured correctly. Thus, "inspecting out" defective items became the chief means of ensuring quality.

Statistical Quality Control (SQC), the forerunner of today's Total Quality Management, had its beginnings in the mid-1920s at the Western Electric plant of the Bell System. Walter Shewhart, a Bell Laboratories physicist, designed the original version of SQC to help eliminate defects in the mass production of telephone exchanges and sets. In 1924, Shewhart made the first sketch of a modern "control chart." The new technique was subsequently developed in various memoranda and articles; and in 1931, Shewhart published a book on statistical quality control which bore the title *Economic Control of Quality of Manufactured Product*. This book provided a precise, measurable definition of quality and promoted statistical techniques, such as control charts, for evaluating production processes and improving quality. In the early thirties, Bell Systems, in collaboration with the American Society for Testing and Materials ATSM), the American Standards Association (ASA), and the American Society for Mechanical Engineers (ASME), undertook to popularize the new statistical methods in the United States. But, despite their support, the rate of adoption of the new techniques proved to be extremely slow.

The initial reluctance of American industry to adopt statistical quality control was rapidly overcome during World War II. The armed services became large consumers of U.S. industrial output, and consequently, influenced the adoption of statistical quality control in two ways. First, the armed services themselves adopted scientifically designed sampling inspection procedures. The initial step in the development of military sampling inspection procedures occurred shortly after America's entry into the war, when a group of Bell Lab engineers were brought to Washington to develop a sampling inspection program for Army Ordnance. Second, the military established educational programs for military and industrial personnel. At the request of the War Department, the American Standards Association developed concise statements of American control-chart practices and published materials for training courses. Between 1943 and 1945, representatives from 810 organizations attended one or more of the 33 intensive courses on SQC offered by the Office of Production Research and Development of the War Production Board. Among those attending were faculty from 43 different educational institutions who wished to prepare themselves to teach quality control.

After World War II, many of the SQC techniques, which had been developed and applied in the production of military goods could have been readily applied in the consumer goods sector. However, many U.S. executives stubbornly resisted the implementation of the new techniques. Deming personally experienced this resistance when he delivered a series of seminars on SQC for auto industry executives at Stanford University in 1945. Because of such experiences, Deming became somewhat bitter toward U.S. management.

In 1946, under the auspices of the Economic and Scientific Section of the U.S. Department of War, Deming spent two months in Japan assisting U.S. occupation forces with studies of nutrition, agricultural production, housing, fisheries, and so forth. Quality control concepts were first introduced to Japanese industry in the form of an occupation forces order to communications equipment manufacturers. During this and many subsequent visits to Japan, Deming discussed his ideas on quality control with members of the Union of Japanese Scientists and Engineers (JUSE). In late 1949, Deming was invited to Japan to teach statistical methods for industry. He preached a very simple message: *controlling the process, rather than inspecting items that come from the process*, is the key to quality; process control depends on a knowledge of statistical process control concepts; and total quality management requires the participation and training of all members of the organization. The Japanese took Deming's message to heart. By the 1980s, many Japanese companies that adopted Deming's approach were outperforming their U.S. competitors.

Eventually, organizations worldwide saw the benefits of implementing Deming's philosophy. While Deming is not solely responsible for quality improvement in Japan or the U.S., he played a key role in increasing awareness of the process of quality control and the need to improve. Few people have changed worldwide business practices as positively and completely as did Deming. Today, he is regarded as a national hero in Japan and has been honored by the Japanese with the world-famous Deming Prize for Quality which recognizes successful efforts in instituting company-wide statistical quality control principles.

STRATEGIC PERSPECTIVE: Many current management concepts can be traced back to Deming's preachings. Quality was his central theme, and most "Total Quality Management" (TQM) programs are based on his beliefs about quality. He emphasized the need to break down barriers between management and labor and to effectively utilize employees to make processes work better. This eventually led to powerful concepts like worker empowerment and cross-functional teams. Highly effective manufacturing systems, such as "Just-in-Time," would not have been possible if not for the quality-based improvements that resulted in reliable processes as well as expectations of high quality products from vendors. And, Deming's philosophy about the need to overhaul and improve business processes laid the foundations for the current trend toward "reengineering." Deming's theories on quality can be found in *Quality, Productivity, and Competitive Position* (1982), *Out of the Crisis* (1982), and *The New Economics for Industry, Government, and Education* (1993).

DEMING'S 14 POINTS FOR MANAGEMENT: Deming's goal was to make companies more competitive and keep them in business. Because he believed 85% of the problems in a system to be the responsibility of management, he offered 14 Points for management as the basis for transforming industry. These can be applied to large or small organizations, service or manufacturing industries, or divisions within a company. The 14 Points are:

1. Create constancy of purpose toward improvement of product and service with a plan to become competitive and stay in business.
2. Adopt the new philosophy of quality.
3. Cease dependence on inspection to achieve quality. Require instead statistical evidence that quality is built in.
4. End the practice of awarding business on the basis of price tag. Consider quality as well as price. Move toward a single supplier for an item, based on a long-term relationship of trust.
5. Improve constantly and forever the system of planning, production, and service, to improve quality and productivity and constantly decrease costs.
6. Institute modern methods of training on the job.
7. Institute modern methods of supervision that emphasize quality rather than sheer numbers.
8. Drive out fear.
9. Break down barriers between departments.
10. Eliminate slogans, exhortations, and targets for the work force.
11. Eliminate numerical quotas for the work force and numerical goals for management.
12. Remove barriers that rob workers (and management) of their right to pride of workmanship.
13. Institute a vigorous program of education and retraining.
14. Put everyone in the company to work to accomplish the transformation.

Deming believed that the consumer is the most important part of the production line. He defined quality as the translation of future needs of the consumer into measurable characteristics so that a product or service can be designed and produced to give satisfaction at a price the user will pay and that will make the user better off in the future. Low quality costs money because it means rework or scrap; wastes material, labor, and machine capacity; increases the cost of the product; and causes a loss of sellable product. As quality improves, costs will decrease and productivity will increase, resulting in greater market share, more jobs, and long-term survival. Deming argued that high quality is achieved by measuring, controlling, and improving the processes involved in the design, production, and delivery of goods or services.

In his 14 Points, Deming highlighted some key business practices that cost companies dearly. These practices include awarding business to the lowest bidder while ignoring quality, creating obstacles that rob workers of pride of quality workmanship, and trying to inspect quality into products. For example, Deming points out that, in most practical cases, "all or none" sampling will minimize inspection costs. For a process in control, cost is minimized with no sampling when proportion defective is less than the break-even proportion which is calculated as cost to inspect divided by cost to repair, and cost is minimized with 100 percent sampling when proportion defective exceeds the break-even proportion. Because properly motivated workers are necessary for the production of high quality products, Deming stressed worker pride and satisfaction rather than slogans and numerical goals. Furthermore, since the system, rather than the worker, is the root cause of most problems, Deming's overall approach focused on improving the system. And this is management's responsibility.

The Deming cycle (Plan, Do, Check, Act) is a methodology for continuous improvement. This methodology was originally developed in *Statistical Method from the Viewpoint of Quality Control* (1939) by Shewhart. Deming suggested that this procedure should be followed for the improvement of any stage of production and as a procedure for finding a special cause detected by statistical signals (see section on Statistical Process Control). The Plan stage involves studying the current situation, gathering data, and planning for improvement. The Do stage consists of implementing the plan on a trial basis. The Check stage is designed to determine if the trial plan is working and to see if any further problems or opportunities have been discovered. The Act stage consists of implementing the final plan. This leads back to the Plan stage for further diagnosis and improvement. The cycle is never ending. Deming used this Plan, Do, Check, Act cycle as the basis for accomplishing his 14th Point: Put everyone in the company to work to accomplish the transformation.

STATISTICAL PROCESS CONTROL: Shewhart (1931) established his criteria for determining when numerical data are in statistical control. Shewhart realized that potential information is generated by all industrial processes. He demonstrated how graphs can be used to identify fluctuations in the process and to determine when a process is exhibiting more than simple random variation. This kind of analysis can be applied to distinguish common causes of trouble that occur due to the natural variability of the system from local sources of trouble, which Shewhart called assignable causes of variation. Local sources of trouble must be eliminated before managerial innovations leading to improved productivity can be achieved through removal of common causes.

Perhaps Deming's most significant contribution was recognizing the tremendous potential of Shewhart's work and the use of statistical tools for the continuous improvement of production processes and the delivery of a quality product. He emphasized that variability in manufacturing and service processes can be traced to either common causes or special causes (Shewhart's assignable causes). Common cause variability is that variability naturally inherent in the system. Common cause variability is not necessarily acceptable, but it cannot be removed by employees attempting to do better. Problems generated by common cause variability can only be addressed by improving the system. On the other hand, special cause variability is that variability beyond the natural variability of the process. Such variability can be addressed by employees. Deming believed that 85% of the variation in a process is due to common causes and only 15% is due to special causes. Deming stressed the importance of clearly identifying the source of the variability. For example, if an employee tries to compensate for variability without any statistical guidance, the employee will probably either overadjust or underadjust to the variability. This will not cure the problem and may actually increase variability. And when employees' best efforts fail to produce improvement, they may become confused and demoralized, eroding their creative spirit and pride of workmanship, and making future process improvements unlikely.

Deming relied on control charts to describe both the natural variability of the system and to detect the existence of a special cause of variation that lies outside the system. A control chart is a time-ordered plot of a sample statistic – usually a sample mean, range, proportion, or number of defects – and upper and lower control limits that reflect the magnitude of common cause variation present in the system. Suppose we want to use a control chart to monitor a process that produces one-inch diameter bolts. Upper and lower control chart limits for the mean diameter of bolts can be developed from previous production runs. These limits describe the normal amount of variability in the bolt diameters. Bolts are then periodically sampled from future production runs and the average bolt diameters from each sample are plotted on the control chart. As long as average bolt diameters fall within the control limits and exhibit no particular pattern, the process is considered to be in statistical control. However, this only means that the process is acting as it has in the past. A process that is in statistical control may still produce defective items; it is stable only in the sense that the amount of variability is predictable. Any sample mean diameter beyond the specified limits can be attributed to special causes that increase the amount of variability in the system. As changes in the system are implemented, the control chart can determine the effectiveness of the changes by monitoring the output and plotting the new sample results on the control chart.

THE SYSTEM OF PROFOUND KNOWLEDGE: Deming described a System of Profound Knowledge (SOPK) which provides insights into the process of effectively transforming organizations. The overriding theme is that the prevailing style of management must first undergo a transformation. There are four interdependent parts of SOPK. The first part of SOPK is an appreciation of the system and systematic thinking. A system is a set of functions or activities within an organization that work together to achieve organizational goals. To run any system, managers must understand the interrelationships among all subsystems and the people that work in them.

The second part of SOPK is knowledge of theory of variation. A production process contains many sources of variation. The complex interaction of the variations in materials, tools, machines, operators, and the environment cannot be understood in isolation. However, the combined effect of all sources of variation can be examined statistically.

The third part of SOPK is theory of knowledge – a branch of philosophy concerned with the presuppositions, nature, and scope of knowledge. Deming emphasized that there is no knowledge without theory and that experience alone cannot establish a theory. Managers thus have a responsibility to learn and apply theory. The final part of SOPK is psychology. Psychology helps managers understand people. Managers must recognize differences between people and use psychology to nurture each individual's positive innate attributes. The 14 Points for management follow naturally as an application of Deming's SOPK.

RESULTS: For over 40 years, Dr. Deming served as a world-renowned consultant in statistics. Companies that have successfully implemented Deming's approach to quality management have experienced reduced inventories, decreased costs, increased profits, and improved worker morale and labor relations. Two examples of companies that have benefited from Deming's philosophies follow.

In *Quality, Productivity, and Competitive Position*, Deming offers the following example. A superintendent at a plant knew there were problems with a certain production line. His only explanation was that the work force of 24 people made a lot of mistakes. The first step of a consulting statistician was to obtain data from inspection and plot the fraction defective day by day over a six-week period. This plot (a run chart) showed statistical control with stable random variation above and below the average. This meant that any substantial improvement had to come from action on the system, the responsibility of management. What could management do? The statistician suggested that perhaps the people on the job, and the inspector, did not fully understand what constituted acceptable work. Operational definitions were developed and posted for everyone to see. The result was a substantial reduction in the proportion defective. The gains were immediate and the costs were essentially zero, all with the same work force and no investment in new machinery. The next step was to reduce the proportion defective through better incoming materials and better maintenance of equipment, both the responsibility of management.

A more recent illustration of a successful implementation of Deming's approach is found in Graphic Enterprises, Incorporated of Detroit, Michigan. The company was on the brink of bankruptcy. Following Deming's principles, the company decided to train and empower action teams to pursue special tasks. These teams implemented many improvements. The company began to monitor and control humidity levels in its storage room to keep paper from becoming sticky or developing static. A customer survey log was instituted to follow up every job. And, internal statistics were used to keep track of the quality of maintenance. As a result, the company became profitable, and sales growth jumped to between 6% and 10% per year.

CASES: Deming visited Ford in 1981 to meet with its president, Donald Petersen, and other company officials. Subsequently, Deming began by giving seminars for top executives and meeting with various employee groups, suggesting changes in accordance with his 14 Points. Ford managers visited Nashua Corporation, the first American company to incorporate Deming's philosophy, to learn how statistical methods were used there, and chief executives from many of Ford's major suppliers visited Japan. Petersen himself took a course on statistical methods. The 14 Points became the basis for a transformation in Ford's management.

Ford's quality commitment was embodied in its "Guiding Principles": quality comes first, customers are the focus of everything we do, continuous improvement is essential to our success, employee involvement is our way of life, dealers and suppliers are our partners, and integrity is never compromised. Management at Ford endeavored to create an environment where everyone could contribute to continuous improvement. Subsequent to the adoption of these ideas, over the period 1982 to 1984, direct labor productivity at Ford increased by 13%, setup time improved by 80%, inventory turns increased from 1.9 to 4.0, and customer service was substantially improved.

Ruan Transportation Management Systems followed Deming's prescription to trim expenses and improve profits. After the federal government deregulated the trucking industry in 1980, Ruan faced new low-cost competitors. To compete, Ruan decided to develop the "mega truck," the first truck designed to last one million miles. Ruan introduced quality circles to help with the development of the truck. At first, the quality circles didn't work because management had no idea what they were supposed to be doing. Then company executives decided to learn about Deming's methods and SPC. It took a long time for the changes to take hold, but top executives made quality integral to the company's mission.

Employees at all levels were trained to effectively investigate and analyze problems, and they began collecting detailed data about how the company worked. For example, Ruan trucks were experiencing an affliction known as engine "pitting." The pitting was caused by insufficient radiator coolant; the manufacturers claimed Ruan's maintenance workers weren't filling the radiators properly and would not honor warranties. Then a team of maintenance workers created a detailed chart of when engine problems were occurring. It turned out that things got worse when the weather changed significantly. The employee team discovered that the company was using a radiator hose that expanded and contracted with changes in temperature, causing coolant leaks. The company purchased a new kind of flexible clamp that kept the hoses properly sealed. The pitting problem was solved.

Deming's ideas also enabled Ruan employees to determine a major cause of excessive maintenance costs. The employees began keeping statistics on maintenance expenditures. Eventually they found a major problem: headlights were burning out too quickly. The company equipped all trucks with long-life halogen headlights, resulting in substantial savings.

The Deming philosophy eventually led to Ruan's "added value" emphasis. By charting every move, employees gained understanding about what they did and what their purpose in the company was. The data showed that many workers were underutilized and could use their time providing additional services to clients. By more effectively utilizing employees, Ruan was able to provide services such as driver training, fuel purchasing, fleet dispatching, and vehicle routing. In 1993, these activities accounted for about half of the $300 million the company was expected to generate.

COLLECTIVE WISDOM: A change in managerial mindset is required in order to successfully implement Deming's ideas. Companies that undertake programs to improve quality without changing the way the company is managed are doomed to failure. Thus, Deming would work with a company only on the invitation of top management, only if top management was willing to actively participate, and only on a long-term basis. Deming noted that the biggest obstacle to improving quality was the lack of constancy of purpose. Improvement takes time, determination, and commitment to change. Workers must perceive that the push for quality is more than just another gimmick or slogan that will have little or no impact on the way the company is run.

Layoffs and labor difficulties are possible symptoms of a failure in the implementation of Deming's concept of quality. Xerox Corporation, considered to be a national leader in total quality, embraced "Leadership Through Quality," a process aimed at providing innovative products and services that fully satisfy customers. The movement depended on employee involvement and empowerment. Employee support was eroded when Xerox decided in 1993 to cut ten percent of the company's document processing work force worldwide. Workers said they feared for their jobs; such fear is disruptive and stifles future improvements. Deming contended that fear is a symptom of failure in hiring, training, supervision, and motivation, and can adversely affect the ability to produce a quality product that the market is willing to pay for.

In the mid-1980s, Briggs and Stratton sent executives and union leaders to a Deming training program. The concepts were never put into practice. Throughout the 1990s, the union and management at the company's Milwaukee plants have been almost continually at odds due to decisions to outsource inefficient operations, sell off nonperforming subsidiaries, and move plants away from Milwaukee to areas with lower labor costs.

Of all the quality gurus, Deming's ideas (the 14 Points and SOPK) are the most philosophical and therefore the most difficult to interpret and implement. Deming desired total transformation of the system of management, but his guidelines for achieving that goal are quite abstract for the typical manager. Managers tend to be practical minded and place little stock in abstract ideas and goals that are difficult to comprehend and carry out. Thus, they often shy away from Deming's approach altogether or abandon it after trying it for a while.

Finally, in order to realize Deming's benefits of quality improvements (the cycle in which improved quality leads to lower costs, resulting in lower prices, increasing market share, leading to more jobs), quality must be related to other business objectives, plans, and functions. This is a point that Deming did not emphasize, but it is very relevant in the context of implementing TQM. Quality must be integrated with other dimensions of business that deliver value to the customer, such as on-time delivery, low cost, and increased flexibility.

Organizations Mentioned: Army Ordnance; ASA; ASME; ASTM; Briggs and Stratton; Ford; Graphic Enterprises; Nashua Corporation; Ryan Transport Management Systems; Stanford University; War Department; War Production Board; Union of Japanese Scientists and Engineers (JUSE); Xerox Corporation.

See **All or none sampling; Assignable causes of variation; Common causes of variations; Deming cycle; Deming prize; Deming's 14 points for management; Integrated quality control; ISO-9000/QS-9000; Malcolm Baldridge national quality award; Natural variability; Performance excellence; Quality assurance; Quality management systems; Sampling in quality control; Special causes of variations; Statistical process control; Statistical quality control using control charts; Total productive management; Variability in manufacturing.**

References

[1] Deming, W. Edwards (1982). *Out of the Crisis*, Center for Advanced Engineering Study, Massachusetts Institute of Technology, Cambridge, Massachusetts.

[2] Deming, W. Edwards (1982). *Quality, Productivity, and Competitive Position*, Center for Advanced Engineering Study, Massachusetts Institute of Technology, Cambridge, Massachusetts.

[3] Deming, W. Edwards (1993). *The New Economics for Industry, Government, and Education*, Center for Advanced Engineering Study, Massachusetts Institute of Technology, Cambridge, Massachusetts.

[4] Shewhart, W. A. (1931). *Economic Control of Quality of Manufactured Product*, E. Van Nostrand Company, New York.

[5] Shewhart, W. A. (1939). *Statistical Method from the Viewpoint of Quality Control*, Graduate School, Department of Agriculture, Washington.

[6] Taylor, Frederick (1911). *Principles of Scientific Management*, Harper and Row, New York.

QUANTITATIVE OVERLOAD

An individual faces quantitative overload when the quantity of work to be performed exceeds his/her capabilities. Basically, it means having too much work to do in a given time frame.

See **Human resource issues and advanced manufacturing technology; Resource issues in manufacturing.**

QUANTITY DISCOUNT MODEL

The quantity to order from supplier to replenish inventory in an important decision in material/inventory management. Economic order quantity (EOQ) model computes the amount to order using the assumptions that cost per unit of purchased items remains fixed regardless of the number of units ordered. But, it is common for suppliers to give discounts when order quantities are high. When discounts are considered the economic order quantity may change. The optional order quantity model investigates the total annual inventory costs with and without discounts, and the optional order quantity is one that minimizes total annual inventory costs. The results is that larger than usual economic order quantities may be justified.

Just-in-time manufacturing systems, or Kanban-based manufacturing systems have somewhat altered the promise behind quantity discount models. In these systems, order quantities tend to be much smaller.

See **Economic order quantity model (EOQ); Inventory flow analysis; Just-in-time manufacturing; Kanban-based manufacturing system; Manufacturing order quantity; Purchasing: Acquiring the best inputs; Simulation of production problems using spreadsheet programs.**

QUEUE LIMITER

In a manufacturing facility, it is common for work or jobs to wait in a queue for processing. Any device which limits the amount of parts or documents from accumulating in front of a work station is called a queue limiter. These devices may act like kanban signals.

See **Just-in-time manufacturing implications; Manufacturing cell design.**

QUEUE TIME

It is common for jobs to wait for processing at work centers. The time a job is made to wait before processing is called queue time. As queue time increases at various work centers in a factory, the manufacturing lead time and work-in process (WIP) inventory increases; both are undesirable. Increased WIP increases inventory costs.

Reducing Manufacturing lead time in an important goal of modern manufacturing systems. Short lead times indicate manufacturing efficiency and effectiveness. Queue times increase when capacity is inadequate and/or production scheduling is inappropriate.

See **Queuing theory; Simulation analysis of manufacturing and logistics systems; Simulation of production problems using spreadsheet programs.**

QUEUEING THEORY

Queueing theory is the mathematical study of waiting lines, which occur whenever the current demand for a service exceeds the current capacity at the work center. The first significant work on queueing was performed by A.K. Erlang of the Danish Telephone Company in 1909 with the objective of determining the optimal number of telephone lines to handle pre-specified call frequencies. Capacity allocation decisions are complicated by the fact that it is often impossible to predict accurately when "customers" will seek service, and how long a service each customer may require. Excessive service capacity leads to increased operating costs. Insufficient capacity causes waiting lines to be unacceptably long, and poor customer service. The objective in capacity design therefore is to achieve a balance between the cost of service and the cost associated with waiting for that service. Queuing theory itself does not solve this problem directly; however, it provides models that furnish key information on the performance characteristics of facilities such as the distribution of the waiting time capacity expansion is justified when waiting time is excessive.

A queueing model can be constructed by parameterizing the following system characteristics: arrival pattern of the customers, service pattern of servers, queue discipline, number of service channels, and number of service stages. The arrival pattern (the input) into a queueing system is often measured in terms of the average number of arrivals per unit of time. In the event that the stream of input is deterministic, then the arrival pattern is fully determined by the mean arrival rate. On the other hand, if there is uncertainty in the arrival pattern, then these mean values only provide measures of central tendency for the input process; additional characterization is required in the form of the probability distribution associated with this random process. Service patterns can also be described by a rate, conditioned on the fact that the system is not empty.

Queue discipline refers to the manner by which customers are selected for service when a queue is formed. While the most common discipline is first-come-first-served, there exist a variety of priority schemes. The number of service channels refers to the number of parallel service stations, which can service customers simultaneously. Stages of service refer to the number of service stations in series that the customer visits in a given sequence to receive complete service.

See **Priority sequencing rules; Queue time; Simulation analysis of manufacturing and logistics systems; Simulation of production problems using spreadsheet programs; Waiting line model.**

QUICK RATIO

This ratio is indicative of a company's ability to meet its short-term cash requirements, (including such expenses as accounts payable, short-term payroll costs, and short-term operating expenses). This ratio is calculated as:

$$\text{Quick ratio} = (\text{cash} + \text{securities} + \text{receivables}) / \text{current liabilities}.$$

The quick ratio is often used as a "red flag," to detect companies is overextended in terms of debt. A supplier with a quick ratio less than 1.0 may be in danger of shutting down the operations due to an inability to meet their short term debts.

In cases when the supplier is a publicly traded company on the stock market, specific data for financial ratios can be obtained from sources such as 10-K reports, Moody's Industrials, and Dunn and Bradstreet Reports. Ratios can also be calculated using information shown in income statements and balance sheets.

See **Performance measurement in manufacturing; Supplier performance measurement.**

QUICK RESPONSE (QR) OR MODULAR MANUFACTURING

See **Agile manufacturing; Lean manufacturing implementation; Manufacturing cells, linked; Mass customization; Mass customization implementation; Safety stocks: Luxury or necessity.**

RADICAL INNOVATION

Innovation is the commercial use of a new idea. Radical innovation involves a large step forward in product, service or organisational design. It may be an entirely new product. The commercialisation of TV is an example It could also be a completely new manufacturing process such as the blast furnace for making steel. In an organisational setting, it could be a new form of activity such as activity based accounting or just-in-time production.

See **Process innovation; Product innovation.**

RANDOMLY GENERATED PROBLEMS

Computer simulation enables the study of complex problems. Test problems are used in computer experiments. Certain parameters of the problem are generated according to a statistical distribution. For example, randomly generated **Line Balancing Problems** have a random number of tasks, random task time, and a random set of precedence relationships.

See **Simulation analysis of manufacturing and logistics systems; Simulation of production problems using spreadsheet programs; U-shaped assembly lines.**

RANGE FLEXIBILITY

A flexible manufacturing system can perform a wide range of activities. Range flexibility defines the total envelope of capability of manufacturing system is capable of achieving in terms of the portfolio of products, processes, and procedures. A manufacturing system with a higher range flexibility has the capability to produce a larger number of products, or it can perform a larger number of different processes.

See **Manufacturing flexibility types and dimension.**

RAPID PROTOTYPING

Rapid prototyping refers to a technology by which physical objects can be directly created from designs or graphical representations. Initially, solid model of a part drawing is created using a computer-aided design (CAD) package, and then the physical part is created by a machine layer by layer by using polymers and lasers. This process of creating physical objects is called stereolithography. The plastic parts made by stereolithography can be used in investment casting to create an actual physical part which will be tested for physical, thermal, electrical, chemical properties. Modifications to the part will be done if required and the entire cycle is repeated until acceptable design is achieved. This technology reduces the time needed for new product development.

See **New product development through supplier integration; Product design; Product innovation; Virtual manufacturing.**

RATE-BASED DECOUPLING OF INVENTORIES

This concept is relevant to process industry scheduling. Rate-based decoupling inventories occur when sequential process stages have different production rates. Two variations exist. First, suppose that two adjacent process steps are continuous flow processes, and that the output of the first process can be immediately processed by the second process. If the process rate of the first process exceeds the process rate of the second process, intermediate inventory must be stored between the two processes, and the first process step must be periodically shut down to help the second process to catch up. Because of the different production rates, the two processes must have separate schedules, and a decoupling inventory between them is required.

A similar situation exists when the production rates are reversed. Suppose a 1,000-liter run of product Q requires 10 hours in the first stage and one hour in the second. In addition, let us assume that the second stage cannot be throttled and must run all 1,000 liters in a single one-hour run. Stage 1 must run enough product Q to begin the second stage. Again, the process units must have different schedules, and a decoupling inventory is required.

See **Process industry scheduling.**

Reference

[1] Taylor, S.G. and S.F. Bolander (1994). *Process Flow Scheduling: A Scheduling Systems Framework for Flow Manufacturing*, APICS, Falls Church, VA.

RATE-BASED MASTER PRODUCTION

Companies that operate in Just-in-Time (JIT) or flow manufacturing environments produce every product they make at a fairly constant rate on a daily or weekly rate. This method of scheduling is more common in high-volume and process industries. It has been extended to job shops using cellular layouts and mixed-model level schedules. In rate-based systems, the production rate is matched to the selling rate and a pull production system becomes a reality.

In JIT systems, rate-based master production scheduling is very common.

See **Just-in-time (JIT) manufacturing implementation; Lean manufacturing implementation; Manufacturing cell design; Manufacturing systems; Master production schedule; MRP; Pull production system; Rate-based material planning; Rate-based production schedule.**

RATE-BASED MATERIAL PLANNING

Examples of firms using rate-based planning can be found in repetitive manufacturing, assembly line, just-in-time, and other flow systems. The primary intent in rate-based scheduling is to establish rates of production for each part or product made in the factory. The rates of production allows the company to move material through the manufacturing system without stops or interruption, in the shortest time possible.

Typically, single-level planning bill of material information is used to convert rate-based master production schedules into material plans that specify the appropriate daily or hourly flow rates for individual components or products. These production systems are characterized by master production schedule (MPS) stability, high rates of material flow, negligible work-in- process inventory levels, short manufacturing lead times, and a relatively small variety of final products in the MPS. Therefore, detailed status information on work-in-process items may be required. This reduces the size of manufacturing data base, the number of transactions, and the number of material planning personnel when compared to time-phased detailed material planning. Rate-based material planning is appropriate for a relatively narrow range of standard products with stable product designs produced in high volume. Rate-based detailed material planning is limited in its ability to cope with changes in product mix. The limited product line permits small changes in the schedule as long as they are within the product design specifications. Enhancements in customer delivery speed are typically accommodated with finished-goods inventories. The use of rate-based material planning and line production yields an opportunity to cut work-in-process inventory and overhead costs, which are important in price-sensitive markets.

See **Bill of materials; Just-in-time manufacturing; Lean manufacturing implementation; Manufacturing cell design; Manufacturing systems; Rate-based master production; Rate-based production schedule.**

RATE-BASED PRODUCTION SCHEDULE

Companies that operate in a flow manufacturing environment will produce products at a fairly constant rate on a weekly, or even daily basis. This steady production greatly simplifies production planning and control. For example, suppose a company makes three different products, A, B, and C. Each day they produce 50 units of A, 100 units of B, and 200 units of C. Consequently, planned production can be determined for any month based on the number of production days in the month. Further, problems concerning sequencing or capacity will have to be solved only once because the production schedule remains constant.

See **Just-in-time manufacturing implementation; Just-in-time manufacturing; Lean manufacturing implementation; Manufacturing cell design; Mixed-model sequencing; rate-based master production; Rate-based material planning; Uniform load scheduling.**

RATED CAPACITY

Each manufacturing facility and production equipment have a rated capacity, which is the expected output capability of the facility or work center. Capacity is often expressed in hours of production capability per unit of time. Rated capacity is reduced by efficiency and the utilization of the equipment.

See **Capacity planning in make-to-order production; Capacity planning: Long-range; Capacity planning: Medium- and short-range.**

R-CHART

The R-chart is a control chart for the range of observations in a sample. The range is found by subtracting the minimum observation from the maximum observation. It is a measure of the variability of the process. Thus, R-charts can be constructed for quality characteristics such as the range of diameters, or the range of tensile strength of parts.

While the range is a simple measure to calculate and understand, it has a limitation. It uses only the two extreme values in a sample for its computation. So, if these two extreme values stay the same, while all of the other values in a sample change drastically because of a major change in the process variability, the range will remain the same and will fail to capture this change in the process. Therefore, other measures such as the standard deviation are used to compute indices of process dispersion. For small sample sizes, say 2 to 10, the range is an acceptable measure of variability. For sample sizes greater than 10, the standard deviation is preferable to the range.

As in X-bar charts, the R-chart is constructed based on at least 25 to 30 subgroups or samples of observations. Each sample is made of items chosen at close proximity with respect to the time of production. This ensures that process conditions remained about the same for observations forming a subgroup. The variation within a subgroup, as indicated by the range of observations, is then a reflection of the variability due to common causes in the system. The control limits on the R-chart are placed at three standard deviations of the range ($3\sigma_R$) from the center line. If the lower control limit is calculated to be negative, it is converted to zero, since the range is never a negative value.

It should be noted that small values of range are desirable. So, if an observation plots below the lower control limit on a range chart, when the lower control limit is greater than zero, the process is out of control indicating the presence of special causes. But, such a case is quite desirable since it indicates unusually small variability within the sample.

See **Statistical process control using control charts; X-bar chart.**

REACTIVE KANBAN CONTROL

Reactive kanban control is a dynamic kanban control (DKC) policy (Takahashi and Nakamura, 1996, 1997a, 1997b), the goal of which is to systematically add and remove kanbans so as to achieve improved system performance. Under a DKC policy, kanbans are added when needed to improve system performance, and are removed when they are no longer needed, or when their presence will result in lowered system performance. In general, we want additional kanbans when the benefit of their presence (e.g., reduced blocking and starvation, improved throughput) outweighs the cost (e.g., increased WIP and operating costs). The goal of the control policy is to achieve a desired order completion time and WIP.

See **Adaptive kanban control; Base kanbans; Capture threshold; Dynamic kanban control for JIT manufacturing; Flexible kanban control; Just-in-time manufacturing implementation; Just-in-time manufacturing; Kanban-based manufacturing systems; Kanbans; Pull production systems; Release threshold.**

References

[1] Takahashi, K., and N. Nakamura (1996). "Reactive JIT ordering systems for unstable demand changes." *Proceedings of Advances in Production Management Systems*, Information Processing Society of Japan.

[2] Takahashi, K., and N. Nakamura (1997a). "Reactive logistics in a JIT environment." *Proceedings of 3rd International Symposium on Logistics*, SGEditoriali.

[3] Takahashi, K., and N. Nakamura (1997b). "A study on reactive kanban systems." *Journal of Japan Industrial Management Association (in Japanese)*, 48, 159–165.

REAL-TIME SYSTEMS

MRP systems are traditionally planed in units of time called buckets, usually a week. In real-time systems, time buckets are discarded. Real-time systems are run daily, without weekly time buckets. In this system plans can be revised whenever necessary. Buckless systems are more recent developments in manufacturing planning and control.

See **Bucketless systems; Just-in-time manufacturing; Manufacturing planning and control; Master production schedule; MRP.**

Reference

[1] Vollmann, T.E., Berry, W.L., Whybark, D.C. *Manufacturing Planning and Control Systems*, Fourth Edition, Irwin/McGraw-Hill, New York, 1997.

RECONCILING ERRORS IN INVENTORY

The integrity of inventory systems depend on the accuracy inventory records. That is, actual or physical inventory agree with inventory records when inventory records are accurate. Periodically, actual inventory is counted (cycle, counting) and the records are reconciled. If inventory records are inaccurate, production schedules based on faulty records cannot be carried out because materials and parts may not be in inventory as expected. It will cause production schedules to fall apart, delay shipment, dissatisfy customers, and confusion will reign in the factory.

See **Cycle counting; Just-in-time manufacturing; MRP implementation; MRP.**

REDUCTION OF SET-UP TIMES

See **Setup reduction; SMED; NOTED.**

REENGINEERING AND THE PROCESS VIEW OF MANUFACTURING

Timothy L. Smunt

Wake Forest University, Winston-Salem, NC, USA

DESCRIPTION: This article considers operations from a process perspective, rather than from the traditional functional point of view. Such a consideration is important as an increasing number of firms are implementing reengineering projects in order to improve their systems' effectiveness. This article first provides a description of the process perspective, then provides implementation guidelines using reengineering principles as a basis for analysis. Finally, examples are given of how firms have made the transition from an operations functional structure to one where a process model is used.

A *process* can be defined as the set of tasks required to transform a set of inputs into a set of outputs for a customer. Customers can be either *internal* or *external*. External customers are ones to which a company sells their final products, sometimes called *end items*. Internal customers are the people within the organization who receive reports or other types of information that are produced by a process but are not given to the external customers.

Most U.S. and Western European companies have been organized around functional expertise, rather than processes. Processes typically require the completion of tasks from a number of functional areas; however, they can become inefficient or ineffective when handoffs must be made across functions. (See Figure 1 for an illustration of the process focus.) As an example, consider the order fulfillment process that takes a customer order as the initial input and ends with the delivery of a product to the customer. In a functionally-oriented organization, the customer order is entered by employees reporting to a sales or marketing manager. The order then proceeds to a production planning department (as depicted in Figure 2) which then schedules the manufacture of the various components and subassemblies in order to complete the end item. Purchasing utilizes the information from production planning to place orders with suppliers for material and parts that are acquired from the firm's vendors. After the end item becomes available in finished goods inventory, the distribution department determines whether the product will be shipped to satellite distribution centers or directly to an external customers.

Note that in this functionally oriented system, responsibilities for the total process are separated among many different departments. Incentives and coordinating mechanisms must be carefully designed and implemented in order to provide smooth completion of tasks and high service levels

Fig. 1. Process focus.

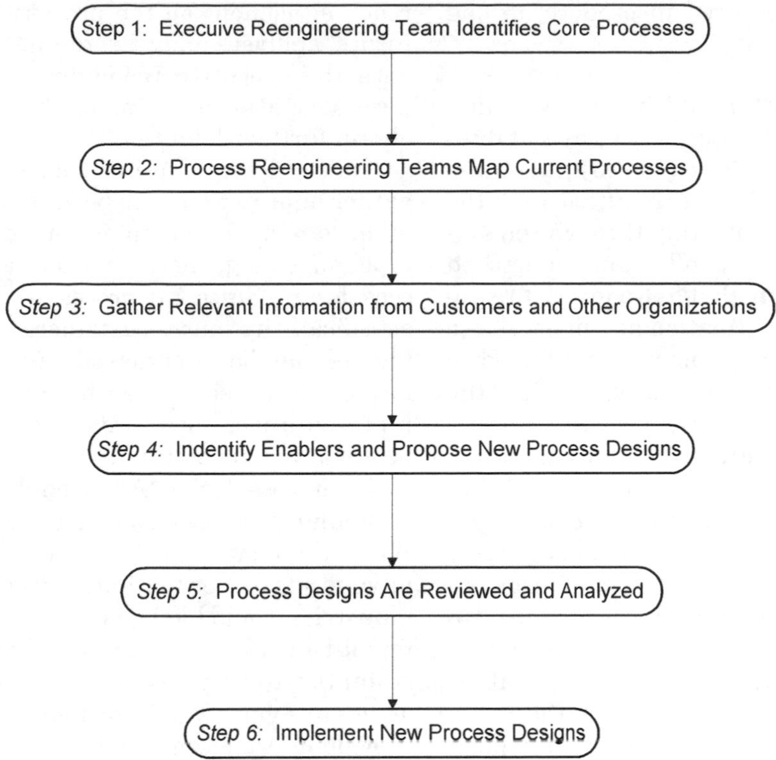

Fig. 2. Major steps in implementing a process focus.

to the external customers. In a manufacturing company, complex information systems are typically implemented to provide a focal point for the various interdepartmental activities. Material Requirements Planning (MRP) systems and their extensions to MRP II or Enterprise Requirements Planning (ERP) systems, which provide additional financial and capacity planning information, are the most common types in use today.

These information systems are often very expensive to either design or purchase and to implement, but are necessary for control in any organization of reasonable size. Even though the information systems provide a link across functional areas, communication between any two departments can remain difficult and may result in delays and errors. A process perspective, on the other hand, attempts to resolve the inherent communication and incentive problems of separate functional departments. These are often referred to as *silos,* since the managers and employees within a department tend to behave narrowly, although they have a deep set of functional skills. The process focus requires that the appropriate personnel from each of the functions required to produce a product be reassigned to work as a team whose sole responsibility is to provide the necessary steps for a particular product or set of products.

HISTORICAL PERSPECTIVE: While the process perspective can be traced to the origins of industrial engineering, it has recently come to light as an important contemporary management issue by Michael Hammer and James Champy through their best selling management book, *Reengineering the Corporation: A Manisfesto for Business Revolution* (Hammer and Champy, 1993). Two other notable books, by James Harrington and Thomas Davenport, have outlined the process approach in fair detail and provide valuable reading in this area (Harrington, 1991 and Davenport, 1993).

The major reasons cited for the recent ability to change from a functional organization structure to a process one is the fact that information systems, including both hardware and software, have finally matured to the extent that they can enable such a change in focus. Previously, communication across functional areas could only be accomplished through a complex reporting structure using various levels of managers reporting up and across a pyramid, with the CEO or president of an organization at the top and the lower level managers and employees at the bottom. The use of email, intranets, electronic white boards, etc. have provided the impetus to consider the dissolution of functional departments and reallocation of appropriate personnel to process teams that handle all, or subsets, of tasks required to produce an end item.

STRATEGIC PERSPECTIVE: The ability to link tasks electronically can both reduce costs and increase customer service levels, mainly due to the fact that fewer handoffs are required and the work-in-process moves smoothly through the system. The work-in-process can be a manufactured item or the information that is required to support the physical transformation activities. These information processes are sometimes referred to as administrative processes and may reflect the majority of expenditures for the firm.

As we find the direct costs shrinking to an average of less than 25% of the cost of goods sold in today's environment, the reduction of costs in administrative processes can often have the largest impact on competitiveness. A number of firms, including **Texas Instruments**, have also used a process focus to address manufacturing capabilities and technology development over time. Rather than delegate the responsibilities for these important issues to more narrowly focused departments such as "research and development" (R&D) or "manufacturing engineering," companies are reorganizing and combining personnel from a number of departments, including manufacturing, engineering, R&D and marketing to work in a process mode on these dynamic, long-term issues.

REENGINEERING: Redesigning the company or parts of it into a set of processes typically requires that a number of steps are followed over a period of one to two years. Recall that Figure 2 shows the major implementation steps.

The first step is to identify core processes of the firm, especially the ones that have the potential for the greatest improvement in competitiveness. A top management team, sometimes called an executive reengineering team, is responsible for the initial identification of the core processes of the firm and for assigning managers and other personnel to process reengineering teams. The executive reengineering team may be specific about the goals or visions for each of the processes or may choose to let the process teams themselves identify the goals after an initial analysis of the current process design or lack thereof.

The second step requires that the process reengineering teams become familiar with the current problems and current practices of their assigned process. One non-trivial analysis here is the determination of process boundaries. It is often unclear where a process starts or stops, and to what degree there is overlap with other processes. However, unless the process teams can reach agreement on the process boundaries, any future analyses may be redundant. Interaction with the executive reengineering team is often necessary at some point in this step in order to obtain further definition.

The third step requires that the process teams gather information from both their customers and other organizations in order to determine the types of change necessary for an effective process design. Since any process should be designed specifically to meet customers' needs, the customers may be interviewed, surveys may be performed, discussions may be had with field sales representatives, or records of warranty work may be reviewed to better understand the direction of the process design. Additionally, benchmarking of similar processes should be made as possible. While the benchmarking of a direct competitor's process may not be feasible, many firms from other industries would likely have similar processes and might be willing to cooperate in this regard. Consultants are often used in this step since they have access to a variety of firms and industries and can provide information on best practices.

The fourth step requires that "enablers of change" are identified and tentative process designs are proposed. Enablers of change include both information technology (IT) enablers and other cultural enablers, including organizational and human resource enablers (Teng *et al.*, 1996). IT enablers could include email, electronic white boards or other file sharing mechanisms, digital imaging technology, laptop computers, local area networks and expert systems. Organizational enablers include the use of case teams, case managers, new budgeting and resource allocation methods, and new management reporting relationships. The intent of the organizational enablers is to enhance employee empowerment so that delays caused by layers of vertical management approvals are drastically reduced. Human resource enablers include compensation systems that are based on team performance, training programs to enhance cross-functional skills, work assignment rotations, and advancement criteria that are related to ability rather than performance. Using "ability" as a criterion provide incentives to employees for increasing the breadth of their skill set rather than just performing well on a narrow set of tasks.

Process designs are then tentatively proposed taking advantage of the most effective enablers available to the design team and company. It should be recognized that constraints often exist in companies with regard to enablers, either due to budget constraints or the current culture. For example, while the use of electronic white boards and shared data bases might be identified as effective enablers for a process design, the current state of IT at a particular company may not allow quick adoption of these enablers. Or, perhaps, the current corporate culture is highly control-oriented and certain changes to the approval process might be met with stiff resistance from key managers. While the optimal process design would necessitate advancements in these areas, process designers may need to accept such constraints in the short-term in order to make any progress toward moving to a process orientation.

The fifth step calls for the tentative process designs to be reviewed by both the executive reengineering team, customers, and the personnel who will be involved in the new design. Analysis of the proposed process design can be made at this step, and revisions to the process will be incorporated as needed.

The last step, implementing the new process design, requires top management support and financial resources for success. The changes that result from moving towards a process focus can be radical, affecting the fundamental precepts on how an organization works. Such change must be fully supported by the top management with both their time and the company's resources. There are a number of case studies that indicate that the lack of such support can be fatal to the successful implementation of a process focus and the specific designs determined by the process teams (Champy, 1995 and Hall *et al.*, 1993).

TECHNOLOGY PERSPECTIVE: The availability of new information technology and other process technologies can greatly enhance the ability for a company to take a process focus. Essentially, the use of such enabling technology allows an organization to rewrite the "rules" of doing work and what is possible in meeting the customers' needs. The use of fast computers, networks, the internet, intranets, group decision-making software, expert systems, shared data bases, video conferencing, bar code scanning, and other technology provides advantages on a number of dimensions. The speed of the process flow increases, timeliness of information improves, work can be accomplished in a number of different locations, authority can be more easily decentralized, and communications between the company with both customers and suppliers can be vastly enhanced. For example, the use of shared data bases allows information to appear simultaneously in many places, reducing delays in communication and improving the accuracy of information reporting. The use of expert systems provide the capability of lower-skilled generalists to do the work of a number of specialists. While it may be possible to move from a functional focus to a process focus without the use of some type of enabling technology, the benefits of doing so will likely be much smaller.

SIGNIFICANT ANALYTICAL MODELS: Process mapping and related tools are the most prevalent methods used in process design. Additionally, computer simulation is rapidly becoming an analysis tool used to determine both the efficiency and efficacy of new process designs for both manufacturing and administrative processes.

A typical process map (sometimes called a process flow chart) is shown in Figure 3. While any symbols can be used to represent tasks, flows of material and flows of information, the ones used here are widely accepted across the world and provide a common vocabulary for discussion of the current and proposed designs. Also shown in Figure 3 are the three typical perspectives of processes, i.e. "what we think it is," "what we want it to be," and "what it really is." In my experience of mapping current processes for organizations, managers are always shocked to see the visual representation of how a process is currently working. Much of the reason for the complexity of the current systems is due to the fact that the processes are run using a functional organizational structure, causing the creation of a number of "work arounds" to deal with the inherent inefficiencies of communicating across functional silos.

An important analysis of the current process design requires the determination of *value-added* activities. A value-added activity or task is one, when viewed by the customer, is necessary to provide the output that the customer expects. Note that the definition of value-added relates to the customer's expectation, not the company's. While some activities may not be value-added to the customer, a company may still choose to keep them for control or communication purposes. The objective, however, is to eliminate as many non-value added activities as possible in order to reduce costs and time required to complete the process. See Figure 4 for an example of a value-added analysis.

Another common tool for analyzing process designs is an enhancement of the simple process map to one that includes both time and function attributes. These process maps are often called time-function maps or deployment flowcharts, and an example is depicted in Figure 5. The major objective is to illustrate the cumulative time required for completing activities in a sequence and to indicate the functional expertise required at each step.

The concept of mapping a process has been extended to user-friendly computer simulation programs. Many new software packages provide for

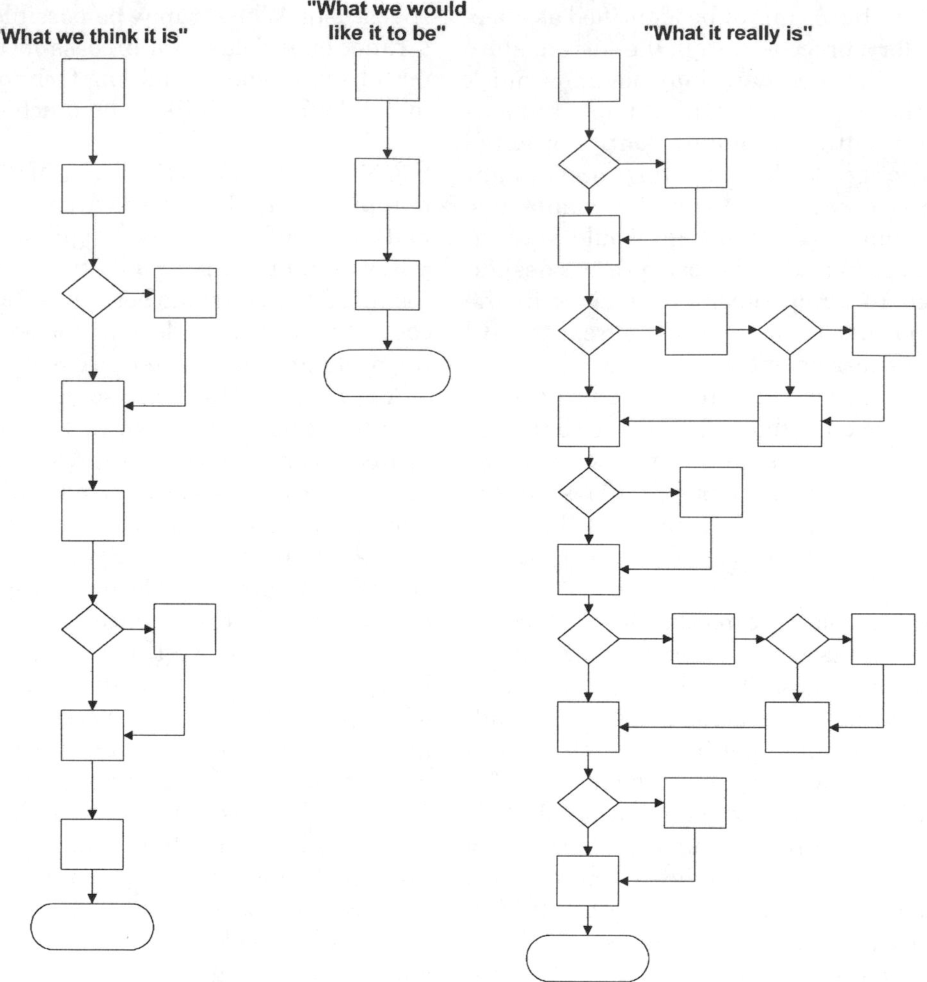

Fig. 3. Process flow charts (process maps).

easy mapping of the process on the screen, with simple menu-driven input windows for customization of task times, assembly tasks, worker assignments and the like. The main advantage of the use of one of these software packages is that more realistic conditions can be rapidly tested without the need for expensive "laboratory" experiments of proposed process designs.

EFFECT: The major effects on performance by moving towards a process focus can be categorized two ways, i.e. in *process efficiency* and in *process effectiveness*. Process efficiency can be measured in terms of *processing time*, including specific measures like cycle time per unit or volume of transactions per employee. Other process efficiency measures include *resources expended*, *value-added*, and *wait time*. Effectiveness measures are those that measure the extent to which outputs of the process meet the needs of the customers. They include measures of *accuracy* (e.g. number of customer complaints or number of errors found in inspection before delivery), measures of *timeliness* (e.g. time to solve problem and closeness to product rollover promise), measures of dependability (e.g. percentage of missed deliveries by promised due date), measures of *responsiveness* (e.g. number of rush orders completed), and measures of *market* and *financial* success, including market share, number of lost customers, return on investment and earnings per share.

RESULTS AND CASES: Companies typically consider the move to a process focus when one or more of the above measures are indicating clearly poor performance, although a few companies have done so as a preemptive action. A number of case studies have been reported over the past few years on successes and results achieved from using a process focus. The Taco Bell and Ford Motor Company Accounts Payable examples are most notable.

The **Taco Bell** case (Hammer and Champy, 1993) is an excellent example of a business that

ACTIVITY (TASK)	TIME	VALUE-ADDED ?
Order taken by salesperson and entered onto order form.	1 hour	Yes
Order form taken back to car and placed into car organizer recently purchased from office super store.	10 minutes	No
Waited in car as salesperson made visits to other customers in her area.	4 days	No
Overnight delivery to central order entry department at headquarters.	1 days	No
Order entered into computer system.	10 minutes	No
Computer system notes missing data.	1 second	No
Order awaits review by supervisor	1 day	No
Supervisor reviews order form and calls saleperson for missing data.	30 minutes	No
Order again entered into computer system with full information; confirmation number assigned.	10 minutes	No

Fig. 4. Value added analysis.

was in danger of losing a market share war, but was turned around by radical changes in its management and operations. While IT was used to provide better information flow within the organization, the major enabler from an operations perspective was the centralization of food preparation. Using what they called a "K-minus" approach for the elimination of the kitchen from each restaurant location, both food preparation costs and timeliness were improved. Today, the food is mostly prepared at central "kitchens" (factories) and shipped to the individual restaurant locations for warming and final assembly to customer order. Less space is then needed at the restaurants for food preparation, allowing more seating area and, thus, larger sales volumes at the same facility cost. There is also less material waste, lower production costs, and higher consistency. Through these and other changes, Taco Bell's growth went from a negative 16% per year to a growth in sales from $500 million to $3 billion in just 10 years.

The **Ford Motor Company** Accounts Payable example (Hammer, 1990) provides an excellent example of how IT can change the rules and provide both reductions in cost and increased effectiveness. In this early implementation of reengineering concepts, Ford found that their accounts payable department used five times more employees than did an equivalent Mazda department. Ford instituted what they called "invoiceless processing" in order to eliminate paperwork that was frequently in error and to place the responsibility of shipment conformation on the receiving dock. The goods were checked on the receiving dock when they arrived, and if the order matched the shipment, the receiving clerk noted such on the computer system. Checks are promptly sent to the vendors. At the same time, inaccurate shipments were sent back to the vendor. In this way, many of the old checks and balances were eliminated along with the need for numerous accounts payable clerks. The use of IT in this situation enabled the process to maintain control at a much lower cost and higher effectiveness.

CAVEATS: There are many other examples of successful reengineering projects that have improved the operations in numerous companies by changing the firm's structure from a functional to process perspective. At the same time, even more failures are reported. The failures can be blamed on a number of factors, including lack of top management commitment, lack of resources for implementation, lack of will power of top and middle managers in making radical changes, and a poor understanding of analyzing and designing processes. This article provides an overview of the issues and tools used in focusing operations towards a process orientation. Further reading and study of the following articles and books are suggested

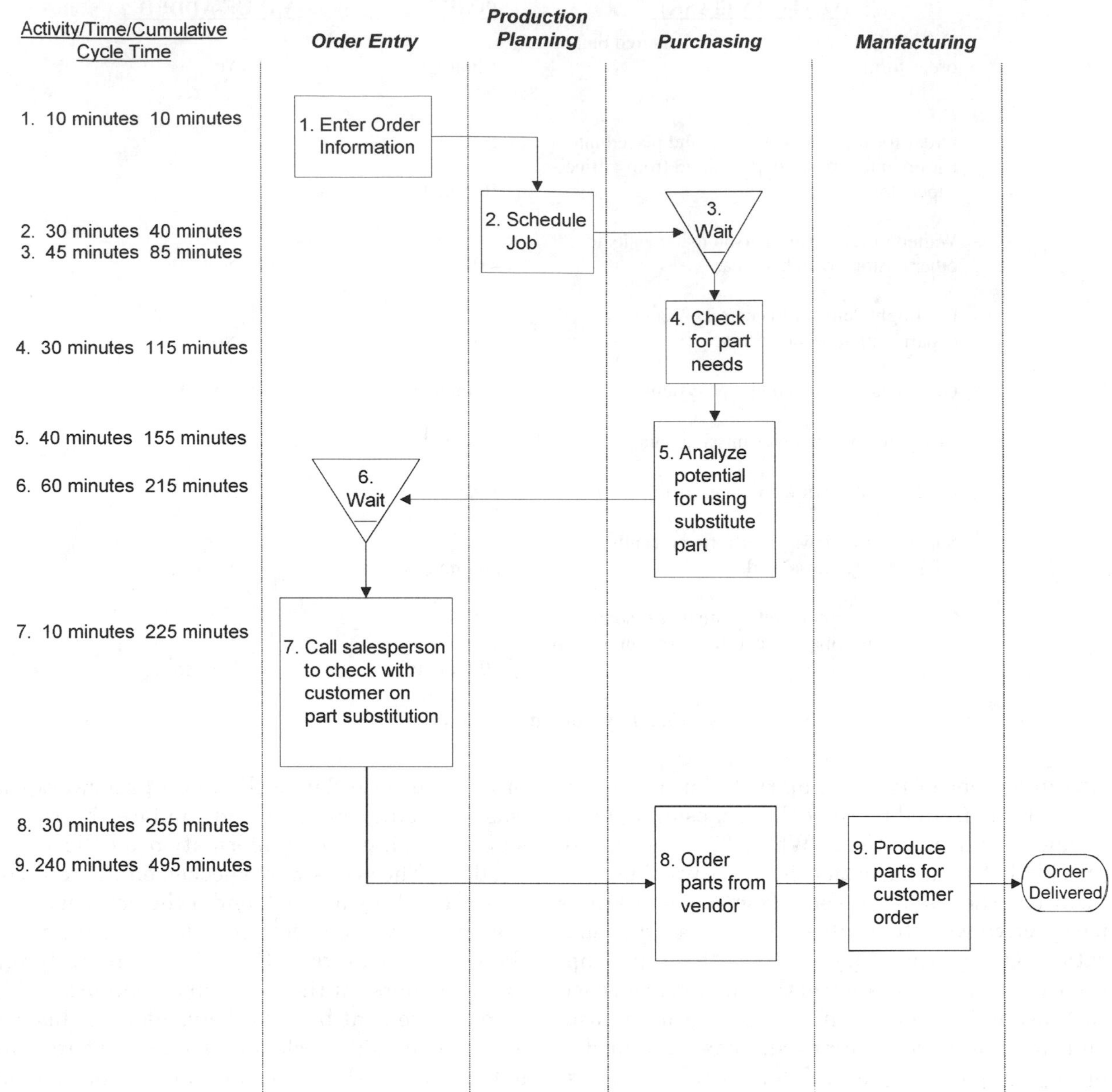

Fig. 5. Time function maps.

to better understand the path to successful implementations.

Reengineering Tools and Implementation:
- Hammer and Stanton, 1995
- Terez, 1990
- Johansson, et. al, 1993
- Shapiro, et. al, 1992
- Majchrzak and Wang, 1996

Case Studies:
- Reeves and Torrey, 1994
- Short and Venkatraman, 1992
- Rohm, 1992/1993

Organizations Mentioned: Ford Motor Company; Taco Bell; Texas Instruments.

See **Business process reengineering; Enterprise resource planning; Resource planning: MRP to MRPII and ERP; MRP; Total quality management.**

References

[1] Champy, J.A. (1995). *Reengineering Management: The Mandate for New Leadership*, Harper Collins, London.

[2] Davenport, T.H. (1993). Process Innovation: *Reengineering Work Through Information Technology*, Harvard Business School Press, Massachusetts.

[3] Hall, G., J. Rosenthal, and J. Wade (1993). "How to Make Reengineering *Really* Work." *Harvard*

Business Review (November–December), Vol. 71, No. 6, 119–131.

[4] Hammer, M. (1990). "Reengineering Work: Don't Automate, Obliterate." *Harvard Business Review* (July-August), Vol. 68, No. 4, 104–102.

[5] Hammer, M. (1996). *Beyond Reengineering*, Harper Business, New York.

[6] Hammer, M. and J. Champy (1993). *Reengineering the Corporation: A Manifestor for Business Revolution*, Harper Business, New York.

[7] Hammer, M. and S. Stanton (1995). *The Reengineering Revolution*, Harper Business, New York.

[8] Harrington, D.H.J. (1991). *Business Process Improvement: The Breakthrough Strategy for Total Quality, Productivity, and Competitiveness*, McGraw-Hill, New York.

[9] Johansson, H.J., P. McHugh, A.J. Pendlebury, and W. A. Wheeler, III (1993). *Business Process Reengineering: BreakPoint Strategies for Market Dominance*, Wiley & Sons, New York.

[10] Majchrzak, A. and Q. Wang (1996). "Breaking the Functional Mind Set in Process Organizations." *Harvard Business Review* (September–October), Vol. 74, No. 5, 93–99.

[11] Reeves, E. and E. Torrey, Editors (1994). *Beyond the Basics of Reengineering: Survival Tactics for the '90s*, Industrial Engineering and Management Press, Georgia.

[12] Rohm, C.E. (Winter 1992/1993), "New England Telephone Opens Customer Service Lines to Change." *National Productivity Review*, Vol. 12, No. 1, 73–82.

[13] Shapiro, B., V.K. Ranga, and J.J. Sviokla (1992). "Staple Yourself to an Order." *Harvard Business Review* (July–August), Vol. 70, No.4, 113–122.

[14] Short, J.E. and N. Venkatraman (1992). "Beyond Business Process Redesign: Redefining Baxter's Business Network." *Sloan Management Review*, (Fall), Vol. 34, No. 1, 7–20.

[15] Tent, J.T.C., V. Grover, and K.D. Fiedler (1996) "Developing Strategic Perspectives on Business Process Reengineering: From Process Reconfiguration to Organizational Change." Omega, *International Journal of Management Science*, Vol. 24, No. 3, 271–294.

[16] Terez, T. (1990). "A Manager's Guidelines for Implementing Successful Operational Changes." *Industrial Management*, Vol. 32, No. 4, 18–20.

REGENERATION

Material Requirements Planning (MRP) records need updating because every day inventory levels change, orders are completed, priorities change, and so on.

MRP records and plans must be brought up to date to reflect daily/weekly changes. If all the MRP records are processed and updated in one computer run it is called regeneration. In regeneration, all part number records are affected and they are completely reconstructed.

See **MRP implementation; MRP regeneration; MRP; Net change processing.**

RELEASE THRESHOLD

During the 1990s, researchers developed a class of kanban control policies, known as dynamic kanban control (DKC) policies. The goal of DKC is to systematically add and remove kanbans so as to achieve improved system performance. Under a DKC policy, kanbans are added when needed to improve system performance and removed when they are no longer needed, or when their presence will result in lowered system performance. In general, we want additional kanbans when the benefit of their presence (e.g., reduced blocking and starvation, improved throughput) outweighs the cost (e.g., increased WIP and operating costs). The point at which kanbans are added is referred to as the *release threshold*; the point at which they are removed is the *capture threshold*. The release threshold is the threshold in a system performance measure at which additional kanbans are released into the system. The capture threshold is the threshold in a system performance measure at which kanbans are captured and removed from the system. The specific release and capture thresholds are functions of the DKC in use.

See **Adaptive kanban control; Base kanbans; Dynamic kanban control for JIT manufacturing; Flexible kanban control; Just-in-time manufacturing; Kanbans; Manufacturing cell design; Pull production systems; Reactive kanban control.**

RELIABILITY

The *reliability* of a piece of equipment is a measure which indicates the extent to which it will adequately perform its specified task for a specified period of time under specified environmental conditions. It is expressed in terms of probability, and therefore its value ranges between 0 and 1. The statement that a bearing has a reliability of 0.98 indicates that for a large number of identical copies of this bearing 98% will fulfill the stated requirements under the specified conditions and 2% will not. *Reliability theory* is concerned with the question of how the reliability of a complex system, composed of many individual components, can be derived, when the reliability of the individual components is given. In a series system, in which the failure of every single component causes the system to fail, the system reliability is the product of the individual component reliabilities.

See **Maintenance management decision models; Total productive maintenance; Total quality management.**

RELIABILITY CENTERED MAINTENANCE

Reliability centered maintenance is a systematic consideration of manufacturing system functions and the way functions can fail. In reliability centered maintenance, the value of maintenance activities is expressed first of all in terms of machine reliability improvement.

See **Maintenance management decision models; Total productive management; Total quality management.**

REMANUFACTURING

Remanufacturing transforms durable products that are worn, defective, or obsolete into functional products through disassembly, cleaning, refurbishment and replacement of components, reassembling, and testing. The range of services can be: repair or replacement of broken or unreliable parts, system upgrade, minor system modification, routine system inspection, and product refurbishment to enhance appearance. A remanufactured product may retain it's old form, or may lose the original identity all together (Ferrer, 1997).

See **Environmental issues and operation management; Environmental issues: Reuse and recycling in manufacturing systems.**

Reference

[1] Ferrer, G. "Communicating Developments in Product Recovery," Working Paper No. 97/30/TM, *Center for the Management of Environmental Resources*, INSEAD, Fontainebleau, France, 1997

REORDER POINT

Replenishing inventory is an important decision. When to order replenishment is sometimes decided by the inventory balance on hand. In such a system, a predetermined reorder level (point) of inventory is used to determine when replenishment orders must be placed. For example, it the reorder point is 15 units of inventory, a replenishment order is generated when inventory level drops to 15 units (on hand inventory and outstanding orders). The orderpoint is based on average usage during the lead time to get replenishment, plus a safety or buffer stock for unexpected demand during the lead time.

See **Inventory flow analysis; Safety stocks: Luxury or necessity.**

REORDER POINT SYSTEM

See **Reorder point.**

REPETITIVE MANUFACTURING

See **Repetitive production.**

REPETITIVE PRODUCTION

A repetitive manufacturing philosophy attempts to combine the low cost advantages of mass production with the advantages of customization or product differentiation. This production system achieves these goals through a modular approach to manufacturing: a great variety of end products may be produced or made from a relatively small number of intermediate assemblies and subassemblies. For example, an automobile manufacturer may offer its customers two different body styles, each with the option of two different engines, with two different transmissions, with two different interior upholstries, and in 6 different colors. This then gives the customer some 96 different cars to choose from just 8 different physical assemblies and 6 colors. The products of repetitive manufacturing may be quasi-customized: they are basically similar but with a set of more-or-less limited options reflect a customer's individual requirements.

Although it might appear that repetitive production is simple, there are many factors that can complicate the picture. For example, dedicated equipment is often used to make products which, require setups when changing from one product to another. Because this setup time can be extensive, companies have tended to prefer long production runs for each product, generating large finished goods inventories, and also risking stock-outs of those products not being produced at a given time. Similarly, large-batch production can create problems for the operations that make component parts, especially if some parts are used only in certain products.

Repetitive production is common in Just-in-Time (JIT) manufacturing because it is the environment in which JIT was originally applied. We now know that JIT concepts can be applied to many different manufacturing environments.

See **Just-in-time manufacturing; Just-in-time manufacturing implementation; Lean manufacturing implementation; Mixed-model sequencing; MRP.**

REPLENISHMENT TIME

Inventory management requires the replenishment of items in inventory when they are consumed. Replenishment time could be either the lead time to replenish an item in inventory. Sometimes, the time between two successive replenishment orders in the replenishment time.

See **Inventory flow analysis; Safety stocks: Luxury or necessity; Simulation analysis of manufacturing and logistics systems.**

RESOURCE PLANNING: MRP TO MRPII AND ERP

V. Sridharan and R. Lawrence LaForge

Clemson University, Clemson, SC, USA

INTRODUCTION: ***Material Requirements Planning*** (MRP) systems have been in vogue since the early 1970s. The MRP approach provided much-needed forward visibility and the ability to coordinate plans for dependent-demand inventory items. MRP-based systems, however, focused mainly on planning materials and organizing shop floor activities. Thus, their scope was limited to the manufacturing function. In addition to its narrow focus, the MRP approach also had several weaknesses including the absence of a formal feedback mechanism, especially to handle situations when the material plan was found to be infeasible due to a shortage of capacity.

Recognition of the weaknesses of the MRP approach led to a cycle of improvements. With these developments came recognition of the need to take a more holistic approach encompassing the entire organization. To some extent, concurrent advances in the area of computing technology (both hardware and software) greatly facilitated the rapid growth of a new integrated approach for manufacturing planning. Over time, the viability of the master production schedule (MPS) was ensured by incorporating the so-called "feedback loop" between shop floor execution and production planning activities. Furthermore, sophisticated approaches for testing the capacity-feasibility of both the master production schedule and MRP-generated material plans were incorporated. This led to the development of rough-cut capacity planning approaches and capacity requirements planning. In turn, these developments resulted in improved execution of the plan and facilitated basing financial planning on the detailed manufacturing plan, leading to a new cadre of systems known as manufacturing resource planning systems.

A ***Manufacturing Resource Planning*** (also known as MRP II) system is essentially a business planning system. In addition to enhancing the planning of manufacturing activities it also integrates information systems across departments. In an enterprise implementing MRP II, manufacturing and marketing managers, financial officers and engineers are linked to a company-wide information system. Each manager has access to information relating to his functional area of management as well as to the information pertaining to all other aspects of the business. In reality, to produce higher quality products and provide excellent customer service this integration is clearly mandatory. For example, the sales department needs the production schedule to promise realistic delivery dates to customers, and finance needs the shipment schedule to project cash flow.

TRANSITION FROM MRP TO MRP II: Manufacturing planning and control activities are closely related to the activities of other functional areas such as accounting and finance, marketing and sales, product/process engineering and design, purchasing, and materials management. Traditionally, each function within an organization had its own way of doing things and its own databases. Consequently communication between the various functional areas has not always been perfect. However, such separation of the activities across functions is artificial. In any business all activities are interrelated, and constitute the whole rather than a collection of different functions.

Therefore, the next logical step to MRP was to combine the manufacturing activities with those of finance, marketing, purchasing and engineering through a common database. This was the leap from MRP to MRP II, known as Manufacturing Resource Planning.

The quality of the major inputs to manufacturing planning, the MPS, bills of material (BOM) and inventory record information is not determined solely by manufacturing. These inputs are prepared, shared and updated by other functions within the organization as well. To start with, the MPS is a statement of planned production and thus is the basis for making delivery promises. It forms the basis for coordinating the activities of sales and production departments. Any changes or updates by sales need to be approved by manufacturing and vice versa. In addition, whereas marketing is charged with creating the demand, manufacturing is responsible to fulfill it. Therefore, any marketing activity that may influence future demand needs to be communicated to and confirmed by manufacturing.

BOM manufacturing and engineering. Any changes in the BOM will have to be agreed upon by both engineering and manufacturing to assure the feasibility of tolerances, impact of product revisions and new product introductions (where marketing also is involved) on the current shop floor system. The impact of changes in the BOM can be seen on material routings and lead times as well, which are used in material and capacity planning. Changes in agreements with suppliers affect purchase quantities and delivery lead times for purchased items.

Finally, the accounting/finance functions should use the same data as manufacturing, converting the materials, units produced, and activities into dollars. The MPS converted to dollars represents the revenue projection; purchase orders converted to dollars represent the cost of materials, shop floor activities (represented in work orders) converted to dollars reflect the labor and overhead costs. Discrepancies in the information used by manufacturing and finance/accounting is therefore unacceptable.

Once we use the MRP II logic, it is easy to realize that there is one physical system in operation in a company, and there is no justification for having more than one information system representing different dimensions of the physical system. Multiple systems lead to discrepancies and errors over time. The information system should also be unique and reflect the actual physical system. Thus, MRP systems evolved into MRP II when a common database was shared by all functions, and any changes and updates by one functional area immediately became visible to the rest of the organization.

With the advent of faster and cheaper computers it became possible to provide MRP II systems with the ability to simulate and thus provide a "what if" capability. Thus, modern MRP II systems can be used to simulate what would happen if various decisions were implemented, without changing the actual database. This makes it possible to see the impact of changes in the schedule on capacity requirements, the impact of schedule changes on material requirements, and/or the impact of design changes on customer responsiveness.

MRP II FEATURES:

Interfunction Coordination: As already mentioned, MRP II systems grew incrementally from material requirements planning (MRP) systems. One of the goals was to take a resource-based view of the firm. That is to say, engineering, marketing, manufacturing, distribution, and finance are all resources that must be managed in a coherent fashion to ensure consistency of purpose over time. This warranted improved coordination among the various functions.

Consider for example the activities of marketing and manufacturing. Recall that the master production schedule is a key input for planning material requirements for manufacturing. Given a master schedule, customer order promising is regulated via the so-called Available-To-Promise (ATP) logic to achieve coordination between production and sales. MRP II systems extend this similar integration and coordination between operations and other functions of the business such as human resource management, finance, and engineering.

Closed Loop Planning: Another enhanced functionality of MRP II is the "closing the loop" featureto ensure capacity-feasibility of material plans. APICS, The Educational Society for Resource Management, defines a closed-loop system as follows.

> *"A system built around material requirements that includes the additional planning functions of sales and operations (production planning, master production scheduling, and capacity requirements planning). . . . The term 'closed-loop' implies that is each of these elements included in the overall system, but also that feedback is provided by the execution functions so that the planning can be kept valid at all times." (APICS Dictionary, 8th ed. 1995)*

This is accomplished at two levels – first at the master production scheduling (MPS) level, and again after the detailed material plans are

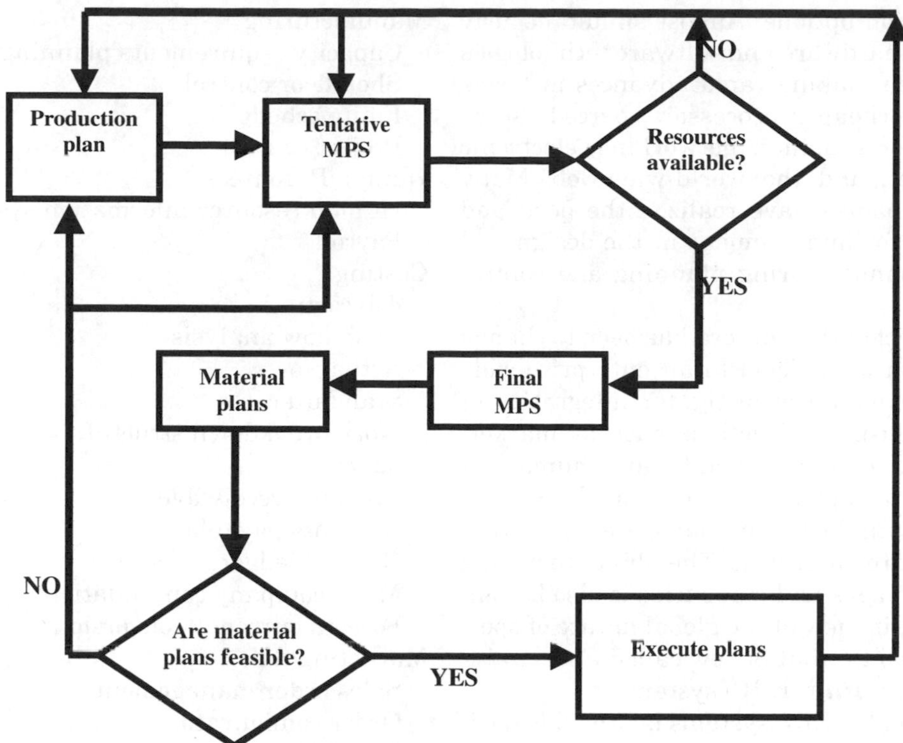

Fig. 1. Closed loop planning.

developed. Figure 1 presents a schematic of a closed- loop system. As indicated by the figure, first a trial MPS is prepared. Then, the feasibility of the trial MPS is verified using the so-called rough-cut capacity planning techniques. The available techniques range from the simple (which use historical load factors to project time-aggregated total capacity requirements at each work center) to the complex (which produce time-phased workloads at each work center using product routing, bill of material, and lead time data). If adequate resources are not available to meet the MPS, the MPS may be revised to achieve feasibility. Thus, feasibility of the final MPS is ensured.

Notice, however, that a feasible MPS cannot guarantee feasibility of the detailed material plans developed by MRP. Hence, MRP II systems also provide users the capability to check the feasibility of the detailed material plans. This is accomplished by means of what is called Capacity Requirements Planning (CRP). The CRP module verifies capacity feasibility taking into account; (1) the time-phased material plan; (2) inventory on-hand; (3) current status of work-in-progress in the shop; and (4) demand for service parts. Obviously the CRP module is highly data and computation intensive. The benefits, however, are realistic plans that can be achieved and better customer service via increased on-time delivery rate. The loop is closed by providing the necessary feed back of the status of plan execution to relevant planning modules.

What-If Analysis Capability: MRP II systems also provide the capability to conduct simulations to study and evaluate alternative decision options. Thus the effect of operations decisions on cash flow or working capital requirements can be easily examined. This capability to examine the consequences of each course of action without actually disturbing the system greatly enhances the quality of the decisions made. In turn, costs can be reduced without sacrificing customer service.

MRP II TO ERP SYSTEMS: Both MRP and MRP II systems focus at the plant production level. The reader may be aware of recent economic trends such as global competition; multiple plant-locations; geographically diverse, international, network of operations; and global network of suppliers. In addition, the markets are also becoming global. These changes in economic and market environments are complicated by the diversity of the political, cultural, and financial conditions in different parts of the globe. One consequence of these developments has been the need for instant access to information pertaining to customer needs, operations capabilities, and sourcing and distribution options. Almost simultaneously the computer hardware and software technologies have also been making rapid advances in terms of faster and cheaper processors, barcode scanners, computer graphics, networking, electronic mail, Internet, and the world-wide-web. Many software companies have realized the need and the potential for improvement in the design and practice of manufacturing planning and control systems.

The net effect of the new era has been to change the focus from plant level to an enterprise-wide view. This means an even tighter integration of the various business functions such as marketing and sales, accounting and finance, human resources, purchasing, distribution, and design and engineering with the traditional operations activities of the entire enterprise. The ability to handle multiple languages and currencies is also becoming important in view of the global nature of operations. This led to what is now called ***Enterprise Resource Planning*** (ERP) systems.

The main goal of ERP systems is to enable business processes that support customer needs. ERP accomplishes this goal by developing and presenting an integrated view of the supply chain. Such a view exposes system bottlenecks and constraints. The core view of ERP systems, however, remains resource-based. That is to say, ERP systems view the business as a collection of interconnected resources, whose optimization produces a superior performance. In turn, this approach enables accurate forecasting and realistic planning.

ERP systems tend to address the present-day needs for distributed planning and control. ERP systems also cater to the unique needs of personnel at different layers of the organization, such as individual divisions, locations, and machine centers. A typical ERP system provides the following functionality:

Engineering:
- Item part number control
- Bill of material control

Engineering Change & Documentation Control
- Routing
- Estimating
- Design engineering

Purchasing
- Vendor performance
- Purchase order management
- Subcontracting

Materials
- Inventory control
- Master production scheduling
- Material requirements planning
- Lot tracking
- Rough-cut capacity planning

Manufacturing
- Capacity requirements planning
- Shop floor control
- Finite scheduling
- Project control

Human Resources
- Human resource information system
- Payroll

Costing
- Job costing
- Cash flow analysis
- Actual costs
- Standard costs
- Work breakdown structure

Finance
- Accounts receivable
- Accounts payable
- General ledger
- Multi-company consolidation
- Foreign currency conversion

Marketing/Sales
- Sales order management
- Order configurator
- Billing/Invoicing
- Full sales analysis
- Commission calculation/reporting
- Sales forecasting /rollups
- Quoting

Today, many ERP systems are run on a network of personal computers. Their features include a client/server architecture supported by a distributed relational database system with query and reporting capabilities, electronic data interchange capability to communicate with both suppliers and customers, decision support systems for managers, a graphical user interface, and standard application programming interfaces.

In its 1998 software survey, APICS identified over one hundred products/vendors, indicating the prevalence of ERP systems. A small but representative sample of software companies (and their products) that market ERP systems include SAP America (R/3), Bann Company (Bann ERP), MAPICS Inc. (MAPICS XA), and Marcam Solutions Inc. (PRISM).

SUMMARY: The science and practice of manufacturing planning and control has evolved over the years from MRP to MRP II to ERP systems. This evolution has been necessary to meet the changing needs of businesses, perhaps due to the changing economic and market conditions, and to bring the state-of-the art technology to the work place. The evolution has also brought about a resource-view of the corporation and a process-based view of its activities. Consequently, it has produced a tight integration of the activities of the various business functions such as engineering, operations, marketing, finance, and human resource management. The net effect of this evolution is to make enterprises more lean and agile.

See **Bill-of-materials (BOM); Capacity planning; Capacity planning in make - to-order production systems; Capacity planning: Medium- and short-range; ERP; Long range; MPS; MRP; MRP implementation; Order release; Schedule stability.**

References

[1] *APICS Dictionery* (1995). 8th Edition, APICS, Falls Church, VA.

[2] *APICS The Performance Advantage* (1998). APICS, Falls Church, VA, June.

[3] *APICS The Performance Advantage* (1998). APICS, Falls Church, VA, September.

[4] Vollmann T.E., W.L. Berry, and D.C. Whybark (1997). *Manufacturing Planning and Control Systems*, 4th Edition, Irwin/McGraw Hill.

RESOURCE PROFILE

The Resource Profiles (RP) method is one of the techniques that can be used for a company's resource requirements planning (RRP, i.e., roughcut capacity planning, RCP). The RP method takes into account the specific timing of the projected workloads at individual work centers. A resource profile includes the time-phased profile of the capacity requirements for each end item. The resource profile indicates standard hours required at each operation (machine or work center) for a master production schedule (MPS) item. Furthermore, this profile shows the lead time offset from the master schedule for each operation. Because the MPS generally shows the week in which production of an item will be *completed*, the lead time offset for each operation indicates how far ahead of that completion time each operation must *begin*. The following table is an example resource profile showing standard hours at work centers, and lead time offsets (in weeks) from the week planned for completion in the MPS for product Z.

Resource profile – product Z.

	Lead time in weeks		
Work center	1 week	2 weeks	3 weeks
Drill (standard hours)	.35		
Mill (standard hours)		.40	
Assembly (std. Hours)			.25

This information can be used in rough-cut capacity planning to estimate capacity requirements at each machine or work center once the MPS is ready. Because lead time information is included in the resource profile, this method for rough-cut capacity planning is generally the most accurate, but also the one that requires the most detailed data.

The resource profile method works as follows. Suppose 100 units of Product Z are planned for production in week 8. The resource profile shows us that 35 standard hours of capacity will be required in the Drill work center during week 6 ($8 - 2 = 6$) to meet that production plan.

See **Aggregate plans and master production schedules linkage; Capacity planning in make-to-stock production; Capacity planning: Medium- and short-range; Capacity planning: Long-range; MRP; Resource requirement planning; Resource requirements planning; Rough-cut capacity planning.**

RESOURCE REQUIREMENTS PLANNING (RRP)

To be feasible and acceptable, a master production schedule (MPS) should meet the resource constraints at key work centers. This feasibility check is called resource requirements planning (RRP) or Rouch-Cut Capacity Planning (RCP). If resource requirements planning indicates potential infeasibility at some critical operations, then the master schedule should be revised to accommodate the capacity limitations of the key facilities. Sometimes, a decision may be made to maintain the master schedule and to implement adjustments to the capacities at relevant work centers. This may require adjustments to the long-term capacity plan or the aggregate production plan. After the feasibility check via resource requirements planning, the feasible and acceptable master schedule becomes an authorized master schedule, and can be further broken down into detailed schedules using systems such as material requirements planning (MRP). There are several techniques that can be employed for RCP. The Overall Factors (OF) method is a relatively simple approach which uses planning factors derived from standards or historical data for end products. When these planning factors are applied to the MPS data, overall labor or machine-hour capacity requirements can be estimated. This overall estimate is thereafter allocated to individual work centers on the basis of historical data of shop workloads.

The Capacity Bills (CB) method provides more direct linkage between individual end products in the MPS and the capacity required at individual work centers. A bill of capacity indicates total standard time required to produce one end product in each work center required in its manufactures. The capacity requirements at individual work centers can be estimated by multiplying the capacity bills by the MPS quantities. The Resource Profiles (RP) method takes into account the specific timing of the projected workloads at individual work centers. A resource profile includes the time-phased profile of the capacity requirements for each end item over its production lead time.

See **Capacity planning: Long-range; Capa city planning: Medium- and short-range; disaggregation in an automated manufacturing environment; Master production schedule; MRP; Resource profile; Resource profile; Rough-cut capacity planning.**

Reference

[1] Berry, W.L., T.G. Schmitt, and T.E. Vollmann (1982). "Capacity planning Techniques for Manufacturing Control Systems: Information Requirements and Operational Features," *Journal of Operations Management*, 3(1), pp. 13–25.

RISK/REWARD SHARING AGREEMENTS

This is an agreement between a purchasing organization and a supplier who are actively involved in the new product development process. In most companies, such an agreement defines acceptable development and production costs (prices) based on projected volumes. Based on research by Ragatz, Handfield, and Scannell (1997), it was found that few companies reach advance agreements, where expectations and actual volumes are likely to differ substantially. Many of the companies indicated that such risks and rewards are shared 50/50 based on a handshake agreement, which highlights the importance of mutual trust.

See **New product development through suppliers integration; Supplier partnership as strategy; Supplier relationships.**

Reference

[1] Ragatz, G., Handfield, R., and Scannell, T. (1997). "Success factors for integrating suppliers into new product development," *Journal of Product Innovation Management*, 7–20.

ROBOT REPEATABILITY

Robots are used in repetitive and monotonous operations. Robot repeatability is the ability of the robot to return to the same point. This ability is essential for the robot to carryout repetitive operations.

See **Programmable automation; Robot selection.**

ROBOT SELECTION

Moutaz Khouja

University of North Carolina – Charlotte, Charlotte, NC, USA

O. Felix Offodile, David E. Booth and Michael Suh

Kent State University, Kent, OH, USA

INTRODUCTION: Industrial robot usage has increased greatly in the past two decades (Nof, 1985). The growth has led to an increase in the number of robot manufacturers with each offering a number of models (Agrawal *et al.*, 1991; Offodile *et al.*, 1987). A decision-maker (DM) that could use an industrial robot is faced with a choice between many robot models with different performance capabilities and costs. Subsequently, considerable effort has been focused on developing models to aid DMs facing a robot selection problem (RSP). RSP models are arranged into five groups:

1. Multi-criteria decision making models (MCDM)
2. Production system performance optimization models
3. Computer-assisted models
4. Statistical models
5. Other approaches

The approaches to the robot selection problem are evaluated in terms of:

1. Robot attribute consideration – what robot attributes are considered in the selection decision?
2. Data requirement – what parameters need to be estimated and how difficult is the estimation?
3. Versatility – can the model be applied to different robotic applications or is it limited to a specific application?

Table 1. Attributes used for evaluating robots.

Engineering attributes	Vendor-related attributes	Cost attributes
Load capacity	Brand	Robot cost
Repeatability	Availability of Training	Installation support
Velocity	Quality of Training	Tooling cost
Programming method	Documentation	Energy consumption
Vertical reach	Installation support	Labor cost
Horizontal reach	Spare parts availability	Maintenance cost
Memory size	Installation leadtime	
Acceleration	Pre-sale services	
Deceleration	Servicing ability	
Degrees of Freedom	Warranty	
Reliability		
Diagnostic capability		
Programming language		
Software		
Control type		
Recovery from error		

4. Built-in assumptions – what stated and unstated assumptions are built into the model?

ROBOT PERFORMANCE AND SELECTION: Three groups of attributes, as shown in Table 1, have been used to evaluate robots (Agrawal *et al.*, 1991; Nnaji and Yannacopoulou, 1989; Seidmann *et al.*, 1984):

1. Engineering Attributes: determine the ability of robots to perform tasks and include load capacity, speed, repeatability, and accuracy.
2. Vendor-related Attributes: determine the attractiveness of robot vendors.
3. Cost Attributes: determine total costs of installing and operating robots.

An implicit assumption in the models is that engineering attributes are satisfying attributes, and hence are only used to find a set of feasible robots and not to make a final selection.

Accuracy and repeatability are among the most important but confused robot engineering attributes (Albertson, 1983). Repeatability is a measure of the ability of a robot to return to the same point.

Robot selection has a critical impact on the performance of the production system. As Figure 1 shows, important system performance measures such as output quality, cycle time, and manufacturing flexibility are all influenced by robot engineering attributes. Also, engineering attributes present a DM with a trade-off. The reason for this trade-off is that large loads exert large force that causes poor accuracy and repeatability. High-speed robots must also have high acceleration and deceleration that impedes their ability to provide good accuracy and repeatability.

The lack of industry-wide standards for specifying robot engineering performance further complicates the RSP (Albertson, 1983). Robot testing prior to selection is frequently prohibitive in terms of both time and cost. Published specifications, however, are frequently mutually exclusive (Offodile and Ugwu, 1991).

RSP models which are solely based on financial estimates of costs/benefits not related to robot engineering attributes (Knott and Getto, 1982) do not provide the level of detail needed to make a selection. A summary of the reviewed models is presented in Table 2. For quick reference, the table shows the application domain, solution approach, robot attributes considered, and the selection criterion for each model.

MULTI-CRITERIA DECISION MAKING (MCDM) MODELS FOR ROBOT SELECTION: Multi-criteria decision making (MCDM) models include multi-attribute decision-making models (MADM) and multi-objective decision-making models (MODM). In MADM, all objectives of the decision makers are unified under a super-function termed the decision maker's utility. In MODM, the decision maker's objectives remain explicit and are assigned weights reflecting their relative importance.

Several MADM-based robot selection models have been developed. Jones, Malmborg, and Agee (1985) present the decision maker with predefined value functions from which he/she chooses the ones that describe his/her preference for attributes. Scaling constants are obtained from the decision maker by asking him/her to rate a set of robots with the robot (hypothetical or actual) that combines the best of all specifications as a reference, and it is assigned a 100 utility rating. The robot that maximizes utility is then selected. Nnaji and Yannacopoulou (1989) followed a similar approach to Jones *et al.* with some refinements (Khouja and Offodile, 1994).

Agrawal, Khohl, and Gupta (1991) used an expert system together with an MADM model to address the robot selection process. The decision maker the type of application for which the robot is

Table 2. A summary of robot selection models.

Author (year)	Application	Solution approach	Robot attributes considered	Selection criterion
Seidmann *et al.* (1984)	General	MODM	Engineering Vendor Related	Cost Quality Utilization
Jones *et al.* (1985)	General	MADM	Engineering	DM's utility
Nof (1985)	General	Robot time and motion	Engineering	Cycle time
Abdel-Malek and Boucher (1985)	Assembly simple geometry	Performanc optimization	Engineeri	Probability of producing a conforming unit
Abdel-Malek (1986)	Assembly, complex geometry	Performance optimization	Engineering	Probability of producing a conforming unit
Nnaji (1986)	General	Computer-assisted	Engineering Vendor related	N/A
Offodile *et al.* (1987)	General	Computer-assisted	Engineering Cost	First year cost or NPV
Abdel-Malek (1987)	Assembly	Performance optimization	Engineering	Profit
Abdel-Malek (1988)	Flexible transfer lines	Performance optimization	Engineering	FTL production rate and % of time it's operational
Nnaji (1988)	General	Math. prog. And MADM	Engineering Vendor related Cost	DM's utility
Nnaji and Yannacopoulou (1988)	Assembly	MADM	Engineering	DM's utility
Fisher and Maimon (1988)	General	Computer-assisted	Engineering Cost	N/A
Fisher and Maimon (1988)	General	Math. prog.	Engineering Cost	Cost
Seidmann and Nof (1989)	Robotic assembly station	Performance optimization	Engineering	Task level Product level System level
Abdel-Malek (1989)	Assembly, conveyor tracking	Performance optimization	Engineering	Cycle time and probability of tracking producing a conforming unit
Imany and Schlesinger (1989)	General	Statistical	Engineering	Minimum absolute regression residual
Boubekri *et al.* (1991)	General	Computer-assisted	Engineering Cost	Payback period
Agrawal *et al.* (1991)	General	MADM	Engineering	DM's utility
Khouja and Booth (1991)	General	Statistical	Engineering	Minimum regression residual
Booth *et al.* (1992)	General	Statistical	Engineering	Maximum robust Mahalanobis distance
Khouja (1995)	General	DEA and MADM	Engineering	DM's utility
Khouja and Booth (1995)	General	Statistical	Engineering	Fuzzy cluster analysis
Baker and Talluri (1997)	General	DEA and CEA	Engineering	DM's utility

needed, and with assistance of an expert system, decides on the key attributes and their minimum acceptable values. A search is then performed to obtain a short list of feasible robots from a data base of available robots. MADM is then used to make a selection among feasible robots. Technique for Order Preference by Similarity to Ideal Solution (TOPSIS), a method of MADM, is used.

The strengths of MADM models are their versatility and ability to consider a large number of robot attributes. Using MADM, the decision maker can take into account, engineering, vendor-related, and cost attributes. The versatility of MADM models is a result of their flexibility in assigning varying weights (importance) to different robot attributes depending on the application. While the data requirements from the decision maker can be reduced by using an expert system (Agrawal *et al.*, 1991), the decision maker is still required to assign weights to each robot attribute where he/she is likely to encounter difficulty (Khouja and Offodile, 1994). MADM models assume accuracy and repeatability of robots to be constant over all weights handled and operating speeds achieved. However, a robot's repeatability deteriorates as loads and speeds are increased (Offodile and Ugwu, 1991) which may lead to a violation of the mutual preferential independence assumption of MADM.

Seidmann, Arbel, and Shapira (1984) solved the robot selection process a two-phase problem using MODM. To solve the problem, the authors used the Analytical Hierarchy Process (AHP). The AHP is attractive for complex problems because it decomposes the problem into a hierarchy with manageable parts at each level. By decomposing the problem, the priorities between elements (objectives) at each level are obtained using pairwise comparison, thus, easing the information processing burden on DMs.

Khouja (1995) used data envelopment analysis (DEA) for identitifying a small set of robots from which the final selection is to be made. Baker and Talluri (1997) observed that Khouja's method of selecting the candidate robots can be improved. In order to improve upon Khouja's approach, they suggest that the MADM be replaced with cross-efficiency analysis (CEA) to make a final robot selection.

Robot Selection Based on Production System Optimization: Models in this category select a robot that optimizes some performance measure(s) of the production system. The work of Abdel-Malek and Boucher (1985), and Abdel-Malek (1985, 1986, 1988, 1989) fall within this category. Abdel-Malek and Boucher's (1985) thesis is that product design specifications and robot repeatability determine the probability of successful assembly. For a given product design and robot repeatability, the authors estimate the probability of successful assembly. To improve that probability, a robot with better repeatability can be used, and/or the product can be redesigned by tightening tolerances and/or increasing allowances.

The ability of a robot to recover from assembly failure and retry the assembly task was analyzed by Abdel-Malek (1985). The decision variable here is the number of assembly trials the robot should attempt for a given assembly in order to maximize the production rate. Abdel-Malek (1986) extended Abdel-Malek and Boucher's model (1985) to parts with complex geometries.

Abdel-Malek (1989) extended the models to robotic assembly with conveyor tracking (RACT). In RACT a robot performs assembly tasks as parts move within the robot's envelop on a conveyor belt. Reach and control type become important attributes in determining robot feasibility.

Abdel-Malek (1988) introduced a model that can be used to select robots to load and unload machine tools in the cells of an assembly line. The product goes through every cell before completion. Abdel-Malek suggests using robots with overlapping envelopes so that, if a robot fails, the adjacent robots can share its tasks. Three robot attributes are critical here: reach, programmability, and speed. Regardless of the type of failure and repair process, using robots with overlapping envelopes is found to improve the line's production rate and efficiency.

Seidmann and Nof (1989) address the robot selection process in unitary robotic assembly stations. The authors evaluate the station at three levels: Task, product, and system. The authors present a numerical example in which robots with different speeds can be selected.

In the above models, engineering and cost attributes are considered. The relationship between robot engineering attributes and performance of production systems is explored in detail. Furthermore, the models sensitize the DM to issues relating to robust product design.

Models in this category are not as versatile as those in MCDM. These models are most suitable for assembly and machine loading applications. However, such applications usually give rise to the most difficult robot selection process due to the critical effect of engineering attributes on system performance. In that sense, the models tackle some of the most complex robot selection process. The above models are developed for a single-product production system. Most production systems that employ robots manufacture a number of items that may complicate the analysis. Data requirements for the models include product design specifications and engineering attributes of available robots. The models require robot cycle time as a parameter of the analysis.

Computer-Assisted Models

Because of the large number of both robot attributes and available robots, computer-assisted methods provide an attractive approach to therobot selection process. In general, the selection process is shown in Figure 2. A decision maker starts by providing information about the robot application. These data are used by an expert system to provide a list of important robot engineering attributes and their desired values, which in turn is used to obtain a list of feasible robots from a descriptive data base of available robots. If more than one robot is feasible, a decision model is used to make a selection.

Nnaji (1986) uses a total of eighty-nine alphanumeric codes to describe robots. The decision maker is asked questions about the application and preferred values of some robot attributes. Based on the decision maker's answers, the expert system builds a structural model of the robot that supposedly will be best able to deliver the required performance. The structural model is then translated into codes that are matched with the codes of available industrial robots to obtain a list of feasible robots. To evaluate the set of feasible robots a model similar to MADM is used.

Offodile, Lambert, and Dudek (1987) developed a coding and classification system to be used for storing specifications of available robots in a database. Fisher and Maimon (1988) propose an robot selection process model similar to Seidmann *et al.* (1984) in the sense that it addresses the problem in two-phases with similar aims.

Boubekri, Sahoui, and Lakrib (1991) developed an expert system for industrial robot selection (ESIR). Based on information provided by the decision maker, ESIR selects robot attributes assigned by an expert for the particular application and the best values for those attributes. ESIR then searches the database for feasible robots. If none is found, the DM is asked to modify the input data. If more than one feasible robot is found, then a robot that minimizes the payback period is selected.

The major advantage of models in this category is that they bring expert knowledge to a difficult problem. Also, since the selection process is automated, a large number of robot engineering and cost attributes can be considered. However, the models require a considerable amount of information gathering and processing. All models require a full description of all available industrial robots. Furthermore, versatility, for a problem as difficult as the robot selection process, requires a large number of rules in the expert system, which may be overwhelming. With a technology that is as quickly changing as robotics, a frequent update of the expert system is needed.

Statistical Models

The models in this category focus on the tradeoff between robot engineering attributes. Multivariate statistical procedures are used to identify robots that will deliver the best engineering performance based on Manufacturers' specifications.

Imany and Schlesinger (1989) used least absolute deviation regression (LAD) as a tool for the robot selection process. In many applications, repeatability is more critical than other attributes because of its critical effect on product quality. Thus, a regression equation relating repeatability to load capacity and speed is fitted. Khouja and Booth (1991) used M-estimator regression because it is more efficient in minimizing variance than LAD.

Booth, Khouja, and Hu (1992) used robustified Mahalanobis distance and principal components analysis to identify better performing robots. Mahalanobis distance is used to identify outlying robots while principal components analysis is used to indicate if a robot is an outlier because it provides better or worse combinations of specifications from the average robot. The authors applied both techniques to the data set in Imany and Schlesinger (1989), and then ranked the robots based on their Mahalanobis distances and position on the principal components plots, thus providing an ordering for potential selection or testing.

Because there are so many options in both performance and cost, the robot selection process can be a difficult problem. Khouja and Booth (1995) suggested the use of fuzzy cluster analysis to solve this problem. In this approach, a robot is considered to be feasible if it satisfies three conditions:

1. Minimal environmental conditions,
2. Minimal performance requirements, and
3. Budget ceiling.

For these feasible robots, the ones with the best combination of specifications on the performance parameters are determined using fuzzy cluster analysis. This method has several advantages.

The models in this category base the selection on robot engineering attributes. However, cost attributes can be easily included in the analysis.

The models are most appropriate in applications, where the tradeoff between engineering attributes is important such as assembly.

The only data requirement of the models are the manufacturers' specifications. Since the specifications are usually mutually exclusive, the selected robot may not deliver the required performance. Thus, actual testing of robots may be required. Also, similar to MADM models, the relationship between robot engineering attributes and production system performance is unexplored.

Other Approaches

Nnaji (1988) proposed a mathematical programming approach combining economic analysis and MADM to the Robot Selection Process. In this research, each robot contributes to certain savings due to increased accuracy and productivity. The savings are used to compute the payback period. However, the author provides no explanation of how the savings are obtained for each robot. Obviously, the savings depend on robot engineering attributes. Nof and Lechtman (1982) and Nof (1985) propose a method for evaluating alternative robot work methods. The authors develop a robot time and motion (RTM) system study to find the best robot work method. Finally, Fisher and Maimon (1988) propose a two-phase mathematical programming approach to the RSP.

DISCUSSION AND SUGGESTED FUTURE DIRECTIONS FOR RESEARCH: Diverse approaches have been used to address the robot selection process. MADM and statistical approaches, while versatile and are able to handle a large number of attributes, ignore the effect of robot engineering attributes on production system performance. Computer-assisted models incorporate expert knowledge to automate the decision process. In that sense, they can handle a large number of engineering attributes. However, these attributes are considered satisfying, and explicit delineation of the relationships governing the performance of production systems is absent.

Production systems performance optimization models provide explicit expressions of the relationships between robot engineering attributes and performance of the production system. Due to the complexity of these relationships and their demanding data requirements, only a limited number of attributes can be considered and the models are not as versatile as those in other categories. However, in most applications, the number of critical attributes to be considered is small. Models in this category sensitize the decision maker to means of improving performance such as the use of robust product design, multiple robots, and compliance mechanisms.

Production system performance optimization models can be further generalized by considering the effect of robot speed on output rates of production systems. Such a generalization leads to a quality and output rate tradeoff. As robot speed is increased, production rate increases; however, repeatability deteriorates (Offodile and Ugwu, 1991) and so does quality (Mehrez and Offodile, 1994).

An issue that still needs to be addressed in the robot selection process is manufacturing flexibility. Flexibility in manufacturing is the ability to reconfigure manufacturing resources so as to produce efficiently different products of acceptable quality.

Many robot selection process models do not consider the uncertainty of future market conditions. Deterministic estimates of costs/benefits are made in Boubekri, Sahoui, and Lakrib (1991), and Nnaji (1988). With rapid technological change that may lead to robot obsolescence, and with volatile economic conditions, the certainty assumption is not justified (Knott and Getto, 1982). The capital asset pricing model (CAPM) or options theory may provide an approach in which uncertainty in the demand, represented by states of the economy, and uncertainty in the discount rate are taken into account in the robot selection process.

Some robot attributes, in spite of their importance, received limited attention. For example, plant compatibility (Seidmann *et al.*, 1984) is not considered in other models. The ability of a robot to interface with existing machines and equipment and future acquisitions is an important factor in the RSP. Also, in-house experience with robots is important in determining the ability of the firm to realize the full potential of acquired robots in a short time. DMs may be inclined to select robots similar to those the firm has experience in operating. Other important attributes include warmup period and acceleration and deceleration times. A warmup is a limited period of movement and motion actions that is carried out to prevent positioning errors whenever the robot has been shut down for any length of time (Nof, 1985). Acceleration and deceleration warrant special attention since in many applications the traveled distance is not sufficient for the robot to reach maximum speed (Nof, 1985).

The establishment of standards for reporting robot performance will facilitate robot selection process research. Developing empirical formulas for the relationship between robot engineering attributes such as speed, acceleration, load, and repeatability will provide useful information to solve the robot selection process. Empirical work has been performed toward that end on different robot models. (Offodile and Ugwu, 1991).

SUMMARY: A DM facing an RSP must decide on the selection approach to use. Production systems performance optimization models are most useful when the selected robot will have high impact on product quality or production cycle time as is the case in many assembly applications. When the robot will not have a critical impact on product quality or production cycle time, a combination of the other approaches can be used. Computer-assisted models can be used first to identify a candidate set of robots for selection from those that are commercially available. MCDM or statistical models can then be used to make the final selection decision. When the DM has a clear idea of his/her preferences for robot attributes, MCDM can be applied. When the DM can't articulate his/her preferences, statistical models can be applied.

See **Capital investment in advanced manufacturing technology; Data envelopment analysis (DEA); Flexible automation; Manufacturing cell design; Manufacturing flexibility; Manufacturing flexibility dimensions; Manufacturing technology; Manufacturing technology use and benefits; Manufacturing with flexible systems; Multi-attribute decision models; Multi-criteria decision models; Multi-objective decision models; Robot engineering attributes; Robot repeatability; Unitary robotic assembly station.**

References

[1] Abdel-Malek, L. (1986). "A framework for the robotic assembly of parts with general geometries." *International Journal of Production Research*, 24, 1025–1041.

[2] Abdel-Malek, L. (1988). "The effect of robots with overlapping envelopes on performance of flexible transfer lines." *IIE Transactions*, 20, 213–222.

[3] Abdel-Malek, L. (1989). "Assessment of the economic feasibility of robotic assembly while conveyor tracking (RACT)." *International Journal of Production Research*, 27, 1209–1224.

[4] Abdel-Malek, L. (1985). "A case study on the use of robots in assembly process." *Engineering Costs and Production Economics*, 9, 99–102.

[5] Abdel-Malek, L. and T.O. Boucher (1985). "A Framework for the economic evaluation of production system and product design alternatives for robot assembly." *International Journal of Production Research*, 23, 197–208.

[6] Agrawal, V.P., V. Khohl, and S. Gupta (1991). "Computer aided robot selection: The multiple attribute decision making approach." *International Journal of Production Research*, 29, 1629–1644.

[7] Albertson, P. (1983). "Verifying robot performance." *Robotics Today*, 33–36.

[8] Baker, R.C. and S. Talluri (1997). "A closer look at the use of data envelopment analysis for technology selection." *Computers and Industrial Engineering*, 32, 101–108.

[9] Booth, D.E., M. Khouja, and M. Hu (1992). "A robust multivariate procedure for evaluation and selection of industrial robots." *International Journal of Operations and Production Management*, 12, 15–24.

[10] Boubekri, N., M. Sahoui, and C. Lakrib (1991). "Development of an expert system for industrial robot selection." *Computers and Industrial Engineering*, 20, 119–127.

[11] Fisher, E.L. and O.Z. Maimon (1988). "Specification and selection of robots." In Kusiak, A., (Ed.) *Artificial Intelligence Implications for CIM*, IFS Publications, Bedford.

[12] Imany, M.M. and R.J. Schlesinger (1989). "Decision models for robot selection: a comparison of ordinary least squares and linear goal programming methods." *Decision Sciences*, 20, 40–53.

[13] Jones, M.S., Malmborg, C.J. and M.H. Agee (1985). "Decision support system used for robot selection." *Industrial Engineering*, 17, 66–73.

[14] Khouja, M. (1995). "The use of data envelopment analysis for technology selection." *Computers and Industrial Engineering*, 28, 123–132.

[15] Khouja, M. and D.E. Booth (1991). "A decision model for the robot selection problem using robust regression." *Decision Sciences*, 22, 656–662.

[16] Khouja, M. and O.F. Offodile (1994). "The industrial robots selection problem: literature review and directions for future research." *IIE Transactions*, 26(4), 50–61.

[17] Khouja, M. and D.E. Booth (1995). "Fuzzy clustering procedure for evaluation and selection of industrial robots." *Journal of Manufacturing Systems*, 14, 244–251.

[18] Knott, K. and R.D. Getto, Jr. (1982). "A Model for evaluating alternative robot systems under uncertainty." *International Journal of Production Research*, 20, 155–165.

[19] Mehrez, A. and O.F. Offodile (1994). "A statistical-economic framework for evaluating the effect of robot repeatability on profit." *IIE Transactions*, 26, 101–110.

[20] Nnaji, B.O. (1986). *Computer-Aided Design, Selection and Evaluation of Robots*, Elsevier, New York.

[21] Nnaji, B.O. (1988). "Evaluation methodology for performance and system economics for robotic devices." *Computers and Industrial engineering*, 14, 27–39.

[22] Nnaji, B.O., and M. Yannacopoulou (1989). "A utility theory based robot selection and evaluation for electronics assembly." *Computers and Industrial Engineering*, 14, 477–493.

[23] Nof, S.Y. (1985). *Handbook of Industrial Robotics*, John Wiley & Sons, Inc., New York.

[24] Nof, S.Y. and H. Lechtman (1982). "Robot time and motion provides means for evaluating alternative work methods." *Industrial Engineering*, 14, 38–48.

[25] Offodile, O.F., B.K. Lambert, and R.A. Dudek (1987). "Development of a computer aided robot selection procedure (CARSP)." *International Journal of Production Research*, 25, 1109–1121.

[26] Offodile, O.F. and U. Ugwu (1991). "Evaluating the effect of speed and payload on robot repeatability." *International Journal of Robotics and Computer Integrated Manufacturing*, 8, 27–33.

[27] Seidmann, A., A. Arbel, and R. Shapira (1984). "A two-phase analytic approach to robotic system design." *Robotics and Computer-Integrated Manufacturing*, 12, 181–190.

[28] Seidmann, A. and S.Y. Nof (1989). "Operational analysis of an autonomous assembly robotic station." *IEEE Transactions on Robotics and Automation*, 5, 4–15.

ROBOTIC CELLS

In *robotic cells*, the microcomputers of the the CNC machine tools and a robot are networked together with a cell host computer. All the machines in the cell are programmable, and therefore this kind of automation is very flexible. The robots serve as a substitute for manual operators for material handling.

It is, however, difficult to realize the mechanical, electrical, and computer engineering interfaces required to arrive at a robotic cell that is fully unmanned, flexible, and autonomous.

See **Manufacturing cell design; Robot selection.**

ROLE AMBIGUITY

Workers face role ambiguity, when the expectations associated with their role, the methods for carrying out their roles, and/or the outcomes of their performance are not clearly defined to reduce ambiguity. Individuals should have a clear understanding of : 1) the boundaries of their authority; 2) the goals and objectives pertaining to their jobs; 3) the extent of their responsibilities; 4) their supervisors' expectations; and 5) the various tasks that need to be performed.

See **Human resource issues and advanced manufacturing technology; Human resource issues in operations management.**

ROLE SHIFTING

The essence of supply chain integration is to establish a team of companies that develop an effective "chemistry" in working together so that an overall outstanding level of performance is attained. For the supply chain team to be successful, each member of the team must have a clear and well-specified role. Equally important is the need for each member of the supply chain to perform its role in an exceptional manner – better than any other firm in the supply chain. When this occurs, the supply chain achieves a high level of competitiveness.

As objectives and performance change, it is possible that a member of the supply chain no longer possesses a capability that enhances the entire supply chain's performance. Perhaps one of the other supply chain members can perform the same task at a lower cost and/or higher quality. May be the task no longer needs to be performed at all. And it is always possible that an outside firm with higher levels of performance should become a member of the supply chain. In any of these circumstances, the firm without a clear value-added benefit would be shifted out of the supply chain; that is, because the firm no longer adds unique value, it loses its place in the supply chain.

See **New product development through supplier integration; Supply chain management: Competing through integration; Supplier relationships.**

ROLLING HORIZON

The practice of rolling horizons or rolling schedules is to routinely update or revise the (production) schedules taking into consideration more reliable and recent data as they become available. The typical use of rolling horizon is as follows: solve the planning model and implement only the first period's decisions; next period, update the model to reflect new information resolve the model, and again implement the imminent decisions. The rolling horizon practices simply reflect the continuity of the firm's business and activities. It should be noted, rolling schedules need not always be updated every period. This is particularly true in a relatively stable environment. In such an environment, when a planning model is solved, several periods' decisions are implemented before the model is resolved with new information. This is done with the assumption that no significant change occurs in the environment between the two planning sessions.

See **Linking aggregate plans and master production schedules; Master production schedule; MRP; MRP implementation; Order release; Schedule stability.**

References

[1] Baker, K.R. (1977), "An Experimental Study of The Effectiveness of Rolling Schedules in Production Planning," *Decision Sciences*, Vol. 8, No. 1, pp. 19–27.

[2] Chung, C.H. and Krajewski, L.J. (1987), "Interfacing Aggregate Plans and Master Production Schedules via a Rolling Horizon Procedure," *OMEGA (The International Journal of Management Science*, Vol. 15, No. 5, pp. 401–409.

ROLLING PLANNING HORIZON

See **Rolling horizon.**

ROLL-UP FORECASTS

This is also called pyramid forecasting. Forecasts are made for individual items which are summed or rolled-up into forecasts for product groups. Product group forecasts are summed to produce total business forecasts in units and/or dollars. Once top management plans are finalized based on these forecasts, they are passed down for implementation. This approach ensures coordination and integration's among forecasts prepared by different parts of the company.

See **Forecasting guidelines and methods; Forecasting in operation management; Master production schedule; MRP.**

ROUGH-CUT CAPACITY PLANNING

See **Resource requirements planning.**

ROUTE SHEET, ROUTING, ROUTING SHEET

See **Process planning.**

RUN CHART

See **Integrated quality control; Statistical process control using charts.**

RUN TIME

Run time is the total time to complete an operation at a machine/work center for the entire batch/lot in the order. It is the machine time for one piece multiplied by the lot size. It does not include set up time.

See **Cycle time; Manufacturing cell design; Setup reduction.**

SAFETY LEAD TIME

Lead time is the time to procure or manufacture an item. In material requirements planning, lead time for all items covered by MRP is part of the MRP database. Due to unforeseen events, it may take more time than the planned lead time to purchase or make an item.

When safety lead time is employed, shop orders or purchase orders are released and scheduled to arrive one or more periods earlier to allow for potential delays. Consequently, purchase orders and production orders are released earlier and are the MRP logic would require. Safety lead time can be used in practice with safety stocks use safety lead time when there is an uncertainty in the timing of delivery from a vendor who misses delivery dates.

See **Inventory flow analysis; MRP; Order release; Safety stocks: Luxury or necessity; Safety time; Schedule stability; Synchronous manufacturing using buffers.**

SAFETY STOCK

See **Buffer stock; Inventory flow analysis; MRP; Order release; Safety lead time; Safety stocks: Luxury or necessity; Safety time; Schedule stability.**

SAFETY STOCKS: LUXURY OR NECESSITY

R. Nat Natarajan

Tennessee Technological University, Cookesville, USA

DESCRIPTION: Safety stocks (also called buffer stocks) are maintained by organizations to cope with demand and supply uncertainties. In retail, wholesale, and distribution environments, they are called safety stocks. People in the factory tend to call them buffer stocks because they are the buffer between one operation and the next (Schonberger and Knod, 1997). In this article these terms are used interchangeably. Safety stock is defined as the average level of the net stock just before replenishment arrives. A positive safety stock provides a cushion or buffer against a larger than average demand or disruptions in supply during the effective replenishment lead time. The level of safety stock depends on how quickly the item can be re-supplied when there is a stockout, and on the level of demand (Silver and Peterson, 1985). Thus, safety stock is analogous to an insurance policy. Like in any other form of insurance, there is a price to be paid. In the case of safety stocks, this price consists of the direct and the hidden costs of carrying the inventory or maintaining the level of customer service.

HISTORICAL PERSPECTIVE: Management of safety stocks is closely related to the management of inventories in general. Though inventory-related problems have existed in industry for a long time, it was only in the early part of this century that analytical methods were used to answer the basic questions of inventory management – when and how much of an item should be ordered. Soon attention turned to how to deal with uncertain demand. This led to the development of inventory policies for spare parts in the military and in businesses. The critical questions were, what is the correct level of inventory and where it should be stored, given the conflicting goals of meeting budgets and having items available when needed. Prompted by the urgency of the second world war, such issues were soon being addressed at the practical as well as research levels. By the 1960s, considerable advances were made on both fronts. A large number of analytical techniques, which take into account the different types of costs and service levels, were developed to establish safety stock policies (Arrow, Karlin, and Scarf, 1958; Hadley and Whitin, 1963). These techniques were applied in both military and nonmilitary environments, and their application resulted in significant benefits (Buchan and Koenigsberg, 1963; Brown, 1967).

Since then, impressive progress has been made in identifying the sources of uncertainties, the use of statistical techniques for quantifying the uncertainties, and adjusting the level of safety stock for those uncertainties (Silver and Peterson, 1985). The use of computers for inventory control made the application of these techniques easier and more widespread. Practitioners now have access to methods which provide the conceptual and analytical underpinnings and guidelines for safety stock decisions, as well as the technological tools for implementation. Since the 1980s, many companies have also developed strategies such as Just-in-Time (JIT) and Total Quality Management (TQM). These strategies are proactive rather than adaptive in identifying and eliminating the root causes of uncertainty. In recent years, information and communication technologies have been put to innovative uses to incorporate customer- and supplier-related information in inventory decisions, both to reduce uncertainty and to substitute information for safety stocks.

STRATEGIC PERSPECTIVE: Safety stock policies have strategic implications. The customer service level, provided by a given level of safety stocks, affects customer satisfaction, customer retention, and consequently market share, while direct and hidden costs of carrying the safety stocks affect the cost side of the bottom line. For example, a lot of money gets locked up in safety stocks of critical service parts in the military, and in heavy machinery, construction, agricultural equipment, airlines, and electric power industries all over the world.

In a plant, the location and quantity of buffer stocks can affect the lead time for production. Safety stocks are an important consideration in the design of distribution systems – decentralized or centralized system of distribution may have very different impacts on safety stocks and service levels. Many of the strategies firms pursue, together with the systems and technologies they use, affect safety stocks. These strategies and systems include partnerships with suppliers, JIT in production and supply arrangements, market orientation – make-to-stock or make-to-order, systems for forecasting, materials and distribution planning, and technologies like Electronic Data Interchange (EDI). The following initiatives, which are part of product and process design strategies, all have an impact on safety stocks: increasing the commonality of parts in the product, minimizing the number of parts, reducing the number of levels in the bill of materials, and improving the capability of manufacturing processes to meet the specifications. Also, to a certain degree, changes in customers' preferences, and short product life cycles also affect safety stocks.

IMPLEMENTATION – ESTABLISHING SAFETY STOCK LEVELS: What is the "right" amount of safety stock? Answering this question is akin to answering what is the right amount of insurance to buy. It depends on the risks that safety stock will cover and what it will cost. To figure out the numerical value of safety stock, one needs to 1) understand the sources of uncertainty, 2) measure or quantify the uncertainty, and 3) establish and define criteria, such as customer service level or cost consequences.

Sources of Uncertainty: There are two major sources of uncertainty. On the demand side, uncertainty can be due to forecasting errors, uncertainty in customer orders with respect to the quantity, timing, and the type of the item in the order. On the supply side, which consists of the internal production and the external supply system, uncertainty can be due to variations in the production and/or supply lead times, the quality and quantity of the items that are produced and/or purchased, equipment breakdown, strikes, and absenteeism.

Measurement of Uncertainty: Statistical concepts provide us with various measures of uncertainty. For instance, in the case of supply or demand, the variations in quantity or timing can be quantified by the range or standard deviation of the distribution of the quantity or lead times respectively. If the demand is being forecast, then the measure of forecast error – such as mean absolute deviation (MAD), which is related to the standard deviation of forecast errors – can be used to quantify the uncertainty. Similar measures can be used for measuring variations in product or process quality if the distribution of the quality characteristic relative to the specification limits is available. The key is the data pertaining to the underlying distributions. If such data are not available, one has to use judgmental factors in estimating the uncertainty.

Service Levels and Cost Factors: Several measures of *customer service* can be defined, e.g., ordering periods without a stock out, number of stock outs in a month or a year, and idle time due to material or component shortages. The relevance and appropriateness of each measure depend on the context (Fogarty, Blackstone, and Hoffman, 1991). Likewise, the *customer* could be: the ultimate consumer at the end of a long distribution chain, the intermediate customer that might be a distribution center, or the immediate customer that might be a factory operator in a plant demanding parts and raw materials from manufacturing inventories. Two commonly used measures of customer service are: 1) specifying the probability of not meeting the demand or the required quantity of supply in a replenishment cycle without considering the number of units short; and 2) specifying the percentage of units of an item that can be filled from available supply. This measure, also called *fill rate*, is related to the expected number of units that are short in each ordering cycle.

The cost factors relevant to the determination of safety stock levels include the following: 1) direct and hidden costs of carrying safety stocks, 2) fixed costs of stockouts, and 3) variable costs of stockouts, based on their amount and duration. In addition, the cost of ordering/setup for the item could also be relevant. The costs of consequences, due to stockouts such as the loss of current and future sales, loss of goodwill, expediting, idle machinery or workers, are very difficult to estimate.

Implementing safety stock policies requires relevant data in the form of forecasts of demand and lead time, standard deviations and measures of forecast errors, order quantity, service levels, and the various cost elements for each item. It also presumes a sound forecasting and inventory management system and practices.

SIGNIFICANT ANALYTICAL MODELS: The analytical approaches for determining safety stock levels will be discussed for two types of demand – independent and dependent.

Independent Demand

The following types of items have independent demand: finished goods in the factory or warehouses, maintenance, repair and operating supplies (MRO), and service parts. These items are not directly used in the manufacture of products and their demand is unrelated to the demand for other items. In a make-to-stock environment, the demand for such items is forecast.

If the inventory of these items are controlled by a lot size-reorder point (Q, R) model, where Q is the fixed order quantity and R is the reorder point, then safety stock S is related to R, which determines the timing of inventory replenishment. Specifically, R and S are related as $R = \overline{d_L} + S$, where $\overline{d_L}$ is the expected demand during the replenishment lead time L.

As shown in Figure 1, there is no variability in the lead time – this assumption will be relaxed later. What happens to demand *during the replenishment lead time* is critical because there will be a stockout if demand exceeds the amount on hand. Therefore, safety stock is expressed as $S = k\sigma_L$, where σ_L is the standard deviation of the demand during the lead time (or its estimate σ_L) and k is known as the safety factor. The value of k can be determined if the probability of stockout during the ordering cycle and the statistical distribution of demand (or the errors in the forecasts of demand) during the lead time are specified.

There are two approaches to the determination of k. In one approach, service level, such as probability of stockout in a ordering cycle P, is specified. For instance, if P is 5 percent and d_L is normally distributed, then from the normal curve area tables, k equals the z value of 1.64. Similarly, if fill rate is given, z value can be found by first determining the unit expected quantity short $E(z)$ for the normal distribution and then using the table that relates z and $E(z)$. This procedure is illustrated in a later section of the article. The other approach optimizes the total expected cost. This method requires estimating the costs of placing orders or setups, carrying costs, and shortage costs. The probability of stockout and consequently the value of k can be determined as a condition for minimizing the total expected cost.

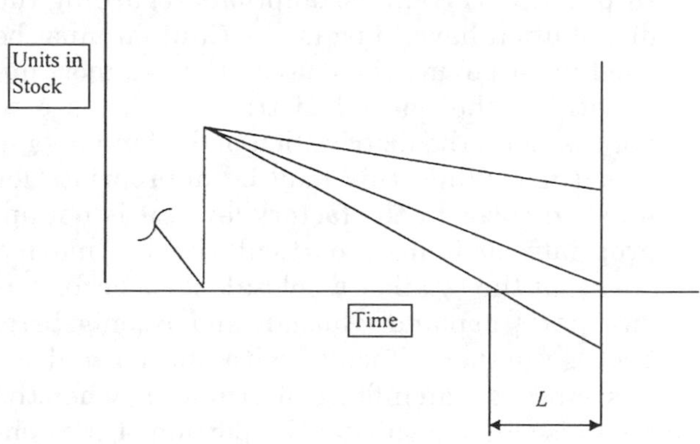

Fig. 1. Varying demand and constant lead time.

If the fixed-time period or periodic review inventory control system is used, the inventory status is known only periodically, e.g., every week or month. This system requires a higher level of safety stocks compared to the (Q, R) system. The longer the review interval T, the greater the uncertainty. In the (Q, R) system, the inventory status is known continually and the order is placed when reorder point is reached. In the fixed-time period model, the order for replenishment can be placed only at the time of review.

It is possible that a large demand can draw the inventory level down to zero immediately after an order is placed, and this stockout would not be noticed until the next review. Then, another order will be placed, which will take lead time L time units to arrive. Thus the item can be stocked out during the review interval T and lead time L, and safety stock must be used to protect against stockouts during $T+L$ time units. Hence, the standard deviation of demand during $T+L$ (not just L, as in the (Q, R) model) must be taken into account in the computation of safety stock.

Statistical Distributions: The general method outlined above applies in the case of any statistical distribution of the demand during lead time.

In practice, certain assumptions regarding the distribution have to be made. Caution must be used in specifying the distribution. In most applications, the normal distribution is used to approximate the distribution of demand during lead time. While this may be appropriate for finished goods at the factory level, it is not appropriate for items – particularly slow moving ones – at the retail or wholesale level or for service parts demand (Buchan and Koenigsberg, 1963). Another difficulty with the normal exists: since it is an infinite distribution, when the mean is small, a substantial portion of the density curve extends into the negative half line. For slow moving items, Poisson distribution is appropriate to use if the actually observed standard deviation is quite close to the square root of the observed mean. The demand uncertainty can also be modeled by a number of other distributions, such as exponential, gamma, logarithmic, pseudo-exponential, and negative binomial (Silver and Peterson, 1985). For the normal distribution, the standard deviation can be approximated as 1.25 times MAD of forecast errors. Most forecasting systems track MAD as a measure of forecast accuracy.

Interaction with Lot Size: Lot size can affect safety stock levels. For instance, the service level is specified in terms of number of stockouts tolerated per year. The number of ordering cycles in a year equals D/Q, where D is the forecast of annual demand and Q is the lot size. Therefore, larger lot sizes imply fewer ordering cycles during the year and hence less exposure to stockouts. Thus, larger lot sizes give protection against stockouts. In fact, there is a trade-off: the bigger the lot size, the smaller the safety stock. In some optimizing models, safety stock – through reorder point calculations – can also affect lot size. Therefore, they have to be determined jointly. For both the service level and the optimizing approaches for calculating safety stock, the exact relationship between lot size and safety stock can be analyzed (Natarajan and Goyal, 1994).

Lead Time Variability: Variability in lead time is quite common in practice. Increased safety stock is required to cope with this additional uncertainty. Generally lead time and demand are correlated. The correlation could be positive: high demand is likely to be associated with longer lead time because of heavy workload placed on the supplier; or it could be negative: low demand can be associated with long lead times because the supplier has to wait longer to accumulate orders to achieve a sufficient run size for the item (Silver and Peterson, 1985). Also, if the successive lead times are assumed to be independent, then the possibility of order crossing (i.e., an order placed later arrives before the one placed earlier) exists. This can be ruled out in the case of a single supplier. However, incorporating these realities into safety stock calculations can be analytically quite complicated. If we assume that lead time and demand are statistically independent and there is no order crossing, then the expected value of demand ($\bar{d}_L$) and the standard deviation of demand (σ_L) during the variable lead time are given by $\bar{L} \times \bar{D}$ and $\sqrt{\bar{L} \times \sigma^2 + \bar{d}^2 \times Var(L)}$, respectively. Here, $\bar{L}$ is the expected lead time, $\bar{d}$ is the expected demand per unit time, and σ is the standard deviation of demand per unit time.

There are two other operationally simpler methods that can be used to deal with lead time variability (Fogarty, Blackstone, and Hoffman, 1991). For most statistical distributions, the maximum of demand and lead time could be infinite. However, actual analysis of past data will reveal that they rarely exceed a finite maximum value denoted as d_m and L_m respectively. Then safety stock can be calculated as $L_m \times d_m - \bar{L} \times \bar{d}$. The other method is to simply record the distribution of the actual joint variation of demand and lead time (see Figure 2). The standard deviation of this empirical distribution can be calculated and used as the estimate of σ_L. An advantage of this method is that it can be used regardless of whether L and d are statistically independent.

Illustrative Example: Suppose the fill rate for a product is 99 percent (fraction short $= .01$). The latest forecast for weekly demand is 200 units, which, given a 50-week year, would translate to 10,000 units annually. The forecast errors are normally distributed and the latest value of MAD is 20. The lead time for the item has a mean of 2.5 weeks with a standard deviation of 0.5 week. The order quantity for the item is 500 units. Using these data, safety stock and reorder point can be calculated.

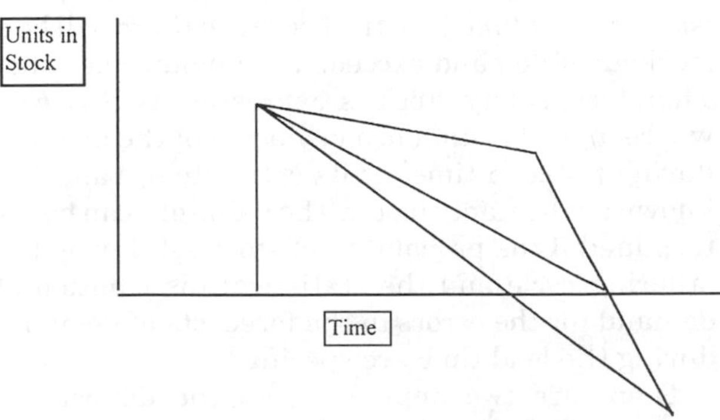

Fig. 2. Varying demand and varying lead time.

First, we calculate the standard deviation of demand during forecast interval as $25 = 1.25 \times$ MAD. Then σ_L is given by $\sqrt{2.5 \times 25^2 + 40{,}000 \times 0.5^2} \approx 107.53$.

To compute safety factor k we use the following relation:

Annual quantity short
= Fraction short per year (0.01)
× Annual demand (10,000)
= Expected quantity short per order ($E(k)$
× 107.53) × Number of orders per year
(10,000/500).

The expected quantity short per order is given by $E(k) \times \sigma_L$, where $E(k)$ is the partial expectation function of k, the safety factor (Brown, 1967). In this case $E(k)$, which is actually $E(z)$ because of normal distribution, works out to be 0.0465. Using the tables, originally provided by R.G. Brown but now widely available in the books on inventory management, we get $z = 1.29$ (Brown, 1967). Hence safety stock $= 1.29 \times 107.53 \approx 139$. The reorder point is $R = \bar{L} \times \bar{d} +$ safety stock $= 2.5 \times 200 + 139 = 639$.

This example illustrates two important points made earlier: 1) Other things remaining the same, for the same fill rate, as order quantity increases, the value of $E(z)$ will increase. This means that corresponding value of z will decrease because z and $E(z)$ are inversely related. Consequently, the safety stock also decreases. For example, if the order quantity is increased to 1,000 units, then $E(z)$ becomes 0.093. This implies a z value of 0.94 and the safety stock is now $0.94 \times 107.53 \approx 101$ units. 2) If there is no variability in lead time, other things remaining the same, σ_L will become $\sqrt{2.5 \times 25^2} = 39.53$. New $E(z)$ and z values are 0.1265 and 0.77 respectively, and safety stock decreases to approximately 31. Another interesting observation is that, in this model, for the given values of the parameters, certain fill rates (e.g., 90 percent) could be supported with negative safety stock (approximately – 14), i.e., reorder point ($R = 500 - 14 = 486$) is less than the expected demand (500) during lead time.

Dependent Demand and Multi-echelon Systems

Dependent demand is directly related to or derived from the bill of material structure for other items or end products. Because these demands are calculated and not forecasted, it may appear that there is no need to maintain safety stocks of these items. However, material requirements planning (MRP) systems for such items have to deal with uncertainty. Here, the two basic sources of uncertainty, supply and demand, can be further classified into uncertainty related to quantity and timing.

Demand quantity uncertainty for a dependent demand item occurs when the requirements fluctuate around a mean value. This could be due to: changes in master production schedule (MPS) to accommodate changes in forecasts or customer requirements; variations in inventory levels; and changes on higher level items which use this item. Supply quantity uncertainty is due to variations in scrap, yield, or quantity delivered by the outside suppliers, which lead to either shortage of lower level items, or production/supply overruns. In either case, the actual quantity varies around planned receipts. Simulation studies indicate that safety stocks are effective in buffering quantity uncertainty while safety lead times – releasing orders earlier than the planned release dates indicated by MRP calculations – is effective in buffering timing uncertainty (Whybark and Williams, 1976).

In principle, the statistical methods outlined earlier can be used to compute safety stock. However, to the extent it is management actions, such as frequently revising the MPS, that are the sources of quantity uncertainty, it becomes difficult to apply statistical measures to quantify the uncertainty. Another important issue is at what level(s) safety stocks should be held in a multi-stage manufacturing system. There is no commonly agreed answer to this.

Generally, safety stock is planned at the MPS level, so that it forms matched sets as it is exploded into the component orders through the MRP process. However, if there is supply-related uncertainty in terms of quantity due to spoilage or unreliable suppliers, safety stock can be included at the *item level* rather than the MPS level. The same would hold when an item is used to satisfy independent demand, e.g., as a service part.

In the production process itself, buffer inventory of dependent demand items is held at strategic points (e.g., before bottlenecks) to protect the throughput of an operation or schedule against the negative effects caused by statistical fluctuations (Fogarty, Blackstone, and Hoffman, 1991). It is typically expressed as a time buffer. Its size should be large enough so that the bottleneck does not run out of work. It can be calculated by examining the variation in each operation and simulating the production sequence. In practice, a trial and error approach based on analysis and experience is used to arrive at the final buffer size.

Safety stocks are also used in the distribution system. The orders placed by the ultimate customers constitute independent demand but they in turn generate a dependent demand for the intermediate stocking points in the distribution system. As in the case of MRP systems, safety stock is more appropriate in the case of quantity-related uncertainty. But the more important issue is *where* in the distribution system safety stocks should be carried? It can be argued that safety stocks should be carried only at the location closest to the customer or at an intermediate location where uncertainty in independent demand exists. This would imply no safety stock in locations where demand is dependent (Vollman, Berry, and Whybark, 1997). However, safety stock amount can be reduced if all of it is held in a central warehouse where uncertainty is aggregated (Brown, 1967).

For example, if a company carrying a safety stock of 500 units in a central warehouse decides to carry the safety stock in four field warehouses, each supplying equivalent part of demand, then for the same level of service, the safety stock at each location is now: Safety Stock for One Location $\div \sqrt{N}$ (where N is the number of locations) $= 500 \div \sqrt{4} = 250$ units. Total units in safety stock is now increased to $4 \times 250 = 1{,}000$. An alternative might be to split the safety stock between the central and local warehouses.

Simulation studies indicate that keeping the safety stocks at the local field warehouse provides higher level of service than the same amount of total safety stock carried at the central warehouse (Vollman, Berry, and Whybark, 1997). Such practice also provides much faster response and adjustment to local conditions. However, there are exceptions to this finding – demand for a service part can arise very infrequently but from practically any part of the world where the product has been in use for some time. It is uneconomical to operate a local warehouse to meet those occasional needs on an individual basis. This again highlights the issue of trade-off between cost and customer service in the location of safety stock in the distribution system. In general, the determination of appropriate safety stocks in a multi-stage manufacturing and distribution system is analytically very complex because of the interaction between various factors, such as the numerous locations for uncertainty, the opportunities for risk pooling, and the manner in which the uncertainty is transmitted in the system (Inderfurth, 1994).

EFFECTS: Experiences of a number of organizations indicate that statistical approaches to determining safety stock policies have resulted in significant improvements in costs and service levels. These approaches are superior to setting safety stocks based on ad hoc rules of thumb such as percent of order quantity or days of supply (Plossl, 1985). Inexpensive and widely available computing power and user-friendly software packages for forecasting and statistical analysis have enabled their application to thousands of items. Yet, while using the above-mentioned tools, one must not forget to incorporate human judgment. For example, to begin with, the items can be classified, and their criticality established, based on multi-criteria ABC analysis (Vollman, Berry, and Whybark, 1997).

In implementation, it is better to try these models on a sample of A type critical items first and then on all A items before extending them to B items. For items with low criticality and value, low safety stock can be maintained and up-dated every six months or a year. For most non-critical C items, the application of these methods is not worthwhile – large size of the orders can provide necessary protection.

A group of items (e.g., provided by the same supplier) can be ordered jointly, with an *aggregate reorder point*, expressed in monetary terms as the total value of order points of each item, and *a group service level*, e.g., total number of stockouts for all the items in the group (Vollman, Berry, and Whybark, 1997; Fogarty, Blackstone, and Hoffman, 1991). In general, joint ordering reduces safety stock requirements because an order typically is placed before all items – except one – reach their reorder point levels. Such joint ordering policies have had significant impact on costs and service levels (Kleijnen and Rens, 1978). In each case, before full-scale implementation, Monte Carlo and other methods of simulation can be used to understand, test and validate the model and what-if-analysis performed to test the sensitivity of the model parameters. Based on safety stock calculations for individual items, exchange curves can be derived (see Figure 3) to show the relationship

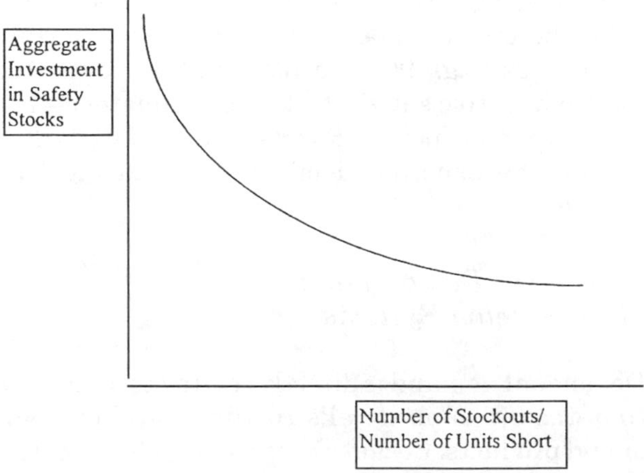

Fig. 3. Aggregate exchange curve.

between aggregate safety stocks (expressed in monetary value) and service level. These exchange curves are very useful to top management for setting overall service levels, and for deciding on how much should be invested in safety stocks (Plossl, 1985). Materials and inventory managers can then decide how to distribute this investment across various products.

COLLECTIVE WISDOM:

Practical Considerations

Understanding the assumptions behind the models is important. For instance, Poisson distribution can be used to describe the number of orders but not the number of units demanded during the lead-time. The average number of units per order must be estimated and checked for constancy across orders before safety stock can be computed (Plossl, 1985). Also, order arrivals must be random and independent. This will not be true if orders are accumulated in batches and then released.

Often lead-time and demand tend to be correlated – lead times increase when the demand is increasing and vice versa – but the formula for σ_L involving variable lead times assumes otherwise. Another important practical concern is the updating of safety stock calculations. Safety stocks have to adjust to changing demand conditions such as seasonal and or trend variations in future demand. For instance, as products approach the end of their life cycle, a fixed safety stock will end up as dead inventory unless it is revised downwards in proportion to the downward trend. Bias in forecasts tends to increase the estimate of standard deviation and therefore overstates the safety stock. Also, if forecasts are consistently optimistic, i.e., biased, then the actual demand will not consume even the replenishment quantity based on the forecast, let alone the inflated safety stock. Hence, along with MAD, forecast error tracking signal should be used to monitor the bias. Then, safety stocks can be modified accordingly. Methods, which address these issues, have been suggested to make safety stocks "flex" with changing conditions (Krupp, 1997).

Nevertheless, just because these methods are available does not mean that they must be used. Their application may be justified only when the demand environment is highly unstable. Here, it is also important to distinguish between variability that is natural and inherent in market demand due to factors such as seasonal and trend patterns, and the variability that is induced by management actions. For example, in an end-of-month push, sales are increased to meet targets. On the flip side, frequent revisions to safety stock amounts can lead to changes in MPS, causing nervousness in downstream replenishment schedules. The implication is that careful managerial judgments are necessary to strike proper balance.

Even understanding the basic concepts without detailed knowledge of the underlying theory of these models will provide very useful insights to practitioners. For example, for many high sales volume items, demand tends to be stable. Hence the required safety stocks will be lower than for low volume items whose demand tends to be more volatile – the maximum demand for such items may be 3 to 4 times the average demand. For the same reason, safety stock requirements invariably go up when product variety is increased by expanding the number of models and options, but the sales volume for each one is lower.

Two items X and Y may have the same average demand, but if X is sold to fewer number of customers than Y, it is likely to have more variability in demand and consequently higher level of safety stock than Y. These insights can be utilized to develop judgment-based safety stock policies in situations where statistical modeling may not be accurate. Such situations can arise with products having many options, volatile demand, long manufacturing lead times – which increase uncertainty – and with new products, which do not have much sales history (Potamianos and Orman, 1996). Later, as more demand data become available, formal models can be used.

Mentioning safety stocks in the context of Just-in-Time appears contrary to the principles of JIT. Under JIT, buffer stocks are reduced so that the underlying sources of uncertainty can be exposed and solutions can be found. However, as progress is being made towards that objective, safety stocks may still be needed. Also, under JIT, lot sizes are reduced, which in turn reduces the protection larger lots offered against stockouts. This increases the emphasis on buffer stocks in JIT environments (Natarajan and Goyal, 1994). However, buffer stock, like Work-in-Process (WIP) inventory, increases throughput time. One approach is to maintain buffer stocks but keep them off line (Schonberger and Knod, 1997). Because they are taken out of regular production stream and WIP, off-line buffer stocks do not inflate throughput time and avoid premium space, handling, and administrative expenses of WIP.

Effects of Information Technology and Business Practices

Both technology and changing business practices are having a significant impact on safety stock policies. Depending on circumstances, they can lead to either an increase or a decrease in safety stocks. To analyze their effects, one must understand how they affect uncertainty. For example, reducing lot sizes in JIT increases stockout risk and hence safety stocks. However, JIT can also reduce lead times, which means decrease in forecast horizon and errors. Same is true with the quick response (QR) systems now commonly used in apparel-fiber-textile and auto industry in the U.S. In QR, all units in the supply chain are electronically linked and downstream demand information collected from point-of-sales terminals in retail stores is simultaneously available to all upstream units. In vendor-managed inventory (VMI) system – which is an advanced version of QR – retailers share sales and inventory information almost on real time basis with the producers, so production and replenishment can be based on the most recent data. In all these cases, information technology (IT) is being used to reduce uncertainty and trade information for inventory. As electronic commerce becomes more prevalent, these practices are likely to spread to other industries and worldwide.

Practices associated with Total Quality Management (TQM), which focuses on proactively reducing variability in processes and cycle times, will tend to reduce safety stocks. On the other hand, global sourcing could mean longer supply lead-time, which increases variability. If global markets are served by local instead of central production and distribution units, then, as mentioned earlier, total safety stocks will increase. However, if these units are able to predict the local demand conditions more accurately because of their closeness to customers, this increase will be mitigated. With businesses placing more emphasis on managing inventories in the entire supply chain, the issue *who* should carry safety stocks becomes critical. Logic would dictate that safety stocks should be carried to counteract uncertainties. However, with JIT supply arrangements, customer firms often shift the burden of carrying inventories, including safety stocks, to their suppliers.

In order to cope with uncertainty and increase flexibility, manufacturers are increasingly using other means, such as safety capacity and safety lead times, in lieu of, or in conjunction with safety stocks. Fragmentation of markets, customization, and short product life cycles will tend to increase the safety stock for individual products. At the same time, this tendency is counteracted by the trend towards make-to-order production, linking the shop floor with the customers, and postponing the differentiation of end products until the latest possible moment. Many such trends are concurrently in effect and are likely to continue. The upshot of all this is that management of safety stocks will be more of a challenge in the years to come. On the positive side, organizations can turn to information technology and analytical tools to meet these challenges.

See **ABC analysis; Buffer stock; Electronic data interchange (EDI); Fill rate; Fixed-quantity inventory model (Q, R model); Fixed-time period inventory model; Information technology (IT); Inventory floor model; Just-in-time manufacturing (JIT); Lean manufacturing implementation; Master production schedule (MPS); Materials requirements planning (MRP); Mean absolute deviation (MAD); Monte Carlo simulation; Multiple criteria ABC analysis; Normal distribution; Poisson distribution; Quick response (QR); Reorder point; Safety stock; Service level; Setup reduction; Simulation; Simulation analysis of manufacturing and logistics systems; Simulation of production problems using spreadsheet programs; Standard deviation; Theory of constraints in manufacturing management; Total quality management (TQM); Tracking signal: Vendor-managed inventory (VMI) system.**

References

[1] Arrow, K.J., S. Karlin, and H.E. Scarf, Editors. (1958). *Studies in the Mathematical Theory of Inventory Management*. Stanford University Press, California.

[2] Brown, R.G. (1967). *Decision Rules for Inventory Management*. Holt, Rinehart and Winston, New York. 2, 3, and 5.

[3] Buchan, J., and E. Koenigsberg (1963). *Scientific Inventory Management*. Prentice-Hall, New Jersey. 1–16, and 19.

[4] Fogarty, D.W., J.H. Blacksotne, and T.R. Hoffman (1991). *Production and Inventory Management*. South-Western, Ohio. 3, 6–8, 17, 18, and 23.

[5] Hadley, G., and T.M. Whitin (1963). *Analysis of Inventory Systems*. Prentice-Hall, New Jersey. 1, 3–5, and 9.

[6] Inderfurth, K. (1994). "Safety Stocks in Multistage Divergent Inventory Systems: A Survey." *International Journal of Production Economics*, 35 321–329.

[7] Kleijnen, J.P.C., and P.J. Rens (1978). "Impact Revisited: A Critical Analysis of IBM's Inventory Package." *Production and Inventory Management*, 19(1) (First Quarter), 71–90.

[8] Krupp, J.A.G. (1997). "Safety Stock Management." *Production and Inventory Management*, 38(3) (Third Quarter), 11–18.

[9] Natarajan, "Nat" R., and S. Goyal (1994). "Safety Stocks in JIT Environments." *International*

Journal of Operations and Production Management, 14(10), 64–71.

[10] Plossl, G.W. (1985). *Production and Inventory Management: Principles and Techniques*, (2nd Edition), Prentice-Hall, New Jersey. 96–125, 200–212.

[11] Potamianos, J., and A.J. Orman (1996). "An Interactive Dynamic Inventory-Production Control System." *Journal of the Operational Research Society*, 47(8), 1016–1028.

[12] Schonberger, R.J., and E.M. Knod Jr. (1997). *Operations Management: Customer Focused Principles*, 6th Edition. Irwin, Illinois. 5, 8, and 10.

[13] Silver, E.A., and R. Peterson (1985). *Decision Systems for Inventory Management and Production Planning*, (2nd Edition), John-Wiley, New York. 3, 4, 7–12, and 17.

[14] Vollman, T.E., W.L. Berry, and D.C. Whybark (1997). *Manufacturing Planning and Control Systems*, (4th Edition), Irwin/McGraw-Hill, New York. 11, 17, and 18.

[15] Whybark, D.C., and J.G. Williams (1976). "Material Requirements Planning Under Uncertainty." *Decision Sciences*, 7(4) (October 1976), 595–606.

SAFETY TIME

In MRP systems, safety time is an element of time added to normal lead time to protect against fluctuations in lead time so that an order can be completed on or before it is actually needed. When used in MRP systems, both order release and order completion will be planned for earlier dates than they otherwise would have been.

In synchronous manufacturing systems, the safety time concept is expressed in terms of time buffers. Time buffers are strategically designed to protect system throughput from the disruptions that occur in any manufacturing process. The different categories of time buffers include constraint, assembly, shipping, and stock replenishment time buffers. Constraint time buffers protect constraint resources from starvation and maintain the integrity of the constraint resource's production schedule. Assembly time buffers are applied to non-constraint parts, and ensure that assembly operations can be performed as soon as – processed parts are available from a constraint resource. Shipping time buffers ensure that customer orders are shipped according to schedule. Stock replenishment time buffers are used to ensure the timely flow of material to internal inventory stock buffers.

See **Synchronous manufacturing using buffers; Theory of constraints (TOC) in manufacturing management.**

References

[1] Cox, J.F., J.H. Blackstone, and M.S. Spencer (eds.) (1995). *APICS Dictionary*: Eighth Edition, American Production and Inventory Control Society, Falls Church, Virginia, 75.

[2] Srikanth, M.L. and M.M. Umble (1997). *Synchronous Management: Profit-Based Manufacturing For The 21st Century, Volume One*, Spectrum Publishing, Guilford, Connecticut, 223.

SALES AND OPERATIONS PLAN

One can logically view production planning activities as beginning with a broad, general plan developed at the upper management levels of an organization, then proceeding to lower levels as the plan is refined into more specific detail. Historically, plans developed at the upper management level have included a sales plan and a production plan, both of which were often developed with very little coordination between the two functional areas. This resulted in a mismatch between what the sales area envisioned it would be selling, and what the operations function planned to produce. Even worse, both plans may have been unrelated to the organization's strategy.

To overcome this problem, many companies have moved to a sales and operations plan that is developed jointly by the two areas. This plan is still developed by upper management and is usually in broad, general terms, often stated in monetary units. However, its main feature is that the plan is agreed upon between the two functional areas so that the operations area plans to produce what the sales area is planning to sell. Further, the plan is consistent with the organization's overall competitive strategy.

See **Aggregate plan and master production schedule linkage; Aggregate planning; Capacity planning in make-to-order production; Capacity planning: Long-range.**

Reference

[1] Sari, J.F. (1989). "Why Don't We Call It Sales and Operations Planning?" *APICS 29th Annual International Conference Proceedings*. American Production and Inventory Control Society, Virginia.

SAMPLING FREQUENCY

See **Sampling in quality control; Statistical process control using control charts.**

SAMPLING IN QUALITY CONTROL

The purpose of quality control is to make sure that processes are performing in an acceptable manner and that the quality of inputs are acceptable. This may be accomplished by monitoring process output using statistical techniques. Sampling in quality control can thus be used for quality assurance. Quality assurance that relies primarily on inspection after production is called acceptance sampling. Acceptance sampling involves testing a random sample of items from a lot and deciding whether to accept or reject the entire lot based on the quality of the random sample. In most cases, acceptance sampling is concerned with incoming purchased parts or final products awaiting shipment to warehouses or customers. Quality assurance efforts that occur during production include statistical process control. Statistical process control involves testing a random sample of output from a process to determine whether the process is producing items within a predetermined acceptable range. When the tested output falls outside the range, this is a signal to adjust the production process to get the process back into the acceptable range.

See **Integrated quality control; Lean manufacturing implementation; Quality: The implications of W. Edwards Deming's approach; Statistical quality control using control charts; Total quality management.**

References

[1] Chase, Richard B. and Nicholas J. Aquilano (1992). *Production and Operations Management: A Life Cycle Approach*, Irwin, Homewood, IL.

[2] Stevenson, William J. (1996). *Production/Operations Management*, Irwin, Homewood, IL.

SASHIMI APPROACH TO PRODUCT DESIGN

"Concurrent engineering," is defined by the Institute for Defense Analysis (IDA) as: "... the systematic approach to the integrated concurrent design of products and related processes including manufacture and support. This approach is to cause the developers, from the outset, to consider all the elements of product life-cycle from conception through disposal including quality, cost, schedule and user requirements."

The above described "sashimi" approach to new product development is in contrast to the "sequential approach", which was developed originally by NASA for large projects for the space program in the 50's and 60's. The disadvantage of sequential systems is that a bottleneck in any one phase can slow down the entire development project, thereby increasing development leadtime. Many Japanese companies such as Fuji, Canon, and Honda began using a new type of system in which activities take place concurrently.

Concurrent engineering involves having events in the product development cycle scheduled in parallel at the interface of different phases (see Handfield, 1994). The benefits of concurrent engineering at the Fuji corporation included faster speed of development, increased flexibility, and greater sharing of information. This system (which has been referred to as "sashimi" development by Fuji, because it resembles the manner in which slices of raw fish overlap one another on a plate) also increases shared responsibility of project members. Members of sales, R&D, manufacturing departments and suppliers all work together as a team to solve the problems occurring in new product design. Some of the disadvantages of the sashimi system include the amplification of ambiguity, tension and conflict in the group.

If we compare the "sashimi" and "sequential" approaches, it is apparent that the former involves the compression of development leadtimes. In order to design a product without defects, all of the stages in product development must be carried out. While the total time (in manhours) spent on development are equal using both approaches, the "sashimi" system compresses development time through overtime and parallel development activities. In addition, concurrent approaches often require that the organization compress its learning into a shorter time period. This means that product development team members must learn to work more as a team, rather than as functions separated by "walls."

North American firms have begun to use concurrent engineering processes more in the last five to ten years. The primary attribute associated with concurrent engineering is scheduling parallel activities, in which functions transfer information in parallel rather than serially (analogous to a "rugby" game as compared to a relay race). Parallel scheduling typically involves the sharing of partial information at critical design interfaces. By sharing "imperfect" information at various nodes, activities which follow can be started before preceding ones are fully complete. Managers must learn how to fully use all available information, consider all of the possibilities, and make decisions on shorter notice.

See **Concurrent engineering; New product development through supplier integration; Product development and concurrent engineering.**

Reference

[1] Handfield, R.B. 1994. "Effects of Concurrent Engineering on Make-to-Order Products", *IEEE Transactions on Engineering Management*, 41(4): 1–11.

SCALE ECONOMICS

A pure economic definition of scale economies would perhaps state, "The lowering of cost per unit as output increases through the allocation of fixed costs over the larger number of units produced." A broader approach to scale economies would incorporate the notion of the learning curve, which also relates production volume to cost. In learning curves, the assumption is that as additional units are produced, a certain amount of learning occurs such that each subsequent unit takes less time to produce.

In a learning curve, assuming a 90 percent learning rate, when the first unit produced requires 100 hour, the second unit requires 90 percent of 100 or 90 hours. At the 90% learning rate each successive doubling of the cumulative production yields a 10 percent improvement. Thus, the fourth unit produced requires 90 percent of the time required for the second unit (90 hours) – $.90 \times 90 = 81$ hours. Further, the eighth unit produced requires 90 percent of the time required for the fourth unit – 72.9 hours. See *Learning Curve Analysis* for a more detailed description.

Much learning occurs with repetitive production. Other sources of learning are: (1) the training of workers, who are responsible for production; and (2) the implementation of new technologies. Since these learning sources require up-front expenditures, they can be justified by high-volume production.

See **Economies of scale; Economies of scope; Global manufacturing rationalization; Industrial robot selection; Learning curve analysis; Manufacturing flexibility; Product-process dynamics.**

SCATTER DIAGRAM

See **Integrated quality control; Statistical process control using control charts.**

SCHEDULE NERVOUSNESS

See **MRP; MRP implementation; Order release; Schedule stability.**

SCHEDULE RECEIPTS

Scheduled receipts is an information included in MRP records. Scheduled receipts indicate when an existing replenishment order (or open orders) for an item due. In an MRP record, the Open Order row shows when to expect these orders to be completed and how much has been ordered. Necessary materials are committed to the open order. The convention in MRP is to show scheduled receipts as available at the beginning of the period to satisfy gross requirements for the period.

See **MRP; MRP implementation; Order release; Schedule stability.**

SCHEDULE STABILITY

R. Lawrence LaForge

Clemson University, South Carolina, USA

Sukran N. Kadipasaoglu

University of Houston, Texas, USA

V. Sridharan

Clemson University, South Carolina, USA

DESCRIPTION: *Schedule stability* occurs when plans for materials management and production activity control in a manufacturing facility do not vary significantly from one scheduling cycle to the next. This very desirable situation reduces the need for expediting, facilitates coordination within and among departments in the firm, and makes it easier to meet the overall manufacturing plan. By contrast, the term *schedule instability* (or *schedule nervousness*) is associated with excessive schedule changes that may result in a chaotic environment in the plant, potentially leading to reduced productivity, low morale, increased costs, and poor customer service.

STRATEGIC PERSPECTIVE: Manufacturing managers constantly face tradeoffs involving the need to adapt to changing market conditions and the need for stability in plant schedules for purchasing and production activity. Flexibility to respond to external market conditions is clearly important and, in some cases, may be critical for the firm to maintain its competitive advantage. However, excessive changes to internal plant schedules can lead to costly production delays, high material costs, reduced productivity and/or quality, and other problems that may ultimately erode the competitive position of the firm.

Schedule instability is very common in manufacturing environments involving uncertain demand for complex products with multiple bill-of-material levels representing major assemblies, subassemblies, fabricated parts, and raw materials. In addition, schedule instability can also result from uncertainty in supply due to scrap losses, production overruns, variations in vendor lead times and shop flow times, machine breakdowns, and/or tooling problems.

Problems associated with schedule instability are most often experienced in manufacturing environments that utilize material requirements planning (MRP) systems. MRP applies a "top down" logic to schedule material flows necessary to meet planned production of end items specified in a master production schedule (MPS). From the material plan developed by MRP, orders are released to production and purchasing to supply needed items to build the products on the master production schedule.

Schedule instability can occur when MRP replanning takes place. This is unfortunate because a regular cycle of replanning is essential for effective materials management and production activity control in an MRP environment. It is through replanning that MRP accounts for the realities of the market and the plant by adjusting plans and schedules to deal with changing conditions. However, these very changes may cause a multitude of problems and costly disruptions in the plant.

Fortunately, not all schedule changes from MRP replanning are caused by uncontrollable or unforeseen circumstances. To a limited extent, schedule stability can be "managed" through master scheduling and MRP practices.

IMPLEMENTATION: A simple but effective measure of schedule instability is a frequency count of the number of order changes resulting from an MRP regeneration. Variations of this measure have been proposed to express the instability index as the average number of order changes per item. In any case, fewer changes (lower instability index) represent greater stability in the schedule.

There is considerable evidence from the research literature that schedule instability is influenced by master scheduling policies and MRP implementation decisions. Master scheduling policies related to *freezing* and *end-item safety stock* have been shown to impact schedule instability. In addition, *MRP lot sizing* is also known to affect the stability of production schedules.

Attempts to reduce schedule instability must be made with full awareness of their effects on customer service and inventory costs. The following sections address management strategies to achieve schedule stability, and the tradeoffs surrounding their implementation.

MPS Freezing

Freezing the master production schedule is almost certain to reduce schedule instability. Freezing means that planned orders in the MPS are firmed and not allowed to change from one planning cycle to another. If the entire MPS is frozen, no schedule changes to any planned or open orders for assemblies, components, or materials will occur as a result of master scheduling. Even with a completely frozen MPS, however, some schedule changes could result from plant actions taken to address scrap losses, production overruns, unexpected downtime, or other problems. Since MPS changes are a major cause of instability, it can be concluded that freezing the entire MPS would usually lead to a major reduction in schedule instability.

Unfortunately, freezing the entire MPS completely eliminates the ability of the planning system to react to changes in market conditions. When demand is uncertain, freezing the entire MPS could lead to excessive inventory costs and/or loss in customer service. In most practical situations, this is not a viable option.

An alternative to freezing the entire MPS is to freeze only a portion of the master schedule. If the MPS is frozen throughout the product cumulative lead time, then no changes to *open orders* for assemblies, parts, and materials will result from master scheduling. However, freezing the cumulative lead time portion of the master schedule does allow for MPS-induced changes to *planned orders* for assemblies, parts, and materials.

Freezing less than the cumulative lead time in the MPS allows for some changes to both open and planned orders of bill-of-material items – which is our definition of schedule instability – but leaves flexibility to accommodate changes in market and plant conditions. Research suggests that longer frozen intervals lead to more stability but at the expense of customer service and inventory-related costs.

Methods Used for Freezing. One final issue related to freezing a portion of the master schedule deals with the method used to implement freezing. In practice, freezing is usually implemented by precluding changes to a certain number of periods in the master schedule. Recent research suggests, however, that there are some advantages to implementing freezing by limiting changes to a specified number of orders in the schedule rather than a specified number of periods. That is, the frozen interval could be defined by the first *n orders* in the schedule, rather than the first *n periods*. Further research is needed on the relative merits of period-based freezing and order-based freezing.

End-Item Safety Stock

Safety stock is extra inventory held to provide protection against uncertainty. It is clear that safety stock generates additional cost associated with carrying the extra buffer of inventory. It is also known that moderate amounts of safety stock can help improve customer service. What is not entirely clear is the effect that end-item safety stock has on schedule stability.

It seems logical that safety stock for master-scheduled items would help absorb unexpected fluctuations in demand, thereby reducing the need for MPS changes that cause schedule instability. To a certain extent, this is true. However, research has shown that there are situations in which safety stock at the end-item level does not help reduce schedule instability.

Evidence from research suggests that small amounts of end-item safety stock do help reduce schedule instability but, at some point, increasing the amount of safety stock serves only to increase costs. That is, the extra inventory created by increasing safety stock levels adds to total cost but does not improve either schedule stability or customer service. In some cases, it has been shown that increasing the amount of end-item safety stock can actually make schedules more unstable. This can occur as MPS lot sizing rules are applied to meet both demand and safety stock requirements.

Safety stock is a viable option for managing schedule stability, but it must be used with caution. Small amounts of end-item safety stock, combined with freezing a portion of the MPS, may be used to balance schedule stability, inventory costs, and customer service.

MRP Lot Sizing

Lot sizing is deciding how much of an item to make or buy, given projections of requirements for the item over time. Economic lot-size models attempt to balance setup (ordering) costs against the costs of carrying inventory. MRP lot sizing refers to successive order-size decisions that are made in translating the master production schedule of end items into a time-phased plan for all lower-level items in the bills of material. In this process, requirements for lower-level items are determined by lot-size decisions for higher-level items. MRP lot-sizing decisions may be made on the basis of company policy or by application of economic lot-size models.

It is well known that schedule stability is affected in part by procedures adopted for MRP lot sizing. When MRP replanning is done with economic lot-size models at successive levels of the bill of material, small changes in requirements and order sizes at higher-levels in the bill of material can be magnified into much larger changes in requirements and orders sizes at lower levels. This obviously increases schedule instability.

Sometimes the adverse effects of economic lot-sizing on schedule stability are made worse by the rolling planning horizon used in MRP. As the planning horizon rolls forward and new requirements for higher-level items are identified at the end of the horizon, the lot-size model may modify previously-determined order sizes, starting the cascading process of altering lower-level requirements and orders.

The savings in inventory-related costs that one hopes to achieve from using economic lot-size models in MRP must be weighed against the potential for reduced schedule stability. Research on the subject suggests that dynamic, economic lot-size models should be used only on a very selective basis.

One option is to apply economic lot-size models only at the end-item level. With this approach, order sizes below the end-item level are made on a lot-for-lot (LFL) basis. LFL simply means that each requirement is treated as a separate order, with no attempt to combine requirements. This eliminates some of the instability problems caused by applying dynamic, economic lot-sizing rules at successive levels in the product structure. Research suggests that the effectiveness of this strategy varies, partly depending on what procedures are used to make order size decisions at the master schedule level.

Some other interesting approaches to deal with schedule instability and MRP lot sizing have been suggested in the research literature, although with less potential for widespread application. One of these approaches is to continue to utilize economic lot-sizing models in MRP, but to incorporate into those models the cost of schedule changes in addition to the cost of ordering and carrying inventory. In theory, when the model is applied during MRP replanning, it attempts to find order sizes that balance traditional inventory-related costs with schedule change costs. While this approach has shown promise in limited studies, there are practical considerations that make implementation difficult.

Finally, it has been suggested that the "end-of-horizon" effect described earlier (in which newly identified requirements at the end of the planning horizon cause changes to previously-determined orders) could be controlled by forecasting demand beyond the normal planning horizon. This would involve extending the forecast horizon to reduce "surprises" that cause changes to orders. The practical difficulty with this approach is that forecast accuracy decreases over longer periods, reducing the effectiveness of the new information for reducing order changes.

CONVENTIONAL WISDOM: Managing schedule stability involves tradeoffs in customer service, inventory-related costs, and schedule nervousness. Balancing these factors usually requires a combination of actions involving freezing a portion of the master production schedule, maintaining limited safety stock for end items, and selectively applying lot-size rules in MRP. While much has been learned from recent research about dealing with schedule instability, it is not possible to prescribe the best combination of these strategies for every manufacturing environment. The relative effectiveness of the strategies, and their combined impact, would be expected to vary from one manufacturing situation to another. Schedule instability is a continuing problem that requires more research.

See **Capacity planning in made-to-order productions; Capacity planning: Medium- and short-range; Economic-order quantity (EOQ) model; End-of-horizon effect; ERP; Freezing the master production schedule; Lot-sizing models; Manufacturing order quantity; MPS; MRP; MRP lot-sizing; MRP to MRPII and ERP: MRP regeneration; MRP replanning; Order-based freezing; Order release; Period-based freezing; Product cumulative lead time; Production activity control; Resource planning; Roll-of-materials (ROM); Rolling planning horizon; Safety stocks: Luxury or necessity; Schedule nervousness; Scheduling cycle.**

References

[1] Blackburn, J.D., D.H. Kropp, and R.A. Millen (1986). "A comparison of strategies to dampen nervousness in MRP systems." *Management Science*, 32(4), 413–429.

[2] Kadipasaoglu, S.N. and V Sridharan (1995). "Alternative approaches for reducing schedule instability in multistage manufacturing under demand uncertainty." *Journal of Operations Management*, 13, 193–211.

[3] Kropp, D.H., R.C. Carlson, and J.V. Jucker (1979). "Use of dynamic lot-sizing to avoid nervousness in material requirements planning systems." *Journal of Production and Inventory Management*, 20(3), 40–58.

[4] Kropp, D.H., R.C. Carlson, and J.V. Jucker (1983). "Heuristic lot-sizing approaches for dealing with MRP system nervousness." *Decision Sciences*, 14(2), 156–169.

[5] Mather, H. (1977). "Reschedule the schedules you just scheduled – Way of life for MRP?" *Journal of Production and Inventory Management*, 18(1), 60–79.

[6] Sridharan, V. and W.L. Berry (1990). "Freezing the master production schedule under demand uncertainty." *Decision Sciences*, 21(1), 97–120.

[7] Sridharan, V and R.L. LaForge (1989). "The impact of safety stock on schedule instability, cost, and service." *Journal of Operations Management*, 8(4), 327–347.

[8] Sridharan, V. and R.L. LaForge (1994). "A model to estimate service levels when a portion of the master production schedule is frozen." *Computers & Operations Research*, 21(5), 477–486.

[9] Sridharan, V, and R.L. LaForge (1994). "Freezing the master production schedule: Implications for fill rate." *Decision Sciences*, 25(3), 461–469.

[10] Sridharan, V., W.L. Berry and V. Udayabanu (1988). "Measuring master production schedule instability under rolling planning horizons." *Decision Sciences*, 19(1), 147–166.

[11] Steele, D.C. (1975). "The nervous MRP system: How to do battle." *Journal of Production and Inventory Management*, 16(4), 83–89.

[12] Zhao, X. and T.S. Lee (1993). "Freezing the master production schedule for material requirements planning systems under demand uncertainty." *Journal of Operations Management*, 11(2), 185–205.

SCHEDULER FOR FMS (DYNAMIC)

Paul Ranky

New Jersey Institute of Technology, Newark, NJ USA

A Flexible Manufacturing System, or FMS, is a highly automated, distributed and feed-back controlled system of data, information and physical processors. A typical system may include computer, and manually controlled machine tools, robots and other equipment's in which decisions have to be taken real-time. This is possible if all information processors (including the human resources of such systems) are "well informed," meaning that they have the right information at the right time, in the format and mode needed to take responsible decisions within given time constraints.

The dynamic scheduler of an FMS is responsible for providing the production system with the sequence in which parts have to be manufactured.

The term "dynamic" refers to the fact that FMSs can react to rapid changes, such as random part arrivals, thus the dynamic scheduler must have the capability of re-scheduling the system typically in less time it takes to change/load or unload a part onto/from a machine. (As an example, in the machine tool business, this represents less than 5–7 seconds, and in the assembly industry it is even less).

The dynamic scheduler must consider in real time not just part processing times and their sequences, but also tool and clamping/fixturing availability at each processor (or machine) of the system, as well as must assure that the selected processor (or machine) is capable of performing the required processes.

The overall specifications of the FMS dynamic scheduler can be summarized as follows:

1. The system should be capable of interpreting and evaluating the production situation from cell controllers by reading FMS system status data via the distributed FMS data processing network in real-time. (It should also display this status in a readily understandable way, using multimedia, if required.)
2. It should read and evaluate the current dynamic (transient) input data which may include breakdowns (e.g., tool breakage at a cell, or the breakdown of the entire cell, or the material handling system, etc.).
3. It should cope with job priority changes, job deletions, and rearrangement of part input queues and buffers. This data is used together with the predetermined off-line operation control input data in order to generate a task for any required instant of time.
4. It should access the variable route part program data base (i.e., the production rule base) along with other manufacturing data to schedule the various jobs from the job list on different cells using a built-in secondary optimization routine.
5. In case of an error in the operation of the FMS or the system, and if either of these is irrecoverable, appropriate messages must be generated. A self-learning, expert diagnostic module could learn from such failures and suggest different ways of changing or updating the rule base.
6. Relevant statistical information and schedule evaluation criteria based on the operations during the particular shift should also be generated as part of this system.
7. The interactive multimedia user interface should be such that even a non-technical shop-floor worker should be able to supervise/operate the real-time operation control system.

The system should be modular in such a way that at each sub-system level of implementation any of the sub-system modules should consist of two layers, these being: (1) The top level layer, which is FMS architecture independent, i.e., allows programming at a generic level that can work in virtually any truly flexible FMS architecture, and (2) The installation specific layer should be such that the FMS could be expanded and modified without affecting the rest of the system.

Furthermore the dynamic FMS scheduling system should incorporate procedures such to delete work piece from any production route and/or the buffer, to provide status of a component, and rearrange any production route in real-time because of a break down or maintenance, and other disruptions.

References

[1] Ranky, P.G. (1998). *Flexible Manufacturing Cells and Systems in CIM,* CIMware Ltd, http://www.cimwareukandusa.com (Book and CD-ROM combo).

[2] Caprihan, Rahul and Wadhwa, Sughash (1977). The Impact of Routing Flexibility on the Performance of an FMS – A simulation study, *The International Journal of Flexible manufacturing Systems* 9 (1977), p. 273–298.

SCHEDULING

See **Just-in-time manufacturing; MRP; Order release; Process industry scheduling; Scheduler for FMS; Schedule stability.**

SCHEDULING BOARD

Before the advent of computers, many companies used a long board on the wall to schedule production. Such a board, called a scheduling board, or control board, is still in use in some situations, but has more often been replaced by computerized scheduling software. The scheduling board usually indicates the jobs (load) scheduled on each machine or work center. However, it may also be used to represent the time line for an entire project. In either case, the scheduling board is just one form of a Gantt chart.

The problem with a manual scheduling board is that scheduling is usually a very complex activity that requires consideration of many options. All these options cannot be represented easily in a scheduling board. Computerized scheduling software, however, can try many options and store those various options, either in memory or on a disk drive with relative ease.

See **Capacity planning: Medium- and short-range; Gantt chart; MRP.**

SCHEDULING CYCLE

Scheduling cycle is the period of time that elapses between successive updates of a plant's schedule. It is also known as rescheduling interval.

See **Just-in-time manufacturing; MRP; Order release; Scheduling framework; Schedule stability.**

SCHEDULING FRAMEWORK

Even a small factory brings together dozens of machines, people, materials and components together in an orderly manner to produce products demanded by customers in a timely manner. The schedule brings machines, operator and materials together at a specified time to finish an operator in the time allotted for the operation.

Scheduling must also address the entire sequence of operations, and integrate the schedule for parts, and subassemblies with final assembly schedules. Material plans ensure material availability to carryout the schedules successfully.

Scheduling is a repetitive task, a schedule is prepared ahead of time, observe actual performance, and update the schedule.

See **MRP; Order release; Scheduling framework; Schedule stability.**

SCIENTIFIC MANAGEMENT

James M. Wilson

Glasgow University, Glasgow, Scotland, UK

HISTORICAL PERSPECTIVE: Scientific management was the first significant body of management thinking. It now "... so permeates ... modern life that we no longer realize it's there" (Kanigel, 1997: 7). The influence of scientific management is pervasive and, although some of its features have been and remain controversial, its general philosophy has been accepted with an impact that would be hard to underestimate. Frederick W. Taylor and his associates developed Scientific Management during the late 1800's with the aim of systematizing managerial practice. Taylor is best known for his work on time study, which was to resolve what he perceived as the central management problem: setting a fair task. Achieving this goal required more than time study, and Taylor's contributions can be categorized under three headings, engineering innovations and practices, personnel management and organizational changes, and production planning and control systems. All were necessary for time study and the organization to function effectively.

TECHNOLOGICAL PERSPECTIVE: ENGINEERING INNOVATIONS AND PRACTICES: Taylor's contributions to engineering are often overlooked. He was prominent in the American Society of Mechanical Engineers and served as its president. Taylor's most significant engineering contribution (Taylor, 1906) was developing high-speed steel, an innovation that promised increased shop productivity. It has been suggested (Aitken, 1960) that the development of high speed steel required a new approach to manufacturing management: it threatened to increase production so significantly that old work practices would have literally exploded under the increased throughput. Although high speed steel did promise to increase productivity Taylor was working on improved engineering practice and management long before (Taylor, 1886, 1895, 1903; Gantt, 1903). These engineering developments were a key prerequisite for the introduction of his management initiatives; management effectiveness was grounded in engineering efficiency. An unreliable, badly designed facility was poorly suited to finely planned and executed management; engineering inadequacies would undermine it.

The key idea underlying Taylor's improvements in engineering management was standardization. Systematic management would be impossible in an organization whose operations were idiosyncratic. All equipment, tooling and procedures used in a company should be standardized under scientific management (Taylor, 1903; Barth, 1912; Gantt, 1916; Copley, 1923). Thus, an employee given a job to do could most easily accomplish it if just one type of machine tool was used for that operation in the factory; this allowed better quality, faster set-up and operation generally, and greater throughput. Where a variety of equipment and tooling existed the staff would seldom use exactly the right machine or tooling for the job; time would be wasted in setting up non-standard operations, looking for or making unique tooling; and workmen would perform the work according to their individual preferences and prejudices.

The standardization of equipment was the essential preliminary stage in scientific management. No work study could create viable standards without first establishing its engineering foundations. Similarly, tooling and work methods had to be standardized – work study could not establish standard times if some staff used poorly designed tools or work practices. Without these prerequisites no standard time for operations would be available for work flow planning, cost accounting, and a variety of other management essentials.

Taylor advocated functional layouts in which equipment of a particular kind was located together. There were two motivations for these layouts. First, the water or steam driven shafting and belting systems that provided power within factories in the late 1800's were not amenable to highly varied loads. It was easier to design, build and maintain a power system that faced sets of similar loads than one that faced more varied and oddly distributed loads. Shaft and belt based power systems were inherently inflexible. It was thus virtually impossible to shift equipment around the factory in response to changing production requirements.

A second motivation for grouping like equipment in one place was to maximize productivity. A specialist then could more effectively supervise machine operators. The idea was not to de-skill staff but to use less skilled staff as effectively as possible. If like equipment was spread throughout the factory so too would be any operational problems. The machine operators would then need to be more skilled, or the supervisor would spend time moving around the factory with delays in responding to problems. Further benefits followed from better use of specialists, who handled set-ups and change-overs between products. Similarly, specialist inspectors and their equipment could more effectively control quality within a department than if they had to move throughout a factory.

Taylor also emphasizes maintenance, an issue that was probably more significant during his time. Repairmen also found that working with equipment conveniently located in one place made regular inspection and general maintenance easier. The supervisors, maintenance staff, set-up people and inspectors could more efficiently move between machines located in a small area, and could more readily cooperate to resolve difficulties. Babcock (1918) describes a process oriented factory under scientific management, but recognizes its atypical operation. Taylor would have been familiar with process oriented layouts, but for factories producing a variety of products a functional layout seemed most appropriate. Scientific management had its origins in Taylor's engineering work. But engineering developments alone were not sufficient to achieve a factory's maximum potential and personnel, organizational and operational improvements were also required.

ORGANIZATIONAL EFFECTS: PERSONNEL MANAGEMENT & ORGANIZATIONAL CHANGES: Taylor's critics saw scientific management as exploiting workmen in two ways. First, it deskilled staff by absorbing their knowledge, analysing it to create jobs unskilled workers could perform. Second, time-study was considered a means for driving workers harder – any standards could be re-set repeatedly until the staff could work no faster. Another criticism (Drucker, 1955) focused on the division between planning and doing, with the view that workers would be reduced to powerless machine tenders. These critiques were extreme, and often promoted inaccurate interpretations of Taylor's approach. In many ways the problems, of managing people were central for Taylor, and his scientific management evolved around maximizing worker's effectiveness. This did not mean that workers would be ruthlessly exploited but instead the system itself should work more effectively. Taylor (Drucker, 1955) recognized that more production did not necessarily require greater resource utilization. Taylor's approach is often portrayed as usurping skilled workmen's discretion in job performance. Taylor's counter-argument (Taylor, 1911, 1917) was that there was no consistent practice among these skilled workmen, his investigations showed which practices were most effective, and deserved to be were more widely and consistently implemented.

Personnel management may be seen as an offshoot of scientific management. Taylor devoted more attention to staffing decisions and policies than had been common; this is evident in the four part explanation provided by Taylor (1917:36):

> First. They develop a science for each element of a man's work, which replaces the old rule of thumb method.
>
> Second. They scientifically select and then train, teach, and develop the workman,

whereas in the past he chose his own work and trained himself as best he could.

Third. They heartily cooperate with the men so as to ensure all of the work being done is in accordance with the principles of the science which has been developed.

Fourth. There is an almost equal division of the work and the responsibility between the management and the workman.

The capstone of Taylor's system was its handling of personnel. The engineering elements could be relatively easily identified, understood and resolved; in fact, had to be implemented before alterations to work routines. The people issues in system design and operation were not so easily identified, understood or resolved, and great effort was needed to appreciate the environment and adapt the system to it most effectively. Taylor's approach assumed people acted on the basis of economic rationality, and, where possible, the social and cultural factors that could disrupt this behaviour were to be minimized.

An employee possessing the right intellectual and physical capabilities, appropriately trained and provided with the proper equipment, tooling and materials, motivated by effective economic incentives and supported by positive management should consistently produce sufficient output on time and to specification. The staff selection and training process should give the shop manager the staff needed. Appropriate incentive payment plans should provide the desired financial motivation, and a management with positive attitudes would facilitate their effective application. But the engineering and personnel developments alone were not enough. Even with well-selected and trained staff using appropriate equipment, tools and materials with properly structured incentives and positive management, efficient operation would be dependent on effective work planning and scheduling to ensure each workman had the right materials at the right time.

OPERATIONAL EFFECTS: PRODUCTION PLANNING AND CONTROL SYSTEMS: Although Taylor (1917) proclaimed "In the past man has been first. In the future the system shall be first," few analysts have approached scientific management as an integrated system. The engineering and personnel activities were linked by Taylor's organizational developments that created new managerial functions. The planning office's (PO) functional foremen provided the shop's planning, organizing and controlling features. It was meant to get the right materials in the proper quantities to the right people at the right times and places. It's functions and behavior are comparable with modern material requirements planning (MRP) systems.

It has been noted that functional foremanship was the least widely implemented element of scientific management. This seems to have largely been a consequence of Taylor's over-promotion of time-study. Significant benefits were ascribed to work study while there was little comment on its engineering and planning foundations. Managers wanted quick solutions to their problems and rapid returns to their investments, neither seemed to follow from laying work-study's essential foundations. The long and costly effort to standardize the engineering equipment and practices, and to set up and operate a planning office, did not seem justified when so much emphasis was laid on work study. Many advocates of scientific management were themselves unfamiliar with these features of Taylor's system, but Taylor too was responsible in not fully promoting them.

The functional foremen and shop clerical staff are identified in Table 1. (Taylor, 1911; Copley, 1923; Sterling, 1914; Thompson, 1914, 1917). Each filled a specific role in managing the production system; and, together they all provided a complex planning and control system.

In the PO, the creation of bills of materials, product specifications and drawings was the province of the engineering office, a sub-division of the PO. The creation and maintenance of routings was the function of the route clerk in the PO. This was a more complex document than seen in MRP. It specified the route to be followed and operations to be performed, and also specified the points at which various components in appropriate quantities would be needed. The routing and bill-of-material were important because they coordinated the men, machines and materials so that production occurred when and as required. The instruction card clerk was responsible for identifying the best way for jobs to be done, and for setting standards for their performance. The routing, bill of materials and instruction cards were then incorporated into the job instructions issued to workmen, when dictated by the production schedule that the "order of work" provided.

The production clerk was the intermediary between the sales department and the manufacturing department. He set the overall production schedule identifying which products were to be produced in each period. This order of work was a master production schedule produced by the production clerk based on sales and other information from the sales department and hi knowledge of the existing workload and capacity. The order of work was a continuously updated (rather than

Table 1. Functional foremen and shop clerks and their functions.

Functional foremen and clerical staff	Functions
Route Clerk	Creates Route Charts showing Components and Quantities, Their Routing and Processing.
Instruction Card Clerk	Creates Instruction Cards Describing the Most Effective Manufacturing Methods Based on Work Study.
Cost and Time Clerk	Accumulates and Records Cost Data from Reports of Manufacturing Activity.
Gang Boss	Trains the Workers in the Methods Shown on the Instruction Card.
Speed Boss	Ensures Machines are Set Up and Run in Accordance with the Instruction Card.
Inspector	Inspects Material to Ensure Quality, Particularly During Production.
Repair Boss	Preventative Maintenance and General Upkeeping.
Shop Disciplinarian	Maintains Discipline.
Production Clerk	Overall Production Schedules.
Route/Tag Clerk	Creates Job and Move Cards for Each Operation.
Balance of Stores Clerk	Maintain and Monitor Inventory Records, Project Requirements and Determine when to Reorder.
Instruction Card and Time Study Clerk	Determines the Detailed Manufacturing Methods.
Order of Work Clerk	Detailed Production Schedules.
Recording Clerk	Monitors the Progress of Work.
Cost Accounting Clerk	Analyses Costs Based on Manufacturing Activities.

Taylor, 1911; Copley, 1923, Vol. 1, p. 296–298; Sterling, 1914; Thompson, 1914; 1917.

regenerated in whole) plan. In the PO there was no forecasting; the order of work was predominantly sales driven. The execution of the order of work was the responsibility of the order of work clerk who released material and work orders to the shop floor as requirement dictated, and as previous jobs were completed.

The function of inventory management in MRP is to use the master production schedule along with the bill of material to identify the production requirements for every component in every period throughout the planning horizon. This function is replicated in Taylor's PO by the balance of stores clerk. Hathaway enumerated the clerk's functions:

> Under the Taylor System it is the practice to keep in the Planning Department "Balance Sheets" or running inventory sheets showing for each article carried in stock: (a) The quantity on hand; that is, actually carried in stores. (b) The quantity on order but not yet received in stores. (c) The quantity required for orders for shipment or manufacture to which they have been apportioned, but not yet issued. (d) The quantity available for future requirements. This information is constantly needed in connection with the planning of work. (Hathaway, 1914: 381)

The clerk must also:

> ... apportion in advance the materials called for on these issues, and subtract then from the quantities available, he compares it [the quantities available] to the minimum quantity [the reorder point]... if it is less than the minimum, he must issue an order for the quantity indicated for replenishment. (Hathaway, 1914: 381)

This is similar to an MRP Tableau in which Gross requirements are netted against On Hand Stocks and Expected Order Receipts to yield projected inventories or Net Requirements with Planned Order Releases. The PO operated on a reorder point basis, whenever stock fell below a specified point a fixed quantity would be reordered. '-f he reorder point was the amount that would be used during the replenishment lead time. The idea of covering requirements was plainly recognized. The need for time phasing the various production activities was also recognized:

> There will be considerable differences in the time necessary for piloting the respective components through their various stages of production and this means that there ought to be all that difference in the time of starting work on the components, if delivery of the whole set or sets is to be synchronized at a given date. (Elbourne, 1918: 170)

The need for careful co-ordination of many production activities across different periods was as well understood under scientific management as it was later in MRP.

The four functional foremen not specifically involved in the planning office acted as its agents on the shop floor. Under scientific management the planning system was augmented by significant control elements that directed and controlled the performance of work in accordance with the established plans. Orders were launched onto the shop floor but this action was backed up with close monitoring and response to problems. The gang boss was responsible for managing a group of workers; he ensured that they worked diligently and acted to resolve any problems that arose. The speed boss ensured that the instruction cards were followed and that the work was performed as instructed at the rates required. The inspector acted to ensure that the product's specifications were met, and periodically inspected each operation while it was in process to ensure that no deviations arose. The ideal of continual inspection and correction of faulty processes was approximated. The disciplinarian kept order in the shop and provided an independent adjudicator for any disputes. The PO made plan execution on the shop floor an integral part of its design and operation.

IMPLEMENTATION CASES: The influence of scientific management and Taylor has been so great that the paucity of implementations is striking. Wrege and Greenwood (1991) maintain that there were only two "full" implementations, and other studies (Nelson, 1992; Whitson, 1996) have noted that few factories used scientific management. Taylor himself was a purist and bitterly denounced the majority of consultants that sprang up to take advantage of his ideas, particularly work study (Copley, 1923). Perhaps the best descriptions of scientific management are to be found in Thompson (1914), a collection of contemporary analyses of the design and implementation issues found in scientific management, and in Thompson (1917), a thorough analysis of Taylor's production planning and control system. Aitken (1960) discusses the implementation of scientific management at Watertown Arsenal that lead to the House of Representatives' investigation of the Taylor and other systems of shop management. Babcock (1918) provides a contemporary discussion of the implementation of scientific management that describes its use by an automobile manufacturer. The best general background reading is Nelson (1980) and the biography by Copley (1923).

STRATEGIC PERSPECTIVE: SCIENTIFIC MANAGEMENT'S IMPACT: The impact of scientific management lies in its promotion and codification of rationalist thinking about management and production activities. Taylor epitomized the nineteenth century ratrionalist approach that saw the world as a system that behaved according to consistent patterns and rules. Identifying and exploiting these rules was possible and would allow people to better understand and control their lives. This was possible not only in the hard sciences but also in social activities, such as management. By thus influencing manager's understanding of their roles and capabilities, scientific management created a management culture that did not exist previously. Scientific management did more than effect a cultural change, it also provided tangible activities within a systematic framework for managers. Management was given new responsibilities in systems design, personnel management and in operations planning and control.

Organizations Mentioned: Watertown Arsenal.

See **Bill-of-material; Capacity planning in made-to-order production; History of manufacturing management; Make-to-order production; Manufacturing systems; Master production schedule.**

References

[1] Aitken, Hugh G.J. (1960). *Taylorism at Watertown Arsenal; Scientific Management in Action*, 1908–1915. Harvard University Press, Cambridge, MA.

[2] Barth, Carl (1912). "Testimony", in *The Taylor and Other Systems of Shop Management, Hearings before the Special Committee of the House of Representatives to Investigate the Taylor and Other Systems of Management*. V 3, Government Printing Office, Washington, D.C., p. 1538–1583.

[3] Babcock, George D. (1918). *The Taylor System in Franklin Management*. The Engineering Magazine Co., New York, New York.

[4] Copley, F.B. (1923). *Frederick W. Taylor: Father of Scientific Management*. Harper & Row, New York, New York.

[5] Drucker, Peter F. (1955). *The Practice of Management*. William Heinemann, Ltd., London.

[6] Elbourne, Edward D. (1918). *Factory Administration and Accounts*, The Library Press Co., Ltd., London, England.

[7] Gantt, H.L. (1903). "A Graphical Daily Balance in Manufacture". *ASME Transactions*, 24, 1322–1336.

[8] Gantt, H.L. (1916) . *Work, Wages and Profits*. The Engineering Magazine Co., New York, New York.

[9] Hathaway, H.K. (1914). "The Planning Department, Its Organization and Function". in *Thompson, C. Bertrand, Scientific Management: A Collection of the More Significant Articles Describing the Taylor System of Management*, Harvard University Press, Cambridge, Mass., p. 366–394.

[10] Kanigel, Robert (1997). *The One Best Way*, Viking Penguin, New York, New York.
[11] Nelson, Daniel (1990). *Frederick W. Taylor and the Rise of Scientific Management*, University of Wisconsin Press, Madison, Wisconsin.
[12] Nelson, Daniel (1992). *A Mental Revolution: Scientific Management Since Taylor*, Ohio State University Press, Columbus, Ohio.
[13] Sterling, Frank W. (1914). "The Successful Operation of a System of Scientific Management". in *Thompson, C. Bertrand, Scientific Management: A Collection of the More Significant Articles Describing the Taylor System of Management*, Harvard University Press, Cambridge, Mass., p. 296–365.
[14] Taylor, F.W. (1886). "Discussion of 'The Shop Order System of Accounts' (by Henry Metcalfe)", *ASME Transactions*, 7, 475–6 & 484–485.
[15] Taylor, Fred. W. (1894). "Notes on Belting", *ASME Transactions*, 15, 204–259.
[16] Taylor, Fred. W. (1895). "A Piece-Rate System, Being a Step Toward Partial Solution of the Labor Problem", *ASME Transactions*, 16, 856–903.
[17] Taylor, Fred. W. (1903). "Shop Management", *ASME Transactions*, 24, 1337–1480.
[18] Taylor, Fred. W. (1906). "On the Art of Cutting Metals", *ASME Transactions*, 27, 31–279, with "Discussions", 280–350.
[19] Taylor, Frederick, Winslow (1911). *Shop Management*, Harper & Brothers Publishers, New York, New York.
[20] Taylor, Frederick, Winslow (1917). *The Principles of Scientific Management*, Harper & Brothers, Publishers, New York, New York.
[21] Thompson, C. Bertrand (1917). *The Taylor System of Scientific Management*, A.W. Shaw Co., Chicago, Illinois.
[22] Thompson, C. Bertrand (1914). *Scientific Management: A Collection of the More Significant Articles Describing the Taylor System of Management*, Harvard University Press, Cambridge, Mass.
[23] Whitson, Kevin (1996). "Scientific Management and Production Management Practice in Britain Between the Wars", *Historical Studies in Industrial Relations*, 1, 47–75.
[24] Wrege, Charles D. and Ronald G. Greenwood (1991). *Frederick W. Taylor, the Father of Scientific Management: Myth* and Reality, Richard D. Irwin, Homewood, Illinois.

SCRAP ALLOWANCES

During production, items that do meet specification and cannot be rectified are rejected as scrap. Therefore, if 100 units of a part are started in the first of many sequential operations, at the end of the final operation, we may be left with less than 100 acceptable units. If 95 units finish all the operations successfully, then the yield in 95 percent and 5 percent is scrap. Therefore, some scrap loss is allowed in plans and it acts in the manner of a buffer or safety stock. To accommodate scrap loss, a larger starting quantity than the expected finish quantity in plans; the difference being the scrap allowance. If the process improves over time, scrap allowance is progressively reduced.

See **Safety stocks: Luxury or necessity; Synchronous manufacturing using buffers.**

SEASONAL INDEX

Seasonal index is used to adjust forecasts when demand exhibits a seasonal pattern. Products such as holiday greeting cards have a seasonal demand. The index captures the extent of seasonal variations. It indicates how much the demand will be higher or lower than the average demand during any given period.

If the average demand is 1000 units, and the demand during the peak period is 1800 units, the index for the peak period would be 1.5.

This index can be used with average demand to make forecasts for any period once the index for the given period (say, June) is known from prior data. Exponential smoothing method of forecasting uses a similar index.

See **Forecasting examples; Forecasting guidelines and methods; Forecasting in operation management; Seasonality.**

SEASONALITY

A seasonal pattern exists (seasonality) when a time series is influenced by seasonal factors (i.e., weekly, monthly, quarterly). Refers to the presence of an underlying process over time that has a seasonal form. The best known product demands that exhibits seasonality are those that are related to winter (snow blowers, Christmas) or summer (lawn mowers). Sometimes the pattern may be daily, as in the case of demand for telephone calls and the response of companies by providing rebates for certain days of the week (usually Sunday) or hours of the day (after 7 p.m.).

See **Forecasting guidelines and methods; Forecasting in operations management.**

SEASONALLY ENHANCED FORECASTS

See **Forecasting examples; Forecasting guidelines and methods; Forecasting in**

operations management; Seasonal index; Seasonality.

SECTIONAL MODULARITY

Sectional modularity is one form of modularity employed in mass customization. Sectional modularity allows a collection of components chosen from a set of component types to be configured in any arbitrary way as long as the components are connected at their interfaces. Each component may have one, two, or more interfaces allowing sequential addition of components. The most basic form of sectional modularly can be demonstrated with Lego blocks. A seemingly infinite number of items can be built from a set of Legos. Another common example of sectional modularity is the sectional sofa system. Many different pieces of the furniture can be combined to form an uniquely shaped sitting area.

See **Mass customization; Mass customization implementation; Product design; Product design for global markets.**

SELF-DIRECTED WORK TEAMS

Self-directed work teams are given complete decision making authority over large segments of the work assigned to them. This authority includes how the work is performed, how resources are selected for the production processes, and what equipment is purchased or utilized. These teams are normally given broad performance goals which are reviewed on a period basis. Self-directed work teams allocate work among team members, and are responsible for managing the performance of individual members and the team as a whole. The team determines work schedules, prioritizes tasks, and manages on-going work operations. Examples may include consulting teams and hospital surgery teams.

See **Issues in operation management; Teams: Design and implementation of manufacturing.**

SELF-ORGANIZING SYSTEMS

A self-organizing systems generate new organizations spontaneously from information and processes wholly internal to the system. Self-organization theory was first developed by Ilya Prigogine and colleagues (Nicolis and Prigogine, 1989), but is today linked to other complex systems theories (Kauffman, 1995). Broadly, self-organization is due to the spontaneous emergence of new patterns as a result of nonlinear interactions between the parts of a system. Examples of self-organizing phenomena range from the periodic aggregation of certain slime molds, to surges and crashes in stock markets, to the emergence of socio-political trends.

See **Manufacturing analysis using Chaos, Fractals, and self-organization.**

References

[1] Kauffman, S. 1995. *At Home in the Universe: The Search for Laws of Self-Organisation and Complexity*. Oxford University Press, New York.

[2] Nicolis, G. & I. Prigogine. 1989. *Exploring Complexity: An Introduction*. W.H. Freeman & Co., San Francisco, CA.

SEMI-VARIABLE COSTS

These costs have a fixed and variable component (and are therefore often called mixed costs). For example, the base salary of an executive is a fixed expense, while an earned bonus is a variable expense. Telephone expenses are another example of a semi-variable cost. Regardless of usage, a customer still receives a fixed charge each month for basic phone service. As long-distance usage increases, costs increase. A third example is a basic equipment rental or lease (such as a car rental or lease). A firm must pay the monthly payment regardless of volume, as equipment usage increases, so do the operating expenses associated with the equipment. A semi-variable cost has a fixed component regardless of volume but the cost also increases as volume or usage increases.

See **Cost analysis for purchasing.**

SENSITIVITY ANALYSIS

One objective in a simulation study might be to understand the relationships between system inputs and system output, namely, the effect of various operating decisions and system performance. For example, one might be interested in examining the impact of lot sizing decisions on average cycle time. Such an investigation, typically referred to as *sensitivity analysis*, may be a preamble to a more detailed design and optimization study.

Techniques for sensitivity analysis have largely been adopted from the theory of experimental design. In experimental design terminology, the input parameters and structural assumptions of a model are called *factors*, and the output performance measures are called *responses*. For example, in a simulation model depicting a flexible manufacturing system, the number of machines, the number of pallets, the release rules, the rate of machine breakdowns are all examples of factors responses of interest may include the total cycle time or capacity utilization. Sensitivity analysis can therefore be defined as an investigation of the impact of input parameters (factors) on system performance (response). If the input parameters vary continuously, this is equivalent to computing the partial derivatives of the expected response function with respect to the input parameters. The vector of these partial derivatives is called the *gradient* of the expected response function. The gradient information not only gives the sensitivity of simulation's expected response to small changes in the input parameters, but it is also fundamental in most mathematical programming approaches for finding optimal settings of the input parameters.

In simulation, experimental design provides a methodology to decide which particular configurations, i.e., settings for factor values, to simulate so that the desired information can be obtained efficiently, that is, with the least amount of computation. This sensitivity information is ultimately used for system design to obtain optimal performance.

See **Manufacturing with flexible systems; Simulation analysis of manufacturing and logistics systems; Simulation of production problems using spreadsheet program.**

SEQUENCING RULES

See **Priority scheduling rules.**

SEQUENTIAL APPROACH TO PRODUCT DESIGN

Some firms use the sequential approach for developing new products, in which various functional groups "hand-off" responsibility after completion of their stage to the next functional group responsible in the process (see Handfield, 1995). The activities performed by each group is typically evaluated at a "gate" to determine whether a set of formal criteria were met. At each gate, the project is measured against a number of clearly defined attributes. The project may be terminated at one of the gates.

In the first stage of new product development, marketing studies provide commercial specifications of the attributes which the future product should possess. At this "breadboard" stage, the principal task is that of examining the feasibility of such a product. The interaction between marketing and engineering departments at this stage is critical in obtaining an accurate portrayal of available technology versus market needs. This may also result in a paper or cardboard model of the product, used for assessing its market potential only (sometimes refered to as a "**Bench model**").

The second gate is a design-driven effort. The goal of design engineers is to design a product which will meet or surpass the attributes determined by marketing in the conception stage without exceeding cost estimates. The design task involves the choice between alternative or existing product technologies which will ensure market success. The result of these activities is the development of an engineering prototype or "**Engineering model**." This prototype embodies the type of "package" to be used, as well as the required components and materials involved.

Between gates 2 and 3, the first manufactured prototypes are run on existing or procured equipment (sometimes refered to as a "**Prototype model**"). Suppliers often prove to be invaluable sources of information in determining simpler and more manufacturable designs. In some cases, suppliers perform the design function for many components and parts, contributing significantly to reduced development leadtimes. For those parts produced in-house, manufacturing personnel are instrumental in matching process capability and design tolerances.

Tests occurring in-house and through field trials determine the performance characteristics of the manufacturing prototypes. These tests provide an assessment of the match between materials, product and process.

The final gate is passed when the product is ready to market. At this stage, all testing has been approved and completed, the customer's suggestions have been integrated into the product, and all process capability studies have been formally agreed upon. If all is in order, the product proceeds into full production. At this point, engineering efforts focus on cost reductions through alternative materials, value analysis, process improvements, and progress down the learning curve.

See **Concurrent engineering; New product development through supplier integration; Product design for global markets; Product development and concurrent engineering.**

Reference

[1] Handfield, R.B. 1995. "Effects of Concurrent Engineering on Make-to-Order Products", *IEEE Transactions on Engineering Management*, 41(4): 1–11.

SEQUENTIAL PROCESSING

Sequential processing in a manufacturing system involves a linear flow of work from its start to its completion with the various transformation operations occurring in a fixed sequence. Although the ordering of operations will generally be based on technical requirements (much as one puts on shoes after socks) there are instances where such ordering is not required. Sequential processing may involve a strictly linear sequence of activities in which each operation has a single predecessor that provides it with materials for processing. Alternatively, a sequential processing on assembly lines may involve multiple sources of inputs.

See **Assembly line design; Manufacturing cell design; Manufacturing systems modeling using petri nets; U-shaped assembly lines.**

SERVER PLANTS

Server plants are built to serve a particular market, i.e. their output is completely absorbed by local markets.

See **International manufacturing.**

SERVICE LEVEL

Service level is a manufacturing performance measure, which expresses entire customer demand that is satisfied on time as a percent of total demand. In a make-to-stock environment it may be stated as the ratio of demand satisfied from stock over the total demand.

See **Capacity management in make-to-order production systems; Performance measurement in manufacturing; Safety stocks: Luxury or necessity.**

SERVICE ORIENTATION

A service orientation refers to a tendency to provide higher levels of service to the customer. For example, a firm with a "service orientation" would choose to carry higher levels of inventory in order to provide higher levels of product availability than would a firm which has a "cost orientation." Thus, a firm with a service orientation tends to make the tradeoff decisions in favor of higher service levels rather than lower cost.

A service orientation could suffer if service levels to be delivered to customers are internally defined based on the firm's internal capabilities rather than related to customer expectations or requirements. In some companies service orientation generally results in a focus on internal measures of performance such as fill-rate percentages and damage ratios, rather than external measurement of customer satisfaction. As a result, a service orientation does not always lead to enhanced customer satisfaction, and can lead to wasted efforts and resources (effort may be expended on efforts that are not valued by the customer). By matching a service orientation with an intimate knowledge of customer needs and expectations, a firm can improve satisfaction levels, customer loyalty, and long-term profitability.

See **Customer service; Performance measurement in manufacturing; Safety stocks: luxury or necessity; Total quality management.**

SERVICE PARTS

When parts fail in the field, while in the custody of the customer, parts are ordered by the customer or the agency performing maintenance or repair service.

The demand for such parts are treated as service parts and are included in the MRP record. Demand for service parts are forecasted separately and added to the gross requirements for the part in MRP records.

Safety or buffer stocks for the part is used to deal with uncertainties in the demand for the part.

See **Maintenance management decision models; MRP; Safety stocks: Luxury or necessity; Total productive maintenance.**

SETUP REDUCTION

John Leschke

University of Virginia, Charlottesville, Virginia USA

INTRODUCTION: A setup is the activity or event related to changing a production process from making one product to another. The time it takes to complete a setup is relevant in manufacturing because it determines how flexible a production process is. For example, faster setups enable a firm to produce a greater variety of products or produce a larger number of units in the same amount of time. Faster setups enable quicker response to customer demands or smaller economical batch sizes.

Therefore, reducing setup times (or costs) is an important operations objective. This article discusses issues related to the process of reducing setup time (or cost), including a brief history, significant analytical and management concepts, and implementation issues. The term "setup" may refer to either setup time or setup cost, however setup reduction typically emphasizes setup time because time is ordinarily the resource that constrains flexibility and setup cost is ordinarily highly correlated with setup time.

HISTORICAL PERSPECTIVE: The growing awareness of the feasibility and value of reduced setups coincided with the Just-in-Time revolution in the United States. As American firms tried to incorporate JIT concepts, they quickly realized that quick changeovers were a critical facilitating factor in achieving the small lot sizes that are the hallmark of JIT production.

Important promoters of small-lot production during this period included Robert Hall (1983) and Richard Schonberger (1982). Their books described examples of dramatic reductions in setup time leading to equally dramatic improvements in work-in-progress inventory, throughput, and leadtime. Omark Industries, a supplier of chains for chainsaws, was one of the most visible early adopters of Japanese manufacturing techniques in North America. Their experience contributed greatly to understanding the critical link between reduced setups and achieving the promise of JIT production.

In 1985, Shigeo Shingo, an industrial engineer and consultant who played a significant role in developing Toyota's just-in-time manufacturing system, formally introduced the single-minute exchange of dies (SMED) setup-reduction methodology. The SMED approach represented a compilation of 35 years of insights and experience with reducing setup. Prior to this time, long setup times were considered a fact of manufacturing life and managers tended to work around setup, rather than try to reduce it. For example, the motivation for research in mathematical lot-sizing models is to optimize the tradeoff between inventory carrying and setup costs. Consider the classic EOQ model; the objective is to determine the optimal lot size *assuming setup cost is fixed*. It is likely that the emphasis on developing more and more sophisticated ways to model or derive an optimal solution to a given situation distracted managers and researchers from recognizing that the situation was, in fact, not given. The SMED methodology, on the other hand, was a breakthrough that represented a systematic approach to reducing setup times and costs that was both inexpensive and effective, that is, a means to change the parameters of the situation rather than to simply make the best of it.

In the meantime, the academic community was examining the theoretical aspects of reduced setups. Ritzman, King and Krajewski (1984) conducted a large-scale factory simulation and concluded that fast setups were the key contributing factor in achieving just-in-time production. Conversely, long setups significantly inhibited achieving the benefits of just-in-time production.

Porteus (1985) represents the seminal analytical piece on the topic of determining the optimal amount to invest in setup reduction. That is to ask, what is the tradeoff between investing in reduced setup and the cost reductions derived. Porteus' work spawned a long stream of research that extended his original model to consider different investment functions and variations on the cost model. More recent research included descriptions of the setup-reduction process and analyses of how managers might allocate a limited amount of setup-reduction resources among competing setup-reduction opportunities.

SIGNIFICANT ANALYTICAL MODELS:

1. ***The Classic Economical Order Quantity Model:*** In his practitioner-oriented book, Schonberger (1983) describes some of the consequences of reducing setup. Beginning with the classical economic order quantity (EOQ) model, let Q be the lot size, D the deterministic, continuous demand rate, A the setup cost, and h the per unit per unit time holding cost, the total relevant inventory and setup cost TRC equals:

   $$\mathrm{TRC}(Q) = \frac{Qh}{2} + \frac{AD}{Q}$$

   It can be further shown that $\mathrm{TRC}(Q^*) = 2AD/h$ for the optimal order quantity Q^*.

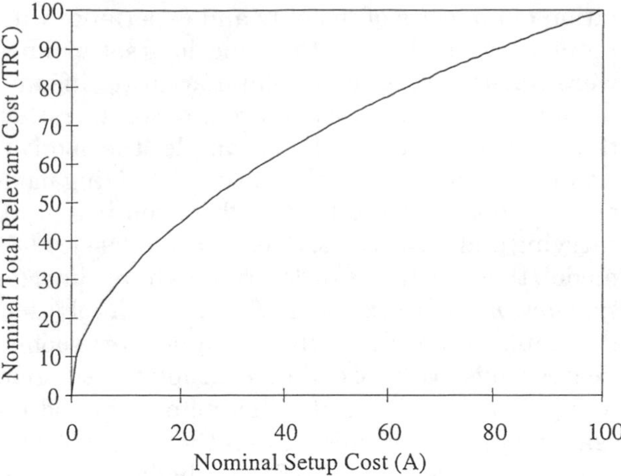

Fig. 1. Total relevant cost as a function of decreasing setup cost.

If, as in Figure 1, one plots total cost as a function of decreasing setup cost it becomes apparent that reducing setup yields marginally *increasing* returns. Note the concave shape of the total cost function as setup decreases from the original setup cost A_0 to 0.

The implications of this behavior are significant. First, it implies *that each marginal improvement in setup is more valuable than the last*, providing incentive for continuing reductions in setup. These benefits arise from the interaction of the lower setup costs and smaller optimal lot sizes. Although smaller lot sizes increase the total number of setups, total setup cost is more than offset by the lower cost of each setup and the smaller cycle stock costs. Moreover, there is a multitude of benefits not accounted for in this simple model. For example, smaller lot sizes imply less work-in-progress inventory which can lead to faster throughput times, lower obsolescence costs, less scrap, lower space requirements, lower material handling costs, and so on. Smaller lot sizes are also linked to higher quality because defects are detected closer to the time of failure, making it more likely the cause of failure can be identified and removed. Finally, smaller economical lot sizes imply greater flexibility in terms of production capacity, product variety and customer responsiveness.

2. ***The Cost-Benefit Tradeoff of Investing in Reduced Setups:*** Porteus (1985) introduced a theoretical framework for investigating the impact of reduced setup costs. Beginning with the economic order quantity (EOQ) model, add $I(A)$, the investment required to reduce the setup cost to a value of A. $I(A)$ is assumed convex and strictly decreasing. The α-factor represents an opportunity cost or discount rate for the investment.

$$\mathrm{TRC}(Q) = \frac{Qh}{2} + \frac{AD}{Q} + \alpha\, I(A).$$

Solving the above equation for the least total cost yields a value for the optimal amount to invest in setup-reduction. By evaluating this model assuming different parameters, Porteus concluded that very large reductions in setup could be economically justified by just the inventory and setup cost savings alone. This supported Shingo's experience and other anecdotal evidence that suggest huge reductions in setup can be justified.

An extension of Porteus' work, Kim, Hayya and Hong (1993) examine the effects of different forms of the investment function. They demonstrated that the optimal amount of setup reduction, indeed whether setup reduction was economically justified at all, was sensitive to the shape of the investment function. For example, if the investments yielded constant reductions in setup (i.e., a linear investment-setup relationship) or the investments yielded marginally increasing reductions (i.e., a concave investment-setup relationship), then the setup should be reduced as much as technologically possible.

In contrast, if there were marginally decreasing reductions in setup with increasing investment (i.e., a convex investment-setup relationship) then the results were less clear. Their research confirmed Porteus' earlier observation that the investment function must be "sufficiently" convex to recommend investing in reduced setups. In some cases, it may be economical to initially invest in reduced setups but because of the marginally increasing costs to reduce setup there would be a point at which it was no longer optimal to invest any more. Thus, both, Porteus and Kim *et al.* concluded that the optimal amount of setup reduction might be significantly higher than what is technologically feasible.

3. ***Allocating a Constrained Budget toward Reduced Setups:*** Leschke and Weiss (1997) consider the question of how to allocate a limited setup-reduction budget among multiple setup-reduction opportunities. They demonstrate that marginal analysis can be applied to the multiple-product problem for a wide range of convex investment functions.

Marginal analysis suggests that managers should identify the setup with the greatest marginal cost savings and allocate the early portion of the budget to this setup. When that setup has been reduced to a point where its marginal benefit is equal to a second product, it is most cost-effective to allocate the budget "simultaneously" between the two setups. Simultaneous allocation is recommended until all setups are being reduced and all marginal benefits remain equal.

In practice, this suggests that it is better for managers to reduce setup across several products rather than to focus on a single setup and reduce it as much as possible before proceeding to the next product's setup. This might be thought of as an example of technology transfer, in that, as a reduction is made in one setup it should be transferred to other setups rather than continuing to reduce the original setup exclusively.

MANAGERIAL (STRATEGIC) AND TECHNOLOGY PERSPECTIVES:

1. ***The SMED Methodology:*** SMED takes its name from the idea of reducing the number of minutes to complete a setup into single digits (i.e., less than ten minutes). In practice, this goal is readily achieved at relatively little cost. The key concept underlying SMED (Shingo, 1985) is recognizing the difference between internal and external setup operations. An internal setup operation is a task that can only be performed while a machine is stopped, such as mounting or removing dies. An external operation can be conducted while the machine is in operation; for example, transporting dies to storage or conveying dies to a machine.

   The SMED methodology consists of four conceptual stages. In the first stage, *internal and external setup operations are not distinguished*. Shingo observed that many firms do not recognize the difference between internal and external operations and often perform inherently external operations while the machine is stopped. Thus, Shingo reasoned, if internal and external operations are identified and separated, then the time the machine must be shut off could be limited to only the time needed for internal setup operations.

   The second stage of SMED involves *identifying and separating internal from external operations*. Shingo estimated that this step could yield reductions between 30 and 50 percent with little or no investment.

   The third stage involves *converting internal setup operations to external operations*. The focus here is to reduce minimize the time the process is not running, that is, maximize capacity and flexibility. Some conversions may require restructuring the setup procedure but typically some sort of process innovation or technological solution is applied. Therefore, the third stage tends to require higher investment.

   The final stage involves *streamlining all aspects of setup operations*. Actions involve reducing or eliminating both internal and external operations so as to reduce the overall time and cost of the setup. The improvements in this stage tend to involve investing in more technology, e.g., process or product redesign, new tooling, or automation.

2. ***The Setup Reduction Process:*** Leschke (1996) describes a field study of the setup-reduction process. Three significant observations emerged. First, there are three basic types of setup-reduction activities, each with different economic implications. Second, these activities tend to be implemented in a logical sequence, suggesting a multistage model of the setup-reduction process. Finally, there is a relationship between the degree of product or process standardization and the stage of setup-reduction.

   A *work center-level* investment is defined as a one-time expenditure that reduces the setups for many or all products in a work center. The distinctive characteristic of a work center-level investment is its economy of scope; that is, one investment reduces setup for many products.

   A *product-level* investment is characterized by a one-time expenditure of time or money that has an effect limited to a single product's setup. Product-level investments typically work to reduce variations in an element of setup common to different products. Thus, although their scope is limited to a single product, product-level activities are often replicated across many products.

   Policy-level investments affect many products, however, they differ from work center- and product-level investments in that they require a continued commitment to sustain the setup reduction. These investments tend to focus on improving the context (e.g., equipment reliability and capability, material quality and availability, and product design) rather than specific elements of the setup process itself.

   How these investment types correspond to different stages of setup reduction, their logical implementation sequence, and how they interact with the product and process technology are summarized in Figure 2.

   The figure indicates that work center-level investments tend to be applied during an initial *organization* stage in which firms focus on coordinating the setup activities of different products made in the same work center (e.g.,

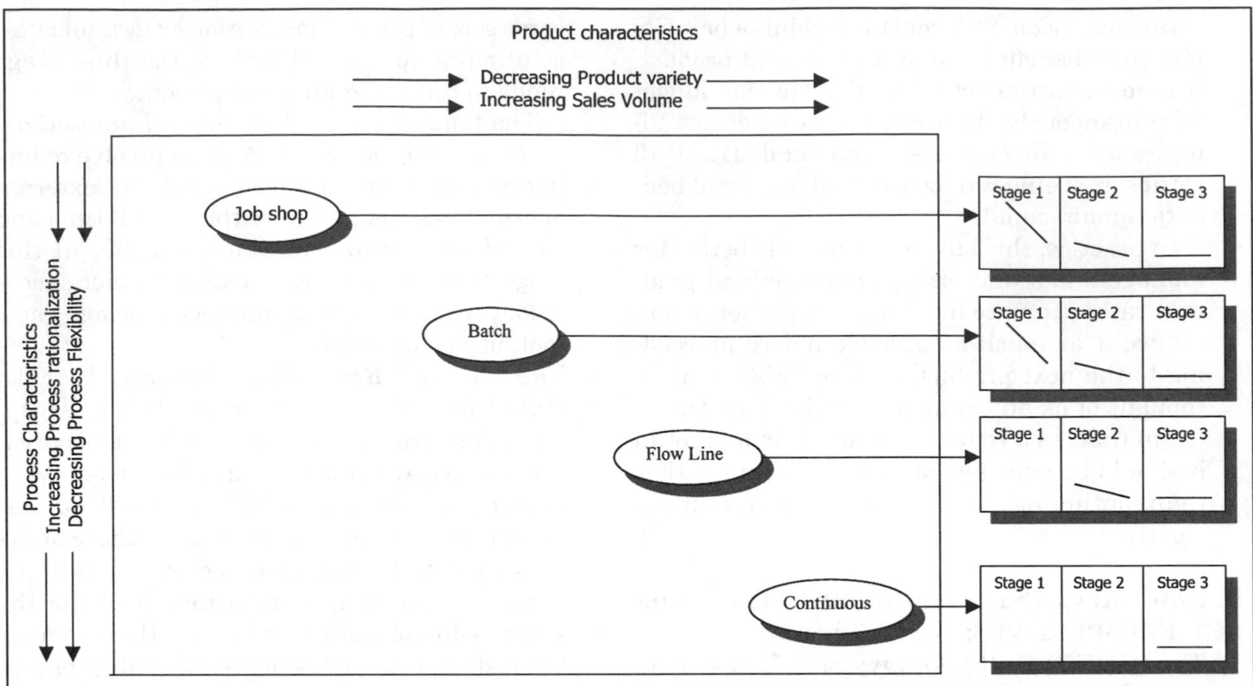

Fig. 2. The setup-reduction process. (Reprinted with permission of the APICS – the Educational Society for Resource Management, Falls Church, VA. *The Journal of Operations Management*, Vol. 38 (1), p. 32–37.)

improving housekeeping, centralizing tools and revising procedures). During the second *standardization* stage, firms tend to make product-specific investments in ensure individual setups conform to a particular standard. Finally, during the *rationalization* stage, work center-level investments tend to dominate because the standardization achieved during the previous stage implies that a change in one procedure will need to be propagated for all setups.

The investment sequence observed here is consistent with other research in product or process development. Specifically, according to the product-process matrix concept, firms with high product variety will tend to use production processes that are more flexible and have a more jumbled process flow. Thus, such firms will likely have high variety in setup procedures and will need to organize common setup elements before beginning to standardize the setup process. In contrast, firms with low product variety will tend to have a more prescribed process flow, therefore, setups are inherently more organized or have significantly fewer product-specific differences. Thus, the need to organize and standardize may be mitigated due to the advanced nature of the product or process technology. Therefore, this study concluded that the nature of the product and process technology influences the point of departure for setup reduction by reducing or even eliminating the early setup-reduction stages for those operations with more advanced process or product technologies.

3. ***Setting Reduction Priorities:*** Leschke (1997) provides some guidance as to how managers should prioritize investments during the standardization stage of the setup-reduction process. Priorities during the organization stage are logically directed toward bottleneck work centers. After this initial stage, however, the nature of setup-reduction activity changes and setting meaningful priorities becomes more complex. Improvements made during this stage are likely to incur greater expense due to the kind of changes that must be made, as well as the fact that only one setup is affected by each investment. Therefore, managers of setup reduction programs must more carefully choose which product's setup to reduce first, second, and so on.

Shingo suggests that those setups that take the longest time should be reduced first. In contrast, the academic literature emphasizes finding the optimal tradeoff between the cost of reducing setup and its benefits; that is, it may not be optimal to drive setup times arbitrarily to single digits. In this study, Leschke compares the relative effectiveness of Shingo's longest setup time "rule" against five other investment priority policies derived from the academic literature.

The specific situation modeled was that of multiple products produced at a single, capacity constrained work center. Setups were assumed to be sequence independent and investments product specific to correspond to the standardization stage. Some of the policies included were longest setup time, uniform allocation, marginal cost-based rules, and random allocation. The last method served as a basis for comparison; any allocation scheme worth considering should be more effective than a random application of the budget.

These results demonstrate that, although, the marginal analysis-based rules performed most effectively, longest setup time and uniform allocation were nearly as effective. The advantage of marginal analysis increased as the variance among demand, setup time, setup cost and product cost among products increased. All rules significantly outperformed a random allocation. The relevance of this research is that managers can use very simple allocation rules to obtain much of the benefit afforded by more sophisticated rules under certain conditions. See also Leschke and Weiss (1998).

IMPLEMENTATION AND APPLICATION: Figure 2 illustrates the relationship between the major process types and combinations of product and process characteristics; the inset shows corresponding setup-reduction processes. For example, setup-reduction processes that include all three stages most likely characterize job shop or small batch processes. Then, as product and process evolve toward less variety and greater flow, respectively, the relevance of the organization and standardization stages diminishes. Thus, Figure 2 offers managers some foresight as to the types of investments they are likely to make and a logical implementation sequence.

This implementation framework is consistent with the specific steps of the SMED methodology. The first step is largely facilitated by work center-level investments to organize the work place and the setup procedure. The second stage is facilitated by the reorganization of the setup procedure and by product-specific investments to standardize setups. The final steps, converting, streamlining and eliminating internal and external setup elements, are facilitated by efforts to standardize and rationalize setups.

In general, this sequence approximates an investment-benefit tradeoff with diminishing marginal returns. Organizational activities are relatively inexpensive, but yield significant setup reductions. During the standardization stage, marginal costs would be somewhat higher because each product's setup is modified individually and there is no economy of scope. Then, during the rationalization stage, marginal costs rise substantially because each action involves every setup and the potential benefits remaining are smaller.

This particular sequence has some practical and theoretical advantages. In general, this sequence will yield a convex investment pattern, hence the major theoretical insights derived by Porteus and others can be applied. Moreover, a sufficiently convex investment pattern will yield a convex total cost function that has the property of maximizing the present value of the setup reduction investment.

1. ***Organization:*** Setup reduction is best delegated to operators or others working most closely with the process. The nature of the setup-reduction process involves many small incremental improvements and much experimentation. A hierarchical or tightly controlled organizational structure can inhibit creativity and speed. Many firms involve production engineering staff in setup reduction, but most firms give operator teams authority and responsibility to make the desired changes. It is, however, important that teams are trained in setup-reduction concepts and problem-solving techniques.

   Other important groups to involve in the process are process and product designers. These people should be involved to ensure future processes and products are designed for quick changeovers and conform to established setup standards.
2. ***Training:*** Training for setup reduction programs typically involves training in SMED concepts, team building, goal setting and problem solving. The theoretical models can be used to supplement SMED training. For example, the three stages of the setup-reduction process model are consistent with SMED, but add a new perspective. Specifically, it suggests economic implications that are not apparent in the steps of the SMED methodology. Useful problem-solving techniques include cause and effect diagrams, brainstorming, process diagramming, flowcharting, and maintaining performance records.
3. ***Implementation:*** In general, setup reduction should be emphasized for bottleneck work centers. Of course, as setup times decrease, bottlenecks, and hence priorities, may shift. The process model of setup reduction outlines a logical investment sequence, hence a general priority sequence. Thus, setup-reduction teams can focus their attention on making the types of investments that are appropriate for the current stage of reduction and stage of product-process development.

   With respect to product-specific investment priorities, when simplicity, amount of information required, effectiveness, robustness over different manufacturing situations, and speed of decision making are considered, the best policy is to emphasize reducing the highest setup times as Shingo suggested. Although other policies may produce slightly lower costs, they require substantially more information to apply.

The idea of setup reduction is most readily associated with batch or repetitive manufacturing. Here, the relatively high variety of products (or components) and the correspondingly high frequency of changeovers make setup reduction an obvious source of benefit. However, there is nothing inherently good about a long setup. Thus, in commodity businesses with relatively fewer products and fewer changeovers, setup is still beneficial. By definition, faster changeover procedures reduce wasted labor, increase capacity, and increase flexibility. Indeed, faster changeovers may transform a commodity supplier into a more specialty-order supplier due to its new capability to economically satisfy smaller demands.

SUMMARY: Quick changeovers are an important source of competitive advantage. With fast setup times come smaller lot sizes, quicker customer response, increased capacity, product and volume flexibility, higher quality, and more. Fast setups make many of the characteristics that define a "world class manufacturing operation" possible. Despite, its huge effect, setup reduction is remarkably simple. It involves good engineering, solid design and discipline. Applying the SMED methodology, for example, has helped many firms reduce setup times by over 90 percent for relatively little cost.

Organizations Mentioned: Omark Industries.

See **Cost-based rules; Economic order quantity (EOQ) model; External setup operation; Implementation; Internal setup operation; Just-in-time manufacturing; Lean manufacturing; Marginal analysis; Marginal mathematical lotsizing models; Obsolescence costs; Optimal lot size; Shigeo Shingo; Theory of constraints in manufacturing management; Total productive maintenance; Uniform allocation.**

References

[1] Hall, R. (1983). *Zero Inventories*. Dow Jones-Irwin. Homewood, Illinois.

[2] Kim, S.L., J.C. Hayya, and J.D. Hong (1992). "Setup Reduction in the Economic Production Quantity Model." *Decision Sciences*, 23(2), 500–508.

[3] Ritzman, L.P., B.E. King and L.J. Krajewski (1984). "Manufacturing Performance – Pulling the Right Levers." *Harvard Business Review*, March–April, 143–152.

[4] Leschke, J. (1996). "An Empirical Study of the Setup-Reduction Process." *Production and Operations Management*, 5(2), 121–131.

[5] Leschke, J. (1997). "The Setup-Reduction Process: Part 2: Setting Reduction Priorities." *Production and Inventory Management Journal*, 38(1), 38–42.

[6] Leschke, J. and E.N. Weiss (1997). "The Uncapacitated Setup-Reduction Investment-Allocation Problem with Continuous Investment-Cost Functions." *Management Science*, 43(6).

[7] Leschke, J. and E.N. Weiss (1998). "A Comparison of Rules for Allocating Setup-Reduction Investments in a Capacitated Environment." *Production and Operations Management, forthcoming*.

[8] Porteus, E.L. (1985). "Investing in Reduced Setups in the EOQ Model." *Management Science*, 31(8), 998–1010.

[9] Schonberger, R.D. (1982). *Japanese Manufacturing Techniques*. The Free Press, New York.

[10] Shingo, S. (1985). *SMED: A Revolution in Manufacturing*, Productivity Press, Cambridge, MA.

SETUP TIME REDUCTION

See **Setup reduction.**

SEVEN "RIGHTS" OF LOGISTICS

For many years, logistics management was viewed as a necessary evil. That is, logistics activities had were needed for a manufacturing or retail firm to stay in business, but these activities were not seen as "strategic. As logistics managers sought to make the value-added contribution of logistics more visible to upper management, the "seven rights" of logistics were coined and publicized. The seven rights are, to deliver the **right** product, in the **right** quantity and the **right** condition, to the **right** place at the **right** time for the **right** customer at the **right** price. These seven rights highlighted the importance of moving and storing materials in an efficient, timely, and reliable manner. The seven rights also linked logistics to the key strategic objectives of cost competitiveness, quality, flexibility, and delivery. The seven rights demonstrate that logistic activities provided the foundation for high levels of customer satisfaction.

See **Just-in-time manufacturing; Logistics: Meeting customer's real needs; Supply chain management; Total quality management.**

SEVEN TOOLS OF QUALITY CONTROL

See **Integrated quality control.**

SHARED COSTS

One of the goals of traditional financial accounting systems is to find a cost for each product. Many costs such as direct labor and materials can be allocated to a specific product. However, other types of costs, overhead for example, are shared among several different products. These shared costs have typically been divided among products based on the proportion of direct labor need to make each of those products. In the past, direct labor was a major component of product cost, so this approach to allocation made sense. Today, however, direct labor is often a very small component of product cost. Thus, allocation of shared costs on the basis of direct labor often leads to distortions in which the cost of manufacturing a product, as determined by the financial accounting system, is not representative of its actual cost.

One approach to overcoming the problem of shared costs is the use of activity-based costing (ABC). With ABC, specific "cost drivers" are identified to determine how the shared costs should be allocated. For example, cost drivers include the number of setups or inspection time required for a given product. Such cost drivers can then be used to allocate shared costs more appropriately.

See **Accounting system implications of TOC; Activity-based costing: An evaluation; Activity-based costing; Cost drivers; Financial accounting system; Performance measurement in manufacturing.**

SHEWHART CONTROL CHARTS

The concept of control charts was first developed by Walter A. Shewhart of Bell Telephone Laboratories in 1924. For this reason, statistical control charts are also known as Shewhart control charts.

A Shewhart control chart is a graphical tool for monitoring the activity of an ongoing process. Such control charts can be constructed for quality characteristics that are variables or attributes. Values of the sample statistics are plotted along the vertical axis as a function of the time sequence in which the samples, sometimes called subgroups, are chosen. Examples of samples statistics that are variables are average diameter, average thickness, average length, average tensile strength, and average resistance.

Examples of sample statistics for which control charts are used are proportion of nonconforming items, number of surface finish defects per unit, number of machine breakdowns per month, and number of defects in 100 square meters of textile.

A typical control chart is shown in Figure 1. Three lines are used in a control chart. The *center*

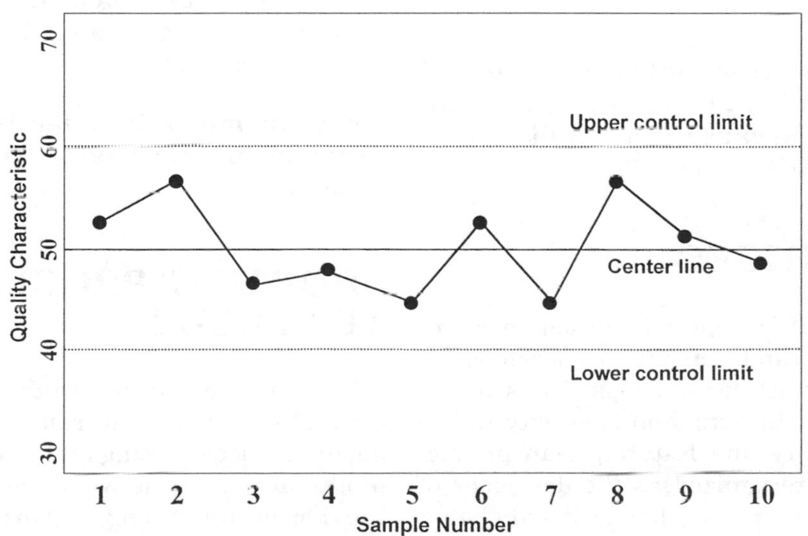

Fig. 1. A typical control chart.

line, which typically represents the average value of the quality characteristic being plotted, is an indication of the process center. The process center should be as close to the target value as possible.

Changes or shifts in the process center are accomplished by changing the speed, feed, or depth of cut, the tool material, the vendor that supplies incoming raw material or components. Two other lines known as the *upper control limit* (UCL) and *lower control limit* (LCL) are also shown in Figure 1. These control limits are so chosen that, if the process is in a state of control, the probability of any sample point falling outside these limits is quite small, or negligible.

Shewhart control charts are constructed by using observations from random samples from the output of the process. Each set of measurements is called a sample or a subgroup. Control limits are based on the variation that occurs within the sampled subgroups, which is a measure of the variation due to common causes. The variation between subgroups is excluded from the computation of the control limits. So, the computations assume that there are no special causes affecting the process. Thus, if a special cause of variation is present, the control chart, based on common causes, is able to detect when and where the special cause has occurred. Shewhart proposed the construction of a control chart for the sample average. He suggested that the range of allowable variation, due to common causes, be plus or minus three times the standard deviation of the sample mean (also known as the standard error). If the distribution of the sample means can be assumed to be normal, then with the control limits placed at three standard errors away from the center line, virtually all of the plotted values (about 99.73%) will fall within the control limits if the process is stable.

See **Statistical process control using control charts; Total quality management; Variability due to common causes; Variability due to special causes; X-Bar chart.**

SHINGO, SHIGEO

Shigeo Shingo (1985), a prominent Japanese industrial engineer and consultant, heralded by some as the "Thomas Edison of Japan," was known for his innovative thinking and creativity in improving productivity and fostering lean production. He is most recognized as the developer of the SMED (single minute exchange of die) system for reducing setup times and costs. His philosophies were instrumental in developing the Toyota production system of just-in-time manufacturing. Specifically, rapid changeover methods developed by him make the small lot of just-in-time manufacturing feasible and economical.

See **Just-in-time manufacturing; NOTED; Setup reduction; SMED.**

Reference

[1] Shingo, S. (1985). *SMED: A Revolution in Manufacturing*, Productivity Press, Cambridge, MA.

SHOP-FLOOR SYSTEM

The data and information system to plan, schedule and control shop floor activities is called the shop-floor system. This system includes the follow capabilities: (1) priority assignment; (2) information on work-in-process; (3) maintains order status data; (4) tracks actual output; (5) collects accounting information; and (6) collects efficiency, utilization, and productivity data on the workforce and machines.

See **MRP; MRP implementation; Order release; Schedule stability.**

SHORTAGE COSTS

When demand exceeds the available inventory for an item, the demand and customer goodwill may be lost. The associate cost is called shortage cost. Some writers estimate shortage costs as equal to the product's contribution margin. If the customer would accept a back order, the cost to the manufacturer is not significant. However, this cost is difficult to estimate.

See **Economic order quantity (EOQ); Inventory flow analysis; Safety stocks: Luxury or necessity.**

SHORTEST PROCESSING TIME RULE

It is common for jobs to await processing at a work center. Several different rules could be used to sequence the jobs waiting at a work center. Shortest processing time rule would require the job with the shortest processing time to be processed ahead of all other waiting jobs. The use of this rule for sequencing will minimize the average time a job spends at the workcenter (waiting time plus processing time), which is called average time in the system.

See **Priority sequencing rules.**

SIGNALING DEVICE

Signaling devices are used in the pull system of production, a system in which production is based on demand or need, not some production plan. For example, if work center B uses a component part produced at work center A, that part will be produced by A only when the need actually exists for it at work center B. A signaling device indicates that need. In many instances, this signaling device will be a card (kanban), but companies have used other signaling devices such as flags, or even colored golf balls (rolled through a tube between work centers). If work center A is a distant supplier, the signaling device could also be a telephone call instead of a physical object. Regardless of what is actually used, the purpose is the same – to indicate the need for more material.

See **C-Kanban; Just-in-time manufacturing implementation; Just-in-time manufacturing; Kanban-based manufacturing systems; Lean manufacturing implementation; P-Kanban.**

SIMULATION

Simulation is the mathematical representation of real-world system whose behavior can be studied through artificial experiments on the mathematical model. Today, computers permit us to build simulation models for small as well as very large real-world systems with accuracy and validity.

A computerized simulation model lets the analyst change input variables and study the effect of the change on the output or other aspects of the whole system.

A simulation model allows relative inexpensive experimentation with a model of real system, which may not permit such experimentation's at all, or may be too expensive an endeavor.

Simulation models can mimic randomness and uncertainty in the real-world system; computer generated probability distributions and random numbers make this possible.

Experimentation's with computer simulation models are used to "test" different decision scenarios until the analyst finds a solution that is satisfactory or appropriate for implementation. Simulations permit the investigation of "what if" scenarios in the search of solutions to problems or improved system design. Plant layout, assembly line, and hundreds of other operations management design problems are routinely solved using simulation.

See **Safety stocks: Luxury or necessity; Simulation analysis manufacturing and logistics systems; Simulation community; Simulation languages; Simulation model; Simulation of production problems using spreadsheet programs; Simulation software selection; Systems, models and simulation.**

SIMULATION ANALYSIS OF MANUFACTURING AND LOGISTICS SYSTEMS

Enver Yücesan

INSEAD, Fontainebleau Cedex, France

John W. Fowler

Arizona State University, Tempe, Arizona, USA

INTRODUCTION: The complexity of modern manufacturing and logistics systems, their high initial cost, and the often lengthy time required to ramp up the production to the designated targets necessitate the use of formal *models* of the system. In such systems, management decisions can be classified according to the time horizon and the extent of the system impacted by the decision.

Strategic decisions are those with a long time horizon (typically over one year), involving major commitment of organizational resources such as the size and the location of a plant, the product portfolio, the choice of technology, etc. *Tactical* decisions have an intermediate time horizon or a narrower scope. Examples include size of the workforce, production rates, factory layout, buffer capacities, etc. *Operational* decisions are made on a daily basis, involving issues such as the release of jobs to the factory floor, the allocation of work to specific work centers, the modification in job routings following a machine breakdown.

To guide the decision making process and to select between alternative courses of action, a set of performance criteria must be in place. While at the strategic level, these criteria are usually based on broad organizational goals (e.g., profitability, growth), a more focused orientation is necessary for tactical and operational decisions. Performance analysis is therefore related to production volume, output quality, cost, and customer service.

A *model* is a stylized representation of a system, reflecting the objectives of the study and embodying simplifying assumptions. Models can be classified into three broad categories: physical, simulation, and analytical. A *physical model* represents the real system by another physical system, typically using a different scale. A *simulation model* is a computer implementation of the logical relationships among the entities in a manufacturing and logistics system. An *analytical model* represents the system through mathematical and/or symbolic relationships.

Since models are intended to support management decisions about the system, a single model will not be capable of supporting all decisions. Rather, different decisions require different models because different aspects of the design and operation of the system will be critical to the model's success in accurately predicting performance. We focus on simulation and analytical models, and show how they can be used in a complementary fashion to design and analyze manufacturing and logistics systems.

The process of modeling involves the following steps with the associated key activities (Chance *et al.*, 1996):

A. *Model Design:*
  1. Identify the issues to be addressed.
  2. Learn about the system.

B. *Model Development*
  3. Choose a modeling approach.
  4. Develop and test the model.
  5. Verify and validate the model.
  6. Develop a model interface for the user.

C. *Model Deployment:*
  7. Experiment with the model.
  8. Implement the results for decision making.

Effective modeling of manufacturing and logistics systems requires both analytical and simulation modeling capability. An analytical model can be developed rather quickly to gain insights into the relationships among the entities in a system, ultimately identifying key parameters affecting the system's performance. Since analytical models will often use approximations, simulation models will be needed to test the robustness of these preliminary conclusions under more realistic operating conditions. Analytical models can be further useful in validating simulation models as well as in reducing the variance of the simulation output, thus minimizing the computational burden.

SIMULATING MANUFACTURING AND LOGISTICS SYSTEMS: Manufacturing and logistics have been one of the earliest simulation application areas (Naylor *et al.*, 1966). Given the inherent complexity of modern manufacturing and logistics systems and the steep performance requirements imposed on them, manufacturing and logistics continues to be one of the principal application areas for simulation. In this section, we identify and discuss the key features of a successful simulation study as a decision support tool. In particular, we focus on model design, model development, and model deployment. Although our discussion of these stages is sequential for ease of exposition, a high degree of overlap (and iteration) is expected – even encouraged – during the execution of a simulation project.

1. ***Model Design:*** Manufacturing and logistics systems are increasingly complex. The natural tendency is to include most, if not all, of the entities of the system in a simulation model. While incorporation of detail may increase the credibility of the model, excessive levels of detail may render a model hard to build, debug, understand, deploy, and maintain. The determination of how much complexity is necessary is the primary goal of the design stage. To this end, one should:
   (a) *Identify project participants:* It is necessary to identify the principal user(s) of the results of the study and their expectations (e.g., materials management, shop floor control, process engineering, warehousing and distribution).
   (b) *Identify project goals:* Once the expectations are determined, they must be operationalized through a project timeline (a planning horizon) and quantifiable performance measures. The planning horizon is an indication of the project's orientation: strategic (e.g., plant location), tactical (e.g., work release policies), or operational (e.g., due date assignment). The orientation, in turn, dictates the level of detail to be incorporated into the model.
   (c) *Produce a Project Specification:* It is highly desirable to produce a "living" document summarizing the above steps, reflecting the project timeline with the associated deliverables.
2. ***Model Development:*** Once the project specification has been finalized, model(s) can be developed. This step entails the selection of a methodology and a tool, the construction and the implementation of the model(s), the collection of the necessary data, and the validation

Table 1. Relative merits of the modeling approaches.

Features	Simulation models	Analytical models
Complexity	Can incorporate virtually unlimited system complexity; a price is paid, however, at run time	Quite limited in depicting complexity; computational complexity increases exponentially with system size
Flexibility	Modern modeling practices (e.g., object orientation) promote reusability of model components	Very easy to change system parameter values; changes in structure may require a new model
Data Requirements	May require large amounts of data	Minimal data requirements due to simple description of system
Output Analysis	Requires sound statistical approach since it is a sampling experiment	Exact solutions typically for the mean and the variance of the performance metric
Efficiency	Modern packages have significantly reduced the model development time; however, experimentation can take a long time	Model development time is highly unpredictable; once the model is built, however, results can be obtained very quickly
User Interface	Modern simulation packages incorporate more and more advanced animation capabilities	Need to be incorporated into a decision support system

and the verification of the model(s). Various sources of uncertainty such as the timing and quantity of customer demand, machine breakdowns, imperfect yields, and other quality problems necessitate a stochastic framework for performance modeling and analysis.

A hybrid (analytic-simulation modeling) approach can be very effective in systems analysis. On the one hand, analytical models may suggest the general form of the relationships between system performance and its parameters; this information can be used in developing and validating the simulation model. On the other hand, it is often necessary to use approximations in analytical models; simulation models can be used to test the robustness of these approximations. Table 1 contrasts the relative merits of the two modeling approaches. Shantikumar and Sargent (1983) illustrate the value of hybrid modeling in effective performance analysis. Pritsker (1989) gives several examples of how analytical and simulation models can be used in a complementary fashion. For useful analytical models for manufacturing and logistics systems see Appendix I.

The most expensive part of modeling is data collection. A simulation model will require detailed data on system characteristics. Data can be used in three ways: in *trace driven simulations*, data can be read into the model from (large) external files. This is a valuable approach for validation and verification. However, it is not possible to replicate the simulation run or to extrapolate the results to more general situations. For instance, if the data do not contain any information on rare events such as a major machine breakdown, the model will never exhibit such behavior either.

Alternatively, an empirical distribution can be fit to the data to significantly reduce the storage requirements and speed up sampling. By fitting an exponential tail to the empirical cumulative distribution function (cdf), one can obtain information about rare events, that were not reflected in the original data set (Bratley *et al.*, 1987). With an empirical cdf, it is also possible to conduct sensitivity analysis through replications.

The most efficient sampling during execution is achieved through the use of a parametric distribution. All modern simulation packages provide utilities for sampling from standard probability distributions. The key challenge in this approach is to select a family of distributions, estimate the parameters of that family, and test the goodness of fit. Law and Kelton (1991), who have a comprehensive list of most widely used parametric distributions, give a detailed discussion on distribution fitting. Software packages such as BestFit (Jankauskas and Lafferty, 1996) or ExpertFit (Law and McComas, 1996) facilitate this task, which was typically ignored by simulation software vendors. The importance of input modeling cannot be overemphasized as Cheng *et al.* (1996) illustrate the contribution of input distributions to the variance in simulation output. For a comprehensive overview of input modeling, see Leemis (1996).

Once the simulation model has been implemented and populated by an appropriate data set, it must be validated and verified. *Model validation* is the process of ensuring that the model accurately represents the actual system for the purposes of the simulation study. *Model verification*, on the other hand, is the process of ensuring that the model produces a correct output given a specified input. Model validation and verification (V&V) is a hard problem (Yücesan and Jacobson, 1996); a large number of heuristic procedures have therefore been developed (Balci, 1994). An excellent illustration of V&V can be found in Krajewski *et al.* (1987), where a large-scale simulation model depicting a "typical" discrete-parts manufacturing environment is built to study the operational factors impacting the performance of manufacturing planning and control policies. The model is validated through a trace-driven experiment, where the simulation is run using historical data from an actual manufacturing site.

3. ***Model Deployment:*** Since stochastic simulation is a statistical sampling experiment, appropriate statistical procedures must be used for output analysis. The discussion in this section is divided into four parts: performance assessment of a single system, comparison among multiple systems, sensitivity analysis, and simulation optimization.

Performance Assessment of a Single System: the first issue in the performance analysis of a single system is to decide whether the simulation run will be of finite horizon or steady state. While steady-state simulations are useful for long-term problems such as capacity planning, most tactical simulations tend to be of finite horizon. For example, Chance (1993) describes a printed circuit board production line that never attains steady-state conditions as new models are continuously introduced into the line. In fact, most manufacturing systems producing a wide variety of products, each in moderate volumes, will exhibit similar behavior. Nuyens *et al.* (1996) caution against using analytical queueing models to assess the performance of such inherently transient systems.

To analyze finite-horizon (transient) simulations, independent replications, i.e., independently seeded runs starting from the same initial condition, are conducted. Let X_j be the performance measure of interest (e.g., average time in system) from the jth replication, $j = 1, 2, \ldots, n$. The sample mean,

$$\overline{X} = \frac{1}{n}\sum_{j=1}^{n} X_j,$$

is an unbiased estimator of the mean performance measure. The sample variance is computed analogously:

$$s^2 = \frac{1}{n-1}\sum_{j=1}^{n}(X_j - \overline{X})^2.$$

An approximate $(1-\alpha)$-percent confidence interval for the performance measure can then be constructed as follows:

$$\overline{X} \pm t_{n-1,1-\alpha/2}\frac{s}{\sqrt{n}},$$

where $t_{n-1,1-\alpha/2}$ is the t-statistic with $n-1$ degrees of freedom. While a large n will ensure the validity of the confidence interval, Schmeiser (1983) suggests that the use of 20 to 30 replications is usually adequate.

Steady-state simulations need further care in analysis, since, in steady-state simulations, the performance measure is defined in the limit as simulated time (or number of simulated entities) goes to infinity, independent of the starting conditions. As simulation runs cannot be infinitely long, steady-state performance measure is approximated through a finite (but sufficiently long) run. The first challenge is to establish the *initial conditions*. The ideal scenario is to start the simulation in steady state; however, steady state is rarely known (otherwise, there would be no need for the simulation study). Hence, initial conditions are set arbitrarily as a matter of convenience, contaminating the data with *initialization bias*.

There are several heuristic ways of dealing with initialization bias. One approach is to overwhelm it by conducting very long runs. For complex manufacturing simulations, this approach may not be feasible. Another approach is to detect when the system has attained steady state and start data collection beyond that point. This is typically referred to as the *truncation* approach, and the observation beyond which data collection is started is called the *truncation point*. There exists a large literature on the determination of the truncation point (Yücesan, 1993).

The next challenge is the serial correlation among the observations from a simulation run, which may lead to an under or over-estimation of the variance. For example, individual customer waiting times in queueing simulations are notoriously positively correlated. Ignoring this positive correlation would lead to an underestimation of the variance, ultimately resulting in smaller, but invalid, confidence intervals. If the underlying system has a *regenerative* structure, it is possible to obtain independent observations by averaging the performance measure over the regenerative cycles (Crane and Iglehart, 1975). For example, in a queueing system with IID arrivals and IID service times with FIFO service, a customer who arrives to find the system completely empty and idle starts a new cycle.

More specifically, assume that, for the output process $Y_1, Y_2, \ldots$, there is a sequence of indices $1 \leq B_1 < B_2 < \ldots$, called *regeneration points*, at which the process starts over probabilistically. The portion of the process between two successive B_j's is called a *regenerative cycle*. Successive cycles are IID replicas of each other; hence, comparable random variables defined over successive cycles are IID. Let $N_j = B_{j+1} - B_j$, $j = 1, 2, \ldots$. Assume $E[N_j] < \infty$. Also let $Z_j = \sum_{i=B_j}^{B_{i+1}-1} Y_i$. The steady state mean is then given by $\mu = \frac{E[Z]}{E[N]}$. Suppose a simulation is run for m cycles generating the following data: $Z_1, Z_2, \ldots, Z_m$, and $N_1, N_2, \ldots, N_m$. Each of these sequences consists of IID random variables. In general, however, Z_j and N_j are not independent. The point estimator $\overline{Z}/\overline{N}$ is not an unbiased estimator for μ. A simple transformation reduces bias (Law and Kelton, 1991: 559).

While this is an elegant theoretical approach, it may not be feasible in practice. First, it may not be possible to identify a regenerative state in the underlying system. Second, even if such a state is identified, the cycle may last too long to collect a sufficient number of observations.

An alternative approach, which provides no guarantee of independence, is to batch the observations and use the batch means in the construction of a confidence interval. Assume that the output process of a simulation is covariance stationary. Suppose we make a simulation run of length m and then divide the resulting observations $Y_1, Y_2, \ldots, Y_m$ into n adjacent batches of size k (i.e., $m = kn$). In other words, we assume that we make a long simulation run and start the data collection after going through the initial transient. Let $\overline{Y}_j(k)$ be the mean of the k observations in the jth batch ($j = 1, 2, \ldots, n$) and let $\overline{\overline{Y}}$ be the grand sample mean, which is the point estimator for μ. If k is sufficiently large, $\overline{Y}(k)$'s will approximately be uncorrelated. Furthermore, for large k, $\overline{y}_j(k)$ will be approximately normally distributed. Under the assumption of a covariance stationary process, $\overline{Y}_j(k)$ are therefore normal random variables with the same mean and variance as the original process. A confidence interval can then be constructed using the batch means in the usual fashion.

Schmeiser (1983) describes the approach and discusses the inherent tradeoffs between the total number of batches and the number of observations per batch in an experiment with a fixed computing budget. Other approaches such as the standardized time series (Schruben, 1983) eliminate altogether the need to estimate the variance by using a ratio estimator, where the variance terms cancel out.

Comparison Among Multiple Systems: The real utility of simulation lies in comparing different scenarios that might represent competing system designs or alternative operating policies. This is a common situation in manufacturing and logistics where system designs arise from choosing among competing machines, schedules, or facilities, subject to constraints on the available budget or technology.

The simplest case arises when two system designs are to be compared (e.g., an existing system and a proposed alternative). A *paired-t confidence interval* can be constructed in this case. For $i = 1, 2$, let $X_{i1}, X_{i2}, \ldots, X_{in}$ be a sample of n IID observations from system i and let $\mu_i = E[X_{ij}]$ be the expected response of interest. A confidence interval for the difference $\delta = \mu_1 - \mu_2$ is desired. Note that a confidence interval for the difference is preferable to a hypothesis test to assess whether the observed difference is significantly different from zero; whereas a hypothesis test results in a "reject" or "fail-to-reject" conclusion, a confidence interval gives us this information (i.e., if the confidence interval contains zero, the difference between the two systems is not statistically significant) and quantifies by how much the two systems differ, if at all.

To construct the confidence interval, we pair up the observations to define $Z_j = X_{1j} - X_{2j}$ for $j = 1, 2, \ldots, n$. Then Z_j are IID random variables with $E[Z_j] = \delta$. Then, let

$$\overline{Z} = \frac{1}{n} \sum_{j=1}^{n} Z_j \text{ and}$$

$$VAR(\overline{Z}) = \frac{1}{n(n-1)} \sum_{j=1}^{n} (Z_j - \overline{Z})^2,$$

which yields a $100(1-\alpha)$ percent confidence interval

$$\bar{Z} \pm t_{n-1,1-\alpha/2}\sqrt{VAR(\bar{Z})}.$$

If Z_j's are normally distributed, then the confidence interval is exact; otherwise, we invoke the central limit theorem for approximate coverage. One could also use nonparametric techniques (Yücesan, 1994) bypassing all distributional assumptions.

In constructing the confidence interval, we did not have to assume that X_{1j} and X_{2j} are independent; nor did we have to assume that $VAR(X_{1j}) = VAR(X_{2j})$. The assumption of common variance is hard to justify in practice, as variance is expected to depend on the alternative designs under consideration, and thus is expected to vary among these alternatives. However, allowing positive correlation between X_{1j} and X_{2j} is generally desirable, since this will reduce the variance of Z_j, yielding a smaller confidence interval. Such correlation induction strategies are known as *variance reduction techniques* (VRT). VRTs can greatly enhance the statistical efficiency of simulation experiments. Positive correlation can be induced through *common random numbers* (CRN). The basic idea is to compare alternative configurations under similar experimental conditions so that any observed differences in performance are due to differences in system configurations rather than the fluctuations in the experimental conditions. In simulation, similar experimental conditions are obtained by driving the replications under different configurations with a common set of random numbers. In the terminology of classical experimental design, CRN is a form of blocking, i.e., comparing the like with the like.

In the above case with two competing system designs,

$$VAR(\bar{Z}) = \frac{VAR(Z_j)}{n}$$
$$= \frac{VAR(X_{1j}) + VAR(X_{2j}) - 2COV(X_{1j}, X_{2j})}{n}.$$

If the simulations of the two systems are conducted independently using different random numbers, the covariance term will vanish. On the other hand, if positive correlation is induced through CRN, then $COV(X_{1j}, X_{2j}) > 0$, resulting in a reduction in the variance of $\bar{Z}$. Note that correlation induction is possible due to the deterministic, reproducible nature of random number generators. For CRN to achieve the maximum possible variance reduction, synchronization among the simulation runs is necessary; see Glasserman and Yao (1992). For a complete discussion of variance reduction techniques, see Bratley *et al.* (1987).

In situations where more than two system configurations are to be compared (e.g., comparisons with a standard or all pairwise comparisons), confidence intervals can be constructed simultaneously. Typically, the Bonferroni inequality is used to ensure that the overall confidence level is at least $1-\alpha$. The *Bonferroni inequality* implies that, for c simultaneous confidence intervals to have an overall confidence level of at least $1-\alpha$, each individual confidence interval should be constructed at the level $1-\alpha/c$. For example, if $c = 10$ and the desired overall confidence level is 90 percent, each individual confidence interval should be at the 99 percent level. Note that for comparisons with a standard system (i.e., $\mu_1 - \mu_k\ \forall k$, where we call the standard system 1 and the other variants systems $2, 3, \ldots, k$), $c = k-1$. For all pairwise comparisons, $c = k(k-1)/2$.

A more ambitious goal would be not only to compare k competing system configurations, but to select the "best" configuration. "Best" is defined in accordance with the performance measure of interest (e.g., the smallest cycle time or the largest throughput). A two-stage procedure is typically used (Bechoffer *et al.*, 1995): in the first stage, a fixed number of replications for each system configuration is performed to estimate the variance of the performance measure of interest. The resulting variance estimates are then used to determine the additional number of replications from each configuration necessary in a second stage of sampling in order to reach a decision with some correct selection guarantee. Chen (1996) has recently introduced a dynamic sampling scheme with the objective of eliminating "inferior" designs as quickly as possible, thus allocating further computing budget to the more promising configurations.

Sensitivity Analysis: Another objective in a simulation study might be to understand the relationships between various operating decisions and system performance. For example, one might be interested in examining the impact of lot sizing decisions on total cycle time. Such an investigation, typically referred to as *sensitivity analysis*, may be a preamble to a more detailed design and optimization study.

Traditional approaches for sensitivity analysis have been adopted from the theory of experimental design. In experimental design terminology, the input parameters and structural assumptions of a model are called *factors*, and the output performance measures are called *responses*. For example, in a simulation model

Table 2. The design matrix for the 2^3 factorial experiment.

Design point	Factor 1	Factor 2	Factor 3	Response
1	−	−	−	R_1
2	+	−	−	R_2
3	−	+	−	R_3
4	+	+	−	R_4
5	−	−	+	R_5
6	+	−	+	R_6
7	−	+	+	R_7
8	+	+	+	R_8

depicting a flexible manufacturing system, the number of machines, the number of pallets, the rate at which jobs are released into the system, the rate of machine breakdowns are all examples of factors; responses of interest may include the total cycle time or capacity utilization.

In simulation, experimental design provides a methodology to decide which particular configurations, i.e., settings for factor values, to simulate so that the desired information can be obtained with the least amount of computation. Suppose the system under study has $k \geq 2$ factors. One economical strategy, which allows for the quantification of main effects as well as interactions, is a 2^k *factorial design*, where two fixed values (typically referred to as *levels*) are chosen for each factor and simulations are conducted at each of the 2^k possible factor-level combinations. The resulting *design matrix* for a 2^3 factorial experiment is presented in Table 2. Typically, a minus sign is associated with a low level, while a plus sign is associated with a high level.

In this setting, the *main effect* of factor j is the average change in the response due to moving factor j from its "−" level to its "+" level while holding all other factors fixed. This average is taken over all combinations of the other factor levels in the design.

For example, the main effect of factor 1 in Table 2 is given by:

$$e_1 = \frac{(R_2 - R_1) + (R_4 - R_3) + (R_6 - R_5) + (R_8 - R_7)}{4},$$

which can be read off the design matrix. A *two-way interaction* between factors i and j is defined as half the difference between the average effect of factor i when factor j is at its "+" level (and all other factors are held constant) and the average effect of factor i when factor j is at its "−" level. In the above illustration, the two-way interaction between factors 1 and 2 is given by:

$$e_{12} = \frac{1}{2}\left[\frac{(R_4 - R_3) + (R_8 - R_7)}{2} - \frac{(R_2 - R_1) + (R_6 - R_5)}{2}\right].$$

Higher-order interactions are defined in the same manner. The principal problem with factorial experiments is the exponential growth of the design in the number of factors, k. For instance, $k = 10$ generates $2^{10} = 1024$ design points. One way to reduce the computational burden is to adopt a *fractional factorial design*. A 2^{k-p} fractional factorial design is constructed by choosing a certain subset of all the 2^k possible design points. We would like p to be large from a computational perspective, but a larger p will also result in less information from the experiment (i.e., we may only observe the main effects and two-way interactions, but none of the higher-order interactions).

Another approach to reduce the effective number of factors is *group screening*, where individual factors would be grouped together in some way. See Krajewski *et al.* (1987) for a good example of grouping in a large-scale simulation. Each group is then regarded as a single "aggregate factor" with all the individual factors within the group set to their levels together. The effects would then be interpreted accordingly as what happens when an entire group moves between levels. The groups that appear unimportant are discarded from further consideration. A recursive process is used to ungroup the individual factors in the important groups until some subset of the original individual factors is finally identified as being important.

Full and fractional factorial designs as well as group screening are discussed in detail in Kleijnen (1987), where other sensitivity analysis approaches such as *response surface methodology* (RSM) are also presented. In RSM, the input-output relationship (i.e., the impact of input factors on the response) is depicted as a polynomial function. RSM can be coupled with a search algorithm for simulation optimization.

An innovative factor screening idea is to change the factor settings *during* the simulation run rather than *between* runs to assess the impact of a change in the value of a factor on the simulation response. Schruben and Cogliano (1987) oscillate the different input parameter values, each at a different frequency, during the course of a simulation. The output is then examined (in the *frequency* domain rather than the *time* domain) to see which input parameters' oscillation frequency can be detected in the output. While an important factor's oscillations will produce noticeable oscillation in the output at the corresponding frequency, those of unimportant factors will not be apparent. This approach may guide the selection of an appropriate fractional factorial design and give an indication of the form of the polynomial function depicting the input-output relationship in RSM. Robinson *et al.* (1993) provide an example of using this technique to analyze a manufacturing system.

Simulation Optimization: Sensitivity analysis was defined as an investigation of the impact of input parameters (factors) on system performance (response). If the input parameters vary continuously, this is equivalent to computing the partial derivatives of the expected response function with respect to the input parameters. The vector of these partial derivatives is called the *gradient* of the expected response function. The gradient information not only gives the sensitivity of simulation's expected response to small changes in the input parameters, but it is also fundamental in most mathematical programming approaches for finding optimal settings of the input parameters. More specifically, gradient information is used in a search algorithm to determine a direction to seek the optimum.

Classical optimization techniques applied to simulation tend to be computationally intensive (Jacobson and Schruben, 1989). One should note, however, that simulation optimization is an extremely difficult problem, since it combines the problems of deterministic mathematical programming and statistical estimation. Deterministic optimization is, in general, a hard problem in itself; this is further complicated by the fact that the inherent randomness of the simulation output makes it impossible even to evaluate exactly the objective function to be optimized.

An innovative approach to this problem has been the technique of *perturbation analysis*. While the technique was introduced in Ho *et al.* (1979), a comprehensive treatment is given in Glasserman (1991) and its application to manufacturing systems is discussed in Glasserman (1994). Perturbation analysis is an efficient method of implementing the idea of finite differencing for derivative estimation, namely $[L(\theta + h) - L(\theta)]/h$, where θ is an input parameter and $L(\bullet)$ is the expected system response. In particular, *infinitesimal* perturbation analysis (IPA) evaluates the derivative

$$\frac{\partial L(\theta)}{\partial \theta} = \lim_{h \to 0} \frac{L(\theta + h) - L(\theta)}{h}$$

exactly through a convergent sequence. The attractive feature of IPA is that it requires a single simulation run to estimate the gradient with p input parameters, while the classical finite-differencing approach would need $2p$ runs for this purpose. Conditions for the applicability of IPA are also presented in Glasserman (1991).

Gradient estimation methods based on likelihood ratios have also been developed (Reiman and Weiss, 1989). Related methods are described by Rubinstein and Shapiro (1993). A unifying framework is given by L'Ecuyer (1990).

SUMMARY: The increasing cost and complexity of manufacturing and logistics systems necessitate accurate performance modeling and analysis of these systems. Both analytical and simulation models can be useful in these endeavors. Analytical models are most useful in understanding basic system performance and in "rough-cut" analysis. However, for precise estimates of system performance, simulation is often necessary.

Simulation has been used successfully as an aid in the design of new production and distribution facilities. It has also been used to evaluate suggested modifications to existing systems. Decision makers have found it useful for evaluating the impact of capital investment in equipment and physical facility as well as the impact of staffing and changes in operating rules. In short, simulation has proven to be a strategic technology for both product and process design.

We conclude this chapter with a warning: the introduction of user-friendly simulation software with impressive graphical interfaces has distracted attention from the statistical aspects of a simulation study. More specifically, the following pitfalls should be avoided in a sound simulation study (Law and McComas, 1989):

1. Failure to have a well-defined set of objectives.
2. Inappropriate level of model detail.
3. Failure to interact with management on a regular basis.
4. Insufficient simulation and statistics training.
5. Inappropriate simulation software.
6. Misuse of animation.
7. Replacement of a distribution with its mean.
8. Use of the wrong probability distribution.
9. Incorrect modeling of machine down times.
10. Misinterpretation of simulation results.

Hybrid modeling, carefully planned simulation experiments, and proper output analysis will not only save considerable time for the decision maker, who typically works under the pressure of tight deadlines, but also improve the accuracy and robustness of the results upon which strategic, tactical, or operational decisions will be based.

APPENDIX I: ANALYTICAL MODELS FOR MANUFACTURING AND LOGISTICS SYSTEMS: Queueing theory has been successfully applied to the modeling and performance assessment of manufacturing and logistics systems. Several recent initiatives primarily focus on stochastic models of manufacturing and logistics systems (e.g., Buzacott and Shantikumar, 1993; Papadopoulos *et al.*, 1993; Yao, 1994; Altiok, 1997). Single-stage queueing models have been proposed for make-to-order systems. Queueing models have also been used in the analysis of asynchronous transfer lines. The key complication here is the interference among the work centers due to finite buffer capacity. Decomposition schemes that usually employ phase-type distributions are invoked to analyze these systems (Perros, 1994).

Queueing networks (Walrand, 1988; Van Dijk, 1993) provide a powerful modeling framework. Job shops are typically modeled through open queueing networks (Jackson, 1963), while closed queueing networks are used extensively to model modern manufacturing facilities such as flexible manufacturing systems (Buzacott and Yao, 1986). Closed queueing network models are also used to model pull-type production systems (Spearman *et al.*, 1990). Recently, sample path-based techniques have been introduced to analyze pull systems (Tayur, 1992).

One obstacle to the widespread application of queueing models in the design and analysis of manufacturing and logistics systems has been the lack of software packages that include a convenient user interface and, more importantly, a robust engine incorporating accurate approximations. Notable exceptions include CAN-Q (Solberg, 1976), QNA (Whitt, 1983), RESQME (Gordon *et al.*, 1986), MPX (Network Dynamics, Inc., 1988), QNAP2 (INRIA, 1984), and Factory Explorer (Wright, Williams & Kelly, 1997). An in-depth discussion of some of these packages can be found in Papadopoulos *et al.* (1993) and in Suri *et al.* (1995).

We close this appendix with approximation formulas that are both very intuitive and powerful. These formulas provide immediate insight into the performance of the system under study. Let t_a represent the mean time between arrivals; $r_a = 1/t_a$ is then the rate of arrivals. In a serial line without yield loss or rework r_a represents the throughput at every station. $c_a^2 = \sigma_a^2/t_a^2$ is the squared coefficient of variation (SCV) of the arrival process with variance σ_a^2. Similarly, let t_e represent the mean processing time, σ_e^2 the processing time variance, and $c_e^2 = \sigma_e^2/t_e^2$ the processing time SCV. Let m denote the number of identical machines at the work center. Finally, let $u = r_a/r_e = r_a t_e/m$ represent the utilization of the work center. For a single-server work center ($m = 1$) with general arrival and processing time distributions, mean time spent in the queue is approximated by (Kingman 1961):

$$CT_q = \left(\frac{c_a^2 + c_e^2}{2}\right)\left(\frac{u}{1-u}\right) t_e.$$

This approximation has several nice properties. First, it is exact for the M/M/1 and M/G/1 queues. Second, the expression separates into three terms: a dimensionless variability term, a utilization term, and a capacity term, which represent three distinct improvement opportunities for reducing the queue time at a work center.

For a work center with m machines, the above expression becomes:

$$CT_q = \left(\frac{c_a^2 + c_e^2}{2}\right)\left(\frac{t_e u^{(\sqrt{2(m+1)}-1)}}{m(1-u)}\right).$$

The total time spent at the work center is then given by $CT = CT_q + t_e$. Applying Little's law, we also obtain the expected WIP level at the work center: $WIP = r_a \times CT$.

In a serial production line, where all the output from work center i becomes input to work center $i+1$, the departure rate from i must equal the arrival rate to $i+1$:

$$t_a(i+1) = t_d(i).$$

Furthermore,

$$c_a(i+1) = c_d(i),$$

where, for $m = 1$,

$$c_d^2 = u^2 c_e^2 + (1-u)c_a^2,$$

and, for $m > 1$,

$$c_d^2 = 1 + (1-u)^2\left(c_a^2 - 1\right) + \frac{u^2}{\sqrt{m}}\left(c_e^2 - 1\right).$$

The simplifying assumptions necessary in building analytical models of manufacturing and logistics systems can limit their utility in performance assessment. A simulation study therefore becomes the next step.

See **Global facility location analysis; Initialization bias in simulation; Inventory floor analysis; Kanban-based manufacturing systems; Maintenance management decision models; Manufacturing analysis using chaos, fractals, and self organizations; Manufacturing system modeling using petri nets;**

Model validation and verificaiton; Queueing theory; Sensitivity analysis; Simulation languages; Simulation model; Simulation of production problems using spreadsheet programs; Simulation software selection; Systems, models, and simulation; Trace driven simulations; Truncation point in simulation.

Short Bibliography/References

[1] Altiok, T. (1997). *Performance Analysis of Manufacturing Systems*. Springer, New York.

[2] Balci, O. (1994). "Model validation, verification, and testing techniques throughout the life cycle of a simulation study." *Annals of Operations Research*, 53, 121–173.

[3] Banks, J., B. Burnette, H. Kozloski, and J. Rose (1995). *Introduction to SIMAN V and CINEMA V*. John Wiley & Sons, New York.

[4] Bechoffer, R., T.J. Santner, and D. Goldsman (1995). *Design and Analysis of Experiments for Statistical Selection, Screening, and Multiple Comparisons*. Wiley, New York.

[5] Bratley, P., B. Fox, and L. Schrage (1987). *A Guide to Simulation* (2nd Ed.), Springer-Verlag. New York.

[6] Buzacott, J.A. and J.G. Shantikumar (1993). *Stochastic Models of Manufacturing Systems*. Prentice-Hall, New Jersey.

[7] Buzacott, J.A. and D.D. Yao (1986). "Flexible manufacturing systems: a review of analytical models." *Management Science*, 32, 890–905.

[8] Chance, F. (1993). "Conjectured upper bounds on transient mean total waiting times in queueing networks." In: *Proceedings of the Winter Simulation Conference*, 414–421.

[9] Chance, F., J. Robinson, and J. Fowler (1996). "Supporting manufacturing with simulation: model design, development, and deployment." In: *Proceedings of the Winter Simulation Conference*, 114–121.

[10] Chen, C.H. (1996). "A lower bound on the correct subset selection probability and its application to discrete-event simulation." *IEEE Transactions on Automatic Control*, 41, 1227–1231.

[11] Cheng, R.H.C., W. Holland, and N.A. Hughes (1996). "Selection of input models using bootstrap goodness of fit." In: *Proceedings of the Winter Simulation Conference*, 199–206.

[12] Conway, R., W.L. Maxwell, J.O. McClain, and S. Worona (1987). *User's Guide to Xcell + Factory Modeling System* (2nd Ed.), Scientific Press, California.

[13] Crane, M.A. and D.L. Iglehart (1975). "Simulating stable stochastic processes, III: regenerative processes and discrete event simulations." *Operations Research*, 23, 33–45.

[14] Glasserman, P. (1991). *Gradient Estimation via Perturbation Analysis*. Kluwer Academic Publishers, Massachusetts.

[15] Glasserman, P. (1994). "Perturbation analysis of production networks." In: *Stochastic Modeling and Analysis of Manufacturing Systems* (Yao, Ed.). Springer-Verlag, New York.

[16] Glasserman, P. and D.D. Yao (1992). "Some guidelines and guarantees for common random numbers." *Management Science*, 38, 884–908.

[17] Gordon, R.F., E.A. MacNair, P.D. Welch, K.J. Gordon, and J.F. Kurose (1986). "Examples of using the research queueing package modeling environment (RESQME)." In: *Proceedings of the Winter Simulation Conference*, 494–503.

[18] Ho, Y.C., M.A. Eyler, and T.T. Chien (1979). "A gradient technique for general buffer storage design in a serial production line." *International Journal of Production Research*, 17, 557–580.

[19] INRIA (1984). *New Users' Introduction to QNAP2*. Sophia Antipolis, France.

[20] Jackson, J.R. (1963). "Jobshop-like queueing systems." *Management Science*, 10, 131–142.

[21] Jacobson, S.H. and L. Schruben (1989). "Techniques for simulation response optimization." *Operations Research Letters*, 8, 1–9.

[22] Jankauskas, L. and S. McLafferty (1996). "BESTFIT, Distribution fitting software by Palisade Corporation." In: *Proceedings of the Winter Simulation Conference*, 551–555.

[23] King, C.B. (1996). "Taylor II manufacturing simulation software." In: *Proceedings of the Winter Simulation Conference*, 569–573.

[24] Kingman, J.F.C. (1961). "The single-server queue in heavy traffic." In: *Proceedings of the Cambridge Philosophical Society*, 57, 902–904.

[25] Kleijnen, J.P.C. (1987). *Statistical Tools for Simulation Practitioners*. Marcel Dekker, New York.

[26] Krajewski, L.J., B.E. King, L.P. Ritzman, and D.S. Wong (1987). "Kanban, MRP, and shaping the manufacturing environment." *Management Science*, 33, 39–57.

[27] Law, A.M. and M.G. McComas (1996). "EXPERTFIT: Total support for simulation input modeling." In: *Proceedings of the Winter Simulation Conference*, 588–593.

[28] Law, A.M. and M.G. McComas "(1989). "Pitfalls to avoid in the simulation of manufacturing systems." *Industrial Engineering*, 31, 28–31, 69.

[29] Law, A.M. and D.W. Kelton (1991). *Simulation Modeling and Analysis* (2nd Ed.), McGraw-Hill, New York.

[30] L'Ecuyer, P. (1990). "A unified view of the IPA, SF, and LR gradient estimation techniques." *Management Science*, 36, 1364–1383.

[31] Leemis, L.M. (1996). "Discrete event simulation input process modeling." In: *Proceedings of the Winter Simulation Conference*, 39–46.

[32] Naylor, T.H., J.L. Balintfy, D.S. Burdick, and K. Chu (1966). *Computer Simulation Techniques*. John Wiley & Sons, New York.

[33] Network Dynamics, Inc. (1991). MPX *User Manual*. Burlington, MA.

[34] Nuyens, R.P.A., N.M. Van Dijk, L. Van Wassenhove, and E. Yücesan (1996). "Transient behavior of simple queueing systems: implications for FMS models." *Simulation: Application and Theory*, 4, 1–29.

[35] Papadopoulos, H.D., C. Heavey, and J. Browne (1993). *Queueing Theory in Manufacturing Systems Analysis and Design*. Chapman & Hall. London, UK.

[36] Perros, H.G. (1994). *Queueing Networks with Blocking*. Oxford University Press. Oxford.

[37] Pritsker, A.A.B. (1989). " Developing analytic models based on simulation results." In: *Proceedings of the Winter Simulation Conference*, 653–660.

[38] Pritsker Corporation (1994). *SLAMSYSTEM User's Guide*. Version 4.5. Indianapolis, IN.

[39] Production Modeling Corporation of Utah (1989). *ProModel User's Manual*. Orem, UT.

[40] Reiman, M.I. and A. Weiss (1989). "Sensitivity analysis for simulations via likelihood ratios." *Operations Research*, 37, 830–844.

[41] Robinson, J., L.W. Schruben, and J.W. Fowler (1993). "Experimenting with large-scale semiconductor manufacturing simulations: a frequency domain approach to factor screening." In: *Proceedings of the Second IE Research Conference*, 112–116.

[42] Rubinstein, R.Y. and A. Shapiro (1993). *Discrete Event Systems: Sensitivity Analysis and Stochastic Optimization by the Score Function Method*. Wiley, New York.

[43] Schmeiser, B.W. (1983). "Batch size effects in the analysis of simulation output." *Operations Research*, 31, 565–568.

[44] Schruben, L. (1983). "Confidence interval estimation using standardized time series." *Operations Research*, 31, 1090–1107.

[45] Schruben, L. and V. J. Cogliano (1987). "An experimental procedure for simulation response surface model identification." *Communications of the ACM*, 30, 716–730.

[46] Shantikumar, J.G. and R.G. Sargent (1983). "A unifying view of hybrid simulation/analytic models and modeling." *Operations Research*, 31, 1030–1052.

[47] Solberg, J.J. (1976). "Optimal design and control of computerized manufacturing systems." In *Proceedings of AIIE Systems Engineering Conference*, 138–144.

[48] Spearman, M.L., D.L. Woodruff, and W.J. Hopp (1990). "CONWIP: a pull alternative to kanban." *International Journal of Production Research*, 28, 879–894.

[49] Suri, R., G.W.W. Diehl, S. De Treville, and M.J. Tomsicek (1995). "From CAN-Q to MPX: evolution of queueing software for manufacturing." *Interfaces*, 25, 128–150.

[50] Tayur, S.R. (1992). "Properties of serial kanban systems." *Queueing Systems*, 12, 297–318.

[51] Van Dijk, N.M. (1993). *Product Forms for Queueing Networks: A System's Approach*. Wiley, New York.

[52] Walrand, J. (1988). *An Introduction to Queueing Networks*. Prentice Hall. New Jersey.

[53] Whitt, W. (1983). "The Queueing Network Analyzer." *The Bell System Technical Journal*, 62, 2779–2815.

[54] Wright, Williams, & Kelly. (1997) *Factory Explorer User's Manual*, Version 2.2. Dublin, California.

[55] Yao, D.D. (1994). *Stochastic Modeling and Analysis of Manufacturing Systems*. Springer-Verlag, New York.

[56] Yücesan, E. (1993). "Randomization tests for initialization bias in simulation output." *Naval Research Logistics*, 40, 643–664.

[57] Yücesan, E. (1994). "Evaluating alternative system configurations using simulation: a non-parametric approach." *Annals of Operations Research*, 53, 471–484.

[58] Yücesan, E. and S.H. Jacobson (1996). "Computational issues for Accessibility in discrete event simulation." *ACM Transactions on Modeling and Computer Simulation*, 6, 53–75.

SIMULATION EXPERIMENTS

See **Simulation; Simulation analysis of manufacturing and logistics systems.**

SIMULATION LANGUAGES

David L. Olson

Texas A&M University, College Station, Texas USA

James R. Evans

University of Cincinnati, Cincinnati, Ohio, USA

DESCRIPTION: Simulation analysis has become and increasingly important operations management tool, especially in recent years. Lane, Mansour and Harpell (1993) reported that simulation ranked third, behind statistics and mathematical programming as the quantitative technique considered most useful by practitioners in a 1988 survey. Over their study period (1973, 1978, 1983, 1988), the importance of simulation increased. The growth of simulation use is due to a number of factors, most notably the proliferation of computers, and the availability of increasingly powerful simulation software tools. Another aspect of simulation that is very valuable is the ability to support evaluation of complex systems before their development.

Simulation provides an extremely useful approach for quickly analyzing complex systems in any decision environment, especially when there is uncertainty regarding system performance or inputs. Simulation is relatively easy to understand and use, and the output can be designed to meet user needs. Simulation is becoming more popular because of the widespread availability and increased performance of personal computers. This trend is enhanced by the development of many simulation software products for modeling specific types of problems, for analyzing input data, for analyzing output, and for animating scenarios.

We will review simulation software by five broad categorical types, and demonstrate how simple spreadsheet models can be applied to useful operations management problems. Specifically, we demonstrate network scheduling, waiting line, and inventory models, and also illustrate the use of risk-analysis add-in software for spreadsheet simulation.

HISTORICAL PERSPECTIVE: A wide variety of language support, ranging from general purpose languages to specialty products, exists for simulation. We briefly review the more popular general-purpose languages (GPPL), general-purpose simulation languages (GPSL), special-purpose simulation systems (SPSS), spreadsheet-based simulation tools (SBST), and adjunct support.

General Purpose Programming Languages: General-purpose programming languages (e.g., FORTRAN) were the first tools for implementing simulation. Developing a simulation using a GPPL involves more work than when using a GPSL or SPSS; however, GPPLs provide the greatest modeling and output flexibility. Cochran, Mackulak, and Savory (1995) conducted a survey of simulation practitioners from industry and research institutes, and found that 82 percent of those surveyed used a general purpose programming language. C and FORTRAN were by far the most, with C rapidly replacing FORTRAN in popularity.

General-Purpose Simulation Languages: GPSLs are designed for large-scale commercial development of a broad variety of simulations, although most focus on queuing situations. One of, if not the, first such languages (early 1960s) was the general purpose simulation system (GPSS). GPSS has evolved over the years, and is available commercially. It is used in many models of factory environments and other problems where waiting lines are present. SLAM and SIMAN are similar products, and also are used widely within industry. Their flexibility allows simulation analysts to model just about any realistic situation. The full scale versions require a substantial investment in both time and money. These products continue to be developed, with stronger features for modeling, animation, and statistical analysis. This is in response to new competitors, such as ARENA and Extend, both of which have graphical modeling environments, animation, and graphical display of output statistics.

Cochran *et al.* (1995) found that 84 percent of the simulation analysts they surveyed used this class of simulation support. SLAM II and SIMAN were the most popular simulation languages. Cochran, *et al.* (1995) noted that SIMAN was the most popular in industry, while SLAM II was the most popular in academic circles. They attributed this difference to the reliance of academics on main frame systems that were already paid for. Note, however, that many features of SIMAN were developed specifically for manufacturing environments.

Special-Purpose Simulation Systems: SPSSs are specifically tailored to be applied to specific problem environments; hence, developing simulation models of those systems is generally easier than with SPSLs. Products in this class include the very powerful SPSLs developed for specific problem types. COMNET III, Best Network, and OPNET Modeler are designed for simulating communication networks. COMNET III's graphical interface allows one to develop a model by using the mouse to drop icons on a drawing window and entering relevant data in predefined forms, a feature available on the newer versions of simulation languages such as SLAM and ARENA. CACI, the company that produces COMNET III, also produces SIMFACTORY for factory organization models, and SIMPROCESS for business process analysis. MOGUL and NETWORK II.5 support simulation of computer systems. Maintsim is a product for analyzing maintenance and repair of equipment. Cellular manufacturing environments are supported by MAST Simulation Environment, and MEModel is designed for simulation of medical decision problems. These specialty products have had a high growth period that should continue into the foreseeable future.

Cochran *et al.* (1995) found much lower use of this class of product (40 percent of those surveyed). PROMODEL was the most popular product, with a market share of 14 percent of those surveyed. The use of this class of product appears to be slowly increasing.

Spreadsheet-Based Simulation Tools: Spreadsheet simulation models have been used for cash flow simulation at least as long ago as the 1970s, when IFPS (Gray, 1983) was developed. IFPS featured the ability to treat interest rates and other financial variables as random variables, and provided the ability to replicate scenarios to generate a distribution of outcomes. Today, IFPS has been superceded by common spreadsheet products such as Lotus, Quattro, and EXCEL, along with add- ins such as @Risk (see Winston, 1996) and Crystal Ball (see Evans and Olson, 1998).

Adjunct Support: Many products that perform specific tasks involved in simulation studies have been developed. The two areas of such support that have received the greatest attention are statistical analysis and animation.

BestFit, CurveFit I, and ExpertFit are examples of software that determine distributions for given data. Simstat 2.0 also provides tools for analysis of simulation output. These products can be used with other systems used for conducting the actual simulation. Another factor is that simulation analysis involves statistical sampling from model output. Sound statistical practice needs to be applied, meaning that sufficient samples must be obtained, requiring repetitive runs of the simulation model. The number of runs to make is a statistical question, but at least 30 runs should be obtained for most applications, and preferably 100 or more runs to allow for more accurate statistical analysis. Add-on products such as @Risk and Crystal Ball make obtaining multiple runs much easier.

Almost all major simulation systems have developed animation support for their products. Animation makes it easier to see how the model is working, a useful feature for debugging, verification, and validation. Animation is even more useful in communicating results to others. Taylor II has excellent animation features. This system has click and drag model building features, as well as three dimensional graphics. Other click and drag products include Hocus Simulation 8000 for processes, and Extend for general purpose simulation (Swain, 1995). FlowChart, ithink, and Process Charter are systems with which analysts can develop simulation models graphically, thereby avoiding the need to program.

STRATEGIC PERSPECTIVE: Today, spreadsheets have replaced the calculator as one of the most useful managerial tools. Few realize the power that spreadsheets have for simulating meaningful problems. The explosive growth of Lotus 1-2-3 and EXCEL have broadened the use of spreadsheets significantly. The use of visual basic (Gray, 1996) to extend spreadsheet applications also advances the potential of spreadsheets as simulation tools. Spreadsheet add-ins such as Crystal Ball and @Risk relieve the user of the burden of trying to replicate spreadsheet models with probabilistic components. They can quickly generate random outcomes for each probabilistic variable, replicate the simulation a sufficient number of times to allow valid statistical results, and compute summary statistics and histograms to analyze risk. They also provide graphical means of displaying the output to make it more easily understood and provide sensitivity analysis of key inputs.

Spreadsheet-based simulation models provide a tool capable of supporting many standard operational management analyses. This is especially true considering the power of add-on software to make implementation of statistically valid results easier. The spreadsheet allows definition of the basic model, and add-on supports the replications and data analysis necessary to obtain statistically valid results. The reasons for using spreadsheet-based simulation models include ease of use, lower cost, and accessibility.

See **Average inventory; Exponential arrivals; Exponential service times; General-purpose language; Holding cost rate; Multi-server problem; PERT; Simulation analysis of manufacturing and logistics systems; Simulation languages; Simulation of production problems using spreadsheet programs; Simulation software selection; Single-server problem; Steady-state behavior; Waiting line model.**

References

[1] Cochran, J.K., G.T. Mackulak, and P.A. Savory (1995). "Simulation project characteristics in industrial settings." *Interfaces* 25:4, 104–113.

[2] Evans, J.R. and D.L. Olson (1998). *Simulation*. Prentice-Hall, Englewood Cliffs, NJ.

[3] Gray, Paul (1983). *Student Guide to IFPS (Interactive Financial Planning System)*. McGraw-Hill, New York.

[4] Gray, Paul (1996). *Visual IFPS/Plus for Business*. Prentice-Hall, Englewood Cliffs, NJ.

[5] Lane, M.S., A.H. Mansour, and J.L. Harpell (1993). "Operations research techniques: A longitudinal update 1973–1988," *Interfaces* 23:2, 63–68.

[6] Law, A.M. and Kelton, W.D. (1991). *Simulation Modeling & Analysis*. McGraw-Hill, Inc., New York.

[7] Swain, J.J. (1995). "Simulation Survey: Tools for process understanding and improvement." *OR/MS Today* 22:4, 64–72 and 74–79.

[8] Winston, W.L. (1996). *Simulation Modeling Using @RISK*. Duxbury Press, Belmont, CA.

SIMULATION MODELS

Simulation models are mathematical representations of real-world systems, which can investigated through simulated experiments on the model. Different simulation models are described below. There is an appropriate simulation model to represent almost nay real-world system we encounter.

Static vs. dynamic simulation models: A static model is one where time plays no role. On the other hand, a dynamic model represents a system as it evolves over time, such as a queueing system.

Deterministic vs. stochastic simulation models: If a simulation model does not contain any random components, it is called deterministic. An analytically intractable system of differential equations representing a chemical reaction might be such a model. Many systems, however, must be modeled with some random components, which are stochastic models. Most queueing and inventory systems are modeled stochastically.

Continuous vs. discrete simulation models: A continuous simulation model is one where state variables change continuously over time. Predator-prey models tracking the population size of certain species are typically continuous. On the other hand, models where the state variables change instantaneously at separated points in time are discrete. Queueing models where customers arrive and depart at discrete points in time are discrete models.

Manufacturing and logistics systems are typically studied using stochastic, dynamic, discrete-event simulation models.

See **Monte Carlo simulation; Production problems using spreadsheet programs; Simulation analysis of manufacturing and logistics systems; Simulation languages; Simulation software selection; Systems, models and simulation.**

SIMULATION OF PRODUCTION PROBLEMS USING SPREADSHEET PROGRAMS

David L. Olson

Texas A&M University, College Station, Texas USA

James R. Evans

University of Cincinnati, Cincinnati, Ohio, USA

INTRODUCTION: To demonstrate the variety of operations management problems that can be simulated with spreadsheets, we present models for network scheduling, waiting line, and inventory problems. The degree of difficulty in modeling these three classes of problems varies. While none are hard, network scheduling models are very easy, as the spreadsheet layout is a natural approach to solving a network scheduling problem. Waiting line models are a bit less obvious, especially when more than one server is present. Inventory models can involve a variety of situations for which spreadsheet models can be adapted.

NETWORK SCHEDULING: Although the critical path method has proven to be very useful in project management planning and control, it is widely recognized that the actual durations of activities often turn out to be different from those assumed in the planning stage. Simulation can be a valuable tool for estimating the *distribution* of outcomes in a project involving interrelated activities.

The Project Evaluation and Review Technique (PERT) addresses the widely recognized uncertainty involved in project management activities, but makes a rigid assumption about the distribution of durations, and the calculation of probability of completion by a specified time disregards non-critical activities. Simulation provides a flexible means to evaluate the probability of projects being completed by specific times. Any distribution of duration can be assumed. The distribution used should be based on empirical data. All activity paths are considered in the simple spreadsheet network. The following simple example demonstrates spreadsheet simulation on a network scheduling model.

Scheduling Example

A firm is implementing a new production operation, requiring coordination of delivery of materials and equipment, as well as development of a new personnel team. A simple critical path model of the operation consists of the activities, predecessor relationships, and durations shown in Table 1. Some of these activity durations involve uncertainty. The best guess of management (based on experience and judgment) includes triangularly distributed data for these uncertain events, measured in weeks.

Development of a spreadsheet simulation model for critical path problems is very easy. An EXCEL example is given in Figure 1. Those durations that are not constant are drawn from the triangular distribution. The full formula is given in cell **F2**. The formula looks imposing, but is simply a contingent formula based on two abutting triangles. The same formula is used in cells **F3**, **F4**, and **F7**, differing only in the row references. The final

Table 1.

Activity	Predecessors	Triangular durations		
		min	mode	max
A select raw materials (by bid)	none	3	3	5
B select & receive equipment (fob plant)	none	4	5	8
C computer system design	none	5	7	10
D select shipping for raw materials	A	1	1	1
E hire & train personnel	B	2	2	2
F computer training program	C, E	3	3	4
G test run	D, F	1	1	1

duration of the project appears in cell **H8**, which is the concluding activity of the project. Columns **G** and **H** implement the network model. Column **G** assigns the start times, equal to the maximum finish time of all predecessor activities. Column **H** calculates activity finish time by adding the duration to the start time. Since the particular observation for each simulation run is determined by the random number drawn in column **E**, different simulation runs may generate different critical paths.

This model can be used to estimate the probability of project completion by any given time, and the distribution of project completion times. Figure 2 shows a sample simulation run; for this case, the completion time is 11.961, or almost 12 weeks.

Although this model represents fractional weeks, durations can be easily converted to integer values.

The results of 100 repetitions can be used to define a distribution of project completion times. When the RAND() function of EXCEL is used, using the data table feature of EXCEL allows generating any number of repetitions. This involves some effort. Add-ins such as @Risk and Crystal Ball do this repetition and statistical analysis with minimal user effort. The calculation of probability is very simple. Divide the number of observed occurrences by the total number of simulation runs. Figure 3 gives the 100 results in tabular form.

We can see that the expected project duration is almost certain to exceed 10 weeks (with a 0.96 probability of exceeding 11 weeks), and will very likely be completed within 14 weeks. But there is a lot of variance between these limits. There is a 0.28 probability that the project will take between 11 and 12 weeks, a 0.38 probability that it will take between 12 and 13 weeks, and a 0.30 probability that it will take over 13 weeks. Simulation provides output that enables interpretation of probability for any particular duration or range. Relying upon sampling theory, the degree of accuracy of these estimates is a matter of the number of samples (runs) taken.

WAITING LINE MODELS: Waiting lines are a common operations management problem to which simulation has been applied. Both single and multi-server problems can be modeled in spreadsheets. To demonstrate, we will compare

	A	B	C	D	E	F	G	H
1	*Activity*	*min*	*mode*	*max*	*random*	*duration*	*start*	*finish*
2	A select raw materials	3	3	5	=RAND()	=IF(E2<=(C2-B2)/(D2-B2), B2+SQRT((C2-B2)*E2),D2-SQRT((D2-C2)*(D2-B2)*(1-E2)))	0	=G2+F2
3	B select & receive equipment	4	5	8	=RAND()	=IF(E3<=(C3-B3)/(D3-B3), B3+SQRT((C3-B3)*E3),D3-SQRT((D3-C3)*(D3-B3)*(1-E3)))	0	=G3+F3
4	C computer system design	5	7	10	=RAND()	=IF(E4<=(C4-B4)/(D4-B4), B4+SQRT((C4-B4)*E4),D4-SQRT((D4-C4)*(D4-B4)*(1-E4)))	0	=G4+F4
5	D select shipping for raw materials	1	1	1		=C5	=H2	=G5+F5
6	E hire & train personnel	2	2	2		=C6	=H3	=G6+F6
7	F computer training program	3	3	4	=RAND()	=IF(E7<=(C7-B7)/(D7-B7), B7+SQRT((C7-B7)*E7),D7-SQRT((D7-C7)*(D7-B7)*(1-E7)))	=MAX(H4,H6)	=G7+F7
8	G test run	1	1	1		=C8	=MAX(H5,H7)	=G8+F8

Fig. 1.

	A	B	C	D	E	F	G	H
1	*Activity*	*min*	*mode*	*max*	*random*	*duration*	*start*	*finish*
2	A select raw materials (by bid)	3	3	5	0.728	3.956	0	3.956
3	B select & receive equipment (fob plant)	4	5	8	0.447	5.423	0	5.423
4	C computer system design	5	7	10	0.583	7.498	0	7.498
5	D select shipping for raw materials		1			1	3.956	4.956
6	E hire & train personnel		2			2	5.423	7.423
7	F computer training program	3	3	4	0.711	3.463	7.498	10.961
8	G test run		1			1	10.961	11.961

Fig. 2.

two standard queuing situations. The first example is a single server model with exponential arrivals and exponential service times. The second is a two-server model with the same assumptions.

Assume that arrivals follow the exponential distribution with a mean of 15 jobs per hour (0.25 jobs per minute). Service is exponentially distributed with a job completed on average every 3 minutes. FIFO priority is used, there are no limits to the number of jobs in the system, and there are a large number of potential jobs. Two alternative systems are being considered. System 1 is an M/M/1 system and system 2 an M/M/2 system. These systems can be analyzed analytically to determine their steady-state behavior (assuming that the system is stable). These systems can also be modeled using spreadsheet-based simulation and multiple runs of the simulation models can be conducted to provide confidence interval estimates for systems performance variables. Simulation is more flexible than the analytic models, in that it allows us to 1) modify the assumptions (e.g., we can assume any arrival and service time distribution), and 2) examine the performance of the system during the transition, as well as steady-state, periods. We now describe the spreadsheet-based waiting line models.

Waiting Line Spreadsheet Model

Implementing a single server waiting line model in a spreadsheet program such as EXCEL is

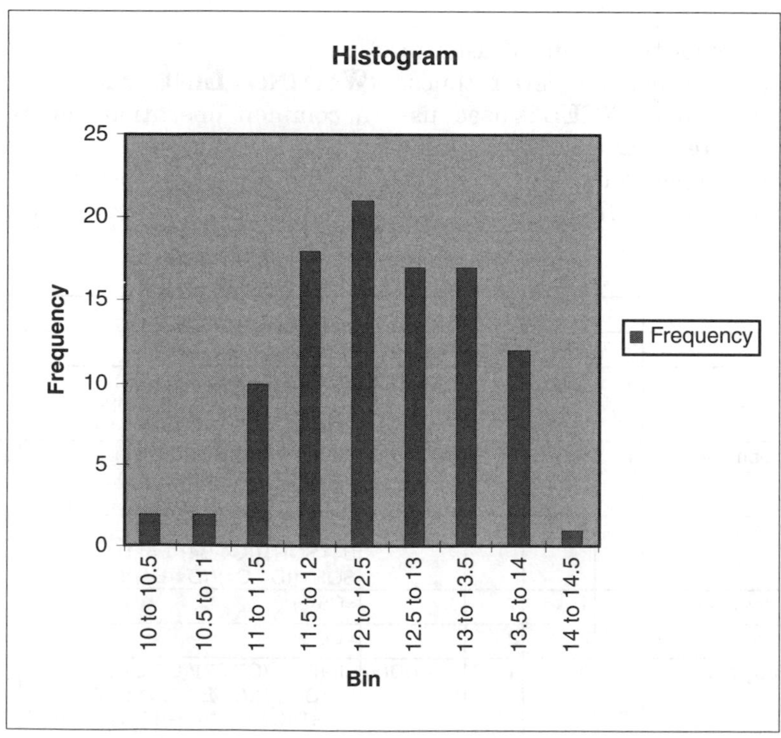

Fig. 3.

	A	B	C	D	E	F	G	H
1	**M/M/1 System**		*Mean arrivals per min*		0.25			***Average***
2			*Mean services per min*		0.333			***Wait Time***
3								=SUM(G7:G106)/100
4								
5			*Arrival*	*Start*	*Service*	*Completion*	*Wait*	*Idle*
6	*Job*	*TBA*	*Time*	*Time*	*Time*	*Time*	*Time*	*Time*
7	=1	=(-1/$E$1)*LN(1-RAND())	=B7	=MAX(0,C7)	=(-1/$E$2)*LN(1-RAND())	=D7+E7	=D7-C7	=D7
8	=A7+1	=(-1/$E$1)*LN(1-RAND())	=B8+C7	=MAX(F7,C8)	=(-1/$E$2)*LN(1-RAND())	=D8+E8	=D8-C8	=D8-F7
9	=A8+1	=(-1/$E$1)*LN(1-RAND())	=B9+C8	=MAX(F8,C9)	=(-1/$E$2)*LN(1-RAND())	=D9+E9	=D9-C9	=D9-F8
10	=A9+1	=(-1/$E$1)*LN(1-RAND())	=B10+C9	=MAX(F9,C10)	=(-1/$E$2)*LN(1-RAND())	=D10+E10	=D10-C10	=D10-F9
11	=A10+1	=(-1/$E$1)*LN(1-RAND())	=B11+C10	=MAX(F10,C11)	=(-1/$E$2)*LN(1-RAND())	=D11+E11	=D11-C11	=D11-F10
12	=A11+1	=(-1/$E$1)*LN(1-RAND())	=B12+C11	=MAX(F11,C12)	=(-1/$E$2)*LN(1-RAND())	=D12+E12	=D12-C12	=D12-F11

Fig. 4.

straightforward. In Figure 4, arrival times and service times are random, and generated by converting a uniform random number R into the exponential distribution with mean $1/\lambda$ by the formula: $Ex = -(1/\lambda)^{*}\ln(1 - R)$ in column **B**. The arrivals in minutes are cumulated to identify when the job entered the system in column **C**. Start time in column **D** is the maximum of arrival for the job, and time of completion of the prior job. Completion time for each job is modeled in column **F** as the sum of start time + duration (generated in column **E**). The waiting time for each job, as well as idle time for the server, can be measured by job, and upon completion of the simulation run, average waiting time (column **G**) and idle time (column **H**) can be calculated.

Figure 5 shows sample results for the first 10 jobs for one run of the model. The exponential distribution is highly variable. In this run, there is no waiting until job #5. Job #4 turned out to be very slow. The slack production capacity was able to catch up again after job #9. Although five of the

	A	B	C	D	E	F	G	H
1	**M/M/1 System**		*Mean arrivals per min*		0.25			***Average***
2			*Mean services per min*		0.333			***Wait Time***
3								2.06
4								
5			*Arrival*	*Start*	*Service*	*Completion*	*Wait*	*Idle*
6	*Job*	*TBA*	*Time*	*Time*	*Time*	*Time*	*Time*	*Time*
7	1	7.494	7.494	7.494	0.345	7.839	0	7.494
8	2	1.118	8.612	8.612	0.081	8.693	0	0.773
9	3	7.691	16.303	16.303	0.194	16.497	0	7.610
10	4	4.457	20.760	20.760	12.604	33.364	0	4.263
11	5	0.113	20.873	33.364	3.741	37.105	12.491	0
12	6	6.828	27.701	37.105	7.486	44.591	9.404	0
13	7	6.061	33.762	44.591	3.909	48.500	10.829	0
14	8	5.617	39.379	48.500	3.107	51.607	9.121	0
15	9	3.310	42.689	51.607	0.119	51.726	8.918	0
16	10	9.526	52.215	52.215	1.602	53.817	0	0.489

Fig. 5.

	A	B	C	D	E	F
1	**M/M/2 System**		*Mean arrivals per min*		0.25	
2			*Mean services per min*		0.333	
3						
4						
5			*Arrival*	*SERVER*	*Start*	*Service*
6	*Job*	*TBA*	*Time*		*Time*	*Time*
7	=1	=(-1/$E$1)*LN(1-RAND())	=B7	=1	=C7	=(-1/$E$2)*LN(1-RAND())
8	=A7+1	=(-1/$E$1)*LN(1-RAND())	=B8+C7	=2	=C8	=(-1/$E$2)*LN(1-RAND())
9	=A8+1	=(-1/$E$1)*LN(1-RAND())	=B9+C8	=IF(H8<I8,1,2)	=MAX(C9,MIN(H8,I8))	=(-1/$E$2)*LN(1-RAND())

	G	H	I	J
1		***Average***		
2		***Wait Time***		
3		=SUM(J7:J106)/100		
4				
5	*Completion*	*server 1*	*server 2*	*Wait*
6	*Time*			*Time*
7	=E7+F7	=IF(D7=1,G7,0)	=IF(D7=2,G7,0)	=E7-C7
8	=E8+F8	=IF(D8=1,G8,H7)	=IF(D8=2,G8,I7)	=E8-C8
9	=E9+F9	=IF(D9=1,G9,H8)	=IF(D9=2,G9,I8)	=E9-C9

Fig. 6.

ten jobs had no waiting time, the average waiting time was over 2 minutes per job.

We now consider a multiple server model. In this case, each customer will be served by the first of the two servers to complete their prior job. As shown in Figure 6, the spreadsheet model for the two server model is similar to the one server model. An extra column (**D**) is added to identify which of the two servers is doing work for the customer. The server is selected with an IF statement, which compares the finish times of the PRIOR customer for each server (**H** and **I**), and selects the server with the lowest finish time. The start time (column **E**) is determined by taking the maximum of arrival time and next available server. The next available server is identified by taking the minimum completion time of the prior customer among the servers. The only other difference is that completion times for both servers need to be identified. The IF statements in columns **H** and **I** replace the prior completion time with the current customer completion time if the server is the appropriate number. Otherwise, the prior completion time for this server is repeated.

Note that the waiting time with the two server model (Figure 7) has disappeared. With enough runs, there will be some waiting time, but for the first ten customers, given the arrival and service times experienced, all waiting was eliminated. The percent idle time could be calculated for the system by adding the service times and dividing by the number of servers times the maximum of clock times for services.

Fifty runs of one 8 hour day each were simulated. Arrivals were cut off after 8 hours. Jobs already in the system at the end of eight hours were completed. The estimates for each of the waiting line system parameters obtained from the simulations appear in Table 2.

Table 3 compares analytic model results for the standard waiting line measures with the 0.95 confidence intervals obtained from the fifty simulation runs. Each spreadsheet has a number of means to obtain multiple runs, including visual

	A	B	C	D	E	F	G	H	I	J
1	**M/M/2 System**		*Mean arrivals per min*		0.25			***Average***		
2			*Mean services permin*		0.333			***Wait Time***		
3								0		
4										
5			*Arrival*	*SERVER*	*Start*	*Service*	*Completion*	*server 1*	*server 2*	*Wait*
6	*Job*	*TBA*	*Time*		*Time*	*Time*	*Time*			*Time*
7	1	7.494	7.494	1	7.494	0.345	7.839	7.839	0.000	0.000
8	2	1.118	8.612	2	8.612	0.081	8.693	7.839	8.693	0.000
9	3	7.691	16.303	1	16.303	0.194	16.497	16.497	8.693	0.000
10	4	4.457	20.760	2	20.760	12.604	33.364	16.497	33.364	0.000
11	5	0.113	27.873	1	20.873	3.741	24.614	24.614	33.364	0.000
12	6	6.828	27.701	1	27.701	7.486	35.187	35.187	33.364	0.000
13	3	6.061	33.762	2	33.762	3.909	37.671	35.187	37.671	0.000
14	4	5.617	39.379	1	39.379	3.107	42.486	42.486	37.671	0.000
15	5	3.310	42.689	2	42.689	0.119	42.808	42.486	42.808	0.000
16	6	9.526	52.215	1	52.215	1.602	53.817	53.817	42.808	0.000

Fig. 7.

basic macros. The results are compatible, in that the analytic model results are within the confidence intervals obtained from simulation (with the exception of System 2 estimates of probabilities), and are close to the middle of those confidence intervals in most cases. The confidence intervals obtained from simulation provide a more descriptive picture of expected results than do the point estimates obtained from analytic formulas.

The ability to model multiple servers makes spreadsheet modeling of waiting line problems a highly effective tool for many operations management environments. Spreadsheet flexibility makes it very easy to analyze the results of simulations as well.

Crystal Ball is an add-in that replicates spreadsheets and provides statistical information about key output variables. Figure 8 shows Crystal Ball output for 1000 replications of the M/M/1 system spreadsheet shown earlier for average wait time. The frequency chart clearly shows that although the average wait time has a mean of about 7.5 minutes, the distribution of waiting times is

Table 2.

		System 1	System 2
Utilization rate	ρ	0.746	0.373
Average number in system	L	2.676	1.332
Average number in queue	L_q	1.930	0.959
Average time in system hours	W	0.177 hours	0.176
Average time in line hours	W_q	0.127 hours	0.126
Average time between arrivals hours	μ	0.05	0.05
Probability system is empty	P_0	0.254	0.256
Probability of one in the system	P_1	0.205	0.205
Probability of two in the system	P_2	0.148	0.148
Probability of more than two in system	$P_{>2}$	0.393	0.391

Table 3.

	System 1		System 2	
	Analytic model	Simulation	Analytic model	Simulation
ρ	0.75	0.585–0.907	0.375	0.293–0.453
L	3	0.128–5.224	0.873	0.058–2.606
L_q	2.25	0–4.360	0.123	0–2.174
W	0.2 hr	0.020–0.334	0.058 hr	0.019–0.333
W_q	0.15 hr	0–0.284	0.008 hr	0–0.283
P_0	0.25	0.093–0.415	0.455	0.093–0.419
P_1	0.187	0.082–0.328	0.341	0.083–0.327
P_2	0.141	0.081–0.215	0.128	0.081–0.215

Forecast: Average Wait Time **Cell: G3**

Summary:
Display Range is from 0.000 to 22.500 Minutes
Entire Range is from 1.181 to 65.072 Minutes
After 1,000 Trials, the Std. Error of the Mean is 0.164

Statistics:	Value
Trials	1000
Mean	7.457
Median	6.236
Mode	---
Standard Deviation	5.172
Variance	26.749
Skewness	3.42
Kurtosis	26.53
Coeff. of Variability	0.69
Range Minimum	1.181
Range Maximum	65.072
Range Width	63.891
Mean Std. Error	0.164

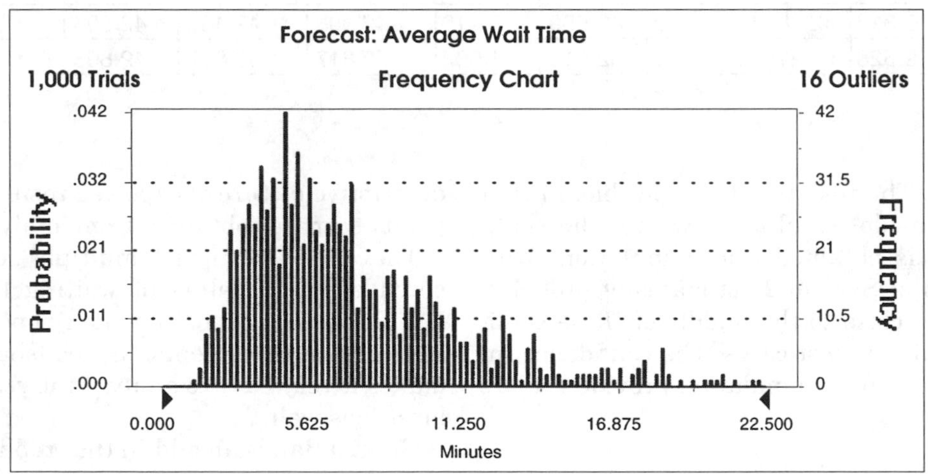

Fig. 8.

skewed to the right, and some jobs might wait over 20 minutes. This type of output can provide more insight than the bad practice of replicating a simulation 3 or 5 times and developing crude confidence intervals.

INVENTORY: Simulation is appropriate for inventory models because of the highly probabilistic nature of many operational demands. Inventory models fit the spreadsheet layout very well. Each time period can be assigned a row, and a variety of inventory complications can be modeled, measuring all appropriate inventory costs.

Inventory Model

Parameters for inventory models can be placed in a convenient locations on the spreadsheet to identify holding cost rates, setup or order cost rates, and stockout rates, as well as initial inventory conditions, reorder point levels, order quantities, lead time conditions, and demand parameters. The model itself consists of one row per time period, which is used to identify the initial inventory, demand input, and ending inventory for the period. Costs can be calculated per time period and aggregated across time periods. Actions taken to generate replacements can also be modeled.

Consider a reorder point model where a product with a normally distributed demand (mean 50 per week, standard deviation 20) is served by a system where a lot size of 200 units is ordered with a lead time of 2 weeks. Products are modeled in Figure 9 to arrive at the end of the week, to allow processing. The reorder point is 160 units. The initial inventory is 100 units. The holding cost rate is \$2 per unit of average inventory (the average of beginning and ending inventory per week), with

	A	B	C	D	E
1		holding rate	2	Reorder Point	160
2		order rate	200	initial	100
3		loss rate	50	lead	2
4				lot size	200
5					
6	*week*	*beginning*	*random*	*demand*	*lost*
7		*inventory*			*sale*
8	1	=E2	=RAND()	=INT(NORMINV(C8,$G$1,$G$2)+0.5)	=IF(D8>B8,INT((D8-B8)*0.8+0.99),0)
9	2	=H8	=RAND()	=INT(NORMINV(C9,$G$1,$G$2)+0.5)+F8	=IF(D9>B9,INT((D9-B9)*0.8+0.99),0)
10	3	=H9	=RAND()	=INT(NORMINV(C10,$G$1,$G$2)+0.5)+F9	=IF(D10>B10,INT((D10-B10)*0.8+0.99),0)
11	4	=H10	=RAND()	=INT(NORMINV(C11,$G$1,$G$2)+0.5)+F10	=IF(D11>B11,INT((D11-B11)*0.8+0.99),0)

	F	G	H	I
1	demand	50		
2	std	20		
3				
4				
5				
6	*backorder*	*arrival*	*ending*	*average*
7			*inventory*	*inventory*
8	=IF(D8>B8,D8-B8-E8,0)	0	=MAX(0,B8-D8+G8,G8)	=(B8+H8)/2
9	=IF(D9>B9,D9-B9-E9,0)	0	=MAX(0,B9-D9+G9,G9)	=(B9+H9)/2
10	=IF(D10>B10,D10-B10-E10,0)	=J8	=MAX(0,B10-D10+G10,G10)	=(B10+H10)/2
11	=IF(D11>B11,D11-B11-E11,0)	=J9	=MAX(0,B11-D11+G11,G11)	=(B11+H11)/2

	J	K	L	M	N
6	order	holding	order	stockout	total
7		cost	cost	cost	cost
8	=IF(H8>$E$1,0,$E$4*(INT(F8/$E$4)+1))	=$C$1*I8	=IF(J8>0,$C$2,0)	=E8*$C$3	=K8+L8+M8
9	=IF(H9>$E$1,0,$E$4*(INT(F9/$E$4)+1)-J8	=$C$1*I9	=IF(J9>0,$C$2,0)	=E9*$C$3	=K9+L9+M9
10	=IF(H10>$E$1,0,$E$4*(INT(F10/$E$4)+1)-J8-J9	=$C$1*I10	=IF(J10>0,$C$2,0)	=E10*$C$3	=K10+L10+M10
11	=IF(H11>$E$1,0,$E$4*(INT(F11/$E$4)+1)-J9-J10	=$C$1*I11	=IF(J11>0,$C$2,0)	=E11*$C$3	=K11+L11+M11

Fig. 9.

an order cost of $200 per order. At least 80% of the shortages are lost with an impact of $50 in lost profit per sale, the remainder are carried forward to the next week as backorders (added to next week's demand).

Demand is generated for each time period in columns **C** and **D**. A uniform random number (between 0 and 1) is generated in column **C** to represent the cumulative normal probability. This is converted in column **D** to a normally distributed number with the specified mean (cell **G1**) and standard deviation (cell **G2**) using the EXCEL function NORMINV (random, mean, standard deviation). Because demand is integer for each time period, the obtained value is rounded to the nearest integer value. Backorders for prior periods (column **F**) are added to the next period's demand.

Lost sales are calculated in column **E**. At least 80% of unmet demand is a lost sale. If demand is greater than beginning inventory, 80% of the unmet demand is calculated, and rounded up. If demand was less than beginning inventory, lost sales are 0. Backorders are the difference between unmet demand and lost sales.

Ending inventory is calculated as beginning inventory (column **B**) minus demand (column **D**) plus arriving inventory (column **G**, which is modeled as being added to inventory at the end of the period, after processing). If demand exceeds supply, the ending inventory is equal to arriving inventory only. The MAX function assures the appropriate value is entered.

Average inventory is the mean of beginning and ending inventory. If another assumption is desired, it can be modeled as well. The quantity ordered is calculated in column **J**. If the ending inventory (column **H**) plus orders en route (the prior two entries in column **J**) is greater than the reorder point (cell **E1**), then no order is placed. If ending inventory plus previous orders is less than or equal to the reorder point, the quantity ordered is the multiple of the lot size (cell **E4**) required to cover backorders plus one lot. Other rules could be implemented as desired.

Columns **K** through **M** calculate the costs by category, and all three costs totaled in column **N**. These costs can be summed over the planning horizon. Alternative policies can be viewed by changing the input parameters at the top of the model.

An example run of the model is shown in Figure 10. The first period will result in a need for placing an order under any demand, as beginning inventory is less than the reorder point. This order will arrive in the third week (but won't be available for issue until week 4). In this case, there was a shortage of 46 units in week 3. Lost sales were 80% (rounded up) of shortages, or 37 units. The remainder of shortages are backorders, added to the demand for week 4 (the random number 0.715737 generated a demand for 61 units, and the backorder of 9 was added to yield a total demand of 70).

No order was place in weeks 2 or 3, because of the order placed in week 1. In week 4, ending inventory plus the quantity due in again fell below the reorder point, so another order was placed.

Spreadsheet simulations provide the opportunity to include a variety of modifications to reflect actual inventory conditions. Through the use of IF statements, and extra random numbers, many complexities of the inventory process can be included.

When and Where to Use Spreadsheet-Based Simulation: Spreadsheet models can be applied to many standard operational management problems. There are obviously operational management problems for which it would be inappropriate. For large scale models of factories with many interrelated waiting lines, GPSLs are clearly the appropriate tool. For the specific problem environments for which they were developed, SPSLs like COMNET III are much easier to use. However, it is difficult to determine the boundary of spreadsheet applicability. Not that long ago the authors argued to their classes that spreadsheet-based simulation models were inappropriate for multiple server waiting line models. There are ways, however, to effectively model such problems, including what was demonstrated in this article.

Spreadsheet models should be considered because of their ready accessibility and lower cost. Another factor is that the output of spreadsheets

	A	B	C	D	E	F	G	H	I
1		holding rate	2	ROP	160	demand	50		
2		order rate	200	initial	100	std	20		
3		loss rate	50	lead	2				
4				lot size	200				
5									
6	*week*	*beginning*	*random*	*demand*	*lost*	*backorder*	*arrival*	*ending*	*average*
7		*inventory*			*sale*			*inventory*	*inventory*
8	1	100	0.082651	22	0	0	0	78	89
9	2	78	0.922128	78	0	0	0	0	39
10	3	0	0.427808	46	37	9	200	200	100
11	4	200	0.715737	70	0	0	0	130	165

	J	K	L	M	N
6	*order*	*holding*	*order*	*stockout*	*total*
7		*cost*	*cost*	*cost*	*cost*
8	200	178	200	0	378
9	0	78	0	0	78
10	0	200	0	1850	2050
11	200	330	200	0	530

Fig. 10.

is easily manipulated to import into other computer tools, such as word processors. However, the cost of modeler time must also be considered. If a simulation analysis involves too much complication, more specialized simulation tools such as GPSLs and SPSLs should be considered.

CONCLUSIONS: Simulation is an extremely valuable analytic method. It has been used in literally every area of scientific endeavor, including operations management. The ability to include any assumption into a simulation model gives it a valuable advantage over closed form analytic formulations. In the past, it took significant amounts of time to encode a simulation model, and to interpret the model's output.

There are more and better products, with the market place providing systems that are ever easier to use, do more powerful things, and make it easier to communicate results. Products have been developed for problem application areas that have proven to be important, such as simulation of plant floors, telecommunications networks, computer systems, and medical environments.

New products are making it much easier to implement simulation. However, there is still a need to look at the basics of simulation, in order that simulation users understand how simulation studies should be set up and designed, and how to apply statistical analysis to interpret the output. As Cochran *et al.* (1995) concluded, successful simulation practice is more dependent on methodology than on software tools used.

See **Average inventory; Exponential arrivals; Exponential service times; General-purpose language; Holding cost rate; Multi-server problem; PERT; Queuing models for manufacturing and logistics; Simulation analysis of manufacturing and logistics systems; Simulation of production problems using spreadsheet programs; Simulation software selection; Single-server problem; Steady-state behavior; Waiting line model.**

References

[1] Cochran, J.K., G.T. Mackulak, and P.A. Savory (1995). "Simulation project characteristics in industrial settings." *Interfaces* 25:4, 104–113.

[2] Evans, J.R. and D.L. Olson (1998). *Simulation*. Prentice-Hall, Englewood Cliffs, NJ.

[3] Gray, Paul (1983). *Student Guide to IFPS (Interactive Financial Planning System)*. McGraw-Hill, New York.

[4] Gray, Paul (1996). *Visual IFPS/Plus for Business*. Prentice-Hall, Englewood Cliffs, NJ.

[5] Lane, M.S., A.H. Mansour, and J.L. Harpell (1993). "Operations research techniques: A longitudinal update 1973–1988," *Interfaces* 23:2, 63–68.

[6] Law, A.M. and Kelton, W.D. (1991). *Simulation Modeling & Analysis*. McGraw-Hill, Inc., New York.

[7] Swain, J.J. (1995). "Simulation Survey: Tools for process understanding and improvement." *OR/MS Today* 22:4, 64–72 and 74–79.

[8] Winston, W.L. (1996). *Simulation Modeling Using @RISK*. Duxbury Press, Belmont, CA.

SIMULATION SOFTWARE SELECTION

Enver Yücesan

INSEAD, Fontainebleau Cedex, France

John W. Fowler

Arizona State University, Tempe, Arizona, USA

INTRODUCTION: As manufacturing and logistics constitute the largest application area of simulation, there are numerous software packages specifically geared to this domain. There are several valuable sources for up-to-date information on simulation packages: the first one is a bi-yearly review that appears in *OR/MS Today*, a publication of the *Institute for Operations Research and Management Science* (www.informs.org). The second is a yearly survey published by the *Society for Computer Simulation* (www.scs.org). The *American Institute of Industrial Engineers* (www.iienet.org) also publish regular surveys of simulation software in their journal, *IE Solutions*.

There are two groups of packages supporting manufacturing simulation. The first group consists of general-purpose languages that also include specific utilities facilitating the modeling of various features of manufacturing systems such as kanban control or material handling systems. A model is developed in a simulation language by writing a program using the languages modeling constructs. SLAMSYSTEM (Pritsker Corporation, 1994) and SIMAN (Banks *et al.*, 1995) are the two most popular examples of such languages.

The second group consists of simulators that are conceived exclusively for manufacturing applications. Xcell+ (Conway *et al.*, 1987), Promod (Production Modeling Corporation, 1989), ARENA (Systems Modeling Corporation, 1995), and Taylor II (King, 1996) are examples of manufacturing simulators.

The emergence of the world-wide web has also had an impact on simulation technology. Several general-purpose simulation libraries written in Java are available over the web (e.g., www.dcs.ed.ac.uk/home/hase/simjava).

Given the long list of commercial simulation software packages, one should consider the following desirable characteristics in selecting a particular package:

General Features. Modeling flexibility, i.e., the capability of incorporating different operating characteristics, is important. One should also keep in mind ease of model development, maximum model size, execution speed, and portability among different hardware platforms.

Animation. This is a capability useful in debugging a simulation model during implementation, in investigating if it is valid, in understanding the dynamic behavior of the system, in explaining the control logic of a system, in training operational personnel, and in presenting the results of the analysis to higher levels of management. Easy-to-use high-quality animation has been one of the key drivers of simulation's increased popularity.

Statistical Capabilities. Since simulation is generally a statistical sampling experiment, the simulation package must be capable of supporting correct experimentation. On the input side, it should provide several streams of random numbers as well as a wide variety of standard distributions. On the output side, it should facilitate the execution of replications or batching. It should provide utilities to specify a warm-up period and to construct confidence intervals. Further support for design of experiments, though rare in commercial packages, is also desirable.

Output Reports. A simulation package should provide time-saving standard reports for commonly occurring performance statistics, but should also allow for tailored reports to be developed easily. Presentation-quality graphical displays are also highly desirable.

Customer Support. The vendor should provide timely technical support for specific modeling problems encountered by the user. Constructs used in a language should not appear as 'black boxes' to the user. Good documentation, including a user's manual and detailed examples, is necessary to make the language more transparent.

See the application of various softwares to simulation problems in one or more articles in the short bibliography at the end of this article.

See **Initialization bias in simulation; Inventory floor analysis; Maintenance management decision models; Manufacturing system modeling using petri nets; Model validation and verification; Queuing theory; Sensitivity analysis; Simulation analysis of manufacturing and logistics systems; Simulation model; Simulation of production problems using spreadsheet programs; Systems, models, and simulation; Trace driven simulations; Truncation point in simulation.**

Short Bibliography

[1] Altiok, T. (1997). *Performance Analysis of Manufacturing Systems*. Springer, New York.

[2] Balci, O. (1994). "Model validation, verification, and testing techniques throughout the life cycle of a simulation study." *Annals of Operations Research*, 53, 121–173.

[3] Banks, J., B. Burnette, H. Kozloskic and J. Rose (1995). *Introduction to SIMAN V and CINEMA V*. John Wiley & Sons, New York.

[4] Bechoffer, R., T.J. Santner and D. Goldsman (1995). *Design and Analysis of Experiments for Statistical Selection, Screening, and Multiple Comparisons*. Wiley, New York.

[5] Bratley, P., B. Fox and L. Schrage (1987). *A Guide to Simulation* (2nd Ed.), Springer-Verlag. New York.

[6] Buzacott, J.A. and J.G. Shantikumar (1993). *Stochastic Models of Manufacturing Systems*. Prentice-Hall, New Jersey.

[7] Buzacott, J.A. and D.D. Yao (1986). "Flexible manufacturing systems: a review of analytical models." *Management Science*, 32, 890–905.

[8] Chance, F. (1993). "Conjectured upper bounds on transient mean total waiting times in queueing networks." In: *Proceedings of the Winter Simulation Conference*, 414–421.

[9] Chance, F., J. Robinson and J. Fowler (1996). "Supporting manufacturing with simulation: model design, development, and deployment." In: *Proceedings of the Winter Simulation Conference*, 114–121.

[10] Chen, C.H. (1996). "A lower bound on the correct subset selection probability and its application to discrete-event simulation." *IEEE Transactions on Automatic Control*, 41, 1227–1231.

[11] Cheng, R.H.C., W. Holland and N.A. Hughes (1996). "Selection of input models using bootstrap goodness of fit." In: *Proceedings of the Winter Simulation Conference*, 199–206.

[12] Conway, R., W.L. Maxwell, J.O. McClain and S. Worona (1987). *User's Guide to Xcell + Factory Modeling System* (2nd Ed.), Scientific Press, California.

[13] Crane, M.A. and D.L. Iglehart (1975). "Simulating stable stochastic processes, III: regenerative processes and discrete event simulations." *Operations Research*, 23, 33–45.

[14] Glasserman, P. (1991). *Gradient Estimation via*

Perturbation Analysis. Kluwer Academic Publishers, Massachusetts.

[15] Glasserman, P. (1994). "Perturbation analysis of production networks." In: *Stochastic Modeling and Analysis of Manufacturing Systems* (Yao, Ed.). Springer-Verlag, New York.

[16] Glasserman, P. and D.D. Yao (1992). "Some guidelines and guarantees for common random numbers." *Management Science*, 38, 884–908.

[17] Gordon, R.F., E.A. MacNair, P.D. Welch, K.J. Gordon and J.F. Kurose (1986). "Examples of using the research queueing package modeling environment (RESQME)." In: *Proceedings of the Winter Simulation Conference*, 494–503.

[18] Ho, Y.C., M.A. Eyler and T.T. Chien (1979). "A gradient technique for general buffer storage design in a serial production line." *International Journal of Production Research*, 17, 557–580.

[19] INRIA (1984). *New Users' Introduction to QNAP2*. Sophia Antipolis, France.

[20] Jackson, J.R. (1963). "Jobshop-like queueing systems." *Management Science*, 10, 131–142.

[21] Jacobson, S.H. and L. Schruben (1989). "Techniques for simulation response optimization." *Operations Research Letters*, 8, 1–9.

[22] Jankauskas, L. and S. McLafferty (1996). "BESTFIT, Distribution fitting software by Palisade Corporation." In: *Proceedings of the Winter Simulation Conference*, 551–555.

[23] King, C.B. (1996). "Taylor II manufacturing simulation software." In: *Proceedings of the Winter Simulation Conference*, 569–573.

[24] Kingman, J.F.C. (1961). "The single-server queue in heavy traffic." In: *Proceedings of the Cambridge Philosophical Society*, 57, 902–904.

[25] Kleijnen, J.P.C. (1987). *Statistical Tools for Simulation Practitioners*. Marcel Dekker, New York.

[26] Krajewski, L.J., B.E. King, L.P. Ritzman and D.S. Wong (1987). "Kanban, MRP, and shaping the manufacturing environment." *Management Science*, 33, 39–57.

[27] Law, A.M. and M.G. McComas (1996). "EXPERTFIT: Total support for simulation input modeling." In: *Proceedings of the Winter Simulation Conference*, 588–593.

[28] Law, A.M. and M.G. McComas (1989). "Pitfalls to avoid in the simulation of manufacturing systems." *Industrial Engineering*, 31, 28–31, 69.

[29] Law, A.M. and D.W. Kelton (1991). *Simulation Modeling and Analysis* (2nd Ed.), McGraw-Hill, New York.

[30] L'Ecuyer, P. (1990). "A unified view of the IPA, SF, and LR gradient estimation techniques." *Management Science*, 36, 1364–1383.

[31] Leemis, L.M. (1996). "Discrete event simulation input process modeling." In: *Proceedings of the Winter Simulation Conference*, 39–46.

[32] Naylor, T.H., J.L. Balintfy, D.S. Burdick and K. Chu (1966). *Computer Simulation Techniques*. John Wiley & Sons, New York.

[33] Network Dynamics, Inc. (1991). *MPX User Manual*. Burlington, MA.

[34] Nuyens, R.P.A., N.M. Van Dijk, L. Van Wassenhove and E. Yücesan (1996). "Transient behavior of simple queueing systems: implications for FMS models." *Simulation: Application and Theory*, 4, 1–29.

[35] Papadopoulos, H.D., C. Heavey and J. Browne (1993). *Queueing Theory in Manufacturing Systems Analysis and Design*. Chapman & Hall. London, UK.

[36] Perros, H.G. (1994). *Queueing Networks with Blocking*. Oxford University Press. Oxford.

[37] Pritsker, A.A.B. (1989). " Developing analytic models based on simulation results." In: *Proceedings of the Winter Simulation Conference*, 653–660.

[38] Pritsker Corporation (1994). SLAMSYSTEM *User's Guide*. Version 4.5. Indianapolis, IN.

[39] Production Modeling Corporation of Utah (1989). *ProModel User's Manual*. Orem, UT.

[40] Reiman, M.I. and A. Weiss (1989). "Sensitivity analysis for simulations via likelihood ratios." *Operations Research*, 37, 830–844.

[41] Robinson, J., L.W. Schruben and J.W. Fowler (1993). "Experimenting with large-scale semiconductor manufacturing simulations: a frequency domain approach to factor screening." In: *Proceedings of the Second IE Research Conference*, 112–116.

[42] Rubinstein, R.Y. and A. Shapiro (1993). *Discrete Event Systems: Sensitivity Analysis and Stochastic Optimization by the Score Function Method*. Wiley, New York.

[43] Schmeiser, B.W. (1983). "Batch size effects in the analysis of simulation output." *Operations Research*, 31, 565–568.

[44] Schruben, L. (1983). "Confidence interval estimation using standardized time series." *Operations Research*, 31, 1090–1107.

[45] Schruben, L. and V.J. Cogliano (1987). "An experimental procedure for simulation response surface model identification." *Communications of the ACM*, 30, 716–730.

[46] Shantikumar, J.G. and R.G. Sargent (1983). "A unifying view of hybrid simulation/analytic models and modeling." *Operations Research*, 31, 1030–1052.

[47] Solberg, J.J. (1976). "Optimal design and control of computerized manufacturing systems." In *Proceedings of AIIE Systems Engineering Conference*, 138–144.

[48] Spearman, M.L., D.L. Woodruff and W.J. Hopp (1990). "CONWIP: a pull alternative to kanban." *International Journal of Production Research*, 28, 879–894.

[49] Suri, R., G.W.W. Diehl, S. De Treville and M.J. Tomsicek (1995). "From CAN-Q to MPX: evolution of queueing software for manufacturing." *Interfaces*, 25, 128–150.

[50] Tayur, S.R. (1992). "Properties of serial kanban systems." *Queueing Systems*, 12, 297–318.

[51] Van Dijk, N.M. (1993). *Product Forms for Queueing Networks: A System's Approach*. Wiley, New York.

[52] Walrand, J. (1988). *An Introduction to Queueing Networks*. Prentice-Hall. New Jersey.

[53] Whitt, W. (1983). "The Queueing Network Analyzer." *The Bell System Technical Journal*, 62, 2779–2815.

[54] Wright, Williams and Kelly (1997). *Factory Explorer User's Manual*, Version 2.2. Dublin, California.

[55] Yao, D.D. (1994). *Stochastic Modeling and Analysis of Manufacturing Systems*. Springer-Verlag, New York.

[56] Yücesan, E. (1993). "Randomization tests for initialization bias in simulation output." *Naval Research Logistics*, 40, 643–664.

[57] Yücesan, E. (1994). "Evaluating alternative system configurations using simulation: a non-parametric approach." *Annals of Operations Research*, 53, 471–484.

[58] Yücesan, E. and S.H. Jacobson (1996). "Computational issues for Accessibility in discrete event simulation." *ACM Transactions on Modeling and Computer Simulation*, 6, 53–75.

SIMULTANEOUS ENGINEERING

See **Concurrent engineering; Product development and concurrent engineering.**

SINGLE CARD SYSTEM

In a Kanban-based manufacturing system either one or two types of Kanbans may be used. In the single card system, each workstation has only outbound inventory, and each container is assigned a Kanban, known as a conveyance Kanban or withdrawal Kanban. When a container is withdrawn by the succeeding stage, the conveyance Kanban is detached and becomes a production order for the preceding stage. The empty container sent to the preceding stage serves as a withdrawal Kanban. This system is therefore appropriate for manufacturers not ready for the strict controls placed on output buffer inventories by the two card system, and is often used as the first stage in the development of a two-card Kanban system. A one-card system is also used when two adjacent workstations are the only source and endpoint of a manufacturing stage.

See **Just-in-time manufacturing; Kanban-based manufacturing systems; Production ordering Kanban.**

SINGLE-SERVER PROBLEM

Queuing or waiting line problems may have one or more servers. A single-server problem is a special case of a waiting line model, where there is precisely one server. Single-server waiting line models are simple enough that widely published analytic models have been published, making it possible to mathematically derive expected waiting time, expected numbers of jobs waiting for service, and time and number in the overall system. The proportion of the time the system is working and idle can also be computed.

See **Queuing models for manufacturing and logistics; Queuing theory; Simulation analysis of manufacturing and logistics systems; Simulation of production problems using spreadsheet programs; Waiting line models.**

SIX-SIGMA

Six-sigma is the name associated with a quality system that perfects its processes to a point that the process produces almost zero defects; precisely 3–4 defects per million. This is accomplished by shrinking process variations to less than one half of design tolerance while allowing the process mean to shift 1.5 times the process standard deviation from the target.

See **Statistical quality control using control charts; Total quality management.**

SKILL OBSOLESCENCE

Much has been written about the displacement of workers created by advanced manufacturing technology. These statements usually refer to operators working in traditional manufacturing environments. New manufacturing technologys have decreased the demand for workers' skills, especially manual skills. Even skilled operators are left behind by rapid changes in manufacturing technology. Even though equipment may be upgraded every five to ten years, software upgrades are much more frequent. It is estimated that one half of a worker's knowledge and skills become obsolete after 3 to 5 years. As technology becomes more sophisticated, training programs should be in place to enable operators to continuously update their skill repertoires.

See **Human resource issues and advanced manufacturing technology; Human resource issues in manufacturing.**

SLACK TIME RULE

Jobs waiting in a line or queue for processing are sequenced using one of many priority sequencing rules; the slack time (ST) rule is one of them. Slack time is computed by the hours from now till the due date minus the unfinished processing for the job. For example, if the due date is 3 days from now and the unfinished processing time on a job is 6 hours, the slack time = $(3 \times 8 \text{ hrs}).) - 6 = 18$ hours (this assumes 8 hours of processing time available each day). The job with least slack time gets the highest priority.

See **Priority sequencing rules.**

SLAM SOFTWARE

See **Simulation languages; Simulation software selection.**

SMALL GROUP IMPROVEMENT ACTIVITIES (SGIA)

Quality circles are called SGIA in Japan where it originated.

See **Integrated quality control; Just-in-time manufacturing; Quality circles; Total quality management.**

SMART

SMART is an acronym for Strategic Measurement and Reporting Technique, an approach to performance measurement developed by Wang Laboratories. SMART clearly defines a procedure for translating corporate objectives into terms that are meaningful for individual departments. These corporate objectives are divided into two categories of measures: market measures and financial measures. Financial measures include process time and cost, while market measures include quality and delivery.

The idea behind SMART is to ensure that all parts of the organization are working together toward the same goals, although those goals may be measured in different terms at different levels of the organization. For example, higher levels of the organization may focus on overall cost reductions while intermediate levels may be concerned with productivity improvement through cost reductions. At the lowest organizational levels, the emphasis may be on finding ways to eliminate waste so that costs can be reduced.

See **Performance measurement in manufacturing; Supplier performance measurement.**

SME

See **Professional associations for manufacturing professionals and managers.**

SMED

JT Black

Auburn University, Auburn, AL, USA

The lean production approach to manufacturing demands that lot sizes be small and tending towards a lot size of one. This is impossible to do if machine setups take hours to accomplish. The economic order quantity (EOQ) formula has traditionally been used in manufacturing to determine batch sizes and to cost-justify a long and costly setup time. The EOQ was a faulty suboptimal approach that accepted long setup times as a given. But, we can reduce setup times thus reducing lot sizes.

Successful setup reduction is easily achieved when approached from a methods engineering perspective. Singeo Shingo (1988) and Taiichi Ohno (1988) have done much of the initial work in this area. Large reductions can be achieved by applying time and motion studies and Shingo's Single Minute Exchange of Dies (SMED) rules for rapid exchange of dies. Setup-time reduction occurs in four steps, see Figure 1. The initial step is to determine what currently is being done in the setup operation. The setup operation is usually videotaped and everyone concerned gets together and reviews the tape to determine the elemental steps in the setup.

The next step is to separate all the elements of the setup activity into two categories, *internal* and *external*. Internal activities can be done only when the machine is not running; external elements can be carried out while the machine is running. This elemental division of setup activities will usually assist in shortening the lead time considerably. Steps three and four focus on reducing the internal time. The idea here is for operators to learn how to reduce setup times by applying simple principles and techniques. If a company must wait for the setup reduction people other than operators to examine every process, lean manufacturing systems will never be achieved.

The similarity in shape and processes needed for a family of parts allows setup time to be reduced

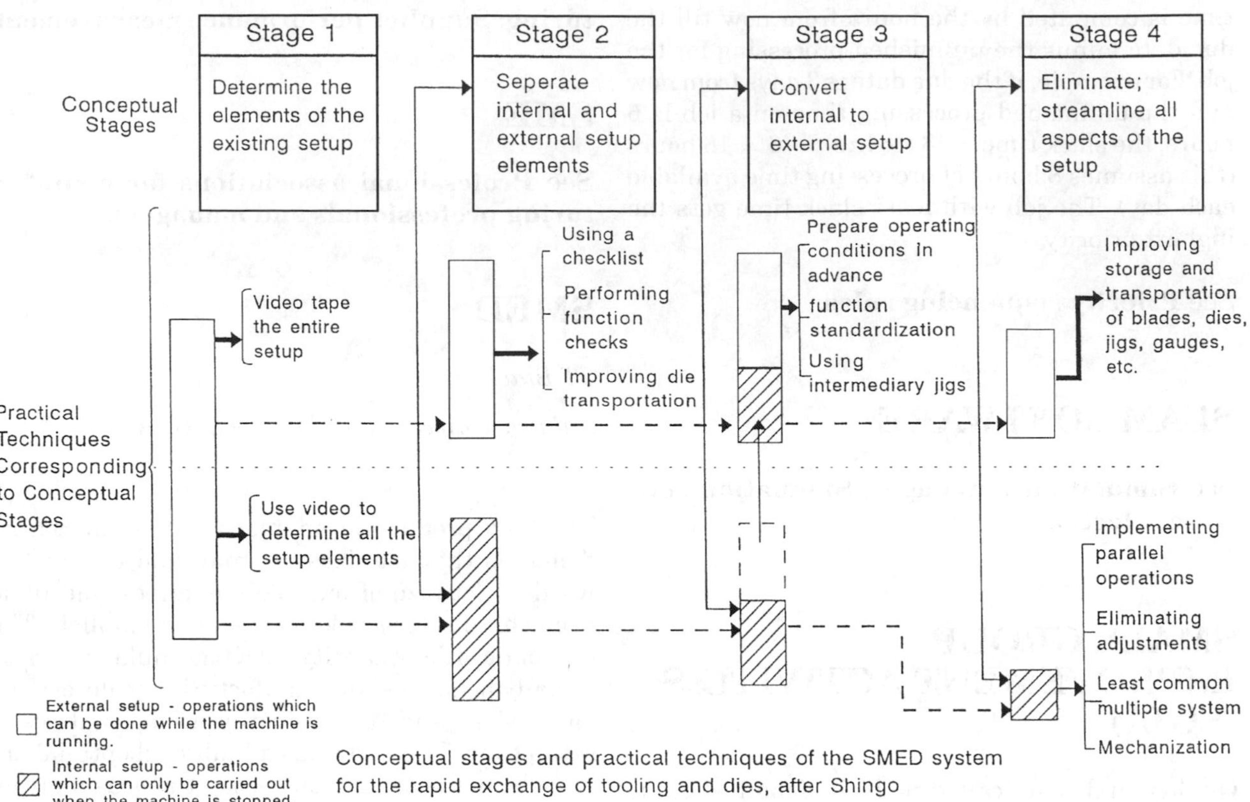

Fig. 1. Conceptual stages of the SMED system for the rapid exchange of tooling and dies.

or even eliminated. Initially, the goal of the setup should be to halve the setup time and then drive it to be less than 10 minutes. This is what Shingo meant by *single-minute exchange of dies*. As the cell manufacturing matures setup times are continually reduced. The final goal is to get them down to zero or at least less than one minute. Shingo called this one touch exchange of dies (OTED).

In the last stages of SMED, it may be necessary to invest capital to drive the setup times below one minute. Automatic positioning of workholders, intermediate jigs and fixtures, and duplicate workholders represent the typical kinds of hardware needed. The result is that the long setup times can be drastically reduced.

The initial objective here is to reduce the setup time until it is equal or less than to the cycle time (one to two minutes) for the cell. This will permit a significant initial reduction in lot sizes. After more reductions, the goal is to get the setup time down to zero or at least about the time an operator spends to each machine to load, unload, inspect, deburr, and so on.

As shown in Table 1 when numerous processes are involved in the cellular manufacturing of a

Table 1. Example of change from part A to part B for four-process cell.

	Four processes in the cell			
Cycles through the cell	Machine 1	Machine 2	Machine 3	Machine 4
1	A	A	A	A
2	A (last A part)	A	A	A
3	Setup change	A (last A part)	A	A
4	B (first B part)	Setup change	A (last A part)	A
5	B	B (first B part)	Setup change	A (last A part)
6	B	B	B (first B part)	Setup change
7	B	B	B	B (first B part)
8	B	B	B	B

Example of OTED

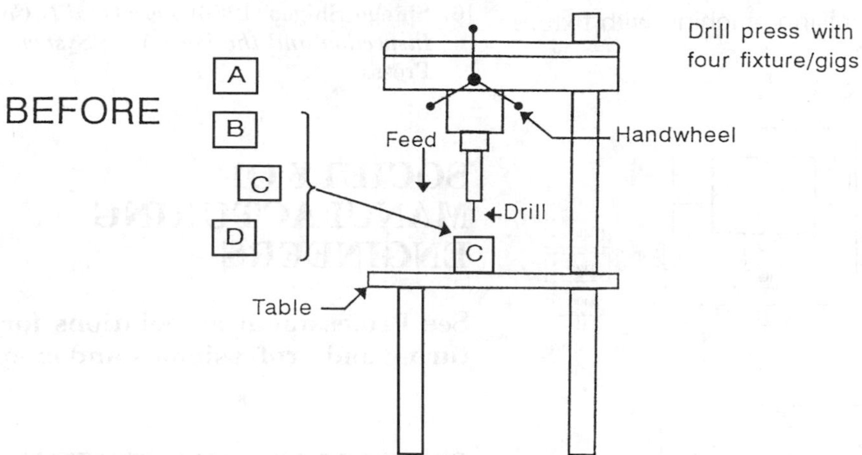

Typically, a machine that has four jobs, with four different fixtures or jigs, would need four different setups, each consisting of changing fixtures or jigs and alignment f each.

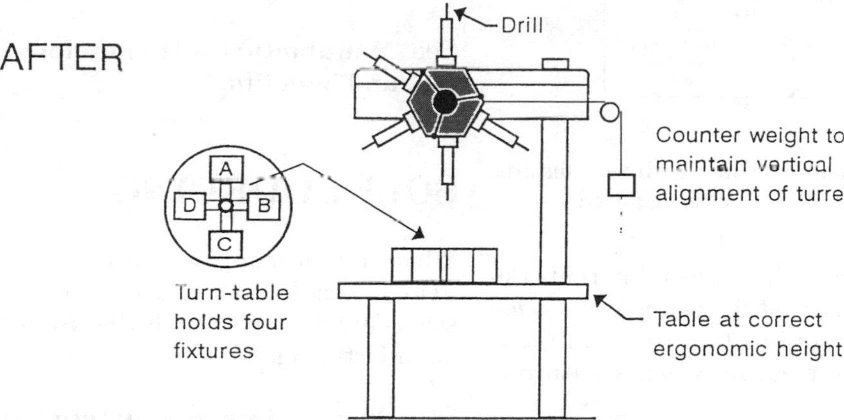

After redesign, the four fixtures are mounted on a turntable and are permanently aligned to the spindle when locked in position. Turret replaces single spindle. Automatic feed can replace handwheel.

Fig. 2. The machines in the interim cell are modified to process a family of parts, reducing setup.

family of parts, sequential or flow through setup changes are utilized. That is, the setups are done sequentially by the operator. After each setup, defect-free products should be made right from the first unit. Ultimately, the ideal condition would be to eliminate setup between parts altogether. Shingo called this No Touch Exchange of Dies (NOTED).

Figure 2 shows how a drilling machine in a cell can be modified for rapid changeover. This internal cell was designed for a family of four parts. To reduce the setup time to a few seconds, or "one touch," the drilling machine was modified to eliminate setup as far as possible. The total cost to modify the drilling machine was around $300. This is an example of one touch exchange of fixtures.

Figure 3 shows a top view of a vertical milling machine equipped with a digital readout device from an interim manufacturing cell. The fixtures for the four parts in the family are never removed from the oversized table. The first part produced with the assistance of these fixtures is always a good part and there was no time spent on fixture exchange. This is an example of no touch exchange of fixtures.

Other setup reduction techniques include the use of group jigs (one jig to accommodate different parts with the use of adapters), training operators in rapid setup techniques, practicing rapid setups, and the use of intermediate jigs and fixtures. *The intermediate jig* is like the cassette for a VCR. The cassette is quickly loaded and locked in place. The tape inside every cassette can be different. To the machine (the tape player), every different fixture (tape) that is placed in the intermediate jig (cassette) looks the same. To the cutting tool (the

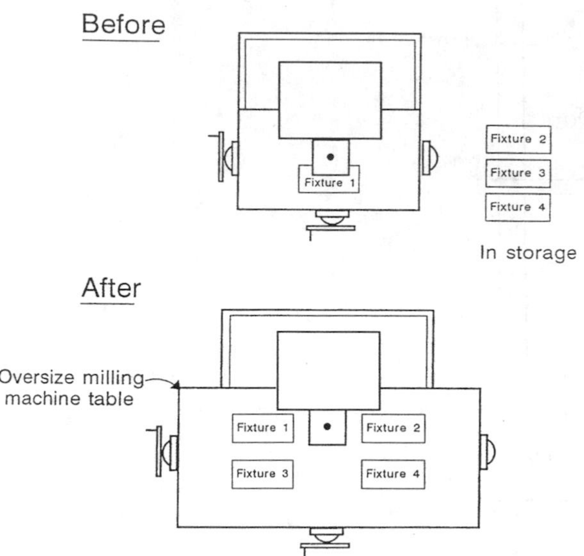

Fig. 3. Top view of milling machine with four fixtures for four parts in the family – no setup required.

tape head), every workholder looks different. Designing workholders so that they appear the same to the machine tool requires the use of an intermediate jig or fixture plate to which a fixture is attached.

In summary, the savings in setup times are used to decrease the lot size, and increase the frequency lot production. The smaller the lot size, the lower will be the inventory level in the system, resulting in shorter throughput times, improved quality and lower operating costs.

See **Economic order quantity; Implementing lean manufacturing systems; Intermediate jig or fixture; Manufacturing cell design; Setup reduction.**

References

[1] Shingo, Shigeo (1985). *A revolution in Manufacturing: The SMED System*. Productivity Press.

[2] Ohno, Taiichi (1988). *Toyota Production System: Beyond Large-Scale Production*. Productivity Press.

[3] Ishikawa, Kaouru (1972). *Guide to Quality Control*. Asian Productivity Organization.

[4] Monden, Yasuhiro (1983). *Toyota Production System*. Industrial Engineering and Management Press, IIE.

[5] Shconberger, Richard J. (1982). *Japanese Manufacturing Techniques: Nine Hidden Lessons in Simplicity*. The Free Press.

[6] Shingo, Shigeo (1986). *Zero Quality Control: Source Inspection and the Poka-Yoke System*. Productivity Press.

SOCIETY OF MANUFACTURING ENGINEERS

See **Professional associations for manufacturing and professionals and managers.**

SOFT MANUFACTURING TECHNOLOGY

See **Manufacturing technology use in the U.S. and benefits.**

SOLE SOURCING

When only one supplier is used for the supply of an item, it is called sole sourcing. The trend towards sole sourcing began with the use of just-in-time-manufacturing.

See **Just-in-time manufacturing; Supplier relationship; Supplier relationship as strategy.**

SOURCE PLANTS

Multinational manufacturers use a network of plants to serve worldwide markets. Ina network of plants, source plants may be located in a region too small to absorb all their outputs but such plants serve as sources for products, process improvements and product innovations for the other plants in the network.

See **International manufacturing.**

SOURCING

Sourcing is the process of locating the right number and mix of suppliers to meet a firm's buying requirements with quality goods and services on time.

See **The future of purchasing; Purchasing: Acquiring the best inputs; Supplier partnership as strategy; Supplier performance measurement; Supplier relationships; Supply chain management.**

SPAN OF PROCESS

The 'Span of Process' of a manufacturing organization specifies the range of activities or processes that the company has under its direct control. This captures the extent of 'Vertical Integration'. Very few manufacturers start with basic raw materials and convert these to products for direct sale to end users. It is more usual for intermediate products to be produced and a supply chain of several manufacturers to exist. For example, a steel producer might produce bar stock, which is sold to a component manufacturer, who produces axles, which are assembled into wheel hubs, which in turn are sold on to another company, which manufactures bicycle wheels, which are sold to the company which assembles bicycles, which are sold by a retail chain to the end consumer.

Each company within the chain has control over a defined span of process. Deciding the extent of the 'span of process' is a key strategic decision for a manufacturing company. In the example above, the hub manufacturer could decide to extend its span *forward* and move into manufacturing complete wheels, or conversely the wheel manufacturer could decide to *integrate backwards* into manufacturing hubs.

'Backward integration,' making instead of buying, allows a company to protect its source of supply, to save the profit margin paid to the supplier, and to gain control over its input materials' quality and delivery. 'Forward integration,' allows a company to protect its markets, to increase its value-added operations, and to move closer to the ultimate customer.

See **Core manufacturing competencies; Manufacturing strategy; A process approach to manufacturing strategy development; Product-process matrix.**

SPC

SPC stands for statistical process control.

See **Statistical process control using control charts; Total quality management.**

SPECIAL CAUSES OF VARIATIONS

W. Edwards Deming elaborated on Walter A. Shewhart's argument that variability in manufacturing and service processes can be traced to either common causes or special causes of variations (Shewhart's assignable causes). Special cause variability is beyond the natural variability of the process. Special cause variability can be identified and addressed by operators. Examples of special causes are operator error, faulty setup, or incoming defective raw material. Deming believed that only about 15% of the variation in a process is due to special causes. Deming relied on control charts to describe both the natural variability of the system, and to detect the existence of a special cause of variation. A process that is operating with special causes of variation is said to be "out of statistical control."

See **Quality: The implications of W. Edwards Deming's approach; Statistical process control using control charts; Total quality management.**

References

[1] Deming, W. Edwards (1982). *Out of the Crisis*, Center for Advanced Engineering Study, Massachusetts Institute of Technology, Cambridge, Massachusetts.

[2] Deming, W. Edwards (1982). *Quality, Productivity, and Competitive Position*, Center for Advanced Engineering Study, Massachusetts Institute of Technology, Cambridge, Massachusetts.

SPECIALIZATION

Specialization in job design assigns a limited number or range of the tasks to operators involved in the production of goods or service. Increasing levels of specialization may allow increased productivity with higher volume and better quality. When workers focus their efforts on doing a limited range of activities they generally become more adept at performing them. There are two theoretical foundations for this phenomenon. The first is based on the economic theory of "comparative advantage." People naturally differ in their aptitudes and abilities, and these differences may be most effectively used by matching them to a particular task or job requirements. Thus, both workers and employers benefit when workers seek, or are given, work that matches their capabilities.

The second argument supporting specialization is based on learning curves; the more experience a worker has in performing a task, they more efficient they become in doing the work.

Thus, a group of workers of equal abilities will be more productive if the specialize. Specialization is one of the oldest industrial concepts. Plato (1987, p. 56–63) comments on specialization leading to the creation of various trades and crafts, and their role in the rise of cities as trading centers. Adam Smith (1976, vol. 1, p. 8) too recognized the benefits of specialization in improving productivity in pin manufacture during the 1700's. In general, specialization is limited by the level of demand – where demand is low the overall production volume may not be sufficient to justify specialization. The rise of mass production stimulated specialization.

See **History of operations management; Human resource issues in operations management; Learning curves; Manufacturing systems; Scientific management.**

References

[1] Babbage, Charles. 1835. *On the Economy of Machinery and Manufactures*. Charles Knight, London, England.

[2] Plato, (Trans. Desmond Lee). 1987. *The Republic*, Penguin Books, London, England.

[3] Smith, Adam. (ed. Edwin Cannan) 1976. *The Wealth of Nations*, University of Chicago Press, Chicago, Illinois.

[4] Stigler, George J. 1951. The Division of Labor is Limited by the Extent of the Market. *The Journal of Political Economy*, Vol. 59, No. 3, p. 185–193.

SQC

SQC stands for Statistical Quality Control.

See **Statistical quality control using control charts; Total quality management.**

STAKEHOLDER GROUPS

Stakeholder groups are people, who have an interest in some aspect of a firm. Typical examples include customers, suppliers, employees, shareholders, creditors, regulators, and the local community.

See **Balanced scorecard.**

STANDARD COST SYSTEM

Standard cost systems make use of standard costs, which are the budgeted or estimated costs deemed to be necessary to manufacture a single unit of product or perform a single service. Standards are traditionally established for each component (material, labor, and overhead) of product cost. Actual costs are compared to standard costs to determine variances (the difference between the actual cost and the cost that was expected to be incurred).In a standard cost system, both standard costs and actual costs are recorded in accounting records. This dual record keeping affords an element of cost control by providing norms against which actual costs operations can be compared.

See **Accounting system implications of TOC; Activity-based costing: An evaluation; Activity-based costing; Standard costing; Standard time; Target costing.**

References

[1] Barfield, Jesse T., Cecily A. Raiborn and Michael R. Kinney (1994). *Cost Accounting: Traditions and Innovations*, West Publishing Company, St. Paul, Minnesota.

[2] Henke, Emerson O. and Charlene W. Spoede (1991). *Cost Accounting: Managerial Use of Accounting Data*, PWS-Kent Publishing Company, Boston, Massachusetts.

STANDARD COSTING

Standard costing is the process of calculating the cost of activities, products or services based on *standard costs*. *Costing* is the British word for *cost accounting*. *Standard costs* are anticipated costs of producing an unit of output, and implies what the costs should be. Firms use standard costs as the benchmark for gauging performance, and usually update these annually, or when technology significantly changes, or when costs of factors of production significantly change.

See **Accounting system implications of TOC; Activity-based costing; Standard cost system; Standard time; Target costing.**

STANDARD TIME

Standard times are frequently used in job design, work flow planning and costing. The basic idea is that the performance of some task involved in making a product will take a consistent time. This time may then be used in a variety of ways. It may be used in planning work flow as is done in line balancing, where the objective is to give every work station an equivalent amount of work to be done. Standard times may also be used in performance evaluation and reward systems. Workers that take longer than the standard times in performing tasks may be regarded as less effective or efficient. If employees are paid on a piece-work or bonus basis, a failure to make or exceed the standard can be problematical, with questions raised about the appropriateness of the standard or external factors that affect productivity. The setting of standards is thus not entirely dictated by technical considerations of the work involved, the tools and equipment provided for its accomplishment, etc.; standards also reflect labor relations and employee motivation.

See **Activity-based costing; History of manufacturing management; Scientific management; Standard cost system; Standard costing; Target costing.**

References

[1] Burawoy, Michael. 1979. *Manufacturing Consent: Changes in the Labor Process Under Monopoly Capitalism*. University of Chicago Press, Chicago, Illinois.

[2] Gilbreth, Frank B.; Lillian M. Gilbreth. 1918. *Applied Motion Study*. George Routledge & Sons, Ltd., London, England.

[3] Gilbreth, Frank B.; L.M. Gilbreth. 1921. *Process Charts. ASME Transactions*, Vol. 43, 1029–1050.

[4] Taylor, Fred. W. 1895. A Piece Rate System. *ASME Transactions*, Vol. 16, 856–903.

STANDARDIZATION

This now typically involves s set of engineering specifications that rigidly define the relevant product characteristics, materials, dimensions, and manufacturing processes. When Eli Whitney and other arms manufacturers first attempted to make standardized muskets with interchangeable parts, they were provided with a set of "pattern" muskets that were to serve as physical standards. The major flaw with this scheme was the failure to issue manufacturers with identical "pattern" muskets – one manufacturer found that the musket parts were not interchangeable between the patterns. Standardization and interchangeability are mutually supportive – interchangeability follows from standardization, and standardization is dependent on interchangeability.

See **History of manufacturing management; Mass customization; Product design for global markets; Product design; Scientific management; Standardized customization; Target costing.**

References

[1] Booker, Peter J. 1979 reprint of 1963 Edition. *A History of Engineering Drawing*. Northgate Publishing Co. Ltd., London, England.

[2] Hounshell, David. 1983. *From the American System to Mass Production*. University of Delaware Press, Wilmington, Delaware.

[3] Howard, Robert A. 1978. Interchangeable Parts Reexamined: The Private Sector of the American Arms Industry on the Eve of the Civil War. *Technology and Culture*, Vol. 19, No. 4, 633–649.

[4] Sinclair, Bruce. 1969. At the Turn of a Screw: William Sellers, the Franklin Institute, and a Standard American Thread. *Technology and Culture*, Vol. 10, No. 1, 20–34.

STANDARDIZED CUSTOMIZATION

In standardized customization, final products are assembled from standardized components. The final product may be built around some central standardized core. Standard customization can be seen in a variety of products that offer the customer a choice of features or options. For example, the desktop computer offers customers a range of choice in the processor speed, amount of memory, type of storage devices (CD ROM, floppy drive, ZIP drive, tape drive), monitors, and keyboards. Customers choice of options configures a unique product using standard options.

See **Mass customization; Mass customization implementation; Product design; Product design for global markets.**

STARVATION AVOIDANCE

Starvation refers to a machine that is idled for want of work. For effective and efficient plant operation, bottleneck resource should not be starved. Starvation avoidance is an order release mechanism. Introduced by Glassey and Resende (1988), starvation avoidance constantly measures the virtual inventory W for the bottleneck workstation and compares that to the bottleneck's replenishment lead time L. If $W > L$, then one expects that a new order released now would reach the bottleneck before the bottleneck operation starved. Otherwise, the bottleneck needs more work.

To account for manufacturing uncertainties, the starvation avoidance order release mechanism has an adjustable threshold; it triggers an order release whenever $W < \alpha L (\alpha > 0)$. L is the total processing time needed for a newly released order to reach the bottleneck. (Ignoring the queue time simplifies the calculations.)

Lozinski and Glassey (1988) describe a graphical tool that is used for order release and bottleneck starvation avoidance. The bottleneck starvation indicators are used daily for controlling high-volume manufacturing. The indicators plot cumulative inventory versus cycle time from the bottleneck and compare that to a target line that shows the inventory needed to achieve desired production levels. If the actual indicator line drops below the target, then the bottleneck resource is in danger of starving at some point in the future. Releasing more work may be necessary. However, if the actual indicator line is above the target, then the system has too much WIP, and no additional order release is necessary.

See **Order release; Synchronous manufacturing with buffers; Theory of constraints in manufacturing management.**

References

[1] Glassey, C. Roger, and Mauricio G.C. Resende, "Closed-loop job release for VLSI circuit manufacturing," *IEEE Transactions on Semiconductor Manufacturing*, Volume 1, Number 1, pages 36–46, 1988. (1988a)

[2] Lozinski, Christopher, and C. Roger Glassey, "Bottleneck starvation indicators for shop floor control," *IEEE Transactions on Semiconductor Manufacturing*, Volume 1, Number 4, pages 147–153, 1988.

STARVED MACHINE

See **Starvation avoidance.**

STATE FLEXIBILITY

State flexibility refers to the capability of a manufacturing system to continue to maintain its state of effective operation despite changes in its internal and external operating environment. For example, if the manufacturing system's labor force is skilled in multiple tasks, and the machines are capable of performing a variety of operations, then the manufacturing system would have a high degree of state flexibility.

See **Manufacturing flexibility; Manufacturing flexibility types and dimension; Manufacturing with flexible systems.**

STATISTICAL FORECASTING PROCEDURES

Statistical forecasting procedures use statistical techniques for forecasting demand using historical data.

See **Forecasting examples; Forecasting guidelines and methods; Forecasting in operations management.**

STATISTICAL PROCESS CONTROL (SPC)

See **Quality: The implications of W. Edwards Deming's approach; Statistical quality control using control charts; Total quality management.**

STATISTICAL PROCESS CONTROL USING CONTROL CHARTS

Amitava Mitra

Auburn University, Auburn, AL, USA

INTRODUCTION: A control chart is a graphical device for monitoring the output of an ongoing manufacturing or production process. Management, while operating under resource constraints, has to ensure that the best quality is achieved at a competitive price. Cost reduction and continuous quality improvement are major objectives of the organization. Elimination of scrap and rework leading to reduction in production costs is a priority for manufacturing. Making things right the first time becomes the motto of production. A process in control is one that is predictable. Of course, being predictable does not imply that customer requirements are satisfied. Continual improvement to meet the needs of the customer is imperative for any organization.

Customers can be internal or external. While the finished product from manufacturing is typically sold to an external customer, the flow of work-in-process inventory from one operation to the next can be regarded as a producer-consumer transaction. For instance, a cylinder casting from the foundry (the producer) goes to the machine shop (the customer) to be turned. Likewise, the turned casting from the machine shop (the producer) proceeds to the drilling unit (the customer) for appropriate holes to be drilled. In this sequence of operations, quality has to be adhered to in each step. For this to occur, the process has to be maintained in control for the various operations that are performed in producing the finished product.

It is not sufficient to just inspect the end product prior to shipment and remove the nonconforming items from the lot. Inspection does not address the root cause of the problem, it is merely a sorting mechanism. A control chart, on the other hand, helps manufacturing to identify when something is wrong with the process, and thereby to take corrective action. It is a mechanism whereby data is collected from the process output, appropriate statistics are calculated and then plotted on a chart that is used to monitor the process.

In any production or manufacturing environment, variability is present and this needs to be understood by management and operating personnel. The causes of variation may be divided into two groups – common causes and special causes. Variability due to *common causes* is part of the designed process and affects all items. It is inherent in the process and is referred to as the natural variation in the process. It is the result of many causes and cannot be totally eliminated. A process that is operating under a stable system of common causes is said to be in statistical control. Examples are the inherent variation in incoming raw material, machines, and equipment used, lack of adequate supervision skills, and changes in working conditions. Management is responsible for such problems. Deming believed that about 85% of all problems are due to common causes and hence can be acted upon by management alone (Deming, 1982).

Variability due to *special causes* are those for which an identifiable reason can be determined. They are not part of the designed process. Examples of such causes are use of a wrong tool, inexperienced operator, improper raw material, and incorrect process settings. In using a control chart, when an observation plots outside the control limits, or if a nonrandom pattern is exhibited, it is assumed that special causes exist and the process is said to be out-of-control. Operating personnel and management can recommend remedial actions to eliminate special causes to bring the process to a state of statistical control. **Thus, while control of a process may be achieved through elimination of special causes, improvement of a process is attained through the reduction of common causes.**

HISTORICAL PERSPECTIVE: The originator of the control chart is Walter A. Shewhart. During the 1920s when Shewhart was with Bell Telephone Laboratories, he introduced the concept in a technical report in 1924. Therefore, control charts are often referred to as Shewhart control charts, based on the founder's name. At that time, however, Shewhart's work did not attract serious attention. In 1931, Shewhart published his classical book (Shewhart, 1931) that outlined statistical methods for use in production, and control chart methods. He was invited to give a series of lectures at University College in London in 1932, where his concepts were warmly received. Thus began the application and utilization of control chart methods in England which further stimulated its interest in the United States.

In the United States, W.E. Deming, another pioneer in the field of quality, was a strong advocate of the use of statistical methods to improve quality. In 1938, Deming invited Shewhart to present seminars on control charts at the U.S. Department of Agriculture. World War II caused an enormous expansion of industries that were involved in supplying war material. Consequently, process control became an important criteria. In 1940, the U.S. War Department published a guide for using control charts to analyze process data. After the end of World War II, early application of statistical process control in the U.S. continued to grow as training courses in this area were given to industry. During the period of 1942–1946, more than 15 quality societies were formed in North America. In 1946, the American Society for Quality Control (ASQC) was formed as a merger of various quality societies, and became a proponent of quality control and its applications.

In the meantime, Deming continued to disseminate his philosophy on quality assurance. But, there was not widespread acceptance by U.S. industry. In 1950, the Union of Japanese Scientists and Engineers (JUSE) invited Deming to address their leading industrialists. He widely taught statistical quality control methods to managers in Japan. The results of such teachings and their adoption lead to dramatic improvements in the quality of Japanese products over the next three to four decades.

Deming developed a set of 14 points for management (Deming, 1986; Deming, 1982). He believed that about 85% of all problems can be solved only by management, because they involved changing the system of operation and are not influenced by the operators. Deming's ideal management style is holistic where the organization is viewed as an integrated entity. Worker's responsibility lies in communicating to management the information they possess regarding the system. Management must use this information as input in making operational and strategic decisions. The idea is to plan for the long run and provide a course of action for the short run. Statistical process control, in this context, is typically used as an on-line control mechanism to verify the state of the process and make changes in it when necessary.

STRATEGIC PERSPECTIVE: The production and manufacturing environment faces the challenges of global competition. Top management has the responsibility of creating a strategic vision for the survival and growth of the organization. Out of the vision will stem the mission statement and thereupon the goals and objectives to be implemented operationally.

Process Improvement

One of the major objectives of management is to first bring the process to a state of control. This is accomplished through the use of control charts whereby special causes are identified and remedial actions are taken for their elimination. From a strategic perspective, attaining a stable and predictive process is important because only then can management take actions to improve the process by focusing on the common causes. Thus, control charts play an important role in creating a precondition before process improvement can take place.

Another contribution of the control chart toward process improvement lies in process centering and variability reduction. Given a target or nominal value for the characteristic under consideration, a control chart can easily determine if the process is off-centered. The calculated center line on the control chart is an estimate of the process center. Adjustments can be made on process parameters, say depth of cut, feed rate, temperature, etc., so as to move the process center toward the target value if there is a deviation between the two.

A control chart is a tool to monitor process variability. As the number of operations performed increases so does variability. The overall variation in product characteristics is a function of the variability in each step and the number of steps required to manufacture the finished product. With proper use of the control chart, measures can be taken to reduce variability in individual operations, which causes a reduction in overall variability. A strategic goal of every organization is to reduce variability and the control chart is a useful aid in accomplishing this goal.

Cost Reduction

The old notion that quality and productivity cannot go hand-in-hand is a myth. In fact, achievement of quality improves productivity. Let us examine this from a cost and profitability point of view.

There are four major categories of quality costs (ASQC, 1971): prevention costs, appraisal costs, internal failure costs, and external failure costs. *Prevention costs* are those incurred for planning, implementing, and maintaining a quality system. *Appraisals costs* are those associated with measuring, evaluating, or auditing products, components, or purchased materials to determine the degree of conformance with respect to some specified standards. Prevention costs are associated with managing the intent or goal, whereas appraisal costs are those associated with managing the outcome. The two other costs are related to failure. *Internal failure costs* are incurred when products fail to meet quality requirements prior to transfer of ownership to the customer. Finally, *external failure costs* are incurred when the product does not perform satisfactorily after ownership is transferred to the customer.

Through the use of control charts, quality control moves up-stream in the process. The emphasis becomes on prevention of problems rather than detection of nonconforming items. Strategic points within the manufacturing cycle of operations are identified for the purpose of process monitoring by control charts. For example, a complex or expensive operation may necessitate the use of a control chart. It is highly undesirable for the process to be out of control because the cost of producing nonconforming items may be excessively high. Collection and analyses of data from such operations by using control charts may increase appraisal costs. However, this may be offset by a reduction in total costs because of the lowering of internal and external failure costs. Availability of user-friendly software in conjunction with hardware that can automatically collect and record data will have an impact on the use of control charts. This will lower appraisal costs as routine tasks are relieved from personnel.

From a strategic perspective, an improvement in quality not only causes a reduction in total quality costs but also improves productivity. As the proportion of scrap and rework is reduced, the proportion of conforming output increases, leading to increased productivity. Another way in which improved quality increases profitability is through the market route. With an increased customer satisfaction level there is an increase in the market share. Also, a lowering of costs leads to an improved competitive position through a reduction in the cost/price ratio. The final impact of all this is increased profitability and a motivation for continual quality improvement in order to maintain and improve the company's competitive position.

Empowerment of People

Another strategic implication is the empowerment of people who are involved in the process. When a control chart indicates an out-of-control signal, operating personnel analyze possible special causes for such an outcome and investigate remedial actions for deployment. By delegating authority to such personnel to take action, it provides a vital input in utilizing their knowledge of the process to identify corrective actions for eliminating special causes. Thus a sense of belonging is created which is important to promoting the spirit of teamwork.

Quality Planning

Statistical process control (SPC) may be used for those quality characteristics that are critical to satisfying the needs of the customer. In order to identify such critical characteristics, the methodology of quality function deployment (QFD) could be employed. QFD analyzes customer requirements and through a process of comparison ranking determines their relative importance. Based on this relative importance ranking, it identifies important product characteristics that meet critical customer requirements. It then goes a step further to determine critical process characteristics that will help achieve the desired product characteristics. So, as part of the planning process for SPC, the QFD framework should be helpful.

PRINCIPLES OF PROCESS CONTROL: In a typical control chart, the value of the appropriate quality characteristic is plotted along the vertical axis while the horizontal axis represents the time sequence of the samples (or subgroups) that are drawn from the process. Samples of a certain size (say 4 or 5) are selected from the process and the quality characteristic (say, average diameter) is calculated. This characteristic is then plotted in the order in which samples are taken.

Quality characteristics can either be variables or attributes. For *variables*, the quality characteristic is such that a numerical value can be obtained in a range or continuum scale. Examples are the average diameter of a shaft, the average strength of a beam, or the average viscosity of a fluid. On the other hand, a characteristic is an *attribute* if discrete measures are used for the purpose of classification and counting the number of units or occurrences. For example, a part can be classified as conforming or nonconforming to stipulated specifications requirement, a product can be classified as defective, adequate, or excellent.

Sometimes, characteristics that are inherently variables, are observed as attributes because of ease of measurement or cost constraints. As an example, the diameter of a bearing is, in principle, a variable. But, if the diameter is measured using a go/no-go gauge as opposed to obtaining a numerical value using a micrometer, it is classified as conforming or not to desired specifications, and is observed as an attribute. Clearly variables type of measurements convey more information than attribute type of measurements. Some companies like Ford require of their suppliers only variables type measurements for control charts.

Three lines are usually used on a control chart. The *center line*, which represents the average value of the quality characteristic, is an indication of the process center. The *upper* and *lower control limits* (typically three standard deviations from the center line of the characteristic that is being plotted) are used to make decisions regarding the process. If the points plot within the control limits and do not exhibit any identifiable pattern, the process is said to be in *statistical control*. If a point plots outside the control limits, or if identifiable non-random patterns (AT&T, 1984; Nelson, 1984) exist , the process is said to be out-of-control. Under these circumstances, production personnel would look for special causes that create such occurrences.

Uses of Control Charts

There are several uses of control charts:

To decide when to take corrective action: A control chart indicates the need for corrective action. This is a situation where special causes may exist in the system and remedial actions must be identified for their removal. When an observation plots outside the control limits, it calls for corrective action.

To decide who should take corrective action: For effective use of control charts, whenever an out-of-control situation is detected through the presence of special causes, it is important that decision making be timely and based on adequate information. This can be accomplished by empowering the operating personnel, who are involved with the process to take corrective action as necessary. These people have an insight on the effect of various process parameters on the quality characteristic in question through their first-hand experience in dealing with process operations. They have good feasible ideas that may eliminate the special causes. Furthermore, by giving them the authority to take remedial action, implementation of such actions take place in a timely manner, which is important. A delay in taking corrective action will result in more undesirable output.

To decide on the type of remedial action: The pattern of the plot may provide some clues on possible special causes that exist in the system. Such information may suggest the type of remedial action that is necessary to eliminate the identified special causes. For example, if there are nine or more consecutive points that plot above the center line of the quality characteristic (say, average length of a part), it may be an indication that remedial action needs to be taken to reduce the part length. Possible steps could be to change the depth of cut or the tool in a machining operation.

To decide when to leave a process alone: It is important to understand that every production process has some variability. When the exhibited variation is due to common causes that are normal and inherent to the process, no remedial action is necessary. In fact, inappropriate over-control of a process through frequent adjustments increases the variability of the process. A control chart can be used to identify when the variation is part of the normal process and hence no corrective actions are needed. For instance, if no identifiable patterns in the plot are observed, or an observation does not fall outside the control limits, even though variability exists in the observations, it is considered to be due to common causes that are an inherent part of the process.

To estimate process capability: Let us first point out the difference between control limits and specification limits. *Specification limits* are determined by the needs of the customer. These limits are placed on a product characteristic by designers and engineers to ensure adequate functioning of the product. They apply to individual observations. *Control limits*, on the other hand, identify the variation that exists between samples or subgroups of measurements. They apply to sample statistics, for example, the average length of a part. They do not apply to individual units except in the case of control charts for individual measurements.

Control limits reflect the variability in the process and have no relationship to specification limits, which are chosen to meet customer needs for the product. Thus, it is possible for a process to be in control, as concluded from the analysis of a control chart, and yet the product may not meet specifications. This implies that while there are no special causes in the system, the inherent variability in the process due to common causes is such that the process is not able to manufacture a product that meets the needs of the customer. The process, though in control, is not capable of satisfying customer expectations. On the contrary, it is also possible for a process to be out of control and yet produce a product that meets customer specifications. Even though this seldom happens, this may represent a situation where the customer specifications are quite loose relative to the process spread. The process may drift out of control either due to a shift in the process mean or an increase in the process variability. However, in this case, since the customer requirements are not very rigid, as evidenced by the large spread between the specification limits, the output from the process may still be acceptable to the customer.

Process capability is a measure of the ability of the process to meet customer specifications. It is a measure of the goodness of the process since the ultimate objective is to satisfy or exceed customer requirements. Once a process is in statistical control, as determined from a control chart through elimination of special causes, its capability can be estimated.

A simple measure of process capability, used for a centered process, is given by the C_p *index* as follows

$$C_p = \frac{USL - LSL}{6\sigma}, \tag{1}$$

where *LSL* and *USL* represent the lower specification limit and upper specification limit on the quality characteristic, and σ is the process standard deviation. The process standard deviation can be estimated from control chart information as shown later, once the process is brought to a state of control.

If the quality characteristic can be assumed to have a normal distribution, and the process is centered, the capability index can be thought of as the ratio of the specification spread to the process spread. Figure 1 demonstrates this concept for a capable process for which $C_p > 1$. The process spread is 6σ. Assuming normality, just about all (99.73%) of the production will be within the lower natural tolerance limit (LNTL)

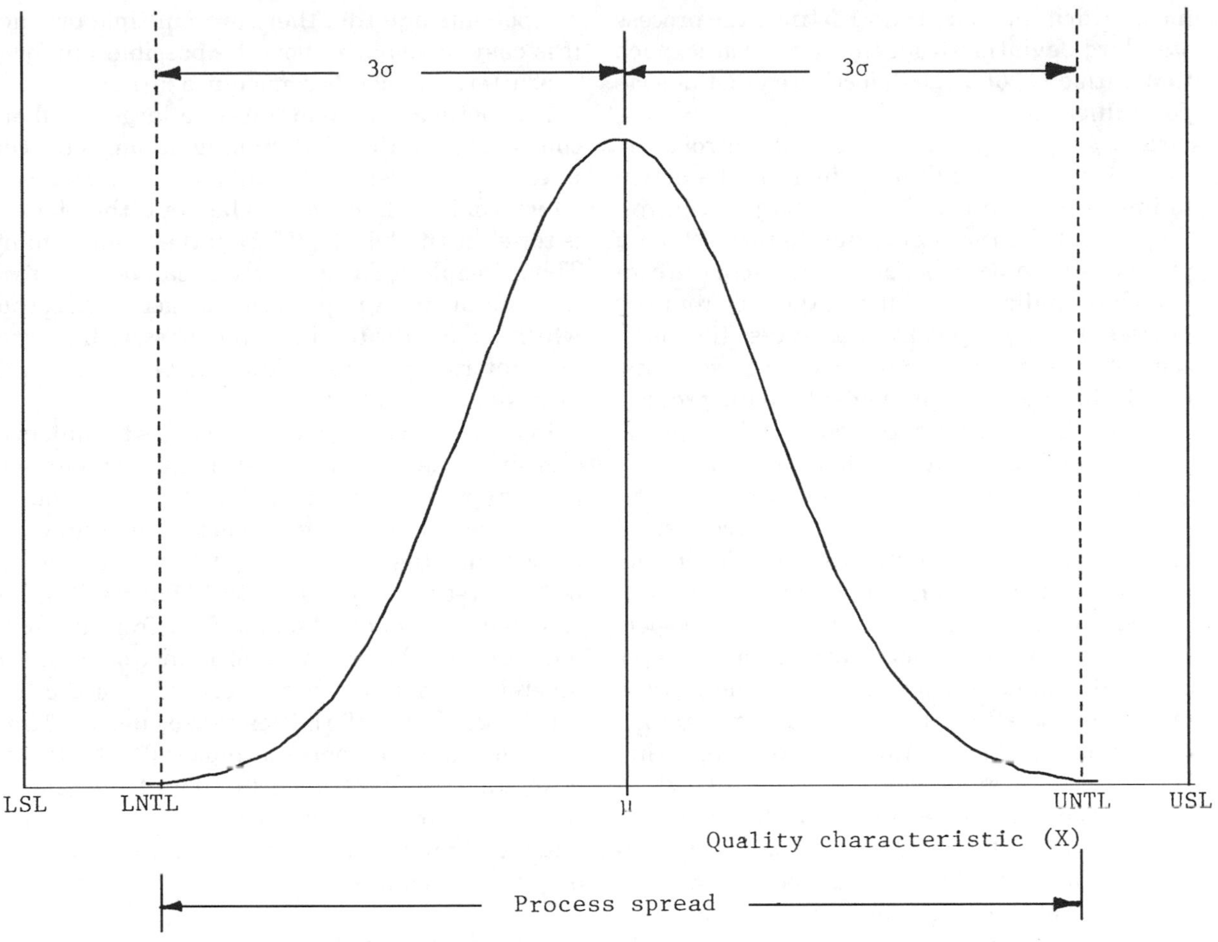

Fig. 1. A capable process with $C_p > 1$.

and the upper natural tolerance limit (UNTL), each of which is three standard deviations away from the process mean (μ). Note that desirable values of C_p are $C_p \geq 1.33$, in which case almost all products will be within the specification limits. When $C_p = 1.3$, the proportion of the product outside two-sided specification limits is 96 parts per million. As C_p increases to 1.5, this proportion decreases to 7 parts per million. In Figure 1, the spread between the specifications exceeds the process spread and so the process is capable.

When the process is not centered between the specifications, that is the process mean is not at the midpoint between the specification limits, the C_p index is not a good measure of process capability. From Equation (1), note that C_p is influenced by the process spread but not by the location of the process, i.e., the process mean. So, if the process spread is less than the spread between the specifications, but the process mean is located at the lower specification limit, the value of C_p would be greater than 1. However, such a process is obviously not capable since, assuming normality, 50% of the process output would be below the lower specification limit.

A general measure of process capability, that may be used for centered or uncentered processes, is the C_{pk} **index** given by:

$$C_{pk} = \min\left[\frac{USL - \mu}{3\sigma}, \frac{\mu - LSL}{3\sigma}\right]. \qquad (2)$$

This index incorporates both the location of the process and its variability. Desirable values of C_{pk} are $C_{pk} \geq 1.33$, in which case almost all of the process output will be conforming. Note that while C_p is a measure of process *potential*, C_{pk} represents the *actual* capability of the process with the existing parameter values. C_{pk} is thus a measure of process performance.

The *six-sigma* concept first introduced by Motorola has a stringent requirement for process capability. Motorola's six-sigma program requires that the process mean will not be closer than six standard deviations from the nearest specification limit. This requires that the C_p index will be at least 2 with a defect rate of about 2 billion. The program also allows for the process mean to drift no more than 1.5 times the process standard deviation from the target value, such that with a C_p of 2 the defect rate is 3.4 defects per million.

Measures for quality improvement: A control chart can assist in determining when to take action to improve a process, i.e., beyond just controlling it, and the types of action to take. When a process is in control, baseline data from the control chart indicates what is expected from the process. If the output from a process that is in control does not meet customer requirements, it calls for quality improvements in the process, without which the process would not be capable.

Consider the situation where a process is in control and has a mean of 20 mm (say the diameter of a component). Suppose the process standard deviation is estimated as 2 mm. Then, the natural tolerance limits of the process, assuming normality, are 14 mm and 26 mm respectively. Assume that specifications on diameter are 20 $\pm$ 4 mm. In this case, since the process mean is at the target value of 20 mm, no action need be taken as far as process centering is concerned. Management must focus their efforts on reducing the process variability, so that the component diameters will lie within specifications. Possible solutions could be through vendor control of incoming raw material or parts for tighter control of variation, operator training to improve consistency in output or reduction of inherent equipment variability through purchase of better machinery.

Continued use of control charts must be made even after the process is capable of meeting customer specifications for the following two reasons. First, as part of continuous improvement, variability should be further reduced and the control chart serves as a monitoring tool. Second, the process being in control is more of an exception rather than the rule. Left to itself, the process will deteriorate and soon go out of control.

PRELIMINARY CONSIDERATIONS PRIOR TO CONSTRUCTION OF CONTROL CHARTS: Several issues, in general, are to be addressed prior to the construction of control chart. These are briefly discussed.

Selection of Quality Characteristics

In most manufacturing situations, there are numerous quality characteristics that could be of interest. A single component can have several quality characteristics such as diameter, thickness, length, hardness, and surface finish. Considering the fact that a product usually consists of several components and that there are multiple products, it is easy to visualize how the possible number of characteristics can become quite large.

It is not feasible to maintain a large number of control charts. Decision-making becomes too complicated. In selecting quality characteristics for which control charts are to be kept, the objective is to select the "vital few" from the "useful many". The principle of Pareto analysis can be used to determine the most important characteristics; those which impact the total cost the most, or have more nonconforming items , or the number of nonconformities is the greatest.

In Pareto analysis, one approach is to rank in decreasing order the total cost or the percentage of nonconforming items as a function of the quality characteristic or type of defect. The characteristic that has the greatest impact on total cost will be the first on which a control chart will be constructed, and so on. Assume that from historical data it is found that nonconformance of outside diameter of a hub leads to defects 25% of the time, hub length nonconformities causes defects 15% of the time, and the nonconformities due to depth of a keyway results in defects 10% of the time. In this case, our preference would be to control the outside diameter of the hub first, if that were the only quality characteristic to be controlled.

Selection of Samples

A fundamental premise in selecting samples (also called subgroups) from the process is that the variation within a sample be due to only common causes. So, if special causes are present, samples are chosen such that their effect will be demonstrated between samples. Therefore, differences within a sample will be minimized, whereas differences between samples will be maximized.

Samples are selected periodically in fixed or random intervals. That way, if the particular computed quality characteristic from the sample plots outside the control limits, indicating a possible presence of special cause(s), manufacturing personnel can look into process parameter settings and conditions at that reference point in time to determine special causes and their remedial actions. Identification of special causes is convenient when one has an idea of the time of its occurrence since logs of process conditions are normally kept as a function of time.

Another concept of selection of rational samples is to be able to identify the effect of several special causes separately in order to better decide on remedial actions. Consider, for example, a job shop with several machines. Samples should be collected at random points in time from each machine in order to study the performance of each machine separately. We would not mix the output of some machines (unless it has been previously determined that there is no difference in their performance, even when they are similar machines such as lathes or boring machines). Control chart for the average value of the quality characteristic, say average outside diameter of hub, for each machine would be plotted separately. If there are two or more operators producing output, a separate control chart should be kept for each operator, unless it has been previously established that there is no difference in the performance of the operators. Likewise, if it is suspected that there is a difference between the output of two shifts, the two outputs should not be mixed and separate charts should be kept for each shift. Such practices will then enable us to determine differences between machines, operators, or shifts.

Sample Size

The number of observations in a sample is known as the sample size. The degree of shift in the process parameter that is expected to take place may influence the choice of the sample size. Large shifts in a process parameter, such as the process mean, can be detected by samples of smaller size than those needed to detect smaller shifts. Other factors that influence the sample size are the unit inspection cost, cost of shipping a nonconforming item, and intangible costs associated with loss of customer goodwill due to unsatisfactory product performance. Sample sizes are normally between 4 and 10. It is quite common in a manufacturing environment to choose small samples (of size 4 or 5) at frequent time intervals. The small sample size helps to minimize variability within samples, while the choice of frequent samples maximizes variability between samples.

Sampling Frequency

Ideally, the sampling scheme that provides the most information for detecting process changes is to choose large samples at very frequent intervals. This, however, may not be feasible due to factors such as resource constraints, production requirements, whether sampling requires the process to be temporarily stopped, and the type of inspection, i.e., destructive or nondestructive. The current state of the process, that is whether it is in control or not, also has an influence. For a stable process one could sample at less frequent intervals. The availability of personnel for taking observations, and the availability of measuring equipment are other factors that have an impact on sampling frequency.

Selection of Measuring Instrument

The accuracy of the data is a function of the accuracy of the gauge or measuring instrument. Measuring instruments should be calibrated and tested for repeatability. The characteristic being measured and the desired degree of precision influence the selection of the measuring instrument. While a micrometer may be acceptable for the bore diameter of a cast iron pipe, an optical sensory equipment may be necessary for measuring the thickness of silicon wafers. Other quality characteristics such as pressure, temperature, specific gravity, strength, and viscosity require different equipment designed specifically for the measurement of such quality characteristics.

Variation always exists in process outcomes but its detection depends on the precision of the measuring instrument. The rule of thumb is that the precision of the measuring instrument should be at least one order of magnitude higher than what is being measured. For example, if a characteristic is being measured to the tenth of a centimeter then the accuracy of the gauge must be at least to the hundredth of a centimeter.

RULES FOR OUT-OF-CONTROL CONDITIONS:

A major objective of a control chart is to determine when a process is out of control. The position of the control limits affects decision making because of its impact on the risks associated with such decisions. There are two types of errors associated with decision making from control charts.

Type I Error: A Type I error occurs when it is inferred that a process is out of control when it is really in control. Suppose we assume a process to be out of control when an observation plots outside the control limits. For a normally distributed quality characteristic, if the control limits are placed at three standard deviations of the characteristic being plotted, on either side of the center line, the probability of a Type I error is small, roughly 0.26%. This is the area under the normal curve outside three standard deviations from the mean. For a process in control, a Type I error represents a false alarm, where we try to look for special causes in the system when there are none. The Type I error can be reduced by moving the control limits further out from the center line, in which case, it is highly unlikely for an observation to plot outside the control limits when the process is in control. However, this has a negative impact on the second type of error that is discussed next.

Type II Error: A Type II error occurs when it is inferred that a process is in control when it is really out of control. Suppose we assume a process to be in control when no observations fall outside the control limits. Now, suppose the process goes out of control because of a change in the process mean as a result of the operator inadvertently changing the depth of cut in a machining operation or a change in the quality level of the raw material from a new vendor. Under these conditions, it is still possible for the sample statistic to fall within the control limits, depending on the degree of change that has taken place. But, if this happens, we would conclude that the process is in control. This is an example of a Type II error, where we should be looking for special causes in the system but do not do so.

The probability of a Type II error is a function of the position of the control limits, and the degree to which the process has gone out of control. The probability of such errors can be reduced by tightening the control limits, i.e., moving them closer to the center line. Doing so, however, will increase the probability of a Type I error. Thus, in production situations, a balance is maintained between the probability of committing these errors. Usually, the probability of a false alarm (Type I error) is maintained at a low level in setting up control limits. For given values of the process parameters, to reduce both the probability of a Type I and a Type II error, the sample size must be increased.

Following are some of the commonly used rules (AT&T, 1984; Mitra, 1998; Nelson, 1984) in analyzing the pattern of the plot in a control chart to determine if the process is out of control.

Rule 1: A process is assumed to be out of control if a single point plots outside the control limits.

This is the most commonly used rule. For a process in control, if control limits are three standard deviations from the mean of the characteristic being plotted, the chance of this happening is quite small (only 0.0026), assuming normality.

Rule 2: A process is assumed to be out of control if two out of three consecutive points fall outside the two-sigma limits on the same side of the center line. The two-sigma limits are those that are two standard deviations from the mean of the quality characteristic.

The two sigma limits are also known as warning limits. For a process in control, the chance of the above pattern happening is very small. Thus, if such a pattern is observed, it is reasonable to assume that the process is out of control.

Rule 3: A process is assumed to be out of control if four out of five consecutive points fall beyond the one-sigma limit on the same side of the center line.

A one-sigma limit is one standard deviation away from the mean of the quality characteristic. Computationally, if the control limits are placed three standard deviations from the mean, the one-sigma limits are merely one-third of that distance away from the mean.

Rule 4: A process is assumed to be out of control if nine or more consecutive points fall to one side of the center line.

For a process in control, roughly an equal number of points should fall above or below the center line with no systematic patterns visible. If, for example, nine or more consecutive points plot above the center line, we would strongly suspect that an upward shift in the process mean has taken place. Some special causes for such a pattern could be change of operator, change in incoming quality of raw material and parts due to a change of vendor, chipped tool, or error in calibration of the measuring instrument.

Rule 5: A process is assumed to be out of control if there is a run of six or more consecutive points steadily increasing or decreasing.

A run is a sequence of like observations. An increasing run as stated above is a good indication of a gradual drift in the process mean. Examples of such trends could be due to causes such as tool wear, gradual deterioration of machinery, or gradual change in pressure or temperature.

CONTROL CHARTS FOR VARIABLES: In this section we discuss some control charts used for quality characteristics measured on a numerical scale. It is assumed that all of the preliminary decisions prior to the construction of control charts, discussed previously, have been taken. So, the steps for the construction of the control charts are described.

Chart for the Mean ($\overline{X}$) and the Range (R)

It is assumed that samples, of size n, are chosen at random points in time from the process and the selected quality characteristic is measured. Both charts are used simultaneously for decision making. The $\overline{X}$-chart monitors process location whereas the R chart monitors process variability. The charts are interrelated since variability affects both charts. Let g denote the number of such samples chosen, each of size n.

Step 1: The sample mean ($\overline{X}$) and range (R) of each sample are found as follows:

$$\overline{X} = \frac{\sum_{i=1}^{n} X_i}{n} \tag{3}$$

$$R = X_{\max} - X_{\min} \tag{4}$$

where X_i represents the ith observation, and X_{max} and X_{min} represent the maximum and minimum observation, respectively, in the sample.

Step 2: The center line and the trial control limits are found for each chart. For the $\bar{X}$-chart, the center line is given by

$$\bar{\bar{X}} = \frac{\sum_{i=1}^{g} \bar{X}_i}{g}. \tag{5}$$

For the R-chart, the center line is

$$\bar{R} = \frac{\sum_{i=1}^{g} R_i}{g}. \tag{6}$$

Assuming that three-sigma limits are to be used, for the $\bar{X}$-chart, the trial control limits are

$$\bar{\bar{X}} \pm 3\sigma_{\bar{x}} \tag{7}$$

where $\sigma_{\bar{x}}$ represents the standard deviation of the sample means.

A relationship exists between the process standard deviation (σ), also known as the standard deviation of the individual values, and the mean of the ranges ($\bar{R}$). It is known that, when sampling from a population that is normally distributed, the distribution of the statistic $w = R/\sigma$, known as relative range, has a mean given by d_2, that is also dependent on the sample size n (Mitra, 1998). So, the process standard deviation is estimated as

$$\hat{\sigma} = \frac{\bar{R}}{d_2}. \tag{8}$$

Since $\sigma_{\bar{x}}$ is estimated as $\hat{\sigma}/\sqrt{n}$, from the Central Limit Theorem that describes the distribution of sample means, we have an estimate of $\sigma_{\bar{x}}$ as

$$\hat{\sigma}_{\bar{x}} = \frac{\bar{R}}{d_2\sqrt{n}}.$$

Now, the trial control limits on an $\bar{X}$ chart can be re-expressed as

$$\bar{\bar{X}} \pm \frac{3\bar{R}}{d_2\sqrt{n}}$$

$$\text{or,} \quad \bar{\bar{X}} \pm A_2\bar{R}, \tag{9}$$

where $A_2 = \frac{3}{d_2\sqrt{n}}$, and is shown in Table 1 (ASTM, 1976) for different values of n.

Equation (9) is the working version of the formula for finding the trial control limits, since $\bar{\bar{X}}$ and $\bar{R}$ can be found from the data.

For the R-chart, conceptually the trial control limits are given by

$$\bar{R} \pm 3\sigma_R,$$

where σ_R represents the standard deviation of the range values. The working version of the formula (ASQC, 1987; Mitra, 1998), using factors based on the sample size, as tabulated in Table 1, is:

$$UCL_R = D_4\bar{R}$$
$$LCL_R = D_3\bar{R} \tag{10}$$

where UCL_R and LCL_R stand for the upper control limit and lower control limit for the R chart, respectively, and factors D_4 and D_3 are shown in Table 1.

Step 3: The sample values of $\bar{X}$ and R are plotted on the respective charts, with the center line and the control limits drawn. The plots are used to determine whether all of the points are in statistical control. If not, special causes associated with the out of control points are investigated and appropriate remedial actions are taken.

The rules for out of control conditions described previously may be used. In most situations Rule 1 is used. Remedial actions that will eliminate the special causes are arrived through a study of the process. An R-chart reflecting process variability is brought to control first.

Note that the control limits on an $\bar{X}$ chart involve process variability and hence $\bar{R}$. The corrective reaction to out-of-control conditions should be timely.

Table 1. Factors for computing center line and three sigma control limits.

	$\bar{X}$ chart	R chart		
Sample size n	A_2	d_2	D_3	D_4
2	1.880	1.128	0	3.267
3	1.023	1.693	0	2.574
4	0.729	2.059	0	2.282
5	0.577	2.326	0	2.114
6	0.483	2.534	0	2.004
7	0.419	2.704	0.076	1.924
8	0.373	2.847	0.136	1.864
9	0.337	2.970	0.184	1.816
10	0.308	3.078	0.223	1.777
11	0.285	3.173	0.256	1.744
12	0.266	3.258	0.283	1.717
13	0.249	3.336	0.307	1.693
14	0.235	3.407	0.328	1.672
15	0.223	3.472	0.347	1.653
16	0.212	3.532	0.363	1.637
17	0.203	3.588	0.378	1.622
18	0.194	3.640	0.391	1.608
19	0.187	3.689	0.403	1.597
20	0.180	3.735	0.415	1.585
21	0.173	3.778	0.425	1.575
22	0.167	3.819	0.434	1.566
23	0.162	3.858	0.443	1.557
24	0.157	3.895	0.451	1.548
25	0.153	3.931	0.459	1.541

Step 4: The revised center line and control limits for the $\overline{X}$ and R charts are determined.

Once special causes associated with the out of control points have been identified and remedial actions taken for their elimination, those out of control points are deleted. The center lines and control limits are calculated using the same formulas as before, except that some of the samples may be omitted from the computations. The cycle of obtaining information, determining trial limits, finding out-of-control points, identifying special causes, taking remedial actions, and constructing revised control limits is continual. The revised control limits will serve as trial control limits for the immediate future until they are revised again. This ongoing process is fundamental to continuous improvement.

Step 5: The control charts are implemented for future use using the revised limits. Production personnel should closely monitor the charts and take prompt action to special causes of variation.

If the process is improved, i.e., fundamental changes as in equipment, operations, and vendors are made, the center line and control limits may have to be recalculated.

Example: Table 2 shows measurements on the outside diameter of a hub (in mm) used in an assembly that were obtained for 25 samples, each of size 4. The sample means ($\overline{X}$) and ranges (R) are also shown in the table. The specifications on the outside diameter are 50 ± 5 mm.

The center line on the $\overline{X}$-chart is

$$\overline{\overline{X}} = \frac{1222.25}{25} = 48.89.$$

The center line on the R-chart is

$$\overline{R} = \frac{97}{25} = 3.88.$$

The trial control limits on the $\overline{X}$ chart are:

$$\overline{\overline{X}} \pm A_2\overline{R} = 48.89 \pm 0.729(3.88) = 48.89 \pm 2.828 = (51.72, 46.06).$$

The trial control limits on the R chart are:

$$UCL_R = D_4\overline{R} = 22.282(3.88) = 8.85$$

$$LCL_R = D_3\overline{R} = 0(3.88) = 0.$$

Figure 2 shows the trial control charts for $\overline{X}$ and R where the *Minitab* (Minitab, 1996) software package is used to construct these charts. The three-sigma limits on each chart are shown. Suppose we use Rule 1 to determine out of control condition. Note that all points on the R-chart are within the control limits. On the $\overline{X}$-chart, the mean for sample number 7 plots above the upper control limit while that for sample number 17 plots below the lower control limit. Special causes should be investigated for the occurrence of these conditions. For sample number 7, it was found that an inexperienced operator was on the job. This can be rectified by assigning only trained operators. For sample number 17, investigation revealed that an improper depth of cut had been used, which was subsequently corrected.

Deleting sample numbers 7 and 17, the revised center line and control limits are calculated for the $\overline{X}$ and R charts. Figure 3 shows the revised control charts using *Minitab* software. All points on both control charts are within the revised control limits. For this stable process, let us estimate its capability.

The process standard derivation is estimated as

$$\hat{\sigma} = \frac{\overline{R}}{d_2} = \frac{3.783}{2.059} = 1.837.$$

The process mean is estimated as $\hat{\mu}$, the revised center line on the $\overline{X}$-chart, which is 48.89. Observe that the process is not centered between the specification limits, since the target value is 50 mm.

Table 2. Measurements on outside diameter of hub.

Sample number	Observations				Sample mean $\overline{X}$	R
1	48	49	50	47	48.50	3
2	50	46	49	48	48.25	4
3	46	48	47	49	47.50	3
4	48	51	47	46	48.00	5
5	51	49	50	50	50.00	2
6	48	51	49	48	49.00	3
7	51	55	50	53	52.25	5
8	47	50	51	51	49.75	4
9	49	49	52	51	50.25	3
10	50	48	50	49	49.25	2
11	46	45	50	46	46.75	5
12	48	52	51	51	50.50	4
13	50	47	50	46	48.25	4
14	49	48	50	51	49.50	3
15	52	50	48	49	49.75	4
16	47	52	46	45	47.50	7
17	43	48	46	45	45.50	5
18	51	50	51	48	50.00	3
19	47	51	48	46	48.00	5
20	47	50	47	50	48.50	3
21	48	52	49	51	50.00	4
22	52	48	47	49	49.00	5
23	48	49	50	47	48.50	3
24	50	46	46	48	47.50	4
25	48	51	50	52	50.25	4

$$\overline{\overline{X}} = \frac{1222.25}{25} = 48.89 \qquad \overline{R} = \frac{97}{25} = 3.88$$

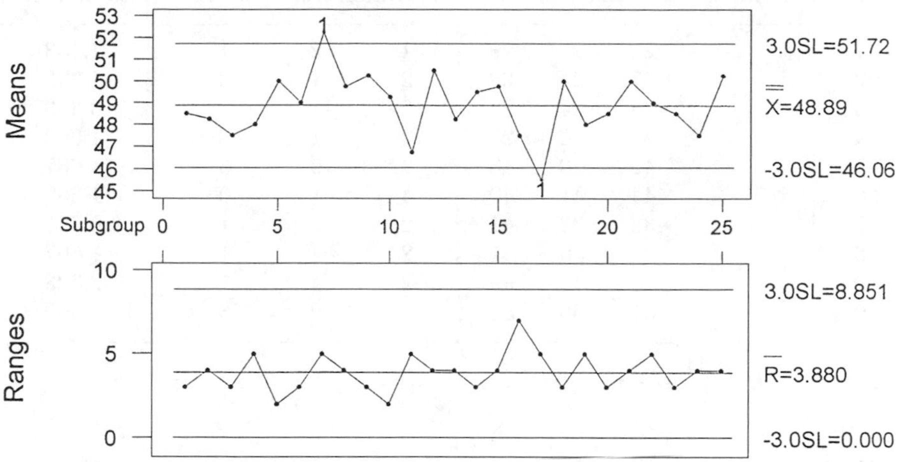

Fig. 2. $\overline{X}$ and R trial control charts for outside diameter of hub.

A measure of the process capability is given by the C_{pk} index which is:

$$C_{pk} = \min\left[\frac{55-48.89}{3(1.837)}, \frac{48.89-45}{3(1.837)}\right]$$
$$= 0.706.$$

Since $C_{pk} < 1$, we have an undesirable situation. The process is not capable; not all of the process output will meet customer specifications. The standardized normal value at the upper specification limit is $Z_1 = (USL - \overline{\overline{X}})/\hat{\sigma} = 3.33$, while that at the lower specification limit is $Z_2 = (LSL - \overline{\overline{X}})/\hat{\sigma} = -2.12$. Using the standard normal distribution tables (Mitra, 1998), the proportion of the output that is above the upper specification limit is 0.0004 or 0.04%, while that below the lower specification limit is 0.0170 or 1.70%. So, the total nonconforming output is 1.74%.

Process Control for Short Production Runs: Companies are faced with short production runs for several reasons, product specialization being one. Organizations have to be responsive to customer needs in order to maintain and improve market share. To produce custom-built items, manufacturing has to change and adapt. This may necessitate changes in equipment, tooling, operations, parts, and components in short time frames. Thus, a long stream of data in a static environment may not be available for statistical process control.

Just-in-time (JIT) manufacturing has also caused a need for short-run production (Alsup and

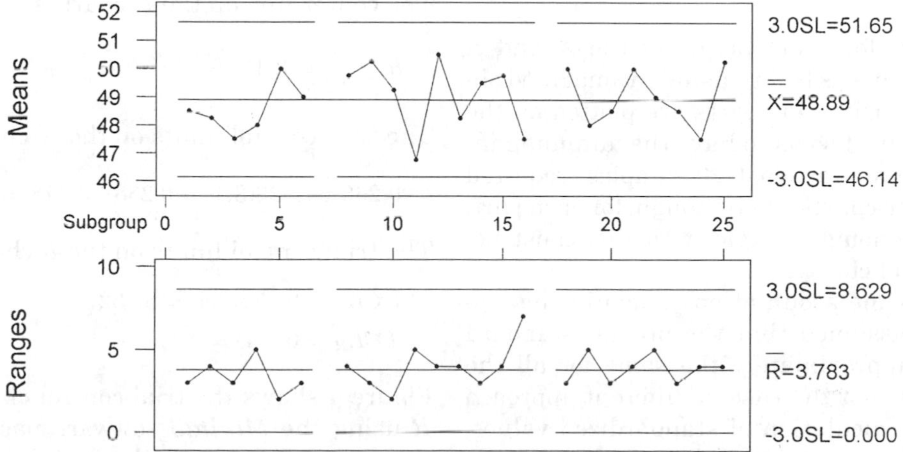

Fig. 3. $\overline{X}$ and R revised control charts for outside diameter of hub.

Table 3. Observations for short run $\overline{X}$ and R charts.

Part number	Sample number	Observations			Deviation from nominal			Sample mean $\overline{X}$	Range R
A	1	26	27	26	1	2	1	1.333	1
A	2	27	24	28	2	−1	3	1.333	4
A	3	25	24	27	0	−1	2	0.333	3
A	4	23	26	23	−2	1	−2	−1.000	3
B	5	42	40	43	2	0	3	1.667	3
B	6	44	41	40	4	1	0	1.667	4
B	7	38	39	41	−2	−1	1	−0.667	3
B	8	38	36	41	−2	−4	1	−1.667	5
C	9	48	46	49	−2	−4	−1	−2.333	3
C	10	51	50	55	1	0	5	2.000	5

$\overline{\overline{X}} = \frac{2.666}{10} = 0.266 \qquad \overline{R} = \frac{34}{10} = 3.4$

Watson, 1993). JIT is a pull-through system which means that the amount of production is driven by the immediate needs of the final assembly. Only as much product as is needed is produced.

So, if a variety of parts is produced for brief time intervals, there would never be enough samples from a particular part to set up control charts for that part alone. What is done is to construct a control chart where different parts are put on the same chart. The justification behind this is that control charts are used for *process* control; so the process that produced these different parts is the one that is being controlled.

$\overline{X}$ *and R Chart for Short Production Runs*

One approach for constructing $\overline{X}$ and R charts in short production runs where different parts may be produced, is to use the deviation from the nominal value (N) as the modified observations. The nominal value may vary from part to part. So, the deviation of the observed value O_i is given by X_i, where

$$X_i = O_i - N. \tag{11}$$

The procedure for construction of the $\overline{X}$ and R charts is the same as before using the modified observations, X_i. Different parts are plotted on the same control chart so as to have the minimum information (usually at least 20 samples) required to construct the charts, even though, for each part, there are not enough samples to justify construction of a control chart.

There are some assumptions made in this approach. It is assumed that the process standard deviation is approximately the same for all the parts. If this is not the case, a different approach that involves calculation of standardized values, is used. That approach is not discussed in this paper and may be found in other references (Mitra, 1998; Montgomery, 1996). Also, the sample size is assumed to be constant for all part numbers.

Example: Table 3 shows measurements (in mm) from three parts A, B, and C. The nominal dimension for part A is 25 mm, while that for parts B and C are 40 mm and 50 mm respectively. The sample size is 3 for each. For part A and part B, 4 samples are chosen while for part C, 2 samples are selected. Even though we do not have at least 20 samples, we demonstrate the procedure as it will be the same had we accumulated more samples.

The nominal values for each part are subtracted from the observed values to obtain the deviation from nominal values. These are shown in Table 3. Sample mean and range values for each sample are computed using these deviation from nominal values.

The center line on the $\overline{X}$-chart is

$$\overline{\overline{X}} = \frac{2.666}{10} = 0.266.$$

The center line on the R-chart is

$$\overline{R} = \frac{34}{10} = 3.4.$$

The trial control limits on the $\overline{X}$-chart are:

$$0.266 \pm 1.023(3.4) = 0.266 \pm 3.478 = (3.744, -3.212).$$

The trial control limits on the R chart are:

$$UCL_R = 2.574(3.4) = 8.752$$
$$LCL_R = 0(3.4) = 0.$$

Figure 4 shows the trial control charts for $\overline{X}$ and R using the *Minitab* software package. None of the points plot outside the control limits on either

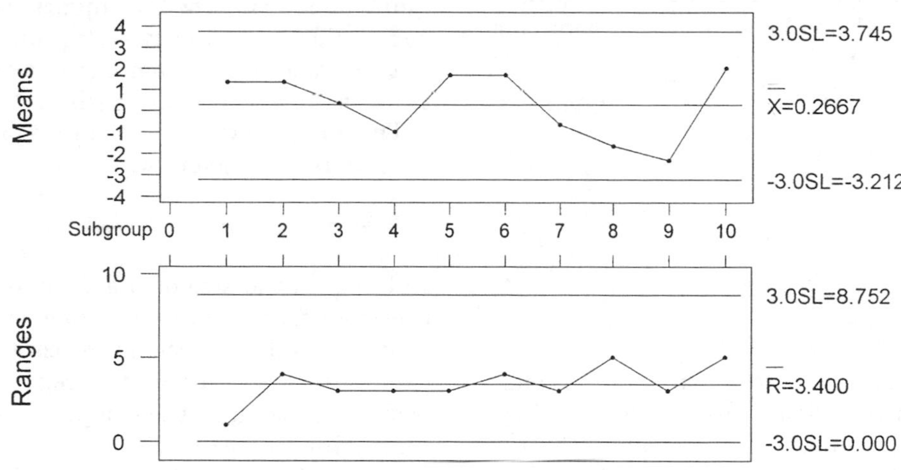

Fig. 4. $\overline{X}$ and R chart for short production run.

chart and no identifiable patterns are observed. The process is in control.

Control Charts for Attributes: In this section, some control charts for attribute data are presented. A quality characteristic that does not meet certain specifications is defined to be a *nonconformity* (or defect), while a product with one or more nonconformities that make it dysfunctional is said to be *nonconforming* (or defective). Three charts for attributes are described. The *p*-chart is used for the proportion of nonconforming items, the *c*-chart for the number of nonconformities, and the *u*-chart for the number of nonconformities per unit.

Control charts for variables usually provide more information than that for attributes. However, resource constraints and feasibility of inspection may dictate the use of attribute charts. For situations requiring summary measures of product and process, attribute charts are preferred. When information is needed for top management, these charts are quite effective. For example, output at the plant level could be best described by a proportion nonconforming chart. On the contrary, for technical problems at the operator or supervisor level, variables charts are more meaningful because they provide specific clues for remedial actions when the process is out of control.

Chart for Proportion Nonconforming (p-Chart)

A *p*-chart may be used to obtain a good overall measure of quality. It may be used to control the acceptability of a single quality characteristic (say, the length of a part), a group of quality characteristics of the same type or on the same part (for example, the length, outside diameter, and thickness of a sleeve), or an entire product. A *p*-chart can also be used to measure the quality of an operator or a machine, a work center, a department, or the entire plant. Another use of the *p*-chart may be to identify where to use X and R charts, since variable charts are more sensitive to variations and are useful in diagnosing special causes.

The computations for a *p*-chart are demonstrated for the case when the sample size (n) is constant. For a given sample number i, the proportion nonconforming (p_i) is estimated as

$$\hat{p}_i = \frac{x_i}{n}, \tag{12}$$

where x_i represents the number of nonconforming items in sample number i. From a total of g samples, expressions for the average proportion nonconforming ($\bar{p}$), representing the center line, and the three sigma control limits are shown in Table 4. If the lower control limit is calculated to be negative, it is set to zero.

Chart for Number of Nonconformities (c-chart)

A c-chart is used to monitor the total number of nonconformities in samples of constant size. The size of the sample is also known as the area of opportunity, which may be single or multiple units of a product. For items produced on a continuous basis, the area of opportunity could be 100 m^2 of paper in a paper mill or 200 m^2 of fabric in a textile plant.

Table 4. Formulas for center line and control limits for attribute charts.

Chart	Center line	Control limits
p	$\bar{p} = \frac{\sum_{i=1}^{g} x_i}{ng}$	$\bar{p} \pm 3\sqrt{\frac{\bar{p}(1-\bar{p})}{n}}$
c	$\bar{c} = \frac{\sum_{i=1}^{g} c_i}{g}$	$\bar{c} \pm \sqrt{\bar{c}}$
u	$\bar{u} = \frac{\sum_{i=1}^{g} c_i}{\sum_{i=1}^{g} n_i}$	$\bar{u} \pm 3\sqrt{\frac{\bar{u}}{n_i}}$

The occurrence of nonconformities is assumed to be modeled by a Poisson distribution. Such a distribution is well-suited to modeling the number of events that happen over a specified amount of time, space, or volume. Certain assumptions must hold to justify the usage of this distribution. First, the opportunity for the occurrence of nonconformities may be large, but the average number of nonconformities observed per sample is small. Consider an example of fabric production. The number of flaws in 100 m^2 of fabric could conceivably be quite large, but the average number of flaws in 100 m^2 is not necessarily a large value. Second, the occurrences of nonconformities must be independent of each other, implying that a nonconformity in a certain segment of the fabric must not influence the occurrence of other nonconformities. Third, each sample should have an equal likelihood of the occurrence of nonconformities. The implication of this is that the prevailing process conditions should be quite similar from sample to sample. A property of the Poisson distribution is that the mean and the variance are equal.

Denoting the observed number of nonconformities in sample i by c_i, and g to be the number of samples, expressions for the center line ($\bar{c}$) and control limits are found in Table 4. If the calculated lower control limit is negative, it is converted to zero.

Chart for Number of Nonconformities per Unit (u-Chart)

When the area of opportunity changes from one sample to another, the center line and the control limits of a *c*-chart will change. For situations in which the sample size varies, a chart for the number of nonconformities per unit (*u*-chart) is used. For example, companies that inspect all items produced, the output may vary because of fluctuations in the availability of raw material, labor, and machinery, affecting the number of items inspected. Even though the control limits on a *u*-chart will change as the sample size varies, the center line remains constant, thus permitting meaningful comparisons to be made between samples of different sizes.

The number of nonconformities per unit for sample number i is given by

$$u_i = \frac{c_i}{n_i}, \tag{13}$$

where n_i is the size of the sample number i. Expressions for the center line and control limits are given in Table 4. If the lower control limit is calculated to be negative, it is converted to zero. It is observed that, as the sample size increases, the control limits are drawn closer to the center line, a behavior similar to that of the *p*-chart. Also, since the limits depend on n_i, they are not necessarily constant from sample to sample.

CONTEMPORARY VIEWS ON PROCESS CONTROL: A control chart, even if used in real time as an on-line tool, is reactive in nature. As Shingo (Shingo, 1986) observes, a control chart does too little too late and does not really prevent defects from occurring. He proposes that defects should be prevented by mistake-proofing the process through a *poka-yoke* system. A poka-yoka is a device that prevents permanently the recurrence of the defect it is designed to eliminate. Such a system is simple, inexpensive, and lends itself to total worker involvement. So, control charts may play a supplementary role in a quality assurance program.

Another tool in a zero-defect environment is successive checking (Shingo, 1986; Robinson and Schroeder, 1990). In this situation, subsequent processes are designed so that they physically cannot operate on defective parts produced by a preceding process. Successive checking shortens the time needed for feedback of problems. The idea of self-checking is similar to successive checking, where the operator checks his/her own output. Feedback time is therefore tightened even further. Poka-yoke devices are critical to the self-checking process.

Source inspection is another approach in a zero-defect program. This is based on the concept that since errors cause defects, elimination of errors will eliminate the occurrence of defects. Actions taken prior to the occurrence of errors could include (Shingo, 1986) a full maintenance program, where scheduled maintenance on machines takes place rather than replacement or repair on failure.

Another alternative deals with the concept of autonomation, where if something goes wrong with a machine it will automatically shut down rather than produce defective output. While it is an attractive alternative, the difficulty is in designing a system whereby the machine shuts down on problem occurence.

While the above tools are viable alternatives, statistical process control (SPC) has played and will continue to play an important in manufacturing and production. In 1984, Bill Conway, then President of Nashua Corporation recalls how he directed the company's hard memory disk division to use SPC and the Deming philosophy. Production volume increased 300 percent without adding employees or floor space. Using about 125 control charts, the division gained control over their business processes and then narrowed their operating ranges. The enormous potential of SPC is illustrated through this example (Amsden, 1998).

For SPC to work, management must develop a culture where everyone understands that they are there to help and not control. They must actively help their associates identify, solve, and implement solutions to problems. Once this social organization is in place, the utility of SPC in improving quality and productivity can be realized. At the heart of statistical process control is the idea of getting most from processes and continually improving these processes and their outcomes.

With the advent of the availability of technological tools for charting, creating control charts is becoming more automated while software becomes more user-friendly. Now, charts can be constructed in real time once parameters such as sample size, sampling frequency and type of chart are specified as input. The development of expert systems with diagnostic capabilities in conjunction with SPC software is a reality. Such systems will identify out of control situations based on prespecified rules. They may also suggest likely special causes and stop the process until remedial actions are taken.

CONCLUSIONS: In any manufacturing environment, stability of a process is vital to producing a product that meets customer specifications. Process centering, and controlling process variability through identification of special causes and remedial actions are fundamental in maintaining a stable process, whose output is predictable. Control charts are such tools that enable us to determine when a process is possibly out of control and remedial actions are necessary to bring it back to control.

Organizations Mentioned: Ford; Motorola; Nashua Corporation.

See **ISO 9000/QS 9000 quality standards; C-chart; Integrated quality control; ISO 9000/QS 9000 quality standards; ISO 9000 and QS 9000; Just-in-time manufacturing; Just-in-time manufacturing implications; Kanban-based manufacturing systems; Linked-cell manufacturing systems; Malcolm baldrige award criteria; Manufacturing cell design; Manufacturing implications; *P*-chart; Performance excellence; Poka-Yoke; Quality: Implications of Deming's approach; Quality management systems; *R*-chart; Sampling frequency; Shewhart control charts; Total quality management; Type I error; Type II error; Variability due to common causes; Variability due to special causes; $\overline{X}$-chart.**

References

[1] American Society for Quality Control (1987). ANSI/ASQC A1-1987. *Definition, Symbols, Formulas, and Tables for Control charts*, ASQC Press, Milwaukee, Wisconsin.

[2] American Society for Quality Control (1971). *Quality Costs – What and How* (2nd. Ed.), Milwaukee, Wisconsin.

[3] American Society for Testing and Materials (1976). *ASTM-STP 150*, Philadelphia, Pennsylvania.

[4] Alsup, F., and R.M. Watson (1993). *Practical Statistical Process Control*, Van Nostrand Reinhold, New York.

[5] Amsden, R.T. (1998). "What's in store for SPC?" *Quality Digest*, February, 37–40.

[6] AT&T (1984). *Statistical Quality Control Handbook*, 10th printing.

[7] Deming, W.E. (1982). *Quality, Productivity, and Competitive Position*, MIT Center for Advanced Engineering Study, Cambridge, Massachusetts.

[8] Deming, W.E. (1986). *Out of the Crisis*, MIT Center for Advanced Engineering Study, Cambridge, Massachusetts.

[9] Grant, E.L. and R.S. Leavenworth (1996). *Statistical Quality Control* (7th. Ed.), McGraw Hill, New York.

[10] Minitab, Inc. (1996). *Minitab Reference Manual*, Release 11, State College, Pennsylvania.

[11] Mitra, A. (1998). *Fundamentals of Quality Control and Improvement* (2nd. Ed.), Prentice Hall, Upper Saddle River, New Jersey.

[12] Montgomery, D.C. (1996). *Introduction to Statistical Quality Control* (3rd. Ed.), John Wiley & Sons, Inc., New York.

[13] Nelson, L.S. (1984). "The Shewhart control chart – tests for special causes." *Journal of Quality Technology*, 16, 237–239.

[14] Robinson, A.G. and D.M. Schroeder (1990). "The limited role of statistical quality control in a zero-defect environment." *Production and Inventory Management Journal* 3rd. Quarter, 60–65.

[15] Shewhart, W.A. (1931). *Economic Control of Quality of Manufactured Product*, D.Van Nostrand Company, Inc., Princeton, New Jersey.

[16] Shingo, S. (1986). *Zero Quality Control: Source Inspection and the Poka-Yoke System*, Productivity Press, Stamford, Connecticut.

STEADY-STATE BEHAVIOR

Artificial simulation of systems is sometimes conducted with the intent steady-state behavior. The simulation can reveal average performance of the system. In this type of simulation, the state of the system (number of jobs in the system, or number of items of inventory on hand) is collected over time. If the initial state of the simulation was not the equilibrium condition, the system state will converge over time to the equilibrium condition. The period required to safely obtain steady-state conditions is referred to as the transient state.

See **Simulation analysis of manufacturing and logistics systems; Simulation of production problems using spreadsheet programs.**

STOCK BUFFER

See **Safety stocks: Luxury or necessity; Synchronous manufacturing using buffers.**

STOCK RATIONING

The process of allocating limited available inventory to the right customer so as to optimize a stated objective.

See **Capacity management in make-to-order production systems.**

STOCKLESS PRODUCTION

Stockless production, a synonym for Just-in-Time (JIT), is rarely used today. It came about because individuals from outside Japan were initially struck by the extremely small amount of inventory they saw in factories using JIT. Another early synonyms was "zero inventories". Today we have come to realize that JIT includes much more than just inventory reduction. In fact, the inventory reduction associated with JIT is not usually viewed as one of its major components. Inventory reduction is a by-product of the reduced lot sizes that are necessary to achieve flow manufacturing. Because unnecessary inventory is viewed as a form of waste in JIT, the reduction that occurs is a natural benefit. In fact, many individuals would argue that the major benefits resulting from JIT are improved customer satisfaction and better relationships with employees and suppliers.

See **History of operations management; Just-in-time manufacturing implementation; Just-in-time manufacturing; Manufacturing cells, linked; Manufacturing systems.**

STOCKOUT PROBABILITY

Stockout occurs when it is impossible to fulfil an order from inventory. The cost of a stockout can be excessive, but it is difficult to assess. Depending on the cost of losing demand due to stockouts, businesses use safety stocks to protect against stockouts. Safety Stocks are computed using an acceptable probability of stockout. Stockout probability may be set by management under ten percent for most items.

See **Safety stocks: Luxury or necessity; Stockouts; Synchronous manufacturing using buffers.**

STOCKOUTS

A stockout is a lack of materials, components, or finished goods that are needed to support production processes or satisfy customer demand. The detrimental effects of stockouts may include lost sales, backorder costs, poor customer service, increased manufacturing lead times, expediting, and additional manufacturing and purchasing costs.

In synchronous manufacturing systems, the likelihood of stockouts are minimized through the combined use of stock and time buffers. Stock buffers are inventories of specific products that are held in finished, partially finished, or raw material form, in order to fill customer orders in less than the normal lead time. Stock buffers are designed to improve the responsiveness of the system to specific market conditions. Time buffers represent additional planned lead time allowed for materials to reach a specified point in the production process.

Strategically placed, time buffers protect the system throughput from the internal disruptions that are inherent in any process.

See **Safety stocks: Luxury or necessity; Synchronous manufacturing using buffers.**

STRATEGIC BUSINESS UNITS

For strategic decision making and planning a company's products may be grouped into strategic business units (SBUs). Typically, an SBU represents a product line. It is common to run each SBU as a profit center, which is responsible for everything including profitability and long-term sustainability of the SBU.

See **Balanced scorecard; Core manufacturing competencies; Manufacturing strategy; Performance measurement in manufacturing.**

STRATEGIC MANAGEMENT OF SUPPLIERS

For most of this century, firms looked at purchasing as a necessary but non-glamorous activity. For a firm to produce a product it had to have materials to use in the transformation process, but materials were widely available and labor represented the most important and most visible cost of doing business. As a result, purchasing was viewed as a clerical activity for order processing to trigger the inflow of materials. Likewise, suppliers were managed reactively and adversarial relationships often developed. Buyers pitted one supplier against another to get a better price. Over time (and especially in the past decade), the purchasing function has evolved into an important and potentially strategic function. Further, with the emergence of just-in-time manufacturers, the role of suppliers has changed. The trend toward developing core competencies has transformed suppliers into primary cost driver as well as a primary source of quality and technology. As a result, many firms have taken a more proactive and strategic approach to managing their supply base. Systematic supplier selection approaches are now used; better performance monitoring is employed; supplier certification is the norm rather than the exception; closer working relationships are developed; and better information is exchanged more frequently as part of strategic supplier management initiatives. The goal is for the firm and its suppliers to work synergistically to produce higher quality, lower cost products and develop new products more frequently.

See **Just-in-time manufacturing; Supplier partnerships as strategy; Supplier relationships; Supply chain management: competing through integration.**

STRATEGIC MANUFACTURING FLEXIBILITY

See **Manufacturing flexibility.**

STRATEGIC QUALITY PLANNING

Strategic quality planning is a systematic approach to quality assurance and improvement plans at the top levels of an organization and linking them with business strategy. In the past, strategic quality planning was viewed as a separate and distinct activity from strategic business planning, as quality planning traditionally took place at low levels of the organization. As customer-driven quality and operational performance excellence have become key business drivers, quality improvement objectives such as increasing customer satisfaction, reducing defects, and reducing cycle times began to receive as much attention as financial and market objectives. Today, strategic quality planning is not thought of as a separate process, but as an integral component of strategic business planning.

See **Core manufacturing competencies; Performance excellence: The Malcolm Baldrige national quality award criteria; Total quality management.**

Reference

[1] James R. Evans and William M. Lindsay, *The Management and Control of Quality, Fourth Edition*, Cincinnati, OH: South-Western Publishing Co., 1999.

STRATEGIC SUPPLIER PARTNERSHIPS

See **Supplier partnerships as strategy.**

STRUCTURE OF THE SUPPLY CHAIN

See **Supplier relationships; Supply chain management: Competing through integration.**

SUPER BILL OF MATERIALS

The super bill of materials is a planning bill of materials which is not used in actually building products. It shows average quantities of component parts and options that make up the finished product. It is prepared by combining the bills of material for more than one but similar products, which share many parts and components. It is a useful input for long-term planning.

See **Bill of materials; Bill of materials structuring; Indented bill of materials; Master production schedule; MRP.**

SUPPLIER ASSOCIATION

Hines (1994, p. 143) describes a *kyoryoku kai*, the Japanese name for a supplier association, as "a mutually benefiting group of a company's most important subcontractors brought together on a regular basis for the purpose of coordination and cooperation." Association members benefit from the knowledge of the sponsor of the association such as an automobile assembler.

Hines (1994) describes a ten stage process that can be used to establish a supplier association. A small group of suppliers, about a dozen, can be selected as the initial members of the association. The suppliers are then benchmarked and opportunities for improvement are jointly identified so that development efforts can be focused. Group activities are then undertaken and performance monitored to see that improvements are delivered. At this point the composition of the association and its goals can be reviewed and a new cycle of activities can be started. The activities include: yearly conferences, monthly seminars, weekly meetings, company visits and social events.

See **Supplier partnership as strategy; Supplier relationships; Supply chain management.**

Reference

[1] Hines, P. (1994). *Creating World Class Suppliers*. Pitman Publishing, London.

SUPPLIER CERTIFICATION PROGRAMS

See **ISO 9000/QS 9000 quality standards; Quality management systems: Baldridge, ISO 9000 and QS 9000.**

SUPPLIER DEVELOPMENT

The single most important task of the purchasing function is to find and/or develop the best possible supplier for any given purchase requirement. In recent years, more and more emphasis has been placed on supplier development. One reason for this added emphasis on developing better suppliers is the fact that companies are required to meet higher levels of performance in the areas of cost, quality, cycle time, responsiveness, and innovation. Given that 50 to 80 percent of the cost-of-goods sold for the typical manufacturer (the number is about 30 percent for service companies) is due to purchased components, the role of suppliers in meeting these higher performance standards is critical. Thus, having good suppliers is not enough; rather, a company needs to have not only the best suppliers possible but also suppliers, who are continually getting better.

To achieve this continual improvement in supplier performance, firms must generally work more closely with their suppliers. Dedicating the time and financial resources needed to work closely with suppliers requires that the supply base be reduced – it is much easier to work with 400 suppliers than with 4,000 suppliers. An initial step in supplier development is therefore an initial screening process that selects the very best suppliers, and consolidates orders with these "preferred" suppliers.

Even after reducing the overall supply base, most firms must categorize their suppliers using an "ABC' classification system. The "A" suppliers are the most important either because of the amount of money spent with these suppliers, or because of the importance of the specific items being purchased from them. The "B" suppliers are also important and typically receive considerable attention. The "C' suppliers represent neither large volumes nor critical components and receive the least amount of attention. After identifying the most important suppliers, the firm invests considerable time working with these suppliers to better understand and improve their key value-added processes.

It is not uncommon for a buyer to send quality experts into a supplier to help the supplier implement statistical process control or some other quality improvement technique. Likewise, buyers often provide training in the areas of cycle time improvement, just-in-time techniques, product development etc. In addition to training, the buyer might help finance the acquisition of new equipment or the development of new processes. The key point is that the buyer invests both time and money to help its most important suppliers improve their performance. By working together, the joint competitiveness of the buyer/supplier team can be enhanced.

See **ABC analysis; Purchasing: Acquiring the best inputs; Supplier partnership as strategy; Supplier relationships; Supply chain management: Competing through integration.**

SUPPLIER PARTNERSHIPS AS STRATEGY

Brian Leavy

Dublin City University, Dublin, Ireland

INTRODUCTION: The partnership approach to supplier relations is now a very common feature in many industries. This is due in large part to the phenomenal success of the 'world class manufacturing' (Schonberger 1985), and 'lean production' crusades (Womack *et al.*, 1990, Womack and Jones, 1994) over the last two decades, which have drawn their inspiration from the proven competitive practices of leading Japanese companies (Dyer and Ouchi, 1993). One of the by-products of this success has been the ever-widening faith in the virtues of supplier partnering.

The purpose of this article is to present a strategic perspective on supplier partnering that will help companies to more critically assess the opportunities and risks associated with the adoption of the partnering approach. The article begins by contrasting the underlying rationale for supplier partnering with the more traditional arm's length approach to buyer-supplier relations. Most companies with aspirations to world class manufacturing standards are now adopting the partnering approach.

A STRATEGIC PERSPECTIVE ON SUPPLIER-PARTNERING: To understand better the nature of the supplier partnering from a strategic perspective, it will be useful in the first instance to examine how it contrasts with the more traditional arm's length approach to buyer-supplier relations.

The Traditional Approach

The traditional approach to buyer-supplier relations is based on two main premises:

(i) *The relationship is best managed through the market mechanism.*
- Suppliers compete with one another for the customers;
- Buyers shop around for the best deal;
- Repeat business is not guaranteed;
- Buyers and suppliers communicate and interact with each other at arm's length;
- The 'discipline of the marketplace' keeps them independently on their toes.

and

(ii) *The buyer and supplier see themselves as essentially competing with each other for margin*
- Both parties see it as a win-lose game;
- The primary focus is on the division of profit margin;
- The lion's share goes to the party with the most economic power.

Under this traditional arm's length perspective any company, at any stage in the industry chain, is seen to be in competition with its upstream suppliers and downstream buyers for profit margin (Porter 1980). For example, the producer of soft-drink cans must compete for margin with the aluminum companies at one end of the supply chain and the soft drink companies at the other. All are ultimately dependent on the dollars of the soft drink consumers for their revenues, and the margin enjoyed by each of these players in the industry is seen to depend on its economic power in the overall market chain.

Within this perspective, company strategists are advised to try to reduce the power of their suppliers by maintaining multiple sources of supply, avoiding any uniqueness in the relationship that might make the cost of switching suppliers high, and searching for readily available substitute materials that would help to keep supplier prices in check. The ability to switch business easily among suppliers and pose a credible threat of bringing component production in-house can be used to keep the most aggressive and ambitious suppliers in line. Moreover, by avoiding long-term commitments and keeping its options continuously open, the buyer will always be able to move his business quickly and easily to those suppliers which, at any time, are the most efficient and technically advanced in their own sectors.

The strategic advice to suppliers in this perspective is the mirror equivalent. Suppliers are advised to keep their options open by not allowing themselves to become over-dependent on any one buyer. At the same time they are also advised to try to increase the switching costs for their customers by building elements of uniqueness into the relationship (in the form of product, service or quality distinctiveness, for example). Strong suppliers can keep their buyers in line, and their demands in check, by posing a credible threat of forward integration. They can also try to increase their influence in the overall market chain by directly developing a brand image for their components in the wider marketplace, as Intel has been able to do with its 'Intel Inside' campaign.

The Partnership Approach

The partnership approach to supplier relations, on the other hand, presents a very different picture, with its own set of strategic implications. Its main premises are:

(i) *The relationship is better managed through direct cooperative agreement rather than through the market mechanism.*
  - Buyers and suppliers seek long-term commitments from each other;
  - Repeat business is guaranteed;
  - Buyers and suppliers communicate with each other directly and share information, and learning/know-how;
  - The supplier is co-opted into the buyer's competitive strategy, and the discipline of the buyer's marketplace keeps the entire partnership on its toes;

  and

(ii) *The buyer and supplier are essentially partners-in-profit*
  - Both parties see it as a win-win (or lose-lose) game;
  - The primary focus is on the creation of profit margin;
  - The partners work together to ensure an equitable share of the spoils;
  - Trust replaces opportunism.

Under this approach, the buyer seeks to improve his competitiveness through the development of close cooperative relationship with a relatively small number of carefully selected suppliers. The emphasis is on the development of long-term partnerships. This strategy encourages a high degree of interdependence between the buyer and the supplier and it is pursued in the belief that it can offer very significant economies of cooperation, which can help to improve the profitability of both parties. The commercial benefits of such an approach are expected to come from the closer coordination of schedules, cooperation on product development and process improvement, and joint-action on cost reduction. Furthermore, the cooperative approach provides some insulation for the supplier from the full winds of competition in its own segment of the marketplace. It also offers the buyer many of the benefits of vertical integration (greater security of supply, and more direct control over cost and quality) without the associated investment or the risks that would normally be involved. This is the approach that has helped Toyota to become the world's most efficient manufacturer of cars, much to the benefit of the company and its suppliers (Womack *et al.* 1990). It is also the approach that has underpinned the successes of Marks and Spencer, sometimes called 'the manufacturer with no factories', and Benetton, 'the retailer with no shops'.

In short, under the partnership approach, the supplier's growth and profitability become less dependent on the competitive forces operating in his own segment of the market chain. Instead, the overall effect is to co-opt the supplier into the competitive strategy of the buyer, and the future growth and profitability of both partners become more closely tied to the evolution of the buyer's market and to his competitive position within it. The fortunes of both parties depend on the ability of the partnership to develop and fully exploit the potential economies of cooperation in ways that will strengthen and secure the growth and competitiveness of the buyer.

SOME RISKS AND IMPLICATIONS: While the partnership approach to buyer-supplier relations can offer many potential advantages to both buyer and supplier, there are a number of implications and limitations that company strategists in both buyer and supplier firms should keep in mind.

1. ***The rationale for partnership may change with industry evolution:*** To begin with, it is important to recognise that the underlying economic rationale can change with industry evolution, as it did in the case of the information technology industry over the last three decades.

   In the 1970's the industry was dominated by a small number of highly integrated manufacturers, like IBM and Digital Equipment Corporation, which controlled the key technologies in both hardware and software. These companies sold very high margin products directly to a relatively small base of industrial and commercial customers. Manufacturing was very labour intensive, with little automation, few economies of scale, and little incentive to buy anything other than very basic components and raw materials. The whole independent sub-supply segment remained very underdeveloped in terms of range, scale, quality and efficiency as companies continued to find it more economical to produce their own printed circuit boards, manufacture their own semiconductors, processors and peripherals, integrate their own systems, write their own software and support their installed base with their own service operations.

The situation then changed dramatically during the 1980s, as the industry experienced the rapid growth that accompanied the increasing convergence of computing and telecommunications and the emergence of a whole new mass-market segment following the arrival of the personal computer. This explosion in the primary market has since allowed the subsupply sector to grow to a size which now supports degrees of specialisation and automation that would have been unthinkable less than two decades ago. Furthermore, many of the industry's strategic technologies have been migrating backwards into the component and subsystems levels, particularly in the printed circuit board and microprocessor areas. The overall effect of these developments has been to increase the opportunities and incentives for a whole variety of alliances and partnerships throughout the industry, as the range of value-adding activities has become too wide and complex for any individual company to be able to excel across the board.

2. ***The partnership approach is always a trade-off, never a panacea:*** Even when the industry context favours the emergence of supplier partnering, we should always remember that the supplier as partner approach involves a fundamental trade-off. It sacrifices some degree of market discipline for the benefits of closer cooperation, and may not always be the best option in all cases. For example, the partnering approach has been very effective in supporting Marks and Spencer, the leading British retailer, in its strategy of providing premium quality store-label products to middle income customers at very competitive prices (Montgomery 1991). On the other hand, however, it is also evident that the traditional arm's length approach has been just as effective in underpinning the Wal-Mart Stores strategic mission of offering consumers branded goods at lowest prices (Huey 1989).

Furthermore, while the security and stability of partnership can be beneficial and help to generate substantial economies of cooperation of the type discussed earlier, there is always a risk of complacency creeping in over time, particularly on the supplier side. The absence of a strong market mechanism may eventually blunt the inventiveness and enterprise of the supply sector.

3. ***The future of both partners is tied to the buyer's market:*** The fact that the partnership approach ties the fortunes of both parties much more closely together, and makes them dependent on what happens in the buyer's marketplace is perhaps its biggest implication. As long as the buyer's industry continues to grow, and the buyer's competitive position remains strong, then both parties will prosper. However, as the industry matures and the growth rate declines, the partners may find themselves facing zero-sum conditions that will really put a strain on the relationship, and test their commitment to each other to the full. The relationship will also come under pressure if the buyer's relative competitive position deteriorates, as suppliers of General Motors discovered a few years ago, when teams of GM "warriors" were sent into their operations to press for price reductions of 20% on previously agreed contracts in order to shave nearly $1 billion from their partner's overall components bill (Lorenz, 1993).

The evolution of the buyer's industry is influenced by many factors external to both partners, and the buyer's competitive position is dependent on many elements that have little direct connection with the quality and effectiveness of the buyer-supplier relationship, as the recent histories of such partnership-oriented companies as Digital Equipment Corporation, Amdahl and Apple can readily testify.

4. ***The partners may hitch their fortunes to the wrong star:*** A further danger for both buyer and supplier in the partnership approach is the risk that they may hitch their fortunes to the wrong star. Will the partners be able to keep pace with each other, and drive each other forward? Or will one partner eventually end up holding the other back?

For the supplier, the risk is that he commits a large portion of his capacity to a company that may be unable to hold or strengthen its competitive position and grow as the industry evolves. For example, Intel has a greater strategic stake in many of its buyer's markets than they have themselves. As the cost of developing each new generation of microprocessor continues to escalate by orders of magnitude (from $2M in 1980 to $1B by 2000), and ever greater volumes are needed to recoup the R&D investment and fill the high-technology fabrication plants to capacity, the company becomes more and more dependent on the ability of its own commercial customers to grow the downstream markets. This is why Andy Grove, Intel's CEO, believes that his company has to be the driving force in the industry, even if it makes its commercial partners uncomfortable and keeps them under pressure (Kirkpatrick 1997).

The risk for the buyer, on the other hand, is that he invests heavily in a partnership where the supplier proves incapable of developing in line with him and ultimately undermines his competitive position. If this happens for either party, the process of unhitching can be expensive because of the switching costs involved. The more traditional perspective on the buyer-supplier relationship serves to remind both parties of such risks, and advises them to always keep some options open.

5. ***A partner may unintentionally mortgage its future:*** One of the major arguments offered by advocates of supplier-partnerships is that they offer firms enhanced opportunities to harness the complementary core competencies of an array of sophisticated suppliers in the creation of value for customers, and to enrich the development of their own competencies through the wider access to learning and new ideas that such relationships can provide. As Quinn and Hilmer (1994; 43) have argued, such strategic partnering can offer a company "the full utilization of external suppliers' investments, innovations and specialized professional capabilities that would be prohibitively expensive or even impossible to duplicate internally".

However, there is also a downside risk to this process. While partnerships may create more learning and innovation than would result from the more traditional arm's length approach, the benefits may not accrue equitably to both partners. As Hamel and Prahalad (1994) have pointed out, competing for the future is closely tied to competing for today's learning opportunities. When a company offers itself as an attractive supplier-partner of key subsystems to a number of major buyers in its industry, it can also be seen as borrowing the customers of its OEMs in order to increase its own production volume of these core components and accelerate its own learning in strategic technologies at the possible expense of its buyer partners. In this way, the hasty and uncritical adoption of the supplier-partner approach may result in the unintended transfer of critical learning opportunities from the buyer segment to the supply segment,which the buyer may regret. Something like this happened to IBM in its relations with Intel and Microsoft in personal computers. It also happened to General Electric in its relationship with Samsung in microwave ovens, and to Bullova in its relationship with Citizen in the watch industry.

CONCLUSION: Perhaps the essence of the partnership approach is best summed up in the old Japanese proverb as two partners "having the same bed, but different dreams". So while partnership clearly offers many strategic advantages, both buyers and suppliers would be well advised to keep the insights offered by the more traditional arm's length approach readily to mind. Buyer-supplier partnerships are rarely marriages among economic equals (Lyons *et al.* 1990). Furthermore, the partners remain separate commercial players with their own goals and aspirations, and the marriage, however close, will always be a conditional one, for better, for richer, and in health, for as long as it continues to make sound economic sense.

It is important, therefore, that buyers and suppliers continue to look out for their own interests, even when the partnering relationship is working well. Few buyer-supplier partnerships can ever be expected to develop to the point where the partners are fully prepared to sink or swim together in times of serious industry downturn or major strategic error. One of the enduring virtues of the traditional perspective on the buyer-supplier relationship is that it serves to remind each of the partners, particularly the economically more dependent one, be prepared for industry downturns or strategic error.

Organizations Mentioned: Amdahl; Apple; Bennetton; Digital Equipment Corporation; General Motors; Intel; Marks and Spencer; Toyota; Wal-Mart.

See **Backward integration; Economics of cooperation; Forward integration; Integrated manufacturer; OEM; Purchasing: Acquiring the best inputs; Purchasing: The future; Supplier relationships; Supplier partnering; Supplier performance management; Supply chain management.**

References

[1] Dyer, J.H. and Ouchi, W.G. (1993). Japanese-style partnerships:giving companies a competitive edge, *Sloan Management Review*, Fall, 51–63.

[2] Hamel, G. and Prahalad, C.K. (1994). *Competing for the Future*, Boston:Harvard Business School Press.

[3] Huey, J. (1989). Wal-Mart, will it take over the world?, *Fortune*, January 30th, 56–62.

[4] Kirkpatrick, D. (1997). Intel's amazing profit machine, *Fortune*, February 17th, 24–30.

[5] Lorenz, A. (1993). Lopez to drive VW on a lean mixture, *The Sunday Times*, London, March 21st, Section 3, p. 9.

[6] Lyons, T., Krachenberg, A.R. and Henke, J.W. (1990). Mixed motive marriages: What's next for buyer-supplier relations?, *Sloan Management Review*, Spring, 29–36.

[7] Mandel, M. (1997). *The new business cycle*, Business Week, March 31st, 48–54.

[8] Montgomery, C. (1991). Marks and Spencer, Ltd. (A), *Harvard Business School*, Case #9-391-089.

[9] Porter, M.E. (1980). *Competitive Strategy*, New York: Free Press.

[10] Quinn, J.B. and Hilmer, F.G. (1994). Strategic outsourcing, *Sloan Management Review*, Summer, 43–55.

[11] Womack, J.P. and Jones, D.T. (1994). From lean production to the lean enterprise, *Harvard Business Review* 72, March–April, 93–103.

[12] Womack, J.P., Jones, D.T. and Roos, D. (1990). *The Machine that Changed the World*, New York: Rawson Associates.

SUPPLIER PERFORMANCE MEASUREMENT

Robert B. Handfield

Michigan State University, East Lansing, MI USA

DESCRIPTION: It is important to distinguish between an on-going evaluation system and a "prior-to-purchase" evaluation process. Once, the purchase decision is made, "on-going" collection of supplier performance data (i.e. delivery, quality, price, service, etc.) will generally be recorded and added to a database every time a purchase is made. This ongoing evaluation is important, because it provides a source of information on whether the supplier's actual performance is meeting expectations. The focus here is the "prior-to-purchase" evaluation of suppliers for a new component or service.

STRATEGIC PERSPECTIVE: Traditionally, most firms relied on competitive price bids from suppliers for awarding a purchase contract. In the past, purchasing managers found it sufficient to obtain three bids and award the contract to the lowest bid. Today, however, purchasers often commit major resources to perform initial supplier evaluations. These evaluations examine a supplier's performance across a number of areas. The supplier selection process has become so important that entire teams of cross-functional personnel are responsible for visiting and evaluating suppliers directly at their facilities. The evaluation and selection process, particularly for important purchase requirements, will continue to receive a great deal of attention now and in the future. A sound supplier selection decision today can reduce or eliminate a host of problems tomorrow.

An effective understanding of a supplier is also a prerequisite for other important supplier-specific strategies. For instance, before considering the development of a strategic alliance with a supplier, organizations will often carry out a detailed evaluation of the supplier. In a buyer-supplier alliance, the buyer is purchasing not only the supplier's products and services, but their capabilities as well. Within the integrated supply chain, trust must be developed not only between a single alliance but among multiple partners located across the supply chain. Moreover, initiating firms must be able to trust not only their supplier, but also their supplier's suppliers. In turn, a supplier cannot simply trust a manufacturer, but must also trust the manufacturer's customers who will dictate demand volumes, pricing, etc. Managing multiple relationships within a supply chain is a challenging task. Thus, the evaluation of the supply chain partner often looks at multiple criteria in detail, in order to truly identify the long-term potential for an alliance relationship.

The extent of an evaluation will vary depending on the nature of the commodity. As shown in Figure 1, the amount of information collected will be greater for those commodities which are considered to be critical to the functioning of the product, or those which are considered "high value-added". Throughout the supplier evaluation and selection process it is important to understand the requirements that are important to the purchaser. These requirements can differ widely from company to company or industry to industry. A rapidly changing industry may need suppliers that are responsive to technological change and are at the leading-edge of new product and process development. Conversely, a slower changing industry, such as household furniture manufacturing, may stress supplier cost competitiveness.

While different requirements can exist for each performance evaluation area, most firms typically evaluate, at a minimum, supplier quality, cost competitiveness, potential delivery performance, and technological capability. This evaluation is typically carried out by different functional personnel using a "check-off form" that covers some of the following ifferent criteria (Dobler and Burt, 1996):

- Purchasing Rating (on-time, delivers at quoted prices, competitive pricing, etc.)

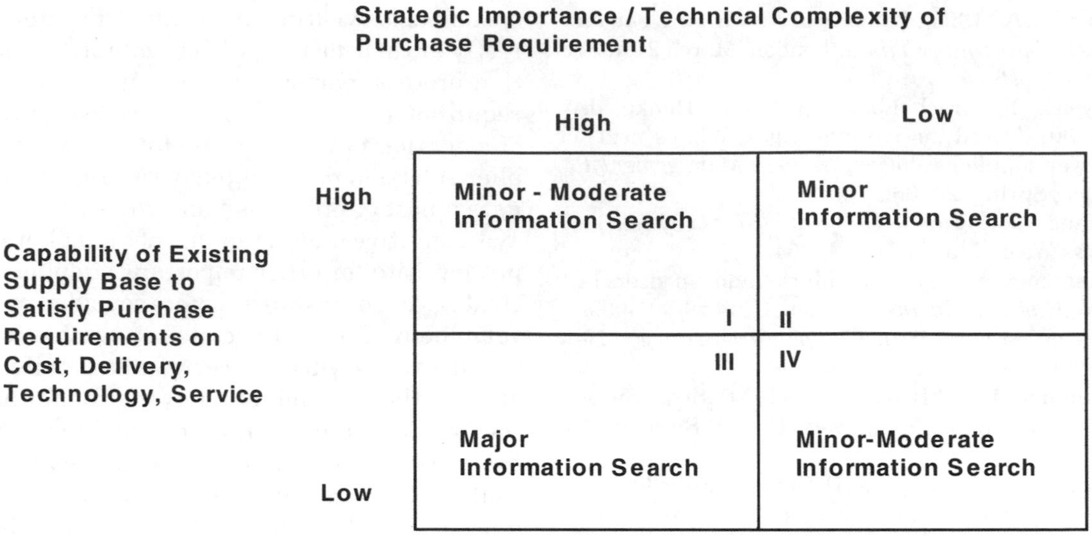

Fig. 1. Information search for potential sources of supply.

- Receiving Rating (good packaging, delivers per routing instructions, good service)
- Accounting Rating (invoices correctly, issues credit memos punctually, does not ask for special financial consideration)
- Engineering Rating (reliability of products, technical ability, accepts responsibility for defects and reacts promptly)
- Quality Rating (quality of material, furnishes certification, replies with corrective action)

Within each area, a buyer must also determine the level of performance expected of suppliers. This may be done by applying a "weighted" approach (such as the one shown in Figure 2), where different capabilities are weighted based on their importance.

IMPLEMENTATION: Many organizations spend a great amount of time and resources evaluating their supply chain partners for critical, important materials or services. This may involve an on-site audit of the supplier. Some of the different criteria that a company may use to assess the potential for integrating a partner into a supply chain include, but are not limited to, the following:

- Management capability
- Organizational ownership
- Major customers
- Major suppliers
- Company history / reputation
- Market share in major product lines
- Overall personnel capabilities
- Cost structure
- Total Quality Management philosophy and programs
- Product and process technological capability
- Environmental regulation compliance
- Financial capability/stability
- Manufacturing operations, including scheduling and control systems
- Performance measurement systems
- Logistical operations
- Information systems capability (e.g., EDI, bar coding, CAD/CAM, etc.)
- Supplier purchasing strategies, policies, and techniques
- Potential and interest in long-term relationship

It is worthwhile to go through each of these areas (Monczka, Trent, and Handfield, 1997). Although it may not be possible to obtain access to information on the potential partner's capabilities in all of these areas, information that can be obtained during the evaluation stage will help to determine the potential for a successful match and can highlight potential problems that need to be addressed in the supply chain relationship.

MANAGEMENT CAPABILITY: It is often important to evaluate the capability of a supplier or customer's management. Different aspects of management capability include (1) management's commitment to continuous process and quality improvement, (2) management's overall professional ability and experience, (3) management's ability to maintain positive relationships with its workforce, (4) management's willingness to make the investments necessary to support future growth, (5) management's willingness to develop a closer working relationship with the buyer, and (6) management's vision of the future. While there is no straightforward "checklist" to follow in evaluating management capabilities, it can be one of the most critical areas to assess. Management capability will an organization's ability to satisfy supply chain requirements, both currently and in the future, and will also determine the level of commitment to the overall supply chain initiative.

Some questions that might be asked in assessing management capability include:

- Does executive management practice strategic long-range planning?
- Has management committed itself to total quality management and continuous improvement?
- Has management developed the commitment of all employee's to continuous performance improvement?
- Is there a high degree of turnover among managers?
- What is the professional managerial experience?
- Does management have a vision about the future direction of the company which is aligned with your company?
- What is the history of management / labor relations?
- Has management prepared the company to withstand the competitive challenges of the next two decades?

Many of these questions are difficult to answer using strict "yes or no" criteria. Organizations undergoing significant re-engineering and change are likely to face a great deal of uncertainty about their future. As such, it may be difficult to discern the true state of affairs within such a company's management team. Nevertheless, asking these questions can help the supply chain manager to develop a "gut feel" for the strategic posture of the managers in the candidate organization, and determine whether this posture is aligned with that of the initiating organization.

PERSONNEL CAPABILITIES: The supplier evaluation process also requires an assessment of non-management personnel. A highly trained and motivated workforce is a critical element of any organization and will be a key factor in determining that organization's performance. An audit team should evaluate: (1) the overall skills and abilities of the workforce, (especially with regard to the level of education and training received), (2) the degree to which employees support the company's quality process and commit to continuous improvement (3) the current state of employee-management relations, as well as a history of labor relations within the organization, (4) workforce turnover, and (5) the willingness of employees to contribute to supply chain objectives. All of these data can provide a general idea of how dedicated the organization's employees are to meeting expectations of the supply chain relationship.

COST STRUCTURE: Evaluating a supply chain partner's cost structure requires an in-depth understanding of their total costs. This includes understanding the partner's, (1) direct labor costs, (2) indirect labor costs, (3) material costs, (4) manufacturing or process operating costs, (5) overhead costs, and (6) sales and general administrative expenses. Understanding the cost structure helps the initiating firm determine the effectiveness and efficiency of the potential partner's operations. A cost analysis also helps identify potential areas of cost improvement.

Collecting this information can be a challenge during the initial evaluation process. A potential partner may not have a detailed understanding of its costs at the level of detail required. Furthermore, many firms view cost data as highly proprietary. Remember, during the initial evaluation, the two firms seeking to develop the relationship usually have not developed any major degree of mutual commitment or trust.

TOTAL QUALITY MANAGEMENT PHILOSOPHY: A major part of the evaluation process focuses on the potential partner's quality management processes, systems, and philosophy. The evaluation team not only evaluates the obvious topics associated with quality (i.e. management commitment, statistical process control, number of defects) but may also evaluate safety, training, and facilities and equipment maintenance.

PROCESS AND TECHNOLOGICAL CAPABILITY: Supply chain evaluation teams often include a member from the manufacturing/industrial engineering or technical staff to evaluate a partner's process and technological capability. Process consists of *the technology, design, methods, and equipment used to manufacture a product or deliver a service*. An organization's production process defines its required technology, human resource skills, and capital equipment requirements.

The ability to produce an item economically at the required quality level is critical. A firm that lacks the ability to meet current and expected future technical requirements has a smaller chance of becoming a supply chain member. An exception may occur if other supply chain members are willing to work directly with the candidate organization to increase their technical ability. In such cases, other organizations may provide additional technical (and possibly even financial) support to further develop the supply chain member, if they believe a significant payback can occur in the form of improved supply chain performance.

The evaluation of an organization's technical and process capability should also focus on *future* process and technical ability. This requires assessing an organization's capital equipment plans and strategy and commitment to research and development.

An assessment of the organization's design capability may also take place during the selection process. For example, buyers often expect suppliers to perform component design *and* production. One way to reduce the time required to develop new products involves using qualified suppliers who are able to perform product design activities. Ford, for example, now requires almost all of its suppliers to have not only production but also design capabilities. The company has transferred most of the design of its component and component system requirements to its suppliers. The trend towards the increased use of external design capabilities makes this area an important part of the organizational evaluation and selection process.

ENVIRONMENTAL REGULATION COMPLIANCE: The 1990s brought about a renewed awareness of the impact that industry has on the environment. Government regulations are becoming increasingly harsh on polluters. The U.S. Clean Air Act of 1990 imposes large fines on producers of ozone-depleting substances and foul-smelling gases, and the U.S. administration has introduced laws regarding recycling content in industrial materials(Business Week, 1993). As a result, an organization's ability to comply with environmental regulations is becoming an important criteria for supply chain alliances. This includes, but is not limited to, the proper disposal of hazardous waste. It is not in the best interest of an organization to do business with a supply chain partner that fails to comply with environmental regulations.

FINANCIAL CAPABILITY/STABILITY: An assessment of a potential partner's financial condition almost always occurs during the initial evaluation process. Some firms view the financial assessment as a screening process or preliminary condition that organizations must pass before a detailed evaluation can begin. A firm can use a financial rating service to help analyze the candidate organization's financial condition. If the organization is publicly held, other financial information found in annual reports and databases is readily available.

Selecting a supply chain partner who is in poor financial condition presents a number of risks. First, there is the risk that the organization will go out of business. This can present serious disruptions in the flow of goods and information occurring in a supply chain. Second, supply chain members who are in poor financial condition may not have the resources to invest in plant, equipment or research which leads to longer-term technological or other performance improvements. Finally, organizations must feel confident that they can commit the resources necessary to support continuous improvement.

Circumstances may exist that support selecting a supply chain members who is in a weaker financial condition. For example, a company may be developing, but has not yet marketed, a leading-edge technology that can provide a market advantage to other supply chain members. A temporarily unstable financial condition may also be a function of uncontrollable or non-repeating circumstances. In such cases, the candidate organization's financial condition should not be the unique reason for elimination.

In cases when the organization is a publicly traded company, specific financial ratios can be obtained from sources such as 10-K reports, Moody's Industrials, and Dunn and Bradstreet to gain insights into the potential partner's financial health. Many of these rations can also be calculated using income statements and balance sheets. Some of the most common ratios and their implications for partner evaluation are as follows:

Liquidity ratios: (How capable is the company of meeting their short-term cash needs?)

- Quick ratio
- Current ratio

Activity ratios: (How effectively is the company managing their assets?)

- Inventory turnover
- Average collection period
- Fixed asset turnover
- Return on net assets
- Installed capacity

Profitability ratios: (Is the company overcharging or undercharging?)

- Gross profit margin
- Net profit margin
- Return on equity
- Return on investment
- Earnings before interest and taxes
- Price earnings ratio

Debt ratios: (Is the company over-leveraged and incapable of paying their long-term debts?)

- Debt ratio
- Times interest earned

In technology intensive industries, the audit team should also look at the company's research and development (R&D) expenditures as a percent of sales, to obtain an indication of how much they are re-investing into product development. These financial ratios can provide quick and valuable insights into a company's financial health. Moreover, managers should track such ratios for possible "red flags" that may signify a potential financial difficulty. If at any point the manager has reason to believe that the candidate organization is facing a financial crisis of some sort, an appropriate investigation should be carried out.

PRODUCTION SCHEDULING AND CONTROL SYSTEMS: Production scheduling includes those systems that release, schedule, and control a production process. The sophistication of the scheduling system can have a major impact on supply chain performance. Some of the criteria to be investigated include the following:

- Computerized material requirements planning (MRP) system to ensure the availability of required components.
- Does the organization track material and production cycle time and compare this against a performance objective or standard?
- Does the production scheduling system support the supply chain requirements?
- How much order lead-time does the production scheduling and control system require?
- What is the on-time delivery performance history?

The purpose behind evaluating the production scheduling and control system is to identify the degree of control the organization has over its scheduling and production process. A well-defined system will use computerized inventory control systems and measure performance against pre-established objectives.

INFORMATION SYSTEMS CAPABILITY (EDI, CAD/CAM, WORLD WIDE WEB, ETC.): The ability to communicate electronically between supply chain members is rapidly becoming a requirement for entering into an alliance. Many companies now insist that their supply chain partners, (1) currently have electronic data interchange capability, or (2) be willing to develop the capability as a precondition. Besides the efficiencies that electronic communication provide, these systems support closer relationships and the exchange of all kinds of information.

Managers should also evaluate other dimensions of the supplier's information technology by asking the following questions:

- Are designers using Computer Aided Design (CAD) and Computer Aided Manufacturing (CAM) technologies?
- Is bar-coding being used when appropriate?
- Is the company able to communicate via the Internet or the World Wide Web?
- Are their managers networked throughout the company?
- Do they use modern systems such as voice mail, CD-ROM, purchasing credit cards, multi-media computing, etc.?

Evidence that the organization is using these technologies can provide reasonable assurance that they are keeping up with the progress in information technology, and will be prepared for the future (Monczka, Nichols, and Handfield, 1994).

SUPPLIER SOURCING STRATEGIES, POLICIES, AND TECHNIQUES: In evaluating a downstream supply chain partner, the performance of a supplier's suppliers deserves consideration.

The concept of managing not just one's suppliers but all of the suppliers in a supply chain is becoming increasingly important. For instance, a recent meeting that took place to discuss the manufacture of a part on an airplane is illustrative. The companies present at the meeting included United Airlines, Boeing (plane manufacturer), Pratt and Whitney (engine manufacturer), and Auto Air (the components manufacturer). In addition, United Airlines was considering bringing along one of their Frequent Flier customers to the meeting. The point of this example is that supply chain management requires an awareness of the sourcing practices of all direct suppliers.

In order to truly integrate the supply chain, a firm must gain insights into the sourcing practices of its first, second, and sometimes even third tier suppliers as they relate to critical purchased items. Evaluating a potential supplier's sourcing strategies, policies, and techniques is one way to gain greater insight and understanding of the supply chain. However, simply evaluating the supplier's sourcing policies is not enough, customers must drive their supplier's sourcing policies through cross-functional team meetings that occur between the two organizations, thereby aligning strategic priorities and requiring improvements in total cost, quality, delivery, and environmental issues (Giunipero and Brewer, 1993).

Because so few firms actually have an understanding of their second and third tier suppliers, those firms that have such an understanding can gain an important advantage over competitors. For instance, firms that integrate their information systems across multiple tiers of suppliers can improve planning and forecasting, reduce lead-time throughout the supply chain, reduce in-transit inventory, and significantly reduce costs. One company that is actively pursuing such multitier integration is Chrysler Corporation, through their "Extended Enterprise" vision.

In summer 1995, the Dodge Ram platform team was tasked with bringing the new model into production by February 1996. The process was managed by using a cross-functional team who paid attention to what materials could be available from suppliers on a timely basis. Time did not permit traditional supplier evaluation, so suppliers were selected based on long-term commodity strategies and were told they needed to begin design and prototype tooling right away. Steve Walukas, Chrysler Manager for Ram Truck Platform Supply noted that "This was not just Chrysler and its first-tier suppliers cooperating. It was Chrysler, first-tier, second-tier and other suppliers all helping to make this vehicle possible. This project included five-tier involvement." (*Supplier*, 1996).

LONGER-TERM RELATIONSHIP POTENTIAL: Firms today increasingly evaluate a supply chain partner's willingness to develop longer-term relationships. A major flaw in managers' conceptualization of supply chains is that all potential supply chain members will want to develop a closer relationships. This may not always be the case. As a result, there remains a desire on the part of management at some companies to remain independent of larger supply chain members. A company seeking a relationship with a smaller organization that requires the sharing of sensitive information should address suppliers' willingness or unwillingness to enter into such an arrangement.

A company that wants to establish a longer-term relationship or partnership with a supplier or customer must go beyond evaluating the performance of the supplier. An initiating organization should ask the following questions when evaluating the possibility of a closer relationship with a potential supply chain candidate's commitment to joint planning and problem solving (Spekman, 1988).

- Has the organization signaled a willingness or commitment to a partnership-type arrangement?
- Is the organization willing to commit resources to this relationship?
- How early in the product design stage is the organization willing or able to participate?
- What does the organization bring to the relationship that is unique?
- Will the organization immediately revert to a negotiated stance if a problem arises?
- Does the organization have a genuine interest in joint problem solving and a win-win agreement?
- Is the organization's senior management committed to the processes inherent in strategic partnerships?
- Will there be free and open exchange of information across functional areas between companies?
- Does the organization have the infrastructure to support such cross-functional interdependence?
- How much future planning is the organization willing to share with us?
- Is the need for confidential treatment of information taken seriously?
- What is the general level of comfort between the companies?
- How well does the organization know our business?
- Will the organization share cost data?
- What will be the organization's commitment to understanding our problems and concerns?
- Will we be special to the organization or just another customer/supplier?

This list provides a framework concerning the types of issues that are important.

CASES:

The Texas Instrument Experience

Texas Instruments, (TI), a high technology electronics company, uses an evaluation worksheet approach to supplier selection similar to the approach shown in Figure 2 (Gregory, 1986). TI's goal was to develop a flexible and reliable method for performing initial supplier assessments and selection. The company wanted the audit worksheet to permit the reviewer(s) to be as objective or systematic as possible in a process that is inherently subjective and involves many performance variables.

As shown in Figure 2, the total score for the quality category for this particular supplier is calculated as follows. The evaluator first evaluated the supplier, and assigned their process control systems a score of 4 out of 5, the supplier's quality commitment a score of 4 out of 5, and their parts-per-million defect rate 5 out of 5 (the maximum possible). Next, the evaluator divided each score by five to obtain a ratio. The ratio was multiplied by the subweight to produce the subweight score. Finally the subweight scores are added to form the category score for this particular supplier, as shown below.

Quality Systems Performance Category (Total = 20) Categories

Process Control Systems

(4 points out of 5 possible points)
$= 0.8 \times 5$ subweight $= 4.0$ points

Total Quality Commitment

(4 points out of 5 possible points)
$= 0.8 \times 8$ subweight $= 6.4$ points

Supplier: Advanced Micro Systems					
CATEGORY	WEIGHT	SUBWEIGHT	SCORE (5 pt. scale)	WEIGHTED SCORE	
Quality Systems	20				
Process Control Systems		5	4	4	
Total Quality Commitment		8	4	6.4	
Parts per mill Defect Performance		7	5	7	
					17.4
Management Capability	10				
Management/Labor Relations		5	4	4	
Management Capability		5	4	4	
					8
Financial	10				
Debt Structure		5	3	3	
Turnover Ratios		5	4	4	
					7
Cost Structure	15				
Costs Relative to Industry		5	5	5	
Understanding of Costs		5	4	4	
Cost Reduction Efforts		5	5	5	
					14
Delivery Performance	15				
Performance to Promise		5	3	3	
Lead Time Requirements		5	3	3	
Responsiveness		5	3	3	
					9
Technical/Process Capability	15				
Product Innovation		5	4	4	
Process Innovation		5	5	5	
Research and Development		5	5	5	
					14
Information Systems Capability	5				
EDI Capability		3	5	3	
CAD/CAM		2	0	0	
					3
General	10				
Support of Minority Suppliers		2	3	1.2	
Environmental Compliance		3	5	3	
Supplier's Supply Base Management		5	4	4	
					8.2
			TOTAL WEIGHTED SCORE		80.6

Fig. 2. Initial supplier evaluation.

PPM Defect Performance

(5 points out of 5 possible points)

$= 1.0 \times 7$ subweight $= 7.0$ points

Total for Category 17.4 points

Which is 87% of total possible points (17.4/20)

In the sample evaluation in Figure 2, the supplier received a total overall evaluation of 80.6 percent.

Professional buyers at TI indicate that a structured approach to initial supplier evaluation and selection has resulted in some definite benefits:

- The formal supplier evaluation supports a concise presentation for review by senior management.
- TI can use the evaluation as a debriefing device with suppliers. This debriefing can highlight both supplier strengths and weaknesses.
- A structured approach to supplier evaluation can result in a higher degree of confidence that TI has made the right sourcing decision. This is particularly true when a cross or multifunctional team conducts the evaluation. The team has a formal document from which to strive for a consensus decision about supplier evaluation scores and final selection. Also, a higher degree of confidence can result when separate team members, using the same evaluation document, reach the same conclusion independently. If team members do not reach the same

conclusion, the evaluation document provides a logical basis for discussing member differences.
- A formal approach to supplier evaluation provides an audit trail supporting the logic behind past decisions.

Collective Wisdom: Purchasing must consider, at least implicitly, a number of important issues throughout the supplier evaluation and selection process. Each of these issues has the potential to affect purchasing's final choice of suppliers.

Size Relationship: A purchaser may decide to select suppliers who are relatively smaller in size. It may be easier to develop closer working relationships when the buyer is somewhat larger than the supplier. Furthermore, a buyer may also have greater influence when it has a relative size advantage over the supplier. For example, Allen-Edmonds Shoe Corporation, a 71-year old maker of premium shoes, tried unsuccessfully to implement just-in-time methods to speed production, boost customer satisfaction, and save money. Why? One of the major reasons it was unsuccessful was because the company had difficulty getting suppliers to agree to the just-in-time requirement of matching delivery to production need. While domestic suppliers of leather soles did agree to make weekly instead of monthly deliveries, European tanneries supplying calf-skin hides refused to cooperate. The reason? Allen-Edmonds Shoe was not a large enough customer to wield much leverage with those suppliers (March, 1993).

Use of International Suppliers: The decision to source internationally can have some important implications during the supplier evaluation and selection process. For one, international sourcing is generally more complex than domestic buying. As a result, the evaluation and selection process can take on added complexity. Sometimes, there simply is no choice but to purchase from an international supplier (Bozarth *et al.*, 1997). This is often true, for example, in the electronics industry where many component suppliers are located in the Far East. In addition, it may be difficult to implement JIT with international suppliers, as leadtimes are frequently twice as long as those for domestic suppliers (Handfield, 1995).

Competitors as Suppliers: Another important issue is the degree to which a buyer is willing to purchase directly from a direct competitor. For example, a major supplier of steering gears to Chrysler is Saginaw Steering Gear, a unit of General Motors. When a buyer purchases from competitors, information sharing between the two parties is limited. The purchase transaction is usually straightforward and the buyer and seller do not develop a working relationship characterized by mutual commitment and confidential information sharing.

Offset Requirements: The need to satisfy counter-trade requirements can also influence or affect the supplier evaluation and selection decision. Boeing, a producer of commercial aircraft, purchases a portion of its supply requirements in markets where it must satisfy the counter-trade requirements of international governments. Supplier selection decisions actually complement the company's marketing effort. A firm involved in extensive worldwide marketing may have to contend with offsetting counter-trade requirements before it can sell to international customers. This can have a direct impact on the supplier evaluation and selection process.

Social Objectives: Many firms are attempting to increase their volume of business with traditionally disadvantaged suppliers. This can include suppliers owned by female, minority, or handicapped members of society. Buyers may also want to conduct business with suppliers that commit to the highest environmental standards. Expect the influence of social objectives on the purchasing function to increase in the future. This trend will affect how buyers approach the supplier selection process.

Organizations Mentioned: Allen Edmonds Shoe Corporation; Auto Air; Boeing; Chrysler; Ford; GM (Saginaw Steering Gear); Pratt and Whitney; Texas Instruments; United Airlines.

See **Current ratio; Debt ratio; Direct labor costs; Fixed asset turnover; Gross profit margin; Indirect labor costs; Net profit margin; Purchasing: Assigning the best inputs; Purchasing rating; Purchasing: The future; Quick ratio; Supplier partnership as strategy; Supplier relationship; Supply chain management; Workforce turnover.**

References

[1] Bozarth, Cecil, Handfield, Robert, and Das, Ajay (1997). "Stages of global sourcing strategy evolution: An exploratory study," Special Issue on Global Manufacturing, *Journal of Operations Management*, in press.

[2] *Business Week*, Nov 1., 1993, 60–61.

[3] Dobler, Donald W., and David Burt (1996). *Purchasing and Supply Management*, McGraw Hill, 1997.

[4] Gregory, R. (1986). "Source selection: A matrix approach," *International Journal of Purchasing and Materials Management*, 24–29.

[5] Handfield, Robert, (1994). "Global sourcing: Patterns of development," *International Journal of Operations and Production Management*, 14(6), 40–51.

[6] March, Barbara (March 5, 1993). "Allen Edmonds Shoe Tries Just-In-Time Production," *Wall Street Journal*, 1.

[7] Monczka, R., Trent, R., and R., Handfield (1997). *Purchasing and Supply Chain Management*, Southwestern Publishing.

[8] Monczka, R.M., E.L., Nichols, and R.B. Handfield, (1994). "Information systems, technology, procurement and supply chain management: A view to the future," *International NAPM Purchasing Conference Proceedings*, Tempe, AZ: NAPM, Inc., 156–160.

[9] Spekman, Robert E. (1988). "Strategic supplier selection: Understanding long-term buyer relationships," *Business Horizons*, 80–81.

[10] *Supplier* (1996). "Extended Enterprise races to make special-edition Dodge Ram a reality," newsletter published by the Chrysler Corporation, March–April, p. 6.

SUPPLIER RATING

See **Purchasing rating; Supplier performance measurement.**

SUPPLIER RELATIONSHIPS

Thomas F. Burgess

University of Leeds, Leeds, United Kingdom

Hasan K. Gules

Selcuk University, Turkey

DESCRIPTION: In a supply transaction, one party requires that goods and/ or services be supplied by a second party, who in return receives monetary benefit or some benefit in kind. From an economic perspective, markets provide the key mechanisms to establish transactions. In this view of supply relationships, the buyer initiates the transaction by selecting the supplier using the criterion of lowest price. This narrow view of supply relationships is no more than fulfilling a commercial contract between the two parties. Further, completing the transaction frees the parties to seek new partnerships without any residual obligations.

A progressive broader view of supply relationships recognizes that social interaction takes place between the two parties, and that the social dimension can be extremely important. From this perspective, individual transactions are considered part of an ongoing dialogue between parties and not as separate, unconnected commercial encounters. In this wider view supplier relationships are perceived as having important social and cultural dimensions as well as their commercial and technical aspects. Such widely defined relations between buyer and seller can take different forms. To many this variation stretches from the pure adversarial relationship to pure collaboration. Pure collaboration ideally consists of long-term relationships between trusting partners, who seek to co-operatively reduce the cost of doing business. The relationship is typified by frequent communications, both informal and formal, between partners whereby information is openly shared.

This wider view also takes into account the surrounding context in which buyers and sellers come together to form a chain, or web, of relationships that parallels the chain along which goods and services flow. The collaborative quality of relationships between individual elements contributes significantly to the performance of the whole supply chain. It is now realized that relationships and structure (topology) of the supply chain are key factors in delivering high performance.

HISTORICAL PERSPECTIVE: Much of the theory and practice concerning supplier relationships have developed in the milieu of the automobile industry. It is therefore fitting that this section reflects that industry's importance. In the early part of the twentieth century Henry Ford sought to control the supply chain for the Model T through complete vertical integration. He imposed his will on the supply chain by owning the links in the chain starting from the plantations that produced rubber for the tires through to the sale of the final vehicle. In this situation, where significant power was exercised over suppliers collaboration was not an issue. The traditional alternative to vertical integration relied on short term supply contracts allocated between competing suppliers on the basis of price; i.e., an adversarial relationship was the norm. For the best part of the century, the mass production paradigm, as epitomized by Henry Ford's assembly line, dominated the industrial landscape. However, attitudes to coordination of supply chains have fluctuated during the century between the extremes of vertical integration, and buyers exercising power over suppliers through adversarial relationships.

The mass production form of industrial organization dominated in the West until challenged by the evolution in Japan of alternative manufacturing approaches. Such methods came to prominence in the eighties. At the core of these approaches was Just in time (JIT) production, also known as the Toyota Production System. These manufacturing approaches depend on buyer-supplier collaboration, a form of relationship often termed Japanese Style Partnership (JSP) (Dyer and Ouchi, 1993). This form of relationship is usually coupled with a revised structuring of the supply chain. The Japanese supply hierarchy is often described as a pyramid with different layers, or tiers, of suppliers. In general, the members of one tier co-ordinate the work of their suppliers, who form the next tier below. Co-operation within and between layers is seen as being more endemic than in the competing Western structure. The number of suppliers in each tier is much reduced in the Japanese arrangement.

Total Quality is one of the cornerstones of JIT but can be applied separately. The Total Quality Management (TQM) approach (Oakland, 1993) recognizes the links with customers and suppliers, both external to the organization and internally. The latter is achieved by identifying quality chains of internal customers and internal suppliers. People within organizations are linked to other people by the processes they carry out, e.g., a secretary is considered an "internal" supplier to his/her principal who is the "internal" customer. TQM places value on collaboration between customer and supplier and is a further impetus in emphasizing supplier relationships.

Hirschman (1970) reported one of the earliest studies of buyer supplier relationships. Helper built on this work to examine the evolving collaborative relationships in the US automotive industry (e.g., see Helper, 1991). The seminal study by Womack, Jones and Roos (1990) into lean production drew attention to the contribution of collaborative relationships to this paradigm. Sriram, Krapfel and Spekman (1992) identified a move toward more collaboration in major American companies such as Xerox and Motorola. An important realization was that relationships between buyer and seller developed and matured over time (Ford, 1980). Additionally, it was realized that stage models could describe the evolving relationships, where whole industries progress through stages of different relationship types. For example, Lamming (1993) outlines a five-stage progression for the Western automobile industry that starts with the traditional approach that dominated before 1975. According to his model, the industry is currently located at the fourth stage reflecting the emulation of Japanese Style Partnerships.

STRATEGIC PERSPECTIVE: Peter Drucker observed about close working suppliers:

> "This is the largest remaining frontier for gaining competitive advantage – nowhere has such a frontier been more neglected" (Drucker, 1982)

The strategic potential of collaboration can be realized in a number of ways. It is argued that establishing collaborative relationships between buyer and supplier improves business performance substantially for both parties. Such improvements are increasingly important given the competitive pressures of globalization, increased customer sophistication and so on. Collaboration can improve all of the competitive capabilities of quality, cost, speed and flexibility. These capabilities are important in gaining and retaining customers. In particular, the importance to a manufacturer of reducing costs is underlined by the point that the majority of costs are not uncommonly tied up in purchased items. For the automotive industry Brown (1996, p. 252) gives examples of Toyota, where 80% of the value comes from suppliers and Chrysler where the figure is 70%. With the continuing growth of outsourcing, the proportion of total costs attributed to external sources will continue to grow.

Apart from improving performance in delivering today's products and services, collaboration leads to strategic improvements in designing and developing new products and services. Collaboration means that the "buyer" company can tap in to the knowledge that is vested in suppliers and can rely on them to speedily and successfully design lower level constituents of the buyer's new products. This frees the buyer to concentrate more on design aspects to meet the needs of their customers.

The more recently implemented approaches to manufacture such as lean production require collaborative approaches. This is also true of more recently projected approaches such as agile manufacturing (Goldman, Nagel and Preiss, 1995). Basically, collaborative relationships are now considered a strategic necessity.

IMPLEMENTATION: In one sense, setting out to increase collaboration simply requires an act of will from one of the parties. However, true collaboration requires the involvement of both parties, i.e. that initiatives are reciprocated. Much of the literature suggests implementation starts from the premise that the buyer initiates the process, and attempts to improve collaboration with the firm's suppliers. This may reflect that power in supply chains often resides with the buyer; although in some sectors suppliers hold the power. Many powerful buyers have long histories of developing their suppliers; for example, many of the large automotive companies have programs to improve supplier performance. What is new is that, instead of simply focussing on restricted aspects such as improving the technical competence of the supplier, improvement practices now target collaboration first, and then, within this cooperative context, the more specialized aspects of the relationship are improved.

Forming a Supplier Association (in Japanese Kyoryoku Kai, see Hines, 1994) is a key method of developing collaboration. A powerful buyer may strongly encourage suppliers to come together in such a manner. However, associations can be in created in other ways. For example, suppliers of a company can come together at their own volition to collaborate with one another. This can involve suppliers who are directly competing with one another. Suppliers who are supplying different companies can also come together to share experiences on improving collaboration. Similarly suppliers located in different supply chains can also learn from one another.

The Purchasing Department occupies a key role in dealing with suppliers and therefore has a major contribution to make in establishing collaboration. Purchasing has to occupy a more strategic role than its more traditional clerical one. For example, Dobler and Burt (1996) describe a four-stage model that reflects how the importance of the Purchasing function has changed over the past few decades. Paradoxically, establishing collaboration also means that other functions have more powerful roles to play since there is a need for wider cross-functional participation both within and between the two parties.

One of the first steps an organization usually takes is to audit or measure the extent of collaboration with their suppliers. Tools such as the Relationship Maturity Grid (Macbeth and Ferguson, 1994) or the method developed by Sako (1992) can be deployed to take stock of the existing situation. These audit measures can then be used to diagnose areas where improvements need to be made. Hines (1994), in chapter nine of his book, treats thoroughly a mechanism for creating world class suppliers that integrates the need to develop close collaboration. The mechanism includes a set of tools, e.g., benchmarking, buyer-supplier relationship questionnaire and quality function deployment.

TECHNOLOGY PERSPECTIVE: Improving collaboration can be seen as a "softer" or managerial technology, and therefore depends less on the presence of a technology infrastructure than other initiatives such as flexible manufacturing systems. However, communication between partners is increasingly a key factor in collaboration, and therefore information technology capability, e.g., Electronic Data Interchange, is seen as a major enabler.

Conversely, it is realized more and more that implementing hard technology successfully relies on collaborative relationships with suppliers. For example, when investing in a new automated machining center the supplier of the machining center could be treated as a collaborator. Otherwise, buyers may not get the best out of the new equipment. Treating the technology as a black box impairs organizational learning and reduces the potential for process improvement. Similarly, suppliers of material to be processed on the machining center also have an important contribution to make and therefore need to be drawn into a close partnership.

EFFECT: With increased collaboration, partners move to longer-term relationships based on mutual trust, on improved quality assurance and on more flexible delivery times. Both parties commit to a future of mutual dependency. This stands in contrast to the traditional approach where partnerships are often short-lived and conducted at arms' length. Traditionally, new partnerships are made, and others set aside, on the criterion of selecting the firm that offers the lowest price from a wide pool of potential suppliers. In such competitive situations, firms guard their trade secrets fiercely and therefore information flows are strictly circumscribed. With collaborative partnerships the amount of information exchanged and the frequency of interchange increases. The barriers between partners become more permeable, and technological and managerial expertise are transferred more freely. Conflict is endemic in the traditional adversarial situation. With increased collaboration, conflicts between partners do not disappear but are resolved in a more constructive manner than hitherto. Price (cost) is not the only, or the major, criteria in decision making but the net result is that major cost reductions can be achieved, and can benefit both partners.

RESULTS: Increased collaboration improves the competitive capabilities (cost, quality, speed and flexibility) of both partners and the performance of the supply chain. Brown (1996, p. 252) credits improved supplier collaboration as the reason behind the turn-around in Chrysler's performance; from a $795 million loss in 1991 to a net profit of $723 million in 1992. Apart from this improvement in operational performance, future potential to develop new products is enhanced, which strengthens the strategic capability. Again, using Chrysler as an example, Goldman, Nagel and Preiss (1995, p. 141) attributes Chrysler's recent product development success story to improved supplier collaboration. The result is new vehicles are brought to market in less than 3.5 years by Chrysler, far faster than the U.S. average.

One of the consequences of longer-term relationships is that the buyer deals with fewer suppliers. A related consequence of increased collaboration may be that the transaction volume between the two partners increases as a result of consolidating business with a few preferred suppliers. Such consolidation progressively reduces the number of suppliers. Other related initiatives such as tiering of the supply chains also contribute to reducing the number of suppliers within the supply chain. The outcomes of these connected changes are closer relationships with fewer suppliers, which reduces the effort required to coordinate the whole supply chain. That is, supply chains become "leaner".

COLLECTIVE WISDOM: Increased collaboration then is generally held to be a "good thing". However, there are some caveats that are considered here. The purchasing of commodities does not require long-term partnerships, and can still be left under the traditional regime of selecting suppliers on the criterion of lowest price. Whether collaboration should be restricted in this way to some suppliers and not others is a debatable point. It can be argued that collaboration is a universally applicable concept.

By its nature, collaboration increases the level of dependency between firms constituting the supply chain. The prospect, and reality, of increased dependency may be difficult for some firms. After decades of perceived exploitation at the hands of a powerful buyer, a supplier may be reluctant to accept that trust can really exist between the partners. Similarly, powerful buyers may find it difficult not to grasp the opportunity to dominate the supplier (Saunders, 1997, p. 277). What this indicates is the existence of deep-seated mindsets that are difficult to change. Management of change is a notoriously intractable area, and therefore many initiatives (e.g., TQM) have high failure rates. Therefore, moving to increased collaboration should not be treated lightly. Like any new initiatives, senior management commitment is a key success factor.

Because of the close-coupled nature of collaborative supply chains, it can be alleged that they are more "fragile" in the sense of being more vulnerable to shocks arising in the environment. This is a criticism that is often made of JIT production and may have some merit.

Organizations Mentioned: Chrysler; Motorola; Toyota; Xerox.

See **Adversarial buyer-seller relationship; Integrated quality control; Internal customer; Internal supplier; Japanese style buyer-supplier collaboration; Just-in-time manufacturing; Kanban-based manufacturing system; Lean manufacturing implementation; Outsourcing; Purchasing: Acquiring the best inputs; Purchasing: The future; Relationship maturity grid; Structure of the supply chain; Supplier association; Supplier partnership as strategy; Supplier performance measurement; Supply chain management; Supplier association; Tiering of the supply chain; Total quality management; Vertical integration; Virtual manufacturing.**

References

[1] Brown, S. (1996). *Strategic Manufacturing for Competitive Advantage: Transforming Operations from Shop Floor to Strategy*. Prentice-Hall, London.

[2] Dobler, D.W. and D.N. Burt (1996). *Purchasing and Supply Management*. McGraw-Hill, London.

[3] Dyer, H.J. and G.W. Ouchi (1993). "Japanese Style Partnerships: Giving Companies a Competitive Edge." *Sloan Management Review*, Fall, 51–63.

[4] Drucker, P. (1982). *The Changing World of the Executive*. Heinemann, London.

[5] Ford, D. (1980). "The development of buyer-seller relationships in industrial markets." *European Journal of Marketing*, 14 (5–6), 339–54.

[6] Goldman, S.L., R.N. Nagel, and K. Preiss, (1995). *Agile Competitors and Virtual Organizations: Strategies for Enriching the Customer*. Van Nostrand Reinhold, New York.

[7] Helper, S.R. (1991). "Strategy and Irreversibility in Supplier Relations: The Case of the US Automobile Industry." *Business History Review*, 65, Winter, 781–824.

[8] Hines, P. (1994). *Creating World Class Suppliers*. Pitman Publishing, London.

[9] Hirschman, A.O. (1970). *Exit, Voice and Loyalty: Responses to Decline in Firms, Organization and Strategies*. Harvard University Press, Cambridge, Massachusetts.

[10] Lamming, R. (1993). *Beyond Partnership: Strategies for Innovation and Lean Supply*. Prentice-Hall, New York.

[11] Macbeth, D.K. and N. Ferguson (1994). *Partnership Sourcing, an Integrated Supply Chain Management Approach*. Pitman Publishing, London.

[12] Oakland, J.S. (1993). *Total Quality Management: The Route to Improving Performance*. Butterworth-Heinemann, Oxford.
[13] Sako, M. (1992). *Prices, Quality and Trust: Inter-Firm Relations in Britain and Japan*. Cambridge University Press, Cambridge.
[14] Saunders, M. (1997). *Strategic Purchasing and Supply Management*. Pitman Publishing, London.
[15] Sriram, V., R. Krapfel, and R. Spekman (1992). "Antecedents to Buyer-Seller Collaboration: An Analysis from the Buyer's Perspective." *Journal of Business Research*, 25, 303–320.
[16] Womack, J., D. Jones, and D. Roos (1990). *The Machine that Changed the World*. Rawson Associates, New York.

SUPPLY CHAIN MANAGEMENT: COMPETING THROUGH INTEGRATION

Stanley E. Fawcett
Brigham Young University, Provo, Utah, USA

DESCRIPTION: The quest to achieve higher levels of customer satisfaction is driving supply chain integration initiatives. To better meet the needs of customers and to increase organization effectiveness, companies have for many years restructured and reorganized their operations. Most of these efforts have failed to achieve their objectives of sustainable competitive advantage and customer loyalty. Part of the problem is that ever higher levels of service and value are needed to retain customers (Jones, *et al.*, 1995). Indeed, managers at world-class firms now believe that, to compete effectively, their firms must develop value-added processes that deliver innovative, high-quality, low-cost products on-time, with shorter cycle times, and greater responsiveness than before. Yet, even as all-around superior performance pursued, managers have come to realize that their firms may lack the resources and the competencies required for success. This realization has led managers to look beyond their firms' organizational boundaries to consider how the resources of their suppliers and customers can be utilized to create the exceptional value that is required for long-term sustainable advantage.

The endeavors to align objectives and integrate resources across organizational boundaries are known as supply chain management initiatives. The typical supply chain involves the firm various tiers of materials suppliers, service providers, and one or more levels of customers (see Figure 1). In the past, materials suppliers and service providers have been managed differently, frequently by different functional areas within the firm. For Example, materials suppliers have been managed by purchasing and production while service providers such as distributors and transportation providers have been managed by logistics, marketing, and at times purchasing. For superior operation, a firm must manage both types of suppliers in a coordinated, seamless manner.

The essence of supply chain management is for the firm to focus on doing exceptionally well a few things for which the firm has unique skills and

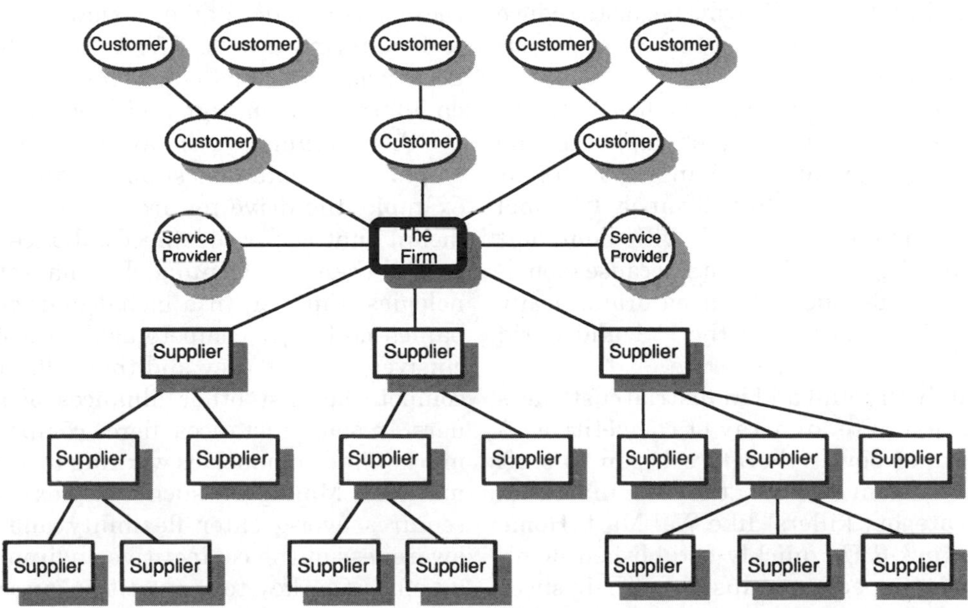

Fig. 1. A simplified supply chain.

advantages. Non-core activities and processes are then obtained from firms that possess superior capabilities in those areas, regardless of the firms' position in the supply chain. Close relationships are formed to assure outstanding and seamless performance levels. In effect, "teams" of firms are formed to create the very best product/service offerings possible. As with other "teams," the most successful supply chain teams are those that not only have the best players but that have established true chemistry – a common understanding of supply chain success factors, an understanding of individual roles, an ability to work together, and a willingness to adjust and adapt in order to create superior value for the customers. These allied teams of companies form a supply chain.

HISTORICAL PERSPECTIVE: The seeds of today's approach to supply chain management in the U.S. were planted in the early 1980s as firms began to seriously consider the threat of Japanese rivals and the competitive implications of just-in-time manufacturing. After several years of focusing on inventory reduction as the key to JIT and improved competitiveness, managers saw the broader aspects of JIT. One of the techniques that emerged from JIT is quite different from traditional management practice, it is the use of long-term partnership relationships with suppliers. Until the advent of Japanese buyer/supplier relationships, U.S. firms had emphasized the practice of multiple sourcing so that suppliers could be leveraged against each other to obtain the lowest possible price on purchased inputs. However, Japanese rivals appeared to achieve significant benefits from supplier certification/development and sole sourcing. Thus, U.S. firms began to reduce their supply bases and emphasize closer buyer/supplier relationships.

Even as greater emphasis was being placed on the strategic management of suppliers, the competitive environment was changing such that channel power began to shift. That is, Channel power, which had long resided with dominant manufacturers, began to dissipate because globalization in recent decades increased dramatically the number of competitors in the U.S. and world markets.

As a result, both retail and industrial customers had greater access to an array of competitive offerings. Channel power therefore began to shift toward the customer from the manufacturer. Moreover, "category killers" like Wal-Mart, Home Depot, and Toys-R-Us quickly established dominance in their respective industries. Their sheer size created the leverage necessary to further shift channel power in a way that put serious pressure on major manufacturers to better manage customer relationships.

Today, power is increasingly concentrated closer to the final consumer, greatly altering the way leading manufacturers operate, and forcing them to pay much greater attention to delivery, timeliness, and cost. For example, a custom-designed delivery system that uses cross-docking, a satellite communications system, and a private trucking fleet enables Wal-Mart to keep its shelves consistently well stocked with low-priced products. Further up the channel, Proctor and Gambles' status as a key Wal-Mart supplier required shorter delivery lead times and a generally much higher level of logistics service than Proctor and Gamble was used to.

The emphasis on closer relationships with both upstream and downstream channel members naturally led to an increasing reliance on strategic alliances. The bottom-line was that only by working as "partners in profit" could firms develop the efficient and effective processes required to produce the high-quality, low-cost products that world consumers required (Schonberger, 1986). This pressure and sentiment continues to persist today, and is exemplified by the adoption of JIT II and customer rationalization strategies by many leading organizations. First, JIT II represents substantial integration of buyers and suppliers in an effort to further reduce inefficiencies and lead times. In JIT II, managers from key suppliers work on-site at the buyer's facilities in order to closely manage important inventories. Second, customer rationalization involves the classification of customers into key accounts and less important customers. Tailored products and services are then offered to those customers that are viewed as important to long-term success. Efforts to customize product/service offerings place tremendous emphasis on information sharing and process integration.

Other circumstances have also led to a greater emphasis on extended supply chain alliances. For example, the drive toward global operations has meant that companies need alliance partners to extend their reach into global markets and technologies. Further, in a global marketplace, companies no longer compete against each other exclusively; rather they and their alliance partners compete against other alliances of global partners. In some instances, fierce competitors in one market find themselves working together in other markets. Managing such complex relationships requires ever-greater flexibility and a constant awareness of the competitive environment. Thus, an emerging key to competitive success is to put together the best team of alliance partners possible, regardless of geographic or channel position.

After all, the supply chain is only as competitive as its weakest link.

To summarize, "As the economy changes, as competition becomes more global, it's no longer company vs. company but supply chain vs. supply chain" (Henkoff, 1994). Thus, meeting the imperatives of today's global marketplace requires that firms and their managers adopt a supply chain management mentality.

STRATEGIC PERSPECTIVE: The primary objective of supply chain management is the elimination of barriers that inhibit communication and cooperation among different members of the entire supply chain. To eliminate these interorganizational barriers, managers must understand and manage the flow of goods and information from the initial source of raw materials all the way to the final customer. The terminology commonly used to describe supply chain management is, "the management of the value-added process from the suppliers' supplier to the customers' customer." Given the difficulties that most firms encounter in their efforts to mitigate the adverse effects of functional barriers that are entirely within the firm, the challenge of supply chain integration is daunting. To achieve synergies among supply chain members companies must find answers to many questions, including:

- Why should I be concerned about how somebody else does business?
- Can we really trust the other supply chain members not to take advantage of us?
- How is our role going to change in the new, integrated supply chain environment?
- Who are the best partners to align our competitive efforts with?
- How many different supply chains can we work with effectively?

Most firms initially lack the answers to these questions, and struggle with the very notion of supply chain management (Elliff, 1996). Many view supply chain integration as a serious threat to independence. Some even view supply chain management as the latest effort of larger companies in the supply chain to squeeze smaller firms' contribution margin and autonomy than before. Despite these serious reservations, and the many unanswered questions, today's changing competitive environment has left most managers feeling that they have no other legitimate options than to participate in integrated supply chain management programs. The fact that key customers request participation while serious competitors are willing to enter into such integrated channel alliances provides strong motivation for adopting a supply chain management perspective. Besides, the competitive improvements that emerge from well-designed and carefully executed supply chain integration are attractive, and create considerable motivation in their own right.

IMPLEMENTATION: The foundation of effective supply chain management is put into place by top management–only top management can provide the direction and the resources needed to build strong supply chains that possess distinctive capabilities (Stalk *et al.*, 1992). Without support from the highest managerial levels, the right people within the firm and across firm boundaries would never come together in a cooperative manner. Without their support the mechanisms needed for successful integration cannot be established. In fact, managerial support is critical to the implementation and management of each of the four primary integrative mechanisms: cross-boundary integration via process change, information systems support, performance measurement, and alliance management systems. Each of these implementation facilitators will be briefly discussed below.

1. ***Process Change:*** Process change has become a core element of competitive strategies of most firms. The focus of many process change activities is on the integration of value-added activities that occur within the firm and throughout the supply chain. Heyer and Lee (1992) noted that greater coordination among operating departments is a critical outcome of process change. Greater supply chain integration has also been identified as an important outcome of process change (Lee and Billington, 1992). Ultimately, most firms believe that some degree of process integration is a prerequisite to competitive success. However, making the transition from reactive, cost-driven relationships to proactive, customer-oriented cooperation requires strong, sustained and targeted emphasis (Hammer, 1990). Several characteristics are identified as important to supply-chain integration initiatives:
   - An increased emphasis on customer input in strategic planning. It is not uncommon today for senior executives to spend upwards of 20% of their time working directly with customers.
   - A high degree of consistency among interdepartmental and interorganizational operating goals (St. John and Young, 1991).
   - A willingness to share information across boundaries. Interestingly, while many firms exhibit a high degree of willingness to share information, they often lack the technological ability to effectively share information.

Bowersox *et al.* (1995) describe this technology compatibility as connectivity.

- The co-locating of employees among supply chain members.
- A greater focus on formal rules and procedures to guide process integration.

A tremendous amount of rhetoric exists regarding these integrative characteristics; yet companies are slow to implement them (Bleakley, 1995).

2. ***Information Capability:*** Information systems play an important role in supply chain management because they link the diverse and often geographically dispersed members of the supply chain. Information is substituted for inventory throughout the supply chain, and is the key to postponement, continuous replenishment, and other time-based competitive strategies (McGrath and Hoole, 1992; Bleakley, 1995). Recent research has shown a wide disparity in the effective use of information system capabilities (Fawcett and Clinton, 1996). Even so, the use of electronic data interchange has become the standard operating procedure for world-class firms. Leading firms have also been more successful at integrating their information applications. Despite this progress in building enhanced information capabilities, much work remains. In fact, while a large majority of leading firms have made significant investments in new information technologies in an effort to keep pace with customer demands and competitive threats, they continue to believe that their information systems are inadequate. Part of the challenge is that information demands in terms of accuracy and timeliness are escalating rapidly in today's information driven and time sensitive marketplace. Thus, continued emphasis and investment is required to develop information systems that can satisfactorily meet the requirements of successful supply chain integration.
3. ***Performance Measurement:*** For the supply chain to establish the right strategic orientation, promote integration among the channel members, and achieve the necessary continuous process renewal, an appropriate performance measurement system must be put in place. Performance measurement's impact on supply chain integration is pervasive since it: 1) provides insight into the real needs of important customers; 2) yields understanding regarding the value-added capability of supply chain members; 3) influences behavior throughout the supply chain; and 4) provides information regarding the results of supply chain activities (Clinton *et al.*, 1996; Kaplan, 1991). In effect, performance measures direct the design of the supply chain as well as assist in monitoring the integration and day-to-day management of supply chain activities.

   In recent years, almost all firms have placed substantial emphasis on improving their measurement systems. Leading firms in particular have expended great efforts to develop measurement systems as a strategic facilitator and supply chain integrator. These firms have truly recognized that without proper alignment among key measures, supply chain activities will not yield the desired product quality, service and competitive advantage. As a result, world-class firms are seeking higher levels of customer input in the measurement of performance, while at the same time using techniques such as performance scorecards to better align supply chain members. Total costing and activity-based costing are also popular integrative tools. Overall, an increasing number of today's firms appear to recognize the importance of performance measurement and have actively sought to enhance their ability to use measurement to direct and mold their supply chain integration efforts.
4. ***Alliance Management:*** Alliance management is the final issue that has received considerable attention in recent years as a key facilitator of supply chain integration. Indeed, alliance management is the essence of supply chain integration. However, in supply chain management the traditional dyadic management of alliances must be successfully extended throughout the entire supply chain. This extension requires the application of best alliance practice and technique. Four practices common to supply chain alliances are: (1) the use of written contracts; (2) the use of clear guidelines and procedures for selecting alliance partners; (3) the use of clear guidelines and procedures for monitoring alliances: and (4) the sharing of risks and rewards.

   Recent research has found that establishing alliances on the principle of shared rewards and risks is perhaps the single most important key to success (Schonberger, 1986; Bowersox *et al.*, 1995). World-class firms are particularly adept at sharing both risks and rewards that emerge from collaboration. These high-performing firms also tend to be active in establishing formal guidelines and procedures for creating and managing alliances. Interestingly, recent research has found that techniques that promote the formalization of alliance management remain relatively little used (Fawcett and Clinton, 1996). This finding suggests that while alliances have been touted as vital for competitive success, most firms have not determined the best way to structure and manage important supply chain relationships.

TECHNOLOGY PERSPECTIVE: Currently conceived supply chain management could not exist without recent technological advances, especially in the areas of computing power and communications. The role of information systems was outlined above. These information systems include many of the following: EDI, barcoding, satellite tracking, data warehouses, and a variety of input/output devices. Together, these technologies allow for accurate information to be collected, stored, processed, and exchanged efficiently and in a timely manner. Without adequate information exchange, supply chain coordination could not occur. It is important to note that the ability and the willingness to exchange information are not the same thing. For many firms, investments in technology have not been matched by a change in mindset to all freer exchange of information that is considered to be proprietary.

EFFECT: In theory, supply chain management enhances communication and coordination among the different members of the entire supply chain or distribution channel. Improvements in coordination lead to more efficient materials management – total supply chain inventory is reduced, cycle times are shortened, and the supply chain is more flexible and responsive. Better communications play an important role in developing innovative products and services, and reduce concept-to-market leadtimes. Role shifting has become an important element of supply chain management since it places value-added activities with the team member best suited to perform a given role based on competency rather than tradition. Of course, role shifting also represents a threat to any firm that does not possess a real and valued competency – such a firm would likely lose its role in the supply chain. When done appropriately, role shifting enhances the entire supply chain's performance.

TIMING AND LOCATION: In reality, only a few firms in any given supply chain possess the "clout" needed to initiate and promote complete supply chain integration. Other firms in the potential supply chain must work to create unique competencies and knowledge bases so that they will possess attractive attributes and skills; that is, they will be strong role players on a supply chain team. Every firm within the supply chain must constantly evaluate its value-added potential from a supply chain perspective and develop competencies that will make it both the supplier and the customer of choice. Firms that fail to add exceptional value risk being left out of key relationships.

RESULTS: Most firms struggle with the implementation of supply chain management because of numerous, inherent information-related intricacies, difficulties encountered in modifying performance measures, and obstacles to developing working relationships within and across firm boundaries. However, some firms have aggressively and judiciously implemented supply chain integration initiatives with dramatic success. At such leading firms, total cycle time reductions of 80 percent have been achieved, inventory has been cut in half, and the cost of purchased materials and services reduced by 10 percent or more. Perhaps the most important fact is that these improvements have enhanced customer service (Elliff, 1996).

CONCLUSION: A fundamental challenge to understanding and achieving effective supply chain integration is that the term is used so differently by different individuals. In many cases, the term supply chain management is used for either simple dyadic relationships between a firm and its supplier or for a slightly more complex three-company relationship that extends forward to include the firm's customers. This frequently used, but narrow, approach to supply chain management has led to a relatively meager understanding of how entire supply chains with their multiple levels and diverse participants should be structured and managed. Supply chain management is one of the most talked about competitive strategies of the 1990s.

Organizations Mentioned: Home Depot; Proctor and Gamble; Toys-R-Us; Wal-Mart.

See **Alliance management; Continuous replenishment; Customer rationalization; Functional barriers; Integrated supply chain; Japanese buyer/supplier relationship; Purchasing: Acquiring the best inputs; Purchasing: The future; Role shifting; Strategic management of suppliers; Supplier development; Supplier partnership as strategy; Supplier performance measurement; Supplier relationships; Total costing.**

References

[1] Bleakley, F. (1995). "Strange Bedfellows." *Wall Street Journal* (January 13), A1, A6.

[2] Bowersox, D., R. Calantone, S. Clinton, D. Closs, M. Cooper, C. Droge, S. Fawcett, R. Frankel, D. Frayer, E. Morash, L. Rinehart, and J. Schmitz (1995). *World Class Logistics: The Challenge of Managing Continuous Change*. Council of Logistics Mgmt, Oak Brook, IL.

[3] Clinton, S.R., D.J. Closs, M.B. Cooper, and S.E. Fawcett (1996). "New Dimensions of World Class Logistics Performance." In *Annual Conference Proceedings Council of Logistics Management*, (pp. 21–33).

[4] Elliff, S. (1996). "Supply Chain Management – New Frontier." *Traffic World* (October 21), 55.

[5] Fawcett, S. and S. Clinton (1996). "Enhancing Logistic Performance to Improve the Competitiveness of Manufacturing Organizations." *Production and Inventory Management Journal*, 37(1), 40–46.

[6] Hammer, M. (1990). "Reengineering Work: Don't Automate, Obliterate." *Harvard Business Review* (July–August), 104–131.

[7] Henkoff, R. (1994). "Delivering the Goods." *Fortune* (November 28), 64–78.

[8] Heyer, S. and R. Lee (1992). "Rewiring the Corporation." *Business Horizons*(May–June), 13–22.

[9] Jones, T.O. and W.E. Sasser, Jr. (1995). "Why Satisfied Customers Defect." *Harvard Business Review* (November–December), 88–99.

[10] Kaplan, R.S. (1991). "New Systems for Measurement and Control." *The Engineering Economist*, 36 (3), 201–218.

[11] Lee, H.L. and C. Billington (1992). "Managing Supply Chain Inventory: Pitfalls and Opportunities." *Sloan Management Review* (Spring), 65–73.

[12] McGrath, M. and R. Hoole (1992). "Manufacturing's New Economies of Scale." *Harvard Business Review* (May–June), 94–102.

[13] Schonberger, R.J. (1986). *World Class Manufacturing*. The Free Press, New York.

[14] St. John, C.H. and S.T. Young (1991). "The Strategic Consistency Between Purchasing and Production." *International Journal of Purchasing and Materials Management* (Spring), 15–20.

[15] Stalk, G., P., Evans, and L.E. Schulman (1992). "Competing on Capabilities: The New Rules of Corporate Strategy." *Harvard Business Review*, 70 (2), 57–69.

[16] Wisner, J.D. and S.E. Fawcett (1991). "Linking Firm Strategy to Operating Decisions Through Performance Measurement." *Production and Inventory Management Journal* 32(3), 5–11.

SUPPLY UNCERTAINTY

Supply uncertainty could be either quantity uncertainty or timing uncertainty. Timing uncertainty occurs when supply lead time is not fixed; the exact time of supply is uncertain. Quantity uncertainty refers to the expected quantity failing to materialize due to supplier's failure, or due to a preceding operation being unable to deliver the needed quantity due to scrap, shortage of raw materials, etc. Supply uncertainty is handled by using safety or buffer stocks.

See **Safety stocks: Luxury or necessity; Synchronous manufacturing using buffers.**

SUSTAINABLE COMPETITIVE ADVANTAGE

Competitive advantage refers to the inherent ability of a firm to meet the real needs of key customers better than the competition. Two types of competitive advantage exist: transient competitive advantages and sustainable competitive advantages. Transient advantages come from one time improvements in the firm's ability to produce higher-quality, lower-cost, or more innovative products. For example, when a company locates a facility in a low-labor-cost country, it can gain a cost advantage over its competitors. However, as soon as the competitor locates in the same country, or in a lower wage country, the labor cost advantage disappears. Likewise, when a company introduces a new and superior product, it possesses an advantage in the marketplace. Unfortunately, most new products are in the market for much less than one year before the competition comes to the market with a better product (often reverse engineered and improved). A third type of transient advantage emerges when a firm implements off-the- shelf technologies that are available to any competitor who wants to buy and implement them. Sustainable advantages result from a critical capability that allows the firm to stay ahead of competition. For example, a firm can gain a sustainable advantage from its global manufacturing if it develops the policies and procedures as well as the organizational culture that allow it to operate a seamless network, and get the most out of its global employees. The fact that most companies do not possess the drive and diplomacy to create this kind of operating culture makes this type of global manufacturing hard to replicate and therefore sustainable. Likewise, in the product development area, a firm can develop a fast-cycle capability that enables it to consistently come to the market with new products faster than the competition. Toyota can readily bring new cars to the market in 18 months and has done so in as few as 15 months (most competitors require 24 months or more).

Finally, firms that can develop their own technologies or implement off-the-shelf technologies rapidly and in an unique manner can gain a sustainable competitive advantage over their competitors. The essence of sustainable advantages is that they are both rare and hard to imitate. They cannot be purchased off-the-shelf but require time, effort, and up-front investments in both people and technologies.

Perhaps the most sustainable advantage is the ability to learn not only as individuals but as an organization. Such a capability positions a company to not only respond to change but to anticipate and even precipitate change in its industry. George Fisher, Chief Executive at Kodak, alluded to the challenge of creating this type of capability, stating, "My biggest job over the next two years is not how we develop the market, not even how we gain growth. It is how we get people to start thinking of systemic, rather than incremental change." Michael Porter has noted that the most important role of governments is to spur this type of capability development, "Governments' proper role is as a catalyst and challenger; it is to encourage – or even push – companies to raise their aspirations and move to higher levels of competitive performance, even though this procedure may be inherently unpleasant and difficult." Ultimately, sustainable advantages are one of the rarest finds in business because complacency almost always leads a firm to stop pushing, decrease learning, and become stagnant at which point the advantages disappear.

Organizations Mentioned: Kodak.

See **Core manufacturing competencies; Global manufacturing rationalization; Industrial robot selection; International manufacturing; Manufacturing strategy.**

SYNCHRONOUS MANUFACTURING USING BUFFERS

M. Michael Umble

Baylor University, Waco, Texas, USA

DESCRIPTION: Synchronous manufacturing (SM) is a less-known manufacturing management philosophy that views the resources and activities of an organization as elements of an interdependent network, and manages them in such a way as to optimize the performance of the entire system (Srikanth and Umble, 1997: 10). Not to be confused with concurrent manufacturing or concurrent engineering, SM builds upon the drum-buffer-rope logistical system (Goldratt, 1990) that emphasizes the identification and management of constraints in an attempt to maximize overall system performance with minimum levels of inventory. SM utilizes two different types of buffers – stock buffers and time buffers – to allow the plant to meet customer requirements with lower levels of inventory, shorter lead times, and less overall manufacturing capacity.

Buffers in a manufacturing process increase inventory. Thus, each buffer must serve a specific, well-conceived purpose. In SM systems, stock buffers are inventories of materials, component parts, or products used to help fill customer orders in less than normal lead time. Stock buffers are required only if manufacturing cannot otherwise respond to customer requirements – that is, if lead times are too long. In SM systems, time buffers represent the additional planned lead time allowed, beyond the minimum required setup and run times, for materials to reach a specific point in the product routing. Time buffers protect the system's throughput from the internal disruptions that are inherent in any process (Umble and Srikanth, 1997). The rationale for establishing stock and time buffers is that they constitute the only planned inventory in the entire system. All other inventories are eliminated. Thus, the total inventory required to run a SM system is often less than for other logistical systems.

BUFFERS IN MRP, JIT, AND SM SYSTEMS: Material Requirements Planning (MRP) is a computer-based information system that systematically determines material requirements and releases jobs to the shop according to a predetermined schedule. These jobs are then "pushed" from one work station to the next where they are processed according to schedule priorities. Jobs are assigned planned lead times based not only on the resources' setup and run times, but also on expected queue times. This expanded definition of planned lead time results in a self- fulfilling prophesy. For example, if all jobs have a six-week lead time, then the system will normally contain six-weeks worth of inventory and jobs will typically require six weeks to process. In most MRP systems, the typical work station inventory queue (buffer) contains multiple jobs.

JIT systems are grounded in the belief that excess inventory is a liability and should be minimized. JIT systems "pull" materials through the process only as needed to replenish materials used in downstream operations. JIT systems generally require less inventory and generate shorter lead times than MRP systems. To make the replenishment system viable, small work-in-process inventories (buffers) are maintained at each work station and materials are moved in small batches.

SM drum-buffer-rope systems are a combination of push and pull systems. Materials are "pulled" into the system only as needed to support the constraint resource schedule (the drum). But once in the system, materials are quickly "pushed" in small transfer batches through the system. Buffers are established only where specifically required to protect the system's throughput. Compared to MRP and JIT systems, SM systems maintain inventories at a relatively few buffer locations. Figure 1 illustrates the push and/or pull nature of MRP, JIT, and SM systems and compares the locations and relative sizes of inventory (buffers) for each system. The size of the buffer containers in the figure represent the relative amounts of planned inventory generally found in each system.

Categories of Stock Buffers:

Finished-Goods Stock Buffer: Finished-goods stock buffers are the most common category of stock buffer. They are needed when products are sold "off the shelf" or when manufacturing lead time is so long that it is extremely difficult or impossible to manufacture products in time to meet customer requirements, even if starting from partially processed materials.

Raw-Material Stock Buffer: Raw-material (or purchased-component) stock buffers are used to ensure that materials are available when needed. This category of stock buffer is especially important when vendor reliability is poor or when manufacturing lead time is relatively long and does not allow sufficient lead time for the vendor to supply parts in order to meet promised ship dates.

Work-in-Process Stock Buffer: Work-in-process stock buffers consist of inventories of partially processed materials, and are appropriate when normal manufacturing lead time is longer than customers will tolerate. There are two categories of work-in-process stock buffers. In processing plants, product stock buffers are composed of partially processed materials that can be fully processed into a finished product. In assembly plants, component stock buffers contain partially processed components that are assembled with other components to form finished products.

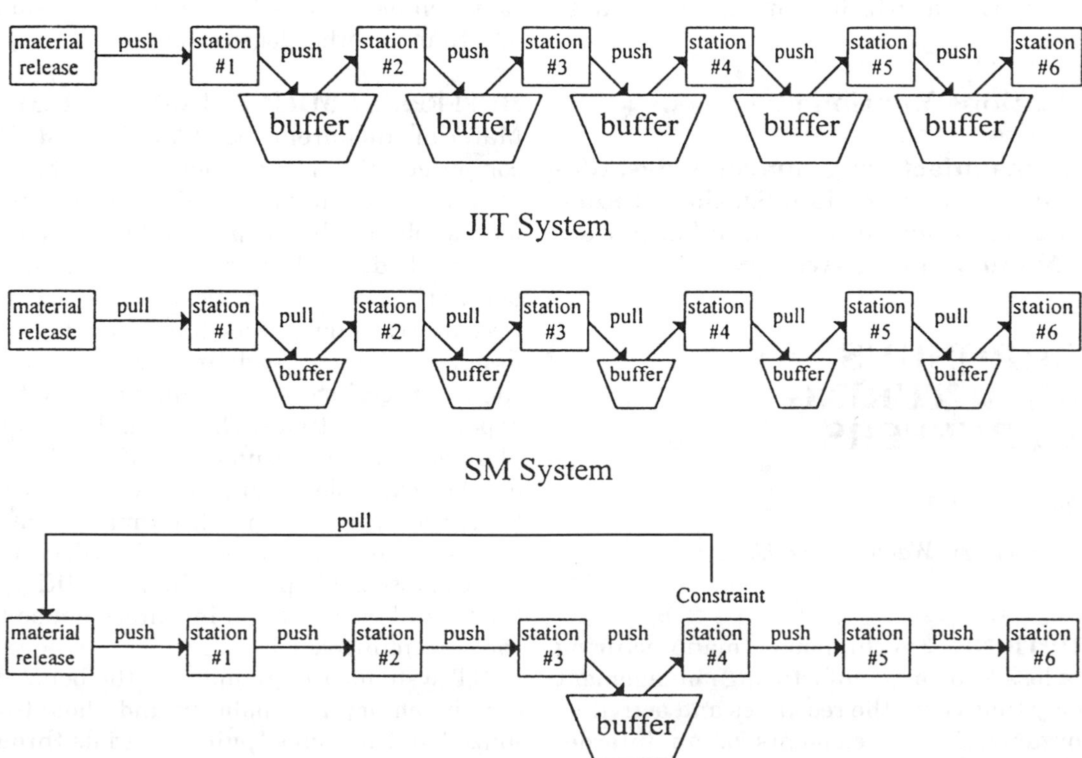

Fig. 1. A comparison of buffers in MRP, JIT, and SM systems.

CATEGORIES OF TIME BUFFERS:

Constraint Time Buffer: When a manufacturing process has a capacity constraint, a constraint time buffer is needed to maintain the integrity of the schedule and protect the constraint resource from starvation. Monitoring the arrival of material at the constraint buffer also keeps managers informed of significant disruptions and their causes, so corrective action to improve the process can be taken.

Assembly Time Buffer: An assembly time buffer is needed at assembly operations for parts that are not processed by a constraint, but are assembled with one or more constraint-processed parts. This buffer protects the assembly operation from disruptions that affect non-constraint parts and ensures that assembly operations can be performed as soon as the constraint-processed parts are available. The assembly and constraint time buffers work together to keep assembly operations on schedule.

Stock Replenishment Time Buffer: Stock buffers should be treated like "internal customers" so that production schedules and customer requirements can be consistently satisfied. Stock replenishment time buffers are used to ensure the timely flow of replenishment material to stock buffers.

Shipping Time Buffer: For products that do not require a finished-goods stock buffer, a shipping time buffer is needed at the end of the process to ensure on-time deliveries and monitor system performance. This time cushion protects shipping due dates from disruptions that occur beyond capacity constraints, at assembly operations, and at final preparation and packaging operations. For example, if an order is scheduled to ship on Friday at 5:00 P.M., and there is a two-day shipping buffer, then final assembly and packaging would be scheduled to be completed at 5:00 P.M. Wednesday.

GUIDELINES FOR ESTABLISHING BUFFERS:

Stock Buffers: The location and size of stock buffers may vary greatly by product and should be designed according to the customer service requirements of specific products. Marketing/sales personnel should help determine whether finished goods and/or work-in-process stock buffers are actually needed for each product. Generally, stock buffers should include only fairly standardized items with high demand. This is especially true for finished-goods stock buffers. In the case of work-in-process stock buffers, the production stage at which the buffer is held must be selected with care. Ideally, it should be at a point where some of the early manufacturing operations can be completed without causing the material to become overly or irreversibly differentiated. The size of a stock buffer must be sufficient to maintain the required level of customer service without carrying excessive stock. This is primarily determined by the product demand rate and the time required to replenish the buffer.

Time Buffers: Time buffers are used to protect critical aspects of manufacturing operations. They are implemented by releasing material from one point in the operation earlier than theoretically needed to arrive at a designated point in the routing. The planned lead time for determining when materials should be released is the sum of the individual operations setup and run times plus the time buffer. For example, suppose that a job has six operations in its routing prior to arriving at a constraint resource. If the sum of setup and run times for the six operations is 10 hours, the earliest that material could arrive at the constraint is 10 hours after it is released. A time buffer helps to ensure that, despite normal processing problems and delays, material will be available at the constraint when it is scheduled to be run. In this case, a time buffer of 15 hours means that material will be released 25 hours before the constraint is scheduled to process that material.

In general, the longer the routing and the greater the frequency and magnitude of delays, the larger should be the size of the time buffer. But the proper sizing of a time buffer is an iterative process that is affected by numerous variables and cannot be optimized by formula. However, a general rule of thumb is to initially set the buffer equal to half of the total lead time stripped from the individual operations. The process is then monitored. If jobs routinely arrive too early or too late at the buffer location, then the buffer size can be reduced or increased. Process improvements that reduce the frequency and magnitude of delays should receive top priority since this allows the time buffer size to be reduced. This, in turn, leads to reductions in lead time and inventory level.

PLANT CLASSIFICATION AND BUFFER REQUIREMENTS: The specific categories of stock and time buffers that are appropriate for a manufacturing process vary greatly from one plant to another. However, based on specific plant characteristics, competitive capabilities, and customer requirements, some generalizations about buffer implementation procedures can be made for plants known as V-plants, A-plants, and T-plants (Umble and Srikanth, 1997).

V-Plants

Processing plants, such as steel rolling mills and textile plants, that convert a small number of materials into a large number of distinct end items are known as V-plants. V-plants generally process all products through the same sequence of operations and are capital intensive. Setup times in V-plants are often quite long, causing relatively long production runs and lengthy manufacturing lead times.

If manufacturing lead time are too long, finished-goods and/or work-in-process product stock buffers are appropriate. Stock replenishment time buffers also must be established to ensure that these stock buffers are replenished in a timely manner. Any product that is not held in finished-goods stock buffers should be protected by a shipping time buffer to make sure that customer orders are shipped on schedule. Finally, V-plants often have a capacity constraint which must be protected with a constraint time buffer.

A-Plants

Plants, such as a General Electric jet engine plant, that assemble a large number of manufactured parts into a relatively small number of distinct end items are called A-plants. In typical A-plants, equipment is quite flexible, components are unique to specific end items, most components have highly dissimilar production routings, and lead times are quite long.

Except in cases of high demand for specific products, finished-goods stock buffers are generally not appropriate for A-plants. Furthermore, work-in-process component stock buffers are not generally used to try to improve market responsiveness because a large number of parts is needed to produce each unique end item, specific component parts are not common to multiple end items, and the volume of demand does not justify carrying the needed parts. However, A-plants often have capacity constraints. In such cases, both capacity constraint and assembly time buffers are necessary to ensure the timely release and control of constraint-processed and non-constraint-processed parts. Shipping time buffers are always needed to protect promised delivery dates.

T-Plants

The critical feature of T-plants is that a very large number of different end items are assembled from a limited number of common components. While production volume in A-plants is usually quite low, it is typically very high in T-plants. Plants that manufacture automobiles, computers, and most household appliances are common examples of T-plants.

Off-the-shelf demand is common in T-plants. But finished-goods stock buffers should be relatively small because carrying large inventories of every configuration is expensive. Instead, work-in-process inventory should be held at component stock buffers located at final-assembly component stores. These buffers contain all components needed at final assembly to replenish finished-goods stock buffers and satisfy customer orders. Stock-replenishment time buffers support the component and finished-goods stock buffers, while shipping buffers protect individual customer orders. True capacity constraints are not typically found in T-plants, so constraint and assembly time buffers are not usually needed.

CASE: To illustrate the proper use of stock and time buffers, an actual Synchronous Manufacturing case is presented. The details of the implementation are disclosed, but the company's identity is withheld.

The Plant Environment

XYZ Textile is a single plant in a large company with several operating divisions. XYZ is a typical V-plant, where material is transformed into a large variety of drapery and upholstery fabrics. The basic production steps are yarn prep, dyeing, weaving, finishing, and cutting/sewing. At yarn prep, cotton and rayon are combined in different percentages to yield a variety of yarn materials. At dyeing, the yarn can be dyed a wide variety of colors and shades. At weaving, yarn is loaded onto looms where it is woven into a number of different fabric types and patterns. At finishing, the fabric is fitted with various thermal backing materials and sprayed with chemicals for durability and appearance. Finally, at cutting/sewing, the finished material is cut and sewn into a variety of sizes. Major product differentiation occurs at the dyeing, weaving, and cutting/sewing operations.

Production and scheduling complexity is created by several factors. (1) There are several thousand distinct end items. (2) Demand for the various products varies greatly. (3) Since all products are processed through the same sequence of operations, products compete for the same resources. (4) There are numerous pieces of equipment for each resource. For example, there are over 300 weaving looms. (5) Setups are generally lengthy (especially at the looms). This complexity created the perception that production could not be scheduled to satisfy customer demand on an order-by-order basis. So XYZ attempted to fill customer orders from finished-goods inventory. But frequent design changes made this a risky and expensive strategy. And despite large inventories, XYZ suffered constant stockouts and poor on-time delivery performance. Increasingly intense competition eventually dictated a change in strategy.

At XYZ, a batch of several thousand yards of fabric could be fully processed through the plant in a few days. Yet the planned lead time was 18 weeks. Each resource had a planned lead time of at least two weeks that included setup, run, and queue times. But even these planned lead times were often inadequate, so additional safety time was added in a futile attempt to maintain the integrity of the schedule. Furthermore, eight weeks' worth of inventory for each product was held in finished goods. And, in an attempt to keep finished-goods inventory well stocked, 10-weeks' worth of work-in-process inventory was kept prior to the cutting/sewing operation. Between the stock buffers and the planned lead time inventory, XYZ maintained about 36 week's worth of inventory.

Policies and performance measures were designed to encourage the efficiency of individual resources. For example, because of the relatively lengthy changeovers, XYZ had a minimum batch size policy of 10,000 yards even though typical order quantities were significantly less than 10,000 yards. This only aggravated the excess inventory and long lead time problems.

Implementing a SM System

Implementing a SM system with appropriate stock and time buffers requires a thorough understanding of the constraints and customer requirements. The plant's primary constraint was the weaving operation, which ran 24 hours a day, while the other resources operated only partial second or third shifts. Another key piece of information was that less than 10 percent of all end items were sold "off the shelf." The remaining 90 percent had delivery due dates of about four weeks.

The first order of business was to start with a clean slate and strategically determine the location and size of stock buffers needed to efficiently satisfy customer requirements. As a general rule, it is best to start with finished-goods stock buffers and work back toward the beginning of the process.

A finished-goods stock buffer was established to service the ten percent of products ordered off-the-shelf. Finished-goods inventory for all other products was eliminated. It was decided that the size of the finished-goods stock buffer for off-the-shelf items should be less than the previous ten weeks. The new size for this buffer would be dependent upon how quickly the buffer stocks could be replenished. It was decided to establish a work-in-process stock buffer prior to the cutting/sewing operation to service all products because that all products could easily be processed through the final cutting/sewing and finishing operations in less than four weeks. This allowed the plant to service the remaining 90 percent of orders within the prescribed four-week lead time. It also mandated that the finished-goods stock buffer for products ordered off-the-shelf need not be larger than four weeks.

The fact that cutting/sewing is a major product differentiation point reinforces it as a location for a work-in-process stock buffer. The cutting/sewing stock buffer was set at six weeks because it was determined that establishing an additional stock buffer prior to the weaving operation would allow all material to be fed to the cutting/sewing stock buffer within six weeks. Weaving was selected since it is a major divergence point and the constraint. A stock buffer at weaving protects and increases scheduling flexibility at the constraint. The weaving stock buffer was set at four weeks, since it was determined that all released materials could easily be processed through yarn prep and dyeing within four weeks.

Once the stock buffers were set, the planned lead times for all operations were stripped down to the actual setup and run times and time buffers were added as necessary to protect due-date performance and ensure the timely replenishment of the stock buffers. The stripped down lead time (based only on changeover times and run times for smaller standard batches) was one day for every operation except weaving, which was two days. Three time buffers were needed to protect the system, mostly to allow time for jobs waiting in queue at the various resources. It was decided that a two-week shipping time buffer allowed sufficient time for materials at cutting/sewing to be scheduled and processed to replenish the finished-goods stock buffer or to be shipped to customers. A four-week stock replenishment time buffer protected materials flowing from weaving to cutting/sewing. And a three-week stock replenishment time buffer allowed for the timely scheduling of materials from yarn prep to weaving.

As the SM system was implemented, a number of key changes reinforced the successful implementation of the buffers. For example, batch sizes were drastically reduced. This helped to drain work-in-process inventory from the system, reduced priority distortions, and virtually eliminated expediting. After the initial implementation phase, highly focused process improvement programs were implemented that were designed to reduce the production delays and disruptions that mandated the need for large stock and time buffers. For example, a setup reduction program significantly reduced setup time at the constraint (weaving looms). This increased scheduling flexibility at the looms and generated extra loom processing capacity. In addition, the further reduction of batch sizes continued to generate a faster and smoother flow of materials throughout the process.

SM Implementation Results

The initial implementation took six months to complete. The total manufacturing lead time – consisting of the individual revised planned lead times plus the time buffers – was ten weeks down from the previous 18 weeks. Moreover, total system inventory – composed of the sum of the time buffers, the work-in-process stock buffers, and a prorated proportion of the finished-goods-stock buffer – was 20 weeks' worth, instead of the previous 36 weeks' worth of inventory.

In the nine months following the initial implementation, the structure of the stock and time buffers radically changed. Due to greatly reduced manufacturing lead times and speedy replenishment times, only two stock buffers and two time buffers were needed. The stock and time buffers at cutting/sewing were eliminated. As system improvements occurred, managers monitored the remaining buffers to determine when and by how much each specific buffer could be reduced. Little by little, each buffer was shrunk. The finished-goods stock buffer for off-the-shelf items was reduced to 20 days and the work-in- process stock buffer at weaving was cut to 12 days. A 10-day time buffer was sufficient to protect the finished-goods stock buffer and customer orders, while a seven-day time buffer was sufficient to protect the weaving stock buffer. The resulting planned lead time was 23 days (six days for setup and run time plus 17 days for time buffers), while total system-wide inventory was limited to 37 days' worth.

In 15 months, XYZ Textile reduced inventory from 36 weeks to 37 days, shortened production lead times from 18 weeks to 23 days, and generated extra capacity at the looms that allowed a larger production and sales volume. This was all accomplished with minimal capital investment. Needless to say, the company's bottom-line financial performance improved sharply.

See **Batch size; Finished goods buffer; Inventory flow analysis; Manufacturing lead time; MRP; Off-the-shelf demand; Order release; Product differentiation; Production run; Safety stocks: Luxury or necessity; Safety time; Setup reduction; Schedule stability; Stock buffer; Stockouts; Time buffer; Theory of constraints in manufacturing management.**

References

[1] Goldratt, E.M. (1990). *The Haystack Syndrome: Sifting Information Out of the Data Ocean*, North River Press, Croton-On-Hudson, New York.

[2] Srikanth, M.L. and M.M. Umble (1997). *Synchronous Management: Profit-Based Manufacturing For The 21st Century – Volume I*, Spectrum Publishing, Guilford, Connecticut.

[3] Umble, M.M. and M.L. Srikanth (1997). *Synchronous Management: Profit-Based Manufacturing For The 21st Century – Volume II, Implementation Issues and Case Studies*, Spectrum Publishing, Guilford, Connecticut.

SYSTEM CORRECTNESS

A correct system design is one in which the system only enters states that are desirable and does not enter states which are undesirable. System correctness is a logical feature of the system design (i.e., it does not refer to system performance issues such as throughput, makespan, and so on). In manufacturing systems, we wish to design systems that are deadlock free, do not have buffer overflows, and do not violate any precedence relationships (e.g., the production steps). (See Desrochers and Al-Jaar, 1995.)

See **Manufacturing systems modeling using petri nets.**

Reference

[1] Desrochers, A.A., and Al-Jaar, R.Y., 1995, *Applications of Petri Nets in Manufacturing Systems: Modeling, Control, and Performance Analysis.* (Piscataway, NJ: IEEE Press).

SYSTEM DYNAMICS

System dynamics studies the behavior of a systems through time. Its earliest roots go way back in physics and mechanics to the study of pendulums and the motion of the planets. Its modern history is often traced to the late 19th and early 20th

centuries in the work of James Maxwell, Henri Poincaré and others (Abraham & Shaw, 1992 provide a fine history and overview that's not too technical; Kaplan & Glass, 1995 are both briefer and more rigorous). Classical dynamics assumed linear system interactions and predictable system behavior as a whole. Modern system dynamics addresses more complex behaviors including regular and irregular cycles, chaos (non- repeating behavior), and self-organization (discontinuous or bifurcating behavior).

See **Manufacturing analysis using Chaos, Fractals, and self-organization; Simulation analysis of manufacturing and logistics systems; Simulation languages; Systems, models and simulation.**

References

[1] Abraham, R.H. & C.D. Shaw. 1992. *Dynamics: The Geometry of Behavior*, 2nd ed. Addison-Wesley, Redwood City, CA.

[2] Kaplan, D. & L. Glass. 1995. *Understanding Nonlinear Dynamics*. Springer Verlag, New York.

SYSTEM OF PROFOUND KNOWLEDGE

See **Quality: The implications of deming's approach.**

SYSTEM'S CONSTRAINT

According to the Theory of Constraints (TOC), every system has a constraint that limits the output of the production system. Organizations using TOC as a process management tool will identify and emphasize the system's constraint in establishing production routine and priorities.

See **Synchronous manufacturing using buffers; Theory of constraints in manufacturing.**

SYSTEM-IMPOSED PACING

Some advanced manufacturing technologies determine the beginning of the work cycle as well as its length. The operator has no ability to control the start or finish of the operation. In this case the machine paces the operator.

See **Human resource issues and advanced manufacturing technology; Human resource issues in manufacturing.**

SYSTEMS, MODELS, AND SIMULATION

A *system* is defined as a collection of entities that act or interact together toward the accomplishment of some logical end (Law and Kelton, 1991). We define a system by its purpose. Thus, we speak of a *communications system, a health care system, a production system, etc*. Using a functional definition helps us avoid thinking of a system as a fixed structure. Consequently, we can view a system in terms of how it ought to function rather than how it has traditionally worked in the past.

Entities making up a system may be either physical or mathematical. A physical entity may be a patient in a hospital or a part in a factory; a mathematical entity might be a variable in an equation. In practice, what is meant by "the system" depends on the objectives of a particular study. The collection of entities that compose a system for one study might be only a subset of the overall system for another.

We study a system to gain some insight into the relationships among its various components or to predict its performance under some new conditions. If it is feasible to alter the system physically and observe its behavior under the new conditions, it would be a valid study. However, it is rarely feasible to do this. Therefore, it is usually necessary to build a *model* as a representation of the system, and study it as a surrogate for the actual system. In certain cases, it may be feasible to build *physical* or *iconic* models such as clay cars in wind tunnels. The vast majority of models built to study engineering and management systems, however, are *mathematical* models, representing the system in terms of logical and quantitative relationships.

Once a mathematical model is built, it can be manipulated to see how the model reacts, and thus learn how the system would react. If the model is sufficiently simple, it may be possible to obtain an exact *analytical* solution. Perhaps the simplest example of an analytical model is the simplest relationship between distance traveled, d, the rate of travel, r, and the time spent traveling, t. Hence, $d = rt$. For instance, if we know the distance to be traveled and the velocity, then we can manipulate the model to get $t = d/r$ as the time that will be required. This is a simple closed-form solution. Some analytical solutions can be extraordinarily complex, requiring vast computing resources; inverting a large nonsparse matrix is a well-known example of a situation in which there is an analytical formula in principle, but obtaining it numerically in a given instance is far from trivial.

If an analytical solution to a mathematical model is available and is computationally efficient, it is desirable to study the model *analytically*. Otherwise, the model must be studied by means of simulation, that is, numerically exercising the model for the inputs in question to see how they affect the output measures of performance. It is also possible to construct *hybrid* (analytical-simulation) models, where the closed-form solution obtained from one of the model components can be used to drive the simulation of the rest of the model. On the one hand, analytical models may suggest the general form of the relationships between system performance and its parameters; this information can be used in developing and validating the simulation model. On the other hand, it is often necessary to use approximations in analytical models; simulation models can be used to test the robustness of these approximations. (Contributed by Yucesan and Fowler; see Simulation Analysis of Manufacturing and Logistics Systems.)

See **Fractals and self organizations; Manufacturing analysis using chaos; Manufacturing systems modeling using petri nets; Simulation; Simulation analysis of manufacturing and logistics systems; Simulation of production problems using spreadsheet programs.**

TACIT SKILLS OF OPERATORS

Tacit skills involve learning to avoid mistakes, and imperfections. Tacit skills are based on implicit knowledge, which is not directly verbalizeable or communicable to others. Implicit knowledge appears to be more salient in complex and uncertain environments. Its development is gained through active involvement in system control rather than through formal training. Tacit skills are similar to a "sixth sense" regarding future mishaps in machine/system operations, their consequences, and the actions needed to prevent them.

See **Human resource issues and advanced manufacturing technology; Human resource issues in operations management.**

TAGUCHI METHOD

It is an off-line quality control method developed by Grenichi Taguchi. It is used in the product and process design stages in the product development cycle. His methods are appropriate for the three phases of product design: system design, parameter design, and tolerance design.

See **Total quality management.**

TAILORED CUSTOMIZATION

In tailored customization, a basic product design is modified to the customer's needs or specifications. The needs of the customer has no effect in the design stage, as products are not designed from scratch. These products are not designed for a specific product, but are altered for each customer at the fabrication stage. Eyeglasses are a good example of tailored customization. Each customer has a unique prescription to correct his/her vision. The eyeglass frames are selected from an assortment of previously designed frames, but the lenses are ground to the unique prescription of the customer. Eyeglasses are tailored to the unique customer, but not designed for the individual; a case of tailored customization.

See **Mass customization.**

TAKT TIME

See **Leveling, balancing and synchronizing the lean manufacturing system.**

TARGET COSTING

Ramachandran Ramanan

University of Notre Dame, Notre Dame, IN, USA

DESCRIPTION: Target costing is the concept of *price-based costing* instead of *cost-based pricing.* A **target price** is the estimated price for a product or service that potential customers will be willing to pay. A target cost is the estimated long-run cost of a product or service that allows the firm to achieve a targeted profit. **Target cost** is derived by subtracting the target profit from the target price.

Target costing is widely used. For example, Mercedes and Toyota in the automobile industry, Panasonic and Sharp in the electronic industry, and Apple and Toshiba in the personal computer industry use target costing (Maher, Stickney and Weil, 1997). This approach is quite different from standard costing. Target costing begins with identifying customers' needs and estimating an acceptable sales price for the product. Working backwards, the full cost of the product is established so as to earn an estimated profit. Target costs serve as goals for research and development, design, production personnel and other departments in the value chain. If the target cost cannot be achieved, management must take a close look at the viability of making the proposed product.

HISTORICAL PERSPECTIVE: In the late 1960's, as personal income in Japan increased rapidly, people divesified their tastes. As consumer needs became diversified, product diversification came about. As a result of these changes, Japanese companies had to develop and produce numerous products with quite different characteristics. An improvement in production methods from mass production was needed to meet these diverse social needs. In particular, the rapid progress of industrial robots and factory automation had made it possible to produce multiple products at low cost. Sakurai (1989) defines factory automation

Exhibit 1.

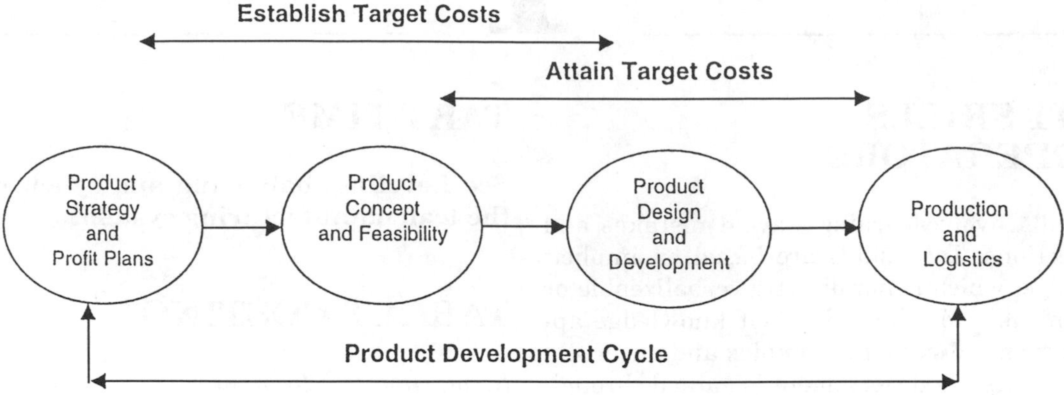

to include *computer-aided design, computer-aided manufacturing, flexible manufacturing systems,* and *office automation*. In this article, Sakurai provides an excellent description of the developments in Japan and the applicability of target costing to various settings.

In addition, product life cycles became shorter as a result of more diversified needs by customers. Therefore, the development, planning, and design phases of a product became critical to cost management. Target costing emerged as a management tool to plan needed features or functions of a product at a cost that would permit the firm to earn a targeted profit. This method has become prevalent in the United States only since the late 80's. The loss of market share to Japanese companies has been a major motivation for the U.S. companies adopting target costing.

STRATEGIC PERSPECTIVE: This approach to "cost management" or "cost planning" over the life of a new product is important because the global manufacturing environment has become very competitive. The increase of sales price has to be evaluated carefully by management in several key industries, taking into account potential losses in market share to competitors offering lower prices. Therefore, in order to be viable, there is a great need for controlling of costs. Further, in a rapidly changing business environment, it becomes difficult to use just one factor, such as quality, to sustain a competitive advantage. The prevailing environment is unforgiving of poor quality, mistakes or delay, and is very demanding since consumers today are better informed about products.

Target costing, as a management initiative, responds to this environment by anticipating costs before they are incurred. It focuses externally on customer requirements and competitive threat, and attempts to systematically link an organization to its suppliers, dealers, customers, and recyclers in a cohesive and integrated profit and cost planning system. Using this approach, firms attempt to continually improve product and process designs at optimal costs.

IMPLEMENTATION: Target costing relies on product design for cost reduction to meet a specified profit, that is, it is applied primarily to new product development efforts. The relationship of target costing activities to the product development cycle, as developed in Ansari *et al.* (1997) is shown in Exhibit 1. Typically, target costing occurs in two stages that correspond roughly to the first and second halves of product development cycle. First is the establishment phase that involves establishing a target cost during the product planning and development stages. The attainment phase involves achieving a target cost and takes place during the design development and production stages of the cycle. The following seven major activities are usually carried out by companies during the establishment phase (see Exhibit 2):

1. Market research is carried out to gain information about unmet needs and wants of customers. This research defines the market and product niche to be exploited.
2. Competitor analysis is done to determine how customers evaluate competitor products, and how they might react to a company's new product introductions.
3. A customer or product niche is then defined by analyzing market and competitor information to decide what particular customer requirements to target.
4. Customer requirements are fine-tuned by introducing an initial product concept and asking customers for their reaction. The products are then designed to meet customer requirements.
5. Product features are defined by setting specific requirements for the features the product will have and the level of performance for those features.

Exhibit 2.

Establishing Target Costs

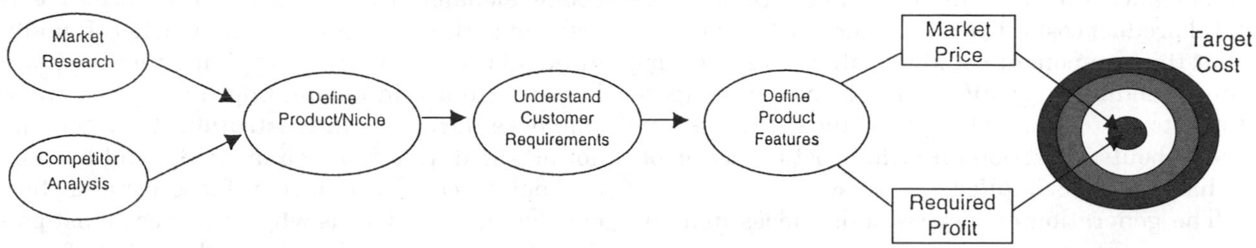

6. A market price is established that is acceptable to customers without affecting the firm's competitive position.
7. The firm then decides on the required profit to be made on the product. This will leave the target, or allowable, cost.

The efforts within the firm to actually attain target costs during the attainment phase can be described by the following three steps:

1. The cost gap is computed. This is the difference between the current estimate of the cost of producing something based on current cost factors or models, and the allowable cost that has been established.
2. The next step is to design costs out of a product. This means reduction of cost through product design, and is the most critical step in attaining target costs.
3. The design is released and the company gets on the continuous improvement path. This is an attempt to reducing costs beyond that which is possible through design alone, and is achieved by improving production yields and eliminating wastes along the way.

TECHNOLOGY PERSPECTIVE: Narrowing the cost gap can be an extremely difficult process though. The initial concept design is the beginning of the cost planning process. Subsequently, the product and the processes are concurrently designed. Common tools used at this stage are Computer Aided Design and Computer Aided Manufacturing. Cost parameters may be built into these computer models so cost impact of changes in design can be simulated concurrently.

Next comes the performance of cost analysis. This involves identifying cost targets for major sub-components and parts of a product as well as cost reduction goals for selected components. The activities include the development of a list of the components, the function(s) of these components and their current estimated costs. The next step is to do a functional cost breakdown. This involves identifying the cost of each of the functions of the product.

The manufacturer's view of the product as a sum of functions needs to be reconciled with the customer's view of the product as a set of important features. A tool called Quality Function Deployment (QFD, borrowed from the quality literature) is used to systematically arrange information about features, functions, and competitive evaluation. This tool highlights the relationships between customer requirements and design parameters. Moreover, a concerted effort is made to include information about how customers evaluate competitor offering on these same features. QFD provides information that allows a manager to convert product feature rankings into functional or component rankings. This analysis provides useful insights that facilitate systematic trade-offs in the design of the various components using value engineering.

VALUE ENGINEERING: Value engineering, which is at the core of target costing, is an organized effort directed at analyzing the functions of the various components of the product for the purpose of achieving these functions at the lowest overall cost without reductions in requires performance, reliability, maintainability, quality, safety, recyclability, and usability. It analyzes product and manufacturing process design and reduces cost by simplifying both.

The first step of value engineering is identifying components for cost reduction. A Value Index is typically computed in order to select components for cost reduction or for enhancements. The value index is the ratio of the degree of importance to the customer (expressed as a percentage) and the cost of each component (also expressed as a percentage of the total product cost). Components with a value index of less than one are targeted for cost reduction while those with a value index of greater than one are targeted for enhancements.

To illustrate this, consider two components A and B. Component A has a relative importance of 35% to customers and costs 46% of the total product cost. The value index is 0.76 and this component is a prime candidate for cost reduction efforts, since it is costing too much relative to its importance to customers. Component B has a relative importance of 10% to customers and it costs 4% of total product costs. The value index of 2.5 suggests that this component could be enhanced since current spending is not sufficient relative to its importance to customers. When the value index is 1.0 or thereabouts, no action for either cost reduction or enhancement is implied.

The generation of cost reduction ideas usually is a brainstorming exercise about what can be reduced, eliminated, combined, substituted, rearranged, or enhanced to provide the same level of functionality from a component at less cost. These ideas are evaluated on technical feasibility and acceptability to customers, prior to implementation.

Estimating achievable cost is an activity that takes place at every iteration of the design and production cycle. Estimates of achievable cost should become more and more accurate as the process moves on and the company gains experience with the product. This not only includes manufacturing costs but also distribution, marketing, and support costs.

EFFECT ON PERFORMANCE: Target costing has definite effects on the performance of employees from different functional areas. Target costing requires all the key players, such as design engineers, process planning engineers, marketing personnel, and management accountants to have early involvement with a product. They are required to provide good estimates for each design iteration, and they must look ahead and provide information about products from incomplete design data. They must help each other on the team. Thus, it calls for an expansion of the traditional role of employees used to functional silos.

Target costing requires different behaviors from all members of an organization. Team members must learn to live with ambiguity because design is an incomplete process, which means hard numbers are not always available. Members of the organization must work closely with other disciplines, as team playing is extremely important. This also calls for more effective communication among department personnel.

A customer focus is achieved because the target costing process work must always focus on how it creates value for a customer, as customer requirements drive these activities. Management must take advantage of these implications of target costing to evaluate employees. The evaluation of employees geared to target costing goalswill insure the involvement of all workers, keep their focus on cross functional teamwork, encourage open sharing of information, and value customer inputs.

Target costing is not without its limitations. Overemphasis on design may lead to longer product development times, which could cause a costly delay in getting a product to the market. Usually, targets for cost, quality, and time must be made. Pressure to attain demanding targets can cause employee burnout and frustration. If targets are not attained despite much overtime and hard effort, employees' future levels of interest and effort may be reduced. This is why employees must participate in target setting. A small amount of slack should also be used to allow an organization to devote extra effort during crisis periods. Also, to reduce stress, the focus should be on continuous improvement and not radical changes.

Another potential problem of target costing is that too much attention to customers can lead to adding features without regard to extra costs involved, and proliferation of product models that can cause market confusion. This can be controlled if engineers are very aware of the costs of new features and marketing does not simply produce a customer "wish list".

If the proper steps to avoid these potential problems are implemented, then target costing can be a great source of competitive advantage. It needs to have the support of all levels of the organization, otherwise it will breakdown somewhere along the line. Ansari *et al.* (1997) point out that target costing focuses on the product as it moves through time, across units, across organizations, and across activities. Further, they argue since target costing is accomplished by cross-functional teams, who have a product and process focus, target costing cannot function in an organization that is not ready to adopt a process orientation.

WHEN IS TARGET COSTING APPROPRIATE?: Target costing is most appropriate in the beginning stages of product development. It needs to start as early as the new product brainstorming stage. If a product is already on the market, target costing to change design features and processes is probably not worth the extra time and effort. It is appropriate in an organization where cross-departmental communication is easily facilitated. It is also better used in an environment where there is a strong understanding of cost drivers and work flows so production process costs can be efficiently estimated. Finally it is applicable mainly for *assembly-oriented* industries that produce a variety of products in medium to small volumes than for *process-oriented* industries.

PRIOR LITERATURE AND CASES: Ansari *et al.* (1997) provide a comprehensive example of target costing with the help of a hypothetical company, Kitchenhelp Inc. They illustrate all the key activities for product concept that combines a coffee grinder and a drip system into a single coffee-maker. These authors also include an excellent discussion of the various facets of target costing as a management tool.

Shank (1995) presents an illustration of target costing for a forest products company, which adopted price-based cost planning initiatives including re-engineering for their continued viability. Cooper (1994) provides another example of target costing practices in the automobile industry in Japan, as it relates to Yokohama Corporation which is a supplier to Tokyo Motors. Cooper and Slagmulder (1997) describe the experiences of several Japanese corporations with regard to target costing and value engineering. In their book, the authors also provide an extensive reference to the literature including an annotated bibliography of selected articles.

Brausch (1994) describes the implementation of target costing by an American textile manufacturer. This practitioner article emphasizes the importance of using a cross functional, team based approach to target costing. Cooper and Chew (1996) emphasize the need for feedforward cost management in the new global economy with intense competition on price, quality and functionality. The article cites the experiences of two Japanese firms, Olympus Optical Company and Komatsu, to illustrate effective application of target costing. In a related vein, Fisher (1995) documents the use of target costing by Matshushita Electric Works and Toyota Motors, Tanaka (1993) describes Toyota Motors' unique differential approach to cost planning, and Hiromota (1988) describes target costing practices at Daihatsu Motors.

Kato (1993) explores the contributions of target costing to cost management in Japanese firms. Information systems necessary to support this process are described, including support systems for market research, target profit computation, R&D, and value engineering. The article also warns against potential disadvantages of target costing, such as the extreme demands it places on the employees. Other potential downsides of target costing, such as longer development cycles, employee burnout, market confusion, and organizational conflicts are very well presented in Kato, Boer and Chow (1995).

Organizations Mentioned: Daihatsu Motors; Komatsu; Matshushita Electric Works; Mercedes-Benz; Olympus Optical Company; Panasonic; Sharp; Toyota; Yokohama Corporation.

See **Activity based costing; Activity-based costing: An evaluation; Concurrent engineering; Cost analysis for purchasing; Cost drivers; Cost management; Cost planning; Cost-based pricing; New product development through supplier integration; Performance measurement in manufacturing; Price-based costing; Product design for global markets; Product development and concurrent engineering; Quality function deployment (QFD); Standard costing; Target price; Value engineering; Value index.**

References

[1] Ansari, S., J. Bell, T. Klammer, and C. Lawrence, "Target Costing." Irwin, 1997.

[2] Brausch, J.M. "Beyond ABC: Target Costing for Profit Enhancement." *Management Accounting*, November 1994, pp. 45–49.

[3] Cooper, R., "Yokohama Corporation Ltd. (A)." Harvard Business School Case 5-195-071, 1994.

[4] Cooper, R., and W.B. Chew, "Control Tomorrow's Costs through Today's Designs." *Harvard Business Review*, January–February 1996, pp. 88–97.

[5] Cooper, R., and R. Slagmulder, *Target Costing and Value Engineering*, Institute of Management Accountants, Productivity Press 1997.

[6] Fisher, J. "Implementing Target Costing." *Journal of Cost Management*, Summer 1995, pp. 50–59.

[7] Hiromoto, T. "Another Hidden Edge – Japanese Management Accounting." *Harvard Business Review*, July–August 1988, pp. 22–26.

[8] Kato, Y. "Target Costing Support Systems: Lessons from Leading Japanese Companies." *Management Accounting Research*, Vol. 4, No. 4, 1993, pp. 33–47.

[9] Kato, Y., G. Boer, and C.W. Chow. "Target Costing: An Integrative Management Process." *Journal of Cost Management*, Spring 1995, pp. 39–51.

[10] Sakurai, M., "Target Costing and How to Use It," *Cost Management*, Summer 1989.

[11] Shank, J. "Strategic Cost Management: A Target Costing Field Study," The Amos Tuck School of Business Administration, Dartmouth College, 1995.

[12] Tanaka, T. "Target Costing at Toyota." *Journal of Cost Management*, Spring, 1993, pp. 4–11.

TARGET PRICE

Target price is selling price based on its value to the customer, constrained by competitors' price for similar items.

See **Target costing.**

TAYLOR, FREDERICK W.

See **Scientific management.**

TEAM

TEAM, which stands for Technologies Enabling Agile Manufacturing is established by the US Department of Energy (DoE). TEAM is a program that leverages the technology strengths of industry, government, and academia to enhance the global competitiveness of US industry and strengthen the nation's defense capability. TEAM brings together dozens of companies, federal organizations, and academic institutions across the nation. Industry members represent many different sectors, including aerospace, automotive, machine tools, consumer electronics, and software. TEAM utilizes and develops state-of-the-art tools used to design, prototype, and manufacture demonstration products. The technologies and methodologies deployed by TEAM in five thrust areas are (1) product design and enterprise concurrency; (2) virtual manufacturing; (3) manufacturing planning and control; (4) intelligent closed-loop processing; and (5) enterprise integration.

See **Agile manufacturing; AMRI; CAM-I; DARPA; MANTECH; NIST.**

TEAMS: DESIGN AND IMPLEMENTATION

John K. McCreery

North Carolina State University, Raleigh, North Carolina, USA

Matthew C. Bloom

University of Notre Dame, Notre Dame, Indiana, USA

The use of teams as a way to organize people and conduct work is not a new phenomenon in business. What is new is the dramatic increase in the use of teams and the variety of purposes for which they are used. Teams are becoming pervasive in manufacturing organizations because they provide a variety of benefits (Katzenbach and Smith, 1993). While teams certainly have their place in the modern manufacturing organization, they should not be formed indiscriminately. If not used properly, teams may fail to provide the benefits hoped for and may even be harmful to manufacturing performance (Campion, Medsker and Higgs; 1993).

HISTORICAL PERSPECTIVE: For many years, managers thought of teams as primarily a productivity tool. The old admonition that "many hands make the work light" served as the guiding principle: teams could simply do more work. This view was later expanded somewhat and viewed teams as a vehicle to induce desireable behaviors. The idea was that people who work as part of a team would be more likely to help each other, offer advice and counsel, or provide informal, on-the-job training to less experienced workers. The central focus of teams was to concentrate more people on a particular task.

In the decades after World War II, conventional wisdom shifted toward seeing teams as a way to increase quality. Quality circles (QCs) first came into use in Japan in the early 1960's (Ishikawa, 1989) and were tremendously successful in fostering quality improvements. Over the next decade, as word of Japan's success with QCs spread, the concept was introduced in the U.S. by managers who realized that a great deal of untapped knowledge resided in the minds of workers. Teams were seen as a mechanism to tap this store of knowledge because the attention of a group of workers could be focused on day-to-day work activities. Since workers know these situations best, it was reasoned, why not get them to think about these situations more broadly?

The success of these early efforts suggested that teams might be a way to increase innovation, better meet customer needs, and facilitate change (Osburn, Moran, Musselwhite and Zenger; 1990). "Teams can solve all your problems" was an often-heard rallying cry. Today, views are moderating as stories of team failures come in (Hackman, 1998). Attention is turning toward seeking to understand when teams can be helpful, and how they should be structured for optimum performance.

STRATEGIC PERSPECTIVE: The decision to use work teams should be placed in the broader context of the organization's competitive and manufacturing strategies. The overall premise is that, for work teams to be truly effective within the organization, they must be geared to provide capabilities or outcomes that are valued by the firm as a whole. In other words, teams should add value to the firm's products and services in ways that are meaningful to the marketplace. This implies an outward-looking view of team design: teams should be empowered to take actions that, either directly or indirectly, support the organization's sources of competitive advantage.

Depending on its charter and design, manufacturing work teams may provide improvements in operational efficiency, quality, delivery, and flexibility. In turn, the organization may exploit these improvements through either low price or product differentiation strategies.

IMPLEMENTATION ISSUES: A number of key implementation issues must be considered when designing teams.

Types of Team

Casual observation tells us that teams come in many shapes and sizes. Although there are many ways to classify teams (Goodman, Devadas and Hughson, 1988; Guzzo and Dickson, 1996), we suggest Table 1 as a useful way to think about teams in a manufacturing context. Table 1 presents what we might call "pure types". In practice, it is likely that teams will exhibit the characteristics of more than one type, but the typology in Table 1 helps suggest what the principal features of teams are, and how they can be used to facilitate manufacturing performance.

Ad Hoc Teams: At one end of the typology are teams that are primarily concerned with solving temporary problems, addressing specific issues, or recommending remedial actions to changing conditions. These teams are characterized by irregular team activities and a focus on temporary versus on-going issues and activities. Ad hoc teams meet sporadically or for short, intense periods of time. Sometimes, the team is created to deal with a specific problem and may disband after a solution has been tendered and implemented. Other times, ad-hoc teams use people from on-going teams, but it may meet occasionally. The normal work responsibilities of the individual team members are usually not highly interdependent; members can perform their routine duties without relying on the team. Team activity is centered around specific, one-time or short-term issues. Team members may come from different functional areas to ensure that the team is complete and balanced.

Consultative Teams: These are somewhat akin to ad hoc teams. The primary responsibility of consultative teams is to provide information for other decision makers. For example, a consultative team may be used as a sounding board to test proposed changes, an example is an internal focus group. Or, managers can use the team

Table 1. Team design continuum.

SCOPE OF RESPONSIBILITY / DEGREE OF AUTONOMY	TEAM TYPE	TEAM CHARACTERISTICS
Increasing ↑	**Autonomous Work Teams**	◆ Team makes all decisions or have veto power regarding the pace and flow of their work. ◆ Team determines the best way to structure and complete all phases of their work.
	Self-directed Work Teams	◆ Employees supervise their own work. ◆ Make decisions about the pace and flow of work. ◆ May make decisions about the best way to get work done.
	Quality of Work Life Initiatives; Problem Solving Teams; Quality Circles	◆ Team identifies work problems or important work issues. ◆ Team makes recommendations or suggestions to management for resolving work problems. ◆ Team does not make final decisions.
	Consultative Teams	◆ As required, management informally seeks worker input and opinions about work issues.
Decreasing ↓	**Ad hoc Teams**	◆ As needed, teams of employees form temporary groups to analyze work issues and recommend remedial action or changes.

to acquire important information, or to get the ideas or perspectives of workers about an issue or problem. Here, the team is a resource for information, data, ideas, opinions, and the like. Again, these teams can be temporary or more permanent. In either case, team activities are irregular and do not comprise the daily or normal work responsibilities of the team members.

Quality of Work Life Initiatives, Problem Solving Teams, or Quality Circles: Quality circles (QC) became popular in the late 1970's and are still around in various forms. They represent a higher level of team functioning in several respects. First, members are usually permanent; temporary QC teams are rare. Second, QC teams meet regularly and team activities are a routine part of team members' work responsibilities. Third, these teams are usually organized around a specific part of the work process or a clearly defined segment of work. For example, an automobile manufacturer might organize a QC around its body painting group, engine assembly group, etc. Members usually come from the same or very similar functional areas. Fourth, QC teams are expressly formed to *identify* work problems or important work issues. This means, for example, they can be used to uncover hidden problems in manufacturing processes or diagnose the root cause of symptoms such as consistent defects. Fifth, QC teams usually focus on a wide range of issues and problems that are common to a particular area. However, QC teams make recommendations to management for resolving work problems; employees do not make final decisions. Finally, QC teams usually have a long-term mandate, and have stable membership (team members do not change often).

Self-directed Work Teams: Self-directed teams represent a key point of demarcation in the typology because they consist of employees who supervise their own work. Here, the team makes decisions about such things as the pace and flow of work. An important distinction between these and the other teams discussed so far is the scope of decision making. Self-directed teams are usually given specific performance targets or final output goals, and then it is left to the team to determine what specific tasks or duties need to be done, when they should be conducted, and how workers, materials, and equipment should be utilized. They may be comprised of members from different functional areas (so-called cross-functional teams), but the work of these members is highly interdependent. Functioning within the team is integral to performing the normal work responsibilities of all members. That is, members rely on the team to perform most of their duties. These teams are permanent and have stable membership.

Autonomous Work Teams: Autonomous work teams are the most complex form and represent the high-end of the typology. These are truly self-contained teams. Organizations set broad performance targets (e.g., meet customer needs), and the team decides not only what must be done to meet the goal, but how to measure the goal itself. Autonomous work teams make all decisions about work, or have veto power regarding the best way to structure and complete all phases of their work. They set interim goals, establish performance standards, develop regulatory policies, and manage resources such as budgets, equipment, and materials. Often, autonomous teams select new members, discipline poorly performing members, and may even set human resource policies. They most often comprise members from a broad spectrum of functional areas, and usually include members from management, professional, or technical levels. That is, team members are usually highly skilled and deal with complex tasks. Autonomous work teams can be accurately thought of as mini-organizations in that most of their activities are self-contained, independent, and not subject to significant external controls.

Motivational Issues

A fact often ignored by managers is that successful implementation of work teams requires paying attention to the human element: teams are made up of people. Issues such as member support, communication, and celebration are important because they are integral to properly motivating people to perform well on teams.

One factor which seems to be more important as one moves up the typology from ad hoc to autonomous work teams is the notion of trust. Sometimes called altruistic attachments, teams seem to function better when the members trust each other. The idea of a combat unit in war times is perhaps one of the best examples of trust and its importance for teams. Team members must trust that others are competent to do the job and also trust the others will be responsible, work hard, and seek the good of the entire team and not act in their own self-interest. Training (often called team building) is an important step during the initial formation of teams because it can help create the basis for ongoing trust among members. Likewise, setting up compensation policies to reward cooperation and trust (e.g., using team versus individual incentive pay plans) can foster trust among members. Developing trust often means managers taking a more laissez faire approach to supervising the team, opting instead to let teams develop their own internal controls consistent with the team's design and objectives. However it is achieved, trust is of central importance.

Another important factor is extensive communication. Centralized information flows may inhibit teams because they cannot get the information they need. Restricted communication flows may also disrupt the natural cooperative process because members cannot create the synergy or rhythm that facilitates good teamwork. Higher performing teams often learn how to do things best "on-the-fly," learning as they go. If communication is blocked, this learning cannot take place. Extensive communication also ensures all team members are apprised of important events, opportunities, threats, and the like. Communication is important for knowledge sharing, mentoring, and other interpersonal activities that strengthen team bonds, create skilled workers, and facilitate progress toward team goals. Training may be important in this area too. High performing teams are characterized by fine-grained information transfers where non-proceduralized or proprietary information is exchanged among members as a course of normal activity. Unless members are good at communicating, this information transfer will be less effective. Finally, communication and trust are connected: if team members do not trust each other, communication is unlikely to occur at all.

The importance of trust and communication point to a third motivational factor: keep the rules imposed from outside the team to a minimum. Especially for self-directed or autonomous teams, rules imposed from the outside cramp team functioning or force the team to operate in ineffective or inefficient ways. As much as is practical, and consistent with the team's design and objectives, the team needs freedom to find its own way, room to improvise, and the opportunity to develop its own performance norms. Consider that, in most societies, the worst form of punishment is not death, but expulsion from the group. This suggests that group or team norms tend to have a much stronger influence on the behavior of members than do rules imposed from outside the group.

The fourth motivational factor critical to successful team implementation is a shared, goal-oriented mindset among team members. Team members need to share a common understanding about what goals the team is pursuing so that their efforts can be directed appropriately. Managers should set goals that are specific, challenging yet attainable, meaningful, and relevant (O'Leary-Kelly, Martocchio, and Frink; 1994). With a clear idea of what the team's purpose is, members tend to perform better. This is probably the best way to exert external control on higher level teams. Set a goal, and let the self-directed or autonomous team take over.

The final motivational factor is that a successful team is intrinsically rewarding to many, if not most, people. People like being part of a successful team. The team needs to celebrate its success. This builds strong member-to-member support, fosters trust, enhances a common mindset, and encourages hard work in the future. Organizational rewards should be tied to behavior that helps the group. Recognition programs are one example, but cash compensation itself should be tied to factors that reward team performance. Team failures can often be attributed, at least in part, to a mismatched reward system that did not support team goals and teamwork (Hackman, 1998). Failing to recognize the satisfying aspects of teamwork per se may leave important motivational potential on the table. Encouraging esprit de corps can be an important precursor to high functioning teams.

Individual Characteristics

Creating a team out of a group of individuals is no simple task. Each potential member comes into the team with his or her own set of particular skills, beliefs, and personality traits. While an overly high level of homogeneity within a team is likely to stifle creativity and diversity of ideas, nonetheless there are certain skills and traits deemed necessary for all team members (Dunphy and Bryant, 1996). At the most fundamental level, all work team members must come into the team with a set of basic skills. These include reading comprehension, rudimentary writing and speaking skills, basic computational ability, and logical thinking. These basic skills are necessary in virtually any modern manufacturing setting, and are needed for members within any type of team design. In addition, certain technical skills are required. These skills depend greatly on the particulars of the production tasks that will be performed by the team. A wide variety of fabrication, process monitoring and control, and product assembly tasks are included in this skill group. These skills build upon the set of basic skills, and are likely to be at least partially product- and process-specific.

A more general set of team management skills or traits are also required by all team members. These include the ability to communicate well with others both within and outside the team, the willingness to trust management and other team members to perform their jobs fairly and well, and the desire to be a cooperative team member. Further, each team member needs a positive attitude toward his or her duties and responsibilities as a member of the team. In essence, regardless of each team member's background, experiences, and beliefs, he or she needs to be willing to be a "team player".

Leadership

Leadership may be exercised within the team by certain members, depending on how the team is designed. This leadership may be formally delineated as part of the team design, or may come about informally as the team matures. Either way, the team management skills and traits mentioned above are also important here. Good communication skills, trustworthiness and a trusting attitude, cooperation, and a positive attitude are all critical for a team member to exhibit effective leadership (Mohrman, Cohen and Mohrman, 1995). In addition, technical skill proficiency, the ability to train team members, and the capability to effectively interface with other organizational units may be necessary, depending on the organizational context within which the team operates.

Leaders outside the boundary of the team have a more difficult role to play. These are often former supervisory and first-line managerial personnel. Their management styles under a more conventional hierarchical organizational structure may be at odds with what is needed in a team-based environment. They need the ability to delegate responsibilities to the team, and then allow the team to perform with minimal interference. The team needs the leeway to find its own solutions to issues and problems as much as is practical. The former supervisor/manager is now more of a coach, giving guidance when requested and providing support and encouragement to the team. The external leader must also be able to effectively manage relationships at the team's boundary, thereby allowing the team to operate without an undue level of influence or interference from the organization or other external constituencies.

Rewards

Performance expectations change when workers are configured into teams. Under more traditional hierarchical reporting structures, individual worker accomplishments are often the primary or sole basis for determining compensation. Typical measures used at the level of the individual worker include the worker's productivity or work output as compared to job standards, the worker's quality performance as measured by scrap and rework levels, and the worker's time-based performance as measured against planned schedules for the worker's station. However, the basis for compensation of individuals organized in teams must be adjusted to reflect their new set of responsibilities and accomplishments (Lawler and Cohen, 1992).

Along with rewards for individual performance, systems for measuring and rewarding team-based performance should be implemented. Using the individual worker performance measure examples from above, analogous team-based performance measures may be appropriate, such as the team's overall level of work output, levels of scrap and rework generated by the team, and the time-based performance of the team versus its production schedule. In addition, depending on the design of the team, other team-based measures may be appropriate. Examples of these are: (1) the number of quality problems solved by the team; (2) the number and impact of manufacturing process and product design changes identified and implemented by the team; and (3) improvements in stated customer satisfaction if the team has a direct effect on customer perceptions.

Whatever performance measures are chosen, for the team-based compensation scheme to be successful over the long term, it must be tied to the stated strategic objectives that first motivated the management to institute the team structures. This, however, is complicated by the fact that strategic objectives are often cross-functional in nature. As such, team performance may not be the sole determinant of whether the strategic objectives are met. Therefore, team performance must be measured against team-specific targets that support these strategic objectives, independent of whether these objectives are in fact attained by the organization as a whole.

Regardless of the particulars of the team-based performance criteria, they should possess a few vital characteristics. First, the performance criteria should be attainable by the team. The performance targets should challenge the team to perform at a high level of competence and effectiveness, but not be so challenging that the team cannot achieve them. Second, the performance criteria should be unambiguous. All team members should have a clear understanding of what is required of them. Third, the criteria should be within the domain and control of the team. Setting performance criteria that forces the team to rely significantly on the actions of other organizational units undermines the motivational aspect of the reward system. And fourth, the team criteria should apply equally to all members of the team. While individual performance criteria may also be utilized with team members, there should be a clear delineation between individual and team performance criteria.

EFFECTS: Work teams, when successful, can enhance manufacturing's ability to support the competitive needs of the organization (see, for example, Katzenbach and Smith, 1993). At the most fundamental level, work teams may increase the quantity of products being built. Teams that share production tasks are often able to coordinate their actions, which increases the total amount of output per unit time. The result can be a dramatic increase in plant-wide efficiency. This in turn may be exploited by the organization that competes on the basis of low price.

Beyond efficiency improvements teams improve efficiency through the discovery of better ways to perform production tasks. Process streamlining, setup reduction, and waste reduction initiatives may all be enhanced in a cooperative team environment.

An addition to achieving higher output levels, teams also improve the quality of the products being built (Banker, Field, Schroeder and Sinha, 1996). Improvements in conformance quality are possible as teams share experience and knowledge of production methods. This may result in lower scrap and rework rates, as well as reductions in manufacturing-related field failures. Teams also have the potential to add to product quality through suggestions for product improvements. At the manufacturing level, the nature of the improvements are likely to be related to changes in fabrication or assembly methods, equipment, tooling, inspection methods, or testing procedures. At the product design level, teams may collectively suggest design changes that make the product easier to produce and test, and improve product durability and reliability.

Third, teams make a strategic impact on time-based competition. With improvements in productivity come the ability for manufacturing products with less internal delays between the receipt of an order and its subsequent shipment. In make-to-order and assemble-to-order environments, reductions in cycle time can yield a significant competitive advantage. In make-to-stock environments the benefit of rapid delivery allows manufacturing to satisfy the changing demands of the distribution channel pipeline while minimizing inventories.

Finally, teams may support the organization that competes on the dimension of flexibility. Specifically, teams enhance the ability of manufacturing to provide a greater mix of products, both existing and newly introduced. Teams that schedule their own work and develop highly cross-trained team members are better able to adjust to market-driven changes in demand. Likewise, empowered teams have the potential to participate in the design of new products, resulting in smoother product releases and shorter product ramp-up times.

Not all types of teams have the breadth of skills and responsibility necessary to deliver all of these strategic benefits to the organization. The next section discusses when different types of work teams should be employed, depending on other characteristics of the manufacturing environment.

LOCATION: Teams can be used in a wide variety of manufacturing settings. While work teams have wide applicability in manufacturing, there is a need for research to identify the proper fit between team design and key characteristics of the manufacturing environment. This section will offer some thoughts and suggestions along these lines.

Referring to the typology of Table 1, it can be seen that team types at the low end of the spectrum – Ad Hoc, Consultative, Quality of Work Life, Problem Solving, QC – can be successfully implemented in virtually any manufacturing setting. The Ad Hoc and Consultative types of teams focus on providing information, identifying problems, and suggesting solutions for a wide variety of manufacturing-related situations. These teams are not usually permanent in nature, and they do not require full time participation from team members. Because of the transitory nature of these teams, there is no pressing long-term need for team members to work together in a highly coordinated manner. Hence, the particular characteristics of team members' normal, non-team based jobs are not of great importance here. Therefore, these types of teams can be successfully deployed in any process design environment, be it job shop, batch manufacturing, line assembly, or continuous flow.

Similarly, the Quality of Work Life, Problem Solving, and QC types of teams have wide applicability in manufacturing. While these types of teams tend to be formed on a permanent basis, they nonetheless focus their efforts on distinct, identifiable issues such as quality improvement and process modification. These teams are not primarily engaged in ongoing, highly interactive sharing of work duties and responsibilities. As with the ad hoc and consultative teams, these team types can be successfully employed in virtually any manufacturing setting.

In contrast to the universal applicability of low-end team types of Table 1, caution needs to be used when implementing team types at the high end of the spectrum, namely Self-Directed and Autonomous Teams. As discussed earlier in this article, these team types are significantly different than the others in that they focus on coordinating and managing ongoing activities directly related to building products. Because they are directly engaged in production tasks on an ongoing basis, successful implementation depends on the nature of the work they perform.

The key issue to examine here is the degree of interdependence among team members. Because these types of teams are designed to use a high degree of coordination among their members, activities must be such that members can perform better as a team than as individuals. To increase the probability of this occurring, there needs to be significant associations and dependencies among tasks to be performed by the team, so that team members can share their knowledge, expertise, or workloads. If the tasks are not dependent or somewhat related in the types of skills and abilities called for, there is little to gain by calling the group of workers a team.

Assuming an acceptable degree of interdependence among the team members, the focus of activities undertaken by self-directed and autonomous teams may vary depending on the particulars of the manufacturing environment. One team activity that should be done with caution is the detailed scheduling of its own work.

In line assembly and continuous flow environments with high levels of capital intensity, well balanced workloads and high process utilization rates are often critical to good performance. In this environment, teams may hinder manufacturing performance if their decentralized, localized work scheduling efforts degrade the workload balance or utilization rates. In contrast, manufacturing environments with lower levels of capital intensity found in batch manufacturing environments are more likely to see performance improvements through teams making work scheduling decisions. Similarly, job shop environments with moderately high levels of capital intensity but without the necessity for well-balanced workloads and high equipment utilization rates are also likely to improve their performance through teams performing their own scheduling of work.

High-end teams Participate in activities related to equipment upkeep, maintenance, and improvement. Again, the ability of the team to successfully undertake these types of activities is somewhat dependent on the nature of the equipment used in the plant. Shops with highly complex, expensive equipment are less likely to have teams heavily involved in upkeep and maintenance. Further, in these environments, process improvements will often include modifications to equipment design or operation, and teams may lack the specialized technical skills necessary to determine and execute these modifications. This is often the case in continuous flow and automated line assembly processes. This situation may also occur in job shops, where equipment tend to be more general purpose than in continuous flow and line assembly plants, but can nonetheless be complex to maintain and improve upon.

In general, manufacturing environments that require centralized scheduling/coordination and high levels of capital intensity may need to use prudence when designing and implementing self-directed and autonomous teams.

RESULTS: As discussed in the Effects section of this article, there are a variety of benefits to be gained through the use of work teams. However, there are also some potential pitfalls to consider. A major concern when making the decision to implement manufacturing work teams is for the management to be realistic in their expectations. Despite what is often written in the popular press, teams will not solve all production-related problems. A mismatch between team design, team membership characteristics, and team objectives will likely result in mediocre or poor team performance.

In addition, a new management style is necessary in the team-based organization. The "command-and-control" approach common in manufacturing firms of the past must give way to a more supportive, flexible, and adaptive management style. To gain the full benefit of teams, they must be empowered to achieve their goals and performance targets without undue constraint from management. A reasonable approach for management is to direct teams by setting performance measures and specific targets, yet allow teams to decide specifically how they will achieve their performance targets.

Even assuming work teams are performing at a high level, there is a team life cycle issue to consider. Teams have a tendency to plateau in their effectiveness over time, as the luster and excitement of working as a coordinated unit starts to wear off. A current problem facing many manufacturing organizations is the need to renew the energy and enthusiasm of the mature team (Harrington-Mackin, 1996). Methods such as changing the team's objectives, increasing or otherwise modifying the team's performance targets, and bringing in new team members are all being attempted by team-based organizations. It remains to be seen if these, or other methods, will consistently revitalize mature teams.

Finally, it must be recognized that there is a cost/benefit tradeoff associated with teams. There are costs in terms of time, money, and effort expended to create a cohesive team out of a group of unique individuals. Further, these costs continue in some fashion throughout the life of the team. For the team to add net value to the organization, the benefits generated from the team must be greater than the team's startup and ongoing costs.

See **Ad hoc teams; Autonomous work teams; Consultative teams; Quality of work life initiatives; Self-directed work teams.**

References

[1] Campion, M.A., G.J. Medsker, and A.C. Higgs (1993). "Relations Between Work Group Characteristics and Effectiveness: Implications for Designing Effective Work Groups", *Personnel Psychology*, 46, 823–850.

[2] Banker, R.D., J.M. Field, R.G. Schroeder, and K.K. Sinha (1996). "Impact of Work Teams on Manufacturing Performance", *Academy of Management Journal*, 39 (4), 867–890.

[3] Cohen, S.G., G.E. Ledford, Jr., and G.M. Spreitzer (1996). "A Predictive Model of Managing Work Team Effectiveness", *Human Relations*, 49 (5), 643–676.

[4] Dunphy, D. and B. Bryant (1996). "Teams: Panaceas or Prescriptions for Improved Performance", *Human Relations*, 49 (5), 677–699.

[5] Goodman, P.S., R. Devadas, and T.L.G. Hughson (1988). "Groups and Productivity: Analyzing the Effectiveness of Self-Managing Teams", in *Productivity in Organizations*, J.P. Campbell *et al.* (ed.), Jossey-Bass, San Francisco, CA.

[6] Guzzo, R.A. and M.W. Dickson (1996). "Teams in Organizations: Recent Research on Performance and Effectiveness", *Annual Review of Psychology*, 47, 307–338.

[7] Hackman, J.R. (1998). "Why Teams Don't Work", in *Theory and Research on Small Groups*, R.S. Tindale *et al.* (ed.), Plenum Press, New York, NY.

[8] Harrington-Mackin, D. (1996). *Keeping the Team Going: A Tool Kit to Renew & Refuel Your Workplace Teams*, AMACOM /American Management Association, New York, NY.

[9] Ishikawa, I. (1989). *Introduction to Quality Control, 3rd edition*, 3A Corporation, Tokyo, Japan.

[10] Katzenbach, J.R. and D.K. Smith (1993). *The Wisdom of Teams: Creating the High-Performance Organization*, Harvard Business School Press, Boston, MA.

[11] Lawler, E.E. and S.G. Cohen (1992). "Designing pay systems for teams." *American Compensation Association Journal*, 1, 6–18.

[12] Mohrman, S.A., S.G. Cohen, and A.M. Mohrman, Jr. (1995). *Designing Team-Based Organizations: New Forms for Knowledge Work*, Jossey-Bass, San Francisco, CA.

[13] O'Leary-Kelly, A.M., J.J. Martocchio, and D.D. Frink (1994). "A Review of the Influence of Group Goals on Group Performance", *Academy of Management Journal*, 37 (5), 1285–1301.

[14] Osburn, J.D., L. Moran, E. Musselwhite, and J.H. Zenger (1990). *Self-directed Work Teams: The New American Challenge*, Irwin Publishing, Chicago, IL.

TEAMWORK REWARD SYSTEM

See **Human resource issues in operations management.**

TECHNOLOGY PUSH MODEL OF INNOVATION

Innovation refers to the commercialization of a new idea. Technology push innovation refers to products and services that result from technological advances. This approach to innovation is less concerned initially with satisfying a known or potential demand but rather with what is technically possible. A purely technology push model may lead to a poor balance between what is at the leading edge of technical feasibility and what customers want.

See **Process innovation; Product innovation.**

TECHNOLOGY ROADMAP

Also known as a "long-term material plan," a technology roadmap refers to the set of performance criteria and undiscovered products and processes an organization intends to develop and/or manufacture within a specified or unspecified time horizon. Many companies define their technology roadmaps in terms of the next decade, while others employ a horizon of fifty years or even a century! While the exact form of a technology roadmap is somewhat industry specific, it is typically defined in terms of:

1. Projected performance specifications for a class of products or processes (e.g., memory size in computers, speed, electrical resistance, temperature, pressure, etc.);
2. An intention to integrate a new material or component (e.g., new form of molecule/chemical);
3. The development of a new product to meet customer requirements that is currently (e.g., new television screen technology);
4. The integration of multiple technologies that results in a radical new product (e.g., combined fax/phone/modem/copier, combining television/cable/computer);
5. A combination of the above, and other possible variations.

While technology roadmaps are often difficult to project and, in fact, may or may not be brought to fruition, the sharing of roadmaps with suppliers is essential for supplier integration.

See **New product development through supplier integration.**

Reference

[1] Ragatz, G., Handfield, R., and Scannell, T. (1997). "Success factors for integrating suppliers into new product development", *Journal of Product Innovation Management*, 7–20.

THEORY OF CONSTRAINTS (TOC)

The concept of the Theory of Constraints (TOC) was introduced in the book, *The Goal* (Goldratt and Cox, 1984). TOC has several synonyms such as synchronous manufacturing, optimized production technology, Drum-Buffer-Rope systems, and constraint, bottleneck, or throughput management. The process of accounting for a TOC operating environment is called Throughput Accounting. Essentially, TOC is a management philosophy that identifies and leverages constraints (sometimes called bottlenecks) in order to maximize the profit potential of a company.

See **Accounting system implications of TOC; Synchronous manufacturing; Theory of constraints in manufacturing management.**

Reference

[1] Goldratt, E.M. and J. Cox (1984). *The Goal: A Process of Ongoing Improvement* (North River Press: Croton-on Hudson, NY).

THEORY OF CONSTRAINTS IN MANUFACTURING MANAGEMENT

Monte R. Swain and Stanley E. Fawcett

Brigham Young University, Provo, Utah

DESCRIPTION: As a philosophical approach to better management practice, the Theory of Constraints (TOC) suggests thatfocusing management attention on a few "constraints" is the key to profitability and success.

In manufacturing, TOC uses constraint management to ensure continuous improvement through synchronized manufacturing (see Figure 1). The basic steps involved in synchronization are to: 1) identify the constraint resources; 2) establish key buffer locations to protect throughput; 3) schedule the constraint resource; 4) release materials at gateway operations to maximize production at the constraint resource; and 5) use forward scheduling of work centers that follow the constraint resource to ensure high levels of due-date performance (Fawcett and Pearson, 1991). This approach, which links all manufacturing resources to the constraint resource, is known as Drum-Buffer-Rope synchronization (Goldratt and Fox, 1986).

The premise of TOC is that only constraint resources should operate at full capacity. That is, a facility can produce finished product no faster than the constraint resource can process the materials that go into the product. To minimize

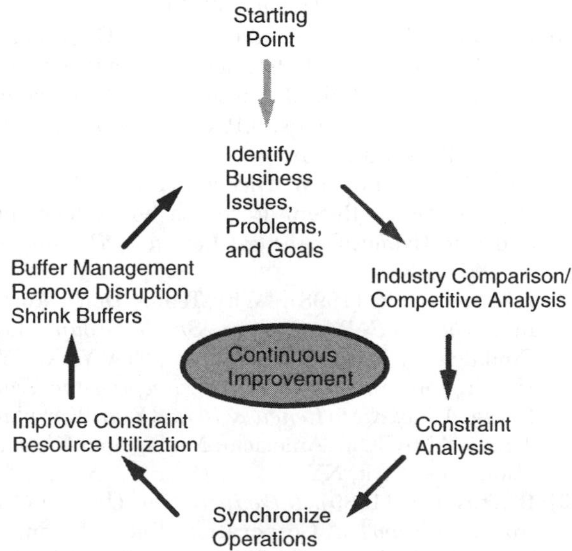

Fig. 1. Constraint management and continuous improvement.

the buildup of work in process (WIP), all non-constraint work centers should operate at less than capacity. The constraint resource acts as the "drum," setting the production pace for the entire operation. "Buffers" are WIP inventory placed in front of the constraint resource and in other strategic locations ensure maximum throughput (sellable finished goods). Simply stated, buffer management assures that constraint resources are never idled, waiting for material to process. Likewise, once material has been processed on the constraint resource, it should not sit idle, but should be kept moving until shipped to the customer. The "rope" is a schedule that ties the production output of upstream processes to the constraint resource. By following a few basic rules, each of which focus managerial attention on the firm's constraining resource, managers can use TOC to maximize the throughput of their production facility. By so doing, the firm is better able to meet its customers' real needs, and maximize its success.

HISTORICAL PERSPECTIVE: To compete against fierce global rivals in the early 1980s, managers across a variety of industries sought to revitalize their manufacturing practice. Optimized Production Technology (OPT), now popularly discussed as the Theory of Constraints (TOC), became one of several competing manufacturing philosophies to receive considerable attention. Optimized Production Technology emerged as the brainchild of Eliyahu Goldratt, an Israeli physicist. Research into heat transfer through bottlenecks combined with a friend's request for help in better managing his business led Goldratt to envision a firm's constrained resources as the key to managerial decision making. Goldratt then developed the algorithms needed to schedule production given bottleneck resources, and packaged this as the commercially available Optimized Production Technology.

Optimized Production Technology gained much of its early visibility through the book *The Goal* (Goldratt and Cox, 1984). It is claimed that constraint management's synchronization was faster, more economical, and more focused than the Japanese JIT production philosophy (Goldratt and Fox, 1986). By using TOC, a firm could develop and maintain a manufacturing-based competitive advantage.

Constraint management can be effectively and profitably utilized in small-firm and capital-constrained manufacturing settings – settings which are often not ideally suited for the implementation of other manufacturing control systems such as JIT production. However, the ability of constraint management to focus management's efforts makes it an ideal tool to use in conjunction with other management practices, including JIT.

STRATEGIC PERSPECTIVE: Constraint management can help a company improve its competitive position through the better scheduling and utilization of a firm's productive resources. Because the constraint resource, which is the operation with the least capacity, dictates the capacity and responsiveness of the entire manufacturing system, it deserves to become the focus of management attention. When the constraint resource becomes the focal point, non-constraint operations will operate below capacity, facilitating several positive outcomes. For example, non-constraint resources will no longer produce excess work-in-process inventory, shop floor clutter and confusion will reduce, production quality will improve, preventive maintenance activities will be enhanced, and more time will be available for employee training. When combined, these outcomes reduce variability in the production process even as the capabilities of the process are enhanced.

By identifying the constraint resource and using it as the reference point for manufacturing decisions, the very nature of key management assumptions concerning product costing, performance measurement, product and process design, and make versus buy (Fox, 1987) is altered. Table 1 outlines more specifically how prior management assumptions are affected by TOC. Of course, the premise is that more accurate assumptions lead to better decisions and enhanced competitiveness.

IMPLEMENTATION: The fact that TOC often requires a dramatic shift in managerial mindset makes its implementation somewhat difficult. The simple reality is that old habits – even when they are counterproductive – are very hard to displace. For example, with the objective of increasing efficiency, managers work ardently to keep both workers and machines busy. Under TOC, idle workers and non-bottleneck machinery might very well be preferred as long the constraint resource is fully utilized. This reality is driven by the TOC view that the use of "excess" capacity at non-constraint work centers simply creates excessive WIP, clutter, and confusion without having any positive impact on overall output.

To the extent that TOC does reduce the "noise and distractions" that complicate manufacturing management, it can help a company streamline and synchronize production so that high-quality, low-cost products are delivered on-time. Indeed, the effective management of constraints means that a firm's resources will be used in a way that enhances the throughput of the most valued

Table 1. The realities of constraint management.

Prior assumption	Current reality
Keeping people busy is the key to making money.	A focus on labor utilization hinders cash flow due to high inventory and the emphasis on keeping people busy.
By keeping utilization high, employees help the company perform well financially.	High utilization of resources does not correlate to profitability.
High labor-utilization rates ensure high levels of customer satisfaction.	High utilization of resources does not necessarily correlate to high customer satisfaction.
If managers release workers to other areas of the operation, they may not get them back when they need them.	Managers will willingly release workers to go where the work is when the right performance measures are used.
Traditional accounting standards tell managers whether they are effective as a total enterprise.	Traditional standards are subjective, inaccurate, and require constant monitoring.
Maximizing the production output per setup and building inventory is key to making money.	Making only what customers order is the key to making money, and on-time delivery is the critical success factor.

(Adapted from Westra *et al.*, 1996).

products. Five basic steps to strategic constraint management have been proposed (Goldratt, 1990):

1. *Understand the system's constraint(s)* – Sometimes the constraint is not on the factory floor or even inside the organization.
2. *Decide how to exploit the system's constraints*– production and pricing decisions should be based on the throughput yield per unit of the constrained resource.
3. *Subordinate everything else to the above decision* – the goal is to coordinate production efforts at non-constraints to keep the constraint operating at optimal capacity.
4. *Elevate the constraint(s)* – elevating the constraint can involve off-loading some work to non-bottlenecks, acquiring more equipment, or outsourcing some bottleneck work.
5. *If a constraint has been broken, go back to step 1* – do not allow inertia to cause a system constraint. Bottlenecks can, and will, shift around within the organization. Not identifying the current constraint because of changes in constraints leads to a return of process inefficiencies and confused management policies. See Figure 1.

1. ***Understand the System's Constraints:*** As managers begin to analyze the production environment looking for constrained resources, it is important to note that not all constraints are found on the factory floor or even inside the organization. In fact, constraints can be classified as either internal constraints or external constraints (Atwater and Gagne, 1997).

   Internal constraints come in several forms. First, process constraints occur when a given process in the company has insufficient capacity to fully satisfy market demand. Second, policy constraints can occur when management or employee unions enforce a rule that limits the company's operation abilities (e.g., a freeze on overtime or hiring, or a restriction on purchasing raw materials, a limit on worker flexibility). Third, logistical constraints occur when specific business methods are used that restrict operations, such as minimum batch sizes or a particular processing sequence that unnecessarily slows down the workflow.

   External constraints include material and market constraints. Material constraints occur when an outside source of material becomes restricted, such as when suppliers cannot meet the organization's needs or when regulations restrict a raw material source. Market constraints occur when the company cannot sell all of the products it can make. Understanding the nature and impact of constraints allows managers to make well-founded, strategic decisions.

   The actual identification of constraint resources begins with the production manager's familiarity with the manufacturing process and focuses on easily made observations and readily available shop floor documentation. In a facility that produces many end items from a relatively small number of raw materials or component parts, inventory accumulates in front of constraint resources. Thus, looking at work-in-process inventory profiles often leads directly to the constraint resource. Some production processes are characterized by a large number of component parts that are assembled into a limited variety of end products (product convergence). In these settings, the key to identifying constraint resources is to look at the late or missing parts list. When component parts are produced at a constraint resource, they will develop a "late-expedite" pattern. Additional verification of constraint resources can be made by checking for long inventory queues, excessive overtime, frequent expediting, assignment of "spare" workers, and broken setups at the suspected constraint. One caveat is that, in a facility overburdened by WIP, many work centers will appear to have problems. By reducing the pace at which material is released to the production floor and working off some of the excess inventory, a better picture of the true constraint will emerge.

2. ***Exploit the System's Constraints:*** When an internal constraint is identified, management must keep the constraint operating as smoothly and efficiently as possible at all times to get the most out of the constraint. After this first step to exploit the constraint, some creative management of the product mix can improve constraint utilization. For example, a strategic analysis of the profitability of each product in the firm's portfolio in terms of profitability per unit of constraint resource consumed should be performed (Fox, 1987). This analysis helps managers determine how the constraint resource should be utilized to maximize customer satisfaction and the firm's profitability. Based on this analysis, management might decide to raise prices to reduce demand and thereby alleviate some of the pressure on the constraint. Since the constraint is producing below market demand, it might make sense to increase margins to generate better cash flow and profitability. However, pricing pressure from industry rivals might preclude this option or make it an unwise long-term decision. Another option is to alter the product mix to increase the firm's revenue per unit of constraint resource. At times, a company might want to abandon or reduce production of a given product because the company's limited resources can be better utilized by focusing on those products that provide the highest profit per constraint resource hour. Again, the combination of market demand and competitor actions will influence product mix decisions, perhaps requiring a more extensive product line than the profitability analysis would suggest as optimal.

3. ***Subordinate Resources to the Constraint Resource:*** After determining the proper product mix at the constraint resource, the key to implementation success is to subordinate all other resources to the constraint resource. The process of resource subordination is managed via Drum-Buffer-Rope synchronization, which links constraint resources to market demand and then ties the remaining work centers responsible for producing the desired output to the constraint resource (Goldratt and Fox, 1986). The schedules produced via DBR control the release of material into the production system and the flow of material among various work centers. Figure 2 highlights the critical issues involved in establishing an effective DBR schedule.

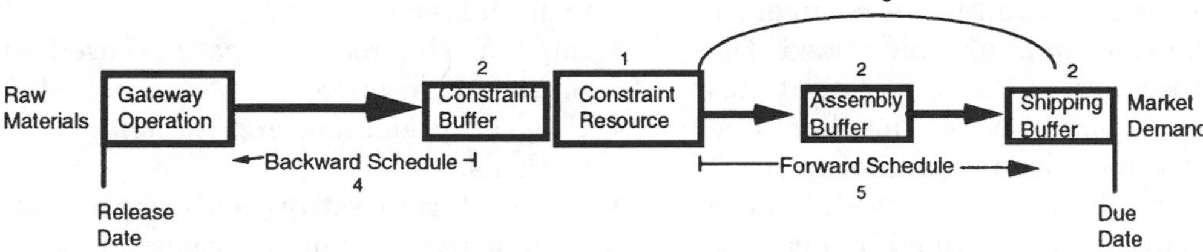

1. Identify constraint resources.
   - Identify work centers where inventory accumulates.
   - Look for a late-expedite pattern in parts produced at constraint resources.
   - Identify consistently late orders.
2. Place time buffers in critical locations to protect throughput.
   - Use buffers in front of constraint resources to keep them running.
   - Use buffers at assembly points of convergence to assure timely output.
   - Use buffers at end of production process to meet specific customer orders.
3. Determine the production schedule for the constraint resource.
   - Link constraint resource to market demand looking at the order lead times between the constraint resource and the shipping buffer.
   - Trade off the constraint resources' limited capacity between process and setup times.
4. Control release of materials to the production floor to feed constraint resources.
   - Backward schedule from the constraint resource to gateway operations.
   - Establish priority rules to determine the sequence jobs are to be processed when conflicts arise.
5. Move material processed on the constraint resource quickly and smoothly through remaining operations.
   - Forward schedule work stations that follow the constraint resource.
   - Allow transfer and process batches to vary to assure a smooth materials flow.

Fig. 2. Steps to synchronize the manufacturing process.

As has been noted earlier, the constraint resource is the Drum that sets the pace for the entire manufacturing process. Once the Drum has been identified, time buffers are placed at key points in the production process to protect throughput. Three points are of particular importance: 1) in front of the constraint resource; 2) at assembly points where parts that have been processed on the constraint resource converge; and 3) at the finished goods/shipping area.

The need for a buffer in front of the constraint resource is obvious – to keep the constraint resource running, when disruptions occur upstream. Buffers are needed at assemblies involving parts that have been processed on the constraint resource to assure that when these parts arrive at the assembly operation, the other parts necessary to begin assembly are also there. These assembly buffers protect against disruptions that occur in the acquisition and manufacture of parts that are not processed on the constraint resource, and assure that, once parts pass through the constraint resource, they move through the remaining work centers in a timely manner. Finally, a buffer is placed at the end of the production process to assure high levels of due-date performance for all specific customer orders. Shipping buffers protect delivery performance against process variability in the operation following the constraint resource.

The final step to DBR synchronization is to schedule the flow of materials through the facility so that the constraint resources operate continually and lead times are minimized. The scheduling procedure is the "rope" that ties the entire production process together, and keeps the operation moving at the pace set by the constraint resource. The schedule for a constraint resource is determined by evaluating the constraint resource's limited capacity, market demand, and the lead time from the constraint resource to the shipping buffer. Material is then released to the production floor in order to feed the constraint resource. Gateway operations are back scheduled from the constraint resource. For all operations with either multiple gateways or multiple product routings, it is important to establish priority rules to determine the sequence in which jobs are to be processed when they compete for the same resource. When DBR synchronization is undertaken, lot sizes are no longer set according to the costs of setups at non-constraining resources; rather, They are based on tradeoffs among throughput, lead times, and setups at the constraint resource. Process and transfer batches at the non-constraining resources are also used to maintain flow, avoid timing constraints, and reduce lead times. This concept is the foundation for related work in lot sizing and sequencing (Jacobs and Bragg, 1988).

4. ***Elevate the Constraint Resource:*** Even as synchronization is undertaken, efforts should be made to elevate the capacity of the constraint resource. Methods for increasing the constraint resource capacity can be relatively easy to implement and are frequently inexpensive. The first of these is to assure that the bottleneck is running during the entire shift. Lunch times and breaks should be scheduled during times when the constraint resource does not need worker supervision. When this is not possible, breaks should be staggered so that the constraint resource is always fully utilized. Poor scheduling of breaks can reduce the capacity of a constraint resource by 20% or more (Umble and Srikanth, 1990). Other approaches would be to use overtime or an additional shift on the constraint resource only. These three options represent low-cost alternatives to increasing overall facility capacity. Other methods of improving the utilization and capacity of constraint resources are (Fawcett and Pearson, 1991):

- Train the constraint resource operator to be more efficient.
- Cross-train workers to prevent idle time due to absenteeism.
- Improve the tools or gauges used at the constraint resource.
- Use setup engineering to simplify setup methods.
- Move internal setups activities to external setup activities whenever possible.
- Reduce major setups by processing families of jobs that require similar setups together.
- Improve constraint resource yield by assuring the quality of the process.
- Eliminate defective parts before they arrive at the constraint resource.
- Make sure parts that have been processed on the constraint resource are not scrapped later.
- Improve preventive maintenance to reduce constraint resource downtime.
- Improve the constraint resource process through process engineering.
- Offload to alternate work centers or manufacturing methods.

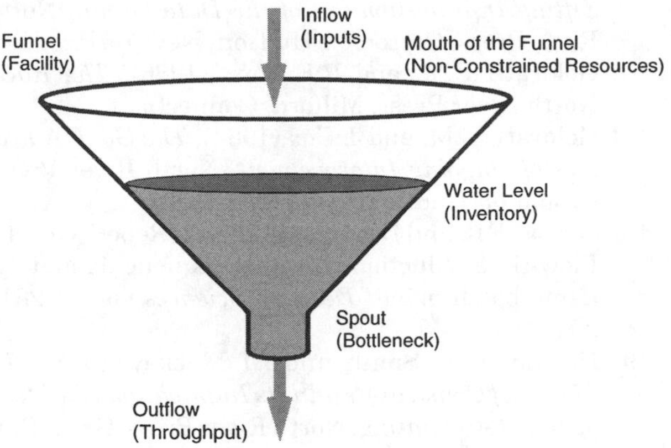

Fig. 3. The funnel analogy.

- Redesign some products to off-load the constraint resource.
- Subcontract bottleneck work.
- Consider capital investment to alleviate the constraint.

EFFECTS: The impact of constraint management on competitiveness is much easier to see when its effect on manufacturing throughput is illustrated. A simple analogy comparing the firm to a funnel helps illustrate the impact of effective constraint management (see Figure 3). The analogy relates the mouth of the funnel to a facility's non-constraining resources, the funnel's spout to a constraint resource, and the water level to the facility's WIP inventory. The flow of water through a funnel compares to the movement of materials through a manufacturing facility. When water is poured into the mouth of the funnel faster than it can pass through the spout, the funnel overflows. This is equivalent to releasing more work to the production floor than can be processed by the constraint resource – the result is excess WIP, confusion, and wasted resources. Similarly, if the person pouring the water into the funnel becomes distracted, it is possible to pour less water into the funnel than can pass through the spout. The result is the inefficient use of the funnel and bottleneck capacity. This allows a constraint resource to sit idle. When this occurs, the throughput of the firm is reduced since the constraint resource's capacity is lost forever. By enlarging the spout and assuring a continuos flow of water into the funnel, the efficiency of transferring water from one container to another can be maximized. This is the objective of constraint management – to maximize the flow of valued products through the facility.

RESULTS AND CASES: A number of case studies report the beneficial results that can be obtained through constraint management (Noreen, Smith and Mackay 1995). The cases also illustrate some key implementation issues regarding performance measurement, worker training, and cost allocation.

1. ***Company J:*** Before TOC, this privately-held metal tool manufacturing company used the ratio of indirect to direct hours as its key performance measure. As predicted by TOC, this performance measure instituted a "keep the workers busy at all costs" attitude, resulting in piles of excess WIP inventory, production inefficiencies, and a poorly focused pricing strategy. Under TOC, non-constrained resources are allowed to sit idle and the company uses this idle time to cross-train employees. Performance is measured via both daily throughput and throughput per constraint hour.
2. ***Western Textile Products:*** Before TOC, this apparel subcontractor focused on keeping people busy. This mindset meant that constraint resources were held in reserve to make sure that workers could be kept busy in case of a machine breakdown. TOC eliminated this counterproductive practice and led to considerably higher throughput and greater profitability.
3. ***Samsonite Europe:*** As a subsidiary of Samsonite, Inc., this luggage manufacturing company makes innovative use of several of TOC-based performance measures, particularly throughput-dollar-days (measures due-date performance) and inventory-dollar-days (measures excess inventory). The physical changes made to the plant layout during TOC implementation reflect the focused factory or manufacturing cell philosophy, suggesting that many of the good ideas in TOC are the same good ideas inherent in other management techniques.

COLLECTIVE WISDOM: Constraint management is designed for practical application and to provide a solid foundation to improve a firm's competitiveness. A key feature of constraint management is that it focuses the firm's finite financial and managerial resources on problems that can be resolved. As these problems are addressed, effective capacity and throughput are both increased. Perhaps constraint management's most valuable selling point is that it emphasizes management's knowledge of its own manufacturing system and the effective use of existing resources. Constraint management harnesses key information to initiate the process of constantly improving the firm's manufacturing processes.

It should be noted that the concepts underlying constraint management are based on sound and proven manufacturing principles. For example, constraint management is a specific application and extension of Pareto's Law, which states that a few items will have a disproportionately larger impact on a system than the remaining majority of items. These few important items – the constraint resources in a manufacturing setting – should become the focus of management attention. Many similarities also exist between constraint management and just-in-time manufacturing. Both advocate continuous improvement utilizing inventory reduction to reveal problems in quality and in the manufacturing system. Also, both constraint management and JIT promote a vigorous problem solving effort to enhance quality and to improve the manufacturing system. Another similarity is that both approaches rely heavily on synchronization to move material smoothly and quickly through the manufacturing system. The objectives in each case are to manage smaller inventories effectively, to increase responsiveness, and in the end, to produce with short lead times what the market demands, and when the market demands.

Organizations Mentioned: Samsonite (Europe); Western Textile Products.

See **Accounting system implications of TOC; Activity-based costing; Just in time manufacturing; Just-in-time manufacturing implications; Synchronous manufacturing using buffers; Constraint management; Constraint resource; Drum-buffer-rope synchronization; External constraints; Internal constraints; Optimized production technology (OPT); System's constraint; Theory of constraints (TOC); Throughput.**

References

[1] Atwater, B. and M.L. Gagne (1997). "The theory of constraints versus contribution margin analysis for product mix decisions," *Journal of Cost Management* (January/February), 6–15.

[2] Chase, R.B. and N.J. Aquilano (1989). *Production and Operations Management* (5th ed.), Irwin, Homewood, Illinois.

[3] Fawcett S.E. and J.N. Pearson (1991). "Understanding and applying constraint management in today's manufacturing environment," *Production and Inventory Management Journal* (Third Quarter), 46–55.

[4] Fox, R.E. (1987). "Theory of Constraints," *NAA Conference Proceedings* (September), 41–52.

[5] Goldratt, E.M. (1990). *The Haystack Syndrome: Sifting Information out of the Data Ocean*, North River Press, Croton-on-Hudson, New York.

[6] Goldratt, E.M. and R.E. Fox. (1986). *The Race*. North River Press, Milford, Connecticut.

[7] Goldratt, E.M. and J. Cox (1984). *The Goal: A Process of Ongoing Improvement*. North River Press, Croton-on-Hudson, New York.

[8] Jacobs, F.R. and D.J. Bragg (1988). "Repetitive lots: Flow-time reductions through sequencing and dynamic batch sizing," *Decision Sciences* Vol. 19, 281–294.

[9] Noreen, E., D. Smith, and J.T. Mackey (1995). *The Theory of Constraints and Its Implications for Management Accounting*. North River Press, Great Barrington, MA.

[10] Umble, M. and M. Srikanth (1990). *Synchronous Manufacturing: Principles for World Class Excellence*. South-Western Publishing, Cincinnati, OH.

[11] D. Westra, D., M.L. Srikanth, and M. Kane (1996). "Measuring operational performance in a throughput world," *Management Accounting* (April), 41–47.

THIRD PARTY LOGISTICS RELATIONSHIP

A third party logistics relationship is one in which a shipper or manufacturer outsources, or contracts with an external organization to handles some or all of their transportation and logistics functions. Thus, for example, a manufacturer may hire outside contractors to handle functions, such as inventory management, which has traditionally been an in-company activity. Functions performed by a third party can encompass the entire logistics process or elected activities within that process. Typically, the services offered by third parties include: (1) logistics re-engineering services such as network planning and site location; (2) transportation management including less-than-truckload (LTL) services; (3) inventory deployment which includes managing cross-dock operations and warehousing operations; and (4) carrier management which includes selection of transportation mode and carrier, payment to carriers and forwarders, and freight cost analysis (Sink and Langley, Jr., 1997).

See **Agile logistics (enterprise logistics); Logistics: Meeting customer's real needs.**

Reference

[1] Sink, H. and C.J. Langley, Jr (1997). "A Managerial Framework for the Acquisition of Third-party Logistics Services," *Journal of Business Logistics*, 18(2), 163–189.

THOMAS REGISTER OF AMERICAN MANUFACTURERS

The first step in the typical supplier selection process is to make a list of every supplier that could possibly meet the firms's needs for the item of interest. The list-making process usually begins with a review of the buying firm's purchasing database, which contains information on all suppliers used in the past several years. Potential supplier names also often come from contacts within (purchasing and sales managers) and outside the firm. Additional sources of information include trade publications and supplier directories. Perhaps the most comprehensive of the supplier directories is the *Thomas Register of American Manufacturers*. The *Thomas Register* includes over 155,000 American and Canadian manufacturers by type of product – the Thomas Register is almost 50,000 pages long. The home page for the Thomas Register (http://www.thomasregister.com) describes the basic role of this directory.

Knowing who produces what and how to contact them is one of the most important types of information in the Register.

The Thomas Register is widely recognized as the most comprehensive database of manufacturers in the United States and Canada. They include information an brand names, multiple SIC codes, geographic location, and so on.

See **Purchasing: Acquiring the best inputs; Supply chain management.**

Reference

ttp://www.thomasregister.com

THROUGHPUT

There are two different definitions for throughput. First, it is the production rate for a given period of time (e.g., the hourly or daily production rate).

Second, Theory of Constraints defines throughput as a dollar measure using the following formula:

$$\text{(Sales Revenue} - \text{Direct Material Costs)} = \text{Throughput}$$

TOC-based throughput (or throughput margin) is similar to the traditional measure of contribution margin: sales revenue less variable costs. However, there are important differences between throughput margin and contribution margin. Throughput margin assumes that the only truly variable costs in the organization are direct materials. The contribution margin measure, on the other hand, typically includes direct labor and some manufacturing overhead as variable costs. The difference between these two measures results from the timeframe relevant to TOC. TOC focuses on maximizing short-term results – typically as measured within a matter of weeks. The contribution margin definition of variable costs assumes a six to twelve month timeframe.

See **Accounting system implications of TOC; Manufacturing systems modeling using Petri nets; Theory of constraints in manufacturing management;**

References

[1] Goldratt, E.M. and J. Cox. (1984). *The Goal: A Process of Ongoing Improvement* (North River Press: Croton-on-Hudson, NY).

[2] Holmen, J.S. (1995). "ABC vs. TOC: It's a matter of time" *Management Accounting* (January), 37–40.

THROUGHPUT ACCOUNTING RATIO (TA)

The Throughput Accounting (TA) ratio is one of a number of performance measures useful in the effort to implement a system of management based on the Theory of Constraints (TOC). This ratio can be calculated per item, or batch, or product line. The TA ratio calculation for an item is as follows:

$$\text{Where: TA Ratio} = \text{(Throughput Return per Factory Hour)/(Cost per Factory Hour)}$$

$$\text{Throughput Return per Factory Hour} = \frac{\text{(Product Sales Price} - \text{Product Material Cost)}}{\text{Time to process the product on Bottleneck Operation (Hours)}}$$

$$\text{Cost per Factory Hour} = \frac{\text{Total Operations Costs of the Factory}}{\text{Total Time Available at the Bottleneck Operation (Hours)}}$$

See **Accounting system implications of TOC; Synchronous manufacturing using buffers; Theory of constraints in manufacturing management.**

References

[1] Holmen, J.S. (1995). "ABC vs. TOC: It's a matter of time," *Management Accounting* (January), 37–40.

[2] Galloway, D., and D. Waldron (1988). "Throughput Accounting: Part 2 – ranking products profitably," *Management Accounting* (UK, December), 34–35 and "Throughput Accounting: Part 3 – a better way to control labour costs," *Management Accounting* (UK, January), 32–33.

THROUGHPUT TIME (TPT)

TPT for a product is the elapsed time between the entry and departure from a manufacturing system upon completion of the product.

See **Manufacturing cell design; Manufacturing systems.**

TIER 1 SUPPLIER

See **Tiering of the supply chain.**

TIERING OF THE SUPPLY CHAIN

A typical supply chain structure is a pyramid-shaped hierarchy. The supply chain for a vehicle assembler is commonly cited as an example of this type of structure. The assembler is at the top of the hierarchy with its direct suppliers at the next level or Tier 1 in the structure. The suppliers to the suppliers are then positioned at the next tier and so on down through the structure. Recently an approach to reconfiguring the organisation of the supply chain has been adopted by Western companies, this new approach is often referred to as "tiering the supply chain". This approach stemming from Japan has a number of elements (see Hines, 1994, pp. 64–74). Customers positioned in one tier are responsible for organising, communicating with and developing the supplier tier below. This is devolution of control in contrast to traditional approaches such as vertical integration. The customers in one tier have few suppliers, conversely this means that in general, individual suppliers have few customers. This makes collaboration much easier to facilitate and co-ordination much easier to achieve. Generally, companies positioned in Western supply chains deal with more suppliers than the comparable Japanese supply chain.

For Western companies, "tiering their supply chains" requires a number of structural alterations. Overall, they have to reduce the number of suppliers by means such as the reduction of the number of components which are multiple sourced. Another mechanism adopted is to outsource assemblies or complete sub-systems to the next tier down (see Saunders, 1997, pp. 162–163). Doing so can remove from the customer tier the need to deal with a whole set of suppliers. This simplified structure enhances collaborative working throughout the supply chain.

See **Supplier relationships; Supply chain management; New product development through supplier integration.**

References

[1] Saunders, M. (1997). *Strategic Purchasing and Supply Chain Management*. Pitman Publising, London.
[2] Hines, P. (1994). *Creating World Class Suppliers*. Pitman Publishing, London.

TIME BUCKETS

Production planning specifies activities for the future. Usually these activities are described in terms of what to do (e.g., how many units of certain products to be produced) in different "time periods" in the future. These time periods are called **time buckets**. The size of the time bucket used for a particular planning activity indicates the degree of detailedness of the plan. Thus, different sizes of time buckets are likely to be used for planning activities at different levels of a company's planning hierarchy. For example, a daily or weekly time bucket may be used for material requirements planning (MRP). A weekly or monthly time bucket may be used for master production scheduling (MPS). A monthly or quarterly time bucket may be used for aggregate production planning. The size of the time bucket for a plan can also affect the length of the planning horizon which is the number of time periods in the plan. For example, if a year's plan is desirable and monthly time buckets are used for planning, then the planning horizon will be 12 periods. If weekly time buckets are used, 52 periods will be needed for a year's plan.

See **Aggregate plans and master production schedules linkage; MRP; Master production schedule; Order release.**

TIME BUFFER

See **Synchronous manufacturing using buffers.**

TIME FENCE

Master production schedule is a manufacturing plan for six months to a year. If the plan is changed or revised, it disrupts production, which increases the cost of production and delays customer orders. In order to minimize and control disruptions, time fences are employed to restrict, limit or bar changes to MPS for selected periods. For example, no changes may be allowed for the first 8 weeks, and minor changes may be allowed in one product between weeks 8 and 16 into the future.

See **Master production schedule (MPS); MRP; Schedule stability.**

TIME SERIES

Time series is a series of historical observations at equally spaced time intervals (days, weeks, months, quarters, years). Monthly sales data for several years into the past is an example of time series data. Many forecasting methods use time series data.

See **Forecasting examples; Forecasting guidelines and methods; Forecasting in operations management.**

TIME STUDY AND MOTION STUDY

Time study is an approach to establish standard times for the performance of a task. There are a number of alternatives to time study. A historical approach may be used in which output over some period of time is observed and productivity or time standard is then inferred. This method is simple and inexpensive but may be inaccurate: it cannot tell if unusual circumstances interfered with production, nor can it tell if the work rate was "normal." Such approaches have probably always been used to gauge whether a "fair day's work" was being done, but the industrial revolution and increasing sub-division of work since the late 19th century required more accurate time and work standards. F.W. Taylor developed the classical form of time study, using stopwatches to measure the time taken by workers on a particular task. Taylor noted that a number of situation-dependent factors could influence the effectiveness of the measurements taken: the worker's innate abilities, their training, the equipment and tooling provided, the work methods used, etc. Once these very important environmental factors were considered and fixed measurements can be taken. Measurements would then focus on a worker that was well trained and suited the pre-defined task to be done. Measurements must be conducted using well designed and maintained equipment and tooling, with adequate materials supplies, lighting etc. The task times with due allowances for fatigue can be converted into useful standards for job design and ork planning.

Taylor's approach was rather basic, and was expanded greatly by Frank Gilbreth. A philosophical difference underlay their two approaches, Taylor was to some extent concerned with the task as a whole, while Gilbreth recognized that all tasks could be broken down into a number of generic constituent movements and acts. Each of these constituents, or elements, could then be measured and a time standard for the task as a whole could be built up from elemental tasks. Gilbreth called his approach Motion Study to differentiate it from Taylor's Time Study. The two terms are now used almost interchangeably but Gilbreth's approach has dominated and is more widely used. His approach allows "synthetic" work standards to be created once the appropriate elemental motions of a task have been identified from past experience. The overall task time is determined from past data on the task's elements rather than by observing workers while performing the task. There is still a need to observe workers to validate such synthetic assessments, but this may reduce the effort and intrusion associated with work study measurement.

See **History of operations management; Scientific management.**

References

[1] Gilbreth, Frank B., Lillian M. Gilbreth. (1918). *Applied Motion Study*. George Routledge & Sons, Ltd., London, England.

[2] Gilbreth, Frank B., L.M. Gilbreth. (1921). *Process Charts. ASME Transactions*, Vol. 43, 1029–1050.

[3] Taylor, Fred. W. 1895. *A Piece Rate System. ASME Transactions*, Vol. 16, 856–903.

[4] Taylor, Frederick W. 1911. *Shop Management*. Harper, New York, New York.

TIME-BASED COMPETITION

Time-based competition is a competitive strategy to gain market share, (a) by compressing the product development time and to introduce new products earlier, or (b) by reducing the manufacturing lead time, thereby delivering the existing products faster than the competitors. This strategy recognizes that firms can succeed by competing on the time dimension much like competing on price or quality.

See **Capital investments in advances manufacturing technologies; Manufacturing strategy; Core manufacturing competencies.**

TIMELINESS

Timelessness is a performance criterion, which assesses the firm's ability to meet the needs of the customer within an expected timeframe, or due date.

See **Reengineering and the process view of manufacturing; Manufacturing strategy; Core manufacturing competencies.**

TIME-PHASED GROSS REQUIREMENTS

In material requirements planning(MRP), gross requirements are the known future demand for an item. In MRP time-phased requirements are stated period-by-period, where each period is typically one week long. Time-phased gross requirements in MRP are satisfied with what is available in inventory. What cannot be satisfied by inventory, is planned for production or purchase. Thus, gross requirements drive the MRP planning process. In Time-phased MRP, gross requirements for week 12 must be met by available inventory, production or purchase on or before week 12. What is available beyond week 12 cannot fulfill the gross requirements for week 12.

See **MRP.**

TIME-PHASED MATERIAL PLANNING

Material requirements planning (MRP) is an example of time-phased material planning, where gross-requirements are presented on a week-by-week basis and inventory, production and purchase are used to meet gross requirements period-by-period (i.e., week-by-week).

See **Capacity planning in make-to-order production; Capacity planning: Medium- and short range; MRP; Order release; Time-phased gross requirements.**

TIME-TO-FAILURE

The *time to failure* of a system is the interval between any given point in time and the next system failure. This is an important measure in maintenance management. In maintenance analysis, the time to failure is usually expressed as a random (or stochastic) variable.

See **Maintenance management decision models; Total productive maintenance.**

TIME-TO-MARKET

Time to market is often defined as the elapsed time between the formulation of a product concept and its introduction into the market. A short time to market is a key factor for product success. It allows being first to the market and thus gives first mover advantages. For example, Japanese automotive suppliers used faster time to market capabilities in the 1980s to update their product lines faster and thus gain market share in the US and Europe.

However, fast time-to-market does not guarantee product success. Even in the automotive industry, a few high end manufacturers continued to thrive with slower product development, but with high product design quality and integrity. Moreover, if short time-to-market is at the expense of design quality, market research, channel buildup, or production ramp up the product is likely to fail. The key is to not force a short time-to-market, but to achieve it with superior process capabilities (Meyer and Utterback, 1995).

In addition, fast time-to-market and the resulting first-to-market advantage may be more appropriate in stable, familiar markets with product line extensions. In new or fast changing markets, and for new products, (Terwiesch *et al.* 1998) leveraging design platforms and the patience to develop a breakthrough design through several failures may be more important than speed.

See **Concurrent engineering; Mass customization; New product development through supplier integration; Product design for global markets; Product design; Product development and concurrent engineering.**

References

[1] Meyer, M.H., and J.M. Utterback (1995). "Product Development Cycle Time and Commercial Success." *IEEE Transactions on Engineering Management*, 42, 297–304.

[2] Terwiesch, C., C.H. Loch, and M. Niederkofler (1998). "When Product Development Performance Makes a Difference: A Statistical Analysis in the Electronics Industry," *Journal of Product Innovation Management*, 15, 3–15.

TITLE VII, CIVIL RIGHTS ACT OF 1964

See **Human resource issues in manufacturing.**

TOC

See **Theory of constraints (TOC) in manufacturing management.**

TOOL MAGAZINE

Tool magazines are the tool holders attached to a computer controlled or CNC machine or placed near the machine for automatic tool change and easy access. Typically, a flexible manufacturing system (FMS) consists of numerically controlled (NC) machine tools, each with automatic tool-changing capabilities. The cutting tools required for all operations performed by the particular machine are stored in that machine's limited-capacity tool magazine. The configurations and the capacities (e.g., the numbers of slots) of the tool magazines become an important consideration (i.e., constraint) in the decision to load jobs on the FMS.

See **Disaggregation in an automated manufacturing environment; Flexible automation; Manufacturing with flexible systems; Tool management for FMS system; Tool planning and management.**

TOOL MANAGEMENT FOR FMS SYSTEM

A Flexible Manufacturing System, or FMS, is a highly automated, distributed and feed-back controlled system of data, information and physical processors, such as computer and manually controlled machine tools, robots and others in which decisions are made in real-time.

In such systems, parts, tools and clamping devices/fixtures must be managed real-time; the term "FMS tool management system" is used to describe this system.

Within this system of real-time executable software packages, tool management is concerned with the fact that tools have to be stored in the tool store, or warehouse, preset to their programmed sizes, and then delivered (in most cases automatically) to the processors (or machines) based on what is described in the FMS part program. (Note, that the FMS part program lists the tools that individual processors, or machines need in the order of processes described by the FMS part programmer).

Furthermore, tools have to be monitored, tested, re-adjusted, maintained and stored for further use. Although several of these activities can be automated, nevertheless this is still an area in which skilled labor and real-time computer systems have to collaborate to avoid system crashes and delay.

See **Disaggregation in automated manufacturing environment; Flexible automation; Manufacturing with flexible systems; Tool magazine; Tool planning and management.**

References

[1] Ranky, P.G. (1998). *Flexible Manufacturing Cells and Systems in CIM*, CIMware Ltd, http://www.cimwareukandusa.com (Book and CD-ROM combo).

[2] Schriber, T. and K.E. Stecke (1986). *Machine utilization and production rates achieved by using balanced aggregate FMS production ratios in a simulated setting.* Proceedings of the Second ORSA/TIMS Conference on Flexible Manufacturing Systems: Operations Research Models and Applications (Amsterdam: Elsevier Science Publishers B.V.), 405–416.

TOOL PLANNING AND MANAGEMENT

In most automated manufacturing systems (AMS) including flexible manufacturing systems (FMS), it becomes a critical issue to properly plan for the tool requirements based on product needs. There is also an on-going need to control and monitor a large number of different tools required by various manufacturing activities. Thus, tool management requires (1) planning strategy to ensure that the appropriate tools are available in the right quantities; (2) a control strategy to coordinate tool transfer between machines and cribs and to see that tools perform properly; and (3) a monitoring strategy to identify and react to unexpected events. At the (computer-aided) process planning level, the number and type of tools needed in producing parts or products are determined. This information provides a basis for projecting and planning for tool requirements for the actual manufacturing activities. Because many auxiliary tools are likely to be used in manufacturing, a control system is essential to avoid excessive duplication and loss or misplacement. In such a system of control, each tool is classified according to type and uses, and is assigned a tool inventory number. These tools are kept in one or more cribs until needed on the production floor. Tool control promotes standardization of tools and assures that tools will be kept in good condition and ready for use when needed. Computer database management systems are often used for this purpose. Other functions of tool management include: tool reduction, tool room support, tool condition monitoring, tool allocation, tool information management, tool transportation, tool replacement, and tool presetting.

See **Capacity planning: Medium- and short-range; Disaggregation in an automated manufacturing environment; Flexible automation; Manufacturing with flexible systems; Tool management for FMS systems.**

References

[1] Gray, A.E., A. Seidmann and K.E. Stecke, "Tool Management in Automated Manufacturing: A Tutorial," *Proceedings of the Third ORSA/TIMS Conference on FMS: Operations Research Models and Applications*, Elsevier, 93–98.

[2] Hankins, S.L. and V.P. Rovito (1984). "The Impact of Tooling in Flexible Manufacturing Systems," Paper presented in International Machine Tool Technical Conference.

TOTAL COST ANALYSIS

The essence of total cost analysis is to identify all relevant costs over the entire life of a product system or project. These costs are then summed to calculate the total cost of a decision. When the total costs are calculated for all attractive options, a proper comparison can be made and the best option selected.

Most decisions have readily visible costs as well as somewhat hidden costs. The visible costs are generally used for decision making while the hidden costs may be overlooked. As a result, many decisions that at first appear to be attractive turn out to be costly and yield an overall negative outcome. For example, Davis (1992) points out that many managers make international sourcing decisions based on the purchase price of the desired items. These managers often overlook many of the costs associated with acquiring, shipping, and working with the lower-priced products. As managers have become more cognizant of managing processes across functional boundaries, the use of total cost analysis has increased. In fact, in a recent study, more than 90 percent of the interviewed managers claimed that their firms use some form of total cost measurement to help manage value-added processes (Fawcett, 1998). However, for many of these firms, total cost analysis is in early – often unsophisticated – stages of use. Many relevant costs are not included in the total cost analysis because the information is either not available or managers have failed to identify them.

The use of total cost analysis should become more prevalent as the information capabilities of firms improve. The adoption of enterprise resource planning (ERP) applications will finally provide managers with the information they need to effectively employ total cost analysis.

See **Activity-based costing; Cost analysis for purchasing; Enterprise resource planning (ERP); Logistics: Meeting customers real needs; Target costing.**

References

[1] Davis, E. (1992). "Global Oursourcing: Have U.S. Managers Thrown the Baby Out With the Bath Water?" *Business Horizons* (July–August), 58–65.

[2] Fawcett, S.E. and M.B. Cooper (1998). "Logistics Performance and Measurement and Customer Success." *Industrial Marketing Management*, 27, 341–357.

TOTAL COST MANAGEMENT

Total Cost Management refers to a broad set of systems that include not just the purchase price of a product or service but the full costs of doing business with a given supplier. These costs are driven primarily by a supplier's quality and delivery performance, as well as the technical performance of the product or service. Additional costs considered as part of total cost management may include some or all of the following: transportation cost, duties and premiums, planning costs, purchasing costs, internal quality control costs, costs of holding inventory (due to leadtime, taxes and duties), warehouse costs (including obsolescence, handling, administration, carrying cost), defective materials costs, costs of field failures, general and administrative costs, factory yield costs, and service costs. Generally speaking, few organizations are able to effectively link these costs to the specific unit of a purchased product, yet the costs are real and are directly or indirectly affected by a supplier's quality and delivery performance.

Transportation cost plays a major role in the decision-making framework. A firm that selects a transportation mode or carrier based solely on the lowest initial cost may suffer for failing to take a total cost perspective. The lowest cost transportation mode or carrier may not provide reliable delivery or other services that exceptional carriers provide. The cost variable, while important, should not be the only variable used to arrive at a transportation decision; cost evaluation should always be in a total cost context. Faster modes of transportation almost always cost more than slower forms of transportation. In factoring other elements such as responsiveness, inventory costs, and customer satisfaction, total costs for faster modes of transportation may be less than the total cost for slower modes of transportation.

See **Activity-based costing; Cost analysis for purchasing; Target costing; Total cost analysis; Total cost of ownership.**

TOTAL COST OF OWNERSHIP (TCO)

Total cost of ownership requires a purchaser to identify and measure costs beyond the standard unit price, transportation, and tooling, when evaluating suppliers' proposals or supplier performance. Formally, total cost of ownership is defined as the sum of all expenses/costs associated with the purchase and use of equipment, materials, and services. To use a total cost approach, a firm must define and measure a purchased item's major cost components. Firms that accurately measure the total cost of ownership have the ability to identify cost variances from planned results and can take corrective action to reduce future problems.

The Total Cost of Ownership (TCO) concept extends beyond purchasing. Moreover, TCO can be viewed as an extension of cost-driven accounting systems such as Activity Based Costing (ABC), which strive to allocate costs to appropriate drivers. In this respect, TCO extends ABC to not only external sources of supply, but also to other value-adding activities both upstream and downstream from the organization (Eccles, 1991; Norton and Kaplan, 1992). Armed with better information about costs, such systems allow managers to make better decisions regarding procurement strategy, market and pricing strategies, and industry trends.

A problem with total cost analysis is that most firms do not have the systems to collect and interpret total cost data. Companies either lack the computerized systems to collect total cost data or they lack the accounting systems to capture cost data in the needed format. Furthermore, an exact definition of total cost requirements can vary widely – even within the same company. Traditional accounting methods simply do not focus on quantifying the costs associated with a total-cost activity system. This information, however, is critical for total cost analysis.

Total cost data must come from a number of different functions within a firm. Activities as diverse as finance, accounting, quality assurance, manufacturing, receiving and inspection, and purchasing all must contribute cost information. It can be difficult to coordinate the collection of data from diverse organizational functions, particularly with a manual data collection system. Purchasing plays a critical role in total cost analysis. It is the logical function to manage a total cost system because of its close interaction with functions inside and outside the firm, and its reliance on total cost data to support the sourcing process.

See **Accounting system implications of TOC; Activity-based costing: An evaluation; Activity-based costing; Cost analysis for purchasing; Target costing; Total cost analysis; Total cost management.**

References

[1] Eccles, Robert G. (1991). "The Performance Measurement Manifesto," *Harvard Business Review*, January–February, 131–137.

[2] Kaplan, Robert S. and David P. Norton (1992). "The Balanced Scorecard – Measures that Drive Performance," *Harvard Business Review*, January–February 71–79.

TOTAL COSTING

See **Accounting systems implications of TOC; Activity-based costing: An evaluation; Activity-based costing; Supply chain management: Competing through integration; Total cost analysis; Total cost management; Total cost of ownership.**

TOTAL ENTERPRISE MANUFACTURING (TEM)

"Total Enterprise Manufacturing is an envisioned state beyond Total Quality Management, Just-In-Time Manufacturing and employee empowerment. It is made possible by advances from many directions" (Hall, 1993: 33). Total Enterprise Manufacturing represents a composite of all characteristics of competitive manufacturers, who display flexibility, supplier partnerships, mass customization, customer interaction, open communication, and continuous improvement. Total Enterprise Manufacturing has the capabilities of virtual organizations, and agile manufacturers. A virtual organization will use teams and teamwork across the boundaries of functional departments within the organization, and across organizations. Agile manufacturers have the ability to respond to all kinds of changes very quickly. TEM have the capabilities of holonic systems, which are autonomous distributed systems that can be computer networks or human systems. To date no enterprise has achieved everything expected of Total Enterprise Manufacturing (Hall, 1993).

See **Customer service through system integration.**

References

[1] Hall, R.W. (1993). "AME's Vision of Total Enterprise Manufacturing." *Target*, 9 (6), 33–38.
[2] Hall, R.W. and J.W. Anderson (1993). "Enterprise Manufacturing." *Target*, 9 (5), 20–27.

TOTAL PREDICTIVE MAINTENANCE

See **Total preventive maintenance; Total productive maintenance.**

TOTAL PREVENTIVE MAINTENANCE

Total preventive maintenance (TPM) is the most rigorous form of equipment maintenance. Maintenance procedures can be categorized into two types: periodic maintenance and irregular maintenance. Both types of maintenance are important to a total preventive maintenance (TPM) program. TPM comprises the following three chief elements:

- Regular maintenance (housekeeping)
- Periodic overhauls (pre-failure replacement)
- Zero tolerance of deficiencies

Through these three guidelines, the philosophy of TPM becomes apparent: quality through readiness. The aim is to have the facility safe and ready for immediate use. In order to achieve this goal, rigorous practices are followed in much the same manner as for Total Quality Management (TQM). In TPM, operators are empowered to maintain their own machines. This takes advantage of the employees' experience and close involvement with their facilities. Maintenance time for minor repairs is reduced, and there is an opportunity for operators to develop preventive and predictive maintenance methods. For more major maintenance, specialists can be used from a maintenance department or cross-functional team. In this way, the role of the maintenance department becomes more clearly defined as back-up and specialists.

See **Just-in-time manufacturing; Lean manufacturing implementation; Maintenance management decision models; Total productive maintenance.**

TOTAL PRODUCTIVE MAINTENANCE (TPM)

Kathleen E. McKone

University of Minnesota, Minneapolis, MN, USA

Elliott N. Weiss

University of Virginia, Charlottesville, VA, USA

INTRODUCTION: Substantial capital investments are required for manufacturing almost all goods of economic significance. The productivity of these investments enable companies and nations to compete. The maintenance of capital investments involves significant recurring expenses. For example, in 1991, DuPont's expenditure on company-wide maintenance was roughly equal to its net income. Maintenance expenses vary depending on the type of industry; however, they are typically 15–40% of production costs. Companies attempt to control maintenance costs by keeping them at a specified budget level, a level often based on the previous year's expenses.

During the last decade, manufacturers found this approach to be insufficient. As companies have invested in programs such as JIT and TQM in an effort to increase organizational capabilities, the benefits from these programs have often been limited by unreliable or inflexible equipment. In the context of JIT & TQM use, rather than being seen simply as an expense that must be controlled, maintenance is now regarded as a strategic competitive tool. Total Productive Maintenance (TPM) has evolved as an effective program for improving equipment performance and increasing organizational capabilities.

TPM has resulted in significant improvements in plant performance. McKone, Schoreder, and Cua (1998) evaluated the impact of TPM practices on manufacturing performance and found that TPM has a positive and significant relationship with low cost (as measured by high inventory turns), high levels of product quality (as measured by higher levels of conformance to specifications), and strong delivery performance (as measured by higher percentage of on-time deliveries and by faster speeds of delivery). Their research indicates that TPM plays a significant role in improving manufacturing performance.

HISTORICAL PERSPECTIVE: TPM originated from the fields of reliability and maintenance – a pair of closely related disciplines that have become standard engineering functions in many industries. The primary objective of these functions is to increase equipment availability and overall effectiveness.

There have been four major periods in the history of maintenance management:

1. The period prior to 1950 was characterized by **reactive maintenance**. During this phase little attention was placed on defining reliability requirements or preventing equipment failures. Typically, equipment specifications included requirements for individual parts without the consideration of the reliability or availability of the entire system.
2. The second period, which saw the growth of **preventive maintenance**, involved an analysis of current equipment to determine the best methods to prevent failure and to reduce repair time. This period resulted from the emergence of the military equipment industry during World War II. Emphasis was placed on the economic efficiency of equipment replacements and repairs as well as on improving equipment reliability to reduce the mean time between failures.
3. The third period, called **productive maintenance**, became well established during the 1960s as the importance of reliability, maintenance, and economic efficiency in plant design was recognized. Productive maintenance has three key elements: maintenance prevention, which is introduced during the equipment-design stages; maintainability improvement, which modifies equipment to prevent breakdowns and facilitate ease of maintenance; and preventive maintenance, which includes periodic inspections and repairs of the equipment. General Electric Corporation is typically credited for initiating productive maintenance in the 1950s but the approach did not gain popularity until the 1960s (Hartmann, 1992). In the late 1950s the concepts of productive maintenance were also promoted in Japan.
4. The most recent period is represented by **Total Productive Maintenance (TPM)**. TPM officially began in the 1970s in Japan. Seiichi Nakajima, vice-chairman of the Japanese Institute of Plant Engineers (JIPE), the predecessor of the Japan Institute of Plant Maintenance (JIPM), promoted TPM throughout Japan and has become known as the father of TPM. In 1971, TPM was described by JIPE as follows:

> TPM is designed to maximize equipment effectiveness (improving overall efficiency) by establishing a comprehensive productive-maintenance system covering the entire life of the equipment, spanning all equipment-related fields (planning, use, maintenance, etc.) and, with the participation of all employees from top management down to shop-floor workers, to promote productive maintenance through motivation management or voluntary small-group activities (Tsuchiya 1992, 4).

TPM provides a comprehensive company-wide approach to maintenance management. In the 1980s, TPM was introduced in the United States. There are several reasons why American companies are utilizing TPM. Dominant, among other reasons, is that many organizations face competitive pressures from firms that have improved plant productivity, often through successful TPM implementation. One particular group of competitors, the Japanese, have demonstrated in many industries that maintenance is a critical component of their success. JIPM has emphasized the importance of maintenance and has awarded over 830 preventive-maintenance prizes to companies that have achieved a high level of success with TPM implementation.

It would, however, be a gross oversimplification to suggest that the Japanese influence is the only factor that has brought about the popularity of TPM. There are also several other changes in the manufacturing environment that have increased the importance of maintenance. JIT, TQM, and Employee Involvement (EI) programs have become more commonplace in industry to address the increasing demands of customers, companies have attempted to reduce costs, shorten production lead-time, and improve quality. These improvement programs require reliable and consistent equipment throughout the entire plant. Moreover, as customers become more demanding and processes become more interrelated, the need for an effective maintenance strategy also

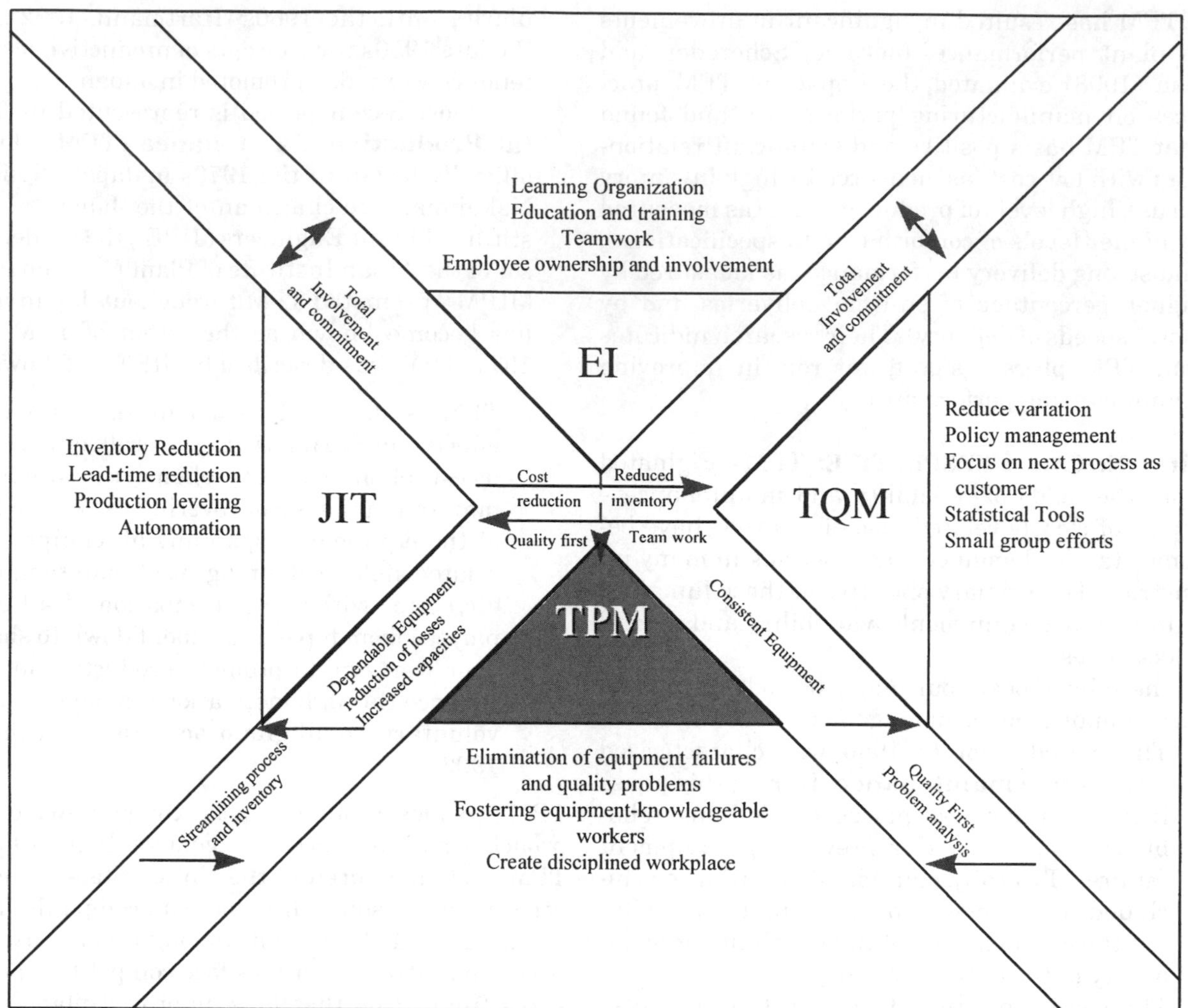

Fig. 1. The interdependence of JIT, TQM, TPM and EI.

increases. Figure 1 shows the interrelationships between TPM, JIT, TQM, and EI. From this diagram, it is clear that proper implementation of TPM will enhance the effectiveness of the other improvement programs. Schonberger (1986) also highlights the importance of JIT, TQM, EI *and* TPM to World Class Manufacturing.

As the technology of production equipment becomes more sophisticated, it is essential that operators and maintenance personnel are provided with the tools and training to support the new demands of the equipment. It is also important that production personnel take advantage of the new maintenance technologies, such as vibration analysis and infrared thermography, that have become useful means for predicting and diagnosing equipment problems.

The trends in manufacturing techniques and processes make it important for organizations to maintain manufacturing equipment in working condition to meet the higher performance criteria, produce maximum returns, and compete aggressively worldwide. Thus TPM has emerged as a result of heightened corporate focus on making better use of available resources.

A DESCRIPTION OF TPM: Figure 2 provides a framework for considering TPM activities. The elements of the framework have been developed based upon popular TPM literature (Hartmann 1992, Nakajima 1988, Shimbun 1995, Suzuki 1992, and Tsuchiya 1992, Gotoh 1991), and a number of site visits and interviews with practicing managers. While maintenance activities primarily focus on cost reduction and equipment effectiveness, TPM, which emphasizes a company-wide approach to maintenance, also plays a vital role in improving other manufacturing performance measures. Therefore, improved cost, quality, delivery, flexibility, and innovativeness are important goals of a TPM program and are represented in the framework.

TPM Values: Critical to successful TPM implementation are its core operating **values**.

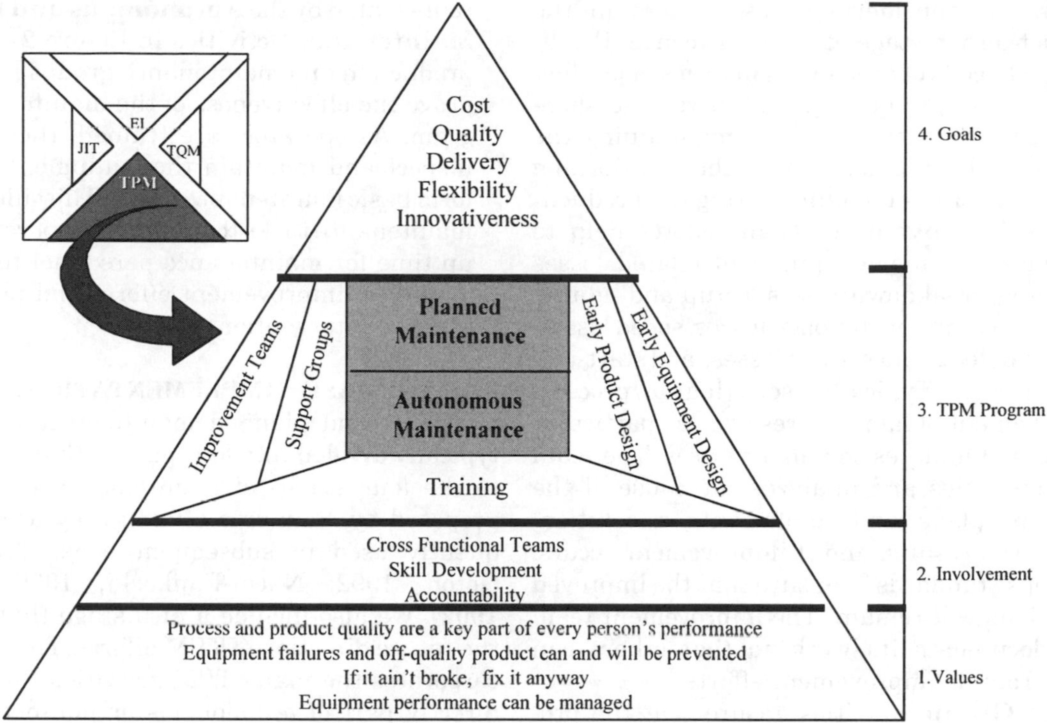

Fig. 2. A model of TPM.

Although these values should be tailored to any given organization, they offer broad guidelines for the management of equipment. The following represent the general values of TPM:

- Process and product quality are a key part of every person's performance.
- Equipment failures and off-quality product can and will be prevented.
- If it ain't broke, fix it anyway.
- Equipment performance can be managed.

Employee Involvement: As indicated by the first value, every employee should be involved in the TPM process. **Accountability** for equipment performance is important to the success of the TPM program. However, employees cannot be accountable for quality and predictable production if they cannot impact equipment performance. Employees' **skills** must be developed to meet the needs of their expanded roles.

The core values and skills of the organization provide the foundation for TPM implementation. Cross-functional teams also encourage employee involvement. Teams help to break down the barriers that are inherent in the traditional approach to maintenance. Teams also help to identify problems and suggest new approaches for elimination of the problems, introduce new skills, initiate training programs, and define TPM processes.

The TPM Program: While TPM is commonly associated with autonomous and planned maintenance activities, the program also includes other activities that help to improve equipment effectiveness over the entire life of the equipment. These activities include: Training, Early Equipment Design, Early Product Design, Focused Improvement Teams, and Support Group Activities (Tajiri and Gotoh 1992, Nakajima 1988, Gotoh 1991). The six major TPM activities are reviewed below.

TPM training should be given to every employee. Although internal capabilities may not be sufficient to achieve the goals of TPM at the start of the program, the capabilities for continuation of TPM must be developed. Also, as needs develop, training should support new TPM activities such as inspection, periodic restoration, and prediction analysis.

Early Equipment Management activities are typically driven by the research and development or engineering functions within the organization. Early equipment management considers the trade-offs between equipment attributes encompassing reliability, maintainability, operability, and safety. The engineering efforts involve the consideration of the life-cycle costing for equipment purchases as well as comprehensive commissioning periods prior to full production.

Early Product Design involves efforts to simplify the manufacturing requirements and improve quality assurance through product design. By considering these factors in the product-design stage, it is easier to meet the diversified needs of consumers in terms of product features, design, quality, and price. The shop-floor employees can focus on maintaining the process and equipment rather than on working out the logistics of manufacturing the product.

Focused Improvement Team efforts help to eliminate the major equipment-related losses including breakdown losses, setup and adjustment losses, minor stoppage losses, speed losses, quality defects and rework losses, and start-up/yield losses. Typically, selecting a process-improvement team to resolve a particular problem eliminates the major losses. The team then identifies and analyzes the cause of the problem, plans and implements a solution, checks the results, and if improvement occurs, develops standards to ensure that the improved conditions will remain. The improvement team also documents its work so that others can learn from its improvement efforts.

Support Group activities ensure that the production department does not produce useless or wasteful products, and that orders are filled on time, at the quality and costs that the development and engineering departments prescribe. This is not the sole responsibility of the production department; it requires a TPM program that embraces the entire company, including the administrative and support departments.

Much of the day-to-day maintenance planning and execution is performed by the production and maintenance personnel. These efforts are represented by the **Autonomous and Planned Maintenance** activities in Figure 2. Together, production and maintenance groups help to improve the effectiveness of the maintenance program. As operators are trained, they begin to inspect and maintain the equipment and perform basic maintenance tasks. This allocation of maintenance tasks to production operators frees up time for maintenance personnel to perform long-term improvement efforts and plan maintenance interventions.

MULTI-PHASE IMPLEMENTATION: The autonomous and planned maintenance efforts are typically divided into four phases that correspond to the four stages of maintenance development proposed by Nakajima (1988); they are also frequently used in subsequent books (Tajiri and Gotoh, 1992; Nachi-Fujikoski, 1990; Suzuki, 1992). We also include a fifth stage that foresees the future direction of TPM efforts. This phase incorporates the many TPM activities that are not directly part of autonomous or planned maintenance roles. Figure 3 shows a five-phase framework for TPM autonomous and planned maintenance development and includes typical steps for implementation. The five phases are described in detail below.

A practitioner's framework for **autonomous and planned maintenance** is shown in Figure 3. It is accepted by practitioners, easy to understand, and closely depicts the managerial decisions made for maintenance on a day-to-day basis within the plant.

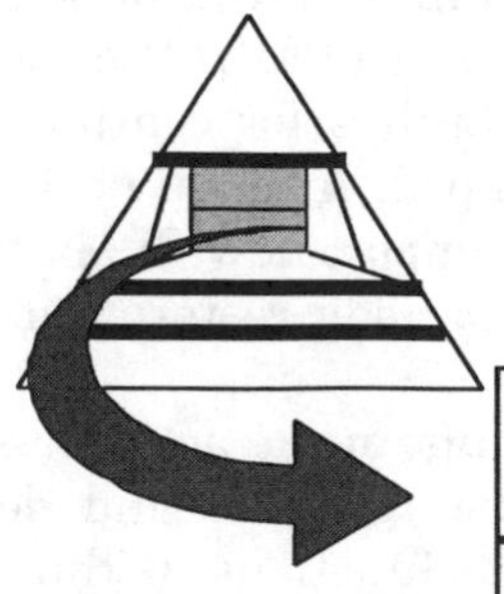

	Phase I	Phase II	Phase III	Phase IV	Phase V
	Reduce Life Span Variability	Lengthen Average Life Span	Estimate Life Span	Predict Life Span	Design Life Span
Planned Maintenance	Step 1: Evaluate Equipment Step 2: Restore Deterioration	Step 3: Correct Design Weaknesses Step 4: Eliminate Unexpected Failures	Step 5: Build Periodic Maintenance System & Identify Symptoms of Deterioration	Step 6: Build Predictive Maintenance System Step 7: Prevent Quality Defects	Step 8: Technical Analysis of Any Deterioration & Design Step 9: Implement in All Support Areas
Autonomous Maintenance	Step 1: Basic Cleaning Step 2: Eliminate Source of Problems Step 3: Set Standards	Step 4: General Inspection of Equipment	Step 5: Autonomous Inspection	Step 6: Maintenance for Quality Step 7: Autonomous Maintenance	Step 8: Process Improvement & Design Team Step 9: Implement in all Support Areas

Fig. 3. Five phases of TPM development.

1. ***Phases I and II: Equipment Improvement.*** The first phase of TPM involves efforts to reduce the variability in equipment life span. This phase involves the removal of assignable causes that reduce equipment life span. The efforts include restoration and cleaning of the equipment. At this stage, practitioners must decide how much time and money to invest in restoring the equipment to a base condition.

   The main objective of Phase II is to lengthen the equipment's life. In this stage, the cleaning and lubrication procedures become standardized and operators are educated to conduct detailed inspection of all equipment. Companies must decide how much time and money to invest in training operators and technicians. Cross-training personnel can make an organization more flexible and more responsive to maintenance needs; however, some tasks will still be too difficult or unsafe for operators to perform. Another Phase II decision is to make investments in eliminating the root causes of contamination and failure, and to simplify the inspection and maintenance tasks. Finally, practitioners must decide how much time to invest in order to sustain the equipment at its base condition.

   Both Phase I and II assume that it is possible to change the distribution of equipment failure occurrences. This assumption is central to the TPM philosophy. In traditional maintenance, the assumption is made that equipment conditions deteriorate over time, leading to failure or the need for replacement. With TPM, investments are made to reduce equipment problems; the assumption is that a goal of zero failures and defects is achievable.

   Most maintenance and reliability literature does not incorporate the important investment considerations present in Phases I and II of TPM development. Recent literature begins to address these improvement phases. McKone and Weiss (1997a) analyze the business decision to reduce the mean and variance of production cycle time through an investment in planned autonomous maintenance.

   Several researchers have investingated the learing benefits of maintenance (Fine, 1988; Marcellus and Dada, 1991; and Dada and Marcellus, 1994). By offering savings in operating costs, these studies demonstrate an economic incentive for gradually improving the maintenance process. Researchers have made some progress in addressing the concerns associated with the early phases of TPM implementation. However, there is still a significant need for research to study the decisions that are being faced by practitioners. In particular, studies are needed to evaluate in investments in time, capital, and employees in support of TPM efforts.
2. ***Phase III:*** After implementing the first two phases, equipment conditions should be dependable and operating conditions should be consistent. Equipment life span can be accurately estimated, and mechanics and operators can plan periodic inspections and renovations. Phase III of TPM development begins to determine the best type and interval of inspections and repairs.

   It is interesting that much of the traditional maintenance literature supports the decisions involved in Phase III of TPM development. Essentially, the failure rate is assumed to be a function of the equipment, and little attention is given to the possibility of improving the equipment or restoring the equipment to a better condition (Phase I and II decisions). See McCall (1965), Pierskalla and Voelker (1976), and Valdez-Flores and Feldman (1989) for a comprehensive reviews of literature relevant to this phase of maintenance management.

   Most of the Phase III literature compares the cost benefits of various maintenance policies: preparedness or preventive maintenance, and periodic- or sequential-replacement policies. More recent work extends the base models to incorporate multi-component equipment and multi-state equipment conditions or relax some of the assumptions of the early models.

   Phase III models have been extremely effective in the development of a philosophy of preventive versus reactive maintenance. Most organizations realize that it can be beneficial to replace equipment prior to failure, when the costs of breakdowns are high. However, in practice, the replacement interval is often chosen based on the equipment manufacturer's recommendations. Therefore, intervals are often established without considering the actual failure distribution (based on the environment in which it operates) or maintenance costs (based on the downtime costs associated with a particular plant). Additional studies are needed to develop better estimates of the actual failure distribution and the associated maintenance costs.
3. ***Phase IV:*** Phase III allows operators and technicians to gain a deeper understanding of the equipment and process. Phase IV permits personnel to use this new knowledge about equipment deterioration, together with diagnostic techniques, to predict failures of equipment, and eliminate equipment-related quality problems. Condition-based maintenance helps

get additional time out of the equipment as well as eliminate unexpected failures. The practitioner must thus decide where, when, and how to use prediction tools. In addition, maintenance policies must be coordinated with quality-control policies in order to reduce product quality problems.

Traditionally, equipment maintenance was treated as a method for increasing equipment availability. The goal of maintenance was to keep the equipment running. With the advent of quality-management efforts, however, the condition of the equipment became important to control the quality of the product. Several recent papers address the relationship between maintenance and quality (see, Tapiero, 1986; Rahim, 1994).

Predictive maintenance is commonly discussed in trade journals; however, very little academic research focuses on this area. Paté-Cornell, Lee, and Tagaras (1987) address some issues of predictive maintenance. They evaluate four maintenance polices: preventive maintenance, scheduled inspection of the process output, maintenance in response to equipment signals, and maintenance in response to output-quality signals. The authors test the various policies under different assumptions of the deterioration process, the signal accuracy, the risk aversion of the decision maker, and the reliability of maintenance. Based on the analysis, they recommend policies that perform best for certain environments. McKone & Weiss (1999) consider the joint use of continuously monitoring prediction tools as well as periodic maintenance policies in order to minimize maintenance costs. They recommend that periodic tools *not* be abandoned for the use of predictive tools, and provide decision rules for selecting the appropriate periodic policy, when predictive tools are available.

The product quality literature clearly provides some support for this stage of TPM development. Tools and techniques that help to define, identify, and eliminate known and/or potential failures, problems, and errors in the system are important to equipment improvement efforts. For example, Failure Model and Effect Analysis(FMEA) (Stamatis, 1995) helps to identify plant equipment and process problems that result in product quality problems. FMEA can help to link both maintenance and quality improvement efforts.

4. ***Phase V:*** The fifth phase involves an organization-wide focus on plant productivity. First, design teams made up of engineers, maintenance personnel, and operators prepare equipment so that cleaning and inspection standards are established and personnel are trained to produce efficiently and effectively. Phase V decisions also consider other non-maintenance systems, such as spare parts, raw materials, and production scheduling, that impact the equipment productivity and quality. Finally, efforts are made to eliminate losses in labor, energy, and materials in addition to equipment efficiency. The decisions in this phase primarily focus on organizational and systemic issues.

   Phase V is comprehensive. At this phase, functional areas within manufacturing must be carefully integrated. Essentially, all TPM activities must be coordinated for effective and efficient equipment operation.

CASE STUDIES: Thilander (1992) conducted a case study of two Swedish firms in an effort to define the benefits of the organizational aspects of TPM. The study shows that well-defined areas of responsibility, one individual who holds the overall responsibility for the maintenance, and direct contact between the operators and maintenance technicians have positive impact on productivity. McKone, Schroeder, and Cua (1999) studied policies at plants from Japan, Italy, and USA, and identified contextual factors that help explain differences in TPM practices.

There are several case studies of TPM that present the TPM development story or examples of TPM improvement and implementation activities in plants (Varughese 1993; Shimbun 1995; Steinbacher and Steinbacher 1993, Chapter 15; Hartmann 1992; Tsuchiya 1992; Suzuki 1992, Chapter 4; and Tajiri and Gotoh 1992).

Organizations Mentioned: DuPont; General Electric; Japanese Institute of Plant Engineers (JIPE); Japan Institute of Plant Maintenance (JIPM).

See **Integrated quality control; Just-in-time manufacturing implications; Kanban-based manufacturing systems; Lean manufacturing implementation; Manufacturing cell design; Total quality management.**

References

[1] Anderson, M.Q. (1981). "Monotone Optimal Preventative Maintenance Polices For Stochastically Failing Equipment," *Naval Research Logistics Quarterly*, 28, 347–358.

[2] Bain, L.J. and M. Engelhardt (1991). *Statistical Analysis Of Reliability And Life-Testing Models*, Marcel Dekker, Inc., New York.

[3] Chikte, S.D., and S.D. Deshmukh (1981). "Preventive Maintenance And Replacement Under Additive Damage," *Naval Research Logistics Quarterly*, 28, 33–46.

[4] Dada, M. and R. Marcellus (1994). "Process Control With Learning," *Operations Research*, 42, 2, 323–336.

[5] Fine, C.H. (1988). "A Quality Control Model With Learning Effects," *Operations Research*, 36 (3), 437–444.

[6] Gotoh, F. (1991). *Equipment Planning For Tpm: Maintenance Prevention Design*, Productivity Press, Cambridge, MA.

[7] Hartmann, E.H. (1992). *Successfully Installing Tpm In A Non-Japanese Plant*. Tpm Press, Inc., Allison Park, PA.

[8] Jorgenson, D.W. and J.J. Mccall (1963). "Optimal Scheduling Of Replacement And Inspection," *Operations Research*, 11, 723–747.

[9] Lee, H.L. and M.J. Rosenblatt (1989). "A Production And Maintenance Planning Model With Restoration Cost Dependent On Detection Delay," *LIE Transactions*, 21 (4), 368–375.

[10] Marcellus, R.L. and M. Dada (1991). "Interactive Process Quality Improvement," *Management Science*, 37 (11), 1365–1376.

[11] Mccall, J.J. (1965). "Maintenance Policies For Stochastically Failing Equipment: A Survey," *Management Science*, 11 (5), 493–524.

[12] Mckone, K., R. Schroeder, and K. Cua (1999). "Total Productive Maintenance: A Contextual View," *Journal Of Operations Management*, 17 (2), 123–144, 1999.

[13] Mckone, K., R. Schroeder, and K. Cua (1998a). "The Impact Of Total Productive Maintenance Practices On Manufacturing Performance," *Carlson School Working Paper*.

[14] Mckone, K. and E. Weiss (1998b). "Total Productive Maintenance: Bridging the Gap Between Practice and Research." *Production Operations Management* 7 (4) (Winter), 335–351.

[15] Mckone, K. and E. Weiss (1997a). "An Autonomous Maintenance Approach To Cycle Time Reduction: Guidelines For Improvement," Darden School Working Paper Series, Dswp-97–30, University Of Virginia, Charlottesville, VA.

[16] Mckone, K. and E. Weiss (1997b). "Analysis Of Investments In Autonomous Maintenance Activities," Carlson School Operations And Management Science Department Working Paper 97-7, University Of Minnesota, Minneapolis, MN.

[17] Mckone, K. and E. Weiss (1999). "Managerial Guidelines For The Use Of Predictive Maintenance," Darden School Working Paper Series, University Of Virginia, Charlottesville, VA.

[18] Nachi-Fujikoshi Corporation (1990). *Training For TPM: A Manufacturing Success Story*, Productivity Press, Cambridge, MA.

[19] Nakajima, S. (1988). *Introduction To TPM*, Productivity Press, Cambridge, MA.

[20] Özekici, S. and S.R. Pliska (1991). "Optimal Scheduling Of Inspections: A Delayed Markov Model With False Positive And Negatives," *Operations Research*, 39 (2), 261–273.

[21] Paté-Cornell, M.E., H.L. Lee and G. Tagaras (1987). "Warning Of Malfunction: The Decision To Inspect And Maintain Production Processes On Schedule Or On Demand," *Management Science*, 33 (10), 1277–1290.

[22] Pierskalla, W.P. and J.A. Voelker (1976). "A Survey Of Maintenance Models: The Control And Surveillance Of Deteriorating Systems," *Naval Research Logistics Quarterly*, 23, 353–388.

[23] Rahim, M.A. (1994). "Joint Determination Of Production Quantity, Inspection Schedule, And Control Chart Design," *IIE Transactions*, 26 (6), 2–11.

[24] Schonberger, R.J. (1986). *World Class Manufacturing: The Lessons Of Simplicity Applied*, Free Press, N.Y.

[25] Shimbun, N.K. (Ed.) (1995). *TPM Case Studies*, Productivity Press, Portland, Oregon.

[26] Steinbacher, H.R. and N.L. Steinbacher (1993). *TPM For America: What It Is And Why You Need It*, Productivity Press, Cambridge, MA.

[27] Stamatis, D.H. (1995). *Failure Mode And Effect Analysis*, Asqc Quality Press, Milwaukee, WI.

[28] Suzuki, T. (1992). *New Directions For TPM*, Productivity Press, Cambridge, MA.

[29] Tajiri, M. and F. Gotoh. (1992). *TPM Implementation: A Japanese Approach*, McGraw-Hill, Inc., New York.

[30] Tapiero, C.S. (1986). "Continuous Quality Production And Machine Maintenance," *Naval Research Logistics Quarterly*, 33, 489–499.

[31] Thilander, M. (1992). "Some Observations Of Operation And Maintenance In Two Swedish Firms," *Integrated Manufacturing Systems*, 3 (2).

[32] Thompson, G. (1968). "Optimal Maintenance Policy And Sale Date Of A Machine," *Management Science*, 14 (9), 543–550.

[33] Tsuchiya, S. (1992). *Quality Maintenance: Zero Defects Through Equipment Management*, Productivity Press, Cambridge, MA.

[34] Valdez-Flores, C. and R.M. Feldman (1989). "A Survey Of Preventative Maintenance Models For Stochastically Deteriorating Single-Unit Systems," *Naval Research Logistics Quarterly*, 36, 419–446.

[35] Varughese, K.K. (1993). "Total Productive Maintenance," University Of Calgary, A Thesis For The Degree Of Master Of Mechanical Engineering.

[36] Waldman, K.H. (1983). "Optimal Replacement Under Additive Damage In Randomly Varying Environments," *Naval Research Logistics Quarterly*, 30, 377–386.

TOTAL QUALITY

Chase and Aquilano (1995, p. 163) define Total Quality Management (TQM) as "managing the entire organization so that it excels on all dimensions of products and services that are important to the customer." Although, total quality movement is traced to advances made in Japanese manufacturing industry, the roots of the approach can be traced to Americans Juran and Deming. In TQM, everybody in the organization, i.e. the total organization, takes on responsibility for quality. The organization is perceived as a large quality chain linking customers and suppliers together by their interactions (Oakland, 1993, p. 8). This view embraces not just external customers and suppliers but recognizes internal ones too.

An organization wishing to implement TQM may use a framework such as that laid down for the Malcolm Baldrige Quality Award Program (Chase and Aquilano, p. 163). The framework lists the extensive set of criteria an organization could use to achieve total quality.

See **Quality management systems: Baldridge, ISO 9000, and QA 9000; Quality: The implications of deming's approach; Statistical quality control using control charts; Supplier relationships; Total quality management.**

References

[1] Chase, R.B. and Aquilano, N.J. (1995). *Production and Operations Management*. Richard D Irwin, Chicago.

[2] Oakland, J.S. (1993). *Total Quality Management: The Route to Improving Performance*. Butterworth-Heinemann, Oxford.

TOTAL QUALITY CONTROL

See: **Total quality management.**

TOTAL QUALITY MANAGEMENT

R. Nat Natarajan

Tennessee Technological University, Cookesville, TN, USA

WHAT IS TQM?: Collis P. Huntington, owner of Newport News Shipbuilding, engraved in 1917 the company's motto on the side of the building: "We shall build good ships here; at a profit, if we can, at a loss if we must, but always build good ships." (Dobyns and Crawford-Mason, 1991: 11). Five thousand years before Huntington, Egyptian inspectors checked the work of masons who dressed the stones of the pyramids, and the ancient Chinese had a department of the government to establish and maintain quality standards (Juran, 1993). These examples show that, meeting the standards, whether they were set by kings or merchants, was important. Ever since businesses have been engaged in the production of goods and services, quality has been a concern to the producer and the user. In that sense, quality has never gone out of fashion and never will. But what has changed over the years is the meaning of the term "quality," its significance to organizations, and how it is managed.

In the present day context, the word "quality" has come to mean more than it did in the past. It means meeting or exceeding the requirements, expectations, and needs of the customers – even those needs which are latent and not articulated by the customers. In recent years, a pragmatic and comprehensive system for managing quality called Total Quality Management (TQM) has evolved. Many organizations have successfully developed and implemented such systems, with dramatic improvements in performance. Many different experts and the practices of diverse organizations have contributed to the principles of TQM. Generally, TOTAL quality management implies performance excellence throughout the total system – including design, production, distribution, service, and the involvement of all categories of employees, customers, and suppliers in the quality initiative. The following are essential characteristics and core values that are common to the most effective TQM systems.

Customer Focus: This puts the customer first for real. The essence of customer focus is identifying the external and internal customers, their needs and expectations, and doing whatever it takes to satisfy them. Under TQM, the scope of the customer extends to those who are not directly involved in using the product but may have a legitimate concern such as the regulators, consumer organizations, the community, and the general public, whose concern over safety, health, the environment, and consumer protection are taken seriously by TQM.

Active Involvement and Support of the Top Management: The top management of the organization must actively demonstrate by their deeds, and not just words, their dedication to total quality. Their leadership must be informed and visible.

Active Involvement of All Employees: Top management must lead everyone, from line workers to clerical employees to professionals and managers, to participate and become actively involved in the TQM process. Such involvement must be supported by policies for training, empowerment, performance measurement, recognition, and reward.

Prevention Emphasis: The emphasis of TQM is on prevention of defects and errors, rather than after-the-fact detection and reaction, such as inspection. Defects are proactively eliminated by designing quality into the product and the process.

Continuous Improvement and Learning: This basic philosophy includes constant efforts to identify and eliminate non-value adding activities, and to continuously improve product/service, processes, and all the inputs. It also includes education, training, and upgrading the skills of human resources.

Management by Fact: This means using facts and data to solve problems, understand sources of variation, and uncover root causes. Analytical tools are used, both intensively and extensively throughout the organization, to collect, communicate, analyze, and share data.

Business Planning and Performance Measurement: Quality plans and objectives are integrated with overall business strategies and other objectives. Performance measurement, which includes comparative and competitive benchmarking, is aligned with quality objectives and organizational goals.

Collaborative Relationships: This refers to partnerships and alliances with suppliers, customers, educational institutions, and other organizations.

It is emphasized that any one of the above characteristics by itself does not imply TQM. They all interact and constitute a total system. Organizations will not realize the full benefits of TQM if they only pick and choose some of the above characteristics and core values for implementation. In order to fully understand what these characteristics imply, we need to consider the evolution of the terms "total," "quality," and "management" in TQM.

HISTORICAL PERSPECTIVE: Before industrial revolution, skilled craftsmen produced products in small quantities customized to meet the needs of individual customers. Generally, a small group of craftsmen were involved in product design, acquisition of the inputs, production process, and interaction with the customer. They were also responsible for the quality of what they produced. There were no quality inspectors as such. All that changed with the advent of mass production in factories to serve the needs of the mass markets. To make mass production more efficient, division of labor and task specialization were introduced. This led to organizations structured along functional specialties like design, production, marketing, and accounting. Consequently, production activities were separated from the task of inspecting the quality of what was produced. Quality meant conformance to certain standards, which were set mostly by the producers themselves. End-of-line inspection, which weeded out the products that did not meet the standards was the primary means for assuring quality.

It is important to note how quality was managed in such a set-up. Of course, it was not economical for cost and time reasons to inspect each and every item that was produced. But, in many instances, very rudimentary inspection procedures or no procedures were followed. This was particularly true of mass production systems, where consumer products were produced. In Ford's Highland Park plant, which had one of the most advanced forms of mass production for its time, finished automobiles were rarely inspected, and no Model T was ever road-tested (Womack, Jones, and Roos, 1991). Any defective product revealed by inspection was sent to a separate rework line, because rectifying the defects on the assembly line sacrificed production. How did customers' needs figure in all of this? According to Henry Ford, "They can have any color they want, as long as it is black." Because mass produced products were competing on price, mass production systems focused on raising productivity and reducing costs by automation and reduction of direct labor. Quality did not receive the same emphasis as cost in mass production.

In contrast, in batch-oriented job-shop type manufacturing systems, product quality was considered important. Even in those organizations, if customers received defective products, the inspectors, rather than the operators, were questioned. The more important question – Why was the product made defective? – was never raised. In other words, the source of chronic waste in the system was not attacked. Such practices also fostered the belief that quality can be inspected in and that inspectors were responsible for quality! In countries where the mass production systems were not widespread because of the smaller size of the markets, quality became a competitive factor in certain niche industries, e.g., optical equipment and cameras in Germany and watches in Switzerland.

However, the situation was quite different when it came to production of military goods, and the customer was the U.S. Department of Defense. Here, exacting standards of performance for the products were applied and were strictly enforced. Often, it was quality attained at any cost. This was true in many industrialized countries of the world including U.S., Japan, and, until recently, the former Soviet Union. Juran observes: "Japanese quality was all military until 1945. Their toys were shoddy, but their torpedoes were superior." (Juran, 1993: 43). Because of the attention given to quality and reliability in the case of hardware and materiel, important techniques were developed for sampling and inspection in production of military goods. World War II provided further stimulus to these efforts, resulting in the development and application of more advanced techniques. In fact, the origins of the now popular ISO 9000 standards could be traced to the standards developed by the U.S. Department of Defense.

From the 1930s onwards, along with advances in methods for process control and inspection, there were significant advances in measurement, gauging, testing, and other technical aspects of quality. Quality control had emerged as a separate function but with a narrow focus and managed by technically oriented specialists. In this era, the word "total" would mean application of quality control techniques across all product lines of a firm. This was the era of quality with a small "q."

STRATEGIC PERSPECTIVE:

From Small "q" to Big "Q"

Things began to change in the 1950s. As the economies of the industrialized countries recovered from the devastation of the war and began to grow, competition among suppliers increased, which increased the choices for the consumers. There were other major forces and trends worldwide: greater complexity and precision of products; product safety and liability litigation; government regulation of quality; and the rise of consumerist movements. In this emerging environment, the firms were forced to address the requirements of the customers and the regulators. Now, satisfying the customers required the coordination of all the activities that had a bearing on customer satisfaction but may take place in separate functional silos. For instance, the impact of design on conformance quality, i.e., meeting the design specifications during production, was recognized. It became imperative that customers' viewpoint was reflected in all the activities and quality had to be managed across the functions from design to marketing.

Dr. A.V. Feingenbaum was one such quality professional who had first-hand experience in dealing with such across-the-company cooperation issues at General Electric Company in the U.S. He originated the concept of Total Quality Control (TQC) to describe the broader scope of quality assurance function. In Western companies, with their functionally and professionally oriented specialization, it was the QC departments that designed and ran the quality programs. Top management was only peripherally involved, and the rank and file in the organization did not play any role in such programs.

By 1960s, many Japanese companies also had TQC or Company Wide Quality Control (CWQC) systems as they were called, but term "total" in TQC had a very different meaning in Japan. It meant involvement of every one in the hierarchy, from top management to the production worker on the shop floor, and the clerical worker in the office. Dr. Juran's lectures on managing quality in 1954 had convinced many Japanese top managers of the importance of top management's responsibility. As a result, they no longer viewed quality as a technical function that could be left to the specialists. Another important development was due to the efforts of Dr. Ishikawa who insisted on involving all employees in studying and promoting QC (Ishikawa, 1985). In 1962, he developed the quality control circles, small groups of Japanese workers who met voluntarily to discuss ways to improve their own work and the system. His approach was to provide easy-to-use analytical tools – including his own innovation, the cause-and-effect or fishbone diagram – that all workers could use to analyze and solve problems. He also persuaded Japanese management to not only support the activities of quality control circles but also incorporate their suggestions in the total quality effort. He also made them think about co-workers and colleagues as internal customers.

In Summary, the momentum generated by the following factors had a profound impact, not only on product quality but on cost as well: (1) top management's involvement and leadership for quality; (2) Lowering the barriers between departments; (3) relying on workers' brainpower for quality improvement; (4) adopting the philosophy of *kaizen* or continuous improvement of the product, the process, and the inputs; and (5) focusing on the needs of external and internal customers. Many Japanese companies, by the 1970s, were able to achieve rapid improvements in the quality of their products, and, more significantly, could deliver them consistently at lower costs to the market (Shiba, Graham, and Walden, 1993).

TQM, Strategy, and Competitiveness

From the 1970s, global competition has intensified due to the revolutions in communication, transportation, and reduction in barriers to trade. Threats to human safety, health and environment have become major concerns for citizens, governments, and corporations.

Quality may be necessary for the survival of a firm without guaranteeing its success. But the TQM perspective can provide the firm with new strategic options for business (Belohlav, 1995). To illustrate, a few years ago, companies like Xerox, Motorola, IBM, and the big three automakers in the U.S. found themselves in serious trouble against competitors from Japan, whose product quality and customer acceptance levels were much higher. For these firms, quality improvement became a necessity, not a choice. In these recovering companies, quality defined their business strategy. At Xerox, quality improvement thrust drove the entire strategy, and major organizational changes that followed. For Motorola, the new strategy meant that high quality would not only differentiate its products in the market place, but it would also make the company a low cost leader in the industry. Top priority given to quality in the overall strategy gave Motorola a significant competitive edge. It had, in fact, set the industry standards for quality, forcing its competitors to play catch up. Sometimes, a perceived quality edge can be sustained even after the competitors have caught up. For instance, General Motor's Geo Prizm and Toyota's Corollas were produced in the same NUMMI plant in California, a joint venture between GM and Toyota. Because of similar designs, same production processes, and workers, the design and conformance quality of these model vehicles were nearly identical. Yet, for many years, the Corollas outsold the Geo Prizms, and commanded a premium price because of the differences in customers' perceived quality.

Other Implications of TQM

Increased expectations of customers regarding not only quality but regarding cost, responsiveness, and flexibility as well, are challenging companies to design, produce, and deliver products better, cheaper, and faster. According to Dr. Curt Reimann, the former Director of the Malcolm Baldrige National Quality Award in the U.S., "Consumers now have choices from around the world. Choices may be made on the basis of price; on the basis of features, variety, and service; on the basis of responsiveness; on the basis of quality. All of the factors in purchase decisions – price, features, variety, services, responsiveness, and quality, not just quality alone – are addressed in an integrated way in total quality management." (Dobyns and Crawford-Mason, 1991: 93).

In some cases, quality improvement becomes a prerequisite for the implementation of strategies that are not necessarily built around quality. Manufacturers trying to implement Just-In-Time (JIT) production have to first achieve stable and capable manufacturing processes. The Pull system of production in JIT treats the next process as the internal customer. In JIT manufacturing, suppliers deliver defect-free parts directly to the line when incoming inspection is eliminated.

IMPLEMENTATION:

Framework of Quality Awards

TQM implementation can be based on many different frameworks. Many companies have used the approaches advocated by one of the quality gurus, i.e., Deming, Juran, Crosby, Feingenbaum, Ishikawa or other experts to launch their TQM efforts. Some companies have been more eclectic, integrating what they thought were the best ideas of the different experts and developing their own approaches to TQM.

In 1951, Japan instituted the Deming Prize to recognize companies that were successful in implementing Company Wide Quality Control, the Japanese term for TQM. In recent years, several countries ranging from Australia to Mauritius, have created national awards to promote quality. In the U.S., apart from the Baldrige award at the national level, many states have their own award programs that were patterned after the Baldrige award. The European Quality Award was developed for recognizing excellence in European companies. The criteria for these awards provide a non-prescriptive framework for TQM. Many companies have used these criteria for self-assessment. In many countries, a lot of prestige is attached to these awards. The winners are recognized in public ceremonies, and the results achieved by these companies and the methods they used to achieve them are widely disseminated. In some cases like the Baldrige, the winners are obligated to share non-proprietary information regarding their quality strategies, systems, and practices with others. Such award-winning companies are viewed as role models. Indeed, they have often stimulated other companies, both large and small, to initiate TQM.

ISO Quality Standards

The requirements of the ISO 9000 series of quality standards can also serve as a framework for developing a quality system. Though compliance with ISO 9000 is voluntary, many customers are requiring that their suppliers be certified. However, the registration process usually deals with only a part of overall business operations and therefore does not represent a system in the sense that most TQM experts like to emphasize. The criteria are largely process focused and do not include many of the essentials of TQM, such as personal leadership by the top management, integration of quality goals with business goals, achieving rapid rates of improvement in quality, emphasis on overall performance and competitive positioning, participation and empowerment of the workforce, and benchmarking (Reimann and Hertz, 1993). As they stand now, the criteria provide only minimum requirements for a quality system and therefore, certification should be viewed only as the beginning and not the end of the evolution towards TQM.

Critical Success Factors and Tools

There is no single correct way to implement TQM. Each organization has to develop a customized system that is tailored to its culture, history, and the industry it is in. However, the companies that have succeeded with TQM share some common traits:

Leadership: Top managers in these firms promote emphasis on quality, establish quality goals, enlarge business plans to include those goals, and provide resources for achieving those goals. They are personally involved in education, training, and recognition. They are accessible and have routine contact with employees, suppliers, and customers. Juran considers these roles of top management in TQM nondelegable (Juran, 1995). A strong leadership system that is not dependent on any one individual is often evident. Written policy, mission, and other documented statements of quality-based values provide clear and consistent communications. Through their personal roles, the senior leaders serve as role models reinforcing the values and expectations.

Systematic processes such as Management-by-Policy (Hoshi-Kanri) or Policy Deployment (PD) are used to deploy company strategies by considering the relevant ends, the means and measures to achieve those ends at each level of management (Shiba, Graham, and Walden, 1993). These processes reassure people that the organization has a strategy for the future, and make clear how they fit into the strategy. Based on leadership team's vision and active participation of all employees, strategy and action plans are generated. The plans cascade down hierarchically with progressively more detailed information on the means to achieve the goals. At each level, priorities (Hoshin) are developed and targets are set to focus on areas identified for improvement. Thus company strategy is made meaningful for all employees in light of their own responsibilities. These methods are oriented towards both results and improvement. At Hewlett-Packard, PD is called a planning and implementation methodology that is driven by data and supported by documentation.

Senior leaders also provide strategic directions in other ways. For instance, dimensions of quality critical to customers are identified. These dimensions are used to set clearly defined customer satisfaction and internal quality objectives and priorities. Aggressive targets are set. These targets go beyond incremental improvements, and look at the possibility of making large gains, and get the workforce to think about different processes. Strong drivers such as cycle time reduction or other targets are used to focus the efforts. Best practices within and outside their industry are benchmarked, and the results are used for improvements. Flatter organizational structures are created that allowed more authority at lower levels. Senior managers act as coaches rather than bosses. Cross-functional management processes and interdepartmental improvement teams are used. Organizational culture with respect to quality practices is changed – prevention is emphasized.

Customer Focus: Customer is the focal point of any TQM effort. Identifying their requirements and expectations is the very important first step in customer satisfaction, retention, and building a relationship with the customer. Successful companies use a variety of strategies and technologies such as market surveys, focus groups, feedback from employees in contact with the customers, trade shows, toll-free lines, electronic mail and bulletin boards for listening to "the voice of the customer." The choice depends on the type and size of customer segments. To be effective, such learning and listening strategies have to be applied continuously and tied to the overall business strategy. Customer-driven quality is more than just meeting specifications. It implies adapting and responding quickly to the changing and emerging customer requirements. It demands an awareness of developments in technology and competitive product offerings.

Quality Function Deployment (QFD) is a methodology designed to ensure that all major requirements of customers are identified and subsequently met or exceeded through the resulting design of the product and its manufacturing. QFD can be viewed as a set of communication and translation tools for making quality customer-driven – it translates the voice of the customer into the relevant characteristics of the product, parts, and process. QFD tries to eliminate the gap between what the customer wants in a new product, and what the product is capable of delivering. "Customer" could be internal, external, present or future, or it can be any set of requirements, such as ISO 9000 standards. QFD acts as a vehicle that facilitates inter-functional communication and simultaneous engineering involving marketing, design, and manufacturing. Through a series of cascading matrices (called House of Quality), the voice of the customer (the "what's") are translated into technical specifications (the "how's") to meet, successively, the requirements of product design, parts, process, and production. Often competitive assessment of "what's" – "where do we stand relative to competitors" – is also taken into account. The "how's" of one stage become the "what's" for the next stage. This can later lead to optimization of the values of the technical parameters ("how much"). QFD can also help speed new products to the market – it enabled Komatsu to introduce eleven new products in two and a half years, shocking its competitor Caterpillar, which managed only one to two per year!

Customer satisfaction is measured and tracked using a mixture of hard and soft measures tailored to different market segments. The results are then compared to key competitors and industry averages. A system for keeping customers satisfied is to provide easy means for their complaints and prompt resolution of complaints. TQM companies accumulate information on customers in a central database so that this intelligence can be used to drive improvements.

Continuous Improvement: Continuous improvement or *Kaizen* is one of the pillars of TQM. Kaizen is the Japanese term for continual improvement involving everyone – both managers and workers. In manufacturing, kaizen relates to finding and eliminating waste in machinery, labor, or production methods. Such an improvement can be either incremental or breakthrough ("reengineering") in nature.

Applied to manufacturing processes, improvements start with controlling the variation in the quality characteristic of the output. In this context, a landmark development took place in the 1920s, when Dr. Walter A. Shewhart of Bell Telephone Laboratories developed the theory of statistical process control (SPC). It was significant because it focused not on the output of the process, but on the monitoring, and, more importantly, on the improvement of the production process. His study of different processes led to the conclusion that all manufacturing processes exhibit variation. He identified two components: a steady component, which appeared to be inherent in the process, and an intermittent component. Shewhart attributed inherent variation, currently called common cause or systemic variation, to chance and undiscoverable causes, and intermittent variation to assignable or special causes. He also developed, based on statistical methodology, a graphical device, which became known as control chart – a misnomer because in reality it does not "control" any aspect of the process – to monitor the variation in the process and signal the presence of special causes.

His major finding was that assignable causes could be economically discovered and removed with a tenacious diagnostic program, but common causes could not be economically discovered or removed without making basic changes in the process (Shainin and Shainin, 1988). To date, this insight remains fundamental to monitoring and improving the quality of manufacturing processes. Deming's lectures to the Japanese industrialists in 1950, which served as a catalyst for the quality revolution in that country, and his theory of management were based on Shewhart's theory (Deming, 1986).

The manufacturing process has to be made stable and predictable, and brought under statistical control by identifying and removing the special causes of variation. Only then does it make sense to find ways to reduce the systemic variation, i.e., to improve the process. Once the process is improved, it is standardized and documented to enable transfer of knowledge and consistent application. Another cycle of improvement, commonly referred to as the Plan, Do, Check, Act (PDCA) Cycle, starts all over again.

Mistake-Proofing. In many modern manufacturing systems using high speed production, the use of SPC is not enough to prevent defects. The signals that control charts provide are often too late to take any preventive action. In such cases, Shigeo Shingo's innovation of mistake-proofing (Poka-Yoke) can be used. Mistake-proofing is a proactive and generally inexpensive technique for building quality into the manufacturing process and preventing problems (Shingo and Robinson, 1990). Examples are manufacturing or setup activities designed to prevent errors that could lead to product defects. For example, in an assembly operation, if every correct part is not used, a sensing device detects that a part was unused and shuts down the operation, thereby preventing advancement of the incomplete assembly to the next station.

Taguchi Method. As the evolution from small "q" to big "Q" took place, some important concepts and tools were developed. Of particular relevance to manufacturing are two methodologies that are used at the design (off-line) stage. One is the set of tools to evaluate and improve the manufacturability of design. For instance, reducing the number of parts and increasing the use of common and standard parts across designs, improve manufacturability and ease of assembly. The other methodology is quality engineering, attributed to Dr. Genichi Taguchi. It consists of off-line quality control methods applied at the product and process design stages in the product development cycle. This concept, developed by Dr. Taguchi, encompasses three phases of product design: system design, parameter design, and tolerance design. The goal is to reduce quality loss by reducing the variability of the product's characteristic during the parameter design phase of product development. The application of such techniques upstream at the design stage prevents of poor quality occurring downstream at the production stage.

Six Sigma. Another powerful concept, pioneered by Motorola, and currently used to drive improvements in many companies, is *six sigma*. Here process variations are reduced to half of the design tolerance, and process mean could shift as much as 1.5 times the process standard deviation from the target of the process mean. This limits the defect rate to 3.4 defects per million. In addition, if the process mean is centered on target, the defect rate is reduced to 2 defects per billion. This lofty goal is attained by reducing process variations due to insufficient product design margin, inadequate process control, and less than optimum parts and materials. Six Sigma could be applied to everything a company does, including administrative activities such as filing, typing, and document preparation.

Motorola launched the *six sigma* program because it found that a four-sigma (defect rate of 6210 per million) manufacturer, who spends an excess of 10% of sales revenues on internal and external repair, cannot compete against a six sigma manufacturer, who spends less than 1%. Motorola also stressed reductions in cycle times in all elements of its business, and set a goal of ten-fold improvements in cycle time to be achieved in five years. They found the goals of six sigma and cycle time reduction to be mutually supportive – cycle times were reduced when fewer mistakes were made.

Benchmarking. Often an important question to be addressed in TQM is *what* should be improved. This can be answered by competitive and comparative benchmarking. Benchmarking is the continuous systematic search for, and implementation of best practices which lead to superior performance. First, key processes of strategic importance are identified. Then, for each key process, the best competitors or the best-in-class companies are benchmarked; gaps in performance are assessed, priorities are developed and improvement initiatives launched to close the gap.

Employee Involvement: Involvement of all employees is absolutely critical to the success of TQM efforts. Continuous improvement and learning cannot take place if employees are not trained and motivated to improve processes. Since the knowledge of production workers plays a crucial role in identifying and eliminating special causes of process variation, training and empowering the workers involved in the process, and creating an organizational climate in which they can apply their knowledge is essential. Systemic variations are beyond the control of workers who work within the system created by the management.

For a long time, managerial implications were not recognized with respect to quality because process control was considered to be the domain of the quality control (QC) specialist. Now, top management in TQM-oriented companies view their employees as internal customers, and consider it their responsibility to provide them with the tools and the authority to take actions to improve customer satisfaction and processes. A systematic approach is used to solicit suggestions from all employees. This approach also recognizes and rewards them for their ideas. Employees are empowered to operate in autonomous, self-directed teams, engage in problem solving, participate in projects like benchmarking and mistake-proofing, and manage their own work. Companies like Motorola, and Dana Corporation have set up corporate universities and made significant investments in employee training and education.

RESULTS: Many studies now provide the evidence that generally TQM works and has a positive impact on: (1) internal measures ranging from productivity to inventory turns; (2) market measures such as stock prices, market share, revenue growth, customer satisfaction; and (3) bottom line measures such as profitability, reduction in costs, and in legal liabilities (Hiam, 1993). Deming Prize winners have demonstrated profit levels twice those of other Japanese companies. Studies conducted by the National Institute of Standards and Technology in the U.S. in 1995 and

1996 showed that, over time, the stocks of Baldrige award-winning companies outperformed the market index by 4 to 1. The studies mentioned above also indicate that TQM has produced better results at some companies than others. This is due to the differences in the approaches used and implementation.

There are a number of challenges faced by companies that implement TQM. As companies learn to make use of self-directed work teams in TQM, they are likely to meet with resistance because it involves transfer of work and responsibility from specialists and supervisors to the work force. Empowerment of workers has to be managed and accompanied by proper training. In the short term, this could be seen as adding to the costs. In many companies, there are a myriad of initiatives such as team building, SPC, and ISO 9000, without any linkage to the overall strategy. Without coordination by top management and strong linkages, they will be ineffective.

Globalilzation creates a problem because of the differences in quality systems in business units dispersed over the globe. Geographic distribution of people also precludes maximum interaction. Such companies have to balance central coordination and standardization of quality management with the local requirements and autonomy. TQM process is vulnerable to sharp downturns in the market and the economy, but TQM puts firms in a far better position to recover because of the superior management systems it creates. During downturns in the market, the commitment of top management to TQM can become weak; layoffs, not quality, may be seen as the way to cut costs, but they often lead to loss of vital knowledge about processes and customers. TQM can complement downsizing if the latter is based on customer and continuous improvement needs and is not a thoughtless reaction to reduce head count. Mergers, acquisitions, outsourcing, and takeovers are often accompanied by leadership and organizational changes and can create uncertainty about the future and direction of TQM.

The following are some predictions about TQM. Quality from the inception of a product will play a critical role in process reengineering to achieve big improvements in performance, in cycle time reduction, and in new product development (Godfrey, 1993). "Driving quality into the business" will be a major issue. As many companies begin to implement TQM, it is not improvement in performance per se but the rate of improvement that will become one of the keys to success. Companies will be looking for ways to achieve improvements, not just incrementally but at a much faster pace through TQM. TQM will increasingly play an enabling role to enhance and accelerate organizational learning through self-assessment, benchmarking, and partnerships with other organizations. Organizations will also focus on gaining a better understanding of the connection between TQM and innovation. The extensions of quality systems to span entire supply chains and virtual organizations will be a major challenge for the constituent organizations. In many companies, environmental issues are being brought under the umbrella of TQM (Wever, 1997). More companies are likely to follow suit.

Organizations Mentioned: Caterpillar; Dana Corporation; General Electric Company; General Motors-Nummi Plant; Hewlett-Packard; Highland Park Plant, Ford; IBM; Komatsu; Motorola; National Institute of Standards and Technology(NIST); Toyota; Xerox.

See **Benchmarking; Control chart; Cycle time; Deming prize; Design for manufacturability; Employee empowerment; Hoshin kanri; Integrated quality control; ISO 9000; ISO 9000/QS 9000 quality standards; Just-in-time manufacturing (JIT); Kaizen; Kanban-based production system; Lean manufacturing implementation; Malcolm baldrige national quality award; Manufacturing; Systems; Mistake-proofing; Plan-do-check-act (PDCA); Poka-yoke; Process capability; Pull system; Quality circles; Quality function deployment (QFD); Quality management systems: Baldridge, ISO 9000, and QS 9000; Quality: The implications of deming's approach; Six sigma; Statistical process control (SPC); Taguchi methodology; Total quality control (TQC).**

References

[1] Belohlav, J.A. (1995). "Quality, Strategy, and Competitiveness." *The Death and Life of the American Quality Movement*. Edited by R.E. Cole. Oxford University Press, New York, 43–58.

[2] Deming, W.E. (1986). *Out of the Crisis*. MIT Center for Advanced Engineering Study. Massachusetts.

[3] Dobyns, L., and C. Crawford-Mason (1991). *Quality or Else: The Revolution in World Business*. Houghton Mifflin, New York.

[4] Godfrey, A.B. (1993). "Ten Areas for Future Research in Total Quality Management." *Quality Management Journal*, October, 47–70.

[5] Hiam, A. (1993). "Does Quality Work?" A Review of Relevant Studies." *Conference Board Report Number 1043*. The Conference Board, Inc., New York, 7–38.

[6] Ishikawa, K. (1985). *What is Total Quality*

Control? The Japanese Way. Translated by David J. Lu. Prentice-Hall, New Jersey.

[7] Juran, J.M. (1993). "Made in U.S.A. A Renaissance in Quality." *Harvard Business Review*, July–August, 42–50.

[8] Juran, J.M. (1995). "Summary, Trends, and Prognosis." *A History of Managing for Quality*. Edited by J.M. Juran, ASQ Quality Press, Wisconsin, 603–655.

[9] Reimann, C.W. and H. Hertz (1993). "The Malcolm Baldrige National Quality Award and ISO 9000 Registration: Understanding Their Many Important Differences." *ASTM Standardization News*, November, 42–51.

[10] Shainin, D. and P.D. Shainin (1988). "Statistical Process Control." *Juran's Quality Control Handbook*. Edited by J.M. Juran and F.M. Gryna (4th ed.). McGraw-Hill, New York. Section 24, 24.1–24.40.

[11] Shiba, S., A. Graham and D. Walden (1993). *A New American TQM: Four Practical Revolutions in Management*. Productivity Press, Oregon, 3 (30), 411–460.

[12] Shingo, S., and A.G. Robinson (1990). *Modern Approaches to Manufacturing Improvements: The Shingo System*. Productivity Press, Massachusetts.

[13] Wever, G.H. (1997). *Strategic Environmental Management: Using TQEM and ISO 14000 for Competitive Advantage*. John Wiley, New York.

[14] Womack, J.P., D.T. Jones and D. Roos. (1991). *The Machine that Changed the World: The Story of Lean Production*. Harper Collins, New York, 37–38, 91–93.

TOYOTA PRODUCTION SYSTEM

Toyota Production System was developed by Toyota after the Second World War and became popular after the oil crisis of the early 1970's. The focal point of the system is to maintain a continuous flow throughout manufacturing. The idea originated with Henry Ford (1924) but was adapted to the specific circumstances faced in Japan. Kiichiro Toyoda introduced these ideas just prior to the War, but under war-time conditions they fell into disuse. Taiichi Ohno revived the system and developed it more fully after the war. Production flow is maintained by using a "pull" philosophy in which the users of material determine the timing and volume of material they receive. This system uses "Kanbans" or cards to manually control the flows of material between work stations, and from suppliers. By obtaining materials "just in time" before their use, Toyota aimed to minimize waste – less inventory is needed, less storage space, less material become obsolete, and overall efficiency increases. This policy also highlighted the importance of quality since defective materials would disrupt the smooth flow. The resulting system emphasized small lot production with great emphasis placed on quality.

See **Just-in-time manufacturing; Kanban-based production system; Lean manufacturing implementation; Manufacturing cells, Linked; Total quality management.**

References

[1] Ohno, Taiichi (1988). *Toyota Production System*. Productivity Press, Cambridge, MA.

[2] Shingo, Shigeo (1989). *A study of the Toyota Production System from an Industrial Engineering Viewpoint*. Productivity Press, Cambridge, MA.

[3] Toyoda, Eiji (1987). *Toyota: Fifty Years in Motion*. Kodansha International Ltd., Tokyo, Japan.

TQM

See **Total quality management.**

TRACE DRIVEN SIMULATIONS

A simulation model requires detailed data on system characteristics. For example, in the simulation of a manufacturing cell, data is needed to characterize the arrival process of the parts, their service requirements (that is, processing times and routings), and machine reliability (that is, frequency of machine breakdowns and repair times). In *trace driven simulations*, historical data can be read into the model from (large) external files to drive the execution of the simulation model. For instance, the arrival process of the parts can be generated via the production orders received during the past six months. This is a valuable approach for validation and verification. However, it may not be possible to replicate the simulation run, or to extrapolate the results to more general situations. For instance, if the data do not contain any information on rare events such as a major machine breakdown, the model will never exhibit such behavior either.

See: **Simulation analysis of manufacturing and logistics systems; Simulation languages; Simulation of production problems using spreadsheet programs; Simulation software selection.**

TRACKING SIGNAL

Tracking signal is an index of the validity of a forecasting model.

See: **Forecasting guidelines and methods; Forecasting in operations management; Safety stocks: Luxury or necessity.**

TRAIN-DO CYCLE

This a training philosophy. In just-in-time manufacturing education a concept, idea, or technique is taught to employees and then quickly followed by on-the-job application. This cycle of training and application are repeated to form the train-do cycle. Once a basic concept or skill is mastered, employees return for additional training. This approach utilizes brief training and educational sessions rather than a lengthy session whereby theory and application are separated by lengthy periods of time. The advantage to this type of approach is that employees can use their new found knowledge immediately while the information is still fresh in their minds.

See **Just-in-manufacturing: Implications; Just-in-time manufacturing; Lean manufacturing implementation; Manufacturing cell design.**

TRANSBORDER CHANNELS

A channel of distribution refers to the physical path that goods take between the source and the point of consumption. A typical channel could include the manufacturer, a wholesaler, a service provider, and one or more types of retailers. When a company enters the global market, its distribution channel becomes more complex because the channel now crosses national borders, and the company must be able to handle all of the problems that emerge in dealing with multiple infrastructures, government agencies, currency fluctuations, and channel intermediaries. These transborder channels are more complex and uncertain, diminish operational control, increase logistics (transportation, inventory, and documentation) costs, and lower the levels of responsiveness and delivery performance.

Because transborder channels are not transparent, it is important to map out the alternative channels, and use total cost analysis to compare their relative cost effectiveness. The need for thorough total costing is highlighted by the large number of costs incurred that are not normally found in domestic channels. Some of the typical costs include the following:

- Buffer Inventory at Production Facility
- Special Packing Costs
- Transportation and Loading Costs (Point of Departure)
- Paperwork Associated with Shipment
- International Carriers Charges
- Insurance Costs
- Pipeline Inventory Costs and Customs Waiting Costs
- Customs Broker Charge and Associated Paperwork
- Import Duties
- Quota and Local Content Laws (Offsetting Exports)
- Repackaging Costs, if Necessary
- Inspection, Point-of-Sale, and Shrinkage or Damage Costs
- Import Deposits

When these costs as well as the performance characteristics of the different available transborder channels are fully understood, the firm can make good decisions regarding market entry, development, and support.

See **Global facilities location; Global manufacturing rationalization.**

TRANSFER LINES

Transfer lines are a complex arrangement of production equipment with automatic parts transfer from one workcenter to the other. The manufacturing of complex components requiring many different machining operations can be automated using transfer lines. An example is the transfer line used in machining an engine block for automobiles.

See **Flexible automation; Manufacturing systems.**

TRANSIT STOCK

Inventory that is in transit between suppliers and customers, between stocking locations or factories is called transit stock. Transit stock is a function of the time the inventory spends in transit. Transit inventories are part of pipeline inventories, which also include inventories in distribution centers, field warehouses, and other locations.

Transit stock can be reduced by changing the method of transportation to a faster one, and by using a supplier located closer to the factory.

See **Bullwhip effect in supply chain management; Purchasing: Acquiring the best inputs; Supply chain management.**

TRANSPARENT CUSTOMIZERS

Transparent customizers provide individual customers with unique goods or services without letting them know explicitly that those products have been customized for them. The transparent approach to customization is appropriate when customers' specific needs are predictable or can easily be deduced, and when customers do not want to state their needs repeatedly. Transparent customizers observe customers without interaction and then inconspicuously customize their offerings within a standard package.

See **Mass customization; Mass customization implementation.**

TREND

A trend is a pattern in data when there is a steady increase or decrease in the data. The sales patterns of many companies, the GNP, stock market, and other economic indicators may reveal trend pattern from time to time. Trend can be short-term or long-term.

See **Forecasting guidelines and methods; Forecasting in operations management.**

TRUNCATION POINT IN SIMULATION

In simulation studies, the starting conditions cause a bias. There are several heuristics dealing with such initialization bias in simulation. One approach is to overwhelm it by conducting very long runs. For complex manufacturing simulations, this approach may not be feasible due to excessive computation cost and/or time. Another approach is to detect when the system has attained steady state conditions and start data collection beyond that point. This is typically referred to as the *truncation* approach, and the observation beyond which data collection is started is called the *truncation point*. The objective is to allow the system to "warm up" and to initiate data collection at the end of the warm-up period. The truncation point therefore signals the end of the warm-up period. There exists a large literature on the determination of the truncation point.

In an example of a manufacturing cell simulation that is initiated in an empty and idle state, the warm-up period ends when the simulation achieves its typical congestion levels. At this point in simulated time, data should be truncated, that is, all simulation output data collected up to this point should be discarded so as to minimize the impact of the initialization bias on the estimates of variables under investigation.

See **Simulation analysis of manufacturing and logistics systems; Simulation of production problems using spreadsheet programs.**

TWO-CARD SYSTEM

Two-card system refers to one kind of Kanban-based manufacturing. In the two card Kanban system, each workstation along the production line is proceeded and succeeded by inventory points: one immediately prior to the production activity storing input materials, and one immediately following the production activity storing the output. The two Kanbans used in this type of system are usually referred to as withdrawal Kanbans and production (ordering) Kanbans. Withdrawal Kanban prescribes the quantity of material that the subsequent process should withdraw from the preceding stage, while production (ordering) Kanban prescribes the quantity of the specific part, which the workstation should manufacture in order to replace those removed. Therefore, each part type must have at least one production ordering Kanban, and one withdrawal Kanban in circulation between each pair of stations.

See **Just-in-time manufacturing implementation; Just-in-time manufacturing; Kanaban-based manufacturing systems.**

TWO-STAGE ALLOCATION PROCESS

Conventional product cost accounting systems that assign indirect costs to jobs or products typically follow a two-stage allocation process. In the first stage, the system identifies indirect costs incurred by support departments (e.g., purchasing, personnel, maintenance, and quality control departments), then allocates these support (i.e., indirect) costs to production departments. In the second stage, the system combines the accumulated indirect costs with production department costs, and using direct labor hours or machine hours as the basis, creates predetermined department overhead application rates (sometimes called burden rates). Overhead costs are allocated to products using the predetermined burden rates.

Activity-based costing (ABC) systems also assign costs to jobs or products in two stages. In the first stage, the ABC system identifies essential activities within the organization and assigns costs to these activities to form activity cost pools. Typically, an ABC system will identify multiple activity cost pools. In the second stage, ABC systems select a unique activity to use as the basis for cost allocation for each activity cost pool, and create a predetermined activity cost application rate (called an activity driver rate). As products actually engage or use activities identified as activity drivers, costs are assigned to products using the predetermined activity driver rates.

Note that both the conventional costing system and the ABC system use a two-stage approach to assigning costs to cost objects such as products. There are two major differences between these two cost assignment systems. First, conventional cost systems use indirect departments in the first stage for pooling costs for assignment to products, while ABC systems use activities as the cost pooling focus. The second difference is that conventional cost systems typically use one cost allocation basis (e.g., direct labor hours) to apply a burden rate in the second stage. On the other hand, ABC systems will identify many different activities as activity drivers for cost allocation purposes. Hence, ABC systems are often much more complex (i.e., involve more cost assignment rates) than conventional cost systems.

See **Process capability; Activity-based costing; Activity-based costing: An evaluation; Unit-based costing.**

TYPE I ERROR (ALPHA ERROR)

In making decisions using control charts, a Type I error is committed when it is inferred that a process is out of control when it is actually in control. A process may be out of control if an observation (measurement) of the output from a process plots outside the control limits of a process control chart. The control limits are a finite distance from the center line, usually three standard deviations of the statistic being plotted. For a control chart for

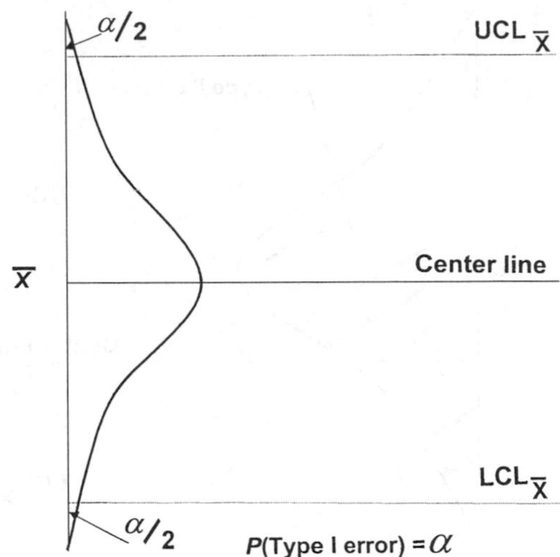

Type I error in control charts.

the sample average (X-bar chart), the control limits are placed at three standard errors (that is, 3 sigma) from the center line of the chart. There is a small chance (a probability of 0.0026) that an observation will fall beyond the three-sigma control charts based on normal distribution theory. When this happens, the inference drawn by the user of the control chart is that the process is out of control. This would be an incorrect conclusion if the process is really in control.

The figure demonstrates this concept of a Type I error. The probability of a Type I error is the sum of the two tail areas outside the control limits. If each tail area is denoted by $\alpha/2$, the probability of a Type I error is given by α. Such errors happen due to the inherent nature of sampling.

A Type I error is also known as a false alarm, which can send operators looking for special causes for the variation. However, no special causes exist. The time and effort expended in looking for special causes when there are none is the cost of a Type I error.

An approach to reduce the chances of a Type I error, as can be seen from the figure, is to widen the control limits. By moving the control limits further out from the center line, there is less chance of an observation falling outside the control limits, when the process is in control. For example, control limits could be placed at four standard errors away on each side of the center line. However, while this does reduce the probability of a Type I error, it increases chance of Type II error where a process is assumed to be in control when it is not.

See **Statistical process control using control charts; Type II error; Variability due to common causes; Variability due to special causes.**

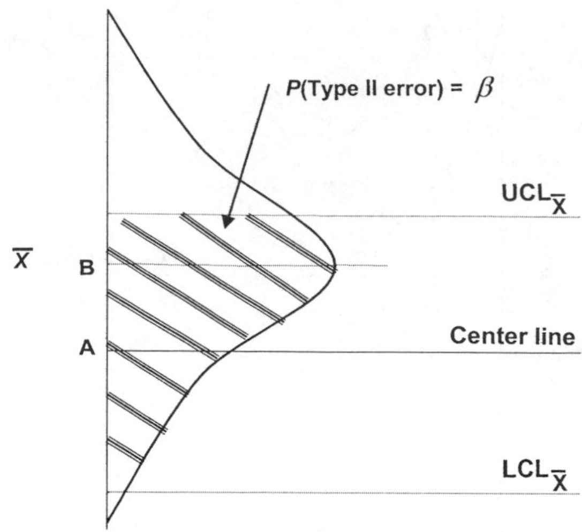

Type II error in control charts.

TYPE II ERROR (BETA ERROR)

A Type II error occurs when it is concluded based on quality control charts that a process is in control, when it really out of control. This is the error in inferring wrongly that there are no special causes prevalent in the system.

Quality control charts are usually interpreted using the rule that if no observations fall outside the control limits, the process is in control. Now suppose the process actually goes out of control. This could happen because of a change in the process mean due to the operator inadvertently changing the feed rate or depth of cut in a machining operation. Under these circumstances, it is still possible for a sample statistic to fall within the control limits, in which case the decision maker would infer that the process is in control. But, in reality, the process is not in control because of the above-mentioned reason. This is a Type II error as illustrated in the figure.

The figure shows the case where a process is out of control as its mean has shifted from point A to B. The sampling distribution of the sample means (X-bar) is normal but centered around the point B. The shaded area indicates the chance of a sample statistic plotting within the control limits. The probability of a Type II error corresponds to this shaded area, which is quite appreciable, and is denoted by β.

From the figure, it is observed that one approach to reduce the chances of a Type II error is to have control limits that are closer to the center line. Rather than three-sigma limits, one could construct two-sigma limits. While this will reduce the probability of a Type II error, it will increase the chance of a Type I error. Thus, a balance should be maintained between the probability of Type I and Type II errors. In most applications, the probability of a Type I error (false alarm) is maintained at a low level (say 0.0026) and control limits are set up based on this selected level. If it is very important for an out-of-control process to be detected as soon as possible because of the high cost of a nonconforming unit being shipped to the customer; control limits will be placed closer to the center line. For given values of the process parameters, to reduce both the probability of a Type I and a Type II error, the sample size must be increased. Increasing the sample size reduces the variance of the sampling distribution of the sample mean, which leads to a reduction of these errors.

See **Process capability; Statistical process control using control charts; Type I error; Variability due to common causes; Variability due to special causes.**

UNCAPACITATED FACILITY LOCATION DECISION

One consideration in facilities location decisions is the determination of the capacity of the new facilities to be built. This has lead to the classification of facilities location decisions (FLD) as capacitated and uncapacitated. In the uncapacitated FLD, the capacity limit of the new facilities is not preset, instead the optimizing model determines the ideal capacity level for the new facilities.

See **Facilities location decision; Global facility location.**

UNIFORM LOAD SCHEDULING

Uniform load scheduling is a synonym for mixed-model sequencing, an approach to scheduling in which each product is made in small batches and in a mix such that daily production of a given product approximately matches average daily demand for that product. The term uniform load scheduling describes this scheduling method for two reasons. First, under this approach, each product is made every day, and in the same number of units. Thus, the production load from one day to the next will be uniform. Second, the demand for components used in those products will be uniform from one day to the next.

For example, suppose a company sells three products, X, Y, and Z, and that the average daily demand is 15 units of X, 10 units of Y, and 5 units of Z. An "ideal" sequence might be XYXYXZ, repeated 5 times each day. This sequence is considered ideal because the production of each product is spread evenly throughout the day, meaning that requirements for the components needed in each product will also be spread evenly throughout the day.

Variations in the demand are handled by varying the frequency of the product/unit mix each day as long as total resource usage remains uniform. Final assembly is the control point for this concept. By establishing a master production schedule (MSP) with a uniform workload in final assembly, a uniform demand is created for all the preceding processes. The uniform workload practice at the macro level is one approach for extending JIT manufacturing to downstream material flow functions. At the micro level, a line-balancing effect is felt throughout the material flow path with the production lot size tending to one. Shingo (1981) refers to this as synchronization of one-piece flow. Uniform workload allows for reduction of inventories, improved utilization of machines, increased effectiveness of labor, and better use of just-in-time delivery of purchased parts (Shingo, 1981; Sugimori *et al.*, 1977).

See **Customer service through system integration; Just-in-time manufacturing; Kanban-based manufacturing; Lean manufacturing implementation; Master production schedule; Mixed-model sequencing.**

References

[1] Shingo, S. (1981). *Study of Toyota Production System From Industrial Engineering Viewpoint.* Japan Management Association, Tokyo.

[2] Sugimori, Y., Kusunoki, F. Cho, and S. Uchikawa (1977). "Toyota production system and kanban system: materialization of just-in-time and respect-for-human system." *International Journal of Production Research*, 15, 553–564.

UNIFORM WORKLOAD

See **Uniform load scheduling.**

UNIT COST

Unit cost is the total cost of material, labor, and overhead incurred in the production or acquisition of a single unit of product or service. Unit cost is calculated by dividing the aggregate cost of producing goods in a given period of time by the number of units turned out in incurring those costs.

Unit cost calculations include both variable and fixed manufacturing costs. Fixed manufacturing costs do not vary with the volume of output of a particular product. Fixed manufacturing costs are typically common costs that are incurred to support the production of many different products and must be arbitrarily allocated to various products according to some allocation system; it is common to allocate fixed costs based on direct labor. Thus, it is generally agreed that because of the necessity of allocating fixed costs, there is no such thing as a true unit cost. Activity-based costing is an alternative to unit cost.

Unit-based costing (UBC) is the traditional accounting method of allocating overhead using direct labor hours, direct labor dollars, or machine hours. Essentially, UBC is a fairly straightforward procedure of allocating the costs of manufacturing overhead to each item produced within the plant. Typically, an allocation basis is selected that is easily tracked within the organization's current information system.

An example of overhead allocation:

1. Let total overhead cost = \$500,000.
2. Let total direct labor hours =10,000 hours.
3. The burden rate then = 500,000/10,000 = \$50/(direct labor hour)
4. Overhead cost for a product with 2.5 hours of direct labor = 2.5×50 = \$125.

See **Activity-based costing; Activity-based costing: An evaluation; Two-stage allocation process.**

References

[1] Barfield, Jesse T., Cecily A. Raiborn, and Michael R. Kinney (1994). *Cost Accounting: Traditions and Innovations*, West Publishing Company, St. Paul, Minnesota.

[2] Henke, Emerson O. and Charlene W. Spoede (1991). *Cost Accounting: Managerial Use of Accounting Data*, PWS-Kent Publishing Company, Boston, Massachusetts.

UNIT-BASED COSTING

See **Unit cost.**

UNITED NATIONS ENVIRONMENT PROGRAMME (UNEP)

Headquartered in Nairobi, Kenya, this organization has evolved from the 1972 Stockholm Conference on the Human Environment. The organization seeks to integrate the environmental efforts of inter-governmental, non-governmental, national and regional bodies. In addition, UNEP fosters the development of environmental science and information, and has sponsored a wide range of studies exploring the sustainable use of natural resources, cleaner production, transborder environmental issues, and global environmental monitoring and outlook.

See **Environmental issues and operations management.**

UNIVERSAL PRODUCT POLICY

A product policy followed by international enterprises under which the company sells a product, which is physically identical in all countries it operates. Such a standardized product is also called uniform or universal product. Minor changes of the standardized product may be adapted to deal with issues such as the language on the package and the manual.

See **Product design for global markets.**

U-SHAPED ASSEMBLY LINES

Gerald Aase

Northern Illinois University, De Kalb, Illinois USA

Robert F. Jacobs

Indiana University, Bloomington, Indiana, USA

INTRODUCTION: Manufacturers often produce high-volume products using assembly lines, where each unit is built incrementally as it moves through a series of workstations. Typically, each workstation is manned by one operator, who repeats a portion of the assembly process on every unit moving on the line. Schonberger (1982) noted that most Western manufacturers arrange assembly lines in a straight-line layout, while their Japanese counterparts prefer U-shaped layouts. In general, this continues to be true, but some evidence suggests this is changing. For instance, a recent article in the *Wall Street Journal* (July 2, 1997) documents how Nokia, a manufacturer of digital cellular phones from Finland, successfully converted their 120-meter-long assembly line into several U-shaped assembly lines.

When using a straight-line layout, operators must work on a contiguous portion of the line. This guideline is followed in both practice and in the academic literature, since it tends to eliminate excessive operator travel and the likelihood of **interference between operators**. When using a U-shaped layout, units on the line follow U-shaped path. However, operators are allowed to work across both "legs" of the line without entering the

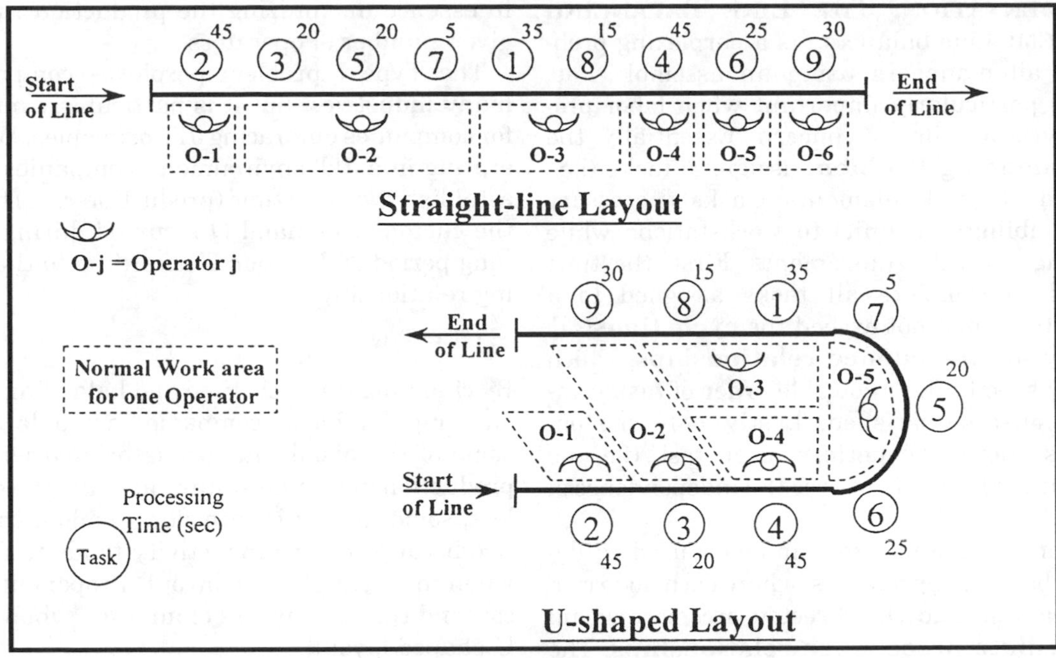

Fig. 1. Balances for straight-line and U-shaped layouts.

normal work area of other operators. Comparable straight-line and U-shaped layouts are shown in Figure 1 to help illustrate this fundamental difference. Notice that operator #2 in the U-shaped layout performs task {3} on the front side of the line, travels to the back side to complete task {9} on a different unit, and then returns to the front side of the line to begin the next cycle.

The remainder of this article provides an overview of U-shaped assembly lines and describes how they differ from lines using the traditional straight-line layout.

Historical and Strategic Perspective:

The assembly line concept was first introduced by Henry Ford in 1913 for assembling engines. However, the benefits for using assembly lines were not fully realized until 1914 when Ford used the moving line concept to reduce the time required to assemble an automobile from 12.75 hours to 1.33 hours (Blackler, 1978). Although other companies quickly adapted Ford's concept for assembling their products, managerial issues associated with the assembly line process were not addressed in the literature until the early 1950s. Since then, assembly lines have been the topic of many academic and professional articles, but most articles consider only the traditional straight-line layout. Literature addressing the U-shaped assembly line is generally written by proponents of just-in-time (JIT) manufacturing, but it is limited and mostly descriptive or anecdotal.

Monden (1983) and Hall (1983) offer the most comprehensive discussion concerning U-shaped layouts while many other books (e.g., Hirano, 1988; Wantuck, 1989; Sekine, 1992; Black, 1996) mention U-shaped layouts in passing. These authors invariably recommend using U-shaped layouts rather than straight-line layouts, asserting that U-shaped layouts support the tenets of JIT better and offer numerous benefits. Several advantages of the U-shaped layout concern the interaction of operators. Overall, communication and teamwork improve when using a U-shaped layout because the clustering of operators fosters more interaction, which creates an environment where continual improvement programs thrive and product quality improves. Another benefit of operator clustering is that more operators are available to offer support when problems occur, which reduces the potential for line disruptions (e.g., stopping the line or removing products from the line).

JIT proponents also claim that U-shaped layouts will often reduce material handling costs because materials enter and finished products leave the plant at the same point. Moreover, they assert that labor productivity will improve because U-shaped layouts will often require fewer workstations (operators) than comparable straight-line layouts. While research concerning total quality management (TQM) and teamwork support most of these benefits, the assertion concerning improvements in labor productivity received little attention until recently.

IMPLEMENTATION: THE LINE BALANCING PROBLEM: Line balancing is a perplexing problem for all managers using an assembly line, but it is particularly important when labor productivity is a primary concern. Essentially, the **Line Balancing Problem** assigns n tasks (i.e., independent work elements (tasks) necessary for assembling one unit) to workstations while satisfying several requirements. First, the time required to complete all tasks assigned to a workstation must not exceed the **cycle time** CT. Secondly, all **precedence relationships**, which partially specify the allowable order of task completion must be satisfied. Lastly, only one operator is allowed to perform each task (i.e., no task splitting), which ensures the independence of tasks.

Network diagrams are commonly used to depict line balancing problems, where each node represents a task and the directed arcs correspond to **immediate precedence relationships**. The network diagram in Figure 2, for example, illustrates a nine-task example problem with ten immediate precedence relationships. This figure also shows the grouping of tasks (henceforth referred to as task-to-workstation assignments) for a feasible balance requiring six workstations when using a straight-line layout and a 50-second cycle time. Notice that this grouping of tasks corresponds to the balance for the straight-line layout shown in Figure 1.

The Line Balancing problem has two variants which are known in the literature as Type I and Type II problems. The Type I problem minimizes the number of workstations (operators) required for a fixed cycle time (production rate). In an alternative formulation, the Type II problem minimizes the cycle time for a fixed number of workstations, in essence maximizing the production rate for a given number of operators.

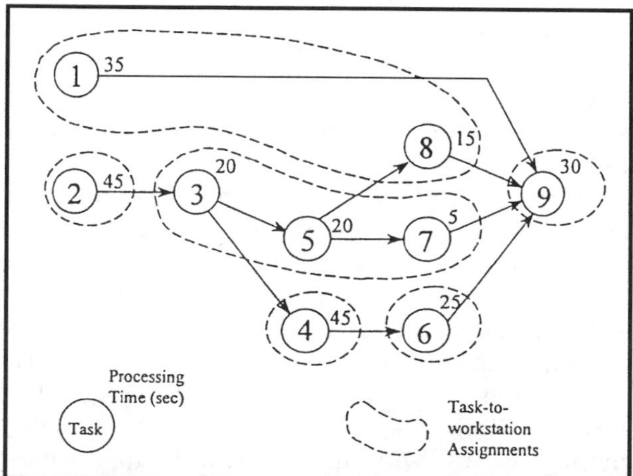

Fig. 2. Network diagram for a nine-task example problem.

The Type I problem perplexes companies in many industries, but it is particularly important for companies embracing JIT principles. When operating in a JIT environment, companies usually establish the cycle time (production rate) based on the customer demand Q incurred during a planning period of duration T according to the following relationship:

$$CT = T/Q \tag{1}$$

By changing the cycle time and balancing the line on a regular basis, companies are able to avert some of the disadvantages associated with overproduction and the resultant inventory. Nevertheless, some companies are slow to adopt this practice because labor productivity tends to decrease when in a straight-line layout. Proponents of JIT contend that this is less of an issue when using a U-shaped layout.

The well-known *Simple Assembly Line Balancing (SALB) Problem* defines the line balancing problem for a single product when using a traditional straight-line layout. The SALB problem simplifies the general line balancing problem further by assuming that each task requires a **deterministic** (i.e., no variance) **processing time** $t_i, i = 1, \ldots, n$. The *Simple U-line Balancing (SULB) Problem* modifies the SALB problem so operators are allowed to perform tasks on both legs of the U-shaped line.

Bryton (1954) was the first to describe the SALB problem, while Salverson (1955), who suggested the topic to Bryton, was the first to publish a mathematical formulation of the problem. Since then, the SALB and other related problems have been the topic of much research as seen in the review paper by Ghosh and Gagnon (1989). Many methods are available for solving the SALB problem including **heuristic** (Helgeson and Bernie, 1961; Hoffmann, 1963) and exact (Jackson, 1956; Held, Karp and Shareshian, 1963; Johnson, 1988; Hoffmann, 1992) procedures. Talbot, Patterson, and Gehrlein (1986) and Baybars (1986) provide comparison studies of heuristic and exact procedures, respectively.

While the fundamental difference between straight-line and U-shaped layouts may seem small, its effect on the line balancing solution procedures is significant. Consequently, the procedures that solve the SALB problem will not solve the SULB problem unless some modifications are made. Miltenburg and Wijngaard (1994) were the first to formulate the SULB problem and offer several solution procedures. By modifying the Held *et al.* (1963) dynamic programming procedure, they were able to solve "small" problems containing up to eleven assembly tasks optimally. They also posed a Modified Ranked Positional Weight heuristic for solving larger problems that are commonly found in practice.

Example Problem

Figure 1 shows optimal solutions for the SALB and SULB problems for the nine-task example problem when using a 50-second cycle time. The straight-line layout requires six operators where the first operator completes task {2}, the second operator performs tasks {3, 5, 7}, and so on. Notice that operators two and three, who were assigned more than one task, work on contiguous portions of the line. By allowing operators to work across both legs of the line, the U-shaped layout in Figure 1 requires only five operators. Since both layouts produce one unit every 50 seconds (i.e., identical production rates), the U-shaped layout yields an increase in labor productivity.

Will U-shaped Layouts Really Improve Labor Productivity?

It is clear that U-shaped layouts provides more alternative tasks-to-workstation assignments than comparable straight-line layouts because operators can handle not only adjacent tasks, but also tasks on opposite sides of the line. However, this flexibility does not necessarily imply that labor productivity will improve. A recent study by Aase (1997) attempts to identify how much labor productivity is expected to improve by switching from a straight-line layout to a U-shaped layout. More important, the author attempts to identify problem characteristics, where the expected improvement in labor productivity is more pronounced by examining the following three questions:

1. Will the improvement in labor productivity be greater when using larger main assembly lines (e.g., Nokia example previously mentioned) or when using smaller subassembly lines?
2. Will the improvement in labor productivity be more pronounced when a company is operating the line during high demand periods (i.e., shorter cycle time, higher production rate) or during low demand periods when the cycle time is longer?
3. Does the strength of the precedence relationships (as measured by **network density**) influence labor productivity improvement? The literature (Talbot *et al.*, 1986) indicates that it is more difficult to develop good, labor efficient balances as the strength of the precedence relationships increases, when using a straight-line layout. Will the U-shaped layout help overcome the difficulties associated with a structured assembly sequence, and allow of a more efficient balance?

Aase (1997) examines these questions by solving a set of **randomly generated problems** that represent a variety of actual applications. Each problem instance was solved as a SALB problem and then again as a SULB problem. By comparing the number of workstations required for the two layouts, the influence that the layout has upon labor productivity is determined.

He used two performance measures to indicate the improvement that a U-shaped layout offers over a straight-line layout.

Reduction in workforce
= (Minimum number of operators for the SALB problem) – (Minimum number of operators for the SULB problem).

Percent Improvement
= **Percent improvement in labor productivity**
= 100 × (Reduction in workforce)/(minimum number of operators for the SALB problem)

Results for the study show that the U-shaped layout requires 0.32 fewer operators than a comparable straight-line layout, representing a average improvement in labor productivity of 2.5%. For the instances where the U-shaped layout required fewer operators than a straight-line layout, the improvement in labor productivity averaged 7.00%. The maximum improvement in labor productivity for some problem instances was nearly 29%. Further analyses revealed that the *Percent Improvement* tends to be more pronounced when the lines are relatively small (fewer than 30 tasks), when the assembly sequence is very structured (network density ≥ 0.50), and during high demand periods when the line is operating at higher production rates.

CONCLUSIONS: JIT proponents assert that U-shaped assembly lines offer many benefits including an improvement in labor productivity relative to assembly systems using the traditional straight-line layout. A recent study by Aase (1997) confirms this assertion, and also found that the improvement was greatest under circumstances where managers find it most difficult to identify a good, labor efficient balance. It is important to note other potential benefits (e.g., teamwork, communication, problem solving) associated with U-shaped layouts.

Organizations Mentioned: Ford; Nokia.

See **Assembly line design; Just-in-time manufacturing; Just-in-time manufacturing implementation; Kanban-based manufacturing; Lean manufacturing implementation; Linked-cell manufacturing systems; Manufacturing cell design; Manufacturing systems.**

References

[1] Aase, Gerald (1997). "Improving Labor Productivity with U-shaped Assembly Lines," *Proceeding for the 26th Annual Meeting of the Decision Science Institute*, San Diego, CA.

[2] Baybars, Ilker (1986). "A Survey of Exact Algorithms for the Simple Assembly Line Balancing Problem," *Management Science*, 32(8), 909–932.

[3] Black, J.T. (1996). *The Design of the Factory With a Future*, McGraw-Hill, New York.

[4] Blackler, F.H.M. (1978). *Job Redesign and Management Control: Studies in British Leyland and Volvo*, Saxon House, Farnborough England.

[5] Bryton, Benjamin (1954). *Balancing of a Continuous Production Line*, Masters Thesis, Northwestern University.

[6] Ghosh and Gagnon (1989). "Assembly Line Research: Historical Roots, Research Life Cycles and Future Directions," *International Journal of Production Research*, 19(5), 381–399.

[7] Hall, Robert W. (1983). *Zero Inventories*, American Production and Inventory Control Society, Dow Jones-Irwin, Homewood, Illinois.

[8] Held, M., R.M. Karp and R. Shareshian (1963). "Assembly Line Balancing – Dynamic Programming with Precedence Constraints," *Operations Research*, 11, 442–459.

[9] Helgeson, W.P. and D.P. Bernie (1961). "Assembly Line Balancing using the Ranked Positional Weight Technique," *J Industrial Engineering*, 12(6), 394–398.

[10] Hirano, Hiroyuki (1988). *JIT Factory Revolution*, Productivity Press, Cambridge, Mass.

[11] Hoffmann, Thomas (1963). "Assembly Line Balancing with a Precedence Matrix," *Management Science*, 9(4), 551–562.

[12] Hoffmann, Thomas (1992). "EUREKA: A Hybrid System for Assembly Line Balancing," *Management Science*, 38(1), 39–47.

[13] Jackson, J.R. (1956). "A Computing Procedure for a Line Balancing Problem," *Management Science*, 2(3), 261–271.

[14] Johnson, Roger (1988). "Optimally Balancing Large Assembly Lines with 'FABLE'," *Management Science*, 34(2), 240–253.

[15] Miltenburg, G.J. and J. Wijngaard (1994). "The U-line Balancing Problem," *Management Science*, 40(10), 1378–1388.

[16] Monden, Yasuhiro (1983). *Toyota Production System*, Institute of Industrial Engineers, Norcross, Georgia.

[17] Salverson, M.E. (1955). "The Assembly Line Balancing Problem," *J Industrial Engineering*, 6(3), 18–25.

[18] Schonberger, Richard J. (1982). *Japanese Manufacturing Techniques*, The Free Press, New York, NY.

[19] Sekine, Kenichi (1992). *One Piece Flow: Cell Design For Transforming the Production Process*, Productivity Press, Cambridge, Mass.

[20] Talbot, Brian and James Patterson (1984). "An Integer Programming Algorithm with Network Cuts for Solving the Assembly Line Balancing Problem," *Management Science*, 30(1), 85–99.

[21] Wantuck, Kenneth (1989). *Just In Time For America*, KWA Media, Southfield, MI.

[22] *Wall Street Journal* (1997). "Nokia bets on Tarzan Yells and Whistles," July 2, B1.

U-SHAPED CELL

U-Shaped cell refers to a general layout of machines in a U-Shape for a manned manufacturing or assembly cell that uses standing, walking workers. This layout results in the input and output of the cell being side by side.

See **Assembly line design; Linked-cell manufacturing systems; Manufacturing cell design; Manufacturing systems; U-shaped assembly lines.**

UTILIZATION

Utilization of a resource such as a machine is typically expressed as a percent of the total available production time.

See **Manufacturing systems modeling using Petri nets; Queing theory; Simulation analysis of manufacturing and logistics systems.**

V

VALUE ANALYSIS

Value analysis is the thorough examination of existing product design specifications with the aim of lowering cost and improving quality, and is typically performed in the purchasing and/or engineering departments. The role of value analysis to ensure that any given product is designed for maximum return for the cost involved. Thus, value analysis implicitly recognizes the cost/performance tradeoff that exists in most product design and use situations. Perhaps the foremost benefit of value analysis is that it is a systematic approach to constantly re-evaluating the value-added of a given product. That is, value analysis forces managers to always look for a better way of doing things. For purchased parts, purchasing department generally leads the value analysis effort in collaboration with engineering. Current and potential suppliers may also be involved in sharing their expertise and viewpoints. For parts made internally, engineering typically drives value analysis initiatives.

See **Purchasing: Acquiring the best inputs.**

References

[1] Leenders, M.R. and Fearon, H.E. (1998). *Purchasing and Materials Management* (10th ed.). Irwin, Inc., Homewood.

[2] Dobler, D.W. and Burt, D.N. (1996). *Purchasing and Supply Management* (6th ed.). McGraw-Hill, New York.

VALUE CHAIN

A value chain represents the sequence of activities performed by a firm to design, produce, market, deliver, and support its product. "The value chain displays total value and consists of value activities and margins. Value activities are the physically and technologically distinct activities a firm performs. These are the building blocks by which a firm creates a product valuable for its buyers" (Porter, 1985: 38).

See **Customer service through system integration; Purchasing: Acquiring the best inputs; Supply chain management.**

Reference

[1] Porter, M. (1985). *Competitive Advantage: Creating and Sustaining Superior Performance*. The Free Press, New York.

VALUE ENGINEERING

Value engineering is an evaluation of the activities in the *value chain* to reduce costs; *value chain* refers to a set of business functions that increase the usefulness of a product or service to the customer, and may typically include research and development, design of products and services, production, marketing, distribution and customer service. See Value Analysis.

See **Target costing; Value analysis; Value chain.**

VALUE INDEX

See **Target costing.**

VALUE PRICING

See **Bullwhip effect in supply chain management.**

VALUE-ADDED ANALYSIS

See **Reengineering and the process view of manufacturing; Value-added tasks; Value analysis.**

VALUE-ADDED TASKS

Manufacturing processes are either value-added or non-value-added tasks. Value-added tasks are tasks that increase the worth of a product or service to a consumer, and a consumer would be willing to pay for it. Value-added tasks contribute to the final usefulness and value of a product, and thus increase the utility of the product.

In most processes, a large majority of tasks – often 85 to 90 percent – are non-value-added tasks. A value-added flow analyses can be designed to carefully analyze processes in order to identify tasks as either value-added tasks or non-value-added tasks. Once identified, management efforts can be undertaken to eliminate or reduce the non-value-added tasks. Material movement is a non-value-added task. In addition, in an attempt to enhance the value of the product or service produced, management's focus should be upon enhancing the value-added tasks involved in the production of goods or the provision of services.

See **Activity-based costing: An evaluation; Business process reengineering; Reengineering and the process view of manufacturing.**

Reference

[1] Barfield, Jesse T., Cecily A. Raiborn, and Michael R. Kinney (1994). *Cost Accounting: Traditions and Innovations*, West Publishing Company, St. Paul, Minnesota.

VARIABILITY DUE TO COMMON CAUSES

Variability is present in any manufacturing process. Part of this variability is due to a set of causes known as common causes or chance causes. Common causes are those that are inherent in a process. Common causes of variability can be reduced but never completely eliminated. Variability due to common causes is due to several sources that are inherent to the process, and it impacts all items processed.

Some examples of common causes of variation are as follows: poor product design, poor process design, unfit operation, unsuitable machine, untrained operators, inherent variability in incoming materials from vendor, lack of adequate supervision skills, poor lighting, poor temperature and humidity, vibration of machinery, inadequate maintenance of equipment, and inadequate environmental conditions due to noise and/or dust.

Variability due to common causes is also known as the natural variation in a process. When a process operates under a stable system of common causes only, it is said to be in statistical control. For a process in control, natural tolerance limits that account for expected variation in the quality characteristics of individual units could be computed. These limits are three standard deviations from the process mean assuming normality of the distribution of the product characteristic. Output from a process in control is predictable since the only variability that remains is due to common causes.

Problems arising due to the variability created by common causes are referred to as system's problems. It is management's responsibility to address these problems. Pioneers in the field of quality management believe that at least 85% of all problems are those due to common causes. Hence, the actions taken by management play a major role in influencing the inherent variability in the process. Operators cannot be held responsible for these problems.

How does one know whether the inherent variability exhibited by the process is acceptable? The process spread is used to answer this question. The process spread is six times the standard deviation of the quality characteristic in question, assuming normality. If the process spread is less than the spread between the upper and lower specification limits, which are based on customer requirements, then a majority of the product manufactured will be acceptable, assuming that the process is centered between the specifications. In this case, the process is deemed as "capable." Management's goal is to ensure that the process spread is less than the specification spread.

Certain actions of the management to accomplish this goal could be: improve product design, streamline an operation, buy a machine that has a better capability (implying less inherent variability), provide training for operators, identify a certified vendor with tighter controls on incoming material, and improve operating and environmental conditions.

Through the use of control charts, how do we identify if only common causes of variation are present? In a control chart, if observations plotted on the chart are within the control limits and no nonrandom pattern is visible, it is assumed that a system of common causes exist and that the process is in a state of statistical control.

See **Process capability; Statistical process control using control charts; Variability due to special causes.**

VARIABILITY DUE TO SPECIAL CAUSES

Special causes, alternatively known as, assignable causes of variability in manufacturing processes, are not part of the process as designed. This variability affects only certain items going through the process. These are causes to which a specific removable reason can be assigned. Hence, variations in the process caused by special causes are not inherent in the process. If special causes are identified and remedial actions are taken to eliminate them, the variability due to special causes can be removed from the system. One of the major objectives of a control charts is to identify special causes as soon as possible.

Examples of variation due to special causes are: the use of a wrong tool, incorrect feed rate, incorrect depth of cut, vendor not capable of supplying raw material or components that meet desired specifications, and fatigued operator. Quite often, detection and rectification of variation due to special causes are the responsibility of the people directly involved with the process. For example, the operator in a certain milling operation can adjust the feed, speed, or depth of cut to the desired level. Sometimes assistance of management is required to eliminate special causes from the system. For instance, if the current vendor is not qualified to deliver raw material and components that meet desirable standards, management must take action to replace the vendor. Management must set policies that prevent these undesirable special causes from recurring.

Special causes can sometimes be desirable. In such cases, management and operating personnel must make an effort to ensure that they do recur and thereby improve the process. Suppose it is found that, in an injection molding operation a new combination of temperature and pressure setting causes much less variation in the output characteristic than current settings. In this case, the new setting should become the standard operating procedure for the particular operation.

The removal of undesirable variability created by special causes is defined as bringing a process into control. When a process is in control, output from the process is stable and predictable. How does one determine if special causes of variation are present? There are several rules for detecting out-of-control processes. These are based on the pattern of the plot in a control chart. Systematic or nonrandom patterns suggest that a process could be out of control. Identification of the special causes of nonrandom patterns requires a knowledge of the process, equipment, and operating methods as well as their impact on the characteristic of interest.

See **Process capability; Statistical process control using control charts; Variability due to common causes.**

VARIABILITY IN MANUFACTURING

Variability is true in nature as well as in every manufacturing operation, no matter how precise and delicate the process is. For example, cans of baked beans vary slightly in net weight from can to can. And, the size of holes drilled by the same machine under exactly identical conditions do vary.

Variability in manufacturing is traceable to two kinds of causes: (1) random or natural variability, which is variability naturally present in a system, and (2) assignable or special variability, which is variability due to specific identifiable causes. Examples of random variability include vibration in machines, slight variations in raw materials, and variability in taking measurements. Causes of special variability include equipment that needs adjustment, defective raw materials, or human error.

See **Quality: The implications of W. Edwards Deming's approach; Statistical process control using control charts; Variability due to common causes; Variability due to special causes.**

Reference

[1] Sinha, M.N. and W.O. Wilborn (1985). *The Management of Quality Assurance*, John Wiley & Sons, New York.

VARIABLE COSTS

Variable costs change in proportion to the quantity of output. As production quantity increases, the cost increases; as production quantity decreases, so do the costs. Most accounting textbooks depict variable costs as varying directly with volume. This also means that the relationship of variable costs to production output is linear. If volume doubles, then variable costs also double.

Larger production volumes tend to generate lower average unit costs (a relationship often referred to as "economies of scale"). Average unit cost is the sum of average fixed cost per unit and the average variable cost per unit. Larger production runs allow a producer to allocate fixed costs over a greater number of units, thereby reducing the average fixed costs associated with each unit.

Examples of variable costs include the materials used directly in production, some types of labor, transportation, material ordering costs, and packaging supplies.

See **Activity based costing; Break point; Cost analysis for purchasing; Unit cost; Target costing.**

VENDOR RELATIONSHIPS

In the traditional manufacturing environment, the purchasing department permits its vendors to make weekly/monthly/semiannual deliveries with long supply lead times. A large safety stock is kept on hand by the customer just in case (JIC) something goes wrong. Quantity variances are large, and late and early deliveries are normal. This chaotic situation requires constant expediting. As a hedge against potential vendor problems, multiple sources are developed. This may happen because one vendor cannot handle all the company's work or the purchasing department likes to pit one vendor against another to reduce the cost of purchased goods.

The lean manufacturing system does it differently, it uses Just-In-Time (JIT) purchasing. JIT purchasing is a program of long-term vendor improvement. The buyer and vendor work together to increase quality, and to reduce lead times, lot sizes, costs and inventory levels. This is a win-win situation, where both companies become more competitive in the world marketplace.

In this environment, longer-term (18 to 24 month) flexible contracts are drawn up with three or four weeks' lead time at the outset. The buyer supplies updated forecasts every month which may be good for 12 months. The buyer commits to long-term quantity, and perhaps even promises to buy out any excess materials from the vendor. Frequent communication between the buyer and the vendor is typical. A Kanban type system controls the material movement between the vendor and the buyer. The vendor facility operates as if it is a remote cell of the buyer.

The buyer moves toward fewer vendors, often going to local, sole sourcing. Frequent visits are made to the vendor by the buyer to provide engineering support (quality, automation, setup reduction, packaging, and the like) to help the vendor become a JIT supplier of parts that need no incoming inspection. This is a form of technology transfer. The vendors learn form the buyer. The buyer and the seller work together to solve material problems.

See **JIT purchasing; Just-in-time manufacturing; Kanban-based manufacturing system; Manufacturing cells, linked; Manufacturing systems; Supplier partnership; Supplier relationships; Supply chain management.**

VENDOR-MANAGED INVENTORY (VMI)

VMI is a new inventory management concept that allows suppliers to manage the inventory at the customer's site, and maintain ownership of all inventory items until they are used by the customer in order to enhance supply chain efficiency. The supplier conducts regular review of on-site inventory (at the customer's site) and inventory is restocked or replenished as agreed with the customer.

See **Bullwhip effect in supply chain management; Just-in-time manufacturing; Purchasing: Acquiring the best inputs; Safety stock: Luxury or necessity.**

VERTICAL INTEGRATION

Vertical integration refers to the extent to which the elements of a supply chain are owned by a single firm. The ownership is meant to ensure that the supply chain elements are tightly coupled together. The classical example of extensive vertical integration is the Ford automobile company in the early part of the 20th century. Ford owned virtually all the supply chain from the assembly of the vehicle back to the plantations where rubber was grown for car tires. Ownership of the supply chain means that the flow of goods and services can be planned and co-ordinated so that disruptions are minimised in the production process.

Hayes and Wheelwright (1994) identify some key aspects of vertical integration strategy. Two major decisions concern the direction and extent of integration. Organisations have the option to vertically integrate "backwards" or "forwards". Backward integration involves owning and controlling the upstream part of the chain, i.e., suppliers and your suppliers' suppliers. Conversely, forward integration is concerned with owning and controlling the downstream part of the supply chain between your company and the end customers. The decision as to how far upstream or downstream the company should control the chain is an important strategic decision.

Current trends where organizations increasingly outsource non-core activities mean that ownership of supply chains can become more fragmented. On the other hand, moves to reduce the number of suppliers and to foster increased collaboration means that improved co-ordination can be achieved by means other than vertical integration.

See **Core manufacturing competencies; Manufacturing strategy; Supplier partnership as strategy; Supplier relationships.**

Reference

[1] Hayes, R.H. and Wheelwright, S.C. (1994). *Restoring Our Competitive Edge*. John Wiley, Chichester.

VIRTUAL CELLULAR MANUFACTURING

The concept of virtual manufacturing cell was proposed by the National Institute of Standards and Technologies (NIST) to address specific control problems in the design of their Automated Manufacturing Research Facility (AMRF). In a "virtual manufacturing cell" the machines that belong to that cell are not physically located together, but are identified as a group only in the system control software. When conditions change, the machines are re-assigned to form different cells. The result is dynamically formed cells, whose compositions evolve in response to changing shop conditions. Unlike traditional cells that are permanent and physical, dynamic cells are temporary and logical, rendering them as "virtual cells." This concept is also labeled as virtual cellular manufacturing.

See **AMRF; NIST; Virtual manufacturing.**

VIRTUAL ENTERPRISE

A virtual enterprise is a seemingly single entity whose vast capabilities result from numerous collaborations assembled as and when they are needed. This concept of "virtual enterprise" can be defined as a rapidly configured multidisciplinary network of small, process-specific firms to meet a window of opportunity to design and produce a specific product.

See **Electronic data interchange in supply chain management; Virtual manufacturing.**

VIRTUAL MANUFACTURING

Pratap S.S. Chinnaiah and Sagar V. Kamarthi
Northeastern University, Boston, MA, USA

STRATEGIC PERSPECTIVE: Business is becoming more global and complex every day. Business climate is changing rapidly and is becoming opportunity-based. Technologies are changing so fast that no single company can meet every market opportunity all alone any more. Customers are also participating in the product specification process helping designers to create and develop new products and services. Even competitors are embracing one another to develop new technologies, enter new markets, and make products that they can't produce on their own. Because of these reasons more and more business partnerships are emerging among companies and entrepreneurs.

Today's joint ventures, strategic alliances, and outsourcing activities represent only a very small beginning of a bigger future for spontaneous partnerships. These partnerships span both manufacturing and service industries. Virtual organization (VO), virtual enterprise, virtual factory, cooperative manufacturing, and networked manufacturing are synonymous terms for virtual manufacturing (Buzzacott, 1995; Camarinha-Matos *et al.*, 1997; Davidow and Malone, 1992). Such partnerships among companies will be less permanent, less formal, and more opportunistic. In VO, companies will band together to meet a specific market opportunity but may go apart once the need evaporates. These partnerships are facilitated by high-speed communication networks, information databases, and common standards for swapping design drawings. Information networks will enable far-flung companies and entrepreneurs link up and work together from start to finish of a project. These relationships make companies far more reliant on one another and require far more trust than ever before. They will share a sense of "co-destiny," in the sense that the fate of each partner is dependent on the others. This new corporate model redefines the traditional boundaries of the company. The closer cooperation among competitors, suppliers, and customers makes it harder to determine where one company ends and another begins. If it becomes widespread, the VO model could become the most important organizational innovation since the 1920s when Pierre Du Pont and Alfred Sloan developed the principles of decentralization to organize giant and complex corporations. There are six strategic reasons, which are listed below, that motivate companies to use a virtual organization model (Goldman *et al.*, 1995).

1. **Sharing infrastructure, R&D efforts, risk, and costs:** The first strategic consideration is the value of sharing risk, infrastructure, R&D, and the cost of resources (human or technological). For a small company, access to specialized manufacturing equipment might be justification enough to join a virtual organization. For a large company, it might be the R&D need that encourages the company seek virtual partnerships. For example, Chrysler, Ford, and GM, together with various government agencies formed the USCAR consortium for cooperatively developing a wide range of generic automotive technologies such as catalytic converters.
2. **Linking complementary core competencies:** The second strategic reason to use the

virtual organization concept is to unite complementary core competencies of different companies in order to serve customers, whom the separate companies could not serve on their own. Each member of the virtual organization brings something unique that is needed to meet a customer need. The joint development of Hewlett-Packard HP-100 palmtop computer with a built-in Motorola pager is an example of this strategy.

3. **Reducing concept-to-cash time through concurrency:** The possibility for multiple companies to operate in parallel and perform many tasks concurrently is the third reason for forming a virtual organization. In this process, the speed of development is increased and concept-to-cash duration is reduced. In the commercial world, concept-to-cash time is an important measure of a company's agility. In a virtual organization, this time is reduced, in part by not having to build new facilities, or find, hire, and train new people, and in part by the concurrent operations of multiple partners. For example, Motorola, Apple, and IBM collaborated to bring out PowerPC chip in the quickest possible time. Another example is Apple and Sony working together to bring out its line of PowerBook products in a short time.
4. **Increasing facilities and apparent size:** The virtual organization is a way of leveraging a company's ability to satisfy and enrich its customers. For small organizations, serving the customers through a virtual organization allows the company to show depth of backup, additional breadth of capability, and the financial ability to deal with what the customers may view as a large company. The need to increase apparent size, facilities, and scope is not limited to small companies competing for larger opportunities. As a truly global economy emerges, with customers scattered all over the world, the ability of a virtual organization to serving the customers anywhere in the world becomes an important competency in its own right. Agile Web in Pennsylvania is an example of cooperative efforts of small companies, who need, on occasion, a critical mass of capability to secure business. This web can form focused virtual organizations quickly out of an amorphous pool of resources.
5. **Gaining access to markets and sharing market or customer loyalty:** Market access and product loyalty are two very valuable core competencies of an organization. Even these values may be shared in a virtual organization. For example, Intel successfully made the public aware of its name by having its customers label their computers "Intel Inside." This idea of the end user being aware of contributing organizations will become prevalent as virtual organizations become more common.
6. **Migrating from selling products to selling solutions:** A product may offer different value to different customers at the same time. The value of a solution is contextually dependent on the significance of what the customer is able to do or avoid doing. Virtual organization is an enabling mechanism for the enrichment of the customer. It is sometimes necessary for skill-based product or service suppliers to join in a virtual organization with other skill-based providers under what one calls the general contractor principle. A general contractor synthesizes the skills of a group of skill-based providers into a solution-based product with known value to the customer. For example, home builders play the role of a general contractor.

DESCRIPTION: The term *virtual* in the phrases *virtual factory, virtual enterprise*, and *virtual organization* has its origins in the computer industry. The interface of Macintosh computer was designed to look and act like a virtual office, with file folders, trashcans, and an area in which one could work simultaneously in several different program *windows*. The so called *Virtual reality*, generates a perception of reality in a human subject. It uses devices that stimulate more than one sense organ, and a dynamic model of real or fictitious environment (Shukla *et al*., 1996). Virtual reality (VR) allows its users to intuitively interact with the virtual environment and its objects as if they were real by immersing them in a 3-dimensional computer generated simulation. This method of conveying information promotes understanding of complex systems even in people with limited past experience or knowledge about the system. Virtual manufacturing (VM) is the name given to an evolving area of research that aims at integrating diverse manufacturing related technologies, under a common umbrella, using VR technology (Shukla *et al*., 1996). The scope can range from integration of the design sub-functions such as drafting, analysis, and prototyping to integration of all the functions within a manufacturing enterprise such as planning, operation, and control.

Virtual manufacturing is an integrated, synthetic manufacturing environment exercised to enhance all levels of decision and control in a manufacturing enterprise (Shukla *et al*., 1996). VM can be described as a simulated model of the actual manufacturing setup, which may or may not exist. It holds all the information relating to the process, the process control and management, and product specific data. It is also possible to have one part of the manufacturing plant to be real and the other to be virtual.

One aspect of VM is the use of computer models and simulations of manufacturing processes to aid in the design and production of manufactured products (Shukla *et al.*, 1996). The concept of virtual manufacturing cell was proposed by the National Bureau of Standards (USA) to address specific control problems in the design of their Automated Manufacturing Research Facility (AMRF). In cellular manufacturing, a *virtual cell* means the machines that belong to that cell are not physically grouped together, but identified as a group only in the system control software. When conditions change the machines are re-assigned to form different cells. The result is dynamically formed cells, whose compositions evolve in response to changing shop conditions. Unlike traditional cells that are permanent and physical, dynamic cells are temporary and logical, rendering them as *virtual cells* (Irani *et al.*, 1993). This concept is also labeled as virtual cellular manufacturing (Kannan and Ghosh, 1996).

Another explanation of VM parrallells the virtual memory concept in computers. *Virtual memory* in the computer terminology refers to a way of making a computer act as if it has more storage capacity than it really possesses. Similarly, the virtual corporation will appear to be a single entity with vast capabilities assembled only when they are needed (Byrne, 1993). This concept of *virtual enterprise* can be defined as a rapidly configured multidisciplinary network of small, process-specific firms configured to meet a window of opportunity to design and produce a specific product.

CHARACTERISTICS: A virtual organization is a temporary network of independent companies, suppliers, customers, even rivals linked by information technology to share skills, resources, costs, and access to one another's markets. It will have neither a central office nor an organization chart. It will have no hierarchy, and no vertical integration. A virtual manufacturing organization will be fluid and flexible, created by a group of collaborators who quickly unite to exploit a specific opportunity. The venture may disband once the opportunity is met. In businesses as diverse as movie making and construction, companies have come together for executing specific projects. Opportunities may last for tens of years as in the case of building Boeing 777 airplanes, or they can be as short as only a few weeks or months. A virtual organization can be defined as a new organizational model that uses technology to dynamically link people, assets, and ideas. The following are the five key characteristics of a virtual organization (Goldman *et al.*, 1995).

1. **Opportunism:** A virtual organization must have a high degree of adaptability. It must be agile in its internal organizational structure, rules, and regulations. A virtual organization is an opportunity-pulled and opportunity-defined integration of core competencies distributed among a number of real organizations.
2. **Excellence:** A virtual organization is assumed to be world-class and excellent in its core competencies. A member of the virtual organization will contribute only what is regarded as its *core competencies*. It will mix and match what it does best with the best of other member companies and entrepreneurs in the virtual organization. In the virtual organization, a capable manufacturer will do manufacturing, an expert design firm will produce product-designs, and a successful marketing company will sell the products.
3. **Technology:** A virtual organization is assumed to offer world-class and state-of-the-art technology in its product as well as service solutions.
4. **Borderlessness:** A virtual organization's competencies can remain dispersed physically and still be synthesized into a coherent productive resource to meet a customer requirement.
5. **Trust:** The members of a virtual organization team must act responsibly toward each other in a trusting and trustworthy manner. If mutual trust does not exist, they cannot succeed in attracting customers to do business with them.

The common benefits of a virtual organization are the reduction in time, costs, and risks; the increase in product service capabilities; and the expansion of resources throughout the concept-to-cash cycle. Rapidly formed virtual corporations composed of the best of several firms have competitive advantage. For instance, the success of Applied Materials, Inc. is based on a collaborative web of suppliers and customers.

Many large corporations are using the virtual concepts to broaden their offerings to their customers or produce sophisticated products less expensively. The virtual concept can also provide muscle and reach for some smaller companies and entrepreneurs. For example, TelePad, Inc. which produces a hand-held pen based computer, is a small company that has limited in-house design talent, a handful of engineers, and no manufacturing plants. The product was designed and co-developed with GVO, Inc.; an Intel Corporation's swat team was brought in to resolve some engineering problems; several other companies developed software for the product; a battery maker developed a portable power supply; and finally spare manufacturing capacity at an IBM plant was used to manufacture the product (Goldman *et al.*, 1995).

PROBLEMS AND PROSPECTUS: While power and flexibility are the obvious benefits of a virtual organization, the virtual model has some real risks too. For starters, a company joining such a network loses control of its functions it cedes to its partners. Proprietary information or technology may be tapped unethically by the other members of the virtual organization. The structure will pose new stiff challenges to managers, who must learn to build trust with outsiders and learn to manage beyond the walls of their company.

One of the big drawbacks of the virtual organization is that it results in loss of total control over technology or operations. For example, To get into the personal computer market quickly, IBM relied on a couple of outsiders for the key technologies: Intel for micro processors and Microsoft for the operating system software. At first, IBM won widespread praise for its unprecedented decision to develop a major product by forming partnerships with teams outside of its corporate walls. Ironically, the approach also meant that IBM's computer system wasn't proprietary, and IBM soon found that it had created a market it could not control. Hundreds of clone makers emerged with lower prices and better products (Goldman *et al.*, 1995).

The three main technologies that companies have employed to create a virtual factory – electronic data interchange (EDI), proprietary groupware such as Lotus Notes, and dedicated wide-area networks – are not in themselves complete solutions. If a large-scale virtual factory has to succeed, it must fulfill the following three demands (Upton and McAfee, 1996).

Sophistication: A virtual factory must be able to accommodate the network of members whose information technology sophistication varies widely – from a small machine shop with a single PC to a large site that operates an array of engineering workstations and mainframes.

Security: While maintaining a high level of security, a virtual factory must be able to cope with a constantly churning pool of suppliers and customers whose relationships fluctuate in closeness and scope.

Functionality: A virtual organization must give its members the capacity to transfer files between computers, the power to access common pools of information, and the capability to access and utilize all the programs on a computer located at a distant site.

EDI, groupware, and wide-area networks can each deal with some of these demands, but none can deal with all of them, nor can combinations of the three technologies (Upton and McAfee, 1996). Nevertheless, the standards and technologies are improving to make the virtual factories possible. For example, AeroTech, a small and young information-services company has built a virtual factory for McDonnell Douglas Aerospace. This networked manufacturing community is open and friendly to even the most unsophisticated users; it provides a very high degree of functionality, and works even though the community's membership is constantly changing. Some technologies such as rapid prototyping, web-aided design (or internet-aided design), virtual machining, virtual assembly & testing, cost & risk analysis models, and product realization process are being developed by TEAM (Technologies Enabling Agile Manufacturing) program established by US Government (Herrin, 1996; Mo *et al.*, 1996; Stevens, 1997).

Apart from technological problems, networked partners face new management challenges. Before companies can more routinely engage in collaboration, they must build a high level of trust in each other. Virtual organizations will demand a different set of skills from their managers. They will have to build relationships, negotiate *win-win* deals, find the right partners with compatible goals and values, and provide the temporary organizations with right balance of freedom and control.

MECHANISMS FOR IMPLEMENTATION: Technologies for VM include all the modules of a manufacturing enterprise such as computer-aided design (CAD) work stations, computer numerical control (CNC) machines, material handling systems, which are linked by either an enterprise network, or internet (Arnst, 1996; Cortese, 1996). A futuristic confederation of National Information Infrastructure that can link computers and machine tools across the US is being envisaged (Byrne, 1993; Stipp, 1996). This communication superhighway would permit far-flung units of different companies to quickly locate suppliers, designers, and manufacturers through an information clearing house. Once connected, they would sign *electronic contracts* to link up without legal headaches. Teams of people in different companies would routinely work together, concurrently rather than sequentially, via computer networks in real time. Artificial intelligence systems and sensing devices would connect engineers directly to the production line. The key is to act like a

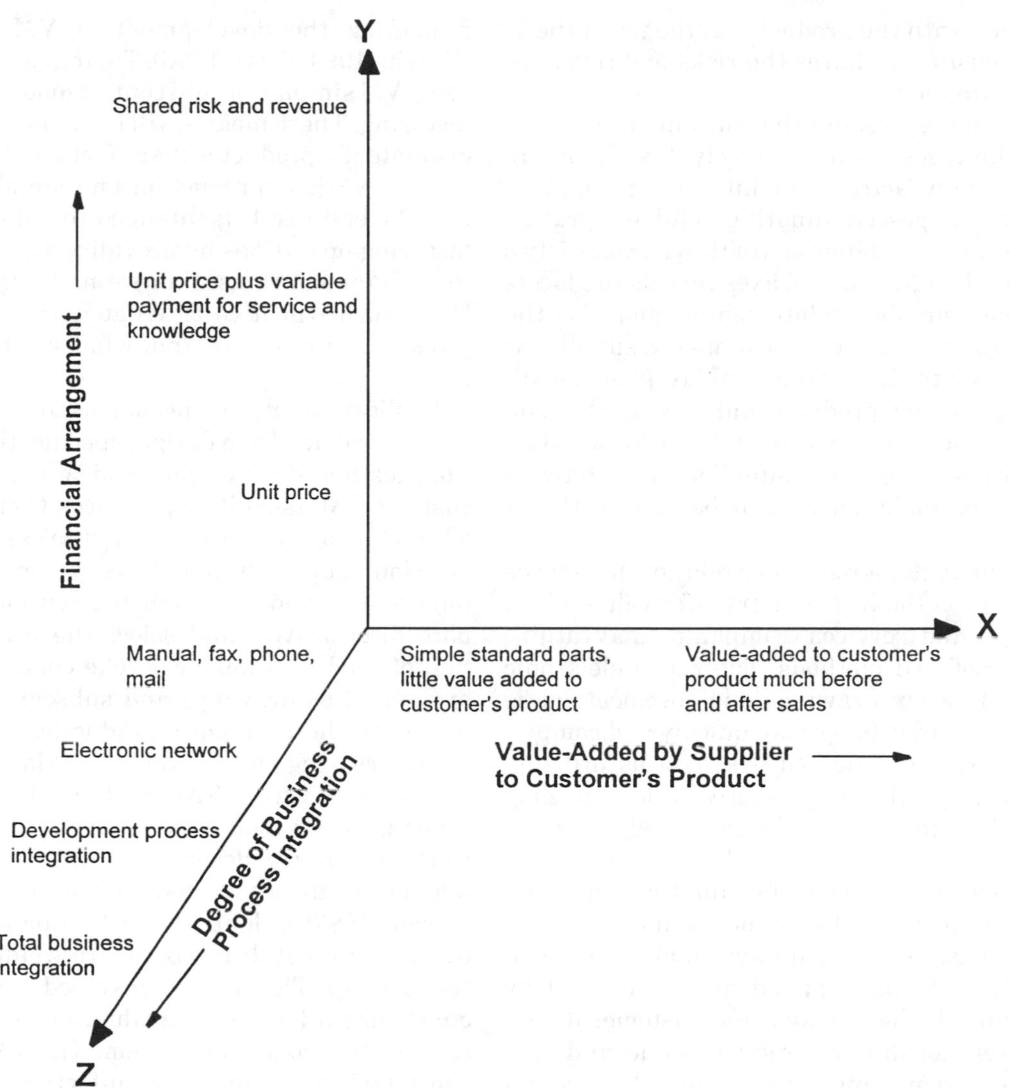

Fig. 1. Agile customer-supplier relationships.

single-source provider, with full accountability to the customer. A virtual organization uses one or more of the existing organizational mechanisms such as partnership, joint venture, strategic alliance, new cooperation, supplier-subcontractor, cooperative agreement, royalty or license, outsourcing contract, or web. Of these mechanisms, the web is a new concept. A web is an open-ended collection of prequalified partners that agree to form a pool of potential members of a virtual organization. To meet individual customer needs, a unique combination of companies is pulled into a virtual relationship because of the distinctive requirements of that customer.

In the world of virtual manufacturing, the types of relationships between supplier-companies (denoted as supplier) and customer-companies (denoted as customer) expand to cover a wide range of possibilities (Preiss *et al.*, 1996; Preiss and Wadsworth, 1995). The scope of the relationship is shown in Figure 1. In the graph, X-axis indicates the value-added by the supplier to the customer's product; Y-axis indicates the financial arrangement between the customer and the supplier; and Z-axis indicates the degree of business process integration. On the X-axis is the total value-added conferred on the customer by the supplier. Simple predefined parts with no value-added service are near the origin of the X-axis.

Moving to the right of the X-axis, the customer receives much more before-sale and after-sale information, knowledge, and value-added service.

The Y-axis represents the financial arrangement employed to control the flow of revenue from customer to supplier. Near the origin of the axis, the customer pays the supplier a fixed unit price. Further up on the Y-axis, the customer pays the supplier a fixed price per unit quantity of products, plus some additional variable price depending on the level of service and knowledge provided in conjunction with the products. At the top of the Y-axis, the customer shares the risks and revenues with the supplier.

The Z-axis represents the continuum of order-process linkages, from passively taking orders to increasingly active and interactive supplier-customer linkages, culminating in full integration of the process. Combinations of the characteristics represented on X, Y, and Z axes reveals the facets of customer-supplier relationships, and also the scope for improvement. For example, a supplier at a low rank on the Z-axis (manual fax, phone, mail) providing complex products and services (high on the X-axis) would be operating slowly. To speed up the business process, the supplier would have to move to electronic network or beyond on the Z-axis.

When the customer-supplier relationship moves to the right on the X-axis to transfer value-added knowledge and services, companies may utilize some type of EDI methods. The use of electronic transfer of design drawings and specifications is necessary in order to operate quickly and competitively. More sophisticated design tools utilizing expert systems, digital interactive video, etc. may be needed in order to provide leading-edge designs quickly.

The basis of payment far up the Y-axis labeled "shared risk and revenue" is undefined in a classical mass production environment in which the products being supplied are pre-defined by the supplier. If the supplier and customer do not have a relationship whereby revenue and risk are shared, then using a more integrated process (shown along Z-axis) for exchange of know-how and services will not be possible.

Other combinations of relationships on the X-Y-Z coordinate system can be interpreted. The specific relationships identified along the axes X, Y, and Z in Figure 1 are just a few examples of the possible combinations in the total spectrum of customer-supplier relationships. Wal-Mart Stores, Inc. implemented EDI to integrate manufacturers, wholesalers, retailers, and customers into a single customer-focused process. Many analysts believe that Wal-Mart has outperformed the information systems (IS) of other retail chains because it was the first to realize that IS is important to success (Preiss and Wadsworth, 1995).

VIRTUAL MANUFACTURING IN PRACTICE: Virtual manufacturing is a relatively new concept. However, many companies already have been practicing the concepts of VM. The US Air Force MANTECH group along with Defense Advanced Research Projects Agency (DARPA) and Defense Modeling and Simulation Office (DMSO) has been promoting the development of VM technology (Herrin, 1994; Scott, 1996). Their objective is to develop VM simulation and control models for manufacturing. These models will have the capability to evaluate if a product is manufacturable. They will analyze various production and supplier scenarios. The end result is intended to enhance manufacturing operations by providing timely answers to such questions as: Can we make the product? How much will it cost? What is the best way to produce the product? And, what are the alternatives?

Applications of VM encompass the entire life cycle of a product, from design, specification, production, testing, distribution, and to delivery to the customer. Matsushita Appliances Company used VR technology to develop a virtual kitchen space decision support system. This system allows customers experience a kitchen environment, evaluate alternatives, and select the best combination of kitchen appliances. The customer choices are stored as drawings and subsequently transmitted to the company's production facilities for manufacturing and delivery (Shukla *et al.*, 1996). Ford worked with C-TAD Systems, Inc. to develop a virtual CAD tool, which helps any designer to work on any car design or to interact with production engineers located anywhere in the world (Owen, 1997b). John Deere Company used VM software to install a robotic arc welding production facility. This project involved a virtual 3-D environment for design, evaluation, and testing of the robotic production system. The VM approach shortened the design-to-manufacturing lead time (Owen, 1997a). Boeing 777 plane was initially developed as a digital plane, on which virtual machining, virtual assembly, and virtual testing were carried out.

A CASE: AeroTech Services Group, which is based in St. Louis, Missouri, has built a highly effective virtual factory for McDonnell Douglas Aerospace (Upton and McAfee, 1996). There are around 400 companies in this manufacturing network. The open and flexible nature of this network accommodates users, whose information technology sophistication and relationships with one another vary greatly. Besides, the network permits the members carry out a wide variety of collaborative tasks very securely. For example, consider the relationship between McDonnell Douglas and UCAR Composites, a $12 million manufacturer of tooling for high-performance composite components based in Irvine, California. To build a new part, CAD files are translated at McDonnell Douglas into the NC machine code needed to operate UCAR's metal cutting machines. Using standard internet protocols over a dedicated high-speed link, McDonnell then translates the CAD file and the metal cutting program to AeroTech's safe network node. AeroTech's system subsequently forwards them to UCAR on normal phone lines. Once information on the job arrives in California, UCAR engineers view it on their own CAD/CAM systems to make last-minute checks on the program. They then run the cutting program on their machines to manufacture parts.

This method of transferring NC programs is also being used by hundreds of small machine shops, many of whose information management systems and expertise are much less sophisticated than UCAR's. Many of these companies, including five- or six-person machine shops, can dial AeroTech using a regular modem, download an NC program or a drawing onto their PCs, and use the data to manufacture the parts.

The virtual factory arrangement also helps its members find the best suppliers much more quickly than before. In the past, McDonnell Douglas would invite representatives of qualified suppliers to come to St. Louis to participate in the bidding process that often took several days. Using the virtual factory's electronic bidding system, both the suppliers and McDonnell Douglas save time and money. AeroTech also helps the manufacturing community coordinate their schedules better by allowing remote members access scheduling software on one another's machines. AeroTech has made it possible for both longtime and casual partners to collaborate easily, securely, and cheaply without having to invest in new and proprietary information technology.

CONCLUSIONS: The existence of a global market, with increasing demands for customized products, is promoting the creation of virtual organizations as dynamic mechanisms to focus on time-based market opportunities. New technologies are enabling simulation of all steps from design to delivery without actually producing a product physically, or without requiring the participating members to be physically near each other. Virtual manufacturing utilizes the enhanced competitive capabilities resulting from cooperation between members of a virtual organization. The success of the virtual manufacturing organization is tied to the ability of the real companies to form the virtual organization rapidly to meet an emerging time-based opportunity. The ability to work intensively with other companies and to be able to trust them from the start of a project is enhanced by prequalification agreements based on company attributes and contractual commitments. The use and role of virtual organizations are subtle and evolving. Once the benefits of virtual manufacturing are understood, many more companies may take advantage of virtual manufacturing.

Organizations Mentioned: Aerotech Services Corp; Apple; Applied Materials, Inc.; Applied Services Corp; AVO Inc.; Chrysler; C-TAD Systems; DARPA; DMSO; Ford; General Motors; Hewlett-Packard; IBM; Intel; John Deere Company; Mantech; Matsushita Appliance Company; McDonnel Douglas Aerospace; Motorola; National Bureau of Standards; Sony; Telepad Inc; UCAR Composites; The U.S. Air Force; USCAR; Wal-Mart Stores.

See **Agile logistics; Agile manufacturing; Lean manufacturing; Mass customization; Mass customization implementation; Supplier partnership and strategy; Supplier relationships; Supply chain management: Competing through integration.**

References

[1] Arnst, C. (1996). "Wiring Small Businesses," *Business Week*, Nov. 25, 164–172.

[2] Buzzacott, J.A. (1995). "A Perspective on New Paradigms in Manufacturing," *Journal of Manufacturing Systems*, 14, 118–125.

[3] Byrne, J.A. (1993). "The Virtual Corporation," *Business Week*, Feb. 8, 98–103.

[4] Camarinha-Matos, L.M., H. Afsarmanesh, C. Garita, and C. Lima (1997). "Towards an Architecture for Virtual Enterprises," In *Proceedings of the Second World Congress on Intelligent Manufacturing Processes & Systems*, Budapest, Hungary, 531–541.

[5] Cortese, A. (1996). "Here comes the Intranet," *Business Week*, Feb. 26, 76–84.

[6] Davidow, W.H. and M.S. Malone (1992). *The Virtual Corporation*, Harper Collins, New York, NY.

[7] Goldman, S.L., R.N. Nagel, and K. Preiss (1995). *Agile Competitors and Virtual Organizations*, Van Nostrand Reinhold, New York, NY.

[8] Herrin, G.E. (1994). "Virtual Manufacturing," *Modern Machine Shop*, 66, 158–159.

[9] Herrin, G.E. (1996). "Industry thrust areas of TEAM," *Modern Machine Shop*, 69, 146–147.

[10] Irani, S.A., T.M. Cavalier, and P.H. Cohen (1993). "Virtual Manufacturing Cells: Exploiting Layout Design and Intercell Flows for the Machine Sharing Problem," *International Journal of Production Research*, 31, 791–810.

[11] Kannan, V.J. and S. Ghosh (1996). "A Virtual Cellular Manufacturing Approach to Batch Production," *Decision Sciences*, 27, 519–539.

[12] Mo, J.P.T., Y. Wang, Y. and C.K. Tang (1996). "The use of the Virtual Manufacturing Device in the

Manufacturing Message Specification Protocol for Robot Task Control," *Computers in Industry*, 28, 123–136.
[13] Owen, J.V. (1997a). "Virtual manufacturing," *Manufacturing Engineering*, 119, 78–83.
[14] Owen, J.V. (1997b). "Virtual Manufacturing," *Manufacturing Engineering*,119, 84–90.
[15] Preiss, K., S.L. Goldman, and R.N. Nagel (1996). *Cooperate to Compete-Building Agile Business Relationships*, Van Nostrand Reinhold, New York.
[16] Preiss, K. and B. Wadsworth (1995). *Agile Customer-Supplier Relations*, Agility Forum, Bethlehem, PA.
[17] Scott, W.B. (1996). "Integrated Software to cut JSF, F-22 Costs," *Aviation Week and Space Technology*, 144, 64–65.
[18] Shukla, C., M. Vazquez, and F. Chen (1996). "Virtual Manufacturing: An Overview," *Computers and Industrial Engineering*, 31, 79–82.
[19] Stevens, T. (1997). "Internet-aided Design," *Industry Week*, 246, 50–55.
[20] Stipp, D. (1996). "The Birth of Digital Commerce," *Fortune*, 134, 159–164.
[21] Upton, D.M. and A. McAfee (1996). "The Real Virtual Factory," *Harvard Business Review*, 74, 123–133.

VIRTUAL MANUFACTURING NETWORK

As firms seek to take advantage of worldwide resources, incorporating low-cost labor and materials into their productive systems, they often look to establish fabrication and assembly facilities in different countries around the world. Unfortunately, such networks are very expensive and hard to manage. As a result, many companies have adopted a philosophy of acquiring worldwide resources through a virtual network of suppliers and alliance partners. These firms emphasize a core competence, and rely on other firms to perform as much of the manufacturing as possible.

Nike is an example of a firm that relies almost exclusively on suppliers for its manufacturing capacity. Nike focuses on research and development as well as marketing to establish a worldwide brand image for excellence in athletic gear and apparel. Nike outsources all of its non R&D production to other companies that are responsible for locating, constructing, and managing efficient manufacturing facilities in different countries around the world. Ikea follows a very similar strategy, designing consumer-friendly household furniture and outsourcing its production. Ikea then sells the unassembled furniture and lets the customer perform the assembly work. Most companies perform more of the actual production of the products that they sell; however, they increasingly look to suppliers for design expertise and manufacturing capacity. After all, it is much less expensive and a lot less risky to allow suppliers to chase low-cost labor and establish manufacturing capacity and expertise. The firm retains the flexibility to buy from new suppliers when technologies or factor input advantages change. The recent emphasis on core competencies by manufactures combined with greater ability to manage geographically dispersed global suppliers should increase the use of virtual manufacturing networks over the next several years.

Organizations Mentioned: Ikea; Nike.

See **Core manufacturing competencies; Global manufacturing rationalization; Virtual manufacturing.**

VIRTUAL ORGANIZATION

See **Electronic data interchange in supply chain management; Virtual enterprise; Virtual manufacturing.**

VIRTUAL REALITY

Virtual reality is a technology that generates a perception of reality in a human subject. It requires the use of devices such as data gloves and head-mounted display systems linked to work stations, that stimulate more than one sensory organ such as eyes, ears, etc. to present a dynamic model of real or fictitious environment. Virtual reality allows its users to intuitively interact with the virtual environment and its objects as if they were real by immersing the users in a 3-dimensional computer generated simulation. Examples of virtual reality are flight simulators, virtual machining and assembly.

See **Agile manufacturing; Virtual manufacturing.**

VOLUNTARY PRODUCT ADAPTATION

Voluntary product adaptations are changes/modifications that the international enterprise willingly makes on the products it sells in overseas markets. These changes are made in order to align with the company's overseas marketing strategy, which strives to meet the needs of consumers in various countries. Such adaptations include the choice of the color used on the package (especially in consumer products) since different colors have various meanings in different countries, the size of the product, and accessories preferred by consumers in the various countries.

See **Mass customization; Product design for global markets.**

WAGNER-WHITIN ALGORITHM

Wagner-Whitin developed an algorithm to find the optimal lot size for each period when net requirements are known for several periods into the future (Gass and Harris, 1996). Generally the algorithm minimizes the cost of ordering by optimally combining the net requirements for one or more periods (weeks).

See **Lot sizing models; MRP; Order release.**

Reference

[1] Gass, S.I. and C.M. Harris (1996). *Encyclopedia of Operations Research and Management Science.* Boston, MA: Kluwer Academic Publishers.

WAITING LINE MODEL

See **Queuing theory; Simulation analysis of manufacturing and logistics systems; Simulation of production problems using spreadsheet programs.**

WASTE ELIMINATION IN JIT

See **Just-in-time manufacturing; Lean manufacturing implementation; Total quality management.**

WEEKLY PLANNED PRODUCTION

The production planning process generally begins with top-level management developing a plan that is stated in broad, general terms, often in monetary units, and usually with one month as the smallest time unit. That plan is later refined into more detail by lower levels of management, usually breaking the time units down to weeks. This statement of weekly planned production is generally referred to as the master production schedule (MPS). The MPS plays several important roles. For one, it is used in rough-cut capacity planning to estimate the load that will be placed on various machines or work centers. Second, the MPS "drives" the material requirements plan (MRP), which is used to determine when, and in what quantities, materials must be ordered so that weekly planned production as indicated by the MPS can occur.

In process industries, such as oil refining, or in companies that have implemented Just-in-Time (JIT), the weekly planned production is often rate-based. In other words, these companies usually produce each product at some fairly constant rate per day or per week, so the production planning process is greatly simplified.

See **Capacity planning: Long-range; Capacity planning: Medium- and short-range; Just-in-time manufacturing; Master production schedule; MRP; Rate-based production schedule; Rough-cut capacity planning.**

WHERE-USED DATA

Some components/parts are used in the assembly of more than one product of subassembly. A where-used list, which can be stored in a computer file or derived from the bill-of-material files, shows all the subassemblies and assemblies that use a lower level part or component. If the lower level part is in short supply, a where-used list helps one to know which assemblies and subassemblies are likely to be affected by the shortage of the said part.

See **Bill-of-materials; MRP.**

WIDE AREA NETWORKS

A Wide area network (WAN) is a information and data system network that spans a large geographic area. WAN technology provides communication connectivity between points on the network. A WAN usually operates at speeds ranging from 9.6 KBPS (kilo bits per second) to 45 MBPS (million bits per second) and beyond to provide connectivity over relatively long distances. Generally, WAN technology interconnects local area networks (LAN).

See **Virtual manufacturing.**

WINTER'S FORECASTING MODEL

Winters' model is one of several models appropriate for forecasting when a seasonal pattern or a pattern containing a combination of trend and seasonality. It is available on most forecasting software packages and is often considered the model of choice for these types of patterns. Winters' model is based on four equations. The first equation is for overall smoothing, the second is for smoothing of trend, and the third for smoothing of seasonality. The final equation combines overall smoothing with trend, and makes an adjustment for seasonality to generate a final forecast. Winters' model is similar to Holt's two parameter exponential smoothing trend model, with an added equation to deal with seasonality. This model uses three parameters, one for overall smoothing, one for trend smoothing, and the third for smoothing of seasonality.

See **Forecasting guidelines and methods; Forecasting in operations management.**

WINTER'S METHOD

See **Forecasting guidelines and methods; Forecasting in operations management; Winter's forecasting model.**

WIP

See **Inventory flow analysis; Just-in-time manufacturing; Work-in-process inventory.**

WITHDRAWAL (OR CONVEYANCE) KANBAN (WLK)

Kanban is a Japanese term, which refers to a card used in the pull inventory and production control system the Kanban system. It is used to withdraw material from the upstream process to the downstream process in the kanban-based just-in-time manufacturing system. It is also called W-kanban.

See **Just-in-time manufacturing; Just-in-time manufacturing implementation; Kanban; Kanban-based manufacturing; Linked-cell manufacturing systems.**

W-KANBAN (WLK)

See **Withdrawal kanban.**

WORK MEASUREMENT

See **Time study.**

WORK STUDY

See **Time study.**

WORKFORCE TURNOVER

Workforce turnover is expressed as a ratio or percentage that indicates the rate at which workers are leaving an organization within a given time period. In most cases, this is calculated as an annual workforce turnover, and expressed as a percentage. Turnover percentage is calculated as: $100 \times$ (Number of New hires in a year)/(Average number of employees in the year).

For example, if the average workforce for an organization has remained at 200 over the last year, yet 100 new employees were hired during this same period, then turnover $= (100/200) \times 100 = 50\%$. That is, 50% of the workforce left for alternative employment or was dismissed, and was replaced by new workers.

Workforce turnover is often used as a surrogate measure for employee satisfaction. Generally speaking, turnover is low in companies with good pay, benefits, and retirement packages, but is high in low-paying jobs that are often considered "boring." For example, turnover in the fast food industry is as high as 200% to 400%.

In evaluating suppliers, workforce turnover is one element of assessment. Ideally, a supplier should have low workforce turnover, indicating that employees are dedicated to the company and are willing to contribute to problem-solving and continuous improvement.

See **Human resource management in manufacturing; Performance measurement in manufacturing; Supplier performance measurement.**

WORKING CAPITAL COSTS

Typically, the cost of working capital is the interest expense of borrowing money or the cost of funding internally. Working capital is the money used to pay for raw material, work-in-process inventory, finished goods inventory, and various other day-to-day activities.

See **Just-in-time manufacturing implications.**

WORK-IN-PROCESS (WIP) INVENTORY

The inventory in a manufacturing system that is awaiting completion is called work-in-process (WIP) inventory. WIP inventory is analogous to the water in a river. A high water river level in the river is equivalent to a high level of inventory in the system. A high water level covers all the rocks in the riverbed. Rocks in the riverbed are equivalent to problems. As the water level drops (i.e., inventory drop), the rocks (problems) in the river bed are exposed. If the exposed problems receive immediate attention, products will flow smoothly through the system with very little WIP inventory. In a river, when all the rocks in the riverbed are removed, the river can run very smoothly with very little water. The goal should be to minimize WIP between manufacturing cells. Within the cell, parts move one at a time resulting in minimum WIP inventory.

The minimum level of inventory that can be achieved is a function of the quality level, the probability of machine breakdown, the length of the setups, the variability in the manual operations, the number of workers in the cell, parts shortages, the transportation distance, etc. It is significant that WIP inventory can be made a controllable variable rather than an uncontrollable variable.

See **Inventory flow analysis; Just-in-time manufacturing implications; Just-in-time manufacturing; Lean manufacturing implementation; Manufacturing cell design; Manufacturing system modeling using Petri nets; Manufacturing systems; Simulation analysis of manufacturing and logistics systems.**

WORLD-CLASS LOGISTICS

World-class logistics is the rapid adoption and effective implementation of best logistics practice. World-class logistics enables managers to exploit logistical practice to attract and retain selected customers by providing superior levels of differentiated service. By using logistics to help customers improve their own competitive position, world-class logistics helps assure the company of short-term competitiveness and long-term prosperity. Within a firm that values logistics as a competitive weapon, the mindset of continuous improvement, constant learning, and understanding the success factors of important customers provides the impetus for logistics strategy evolution. Some general propositions accompany the concept of world-class logistics:

- Logistics practices must not only provide high levels of service to all customers but also must be finely tuned to the specific requirements of the most important customers.
- Logistics must facilitate the establishment and maintenance of strong supply-chain relationships with suppliers and customer.
- Logistics must increasingly rely on information technology to replace inventory and increase customer service. Logistics information is a key to seamless logistics performance throughout the supply chain.

See **Logistics: Meeting customers' real needs; Supply chain management.**

Reference

[1] Bowersox, D.J., Calantone, R.J., Clinton, S.R., Closs, D.J., Cooper, M.B., Droge, C.L., Fawcett, S.E., Frankel, R., Frayer, D.J., Morash, E.A., Rinehart, L.M., and Schmitz, J.M. (1995). *World Class Logistics: The Challenge of Managing Continuous Change*. Council of Logistics Management, Oak Brook, IL.

WORLD-CLASS MANUFACTURING (WCM)

This is either a term used to indicate a standard of excellence among manufacturers, or a term used to describe just-in-manufacturing and its variation practiced in the United States.

See **Just-in-time manufacturing; Just-in-time manufacturing implementation; Lean manufacturing implementation; Manufacturing cells, linked.**

Reference

[1] Schonberger, R.J. (1996). *World Class Manufacturing: The Next Decade*. The Free Press, New York.

WTO (WORLD TRADE ORGANIZATION)

WTO (World Trade Organization) is an umbrella organization that controls international trade. As of 1996, the WTO consists of over 119 member countries, including the US. WTO is empowered to establish international panels to hear disputes about the decisions and laws of member countries. If the country's laws do not comply with those of the WTO, the WTO has the authority to place sanctions against the member country or require the violating member to pay fines.

The function of the WTO is to facilitate the implementation, operation and administration of all multilateral trade agreements. The WTO is also required to provide a forum for negotiations and development of agreements and associated legal instruments among its members on all international trade policy issues. The WTO also helps to achieve greater coherence in global economic policy making by working with the International Monetary Fund and International Bank for Reconstruction and Development. The longer-term objective of the WTO organization is to eliminate all trade barriers.

See **Global facility location analysis.**

Reference

[1] World Trade Organization @ ITL.iru.uit.no/trade_law/documents/freetrade/wat-94/

X-BAR CHART

One of the most commonly used control charts for variables is the X-bar chart, also known as the chart for the sample means. Walter Shewhart, the originator of control charts, proposed a plan for the construction of control charts for the sample mean, and used it as a foundation for the development of the theory of control charts. X-bar-charts may be used for statistics such as average length, average thickness, average diameter, average temperature, average pressure, or average viscosity.

The mean of the sample means, which is the center line on of the X-bar control chart, is a good measure of the process mean. If a target value for the quality characteristic is specified, it is desirable for the center line to be on this target value, or in close proximity to it. A process is said to be biased if the center line deviates from the target value. In such instances, remedial measures are taken to move the center line to the target value.

The control limits on an X-bar chart are usually placed at three standard deviations of the sample mean from the center line. These are known as three-sigma control limits. An underlying assumption in the choice of these control limits is the normality of the distribution of the sample mean. It is quite reasonable to justify this assumption using the Central Limit Theorem.

The Central Limit Theorem states that for random samples of size n selected from a population with mean μ and standard deviation σ:

(a) the sampling distribution of the sample mean will be approximately normal;

(b) the mean of the sampling distribution of the sample mean will be equal to the population mean μ;

(c) the standard deviation of the sample mean (also known as the standard error) is given by (sigma/square root of n). The approximation becomes better as the sample size n increases.

Typically, the Central Limit Theorem may be a good approximation for samples of size 30 or more. However, if the distribution of the quality characteristic is symmetric and unimodal, then the approximation of the distribution of sample means approaching normality may be valid for much smaller samples of size 4 or 5. In most quality control activities for variables, samples of size 4 or 5 are selected and their average X-bar is computed. About 25 to 30 samples are used in the construction of control charts.

From the conclusions of the Central Limit Theorem, it is found that the standard error decreases as the sample size n increases. For a four-fold increase in the sample size, the standard error reduces by half. As the standard error decreases, the control limits are drawn closer to the center line. This implies that if the sample size is large, it is easier to detect changes in the process due to special causes.

Various patterns in an X-bar chart indicate likely special causes that may be present in the system. For example, a sudden shift in the pattern level could occur due to changes in process settings such as depth of cut, pressure, or temperature. New equipment, new measuring instruments, or new methods of processing are other reasons for sudden shifts on X-bar charts.

A gradual shift in the pattern level may occur because of the changes in incoming quality of raw materials or components. A steady increase or decrease in observations may be because of tool wear, die wear, or gradual deterioration in equipment. Cyclic patterns in an X-bar chart may be caused by periodic changes in temperature, rotation of operators, or periodicity in the mechanical or chemical properties of the material.

See **Statistical process control using control charts; Shewhart control charts; Total quality management.**

YO-I-DON

Yo-I-Don is a Japanese term, which literally means ready-set-go. It refers to all the synchronized portions of the plant starting at the same time after production stoppage due to any reason including quality problems.

See **Integrated quality control; Just-in-time manufacturing; Just-in-time manufacturing implementation; Lean-manufacturing implementation.**

ZERO INVENTORY PRODUCTION SYSTEMS (ZIPS)

See **Just-in-time manufacturing; Lean manufacturing; Lean manufacturing implementation; Manufacturing cells, linked.**

Appendix I

Alphabetical Bibliography

A

Aase, Gerald (1997). "Improving Labor Productivity with U-shaped Assembly Lines." Proceeding for the 26th Annual Meeting of the Decision Science Institute, San Diego, CA.

Abdel-Malek, L. (1985). "A case study on the use of robots in assembly process." *Engineering Costs and Production Economics*, 9, 99–102.

Abdel-Malek, L. (1986). "A framework for the robotic assembly of parts with general geometries." *International Journal of Production Research*, 24, 1025–1041.

Abdel-Malek, L. (1988). "The effect of robots with overlapping envelopes on performance of flexible transfer lines." *IIE Transactions*, 20, 213–222.

Abdel-Malek, L. (1989). "Assessment of the economic feasibility of robotic assembly while conveyor tracking (RACT)." *International Journal of Production Research*, 27, 1209–1224.

Abdel-Malek, L. and T.O. Boucher (1985). "A Framework for the economic evaluation of production system and product design alternatives for robot assembly." *International Journal of Production Research*, 23, 197–208.

Abegglen, J.C. and G. Stalk (1985). *The Japanese Corporation*, New York: Basic Books, Inc.

Abernathy, William J. and Kim B. Clark (1985). "Innovation Mapping the Winds of Creative Destruction." *Research Policy*, North-Holland, 14, 3–22.

Abernathy, William J. and Phillip L. Townsend (1975). "Technology, Productivity, and Process Change." *Technology Forecasting and Social Change*, 7, 379–396.

Abraham, R.H. and C.D. Shaw (1992). *Dynamics: The Geometry of Behavior* (2nd ed.). Addison-Wesley, Redwood City, CA.

Aburdene, M.F. (1988). *Computer Simulation of Dynamic Systems*. Brown, Dubuque, IA.

Adam, E.E. and P.M. Swamidass (1989). "Assessing operations management from a strategic perspective." *Journal of Management*, 15 (2), 181–203.

Adkins, A.C., Jr. (1984). "EOQ in the Real World." *Production and Inventory Management*, 25 (4), 50–54.

Adler, P.S. (1990). "Shared Learning." *Management Science*, 938–957.

Adler, P.S., A. Mandelbaum, V. Nguyen, and E. Schwerer (?). "From Project to Process Management: An Empirically Based Framework for Analyzing Product Development Time." *Management Science*, 41, 1995, 458–484.

Adler, P.S. (1993). "Time-and-Motion Regained." *Harvard Business Review*, January–February, 97–108.

Aggarwal, S.C. (1985). "MRP, JIT, OPT, FMS?" *Harvard Business Review* (September–October).

Agrawal, V.P., V. Khohl, and S. Gupta (1991). "Computer aided robot selection: The multiple attribute decision making approach." *International Journal of Production Research*, 29, 1629–1644.

Aigbedo, H. and Y. Monden (1997). "Scheduling for Just-in-Time Mixed Model Assembly Lines – A Parametric Procedure for Multi-Criterion Sequence." *International Journal of Production Research*, 35, 2543–2564.

Aikens, C.H. (1985). "Facility Location Models of Distribution Planning." European *Journal of Operations Research*, 263–279.

Aitken, Hugh G.J. (1960). *Taylorism at Watertown Arsenal; Scientific Management in Action*, 1908–1915. Harvard University Press, Cambridge, Mass.

Akinc, U. and B.M. Khumawala (1977). "An Efficient Branch and Bound Algorithm for Capacitated Warehouse Location Problem." *Management Science*, 23, 585–594.

Alberech, K. and R. Zemke (1985). *Service America: Doing Business in the New Economy*. Dow Jones-Irwin, Homewood, Ill.

Albertson, P. (1983). "Verifying robot performance." *Robotics Today*, 33–36.

Albright, T.L. and T. Smith (1996). "Software for Activity-Based Costing." *Journal of Cost Management*, Summer, 47–53.

Alexander, D.C. (1985). "A Business Approach to Ergonomics." *Industrial Engineering*, 17 (7) (July), 32–39.

Allen, S.J. and E.W. Schuster (1994). "Practical Production Scheduling with Capacity Constraints and Dynamic Demand: Family Planning and Disaggregation." *Production and Inventory Management*, Fourth quarter, 15–21.

Allen, T. J. (1986). "Organizational Structure, Information Technology, and R&D Productivity." *IEEE Transactions on Engineering Management*, EM 33, 212–217.

Alsup, F. and R.M. Watson (1993). *Practical Statistical Process Control*, Van Nostrand Reinhold, New York.

Alting, L. (1982). *Manufacturing Engineering Processes*. Marcel Dekker, New York.

Alting, L. (1993). "Life-Cycle Design of Products: A New Opportunity for Manufacturing Enterprises." in Concurrent Engineering, A. Kusiak (ed.). Wiley, New York.

Altiok, T. (1997). *Performance Analysis of Manufacturing Systems*. Springer, New York.

American Production and Inventory Control Society (1986). *APICS Shop Floor Control Reprints*. American Production and Inventory Control Society, Falls Church, VA.

American Production and Inventory Control Society (1986). *Material Requirements Planning Reprints*. American Production and Inventory Control Society, Falls Church, VA.

American Production and Inventory Control Society (1996). "1996 finite capacity scheduling software and vendor directory." *APICS – the Performance Advantage*, 6, 72–82.

American Productivity Center (1987). *Productivity Perspectives*. Houston, TX. December, p. 12.

American Society for Quality Control (1971). *"Quality Costs – What and How"* (2nd ed.). Milwaukee, Wisconsin.

American Society for Quality Control (1987). ANSI/ASQC A1-1987. *"Definition, Symbols, Formulas, and Tables for Control Charts."* ASQC Press, Milwaukee, Wisconsin.

Ammer, D.S. (1980). *Materials Management and Purchasing* (4th ed.). Irwin, Homewood, IL.

Amsden, R.T. (1998). "What's in store for SPC?" *Quality Digest*, February, 37–40.

Anderson, D., D. Sweeney, and T. Williams (1988). *An Introduction to Management Science: Quantitative Approaches to Decision Making* (5th ed.). West, St. Paul, Minn.

Anderson, D., D. Sweeney, and T. Williams (1989). *Quantitative Methods for Business*. West, St. Paul.

Anderson, D., D. Sweeney, and T. Williams (1990). *Statistics for Business and Economics* (4th ed.). West, Minneapolis, MN.

Anderson, D.A. (1998). *Agile Product Development for Mass Customization, Niche Markets, JIT, Built-to-Order and Flexible Manufacturing*, Irwin, Chicago, IL.

Anderson, D.R. (1992). Motor Carrier/Shipper Electronic Data Interchange, The American Trucking Associations, Inc., Alexandria, VA.

Anderson, J.C., Roger G. Schroeder, Sharon E. Tupy, and Edna M. White (1982). "Material Requirements Planning Systems: The State of the Art." *Production and Inventory Management*, 23 (4), 51–67.

Anderson, M.Q. (1981). "Monotone Optimal Preventative Maintenance Polices for Stochastically Failing Equipment." *Naval Research Logistics Quarterly*, 28, 347–358.

Anderson, P.W., K.J. Arrow, and D. Pines, eds. (1988). *The Economy as an Evolving Complex System*. Addison-Wesley, Redwood City, CA.

Andrew, C.G., *et al.* (1982). "Critical Importance of Production and *Operations Management." Academy of Management Review*, 7:143–147.

Andrews, C.G., "Leadership and motivation: the people side of systems implementation." In Proceedings of the American *Production and Inventory Control* Society Conference, 1978, 352–387.

Andrews, K.R. (1971). *The Concepts of Corporate Strategy*. Dow Jones-Irwin, Homewood, Illinois.

Ang, J.S.K., C.C. Sum, and W.F. Chung. 1995. Critical success factors in implementing MRP and government assistance: A Singapore Context, *Information and Management*, 29, 63–70.

Anonymous (1984). "Manufacturing Cells: Blueprint for Versatility." *Production*, December, 38–40.

Ansari, A. (1986). "Survey Identifies Critical Factors in Successful Implementation of Just-in-Time Purchasing Techniques." *Industrial Engineering*, 18 (10), 44–50.

Ansari, A. and B. Modarres (1986). "Just-in-Time Purchasing: Problems and Solutions." *Journal of Purchasing and Materials Management*, Summer, 11.

Ansari, A. and B. Modarress (1987). "The Potential Benefits of Just-In-Time Purchasing for U.S. Manufacturing." *Production and Inventory Management*, 28 (2) (Second Quarter), 30–35.

Ansari, A. and Jim Heckel (1987). JIT Purchasing: Impact of freight and Inventory Costs. *Journal of Purchasing and Material Management*, Summer, 24.

Ansari, S., J. Bell, T. Klammer, and C. Lawrence, *Target Costing*, Irwin, 1997.

Ansoff, H.I. (1965). *Corporate Strategy*. McGraw-Hill, New York, NY.

Anthony, Robert N. (1965). Planning and Control Systems: A Framework for Analysis. Cambridge, MA, Harvard University Graduate School of Business Administration, Studies in Management Control, 26–47.

APICS (1980). Certification Study Guide: Capacity Management. *American Production and Inventory Control* Society, Washington, D.C.

APICS (1996). *1997 CPIM Exam Content Manual*. APICS, the Educational Society for Resource Management, Virginia.

APICS Dictionary, 8th Edition, APICS, Falls Church, VA, 1995.

Apple, James M. (1977). *Plant Layout and Material Handling*. Wiley, New York.

Appleton, E.L., "How to survive ERP." *Datamation*, 43, 3, 1997, 50–53.

Archetti, F. and A. Sciomachen (1989). "Development, analysis and simulation of Petri net models: an application to AGV systems." F. Archetti, M. Lucertini, P. Serafini (eds.). *Operations Research Models in Flexible Manufacturing Systems*, 91–113, Springer-Verlag, New York.

Argote, L., S. Beckman and D. Epple (1990). "The Persistence and Transfer of Learning in Industrial Settings." *Management Science* (February), 36, 2, 140–154.

Armstrong, J.S. and F. Collopy (1992). "Error measures for generalizations about forecasting methods:

Empirical comparisons with discussion." *International Journal of Forecasting*, 8, 69–80.

Armstrong, J.S. (1978). *Long-Range Forecasting*. John Wiley and Sons, New York.

Armstrong, J.S. (1995). *Long-range Forecasting: From Crystal Ball to Computer*, New York: John Wiley and Sons.

Armstrong, J.S. and F. Collopy (1993). "Causal Forces: Structuring Knowledge for Time Series Extrapolation." *Journal of Forecasting*, 12, 103–115.

Armstrong, J.S. (1984). "Forecasting by Extrapolation: Conclusions from 25 Years of Research." *Interfaces*, November–December, 52–66.

Arn, E.A. (1975). *Group Technology*. Springer-Verlag, West Berlin.

Arntzen, B.C., G.G. Brown, T.P. Harrison, and L.L. Trafton (1995). "Global Supply Chain Management At Digital Equipment Corporation." *Interfaces*, 25(1), 69–93, Jan.–Feb.

Arogyaswamy, B. and R.P. Simmons (1991). "Thriving on independence: The Key to JIT Implementation." *Production and Inventory Management Journal*, 32, 3, 56–60.

Arrow, K.J., S. Karlin, and H. Scarf (1982). *Studies in the Mathematical Theory of Inventory and Production*. Stanford University Press, Stanford, CA.

Arthur, W.B. 1989. "Competing technologies, increasing returns, and lock-in by historical events." *The Economic J.*, 99: 116–31.

Artzt, E.L. (1993). "Customers Want Performance, Price, and Value." *Transportation and Distribution*, 32–34.

Ashby, James R., and Reha Uzsoy (1995). "Scheduling and order release in a single-stage production system." *Journal of Manufacturing Systems*,s 290–306.

Ashton, David, and Leslie Simister (1970). *The Role of Forecasting in Corporate Planning*. Staples Press, London.

AT&T (1984). *Statistical Quality Control Handbook*, 10th printing.

AT&T Corporate Quality Office (1994). *Using ISO 9000 to Improve Business Processes*.

Atwater, B. and M.L. Gagne (1997). "The theory of constraints versus contribution margin analysis for product mix decisions." *Journal of Cost Management* (January/February), 6–15.

Auguston, K. (1995). "How EDI Helps us Clear the Dock Every Day." *Modern Material Handling*, 46–47.

Austin, T.A., H.L. Lee, and L. Kopczak (1997). Unlocking Hidden Value in the Personal Computer Supply Chain, Andersen Consulting Report.

Automotive Industry Action Group (AIAG) (1983). *The Japanese Approach to Productivity*. Videotape Series.

Aven, T. and R. Dekker (1997). "A useful framework for optimal replacement models." *Reliability Engineering and System Safety*, 58, 61–67.

Ayres, R.U. (1969). *Technological Forecasting and Long-Range Planning*. McGraw-Hill, New York.

Ayres, R.U. (1991). *Computer Integrated Manufacturing*. Chapman and Hall.

B

Babbage, Charles (1835). *On the Economy of Machinery and Manufactures*, Charles Knight, London, England.

Babcock, George D. (1918). *The Taylor System in Franklin Management*. The Engineering Magazine Co., New York, New York.

Back, H. (1985). "Decentralized Integration – Advantage or Disadvantage of Logistics and JIT – Concepts." *Computers in Industry* (The Netherlands), 6 (6) (December), 529–541.

Badiru, A.B. (1988). *Project Management in Manufacturing and High Technology Operations*. John Wiley, New York.

Bagchhi, P.K., T.S. Raghunathan, and E.J. Bardi (1987). "The Implication of Just-in-Time Inventory Policy on Carrier Selection." *Logistics and Transportation Review* (Canada), 23 (4) (December), 373–384.

Bahl, H.C. and Larry P. Ritzman (1983). "An Empirical Investigation of Different Strategies for Material Requirements Planning." *Journal of Operations Management*, 3 (2), 67–77.

Bain, L.J. and M. Engelhardt (1991). *Statistical Analysis of Reliability and Life-Testing Models*, Marcel Dekker, Inc., New York.

Baker Kenneth R. (1984). *Introduction to Sequencing and Scheduling*, New York, Wiley.

Baker, J.T. (1980). "Automated Prevention Maintenance Program for Service Industries and Public Institutions." *Industrial Engineering*, 12 (2) (February), 18–21.

Baker, K.R. (1977). "An Experimental Study of the Effectiveness of Rolling Schedules in Production Planning." *Decision Sciences*, 8 (1), 19–27.

Baker, R.C. and S. Talluri (1997). "A closer look at the use of data envelopment analysis for technology selection." *Computers and Industrial Engineering*, 32, 101–108.

Baker, R.D. and A.H. Christer (1994). "Review of delay-time or modeling of engineering aspects of maintenance." *European Journal of Operational Research*, 73, 407–422.

Bakos, J.Y. (1991). "A Strategic Analysis of Electronic Marketplaces." *MIS Quarterly*, 15, 295–310.

Balakrishnan, N., V. Sridharan, and J.W. Patterson (1995). "Rationing Capacity Between Two Product Classes." *Decision Sciences*, 27–2, 185–215.

Balci, O. (1994). "Model validation, verification, and testing techniques throughout the life cycle of a simulation study." *Annals of Operations Research*, 53, 121–173.

Ball, Donald A., and Wendell H. McCulloch (1985). *International Business: Introduction and Essentials*. Business Publications, Plano, TX.

Ballou, R.H. (1985). *Business Logistics Management: Planning and Control* (2nd ed.). Prentice-Hall, Englewood Cliffs, NJ.

Ballou, R.H. (1992). *Business Logistics Management* (3rd ed.). Prentice-Hall, Englewood Cliffs, New Jersey.

Banker, R.D., J.M. Field, R.G. Schroeder, and K.K. Sinha (1996). "Impact of Work Teams on Manufacturing Performance." *Academy of Management Journal*, 39 (4), 867–890.

Banks, J. and J.S. Carson (1984). *Discrete Event System Simulation*. Prentice-Hall, Englewood Cliffs, NJ.

Banks, J., B. Burnette, H. Kozloski, and J. Rose (1995). *Introduction to SIMAN V and CINEMA V*. John Wiley and Sons, New York.

Barczak, G. (1995). New Product Strategy, Structure, Process and Performance in the Telecommunications Industry, *Journal of Product Innovation Management*, pp. 374–391.

Barlow, R.E. and F. Proschan (1965). *Mathematical theory of Reliability*. John Wiley, New York.

Barnes, R. (1980). *Motion and Time Study: Design and Measurement of Work* (8th ed.). John Wiley, New York.

Barney, J. (1991). "Firm Resources and Sustained Competitive Advantage." *Journal of Management*, 19, 1.

Barth, Carl (1912). "Testimony." in The Taylor and Other Systems of Shop Management, Hearings before the Special Committee of the House of Representatives to Investigate the Taylor and Other Systems of Management. 3, Government Printing Office, Washington, D.C., p. 1538–1583.

Bartmess, A. and K. Cerny (1993). "Building Competitive Advantage Through a Global Network of Capabilities." *California Management Review*, 78–102, Winter, 1993.

Batdorf, L. and J.A. Vora (1983). "Use of Analytical Techniques in Purchasing." *Journal of Purchasing and Materials Management* (Spring), 25–29.

Baulkin (1988). "Reward Policies That Support Entrepreneurship." *Compensation Benefits Review*, 20.

Baybars, I. (1986). "A Survey of Exact Algorithms for the Simple Assembly Line Balancing Problem." *Management Science*, 32, 909–932.

Bazarra, M.S. and J.J. Jarvis (1990). *Linear Programming and Network Flow* (2nd ed.). John Wiley, New York.

Bechoffer, R., T.J. Santner, and D. Goldsman (1995). *Design and Analysis of Experiments for Statistical Selection, Screening, and Multiple Comparisons*. Wiley, New York.

Bechte, Wolfgang (1988). "Theory and Practice of Load Oriented Manufacturing Control." *International Journal of Production Research*, 26, 3, 375–395.

Bechtold, S.E. (1981). "Work-Force Scheduling for Arbitrary Cyclic Demands." *Journal of Operations Management*, 1 (4) (May), 205–214.

Beddick, J.F. (1983). "Elements of success – MRP implementation." *Production and Inventory Management*, 24, 2, 26–30.

Bedworth, D. and J. Bailey (1987). *Integrated Production Control Systems* (2nd ed.). John Wiley, New York.

Behara, R.S., D.E. Gundersen, and E.A. Capozzoli (1995). "Trends in Information Systems Outsourcing." *International Journal of Purchasing and Materials Management*, 31 (2), 46–51.

Bell, Robert R. and John M. Burnham (1987). "Managing Change in Manufacturing." *Production and Inventory Management*, 28 (1), 106–115.

Bell, Robert R. and John M. Burnham (1987). "The Politics of Change in Manufacturing." *Production and Inventory Management*, 28 (2), 71–78.

Belohlav, J.A. (1995). "Quality, Strategy, and Competitiveness." *The Death and Life of the American Quality Movement*. Edited by R.E. Cole. Oxford University Press, New York, 43–58.

Below, L.J. (1987). "The Meaning of Integration." Proceedings of the 4th European Conference on Automated Manufacturing, Birmingham, UK, May, 345–352.

Bemowski, K. (1996). "Baldrige Award Celebrates Its 10th Birthday With a New Look." *Quality Progress*, 29, 49–54.

Ben-Arieh, D., C.A. Fritsch, and K. Mandel (1986). "Competitive Product Realization in Today's Electronic Industries." *Industrial Engineering*, February, 34–42.

Bender, P.S., W.D. Northup, and J.F. Shapiro (1981). "Practical Modeling for Resource Management." *Harvard Business Review*, 59 (2), 163–173.

Bennet, J. (1994). "Detroit Struggles to Learn Another Lesson From Japan." *The New York Times*, June 19, 1994, section F, 5.

Bergamaschi, D., R. Cigolini, M. Perona, and A. Portioli (1997). "Order review and release strategies in a job shop environment: a review and a classification." *International Journal of Production Research*, 35, 2, 399–420.

Berger, J. (1987). "Productivity: Why It's the Number 1 Underachiever." *Business Week*, April 20, 54–55.

Berkley, B. (1992). "A review of the kanban production control research literature." *Production and Operations Management*, 1 (4), 393–411.

Berliner, C. and J.A. Brimson (1988). *Cost Management for Today's Advanced Manufacturing*. Harvard Business School Press, Boston.

Berry W.L., T.E. Vollman, and D.C. Whybark (1979). Master Production Scheduling. American *Production and Inventory Control* Society, Falls Church, VA.

Berry, L., T.E. Vollman, and D.C. Whybark (1979). Master Production Scheduling: Principles and Practice. American *Production and Inventory Control* Society, Falls Church, VA.

Berry, L.L. (1983). "Relationship Marketing." in L.L. Berry, G.L. Shostack and G.D. Upah (eds.). *Emerging Perspectives on Services Marketing*. AMA, Chicago, 25–28.

Berry, L.L., V.A. Zeithmal, and A. Parasuraman (1985). "Quality Counts in Services, Too." *Business Horizons*, 28 (3) (May–June), 44–52.

Berry, W.L., T. Schmitt, and T.E. Vollmann (1982). "Capacity Planning Techniques for Manufacturing Control Systems: Information Requirements and Operational Features." *Journal of Operations Management*, 3 (10), 13–25.

Bertain, Leonard (1993). *The New Turnaround: A Breakthrough for People, Profits, and Change*, North River Press, Croton-on-Hudson, New York.

Bertrand, J.W.M. and V. Sridharan (1998). A Study of Simple Rules for Subcontracting in Make-to-Order Manufacturing. *European Journal of Operational Research*, Forthcoming.

Best, M.J. and K. Ritter (1985). *Linear Programming: Action Set Analysis and Computer Programs*, Prentice-Hall, Englewood Cliffs, NJ.

Besterfield, D.H. (1990). *Quality Control* (3rd ed.). Prentice-Hall, Englewood Cliffs, NJ.

Betz F. *Strategic Technology Management*, New York, McGraw-Hill 1993.

Bevis, G.E. (1976). "A management viewpoint on the implementation of a MRP system." *Production and Inventory Management*, 16, 1, 1976, 105–116.

Billesbach, Thomas J. (1994). "Applying Lean Production Principles to A Process Facility." *Production and Inventory Management Journal*, 35, 3, 40–44.

Billesbach, Thomas J. (1994). "Simplified Flow Control Using Kanban Signals." *Production and Inventory Management Journal*, 35, 2, 72–75.

Billesbach, Thomas J. and R. Hayen (1994). "Long-Term Impact of Just-In-Time on Inventory Performance Measures." *Production and Inventory Management Journal*, V. 35, 1, 62–67.

Billesbach, Thomas J., A. Harrison, and S. Croom-Morgan (1991). "Just-in-Time: A United States-United Kingdom Comparison." *International Journal of Operations and Production Management*, 11, 10, 44–57.

Billesbach, Thomas J. and Marc J. Schniederjans (1989). "Applicability of Just-In-Time Techniques in Administration." *Production and Inventory Management Journal*, 30, 3, 40–45.

Birou, L.M. and S.E. Fawcett (1993). "International Purchasing: Benefits, Requirements, and Challenges." *International Journal of Purchasing and Materials Management*, 29 (2), 28–37.

Bitran, G.D., E.A. Haas, and A.C. Hax (1982). "Hierarchical Production Planning: A Two-Stage System." *Operations Research*, March–April, 232–251.

Bitran, G.R., D.M. Marini, H. Matsuo, and J.W. Noonan (1984). "Multiplant MRP." *Journal of Operations Management*, 5 (2) (February), 183–204.

Black, J.T. (1983). "Cellular Manufacturing Systems Reduce Setup Time, Make Small Lot Production Economical." *Industrial Engineering*, 15 (11) (November), 36–48.

Black, J.T. (1991). *The Design of The Factory With A Future*. McGraw-Hill.

Black, J.T., B.C Jiang, and G.J. Wiens (1991). *Design, Analysis and Control of Manufacturing Cells*. PED-53, ASME.

Blackburn, J.D. (1991). "The Quick-Response Movement in the Apparel Industry: A Case Study Time-Compressing Supply Chains." edited by J.D. Blackburn, *Time-Based Competition: The Next Battle Ground in American Manufacturing*, Irwin, Homewood, IL, 246–269.

Blackburn, J.D. (1991). *Time Based Competition*, Homewood: Business One Irwin.

Blackburn, J.D., D.H. Kropp, and R.A. Millen (1986). "A comparison of strategies to dampen nervousness in MRP systems." *Management Science*, 32–4, 413–429.

Blackler, F.H.M. (1978). *Job Redesign and Management Control: Studies in British Leyland and Volvo*, Saxon House, Farnborough England.

Blackstone, J.H. (1989). *Capacity Management*, South-Western Publishing, Cincinnati.

Blackstone, J.H., D.T. Phillips, and G.L. Hogg (1982). "A State-of-the-Art Survey of Dispatching Rules for Manufacturing Job Shop Operations." *International Journal of Production Research*, 20, 27–45.

Blackstone, John H., Jr., and James F. Cox (1988). "MRP Design and Implementation Issues for Small Manufacturers." *Production and Inventory Management*, 3rd Quarter, 65–76.

Blasingame, J.W. and J.K. Weeks (1981). "Behavioral dimensions of MRP change; assessing your organisation's strengths and weaknesses." *Production and Inventory Management*, 22, 1, 81–95.

Blattberg, R.C. and S.J. Hoch (1990). "Database models and managerial intuition: 50% model and 50% manager." *Management Science*, 36, 889–899.

Bleakley, F. (1995). "Strange Bedfellows." *Wall Street Journal* (January 13), A1, A6.

Blois, K.J. (1986). "Marketing Strategies and the New Manufacturing Technologies." *International Journal of Operations and Production Management*, 6 (1), 34–41.

Bluemenstein, R. (1997). "Global strategy: GM is building plants in developing nations to woo new markets." *Wall Street Journal*, August 4, A1, A4.

Blumberg, D.F. (1980). "Factors Affecting the Design of a Successful MRP System." *Production and Inventory Management*, 5 (2) (Fourth Quarter), 50–62.

Blumenthal, K. (1996). "Compaq is segmenting the home-computer market." *Wall Street Journal*, July 16, A3.

Boddewyn, J.J., R. Soehl, and J. Picard (1986). "Standardization in international marketing: Is Ted Levitt in fact right?" *Business Horizons*, November–December, 69–75.

Boer, G. (1994). "Five modern management accounting myths." *Management Accounting* (USA, January), 22–27.

Bohn, Roger E. (1995). "Noise and Learning in Semiconductor Manufacturing." *Management Science*, 41, 1, 31–42.

Bolander, Steven F., Richard C. Heard, Samuel M. Steward, and Sam G. Taylor (1981). Manufacturing Planning and Control in Process Industries. American *Production and Inventory Control* Society, Falls Church, VA.

Bonvik, A.M., C.E. Couch and S.B. Gershwin (1997). "A Comparison of Production Line Control

Mechanisms." *International Journal of Production Research*, 35, 789–804.

Booth, D.E., M. Khouja, and M. Hu (1992). "A robust multivariate procedure for evaluation and selection of industrial robots." *International Journal of Operations and Production Management*, 12, 15–24.

Bornemann, A.H. (1974). *Essentials of Purchasing*. Grid, Columbus, OH.

Botta, Elwood (1984). *Meeting the Competitive Challenge*. Richard D. Irwin, Homewood, IL.

Boubekri, N., M. Sahoui, and C. Lakrib (1991). "Development of an expert system for industrial robot selection." *Computers and Industrial Engineering*, 20, 119–127.

Bourke, Richard W. (1980). "Surveying the Software." *Datamation*, October, 101–107.

Bower J. and T. Hout (1988). "Fast Cycle Capability For Competitive Power." *Harvard Business Review*, Nov./Dec.

Bowerman, Bruce L., and Richard T. O'Connell (1979). *Time Series and Forecasting*. Duxbury Press, North Scituate, MA.

Bowersox, D.J., R.J. Calantone, S.R. Clinton, D.J. Closs, M.B. Cooper, C.L. Droge, S.E. Fawcett, R. Frankel, D.J.Frayer, E.A. Morash, L.M. Rinehart, and J.M. Schmitz (1995). *World Class Logistics: The Challenge of Managing Continuous Change*. Council of Logistics Management, Oak Brook, IL.

Bowersox, D.J. and P.J. Daugherty (1995). "Logistics Paradigms: The Impact of Information Technology." *Journal of Business Logistics*, 16 (1), 65–80.

Bowersox, D.J., D.J. Closs, and O.K. Helferich (1986). *Logistical Management: A Systems Integration of Physical Distribution, Manufacturing Support, and Materials Procurement* (3rd ed.). Macmillan, New York.

Bowles, J.G. (1986). "The Quality Imperative." *Fortune* (September 29), 61–96.

Bowman, E.H. (1956). "Production Scheduling by the Transportation Method of Linear Programming." *Operations Research*, February, 100–103.

Box, George E. and G.M. Jenkins (1970). *Time Series Analysis: Forecasting and Control*. Holden Day, Oakland, CA.

Boxoma, O.J. and R. Syski, eds. (1988). *Queuing Theory and Its Applications*. North Holland, Amsterdam.

Boyer, K.K. and G.K. Leong (1996). "Manufacturing Flexibility at the Plant Level." Omega: *The International Journal of Management Science*, 495–510.

Boyer, K.K., G.K. Leong, P.T. Ward, and Krajewski, L. (1997). "Unlocking the Potential of Advanced Manufacturing Technologies." *Journal of Operations Management*, 15 (4), 331–347.

Bozarth, R. Handfield, and A. Das (1997). "Stages of global sourcing strategy evolution: An exploratory study." Secial Issue on Global Manufacturing, *Journal of Operations Management*, in press.

Bradley, S.P., A.C. Hax, and T.L. Magnanti (1977). *Applied Mathematical Programming*. Addison-Wesley, Reading, Massachusetts.

Branch, A.E. (1996). *Elements of Shipping*, Seventh edition, Chapman and Hall, London, UK.

Brandeau, M.L. and S.S. Chui (1989). "An overview of representative problems on location research." *Management Science*, 35:645–74.

Bratley, P., B. Fox and L. Schrage (1987). *A Guide to Simulation* (2nd ed.). Springer-Verlag. New York.

Brausch, J.M. (1994). "Beyond ABC: Target Costing for Profit Enhancement." *Management Accounting*, November, 45–49.

Brennan, L., S.M. Gupta and K. Taleb (1994). "Operations Planning Issues in an Assembly/Disassembly Environment." *International Journal of Operations and Production Management*, 14 (9), 57–67.

Bridger, R.S. (1995). *Introduction to Ergonomics*, McGraw-Hill, NY, p. 1–30.

Britan, Gabriel R. and Arnoldo C. Hax (1977). "On the Design of Hierarchical Planning Systems." *Decision Sciences*, 8, January.

Briuand D.A., H. Garcia, D. Gateau, L. Molloholli, and B. Mutel (1989). "Part families and manufacturing cell formation". *Prolamat's Proceedings*, Dresden, GDR.

Brown, A.D. (1994). "Implementing MRP: leadership, rites and cognitive change." *Logistics Information Management*, 7, 2, 6–11.

Brown, Jimmie, John Harhen, and James Shivnan (1988). *Production Management Systems: A CIM Perspective*. Addison Wesley.

Brown, R.G. (1963). *Smoothing, Forecastings and Prediction of Discrete Time Series*. Prentice-Hall, Englewood Cliffs, N.S.

Brown, R.G. (1967). *Decision Rules for Inventory Management*, Holt, Rinehart and Winston, New York.

Brown, R.G. (1971). *Management Decisions for Production Operations*. The Dryden Press, Hinsdale, IL.

Brown, R.G. (1977). *Materials Management Systems*. John Wiley & Sons, New York.

Brown, R.G. (1982). *Advanced Service Parts Inventory Control. Materials Management Systems*, Inc., Norwich, VT.

Brown, S. (1996). *Strategic Manufacturing for Competitive Advantage: Transforming Operations from Shop Floor to Strategy*. Prentice-Hall, London.

Brown, S.L. and K.M. Eisenhardt (1997). "The art of continuous change: linking complexity theory and time-paced evolution in relentlessly shifting organizations." *Administrative Science Quarterly*, 42: 1–34.

Browne J. (1988). "Production Activity Control – A Key Aspect of Production Control." *International Journal of Production Research*, 26, no. 3, 415–427.

Browne, J., D. Dubois, K. Rathmill, S. Sethi, and K. Stecke (1984). "Classification of flexible manufacturing systems." *The FMS Magazine*, April, 114–117.

Bruns, W.J., Jr. and R.S. Kaplan, eds. (1987). *Accounting and Management: Field Study Perspectives*. Harvard Business School Press, Boston, 204–228.

Buchan, J. and E. Koenigsberg (1963). *Scientific Inventory Management*. Prentice-Hall, New Jersey. Chapters 1–16 and 19.

Buffa, E.S. (1984). *Meeting the Competitive Challenge: Manufacturing Strategy for U.S. Companies*. Dow Jones-Irwin, Homewood, IL.

Buffa, E.S. and Jeffrey G. Miller (1979). *Production-Inventory Systems: Planning and Control* (3rd ed.). Richard D. Irwin, Homewood, IL.

Buggie, Frederick D. (1981). *New Product Development Strategies*. AMACOM, New York.

Bulgren, W. (1982). *Discrete System Simulation*. Prentice-Hall, Englewood Cliffs, NJ.

Bunday, B.D. (1986). *Basic Queuing Theory*. Edward Arnold, London.

Burbidge, J.L. and B.G. Dale (1984). "Planning the Introduction and Predicting the Benefits of Group Technology." *Engineering Costs and Production Economics*, 8 (1), 117–128.

Burgess, A.R. and James B. Killebrew (1962). "Variation in Activity Level on a Cyclical Arrow Diagram." *Journal of Industrial Engineering*, 13, 2 (March–April), 76–83.

Burlingame, J.W. and J.K. Weeks (1981). "Behavioral Dimensions of MRP Change: Assessing Your Organization's Strengths and Weaknesses." *Production and Inventory Management*, 22 (1) (First Quarter), 81–95.

Burnham, J.M. (1980). "The Operating Plant – Case Studies and Generalizations." *Production and Inventory Management*, 21, 4, 63–94.

Burnham, J.M. (1985). "Improving Manufacturing Performance: A Need for Integration." *Production and Inventory Management*, 26, 4, 51–59.

Burnham, John M. (1987). "Some Conclusions About JIT Manufacturing." *Production and Inventory Management*, 28 (3), 7–11.

Burns T. and G. Stalker (1960). *The Management Of Innovation*. London, Tavistock.

Burns, O.M. and D. Turnipseed (1991). "Critical success factors in manufacturing resource planning implementation." *International Journal of Operations and Production Management*, 11 (4), 5–19.

Burt, D. (1984). *Proactive Purchasing*. Prentice-Hall, Englewood Cliffs, NJ.

Burt, D.N. (1989). "Managing Suppliers Up To Speed." *Harvard Business Review*, 67(4).

Burt, D.N., W. Norquist, and J. Anklesaria (1990). *Zero Base Pricing: Achieving World Class Competitiveness Through Reduced All-in-Cost*. Probus Publishing, Chicago.

Buzacott, J. (1982). "The fundamental principles of flexibility in manufacturing systems." Proceedings of 1st Intl. Conference on Flexible Manufacturing Systems, North Holland, UK.

Buzacott, J.A. and D.D. Yao (1986). "Flexible manufacturing systems: a review of analytical models." *Management Science*, 32, 890–905.

Buzacott, J.A. and J.G. Shantikumar (1993). *Stochastic Models of Manufacturing Systems*. Prentice-Hall, New Jersey.

Buzzell, R.D. and B.T. Gale (1987). *The PIMS Principles: Linking Strategy to Performance*. Free Press, New York.

Buzzell, R.D., B.T. Gale, and R.G. Sultan (1975). "Market share: a key to profitability." *Harvard Business Review*, 54 (1), 97–106.

Bylinski, G. (1983). "The Race to the Automatic Factory." *Fortune*, February 21, 52–60.

Byrett, Donald L., Mufit H. Ozden, and Jon M. Patton (1988). "Integrating Flexible Manufacturing Systems with Traditional Manufacturing, Planning, and Control." *Production and Inventory Management Journal*, Third Quarter, 15–21.

C

Calvert, J.E., *et al.* (1989). *Linear Programming*. Harcourt Brace Jovanovich, New York.

Camm, J.D., T.E. Chorman, F.A. Dill, J.R. Evans, D.J. Sweeney, and G.W. Wegryn (1997). "Blending OR/MS, Judgment, and GIS: Restructuring P&G's Supply Chain." *Interfaces*, 27 (1), 128–142.

Campbell, Kenneth L. (1975). "Inventory Turns and ABC-Analysis-Outmoded Textbook Concepts?" *American Production and Inventory Control Conference Proceedings*, 420.

Campion, M.A., G.J. Medsker, and A.C. Higgs (1993). "Relations Between Work Group Characteristics and Effectiveness: Implications for Designing Effective Work Groups." *Personnel Psychology*, 46, 823–850.

Canel, T.C. and B.M. Khumawala (1997). "Multi-Period International Facility Location: An Algorithm and Application." *International Journal of Production Research*, 35 (7), 1891–1910.

Cannon, Joseph J. and Gary Kapusta (1989). "Just-in-Time at AT&T's Reading Works." Just-in-Time Seminar Proceedings, July, 24–26.

Cantwell, J. (1985). "The How and Why of Cycle Counting: The ABC Method." *Production and Inventory Control*, 26 (2) (Second Quarter), 50–54.

Cao, T. and A.C. Sanderson (1994). "Task decomposition and analysis of robotic assembly task plans using Petri nets." *IEEE Transactions on Industrial Electronics*, 41, 620–630.

Carbone, J. (1995). "Lessons from Detroit." *Purchasing* (October 19), 38–42.

Carlisle, Brian (1983). "Job Design Implications for Operations Managers." *International Journal of Operations and Production Management*, 3 (3), 40–48.

Carson, J.W. and T. Richards (1979). *Industrial New Product Development: A Manual for the 1980's*. Wiley, New York.

Carter, Phillip L., and Chrwan-Jyh Ho (1984). "Vendor Capacity Planning: An Approach to Vendor Scheduling." *Production and Inventory Management*, Fourth Quarter, 63–73.

Casti, J.L. (1997). *Would-Be Worlds: How Simulation is Changing the Frontiers of Science*. John Wiley and Sons, New York.

Cavinato, J.L. "A Total Cost/Value Model for Supply Chain Competitiveness, *Journal of Business Logistics*, 13, 2, 285–301.

Cavinato, J.L. (1984). *Purchasing and Materials Management*. West, St. Paul, MN.

Celley, Albert F., William H. Clegg, Arthur W. Smith, and Mark A. Vonderembse (1986). "Implementation of JIT in the United States." *Journal of Purchasing and Materials Management*, Winter, 9.

Chaffee, R.H. (1995). "Creating Fulfillment Agility in the Face of Product Proliferation." *Logistics: Mercer Management Consulting*, 20–23.

Chakravarty, A.K. and A. Shtub (1984). "Selecting Parts and Loading Flexible Manufacturing Systems." Proceedings of The 1st ORSA/TIMS FMS Conference, Ann Arbor, MI.

Champy, J.A. (1995). *Reengineering Management: The Mandate for New Leadership*, Harper Collins, London.

Chance, F. (1993). "Conjectured upper bounds on transient mean total waiting times in queueing networks." In: Proceedings of the Winter Simulation Conference, 414–421.

Chance, F., J. Robinson, and J. Fowler (1996). "Supporting manufacturing with simulation: model design, development, and deployment." In: Proceedings of the Winter Simulation Conference, 114–121.

Chandler, Alfred (1977). *The Visible Hand: The Managerial Revolution in American Business*. Belknap Press, Cambridge, Mass.

Chandrasekharan, M.P. and R. Rajagoplan (1986). "An ideal-seed non-hierarchical clustering algorithm for cellular manufacturing." *International Journal of Production Research*, 24, 451–554.

Chang, T.C. and R.A. Wysk (1985). *An Introduction to Automated Process Planning Systems*. Prentice-Hall, Englewood Cliffs, N.J.

Chang, T.C., R.A. Wysk, and H.-P. Wang (1998). *Computer-Aided Manufacturing* (2nd ed.). Prentice-Hall, New Jersey.

Chase, R.B (1978). "Where Does the Customer Fit in a Service Operation?" *Harvard Business Review*, November–December, 137–142.

Chase, R.B. and D.A. Garvin (1989). "The Service Factory." *Harvard Business Review*, 67 (4) (July–August), 61–69.

Chase, R.B. and D.A. Tansik (1983). "The Customer Contact Model for Organization and Design." *Management Science*, 29, 9 (September) 1037–1050.

Chase, R.B. and E.L. Prentis (1987). "*Operations Management*: A Field Rediscovered." Journal of Management 13, 2, October, 351–366.

Chase, R.B. and N.J. Aquilano (1989). *Production and Operations Management* (5th ed.). Irwin, Homewood, Illinois.

Chase, R.B. and W.K. Erikson (1988). "The Service Factory." *Academy of Management of Executive* (August), 191–196.

Chase, R.B.K., R. Kumar, and W.E. Youngdahl (1992). "Service-Based Manufacturing: The Service Factory." *Production and Operations Management*, 1 (2), 175–184.

Chen, C.H. (1996). "A lower bound on the correct subset selection probability and its application to discrete-event simulation." *IEEE Transactions on Automatic Control*, 41, 1227–1231.

Chen, I.J. and Chung, C.H. (1991). "Effects of loading and routing decisions on performance of flexible manufacturing systems." *International Journal of Production Research*, 29, 2209–2225.

Cheng, R.H.C., W. Holland, and N.A. Hughes (1996). "Selection of input models using bootstrap goodness of fit." In: Proceedings of the Winter Simulation Conference, 199–206.

Cherukuri, S.S., R.G. Nierman, and N.C. Sirianni (1995). "Cycle Time and The Bottom Line." *Industrial Engineering*, March.

Chikte, S.D. and D. Deshmukh (1981). "Preventive Maintenance and Replacement Under Additive Damage." *Naval Research Logistics Quarterly*, 28, 33–46.

Chiola, G., G. Franceschinis, R. Gaeta, and M. Ribaudo (1995). "GreatSPN 1.7: graphical editor and analyzer for timed and stochastic Petri nets." *Performance Evaluation*, 24, 47–68.

Cho, H. (1998). "Petri net models for message manipulation and event monitoring in an FMS cell." *International Journal of Production Research*, 36, 231–250.

Choobineh, F. (1984). "Optimum Loading for GT/MRP Manufacturing Systems." *Computer and Industrial Engineering* **8**, 3/4, 197–206.

Chorafas, D.N. (1965). *Systems and Simulation*. Academic Press, New York.

Chowdhury, A.R. (1985). "Reliability as It Relates to QA/QC." *Quality Progress*, 18 (12) (December), 27–30.

Christer, A.H. and W.M. Waller (1984). "Reducing production downtime using delay-time analysis." *Journal of the Operational Research Society*, 35, 499–512.

Christy, David P. and Hugh J. Watson (1983). "The Application of Simulation: A Survey of Industry Practice." *Interfaces*, 13, 5, 47–52.

Chu, C.H. and J.C. Hayya (1991). A fuzzy clustering approach to manufacturing cell formation, *International Journal of Production Research*, 29, 1475–1487.

Chung, C.H. (1991). "Planning Tool Requirements for Flexible Manufacturing Systems." *Journal of Manufacturing Systems*, 10 (6), 476–483.

Chung, C.H. and L.J. Krajewski (1984). "Planning Horizons for Master Production Schedules." *Journal of Operations Management*, (4), 389–406.

Chung, C.H. and L.J. Krajewski (1986). "Replanning Frequencies for Master Production Schedules." *Decision Sciences*, 17 (2), 263–273.

Chung, C.H. and L.J. Krajewski (1986). "Integrating Resource Requirements Planning with Master Production Scheduling." Lee, S.M., L.D. Digman and M.J. Schniederjans, eds., Proceedings of the 18th Decision Sciences Institute National Meeting, Hawaii, 1165–1167.

Chung, C.H. and L.J. Krajewski (1987). "Interfacing Aggregate Plans and Master Production Schedules via a Rolling Horizon Procedure." *OMEGA*, 15 (5), 401–409.

Chung, C.H. and W.H. Luo (1996). "Computer Support of Decentralized Production Planning." *The Journal of Computer Information Systems*, 36 (2), 53–59.

Clark J.T. (1984). "Inventory Flow Models." Proceedings, 27th Annual Conference, American *Production and Inventory Control* Society.

Clark, Charles T. and Eleanor W. Jordan (1985). *Introduction to Business and Economic Statistics*. Southwestern, Cincinatti.

Clark, K.B. (1989). "Project Scope and Project Performance: The Effect of Parts Strategy and Supplier Involvement on Product Development." *Management Science*, 35 (10), 1247–1263

Clark, K.B. (1992). "Design for Manufacturability at Midwest Industries: Teaching Note." Note no. 5-693-007, Harvard Business School Press, Boston.

Clark, K.B. and T. Fujimoto (1991). *Product Development Performance: Strategy, Organization and Management in the World Auto Industry*. Harvard Business School Press, Cambridge.

Cleland, D.I. and W.R. King (1983). *Project Management Handbook*. Van Nostrand Reinhold, New York.

Clemen, R.T. (1989). "Combining forecasts: A review and annotated bibliography." *International Journal of Forecasting*, 559–584.

Clingen, C.T. (1964). "A Modification of Fulkerson's PERT Algorithm." *Operations Research*, 12, 4 (July–August), 629–631.

Clinton, S.R., D.J. Closs, M.B. Cooper, and S.E. Fawcett (1996). "New Dimensions of World Class Logistics Performance." In Annual Conference Proceedings Council of Logistics Management (21–33).

Close, Arthur C. (1970). "Projected Inventory Velocity Measurement." *Production and Inventory Management*, 11 (3), 66.

Cochran, J.K., G.T. Mackulak, and P.A. Savory (1995). "Simulation project characteristics in industrial settings." *Interfaces*, 25 (4), 104–113.

Cohen, M.A. and H.L. Lee (1989). "Resource deployment analysis of global manufacturing and distribution networks." *Journal of Manufacturing & Operations Management*, 2 (81)–104.

Cohen, M.A. and U.M. Apte (1997). *Manufacturing Automation*, Irwin, Illinois.

Cohen, Oded (1988). "The Drum-Buffer-Rope (DBR) Approach to Logistics." In *Computer-Aided Production Management*, A. Rolstadas, ed. Springer-Verlag, New York.

Cohen, S.G., G.E. Ledford, Jr., and G.M. Spreitzer (1996). "A Predictive Model of Managing Work Team Effectiveness." *Human Relations*, 49 (5), 643–676.

Cohen, S.S. and J. Zysman (1987). *Manufacturing Matters: The Myth of the Post Industrial Economy*, Basic Books, New York.

Colley, J.L. (1977). "Inventory Flow Models." Technical Note OM-120, Darden Graduate School of Business, University of Virginia, Charlottesville, Virginia.

Collier, D.A. (1980). "The Interaction of Single-Stage Lot Size Models in a Material Requirements Planning System." *Production and Inventory Management*, 21 (4), 11–20.

Collier, D.A. (1985). *Service Management: The Automation of Services*. Reston Publishing, Reston, VA.

Collier, D.A. (1987). *Service Management: Operating Decisions*. Prentice-Hall, Englewood Cliffs, NJ.

Collins, R. and R.W. Schmenner (1995). "Taking manufacturing advantage of Europe's single market." *European Management Journal*, 13:257–68.

Collopy, F. and J.S. Armstrong (1992). "Rule Based Forecasting: Development and Validation of an Expert System Approach to Combining Time Series Extrapolations." *Management Science*, 38, 10, 1394-1414, 1992.

Colton, R.R. and W.F. Rohrs (1985). *Industrial Purchasing and Effective Materials Management*. Reston Publishing, Reston, VA.

Combs, P.M. (1976). *Handbook of International Purchasing*. Cahners, Boston.

Conlon, J.R. (1977). "Is Your Master Schedule Feasible?" APICS Master Production Scheduling Reprints. American *Production and Inventory Control* Society, Falls Church, VA.

Connell, B.C., E.E. Adam, Jr., and A.N. Moore (1984). "Aggregate Planning in a Health Care Foodservice System with Varying Technologies." *Journal of Operations Management*, 5, 1, 41–55.

Conover, W.J. (1980). *Practical Nonparametric Statistics* (2nd ed.). John Wiley, New York.

Conroy, Paul G. (1977). "Data General ABC Inventory Management." *Production and Inventory Management*, 18, 4, 63.

Constantinides, K. and J.K. Shank (1994). "Matching accounting to strategy: One mill's experience." *Management Accounting* (USA, September), 32–36.

Conway, R.W. and Andrew Schultz, Jr. (1959). "The Manufacturing Progress Function." *Journal of Industrial Engineering*, 10, 1, 39–54.

Conway, R., W.L. Maxwell, J.O. McClain, and S. Worona (1987). *User's Guide to Xcell + Factory Modeling System* (2nd ed.). Scientific Press, California.

Cooper, Carolyn C. (1981–2). "The Production Line at Portsmouth Block Mill." *Industrial Archaeology Review*. 6, 28–44.

Cooper, Carolyn C. (1984). The Portsmouth System of Manufactures. *Technology and Culture*, 25, 182–225.

Cooper, L.K., D.J. Duncan, and J. Whetstone (1996). "Is Electronic Commerce Ready for The Internet?: A Case Study of a Financial Service Provider." *Information Systems Management*, 13, 25–36.

Cooper, M.C., D.M. Lambert, and J.D. Pagh (1997). "Supply Chain Management: More Than A New Name for Logistics." *International Journal of Logistics Management*, 1, 1–13.

Cooper, M.C. and Lisa M. Ellram (1993). "Characteristics of Supply Chain Management and the Implications for Purchasing and Logistics Strategy." *International Journal of Logistics Management*, 4, 2, 10–15.

Cooper, R. (1987). "Does your company need a new cost system?" *Journal of Cost Management* (Spring), 45–49.

Cooper, R. (1988). "The rise of activity-based costing – part one: What is an activity-based cost system?" *Journal of Cost Management* (Summer), 45–54.

Cooper, R. (1988). "The rise of activity-based costing – part two: When do I need an activity-based cost system?" *Journal of Cost Management* (Fall), 41–48.

Cooper, R. (1989). "You Need a New Cost System When...." *Harvard Business Review*, 77 (1) (January–February), 77–82.

Cooper, R. (1989). "The rise of activity-based costing – part three: How many cost drivers do you need, and how do you select them?" *Journal of Cost Management* (Winter), 34–46.

Cooper, R. (1990). "Cost classifications in unit-based and activity-based manufacturing cost systems." *Journal of Cost Management for the Manufacturing Industry* (Fall), 4–14.

Cooper, R. (1994). "Yokohama Corporation Ltd. (A)." *Harvard Business School Case* 5-195-071, 1994.

Cooper, R. and R.S. Kaplan (1988). "Measuring costs right make the right decisions." *Harvard Business Review*, 66, 96–103.

Cooper, R. and R.S. Kaplan (1991). "Profit Priorities from Activity-Based Costing." *Harvard Business Review*, May–June, 130–135.

Cooper, R. and R.S. Kaplan (1991). *The Design of Cost Management Systems*, Prentice-Hall, Englewood Cliffs.

Cooper, R. and R. Slagmulder (1997). *Target Costing and Value Engineering, Institute of Management Accountants*. Productivity Press.

Cooper, R. and W.B. Chew. "Control Tomorrow's Costs through Today's Designs." *Harvard Business Review*, January–February, 1996, 88–97.

Cooper, R., R.S. Kaplan, L.S. Maisel, E. Morrissey, and R.M. Oehm (1992). *Activity-Based Cost Management: Moving from Analysis to Action*. The Institute of Management Accountants, New Jersey.

Cooper, R.B. (1981). *Introduction to Queuing Theory* (2nd ed.). Elsevier North-Holland, New York.

Cooper, R.G., E.J. Kleinschmidt (1987). New Products: What Separates Winners from Losers? *Journal of Product Innovation Management*, 4, 3, 69–184.

Cooper, R.B. and R.W. Zmud (1989). "Materials requirements planning system infusion." *OMEGA International Journal of Management Sciences*, 17, 5, 1989, 471–481.

Cooper, R.B. and R.W. Zmud (1990). IT implementation research: a technological diffusion approach." *Management Science*, 36, 2, 1990, 123–139.

Copley, Frank Barkley (1923). *Frederick W. Taylor: Father of Scientific Management*. Reprinted 1969, Augustus Kelley, New York, New York.

Corbett, C. and L. Van Wassehnove (1993). "Trade-Offs? What Trade-Offs? Competence and Competitiveness in Manufacturing Strategy." *California Management Review*, Summer, 107–122.

Cordero, S.T. (1987). *Maintenance Management*. Fairmont Press, Englewood Cliffs, NJ.

Council of Logistics Management (1995). *World Class Logistics: The Challenge of Managing Continuous Change*, Council of Logistics Management, IL.

Courtis, John K. (1995). "JIT's Impact on a Firm's Financial Statements." *International Journal of Purchasing and Materials Management*, 31, 1, Winter, 46–51.

Cox, T. (1994). *Cultural Diversity in Organizations: Theory, Research and Practice*. Berret-Koehler, San Francisco.

Coyle, J.J. and E.J. Bardi (1988). *The Management of Business Logistics*. St. Paul, West.

Craighead, Thomas G. (1989). "EDI Impact Grows for Cost Efficient Manufacturing." *PIM Review*, July, 32.

Crane, M.A. and D.L. Iglehart (1975). "Simulating stable stochastic processes, III: regenerative processes and discrete event simulations." *Operations Research*, 23, 33–45.

Crawford, K.M., J.F. Cox, and J.H. Blackstone, Jr. (1988). Performance Measurement Systems and the JIT Philosophy: Principles and Cases. *American Production and Inventory Control Society*, Falls Church, VA.

Crawford, K.M., J.H. Blackstone, and J.F. Cox (1988). "A Study of JIT Implementation and Operating Problems." *International Journal of Production Research*, 26, 1561–1568.

Crosby, L.B. (1984). "The Just-in-Time Manufacturing Process: Control of Quality and Quantity." *Production and Inventory Management*, 25, 4, 21–33.

Crosby, P.B. (1979). *Quality is Free: The Art of Making Quality Certain*. McGraw-Hill, New York.

Crosby, P.B. (1984). *Quality Without Tears*. McGraw-Hill, New York.

Cryer, Jonathan D. (1986). *Time Series Analysis*. Duxbury Press, Boston.

Cyert, R. (?). "Saving the Smokestack Industries." *Bell Atlantic Quarterly* 2 (3) (Autumn), 17–24.

Czinkota, M.R. and I.A. Ronkainen (1993). *International Marketing*. Dryden Press, Fort Worth.

D

D'Souza, K.A. and S.K. Khator (1994). "A survey of Petri net applications in modeling controls for manufacturing systems." *Computers in Industry*, 24, 5–16.

Dada, M. and R. Marcellus (1994). "Process Control with Learning." *Operations Research*, 42, 2, 323–336.

Dalrymple, D.J. (1984). "Sales Forecasting Methods and Accuracy." *Business Horizons*, December, 69–73.

Dalrymple, D.J. (1987). "Sales Forecasting Practices – The Results of a US Survey." *International Journal of Forecasting*, 3, 3, 379–391.

Daniels, J.D. and L.H. Radebaugh (1998). *International Business Environments and Operations* (8th ed.). Addison Wesley, Reading, Mass.

Dapiran, P. (1992). "Benetton – Global Logistics in Action." *International Journal of Physical Distribution and Logistics Management*, 22(6).

Daskin, M.S. (1985). "Logistics: An Overview of the State of the Art and Perspectives on Future Research." *Transportation Journal*, 19A(5/6), 383–398.

Daugherty P.J., R.E. Sabath and D.S. Rogers (1992). "Competitive Advantage Through Customer Responsiveness." *The Logistics and Transportation Review*, 28(3), 257–271.

Davenport, T.H. (1993). *Process Innovation: Reengineering Work Through Information Technology*, Harvard Business School Press, Boston, Massachusetts, p. 5–9.

Davenport, T.H. and J.E. Short (1990). "The New Industrial Engineering: Information Technology and Business Process Redesign." *Sloan Management Review*, Summer, 11–27.

Davidow, W.H. and B. Uttal (1989). "Service Companies: Focus or Falter." *Harvard Business Review*, 67(4), 77–85.

Davis, P.S. and T.L. Ray (1969). "A Branch and Bound Algorithm for the Capacitated Facilities Location Problems." *Naval Research Logistics Quarterly*, 16(3), Sept.

Davis, S.M. (1987). *Future Perfect, Reading, Massachusetts*, Addison-Wesley Publishing Company.

De Meyer, A. (1991). "Tech talk: How managers are stimulating global R&D communication." *Sloan Management Review*, Spring, 49–58.

De Meyer, A. and A. Vereecke (1996). "International Operations" in *International Encyclopedia of Business and Management*. Editor M. Werner. Routledge.

De Vera, Dennis, Tom Glennon, Andrew A. Kenney, Mohammad A.H. Khan, and Mike Mayer (1988). "An Automotive Case Study." *Quality Progress*, June, 35–38.

Dean, J.H., Jr. and G.I. Susman (1989). "Organizing for Manufacturable Design." *Harvard Business Review*, January–February, 1989, 28–36.

Deane, R.H., P.P. McDougall, and V.B. Gargeya (1991). "Manufacturing and Marketing Interdependence in the New Venture Firm: An Empirical Study." *Journal of Operations Management*, 10 (3), 329–343.

DeCarlo, N.J. and W.K. Sterret (1990). "History of the Malcolm Baldrige National Quality Award." *Quality Progress*, 23, 21–27.

Defoe, Daniel (1697). *An Essay Upon Projects*. Reprinted 1887. Cassell & Co., London, England.

DeGarmo, E. Paul, JT. Black, A. Kohser, A. Ronald (1997). *Materials and Processes in Manufacturing* – 8th Edition. Prentice-Hall, Inc.

Dekker, R., C.F.H. Van Rijn, and A.C.J.M. Smit (1994). "Mathematical models for the optimization of maintenance and their application in practice." *Maintenance*, 9, 22–26.

Dekker, R., F.A. Van der Duyn Schouten and R.E. Wildeman (1997). "A review of multicomponent maintenance models with economic dependence." *Mathematical Methods of Operations Research*, 45, 411–435.

Deleersnyder, J.L., T.J. Hodgson, H. Muller and P.J. O'Grady (1989). "Controlled Pull Systems: An Analytical Approach." *Management Science*, 35, 1079–1091.

DelMar, D. and G. Sheldon (1988). *Introduction to Quality Control*. West, St. Paul, MN.

DeLurgio, S.A. and C.D. Bahme (1991). *Forecasting Systems for Operations Management*, Business One Irwin, Homewood, IL.

DeLurgio, S. (1998). *Forecasting Principles and Applications*, Irwin/McGraw-Hill.

Dematheis, J.J. (1968). "An Economic Lot-Sizing Technique: The Part Period Algorithms." *IBM Systems Journal*, 7, 1, 50–51.

DeMeyer, A., J. Nakane, J.G. Miller, and K. Ferdows (1987). Flexibility: The Next Competitive Battle." Manufacturing Roundtable Research Report Series. School of Management, February. Reprinted in *Strategic Management Journal*, 10 (1989), 135–144.

Deming, W.E. (1986). *Out of the Crisis*. MIT Center for Advanced Engineering Study, Cambridge, MA.

Deming, W.E. (1982). *Quality, Productivity, and Competitive Position*, Center for Advanced Engineering Study, Massachusetts Institute of Technology, Cambridge, Massachusetts.

Deming, W.E. (1993). *The New Economics for Industry, Government, and Education*, Center for Advanced Engineering Study, Massachusetts Institute of Technology, Cambridge, Massachusetts.

Denison, R.A. and J. Ruston (1990). *Recycling and Incineration*. Island Press, Washington, D.C.

Denna, E.L., J.O. Cherrington, D.P. Andros, and A.S. Hollander (1993). *Event-Driven Business Solutions*, Business One Irwin, Homewood, IL.

Dertouzos, Michael L., Richard K. Lester, Robert M. Solow (1989) and the MIT Commission on Industrial Productivity. *Made in America: Regaining the Productive Edge*. The MIT Press, Cambridge, MA.

DeSantis, G.F. (1977). "Implementation considerations in system design." In Proceedings of the American and Production Inventory Control Society Conference, 210–217.

Desrochers, A.A. and R.Y. Al-Jaar (1995). *Applications of Petri Nets in Manufacturing Systems: Modeling, Control, and Performance Analysis*. IEEE Press, Piscataway, NJ.

Dess, G.G. and P.S. Davis (1984). "Porter's (1980) generic strategies as determinants of strategic group membership and organizational performance." *Academy of Management Journal*, 27 (3), 467–488.

DeVor, R., R. Graves, and J.J. Mills (1997). "Agile Manufacturing Research: Accomplishments and Opportunities." *IIE Transactions*, 29, 813–823.

Dhavale D.G. and G.L. Otterson, Jr. (1980). "Maintenance by Priority." *Industrial Engineering*. 12 (2) (February), 24–27.

Dilworth, J.D. (1987). *Information Systems for JIT Manufacturing*. Association for Manufacturing Excellence, Wheeling, IL.

Dilworth, J.B. (1992). *Operations Management:* Design, Planning and Control for Manufacturing and Services. McGraw-Hill.

Diprimio, A. (1987). *Quality Assurance in Service Organizations*. Radnor, PA, Chilton.

Disckson, Gary W., James A. Senn, and Normal L. Chervany (1977). "Research in Management Information Systems: The Minnesota Experiments." *Management Science*, May.

Diskie, H. Ford (1951). "ABC Inventory Analysis Shoots for Dollars." *Factory Management and Maintenance*, July, 92.

Dixon, J.R., A.J. Nanni, Jr., and T.E. Vollmann (1990). *The New Performance Challenge: Measuring Operations for World-Class Competition*. Business One Irwin, IL.

Dixon, Lance and Anne Miller Porter (1994). *JIT II – Revolution in Buying and Selling*, Cahner's Publishing Company, Newton, MA.

Dobler, D.W. and Burt, D.N. (1996). *Purchasing and Materials Management*. McGraw-Hill Book Company, New York.

Dobler, D.W., L. Lee, Jr., and D.N. Burt (1984). *Purchasing and Materials Management: Text and Cases*. McGraw-Hill, New York.

Doblin, J. (1987). "A Short Grandiose Theory of Design." *STA Design Journal*.

Dobyns, L. and C. Crawford-Mason (1991). *Quality or Else: The Revolution in World Business*. Houghton Mifflin, New York.

Doll, William, and Mark Vonderembse (1987). "Forging a Partnership to Achieve Competitive Advantage: The CIM Challenge." *MIS Quarterly*, 11 (2), 205–220.

Donlon, J.P. (1993). "The Six Sigma Encore; Quality Standards at Motorola." *Chief Executive*, November, 90.

Doos, B.R. (1994). "The State of the Global Environment: Is there a need for 'An orderly retreat?" *Environment Cooperation in Europe: The political dimension*, Holl, O. (eds.), Westview Press, Boulder, Colorado, 33–50.

Dorf, R.C. and A. Kusiak (1994). *Handbook of Design, Manufacturing, and Automation*. Wiley, New York.

Douglas, S.P. and Y. Wind (1987). "The myth of globalization." *Columbia Journal of World Business*, Winter, 19–29.

Dove, R. (1995). "The Challenges of Change." *Production*, 107 (2), 16–17.

Dove, R. (1996). *Tools for Analyzing and Constructing Agile Capabilities*, Agility Forum, Bethlehem, PA.

Dove, R., S. Hartman, and S. Benson (1996). *An Agile Enterprise Reference Model with a Case Study of Remmele Engineering*, Agile Forum, Bethlehem, PA.

Doz, Y. (1986). *Strategic Management in Multinational Companies*. Pergamon Press, Oxford.

Draper, C. (1984). *Flexible Manufacturing Systems Handbook*, Automation and Management Systems Division, The Charles Stark Draper Laboratory Inc., Noyes Publication.

Dreifus, S.B., ed. (1992). *Business International's Global Management Desk Reference*, McGraw-Hill, New York.

Droy, John B. (1984). "It's Time to "Cell' Your Factory." *Production Engineering*, December, 50–52.

Drucker, P. (1954). *The Practice of Management*, Harper and Brothers, New York, 195–196.

Drucker, P. (1963). "Managing for business effectiveness." *Harvard Business Review* (May–June), 59–66.

Drucker, P. (1974). *Management: Tasks, Responsibilities*, Practices. New York, Harper and Rowe.

Drucker, P. (1982). *The Changing World of the Executive*. Heinemann, London.

Drucker, P. (1986). "The Changed World Economy." *Foreign Affairs* (Spring), 768–791.

DTI (1988). *Competitive Manufacturing: A Practical Approach to the Development of a Manufacturing Strategy*. IFS, Bedford.

Dubois, D. and K.E. Stecke (1983). "Using Petri nets to represent production processes." Proceedings of 22nd IEEE Conf. Decision and Control, IEEE Press.

Duchessi, P., C.M. Schaninger, D.R. Hobbs, and L.P. Pentak (1988). "Determinants of success in implementing materials requirements planning (MRP)." *Journal of Manufacturing and Operations Management*, 1 (3), 263–304.

Dugdale, D. and D. Jones (1996). "Accounting for throughput: Part 1 – the theory." *Management Accounting* (UK, April), 24–29.

Duncan, A.J. (1986). *Quality Control and Industrial Statistics* (5th ed.). Irwin, Homewood, IL.

Dunphy, D. and B. Bryant (1996). "Teams: Panaceas or Prescriptions for Improved Performance." *Human Relations*, 49 (5), 677–699.

Dyer, J.H. and Ouchi, W.G. (1993). Japanese-style partnerships: giving companies a competitive edge. *Sloan Management Review*, 35 (1): 51–63.

E

Ealey, L.A. (1988). *Quality by Design: Tagauchi Methods and U.S. Industry*. ASI Press, Dearborn, MI.

Eccles, R.G. (1991). "The performance measurement manifesto." *Harvard Business Review*.

Eckstein, A.L.H. and J. Balakrishnan (1993). "The ISO 9000 Series: Quality Management Systems for the Global Economy." *Production and Inventory Management Journal* 34, 66–71.

Edmundson, R., M. Lawrence, and M. O'Connor (1988). "The use of non-time series information in sales forecasting: a case study." *Journal of forecasting*, 7, 201–211.

Eilon, S., *et al.* (1967). Job shop scheduling with due date. *International Journal of Production Research* 6 (1), 1–13.

Eisenhardt, K.M. and B. N. Tabrizi (1995). "Accelerating Adaptive Processes: Product Innovation in the Global Computer Industry." *Administrative Science Quarterly*, 40, 84–110.

Elantora, E. (1992). "The Future Factory: Challenge for One-of-a-Kind Production." *International Journal of Production Economics*, 28, 131–142.

Elbourne, Edward D. (1918). *Factory Administration and Accounts*, The Library Press Co., Ltd., London, England.

Elfring, T. and G. Baven (1994). "Outsourcing Technical Services: Stages of Development." *Long Range Planning*, 27 (5), 42–51.

Elliff, S. (1996). "Supply Chain Management-New Frontier." *Traffic World* (October 21), 55.

Ellram, L.M. (1994). "The 'What' of Supply Chain Management." *NAPM Insights*, 5 (3), 26–27.

Elmagrhaby, Salah E. "The Theory of Networks and Management Science, II." *Management Science*, 13 (5) (January), 299–306.

Elmagrhaby, Salah E. (1966). "On Generalized Activity Networks." *Journal of Industrial Engineering*, 17, 11 (November), 621–631.

Emmelhorne, Margaret A. (1971). *Guide to Purchasing: Electronic Data Interchange. National Association of Purchasing Management*, John Wiley & Sons, New York.

Engelberger, J.F. (1980). *Robotics in Practice*. AMACOM, New York.

Enrick, N.L. (1985). *Quality, Reliability, and Process Improvement* (8th ed.). Industrial Press, New York.

Ephremides, A., P. Varauoa, and J. Walrand (1980). A simple dynamic routing problem, *IIEE Transactions on Automatic Control*, 690–693.

Eppen Gary D., J.F. Gould, and C.P. Schmidt (1987). *Introductory Management Science* (2nd ed.). Prentice-Hall, Englewood Cliffs, N.J.

Eppinger, S.D., D.E. Whitney, R.P. Smith, and D.A. Gebala (1994). "A Model-Based Method for Organizing Tasks in Product Development." *Research in Engineering Design*, 6, 1–13.

Epple, Dennis, Linda Argote, and Kenneth Murphy (1996). "An Empirical Investigation of the Microstructure of Knowledge Acquisition and Transfer Through Learning by Doing." *Operations Research*, 44 (1), 77–86.

Erens, F.J. and Hegge, H.M.H. (1994). "Manufacturing and Sales Coordination for Product Variety." *International Journal of Production Economics*, 37–1, 83–100.

Erikson, W.J. and O.P. Hall, Jr. (1983). *Computer Models for Management Science*. Addison-Wesley, Reading, MA.

Esparrago, R.A., Jr. (1988). "Kanban." *Production and Inventory Management*, 29 (1), 6–10.

Etienne, E.C. (1983). "MRP May Not Be Right for You, At Least not Yet." *Production and Inventory Management*, 24 (3) (Third Quarter), 33–46.

Ettlie, J.E. (1990). "What makes a Manufacturing Firm Innovative?" *Academy of Management Executive*, 4 (4).

Ettlie, J.E. and J.D. Penner-Hahn (1994). "Flexibility ratios and manufacturing strategy." *Management Science*, 40, 1444–1454.

Ettlie, J.E. (1988). *Taking Charge of Manufacturing*, Jossey-Bass Publishers, San Francisco.

Ettlie, J.E., M.C. Burstein, and A. Fiegenbaum (1990). *Manufacturing Strategy*. Kluwer, Boston.

Evans, G.N., M.M. Naim, and D.R. Towill (1993). "Dynamic Supply Chain Performance: Assessing the Impact of Information Systems." *Logistics Information Management*, 6 (4), 15–25.

Evans, J.R. (1996). "Something old, something new: a process view of the Baldrige criteria." *International Journal of Quality Science*, 1, 62–68.

Evans, J.R. (1997). "Beyond QS-9000." *Production and Inventory Management Journal*, 38, 3, Third Quarter, 72–76.

Evans, J.R. and D.L. Olson (1998). *Simulation*. Prentice-Hall, Englewood Cliffs, NJ.

Evans, J.R. and M.W. Ford (1997). "Value-Driven Quality." *Quality Management Journal*, 4 (4), 19–31.

Evans, J.R. and W.M. Lindsay (1998). *The Management and Control of Quality* (4th ed.). South-Western, Cincinnati, OH.

Evans, Oliver (1834). *The Young Mill-wright and Miller's Guide*. Reprinted 1989. Arno Press. New York, New York.

Everdell, Romeyn (1987). Master Scheduling: APICS Training Aid. American *Production and Inventory Control* Society, Falls Church, Virginia.

Eyman, Earl D. (1988). *Modeling, Simulation and Control*. West, St. Paul.

Ezpeleta, J. and J. Martinez (1992). "Petri nets as a specification language for manufacturing systems". In J.-C. Gentina, S.G. Tzafestas (eds.), *Robotics and Flexible Manufacturing Systems*, 427–436, North-Holland, New York.

F

Fallon, Carlos (1971). *Value Analysis to Improve Productivity*. John Wiley and Sons, New York.

Faurote, Fay Leone and Harold Lucien Arnold (1916). The Ford Methods and the Ford Shops. Engineering Magazine Co., New York, New York.

Fava, J.A., R. Denison, B. Jones, M.A. Curran, B. Vigon, S. Selke, and J. Barnum (ed.) (1991). *A Technical Framework for Life-Cycle Assessments*. Washington, DC: Society of Environmental Toxicology and Chemistry (SETEC).

Fawcett S.E. and J.N. Pearson (1991). "Understanding and applying constraint management in today's manufacturing environment." *Production and Inventory Management Journal* (Third Quarter), 46 55.

Fawcett, S.E. (1990). "Logistics and Manufacturing Issues in Maquiladora Operations." *International Journal of Physical Distribution and Logistics Management*, 20 (4), 13–21.

Fawcett, S.E. and S.A. Fawcett (1995). "The Firm as a Value-Added System: Integrating Logistics, Operations, and Purchasing." *International Journal of Physical Distribution and Logistics Management*, 25 (3), 24–42.

Fawcett, S.E. and S. Clinton (1996). "Enhancing Logistic Performance to Improve the Competitiveness of Manufacturing Organizations." *Production and Inventory Management Journal*, 37 (1), 40–46.

Fawcett, S.E. and S.R. Smith. (1995). "Logistics Measurement and Performance for United States – Mexican Operations under NAFTA." *Transportation Journal*, 34 (3), 25–34.

Fawcett, S.E., L.E. Stanley, and S.R. Smith (1997). "Developing a Logistics Capability to Improve the Performance of International Operations." *Journal of Business Logistics*, 18 (2), 101–128.

Fazar, Willard (1962). The Origins of PERT. *The Controller*, 8, 598–602, 618–621.

Feigenbaum, A.V. (1986). *Total Quality Control* (3rd ed.). McGraw-Hill, New York.

Feitzinger, E. and L.H. Lee (1997). "Mass Customization at Hewlett-Packard: The Power of Postponement." *Harvard Business Review*, 75, 116–121.

Feldman, E., F.A. Lehrer, and T.L. Ray (1966). "Warehouse Locations Under Continuous Economies of Scale." *Management Science*, 12, May.

Ferdows, K. (1997). "Making the most of foreign factories." *Harvard Business Review*, 73–88.

Ferdows, K. and A. De Meyer (1990). "Lasting Improvements in Manufacturing Performance: In Search of a New Theory." *Journal of Operations Management*, 9 (2), 168–184.

Ferdows, K. (1989). Editor. *Managing International Manufacturing*. North-Holland, New York.

Fiegenbaum, A.V. (1961). *Total Quality Control: Engineering and Management*. McGraw-Hill, New York.

Fierman, J. (1995). "Americans Can't Get No Satisfaction." *Fortune* (December 11), 186–194.

Finch, Byron J., and James F. Cox (1987). Planning and Control System Design: Principles and Cases For Process Manufacturers. American *Production and Inventory Control* Society, Falls Church, VA.

Fine, C.H. (1988). "A Quality Control Model with Learning Effects." *Operations Research*, 36, 3, 437–444.

Fine, C.H. and A.C. Hax (1985). "Manufacturing Strategy: A Methodology and an Illustration." *Interfaces*, 15 (6), 28–46.

Fine, C.H. and R.M. Freund (1990). "Optimal Investment in Product-Flexible Manufacturing Capacity." *Management Science*, 36 (4), 449–466.

Fisher, E.L. and O.Z. Maimon (1988). "Specification and selection of robots." In Kusiak, A. (ed.). *Artificial Intelligence Implications for CIM*, IFS Publications, Bedford.

Fisher, J. (1995). "Implementing Target Costing." *Journal of Cost Management*, Summer, 50–59.

Fisher, M.L., J.H. Hammond, W.R. Obermeyer, A. Raman (1994). "Making Supply Meet Demand in an Uncertain World." *Harvard Business Review*, 83–93.

Fishman, G.S. (1971). Estimating Sample Size in Computer Simulation Experiments." *Management Science*, 18 (1), 21–38.

Fitzsimmons, James A., P. Kouvelis, and D.N. Mallick (1991). "Design Strategy and Its Interface with Manufacturing and Marketing: A Conceptual Framework." *Journal of Operations Management*, 10 (3) August, p 398–415.

Flaherty, M.T. (1996). Global *Operations Management*, McGraw-Hill, New York.

Flores, B.E. and D.C. Whybark (1987). "Implementing Multiple Criteria ABC Analysis." Journal of *Operations Management*, 7 (1–2) (October), 79–85.

Florin, G. and S. Natkin (1982). "Evaluation based upon stochastic Petri nets of the maximum throughput of a full duplex protocol". C. Girault and W. Reisig (eds.), *Application and Theory of Petri Nets*, 280–288, Springer-Verlag, New York.

Flynn, Barbara B. and F. Robert Jacobs (1987). "An Experimental Comparison of Cellular (Group Technology) Layout with Process Layout." *Decision Sciences*, 18, 562–581.

Fogarty, D.W. (1987). "Material Flow in Manufacturing Cell and Material Requirements Planning Environment: Problems and a Solution." *Material Flow*, 4, 139–146.

Fogarty, D.W. and Robert L. Barringer (1987). "Joint Order Releases Under Dependent Demand." *Production and Inventory Management*, First Quarter, 55–61.

Fogarty, D.W. and T.R. Hoffman (1980). "Customer Service." *Production and Inventory Management*, Fall.

Fogarty, D.W. and T.R. Hoffman (1983). *Production and Inventory Management*. South-Western, Cincinnati, OH.

Fogarty, D.W., T.R. Hoffman, and P.W. Stonebraker (1989). *Production and Operations Management*. South-Western, Cincinnati.

Foley, M.J. (1988). "Post-MRP II: What Comes Next." *Datamation*, December, 1, 24–36.

Ford, D. (1980). "The development of buyer-seller relationships in industrial markets." *European Journal of Marketing*, 14 (5–6), 339–54.

Ford, Henry with Samuel Crowther (1926). *My Life and Work*. William Heineman, Ltd., London.

Ford, Henry with Samuel Crowther (1928). *Today and Tomorrow*. William Heineman, Ltd., London.

Fordyce, J.M. and F.M. Webster (1984). "The Wagner-Whitin Algorithm Made Simple." *Production and Inventory Management*, 25 (2), 21–30.

Forest, J. (1991). "Models For The Process Of Technological Innovation." *Technology Analysis And Strategic Management*, 3 (4).

Forrester, J.W. (1988). *Industrial Dynamics*. MIT Press, Cambridge, Mass.

Fortuna, Ronald M. (1988). "Beyond Quality: Taking SPC Upstream." *Quality Progress*, June, 23–28.

Fox, Edward A. (1989). "The Coming Revolution in Interactive Digital Video." *Communications of the ACM*, July, 794.

Francis, R.L. and J.A. White (1987). *Facilities Layout and Location; An Analytical Approach*. Prentice-Hall, Englewood Cliffs, NJ.

Fransoo, J.C., V. Sridharan, and J.W.M. Bertrand (1995). "A Hierarchical Approach for Capacity Coordination in Multiple Products Single-Machine Production Systems With Stationary Stochastic Demands." *European Journal of Operational Research*, 86 (1), 57–72.

Freeland, J. and R. Landel (1984). *Aggregate Production Planning: Text and Cases*. Reston Publishing, Reston, VA.

Freidenfelds, J. (1981). *Capacity Expansions: Analysis of Simple Models with Applications*. Elsevier North-Holland, New York.

Freund, J.D. and R.E. Walpole (1980). *Mathematical Statistics* (3rd ed.). Prentice-Hall, Englewood Cliffs, NJ.

Frohman, A.L. (1982). "Technology as a Competitive Weapon." *Harvard Business Review*, 60 (1) (January–February), 97–104.

Fry, Timothy D. and Allen E. Smith (1987). "A procedure for implementing input/output control: a case study." *Production and Inventory Management Journal*, 28 (3), 50–52.

Fuller, J.B., J. O'Conor, and R. Rawlinson (1993). "Tailored Logistics: The Next Advantage." *Harvard Business Review*, 71, 87–98.

G

Gabel, D. (1985). "Project Management Software." *PC Week*. September 10, 59–67.

Gaither, N. and G.V. Frazier (1998). *Production and Operations Management* (Eighth Edition), Southwestern Publishing, New York.

Galbraith, C. and D. Schendel (1983). "An empirical analysis of strategy types." *Strategic Management Journal*, 4 (2), 153–173.

Galbraith, J. (1973). *Designing Complex Organizations*. Addison-Wesley, Reading, Mass.

Galdwin, C.H. (1989). *Ethnographic Decision Tree Modeling*. Sage, Beverly Hills, CA.

Gale, B.T. (1985). *Quality as a Strategic Weapon*. Strategic Planning Institute, Cambridge, MA.

Gale, Bradley T. (1980). "Can More Capital Buy Higher Productivity?" *Harvard Business Review*, 59 (4), 78–86.

Gallagher, Colin (1978). Batch Production and Functional Layout. *Industrial Archaeology*. 13, 157–161.

Gallagher, G.R. (1980). "How to Develop a Realistic Master Schedule." *Management Review* (April), 19–25.

Galloway, D. and D. Waldron (1988). "Throughput Accounting: Part 2 – ranking products profitably." *Management Accounting* (UK, December), 34–35.

Galloway, D. and D. Waldron (1989). "Throughput Accounting: Part 3 – a better way to control labor costs." *Management Accounting* (UK, January), 34–35.

Galloway, D. and D. Waldron (1989). "Throughput Accounting: Part 4 – moving on to complex products." *Management Accounting* (UK, February), 34–35.

Galloway, D. and D. Waldron (1989). "The rise of activity-based costing – part four: What do activity-based cost systems look like?" *Journal of Cost Management* (Spring), 38–49.

Gantt, H.L. (1903). A Graphical Daily Balance in Manufacture. *ASME Transactions*, 24, 1322–1336.

Gantt, H.L. (1916). *Work, Wages and Profits*. The Engineering Magazine Co., New York, New York.

Gardner, E. (1985). "Exponential Smoothing: The State of the Art." *Journal of Forecasting*, 4, 1–38.

Gardner, E. and E. McKenzie (1985). "Forecasting Trends in Time Series." *Management Science*, 31 (10), 1985, 1237–1246.

Gardner, E.S. (1980). "Inventory Theory and the Gods of Olympus." *Interfaces*, 10 (4), 42–45.

Garud, R. and A. Kumaraswamy (1993). "Technological and Organizational Designs for Realizing Economies of Substitution." *Strategic Management Journal*, 16, 93–109.

Garvin, D.A. (1983). "Quality on the Line." *Harvard Business Review* (September–October), 64–75.

Garvin, D.A. (1984). "What Does 'Product Quality' Really Mean?" *Sloan Management Review*. Fall, 25–43.

Garvin, D.A. (1987). "Competing on the Eight Dimensions of Quality." *Harvard Business Review*. November–December, 101–109.

Garvin, D.A. (1988). *Managing Quality: The Strategic and Competitive Edge*. Free Press, New York.

Garvin, D.A. (1993). "Manufacturing Strategic Planning." *California Management Review*, Summer, 85–106.

Garwood, Dave (1971). "Delivery as Promised." *Production and Inventory Management*, 12, 3.

Garwood, Dave, and Michael Bane (1988). *A Jumpstart to World Class Performance*. Dogwood Publishing, Marietta, GA.

Gass, S.I. and C.M. Harris (1996). *Encyclopedia of Operations Research and Management Science*, Kluwer, Boston.

Gates, S. (1994). *The Changing Global Role of the Foreign Subsidiary Manager*, The Conference Board, New York.

Gecowets, G.A. (1979). "Physical Distribution Management." *Defense Transportation Journal*, 35, 11.

Gelb, B.D. and B.M. Khumawala (1984). "Reconfiguration of an Insurance Company's Sales Regions." *Interfaces*, 14, 87–94.

Gen, M. and R. Cheng (1997). *Genetic Algorithms and Engineering Design*, Wiley and Sons, New York.

Geoffrion, A.M. (1974). "Multicommodity Distribution System Design by Benders Decomposition." *Management Science*, January, 822.

Geoffrion, A.M. (1976). "Better Distribution Planning With Computer Models." *Harvard Business Review*, July–August, 92.

Geoffrion, A.M. and G.W. Graves (1974). "Multicomodity Distribution System Design by Bender's Decomposition." *Management Science*, 12, 822–844.

Geoffrion, A.M. and R.F. Powers (1980). "Facility Location Analysis Is Just the Beginning." *Interfaces*, 10, 2 (April), 22–30.

George, J.F. (1991). The Conceptual and Development of Organizational Decision Support System." *Journal of Management Information Systems*, 8 (3), 109–125.

Georgoff, D.M. and R.G. Murdick (1986). "Manager's Guide to Forecasting." *Harvard Business Review*, 64 (January–February), 110–120.

Georgoff, David M., and Robert G. Murdick (1986). Manager's Guide to Forecasting, *Harvard Business Review*, Jan.–Feb., 110–120.

Gershwin, S.B. (1986). "An Approach to Hierarchical Production Planning and Scheduling." in Jackson, R.H.F. and A.W.T. Jones, eds., *Real-Time Optimization in Automated Manufacturing Facilities*, U.S. Department of Commerce, National Bureau of Standards, 1–14.

Gerwin, D. (1981). "Control and Evaluation in the Innovation Process: The Case of Flexible Manufacturing Systems." *IEEE Transactions on Engineering Management*, EM-28, 3 (August), 62–70.

Gerwin, D. (1982). "Do's and Don'ts of Computerized Manufacturing." *Harvard Business Review* (March–April), 107–116.

Gerwin, D. (1987). "An Agenda for Research on the Flexibility of Manufacturing Processes." *International Journal of Operations and Production Management*, 7, 38–49.

Gerwin, D. (1993). Manufacturing flexibility: A Strategic Perspective. *Management Science*, 39 (4), 395–409.

Gerwin, D. (1993). "Integrating Manufacturing into the Strategic Phases of New Product Development." *California Management Review*, Summer, 123–136.

Gerwin, D. and G. Susman (1996). "Special Issue on Concurrent Engineering." *IEEE Transactions on Engineering Management*, 43, 118–123.

Gerwin, D. and H. Kolodny (1992). *Management of Advanced Manufacturing Technology*, New York: Wiley-Interscience.

Gerwin, D. and J.C. Tarondeau (1982). "Case Studies of Computer Integrated Manufacturing Systems: A View of Uncertainty and Innovation Process." *Journal of Operations Managemet*, 2, 2 (February), 87–89.

Ghosh and Gagnon (1989). "Assembly Line Research: Historical Roots, Research Life Cycles and Future Directions." *International Journal of Production Research*, 19 (5), 381–399.

Gido, J. (1985). *An Introduction to Project Planning* (2nd ed.). Industrial Press, New York.

Gilbert, J.P. (1990). "The state of JIT implementation and development in the USA." *International Journal of Production Research*, 28, 1099–1109.

Gilbert, K.R. (1965). *The Portsmouth Block-Making Machinery*. Her Majesty's Stationery Office. London, England.

Gilbreath, R.D. (1986). *Winning at Project Management*. John Wiley, New York.

Gilmore, H.J. and J.B. Pine II (1997). "The Four Faces of Mass Customization." *Harvard Business Review*, 75, 91–101.

Ginsburg, I. and N. Miller (1992). "Value-Driven Management." *Business Horizons* (May–June), 23–27.

Gitlow, H.S. and S.J. Gitlow (1987). *The Deming Guide to Quality and Competitive Position*. Prentice-Hall, Englewood Cliffs, NJ.

Gitlow, H.S. and P.T. Hertz (1983). "Product Defects and Productivity." *Harvard Business Review* (September–October), 131–141.

Giunipero, L. and J. Vogt (1997). "Empowering the Purchasing Function: Moving to Team Decisions." *International Journal of Purchasing and Materials Management*, Winter, 8–15.

Giunipero, L.C. and R.R. Brand (1996). "Purchasing's Role in Supply Chain Management." *International Journal of Logistics Management*, 7 (1), 25–38.

Glasserman, P. (1991). *Gradient Estimation via Perturbation Analysis*. Kluwer Academic Publishers, MA.

Glasserman, P. (1994). "Perturbation analysis of production networks." In: *Stochastic Modeling and Analysis of Manufacturing Systems* (Yao, ed.). Springer-Verlag, New York.

Glasserman, P. and D.D. Yao (1992). "Some guidelines and guarantees for common random numbers." *Management Science*, 38, 884–908.

Glassey, C. Roger and Mauricio G.C. Resende (1988a). "Closed-loop job release for VLSI circuit manufacturing." *IEEE Transactions on Semiconductor Manufacturing*, 1 (1), 36–46.

Gleick, J. (1987). *Chaos: Making a New Science*. Viking, New York.

Gobeli, D.H. and E.W. Larson (1987). "Relative Effectiveness of Different Project Structures." *Project Management Journal*. 18, 2 (June), 81–85.

Goddard, Walter, and Associates (1986). *Just-in-Time: Surviving by Breaking Tradition*. Oliver Wight Publications, Essex Junction, VT.

Godfrey, A.B. (1993). "Ten Areas for Future Research in Total Quality Management." *Quality Management Journal*, October, 47–70.

Godrad, Eliyahu M. and Jeff Cox (1986). *The Goal: A Process of Ongoing Improvement*. North River Press, Croton-on-Hudson, NY.

Gold, B. (1980). "CAMSets New Rules for Production." *Harvard Business Review*, 60 (6) (November–December), 78–86.

Gold, B. (1981). "Changing Perspectives on Size, Scale, and Returns: An Interpretive Survey." *Journal of Economic Literature*, 19, March.

Goldhar, J.D. and M. Jelinek (1983). "Plan for Economics of Scope." *Harvard Business Review*, 61, 6 (November–December), 141–148.

Goldhar, J.D. and M. Jelinek (1983). "Plan for Economies of Scope." *Harvard Business Review*, 61 (6), 141–148.

Goldman, S., R. Nagel and K. Preiss (1995). *Agile Competitors and Virtual Organizations*, Van Nostrand Reinhold, New York.

Goldratt, E.M. (1983). "Cost Accounting is Enemy Number One of Productivity." International Conference Proceedings, American *Production and Inventory Control* Society, October.

Goldratt, E.M. (1990). *The Haystack Syndrome: Sifting Information out of the Data Ocean*, North River Press, Croton-on-Hudson, New York.

Goldratt, E.M. (1990). *Theory of Constraints*. North River Press, New York.

Goldratt, E.M. and J. Cox (1986). *The Goal: A Process of Ongoing Improvement*. North River Press, Croton-on-Hudson, N.Y.

Goldratt, E.M. and Robert Fox (1986). *The Race For a Competitive Edge*. North River Press, Croton-on-Hudson, NY.

Goodman, P.S., R. Devadas, and T.L.G. Hughson (1988). "Groups and Productivity: Analyzing the Effectiveness of Self-Managing Teams." in *Productivity in Organizations*, J.P. Campbell *et al.* (ed.). Jossey-Bass, San Francisco, CA.

Goodman, Stephen H., Stanley T. Hardy, and Joseph R. Biggs (1977). "Lot Sizing Rules in a Hierarchical Multistage Inventory System." *Production and Inventory Management*, 18, 1, 104–116.

Goodwin, P. and G. Wright (1993). "Improving judgmental time series forecasting: A review of the guidance provided by research." *International Journal of Forecasting*, 9, 147–161.

Gordon, Geoffrey (1975). *The Application of GPSS V to Discrete System Simulation*. Prentice-Hall, Englewood Cliffs, N.J.

Gordon, R.F., E.A. MacNair, P.D. Welch, K.J. Gordon, and J.F. Kurose (1986). "Examples of using the research queueing package modeling environment (RESQME)." In: Proceedings of the Winter Simulation Conference, 494–503.

Gotoh, F. (1991). *Equipment Planning for TPM: Maintenance Prevention Design*, Productivity Press, Cambridge, MA.

Goyal, S.K. and S.G. Deshmukh (1992). "A critique of the literature on just-in-time Manufacturing." *International Journal of Operations and Production Management*, 12, 1, 18–28.

Graedel, T.E. and B.R. Allenby (1995). *Industrial Ecology*. Prentice-Hall, New Jersey.

Granger, C.W.J. (1980). *Forecasting in Business and Economics*. Academic Press, New York.

Grant, E.L. and R.S. Leavenworth (1996). *Statistical Quality Control* (7th ed.). McGraw Hill, New York.

Gravel, M. and W.L. Price (1988). "Using the Kanban in a job-shop Environment." *International Journal of Production Research*, 26, 1105–1118.

Gray, Allison F. (1984). "Educating and Scheduling the Outside Shop: A Case Study." In Readings in *Production and Inventory Control Interfaces*, 48–50. American Production and Inventory Control Society, Falls Church VA.

Gray, C. (1987). *The Right Choice: A Complete Guide to Evaluating, Selecting and Installing MRP II Software*. Essex Junction, VT., Oliver Wight Ltd.

Graybeal, W. and U.W. Pooch (1980). *Simulation: Principles and Methods*. Cambridge, MA, Winthrop.

Green, Donald R. (1987). "Direct Pegging, MRP/JIT Bridge." *P&IM Review*, April.

Green, G.I. and L.B. Appel (1981). "An Empirical Analysis of Job Shop Dispatch Rule Selection." *Journal of Operations Management*, 1, 4 (May), 197–204.

Green, T.J. and R.P. Sadowski (1984). "A Review of Cellular Manufacturing Assumptions, Advantages, and Design Techniques." *Journal of Operations Management*, 4, 2 (February), 85–97.

Greene, J.H. (1987). Editor. *Production and Inventory Control Handbook* (2nd ed.). McGraw-Hill, New York.

Greene, James H. (1974). *Production and Inventory Control* Systems and Decisions (2nd ed.). Richard D. Irwin, Homewood, IL.

Greene, T.J. and R.P. Sadowski (1983). "A Review of Cellular Manufacturing Assumptions, Advantages, and Design Techniques." *Journal of Operations Management*, 4, 2 (February), 85–97.

Greeson, C.B. and M.C. Kocakulah (1997). "Implementing and ABC Pilot at Whirlpool." *Journal of Cost Management*, March–April, 16–21.

Gregory, R. (1986). "Source selection: A matrix approach." *International Journal of Purchasing and Materials Management*, 24–29.

Greis, N.P. and J.D. Kasarda (1997). "Enterprise Logistics in the Information Era." *California Management Review*, 39 (4), 55–78.

Griffin, R.W. (1982). *Task Design: An Integrative Approach*. Glenview, IL, Scott Foresman.

Grönroos, C. (1990). "Relationship Approach to Marketing in Service Contexts: The Marketing and Organizational Behaviour Interface." *Journal of Business Research*, 20, 3–11.

Groover, M.P. (1980). *Automation, Production Systems, and Computer-Aided Manufacturing*. Prentice-Hall, Englewood Cliffs, NJ.

Groover, M.P. (1987). *Automation, Production Systems, and Computer-Aided Manufacturing*. Englewood Cliffs, N.J., Prentice-Hall, 119–128.

Groover, M.P. and E.W. Zimmers, Jr. (1984). *CAD/CAM: Computer Aided Design and Manufacturing*. Prentice-Hall, Englewood Cliffs, NJ.

Gross, Charles W. and Peterson T. Robin (1983). *Business Forecasting* (2nd ed.). Houghton Mifflin, Boston.

Grubbs, S.C. (1994). "MRPII: successful implementation the hard way." *Hospital Material Management Quarterly*, 15 (4), 40–47.

Gstettner, S. and H. Kuhn (1996). "Analysis of Production Control Systems Kanban and CONWIP." *International Journal of Production Research*, 34, 3253–3273.

Gubman, E.L. (1998). *The Talent Solution: Aligning Strategy and People to Achieve Extraordinary Results*. New York: McGraw-Hill.

Guerrero, Hector H. (1987). "Group Technology: I. The Essential Concepts." *Production and Inventory Management*, First Quarter, 62–70.

Guest, R.H. (1982). *Innovation Work Practices*. Elmsford, NY, Pergamon Press.

Gulati, R.K. and S.D. Eppinger (1996). "The Coupling of Product Architecture and Organizational Structure Decisions." MIT Working Paper 3906.

Gunasingh, K.R. and R.S. Lashkari (1989). "Machine grouping problem in cellular manufacturing systems: An integer programming approach." *International Journal of Production Research*, 27, 1465–1473.

Gunnm, Thomas G. (1981). *Computer Applications in Manufacturing*. Industrial Press, New York.

Gunther, T.J. and R.P. Sadowski (1984). "Currently Practiced Formulations of the Assembly Line Balance Problem." *Journal of Operations Management*, 3 (3) (August), 209–221.

Gupta, M., S. Baxendale, and K. McNamara (1997). "Integrating TOC and ABCM in a health care company." *Journal of Cost Management*, 11, 23–33.

Gupta, S.M. and C.R. McLean (1996). "Disassembly of Products." *Computers and Industrial Engineering*, 31, 225–228.

Gupta, S.M. and K. Taleb (1994). "Scheduling Disassembly." *International Journal of Production Research*, 32 (8), 1857–1866.

Gupta, S.M. and Y.A.Y. Al-Turki (1998). "Adapting just-in-time manufacturing systems to preventive maintenance interruptions." *Production Planning and Control*, 9 (4).

Gupta, S.M. and Y.A.Y. Al-Turki (1997). "An Algorithm to Dynamically Adjust the Number of Kanbans in Stochastic Processing Times and Variable Demand Environment." *Production Planning and Control*, 8, 133–141.

Gupta, S.M. and Y.A.Y. Al-Turki (1997). "An algorithm to dynamically adjust the number of kanbans in stochastic processing times and variable demand environment." *Production Planning and Control*, 8, 133–141.

Gupta, S.M. and Y.A.Y. Al-Turki (1998). "Adapting just-in-time manufacturing systems to preventive maintenance interruptions." *Production Planning and Control*, 9 (4), 349–359.

Gupta, S.M. and Y.A.Y. Al-Turki (1998). "The effect of sudden material handling system breakdown on the performance of a JIT system." *International Journal of Production Research*, 36 (7), 1935–1960.

Gupta, S.M., Y.A.Y. Al-Turki, and R.F. Perry (1995). "Coping with processing times variation in a JIT environment." Proceedings of Northeast Decision Sciences Institute Conference.

Gupta, Y.P. and P. Bagchi (1987). "Inbound Freight Consolidation Under Just-In-Time Procurement, Application of Clearing Models." *Journal of Business Logistics*, 8, 2, 74–94.

Guzzo, R.A. and M.W. Dickson (1996). "Teams in Organizations: Recent Research on Performance and Effectiveness." *Annual Review of Psychology*, 47, 307–338.

H

Ha, A.Y. and E.L. Porteus (1995). "Optimal Timing of Reviews in Concurrent Design for Manufacturability." *Management Science*, 41, 1431–1447.

Haas, E.A. (1987). "Breakthrough Manufacturing." *Harvard Business Review* (March–April), 75–81.

Hackman, J.R. (1998). "Why Teams Don't Work." in Theory and Research on small Groups, R.S. Tindale *et al.* (ed.). Plenum Press, New York, NY.

Hackman, J.R. and G.R. Oldham (1980). *Work Redesign*. Addision-Wesley, Reading, MA.

Haddad, C.J. (1996). "Operationalizing the Concept of Concurrent Engineering: A Case Study from the U.S. Auto Industry." *IEEE Transactions on Engineering Management*, 43 (2), 124–132.

Hadjinicola, G.C. and K.R. Kumar (1997). "Factors affecting international product design." *Journal of the Operational Research Society*, 48, 1131–1143.

Hadley, G. and T.M. Within (1963). *Analysis of Inventory Systems*, 1, 3–5, and 9. Prentice-Hall, New Jersey.

Hahn, C.K., E.A. Duplaga, and K.Y. Kim (1994). "Production/Sales Interface: MPS at Hyundai Motor." *International Journal of Production Economics*, 37–1, 5–18.

Hahn, C.K., K.H. Kim, and J.S. Kim (1986). "Costs of Competition, Implications for Purchasing Strategy." *Journal of Purchasing and Materials Management* (Fall), 2–7.

Haider, S.W. and J. Banks (1986). "Simulation Software Products for Analyzing Manufacturing Systems." *Industrial Engineering*, 18, 7 (July), 98–103.

Hajek, B. (1984). Optimal control of two interacting service stations, *IEEE Transactions on Automatic Control*, 29, 491–499.

Halberstam, D. (1986). *The Reckoning*. Morrow Publishers, New York.

Hall, G., J. Rosenthal, and J. Wade (1993). "How to Make Reengineering Really Work." *Harvard Business Review* (November–December), 71 (6), 119–131.

Hall, R.W. (1982). "Repetitive Manufacturing." *Production and Inventory Management*, Second Quarter, 78–86.

Hall, R.W. (1982). *Kawasaki U.S.A.: A Case Study*. APICS.

Hall, R.W. (1983). Zero Inventories. Dow Jones-Irwin. Homewood, Illinois.

Hall, R.W. (1993). "The Challenges of the Three-Day Car." Target, 9 (2), March/April, 21–29.

Hall, R.W. (1993). "AME's Vision of Total Enterprise Manufacturing." *Target*, 9, 6, November/December, 33–38.

Hall, R.W., N.P. Galambos, and M. Karlsson (1997). "Constraint-based profitability analysis: stepping beyond the theory of constraints." *Journal of Cost Management*, 11, 6–10.

Hall, R.W. (1987). *Attaining Manufacturing Excellence*. Homewood, Ill., Dow Jones-Irwin.

Hall, R.W. and T.E. Vollman (1978). "Planning your requirements planning." *Harvard Business Review*, 56, 2, 105–112.

Hall, R.W. and Y. Yamada (1993). "Sekisui's Three-Day House." *Target*, 9, 4, July/August, 6–11.

Halley, W.H., Jr. and K.M. Jennings (1984). *The Labor Relations Process* (2nd ed.). Hinsdale, IL, Dryden Press.

Hallihan, A.P. Sackett, and G.M. Williams (1997). "JIT manufacturing: The evolution to an implementation model founded in current practice." *International Journal of Production Research*, 35 (4), 901–920.

Ham, I., K. Hitomi, and T. Yoshida (1985). *Group Technology Application to Production Management*. Boston, Kluwer-Nijhoff.

Hambrick, D.C. (1983). "High Profit Strategies in Mature Capital Goods Industries: A Contingency Approach." *Academy of Management Journal*, 26 (4), 687–707.

Hamel, G. and Prahalad, C.K. (1994). *Competing for the Future*, Boston: Harvard Business School Press.

Hammer M. and J. Champy (1993). "Reengineering the Corporation: A Manifesto for Business Revolution." *Harper*, New York, NY.

Hammer, M. (1990). "Reengineering Work: Don't Automate, Obliterate." *Harvard Business Review* (July–August), 68 (4), 104–102.

Hammer, M. (1996). *Beyond Reengineering*, Harper Business, New York.

Hammer, M. and J. Champy (1993). *Reengineering the Corporation: A Manifestor for Business Revolution*, Harper Business, New York.

Hammer, M. and S. Stanton (1995). *The Reengineering Revolution*, Harper Business, New York.

Hampton, W.J. (1988). "How Does Japan, Inc. Pick Its American Workers?" *Business Week*, October 3, 84, 88.

Han, Min-Hong and Leon F. McGinnis (1988). "Throughput Rate Maximization in Flexible Manufacturing Cells." *IIE Transactions* 20, 4 (December), 409–417.

Handfield, R.B. (1994). "Global sourcing: Patterns of development." *International Journal of Operations and Production Management*, 14 (6), 40–51.

Handfield, R.B. (1994). Effects of concurrent engineering on make-to-order products. *IEEE Transactions on Engineering Management*, 41 (4), 1–11.

Handfield, R.B. and R. Pannesi (1994). "Managing component life cycles in dynamic technological environments." *International Journal of Purchasing and Materials Management* (Spring), 20–27.

Handfield, R.B., S.V. Walton, L.K. Seegers, and S.A. Melnyk (1997). "'Green' Value Chain Practices in the Furniture Industry." *Journal of Operations Management*, 15 (4), 293–315.

Handfield, R.B. (1995). *Re-engineering for Time-based Competition-Benchmarks and Best Practices for Production, R&D, and Purchasing*, Quorum Books, Westport, CT.

Hanke, John E. and Arthur G. Reitsch, 1995, *Business Forecasting*, Fifth Edition, Prentice-Hall.

Hannah, K.H. (1987). "Just-in-Time: Meeting the Competitive Challenge." *Production and Inventory Management*, 28, 3, 1–3.

Hansen, B.L. and P.M. Ghare (1987). *Quality Control and Application*. Prentice-Hall, Englewood CLiffs, NJ.

Harding, C.F. (1980). "Your Business: Right Ingredients, Wrong Location." *INC* (February), 20–22.

Harl, J.E. (1983). "Reducing Capacity Problems in Material Requirements Planning Systems." *Production and Inventory Management* (Third Quarter), 52–60.

Harmon, P. and D. King (1985). *Expert Systems: Artificial Intelligence in Business*. Wiley, New York.

Harmon, Roy L. and Peterson, Leroy D. (1990). *Reinventing The Factory: Productivity Breakthroughs In Manufacturing Today*. The Free Press.

Harrigan, Kenneth W. (1984). "Making Quality Job One – A Cultural Revolution." *Business Quarterly*, Winter, 68–71.

Harrington, D.H.J. (1991). *Business Process Improvement: The Breakthrough Strategy for Total Quality, Productivity, and Competitiveness*, McGraw-Hill, New York.

Harrington-Mackin, D. (1996). *Keeping the Team Going: A Tool Kit to Renew and Refuel Your Workplace Teams*, AMACOM/American Management Association, New York, NY.

Harrison, G.C. (1921). "What is wrong with cost accounting?" *N.A.C.A. Bulletin*.

Hartley, John (1984). *FMS at Work* IFS Publications, Bedford, UK.

Hartmann, E.H. (1992). *Successfully Installing TPM in a Non-Japanese Plant*. TPM Press, Inc., Allison Park, PA.

Hathaway, H.K. (1914). "The Planning Department, Its Organization and Function." in Thompson, C. Bertrand, *Scientific Management: A Collection of the More Significant Articles Describing the Taylor System of Management*, Harvard University Press, Cambridge, MA., p. 366–394.

Haug, P. (1985). "A Multiple Period, Mixed Integer Programming Model for Multinational Facility Location." *Journal of Management*, 11 (3), 83–96.

Hauptman, O. and K.K. Hirji (1996). "The Influence of Process Concurrency on Project Outcomes in Product Development: an Empirical Study with Cross-Functional Teams." *IEEE Transactions in Engineering Management*, 43, 153–164.

Hauser, J.R. and Don Clausing (1988). "The House of Quality." *Harvard Business Review*, 66 (3) (May–June), 63–73.

Hax, A.C. and D. Candea (1984). *Production and Inventory Management*, Prentice-Hall, Englewood Cliffs, NJ.

Hax, A.C. and H.C. Meal (1973). "Hierarchical Integration of Production Planning and Scheduling." *Studies in Management Sciences, Logistics 1*. Edited by M.A. Geisler, North Holland, New York.

Hay, E.J. (1988). *The Just-In-Time Breakthrough: Implementing the New Manufacturing Basics*. John Wiley and Sons, New York: NY.

Hayes, R. (1981). "Why Japanese Factories Work." *Harvard Business Review* (July–August).

Hayes, R.H. and G.P. Pisano (1994). "Beyond World Class: The New Manufacturing Strategy." *Harvard Business Review*, January–February, 77–86.

Hayes, R.H. and G.P. Pisano (1996). "Manufacturing strategy: at the intersection of two paradigm shifts." *Production and Operations Management*, 5 (1) 25–41.

Hayes, R.H. and J. Jaikumar (1988). Manufacturing's crisis: new technologies, obsolete organizations, *Harvard Business Review*, September, 77–85.

Hayes, R.H. and K.B. Clark (1986). "Why Some Factories are More Productive Than Others." *Harvard Business Review*, September–October, 66–73.

Hayes, R.H. and R.W. Schmenner (1978). "How should you organize manufacturing?" *Harvard Business Review* (January–February), 105–117.

Hayes, R.H. and R.W. Schmenner (1978). "How should you organize manufacturing?" *Harvard Business Review*, 105–19.

Hayes, R.H. and S.C. Wheelwright (1979). "Link Manufacturing Process and Product Life Cycles. *Harvard Business Review* (January–February), 133–140.

Hayes, R.H. and S.C. Wheelwright (1979b). "The dynamics of process-product life cycles." *Harvard Business Review* (March–April), 127–136.

Hayes, R.H. and S.C. *Wheelwright, 1984. Restoring Our Competitive Edge: Competing Through Manufacturing*, John Wiley and Sons, New York.

Hayes, R.H. and W.J. Abernathy (1980). "Managing Our Way to Economic Decline." *Harvard Business Review*, 58, July–August, 67–77.

Hayes, R.H., G.P. Pisano, and D.M. Upton (1996). *Strategic Operations*, Free Press, New York.

Hayes, R.H., S. Wheelwright, and K. Clark (1988). *Dynamic Manufacturing: Creating the Learning Organization*. The Free Press, New York, 161–184.

Hayhurst, G. (1987). *Mathematical Programming Applications*. MacMillan, New York.

Heinritz, S.F., P.V. Farrell, and C.L. Smith (1986). *Purchasing Principles and Applications*. Prentice-Hall, Englewood Cliffs, NJ.

Held, M., R.M. Karp, and R. Shareshian (1963). "Assembly Line Balancing – Dynamic Programming with Precedence Constraints." *Operations Research*, 11, 442–459.

Helgeson, W.P. and D.P. Bernie (1961). "Assembly Line Balancing using the Ranked Positional Weight Technique." *Journal of Industrial Engineering*, 12, 6, 394–398.

Helming, B. and Mistry, P. (1998). "Making Your Supply Chain Strategic." *APICS: The Performance Advantage*, 8, 54–56.

Helper, S.R. (1991). "Strategy and Irreversibility in Supplier Relations: The Case of the US Automobile Industry." *Business History Review*, 65, Winter, 781–824.

Hendricks, K.B. and V.R. Singhal (1997). "Delays in New Product Introductions and the Market Value of the Firm: The Consequences of Being Late to the Market." *Management Science*, 43 (4), 422–436.

Hendry, L.C. and B.G. Kingsman (1991). "Job release: part of a hierarchical system to manage manufacturing lead times in make-to-order companies." OR: *Journal of the Operational Research Society*, 42 (10), pages 871–883.

Henkoff, R. (1994). "Delivering the Goods." *Fortune* (November 28), 64–78.

Henley, E.J. and H. Kavmamoto (1981). *Reliability Engineering and Risk Assessment*. Prentice-Hall, Englewood Cliffs, NJ.

Hentschel, D.C. (1993). "The Greening of Products and Production – A New Challenge for Engineers." *Advances in Production Management Systems*, Elsevier Science, North-Holland, 39–46.

Herrin, G.E. (1996). "Industry thrust areas of TEAM." *Modern Machine Shop*, 69, 146–147.

Herrmann, Jeffrey W., Ioannis Minis, Murali Narayanaswamy, and Philippe Wolff (1996). "Work order release in job shops." 733–736, Proceedings of the 4th IEEE Mediterranean Symposium on New Directions in Control and Automation, Maleme, Krete, Greece.

Herron, David P. (1979). "Managing Physical Distribution for Profit." *Harvard Business Review*, May–June, 121–132.

Hertz, D.D. (1982). *Risk Analysis and Its Applications*. John Wiley, New York.

Hertz, D.D. (1983). *Practical Risk Analysis*. John Wiley, New York.

Herzberg, F. (1987). "One More Time: How Do You Motivate Employees?" (with retrospective commentary). *Harvard Business Review* (September–October), 109–120.

Herzog, D.R. (1985). *Industrial Engineering Methods and Controls*. Reston Publishing, Reston, VA.

Heskett, J.L. (1977). "Logistics-Essential to Strategy." *Harvard Business Review*, November–December, 85–96.

Heskett, J.L. (1986). *Managing in the Service Economy*. Harvard Business School Press, Boston.

Heyer, S. and Lee, R. (1992). "Rewiring the Corporation." *Business Horizons* (May–June), 13–22.

Hill, Arthur D., J.D. Naumann, and Norman L. Chervany (1983). "SCAT and SPAT: Large-Scale Computer-Based Optimization Systems for the Personnel Assignment Problem." *Decision Sciences*, 14 (2) (April), 207–220.

Hill, C. (1988). "Differentiation versus low cost or differentiation and low cost: A contingency framework." *Academy of Management Journal*, 13 (3), 401–412.

Hill, J.S. and R. R. Still (1984). "Adapting products to LDC tastes." *Harvard Business Review*, March–April, 92–101.

Hill, T. (1989). *Manufacturing Strategy: Text and Cases*. Addison-Wesley, Wokingham, England.

Hill, T. (1994). *Manufacturing Strategy, Text and Cases*, Richard D. Irwin Inc., Burr Ridge, Illinois.

Hill, T. (1983). "Manufacturing's Strategic Role." *Journal of the Operational Research Society*, 34 (9), 853–860.

Hiller, Randall S. and Jeremy F. Shapiro (1986). "Optimal Capacity Expansion Planning When There Are Learning Effects." *Management Science*, 32 (9), 1153–1163.

Hillier, F.S. and G.J. Lieberman (1990). *Introduction to Operations Research*. McGraw-Hill, New York.

Hillier, F.S. and O.S. Yu (1981). *Queuing Tables and Graphs*. Elsevier North-Holland, New York.

Hillion, H.P. and J.M. Proth (1989). "Performance evaluation of job-shop systems using timed event-graphs." *IEEE Transactions on Automatic Control*, 34.

Hines, P. (1994). *Creating World Class Suppliers*. Pitman Publishing, London.

Hirano, Hiroyuki (Black, J.T.,/English Edition). (1988). *JIT Factory Revolution*. Productivity Press.

Hiromoto, T. "Another Hidden Edge-Japanese Management Accounting." *Harvard Business Review*, July–August, 1988, 22–26.

Hirschman, A.O. (1970). *Exit, Voice and Loyalty: Responses to Decline in Firms, Organization and Strategies*. Harvard University Press, Cambridge, MA.

Hirst, Mark (1990). "Accounting education and advanced cost management systems." *Australian Accountant*, 60 (11), 59–65.

Ho, Y.C., M.A. Eyler, and T.T. Chien (1979). "A gradient technique for general buffer storage design in a serial production line." *International Journal of Production Research*, 17, 557–580.

Hoaglin, D.C., F. Mosteller, and J.W. Tukey (1983). *Understanding Robust and Exploratory Data Analysis*. John Wiley, New York.

Hodder, J.E. and H.E. Riggs (1984). "Pitfalls in evaluating risky projects." *Harvard Business Review*, 62, 26–30.

Hodder, J.E. and J.V. Jucker (1985). "International Plant Location under Price and Exchange Rate Uncertainty." *Engineering Costs and Production Economics*, 9, 225–229.

Hofer, C. and D. Schendel (1978). *Strategy Formulation: Analytical Concepts*, West, St. Paul.

Hoffman, J.J. and M.J. Schniederjans (1990). "An International Strategic Management/Goal Programming Model for Structuring Global Expansion Decisions in the Hospitality Industry." *International Journal of Hospitality Management*, 9 (3), 175–190.

Hoffman, J.J. and M.J. Schniederjans (1994). "A Two-Stage Model for Structuring Global Facility Site Selection Decisions: The Case of the Brewing Industry." *International Journal of Operations and Production Management*, 14 (4), 79–96.

Hoffman, J.J. and M.J. Schniederjans, and G.S. Sirmans (1991). "A Strategic Value Model for International Property Appraisal." *The Journal of Real Estate Appraisal and Economics*, 5 (1), 15–19.

Hoffman, K.L. and M. Padberg (1996). "Traveling Salesman Problem." In *Encyclopedia of Operations Research and Management Science*, S.I. Gass and C.M. Harris, ed., Kluwer, Boston.

Hoffman, Thomas R. and Gary D. Scudder (1983). "Priority Scheduling With cost Considerations." *International Journal of Production Research*, 21, 6, 881–889.

Hoffmann, Thomas (1963). "Assembly Line Balancing with a Precedence Matrix." *Management Science*, 9 (4), 551–562.

Hoffmann, Thomas (1992). "EUREKA: A Hybrid System for Assembly Line Balancing." *Management Science*, 38 (1), 39–47.

Hogarth, R.M. and S.G. Makridakis (1981). "Forecasting and Planning: An Evaluation." *Management Science*, 27 (February), 115–138.

Holland, J.H. (1995). *Hidden Order: How Adaptation Builds Complexity*. Addison-Wesley, Reading, MA 185.

Holloway, C.A. (1979). *Decisions Making Under Uncertainty*. Prentice-Hall, Englewood Cliffs, NJ.

Holmen, J.S. (1995). "ABC vs. TOC: It's a matter of time." *Management Accounting* (USA, January), 37–40.

Holt, C.C., F. Modigliani, and H. Simon (1955). "A Linear Decision Rule for Production and Employment Scheduling." *Management Science*, 2 (1) (October), 1–30.

Holt, C.C., F. Modigliani, J.F. Muth, and H.A. Simon (1960). *Planning Production, Inventories, and Workforce*. Prentice-Hall, New York.

Holusha, J. (1996). Producing Custom-Made Clothes for the Masses. *New York Times*, February 19.

Hooker, J.N. (1986, 1987). "Karmarkar's Linear Programming Algorithm." *Interfaces*, 16 (4) (July–August), 75–90 and 17 (1) (January–February), 128.

Hoover, S.V. and R.F. Perry (1989). *Simulation: A Problem Approach*. Addison-Wesley, Reading, MA.

Hopp, W.J. and M.L. Spearman (1996). *Factory Physics*. Irwin, IL.

Hora, M.E. (1987). "The Unglamorous Game of Managing Maintenance." *Business Horizons* 30 (3) (May–June), 67–75.

Horn, R.E. and A. Clearas (1980). *The Guide to Simulation and Games* (4th ed.). Beverly Hills, CA, Sage.

Hostack, G.L. (1984). "Designing Services That Deliver." *Harvard Business Review*, January–February, 133–139.

Hounshell, David W. (1984). *From the American System to Mass Production*, 1800–1932. The John Hopkins University Press. Baltimore, Maryland.

Howell, R.A. and S.R. Soucy (1987). "Major trends for management accounting." *Management Accounting*, 68, 21–27.

Howell, R.A., J.D. Brown, S.R. Soucy, and A.H. Reed (1988). *Management Accounting in the New Environment*, National Association of Accountants, USA.

Huang, C.C. and A. Kusiak (1996). "Overview of Kanban Systems." *International Journal of Computer Integrated Manufacturing*, 9, 169–189.

Huey, J. (1989). "Wal-Mart, will it take over the world?" *Fortune*, January 30th, 56–62.

Hughes, M.W. (1986). "Why Projects Fail: The Effects of Ignoring the Obvious." *Industrial Engineering*, 18, 4 (April), 14–18.

Hutchins, D. (1986). "Having a Hard Time with Just-in-Time." *Fortune*, June 9, 64–66.

Hutchinson, R.D. (1981). *New Horizons for Human Factors in Job Design*. McGraw-Hill, New York.

Hyer, N.L. (1984). *Group Technology at Work. Society of Manufacturing Engineers*, Dearborn, Mich.

Hyer, N.L. and U. Wemmerlov (1984). "Group Technology and Productivity." *Harvard Business Review*, 62 (July–August), 140–149.

I

Iansiti, M. (1995). "Shooting the rapids: managing product development in turbulent environments."

California Management Review, 38 (1), 37–58.

Ignizio, J.P. (1982). *Linear Programming in Single and Multiple Objective Systems*. Prentice-Hall, Englewood Cliffs, NJ.

Imai, K., I. Nonaka, and H. Takeuchi (1985). "Managing the New Product Development Process: How the Japanese Companies Learn and Unlearn." In: Clark, K.B., R.H. Hayes, and C. Lorenz (eds.). *The Uneasy Alliance*. Harvard Business School Press, Boston.

Imai, Masaaki (1986). Kaizen: The Key to Japan's Competitive Success. Random House, New York.

Imany, M.M. and R.J. Schlesinger (1989). "Decision models for robot selection: a comparison of ordinary least squares and linear goal programming methods." *Decision Sciences*, 20, 40–53.

Inderfurth, K. (1994). "Safety Stocks in Multistage Divergent Inventory Systems: A Survey." *International Journal of Production Economics*, 35, 321–329.

Ingle, S. (1985). *In Search of Perfection: How to Create/Maintain/Improve Quality*. Prentice-Hall, Englewood Cliffs, NJ.

Ishikawa, I. (1989). *Introduction to Quality Control* (3rd ed.). 3A Corporation, Tokyo, Japan.

Ishikawa, K. (1972). *Guide to Quality Control*. Asian Productivity Organization.

Ishikawa, K. (1985). *What is Total Quality Control? The Japanese Way*. Translated by David J. Lu. Prentice-Hall, NJ.

J

Jackson, G.C. (1985). "A Survey of Freight Consolidation Practices." *Journal of Business Logistics*, 6, 13–34.

Jackson, J.R. (1956). "A Computing Procedure for a Line Balancing Problem." *Management Science*, 2, 3, 261–271.

Jackson, J.R. (1963). "Jobshop-like queueing systems." *Management Science*, 10, 131–142.

Jacobs, F.R. (1983). "The OPT Scheduling System: A Review of a New Production Scheduling System." *Production and Inventory Management*, 24 (3), 47–51.

Jacobs, F.R. (1984). "OPT Uncovered: Many Production Planning and *Scheduling Concepts Can Be Applied with or without the Software*." Industrial Engineering (October), 32–41.

Jacobs, F.R., J.W. Bradford, and L.P. Ritzman (1980). "Computerized Layout: An Integrated Approach to Spatial Planning and Communication Requirement." *Industrial Engineering* (July), 56–61.

Jacobs, F.R. and D.J. Bragg (1988). "Repetitive lots: Flow-time reductions through sequencing and dynamic batch sizing." *Decision Sciences*, 19, 281–294.

Jacobson, S.H. and L. Schruben (1989). "Techniques for simulation response optimization." *Operations Research Letters*, 8, 1–9.

Jaikumar, R. (1986). "Postindustrial Manufacturing." *Harvard Business Review*, November–December, 69–76.

Jaikumar, R. and R. Bohn (1992). "A Dynamic Approach to *Operations Management*: An Alternative to Static Optimization." *International Journal of Production Economics*.

Jain, S.J. (1989). "Standardization of international marketing strategy: Some research hypotheses." *Journal of Marketing*, 53, 70–79.

Jankauskas, L. and S. McLafferty (1996). "BESTFIT, Distribution fitting software by Palisade Corporation." In: Proceedings of the Winter Simulation Conference, 551–555.

Janson, R.L. (1989). *Handbook of Inventory Management*. Prentice-Hall, Englewood Cliffs, NJ.

Japan Management Association (1985). *Kanban: Just-In-Time at Toyota*. Productivity Press.

Jayson, Susan (1987). "Goldratt and Fox: Revolutionizing the Factory Floor." *Management Accounting*, May, 18–22.

Jelinek, M. (1977). "Technology, Organizations, and Contingency." *Academy of Management Review*, January, 17–26.

Jelinek, M. and J. Goldhar (1983). "The Interface Between Strategy and Manufacturing Technology." *Columbia Journal of World Business*, Spring.

Jensen, K. and G. Rozenberg (eds.) (1991). *High-Level Petri Nets: Theory and Application*. Springer-Verlag, New York.

Johansson, H.J., P. McHugh, A.J. Pendlebury, and W.A. Wheeler, III (1993). *Business Process Reengineering: BreakPoint Strategies for Market Dominance*, Wiley & Sons, New York.

Johnson, H.T. (1992). *Relevance Regained: From Top-Down Control to Bottom-Up Empowerment*, Free Press, New York.

Johnson, H.T. and R.S. Kaplan (1987). *Relevance Lost! The Rise and Fall of Management Accounting*, Harvard Business School Press, Boston, MA.

Johnson, J.C. and D.F. Wood (1986). *Contemporary Physical Distribution and Logistics* (3rd ed.). MacMillan, New York.

Johnson, Linwood and Douglas C. Montgomery (1974). *Operatons Research in Production Planning, Scheduling and Inventory Control*. John Wiley & Sons, New York.

Johnson, R. (1982). "SPACECRAFT for Multi-Floor Layout Planning." *Management Science*, 28 (4) (April), 407–417.

Johnson, R. (1988). "Optimally Balancing Large Assembly Lines with 'FABLE'" *Management Science*, 34 (2), 240–253.

Jonas, N. (1986). "The Hollow Corporation." *Business Week*. March 3, 57–85.

Jones, M.S., C.J. Malmborg, and M.H. Agee (1985). "Decision support system used for robot selection." *Industrial Engineering*, 17, 66–73.

Jones, T.O. and W.E. Sasser, Jr. (1995). "Why Satisfied Customers Defect." *Harvard Business Review* (November–December), 88–99.

Jones, Thomas C. and D.W. Riley. "Using Inventory for Competitive Advantage Through Supply Chain

Management." *International Journal of Physical Distribution and Materials Management*, 15, 5, 16–26, 1985.

Jordan, Henry (1974). "Relating customer Service to Inventory Control." *American Management Association's Advanced Management Journal*, 39 (4).

Jordan, J. (1997). "Enablers for Agile Virtual Enterprise Integration." *Agility and Global Competition*, 1 (3), 26–46.

Jordan, W.C. and S.C. Graves (1995). "Principles on the benefits of manufacturing process flexibility." *Management Science*, 41, 577–594.

Jorgenson, D.W. and J.J. McCall (1963). "Optimal Scheduling of Replacement and Inspection." *Operations Research*, 11, 723–747.

Juran, J.M. (1993). "Made in U.S.A. A Renaissance in Quality." *Harvard Business Review*, July–August, 42–50.

Juran, J.M. (1995). "Summary, Trends, and Prognosis." *A History of Managing for Quality*. Edited by J.M. Juran, ASQ Quality Press, Wisconsin, 603–655.

Juran, J.M. and F.M. Gryna (1980). *Quality Planning and Analysis* (2nd ed.). McGraw-Hill, New York.

Juran, J.M., F.M. Gryna, Jr., and R.S. Bingham, Jr. eds. (1988). *Quality Control Handbook* (4th ed.). McGraw-Hill, New York.

K

Kadipasaoglu, S.N. and V. Sridharan (1995). "Alternative approaches for reducing schedule instability in multistage manufacturing under demand uncertainty." *Journal of Operations Management*, 13, 193–211.

Kanet, J.K. and J.C. Hayya (1982). "Priority Dispatching with Operations Due Dates in a Job Shop." *Journal of Operations Management*, 2 (3) (May), 167–175.

Kanigel, Robert (1997). *The One Best Way*, Viking Penguin, New York, New York.

Kanter, R. (1985). *The Change Masters: Innovation For Productivity In The American Corporation*. New York, Simon and Schuster.

Kanter, R.M. (1995). *World Class: Thriving Locally in the Global Economy*, Simon and Shuster, New York.

Kantrow, A.M. (1980). "Keping Informed: The Strategy Technology Connection." *Harvard Business Review* 58, July–August, 6–21.

Kantrow, A.M. (1980). "The Strategy-Technology Connection." *Harvard Business Review*, July–August, 12, 6–8.

Kaplan, G. (1993). "Manufacturing S La Carte – Agile Assembly Lines, Faster Development Cycles." *IEEE Spectrum*, 30, 24–34.

Kaplan, R.S. (1986). "Must CIM be justified by faith alone." *Harvard Business Review*, 64, 87–95.

Kaplan, R.S. (1986). "Must CIM be Justified by Faith Alone?" *Harvard Business Review*, March–April, 87–95.

Kaplan, R.S. (1986). "Must CIM be Justified by Faith Alone?" *Harvard Business Review*, March–April, 86 (2), 87–95.

Kaplan, R.S. (1988). "One Cost System Isn't Enough." *Harvard Business Review*, January/February, 61–68.

Kaplan, R.S. (1988). "One cost system isn't enough." *Harvard Business Review* (January–February), 61–66.

Kaplan, R.S. (1991). "New Systems for Measurement and Control." *The Engineering Economist*, 36 (3), 201–218.

Kaplan, R.S. and D.P. Norton (1992). "The balanced scorecard – measures that drive performance." *Harvard Business Review*, 70, 71–79.

Kaplan, R.S. and D.P. Norton (1993). "Putting the balanced scorecard to work." *Harvard Business Review*, September–October.

Kaplan, R.S. and D.P. Norton (1996). "Using the balanced scorecard as a strategic management system." *Harvard Business Review*, 74, 79–85.

Kaplan, R.S. and H.T. Johnson (1987). *Relevance Lost: The Rise and Fall of Cost Accounting*, Harvard Business School Press, Boston, MA.

Karger, D.W. (1982). *Advanced Work Measurement*. Industrial Press, New York.

Karmarker, N. (1984). "A New Polynomial-Time Algorithm for Linear Programming." *Combinatorica*, 4 (4), 373–395.

Karni, R. (1982). "Capacity Requirements Planning – A Systematization." *International Journal of Production Research*, 20 (6), 715–739.

Kasarda J.D. and D.A. Rondinelli (1998). "Innovative Infrastructure for Agile Manufacturing." *Sloan Management Review*, 39 (2), 73–82.

Kashani, K. (1989). "Beware of the pitfalls of global marketing." *Harvard Business Review*, September–October, 91–98.

Kate, H.A. ten (1994). "Towards a Better Understanding of Order Acceptance." *International Journal of Production Economics*, 37 (1), 139–152.

Kato, Y. (1993). "Target Costing Support Systems: Lessons from Leading Japanese Companies." *Management Accounting Research*, 4 (4), 33–47.

Kato, Y., G. Boer, and C.W. Chow (1995). "Target Costing: An Integrative Management Process." *Journal of Cost Management*, Spring, 39–51.

Katzel, J. (1987). "Maintenance Management Software." *Plant Engineering*, 41 (12) (June 18), 124–170.

Katzenbach, J.R. and D.K. Smith (1993). *The Wisdom of Teams: Creating the High-Performance Organization*, Harvard Business School Press, Boston, MA.

Kauffman, S. (1995). *At Home in the Universe: The Search for Laws of Self-Organization and Complexity*. Oxford University Press, New York.

Keating, Giles (1985). *The Production and Use of Economic Forecasts*. Methuen, London.

Keaton M. (1995). "A New look at the Kanban Production Control System." *Production and Inventory Management Journal*, Third Quarter, 71–78.

Kekre, S. and T. Mukhopadhyay (1992). "Impact of Electronic Data Interchange Technology on Quality

Improvement and Inventory Reduction Programs: A Field Study." *International Journal of Production Economics*, 28, 265–282.

Keller, A.Z. and A. Kazazi (1993). "Just-in-time manufacturing systems: A literature review." *Industrial Management and Data Systems*, 93 (7), 1–32.

Kelley, James E., Jr. and Morgan R. Walker (1989). *The Origins of CPM: A Personal History*. PMNetwork, 3, 7–22.

Kelly P. and M. Krantzberg (1978). *Technological Innovation: A Critical Review Of Current Knowledge*. San Francisco Press, USA.

Kendrick, J.W. (1984). *Improving Company Productivity*, John Hopkins University Press, Baltimore.

Kenny, Andrew A. (1988). "A New Paradigm for Quality Assurance." Quality Progress, June, 30–32.

Keoleian, G.A. and D. Menerey (1994). "Sustainable Development by Design: Review of Life Cycle Design and Related Approaches." *Journal of the Air and Waste Management Association*, 44 (5), 645–668.

Kerzner, H. (1984). *Project Management: A Systems Approach to Planning, Scheduling, and Controlling*. New York, Van Nostrand.

Kester, W.C. (1984). "Today's options for tomorrow's growth." *Harvard Business Review*, 62, 153–160.

Keuhn, A.A. and M.J. Hamburger (1963). "A Heuristic Program for Locating Warehouses." *Management Science*, 9, July.

Kezbom, D.S., D.L. Schilling, and V.A. Edward (1989). *Dynamic Project Management*. John Wiley, New York.

Khouja, M. (1995). "The use of data envelopment analysis for technology selection." *Computers and Industrial Engineering*, 28, 123–132.

Khouja, M. and D.E. Booth (1991). "A decision model for the robot selection problem using robust regression." *Decision Sciences*, 22, 656–662.

Khouja, M. and D.E. Booth (1995). "Fuzzy clustering procedure for evaluation and selection of industrial robots." *Journal of Manufacturing Systems*, 14, 244–251.

Khouja, M. and O.F. Offodile (1994). "The industrial robots selection problem: literature review and directions for future research." *IIE Transactions*, 26 (4), 50–61.

Khumawala, B.M. (1972). "An Efficient Branch and Bound Algorithm for the Warehouse Location Problem." *Management Science*, 18 (12), 718–731.

Khumawala, B.M. (1974). "An Efficient Heuristic Procedure for the Capacitated Warehouse Location Problem." *Naval Research Logistics Quarterly*, 21, 609–623.

Khumawala, B.M. and D.C. Whybark (1971). "A Comparison of Some Recent Warehouse Location Techniques." *The Logistics Review*, 7 (31), 3–19.

Khumawala, B.M., C. Hixon, and J.S. Law (1986). "MRP II in the Services Industries." *Production and Inventory Management* 27 (3) (Third Quarter), 57–63.

Kidd, P.T. (1994). *Agile Manufacturing-Forging New Frontiers*, Addison-Wesley Publishing Company, Wokingham, England.

Kiel, L.D. and E. Elliott, eds. (1996). *Chaos and the Social Sciences: Foundations and Applications*. University of Michigan Press, Ann Arbor.

Kim, L. and Y. Lim (1988). "Environment, Generic Strategies, and Performance in a Rapidly Developing Country: A Taxonomic Approach." *Academy of Management Journal*, 31 (4), 802–826.

Kim, S.L., J.C. Hayya, and J.D. Hong (1992). "Setup Reduction in the Economic Production Quantity Model." *Decision Sciences*, 23 (2), 500–508.

Kim, W.C. and R. Mauborgne (1997). "Fair Process: Managing in the Knowledge Economy." *Harvard Business Review* (Jul./Aug.).

Kim, Y. and J. Lee (1993). "Manufacturing Strategy and Production Systems: An Integrated Framework." *Journal of Operations Management*, 11 (1), 3–15.

King, C.B. (1996). "Taylor II manufacturing simulation software." In: Proceedings of the Winter Simulation Conference, 569–573.

King, C.A. (1985). "Service Quality Assurance Is Different." *Quality Progress*, 18 (6), 14–18.

King, J.R. (1980). "Machine-component grouping in production flow analysis: an approach using a rank order clustering algorithm." *International Journal of Production Research*, 18, 231–237.

King, J.R. and V. Nakornchai (1982). "Machine-Component Group Formation in Group Technology: Review and Extension." *International Journal of Production Research*, 20 (2), 117–133.

King, J.R. and V. Nakornchai (1982). Machine-component group formation in group technology – review and extension, *International Journal of Production Research*, 20, 117–133.

King, R.H. and R.R. Love, Jr. (1980). "Coordinating Decisions for Increased Profit." *Interfaces*, 10 (6), 4–19.

Kingman, J.F.C. (1961). "The single-server queue in heavy traffic." In: Proceedings of the Cambridge Philosophical Society, 57, 902–904.

Kinney, H.D., Jr., and L.F. McGinnis (1987). "Design and Control of Manufacturing Cells." *Industrial Engineering* (October), 28–38.

Kinney, Hugh D. and Leon F. McGinnis (1987). "Design and Control of Manufacturing Cells." *Industrial Engineering* (October), 28–38.

Kirkpatrick, D. (1997). "Intel's amazing profit machine." *Fortune*, February 17th, 24–30.

Kiviat, P.J.R. Villanueva and H.M. Markowitz (1969). *The SIMSCRIPT II. 5 Programming Language*. Consolidated Analysis centers, Los Angeles.

Klassen, R.D. and N.P. Greis (1993). "Managing Environmental Improvement through Product and Process Innovation: Implications of Environmental Life Cycle Assessment." *Industrial and Environmental Crisis Quarterly*, 7 (4), 293–318.

Kleijnen, J.P.C. and P.J. Rens (1978). "Impact Revisited: A Critical Analysis of IBM's Inventory Package." *Production and Inventory Management*, 19 (1) (First Quarter), 71–90.

Kleijnen, J.P.C. (1987). *Statistical Tools for Simulation Practitioners*. Marcel Dekker, New York.

Klein, J.A. and P.A. Posey (1986). "Good Supervisors Are Good Supervisors – Anywhere." *Harvard Business Review* (November–December), 125–128.

Kleuthgen, P.O. and J.C. McGee (1985). "Development and Implementation of an Integrated Inventory Management Program at Pfizer Pharmaceuticals." *Interfaces* 15 (1) (January–February), 69–87.

Klingel, A.R., Jr. (1966). "Bias in PERT Project Completion Time Calculations for a Real Network." *Management Science*, 13 (4), 194–201.

Knight, Donald O. and Michael L. Wall (1989). "Using Group Technology for Improving Communication and Coordination Among Teams of Workers in Manufacturing Cells." *Industrial Engineering* (January), 32–34.

Knights, D., H. Willmott, and D. Collisons, eds. (1985). Job Redesign. Gower, Hants, England.

Knott, K. and R.D. Getto, Jr. (1982). "A Model for evaluating alternative robot systems under uncertainty." *International Journal of Production Research*, 20, 155–165.

Knutson, J. (1989). "How to be a Successful Project Manager." *American Management Association*, Saranac Lake, NY.

Kogut, B. (1985). "Designing Global Strategies: Comparative and Competitive Value-Added Chains." *Sloan Management Review*, 26 (4), 15–28.

Kogut, B. (1985). "Designing Global Strategies: Profiting from Operational Flexibility." *Sloan Management Review*, 27 (1), 27–38.

Kolata, G. (1984). "A Fast Way to Solve Hard Problems." *Science*, September 21, 1379–1380.

Kolchin, M. and L. Giunipero (1993). *Purchasing Education and Training*, Center for Advanced Purchasing Studies, Tempe, AZ.

Kolman, B. and R.E. Beck (1980). *Elementary Linear Programming with Applications*. Academic Press, New York.

Kontz, Stephan (1983). *Work Design: Industrial Ergonomics* (2nd ed.). Wiley, New York.

Koren, Y. and G. Ulsoy (1997). "Reconfigurable Manufacturing Systems." ERC Technical Report #1, The University of Michigan, Ann Arbor, Michigan.

Kotha, S. (1995). "Mass Customization: Implementing the Emerging Paradigm for the Competitive Advantage." *Strategic Management Journal*, 16, 21–42.

Kotha, S. and D. Orne (1989). "Generic Manufacturing Strategies: A Conceptual Synthesis." *Strategic Management Journal*, 10 (3), 211–231.

Kotha, S. and P.M. Swamidass (1998). "Advanced manufacturing technology use: Exploring the effect of the nationality varible." *International Journal of Production Research*, 36 (11), 3135–3146.

Kotler, P. (1984). *Marketing Management: Analysis, Planning, and Control* (5th ed.). Prentice-Hall, New Jersey.

Krajewski, L.J. and L.P. Ritzman (1977). "Disaggregation in Manufacturing and Service Organizations: Survey of Problems and Research." *Decision Sciences*, 8 (1).

Krajewski, L.J., B. King, L.P. Ritzman, and D.S. Wong (1987). "Kanban, MRP, and Shaping the Manufacturing Environment." *Management Science*, 33 (1). (January), 39–57.

Kraus, R.L. (1980). "A Strategic Planning Approach to Facility Site Selection." *Dun's Review* (November), 14–16.

Krishnan, V., S.D. Eppinger, and D.E. Whitney (1997). "A Model-Based Framework to Overlap Product Development Activities." *Management Science*, 43, 437–451.

Kropp, D.H., R.C. Carlson, and J.V. Jucker (1979). "Use of dynamic lot-sizing to avoid nervousness in material requirements planning systems." *Journal of Production and Inventory Management*, 20 (3), 40–58.

Kropp, D.H., R.C. Carlson, and J.V. Jucker (1983). "Heuristic lot-sizing approaches for dealing with MRP system nervousness." *Decision Sciences*, 14 (2), 156–169.

Krupp, J.A. (1981). "Inventory Turn Ratio, as a Management Control Tool." *Inventories and Production* 1, 4 (September/October), 18–21.

Krupp, J.A.G. (1984). "Why MRP Systems Fail: Traps to Avoid." *Production and Inventory Management*, 25, no. 3, 48–53.

Krupp, J.A.G. (1997). "Safety Stock Management." *Production and Inventory Management*, 38, 3 (Third Quarter), 11–18.

Krupp, J.A.G. (1984). "Why MRP Systems Fail: Traps to Avoid." *Production and Inventory Management*, 25, 3 (Third Quarter), 49–53.

Kubiak, J. (1993). "A Joint Venture in Mass Customization." *Planning Review*, July 21 (4), 25.

Kulatilaka, N. (1988). "Valuing the flexibility of flexible manufacturing systems." *IEEE Transactions on Engineering Management*, 35, 250–257.

Kumar, V. (1987). "Entropy measures of measuring manufacturing flexibility." *International Journal of Production Research*, 25, 957–966.

Kupanhy, Lumbidi (1995). "Classification of JIT Techniques and Their Implications." *Industrial Engineering*, 27 (2), Feb., 62–66.

Kusiak, A. (1985). "The Part Families Problem in Flexible Manufacturing Systems." *Annals of Operations Research*, 3, 279–300.

Kusiak, A. (1987). The production equipment requirements problem, International *Journal of Production Research*, 25, 319–325.

Kusiak, A. (1987). "An Expert System for Group Technology." *Industrial Engineering* (October), 56–61.

Kusiak, A. (1992). *Intelligent Design and Manufacturing*. Wiley, New York.

Kusiak, A. (1993). *Concurrent Engineering: Automation, Tools, and Techniques*. Wiley, New York.

Kusiak, A. and U. Belhe (1992). "Concurrent engineering: A design process perspective." Proceedings of the American Society of Mechanical Engineers, PED 59, 387–401.

L

L'Ecuyer, P. (1990). "A unified view of the IPA, SF, and LR gradient estimation techniques." *Management Science*, 36, 1364–1383.

Lachica, E. (1997). "East, Southeast Asia Make Huge Strides In Reducing Poverty, World Bank Says." *Wall Street Journal*, A8, August 27.

LaLonde, B.J., M.C. Cooper, and T.G. Noordewier (1988). *Customer Service: A Management Perspective*. Council of Logistics Management, Oak Brook, IL.

Lambert, D.M. and J.R. Stock (1993). *Strategic Logistics Management* (3rd ed.). Richard D. Irwin, Inc., Homewood, IL.

Lamming, R. (1993). *Beyond Partnership: Strategies for Innovation and Lean Supply*. Prentice-Hall, London.

Landel, Robert D. (1986). *Managing Productivity Through People: An Operations Perspective*. Reston, Prentice-Hall, Englewood Cliffs, N.J.

Lane, M.S., A.H. Mansour, and J.L. Harpell (1993). "Operations research techniques: A longitudinal update 1973–1988." *Interfaces*, 23 (2), 63–68.

Lankford, Raymond L. (1978). "Short Term Planning of a Manufacturing Capacity." APICS Conference Proceedings, 37–68.

Lau, R.S.M. (1995). "Mass Customization: The Next Industrial Revolution." *Industrial Management*, 37, 18–19.

Law, A.M. and M.G. McComas (1989). "Pitfalls to avoid in the simulation of manufacturing systems." *Industrial Engineering*, 31, 28–31, 69.

Law, A.M. and M.G. McComas (1996). "EXPERTFIT: Total support for simulation input modeling." In: Proceedings of the Winter Simulation Conference, 588–593.

Law, A.M. and W.D. Kelton (1982). "Computer Simulation of Manufacturing Systems: Part I." *Industrial Engineering*, 18, no. 5 (May), 46–63.

Law, A.M. and W.D. Kelton (1982). *Simulation Modeling and Analysis*. McGraw-Hill, New York.

Lawler, E.E. and S.G. Cohen (1992). "Designing pay systems for teams." *American Compensation Association Journal*, 1, 6–18.

Lawler, E.E., III (1986). *High-Involvement Management*. Jossey-Bass, San Francisco.

Lawler, E.E., III, and S. Mohrman (1985). "Quality Circles After the Fad." *Harvard Business Review* (January–February), 65–71.

Lawrence, M.J., R.H. Edmundson, and M.J. O'Conner (1986). "The Accuracy of Combining Judgmental and Statistical Forecasts." *Management Science*, 32 (December), 1521–1532.

Ledvinka, J. and V.G. Scarpello (1991). *Federal Regulation of Personnel and Human Resource Management* (2nd ed.). PWS-Kent, Boston.

Lee, H.L. and C. Billington (1992). "Managing Supply Chain Inventory: Pitfalls and Opportunities." *Sloan Management Review* (Spring), 65–73.

Lee, H.L. and M.J. Rosenblatt (1989). "A Production and Maintenance Planning Model with Restoration Cost Dependent on Detection Delay." *IIE Transactions*, 21 (4), 368–375.

Lee, H.L., C. Billington, and B. Carter (1993). "Hewlett-Packard gains control of inventory and service through design for localization." *Interfaces*, 23, 1–11.

Lee, H.L., V. Padmanabhan, and S. Whang (1997). "Information Distortion in a Supply Chain: The Bullwhip Effect." *Management Science*, 43, 546–558.

Lee, H.L., V. Padmanabhan, and S. Whang (1997). "The Bullwhip Effect in Supply Chains." *Sloan Management Review*, 93–102.

Lee, S.M. (1972). *Goal Programming for Decision Analysis*. Auerbach Publishers, Pennsauken, N.J.

Lee, S.M. and G. Schwediman, eds. (1982). *Management by Japanese Systems*. Praeger, New York.

Lee, S.M. and L.J. Moore (1974). "A Practical Approach to Production Scheduling." *Production and Inventory Management*, 15 (1), 79–92.

Lee, S. and M. Ebrahimpour (1984). "Just-In-Time Production System: Some Requirements for Implementation." *International Journal of Operations and Production Management*, 4 (1), 3–11.

Lee, W.B. and B.M. Khumawala (1974). "Simulation Testing of Aggregate Production Planning Models in an Implementation Methodology." *Management Science*, 20 (6) (February), 903–911.

Lee, W.B. and C.P. McLaughlin (1974). "Corporate Simulation Models for Aggregate Materials Management." *Production and Inventory Management*, 15 (1), 55–67.

Lee-Kwang, H. and J. Favrel (1985). "Hierarchical reduction method for analysis and decomposition of Petri nets." *IEEE Transactions on Systems, Man, and Cybernetics*, SMC-15, 272–280.

Lee-Kwang, H., J. Favrel, and P. Baptiste (1987). "Generalized Petri net reduction method." *IEEE Transactions on Systems, Man and Cybernetics*, SMC-17, 297–303.

Leemis, L.M. (1996). "Discrete event simulation input process modeling." *In: Proceedings of the Winter Simulation Conference*, 39–46.

Leenders, M. and D. Blenkhorn (1988). *Reverse Marketing – The New Buyer Seller Relationship*. The Free Press, New York.

Leenders, M. and H. Fearon (1997). *Purchasing and Supply Management*. R.D. Irwin, Homewood, IL.

Leenders, M.R., H.E. Gearon, and W.B. England (1985). *Purchasing and Materials Management* (8th ed.). Irwin, Homewood, IL.

Leonard, F.S. and W.E. Sasser (1982). "The Incline of Quality." *Harvard Business Review* (September–October), 163–171.

Leone, R.A. and J.R. Meyer (1980). "Capacity Strategies for the 1980s." *Harvard Business Review*, 58, 6 (November–December), 133–140.

Leong, G.K., D. Snyder and P. Ward, 1990. "Research in the Process and Content of Manufacturing Strategy." *Omega*, 18, 109–122.

LeRose, J.W. (1993). "Intermodal EDI." *EDI World* 3, 22–24.

Leschke, J. (1996). "An Empirical Study of the Setup-Reduction Process." *Production and Operations Management*, 5 (2), 121–131.

Leschke, J. (1997). "The Setup-Reduction Process: Part 2: Setting Reduction Priorities." *Production and Inventory Management Journal*, 38 (1), 38–42.

Leschke, J. and E.N. Weiss (1997). "The Uncapacitated Setup-Reduction Investment-Allocation Problem with Continuous Investment-Cost Functions." *Management Science*, 43 (6).

Leschke, J. and E.N. Weiss (1998). "A Comparison of Rules for Allocating Setup-Reduction Investments in a Capacitated Environment." *Production and Operations Management*.

Lev, B. and H.J. Weiss (1982). *Introduction to Mathematical Programming*. Elsevier, North-Holland, New York.

Levin, D.P. (1994). "Compaq storms the PC heights from its factory floor." *New York Times*, November 13, C5.

Levin, Richard I. and Charles A. Kirkpatrick (1966). *Planning and Control with PERT/CPM*. McGraw-Hill, New York.

Levine, H.A. (1986). *Project Management Using Microcomputers*. Osborne McGraw-Hill, Berkeley, CA.

Levitt, T. (1983). *"The Globalization of Markets."* Harvard Business Review, 61 (3), 92–102.

Levy, D. (1994). "Chaos theory and strategy: theory, application, and managerial implications." *Strategic Management Journal*, 15, 167–78.

Levy, D. (1997). "Lean production in an international supply chain." *Sloan Management Review*, 38 (2): 94–102.

Levy, Ferdinand K., Gerald L. Thompson, and Jerome D. Wiest (1963). "The ABC's of the Critical Path Method." *Harvard Business Review*, September–October, 98–108.

Lewin, Arie and John W. Minton (1986). "Determining Organizational Effectiveness: Another Look, and an Agenda for Research." *Management Science*, 32 (5) (May), 514–538.

Lewin, K. (1947). "Group decision and social change." In *Readings in Social Psychology*, New York; Holt, Rinehart and Winston, 197–211.

Lewin, K. (1951). "Field Theory in Social Sciences." *In Field Theory*, by D. Cartwright. Harper Brothers, New York.

Lewis, J. and A. Wiegert (1985). *Trust as a social reality. Social Forces*, 63, 967–988.

Lewis, James P. (1993). *The project manager's desk reference*. New York: McGraw-Hill.

Lewis, James P. (1995). *Project planning, scheduling, and control*, Revised Edition. New York: McGraw-Hill.

Lewis, James P. (1997). *Team-based project management*. New York: AMACOM.

Lewis, James P. (1998). *Mastering project management*. New York: McGraw-Hill.

Lewis, W.P. and T.E. Block (1980). "On the Application of Computer Aids to Plant Layout." *International Journal of Production Research*, 18 (1), 11–20.

Liker, J.K., R.R. Kamath, S.N. Wasti, and M. Nagamachi (1995). "Integrating Suppliers into Fast-Cycle Product Development." In J.K. Liker, J, J.E. Ettlie, and J.C. Campbell (eds.). *Engineered in Japan: Japanese Technology Management Practices*. Oxford University Press, Oxford.

Lilien, G. and P. Kotler (1983). *Marketing Decision Making – A Model-Building Approach*, Harper and Row Publishers, New York.

Limprecht, Joseph A., and Robert H. Hayes (1980). "Germany's World-Class Manufacturers." *Harvard Business Review*, 63, September–October, 142–150.

Lin, J.T. and C.C. Lee (1992). "A modular approach for the modelling of a class of zone-control conveyor system using timed Petri nets." *International Journal of Computer Integrated Manufacturing*, 5, 277–289.

Lin, Li, and David D. Bedworth (1988). "A Semi – Generative Approach to Computer Aided Process Planning Using Group Technology." *Computers and Industrial Engineering*, 14 (2), 127–137.

Lindemann, C. (1998). *Performance Modeling with Deterministic and Stochastic Petri Nets*. John Wiley.

Lingayat, Sunil, John Mittenthal, and Robert M. O'Keefe (1995). "Order release in automated manufacturing systems." *Decision Sciences*, 26 (2), 175–205.

Littler, D., F. Leverick, and M. Bruce (1995). Factors affecting the process of collaborative product development: a study of UK manufacturers of information and communications technology products. *The Journal of Product Innovation Management*, 12 (1), 16–23.

Loch, C.H. and C. Terwiesch (1998). "Communication and Uncertainty in Concurrent Engineering." *Management Science*, 44, August.

Lock, D. (1987). *Project Management* (4th ed.). Gower, Brookfield, VT.

Lockamy, A. III and J.F. Cox III (1994). *Reengineering Performance Measurement*. Irwin, New York.

Locke, D. (1996). *Global Supply Management*, Irwin, Chicago.

Logue, J. (1982). "Semi-Autonomous Groups at Saab: More Freedom, High Output." *Management Review*, 71, 9 (September), 32–33.

Lorenz, A. (1993). *Lopez to drive VW on a lean mixture. The Sunday Times, London*, March 21st, Section 3, 9.

Lorenz, C. (1986). *The Design Dimension: The New Competitive Weapon for Business*, Basil Blackwell, N.Y.

Lovelock, C. (1985). *Managing Services*. Prentice-Hall, New York.

Lowe, P.H. and J.E. Eguren (1980). "The Determination of Capacity Expansion Programmes with Economies of Scale." *International Journal of Production Research*, 18, 3, 379–390.

Lozinski, Christopher, and C. Roger Glassey (1988). "Bottleneck starvation indicators for shop floor control." *IEEE Transactions on Semiconductor Manufacturing*, 1 (4), 147–153.

Lubben, Richard T. (1988), *Just-In-Time: An Aggressive Manufacturing Strategy*, McGraw-Hill, New York.

Luenberger, D.G. (1985). *Linear and Nonlinear Programming* (2nd ed.). Addison-Wesley, Reading, MA.

Lynch, L.L. and K.F. Cross (1995). *Measure up: Yardsticks for Continuous Improvement*. Basil Blackwell, Massachusetts.

Lyons, T., A.R. Krachenberg, and J.W. Henke (1990). Mixed motive marriages: What's next for buyer-supplier relations? *Sloan Management Review*, Spring, 29–36.

M

Mabert, V.A. (1982). "Static vs. Dynamic Priority Rules for Check Processing in Multiple Dispatch-Multiple Branch Banking." *Journal of Operations Management*, 2 (1) (May), 187–196.

Mabert, V.A. and C.A. Watts (1982). "Simulation Analysis of Tour-Shift Construction Procedures." *Management Science* (May), 520–532.

Mabert, V.A. and M.J. Showalter (1980). "Priority Rules for Check Processing in Multiple Branch Banking: An Experimental Analysis." *Journal of Operations Management*, 1, 1 (August), 15–22.

Mabert, V.A., J.F. Muth, and R.W. Schmenner (1992). "Collapsing New Product Development Times: Six Case Studies." *Journal of Product Innovation Management*, 9, 200–212.

Macbeth, D.K. and N. Ferguson (1994). *Partnership Sourcing an Integrated Supply Chain Management Approach*. Pitman Publishing, London.

MacCarthy, B.L. and J. Liu (1993). A new classification scheme for flexible manufacturing systems, *International Journal of Production Research*, 31, 299–309.

MacCormack, A.D., L.J. Newmann, III, and D.B. Rosenfield (1994). "The New Dynamics of Global Manufacturing Site Location." *Sloan Management Review*, 35 (4), 69–80, Summer.

MacCrimmon, Kenneth R. and Charles A. Ryavec (1964). "An Analytical Study of the PERT Assumptions." *Operations Research*, January–February, 16–37.

Mackey, J.T. and C.J. Davis (1993). "The Drum-Buffer-Rope application of the theory of constraints in manufacturing: What is it and what does it do to your accounting system?" Proceedings of the 1993 American Accounting Association Annual Convention, San Francisco, CA.

Magee, John R. and David M. Boodman (1967). *Production Planning and Inventory Control* (2nd ed.). McGraw-Hill, New York.

Mahmoud, E. (1984). "Accuracy in forecasting: A Survey." *Journal of Forecasting*, 3 (139–159).

Maimon, O.Z. and Y.F. Choong (1987). Dynamic routing in reentrant flexible manufacturing systems, *Robotics and Computer-Integrated Manufacturing*, 3, 295–300.

Main, J. (1990). "How to Win the Baldridge Award." *Fortune*, April 25, 101–116.

Maisel, Herbert and Guiliano Gnugnoli (1972). *Simulation of Discrete Stochastic Systems*. Science Research Associates, Chicago.

Majchrzak, A. and Q. Wang (1996). "Breaking the Functional Mind Set in Process Organizations." *Harvard Business Review* (September–October), 74 (5), 93–99.

Maki, Daniel P. and Maynard Thompson (1973). *Mathematical Models and Applications*. Prentice-Hall, Englewood Cliffs, N.J.

Makridakis, S. (1986). "The Art and Science of Forecasting." *International Journal of Forecasting*, 2, 5–39.

Makridakis, S.G., S.C. Wheelwright, and V.E. McGee (1983). *Forecasting: Methods and Applications* (2nd ed.). John Wiley, New York.

Makridakis, S., A. Andersen, R. Carbone, R. Fildes, M. Hibon, R. Lewandowski, J. Newton, E. Parzen, and R. Winkler (1982). "The Accuracy of Extrapolation (time series) Methods: Results of a Forecasting Competition." *Journal of Forecasting*, 1, 111–153.

Makridakis, S., S. Wheelwright and V. McGee (1983). *Forecasting: Methods and Applications* (2nd ed.). New York: John Wiley & Sons.

Malcom, D.G., J.H. Roseboom, C.E. Clark, and W. Fazar (1959). "Application of a Technique for Research and Development Program Evaluation." *Operations Research*, 7 (5), September–October, 98–108.

Mallick, D.N. and P. Kouvelis (1992). "Management of Product Design: A Strategic Approach." *In Intelligent Design and Manufacturing*, ed. A. Kusiak, Wiley, New York.

Malott, R.H. (1983). "Let's Restore Balance to Product Liability Law." *Harvard Business Review*, May–June, 66–74.

Maltz, Arnald, and Lisa Ellram, "Outsourcing: Implications for Supply Management" *CAPS*, p. 140, 1997.

Mandel, M. (1997). The new business cycle, *Business Week*, March 31st, 48–54.

Mandelbrot, B.B. (1983). *The Fractal Geometry of Nature*, W.H. Freeman & Co., San Francisco, CA.

Mangiameli, P. and L. Krajewski (1983). "The Effects of Workforce Strategies on Manufacturing Operations." *Journal of Operations Management*, 3 (4) (August), 183–196.

Manivannan S. and D. Chudhuri (1984). "Computer-Aided Facility layout Algorithm Generates Alternatives to Increase Firm's Productivity." *Industrial Engineering* (May), 81–84.

Mann, L., Jr. (1983). *Maintenance Management*. Rev. ed. Lexington Books, Lexington MA.

Mantel, Samuel J. and Jack R. Meridith (1986). "IEs Are Best Suited to Challenging Role of Project Manager." *Industrial Engineering*, 18 (4), 54–60.

Marcellus, R.L. and M. Dada (1991). "Interactive Process Quality Improvement." *Management Science*, 37 (11), 1365–1376.

March, A. and D.A. Garvin (1986). Copeland Corporation: Evolution of a Manufacturing Strategy, Harvard Business School Case 9-686-088.

March, Barbara (1993). "Allen Edmonds Shoe Tries Just-In-Time Production." *Wall Street Journal*, March 5, 1.

Marquardt, M. and A. Reynolds (1994). *The Global Learning Organization*. Irwin, Burr Ridge, IL.

Marshal, J. and P. Lawrence (1991). "Design as A Strategic Resource: A Business Perspective." Proceedings of the Design Leadership Symposium, Corporate Design Foundation, Boston, MA.

Martin, A.J. (1983). *Distribution Resource Planning*. Prentice-Hall, Englewood Cliffs, NJ; Oliver Wight Ltd., Essex Junction, VT.

Martinez, J.I. and J.C. Jarillo (1989). "The evolution of research on coordination mechanisms in multinational corporations." *Journal of International Business Studies*, 20, 489–514.

Masaracchia, Philip (1987). "TQC – The 'Quality' Component of J-I-T." *P&IM Review*, April, 44.

Massimilian, R.D. and L. Pedro (1990). "Back from the future: Gearing up for the productivity challenge." 79 (2), *Management Review*, 41–43.

Mather, H. (1977). "Reschedule the schedules you just scheduled – Way of life for MRP?" *Journal of Production and Inventory Management*, 18 (1), 60–79.

Mather, H. (1987). *Bills of Material*. Dow Jones-Irwin, Homewood, IL.

Mazzola, Joseph B. and Kevin F. McCardle (1996). "A Bayesian Approach to Managing Learning-Curve Uncertainty." *Management Science*, 42 (5), 680–692.

McAllister, D. (1995). Affect- and cognition-based trust as foundations for interpersonal cooperation in organizations. *Academy of Management Journal*, 38 (1), 24–59.

McCall, J.J. (1965). "Maintenance Policies for Stochastically Failing Equipment: A Survey." *Management Science*, 11 (5), 493–524.

McCormick Ernest J. (1982). *Human Factors in Engineering and Design*. McGraw-Hill, New York.

McCuley, J. (1972). "Machine grouping for efficient production." *Production Engineering*, 2, 53–57.

McDonald, A.L. (1986). "Of Floating Factories and Mating Dinosaurs." *Harvard Business Review* (November–December), 81–86.

McDougall, P.P., R.H. Deane, and D.E. D'Souza (1992). "Manufacturing Strategy and Business Origin of New Venture Firms in the Computer and Communications Equipment Industries." *Production and Operations Management*, 1 (1), 53–69.

McGrath, M. and R. Hoole (1992). "Manufacturing's New Economies of Scale." *Harvard Business Review* (May–June), 94–102.

McGrath, R.G., M. Tsai, S. Venkataraman, and I.C. MacMillan (1996). "Innovation, Competitive Advantage and Rent: A Model and Test." *Management Science*, 42 (3), 389–403.

McGraw, D. (1997). "Shootout at PC Corral." U.S. *News and World Report*, June 23.

McGuire, Kenneth J. (1984). "Impressions From Our Most Worthy Competitor." American *Production and Inventory Control* Society, Falls Church, VA.

McKenzie, J., R. Schaefer, and E. Farber (1995). *The Student Edition of MINITAB for Windows*. Addison-Wesley, Reading, Mass, 4.

McKeown, R. (1990). "New products from existing technologies." *The Journal of Business and Industrial Marketing*, 5 (1).

McKone, K. and E. Weiss (1999). "Managerial Guidelines for the Use of Predictive Maintenance." Darden School Working Paper Series, University of Virginia, Charlottesville, VA.

McKone, K., R. Schroeder, and K. Cua (1998). "The Impact of Total Productive Maintenance Practices on Manufacturing Performance." Carlson School Working Paper.

McKone, K., R. Schroeder, and K. Cua (1999). "Total Productive Maintenance: A Contextual View." *Journal of Operations Management*, 17 (2), p. 123–144.

McLeavey, D.W. and S.L. Narasimhan (1985). *Production Planning and Inventory Control*. Allyn and Bacon, Boston.

McMillan, Claude, Jr. and Richard F. Gonzales (1968). *Systems Analysis: A Computer Approach to Decision Models* (rev. ed.). Richard D. Irwin, Homewood, Il.

McMullen, P.R. and G.V. Frazier (1997). "A Heuristic Solution for Stochastic Assembly Line Balancing Problems with Parallel Processing for Mixed-Models." *International Journal of Production Economics*, 51, 177–190.

McMullen, P.R. and G.V. Frazier (1998). "Using Simulated Annealing to Solve a Multiobjective Assembly Line Balancing Problem with Parallel Work Stations." *International Journal of Production Research*, 10, 2717–2741.

McQuade, W. (1984). "Easing Tensions Between Man and Machine." *Fortune*, March 19, 58–66.

McQuater, R.E., A.J. Peters, B.G. Dale, M. Spring, J.H. Rogerson, E.M. Rooney (1998). "The Management and Organisational Context of New Product Development." *International Journal of Production Economics*, 55, 121–131.

Meadows, D.H., D.L. Meadows, J. Randers and W.W. Behrens, III (1974). *The Limits to Growth*. Universe Books. New York.

Meal, C.H. (1984). "Putting Production Decisions Where They Belong." *Harvard Business Review* (March–April), 102–111.

Mecklenburg, J.C. (1983). *Plant Layout: A Guide to the Layout of Process Plants and Sites*. Wiley, New York.

Mehra, Satish and M.J. Reid (1982). "MRP Implementation Using an Action Plan." *Interfaces* 12 (1) (February), 69–73.

Mehrez, A. and O.F. Offodile (1994). "A statistical-economic framework for evaluating the effect of robot repeatability on profit." *IIE Transactions*, 26, 101–110.

Melnyk, S.A. (1987). *Production Activity Control* (Shop Floor Control). Dow Jones-Irwin, Homewood, IL.

Melnyk, S.A. and R.F. Gonzalez (1985). "MRP II: The Early Returns are In." *Production and Inventory Management*, 26 (1) (First Quarter), 124–137.

Melnyk, Steven A. and Gary L. Ragatz (1989). "Order Review/Release and Its Impact on the Shop Floor." *Production and Inventory Management*, 29 (2), 13–17.

Melnyk, Steven A. and Gary L. Ragatz (1989). "Order review/release: research issues and perspectives." *International Journal of Production Research*, 27 (7), 1081–1096.

Melnyk, Steven A. and Phillip L. Carter (1986). "Identifying the Principles of Effective Production Activity Control." APICS Conference Proceedings, 227–231.

Mendenhall, W., R.L. Schaeffer, and D. Wackerly (1986). *Mathematical Statistics with Applications* (3rd ed.). PWS-Kent, Boston.

Mentzer, J., K. Kahn, and C. Bienstock (1996). *Sales Forecasting Benchmarking Study*. The University of Tennessee, Knoxville.

Mentzer, J. and J. Cox (1984). "Familiarity, Application and Performance of Sales Forecasting Techniques." *Journal of Forecasting*, 3 (27–36).

Mentzer, J. and K. Kahn (1995). "Forecasting techniques familiarity, satisfaction, usage and application." *Journal of Forecasting*, 14, 465–476.

Meredith, J.R. (1992). *The Management of Operations: A Conceptual Emphasis*. Wiley, New York.

Meredith, J.R. and S.J. Mantel, Jr. (1989). Project Management: A Managerial Approach (2nd ed.). John Wiley, New York.

Messina, W.S. (1987). *Statistical Quality Control for Manufacturing Managers*. John Wiley, New York.

Messmer, E. (1992). "EDI Suppliers to Test Fed Purchasing Net." *Network World*, 9, 43–44.

Messner, W.A. (1982). *Profitable Purchasing Management*. AMACOM, New York.

META (1993). *Design CPN Tutorial for X-Windows*. Meta Software Corporation, Cambridge, MA.

Metters, R. (1997). "Quantifying the Bullwhip Effect in Supply Chains." *Journal of Operations Management*, 15, 89–100.

Meyer, M.H. (1997). "Revitalize your Product Lines Through Continuous Platform Renewal." *Research Technology Management*, 40, 17–28.

Micro-*CRAFT* (1986). *Plant Layout Software*. Industrial Engineering and Management Press, IIE, Atlanta, GA.

Milbrandt, B. (1987). *Electronic Data Interchange: Making Business More Efficient, Automated Graphic Systems*, White Plains, MD.

Miles, R. and C. Snow (1978). *Organizational Strategy, Structure, and Process*. McGraw-Hill, New York.

MIL-HDBK-59A-CALS (1995). *Military Handbook on Computer-Aided Acquisition and Logistics Support*.

Miller, D. (1986). "Configurations of strategy and structure: Towards a synthesis." *Strategic Management Journal*, 7(3), 233–249.

Miller, D. (1987). "The structural and environmental correlates of business strategy." *Strategic Management Journal*, 8 (1), 55–76.

Miller, D. (1988). "Relating Porter's business strategies to environment and structure: Analysis and performance implications." *Academy of Management Journal*, 31 (3), 280–308.

Miller, D. and P.H. Friesen (1984). *Organizations: A quantum view*, Prentice-Hall, Englewood Cliffs, NJ.

Miller, J. (1995). "Innovative cost savings." *NAPM Insights* (September), 30–32.

Miller, J.G., A. De Meyer, and J. Nakane (1992). *Benchmarking Global Manufacturing*, Irwin, Homewood, IL.

Miller, J.G. (1981). "Fit Production Systems to the Task." *Harvard Business Review* (January–February), 145–154.

Miller, J.G. and T.E. Vollmann (1985). "The Hidden Factory." *Harvard Business Review* (September–October), 142–150.

Miltenburg, G.J. and J. Wijngaard (1994). "The U-line Balancing Problem." *Management Science*, 40 (10), 1378–1388.

Min, H. and W.P. Galle (1991). "International Purchasing Strategies of Multinational U.S. Firms." *International Journal of Purchasing and Materials Management* (Summer), 9–18.

Minitab, Inc. (1996). *Minitab Reference Manual*, Release 11, State College, Pennsylvania.

Minoux, M. (1986). *Mathematical Programming*. John Wiley, New York.

Mintzberg, H. (1973). "Strategy-making in three modes." *California Management Review*, 16 (2), 44–53.

Mintzberg, H. (1978). "Patterns in strategy formation." *Management Science*, 24 (9), 934–48.

Mintzberg, H. (1988). "Generic Strategies: Toward a Comprehensive Framework." *Advances in Strategic Management*, 5, 1–67.

Mitchell, R. (1986). "How Ford Hit the Bull's-Eye with Taurus." *Business Week*, June 30, 69–70.

Mitra, A. (1998). *Fundamentals of Quality Control and Improvement* (2nd ed.). Prentice-Hall, Upper Saddle River, New Jersey.

Mize, J.H. and J.G. Cox (1968). *Essentials of Simulation*. Prentice-Hall, Englewood Cliffs, N.J.

Moder, J.J., C.R. Phillips, and E.W. Davis (1983). *Project Management with CPM, PERT, and Precedence Diagramming* (3rd ed.). Van Nostrand Reinhold, New York.

Moder, Joseph, J. and Cecil R. Phillips (1970). *Project Management with CPM and PERT* (2nd ed.). Littleton Educational Publishing, Van Nostrand Reinhold, New York.

Modic, S. (1987). "Myths about Japanese Management." *Industry Week*, October 5.

Moffat, S. (1990). "Japan's New Personalized Production." *Fortune*, 122, 132–135.

Mohrman, S.A., S.G. Cohen, and A.M. Mohrman, Jr. (1995). *Designing Team-Based Organizations: New Forms for Knowledge Work*. Jossey-Bass, San Francisco, CA.

Molloy, M.K. (1982). "Performance analysis using stochastic Petri nets." *IEEE Transactions on Computers*, C-31, 913–917.

Monahan, G.E. and T.L. Smunt (1987). "A multilevel decision support system for the financial justification of automated flexible manufacturing systems." *Interfaces*, 17, 29–40.

Monczka, R. and L.C. Giunipero (1984). "International Purchasing: Characteristics and Implementation." *Journal of Purchasing and Materials Management* (Autumn), 2–9.

Monczka, R. and S. Trecha (1988). "Cost-based supplier performance evaluation." *International Journal of Purchasing and Materials Management* (Spring), 2–7.

Monczka, R. and Robert J. Trent (1995). "Global sourcing: a development approach." *International Journal of Purchasing and Materials Management* 11 (2), 2–8.

Monczka, R. and R.J. Trent (1997). *Purchasing and sourcing 1997: trends and implications*, Center for Advanced Purchasing Studies (CAPS), Tempe, Arizona.

Monczka, R. and R.J. Trent (1993). *Cross-Functional Sourcing, Team Effectiveness*, Center for Advanced Purchasing Studies, Tempe, AZ.

Monczka, R., Frayer R., Handfield, G. Ragatz, and R. Trent (1998). "*Supplier Integration into New Product/Process Development: Best Practices*." Milwaukee: ASQC.

Monczka, R., R. Trent, and R. Handfield (1997). *Purchasing and Supply Chain Management*, Cincinnati, OH: Southwestern Publishing, College Division.

Monden, Y. (1981). "What makes the Toyota production system really tick?" *Industrial Engineering*, 13, 36–46.

Monden, Y. (1983). *Toyota Production System: A Practical Approach to Production Management*. Industrial Engineers and Management Press, Norcross, GA.

Montgomery, D.C. (1996). *Introduction to Statistical Quality Control* (3rd ed.). John Wiley & Sons, Inc., New York.

Moore, Harry D. and Donald R. Kibbey (1975). "Manufacturing Materials and Processes." *Grid*, Columbus Ohio.

Moore, J.M. (1980). "The Zone of Compromise for Evaluating Layout Arrangements." *International Journal of Production Research*. 18 (1), 1–10.

Moore, K.E. and S.M. Gupta (1996). "Petri net models of flexible and automated manufacturing systems: a survey." *International Journal of Production Research*, 34, 3001–3035.

Morelli, M.D., S.D. Eppinger, and R.K. Gulati (1995). "Predicting Technical Communication in Product Development Organizations." *IEEE Transactions on Engineering Management*, 42, 215–222.

Morris, William T. (1967). *The Capacity Decision System*. Irwin, Homewood Ill.

Mosier, C.T., D.A. Evers, and D. Kelly (1984). "Analysis of Group Technology Scheduling Heuristics." *International Journal of Production Research*, 22 (5), 857–875.

Muhlemann, A.P., A.G. Lockett, and C.I. Farn (1982). "Job Shop Scheduling Heuristics and Frequency of Scheduling." *International Journal of Production Research*, 20 (2), 227–241.

Muller, T. (1983). *Automated Guided Vehicles*. IFS Publications, Bedford, UK.

Muramatsu, M., H. Miyazaki, and K. Ishii (1987). "A Successful Application of Job Enlargement/Enrichment at Toyota." *IIE Transactions*, 19 (4), 451–459.

Murata, T. (1989). "Petri nets: properties, analysis and applications." Proceedings of the IEEE, 77, 541–580.

Murray, Jr., J.E. (1996). "Electronic Purchasing and Purchasing Law." *Purchasing* (February 15), 29–30.

Murty, K. (1983). *Linear Programming* (2nd ed.). John Wiley, New York.

Muther, Richard, and K. McPherson (1970). "Four Approaches to Computerized Layout." *Industrial Engineering*, 2 (2), 39–42.

Myers, S.C. (1984). "Finance theory and finance strategy." *Interfaces*, 114, 126–137.

N

Nachi-Fujikoshi Corporation (1990). *Training for TPM: A Manufacturing Success Story*, Productivity Press, Cambridge, MA.

Nadler, G. and D. Smith (1963). "Manufacturing Progress Functions for Types of Processes." *International Journal of Production Research*, 2 (2), 115–135.

Nakajima, S. (1986). "TPM-challenge to the improvement of productivity by small group activities." *Maintenance Management International*. 6, 73–83.

Nakajima, Seiichi (1988). *Introduction to TPM: Total Productive Maintenance*. Productivity Press, Cambridge, MA.

Nakane, J. and R.W. Hall (1983). "Management Specs for Stockless Production." *Harvard Business Review* (May–June), 84–91.

Narasimhan, R. (1983). "An Analytical Approach to Supplier Selection." *Journal of Purchasing and Materials Management* (Winter), 27–32.

Natarajan, R. and S. Goyal (1994). "Safety Stocks in JIT Environments." *International Journal of Operations and Production Management*, 14 (10).

Naylor, T.J., J.L. Balintfy, D.S. Burdick, and Cong Chu (1966). *Computer Simulation Techniques*. John Wiley & Sons, New York.

Neelamkavil, F. (1987). *Computer Simulation and Modeling*. John Wiley, New York.

Nelson, Daniel (1990). *Frederick W. Taylor and the Rise of Scientific Management*. University of Wisconsin Press, Madison, Wisconsin.

Nelson, Daniel (1992). *A Mental Revolution: Scientific Management Since Taylor*. Ohio State University Press, Columbus, Ohio.

Nelson, L.S. (1984). "The Shewhart control chart – tests for special causes." *Journal of Quality Technology*, 16, 237–239.

Nelson, N.S. (1981). "MRP and Inventory and Production Control in Process Industries." *Production and Inventory Control*, 22 (4) (Fourth Quarter), 15–22.

Neter, J., W. Wasserman, and G.A. Whitmore (1987). *Applied Statistics* (3rd ed.). Boston, Allyn and Bacon.

Network Dynamics, Inc. (1991). *MPX User Manual*. Burlington, MA.

Neves, J.S. and B. Nakhai (1994). "The Evolution of the Baldrige Award." *Quality Progress*, 27, 65–70.

Nevins, J.L. and D.E. Whitney (1989). *Concurrent Design of Products and Processes*. McGraw-Hill, New York.

New, C. Colin (1977). "MRP & GT, A New Strategy for Component Production." *Production and Inventory Management*, 18, 3, 50–62.

Newberry Thomas L., ed. (1978). *Inventory Management Systems*. American Software, Atlanta, GA., 8-1 and 8-2.

Newberry, Thomas L. and Carl D. Bhame (1981). "How Management Should Use and Interact with Sales Forecasts." *Inventories and Production Management* 1 (3) (July–August), 1–13.

Newbold, Paul, and Theodore Bos (1990). *Introductory Business Forecasting*. South-Western, Cincinnati.

Newell, Gordon F. (1982). *Applications of Queuing Theory*. Chapman and Hall, New York.

Nicolis, G. & I. Prigogine (1989). *Exploring Complexity: An Introduction*. W.H. Freeman & Co., San Francisco, CA.

Niebel, B.W. (1988). *Motion and Time Study* (8th ed.). Irwin, Homewood IL.

NKS/Factory Magazine (Introduction by Shigeo Shingo) (1988). *Poka-Yoke: Improving Quality by Preventing Defects*. Productivity Press.

Nnaji, B.O. (1986). *Computer-Aided Design, Selection and Evaluation of Robot*. Elsevier, New York.

Nnaji, B.O. (1988). "Evaluation methodology for performance and system economics for robotic devices." *Computers and Industrial Engineering*, 14, 27–39.

Nnaji, B.O. and M. Yannacopoulou (1989). "A utility theory based robot selection and evaluation for electronics assembly." *Computers and Industrial Engineering*, 14, 477–493.

Nobeoka, K. and M.A. Cusumano (1995). "Multiproject Strategy, Design Transfer, and Project Performance." *IEEE Transactions on Engineering Management*, 42, 397–409.

Noe, R.A., J.R. Hollenbeck, B. Gerhart, and P.M. Wright (1997). *Human Resource Management: Gaining a Competitive Advantage* (2nd ed.). Irwin, Chicago.

Nof, S.Y. (1985). *Handbook of Industrial Robotics*, John Wiley & Sons, Inc., New York.

Nof, S.Y., M.M. Barash, and J.J. Solberg (1979). "Operational control of item flow in versatile manufacturing systems." *International Journal of Production Research*, 17, 479–489.

Nof, S.Y. and H. Lechtman (1982). "Robot time and motion provides means for evaluating alternative work methods." *Industrial Engineering*, 14, 38–48.

Nonaka, I. and H. Takeuchi (1995). *The Knowledge Creating Company*. Oxford University Press, New York.

Norbom, Rolf (1973). "The Simple ABC's (A Loose Piece Float System)." *Production and Inventory Management*, 14, 1, 16.

Noreen, E., D. Smith, and J.T. Mackay (1995). *The Theory of Constraints and its Implications for Management Accounting*. North River Press, Great Barrington, MA.

Normann, R. (1984). Service Management: Strategy and Leadership in Service Businesses. John Wiley, New York.

Nuyens, R.P.A., N.M. Van Dijk, L. Van Wassenhove, and E. Yücesan (1996). "Transient behavior of simple queueing systems: implications for FMS models." *Simulation: Application and Theory*, 4, 1–29.

O

O'Donnell, Merle, and Robert J. O'Donnell (1984). "Quality Circles – The Latest Fad or a Real Winner?" *Business Horizons* (May–June), 113–121.

O'Leary-Kelly, A.M., J.J. Martocchio, and D.D. Frink (1994). "A Review of the Influence of Group Goals on Group Performance." *Academy of Management Journal*, 37 (5), 1285–1301.

O'Neal, Charles R. (1987). "The Buyer-Seller Linkage in a Just-in-time Environment." *Journal of Purchasing and Materials Management*, Spring, 7.

O'Neal, K. (1987). "Project Management Computer Buyer's Guide." *Industrial Engineering* 19 (1) (January), 53–63.

Oakland, J.S. (1993). *Total Quality Management: The Route to Improving Performance*. Butterworth-Heinemann, Oxford.

Offodile, O.F. and U. Ugwu (1991). "Evaluating the effect of speed and payload on robot repeatability." *International Journal of Robotics and Computer Integrated Manufacturing*, 8, 27–33.

Offodile, O.F., B.K. Lambert, and R.A. Dudek (1987). "Development of a computer aided robot selection procedure (CARSP)." *International Journal of Production Research*, 25, 1109–1121.

Ohmae, K. (1987). "The Triad World View." *Journal of Business Strategy* (Spring), 8–19.

Ohmae, K. (1988). "Getting Back to Strategy." *Harvard Business Review*, 66 (6), 149–156.

Ohmae, K. (1989). "Managing in a Borderless World." *Harvard Business Review*, 67 (3), 151–161.

Ohno, Taiichi (1988). *Toyota Production System: Beyond Large-Scale Production*. Productivity Press, Cambridge, MA.

Oliff, Michael, and Earl Burch (1985). "Multiproduct Production Scheduling at Owens-Corning Fiberglass." *Interfaces* 15 (5) (September–October), 25–34.

Olins, W. (1989). *Corporate Identity: Making Business Strategy Visible through Design*, Harvard Business School Press, Boston, MA.

Orlicky, Joseph (1975). *Material Requirements Planning*. McGraw-Hill, New York.

Ormsby, J.G., S.Y. Ormsby, and C.R. Ruthstrom (1990). "MRPII implementation: a case study." *Production and Inventory Management Journal*, 31 (4), 77–81.

Osburn, J.D., L. Moran, E. Musselwhite, and J.H. Zenger (1990). *Self-directed Work Teams: The New American Challenge*. Irwin Publishing, Chicago, IL.

Ouchi, W. (1981). *Theory Z*. Addison-Wesley, Reading, MA.

Özekici, S. (1996). Editor. *Reliability and maintenance of complex systems. NATO ASI Series*, vol. 154, Springer, Berlin.

Ozekici, S. and S.R. Pliska (1991). "Optimal Scheduling of Inspections: A Delayed Markov Model with False Positive and Negatives." *Operations Research*, 39 (2), 261–273.

P

Padulo, Louis and Michael A. Arbib (1974). *Systems Theory*. Hemisphere, Washington, D.C.

Papadopoulos, H.D., C. Heavey, and J. Browne (1993). *Queueing Theory in Manufacturing Systems Analysis and Design*. Chapman & Hall, London, UK.

Parasuraman, A., V.A. Zeithaml, and L.L. Berry (1985). "A Conceptual Model of Service Quality and Its Implications for Future Research." *Journal of Marketing*, 49 (4), 41–50.

Pascale, Richard T. and Anthony G. Athos (1981). *The Art of Japanese Management*. Simon & Schuster, New York.

Pate-Cornell, M.E., H.L. Lee, and G. Tagaras (1987), "Warning of Malfunction: The Decision to Inspect and Maintain Production Processes on Schedule or On Demand." *Management Science*, 33 (10), 1277–1290.

Patton, J.D., Jr. (1983). *Preventive Maintenance. Instrument Society of America*. Research Triangle Park, NC.

Payne, J.A. (1982). *An Introduction to Simulation*. McGraw-Hill, New York.

Peek, L.E. and John H. Blackstone, Jr. (1987). "Developing a Time-phased Order-point Template Using A Spreadsheet." *Production and Inventory Control Management Journal*, 28 (4), 6–10.

Pegden, C. Dennis (1987). *Introduction to SIMAN*. Systems Modeling Corp, State College, Pa.

Pennente, Ernest and Ted Levy (1982). "MRP on Microcomputers." *Production and Inventory Management Review*, May, 20–25.

Penrose, E. (1959). *The Theory of the Growth of the Firm*, Wiley, New York.

Perros, H.G. (1994). *Queueing Networks with Blocking*. Oxford University Press. Oxford.

Peters, T. (1987). *Thriving on Chaos: Handbook for a Management Revolution*. Knopf, New York.

Peters, T.J. and R.H. Waterman, Jr. (1982). *In Search of Excellence*. Harper & Rowe, New York.

Peterson, R. and E.A. Silver (1984). *Decision Systems for Inventory Management and Production Planning*. (2nd ed.). Wiley, New York.

Pfeffer, J. (1995). "Producing sustainable competitive advantage through the effective management of people." *Academy of Management Executive*, 9, 55–69.

Philipoom, P.R. and T.D. Fry (1992). "Capacity-based order review/release strategies to improve manufacturing performance." *International Journal of Production Research*, 30 (11), 2559–2572.

Philipoom, P.R., L.P. Rees, and B.W. Taylor (1996). "Simultaneously Determining the Number of Kanbans, Container Sizes and the Final Assembly Sequence of Products in a Just-in-Time Shop." *International Journal of Production Research*, 34, 51–69.

Phillips, F. and N. Kim (1996). "Implications of chaos research for new product forecasting." *Technological Forecasting and Social Change*. 53: 239–61.

Pidd, M. (1984). *Computer Simulation in Management Science*. John Wiley, New York.

Pierskalla, W.P. and J.A. Voelker (1976). "A Survey of Maintenance Models: The Control and Surveillance of Deteriorating Systems." *Naval Research Logistics Quarterly*, 23, 353–388.

Pine II, B.J. (1993). "Mass Customizing Products and Services." *Planning Review*, July, 21 (4), 6.

Pine II, B.J. and T.W. Pietrocini (1993). "Standard Modules Allow Mass Customization at Bally Engineered Structures." *Planning Review*, July, 21 (4), 20.

Pine II, B.J., B. Victor, and A.C. Boyton (1993). "Making Mass Customization Work." *Harvard Business Review*, 71, 108–119.

Pine II, J.B. (1993). *Mass Customization – The New Frontier in Business Competition*, Harvard Business School Press, Boston, MA.

Pine II, J.B., D. Peppers, and M. Rogers (1995). "Do You Want to Keep Your Customers Forever?" *Harvard Business Review*. 73, 103–114.

Pinella, Paul, and Eric Cheathan (1981). "In-House Inventory Control." *ICP INTERFACE Manufacturing and Engineering*, Winter, 18–21.

Pintelon, L.M. and L.F. Gelders (1992). "Maintenance management decision making." *European Journal of Operational Research*, 58, 301–317.

Pinto, J.K. (1987). "Strategy and Tactics in a Process Model of Project Implementation." *Interfaces* 17 (3) (May–June), 34–46.

Piore, M.J. and C.F. Sabel (1984). *The Second Industrial Divide*, Basic Books, New York, N.Y.

Pisano, G. (1997). *The Development Factory*. Harvard Business School Press, Boston.

Plane, Donald, and Gary Cochenberger (1972). *Operations Research for Mangerial Decisions*. Richard D. Irwin, Inc., Homewood, IL.

Platts, K.W. (1994). "Characteristics of Methodologies for Manufacturing Strategy Formulation." *Computer Integrated Manufacturing Systems*, 7, 2, 93–9.

Player, R.S. and D.E. Keys (1995). "Lessons from the ABM battlefield: Getting off to the right start." *Journal of Cost Management* (Spring), 26–38.

Player, R.S. and D.E. Keys (1995). "Lessons from the ABM battlefield: Developing the pilot." *Journal of Cost Management* (Summer), 20–35.

Player, R.S. and D.E. Keys (1995). "Lessons from the ABM battlefield: Moving from pilot to mainstream." *Journal of Cost Management* (Fall): 31–41.

Player, S. and D. Keys (1995). Editors. *Activity-Based Management: Arthur Andersen's Lessons From the ABM Battlefield*, MasterMedia Limited, New York.

Plenert, G. and T.D. Best (1986). "MRP, JIT, and OPT: What's Best?" *Production and Inventory Management*, 27 (2), 22–29.

Plossl, G.W. (1985). *Just-in-Time: A Special Round table*. George Plossl Educational Services, Georgia.

Plossl, G.W. (1973). "How Much Inventory is Enough?" *Production and Inventory Management*. Second Quarter.

Plossl, G.W. (1973). *Manufacturing Control: The Last Frontier for Profits*. Reston, VA., Reston Publishing.

Plossl, G.W. (1983). *Production and Inventory Control*: Applications. George Plossl Education Services, Atlanta.

Plossl, G.W. (1985). *Production and Inventory Management: Principles and Techniques*, (2nd ed.). Prentice-Hall, New Jersey.

Plossl, G.W. (1994). *Orlicky's Material Requirements Planning* (2nd ed.). McGraw-Hill.

Plossl, G.W. and O.W. Wight (1967). *Production and Inventory Control*. Prentice-Hall, Englewood Cliffs, NJ.

Plossl, G.W. and O.W. Wight (1975). "Capacity Planning and Control." *Capacity Planning and Control*. 50–86. American Production and Inventory Control Society, Falls Church, VA.

Plossl, G.W. and W.E. Welch (1979). *The Role of Top Management in the Control of Inventory*. Reston, Reston, VA.

Poist, R.F. (1986). "Evolution of Conceptual Approaches to Designing Business Logistics Systems." *Transportation Journal*, Fall, 55–64.

Poppel, H.L. (1982). "Who Needs the Office of the Future?" *Harvard Business Review* (November–December), 146–155.

Porter, M.E. (1980). *Competitive Strategy – Techniques for Analysing Industries and Competitors*. Macmillan Free Press, New York.

Porter, M.E. (1985). *Competitive Advantage: Creating and Sustaining Superior Performance*. Free Press, New York.

Porter, M.E. (1986). "Changing Patterns of International Competition." *California Management Review* (Winter), 9–40.

Porter, M.E. (1986). Editor. *Competition in Global Industries*. Harvard Business School Press, Boston.

Porter, M.E. (1990). "The Competitive Advantage of Nations." *Harvard Business Review*, 68 (2), 73–93.

Porteus, E.L. (1985). "Investing in Reduced Setups in the EOQ Model." *Management Science*, 31 (8), 998–1010.

Posner, B.Z. (1987). "What It Takes to Be a Good Project Manager." *Project Management Journal* 18 (1) (March), 51–54.

Potamianos, J. and A.J. Orman (1996). "An Interactive Dynamic Inventory-Production Control System." *Journal of the Operational Research Society*, 47 (8), 1016–1028.

Powell, C. (1978). "Systems Planning/Systems Control." *Production and Inventory Management* 18 (3), 15–25.

Powell, C. (1987). "Workshop Report: Cincinnati Milacron-Electronics System Division." *Target*, Summer, 28–33.

Powers, T.L., J.U. Sterling, and J.F. Wolter (1988). "Marketing and Manufacturing Conflict: Sources and Resolution." *Production and Inventory Management Journal*, 29 (4), 56–60.

Poza, Ernesto J. (1983). "Twelve Actions to Build Strong U.S. Factories." *Sloan Management Review*, 21, Winter, 47–58.

Pratsini, Eleni, Jeffrey D. Camm, and Amitabh S. Raturi (1993). "Effect of Process Learning on Manufacturing Schedules." *Computers and Operation Research*, 20 (1), 15–24.

Preiss K. (1997). "A Systems Perspective of Lean and Agile Manufacturing." *Agility and Global Competition*, 1 (1), 59–75.

Preiss K., S. Goldman, and R. Nagel (1996). *Cooperate to Compete: Building Agile Business Relationships*. Van Nostrand Reinhold, New York.

Preiss, K. (1995). "Mass, Lean, and Agile as Static and Dynamic Systems." *Agility Forum*, Bethlehem, PA.

Preiss, K. (1995). "Models of the Agile Competitive Environment." *Agility Forum*, Bethlehem, PA.

Premkumar, G., K. Ramamurthy, and S. Nilakanta (1994). "Implementation of Electronic Data Interchange: An Innovation Diffusion Perspective." *Journal of Management Information Systems*, 11, 157–186.

Pritsker, A.A.B. (1986). *Introduction to Simulation and SLAM II*. John Wiley & Sons, New York.

Pritsker, A.A.B. (1989). "Developing analytic models based on simulation results." In: Proceedings of the Winter Simulation Conference, 653–660.

Production Modeling Corporation of Utah (1989). *ProModel User's Manual*. Orem, UT.

Proud, J.F. (1981). "Controlling the Master Schedule." *Production and Inventory Management* 22 (2) (Second Quarter), 78–90.

Ptak, C.A.M. (1997). *MRP and Beyond*, Richard D. Irwin.

Putnam, Arnold O., R. Everdell, D.H. Dorman, R.R. Cronan, and L.H. Lindgren (1971). "Updating Critical Ratio and Slack-Time Priority Scheduling Rules." *Production and Inventory Management*, 12, 4.

Pye, C. (1997). "Another Option for Internet-Based EDI: Electronic Communication." *Purchasing Today*, 8, 38–40.

Q

Quality System Requirements QS-9000, International Organization for Standardization, Chrysler Corporation, ford Motor Company, and General Motors Corporation, August 1994.

Quelch, J.A. and E.J. Hoff (1986). "Customizing global marketing." *Harvard Business Review*, May–June, 59–68.

Quinlan, J. (1982). "Just-in-Time at the Tractor Works." *Material Handling Engineering*, June, 62–65.

Quinn, J.B. (1978) "Strategic change: logical incrementalism." *Sloan Management Review*, 1, 20.

Quinn, J.B. (1985) "Managing Innovation: Controlled Chaos." *Harvard Business Review*, May/June.

Quinn, J.B. (1992). *Intelligent Enterprise*. Free Press, New York.

Quinn, J.B. and F.G. Hilmer (1994). "Strategic outsourcing." *Sloan Management Review*, Summer, 43–55.

R

Ragatz, G., R. Handfield, and T. Scannell (1997). "Success factors for integrating suppliers into new product development." *Journal of Product Innovation Management*, 7 (20).

Ragatz, G.L. and V.A. Mabert (1984). "A Simulation Analysis of Due Date Assignment Rules." *Journal of Operations Management*, 5 (1) (November), 27–40.

Rahim, M.A. (1994). "Joint Determination of Production Quantity, Inspection Schedule, and Control Chart Design." *IIE Transactions*, 26 (6), 2–11.

Rajagopalan, R. and J.L. Batra (1975). "Design of cellular production systems: a graphtheoretic approach." *International Journal of Production Research*, 13. 567–579.

Ramamoorthy, C.V. and G.S. Ho (1980). "Performance evaluation of asynchronous concurrent systems using Petri nets." *IEEE Transactions on Software Engineering*, SE-6, 440–449.

Ramasesh, R.V. and M.D. Jayakumar (1991). "Measurement of manufacturing flexibility: a value-based approach." *Journal of Operations Management*, 10, 1–8.

Ramasesh, R.V. and M.D. Jayakumar (1993). "Economic justification of advanced manufacturing technology." *OMEGA*, 21, 2819–306.

Ramstad (1997). "Shift at Compaq May Push Down Prices for PCS." *Wall Street Journal*, March 31.

Ranky, P.G. (1982). *The Design and Operation of FMS* (IFS (Publications) Ltd. and North Holland).

Ranky, P.G. (1985). "FMS in CIM" (Flexible Manufacturing Systems in Computer Integrated Manufacturing), *ROBOTICA*, Cambridge University Press, 3, 205–214.

Ranky, P.G. (1985). *Computer Integrated Manufacturing*. Prentice-Hall International.

Ranky, P.G. (1998). "An Introduction to Concurrent/Simultaneous Engineering." CIMware Ltd., http://www.cimwareukandusa.com (Book and CD-ROM combo).

Ranky, P.G. (1998). "Flexible Manufacturing Cells and Systems in CIM." CIMware Ltd., http://www.cimwareukandusa.com (Book and CD-ROM combo).

Reddy, J. and A. Berger (1983). "Three Essentials of Product Quality." *Harvard Business Review* (July–August), 153–159.

Reed, Ruddell (1966). *Plant Layout: Factors, Principles, and Techniques*. Irwin, Homewood, IL.

Reeves, E. and E. Torrey, Editors (1994). *Beyond the Basics of Reengineering: Survival Tactics for the '90s*. Industrial Engineering and Management Press, Georgia.

Reiman, M.I. and A. Weiss (1989). "Sensitivity analysis for simulations via likelihood ratios." *Operations Research*, 37, 830–844.

Reimann, C.W. and H. Hertz (1993). "The Malcolm Baldrige National Quality Award and ISO 9000 Registration: Understanding Their Many Important Differences." *ASTM Standardization News*, November, 42–51.

Reinfeld, N.V. (1987). *Handbook of Production and Inventory Control*. Prentice-Hall, Englewood Cliffs, NJ.

Render, B. and J. Heizer (1997). *Principles of Operations Management*. Prentice-Hall, Upper Saddle River, New Jersey.

Rhodes, D.J. (1991). "The facilitator – an organizational necessity for the successful implementation of IT and operations strategies." *Computer Integrated Manufacturing Systems*, 4, 2, 109–13.

Rice, R.S. (1977). "Survey of Work Measurement and Wage Incentives." *Industrial Engineering* 9 (7) (July), 18–31.

Richards, C.W. (1996). "Agile Manufacturing: Beyond Lean?" *Production and Inventory Mangement Journal*, 37, 60–64.

Richardson, P.R., A.J. Taylor, and J.R.M. Gordon (1985). "A Strategic approach to Evaluating Manufacturing Performance." *Interfaces*, November–December, 15–27.

Riopel, Robert J. (1986). "JIT: Evolutionary Revolution." *Manufacturing Systems*, July, 44.

Ritzman, L.P., B.E. King and L.J. Krajewski (1984). "Manufacturing Performance – Pulling the Right Levers." *Harvard Business Review*, March–April, 143–152.

Roberts, M.W. and K.J. Silvester (1996). "Why ABC Failed and How It May Yet Succeed." *Journal of Cost Management*, Winter 23–35.

Robertson, T. (1971). *Innovative Behaviour and Communication*. New York, Holt, Rinehart and Winston.

Robinson, A.G. and D.M. Schroeder (1990). "The limited role of statistical quality control in a zero-defect environment." *Production and Inventory Management Journal* 3rd. Quarter, 60–65.

Robinson, J., L.W. Schruben, and J.W. Fowler (1993). "Experimenting with large-scale semiconductor manufacturing simulations: a frequency domain approach to factor screening." In: Proceedings of the Second IE Research Conference, 112–116.

Robinson, Richard D. (1984). *Internationalization of Business: An Introduction*. Free Press, Chicago.

Roderick, Larry M., Don T. Phillips, and Gary L. Hogg (1992). "A comparison of order release strategies in production control systems." *International Journal of Production Research*, 30 (3), 611–626.

Rogers, E.M. (1983). *Diffusion of Innovation*, Free Press, New York, NY.

Rohm, C.E. (Winter 1992/1993). "New England Telephone Opens Customer Service Lines to Change." *National Productivity Review*, 12 (1), 73–82.

Ronen, D. (1983). "Inventory Service Measures – A Comparison of Measures." *International Journal of Operations and Production Management*, 3 (2), 37–45.

Roos, D. (1995). "Agile/Lean: A Common Strategy for Success." *Agility Forum*, Bethlehem, PA.

Rosander, A.C. (1985). *Applications of Quality Control in the Service Industries*. New York, Marcel Dekker.

Rosenberg, Nathan (1969). *The American System of Manufactures*. Edinburgh University Press, Edinburgh, Scotland.

Rosenthal, S.R. (1984). "Progress Towards the 'Factory of the Future." *Journal of Operations Management*, 4 (3) (May), 203–229.

Rosenthal, S.R. (1992). *Effective Product Design and Development*. Business One Irwin, Homewood.

Rosow, I.M. and R. Zager (1988). *Training: The Competitive Edge*. San Francisco, Jossey-Bass.

Ross, Phillip J. (1988). "The Role of Taguchi Methods and Design of Experiemts in QFD." *Quality Progress*, June, 41–47.

Rubin, P.A. and G.L. Ragatz (1995). "Scheduling in a Sequence Dependent Setup Environment with Genetic Search." *Computers and Operations Research*, 22 (1), 85–99.

Rubinstein, R.Y. and A. Shapiro (1993). *Discrete Event Systems: Sensitivity Analysis and Stochastic Optimization by the Score Function Method*. Wiley, New York.

Ruch, W.A., H.E. Fearon, and C.D. Wieters (1992). *Fundamental of Production/Operations Management* (5th ed.). West Publishing Company, St. Paul.

Ruefli, T. (1971). "A Generalized Goal Decomposition Model." *Management Science*, 17 (8), 505–518.

Russell, S. and P. Norvig (1995). *Artificial Intelligence: A Modern Approach*, Prentice-Hall, Englewood Cliffs, NJ.

Ryan, T.A., B.L. Joiner, and B.F. Ryan (1985). *Minitab Handbook* (2nd ed.). Boston, PWS-Kent.

Rycroft, R. (1993). "Microcomputer Software of Interest to Forecasters in Comparative Review: An Update." *International Journal of Forecasting*, 9 (4), December, 531–575.

S

Sadowski, R.P. (1984). "Computer-Integrated Manufacturing Series Will Apply Systems Approach to Factory of the Future." *Industrial Engineering*, 16 (1) (January), 35–40.

Sakakibara, S., B.B. Flynn, and R.G. Schroeder (1993). "A framework and measurement instrument for just-in-time manufacturing." *Production and Operations Management*, 2, 177–194.

Sako, M. (1992). Prices, Quality and Trust: Inter-Firm Relations in Britain and Japan. Cambridge University Press, Cambridge.

Sakurai, M. (1989). "Target Costing and How to Use It." *Cost Management*, Summer.

Saleh S. and C. Wang (1993). "The Management Of Innovation: Strategy Structure and Organizational Climate" *IEEE Transactions On Engineering Management*. 40 (1).

Salomon, Daniel P., and John E. Biegel (1984). "Assessing Economic Attractiveness of FMS Applications in Small Batch Manufacturing." *Industrial Engineering* (June), 88–96.

Salvador, Michael S. *et al.* (1975). "Mathematical Modeling Optimization and Simulation Improve Large-Scale Finished Goods Inventory Management." *Production and Inventory Control Management*, Second Quarter.

Salverson, M.E. (1955). "The Assembly Line Balancing Problem." *Journal of Industrial Engineering*, 6 (3), 18–25.

Salveson, M.E. (1955). "The Assembly Line Balancing Problem." *Journal of Industrial Engineering*, 6, 18–25.

Sampson, R.J., M.T. Farris, and D.L. Schrock (1990). *Domestic Transportation: Practice, Theory, and Policy* (6th ed.). Houghton Mifflin Company, Boston.

Sanders, M.A. and E.J. McCormick (1987). *Human Factors in Engineering and Design* (6th ed.). McGraw-Hill, New York.

Sanders, N. (1997). "The Status of Forecasting in Manufacturing Firms." *Production and Inventory Mangement Journal*, 38 (2), 2nd Quarter, 32–35.

Sanders, N.R. and L.P. Ritzman (1992). "The need for contextual and technical knowledge in Judgmental forecasting." *Journal of Behavioral Decision Making*, 5, 39–52.

Sandras, William A. Jr. "JIT/TQC Changes in Thinking for Information Systems." *P&IM Review*, April, 33.

Saporito, B. (1994). "Behind the Tumult at P & G." *Fortune*, 74–82.

Saren A. (1984). "A Classification And Review of Models of The Intra-Firm Innovation Process." *R&D Management*, 14.

Sarin, S.C. and S.K. Das (1994). "Computer Integrated Production Planning and Control." in Dorf, R.C. and A. Kusiak, eds., *Handbook of Design, Manufacturing, and Automation*, John Wiley & Sons, New York.

Saunders, M. (1997). *Strategic Purchasing and Supply Management*. Pitman Publishing, London.

Savsar, M. (1996). "Effects of Kanban Withdrawal Policies and other Factors on the Performance of JIT Systems – A Simulation Study." *International Journal of Production Research*, 34, 2879–2899.

Scarpello, V., W.R. Boulton, and C.R. Hofer (1986). "Reintegrating R&D into Business Strategy." *Journal of Business Strategy*, Spring, 49–56.

Schein, E.H. (1964). "The mechanism of change." *Interpersonal Dynamics*, The Dorsey Press: Homewood, IL. 199–236.

Schendel, D. and C. Hofer (1979). *Strategic Management: A new view of Business Policy and Planning*. Little, Brown & Co., Boston, MA.

Schewe, G. (1994). "Successful Innovation Management: An Integrative Perspective." *Journal Of Engineering and Technology Management*, 11.

Schilling, D.A. (1980). "Dynamic Location Modeling for Public Sector Facilities: A Multi-Criteria Approach." *Decision Sciences* 11 (4) (October), 714–724.

Schmeiser, B.W. (1983). "Batch size effects in the analysis of simulation output." *Operations Research*, 31, 565–568.

Schmenner, R.W. (1976). "Before You Build a Big Factory." *Harvard Business Review*, 54 (4) (July–August), 71–81.

Schmenner, R.W. (1982). "Multiple Manufacturing Strategies Among the Fortune 500." *Journal of Operations Management*, 2 (2) (February), 77–86.

Schmenner, R.W. (1979). "Look Beyond the Obvious in Plant Location." *Harvard Business Review*. (January–February), 126–132.

Schmenner, R.W. (1982). *Making Business Location Decisions*, Prentice-Hall, Englewood Cliffs, NJ.

Schmenner, R.W. (1983). "Every Factory Has a Cycle." *Harvard Business Review* (March–April), 121–129.

Schmenner, R.W. (1986). "How Can Service Businesses Survive and Prosper?" *Sloan Management Review*, Spring, 21–32.

Schmidgall, R.S. (1986). *Managerial Accounting*, The Educational Institute, East Lansing.

Schneider, J.D. and M.A. Leatherman (1992). "Integrated Just-In-Time: A Total Business Approach." *Production and Inventory Management Journal*, 33 (1), 78–82.

Schniederjans, M.J. (1998). *Operations Management: A Global Context*, Quorum Books, New York.

Schniederjans, M.J. (1999). *Global Facility Location and Acquisition*, Quorum Books, New York.

Schniederjans, M.J. and J.J. Hoffman (1992). "Multinational Acquisition Analysis: A Zero-One Goal Programming Model." *European Journal of Operational Research*, 62, 174–185.

Schnor, John E., and Thomas F. Wallace (1986). *High Performance Purchasing*, Oliver Wight Publications, Essex Junction, VT.

Schonberger, R.J. (1982). "Some Observations on the Advantages and Implementation Issues of Just-in-Time Production Systems." *Journal of Operations Management*, 2 (1) (November), 1–12.

Schonberger, R.J. (1982). *Japanese Manufacturing Techniques: Nine Hidden Lessons in Simplicity*. The Free Press, New York.

Schonberger, R.J. (1983). "Applications of Single-Card and Dual-Card Kanban." *Interfaces*, 13 (4) (August), 56–67.

Schonberger, R.J. (1983). "Selecting the Right Manufacturing Inventory Systems: Western and Japanese Approaches." *Production and Inventory Control*, 24 (2) (Second Quarter), 33–44.

Schonberger, R.J. (1984). "Just-In-Time Production System: Replacing Complexity with Simplicity in Manufacturing Management." *Industrial Engineering*, Oct, 52–63.

Schonberger, R.J. (1986). *World Class Manufacturing: The Lessons of Simplicity Applied*. Free Press, New York.

Schonberger, R.J. (1987). *World Class Manufacturing Casebook: Implementing JIT and TQC*. Free Press, New York.

Schonberger, R.J. (1996). *World Class Manufacturing: The Next Decade*. The Free Press.

Schonberger, R.J. and E.M. Knod, Jr. (1997). *Operations Management: Customer Focused Principles* (6th ed.). Chapters 5, 8, and 10. Irwin, IL.

Schonberger, R.J. and J. Gilbert (1983). "Just-in-Time Purchasing: A Challenge for U.S. Industry." *California Management Review*, 26 (1) (Fall), 54–68.

Schonberger, R.J. and M.J. Schniederjans (1984). "Reinventing Inventory Control." *Interfaces* 14 (3) (May–June), 76–83.

Schrage, L. (1986). *Linear, Integer, and Quadratic Programming with LINDO* (3rd ed.). Scientific Press, Palo alto, CA.

Schroeder, R.G. and P. Larson (1986). "A Reformulation of the Aggregate Planning Problem." *Journal of Operations Management*, 6 (3) (May), 245–256.

Schroeder, R.G., G.D Scudder, and D.R. Elmm (1989). "Innovation in Manufacturing." *Journal of Operations Management*, 8 (1), 1–15.

Schroeder, R.G., J.C. Anderson, and G. Cleveland (1986). "The Content of Manufacturing Strategy: An Empirical Study." *Journal of Operations Management*, 6 (3–4), 405–415.

Schroeder, R.G., J.C. Anderson, S.E. Tupy, and E.M. White (1981). "A Study of MRP Benefits and Costs." *Journal of Operations Management*, 2 (1) (October), 1–9.

Schruben, L. (1983). "Confidence interval estimation using standardized time series." *Operations Research*. 31, 1090–1107.

Schruben, L. and V.J. Cogliano (1987). "An experimental procedure for simulation response surface model indentification." *Communications of the ACM*, 30, 716–730.

Schuler, R.S., L.P. Ritzman, and V.L. Davis (1981). "Merging Prescriptive and Behavioral Approaches for Office Layout." *Journal of Operations Management*, 1 (3) (February), 131–142.

Schumpeter, J. (1934). *The Theory Of Economic Development*. Harvard University Press. Cambridge, MA.

Schwarz, Leroy B., and Robert E. Johnson (1978). "An Appraisal of the Empirical Performance of the Linear Decision Rule for Aggregate Planning." *Management Science*, 24 (8) (April), 844–849.

Schwarzendahl, R. (1996). "The Introduction of Kanban Principles to Material Management in EWSD Production." *Production Planning and Control*, 7, 212–221.

Scott, B. (1994). *Manufacturing Planning Systems*. McGraw Hill.

Scully, J.I. and S.E. Fawcett (1993). "Comparative Logistics and Production Costs for Global Manufacturing Strategy." *International Journal of Operations and Production Management*, 13 (12), 62–78.

Scully, J.I. and S.E. Fawcett (1994). "International Procurement Strategies: Opportunities and Challenges for the Small Firm." *Production and Inventory Mangement Journal*, 35 (2), 39–46.

Seglund, R., and S. Ibarreche (1984). "Just-in-Time: The Accounting Implications." *Management Accounting* (August), 43–45.

Seidmann, A. and S.Y. Nof (1989). "Operational analysis of an autonomous assembly robotic station."

IEEE Transactions on Robotics and Automation, 5, 4–15.

Seidmann, A., A. Arbel, and R. Shapira (1984). "A two-phase analytic approach to robotic system design." *Robotics and Computer-Integrated Manufacturing*, 12, 181–190.

Seifoddini, H.K. (1989). "Single linkage versus average linkage clustering in machine cells formation." *Computers Industrial Engineering*, 16, 419–426.

Seifoddini, H.K. and P. Wolfe (1986). "Application of the similarity coefficient method in group technology." *AIIE Transactions*. 18, 271–277.

Sekine, Kenichi (1990). *One-Piece Flow: Cell Design for Transforming the Production Process*. Productivity Press.

Sekine, Kenichi (1992). *One Piece Flow: Cell Design For Transforming the Production Process*. Productivity Press, Cambridge, Mass.

Sekine, Kenichi and Arai, Keisuke (1992). *Kaizen for Quick Changeover*. Productivity Press.

Senge, P.M. (1990). *The Fifth Discipline: The Art and Practice of the Learning Organization*. Doubleday Currency, New York: NY.

Sepehri, M. (1986). *Just-in-Time, Not Just in Japan*. American Production and Inventory Control Society, Virginia.

Sethi, A.K. and S.P. Sethi (1990). "Flexibility in Manufacturing: A Survey." *International Journal of Flexible Manufacturing Systems*, 2, 289–328.

Shaffer, L.R., J.B. Ritter, and W.L. Meyer (1965). *The Critical Path Method*, McGraw Hill, New York.

Shainin, D. and P.D. Shainin (1988). "Statistical Process Control." *Juran's Quality Control Handbook*. Edited by J.M. Juran and F.M. Gryna (4th ed.). Section 24, 24.1–24.40. McGraw-Hill, New York.

Shank, J. (1995). "Strategic Cost Management: A Target Costing Field Study." *The Amos Tuck School of Business Administration*, Dartmouth College.

Shannon, Robert E. (1975). *System Simulation – the Art and Science*. Prentice-Hall, Englewood Cliffs, N.J.

Shantikumar, J.G. and R.G. Sargent (1983). "A unifying view of hybrid simulation/analytic models and modeling." *Operations Research*, 31, 1030–1052.

Shapiro, B., V.K. Ranga, and J.J. Sviokla (1992). "Staple Yourself to an Order." *Harvard Business Review* (July–August), 70 (4), 113–122.

Shapiro, B.P. (1977). "Can Marketing and Manufacturing coexist?" *Harvard Business Review*, 55 (5), 104–114.

Shapiro, R.D. and J.L. Heskett (1985). *Logistics Strategy: Cases and Concepts*. West, St. Paul, MN.

Sharman, G. (1984). "The Rediscovery of Logistics." *Harvard Business Review* (September–October), 71–79.

Shaw, M.J. and A.B. Whinston (1988). "A Distributed Knowledge-Based Approach to Flexible Automation: The Contract Net Framework." *Flexible Manufacturing Systems*, 85–104.

Sheldon, D. (1991). "MRPII – What it really is." *Production and Inventory Management Journal*, 32, 3, 12–15.

Sheldon, D. (1994). "MRPII implementation: a case study." *Hospital Material Management Quarterly*, 15 (4), 48–52.

Sheridan, J.H. (1993). "Agile Manufacturing: Beyond Lean Production." *Industry Week*, 242, 34–36.

Shewhart, W.A. (1931). *Economic Control of Quality of Manufactured Product*, E. Van Nostrand Company, New York.

Shewhart, W.A. (1939). *Statistical Method from the Viewpoint of Quality Control*, Graduate School, Department of Agriculture, Washington.

Shiba, S., A. Graham, and D. Walden (1993). *A New American TQM: Four Practical Revolutions in Management*, 3 (3)0, 411–460. Productivity Press, Oregon.

Shimbun, N.K. (ed.). (1995). *TPM Case Studies*, Productivity Press, Portland, OR.

Shingo, S. (1981). *Study of Toyota Production System From Industrial Engineering Viewpoint*. Japan Management Association, Tokyo.

Shingo, S. (1985). *SMED: A Revolution in Manufacturing*, Productivity Press, Cambridge, MA.

Shingo, S. (1986). *Zero Quality Control: Source Inspection and the Poka-Yoke System*, Productivity Press, Stamford, Connecticut.

Shingo, S. (1988). *Non-Stock Production: The Shingo System for Continuous Improvement*, Productivity Press.

Shingo, S. and A.G. Robinson (1990). *Modern Approaches to Manufacturing Improvements: The Shingo System*, Productivity Press, MA.

Shobrys, D.E. (1995). "Scheduling." *Planning, Scheduling, and Control Integration in the Process Industries*, C.E. Bodington, ed., McGraw-Hill, New York.

Shore, Barry (1973). *Operations Management*. McGraw-Hill, New York.

Short, J.E. and N. Venkatraman (1992). "Beyond Business Process Redesign: Redefining Baxter's Business Network." *Sloan Management Review*, (Fall), 34 (1), 7–20.

Shostack, G.L. (1984). "Designing Services That Deliver." *Harvard Business Review*, January–February, 133–139.

Shtub, A. (1989). "Estimating the Effects of Conversion to a Group Technology Layout on the Cost of Material Handling." *Engineering Costs and Production Economics*, 16, 103–109.

Shtub, A. (1989). "Modeling group technology cell formation as a generalized assignment problem." *International Journal of Production Research*, 27, 775–782.

Shultz, Terry (1982). "BRP: The Journey to Excellence." *The Forum*, Milwaukee, WI.

Shycon, H.N. and C.R. Sprague (1975). "Put a Price Tag on Your Customer Servicing Levels." *Harvard Business Review*, 53 (July–August), 71–78.

Siha, S. (1994). "The Pull Production System: Modelling and Characteristics." *International Journal of Production Research*, 32, 933–950.

Silver, E.A. (1974). "A Control System for Coordinated Inventory Replenishment." *International Journal of Production Research*, 12 (6), 647–671.

Silver, E.A. (1975). "Modifying the Economic Order Quantity (EOQ) to Handle Coordinated Replenishment of Two or More Items." *Production and Inventory Management*, 16 (3), 26–38.

Silver, E.A. and Peter Kelle (1988). "More on Joint Order Releases Under Dependent Demand." *Production and Inventory Management*, First Quarter.

Silver, E.A. and R. Peterson (1985). *Decision Systems for Inventory Management and Production Planning*, (2nd ed.). Chapters 3, 4, 7–12, and 17. John-Wiley, New York.

Silverberg, G., G. Dosi and L. Orsenigo (1988). "Innovation, diversity and diffusion: a self-organization model." *The Economic Journal*, 98, 1032–54.

Simpson, J.A. and E.S.C. Weiner (1989). *The Oxford English Dictionary*, Second Edition, Clarendon Press, Oxford, UK, 519.

Sinclair, J.M. (1984). "Is the Matrix Really Necessary?" *Project Management Journal*, 15 (1) (March), 49–52.

Singhal, K., C. Fine, J.R. Meredith, and R. Suri (1987). "Research and Models for Automated Manufacturing." *Interfaces*, 17 (6), November–December, 5–14.

Sinha, M.N. and W.O. Wilborn (1985). *The Management Quality Assurance*, John Wiley, New York.

Sinha, S.K. and B.K. Kale (1980). *Life Testing and Reliability Estimation*, John Wiley, New York.

Sirianni, N.C. (1979). "Inventory: How Much Do You Really Need?." *Production Magazine*, November.

Sirianni, N.C. (1981). "Right Inventory Level: Top Down or Bottom UP." *Inventories and Production Magazine*, May–June.

Sisk, G.A. (1986). "Certainty equivalent approach to capital budgeting." *Financial Management*, Winter, 23–32.

Skinner, W. (1969). "Manufacturing – Missing Link in Corporate Strategy." *Harvard Business Review*, 47 (3) (May–June), 136–145.

Skinner, W. (1974). "The Focused Factory." *Harvard Business Review*, 52 (3), 113–121.

Skinner, W. (1978). *Manufacturing in Corporate Strategy*, John Wiley and Sons. New York.

Skinner, W. (1983). "Getting Physical: New strategic Leverage from Operations." *Journal of Business Strategy*, 3, Spring, 74–79.

Skinner, W. (1983). "Wanted: Managers for the Factory of the Future." *Annals of the American Academy of Political and Social Science*, November, 102–114.

Skinner, W. (1984). "Operations Technology: Blind Spot in Strategic Management." *Interfaces*, January–February, 116–125.

Skinner, W. (1985). *Manufacturing the Formidable Competitive Weapon*, John Wiley, New York.

Skinner, W. (1986). "The Productivity Paradox." *Harvard Business Review*, 64 (4), July–August, 55–59.

Skinner, W. (1992). "Missing the Links in Manufacturing Strategy." in: C. Voss (Ed.), *Manufacturing Strategy: Process and Content*, Chapman-Hall, London, 13–25.

Skinner, W. (1996). "Manufacturing strategy on the "S" curve." *Production and Operations Management*, 5 (1) 3–14.

Slack, N. (1987). "The flexibility of manufacturing systems." *International Journal of Operations and Production Management*, 7, 35–45.

Slautterback, W.H. (1985). "Manufacturing in the Year 2000." *Manufacturing Engineering*, August, 57–60.

Slautterback, W.H. and W.B. Werther (1984). "The Third Revolution: Computer Integrated Manufacturing." *National Productivity Review*, Autumn, 367–374.

Sleigh, P. (1993). *The world automotive components industry: A review of leading manufacturers and trends*, Research Report, 1–4, Economist Intelligence Unit, London.

Slocombe, D.S. (1997). *Chaos, Complexity and Self-Organization – An Annotated Bibliography and Review Essay on Management and Design in Complex Systems from Organizations to Ecosystems*, WLU Dept. of Geography & School of Business, Waterloo.

Smith, A.M. (1993). *Reliability centered maintenance*, McGraw-Hill, New York.

Smith, D.J. (1985). *Reliability and Maintainability in Perspective* (2nd ed.). John Wiley, New York.

Smith, L.A. and Joan Mills (1982). "Project Management Network Programs." *Project Management Quarterly*, 18–29.

Smith, L.A. and Sushil Gupta (1985). "*Project Management Software in P&IM." P&IM Review*, 5 (6), 66–68.

Smith, R. (1997). "The Historical Roots of Concurrent Engineering Fundamentals." *IEEE Transactions on Engineering Management*, 44 (1), 67–78.

Smith, R.P. and S.D. Eppinger (1997). "Identifying Controlling Features of Engineering Design Iteration." *Management Science*, 43, 276–293.

Smith, S.B. (1989). *Computer Based Production and Inventory Control*. Prentice-Hall, Englewood Cliffs, NJ.

Smolik, D.P. (1983). *Material Requirements of Manufacturing*. New York, Van Nostrand Reinhold.

Smunt, Timothy L. (1986). "A Comparison of Learning Curve Analysis and Moving Average Ratio Analysis for Detailed Operational Planning." *Decision Sciences*, 17 (4), 475–495.

Smunt, Timothy L. (1986). "Incorporating Learning Curve Analysis into Medium-Term Capacity Planning Procedures: A Simulation Experiment." *Management Science*, 17 (4), 475–495.

Smunt, Timothy L. (1996). "Rough Cut Capacity Planning in a Learning Environment." *IEEE Transactions on Engineering Management*, 43 (3), 334–341.

Smunt, Timothy L. and Thomas E. Morton (1985). "The Effects of Learning on Optimal Lot Sizes: Further Developments on the Single Product Case." *IIE Transactions*, 17 (1), 3–37.

Sohal, A.S., L. Ramsay, and D. Samson (1993). "JIT manufacturing: Industry analysis and a methodology for implementation." *International Journal of Operations and Production Management*, 13 (7), 22–56.

Solberg, J.J. (1976). "Optimal design and control of computerized manufacturing systems." In Proceedings of AIIE Systems Engineering Conference, 138–144.

Solomon, S.L. (1983). *Simulation of Waiting Line Systems*. Prentice-Hall, Englewood Cliffs, NJ.

Son, Y.K. (1992). "A Comprehensive Bibliography on Justification of Advanced Manufacturing Technologies." *Engineering Economist*, 38 (1), 59–71.

Sorenson, R.Z. and U.E. Wiechmann (1975). "How multinationals view marketing standardization." *Harvard Business Review*, May–June, 38–54.

Souder, W. (1987). *Managing New Product Innovations*. Lexington Lexington Books, MA.

Spearman, M.L., D.L. Woodruff, and W.J. Hopp (1990). "CONWIP: a pull alternative to kanban." *International Journal of Production Research*, 28 (5), 879–894.

Spekman, Robert E. (1988). "Strategic supplier selection: Understanding long-term buyer relationships." *Business Horizons*, 80–81.

Spencer, M.S. (1980). "Scheduling Components for Group Technology Lines." *Production and Inventory Management*, Fourth Quarter, 43–49.

Spencer, M.S. and J.F. Cox (1995). "An analysis of the product-process matrix and repetitive manufacturing." *International Journal of Production Research*, 33 (5), 1275–1294.

Spira, J.S. and B.J. Pine II (1993). "Mass Customization." *Chief Executive*, March, 83, 26.

Squires, F.H. (1982). "What Do Quality Control Charts Control?" *Quality*. (November), 63.

Sridharan, V. and R.L. LaForge (1989). "The impact of safety stock on schedule instability, cost, and service." *Journal of Operations Management*, 8 (4), 327–347.

Sridharan, V. and R.L. LaForge (1994). "A model to estimate service levels when a portion of the master production schedule is frozen." *Computers & Operations Research*, 21 (5), 477–486.

Sridharan, V. and R.L. LaForge (1994). "Freezing the master production schedule: Implications for fill rate." *Decision Sciences*, 25 (3), 461–469.

Sridharan, V. and W.L. Berry (1990). "Freezing the master production schedule under demand uncertainty." *Decision Sciences*, 21 (1), 97–120.

Sridharan, V., W.L. Berry and V. Udayabanu (1988). "Measuring master production schedule instability under rolling planning horizons." *Decision Sciences*, 19 (1), 147–166.

Srikanth, M.L. and M.M. Umble (1997). *Synchronous Management: Profit-Based Manufacturing For The 21st Century* – I, Spectrum Publishing, Guilford, CT.

Sriram, V., R. Krapfel, and R. Spekman (1992). "Antecedents to Buyer-Seller Collaboration: An Analysis from the Buyer's Perspective." *Journal of Business Research*, 25, 303–320.

St. John, C.H. and S.T. Young (1991). "The Strategic Consistency Between Purchasing and Production." *International Journal of Purchasing and Materials Management*, (Spring), 15–20.

St. Onge, A. (1997). "Personalized Production: Riding the Information Superhighway Back to the Future." *Supply Chain Management Review*, 1 (1), 44–52.

Stacey, R.D. (1996). *Complexity and Creativity in Organizations*, Berrett-Koehler, San Francisco.

Stalk, G., Jr. (1988). "Time – The Next Source of Competitive Advantage." *Harvard Business Review*, July–August, 41–51.

Stalk, G., Jr. (1988). "Time – The Next Source of Competitive Advantage." *Harvard Business Review*. (July–August), 41–51.

Stalk, G., Jr. and T. Hout (1990). *Competing Against Time*, The Free Press, New York.

Stalk, G., Jr., P. Evans, and L.E. Schulman (1992). "Competing on Capabilities: The New Rules of Corporate Strategy." *Harvard Business Review*, 70 (2), 57–69.

Stamatix, D.H. (1995). *Failure Mode and Effect Analysis, ASQC Quality Press*. Milwaukee, WI.

Starr, M.K. (1965). "Modular production – a new concept." *Harvard Business Review*, November–December, 137–145.

Starr, M.K. (1988). *Global Competitiveness: Getting the U.S. Back on Track*. Norton, New York.

Stecke, K.E. (1983). "Formulation and Solution of Nonlinear Integer Production Planning Problems for Flexible Manufacturing Systems." *Management Science*, 29 (3), 273–288.

Stecke, K.E. (1986). "A Hierarchical Approach to Solving Machine Grouping and Loading Problems of Flexible Manufacturing Systems." *European Journal of Operational Research*, 24, 369–378.

Stecke, K.E. and I. Kim (1991). "A Flexible Approach to Part Type Selection in Flexible Flow Systems Using Part Mix Ratios." *International Journal of Production Research*, 29 (1), 53–75.

Steele L. (1989). "Managing Technology: The Strategic View." New York, McGraw-Hill.

Steele, D.C. (1975). "The nervous MRP system: How to do battle." *Journal of Production and Inventory Management*, 16 (4), 83–89.

Steinbacher, H.R. and N.L. Steinbacher (1993). *TPM for America: What It Is and Why You Need It*. Productivity Press, Cambridge, MA.

Steinberg, E.E., B. Khumawala, and R. Scamell (1982). "Requirements Planning Systems in the Health Care Environment." *Journal of Operations Management*, 2 (4) (August), 251–259.

Steinberg, E.E., W.B. Lee, and B. Khumawala (1980). "A Requirements Planning System for the Space Shuttle Operations Schedule." *Journal of Operations Management*, 1 (2) (November), 69–76.

Stenger, Alan J. and Joseph L. Cavinato (1979). "Adapting MRP to the Outbound Side Distribution Requirements Planning." *Production and Inventory Management*, 20 (4), 1–13.

Sterling, Frank W. (1914) "The Successful Operation of a System of Scientific Management." in Thompson, C. Bertrand, *Scientific Management: A Collection of the More Significant Articles Describing the Taylor System of Management*. Harvard University Press, Cambridge, MA, 296–365.

Stevenson, W.J. (1989). *Introduction to Management Science*. Irwin, Homewood, IL.

Stevenson, W.J. (1996). *Production/Operations Management* (5th ed.). Irwin, Homewood, IL.

Steward, D.V. (1981). *Systems Analysis and Management: Structure, Strategy and Design*. Petrocelli Books, New York.

Stewart, T. (1995). "After all you've done for you customers, why are they still NOT HAPPY?" *Fortune* (December 11), 178–818.

Stewart, T.A. (1997). "A Satisfied Customer Isn't Enough." *Fortune*, 136 (July 21), 112–113.

Stipp, D. (1996). "The Birth of Digital Commerce." *Fortune*. 134, 159–164.

Stobaugh, R. and P. Telesio (1983). "Match Manufacturing Policies and Product Strategies." *Harvard Business Review*, March–April, 113–120.

Stock, J. and Lambert, D. (1992). "Becoming a "World Class" Company With Logistics Service Quality." *International Journal of Logistics Management*, 3 (1), 73–80.

Stone, P.J. and R. Luchetti (1985). "Your Office Is Where You Are." *Harvard Business Review* (March/April), 102–117.

Storhagen, N.G. and R. Hellberg (1987). "Just-in-Time from a Business Logistics Perspective." *Engineering Cost and Production Economics* (The Netherlands), 12 (1–4) (July), 117–121.

Strayer, J.K. (1989). *Linear Programming and Its Applications*. Springer-Verlag, New York.

Sugimori, Y., K. Kusunoki, F. Cho, and S. Uchikawa (1977). "Toyota production system and kanban system materialization of just-in-time and respect-for-human system." *International Journal of Production Research*, 15, 553–564.

Sullivan, L. (1995). "Custom-made" *Forbes*, 156, 124–125.

Sullivan, L.P. (1986). "The Seven Stages in Company-Wide Quality Control." *Quality Progress*, May, 77–83.

Sullivan, L.P. (1986). "Quality Function Deployment." *Quality Progress*, 19 (6) (June), 36–50.

Sullivan, L.P. (1988). "Policy Management Through Quality Functional Deployment." *Quality Progress*. June, 18–20.

Sum, C.C., J.S.K. Ang, and L.N. Yeo (1997). "Contextual elements of critical success factors in MRP implementation." *Production and Inventory Management Journal*, 38 (3), 77–83.

Sum, C.C., K.K. Yang, J.S.K. Ang, and S.A. Quek (1995). "An analysis of material requirements planning (MRP) benefits using alternating conditional expectation (ACE)." *Journal of Operations Management*. 13 (1), 35–58.

Supplier (1996). "Extended Enterprise races to make special-edition Dodge Ram a reality." newsletter published by the Chrysler Corporation, March–April, 6.

Suresh, N.C. (1990). "Towards an integrated evaluation of flexible automation investments." *International Journal of Production Research*, 28, 1657–1672.

Suresh, N.C. and J.R. Meredith (1985). "Achieving Factory Automation Through Group Technology Principles." *Journal of Operations Management*, 5 (3) (February), 151–167.

Suri, R. and R. Desiraju (1977). "Performance analysis of Flexible Manufacturing Systems with a Single Discrete Material – Handling Device." *The International Journal of Flexible manufacturing Systems*. 9, 223–249.

Suri, R., G.W.W. Diehl, S. De Treville, and M.J. Tomsicek (1995). "From CAN-Q to MPX: evolution of queueing software for manufacturing." *Interfaces*. 25, 128–150.

Susman, G.I. (1992). *Integrating Design and Manufacturing for Competitive Advantage*. Oxford University Press, New York.

Susman, G.I. and R.B. Chase (1986). "A Sociotechnical Analysis of the Integrated Factory." *Journal of Applied Behavioral Science* 22 (3), 257–270.

Suzaki, K. (1985). "Japanese Manufacturing Techniques, Their Importance to U.S. Manufacturers." *Journal of Business Strategy*, 5 (3), 10–20.

Suzaki, K. (1987). *The New Manufacturing Challenge: Techniques for Continuous Improvement*. The Free Press, New York.

Suzuki, T. (1992). *New Directions for TPM*, Productivity Press, Cambridge, MA:

Swain, J.J. (1995). "Simulation Survey: Tools for process understanding and improvement." *OR/MS Today*. 22 (4), 64–72 and 74–79.

Swain, M.R. and E.L. Denna (1998). "Making new wine: Useful management accounting systems." *International Journal of Applied Quality Management*, 1, 25–44.

Swamidass, P.M. (1987). "Manufacturing Strategy: Its assessment and practice." *Journal of Opeations Management*, 6 (4), 471–484.

Swamidass, P.M. (1987). "Planning for manufacturing technology." *Long Range Planning*, 20 (5), 125–133.

Swamidass, P.M. (1988). *Manufacturing Flexibility. Operations Management* Association Monograph 2, Naman and Schneider Associates Group, Texas.

Swamidass, P.M. (1990). "A Comparison of Plant Location Strategies of Foreign and Domestic Manufacturers in the U.S." *Journal of International Business Studies*, 8 (3), 263–277.

Swamidass, P.M. (1991). "Empirical Science: The New Frontier in Operations Management." *Academy of Management Review*, 16 (4), 793–814.

Swamidass, P.M. (1993). "Import Sourcing Dynamics: An Integrative Perspective." *Journal of International Business Studies*, 24 (4).

Swamidass, P.M. (1995). "Making sense out of manufacturing innovations." *Design Management Journal*, 6 (2), 52–56.

Swamidass, P.M. (1996). "Benchmarking Manufacturing Technology Use in the United States." *Handbook of Technology Management*, Gerard H. "Gus" Gaynor (ed.). 37.1–37.40, McGraw-Hill Inc., New York.

Swamidass, P.M. (1998). *Technology on the Factory Floor III*. Manufacturing Institute of the National Association of Manufacturers, Washington, D.C.

Swamidass, P.M. and C. Majerus (1991). "Statistical Control of Cycle Time and Project Time: Lessons from stratistical process control." *International Journal of Production Research*, 29 (3), 551–564.

Swamidass, P.M. and M. Waller (1990). "A classification of approaches to planning and justifying new manufacturing technologies." *Journal of Manufacturing Systems*, 9 (3), 181–193.

Swamidass, P.M. and S. Kotha (1998). "Explaining manufacturing technology use, firm size and performance using a mulitdimensional view of technology." *Journal of Operations Management*, 17, 23–37.

Swamidass, P.M. and W.T. Newell (1987). "Manufacturing Strategy, Environmental Uncertainty and Performance: A Path Analytic Model." *Management Science*, 33 (4), 509–524.

Swamidass, P.M., S.S. Nair, and S.I. Mistry (1999). "The use of a nueral factory to investigate the effect of product line width on manufacturing performance." *Management Science*, forthcoming.

Swann, Don (1989). "What is CIM, and Why Does it Cost $40 Million." *P&IM Review*, July, 34.

Swink, M.L. and M. Way (1995). "Manufacturing Strategy: Propositions, Current Research, Renewed Directions." *International Journal of Operations and Production Management*, 15 (7), 4–27.

Swink, M.L., J.C. Sandvig, and V.A. Mabert (1996). "Adding Zip to Product Development: Concurrent Engineering Methods and Tools." *Business Horizons*, 39 (2), 41–49.

Swink, M.L., J.C. Sandvig, and V.A. Mabert (1996). "Customizing Concurrent Engineering Processes: Five Case Studies." *Journal of Product and Innovation Management*, 13 (3), 229–244.

Szilagyi, A.D. and M.J. Wallace (1987). *Organizational Behavioral and Performance*. Scott, Foresman, Glenview IL.

T

Tadjer, R. (1988). "Enterprise resource planning." *Internet Week*, 710, 40–41.

Taguchi, Genichi (1986). *Introduction to Quality Engineering: Designing Quality into Products and Processes*. Asian Productivity Organization, White Plains, NY.

Taha, H.A. (1987). *Operations Research: An Introduction* (4th ed.). Prentice-Hall, New York.

Tajiri, M. and F. Gotoh (1992). *TPM Implementation: A Japanese Approach*. McGraw-Hill, Inc., New York.

Takahashi, K. and N. Nakamura (1996). "Reactive JIT ordering systems for unstable demand changes." *Proceedings of Advances in Production Management Systems*, Information Processing Society of Japan.

Takahashi, K. and N. Nakamura (1997). "A study on reactive kanban systems." *Journal of Japan Industrial Management Association* (in Japanese). 48, 159–165.

Takeuchi, H. and M.E. Porter (1986). "Three roles on international marketing global strategy." In M.E. Porter (ed.). *Competition in Global Industries*. Harvard Business School Press, Boston, 111–146.

Takeuchi, H. and I. Nonaka (1986). "The New Product Development Game." *Harvard Business Review*, 64, 137–146.

Talbot, Braian and James Patterson (1984). "An Integer Programming Algorithm with Network Cuts for Solving the Assembly Line Balancing Problem." *Management Science*, 30 (1), 85–99.

Taleb, K., S.M. Gupta, and L. Brennan (1997). "Disassembly of Complex Products with Parts and Materials Commonality." *Production Planning and Control*, 8 (3), 255–269.

Tanaka, T. (1993). "Target Costing at Toyota." *Journal of Cost Management*, Spring, 4–11.

Tapiero, C.S. (1986). "Continuous Quality Production and Machine Maintenance." *Naval Research Logistics Quarterly*, 33, 489–499.

Tardif, V. and L. Maaseidvaag (1997). *Adaptive single-stage pull control systems*. University of Texas at Austin, Graduate Program in Operations Research and Industrial Engineering, Austin, TX.

Taylor Bernard III, Frederick S. Hills, and K. Roscoe Davis (1979). "The Effects of Change Factors on the Production Operation and the Production Manager." *Production and Inventory Management*, 20 (3), 18–32.

Taylor, Fred. W. (1886). "Discussion of 'The Shop Order System of Accounts" (by Henry Metcalfe). *ASME Transactions*, 7, 475–6 and 484–485.

Taylor, Fred. W. (1894). "Notes on Belting." *ASME Transactions*, 15, 204–259.

Taylor, Fred. W. (1895). "A Piece-Rate System, Being a Step Toward Partial Solution of the Labor Problem." *ASME Transactions*, 16, 856–903.

Taylor, Fred. W. (1903). "Shop Management." *ASME Transactions*, 24, 1337–1480.

Taylor, Fred. W. (1906). "On the Art of Cutting Metals." *ASME Transactions*, 27, 31–279, with "Discussions." 280–350.

Taylor, Fred. W. (1911). *Principles of Scientific Management*. Harper and Row, New York.

Taylor, Fred. W. (1911). *Shop Management*. Harper & Brothers Publishers, New York, New York.

Taylor, S.G. and S.F. Bolander (1994). *Process Flow Scheduling: A Scheduling Systems Framework for Flow Manufacturing*. APICS, Falls Church, VA.

Taylor, S.G. and S.F. Bolander (1997). "Process Flow Scheduling: Past, Present, and Future." *Production and Inventory Mangement Journal*, 38 (2), 21–25.

Tayur, S.R. (1992). "Properties of serial kanban systems." *Queueing Systems*, 12, 297–318.

Tayur, S.R. (1993). "Structural Properties and a Heuristic for Kanban-Controlled Serial Lines." *Management Science*, 39, 1347–1368.

Tent, J.T.C., V. Grover, and K.D. Fiedler (1996). "Developing Strategic Perspectives on Business Process Reengineering: From Process Reconfiguration to Organizational Change." Omega, *International Journal of Management Science*, 24 (3), 271–294.

Terez, T. (1990). "A Manager's Guidelines for Implementing Successful Operational Changes." *Industrial Management*, 32 (4), 18–20.

Tersine, R.J. (1987). *Principles of Inventory and Materials Management* (3rd ed.). North-Holland, New York.

Terwiesch, C. and C.H. Loch (1997). "Management of Overlapping Development Activities: a Framework for Exchanging Preliminary Information." Proceedings of the 4th EIASM Conference on Product Development, 797–812.

Terwiesch, C., C.H. Loch, and M. Niederkofler (1996). "Managing Uncertainty in Concurrent Engineering." Proceedings of the 3rd EIASM Conference on Product Development, 693–706.

Tharumarajah, A. and A.J. Wells (1997). "A behaviour-based approach to scheduling in distributed manufacturing systems." *Integrated Computer Aided Engineering*, 4, 235–49.

Thilander, M. (1992). "Some Observations of Operation and Maintenance in Two Swedish Firms." *Integrated Manufacturing Systems*, 3 (2).

Thomas, D.R.E. (1978). "Strategy Is Different in Service Businesses." *Harvard Business Review*. July–August, 158–165.

Thompson, C. Bertrand (1914). *Scientific Management: A Collection of the More Significant Articles Describing the Taylor System of Management*, Harvard University Press, Cambridge, MA.

Thompson, C. Bertrand (1917). *The Taylor System of Scientific Management*. A.W. Shaw Co., Chicago, IL.

Thompson, G. (1968). "Optimal Maintenance Policy and Sale Date of a Machine." *Management Science*. 14 (9), 543–550.

Thompson, K. (1983). "MRP II in the Repetitive Manufacturing Environment." *Production and Inventory Management*, 24 (4) (Fourth Quarter), 1–14.

Thurow, L.C. (1992). "Who Owns the Twenty-First Century?" *Sloan Management Review*, 5–17, Spring.

Timbers, M.J. (1992). "ISO 9000 and Europe's Attempts to Mandate Quality." *The Journal of European Business*, 14–25.

Timm, P.R. and B.D. Peterson (1986). *People at Work: Human Relations in Organizations*, West, St. Paul, MN.

Toffler, Alvin (1971). *Future Shock*. Bantam Books, New York.

Tombari, H. (1982). "Designing a Maintenance Management System." *Production and Inventory Management*, 23 (4), 139–147.

Tomlinson, P.D. (1987). "Organizing for Productive Maintenance." *Production Engineering*, 34 (10) (October), 38–40.

Tompkins, J.A., J.A. White, Y.A. Bozer, E.H. Frazelle, J.M.A. Tanchoco, and J. Trevino (1996). *Facilities Planning*, Wiley, NY, p. 1–17.

Tong, H.M. and C.K. Walter (1980). "An Empirical Study of Plant Location Decisions of Foreign Manufacturing Investors in the United States." *Columbia Journal of World Business*, (Spring), 66–73.

Toomey, J.W. (1996). *MRP II: Planning for Manufacturing Excellence*. Chapman and Hall, New York.

Townsend, P.L. (1986). *Commit to Quality*. John Wiley, New York.

Toyoda, Eiji (1987). *Toyota: Fifty Years in Motion*. Kodansha International, Tokyo, Japan.

Triantis, A.J. and J.E. Hodder (1990). "Valuing flexibility as a complex option." *Journal of Finance*, XLV, 549–565.

Trigeorgis, L. (1996). *Real Options: Managerial Flexibility and Strategy in Resource Allocation*. The MIT Press, MA.

Tsuchiya, S. (1992). *Quality Maintenance: Zero Defects Through Equipment Management*, Productivity Press, Cambridge, MA.

Tully, S. (1995). "Purchasing's New Muscle." *Fortune* (February 20), 75–83.

Turban, E. (1990). *Decision Support and Expert Systems: Management Support Systems*. MacMillan, New York.

Turban, E. and J. Aronson (1998). *Decision Support Systems and Intelligent Systems*, Fifth edition, Prentice-Hall Inc., Upper Saddle River, New Jersey.

Turban, E. and J.R. Meredith (1991). *Fundamentals of Management Science* (5th ed.). Irwin, Homewood, IL.

Turbide, D.A. (1993). "Total Productive Maintenance Finds Its Place in the Sun." *APICS – The Performance Advantage*, August, 32–34.

Turney, P.B.B. (1992). *Common Cents: The ABC Performance Breakthrough*. Cost Technology, Portland, OR.

Twigg, D. and N. Slack (1998). "Lessons from Using Supplier Guest Engineers in the Automotive Industry." *International Journal of Logistics, Research and Applications*, 1 (2), 181–192.

Twiss, B.C. (1980). *Managing Technological Innovation*, 2nd ed., London, Longman Group.

U

U.S. Department of Commerce, International Trade Administration. Office of Capital Goods and International Construction Sector Group (1985). *A Competitive Assessment of the U.S. Flexible Manufacturing System Industry*. U.S. Government Printing Office, Washington, D.C.

U.S. Department of Commerce (1987). Washington, D.C., *A Competitive Assessment of the U.S. Flexible Manufacturing System Industry*, July.

U.S. Department of Commerce (1998). Malcolm Baldrige National Quality Award 1998 Criteria for Performance Excellence, Washington, D.C., GPO.

U.S. Department of Commerce (undated). Malcolm Baldrige National Quality Award, Profiles of Winners.

Ulrich, K.T. (1995). "The Role of Product Architecture in the Manufacturing Firm." *Research Policy*, 24, 419–440.

Ulrich, K.T. and K. Tung (1991). "Fundamentals of Product Modularity." Proceedings of the 1991 ASME Winter Annual Meeting Symposium on Issues in Design/Manufacturing Integration, Atlanta.

Ulrich, K.T. and S.D. Eppinger (1995). *Product Design and Development*, McGraw-Hill, NY, 77–127.

Ulvill, J. and R.V. Brown (1982). "Decision Analysis Comes of Age." *Harvard Business Review* (September–October), 130–141.

Umble, M. and M.L. Srikanth (1990). *Synchronous Manufacturing: Principles for World Class Excellence*. South-Western Publishing, Cincinnati, OH.

Umble, M. and M.L. Srikanth (1995). *Synchronous Manufacturing*, Spectrum Publishing, CT.

Umble, M. and M.L. Srikanth (1997). *Synchronous Management: Profit-Based Manufacturing For The 21st Century* – Volume II, Implementation Issues and Case Studies, Spectrum Publishing, Guilford, Connecticut.

Upton, D.M. (1994). "The Management of Manufacturing Flexibility." *California Management Review*. Winter, 72–89.

Upton, D.M. (1995). "Flexibility as process mobility: The management of plant capabilities for quick response manufacturing." *Journal of Operations Management*, 12 (3&4), June, 205–224.

Upton, D.M. (1995). "What Really Makes Factories Flexible." *Harvard Business Review*, July–August, 74–79.

Upton, D.M. (1997). "Process range in manufacturing: An empirical study of flexibility." *Management Science*, 4 (8), August, 1079–1092.

Uttal, B. (1987). "Speeding New Ideas to Market." *Fortune*, March 2, 62–66.

Utterback J. (1972). "The Process Of Technological Innovation Within Firms." *Academy Of Management Journal*.

Utterback, J.M. and W.J. Abernathy (1975). "A dynamic model of process and product innovation." *OMEGA*, 3 (6), 639–656.

Utterback, J.M. and W.J. Abernathy (1981). "Multivariant Models for Innovation/Looking at the Abernathy and Utterback Model with Other Data." *OMEGA*, 9.

V

Vakharia, Asoo J. (1986). "Methods of Cell Formation in Group Technology: A Framework for Evaluation." *Journal of Operations Management*, 5 (3) (February), 151–167.

Valdez-Flores, C. and R.M. Feldman (1989). "A Survey of Preventative Maintenance Models for Stochastically Deteriorating Single-Unit Systems." *Naval Research Logistics Quarterly*, 36, 419–446.

Valentine, Lloyd M. and Carl A. Dauten (1983). *Cycles and Forecasting*. South-Western, Cincinnati.

Van De Ven, Andrew H. (1986). "Central Problems in the Management of Innovation." *Management Science*, 32 (5) (May), 590–607.

Van Dierdonck, R. and J.G. Miller (1980). "Designing Production Planning and Control Systems." *Journal of Operations Management*, 1 (1) (August), 37–46.

Van Dijk, N.M. (1993). *Product Forms for Queueing Networks: A System's Approach*. Wiley, New York.

Van Gigch, John P. (1978). *Applied General Systems Theory*. Harper & Rowe, New York.

Van Slyke, Richard M. (1963). "Monte Carlo Methods and the PERT Problem." *Operations Research*, 11 (5) (September–October), 129–143.

Vanneste, S.G. and L.M. Van Wassenhove (1995). "An integrated and structured approach to improve maintenance." *European Journal of Operational Research*, 82, 241–257.

Varughese, K.K. (1993). "Total Productive Maintenance." University of Calgary, A Thesis for the Degree of Master of Mechanical Engineering.

Veerakamolmal, P. and S.M. Gupta (1998). "Disassembly Process Planning." *Engineering Design and Automation*. (forthcoming).

Veeramani, D., J.J. Bernardo, C.H. Chung, and Y.P. Gupta (1995). "Computer-Integrated Manufacturing: A Taxonomy of Integration and Research Issues." *Production and Operations Management*, 4 (4), 360–380.

Venkatraman, N. and Zaheer, A. (1990). "Electronic Integration and Strategic Advantage: A Quasi-experimental Study in the Insurance Industry." *Information Systems Research*, 1, 377–393.

Venketesan, R. (1990). "Cummins flexes its Factory. *Harvard Business Review*, 90 (2), March–April, 120–127.

Vinrod, B. and T. Altiok. (1986). "Approximating Unreliable Quewing Networks Under the Assumption of Exponentiality." *Journal of the Operational Research Society* (March), 309–316.

Viswanadham, N. and Y. Narahari (1992). *Performance Modeling of Automated Manufacturing Systems*. Prentice-Hall, New Jersey.

Vollman, T.E. (1986). "OPT as an Enhancement to MRP II." *Production and Inventory Management*, 27 (2), 38–47.

Vollman, T.E., W.L. Berry, and D.C. Whybark (1997). *Manufacturing Planning and Control Systems* (4th ed.). Chapters 11, 17, and 18, Irwin/McGraw-Hill, New York.

Vollmann T.E., W.L. Berry, and D.C. Whybark (1997). *Manufacturing Planning and Control Systems* (4th ed.). Irwin/McGraw Hill.

Vollmann, T.E. (1986). "OPT as an Enhancement to MRP II." *Production and Inventory Management*, 27 (2), 38–47.

Von Winterfeldt, D. and W. Edwards (1986). *Decision Analysis and Behavioral Research*, Cambridge University Press, New York.

Vonderembse, Mark A. and Gregory S. Wobser (1987). "Steps for Implementing a Flexible Manufacturing System." *Industrial Engineering*, 15 (11) (November), 58–64.

Voss, C.A. (1986). Managing New Manufacturing Technologies, Monograph 1, *Operations Management* Association, September.

W

Wagner, Harvey (1974). *Operations Research* (2nd ed.). Prentice-Hall, Englewood Cliffs, N.J.

Wagner, Harvey and Thompson M. Whitlin (1958). "Dynamic Version of the Economic Lot Size Model." *Management Science*, 5, 89–96.

Waldman, K.H. (1983). "Optimal Replacement Under Additive Damage in Randomly Varying Environments." *Naval Research Logistics Quarterly*, 30, 377–386.

Waldron, D. (1988). "Accounting for CIM: The new yardsticks." *EMAP Business and Computing Supplement* (February), 1–2.

Wallace, T.F. (1990). *MRPII: Making It Happen*. Oliver Wight Limited Publications, Inc.

Walleigh, R.C. (1986). "Getting Things Done: What's Your Excuse for Not Using JIT?" *Harvard Business Review* (March–April), 36–54.

Walrand, J. (1988). *An Introduction to Queuing Networks*. Prentice-Hall, Englewood Cliffs, NJ.

Walters, P.G. (1986). "International marketing policy, a discussion of the standardization construct and its relevance for corporate policy." *Journal of International Business Studies*, 17, 55–69.

Walters, P.G. and B. Toyne (1989). "Product modification and standardization in international markets: Strategic options and facilitating policies." *Columbia Journal of World Business*, Winter, 37–44.

Walton, L.W. (1996). "The ABC's of EDI: The Role of Activity-Based Costing (ABC) in Determining EDI Feasibility in Logistics Organizations." *Transportation Journal*, 36, 43–50.

Walton, R.E. and G.I. Susman (1987). "People Policies for the New Machines." *Harvard Business Review* (March–April), 98–106.

Wantuck, K. (1989). *Just In Time For America*, KWA Media, Southfield, MI.

Wantuck, K. (1978). "Calculating Optimum Inventories." Proceedings 21st Annual Conference, American *Production and Inventory Control* Society.

Ward, A., J.K. Liker, J.J. Cristiano, and D.K. Sobek II (1995). "The Second Toyota Paradox: How Delaying Decisions Can Make Better Cars Faster." *Sloan Management Review*, Spring, 43–61.

Warnecke, H.J. (1993). *The Fractal Company*. Springer-Verlag, New York, N.Y.

Wasserman, A.S. (1989). "Redesigning Xerox: A Design Strategy Based on Operability" in *Ergonomics: Harness the Power of Human Factors in Your Business*, ed Klemmer, E.T., Norwood, N.J.

Wassweiler, William R. (1980). "Fundamentals of Shop Floor Control." *APICS Conference Proceedings*, 352–354.

Waterman, R.H., Jr. (1987). *The Renewal Factor: How to Get and Keep the Competitive Edge*. Bantam, Toronto.

Watson Hugh J. (1981). *Computer Simulation in Business*. Wiley, New York.

Watson, H.J. and J.H. Blackstone, Jr. (1989). *Computer Simulation* (2nd ed.). John Wiley, New York.

Watson, S.R. and D.M. Buede (1987). *Decision Synthesis*. Cambridge University Press, New York.

Weatherford, L.R. and S.E. Bodily (1992). "A taxonomy and research overview of perishable-asset revenue management: Yield management, overbooking, and pricing." *Operations Research*, 40 (5), 831–844.

Weatherston, D. (1986). "Transportation Makes the JIT Connection." *Industrial Management* (April), 40–44.

Webby, R. and M. O'Connor (1996). "Judgmental and statistical time series forecasting: A review of the literature." *International Journal of Forecasting*, 12, 91–118.

Webster, S.E. (1997). "ISO 9000 Certification, A Success Story at Nu Visions Manufacturing." *IIE Solutions*. 18–21.

Wein, Lawrence M. (1988). "Scheduling semiconductor wafer fabrication." *IEEE Transactions on Semiconductor Manufacturing*, 1 (3), 115–130.

Wein, Lawrence M. and Philippe B. Chevalier (1992). "A broader view of the job shop scheduling problem." *Management Science*, 38 (7), 1018–1033.

Weinberg, G.M. (1975). *An Introduction to General Systems Theory*. Wiley, New York.

Weiss, A. (1984). "Simple Truths of Japanese Manufacturing." *Harvard Business Review* (July–August), 119–125.

Weist, Jerome D. and K. Levy Ferdinand (1977). *A Management Guide to PERT/CPM* (2nd ed.). Prentice-Hall, Englewood Cliffs, NJ.

Wellstead, P.E. (1979). *Introduction to Physical System Modeling*. Academic Press, London.

Wemmerlov, U. (1984). *Case Studies in Capacity Management. American Production and Inventory Control Society*, Virginia.

Wemmerlov, U. and N.L. Hyer (1986). "The Part Family/ Machine Group Identification Problem in Cellular Manufacturing." *Journal of Operations Management*. 6(2), 125–147.

Wen, H.J., C.H. Smith, and E.D. Minor (1996). "Formation and dynamic routing of part families among flexible manufacturing cells." *International Journal of Production Research*, 34, 2229–2245.

Wernerfelt, B. (1984). "A Resource-based View of the Firm." *Strategic Management Journal*, 5, 171–180.

Weston, Frederick C. Jr. (1980). "A Simulation Approach to Examining Traditional EOQ/EOP and Single Order Exponential Smoothing Efficiency Adopting a Small Business Perspective." *Production and Inventory Management*, 21 (2), 67–83.

Westra, D., M.L. Srikanth, and M. Kane (1996). "Measuring operational performance in a throughput world." *Management Accounting* (April), 41–47.

Wever, G.H. (1997). *Strategic Environmental Management: Using TQEM and ISO 14000 for Competitive Advantage*. John Wiley, New York.

Wexley, K.N. and G.P. Latham (1981). *Developing and Training Human Resources in Organizations*. Scott, Foresman, Glenview IL.

Wheelright, S.C. (1978). "Reflecting Corporate Strategy in Manufacturing Decisions." *Business Horizons*, 21, February, 57–66.

Wheelwright, S.C. (1981). "Japan – Where Operations Really Are Strategic." *Harvard Business Review*. July–August, 67–74.

Wheelwright, S.C. (1984). "Manufacturing Strategy: Defining the Missing Link." *Strategic Management Journal*, 5 (1), 77–91.

Wheelwright, S.C. and R.H. Hayes (1985). "Competing Through Manufacturing." *Harvard Business Review*, January–February, 99–109.

Wheelwright, S.C. and S.G. Makridakis (1985). *Forecasting Methods for Management* (4th ed.) John Wiley, New York.

Wheelwright, S.C. and K.B. Clark (1992). *Revolutionizing Product Development*. The Free Press, New York.

White, E.M. (1980). "Implementing an MRP system using the Lewin-Schein theory of change." *Production and Inventory Management*, 21 (1), 1–2.

White, E.M., J.C. Anderson, R.G. Schroeder, and S.E. Tupy (1982). "A study of the MRP implementation process." *Journal of Operations Management*, 2 (3), 145–153.

White, G.P. and J.A. Ponce de Leon (1998). *Performance Measurement Systems: A Model and Seven Case Studies*. APICS Education & Research Foundation, VA.

White, Jeffrey (1995). "Learning Pricing Moves and Strategies." *NAPM Insights*, September, 16–19.

White, R.E. (1993). "An empirical assessment of JIT in U.S. manufacturers." *Production and Inventory Management Journal*, 34 (2), 38–42.

White, R.E. and W.A. Ruch (1990). "The composition and scope of JIT." *Operations Management Review*. 7 (3&4), 9–18.

Whitney, D.E. (1988). "Manufacturing by Design." *Harvard Business Review*, July–August, 83–91.

Whitson, Kevin (1996). "Scientific Management and Production Management Practice in Britain Between the Wars." *Historical Studies in Industrial Relations*, 1, 47–75.

Whitt, W. (1983). "The Queueing Network Analyzer." *The Bell System Technical Journal*, 62, 2779–2815.

Whybark, C. and J.G. Williams (1976). "Material Requirements Planning under Uncertainty." *Decision Sciences*, 7 (1), 595–606.

Wiersema, W.H. (1996). "Implementing Activity-Based Management: Overcoming the Data Barrier." *Journal of Cost Management*, Summer, 17–21.

Wiest, Jerome D. (1966). "Heuristic Programs for Decision Making." *Harvard Business Review*, September–October, 129–143.

Wight, O.W. (1974). *Production and Inventory Management in the Computer Age*. Cahner Books, Boston, MA.

Wight, O.W. (1970). "Input-output control, a real handle on lead time." *Production and Inventory Management*, 11, 9–31.

Wight, O.W. (1982). *The Executive's Guide to Successful MRP II*. Oliver Wight Publishing, Essex VT.

Wight, O.W. (1984). *MRP II: Unlocking America's Productivity Potential*. CBI, Boston, MA.

Wild, Ray (1974). The Origins and Development of Flow-Line Production. *Industrial Archaeology Review*, 11, 43–55.

Wildemann, Horst (1987). "JIT Progress in West Germany." *Target*, Summer, 23–27.

Williams, Vearl A. (1985). "FMS in Action." *Production*, January, 41–43.

Willis, R.E. (1987). *A Guide to Forecasting for Planners and Managers*. Prentice-Hall, Englewood Cliffs, NJ.

Wilson, James M. (1995). "An Historical Perspective on Operations Management." *Production and Inventory Management Journal*, 36, 61–66.

Wilson, James M. (1996). "Henry Ford: A Just-in-Time Pioneer. "*Production and Inventory Management* Journal, 37, 26–31.

Winston, W.L. (1996). *Simulation Modeling Using @RISK*. Duxbury Press, Belmont, CA.

Winters, P.R. (1960). "Forecasting Sales by Exponentially Weighted Moving Average." *Management Science*, 6, April, 324–342.

Wireman, T. (1984). *Preventive Maintenance*. Englewood Cliffs, NJ, Reston.

Wisner, J.D. and S.E. Fawcett (1991). "Linking Firm Strategy to Operating Decisions Through Performance Measurement." *Production and Inventory Management Journal*, 32 (3), 5–11.

Womack, J.P. and D.T. Jones (1994). "From lean production to the lean enterprise." *Harvard Business Review*, 72, March–April, 93–103.

Womack, J.P., D.T. Jones, and D. Roos (1991). *The Machine that Changed the World: The Story of Lean Production*. Harper Collins, New York, 37–38, 91–93.

Woodhouse, J. (1997). "The MACRO project: cost/risk evaluation of engineering and management decisions." In: C. Guedes Soares (ed.). Advances in Safety and Reliability. Pergamon Press, Oxford, UK, 237–246

Works, M. (1998). "GenCorp Aerojet says time is right for SAP in the aerospace and defense industries." *Automatic I.D. News*, 14 (2), 50–52.

Wortmann, J.C. (1992). "Production Management Systems for One-of-a-Kind Products." *Computers in Industry*, 19, 79–88.

Wrege, Charles D., and Ronald G. Greenwood (1991). *Frederick W. Taylor, the Father of Scientific Management: Myth and Reality*, Richard D. Irwin, Homewood, IL.

Wren, Daniel (1994). *The Evolution of Management Thought*. John Wiley & Sons, New York, New York.

Wright, T.P. (1936). "Factors Affecting the Cost of Airplanes." *Journal of Aeronautical Sciences*, 3 (4), 122–128.

Wright, Williams and Kelly (1997). *Factory Explorer User's Manual*, Version 2.2. Dublin, CA.

Wygant, Robert M. (1987). "Improving Productivity with Financial Incentives." *Engineering Management International*, 4, 87–93.

Y

Yacura, J.A. (1994). "The Expert System: Purchasing's Tool of the Future." *NAPM Insights* (October), 32–33.

Yao, D.D. (1985). "Material and information flows in flexible manufacturing systems." *Material Flow*, 2, 143–149.

Yao, D.D. (1994). *Stochastic Modeling and Analysis of Manufacturing Systems*. Springer-Verlag, New York.

Yelle, Y.E. (1979). "The Learning Curve: Historical Review and Comprehensive Survey." *Decision Sciences* 10 (2) (April), 302–328.

Yip, G.S. (1989). "Global strategy.... in a world of nations?" *Sloan Management Review*, 31, 29–41.

Yücesan, E. (1993). "Randomization tests for initialization bias in simulation output." *Naval Research Logistics*, 40, 643–664.

Yücesan, E. and S.H. Jacobson (1996). "Computational issues for Accessibility in discrete event simulation." *ACM Transactions on Modeling and Computer Simulation*, 6, 53–75.

Yurkiewicz, J. (1993). "Forecasting software: Clearing up a cloudy picture." *OR/MS Today*, 2, 64–75.

Z

Zandin, K. (1980). *MOST Work Measurement Systems*. Marcel Dekker, New York.

Zeithaml, V.A., L.L. Berry, and A. Parasuraman (1996). "The Behavioral Consequences of Service Quality." *Journal of Marketing*, 60, 31–46.

Zelenovic, D.M. (1982). "Flexibility – a condition for effective production systems." *International Journal of Production Research*, 20, 319–337.

Zhao, X. and T.S. Lee (1993). "Freezing the master production schedule for material requirements planning systems under demand uncertainty." *Journal of Operations Management*, 11 (2), 185–205.

Zhou, M. and F. DiCesare (1993). *Petri Net Synthesis for Discrete Event Control of Manufacturing Systems*. Kluwer, Boston, MA.

Zimmerman, Gary (1975). "The ABC's of Wilfredo Pareto." *Production Inventory Management*, 16 (3), 1–9.

Zisk, B.I. (1983). "Flexibility is Key to Automated Material Transport System for Manufacturing Cells." *Industrial Engineering*, 15 (11) (November), 58–64.

Zuckerman, A. (1997). "ISO/QS-9000 Registration Issues Heating Up Worldwide." *The Quality Observer*, June, 21–23.

Zurawski, R. (1994). "Systematic construction of functional abstractions of Petri net models of flexible manufacturing systems." *IEEE Transactions on Industrial Electronics*, 41, 584–591.

Zurawski, R. and M. Zhou (1994). "Petri nets and industrial applications: a tutorial." *IEEE Transactions on Industrial Electronics*, 41, 567–583.

Appendix II

Topical Bibliography

List of Topics

(Alphabetical Order)

1. **Capacity planning and forecasting**
2. **Costing, finance and investment**
3. **Flexible manufacturing**
4. **History and overview of manufacturing systems**
5. **Human issues and teams**
6. **Layout and manufacturing cells**
7. **Manufacturing analysis**
8. **Manufacturing performance**
9. **Manufacturing technology**
10. **Material and inventory management**
11. **Modern manufacturing ideas: Reengineering, JIT, lean, agile manufacturing**
12. **Plant location**
13. **Plant maintenance**
14. **Product development and design**
15. **Production scheduling, and control**
16. **Project management**
17. **Quality**
18. **Service**
19. **Software and computer integration**
20. **Strategic, environmental and international issues**
21. **Supply chain and logistics**

Capacity Planning and Forecasting

Allen, S.J. and E.W. Schuster (1994). "Practical Production Scheduling with Capacity Constraints and Dynamic Demand: Family Planning and Disaggregation." *Production and Inventory Management*, Fourth quarter, 15–21.

Anthony, Robert N. (1965). *Planning and Control Systems: A Framework for Analysis*. Cambridge, MA, Harvard University Graduate School of Business Administration, Studies in Management Control, 26–47.

APICS (1980). *Certification Study Guide: Capacity Management*. American Production and Inventory Control Society, Washington, D.C.

Armstrong, J.S. and F. Collopy (1992). "Error measures for generalizations about forecasting methods: Empirical comparisons with discussion." *International Journal of Forecasting*, 8, 69–80.

Armstrong, J.S. (1978). *Long-Range Forecasting*, John Wiley & Sons, New York.

Armstrong, J.S. (1995). *Long-Range Forecasting: From Crystal Ball to Computer*, New York: John Wiley and Sons.

Armstrong, J.S. and F. Collopy (1993). "Causal Forces: Structuring Knowledge for Time Series Extrapolation." *Journal of Forecasting*, 12, 103–115.

Armstrong, J.S. (1984). "Forecasting by Extrapolation: Conclusions from 25 Years of Research." *Interfaces*, November–December, 52–66.

Ashton, David and Leslie Simister (1970). *The Role of Forecasting in Corporate Planning*. Staples Press, London.

Ayres, R.U. (1969). *Technological Forecasting and Long-Range Planning*. McGraw-Hill, New York.

Balakrishnan, N., V. Sridharan, and J.W. Patterson (1995). "Rationing Capacity Between Two Product-Classes." *Decision Sciences*, 27–2, 185–215.

Berry, W.L., T.G. Schmitt, and T.E. Vollman (1982). "Capacity Planning Techniques for Manufacturing Control Systems: Information Requirements and Operational Features." *Journal of Operations Management*, 3, 1,13–26.

Bitran, G.D., E.A. Haas, and A.C. Hax (1982). "Hierarchical Production Planning: A Two-Stage System." *Operations Research*, March–April, 232–51.

Blackstone, J.H. (1989). *Capacity Management*, South-Western Publishing, Cincinnati.

Bolander, Steven F., Richard C. Heard, Samuel M. Steward, and Sam G. Taylor (1981). *Manufacturing Planning and Control in Process Industries*. American Production and Inventory Control Society, Falls Church, VA.

Bowerman, Bruce L. and Richard T. O'Connell (1979). *Time Series and Forecasting*. Duxbury Press, North Scituate, MA.

Box, George E. and G.M. Jenkins (1970). *Time Series Analysis: Forecasting and Control*. Holden Day, Oakland, CA.

Britan, Gabriel R. and Arnoldo C. Hax (1977). "On the Design of Hierarchical Planning Systems." *Decision Sciences*, 8, January.

Brown, R.G. (1963). *Smoothing, Forecastings and Prediction of Discrete Time Series*. Prentice-Hall, Englewood Cliffs, N.S.

Chung, C.H. and L.J. Krajewski (1986b). "Integrating Resource Requirements Planning with Master Production Scheduling." Lee, S.M., L.D. Digman and M.J. Schniederjans, eds., *Proceedings of the 18th Decision Sciences Institute National Meeting*, Hawaii, 1165–1167.

Chung, C.H. and L.J. Krajewski (1987). "Interfacing Aggregate Plans and Master Production Schedules via a Rolling Horizon Procedure." *OMEGA*, 15 (5), 401–409.

Clemen, R.T. (1989). "Combining forecasts: A review and annotated bibliography." *International Journal of Forecasting*, 559–584.

Collopy, F. and J.S. Armstrong (1992). "Rule Based Forecasting: Development and Validation of an Expert System Approach to Combining Time Series Extrapolations." *Management Science*, 38, 10, 1394–1414.

Connell, B.C., E.E. Adam, Jr., and A.N. Moore (1984). "Aggregate Planning in a Health Care Foodservice System with Varying Technologies." *Journal of Operations Management*, 5, 1, 41–55.

Cryer, Jonathan D. (1986). *Time Series Analysis*. Duxbury Press, Boston.

Dalrymple, D.J. (1984). "Sales Forecasting Methods and Accuracy." *Business Horizons*, December, 69–73.

Dalrymple, D.J. (1987). "Sales Forecasting Practices – The Results of a US Survey." *International Journal of Forecasting*, 3, 3, 379–391.

Davis, S.M. (1987). *Future Perfect, Reading*, Massachusetts, Addison-Wesley Publishing Company.

DeLurgio, S.A. and C.D. Bahme (1991). *Forecasting Systems for Operations Management*, Business One Irwin, Homewood, IL.

DeLurgio, S. (1998). *Forecasting Principles and Applications*, Irwin/McGraw-Hill.

Edmundson, R., M. Lawrence, and M. O'Connor (1988). "The use of non-time series information in sales forecasting: a case study." *Journal of forecasting*, 7, 201–211.

Finch, Byron J. and James F. Cox (1987). *Planning and Control System Design: Principles and Cases For Process Manufacturers*. American Production and Inventory Control Society, Falls Church, VA.

Fine, C.H. and R.M. Freund (1990). "Optimal Investment in Product-Flexible Manufacturing Capacity." *Management Science*, 36 (4), 449–466.

Freeland, J. and R. Landel (1984). *Aggregate Production Planning: Text and Cases.* Reston Publishing, Reston, VA.

Freidenfelds, J. (1981). *Capacity Expansions: Analysis of Simple Models with Applications*. Elsevier North-Holland, New York.

Gardner, E. (1985). "Exponential Smoothing: The State of the Art." *Journal of Forecasting*, 4, 1–38.

Gardner, E. and E. McKenzie (1985). "Forecasting Trends in Time Series." *Management Science*, 31 (10), 1237–1246.

Garvin, D.A. (1993). "Manufacturing Strategic Planning." *California Management Review*, Summer, 85–106.

Georgoff, D.M. and R.G. Murdick (1986). "Manager's Guide to Forecasting." *Harvard Business Review*, 64 (January–February), 110–120.

Gershwin, S.B. (1986). "An Approach to Hierarchical Production Planning and Scheduling." in Jackson, R.H.F. and A.W.T. Jones, eds., *Real-Time Optimization in Automated Manufacturing Facilities*, U.S. Department of Commerce, National Bureau of Standards, 1–14.

Goodwin, P. and G. Wright (1993). "Improving judgmental time series forecasting: A review of the guidance provided by research." *International Journal of Forecasting*, 9, 147–161.

Granger, C.W.J. (1980). *Forecasting in Business and Economics*. Academic Press, New York.

Gross, Charles W. and Peterson T. Robin (1983). *Business Forecasting* (2nd ed.). Houghton Mifflin, Boston.

Hahn, C.K., E.A. Duplaga, and K.Y. Kim (1994). "Production/Sales Interface: MPS at Hyundai Motor." *International Journal of Production Economics*, 37–1, 5–18.

Hall, R.W. and T.E. Vollman (1978). "Planning your requirements planning." *Harvard Business Review*, 56 (2), 105–112.

Hanke, John E. and Arthur G. Reitsch (1995). *Business Forecasting*, Fifth Edition, Prentice-Hall.

Harl, J.E. (1983). "Reducing Capacity Problems in Material Requirements Planning Systems." *Production and Inventory Management* (Third Quarter), 52–60.

Hax, A.C. and H.C. Meal (1973). "Hierarchical Integration of Production Planning and Scheduling." *Studies in Management Sciences, Logistics 1*. Edited by M.A. Geisler, North Holland, New York.

Hiller, Randall S. and Jeremy F. Shapiro (1986). "Optimal Capacity Expansion Planning When There Are Learning Effects." *Management Science*, 32 (9), 1153–1163.

Hogarth, R.M. and S.G. Makridakis (1981). "Forecasting and Planning: An Evaluation." *Management Science*, 27 (February), 115–138.

Karni, R. (1982). "Capacity Requirements Planning – A Systematization." *International Journal of Production Research*, 20 (6), 715–739.

Kate, H.A. ten (1994). "Towards a Better Understanding of Order Acceptance." *International Journal of Production Economics*, 37 (1), 139–152.

Keating, Giles (1985). *The Production and Use of Economic Forecasts*. Methuen, London.

Kraus, R.L. (1980). "A Strategic Planning Approach to Facility Site Selection." *Dun's Review* (November), 14–16.

Lankford, Raymond L. (1978). "Short Term Planning of a Manufacturing Capacity." *APICS Conference Proceedings*, 37–68.

Lawrence, M.J., R.H. Edmundson, and M.J. O'Conner (1986). "The Accuracy of Combining Judgmental and Statistical Forecasts." *Management Science*, 32 (December), 1521–1532.

Lee, W.B. and B.M. Khumawala (1974). "Simulation Testing of Aggregate Production Planning Models in an Implementation Methodology." *Management Science*, 20 (6) (February), 903–911.

Lee, W.B. and C.P. McLaughlin (1974). "Corporate Simulation Models for Aggregate Materials Management." *Production and Inventory Management*, 15 (1), 55–67.

Leone, R.A. and J.R. Meyer (1980). "Capacity Strategies for the 1980s." *Harvard Business Review*, 58, 6 (November–December), 133–140.

Lowe, P.H. and J.E. Eguren (1980). "The Determination of Capacity Expansion Programmes with Economies of Scale." *International Journal of Production Research*, 18 (3), 379–390.

Mahmoud, E. (1984). "Accuracy in forecasting: A Survey." *Journal of Forecasting*, 3, 139–159.

Makridakis, S. (1986). "The Art and Science of Forecasting." *International Journal of Forecasting*, 2, 5–39.

Makridakis, S.G., S.C. Wheelwright, and V.E. McGee (1983). *Forecasting: Methods and Applications* (2nd ed.). John Wiley, New York.

Makridakis, S., A. Andersen, R. Carbone, R. Fildes, M. Hibon, R. Lewandowski, J. Newton, E. Parzen and R. Winkler (1982). "The Accuracy of Extrapolation (time series) Methods: Results of a Forecasting Competition." *Journal of Forecasting*, 1, 111–153.

Makridakis, S., S. Wheelwright, and V. McGee (1983). *Forecasting: Methods and Applications* (2nd ed.), New York: John Wiley & Sons.

McLeavey, D.W. and S.L. Narasimhan (1985). *Production Planning and Inventory Control*. Allyn and Bacon, Boston.

Melnyk, S.A. and R.F. Gonzalez (1985). "MRP II: The Early Returns are In." *Production and Inventory Management*, 26 (1) (First Quarter), 124–137.

Mentzer, J., K. Kahn, and C. Bienstock (1996). *Sales Forecasting Benchmarking Study,* The University of Tennessee, Knoxville.

Mentzer, J. and J. Cox (1984). "Familiarity, Application and Performance of Sales Forecasting Techniques." *Journal of Forecasting*, 3, 27–36.

Mentzer, J. and K. Kahn (1995). "Forecasting techniques familiarity, satisfaction, usage and application." *Journal of Forecasting*, 14, 465–476.

Morris, William T. (1967). *The Capacity Decision System*. Irwin, Homewood Ill.

Newberry, Thomas L. and Carl D. Bhame (1981). "How Management Should Use and Interact with Sales Forecasts." *Inventories and Production Management*, 1 (3) (July–August), 1–13.

Newbold, Paul and Theodore Bos (1990). *Introductory Business Forecasting*. South-Western, Cincinnati.

Phillips, F. and N. Kim (1996). "Implications of chaos research for new product forecasting." *Technological Forecasting and Social Change*, 53, 239–61.

Plossl, G.W. and O.W. Wight (1975). "Capacity Planning and Control." *Capacity Planning and Control*, 50–86. American Production and Inventory Control Society, Falls Church, VA.

Proud, J.F. (1981). "Controlling the Master Schedule." *Production and Inventory Management*, 22 (2) (Second Quarter), 78–90.

Rycroft, R. (1993). "Microcomputer Software of Interest to Forecasters in Comparative Review: An Update." *International Journal of Forecasting*, 9 (4), December, 531–575.

Sanders, N. (1997). "The Status of Forecasting in Manufacturing Firms." *Production and Inventory Management Journal*, 38 (2) (2nd Quarter), 32–35.

Sanders, N.R. and L.P. Ritzman (1992). "The need for contextual and technical knowledge in Judgmental forecasting." *Journal of Behavioral Decision Making*, 5, 39–52.

Schroeder, R.G. and P. Larson (1986). "A Reformulation of the Aggregate Planning Problem." *Journal of Operations Management*, 6 (3) (May), 245–256.

Schruben, L. (1983). "Confidence interval estimation using standardized time series." *Operations Research*, 31, 1090–1107.

Schwarz, Leroy B. and Robert E. Johnson (1978). "An Appraisal of the Empirical Performance of the Linear Decision Rule for Aggregate Planning." *Management Science*, 24 (8) (April), 844–849.

Scott, B. (1994). *Manufacturing Planning Systems*. McGraw-Hill.

Smunt, Timothy L. (1986). "Incorporating Learning Curve Analysis into Medium-Term Capacity Planning Procedures: A Simulation Experiment." *Management Science*, 17 (4), 475–495.

Smunt, Timothy L. (1996). "Rough Cut Capacity Planning in a Learning Environment." *IEEE Transactions on Engineering Management*, 43 (3), 334–341.

Sridharan, V. and R.L. LaForge (1994). "A model to estimate service levels when a portion of the master production schedule is frozen." *Computers and Operations Research*, 21 (5), 477–486.

Sridharan, V. and R.L. LaForge (1994). "Freezing the master production schedule: Implications for fill rate." *Decision Sciences*, 25 (3), 461–469.

Sridharan, V. and W.L. Berry (1990). "Freezing the master production schedule under demand uncertainty." *Decision Sciences*, 21 (1), 97–120.

Sridharan, V., W.L. Berry and V. Udayabanu (1988). "Measuring Master production schedule instability under rolling planning horizons." *Decision Sciences*, 19 (1), 147–166.

Taleb, K., S.M. Gupta, and L. Brennan (1997). "Disassembly of Complex Products with Parts and Materials Commonality." *Production Planning and Control*, 8 (3), 255–269.

Tompkins, J.A., J.A. White, Y.A. Bozer, E.H. Frazelle, J.M.A. Tanchoco, and J. Trevino (1996). *Facilities Planning*, Wiley, NY, p. 1–17.

Valentine, Lloyd M. and Carl A. Dauten (1983). *Cycles and Forecasting*. South-Western, Cincinnati.

Webby, R. and M. O'Connor (1996). "Judgmental and statistical time series forecasting: A review of the literature." *International Journal of Forecasting*, 12, 91–118.

Wemmerlov, U. (1984). *Case Studies in Capacity Management*. American Production and Inventory Control Society, Virginia.

Wheelwright, S.C. and S.G. Makridakis (1985). *Forecasting Methods for Management* (4th ed.). John Wiley, New York.

Willis, R.E. (1987). *A Guide to Forecasting for Planners and Managers*. Prentice-Hall, Englewood Cliffs, NJ.

Winters, P.R. (1960). "Forecasting Sales by Exponentially Weighted Moving Average." *Management Science*, 6, April, 324–342.

Yurkiewicz, J. (1993). "Forecasting software: Clearing up a cloudy picture." *OR/MS Today*, 2, 64–75.

COSTING AND INVESTMENT

Abdel-Malek, L. (1989). "Assessment of the economic feasibility of robotic assembly while conveyor tracking (RACT)." *International Journal of Production Research,* 27, 1209–1224.

Abdel-Malek, L. and T.O. Boucher (1985). "A Framework for the economic evaluation of production system and product design alternatives for robot assembly." *International Journal of Production Research*, 23,197–208.

Agrawal, V.P., V. Khohl, and S. Gupta (1991). "Computer aided robot selection: The multiple attribute decision making approach." *International Journal of Production Research*, 29, 1629–1644.

Albright, T.L. and T. Smith (1996). "Software for Activity-Based Costing." *Journal of Cost Management*, Summer, 47–53.

American Society for Quality Control (1971). "*Quality Costs – What and How*" (2nd ed.), Milwaukee, Wisconsin.

Ansari, S., J. Bell, T. Klammer, and C. Lawrence, *Target Costing*, Irwin, 1997.

Baker, R.C. and S. Talluri (1997). "A closer look at the use of data envelopment analysis for technology selection." *Computers and Industrial Engineering*, 32, 101–108.

Barnes, R. (1980). *Motion and Time Study: Design and Measurement of Work* (8th ed.). John Wiley, New York.

Berliner, C. and J.A. Brimson (1988). *Cost Management for Today's Advanced Manufacturing*. Harvard Business School Press, Boston.

Boer, G. (1994). "Five modern management accounting myths." *Management Accounting* (USA, January), 22–27.

Booth, D.E., M. Khouja, and M. Hu (1992). "A robust multivariate procedure for evaluation and selection of industrial robots." *International Journal of Operations and Production Management*, 12, 15–24.

Boubekri, N., M. Sahoui, and C. Lakrib (1991). "Development of an expert system for industrial robot selection." *Computers and Industrial Engineering*, 20, 119–127.

Brausch, J.M. (1994). "Beyond ABC: Target Costing for Profit Enhancement." *Management Accounting*, November, 45–49.

Bruns, W.J., Jr., and R.S. Kaplan, eds. (1987). *Accounting and Management: Field Study Perspectives*. Harvard Business School Press, Boston, 204–228.

Burt, D.N., W. Norquist, and J. Anklesaria (1990). *Zero Base Pricing: Achieving World Class Competitiveness Through Reduced All-in-Cost*. Probus Publishing, Chicago.

Constantinides, K. and J.K. Shank (1994). "Matching accounting to strategy: One mill's experience." *Management Accounting* (USA, September), 32–36.

Cooper, R. (1987). "Does your company need a new cost system?" *Journal of Cost Management* (Spring), 45–49.

Cooper, R. (1988). "The rise of activity-based costing – part one: What is an activity-based cost system?" *Journal of Cost Management* (Summer), 45–54.

Cooper, R. (1988). "The rise of activity-based costing – part two: When do I need an activity-based cost system?" *Journal of Cost Management* (Fall), 41–48.

Cooper, R. (1989). "You Need a New Cost System When...." *Harvard Business Review,* 77 (1) (January–February), 77–82.

Cooper, R. (1989). "The rise of activity-based costing – part three: How many cost drivers do you need, and how do you select them?" *Journal of Cost Management* (Winter), 34–46.

Cooper, R. (1990). "Cost classifications in unit-based and activity-based manufacturing cost systems." *Journal of Cost Management for the Manufacturing Industry* (Fall), 4–14.

Cooper, R. and R.S. Kaplan (1988). "Measuring costs right make the right decisions." *Harvard Business Review*, 66, 96–103.

Cooper, R. and R.S. Kaplan (1991). "Profit Priorities from Activity-Based Costing." *Harvard Business Review*, May–June, 130–135.

Cooper, R. and R.S. Kaplan (1991). *The Design of Cost Management Systems*, Prentice-Hall, Englewood Cliffs.

Cooper, R. and R. Slagmulder (1997). *Target Costing and Value Engineering, Institute of Management Accountants*. Productivity Press.

Cooper, R. and W.B. Chew (1996). "Control Tomorrow's Costs through Today's Designs." *Harvard Business Review*, January–February, 88–97.

Cooper, R., R.S. Kaplan, L.S. Maisel, E. Morrissey, and R.M. Oehm (1992). *Activity-Based Cost Management: Moving from Analysis to Action*. The Institute of Management Accountants, New Jersey.

Dada, M. and R. Marcellus (1994). "Process Control with Learning." *Operations Research*, 42, 2, 323–336.

Dugdale, D. and D. Jones (1996). "Accounting for throughput: Part 1 – the theory." *Management Accounting* (UK, April), 24–29.

Elbourne, Edward D. (1918). *Factory Administration and Accounts*, The Library Press Co., Ltd., London, England.

Fisher, J. (1995). "Implementing Target Costing." *Journal of Cost Management*, Summer, 50–59.

Galloway, D. and D. Waldron (1988). "Throughput Accounting: Part 2 – ranking products profitably." *Management Accounting* (UK, December), 34–35.

Galloway, D. and D. Waldron (1989). "Throughput Accounting: Part 3 – a better way to control laborcosts." *Management Accounting* (UK, January), 34–35.

Galloway, D. and D. Waldron (1989). "Throughput Accounting: Part 4 – moving on to complex products." *Management Accounting* (UK, February), 34–35.

Galloway, D. and D. Waldron (1989). "The rise of activity-based costing – part four: What do activity-based cost systems look like?" *Journal of Cost Management* (Spring), 38–49.

Goldratt, E.M. (1983). "Cost Accounting is Enemy Number One of Productivity." International Conference Proceedings, American *Production and Inventory Control* Society, October.

Greeson, C.B. and M.C. Kocakulah (1997). "Implementing an ABC Pilot at Whirlpool." *Journal of Cost Management*, March–April, 16–21.

Hahn, C.K., K.H. Kim, and J.S. Kim (1986). "Costs of Competition, Implications for Purchasing Strategy." *Journal of Purchasing and Materials Management* (Fall), 2–7.

Hall, R.W., N.P. Galambos, and M. Karlsson (1997). "Constraint-based profitability analysis: stepping beyond the theory of constraints." *Journal of Cost Management*, 11, 6–10.

Harrison, G.C. (1921). "What is wrong with cost accounting?" *N.A.C.A. Bulletin*.

Hill, C. (1988). "Differentiation versus low cost or differentiation and low cost: A contingency framework." *Academy of Management Journal*, 13 (3), 401–412.

Hiromoto, T. (1988). "Another Hidden Edge-Japanese Management Accounting." *Harvard Business Review*, July–August, 22–26.

Hirst, Mark (1990). "Accounting education and advanced cost management systems." *Australian Accountant*, 60 (11), 59–65.

Hodder, J.E. and H.E. Riggs (1984). "Pitfalls in evaluating risky projects." *Harvard Business Review*, 62, 26–30.

Holmen, J.S. (1995). "ABC vs. TOC: It's a matter of time." *Management Accounting* (USA, January), 37–40.

Howell, R.A. and S.R. Soucy (1987). "Major trends for management accounting." *Management Accounting*, 68, 21–27.

Howell, R.A., J.D. Brown, S.R. Soucy, and A.H. Reed (1988). *Management Accounting in the New Environment*, National Association of Accountants, USA.

Jayson, Susan (1987). "Goldratt and Fox: Revolutionizing the Factory Floor." *Management Accounting*, May, 18–22.

Johnson, H.T. and R.S. Kaplan (1987). *Relevance Lost! The Rise and Fall of Management Accounting*, Harvard Business School Press, Boston, MA.

Kaplan, R.S. (1986). "Must CIM be justified by faith alone." *Harvard Business Review*, 64, 87 95.

Kaplan, R.S. (1988). "One Cost System Isn't Enough." *Harvard Business Review*, January/February, 61–68.

Kaplan, R.S. (1991). "New Systems for Measurement and Control." *The Engineering Economist*, 36 (3), 201–218.

Kaplan, R.S. and D.P. Norton (1992). "The balanced scorecard – measures that drive performance." *Harvard Business Review*, 70, 71–79.

Kaplan, R.S. and D.P. Norton (1993). "Putting the balanced scorecard to work" *Harvard Business Review*, September–October.

Kaplan, R.S. and D.P. Norton (1996). "Using the balanced scorecard as a strategic management system." *Harvard Business Review*, 74, 79–85.

Kaplan, R.S. and H.T. Johnson (1987). *Relevance Lost: The Rise and Fall of Cost Accounting*, Harvard Business School Press, Boston, MA.

Kato, Y. (1993). "Target Costing Support Systems: Lessons from Leading Japanese Companies." *Management Accounting Research*, 4 (4), 33–47.

Kato, Y., G. Boer, and C.W. Chow (1995). "Target Costing: An Integrative Management Process." *Journal of Cost Management*, Spring, 39–51.

Leschke, J. and E.N. Weiss (1998). "A Comparison of Rules for Allocating Setup-Reduction Investments in a Capacitated Environment." *Production and Operations Management*.

Malcom, D.G., J.H. Roseboom, C.E. Clark, and W. Fazar (1959). "Application of a Technique for Research and Development Program Evaluation." *Operations Research*, 7 (5), September–October, 98–108.

Mallick, D.N. and P. Kouvelis (1992). "Management of Product Design: A Strategic Approach." *In Intelligent Design and Manufacturing*, ed. A. Kusiak, Wiley, New York.

Mazzola, Joseph B. and Kevin F. McCardle (1996) "A Bayesian Approach to Managing Learning-Curve Uncertainty." *Management Science*, 42 (5), 680–692.

Mehrez, A. and O.F. Offodile (1994). "A statistical-economic framework for evaluating the effect of robot repeatability on profit." *IIE Transactions*, 26, 101–110.

Miller, J. (1995). "Innovative cost savings." *NAPM Insights* (September), 30–32.

Monahan, G.E. and T.L. Smunt (1987). "A multilevel decision support system for the financial justification of automated flexible manufacturing systems." *Interfaces*, 17, 29–40.

Monczka, R. and S. Trecha (1988). "Cost-based supplier performance evaluation." *International Journal of Purchasing and Materials Management* (Spring), 2–7.

Myers, S.C. (1984). "Finance theory and finance strategy." *Interfaces*, 114, 126–137.

Nnaji, B.O. (1986). *Computer-Aided Design, Selection and Evaluation of Robot.*, Elsevier, New York.

Nnaji, B.O. (1988). "Evaluation methodology for performance and system economics for robotic devices." *Computers and Industrial Engineering*, 14, 27–39.

Nnaji, B.O. and M. Yannacopoulou (1989). "A utility theory based robot selection and evaluation for electronics assembly." *Computers and Industrial Engineering*, 14, 477–493.

Noreen, E., D. Smith, and J.T. Mackay (1995). *The Theory of Constraints and its Implications for Management Accounting*. North River Press, Great Barrington, MA.

Offodile, O.F., B.K. Lambert, and R.A. Dudek (1987). "Development of a computer aided robot selection procedure (CARSP)." *International Journal of Production Research*, 25, 1109–1121.

Player, R.S. and D.E. Keys (1995). "Lessons from the ABM battlefield: Getting off to the right start." *Journal of Cost Management* (Spring), 26–38.

Player, R.S. and D.E. Keys (1995). "Lessons from the ABM battlefield: Developing the pilot." *Journal of Cost Management* (Summer), 20–35.

Player, R.S. and D.E. Keys (1995). "Lessons from the ABM battlefield: Moving from pilot to mainstream." *Journal of Cost Management* (Fall), 31–41.

Player, S. and D. Keys (1995). Editors. *Activity-Based Management: Arthur Andersen's Lessons From the ABM Battlefield*, MasterMedia Limited, New York.

Ramasesh, R.V. and M.D. Jayakumar (1991). "Measurement of manufacturing flexibility: a value-based approach." *Journal of Operations Management*, 10, 1–8.

Ramasesh, R.V. and M.D. Jayakumar (1993). "Economic justification of advanced manufacturing technology." *OMEGA*, 21, 2819–306.

Roberts, M.W. and K.J. Silvester (1996). "Why ABC Failed and How It May Yet Succeed." *Journal of Cost Management*, Winter, 23–35.

Sakurai, M. (1989). "Target Costing and How to Use It." *Cost Management*, Summer.

Salomon, Daniel P. and John E. Biegel (1984). "Assessing Economic Attractiveness of FMS Applications in Small Batch Manufacturing." *Industrial Engineering* (June), 88–96.

Schmidgall, R.S. (1986). *Managerial Accounting*, The Educational Institute, East Lansing.

Schroeder, R.G., J.C. Anderson, S.E. Tupy, and E.M. White (1981). "A Study of MRP Benefits and Costs." *Journal of Operations Management*, 2 (1) (October), 1–9.

Scully, J. and Fawcett, S.E. (1993). "Comparative Logistics and Production Costs for Global Manufacturing Strategy." *International Journal of Operations and Production Management*, 13 (12), 62–78.

Seglund, R. and S. Ibarreche (1984). "Just-in-Time: The Accounting Implications." *Management Accounting* (August), 43–45.

Shank, J. (1995). "Strategic Cost Management: A Target Costing Field Study." The *Amos Tuck School of Business Administration*, Dartmouth College.

Shtub, A. (1989). "Estimating the Effects of Conversion to a Group Technology Layout on the Cost of Material Handling." *Engineering Costs and Production Economics*, 16, 103–109.

Shycon, H.N. and C.R. Spraque (1975). "Put a Price Tag on Your Customer Servicing Levels." *Harvard Business Review*, 53 (July–August), 71–78.

Sisk, G.A. (1986). "Certainty equivalent approach to capital budgeting." *Financial Management*, Winter, 23–32.

Smunt, Timothy L. (1986). "A Comparison of Learning Curve Analysis and Moving Average Ratio Analysis for Detailed Operational Planning." *Decision Sciences*, 17 (4), 475–495.

Smunt, Timothy L. (1986). "Incorporating Learning Curve Analysis into Medium-Term Capacity Planning Procedures: A Simulation Experiment." *Management Science*, 17 (4), 475–495.

Smunt, Timothy L. and Thomas E. Morton (1985). "The Effects of Learning on Optimal Lot Sizes: Further Developments on the Single Product Case." *IIE Transactions*, 17 (1), 3–37.

Son, Y.K. (1992). "A Comprehensive Bibliography on Justification of Advanced Manufacturing Technologies." *Engineering Economist*, 38 (1), 59–71.

Swain, M.R. and E.L. Denna (1998). "Making new wine: Useful management accounting systems." *International Journal of Applied Quality Management*, 1, 25–44.

Swann, Don (1989). "What is CIM, and Why Does it Cost $40 Million." *P&IM Review*, July, 34.

Tanaka, T. (1993). "Target Costing at Toyota." *Journal of Cost Management*, Spring, 4–11.

Waldron, D (1988). "Accounting for CIM: The new yardsticks." *EMAP Business and Computing Supplement* (February), 1–2.

Walton, L.W. (1996). "The ABC's of EDI: The Role of Activity-Based Costing (ABC) in Determining EDI Feasibility in Logistics Organizations." *Transportation Journal*, 36, 43–50.

Wiersema, W.H. (1996). "Implementing Activity-Based Management: Overcoming the Data Barrier." *Journal of Cost Management*, Summer, 17–21.

Woodhouse, J. (1997). "The MACRO project: cost/risk evaluation of engineering and management decisions." In: C. Guedes Soares (ed.), *Advances in Safety and Reliability*. Pergamon Press, Oxford, UK, 237–246.

Wright, T.P. (1936). "Factors Affecting the Cost of Airplanes." *Journal of Aeronautical Sciences*, 3 (4), 122–128.

Wygant, Robert M. (1987). "Improving Productivity with Financial Incentives." Engineering Management International, 4, 87–93.

Yelle, Y.E. (1979). "The Learning Curve: Historical Review and Comprehensive Survey." *Decision Sciences*, 10 (2) (April), 302–328.

Flexible Manufacturing

Blackburn, J.D. (1991). *Time Based Competition*, Homewood: Business One Irwin.

Blackburn, J.D. (1991). "The Quick-Response Movement in the Apparel Industry: A Case Study Time-Compressing Supply Chains." edited by J.D. Blackburn, *Time-Based Competition: The Next Battle Ground in American Manufacturing*, Irwin, Homewood, IL, 246–269.

Boyer, K.K. and G.K. Leong (1996). "Manufacturing Flexibility at the Plant Level", Omega: *The*

International Journal of Management Science, 495–510.

Browne, J., D. Dubois, K. Rathmill, S. Sethi, and K. Stecke (1984). "Classification of flexible manufacturing systems." *The FMS Magazine*, April, 114–117.

Buzacott, J. (1982). "The fundamental principles of flexibility in manufacturing systems." *Proceedings of 1st Intl. Conference on Flexible Manufacturing Systems*, North Holland, UK.

Buzacott, J.A. and D.D. Yao (1986). "Flexible manufacturing systems: a review of analytical models." *Management Science*, 32, 890–905.

Bylinski, G. (1983). "The Race to the Automatic Factory." *Fortune*, February 21, 52–60.

Byrett, Donald L., Mufit H. Ozden, and Jon M. Patton (1988). "Integrating Flexible Manufacturing Systems with Traditional Manufacturing, Planning, and Control." *Production and Inventory Management Journal*, Third Quarter, 15–21.

Chakravarty, A.K. and A. Shtub (1984). "Selecting Parts and Loading Flexible Manufacturing Systems." Proceedings of The 1st ORSA/TIMS FMS Conference, Ann Arbor, MI.

Chang, T.C., R.A. Wysk, and H.-P. Wang (1998). *Computer-Aided Manufacturing* (2nd ed.). Prentice-Hall, New Jersey.

Chen, I.J. and C.H. Chung (1991). "Effects of loading and routing decisions on performance of flexible manufacturing systems". *International Journal of Production Research*, 29, 2209–2225.

Cherukuri, S.S., R.G. Nierman, and N.C. Sirianni (1995). "Cycle Time and The Bottom Line." *Industrial Engineering*, March, 1995.

Chung, C.H. (1991). "Planning Tool Requirements for Flexible Manufacturing Systems." *Journal of Manufacturing systems*, 10 (6), 476–483.

Cohen, M.A. and U.M. Apte (1997). *Manufacturing Automation*, Irwin, Illinois.

DeMeyer, A., J. Nakane, J.G. Miller, and K. Ferdows (1987). "Flexibility: The Next Competitive Battle." Manufacturing Roundtable Research Report Series. School of Management, February. Reprinted in *Strategic Management Journal*, 10 (1989), 135–144.

Doll, William, and Mark Vonderembse (1987). "Forging a Partnership to Achieve Competitive Advantage:The CIM Challenge." *MIS Quarterly*, 11 (2), 205–220.

Draper, C. (1984). *Flexible Manufacturing Systems Handbook*, Automation and Management Systems Division, The Charles Stark Draper Laboratory Inc., Noyes Publication.

Engelberger, J.F. (1980). *Robotics in Practice.* AMACOM, New York.

Ettlie, J.E. and J.D. Penner-Hahn (1994). "Flexibility ratios and manufacturing strategy." *Management Science*, 40, 1444–1454.

Fisher, E.L. and O.Z. Maimon (1988). "Specification and selection of robots." In Kusiak, A. (Ed.) *Artificial Intelligence Implications for CIM*, IFS Publications, Bedford.

Gerwin, D. (1981). "Control and Evaluation in the Innovation Process: The Case of Flexible Manufacturing Systems." *IEEE Transactions on Engineering Management*, EM-28, 3 (August), 62–70.

Gerwin, D. (1982). "Do's and Don'ts of Computerized Manufacturing." *Harvard Business Review* (March–April), 107–116.

Gerwin, D. (1987). "An Agenda for Research on the Flexibility of Manufacturing Processes." *International Journal of Operations and Production Management*, 7, 38–49.

Gerwin, D. (1993). Manufacturing flexibility: A Strategic Perspective. *Management Science*, 39 (4), 395–409.

Gerwin, D. and J.C. Tarondeau (1982). "Case Studies of Computer Integrated Manufacturing Systems: A View of Uncertainty and Innovation Process." *Journal of Operations Management*, 2, 2 (February), 87–89.

Gold, B. (1980). "CAM Sets New Rules for Production." *Harvard Business Review*, 60 (6) (November–December), 78–86.

Goldhar, J.D. and M. Jelinek (1983). "Plan for Economics of Scope." *Harvard Business Review*, 61 (6) (November–December), 141–148.

Groover, M.P. (1980). *Automation, Production Systems, and Computer-Aided Manufacturing*. Prentice-Hall, Englewood Cliffs, NJ.

Groover, M.P. (1987). *Automation, Production Systems, and Computer-Aided Manufacturing.* Englewood Cliffs, N.J., Prentice-Hall, 119–128.

Groover, M.P. and E.W. Zimmers, Jr. (1984). *CAD/CAM: Computer Aided Design and Manufacturing*. Prentice-Hall, Englewood Cliffs, NJ.

Han, Min-Hong and Leon F. McGinnis (1988). "Throughput Rate Maximization in Flexible Manufacturing Cells." *IIE Transactions*, 20, 4 (December), 409–417.

Hartley, John (1984). *FMS at Work.* IFS Publications, Bedford, UK.

Jordan, W.C. and S.C. Graves (1995). "Principles on the benefits of manufacturing process flexibility." *Management Science*, 41, 577–594.

Koren, Y. and G. Ulsoy (1997). "Reconfigurable Manufacturing Systems." ERC Technical Report #1, The University of Michigan, Ann Arbor, Michigan.

Kulatilaka, N. (1988). "Valuing the flexibility of flexible manufacturing systems." *IEEE Transactions on Engineering Management*, 35, 250–257.

Kumar, V. (1987). "Entropy measures of measuring manufacturing flexibility." *International Journal of Production Research*, 25, 957–966.

Kusiak, A. (1985). "The Part Families Problem in Flexible Manufacturing Systems." *Annals of Operations Research*, 3, 279–300.

Lingayat, Sunil, John Mittenthal, and Robert M. O'Keefe (1995). "Order release in automated manufacturing systems." *Decision Sciences*, 26 (2), 175–205.

MacCarthy, B.L. and L.Liu (1993). A new classification scheme for flexible manufacturing systems,

International Journal of Production Research, 31, 299–309.

Maimon, O.Z. and Y.F. Choong (1987). Dynamic routing in reentrant flexible manufacturing systems, *Robotics and Computer-Integrated Manufacturing*, 3, 295–300.

Ramasesh, R.V. and M.D. Jayakumar (1991). "Measurement of manufacturing flexibility: a value-based approach." *Journal of Operations Management*, 10, 1–8.

Ranky, P.G. (1982). *The Design and Operation of FMS* (IFS (Publications) Ltd. and North Holland).

Ranky, P.G. (1985). "FMS in CIM" (Flexible Manufacturing Systems in Computer Integrated Manufacturing), *ROBOTICA*, Cambridge University Press, 3, 205–214.

Ranky, P.G. (1985). *Computer Integrated Manufacturing*. Prentice-Hall International.

Ranky, P.G. (1998). "Flexible Manufacturing Cells and Systems in CIM." CIMware Ltd, http://www.cimwareukandusa.com (Book and CD-ROM combo).

Sethi, A.K. and S.P. Sethi (1990). "Flexibility in Manufacturing: A Survey." *International Journal of Flexible Manufacturing Systems*, 2, 289–328.

Shaw, M.J. and A.B. Whinston (1988). "A Distributed Knowledge-Based Approach to Flexible Automation: The Contract Net Framework." *Flexible Manufacturing Systems*, 85–104.

Slack, N. (1987). "The flexibility of manufacturing systems." *International Journal of Operations and Production Management*, 7, 35–45.

Stecke, K.E. (1983). "Formulation and Solution of Nonlinear Integer Production Planning Problems for Flexible Manufacturing Systems." *Management Science*, 29 (3), 273–288.

Stecke, K.E. (1986). "A Hierarchical Approach to Solving Machine Grouping and Loading Problems of Flexible Manufacturing Systems." *European Journal of-Operational Research*, 24, 369–378.

Suresh, N.C. (1990). "Towards an integrated evaluation of flexible automation investments." *International Journal of Production Research*, 28, 1657–1672.

Swamidass, P.M. (1988). *Manufacturing Flexibility. Operations Management*, Association Monograph 2, Naman and Schneider Associates Group, Texas.

Triantis, A.J. and J.E. Hodder (1990). "Valuing flexibility as a complex option." *Journal of Finance*, XLV, 549–565.

U.S. Department of Commerce (1987). Washington, D.C., *A Competitive Assessment of the U.S. Flexible Manufacturing System Industry*, July.

Upton, D.M. (1994). "The Management of Manufacturing Flexibility." *California Management Review*, Winter, 72–89.

Upton, D.M. (1995). "Flexibility as process mobility: The management of plant capabilities for quick response manufacturing." *Journal of Operations Management*, 12 (3&4), June, 205–224.

Upton, D.M. (1997). "Process range in manufacturing: An empirical study of flexibility." *Management Science*, 4 (8), August, 1079–1092.

Venketesan, R. (1990). "Cummins flexes its Factory. *Harvard Business Review*, 90 (2), March–April, 120–127.

Vonderembse, Mark A. and Gregory S. Wobser (1987). "Steps for Implementing a Flexible Manufacturing Systems." *Industrial Engineering*, 15 (11) (November), 58–64.

Williams, Vearl A. (1985). "FMS in Action." *Production*. January, 41–43.

Yao, D.D. (1985). "Material and information flows in flexible manufacturing systems." *Material Flow*, 2, 143–149.

Zelenovic, D.M. (1982). "Flexibility – a condition for effective production systems." *International Journal of Production Research*, 20, 319–337.

HISTORY AND OVERVIEW

Aitken, Hugh G.J. (1960). *Taylorism at Watertown Arsenal; Scientific Management in Action*, 1908–1915. Harvard University Press, Cambridge, Mass.

APICS (1996). *1997 CPIM Exam Content Manual*. APICS, the Educational Society for Resource Management, Virginia.

Babbage, Charles (1835). *On the Economy of Machinery and Manufactures*, Charles Knight, London, England.

Babcock, George D. (1918). *The Taylor System in Franklin Management*. The Engineering Magazine Co., New York. New York.

Barth, Carl (1912). "Testimony." in *The Taylor and Other Systems of Shop Management, Hearings before the Special Committee of the House of Representatives to Investigate the Taylor and Other Systems of Management*, 3, Government Printing Office, Washington, D.C., p. 1538–1583.

Brown, R.G. (1971). *Management Decisions for Production Operations*. The Dryden Press, Hinsdale, IL.

Buchan, J. and E. Koenigsberg (1963). *Scientific Inventory Management*. Prentice-Hall, New Jersey. Chapters 1–16, and 19.

Burnham, J.M. (1980). "The Operating Plant – Case Studies and Generalizations." *Production and Inventory Management*, 21, 4, 63–94.

Chandler, Alfred (1977). *The Visible Hand: The Managerial Revolution in American Business*. Belknap Press, Cambridge, Mass.

Chase, R.B. and D.A. Garvin (1989). "The Service Factory." *Harvard Business Review*, 67 (4) (July–August), 61–69.

Chase, R.B. and D.A. Tansik (1983). "The Customer Contact Model for Organization and Design." *Management Science*, 29, 9 (September), 1037–1050.

Chase, R.B. and E.L. Prentis (1987). "*Operations Management*, A Field Rediscovered." *Journal of Management*, 13, 2, October, 351–366.

Chase, R.B. and N.J. Aquilano (1989). *Production and Operations Management* (5th ed.), Irwin, Homewood, Illinois.

Chase, R.B. and W.K. Erikson (1988). "The Service Factory." *Academy of Management of Executive*, (August), 191–196.

Chase, R.B.K., R. Kumar, and W.E. Youngdahl (1992). "Service-Based Manufacturing: The Service Factory." *Production and Operations Management*, 1(2), 175–184.

Cohen, S.S. and J. Zysman (1987). *Manufacturing Matters: The Myth of the Post Industrial Economy*, Basic Books, New York.

Conway, R.W. and Andrew Schultz, Jr. (1959). "The Manufacturing Progress Function." *Journal of Industrial Engineering*, 10, 1, 39–54.

Cooper, Carolyn C. (1981–2). "The Production Line at Portsmouth Block Mill." *Industrial Archaeology Review*, 6, 28–44.

Cooper, Carolyn C. (1984). The Portsmouth System of Manufactures. *Technology and Culture*, 25, 182–225.

Copley, Frank Barkley (1923). *Frederick W. Taylor: Father of Scientific Management*, Reprinted 1969, Augustus Kelley, New York, New York.

Cyert, R. (?). "Saving the Smokestack Industries." *Bell Atlantic Quarterly*, 2 (3) (Autumn), 17–24.

DeGarmo, E. Paul, J.T. Black, A. Kohser, A. Ronald (1997). *Materials and Processes in Manufacturing* (8th ed.). Prentice-Hall, Inc.

Deming, W.E. (1986). *Out of the Crisis*. MIT Center for Advanced Engineering Study, Cambridge, MA.

Deming, W.E. (1993). *The New Economics for Industry, Government, and Education*, Center for Advanced Engineering Study, Massachusetts Institute of Technology, Cambridge, Massachusetts.

Dilworth, J.B. (1992). *Operations Management*: Design, Planning and Control for Manufacturing and Services. McGraw-Hill.

Dove, R. (1995). "The Challenges of Change." *Production*, 107 (2), 16–17.

Drucker, P. (1954). *The Practice of Management*, Harper and Brothers, New York, 195–196.

Drucker, P. (1963). "Managing for business effectiveness." *Harvard Business Review* (May–June), 59–66.

Drucker, P. (1974). *Management: Tasks, responsibilities*, Practices. New York, Harper and Rowe.

Drucker, P. (1982). *The Changing World of the Executive*. Heinemann, London.

Ettlie, J.E. (1988). *Taking Charge of Manufacturing*, Jossey-Bass Publishers, San Francisco.

Evans, Oliver (1834). *The Young Mill-wright and Miller's Guide*. Reprinted 1989. Arno Press. New York, New York.

Faurote, Fay Leone and Harold Lucien Arnold (1916). *The Ford Methods and the Ford Shops*. Engineering Magazine Co., New York, New York.

Fierman, J. (1995). "Americans Can't Get No Satisfaction." *Fortune* (December 11), 186–194.

Fogarty, D.W., T.R. Hoffman, and P.W. Stonebraker (1989). *Production and Operations Management*. South-Western, Cincinnati.

Ford, Henry with Samuel Crowther (1926). *My Life and Work*. William Heineman, Ltd., London.

Ford, Henry with Samuel Crowther (1928). *Today and Tomorrow*. William Heineman,Ltd., London.

Gaither, N. and G.V. Frazier (1998). *Production and Operations Management* (8th ed.). South western Publishing, New York.

Galbraith, J. (1973). *Designing Complex Organizations*. Addison-Wesley, Reading, Mass.

Gantt, H.L. (1903). A Graphical Daily Balance in Manufacture. *ASME Transactions*, 24, 1322–1336.

Gilbert, K.R. (1965). *The Portsmouth Block-Making Machinery*. Her Majesty's Stationery Office. London, England.

Ginsburg, I. and N. Miller (1992). "Value-Driven Management." *Business Horizons* (May–June), 23–27.

Godrad, Eliyahu M. and Jeff Cox (1986). *The Goal: A Process of Ongoing Improvement*. North River Press, Croton-on-Hudson, NY.

Gold, B. (1981). "Changing Perspectives on Size, Scale, and Returns: An Interpretive Survey." *Journal of Economic Literature*, 19, March.

Goldratt, E.M. (1990). *Theory of Constraints*. North River Press, New York.

Goldratt, E.M. and J. Cox (1986). *The Goal: A Process of Ongoing Improvement*. North River Press, Croton-on-Hudson, N.Y.

Greene, J.H. (1987). Editor. *Production and Inventory Control Handbook* (2nd ed.). McGraw-Hill, New York.

Greene, James H. (1974). *Production and Inventory Control Systems and Decisions* (2nd ed.). Richard D. Irwin, Homewood, IL.

Haas, E.A. (1987). "Breakthrough Manufacturing." *Harvard Business Review* (March–April), 75–81.

Halberstam, D. (1986). *The Reckoning*. Morrow Publishers, New York.

Hall, R.W. (1993). "AME's Vision of Total Enterprise Manufacturing." *Target*, 9, 6, November/December, 33–38.

Hall, R.W. (1987). *Attaining manufacturing Excellence*. Homewood, Ill., Dow Jones-Irwin.

Hall, R.W. and Y. Yamada (1993). "Sekisui's Three-Day House." *Target*, 9, 4, July/August, 6–11.

Hammer, M. (1996). *Beyond Reengineering*, Harper Business, New York.

Hammer, M. and J. Champy (1993). *Reengineering the Corporation: A Manifestor for Business Revolution*, Harper Business, New York.

Hammer, M. and S. Stanton (1995). *The Reengineering Revolution*, Harper Business, New York.

Hathaway, H.K. (1914). "The Planning Department, Its Organization and Function." in Thompson, C. Bertrand, *Scientific Management: A Collection of the More Significant Articles Describing the Taylor System of Management*, Harvard University Press, Cambridge, MA., p. 366–394.

Hax, A.C. and D. Candea (1984). *Production and Inventory Management*, Prentice-Hall, Englewood Cliffs, NJ.

Hayes, R. (1981). "Why Japanese Factories Work." *Harvard Business Review* (July–August).

Hayes, R.H. and J. Jaikumar (1988). Manufacturing's crisis: new technologies, obsolete organizations, *Harvard Business Review*, September, 77–85.

Hayes, R.H. and R.W. Schmenner (1978). "How should you organize manufacturing?" *Harvard Business Review* (January–February), 105–117.

Hayes, R.H. and W.J. Abernathy (1980). "Managing Our Way to Economic Decline." *Harvard Business Review*, 58, July–August, 67–77.

Hayes, R.H., S. Wheelwright, and K. Clark (1988). *Dynamic Manufacturing: Creating the Learning Organization*, The Free Press, New York, 161–184.

Henkoff, R. (1994). "Delivering the Goods." *Fortune* (November 28), 64–78.

Heyer, S. and Lee, R. (1992). "Rewiring the Corporation." *Business Horizons* (May–June),13–22.

Hirschman, A.O. (1970). *Exit, Voice and Loyalty: Responses to Decline in Firms, Organization and Strategies*. Harvard University Press, Cambridge, MA.

Holland, J.H. (1995). *Hidden Order: How Adaptation Builds Complexity*. Addison-Wesley, Reading, MA 185.

Hounshell, David W. (1984). *From the American System to Mass Production*, 1800–1932. The John Hopkins University Press. Baltimore, Maryland.

Jaikumar, R. (1986). "Postindustrial Manufacturing." *Harvard Business Review*, November–December, 69–76.

Jelinek, M. (1977). "Technology, Organizations, and Contingency." *Academy of Management Review*, January, 17–26.

Johnson, H.T. (1992). *Relevance Regained: From Top-Down Control to Bottom-Up Empowerment*, Free Press, New York.

Jonas, N. (1986). "The Hollow Corporation." *Business Week*. March 3, 57–85.

Kanigel, Robert (1997). *The One Best Way*, Viking Penguin, New York, New York.

Kanter R. (1985). *The Change Masters: Innovation For Productivity In the American Corporation*. New York, Simon and Schuster.

Kelley, James E., Jr. and Morgan R. Walker (1989). *The Origins of CPM: A Personal History*. PMNetwork, 3, 7–22.

Lau, R.S.M. (1995). "Mass Customization: The Next Industrial Revolution." *Industrial Management*, 37, 18–19.

Lawler, E.E., III (1986). *High-Involvement Management*. Jossey-Bass, San Francisco.

Lee, S.M. and G. Schwediman, eds. (1982). *Management by Japanese Systems*. Praeger, New York.

Limprecht, Joseph A. and Robert H. Hayes (1980). "Germany's World-Class Manufacturers." *Harvard Business Review*, 63, September–October, 142–150.

Marquardt, M. and A. Reynolds (1994). *The Global Learning Organization*. Irwin, Burr Ridge, IL.

Meadows, D.H., D.L. Meadows, J. Randers, and W.W. Behrens, III (1974). *The Limits to Growth*. Universe Books, New York.

Meredith, J.R. (1992). *The Management of Operations: A Conceptual Emphasis*. Wiley, New York.

Miller, J.G. (1981). "Fit Production Systems to the Task." *Harvard Business Review* (January–February), 145–154.

Miller, J.G. and T.E. Vollmann (1985). "The Hidden Factory." *Harvard Business Review* (September–October), 142–150.

Moore, Harry D. and Donald R. Kibbey (1975). "Manufacturing Materials and Processes." *Grid*, Columbus Ohio.

Nadler, G. and D. Smith (1963). "Manufacturing Progress Functions for Types of Processes." *International Journal of Production Research*, 2 (2), 115–135.

Nelson, Daniel (1990). *Frederick W. Taylor and the Rise of Scientific Management*. University of Wisconsin Press, Madison, Wisconsin.

Nelson, Daniel (1992). *A Mental Revolution: Scientific Management Since Taylor*. Ohio State University Press, Columbus, Ohio.

Nonaka, I. and H. Takeuchi (1995). *The Knowledge Creating Company*. Oxford University Press, New York.

Ohmae, K. (1987). "The Triad World View." *Journal of Business Strategy* (Spring), 8–19.

Ouchi, W. (1981). *Theory Z*. Addison-Wesley, Reading, MA.

Peters, T.J. and R.H. Waterman, Jr. (1982). *In Search of Excellence*. Harper and Rowe, New York.

Piore, M.J. and C.F. Sabel (1984). *The Second Industrial Divide*, Basic Books, New York, N.Y.

Plossl, G.W. (1985). *Production and Inventory Management: Principles and Techniques* (2nd ed.). Prentice-Hall, New Jersey, 96–125, 200–212.

Poza, Ernesto J. (1983). "Twelve Actions to Build Strong U.S. Factories." *Sloan Management Review*, 21, Winter, 47–58.

Render, B. and J. Heizer (1997). *Principles of Operations Management*. Prentice-Hall, Upper Saddle River, New Jersey.

Rosenberg, Nathan (1969). *The American System of Manufactures*. Edinburgh University Press, Edinburgh, Scotland.

Rosenthal, S.R. (1984). "Progress Towards the 'Factory of the Future." *Journal of Operations Management*, 4 (3) (May), 203–229.

Ruch, W.A., H.E. Fearon, and C.D. Wieters (1992). *Fundamental of Production/Operations Management* (5th ed.). West Publishing Company, St.Paul.

Schniederjans, M.J. (1998). *Operations Management: A Global Context*, New York, Qurum Books.

Schonberger, R.J. and E.M. Knod Jr. (1997). *Operations Management: Customer Focused Principles* (6th ed.). Irwin, IL. Chapters 5, 8, and 10.

Senge, P.M. (1990). *The Fifth Discipline: The Art and Practice of the Learning Organization*. Doubleday Currency, New York: NY.

Shore, Barry (1973). *Operations Management*. McGraw-Hill, New York.

Slautterback, W.H. (1985). *"Manufacturing in the Year 2000." Manufacturing Engineering*, August, 57–60.

Sleigh, P. (1993). *The world automotive components industry: A review of leading manufacturers and trends*, Research Report, 1–4, Economist Intelligence Unit, London.

Sterling, Frank W. (1914). "The Successful Operation of a System of Scientific Management." in Thompson, C. Bertrand, *Scientific Management: A Collection of the More Significant Articles Describing the Taylor System of Management*. Harvard University Press, Cambridge, MA, 296–365.

Stevenson, W.J. (1996). *Production/Operations Management* (5th ed.). Irwin, Homewood, IL.

Steward, D.V. (1981). *Systems Analysis and Management: Structure, Strategy and Design*. Petrocelli Books, New York.

Suzaki, K. (1987). *The New Manufacturing Challenge: Techniques for Continuous Improvement*. The Free Press, New York.

Swamidass, P.M. (1991). "Empirical Science: The New Frontier in Operations Management." *Academy of Management Review*, 16 (4), 793–814.

Taylor, Fred. W. (1886). "Discussion of The Shop Order System of Accounts" (by Henry Metcalfe). *ASME Transactions*, 7, 475–6 and 484–485.

Taylor, Fred. W. (1894). "Notes on Belting." *ASME Transactions*, 15, 204–259.

Taylor, Fred. W. (1895). "A Piece-Rate System, Being a Step Toward Partial Solution of the Labor Problem." *ASME Transactions*, 16, 856–903.

Taylor, Fred. W. (1903). "Shop Management." *ASME Transactions*, 24, 1337–1480.

Taylor, Fred. W. (1906). "On the Art of Cutting Metals." *ASME Transactions*, 27, 31–279, with "Discussions." 280–350.

Taylor, Fred. W. (1911). *Principles of Scientific Management*. Harper & Row, New York.

Taylor, Fred. W. (1911). *Shop Management*. Harper & Brothers Publishers, New York, New York.

Terez, T. (1990). "A Manager's Guidelines for Implementing Successful Operational Changes." *Industrial Management*, 32 (4), 18–20.

Tharumarajah, A. and A.J. Wells (1997). "A behaviour-based approach to scheduling in distributed manufacturing systems." *Integrated Computer Aided Engineering*, 4, 235–249.

Thompson, C. Bertrand (1914). *Scientific Management: A Collection of the More Significant Articles Describing the Taylor System of Management*, Harvard University Press, Cambridge, MA.

Thompson, C. Bertrand (1917). *The Taylor System of Scientific Management*. A.W. Shaw Co., Chicago, IL.

Toffler, Alvin (1971). *Future Shock*. Bantam Books, New York.

Toyoda, Eiji (1987). *Toyota: Fifty Years in Motion*. Kodansha International, Tokyo, Japan.

Weinberg, G.M. (1975). *An Introduction to General Systems Theory*. Wiley, New York.

Wernerfelt, B. (1984). "A Resource-based View of the Firm." *Strategic Management Journal*, 5, 171–180.

Whitson, Kevin (1996). "Scientific Management and Production Management Practice in Britain Between the Wars." *Historical Studies in Industrial Relations*, 1, 47–75.

Wild, Ray (1974). The Origins and Development of Flow-Line Production. *Industrial Archaeology Review*, 11, 43–55.

Wilson, James M. (1995). "An Historical Perspective on Operations Management." *Production and Inventory Management Journal*, 36, 61–66.

Wilson, James M. (1996). "Henry Ford: A Just-in-Time Pioneer." *Production and Inventory Management Journal*, 37, 26–31.

Womack, J.P., D.T. Jones, and D. Roos (1991). *The Machine that Changed the World: The Story of Lean Production*. Harper Collins, New York, 37–38, 91–93.

Wrege, Charles D. and Ronald G. Greenwood (1991). *Frederick W. Taylor, the Father of Scientific Management: Myth and Reality*, Richard D. Irwin, Homewood, IL.

Wren, Daniel (1994). *The Evolution of Management Thought*. John Wiley & Sons, New York, New York.

Zimmerman, Gary (1975). "The ABC's of Wilfredo Pareto." *Production Inventory Management*, 16 (3), 1–9.

Human Issues and Teams

Adler, P.S. (1990). "Shared Learning." *Management Science*, 938–957.

Andrews, C.G. (1978). "Leadership and motivation: the people side of systems implementation." In *Proceedings of the American Production and Inventory Control Society Conference*, 352–387.

Argote, L., S. Beckman and D. Epple (1990). "The Persistence and Transfer of Learning in Industrial Settings." *Management Science* (February), 36, 2, 140–154.

Banker, R.D., J.M. Field, R.G. Schroeder, and K.K. Sinha (1996). "Impact of Work Teams on Manufacturing Performance." *Academy of Management Journal*, 39 (4), 867–890.

Baulkin (1988). "Reward Policies That Support Entrepreneurship." *Compensation Benefits Review*, 20.

Blackler, F.H.M. (1978). *Job Redesign and Management Control: Studies in British Leyland and Volvo*, Saxon House, Farnborough England.

Blasingame, J.W. and J.K Weeks (1981). "Behavioral dimensions of MRP change; assessing your organisation's strengths and weaknesses." *Production and Inventory Management*, 22, 1, 81–95.

Bohn, Roger E. (1995). "Noise and Learning in Semiconductor Manufacturing." *Management Science*, 41, 1, 31–42.

Bridger, R.S. (1995). *Introduction to Ergonomics*, McGraw-Hill, NY, p. 1–30.

Burlingame, J.W. and J.K. Weeks (1981). "Behavioral Dimensions of MRP Change: Assessing Your Organization's Strengths and Weaknesses." *Production and Inventory Management*, 22 (1) (First Quarter), 81–95.

Campion, M.A., G.J. Medsker, and A.C. Higgs (1993). "Relations Between Work Group Characteristics and Effectiveness: Implications for Designing Effective Work Groups." *Personnel Psychology*, 46, 823–850.

Carlisle, Brian (1983). "Job Design Implications for Operations Managers." *International Journal of Operations and Production Management*, 3 (3), 40–48.

Cohen, S.G., G.E. Ledford, Jr., and G.M. Spreitzer (1996). "A Predictive Model of Managing Work Team Effectiveness." *Human Relations*, 49(5), 643–676.

Cox, T. (1994). *Cultural Diversity in Organizations: Theory, Research and Practice*. Berret-Koehler, San Francisco.

Dada, M. and R. Marcellus (1994). "Process Control with Learning." *Operations Research*, 42, 2, 323–336.

Dunphy, D. and B. Bryant (1996). "Teams: Panaceas or Prescriptions for Improved Performance." *Human Relations*, 49 (5), 677–699.

Epple, Dennis, Linda Argote and Kenneth Murphy (1996). "An Empirical Investigation of the Microstructure of Knowledge Acquisition and Transfer Through Learning by Doing." *Operations Research*, 44 (1), 77–86.

Fine, C.H. (1988). "A Quality Control Model with Learning Effects." *Operations Research*, 36, 3, 437–444.

Gantt, H.L. (1916). *Work, Wages and Profits*. The Engineering Magazine Co., New York, New York.

Goodman, P.S., R. Devadas, and T.L.G. Hughson (1988). "Groups and Productivity: Analyzing the Effectiveness of Self-Managing Teams." In *Productivity in Organizations*, J.P. Campbell *et al.* (ed.), Jossey-Bass, San Francisco, CA.

Griffin, R.W. (1982). *Task Design: An Integrative Approach*. Glenview, IL, Scott Foresman.

Gubman, E.L. (1998). *The Talent Solution: Aligning Strategy and People to Achieve Extraordinary Results*. New York: McGraw-Hill.

Guest, R.H. (1982). *Innovation Work Practices*. Elmsford, NY, Pergamon Press.

Guzzo, R.A. and M.W. Dickson (1996). "Teams in Organizations: Recent Research on Performance and Effectiveness" *Annual Review of Psychology*, 47, 307–338.

Ha, A.Y. and E.L. Porteus (1995). "Optimal Timing of Reviews in Concurrent Design for Manufacturability." *Management Science*, 41, 1431–1447.

Hackman, J.R. (1998). "Why Teams Don't Work." In *Theory and Research on Small Groups*, R.S. Tindale *et al.* (ed.). Plenum Press, New York, NY.

Hackman, J.R. and G.R. Oldham (1980). *Work Redesign*. Addison-Wesley, Reading, MA.

Halley, W.H., Jr. and K.M. Jennings (1984). *The Labor Relations Process* (2nd ed.). Hinsdale, IL, Dryden Press.

Hampton, W.J. (1988). "How Does Japan, Inc. Pick Its American Workers?" *Business Week*, October 3, 84, 88.

Harrington-Mackin, D. (1996). *Keeping the Team Going: A Tool Kit to Renew and Refuel Your Workplace Teams*, AMACOM/American Management Association, New York, NY.

Hauptman, O. and K.K. Hirji (1996). "The Influence of Process Concurrency on Project Outcomes in Product Development: an Empirical Study with Cross-FunctionalTeams." *IEEE Transactions in Engineering Management*, 43, 153–164.

Herzberg, F. (1987). "One More Time: How Do You Motivate Employees?" (with retrospective commentary). *Harvard Business Review* (September–October), 109–120.

Hiller, Randall S. and Jeremy F. Shapiro (1986). "Optimal Capacity Expansion Planning When There Are Learning Effects." *Management Science*, 32 (9), 1153–1163.

Hutchinson, R.D. (1981). *New Horizons for Human Factors in Job Design*. McGraw-Hill, New York.

Imai, K., I. Nonaka, and H. Takeuchi (1985). "Managing the New Product Development Process: How the Japanese Companies Learn and Unlearn." In: Clark, K.B., R.H. Hayes, and C. Lorenz (eds.). *The Uneasy Alliance*. Harvard Business School Press, Boston.

Johnson, H.T. (1992). *Relevance Regained: From Top-Down Control to Bottom-Up Empowerment*, Free Press, New York.

Karger, D.W. (1982). *Advanced Work Measurement*. Industrial Press, New York.

Katzenbach, J.R. and D.K. Smith (1993). *The Wisdom of Teams: Creating the High-Performance Organization*, Harvard Business School Press, Boston, MA.

Klein, J.A. and P.A. Posey (1986). "Good Supervisors Are Good Supervisors – Anywhere." *Harvard Business Review* (November–December), 125–128.

Knight, Donald O. and Michael L. Wall (1989). "Using Group Technology for Improving Communication and Coordination Among Teams of Workers in Manufacturing Cells." *Industrial Engineering* (January), 32–34.

Knights, D., H. Willmott, and D. Collisons, eds. (1985). *Job Redesign*. Gower, Hants, England.

Kontz, Stephan (1983). Work Design: *Industrial Ergonomics* (2nd ed.). Wiley, New York.

Lawler, E.E. and S.G. Cohen (1992). "Designing pay systems for teams." *American Compensation Association Journal*, 1, 6–18.

Ledvinka, J. and V.G. Scarpello (1991). *Federal Regulation of Personnel and Human Resource Management* (2nd ed.). PWS-Kent, Boston.

Lewin, K. (1947). "Group decision and social change." In *Readings in Social Psychology*, New York; Holt, Rinehart and Winston, 197–211.

Lewis, J. and Wiegert, A. (1985). Trust as a social reality. *Social Forces*, 63, 967–988.

Lewis, James P. (1997). *Team-based project management*. New York: AMACOM.

Logue, J. (1982). "Semi-Autonomous Groups at Saab: More Freedom, High Output." *Management Review*, 71, 9 (September), 32–33.

Mangiameli, P. and L. Krajewski (1983). "The Effects of Workforce Strategies on Manufacturing Operations." *Journal of Operations Management*, 3 (4) (August), 183–196.

Mazzola, Joseph B. and Kevin F. McCardle (1996). "A Bayesian Approach to Managing Learning-Curve Uncertainty." *Management Science*, 42 (5), 680–692.

McAllister, D. (1995). Affect- and cognition-based trust as foundations for interpersonal cooperation in organizations. *Academy of Management Journal*, 38 (1), 24–59.

McCormick Ernest J. (1982). *Human Factors in Engineering and Design*. McGraw-Hill, New York.

McQuade, W. (1984). "Easing Tensions Between Man and Machine." *Fortune*, March 19, 58–66.

Mohrman, S.A., S.G. Cohen, and A.M. Mohrman, Jr. (1995). *Designing Team-Based Organizations: New Forms for Knowledge Work*, Jossey-Bass, San Francisco, CA.

Monczka, R.M. and R.J. Trent (1993). *Cross-Functional Sourcing, Team Effectiveness*, Center for Advanced Purchasing Studies, Tempe, AZ.

Muramatsu, M., H. Miyazaki, and K. Ishii (1987). "A Successful Application of Job Enlargement/Enrichment at Toyota." *IIE Transactions*, 19 (4), 451–459.

Niebel, B.W. (1988). *Motion and Time Study* (8th ed.). Irwin, Homewood IL.

Noe, R.A., J.R. Hollenbeck, B. Gerhart, and P.M. Wright (1997). *Human Resource Management: Gaining a Competitive Advantage* (2nd ed.). Irwin, Chicago.

Nof, S.Y. and H. Lechtman (1982). "Robot time and motion provides means for evaluating alternative work methods." *Industrial Engineering*, 14, 38–48.

O'Leary-Kelly, A.M., J.J. Martocchio, and D.D. Frink (1994). "A Review of the Influence of Group Goals on Group Performance." *Academy of Management Journal*, 37 (5), 1285–1301.

Osburn, J.D., L. Moran, E. Musselwhite, and J.H. Zenger (1990). *Self-directed Work Teams: The New American Challenge*. Irwin Publishing, Chicago, IL.

Pfeffer, J. (1995). "Producing sustainable competitive advantage through the effective management of people." *Academy of Management Executive*, 9, 55–69.

Pratsini, Eleni, Jeffrey D. Camm, and Amitabh S. Raturi (1993). "Effect of Process Learning on Manufacturing Schedules." *Computers and Operation Research*, 20 (1), 15–24.

Rice, R.S. (1977). "Survey of Work Measurement and Wage Incentives." *Industrial Engineering*, 9 (7) (July), 18–31.

Rosow, I.M. and R. Zager (1988). *Training: The Competitive Edge*. San Francisco, Jossey-Bass.

Sako, M. (1992). *Prices, Quality and Trust: Inter-Firm Relations in Britain and Japan*. Cambridge University Press, Cambridge.

Sanders, M.A. and E.J. McCormick (1987). *Human Factors in Engineering and Design* (6th ed.). McGraw-Hill, New York.

Schein, E.H. (1964). "The mechanism of change." *Interpersonal Dynamics*, 199–236. Homewood, IL: The Dorsey Press.

Schuler, R.S., L.P. Ritzman, and V.L. Davis (1981). "Merging Prescriptive and Behavioral Approaches for Office Layout." *Journal of Operations Management*, 1 (3) (February), 131–142.

Senge, P.M. (1990). *The Fifth Discipline: The Art and Practice of the Learning Organization*. Doubleday Currency, New York: NY.

Sugimori, Y., K. Kusunoki, F. Cho, and S. Uchikawa (1977). "Toyota production system and kanban system materialization of just-in-time and respect-for-human system." *International Journal of Production Research*, 15, 553–564.

Taylor, Fred. W. (1895). "A Piece-Rate System, Being a Step Toward Partial Solution of the Labor Problem." *ASME Transactions*, 16, 856–903.

Timm, P.R. and B.D. Peterson (1986). *People at Work: Human Relations in Organizations*, West, St. Paul, MN.

Walton, R.E. and G.I. Susman (1987). "People Policies for the New Machines." *Harvard Business Review* (March–April), 98–106.

Wexley, K.N. and G.P. Latham (1981). *Developing and Training Human Resources in Organizations*. Scott, Foresman, Glenview IL.

Wygant, Robert M. (1987). "Improving Productivity with Financial Incentives." *Engineering Management International*, 4, 87–93.

Yelle, Y.E. (1979). "The Learning Curve: Historical Review and Comprehensive Survey." *Decision Sciences*, 10 (2) (April), 302–328.

Zandin, K. (1980). *MOST Work Measurement Systems*. Marcel Dekker, New York.

Layout and Manufacturing Cells

Aase, Gerald (1997). "Improving Labor Productivity with U-shaped Assembly Lines." Proceeding for the 26th Annual Meeting of the Decision Science Institute, San Diego, CA.

Abdel-Malek, L. (1985). "A case study on the use of robots in assembly process." *Engineering Costs and Production Economics*, 9, 99–102.

Aigbedo, H. and Y. Monden (1997). "Scheduling for Just-in-Time Mixed Model Assembly Lines – A Parametric Procedure for Multi-Criterion Sequence." *International Journal of Production Research*, 35, 2543–2564.

Anonymous (1984). "Manufacturing Cells: Blueprint for Versatility." *Production*, December, 38–40.

Apple, James M. (1977). *Plant Layout and Material Handling*. Wiley, New York.

Arn, E.A. (1975). *Group Technology*. Springer-Verlag, West Berlin.

Baybars, I. (1986). "A Survey of Exact Algorithms for the Simple Assembly Line Balancing Problem." *Management Science*, 32, 909–932.

Black, J.T. (1983). "Cellular Manufacturing Systems Reduce Setup Time, Make Small Lot Production

Economical." *Industrial Engineering*, 15 (11) (November), 36–48.

Black, J.T. (1991). *The Design of The Factory With A Future*. McGraw-Hill.

Black, J.T., B.C. Jiang, and G.J. Wiens (1991). *Design, Analysis and Control of Manufacturing Cells*. PED-53, ASME.

Briuand, D.A., H.Garcia, D. Gateau, L. Molloholli, and B. Mutel (1989). "Part families and manufacturing cell formation." *Prolamat's Proceedings*, Dresden, GDR.

Burbidge, J.L. and B.G. Dale (1984). "Planning the Introduction and Predicting the Benefits of Group Technology." *Engineering Costs and Production Economics*, 8 (1), 117–128.

Chu, C.H. and J.C. Hayya (1991). A fuzzy clustering approach to manufacturing cell formation, *International Journal of Production Research*, 29, 1475–1487.

Cooper, Carolyn C. (1981–2). "The Production Line at Portsmouth Block Mill." *Industrial Archaeology Review*, 6, 28–44.

Droy, John B. (1984). "It's Time to "Cell' Your Factory." *Production Engineering*, December, 50–52.

Flynn, Barbara B. and F. Robert Jacobs (1987). "An Experimental Comparison of Cellular (Group Technology) Layout with Process Layout." *Decision Sciences* 18, 562–581.

Fogarty, D.W. (1987). "Material Flow in Manufacturing Cell and Material Requirements Planning Environment: Problems and a Solution." *Material Flow*, 4, 139–146.

Gallagher, Colin (1978). Batch Production and Functional Layout. *Industrial Archaeology*, 13, 157–161.

Ghosh and Gagnon (1989). "Assembly Line Research: Historical Roots, Research Life Cycles and Future Directions." *International Journal of Production Research*, 19 (5), 381–399.

Green, T.J. and R.P. Sadowski (1984). "A Review of Cellular Manufacturing Assumptions, Advantages, and Design Techniques." *Journal of Operations Management*, 4, 2 (February), 85–97.

Guerrero, Hector H. (1987). "Group Technology: I. The Essential Concepts." *Production and Inventory Management*, First Quarter, 62–70.

Gunasingh, K.R. and R.S. Lashkari (1989). "Machine grouping problem in cellular manufacturing systems: An integer programming approach." *International Journal of Production Research*, 27, 1465–1473.

Gunther, T.J. and R.P. Sadowski (1984). "Currently Practiced Formulations of the Assembly Line Balance Problem." *Journal of Operations Management*, 3 (3) (August), 209–221.

Gupta, S.M. and Y.A.Y. Al-Turki (1997). "An Algorithm to Dynamically Adjust the Number of Kanbans in Stochastic Processing Times and Variable Demand Environment." *Production Planning and Control*, 8, 133–141.

Ham, I., K. Hitomi, and T. Yoshida (1985). *Group Technology Application to Production Management*. Boston, Kluwer-Nijhoff.

Held, M., R.M. Karp, and R. Shareshian (1963). "Assembly Line Balancing – Dynamic Programming with Precedence Constraints." *Operations Research*, 11, 442–459.

Helgeson, W.P. and D.P. Bernie (1961). "Assembly Line Balancing using the Ranked Positional Weight Technique." *Journal of Industrial Engineering*, 12, 6, 394–398.

Ho, Y.C., M.A. Eyler, and T.T. Chien (1979). "A gradient technique for general buffer storage design in a serial production line." *International Journal of Production Research*, 17, 557–580.

Hoffmann, Thomas (1963). "Assembly Line Balancing with a Precedence Matrix." *Management Science*, 9 (4), 551–562.

Hoffmann, Thomas (1992). "EUREKA: A Hybrid System for Assembly Line Balancing." *Management Science*, 38 (1), 39–47.

Hyer, N.L. (1984). *Group Technology at Work*. Society of Manufacturing Engineer, Dearborn, Mich.

Hyer, N.L. and U. Wemmerlov (1984). "Group Technology and Productivity." *Harvard Business Review*, 62 (July–August), 140–149.

Jackson, J.R. (1956). "A Computing Procedure for a Line Balancing Problem." *Management Science*, 2, 3, 261–271.

Jacobs, F.R., J.W. Bradford, and L.P. Ritzman (1980). "Computerized Layout: An Integrated Approach to Spatial Planning and Communication Requirement." *Industrial Engineering* (July), 56–61.

Johnson, R. (1982). "SPACECRAFT for Multi-Floor Layout Planning." *Management Science*, 28 (4) (April), 407–417.

Johnson, R. (1988). "Optimally Balancing Large Assembly Lines with 'FABLE'." *Management Science*, 34 (2), 240–253.

King, J.R. (1980). "Machine-component grouping in production flow analysis: an approach using a rank order clustering alogrithm." *International Journal of Production Research*, 18 (231–237).

King, J.R. and V. Nakornchai (1982). "Machine-Component Group Formation in Group Technology: Review and Extension." *International Journal of Production Research*, 20 (2), 117–133.

Kinney, H.D. Jr. and L.F. McGinnis (1987). "Design and Control of Manufacturing Cells." *Industrial Engineering* (October), 28–38.

Knight, Donald O. and Michael L. Wall (1989). "Using Group Technology forImproving Communication and Coordination Among Teams of Workers in Manufacturing Cells." *Industrial Engineering* (January), 32–34.

Kusiak, A. (1985). "The Part Families Problem in Flexible Manufacturing Systems." *Annals of Operations Research*, 3, 279–300.

Kusiak, A. (1987). "An Expert System for Group Technology." *Industrial Engineering* (October), 56–61.

Lewis, W.P. and T.E. Block (1980). "On the Application of Computer Aids to Plant Layout." *International Journal of Production Research*, 18 (1), 11–20.

Lin, Li, and David D. Bedworth (1988). "A Semi-Generative Approach to Computer Aided Process Planning Using Group Technology." *Computers and Industrial Engineering*, 14 (2), 127–137.

McCuley, J. (1972). "Machine grouping for efficient production." *Production Engineering*, 2, 53–57.

McMullen, P.R. and G.V. Frazier (1997). "A Heuristic Solution for Stochastic Assembly Line Balancing Problems with Parallel Processing for Mixed–Models." *International Journal of Production Economics*, 51, 177–190.

McMullen, P.R. and G.V. Frazier (1998). "Using Simulated Annealing to Solve a Multiobjective Assembly Line Balancing Problem with Parallel Work Stations." *International Journal of Production Research*, 10, 2717–2741.

Mecklenburg, J.C. (1983). *Plant Layout: A Guide to the Layout of Process Plants and Sites*. Wiley, New York.

Micro-CRAFT (1986). *Plant Layout Software*. Industrial Engineering and Management Press, IIE, Atlanta, GA.

Miltenburg, G.J. and J. Wijngaard (1994). "The U-line Balancing Problem." *Management Science*, 40 (10), 1378–1388.

Moore, J.M. (1980). "The Zone of Compromise for Evaluating Layout Arrangements." *International Journal of Production Research*, 18 (1), 1–10.

Mosier, C.T., D.A. Evers, and D. Kelly (1984). "Analysis of Group Technology Scheduling Heuristics." *International Journal of Production Research*, 22 (5), 857–875.

Muther, Richard, and K. McPherson (1970). "Four Approaches to Computerized Layout." *Industrial Engineering*, 2 (2), 39–42.

Rajagopalan, R. and J.L. Batra (1975). "Design of cellular production systems: a graphtheoretic approach." *International Journal of Production Research*, 13, 567–579.

Reed, Ruddell (1966). *Plant Layout: Factors, Principles, and Techniques*. Irwin, Homewood, IL.

Salverson, M.E. (1955). "The Assembly Line Balancing Problem." *Journal of Industrial Engineering*, 6 (3), 18–25.

Seifoddini, H.K. (1989). "Single linkage versus average linkage clustering in machine cells formation." *Computers Industrial Engineering*, 16, 419–426.

Seifoddini, H.K. and Wolfe, P. (1986). "Application of the similarity coefficient method in group technology." *AIIE Transactions*, 18, 271–277.

Sekine, Kenichi (1990). *One-Piece Flow: Cell Design for Transforming the Production Process*. Productivity Press.

Sekine, Kenichi (1992). *One Piece Flow: Cell Design For Transforming the Production Process*. Productivity Press, Cambridge, Mass.

Shtub, A. (1989). "Estimating the Effects of Conversion to a Group Technology Layout on the Cost of Material Handling." *Engineering Costs and Production Economics*, 16, 103–109.

Shtub, A. (1989). "Modeling group technology cell formation as a generalized a ssignment problem." *International Journal of Production Research*, 27, 775–782.

Spencer, M.S. (1980). "Scheduling Components for Group Technology Lines." *Production and Inventory Management* (Fourth Quarter), 43–49.

Stecke, K.E. and I. Kim (1991). "A Flexible Approach to Part Type Selection in Flexible Flow Systems Using Part Mix Ratios." *International Journal of Production Research*, 29 (1), 53–75.

Suresh, N.C. and J.R. Meredith (1985). "Achieving Factory Automation Through Group Technology Principles." *Journal of Operations Management*, 5 (3) (February), 151–167.

Talbot, Brian and James Patterson (1984). "An Integer Programming Algorithm with Network Cuts for Solving the Assembly Line Balancing Problem." *Management Science*, 30 (1), 85–99.

Tompkins, J.A., J.A. White, Y.A. Bozer, E.H. Frazelle, J.M.A. Tanchoco, and J. Trevino (1996). *Facilities Planning*, Wiley, NY, p. 1–17.

Vakharia, Asoo J. (1986). "Methods of Cell Formation in Group Technology: A Framework for Evaluation." *Journal of Operations Management*, 5 (3) (February), 151–167.

Wemmerlov, U. and N.L. Hyer (1986). "The Part Family/Machine Group Identification Problem in Cellular Manufacturing." *Journal of Operations Management*, 6 (2), 125–147.

Wen, H.J., C.H. Smith, and E.D. Minor (1996). "Formation and dynamic routing of part families among flexible manufacturing cells." *International Journal of Production Research*, 34, 2229–2245.

Zisk, B.I. (1983). "Flexibility is Key to Automated Material Transport System for Manufacturing Cells." *Industrial Engineering*, 15 (11) (November), 58–64.

MANUACTURING ANALYSIS

Aburdene, M.F. (1988). *Computer Simulation of Dynamic Systems*. Brown, Dubuque, IA.

Akinc, U. and B.M. Khumawala (1977). "An Efficient Branch and Bound Algorithm for Capacitated Warehouse Location Problem." *Management Science*, 23, 585–594.

Alsup, F. and R.M. Watson (1993). *Practical Statistical Process Control*, Van Nostrand Reinhold, New York.

Anderson, D., D. Sweeney, and T. Williams (1988). *An Introduction to Management Science: Quantitative Approaches to Decision Making* (5th ed.). West, St. Paul, Minn.

Anderson, D., D. Sweeney, and T. Williams (1990). *Statistics for Business and Economics* (4th ed.). West, Minneapolis, MN.

Anderson, M.Q. (1981). "Monotone Optimal Preventative Maintenance Polices for Stochastically Failing Equipment." *Naval Research Logistics Quarterly*, 28, 347–358.

Archctti, F. and A. Sciomachen (1989). "Development, analysis and simulation of Petri net models: an application to AGV systems." F. Archetti, M. Lucertini,

P. Serafini (eds.) *Operations Research Models in Flexible Manufacturing Systems*, 91–113, Springer-Verlag, New York.

Arrow, K.J., S. Karlin, and H. Scarf (1982). *Studies in the Mathematical Theory of Inventory and Production*. Stanford University Press, Stanford, CA.

Aven, T. and R. Dekker (1997). "A useful framework for optimal replacement models." *Reliability Engineering and System Safety*, 58, 61–67.

Bain, L.J. and M. Engelhardt (1991). *Statistical Analysis of Reliability and Life-Testing Models*, Marcel Dekker, Inc., New York.

Balakrishnan, N., V. Sridharan, and J.W. Patterson (1995). "Rationing Capacity Between Two Product Classes." *Decision Sciences*, 27–2, 185–215.

Balci, O. (1994). "Model validation, verification, and testing techniques throughout the life cycle of a simulation study." *Annals of Operations Research*, 53, 121–173.

Banks, J. and J.S. Carson (1984). *Discrete Event System Simulation*. Prentice-Hall, Englewood Cliffs, NJ.

Banks, J., B. Burnette, H. Kozloski, and J. Rose (1995). *Introduction to SIMAN V and CINEMA V*. John Wiley & Sons, New York.

Barlow, R.E. and F. Proschan (1965). *Mathematical theory of reliability*. John Wiley, New York.

Batdorf, L. and J.A. Vora (1983). "Use of Analytical Techniques in Purchasing." *Journal of Purchasing and Materials Management* (Spring), 25–29.

Baybars, I. (1986). "A Survey of Exact Algorithms for the Simple Assembly Line Balancing Problem." *Management Science*, 32, 909–932.

Bazarra, M.S. and J.J. Jarvis (1990). *Linear Programming and Network Flow* (2nd ed.). John Wiley, New York.

Bechoffer, R., T.J. Santner, and D. Goldsman (1995). *Design and Analysis of Experiments for Statistical Selection, Screening, and Multiple Comparisons*. Wiley, New York.

Bender, P.S., W.D. Northup, and J.F. Shapiro (1981). "Practical Modeling for Resource Management." *Harvard Business Review*, 59 (2), 163–173.

Best, M.J. and K. Ritter (1985). *Linear Programming: Action Set Analysis and Computer Programs*, Prentice-Hall, Englewood Cliffs, NJ.

Bowman, E.H. (1956). "Production Scheduling by the Transportation Method of Linear Programming." *Operations Research*, February, 100–103.

Boxoma, O.J. and R. Syski, eds. (1988). *Queuing Theory and Its Applications*. North Holland, Amsterdam.

Bradley, S.P., A.C. Hax, and T.L. Magnanti (1977). *Applied Mathematical Programming*. Addison-Wesley, Reading, Massachusetts.

Bratley, P., B. Fox, and L. Schrage (1987). *A Guide to Simulation* (2nd ed.). Springer-Verlag. New York.

Brown, S.L. and K.M. Eisenhardt (1997). "The art of continuous change: linking complexity theory and time-paced evolution in relentlessly shifting organizations." *Administrative Science Quarterly*, 42, 1–34.

Bulgren, W. (1982). *Discrete System Simulation*. Prentice-Hall, Englewood Cliffs, NJ.

Bunday, B.D. (1986). *Basic Queuing Theory*. Edward Arnold, London.

Buzacott, J.A. and D.D. Yao (1986). "Flexible manufacturing systems: a review of analytical models." *Management Science*, 32, 890–905.

Buzacott, J.A. and J.G. Shantikumar (1993). *Stochastic Models of Manufacturing Systems*. Prentice-Hall, New Jersey.

Calvert, J.E., *et al.* (1989). *Linear Programming*. Harcourt Brace Jovanovich, New York.

Canel, T.C. and B.M. Khumawala (1997). "Multi-Period International Facility Location: An Algorithm and Application." *International Journal of Production Research*, 35 (7), 1891–1910.

Cao, T. and A.C. Sanderson (1994). "Task decomposition and analysis of robotic assembly task plans using Petri nets." *IEEE Transactions on Industrial Electronics*, 41, 620–630.

Casti, J.L. (1997). *Would-Be Worlds: How Simulation is Changing the Frontiers of Science*. John Wiley & Sons, New York.

Chance, F. (1993). "Conjectured upper bounds on transient mean total waiting times in queueing networks." In: Proceedings of the Winter Simulation Conference, 414–421.

Chance, F., J. Robinson, and J. Fowler (1996). "Supporting manufacturing with simulation: model design, development, and deployment." In: Proceedings of the Winter Simulation Conference, 114–121.

Chandrasekharan, M.P. and R. Rajagoplan (1986). "An ideal-seed non-hierarchical clustering algorithm for cellular manufacturing." *International Journal of Production Research*, 24, 451–554.

Chen, C.H. (1996). "A lower bound on the correct subset selection probability and its application to discrete-event simulation." *IEEE Transactions on Automatic Control*, 41, 1227–1231.

Cheng, R.H.C., W. Holland, and N.A. Hughes (1996). "Selection of input models using bootstrap goodness of fit." In: Proceedings of the Winter Simulation Conference, 199–206.

Chiola, G., G. Franceschinis, R. Gaeta, and M. Ribaudo (1995). "GreatSPN 1.7: graphical editor and analyzer for timed and stochastic Petri nets." *Performance Evaluation*, 24, 47–68.

Cho, H. (1998). "Petri net models for message manipulation and event monitoring in an FMS cell." *International Journal of Production Research*, 36, 231–250.

Chorafas, D.N. (1965). *Systems and Simulation*. Academic Press, New York.

Christy, David P. and Hugh J. Watson (1983). "The Application of Simulation: A Survey of Industry Practice." *Interfaces*, 13, 5, 47–52.

Chu, C.H. and J.C. Hayya (1991). A fuzzy clustering approach to manufacturing cell formation, *International Journal of Production Research*, 29, 1475–1487.

Clark, Charles T. and Eleanor W. Jordan (1985). *Introduction to Business and Economic Statistics*. South-western, Cincinatti.

Clingen, C.T. (1964). "A Modification of Fulkerson's PERT Algorithm." *Operations Research*, 12, 4 (July–August), 629–631.

Cochran, J.K., G.T. Mackulak, and P.A. Savory (1995). "Simulation project characteristics in industrial settings." *Interfaces*, 25 (4), 104–113.

Conover, W.J. (1980). *Practical Nonparametric Statistics* (2nd ed.). John Wiley, New York.

Conway, R.W. and Andrew Schultz, Jr. (1959). "The Manufacturing Progress Function." *Journal of Industrial Engineering*, 10, 1, 39–54.

Cooper, R.B. (1981). *Introduction to Queuing Theory*. (2nd ed.). Elsevier North-Holland, New York.

Crane, M.A. and D.L. Iglehart (1975). "Simulating stable stochastic processes, III: regenerative processes and discrete event simulations." *Operations Research*, 23, 33–45.

D'Souza, K.A. and S.K. Khator (1994). "A survey of Petri net applications in modeling controls for manufacturing systems." *Computers in Industry*, 24, 5–16.

Davis, P.S. and T.L. Ray (1969). "A Branch and Bound Algorithm for the Capacitated Facilities Location Problems." *Naval Research Logistics Quarterly*, 16 (3), Sept.

Dekker, R., C.F.H. Van Rijn, and A.C.J.M. Smit (1994). "Mathematical models for the optimization of maintenance and their application in practice." *Maintenance*, 9, 22–26.

Dematheis, J.J. (1968). "An Economic Lot-Sizing Technique: The Part Period Algorithms." *IBM Systems Journal*, 7, 1, 50–51.

Desrochers, A.A. and R.Y. Al-Jaar (1995). *Applications of Petri Nets in Manufacturing Systems: Modeling, Control, and Performance Analysis*. IEEE Press, Piscataway, NJ.

Dubois, D. and K.E. Stecke (1983). "Using Petri nets to represent production processes." Proceedings of 22nd IEEE Conf. Decision and Control, IEEE Press.

Elmagrhaby, Salah E. "The Theory of Networks and Management Science, II." *Management Science*, 13 (5) (January), 299–306.

Elmagrhaby, Salah E. (1966). "On Generalized Activity Networks." *Journal of Industrial Engineering*, 17, 11 (November), 621–631.

Eppen Gary D., J.F. Gould, and C.P. Schmidt (1987). *Introductory Management Science* (2nd ed.). Prentice-Hall, Englewood Cliffs, N.J.

Erikson, W.J. and O.P. Hall, Jr. (1983). *Computer Models for Management Science*. Addison-Wesley, Reading, MA.

Evans, J.R. and D.L. Olson (1998). *Simulation*. Prentice-Hall, Englewood Cliffs, NJ.

Eyman, Earl D. (1988). Modeling, *Simulation and Control*. West, St. Paul.

Ezpeleta, J. and J. Martínez (1992). "Petri nets as a specification language for manufacturing systems." In J.-C. Gentina, S. G. Tzafestas (eds.). *Robotics and Flexible Manufacturing Systems*, 427–436, North-Holland, New York.

Fava, J.A., R. Denison, B. Jones, M.A. Curran, B. Vigon, S. Selke, and J. Barnum (Ed.) (1991). *A Technical Framework for Life-Cycle Assessments*. Washington, DC: Society of Environmental Toxicology and Chemistry (SETEC).

Feldman E., F.A. Lehrer, and T.L. Ray (1966). "Warehouse Locations Under Continuous Economies of Scale." *Management Science*, 12, May.

Fishman, G.S. (1971). "Estimating Sample Size in Computer Simulation Experiments." *Management Science*, 18 (1), 21–38.

Florin, G. and S. Natkin (1982). "Evaluation based upon stochastic Petri nets of the maximum throughput of a full duplex protocol." C. Girault, W. Reisig (eds.). *Application and Theory of Petri Nets*, 280–288, Springer-Verlag, New York.

Fordyce, J.M. and F.M. Webster (1984). "The Wagner-Whitin Algorithm Made Simple." *Production and Inventory Management*, 25 (2), 21–30.

Forrester, J.W. (1988). *Industrial Dynamics*. MIT Press, Cambridge, Mass.

Francis, R.L. and J.A. White (1987). *Facilities Layout and Location; An Analytical Approach*. Prentice-Hall, Englewood Cliffs, NJ.

Fransoo, J.C., V. Sridharan, and J.W.M. Bertrand (1995). "A Hierarchical Approach for Capacity Coordination in Multiple Products Single-Machine Production Systems With Stationary Stochastic Demands." *European Journal of Operational Research*, 86 (1), 57–72.

Freund, J.D. and R.E. Walpole (1980). *Mathematical Statistics* (3rd ed.). Prentice-Hall, Englewood Cliffs, NJ.

Galdwin, C.H. (1989). *Ethnographic Decision Tree Modeling*. Sage, Beverly Hills, CA.

Gass, S.I. and C.M. Harris (1996). *Encyclopedia of Operations Research and Management Science*, Kluwer, Boston.

Gen, M. and R. Cheng (1997). *Genetic Algorithms and Engineering Design*, Wiley and Sons, New York.

Geoffrion, A.M. and G.W. Graves (1974). "Multicomodity Distribution System Design by Bender's Decomposition." *Management Science*, 12, 822–844.

Glasserman, P. (1991). *Gradient Estimation via Perturbation Analysis*. Kluwer Academic Publishers, MA.

Glasserman, P. (1994). "Perturbation analysis of production networks." In: *Stochastic Modeling and Analysis of Manufacturing Systems* (Yao, Ed.). Springer-Verlag, New York.

Glasserman, P. and D.D. Yao (1992). "Some guidelines and guarantees for common random numbers." *Management Science*, 38, 884–908.

Gleick, J. (1987). *Chaos: Making a New Science*. Viking, New York.

Graybeal, W. and U.W. Pooch (1980). *Simulation: Principles and Methods*. Cambridge, MA, Winthrop.

Gunasingh, K.R. and R.S. Lashkari (1989). "Machine grouping problem in cellular manufacturing systems: An integer programming approach." *International Journal of Production Research*, 27, 1465–1473.

Haider, S.W. and J. Banks (1986). "Simulation Software Products for Analyzing Manufacturing

Systems." *Industrial Engineering*, 18, 7 (July), 98–103.

Hajek, B. (1984). Optimal control of two interacting service stations, *IEEE Transactions on Automatic Control*, 29, 491–499.

Harmon, P. and D. King (1985). *Expert Systems: Artificial Intelligence in Business*. Wiley, New York.

Haug, P. (1985). "A Multiple Period, Mixed Integer Programming Model for Multinational Facility Location." *Journal of Management*, 11 (3), 83–96.

Hayhurst, G. (1987). *Mathematical Programming Applications*. MacMillan, New York.

Held, M., R.M. Karp, and R. Shareshian (1963). "Assembly Line Balancing – Dynamic Programming with Precedence Constraints." *Operations Research*, 11, 442–459.

Helgeson, W.P. and D.P. Bernie (1961). "Assembly Line Balancing using the Ranked Positional Weight Technique." *Journal of Industrial Engineering*, 12, 6, 394–398.

Hertz, D.D. (1982). *Risk Analysis and Its Applications*. John Wiley, New York.

Hertz, D.D. (1983). *Practical Risk Analysis*. John Wiley, New York.

Herzog, D.R. (1985). *Industrial Engineering Methods and Controls*. Reston Publishing, Reston, VA.

Hill, Arthur D., J.D. Naumann, and Norman L. Chervany (1983). "SCAT and SPAT: Large-Scale Computer-Based Optimization Systems for the Personnel Assignment Problem." *Decision Sciences*, 14 (2), (April), 207–220.

Hillier, F.S. and G.J. Lieberman (1990). *Introduction to Operations Research*. McGraw-Hill, New York.

Hillier, F.S. and O.S. Yu (1981). *Queuing Tables and Graphs*. Elsevier North-Holland, New York.

Hoaglin, D.C., F. Mosteller, and J.W. Tukey (1983). *Understanding Robust and Exploratory Data Analysis*. John Wiley, New York.

Hodder, J.E. and J.V. Jucker (1985). "International Plant Location under Price and Exchange Rate Uncertainty." *Engineering Costs and Production Economics*, 9, 225–229.

Hoffman, J.J. and M.J. Schniederjans (1990). "An International Strategic Management/Goal Programming Model for Structuring Global Expansion Decisions in the Hospitality Industry." *International Journal of Hospitality Management*, 9 (3), 175–190.

Hoffman, J.J. and M.J. Schniederjans (1994). "A Two-Stage Model for Structuring Global Facility Site Selection Decisions: The Case of the Brewing Industry." *International Journal of Operations and Production Management*, 14(4), 79–96.

Hoffman, K.L. and M. Padberg (1996). "Traveling Salesman Problem." In *Encyclopedia of Operations Research and Management Science*, S.I. Gass and C.M. Harris, ed., Kluwer, Boston.

Hoffman, Thomas R. and Gary D. Scudder (1983). "Priority Scheduling With Cost Considerations." *International Journal of Production Research*, 21, 6, 881–889.

Holloway, C.A. (1979). *Decision Making Under Uncertainty*. Prentice-Hall, Englewood Cliffs, NJ.

Hooker, J.N. (1986, 1987). "Karmarkar's Linear Programming Algorithm." *Interfaces*, 16 (4) (July–August), 75–90 and 17 (1) (January–February), 128.

Hoover, S.V. and R.F. Perry (1989). *Simulation: A Problem Approach*, Addison-Wesley, Reading, MA.

Hopp, W.J. and M.L. Spearman (1996). *Factory Physics*, Irwin, IL.

Horn, R.E. and A. Clearas (1980). *The Guide to Simulation and Games* (4th ed.) Beverly Hills, CA, Sage.

Ignizio, J.P. (1982). *Linear Programming in Single and Multiple Objective Systems*. Prentice-Hall, Englewood Cliffs, NJ.

Imany, M.M. and R.J. Schlesinger (1989). "Decision models for robot selection: a comparison of ordinary least squares and linear goal programming methods." *Decision Science*, 20, 40–53.

Jackson, J.R. (1963). "Jobshop-like queueing systems." *Management Science*, 10, 131–142.

Jacobson, S.H. and L. Schruben (1989). "Techniques for simulation response optimization." *Operations Research Letters*, 8, 1–9.

Jaikumar, R. and R. Bohn (1992). "A Dynamic Approach to *Operations Management*: An Alternative to Static Optimization." *International Journal of Production Economics*.

Jankauskas, L. and S. McLafferty (1996). "BESTFIT, Distribution fitting software by Palisade Corporation." In: Proceedings of the Winter Simulation Conference, 551–555.

Jensen, K. and G. Rozenberg (eds.) (1991). *High-Level Petri Nets: Theory and Application*. Springer-Verlag, New York.

Johnson, Linwood and Douglas C. Montgomery (1974). *Operatons Research in Production Planning, Scheduling and Inventory Control*. John Wiley & Sons, New York.

Karmarker, N. (1984). "A New Polynomial-Time Algorithm for Linear Programming." *Combinatorica*, 4 (4), 373–395.

Kauffman, S. (1995). *At Home in the Universe: The Search for Laws of Self-Organization and Complexity*. Oxford University Press, New York.

Keuhn, A.A. and M.J. Hamburger (1963). "A Heuristic Program for Locating Warehouses." *Management Science*, 9, July.

Khouja, M. (1995). "The use of data envelopment analysis for technology selection." *Computers and Industrial Engineering*, 28, 123–132.

Khouja, M. and D.E. Booth (1991). "A decision model for the robot selection problem using robust regression." *Decision Sciences*, 22, 656–662.

Khouja, M. and D.E. Booth (1995). "Fuzzy clustering procedure for evaluation and selection of industrial robots." *Journal of Manufacturing Systems*, 14, 244–251.

Khouja, M. and O.F. Offodile (1994). "The industrial robots selection problem: literature review and directions for future research." *IIE Transactions*, 26 (4), 50–61.

Khumawala, B.M. (1972). "An Efficient Branch and Bound Algorithm for the Warehouse Location Problem." *Management Science*, 18 (12), 718–731.

Khumawala, B.M. (1974). "An Efficient Heuristic Procedure for the Capacitated Warehouse Location Problem." *Naval Research Logistics Quarterly*, 21, 609–623.

Khumawala, B.M. and D.C. Whybark (1971). "A Comparison of Some Recent Warehouse Location Techniques." *The Logistics Review*, 7 (31), 3–19.

Kiel, L.D. and E. Elliott, eds. (1996). *Chaos and the Social Sciences: Foundations and Applications*. University of Michigan Press, Ann Arbor.

King, C.B. (1996). "Taylor II manufacturing simulation software." In: Proceedings of the Winter Simulation Conference, 569–573.

King, J.R. (1980). "Machine-component grouping in production flow analysis: an approach using a rank order clustering algorithm." *International Journal of Production Research*, 18 (231–237).

Kingman, J.F.C. (1961). "The single-server queue in heavy traffic." In: *Proceedings of the Cambridge Philosophical Society*, 57, 902–904.

Kiviat, P.J.R. Villanueva, and H.M. Markowitz (1969). *The SIMSCRIPT II.5 Programming Language*. Consolidated Analysis centers, Los Angeles.

Kleijnen, J.P.C. (1987). *Statistical Tools for Simulation Practitioners*. Marcel Dekker, New York.

Knott, K. and R.D. Getto, Jr. (1982). "A Model for evaluating alternative robot systems under uncertainty." *International Journal of Production Research*, 20, 155–165.

Kolata, G. (1984). "A Fast Way to Solve Hard Problems." *Science*, September 21, 1379–1380.

Kolman, B. and R.E. Beck (1980). *Elementary Linear Programming with Applications*. Academic Press, New York.

Kusiak, A. (1987). "An Expert System for Group Technology." *Industrial Engineering* (October), 56–61.

L'Ecuyer, P. (1990). "A unified view of the IPA, SF, and LR gradient estimation techniques." *Management Science*, 36, 1364–1383.

Lane, M.S., A.H. Mansour, and J.L. Harpell (1993). "Operations research techniques: A longitudinal update 1973–1988." *Interfaces*, 23 (2), 63–68.

Law, A.M. and M.G. McComas (1989). "Pitfalls to avoid in the simulation of manufacturing systems." *Industrial Engineering*, 31, 28–31, 69.

Law, A.M. and M.G. McComas (1996). "EXPERTFIT: Total support for simulation input modeling." In: Proceedings of the Winter Simulation Conference, 588–593.

Law, A.M. and W.D. Kelton (1982). "Computer Simulation of Manufacturing Systems: Part I." *Industrial Engineering*, 18, no. 5 (May), 46–63.

Law, A.M. and W.D. Kelton (1982). *Simulation Modeling and Analysis*. McGraw-Hill, New York.

Lee, S.M. (1972). *Goal Programming for Decision Analysis*. Auerbach Publishers, Pennsauken, N.J.

Lee, W.B. and B.M. Khumawala (1974). "Simulation Testing of Aggregate Production Planning Models in an Implementation Methodology." *Management Science*, 20 (6) (February), 903–911.

Lee, W.B. and C.P. McLaughlin (1974). "Corporate Simulation Models for Aggregate Materials Management." *Production and Inventory Management*, 15 (1), 55–67.

Lee-Kwang, H. and J. Favrel (1985). "Hierarchical reduction method for analysis and decomposition of Petri nets." *IEEE Transactions on Systems, Man, and Cybernetics*, SMC-15, 272–280.

Lee-Kwang, H., J. Favrel, and P. Baptiste (1987). "Generalized Petri net reduction method." *IEEE Transactions on Systems, Man and Cybernetics*, SMC-17, 297–303.

Leemis, L.M. (1996). "Discrete event simulation input process modeling." *In: Proceedings of the Winter Simulation Conference*, 39–46.

Lev, B. and H.J. Weiss (1982). *Introduction to mathematical Programming*. Elsevier North-Holland, New York.

Levy, D. (1994). "Chaos theory and strategy: theory, application, and managerial implications." *Strategic Management Journal*, 15, 167–78.

Lilien, G. and P. Kotler (1983). *Marketing Decision Making – A Model-Building Approach*, Harper & Row Publishers, New York.

Lin, J.T. and C.C. Lee (1992). "A modular approach for the modelling of a class of zone-control conveyor system using timed Petri nets." *International Journal of Computer Integrated Manufacturing*, 5, 277–289.

Lindemann, C. (1998). *Performance Modeling with Deterministic and Stochastic Petri Nets*. John Wiley.

Luenberger, D.G. (1985). *Linear and Nonlinear Programming* (2nd ed.). Addison-Wesley, Reading, MA.

Mabert, V.A. and C.A. Watts (1982). "Simulation Analysis of Tour-Shift Construction Procedures." *Management Science* (May), 520–532.

Maisel, Herbert, and Guiliano Gnugnoli (1972). *Simulation of Discrete Stochastic Systems*. Science Research Associates, Chicago.

Maki, Daniel P. and Maynard Thompson (1973). *Mathematical Models and Applications*. Prentice-Hall, Englewood Cliffs, N.J.

Mandelbrot, B.B. (1983). *The Fractal Geometry of Nature*, W.H. Freeman & Co., San Francisco, CA.

McCall, J.J. (1965). "Maintenance Policies for Stochastically Failing Equipment: A Survey." *Management Science*, 11 (5), 493–524.

McMillan, Claude, Jr., and Richard F. Gonzales (1968). *Systems Analysis: A Computer Approach to Decision Models* (rev. ed.). Richard D. Irwin, Homewood, Il.

McMullen, P.R. and G.V. Frazier (1997). "A Heuristic Solution for Stochastic Assembly Line Balancing Problems with Parallel Processing for Mixed-Models." *International Journal of Production Economics*, 51, 177–190.

McMullen, P.R. and G.V. Frazier (1998). "Using Simulated Annealing to Solve a Multiobjective Assembly Line Balancing Problem with Parallel Work

Stations." *International Journal of Production Research*, 10, 2717–2741.

Mendenhall, W., R.L. Schaeffer, and D. Wackerly (1986). *Mathematical Statistics with Applications* (3rd ed.). PWS-Kent, Boston.

Minoux, M. (1986). *Mathematical Programming*. John Wiley, New York.

Mize, J.H. and J.G. Cox (1968). *Essentials of Simulation*. Prentice-Hall, Englewood Cliffs, N.J.

Molloy, M.K. (1982). "Performance analysis using stochastic Petri nets." *IEEE Transactions on Computers*, C-31, 913–917.

Moore, K.E. and S.M. Gupta (1996). "Petri net models of flexible and automated manufacturing systems: a survey." *International Journal of Production Research*, 34, 3001–3035.

Murata, T. (1989). "Petri nets: properties, analysis and applications." Proceedings of the IEEE, 77, 541–580.

Murty, K. (1983). *Linear Programming* (2nd ed.). John Wiley, New York.

Narasimhan, R. (1983). "An Analytical Approach to Supplier Selection." *Journal of Purchasing and Materials Management* (Winter), 27–32.

Naylor, T.J., J.L. Balintfy, D.S. Burdick, and Cong Chu (1966). *Computer Simulation Techniques*. John Wiley & Sons, New York.

Neelamkavil, F. (1987). *Computer Simulation and Modeling*. John Wiley, New York.

Neter, J., W. Wasserman, and G.A. Whitmore (1987). *Applied Statistics* (3rd ed.). Boston, Allyn and Bacon.

Newell, Gordon F. (1982). *Applications of Queuing Theory*. Chapman and Hall, New York.

Nicolis, G. and I. Prigogine (1989). *Exploring Complexity: An Introduction*. W.H. Freeman & Co., San Francisco, CA.

Niebel, B.W. (1988). *Motion and Time Study* (8th ed.). Irwin, Homewood IL.

Nuyens, R.P.A., N.M. Van Dijk, L. Van Wassenhove, and E. Yücesan (1996). "Transient behavior of simple queueing systems: implications for FMS models." *Simulation: Application and Theory*, 4, 1–29.

Özekici, S. (1996). Editor. *Reliability and maintance of complex systems. NATO ASI Series*, vol. 154, Springer, Berlin.

Ozekici, S. and S.R. Pliska (1991). "Optimal Scheduling of Inspections: A Delayed Markov Model with False Positive and Negatives." *Operations Research*, 39 (2), 261–273.

Padulo, Louis, and Michael A. Arbib (1974). *Systems Theory*. Hemisphere, Washington, D.C.

Papadopoulos, H.D., C. Heavey, and J. Browne (1993). *Queueing Theory in Manufacturing Systems Analysis and Design*. Chapman and Hall. London, UK.

Payne, J.A. (1982). *An Introduction to Simulation*. McGraw-Hill, New York.

Perrors, H.G. (1994). *Queueing Networks with Blocking*. Oxford University Press Oxford.

Peters, T. (1987). *Thriving on Chaos: Handbook for a Management Revolution*. Knopf, New York.

Phillips, F. and N. Kim (1996). "Implications of chaos research for new product forecasting." *Technological Forecasting and Social Change*, 53, 239–61.

Pidd, M. (1984). *Computer Simulation in Management Science*. John Wiley, New York.

Plane, Donald, and Gray Cochenberger (1972). *Operations Research for Managerial Decisions*. Richard D. Irwin, Inc., Homewood, IL.

Pritsker, A.A.B. (1986). *Introduction to Simulation and SLAM II*. John Wiley & Sons, New York.

Pritsker, A.A.B. (1989). "Developing analytic models based on simulation results." In: Proceedings of the Winter Simulation Conference, 653–660.

Production Modeling Corporation of Utah (1989). *ProModel User's Manual*. Orem, UT.

Ragatz, G.L. and V.A. Mabert (1984). "A Simulation Analysis of Due Date Assignment Rules." *Journal of Operations Management*, 5 (1) (November), 27–40.

Rajagopalan, R. and J.L. Batra (1975). "Design of cellular production systems: a graphtheoretic approach." *International Journal of Production Research*, 13, 567–579.

Ramamoorthy, C.V. and G.S. Ho (1980). "Performance evaluation of asynchronous concurrent systems using Petri nets." *IEEE Transactions on Software Engineering*, SE-6, 440–449.

Reiman, M.I. and A. Weiss (1989). "Sensitivity analysis for simulations via likelihood ratios." *Operations Research*, 37, 830–844.

Robinson, J., L.W. Schruben, and J.W. Fowler (1993). "Experimenting with large-scale semiconductor manufacturing simulations: a frequency domain approach to factor screening." In: *Proceedings of the Second IE Research Conference*, 112–116.

Rubin, P.A. and G.L. Ragatz (1995)."Scheduling in a Sequence Dependent Setup Environment with Genetic Search." *Computers and Operations Research*, 22 (1), 85–99.

Rubinstein, R.Y. and A. Shapiro (1993). *Discrete Event Systems: Sensitivity Analysis and Stochastic Optimization by the Score Function Method*. Wiley, New York.

Ruefli, T. (1971). "A Generalized Goal Decomposition Model." *Management Science*, 17 (8), 505–518.

Russell, S. and P. Norvig (1995). *Artificial Intelligence: A Modern Approach*, Prentice-Hall, Englewood Cliffs, NJ.

Salvador, Michael S., *et al*. (1975). "Mathematical Modeling Optimization and Simulation Improve Large-Scale Finished Goods Inventory Management." *Production and Inventory Control Management*, Second Quarter.

Savsar, M. (1996). "Effects of Kanban Withdrawal Policies and other Factors on the Performance of JIT Systems – A Simulation Study." *International Journal of Production Research*, 34, 2879–2899.

Schilling, D.A. (1980). "Dynamic Location Modeling for Public Sector Facilities: A Multi-Criteria Approach." *Decision Sciences*, 11 (4) (October), 714–724.

Schmeiser, B.W. (1983). "Batch size effects in the analysis of simulation output." *Operations Research*, 31, 565–568.

Schniederjans, M.J. and J.J. Hoffman (1992). "Multinational Acquisition Analysis: A Zero-One Goal Programming Model." *European Journal of Operational Research*, 62, 174–185.

Schrage, L. (1986). *Linear, Integer, and Quadratic Programming with LINDO* (3rd ed.) Scientific Press, Palo alto, CA.

Schruben, L. and V.J. Cogliano (1987). "An experimental procedure for simulation response surface model identification." *Communications of the ACM*, 30, 716–730.

Seidmann, A. and S.Y. Nof (1989). "Operational analysis of an autonomous assembly robotic station." *IEEE Transactions on Robotics and Automation*, 5, 4–15.

Seidmann, A., A. Arbel, and R. Shapira (1984). "A two-phase analytic approach to robotic system design." *Robotics and Computer-Integrated Manufacturing*, 12, 181–190.

Shannon, Robert E. (1975). *System Simulation – the Art and Science*. Prentice-Hall, Englewood Cliffs, N.J.

Shantikumar, J.G. and R.G. Sargent (1983). "A unifying view of hybrid simulation/analytic models and modeling." *Operations Research*, 31, 1030–1052.

Shtub, A. (1989). "Modeling group technology cell formation as a generalized assignment problem." *International Journal of Production Research*, 27, 775–782.

Siha, S. (1994). "The Pull Production System: Modelling and Characteristics." *International Journal of Production Research*, 32, 933–950.

Sinha, S.K. and B.K. Kale (1980). *Life Testing and Reliability Estimation*. John Wiley, New York.

Smith, L.A. and Joan Mills (1982). "*Project Management Network Programs.*" *Project Management Quarterly*, 18–29.

Solberg, J.J. (1976). "Optimal design and control of computerized manufacturing systems." In Proceedings of AIIE Systems Engineering Conference, 138–144.

Solomon, S.L. (1983). *Simulation of Waiting Line Systems*. Prentice-Hall, Englewood Cliffs, NJ.

Stecke, K.E. (1983). "Formulation and Solution of Nonlinear Integer Production Planning Problems for Flexible Manufacturing Systems." *Management Science*, 29 (3), 273–288.

Stecke, K.E. (1986). "A Hierarchical Approach to Solving Machine Grouping and Loading Problems of Flexible Manufacturing Systems." *European Journal of Operational Research*, 24, 369–378.

Stecke, K.E. and I. Kim (1991). "A Flexible Approach to Part Type Selection in Flexible Flow Systems Using Part Mix Ratios." *International Journal of Production Research*, 29 (1), 53–75.

Stevenson, W.J. (1989). *Introduction to Management Science*. Irwin, Homewood, IL.

Strayer, J.K. (1989). *Linear Programming and Its Applications*. Springer-Verlag, New York.

Swain, J.J. (1995). "Simulation Survey: Tools for process understanding and improvement." *OR/MS Today*, 22 (4), 64–72 and 74–79.

Swamidass, P.M., S.S. Nair, and S.I. Mistry (1999). "The use of a nueral factory to investigate the effect of product line width on manufacturing performance." *Management Science*, forthcoming.

Taha, H.A. (1987). *Operations Research: An Introduction* (4th ed.). Prentice-Hall, New York.

Talbot, Brian and James Patterson (1984). "An Integer Programming Algorithm with Network Cuts for Solving the Assembly Line Balancing Problem." *Management Science*, 30 (1), 85–99.

Turban, E. and J.R. Meredith (1991). *Fundamentals of Management Science* (5th ed.). Irwin, Homewood, IL.

Ulvill, J. and R.V. Brown (1982). "Decision Analysis Comes of Age." *Harvard Business Review* (September–October), 130–141.

Van Dijk, N.M. (1993). *Product Forms for Queueing Networks: A System's Approach*, Wiley, New York.

Van Gigch, John P. (1978). *Applied General Systems Theory*. Harper and Rowe, New York.

Van Slyke, Richard M. (1963). "Monte Carlo Methods and the PERT Problem." *Operations Research*, 11 (5) (September–October), 129–143.

Vinrod, B. and T. Altiok (1986). "Approximating Unreliable Quewing Networks Under the Assumption of Exponentiality." *Journal of the Operational Research Society* (March), 309–316.

Viswanadham, N. and Y. Narahari (1992). *Performance Modeling of Automated Manufacturing Systems*. Prentice-Hall, New Jersey.

Von Winterfeldt, D. and W. Edwards (1986). *Decision Analysis and Behavioral Research*. Cambridge University Press, New York.

Wagner, Harvey (1974). *Operations Research* (2nd ed.). Prentice-Hall, Englewood Cliffs, N.J.

Waldman, K.H. (1983). "Optimal Replacement Under Additive Damage in Randomly Varying Environments." *Naval Research Logistics Quarterly*, 30, 377–386.

Walrand, J. (1988). *An Introduction to Queuing Networks*. Prentice-Hall, Englewood Cliffs, NJ.

Warnecke, H.J. (1993). *The Fractal Company*. Springer-Verlag, New York, N.Y.

Watson Hugh J. (1981). *Computer Simulation in Business*. Wiley, New York.

Watson, H.J. and J.H. Blackstone, Jr. (1989). *Computer Simulation* (2nd ed.) John Wiley, New York.

Watson, S.R. and D.M. Buede (1987).*Decision Synthesis*. Cambridge University Press, New York.

Weatherford, L.R. and S.E. Bodily (1992). "A taxonomy and research overview of perishable-asset revenue management: Yield management, overbooking, and pricing." *Operations Research*, 40 (5), 831–844.

Wellstead, P.E. (1979). *Introduction to Physical System Modeling*. Academic Press, London.

Weston, Frederick C. Jr. (1980). "A Simulation Approach to Examining Traditional EOQ/EOP and Single Order Exponential Smoothing Efficiency Adopting a Small Business Perspective." *Production and Inventory Management*, 21 (2), 67–83.

Whitt, W. (1983). "The Queueing Network Analyzer." *The Bell System Technical Journal*, 62, 2779–2815.

Whybark, C. and J.G. Williams (1976). "Material Requirements Planning under Uncertainty." *Decision Sciences*, 7 (1), 595–606.

Wiest, Jerome D. (1966). "Heuristic Programs for Decision Making." *Harvard Business Review*, September–October, 129–143.

Winston, W.L. (1996). *Simulation Modeling Using @RISK*. Duxbury Press, Belmont, CA.

Yao, D.D. (1994). *Stochastic Modeling and Analysis of Manufacturing Systems*. Springer-Verlag, New York.

Ýucesan, E. (1993). "Randomization tests for initialization bias in simulation output." *Naval Research Logistics*, 40, 643–664.

Yücesan, E. and S.H. Jacobson (1996). "Computational issues for Accessibility in discrete event simulation." *ACM Transactions on Modeling and Computer Simulation*, 6, 53–75.

Zhou, M. and F. DiCesare (1993). *Petri Net Synthesis for Discrete Event Control of Manufacturing Systems*. Kluwer, Boston, MA.

Zurawski, R. (1994). "Systematic construction of functional abstractions of Petri net models of flexible manufacturing systems." *IEEE Transactions on Industrial Electronics*, 41, 584–591.

Zurawski, R. and M. Zhou (1994). "Petri nets and industrial applications: a tutorial." *IEEE Transactions on Industrial Electronics*, 41, 567–583.

MANUFACTURING PERFORMANCE

Aase, Gerald (1997). "Improving Labor Productivity with U-shaped Assembly Lines." *Proceeding for the 26th Annual Meeting of the Decision Science Institute*, San Diego, CA.

Abernathy, William J., and Phillip L. Townsend (1975). "Technology, Productivity, and Process Change." *Technology Forecasting and Social Change*, 7, 379–396.

Adler, P.S. (1990). "Shared Learning." *Management Science*, 938–957.

Adler, P.S. (1993). "Time-and-Motion Regained." *Harvard Business Review*, January–February, 97–108.

Aitken, Hugh G.J. (1960). *Taylorism at Watertown Arsenal; Scientific Management in Action*, 1908–1915. Harvard University Press, Cambridge, Mass.

Albertson, P. (1983). "Verifying robot performance." *Robotics Today*, 33–36.

Allen, Th. J. (1986). "Organizational Structure, Information Technology, and R&D Productivity." *IEEE Transactions on Engineering Management*, EM 33, 212–217.

Altiok, T. (1997). *Performance Analysis of Manufacturing Systems*. Springer, New York.

American Productivity Center (1987). *Productivity Perspectives*. Houston, TX. December, p. 12.

Argote, L., S. Beckman, and D. Epple (1990). "The Persistence and Transfer of Learning in Industrial Settings." *Management Science* (February), 36, 2, 140–154.

AT&T Corporate Quality Office (1994). *Using ISO 9000 to Improve Business Processes*.

Automotive Industry Action Group (AIAG) (1983). *The Japanese Approach to Productivity*. Videotape Series.

Babcock, George D. (1918). *The Taylor System in Franklin Management*. The Engineering Magazine Co., New York, New York.

Banker, R.D., J.M. Field, R.G. Schroeder, and K.K. Sinha (1996). "Impact of Work Teams on Manufacturing Performance." *Academy of Management Journal*, 39 (4), 867–890.

Barnes, R. (1980). *Motion and Time Study: Design and Measurement of Work* (8th ed.). John Wiley, New York.

Bemowski, K. (1996). "Baldrige Award Celebrates Its 10th Birthday With a New Look." *Quality Progress*, 29, 49–54.

Berger, J. (1987). "Productivity: Why It's the Number 1 Underachiever." *Business Week*, April 20, 54–55.

Billesbach, Thomas J. and Hayen, Roger (1994). "Long-Term Impact of Just-In-Time on Inventory Performance Measures." *Production and Inventory Management Journal*, V. 35, 1, 62–67.

Burnham, J.M. (1985). "Improving Manufacturing Performance: A Need for Integration." *Production and Inventory Management*, 26, 4, 51–59.

Cooper, R. (1987). "Does your company need a new cost system?" *Journal of Cost Management* (Spring), 45–49.

Cooper, R. (1988). "The rise of activity-based costing – part one: What is an activity-based cost system?" *Journal of Cost Management* (Summer), 45–54.

Cooper, R. (1988). "The rise of activity-based costing – part two: When do I need an activity-based cost system?" *Journal of Cost Management* (Fall), 41–48.

Cooper, R. (1989). "You Need a New Cost System When...." *Harvard Business Review*, 77 (1) (January–February), 77–82.

Cooper, R. (1989). "The rise of activity-based costing – part three: How many cost drivers do you need, and how do you select them?" *Journal of Cost Management* (Winter), 34–46.

Cooper, R. (1990). "Cost classifications in unit-based and activity-based manufacturing cost systems." *Journal of Cost Management for the Manufacturing Industry* (Fall), 4–14.

Cooper, R. and R.S. Kaplan (1988). "Measuring costs right make the right decisions." *Harvard Business Review*, 66, 96–103.

Cooper, R. and R.S. Kaplan (1991). "Profit Priorities from Activity-Based Costing." *Harvard Business Review*, May–June, 130–135.

Cooper, R. and R.S. Kaplan (1991). *The Design of Cost Management Systems*, Prentice-Hall, Englewood Cliffs.

Cooper, R. and R. Slagmulder (1997). *Target Costing and Value Engineering, Institute of Management Accountants*. Productivity Press.

Cooper, R., and W.B. Chew (1996). "Control Tomorrow's Costs through Today's Designs." *Harvard Business Review*, January–February, 88–97.

Cooper, R., R.S. Kaplan, L.S. Maisel, E. Morrissey, and R.M. Oehm (1992). *Activity-Based Cost Management: Moving from Analysis to Action*. The Institute of Management Accounts, New Jersey.

Courtis, John K. (1995). "JIT's Impact on a Firm's Financial Statements." *International Journal of Purchasing and Materials Management*, 31, 1, Winter, 46–51.

Craighead, Thomas G. (1989). "EDI Impact Grows for Cost Efficient Manufacturing." *PIM Review*, July, 32.

Crawford, K.M., J.F. Cox, and J.H. Blackstone, Jr. (1988). Performance Measurement Systems and the JIT Philosophy: Principles and Cases. *American Production and Inventory Control Society*, Falls Church, VA.

Dixon, J.R., A.J. Nanni, Jr., and T.E. Vollmann (1990). *The New Performance Challenge: Measuring Operations for World-Class Competition*. Business One Irwin, IL.

Eccles, R.G. (1991). "The performance measurement manifesto." *Harvard Business Review*.

Fallon, Carlos (1971). *Value Analysis to Improve Productivity*. John Wiley & Sons, New York.

Fawcett, S.E., L.E. Stanley, and S.R. Smith (1997). "Developing a Logistics Capability to Improve the Performance of International Operations." *Journal of Business Logistics*, 18 (2), 101–128.

Ferdows, K. and A. De Meyer (1990). "Lasting Improvements in Manufacturing Performance: In Search of a New Theory." *Journal of Operations Management*, 9 (2), 168–184.

Gale, Bradley T. (1980). "Can More Capital Buy Higher Productivity?" *Harvard Business Review*, 59 (4), 78–86.

Gantt, H.L. (1916). *Work, Wages and Profits*. The Engineering Magazine Co., New York, New York.

Garwood, Dave, and Michael Bane (1988). *A Jumpstart to World Class Performance*. Dogwood Publishing, Marietta, GA.

Gitlow, H.S. and P.T. Hertz (1983). "Product Defects and Productivity." *Harvard Business Review* (September–October), 131–141.

Goldratt, E.M. (1983). "Cost Accounting is Enemy Number One of Productivity." International Conference Proceedings, American *Production and Inventory Control Society*, October.

Hall, R.W., N.P. Galambos, and M. Karlsson (1997). "Constraint-based profitability analysis: stepping beyond the theory of constraints." *Journal of Cost Management*, 11, 6–10.

Harmon, Roy L. and Peterson, D. Leroy (1990). *Reinventing The Factory: Productivity Breakthroughs In Manufacturing Today*. The Free Press.

Hayes, R.H. and K.B. Clark (1986). "Why Some Factories are More Productive Than Others." *Harvard Business Review*, September–October, 66–73.

Hillion, H.P. and J.M. Proth (1989). "Performance evaluation of job-shop systems using timed event-graphs." *IEEE Transactions on Automatic Control*, 34.

Hughes, M.W. (1986). "Why Projects Fail: The Effects of Ignoring the Obvious." *Industrial Engineering*, 18, 4 (April), 14–18.

Kanter R. (1985). *The Change Masters: Innovation For Productivity In The American Corporation*. New York, Simon and Schuster.

Kaplan, R.S. (1988). "One Cost System Isn't Enough." *Harvard Business Review*, January/February, 61–68.

Kaplan, R.S. (1991). "New Systems for Measurement and Control." *The Engineering Economist*, 36 (3), 201–218.

Kaplan, R.S. and D.P. Norton (1992). "The balanced scorecard – measures that drive performance." *Harvard Business Review*, 70, 71–79.

Kaplan, R.S. and D.P. Norton (1993). "Putting the balanced scorecard to work" *Harvard Business Review*, September–October.

Kaplan, R.S. and D.P. Norton (1996). "Using the balanced scorecard as a strategic management system." *Harvard Business Review*, 74, 79–85.

Kaplan, R.S. and H.T. Johnson (1987). *Relevance Lost: The Rise and Fall of Cost Accounting*, Harvard Business School Press, Boston, MA.

Karger, D.W. (1982). *Advanced Work Measurement*. Industrial Press, New York.

Kato, Y. (1993). "Target Costing Support Systems: Lessons from Leading Japanese Companies." *Management Accounting Research*, 4 (4), 33– 47.

Kato, Y., G. Boer, and C.W. Chow (1995). "Target Costing: An Integrative Management Process." *Journal of Cost Management*, Spring, 39–51.

Kendrick, J.W. (1984). *Improving Company Productivity*, John Hopkins University Press, Baltimore.

Krupp, J.A. (1981). "Inventorry Turn Ratio, as a Management Control Tool." *Inventories and Production*, 1, 4 (September/October), 18–21.

Landel, Robert D. (1986). *Managing Productivity Through People: An Operations Perspective*. Reston, Prentice-Hall, Englewood Cliffs, N.J.

Lewin, Arie, and John W. Minton (1986). "Determining Organizational Effectiveness: Another Look, and an Agenda for Research." *Management Science*, 32 (5) (May), 514–538.

Lynch, L.L. and K.F. Cross (1995). *Measure up: Yardsticks for Continuous Improvement*. Basil Blackwell, Massachusetts.

Main, J. (1990). "How to Win the Baldridge Award." *Fortune*, April 25, 101–116.

Massimilian, R.D. and L. Pedro (1990). "Back from the future: Gearing up for the productivity challenge." 79 (2), *Management Review*, 41–43.

McCuley, J. (1972). "Machine grouping for efficient production." *Production Engineering*, 2, 53–57.

Miller, J.G. and T.E. Vollmann (1985). "The Hidden Factory." *Harvard Business Review* (September–October), 142–150.

Monczka, R. and S. Trecha (1988). "Cost-based supplier performance evaluation." *International Journal of Purchasing and Materials Management* (Spring), 2–7.

Nobeoka, K. and M.A. Cusumano (1995). "Multi-project Strategy, Design Transfer, and Project

Performance." *IEEE Transactions on Engineering Management*, 42, 397–409.

Oakland, J.S. (1993). *Total Quality Management: The Route to Improving Performance*. Butterworth-Heinemann, Oxford.

Philipoom, P.R. and T.D. Fry (1992). "Capacity-based order review/release strategies to improve manufacturing performance." *International Journal of Production Research*, 30 (11), 2559–2572.

Richardson, P.R., A.J. Taylor, and J.R.M. Gordon (1985). "A Strategic approach to Evaluating Manufacturing Performance." *Interfaces*, November–December, 15–27.

Ritzman, L.P., B.E. King, and L.J. Krajewski (1984). "Manufacturing Performance – Pulling the Right Levers." *Harvard Business Review*, March–April, 143–152.

Roberts, M.W. and K.J. Silvester (1996). "Why ABC Failed and How It May Yet Succeed." *Journal of Cost Management*, Winter, 23–35.

Ronen, D. (1983). "Inventory Service Measures – A Comparison of Measures." *International Journal of Operations and Production Management*, 3 (2), 37–45.

Sakakibara, S., B.B. Flynn, and R.G. Schroeder (1993). "A framework and measurement instrument for just-in-time manufacturing." *Production and Operations Management*, 2, 177–194.

Skinner, W. (1986). "The Productivity Paradox." *Harvard Business Review*, 64 (4), July–August, 55–59.

Szilagyi, A.D. and M.J. Wallace (1987). *Organizational Behavioral and Performance*. Scott, Foresman, Glenview IL.

Turney, P.B.B. (1992). *Common Cents: The ABC Performance Breakthrough*. Cost Technology, Portland, OR.

Viswanadham, N. and Y. Narahari (1992). *Performance Modeling of Automated Manufacturing Systems*. Prentice-Hall, New Jersey.

Westra, D., M.L. Srikanth, and M. Kane (1996). "Measuring operational performance in a throughput world." *Management Accounting* (April), 41–47.

White, G.P. and J.A. Ponce de Leon (1998). *Performance Measurement Systems: A Model and Seven Case Studies*. APICS Education and Research Foundation, VA.

MANUFACTURING TECHNOLOGY

Abdel-Malek, L. (1985). "A case study on the use of robots in assembly process." *Engineering Costs and Production Economics*, 9, 99–102.

Abdel-Malek, L. (1986). "A framework for the robotic assembly of parts with general geometries." *International Journal of Production Research*, 24, 1025–1041.

Abdel-Malek, L. (1988). "The effect of robots with overlapping envelopes on performance of flexible transfer lines." *IIE Transactions*, 20, 213–222.

Abdel-Malek, L. (1989). "Assessment of the economic feasibility of robotic assembly while conveyor tracking (RACT)." *International Journal of Production Research*, 27, 1209–1224.

Abernathy, William J. and Phillip L. Townsend (1975). "Technology, Productivity, and Process Change." *Technology Forecasting and Social Change*, 7, 379–396.

Agrawal, V.P., V. Khohl, and S. Gupta (1991). "Computer aided robot selection: The multiple attribute decision making approach." *International Journal of Production Research*, 29, 1629–1644.

Albertson, P. (1983). "Verifying robot performance." *Robotics Today*, 33–36.

Archetti, F. and A. Sciomachen (1989). "Development, analysis and simulation of Petri net models: an application to AGV systems." F. Archetti, M. Lucertini, P. Serafini (eds.). *Operations Research Models in Flexible Manufacturing Systems*, 91–113, Springer–Verlag, New York.

Arthur, W.B. 1989. "Competing technologies, increasing returns, and lock-in by historical events." *The Economic J.*, 99, 116–31.

Babbage, Charles (1835). *On the Economy of Machinery and Manufactures*, Charles Knight, London, England.

Baker, R.C. and S. Talluri (1997). "A closer look at the use of data envelopment analysis for technology selection." *Computers and Industrial Engineering*, 32, 101–108.

Berliner, C. and J.A. Brimson (1988). *Cost Management for Today's Advanced Manufacturing*. Harvard Business School Press, Boston.

Betz F. (1993). *Strategic Technology Management*, McGraw-Hill, New York.

Blois, K.J. (1986). "Marketing Strategies and the New Manufacturing Technologies." *International Journal of Operations and Production Management*, 6 (1), 34–41.

Booth, D.E., M. Khouja, and M. Hu (1992). "A robust multivariate procedure for evaluation and selection of industrial robots." *International Journal of Operations and Production Management*, 12, 15–24.

Boubekri, N., M. Sahoui, and C. Lakrib (1991). "Development of an expert system for industrial robot selection." *Computers and Industrial Engineering*, 20, 119–127.

Boyer, K.K., G.K. Leong, P.T. Ward, and Krajewski, L. (1997). "Unlocking the Potential of Advanced Manufacturing Technologies." *Journal of Operations Management*, 15 (4), 331–347.

Browne, J., D. Dubois, K. Rathmill, S. Sethi, and K. Stecke (1984). "Classification of flexible manufacturing systems." *The FMS Magazine*, April, 114–117.

Burbidge, J.L. and B.G. Dale (1984). "Planning the Introduction and Predicting the Benefits of Group Technology." *Engineering Costs and Production Economics*, 8 (1), 117–128.

Burns T. and G. Stalker (1960). *The Management Of Innovation*. London, Tavistock.

Buzacott, J. (1982). "The fundamental principles of flexibility in manufacturing systems." Proceedings of 1st Intl. Conference on Flexible Manufacturing Systems, North Holland, UK.

Bylinski, G. (1983). "The Race to the Automatic Factory." *Fortune*. February 21, 52–60.

Cao, T. and A.C. Sanderson (1994). "Task decomposition and analysis of robotic assembly task plans using Petri nets." *IEEE Transactions on Industrial Electronics*, 41, 620–630.

Chang, T.C. and R.A. Wysk (1985). *An Introduction to Automated Process Planning Systems*. Prentice-Hall, Englewood Cliffs, N.J.

Chang, T.C., R.A. Wysk, and H.-P. Wang (1998). *Computer-Aided Manufacturing* (2nd ed.). Prentice-Hall, New Jersey.

Cho, H. (1998). "Petri net models for message manipulation and event monitoring in an FMS cell." *International Journal of Production Research*, 36, 231–250.

Choobineh, F. (1984). "Optimum Loading for GT/MRP Manufacturing Systems." *Computer and Industrial Engineering*, 8, 3/4, 197–206.

Cohen, M.A. and U.M. Apte (1997). *Manufacturing Automation*, Irwin, Illinois.

DeGarmo, E. Paul, J.T. Black, A. Kohser, A. Ronald (1997). *Materials and Processes in Manufacturing* (8th ed.). Prentice-Hall, Inc.

Dorf, R.C. and A. Kusiak (1994). *Handbook of Design, Manufacturing, and Automation*. Wiley, New York.

Draper, C. (1984). *Flexible Manufacturing Systems Handbook*, Automation and Management Systems Division, The Charles Stark Draper Laboratory Inc., Noyes Publication.

Engelberger, J.F. (1980). *Robotics in Practice*. AMACOM, New York.

Ettlie J.E. (1990). "What makes a Manufacturing Firm Innovative?" *Academy Of Management Executive*, 4 (4).

Fisher, E.L. and O.Z. Maimon (1988). "Specification and selection of robots." In Kusiak, A. (ed.) *Artificial Intelligence Implications for CIM*, IFS Publications, Bedford.

Forest J. (1991). "Models For The Process Of Technological Innovation." *Technology Analysis And Strategic Management*, 3 (4).

Frohman, A.L. (1982). "Technology as a Competitive Weapon." *Harvard Business Review*, 60 (1) (January–February), 97–104.

Gale, Bradley T. (1980). "Can More Capital Buy Higher Productivity?" *Harvard Business Review*, 59 (4), 78–86.

Gerwin, D. (1981). "Control and Evaluation in the Innovation Process: The Case of Flexible Manufacturing Systems." *IEEE Transactions on Engineering Management*, EM-28, 3 (August), 62–70.

Gerwin, D. (1982). "Do's and Don'ts of Computerized Manufacturing." *Harvard Business Review*, (March–April), 107–116.

Gerwin, D. (1987). "An Agenda for Research on the Flexibility of Manufacturing Processes." *International Journal of Operations and Production Management*, 7, 38–49.

Gerwin, D. and H. Kolodny (1992). *Management of Advanced Manufacturing Technology*, New York: Wiley-Interscience.

Gerwin, D. and J.C. Tarondeau (1982). "Case Studies of Computer Integrated Manufacturing Systems: A View of Uncertainty and Innovation Process." *Journal of Operations Management*, 2, 2 (February), 87–89.

Gold, B. (1980). "CAM Sets New Rules for Production." *Harvard Business Review*, 60 (6) (November–December), 78–86.

Gold, B. (1981). "Changing Perspectives on Size, Scale, and Returns: An Interpretive Survey." *Journal of Economic Literature*, 19, March.

Goldhar, J.D. and M. Jelinek (1983). "Plan for Economics of Scope." *Harvard Business Review*, 61 (6) (November–December), 141–148.

Groover, M.P. (1980). *Automation, Production Systems, and Computer-Aided Manufacturing*. Prentice-Hall, Englewood Cliffs, NJ.

Groover, M.P. and E.W. Zimmers, Jr. (1984). *CAD/CAM: Computer Aided Design and Manufacturing*. Prentice-Hall, Englewood Cliffs, NJ.

Groover, Mikell P. (1987). *Automation, Production Systems, and Computer-Aided Manufacturing*. Englewood Cliffs, N.J., Prentice-Hall, 119–128.

Hammer, M. (1990). "Reengineering Work: Don't Automate, Obliterate." *Harvard Business Review* (July–August), 68 (4), 104–102.

Hartley, John (1984). *FMS at Work*. IFS Publications, Bedford, UK.

Hayes, R.H. and J. Jaikumar (1988). Manufacturing's crisis: new technologies, obsolete organizations, *Harvard Business Review*, September, 77–85.

Imany, M.M. and R.J. Schlesinger (1989). "Decision models for robot selection: a comparison of ordinary least squares and linear goal programming methods." *Decision Sciences*, 20, 40–53.

Jelinek, M. (1977). "Technology, Organizations, and Contingency." *Academy of Management Review*, January, 17–26.

Jelinek, M. and J. Goldhar (1983). "The Interface Between Strategy and Manufacturing Technology." *Columbia Journal of World Business*, Spring.

Jones, M.S., C.J. Malmborg, and M.H. Agee (1985). "Decision support system used for robot selection." *Industrial Engineering*, 17, 66–73.

Kantrow, A.M. (1980). "The Strategy-Technology Connection." *Harvard Business Review*, July–August, 6–8, 12.

Kaplan, R.S. (1986). "Must CIM be justified by faith alone." *Harvard Business Review*, 64, 87–95.

Kelly P. and M. Krantzberg (1978). *Technological Innovation: A Critical Review Of Current Knowledge*. San Francisco Press, USA.

Khouja, M. (1995). "The use of data envelopment analysis for technology selection." *Computers and Industrial Engineering*, 28, 123–132.

Khouja, M. and D.E. Booth (1991). "A decision model for the robot selection problem using robust regression." *Decision Sciences*, 22, 656–662.

Khouja, M. and D.E. Booth (1995). "Fuzzy clustering procedure for evaluation and selection of industrial robots." *Journal of Manufacturing Systems*, 14, 244–251.

Khouja, M. and O.F. Offodile (1994). "The industrial robots selection problem: literature review and directions for future research." *IIE Transactions*, 26 (4), 50–61.

Knott, K. and R.D. Getto, Jr. (1982). "A Model for evaluating alternative robot systems under uncertainty." *International Journal of Production Research*, 20, 155–165.

Koren, Y. and G. Ulsoy (1997). "Reconfigurable Manufacturing Systems." ERC Technical Report #1, The University of Michigan, Ann Arbor, Michigan.

Kotha, S. and P.M. Swamidass (1998). "Advanced manufacturing technology use: Exploring the effect of the nationality varible." *International Journal of Production Research*, 36 (11), 3135–3146.

Kusiak, A. (1987). The production equipment requirements problem, *International Journal of Production Research*, 25, 319–325.

MacCarthy, B.L. and J. Liu (1993). A new classification scheme for flexible manufacturing systems, *International Journal of Production Research*, 31, 299–309.

Malcom, D.G., J.H. Roseboom, C.E. Clark, and W. Fazar (1959). "Application of a Technique for Research and Development Program Evaluation." *Operations Research*, 7 (5), September–October, 98–108.

McGrath, R.G., M. Tsai, S. Venkataraman and I.C. MacMillan (1996). "Innovation, Competitive Advantage and Rent: A Model and Test." *Management Science*, 42 (3), 389–403.

Mehrez, A. and O.F. Offodile (1994). "A statistical-economic framework for evaluating the effect of robot repeatability on profit." *IIE Transactions*, 26, 101–110.

Monahan, G.E. and T.L. Smunt (1987). "A multilevel decision support system for the financial justification of automated flexible manufacturing systems." *Interfaces*, 17, 29–40.

Moore, K.E. and S.M. Gupta (1996). "Petri net models of flexible and automated manufacturing systems: a survey." *International Journal of Production Research*, 34, 3001–3035.

Muller, T. (1983). *Automated Guided Vehicles*. IFS Publications, Bedford, UK.

Nnaji, B.O. (1986). *Computer-Aided Design, Selection and Evaluation of Robot*, Elsevier, New York.

Nnaji, B.O. (1988). "Evaluation methodology for performance and system economics for robotic devices." *Computers and Industrial Engineering*, 14, 27–39.

Nnaji, B.O. and M. Yannacopoulou (1989). "A utility theory based robot selection and evaluation for electronics assembly." *Computers and Industrial Engineering*, 14, 477–493.

Nof, S.Y. (1985). *Handbook of Industrial Robotics*, John Wiley & Sons, Inc., New York.

Offodile, O.F. and U. Ugwu (1991). "Evaluating the effect of speed and payload on robot repeatability." *International Journal of Robotics and Computer Integrated Manufacturing*, 8, 27–33.

Offodile, O.F., B.K. Lambert, and R.A. Dudek (1987). "Development of a computer aided robot selection procedure (CARSP)." *International Journal of Production Research*, 25, 1109–1121.

Quinn J.B. (1985). "Managing Innovation: Controlled Chaos." *Harvard Business Review*, May/June.

Quinn, J.B. (1992). *Intelligent Enterprise*. Free Press, New York.

Ramasesh, R.V. and M.D. Jayakumar (1993). "Economic justification of advanced manufacturing technology." *OMEGA*, 21, 2819–306.

Ranky, P.G. (1982). *The Design and Operation of FMS* (IFS (Publications) Ltd. And North Holland).

Ranky, P.G. (1985). "FMS in CIM" (Flexible Manufacturing Systems in Computer Integrated Manufacturing), *ROBOTICA*, Cambridge University Press, 3, 205–214.

Ranky, P.G. (1985). *Computer Integrated Manufacturing*. Prentice-Hall International.

Rogers, E.M. (1983). *Diffusion of Innovation*, Free Press, New York, NY.

Rosenthal, S.R. (1984). "Progress Towards the 'Factory of the Future." *Journal of Operations Management*, 4 (3) (May), 203–229.

Sadowski, R.P. (1984). "Computer-Integrated Manufacturing Series Will Apply Systems Approach to Factory of the Future." *Industrial Engineering*, 16 (1) (January), 35–40.

Saleh, S. and C. Wang (1993). "The Management Of Innovation: Strategy Structure and Organizational Climate" *IEEE Transactions On Engineering Management*, 40 (1).

Salomon, Daniel P. and John E. Biegel (1984). "Assessing Economic Attractiveness of FMS Applications in Small Batch Manufacturing." *Industrial Engineering*, (June), 88–96.

Saren, A. (1984). "A Classification And Review of Models of The Intra-Firm Innovation Process." *R&D Management*, 14.

Sarin, S.C. and S.K. Das (1994). "Computer Integrated Production Planning and Control." in Dorf, R.C. and A. Kusiak, eds., *Handbook of Design, Manufacturing, and Automation*, New York: John Wiley & Sons.

Scarpello, V., W.R. Boulton, and C.R. Hofer (1986). "Reintegrating R&D into Business Strategy." *Journal of Business Strategy*, Spring, 49–56.

Schewe, G. (1994). "Successful Innovation Management: An Integrative Perspective." *Journal Of Engineering and Technology Management*, 11.

Schroeder, R.G., G.D Scudder, and D.R. Elmm, (1989). "Innovation in Manufacturing." *Journal of Operations Management*, 8 (1), 1–15.

Seidmann, A. and S.Y. Nof (1989). "Operational analysis of an autonomous assembly robotic station." *IEEE Transactions on Robotics and Automation*, 5, 4–15.

Seidmann, A., A. Arbel, and R. Shapira (1984). "A two-phase analytic approach to robotic system design." *Robotics and Computer-Integrated Manufacturing*, 12, 181–190.

Shaw, M.J. and A.B. Whinston (1988). "A Distributed Knowledge-Based Approach to Flexible Automation: The Contract Net Framework." *Flexible Manufacturing Systems*, 85–104.

Silverberg, G., G. Dosi, and L. Orsenigo (1988). "Innovation, diversity and diffusion: a self-organization model." *The Economic Journal*. 98, 1032–54.

Singhal, K., C. Fine, J.R. Meredith, and R. Suri (1987).

"Research and Models for Automated Manufacturing." *Interfaces*, 17 (6), November–December, 5–14.

Skinner, W. (1984). "Operations Technology: Blind Spot in Strategic Management." *Interfaces*. January–February, 116–125.

Slack, N. (1987). "The flexibility of manufacturing systems." *International Journal of Operations and Production Management*, 7, 35–45.

Slautterback, W.H. and W.B. Werther (1984). "The Third Revolution: Computer Integrated Manufacturing." *National Productivity Review*, Autumn, 367–374.

Solberg, J.J. (1976). "Optimal design and control of computerized manufacturing systems." In *Proceedings of AIIE Systems Engineering Conference*, 138–144.

Son, Y.K. (1992). "A Comprehensive Bibliography on Justification of Advanced Manufacturing Technologies." *Engineering Economist*, 38 (1), 59–71.

Stacey, R.D. (1996). *Complexity and Creativity in Organizations*. Berrett-Koehler, San Francisco.

Steele, L. (1989). "Managing Technology: The Strategic View." New York, McGraw-Hill.

Suresh, N.C. (1990). "Towards an integrated evaluation of flexible automation investments." *International Journal of Production Research*, 28, 1657–1672.

Suresh, N.C. and J.R. Meredith (1985). "Achieving Factory Automation Through Group Technology Principles." *Journal of Operations Management*, 5 (3), (February), 151–167.

Swamidass, P.M. (1987). "Planning for manufacturing technology." *Long Range Planning*, 20 (5), 125–133.

Swamidass, P.M. (1995). "Making sense out of manufacturing innovations." *Design Management Journal*, 6 (2), 52–56.

Swamidass, P.M. (1996). "Benchmarking Manufacturing Technology Use in the United States." *Handbook of Technology Management*, Gerard H. "Gus" Gaynor (ed.), 37.1–37.40, McGraw-Hill Inc., New York.

Swamidass, P.M. (1998). *Technology on the Factory Floor III*. Manufacturing Institute of the National Association of Manufacturers, Washington, D.C.

Swamidass, P.M. and M. Waller (1990). "A classification of approaches to planning and justifying new manufacturing technologies." *Journal of Manufacturing Systems*, 9 (3), 181–193.

Swamidass, P.M. and S. Kotha (1998). "Explaining manufacturing technology use, firm size and performance using a multidimensional view of technology." *Journal of Operations Management*, 17, 23–37.

Swann, Don (1989). "What is CIM, and Why Does it Cost $40 Million." *P&IM Review*, July, 34.

Twiss, B.C. (1980). *Managing Technological Innovation* (2nd ed.). London, Longman Group.

U.S. Department of Commerce, International Trade Administration. Office of Capital Goods and International Construction Sector Group (1985). *A Competitive Assessment of the U.S. Flexible Manufacturing System Industry*. U.S. Government Printing Office, Washington, D.C.

U.S. Department of Commerce (1987). Washington, D.C., *A Competitive Assessment of the U.S. Flexible Manufacturing System Industry*, July.

Utterback, J. (1972). "The Process Of Technological Innovation Within Firms." *Academy Of Management Journal*.

Utterback, J.M. and W.J. Abernathy (1975). "A dynamic model of process and product innovation." *OMEGA*, 3 (6), 639–656.

Utterback, J.M. and W.J. Abernathy (1981). "Multivariant Models for Innovation/Looking at the Abernathy and Utterback Model with Other Data." *OMEGA*, 9.

Van De Ven, Andrew H. (1986). "Central Problems in the Management of Innovation." *Management Science*, 32 (5) (May), 590–607.

Veeramani, D., J.J. Bernardo, C.H. Chung, and Y.P. Gupta (1995). "Computer-Integrated Manufacturing: A Taxonomy of Integration and Research Issues." *Production and Operations Management*, 4 (4), 360–380.

Vonderembse, Mark A., and Gregory S. Wobser (1987). "Steps for Implementing a Flexible Manufacturing System." *Industrial Engineering*, 15 (11) (November), 58–64.

Voss, C.A. (1986). Managing New Manufacturing Technologies, Monograph 1, *Operations Management* Association, September.

Waldron, D. (1988). "Accounting for CIM: The new yardsticks." *EMAP Business and Computing Supplement* (February), 1–2.

Williams, Vearl A. (1985). "FMS in Action." *Production*. January, 41–43.

Material and Inventory Management

Adkins, A.C., Jr. (1984). "EOQ in the Real World." *Production and Inventory Management*, 25 (4), 50–54.

American Production and Inventory Control Society (1986). *Material Requirements Planning Reprints*. American Production and Inventory Control Society, Falls Church, VA.

American Production and Inventory Control Society (1996). "1996 finite capacity scheduling software and vendor directory." *APICS – the Performance Advantage*, 6, 72–82.

Ammer, D.S. (1980). *Materials Management and Purchasing* (4th ed.). Irwin, Homewood, IL.

Anderson, J.C., Roger G. Schroeder, Sharon E. Tupy, and Edna M. White (1982). "Material Requirements Planning Systems: The State of the Art." *Production and Inventory Management*, 23 (4), 51–67.

Arrow, K.J., S. Karlin, and H. Scarf (1982). *Studies in the Mathematical Theory of Inventory and Production*. Stanford University Press, Stanford, CA.

Bahl, H.C. and Larry P. Ritzman (1983). "An Empirical Investigation of Different Strategies for Material Requirements Planning." *Journal of Operations Management*, 3 (2), 67–77.

Beddick, J.F. (1983). "Elements of success-MRP implementation." *Production and Inventory Management*, 24, 2, 26–30.

Bevis, G.E. (1976). "A management viewpoint on the implementation of a MRP system." *Production and Inventory Management*, 16, 1, 105–116.

Bitran, G.R., D.M. Marini, H. Matsuo, and J.W. Noonan (1985). "Multiplant MRP." *Journal of Operations Management*, 5 (2) (February). 183–204.

Blackburn, J.D., D.H. Kropp, and R.A. Millen (1986). "A comparison of strategies to dampen nervousness in MRP systems." *Management Science*, 32–4, 413–429.

Blackstone, John H., Jr., and James F. Cox (1988). "MRP Design and Implementation Issues for Small Manufacturers." *Production and Inventory Management* (3rd Quarter), 65–76.

Blumberg, D.F. (1980). "Factors Affecting the Design of a Successful MRP System." *Production and Inventory Management*, 5 (2) (Fourth Quarter), 50–62.

Brown, A.D. (1994). "Implementing MRP: leadership, rites and cognitive change." *Logistics Information Management*, 7, 2, 6–11.

Brown, Jimmie, John Harhen, and James Shivnan (1988). *Production Management Systems: A CIM Perspective*. Addison Wesley.

Brown, R.G. (1967). *Decision Rules for Inventory Management*, Holt, Rinehart and Winston, New York.

Brown, R.G. (1977). *Materials Management Systems*. John Wiley & Sons, New York.

Brown, R.G. (1982). *Advanced Service Parts Inventory Control*. Materials Management Systems, Inc., Norwich, VT.

Buchan, J. and E. Koenigsberg (1963). *Scientific Inventory Management*. Prentice-Hall, New Jersey. Chapters 1–16, and 19.

Buffa, E.S. and Jeffrey G. Miller (1979). *Production-Inventory Systems: Planning and Control* (3rd ed.). Richard D. Irwin, Homewood, IL.

Burlingame, J.W. and J.K. Weeks (1981). "Behavioral Dimensions of MRP Change: Assessing Your Organization's Strengths and Weaknesses." *Production and Inventory Management*, 22 (1) (First Quarter), 81–95.

Burns, O.M. and D. Turnipseed (1991). "Critical success factors in manufacturing resource planning implementation." *International Journal of Operations and Production Management*, 11 (4), 5–19.

Campbell, Kenneth L. (1975). "Inventory Turns and ABC-Analysis-Outmoded Textbook Concepts?" *American Production and Inventory Control Conference Proceedings*, 420.

Cantwell, J. (1985). "The How and Why of Cycle Counting: The ABC Method." *Production and Inventory Control*, 26 (2) (Second Quarter), 50–54.

Cavinato, J.L. (1984). *Purchasing and Materials Management*. West, St. Paul, MN.

Choobineh, F. (1984). "Optimum Loading for GT/MRP Manufacturing Systems." *Computer and Industrial Engineering*, 8, 3/4, 197–206.

Chung, C.H. and L.J. Krajewski (1984). "Planning Horizons for Master Production Schedules." *Journal of Operations Management*, (4), 389–406.

Chung, C.H. and L.J. Krajewski (1986a). "Replanning Frequencies for Master Production Schedules." *Decision Sciences*, 17 (2), 263–273.

Chung, C.H. and L. Krajewski (1984). "Planning Horizons for Master Production Scheduling." *Journal of Operations Management*, 4 (4) (August), 389–405.

Clark, J.T. (1984). "Inventory Flow Models." Proceedings, 27th Annual Conference, *American Production and Inventory Control Society*.

Close, Arthur C. (1970). "Projected Inventory Velocity Measurement." *Production and Inventory Management*, 11 (3), 66.

Colley, J.L. (1977). "Inventory Flow Models." Technical Note OM-120, Darden Graduate School of Business, University of Virginia, Charlottesville, Virginia.

Collier, D.A. (1980). "The Interaction of Single-Stage Lot Size Models in a Material Requirements Planning System." *Production and Inventory Management*, 21 (4), 11–20.

Conroy, Paul G. (1977). "Data General ABC Inventory Management." *Production and Inventory Management*, 18, 4, 63.

Cooper, R.B. and R.W. Zmud (1989). "Materials requirements planning system infusion." *OMEGA International Journal of Management Sciences*, 17, 5, 471–481.

Diskie, H. Ford (1951). "ABC Inventory Analysis Shoots for Dollars." *Factory Management and Maintenance*, July, 92.

Duchessi, P., C.M. Schaninger, D.R. Hobbs, and L.P. Pentak (1988). "Determinants of success in implementing materials requirements planning (MRP)." *Journal of Manufacturing and Operations Management*, 1 (3), 263–304.

Etienne, E.C. (1983). "MRP May Not Be Right for You, At Least not Yet." *Production and Inventory Management*, 24 (3) (Third Quarter), 33–46.

Flores, B.E. and D.C. Whybark (1987). "Implementing Multiple Criteria ABC Analysis." Journal of *Operations Management*, 7 (1–2) (October), 79–85.

Fogarty, D.W. (1987). "Material Flow in Manufacturing Cell and Material Requirements Planning Environment: Problems and a Solution." *Material Flow*, 4, 139–146.

Gallagher, G.R. (1980). "How to Develop a Realistic Master Schedule." *Management Review* (April), 19–25.

Gardner, E.S. (1980). "Inventory Theory and the Gods of Olympus." *Interfaces*, 10 (4), 42–45.

Gray, C. (1987). *The Right Choice: A Complete Guide to Evaluating, Selecting and Installing MRP II Software*. Essex Junction, VT., Oliver Wight Ltd.

Green, Donald R. (1987). "Direct Pegging, MRP/JIT Bridge." *P&IM Review*, April.

Greene J.H. (1987). Editor. *Production and Inventory Control Handbook* (2nd ed.). McGraw-Hill, New York.

Greene, James H. (1974). *Production and Inventory Control Systems and Decisions* (2nd ed.). Richard D. Irwin, Homewood, IL.

Gunnm, Thomas G. (1981). *Computer Applications in Manufacturing*. Industrial Press, New York.

Hadley, G. and T.M. Whitin (1963). *Analysis of Inventory Systems*. Prentice-Hall, New Jersey, 1, 3–5, and 9.

Hall, R.W. and T.E. Vollman (1978) "Planning your

requirements planning." *Harvard Business Review*, 56, 2, 105–112.

Harl, J.E. (1983). "Reducing Capacity Problems in Material Requirements Planning Systems." *Production and Inventory Management* (Third Quarter), 52–60.

Hax, A.C. and D. Candea (1984). *Production and Inventory Management*, Prentice-Hall, Englewood Cliffs, NJ.

Holt, C.C., F. Modigliani, J.F. Muth, and H.A. Simon (1960). *Planning Production, Inventories, and Workforce*. Prentice-Hall, New York.

Inderfurth, K. (1994). "Safety Stocks in Multistage Divergent Inventory Systems: A Survey." *International Journal of Production Economics*, 35, 321–329.

Janson, R.L. (1989). *Handbook of Inventory Management*. Prentice-Hall, Englewood Cliffs, NJ.

Jones, Thomas C. and D.W. Riley (1985). "Using Inventory for Competitive Advantage Through Supply Chain Management." *International Journal of Physical Distribution and Materials Management*, 15, 5, 16–26.

Jordan, Henry (1974). "Relating customer Service to Inventory Control." *American Management Association's Advanced Management Journal*, 39 (4).

Kim, S.L., J.C. Hayya, and J.D. Hong (1992). "Setup Reduction in the Economic Production Quantity Model." *Decision Sciences*, 23 (2), 500–508.

Kleuthgen, P.O. and J.C. McGee (1985). "Development and Implementation of an Integrated Inventory Management Program at Pfizer Pharmaceuticals." *Interfaces*, 15 (1) (January–February), 69–87.

Kropp, D.H., R.C. Carlson and J.V. Jucker (1979). "Use of dynamic lot-sizing to avoid nervousness in material requirements planning systems." *Journal of Production and Inventory Management*, 20 (3), 40–58.

Kropp, D.H., R.C. Carlson and J.V. Jucker (1983). "Heuristic lot-sizing approaches for dealing with MRP system nervousness." *Decision Sciences*, 14 (2), 156–169.

Krupp, J.A. (1981). "Inventorry Turn Ratio, as a Management Control Tool." *Inventories and Production*, 1, 4 (September/October), 18–21.

Krupp, J.A.G. (1984). "Why MRP Systems Fail: Traps to Avoid." *Production and Inventory Management*, 25, no. 3, 48–53.

Krupp, J.A.G. (1997). "Safety Stock Management." *Production and Inventory Management*, 38, 3 (Third Quarter), 11–18.

Lee, H.L. and C. Billington (1992). "Managing Supply Chain Inventory: Pitfalls and Opportunities." *Sloan Management Review* (Spring), 65–73.

Lee, H.L., C. Billington, and B. Carter (1993). "Hewlett-Packard gains control of inventory and service through design for localization." *Interfaces*, 23, 1–11.

Magee, John R., and David M. Boodman (1967). *Production Planning and Inventory Control* (2nd ed.). McGraw-Hill, New York.

Mather, H. (1977). "Reschedule the schedules you just scheduled-Way of life for MRP?" *Journal of Production and Inventory Management*, 18 (1), 60–79.

Mather, H. (1987). *Bills of Material*. Dow Jones-Irwin, Homewood, IL.

McLeavey, D.W. and S.L. Narasimhan (1985). *Production Planning and Inventory Control*. Allyn and Bacon, Boston.

Mehra, Satish, and M.J. Reid (1982). "MRP Implementation Using an Action Plan." *Interfaces*, 12 (1) (February), 69–73.

Nakane, J. and R.W. Hall (1983). "Management Specs for Stockless Production." *Harvard Business Review* (May–June), 84–91.

Natarajan, R. and S. Goyal (1994). "Safety Stocks in JIT Environments." *International Journal of Operations and Production Management*, 14 (10).

Nelson, N.S. (1981). "MRP and Inventory and Production Control in Process Industries." *Production and Inventory Control*, 22 (4) (Fourth Quarter), 15–22.

New, C. Colin (1977). "MRP & GT, A New Strategy for Component Production." *Production and Inventory Management*, 18, 3, 50–62.

Newberry Thomas L., ed. (1978). *Inventory Management Systems*. American Software, Atlanta, GA., 8–1 and 8–2.

Orlicky, Joseph (1975). *Material Requirements Planning*. McGraw-Hill, New York.

Ormsby, J.G., S.Y. Ormsby, and C.R. Ruthstrom (1990). "MRP II implementation: a casestudy." *Production and Inventory Management Journal*, 31 (4), 77–81.

Peek, L.E. and John H. Blackstone, Jr. (1987). "Developing a Time-phased Order-point Template Using A Spred sheet." *Production and Inventory Control Management Journal*, 28 (4), 6–10.

Pennente, Ernest and Ted Levy (1982). "MRP on Microcomputers." *Production and Inventory Management Review*, May, 20–25.

Peterson, R. and E.A. Silver (1984). *Decision Systems for Inventory Management and Production Planning* (2nd ed.). Wiley, New York.

Pinella, Paul, and Eric Cheathan (1981). "In-House Inventory Control." *ICP INTERFACE Manufacturing and Engineering*, Winter, 18–21.

Plenert, G. and T.D. Best (1986). "MRP, JIT, and OPT: What's Best?" *Production and Inventory Management* 27, (2), 22–29.

Plossl, G.W. (1973). "How Much Inventory is Enough?" *Production and Inventory Management*. Second Quarter.

Plossl, G.W. (1983). *Production and Inventory Control*: Applications. George Plossl Education Services, Atlanta.

Plossl, G.W. (1985). *Production and Inventory Management: Principles and Techniques* (2nd ed.). Prentice-Hall, New Jersey.

Plossl, G.W. (1994). *Orlicky's Material Requirements Planning* (2nd ed.). McGraw-Hill.

Plossl, G.W. and O.W. Wight (1967). *Production and Inventory Control*. Prentice-Hall, Englewood Cliffs, NJ.

Plossl, G.W. and W.E. Welch (1979). *The Role of Top Management in the Control of Inventory*. Reston, Reston, VA.

Porteus, E.L. (1985). "Investing in Reduced Setups in the EOQ Model." *Management Science*, 31 (8), 998–1010.

Potamianos, J. and A.J. Orman (1996). "An Interactive Dynamic Inventory-Production Control System." *Journal of the Operational Research Society*, 47 (8), 1016–1028.

Ptak, C.A.M. (1997). *MRP and Beyond*, Richard D. Irwin.

Reinfeld, N.V. (1987). *Handbook of Production and Inventory Control*. Prentice-Hall, Englewood Cliffs, NJ.

Ronen, D. (1983). "Inventory Service Measures – A Comparison of Measures." *International Journal of Operations and Production Management*, 3 (2), 37–45.

Salvador, Michael S., *et al*. (1975). "Mathematical Modeling Optimization and Simulation Improve Large-Scale Finished Goods Inventory Management." *Production and Inventory Control Management*, Second Quarter.

Schonberger, R.J. (1983). "Selecting the Right Manufacturing Inventory Systems: Western and Japanese Approaches." *Production and Inventory Control*, 24 (2) (Second Quarter), 33–44.

Schonberger, R.J. (1984). "Just-In-Time Production System: Replacing Complexity with Simplicity in Manufacturing Management." *Industrial Engineering*, Oct, 52–63.

Schonberger, R.J. and M.J. Schniederjans (1984). "Reinventing Inventory Control." *Interfaces*, 14 (3) (May–June), 76–83.

Schroeder, R.G., J.C. Anderson, S.E. Tupy, and E.M. White (1981). "A Study of MRP Benefits and Costs." *Journal of Operations Management*, 2 (1) (October), 1–9.

Schwarzendahl, R. (1996). "The Introduction of Kanban Principles to Material Management in EWSD Production." *Production Planning and Control*, 7, 212–221.

Sheldon, D. (1991). "MRP II – What it really is." *Production and Inventory Management Journal*, 32, 3, 12–15.

Sheldon, D. (1994). "MRP II implementation: a case study." Hospital Material Management Quarterly, 15 (4), 48–52.

Shingo, S. (1988). *Non-Stock Production: The Shingo System for Continuous Improvement*. Productivity Press.

Silver, E.A. (1974). "A Control System for Coordinated Inventory Replenishment." *International Journal of Production Research*, 12 (6), 647–671.

Silver, E.A. (1975). "Modifying the Economic Order Quantity (EOQ) to Handle Coordinated Replenishment of Two or More Items." *Production and Inventory Management*, 16 (3), 26–38.

Silver, E.A. and Peter Kelle (1988). "More on Joint Order Releases Under Dependent Demand." *Production and Inventory Management*, First Quarter.

Silver, E.A. and R. Peterson (1985). *Decision Systems for Inventory Management and Production Planning* (2nd ed.). John–Wiley, New York. Chapters 3, 4, 7–12, and 17.

Sirianni, N.C. (1979). "Inventory: How Much Do You Really Need?" *Production Magazine*, November.

Sirianni, N.C. (1981). "Right Inventory Level: Top Down or Bottom UP." *Inventories and Production Magazine*, May–June.

Smith, S.B. (1989). *Computer Based Production and Inventory Control*. Prentice-Hall, Englewood Cliffs, NJ.

Smolik, D.P. (1983). *Material Requirements of Manufacturing*. New York, Van Nostrand Reinhold.

Smunt, Timothy L. and Thomas E. Morton (1985). "The Effects of Learning on Optimal Lot Sizes: Further Development on the Single Product Case." *IIE Transactions*, 17 (1), 3–37.

Sridharan, V. and R.L. LaForge (1989). "The impact of safety stock on schedule instability, cost, and service." *Journal of Operations Management*, 8 (4), 327–347.

Steele, D.C. (1975). "The nervous MRP system: How to do battle." *Journal of Production and Inventory Management*, 16 (4), 83–89.

Steinberg, E.E., B. Khumawala, and R. Scamell (1982). "Requirements Planning Systems in the Health Care Environment." *Journal of Operations Management*, 2 (4) (August), 251–259.

Steinberg, E.E., W.B. Lee, and B. Khumawala (1980). "A Requirements Planning System for the Space Shuttle Operations Schedule." *Journal of Operations Management*, 1 (2) (November), 69–76.

Stenger, Alan J. and Joseph L. Cavinato (1979). "Adapting MRP to the Outbound Side Distribution Requirements Planning." *Production and Inventory Management*, 20 (4), 1–13.

Sum, C.C., J.S.K. Ang, and L.N. Yeo (1997). "Contextual elements of critical success factors in MRP implementation." *Production and Inventory Management Journal*, 38 (3), 77–83.

Sum, C.C., K.K. Yang, J.S.K. Ang, and S.A. Quek (1995). "An analysis of material requirements planning (MRP) benefits using alternating conditional expectation (ACE)." *Journal of Operations Management*, 13 (1), 35–58.

Tadjer, R. (1988). "Enterprise resource planning." *Internet Week*, 710, 40–41.

Takahashi, K. and N. Nakamura (1996). "Reactive JIT ordering systems for unstable demand changes." *Proceedings of Advances in Production Management Systems*, Information Processing Society of Japan.

Tersine, R.J. (1987). *Principles of Inventory and Materials Management* (3rd ed.). North-Holland, New York.

Thompson, K. (1983). "MRP II in the Repetitive Manufacturing Environment." *Production and Inventory Management*, 24 (4) (Fourth Quarter), 1–14.

Toomey, J.W. (1996). *MRP II: Planning for Manufacturing Excellence*. Chapman and Hall, New York.

Vollman, T.E. (1986). "OPT as an Enhancement to MRP II." *Production and Inventory Management*, 27 (2), 38–47.

Vollmann T.E., W.L. Berry, and D.C. Whybark (1997).

Manufacturing Planning and Control Systems (4th ed.). Irwin/McGraw-Hill.

Wagner, Harvey and Thompson M. Whitlin (1958). "Dynamic Version of the Economic Lot Size Model." *Management Science*, 5, 89–96.

Waldman, K.H. (1983). "Optimal Replacement Under Additive Damage in Randomly Varying Environments." *Naval Research Logistics Quarterly*, 30, 377–386.

Wallace, T.F. (1990). *MRPII: Making It Happen*. Oliver Wight Limited Publications, Inc.

Wantuck, K. (1979). "Calculating Optimum Inventories." Proceedings 21st Annual Conference, *American Production and Inventory Control Society*.

White, E.M. (1980). "Implementing an MRP system using the Lewin-Schein theory of change." *Production and Inventory Management*, 21 (1), 1–12.

White, E.M., J.C. Anderson, R.G. Schroeder, and S.E. Tupy (1982). "A study of the MRP implementation process." *Journal of Operations Management*, 2 (3), 145–153.

Whybark, C. and J.G. Williams (1976). "Material Requirements Planning under Uncertainty." *Decision Sciences*, 7 (1), 595–606.

Wight, O.W. (1974). *Production and Inventory Management in the Computer Age*. Cahner Books, Boston, MA.

Wight, O.W. (1982). *The Executive's Guide to Successful MRP II*. Oliver Wight Publishing, Essex VT.

Wight, O.W. (1984). *MRP II: Unlocking America's Productivity Potential*. CBI, Boston, MA.

Works, M. (1998). "GenCorp Aerojet says time is right for SAP in the aerospace and defense industries." *Automatic I.D. News*, 14 (2), 50–52.

Zhao, X. and T.S. Lee (1993). "Freezing the master production schedule for material requirements planning systems under demand uncertainty." *Journal of Operations Management*, 11 (2), 185–205.

New Ideas: Reengineering, JIT, Lean, and Agile Manufacturing

Abegglen, J.C. and G. Stalk (1985). *The Japanese Corporation*, New York: Basic Books, Inc.

Aggarwal, S.C. (1985). "MRP, JIT, OPT, FMS?" *Harvard Business Review* (September–October).

Ansari, A. (1986). "Survey Identifies Critical Factors in Successful Implementation of Just-in-Time Purchasing Techniques." *Industrial Engineering*, 18 (10), 44–50.

Ansari, A. and B. Modarres (1986). "Just-in-Time Purchasing: Problems and Solutions." *Journal of Purchasing and Materieals Management*, Summer, 11.

Ansari, A. and B. Modarress (1987). "The Potential Benefits of Just-In-Time Purchasing for U.S. Manufacturing." *Production and Inventory Management*, 28 (2) (SecondQuarter), 30–35.

Ansari, A. and Jim Heckel (1987). JIT Purchasing: Impact of freight and Inventory Costs. *Journal of Purchasing and Material Management*, Summer, 24.

Arogyaswamy, B. and R.P. Simmons (1991). "Thriving on independence: The Key to JIT Implementation." *Production and Inventory Management Journal*, 32, 3, 56–60.

Automotive Industry Action Group (AIAG) (1983). *The Japanese Approach to Productivity*. Videotape Series.

Back, H. (1985). "Decentralized Integration – Advantage or Disadvantage of Logistics and JIT – Concepts." *Computers in Industry* (The Netherlands) 6 (6) (December), 529–541.

Bagchhi, P.K., T.S. Raghunathan, and E.J. Bardi (1987). "The Implication of Just-in-Time Inventory Policy on Carrier Selection." *Logistics and Transportation Review* (Canada) 23 (4) (December), 373–384.

Bennet, J. (1994). "Detroit Struggles to Learn Another Lesson From Japan." *The New York Times*, June 19, 1994, section F, 5.

Berkley, B. (1992). "A review of the kanban production control research literature." *Production and Operations Management*, 1 (4), 393–411.

Billesbach, Thomas J., (1994). "Applying Lean Production Principles to A Process Facility." *Production and Inventory Management Journal*, 35, 3, 40–44.

Billesbach, Thomas J. (1994). "Simplified Flow Control Using Kanban Signals." *Production and Inventory Management Journal*, 35, 2, 72–75.

Billesbach, Thomas J. and R. Hayen (1994). "Long-Term Impact of Just-In-Time on Inventory Performance Measures." *Production and Inventory Management Journal*, V. 35, 1, 62–67.

Billesbach, Thomas J., A. Harrison, and S. Croom-Morgan (1991). "Just-in-Time: A United States-United Kingdom Comparison." *International Journal of Operations and Production Management*, 11, 10, 44–57.

Billesbach, Thomas J. and Marc J. Schniederjans (1989). "Applicability of Just-In-Time Techniques in Administration." *Production and Inventory Management Journal*, 30, 3, 40–45.

Blackburn, J.D. (1991). *Time Based Competition*, Homewood: Business One Irwin.

Burnham, J.M. (1987). "Some Conclusions About JIT Manufacturing." *Production and Inventory Management*, 28 (3), 7–11.

Cannon, Joseph J. and Gary Kapusta (1989). "Just-in-Time at AT&T's Reading Works." Just-in-Time Seminar Proceedings, July 24–26.

Celley, Albert F., William H. Clegg, Arthur W. Smith, and Mark A. Vonderembse (1986). "Implementation of JIT in the United States." *Journal of Purchasing and Materials Management*, Winter, 9.

Chaffee, R.H. (1995). "Creating Fulfillment Agility in the Face of Product Proliferation." *Logistics: Mercer Management Consulting*, 20–23.

Champy, J.A. (1995). *Reengineering Management: The Mandate for New Leadership*, Harper Collins, London.

Cherukuri, S.S., R.G. Nierman, and N.C. Sirianni (1995). "Cycle Time and The Bottom Line." *Industrial Engineering*, March.

Christer, A.H. and W.M. Waller (1984). "Reducing production downtime using delay-time analysis."

Journal of the Operational Research Society, 35, 499–512.

Clark, K.B. (1992). "Design for Manufacturability at Midwest Industries: Teaching Note." Note no. 5-693-007, Harvard Business School Press, Boston.

Courtis, John K. (1995). "JIT's Impact on a Firm's Financial Statements." *International Journal of Purchasing and Materials Management*, 31, 1, Winter, 46–51.

Crawford, K.M., J.F. Cox, and J.H. Blackstone, Jr. (1988). Performance Measurement Systems and the JIT Philosophy: Principles and Cases. *American Production and Inventory Control Society*, Falls Church, VA.

Crawford, K.M., J.H. Blackstone, and J.F. Cox (1988). "A Study of JIT Implementation and Operating Problems." *International Journal of Production Research*, 26, 1561–1568.

Crosby, L.B. (1984). "The Just-in-Time Manufacturing Process: Control of Quality and Quantity." *Production and Inventory Management*, 25, 4, 21–33.

Davenport, T.H. (1993). *Process Innovation: Reengineering Work Through Information Technology*, Harvard Business School Press, Boston, Massachusetts, p. 5–9.

Davenport, T.H. and J.E. Short (1990). "The New Industrial Engineering: Information Technology and Business Process Redesign." *Sloan Management Review*, Summer, 11–27.

Deleersnyder, J.L., T.J. Hodgson, H. Muller, and P.J. O'Grady (1989). "Controlled Pull Systems: An Analytical Approach." *Management Science*, 35, 1079–1091.

Dertouzos, Michael L., Richard K. Lester, and Robert M. Solow (1989) and the MIT Commission on Industrial Productivity. *Made in America: Regaining the Productive Edge*. The MIT Press, Cambridge, MA.

DeVor, R., R. Graves, and J.J. Mills (1997). "Agile Manufacturing Research: Accomplishments and Opportunities." *IIE Transactions*, 29, 813–823.

Dilworth, J.D. (1987). *Information Systems for JIT Manufacturing*. Association for Manufacturing Excellence, Wheeling, IL.

Dixon, Lance and Anne Miller Porter (1994). *JIT II – Revolution in Buying and Selling*, Cahner's Publishing Company, Newton, MA.

Dove, R. (1996). *Tools for Analyzing and Constructing Agile Capabilities*, Agility Forum, Bethlehem, PA.

Dove, R., S. Hartman, and S. Benson (1996). *An Agile Enterprise Reference Model with a Case Study of Remmele Engineering*, Agile Forum, Bethlehem, PA.

DTI (1988). *Competitive Manufacturing: A Practical Approach to the Development of a Manufacturing Strategy*. IFS, Bedford.

Esparrago, R.A. Jr. (1988). "Kanban." *Production and Inventory Management*, 29 (1), 6–10.

Fiegenbaum, A.V. (1961). *Total Quality Control: Engineering and Management*. McGraw-Hill, New York.

Garwood, Dave and Michael Bane (1988). *A Jumpstart to World Class Performance*. Dogwood Publishing, Marietta, GA.

Gilbert, J.P. (1990). "The state of JIT implementation and development in the USA." *International Journal of Production Research*, 28, 1099–1109.

Goddard, Walter and Associates (1986). *Just-in-Time: Surviving by Breaking Tradition*. Oliver Wight Publications, Essex Junction, VT.

Goldman, S., R. Nagel, and K. Preiss (1995). *Agile Competitors and Virtual Organizations*, Van Nostrand Reinhold, New York.

Goyal, S.K. and S.G. Deshmukh (1992). "A critique of the literature on just-in-time Manufacturing." *International Journal of Operations and Production Management*, 12, 1, 18–28.

Gravel, M. and W.L. Price (1988). "Using the Kanban in a Job-shop Environment." *International Journal of Production Research*, 26, 1105–1118.

Gupta, S.M. and Y.A.Y. Al-Turki (1998). "Adapting just-in-time manufacturing systems to preventive maintenance interruptions." *Production Planning and Control*, 9 (4).

Gupta, S.M. and Y.A.Y. Al-Turki (1998). "The effect of sudden material handling system breakdown on the performance of a JIT system." *International Journal of Production Research*, 36 (7), 1935–1960.

Gupta, S.M., Y.A.Y. Al-Turki, and R.F. Perry (1995). "Coping with processing times variation in a JIT environment." *Proceedings of Northeast Decision Sciences Institute Conference*.

Gupta, Y.P. and P. Bagchi (1987). "Inbound Freight Consolidation Under Just-In-Time Procurement, Application of Clearing Models." *Journal of Business Logistics*, 8, 2, 74–94.

Hall, G., J. Rosenthal, and J. Wade (1993). "How to Make Reengineering Really Work." *Harvard Business Review* (November–December), 71 (6), 119–131.

Hall, R.W. (1982). *Kawasaki U.S.A.: A Case Study*. APICS.

Hall, R.W. (1983). *Zero Inventories*. Dow Jones-Irwin. Homewood, Illinois.

Hall, R.W. (1993). "The Challenges of the Three-Day Car." *Target*, 9 (2), March/April, 21–29.

Hall, R.W. (1993). "AME's Vision of Total Enterprise Manufacturing." *Target*, 9, 6, November/December, 33–38.

Hallihan, A.P. Sackett, and G. M. Williams (1997). "JIT manufacturing: The evolution to an implementation model founded in current practice." *International Journal of Production Research*, 35 (4), 901–920.

Hammer, M. (1996). *Beyond Reengineering*, Harper Business, New York.

Hammer, M. and J. Champy (1993). *Reengineering the Corporation: A Manifestor for Business Revolution*, Harper Business, New York.

Hammer, M. and S. Stanton (1995). *The Reengineering Revolution*, Harper Business, New York.

Handfield, R.B. (1995). *Re-engineering for Time-based Competition-Benchmarks and Best Practices for Production, R&D, and Purchasing*, Quorum Books, Westport, CT.

Hannah, K.H. (1987). "Just-in-Time: Meeting the Competitive Challenge." *Production and Inventory Management*, 28, 3, 1–3.

Harrington, D.H.J. (1991). *Business Process Improvement: The Breakthrough Strategy for Total Quality, Productivity, and Competitiveness*, McGraw-Hill, New York.

Hartmann, E.H. (1992). *Successfully Installing TPM in a Non-Japanese Plant*. TPM Press, Inc., Allison Park, PA.

Hay, E.J. (1988). *The Just-In-Time Breakthrough: Implementing the New Manufacturing Basics*. John Wiley & Sons, New York: NY.

Hayes, R.H. and G.P. Pisano (1994). "Beyond World Class: The New Manufacturing Strategy." *Harvard Business Review*, January–February, 77–86.

Hirano, Hiroyuki (Black, J.T.,/English Edition). (1988). *JIT Factory Revolution*. Productivity Press.

Huang, C.C. and A. Kusiak (1996). "Overview of Kanban Systems." *International Journal of Computer Integrated Manufacturing*, 9, 169–189.

Hutchins, D. (1986). "Having a Hard Time with Just-in-Time." *Fortune*, June 9, 64–66.

Imai, Masaaki (1986). *Kaizen: The Key to Japan's Competitive Success*. Random House, New York.

Japan Management Association (1985). *Kanban: Just-In-Time at Toyota*. Productivity Press.

Johansson, H.J., P. McHugh, A.J. Pendlebury, and W. A. Wheeler, III (1993). *Business Process Reengineering: BreakPoint Strategies for Market Dominance*, Wiley & Sons, New York.

Jordan, J. (1997). "Enablers for Agile Virtual Enterprise Integration." *Agility and Global Competition*, 1 (3), 26–46.

Kasarda, J.D. and D.A. Rondinelli (1998). "Innovative Infrastructure for Agile Manufacturing." *Sloan Management Review*, 39 (2), 73–82.

Keaon, M. (1995). "A New look at the Kanban Production Control System." *Production and Inventory Management Journal*, Third Quarter, 71–78.

Keller, A.Z. and A. Kazazi (1993). "Just-in-time manufacturing systems: A literature review." *Industrial Management and Data Systems*, 93 (7), 1–32.

Kidd, P.T. (1994). *Agile Manufacturing-Forging New Frontiers*, Addison-Wesley Publishing Company, Wokingham, England.

Krajewski, L.J., B. King, L.P. Ritzman, and D.S. Wong (1987). "Kanban, MRP, and Shaping the Manufacturing Environment." *Management Science*, 33 (1) (January), 39–57.

Kupanhy, Lumbidi (1995). "Classification of JIT Techniques and Their Implications." *Industrial Engineering*, 27 (2), Feb., 62–66.

Lee, S. and M. Ebrahimpour (1984). "Just-In-Time Production System: Some Requirements for Implementation." *International Journal of Operations and Production Management*, 4 (1), 3–11.

Levy, D. (1997). "Lean production in an international supply chain." *Sloan Management Review*, 38 (2), 94–102.

Lubben, Richard T. (1988). *Just-In-Time: An Aggressive Manufacturing Strategy*, McGraw-Hill, New York.

March, Barbara (1993). "Allen Edmonds Shoe Tries Just-In-Time Production." *Wall Street Journal*, March 5, 1.

Masaracchia, Philip (1987). "TQC – The 'Quality' Component of J.I.T." *P&IM Review*, April, 44.

McGuire, Kenneth J. (1984). "Impressions From Our Most Worthy Competitor." American *Production and Inventory Control Society*, Falls Church, VA.

McKone, K. and E. Weiss (1999). "Managerial Guidelines for the Use of Predictive Maintenance." Darden School Working Paper Series, University of Virginia, Charlottesville, VA.

McKone, K., R. Schroeder, and K. Cua (1998). "The Impact of Total Productive Maintenance Practices on Manufacturing Performance." Carlson School Working Paper.

McKone, K., R. Schroeder, and K. Cua (1999). "Total Productive Maintenance: A Contextual View." *Journal of Operations Management*, 17 (2), p. 123–144.

Modic, S. (1987). "Myths about Japanese Management." *Industry Week*, October 5.

Moffat, S. (1990). "Japan's New Personalized Production." *Fortune*, 122, 132–135.

Monden, Y. (1981). "What makes the Toyota production system really tick?" *Industrial Engineering*, 13, 36–46.

Monden, Y. (1983). *Toyota Production System: A Practical Approach to Production Management*. Industrial Engineers and Management Press, Norcross, GA.

Nachi-Fujikoshi Corporation (1990). *Training for TPM: A Manufacturing Success Story*, Productivity Press, Cambridge, MA.

Nakajima, S. (1986). "TPM-challenge to the improvement of productivity by small group activities." *Maintenance Management International*, 6, 73–83.

Nakajima, Seiichi (1988). *Introduction to TPM: Total productive Maintenance*. Productivity Press.

Natarajan, R. and S. Goyal (1994). "Safety Stocks in JIT Environments." *International Journal of Operations and Production Management*, 14 (10).

NKS/Factory Magazine (Introduction by Shigeo Shingo) (1988). *Poka-Yoke: Improving Quality by Preventing Defects*. Productivity Press.

O'Neal, Charles R. (1987). "The Buyer-Seller Linkage in a Just-in-time Environment." *Journal of Purchasing and Materials Management*, Spring, 7.

Ohno, Taiichi (1988). *Toyota Production System: Beyond Large-Scale Production*. Productivity Press, Cambridge, MA.

Pascale, Richard T. and Anthony G. Athos (1981). *The Art of Japanese Management*. Simon and Schuster, New York.

Philipoom, P.R., L.P. Rees, and B.W. Taylor (1996). "Simultaneously Determining the Number of Kanbans, Container Sizes and the Final Assembly Sequence of Products in a Just-in-Time Shop." *International Journal of Production Research*, 34, 51–69.

Plenert, G. and T.D. Best (1986). "MRP, JIT, and OPT: What's Best?" *Production and Inventory Management*, 27 (2), 22–29.

Plossl, G.W. (1985). *Just in-Time: A Special Roundtable*. George Plossl Educational Services, Georgia.

Preiss, K. (1997). "A Systems Perspective of Lean and Agile Manufacturing." *Agility and Global Competition*, 1 (1), 59–75.

Preiss, K., S. Goldman, and R. Nagel (1996). *Cooperate to Compete: Building Agile Business Relationships*. Van Nostrand Reinhold, New York.

Preiss, K. (1995). "Mass, Lean, and Agile as Static and Dynamic Systems." *Agility Forum*, Bethlehem, PA.

Preiss, K. (1995). "Models of the Agile Competitive Environment." *Agility Forum*, Bethlehem, PA.

Quinlan, J. (1982). "Just-in-Time at the Tractor Works." *Material Handling Engineering*, June, 62–65.

Reeves, E. and E. Torrey, Editors (1994). *Beyond the Basics of Reengineering: Survival Tactics for the '90s*. Industrial Engineering and Management Press, Georgia.

Richards, C.W. (1996). "Agile Manufacturing: Beyond Lean?" *Production and Inventory Management Journal*, 37, 60–64.

Riopel, Robert J. (1986). "JIT: Evolutionary Revolution." *Manufacturing Systems*, July, 44.

Roos, D. (1995). "Agile/Lean: A Common Strategy for Success." *Agility Forum*, Bethlehem, PA.

Sakakibara, S., B.B. Flynn, and R.G. Schroeder (1993). "A framework and measurement instrument for just-in-time manufacturing." *Production and Operations Management*, 2, 177–194.

Savsar, M. (1996). "Effects of Kanban Withdrawal Policies and other Factors on the Performance of JIT Systems – A Simulation Study." *International Journal of Production Research*, 34, 2879–2899.

Schneider, J.D. and M.A. Leatherman (1992). "Integrated Just-In-Time: A Total Business Approach." *Production and Inventory Management Journal*, 33 (1), 78–82.

Schonberger, R.J. (1982). "Some Observations on the Advantages and Implementation Issues of Just-in-Time Production Systems." *Journal of Operations Management*, 2 (1) (November), 1–12.

Schonberger, R.J. (1982). *Japanese Manufacturing Techniques: Nine Hidden Lessons in Simplicity*. The Free Press, New York.

Schonberger, R.J. (1983). "Applications of Single-Card and Dual-Card Kanban." *Interfaces*, 13 (4) (August), 56–67.

Schonberger, R.J. (1983). "Selecting the Right Manufacturing Inventory Systems: Western and Japanese Approaches." *Production and Inventory Control*, 24 (2) (Second Quarter), 33–44.

Schonberger, R.J. (1984). "Just-In-Time Production System: Replacing Complexity with Simplicity in Manufacturing Management." *Industrial Engineering*, Oct, 52–63.

Schonberger, R.J. (1986). *World Class Manufacturing: The Lessons of Simplicity Applied*. Free Press, New York.

Schonberger, R.J. (1987). *World Class Manufacturing Casebook: Implementing JIT and TQC*. Free Press, New York.

Schonberger, R.J. (1996). *World Class Manufacturing: The Next Decade*. The Free Press.

Schonberger, R.J. and J. Gilbert (1983). "Just-in-Time Purchasing: A Challenge for U.S. Industry." *California Management Review*, 26 (1) (Fall), 54–68.

Schwarzendahl, R. (1996). "The Introduction of Kanban Principles to Material Management in EWSD Production." *Production Planning and Control*, 7, 212–221.

Seglund, R. and S. Ibarreche (1984). "Just-in-Time: The Accounting Implications." *Management Accounting* (August), 43–45.

Sepehri, M. (1986). *Just-in-Time, Not Just in Japan*. American Production and Inventory Control Society, Virginia.

Sheridan, J.H. (1993). "Agile Manufacturing: Beyond Lean Production." *Industry Week*, 242, 34–36.

Shiba, S., A. Graham, and D. Walden (1993). *A New American TQM: Four Practical Revolutions in Management*. Productivity Press, Oregon, 3 (3)0, 411–460.

Shimbun, N.K. (ed.) (1995). *TPM Case Studies*. Productivity Press, Portland, OR.

Shingo, S. (1981). *Study of Toyota Production System From Industrial Engineering Viewpoint*. Japan Management Association, Tokyo.

Shingo, S. and A.G. Robinson (1990). *Modern Approaches to Manufacturing Improvements: The Shingo System*. Productivity Press, MA.

Short, J.E. and N. Venkatraman (1992). "Beyond Business Process Redesign: Redefining Baxter's Business Network." *Sloan Management Review* (Fall), 34 (1), 7–20.

Shultz, Terry (1982). "BRP: The Journey to Excellence." *The Forum*. Milwaukee, WI.

Siha, S. (1994). "The Pull Production System: Modelling and Characteristics." *International Journal of Production Research*, 32, 933–950.

Sohal, A.S., L. Ramsay, and D. Samson (1993). "JIT manufacturing: Industry analysis and a methodology for implementation." *International Journal of Operations and Production Management*, 13 (7), 22–56.

Spearman, M.L., D.L. Woodruff, and W.J. Hopp (1990). "CONWIP: a pull alternative to kanban." *International Journal of Production Research*, 28 (5), 879–894.

Steinbacher, H.R. and N.L. Steinbacher (1993). *TPM for America: What It Is and Why You Need It*. Productivity Press, Cambridge, MA.

Storhagen, N.G. and R. Hellberg (1987). "Just-in-Time from a Business Logistics Perspective." *Engineering Cost and Production Economics* (The Netherlands) 12 (1–4) (July), 117–121.

Sugimori, Y., K. Kusunoki, F. Cho, and S. Uchikawa (1977). "Toyota production system and kanban system materialization of just-in-time and respect-for-human system." *International Journal of Production Research*, 15, 553–564.

Suzaki, K. (1985). "Japanese Manufacturing Techniques, Their Importance to U.S. Manufacturers." *Journal of Business Strategy*, 5 (3), 10–20.

Suzaki, K. (1987). *The New Manufacturing Challenge: Techniques for Continuous Improvement*. The Free Press, New York.

Suzuki, T. (1992). *New Directions for TPM*, Productivity Press, Cambridge, MA.

Tajiri, M. and F. Gotoh (1992). *TPM Implementation: A Japanese Approach*. McGraw-Hill, Inc., New York.

Tent, J.T.C., V. Grover, and K.D. Fiedler (1996). "Developing Strategic Perspectives on Business Process Reengineering: From Process Reconfiguration to Organizational Change." *Omega, International Journal of Management Science*, 24 (3), 271–294.

Walleigh, R.C. (1986). "Getting Things Done: What's Your Excuse for Not Using JIT?" *Harvard Business Review* (March–April), 36–54.

Wantuck, K. (1989). *Just In Time For America*, KWA Media, Southfield, MI.

Weatherston, D. (1986). "Transportation Makes the JIT Connection." *Industrial Management* (April), 40–44.

Weiss, A. (1984). "Simple Truths of Japanese Manufacturing." *Harvard Business Review* (July–August), 119–125.

White, R.E. (1993). "An empirical assessment of JIT in U.S. manufacturers." *Production and Inventory Management Journal*, 34 (2), 38–42.

White, R.E. and W.A. Ruch (1990). "The composition and scope of JIT." *Operations Management Review*, 7 (3&4), 9–18.

Wildemann, Horst (1987). "JIT Progress in West Germany." *Target*, Summer, 23–27.

Womack, J.P. and D.T. Jones (1994). "From lean production to the lean enterprise." *Harvard Business Review*, 72, March–April, 93–103.

Womack, J.P., D.T. Jones, and D. Roos. (1991). *The Machine that Changed the World: The Story of Lean Production*. Harper Collins, New York, 37–38, 91–93.

PLANT LOCATION

Cohen, M.A. and H.L. Lee (1989). "Resource deployment analysis of global manufacturing and distribution networks." *Journal of Manufacturing and Operations Management*, 2, 81–104.

Collins, R. and R.W. Schmenner (1995). "Taking manufacturing advantage of Europe's single market." *European Management Journal*, 13, 257–68.

Daniels, J.D. and L.H. Radebaugh (1998). *International Business Environments and Operations* (8th ed.). Addison Wesley, Reading, Mass.

De Meyer, A. and A. Vereecke (1996). "International Operations" in *International Encyclopedia of Business and Management* Editor M. Werner. Routledge.

Dreifus, S.B., ed. (1992). *Business International's Global Management Desk Reference*, McGraw-Hill, New York.

Fawcett, S.E. (1990). "Logistics and Manufacturing Issues in Maquiladora Operations." *International Journal of Physical Distribution and Logistics Management*, 20 (4), 13–21.

Feldman E., F.A. Lehrer, and T.L. Ray (1966). "Warehouse Locations Under Continuous Economies of Scale." *Management Science*, 12, May.

Ferdows, K. (1997). "Making the most of foreign factories." *Harvard Business Review*, 73–88.

Flaherty, M.T. (1996). *Global Operations Management*, McGraw-Hill, New York.

Handfield, R.B. (1994). "Global sourcing: Patterns of development." *International Journal of Operations and Production Management*, 14 (6), 40–51.

Harding, C.F. (1980). "Your Business: Right Ingredients, Wrong Location." *INC* (February), 20–22.

Hodder, J.E. and J.V. Jucker (1985). "International Plant Location under Price and Exchange Rate Uncertainty." *Engineering Costs and Production Economics*, 9, 225–229.

Hoffman, J.J. and M.J. Schniederjans (1990). "An International Strategic Management/Goal Programming Model for Structuring Global Expansion Decisions in the Hospitality Industry." *International Journal of Hospitality Management*, 9 (3), 175–190.

Jain, S.J. (1989). "Standardization of international marketing strategy: Some research hypotheses." *Journal of Marketing*, 53, 70–79.

Kanter, R.M. (1995). *World Class: Thriving Locally in the Global Economy*, Simon and Shuster, New York.

Kashani, K. (1989). "Beware of the pitfalls of global marketing." *Harvard Business Review*, September–October, 91–98.

Kraus, R.L. (1980). "A Strategic Planning Approach to Facility Site Selection." *Dun's Review* (November), 14–16.

Levitt, T. (1983). *"The Globalization of Markets." Harvard Business Review*, 61 (3), 92–102.

MacCormack, A.D., L.J. Newmann, III, and D.B. Rosenfield (1994). "The New Dynamics of Global Manufacturing Site Location." *Sloan Management Review*, 35 (4), 69–80, Summer.

Martinez, J.I. and J.C. Jarillo (1989). "The evolution of research on coordination mechanisms in multinational corporations." *Journal of International Business Studies*, 20 (489–514).

Miller, J.G., A. De Meyer, and J. Nakane (1992). *Benchmarking Global Manufacturing*, Irwin, Homewood, IL.

Porter, M.E. (1986). Editor. *Competition in Global Industries*. Harvard Business School Press, Boston.

Porter, M.E. (1990). *"The Competitive Advantage of Nations." Harvard Business Review*, 68 (2), 73–93.

Robinson, Richard D. (1984). *Internationalization of Business: An Introduction*. Free Press, Chicago.

Schilling, D.A. (1980). "Dynamic Location Modeling for Public Sector Facilities: A Multi-Criteria Approach." *Decision Sciences*, 11 (4) (October), 714–724.

Schniederjans, M.J. and J.J. Hoffman (1992). "Multinational Acquisition Analysis: A Zero-One Goal Programming Model." *European Journal of Operational Research*, 62, 174–185.

Starr, M.K. (1988). *Global Competitiveness: Getting the U.S. Back on Track*. Norton, New York.

Tong, H.M. and C.K. Walter (1980). "An Empirical Study of Plant Location Decisions of Foreign Manufacturing Investors in the United States." *Columbia Journal of World Business* (Spring), 66–73.

PLANT MAINTANANCE

Anderson, M.Q. (1981). "Monotone Optimal Preventative Maintenance Polices for Stochastically Failing Equipment." *Naval Research Logistics Quarterly*, 28, 347–358.

Aven, T. and R. Dekker (1997). "A useful framework for optimal replacement models." *Reliability Engineering and System Safety*, 58, 61–67.

Christer, A.H. and W.M. Waller (1984). "Reducing production downtime using delay-time analysis." *Journal of the Operational Research Society*, 35, 499–512.

Cordero, S.T. (1987). *Maintenance Management*. Fairmont Press, Englewood Cliffs, NJ.

Dekker, R., C.F.H. Van Rijn, and A.C.J.M. Smit (1994). "Mathematical models for the optimization of maintenance and their application in practice." *Maintenance*, 9, 22–26.

Dekker, R., F.A. Van der Duyn Schouten, and R.E. Wildeman (1997). "A review of multi-component maintenance models with economic dependence." *Mathematical Methods of Operations Research*, 45, 411–435.

Dhavale D.G. and G.L. Otterson, Jr. (1980). "Maintenance by Priority." *Industrial Engineering*, 12 (2) (February), 24–27.

Gotoh, F. (1991). *Equipment Planning for TPM: Maintenance Prevention Design*, Productivity Press, Cambridge, MA.

Gupta, S.M. and Y.A.Y. Al-Turki (1998). "Adapting just-in-time manufacturing systems to preventive maintenance interruptions." *Production Planning and Control*, 9 (4).

Hartmann, E.H. (1992). *Successfully Installing TPM in a Non-Japanese Plant*. TPM Press, Inc., Allison Park, PA.

Henley, E.J. and H. Kavmamoto (1981). *Reliability Engineering and Risk Assessment*. Prentice-Hall, Englewood Cliffs, NJ.

Hertz, D.D. (1982). *Risk Analysis and Its Applications*. John Wiley, New York.

Hertz, D.D. (1983). *Practical Risk Analysis*. John Wiley, New York.

Hora, M.E. (1987). "The Unglamorous Game of Managing Maintenance." *Business Horizons*, 30 (3) (May–June), 67–75.

Katzel, J. (1987). "Maintenance Management Software." *Plant Engineering*, 41 (12) (June 18), 124–170.

Lee, H.L. and M.J. Rosenblatt (1989). "A Production and Maintenance Planning Model with Restoration Cost Dependent on Detection Delay." *IIE Transactions*, 21 (4), 368–375.

Mann, L., Jr. (1983). *Maintenance Management*. Rev. ed. Lexington Books, Lexington, MA.

McCall, J.J. (1965). "Maintenance Policies for Stochastically Failing Equipment: A Survey." *Management Science*, 11 (5), 493–524.

McKone, K. and E. Weiss (1999). "Managerial Guidelines for the Use of Predictive Maintenance." Darden School Working Paper Series, University of Virginia, Charlottesville, VA.

McKone, K., R. Schroeder, and K. Cua (1998). "The Impact of Total Productive Maintenance Practices on Manufacturing Performance." Carlson School Working Paper.

McKone, K., R. Schroeder, and K. Cua (1999). "Total Productive Maintenance: A Contextual View." *Journal of Operations Management*, 17 (2), p. 123–144.

Nachi-Fujikoshi Corporation (1990). *Training for TPM: A Manufacturing Success Story*, Productivity Press, Cambridge, MA.

Nakajima, S. (1986). "TPM-challenge to the improvement of productivity by small group activities." *Maintenance Management International*, 6, 73–83.

Nakajima, S. (1988). *Introduction to TPM*, Productivity Press, Cambridge, MA.

Özekici, S. (1996). Editor. *Reliability and maintenance of complex systems. NATO ASI Series*, vol. 154, Springer, Berlin.

Pate-Cornell, M.E., H.L. Lee, and G. Tagaras (1987), "Warning of Malfunction: The Decision to Inspect and Maintain Production Processes on Schedule or On Demand." *Management Science*, 33 (10), 1277–1290.

Patton, J.D., Jr. (1983). *Preventive Maintenance. Instrument Society of America*. Research Triangle Park, NC.

Pierskalla, W.P. and J.A. Voelker (1976). "A Survey of Maintenance Models: The Control and Surveillance of Deteriorating Systems" *Naval Research Logistics Quarterly*, 23, 353–388.

Pintelon, L.M. and L.F. Gelders (1992). "Maintenance management decision making." *European Journal of Operational Research*, 58, 301–317.

Shimbun, N.K. (ed.) (1995). *TPM Case Studies*. Productivity Press, Portland, OR.

Sinha, S.K. and B.K. Kale (1980). *Life Testing and Reliability Estimation*. John Wiley, New York.

Smith, A.M. (1993). *Reliability centered maintenance*. Mc. Graw-Hill, New York.

Smith, D.J. (1985). *Reliability and Maintainability in Perspective* (2nd ed.). John Wiley, New York.

Steinbacher, H.R. and N.L. Steinbacher (1993). *TPM for America: What It Is and Why You Need It*. Productivity Press, Cambridge, MA.

Suzuki, T. (1992). *New Directions for TPM*, Productivity Press, Cambridge, MA.

Tajiri, M. and F. Gotoh (1992). *TPM Implementation: A Japanese Approach*. McGraw-Hill, Inc., New York.

Tapiero, C.S. (1986). "Continuous Quality Production and Machine Maintenance." *Naval Research Logistics Quarterly*, 33, 489–499.

Thilander, M. (1992). "Some Observations of Operation and Maintenance in Two Swedish Firms." *Integrated Manufacturing Systems*, 3 (2).

Thompson, G. (1968). "Optimal Maintenance Policy and Sale Date of a Machine." *Management Science*, 14 (9), 543–550.

Tombari, H. (1982). "Designing a Maintenance Management System." *Production and Inventory Management*, 23 (4), 139–147.

Tomlinson, P.D. (1987). "Organizing for Productive Maintenance." *Production Engineering*, 34 (10) (October), 38–40.

Tsuchiya, S. (1992). *Quality Maintenance: Zero Defects Through Equipment Management*, Productivity Press, Cambridge, MA.

Turbide, D.A. (1993). "Total Productive Maintenance Finds Its Place in the Sun." *APICS – The Performance Advantage*, August, 32–34.

Valdez-Flores, C. and R.M. Feldman (1989). "A Survey of Preventative Maintenance Models for Stochastically Deteriorating Single-Unit Systems." *Naval Research Logistics Quarterly*, 36, 419–446.

Vanneste, S.G. and L.M. Van Wassenhove (1995). "An integrated and structured approach to improve maintenance." *European Journal of Operational Research*, 82, 241–257.

Veeramani, D., J.J. Bernardo, C.H. Chung, and Y.P. Gupta (1995). "Computer-Integrated Manufacturing: A Taxonomy of Integration and Research Issues." *Production and Operations Management*, 4 (4), 360–380.

Wireman, T. (1984). *Preventive Maintenance*. Englewood Cliffs, NJ, Reston.

PRODUCT DEVELOPMENT AND DESIGN

Adler, P.S., A. Mandelbaum, V. Nguyen, and E. Schwerer (1995). "From Project to Process Management: An Empirically Based Framework for Analyzing Product Development Time." *Management Science*, 41, 458–484.

Alexander, D.C. (1985). "A Business Approach to Ergonomics." *Industrial Engineering*, 17 (7) (July), 32–39.

Alting, L. (1982). *Manufacturing Engineering Processes*. Marcel Dekker, New York.

Alting, L. (1993). "Life-Cycle Design of Products: A New Opportunity for Manufacturing Enterprises." In *Concurrent Engineering*, A. Kusiak (ed.). Wiley, New York.

Ben-Arieh, D., C.A. Fritsch, and K. Mandel (1986). "Competitive Product Realization in Today's Electronic Industries." *Industrial Engineering*, February, 34–42.

Bridger, R.S. (1995). *Introduction to Ergonomics*, McGraw-Hill, NY, p. 1–30.

Buggie, Frederick D. (1981). *New Product Development Strategies*. AMACOM, New York.

Carson, J.W. and T. Richards (1979). *Industrial New Product Development: A Manual for the 1980's*. Wiley, New York.

Clark, K.B. (1989). "Project Scope and Project Performance: The Effect of Parts Strategy and Supplier Involvement on Product Development." *Management Science*, 35 (10), 1247–1263.

Clark, K.B. (1992). "Design for Manufacturability at Midwest Industries: Teaching Note." Note no. 5-693-007, Harvard Business School Press, Boston.

Clark, K.B. and T. Fujimoto (1991). *Product Development Performance: Strategy, Organization and Management in the World Auto Industry*. Harvard Business School Press, Cambridge.

Cooper, R.G., E.J. Kleinschmidt (1987). New Products: What Separates Winners from Losers? *Journal of Product Innovation Management*, 4, 3, 69–184.

Dean, J.H., Jr. and G.I. Susman (1989). "Organizing for Manufacturable Design." *Harvard Business Review*, January–February, 28–36.

Diprimio, A. (1987). *Quality Assurance in Service Organizations*. Radnor, PA, Chilton.

Doblin, J. (1987). "A Short Grandiose Theory of Design." *STA Design Journal*.

Dorf, R.C. and A. Kusiak (1994). *Handbook of Design, Manufacturing, and Automation*. Wiley, New York.

Eisenhardt, K.M. and B.N. Tabrizi (1995). "Accelerating Adaptive Processes: Product Innovation in the Global Computer Industry." *Administrative Science Quarterly*, 40, 84–110.

Eppinger, S.D., D.E. Whitney, R.P. Smith, and D.A. Gebala (1994). "A Model-Based Method for Organizing Tasks in Product Development." *Research in Engineering Design*, 6, 1–13.

Erens, F.J. and Hegge, H.M.H. (1994). "Manufacturing and Sales Coordination for Product Variety." *International Journal of Production Economics*, 37–1, 83–100.

Feitzinger, E. and L.H. Lee (1997). "Mass Customization at Hewlett-Packard: The Power of Postponement." *Harvard Business Review*, 75, 116–121.

Fitzsimmons, James A., P. Kouvelis, and D.N. Mallick (1991). "Design Strategy and Its Interface with Manufacturing and Marketing: A Conceptual Framework." *Journal of Operations Management*, 10 (3) August, p. 398–415.

Forest, J. (1991). "Models For The Process Of Technological Innovation." *Technology Analysis And Strategic Management*, 3 (4).

Gerwin, D. (1993). "Integrating Manufacturing into the Strategic Phases of New Product Development." *California Management Review*, Summer, 123–136.

Gerwin, D. and G. Susman (1996). "Special Issue on Concurrent Engineering." *IEEE Trans-actions on Engineering Management*, 43, 118–123.

Gilmore, H.J. and J.B. Pine II (1997). "The Four Faces of Mass Customization." *Harvard Business Review*, 75, 91–101.

Gulati, R.K. and S.D. Eppinger (1996). "The Coupling of Product Architecture and Organizational Structure Decisions." MIT Working Paper 3906.

Gupta, S.M. and C.R. McLean (1996). "Disassembly of Products." *Computers and Industrial Engineering*, 31, 225–228.

Haddad, C.J. (1996). "Operationalizing the Concept of Concurrent Engineering: A Case Study from the U.S. Auto Industry." *IEEE Transactions on Engineering Management*, 43 (2), 124–132.

Hadjinicola, G.C. and K.R. Kumar (1997). "Factors affecting international product design." *Journal of the Operational Research Society*, 48, 1131–1143.

Handfield, R.B. (1994). Effects of concurrent engineering on make-to-order products. *IEEE Transactions on Engineering Management*, 41 (4), 1–11.

Handfield, R.B. and R. Pannesi (1994). "Managing component life cycles in dynamic technological environments." *International Journal of Purchasing and Materials Management* (Spring), 20–27.

Hauptman, O. and K.K. Hirji (1996). "The Influence of Process Concurrency on Project Outcomes in Product Development: an Empirical Study with Cross-Functional Teams." *IEEE Transactions in Engineering Management*, 43, 153–164.

Hendricks, K.B. and V.R. Singhal (1997). "Delays in New Product Introductions and the Market Value of the Firm: The Consequences of Being Late to the Market." *Management Science*, 43 (4), 422–436.

Hill, J.S. and R.R. Still (1984). "Adapting products to LDC tastes." *Harvard Business Review*, March–April, 92–101.

Iansiti, M. (1995). "Shooting the rapids: managing product development in turbulent environments." *California Management Review*, 38 (1), 37–58.

Imai, K., I. Nonaka, and H. Takeuchi (1985). "Managing the New Product Development Process: How the Japanese Companies Learn and Unlearn." In: Clark, K. B., R.H. Hayes, and C. Lorenz (eds.). *The Uneasy Alliance*. Harvard Business School Press, Boston.

Keoleian, G.A. and D. Menerey (1994). "Sustainable Development by Design: Review of life cycle Design and Related Approaches." *Journal of the Air and Waste Management Association*, 44 (5), 645–668.

Klassen, R.D. and N.P. Greis (1993). "Managing Environmental Improvement through Product and Process Innovation: Implications of Environmental Life Cycle Assessment." *Industrial and Environmental Crisis Quarterly*, 7 (4), 293–318.

Kotha, S. (1995). "Mass Customization: Implementing the Emerging Paradigm for the Competitive Advantage." *Strategic Management Journal*, 16, 21–42.

Krishnan, V., S.D. Eppinger, and D.E. Whitney (1997). "A Model-Based Framework to Overlap Product Development Activities." *Management Science*, 43, 437–451.

Kubiak, J. (1993). "A Joint Venture in Mass Customization." *Planning Review*, July 21 (4), 25.

Kusiak, A. (1992). *Intelligent Design and Manufacturing*. Wiley, New York.

Kusiak, A. (1993). *Concurrent Engineering: Automation, Tools, and Techniques*. Wiley, New York.

Kusiak, A. and U. Belhe (1992). "Concurrent engineering: A design process perspective." Proceedings of the American Society of Mechanical Engineers, PED 59, 387–401.

Liker, J.K., R.R. Kamath, S.N. Wasti, and M. Nagamachi (1995). "Integrating Suppliers into Fast-Cycle Product Development." In J.K. Liker, J, J.E. Ettlie, and J.C. Campbell (eds.). *Engineered in Japan: Japanese Technology Management Practices*. Oxford University Press, Oxford.

Littler, D., F. Leverick, and M. Bruce (1995). Factors affecting the process of collaborative product development: a study of UK manufacturers of information and communications technology products. *The Journal of Product Innovation Management*, 12 (1), 16–23.

Loch, C.H., C. Terwiesch (1998). "Communication and Uncertainty in Concurrent Engineering." *Management Science*, 44, August.

Lockamy, A. III and J.F. Cox III (1994). *Reengineering Performance Measurement*. Irwin, New York.

Lorenz, C. (1986). *The Design Dimension: The New Competitive Weapon for Business*, Basil Blackwell, NY.

Mabert, V.A., J.F. Muth, and R.W. Schmenner (1992). "Collapsing New Product Development Times: Six Case Studies." *Journal of Product Innovation Management*, 9, 200–212.

Mallick, D.N. and P. Kouvelis (1992). "Management of Product Design: A Strategic Approach." In *Intelligent Design and Manufacturing*, ed. A. Kusiak, Wiley, New York.

Malott, R.H. (1983). "Let's Restore Balance to Product Liability Law." *Harvard Business Review*, May–June, 66–74.

McCormick Ernest J. (1982). *Human Factors in Engineering and Design*. McGraw-Hill, New York.

McKeown, R. (1990). "New products from existing technologies." *The Journal of Business and Industrial Marketing*, 5 (1).

McQuater, R.E., A.J. Peters, B.G. Dale, M. Spring, J.H. Rogerson, E.M. Rooney (1998). "The Management and Organisational Context of New Product Development." *International Journal of Production Economics*, 55, 121–131.

Meyer, M.H. (1997). "Revitalize your Product Lines Through Continuous Platform Renewal." *Research Technology Management*, 40, 17–28.

Monczka, R.D. Frayer, R. Handfield, G. Ragatz and R. Trent (1998). "*Supplier Integration into New Product/Process Development: Best Practices*." Milwaukee: ASQC.

Morelli, M.D., S.D. Eppinger, and R.K. Gulati (1995). "Predicting Technical Communication in Product Development Organizations." *IEEE Transactions on Engineering Management*, 42, 215–222.

Muhlemann, A.P., A.G. Lockett, and C.I. Farn (1982). "Job Shop Scheduling Heuristics and Frequency of Scheduling." *International Journal of Production Research*, 20 (2), 227–241.

Nevins, J.L. and D.E. Whitney (1989). *Concurrent Design of Products and Processes*. McGraw-Hill, New York.

Pine II, B.J. (1993). "Mass Customizing Products and Services." *Planning Review*, July 21 (4), 6.

Pine II, B.J. and T.W. Pietrocini (1993). "Standard Modules Allow Mass Customization at Bally Engineered Structures." *Planning Review*, July 21 (4), 20.

Pine II, B.J., B. Victor, and A.C. Boyton (1993). "Making Mass Customization Work." *Harvard Business Review*, 71, 108–119.

Pine II, J.B. (1993). *Mass Customization – The New Frontier in Business Competition*, Harvard Business School Press, Boston, MA.

Pisano, G. (1997). *The Development Factory*. Harvard Business School Press, Boston.

Ragatz, G., R. Handfield, and T. Scannell (1997). "Success factors for integrating suppliers into new product development." *Journal of Product Innovation Management*, 7 (20).

Rosenthal, S.R. (1992). *Effective Product Design and Development*. Business One Irwin, Homewood.

Ross, Phillip J. (1988). "The Role of Taguchi Methods and Design of Experiemts in QFD." *Quality Progress*, June, 41–47.

Smith, R. (1997). "The Historical Roots of Concurrent Engineering Fundamentals." *IEEE Transactions on Engineering Management*, 44 (1), 67–78.

Smith, R.P. and S.D. Eppinger (1997). "Identifying Controlling Features of Engineering Design Iteration." *Management Science*, 43, 276–293.

Souder, W. (1987). *Managing New Product Innovations*. Lexington MA, Lexington Books.

Spira, J.S. and Pine II, B.J. (1993). "Mass Customization." *Chief Executive*, March, 83, 26.

Starr, M.K. (1965). "Modular production – a new concept." *Harvard Business Review*, November–December, 137–145.

Stobaugh, R. and P. Telesio (1983). "Match Manufacturing Policies and Product Strategies." *Harvard Business Review*, March–April, 113–120.

Sullivan, L. (1995). "Custom-made." *Forbes*, 156, 124–125.

Sullivan, L.P. (1986). "Quality Function Deployment." *Quality Progress*, 19 (6) (June), 36–50.

Sullivan, L.P. (1988). "Policy Management Through Quality Functional Deployment." *Quality Progress*, June, 18–20.

Susman, G.I. (1992). *Integrating Design and Manufacturing for Competitive Advantage*, Oxford University Press, New York.

Swink, M.L., J.C. Sandvig, and V.A. Mabert (1996). "Adding Zip to Product Development: Concurrent Engineering Methods and Tools." *Business Horizons*,39 (2), 41–49.

Swink, M.L., J.C. Sandvig, and V.A. Mabert (1996). "Customizing Concurrent Engineering Processes: Five Case Studies." *Journal of Product and Innovation Management*, 13 (3), 229–244.

Takeuchi, H. and I. Nonaka (1986). "The New Product Development Game." *Harvard Business Review*, 64, 137–146.

Terwiesch, C. and C.H. Loch (1997). "Management of Overlapping Development Activities: a Framework for Exchanging Preliminary Information." *Proceedings of the 4th EIASM Conference on Product Development*, 797–812.

Terwiesch, C., C.H. Loch, and M. Niederkofler (1996). "Managing Uncertainty in Concurrent Engineering." *Proceedings of the 3rd EIASM Conference on Product Development*, 693–706.

Ulrich, K.T. (1995). "The Role of Product Architecture in the Manufacturing Firm." *Research Policy*, 24, 419–440.

Ulrich, K.T. and K. Tung (1991). "Fundamentals of Product Modularity." *Proceedings of the 1991 ASME Winter Annual Meeting Symposium on Issues in Design/Manufacturing Integration*. Atlanta.

Ulrich, K.T. and S.D. Eppinger (1995). *Product Design and Development*. McGraw-Hill, NY, 77–127.

Uttal, B. (1987). "Speeding New Ideas to Market." *Fortune*, March 2, 62–66.

Utterback, J.M. and W.J. Abernathy (1975). "A dynamic model of process and product innovation." *OMEGA*, 3 (6), 639–656.

Utterback, J.M. and W.J. Abernathy (1981). "Multivariant Models for Innovation/Looking at the Abernathy and Utterback Model with Other Data." *OMEGA*, 9.

Walters, P.G. and B. Toyne (1989). "Product modification and standardization in international markets: Strategic options and facilitating policies." *Columbia Journal of World Business* Winter, 37–44.

Wheelwright, S.C. and K.B. Clark (1992). *Revolutionizing Product Development*. The Free Press, New York.

Production Scheduling and Control

Aigbedo, H. and Y. Monden (1997). "Scheduling for Just-in-Time Mixed Model Assembly Lines – A Parametric Procedure for Multi-Criterion Sequence." *International Journal of Production Research*, 35, 2543–2564.

American Production and Inventory Control Society (1986). *APICS Shop Floor Control Reprints*. American Production and Inventory Control Society, Falls Church, VA.

Anderson, J.C., Roger G. Schroeder, Sharon E. Tupy, and Edna M. White (1982). "Material Requirements Planning Systems: The State of the Art." *Production and Inventory Management*, 23 (4), 51–67.

Ang, J.S.K., C.C., Sum, and W.F., Chung (1995). "Critical success factors in implementing MRP and government assistance: A Singapore Context." *Information and Management*, 29, 63–70.

Ashby, James R. and Reha Uzsoy (1995). "Scheduling and order release in a single-stage production system." *Journal of Manufacturing Systems*, 290–306.

Atwater, B. and M.L. Gagne (1997). "The theory of constraints versus contribution margin analysis for product mix decisions." *Journal of Cost Management*, (January/February), 6–15.

Baker, Kenneth R. (1984). *Introduction to Sequencing and Scheduling*, New York, Wiley.

Baker, K.R. (1977). "An Experimental Study of the Effectiveness of Rolling Schedules in Production Planning." *Decision Sciences*, 8 (1), 19–27.

Bechte, Wolfgang (1988). "Theory and Practice of Load Oriented Manufacturing Control." *International Journal of Production Research*, 26, 3, 375–395.

Bechtold, S.E. (1981). "Work-Force Scheduling for Arbitrary Cyclic Demands." *Journal of Operations Management*, 1 (4) (May), 205–214.

Bedworth, D. and J. Bailey (1987). *Integrated Production Control Systems* (2nd ed.). John Wiley, New York.

Bergamaschi, D., R. Cigolini, M. Perona, and A. Portioli, (1997). "Order review and release strategies in a job shop environment: a review and a classification." *International Journal of Production Research*, 35 (2), 399–420.

Berry, W.L., T.E. Vollman, and D.C. Whybark (1979). *Master Production Scheduling*. American Production and Inventory Control Society, Falls Church, VA.

Bertrand, J.W.M. and V. Sridharan (1998). A Study of Simple Rules for Subcontracting in Make-to-Order Manufacturing. *European Journal of Operational Research*, Forthcoming.

Bevis, G.E. (1976). "A management viewpoint on the implementation of a MRP system." *Production and Inventory Management*, 16 (1), 105–116.

Billesbach, Thomas J. (1994) "Simplified Flow Control Using Kanban Signals." *Production and Inventory Management Journal*, 35 (2), 72–75.

Billesbach, Thomas J., A. Harrison, and S. Croom-Morgan (1991). "Just-in-Time: A United States-United Kingdom Comparison." *International Journal of Operations and Production Management*, 11 (10), 44–57.

Bitran, G.D., E.A. Haas, and A.C. Hax (1982). "Hierarchical Production Planning: A Two-Stage System." *Operations Research*, March–April, 232–51.

Bitran, G.R., D.M. Marini, H. Matsuo, and J.W. Noonan (1984). "Multiplant MRP." *Journal of Operations Management*, 5 (2) (February), 183–204.

Black, J.T. (1983). "Cellular Manufacturing Systems Reduce Setup Time, Make Small Lot Production Economical." *Industrial Engineering*, 15 (11) (November), 36–48.

Black, J.T., B.C. Jiang, G.J. Wiens (1991). *Design, Analysis and Control of Manufacturing Cells*. PED-53, ASME.

Blackburn, J.D., D.H. Kropp, and R.A. Millen (1986). "A comparison of strategies to dampen nervousness in MRP systems." *Management Science*, 32 (4), 413–429.

Blackstone, J.H., D.T. Phillips, and G.L. Hogg (1982). "A State-of-the-Art Survey of Dispatching Rules for Manufacturing Job Shop Operations." *International Journal of Production Research*, 20, 27–45.

Blumberg, D.F. (1980). "Factors Affecting the Design of a Successful MRP System." *Production and Inventory Management*, 5 (2) (Fourth Quarter), 50–62.

Bonvik, A.M., C.E. Couch, and S.B. Gershwin (1997). "A Comparison of Production Line Control Mechanisms." *International Journal of Production Research*, 35, 789–804.

Bowman, E.H. (1956). "Production Scheduling by the Transportation Method of Linear Programming." *Operations Research*, February, 100–103.

Brennan, L., S.M. Gupta, and K. Taleb (1994). "Operations Planning Issues in an Assembly/Disassembly Environment." *International Journal of Operations and Production Management*, 14 (9), 57–67.

Britan, Gabriel R. and Arnoldo C. Hax (1977). "On the Design of Hierarchical Planning Systems." *Decision Sciences*, 8, January.

Brown, A.D. (1994). "Implementing MRP: leadership, rites and cognitive change." *Logistics Information Management*, 7 (2), 6–11.

Brown, Jimmie, John Harhen, and James Shivnan (1988). *Production Management Systems: A CIM Perspective*. Addison Wesley.

Browne, J. (1988). "Production Activity Control – A Key Aspect of Production Control." *International Journal of Production Research*, 26 (3), 415–427.

Buffa, E.S. and Jeffrey G. Miller (1979). *Production-Inventory Systems: Planning and Control* (3rd ed.). Richard D. Irwin, Homewood, IL.

Byrett, Donald L., Mufit H. Ozden, and Jon M. Patton (1988). "Integrating Flexible Manufacturing Systems with Traditional Manufacturing, Planning, and Control." *Production and Inventory Management Journal* (Third Quarter), 15–21.

Celley, Albert F., William H. Clegg, Arthur W. Smith, and Mark A. Vonderembse (1986). "Implementation of JIT in the United States." *Journal of Purchasing and Materials Management*, Winter, 9.

Chakravarty, A.K. and A. Shtub (1984). "Selecting Parts and Loading Flexible Manufacturing Systems." *Proceedings of The 1st ORSA/TIMS FMS Conference*, Ann Arbor, MI.

Chang, T.C. and R.A. Wysk (1985). *An Introduction to Automated Process Planning Systems*. Prentice-Hall, Englewood Cliffs, N.J.

Chung, C.H. and L.J. Krajewski (1984). "Planning Horizons for Master Production Schedules." *Journal of Operations Management*, (4), 389–406.

Chung, C.H. and L.J. Krajewski (1986). "Replanning Frequencies for Master Production Schedules." *Decision Sciences*, 17 (2), 263–273.

Chung, C.H. and L.J. Krajewski (1986). "Integrating Resource Requirements Planning with Master Production Scheduling." Lee, S.M., L.D. Digman and M.J. Schniederjans, eds., *Proceedings of the 18th Decision Sciences Institute National Meeting*, Hawaii, 1165–1167.

Chung, C.H. and L.J. Krajewski (1987). "Interfacing Aggregate Plans and Master Production Schedules via a Rolling Horizon Procedure." *OMEGA*, 15 (5), 401–409.

Chung, C.H. and W.H. Luo (1996). "Computer Support of Decentralized Production Planning." *The Journal of Computer Information Systems*, 36 (2), 53–59.

Cohen, Oded (1988). "The Drum-Buffer-Rope (DBR) Approach to Logistics." In *Computer-Aided Production Management*, A. Rolstadas ed. Springer-Verlag, New York.

Colley, J.L. (1977). "Inventory Flow Models." Technical Note OM-120, Darden Graduate School of Business, University of Virginia, Charlottesville, Virginia.

Collier, D.A. (1980). "The Interaction of Single-Stage Lot Size Models in a Material Requirements Planning System." *Production and Inventory Management*, 21 (4), 11–20.

Conlon, J.R. (1977). "Is Your Master Schedule Feasible?" *APICS Master Production Scheduling Reprints*. American Production and Inventory Control Society, Falls Church, VA.

Cooper, R.B. and R.W. Zmud (1989). "Materials requirements planning system infusion." *OMEGA International Journal of Management Sciences*, 17 (5), 471–481.

Crawford, K.M., J.H. Blackstone, and J.F. Cox (1988). "A Study of JIT Implementation and Operating Problems." *International Journal of Production Research*, 26, 1561–1568.

Deleersnyder, J.L., T.J. Hodgson, H. Muller, and P.J. O'Grady (1989). "Controlled Pull Systems: An Analytical Approach." *Management Science*, 35, 1079–1091.

Dematheis, J.J. (1968). "An Economic Lot-Sizing Technique: The Part Period Algorithms." *IBM Systems Journal*, 7 (1), 50–51.

Duchessi, P., C.M. Schaninger, D.R. Hobbs, and L.P. Pentak (1988). "Determinants of success in implementing materials requirements planning (MRP)." *Journal of Manufacturing and Operations Management*, 1 (3), 263–304.

Eilon, S., *et al.* (1967). "Job shop scheduling with due date." *International Journal of Production Research* 6 (1), 1–13.

Elantora, E. (1992). "The Future Factory: Challenge for One-of-a-Kind Production." *International Journal of Production Economics*, 28, 131–142.

Ephremides, A., P. Varauoa, and J. Walrand (1980). A simple dynamic routing problem. *IIEE Transactions on Automatic Control*, 690–693.

Esparrago, R.A., Jr. (1988). "Kanban." *Production and Inventory Management*, 29 (1), 6–10.

Etienne, E.C. (1983). "MRP May Not Be Right for You, At Least not Yet." *Production and Inventory Management*, 24 (3) (Third Quarter), 33–46.

Everdell, Romeyn (1987). *Master Scheduling: APICS Training Aid*. American Production and Inventory Control Society, Falls Church, Virginia.

Fawcett, S.E. and J.N. Pearson (1991). "Understanding and applying constraint management in today's manufacturing environment." *Production and Inventory Management Journal* (Third Quarter), 46–55.

Finch, Byron J. and James F. Cox (1987). *Planning and Control System Design: Principles and Cases For Process Manufacturers*. American Production and Inventory Control Society, Falls Church, VA.

Fogarty, D.W. and Robert L. Barringer (1987). "Joint Order Releases Under Dependent Demand." *Production and Inventory Management*, First Quarter, 55–61.

Foley, M.J. (1988). "Post-MRP II: What Comes Next." *Datamation*, December 1, 24–36.

Fry, Timothy D., and Allen E. Smith (1987). "A procedure for implementing input/output control: a case study." *Production and Inventory Management Journal*, 28 (3), 50–52.

Gallagher, Colin (1978). "Batch Production and Functional Layout." *Industrial Archaeology*, 13, 157–161.

Gallagher, G.R. (1980). "How to Develop a Realistic Master Schedule." *Management Review* (April), 19–25.

Gershwin, S.B. (1986). "An Approach to Hierarchical Production Planning and Scheduling." in Jackson, R.H.F. and A.W.T. Jones, eds., *Real-Time Optimization in Automated Manufacturing Facilities*, U.S. Department of Commerce, National Bureau of Standards, 1–14.

Gilbert, J.P. (1990). "The state of JIT implementation and development in the USA." *International Journal of Production Research*, 28, 1099–1109.

Glassey, C. Roger, and Mauricio G.C. Resende (1988a). "Closed-loop job release for VLSI circuit manufacturing." *IEEE Transactions on Semiconductor Manufacturing*, 1 (1), 36–46.

Goldratt, E.M. (1990). *Theory of Constraints*. North River Press, New York.

Goldratt, E.M. and J. Cox (1986). *The Goal: A Process of Ongoing Improvement*. North River Press, Croton-on-Hudson, N.Y.

Goodman, Stephen H., Stanley T. Hardy, and Joseph R. Biggs (1977). "Lot Sizing Rules in a Hierarchical Multistage Inventory System." *Production and Inventory Management*, 18, 1, 104–116.

Goyal, S.K. and S.G. Deshmukh (1992). "A critique of the literature on just-in-time Manufacturing." *International Journal of Operations & Production Management*, 12 (1), 18–28.

Gravel, M. and W.L. Price (1988). "Using the Kanban in a Job-shop Environment." *International Journal of Production Research*, 26, 1105–1118.

Gray, Allison F. (1984). "Educating and Scheduling the Outside Shop: A Case Study." In *Readings in Production and Inventory Control Interfaces*, 48–50. American Production and Inventory Control Society, Falls Church VA.

Green, Donald R. (1987). "Direct Pegging, MRP/JIT Bridge." *P&IM Review*, April.

Green, G.I. and L.B. Appel (1981). "An Empirical Analysis of Job Shop Dispatch Rule Selection." *Journal of Operations Management*, 1, 4 (May), 197–204.

Grubbs, S.C. (1994). "MRPII: successful implementation the hard way." *Hospital Material Management Quarterly*, 15 (4), 40–47.

Gstettner, S. and H. Kuhn (1996). "Analysis of Production Control Systems Kanban and CONWIP." *International Journal of Production Research*, 34, 3253–3273.

Gupta, S.M. and C.R. McLean (1996). "Disassembly of Products." *Computers and Industrial Engineering*, 31, 225–228.

Gupta, S.M. and K. Taleb (1994). "Scheduling Disassembly." *International Journal of Production Research*, 32 (8), 1857–1866.

Gupta, S.M. and Y.A.Y. Al-Turki (1997). "An Algorithm to Dynamically Adjust the Number of Kanbans in Stochastic Processing Times and Variable Demand Environment." *Production Planning and Control*, 8, 133–141.

Gupta, S.M. and Y.A.Y. Al-Turki (1998). "The effect of sudden material handling system breakdown on the performance of a JIT system." *International Journal of Production Research*, 36 (7), 1935–1960.

Gupta, S.M., Y.A.Y. Al-Turki, and R.F. Perry (1995). "Coping with processing times variation in a JIT environment." *Proceedings of Northeast Decision Sciences Institute Conference*.

Hall, R.W. (1982). "Repetitive Manufacturing." *Production and Inventory Management*, Second Quarter, 78–86.

Hall, R.W. (1982). *Kawasaki U.S.A.: A Case Study*. APICS.

Hall, R.W. (1983). *Zero Inventories*. Dow Jones-Irwin. Homewood, Illinois.

Hallihan, A.P. Sackett. and G.M. Williams (1997). "JIT manufacturing: The evolution to an implementation model founded in current practice." *International Journal of Production Research*, 35 (4), 901–920.

Han, Min-Hong and Leon F. McGinnis (1988). "Throughput Rate Maximization in Flexible Manufacturing Cells." *IIE Transactions*, 20 (4) (December), 409–417.

Hannah, K.H. (1987). "Just-in-Time: Meeting the Competitive Challenge." *Production and Inventory Management*, 28 (3), 1–3.

Hay, E.J. (1988). *The Just-In-Time Breakthrough: Implementing the New Manufacturing Basics*. John Wiley & Sons, New York: NY.

Hendry, L.C. and B.G. Kingsman (1991). "Job release: part of a hierarchical system to manage manufacturing lead times in make-to-order companies." *OR: Journal of the Operational Research Society*, 42 (10), 871–883.

Herrmann, Jeffrey W., Ioannis Minis, Murali Narayanaswamy, and Philippe Wolff (1996). "Work order release in job shops." 733–736, *Proceedings of the 4th IEEE Mediterranean Symposium on New Directions in Control and Automation*, Maleme, Krete, Greece.

Hirano, Hiroyuki (Black, J.T., English Edition). (1988). *JIT Factory Revolution*. Productivity Press.

Ho, Y.C., M.A. Eyler, and T.T. Chien (1979). "A gradient technique for general buffer storage design in a serial production line." *International Journal of Production Research*, 17, 557–580.

Hoffman, Thomas R. and Gary D. Scudder (1983). "Priority Scheduling With cost Considerations." *International Journal of Production Research* 21, 6, 881–889.

Holmen, J.S. (1995). "ABC vs. TOC: It's a matter of time." *Management Accounting* (USA, January), 37–40.

Holt, C.C., F. Modigliani, and H. Simon (1955). "A Linear Decision Rule for Production and Employment Scheduling." *Management Science*, 2 (1) (October), 1–30.

Holt, C.C., F. Modigliani, J.F. Muth, and H.A. Simon (1960). *Planning Production, Inventories, and Workforce*. Prentice-Hall, New York.

Huang, C.C. and A. Kusiak (1996). "Overview of Kanban Systems." *International Journal of Computer Integrated Manufacturing*, 9, 169–189.

Hutchins, D. (1986). "Having a Hard Time with Just-in-Time." *Fortune*, June 9, 64–66.

Jacobs, F.R. (1983). "The OPT Scheduling System: A Review of a New Production Scheduling System." *Production and Inventory Management*, 24 (3), 47–51.

Jacobs, F.R. (1984). "OPT Uncovered: Many Production Planning and Scheduling Concepts Can Be Applied with or without the Software." *Industrial Engineering* (October), 32–41.

Jacobs, F.R. and D.J. Bragg (1988). "Repetitive lots: Flow-time reductions through sequencing and dynamic batch sizing." *Decision Sciences*, 19, 281–294.

Japan Management Association (1985). *Kanban: Just-In-Time at Toyota*. Productivity Press.

Jayson, Susan (1987). "Goldratt & Fox: Revolutionizing the Factory Floor." *Management Accounting*, May, 18–22.

Johnson, Linwood and Douglas C. Montgomery (1974). *Operatons Research in Production Planning, Scheduling and Inventory Control*. John Wiley & Sons, New York.

Jorgenson, D.W. and J.J. McCall (1963). "Optimal Scheduling of Replacement and Inspection." *Operations Research*, 11, 723–747.

Kadipasaoglu, S.N. and V. Sridharan (1995). "Alternative approaches for reducing schedule instability in multistage manufacturing under demand uncertainty." *Journal of Operations Management*, 13, 193–211.

Kanet, J.K. and J.C. Hayya (1982). "Priority Dispatching with Operations Due Dates in a Job Shop." *Journal of Operations Management*, 2 (3) (May), 167–175.

Kate, H.A. ten (1994). "Towards a Better Understanding of Order Acceptance." *International Journal of Production Economics*, 37 (1), 139–152.

Keaton M. (1995). "A New look at the Kanban Production Control System." *Production and Inventory Management Journal*, Third Quarter, 71–78.

Keller, A.Z. and A. Kazazi (1993). "Just-in-time manufacturing systems: A literature review." *Industrial Management and Data Systems*, 93 (7), 1–32.

Kim, S.L., J.C. Hayya, and J.D. Hong (1992). "Setup Reduction in the Economic Production Quantity Model." *Decision Sciences*, 23 (2), 500–508.

Krajewski, L.J. and L.P. Ritzman (1977). "Disaggregation in Manufacturing and Service Organizations: Survey of Problems and Research." *Decision Sciences*, 8 (1).

Krajewski, L.J., B. King, L.P. Ritzman, and D.S. Wong (1987). "Kanban, MRP, and Shaping the Manufacturing Environment." *Management Science*, 33 (1) (January), 39–57.

Kropp, D.H., R.C. Carlson, and J.V. Jucker (1979). "Use of dynamic lot-sizing to avoid nervousness in material requirements planning systems." *Journal of Production and Inventory Management*, 20 (3), 40–58.

Kropp, D.H., R.C. Carlson, and J.V. Jucker (1983). "Heuristic lot-sizing approaches for dealing with

MRP system nervousness." *Decision Sciences*, 14 (2), 156–169.

Krupp, J.A.G. (1997). "Safety Stock Management." *Production and Inventory Management*, 38 (3) (Third Quarter), 11–18.

Kupanhy, Lumbidi (1995). "Classification of JIT Techniques and Their Implications." *Industrial Engineering*, 27 (2), Feb., 62–66.

Lee, S.M. and L.J. Moore (1974). "A Practical Approach to Production Scheduling." *Production and Inventory Management*, 15 (1), 79–92.

Lee, S. and M. Ebrahimpour (1984). "Just-In-Time Production System: Some Requirements for Implementation." *International Journal of Operations and Production Management*, 4 (1), 3–11.

Leschke, J. (1996). "An Empirical Study of the Setup-Reduction Process." *Production and Operations Management*, 5 (2), 121–131.

Leschke, J. (1997). "The Setup-Reduction Process: Part 2: Setting Reduction Priorities." *Production and Inventory Management Journal*, 38 (1), 38–42.

Leschke, J. and E.N. Weiss (1997). "The Uncapacitated Setup-Reduction Investment-Allocation Problem with Continuous Investment-Cost Functions." *Management Science*, 43 (6).

Leschke, J. and E.N. Weiss (1998). "A Comparison of Rules for Allocating Setup-Reduction Investments in a Capacitated Environment." *Production and Operations Management*.

Lingayat, Sunil, John Mittenthal, and Robert M. O'Keefe (1995). "Order release in automated manufacturing systems." *Decision Sciences*, 26 (2), 175–205.

Lozinski, Christopher and C. Roger Glassey (1988). "Bottleneck starvation indicators for shop floor control." *IEEE Transactions on Semiconductor Manufacturing*, 1 (4), 147–153.

Mackey, J.T. and C.J. Davis (1993). "The Drum-Buffer-Rope application of the theory of constraints in manufacturing: What is it and what does it do to your accounting system?" *Proceedings of the 1993 American Accounting Association Annual Convention*, San Francisco, CA.

Maimon, O.Z. and Y.F. Choong (1987). Dynamic routing in reentrant flexible manufacturing systems." *Robotics and Computer-Integrated Manufacturing*, 3, 295–300.

Mather, H. (1977). "Reschedule the schedules you just scheduled – Way of life for MRP?" *Journal of Production and Inventory Management*, 18 (1), 60–79.

Mehra, Satish and M.J. Reid (1982). "MRP Implementation Using an Action Plan." *Interfaces*, 12 (1) (February), 69–73.

Melnyk, S.A. (1987). *Production Activity Control (Shop Floor Control)*. DowJones-Irwin, Homewood, IL.

Melnyk, S.A. and R.F. Gonzalez (1985). "MRP II: The Early Returns are In." *Production and Inventory Management*, 26 (1) (First Quarter), 124–137.

Melnyk, Steven A. and Gary L. Ragatz (1989). "Order Review/Release and Its Impact on the Shop Floor." *Production and Inventory Management*, 29 (2), 13–17.

Melnyk, Steven A. and Gary L. Ragatz (1989). "Order review/release: research issues and perspectives." *International Journal of Production Research*, 27 (7), 1081–1096.

Melnyk, Steven A. and Phillip L. Carter (1986). "Identifying the Principles of Effective Production Activity Control." *APICS Conference Proceedings*, 227–231.

Mosier, C.T., D.A. Evers, and D. Kelly (1984). "Analysis of Group Technology Scheduling Heuristics." *International Journal of Production Research*, 22 (5), 857–875.

New, C. Colin (1977). "MRP & GT, A New Strategy for Component Production." *Production and Inventory Management*, 18 (3), 50–62.

Nof, S.Y., M.M. Barash, and J.J. Solberg (1979). "Operational control of item flow in versatile manufacturing systems." *International Journal of Production Research*, 17, 479–489.

Norbom, Rolf (1973). "The Simple ABC's (A Loose Piece Float System)." *Production and Inventory Management*, 14 (1), 16.

Noreen, E., D. Smith, and J.T. Mackay (1995). *The Theory of Constraints and its Implications for Management Accounting*. North River Press, Great Barrington, MA.

Oliff, Michael and Earl Burch (1985). "Multiproduct Production Scheduling at Owens-Corning Fiberglass." *Interfaces*, 15 (5) (September–October), 25–34.

Orlicky, Joseph (1975). *Material Requirements Planning*. McGraw-Hill, New York.

Ormsby, J.G., S.Y. Ormsby, and C.R. Ruthstrom (1990). "MRP II implementation: a casestudy." *Production and Inventory Management Journal*, 31 (4), 77–81.

Ozekici, S. and S.R. Pliska (1991). "Optimal Scheduling of Inspections: A Delayed Markov Model with False Positive and Negatives." *Operations Research*, 39 (2), 261–273.

Peterson, R. and E.A. Silver (1984). *Decision Systems for Inventory Management and Production Planning* (2nd ed.). Wiley, New York.

Philipoom, P.R. and T.D. Fry (1992). "Capacity-based order review/release strategies to improve manufacturing performance." *International Journal of Production Research*, 30 (11), 2559–2572.

Philipoom, P.R., L.P. Rees, and B.W. Taylor (1996). "Simultaneously Determining the Number of Kanbans, Container Sizes and the Final Assembly Sequence of Products in a Just-in-Time Shop." *International Journal of Production Research*, 34, 51–69.

Plossl, G.W. (1973). *Manufacturing Control: The Last Frontier for Profits*. Reston, VA, Reston Publishing.

Plossl, G.W. (1983). *Production and Inventory Control: Applications*. George Plossl Education Services, Atlanta.

Plossl, G.W. (1985). *Production and Inventory Control: Principles and Techniques* (2nd ed.). Prentice-Hall, Englewood Cliffs, NJ.

Plossl, G.W. (1994). *Orlicky's Material Requirements Planning* (2nd ed.). McGraw-Hill.

Plossl, G.W. and O.W. Wight (1967). *Production and Inventory Control*. Prentice-Hall, Englewood Cliffs, NJ.

Plossl, G.W. and O.W. Wight (1975). *Capacity Planning and Control*, 50–86. American Production and Inventory Control Society, Falls Church, VA.

Potamianos, J. and A.J. Orman (1996). "An Interactive Dynamic Inventory-Production Control System." *Journal of the Operational Research Society*, 47 (8), 1016–1028.

Powell, C. (1978). "Systems Planning/Systems Control." *Production and Inventory Management*, 18 (3), 15–25.

Pratsini, Eleni, Jeffrey D. Camm, and Amitabh S. Raturi (1993). "Effect of Process Learning on Manufacturing Schedules." *Computers and Operation Research*, 20 (1), 15–24.

Proud, J.F. (1981). "Controlling the Master Schedule." *Production and Inventory Management*, 22 (2) (Second Quarter), 78–90.

Ptak, C.A.M. (1997). *MRP and Beyond*, Richard D. Irwin.

Putnam, Arnold O., R. Everdell, D.H. Dorman, R.R. Cronan, and L.H. Lindgren (1971). "Updating Critical Ratio and Slack-Time Priority Scheduling Rules." *Production and Inventory Mangement*, 12, 4.

Quinlan, J. (1982). "Just-in-Time at the Tractor Works." *Material Handling Engineering*, June, 62–65.

Ragatz, G.L. and V.A. Mabert (1984). "A Simulation Analysis of Due Date Assignment Rules." *Journal of Operations Management*, 5 (1) (November), 27–40.

Rahim, M.A. (1994). "Joint Determination of Production Quantity, Inspection Schedule, and Control Chart Design." *IIE Transactions*, 26 (6), 2–11.

Reinfeld, N.V. (1987). *Handbook of Production and Inventory Control*. Prentice-Hall, Englewood Cliffs, NJ.

Roderick, Larry M., Don T. Phillips, and Gary L. Hogg (1992). "A comparison of order release strategies in production control systems." *International Journal of Production Research*, 30 (3), 611–626.

Rubin, P.A. and G.L. Ragatz (1995). "Scheduling in a Sequence Dependent Setup Environment with Genetic Search." *Computers and Operations Research*, 22 (1), 85–99.

Sarin, S.C. and S.K. Das (1994). "Computer Integrated Production Planning and Control." In Dorf, R.C. and A. Kusiak, eds., *Handbook of Design, Manufacturing, and Automation* New York: John Wiley & Sons.

Schmeiser, B.W. (1983). "Batch size effects in the analysis of simulation output." *Operations Research*, 31, 565–568.

Schonberger, R.J. (1982). "Some Observations on the Advantages and Implementation Issues of Just-in-Time Production Systems." *Journal of Operations Management*, 2 (1) (November), 1–12.

Schonberger, R.J. (1982). *Japanese Manufacturing Techniques: Nine Hidden Lessons in Simplicity*. The Free Press, New York.

Schonberger, R.J. (1983). "Applications of Single-Card and Dual-Card Kanban." *Interfaces*, 13 (4) (August), 56–67.

Schonberger, R.J. (1986). *World Class Manufacturing: The Lessons of Simplicity Applied*. Free Press, New York.

Schonberger, R.J. (1987). *World Class Manufacturing Casebook: Implementing JIT and TQC*. Free Press, New York.

Schonberger, R.J. (1996). *World Class Manufacturing: The Next Decade*. The Free Press.

Sekine, Kenichi and Arai, Keisuke (1992). *Kaizen for Quick Changeover*. Productivity, Press.

Sheldon, D. (1991). "MRP II – What it really is." *Production and Inventory Mangement Journal*, 32 (3), 12–15.

Sheldon, D. (1994). "MRP II implementation: a case study." *Hospital Material Management Quarterly*, 15 (4), 48–52.

Shingo, S. (1985). SMED: *A Revolution in Manufacturing*. Productivity Press, Cambridge, MA.

Shingo, S. (1988). *Non-Stock Production: The Shingo System for Continuous Improvement*. Productivity Press.

Shingo, S. and A.G. Robinson (1990). *Modern Approaches to Manufacturing Improvements: The Shingo System*. Productivity Press, MA.

Shobrys, D.E. (1995). "Scheduling." *Planning, Scheduling, and Control Integration in the Process Industries*. C.E. Bodington, ed., McGraw-Hill, New York.

Silver, E.A. (1974). "A Control System for Coordinated Inventory Replenishment." *International Journal of Production Research*, 12 (6), 647–671.

Silver, E.A. (1975). "Modifying the Economic Order Quantity (EOQ) to Handle Coordinated Replenishment of Two or More Items." *Production and Inventory Management*, 16 (3), 26–38.

Silver, E.A. and Peter Kelle (1988). "More on Joint Order Releases Under Dependent Demand." *Production and Inventory Management*, First Quarter.

Silver, E.A. and R. Peterson (1985). *Decision Systems for Inventory Management and Production Planning*. (2nd ed.). John-Wiley, New York. Chapters 3, 4, 7–12, and 17.

Sirianni, N.C. (1979). "Inventory: How Much Do You Really Need?" *Production Magazine*, November.

Sirianni, N.C. (1981). "Right Inventory Level: Top Down or Bottom UP." *Inventories and Production Magazine*, May–June.

Smunt, Timothy L. (1986). "A Comparison of Learning Curve Analysis and Moving Average Ratio Analysis for Detailed Operational Planning." *Decision Sciences*, 17 (4), 475–495.

Sohal, A.S., L. Ramsay, and D. Samson (1993). "JIT manufacturing: Industry analysis and a methodology for implementation." *International Journal of Operations and Production Management*, 13 (7), 22–56.

Spearman, M.L., D.L. Woodruff, and W.J. Hopp (1990). "CONWIP: a pull alternative to kanban."

International Journal of Production Research, 28 (5), 879–894.

Spencer, M.S. (1980). "Scheduling Components for Group Technology Lines." *Production and Inventory Management* (Fourth Quarter), 43–49.

Spencer, M.S. and J.F. Cox (1995). "An analysis of the product-process matrix and repetitive manufacturing." *International Journal of Production Research*, 33 (5), 1275–1294.

Sridharan, V. and R.L. LaForge (1989). "The impact of safety stock on schedule instability, cost, and service." *Journal of Operations Management*, 8 (4), 327–347.

Sridharan, V. and R.L. LaForge (1994). "A model to estimate service levels when a portion of the master production schedule is frozen." *Computers & Operations Research*, 21 (5), 477–486.

Sridharan, V. and R.L. LaForge (1994). "Freezing the master production schedule: Implications for fill rate." *Decision Sciences*, 25 (3), 461–469.

Sridharan, V. and W.L. Berry (1990). "Freezing the master production schedule under demand uncertainty." *Decision Sciences*, 21 (1), 97–120.

Sridharan, V., W.L. Berry, and V. Udayabanu (1988). "Measuring master production schedule instability under rolling planning horizons." *Decision Sciences*, 19 (1), 147–166.

Srikanth, M.L. and M.M. Umble (1997). *Synchronous Management: Profit-Based Manufacturing For The 21st Century-I*, Spectrum Publishing, Guilford, CT.

Starr, M.K. (1965). "Modular production – a new concept." *Harvard Business Review*, November–December, 137–145.

Steele, D.C. (1975). "The nervous MRP system: How to do battle." *Journal of Production and Inventory Management*, 16 (4), 83–89.

Sum, C.C., J.S.K. Ang, and L.N. Yeo (1997). "Contextual elements of critical success factors in MRP implementation." *Production and Inventory Mangement Journal*, 38 (3), 77–83.

Tadjer, R. (1988). "Enterprise resource planning." *Internet Week*, 710, 40–41.

Takahashi, K. and N. Nakamura (1996). "Reactive JIT ordering systems for unstable demand changes." *Proceedings of Advances in Production Management Systems*, Information Processing Society of Japan.

Takahashi, K. and N. Nakamura (1997). "A study on reactive kanban systems." *Journal of Japan Industrial Management Association* (in Japanese). 48, 159–165.

Taleb, K., S.M. Gupta, and L. Brennan (1997). "Disassembly of Complex Products with Parts and Materials Commonality." *Production Planning and Control*, 8 (3), 255–269.

Taylor, S.G. and S.F. Bolander (1994). *Process Flow Scheduling: A Scheduling Systems Framework for Flow Manufacturing*. APICS, Falls Church, VA.

Taylor, S.G. and S.F. Bolander (1997). "Process Flow Scheduling: Past, Present, and Future." *Production and Inventory Mangement Journal*, 38 (2), 21–25.

Tayur, S.R. (1992). "Properties of serial kanban systems." *Queueing Systems*, 12, 297–318.

Tayur, S.R. (1993). "Structural Properties and a Heuristic for Kanban-Controlled SerialLines." *Management Science*, 39, 1347–1368.

Tersine, R.J. (1987). *Principles of Inventory and Materials Management* (3rd ed.). North-Holland, New York.

Tharumarajah, A. and A.J. Wells (1997). "A behaviour-based approach to scheduling in distributed manufacturing systems." *Integrated Computer Aided Engineering*, 4, 235–249.

Thompson, K. (1983). "MRP II in the Repetitive Manufacturing Environment." *Production and Inventory Management*, 24 (4) (Fourth Quarter), 1–14.

Toomey, J.W. (1996). *MRP II: Planning for Manufacturing Excellence*. Chapman and Hall, New York.

Umble, M. and M.L. Srikanth (1990). *Synchronous Manufacturing: Principles for World Class Excellence*. South–Western Publishing, Cincinnati, OH.

Umble, M. and M.L. Srikanth (1995). *Synchronous Manufacturing*, Spectrum Publishing, CT.

Umble, M. and M.L. Srikanth (1997). *Synchronous Management: Profit-Based Manufacturing For The 21st Century* – Volume II, Implementation Issues and Case Studies, Spectrum Publishing, Guilford, Connecticut.

Van Dierdonck, R. and J.G. Miller (1980). "Designing Production Planning and Control Systems." *Journal of Operations Management*, 1 (1) (August), 37–46.

Veerakamolmal, P. and S.M. Gupta (1998). "Disassembly Process Planning." *Engineering Design and Automation* (forthcoming).

Vollman, T.E. (1986). "OPT as an Enhancement to MRP II." *Production and Inventory Management*, 27 (2), 38–47.

Vollmann T.E., W.L. Berry, and D.C. Whybark (1997). *Manufacturing Planning and Control Systems* (4th ed.). Irwin/McGraw-Hill.

Wallace, T.F. (1990). *MRPII: Making It Happen*. Oliver Wight Limited Publications, Inc.

Wantuck, K. (1978). "Calculating Optimum Inventories." *Proceedings 21st Annual Conference*, American Production and Inventory Control Society.

Wassweiler, William R. (1980). "Fundamentals of Shop Floor Control." *APICS Conference Proceedings*, 352–354.

Wein, Lawrence M. (1988). "Scheduling semiconductor wafer fabrication." *IEEE Transactions on Semiconductor Manufacturing*, 1 (3), 115–130.

Wein, Lawrence M. and Philippe B. Chevalier (1992). "A broader view of the job shop scheduling problem." *Management Science*, 38 (7), 1018–1033.

Weiss, A. (1984). "Simple Truths of Japanese Manufacturing." *Harvard Business Review*, (July-August), 119–125.

Weston, Frederick C., Jr. (1980). "A Simulation Approach to Examining Traditional EOQ/EOP and Single Order Exponential Smoothing Efficiency Adopting a Small Business Perspective." *Production and Inventory Management*, 21 (2), 67–83.

Westra, D., M.L. Srikanth, and M. Kane (1996). "Measuring operational performance in a throughput world." *Management Accounting* (April), 41–47.

Wight, O.W. (1970). "Input–output control, a real handle on lead time." *Production and Inventory Management*, 11, 9–31.

Wight, O.W. (1982). *The Executive's Guide to Successful MRP II*. Oliver Wight Publishing, Essex VT.

Wight, O.W. (1984). *MRP II: Unlocking America's Productivity Potential*. CBI, Boston, MA.

Works, M. (1998). "GenCorp Aerojet says time is right for SAP in the aerospace and defense industries." *Automatic I.D. News*, 14 (2), 50–52.

Wortmann, J.C. (1992). "Production Management Systems for One-of-a-Kind Products." *Computers in Industry*, 19, 79–88.

Zhao, X. and T.S. Lee (1993). "Freezing the master production schedule for material requirements planning systems under demand uncertainty." *Journal of Operations Management*, 11 (2), 185–205.

PROJECT MANAGEMENT

Bertain, Leonard (1993). *The New Turnaround: A Breakthrough for People, Profits, and Change*. North River Press, Croton-on-Hudson, New York.

Defoe, Daniel (1697). *An Essay Upon Projects*. Reprinted 1887. Cassell & Co., London, England.

Fazar, Willard (1962). The Origins of PERT. *The Controller*, 8, 598–602, 618–621.

Gabel, D. (1985). "Project Management Software." *PC Week*. September 10, 59–67.

Gido, J. (1985). *An Introduction to Project Planning* (2nd ed.) Industrial Press, New York.

Gilbreath, R.D. (1986). *Winning at Project Management*. John Wiley, New York.

Gobeli, D.H. and E.W. Larson (1987). "Relative Effectiveness of Different Project Structures." *Project Management Journal*, 18, 2 (June), 81–85.

Hughes, M.W. (1986). "Why Projects Fail: The Effects of Ignoring the Obvious." *Industrial Engineering*, 18, 4 (April), 14–18.

Kerzner, H. (1984). *Project Management: A Systems Approach to Planning, Scheduling, and Controlling*. New York, Van Nostrand.

Kezbom, D.S., D.L. Schilling, and V.A. Edward (1989). *Dynamic Project Management*. John Wiley, New York.

Klingel, A.R., Jr. (1966). "Bias in PERT Project Completion Time Calculations for a Real Network." *Management Science*, 13 (4), 194–201.

Knutson, J. (1989). *How to be a Successful Project Manager*. American Management Association, Saranac Lake, NY.

Levin, Richard I. and Charles A. Kirkpatrick (1966). *Planning and Control with PERT/CPM*. McGraw-Hill, New York.

Levine, H.A. (1986). *Project Management Using Microcomputers*. Osborne McGraw-Hill, Berkeley, CA.

Levy, Ferdinand K., Gerald L. Thompson, and Jerome D. Wiest (1963). "The ABC's of the Critical Path Method." *Harvard Busines Review*, September–October, 98–108.

Lewis, James P. (1993). *The project manager's desk reference*. New York: McGraw-Hill.

Lewis, James P. (1995). *Project planning, scheduling, and control*, Revised Edition. New York: McGraw-Hill.

Lewis, James P. (1997). *Team-based project management*. New York: AMACOM.

Lewis, James P. (1998). *Mastering project management*. New York: McGraw-Hill.

Lock, D. (1987). *Project Management* (4th ed.). Gower, Brookfield, VT.

MacCrimmon, Kenneth R. and Charles A. Ryavec (1964). "An Analytical Study of the PERT Assumptions." *Operations Research*, January–February, 16–37.

Mantel, Samuel J. and Jack R. Meridith (1986). "IEs Are Best Suited to Challenging Role of Project Manager." *Industrial Engineering*, 18 (4), 54–60.

Meredith, J.R. and S.J. Mantel, Jr. (1989). *Project Management*: A Managerial Approach (2nd ed.). John Wiley, New York.

Moder, J.J., C.R. Phillips, and E.W. Davis (1983). *Project Management with CPM, PERT, and Precedence Diagramming* (3rd ed.). Van Nostrand Reinhold, New York.

Nobeoka, K. and M.A. Cusumano (1995). "Multiproject Strategy, Design Transfer, and Project Performance." *IEEE Transactions on Engineering Management*, 42, 397–409.

O'Neal, K. (1987). "Project Management Computer Buyer's Guide." *Industrial Engineering* 19 (1) (January), 53–63.

Pinto, J.K. (1987). "Strategy and Tactics in a Process Model of Project Implementation." *Interfaces*, 17 (3) (May–June), 34–46.

Posner, B.Z. (1987). "What It Takes to Be a Good Project Manager." *Project Management Journal*, 18 (1) (March), 51–54.

Shaffer, L.R., J.B. Ritter, and W.L. Meyer (1965). *The Critical Path Method*, McGraw-Hill, New York.

Smith, L.A. and Joan Mills (1982). "Project Management Network Programs." *Project Management Quarterly*, 18–29.

Smith, L.A. and Sushil Gupta (1985). "Project Management Software in P&IM." *P&IM Review*, 5 (6), 66–68.

Swamidass, P.M. and C. Majerus (1991). "Statistical Control of Cycle Time and Project Time: Lessons from statistical process control." *International Journal of Production Research*, 29 (3), 551–564.

Van Slyke, Richard M. (1963). "Monte Carlo Methods and the PERT Problem." *Operations Research*, 11 (5) (September–October), 129–143.

Weist, Jerome D. and K. Levy Ferdinand (1977). *A Management Guide to PERT/CPM* (2nd ed.). Prentice-Hall, Englewood Cliffs, NJ.

Quality

American Society for Quality Control (1987). ANSI/ ASQC A1-1987. *Definition, Symbols, Formulas, and Tables for Control charts*. ASQC Press, Milwaukee, Wisconsin.

Amsden, R.T. (1998). "What's in store for SPC?" *Quality Digest*, February, 37–40.

AT&T (1984). *Statistical Quality Control Handbook*, 10th printing.

AT&T Corporate Quality Office (1994). *Using ISO 9000 to Improve Business Processes*.

Belohlav, J.A. (1995). "Quality, Strategy, and Competitiveness." *The Death and Life of the American Quality Movement*. Edited by R.E. Cole. Oxford University Press, New York, 43–58.

Bemowski, K. (1996). "Baldrige Award Celebrates Its 10th Birthday With a New Look." *Quality Progress*, 29, 49–54.

Berry, L.L., V.A. Zeithmal, and A. Parasuraman (1985). "Quality Counts in Services, Too." *Business Horizons*, 28 (3) (May–June), 44–52.

Besterfield, D.H. (1990). *Quality Control* (3rd ed.). Prentice-Hall, Englewood Cliffs, NJ.

Bowles, J.G. (1986). "The Quality Imperative." *Fortune* (September 29), 61–96.

Chowdhury, A.R. (1985). "Reliability as It Relates to QA/QC." *Quality Progress*, 18 (12) (December), 27–30.

Crosby, L.B. (1984). "The Just-in-Time Manufacturing Process: Control of Quality and Quantity." *Production and Inventory Management*, 25, 4, 21–33.

Crosby, P.B. (1979). *Quality is Free: The Art of Making Quality Certain*. McGraw-Hill, New York.

Crosby, P.B. (1984). *Quality Without Tears*. McGraw-Hill, New York.

De Vera, Dennis, Tom Glennon, Andrew A. Kenney, Mohammad A.H. Khan, and Mike Mayer (1988). "An Automotive Case Study." *Quality Progress*, June, 35–38.

DeCarlo, N.J. and W.K. Sterret (1990). "History of the Malcolm Baldrige National Quality Award." *Quality Progress*, 23, 21–27.

DelMar, D. and G. Sheldon (1988). *Introduction to Quality Control*. West, St. Paul, MN.

Deming, W.E. (1982). *Quality, Productivity, and Competitive Position*, Center for Advanced Engineering Study, Massachuseths Institute of Technology, Cambridge, Massachusetts.

Dobyns, L. and C. Crawford-Mason (1991). *Quality or Else: The Revolution in World Business*. Houghton Mifflin, New York.

Donlon, J.P. (1993). "The Six Sigma Encore; Quality Standards at Motorola." *Chief Executive*, November, 90.

Duncan, A.J. (1986). *Quality Control and Industrial Statistics* (5th ed.). Irwin, Homewood IL.

Ealey, L.A. (1988). *Quality by Design: Tagauchi Methods and U.S. Industry*. ASI Press, Dearborn, MI.

Eckstein, A.L.H. and J. Balakrishnan (1993). "The ISO 9000 Series: Quality Management Systems for the Global Economy." *Production and Inventory Management Journal*, 34, 66–71.

Enrick, N.L. (1985). *Quality, Reliability, and Process Improvement* (8th ed.). Industrial Press, New York.

Evans, J.R. (1996). "Something old, something new: a process view of the Baldrige criteria." *International Journal of Quality Science*, 1, 62–68.

Evans, J.R. (1997). "Beyond QS-9000." *Production and Inventory Management Journal*, 38, 3 (Third Quarter), 72–76.

Evans, J.R. and M.W. Ford (1997). "Value-Driven Quality." *Quality Management Journal*, 4 (4), 19–31.

Evans, J.R. and W.M. Lindsay (1998). *The Management and Control of Quality*, (4th ed.). South-Western, Cincinnati, OH.

Fiegenbaum, A.V. (1961). *Total Quality Control: Engineering and Management*. McGraw-Hill, New York.

Feigenbaum, A.V. (1986). *Total Quality Control* (3rd ed.). McGraw-Hill, New York.

Fine, C.H. (1988). "A Quality Control Model with Learning Effects." *Operations Research*, 36, 3, 437–444.

Fortuna, Ronald M. (1988). "Beyond Quality: Taking SPC Upstream." *Quality Progress*, June, 23–28.

Gale, B.T. (1985). *Quality as a Strategic Weapon*. Strategic Planning Institute, Cambridge, MA.

Garvin, D.A. (1983). "Quality on the Line." *Harvard Business Review* (September–October), 64–75.

Garvin, D.A. (1984). "What Does 'Product Quality' Really Mean?" *Sloan Management Review*. Fall, 25–43.

Garvin, D.A. (1987). "Competing on the Eight Dimensions of Quality." *Harvard Business Review*. November–December, 101–109.

Garvin, D.A. (1988). *Managing Quality: The Strategic and Competitive Edge*. Free Press, New York.

Gitlow, H.S. and S.J. Gitlow (1987). *The Deming Guide to Quality and Competitive Position*. Prentice-Hall, Englewood Cliffs, NJ.

Gitlow, H.S. and P.T. Hertz (1983). "Product Defects and Productivity." *Harvard Business Review* (September–October), 131–141.

Godfrey, A.B. (1993). "Ten Areas for Future Research in Total Quality Management." *Quality Management Journal*, October, 47–70.

Grant, E.L. and R.S. Leavenworth (1996). *Statistical Quality Control* (7th ed.). McGraw-Hill, New York.

Hansen, B.L. and P.M. Ghare (1987). *Quality Control and Application*. Prentice-Hall, Englewood CLiffs, NJ.

Harrigan, Kenneth W. (1984). "Making Quality Job One – A Cultural Revolution." *Business Quarterly*, Winter, 68–71.

Harrington, D.H.J. (1991). *Business Process Improvement: The Break through Strategy for Total Quality, Productivity, and Competitiveness*, McGraw-Hill, New York.

Hauser, J.R. and Don Clausing (1988). "The House of Quality." *Harvard Business Review*, 66 (3) (May–June), 63–73.

Ingle, S. (1985). *In Search of Perfection: How to Create/Maintain/Improve Quality*. Prentice-Hall, Englewood Cliffs, NJ.

Ishikawa, I. (1989). *Introduction to Quality Control* (3rd ed.). 3A Corporation, Tokyo, Japan.

Ishikawa, K. (1972). *Guide to Quality Control*. Asian Productivity Organization.

Ishikawa, K. (1985). *What is Total Quality Control? The Japanese Way*. Translated by David J. Lu. Prentice-Hall, NJ.

Jorgenson, D.W. and J.J. McCall (1963). "Optimal Scheduling of Replacement and Inspection." *Operations Research*, 11, 723–747.

Juran, J.M. (1993). "Made in U.S.A.: A Renaissance in Quality." *Harvard Business Review*, July–August, 42–50.

Juran, J.M. (1995). "Summary, Trends, and Prognosis." *A History of Managing for Quality*. Edited by J.M. Juran, ASQ Quality Press, Wisconsin, 603–655.

Juran, J.M. and F.M. Gryna (1980). *Quality Planning and Analysis* (2nd ed.). McGraw-Hill, New York.

Juran, J.M., F.M. Gryna, Jr., and R.S. Bingham, Jr. eds. (1988). *Quality Control Handbook* (4th ed.). McGraw-Hill, New York.

Kekre, S. and T. Mukhopadhyay (1992). "Impact of Electronic Data Interchange Technology on Quality Improvement and Inventory Reduction Programs: A Field Study." *International Journal of Production Economics*, 28, 265–282.

Kenny, Andrew A. (1988). "A New Paradigm for Quality Assurance." *Quality Progress*, June, 30–32.

Lawler, E.E., III and S. Mohrman (1985). "Quality Circles After the Fad." *Harvard Business Review*, (January–February), 65–71.

Leonard, F.S. and W.E. Sasser (1982). "The Incline of Quality." *Harvard Business Review* (September–October), 163–171.

Main, J. (1990). "How to Win the Baldridge Award." *Fortune*, April 25, 101–116.

Marcellus, R.L. and M. Dada (1991). "Interactive Process Quality Improvement." *Management Science*, 37 (11), 1365–1376.

Masaracchia, Philip (1987). "TQC – The 'Quality' Component of J. I. T." *P&IM Review*, April, 44.

Messina, W.S. (1987). *Statistical Quality Control for Manufacturing Managers*. John Wiley, New York.

Mitra, A. (1998). *Fundamentals of Quality Control and Improvement* (2nd ed.). Prentice-Hall, Upper Saddle River, New Jersey.

Montgomery, D.C. (1996). *Introduction to Statistical Quality Control* (3rd ed.). John Wiley & Sons, Inc., New York.

Nelson, L.S. (1984). "The Shewhart control chart – tests for special causes." *Journal of Quality Technology*, 16, 237–239.

Neves, J.S. and B. Nakhai (1994). "The Evolution of the Baldrige Award." *Quality Progress*, 27, 65–70.

NKS/Factory Magazine (Introduction by Shigeo Shingo) (1988). *Poka-Yoke: Improving Quality by Preventing Defects*. Productivity Press.

O'Donnell, Merle and Robert J. O'Donnell (1984). "Quality Circles – The Latest Fad or a Real Winner?" *Business Horizons* (May–June), 113–121.

Oakland, J.S. (1993). *Total Quality Management: The Route to Improving Performance*. Butterworth-Heinemann, Oxford.

Quality System Requirements QS-9000, International Organization for Standardization, Chrysler Corporation, ford Motor Company, and General Motors Corporation, August 1994.

Rahim, M.A. (1994). "Joint Determination of Production Quantity, Inspection Schedule, and Control Chart Design." *IIE Transactions*, 26 (6), 2–11.

Reddy, J. and A. Berger (1983). "Three Essentials of Product Quality." *Harvard Business Review*, (July–August), 153–159.

Reimann, C.W. and H. Hertz (1993). "The Malcolm Baldrige National Quality Award and ISO 9000 Registration: Understanding Their Many Important Differences." *ASTM Standardization News*, November, 42–51.

Robinson, A.G. and D.M. Schroeder (1990). "The limited role of statistical quality control in a zero-defect environment." *Production and Inventory Management Journal*, 3rd. Quarter, 60–65.

Rosander, A.C. (1985). *Applications of Quality Control in the Service Industries*. New York, Marcel Dekker.

Sako, M. (1992). *Prices, Quality and Trust: Inter-Firm Relations in Britain and Japan*. Cambridge University Press, Cambridge.

Shainin, D. and P.D. Shainin (1988). "Statistical Process Control." *Juran's Quality Control Handbook*. Edited by J.M. Juran and F.M. Gryna (4th ed.). McGraw-Hill, New York. Section 24, 24.1–24.40.

Shewhart, W.A. (1931). *Economic Control of Quality of Manufactured Product*. E. Van Nostrand Company, New York.

Shewhart, W.A. (1939). *Statistical Method from the Viewpoint of Quality Control*. Graduate School, Department of Agriculture, Washington.

Shiba, S., A. Graham, and D. Walden (1993). *A New American TQM: Four Practical Revolutions in Management*. Productivity Press, Oregon, 3 (3) 0, 411–460.

Shingo, S. (1986). *Zero Quality Control: Source Inspection and the Poka-Yoke System*, Productivity Press, Stamford, Connecticut.

Sinha, M.N. and W.O. Wilborn (1985). *The Management Quality Assurance*. John Wiley, New York.

Squires, F.H. (1982). "What Do Quality Control Charts Control?" *Quality* (November), 63.

Stamatix, D.H. (1995). *Failure Mode and Effect Analysis, ASQC Quality Press*. Milwaukee, WI.

Sullivan, L.P. (1986). "The Seven Stages in Company-Wide Quality Control." *Quality Progress*. May, 77–83.

Sullivan, L.P. (1986). "Quality Function Deployment." *Quality Progress*, 19 (6) (June), 36–50.

Swamidass, P.M. and C. Majerus (1991). "Statistical Control of Cycle Time and Project Time: Lessons from stratistical process control." *International Journal of Production Research*, 29 (3), 551–564.

Taguchi, Genichi (1986). *Introduction to Quality Engineering: Designing Quality into Products and*

Processes. Asian Productivity Organization, White Plains, NY.

Timbers, M.J. (1992). "ISO 9000 and Europe's Attempts to Mandate Quality." *The Journal of European Business*, 14–25.

Townsend, P.L. (1986). *Commit to Quality*. John Wiley, New York.

U.S. Department of Commerce (1998). Malcolm Baldrige National Quality Award 1998 Criteria for Performance Excellence, Washington, D.C., GPO.

U.S. Department of Commerce (undated). Malcolm Baldrige National Quality Award, Profiles of Winners.

Webster, S.E. (1997). "ISO 9000 Certification, A Success Story at Nu Visions Manufacturing." *IIE Solutions*, 18–21.

Zeithaml, V.A., L.L. Berry, and A. Parasuraman (1996). "The Behavioral Consequences of Service Quality." *Journal of Marketing*, 60, 31–46.

Zuckerman, A. (1997). "ISO/QS-9000 Registration Issues Heating Up Worldwide." *The Quality Observer*, June, 21–23.

SERVICE

Berry, L.L., V.A. Zeithmal, and A. Parasuraman (1985). "Quality Counts in Services, Too." *Business Horizons*, 28 (3) (May–June), 44–52.

Chase, R.B. (1978). "Where Does the Customer Fit in a Service Operation?" *Harvard Business Review*, November–December, 137–142.

Chase, R.B. and D.A. Garvin (1989). "The Service Factory." *Harvard Business Review*, 67 (4) (July–August), 61–69.

Chase, R.B. and D.A. Tansik (1983). "The Customer Contact Model for Organization and Design." *Management Science*, 29, 9 (September), 1037–1050.

Chase, R.B. and W.K. Erikson (1988). "The Service Factory." *Academy of Management of Executive*, (August), 191–196.

Chase, R.B.K., R. Kumar, and W.E. Youngdahl (1992), "Service-Based Manufacturing: The Service Factory." *Production and Operations Management*, 1(2), 175–184.

Collier, D.A. (1985). *Service Management: The Automation of Services*. Reston Publishing, Reston, VA.

Collier, D.A. (1987). *Service Management: Operating Decisions*. Prentice-Hall, Englewood Cliffs, NJ.

Davidow, W.H. and B. Uttal (1989). "Service Companies: Focus or Falter." *Harvard Business Review*, 67(4), 77–85.

Diprimio, A. (1987). *Quality Assurance in Service Organizations*. Radnor, PA, Chilton.

Fogarty, D.W. and T.R. Hoffman (1980). "Customer Service." *Production and Inventory Management*, Fall.

Grönroos, C. (1990). "Relationship Approach to Marketing in Service Contexts: The Marketing and Organizational Behaviour Interface." *Journal of Business Research*, 20, 3–11.

Heskett, J.L. (1986). *Managing in the Service Economy*. Harvard Business School Press, Boston.

Hostack, G.L. (1984). "Designing Services That Deliver." *Harvard Business Review*, January–February, 133–139.

Jones, T.O. and W.E. Sasser, Jr. (1995). "Why Satisfied Customers Defect." *Harvard Business Review* (November–December), 88–99.

Kantrow, A.M. (1980). "Keping Informed: The Strategy Technology Connection." *Harvard Business Review*, 58, July–August, 6–21.

King, C.A. (1985). "Service Quality Assurance Is Different." *Quality Progress*, 18 (6), 14–18.

LaLonde, B.J., M.C. Cooper, and T.G. Noordewier (1988). *Customer Service: A Management Perspective*. Council of Logistics Management, Oak Brook, IL.

Lovelock, C. (1985). *Managing Services*. Prentice-Hall, New York.

Normann, R. (1984). *Service Management: Strategy and Leadership in Service Businesses*. John Wiley, New York.

Parasuraman, A., V.A. Zeithaml, and L.L. Berry (1985). "A Conceptual Model of Service Quality and Its Implications for Future Research." *Journal of Marketing*, 49 (4), 41–50.

Pine II, J.B., D. Peppers, and M. Rogers (1995). "Do You Want to Keep Your Customers Forever?" *Harvard Business Review*, 73, 103–114.

Schmenner, R.W. (1986). "How Can Service Business Survive and Prosper?" *Sloan Management Review*, Spring, 21–32.

Shostack, G.L. (1984). "Designing Service That Deliver." *Harvard Business Review*, January–February, 133–139.

Shycon, H.N. and C.R. Spraque (1975). "Put a Price Tag on Your Customer Servicing Levels." *Harvard Business Review*, 53 (July–August), 71–78.

Stewart, T. (1995). "After all you've done for you customers, why are they still NOT HAPPY?" *Fortune*, (December 11), 178–18.

Stewart, T.A. (1997). "A Satisfied Customer Isn't Enough." *Fortune*, 136 (July 21), 112–113.

Thomas, D.R.E. (1978). "Strategy Is Different in Service Businesses." *Harvard Business Review*. July–August, 1.

SOFTWARE AND COMPUTER INTEGRATION

Albright, T.L. and T. Smith (1996). "Software for Activity-Based Costing." *Journal of Cost Management*, Summer, 47–53.

American Production and Inventory Control Society (1996). "1996 finite capacity scheduling software and vendor directory." *APICS – the Performance Advantage*, 6, 72–82.

Banks, J., B. Burnette, H. Kozloski, and J. Rose (1995). *Introduction to SIMAN V and CINEMA V*. John Wiley & Sons, New York.

Berry, W.L., T.G. Schmitt, and T.E. Vollman (1982). "Capacity Planning Techniques for Manufacturing

Control Systems: Information Requirements and Operational Features." *Journal of Operations Management*, 3 (1), 13–26.

Best, M.J. and K. Ritter (1985). *Linear Programming: Action Set Analysis and Computer Programs*, Prentice-Hall, Englewood Cliffs, NJ

Bowersox, D.J. and P.J. Daugherty (1995). "Logistics Paradigms: The Impact of Information Technology." *Journal of Business Logistics*, 16 (1), 65–80.

Bratley, P., B. Fox, and L. Schrage (1987). *A Guide to Simulation* (2nd ed.). Springer-Verlag. New York.

Bulgren, W. (1982). *Discrete System Simulation*. Prentice-Hall, Englewood Cliffs, NJ.

Burnham, J.M. (1985). "Improving Manufacturing Performance: A Need for Integration." *Production and Inventory Management*, 26 (4), 51–59.

Camm, J.D., T.E. Chorman, F.A. Dill, J.R. Evans, D.J. Sweeney, and G.W. Wegryn (1997). "Blending OR/MS, Judgment, and GIS: Restructuring P&G's Supply Chain." *Interfaces*, 27(1), 128–142.

Chung, C.H. and W.H. Luo (1996). "Computer Support of Decentralized Production Planning." *The Journal of Computer Information Systems*, 36 (2), 53–59.

Craighead, Thomas G. (1989). "EDI Impact Grows for Cost Efficient Manufacturing." *PIM Review*, July, 32.

Davenport, T.H. (1993). *Process Innovation: Reengineering Work Through Information Technology*, Harvard Business School Press, Boston, Massachusetts, p. 5–9.

Davenport, T.H. and J.E. Short (1990). "The New Industrial Engineering: Information Technology and Business Process Redesign." *Sloan Management Review*, Summer, 11–27.

Dilworth, J.D. (1987). *Information Systems for JIT Manufacturing*. Association for Manufacturing Excellence, Wheeling, IL.

Emmelhorne, Margaret A. (1971). *Guide to Purchasing: Electronicdate Interchange. National Association of Purchasing Management*, John Wiley & Sons, New York.

Erikson, W.J. and O.P. Hall, Jr. (1983). *Computer Models for Management Science*. Addison-Wesley, Reading, MA.

Evans, G.N., M.M. Naim, and D.R. Towill (1993). "Dynamic Supply Chain Performance: Assessing the Impact of Information Systems." *Logistics Information Management*, 6 (4), 15–25.

Foley, M.J. (1988). "Post-MRP II: What Comes Next." *Datamation*, December 1, 24–36.

Gabel, D. (1985). "Project Management Software." *PC Week*. September 10, 59–67.

Geoffrion, A.M. (1976). "Better Distribution Planning with Computer Models." *Harvard Business Review*, July–August, 92.

Gray, C. (1987). *The Right Choice: A Complete Guide to Evaluating, Selecting and Installing MRP II Software*. Essex Junction, VT., Oliver Wight Ltd.

Grubbs, S.C. (1994). "MRPII: successful implementation the hard way." *Hospital Material Management Quarterly*, 15 (4), 40–47.

Haider, S.W. and J. Banks (1986). "Simulation Software Products for Analyzing Manufacturing Systems." *Industrial Engineering*, 18, 7 (July), 98–103.

Harmon, P. and D. King (1985). *Expert Systems: Artificial Intelligence in Business*. Wiley, New York.

Hoffmann, Thomas (1992). "EUREKA: A Hybrid System for Assembly Line Balancing." *Management Science*, 38 (1), 39–47.

Jackson, J.R. (1956). "A Computing Procedure for a Line Balancing Problem." *Management Science*, 2, 3, 261–271.

Jankauskas, L. and S. McLafferty (1996). "BESTFIT, Distribution fitting software by Palisade Corporation." In: *Proceedings of the Winter Simulation Conference*, 551–555.

Johnson, R. (1982). "SPACECRAFT for Multi-Floor Layout Planning." *Management Science*, 28 (4) (April), 407–417.

Johnson, R. (1988). "Optimally Balancing Large Assembly Lines with 'FABLE'." *Management Science*, 34 (2), 240–253.

Jones, M.S., C.J. Malmborg, and M.H. Agee (1985). "Decision support system used forrobot selection." *Industrial Engineering*, 17, 66–73.

Katzel, J. (1987). "Maintenance Management Software." *Plant Engineering*, 41 (12) (June 18), 124–170.

Kekre, S. and T. Mukhopadhyay (1992). "Impact of Electronic Data Interchange Technology on Quality Improvement and Inventory Reduction Programs: A Field Study." *International Journal of Production Economics*, 28, 265–282.

King, C.B. (1996). "Taylor II manufacturing simulation software." In: *Proceedings of the Winter Simulation Conference*, 569–573.

Law, A.M. and M.G. McComas (1996). "EXPERTFIT: Total support for simulation input modeling." In: *Proceedings of the Winter Simulation Conference*, 588–593.

LeRose, J.W. (1993). "Intermodal EDI." *EDI World* 3, 22–24.

Lewis, W.P. and T.E. Block (1980). "On the Application of Computer Aids to Plant Layout." *International Journal of Production Research*, 18 (1), 11–20.

Lin, Li and David D. Bedworth (1988). "A Semi-Generative Approach to Computer Aided Process Planning Using Group Technology." *Computers and Industrial Engineering*, 14 (2), 127–137.

Manivannan, S. and D. Chudhuri (1984). "Computer-Aided Facility layout Algorithm Generates Alternatives to Increase Firm's Productivity." *Industrial Engineering* (May), 81–84.

McMillan, Claude Jr. and Richard F. Gonzales (1968). *Systems Analysis: A Computer Approach to Decision Models* (Rev. ed.). Richard D. Irwin, Homewood, IL.

Messmer, E. (1992). "EDI Suppliers to Test Fed Purchasing Net." *Network World* 9 (43–44).

Micro-Craft (1986). *Plant Layout Software*. Industrial Engineering and Management Press, IIE, Atlanta, GA.

Muther, Richard and K. McPherson (1970). "Four Approaches to Computerized Layout." *Industrial Engineering*, 2 (2), 39–42.

O'Neal, K. (1987). "Project Management Computer Buyer's Guide." *Industrial Engineering*, 19 (1) (January), 53–63.

Premkumar, G., K. Ramamurthy, and S. Nilakanta (1994). "Implementation of Electronic Data Interchange: An Innovation Diffusion Perspective." *Journal of Management Information Systems*, 11, 157–186.

Rhodes, D.J. (1991). "The facilitator – an organizational necessity for the successful implementation of IT and operations strategies." *Computer Integrated Manufacturing Systems*, 4 (2), 109–113.

Rycroft, R. (1993). "Microcomputer Software of Interest to Forecasters in Comparative Review: An Update." *International Journal of Forecasting*, 9 (4), December, 531–575.

Schrage, L. (1986). *Linear, Integer, and Quadratic Programming with LINDO* (3rd ed.). Scientific Press, Palo alto, CA.

Smith, L.A. and Sushil Gupta (1985). "Project Management Software in P&IM." *P&IM Review*, 5 (6), 66–68.

Smith, S.B. (1989). *Computer Based Production and Inventory Control*. Prentice-Hall, Englewood Cliffs, NJ.

Swain, J.J. (1995). "Simulation Survey: Tools for process understanding and improvement." *OR/MS Today*. 22(4), 64–72 and 74–79.

Wiest, Jerome D. (1966). "Heuristic Programs for Decision Making." *Harvard Business Review*. September–October, 129–143.

Winston, W.L. (1996). *Simulation Modeling Using @RISK*. Duxbury Press, Belmont, CA.

Yacura, J.A. (1994). "The Expert System: Purchasing's Tool of the Future." *NAPM Insights* (October), 32–33.

Strategic, Environmental and International Issues

Abegglen, J.C. and G. Stalk (1985). *The Japanese Corporation*, New York: Basic Books, Inc.

Belohlav, J.A. (1995). "Quality, Strategy, and Competitiveness." *The Death and Life of the American Quality Movement*. Edited by R.E. Cole. Oxford University Press, New York, 43–58.

Betz, F. (1993). *Strategic Technology Management*, New York, Mc Graw-Hill.

Blois, K.J. (1986). "Marketing Strategies and the New Manufacturing Technologies." *International Journal of Operations and Production Management*, 6 (1), 34–41.

Bozarth, R. Handfield and A. Das (1997). "Stages of global sourcing strategy evolution: An exploratory study." Secial Issue on Global Manufacturing, *Journal of Operations Management*, in press.

Clark, K.B. and T. Fujimoto (1991). *Product Development Performance: Strategy, Organization and Management in the World Auto Industry*. Harvard Business School Press, Cambridge.

Cooper, R.G. E.J. Kleinschmidt (1987). New Products: What Separates Winners from Losers?" *Journal of Product Innovation Management*, 4 (3), 69–184.

Cyert, R. (?). "Saving the Smoke stack Industries." *Bell Atlantic Quarterly*, 2 (3) (Autumn), 17–24.

Daugherty, P.J., R.E. Sabath, and D.S. Rogers (1992). "Competitive Advantage Through Customer Responsiveness." *The Logistics and Transportation Review*, 28 (3), 257–271.

Deming, W.E. (1982). *Quality, Productivity, and Competitive Position*, Center for Advanced Engineering Study, Massachusetts Institute of Technology, Cambridge, Massachusetts.

Dertouzos, Michael L., Richard K. Lester, and Robert M. Solow (1989), and the MIT Commission on Industrial Productivity. *Made in America: Regaining the Productive Edge*. The MIT Press, Cambridge, MA.

Ettlie, J.E. and J.D. Penner-Hahn (1994). "Flexibility ratios and manufacturing strategy." *Management Science*, 40, 1444–1454.

Frohman, A.L. (1982). "Technology as a Competitive Weapon." *Harvard Business Review*, 60 (1) (January–February), 97–104.

Galbraith, C. and D. Schendel (1983). "An empirical analysis of strategy types." *Strategic Management Journal*, 4 (2), 153–173.

Gale, B.T. (1985). *Quality as a Strategic Weapon*. Strategic Planning Institute, Cambridge, MA.

Garud, R. and A. Kumaraswamy (1993). "Technological and Organizational Designs for Realizing Economies of Substitution." *Strategic Management Journal*, 16, 93–109.

Garvin, D.A. (1993). "Manufacturing Strategic Planning." *California Management Review*, Summer, 85–106.

Gates, S. (1994). *The Changing Global Role of the Foreign Subsidiary Manager*. The Conference Board, New York.

Gerwin, D. (1993). Manufacturing flexibility: A Strategic Perspective. *Management Science*, 39 (4), 395–409.

Gerwin, D. (1993). "Integrating Manufacturing into the Strategic Phases of New Product Development." *California Management Review*, Summer, 123–136.

Gerwin, D. and H. Kolodny (1992). *Management of Advanced Manufacturing Technology*, New York: Wiley-Interscience.

Goldratt, E.M. and Robert Fox (1986). *The Race for a Competitive Edge*. North River Press, Croton-on-Hudson, NY.

Graedel, T.E. and B.R. Allenby (1995). *Industrial Ecology*. Prentice-Hall, New Jersey.

Hadjinicola, G.C. and K.R. Kumar (1997). "Factors affecting international product design." *Journal of the Operational Research Society*, 48, 1131–1143.

Hahn, C.K., E.A. Duplaga, and K.Y. Kim (1994). "Production/Sales Interface: MPS at Hyundai Motor." *International Journal of Production Economics*, 37–1, 5–18.

Hambrick, D.C. (1983). "High Profit Strategies in Mature Capital Goods Industries: A Contingency Approach." *Academy of Management Journal*, 26 (4), 687–707.

Hamel, G. and C.K. Prahalad (1994). *Competing for the Future*, Boston: Harvard Business School Press.

Handfield, R.B. (1995). *Re-engineering for Time-based Competition-Benchmarks and Best Practices for Production, R&D, and Purchasing*, Quorum Books, Westport, CT.

Harding, C.F. (1980). "Your Business: Right Ingredients, Wrong Location." *INC* (February), 20–22.

Hayes, R.H. and G.P. Pisano (1996). "Manufacturing strategy: at the intersection of two paradigm shifts." *Production and Operations Management*, 5 (1), 25–41.

Hayes, R.H. and G.P. Pisano (1994). "Beyond World Class: The New Manufacturing Strategy." *Harvard Business Review*, January–February, 77–86.

Hayes, R.H. and S.C. Wheelwright (1979). "Link Manufacturing Process and Product Life Cycles." *Harvard Business Review*, (January–February), 133–140.

Hayes, R.H. and S.C. Wheelwright (1979). "The dynamics of process-product lifecycles." *Harvard Business Review* (March–April), 127–136.

Hayes, R.H. and S.C. Wheelwright (1984). *Restoring Our Competitive Edge: Competing Through Manufacturing*, John Wiley & Sons, New York.

Hayes, R.H. and W.J. Abernathy (1980). "Managing Our Way to Economic Decline." *Harvard Business Review*, 58, July–August, 67–77.

Hayes, R.H., G.P. Pisano, and D.M. Upton (1996). *Strategic Operations*, Free Press, New York.

Hayes, R.H., S. Wheelwright, and K. Clark (1988). *Dynamic Manufacturing: Creating the Learning Organization*, The Free Press, New York, 161–184.

Helming, B. and Mistry, P. (1998). "Making Your Supply Chain Strategic." *APICS: The Performance Advantage*, 8, 54–56.

Helper, S.R. (1991). "Strategy and Irreversibility in Supplier Relations: The Case of the US Automobile Industry." *Business History Review*, 65, Winter, 781–824.

Hentschel, D.C. (1993). "The Greening of Products and Production – A New Challenge for Engineers." *Advances in Production Management Systems*, Elsevier Science, North-Holland, 39–46.

Hill, C. (1988). "Differentiation versus low cost or differentiation and low cost: A contingency framework." *Academy of Management Journal*, 13 (3), 401–412.

Hill, J.S. and R.R. Still (1984). "Adapting products to LDC tastes." *Harvard Business Review*, March–April, 92–101.

Hill, T. (1989). *Manufacturing Strategy: Text and Cases*. Addison-Wesley, Wokingham, England.

Hill, T. (1994). *Manufacturing Strategy, Text and Cases*, Richard D. Irwin Inc., Burr Ridge, Illinois.

Hill, T. (1983). "Manufacturing's Strategic Role." *Journal of the Operational Research Society*, 34 (9), 853–860.

Hofer, C. and D. Schendel (1978). *Strategy Formulation: Analytical Concepts*, West, St. Paul.

Jain, S.J. (1989). "Standardization of international marketing strategy: Some research hypotheses." *Journal of Marketing*, 53, 70–79.

Jelinek, M. and J. Goldhar (1983). "The Interface Between Strategy and Manufacturing Technology." *Columbia Journal of World Business*, Spring.

Jordan, W.C. and S.C. Graves (1995). "Principles on the benefits of manufacturing process flexibility." *Management Science*, 41, 577–594.

Kanter, R.M. (1995). *World Class: Thriving Locally in the Global Economy*, Simon and Shuster, New York.

Kantrow, A.M. (1980). "The Strategy-Technology Connection." *Harvard Business Review*, July–August, 6–8, 12.

Kashani, K. (1989). "Beware of the pitfalls of global marketing." *Harvard Business Review*, September–October, 91–98.

Kim, L. and Y. Lim (1988). "Environment, Generic Strategies, and Performance in a Rapidly Developing Country: A Taxonomic Approach." *Academy of Management Journal*, 31 (4), 802–826.

Kim, Y. and J. Lee (1993). "Manufacturing Strategy and Production Systems: An Integrated Framework." *Journal of Operations Management*, 11 (1), 3–15.

Klassen, R.D. and N.P. Greis (1993). "Managing Environmental Improvement through Product and Process Innovation: Implications of Environmental Life Cycle Assessment." *Industrial and Environmental Crisis Quarterly*, 7 (4), 293–318.

Kogut, B. (1985). "Designing Global Strategies: Comparative and Competitive Value-Added Chains." *Sloan Management Review*, 26 (4), 15–28.

Kogut, B. (1985). "Designing Global Strategies: Profiting from Operational Flexibility." *Sloan Management Review*, 27 (1), 27–38.

Kotha, S. and D. Orne (1989). "Generic Manufacturing Strategies: A Conceptual Synthesis." *Strategic Management Journal*, 10 (3), 211–231.

Kotler, P. (1984). *Marketing Management: Analysis, Planning, and Control* (5th ed.). Prentice-Hall, New Jersey.

Kubiak, J. (1993). "A Joint Venture in Mass Customization." *Planning Review*, July 21 (4), 25.

Kulatilaka, N. (1988). "Valuing the flexibility of flexible manufacturing systems." *IEEE Transactions on Engineering Management*, 35, 250–257.

Lamming, R. (1993). *Beyond Partnership: Strategies for Innovation and lean Supply*. Prentice-Hall, London.

Leong, G.K., D. Snyder, and P. Ward (1990). "Research in the Process and Content of Manufacturing Strategy." *Omega*, 18, 109–122.

Levitt, T. (1983). *"The Globalization of Markets." Harvard Business Review*, 61 (3), 92–102.

Limprecht, Joseph A. and Robert H. Hayes (1980). "Germany's World-Class Manufacturers." *Harvard Business Review*, 63, September–October, 142–150.

Lorenz, C. (1986). *The Design Dimension: The New Competitive Weapon for Business*, Basil Blackwell, NY.

Lubben, Richard T. (1988). *Just-In-Time: An Aggressive Manufacturing Strategy*, McGraw-Hill, New York.

MacCormack, A.D., L.J. Newmann, III, and D.B. Rosenfield (1994). "The New Dynamics of Global Manufacturing Site Location." *Sloan Management Review*, 35 (4), 69–80, Summer.

March, A. and D.A. Garvin (1986). Copeland Corporation: Evolution of a Manufacturing Strategy, Harvard Business School Case 9-686-G088.

Marquardt, M. and A. Reynolds (1994). *The Global Learning Organization*. Irwin, Burr Ridge, IL.

Marshal, J. and P. Lawrence (1991). "Design as A Strategic Resource: A Business Perspective." Proceedings of the Design Leadership Symposium, Corporate Design Foundation, Boston, MA.

Martinez, J.I. and J.C. Jarillo (1989). "The evolution of research on coordination mechanisms in multinational corporations." *Journal of International Business Studies*, 20 (489–514).

McDougall, P.P., R.H. Deane, and D.E. D'Souza, (1992). "Manufacturing Strategy and Business Origin of New Venture Firms in the Computer and Communications Equipment Industries." *Production and Operations Management*, 1 (1), 53–69.

McGrath, M. and R. Hoole (1992). "Manufacturing's New Economies of Scale."*Harvard Business Review* (May–June), 94–102.

Miles, R. and C. Snow (1978). *Organizational Strategy, Structure, and Process*. McGraw-Hill, New York.

Miller, D. (1986). "Configurations of strategy and structure: Towards a synthesis." *Strategic Management Journal*, 7 (3), 233–249.

Miller, D. (1987). "The structural and environmental correlates of business strategy." *Strategic Management Journal*, 8 (1), 55–76.

Miller, D. (1988). "Relating Porter's business strategies to environment and structure: Analysis and performance implications." *Academy of Management Journal*, 31 (3), 280–308.

Miller, J.G., A. De Meyer, and J. Nakane (1992). *Benchmarking Global Manufacturing*, Irwin, Homewood, IL.

Min, H. and W.P. Galle (1991). "International Purchasing Strategies of Multinational U.S. Firms." *International Journal of Purchasing and Materials Management* (Summer), 9–18.

Mintzberg, H. (1973). "Strategy-making in three modes." *California Management Review*, 16 (2), 44–53.

Mintzberg, H. (1978). "Patterns in strategy formation." *Management Science*, 24 (9), 934–948.

Mintzberg, H. (1988). "Generic Strategies: Toward a Comprehensive Framework." *Advances in Strategic Management*, 5, 1–67.

Monczka, R. and Robert J. Trent (1995). "Global sourcing: a development approach." *International Journal of Purchasing and Materials Management*, 11 (2), 2–8.

Noe, R.A., J.R. Hollenbeck, B. Gerhart, and P.M. Wright (1997). *Human Resource Management: Gaining a Competitive Advantage* (2nd ed.). Irwin, Chicago.

Ohmae, K. (1988). "Getting Back to Strategy." *Harvard Business Review*, 66 (6), 149–156.

Ohmae, K. (1989). "Managing in a Borderless World." *Harvard Business Review*, 67 (3),151–161.

Olins, W. (1989). *Corporate Identity: Making Business Strategy Visible through Design*, Harvard Business School Press, Boston, MA.

Penrose, E. (1959). *The Theory of the Growth of the Firm*, Wiley, New York.

Pfeffer, J. (1995). "Producing sustainable competitive advantage through the effective management of people." *Academy of Management Executive*, 9, 55–69.

Pine II, J.B. (1993). *Mass Customization – The New Frontier in Business Competition*, Harvard Business School Press, Boston, MA.

Platts, K.W. (1994). "Characteristics of Methodologies for Manufacturing Strategy Formulation." *Computer Integrated Manufacturing Systems*, 7, 2, 93–9.

Porter, M.E. (1980). *Competitive Strategy – Techniques for Analysing Industries and Competitors*. Macmillan Free Press, New York.

Porter, M.E. (1985). *Competitive Advantage: Creating and Sustaining Superior Performance*. Free Press, New York.

Porter, M.E. (1986). "Changing Patterns of International Competition." *California Management Review* (Winter), 9–40.

Porter, M.E. (1986). Editor. *Competition in Global Industries*. Harvard Business School Press, Boston.

Porter, M.E. (1990). "*The Competitive Advantage of Nations.*" *Harvard Business Review*, 68 (2), 73–93.

Powers, T.L., J.U. Sterling, and J.F. Wolter (1988). "Marketing and Manufacturing Conflict: Sources and Resolution." *Production and Inventory Management Journal*, 29 (4), 56–60.

Quelch, J.A. and E.J. Hoff (1986). "Customizing global marketing." *Harvard Business Review*, May–June, 59–68.

Quinn, J.B. (1978). "Strategic change: logical incrementalism." *Sloan Management Review*, 1, 20.

Quinn, J.B. (1992). *Intelligent Enterprise*. Free Press, New York.

Quinn, J.B. and F.G. Hilmer (1994). "Strategic outsourcing." *Sloan Management Review*, Summer, 43–55.

Rhodes, D.J. (1991). "The facilitator – an organizational necessity for the successful implementation of IT and operations strategies." *Computer Integrated Manufacturing Systems*, 4, 2, 109–13.

Richardson, P.R., A.J. Taylor, and J.R.M. Gordon (1985). "A Strategic approach to Evaluating Manufacturing Performance." *Interfaces*, November–December, 15–27.

Robinson, Richard D. (1984). *Internationalization of Business: An Introduction*, FreePress, Chicago.

Rosow, I.M. and R. Zager (1988). *Training: The Competitive Edge*, San Francisco, Jossey-Bass.

Scarpello, V., W.R. Boulton, and C.R. Hofer (1986). "Reintegrating R&D into Business Strategy." *Journal of Business Strategy*, Spring, 49–56.

Schendel, D. and C. Hofer (1979). *Strategic Management: A new view of Business Policy and Planning*. Little, Brown & Co., Boston, MA.

Schmenner, R.W. (1982). "Multiple Manufacturing Strategies Among the Fortune 500." *Journal of Operations Management*, 2 (2) (February), 77–86.

Schniederjans, M.J. (1999). *Global Facility Location and Acquisition*, New York, Quorum Books.

Schroeder, R.G., J.C. Anderson, and G. Cleveland (1986). "The Content of Manufacturing Strategy: An Empirical Study." *Journal of Operations Management*, 6 (3–4), 405–415.

Scully, J.I. and S.E. Fawcett (1993). "Comparative Logistics and Production Costs for Global Manufacturing Strategy." *International Journal of Operations and Production Management*, 13 (12), 62–78.

Scully, J.I. and S.E. Fawcett (1994). "International Procurement Strategies: Opportunities and Challenges for the Small Firm." *Production and Inventory Mangement Journal*, 35 (2), 39–46.

Sethi, A.K. and S.P. Sethi (1990). "Flexibility in Manufacturing: A Survey." *International Journal of Flexible Manufacturing Systems*, 2, 289–328.

Shapiro, B.P. (1977). "Can Marketing and Manufacturing coexist?" *Harvard Business Review*, 55 (5), 104–114.

Skinner, W. (1969). "Manufacturing – Missing Link in Corporate Strategy." *Harvard Business Review*, 47 (3) (May–June), 136–145.

Skinner, W. (1974). "The Focused Factory." *Harvard Business Review*, 52 (3), 113–121.

Skinner, W. (1978). *Manufacturing in Corporate Strategy*, New York: John Wiley and Sons.

Skinner, W. (1983). "Getting Physical: New strategic Leverage from Operations." *Journal of Business Strategy*, 3, Spring, 74–79.

Skinner, W. (1983). "Wanted: Managers for the Factory of the Future." *Annals of the American Academy of Political and Social Science*. November, 102–114.

Skinner, W. (1984). "Operations Technology: Blind Spot in Strategic Management." *Interfaces*, January–February, 116–125.

Skinner, W. (1985). *Manufacturing the Formidable Competitive Weapon*. John Wiley, New York.

Skinner, W. (1992). "Missing the Links in Manufacturing Strategy." in: C. Voss (Ed.), *Manufacturing Strategy: process and Content*, Chapman-Hall, London, 13–25.

Skinner, W. (1996). "Manufacturing strategy on the "S" curve." *Production and Operations Management*, 5 (1), 3–14.

Slautterback, W.H. and W.B. Werther (1984). "The Third Revolution: Computer Integrated Manufacturing." *National Productivity Review*, Autumn, 367–374.

Sorenson, R.Z. and U.E. Wiechmann (1975). "How multinationals view marketing standardization." *Harvard Business Review*, May–June, 38–54.

Spencer, M.S. and J.F. Cox (1995). "An analysis of the product-process matrix and repetitive manufacturing." *International Journal of Production Research*, 33 (5), 1275–1294.

Stalk, G., Jr., (1988). "Time – The Next Source of Competitive Advantage." *Harvard Business Review*, July–August, 41–51.

Stalk, G., Jr., and T. Hout (1990). *Competing Against Time*, The Free Press, New York.

Stalk G., Jr., P. Evans, and L. E. Schulman (1992). "Competing on Capabilities: The New Rules of Corporate Strategy." *Harvard Business Review*, 70 (2), 57–69.

Starr, M.K. (1988). *Global Competitiveness: Getting the U.S. Back on Track*. Norton, New York.

Steele L. (1989). "Managing Technology: The Strategic View." New York, McGraw-Hill.

Stobaugh, R. and P. Telesio (1983). "Match Manufacturing Policies and Product Strategies." *Harvard Business Review*, March–April, 113–120.

Susman, G.I. (1992). *Integrating Design and Manufacturing for Competitive Advantage*. Oxford University Press, New York.

Swamidass, P.M. (1987). "Manufacturing Strategy: Its assessment and practice." *Journal of Opeations Management*, 6 (4), 471–484.

Swamidass, P.M. (1988). *Manufacturing Flexibility. Operations Management* Association Monograph 2, Naman and Schneider Associates Group, Texas.

Swamidass, P.M. (1990). "A Comparison of Plant Location Strategies of Foreign and Domestic Manufacturers in the U.S." *Journal of International Business Studies*, 8 (3), 263–277.

Swamidass, P.M. and W.T. Newell (1987). "Manufacturing Strategy, Environmental Uncertainty and Performance: A Path Analytic Model." *Management Science*, 33 (4), 509–524.

Swamidass, P.M., S.S. Nair, and S.I. Mistry (1999). "The use of a neural factory to investigate the effect of product line width on manufacturing performance." *Management Science*, forthcoming.

Swink, M.L. and M. Way (1995). "Manufacturing Strategy: Propositions, Current Research, Renewed Directions." *International Journal of Operations and Production Management*, 15 (7), 4–27.

Takeuchi, H. and M.E. Porter (1986). "Three roles on international marketing global strategy." In M.E. Porter (ed.). *Competition in Global Industries*. Harvard Business School Press, Boston, 111–146.

Tong, H.M. and C.K. Walter (1980). "An Empirical Study of Plant Location Decisions of Foreign Manufacturing Investors in the United States." *Columbia Journal of World Business* (Spring), 66–73.

Trigeorgis, L. (1996). *Real Options: Managerial Flexibility and Strategy in Resource Allocation*. The MIT Press, MA.

Upton, D.M. (1994). "The Management of Manufacturing Flexibility." *California Management Review*, Winter, 72–89.

Upton, D.M. (1995). "Flexibility as process mobility: The management of plant capabilities for quick response manufacturing." *Journal of Operations Management*, 12 (3&4), June, 205–224.

Upton, D.M. (1995). "What Really Makes Factories Flexible." *Harvard Business Review*, July–August, 74–79.

Upton, D.M. (1997). "Process range in manufacturing: An empirical study of flexibility." *Management Science*, 4 (8), August, 1079–1092.

Venketesan, R. (1990). "Cummins flexes its Factory. *Harvard Business Review*, 90 (2), March–April, 120–127.

Walters, P.G. (1986). "International marketing policy, a discussion of the standardization construct and its relevance for corporate policy." *Journal of International Business Studies*, 17, 55–69.

Walters, P.G. and B. Toyne (1989). "Product modification and standardization in international markets: Strategic options and facilitating policies." *Columbia Journal of World Business*, Winter, 37–44.

Walton, L.W. (1996). "The ABC's of EDI: The Role of Activity-Based Costing (ABC) in Determining EDI Feasibility in Logistics Organizations." *Transportation Journal*, 36, 43–50.

Waterman, R.H., Jr. (1987). *The Renewal Factor: How to Get and Keep the Competitive Edge*, Bantam, Toronto.

Wernerfelt, B. (1984). "A Resource-based View of the Firm." *Strategic Management Journal*, 5, 171–180.

Wever, G.H. (1997). *Strategic Environmental Management: Using TQEM and ISO14000 for Competitive Advantage*. John Wiley, New York.

Wheelright, S.C. (1978). "Reflecting Corporate Strategy in Manufacturing Decisions." *Business Horizons*. 21, February, 57–66.

Wheelwright, S.C. (1981). "Japan – Where Operations Really Are Strategic." *Harvard Business Review*. July–August, 67–74.

Wheelwright, S.C. (1984). "Manufacturing Strategy: Defining the Missing Link." *Strategic Management Journal*. 5 (1), 77–91.

Wheelwright, S.C. and R.H. Hayes (1985). "Competing Through Manufacturing." *Harvard Business Review*, January–February, 99–109.

Wisner, J.D. and S.E. Fawcett (1991). "Linking Firm Strategy to Operating Decisions Through Performance Measurement." *Production and Inventory Mangement Journal*, 32 (3), 5–11.

Yip, G.S. (1989). "Global strategy . . . in a world of nations?" *Sloan Management Review*, 31, 29–41.

SUPPLY CHAIN AND LOGISTICS

Aikens, C.H. (1985). "Facility Location Models of Distribution Planning." *European Journal of Operations Research*, 263–279.

Akinc, U. and B.M. Khumawala (1977). "An Efficient Branch and Bound Algorithm for Capacitated Warehouse Location Problem." *Management Science*, 23, 585–594.

Ammer, D.S. (1980). *Materials Management and Purchasing* (4th ed.). Irwin, Homewood, IL.

Birou, L.M. and S.E. Fawcett (1993). "International Purchasing: Benefits, Requirements, and Challenges." *International Journal of Purchasing and Materials Management*, 29 (2), 28–37.

Blackburn, J.D. (1991). "The Quick-Response Movement in the Apparel Industry: A Case Study Time-Compressing Supply Chains." edited by J.D. Blackburn, *Time-Based Competition: The Next Battle Ground in American Manufacturing*, Irwin, Homewood, IL, 246–269.

Bornemann, A.H. (1974). *Essentials of Purchasing*. Grid, Columbus, OH.

Bowersox, D.J., R.J. Calantone, S.R. Clinton, D.J. Closs, M.B. Cooper, C.L. Droge, S.E. Fawcett, R. Frankel, D.J. Frayer, E.A. Morash, L.M. Rinehart, and J.M. Schmitz (1995). *World Class Logistics: The Challenge of Managing Continuous Change*. Council of Logistics Management, Oak Brook, IL.

Bowersox, D.J. and P.J. Daugherty (1995). "Logistics Paradigms: The Impact of Information Technology." *Journal of Business Logistics*, 16 (1), 65–80.

Bowersox, D.J., D.J. Closs, and O.K. Helferich (1986). *Logistical Management: A Systems Integration of Physical Distribution, Manufacturing Support, and Materials Procurement* (3rd ed.). Macmillan, New York.

Bozarth, R. Handfield, and A. Das (1997). "Stages of global sourcing strategy evolution: An exploratory study." Secial Issue on Global Manufacturing, *Journalof Operations Management*, in press.

Branch, A.E. (1996). *Elements of Shipping*, Seventh edition, Chapman and Hall, London, UK.

Burt, D. (1984). *Proactive Purchasing*. Prentice-Hall, Englewood Cliffs, NJ.

Burt, D.N. (1989). "Managing Suppliers Up To Speed." *Harvard Business Review*, 67 (4).

Camm, J.D., T.E. Chorman, F.A. Dill, J.R. Evans, D.J. Sweeney, and G.W. Wegryn (1997). "Blending OR/MS, Judgment, and GIS: Restructuring P&G's Supply Chain." *Interfaces*, 27 (1), 128–142.

Carbone, J. (1995). "Lessons from Detroit." *Purchasing* (October 19), 38–42.

Carter, Phillip L., and Chrwan-Jyh Ho (1984). "Vendor Capacity Planning: An Approach to Vendor Scheduling." *Production and Inventory Management*, Fourth Quarter, 63–73.

Cavinato, J.L. (1992). "A Total Cost/Value Model for Supply Chain Competitiveness, *Journalof Business Logistics*, 13, 2, 285–301.

Cavinato, J.L. (1984). *Purchasing and Materials Management*, West, St. Paul, MN.

Clinton, S.R., D.J. Closs, M.B. Cooper, and S.E. Fawcett (1996). "New Dimensions of World Class Logistics Performance." In *Annual Conference Proceedings Council of Logistics Management*, 21–33.

Colton, R.R. and W.F. Rohrs (1985). *Industrial Purchasing and Effective Materials Management*, Reston Publishing, Reston, VA.

Combs, P.M. (1976). *Handbook of International Purchasing*. Cahners, Boston.

Cooper, M.C., D.M. Lambert, and J.D. Pagh (1997). "Supply Chain Management: More Than A New Name for Logistics." *International Journal of Logistics Management*, 1, 1–13.

Cooper, M.C. and Lisa M. Ellram (1993). "Characteristics of Supply Chain Management and the Implications for Purchasing and Logistics Strategy." *International Journal of Logistics Management*, 4, 2, 10–15.

Council of Logistics Management (1995). *World Class Logistics: The Challenge of Managing Continuous Change*, Council of Logistics Management, IL.

Coyle, J.J. and E.J. Bardi (1988). *The Management of Business Logistics*. St. Paul, West.

Daskin, M.S. (1985). "Logistics: An Overview of the State of the Art and Perspectives on Future Research." *Transportation Journal*, 19A(5/6), 383–398.

Denison, R.A. and J. Ruston (1990). *Recycling and Incineration*. Island Press, Washington, D.C.

Dixon, Lance and Anne Miller Porter (1994). *JIT II – Revolution in Buying and Selling*, Cahner's Publishing Company, Newton, MA.

Dobler, D.W. and D.N. Burt (1996). *Purchasing and Materials Management*. McGraw-Hill Book Company, New York.

Dobler, D.W., L. Lee, Jr., and D.N. Burt (1984). *Purchasing and Materials Management: Text and Cases*. McGraw-Hill, New York.

Elfring, T. and G. Baven (1994). "Outsourcing Technical Services: Stages of Development." *Long Range Planning*, 27 (5), 42–51.

Elliff, S. (1996). "Supply Chain Management – New Frontier." *Traffic World* (October 21), 55.

Ellram, L.M. (1994). "The 'What' of Supply Chain Management." *NAPM Insights*, 5 (3), 26–27.

Emmelhorne, Margaret A. (1971). *Guide to Purchasing: Electronic Data Interchange. National Association of Purchasing Management*, John Wiley & Sons, New York.

Evans, G.N., M.M. Naim, and D.R. Towill (1993). "Dynamic Supply Chain Performance: Assessing the Impact of Information Systems." *Logistics Information Management*, 6 (4), 15–25.

Fawcett, S.E. and S.A. Fawcett (1995). "The Firm as a Value-Added System: Integrating Logistics, Operations, and Purchasing." *International Journal of Physical Distribution and Logistics Management*, 25 (3), 24–42.

Fawcett, S.E. and S.R. Smith (1995). "Logistics Measurement and Performance for United States – Mexican Operations under NAFTA." *Transportation Journal*, 34 (3), 25–34.

Fawcett, S.E., L.E. Stanley, and S.R. Smith (1997). "Developing a Logistics Capability to Improve the Performance of International Operations." *Journal of Business Logistics*, 18 (2), 101–128.

Fisher, M.L., J.H. Hammond, W.R. Obermeyer, and A. Raman (1994). "Making Supply Meet Demand in an Uncertain World." *Harvard Business Review*, 83–93.

Ford, D. (1980). "The development of buyer-seller relationships in industrial markets." *European Journal of Marketing*, 14 (5–6), 339–54.

Fuller, J.B., J. O'Conor, and R. Rawlinson (1993). "Tailored Logistics: The Next Advantage." *Harvard Business Review*, 71, 87–98.

Garwood, Dave (1971). "Delivery as Promised." *Production and Inventory Management*, 12, 3.

Gecowets, G.A. (1979). "Physical Distribution Management." *Defense Transportation Journal*, 35, 11.

Geoffrion, A.M. (1974). "Multicommodity Distribution System Design by Benders Decomposition." *Management Science*, January, 822.

Geoffrion, A.M. (1976). "Better Distribution Planning with Computer Models." *Harvard Business Review*, July–August, 92.

Geoffrion, A.M. and G.W. Graves (1974). "Multicomodity Distribution System Design by Bender's Decomposition." *Management Science*, 12, 822–844.

Giunipero, L. and J. Vogt (1997). "Empowering the Purchasing Function: Moving to Team Decisions." *International Journal of Purchasing and Materials Management*, Winter, 8–15.

Giunipero, L.C. and R.R. Brand (1996). "Purchasing's Role in Supply Chain Management." *International Journal of Logistics Management*, 7 (1), 25–38.

Gregory, R. (1986). "Source selection: A matrix approach." *International Journal of Purchasing and Materials Management*, 24–29.

Greis, N.P. and J.D. Kasarda (1997). "Enterprise Logistics in the Information Era." *California Management Review*, 39 (4), 55–78.

Gupta, Y.P. and P. Bagchi (1987). "Inbound Freight Consolidation Under Just-In-Time Procurement, Application of Clearing Models." *Journal of Business Logistics*, 8 (2), 74–94.

Hahn, C.K., K.H. Kim, and J.S. Kim (1986). "Costs of Competition, Implications for Purchasing Strategy." *Journal of Purchasing and Materials Management* (Fall), 2–7.

Handfield, R.B. (1994). "Global sourcing: Patterns of development." *International Journal of Operations and Production Management*, 14 (6), 40–51.

Handfield, R.B., S.V. Walton, L.K. Seegers, and S.A. Melnyk (1997). "'Green' Value Chain Practices in the Furniture Industry." *Journal of Operations Management*,15 (4), 293–315.

Heinritz, S.F., P.V. Farrell, and C.L. Smith (1986). *Purchasing Principles and Applications*. Prentice-Hall, Englewood Cliffs, NJ.

Helming, B. and P. Mistry (1998). "Making Your Supply Chain Strategic." *APICS: The Performance Advantage*, 8, 54–56.

Helper, S.R. (1991). "Strategy and Irreversibility in Supplier Relations: The Case of the U.S. Automobile Industry." *Business History Review*, 65, Winter, 781–824.

Herron, David P. (1979). "Managing Physical Distribution for Profit." *Harvard Business Review*, May–June, 121–132.

Heskett, J.L. (1977). "Logistics-Essential to Strategy." *Harvard Business Review*, November–December, 85–96.

Hines, P. (1994). *Creating World Class Suppliers*. Pitman Publishing, London.

Jackson, G.C. (1985). "A Survey of Freight Consolidation Practices." *Journal of Business Logistics*, 6, 13–34.

Johnson, J.C. and D.F. Wood (1986). *Contemporary Physical Distribution and Logistics* (3rd ed.). MacMillan, New York.

Jones, Thomas C. and D.W. Riley (1985). "Using Inventory for Competitive Advantage Through Supply

Chain Management." *International Journal of Physical Distribution and Materials Management*, 15, 5, 16–26.

Jordan, Henry (1974). "Relating customer Service to Inventory Control." *American Management Association's Advanced Management Journal*, 39 (4).

Keuhn, A.A. and M.J. Hamburger (1963). "A Heuristic Program for Locating Warehouses." *Management Science*, 9, July.

Khumawala B.M. (1972). "An Efficient Branch and Bound Algorithm for the Warehouse Location Problem." *Management Science*, 18 (12), 718–731.

Khumawala, B.M. (1974). "An Efficient Heuristic Procedure for the Capacitated Warehouse Location Problem." *Naval Research Logistics Quarterly*, 21, 609–623.

Khumawala, B.M. and D.C. Whybark (1971). "A Comparison of Some Recent Warehouse Location Techniques." *The Logistics Review*, 7 (31), 3–19.

Kolchin, M. and L. Giunipero (1993). *Purchasing Education and Training*, Center for Advanced Purchasing Studies, Tempe, AZ.

Lambert, D.M. and J.R. Stock (1993). *Strategic Logistics Management* (3rd ed.). Richard D. Irwin, Inc., Homewood, IL.

Lamming, R. (1993). *Beyond Partnership: Strategies for Innovation and lean Supply*. Prentice-Hall, London.

Lee, H.L. and C. Billington (1992). "Managing Supply Chain Inventory: Pitfalls and Opportunities." *Sloan Management Review* (Spring), 65–73.

Lee, H.L., V. Padmanabhan, and S. Whang (1997). "Information Distortion in a Supply Chain: The Bullwhip Effect." *Management Science*, 43, 546–558.

Lee, H.L., V. Padmanabhan, and S. Whang (1997). "The Bullwhip Effect in Supply Chains." *Sloan Management Review*, 93–102.

Leenders, M. and D. Blenkhorn (1988). *Reverse Marketing – The New Buyer Seller Relationship*. The Free Press, New York.

Leenders, M. and H. Fearon (1997). *Purchasing and Supply Management*. R.D. Irwin, Homewood, IL.

Leenders, M.R., H.E. Gearon, and W.B. England (1985). *Purchasing and Materials Management* (8th ed.). Irwin, Homewood, IL.

LeRose, J.W. (1993). "Intermodal EDI." *EDI World*, 3, 22–24.

Levy, D. (1997). "Lean production in an international supply chain." *Sloan Management Review*, 38 (2), 94–102.

Liker, J.K., R.R. Kamath, S.N. Wasti, and M. Nagamachi (1995). "Integrating Suppliers into Fast-Cycle Product Development." In J.K. Liker, J, J.E. Ettlie, and J.C. Campbell (eds.). *Engineered in Japan: Japanese Technology Management Practices*. Oxford University Press, Oxford.

Locke, D. (1996). *Global Supply Management*, Irwin, Chicago.

Lyons, T., A.R. Krachenberg, and J.W. Henke (1990). Mixed motive marriages: What's next for buyer-supplier relations? *Sloan Management Review*, Spring, 29–36.

Macbeth, D.K. and N. Ferguson (1994). *Partnership Sourcing, an Integrated Supply Chain Management Approach*. Pitman Publishing, London.

Maltz, Arnald and Lisa Ellram (1997). "Outsourcing: Implications for Supply Management." *CAPS*, p. 140.

Martin, A.J. (1983). *Distribution Resource Planning*. Prentice-Hall, Englewood Cliffs, NJ; Oliver Wight Ltd., Essex Junction, VT.

Messmer, E. (1992). "EDI Suppliers to Test Fed Purchasing Net." *Network World*, 9, 43–44.

Messner, W.A. (1982).*Profitable Purchasing Management*. AMACOM, New York.

Metters, R. (1997). "Quantifying the Bullwhip Effect in Supply Chains." *Journal of Operations Management*, 15, 89–100.

MIL-HDBK-59A-CALS (1995). *Military Handbook on Computer-Aided Acquisition and Logistics Support*.

Miller, J. (1995). "Innovative cost savings." *NAPM Insights* (September), 30–32.

Min, H. and W.P. Galle (1991). "International Purchasing Strategies of Multinational U.S. Firms." *International Journal of Purchasing and Materials Management* (Summer), 9–18.

Monczka, R. and Robert J. Trent (1995). "Global sourcing: a development approach." *International Journal of Purchasing and Materials Management*, 11 (2), 2–8.

Monczka, R. and L.C. Giunipero (1984). "International Purchasing: Characteristics and Implementation." *Journal of Purchasing and Materials Management* (Autumn), 2–9.

Monczka, R. and R.J. Trent (1997). *Purchasing and sourcing 1997: trends and implications*, Center for Advanced Purchasing Studies (CAPS), Tempe, Arizona.

Monczka, R., R. Trent, and R. Handfield (1997). *Purchasing and Supply Chain Management*, Cincinnati, OH: Southwestern Publishing, College Division.

Monczka, R., Frayer, R., Handfield, G. Ragatz, and R. Trent (1998). "*Supplier Integration into New Product/Process Development: Best Practices*." Milwaukee: ASQC.

Murray, Jr., J.E. (1996). "Electronic Purchasing and Purchasing Law." *Purchasing* (February, 15), 29–30.

Narasimhan, R. (1983). "An Analytical Approach to Supplier Selection." *Journal of Purchasing and Materials Management* (Winter), 27–32.

O'Neal, Charles R. (1987). "The Buyer-Seller Linkage in a Just-in-time Environment." *Journal of Purchasing and Materials Management*, Spring, 7.

Poist, R.F. (1986). "Evolution of Conceptual Approaches to Designing Business Logistics Systems." *Transportation Journal*, Fall, 55–64.

Premkumar, G., K. Ramamurthy, and S. Nilakanta (1994). "Implementation of Electronic Data Interchange: An Innovation Diffusion Perspective." *Journal of Management Information Systems*, 11, 157–186.

Pye, C. (1997). "Another Option for Internet-Based EDI: Electronic Communication." *Purchasing Today*, 8, 38–40.

Quinn, J.B. and F.G. Hilmer (1994). "Strategic outsourcing." *Sloan Management Review*, Summer, 43–55.

Ragatz, G., R. Handfield, and T. Scannell (1997). "Success factors for integrating suppliers into new product development." *Journal of Product Innovation Management*, 7 (20).

Sampson, R.J., M.T. Farris, and D.L. Schrock (1990). *Domestic Transportation: Practice, Theory, and Policy* (6th ed.). Houghton Mifflin Company, Boston.

Saunders, M. (1997). *Strategic Purchasing and Supply Management*. Pitman Publishing, London.

Schnor, John E., and Thomas F. Wallace (1986). *High Performance Purchasing*, Oliver Wight Publications, Essex Junction, VT.

Schonberger, R.J. and J. Gilbert (1983). "Just-in-Time Purchasing: A Challenge for U.S. Industry." *California Management Review*, 26 (1) (Fall), 54–68.

Shapiro, R.D. and J.L. Heskett (1985). *Logistics Strategy: Cases and Concepts*. West, St. Paul, MN.

Sharman, G. (1984). "The Rediscovery of Logistics." *Harvard Business Review* (September–October), 71–79.

Spekman, Robert E. (1988). "Strategic supplier selection: Understanding long-term buyer relationships." *Business Horizons*, 80–81.

Sriram, V., R. Krapfel, and R. Spekman (1992). "Antecedents to Buyer-Seller Collaboration: An Analysis from the Buyer's Perspective." *Journal of Business Research*, 25, 303–320.

St. John, C.H. and S.T. Young (1991). "The Strategic Consistency Between Purchasing and Production." *International Journal of Purchasing and Materials Management*, (Spring), 15–20.

Stenger, Alan J. and Joseph L. Cavinato (1979). "Adapting MRP to the Out Bound Side Distribution Requirements Planning." *Production and Inventory Management*, 20 (4), 1–13.

Stock, J. and D. Lambert (1992). "Becoming a "World Class" Company With Logistics Service Quality." *International Journal of Logistics Management*, 3 (1), 73–80.

Storhagen, N.G. and R. Hellberg (1987). "Just-in-Time from a Business Logistics Perspective." *Engineering Cost and Production Economics* (The Netherlands), 12 (1–4) (July), 117–121.

Swamidass, P.M. (1993). "Import Sourcing Dynamics: An Integrative Perspective." *Journal of International Business Studies*, 24 (4).

Tully, S. (1995). "Purchasing's New Muscle." *Fortune*, (February 20), 75–83.

Weatherston, D. (1986). "Transportation Makes the JIT Connection." *Industrial Management* (April), 40–44.

Yacura, J.A. (1994). "The Expert System: Purchasing's Tool of the Future."*NAPM Insights* (October), 32–33.

Appendix III

Production Management Journals and Magazines

Without being comprehensive, the following is a selection of journals and magazines widely used by production management practitioners and researchers.

APICS: The Performance Advantage
Computers and Industrial Engineering
Computers and Inventory Management Journal
Computers and Operations Research
Decision Sciences
European Management Journal
Harvard Business Review
IEEE Transactions in Engineering Management
IIE Transactions
Integrated Manufacturing Systems
Interfaces
International Journal of Industrial Engineering
International Journal of Operations and Production Management
International Journal of Production Economics
International Journal of Production Research
International Journal of Purchasing and Materials Management
International Transactions in Operational Research
Journal of Business and Management
Journal of Manufacturing Systems
Journal of Operations Management
Journal of Productivity Analysis
Journal of the Operational Research Society
Management Science
Naval Research Logistics
Omega
Operations Research
Production and Inventory Management Journal
Production and Operations Management Society Journal
Purchasing
Quality and Reliability Management
Quality Journal
Quality Progress
Sloan Management Review
TQM Magazine

Index

B

D

E

F

J

K

L

M

N

O

P

Q

R

S

T

U

V

W

X

Y

Z